World Tables

1988–89 Edition

FROM THE DATA FILES OF THE WORLD BANK

Published for The World Bank
The Johns Hopkins University Press
Baltimore and London

CONTENTS

SYMBOLS AND CONVENTIONS

..	data not available, or nonexistant
Scientific notation *e.g. .63E-2*	data with insignificant digits, using units appropriate in 1986 -- see *Introduction*
0, 0.0, .0E 0	zero, or less than half the unit shown and not known more precisely
A,B,C, etc.	general issue of methodology, see *General Notes*
f	country issue of methodology, see *Country Notes*
Billion	1,000,000,000

1980 is the base year for constant series data.

The cut-off date for all data is February 14, 1989.

INTRODUCTION

This edition of *World Tables* (*WT*) updates *core* socio-economic indicators given on country pages in the 1987 edition; it adds topical pages and a related explanatory text. This edition covers 137 Bank members plus Switzerland, adding 12 economies not previously covered but dropping Burma due to data inadequacies.

WT disseminates, with little delay, country estimates used by the Bank in its dialogue about economic and social trends in developing countries, which emphasizes Bank borrowers. To make it a more useful resource, other economies are covered, to the extent that they provide internationally comparable measures in readily usable form; these are not subject to detailed scrutiny by Bank staff. Data for industrial market economies are based on reports of the Organisation for Economic Co-operation and Development (OECD) and the International Monetary Fund (IMF). National publications are used for the remaining Bank members. Such sources, generally specified in *WT* Country Notes, should be consulted by readers wishing to be assured of the most timely and complete reports.

Country time series are also used by the Bank to measure trends in groups of countries, notably for the Bank's *World Development Report* (*WDR*) and *World Development Indicators* (*WDI*). *WT* topical pages fill a gap between *WT* country pages and measures for analytical groupings of countries given in *WDR* and *WDI*: the derivation of group indicators from country time series is more evident in the form of *WT* topical pages, and is explained more fully, below.

WT is issued annually, in machine-readable as well as book form; this edition will also be available to the public on diskette. The tables report annual time series, 1967-87 in book and diskette form, but extended back to 1950, where possible, on the computer tape. Diskettes and Tapes may be obtained by contacting the Publications Sales Unit, The World Bank, Washington D.C., 20433; (202) 473-2943. Books may be obtained by contacting World Bank Publications, P.O. Box 7247-8619, Philadelphia, PA 19170-8619 USA; or call (201) 225-2165.

TOPICAL PAGES

Like country pages, topical pages cover two double-page spreads and report annual data for 1967-87. Topical pages recast some indicators from country pages as a global backdrop for analyzing how low- and middle-income economies have fared in recent decades. These are subdivided first by geographic region, and in a sub-table, by some of the analytical groupings used in *WDI*. The last few lines of each sub-table indicate trends for high-income economies.

Since only sparse data are available on the selected topics for the countries referred to as *nonreporting nonmembers* in *WDI*, this category is omitted from the sub-table on *WT* topical pages, and no attempt is made to report global indicators. With this exception, all economies of the world are covered by indicators in the sub-tables, including those not actually selected for separate presentation on the topical pages.

Simple addition is used when a variable is expressed in reasonably comparable units of account, say *imports in current dollars*. Indicators that do not seem naturally additive, say *quantities of diverse imports*, are usually combined by a price-weighting scheme, say *dollar values of the diverse imports in 1980*. Fixed (Laspeyers) weights are normally used for quantities, with moving (Paasche) weights for prices to fulfill the expectation that Value = Price x Quantity. The identity also means that one part of the identity can be derived from the other two; the one derived being described as implicit, e.g., *implicit deflators in national accounts*.

The weighting scheme can be incorporated into the indicator, by scaling the volume indicator to match the base period weight, which means reporting imports in 1980 dollars in this example. The ratio of the sums of imports in current and constant dollars can then provide the implicit deflator for imports. It should be emphasized, however, that use of a single base year raises problems over a period encompassing profound structural changes and significant changes in relative prices, as certainly occurred during 1967-87. The problems exist in indicators for a single country and are simply more apparent when country indicators are agglomerated.

It is debatable whether an analytical ratio, say *imports to GDP*, should be a weighted average of country ratios or a ratio of separately aggregated numerator and denominator. The results will be different, sometimes significantly so. *WT* first estimates the numerator and denominator for the group and then calculates the group ratio.

The *WDI*-style groupings given in *WT* sub-tables are more comprehensive than those given thus far in the *WDI*, although the next *WDI* will adopt *WT* methods. *WT* strives for group time series that retain the same country composition over time and across topics. It does so by permitting group measures to be compiled only if the country data available for a given year account for *at least two-thirds of the full group*, as defined by 1980 benchmarks. So long as that criterion is met, uncurrent reporters (and those not providing ample history) are assumed to behave like the sample of the group that does provide estimates.

The same technique applies to regional groupings in the main part of *WT* topical pages; for these, however, South Africa is omitted from Sub-Saharan Africa while Iran and Iraq are omitted from Middle East and North Africa (since limited data would otherwise prevent calculation of most group measures for recent years).

The benchmarking procedure underlying *WT* group measures requires that some weight be assigned to each economy within the group. For an economy reporting inadequate data in every year, Bank staff must choose some arbitrary 1980 base value within a broad range of plausible estimates. Readers should keep in mind that the purpose is to maintain an appropriate relationship across topics, despite myriad country problems, and that nothing meaningful can be deduced about behavior at the country level by working back from group indicators. In addition, the weighting process may result in discrepancies between summed sub-group figures and overall totals.

CONTRIBUTIONS TO GDP
Several topical pages express sources and uses of gross domestic production (or expenditure) as "contributions" to GDP. The term, contribution, combines information about growth rates and percentage shares of GDP components. This form of presentation, sometimes referred to as "percentage points of GDP," shows by how much GDP would have changed if other GDP components were unchanged.

For example, if agriculture has a 9% growth rate and accounts for a third of GDP, it contributes 2 percentage points to GDP growth. If industry and services remained level, GDP's growth rate would have been 2%. In practice, contributions are obtained by expressing the year-to-year change in a component, for instance, agriculture, as a percentage of GDP in the earlier year.

National accounts in constant 1980 prices are used. Regional aggregation requires that country estimates be expressed in 1980 U.S. dollar terms and then added. Even so, regional aggregation raises an index number problem closely related to the partial rebasing issue arising at the country level (see *Notes on Sources and Methods*), which is resolved in a similar manner. It should be noted, moreover, that regional measures of the contribution to GDP of the resource balance are actually the difference between separately compiled measures for export and imports (of goods and nonfactor services); and regional measures for private consumption are then derived as a residual (GDP less government consumption, investment, and the resource balance).

COVERAGE OF COUNTRY PAGES
Most time series selected for the country pages are concerned with national accounts and international transactions (*foreign trade, balance of payments,* and *external debt*). The Bank uses methodologies and historical data files developed by other international organizations, but adds information obtained by Bank staff directly from national sources, when it can be fitted to internationally agreed classification schemes.

For other data sets (*manufacturing, monetary and fiscal,* and *social indicators*), *WT* recasts a few of the time series made available to the Bank by other international agencies, but readers should refer to publications of the agencies concerned for fuller information. A more diverse selection of indicators, not in time series form, is given in the *WDI.*

WT provides limited time series on external debt, since the Bank's *World Debt Tables* is the authoritative source. Similarly, a wider range of social indicators is available in the Bank's *Social Indicators of Development.* These may also be obtained from the Publication Sales Unit (see above).

BASIC CONCEPTS

As far as is practicable, the *WT* economic indicators conform to the UN *System of National Accounts* (*SNA*), and its social indicators conform to methodologies of UN specialized agencies. In some areas of economic statistics, additional guidelines have been developed that are broadly in line with *SNA* but differ in some respects. For example, the IMF has played a leading role in helping national compilers elaborate balance of payments, monetary, and government finance statistics. While further revision of *SNA*, already under study, may reduce definitional and classification differences, few are resolved in the *WT* amalgam of statistical sources.

Where possible, however, the Bank does harmonize related data sets drawn from diverse sources. This conceptual process, perhaps as much as national estimates obtained by Bank staff, imparts a unique quality to the national accounts and international transactions data reported in *WT*. The Bank has devised for its analytical purposes certain methods and concepts such as partial rebasing, gross domestic income (GDY) and gross national income (GNY). It has also developed a foreign trade data system that offers coherent indicators in nominal and constant US dollars, with related deflators.

The concepts and methods are described in *Sources and Methods* and in the *Glossary of Terms*. It should be emphasized, however, that *WT* notes, like *WT* time series, are selective rather than encyclopedic. Readers interested in more comprehensive and technically precise descriptions should consult the basic references noted in *Sources and Methods*.

STATISTICAL ISSUES

A concerted effort has been made to standardize data and to note exceptions to standards. However, full comparability cannot be ensured and care must be taken in interpreting the indicators. The data are drawn from sources thought to be most authoritative, but many of them are subject to considerable margins of error. This is particularly true for the most recent year or two, since conventional statistical reports take time to digest. Moreover, intercountry and intertemporal comparisons always involve complex technical problems, which have no full and unequivocal solution.

The statistical systems in many developing economies are weak, and this affects the availability and reliability of the data. Readers are urged to take these limitations into account in interpreting the indicators, particularly when making comparisons across economies. *WT* addresses issues of data reliability partly by omitting questionable estimates but also by flagging methodological issues that can influence analysis.

The *Notes* column on country pages flags statistical issues known to affect the data shown. Readers are advised to refer to the *General and Country Notes* sections when using time series with such flags: that is, uppercase letters for general notes, and a lower case *f* to indicate a country-specific footnote.

Country pages for about a dozen Bank members have been omitted for lack of adequate data; some time series on other country pages are blank or uncertain.

Unless otherwise stated, data are reported for calendar years.

SCIENTIFIC NOTATION

For some countries such as Argentina and Bolivia that have experienced rapid inflation, national currency figures and price indexes for early years are very small when shown in current values. Scientific notation is used in these cases. These data are denoted by an *E* followed by a digit indicating the power of 10 by which the number preceding the *E* must be multiplied to restore it to its original form. For example,

$$.63E\text{-}2 = 0.63 \times 10^{-2} = 0.63 \times \frac{1}{100} = 0.0063.$$

For figures in scientific notation, the number of digits shown depends on the level of precision with which data are on file. Hence, a figure of *.1E 0* indicates that no additional digits are maintained, and trailing zeros (to the right of the figure shown) are omitted.

STRUCTURE

The country pages are in alphabetical order with two double pages for each country. A list of countries covered on page 653 also notes the original base years for constant price national accounts prior to partial rebasing (see *Sources and Methods*).

3

SOURCES AND METHODS

The *GNP per capita* figures are calculated according to the *World Bank Atlas* method of converting data in national currency to US dollars. In this method, the conversion factor for any year is the average exchange rate for that year and the two preceding years, adjusted for differences in rates of inflation between the country and the United States. This averaging smooths fluctuations in prices and exchange rates. The resulting estimate of GNP in US dollars is divided by the midyear population to obtain the per capita GNP in current US dollars. The *conversion factor* line on a country page reports the underlying annual observations used for this *Atlas* method. As a rule, it is the official exchange rate reported in the International Monetary Fund's *International Financial Statistics* (IFS)—line *rf*.

Where multiple exchange rate practices are officially maintained, and the spread between rates is analytically significant, a transactions-weighted average is given, if possible, in the *WT*. When the official exchange rate, including any multiple rate, is judged to diverge by an exceptionally large margin from the rate effectively applied to international transactions, a different conversion factor is used by the Bank. Where national compilers used an official exchange rate to assign a national currency value to international transactions, however, that rate must be used to convert the same items to dollars, regardless of whether it was the rate actually applied to international transactions. In these cases, country pages report both the *conversion factor*, underlying the *Atlas* method to convert values of international transactions;, and an *additional conversion factor*, underlying the *Atlas* method to convert the remaining components of GNP.

Population estimates are based on data from the UN Population Division with some adjustments by the World Bank. In many cases, the data take into account the results of recent population censuses. (Note that refugees not permanently settled in the country of asylum are generally considered to be part of the population of their country of origin.)

Statistical background for the estimates is available in the UN annual *Population and Vital Statistics Report* and the Bank's annual *World Population Projections*.

ORIGIN AND USE OF RESOURCES

The time series are based mainly on national sources, as collected by World Bank regional country economists. They generally accord with the *System of National Accounts (SNA)*. Most definitions of indicators given below are those of the UN *SNA, series F, No. 2. revision 3*.

For most countries, GDP by industrial origin is the sum of value added by factors of production measured at their producer prices. When GDP components are available only at producer values, they do not add up to total GDP. When net indirect taxes are added to GDP at producer prices, the result is equal to GDP at market prices.

Data are expressed in national currency units and shown at current prices and constant 1980 prices. To facilitate international comparisons, constant price data series based on years other than 1980 have been partially rebased to the 1980 base. This is accomplished by rescaling, which moves the year in which current and constant price versions of the same time series have the same value, without altering the trend of either. Components of GDP are individually rescaled, and summed up to provide GDP and its sub aggregates. In this process, a rescaling deviation may occur between constant price GDP by industrial origin and constant price GDP by expenditure. Such rescaling deviations are absorbed in private consumption, etc. on the assumption that GDP by industrial origin is a more reliable estimate than GDP by expenditure.

This approach takes into account the effects of changes in intersectoral relative prices between the original and new base period (original base periods are noted in the country list). Because private consumption is calculated as a residual the national accounting identities are maintained. This method of accounting does, however, involve "burying" in private consumption whatever statistical discrepancies arise on the expenditures side in the rebasing process. Large discrepancies are flagged in the *General Notes*.

Partial rebasing requires constant price estimates by industrial origin at the rather aggregated level shown in *WT*. Where less detail is available, the original constant price estimates of GDP are directly rescaled to 1980 prices. This procedure is also used in exceptional cases (see *General Notes*) where partial rebasing would greatly overstate the importance of exports in early years (with a consequential understating of private consumption, derived as a residual).

DOMESTIC PRICES

These data are based on national accounts data,

discussed above. **Overall (GDP)** and **domestic absorption** prices are implicit deflators, that is, they are ratios of current and constant price estimates of relevant aggregates. **Agricultural, industrial,** and **manufacturing** domestic prices are, in principle, price indexes of value added. For most countries, price indexes are implicit deflators derived from volume and value estimates.

MANUFACTURING ACTIVITY

The primary source is the United Nations Industrial Development Organization (UNIDO) database. To improve cross-country comparability, UNIDO has standardized the coverage of establishments to those with five or more employees.

The concepts and definitions are in accordance with the *International Recommendations for Industrial Statistics*, published by the United Nations. The term *employees* refers to two categories defined by the UN: *employees* and *persons engaged*. The term *employees* excludes working proprietors, active business partners, and unpaid family workers, in contrast to the term *persons engaged*, which includes them. Most countries report data on *employees*, but some, as indicated in the *Notes* column, report data on *persons engaged*. Both terms exclude homeworkers. The number of employees or persons engaged usually refers to the average number employed throughout the year.

Real earnings per employee (wages and salaries) cover all payments in cash or kind made by the employer during the year, in connection with work done. The payments include (a) all regular and overtime cash payments, bonuses, and cost of living allowances; (b) wages and salaries paid during vacation and sick leave; (c) taxes and social insurance contributions and the like, payable by the employees and deducted by the employer and (d) payments in kind.

The value of gross **real output per employee** is estimated on the basis of either production or shipments. On the production basis, it consists of (a) the value of all products of the establishment; (b) the value of industrial services rendered to others; (c) the value of goods shipped in the same condition as received; (d) the value of electricity sold; and (e) the net change between the value of work-in-progress at the beginning and the end of the reference period. In the case of estimates compiled on a shipment basis, the net change between the beginning and the end of the reference period in the value of stocks of finished goods is also included. **Value added** is defined as the current value of gross output less the current cost

of (a) materials, fuels, and other supplies consumed; (b) contract and commission work done by others; (c) repair and maintenance work done by others; (d) goods shipped in the same condition as received.

MONETARY HOLDINGS

The primary source is the International Monetary Fund (IMF) *International Financial Statistics* (IFS) database. The concepts and definitions employed are given by the Fund in *A Guide to Money and Banking Statistics in IFS*.

For most countries, the **money supply broadly defined** comprises **money** (IFS line 34) and **quasi-money** (IFS line 35), the normal forms of financial liquidity that economic transactors hold in the monetary system. By definition, holdings of nonresidents and the central goverment are excluded.

In some countries, other (nonmonetary) financial institutions may also incur **quasi-monetary liabilities,** that is, they may issue financial instruments on terms similar to those for **quasi-money.** Where these are significant, **money supply broadly defined** is a measure of liquid liabilities comprising the monetary and **quasi-monetary liabilities** of both monetary and nonmonetary financial institutions.

CENTRAL GOVERNMENT FINANCES

The primary source is the IMF *Government Finance Statistics Yearbook (GFSY). GFSY* data are reported by countries using the system of common definitions and classifications found in the IMF *Manual on Government Finance Statistics (1986)*.

The inadequate statistical coverage of state, provincial, and local governments has dictated the use of central government data only. This may seriously understate or distort the role of government, especially in large countries where lower levels of government are important. A *general note* (E) indicates instances where this is thought to be likely.

Grants, reported separately in *GFSY*, are here included in current revenue; government lending operations (*GFSY* lending minus repayments) are classified as capital expenditure.

FOREIGN TRADE (CUSTOMS BASIS)

Exports and imports cover international movements of goods across customs borders; generally, exports are valued f.o.b. (free on board), and imports c.i.f. (cost, insurance, and freight).

The primary source is the UN trade data system made available through the UN International Computing Center (Geneva). Apart from greater currentness, the UN trade data system accords with the UN *Yearbook of International Trade Statistics,* that is, the data are based on countries' customs returns.

For the most recent years, secondary sources are often used, notably the UN Conference on Trade and Development (UNCTAD) and the IMF. Estimates are also obtained directly from national sources by World Bank staff. Secondary sources are based on aggregated reports from national authorities that become available before the detailed reports submitted to the UN are released.

Exports and imports price indexes for developing countries are World Bank estimates. The indexes are based on international prices for primary commodities and unit value indexes for manufactures. They are aggregated by broad commodity groups for each country, to ensure consistency between data for a group of countries and those for individual countries. For industrial economies, the indexes are from the UN *Yearbook of International Trade Statistics* and *Monthly Bulletin of Statistics,* and the IMF *IFS.*

BALANCE OF PAYMENTS

The primary source is the files producing the IMF *Balance of Payments Statistics Yearbook.* The methodology is described in the Fund's *Balance of Payments Manual* (Fourth Edition). Supplementary data, usually most recent estimates, are obtained from national sources or estimated by World Bank staff.

For **long-term loans,** (net disbursements, gross disbursements, and repayments) data are reported to the World Bank's Debtor Reporting System (DRS). Any difference between the IMF Balance of Payments long-term capital and those in the DRS is shown as **other long-term inflows (net).**

It is not yet possible to reconcile related measures in *WT* sections on balance of payments, foreign trade, and national accounts. This reflects differences in definitions used, timing, recording, and valuation of transactions as well as the nature of basic data sources. Both general and country notes indicate classification and coverage issues that produce discrepancies. The *WT* country pages report related time series on international transactions in each section in order to help readers make their own judgements about how such discrepancy may affect the analytical purpose at hand.

EXTERNAL DEBT, ETC.

The source of debt data is the World Bank DRS, supplemented by World Bank estimates. DRS is concerned solely with developing economies and does not collect data on external debt for other groups of borrowers, nor from economies that are not members of the World Bank. The figures on debt refer to amounts disbursed and outstanding, expressed in US dollars converted at official exchange rates. Total disbursements and total repayments are also reported as separate items of the balance of payments. Valuation adjustments explain differences between the change in debt outstanding and the net movements shown in the balance of payment items.

Data on **international reserves** and **gold holdings** are from the IMF *IFS* data files.

SOCIAL INDICATORS

The primary sources of social indicators data are the data files and publications of specialized international agencies, such as FAO, ILO, Unesco, WHO, and the UN Statistical Office. Supplementary sources are the Population Council, UN Research Institute for Social Development (UNRISD), and World Bank data files. Note that some demographic and labor force indicators are estimated by interpolating census observations.

The index of **food production per capita** shows the average annual quantity of food produced per capita. For this index, food is defined as comprising cereals, starchy roots, sugar cane, sugar beet, pulses, edible oils, nuts, fruits, vegetables, livestock, and livestock products. Quantities of food are measured net of animal feed, seeds for use in agriculture, and food lost in processing and distribution.

The data on **primary school enrollment** are estimates of children of all ages enrolled in primary school. Figures are expressed as the ratio of pupils to the population of children in the country's school age-group. While many countries consider primary school age to be 6-11 years, others do not. For some countries with universal primary education, the gross enrollment ratios may exceed 100 percent because some pupils are younger or older than the country's standard primary-school age.

The data on **secondary school enrollment** are calculated in the same manner and the definition of secondary school age also differs among countries. It is most commonly considered to be 12-17 years.

Many indicators are based on census or household surveys which occur infrequently. Thus some reported figures are interpolated or extrapolated estimates.

GLOSSARY OF TERMS

(in order of appearance on *country pages*)

Current GNP Per Capita
GNP per capita estimates at current purchaser values (market prices), in US dollars, calculated according to *World Bank Atlas* methodology.

Population
Total population – mid-year estimates.

ORIGIN AND USE OF RESOURCES
(current and constant prices)

Gross National Product (GNP)
Gross domestic product (GDP) at purchaser values (market prices) plus net factor income from abroad.

Net Factor Income from Abroad
Includes the net compensation of employees (with less than one year of residence in the host country) and the net property and entrepreneurial income components of the System of National Accounts (SNA). The major components of the latter are investment income and interest on short- and long-term capital.

Gross Domestic Product (GDP)
Gross Domestic Product at purchaser values (market prices) is the sum of GDP at factor cost and indirect taxes, less subsidies.

Indirect Taxes, net
Equals total indirect taxes less subsidies.

GDP at Factor Cost (producer prices)
Derived as the sum of the value added in the Agriculture, Industry, and Services sectors. If the value added of these sectors is calculated at purchaser values (market prices), GDP at factor cost is derived by subtracting the net indirect taxes from the GDP at purchaser values (market prices).

Agriculture (value added)
Comprises agricultural and livestock production and services, hunting, fishing, logging, and forestry.

Industry (value added)
Comprises mining and quarrying; manufacturing; construction; and electricity, gas, and water.

Services (value added)
Includes all service activities, that is, transport, storage, and communications; wholesale and retail trade; banking, insurance, and real estate, ownership of dwellings; public administration and defense; and other services. The *WT* line is called **Services, etc.,** because it includes any statistical discrepancy in the origin of resources.

Resource Balance
Equals exports of goods and nonfactor services less imports of goods and nonfactor services.

Exports/Imports of Goods and Nonfactor Services
Consists of transactions of residents of a given country with the rest of the world, and covers merchandise, transportation, travel, insurance, and other nonfactor services such as government transactions and various fees but excludes dividends, interest, and other investment income receipts or payments, as well as labor income.

Domestic Absorption
Equals private consumption, general government consumption, plus gross domestic investment.

Private Consumption
Equals the market value of all goods and services purchased or received as income in kind by individuals and nonprofit institutions. It excludes purchases of dwellings, but includes the imputed rent of owner-occupied dwellings. The *WT* line is called **Private consumption, etc.** because it includes any statistical discrepancy in the use of resources. At constant prices, it also includes the rescaling deviation from partial rebasing (see *Sources and Methods*.)

General Government Consumption
Equals the sum of (i) purchases, less sales, of consumer goods and services, reduced by the value of the own-account production of fixed assets, (ii) compensation of employees, (iii) consumption of fixed assets, and (iv) any payments of indirect taxes.

Gross Domestic Investment
The sum of gross domestic fixed investment and the change in stocks.

Fixed Investment
Made up of all outlays (purchases and own-account production) of industries, producers of government services, and producers of private nonprofit services on additions of new and imported durable goods to their stocks of fixed assets, reduced by the proceeds of net sales (sales less purchases) of

similar secondhand and scrapped goods. Excluded is the outlay of producers of government services on durable goods primarily for military purposes, which is classified by the UN *SNA* as current consumption.

Gross Domestic Saving (at current prices)
Equals gross domestic product minus total consumption, etc. (or gross domestic investment plus the resource balance). *Gross Domestic Saving* (at constant prices) equals gross domestic income minus total consumption, etc. (or gross domestic investment plus the resource balance plus the terms of trade adjustment).

Gross National Saving (at current prices)
Equals gross domestic saving plus net factor income and net current transfers from abroad.

Capacity to Import
Value of exports of goods and nonfactor services deflated by the import price index.

Terms of Trade Adjustment
Equals capacity to import less exports of goods and nonfactor services in constant prices.

Gross Domestic Income
Derived as the sum of GDP and the terms of trade adjustment.

Gross National Income
Derived as the sum of GNP and the terms of trade adjustment.

DOMESTIC PRICES (DEFLATORS)

Overall (GDP)
The deflator is derived by dividing current price estimates of GDP at purchaser values (market prices) by constant price estimates.. Also called the implicit GDP deflator.

Domestic Absorption
The deflator is derived by dividing current price domestic absorption estimates by constant price estimates.

Agriculture, Industry, Manufacturing
Price indexes are mostly implicit deflators derived from volume and value estimates.

MANUFACTURING ACTIVITY

Employment
The average number of employees or persons engaged during the year. For further information about these classifications, see *Sources and Methods*.

Real Earnings per Employee
Derived by deflating nominal earnings per employee data from UNIDO by the consumer price index (CPI).

Real Output per Employee
Obtained as UNIDO data on gross output per employee in current prices deflated by price indexes of value added in manufacturing, where available, or in industry, where not.

Earnings as % of Value Added
Derived by dividing total nominal earnings of employees by nominal value added, to show labor's share in income generated in the manufacturing sector.

MONETARY HOLDINGS

Money Supply, Broadly Defined
Comprises the monetary and quasi-monetary liabilities of a country's financial institutions to residents other than the central government.

Money as Means of Payment
The sum of currency outside banks plus demand deposits held in the financial system by the rest of the domestic economy, other than central government.

Currency Outside Banks
Comprises bank notes and coin accepted as legal tender in the domestic economy, excluding amounts held by the monetary system, central government, and nonresidents.

Demand Deposits
Deposits payable on demand. Typically comprises accounts transferable by checks, and any alternative instrument, forms, and mechanisms for transferring money.

Quasi-Monetary Liabilities
Comprises time and savings deposits, and similar bank accounts that the issuer will readily exchange for money. Where nonmonetary financial institutions are important issuers of quasi-monetary liabilities, these are also included in the measure of monetary holdings.

CENTRAL GOVERNMENT FINANCES

Government Deficit or Surplus
Defined as the sum of current and capital revenue and all grants received, less the sum of current and

capital expenditure and government lending minus repayments.

Current Revenue
Comprises tax revenue and nontax revenue. Capital receipts are excluded. Tax revenue covers tax on income, profits, social security contributions, taxes on property, domestic taxes on goods and services, etc. Nontax revenue consists of grants, property income and operating surpluses of departmental enterprises, receipts from public enterprises, administrative fees and charges, fines, etc..

Current Expenditure
Expenditure for goods and services, interest payments, and subsidies and other current transfers. Excludes capital payments.

Current Budget Balance
The excess of current revenue over current expenditure.

Capital Receipts
Proceeds from the sale of nonfinancial capital assets, including land, intangible assets, stocks, and fixed capital assets of buildings, construction, and equipment of more than a minimum value and usable for more than one year in the process of production, and receipts of unrequited transfers for capital purposes from nongovernmental sources.

Capital Payments
Expenditure for acquisition of land, intangible assets, government stocks, and nonmilitary and nonfinancial assets; also for capital grants and lending minus repayments.

FOREIGN TRADE (CUSTOMS BASIS)

Value of Exports, fob/ Imports, cif
With some exceptions, cover international movements of goods only across customs borders. Exports are valued fob (free on board), imports cif (cost, insurance, and freight), unless otherwise specified.

Nonfuel Primary Products
Comprises commodities in SITC Revision 1, Sections 0, 1, 2, 4, and 68 (food and live animals, beverages and tobacco, inedible crude materials, oils, fats, waxes, and nonferrous metals).

Fuels
Comprises commodities in SITC Revision 1, Section 3 (mineral fuels and lubricants and related materials).

Manufactures
Comprises commodities in SITC Revision 1, Sections 5 through 9 (chemicals and related products, basic manufactures, machinery and transport equipment, other manufactured articles and goods not elsewhere classified) excluding Division 68 (nonferrous metals).

Terms of Trade
The relative level of export prices compared to import prices, calculated as the ratio of a country's index of average export price to the average import price index.

Export Price, fob/Import Price, cif
Price index measuring changes in the aggregate price level of a country's merchandise exports and imports over time.

Trade at Constant 1980 Prices (exports/imports)
Derived by deflating the current values of exports and imports data by relevant price indexes.

BALANCE OF PAYMENTS

Exports /Imports of Goods and Services
Comprises all transactions involving change of ownership of goods and services between residents of a country and the rest of the world. It includes merchandise, nonfactor services, and factor services.

Merchandise, fob
Comprises the market value of movable goods, including nonmonetary gold. It also includes the market value of related distributive services up to the customs frontier of the exporting economy, that is, fob (free on board) value. The few types of goods that are not covered by the merchandise account include travellers' purchases abroad (which are included in travel) and purchases of goods by diplomatic and military personnel, which are classified under other official goods, services, and income.

Nonfactor Services
Comprises shipment, passenger and other transport services, and travel, as well as current account transactions not separately reported (e.g., not classified as merchandise, nonfactor services, or transfers). These include transactions with nonresidents by government agencies and their personnel abroad, and also transactions by private residents with foreign governments and government personnel stationed in the reporting country.

Factor Services

Comprises services of labor and capital, thus covering income from direct investment abroad, interest, dividends, and property and labor income.

Long-term Interest

Comprises interest on the disbursed portion of outstanding public and private loans repayable in foreign currencies, goods or services. It may include commitment charges on undisbursed loans.

Current Transfers, net

Comprises net transfer payments—between private persons, nonofficial organizations, and governments of the reporting country and the rest of the world—that carry no provisions for repayments. Included are workers' remittances, transfers by migrants, gifts, dowries, and inheritances, alimony and other support remittances, and government grants of real resources and financial items such as subsidies to current budgets, grants of technical assistance, and government contributions to international organizations for administrative expenses.

Workers' Remittances

Comprises remittances of income by those migrants who have come to an economy and who stay or are expected to stay for a year or more, and who are employed by their new economy, where they are considered to be residents.

Totals to be Financed

The sum of the net exports of goods and nonfactor services, net factor service income, and net official private transfers. In some cases, it includes official grants other than those for capital formation.

Official Capital Grants

Comprises unrequited interofficial transfers that are designed to finance capital formation, other forms of accumulation, or long-term expenditure, which is usually nonrecurrent.

Current Account Balance

The sum of the net exports of goods and nonfactor services, net factor service income, and net transfers. Official capital grants are always included.

Long-Term Capital, net

Comprises changes, apart from valuation adjustments in residents' long-term foreign liabilities less their long-term assets, excluding any long-term items classified as reserves.

Direct Investment

Comprises all capital transactions that are made to acquire a lasting interest in an enterprise operating in an economy other than that of the investor, where the investor's purpose is to have an effective voice in the management of the enterprise. Direct investment includes items such as equity capital, reinvestment of earnings, and other long-term and short-term capital.

Long-Term Loans

Comprises all public, publicly guaranteed, and private nonguaranteed loans that have an original or extended maturity of more than a years, and which are repayable in foreign currencies, goods, or services. These data are as reported in the Bank's Debt Reporting System and accord with the stock data on external debt, discussed below.

Disbursements

Comprises the total amounts drawn on public, publicly guaranteed, and private nonguaranteed loans, net of commitment cancellations..

Repayments

Comprises repayments (amortization) of the principal of public, publicly guaranteed, and private nonguaranteed loans.

Other Long-Term Capital

Comprises the difference between long-term as defined above, and the similar item reported in IMF Balance of Payments statistics.

Other Capital, net

Comprises the sum of short-term capital, net errors and omissions, and capital transactions not included elsewhere.

Change in Reserves

Comprises the net change in a country's holdings of international reserves resulting from transactions on the current and capital accounts. These include changes in holdings of monetary gold, SDRs, reserve position in the Fund, foreign exchange assets, and other claims on nonresidents that are available to the central authority. The measure is net of liabilities constituting foreign authorities' reserves, and counterpart items for valuation of monetary gold and SDRs, which are reported separately in IMF sources.

Conversion Factors (Annual Average)

Annual averages of market exchange rates for countries quoting rates in units of national currency per US dollar. (See *Sources and Methods* for details of *additional conversion factors*.)

EXTERNAL DEBT

Public/Publicly Guaranteed Long-Term

Public debt is all external obligations of public debtors, including the national government, its agencies and autonomous public bodies. Publicly guaranteed debt is all external obligations of private debtors that are guaranteed for repayment by a public entity. Long- term debt has an original or extended maturity of more than a year.

Official Creditors

Debt from official creditors comprises loans and credits from the World Bank (IBRD and IDA), regional development banks, and other multilateral and intergovernmental agencies. Excluded are loans from funds administered by an international organization on behalf of a single donor government; such loans are classified as loans from governments.

Private Creditors

Debt from private creditors comprises credits from manufacturers, exporters, or other suppliers of goods; loans from private banks and other private financial institutions as well as publicly issued and privately placed bonds; and external liabilities on account of nationalized properties and unclassified debts to private creditors.

Private Nonguaranteed Long-Term

Contractual obligations of a direct-investment enterprise to a foreign parent company or its affiliate.

Use of Fund Credit

This measures the net use by a member of its conditional drawing rights in the IMF, excluding those resulting from drawings in the reserve tranche and on the IMF Trust Fund.

Short-Term Debt

The sum of public and private external obligations with original or extended maturity of a year or less.

International Reserves Excluding Gold

Comprises a country's monetary authorities' (central banks, currency boards, exchange stabilization funds, and treasuries) holdings of SDRs, reserve position in the Fund, and foreign exchange.

Gold Holdings (at Market Prices)

Official holdings of gold (fine troy ounces) valued at end-year London market prices.

SOCIAL INDICATORS

Total Fertility Rate

The average number of children that would be born alive to a woman during her lifetime if she were to bear children at each age in accordance with prevailing age-specific fertility rates.

Crude Birth Rate

Annual live births per thousand population.

Infant Mortality Rate

Number of infants per thousand live births, in a given year, who die before reaching one year of age.

Life Expectancy at Birth

Number of years a newborn infant would live if prevailing patterns of mortality for all people at the time of his or her birth were to stay the same throughout his or her life.

Food Production per capita

Index of annual production of all food commodities adjusted for population growth. Production excludes animal feed and seed for agriculture.

Labor Force, Agriculture (%)

Labor force in farming, forestry, hunting, and fishing as a percentage of total labor force, which comprises economically active persons, including armed forces and unemployed but excluding housewives and students.

Labor Force, Female (%)

Female labor force as a percentage of total labor force.

School Enrollment Ratio, Primary

Gross enrollment of all ages at primary level as a percentage of children in the country's primary school age-group.

School Enrollment Ratio, Secondary

Computed in a similar manner, but includes pupils enrolled in vocational, or teacher-training secondary schools.

Topical Pages

Table 1. Gross national product per capita

US dollars, Atlas methodology	1967	1968	1969	1970	1971	1972	1973	1974	1975	1976	1977
SUB-SAHARAN AFRICA	*120*	*120*	*130*	*140*	*160*	*180*	*200*	*250*	*280*	*340*	*380*
Excluding Nigeria	*130*	*130*	*140*	*140*	*150*	*160*	*170*	*210*	*240*	*260*	*280*
Benin	100	110	110	110	100	110	130	150	170	180	200
Botswana	100	100	120	130	150	180	240	340	350	400	420
Burkina Faso	60	60	60	60	60	60	70	90	100	110	120
Burundi	50	50	50	70	80	70	80	90	100	120	140
Cameroon	130	150	170	180	180	180	210	270	310	360	410
Cape Verde	..	..	..	..	..	..	..	..	270	320	360
Central African Republic	90	100	110	100	100	110	120	150	170	200	230
Chad	90	90	100	100	90	100	100	120	150	150	160
Comoros	100	90	90	100	110	120	140	150	170	190	200
Congo, People's Republic	200	220	240	240	260	290	340	430	520	510	500
Cote d'Ivoire	210	240	260	270	270	280	320	390	500	580	670
Ethiopia	50	60	60	60	70	70	70	80	90	90	90
Gabon	480	520	610	670	690	780	890	1,710	2,660	4,170	3,750
Gambia, The	110	100	110	100	120	130	140	200	210	230	250
Ghana	240	220	230	250	270	260	260	290	280	280	300
Guinea-Bissau	..	..	..	..	..	160	160	170	190	170	160
Kenya	120	120	130	130	160	180	180	210	230	240	270
Lesotho	80	80	80	100	110	110	150	200	230	260	290
Liberia	270	280	290	310	320	320	320	370	410	470	500
Madagascar	120	130	130	140	140	140	160	190	230	230	240
Malawi	70	60	60	60	80	80	90	110	120	120	130
Mali	..	..	60	70	70	80	80	90	120	140	170
Mauritania	160	170	160	180	170	190	200	260	300	350	360
Mauritius	300	280	280	280	300	340	430	590	720	860	930
Mozambique	..	..	..	..	..	..	..	..	..	..	..
Niger	180	170	160	160	170	160	150	200	230	220	240
Nigeria	90	90	110	150	190	230	260	360	430	590	700
Rwanda	50	50	60	60	60	60	70	70	90	120	160
Senegal	200	210	200	210	200	220	230	270	340	390	380
Seychelles	340	350	340	350	410	470	580	710	800	840	980
Sierra Leone	150	140	150	160	160	160	170	190	220	210	220
Somalia	90	90	90	90	100	110	120	110	160	180	220
Sudan	110	120	120	130	150	150	150	190	250	320	380
Swaziland	190	200	260	270	320	330	370	460	590	580	590
Tanzania	..	..	..	..	..	..	..	..	..	..	..
Togo	130	130	140	140	140	150	170	210	260	270	310
Uganda	160	160	180	190	210	210	210	220	220	270	280
Zaire	170	170	180	180	200	200	240	270	300	300	330
Zambia	320	360	400	450	430	420	420	520	550	540	500
Zimbabwe	250	240	270	280	320	370	420	530	570	560	540
SOUTH ASIA	*90*	*90*	*100*	*110*	*110*	*110*	*120*	*130*	*150*	*150*	*160*
Bangladesh	80	90	100	100	100	70	80	90	130	130	110
India	90	90	90	100	100	110	120	140	160	160	160
Nepal	80	80	80	80	80	80	80	100	110	120	120
Pakistan	130	140	150	170	170	160	130	130	140	170	190
Sri Lanka	170	170	170	170	170	160	170	190	220	230	220
EAST ASIA AND PACIFIC	*100*	*100*	*120*	*140*	*140*	*140*	*170*	*200*	*230*	*240*	*270*
China	100	90	100	120	120	130	150	160	180	170	180
Fiji	340	350	350	400	440	520	660	850	1,030	1,190	1,290
Hong Kong	680	700	800	900	1,020	1,220	1,560	1,890	2,160	2,700	3,200
Indonesia	50	60	70	90	90	100	120	150	210	280	330
Korea, Republic of	140	170	220	260	300	210	460	570	580	750	910
Malaysia	350	370	380	390	410	450	550	700	820	920	1,000
Papua New Guinea	190	200	230	260	280	310	370	460	530	530	550
Philippines	210	230	250	230	220	220	250	300	360	420	450
Singapore	660	740	840	950	1,080	1,270	1,580	2,020	2,540	2,760	2,940
Solomon Islands	..	..	..	..	..	..	..	..	270	260	300	320
Thailand	170	190	200	210	210	210	250	300	360	410	460
Vanuatu	..	..	..	..	..	..	..	..	..	..	..

NOTE: See glossary for definitions.

1978	1979	1980	1981	1982	1983	1984	1985	1986	1987	
400	*470*	*540*	*570*	*570*	*510*	*470*	*470*	*410*	*330*	*SUB-SAHARAN AFRICA*
300	*340*	*390*	*410*	*390*	*360*	*330*	*310*	*320*	*330*	*Excluding Nigeria*
210	260	320	350	330	280	260	260	270	300	Benin
490	600	790	950	1,000	1,040	980	870	910	1,050	Botswana
140	170	200	210	200	170	150	150	170	200	Burkina Faso
150	170	200	250	240	230	230	240	240	250	Burundi
500	600	760	900	900	870	820	840	920	960	Cameroon
360	380	500	460	450	430	430	420	460	500	Cape Verde
260	290	320	330	340	290	280	280	290	330	Central African Republic
170	150	160	160	150	130	110	140	140	150	Chad
210	260	340	380	350	330	310	300	320	370	Comoros
540	650	880	1,190	1,300	1,210	1,170	1,070	930	870	Congo, People's Republic
820	980	1,170	1,130	960	760	660	630	700	750	Cote d'Ivoire
100	110	120	120	120	120	110	110	120	130	Ethiopia
2,770	3,110	3,830	4,260	4,410	3,950	3,440	3,330	3,190	2,760	Gabon
260	340	380	340	320	290	250	230	210	220	Gambia, The
350	370	400	400	370	340	360	370	390	390	Ghana
170	170	140	170	190	190	180	180	170	160	Guinea-Bissau
310	360	410	420	390	340	310	300	310	330	Kenya
330	360	420	480	540	510	490	410	360	370	Lesotho
540	590	590	600	550	500	470	480	460	450	Liberia
250	300	350	330	320	300	270	240	230	210	Madagascar
160	170	180	180	190	190	180	170	160	170	Malawi
170	210	240	250	230	180	160	150	180	210	Mali
360	400	440	480	450	460	420	410	420	440	Mauritania
1,030	1,210	1,190	1,240	1,230	1,140	1,080	1,100	1,230	1,490	Mauritius
..	..	..	..	190	160	170	180	220	170	Mozambique
300	360	430	440	380	320	250	240	250	260	Niger
720	850	1,020	1,090	1,110	980	900	950	700	370	Nigeria
180	210	240	270	260	270	260	280	300	300	Rwanda
370	450	490	470	490	430	370	370	410	510	Senegal
1,200	1,610	2,070	2,310	2,400	2,330	2,340	2,560	2,740	3,170	Seychelles
230	270	320	370	380	390	370	350	320	300	Sierra Leone
260	260	200	220	240	250	250	270	280	290	Somalia
410	430	430	430	400	380	320	280	300	330	Sudan
600	720	820	1,010	1,010	930	880	740	690	700	Swaziland
..	..	..	..	..	..	..	..	..	..	Tanzania
360	360	430	400	340	270	240	230	250	290	Togo
280	290	280	220	240	220	220	230	230	260	Uganda
360	410	430	400	340	290	200	160	150	150	Zaire
500	510	600	720	660	570	470	410	270	250	Zambia
530	580	710	860	900	840	730	640	590	590	Zimbabwe
180	*190*	*220*	*260*	*270*	*270*	*270*	*270*	*280*	*290*	*SOUTH ASIA*
100	120	140	160	160	150	140	150	160	160	Bangladesh
180	190	230	270	280	280	280	290	290	300	India
120	130	140	160	160	160	170	170	170	160	Nepal
220	250	290	330	350	350	350	340	350	350	Pakistan
220	230	260	300	320	330	360	380	390	400	Sri Lanka
320	*380*	*440*	*480*	*500*	*500*	*510*	*510*	*500*	*520*	*EAST ASIA AND PACIFIC*
210	250	290	310	310	310	320	320	300	290	China
1,340	1,580	1,750	2,000	1,870	1,700	1,770	1,630	1,730	1,570	Fiji
3,750	4,260	5,210	6,110	6,390	6,230	6,450	6,230	7,030	8,260	Hong Kong
380	410	480	560	620	580	560	520	500	450	Indonesia
1,190	1,510	1,620	1,830	1,880	2,020	2,120	2,160	2,370	2,690	Korea, Republic of
1,140	1,400	1,680	1,890	1,910	1,900	2,040	1,970	1,830	1,810	Malaysia
660	720	780	810	800	760	710	710	710	740	Papua New Guinea
500	590	680	770	800	750	640	570	560	590	Philippines
3,310	3,880	4,550	5,450	6,160	6,930	7,730	7,620	7,450	7,940	Singapore
350	450	440	550	570	570	560	510	530	420	Solomon Islands
530	590	670	750	780	810	840	810	800	840	Thailand
..	..	..	..	..	..	..	960	..	..	Vanuatu

Table 1. Gross national product per capita (cont'd)

US dollars, Atlas methodology	1967	1968	1969	1970	1971	1972	1973	1974	1975	1976	1977
LATIN AMERICA AND CARIBBEAN	*460*	*500*	*520*	*570*	*620*	*670*	*780*	*960*	*1,120*	*1,260*	*1,350*
Argentina	910	900	900	910	1,000	1,040	1,240	1,630	1,810	1,680	1,610
Bahamas	2,240	2,480	2,780	2,700	2,900	2,900	3,260	2,950	2,730	3,000	3,330
Barbados	590	630	680	750	830	900	1,040	1,220	1,520	1,740	1,960
Belize	420	430	430	440	480	530	590	670	800	850	870
Bolivia	210	220	230	230	250	270	290	320	360	400	410
Brazil	310	370	410	450	500	570	700	880	1,070	1,300	1,470
Chile	770	790	820	850	1,000	1,090	1,090	1,160	870	890	1,040
Colombia	310	310	320	340	360	390	430	510	560	620	710
Costa Rica	430	470	500	560	600	650	750	860	950	1,070	1,320
Dominican Republic	260	260	290	320	330	370	440	510	600	690	770
Ecuador	240	260	280	290	290	300	360	430	550	680	790
El Salvador	280	280	290	290	300	310	330	380	440	500	590
Grenada	..	..	..	..	..	..	..	..	..	..	..
Guatemala	300	330	340	360	380	390	430	500	570	660	790
Guyana	..	..	..	380	420	400	410	500	640	660	620
Haiti	80	80	90	90	100	100	110	130	150	170	180
Honduras	240	260	260	270	280	290	310	330	360	400	460
Jamaica	580	620	650	720	760	900	950	1,080	1,260	1,330	1,450
Mexico	530	580	620	710	750	820	940	1,120	1,360	1,500	1,490
Nicaragua	350	350	370	380	390	390	430	550	650	710	750
Panama	560	590	640	680	740	770	840	920	1,030	1,080	1,120
Paraguay	230	240	240	260	270	300	350	440	550	620	720
Peru	490	480	490	520	580	620	700	850	1,000	1,040	970
St. Vincent and the Grenadines	180	190	200	220	230	290	290	300	340	370	380
Trinidad and Tobago	780	790	810	830	850	970	1,110	1,280	2,000	2,470	2,760
Uruguay	470	550	650	780	910	910	990	1,140	1,370	1,440	1,450
Venezuela	1,060	1,140	1,160	1,240	1,270	1,340	1,550	1,920	2,380	2,900	3,160
MIDDLE EAST AND NORTH AFRICA	*210*	*220*	*250*	*260*	*280*	*320*	*350*	*420*	*520*	*620*	*680*
Algeria	260	300	330	360	350	460	530	680	880	1,070	1,190
Egypt, Arab Republic of	180	190	210	230	240	250	270	270	310	350	390
Jordan	..	..	..	..	..	370	380	410	530	690	780
Morocco	220	240	250	260	280	300	340	410	500	570	610
Oman	170	250	310	360	370	390	390	550	1,280	2,850	2,870
Syrian Arab Republic	270	290	370	350	390	460	460	650	890	1,020	1,010
Tunisia	220	240	260	280	320	390	430	560	710	800	840
Yemen Arab Republic	..	..	..	..	..	70	90	120	140	180	220
Yemen, PDR	..	..	..	..	..	..	..	..	..	..	330
EUROPE AND MEDITERRANEAN	..	..	..	..	..	..	..	..	..	..	*1,480*
Cyprus	..	..	..	..	..	..	..	..	..	..	..
Greece	840	930	1,060	1,170	1,300	1,460	1,720	1,940	2,370	2,660	2,860
Hungary	..	..	..	..	..	..	..	..	..	..	1,310
Israel	1,430	1,630	1,770	1,830	2,100	2,450	2,680	3,310	3,890	4,080	4,080
Malta	610	650	680	760	820	910	1,030	1,220	1,540	1,730	1,930
Poland	..	..	..	..	..	..	..	..	..	..	..
Portugal	500	570	610	700	790	910	1,150	1,380	1,480	1,650	1,780
Turkey	330	370	400	400	400	410	470	630	830	1,000	1,110
Yugoslavia	520	510	600	650	740	780	870	1,150	1,380	1,620	1,970
Developing countries	170	170	190	210	220	230	270	320	380	430	470
Highly indebted	380	400	430	470	510	560	650	810	950	1,090	1,180
Low-income countries	100	90	100	110	120	120	130	150	170	160	180
Low-income Africa	120	120	130	130	140	150	160	180	200	220	240
China and India	90	90	100	110	120	120	130	150	170	160	170
Other low-income Asia	100	110	120	120	120	110	100	110	130	150	150
Middle-income countries	300	320	350	380	420	450	540	670	810	960	1,060
High-income oil exporters	1,000	1,130	1,240	1,270	1,370	1,510	1,800	2,700	4,220	7,300	8,480
Industrialized countries	2,530	2,770	3,010	3,200	3,480	3,890	4,540	5,170	5,810	6,460	7,140
Japan	1,190	1,430	1,680	1,930	2,130	2,540	3,230	3,820	4,490	4,970	5,690
United States	4,080	4,450	4,780	4,970	5,320	5,780	6,410	6,890	7,400	8,180	9,040
European Community	1,900	2,050	2,220	2,390	2,630	2,990	3,620	4,290	4,930	5,460	6,020

NOTE: See glossary for definitions.

1978	1979	1980	1981	1982	1983	1984	1985	1986	1987	
1,470	*1,680*	*1,930*	*2,150*	*2,120*	*1,890*	*1,790*	*1,750*	*1,740*	*1,780*	*LATIN AMERICA AND CARIBBEAN*
1,570	1,790	1,960	1,930	1,860	1,950	2,140	2,120	2,360	2,360	Argentina
3,680	5,070	5,790	5,660	6,620	7,170	8,300	9,000	9,550	10,320	Bahamas
2,250	2,670	3,130	3,520	3,760	4,010	4,410	4,640	5,140	5,350	Barbados
920	1,090	1,190	1,250	1,160	1,070	1,100	1,120	1,200	1,250	Belize
440	460	490	550	530	490	490	490	510	570	Bolivia
1,660	1,890	2,080	2,060	2,060	1,820	1,710	1,670	1,840	2,020	Brazil
1,340	1,690	2,110	2,610	2,230	1,930	1,700	1,450	1,310	1,310	Chile
870	1,050	1,220	1,370	1,440	1,440	1,430	1,350	1,280	1,230	Colombia
1,560	1,790	1,950	1,530	1,140	1,060	1,260	1,400	1,500	1,550	Costa Rica
810	890	1,000	1,120	1,170	1,130	950	760	710	730	Dominican Republic
930	1,080	1,260	1,490	1,490	1,310	1,150	1,150	1,130	1,040	Ecuador
700	760	740	740	730	760	810	830	820	850	El Salvador
..	..	740	870	980	1,030	1,080	1,170	1,240	1,340	Grenada
910	1,030	1,120	1,200	1,190	1,170	1,180	1,190	1,050	950	Guatemala
610	650	720	730	620	550	500	480	470	380	Guyana
200	220	250	270	270	270	290	310	340	360	Haiti
530	600	640	690	690	700	720	740	770	810	Honduras
1,420	1,280	1,160	1,250	1,310	1,370	1,150	920	870	960	Jamaica
1,580	1,820	2,320	3,000	2,820	2,290	2,130	2,200	1,900	1,820	Mexico
730	580	690	760	800	820	790	770	730	830	Nicaragua
1,290	1,420	1,680	1,880	2,000	1,980	2,010	2,100	2,180	2,240	Panama
830	1,030	1,350	1,680	1,730	1,660	1,500	1,190	1,000	1,000	Paraguay
840	850	990	1,240	1,380	1,140	1,110	1,010	1,150	1,430	Peru
430	480	540	640	740	830	900	970	1,030	1,070	St. Vincent and the Grenadines
3,350	3,690	4,690	5,980	6,970	6,450	6,060	6,180	5,360	4,220	Trinidad and Tobago
1,660	2,120	2,810	3,600	3,420	2,460	1,960	1,730	1,920	2,160	Uruguay
3,390	3,730	4,070	4,730	4,920	4,790	4,140	3,800	3,550	3,230	Venezuela
740	*860*	*990*	*1,120*	*1,190*	*1,170*	*1,170*	*1,190*	*1,200*	*1,220*	*MIDDLE EAST AND NORTH AFRICA*
1,380	1,680	1,960	2,250	2,400	2,430	2,420	2,550	2,590	2,680	Algeria
390	410	480	530	580	590	610	640	650	670	Egypt, Arab Republic of
950	1,090	1,430	1,640	1,720	1,630	1,580	1,530	1,550	1,560	Jordan
650	760	880	840	820	710	620	560	580	620	Morocco
2,800	3,150	3,140	6,120	7,560	7,090	7,230	7,510	6,510	5,830	Oman
1,120	1,240	1,460	1,660	1,670	1,640	1,580	1,610	1,630	1,640	Syrian Arab Republic
930	1,080	1,290	1,390	1,320	1,250	1,220	1,190	1,130	1,210	Tunisia
290	360	420	470	640	670	690	670	630	590	Yemen Arab Republic
370	410	420	490	510	500	540	490	420	420	Yemen, PDR
1,670	*1,920*	*2,100*	*2,240*	*2,190*	*2,110*	*2,040*	*1,970*	*2,040*	*2,150*	*EUROPE AND MEDITERRANEAN*
..	..	..	..	..	..	4,070	4,110	4,450	5,200	Cyprus
3,270	3,870	4,370	4,470	4,330	3,970	3,810	3,620	3,670	4,020	Greece
1,500	1,680	1,930	2,140	2,270	2,160	2,060	1,940	2,020	2,240	Hungary
4,250	4,730	5,320	6,090	6,250	6,480	6,390	6,210	6,220	6,810	Israel
2,170	2,600	3,150	3,590	3,800	3,480	3,360	3,250	3,440	4,020	Malta
..	..	..	..	1,570	1,830	2,100	2,120	2,080	1,920	Poland
1,850	2,050	2,350	2,470	2,480	2,230	1,990	1,970	2,250	2,810	Portugal
1,210	1,370	1,400	1,450	1,300	1,180	1,100	1,080	1,110	1,200	Turkey
2,400	2,870	3,250	3,450	3,230	2,640	2,270	2,060	2,300	2,480	Yugoslavia
520	600	680	750	770	740	720	720	710	720	Developing countries
1,300	1,490	1,720	1,890	1,860	1,640	1,520	1,480	1,440	1,430	Highly indebted
200	220	260	280	290	290	290	290	280	290	Low-income countries
260	290	320	330	310	290	260	250	250	260	Low-income Africa
200	220	270	290	300	300	300	310	290	300	China and India
160	180	200	230	240	240	240	240	250	250	Other low-income Asia
1,160	1,340	1,500	1,670	1,700	1,620	1,570	1,540	1,540	1,570	Middle-income countries
9,010	10,520	12,390	15,090	15,540	13,470	11,510	9,850	8,290	..	High-income oil exporters
8,100	9,400	10,650	11,440	11,380	11,350	11,730	12,100	12,990	14,670	Industrialized countries
7,020	8,620	9,870	10,390	10,280	10,320	10,580	11,270	12,840	15,760	Japan
10,100	11,150	12,000	13,270	13,620	14,510	15,910	16,800	17,560	18,530	United States
6,790	8,220	9,760	10,240	9,860	9,090	8,720	8,590	9,390	11,210	European Community

Table 2. Gross national income per capita

1980 US dollars	1967	1968	1969	1970	1971	1972	1973	1974	1975	1976	1977
SUB-SAHARAN AFRICA	*430*	*420*	*460*	*490*	*510*	*510*	*530*	*590*	*540*	*560*	*570*
Excluding Nigeria	*400*	*400*	*410*	*410*	*410*	*400*	*410*	*420*	*380*	*380*	*390*
Benin	350	340	360	350	340	340	360	340	310	310	320
Botswana	270	310	340	370	380	420	530	660	570	630	620
Burkina Faso	160	170	180	180	180	170	170	190	180	200	190
Burundi	170	170	160	220	210	200	210	200	190	220	260
Cameroon	500	540	560	580	560	530	540	600	570	570	640
Cape Verde	..	..	..	..	..	..	220	240	230	260	320
Central African Republic	320	370	390	390	390	380	380	370	330	380	410
Chad	270	260	270	270	260	270	240	260	250	250	240
Comoros	..	..	..	..	..	..	..	..	..	..	..
Congo, People's Republic	710	700	710	710	750	720	850	1,030	870	850	720
Cote d'Ivoire	810	910	950	970	990	990	1,020	1,080	1,090	1,170	1,260
Ethiopia	100	100	100	110	110	110	110	110	100	110	110
Gabon	..	..	..	..	..	..	..	..	..	..	..
Gambia, The	330	300	340	330	350	360	350	420	400	390	430
Ghana	440	420	450	470	470	450	460	480	410	380	400
Guinea-Bissau	..	..	..	210	190	200	190	190	200	180	160
Kenya	320	330	350	330	370	420	410	400	370	380	440
Lesotho	220	240	240	240	250	250	320	350	380	420	480
Liberia	800	890	1,000	760	800	790	700	670	700	740	700
Madagascar	440	450	480	490	470	460	440	440	420	400	400
Malawi	160	150	160	160	190	190	190	200	210	200	210
Mali	210	210	210	210	210	210	200	190	210	240	250
Mauritania	450	480	440	490	470	480	440	530	500	540	470
Mauritius	810	830	680	730	750	850	890	1,260	1,330	1,270	1,290
Mozambique	..	..	..	..	..	..	..	..	..	..	..
Niger	540	530	490	490	510	460	360	410	330	340	370
Nigeria	550	530	620	770	860	860	900	1,130	1,050	1,130	1,150
Rwanda	170	180	180	200	200	200	190	190	190	190	200
Senegal	600	570	510	560	550	570	520	540	560	580	570
Seychelles	..	..	..	..	..	..	..	..	..	..	..
Sierra Leone	310	280	300	330	330	300	310	350	310	280	290
Somalia	130	130	130	120	130	130	120	100	130	120	160
Sudan	350	350	330	330	340	320	280	300	330	380	430
Swaziland	710	740	900	900	970	990	1,000	1,040	1,100	1,060	1,000
Tanzania	280	290	290	300	300	300	300	300	290	300	300
Togo	330	330	350	350	350	350	350	460	360	400	430
Uganda	220	220	230	230	230	220	220	210	200	200	200
Zaire	540	500	550	530	530	510	570	550	460	430	430
Zambia	1,280	1,310	1,740	1,390	1,070	1,110	1,190	1,170	760	810	710
Zimbabwe	..	..	..	..	..	..	..	..	..	800	730
SOUTH ASIA	*210*	*210*	*220*	*230*	*220*	*220*	*220*	*210*	*220*	*220*	*240*
Bangladesh	150	160	160	160	140	120	120	130	130	140	140
India	220	220	230	240	240	230	230	220	240	240	260
Nepal	..	..	..	..	..	..	..	..	..	..	..
Pakistan	220	230	240	260	240	230	240	250	240	240	250
Sri Lanka	180	200	200	210	210	200	210	210	210	230	260
EAST ASIA AND PACIFIC	*220*	*220*	*250*	*280*	*290*	*280*	*320*	*320*	*340*	*340*	*370*
China	160	150	170	200	210	210	220	220	240	220	240
Fiji	1,030	1,070	1,100	1,220	1,250	1,330	1,470	1,670	1,760	1,740	1,810
Hong Kong	2,250	2,300	2,520	2,730	2,960	3,250	3,530	3,380	3,400	4,040	4,430
Indonesia	190	220	230	240	250	260	290	330	340	350	380
Korea, Republic of	700	760	840	890	950	990	1,120	1,170	1,210	1,390	1,530
Malaysia	880	870	860	930	920	940	1,110	1,180	1,090	1,270	1,370
Papua New Guinea	630	640	660	690	730	740	760	850	810	720	760
Philippines	480	500	510	530	540	550	600	610	610	620	650
Singapore	..	..	..	..	..	..	..	..	..	..	..
Solomon Islands	..	..	..	..	..	..	..	..	..	..	..
Thailand	430	450	480	490	480	490	560	550	550	580	610
Vanuatu	..	..	..	..	..	..	..	..	..	..	..

NOTE: See glossary for definitions.

1978	1979	1980	1981	1982	1983	1984	1985	1986	1987	
550	*540*	*570*	*560*	*540*	*510*	*480*	*490*	*470*	*440*	*SUB-SAHARAN AFRICA*
380	*380*	*380*	*370*	*370*	*360*	*350*	*350*	*350*	*340*	*Excluding Nigeria*
320	320	330	330	340	330	320	340	320	310	Benin
630	780	940	780	850	860	860	1,030	1,180	..	Botswana
190	200	210	210	210	210	210	220	240	240	Burkina Faso
220	230	220	230	220	220	220	220	240	210	Burundi
680	710	780	850	880	900	930	1,060	1,050	950	Cameroon
320	340	450	400	370	390	400	420	..	..	Cape Verde
400	380	350	330	350	310	330	330	340	320	Central African Republic
240	180	160	160	160	170	160	210	190	190	Chad
..	..	..	..	..	..	..	..	..	..	Comoros
690	760	990	1,210	1,420	1,410	1,490	1,410	910	..	Congo, People's Republic
1,290	1,220	1,200	1,130	1,090	1,030	990	1,010	1,050	940	Cote d'Ivoire
100	110	110	110	110	110	100	100	100	..	Ethiopia
..	..	..	..	..	..	..	..	..	..	Gabon
390	390	380	350	350	370	..	..	..	..	Gambia, The
430	410	410	380	340	320	340	340	340	340	Ghana
170	170	130	160	160	150	160	160	160	160	Guinea-Bissau
420	420	410	400	380	370	370	360	370	360	Kenya
500	480	480	480	520	490	..	..	..	..	Lesotho
670	660	590	540	470	460	450	450	430	..	Liberia
370	380	370	320	310	300	300	290	280	270	Madagascar
220	200	190	180	180	180	190	180	170	200	Malawi
240	260	250	250	260	250	250	240	270	270	Mali
440	430	430	470	430	440	420	450	480	470	Mauritania
1,320	1,340	1,160	1,120	1,140	1,180	1,210	1,280	1,510	1,770	Mauritius
..	..	200	190	170	150	150	130	130	..	Mozambique
400	460	450	460	450	440	370	370	390	360	Niger
1,090	1,060	1,170	1,160	1,090	980	880	910	840	790	Nigeria
200	220	230	230	230	240	230	230	230	210	Rwanda
520	540	500	490	550	540	500	500	510	520	Senegal
..	..	..	..	..	..	..	..	..	..	Seychelles
310	330	330	350	330	320	320	300	290	290	Sierra Leone
160	150	140	160	160	140	120	140	130	140	Somalia
410	360	350	340	360	360	330	290	280	280	Sudan
930	990	950	1,000	840	660	750	..	..	..	Swaziland
290	280	270	260	250	240	240	240	240	240	Tanzania
430	410	430	380	360	320	320	330	310	310	Togo
180	150	140	..	..	..	..	..	..	..	Uganda
390	390	380	370	330	320	300	300	290	290	Zaire
630	680	640	620	510	510	510	520	410	430	Zambia
690	690	750	810	800	780	740	750	690	660	Zimbabwe
250	*230*	*240*	*250*	*250*	*270*	*270*	*280*	*280*	*290*	*SOUTH ASIA*
140	140	140	150	140	150	150	150	150	160	Bangladesh
270	250	250	260	270	280	280	300	300	300	India
..	..	..	..	..	..	..	..	..	..	Nepal
260	270	280	290	290	300	310	320	330	340	Pakistan
270	270	270	270	280	300	330	310	320	320	Sri Lanka
400	*430*	*440*	*460*	*480*	*520*	*570*	*600*	*630*	*680*	*EAST ASIA AND PACIFIC*
260	280	290	300	330	360	410	450	470	510	China
1,790	1,920	1,870	1,860	1,680	1,570	1,670	1,560	1,660	1,600	Fiji
4,760	4,930	5,470	5,820	6,010	6,270	6,900	6,990	7,470	8,320	Hong Kong
400	440	510	560	550	550	570	570	570	580	Indonesia
1,680	1,760	1,580	1,650	1,730	1,900	2,050	2,120	2,430	2,750	Korea, Republic of
1,420	1,580	1,720	1,710	1,720	1,770	1,920	1,770	1,540	1,670	Malaysia
760	780	820	690	670	710	670	690	680	690	Papua New Guinea
670	710	710	710	690	710	640	580	610	630	Philippines
..	..	..	..	..	..	..	..	..	..	Singapore
..	..	510	480	460	480	580	500	460	..	Solomon Islands
650	670	680	690	690	740	770	760	790	830	Thailand
..	..	..	..	..	..	..	..	..	..	Vanuatu

Table 2. Gross national income per capita (cont'd)

1980 US dollars	1967	1968	1969	1970	1971	1972	1973	1974	1975	1976	1977
LATIN AMERICA AND CARIBBEAN	*1,210*	*1,260*	*1,320*	*1,410*	*1,460*	*1,530*	*1,630*	*1,690*	*1,700*	*1,770*	*1,830*
Argentina	1,600	1,640	1,750	1,810	1,860	1,870	1,910	1,970	1,930	1,880	1,960
Bahamas	..	..	..	..	..	..	..	..	..	..	4,170
Barbados	..	..	..	..	..	..	..	..	..	..	..
Belize	..	..	..	..	..	..	..	..	..	..	..
Bolivia	400	410	420	450	440	460	480	520	510	530	530
Brazil	880	950	1,020	1,110	1,200	1,310	1,480	1,550	1,580	1,710	1,760
Chile	2,340	2,400	2,540	2,500	2,660	2,550	2,410	2,380	1,780	1,850	1,990
Colombia	770	790	820	880	910	970	1,040	1,060	1,050	1,120	1,170
Costa Rica	1,450	1,510	1,550	1,650	1,640	1,690	1,800	1,780	1,790	1,920	2,140
Dominican Republic	660	650	710	760	750	810	870	940	1,020	970	990
Ecuador	740	740	750	790	790	790	860	1,020	1,060	1,140	1,240
El Salvador	710	670	670	730	720	740	780	770	770	900	1,040
Grenada	..	..	..	..	..	..	..	..	..	..	..
Guatemala	820	870	880	930	930	950	1,000	1,010	990	1,050	1,170
Guyana	..	..	..	710	780	710	640	860	1,060	850	800
Haiti	210	210	210	210	230	220	220	220	220	240	240
Honduras	530	550	540	520	510	530	560	540	520	590	660
Jamaica	1,280	1,320	1,360	1,520	1,480	1,630	1,600	1,620	1,700	1,470	1,470
Mexico	1,590	1,660	1,710	1,880	1,890	1,980	2,070	2,140	2,200	2,220	2,220
Nicaragua	970	970	970	950	950	960	950	1,010	960	1,070	1,110
Panama	1,250	1,290	1,380	1,430	1,550	1,600	1,620	1,600	1,590	1,540	1,480
Paraguay	750	760	760	790	810	850	920	960	950	1,040	1,180
Peru	1,010	1,010	1,050	1,070	1,060	1,040	1,120	1,180	1,120	1,100	1,070
St. Vincent and the Grenadines	..	..	..	..	..	..	..	..	..	..	510
Trinidad and Tobago	1,990	1,890	1,810	1,860	2,360	2,190	2,160	2,730	3,470	3,750	4,010
Uruguay	2,430	2,450	2,610	2,770	2,790	2,770	2,860	2,750	2,850	2,960	2,950
Venezuela	..	..	..	..	..	..	..	3,970	4,120	4,430	4,650
MIDDLE EAST AND NORTH AFRICA	*510*	*530*	*560*	*580*	*610*	*650*	*660*	*790*	*800*	*860*	*900*
Algeria	690	710	760	840	940	1,080	1,120	1,560	1,490	1,610	1,690
Egypt, Arab Republic of	310	320	340	350	350	340	350	350	350	390	430
Jordan	..	..	..	1,030	980	1,010	960	950	1,010	1,180	1,230
Morocco	640	680	710	720	750	730	740	790	810	850	870
Oman	..	..	..	..	..	..	..	..	..	..	..
Syrian Arab Republic	580	590	690	610	690	760	740	1,090	1,270	1,320	1,250
Tunisia	690	730	740	790	840	980	950	1,070	1,100	1,090	1,100
Yemen Arab Republic	..	..	..	220	260	280	310	320	330	360	380
Yemen, PDR	..										
EUROPE AND MEDITERRANEAN	*..*	*..*	*..*	*..*	*1,560*	*1,650*	*1,710*	*1,800*	*1,830*	*1,890*	*1,960*
Cyprus	..	..	..	..	..	..	..	..	..	..	..
Greece	2,440	2,570	2,810	3,020	3,240	3,490	3,760	3,580	3,710	3,890	3,990
Hungary	..	..	..	..	1,490	1,570	1,670	1,780	1,780	1,870	1,950
Israel	3,330	3,820	4,230	4,370	4,740	5,250	5,120	5,240	5,240	5,250	5,280
Malta	1,400	1,550	1,690	1,770	1,810	1,910	1,950	2,040	2,400	2,730	2,920
Poland	..	..	..	..	..	..	..	..	..	..	..
Portugal	1,350	1,430	1,700	1,790	1,920	2,070	2,270	2,530	2,420	2,130	2,210
Turkey	930	960	990	1,020	1,070	1,130	1,160	1,200	1,250	1,340	1,360
Yugoslavia	1,690	1,780	1,970	2,030	2,200	2,230	2,270	2,540	2,540	2,650	2,840
Developing countries	430	440	470	500	520	520	560	590	600	620	660
Highly indebted	1,060	1,100	1,170	1,260	1,320	1,370	1,450	1,540	1,530	1,600	1,650
Low-income countries	200	190	210	230	230	230	230	230	230	230	240
Low-income Africa	360	350	370	360	360	360	360	360	320	320	330
China and India	180	180	190	220	220	220	230	220	240	230	240
Other low-income Asia	170	180	180	190	180	160	170	170	170	180	180
Middle-income countries	880	920	980	1,040	1,080	1,120	1,220	1,320	1,330	1,410	1,500
High-income oil exporters	..	..	..	..	..	..	..	..	..	..	..
Industrialized countries	7,840	8,190	8,550	8,810	9,050	9,430	9,910	9,700	9,590	9,950	10,220
Japan	5,100	5,620	6,220	6,910	7,220	7,760	8,180	7,750	7,750	8,050	8,420
United States	9,980	10,340	10,540	10,430	10,590	10,980	11,520	11,160	10,940	11,380	11,760
European Community	7,240	7,550	7,950	8,310	8,590	8,900	9,320	9,230	9,130	9,480	9,680

NOTE: See glossary for definitions.

1978	1979	1980	1981	1982	1983	1984	1985	1986	1987	
1,840	*1,920*	*2,010*	*1,970*	*1,860*	*1,740*	*1,780*	*1,800*	*1,810*	*1,830*	*LATIN AMERICA AND CARIBBEAN*
1,860	1,970	1,990	1,810	1,630	1,650	1,670	1,560	1,630	1,630	Argentina
4,200	5,510	5,820	5,400	5,730	5,970	6,640	7,160	7,320	7,410	Bahamas
..	..	..	..	..	..	..	..	..	..	Barbados
..	..	..	..	..	..	..	..	..	..	Belize
520	510	520	510	450	410	400	380	340	340	Bolivia
1,770	1,830	1,910	1,770	1,720	1,630	1,690	1,790	1,950	1,980	Brazil
2,110	2,290	2,410	2,420	1,930	1,910	1,910	1,940	2,030	2,190	Chile
1,230	1,250	1,290	1,260	1,250	1,240	1,270	1,270	1,360	1,390	Colombia
2,140	2,130	2,070	1,840	1,630	1,700	1,830	1,870	1,980	1,910	Costa Rica
960	960	1,050	1,060	1,040	1,060	1,030	950	960	..	Dominican Republic
1,250	1,330	1,370	1,350	1,290	1,220	1,190	1,180	1,090	1,010	Ecuador
940	890	780	660	610	590	600	610	650	610	El Salvador
..	..	..	..	..	..	..	..	..	..	Grenada
1,150	1,140	1,130	1,080	1,010	960	940	890	890	880	Guatemala
760	690	720	670	550	480	430	430	460	420	Guyana
240	250	270	250	240	240	230	230	230	230	Haiti
680	670	650	610	550	540	540	540	550	540	Honduras
1,440	1,260	1,170	1,120	1,140	1,130	1,120	1,000	1,110	1,160	Jamaica
2,340	2,500	2,690	2,840	2,680	2,410	2,460	2,500	2,260	2,290	Mexico
970	640	740	680	630	640	600	550	490	..	Nicaragua
1,630	1,620	1,700	1,740	1,740	1,700	1,740	1,770	1,830	..	Panama
1,230	1,160	1,470	1,520	1,440	1,360	1,360	1,360	1,300	1,320	Paraguay
1,020	1,100	1,140	1,140	1,110	930	940	920	1,010	1,080	Peru
570	540	550	600	660	710	760	850	910	..	St. Vincent and the Grenadines
4,480	4,820	5,410	5,400	5,140	4,290	3,560	3,380	2,330	..	Trinidad and Tobago
3,100	3,300	3,450	3,510	3,110	2,780	2,670	2,690	2,970	3,120	Uruguay
4,460	4,460	4,640	4,610	4,200	3,850	3,980	3,720	3,540	3,400	Venezuela
910	*1,000*	*1,080*	*1,120*	*1,150*	*1,170*	*1,180*	*1,200*	*1,110*	*1,080*	*MIDDLE EAST AND NORTH AFRICA*
1,720	1,970	2,200	2,340	2,320	2,380	2,370	2,440	2,070	2,050	Algeria
430	470	510	510	540	560	580	580	570	540	Egypt, Arab Republic of
1,370	1,360	1,500	1,490	1,600	1,540	1,580	1,560	1,680	1,520	Jordan
870	890	890	820	860	860	850	850	920	900	Morocco
4,380	4,810	5,420	6,840	6,970	7,400	8,240	8,780	6,710	..	Oman
1,310	1,390	1,490	1,610	1,570	1,530	1,440	1,440	1,330	1,160	Syrian Arab Republic
1,160	1,240	1,340	1,380	1,350	1,380	1,410	1,410	1,300	1,340	Tunisia
410	430	430	430	550	540	540	550	580	590	Yemen Arab Republic
..	..	..	..	..	..	..	..	..	..	Yemen, PDR
2,050	*2,080*	*2,070*	*2,030*	*2,020*	*2,020*	*2,040*	*2,080*	*2,180*	*..*	*EUROPE AND MEDITERRANEAN*
..	..	..	..	..	..	..	..	..	..	Cyprus
4,180	4,250	4,300	4,290	4,230	4,180	4,180	4,250	4,250	4,190	Greece
2,020	2,030	2,030	2,040	2,100	2,100	2,120	2,100	2,100	2,180	Hungary
5,410	5,480	5,580	5,730	5,740	5,880	5,660	5,700	5,870	..	Israel
3,090	3,210	3,360	3,510	3,700	3,500	3,590	3,660	3,740	3,940	Malta
..	..	1,530	1,350	1,300	1,350	1,420	1,490	..	..	Poland
2,270	2,390	2,420	2,360	2,360	2,340	2,260	2,330	2,570	2,740	Portugal
1,370	1,320	1,260	1,260	1,270	1,280	1,340	1,380	1,480	1,550	Turkey
3,090	3,200	3,240	3,230	3,220	3,150	3,140	3,140	3,300	3,180	Yugoslavia
670	690	710	720	720	720	740	760	780	790	Developing countries
1,670	1,720	1,800	1,770	1,680	1,570	1,570	1,590	1,600	1,590	Highly indebted
260	260	270	270	290	300	330	350	360	370	Low-income countries
320	310	310	300	290	280	270	270	270	270	Low-income Africa
270	270	280	290	300	330	360	380	400	420	China and India
190	200	200	210	210	220	220	230	230	240	Other low-income Asia
1,500	1,540	1,580	1,590	1,570	1,540	1,560	1,560	1,580	1,600	Middle-income countries
..	..	..	..	..	..	..	..	..	..	High-income oil exporters
10,610	10,850	10,750	10,800	10,720	10,960	11,410	11,730	12,160	12,490	Industrialized countries
8,910	9,060	9,070	9,330	9,550	9,770	10,210	10,670	11,350	11,770	Japan
12,220	12,340	12,000	12,130	11,760	12,120	12,850	13,140	13,400	13,650	United States
10,040	10,360	10,330	10,180	10,260	10,420	10,630	10,900	11,420	11,760	European Community

Table 3. Private consumption per capita

1980 US dollars	1967	1968	1969	1970	1971	1972	1973	1974	1975	1976	1977
SUB-SAHARAN AFRICA	*352*	*340*	*352*	*375*	*401*	*392*	*393*	*420*	*392*	*395*	*380*
Excluding Nigeria	*292*	*279*	*282*	*275*	*286*	*289*	*292*	*292*	*280*	*281*	*281*
Benin	361	361	346	343	333	331	339	321	306	296	327
Botswana	227	313	320	319	264	247	324	351	372	386	408
Burkina Faso	167	165	165	148	147	133	126	130	134	141	161
Burundi	157	154	145	190	191	179	181	171	178	190	205
Cameroon	427	463	461	456	485	465	450	479	474	482	529
Cape Verde	..	..	..	..	..	..	260	202	210	230	238
Central African Republic	258	262	277	277	271	271	258	266	275	298	345
Chad	178	172	169	174	163	166	205	166	171	172	171
Comoros	..	..	..	..	..	..	..	..	..	..	..
Congo, People's Republic	408	440	470	482	475	468	460	449	460	486	409
Cote d'Ivoire	657	677	708	730	766	768	823	769	834	906	958
Ethiopia	80	79	80	86	87	87	86	85	84	78	81
Gabon	1,105	1,103	1,094	1,121	1,178	1,246	1,468	1,189	1,139	1,419	1,391
Gambia, The	316	325	336	293	376	351	356	399	303	409	408
Ghana	413	373	405	393	400	358	392	425	345	325	319
Guinea-Bissau	..	..	..	181	177	192	189	187	161	148	136
Kenya	235	206	205	179	251	274	247	274	259	237	247
Lesotho	208	203	198	208	219	236	282	321	340	377	390
Liberia	383	406	411	312	379	271	387	393	362	405	439
Madagascar	317	325	330	358	363	365	345	329	318	302	309
Malawi	134	133	139	122	152	146	141	138	143	144	153
Mali	178	171	163	168	160	177	168	186	196	214	203
Mauritania	336	364	345	362	351	357	314	358	332	377	332
Mauritius	607	471	434	509	536	520	468	682	803	830	897
Mozambique	..	..	..	..	..	..	..	..	..	..	..
Niger	387	377	358	335	345	317	234	291	241	238	241
Nigeria	545	539	578	686	753	715	709	816	739	753	693
Rwanda	152	164	180	178	174	169	156	163	146	144	153
Senegal	449	448	387	409	405	396	382	375	399	422	402
Seychelles	..	..	..	..	..	..	..	..	..	..	..
Sierra Leone	..	..	..	..	..	..	..	..	..	..	..
Somalia	119	114	109	111	118	90	95	72	102	98	152
Sudan	227	237	216	203	230	227	185	207	261	272	339
Swaziland	546	497	649	544	526	531	489	321	525	448	507
Tanzania	..	..	..	..	..	..	..	..	..	188	..
Togo	203	207	210	209	199	208	216	228	194	244	253
Uganda	..	..	..	..	..	..	..	..	..	..	..
Zaire	513	433	468	418	428	504	548	492	469	444	367
Zambia	633	616	544	484	441	442	395	434	475	436	386
Zimbabwe	379	349	423	423	462	484	476	525	522	520	447
SOUTH ASIA	*155*	*158*	*159*	*165*	*164*	*156*	*156*	*157*	*163*	*161*	*171*
Bangladesh	125	129	129	136	126	112	99	117	118	131	123
India	160	163	163	167	168	160	161	159	167	163	176
Nepal	..	..	..	..	..	..	..	..	..	..	..
Pakistan	170	177	185	207	197	183	189	207	200	203	207
Sri Lanka	184	196	209	199	197	182	209	200	213	206	228
EAST ASIA AND PACIFIC	*153*	*149*	*163*	*174*	*176*	*180*	*190*	*197*	*202*	*201*	*212*
China	104	93	107	118	117	121	127	128	132	124	132
Fiji	699	664	716	818	860	896	1,056	1,169	1,153	1,179	1,199
Hong Kong	1,347	1,458	1,556	1,691	1,873	1,961	2,177	2,087	2,103	2,230	2,588
Indonesia	157	175	181	180	179	176	185	208	219	220	222
Korea, Republic of	524	562	601	646	693	714	765	809	841	898	944
Malaysia	522	516	535	540	558	573	607	646	627	648	698
Papua New Guinea	413	430	460	489	495	483	463	457	458	461	486
Philippines	392	398	400	395	399	403	414	423	431	441	452
Singapore	1,149	1,244	1,326	1,468	1,612	1,727	1,863	1,969	1,999	2,077	2,161
Solomon Islands	..	..	..	..	..	..	..	..	..	..	..
Thailand	295	315	321	324	320	332	351	359	370	389	410
Vanuatu	..	..	..	..	..	..	..	..	..	..	..

NOTE: See glossary for definitions.

1978	1979	1980	1981	1982	1983	1984	1985	1986	1987	
388	*370*	*387*	*406*	*403*	*383*	*361*	*372*	*368*	*342*	*SUB-SAHARAN AFRICA*
282	*283*	*272*	*276*	*271*	*266*	*255*	*259*	*261*	*255*	*Excluding Nigeria*
328	320	317	333	353	309	329	336	327	303	Benin
419	492	512	492	510	514	501	534	517	..	Botswana
156	162	168	171	169	155	141	164	171	171	Burkina Faso
187	192	194	192	194	178	189	190	186	..	Burundi
521	593	651	651	636	610	575	683	766	747	Cameroon
295	293	333	254	294	385	363	408	..	..	Cape Verde
347	321	329	282	306	259	277	280	301	299	Central African Republic
169	124	118	119	130	141	134	156	..	189	Chad
..	..	..	..	..	..	..	..	..	..	Comoros
428	460	436	552	630	630	657	661	595	..	Congo, People's Republic
1,009	976	758	865	819	784	813	797	857	736	Cote d'Ivoire
83	86	87	84	86	86	80	79	79	..	Ethiopia
1,509	1,474	1,406	1,454	1,519	1,560	1,180	1,173	1,105	934	Gabon
437	310	280	280	258	290	..	..	..	..	Gambia, The
356	342	347	307	258	254	279	288	288	285	Ghana
132	126	100	121	133	124	136	140	133	143	Guinea-Bissau
272	272	260	240	241	224	230	230	228	231	Kenya
414	408	375	414	440	431	..	..	..	..	Lesotho
424	391	340	312	295	335	326	295	281	..	Liberia
285	291	286	253	252	246	239	234	227	212	Madagascar
147	151	144	139	134	133	135	139	138	132	Malawi
218	236	226	229	235	222	217	241	250	250	Mali
297	301	310	332	361	442	386	365	359	355	Mauritania
930	953	884	861	833	841	871	908	952	1,081	Mauritius
..	..	163	158	155	122	106	111	84	..	Mozambique
258	288	309	304	314	291	264	257	263	237	Niger
723	648	750	818	821	754	696	729	704	621	Nigeria
160	170	189	183	181	192	175	174	179	166	Rwanda
393	413	407	409	442	429	380	397	399	404	Senegal
..	..	..	..	..	..	..	..	..	..	Seychelles
..	..	..	..	..	..	..	..	..	..	Sierra Leone
114	150	129	180	181	165	149	158	146	166	Somalia
347	301	284	305	319	313	284	237	250	238	Sudan
508	692	653	731	585	505	531	..	..	..	Swaziland
..	..	211	199	193	193	195	208	221	226	Tanzania
251	246	263	228	218	200	187	195	186	201	Togo
..	..	..	..	..	..	..	..	..	..	Uganda
326	316	302	304	280	280	268	274	247	259	Zaire
379	406	380	391	350	343	334	372	385	370	Zambia
468	512	493	557	542	582	446	371	346	352	Zimbabwe
176	*168*	*180*	*185*	*186*	*193*	*196*	*198*	*204*	*203*	*SOUTH ASIA*
132	134	132	136	119	122	125	128	126	129	Bangladesh
180	168	183	187	190	198	200	200	207	207	India
..	..	..	..	..	..	..	..	..	..	Nepal
217	227	238	246	246	249	260	272	273	271	Pakistan
221	215	219	242	252	274	291	320	325	322	Sri Lanka
226	*238*	*256*	*266*	*275*	*293*	*310*	*315*	*322*	*330*	*EAST ASIA AND PACIFIC*
139	144	161	169	176	191	207	212	216	217	China
1,202	1,241	1,132	1,132	1,076	1,095	1,106	1,073	1,055	1,105	Fiji
3,026	3,053	3,400	3,616	3,780	4,032	4,214	4,333	4,687	5,129	Hong Kong
242	257	279	300	312	324	332	333	337	348	Indonesia
1,020	1,095	1,070	1,087	1,122	1,188	1,241	1,285	1,354	1,429	Korea, Republic of
753	817	899	923	931	936	970	946	829	815	Malaysia
499	513	521	494	471	480	481	506	496	502	Papua New Guinea
463	473	481	487	490	502	494	473	467	482	Philippines
2,279	2,386	2,497	2,582	2,650	2,742	2,852	2,794	2,892	3,103	Singapore
..	..	357	368	289	311	342	328	320	..	Solomon Islands
425	447	459	459	461	484	507	507	518	542	Thailand
..	..	..	..	..	..	..	..	..	..	Vanuatu

Table 3. Private consumption per capita (cont'd)

1980 US dollars	1967	1968	1969	1970	1971	1972	1973	1974	1975	1976	1977
LATIN AMERICA AND CARIBBEAN	*856*	*893*	*913*	*961*	*1,013*	*1,061*	*1,104*	*1,146*	*1,145*	*1,190*	*1,221*
Argentina	1,053	1,075	1,138	1,148	1,189	1,193	1,223	1,287	1,265	1,135	1,138
Bahamas	..	..	..	..	..	..	..	..	..	..	2,587
Barbados	..	..	..	..	..	..	..	..	..	..	..
Belize	..	..	..	..	..	..	..	..	..	..	..
Bolivia	318	307	313	314	317	313	330	332	338	355	364
Brazil	631	677	682	766	849	924	1,010	1,110	1,075	1,182	1,224
Chile	1,730	1,762	1,820	1,776	1,972	2,086	1,914	1,544	1,345	1,326	1,512
Colombia	566	588	618	651	678	714	739	752	760	797	813
Costa Rica	1,066	1,103	1,113	1,197	1,166	1,191	1,220	1,251	1,248	1,273	1,407
Dominican Republic	473	480	510	556	595	604	645	694	727	768	773
Ecuador	543	554	562	574	588	591	615	654	701	744	779
El Salvador	557	541	530	544	550	555	580	583	588	631	691
Grenada	..	..	..	..	..	..	..	..	..	..	..
Guatemala	666	677	695	715	735	762	783	790	791	821	862
Guyana	..	..	..	378	367	405	433	457	429	492	497
Haiti	186	184	184	179	188	189	183	193	181	199	203
Honduras	388	408	399	403	380	391	420	397	400	446	491
Jamaica	903	906	929	1,009	1,060	1,225	1,021	1,088	1,136	1,042	1,024
Mexico	1,231	1,299	1,316	1,335	1,359	1,404	1,449	1,477	1,515	1,540	1,530
Nicaragua	815	814	787	759	761	724	840	883	811	816	852
Panama	774	766	848	853	885	861	862	932	885	867	907
Paraguay	612	638	639	632	667	673	698	758	754	794	880
Peru	640	634	650	766	772	775	784	815	825	818	798
St. Vincent and the Grenadines	..	..	..	..	..	..	..	..	..	..	463
Trinidad and Tobago	1,429	1,435	1,596	1,461	1,387	1,584	1,532	1,591	1,707	1,956	2,145
Uruguay	1,937	1,919	2,070	2,291	2,299	2,313	2,323	2,280	2,355	2,292	2,273
Venezuela	..	..	..	..	..	..	..	1,926	2,125	2,287	2,428
MIDDLE EAST AND NORTH AFRICA	*..*	*..*	*..*	*365*	*396*	*416*	*421*	*485*	*500*	*515*	*562*
Algeria	313	325	340	383	468	500	477	640	699	731	828
Egypt, Arab Republic of	..	..	..	..	..	..	..	285	258	251	290
Jordan	..	..	..	1,092	1,028	1,052	958	862	949	995	1,075
Morocco	485	506	520	526	536	537	547	553	543	570	601
Oman	..	..	..	..	..	..	..	..	..	..	..
Syrian Arab Republic	460	428	527	465	515	531	559	744	865	870	835
Tunisia	390	393	419	463	488	560	587	612	669	686	726
Yemen Arab Republic	..	..	..	210	221	237	286	307	323	371	426
Yemen, PDR	..	..	..	..	..	..	..	..	..	..	..
EUROPE AND MEDITERRANEAN	*..*	*..*	*..*	*928*	*992*	*1,050*	*1,081*	*1,178*	*1,179*	*1,214*	*1,266*
Cyprus	..	..	..	..	..	..	..	..	..	..	..
Greece	1,557	1,659	1,755	1,905	2,002	2,128	2,280	2,287	2,390	2,484	2,558
Hungary	..	..	..	911	961	992	1,030	1,087	1,132	1,147	1,198
Israel	2,091	2,288	2,456	2,447	2,509	2,684	2,790	2,917	2,858	2,930	3,004
Malta	880	1,005	1,187	1,299	1,306	1,376	1,377	1,462	1,529	1,701	1,936
Poland	..	..	..	..	..	..	..	..	..	..	..
Portugal	877	1,109	1,161	1,199	1,251	1,416	1,586	1,733	1,642	1,657	1,670
Turkey	720	749	772	763	818	874	858	902	927	990	1,010
Yugoslavia	731	781	926	988	1,100	1,111	1,108	1,389	1,309	1,276	1,405
Developing countries	297	303	317	338	349	356	369	386	389	398	412
Highly indebted	756	784	813	868	922	948	977	1,038	1,021	1,053	1,071
Low-income countries	141	135	142	149	149	147	150	151	154	150	157
Low-income Africa	279	261	261	251	260	264	265	264	247	242	241
China and India	127	121	129	138	137	137	140	140	146	140	150
Other low-income Asia	139	144	149	159	152	140	139	150	149	156	156
Middle-income countries	604	634	662	714	749	773	807	856	860	894	919
High-income oil exporters	..	..	..	..	..	..	..	..	..	..	..
Industrialized countries	4,506	4,699	4,893	5,049	5,213	5,457	5,698	5,673	5,756	5,968	6,130
Japan	2,972	3,207	3,458	3,654	3,814	4,112	4,453	4,356	4,476	4,598	4,747
United States	5,832	6,071	6,226	6,302	6,418	6,692	6,906	6,780	6,865	7,166	7,408
European Community	4,086	4,240	4,443	4,650	4,855	5,061	5,284	5,322	5,384	5,568	5,688

1978	1979	1980	1981	1982	1983	1984	1985	1986	1987	
1,254	*1,320*	*1,374*	*1,357*	*1,311*	*1,263*	*1,261*	*1,260*	*1,293*	*1,302*	*LATIN AMERICA AND CARIBBEAN*
1,098	1,245	1,332	1,243	1,086	1,112	1,166	1,068	1,141	1,127	Argentina
2,909	4,075	3,923	3,700	3,610	3,619	4,658	4,799	4,855	4,871	Bahamas
..	..	..	..	..	..	..	..	..	..	Barbados
..	..	..	..	..	..	..	..	..	..	Belize
376	369	368	418	341	333	327	338	341	331	Bolivia
1,246	1,330	1,388	1,297	1,302	1,269	1,269	1,302	1,394	1,469	Brazil
1,598	1,673	1,756	1,912	1,652	1,569	1,563	1,514	1,561	1,609	Chile
865	885	906	914	910	897	906	907	917	935	Colombia
1,489	1,489	1,428	1,276	1,127	1,162	1,238	1,319	1,321	1,369	Costa Rica
750	738	847	829	836	828	773	733	..	..	Dominican Republic
800	826	861	877	868	823	821	826	810	808	Ecuador
692	626	565	505	466	460	472	483	473	468	El Salvador
..	..	..	..	..	..	..	..	..	..	Grenada
881	893	898	885	835	800	786	762	747	751	Guatemala
414	369	403	483	436	414	376	415	434	423	Guyana
201	201	221	215	189	192	183	184	189	192	Haiti
473	463	482	461	453	427	416	413	410	416	Honduras
941	868	825	799	822	874	885	857	826	906	Jamaica
1,612	1,713	1,799	1,889	1,802	1,669	1,687	1,705	1,622	1,566	Mexico
759	538	659	523	450	415	381	328	317	..	Nicaragua
988	1,026	1,027	999	1,030	1,064	1,163	1,158	1,083	..	Panama
898	857	1,101	1,095	1,084	1,025	1,072	1,049	1,045	1,039	Paraguay
718	716	732	753	741	656	652	649	717	743	Peru
385	426	439	440	547	546	513	511	623	..	St. Vincent and the Grenadines
2,360	2,576	2,612	2,678	3,472	3,280	2,629	2,262	1,688	..	Trinidad and Tobago
2,338	2,448	2,640	2,684	2,409	2,167	2,023	2,075	2,229	2,407	Uruguay
2,596	2,614	2,534	2,517	2,438	2,479	2,326	2,206	2,390	2,142	Venezuela
565	*623*	*654*	*697*	*704*	*705*	*706*	*722*	*714*	*696*	*MIDDLE EAST AND NORTH AFRICA*
797	916	980	1,112	1,109	1,096	1,118	1,152	1,117	1,104	Algeria
294	347	375	370	376	375	396	411	410	401	Egypt, Arab Republic of
1,144	1,396	1,277	1,315	1,441	1,530	1,520	1,467	1,445	1,336	Jordan
607	623	626	611	628	620	626	617	660	636	Morocco
..	..	..	..	..	..	..	..	..	..	Oman
908	926	988	1,199	1,062	1,069	931	951	875	789	Syrian Arab Republic
748	764	845	885	890	910	942	945	932	923	Tunisia
369	404	366	381	512	531	473	509	494	533	Yemen Arab Republic
..	..	..	..	..	..	..	..	..	..	Yemen, PDR
1,343	*1,340*	*1,299*	*1,289*	*1,277*	*1,290*	*1,302*	*1,308*	*1,373*	*..*	*EUROPE AND MEDITERRANEAN*
..	..	..	..	..	..	..	..	..	..	Cyprus
2,668	2,705	2,662	2,713	2,800	2,794	2,826	2,929	2,915	2,919	Greece
1,238	1,259	1,266	1,303	1,340	1,351	1,368	1,389	1,419	1,465	Hungary
3,178	3,343	3,009	3,304	3,474	3,686	3,363	3,300	3,636	..	Israel
1,973	1,992	2,016	2,083	2,167	2,197	2,291	2,398	2,465	2,588	Malta
..	..	1,066	1,018	883	922	952	969	..	..	Poland
1,641	1,646	1,659	1,683	1,712	1,678	1,622	1,634	1,744	1,861	Portugal
1,056	1,015	940	910	933	950	1,002	990	1,049	1,079	Turkey
1,607	1,597	1,598	1,570	1,557	1,520	1,494	1,483	1,582	1,554	Yugoslavia
425	432	451	459	458	462	468	472	481	483	Developing countries
1,110	1,145	1,193	1,193	1,160	1,115	1,103	1,104	1,129	1,116	Highly indebted
163	161	174	179	182	192	199	202	206	206	Low-income countries
238	235	230	227	224	219	213	215	214	214	Low-income Africa
156	154	170	176	182	194	204	207	212	213	China and India
163	167	172	178	174	178	186	193	194	194	Other low-income Asia
947	969	998	1,009	998	990	990	993	1,012	1,016	Middle-income countries
..	..	..	..	..	..	..	..	..	..	High-income oil exporters
6,323	6,491	6,495	6,511	6,559	6,712	6,879	7,082	7,311	7,504	Industrialized countries
4,960	5,235	5,269	5,297	5,478	5,617	5,734	5,855	6,005	6,206	Japan
7,630	7,714	7,609	7,618	7,638	7,916	8,219	8,523	8,802	8,957	United States
5,880	6,073	6,127	6,130	6,159	6,236	6,320	6,464	6,696	6,922	European Community

Table 4. Gross domestic investment per capita

1980 US dollars	1967	1968	1969	1970	1971	1972	1973	1974	1975	1976	1977
SUB-SAHARAN AFRICA	*64*	*64*	*63*	*84*	*95*	*94*	*103*	*112*	*128*	*147*	*157*
Excluding Nigeria	*63*	*65*	*62*	*72*	*77*	*74*	*77*	*89*	*84*	*83*	*86*
Benin	22	24	24	22	21	30	29	42	41	40	36
Botswana	66	98	137	189	249	323	388	381	322	299	214
Burkina Faso	20	21	24	25	34	43	49	53	52	51	46
Burundi	14	17	16	11	20	6	12	19	14	20	29
Cameroon	61	60	56	72	86	90	93	84	95	85	114
Cape Verde	..	..	..	..	..	..	76	81	69	104	148
Central African Republic	73	78	79	86	87	86	89	54	44	46	46
Chad	56	53	62	59	61	55	-13	57	66	58	65
Comoros	..	..	..	..	..	..	..	..	..	..	..
Congo, People's Republic	378	347	318	291	332	379	433	493	566	421	329
Cote d'Ivoire	145	143	165	191	183	168	203	191	189	229	315
Ethiopia	16	16	15	13	14	15	14	12	12	15	14
Gabon	655	678	740	820	895	1,022	1,123	3,010	3,553	4,847	3,618
Gambia, The	25	26	23	13	16	21	18	19	50	50	65
Ghana	30	31	35	45	46	23	30	42	35	27	39
Guinea-Bissau	..	..	..	92	75	62	49	37	34	35	39
Kenya	74	100	100	133	130	115	134	123	81	87	114
Lesotho	21	19	19	19	24	27	38	36	46	68	97
Liberia	151	106	118	159	145	227	92	114	208	204	189
Madagascar	80	95	100	90	104	78	80	83	77	63	58
Malawi	26	34	40	60	50	63	54	61	69	53	51
Mali	39	46	46	40	41	41	38	29	38	36	44
Mauritania	37	51	38	43	44	61	44	137	201	251	218
Mauritius	214	150	187	126	188	213	348	412	421	400	389
Mozambique	..	..	..	..	..	..	..	..	..	..	..
Niger	119	123	76	124	91	120	171	159	127	127	138
Nigeria	69	59	66	121	154	157	183	183	268	350	383
Rwanda	15	17	15	17	21	21	21	23	30	31	35
Senegal	76	81	86	102	102	110	110	117	101	113	106
Seychelles	..	..	..	..	..	..	..	..	..	..	..
Sierra Leone	58	55	65	82	55	40	41	56	56	39	38
Somalia	26	26	25	24	23	30	29	29	31	39	62
Sudan	44	45	43	47	42	31	35	59	62	93	77
Swaziland	178	155	180	207	233	268	304	353	234	363	324
Tanzania	55	58	50	73	85	66	67	73	65	70	78
Togo	76	65	78	83	90	96	98	98	141	112	194
Uganda	30	29	30	31	35	25	18	23	15	12	11
Zaire	25	26	35	36	46	45	47	50	44	33	62
Zambia	413	417	297	426	460	460	423	517	433	225	196
Zimbabwe	197	211	144	168	193	205	218	290	253	177	140
SOUTH ASIA	*39*	*38*	*39*	*41*	*42*	*38*	*42*	*42*	*43*	*44*	*45*
Bangladesh	20	27	24	22	16	7	13	12	10	11	13
India	41	38	41	43	46	42	48	48	49	50	51
Nepal	..	..	..	..	..	..	..	..	..	..	..
Pakistan	51	57	45	52	50	46	46	42	46	45	49
Sri Lanka	21	24	29	32	29	31	24	37	33	43	41
EAST ASIA AND PACIFIC	*42*	*44*	*51*	*70*	*75*	*73*	*85*	*91*	*95*	*94*	*105*
China	27	26	30	52	56	53	61	60	67	60	66
Fiji	267	375	352	345	382	419	442	389	415	411	409
Hong Kong	710	637	653	734	887	944	1,030	1,015	1,000	1,276	1,445
Indonesia	23	28	34	45	53	62	71	83	92	96	108
Korea, Republic of	121	160	222	214	226	208	259	334	318	360	447
Malaysia	194	197	180	240	237	250	310	399	304	332	397
Papua New Guinea	188	193	274	429	451	299	186	192	186	164	174
Philippines	116	123	125	117	118	112	120	144	173	194	189
Singapore	488	596	737	1,044	1,202	1,288	1,357	1,605	1,458	1,527	1,451
Solomon Islands	..	..	..	..	..	..	..	..	..	..	..
Thailand	104	112	129	134	126	114	152	143	143	139	171
Vanuatu	..	..	..	..	..	..	..	..	..	..	..

NOTE: See glossary for definitions

1978	1979	1980	1981	1982	1983	1984	1985	1986	1987	
125	*110*	*120*	*124*	*108*	*86*	*63*	*69*	*65*	*60*	*SUB-SAHARAN AFRICA*
77	*74*	*80*	*81*	*77*	*65*	*59*	*61*	*58*	*52*	*Excluding Nigeria*
33	46	62	68	70	32	24	27	25	26	Benin
283	334	415	409	381	255	242	373	323	..	Botswana
46	42	37	36	43	40	38	43	38	35	Burkina Faso
33	33	31	42	30	54	41	34	41	..	Burundi
163	151	163	222	220	227	207	218	216	169	Cameroon
129	166	127	195	199	207	184	189	..	..	Cape Verde
43	36	24	29	40	58	63	64	55	48	Central African Republic
50	38	30	17	13	11	11	11	..	45	Chad
..	..	..	..	..	..	..	..	..	..	Comoros
253	263	380	666	882	587	506	465	300	..	Congo, People's Republic
352	314	356	269	222	194	89	127	97	103	Cote d'Ivoire
9	10	11	12	14	14	14	11	11	..	Ethiopia
1,316	1,555	1,523	2,077	1,813	1,623	1,659	1,764	1,464	847	Gabon
129	95	102	96	77	80	..	..	..	..	Gambia, The
26	26	23	21	15	15	16	20	18	22	Ghana
49	58	39	32	40	38	39	42	42	38	Guinea-Bissau
136	99	128	119	88	74	79	69	78	88	Kenya
84	97	95	93	91	78	..	..	..	..	Lesotho
171	183	164	97	80	63	52	42	45	..	Liberia
57	95	88	60	47	45	46	46	44	44	Madagascar
85	60	50	35	41	44	25	33	16	18	Malawi
42	45	42	44	47	43	48	46	57	43	Mali
147	122	166	199	245	94	128	129	123	110	Mauritania
415	427	243	293	241	235	280	296	389	535	Mauritius
..	..	38	38	43	38	25	7	9	..	Mozambique
169	195	169	170	139	128	9	92	58	64	Niger
274	223	250	262	208	150	76	93	86	85	Nigeria
33	26	37	34	48	39	51	54	49	50	Rwanda
98	83	81	81	83	85	81	69	76	74	Senegal
..	..	..	..	..	..	..	..	..	..	Seychelles
39	44	55	65	48	45	45	39	32	31	Sierra Leone
78	54	61	44	50	43	59	55	49	51	Somalia
62	50	53	52	73	66	59	50	40	34	Sudan
543	445	391	285	272	285	254	..	..	..	Swaziland
73	77	63	58	61	38	42	46	39	31	Tanzania
222	205	132	123	99	78	76	83	83	57	Togo
12	9	8	..	..	..	..	..	..	..	Uganda
28	46	58	57	50	36	47	41	54	51	Zaire
184	100	160	138	102	75	77	80	82	53	Zambia
84	89	144	203	169	119	115	140	144	111	Zimbabwe
51	*50*	*56*	*60*	*56*	*57*	*57*	*63*	*61*	*61*	*SOUTH ASIA*
16	16	22	22	28	26	24	24	24	23	Bangladesh
57	54	61	67	61	61	63	71	68	68	India
..	..	..	..	..	..	..	..	..	..	Nepal
50	51	53	50	56	58	59	62	64	69	Pakistan
59	77	92	80	83	82	76	59	60	59	Sri Lanka
128	*136*	*135*	*147*	*154*	*166*	*186*	*219*	*228*	*255*	*EAST ASIA AND PACIFIC*
86	88	88	85	99	112	143	195	207	234	China
433	566	605	629	443	378	333	313	348	196	Fiji
1,564	1,686	1,966	2,105	2,011	1,904	1,940	1,791	1,943	2,213	Hong Kong
122	131	130	232	215	211	180	197	198	193	Indonesia
571	652	509	533	526	608	711	713	780	886	Korea, Republic of
420	466	541	620	692	726	745	579	465	462	Malaysia
174	209	213	200	222	227	200	151	152	160	Papua New Guinea
199	216	220	219	206	195	108	81	73	85	Philippines
1,667	1,956	2,249	2,381	2,702	2,972	3,219	2,779	2,492	2,728	Singapore
..	..	186	177	152	167	159	170	165	..	Solomon Islands
197	184	182	196	172	206	216	202	188	203	Thailand
..	..	..	..	..	..	..	..	..	..	Vanuatu

Table 4. Gross domestic investment per capita (cont'd)

1980 US dollars	1967	1968	1969	1970	1971	1972	1973	1974	1975	1976	1977
LATIN AMERICA AND CARIBBEAN	*243*	*252*	*289*	*300*	*311*	*321*	*364*	*419*	*429*	*432*	*452*
Argentina	279	300	363	365	399	396	371	369	370	389	462
Bahamas	..	..	..	..	..	..	..	..	..	..	593
Barbados	..	..	..	..	..	..	..	..	..	..	..
Belize	..	..	..	..	..	..	..	..	..	..	..
Bolivia	78	105	100	95	103	123	106	103	144	127	127
Brazil	149	173	236	230	256	291	354	403	434	426	417
Chile	394	421	442	470	447	317	278	498	232	228	262
Colombia	139	162	160	183	187	178	191	224	181	194	221
Costa Rica	256	245	274	305	363	328	383	406	361	447	534
Dominican Republic	98	95	123	144	164	180	222	258	272	248	262
Ecuador	194	199	183	188	230	195	209	281	320	295	348
El Salvador	105	79	86	92	109	99	125	154	135	146	205
Grenada	..	..	..	..	..	..	..	..	..	..	..
Guatemala	132	165	123	144	162	133	153	201	164	222	240
Guyana	..	..	..	282	236	193	284	250	427	364	253
Haiti	10	10	12	18	18	20	23	25	30	36	37
Honduras	129	126	123	126	100	95	116	150	114	118	153
Jamaica	412	502	555	554	547	473	625	459	507	352	226
Mexico	340	346	364	429	395	422	471	535	546	519	503
Nicaragua	187	154	173	170	170	124	207	263	178	181	273
Panama	318	349	389	445	532	592	598	554	521	504	342
Paraguay	112	103	108	104	113	125	169	183	191	227	247
Peru	443	302	303	184	208	179	271	364	335	284	236
St. Vincent and the Grenadines	..	..	..	..	..	..	..	..	..	..	156
Trinidad and Tobago	356	407	320	479	742	648	578	826	988	1,114	1,133
Uruguay	213	187	237	266	281	235	224	232	294	354	413
Venezuela	..	..	..	..	..	..	..	1,204	1,439	1,671	2,045
MIDDLE EAST AND NORTH AFRICA	*115*	*122*	*140*	*154*	*154*	*165*	*178*	*243*	*290*	*309*	*345*
Algeria	236	277	334	407	389	440	532	710	747	761	852
Egypt, Arab Republic of	53	46	49	51	43	44	56	81	124	120	132
Jordan	..	..	..	123	168	188	159	240	271	384	450
Morocco	101	108	130	151	163	129	131	153	247	283	320
Oman	..	..	..	..	..	..	..	..	..	..	..
Syrian Arab Republic	119	142	175	133	154	202	119	279	296	377	426
Tunisia	305	294	283	269	269	324	294	340	354	368	366
Yemen Arab Republic	..	..	..	18	35	38	36	46	46	56	64
Yemen, PDR	..	..	..	..	..	..	..	..	..	..	..
EUROPE AND MEDITERRANEAN	*330*	*346*	*373*	*430*	*462*	*466*	*521*	*526*	*547*	*568*	*614*
Cyprus	..	..	..	..	..	..	..	..	..	..	..
Greece	559	618	771	857	907	1,004	1,272	972	950	957	953
Hungary	352	356	372	432	526	459	459	562	616	625	674
Israel	699	1,014	1,246	1,367	1,613	1,801	1,826	1,693	1,742	1,530	1,355
Malta	475	580	637	638	564	519	454	506	503	625	666
Poland	..	..	..	..	..	..	..	..	..	..	..
Portugal	472	457	481	592	614	703	808	742	511	556	670
Turkey	153	165	166	198	194	204	225	275	326	333	360
Yugoslavia	652	655	669	768	832	761	840	903	985	1,011	1,175
Developing countries	87	89	98	117	124	123	138	152	163	166	176
Highly indebted	226	231	259	281	297	298	335	376	409	429	457
Low-income countries	34	33	36	48	50	47	52	53	55	52	56
Low-income Africa	53	56	53	63	66	61	62	67	59	56	61
China and India	33	31	35	49	52	49	55	55	60	56	60
Other low-income Asia	31	36	31	33	29	24	26	25	26	26	29
Middle-income countries	191	198	221	253	270	275	309	351	379	395	417
High-income oil exporters	1,013	1,142	1,125	826	767	871	944	1,290	1,888	2,484	3,209
Industrialized countries	1,738	1,877	2,005	2,061	2,095	2,206	2,431	2,321	1,993	2,183	2,266
Japan	1,599	1,918	2,185	2,535	2,498	2,696	3,032	2,769	2,535	2,609	2,698
United States	1,657	1,742	1,762	1,595	1,731	1,901	2,107	1,930	1,522	1,783	2,030
European Community	1,759	1,892	2,053	2,134	2,117	2,187	2,390	2,303	1,979	2,191	2,193

1978	1979	1980	1981	1982	1983	1984	1985	1986	1987	
457	*457*	*499*	*487*	*406*	*294*	*309*	*320*	*326*	*326*	*LATIN AMERICA AND CARIBBEAN*
383	415	446	336	266	234	204	161	187	215	Argentina
565	781	1,163	1,152	1,371	1,354	1,218	1,527	1,475	1,554	Bahamas
..	..	..	..	..	..	..	..	..	..	Barbados
..	..	..	..	..	..	..	..	..	..	Belize
125	105	78	75	55	31	23	24	16	19	Bolivia
420	411	452	380	342	256	286	312	370	327	Brazil
318	404	521	625	213	173	298	273	306	379	Chile
231	226	246	273	282	271	251	222	230	237	Colombia
519	554	579	352	257	333	362	375	425	388	Costa Rica
262	277	289	252	228	228	235	244	..	..	Dominican Republic
378	357	377	318	348	232	222	233	231	227	Ecuador
207	156	104	99	89	80	81	76	91	103	El Salvador
..	..	..	..	..	..	..	..	..	..	Grenada
258	218	181	203	159	129	133	104	102	130	Guatemala
174	237	232	211	143	122	142	150	164	132	Guyana
40	46	46	45	41	43	44	37	31	32	Haiti
185	186	170	138	79	93	127	121	109	104	Honduras
249	238	197	254	240	244	215	212	170	205	Jamaica
549	630	751	842	624	446	463	508	395	405	Mexico
109	-47	121	175	149	156	152	148	142	..	Nicaragua
418	441	505	556	505	391	341	356	392	401	Panama
305	343	460	510	406	315	301	297	305	309	Paraguay
216	247	328	386	350	211	185	162	233	256	Peru
180	212	226	212	203	187	219	241	267	..	St. Vincent and the Grenadines
1,427	1,447	1,743	1,427	1,478	1,296	943	976	456	..	Trinidad and Tobago
469	577	604	547	444	280	285	234	229	282	Uruguay
1,950	1,499	1,217	1,180	1,222	586	671	693	735	861	Venezuela
360	*351*	*353*	*360*	*376*	*373*	*382*	*378*	*332*	*286*	*MIDDLE EAST AND NORTH AFRICA*
992	894	887	889	916	965	919	885	808	718	Algeria
143	156	149	157	172	173	186	195	166	133	Egypt, Arab Republic of
452	492	622	836	821	641	579	550	488	415	Jordan
232	230	208	180	188	159	158	166	150	143	Morocco
1,212	1,373	1,355	1,612	2,074	2,157	3,109	3,213	2,365	..	Oman
358	352	409	404	403	404	398	409	372	270	Syrian Arab Republic
395	410	403	453	440	421	447	382	305	269	Tunisia
132	142	167	137	125	83	84	68	62	73	Yemen Arab Republic
										Yemen, PDR
607	*647*	*631*	*588*	*573*	*560*	*548*	*552*	*564*	*574*	*EUROPE AND MEDITERRANEAN*
..	..	..	..	..	..	..	..	..	..	Cyprus
1,017	1,103	1,042	886	824	813	786	832	750	640	Greece
792	658	635	621	600	557	544	526	573	584	Hungary
1,362	1,487	1,248	1,153	1,287	1,395	1,268	1,081	1,136	..	Israel
672	725	768	925	1,247	1,063	1,019	997	925	913	Malta
..	..	420	324	310	323	341	359	..	..	Poland
711	767	848	867	914	714	582	578	662	771	Portugal
268	263	281	295	275	266	264	292	326	331	Turkey
1,128	1,372	1,307	1,313	1,263	1,246	1,270	1,240	1,210	1,245	Yugoslavia
180	184	194	199	187	181	186	199	200	210	Developing countries
436	440	468	459	391	300	287	294	293	294	Highly indebted
66	67	69	69	74	78	92	117	120	132	Low-income countries
57	54	57	53	51	43	41	41	41	38	Low-income Africa
74	74	77	78	83	91	109	142	148	163	China and India
33	35	40	38	43	42	41	41	41	43	Other low-income Asia
406	415	440	453	410	381	369	359	353	358	Middle-income countries
3,447	3,049	3,058	3,554	3,163	3,574	3,413	..	..	..	High-income oil exporters
2,345	2,458	2,371	2,314	2,166	2,213	2,475	2,497	2,573	2,720	Industrialized countries
2,866	3,043	2,994	3,059	3,057	2,994	3,163	3,364	3,500	3,801	Japan
2,222	2,192	1,918	2,032	1,649	1,841	2,382	2,283	2,284	2,373	United States
2,202	2,354	2,338	2,084	2,094	2,089	2,168	2,194	2,301	2,432	European Community

Table 5. Gross national income

Average annual growth (percent)	1967	1968	1969	1970	1971	1972	1973	1974	1975	1976	1977
SUB-SAHARAN AFRICA	*-4.0*	*1.5*	*11.5*	*10.4*	*6.8*	*2.5*	*5.7*	*15.4*	*-6.7*	*7.4*	*4.9*
Excluding Nigeria	*2.2*	*2.5*	*7.3*	*1.8*	*1.4*	*2.2*	*4.8*	*5.4*	*-8.3*	*3.9*	*6.2*
Benin	3.8	1.6	6.4	1.5	-1.1	2.8	8.4	-2.3	-5.7	0.2	7.5
Botswana	8.7	16.4	13.4	11.1	5.4	15.2	31.0	30.7	-10.3	13.4	3.5
Burkina Faso	1.8	8.9	4.7	1.0	1.2	0.1	2.5	9.2	-1.2	12.3	0.4
Burundi	7.1	-2.0	-2.2	35.3	-0.6	-7.2	10.3	-2.2	-1.9	17.5	20.3
Cameroon	-11.8	9.1	6.8	5.3	-0.9	-2.0	5.0	14.4	-2.8	2.1	16.4
Cape Verde	..	..	..	..	..	..	..	8.9	-5.5	18.7	22.6
Central African Republic	-10.5	19.5	5.1	3.5	0.4	-0.2	0.9	-2.0	-7.6	16.9	9.3
Chad	-1.4	-0.9	3.9	3.5	-2.3	4.7	-7.4	7.8	-1.5	1.7	-0.5
Comoros	..	..	..	..	..	..	..	..	..	..	..
Congo, People's Republic	6.0	1.0	3.3	3.8	7.0	-0.9	21.5	24.3	-12.9	0.2	-12.4
Cote d'Ivoire	4.7	17.2	8.3	6.9	6.1	3.8	8.0	9.6	5.7	11.5	12.8
Ethiopia	2.3	0.9	2.7	10.1	2.1	4.3	3.4	2.2	-2.1	5.0	2.7
Gabon	..	..	..	..	..	..	..	..	..	..	..
Gambia, The	-2.6	-5.7	16.0	-1.2	9.8	7.2	-1.2	23.3	-2.4	1.7	14.4
Ghana	8.8	-1.7	8.3	6.5	2.6	-2.5	6.6	8.0	-12.9	-5.6	5.9
Guinea-Bissau	..	..	..	..	-4.9	7.0	1.3	2.7	8.3	-5.3	-3.3
Kenya	1.6	6.4	8.0	-0.5	15.2	18.0	0.8	0.6	-2.5	6.0	19.3
Lesotho	15.9	9.3	1.0	1.8	9.1	2.1	28.0	14.6	11.1	11.8	16.2
Liberia	14.7	14.5	15.6	-21.8	8.0	1.9	-8.5	-1.6	7.6	9.6	-3.1
Madagascar	5.4	5.1	8.6	4.6	-0.5	0.1	-2.0	0.7	-0.5	-3.2	2.2
Malawi	13.2	0.8	8.5	4.9	18.5	3.0	5.8	8.6	3.3	-0.7	8.1
Mali	..	5.1	-0.7	5.1	0.4	4.5	-4.5	-4.3	13.4	16.9	7.6
Mauritania	6.2	7.4	-5.8	13.8	-2.2	4.8	-6.6	23.0	-3.0	10.1	-9.4
Mauritius	-1.5	4.9	-16.7	8.7	3.9	15.6	5.3	43.7	6.9	-3.0	3.5
Mozambique	..	..	..	..	..	..	..	..	..	..	..
Niger	-2.6	-0.1	-6.0	1.3	7.1	-5.6	-20.7	16.6	-15.3	4.4	11.0
Nigeria	-14.3	-0.6	20.1	26.5	15.0	3.0	6.9	28.4	-4.9	11.1	3.7
Rwanda	8.7	6.8	6.2	11.1	5.9	1.6	-0.1	3.6	-0.2	3.3	12.0
Senegal	-2.4	-1.8	-8.9	13.2	0.0	5.1	-5.5	5.8	5.6	7.9	-0.2
Seychelles	..	..	..	..	..	..	..	..	..	..	..
Sierra Leone	1.6	-6.0	7.2	13.6	2.0	-8.0	4.2	15.1	-8.7	-8.6	4.8
Somalia	8.3	5.8	2.3	-1.5	4.1	7.0	-3.1	-20.0	36.2	0.8	27.5
Sudan	-5.9	1.6	-3.1	4.2	6.2	-3.9	-9.7	12.1	13.5	16.3	18.3
Swaziland	-4.8	6.9	26.1	2.3	10.5	4.5	3.8	7.5	9.1	-0.5	-3.7
Tanzania	2.6	6.4	3.2	6.2	2.3	4.8	4.5	1.1	1.2	8.0	2.9
Togo	5.6	4.4	10.4	4.5	0.8	3.4	4.0	33.9	-19.7	13.4	9.6
Uganda	4.7	3.9	11.5	1.8	1.0	0.2	1.4	0.7	-5.1	3.2	4.3
Zaire	3.3	-5.1	11.8	-1.8	2.3	-0.2	15.5	-0.5	-14.1	-3.5	3.8
Zambia	6.0	5.1	36.5	-17.4	-21.0	6.9	11.0	1.3	-33.3	9.9	-9.4
Zimbabwe	..	..	..	..	..	..	..	..	..	..	-6.1
SOUTH ASIA	*7.6*	*4.3*	*5.9*	*6.2*	*1.5*	*-0.8*	*2.6*	*0.0*	*7.5*	*3.7*	*8.5*
Bangladesh	-0.6	8.0	1.8	5.5	-7.3	-15.1	3.3	9.3	2.6	11.9	1.6
India	9.2	3.2	6.4	5.9	2.6	0.2	2.0	-1.4	9.1	2.7	9.1
Nepal	..	..	..	..	..	..	..	..	..	..	..
Pakistan	5.1	8.0	5.6	9.8	-1.8	-0.2	7.6	5.1	-1.4	5.0	6.5
Sri Lanka	1.9	10.8	3.7	7.0	0.2	-3.0	7.2	1.8	0.0	12.8	15.3
EAST ASIA AND PACIFIC	*-1.8*	*1.1*	*14.2*	*16.3*	*6.7*	*0.4*	*15.8*	*3.5*	*5.4*	*2.9*	*9.3*
China	-7.3	-6.4	19.8	23.5	6.7	2.9	8.2	1.1	7.7	-4.8	8.5
Fiji	5.4	5.9	5.5	13.9	4.7	8.2	12.7	15.9	7.5	1.1	5.7
Hong Kong	2.3	4.1	11.0	9.8	10.9	12.1	10.9	-2.1	3.0	20.2	11.3
Indonesia	3.1	17.4	6.5	9.0	5.9	6.7	11.9	19.5	3.5	5.7	10.7
Korea, Republic of	6.2	10.7	13.2	8.2	9.4	6.2	14.9	6.5	4.7	16.8	11.9
Malaysia	5.3	1.4	2.2	10.4	2.0	4.8	21.1	8.3	-4.9	18.9	10.4
Papua New Guinea	5.4	4.7	4.9	7.8	8.2	2.7	5.1	14.5	-2.7	-9.1	7.3
Philippines	3.1	7.1	6.0	5.5	5.2	4.4	13.6	4.3	3.0	3.5	7.3
Singapore	..	..	..	..	..	..	..	..	..	..	..
Solomon Islands	..	..	..	..	..	..	..	..	..	..	..
Thailand	7.7	9.4	8.7	6.3	-0.1	5.7	16.2	0.9	2.1	8.5	8.3
Vanuatu	..	..	..	..	..	..	..	..	..	..	..

NOTE: See glossary for definitions.

1978	1979	1980	1981	1982	1983	1984	1985	1986	1987	
-1.2	*0.8*	*8.4*	*1.8*	*-1.0*	*-3.5*	*-3.1*	*5.3*	*-1.0*	*-1.6*	*SUB-SAHARAN AFRICA*
0.0	*2.0*	*3.5*	*1.6*	*0.6*	*0.3*	*0.7*	*3.9*	*1.9*	*0.3*	*Excluding Nigeria*
1.8	5.0	4.8	4.6	5.7	-2.1	1.7	7.8	-1.9	-0.9	Benin
3.8	28.6	24.8	-13.3	12.2	4.9	3.8	23.9	18.1	..	Botswana
1.5	8.7	3.5	4.4	4.1	-0.8	2.4	10.2	11.3	1.8	Burkina Faso
-11.2	3.7	-0.7	7.3	-3.6	3.9	2.7	3.2	10.7	-8.0	Burundi
9.7	7.4	14.6	12.1	6.5	6.2	6.1	18.1	1.9	-6.0	Cameroon
1.0	6.7	34.7	-8.9	-4.6	6.2	6.1	5.9	..	..	Cape Verde
2.1	-4.3	-5.4	-2.9	7.1	-7.8	9.9	2.8	3.0	-1.8	Central African Republic
0.3	-21.1	-8.6	1.4	3.5	8.5	-4.9	32.1	-6.3	0.1	Chad
..	..	..	..	..	..	..	..	..	..	Comoros
-1.2	13.9	33.4	26.8	21.2	2.6	8.7	-2.4	-33.3	..	Congo, People's Republic
6.8	-1.7	2.3	-1.5	0.3	-1.0	0.0	6.2	8.0	-6.9	Cote d'Ivoire
-0.8	6.3	5.9	0.4	2.0	4.2	-6.0	1.6	4.7	..	Ethiopia
..	..	..	..	..	..	..	..	..	..	Gabon
-5.3	2.6	0.6	-4.8	3.1	11.3	..	..	..	..	Gambia, The
9.1	-1.7	1.7	-3.5	-9.4	-2.8	10.0	4.3	4.5	4.0	Ghana
11.4	3.9	-20.3	25.2	2.6	-3.7	7.9	2.0	0.3	2.4	Guinea-Bissau
-0.5	2.9	3.3	0.2	0.6	0.1	4.7	0.8	7.4	2.8	Kenya
8.9	-1.4	0.9	4.1	10.0	-2.2	..	..	..	..	Lesotho
-0.3	1.1	-7.7	-4.9	-10.5	0.9	0.9	4.6	-2.6	..	Liberia
-3.9	6.5	-1.0	-11.0	-0.1	1.3	1.4	0.9	-0.5	1.5	Madagascar
9.0	-4.9	-3.9	-0.9	3.1	3.2	7.7	-1.4	-1.4	20.1	Malawi
-2.9	10.7	-1.4	3.9	7.0	-3.6	3.3	-2.6	15.4	3.6	Mali
-5.8	0.3	4.5	10.2	-4.6	4.3	-2.2	11.0	7.5	1.7	Mauritania
3.7	3.9	-11.7	-1.8	2.6	4.6	3.7	6.0	19.4	18.7	Mauritius
..	..	..	-0.5	-7.2	-11.8	3.4	-9.5	1.4	..	Mozambique
13.7	16.9	1.8	4.4	1.4	1.3	-14.8	3.8	7.8	-4.8	Niger
-2.6	-0.5	13.8	2.1	-2.7	-7.4	-7.3	6.9	-4.1	-3.6	Nigeria
3.1	14.6	4.4	6.5	2.2	6.1	-0.4	2.8	5.2	-6.6	Rwanda
-6.3	8.0	-4.8	1.0	13.5	1.7	-3.7	2.7	3.6	6.2	Senegal
..	..	..	..	..	..	..	..	..	..	Seychelles
12.5	6.3	1.8	10.2	-3.0	-1.1	1.5	-2.7	-2.4	3.7	Sierra Leone
8.0	-8.2	-3.3	17.4	5.7	-13.3	-6.1	14.7	-5.0	11.5	Somalia
-1.9	-10.2	-0.2	1.1	8.0	3.3	-6.7	-11.2	2.0	2.1	Sudan
-3.6	9.6	-0.6	8.7	-13.3	-18.6	18.4	..	..	..	Swaziland
-2.3	0.3	1.2	-1.3	0.5	-1.7	4.9	1.8	5.2	2.4	Tanzania
2.3	-2.2	7.7	-7.8	-3.6	-7.1	3.9	4.3	-0.4	1.1	Togo
-8.5	-15.6	-5.7	..	..	..	..	..	..	..	Uganda
-7.9	5.1	0.1	-0.3	-6.2	-2.0	-3.1	4.1	0.4	2.3	Zaire
-7.8	10.3	-3.1	1.4	-16.0	4.5	3.9	4.6	-17.1	6.5	Zambia
-2.6	2.9	12.0	11.1	2.1	1.1	-0.7	5.4	-5.9	-1.1	Zimbabwe
6.8	*-3.4*	*4.7*	*6.4*	*3.6*	*7.6*	*3.5*	*6.0*	*4.3*	*3.2*	*SOUTH ASIA*
6.4	4.5	2.3	5.6	-1.0	4.1	5.5	5.6	2.8	4.1	Bangladesh
6.7	-5.2	4.1	6.8	3.8	8.4	2.6	6.2	4.5	2.6	India
..	..	..	..	..	..	..	..	..	..	Nepal
7.8	4.4	9.9	4.6	5.3	5.0	6.8	7.7	5.0	7.6	Pakistan
5.1	2.0	2.6	1.8	3.8	9.5	12.9	-5.0	2.5	4.3	Sri Lanka
11.5	*7.9*	*5.9*	*5.4*	*6.4*	*8.8*	*10.7*	*6.7*	*7.4*	*9.4*	*EAST ASIA AND PACIFIC*
13.5	6.6	6.5	4.4	10.5	11.0	14.5	10.9	6.7	8.9	China
1.1	9.0	-0.5	1.1	-7.8	-4.4	8.4	-4.5	8.6	-2.7	Fiji
9.4	12.3	12.2	7.6	4.5	5.6	11.2	2.5	8.3	12.8	Hong Kong
8.7	13.2	18.2	13.0	-0.2	1.8	5.7	2.1	1.8	5.0	Indonesia
12.0	5.9	-8.5	5.6	6.4	11.7	9.6	4.8	16.0	14.4	Korea, Republic of
6.2	13.7	10.7	2.2	3.2	5.4	11.2	-4.6	-10.8	11.5	Malaysia
3.3	3.8	7.7	-14.0	-0.3	7.1	-3.2	4.6	1.6	3.8	Papua New Guinea
6.5	9.2	3.1	2.0	0.0	2.9	-7.8	-4.7	7.5	5.6	Philippines
..	..	..	..	..	..	..	..	..	..	Singapore
..	..	..	-1.2	-0.4	6.1	25.3	-9.6	-5.6	..	Solomon Islands
9.6	5.1	4.4	2.4	2.0	10.6	5.5	1.0	6.1	6.6	Thailand
..	..	..	..	..	..	..	..	..	..	Vanuatu

Table 5. Gross national income (cont'd)

Average annual growth (percent)	1967	1968	1969	1970	1971	1972	1973	1974	1975	1976	1977
LATIN AMERICA AND CARIBBEAN	*3.9*	*7.1*	*7.7*	*9.4*	*6.6*	*7.3*	*9.2*	*6.1*	*3.0*	*6.8*	*5.6*
Argentina	3.4	3.6	8.4	5.0	4.8	2.0	3.9	4.9	-0.5	-0.6	5.7
Bahamas	..	..	..	..	..	..	..	..	..	..	..
Barbados	..	..	..	..	..	..	..	..	..	..	..
Belize	..	..	..	..	..	..	..	..	..	..	..
Bolivia	0.9	4.7	3.9	9.1	2.2	5.2	6.9	12.1	-0.2	6.6	3.3
Brazil	4.5	11.3	10.2	11.0	11.0	12.4	15.5	6.8	4.4	10.8	5.8
Chile	0.4	4.5	7.9	0.1	8.3	-2.1	-3.9	0.1	-23.9	5.8	9.4
Colombia	3.0	6.2	6.8	9.8	5.5	8.8	8.8	4.5	0.8	8.3	7.3
Costa Rica	4.9	7.0	5.7	9.2	3.3	5.9	8.0	1.4	2.9	9.6	15.0
Dominican Republic	2.3	1.3	11.6	10.2	2.6	9.8	11.0	11.3	10.9	-2.6	4.6
Ecuador	7.4	3.0	3.6	9.1	3.8	2.9	11.6	22.1	7.5	10.5	11.9
El Salvador	3.6	-2.0	3.4	11.6	2.3	6.2	7.4	1.0	2.6	19.2	18.3
Grenada	..	..	..	..	..	..	..	..	..	..	..
Guatemala	1.9	9.2	4.4	8.9	2.6	5.2	7.8	4.0	1.0	8.3	14.6
Guyana	13.7	5.6	11.3	-1.0	11.6	-7.4	-7.3	35.8	25.3	-19.8	-4.5
Haiti	-4.7	2.4	2.3	0.9	8.2	0.5	0.2	4.9	-1.3	10.8	2.5
Honduras	5.7	6.8	0.5	0.7	0.4	7.2	8.8	0.0	-0.8	17.1	16.9
Jamaica	1.7	4.1	4.6	13.6	-1.1	12.3	0.1	2.7	5.6	-12.5	1.1
Mexico	6.8	8.0	6.0	13.9	4.1	8.1	8.1	6.6	5.9	3.8	2.7
Nicaragua	-1.7	2.9	3.8	0.8	3.5	4.1	2.3	10.5	-2.0	14.5	6.7
Panama	9.1	6.9	9.8	6.8	11.4	5.8	3.8	1.3	2.4	-1.4	-1.3
Paraguay	7.3	4.3	3.1	5.6	6.4	6.7	11.3	7.3	2.4	12.7	17.1
Peru	0.3	2.5	7.0	4.7	2.4	1.2	10.3	7.9	-1.8	1.0	-0.3
St. Vincent and the Grenadines	..	..	..	..	..	..	..	..	..	..	..
Trinidad and Tobago	4.3	-3.7	-3.0	4.1	27.9	-6.4	0.0	27.5	28.8	9.8	8.5
Uruguay	-8.4	1.8	7.3	6.7	0.8	-0.5	3.2	-3.7	4.0	4.1	0.1
Venezuela	..	..	..	..	..	..	..	..	7.6	11.3	8.6
MIDDLE EAST AND NORTH AFRICA	*9.4*	*6.8*	*8.7*	*5.7*	*9.3*	*8.7*	*4.2*	*22.6*	*4.4*	*10.0*	*6.8*
Algeria	16.8	5.9	9.4	14.4	14.7	18.5	7.5	43.6	-1.5	11.4	7.8
Egypt, Arab Republic of	2.9	5.6	6.8	4.8	2.7	0.6	4.3	2.5	3.7	14.0	13.2
Jordan	..	..	..	..	-0.6	7.1	-1.9	2.6	9.8	23.7	9.3
Morocco	10.2	9.0	7.1	3.3	7.7	0.4	2.6	10.3	4.9	7.1	5.2
Oman	..	..	..	..	..	..	..	..	..	..	..
Syrian Arab Republic	19.0	5.1	22.2	-9.4	17.9	13.5	0.7	52.0	21.2	7.4	-1.7
Tunisia	1.0	7.9	3.8	8.8	9.0	17.9	-0.6	14.3	4.6	1.5	3.6
Yemen Arab Republic	..	..	..	..	21.5	10.9	16.2	5.2	6.2	12.4	7.8
Yemen, PDR	..	..	..	..	..	..	..	..	..	..	..
EUROPE AND MEDITERRANEAN	..	..	..	..	..	*6.8*	*5.2*	*6.5*	*3.2*	*4.6*	*5.3*
Cyprus	..	..	..	..	..	..	..	..	..	24.5	13.9
Greece	5.4	5.8	9.6	7.9	7.6	8.7	8.2	-4.6	4.8	6.1	4.2
Hungary	..	..	..	..	..	5.8	7.0	6.8	0.9	5.2	5.1
Israel	2.1	17.1	13.5	6.8	12.0	13.5	1.5	5.4	2.4	2.4	2.8
Malta	4.4	10.3	10.6	5.6	1.9	3.5	2.9	5.4	19.1	14.2	7.9
Poland	..	..	..	..	..	..	..	..	..	..	..
Portugal	1.5	6.2	19.0	4.8	6.2	7.7	10.0	11.5	0.3	-9.8	4.3
Turkey	3.1	6.3	5.7	5.0	9.2	8.2	4.6	5.8	7.0	9.1	3.9
Yugoslavia	3.2	6.3	11.9	3.8	9.4	2.1	2.6	13.2	0.9	5.5	8.3
Developing countries	2.6	4.7	9.4	10.1	5.7	4.0	10.0	7.7	3.3	6.1	8.7
Highly indebted	2.0	6.7	9.2	10.3	7.8	6.1	8.5	9.2	1.8	7.1	5.5
Low-income countries	0.6	-0.4	11.1	12.0	3.6	1.3	5.3	1.1	4.6	-0.3	8.1
Low-income Africa	2.7	0.9	7.4	1.6	0.8	2.2	4.4	3.8	-10.4	3.2	6.3
China and India	0.0	-1.8	13.0	15.2	4.9	1.8	5.5	0.1	8.3	-1.8	8.8
Other low-income Asia	1.5	8.8	3.9	7.3	-3.0	-5.0	5.3	5.7	0.9	7.7	5.7
Middle-income countries	3.6	7.2	8.6	9.2	6.7	5.3	12.1	10.5	2.8	8.5	8.8
High-income oil exporters	..	..	..	..	..	..	..	..	..	..	..
Industrialized countries	3.8	5.3	5.5	4.1	3.9	5.2	5.9	-1.2	-0.4	4.4	3.4
Japan	10.6	10.5	12.9	12.4	5.9	9.0	6.4	-3.5	1.7	4.6	5.6
United States	2.3	4.6	3.0	0.1	2.8	4.8	6.0	-2.3	-1.0	5.1	4.3
European Community	3.5	4.9	6.0	5.4	4.3	4.3	5.4	-0.6	-0.8	4.1	2.5

NOTE: See glossary for definitions.

1978	1979	1980	1981	1982	1983	1984	1985	1986	1987	
3.4	*6.5*	*7.2*	*0.2*	*-3.5*	*-4.6*	*4.5*	*3.3*	*3.1*	*3.2*	*LATIN AMERICA AND CARIBBEAN*
-3.7	7.8	2.5	-7.7	-8.2	2.6	3.1	-5.2	6.3	1.3	Argentina
2.9	34.3	7.7	-5.4	8.1	6.2	13.2	10.2	4.5	3.4	Bahamas
..	..	..	..	..	..	..	..	..	..	Barbados
..	..	..	..	..	..	..	..	..	..	Belize
0.7	1.4	3.7	1.7	-9.5	-6.9	-0.5	-1.7	-7.0	1.7	Bolivia
2.7	5.8	7.0	-5.5	-0.1	-3.6	6.5	8.1	11.4	3.4	Brazil
7.8	10.1	7.2	2.3	-18.9	0.3	2.1	3.5	6.2	9.7	Chile
6.9	3.2	5.3	0.2	0.8	1.5	3.6	2.2	9.0	3.9	Colombia
2.2	1.7	-0.2	-9.3	-9.4	7.2	10.1	4.1	12.7	-1.2	Costa Rica
-1.1	3.2	11.1	4.1	-0.2	5.1	-1.2	-5.7	4.3	..	Dominican Republic
4.0	9.2	6.3	1.5	-1.9	-2.7	0.4	2.0	-5.2	-4.3	Ecuador
-7.5	-3.0	-11.7	-13.9	-6.4	-2.8	2.9	3.0	8.4	-4.0	El Salvador
..	..	..	..	..	..	..	..	..	..	Grenada
1.4	1.8	1.7	-1.3	-4.5	-2.1	0.8	-2.2	2.8	2.0	Guatemala
-5.1	-8.2	6.3	-7.3	-16.3	-12.1	-9.1	-1.3	8.7	-8.6	Guyana
4.2	4.9	8.0	-3.0	-5.1	1.3	0.4	0.7	1.9	0.6	Haiti
5.2	2.4	1.2	-3.3	-6.2	1.3	3.9	4.2	4.6	2.2	Honduras
-0.7	-11.4	-6.8	-3.1	3.9	1.4	0.4	-9.2	12.1	4.7	Jamaica
8.3	9.5	9.8	8.3	-3.5	-8.4	4.3	4.1	-7.7	3.7	Mexico
-10.2	-31.9	17.7	-5.4	-3.4	3.5	-2.2	-5.3	-7.6	..	Nicaragua
12.5	1.9	7.1	4.1	2.4	0.2	4.6	3.9	5.5	..	Panama
7.4	-2.8	31.0	7.5	-2.6	-2.1	3.0	3.2	-1.7	5.1	Paraguay
-2.6	11.6	5.8	2.4	-0.5	-13.7	4.4	0.7	12.4	9.2	Peru
14.1	-3.4	3.2	10.2	10.9	9.6	8.1	13.5	7.9	..	St. Vincent and the Grenadines
13.6	9.4	14.3	1.6	-3.1	-15.1	-15.8	-4.0	-30.2	..	Trinidad and Tobago
6.0	7.1	5.1	2.5	-10.8	-10.1	-3.9	-0.6	12.5	6.5	Uruguay
-0.6	3.3	7.5	2.5	-6.3	-5.7	6.3	-4.0	-2.3	-1.3	Venezuela
4.9	*12.1*	*11.2*	*6.9*	*5.0*	*4.7*	*4.0*	*4.8*	*-5.2*	*-0.6*	*MIDDLE EAST AND NORTH AFRICA*
5.4	17.9	15.3	9.4	2.5	4.9	4.1	6.2	-12.7	1.6	Algeria
1.5	12.0	11.8	3.9	8.0	7.5	5.2	3.3	-0.1	-2.1	Egypt, Arab Republic of
16.4	-0.2	14.6	3.3	11.2	0.2	6.4	2.6	11.4	-6.5	Jordan
1.5	5.5	1.8	-5.1	7.1	1.8	1.4	3.3	10.4	0.0	Morocco
..	15.7	18.9	32.9	7.2	11.3	16.4	10.8	-20.4	..	Oman
7.9	9.8	10.5	11.8	1.0	1.1	-2.5	3.3	-4.4	-9.2	Syrian Arab Republic
8.0	10.0	10.4	5.6	0.0	5.0	4.0	2.8	-5.9	5.4	Tunisia
11.4	7.7	3.4	3.8	30.1	1.2	2.2	4.1	8.9	4.2	Yemen Arab Republic
..	..	..	..	..	..	..	..	..	..	Yemen, PDR
5.7	*2.4*	*0.9*	*-1.0*	*0.6*	*1.3*	*2.5*	*2.9*	*5.9*	*..*	*EUROPE AND MEDITERRANEAN*
5.3	9.7	5.1	2.5	8.5	2.4	11.9	3.7	5.1	8.0	Cyprus
6.2	3.0	2.1	0.7	-0.7	-0.8	0.6	1.9	0.7	-1.0	Greece
4.0	0.4	0.5	0.6	2.7	0.0	0.6	-1.4	-0.1	3.7	Hungary
4.7	4.1	4.2	4.8	2.0	4.2	-2.3	2.4	4.7	..	Israel
8.2	6.0	10.0	4.4	4.3	-5.4	2.2	2.2	2.1	5.5	Malta
..	..	..	-10.7	-3.1	5.3	5.6	5.6	..	..	Poland
3.5	6.1	1.7	-2.3	0.6	0.1	-2.9	3.4	10.3	7.1	Portugal
2.8	-1.7	-3.0	2.7	3.4	3.3	7.7	5.5	9.9	6.9	Turkey
9.6	4.4	2.0	0.4	0.4	-1.6	0.5	0.6	6.0	-3.2	Yugoslavia
4.1	4.1	5.0	3.7	2.4	2.2	5.1	4.2	4.1	4.5	Developing countries
3.4	5.8	7.1	0.4	-2.7	-4.3	2.5	3.3	3.2	1.9	Highly indebted
8.6	1.9	5.2	4.6	6.5	8.4	8.9	8.3	5.5	6.5	Low-income countries
-1.4	1.6	1.9	0.1	-0.3	0.0	-0.3	2.0	2.6	2.5	Low-income Africa
10.6	1.8	5.6	5.3	7.9	10.0	10.2	9.3	6.0	6.9	China and India
7.0	4.3	6.9	4.9	3.2	4.7	6.9	5.2	3.6	5.7	Other low-income Asia
2.6	4.9	4.9	3.4	1.0	-0.1	3.6	2.5	3.4	3.6	Middle-income countries
..	..	..	..	..	..	..	..	..	..	High-income oil exporters
4.6	2.9	-0.2	1.1	-0.1	2.9	4.7	3.4	4.2	3.3	Industrialized countries
6.7	2.5	0.9	3.7	3.0	3.1	5.1	5.2	7.0	4.2	Japan
5.1	2.1	-1.6	2.2	-2.1	4.0	7.1	3.3	2.9	2.8	United States
4.1	3.5	0.1	-1.1	0.9	1.9	2.2	2.8	5.0	3.3	European Community

Table 6. Gross domestic product

Average annual growth (percent)	1967	1968	1969	1970	1971	1972	1973	1974	1975	1976	1977
SUB-SAHARAN AFRICA	*-4.7*	*1.9*	*11.3*	*13.4*	*9.4*	*4.1*	*4.1*	*8.0*	*-2.6*	*6.6*	*3.9*
Excluding Nigeria	*2.8*	*3.5*	*4.4*	*5.3*	*5.4*	*3.8*	*2.7*	*5.2*	*-2.0*	*3.9*	*1.8*
Benin	3.6	1.1	3.8	2.9	-1.7	6.3	3.5	3.0	-5.0	1.4	4.7
Botswana	8.1	6.3	15.0	12.1	18.6	32.3	21.9	19.9	-1.3	15.0	2.3
Burkina Faso	0.5	8.8	3.1	2.0	1.4	2.2	0.3	8.5	2.8	8.5	0.4
Burundi	9.2	-1.1	-1.3	24.1	2.9	-6.3	6.8	-0.7	0.8	7.9	11.5
Cameroon	-10.9	6.3	4.9	3.1	3.5	2.7	5.4	10.7	-0.8	4.3	8.5
Cape Verde	..	..	..	..	-7.0	-4.6	-0.7	-2.8	3.2	0.0	0.7
Central African Republic	4.7	2.3	6.5	2.2	1.2	1.1	2.2	5.9	0.3	4.7	3.6
Chad	-1.8	-0.5	6.9	1.9	-2.2	1.1	-8.4	5.0	9.0	3.0	2.2
Comoros	..	..	..	..	..	..	..	..	..	..	..
Congo, People's Republic	4.0	7.2	7.4	9.5	7.1	7.9	8.7	10.0	10.4	1.8	-10.1
Cote d'Ivoire	4.7	16.0	7.6	7.6	11.7	6.6	4.8	6.2	10.2	5.8	-0.2
Ethiopia	4.1	1.3	3.4	6.4	4.2	5.1	3.2	1.3	0.2	2.7	2.7
Gabon	4.1	2.5	8.1	8.7	10.3	11.3	10.2	39.5	19.2	35.6	-12.6
Gambia, The	3.5	1.4	10.8	-9.9	11.5	7.2	4.6	19.8	-2.2	9.9	3.1
Ghana	3.0	0.0	5.9	10.0	5.4	-1.8	1.9	7.3	-13.4	-3.5	1.8
Guinea-Bissau	..	..	..	..	-4.6	6.7	0.7	4.3	6.9	4.8	-7.7
Kenya	3.4	7.9	7.9	-4.8	22.0	17.0	5.7	3.6	1.3	2.2	9.4
Lesotho	-0.4	10.9	-0.4	1.5	5.1	-0.2	26.4	11.0	-13.5	11.0	21.8
Liberia	6.2	5.5	6.6	6.5	5.1	4.0	-2.6	4.5	-2.3	5.9	2.6
Madagascar	2.1	5.5	6.8	3.7	3.3	-0.9	-2.2	2.8	1.0	-3.7	2.5
Malawi	13.2	-1.8	6.0	0.6	16.5	5.8	2.8	7.4	6.1	4.8	4.7
Mali	3.3	3.4	-0.9	6.6	1.9	5.1	-3.3	-2.4	13.9	13.6	6.8
Mauritania	3.6	8.9	-2.1	9.7	0.0	3.7	-6.8	10.3	-6.4	8.0	-1.3
Mauritius	-3.6	4.4	-6.9	4.9	4.1	8.1	12.5	9.6	2.8	12.3	8.2
Mozambique	..	..	..	..	..	..	..	..	..	..	..
Niger	0.3	0.1	-3.9	2.7	6.0	-5.7	-18.3	8.8	-2.8	0.7	7.8
Nigeria	-15.5	-0.9	23.9	26.1	14.7	4.4	5.7	11.4	-3.3	9.5	6.1
Rwanda	7.0	7.0	7.0	11.0	6.0	1.2	0.3	3.4	0.8	2.7	5.0
Senegal	-1.4	6.4	-6.7	8.6	15.8	6.1	-5.4	3.9	7.6	8.8	-2.6
Seychelles	0.5	7.6	0.0	8.9	15.8	6.3	9.1	1.3	3.1	3.7	14.2
Sierra Leone	-2.9	-1.0	11.7	10.3	2.4	-1.2	2.9	3.6	3.1	-3.1	1.2
Somalia	9.1	5.9	2.0	-1.7	1.9	9.5	-2.3	-21.3	36.3	-0.2	25.7
Sudan	-3.2	6.2	-2.0	4.9	6.9	-2.1	-8.1	10.1	12.7	18.4	15.2
Swaziland	-4.8	4.2	29.5	1.7	12.2	5.5	11.2	3.3	2.9	-2.3	1.3
Tanzania	4.6	5.6	2.8	6.4	3.0	7.3	5.3	2.1	4.0	3.3	0.4
Togo	5.8	4.8	10.9	2.7	0.1	7.9	4.0	5.0	2.4	-1.8	7.7
Uganda	5.1	3.2	11.7	0.8	1.6	1.0	1.2	1.2	-5.1	2.3	1.7
Zaire	6.8	-1.0	9.3	-0.2	6.0	0.2	8.1	3.1	-5.0	-5.3	0.8
Zambia	7.8	1.2	-0.1	4.4	0.0	9.4	-0.9	6.6	-2.3	6.0	-4.7
Zimbabwe	1.5	8.4	2.0	21.7	8.8	8.5	3.0	6.4	-2.1	0.3	-7.1
SOUTH ASIA	*7.0*	*4.5*	*5.9*	*6.3*	*1.7*	*-1.3*	*3.9*	*1.3*	*8.8*	*2.3*	*7.2*
Bangladesh	-2.0	9.5	1.2	5.6	-5.5	-14.0	3.4	12.1	3.4	12.2	1.3
India	8.4	3.5	6.4	5.8	2.3	-0.7	3.7	0.2	10.0	1.1	8.2
Nepal	..	..	..	..	..	..	..	..	..	..	..
Pakistan	5.5	7.2	5.5	11.4	0.5	0.7	7.1	3.4	4.1	5.3	4.0
Sri Lanka	6.9	5.5	8.3	3.9	1.4	-1.2	7.6	3.8	6.5	3.5	5.1
EAST ASIA AND PACIFIC	*-1.5*	*1.0*	*14.0*	*15.9*	*7.3*	*5.4*	*10.0*	*3.5*	*6.5*	*2.5*	*8.6*
China	-7.3	-6.5	19.2	23.4	7.0	2.9	8.4	1.1	8.3	-5.4	7.9
Fiji	13.5	7.9	2.6	12.7	6.9	7.5	11.6	2.6	0.5	2.7	5.9
Hong Kong	1.9	3.2	11.8	9.4	7.3	11.0	12.7	2.2	0.2	17.1	12.5
Indonesia	1.6	13.9	8.8	9.1	6.5	9.5	11.3	7.7	5.0	6.9	9.0
Korea, Republic of	5.5	10.6	13.3	8.3	9.8	5.6	14.3	8.6	7.5	13.2	10.9
Malaysia	7.8	3.9	8.0	4.9	5.9	9.4	11.7	8.2	0.8	11.7	7.8
Papua New Guinea	5.8	4.1	4.4	8.3	10.8	6.3	5.6	6.5	2.6	-0.9	-3.4
Philippines	6.2	5.5	4.9	4.7	4.9	5.6	8.5	5.0	6.6	7.8	6.2
Singapore	13.0	14.2	13.3	13.3	12.5	13.3	11.2	6.8	4.0	7.1	7.8
Solomon Islands	..	4.7	-2.9	5.1	2.4	-28.3	11.6	26.3	-11.9	13.8	14.4
Thailand	7.3	8.5	8.0	9.2	5.0	3.9	9.8	4.3	4.8	9.3	9.6
Vanuatu	..	..	..	..	..	..	..	..	..	..	..

NOTE: See glossary for definitions.

34

1978	1979	1980	1981	1982	1983	1984	1985	1986	1987	
-1.9	*4.5*	*3.4*	*-1.1*	*1.2*	*-2.2*	*-2.8*	*5.1*	*3.1*	*-1.3*	*SUB-SAHARAN AFRICA*
1.7	*2.1*	*3.4*	*3.3*	*2.3*	*1.1*	*0.9*	*3.4*	*3.8*	*1.3*	*Excluding Nigeria*
1.7	6.4	6.5	9.0	6.8	-2.0	2.3	6.6	0.0	-2.5	Benin
18.4	12.1	14.1	8.6	-1.0	23.8	19.9	8.2	13.9	14.7	Botswana
4.6	4.4	0.0	4.3	6.8	-1.1	1.6	14.8	11.7	3.4	Burkina Faso
-0.9	1.7	1.2	13.4	-2.9	1.1	2.7	4.2	4.4	2.3	Burundi
14.7	13.3	15.6	12.9	2.6	7.8	5.8	11.7	9.1	-2.7	Cameroon
10.6	11.0	5.0	5.2	14.6	5.7	3.5	6.3	5.4	6.0	Cape Verde
2.5	-2.7	-4.6	-2.2	7.5	-6.7	8.8	3.8	1.4	1.0	Central African Republic
-0.5	-21.4	-6.0	1.0	5.3	5.6	-6.0	33.0	-5.3	0.5	Chad
..	..	..	..	..	..	..	..	..	..	Comoros
5.8	12.0	18.8	24.9	17.1	7.7	6.2	-0.7	-4.8	-4.2	Congo, People's Republic
13.9	2.9	-0.8	4.3	1.6	-1.2	-4.4	9.2	3.6	-2.9	Cote d'Ivoire
-1.1	6.4	4.4	2.1	1.6	5.1	-2.2	-7.1	6.7	8.0	Ethiopia
-24.1	0.5	2.5	-4.0	2.7	0.9	6.4	6.3	-5.6	-12.0	Gabon
-3.6	4.5	1.8	-6.5	9.2	13.4	1.9	0.1	5.4	5.7	Gambia, The
9.8	-1.7	0.6	-2.9	-6.5	-4.5	8.7	4.5	5.0	4.4	Ghana
13.6	0.5	-18.9	18.9	4.4	-3.2	5.6	4.6	-0.7	5.7	Guinea-Bissau
6.8	5.2	5.3	4.0	2.6	1.0	2.0	4.2	5.8	4.9	Kenya
18.3	2.9	0.7	-1.4	1.0	-4.4	8.2	1.6	5.8	1.0	Lesotho
5.1	3.2	-4.5	-1.2	-2.0	-1.6	-1.6	-0.7	-1.7	-1.0	Liberia
-3.1	9.6	1.0	-9.2	-1.2	1.1	1.9	2.3	1.0	2.2	Madagascar
10.1	5.6	-0.1	-5.2	2.3	4.0	5.3	4.6	0.9	0.4	Malawi
-2.7	10.9	-1.3	4.6	6.7	-4.5	1.7	-0.3	17.6	3.8	Mali
-0.4	4.6	4.0	3.8	-2.1	4.9	-4.2	3.0	5.4	2.7	Mauritania
5.9	5.7	-9.3	4.8	5.3	0.9	4.8	6.8	9.9	11.1	Mauritius
..	..	..	0.2	-3.4	-12.6	1.7	-9.1	1.5	13.3	Mozambique
13.5	7.1	4.8	1.2	-1.2	-1.8	-14.7	5.9	6.5	-4.9	Niger
-5.5	7.2	3.4	-5.7	-0.2	-6.2	-7.4	7.6	2.3	-4.6	Nigeria
9.2	10.0	10.2	8.8	1.7	6.2	-4.3	4.7	4.1	-2.8	Rwanda
-6.0	9.4	-3.1	-0.6	15.0	2.7	-4.3	3.7	4.4	4.7	Senegal
10.9	9.3	1.3	-3.9	-2.1	-0.6	3.5	9.4	2.4	4.1	Seychelles
-0.3	7.3	3.2	5.6	1.8	-1.6	1.4	-2.4	-3.4	0.5	Sierra Leone
6.3	-6.4	2.3	8.3	2.5	-12.9	5.2	10.3	-3.1	11.0	Somalia
-1.6	-10.4	1.0	2.1	7.5	3.4	-3.7	-13.1	5.4	1.1	Sudan
13.2	-0.3	-4.8	6.7	1.2	-0.8	3.9	3.5	9.0	2.5	Swaziland
1.9	2.8	2.6	-0.4	0.6	-2.1	3.4	2.9	3.8	3.9	Tanzania
10.5	-4.9	14.5	-3.4	-3.3	-5.7	0.5	3.5	3.3	2.0	Togo
-6.6	-15.5	-5.2	9.5	13.1	7.1	-6.2	-6.0	-6.0	4.0	Uganda
-5.3	0.4	2.2	2.9	-2.6	0.8	2.7	2.5	2.7	2.7	Zaire
0.8	-3.1	3.0	6.1	-2.7	-2.1	-0.5	1.4	0.4	-0.2	Zambia
-2.8	2.8	11.4	12.6	3.4	1.3	-1.7	4.7	2.6	-0.1	Zimbabwe
6.8	*-3.0*	*6.6*	*6.8*	*4.1*	*7.2*	*3.7*	*5.9*	*4.6*	*3.1*	*SOUTH ASIA*
6.5	4.5	1.3	6.8	0.8	3.6	4.2	3.6	4.9	4.0	Bangladesh
6.6	-4.8	6.5	6.6	4.0	7.7	3.3	6.1	4.4	2.5	India
..	..	..	..	..	..	..	..	..	..	Nepal
8.1	3.7	10.4	7.9	6.0	6.5	5.6	7.7	7.0	7.0	Pakistan
5.4	6.4	5.8	5.6	3.8	5.8	5.9	1.0	2.5	2.8	Sri Lanka
10.7	*7.3*	*5.7*	*5.7*	*5.8*	*8.1*	*10.3*	*7.5*	*7.4*	*8.8*	*EAST ASIA AND PACIFIC*
12.5	6.9	6.6	4.9	8.8	10.2	14.5	12.7	7.9	9.4	China
1.9	12.2	-1.6	6.3	-6.0	-4.2	8.4	-4.0	7.7	-7.9	Fiji
9.5	11.7	10.9	9.4	3.0	6.5	9.5	-0.1	11.2	13.6	Hong Kong
7.7	6.2	7.9	7.4	-0.4	3.3	6.1	2.4	4.1	3.6	Indonesia
10.9	7.4	-3.0	7.4	5.7	10.9	8.6	5.4	11.7	11.1	Korea, Republic of
6.8	9.5	7.4	6.9	6.0	6.4	7.9	-1.1	1.3	5.5	Malaysia
0.8	8.5	1.8	-0.4	0.3	3.5	1.5	5.2	4.6	5.6	Papua New Guinea
5.5	6.3	5.3	3.8	2.9	1.1	-6.3	-4.5	1.5	5.3	Philippines
8.6	9.3	9.7	9.6	6.9	8.2	8.3	-1.6	1.8	8.8	Singapore
8.9	24.8	-6.0	13.4	2.9	13.9	1.2	0.8	19.8	-5.0	Solomon Islands
10.6	5.0	4.7	6.3	4.0	7.2	7.1	3.5	4.6	6.8	Thailand
..	..	-11.2	3.6	11.6	3.4	6.5	4.3	-1.6	..	Vanuatu

Table 6. Gross domestic product (cont'd)

Average annual growth (percent)	1967	1968	1969	1970	1971	1972	1973	1974	1975	1976	1977
LATIN AMERICA AND CARIBBEAN	*4.4*	*7.3*	*6.3*	*7.0*	*6.0*	*6.8*	*8.3*	*6.1*	*3.3*	*6.0*	*4.7*
Argentina	3.2	4.2	9.2	6.1	3.8	2.0	3.5	5.5	-0.5	-0.2	6.5
Bahamas	9.6	8.4	9.0	-5.6	1.6	-3.5	7.6	-16.8	-14.8	5.2	9.2
Barbados	10.6	6.9	7.5	9.5	3.9	1.3	1.4	-4.7	4.0	0.0	4.3
Belize	4.8	4.9	4.9	4.9	4.8	3.9	10.2	5.4	8.1	5.0	3.6
Bolivia	6.9	5.8	3.1	-0.5	4.9	5.8	6.7	5.1	6.6	6.1	4.2
Brazil	4.9	11.4	9.8	8.8	11.3	12.1	14.0	9.0	5.2	9.8	4.6
Chile	3.2	3.6	3.9	2.0	9.1	-1.2	-5.5	0.8	-13.2	3.6	9.8
Colombia	4.1	6.4	6.5	7.2	6.2	7.8	7.1	5.6	2.1	4.7	4.1
Costa Rica	5.5	8.2	5.4	7.5	6.7	8.0	7.7	5.4	2.0	5.4	8.9
Dominican Republic	3.0	0.4	10.8	10.4	10.7	10.0	12.5	5.8	4.8	6.7	4.8
Ecuador	7.0	3.8	2.6	6.4	6.6	16.2	28.1	5.8	5.3	9.7	6.4
El Salvador	5.4	3.3	3.5	3.0	4.7	5.3	5.0	6.5	3.7	3.7	5.8
Grenada	1.3	1.1	17.6	0.5	6.9	11.1	-13.9	-20.9	11.0	9.1	5.9
Guatemala	4.1	8.8	4.7	5.7	5.6	7.3	6.8	6.4	1.9	7.4	7.8
Guyana	4.3	1.1	7.1	4.1	3.1	-3.5	0.9	7.9	7.9	1.4	-2.7
Haiti	-4.0	1.8	2.9	-0.3	8.4	1.4	1.2	6.3	-2.3	8.6	0.5
Honduras	6.0	6.6	0.9	4.0	3.8	6.0	7.9	-0.9	2.5	10.5	10.5
Jamaica	1.0	5.3	3.8	12.0	3.2	8.8	0.2	-3.3	-0.8	-7.0	-2.8
Mexico	6.4	8.1	6.3	7.4	4.1	8.5	8.3	6.1	5.7	4.2	3.2
Nicaragua	6.5	1.4	6.1	0.6	4.9	3.2	5.1	12.7	2.2	5.0	0.9
Panama	8.6	7.0	8.5	7.1	9.6	4.7	5.4	2.6	1.7	1.6	1.1
Paraguay	8.0	3.0	4.0	5.0	5.2	6.3	7.1	8.3	6.4	6.8	10.7
Peru	3.8	0.4	3.9	5.7	4.2	2.9	5.4	9.3	3.5	1.9	0.4
St. Vincent and the Grenadines	0.0	-9.5	6.5	2.9	10.7	3.0	25.8	-11.1	-8.8	-7.6	5.5
Trinidad and Tobago	7.1	6.0	2.4	3.7	1.8	6.2	1.3	3.6	0.7	8.3	6.7
Uruguay	-6.3	2.7	6.0	6.3	0.1	-1.5	0.3	3.1	6.0	4.0	1.3
Venezuela	2.9	7.4	0.7	7.7	1.5	1.3	7.1	2.1	2.9	7.7	6.3
MIDDLE EAST AND NORTH AFRICA	*4.7*	*9.7*	*10.0*	*6.7*	*2.3*	*12.6*	*2.2*	*6.8*	*6.2*	*7.2*	*9.1*
Algeria	9.5	10.8	8.4	9.7	-8.0	25.5	3.3	9.7	5.2	9.1	6.4
Egypt, Arab Republic of	0.6	2.7	7.0	5.8	3.6	1.8	0.8	2.7	9.1	15.3	13.5
Jordan	..	..	..	..	1.8	10.5	-1.8	-2.0	9.8	23.3	12.7
Morocco	6.9	11.2	7.0	4.6	6.2	2.3	2.6	4.1	5.9	11.8	6.2
Oman	..	..	..	..	..	..	..	..	..	..	..
Syrian Arab Republic	6.9	5.7	20.0	-5.9	9.5	16.6	-5.3	26.6	22.6	8.9	-1.4
Tunisia	0.0	9.6	5.6	6.9	10.6	17.7	-0.6	8.1	7.1	7.9	3.4
Yemen Arab Republic	..	..	..	..	23.0	7.0	13.9	5.2	11.7	11.2	6.7
Yemen, PDR	..	..	..	..	..	..	..	..	..	24.5	21.3
EUROPE AND MEDITERRANEAN	*4.5*	*7.2*	*9.2*	*5.4*	*8.6*	*6.2*	*4.9*	*7.9*	*4.4*	*4.8*	*6.2*
Cyprus	..	..	..	..	..	..	..	..	..	20.3	15.8
Greece	5.2	5.9	9.7	8.0	7.0	8.8	7.1	-3.2	6.0	6.0	3.0
Hungary	7.6	5.0	6.8	4.7	6.2	6.1	6.9	5.9	6.2	3.5	6.7
Israel	3.0	16.2	13.6	7.3	11.2	13.7	3.3	6.8	3.3	1.3	0.0
Malta	7.0	10.2	6.4	12.6	2.5	5.8	9.8	9.9	19.6	17.0	12.2
Poland	..	..	..	..	..	..	..	..	..	..	..
Portugal	4.1	7.6	8.9	2.1	9.1	6.6	8.0	11.2	1.2	-4.3	6.9
Turkey	4.5	7.1	5.6	5.0	8.9	6.5	4.6	8.6	8.9	8.8	4.7
Yugoslavia	3.6	6.4	11.6	4.9	9.1	1.8	2.4	14.8	0.7	5.3	8.3
Developing countries	2.9	5.5	9.0	9.7	5.7	5.7	7.2	5.3	4.2	5.4	6.1
Highly indebted	2.1	6.6	8.3	8.8	7.3	6.0	7.5	7.4	2.6	6.6	5.1
Low-income countries	0.3	-0.2	10.3	12.5	4.4	1.2	5.6	1.5	6.3	-1.4	7.0
Low-income Africa	3.6	2.1	4.0	4.1	4.4	2.7	1.8	3.1	-4.9	1.6	3.8
China and India	-0.6	-1.9	12.9	15.2	5.0	1.4	6.5	0.7	9.0	-2.8	8.0
Other low-income Asia	1.6	8.5	4.0	8.0	-0.9	-3.8	4.9	6.1	4.1	7.2	3.5
Middle-income countries	..	..	..	..	..	..	..	..	..	..	..
High-income oil exporters	10.5	12.7	9.6	6.4	7.2	6.1	9.0	2.9	0.3	11.1	11.2
Industrialized countries	3.6	5.1	5.3	3.8	3.5	4.9	5.8	0.6	-0.3	4.5	3.6
Japan	10.5	10.4	12.6	12.1	4.3	8.4	7.9	-1.2	2.6	4.8	5.3
United States	2.3	4.2	2.9	-0.1	2.7	4.9	4.9	-0.7	-1.0	4.8	4.6
European Community	3.3	4.9	5.7	4.9	3.7	4.0	6.0	1.7	-1.0	4.4	2.6

1978	1979	1980	1981	1982	1983	1984	1985	1986	1987	
4.9	*6.5*	*6.0*	*1.4*	*-1.2*	*-2.8*	*3.6*	*3.7*	*3.6*	*2.6*	*LATIN AMERICA AND CARIBBEAN*
-3.4	7.2	1.9	-6.9	-5.5	2.9	2.5	-4.5	5.7	1.6	Argentina
14.2	26.1	6.5	-9.2	6.7	3.6	14.2	4.8	1.8	3.0	Bahamas
6.1	7.7	4.7	-3.2	-5.0	0.2	3.4	0.4	9.0	1.4	Barbados
6.5	7.0	3.6	1.6	-2.1	-3.5	5.4	0.3	5.8	..	Belize
3.4	0.0	-0.9	0.9	-4.4	-6.5	-0.3	-0.1	-2.9	2.4	Bolivia
4.8	7.2	9.1	-3.1	1.1	-2.8	5.7	8.4	8.0	3.0	Brazil
8.4	8.3	7.8	5.6	-14.2	-0.7	6.3	2.4	5.6	5.7	Chile
8.2	5.2	4.4	2.2	1.0	1.6	3.5	3.3	5.4	5.4	Colombia
6.2	5.0	0.7	-2.3	-7.0	2.8	7.8	1.0	4.6	3.0	Costa Rica
2.3	4.3	5.9	4.0	1.7	4.6	-1.9	-1.2	0.8	..	Dominican Republic
7.0	5.4	4.6	4.1	1.1	-2.1	4.0	4.2	3.0	-6.4	Ecuador
6.7	-1.6	-8.5	-8.2	-5.5	0.6	2.4	1.8	0.5	2.6	El Salvador
8.1	2.3	-3.1	8.0	0.6	3.5	4.7	7.3	3.8	6.0	Grenada
5.0	4.7	3.7	0.7	-3.5	-2.6	0.5	-0.6	0.1	3.1	Guatemala
-1.5	-2.1	1.8	1.8	-13.6	-6.6	0.2	0.9	1.5	0.4	Guyana
4.8	7.3	7.6	-2.8	-3.4	0.8	0.3	0.2	0.6	0.5	Haiti
8.3	6.3	1.3	1.5	-2.0	-0.3	2.8	3.2	2.8	4.1	Honduras
0.4	-2.0	-6.2	2.5	0.8	2.4	-1.1	-4.9	2.1	5.3	Jamaica
8.2	9.3	8.4	8.8	-0.6	-4.2	3.6	2.6	-4.0	1.4	Mexico
-7.0	-23.4	10.0	0.1	-0.8	4.6	-1.6	-4.1	-0.6	1.7	Nicaragua
9.8	4.6	15.4	4.2	5.6	0.4	-0.4	4.7	3.0	2.8	Panama
10.9	11.1	14.5	8.6	-1.5	-3.6	3.5	3.1	0.0	4.0	Paraguay
0.2	5.8	4.5	4.5	0.2	-12.3	4.7	2.3	9.5	6.9	Peru
10.8	3.4	2.0	8.0	8.9	5.6	6.1	6.2	4.1	2.7	St. Vincent and the Grenadines
10.9	2.4	7.6	3.8	-4.7	-7.1	-9.5	-5.0	-2.0	-2.3	Trinidad and Tobago
5.4	6.2	5.9	1.8	-9.5	-6.0	-1.4	0.2	6.7	5.1	Uruguay
2.4	0.8	-4.5	-0.3	-2.1	-3.8	-1.1	1.3	6.8	3.0	Venezuela
6.3	*7.0*	*4.8*	*4.3*	*6.3*	*5.4*	*4.4*	*5.9*	*2.3*	*0.8*	*MIDDLE EAST AND NORTH AFRICA*
9.3	10.6	0.9	3.6	4.0	5.4	4.1	5.2	0.9	1.3	Algeria
5.9	6.2	10.3	3.8	10.1	7.6	6.2	6.7	2.7	2.5	Egypt, Arab Republic of
15.4	0.7	19.0	4.7	7.4	2.0	8.6	3.5	5.4	-1.4	Jordan
2.9	4.6	3.3	-2.6	7.2	2.0	1.7	4.7	6.4	0.2	Morocco
1.0	1.1	3.8	16.7	10.6	18.1	16.1	15.0	5.2	2.3	Oman
7.9	4.2	8.4	9.5	2.9	1.8	-3.6	3.0	-1.2	-9.3	Syrian Arab Republic
6.4	6.6	7.4	5.5	-0.5	4.7	5.7	5.7	-1.6	5.8	Tunisia
10.8	6.5	5.2	7.5	33.0	2.1	3.3	4.6	9.4	4.7	Yemen Arab Republic
8.9	4.2	-7.3	6.6	13.4	0.4	8.8	-6.6	-11.8	3.2	Yemen, PDR
6.2	*3.7*	*2.2*	*0.2*	*1.2*	*2.0*	*3.3*	*2.9*	*4.2*	*2.9*	*EUROPE AND MEDITERRANEAN*
7.6	9.9	5.9	3.1	6.3	5.3	8.9	4.7	4.1	7.2	Cyprus
6.8	3.4	2.1	0.0	0.5	0.1	2.9	3.1	1.2	-0.5	Greece
4.6	1.6	0.0	2.9	2.8	0.7	2.6	-0.3	1.5	3.4	Hungary
4.8	6.3	6.0	4.3	1.2	3.0	0.5	2.4	2.0	4.6	Israel
11.2	10.5	7.0	3.3	2.3	-0.6	0.9	2.6	4.0	4.5	Malta
..	..	..	-10.0	-4.8	5.6	5.6	4.6	4.9	2.2	Poland
3.5	5.9	4.5	1.1	3.1	-0.1	-1.7	3.3	4.3	5.0	Portugal
3.3	-0.9	-0.7	4.2	4.9	3.8	5.9	4.9	8.1	7.2	Turkey
9.2	5.3	2.6	1.2	0.6	-1.1	1.8	1.0	3.4	-0.5	Yugoslavia
4.4	4.2	4.2	2.9	2.7	2.9	4.9	4.8	4.8	4.4	Developing countries
4.0	6.6	5.3	0.6	-0.6	-2.8	1.9	3.8	3.5	1.7	Highly indebted
8.5	2.1	6.0	5.1	6.1	7.9	8.9	9.1	6.4	6.7	Low-income countries
1.1	0.8	2.4	1.5	2.1	0.6	0.3	1.6	3.9	2.9	Low-income Africa
10.1	2.2	6.5	5.6	7.0	9.3	10.4	10.4	6.8	7.2	China and India
7.2	4.3	7.1	7.2	4.4	5.4	5.2	5.5	5.4	5.3	Other low-income Asia
3.0	4.9	3.6	2.2	1.6	1.2	3.4	3.1	4.1	3.3	Middle-income countries
4.0	10.6	3.7	-2.0	-3.0	-9.7	-0.3	-7.2	-8.6	4.2	High-income oil exporters
4.1	3.2	1.3	1.6	-0.2	2.6	4.7	3.4	2.7	3.3	Industrialized countries
5.1	5.2	4.5	3.9	2.8	3.2	5.0	4.7	2.4	4.3	Japan
5.2	2.1	-0.1	2.0	-2.5	3.7	7.0	3.6	3.1	3.6	United States
3.1	3.5	1.0	0.2	1.0	1.6	2.4	2.4	2.7	2.8	European Community

Table 7. Agriculture: contribution to gross domestic product growth

Percentage points	1967	1968	1969	1970	1971	1972	1973	1974	1975	1976	1977
SUB-SAHARAN AFRICA	*-3.3*	*0.0*	*3.2*	*4.2*	*1.7*	*-0.8*	*1.1*	*2.0*	*-2.3*	*-0.3*	*1.4*
Excluding Nigeria	*1.0*	*0.5*	*0.5*	*1.3*	*1.2*	*1.1*	*-0.7*	*0.4*	*-1.0*	*-0.2*	*0.8*
Benin	..	..	..	..	-1.9	1.8	0.9	-2.0	-2.4	4.9	-0.2
Botswana	2.9	2.4	6.7	-5.2	3.9	6.8	4.6	8.2	-2.7	0.3	0.9
Burkina Faso	..	..	..	..	0.6	-0.2	-2.8	7.2	-0.4	1.9	-2.8
Burundi	-2.9	-3.0	-0.2	24.3	2.2	-6.8	8.0	-2.8	3.1	2.5	2.8
Cameroon	1.9	1.9	2.0	1.8	0.5	2.1	1.5	1.8	0.4	0.5	0.5
Cape Verde	..	..	..	..	..	..	..	..	..	..	..
Central African Republic	0.9	1.7	0.6	0.7	1.3	0.2	0.3	1.7	0.9	1.4	0.7
Chad	..	-0.7	2.9	-0.3	-1.1	1.5	-6.2	1.4	3.8	-0.5	0.5
Comoros	..	..	..	..	..	..	..	..	..	..	..
Congo, People's Republic	1.2	0.9	-0.6	1.5	0.6	0.6	0.6	0.5	0.5	-0.6	0.5
Cote d'Ivoire	-0.2	7.4	0.2	1.8	3.3	0.2	2.3	-4.0	3.0	1.8	-1.5
Ethiopia	2.0	0.5	1.3	1.4	1.2	2.0	0.4	-0.3	-1.0	1.5	-0.2
Gabon	..	..	..	..	..	..	..	..	..	..	..
Gambia, The	1.2	0.5	3.8	-3.4	4.0	2.4	1.6	6.8	-2.1	0.5	-0.1
Ghana	2.7	-0.7	3.8	6.2	2.6	2.1	-1.2	5.0	-11.4	-0.9	-2.9
Guinea-Bissau	..	..	..	..	-5.6	5.6	-0.6	2.4	1.0	5.0	-8.5
Kenya	0.7	2.0	3.5	-3.2	6.3	4.3	1.0	-0.9	2.5	0.7	3.5
Lesotho	..	..	..	..	1.7	-11.4	7.6	9.7	-11.5	-2.2	9.0
Liberia	1.4	1.7	2.2	2.7	1.3	0.7	2.7	1.8	0.5	2.9	0.9
Madagascar	..	..	..	..	-0.8	0.7	0.4	3.4	-0.2	-3.4	1.2
Malawi	..	-2.8	1.8	-0.5	6.9	4.0	-2.7	0.8	0.3	4.2	4.2
Mali	1.5	1.6	-3.4	4.9	1.5	1.0	-6.7	-3.6	13.2	8.5	4.7
Mauritania	1.8	2.0	-8.3	-1.6	-4.6	12.7	-7.7	-0.1	-5.5	0.8	1.7
Mauritius	..	..	..	..	2.6	4.3	1.6	-0.5	-8.5	2.1	0.0
Mozambique	..	..	..	..	..	..	..	..	..	..	..
Niger	0.5	-1.6	-5.3	3.9	3.6	-6.5	-21.5	11.5	-11.9	-4.4	7.0
Nigeria	-7.6	-0.7	7.4	8.0	2.2	-2.8	3.1	3.6	-3.6	-0.5	2.0
Rwanda	..	..	..	3.7	1.1	-2.0	1.8	6.9	-2.5	11.8	0.4
Senegal	2.0	0.1	0.4	1.5	-4.0	4.8	-3.8	4.8	0.9	3.6	-1.7
Seychelles	..	..	..	..	..	..	..	..	..	..	-0.7
Sierra Leone	1.1	-0.5	2.1	0.0	0.7	-0.3	0.6	-0.3	1.3	1.7	1.7
Somalia	..	..	..	..	-1.1	4.0	-2.8	-22.0	39.2	2.3	23.4
Sudan	-0.5	1.0	-4.5	10.0	4.3	-0.7	-4.4	8.0	2.5	1.7	7.4
Swaziland	-7.6	-2.7	9.9	2.2	6.3	-1.4	2.4	-0.1	-0.9	-0.6	2.1
Tanzania	0.0	2.3	0.2	2.1	-0.7	4.2	0.5	-2.1	4.1	-2.6	0.5
Togo	1.7	2.7	2.4	-2.5	-0.2	1.8	1.0	0.5	0.9	-1.5	-1.8
Uganda	1.4	1.0	9.4	0.8	-1.6	2.5	3.5	-0.7	-0.2	0.5	1.9
Zaire	..	..	0.2	-4.0	0.7	0.4	1.0	0.5	-0.4	1.3	-0.9
Zambia	-0.1	0.1	0.2	0.6	0.3	0.6	-0.2	0.6	0.6	1.0	0.2
Zimbabwe	..	..	..	-1.4	3.9	2.2	-3.4	2.5	-0.9	1.7	-3.5
SOUTH ASIA	*4.8*	*1.3*	*2.3*	*3.5*	*-0.4*	*-2.4*	*2.4*	*0.1*	*4.0*	*-1.4*	*3.7*
Bangladesh	-1.6	5.9	0.6	3.1	-2.6	-6.2	0.2	6.1	-0.6	6.5	-1.7
India	6.2	0.3	2.6	3.7	-0.2	-2.7	3.0	-0.8	5.2	-2.6	4.6
Nepal	..	..	..	..	..	..	..	..	..	..	..
Pakistan	2.0	4.3	1.7	3.6	-1.1	1.2	0.6	1.5	-0.7	1.4	0.8
Sri Lanka	2.8	2.0	0.4	1.2	-0.8	1.0	-0.2	1.7	-0.7	0.4	3.0
EAST ASIA AND THE PACIFIC	*0.2*	*0.5*	*0.7*	*2.9*	*1.0*	*0.0*	*2.7*	*1.3*	*0.6*	*1.0*	*-0.1*
China	0.7	-0.8	0.0	4.7	0.9	-0.5	3.3	1.5	0.8	0.4	-0.8
Fiji	0.2	2.2	-1.0	2.5	-1.6	-0.2	1.5	-0.8	0.1	0.8	3.4
Hong Kong	..	..	..	..	..	..	..	..	..	..	..
Indonesia	-0.4	4.1	0.9	1.8	2.0	0.5	2.8	1.1	0.0	1.3	0.3
Korea, Republic of	-2.4	0.5	3.5	-0.5	1.0	0.6	1.9	1.8	1.0	2.4	0.7
Malaysia	..	..	..	..	0.4	2.2	3.3	1.9	-0.8	3.2	0.6
Papua New Guinea	1.3	0.9	1.6	0.8	0.7	2.0	1.2	1.8	1.3	0.4	0.3
Philippines	0.9	1.9	0.8	0.6	1.3	1.1	1.6	0.7	1.0	2.0	1.2
Singapore	0.1	0.2	0.1	0.1	0.3	0.2	-0.1	-0.1	0.0	0.2	0.0
Solomon Islands	..	..	..	..	..	..	..	..	..	..	..
Thailand	-1.0	3.0	2.4	1.3	1.3	-0.5	2.7	0.8	1.2	1.7	0.6
Vanuatu	..	..	..	..	..	..	..	..	..	..	..

NOTE: See glossary for definitions.

1978	1979	1980	1981	1982	1983	1984	1985	1986	1987	
-0.9	*-0.3*	*0.9*	*-1.9*	*1.0*	*-0.4*	*-0.8*	*1.9*	*2.4*	*-0.8*	*SUB-SAHARAN AFRICA*
1.0	*0.2*	*0.6*	*0.6*	*1.2*	*-0.7*	*0.0*	*0.6*	*2.3*	*0.6*	*Excluding Nigeria*
4.9	2.4	1.5	-3.4	0.8	0.4	0.6	5.4	2.5	-3.9	Benin
-1.3	0.0	-0.5	-1.2	-0.4	-1.5	-1.0	-0.4	0.6	-0.4	Botswana
2.5	0.8	-2.6	3.3	1.0	0.0	0.0	9.1	5.9	-1.1	Burkina Faso
-1.8	0.9	1.1	10.9	-6.2	2.4	-3.2	4.7	1.9	2.1	Burundi
1.2	4.4	2.5	3.7	1.1	-2.1	2.1	-1.1	1.7	0.7	Cameroon
..	..	..	..	..	..	..	..	..	..	Cape Verde
1.2	-0.6	-0.6	-0.7	0.6	-1.6	3.7	1.5	1.2	1.5	Central African Republic
0.6	-4.3	2.0	-5.9	2.4	2.3	-8.3	16.3	0.8	-3.1	Chad
..	..	..	..	..	..	..	..	..	..	Comoros
-0.1	0.9	1.0	-0.3	0.4	0.4	-0.1	0.1	0.4	-0.3	Congo, People's Republic
3.3	0.1	3.8	0.7	-0.4	-3.4	0.8	2.9	3.5	-0.3	Cote d'Ivoire
-0.7	1.3	2.5	1.2	-0.6	2.3	-4.8	-7.4	3.8	4.7	Ethiopia
..	..	..	..	..	..	..	..	..	..	Gabon
-4.5	9.2	-5.1	-3.5	6.6	6.7	-2.0	1.4	3.8	1.1	Gambia, The
9.5	2.0	1.2	-1.5	-3.2	-4.1	5.6	0.4	1.8	0.0	Ghana
9.1	-5.0	-21.6	12.6	2.7	-1.0	1.6	3.2	2.1	4.6	Guinea-Bissau
1.4	-0.2	0.4	1.9	3.5	0.6	-1.2	1.3	1.6	1.3	Kenya
4.2	1.3	-9.1	-0.6	-3.3	2.6	2.1	1.0	-3.3	1.2	Lesotho
1.3	0.9	0.5	-1.4	-0.3	1.1	1.1	0.9	0.1	..	Liberia
-2.5	2.6	0.9	-1.6	1.5	1.0	1.3	0.8	1.2	0.9	Madagascar
1.2	1.2	-2.5	-2.9	2.2	-0.2	3.7	0.2	0.4	0.6	Malawi
-4.5	9.1	-1.9	-1.0	4.3	-6.9	-4.1	-2.7	13.9	3.0	Mali
-0.9	0.5	3.3	2.7	0.5	5.1	-6.9	0.8	2.9	0.9	Mauritania
0.0	0.7	-5.8	2.7	2.8	-2.1	0.1	1.5	1.5	-0.3	Mauritius
..	..	..	-1.4	-3.0	-8.2	-4.7	-8.3	..	..	Mozambique
6.6	-2.8	2.2	-0.4	-1.1	4.7	-4.7	7.3	4.2	..	Niger
-2.5	-0.9	1.3	-4.3	0.7	0.0	-1.7	3.6	2.5	-2.2	Nigeria
1.4	10.0	-0.8	-2.9	5.7	1.9	-3.9	1.0	0.2	-1.3	Rwanda
-5.7	5.3	-4.1	-1.1	4.5	1.0	-3.8	1.4	4.3	0.3	Senegal
-0.6	1.1	0.0	1.9	-2.1	0.6	-0.8	-0.3	0.2	..	Seychelles
0.7	3.2	-1.1	0.3	1.0	-0.3	0.1	0.0	3.4	-1.3	Sierra Leone
3.6	-8.1	4.8	10.2	1.3	-12.8	9.8	9.0	-5.0	11.0	Somalia
0.8	-6.9	-2.4	-1.5	10.3	-2.9	-0.9	-6.7	6.2	1.8	Sudan
3.8	-1.0	1.9	3.1	-1.3	-0.3	1.2	2.3	2.7	..	Swaziland
-0.8	0.3	1.7	0.4	0.6	1.3	1.9	2.9	2.8	2.2	Tanzania
4.4	-1.3	4.5	-0.4	-2.5	0.2	3.6	0.6	-1.5	0.6	Togo
-0.4	-10.6	-4.6	4.9	7.4	3.8	-7.9	-2.6	-3.4	2.0	Uganda
0.0	0.9	0.8	0.8	0.6	0.6	0.9	1.0	1.9	1.0	Zaire
0.1	-0.8	-0.3	1.2	-1.7	1.1	0.8	0.5	1.4	-0.2	Zambia
1.4	0.0	0.2	2.0	-1.0	-2.1	2.7	3.3	-0.8	-2.8	Zimbabwe
1.4	*-3.9*	*3.7*	*2.2*	*-0.2*	*3.4*	*-0.5*	*0.5*	*-0.3*	*-1.5*	*SOUTH ASIA*
4.6	-0.8	0.1	2.7	0.5	2.3	0.8	0.5	1.9	0.1	Bangladesh
1.2	-5.1	4.4	2.2	-0.6	3.9	-0.7	0.1	-0.9	-2.1	India
..	..	..	..	..	..	..	..	..	..	Nepal
0.9	0.9	2.0	1.1	1.0	1.1	-1.6	3.0	1.5	0.5	Pakistan
1.6	0.6	0.9	1.9	0.7	1.4	-0.1	2.2	0.7	-1.5	Sri Lanka
1.1	*2.4*	*-0.4*	*1.7*	*2.0*	*1.5*	*2.3*	*0.9*	*0.8*	*1.0*	*EAST ASIA AND THE PACIFIC*
1.9	3.8	-0.6	2.0	3.6	2.6	4.0	1.1	1.0	1.7	China
-0.3	3.8	-1.5	2.9	0.4	-4.4	5.3	-3.3	4.1	-1.4	Fiji
..	..	..	..	..	..	..	..	..	..	Hong Kong
1.3	1.6	1.7	1.2	0.3	0.4	1.0	1.0	0.6	0.6	Indonesia
-2.1	1.3	-3.5	3.3	0.5	1.1	0.0	0.7	0.6	-0.6	Korea, Republic of
0.4	1.4	0.3	1.1	1.4	-0.1	0.6	0.5	0.8	1.5	Malaysia
0.8	0.2	0.8	2.6	-0.5	1.1	-1.0	0.9	1.4	3.0	Papua New Guinea
0.9	1.1	1.1	0.9	0.8	-0.5	0.5	0.8	1.0	0.1	Philippines
0.0	0.0	0.0	0.0	-0.1	0.0	0.0	-0.1	-0.1	-0.1	Singapore
..	..	..	..	..	..	..	..	..	..	Solomon Islands
3.2	-0.5	0.4	1.2	0.7	1.0	1.2	1.3	0.3	-0.5	Thailand
..	..	-3.7	4.2	-2.6	14.3	1.0	-1.1	1.0	..	Vanuatu

Table 7. Agriculture: contribution to gross domestic product growth (cont'd)

Percentage points	1967	1968	1969	1970	1971	1972	1973	1974	1975	1976	1977
LATIN AMERICA AND CARIBBEAN	*0.7*	*0.4*	*0.4*	*0.3*	*0.7*	*0.2*	*0.3*	*0.4*	*0.4*	*0.2*	*0.8*
Argentina	0.3	-0.5	0.5	-1.0	0.1	0.2	0.9	0.3	-0.3	0.4	0.2
Bahamas	..	..	..	..	..	..	..	..	..	..	..
Barbados	3.3	-4.0	-1.8	0.9	-1.5	-2.1	0.5	-0.6	-0.4	0.7	0.2
Belize	..	..	..	..	2.1	1.1	1.7	2.8	1.1	-3.1	9.5
Bolivia	-0.3	1.3	-0.2	0.7	1.1	1.1	0.9	0.7	1.4	0.9	-0.1
Brazil	1.6	0.9	0.7	0.2	1.6	0.6	0.0	0.1	0.8	0.3	1.3
Chile	0.2	0.4	-1.0	0.2	-0.1	-0.6	-0.6	1.6	0.3	-0.1	0.9
Colombia	1.3	1.7	0.9	0.6	0.2	1.8	0.5	1.2	1.3	0.6	0.7
Costa Rica	1.8	2.1	2.4	1.0	1.1	1.3	1.3	-0.4	0.6	0.1	0.4
Dominican Republic	-0.1	1.1	2.4	1.8	2.3	0.9	1.9	0.0	-0.5	1.4	0.3
Ecuador	1.5	0.9	-0.1	1.6	1.1	0.9	0.2	1.4	0.4	0.4	0.3
El Salvador	1.6	0.5	1.0	1.8	1.1	0.4	0.5	2.7	1.7	-2.2	0.9
Grenada	..	..	..	..	..	..	..	..	..	..	..
Guatemala	..	..	..	..	..	..	..	..	..	..	..
Guyana	1.2	-0.3	2.1	-0.1	1.5	-2.3	0.0	3.7	-0.9	0.8	-0.7
Haiti	..	..	..	..	..	..	..	..	..	..	..
Honduras	1.4	2.5	-1.4	-1.5	2.9	0.6	1.8	-2.8	-1.9	2.9	1.6
Jamaica	-0.3	-0.3	-0.4	0.4	1.0	0.1	-1.0	0.3	0.1	0.0	0.2
Mexico	0.4	0.4	0.1	0.6	0.6	0.1	0.4	0.3	0.2	0.1	0.7
Nicaragua	1.6	-0.9	2.7	-1.9	1.6	0.2	1.2	1.8	1.3	0.3	0.8
Panama	0.8	0.9	0.8	-0.1	1.0	-0.4	0.1	-0.7	0.8	0.6	0.6
Paraguay	0.8	0.4	0.8	2.3	0.5	1.4	2.2	3.4	2.9	1.3	2.3
Peru	0.8	-0.4	0.7	1.5	-0.3	-0.9	-0.1	0.6	-0.1	0.3	-0.1
St. Vincent and the Grenadines	..	..	..	..	..	..	..	..	..	..	..
Trinidad and Tobago	-0.3	0.4	-0.3	0.6	-0.2	0.5	-0.4	-0.1	0.2	-0.1	0.0
Uruguay	-2.0	-0.2	1.7	1.4	-0.9	-1.4	0.3	0.4	0.7	0.2	0.4
Venezuela	0.2	0.2	0.4	0.2	0.1	-0.1	0.2	0.3	0.3	-0.2	0.3
MIDDLE EAST AND NORTH AFRICA	*1.8*	*2.2*	*0.0*	*0.1*	*1.4*	*1.6*	*-1.6*	*2.2*	*0.7*	*1.0*	*-1.2*
Algeria	2.0	1.5	-0.7	0.6	0.4	0.0	-0.6	1.6	2.1	-0.2	-0.6
Egypt, Arab Republic of	0.2	0.7	1.2	0.8	0.4	2.4	0.9	0.1	1.7	0.4	-0.7
Jordan	..	..	..	..	4.2	0.8	-5.4	2.3	-2.0	1.8	-0.1
Morocco	2.5	7.2	-1.2	1.1	2.5	-0.6	-2.3	2.5	-3.5	2.2	-1.7
Oman	..	..	..	..	..	..	..	..	..	..	..
Syrian Arab Republic	6.1	-3.1	3.9	-5.7	1.3	9.5	-9.9	11.0	0.2	3.3	-2.5
Tunisia	-2.1	4.3	-0.7	1.3	4.0	4.8	-2.2	2.9	1.0	1.4	-2.1
Yemen Arab Republic	..	..	..	..	13.4	0.1	6.0	-2.5	7.1	1.3	-3.1
Yemen, PDR	..	..	..	..	..	..	..	..	..	5.9	0.2
EUROPE AND MEDITERRANEAN	..	..	..	..	..	..	..	..	..	..	..
Cyprus	..	..	..	..	..	..	..	..	..	1.3	0.0
Greece	0.4	-2.2	1.4	2.0	0.8	1.3	-0.1	0.9	1.2	-0.3	-1.4
Hungary	0.7	0.7	2.8	-2.7	1.3	0.5	1.1	0.5	0.4	-0.7	2.1
Israel	..	..	..	..	..	..	..	..	..	..	..
Malta	..	..	..	..	..	..	..	..	..	..	..
Poland	..	..	..	..	..	..	..	..	..	..	..
Portugal	..	..	..	..	..	..	..	..	..	..	..
Turkey	0.0	0.5	0.4	0.6	3.6	-0.5	-2.7	2.4	2.5	1.8	-0.3
Yugoslavia	-0.2	-0.6	1.6	-1.0	1.0	0.6	1.3	0.9	-0.4	0.9	0.8
Developing countries	0.7	0.5	1.0	1.6	0.9	-0.1	1.0	1.0	0.6	0.3	0.7
Highly indebted	-0.4	0.5	1.1	1.0	0.9	-0.1	0.7	0.8	-0.2	0.3	0.9
Low-income countries	2.4	0.2	1.1	3.7	0.4	-1.0	2.3	0.8	1.6	-0.5	1.2
Low-income Africa	1.0	-0.3	0.4	1.3	0.9	1.0	-1.1	0.5	-1.4	-0.7	1.4
China and India	2.9	-0.3	1.3	4.2	0.5	-1.4	3.2	0.7	2.5	-0.7	1.3
Other low-income Asia	0.1	4.9	1.1	3.0	-1.0	-1.2	-0.2	3.2	-0.2	2.9	0.2
Middle-income countries	-0.1	0.7	1.0	0.8	1.1	0.2	0.5	1.0	0.2	0.6	0.5
High-income oil exporters	0.0	0.1	0.0	0.0	0.1	0.1	0.1	0.0	0.1	0.1	0.0
Industrialized countries	..	..	..	..	0.1	0.1	0.0	0.0	0.0	-0.1	0.0
Japan	..	..	..	..	-0.3	0.7	0.4	-0.1	0.0	-0.2	-0.1
United States	..	..	..	..	0.1	0.0	0.0	0.0	0.1	-0.1	0.0
European Community	..	..	..	..	..	..	..	..	..	..	..

NOTE: See glossary for definitions.

1978	1979	1980	1981	1982	1983	1984	1985	1986	1987	
0.2	*0.3*	*0.5*	*0.6*	*0.0*	*0.0*	*0.4*	*0.6*	*-0.3*	*0.6*	*LATIN AMERICA AND CARIBBEAN*
0.3	0.3	-0.5	0.2	0.7	0.2	0.3	-0.1	-0.3	0.2	Argentina
..	..	..	..	..	..	..	..	..	..	Bahamas
0.0	0.8	0.6	-1.8	-0.2	0.3	0.9	-0.1	0.4	-1.6	Barbados
1.9	-0.5	0.3	1.0	2.3	-2.6	0.5	-1.2	..	..	Belize
0.4	0.6	0.3	-0.2	1.3	-3.3	4.1	2.0	-1.1	0.0	Bolivia
-0.4	0.5	1.0	0.9	0.0	0.0	0.4	1.2	-1.0	1.2	Brazil
-0.3	0.5	0.3	0.3	-0.1	-0.2	0.6	0.5	0.7	0.3	Chile
1.8	1.1	0.4	0.7	-0.4	0.6	0.5	0.3	0.7	1.2	Colombia
1.2	0.1	-0.1	0.9	-0.9	0.8	2.0	-0.6	0.6	-1.0	Costa Rica
0.8	0.2	0.8	1.0	0.8	0.5	0.0	-0.6	-0.6	..	Dominican Republic
-0.5	0.4	0.6	0.8	0.2	-1.7	1.2	1.1	1.0	1.3	Ecuador
3.3	0.9	-1.4	-1.8	-1.3	-0.9	0.9	-0.3	-0.8	0.7	El Salvador
..	..	..	..	..	..	..	..	..	..	Grenada
..	..	..	..	..	..	..	..	..	..	Guatemala
1.9	-1.7	0.1	0.5	-0.3	-1.0	0.9	0.0	1.0	-1.7	Guyana
..	..	..	..	..	..	..	..	..	..	Haiti
1.8	2.1	-0.2	1.0	-0.1	-0.2	0.3	0.7	0.4	1.8	Honduras
0.8	-0.9	-0.3	0.2	-0.6	0.5	0.8	-0.3	-0.2	0.2	Jamaica
0.6	-0.2	0.6	0.5	-0.2	0.2	0.2	0.3	-0.2	0.0	Mexico
1.8	-0.9	-4.9	1.5	0.7	1.4	-1.4	-1.2	-1.3	..	Nicaragua
0.9	-0.5	-0.4	0.7	-0.1	0.3	0.2	0.5	-0.2	0.7	Panama
1.7	2.1	2.6	2.9	0.1	-0.7	1.8	1.4	-1.9	2.0	Paraguay
0.0	0.5	-0.7	0.9	0.3	-1.2	1.3	0.4	0.7	0.5	Peru
3.6	-3.4	-2.6	6.3	1.1	0.9	1.3	2.5	0.1	..	St. Vincent and the Grenadines
-0.1	-0.1	-0.2	0.0	0.1	0.4	0.0	0.2	0.1	0.1	Trinidad and Tobago
-0.8	0.0	1.6	0.6	-0.8	0.2	-0.9	0.6	0.9	0.1	Uruguay
0.1	0.1	0.1	-0.1	0.2	0.0	0.0	0.4	0.5	0.2	Venezuela
1.5	*0.1*	*1.6*	*-0.4*	*0.2*	*-0.1*	*0.3*	*1.4*	*0.8*	*-0.1*	*MIDDLE EAST AND NORTH AFRICA*
0.8	0.7	0.9	0.1	-0.7	-0.1	0.6	1.8	0.4	0.5	Algeria
1.2	0.8	0.7	0.3	0.7	0.5	0.3	0.5	0.3	0.3	Egypt, Arab Republic of
3.9	-2.6	4.1	-0.8	0.1	-0.3	0.6	0.6	1.0	0.1	Jordan
2.9	-0.3	1.1	-4.2	2.9	-0.6	0.0	1.8	3.7	-2.4	Morocco
..	0.5	0.5	0.0	0.2	0.4	0.2	0.4	-0.2	..	Oman
3.9	-2.8	6.3	0.8	-0.4	-0.1	-1.5	1.1	1.0	-2.4	Syrian Arab Republic
1.1	-0.9	1.6	1.1	-1.7	0.4	1.9	2.7	-2.1	2.7	Tunisia
-3.6	4.5	1.3	2.4	1.4	-2.6	0.0	1.7	2.3	0.5	Yemen Arab Republic
-3.6	1.4	2.7	-1.8	..	..	..	..	..	..	Yemen, PDR
0.3	*0.3*	*0.5*	*0.0*	*1.0*	*-0.4*	*0.6*	*-0.1*	*0.9*	*0.1*	*EUROPE AND MEDITERRANEAN*
-0.4	0.4	0.6	-0.1	0.0	-0.1	0.8	0.0	-0.2	0.4	Cyprus
1.8	-1.1	2.1	-0.3	0.4	-1.6	1.1	0.3	0.1	-0.7	Greece
0.2	-0.2	0.6	0.6	2.0	0.0	0.9	-0.8	0.6	0.0	Hungary
..	..	..	..	..	..	..	..	..	..	Israel
..	..	..	..	..	..	..	..	..	..	Malta
..	..	..	..	..	..	..	..	..	..	Poland
0.6	1.9	0.0	-1.4	0.5	-0.5	0.2	0.3	..	..	Portugal
0.6	0.6	0.4	0.0	1.4	0.0	0.8	0.5	1.6	0.4	Turkey
-0.8	0.5	0.0	0.3	0.9	-0.1	0.3	-0.9	0.8	0.2	Yugoslavia
0.5	0.3	0.6	0.7	0.8	0.6	0.8	0.8	0.5	0.3	Developing countries
-0.1	0.2	0.6	-0.1	0.2	-0.1	0.1	0.8	0.3	0.2	Highly indebted
1.6	0.4	1.0	1.9	2.0	2.6	1.9	0.8	0.7	0.6	Low-income countries
0.7	-0.1	0.2	0.2	1.9	0.1	-0.4	0.6	2.7	1.2	Low-income Africa
1.6	0.5	1.1	2.1	2.1	3.1	2.4	0.8	0.4	0.6	China and India
2.3	0.6	1.5	2.1	1.1	1.5	-0.1	1.9	1.5	0.2	Other low-income Asia
0.2	0.3	0.4	0.3	0.3	-0.1	0.3	0.8	0.4	0.2	Middle-income countries
0.2	0.1	0.1	0.1	0.1	0.2	0.1	0.2	0.3	..	High-income oil exporters
0.1	0.1	-0.1	0.2	0.1	-0.2	0.2	0.1	0.1	..	Industrialized countries
0.0	0.1	-0.3	-0.1	0.2	0.0	0.1	0.0	-0.1	..	Japan
0.0	0.2	-0.1	0.4	0.0	-0.6	0.3	0.4	0.2	..	United States
..	..	..	..	..	..	..	..	..	..	European Community

Table 8. Industry: contribution to gross domestic product growth

Percentage points	1967	1968	1969	1970	1971	1972	1973	1974	1975	1976	1977
SUB-SAHARAN AFRICA	*-1.8*	*-1.6*	*6.3*	*7.3*	*5.2*	*4.0*	*0.4*	*3.8*	*-3.1*	*4.0*	*1.2*
Excluding Nigeria	*..*	*..*	*1.6*	*2.9*	*1.4*	*1.6*	*1.3*	*1.4*	*-1.0*	*0.0*	*0.5*
Benin	..	..	..	..	-1.2	2.0	-0.3	3.8	-2.2	-1.5	1.6
Botswana	2.1	1.2	-3.7	29.2	4.8	10.7	7.3	6.0	0.6	8.1	-1.3
Burkina Faso	..	..	..	..	0.8	0.4	2.8	-2.9	4.2	2.5	-4.1
Burundi	4.6	0.1	0.0	1.0	0.4	-0.2	0.2	0.3	0.1	0.9	2.3
Cameroon	0.2	1.9	1.1	-0.2	2.0	-0.6	0.7	0.9	0.6	1.3	3.1
Cape Verde	..	..	..	..	..	..	..	..	..	..	..
Central African Republic	1.3	1.8	0.9	1.4	-0.2	2.2	0.7	1.8	-0.1	-1.0	1.7
Chad	..	0.4	0.1	0.4	0.2	0.2	0.1	0.9	1.4	1.5	0.2
Comoros	..	..	..	..	..	..	..	..	..	..	..
Congo, People's Republic	7.1	0.2	1.1	6.9	0.2	2.0	4.4	8.0	9.6	4.2	-9.2
Cote d'Ivoire	2.1	-0.1	0.4	3.7	3.0	1.1	-0.6	-0.3	1.8	2.0	1.5
Ethiopia	1.4	-0.7	1.0	1.6	1.4	0.7	0.5	-0.1	-0.2	-0.6	0.2
Gabon	..	..	..	..	..	..	..	..	..	..	..
Gambia, The	0.6	0.2	2.0	-1.9	2.2	1.4	0.8	3.6	-1.2	0.2	0.5
Ghana	1.0	0.8	1.4	1.0	0.2	-1.4	2.5	-0.1	-0.1	-0.4	0.6
Guinea-Bissau	..	..	..	..	-1.4	1.4	-0.6	0.2	3.0	-6.9	3.4
Kenya	0.8	1.9	1.2	-0.9	4.1	7.5	1.6	0.2	0.1	-0.1	2.5
Lesotho	..	..	..	..	-1.8	4.2	4.0	2.0	-0.7	4.5	8.7
Liberia	3.6	-0.3	4.1	2.5	1.1	1.7	0.1	2.3	-4.3	0.3	-2.0
Madagascar	..	..	..	..	1.0	0.2	-0.4	1.2	0.0	-1.6	0.5
Malawi	..	0.8	1.0	0.7	0.9	2.8	-0.2	1.3	3.0	-0.6	0.9
Mali	0.7	0.1	0.4	1.4	-3.1	1.5	0.3	0.0	1.2	0.7	1.4
Mauritania	1.7	3.4	0.3	2.9	1.7	-3.0	-0.9	4.4	-3.0	3.4	-3.3
Mauritius	..	..	..	..	1.2	2.6	3.5	5.7	-0.3	5.0	2.3
Mozambique	..	..	..	..	..	..	..	..	..	..	..
Niger	0.1	0.5	4.8	-1.7	1.6	0.1	1.8	-2.0	2.1	2.1	1.8
Nigeria	-4.4	-4.2	13.6	13.5	9.9	6.7	-0.5	6.4	-5.3	8.1	1.9
Rwanda	..	..	..	..	..	..	..	..	..	..	1.9
Senegal	0.8	0.7	0.4	1.8	0.6	0.5	-0.8	2.3	1.4	1.7	3.9
Seychelles	..	..	..	..	..	..	..	..	..	..	0.1
Sierra Leone	-4.0	-1.6	3.1	5.9	-0.9	-1.5	0.1	0.0	1.1	-3.1	-2.3
Somalia	..	..	..	..	0.6	0.2	-1.8	0.3	-0.4	1.0	0.8
Sudan	-1.6	0.6	-0.7	-0.6	-0.1	-0.7	-1.1	2.0	1.7	2.8	1.0
Swaziland	3.1	-2.3	2.6	-1.6	1.6	3.7	2.3	0.6	-1.5	0.5	1.1
Tanzania	2.0	0.2	0.8	0.8	2.4	0.9	0.4	0.2	-0.3	1.7	-0.4
Togo	3.3	0.1	1.5	-1.3	0.5	3.1	2.0	1.8	0.5	1.1	3.8
Uganda	0.4	0.6	0.9	0.2	0.1	-0.3	-0.8	-0.1	-1.0	-0.4	-0.3
Zaire	..	..	2.8	9.1	1.8	0.2	2.6	1.6	-1.1	-2.6	0.3
Zambia	2.0	-0.3	2.6	-0.9	0.8	5.3	0.3	3.9	-1.3	1.2	-2.8
Zimbabwe	..	..	..	4.9	2.4	4.1	4.1	1.0	-0.3	-1.8	-3.0
SOUTH ASIA	*0.7*	*1.2*	*2.0*	*0.5*	*0.4*	*0.3*	*0.9*	*0.6*	*1.5*	*2.0*	*1.7*
Bangladesh	-2.3	2.2	0.9	0.0	-2.5	-4.7	4.4	1.5	4.3	0.7	1.0
India	1.0	1.1	2.1	0.2	0.7	0.9	0.4	0.4	1.4	2.2	1.8
Nepal	..	..	..	..	..	..	..	..	..	..	..
Pakistan	0.8	1.1	2.5	3.0	0.7	-1.1	2.4	2.1	0.5	1.1	1.6
Sri Lanka	2.2	4.1	2.9	2.7	0.1	-0.3	1.8	-1.1	0.5	2.2	-1.1
EAST ASIA	*-1.4*	*0.6*	*7.3*	*8.4*	*4.1*	*3.3*	*4.6*	*1.3*	*3.7*	*1.9*	*5.3*
China	-4.3	-2.7	9.9	11.8	4.9	2.0	3.4	0.4	5.4	-1.6	5.7
Fiji	9.2	2.7	-2.8	2.5	0.0	1.8	2.2	-1.0	-1.0	0.6	2.2
Hong Kong	..	..	..	..	..	..	..	..	..	..	..
Indonesia	1.4	7.8	6.2	5.3	1.5	7.5	8.3	3.0	0.4	5.2	5.7
Korea, Republic of	3.4	5.0	5.3	3.6	3.0	2.4	6.9	4.3	3.7	6.6	5.9
Malaysia	..	..	..	..	3.5	3.0	3.7	1.6	0.1	6.0	2.8
Papua New Guinea	..	..	..	..	..	..	..	..	..	..	..
Philippines	2.6	1.5	1.6	1.9	2.3	3.4	4.1	1.8	3.0	3.7	2.9
Singapore	4.4	5.8	6.0	7.0	6.0	6.5	3.6	1.6	0.9	4.2	2.5
Solomon Islands	..	..	..	..	..	..	..	..	..	..	..
Thailand	3.3	1.7	2.3	2.4	2.2	2.2	2.8	1.2	1.3	4.4	4.5
Vanuatu	..	..	..	..	..	..	..	..	..	..	..

NOTE: See glossary for definitions.

1978	1979	1980	1981	1982	1983	1984	1985	1986	1987	
-0.6	*4.2*	*0.3*	*-1.4*	*-0.2*	*-2.2*	*-0.6*	*2.0*	*-0.1*	*-0.5*	*SUB-SAHARAN AFRICA*
0.3	*1.0*	*1.7*	*1.5*	*0.4*	*1.0*	*0.8*	*0.9*	*0.7*	*0.5*	*Excluding Nigeria*
-0.8	1.4	0.0	2.6	4.3	0.1	1.6	2.1	-1.9	0.6	Benin
17.0	2.5	6.0	7.6	-3.1	20.5	16.8	1.7	10.4	11.5	Botswana
1.6	0.3	-0.1	-0.3	-1.4	-1.5	0.0	1.7	1.3	0.9	Burkina Faso
1.3	0.1	1.3	1.0	0.1	0.6	0.9	0.7	0.5	0.9	Burundi
1.9	7.7	7.5	7.9	4.9	7.0	1.6	4.3	0.8	-1.4	Cameroon
..	..	..	..	..	..	..	..	..	..	Cape Verde
2.6	-1.2	0.2	-1.6	-0.1	1.6	0.6	0.7	0.6	-0.4	Central African Republic
-0.7	-6.5	-2.8	4.0	1.7	2.0	4.2	-0.7	-0.6	0.9	Chad
..	..	..	..	..	..	..	..	..	..	Comoros
1.5	9.8	11.1	14.5	18.1	9.0	1.8	0.5	1.4	-0.5	Congo, People's Republic
3.5	1.4	1.6	2.0	-3.1	0.5	0.5	-0.8	-0.9	-0.8	Cote d'Ivoire
-0.5	2.0	1.4	0.5	0.5	0.8	0.8	0.0	0.7	1.2	Ethiopia
..	..	..	..	..	..	..	..	..	..	Gabon
5.2	-10.8	1.4	6.4	1.5	-0.8	-0.4	0.1	2.4	1.2	Gambia, The
-1.2	-1.9	0.0	-1.9	-1.7	-1.1	0.7	1.5	0.7	1.1	Ghana
5.6	0.0	-2.4	-0.8	1.4	-0.3	5.1	1.1	-3.5	-0.5	Guinea-Bissau
2.2	1.5	1.1	1.0	-0.1	0.4	0.4	1.3	0.8	1.1	Kenya
3.2	3.6	2.3	-3.8	5.2	-9.5	2.6	1.5	4.3	-0.5	Lesotho
0.6	1.2	0.0	-2.9	-1.8	-1.3	-1.8	-0.7	-0.4	..	Liberia
0.7	2.3	-0.5	-4.1	-2.2	0.2	0.7	0.6	-0.4	1.3	Madagascar
2.2	0.1	0.0	-0.3	0.0	0.6	-0.1	1.2	1.2	-1.3	Malawi
-0.2	-1.9	-0.1	2.7	0.6	0.7	2.4	0.9	1.6	-0.5	Mali
1.1	2.9	-1.9	0.0	-1.0	-1.5	7.6	1.3	1.1	-0.3	Mauritania
1.8	0.3	-2.5	1.1	1.2	0.1	2.4	3.6	4.8	3.7	Mauritius
..	..	..	0.8	-1.4	-2.0	-0.8	-1.8	..	..	Mozambique
3.4	6.5	0.9	1.0	-0.8	-2.4	-1.3	-0.6	0.2	..	Niger
-1.4	7.4	-1.0	-4.2	-0.9	-5.8	-2.2	3.4	-1.0	-1.6	Nigeria
1.2	0.4	4.8	0.6	-0.5	2.7	0.7	0.8	1.4	..	Rwanda
-2.3	2.7	-0.7	1.4	4.0	0.6	-0.5	0.6	1.8	1.4	Senegal
1.3	3.8	-0.2	-1.4	-2.0	0.0	1.3	2.2	0.4	..	Seychelles
-4.1	1.0	2.6	-0.6	-0.2	-2.5	1.5	0.0	-2.8	2.4	Sierra Leone
-2.7	1.2	0.8	-0.6	0.4	-0.9	0.3	0.5	0.6	0.2	Somalia
-1.3	-0.4	1.2	1.2	-0.1	2.1	-1.0	-0.3	0.4	0.3	Sudan
6.1	-0.5	-0.1	3.4	0.6	-0.9	-0.1	-0.5	4.2	..	Swaziland
-0.2	1.0	-0.1	-1.3	-0.1	-2.4	0.8	-0.5	0.6	0.5	Tanzania
1.1	-0.3	3.7	-1.1	0.2	-2.1	-2.0	1.6	0.7	0.4	Togo
-1.1	-1.6	-0.1	0.1	0.4	0.0	0.2	-0.5	-0.1	0.6	Uganda
-2.5	-2.0	1.5	1.6	-0.9	1.4	3.0	0.8	1.0	0.3	Zaire
1.8	-2.2	0.9	2.1	-0.2	-1.3	-0.9	0.3	-0.6	0.3	Zambia
-1.6	2.1	3.6	2.7	-0.3	-0.9	-1.1	2.2	1.3	1.3	Zimbabwe
1.9	*0.0*	*0.6*	*2.0*	*1.3*	*2.0*	*1.8*	*2.0*	*2.2*	*2.0*	*SOUTH ASIA*
0.3	3.1	-0.8	1.1	0.5	0.1	1.2	1.0	0.4	1.1	Bangladesh
2.0	-0.6	0.4	2.1	1.2	2.2	1.8	2.2	2.4	2.1	India
..	..	..	..	..	..	..	..	..	..	Nepal
2.2	1.8	2.7	2.4	2.8	2.1	2.8	2.2	2.6	2.2	Pakistan
4.0	2.9	1.5	0.7	0.8	0.5	1.9	1.0	1.7	1.8	Sri Lanka
5.8	*3.1*	*3.8*	*1.6*	*1.9*	*4.1*	*5.6*	*5.6*	*4.8*	*7.0*	*EAST ASIA*
7.1	3.3	5.6	1.1	3.6	5.2	7.6	10.3	5.8	9.4	China
0.0	3.3	0.3	1.4	-1.3	-1.5	0.4	-1.3	3.1	-3.0	Fiji
..	..	..	..	..	..	..	..	..	..	Hong Kong
1.5	1.7	3.2	2.1	-3.2	0.9	3.0	0.2	1.8	0.7	Indonesia
7.9	3.4	-0.3	2.3	2.7	5.9	5.7	2.2	6.6	7.0	Korea, Republic of
3.6	4.8	2.6	1.3	2.5	3.8	4.4	-1.3	1.8	2.7	Malaysia
..	..	..	-3.1	0.2	2.0	0.1	4.5	2.3	1.8	Papua New Guinea
2.2	2.8	1.7	1.8	0.7	0.3	-3.7	-3.6	-0.7	2.5	Philippines
2.7	4.5	3.8	4.1	1.9	3.8	3.8	-3.6	-0.3	3.5	Singapore
..	..	..	..	..	..	..	..	..	..	Solomon Islands
3.5	2.0	1.1	1.8	0.9	2.5	2.6	-0.1	2.0	3.0	Thailand
..	..	-1.0	-0.8	3.9	-9.2	1.9	0.1	0.5	..	Vanuatu

Table 8. Industry: contribution to gross domestic product growth (cont'd)

Percentage points	1967	1968	1969	1970	1971	1972	1973	1974	1975	1976	1977
LATIN AMERICA AND CARIBBEAN	*1.6*	*3.4*	*2.3*	*3.1*	*2.0*	*2.7*	*3.9*	*1.9*	*0.2*	*2.6*	*1.3*
Argentina	1.6	4.0	4.4	-0.5	2.5	1.3	0.5	2.3	-0.3	0.2	3.3
Bahamas	..	..	..	..	..	..	..	..	..	..	..
Barbados	0.6	1.8	1.4	1.0	0.2	1.0	1.4	-1.0	-0.2	3.3	-1.1
Belize	..	..	..	..	0.2	2.0	1.0	1.9	1.3	0.8	1.2
Bolivia	4.0	2.6	2.1	-0.9	2.0	3.4	6.6	-0.2	0.4	1.6	1.0
Brazil	1.1	5.1	4.6	3.5	4.6	5.6	6.8	3.4	1.9	4.8	1.3
Chile	1.1	1.7	1.6	0.8	4.1	-0.9	-3.0	2.3	-9.3	1.5	2.3
Colombia	1.6	1.6	2.6	2.1	1.9	3.4	3.9	0.5	-0.2	1.0	0.0
Costa Rica	1.4	2.7	1.2	1.9	2.2	2.8	2.0	2.7	0.9	2.2	2.7
Dominican Republic	2.6	-0.7	3.7	4.3	4.5	4.5	5.9	1.6	2.4	2.4	1.8
Ecuador	2.3	-0.7	1.5	1.6	3.2	13.6	22.7	-0.9	-0.4	5.7	0.8
El Salvador	1.2	0.4	0.4	0.7	1.5	1.5	0.9	1.2	2.1	1.2	2.2
Grenada	..	..	..	..	..	..	..	..	..	..	..
Guatemala	..	..	..	..	..	..	..	..	..	..	..
Guyana	0.8	0.9	3.3	1.8	1.3	-2.6	-1.0	3.3	3.6	-1.6	-1.3
Haiti	..	..	..	..	..	..	..	..	..	..	..
Honduras	3.6	1.5	0.1	0.9	0.7	1.6	3.0	1.2	0.1	1.2	3.1
Jamaica	-0.3	3.4	-0.3	6.6	1.3	2.7	-0.2	-0.6	-1.8	-5.1	-2.5
Mexico	2.5	2.6	2.6	2.8	0.6	3.0	3.3	2.2	1.6	1.6	0.7
Nicaragua	2.2	0.5	1.4	1.5	1.1	1.8	1.1	5.0	0.3	1.8	3.6
Panama	2.4	1.9	1.8	1.9	2.7	0.8	2.5	-0.7	0.7	0.8	-1.7
Paraguay	1.8	0.4	1.2	1.6	1.5	1.4	2.0	1.7	0.5	1.8	4.0
Peru	1.1	0.2	0.5	3.3	1.4	1.6	2.5	3.8	0.3	1.4	0.5
St. Vincent and the Grenadines	..	..	..	..	..	..	..	..	..	..	..
Trinidad and Tobago	4.1	4.6	1.3	2.0	2.0	4.3	1.3	4.3	-0.4	8.1	1.5
Uruguay	-1.2	1.3	1.7	0.4	-0.2	0.5	-0.7	0.7	2.8	1.2	1.5
Venezuela	1.4	5.2	-2.3	7.5	-0.8	-1.3	4.1	-2.3	-3.7	2.6	1.8
MIDDLE EAST AND NORTH AFRICA	*3.1*	*3.5*	*4.2*	*3.2*	*-2.1*	*9.0*	*1.8*	*1.0*	*3.5*	*4.8*	*4.0*
Algeria	7.8	5.4	4.9	5.2	-8.8	21.4	3.4	0.2	1.6	6.7	4.8
Egypt, Arab Republic of	-1.2	1.5	3.8	2.2	1.4	0.4	-0.6	-1.0	4.6	4.6	6.7
Jordan	..	..	..	..	-1.8	3.1	5.5	3.5	3.9	9.8	1.2
Morocco	1.4	1.3	2.2	1.0	2.3	1.9	1.9	1.5	3.6	3.3	3.1
Oman	..	..	..	..	..	..	..	..	..	..	..
Syrian Arab Republic	-0.5	1.8	9.2	2.2	5.5	2.8	-0.1	6.1	7.8	2.6	-1.8
Tunisia	1.4	7.7	0.4	2.6	2.1	4.2	0.3	1.2	4.7	-0.3	2.4
Yemen Arab Republic	..	..	..	..	1.9	1.1	1.4	1.3	-0.4	2.9	2.5
Yemen, PDR	..	..	..	..	..	..	..	..	..	4.9	8.1
EUROPE AND MEDITERRANEAN	*..*	*..*	*..*	*..*	*..*	*..*	*..*	*..*	*..*	*..*	*..*
Cyprus	..	..	..	..	..	..	..	..	..	9.4	7.8
Greece	1.4	3.9	4.1	2.4	3.6	3.8	3.9	-3.7	1.7	2.8	1.5
Hungary	3.3	1.8	1.6	2.6	2.6	2.5	3.0	3.2	2.8	2.3	2.6
Israel	..	..	..	..	..	..	..	..	..	..	..
Malta	..	..	..	..	..	..	..	..	..	..	..
Poland	..	..	..	..	..	..	..	..	..	..	..
Portugal	..	..	..	..	..	..	..	..	..	..	..
Turkey	2.0	3.5	2.6	1.2	1.5	1.7	3.1	2.3	2.7	3.0	2.7
Yugoslavia	0.6	2.0	3.4	2.6	2.6	7.4	1.5	4.1	3.0	1.7	4.1
Developing countries	1.1	2.4	4.6	5.0	2.0	3.5	3.3	1.7	1.0	2.8	2.4
Highly indebted	0.8	2.4	3.4	4.1	3.0	3.6	3.1	2.6	-0.2	3.3	1.7
Low-income countries	-1.6	-0.5	5.0	5.6	2.6	1.3	2.1	0.6	2.9	-0.1	3.4
Low-income Africa	..	0.7	1.7	2.8	1.1	1.6	1.2	1.5	-1.7	-0.5	0.7
China and India	-2.2	-1.0	6.2	6.7	3.2	1.6	2.2	0.4	3.9	-0.2	4.2
Other low-income Asia	-0.5	1.9	1.6	1.6	-0.4	-2.1	2.6	1.3	1.7	1.0	1.1
Middle-income countries	2.3	3.7	4.3	4.7	1.7	4.4	3.8	2.1	0.3	3.9	2.0
High-income oil exporters	8.5	19.6	9.9	7.8	4.5	3.1	6.8	-0.8	-2.8	8.5	8.4
Industrialized countries	..	..	..	..	0.5	2.0	5.4	-0.9	-1.8	2.2	1.3
Japan	..	..	..	..	2.6	3.5	4.4	-1.1	-0.6	2.6	1.6
United States	..	..	..	..	0.0	2.2	2.9	-1.4	-2.3	2.3	2.0
European Community	..	..	..	..	..	..	..	..	..	..	..

NOTE: See glossary for definitions.

1978	1979	1980	1981	1982	1983	1984	1985	1986	1987	
2.0	2.6	2.2	-0.7	-1.0	-2.0	1.3	1.5	1.8	0.6	*LATIN AMERICA AND CARIBBEAN*
-3.1	3.0	-0.5	-4.7	-2.4	2.0	0.3	-2.7	3.5	0.6	Argentina
..	..	..	..	..	..	..	..	..	..	Bahamas
1.2	2.9	1.0	-0.2	-1.5	0.7	0.6	-1.1	1.3	-0.1	Barbados
1.9	1.4	1.6	-1.1	-1.2	0.2	1.5	-0.1	..	..	Belize
-0.9	-2.3	0.1	-0.9	-2.2	-1.5	-4.0	-2.8	-1.7	0.8	Bolivia
2.6	2.7	3.8	-3.8	0.0	-2.4	2.4	3.3	4.4	0.3	Brazil
2.7	3.4	3.0	2.3	-5.6	0.3	2.8	1.4	2.1	1.9	Chile
0.1	0.3	3.4	0.2	-0.3	1.1	1.8	2.0	3.3	1.9	Colombia
2.0	1.6	0.3	-1.1	-3.6	1.0	2.8	0.4	1.6	1.6	Costa Rica
-0.8	2.4	0.8	0.9	-1.1	2.3	0.4	-1.4	0.8	..	Dominican Republic
3.8	2.7	-0.2	2.8	0.1	2.6	1.3	1.6	0.9	-8.6	Ecuador
0.7	-0.6	-2.5	-2.1	-1.4	0.5	0.1	0.8	0.5	0.9	El Salvador
..	..	..	..	..	..	..	..	..	..	Grenada
..	..	..	..	..	..	..	..	..	..	Guatemala
-2.7	-0.6	1.2	-0.7	-6.7	-6.7	0.6	1.1	-0.8	..	Guyana
..	..	..	..	..	..	..	..	..	..	Haiti
3.2	4.0	-1.8	-2.1	1.2	0.5	1.5	-0.3	0.0	0.1	Honduras
-0.5	-1.3	-3.7	0.4	-0.8	1.1	-1.3	-2.4	1.6	2.5	Jamaica
3.3	3.6	3.1	2.9	-0.7	-2.9	1.5	1.5	-1.9	0.6	Mexico
-2.6	-10.1	5.9	-0.4	-0.9	1.7	0.3	-0.9	0.7	..	Nicaragua
2.6	1.6	2.2	0.1	1.8	-2.4	-1.1	0.4	0.8	0.4	Panama
3.5	3.5	7.4	1.9	-1.2	-1.8	0.9	0.0	0.5	0.4	Paraguay
1.2	3.3	2.3	0.5	0.1	-6.3	1.8	1.1	4.1	3.3	Peru
2.4	4.5	1.3	0.5	1.1	1.0	0.4	0.3	0.9	..	St. Vincent and the Grenadines
8.8	1.9	3.1	1.1	-6.9	-5.5	-5.7	-6.5	-2.1	-2.5	Trinidad and Tobago
3.0	2.6	1.0	-0.9	-4.2	-3.6	0.1	-1.5	2.7	3.0	Uruguay
1.6	1.2	-2.4	-0.9	-1.8	-2.4	-1.2	0.2	3.2	1.1	Venezuela
4.1	3.2	0.7	1.2	2.7	3.0	2.4	2.5	1.4	0.2	*MIDDLE EAST AND NORTH AFRICA*
6.0	4.9	-0.9	0.2	4.5	3.1	2.9	2.0	0.9	0.8	Algeria
5.2	3.2	5.2	1.4	1.0	2.3	3.1	2.7	0.5	0.8	Egypt, Arab Republic of
4.3	7.1	5.6	6.5	3.6	0.2	4.1	-1.0	0.0	-1.0	Jordan
-0.3	2.6	-0.8	0.0	0.7	0.9	-0.4	0.7	-0.1	0.9	Morocco
..	-3.1	-0.2	11.0	6.0	14.2	10.2	11.8	7.6	..	Oman
0.8	0.6	-0.3	-0.2	1.3	1.2	-1.9	1.8	2.7	-4.7	Syrian Arab Republic
3.0	3.7	3.2	2.5	0.0	2.4	1.4	0.6	-0.1	0.1	Tunisia
5.5	2.7	-0.9	0.2	6.3	2.2	1.3	0.7	1.5	1.8	Yemen Arab Republic
8.9	-3.8	-1.9	2.9	..	..	..	..	..	..	Yemen, PDR
2.7	1.4	0.1	1.0	0.3	0.4	1.2	1.0	..	..	*EUROPE AND MEDITERRANEAN*
3.7	3.7	1.9	-0.9	0.8	0.5	1.5	0.4	0.2	1.8	Cyprus
2.0	1.9	-0.8	-0.5	-0.7	0.1	0.4	1.1	0.1	-0.4	Greece
2.2	2.1	-0.7	1.8	1.6	0.8	0.5	-1.0	-0.2	1.5	Hungary
..	..	..	..	..	..	..	..	..	..	Israel
..	..	..	..	..	..	..	..	..	..	Malta
..	..	..	..	..	..	..	..	..	..	Poland
2.7	2.0	1.9	0.8	1.2	0.2	-1.0	1.3	..	..	Portugal
2.0	-1.1	-1.4	1.7	1.2	1.9	2.6	1.8	2.8	2.9	Turkey
4.0	3.0	1.4	1.0	-0.8	-0.7	1.5	0.9	1.7	1.0	Yugoslavia
1.8	1.8	1.2	0.5	1.0	0.9	2.6	2.9	2.7	3.1	Developing countries
1.7	3.3	1.7	-0.8	-0.8	-2.1	0.8	1.5	1.4	0.5	Highly indebted
4.1	1.6	3.2	1.3	2.4	3.4	4.8	6.6	4.2	6.3	Low-income countries
-0.2	0.1	0.9	0.3	-0.1	0.1	0.7	0.6	0.6	0.7	Low-income Africa
5.1	1.8	3.8	1.4	2.8	4.2	5.6	7.8	4.8	7.3	China and India
1.6	2.2	1.3	1.7	1.7	1.2	2.1	1.5	1.6	1.7	Other low-income Asia
1.0	1.9	0.5	0.2	0.5	0.0	1.7	1.2	1.9	1.4	Middle-income countries
1.8	7.5	-1.3	-3.5	-5.1	-11.5	-1.3	-6.8	-7.0	..	High-income oil exporters
1.5	1.3	-0.1	0.1	-0.6	1.0	2.1	1.3	0.5	..	Industrialized countries
3.0	2.3	2.4	1.6	1.6	1.7	3.5	2.6	0.7	..	Japan
1.9	0.5	-1.2	0.2	-1.7	1.3	3.4	1.1	0.4	..	United States
..	..	..	..	..	..	..	..	..	..	European Community

Table 9. Services: contribution to gross domestic product growth

Percentage points	1967	1968	1969	1970	1971	1972	1973	1974	1975	1976	1977
SUB-SAHARAN AFRICA	*0.4*	*3.5*	*1.8*	*1.9*	*2.5*	*0.9*	*2.6*	*2.2*	*2.8*	*2.9*	*1.3*
Excluding Nigeria	..	*3.0*	*2.3*	*1.1*	*2.8*	*1.1*	*2.1*	*3.4*	*0.0*	*4.1*	*0.5*
Benin	..	..	..	..	0.8	2.9	3.8	3.9	-1.4	-3.3	2.8
Botswana	3.1	2.6	12.0	-11.9	10.0	14.7	10.0	5.7	0.8	6.6	2.8
Burkina Faso	..	..	..	..	-0.1	2.8	-0.8	4.4	-1.3	2.5	6.4
Burundi	7.4	0.7	-0.8	-1.9	1.3	0.3	-1.2	1.3	-1.7	2.8	1.2
Cameroon	-13.0	2.5	1.8	1.5	1.0	1.2	3.2	8.1	-1.8	2.6	4.9
Cape Verde											
Central African Republic	2.4	-2.0	5.1	1.8	-0.4	-2.3	0.0	4.1	0.1	3.5	2.8
Chad	..	-0.2	4.0	1.7	-1.4	-0.6	-2.3	2.7	3.8	2.0	1.6
Comoros	..		..					..			
Congo, People's Republic	-4.4	6.2	6.8	1.1	6.3	5.2	3.7	1.4	0.3	-1.8	-1.4
Cote d'Ivoire	1.9	8.5	4.5	0.8	9.6	5.1	2.1	11.9	9.2	-7.8	-8.5
Ethiopia	0.9	1.9	1.5	2.7	1.7	1.8	1.6	1.8	1.2	1.5	0.2
Gabon	..	..	..	..	..	..	..	..	..	..	..
Gambia, The	1.7	0.6	5.1	-4.6	5.4	3.5	2.2	9.4	-2.9	4.9	1.1
Ghana	-0.7	0.0	0.7	2.8	2.6	-2.6	0.6	2.5	-1.9	-2.2	4.1
Guinea-Bissau	..	..	..		2.3	-0.3	1.9	1.8	3.0	6.6	-2.6
Kenya	2.3	3.5	2.9	-1.5	10.6	4.1	1.7	1.8	1.4	1.8	2.7
Lesotho	..	..	..	..	6.2	6.5	11.0	-1.7	-0.3	9.4	1.1
Liberia	0.8	3.9	1.1	1.8	2.5	1.0	-5.6	0.9	1.9	2.0	2.2
Madagascar	..	..	..	..	3.1	-1.8	-2.3	-1.8	1.2	1.3	0.7
Malawi	..	0.2	2.4	-0.9	8.3	-0.8	6.4	5.6	2.1	2.7	-1.1
Mali	1.0	1.6	2.2	0.3	3.5	2.6	3.1	1.2	-0.5	4.4	0.7
Mauritania	-0.7	0.4	5.5	11.1	2.9	-7.1	0.5	2.9	4.8	3.6	1.6
Mauritius	..	..	..	..	1.0	3.4	7.2	7.2	10.8	8.0	5.0
Mozambique	..	..	..	..	..	..	..	..	..	..	..
Niger	-0.4	1.2	-3.3	0.6	0.8	0.7	1.4	-0.8	7.0	2.9	-1.0
Nigeria	-3.5	2.8	2.7	6.1	2.4	0.9	3.5	2.3	5.5	1.8	2.3
Rwanda	..	..	..	..	..	..	..	..	..	..	2.7
Senegal	-4.1	5.6	-7.5	5.3	3.6	0.9	-0.8	-3.2	5.3	3.6	-4.7
Seychelles	..	..	..	..	..	..	..	..	..	..	8.4
Sierra Leone	-0.9	1.2	4.5	4.6	2.4	0.6	3.0	2.8	1.1	-0.8	2.8
Somalia	..	..	..	..	1.0	4.1	1.7	-3.3	5.1	-2.8	2.9
Sudan	-1.7	3.5	-1.1	-3.1	2.5	-1.5	-1.6	2.4	7.7	12.6	8.4
Swaziland	-0.2	10.5	9.0	3.6	4.5	1.7	5.0	0.8	3.5	2.5	0.6
Tanzania	1.7	2.5	0.7	2.8	2.0	1.7	1.9	3.9	2.3	2.4	0.3
Togo	0.7	2.0	7.1	6.6	-0.2	3.0	1.0	2.8	1.0	-1.3	5.6
Uganda	0.7	0.6	1.7	0.4	2.3	-0.5	-1.2	0.5	-0.2	0.6	0.3
Zaire	..	..	6.4	-5.3	3.5	-0.4	4.6	1.1	-3.5	-4.0	1.3
Zambia	5.9	1.3	-2.8	4.8	-1.1	3.5	-1.0	2.1	-1.7	3.8	-2.0
Zimbabwe	..	..	..	18.2	2.8	2.3	1.7	2.5	1.1	-0.6	-0.7
SOUTH ASIA	*1.5*	*2.0*	*1.6*	*2.3*	*1.7*	*0.8*	*0.6*	*0.6*	*3.3*	*1.7*	*1.8*
Bangladesh	1.9	1.5	-0.2	2.5	-0.4	-3.1	-1.1	4.5	-0.4	5.1	1.9
India	1.2	1.3	1.4	1.5	1.1	0.9	1.3	1.3	2.8	1.2	2.0
Nepal	..	..	..	..	..	..	..	..	..	..	..
Pakistan	1.0	1.6	2.3	3.8	0.7	1.0	4.0	2.7	5.1	0.9	1.7
Sri Lanka	0.3	2.3	1.6	0.5	0.8	2.4	2.1	2.4	2.8	0.5	2.2
EAST ASIA AND THE PACIFIC	*-0.3*	*-0.1*	*6.0*	*4.6*	*2.2*	*2.1*	*2.7*	*0.9*	*2.2*	*-0.4*	*3.4*
China	-3.7	-3.0	9.3	6.9	1.2	1.5	1.7	-0.9	2.1	-4.1	3.0
Fiji	3.2	2.0	5.3	7.5	8.1	6.3	9.1	4.3	1.0	1.3	0.4
Hong Kong	..	..	..	..	..	..	..	..	..	..	..
Indonesia	0.6	2.0	1.7	2.0	3.0	1.5	0.2	3.6	4.6	0.4	2.9
Korea, Republic of	4.5	5.1	4.5	5.2	5.8	2.6	5.5	2.6	2.8	4.2	4.4
Malaysia	..	..	..	..	1.9	4.2	4.7	4.7	1.5	2.5	4.4
Papua New Guinea	..	..	..	..	..	..	..	..	..	..	..
Philippines	2.7	2.1	2.4	2.1	1.3	1.1	2.8	2.4	2.6	2.1	2.1
Singapore	8.5	8.2	7.1	6.2	6.2	6.6	7.7	5.3	3.0	2.8	5.3
Solomon Islands	..	..	..	..	..	..	..	..	..	..	..
Thailand	5.0	3.8	3.3	5.5	1.4	2.2	4.3	2.3	2.3	3.2	4.5
Vanuatu	..	..	..	..	..	..	..	..	..	..	..

NOTE: See glossary for definitions.

1978	1979	1980	1981	1982	1983	1984	1985	1986	1987	
-0.4	*0.6*	*2.2*	*2.2*	*0.4*	*0.4*	*-1.4*	*1.2*	*0.8*	*0.0*	*SUB-SAHARAN AFRICA*
0.4	*0.9*	*1.1*	*1.2*	*0.7*	*0.8*	*0.1*	*1.9*	*0.8*	*0.2*	*Excluding Nigeria*
-1.4	3.2	4.1	8.2	2.0	-2.5	0.2	0.3	0.1	-0.5	Benin
2.7	9.6	8.7	2.2	2.5	4.8	4.1	6.9	2.8	3.6	Botswana
1.6	4.3	2.2	1.8	7.7	1.6	2.4	3.8	4.7	3.4	Burkina Faso
1.4	1.0	1.5	2.7	1.1	-0.5	1.3	0.3	0.9	2.1	Burundi
11.5	1.2	5.6	1.3	-3.3	2.8	2.1	8.4	6.6	-2.0	Cameroon
..	..	..	..	..	..	..	..	..	..	Cape Verde
-2.6	-0.7	-1.4	1.0	4.0	-6.8	4.6	1.7	-0.2	1.7	Central African Republic
-0.3	-10.7	-5.2	2.8	1.3	1.2	-2.0	17.4	-5.6	2.7	Chad
..	..	..	..	..	..	..	..	..	..	Comoros
4.4	1.3	6.7	10.8	-1.4	-1.7	4.6	-1.3	-6.6	-3.4	Congo, People's Republic
17.0	3.5	-5.6	3.6	5.2	2.9	-8.4	13.8	1.1	-2.8	Cote d'Ivoire
-0.2	2.4	1.7	1.3	1.3	2.2	0.3	0.3	2.2	2.1	Ethiopia
..	..	..	..	..	..	..	..	..	..	Gabon
-2.1	4.0	4.9	-4.1	3.2	6.8	2.0	-2.5	2.8	2.7	Gambia, The
1.5	-1.8	-0.7	0.4	-1.6	0.7	2.4	2.6	2.5	3.2	Ghana
-1.0	5.5	5.1	7.1	0.3	-2.0	-1.1	0.3	0.8	1.6	Guinea-Bissau
2.6	3.0	2.5	3.1	1.4	1.3	1.6	2.2	3.0	2.4	Kenya
7.9	-1.1	7.7	2.0	-1.4	2.5	4.6	0.2	6.0	0.2	Lesotho
1.4	1.7	-4.4	2.9	-0.3	-1.1	-0.3	-0.6	-1.7	..	Liberia
-1.3	4.6	0.6	-3.5	-0.5	-0.1	-0.1	0.9	0.2	0.0	Madagascar
5.2	2.1	2.4	-2.1	0.4	3.4	0.7	3.0	1.1	1.5	Malawi
2.1	3.7	0.7	2.9	1.8	1.7	3.4	1.6	2.1	1.3	Mali
0.9	0.4	4.1	-0.5	-3.1	1.6	-4.7	0.9	1.8	1.9	Mauritania
2.3	2.6	-0.7	2.1	1.5	3.0	2.3	1.5	2.5	5.1	Mauritius
..	..	..	1.2	1.9	-3.3	9.3	5.6	..	..	Mozambique
3.5	3.5	1.8	0.6	0.7	-4.1	-8.7	-0.8	2.2	..	Niger
-1.8	0.8	1.6	2.1	0.1	0.5	-2.7	0.1	0.6	-0.1	Nigeria
6.6	-0.4	6.2	11.1	-3.5	1.6	-1.2	2.9	2.5	..	Rwanda
2.0	1.4	1.8	-0.9	6.5	1.1	0.0	1.8	-1.8	3.0	Senegal
6.3	11.4	-2.4	-7.5	2.5	1.5	3.0	7.5	1.8	..	Seychelles
3.0	1.8	2.6	3.8	3.6	3.8	-2.1	0.0	-3.1	-0.6	Sierra Leone
0.8	0.8	2.3	-2.1	1.4	-0.3	-0.1	0.5	0.2	1.3	Somalia
-0.7	-3.5	4.9	1.7	-0.2	1.3	-1.0	-4.2	-0.6	1.0	Sudan
1.1	1.7	0.5	2.0	3.4	0.9	2.1	0.6	2.0	..	Swaziland
2.8	1.4	0.9	0.4	0.2	-0.9	0.7	0.5	0.4	1.2	Tanzania
5.0	-3.3	6.3	-1.8	-1.0	-3.8	-1.1	1.3	4.0	0.9	Togo
-1.6	-0.8	0.1	0.1	1.2	0.8	0.2	0.9	0.9	0.4	Uganda
-2.9	1.5	-0.1	0.5	-2.3	-1.1	-1.2	0.7	-0.3	1.4	Zaire
-1.1	-0.1	2.3	2.9	-0.7	-1.9	-0.4	0.5	-0.4	-0.3	Zambia
-0.6	-0.4	6.8	5.1	2.7	-0.6	0.7	2.1	1.9	1.7	Zimbabwe
3.5	*0.9*	*2.3*	*2.6*	*3.0*	*1.8*	*2.4*	*3.4*	*2.7*	*2.6*	*SOUTH ASIA*
1.7	2.2	1.9	2.9	-0.1	1.2	2.1	2.2	2.6	2.8	Bangladesh
2.6	0.9	1.9	1.9	2.5	1.8	2.1	2.5	2.5	2.5	India
..	..	..	..	..	..	..	..	..	..	Nepal
4.6	2.8	2.8	2.9	3.4	3.2	3.2	3.9	2.9	3.0	Pakistan
2.7	2.8	3.4	3.0	3.4	3.0	3.2	1.8	1.9	1.2	Sri Lanka
3.8	*1.8*	*2.3*	*2.4*	*1.9*	*2.5*	*2.4*	*1.0*	*1.8*	*0.8*	*EAST ASIA AND THE PACIFIC*
3.5	-0.2	1.6	1.8	1.5	2.4	2.9	1.3	1.1	-1.6	China
2.1	4.9	-0.4	1.7	-0.2	2.0	2.5	-0.3	1.9	-3.5	Fiji
..	..	..	..	..	..	..	..	..	..	Hong Kong
5.0	2.9	3.0	4.1	2.6	1.9	2.1	1.2	1.8	2.3	Indonesia
5.1	2.7	0.8	1.8	2.4	4.0	3.0	2.5	4.6	4.7	Korea, Republic of
2.8	3.3	4.6	4.4	2.1	2.7	2.9	-0.3	-1.3	1.2	Malaysia
..	..	..	0.1	0.6	0.4	2.4	-0.2	0.9	0.9	Papua New Guinea
2.4	2.4	2.4	1.2	1.4	1.3	-3.0	-1.7	1.2	2.6	Philippines
6.0	4.7	5.8	5.5	5.0	4.4	4.5	2.2	2.2	5.4	Singapore
..	..	..	..	..	..	..	..	..	..	Solomon Islands
4.0	3.5	3.1	3.3	2.4	3.7	3.2	2.3	2.3	4.3	Thailand
..	..	-6.4	0.3	10.3	-1.7	3.7	5.3	-3.2	..	Vanuatu

Table 9. Services: contribution to gross domestic product growth (cont'd)

Percentage points	1967	1968	1969	1970	1971	1972	1973	1974	1975	1976	1977
LATIN AMERICA AND CARIBBEAN	*2.1*	*3.5*	*3.6*	*3.6*	*3.3*	*3.9*	*4.1*	*3.8*	*2.7*	*3.2*	*2.6*
Argentina	1.3	1.6	4.2	8.0	1.2	0.5	2.0	3.0	0.1	-0.9	3.0
Bahamas	..	..	..	..	..	..	..	..	..	..	..
Barbados	7.1	9.1	6.0	7.1	5.2	2.3	1.0	-0.7	-1.5	0.6	4.5
Belize	..	..	..	..	1.4	3.2	2.5	4.3	1.8	4.0	-0.7
Bolivia	3.2	1.9	1.1	-0.2	1.8	1.3	-0.8	4.7	4.8	3.5	3.4
Brazil	2.2	5.2	4.6	5.8	5.1	5.8	7.1	5.5	2.4	4.8	2.0
Chile	1.9	1.4	3.3	1.0	5.0	0.3	-2.0	-3.1	-4.2	2.2	6.6
Colombia	1.7	2.9	3.0	4.1	4.4	3.5	3.8	3.0	0.9	2.7	3.0
Costa Rica	2.4	3.4	1.7	4.6	3.4	4.0	4.5	3.1	0.5	3.1	5.8
Dominican Republic	0.5	0.1	4.7	4.3	3.9	4.6	4.8	4.2	2.9	3.0	2.7
Ecuador	3.2	3.6	1.2	3.3	2.2	1.8	5.2	5.4	5.3	3.5	5.2
El Salvador	2.6	2.3	2.1	0.5	2.1	3.4	3.6	2.6	1.5	4.7	2.8
Grenada	..	..	..	..	..	..	..	..	..	..	..
Guatemala	..	..	..	..	..	..	..	..	..	..	..
Guyana	2.1	0.0	1.4	2.5	0.8	0.9	3.0	1.2	6.5	2.3	-0.9
Haiti	..	..	..	..	..	..	..	..	..	..	..
Honduras	1.2	2.2	2.1	3.1	0.6	4.2	2.7	0.8	4.2	5.4	3.8
Jamaica	1.6	2.2	4.5	5.1	0.9	5.9	1.4	-3.0	1.0	-2.0	-0.5
Mexico	3.6	5.1	3.6	4.0	2.9	5.5	4.5	3.6	3.9	2.5	1.9
Nicaragua	2.7	1.8	2.0	1.0	2.2	1.2	2.8	5.9	0.6	2.9	-3.5
Panama	5.4	4.3	5.9	5.3	5.9	4.3	2.8	4.0	0.2	0.2	2.2
Paraguay	5.5	2.2	2.0	1.1	3.1	3.4	2.9	3.3	3.1	3.7	4.4
Peru	2.0	0.6	2.6	0.9	3.2	2.2	3.0	4.8	3.2	0.3	0.0
St. Vincent and the Grenadines	..	..	..	..	..	..	..	..	..	..	..
Trinidad and Tobago	3.4	1.0	1.3	0.7	-0.3	1.3	1.4	2.9	1.1	1.1	4.7
Uruguay	-0.4	0.6	2.3	0.6	1.2	-0.5	0.7	1.9	2.5	2.6	-0.7
Venezuela	1.2	2.0	2.4	0.2	2.1	2.5	2.8	4.0	5.9	5.2	4.0
MIDDLE EAST AND NORTH AFRICA	*-0.2*	*4.0*	*5.8*	*3.4*	*3.0*	*2.0*	*2.0*	*3.6*	*2.0*	*1.4*	*6.3*
Algeria	-0.4	4.0	4.2	3.9	0.4	4.1	0.5	7.9	1.5	2.6	2.1
Egypt, Arab Republic of	0.4	-0.2	1.6	2.3	2.1	1.8	7.1	6.4	4.0	8.9	4.1
Jordan	..	..	..	..	-1.3	6.0	-3.0	5.5	6.9	6.5	3.7
Morocco	3.0	2.7	6.0	2.5	1.3	1.0	3.0	0.2	5.7	6.2	4.9
Oman	..	..	..	..	..	..	..	..	..	..	..
Syrian Arab Republic	1.3	7.0	6.9	-2.5	2.7	4.4	4.7	9.6	14.6	3.0	2.9
Tunisia	0.8	-1.1	4.5	3.1	5.1	6.5	0.4	3.4	4.5	2.9	1.1
Yemen Arab Republic	..	..	..	..	7.6	5.8	6.5	6.3	5.1	7.1	7.3
Yemen, PDR	..	..	..	..	..	..	..	..	..	13.2	12.0
EUROPE AND MEDITERRANEAN	..	..	..	..	..	..	..	..	..	..	..
Cyprus	..	..	..	..	..	..	..	..	..	9.1	8.0
Greece	2.7	3.2	3.6	3.9	3.4	3.9	4.2	1.2	2.3	3.2	2.4
Hungary	3.5	2.5	2.4	4.9	2.4	3.1	2.8	2.2	3.0	1.9	2.1
Israel	..	..	..	..	..	..	..	..	..	..	..
Malta	..	..	..	..	..	..	..	..	..	..	..
Poland	..	..	..	..	..	..	..	..	..	..	..
Portugal	..	..	..	..	..	..	..	..	..	..	..
Turkey	1.9	3.7	2.8	3.6	3.6	4.9	4.0	4.1	3.7	4.2	2.6
Yugoslavia	2.2	4.5	6.9	2.7	5.3	-5.1	0.9	4.5	0.3	2.4	2.9
Developing countries	1.1	2.6	3.4	3.0	2.8	2.3	2.9	2.6	2.6	2.3	3.0
Highly indebted	1.6	3.6	3.7	3.9	3.4	2.7	3.8	3.6	3.1	2.8	2.4
Low-income countries	-0.5	0.1	4.2	3.2	1.4	0.9	1.2	0.1	1.8	-0.8	2.4
Low-income Africa	..	1.7	1.9	0.0	2.4	0.1	1.7	1.1	-1.8	2.8	1.7
China and India	-1.3	-0.6	5.4	4.3	1.3	1.2	1.1	-0.4	2.6	-1.9	2.5
Other low-income Asia	2.0	1.7	1.3	3.4	0.5	-0.5	2.5	1.6	2.6	3.3	2.2
Middle-income countries	1.8	3.4	3.2	3.0	3.4	2.8	3.5	3.5	2.9	3.2	3.4
High-income oil exporters	2.0	-7.0	-0.3	-1.4	2.6	2.9	2.1	3.7	3.0	2.5	2.8
Industrialized countries	..	..	..	..	2.9	2.9	0.3	1.7	1.3	2.4	2.3
Japan	..	..	..	..	2.0	4.1	3.1	0.0	3.2	2.4	3.8
United States	..	..	..	..	2.6	2.7	2.0	0.8	1.1	2.6	2.6
European Community	..	..	..	..	..	..	..	..	..	..	..

NOTE: See glossary for definitions.

1978	1979	1980	1981	1982	1983	1984	1985	1986	1987	
2.7	*3.6*	*3.3*	*1.5*	*-0.2*	*-0.8*	*1.9*	*1.6*	*2.1*	*1.4*	*LATIN AMERICA AND CARIBBEAN*
-0.6	4.0	3.0	-2.4	-3.8	0.8	1.8	-1.6	2.6	0.9	Argentina
..	..	..	..	..	..	..	..	..	..	Bahamas
3.6	4.3	2.8	0.1	-3.2	-0.6	2.1	2.1	3.0	3.7	Barbados
1.9	3.0	1.4	0.4	-1.0	1.4	1.9	1.0	..	..	Belize
3.9	1.8	-1.3	1.9	-3.4	-1.7	-0.4	0.6	-0.1	1.6	Bolivia
2.5	4.0	4.6	-0.4	1.2	-0.3	3.0	3.9	4.6	1.5	Brazil
6.0	4.5	4.5	2.9	-8.5	-0.7	2.9	0.6	2.8	3.6	Chile
4.1	2.8	2.1	1.8	1.2	0.2	0.7	0.7	1.2	2.0	Colombia
3.0	3.3	0.5	-2.0	-2.5	1.0	3.0	1.1	2.4	2.3	Costa Rica
2.3	1.7	4.3	2.1	2.0	1.8	-2.3	0.8	0.7	..	Dominican Republic
3.7	2.4	4.2	0.5	0.8	-2.9	1.6	1.5	1.2	0.8	Ecuador
2.6	-1.9	-4.6	-4.3	-2.8	1.1	1.3	1.4	0.8	1.1	El Salvador
..	..	..	..	..	..	..	..	..	..	Grenada
..	..	..	..	..	..	..	..	..	..	Guatemala
-1.9	0.7	0.4	-0.1	-3.4	-1.8	0.6	-0.1	0.1	..	Guyana
..	..	..	..	..	..	..	..	..	..	Haiti
4.6	3.2	1.3	2.5	-2.2	-0.8	0.2	1.4	1.9	2.5	Honduras
0.1	0.2	-2.1	1.9	2.2	0.8	-0.5	-2.2	0.7	2.7	Jamaica
4.4	5.8	4.7	5.3	0.2	-1.5	1.9	0.8	-1.8	0.8	Mexico
-6.1	-12.3	9.0	-1.1	-0.6	1.4	-0.5	-2.0	0.0	..	Nicaragua
6.3	3.6	13.6	3.3	3.9	2.5	0.5	3.9	2.4	1.7	Panama
5.8	5.5	4.5	3.8	-0.3	-1.1	0.8	1.7	1.4	1.5	Paraguay
-1.0	1.9	3.0	3.1	-0.2	-4.8	1.6	0.8	4.8	3.1	Peru
4.9	2.1	5.2	1.4	3.6	3.1	2.5	2.6	2.0	..	St. Vincent and the Grenadines
3.3	3.3	6.7	2.1	3.5	-2.9	-6.4	0.8	-3.7	-0.6	Trinidad and Tobago
3.3	3.7	3.3	2.1	-4.5	-2.7	-0.6	1.1	3.2	2.0	Uruguay
0.7	-0.5	-2.1	0.6	-0.5	-1.4	0.0	0.6	3.2	1.7	Venezuela
0.7	*3.7*	*2.5*	*3.5*	*3.4*	*2.5*	*1.7*	*2.0*	*0.1*	*0.7*	*MIDDLE EAST AND NORTH AFRICA*
2.4	5.1	0.9	3.4	0.1	2.3	0.7	1.4	-0.4	-0.1	Algeria
2.7	5.5	4.6	2.3	9.8	4.7	2.8	3.5	1.8	1.5	Egypt, Arab Republic of
12.4	5.4	11.2	0.1	5.2	2.3	3.5	2.1	2.6	-0.6	Jordan
0.3	2.3	3.1	1.6	3.5	1.7	2.1	2.2	2.8	1.7	Morocco
..	3.7	3.5	5.7	4.4	3.5	5.7	2.9	-2.2	..	Oman
3.1	6.5	2.3	8.9	2.1	0.7	-0.2	0.1	-5.0	-2.1	Syrian Arab Republic
2.7	4.0	3.7	3.1	1.1	1.9	2.5	2.4	0.6	3.0	Tunisia
8.8	-0.6	4.9	4.8	25.3	2.5	2.0	2.2	5.6	2.4	Yemen Arab Republic
1.5	6.3	-8.1	1.9	..	..	..	..	..	..	Yemen, PDR
2.9	*1.6*	*1.4*	*-1.1*	*-0.3*	*1.7*	*1.4*	*2.0*	*..*	*..*	*EUROPE AND MEDITERRANEAN*
4.4	5.8	3.5	4.0	5.4	4.8	6.5	4.3	4.1	5.0	Cyprus
2.8	2.4	1.2	0.9	0.9	1.6	1.6	1.9	1.1	0.6	Greece
2.3	-0.3	0.1	0.5	-0.8	-0.1	1.2	1.5	1.0	1.9	Hungary
..	..	..	..	..	..	..	..	..	..	Israel
..	..	..	..	..	..	..	..	..	..	Malta
..	..	..	..	..	..	..	..	..	..	Poland
0.8	2.4	2.3	1.9	1.4	0.4	-0.8	1.6	..	..	Portugal
1.9	-0.2	0.4	2.0	1.9	2.1	2.7	1.9	2.8	3.2	Turkey
6.0	1.3	2.8	-0.4	1.3	-1.4	0.5	3.0	0.8	0.7	Yugoslavia
2.1	2.1	2.4	1.7	0.9	1.4	1.5	1.1	1.6	1.0	Developing countries
2.3	3.0	3.0	1.6	0.1	-0.6	1.0	1.6	1.7	1.2	Highly indebted
2.8	0.1	1.8	1.9	1.7	1.9	2.2	1.7	1.5	-0.2	Low-income countries
0.6	0.8	1.3	1.0	0.3	0.4	0.0	0.4	0.6	1.0	Low-income Africa
3.4	-0.1	1.6	2.1	2.1	2.0	2.4	1.8	1.6	-0.7	China and India
3.3	1.5	4.3	3.4	1.6	2.7	3.2	2.1	2.3	3.4	Other low-income Asia
1.8	2.7	2.7	1.7	0.8	1.3	1.4	1.1	1.8	1.7	Middle-income countries
2.0	3.0	4.9	1.4	2.0	1.6	0.9	-0.6	-1.9	..	High-income oil exporters
2.4	1.8	1.5	1.2	0.3	1.8	2.4	1.9	2.2	..	Industrialized countries
2.2	2.8	2.4	2.4	1.1	1.4	1.4	2.2	1.8	..	Japan
3.3	1.3	1.0	1.4	-0.9	3.0	3.3	1.9	2.5	..	United States
..	..	..	..	..	..	..	..	..	..	European Community

Table 10. Total consumption: contribution to growth of gross domestic product expenditure

Percventage points	1967	1968	1969	1970	1971	1972	1973	1974	1975	1976	1977
SUB-SAHARAN AFRICA	*-3.2*	*1.8*	*7.2*	*6.9*	*7.9*	*1.3*	*1.6*	*7.5*	*-1.9*	*2.8*	*2.4*
Excluding Nigeria	*2.2*	*1.6*	*3.1*	*2.1*	*5.7*	*1.6*	*1.9*	*3.6*	*-0.9*	*2.3*	*4.2*
Benin	8.6	3.3	2.9	-1.3	-0.2	5.3	4.5	-0.4	-2.9	-0.7	7.9
Botswana	14.4	35.2	4.4	2.7	-10.8	0.9	16.8	18.0	5.2	7.3	9.2
Burkina Faso	0.8	14.2	4.2	1.8	1.0	-4.7	-1.6	5.4	10.1	5.4	11.6
Burundi	6.2	0.4	-4.0	30.8	2.2	-3.5	4.0	-2.6	6.2	5.0	13.0
Cameroon	-11.6	9.7	3.0	0.6	7.4	0.1	0.6	9.2	0.7	5.2	9.0
Cape Verde	..	..	..	..	..	..	..	-17.1	7.6	11.7	9.9
Central African Republic	5.9	5.4	5.3	1.8	-1.3	0.5	-0.6	2.0	3.1	6.7	11.7
Chad	-4.2	-0.8	2.1	6.4	-4.2	5.9	28.0	-27.6	1.1	3.2	1.2
Comoros	..	..	..	..	..	..	..	..	..	..	..
Congo, People's Republic	-1.4	11.1	9.1	8.2	-1.7	-1.5	-1.2	-0.5	4.5	10.7	-11.3
Cote d'Ivoire	5.4	8.1	6.9	6.0	10.2	3.6	7.3	3.3	10.4	4.7	3.0
Ethiopia	0.5	0.5	3.9	8.5	3.2	3.6	1.0	0.9	4.2	-1.9	5.1
Gabon	..	..	..	..	..	..	..	..	..	..	..
Gambia, The	-3.2	6.4	4.4	-9.3	27.7	-1.9	5.0	7.2	-14.9	26.4	4.3
Ghana	9.5	-3.6	6.9	4.3	3.6	-3.1	3.8	10.1	-13.6	-3.0	3.3
Guinea-Bissau	..	..	..	..	3.7	19.2	6.5	6.3	-10.6	-1.6	-12.1
Kenya	5.5	-3.6	4.0	-7.5	30.7	13.9	-4.3	9.2	2.9	-2.0	7.3
Lesotho	11.1	8.7	0.0	-1.8	12.5	16.3	31.9	24.2	16.1	23.5	11.7
Liberia	1.9	4.6	2.9	-13.2	12.9	-14.4	15.6	3.0	0.7	12.0	9.6
Madagascar	1.8	1.4	5.8	3.3	3.3	0.1	-3.0	-2.7	1.0	-4.4	4.7
Malawi	11.4	2.3	5.6	-8.4	22.5	-0.8	1.2	2.1	5.5	3.6	5.3
Mali	9.2	-1.7	-0.8	4.9	-0.4	9.2	-2.2	10.3	6.9	9.3	-0.5
Mauritania	4.4	5.6	-0.8	9.5	-0.5	7.0	-5.8	13.6	-2.4	33.2	-7.6
Mauritius	6.2	4.5	-20.3	-4.6	3.4	2.0	-5.1	22.7	13.3	9.3	8.5
Mozambique	..	..	..	..	..	..	..	..	..	..	..
Niger	0.0	-0.8	3.0	-10.2	6.9	-7.6	-28.5	30.8	-17.7	1.8	4.2
Nigeria	-11.3	1.7	15.1	14.8	10.9	1.0	1.3	12.0	-3.5	3.1	0.8
Rwanda	10.4	5.1	7.1	12.2	2.3	1.9	1.1	-2.7	6.7	-1.3	7.7
Senegal	0.3	1.7	-10.5	6.4	1.4	0.2	-1.3	1.2	7.5	6.2	2.1
Seychelles	..	..	..	..	..	..	..	..	..	..	5.1
Sierra Leone	4.8	-5.3	-2.6	11.4	4.7	-2.0	-1.5	14.5	-2.6	-5.0	3.2
Somalia	12.0	5.6	-0.6	-1.1	7.1	-1.5	2.3	-17.7	33.9	-5.3	45.3
Sudan	-10.2	6.6	-4.3	-0.5	11.2	-1.4	-12.3	6.4	19.7	6.5	19.9
Swaziland	10.8	-1.7	24.7	-6.8	-3.0	7.2	-2.4	-12.6	17.6	-5.4	10.3
Tanzania	1.7	9.2	2.0	0.6	0.9	10.1	6.9	5.0	2.1	-4.0	0.7
Togo	8.6	5.6	5.0	3.4	1.2	4.9	1.9	5.2	-6.2	16.2	4.7
Uganda	2.1	3.3	10.5	0.0	1.4	2.0	3.7	-1.0	-1.6	3.9	4.1
Zaire	9.3	-2.8	9.2	0.1	4.9	1.5	9.2	1.7	-4.8	-5.7	1.5
Zambia	15.2	2.7	-5.0	-7.6	2.3	2.7	-8.4	1.8	2.4	3.4	-1.1
Zimbabwe	1.4	3.0	-1.9	13.8	17.9	8.3	0.0	1.9	0.2	2.1	-1.4
SOUTH ASIA	*7.3*	*2.5*	*3.8*	*6.9*	*0.9*	*-0.9*	*-1.3*	*0.9*	*5.9*	*1.7*	*7.7*
Bangladesh	-0.7	3.4	3.5	8.0	-4.0	-6.7	-12.9	22.3	3.0	14.1	-5.2
India	8.9	1.8	3.8	5.7	1.8	0.3	-1.2	-1.6	7.1	0.3	8.9
Nepal	..	..	..	..	..	..	..	..	..	..	..
Pakistan	6.7	5.4	2.7	19.0	-3.5	-6.0	3.0	9.7	-0.5	4.6	6.1
Sri Lanka	0.2	9.8	9.9	-3.2	-1.1	-5.0	9.2	1.7	1.2	3.0	8.4
EAST ASIA AND THE PACIFIC	*2.6*	*0.4*	*10.1*	*7.6*	*3.5*	*3.5*	*5.7*	*3.0*	*4.0*	*1.0*	*5.1*
China	-0.3	-6.1	15.1	10.8	2.7	3.6	4.6	1.9	3.9	-2.7	5.1
Fiji	3.3	-0.8	6.4	11.0	5.4	5.6	11.6	8.2	1.7	5.8	3.4
Hong Kong	1.3	6.2	5.8	6.5	7.9	4.9	8.8	-0.6	2.3	4.8	10.5
Indonesia	5.5	9.8	3.9	1.7	1.6	1.3	5.7	6.0	4.8	2.7	2.8
Korea, Republic of	5.4	8.5	6.5	8.6	10.1	2.7	8.3	6.5	5.1	7.1	4.1
Malaysia	3.7	0.7	0.9	3.6	4.6	5.2	4.6	6.1	0.7	3.5	6.4
Papua New Guinea	6.0	4.3	4.9	5.5	2.1	3.1	5.2	2.0	2.0	1.6	-1.0
Philippines	7.7	7.9	5.0	1.6	3.6	5.1	4.5	5.8	2.3	0.7	4.7
Singapore	..	..	..	..	..	..	..	..	..	..	..
Solomon Islands	..	..	..	..	..	..	..	..	..	..	..
Thailand	8.5	10.4	5.4	6.0	-1.7	5.6	8.8	2.0	4.6	6.8	6.2
Vanuatu	..	..	..	..	..	..	..	..	..	..	..

NOTE: See glossary for definitions.

1978	1979	1980	1981	1982	1983	1984	1985	1986	1987	
6.0	*-1.7*	*4.4*	*6.6*	*2.4*	*-1.9*	*-2.3*	*2.9*	*1.3*	*-0.6*	*SUB-SAHARAN AFRICA*
3.5	*1.7*	*2.0*	*2.4*	*1.5*	*0.5*	*-0.2*	*3.2*	*2.6*	*..*	*Excluding Nigeria*
2.7	5.2	6.9	8.9	9.2	-12.7	6.2	6.1	-0.4	-3.2	Benin
5.7	9.8	7.8	3.2	8.0	3.1	3.1	10.3	-0.5	..	Botswana
-0.9	7.0	3.7	5.5	4.6	-2.9	-3.9	16.3	9.0	3.7	Burkina Faso
-4.9	4.1	1.9	1.6	4.9	-5.6	7.5	2.3	1.4	-2.6	Burundi
2.9	13.6	10.8	3.8	0.9	0.1	-1.0	13.3	12.8	1.0	Cameroon
20.6	1.2	23.9	-23.7	15.4	27.7	-1.3	18.0	..	..	Cape Verde
4.4	-3.3	2.7	-10.9	7.2	-13.7	8.2	3.6	7.3	1.7	Central African Republic
2.4	-20.9	-2.7	-0.5	4.3	7.5	-6.3	31.3	..	..	Chad
..	..	..	..	..	..	..	..	..	..	Comoros
7.5	-7.2	12.9	19.1	0.7	3.7	3.7	4.6	-2.8	..	Congo, People's Republic
13.6	2.8	-6.6	5.9	0.0	0.2	-1.9	3.6	6.9	-2.6	Cote d'Ivoire
6.4	4.6	3.8	0.2	4.6	4.3	-2.6	-0.9	10.0	..	Ethiopia
..	..	..	..	..	..	..	..	..	..	Gabon
17.5	-24.8	6.0	4.7	-3.1	5.1	..	..	..	..	Gambia, The
12.6	-3.2	3.2	-2.9	-9.6	-0.5	7.8	4.1	3.3	4.7	Ghana
6.1	-5.3	-11.7	23.1	12.5	-4.0	10.0	4.9	-1.9	4.5	Guinea-Bissau
11.0	5.2	-0.5	-2.1	3.3	-0.4	3.0	2.4	2.7	4.6	Kenya
18.9	2.8	7.4	13.6	6.0	4.4	..	..	..	..	Lesotho
0.0	-3.6	-8.3	3.2	0.7	6.3	-2.2	-0.8	-6.0	..	Liberia
-1.7	6.7	2.1	-9.4	2.1	-0.6	-0.7	1.4	0.6	1.8	Madagascar
3.6	7.3	2.2	-2.3	-1.2	1.6	7.2	5.6	3.1	-0.7	Malawi
8.7	10.3	-0.7	2.9	4.7	-0.9	1.5	12.2	6.0	3.3	Mali
-8.0	-0.1	-7.7	2.4	5.5	23.6	-14.0	-2.5	1.6	4.0	Mauritania
4.2	2.6	-2.1	-2.6	-1.5	2.0	3.1	2.9	2.0	11.5	Mauritius
..	..	..	1.0	-4.5	-11.3	8.8	-2.6	11.4	..	Mozambique
0.8	12.0	10.4	1.3	10.8	-2.7	9.9	-11.3	7.8	-5.4	Niger
8.2	-4.7	6.7	10.3	3.3	-4.8	-4.9	2.8	-0.3	-5.1	Nigeria
11.1	9.6	10.2	8.8	0.4	6.7	-6.5	4.5	6.2	-3.8	Rwanda
1.3	8.0	1.3	3.2	10.3	0.7	-5.8	6.5	-0.6	4.4	Senegal
15.1	11.4	-1.5	0.0	6.4	..	..	..	..	..	Seychelles
8.6	8.3	7.8	6.5	-3.8	1.6	-10.2	-1.4	-3.3	0.9	Sierra Leone
-11.3	23.2	-7.5	29.5	2.7	-8.3	-11.4	10.6	-9.2	19.1	Somalia
3.1	-9.2	3.2	7.8	6.5	-0.5	-4.0	-13.7	5.4	-1.4	Sudan
5.3	18.6	0.7	15.1	-16.9	-11.6	8.2	..	..	..	Swaziland
12.7	-6.3	6.3	-2.9	1.4	2.1	3.3	5.4	7.7	5.3	Tanzania
4.5	-2.0	14.6	-4.0	0.2	-6.4	-1.5	4.9	2.3	7.9	Togo
-7.7	-15.7	-3.6	..	..	..	..	..	..	..	Uganda
-8.6	-0.1	-0.3	3.0	-5.4	1.0	-2.5	3.7	-5.3	5.8	Zaire
-4.3	9.1	0.3	6.7	-11.3	1.1	0.7	5.0	-1.4	3.4	Zambia
2.9	6.1	6.2	10.3	6.5	3.2	-6.5	1.6	-9.9	2.6	Zimbabwe
6.5	*-2.2*	*2.7*	*3.9*	*5.0*	*7.0*	*2.8*	*4.4*	*4.0*	*2.2*	*SOUTH ASIA*
10.1	3.6	1.5	4.6	-4.6	4.1	6.7	6.4	1.5	5.1	Bangladesh
6.1	-4.1	1.9	3.9	5.7	7.7	1.8	4.0	4.4	1.9	India
..	..	..	..	..	..	..	..	..	..	Nepal
7.4	7.5	7.7	2.5	3.7	3.9	6.9	7.1	3.0	3.1	Pakistan
6.8	2.9	7.4	2.5	7.9	8.6	4.0	8.7	4.1	2.5	Sri Lanka
6.5	*6.5*	*5.1*	*2.7*	*3.4*	*5.5*	*4.6*	*2.0*	*2.3*	*2.4*	*EAST ASIA AND THE PACIFIC*
5.8	5.9	6.3	3.7	3.2	6.9	6.3	2.3	2.0	1.0	China
0.4	3.5	-3.8	7.3	-3.5	-2.0	6.1	-4.0	2.4	4.7	Fiji
11.7	6.4	8.3	6.1	3.9	5.4	3.8	2.6	6.6	7.2	Hong Kong
6.1	4.9	7.6	0.0	4.4	3.4	5.1	1.0	2.5	3.2	Indonesia
7.7	6.9	-0.5	2.2	4.1	4.5	3.0	3.4	5.2	4.4	Korea, Republic of
5.9	5.0	10.4	5.2	2.9	2.0	1.7	-0.2	-6.2	-0.1	Malaysia
3.0	5.3	0.7	-1.6	-2.8	1.2	0.8	4.4	-0.5	3.3	Papua New Guinea
5.2	5.7	2.4	2.2	4.9	3.7	-0.9	-4.8	2.6	9.2	Philippines
..	..	..	..	..	..	..	..	..	..	Singapore
..	..	..	5.2	-12.7	7.9	8.5	1.2	2.0	..	Solomon Islands
4.6	9.9	2.4	0.7	0.9	9.6	1.6	1.5	2.7	6.5	Thailand
..	..	..	..	..	..	..	..	..	..	Vanuatu

Table 10. Total consumption: contribution to growth of gross domestic product expenditure (cont'd)

Percentage points	1967	1968	1969	1970	1971	1972	1973	1974	1975	1976	1977
LATIN AMERICA AND CARIBBEAN	*4.1*	*5.9*	*2.6*	*6.7*	*5.7*	*4.3*	*4.9*	*4.3*	*3.2*	*5.3*	*3.6*
Argentina	2.6	3.5	5.1	4.5	2.7	1.5	3.8	5.5	-0.3	-3.9	1.9
Bahamas	..	..	..	..	..	..	..	..	..	..	..
Barbados	..	..	..	..	..	..	..	..	..	..	..
Belize	..	..	..	..	..	..	..	..	..	..	..
Bolivia	2.0	-0.4	3.0	4.8	2.9	0.8	6.4	2.9	4.3	5.3	3.9
Brazil	6.9	9.0	2.2	10.6	10.6	9.1	10.0	8.8	0.9	9.4	3.6
Chile	2.8	3.5	4.9	0.4	11.4	6.1	-5.5	-13.1	-10.4	0.3	11.2
Colombia	1.2	4.5	6.3	6.6	6.9	5.4	5.3	2.6	2.3	5.1	3.1
Costa Rica	4.6	6.4	4.2	8.9	2.5	5.3	4.2	4.6	2.0	3.9	10.0
Dominican Republic	-0.3	2.5	7.1	9.7	6.6	2.4	6.5	10.8	3.1	3.8	2.9
Ecuador	4.9	4.5	5.6	5.2	3.2	2.2	4.4	8.7	8.3	7.6	7.2
El Salvador	3.6	-0.4	2.8	6.3	2.2	3.7	8.1	1.5	1.9	10.6	12.3
Grenada	..	..	..	..	..	..	..	..	..	..	..
Guatemala	4.4	2.7	6.2	5.4	4.0	5.7	4.6	3.1	3.0	7.2	7.0
Guyana	-1.4	6.0	8.1	3.8	-0.5	6.8	8.3	3.3	4.7	16.1	-5.5
Haiti	-5.7	1.4	2.1	-0.7	8.0	0.8	-1.5	5.6	-3.9	9.1	2.8
Honduras	1.9	5.8	1.5	5.0	-2.1	4.6	6.7	0.7	3.5	12.6	11.0
Jamaica	3.0	2.2	3.0	9.5	5.3	13.2	-7.7	7.9	5.1	-3.0	-1.3
Mexico	5.5	7.2	3.6	4.7	4.4	5.7	5.2	4.0	5.4	3.5	1.0
Nicaragua	-1.7	2.3	0.3	0.3	3.0	-0.4	14.0	8.0	-2.9	3.6	5.4
Panama	7.0	2.0	8.9	4.7	6.0	2.7	1.5	8.1	-0.2	0.2	5.9
Paraguay	5.7	6.0	3.1	1.7	6.1	2.6	4.2	8.1	3.0	6.9	11.6
Peru	3.6	8.8	4.1	12.5	3.2	3.0	3.2	5.3	4.0	1.8	1.7
St. Vincent and the Grenadines	..	..	..	..	..	..	..	..	..	..	..
Trinidad and Tobago	6.8	-3.6	2.9	-1.6	9.2	-1.4	-3.8	0.9	8.6	6.1	9.3
Uruguay	-4.9	1.3	6.6	8.7	0.5	-0.9	2.3	-0.2	3.0	-0.4	-0.5
Venezuela	..	..	..	..	..	..	..	..	11.3	7.8	5.9
MIDDLE EAST AND NORTH AFRICA	*3.4*	*5.9*	*5.9*	*5.5*	*7.9*	*3.6*	*2.7*	*9.7*	*5.8*	*2.6*	*9.7*
Algeria	0.3	2.0	1.7	4.3	7.3	4.8	-0.5	10.9	7.1	3.8	7.4
Egypt, Arab Republic of	5.8	8.5	5.3	6.9	5.3	1.7	0.2	7.7	0.0	3.9	9.0
Jordan	..	..	..	..	-2.2	10.8	-2.7	-3.0	12.6	29.4	17.7
Morocco	7.4	7.5	2.6	3.8	4.2	1.7	2.8	5.3	3.8	13.4	4.9
Oman	..	..	..	..	..	..	..	..	..	..	..
Syrian Arab Republic	11.7	0.2	14.2	-4.7	11.6	5.0	6.6	32.2	23.8	2.8	0.0
Tunisia	-0.1	1.1	5.7	10.5	4.5	10.2	4.3	4.5	8.5	4.4	6.0
Yemen Arab Republic	..	..	..	..	15.1	11.0	23.3	10.9	12.0	20.6	19.5
Yemen, PDR	..	..	..	..	..	..	..	..	..	..	..
EUROPE AND MEDITERRANEAN	..	..	..	..	*6.8*	*3.5*	*2.4*	*9.6*	*1.7*	*1.6*	*3.1*
Cyprus	..	..	..	..	..	..	..	..	..	12.6	8.6
Greece	5.7	5.7	5.4	4.9	5.6	6.4	3.8	0.1	6.4	4.4	4.5
Hungary	-0.9	-1.4	15.9	-6.2	4.3	3.1	3.3	4.5	3.5	1.5	3.3
Israel	11.6	10.5	11.7	10.8	3.8	4.5	18.6	5.4	4.5	-1.9	-2.8
Malta	9.9	16.9	23.5	17.0	-1.1	0.7	0.6	7.4	4.3	11.3	11.5
Poland	..	..	..	..	..	..	..	..	..	..	..
Portugal	-1.6	3.1	22.0	2.7	3.5	8.5	3.0	10.3	10.9	-2.5	5.4
Turkey	2.7	7.2	5.1	2.6	11.6	7.3	0.8	8.4	6.3	7.9	4.3
Yugoslavia	3.0	5.9	10.6	4.3	7.3	0.6	0.0	16.1	-3.5	-0.6	6.0
Developing countries	3.1	4.1	5.4	5.9	4.5	2.5	4.0	5.5	3.4	3.2	5.0
Highly indebted	2.5	6.1	4.2	7.4	6.4	4.4	4.7	-0.1	2.0	4.2	3.1
Low-income countries	3.3	-1.2	7.6	7.3	2.3	1.5	1.7	1.9	3.4	-0.5	6.2
Low-income Africa	3.0	0.7	2.5	0.9	5.1	1.5	1.1	3.7	-3.5	0.3	4.9
China and India	3.6	-2.4	9.5	8.5	2.3	2.2	2.3	0.5	5.2	-1.4	6.7
Other low-income Asia	1.1	5.4	4.1	11.7	-2.8	-5.3	-1.3	11.6	1.3	7.2	3.0
Middle-income countries	2.9	6.2	4.5	5.3	5.5	2.8	4.8	6.7	3.5	4.5	4.7
High-income oil exporters	8.7	-8.6	-1.3	-1.6	6.6	0.8	1.7	10.6	3.3	6.4	8.8
Industrialized countries	..	..	..	..	2.9	4.0	3.4	0.7	1.9	2.9	2.4
Japan	..	..	..	..	3.7	5.9	5.8	0.1	3.3	2.6	3.0
United States	..	..	..	..	1.7	3.6	2.4	-0.3	1.7	3.5	3.2
European Community	..	..	..	..	..	3.9	2.9	2.3	3.8	0.1	2.6

NOTE: See glossary for definitions.

1978	1979	1980	1981	1982	1983	1984	1985	1986	1987	
3.8	*5.3*	*5.0*	*1.3*	*0.1*	*-1.9*	*1.6*	*2.1*	*4.0*	*1.2*	*LATIN AMERICA AND CARIBBEAN*
-0.6	7.8	3.3	-3.0	-5.9	3.6	4.3	-4.0	5.6	0.8	Argentina
10.1	25.1	-0.7	-1.0	-0.5	2.9	20.0	4.7	2.4	1.8	Bahamas
..	..	..	..	..	..	..	..	..	..	Barbados
..	..	..	..	..	..	..	..	..	..	Belize
5.0	2.4	1.5	8.0	-9.6	-1.2	-0.1	3.6	-0.1	2.6	Bolivia
3.7	7.4	4.7	-3.2	3.1	-1.6	0.8	5.6	7.8	4.0	Brazil
6.7	6.1	4.0	7.2	-10.1	-2.4	1.6	-0.6	3.1	3.2	Chile
6.1	4.1	4.2	2.4	1.3	0.7	2.8	2.2	2.5	3.0	Colombia
6.5	2.9	-1.1	-4.9	-6.0	2.6	5.9	4.4	4.5	4.8	Costa Rica
0.6	1.8	14.3	2.4	2.3	1.5	-5.8	-2.2	..	..	Dominican Republic
3.8	4.8	5.4	4.0	1.2	-0.9	0.7	1.0	-0.1	0.6	Ecuador
2.4	-8.5	-6.7	-6.2	-4.9	-0.8	4.1	3.9	0.9	-2.2	El Salvador
..	..	..	..	..	..	..	..	..	..	Grenada
4.1	3.3	3.0	0.4	-3.4	-1.9	1.4	-0.7	0.2	4.7	Guatemala
-14.3	-6.0	8.0	8.0	-13.9	-2.1	-16.5	7.9	5.0	-0.3	Guyana
2.5	2.5	10.5	-1.1	-7.9	1.5	-1.6	3.0	2.7	1.6	Haiti
0.8	1.7	6.3	-0.8	-0.1	-1.0	1.6	2.5	2.4	4.0	Honduras
-6.0	-6.3	-2.8	-1.1	6.4	0.7	2.6	-1.6	-1.4	4.8	Jamaica
6.2	6.7	5.8	5.8	-1.4	-3.1	2.8	2.1	-1.5	-1.0	Mexico
-4.5	-21.3	24.3	-12.4	-5.1	2.3	3.3	1.2	-1.2	..	Nicaragua
8.1	4.3	3.4	1.9	5.0	1.8	7.5	1.3	-1.8	2.7	Panama
4.6	-5.2	26.4	3.3	1.3	-1.3	5.8	0.3	1.8	1.7	Paraguay
-6.3	0.7	4.7	3.0	1.8	-6.3	1.1	1.8	8.6	4.4	Peru
-11.8	7.9	4.0	3.1	24.5	2.6	-3.5	1.9	14.1	..	St. Vincent and the Grenadines
7.8	8.0	0.4	6.3	8.8	-5.5	-14.8	-12.4	-5.8	..	Trinidad and Tobago
4.0	5.5	6.0	2.6	-7.8	-7.2	-4.2	1.2	7.1	7.0	Uruguay
3.6	0.7	4.9	3.2	1.8	-1.7	-1.3	-1.7	7.1	-4.6	Venezuela
2.8	*7.8*	*5.9*	*7.5*	*3.5*	*2.8*	*3.2*	*3.8*	*1.4*	*-0.6*	*MIDDLE EAST AND NORTH AFRICA*
1.1	7.9	6.1	7.9	1.9	0.9	4.1	4.4	-1.6	0.4	Algeria
4.1	10.4	9.4	4.3	3.8	4.3	6.6	4.5	2.0	1.1	Egypt, Arab Republic of
18.9	16.8	-7.8	-1.2	15.4	12.8	6.9	1.9	16.4	-7.2	Jordan
4.2	5.2	1.2	0.5	5.8	-0.2	2.0	2.5	9.1	-1.5	Morocco
..	-2.0	10.9	11.5	9.1	8.6	3.5	7.3	6.4	..	Oman
9.5	8.9	6.9	18.2	-5.2	4.1	-4.8	1.8	-5.5	-7.5	Syrian Arab Republic
4.0	3.0	9.2	5.6	2.9	3.9	4.8	2.3	0.5	1.2	Tunisia
-4.6	15.5	-2.8	11.4	37.8	4.5	-5.6	5.4	3.1	10.0	Yemen Arab Republic
..	..	..	..	..	..	..	..	..	..	Yemen, PDR
4.9	*-0.2*	*-7.2*	*1.0*	*-0.3*	*1.7*	*1.7*	*2.1*	*10.6*	*..*	*EUROPE AND MEDITERRANEAN*
3.8	5.6	3.2	1.7	9.3	4.9	5.1	3.9	1.4	3.6	Cyprus
3.9	1.6	-0.4	5.8	5.2	0.7	0.2	5.5	1.1	2.0	Greece
3.1	1.9	0.4	2.1	0.9	0.4	0.8	1.2	1.6	2.0	Hungary
7.5	1.5	4.4	8.5	1.3	3.0	-2.5	0.9	3.3	..	Israel
2.0	3.3	4.9	1.2	1.8	0.1	2.8	5.5	2.2	6.3	Malta
..	..	..	-5.1	-9.3	3.2	3.1	4.9	..	..	Poland
-0.5	1.3	1.7	2.7	1.8	-1.3	-2.4	0.5	4.5	4.7	Portugal
1.4	-1.3	-2.4	-0.8	2.6	4.6	7.3	1.8	8.0	5.3	Turkey
10.2	-0.6	0.8	-1.4	0.2	-1.3	-0.5	0.7	5.5	-2.8	Yugoslavia
4.6	2.7	4.6	2.9	1.6	1.7	2.6	2.1	3.0	1.6	Developing countries
5.0	3.7	4.4	2.3	0.8	-1.9	0.6	2.0	3.7	0.7	Highly indebted
5.6	2.0	4.4	3.4	3.6	6.1	4.3	3.1	2.7	1.5	Low-income countries
2.2	0.8	2.3	1.9	1.5	0.5	0.0	1.9	2.2	3.5	Low-income Africa
6.0	1.9	4.6	3.8	4.2	7.2	4.6	2.8	2.8	1.3	China and India
7.7	5.0	5.7	3.6	2.2	4.2	6.7	6.4	2.5	3.4	Other low-income Asia
4.1	2.9	4.6	2.8	1.0	0.1	2.0	1.6	3.3	1.7	Middle-income countries
5.2	7.5	8.6	2.6	12.1	-1.4	..	..	..	..	High-income oil exporters
2.9	2.2	0.9	0.9	1.1	2.1	2.6	3.0	3.1	2.5	Industrialized countries
3.7	4.3	1.1	1.2	2.5	2.1	1.8	1.7	2.4	2.1	Japan
3.2	1.6	0.2	1.1	1.2	3.3	4.1	4.4	3.5	1.7	United States
2.7	2.6	1.2	0.6	0.7	1.1	1.1	1.8	2.8	2.6	European Community

Table 11. Gross domestic investment: contribution to growth of GDP expenditure

Percentage points	1967	1968	1969	1970	1971	1972	1973	1974	1975	1976	1977
SUB-SAHARAN AFRICA	*0.1*	*0.3*	*0.2*	*4.9*	*2.6*	*0.3*	*1.9*	*2.1*	*3.3*	*4.0*	*2.4*
Excluding Nigeria	*1.3*	*1.2*	*-0.4*	*3.1*	*1.6*	*-0.1*	*1.1*	*3.6*	*-0.7*	*0.3*	*1.4*
Benin	1.8	0.6	0.4	-0.5	-0.2	2.9	0.1	4.1	0.0	0.2	-1.1
Botswana	6.9	12.1	14.3	16.8	19.3	21.0	15.1	1.2	-6.5	-1.8	-10.3
Burkina Faso	-0.9	0.7	2.6	0.6	6.1	5.8	4.4	3.3	-0.3	0.3	-2.4
Burundi	0.8	1.4	0.0	-2.9	4.1	-6.1	3.0	3.8	-2.6	3.3	4.4
Cameroon	-1.2	0.1	-0.5	3.3	2.9	1.1	1.1	-1.4	2.3	-1.2	5.3
Cape Verde	..	..	..	..	..	..	..	1.8	-4.2	14.9	18.4
Central African Republic	-4.0	1.8	0.5	2.2	0.9	0.0	1.0	-9.0	-2.4	0.8	0.1
Chad	3.8	-0.8	4.1	-0.5	1.0	-2.1	-28.8	33.4	4.8	-2.7	3.3
Comoros	..	..	..	..	..	..	..	..	..	..	..
Congo, People's Republic	17.9	-3.7	-3.3	-2.9	7.0	7.5	8.2	8.8	9.9	-13.7	-8.6
Cote d'Ivoire	-1.0	0.4	2.9	3.4	-0.1	-0.6	3.7	-0.4	0.5	3.9	7.8
Ethiopia	0.9	0.1	-0.1	-1.5	0.8	1.8	-1.0	-1.5	0.7	2.4	0.3
Gabon	-8.6	0.7	1.9	2.2	2.9	4.1	3.4	42.5	11.2	21.1	-11.3
Gambia, The	1.5	0.4	-0.7	-2.8	1.2	1.6	-0.5	0.4	7.7	0.5	4.3
Ghana	-2.8	0.5	1.1	2.3	0.4	-4.3	1.5	2.8	-1.2	-1.7	3.1
Guinea-Bissau	..	..	..	..	-7.4	-5.8	-6.1	-6.0	-1.0	1.8	2.9
Kenya	1.6	10.2	0.8	11.8	0.6	-3.1	6.2	-1.8	-9.7	2.5	8.2
Lesotho	3.0	-0.6	0.0	0.4	3.0	2.2	7.3	-0.9	5.1	11.8	15.4
Liberia	5.3	-6.8	2.4	7.2	-1.6	13.1	-19.2	3.9	15.2	0.5	-1.5
Madagascar	2.2	4.4	1.7	-1.9	3.7	-5.3	0.8	1.2	-0.8	-3.1	-1.0
Malawi	-4.2	5.7	5.1	14.0	-5.5	8.5	-3.9	4.7	5.6	-7.9	0.2
Mali	0.9	4.2	0.5	-2.9	0.9	0.3	-1.0	-4.1	5.0	-0.4	3.9
Mauritania	3.2	3.3	-2.5	1.2	0.6	3.7	-3.2	21.1	14.0	12.4	-5.8
Mauritius	8.4	-7.6	4.9	-7.9	8.4	3.4	16.3	7.3	1.6	-1.4	-0.4
Mozambique	..	..	..	..	..	..	..	..	..	..	..
Niger	2.0	1.3	-8.5	9.9	-5.8	6.1	11.4	-1.8	-7.0	0.8	4.2
Nigeria	-1.7	-1.3	1.4	7.6	3.9	0.7	2.8	0.4	7.7	8.0	3.4
Rwanda	-2.2	1.8	-1.0	1.1	2.6	0.5	0.2	1.3	4.2	1.2	2.5
Senegal	1.1	1.3	1.2	3.5	0.5	1.9	0.4	1.7	-2.5	2.8	-0.8
Seychelles	..	..	..	..	..	..	..	..	..	..	5.7
Sierra Leone	-4.4	-0.7	3.8	5.6	-7.3	-4.2	0.7	4.6	0.5	-4.8	-0.1
Somalia	2.3	0.7	-0.1	-0.1	-0.1	6.4	-0.4	0.6	2.9	7.8	20.2
Sudan	-0.4	0.8	-0.3	1.5	-1.1	-2.9	1.7	8.9	1.4	10.3	-3.7
Swaziland	-4.8	-2.8	4.1	3.8	3.7	4.6	4.5	5.6	-10.6	13.3	-2.9
Tanzania	2.8	1.7	-2.2	9.4	5.2	-6.0	1.1	2.9	-2.0	2.6	3.7
Togo	-5.7	-2.3	4.7	2.4	2.4	2.5	1.2	0.6	11.7	-6.5	23.1
Uganda	2.5	0.2	1.1	0.7	2.4	-4.3	-2.9	2.7	-3.5	-1.7	-0.2
Zaire	-0.6	0.2	2.2	0.2	2.4	-0.1	0.8	0.7	-0.7	-2.0	6.7
Zambia	8.0	1.9	-13.4	17.6	6.0	1.8	-2.9	13.7	-8.5	-25.5	-2.8
Zimbabwe	9.1	3.6	-9.2	4.5	3.9	2.2	2.2	9.3	-3.3	-8.3	-4.0
SOUTH ASIA	*0.8*	*0.0*	*1.0*	*1.3*	*1.0*	*-1.4*	*2.8*	*0.5*	*1.0*	*1.0*	*0.8*
Bangladesh	1.6	5.1	-1.4	-1.2	-3.5	-6.4	5.7	-0.6	-1.1	0.5	1.5
India	0.2	-1.0	1.9	1.2	1.6	-1.2	3.2	0.5	1.1	1.0	0.6
Nepal	..	..	..	..	..	..	..	..	..	..	..
Pakistan	2.7	3.1	-4.5	3.5	0.0	-1.1	0.7	-0.9	2.0	0.3	2.2
Sri Lanka	0.9	1.7	2.8	1.9	-1.2	1.2	-3.2	6.3	-1.4	4.3	-0.6
EAST ASIA AND THE PACIFIC	*-3.4*	*1.2*	*3.7*	*8.4*	*2.6*	*-0.1*	*4.3*	*2.7*	*1.7*	*0.3*	*3.5*
China	-7.0	-0.4	3.7	13.8	2.6	-0.8	4.1	0.5	3.3	-2.5	3.2
Fiji	1.1	9.5	-1.1	0.1	3.1	3.2	2.0	-2.6	2.0	0.2	0.3
Hong Kong	-6.0	-2.6	1.1	3.5	6.0	2.6	3.3	0.2	0.2	8.3	4.7
Indonesia	-2.1	2.0	2.7	3.8	3.1	3.1	3.0	3.5	3.0	1.3	3.5
Korea, Republic of	2.3	6.4	9.1	-0.5	2.0	-1.5	5.8	7.4	-0.9	3.9	6.9
Malaysia	1.4	0.9	-1.2	6.4	0.3	1.8	5.8	7.9	-6.6	2.7	5.2
Papua New Guinea	1.9	1.5	13.0	24.1	4.5	-18.5	-13.4	1.2	-0.2	-2.0	1.5
Philippines	3.4	2.4	1.2	-0.9	0.8	-0.7	2.3	4.9	5.7	4.4	0.1
Singapore	4.1	6.9	7.9	15.1	7.5	4.2	3.2	8.6	-3.7	2.6	-1.5
Solomon Islands	..	..	..	..	..	..	..	..	..	..	..
Thailand	2.4	3.0	5.2	1.9	-0.8	-1.8	8.7	-1.1	0.7	-0.1	6.2
Vanuatu	..	..	..	..	..	..	..	..	..	..	..

NOTE: See glossary for definitions.

1978	1979	1980	1981	1982	1983	1984	1985	1986	1987	
-4.8	*-2.0*	*2.4*	*1.3*	*-2.2*	*-3.6*	*-3.9*	*1.5*	*-0.4*	*-0.6*	*SUB-SAHARAN AFRICA*
-1.5	*-0.2*	*2.0*	*0.8*	*-0.5*	*-2.4*	*-1.2*	*1.0*	*-0.3*	*-1.3*	*Excluding Nigeria*
-0.6	4.6	5.3	2.6	1.4	-10.4	-1.9	1.0	-0.3	0.5	Benin
10.7	7.4	10.5	0.7	-1.3	-11.5	-0.4	10.2	-2.6	..	Botswana
0.9	-1.9	-1.8	0.0	4.1	-0.8	-0.9	3.7	-2.3	-0.9	Burkina Faso
2.1	0.3	-0.6	5.2	-4.5	11.1	-5.4	-2.6	3.4	..	Burundi
8.6	-1.0	2.2	7.8	0.5	1.6	-1.4	1.8	0.5	-3.5	Cameroon
-7.7	15.1	-12.5	23.9	2.7	3.4	-5.5	2.4	..	..	Cape Verde
-0.3	-1.6	-2.9	1.5	3.7	5.5	1.9	0.7	-2.1	-1.7	Central African Republic
-5.8	-4.6	-4.7	-7.2	-2.3	-1.1	-0.4	0.5	..	..	Chad
..	..	..	..	..	..	..	..	..	..	Comoros
-8.2	2.2	14.0	29.0	19.1	-18.9	-4.3	-1.6	-10.3	..	Congo, People's Republic
4.2	-1.8	4.4	-6.0	-2.9	-1.6	-8.7	4.0	-2.3	0.8	Cote d'Ivoire
-5.3	1.5	1.3	1.1	2.0	0.7	0.5	-3.2	0.9	..	Ethiopia
-28.6	5.4	0.6	12.0	-3.7	-2.4	2.4	4.1	-5.5	-13.4	Gabon
16.4	-7.8	2.8	-0.6	-4.9	1.6	..	..	..	..	Gambia, The
-3.2	0.2	-0.5	-0.5	-1.2	0.0	0.7	1.2	-0.2	1.1	Ghana
7.7	6.6	-10.1	-4.9	5.8	-0.8	1.4	2.6	0.0	-1.6	Guinea-Bissau
6.7	-8.0	8.1	-1.0	-6.5	-2.6	2.1	-1.6	3.1	3.2	Kenya
-4.5	5.5	0.3	-0.1	0.3	-4.3	..	..	..	..	Lesotho
-2.0	2.7	-2.0	-10.7	-2.5	-2.7	-1.8	-1.6	0.8	..	Liberia
0.1	11.6	-1.2	-7.1	-3.3	-0.2	0.7	0.3	-0.1	0.5	Madagascar
18.7	-11.3	-3.7	-7.2	4.2	2.3	-9.9	5.2	-8.4	1.4	Malawi
-0.5	1.4	-0.7	1.4	1.3	-0.9	2.3	-0.1	4.9	-4.6	Mali
-14.7	-5.0	10.6	8.5	11.3	-33.7	8.2	1.2	-0.7	-2.4	Mauritania
2.8	1.6	-13.6	4.6	-4.0	-0.3	3.8	1.4	7.0	10.1	Mauritius
..	..	..	0.6	3.0	-1.8	-8.2	-11.2	1.6	..	Mozambique
9.1	7.3	-4.7	1.3	-5.8	-1.8	-28.7	25.0	-9.0	2.0	Niger
-8.1	-4.0	2.8	1.7	-4.2	-5.0	-7.3	2.3	-0.5	0.3	Nigeria
-0.4	-3.1	5.7	-0.6	6.3	-2.9	5.6	2.1	-1.3	0.9	Rwanda
-0.9	-2.4	0.1	0.4	0.9	0.8	-0.4	-1.8	1.8	0.1	Senegal
6.1	-1.7	3.4	-3.9	7.0	..	..	..	..	..	Seychelles
0.7	1.7	3.7	3.4	-4.5	-0.6	0.3	-1.6	-1.8	0.0	Sierra Leone
12.4	-15.0	6.2	-11.5	5.0	-4.0	15.0	-2.6	-3.3	2.8	Somalia
-3.0	-2.6	1.4	0.2	6.4	-1.3	-1.5	-2.2	-3.1	-1.9	Sudan
24.0	-7.7	-4.0	-10.0	-0.4	2.3	-2.4	..	..	..	Swaziland
-1.0	2.5	-4.4	-1.2	2.1	-8.6	2.2	2.2	-2.0	-3.1	Tanzania
8.4	-2.7	-17.8	-1.2	-4.9	-4.9	0.2	2.6	1.0	-7.0	Togo
0.5	-1.1	-0.6	..	..	..	..	..	..	..	Uganda
-7.4	4.8	3.5	0.3	-1.5	-3.5	3.6	-1.5	4.2	-0.4	Zaire
-0.8	-11.0	9.4	-2.6	-4.6	-3.7	0.8	1.0	0.8	-4.8	Zambia
-7.1	1.1	8.3	8.8	-3.5	-5.4	0.0	3.9	1.2	-3.9	Zimbabwe
3.0	*-0.1*	*3.1*	*2.4*	*-0.9*	*0.6*	*0.7*	*2.8*	*-0.3*	*0.6*	*SOUTH ASIA*
2.5	0.5	4.4	0.5	4.6	-1.1	-0.9	0.8	-0.1	0.1	Bangladesh
3.1	-0.5	3.2	3.1	-1.9	0.8	0.9	3.3	-0.5	0.4	India
..	..	..	..	..	..	..	..	..	..	Nepal
0.8	0.9	1.5	-0.7	2.8	1.2	0.8	1.6	1.1	2.0	Pakistan
8.1	7.5	6.5	-3.9	1.5	-0.2	-1.5	-5.1	0.6	0.1	Sri Lanka
6.5	*2.7*	*0.3*	*3.2*	*1.9*	*3.0*	*4.3*	*6.4*	*2.1*	*5.0*	*EAST ASIA AND THE PACIFIC*
8.7	1.3	0.4	-0.5	5.0	4.4	9.2	13.4	3.4	6.3	China
1.8	8.1	2.5	1.9	-8.9	-3.1	-2.2	-0.7	2.5	-8.3	Fiji
3.3	5.4	6.1	3.0	-1.2	-1.4	0.9	-1.9	2.6	4.0	Hong Kong
3.5	2.5	0.4	20.2	-2.1	0.1	-4.9	3.7	0.8	-0.1	Indonesia
9.0	5.6	-7.9	2.0	0.0	5.1	5.8	0.5	3.4	4.9	Korea, Republic of
2.1	3.6	5.2	5.2	4.8	2.7	2.0	-7.1	-5.0	0.5	Malaysia
0.6	4.9	1.0	-1.1	3.2	1.2	-2.8	-5.7	0.6	1.4	Papua New Guinea
2.3	3.4	1.3	0.7	-1.0	-1.4	-11.5	-3.5	-1.1	2.4	Philippines
6.1	7.6	7.2	3.3	6.7	5.5	4.7	-6.4	-4.2	4.2	Singapore
..	..	..	-0.5	-3.4	3.8	-0.5	2.9	0.3	..	Solomon Islands
5.2	-1.4	0.4	2.7	-3.0	5.3	1.8	-1.2	-1.3	2.2	Thailand
..	..	..	..	..	..	..	..	..	..	Vanuatu

Table 11. Gross domestic investment: contribution to growth of GDP expenditure (cont'd)

Percentage points	1967	1968	1969	1970	1971	1972	1973	1974	1975	1976	1977
LATIN AMERICA AND CARIBBEAN	*0.2*	*1.2*	*3.2*	*1.3*	*1.3*	*1.2*	*3.3*	*3.9*	*1.2*	*0.8*	*1.7*
Argentina	0.7	1.5	4.2	0.5	2.2	0.2	-1.0	0.2	0.3	1.3	4.3
Bahamas	..	..	..	..	..	..	..	..	..	..	..
Barbados	..	..	..	..	..	..	..	..	..	..	..
Belize	..	..	..	..	..	..	..	..	..	..	..
Bolivia	-1.2	6.3	-0.4	-0.6	2.2	4.8	-2.8	-0.2	8.3	-2.5	0.7
Brazil	-1.1	3.3	7.2	0.0	3.0	3.6	5.5	4.1	2.7	0.2	0.0
Chile	-1.9	1.6	1.3	1.6	-0.6	-5.0	-1.5	10.2	-11.9	0.0	2.0
Colombia	-1.8	3.4	0.3	3.1	0.9	-0.5	1.6	3.6	-3.7	1.5	2.8
Costa Rica	1.2	-0.3	2.5	2.4	4.5	-1.6	3.5	1.8	-1.9	5.0	5.2
Dominican Republic	0.8	0.0	5.0	3.8	3.4	2.6	5.8	4.6	2.2	-1.8	2.0
Ecuador	4.2	1.4	-1.2	1.4	6.2	-3.6	2.2	7.0	4.1	-1.4	4.9
El Salvador	-2.2	-3.2	1.3	1.3	2.7	-0.9	3.9	4.2	-2.0	1.7	7.6
Grenada	..	..	..	..	..	..	..	..	..	..	..
Guatemala	3.1	4.8	-4.7	3.0	2.6	-2.9	2.6	5.6	-3.2	6.5	2.4
Guyana	6.5	-8.3	-0.8	3.4	-5.3	-4.9	12.8	-3.9	23.1	-7.2	-12.8
Haiti	0.1	0.4	0.9	2.7	0.2	1.4	1.5	1.1	2.4	3.0	0.6
Honduras	5.8	0.0	0.2	1.2	-4.2	-0.4	4.4	6.4	-5.6	1.4	6.6
Jamaica	2.1	7.1	4.3	0.5	0.1	-4.0	9.5	-9.3	3.1	-9.5	-8.5
Mexico	1.8	1.0	1.6	4.2	-1.1	2.1	3.1	3.8	1.2	-0.5	-0.1
Nicaragua	0.3	-2.8	2.5	0.3	0.6	-4.3	9.1	6.4	-7.2	0.7	9.0
Panama	1.5	3.5	4.1	5.3	7.5	5.3	1.5	-2.0	-1.3	-0.3	-10.3
Paraguay	2.3	-0.8	1.0	-0.1	1.3	1.8	5.3	2.1	1.4	4.1	2.6
Peru	-1.0	-12.4	1.0	-10.8	2.8	-2.3	9.2	9.1	-1.6	-3.6	-3.5
St. Vincent and the Grenadines	..	..	..	..	..	..	..	..	..	..	..
Trinidad and Tobago	-3.0	1.5	-2.1	4.2	6.7	-2.2	-1.5	6.0	4.0	3.3	0.8
Uruguay	0.6	-1.0	2.1	1.2	0.6	-1.7	-0.4	0.3	2.3	2.1	2.1
Venezuela	..	..	..	..	..	..	..	..	5.9	6.0	8.9
MIDDLE EAST AND NORTH AFRICA	*2.5*	*1.6*	*3.2*	*2.4*	*0.5*	*2.0*	*2.1*	*8.6*	*6.2*	*3.1*	*4.7*
Algeria	6.4	3.7	4.8	5.8	-0.4	4.6	6.3	11.6	3.3	2.0	5.9
Egypt, Arab Republic of	-2.9	-1.6	1.3	0.7	-1.7	0.4	3.8	7.4	12.8	-0.1	3.6
Jordan	..	..	..	..	5.4	2.9	-2.3	9.5	4.5	13.9	7.9
Morocco	4.5	1.6	3.8	3.6	2.3	-4.2	0.7	3.5	13.2	5.5	5.2
Oman	..	..	..	..	..	..	..	..	..	..	..
Syrian Arab Republic	-0.6	3.5	4.8	-4.0	3.1	6.1	-7.8	18.4	2.4	7.0	4.5
Tunisia	1.8	-0.6	-0.7	-1.1	0.5	6.7	-2.3	5.0	1.9	2.0	0.6
Yemen Arab Republic	..	..	..	..	9.8	1.7	-0.4	4.3	0.8	4.0	3.0
Yemen, PDR	..	..	..	..	..	..	..	..	..	..	..
EUROPE AND MEDITERRANEAN	*0.9*	*2.4*	*3.4*	*6.2*	*3.8*	*0.9*	*5.3*	*1.0*	*2.5*	*1.2*	*4.4*
Cyprus	..	..	..	..	..	..	..	..	..	8.5	13.9
Greece	0.5	2.6	6.4	3.3	1.9	3.3	8.2	-8.3	-0.4	0.6	0.3
Hungary	5.5	0.4	1.5	4.7	7.1	-4.6	0.1	6.6	3.4	0.7	2.8
Israel	-6.2	10.0	7.0	4.0	6.8	5.0	1.9	-1.6	1.7	-3.3	-2.7
Malta	8.3	10.6	5.9	0.6	-5.8	-4.1	-4.3	3.5	0.2	6.2	2.0
Poland	..	..	..	..	..	..	..	..	..	..	..
Portugal	1.2	-1.1	1.6	6.7	1.1	4.8	5.4	-3.0	-9.0	2.6	5.7
Turkey	0.2	1.8	0.5	4.0	0.2	1.5	2.5	5.1	5.1	1.2	2.6
Yugoslavia	1.3	0.6	1.1	5.4	3.6	-2.9	3.9	3.2	3.6	1.4	6.6
Developing countries	0.0	0.8	2.5	3.8	1.9	0.4	3.2	2.9	2.4	1.1	2.1
Highly indebted	0.2	0.9	2.8	2.2	1.7	0.6	3.0	5.6	2.8	2.0	2.5
Low-income countries	-2.5	0.1	1.7	6.6	1.8	-1.1	3.1	0.7	1.6	-0.9	2.1
Low-income Africa	1.2	1.4	-0.4	3.2	1.3	-0.8	0.7	2.2	-1.8	-0.6	2.2
China and India	-3.9	-0.7	2.8	7.9	2.2	-1.0	3.7	0.5	2.4	-1.1	2.1
Other low-income Asia	3.0	4.1	-2.6	1.5	-1.3	-2.4	1.2	0.2	0.6	0.8	1.8
Middle-income countries	1.1	1.1	2.9	2.7	1.9	1.0	3.3	3.7	2.6	1.8	2.1
High-income oil exporters	1.3	1.6	0.3	-2.0	-0.1	1.2	1.0	3.1	5.3	5.8	6.8
Industrialized countries	..	..	..	..	0.7	1.5	2.7	-0.9	-3.3	2.2	1.0
Japan	..	..	..	..	-0.1	3.6	5.1	-2.8	-2.6	1.3	1.5
United States	..	..	..	..	1.6	1.9	2.1	-1.5	-3.6	2.6	2.4
European Community	..	..	..	..	..	1.0	2.5	-0.9	-3.5	2.4	0.1

NOTE: See glossary for definitions.

1978	1979	1980	1981	1982	1983	1984	1985	1986	1987	
0.8	*0.6*	*2.7*	*-0.1*	*-3.5*	*-5.4*	*1.2*	*1.0*	*0.6*	*0.4*	*LATIN AMERICA AND CARIBBEAN*
-3.7	2.1	1.9	-5.2	-3.6	-1.6	-1.5	-2.3	1.8	1.8	Argentina
-0.4	4.8	6.6	0.2	4.3	0.1	-1.9	5.1	-0.3	1.6	Bahamas
..	..	..	..	..	..	..	..	..	..	Barbados
..	..	..	..	..	..	..	..	..	..	Belize
0.2	-2.9	-4.4	-0.2	-3.4	-4.6	-1.5	0.4	-1.8	0.8	Bolivia
0.8	0.1	2.8	-3.2	-1.6	-4.3	2.0	1.8	3.4	-1.8	Brazil
3.0	4.2	5.4	4.6	-15.8	-1.7	6.1	-0.9	1.7	3.4	Chile
1.3	-0.1	2.0	2.5	1.1	-0.5	-1.3	-1.9	0.9	0.9	Colombia
-0.1	2.2	1.8	-10.0	-4.3	4.5	1.9	1.1	3.8	-1.4	Costa Rica
0.7	2.1	1.8	-2.9	-1.6	0.5	1.1	1.4	..	..	Dominican Republic
3.1	-0.8	2.2	-3.5	2.7	-7.6	-0.2	1.3	0.3	0.1	Ecuador
0.7	-5.2	-5.7	-0.5	-1.3	-1.2	0.4	-0.7	2.4	2.1	El Salvador
..	..	..	..	..	..	..	..	..	..	Grenada
2.3	-3.1	-2.8	2.4	-3.5	-2.6	0.8	-2.6	0.0	3.5	Guatemala
-9.6	8.2	-0.4	-2.4	-8.5	-3.0	3.4	1.5	2.6	-5.0	Guyana
1.7	2.8	0.1	0.1	-1.2	0.9	0.8	-2.5	-2.3	0.6	Haiti
5.8	1.0	-1.3	-4.0	-8.2	2.6	6.2	-0.1	-1.4	-0.2	Honduras
1.9	-0.6	-2.9	4.9	-0.8	0.7	-2.0	0.0	-3.5	3.1	Jamaica
2.6	3.9	5.3	4.0	-7.0	-5.9	1.0	2.0	-3.8	0.7	Mexico
-14.8	-16.0	23.3	7.6	-2.9	1.8	0.2	0.2	-0.3	..	Nicaragua
5.8	2.1	4.6	3.5	-2.1	-5.5	-2.3	1.3	2.4	0.9	Panama
6.0	4.1	10.1	4.7	-5.9	-5.6	-0.3	0.4	1.2	1.0	Paraguay
-1.2	3.3	7.6	5.6	-2.3	-11.1	-2.1	-1.9	7.5	2.6	Peru
5.5	6.5	3.1	-2.2	-0.9	-2.1	5.1	3.6	4.0	..	St. Vincent and the Grenadines
6.5	0.8	6.1	-5.1	1.3	-2.9	-6.8	1.0	-12.4	..	Trinidad and Tobago
2.0	3.6	0.9	-1.5	-2.8	-5.1	0.2	-1.8	0.0	1.9	Uruguay
-0.5	-7.8	-4.8	0.0	1.8	-14.6	2.6	1.1	1.7	3.8	Venezuela
2.5	*0.1*	*1.1*	*1.5*	*2.4*	*0.7*	*1.7*	*0.6*	*-3.0*	*-3.2*	*MIDDLE EAST AND NORTH AFRICA*
8.4	-3.2	0.9	1.3	2.4	3.2	-0.3	-0.2	-2.3	-3.1	Algeria
3.2	3.4	-0.4	2.3	3.6	1.1	2.9	2.2	-3.7	-4.5	Egypt, Arab Republic of
1.9	3.4	11.5	16.3	1.2	-9.8	-2.5	-0.4	-2.8	-3.6	Jordan
-9.3	0.3	-1.8	-2.6	1.4	-2.7	0.3	1.4	-1.3	-0.4	Morocco
..	3.7	0.9	5.6	8.5	2.7	13.7	2.6	-7.6	..	Oman
-4.1	0.4	5.0	0.6	0.8	1.0	0.5	1.8	-1.6	-6.8	Syrian Arab Republic
3.3	2.0	0.3	4.4	-0.2	-0.7	2.6	-3.9	-4.7	-2.0	Tunisia
22.5	4.1	8.5	-7.1	-2.5	-7.9	0.4	-2.8	-0.9	2.4	Yemen Arab Republic
..	..	..	..	..	..	..	..	..	..	Yemen, PDR
-1.0	*3.3*	*7.0*	*-1.8*	*-0.7*	*-0.8*	*-0.1*	*0.7*	*-4.8*	*-1.3*	*EUROPE AND MEDITERRANEAN*
4.8	6.0	1.7	-3.5	1.1	-0.1	6.8	-1.8	-3.2	0.5	Cyprus
2.0	2.5	-1.2	-3.6	-1.4	-0.2	-0.6	1.1	-1.8	-2.5	Greece
6.2	-6.5	-1.1	-0.7	-1.0	-2.0	-0.6	-0.8	2.0	0.5	Hungary
0.7	3.0	-3.7	-1.2	2.7	2.3	-1.8	-2.9	1.2	..	Israel
0.9	2.4	2.6	5.0	9.6	-5.5	-1.4	-0.6	-2.1	-0.4	Malta
..	..	..	-5.9	-0.8	1.2	1.5	1.4	..	..	Poland
2.0	2.7	3.6	0.9	2.0	-7.5	-5.1	-0.1	3.3	4.2	Portugal
-6.5	0.0	1.8	1.6	-1.1	-0.1	0.3	2.6	2.9	0.9	Turkey
-1.3	8.3	-1.8	0.5	-1.2	-0.2	1.0	-0.7	-0.6	1.2	Yugoslavia
1.1	1.0	2.0	1.3	-1.0	-0.4	1.2	2.3	0.7	1.7	Developing countries
-0.6	0.9	2.3	0.1	-3.4	-5.0	-0.3	0.9	0.4	0.5	Highly indebted
5.0	0.5	1.6	0.6	2.1	2.2	5.0	8.3	1.8	3.8	Low-income countries
-0.7	-0.5	1.4	-0.6	-0.4	-2.1	-0.1	0.3	0.3	-0.4	Low-income Africa
6.3	0.6	1.5	0.8	2.4	3.1	6.2	10.0	2.1	4.4	China and India
2.5	1.7	2.8	-0.2	2.9	-0.1	-0.2	0.7	0.4	1.3	Other low-income Asia
-0.1	1.2	2.2	1.5	-2.1	-1.3	-0.2	-0.1	0.1	0.7	Middle-income countries
3.0	-1.7	1.1	4.7	-1.8	4.6	-0.2	..	..	..	High-income oil exporters
0.9	1.3	-0.7	-0.4	-1.2	0.6	2.5	0.3	0.9	1.4	Industrialized countries
2.4	2.4	-0.3	1.0	0.2	-0.4	1.9	2.2	1.5	2.9	Japan
1.9	-0.1	-2.1	1.2	-3.1	1.8	4.8	-0.6	0.5	0.9	United States
0.2	1.6	-0.1	-2.4	0.2	0.0	0.8	0.3	1.0	1.2	European Community

Table 12. Net exports of goods and nonfactor services: contribution to growth of GDP expenditure

Percentage points	1967	1968	1969	1970	1971	1972	1973	1974	1975	1976	1977
SUB-SAHARAN AFRICA	*-1.6*	*-0.2*	*3.9*	*1.6*	*-1.1*	*2.5*	*0.6*	*-1.6*	*-4.0*	*-0.2*	*-0.9*
Excluding Nigeria	*-0.7*	*0.7*	*1.7*	*0.1*	*-1.9*	*2.3*	*-0.3*	*-2.0*	*-0.4*	*1.3*	*-3.8*
Benin	-4.5	-0.2	2.2	-0.3	-1.2	-1.9	-1.1	-0.6	-2.0	1.9	-2.1
Botswana	-13.2	-41.0	-3.7	-7.4	10.1	10.3	-10.0	0.7	0.1	9.5	3.4
Burkina Faso	-5.4	-2.1	-2.5	5.4	-5.8	1.2	-2.6	-0.1	-7.0	2.8	-8.8
Burundi	2.3	-2.9	2.7	-3.8	-3.4	3.4	-0.2	-1.9	-2.8	-0.5	-5.8
Cameroon	1.9	-3.4	2.4	-0.8	-6.8	1.4	3.7	2.9	-3.8	0.2	-5.9
Cape Verde	..	..	..	..	..	..	..	12.5	-0.2	-26.6	-27.7
Central African Republic	2.7	-4.9	0.7	-1.8	1.7	0.5	1.8	12.9	-0.4	-2.7	-8.2
Chad	0.1	1.1	0.7	-4.0	0.9	-2.7	-7.6	-0.8	3.1	2.4	-2.3
Comoros	..	..	..	..	..	..	..	..	..	..	..
Congo, People's Republic	-12.5	-0.3	1.5	4.2	1.8	1.8	1.7	1.6	-4.0	4.8	9.8
Cote d'Ivoire	0.2	7.5	-2.2	-1.7	1.6	3.6	-6.2	3.2	-0.7	-2.9	-10.9
Ethiopia	2.7	0.6	-0.4	-0.6	0.1	-0.3	3.3	1.8	-4.7	2.2	-2.7
Gabon	..	..	..	..	..	..	..	..	..	..	..
Gambia, The	5.1	-5.3	7.2	2.2	-17.3	7.5	0.1	12.2	5.0	-17.0	-5.4
Ghana	-3.8	3.2	-2.1	3.3	1.4	5.6	-3.4	-5.6	1.5	1.3	-4.5
Guinea-Bissau	..	..	..	..	-0.9	-6.7	0.4	4.0	18.5	4.6	1.4
Kenya	-3.8	1.3	3.1	-9.1	-9.4	6.2	3.8	-3.7	8.1	1.7	-6.1
Lesotho	-0.5	0.2	3.2	-6.6	-10.4	-18.6	-12.8	-12.4	-34.7	-24.2	-5.3
Liberia	-1.0	7.8	1.2	12.6	-6.2	5.3	1.0	-2.5	-18.2	-6.6	'-5.5
Madagascar	3.1	-4.1	-1.2	0.3	-3.8	4.3	0.0	4.3	0.8	3.7	-1.2
Malawi	6.7	-9.8	-4.6	-5.0	-0.6	-1.9	5.5	0.6	-4.9	9.0	-0.8
Mali	-6.8	0.9	-0.6	4.6	1.5	-4.4	-0.1	-8.6	2.0	4.8	3.3
Mauritania	-4.1	-0.1	1.1	-1.0	0.0	-7.1	2.2	-24.5	-18.0	-37.6	12.1
Mauritius	-6.7	15.0	3.6	-5.1	-7.7	2.8	1.3	-20.4	-12.1	4.4	0.0
Mozambique	..	..	..	..	..	..	..	..	..	..	..
Niger	-1.8	-0.5	1.6	3.0	4.8	-4.2	-1.3	-20.2	21.9	-1.9	-0.6
Nigeria	-2.6	-1.4	7.5	3.7	-0.1	2.8	1.6	-1.1	-7.4	-1.6	1.9
Rwanda	4.3	-2.3	-1.2	3.3	-3.6	-1.5	6.8	-8.9	0.6	3.7	-5.2
Senegal	-2.7	3.4	2.6	-1.3	-1.7	4.1	-4.5	1.0	2.6	-0.2	-3.9
Seychelles	..	..	..	..	..	..	..	..	..	..	-2.9
Sierra Leone	-3.3	5.0	10.5	-6.6	5.0	5.0	3.8	-15.5	5.3	6.7	-1.9
Somalia	-1.3	1.3	-1.0	2.7	-5.1	4.6	-4.2	-4.1	-0.4	-2.7	-39.8
Sudan	7.5	-1.3	2.6	3.9	-3.2	2.1	2.6	-5.2	-8.5	1.6	-1.0
Swaziland	-10.8	8.7	0.7	4.8	11.5	-6.2	9.1	10.3	-4.2	-10.2	-6.1
Tanzania	0.1	-5.2	3.0	-3.6	-3.1	3.3	-2.6	-5.8	4.0	4.7	-3.9
Togo	2.9	1.5	1.2	-3.0	-3.5	0.5	0.9	-0.8	-3.1	-11.5	-20.2
Uganda	0.4	-0.3	0.1	0.0	-2.2	3.3	0.4	-0.5	0.0	0.1	-2.2
Zaire	2.5	2.3	-2.1	-0.6	-1.2	-1.3	-1.9	0.7	0.5	2.4	-7.5
Zambia	-15.4	-3.4	18.3	-5.5	-8.3	4.9	10.5	-9.0	3.8	28.1	-0.8
Zimbabwe	..	..	..	..	..	..	..	..	..	..	-1.7
SOUTH ASIA	*-1.1*	*2.0*	*1.1*	*-1.9*	*-0.2*	*1.0*	*2.4*	*-0.1*	*1.9*	*-0.4*	*-1.3*
Bangladesh	-2.9	1.1	-0.9	-1.2	1.9	-0.9	10.5	-9.6	1.5	-2.3	4.9
India	-0.7	2.7	0.7	-1.1	-1.0	0.3	1.7	1.3	1.8	-0.3	-1.4
Nepal	..	..	..	..	..	..	..	..	..	..	..
Pakistan	-3.8	-1.3	7.3	-11.1	3.9	7.8	3.4	-5.3	2.6	0.3	-4.3
Sri Lanka	5.8	-6.0	-4.4	5.2	3.7	2.7	1.6	-4.2	6.8	-3.8	-2.6
EAST ASIA AND THE PACIFIC	*-0.7*	*-0.6*	*0.2*	*-0.1*	*1.2*	*2.0*	*0.0*	*-2.6*	*0.8*	*1.2*	*0.0*
China	0.0	0.0	0.5	-1.2	1.6	0.2	-0.3	-1.3	1.1	-0.3	-0.5
Fiji	9.1	-0.8	-2.7	1.6	-1.6	-1.2	-2.0	-3.0	-3.1	-3.3	2.2
Hong Kong	6.6	-0.4	4.9	-0.5	-6.6	3.5	0.6	2.6	-2.3	4.1	-2.7
Indonesia	-1.7	2.1	2.2	3.6	1.9	5.2	2.7	-1.8	-2.8	2.9	2.7
Korea, Republic of	-2.2	-4.3	-2.4	0.2	-2.2	4.4	0.3	-5.2	3.3	2.3	-0.1
Malaysia	2.1	7.0	2.5	-3.9	1.0	2.4	1.2	-5.8	6.7	5.6	-3.8
Papua New Guinea	-1.9	-1.8	-10.5	-15.4	-0.2	22.0	21.2	0.7	-2.6	-1.0	-3.3
Philippines	-4.8	-4.8	-1.3	3.9	0.5	1.1	1.8	-5.7	-1.4	2.8	1.4
Singapore	..	..	..	..	..	..	..	..	..	..	..
Solomon Islands	..	..	..	..	..	..	..	..	..	..	..
Thailand	-3.5	-4.9	-2.6	1.3	7.5	0.1	-7.7	3.4	-0.5	2.6	-2.8
Vanuatu	..	..	..	..	..	..	..	..	..	..	..

NOTE: See glossary for definitions.

1978	1979	1980	1981	1982	1983	1984	1985	1986	1987	
-3.1	*8.2*	*-3.4*	*-9.0*	*1.0*	*3.3*	*3.4*	*0.7*	*2.2*	*-0.1*	*SUB-SAHARAN AFRICA*
-0.3	*0.6*	*-0.6*	*0.1*	*1.3*	*3.0*	*2.3*	*-0.8*	*1.5*	*..*	*Excluding Nigeria*
-0.3	-3.4	-5.7	-2.4	-3.8	21.0	-1.9	-0.5	0.7	0.1	Benin
2.0	-5.2	-4.2	4.6	-7.7	32.2	17.1	-12.3	17.0	..	Botswana
4.7	-0.8	-1.9	-1.1	-1.9	2.6	6.5	-5.1	5.0	0.5	Burkina Faso
1.8	-2.8	-0.1	6.6	-3.3	-4.5	0.5	4.5	-0.4	2.6	Burundi
3.2	0.6	2.6	1.4	1.3	6.0	8.2	-3.4	-4.2	-0.2	Cameroon
-2.2	-5.3	-6.5	5.0	-3.5	-25.4	10.3	-14.1	..	..	Cape Verde
-1.6	2.3	-4.4	7.1	-3.5	1.5	-1.3	-0.5	-3.9	0.9	Central African Republic
3.0	4.1	1.4	8.7	3.4	-0.8	0.6	1.2	-28.0	-3.5	Chad
..	..	..	..	..	..	..	..	..	..	Comoros
6.5	17.0	-8.1	-23.1	-2.6	22.9	6.8	-3.7	8.3	-15.9	Congo, People's Republic
-3.9	1.9	1.5	4.4	4.5	0.2	6.2	1.6	-1.0	-1.2	Cote d'Ivoire
-2.3	0.3	-0.8	0.8	-5.1	0.2	-0.1	-3.0	-4.2	..	Ethiopia
..	..	..	..	..	..	..	..	..	..	Gabon
-37.4	37.1	-7.0	-10.6	17.2	6.7	7.9	..	..	..	Gambia, The
0.4	1.4	-2.1	0.5	4.3	-3.9	0.3	-0.8	2.0	-1.5	Ghana
-0.2	-0.8	2.9	0.7	-13.9	1.6	-5.9	-2.9	1.3	2.9	Guinea-Bissau
-10.9	8.0	-2.3	7.2	5.7	4.0	-3.1	3.4	0.0	-2.9	Kenya
3.9	-5.4	-6.9	-14.9	-5.3	-4.4	131.0	..	..	..	Lesotho
7.1	4.1	5.8	6.3	-0.1	-5.2	2.4	1.8	3.5	..	Liberia
-1.5	-8.8	0.1	7.3	0.0	1.9	1.9	0.6	0.6	0.0	Madagascar
-12.1	9.5	1.4	4.2	-0.7	0.1	8.0	-6.1	6.2	-0.3	Malawi
-10.9	-0.8	0.1	0.4	0.7	-2.6	-2.1	-12.4	6.7	5.1	Mali
22.2	9.6	1.1	-7.0	-18.8	14.9	1.6	4.3	4.6	1.1	Mauritania
-1.1	1.5	6.4	2.8	10.8	-0.9	-2.1	2.4	0.9	-10.5	Mauritius
..	..	..	-1.4	-1.8	0.6	1.0	4.7	-11.5	31.6	Mozambique
3.6	-12.2	-0.9	-1.5	-6.2	2.7	4.2	-7.9	7.7	-0.2	Niger
-5.6	15.8	-6.0	-17.8	0.8	3.6	4.7	2.6	3.1	0.3	Nigeria
-1.5	3.5	-5.7	0.6	-5.0	2.5	-3.5	-1.8	-0.8	0.4	Rwanda
-6.4	3.9	-4.4	-4.2	3.8	1.2	1.8	-1.0	3.2	0.2	Senegal
-14.2	6.5	-4.4	-3.1	-14.9	30.7	..	..	..	..	Seychelles
-9.6	-2.7	-8.3	-4.3	10.1	-2.6	11.3	0.5	1.8	-0.5	Sierra Leone
5.2	-14.7	3.5	-9.8	-5.3	-0.6	1.6	2.3	9.4	-11.0	Somalia
-1.6	1.4	-3.7	-6.0	-5.3	5.2	1.7	2.7	3.1	4.3	Sudan
-16.1	-11.2	-1.5	1.5	18.5	8.4	-1.9	..	..	..	Swaziland
-9.9	6.6	0.6	3.7	-2.9	4.4	-2.1	-4.8	-1.9	1.8	Tanzania
-2.4	-0.3	17.7	1.7	1.4	5.6	1.8	-4.1	-0.1	1.0	Togo
0.5	1.3	-0.9	..	..	..	..	..	..	..	Uganda
10.6	-4.2	-1.0	-0.4	4.2	3.3	1.6	0.3	3.8	-2.8	Zaire
5.9	-1.3	-6.7	2.0	13.2	0.6	-1.9	-4.6	0.9	1.3	Zambia
1.3	-4.4	-3.1	-6.5	0.4	3.5	4.8	-0.8	11.3	1.2	Zimbabwe
-2.7	*-0.7*	*0.8*	*0.5*	*0.0*	*-0.4*	*0.2*	*-1.3*	*0.9*	*0.3*	*SOUTH ASIA*
-6.0	0.4	-4.6	1.7	0.8	0.6	-1.6	-3.6	3.4	-1.3	Bangladesh
-2.6	-0.3	1.3	-0.3	0.2	-0.8	0.6	-1.2	0.5	0.2	India
..	..	..	..	..	..	..	..	..	..	Nepal
-0.1	-4.8	1.2	6.1	-0.5	1.4	-2.0	-1.0	2.9	1.9	Pakistan
-9.5	-4.0	-8.2	7.0	-5.6	-2.6	3.3	-2.6	-2.2	0.2	Sri Lanka
-2.3	*-1.9*	*0.3*	*-0.2*	*0.5*	*-0.4*	*1.4*	*-0.9*	*3.0*	*1.4*	*EAST ASIA AND THE PACIFIC*
-2.0	-0.3	-0.1	1.7	0.5	-1.1	-0.9	-3.1	2.5	2.1	China
-0.3	0.6	-0.4	-2.9	6.4	0.9	4.6	0.7	2.8	-4.3	Fiji
-5.5	-0.2	-3.5	0.3	0.3	2.6	4.7	-0.8	1.9	2.4	Hong Kong
-1.9	-1.3	-0.1	-12.8	-2.6	-0.3	5.9	-2.3	0.8	0.6	Indonesia
-5.7	-5.1	5.3	3.2	1.5	1.4	-0.1	1.5	3.0	1.8	Korea, Republic of
-1.2	0.9	-8.2	-3.6	-1.8	1.6	4.2	6.2	12.6	5.0	Malaysia
3.1	-4.0	-4.9	2.2	-0.1	1.1	3.5	6.5	4.5	0.9	Papua New Guinea
-2.0	-2.8	1.6	0.9	-1.0	-1.2	6.2	3.8	0.1	-6.3	Philippines
..	..	..	..	..	..	..	..	..	..	Singapore
..	..	..	8.7	19.0	2.1	-6.8	-3.3	17.5	..	Solomon Islands
0.8	-3.5	1.9	2.9	6.1	-7.7	3.6	3.2	3.1	-1.8	Thailand
..	..	..	..	..	..	..	..	..	..	Vanuatu

Table 12. Net exports of goods and nonfactor services: contribution to growth of GDP expenditure (cont'd)

Percentage points	1967	1968	1969	1970	1971	1972	1973	1974	1975	1976	1977
LATIN AMERICA AND CARIBBEAN	*0.1*	*0.2*	*0.5*	*-1.0*	*-1.0*	*1.3*	*0.1*	*-2.1*	*-1.1*	*-0.1*	*-0.6*
Argentina	0.0	-0.9	-0.2	1.2	-1.1	0.3	0.7	-0.2	-0.5	2.4	0.4
Bahamas	..	..	..	..	..	..	..	..	..	..	..
Barbados	..	..	..	..	..	..	..	..	..	..	..
Belize	..	..	..	..	..	..	..	..	..	..	..
Bolivia	6.1	-0.1	0.4	-4.7	-0.2	0.2	3.1	2.5	-6.1	3.2	-0.4
Brazil	-0.9	-0.9	0.4	-1.8	-2.3	-0.7	-1.5	-3.8	1.7	0.2	1.0
Chile	2.3	-1.5	-2.4	0.0	-1.7	-2.2	1.4	3.6	9.1	3.3	-3.4
Colombia	4.8	-1.6	0.0	-2.5	-1.5	2.9	0.2	-0.6	3.5	-1.9	-1.8
Costa Rica	-0.3	2.2	-1.3	-3.8	-0.3	4.3	0.0	-1.0	1.9	-3.5	-6.2
Dominican Republic	2.5	-2.0	-1.2	-3.1	0.7	4.9	0.2	-9.6	-0.5	4.7	0.0
Ecuador	-2.1	-2.2	-1.8	-0.1	-2.8	17.6	21.4	-9.8	-7.1	3.5	-5.8
El Salvador	4.0	6.9	-0.6	-4.6	-0.3	2.6	-6.9	0.8	5.4	-8.6	-14.0
Grenada	..	..	..	..	..	..	..	..	..	..	..
Guatemala	-3.4	1.2	3.3	-2.7	-1.0	4.5	-0.5	-2.3	2.1	-6.4	-1.6
Guyana	-0.8	3.4	-0.2	-3.1	9.0	-5.4	-20.3	8.5	-19.9	-7.4	15.7
Haiti	1.7	0.0	-0.1	-2.4	0.2	-0.8	1.2	-0.5	-0.9	-3.6	-2.9
Honduras	-1.6	0.8	-0.9	-2.1	10.1	1.7	-3.1	-8.0	4.6	-3.5	-7.2
Jamaica	-4.1	-4.1	-3.5	2.0	-2.2	-0.4	-1.7	-1.8	-8.9	5.4	7.0
Mexico	-0.9	-0.1	1.0	-1.5	0.8	0.7	0.0	-1.8	-0.9	1.3	2.3
Nicaragua	8.0	2.0	3.3	0.0	1.3	7.9	-17.9	-1.7	12.3	0.7	-13.5
Panama	0.1	1.4	-4.5	-3.0	-3.8	-3.3	2.3	-3.5	3.2	1.8	5.6
Paraguay	-0.1	-2.2	-0.1	3.3	-2.3	1.9	-2.4	-1.8	2.1	-4.2	-3.6
Peru	1.2	4.0	-1.2	4.0	-1.7	2.1	-7.0	-5.1	1.2	3.7	2.2
St. Vincent and the Grenadines	..	..	..	..	..	..	..	..	..	..	..
Trinidad and Tobago	3.4	8.1	1.6	1.1	-14.1	9.8	6.6	-3.4	-11.9	-1.1	-3.4
Uruguay	-2.0	2.4	-2.8	-3.7	-1.0	1.2	-1.5	3.0	0.7	2.3	-0.3
Venezuela	..	..	..	..	..	..	..	..	-15.1	-7.0	-9.3
MIDDLE EAST AND NORTH AFRICA	*-1.2*	*2.2*	*0.9*	*-1.2*	*-6.1*	*7.0*	*-2.6*	*-11.5*	*-5.8*	*1.5*	*-5.3*
Algeria	2.8	5.1	1.9	-0.3	-14.9	16.1	-2.5	-12.8	-5.2	3.4	-6.9
Egypt, Arab Republic of	-2.3	-4.2	0.4	-1.9	0.0	-0.2	-3.2	-12.4	-3.6	11.5	0.9
Jordan	..	..	..	..	-1.4	-3.2	3.2	-8.5	-7.3	-20.1	-12.8
Morocco	-5.1	2.1	0.6	-2.8	-0.3	4.7	-0.9	-4.7	-11.1	-7.1	-3.8
Oman	..	..	..	..	..	..	..	..	..	..	..
Syrian Arab Republic	-4.1	1.9	0.9	2.7	-5.2	5.5	-4.2	-24.0	-3.5	-0.9	-5.9
Tunisia	-1.8	9.1	0.6	-2.6	5.6	0.8	-2.7	-1.4	-3.3	1.5	-3.3
Yemen Arab Republic	..	..	..	..	-1.9	-5.7	-9.0	-10.0	-1.1	-13.4	-15.9
Yemen, PDR	..	..	..	..	..	..	..	..	..	..	..
EUROPE AND MEDITERRANEAN	*..*	*..*	*..*	*..*	*-2.0*	*1.8*	*-2.8*	*-2.7*	*0.2*	*2.0*	*-2.0*
Cyprus	..	..	..	..	..	..	..	..	..	-0.8	-6.7
Greece	-1.0	-2.3	-2.0	-0.1	-0.5	-1.0	-5.0	5.0	0.0	1.0	-1.8
Hungary	0.0	0.0	0.0	-10.0	-5.1	7.6	3.5	-5.2	-0.6	1.3	0.6
Israel	-2.4	-4.3	-5.1	-7.4	0.7	4.2	-17.2	3.1	-2.8	6.5	5.5
Malta	-11.2	-17.3	-22.9	-5.0	9.4	9.2	13.4	-1.0	15.1	-0.4	-1.3
Poland	..	..	..	..	..	..	..	..	..	..	..
Portugal	4.2	-14.7	-2.6	-1.0	-3.7	0.3	-5.1	-8.5	8.8	-1.3	-4.5
Turkey	1.6	-1.9	0.0	-1.6	-2.8	-2.3	1.4	-5.0	-2.5	-0.4	-2.3
Yugoslavia	-0.6	-0.1	-0.2	-4.8	-1.7	4.0	-1.5	-4.5	0.6	4.5	-4.3
Developing countries	-0.2	0.6	1.1	-0.1	-0.7	2.8	0.0	-3.1	-1.6	1.1	-1.0
Highly indebted	-0.7	-0.5	1.2	-0.6	-0.8	1.2	-0.1	1.5	-2.1	0.2	-0.6
Low-income countries	-0.5	0.9	1.0	-1.4	0.3	0.8	0.8	-1.1	1.3	0.0	-1.3
Low-income Africa	-0.6	0.0	1.9	0.0	-2.0	2.0	0.0	-2.8	0.4	1.9	-3.3
China and India	-0.3	1.2	0.6	-1.2	0.5	0.2	0.5	-0.3	1.4	-0.3	-0.8
Other low-income Asia	-2.5	-1.0	2.5	-5.2	3.2	3.9	5.0	-5.7	2.2	-0.8	-1.3
Middle-income countries	0.0	0.5	1.1	0.5	-1.2	3.6	-0.3	-3.8	-2.7	1.4	-0.9
High-income oil exporters	0.5	19.7	10.6	10.0	0.7	4.1	6.3	-10.8	-8.3	-1.1	-4.4
Industrialized countries	..	..	..	..	-0.1	-0.5	-0.4	1.0	0.9	-0.6	0.2
Japan	..	..	..	..	0.6	-0.9	-3.0	1.2	1.9	1.0	0.9
United States	..	..	..	..	-0.5	-0.5	0.6	1.3	0.8	-1.3	-1.0
European Community	..	..	..	..	..	-0.5	-0.3	1.5	0.6	-0.5	0.7

NOTE: See glossary for definitions.

1978	1979	1980	1981	1982	1983	1984	1985	1986	1987	
0.3	*0.6*	*-1.7*	*0.2*	*2.2*	*4.5*	*0.8*	*0.6*	*-1.0*	*1.0*	*LATIN AMERICA AND CARIBBEAN*
0.9	-2.6	-3.3	1.2	4.0	0.9	-0.4	1.8	-1.7	-0.9	Argentina
4.5	-3.8	0.5	-8.4	2.9	0.6	-4.0	-5.0	-0.3	-0.4	Bahamas
..	..	..	..	..	..	..	..	..	..	Barbados
..	..	..	..	..	..	..	..	..	..	Belize
-1.8	0.5	2.0	-6.9	8.6	-0.7	1.4	-4.1	-1.0	-1.1	Bolivia
0.4	-0.3	1.7	3.3	-0.4	3.1	2.8	0.9	-3.1	0.7	Brazil
-1.3	-2.0	-1.5	-6.3	11.7	3.5	-1.4	3.9	0.8	-0.8	Chile
0.7	1.2	-1.7	-2.7	-1.5	1.4	2.0	3.0	1.9	1.4	Colombia
-0.2	-0.2	0.1	12.6	3.4	-4.3	0.0	-4.5	-3.8	-0.4	Costa Rica
1.1	0.4	-10.2	4.5	1.0	2.6	2.8	-0.4	-2.1	..	Dominican Republic
0.0	1.4	-3.1	3.5	-2.8	6.3	3.6	2.0	2.9	-7.2	Ecuador
3.6	12.1	3.9	-1.5	0.8	2.7	-2.1	-1.4	-2.8	2.6	El Salvador
..	..	..	..	..	..	..	..	..	..	Grenada
-1.4	4.5	3.6	-2.2	3.4	1.9	-1.7	2.7	-0.2	-5.1	Guatemala
22.4	-4.3	-5.9	-3.9	8.9	-1.5	13.2	-8.4	-6.1	6.0	Guyana
0.6	2.0	-3.0	-1.8	5.6	-1.7	1.1	-0.3	0.2	-1.7	Haiti
1.7	3.6	-3.6	6.2	6.2	-1.9	-5.0	0.8	1.8	0.3	Honduras
4.6	4.9	-0.5	-1.2	-4.8	1.0	-1.6	-3.2	7.0	-2.6	Jamaica
-0.5	-1.4	-2.7	-1.1	7.7	4.8	-0.2	-1.6	1.3	1.7	Mexico
12.3	13.9	-37.6	5.0	7.2	0.6	-5.0	-5.6	0.8	..	Nicaragua
-4.1	-1.8	7.3	-1.3	2.7	4.1	-5.7	2.2	2.4	..	Panama
0.3	12.2	-22.0	0.7	3.1	3.2	-2.0	2.4	-3.1	1.2	Paraguay
7.7	1.7	-7.8	-4.2	0.7	5.2	5.7	2.4	-6.6	-0.1	Peru
17.1	-11.0	-5.1	7.1	-14.7	5.1	4.5	0.7	-14.0	..	St. Vincent and the Grenadines
-3.3	-6.5	1.2	2.6	-14.8	1.4	12.1	6.4	16.2	..	Trinidad and Tobago
-0.6	-2.8	-1.0	0.7	1.1	6.3	2.6	0.8	-0.4	-3.8	Uruguay
-0.6	8.7	-5.2	-4.1	-6.6	14.6	-3.9	2.6	-2.5	5.0	Venezuela
1.0	*-0.9*	*-2.2*	*-4.7*	*0.4*	*1.9*	*-0.5*	*1.5*	*3.9*	*4.6*	*MIDDLE EAST AND NORTH AFRICA*
-0.3	5.9	-6.1	-5.6	-0.4	1.3	0.3	1.0	4.8	3.9	Algeria
-1.3	-7.6	1.3	-2.7	2.7	2.2	-3.3	0.0	4.4	5.9	Egypt, Arab Republic of
-5.4	-19.5	15.3	-10.4	-9.2	-1.0	4.3	2.0	-8.2	9.5	Jordan
8.0	-1.0	4.0	-0.5	0.0	4.9	-0.6	0.8	-1.4	2.1	Morocco
..	-0.6	-8.0	-0.4	-6.9	6.8	-1.1	5.2	6.4	..	Oman
2.5	-5.1	-3.6	-9.2	7.3	-3.2	0.6	-0.6	5.8	5.1	Syrian Arab Republic
-0.8	1.6	-2.1	-4.5	-3.2	1.5	-1.7	7.2	2.5	6.6	Tunisia
-7.2	-13.1	-0.5	3.2	-2.4	5.5	8.5	2.0	7.2	-7.8	Yemen Arab Republic
..	..	..	..	..	..	..	..	..	..	Yemen, PDR
2.0	*0.2*	*2.2*	*0.7*	*1.4*	*0.8*	*1.6*	*0.1*	*-1.6*	*..*	*EUROPE AND MEDITERRANEAN*
-1.0	-1.8	1.0	4.8	-4.1	0.5	-3.0	2.6	5.9	3.2	Cyprus
0.9	-0.7	3.7	-2.2	-3.3	-0.5	3.3	-3.6	1.9	0.0	Greece
-4.7	6.2	0.7	1.4	2.9	2.4	2.4	-0.6	-2.1	0.9	Hungary
-3.4	1.7	5.4	-3.0	-2.8	-2.2	4.8	4.3	-2.5	..	Israel
8.3	4.9	-0.5	-2.9	-9.1	4.8	-0.4	-2.4	3.9	-1.5	Malta
..	..	..	1.0	5.3	1.1	1.1	-1.7	..	..	Poland
2.0	1.9	-0.7	-2.5	-0.8	8.7	5.8	2.9	-4.5	-4.9	Portugal
8.4	0.4	-0.1	3.3	3.4	-0.7	-1.8	0.5	-2.9	1.0	Turkey
0.3	-2.4	3.6	2.2	1.7	0.4	1.3	0.9	-1.4	1.0	Yugoslavia
-1.3	0.5	-2.4	-1.3	2.1	1.6	1.1	0.4	1.1	1.1	Developing countries
-0.5	1.9	-1.4	-1.7	2.1	4.1	1.6	1.0	-0.7	0.7	Highly indebted
-2.1	-0.4	0.0	1.1	0.4	-0.4	-0.4	-2.3	1.9	1.4	Low-income countries
-0.4	0.5	-1.3	0.2	1.0	2.2	0.4	-0.6	1.4	-0.2	Low-income Africa
-2.2	-0.3	0.4	1.0	0.4	-1.0	-0.4	-2.4	1.9	1.5	China and India
-3.0	-2.4	-1.4	3.8	-0.7	1.3	-1.3	-1.6	2.5	0.6	Other low-income Asia
-1.0	0.8	-3.2	-2.1	2.7	2.4	1.6	1.6	0.7	0.9	Middle-income countries
-4.2	4.8	-6.0	-9.3	-13.3	-12.9	..	..	..	..	High-income oil exporters
0.2	-0.3	1.1	1.0	-0.1	-0.1	-0.4	0.0	-1.2	-0.5	Industrialized countries
-0.9	-1.4	3.4	1.5	0.3	1.5	1.3	1.1	-1.4	-0.7	Japan
0.3	1.0	1.8	-0.3	-0.7	-1.6	-2.3	-0.8	-1.2	0.2	United States
0.0	-1.1	0.1	2.0	-0.1	0.4	0.6	0.3	-1.2	-1.1	European Community

Table 13. Gross domestic investment: percentage of gross domestic product

Percentage points	1967	1968	1969	1970	1971	1972	1973	1974	1975	1976	1977
SUB-SAHARAN AFRICA	*15.0*	*15.0*	*14.0*	*15.7*	*17.7*	*18.0*	*18.3*	*18.0*	*21.3*	*23.2*	*23.3*
Excluding Nigeria	*15.7*	*16.0*	*14.9*	*17.0*	*18.4*	*17.8*	*17.8*	*20.2*	*20.9*	*20.1*	*20.5*
Benin	14.5	15.2	15.5	14.2	12.6	16.4	16.0	22.3	23.7	20.4	21.8
Botswana	23.3	27.4	33.6	42.2	47.7	52.5	49.9	51.7	47.7	42.0	31.3
Burkina Faso	10.8	11.1	12.9	13.8	20.0	23.3	27.8	34.4	31.1	30.8	26.4
Burundi	6.9	8.2	8.3	4.5	7.6	3.2	5.3	4.1	7.6	9.1	11.1
Cameroon	13.8	13.4	10.7	16.0	16.6	18.2	19.9	17.1	20.0	17.6	21.8
Cape Verde	..	..	..	..	..	..	23.4	25.5	21.6	33.2	47.1
Central African Republic	21.8	17.4	17.1	18.9	19.7	19.2	22.1	14.6	13.9	12.1	11.7
Chad	17.6	16.9	18.9	18.1	17.8	16.5	-3.8	20.7	23.0	19.1	20.8
Comoros	..	..	..	..	..	..	..	..	..	..	..
Congo, People's Republic	31.5	29.6	26.7	24.2	26.6	29.1	32.0	35.1	39.1	31.0	26.6
Cote d'Ivoire	19.8	17.8	19.5	22.5	21.8	20.9	23.2	22.0	22.4	23.0	27.3
Ethiopia	14.3	14.0	13.4	11.5	11.8	12.7	11.4	9.9	10.4	9.6	8.8
Gabon	33.8	34.0	34.5	34.6	36.4	34.5	46.6	54.7	66.7	64.4	55.6
Gambia, The	9.7	10.5	8.8	4.6	6.1	7.6	7.9	4.2	11.6	10.9	14.1
Ghana	10.3	11.1	11.8	14.2	14.1	7.1	9.0	13.0	12.7	8.9	11.1
Guinea-Bissau	..	..	..	29.6	28.0	22.8	20.9	16.8	15.4	16.3	19.8
Kenya	20.2	20.0	19.4	24.4	23.9	22.3	25.8	25.8	18.1	20.2	23.7
Lesotho	10.5	11.4	11.1	11.4	15.8	15.2	18.6	17.7	18.7	36.1	25.0
Liberia	23.1	15.8	16.4	22.2	20.1	21.2	14.5	19.7	34.0	31.5	32.8
Madagascar	14.6	16.2	17.0	15.6	17.7	13.9	14.3	13.6	12.8	12.8	12.7
Malawi	13.5	16.1	18.4	25.7	19.2	24.4	22.4	27.3	33.7	26.3	24.7
Mali	18.4	18.1	18.4	16.2	17.0	18.1	17.9	14.3	16.5	14.2	16.5
Mauritania	21.4	27.5	22.4	22.0	22.8	30.4	8.6	16.1	34.5	42.4	39.1
Mauritius	16.5	11.9	13.9	9.9	14.5	15.3	24.9	25.2	26.3	31.8	30.0
Mozambique	..	..	..	..	..	..	..	..	..	..	..
Niger	8.7	9.7	6.2	9.8	7.8	11.1	14.9	15.1	14.3	14.2	19.7
Nigeria	12.0	10.9	10.7	13.0	16.3	18.3	19.4	14.6	21.8	27.1	26.7
Rwanda	7.3	8.2	6.6	7.0	9.1	9.6	9.4	10.5	13.7	13.8	15.1
Senegal	11.2	11.5	13.4	15.7	16.2	17.0	18.9	22.2	17.8	16.5	17.5
Seychelles	..	..	..	..	..	..	..	..	..	..	..
Sierra Leone	12.5	12.2	14.7	16.6	15.3	11.7	11.7	15.8	15.7	11.3	13.1
Somalia	12.0	12.2	12.4	11.7	11.9	13.6	14.3	26.9	22.5	29.9	39.1
Sudan	12.9	12.9	12.9	13.6	11.8	9.2	11.7	18.4	17.5	23.2	17.1
Swaziland	21.1	18.2	16.6	19.2	19.9	22.3	23.4	26.9	17.8	29.6	27.0
Tanzania	17.9	17.7	15.5	22.5	26.4	21.8	21.1	22.0	21.1	22.9	26.1
Togo	12.6	11.0	13.2	15.1	18.9	20.7	21.7	16.4	27.7	24.9	34.3
Uganda	13.3	13.0	14.0	13.3	15.2	11.0	8.2	10.7	7.6	5.8	5.5
Zaire	10.1	10.1	11.7	13.4	16.9	17.1	15.4	15.7	16.4	13.3	19.7
Zambia	34.2	34.0	19.2	28.2	37.1	35.3	29.2	36.7	40.6	23.7	24.7
Zimbabwe	19.4	23.5	18.8	20.4	22.4	26.4	25.4	27.5	26.3	17.9	19.1
SOUTH ASIA	*16.5*	*16.4*	*16.5*	*17.1*	*17.6*	*16.4*	*17.7*	*18.6*	*19.0*	*20.2*	*19.7*
Bangladesh	10.7	14.2	12.5	11.3	8.2	4.7	8.9	7.6	6.2	10.2	12.0
India	17.7	16.6	17.6	18.2	19.4	17.8	19.2	20.8	22.1	22.1	20.7
Nepal	5.0	5.7	4.8	6.0	7.8	7.3	9.2	8.8	14.5	15.1	16.0
Pakistan	16.7	18.0	14.8	15.8	15.6	14.2	12.9	13.4	16.2	17.2	19.3
Sri Lanka	14.4	15.2	18.2	18.9	17.1	17.3	13.7	15.7	15.6	16.2	14.4
EAST ASIA AND PACIFIC	*19.9*	*20.3*	*20.9*	*27.0*	*27.6*	*26.2*	*27.9*	*29.0*	*29.3*	*27.9*	*28.2*
China	19.9	20.0	20.4	29.2	29.7	27.8	29.7	29.7	31.1	29.2	29.7
Fiji	20.0	25.9	24.9	22.2	24.8	24.0	22.2	18.9	20.6	21.5	23.2
Hong Kong	21.0	17.5	17.2	21.4	25.6	24.6	24.0	25.4	24.2	26.8	27.9
Indonesia	9.3	10.2	13.6	15.8	18.4	21.8	20.8	19.5	23.7	24.1	23.4
Korea, Republic of	22.3	26.2	29.2	24.7	25.1	20.8	24.5	31.6	27.1	25.3	27.3
Malaysia	20.3	20.3	17.3	22.4	22.4	23.3	25.5	30.8	25.3	23.6	25.8
Papua New Guinea	21.9	21.7	29.1	41.6	42.8	27.8	15.5	17.4	20.0	18.6	21.6
Philippines	20.9	21.1	20.4	21.2	20.9	19.3	20.2	25.2	29.5	31.0	28.8
Singapore	22.2	24.9	28.6	38.7	40.6	41.4	39.4	45.3	39.9	40.8	36.2
Solomon Islands	..	..	..	..	..	..	..	..	..	..	..
Thailand	23.2	24.6	25.8	25.6	24.2	21.7	27.0	26.6	26.7	24.0	26.9
Vanuatu	..	..	..	..	..	..	..	..	..	..	..

NOTE: See glossary for definitions.

1978	1979	1980	1981	1982	1983	1984	1985	1986	1987	
21.6	*19.3*	*20.4*	*20.8*	*18.3*	*14.6*	*11.4*	*12.2*	*14.3*	*16.1*	*SUB-SAHARAN AFRICA*
19.2	*18.6*	*20.4*	*20.1*	*19.4*	*16.7*	*16.2*	*16.0*	*15.8*	*15.8*	*Excluding Nigeria*
19.7	23.3	18.6	19.2	34.1	11.6	7.4	13.6	12.7	13.9	Benin
39.6	38.3	41.0	43.6	44.2	29.4	24.4	30.1	23.1	..	Botswana
24.7	22.1	20.3	19.0	23.5	23.7	23.3	19.9	19.0	15.7	Burkina Faso
14.0	14.9	13.9	17.0	14.5	22.8	18.0	13.9	10.8	19.9	Burundi
22.7	21.7	18.9	24.7	23.4	24.4	20.8	19.5	19.8	18.2	Cameroon
42.1	56.8	42.3	62.9	57.3	57.4	50.1	51.6	..	..	Cape Verde
11.0	9.7	7.0	8.7	6.7	11.8	12.3	14.5	15.4	13.9	Central African Republic
18.2	20.4	18.2	11.3	9.1	7.3	7.5	6.4	..	18.4	Chad
..	..	..	..	..	29.0	45.8	29.7	23.6	12.3	Comoros
27.3	26.0	35.8	48.2	59.7	38.4	30.4	30.0	29.5	23.6	Congo, People's Republic
29.8	28.0	28.2	25.9	23.2	20.6	10.9	12.6	11.1	13.0	Cote d'Ivoire
7.5	8.7	10.0	10.4	11.8	11.2	12.3	11.4	11.2	14.5	Ethiopia
35.0	32.9	28.4	36.3	34.6	33.6	33.5	37.3	45.9	31.5	Gabon
30.1	26.5	26.7	26.1	23.3	21.8	19.3	16.2	23.7	19.0	Gambia, The
5.4	6.5	5.6	4.6	3.4	3.7	6.9	9.6	9.7	10.8	Ghana
21.7	21.8	29.6	25.7	28.3	22.7	30.0	32.0	31.6	18.8	Guinea-Bissau
29.8	22.6	30.0	28.4	22.1	21.2	23.5	20.1	22.6	24.6	Kenya
25.9	34.6	33.9	35.4	38.1	30.5	34.1	32.1	33.2	25.2	Lesotho
29.7	30.1	27.3	16.3	14.3	11.8	10.2	8.7	9.7	..	Liberia
15.1	25.3	23.5	18.1	13.4	13.2	13.6	14.0	13.8	13.8	Madagascar
38.4	31.3	24.5	17.7	21.4	22.8	12.9	18.6	11.1	14.0	Malawi
16.9	16.4	17.0	17.5	18.1	18.2	19.4	19.6	22.2	16.5	Mali
31.0	25.5	36.2	41.9	47.1	17.8	25.1	24.3	23.0	20.5	Mauritania
30.7	31.2	20.7	25.3	18.2	17.5	22.0	23.5	22.1	26.1	Mauritius
..	..	18.9	20.0	19.3	10.0	10.5	6.9	9.7	22.4	Mozambique
23.0	25.8	36.6	20.3	18.2	12.8	3.2	12.7	10.1	9.2	Niger
24.8	20.1	20.5	21.7	17.0	12.3	6.4	7.9	10.0	15.8	Nigeria
16.6	12.0	16.1	13.3	17.8	13.5	15.8	17.3	16.0	16.7	Rwanda
17.4	14.4	15.5	16.4	15.4	16.2	15.8	13.7	13.7	13.2	Senegal
..	..	..	32.4	32.3	21.2	21.7	22.7	22.8	..	Seychelles
11.3	13.4	16.2	19.1	13.4	14.3	12.7	11.3	11.1	8.8	Sierra Leone
50.7	39.2	44.2	26.4	30.0	27.4	35.6	29.5	33.0	34.8	Somalia
14.4	13.2	15.1	14.3	19.0	17.4	16.5	16.2	13.1	11.2	Sudan
43.9	39.7	40.7	31.0	32.2	35.0	32.0	32.0	24.5	..	Swaziland
25.2	26.1	23.0	20.6	21.0	13.7	15.6	15.8	16.4	17.0	Tanzania
46.3	48.8	29.9	30.2	26.3	22.5	18.5	20.0	24.7	17.3	Togo
7.8	6.4	6.1	7.0	7.3	8.7	16.3	8.6	8.1	12.0	Uganda
11.0	11.7	14.9	15.0	14.4	10.9	13.9	13.5	11.2	13.0	Zaire
23.9	14.1	23.3	19.3	16.8	13.8	14.7	14.9	14.7	14.9	Zambia
11.9	12.7	18.8	23.1	21.3	16.1	17.4	21.7	22.5	18.5	Zimbabwe
21.8	*22.6*	*23.1*	*24.3*	*22.3*	*22.1*	*22.0*	*23.9*	*22.6*	*22.4*	*SOUTH ASIA*
11.6	11.3	15.1	16.0	15.0	13.6	12.2	12.5	11.8	11.5	Bangladesh
23.5	24.3	24.2	26.1	23.4	23.2	23.7	26.0	24.5	24.2	India
18.3	15.8	18.3	17.6	16.5	20.1	18.8	20.5	21.0	21.0	Nepal
17.9	17.9	18.5	17.1	17.7	17.5	17.0	16.8	16.4	17.0	Pakistan
20.0	25.8	33.8	27.8	30.8	28.9	25.8	23.8	23.7	23.3	Sri Lanka
31.2	*32.1*	*30.1*	*29.4*	*28.9*	*29.3*	*29.3*	*30.8*	*29.7*	*29.7*	*EAST ASIA AND PACIFIC*
34.2	32.5	30.0	27.4	29.1	29.6	32.2	38.8	38.8	37.6	China
22.8	30.1	31.8	34.3	25.6	21.1	19.1	17.5	21.1	12.5	Fiji
30.0	34.0	36.0	36.0	31.8	27.1	24.7	21.7	23.3	25.3	Hong Kong
23.9	26.6	24.3	29.6	27.5	29.4	25.5	26.4	24.6	26.3	Indonesia
31.4	35.5	31.1	29.1	27.4	28.9	30.8	30.0	28.8	29.1	Korea, Republic of
26.7	28.9	30.4	35.0	37.3	36.1	33.6	27.5	25.1	23.3	Malaysia
22.6	23.5	25.2	27.2	32.2	31.7	28.4	21.9	21.4	22.1	Papua New Guinea
28.9	31.1	30.7	30.6	28.3	26.7	17.0	13.9	12.9	14.5	Philippines
39.0	43.4	46.3	46.3	47.9	47.9	48.5	42.5	38.2	39.4	Singapore
..	..	36.4	33.9	30.0	36.8	26.0	32.4	36.3	..	Solomon Islands
28.2	27.2	26.4	26.3	23.1	25.9	24.9	24.0	22.4	23.8	Thailand
..	..	..	..	..	29.2	28.6	29.0	..	..	Vanuatu

Table 13. Gross domestic investment: percentage of gross domestic product (cont'd)

	1967	1968	1969	1970	1971	1972	1973	1974	1975	1976	1977
LATIN AMERICA AND CARIBBEAN	*19.5*	*20.0*	*21.2*	*21.3*	*20.8*	*20.5*	*21.1*	*23.1*	*25.0*	*23.6*	*24.1*
Argentina	18.3	18.8	21.0	21.6	20.8	20.8	18.0	19.3	25.9	26.8	27.2
Bahamas	..	..	..	..	..	..	..	..	..	..	12.5
Barbados	20.5	23.7	24.0	26.1	23.1	22.3	22.2	22.8	19.2	27.0	19.5
Belize	..	..	..	..	..	..	..	..	..	..	..
Bolivia	19.3	24.0	23.9	23.8	24.1	29.6	29.1	21.4	36.9	28.3	27.3
Brazil	17.0	18.9	22.4	20.5	21.1	21.2	23.2	25.4	26.8	23.1	22.1
Chile	16.1	16.3	15.2	16.5	14.5	12.2	7.9	21.2	13.1	12.8	14.4
Colombia	17.0	19.5	18.8	20.2	19.4	18.1	18.3	21.5	17.0	17.6	18.8
Costa Rica	19.4	18.0	20.0	20.5	24.3	22.0	24.0	26.7	21.6	23.7	24.3
Dominican Republic	15.5	15.4	18.1	20.6	19.2	21.2	23.8	25.1	26.4	24.0	23.5
Ecuador	17.4	18.0	17.5	18.2	23.2	20.0	19.5	22.5	26.7	23.8	26.5
El Salvador	14.4	11.1	12.7	12.9	15.6	14.2	18.3	22.6	22.1	19.6	23.4
Grenada	..	..	..	..	..	..	..	..	..	..	..
Guatemala	12.9	15.1	11.4	12.8	14.4	12.1	13.7	18.6	16.1	21.4	20.0
Guyana	25.8	22.3	20.8	22.8	18.6	19.8	27.2	26.4	33.1	37.4	29.1
Haiti	6.7	6.8	8.8	11.4	11.4	13.9	15.9	16.5	14.7	15.8	15.5
Honduras	19.7	18.5	19.4	20.9	17.1	15.9	19.1	26.2	19.0	19.2	23.1
Jamaica	29.3	34.9	35.2	31.5	32.1	27.3	31.5	24.3	25.8	18.2	12.2
Mexico	20.4	19.6	19.8	21.3	18.9	19.0	20.0	21.8	22.3	21.0	21.6
Nicaragua	21.0	17.7	19.0	18.6	17.7	12.8	24.0	31.5	21.4	17.3	27.2
Panama	21.0	22.3	23.6	27.8	30.4	31.8	33.6	33.6	30.8	31.6	23.7
Paraguay	16.5	15.9	16.0	14.7	14.6	15.1	19.0	21.0	24.1	24.6	24.7
Peru	35.1	26.1	25.1	15.5	17.9	15.2	20.3	25.9	24.7	21.9	19.2
St. Vincent and the Grenadines	..	..	..	..	..	..	..	..	..	..	28.1
Trinidad and Tobago	17.6	19.8	17.2	25.9	34.0	31.3	26.0	21.8	27.3	24.6	26.7
Uruguay	13.8	10.2	10.9	11.5	12.6	11.8	12.6	11.5	13.5	14.8	15.2
Venezuela	23.9	28.7	29.6	32.9	32.2	33.2	31.0	26.1	32.8	36.2	43.1
MIDDLE EAST AND NORTH AFRICA	*16.3*	*16.3*	*17.6*	*19.6*	*19.6*	*20.0*	*21.5*	*27.8*	*33.4*	*34.7*	*37.1*
Algeria	23.5	27.8	32.1	36.4	35.7	34.3	40.3	39.7	45.2	43.1	46.8
Egypt, Arab Republic of	14.4	12.6	13.0	13.9	13.2	12.3	13.1	22.5	33.4	28.4	29.2
Jordan	..	..	..	12.8	19.1	20.6	18.0	26.5	28.5	35.6	39.4
Morocco	13.3	12.6	14.2	18.5	17.9	15.3	16.9	20.6	25.2	28.1	34.2
Oman	31.2	12.6	9.3	13.8	28.5	29.8	26.2	30.6	35.6	35.9	30.6
Syrian Arab Republic	11.5	13.0	14.9	13.7	14.8	20.1	12.4	24.8	25.0	31.4	35.5
Tunisia	24.8	23.0	21.7	21.1	21.5	22.7	21.3	25.8	28.0	30.7	30.5
Yemen Arab Republic	..	..	..	0.4	16.4	17.6	16.8	20.2	17.0	18.4	18.8
Yemen, PDR	..	..	..	..	..	..	..	..	..	..	..
EUROPE AND MEDITERRANEAN	..	*24.8*	*25.4*	*28.5*	*29.0*	*28.2*	*29.1*	*28.9*	*29.3*	*29.2*	*30.1*
Cyprus	..	..	..	..	..	..	..	..	22.5	26.6	34.9
Greece	22.3	23.1	25.9	28.1	27.9	29.6	35.8	29.3	27.0	26.3	26.4
Hungary	..	30.9	30.6	33.6	37.7	31.7	29.7	35.8	37.8	35.9	37.2
Israel	16.9	22.5	25.5	27.4	29.9	30.0	30.9	29.1	29.2	24.6	22.6
Malta	29.2	32.3	32.7	32.7	29.0	25.0	22.1	27.1	23.5	27.5	26.0
Poland	..	..	..	..	..	..	..	..	..	..	..
Portugal	26.9	23.7	23.5	26.2	26.3	28.9	29.7	28.5	24.3	25.9	29.0
Turkey	17.4	18.0	17.5	19.8	17.8	20.9	19.0	21.6	23.2	25.1	25.2
Yugoslavia	30.0	29.5	29.2	32.3	32.8	28.1	28.4	30.2	33.5	31.9	35.8
Developing countries	19.6	19.7	20.4	22.7	23.1	22.4	23.5	24.6	26.6	26.0	26.3
Highly indebted	20.0	20.4	21.4	21.6	21.3	20.8	21.4	22.8	25.5	24.9	25.8
Low-income countries	17.8	17.8	17.9	22.6	23.3	22.0	23.6	23.7	24.6	23.8	24.0
Low-income Africa	14.8	15.3	14.2	16.1	17.5	16.3	15.7	17.8	17.4	16.5	17.4
China and India	19.1	18.7	19.3	25.1	25.8	24.2	26.0	26.3	27.9	26.5	26.2
Other low-income Asia	13.6	15.8	13.8	14.2	12.9	11.6	11.2	10.9	11.2	14.3	16.2
Middle-income countries	20.6	20.8	21.7	22.6	22.9	22.5	23.4	24.9	27.3	26.6	27.0
High-income oil exporters	23.6	23.9	23.8	16.3	14.3	16.8	19.0	13.3	18.7	22.9	27.4
Industrialized countries	20.2	19.9	19.7	21.9	21.8	22.4	24.0	23.7	20.5	21.7	22.0
Japan	29.8	31.0	31.6	39.7	36.5	36.1	38.6	37.9	33.3	32.4	31.5
United States	17.6	16.8	15.7	14.7	15.8	16.8	17.7	16.6	13.9	15.8	17.6
European Community	21.2	21.5	22.1	25.9	24.6	24.3	25.5	25.2	21.6	23.1	22.4

NOTE: See glossary for definitions.

1978	1979	1980	1981	1982	1983	1984	1985	1986	1987	
24.5	*23.8*	*24.2*	*23.6*	*21.4*	*16.8*	*16.7*	*17.5*	*17.4*	*18.2*	*LATIN AMERICA AND CARIBBEAN*
24.4	22.7	22.2	18.8	15.9	17.3	11.3	8.5	8.8	9.9	Argentina
11.3	12.8	18.3	18.8	21.5	20.3	16.3	19.2	18.4	18.8	Bahamas
22.9	23.5	25.3	27.6	22.6	19.9	16.2	15.4	15.8	14.8	Barbados
..	29.9	25.3	23.1	26.6	19.7	21.1	17.5	..	..	Belize
24.0	20.0	14.1	12.1	9.7	3.5	5.7	-0.8	8.0	9.5	Bolivia
23.0	22.8	22.9	21.1	20.0	14.7	15.5	16.7	18.5	19.7	Brazil
17.8	17.8	21.0	22.7	11.3	9.8	13.6	13.7	14.6	16.9	Chile
18.3	18.2	19.1	20.6	20.5	19.9	19.0	19.0	18.0	19.0	Colombia
23.5	25.3	26.6	29.0	24.7	24.2	21.1	20.7	21.9	21.2	Costa Rica
25.7	27.3	26.7	25.1	22.1	22.2	21.5	17.8	..	..	Dominican Republic
28.4	25.3	26.1	23.2	25.2	17.6	17.2	18.2	20.5	23.4	Ecuador
23.8	18.1	13.3	14.2	13.2	12.1	12.0	10.8	13.3	14.0	El Salvador
..	..	..	..	..	..	..	..	..	..	Grenada
21.6	18.7	15.9	17.0	14.1	11.1	11.6	11.5	10.3	13.8	Guatemala
20.5	31.0	29.8	29.5	24.2	21.0	27.4	29.8	33.8	31.9	Guyana
16.8	18.7	16.9	16.9	16.6	16.3	15.9	14.2	10.9	12.5	Haiti
27.2	26.5	24.5	20.7	13.5	14.9	19.0	18.1	15.7	15.4	Honduras
15.0	19.2	15.7	20.4	20.9	22.3	23.0	25.0	18.5	22.6	Jamaica
22.3	24.7	27.2	27.4	22.9	20.8	19.9	21.9	18.5	15.5	Mexico
13.1	-5.7	15.4	23.6	18.8	22.5	22.2	23.1	18.3	..	Nicaragua
26.6	28.1	27.7	30.1	27.7	21.4	16.7	15.4	17.4	..	Panama
27.2	28.6	31.7	31.6	27.9	23.1	24.8	22.9	24.1	24.9	Paraguay
19.4	21.7	27.5	31.4	31.4	23.7	22.7	22.4	24.2	24.6	Peru
28.8	35.0	40.3	33.2	28.8	25.5	27.9	29.0	30.4	..	St. Vincent and the Grenadines
30.2	29.1	30.6	27.6	28.3	26.0	21.6	20.6	22.2	..	Trinidad and Tobago
16.0	17.3	17.3	15.4	14.4	10.0	9.9	8.2	7.4	9.3	Uruguay
43.9	33.2	26.4	24.4	27.7	12.2	15.7	17.2	20.2	24.3	Venezuela
37.8	*34.3*	*32.2*	*32.1*	*32.0*	*30.9*	*30.7*	*29.1*	*27.3*	*23.6*	*MIDDLE EAST AND NORTH AFRICA*
52.1	42.5	39.1	37.0	37.3	37.6	36.4	33.7	32.4	28.9	Algeria
31.7	32.8	27.5	29.5	30.1	28.7	27.5	26.7	23.7	19.3	Egypt, Arab Republic of
35.3	37.2	41.1	50.5	46.9	35.9	32.4	30.5	29.6	26.4	Jordan
25.4	24.5	22.6	22.4	23.3	20.9	21.8	22.9	20.3	19.1	Morocco
28.9	26.0	22.3	23.0	26.6	26.5	30.0	27.6	27.6	..	Oman
27.4	26.2	27.5	23.2	23.7	23.6	23.7	23.8	23.3	18.6	Syrian Arab Republic
30.8	29.4	29.4	32.3	31.6	29.3	32.0	26.6	23.5	21.1	Tunisia
39.1	39.2	45.8	46.7	29.8	19.4	18.3	14.4	13.0	13.3	Yemen Arab Republic
..	..	..	..	..	..	..	..	..	..	Yemen, PDR
29.3	*29.6*	*30.2*	*28.4*	*28.2*	*26.7*	*26.1*	*26.7*	*27.8*	*28.3*	*EUROPE AND MEDITERRANEAN*
37.2	38.7	37.8	33.8	31.7	30.1	33.6	30.4	25.8	24.1	Cyprus
27.7	30.1	25.0	25.4	21.1	21.9	20.4	21.3	19.2	17.1	Greece
41.3	34.0	30.7	29.7	28.5	26.5	25.7	25.0	26.9	26.8	Hungary
24.8	25.5	21.5	19.9	21.7	21.9	20.2	16.5	17.5	..	Israel
23.4	25.5	24.6	27.1	31.5	30.0	28.9	28.1	25.5	25.9	Malta
..	..	26.3	18.5	28.0	25.0	26.3	27.7	28.6	..	Poland
30.5	29.5	34.1	36.1	37.0	29.1	22.8	21.8	22.4	24.0	Portugal
18.7	18.6	21.9	22.0	20.6	19.6	19.9	21.1	24.7	25.5	Turkey
34.3	38.3	39.9	38.9	35.6	35.7	37.2	38.8	38.4	38.6	Yugoslavia
27.1	26.8	27.0	26.7	25.2	24.1	23.5	24.0	23.8	24.1	Developing countries
25.7	25.1	25.3	24.8	22.3	17.8	16.6	17.6	18.6	19.4	Highly indebted
27.0	26.7	25.9	24.7	24.7	24.6	26.0	29.3	28.2	27.9	Low-income countries
16.7	16.1	18.2	16.5	16.1	13.9	14.6	14.1	13.8	14.4	Low-income Africa
30.3	29.7	27.8	26.8	26.8	27.0	28.8	33.3	32.3	31.5	China and India
16.4	17.2	19.2	18.3	18.6	17.9	16.6	16.4	15.8	16.1	Other low-income Asia
27.1	26.9	27.3	27.3	25.3	23.9	22.6	22.1	22.4	23.0	Middle-income countries
31.5	26.1	21.2	23.1	22.6	27.5	28.5	26.3	25.6	23.8	High-income oil exporters
22.3	22.9	22.2	21.4	19.6	19.2	20.6	19.7	20.1	20.7	Industrialized countries
31.6	33.2	33.0	32.0	30.9	29.1	29.1	29.3	28.9	29.9	Japan
18.8	18.5	16.3	17.2	14.4	15.0	17.8	16.1	15.8	15.9	United States
21.5	22.5	22.8	20.4	20.0	19.4	19.7	19.4	19.4	19.8	European Community

Table 14. Gross domestic savings: percentage of GDP

	1967	1968	1969	1970	1971	1972	1973	1974	1975	1976	1977
SUB-SAHARAN AFRICA	*14.0*	*14.3*	*15.7*	*16.3*	*15.7*	*17.8*	*19.7*	*23.9*	*18.2*	*21.4*	*21.0*
Excluding Nigeria	*14.9*	*15.4*	*17.0*	*17.9*	*15.2*	*16.3*	*17.5*	*19.8*	*15.2*	*17.0*	*17.1*
Benin	0.1	4.7	6.0	6.5	4.7	3.9	7.1	5.9	2.1	-0.5	1.2
Botswana	-8.4	-12.8	-6.1	1.7	14.7	27.5	28.0	27.6	22.5	22.9	14.3
Burkina Faso	-0.4	0.3	-0.4	0.4	1.0	3.9	5.9	13.7	0.6	6.4	-3.5
Burundi	5.8	5.1	5.9	3.6	1.9	-0.2	3.0	-1.7	-3.1	6.2	11.4
Cameroon	11.8	10.8	11.2	17.6	12.5	12.6	16.6	19.7	17.1	12.9	19.1
Cape Verde	..	..	..	..	..	..	..	-23.7	-15.8	-13.7	-21.9	-25.6
Central African Republic	1.2	4.5	2.1	3.6	6.5	5.6	7.9	5.8	-5.7	4.1	1.5
Chad	12.2	11.6	12.5	9.8	10.5	9.7	-14.8	8.8	6.8	4.5	2.3
Comoros	..	..	..	..	..	..	..	..	..	..	..
Congo, People's Republic	6.2	2.7	1.9	1.2	6.5	5.6	20.2	28.8	11.3	4.8	6.9
Cote d'Ivoire	27.2	30.5	29.2	29.2	25.6	25.4	25.7	29.3	22.6	28.5	33.6
Ethiopia	11.7	12.1	11.3	11.2	9.9	10.8	13.4	13.3	7.5	8.7	5.7
Gabon	33.8	33.2	35.1	38.1	41.1	34.7	41.5	68.4	67.3	70.0	62.6
Gambia, The	8.0	-1.5	7.0	9.4	-0.8	7.8	1.2	8.8	21.5	-1.6	8.0
Ghana	7.8	10.6	11.2	12.8	9.6	12.6	14.1	9.6	13.7	8.5	10.0
Guinea-Bissau	..	..	..	3.4	-2.2	-8.5	-13.8	-17.4	-5.4	-6.4	-3.5
Kenya	19.3	20.2	20.8	23.6	17.4	20.2	24.5	18.5	13.5	20.9	27.0
Lesotho	-30.1	-28.5	-24.9	-32.2	-42.6	-49.9	-45.9	-66.0	-70.4	-83.3	-74.5
Liberia	40.5	43.7	50.0	42.0	38.1	40.6	32.6	31.9	38.4	34.2	25.6
Madagascar	11.6	12.1	12.9	14.5	11.5	10.5	10.3	9.7	8.5	11.4	10.1
Malawi	2.7	0.8	2.3	10.8	7.1	9.7	12.4	16.4	17.0	17.8	20.1
Mali	5.2	9.8	10.1	10.2	9.7	8.0	4.5	-15.5	-4.4	3.5	9.3
Mauritania	32.1	36.4	30.1	29.7	30.4	27.5	9.8	5.8	13.0	2.9	-1.3
Mauritius	11.4	9.5	16.1	11.2	10.1	17.5	22.8	31.1	27.3	24.9	19.3
Mozambique	..	..	..	..	..	..	..	..	..	..	..
Niger	2.7	3.4	-0.9	2.5	4.5	4.8	6.7	-1.6	2.6	2.1	10.0
Nigeria	9.4	8.4	10.2	12.8	16.7	20.5	24.1	30.4	22.7	26.9	26.0
Rwanda	2.4	1.8	-0.1	3.3	2.5	1.6	7.6	1.3	5.2	9.0	11.7
Senegal	8.0	4.2	5.3	11.1	8.9	13.3	8.8	15.6	12.5	8.4	11.3
Seychelles	..	..	..	..	..	..	..	..	..	..	..
Sierra Leone	8.5	8.9	14.3	13.8	14.2	10.3	13.2	13.5	5.7	3.3	5.7
Somalia	3.4	5.5	4.7	6.5	5.3	8.8	1.4	4.3	8.1	16.1	0.2
Sudan	14.5	10.6	11.7	15.2	10.2	7.6	11.9	14.5	6.1	13.2	10.7
Swaziland	11.5	17.9	18.1	25.1	31.4	30.2	38.1	49.5	33.9	37.7	23.7
Tanzania	18.2	15.2	15.8	20.5	21.7	18.3	14.9	7.4	8.6	20.7	22.8
Togo	13.3	13.0	14.9	13.8	15.0	13.2	15.0	33.4	10.2	15.7	16.7
Uganda	14.2	14.4	15.2	16.4	11.3	13.4	11.4	10.4	5.5	7.4	6.8
Zaire	19.3	19.3	19.8	21.5	19.9	18.4	19.9	21.2	14.4	12.4	14.1
Zambia	40.4	40.6	50.9	45.1	35.0	36.9	45.0	46.1	21.0	28.8	22.1
Zimbabwe	19.0	19.5	21.5	21.4	20.7	29.2	27.4	27.0	25.1	21.8	21.5
SOUTH ASIA	*13.4*	*14.5*	*15.3*	*15.0*	*15.5*	*14.9*	*17.1*	*16.3*	*16.8*	*18.8*	*18.7*
Bangladesh	3.5	10.9	9.0	7.2	3.8	-3.4	9.9	0.6	1.0	-2.8	6.9
India	15.5	15.7	17.3	17.4	18.4	17.3	18.7	19.7	21.7	22.9	21.3
Nepal	4.9	5.0	1.6	2.6	4.0	5.1	5.4	3.7	10.0	11.7	13.5
Pakistan	10.3	12.2	12.0	9.0	8.7	9.2	10.1	6.4	4.9	7.9	9.5
Sri Lanka	10.3	10.9	9.2	15.8	15.1	15.7	12.5	8.2	8.1	13.9	18.1
EAST ASIA AND PACIFIC	*18.9*	*18.9*	*19.9*	*25.8*	*26.8*	*26.4*	*28.6*	*27.8*	*27.8*	*28.3*	*29.0*
China	20.3	20.5	21.2	29.2	30.4	28.6	30.2	29.1	30.9	29.4	29.9
Fiji	17.3	21.5	20.6	18.9	17.0	14.7	8.7	13.7	20.0	16.8	20.4
Hong Kong	21.7	18.5	21.9	25.0	25.4	28.5	26.4	26.1	25.3	33.6	30.3
Indonesia	2.3	6.5	8.7	13.9	16.8	22.6	22.5	27.9	25.9	26.3	28.2
Korea, Republic of	11.6	13.5	17.2	15.0	14.8	16.1	21.6	20.6	18.5	24.1	27.5
Malaysia	22.7	23.6	26.7	26.6	23.9	22.0	31.0	31.1	25.8	33.7	33.1
Papua New Guinea	-0.3	0.5	2.1	6.1	5.9	10.8	25.5	24.5	13.9	19.6	20.3
Philippines	19.0	17.9	16.9	20.9	20.1	18.6	23.7	22.1	22.7	24.7	25.0
Singapore	13.7	18.3	18.0	18.4	18.9	24.6	29.3	29.1	29.4	32.6	33.5
Solomon Islands	..	..	..	4.2	6.2	6.8	1.5	5.2	-15.0	3.7	21.6
Thailand	21.2	20.9	22.0	21.2	21.3	20.7	25.6	24.3	22.1	21.5	21.5
Vanuatu	..	..	..	..	..	..	..	..	..	..	..

NOTE: See glossary for definitions.

66

1978	1979	1980	1981	1982	1983	1984	1985	1986	1987	
16.7	*18.7*	*21.8*	*15.3*	*11.6*	*10.6*	*11.6*	*12.8*	*11.2*	*13.0*	*SUB-SAHARAN AFRICA*
13.4	*13.3*	*14.5*	*12.7*	*12.3*	*12.1*	*14.9*	*14.3*	*12.5*	*11.3*	*Excluding Nigeria*
0.0	1.9	-6.3	-14.9	-2.6	0.1	-6.4	-0.8	0.4	4.4	Benin
12.3	22.6	28.5	22.4	7.8	16.8	23.8	28.4	38.7	..	Botswana
-3.0	-5.9	-9.0	-10.0	-10.0	-7.3	0.6	-8.7	-0.5	-7.7	Burkina Faso
5.5	4.4	0.3	5.6	-0.9	8.1	4.2	5.0	6.9	7.7	Burundi
18.2	14.5	15.7	22.3	23.1	28.8	32.7	30.2	20.1	14.6	Cameroon
-32.1	-20.8	-34.4	-22.7	-15.5	-15.4	-9.6	-11.7	..	..	Cape Verde
-0.7	-1.7	-9.6	-1.9	-6.8	-1.2	-0.3	-0.3	0.9	-2.5	Central African Republic
1.8	0.1	-6.4	-4.7	-5.8	-2.8	-1.9	0.2	..	-12.3	Chad
..	..	..	..	..	2.8	-10.8	-9.2	-3.5	-10.8	Comoros
10.0	24.7	35.7	30.1	46.5	44.6	46.4	41.7	15.6	21.2	Congo, People's Republic
28.8	24.9	22.2	18.8	20.3	18.9	22.4	25.8	23.0	18.6	Côte d'Ivoire
1.9	3.4	4.9	5.3	3.2	2.9	2.5	2.8	1.1	3.3	Ethiopia
49.9	54.0	60.6	57.0	53.2	54.3	53.5	51.0	30.3	34.2	Gabon
-13.3	13.4	5.2	-6.2	3.5	9.7	-4.8	7.7	6.2	6.4	Gambia, The
4.0	6.6	4.9	4.0	3.7	0.6	6.6	7.6	11.0	7.8	Ghana
-5.8	-10.7	-6.0	1.2	-5.2	-3.6	-8.3	-11.0	-4.0	5.2	Guinea-Bissau
20.0	16.6	18.7	19.5	18.3	20.1	22.1	19.0	22.6	19.6	Kenya
-47.4	-66.2	-66.2	-77.3	-89.3	-104.0	-100.2	-88.6	-86.4	-73.3	Lesotho
24.6	26.9	27.3	18.8	12.5	10.0	14.4	14.3	18.4	..	Liberia
8.8	9.0	6.5	6.5	3.8	6.3	9.2	8.7	8.8	7.2	Madagascar
20.5	10.9	10.6	11.9	15.1	15.3	14.8	13.8	9.0	11.6	Malawi
-2.5	-2.0	-1.9	-0.3	1.0	-1.2	0.8	-14.3	-2.1	-0.4	Mali
0.0	-0.6	6.9	12.6	3.0	-12.1	-1.9	8.8	15.4	13.7	Mauritania
18.4	19.5	10.5	14.8	15.4	17.1	18.7	21.6	28.7	28.8	Mauritius
..	..	0.5	-0.1	-3.4	-10.7	-6.1	-3.0	-2.1	-9.5	Mozambique
17.7	17.5	22.6	8.7	7.8	7.5	-1.6	2.4	5.6	4.5	Niger
21.1	25.3	29.5	18.2	10.9	8.9	8.0	11.2	8.1	19.9	Nigeria
7.6	10.3	4.2	1.4	3.8	2.9	7.1	7.1	6.4	5.1	Rwanda
4.7	4.2	-0.5	-4.6	0.5	2.0	4.7	1.2	6.0	6.0	Senegal
..	..	..	13.5	-0.2	-1.0	8.2	7.2	5.6	..	Seychelles
6.4	5.2	0.9	2.4	3.2	3.3	10.9	10.4	9.6	9.7	Sierra Leone
20.5	-12.1	-7.6	-13.7	-9.5	-18.2	-7.0	-1.3	3.4	0.5	Somalia
6.3	6.6	3.4	1.0	1.2	4.0	6.4	7.9	4.5	6.2	Sudan
22.1	6.6	7.4	1.4	2.0	-1.1	-4.7	-3.5	12.6	..	Swaziland
10.0	13.3	9.8	12.1	10.9	7.4	7.9	5.8	1.9	-5.9	Tanzania
20.2	16.8	15.0	14.8	12.2	13.9	12.7	10.6	12.4	5.5	Togo
6.1	6.6	4.6	0.9	3.8	5.3	16.1	5.3	5.5	4.8	Uganda
11.8	13.1	13.7	9.4	8.7	8.2	17.4	18.1	12.2	9.8	Zaire
20.5	23.1	19.3	6.8	8.0	12.6	18.5	17.2	10.3	19.5	Zambia
15.4	12.5	15.8	15.8	15.3	12.9	17.9	22.8	26.5	21.9	Zimbabwe
18.9	*18.4*	*17.8*	*19.1*	*17.2*	*17.9*	*17.8*	*19.0*	*18.9*	*19.3*	*SOUTH ASIA*
2.2	1.5	2.1	3.2	0.4	0.3	1.1	2.3	2.4	2.2	Bangladesh
22.2	21.9	20.6	22.8	20.7	20.9	21.3	22.9	22.4	22.3	India
13.4	11.6	11.1	10.9	9.2	9.1	10.0	11.5	11.9	11.4	Nepal
8.6	5.8	6.9	6.3	5.4	6.3	5.3	4.7	7.0	10.5	Pakistan
15.3	13.8	11.2	11.7	11.9	13.8	19.9	11.9	12.0	12.8	Sri Lanka
30.7	*30.9*	*29.2*	*28.3*	*28.6*	*28.9*	*30.8*	*31.0*	*32.6*	*34.7*	*EAST ASIA AND PACIFIC*
33.6	31.4	28.7	27.7	31.0	30.6	32.3	34.3	36.0	37.7	China
18.4	24.7	26.9	19.9	19.2	15.7	18.0	17.2	23.3	17.1	Fiji
26.4	31.1	31.4	30.4	28.2	25.1	28.9	27.3	27.9	30.7	Hong Kong
26.3	33.6	37.1	33.7	27.9	28.3	30.5	28.3	24.9	29.1	Indonesia
28.5	28.2	23.3	23.7	24.9	27.6	30.5	30.5	34.6	37.6	Korea, Republic of
32.2	37.8	32.9	28.8	28.6	30.8	35.5	32.7	31.5	36.9	Malaysia
19.1	23.4	15.1	6.7	8.5	13.3	13.6	11.5	13.7	16.6	Papua New Guinea
23.9	25.6	25.0	25.1	21.6	19.9	16.9	16.9	19.1	15.4	Philippines
34.0	36.3	37.5	40.7	43.5	46.1	45.7	40.1	38.7	39.9	Singapore
16.9	35.4	6.8	1.7	15.5	15.3	19.3	10.4	2.8	..	Solomon Islands
23.9	20.5	20.1	20.0	21.4	18.7	20.6	21.2	24.5	23.9	Thailand
..	..	..	..	..	12.3	20.9	4.4	..	..	Vanuatu

67

Table 14. Gross domestic savings: percentage of gross domestic product (cont'd)

	1967	1968	1969	1970	1971	1972	1973	1974	1975	1976	1977
LATIN AMERICA AND CARIBBEAN	*19.8*	*19.8*	*21.0*	*20.6*	*19.7*	*19.5*	*21.1*	*22.1*	*22.5*	*22.3*	*22.9*
Argentina	21.7	20.3	21.0	21.6	21.6	21.7	20.8	20.4	25.5	31.2	30.3
Bahamas	..	..	..	..	..	..	..	..	..	..	29.0
Barbados	11.0	12.0	7.7	7.0	10.4	9.0	10.6	15.8	9.2	8.3	4.8
Belize	..	..	..	..	..	..	..	..	..	..	..
Bolivia	16.3	20.4	21.5	24.2	21.3	26.9	28.7	30.1	30.8	26.3	24.6
Brazil	16.6	18.5	22.3	20.1	19.4	19.6	22.0	19.5	22.9	20.7	21.4
Chile	17.3	17.1	17.5	17.1	13.5	8.7	6.1	21.8	11.1	17.1	12.6
Colombia	17.5	18.2	17.7	18.4	15.4	17.7	19.3	20.4	18.8	20.7	22.4
Costa Rica	13.2	13.5	14.8	13.8	13.9	15.6	17.9	11.9	13.2	17.7	18.8
Dominican Republic	8.6	7.2	10.5	10.6	9.4	17.6	19.3	15.5	23.9	17.3	16.8
Ecuador	13.5	12.4	11.5	13.6	13.7	16.4	22.7	27.6	20.3	22.4	23.5
El Salvador	11.4	9.4	10.1	13.2	13.7	15.3	15.5	14.2	17.0	20.0	24.3
Grenada	..	..	..	..	..	..	..	..	..	..	..
Guatemala	9.6	13.5	11.8	13.6	13.0	12.5	14.4	15.3	14.3	15.4	18.2
Guyana	23.5	25.4	24.6	22.2	22.4	18.5	10.5	28.3	33.0	12.5	13.5
Haiti	3.1	4.7	5.8	7.5	8.0	9.0	11.2	9.3	6.5	6.9	6.6
Honduras	17.6	17.3	16.5	14.7	16.3	17.0	18.2	14.3	10.0	14.8	18.6
Jamaica	28.5	29.6	28.5	27.4	25.0	19.0	21.9	14.1	15.4	9.4	10.8
Mexico	18.2	17.2	18.1	18.7	17.1	17.4	18.1	18.9	19.0	18.8	20.6
Nicaragua	13.2	14.7	16.6	16.1	15.5	18.6	14.4	18.0	12.5	18.5	21.6
Panama	20.2	23.2	22.5	24.5	26.6	26.9	29.6	24.2	23.5	24.6	20.1
Paraguay	13.1	11.4	11.4	13.5	12.0	15.0	19.7	18.7	18.6	21.2	21.7
Peru	30.3	25.3	25.8	17.4	17.0	14.6	18.3	18.4	13.6	13.7	12.4
St. Vincent and the Grenadines	..	..	..	..	..	..	..	..	..	..	-15.4
Trinidad and Tobago	26.1	29.2	21.4	27.0	28.8	24.7	31.7	46.3	45.1	40.3	39.3
Uruguay	15.0	13.0	12.0	10.1	11.4	12.2	13.8	8.9	9.9	14.1	12.3
Venezuela	32.7	33.8	33.6	37.0	37.7	36.1	39.3	48.1	39.7	36.9	36.2
MIDDLE EAST AND NORTH AFRICA	*13.1*	*13.4*	*14.4*	*14.9*	*14.2*	*15.4*	*16.6*	*23.8*	*21.1*	*23.5*	*22.0*
Algeria	25.1	26.3	27.7	29.3	26.3	29.1	34.0	43.0	36.0	39.0	35.6
Egypt, Arab Republic of	11.3	8.8	9.7	9.4	8.1	6.6	8.0	5.7	12.3	16.7	18.5
Jordan	..	..	..	-21.4	-19.5	-18.8	-20.5	-20.3	-29.2	-23.0	-22.0
Morocco	11.0	10.9	13.1	14.5	15.2	14.6	15.6	20.0	14.3	8.7	13.8
Oman	42.5	68.0	72.0	67.8	62.1	45.5	38.8	56.6	52.5	52.2	44.9
Syrian Arab Republic	8.1	9.9	10.4	10.1	9.9	14.6	8.7	15.2	12.5	16.6	13.0
Tunisia	15.3	20.1	17.3	16.6	19.5	21.6	18.8	26.2	23.2	22.7	20.3
Yemen Arab Republic	..	..	..	-25.9	-7.3	-7.3	-10.6	-7.9	-12.1	-15.4	-31.3
Yemen, PDR	..	..	..	..	..	..	..	..	..	..	..
EUROPE AND MEDITERRANEAN	*..*	*22.8*	*23.6*	*24.4*	*24.1*	*26.4*	*25.1*	*20.4*	*20.7*	*23.4*	*23.5*
Cyprus	..	..	..	..	..	..	..	..	1.3	13.7	14.9
Greece	14.9	14.3	16.9	19.7	19.8	21.2	24.8	19.8	17.0	18.0	18.0
Hungary	..	30.1	32.7	31.3	31.4	33.0	34.2	31.3	30.3	31.8	32.7
Israel	4.7	8.4	9.5	7.9	14.0	18.8	7.8	8.4	4.9	7.0	10.8
Malta	10.4	8.3	5.4	2.7	3.3	2.0	2.5	-1.7	10.0	15.8	11.6
Poland	..	..	..	..	..	..	..	..	..	..	..
Portugal	24.6	18.9	19.3	19.6	19.2	24.2	22.7	13.2	11.9	12.5	14.0
Turkey	16.4	16.2	16.1	17.5	14.3	18.1	16.8	15.2	15.0	18.3	17.1
Yugoslavia	28.6	27.9	27.5	27.1	26.5	25.6	24.8	20.5	25.7	29.3	29.6
Developing countries	18.6	18.7	19.7	21.2	21.2	21.5	23.6	24.4	23.7	24.8	24.8
Highly indebted	20.2	20.0	21.0	20.7	20.1	20.0	21.7	22.8	22.7	23.2	23.9
Low-income countries	16.5	17.0	17.9	21.9	22.3	21.6	23.5	22.3	22.7	22.7	22.9
Low-income Africa	14.2	14.5	16.2	17.1	14.1	14.8	15.8	16.5	10.9	12.4	13.0
China and India	18.5	18.6	19.6	24.8	26.0	24.5	26.1	25.4	27.6	26.9	26.6
Other low-income Asia	7.9	11.5	10.2	8.9	7.7	6.7	10.2	4.7	4.2	6.1	9.8
Middle-income countries	19.8	19.7	20.6	20.7	20.5	21.4	23.5	25.3	24.0	25.5	25.4
High-income oil exporters	51.7	54.8	52.9	49.5	56.3	57.6	60.7	73.7	67.9	64.0	60.0
Industrialized countries	20.7	20.6	20.3	22.6	22.7	23.3	24.8	23.6	21.4	22.0	22.3
Japan	29.8	31.9	33.0	40.8	39.1	38.5	38.7	37.0	33.2	33.1	33.1
United States	18.4	17.3	16.2	15.5	16.4	17.1	19.0	17.8	15.8	16.9	17.6
European Community	21.6	22.2	22.6	26.4	25.6	25.3	25.9	24.5	22.1	22.9	22.9

NOTE: See glossary for definitions.

1978	1979	1980	1981	1982	1983	1984	1985	1986	1987	
22.6	*22.5*	*22.8*	*22.0*	*21.3*	*20.8*	*21.8*	*22.1*	*19.7*	*19.8*	*LATIN AMERICA AND CARIBBEAN*
28.3	23.0	20.0	18.3	19.1	22.1	15.5	15.2	11.1	10.1	Argentina
26.5	21.5	25.8	23.9	27.7	28.9	21.6	24.7	24.3	24.3	Bahamas
15.0	13.9	19.3	12.6	16.4	14.6	16.9	19.3	16.0	15.0	Barbados
..	14.1	11.6	8.3	3.4	3.1	6.4	3.7	..	..	Belize
16.6	16.8	19.3	11.5	14.1	6.3	8.6	0.1	3.0	1.5	Bolivia
21.8	20.7	20.7	20.8	19.3	17.1	21.0	21.7	21.1	22.7	Brazil
14.5	15.0	16.8	12.4	9.4	12.5	12.6	16.5	18.4	21.0	Chile
21.1	19.9	19.7	17.1	16.2	17.1	18.4	20.3	24.7	25.8	Colombia
15.6	15.0	16.2	24.1	27.6	23.4	21.6	19.0	23.3	18.0	Costa Rica
16.2	18.1	13.7	17.9	13.5	16.6	18.3	11.6	..	..	Dominican Republic
22.8	25.9	25.9	24.2	22.9	21.6	23.7	24.1	20.8	16.8	Ecuador
14.6	17.9	14.2	8.1	8.2	6.6	5.2	3.3	10.1	7.8	El Salvador
..	..	..	..	..	..	..	..	..	..	Grenada
15.8	14.2	13.1	10.5	10.2	9.5	9.4	9.9	11.8	7.3	Guatemala
20.5	21.6	19.3	7.6	9.1	2.5	18.2	12.9	20.8	21.7	Guyana
7.9	8.8	8.1	1.7	6.4	5.9	6.8	5.8	5.1	4.8	Haiti
22.6	22.4	17.3	14.4	11.6	11.2	12.3	13.5	14.3	12.7	Honduras
16.4	17.7	13.6	10.0	9.3	10.2	16.4	14.7	20.7	22.9	Jamaica
20.7	22.3	24.9	24.8	27.9	30.3	27.7	26.7	23.2	17.5	Mexico
16.5	8.2	-2.7	3.3	3.8	6.7	4.8	-1.2	-2.1	..	Nicaragua
21.9	19.3	24.4	24.7	23.5	21.8	15.0	15.9	20.8	..	Panama
23.5	23.1	18.3	19.9	18.2	18.2	16.9	16.8	12.5	17.9	Paraguay
20.0	30.3	27.5	25.4	25.7	22.7	26.4	27.0	22.1	22.9	Peru
7.8	-1.7	-2.5	6.1	-2.1	3.9	14.4	22.7	18.5	..	St. Vincent and the Grenadines
34.6	34.8	42.1	37.3	20.9	16.0	19.9	24.5	18.4	..	Trinidad and Tobago
13.5	12.8	11.7	11.4	11.3	11.6	14.0	12.1	12.7	11.3	Uruguay
32.7	34.5	33.3	29.3	25.1	20.5	27.0	26.8	20.3	25.1	Venezuela
22.1	*23.0*	*25.5*	*23.8*	*23.0*	*24.0*	*23.7*	*23.7*	*19.0*	*17.0*	*MIDDLE EAST AND NORTH AFRICA*
37.5	40.8	43.1	40.7	39.2	39.7	38.8	37.0	28.9	28.7	Algeria
16.4	14.2	15.2	14.1	15.2	17.8	14.0	14.5	13.8	8.4	Egypt, Arab Republic of
-18.6	-27.3	-9.0	-15.1	-17.1	-19.2	-16.8	-12.5	-3.0	-3.3	Jordan
11.7	11.6	11.5	7.4	8.4	11.6	10.4	13.2	13.0	14.1	Morocco
38.5	46.4	47.3	49.6	42.0	42.1	42.7	40.2	13.4	..	Oman
11.1	9.3	10.3	5.9	12.6	10.2	12.1	11.2	13.3	9.9	Syrian Arab Republic
21.1	24.4	24.0	23.9	21.1	20.7	20.3	20.4	16.2	19.9	Tunisia
-12.5	-22.7	-19.7	-18.2	-16.7	-19.4	-13.4	-12.9	-8.9	-20.5	Yemen Arab Republic
..	..	..	..	..	..	..	..	..	..	Yemen, PDR
23.6	*24.1*	*24.4*	*23.3*	*25.1*	*24.2*	*25.7*	*26.5*	*28.2*	*..*	*EUROPE AND MEDITERRANEAN*
16.6	19.6	20.0	20.7	18.4	16.5	21.2	20.3	22.5	24.3	Cyprus
20.6	22.4	19.7	18.9	10.8	11.6	12.1	9.7	10.7	8.2	Greece
32.1	30.7	28.5	28.6	29.3	28.4	28.9	27.1	25.5	26.3	Hungary
8.4	10.7	10.0	6.5	9.2	10.4	10.7	9.7	11.4	..	Israel
16.3	20.3	19.2	18.7	15.3	15.0	13.7	12.3	14.6	15.1	Malta
..	..	23.4	16.3	30.0	26.8	28.3	29.1	30.3	..	Poland
18.1	18.7	19.0	15.9	16.8	16.8	15.9	18.3	19.7	18.5	Portugal
15.1	15.4	14.1	16.7	17.3	15.3	15.7	17.9	21.8	23.5	Turkey
29.5	31.4	35.6	36.4	35.5	35.9	38.0	40.6	40.1	39.7	Yugoslavia
24.5	25.7	25.0	23.7	23.4	23.8	24.5	24.8	24.4	25.3	Developing countries
23.1	23.6	24.6	22.7	21.4	20.6	21.3	22.2	21.0	21.2	Highly indebted
24.6	23.7	22.1	21.7	22.2	22.5	23.7	24.6	24.7	25.6	Low-income countries
9.9	9.9	10.3	8.1	7.6	7.3	9.5	8.7	8.6	7.8	Low-income Africa
29.5	28.1	25.7	25.7	26.7	26.6	27.9	29.4	29.8	30.6	China and India
7.9	7.0	7.4	7.2	6.0	6.7	6.7	5.8	7.0	8.8	Other low-income Asia
24.5	26.3	25.9	24.2	23.8	24.2	24.7	24.8	24.3	25.2	Middle-income countries
53.1	51.2	56.4	58.2	49.2	39.0	31.7	28.3	21.9	..	High-income oil exporters
23.2	23.1	22.1	22.0	20.2	19.9	20.8	20.0	20.7	21.1	Industrialized countries
33.3	32.4	32.1	32.6	31.6	31.0	32.0	33.1	33.3	33.7	Japan
18.9	19.3	17.5	18.4	15.2	14.8	16.2	14.2	13.3	13.2	United States
22.8	22.5	21.8	20.3	20.2	20.3	20.8	21.0	21.7	21.5	European Community

Table 15. Private consumption: percentage of GDP

	1967	1968	1969	1970	1971	1972	1973	1974	1975	1976	1977
SUB-SAHARAN AFRICA	*70.1*	*68.9*	*67.2*	*68.6*	*68.9*	*67.5*	*65.5*	*63.0*	*67.5*	*65.5*	*64.9*
Excluding Nigeria	*67.1*	*65.8*	*64.2*	*62.8*	*64.7*	*64.9*	*64.3*	*62.7*	*66.8*	*65.9*	*65.8*
Benin	88.7	84.3	83.0	81.6	82.9	83.6	80.7	83.5	86.6	90.0	88.7
Botswana	82.4	86.3	84.0	78.0	66.9	57.0	59.7	56.1	60.1	56.7	61.6
Burkina Faso	92.4	88.5	89.5	86.8	85.8	82.7	79.2	71.0	75.7	74.9	85.0
Burundi	87.0	87.1	85.9	86.6	87.5	87.2	85.4	89.1	91.6	85.1	78.1
Cameroon	74.3	75.9	75.2	70.5	75.3	75.3	71.8	69.2	72.0	76.4	71.1
Cape Verde	..	..	..	..	..	..	115.5	101.1	96.8	103.0	101.2
Central African Republic	76.7	74.0	77.0	75.3	73.6	74.6	71.5	75.4	88.5	82.2	86.0
Chad	65.5	65.2	62.3	63.6	62.1	62.6	84.0	63.7	67.9	71.8	73.7
Comoros	..	..	..	..	..	..	..	..	..	..	..
Congo, People's Republic	77.7	79.3	80.4	81.8	76.3	77.0	62.3	53.4	70.7	77.5	73.7
Cote d'Ivoire	61.8	58.3	58.6	57.3	58.5	59.0	58.5	54.5	60.4	55.3	52.7
Ethiopia	77.7	77.4	78.6	78.9	80.3	78.5	75.9	76.1	79.4	76.9	80.2
Gabon	52.5	52.8	50.9	48.3	41.0	49.4	42.9	22.4	20.5	19.4	23.5
Gambia, The	78.1	86.3	79.7	78.6	86.6	78.1	84.0	77.5	63.3	86.0	75.8
Ghana	77.2	72.6	74.6	74.4	77.4	74.8	75.0	78.2	73.3	79.2	77.4
Guinea-Bissau	..	..	..	76.7	80.5	85.7	89.3	92.2	82.5	81.5	85.0
Kenya	66.6	64.6	62.8	60.2	64.7	62.2	59.0	64.4	68.2	61.6	55.6
Lesotho	115.5	115.4	112.7	120.4	126.1	133.1	132.1	149.9	151.2	163.9	157.2
Liberia	47.0	44.2	38.3	46.9	49.5	47.5	56.0	57.7	51.5	54.1	60.7
Madagascar	70.6	69.6	71.1	67.0	67.6	70.2	72.4	74.5	76.2	72.8	74.4
Malawi	80.3	81.8	81.2	72.8	78.0	76.4	74.2	69.4	68.9	68.1	66.4
Mali	84.4	80.0	79.4	79.9	80.2	80.4	87.1	103.8	94.1	88.7	81.5
Mauritania	52.7	50.6	55.0	56.2	55.1	53.2	71.5	76.9	67.3	65.2	65.3
Mauritius	74.2	76.9	71.1	75.3	76.6	69.4	66.4	59.4	61.6	62.8	67.2
Mozambique	..	..	..	..	..	..	..	..	..	..	..
Niger	89.2	88.1	92.2	88.8	87.2	85.9	84.7	93.5	86.6	84.9	75.7
Nigeria	85.2	85.3	81.6	81.4	77.5	72.6	68.6	63.5	68.5	64.9	63.7
Rwanda	86.3	89.6	91.3	87.9	87.6	87.7	81.0	86.7	78.2	74.3	71.0
Senegal	74.4	79.7	79.2	74.0	75.4	71.7	75.5	69.7	72.2	75.9	72.6
Seychelles	..	..	..	..	..	..	..	..	..	..	..
Sierra Leone	84.2	83.2	73.7	74.5	76.8	80.7	76.2	76.2	83.2	86.2	84.8
Somalia	87.3	84.6	85.2	83.2	85.0	70.8	79.1	71.1	71.6	67.0	85.7
Sudan	65.9	70.1	65.6	63.8	68.8	74.8	69.6	71.0	80.1	74.4	77.4
Swaziland	68.6	59.7	64.3	54.2	52.2	49.7	43.3	30.2	49.4	44.5	55.7
Tanzania	71.2	72.1	70.5	68.8	66.7	69.2	70.5	75.4	74.1	63.0	62.3
Togo	79.8	80.1	78.7	75.6	72.4	73.2	72.7	57.2	75.3	69.3	68.3
Uganda	75.1	74.9	74.2	..	..	..	..	..	..	..	..
Zaire	59.0	53.9	53.1	46.8	51.7	59.0	57.3	54.9	66.8	76.8	75.7
Zambia	43.6	44.4	36.3	39.4	41.4	39.7	33.3	34.9	51.4	44.8	51.5
Zimbabwe	68.5	67.6	66.6	66.9	67.8	59.8	61.0	61.1	62.1	63.5	61.1
SOUTH ASIA	*81.3*	*78.9*	*78.1*	*76.1*	*77.4*	*76.4*	*74.8*	*77.5*	*76.2*	*72.1*	*72.7*
Bangladesh	81.4	77.1	77.8	79.5	96.2	103.4	86.0	94.6	95.8	98.6	87.8
India	81.3	78.8	77.4	74.1	74.0	73.4	72.8	74.8	71.4	68.0	70.3
Nepal	95.1	95.0	98.4	97.4	96.0	94.9	94.6	96.3	82.4	80.8	79.2
Pakistan	78.6	77.6	78.0	80.9	80.9	78.8	78.3	83.8	84.4	81.3	79.3
Sri Lanka	77.1	76.8	79.2	72.3	72.4	71.8	76.5	80.2	82.6	76.1	73.3
EAST ASIA AND PACIFIC	*67.7*	*67.4*	*65.7*	*60.2*	*58.7*	*59.3*	*58.1*	*59.6*	*58.9*	*58.7*	*58.0*
China	64.4	63.9	62.3	55.3	53.6	55.4	54.5	55.8	53.7	55.1	54.6
Fiji	70.5	67.6	69.1	65.0	69.5	69.1	77.2	74.3	68.0	69.6	64.1
Hong Kong	70.1	73.2	70.4	67.5	67.7	64.6	66.9	66.7	67.1	59.6	62.9
Indonesia	91.0	86.7	84.6	78.1	74.7	69.1	67.8	65.0	65.1	64.3	61.9
Korea, Republic of	79.8	75.5	71.8	74.8	75.7	73.8	69.7	70.4	70.9	66.0	62.9
Malaysia	61.8	61.1	58.8	57.7	59.2	58.5	53.2	53.4	56.5	50.9	50.2
Papua New Guinea	65.6	66.2	66.0	63.6	62.7	58.2	47.8	47.3	53.7	52.2	53.4
Philippines	75.1	74.3	73.3	69.6	71.0	70.8	66.8	67.5	66.4	64.4	66.6
Singapore	76.1	73.7	68.5	67.5	66.2	61.9	61.8	60.7	60.4	58.7	57.8
Solomon Islands	..	..	..	..	..	..	..	..	..	..	..
Thailand	71.6	71.4	69.4	70.0	68.3	69.1	67.1	68.1	69.7	68.6	67.4
Vanuatu	..	..	..	..	..	..	..	..	..	..	..

NOTE: See glossary for definitions.

1978	1979	1980	1981	1982	1983	1984	1985	1986	1987	
68.7	*67.9*	*65.7*	*71.8*	*74.7*	*74.9*	*74.0*	*74.4*	*74.1*	*70.7*	*SUB-SAHARAN AFRICA*
70.0	*70.7*	*69.5*	*71.4*	*71.2*	*71.2*	*68.5*	*69.7*	*70.9*	*71.1*	*Excluding Nigeria*
89.8	88.3	95.7	103.6	92.2	89.0	96.2	92.2	90.3	86.1	Benin
62.3	57.0	50.5	51.9	61.6	54.4	48.3	44.9	37.2	..	Botswana
84.5	85.9	92.0	92.9	88.4	85.2	77.6	91.7	81.5	86.9	Burkina Faso
80.7	82.8	86.5	78.4	85.9	78.6	82.7	82.3	78.9	75.5	Burundi
72.0	76.3	75.5	68.9	68.2	62.3	57.9	61.2	69.5	74.3	Cameroon
110.8	98.9	110.4	99.5	95.0	96.5	88.3	91.8	..	..	Cape Verde
86.4	86.2	94.5	86.9	90.9	84.3	86.0	86.5	85.8	89.0	Central African Republic
72.8	68.5	72.5	72.7	78.2	74.5	80.7	77.9	..	104.1	Chad
..	..	..	..	..	68.7	82.5	79.4	75.9	84.2	Comoros
59.5	55.1	41.0	40.7	40.0	39.8	38.8	41.8	59.4	58.3	Congo, People's Republic
54.9	56.9	60.0	63.8	62.5	64.2	62.3	60.3	61.8	64.6	Cote d'Ivoire
81.1	82.0	79.9	79.0	80.5	79.8	79.1	77.7	80.0	77.5	Ethiopia
36.5	33.9	26.2	28.8	32.6	27.9	28.1	30.4	46.9	42.6	Gabon
91.3	70.2	73.3	83.2	72.9	70.4	86.7	75.9	71.1	73.8	Gambia, The
84.7	83.1	83.9	87.2	89.8	90.8	86.1	83.0	77.9	81.6	Ghana
84.7	88.5	77.0	70.3	75.7	78.1	83.6	86.8	86.6	87 2	Guinea-Bissau
60.5	63.5	61.0	61.5	63.1	60.5	60.3	63.1	59.2	61.1	Kenya
129.7	144.8	133.0	144.2	162.5	175.8	173.7	159.1	156.3	157.5	Lesotho
60.7	58.4	56.4	61.9	65.9	70.9	68.6	64.7	64.5	..	Liberia
74.4	73.7	76.4	77.2	81.2	80.1	77.3	77.8	76.4	76.1	Madagascar
62.7	69.5	70.4	70.1	67.4	68.3	69.5	68.5	71.5	70.4	Malawi
92.2	92.1	91.5	90.4	89.2	90.3	88.4	102.4	91.4	89.9	Mali
65.3	70.9	67.8	65.7	79.1	93.7	84.7	76.4	71.0	73.1	Mauritania
67.9	67.3	75.5	71.3	70.8	69.5	68.5	66.9	61.3	60.9	Mauritius
..	..	81.8	79.6	82.9	87.5	85.5	86.7	84.9	89.6	Mozambique
70.7	71.1	67.1	80.4	80.5	81.1	91.4	86.8	82.9	83.8	Niger
66.8	64.5	61.5	72.1	78.5	78.9	79.6	79.6	81.0	68.9	Nigeria
76.6	76.6	83.3	78.6	83.2	85.3	82.7	81.6	81.6	82.7	Rwanda
76.9	76.7	78.1	82.1	79.5	78.1	74.9	80.3	76.8	76.6	Senegal
..	..	..	52.7	62.5	68.2	61.2	58.2	56.4	..	Seychelles
84.6	85.5	90.7	90.7	88.2	87.8	82.1	83.9	85.7	83.3	Sierra Leone
60.7	89.9	93.8	102.0	97.7	104.8	95.9	91.9	86.4	88.6	Somalia
82.2	80.9	80.6	83.0	82.7	82.0	78.7	75.5	81.0	79.0	Sudan
56.1	73.3	68.0	71.9	72.2	77.7	76.9	79.5	62.3	..	Swaziland
72.6	70.3	77.1	75.4	75.4	79.0	79.7	84.7	90.3	98.1	Tanzania
65.3	66.2	59.6	70.4	72.3	71.6	72.7	74.4	67.9	73.8	Togo
..	..	..	..	..	85.6	76.7	90.2	85.2	89.6	Uganda
78.4	77.0	77.4	81.2	75.2	79.5	67.9	67.1	73.3	72.8	Zaire
55.6	53.1	55.2	64.9	64.3	63.3	56.4	58.8	66.1	55.1	Zambia
65.4	68.5	64.5	67.0	65.9	68.7	61.2	57.3	54.2	58.5	Zimbabwe
73.7	*73.2*	*75.2*	*74.1*	*73.9*	*73.0*	*72.9*	*70.4*	*70.5*	*68.3*	*SOUTH ASIA*
92.9	92.1	91.5	90.1	93.6	93.9	92.1	90.6	89.8	89.8	Bangladesh
70.9	69.9	73.0	71.2	70.6	70.3	69.8	66.6	67.1	65.0	India
79.7	79.9	82.2	82.1	82.3	80.8	80.8	78.6	78.1	77.9	Nepal
80.6	83.8	83.1	83.2	83.5	82.0	82.3	83.1	80.9	76.9	Pakistan
75.2	77.1	80.3	80.9	79.8	78.1	72.3	77.9	77.7	77.2	Sri Lanka
56.2	*55.5*	*56.8*	*58.0*	*58.1*	*57.6*	*56.4*	*56.2*	*55.0*	*53.1*	*EAST ASIA AND PACIFIC*
50.9	51.4	55.0	56.4	54.4	54.6	53.3	51.8	50.4	49.1	China
65.3	61.0	59.6	62.5	61.5	65.5	62.3	64.2	60.0	64.5	Fiji
66.7	62.5	62.2	62.1	63.8	66.9	63.8	65.0	64.4	62.2	Hong Kong
63.0	56.8	52.3	55.3	60.5	60.7	59.0	60.2	63.2	60.6	Indonesia
61.5	62.6	65.4	64.9	64.3	61.1	59.2	58.4	54.8	52.2	Korea, Republic of
51.7	48.3	50.5	53.1	53.1	51.7	49.8	52.0	51.4	47.5	Malaysia
55.9	54.6	61.5	65.7	63.9	63.1	63.0	64.8	64.3	61.6	Papua New Guinea
66.9	67.4	67.3	67.8	68.8	69.8	74.6	76.6	75.1	73.3	Philippines
56.9	54.8	51.5	48.8	46.8	44.1	43.9	45.1	47.4	47.2	Singapore
..	..	70.1	74.7	61.2	60.8	59.6	65.0	70.1	..	Solomon Islands
64.5	65.1	66.6	65.7	65.9	65.8	65.3	65.3	64.1	62.7	Thailand
..	..	..	..	..	60.0	55.7	56.9	..	..	Vanuatu

Table 15. Private consumption: percentage of gross domestic product (cont'd)

	1967	1968	1969	1970	1971	1972	1973	1974	1975	1976	1977
LATIN AMERICA AND CARIBBEAN	*67.7*	*67.5*	*66.0*	*68.8*	*69.4*	*69.5*	*67.9*	*67.0*	*66.2*	*66.7*	*66.6*
Argentina	68.3	69.6	69.1	68.2	68.0	68.6	67.3	66.0	60.7	58.8	60.3
Bahamas	..	..	..	..	..	..	..	..	..	..	55.7
Barbados	62.4	63.6	66.8	67.9	64.2	65.9	62.8	59.8	74.2	74.4	77.9
Belize	..	..	..	..	..	..	..	..	..	..	..
Bolivia	77.4	72.8	72.8	70.5	72.4	69.8	67.5	65.0	67.7	67.3	67.1
Brazil	62.0	60.8	56.3	68.6	69.6	69.7	67.5	70.7	66.5	68.8	69.2
Chile	71.6	71.9	71.4	70.1	71.3	75.2	80.8	62.4	73.2	68.9	72.8
Colombia	74.2	73.6	73.7	72.4	73.7	72.8	71.3	70.9	72.3	71.1	69.9
Costa Rica	74.2	74.1	72.7	73.7	72.2	70.0	68.2	73.8	71.6	66.3	65.2
Dominican Republic	80.3	81.2	78.8	79.2	81.7	75.4	74.2	75.5	71.7	80.6	80.9
Ecuador	78.4	78.3	78.2	75.3	76.0	73.5	67.0	59.8	65.2	63.6	61.7
El Salvador	79.5	81.2	79.4	76.1	76.1	74.1	74.1	74.9	71.8	68.0	64.5
Grenada	..	..	..	..	..	..	..	..	..	..	..
Guatemala	82.6	79.4	80.4	78.4	80.0	80.0	79.2	78.1	78.9	77.8	75.3
Guyana	60.9	59.4	59.5	60.8	59.6	62.0	64.7	54.7	47.4	59.3	60.7
Haiti	87.4	86.0	84.9	83.0	81.8	81.8	80.3	83.3	84.5	84.7	85.1
Honduras	72.3	72.8	72.4	73.8	71.7	71.0	71.7	74.0	77.7	72.3	68.9
Jamaica	62.7	60.9	61.2	60.9	62.6	67.3	61.8	68.0	66.2	69.8	68.5
Mexico	74.0	74.8	73.6	73.2	74.4	72.9	71.6	71.3	70.3	69.6	68.2
Nicaragua	77.5	76.3	74.7	74.3	74.8	72.1	77.5	74.3	78.5	72.2	69.0
Panama	67.0	64.1	65.0	60.6	57.7	55.2	53.1	57.7	57.3	55.7	60.0
Paraguay	79.1	80.2	79.6	77.5	79.5	77.0	73.8	75.8	75.1	72.5	72.1
Peru	63.9	65.6	65.4	70.4	70.6	72.3	68.8	70.1	74.1	74.1	73.5
St. Vincent and the Grenadines	..	..	..	..	..	..	..	..	..	..	90.9
Trinidad and Tobago	61.7	59.3	66.6	60.0	55.3	59.3	54.0	42.3	42.6	47.5	47.9
Uruguay	70.9	73.9	73.1	74.5	72.3	75.5	72.0	76.2	76.5	72.1	75.4
Venezuela	56.5	55.5	55.2	51.9	50.6	52.2	49.6	42.0	48.9	50.8	51.5
MIDDLE EAST AND NORTH AFRICA	*66.9*	*65.8*	*64.7*	*63.6*	*63.6*	*62.8*	*61.5*	*58.2*	*59.5*	*59.3*	*61.0*
Algeria	58.9	58.4	57.2	55.8	57.3	55.3	51.4	46.3	51.1	48.5	51.1
Egypt, Arab Republic of	68.1	68.2	66.4	65.5	65.9	66.6	63.8	73.5	62.9	58.5	62.0
Jordan	..	..	..	87.5	86.7	85.5	83.9	80.8	94.6	86.0	91.5
Morocco	76.8	76.6	75.0	73.4	72.5	73.0	72.7	68.0	69.4	68.9	65.6
Oman	53.1	27.2	22.9	19.4	17.1	24.9	24.0	8.7	15.9	20.5	26.0
Syrian Arab Republic	76.8	72.2	73.3	72.5	72.4	67.8	70.0	67.1	66.4	63.3	67.4
Tunisia	67.2	62.6	66.2	66.5	64.8	63.9	66.0	60.6	62.2	62.1	63.5
Yemen Arab Republic	..	..	..	116.5	96.7	94.4	96.6	95.1	97.8	101.0	117.6
Yemen, PDR	..	..	..	..	..	..	..	..	..	..	..
EUROPE AND MEDITERRANEAN	*..*	*..*	*..*	*63.3*	*63.7*	*62.5*	*62.7*	*66.3*	*65.4*	*63.0*	*63.5*
Cyprus	..	..	..	..	..	..	..	..	81.2	75.7	75.7
Greece	72.4	71.9	69.2	69.2	68.0	65.7	63.4	67.7	67.5	65.8	65.9
Hungary	..	..	..	58.4	58.1	57.2	56.3	58.2	59.2	58.1	57.4
Israel	66.0	63.2	61.3	57.8	54.0	52.6	53.1	54.9	54.9	56.4	56.6
Malta	72.1	74.6	77.9	78.0	76.8	78.7	77.8	81.3	71.6	66.6	71.9
Poland	..	..	..	..	..	..	..	..	..	..	..
Portugal	65.4	68.5	69.1	65.9	68.3	64.2	64.8	72.6	77.1	75.0	72.0
Turkey	71.2	71.2	71.5	69.7	72.3	69.9	70.7	73.3	72.7	69.0	69.4
Yugoslavia	54.4	54.2	54.9	55.2	56.9	57.5	59.2	62.1	56.1	53.2	52.9
Developing countries	69.0	68.4	67.0	66.0	66.0	65.7	63.8	63.6	63.5	62.4	62.6
Highly indebted	67.7	67.5	66.1	68.7	69.1	69.0	67.3	66.3	65.6	65.2	64.9
Low-income countries	71.4	70.3	68.9	64.4	64.2	64.4	63.0	65.4	64.3	63.7	63.7
Low-income Africa	67.1	65.7	63.9	62.4	64.6	65.4	65.0	64.8	69.8	69.4	69.2
China and India	70.8	69.9	68.2	62.3	61.2	61.9	61.1	63.1	60.0	60.0	60.7
Other low-income Asia	81.2	79.1	80.0	81.2	86.2	86.1	83.0	86.9	88.5	84.9	80.7
Middle-income countries	67.5	67.3	65.9	67.0	67.0	66.3	64.2	62.8	63.2	62.1	62.3
High-income oil exporters	31.2	28.9	29.5	33.1	27.7	26.4	24.0	13.5	18.0	18.8	20.9
Industrialized countries	60.8	60.5	60.1	60.1	60.1	59.9	59.0	59.4	60.8	60.6	60.7
Japan	55.9	53.9	52.8	51.5	52.9	53.4	53.0	53.7	56.6	56.9	57.0
United States	62.1	62.4	62.5	63.5	63.3	63.1	62.3	63.1	64.1	64.1	64.0
European Community	60.3	59.8	59.1	58.5	58.7	58.8	58.2	58.7	60.0	59.4	59.5

NOTE: See glossary for definitions.

1978	1979	1980	1981	1982	1983	1984	1985	1986	1987	
66.7	*66.8*	*66.6*	*67.0*	*67.3*	*68.3*	*68.1*	*67.5*	*69.4*	*69.4*	*LATIN AMERICA AND CARIBBEAN*
60.6	65.7	66.2	68.5	68.7	64.7	73.5	73.8	77.1	84.3	Argentina
58.1	65.9	61.7	62.3	59.1	57.4	65.4	62.9	63.5	63.7	Bahamas
68.0	70.5	63.3	70.6	67.1	69.0	66.2	61.9	66.5	66.2	Barbados
..	68.5	69.2	71.6	71.1	71.1	68.5	70.9	..	..	Belize
70.6	68.0	66.7	76.7	70.3	76.9	70.9	84.4	85.8	84.3	Bolivia
68.6	69.5	70.3	69.9	70.3	73.4	70.8	68.7	68.7	65.1	Brazil
71.1	70.7	70.7	74.5	75.3	73.3	73.0	69.3	68.9	67.6	Chile
70.3	70.8	70.2	72.5	72.9	71.9	70.6	69.0	65.4	64.6	Colombia
67.6	66.9	65.5	60.1	57.8	61.5	63.8	66.3	62.4	67.3	Costa Rica
79.8	75.8	78.2	72.4	77.0	73.9	73.0	80.1	..	..	Dominican Republic
63.4	61.2	59.6	61.6	63.1	65.9	64.1	64.5	67.0	71.0	Ecuador
72.5	68.9	71.8	76.0	76.1	77.5	78.8	81.2	77.9	80.8	El Salvador
..	..	..	..	..	..	..	..	..	..	Grenada
77.0	78.7	78.9	81.6	82.0	82.9	83.0	83.1	81.1	84.9	Guatemala
56.2	52.2	51.8	63.4	63.8	66.3	63.7	69.0	60.7	62.6	Guyana
82.7	81.7	81.9	87.1	82.2	83.4	82.4	82.3	84.5	85.4	Haiti
65.7	65.9	69.4	72.0	74.5	74.3	73.0	71.5	70.6	71.3	Honduras
63.6	63.1	66.0	69.3	68.8	69.7	67.1	69.6	63.9	62.5	Jamaica
68.0	66.6	65.1	64.4	61.6	60.9	63.1	64.0	67.7	72.7	Mexico
71.0	74.0	84.0	74.4	70.9	62.7	59.8	56.3	62.3	..	Nicaragua
58.4	60.5	56.5	54.3	54.0	56.7	63.0	62.8	57.3	..	Panama
69.8	71.2	75.7	73.4	74.9	74.8	77.2	81.5	81.0	75.7	Paraguay
68.8	61.0	61.3	63.2	61.7	64.8	62.5	61.8	66.8	66.6	Peru
67.5	77.5	78.3	70.6	78.6	73.0	63.7	56.9	64.8	..	St. Vincent and the Grenadines
52.0	51.3	45.9	49.9	62.8	66.7	61.2	57.2	62.4	..	Trinidad and Tobago
74.1	75.4	75.8	74.4	73.1	74.5	73.5	74.8	74.1	75.9	Uruguay
55.6	54.3	54.9	57.9	62.4	67.6	62.4	62.8	68.6	64.9	Venezuela
60.8	*60.4*	*58.7*	*59.6*	*59.8*	*59.2*	*59.1*	*59.3*	*63.4*	*66.1*	*MIDDLE EAST AND NORTH AFRICA*
48.5	45.6	43.2	45.5	46.0	45.4	45.5	47.1	54.7	54.8	Algeria
63.1	68.7	69.2	66.9	67.0	65.0	67.9	68.2	69.6	77.2	Egypt, Arab Republic of
88.6	96.0	84.3	90.5	92.4	94.8	91.7	87.3	75.5	76.0	Jordan
67.5	67.0	68.1	70.8	70.3	68.6	71.4	69.4	69.6	68.4	Morocco
32.8	26.1	27.7	25.8	31.9	31.4	30.8	33.9	38.5	..	Oman
69.0	69.1	66.5	73.4	65.4	67.8	65.2	65.1	65.4	71.9	Syrian Arab Republic
62.5	60.4	61.5	61.3	62.4	62.5	63.2	63.0	66.5	63.8	Tunisia
97.0	104.6	100.4	93.3	93.5	96.4	93.8	95.0	91.0	100.7	Yemen Arab Republic
..	..	..	..	..	..	..	..	..	..	Yemen, PDR
63.2	*63.5*	*62.1*	*62.3*	*59.3*	*61.3*	*61.3*	*60.6*	*62.0*	*..*	*EUROPE AND MEDITERRANEAN*
74.4	68.2	66.9	65.5	66.2	67.4	63.7	63.9	61.9	60.8	Cyprus
65.2	63.3	63.9	67.5	67.4	66.7	64.7	65.5	66.5	67.1	Greece
57.4	58.8	61.2	61.3	60.8	61.5	61.4	62.8	63.9	63.4	Hungary
57.4	58.6	51.9	53.7	54.5	56.1	52.3	55.0	57.9	..	Israel
67.1	63.2	64.7	64.0	66.2	67.0	68.9	70.0	68.0	66.9	Malta
..	..	66.9	74.1	62.7	64.8	63.3	61.7	..	..	Poland
68.0	67.5	66.6	69.3	68.7	68.6	69.7	67.5	66.6	62.9	Portugal
71.4	71.0	73.4	72.4	71.8	74.5	75.4	73.7	69.5	67.4	Turkey
53.0	51.5	48.8	48.6	49.4	49.8	48.3	46.1	46.3	46.7	Yugoslavia
62.9	62.0	62.8	64.2	63.8	63.9	63.4	63.2	63.4	62.6	Developing countries
65.3	65.0	64.5	66.1	67.0	68.2	68.4	67.6	68.3	67.6	Highly indebted
62.6	62.4	65.3	66.3	65.4	65.0	63.9	62.9	62.9	61.2	Low-income countries
73.5	74.0	73.7	76.3	75.6	76.0	73.8	75.0	74.8	74.7	Low-income Africa
58.2	57.9	61.7	62.3	61.0	61.1	59.9	58.2	58.1	56.4	China and India
82.9	83.7	83.4	83.4	84.5	83.1	82.8	83.4	81.9	79.9	Other low-income Asia
63.0	61.9	61.9	63.6	63.4	63.4	63.3	63.3	63.6	63.2	Middle-income countries
26.0	25.4	24.6	23.9	27.7	35.4	39.1	40.1	42.0	..	High-income oil exporters
60.2	60.5	60.8	60.8	62.0	62.5	61.8	62.2	61.8	61.6	Industrialized countries
57.1	58.0	58.1	57.3	58.5	59.1	58.3	57.6	57.2	57.0	Japan
63.2	63.6	64.5	63.8	65.8	66.6	65.3	66.2	66.8	67.0	United States
59.4	59.6	59.7	60.5	60.5	60.5	60.3	60.3	59.8	59.9	European Community

Table 16. Balance of payments: total to be financed

Percentage of GDP	1967	1968	1969	1970	1971	1972	1973	1974	1975	1976	1977
SUB-SAHARAN AFRICA	..	..	..	*-3.5*	*-5.7*	*-4.5*	*-3.2*	*2.1*	*-5.8*	*-4.3*	*-5.0*
Excluding Nigeria	..	..	..	*-3.1*	*-6.7*	*-5.2*	*-4.1*	*-5.5*	*-9.5*	*-6.5*	*-7.0*
Benin	..	..	..	..	..	..	..	..	..	..	..
Botswana	..	..	..	-42.9	-41.3	-43.8	-31.3	-2.1	-21.2	-9.2	-14.2
Burkina Faso	..	..	..	-6.9	-10.8	-11.8	-13.6	-19.5	-24.8	-18.2	-24.5
Burundi	..	..	..	..	..	..	..	..	..	..	..
Cameroon	..	..	..	-4.1	-5.0	-7.6	-2.0	-1.7	-7.2	-4.8	-3.7
Cape Verde	..	..	..	..	..	..	..	..	..	..	..
Central African Republic	..	..	..	-13.4	-12.1	-12.2	-11.6	-18.0	-22.7	-9.7	-11.3
Chad	..	..	..	-10.0	-9.6	-9.0	-13.4	-13.7	-22.0	-14.9	-19.9
Comoros	..	..	..	..	..	..	..	..	..	..	..
Congo, People's Republic	..	..	..	-17.8	-24.5	-28.1	-16.6	-19.8	-37.2	-35.6	-28.8
Cote d'Ivoire	..	..	..	-5.1	-9.2	-8.5	-10.9	-3.5	-10.8	-6.0	-3.5
Ethiopia	..	..	..	-2.4	-3.0	-0.3	2.6	0.8	-3.0	-0.3	-2.7
Gabon	..	..	..	-4.4	2.3	-6.1	-9.1	12.7	1.4	0.3	2.5
Gambia, The	..	..	..	0.1	-4.7	-2.2	-7.0	-0.9	6.1	-15.1	-7.2
Ghana	..	..	..	..	..	..	..	..	..	..	..
Guinea-Bissau	..	..	..	..	..	..	..	..	..	..	..
Kenya	..	..	..	-5.4	-10.3	-5.1	-6.6	-11.9	-8.6	-4.6	-0.8
Lesotho	..	..	..	-42.2	-51.0	-59.8	-65.6	-81.1	-14.1	-27.4	-20.3
Liberia	..	..	..	-3.4	-1.7	-1.1	1.3	-5.4	6.3	4.3	-5.4
Madagascar	..	..	..	-4.7	-8.2	-4.7	-5.0	-5.9	-7.6	-5.6	-4.6
Malawi	..	..	..	..	..	..	..	..	..	..	..
Mali	..	..	..	..	..	..	..	..	..	..	..
Mauritania	..	..	..	..	..	..	..	..	..	..	..
Mauritius	..	..	..	..	..	..	..	..	..	..	..
Mozambique	..	..	..	..	..	..	..	..	..	..	..
Niger	..	..	..	-5.0	-2.6	-4.4	-6.3	-15.2	-11.6	-12.4	-13.2
Nigeria	..	..	..	-3.0	-2.7	-2.0	0.0	14.2	0.1	-0.7	-1.7
Rwanda	..	..	..	..	..	..	..	..	..	..	..
Senegal	..	..	..	..	..	..	..	..	..	..	..
Seychelles	..	..	..	1.1	-18.2	-23.8	-23.6	-14.1	-17.8	-20.8	-18.9
Sierra Leone	..	..	..	-4.7	-5.7	-2.9	-7.4	-14.5	-11.4	-10.6	-10.0
Somalia	..	..	..	-5.7	-4.9	-5.4	-12.7	-21.4	-14.1	-13.5	-27.8
Sudan	..	..	..	..	..	..	..	..	..	-8.1	-4.5
Swaziland	..	..	..	..	..	..	..	10.4	11.4	7.9	-0.7
Tanzania	..	..	..	..	..	..	..	..	..	..	..
Togo	..	..	..	-5.4	-8.6	-13.1	-10.1	16.5	-19.1	-9.7	-16.7
Uganda	..	..	..	1.0	-4.2	0.6	1.8	-1.1	-2.9	1.3	2.1
Zaire	..	..	..	..	..	..	..	..	..	..	..
Zambia	..	..	..	..	..	..	..	..	..	..	..
Zimbabwe	..	..	..	-0.8	-3.1	0.2	-0.4	-2.4	-4.2	1.6	-0.4
SOUTH ASIA	..	..	..	*-2.3*	*-2.2*	*-1.6*	*-1.2*	*-2.0*	*-2.1*	*-0.5*	*0.1*
Bangladesh	..	..	..	-3.5	-4.8	-9.5	-6.4	-6.1	-7.1	-12.2	-6.4
India	..	..	..	-1.1	-1.3	-0.7	-0.7	-1.1	-0.1	1.4	1.5
Nepal	..	..	..	-2.9	-3.2	-1.2	-2.7	-4.7	-3.4	-1.1	0.0
Pakistan	..	..	..	-7.0	-5.6	-3.2	-1.3	-5.6	-9.4	-6.2	-6.0
Sri Lanka	..	..	..	-3.6	-2.7	-2.5	-1.7	-6.4	-6.3	-2.3	3.0
EAST ASIA AND PACIFIC	..	..	..	*-1.3*	*-1.5*	*-0.4*	*0.6*	*-1.6*	*-1.8*	*0.0*	*0.5*
China	..	..	..	-0.1	-0.1	-0.1	0.4	-0.5	-0.2	0.2	0.3
Fiji	..	..	..	-8.2	-12.7	-12.5	-15.2	-5.9	-4.1	-7.5	-4.0
Hong Kong	..	..	..	6.2	2.4	6.6	5.1	3.7	6.0	10.2	5.7
Indonesia	..	..	..	-3.9	-4.2	-3.3	-3.1	2.0	-3.5	-2.4	-0.1
Korea, Republic of	..	..	..	-7.9	-9.3	-3.9	-2.5	-11.0	-9.1	-1.6	-0.1
Malaysia	..	..	..	0.0	-2.8	-4.8	1.1	-5.6	-5.3	5.0	3.1
Papua New Guinea	..	..	..	-37.0	-41.3	-17.9	2.4	-0.1	-10.2	-6.9	-7.9
Philippines	..	..	..	-1.9	-1.3	-1.2	3.2	-2.5	-6.8	-6.7	-4.2
Singapore	..	..	..	-30.8	-32.8	-17.1	-12.7	-19.8	-10.3	-9.5	-4.4
Solomon Islands	..	..	..	..	..	..	..	..	-39.9	-17.0	-8.7
Thailand	..	..	..	-4.2	-2.9	-1.0	-0.7	-0.8	-4.4	-2.7	-5.6
Vanuatu	..	..	..	..	..	..	..	..	..	..	..

NOTE: See glossary for definitions

1978	1979	1980	1981	1982	1983	1984	1985	1986	1987	
-7.7	*-3.5*	*-3.0*	*-9.6*	*-10.0*	*-7.4*	*-3.3*	*-2.9*	*-6.1*	*-7.3*	*SUB-SAHARAN AFRICA*
-8.3	*-7.6*	*-10.5*	*-11.8*	*-11.4*	*-9.0*	*-6.0*	*-6.2*	*-8.3*	*-8.0*	*Excluding Nigeria*
..	..	-17.2	-26.4	-37.9	-23.0	-12.6	-14.6	-13.2	..	Benin
-29.0	-14.0	-22.7	-36.5	-24.5	-16.0	-12.6	1.2	2.2	30.2	Botswana
-22.1	-21.0	-20.3	-19.0	-25.7	-22.6	-18.2	-6.2	-8.5	..	Burkina Faso
..	..	-15.2	-15.2	-21.0	-18.6	-13.7	-9.3	..	..	Burundi
-4.5	-2.7	-5.3	-5.3	-4.5	-0.7	2.5	3.9	-5.3	-8.8	Cameroon
..	..	..	..	..	..	..	..	..	..	Cape Verde
-12.6	-14.5	-17.6	-11.2	-15.1	-16.0	-15.3	-18.1	-18.0	-19.9	Central African Republic
-20.8	-6.2	-2.2	-3.3	-7.6	-12.9	-15.3	-33.1	-30.4	-33.2	Chad
..	..	..	..	-26.3	-27.9	-59.7	-40.4	-29.1	-30.9	Comoros
-26.4	-11.9	-13.5	-25.3	-16.7	-20.9	7.7	-9.3	-35.6	..	Congo, People's Republic
-11.1	-15.2	-17.5	-16.9	-13.8	-14.1	-1.3	0.6	-2.2	-2.5	Cote d'Ivoire
-5.1	-4.8	-4.5	-4.6	-7.7	-6.8	-8.0	-6.3	-6.7	-8.8	Ethiopia
1.6	7.0	8.2	9.7	7.8	1.9	2.5	-4.8	-31.8	-6.6	Gabon
-32.6	-25.6	-46.1	-51.9	-29.5	-28.9	1.3	4.6	3.0	1.1	Gambia, The
..	..	..	..	..	-6.1	-4.1	-5.4	..	-8.4	Ghana
..	..	..	..	-48.0	-44.8	-49.2	-49.3	-44.8	-49.0	Guinea-Bissau
-13.9	-9.5	-14.2	-10.1	-5.6	-2.9	-4.0	-3.5	-2.7	-8.0	Kenya
-16.4	-27.2	-30.6	-34.3	-23.8	-24.9	-27.5	-28.3	-28.7	-4.3	Lesotho
-3.9	-1.7	0.9	0.6	-8.0	-20.3	-9.6	-3.2	-1.0	..	Liberia
-7.2	-17.4	-19.5	-16.9	-13.6	-10.7	-11.1	-12.0	..	..	Madagascar
..	..	-25.2	-16.2	-13.5	-14.3	..	..	..	..	Malawi
..	..	..	..	-16.6	-18.8	-19.3	..	..	..	Mali
..	..	..	..	..	..	..	..	-18.5	-15.6	Mauritania
..	..	..	..	..	..	..	..	5.1	2.6	Mauritius
..	..	-17.5	-20.1	-23.5	-22.2	-18.5	-12.9	-14.9	-42.5	Mozambique
-18.0	-16.9	-16.9	-17.6	-21.9	-11.2	-10.8	-15.2	-8.3	..	Niger
-6.0	2.2	5.1	-6.5	-7.8	-4.8	0.2	1.4	0.8	..	Nigeria
..	..	..	-13.1	-13.6	-10.7	-8.9	-10.3	-9.6	..	Rwanda
..	..	..	-25.9	-17.4	-18.1	-17.7	..	..	..	Senegal
-15.4	-17.8	-20.3	-20.6	-34.0	-27.4	-18.7	-19.5	-22.5	..	Seychelles
-16.0	-20.7	-19.0	-13.9	-15.4	-3.4	-4.4	-2.6	-0.8	..	Sierra Leone
-16.4	-44.7	-43.3	-27.7	-41.3	-38.3	-40.0	-32.4	-37.8	-5.8	Somalia
-6.0	-4.8	-9.6	-12.2	-17.1	-12.1	-11.1	-9.0	-11.4	-8.1	Sudan
-25.8	-37.9	-38.9	-28.1	-34.2	-32.4	-31.6	-28.0	-13.1	-0.4	Swaziland
..	..	-11.8	-8.5	-10.2	-6.4	-7.0	-6.8	-9.7	-17.3	Tanzania
-30.3	-33.4	-15.9	-13.4	-18.9	-15.1	-7.3	-11.5	-18.0	-11.9	Togo
-5.5	0.5	-7.0	-8.6	-9.4	-4.7	-1.5	-3.5	-0.7	-5.2	Uganda
..	..	-5.4	-9.4	-8.3	-6.8	-10.8	-9.4	-10.4	..	Zaire
..	..	..	..	..	..	..	..	..	..	Zambia
1.1	-2.6	-5.6	-11.3	-11.0	-8.1	-3.6	-3.3	-1.0	-0.3	Zimbabwe
-1.3	*-2.0*	*-2.9*	*-2.8*	*-2.9*	*-2.0*	*-2.1*	*-3.2*	*-2.3*	*-2.1*	*SOUTH ASIA*
-8.0	-7.3	-11.2	-10.0	-12.0	-9.1	-6.8	-8.2	-7.0	-5.5	Bangladesh
-0.3	-0.7	-1.6	-1.8	-1.5	-1.3	-1.5	-2.5	-1.7	-1.6	India
-2.9	-2.1	-4.7	-4.2	-5.0	-8.9	-6.9	-7.2	-7.2	-7.1	Nepal
-2.7	-4.2	-3.7	-2.7	-3.6	-0.7	-2.2	-4.0	-2.3	-2.0	Pakistan
-4.5	-11.0	-19.7	-13.6	-14.9	-12.3	-3.3	-9.9	-9.2	-8.6	Sri Lanka
-0.7	*-1.2*	*-1.8*	*-1.8*	*-1.2*	*-1.0*	*0.8*	*-0.2*	*2.6*	*5.0*	*EAST ASIA AND PACIFIC*
-0.6	-0.9	-1.1	0.5	2.2	1.5	0.8	-4.1	-2.6	0.1	China
-4.6	-7.8	-5.0	-16.0	-9.4	-8.1	-3.8	-4.0	-1.0	-1.0	Fiji
-0.4	0.9	-4.6	-5.5	-3.5	-1.9	4.4	5.6	4.0	..	Hong Kong
-2.6	1.7	3.6	-0.9	-5.8	-7.9	-2.3	-2.3	-5.5	-3.0	Indonesia
-2.2	-6.5	-8.6	-6.8	-3.7	-2.1	-1.6	-1.0	4.7	8.1	Korea, Republic of
0.5	4.2	-1.3	-10.0	-13.5	-11.8	-5.0	-2.3	-0.1	6.9	Malaysia
-15.8	-13.1	-21.0	-32.2	-31.3	-27.0	-25.4	-16.8	-12.4	-17.5	Papua New Guinea
-5.0	-5.5	-5.8	-5.8	-8.4	-8.6	-4.7	-0.7	2.5	-2.1	Philippines
-5.7	-7.8	-13.3	-10.5	-8.4	-3.5	-2.0	0.0	3.2	2.8	Singapore
-18.1	-7.0	-27.0	-34.1	-16.9	-17.0	-6.1	-23.5	..	..	Solomon Islands
-5.0	-7.7	-6.9	-7.7	-3.2	-7.6	-5.4	-4.5	0.2	-1.5	Thailand
..	..	..	..	-23.3	-17.4	-7.1	-18.7	-19.2	..	Vanuatu

Table 16. Balance of payments: total to be financed (cont'd)

Percentage of gross domestic product	1967	1968	1969	1970	1971	1972	1973	1974	1975	1976	1977
LATIN AMERICA AND CARIBBEAN	..	..	..	-2.3	-2.8	-2.3	-1.5	-2.3	-4.0	-2.8	-2.7
Argentina	..	..	..	-0.7	-1.5	-0.9	1.8	0.2	-3.3	1.7	2.7
Bahamas	..	..	..	..	..	..	-22.1	-19.8	6.5	12.4	8.5
Barbados	..	..	..	..	..	..	..	..	..	..	..
Belize	..	..	..	..	..	..	..	..	..	..	..
Bolivia	..	..	..	0.2	-0.7	-1.0	0.2	7.7	-9.8	-3.4	-5.9
Brazil	..	..	..	-2.0	-3.4	-2.9	-2.7	-7.2	-5.7	-4.3	-2.9
Chile	..	..	..	-1.2	-2.0	-3.9	-2.8	-2.7	-6.9	1.3	-4.2
Colombia	..	..	..	-4.6	-6.2	-2.5	-0.8	-3.1	-1.4	1.0	1.9
Costa Rica	..	..	..	..	..	..	..	..	..	..	..
Dominican Republic	..	..	..	-7.4	-8.4	-2.6	-4.5	-8.9	-2.3	-3.6	-3.1
Ecuador	..	..	..	-7.3	-10.2	-4.5	-0.5	0.6	-5.5	-0.6	-5.7
El Salvador	..	..	..	0.7	-1.4	0.8	-3.4	-8.6	-5.3	0.8	0.8
Grenada	..	..	..	..	..	..	..	..	..	..	1.4
Guatemala	..	..	..	-0.4	-2.4	-0.5	0.3	-3.2	-1.8	-1.8	-0.6
Guyana	..	..	..	-8.1	-2.7	-5.3	-21.1	-2.2	-4.6	-31.7	-22.0
Haiti	..	..	..	..	..	..	..	..	..	..	..
Honduras	..	..	..	-9.3	-3.6	-2.0	-4.2	-12.0	-11.1	-8.5	-8.3
Jamaica	..	..	..	-10.6	-10.7	-10.6	-12.7	-3.5	-10.1	-10.3	-1.4
Mexico	..	..	..	-2.9	-2.0	-2.0	-2.4	-3.8	-4.4	-3.7	-2.2
Nicaragua	..	..	..	-5.5	-5.7	-0.4	-8.8	-17.7	-12.5	-2.6	-9.2
Panama	..	..	..	-7.7	-7.8	-9.2	-9.0	-16.1	-10.6	-10.2	-8.8
Paraguay	..	..	..	-3.2	-3.9	-1.1	-1.8	-3.9	-5.6	-4.2	-2.8
Peru	..	..	..	2.0	-0.9	-0.7	-2.8	-5.6	-9.6	-8.0	-6.9
St. Vincent and the Grenadines	..	..	..	..	..	..	..	..	..	..	..
Trinidad and Tobago	..	..	..	-12.7	-18.1	-10.9	-2.2	13.5	14.0	11.0	6.0
Uruguay	..	..	..	-2.3	-2.6	2.0	0.6	-3.6	-5.5	-2.2	-4.0
Venezuela	..	..	..	-0.7	0.0	-0.6	4.4	19.4	6.8	0.8	-7.1
MIDDLE EAST AND NORTH AFRICA	..	..	..	..	..	..	..	..	..	..	..
Algeria	..	..	..	-3.3	-3.5	-2.5	-5.1	4.3	-10.8	-5.1	-11.1
Egypt, Arab Republic of	..	..	..	..	..	..	..	-20.3	-25.2	-16.5	-16.5
Jordan	..	..	..	-26.9	-31.5	-31.0	-26.1	-32.2	-37.4	-25.0	-0.7
Morocco	..	..	..	-4.1	-2.2	0.3	1.0	2.5	-6.3	-15.1	-16.9
Oman	..	..	..	..	..	..	..	9.4	-7.2	-0.6	0.3
Syrian Arab Republic	..	..	..	..	..	..	..	..	..	..	..
Tunisia	..	..	..	-6.1	-1.2	-1.6	-3.7	0.6	-4.9	-12.4	-11.1
Yemen Arab Republic	..	..	..	-40.2	-30.2	-30.9	-38.0	-9.4	1.6	18.0	20.7
Yemen, PDR	..	..	..	..	..	..	..	..	..	..	..
EUROPE AND MEDITERRANEAN	..	..	..	..	-4.3	-0.7	-2.0	-6.4	-6.4	-4.9	-5.1
Cyprus	..	..	..	..	..	..	..	..	..	..	..
Greece	..	..	..	-4.3	-3.1	-3.2	-7.3	-6.1	-4.2	-4.2	-4.1
Hungary	..	..	..	..	..	..	..	..	..	..	..
Israel	..	..	..	-13.7	-10.1	-5.1	-14.3	-18.2	-23.1	-16.4	-10.3
Malta	..	..	..	-12.9	-4.9	3.7	7.9	-3.4	11.3	9.9	5.9
Poland	..	..	..	..	..	..	..	..	..	..	..
Portugal	..	..	..	-10.4	-7.1	4.1	3.0	-6.2	-4.9	-8.3	-5.9
Turkey	..	..	..	-0.5	-0.2	1.2	3.1	-2.0	-4.7	-4.9	-6.6
Yugoslavia	..	..	..	-2.8	-2.7	2.7	2.4	-3.6	-2.1	0.5	-2.9
Developing countries	..	..	..	-2.8	-3.0	-1.7	-1.0	-0.8	-3.4	-1.8	-1.9
Highly indebted	..	..	..	-2.2	-2.6	-1.8	-0.8	-0.7	-3.7	-2.8	-3.0
Low-income countries	..	..	..	-1.4	-1.8	-1.3	-0.7	-2.1	-2.4	-1.3	-1.1
Low-income Africa	..	..	..	-1.8	-5.9	-4.0	-2.6	-6.0	-9.7	-6.6	-7.3
China and India	..	..	..	-0.5	-0.6	-0.3	0.0	-0.7	-0.1	0.7	0.7
Other low-income Asia	..	..	..	-5.1	-4.7	-4.5	-3.4	-5.2	-7.2	-6.5	-4.5
Middle-income countries	..	..	..	-3.6	-3.7	-1.8	-1.2	-0.3	-3.8	-2.0	-2.2
High-income oil exporters	..	..	..	15.9	25.1	22.7	22.5	64.1	37.7	34.6	23.4
Industrialized countries	..	..	..	0.5	0.6	0.5	0.6	-0.3	0.5	0.0	0.0
Japan	..	..	..	1.1	2.6	2.3	0.0	-1.0	-0.1	0.7	1.6
United States	..	..	..	0.5	0.1	-0.2	0.7	0.6	1.4	0.5	-0.5
European Community	..	..	..	0.7	1.1	1.0	0.7	-0.6	0.5	-0.1	0.6

NOTE: See glossary for definitions.

1978	1979	1980	1981	1982	1983	1984	1985	1986	1987	
-3.7	*-3.4*	*-4.0*	*-5.0*	*-5.7*	*-1.4*	*-0.3*	*-0.5*	*-2.3*	*-1.4*	*LATIN AMERICA AND CARIBBEAN*
4.3	-1.0	-8.4	-8.2	-4.1	-3.8	-3.1	-1.4	-3.6	-5.3	Argentina
4.3	0.4	-2.6	-6.9	-5.6	-3.1	-3.2	-2.3	-1.3	..	Bahamas
..	..	-3.1	-11.9	-4.2	-4.4	1.8	4.1	0.1	..	Barbados
..	..	..	..	..	..	-9.1	-4.2	-6.6	..	Belize
-15.1	-16.1	-1.7	-14.6	-6.1	-6.5	-7.5	-9.8	-11.2	-13.4	Bolivia
-3.5	-4.7	-5.4	-4.5	-6.0	-3.3	0.0	-0.1	-1.6	-0.4	Brazil
-7.2	-5.8	-7.3	-14.7	-9.7	-5.9	-11.0	-8.4	-7.0	-4.6	Chile
1.0	1.6	-0.6	-5.4	-7.8	-7.8	-3.7	-5.2	0.9	0.7	Colombia
..	..	-13.6	-15.7	-10.7	-10.6	-7.2	-8.0	-4.6	..	Costa Rica
-7.2	-7.1	-11.7	-5.9	-6.7	-6.9	-4.6	-5.0	-2.8	..	Dominican Republic
-9.6	-7.0	-6.0	-7.4	-10.0	-0.3	-1.6	0.6	-5.8	-10.3	Ecuador
-9.5	0.4	0.0	-7.9	-7.8	-5.7	-5.9	-6.2	-2.0	-4.1	El Salvador
-0.9	-10.9	-16.1	-30.5	-37.5	-30.7	-20.3	-23.9	-27.8	..	Grenada
-4.4	-3.0	-2.1	-6.9	-4.6	-2.5	-4.0	-2.5	-0.5	-7.9	Guatemala
-4.6	-16.0	-21.4	-31.6	-28.9	-32.9	-22.6	-20.4	..	..	Guyana
..	..	-9.6	-16.4	-12.9	-12.7	-11.2	-9.5	..	-7.0	Haiti
-9.0	-9.3	-13.0	-11.6	-8.7	-8.4	-12.0	-9.2	-7.2	-8.2	Honduras
-2.3	-6.1	-6.6	-12.3	-14.2	-11.1	-15.7	-18.4	-3.1	-5.6	Jamaica
-3.0	-3.9	-4.3	-5.6	-3.7	3.5	2.3	0.3	-1.4	2.5	Mexico
-2.1	5.7	-24.5	-26.6	-23.0	-25.5	-24.2	-28.7	-29.2	..	Nicaragua
-9.7	-13.0	-10.6	-0.6	-3.5	7.1	1.6	3.3	6.3	4.2	Panama
-4.6	-6.2	-13.5	-11.4	-9.2	-5.1	-8.8	-11.7	-14.4	..	Paraguay
-2.0	3.9	-1.2	-7.5	-6.9	-5.5	-1.8	0.0	-4.2	-3.1	Peru
..	..	-23.2	-8.0	-17.3	-6.1	-3.5	0.3	-2.5	..	St. Vincent and the Grenadines
1.7	-0.3	5.7	5.8	-7.3	-12.5	-6.4	-1.1	-8.4	..	Trinidad and Tobago
-2.6	-4.9	-7.1	-4.1	-2.6	-1.3	-2.7	-2.2	1.0	-1.8	Uruguay
-11.9	0.6	6.8	5.2	-5.3	5.5	8.1	6.1	-2.4	-2.0	Venezuela
..	..	..	..	..	..	..	..	..	..	*MIDDLE EAST AND NORTH AFRICA*
-13.5	-4.9	0.8	1.1	0.0	0.1	0.5	2.1	-3.5	-1.5	Algeria
										Egypt, Arab Republic of
-13.6	-0.3	11.3	-1.2	-9.0	-9.9	-6.9	-6.2	-1.0	-7.0	Jordan
-10.5	-9.8	-8.6	-13.1	-13.6	-7.4	-9.1	-8.4	-2.5	1.0	Morocco
-2.5	9.9	..	..	..	..	..	..	..	..	Oman
..	..	..	..	..	..	..	..	..	-5.7	Syrian Arab Republic
-9.8	-4.8	-4.0	-7.7	-9.4	-7.4	-10.9	-7.1	-8.0	-1.0	Tunisia
1.2	-10.8	-26.8	-23.0	-13.5	-11.4	-6.8	-7.5	-2.3	-14.3	Yemen Arab Republic
..	..	-32.1	-28.3	-25.8	-23.1	-26.3	..	..	..	Yemen, PDR
-4.3	*-4.9*	*-5.5*	*-5.2*	*-3.7*	*-3.1*	*-2.6*	*-2.4*	*-1.7*	*..*	*EUROPE AND MEDITERRANEAN*
..	..	-13.1	-8.8	-9.4	-11.0	-10.0	-7.2	..	..	Cyprus
-3.0	-4.9	-5.5	-7.0	-6.4	-7.8	-8.4	-12.4	-7.7	-6.3	Greece
..	..	..	..	..	..	..	..	..	..	Hungary
-15.3	-13.9	-12.3	-13.2	-14.0	-13.8	-15.0	-12.4	-9.4	-12.8	Israel
7.6	2.7	1.8	3.0	-1.9	-3.3	-2.2	-4.1	-1.8	-0.8	Malta
..	..	-6.0	-8.5	-3.9	-2.1	-1.7	-1.7	-1.3	-0.9	Poland
-2.6	-0.3	-4.4	-10.8	-13.8	-4.8	-2.8	1.3	3.2	0.9	Portugal
-2.4	-2.1	-6.0	-3.4	-2.0	-4.2	-3.3	-2.4	-3.0	-2.0	Turkey
-2.3	-5.3	-3.2	-1.3	-0.7	0.6	1.1	1.8	1.7	..	Yugoslavia
-2.9	-1.9	-2.9	-4.1	-3.5	-1.8	-0.7	-1.0	-0.8	0.5	Developing countries
-4.0	-3.3	-3.1	-5.0	-5.8	-1.9	-0.1	0.0	-1.5	..	Highly indebted
-2.2	-2.4	-3.2	-2.6	-1.8	-1.3	-1.5	-4.4	-3.5	-2.3	Low-income countries
-7.9	-6.8	-10.3	-11.6	-11.6	-8.5	-7.4	-7.7	-8.0	-9.2	Low-income Africa
-0.5	-0.8	-1.3	-0.4	0.7	0.4	-0.1	-3.4	-2.2	-0.7	China and India
-4.7	-6.2	-7.3	-6.2	-7.2	-4.6	-3.8	-5.8	-4.4	-3.9	Other low-income Asia
-3.2	-1.7	-2.9	-4.6	-4.1	-2.0	-0.4	0.2	0.1	1.4	Middle-income countries
10.1	25.9	39.6	28.0	11.1	2.9	3.9	7.7	3.4	..	High-income oil exporters
0.6	0.0	-0.4	0.1	0.1	0.0	-0.4	-0.3	0.1	-0.1	Industrialized countries
1.8	-0.8	-0.9	0.5	0.8	1.9	2.9	3.8	4.5	3.8	Japan
-0.5	0.2	0.3	0.5	0.0	-1.1	-2.6	-2.6	-3.0	-3.2	United States
1.6	0.5	-0.7	0.1	0.1	0.8	1.2	1.5	2.1	1.5	European Community

Table 17. Value of merchandise imports

Millions of US dollars	1967	1968	1969	1970	1971	1972	1973	1974	1975	1976	1977
SUB-SAHARAN AFRICA	*5,371*	*5,667*	*6,220*	*7,326*	*8,715*	*9,136*	*11,393*	*15,942*	*21,147*	*22,901*	*28,752*
Excluding Nigeria	*4,797*	*5,187*	*5,586*	*6,329*	*7,267*	*7,700*	*9,618*	*13,275*	*15,023*	*14,452*	*17,340*
Benin	43	49	55	64	76	93	112	164	197	219	246
Botswana	31	33	43	49	49	109	166	185	218	209	276
Burkina Faso	45	51	62	57	73	94	130	180	216	195	260
Burundi	20	23	22	22	30	31	32	40	63	58	91
Cameroon	178	178	207	242	250	299	335	437	598	595	747
Cape Verde	..	..	..	..	..	..	..	..	..	..	..
Central African Republic	60	54	53	57	51	56	69	84	110	100	136
Chad	59	54	54	62	62	61	82	87	133		
Comoros	..	..	..	..	..	..	..	..	..		
Congo, People's Republic	82	84	79	57	79	90	125	123	165	168	183
Cote d'Ivoire	263	307	334	387	398	453	714	967	1,127	1,295	1,752
Ethiopia	143	173	155	173	188	189	213	273	294	353	391
Gabon	67	64	78	80	91	133	190	332	469	503	716
Gambia, The	19	21	22	17	21	23	32	38	49	65	73
Ghana	307	308	347	410	434	292	448	818	788	862	1,144
Guinea-Bissau	..	..	..	..	..	..	..	..	..	..	..
Kenya	336	356	360	442	560	554	655	1,026	987	923	1,286
Lesotho	33	34	34	32	39	56	87	120	160	207	229
Liberia	125	107	114	150	162	179	194	288	331	399	464
Madagascar	145	170	183	170	213	202	203	281	367	290	353
Malawi	70	70	74	86	109	130	142	188	251	206	232
Mali	52	34	39	45	59	74	126	179	190	150	159
Mauritania	37	35	45	56	63	85	135	188	235	304	334
Mauritius	78	76	75	76	83	120	170	309	331	358	445
Mozambique	199	234	260	323	335	327	465	464	417	300	328
Niger	46	42	49	58	53	66	87	96	99	127	196
Nigeria	626	541	696	1,059	1,511	1,505	1,862	2,781	6,041	8,195	11,020
Rwanda	20	23	24	29	33	35	41	72	102	128	125
Senegal	157	181	199	192	218	279	359	498	582	644	762
Seychelles	5	6	7	10	15	21	25	28	32	44	46
Sierra Leone	90	91	112	117	113	121	158	222	159	166	181
Somalia	44	47	56	46	56	72	112	154	162	176	206
Sudan	214	258	266	311	355	354	480	656	957	952	983
Swaziland	..	..	..	..	..	..	..	..	..	..	..
Tanzania	182	214	199	272	338	363	447	760	718	645	730
Togo	51	56	68	76	85	94	95	116	269	218	305
Uganda	159	165	175	172	250	162	162	213	206	172	241
Zaire	256	310	452	533	619	766	782	940	933	840	852
Zambia	429	456	436	477	559	563	532	787	929	655	671
Zimbabwe	262	290	279	329	397	416	549	753	814	612	618
SOUTH ASIA	*4,406*	*4,108*	*3,804*	*3,914*	*3,930*	*3,705*	*5,691*	*8,979*	*10,954*	*10,053*	*11,424*
Bangladesh	..	..	..	..	..	189	835	1,032	1,200	1,137	1,058
India	2,722	2,507	2,118	2,094	2,406	2,230	3,146	5,167	6,385	5,710	6,601
Nepal	58	50	74	85	70	88	96	110	171	214	244
Pakistan	1,101	996	1,011	1,171	926	682	966	1,729	2,153	2,183	2,455
Sri Lanka	359	365	427	387	328	336	422	688	745	552	701
EAST ASIA AND PACIFIC	*12,061*	*13,331*	*15,136*	*17,310*	*19,424*	*22,690*	*34,626*	*53,803*	*55,010*	*62,632*	*72,570*
China	2,020	1,950	1,830	2,330	2,210	2,860	5,160	7,620	7,486	6,578	7,214
Fiji	70	79	89	104	128	158	203	273	267	263	306
Hong Kong	1,814	2,058	2,457	2,905	3,387	3,895	5,631	6,710	6,757	8,909	10,457
Indonesia	891	923	1,122	1,258	1,379	1,611	2,969	5,231	6,196	7,700	8,495
Korea, Republic of	996	1,468	1,823	1,983	2,394	2,522	4,240	6,844	7,271	8,765	10,803
Malaysia	1,106	1,184	1,162	1,390	1,485	1,674	2,517	4,297	3,851	4,177	5,004
Papua New Guinea	178	178	220	300	356	336	356	503	556	502	642
Philippines	1,184	1,280	1,256	1,210	1,319	1,388	1,790	3,468	3,776	3,953	4,270
Singapore	1,440	1,661	2,040	2,461	2,827	3,383	5,070	8,344	8,135	9,070	10,472
Solomon Islands	11	12	11	13	15	17	18	27	33	30	33
Thailand	1,060	1,150	1,286	1,293	1,287	1,484	2,073	3,156	3,279	3,572	4,613
Vanuatu	..	..	..	..	..	..	..	..	..	..	..

NOTE: See glossary for definitions.

1978	1979	1980	1981	1982	1983	1984	1985	1986	1987	
33,662	*33,470*	*46,724*	*50,035*	*42,982*	*36,103*	*32,088*	*31,473*	*32,190*	*32,274*	*SUB-SAHARAN AFRICA*
20,454	*22,993*	*29,593*	*28,481*	*26,436*	*23,460*	*22,466*	*22,350*	*24,621*	*26,202*	*Excluding Nigeria*
312	320	331	543	464	283	285	312	370	346	Benin
353	521	691	799	686	735	707	583	684	783	Botswana
296	370	446	422	436	377	319	334	426	513	Burkina Faso
98	153	167	161	214	194	187	189	207	225	Burundi
1,009	1,271	1,538	1,421	1,243	1,250	1,213	1,163	1,704	1,922	Cameroon
..	..	68	..	..	..	..	..	..	..	Cape Verde
156	159	221	171	166	150	153	170	182	193	Central African Republic
..	85	74	108	109	157	171	..	..	..	Chad
..	..	55	52	50	50	60	64	71	..	Comoros
242	266	500	828	807	746	737	752	629	570	Congo, People's Republic
2,310	2,389	3,015	2,393	2,184	1,814	1,497	1,749	1,969	2,245	Cote d'Ivoire
505	567	721	737	785	875	928	993	1,214	1,202	Ethiopia
682	673	686	843	799	852	724	861	924	663	Gabon
100	141	132	145	151	118	100	93	105	113	Gambia, The
1,003	853	1,129	1,106	705	600	640	731	783	836	Ghana
..	..	53	55	52	58	57	63	72	..	Guinea-Bissau
1,706	1,656	2,590	2,081	1,603	1,390	1,529	1,455	1,649	1,866	Kenya
273	373	490	564	514	561	504	377	425	483	Lesotho
481	507	534	477	422	412	369	284	235	208	Liberia
460	698	677	473	439	411	412	465	374	386	Madagascar
338	398	440	350	311	311	270	287	260	281	Malawi
286	361	439	365	332	353	368	432	438	447	Mali
302	323	363	436	482	427	341	358	345	353	Mauritania
498	566	619	554	464	440	471	529	676	831	Mauritius
521	571	800	801	836	636	540	424	543	..	Mozambique
306	462	608	510	442	361	331	388	466	500	Niger
12,763	10,274	16,642	20,877	16,060	12,254	9,364	8,890	7,466	6,134	Nigeria
202	224	277	279	297	277	282	313	348	463	Rwanda
756	931	1,038	1,076	992	1,025	1,035	898	933	1,073	Senegal
53	90	99	93	98	88	88	99	106	137	Seychelles
278	316	425	311	298	160	157	149	141	132	Sierra Leone
275	394	461	426	542	416	535	380	402	426	Somalia
878	916	1,499	1,519	1,754	1,534	1,370	1,114	1,055	832	Sudan
..	..	538	506	440	464	381	281	352	425	Swaziland
1,141	1,077	1,227	1,176	1,134	795	847	1,028	1,048	1,165	Tanzania
497	558	638	504	487	350	310	288	359	350	Togo
255	197	293	345	377	377	344	327	350	417	Uganda
797	826	1,117	1,019	913	842	877	919	1,008	1,149	Zaire
628	750	1,111	1,062	831	690	730	654	663	745	Zambia
592	937	1,448	1,696	1,639	1,205	955	1,031	1,099	1,055	Zimbabwe
14,214	*18,593*	*25,000*	*26,889*	*25,421*	*25,436*	*26,575*	*28,874*	*27,602*	*30,843*	*SOUTH ASIA*
1,338	1,711	2,254	2,649	2,581	2,246	2,353	2,647	2,364	2,620	Bangladesh
7,854	10,142	14,090	15,654	14,387	14,782	15,424	17,295	16,801	18,985	India
339	379	342	369	395	447	416	453	459	569	Nepal
3,161	4,061	5,351	5,413	5,233	5,341	5,873	5,890	5,377	5,822	Pakistan
942	1,449	2,035	1,804	1,770	1,789	1,847	1,988	1,948	2,085	Sri Lanka
92,787	*121,991*	*155,061*	*173,265*	*168,420*	*177,093*	*190,234*	*195,496*	*203,097*	*252,496*	*EAST ASIA AND PACIFIC*
10,893	15,675	19,550	19,482	18,546	21,346	27,350	42,526	43,172	43,392	China
366	470	561	631	514	483	450	442	437	435	Fiji
13,451	17,137	22,027	24,671	23,461	24,010	28,567	29,580	35,366	48,462	Hong Kong
9,493	10,364	14,139	18,527	19,996	19,853	16,853	14,230	13,103	14,453	Indonesia
14,966	20,296	22,228	26,028	24,236	26,174	30,609	31,119	31,518	40,934	Korea, Republic of
6,364	8,708	11,602	13,052	14,042	14,669	14,849	12,746	11,447	12,506	Malaysia
770	903	1,176	1,261	1,170	1,120	1,110	1,008	1,130	1,222	Papua New Guinea
5,143	6,613	8,295	8,478	8,262	7,977	6,424	5,445	5,393	7,144	Philippines
13,049	17,638	24,003	27,572	28,168	28,158	28,656	26,250	25,461	32,480	Singapore
42	70	89	79	71	74	79	83	80	..	Solomon Islands
5,314	7,132	9,450	10,055	8,527	10,279	10,518	9,239	9,124	13,006	Thailand
..	..	70	60	51	52	55	56	56	68	Vanuatu

Table 17. Value of merchandise imports (cont'd)

Millions of US dollars	1967	1968	1969	1970	1971	1972	1973	1974	1975	1976	1977
LATIN AMERICA AND CARIBBEAN	*11,313*	*12,480*	*13,635*	*15,730*	*17,758*	*20,185*	*26,505*	*45,456*	*48,810*	*49,938*	*55,698*
Argentina	1,096	1,169	1,571	1,689	1,846	1,904	2,235	3,635	3,945	3,029	4,158
Bahamas	162	176	296	332	511	485	764	2,112	2,482	2,893	2,871
Barbados	77	84	97	117	122	142	168	204	216	237	272
Belize	..	..	..	..	..	..	..	..	..	..	..
Bolivia	151	153	165	159	170	179	230	366	574	594	586
Brazil	1,667	2,129	2,263	2,845	3,696	4,776	6,992	14,163	13,578	13,714	13,254
Chile	738	833	891	980	1,049	1,168	1,562	2,137	1,733	1,672	2,418
Colombia	497	643	685	843	929	859	1,062	1,597	1,495	1,708	2,028
Costa Rica	191	214	245	317	350	373	455	720	694	800	1,059
Dominican Republic	197	222	246	304	358	388	489	808	889	878	975
Ecuador	213	253	239	273	339	325	396	781	985	952	1,165
El Salvador	224	214	209	214	247	278	374	563	598	718	943
Grenada	..	..	..	..	..	..	..	..	..	..	..
Guatemala	247	249	250	284	303	328	427	701	733	838	1,053
Guyana	130	110	118	134	135	143	177	255	344	364	315
Haiti	38	37	45	59	66	85	113	166	196	246	287
Honduras	165	185	184	221	193	193	262	380	404	453	579
Jamaica	349	387	436	525	551	620	664	936	1,123	911	859
Mexico	1,746	1,962	2,078	2,461	2,407	2,935	4,146	6,057	6,572	6,033	5,589
Nicaragua	202	183	176	198	210	218	326	559	517	532	762
Panama	251	267	294	357	396	441	502	822	892	848	859
Paraguay	..	..	..	..	..	..	..	..	..	..	..
Peru	820	630	600	622	750	796	1,024	1,595	2,380	1,798	1,598
St. Vincent and the Grenadines	..	..	..	..	..	..	..	..	..	..	..
Trinidad and Tobago	418	420	483	543	663	762	792	1,847	1,489	1,976	1,809
Uruguay	170	159	197	233	222	187	285	461	517	599	669
Venezuela	1,445	1,666	1,719	1,869	2,079	2,430	2,806	4,186	6,004	7,663	10,938
MIDDLE EAST AND NORTH AFRICA	*3,243*	*3,369*	*3,690*	*4,232*	*4,540*	*5,322*	*7,355*	*13,707*	*18,830*	*19,042*	*25,094*
Algeria	639	815	1,009	1,257	1,227	1,492	2,259	4,036	5,974	5,307	7,102
Egypt, Arab Republic of	792	666	638	787	920	898	914	2,351	3,934	3,862	4,815
Jordan	154	161	190	185	215	267	328	487	731	1,022	1,381
Morocco	517	550	560	684	697	764	1,098	1,901	2,547	2,593	3,169
Oman	10	10	13	18	33	49	117	393	671	667	875
Syrian Arab Republic	264	311	368	350	438	539	613	1,229	1,669	1,959	2,656
Tunisia	260	217	255	305	342	459	606	1,120	1,418	1,526	1,821
Yemen Arab Republic	30	32	36	32	37	80	123	190	293	410	1,040
Yemen, PDR	210	203	218	200	156	149	171	419	323	412	544
EUROPE AND MEDITERRANEAN	*11,315*	*12,402*	*13,948*	*16,992*	*19,216*	*22,178*	*32,010*	*45,743*	*50,503*	*54,236*	*60,890*
Cyprus	168	170	202	235	259	315	449	406	306	430	619
Greece	1,186	1,393	1,594	1,958	2,098	2,346	3,473	4,385	5,321	6,051	6,853
Hungary	1,330	1,350	1,444	1,877	2,248	2,356	3,018	4,453	5,400	5,533	6,531
Israel	1,045	1,351	1,646	2,084	2,375	2,447	4,235	5,380	5,975	5,645	5,745
Malta	111	123	147	161	156	174	240	357	375	421	513
Poland	2,353	2,539	2,856	3,210	3,593	4,745	6,953	9,327	11,155	12,898	13,420
Portugal	1,059	1,178	1,298	1,590	1,824	2,227	3,073	4,641	3,863	4,316	4,964
Turkey	685	764	747	886	1,088	1,508	2,049	3,720	4,640	4,993	5,694
Yugoslavia	1,707	1,797	2,134	2,874	3,297	3,233	4,783	7,520	7,699	7,367	8,973
Developing countries	52,377	56,229	61,882	71,849	81,023	90,813	127,834	201,456	230,577	243,819	280,325
Highly indebted	13,379	14,671	16,069	19,030	21,607	23,894	32,104	53,270	60,788	62,874	73,506
Low-income countries	10,603	10,480	10,518	11,579	12,487	13,066	18,569	26,811	30,554	28,150	32,239
Low-income Africa	3,384	3,676	3,967	4,464	5,225	5,355	6,433	9,042	10,118	9,566	11,261
China and India	4,742	4,457	3,948	4,424	4,616	5,090	8,306	12,787	13,871	12,288	13,815
Other low-income Asia	1,684	1,600	1,686	1,821	1,523	1,474	2,545	3,811	4,568	4,343	4,823
Middle-income countries	41,754	45,733	51,351	60,259	68,526	77,740	109,255	174,624	199,995	215,568	247,968
High-income oil exporters	2,150	2,401	2,699	2,660	3,257	4,331	6,731	11,540	15,542	22,214	33,303
Industrialized countries	150,993	169,544	194,873	224,118	248,626	295,404	405,484	575,852	577,179	665,382	753,470
Japan	11,664	12,988	15,025	18,883	19,712	23,471	38,313	62,094	57,865	64,505	70,561
United States	26,816	33,089	36,043	39,952	45,563	55,563	69,476	100,997	96,904	121,795	147,863
European Community	80,824	88,926	104,938	120,961	134,428	159,937	223,207	306,603	313,311	357,913	401,188

1978	1979	1980	1981	1982	1983	1984	1985	1986	1987	
63,282	*80,287*	*104,757*	*111,664*	*90,102*	*69,069*	*69,519*	*69,012*	*70,327*	*77,488*	*LATIN AMERICA AND CARIBBEAN*
3,832	6,692	10,539	9,430	5,337	4,504	4,585	3,814	4,724	6,119	Argentina
2,482	3,985	3,479	4,124	3,615	3,342	3,524	3,265	2,615	..	Bahamas
312	421	522	572	551	621	659	602	593	520	Barbados
..	..	228	240	234	237	252	240	252	..	Belize
770	842	665	917	578	589	492	552	716	776	Bolivia
15,016	19,731	24,900	24,100	21,100	16,800	15,200	14,300	15,555	16,581	Brazil
3,276	4,777	6,185	6,331	4,117	3,186	3,770	3,300	3,471	4,622	Chile
2,836	3,233	4,660	5,200	5,460	4,970	4,490	4,130	3,850	4,230	Colombia
1,212	1,446	1,597	1,274	945	993	1,091	1,098	1,148	1,377	Costa Rica
987	1,213	1,640	1,668	1,444	1,471	1,446	1,487	1,433	1,783	Dominican Republic
1,499	1,986	2,215	1,907	1,737	1,487	1,616	1,767	1,810	2,250	Ecuador
1,024	1,012	976	1,045	945	892	977	961	903	1,035	El Salvador
..	..	81	65	80	78	90	86	89	..	Grenada
1,261	1,362	1,559	2,009	1,388	1,135	1,277	1,175	960	1,479	Guatemala
279	318	365	438	282	246	214	248	217	226	Guyana
311	368	492	527	448	412	435	449	367	378	Haiti
699	826	1,009	945	690	823	893	888	875	895	Honduras
880	992	1,178	1,487	1,373	1,531	1,130	1,124	964	1,207	Jamaica
8,053	12,587	19,517	24,161	15,041	8,023	11,788	13,994	12,032	12,731	Mexico
780	415	882	994	775	799	848	964	857	923	Nicaragua
942	1,184	1,448	1,562	1,568	1,411	1,412	1,383	1,229	1,248	Panama
..	..	803	915	822	629	741	585	835	1,202	Paraguay
1,601	1,951	3,062	3,803	3,721	2,722	2,212	2,023	2,909	3,435	Peru
..	..	99	93	98	88	87	101	118	..	St. Vincent and the Grenadines
1,980	2,105	3,180	3,120	3,700	2,582	1,919	1,533	1,370	1,219	Trinidad and Tobago
716	1,173	1,650	1,630	1,110	788	776	708	870	1,140	Uruguay
11,766	10,670	11,827	13,106	12,944	8,710	7,594	8,234	9,565	8,725	Venezuela
29,246	*33,631*	*41,537*	*45,416*	*43,893*	*44,582*	*44,581*	*41,247*	*35,790*	*36,614*	*MIDDLE EAST AND NORTH AFRICA*
8,667	8,407	10,525	11,302	10,679	10,332	10,263	9,814	9,234	9,263	Algeria
6,727	7,073	8,047	8,840	9,078	10,766	11,594	10,581	8,453	10,586	Egypt, Arab Republic of
1,499	1,949	2,394	3,146	3,218	3,016	2,784	2,733	2,432	2,339	Jordan
2,950	3,674	4,182	4,353	4,315	3,597	3,907	3,850	3,790	4,229	Morocco
947	1,246	1,732	2,288	2,683	2,492	2,746	3,153	2,375	0	Oman
2,444	3,324	4,118	5,040	4,014	4,542	4,116	3,487	2,325	2,465	Syrian Arab Republic
2,158	2,842	3,509	3,771	3,396	3,100	3,115	2,757	2,890	2,955	Tunisia
1,358	1,492	1,853	1,758	1,521	1,593	1,565	1,360	982	1,072	Yemen Arab Republic
575	925	1,527	1,419	1,599	1,483	1,543	1,311	1,105	1,450	Yemen, PDR
67,984	*81,880*	*93,884*	*90,834*	*84,004*	*80,408*	*82,137*	*84,555*	*88,498*	*..*	*EUROPE AND MEDITERRANEAN*
748	1,000	1,195	1,101	1,207	1,208	1,351	1,234	1,263	1,463	Cyprus
7,655	9,594	10,531	8,781	10,012	9,500	9,611	10,138	11,241	12,908	Greece
7,990	8,682	9,245	9,139	8,819	8,509	8,091	8,224	9,599	..	Hungary
7,162	8,428	9,512	10,145	9,506	9,471	9,493	9,752	10,366	12,704	Israel
569	753	938	855	789	733	717	759	887	1,002	Malta
14,744	16,142	16,690	12,792	10,648	10,927	10,985	11,855	11,535	..	Poland
5,229	6,509	9,293	9,946	9,605	8,257	7,975	7,650	9,393	13,438	Portugal
4,479	4,946	7,573	8,864	8,794	9,235	10,757	11,515	11,027	14,008	Turkey
9,770	14,037	15,064	15,757	14,100	12,155	11,996	12,163	11,749	13,114	Yugoslavia
327,891	397,957	515,016	556,609	509,460	481,553	491,190	487,444	493,087	578,713	Developing countries
84,392	103,067	135,193	145,204	118,384	92,099	87,932	87,140	87,980	96,059	Highly indebted
40,868	51,880	67,063	68,042	64,563	65,382	73,180	90,725	91,414	96,190	Low-income countries
13,101	14,492	18,755	17,687	16,450	14,420	14,429	14,227	15,258	16,323	Low-income Africa
18,747	25,817	33,640	35,136	32,933	36,128	42,774	59,821	59,973	62,377	China and India
6,361	8,453	10,910	11,235	11,034	10,654	11,151	11,579	10,801	11,858	Other low-income Asia
286,960	346,055	447,923	488,456	444,843	416,162	418,155	397,335	402,288	484,255	Middle-income countries
40,044	49,481	61,067	72,458	75,464	69,383	61,406	45,888	40,373	42,803	High-income oil exporters
882,850	1,134,660	1,357,197	1,292,886	1,216,333	1,200,217	1,306,527	1,356,999	1,515,818	..	Industrialized countries
78,731	110,108	139,892	140,830	130,319	125,017	134,257	127,512	119,424	..	Japan
182,196	217,387	250,280	271,213	253,033	267,971	338,190	358,705	381,363	..	United States
474,687	623,130	748,700	662,411	633,026	608,797	620,038	648,135	760,295	..	European Community

Table 18. Value of merchandise exports

Millions of US dollars	1967	1968	1969	1970	1971	1972	1973	1974	1975	1976	1977
SUB-SAHARAN AFRICA	*5,289*	*5,738*	*6,851*	*7,788*	*8,014*	*9,485*	*13,422*	*22,815*	*20,386*	*25,119*	*29,238*
Excluding Nigeria	*4,616*	*5,141*	*5,938*	*6,548*	*6,200*	*7,299*	*9,932*	*13,496*	*12,287*	*14,327*	*17,187*
Benin	23	33	43	58	70	67	93	93	116	86	129
Botswana	13	10	18	22	28	58	85	121	142	176	180
Burkina Faso	24	28	27	25	25	36	44	66	74	83	95
Burundi	14	16	12	24	19	26	30	30	32	54	89
Cameroon	146	186	232	231	213	225	362	491	470	535	704
Cape Verde	..	..	..	..	..	..	..	..	..	..	..
Central African Republic	36	44	37	43	44	41	55	57	54	70	105
Chad	27	28	31	25	26	34	36	35	40	63	107
Comoros	..	..	..	..	..	..	..	..	..	..	..
Congo, People's Republic	40	40	35	28	35	46	79	208	179	190	174
Cote d'Ivoire	325	422	458	469	456	553	861	1,214	1,182	1,631	2,155
Ethiopia	101	103	119	122	126	168	237	267	229	278	334
Gabon	125	129	142	140	178	228	330	768	943	1,135	1,343
Gambia, The	18	14	17	16	15	19	21	40	49	36	48
Ghana	278	307	302	433	341	393	565	647	737	765	951
Guinea-Bissau	..	..	..	..	..	..	..	..	..	..	..
Kenya	240	250	272	305	314	372	516	661	647	824	1,195
Lesotho	6	6	7	6	4	8	13	14	17	14	32
Liberia	159	168	195	217	222	244	324	400	394	463	454
Madagascar	104	116	113	145	147	164	203	244	301	283	347
Malawi	57	48	53	59	72	81	99	121	139	166	200
Mali	17	11	17	36	36	37	58	64	37	85	125
Mauritania	72	72	78	89	94	119	155	182	174	178	157
Mauritius	63	63	58	69	65	107	138	313	298	265	310
Mozambique	122	155	142	157	160	175	227	296	202	150	149
Niger	26	29	24	32	38	54	63	53	91	134	160
Nigeria	667	592	905	1,228	1,793	2,161	3,448	9,195	7,992	10,569	11,848
Rwanda	14	15	14	25	22	20	51	54	58	114	127
Senegal	137	151	124	161	125	216	195	391	462	485	623
Seychelles	2	3	3	2	2	3	4	7	6	9	11
Sierra Leone	63	91	105	102	99	118	131	146	146	107	131
Somalia	28	30	33	31	38	57	57	64	89	81	71
Sudan	215	233	248	295	332	361	417	441	429	578	661
Swaziland	..	..	..	..	..	..	..	..	..	..	..
Tanzania	217	221	233	238	251	300	344	388	349	490	559
Togo	50	60	73	68	73	70	72	215	141	159	199
Uganda	182	183	196	246	235	261	300	316	264	352	588
Zaire	435	505	679	735	687	738	1,001	1,382	865	809	1,110
Zambia	658	757	1,056	1,001	679	758	1,142	1,407	812	1,041	897
Zimbabwe	272	263	325	370	404	515	693	864	936	974	877
SOUTH ASIA	*2,785*	*2,970*	*3,025*	*3,247*	*3,215*	*3,951*	*4,857*	*6,168*	*6,518*	*7,920*	*9,043*
Bangladesh	..	..	..	..	..	321	326	368	265	362	435
India	1,614	1,754	1,834	2,026	2,043	2,422	2,968	3,906	4,365	5,526	6,355
Nepal	55	40	53	48	48	58	63	66	100	98	81
Pakistan	645	720	682	723	666	698	947	1,102	1,049	1,167	1,174
Sri Lanka	345	342	322	338	327	326	405	523	559	567	763
EAST ASIA AND PACIFIC	*9,417*	*10,221*	*11,870*	*13,401*	*15,432*	*19,663*	*32,514*	*47,323*	*46,499*	*59,567*	*71,435*
China	2,140	2,100	2,200	2,260	2,640	3,440	5,820	6,950	7,260	6,860	7,590
Fiji	49	53	57	68	68	73	78	144	159	127	173
Hong Kong	1,524	1,744	2,178	2,514	2,871	3,478	5,051	5,907	6,019	8,522	9,624
Indonesia	666	731	800	1,055	1,200	1,778	3,211	7,426	7,130	8,556	10,853
Korea, Republic of	320	455	623	835	1,068	1,624	3,225	4,460	5,081	7,715	10,016
Malaysia	1,216	1,347	1,650	1,687	1,639	1,722	3,040	4,234	3,847	5,295	6,079
Papua New Guinea	59	65	72	79	113	149	294	710	580	456	613
Philippines	800	816	823	1,060	1,116	1,029	1,797	2,651	2,218	2,508	3,080
Singapore	1,140	1,271	1,549	1,554	1,755	2,181	3,610	5,785	5,377	6,586	8,242
Solomon Islands	..	..	..	..	..	..	..	..	..	..	..
Thailand	680	654	702	710	827	1,067	1,566	2,449	2,195	2,978	3,490
Vanuatu	..	..	..	..	..	..	..	..	..	..	..

NOTE: See glossary for definitions.

1978	1979	1980	1981	1982	1983	1984	1985	1986	1987	
27,070	*39,803*	*52,751*	*42,243*	*34,122*	*31,668*	*34,992*	*35,514*	*28,959*	*30,401*	*SUB-SAHARAN AFRICA*
16,375	*21,768*	*26,303*	*24,013*	*21,658*	*21,010*	*22,732*	*22,512*	*22,267*	*22,958*	*Excluding Nigeria*
126	133	223	184	144	122	147	152	94	121	Benin
222	436	503	413	456	637	674	745	858	1,010	Botswana
108	133	161	159	126	130	138	112	143	202	Burkina Faso
69	105	65	71	88	77	98	112	167	140	Burundi
805	1,170	1,447	1,764	1,616	1,830	2,080	2,322	1,791	1,770	Cameroon
..	..	9	6	3	2	3	3	3	..	Cape Verde
110	122	147	118	112	115	115	132	145	154	Central African Republic
99	72	72	83	58	74	111	..	..	..	Chad
..	..	11	16	20	20	7	15	19	..	Comoros
161	524	980	1,115	1,213	1,097	1,265	1,145	718	884	Congo, People's Republic
2,323	2,507	3,142	2,535	2,288	2,092	2,707	2,972	3,179	2,961	Cote d'Ivoire
310	422	425	378	404	403	417	333	453	402	Ethiopia
1,107	1,848	2,173	2,201	2,161	1,975	2,011	1,974	1,114	1,285	Gabon
39	58	31	27	44	48	49	43	48	35	Gambia, The
992	1,041	1,257	1,063	873	536	575	663	914	1,056	Ghana
..	..	11	14	12	9	17	12	..	..	Guinea-Bissau
1,023	1,107	1,389	1,183	1,045	979	1,083	988	1,217	985	Kenya
45	58	51	50	35	35	28	21	24	28	Lesotho
496	547	597	529	477	438	452	436	404	385	Liberia
399	408	402	324	330	310	340	287	320	310	Madagascar
185	223	285	274	246	229	314	250	243	264	Malawi
112	148	205	154	146	167	192	181	192	217	Mali
123	147	196	270	240	315	294	374	422	403	Mauritania
326	376	431	324	367	368	372	435	662	817	Mauritius
163	254	281	359	306	240	..	..	..	..	Mozambique
283	448	580	455	333	311	228	223	228	214	Niger
10,560	17,713	25,968	17,846	12,154	10,370	11,891	12,566	6,769	7,475	Nigeria
112	203	134	113	109	124	143	126	167	133	Rwanda
421	533	477	500	548	581	613	498	611	723	Senegal
15	22	21	17	15	20	26	28	19	24	Seychelles
178	194	207	136	111	120	133	130	127	120	Sierra Leone
110	106	133	175	171	98	55	91	105	118	Somalia
483	583	594	538	432	581	722	595	497	482	Sudan
..	..	368	388	325	304	237	176	267	363	Swaziland
472	511	537	564	413	379	340	255	348	348	Tanzania
262	291	476	378	345	274	240	243	275	272	Togo
350	436	345	243	349	372	399	394	443	222	Uganda
899	2,004	2,507	2,030	1,530	1,388	1,558	1,526	1,521	1,594	Zaire
869	1,376	1,360	998	932	997	884	815	689	869	Zambia
891	1,217	1,423	1,406	1,273	1,128	1,154	1,113	1,254	1,358	Zimbabwe
9,858	*12,052*	*13,351*	*13,801*	*13,812*	*15,112*	*15,579*	*14,984*	*16,344*	*19,586*	*SOUTH ASIA*
553	662	759	662	671	705	811	934	819	1,074	Bangladesh
6,650	7,850	8,332	8,698	9,225	9,770	10,192	9,465	10,460	12,548	India
91	108	109	134	115	94	128	160	142	151	Nepal
1,475	2,056	2,618	2,779	2,377	3,075	2,592	2,739	3,384	4,172	Pakistan
846	981	1,049	1,036	1,014	1,066	1,454	1,333	1,215	1,393	Sri Lanka
86,106	*113,883*	*148,377*	*160,763*	*157,075*	*167,260*	*192,968*	*192,501*	*209,235*	*272,620*	*EAST ASIA AND PACIFIC*
9,115	12,525	18,265	21,554	21,820	22,175	25,922	27,326	31,148	39,542	China
216	249	361	298	284	240	256	237	274	314	Fiji
11,499	15,156	19,703	21,737	20,964	21,951	28,318	30,185	35,440	48,475	Hong Kong
11,643	15,590	21,909	22,260	19,747	20,961	20,345	18,711	13,567	17,651	Indonesia
12,695	15,052	17,483	21,250	21,850	24,437	29,248	30,283	34,702	47,172	Korea, Republic of
7,387	11,075	12,939	11,734	12,027	14,100	16,484	15,632	13,830	17,865	Malaysia
756	877	1,033	864	791	820	892	912	1,033	1,172	Papua New Guinea
3,349	4,601	5,788	5,722	5,021	5,001	5,391	4,629	4,842	5,649	Philippines
10,134	14,233	19,376	20,968	20,788	21,833	24,055	22,815	22,428	28,592	Singapore
..	..	74	66	58	62	93	70	66	72	Solomon Islands
4,085	5,297	6,505	7,035	6,957	6,368	7,413	7,121	8,876	11,665	Thailand
..	..	16	19	11	18	32	17	15	17	Vanuatu

Table 18. Value of merchandise exports (cont'd)

Millions of US dollars	1967	1968	1969	1970	1971	1972	1973	1974	1975	1976	1977
LATIN AMERICA AND CARIBBEAN	*11,936*	*12,507*	*13,651*	*15,167*	*15,125*	*17,764*	*25,761*	*41,813*	*38,955*	*44,892*	*52,125*
Argentina	1,465	1,368	1,612	1,773	1,740	1,941	3,266	3,931	2,961	3,912	5,642
Bahamas	32	51	53	88	267	343	530	1,795	2,216	2,616	2,409
Barbados	41	40	37	39	42	45	53	86	107	86	96
Belize	..	..	..	..	..	..	..	..	..	..	..
Bolivia	150	152	172	190	181	201	261	557	444	568	632
Brazil	1,654	1,881	2,311	2,739	2,904	3,991	6,199	7,951	8,670	10,128	12,120
Chile	908	936	1,068	1,234	961	856	1,249	2,481	1,649	2,209	2,138
Colombia	510	558	603	728	689	863	1,176	1,417	1,465	1,745	2,443
Costa Rica	144	171	190	231	225	281	345	440	494	600	840
Dominican Republic	156	164	183	214	241	353	443	649	894	716	780
Ecuador	158	195	153	190	199	326	532	1,124	974	1,258	1,436
El Salvador	207	212	202	236	243	302	358	463	531	743	973
Grenada	..	..	..	..	..	..	..	..	..	..	..
Guatemala	198	228	255	290	283	328	436	572	623	760	1,160
Guyana	110	104	117	133	149	147	137	270	364	268	261
Haiti	33	37	39	48	53	63	95	130	143	176	216
Honduras	156	179	166	170	183	193	247	253	293	392	511
Jamaica	229	215	247	335	330	366	390	731	784	633	778
Mexico	1,026	1,110	1,254	1,205	1,321	1,845	2,632	2,957	2,993	3,468	4,284
Nicaragua	147	157	155	175	183	246	275	377	372	539	633
Panama	85	94	109	110	117	123	138	211	287	237	250
Paraguay	..	..	..	..	..	..	..	..	..	..	..
Peru	755	866	866	1,044	893	944	1,050	1,517	1,291	1,360	1,726
St. Vincent and the Grenadines	..	..	..	..	..	..	..	..	..	..	..
Trinidad and Tobago	441	466	475	482	521	558	696	2,038	1,773	2,219	2,180
Uruguay	159	179	200	233	206	214	322	381	381	536	599
Venezuela	3,108	3,078	3,113	3,197	3,110	3,126	4,773	11,258	9,010	9,466	9,627
MIDDLE EAST AND NORTH AFRICA	*2,344*	*2,708*	*3,152*	*3,328*	*3,303*	*4,165*	*5,982*	*12,347*	*12,110*	*12,318*	*13,822*
Algeria	724	830	934	1,009	857	1,306	1,906	4,260	4,291	4,972	5,809
Egypt, Arab Republic of	566	656	824	870	927	887	1,117	1,588	1,596	1,785	2,015
Jordan	32	34	33	34	32	48	58	155	153	209	249
Morocco	424	450	485	488	499	634	877	1,706	1,543	1,262	1,300
Oman	36	144	194	206	221	242	344	1,211	1,437	1,566	1,573
Syrian Arab Republic	155	176	207	203	195	287	351	784	930	1,055	1,063
Tunisia	149	158	166	183	216	311	386	914	856	789	929
Yemen Arab Republic	3	4	4	3	4	4	8	13	11	8	11
Yemen, PDR	137	110	134	135	96	96	100	228	172	177	181
EUROPE AND MEDITERRANEAN	*8,818*	*9,405*	*10,774*	*11,907*	*13,174*	*16,568*	*22,910*	*28,922*	*32,100*	*36,225*	*40,238*
Cyprus	75	82	88	109	115	134	173	152	151	257	318
Greece	495	468	554	643	663	871	1,454	2,030	2,278	2,558	2,757
Hungary	1,268	1,334	1,553	1,726	1,847	2,403	3,354	3,942	4,519	4,927	5,834
Israel	555	640	724	776	960	1,149	1,509	1,825	1,941	2,416	3,083
Malta	27	34	38	39	45	67	98	134	167	228	289
Poland	2,527	2,858	3,320	3,548	3,872	4,927	6,432	8,321	10,289	11,024	12,273
Portugal	701	761	853	949	1,053	1,294	1,862	2,302	1,940	1,820	2,013
Turkey	522	496	537	589	677	885	1,317	1,538	1,401	1,960	1,753
Yugoslavia	1,252	1,264	1,475	1,679	1,836	2,237	3,020	3,805	4,072	4,896	4,896
Developing countries	43,922	47,148	52,964	58,459	62,120	76,409	113,116	185,261	182,988	212,962	246,378
Highly indebted	13,731	14,255	15,933	18,022	18,459	21,568	32,195	53,315	48,122	56,747	65,543
Low-income countries	8,517	9,034	9,777	10,561	10,508	12,777	17,864	22,373	21,896	24,291	28,307
Low-income Africa	3,267	3,612	4,225	4,726	4,327	5,016	6,656	8,577	7,322	8,454	10,421
China and India	3,753	3,854	4,034	4,286	4,683	5,862	8,788	10,856	11,625	12,386	13,945
Other low-income Asia	1,172	1,216	1,191	1,220	1,172	1,529	1,889	2,262	2,153	2,392	2,686
Middle-income countries	35,596	38,312	43,390	48,109	51,798	63,853	95,521	162,950	161,142	188,602	218,013
High-income oil exporters	4,839	6,051	6,519	7,811	10,819	13,582	20,420	65,974	56,381	71,389	79,669
Industrialized countries	145,748	163,738	188,712	218,607	245,224	289,742	395,189	528,174	564,723	627,145	710,521
Japan	10,442	12,973	15,991	19,319	24,019	28,591	36,931	55,538	55,754	67,203	80,470
United States	31,534	34,389	38,006	43,224	44,130	49,778	71,339	98,507	107,592	114,992	120,133
European Community	75,199	84,309	98,145	113,685	131,068	156,811	213,409	279,711	303,276	333,066	388,632

84

1978	1979	1980	1981	1982	1983	1984	1985	1986	1987	
55,928	*74,634*	*97,290*	*105,082*	*94,285*	*93,894*	*102,968*	*98,356*	*82,936*	*93,963*	*LATIN AMERICA AND CARIBBEAN*
6,394	7,808	8,021	9,143	7,625	7,836	8,107	8,396	6,852	6,360	Argentina
2,118	3,784	4,906	6,178	4,520	3,938	3,911	3,945	3,974	..	Bahamas
130	152	196	244	263	321	391	352	275	156	Barbados
..	..	82	75	60	65	73	64	73	..	Belize
629	760	942	912	828	755	725	623	564	566	Bolivia
12,659	15,244	20,100	23,300	20,200	21,900	27,000	25,600	22,396	26,225	Brazil
2,462	3,894	4,705	3,836	3,710	3,836	3,650	3,823	4,222	5,091	Chile
3,038	3,300	3,950	3,160	3,110	3,080	3,480	3,550	5,110	5,024	Colombia
919	934	1,032	1,011	877	867	967	976	1,125	1,155	Costa Rica
676	869	962	1,188	768	782	868	735	718	711	Dominican Republic
1,558	2,104	2,480	2,168	2,291	2,348	2,621	2,905	2,172	2,021	Ecuador
801	1,131	1,074	797	699	736	725	679	757	577	El Salvador
..	..	17	19	19	19	18	22	28	..	Grenada
1,112	1,161	1,486	1,115	1,120	1,159	1,127	1,060	1,103	1,084	Guatemala
291	290	389	346	241	189	210	206	229	243	Guyana
241	240	341	273	296	272	313	333	285	261	Haiti
602	721	813	713	656	672	725	765	854	827	Honduras
747	817	965	985	739	713	714	549	596	649	Jamaica
6,301	8,817	15,308	20,041	21,230	22,312	24,196	22,108	16,237	20,887	Mexico
646	633	414	476	391	435	387	302	247	300	Nicaragua
244	292	353	319	310	303	258	335	350	357	Panama
..	..	400	399	396	326	361	324	573	952	Paraguay
1,941	3,491	3,898	3,249	3,293	3,015	3,147	2,966	2,531	2,605	Peru
..	..	22	30	34	42	54	62	68	..	St. Vincent and the Grenadines
2,043	2,610	4,080	3,760	3,090	2,353	2,173	2,161	1,386	1,462	Trinidad and Tobago
682	787	1,060	1,220	1,020	1,050	925	855	1,090	1,190	Uruguay
9,270	14,318	19,293	20,125	16,499	14,571	15,841	14,660	9,122	10,567	Venezuela
15,130	*22,317*	*31,414*	*33,320*	*31,014*	*30,601*	*31,117*	*30,864*	*22,200*	*21,875*	*MIDDLE EAST AND NORTH AFRICA*
6,126	9,863	13,871	14,396	13,144	12,583	12,795	13,034	8,066	9,242	Algeria
2,458	3,320	4,759	5,048	5,033	5,936	6,076	5,193	4,040	4,482	Egypt, Arab Republic of
297	402	561	732	739	579	752	789	733	669	Jordan
1,511	1,959	2,403	2,320	2,059	2,062	2,172	2,165	2,428	2,807	Morocco
1,598	2,280	3,748	4,696	4,421	4,248	4,413	4,970	2,889	0	Oman
1,060	1,645	2,108	2,103	2,026	1,923	1,854	1,759	1,368	1,644	Syrian Arab Republic
1,126	1,791	2,234	2,504	1,984	1,872	1,796	1,738	1,759	2,109	Tunisia
7	14	23	48	39	27	32	53	25	19	Yemen Arab Republic
192	466	777	607	795	674	645	681	393	409	Yemen, PDR
46,838	*55,754*	*65,091*	*65,297*	*62,860*	*62,586*	*67,448*	*68,444*	*73,095*	*..*	*EUROPE AND MEDITERRANEAN*
344	456	533	559	554	494	575	476	500	579	Cyprus
3,375	3,877	5,142	4,250	4,297	4,412	4,864	4,536	5,660	6,489	Greece
6,408	7,930	8,672	8,707	8,773	8,702	8,563	8,538	9,645	..	Hungary
3,924	4,553	5,540	5,664	5,280	5,112	5,804	6,256	7,135	8,750	Israel
342	424	483	448	411	363	395	400	497	557	Malta
14,114	16,249	16,997	13,249	11,174	11,572	11,750	11,489	12,074	..	Poland
2,411	3,480	4,629	4,180	4,171	4,602	5,208	5,685	7,160	9,167	Portugal
2,288	2,261	2,910	4,702	5,890	5,905	7,389	8,255	7,583	10,322	Turkey
5,546	6,800	8,978	10,929	10,752	9,914	10,255	10,642	10,298	11,397	Yugoslavia
277,496	373,408	473,842	462,645	440,232	447,060	483,745	479,391	474,464	576,936	Developing countries
69,889	95,854	128,032	128,503	113,695	111,722	123,788	119,985	99,532	112,628	Highly indebted
29,961	38,636	48,325	50,093	48,720	50,104	55,028	55,329	61,768	74,334	Low-income countries
9,431	12,320	14,587	12,839	11,060	10,791	11,342	10,650	11,705	11,729	Low-income Africa
15,765	20,375	26,597	30,252	31,045	31,945	36,114	36,791	41,608	52,090	China and India
3,207	4,201	5,019	5,103	4,587	5,342	5,387	5,519	5,884	7,038	Other low-income Asia
247,483	334,770	425,517	412,549	391,507	396,952	428,731	424,093	412,642	502,609	Middle-income countries
76,236	120,215	186,217	186,133	133,046	93,291	86,646	71,992	49,628	55,459	High-income oil exporters
850,460	1,043,171	1,226,368	1,212,054	1,150,347	1,135,998	1,212,314	1,254,721	1,448,274	..	Industrialized countries
97,502	102,965	129,542	151,910	138,584	146,804	170,038	175,858	209,081	..	Japan
142,536	176,788	216,916	230,506	210,929	199,144	216,008	211,419	211,897	..	United States
471,215	591,274	678,076	625,496	604,548	589,542	602,996	640,817	780,544	..	European Community

Table 19. Growth of merchandise imports

Average annual growth (percent)	1967	1968	1969	1970	1971	1972	1973	1974	1975	1976	1977
SUB-SAHARAN AFRICA	*2.3*	*9.0*	*9.6*	*6.6*	*3.3*	*-1.8*	*-3.1*	*-2.9*	*26.0*	*9.1*	*14.7*
Excluding Nigeria	*2.7*	*9.7*	*8.6*	*4.1*	*1.3*	*-0.6*	*-4.1*	*-6.7*	*8.7*	*-2.7*	*9.6*
Benin	29.9	15.6	0.8	11.9	9.3	15.7	2.0	1.8	10.4	4.5	-1.5
Botswana	20.7	7.3	26.1	34.6	22.9	16.9	17.5	-31.5	12.3	-7.6	22.1
Burkina Faso	-0.7	16.4	18.7	-4.9	15.9	14.5	0.4	-16.5	21.7	-8.9	23.8
Burundi	6.2	15.0	-10.7	17.5	-4.0	-6.9	-22.1	-16.4	47.1	-6.5	45.2
Cameroon	36.5	3.5	10.6	10.9	-5.1	7.6	-9.9	-13.0	29.5	-2.4	14.8
Cape Verde	..	..	..	..	..	..	..	..	..	..	..
Central African Republic	25.6	-10.5	12.5	-28.8	-14.9	2.2	-4.2	-4.5	19.4	-10.4	24.7
Chad	..	..	..	..	..	..	..	..	..	..	..
Comoros	..	..	..	..	..	..	..	..	..	..	..
Congo, People's Republic	17.4	-11.9	-8.2	-35.4	30.2	4.4	14.8	-25.0	24.7	1.0	-4.5
Cote d'Ivoire	3.0	17.4	4.5	9.3	-7.4	10.7	15.5	-0.4	12.2	12.7	22.4
Ethiopia	-3.0	9.2	-8.6	10.4	1.1	-12.6	-7.9	-19.7	3.7	15.8	2.4
Gabon	4.7	-22.8	14.1	-4.1	5.7	34.1	18.4	39.0	30.3	3.9	28.8
Gambia, The	-3.0	11.6	2.0	-24.4	12.3	0.6	16.0	-21.8	27.4	34.4	1.3
Ghana	-4.6	4.9	7.6	10.1	-4.8	-25.0	6.7	17.4	-6.6	6.0	22.4
Guinea-Bissau	..	..	..	..	..	..	..	..	..	..	..
Kenya	0.8	7.8	-3.1	14.7	4.5	-1.8	-9.7	-2.5	-4.4	-11.1	24.7
Lesotho	8.9	19.3	0.5	-5.8	7.1	26.3	20.0	-6.0	26.8	23.1	1.8
Liberia	-1.5	-9.2	-13.0	43.4	0.2	-2.5	-13.3	0.8	7.3	20.3	6.2
Madagascar	-2.4	20.1	6.2	-6.2	6.9	-6.5	-21.8	-16.1	33.7	-22.1	8.9
Malawi	-5.5	3.3	2.2	9.2	32.4	6.7	-14.1	-18.1	24.5	-17.5	3.7
Mali	52.4	-32.9	17.5	12.1	18.0	17.1	18.9	-21.0	12.5	-17.0	-1.1
Mauritania	45.8	16.5	12.8	32.0	-6.0	20.4	19.9	-3.3	11.0	32.7	3.8
Mauritius	24.4	5.9	-8.5	-2.7	-5.1	37.7	3.7	14.2	7.4	9.1	16.0
Mozambique	..	..	..	..	..	..	..	..	..	..	..
Niger	-9.8	-10.4	6.4	13.6	-0.4	14.7	4.0	-33.4	-2.5	29.6	37.3
Nigeria	-0.6	3.3	17.9	24.2	15.5	-8.3	2.9	17.6	99.7	36.4	23.0
Rwanda	-5.4	12.4	1.7	14.6	4.0	3.5	-3.7	8.5	35.2	22.6	-11.7
Senegal	-28.9	37.5	32.1	-15.8	8.3	13.0	-2.9	-8.4	17.4	12.6	8.4
Seychelles	21.4	26.5	2.1	17.6	33.9	18.0	4.5	-30.2	12.9	36.2	-4.1
Sierra Leone	-12.0	2.1	12.0	-5.9	-3.4	-3.3	-4.0	-5.0	-31.2	4.9	1.2
Somalia	1.2	1.5	21.6	-16.0	0.2	20.0	18.8	-4.5	-2.4	11.7	7.5
Sudan	8.5	30.3	19.3	8.6	-4.0	-10.6	1.9	-12.5	32.1	1.1	-4.3
Swaziland	..	..	..	..	..	..	..	..	..	..	..
Tanzania	-24.0	18.9	-5.7	24.5	10.3	-0.7	-0.1	-2.3	-12.6	-4.1	1.3
Togo	-2.8	9.2	16.6	6.9	6.9	2.3	-17.8	-15.2	104.4	-18.5	25.7
Uganda	-3.2	3.1	1.8	-4.8	27.8	-40.8	-16.7	4.6	-9.9	-1.6	23.5
Zaire	-21.0	24.3	73.0	-10.7	-0.1	10.4	-20.7	-18.2	-4.8	-10.1	-7.2
Zambia	36.4	7.2	-2.6	1.5	-7.1	-13.4	-17.6	-9.6	13.1	-30.3	-6.2
Zimbabwe	18.2	11.1	-4.7	9.7	1.6	-6.0	3.9	-7.5	4.3	-26.1	-8.4
SOUTH ASIA	*4.7*	*0.3*	*-13.1*	*7.3*	*-1.9*	*-15.9*	*21.7*	*-9.4*	*18.5*	*-5.8*	*5.5*
Bangladesh	..	..	..	..	..	..	20.2	-37.1	18.6	2.1	-8.2
India	3.4	0.2	-21.4	12.6	12.9	-10.7	10.1	-7.1	17.9	-9.0	7.0
Nepal	-30.2	-10.1	42.4	35.3	-22.3	23.8	-16.7	-32.6	63.6	24.0	3.0
Pakistan	20.1	-2.8	-3.0	10.3	-23.1	-35.0	8.9	12.7	25.3	2.0	0.8
Sri Lanka	-14.9	13.1	2.0	-25.4	-25.2	-9.0	24.6	-7.0	2.3	-15.0	25.3
EAST ASIA AND PACIFIC	*8.2*	*11.2*	*8.5*	*10.8*	*2.4*	*5.9*	*11.5*	*10.2*	*-1.2*	*11.6*	*8.0*
China	-9.4	-2.4	-9.4	24.7	-8.9	19.5	25.7	23.4	1.1	-15.1	6.4
Fiji	9.2	11.8	11.0	14.1	3.8	9.7	-4.1	-14.6	-4.0	-4.7	8.4
Hong Kong	4.7	12.6	13.8	11.9	9.0	3.6	14.0	-7.5	-3.8	26.5	7.9
Indonesia	33.8	-4.6	29.0	3.4	7.9	5.7	43.3	33.3	12.9	26.3	5.8
Korea, Republic of	38.8	40.7	24.8	10.3	12.4	-2.1	18.3	9.7	7.4	16.8	15.2
Malaysia	0.4	10.8	-6.1	10.9	-5.2	-14.4	9.4	14.1	-13.8	8.0	10.2
Papua New Guinea	35.0	-31.5	-6.3	25.5	7.2	-8.7	-16.1	-3.0	7.1	-8.3	21.1
Philippines	21.6	10.9	-3.8	0.5	-3.1	-6.5	-2.0	13.8	5.1	3.1	0.3
Singapore	15.1	17.3	13.7	10.2	1.3	9.6	6.6	-1.1	-4.3	8.4	4.1
Solomon Islands	..	..	..	..	..	..	..	..	..	..	..
Thailand	-19.2	14.7	1.7	5.7	-7.9	3.8	11.7	-6.1	0.2	5.7	18.3
Vanuatu	..	..	..	..	..	..	..	..	..	..	..

NOTE: See glossary for definitions.

1978	1979	1980	1981	1982	1983	1984	1985	1986	1987	
3.4	*-14.1*	*20.9*	*5.6*	*-10.9*	*-13.6*	*-9.8*	*1.3*	*-4.6*	*-6.7*	*SUB-SAHARAN AFRICA*
4.7	*-4.4*	*9.6*	*-6.2*	*-2.3*	*-8.7*	*-2.6*	*3.5*	*4.9*	*-1.0*	*Excluding Nigeria*
17.9	-11.6	-7.6	66.0	-13.0	-36.8	4.0	11.6	18.3	0.3	Benin
15.4	26.5	14.1	15.6	-9.7	9.3	-2.7	-15.1	11.8	6.1	Botswana
1.8	8.0	3.7	-8.7	8.9	-10.7	-13.5	6.9	14.1	11.5	Burkina Faso
-2.5	32.8	-4.8	-5.6	39.1	-7.0	-2.2	2.2	1.3	2.0	Burundi
18.9	9.8	4.9	-8.7	-9.4	2.9	-1.0	-4.2	26.9	0.4	Cameroon
..	..	..	..	..	..	..	..	..	..	Cape Verde
3.5	-11.1	25.9	-23.0	-0.6	-7.5	3.5	11.9	-7.5	0.7	Central African Republic
..	..	..	..	..	..	..	..	..	..	Chad
..	..	..	..	..	..	..	..	..	..	Comoros
23.3	-7.3	65.6	63.5	1.9	-4.6	0.9	3.0	-21.5	-15.4	Congo, People's Republic
15.9	-11.4	8.5	-22.6	-2.6	-14.0	-16.5	19.4	16.2	3.5	Cote d'Ivoire
12.8	-2.6	6.1	-0.9	10.6	15.8	7.6	7.8	10.0	3.0	Ethiopia
-15.6	-14.2	-6.4	23.3	-2.7	9.5	-13.8	18.9	-5.8	-31.6	Gabon
22.9	22.9	-18.6	8.9	16.3	-23.8	-13.2	9.9	7.2	6.4	Gambia, The
-21.9	-26.6	9.1	-5.3	-32.5	-11.3	8.6	16.3	10.6	-4.7	Ghana
..	..	..	..	..	..	..	..	..	..	Guinea-Bissau
17.5	-17.7	28.2	-22.9	-18.4	-9.9	9.9	-1.7	24.0	1.2	Kenya
7.9	15.7	19.3	16.1	-4.4	9.9	-11.8	-18.4	-12.4	3.7	Lesotho
-5.1	-11.1	-13.3	-14.1	-3.7	1.6	-7.9	-20.2	-12.1	-19.2	Liberia
15.7	28.5	-17.9	-31.3	-1.6	-3.3	0.5	15.0	-5.9	-13.9	Madagascar
30.2	0.1	-5.9	-22.7	-8.0	3.1	-11.3	6.6	-14.5	-1.3	Malawi
66.0	3.8	1.7	-19.1	-4.5	8.9	4.4	24.0	0.7	-5.2	Mali
-19.4	-9.2	-4.5	21.3	18.0	-9.2	-19.9	13.5	-8.8	5.6	Mauritania
-1.2	-2.4	-5.8	-12.9	-10.2	-2.0	9.5	14.0	22.0	12.1	Mauritius
..	..	..	..	..	..	..	..	..	..	Mozambique
47.4	28.5	9.7	-17.9	-7.1	-16.8	-8.0	21.4	15.1	2.0	Niger
1.5	-29.0	44.0	24.0	-21.1	-20.8	-21.9	-3.4	-26.1	-24.7	Nigeria
42.4	-5.7	8.6	-0.2	10.4	-4.3	4.6	11.9	13.6	26.5	Rwanda
-9.8	5.4	-6.9	-0.5	1.8	7.1	2.7	-6.7	-3.0	12.3	Senegal
4.7	42.6	-9.2	-7.2	11.6	-8.3	3.8	21.4	4.0	14.6	Seychelles
39.0	-5.8	14.5	-27.9	-0.1	-44.7	0.7	-1.9	-10.9	-12.0	Sierra Leone
21.6	21.5	3.8	-8.8	33.2	-21.8	31.4	-27.2	-5.3	0.0	Somalia
-21.1	-8.5	44.5	-0.1	23.6	-11.3	-9.7	-7.7	-10.5	-18.3	Sudan
..	..	..	..	..	..	..	..	..	..	Swaziland
37.4	-19.2	-3.1	-7.7	2.7	-27.6	8.5	23.3	-0.1	5.1	Tanzania
55.8	-3.8	-3.4	-20.4	2.0	-26.8	-10.8	-1.6	14.1	-7.1	Togo
-3.6	-37.0	21.7	14.2	13.4	4.0	-6.8	-4.1	-1.0	8.5	Uganda
-17.6	-11.8	18.2	-10.0	-6.7	-6.0	6.9	6.0	-0.1	3.8	Zaire
-13.9	-1.5	24.3	-6.2	-18.4	-14.7	8.0	-10.4	3.7	-1.8	Zambia
-14.9	35.2	32.2	13.4	2.6	-24.3	-18.9	9.9	-0.4	-12.5	Zimbabwe
14.3	*8.6*	*9.3*	*4.7*	*1.1*	*2.8*	*4.3*	*12.4*	*-1.8*	*0.8*	*SOUTH ASIA*
13.9	5.7	12.3	15.3	3.6	-11.0	5.1	19.6	-5.3	-6.0	Bangladesh
10.4	8.1	8.9	7.5	-0.8	5.5	4.4	14.7	-1.9	1.7	India
28.8	-9.1	-28.0	5.3	16.1	12.0	-6.9	13.7	-0.7	9.8	Nepal
20.6	4.4	13.0	-0.6	2.0	4.7	7.7	5.2	-0.3	0.6	Pakistan
14.0	27.8	14.7	-14.1	4.2	5.4	6.0	12.3	-0.1	-1.1	Sri Lanka
14.4	*9.5*	*8.4*	*10.5*	*2.5*	*7.4*	*9.3*	*5.2*	*1.5*	*10.6*	*EAST ASIA AND PACIFIC*
33.8	22.9	12.5	1.5	2.4	12.2	30.6	57.5	-11.0	-8.4	China
7.3	6.1	0.2	9.6	-14.0	-2.1	-5.2	-1.0	0.2	-9.4	Fiji
13.5	9.3	15.0	11.9	-1.4	4.0	20.9	5.4	7.9	21.9	Hong Kong
-3.2	-5.2	15.6	29.2	13.2	2.8	-13.3	-12.9	-7.1	-3.7	Indonesia
23.5	10.8	-9.0	15.7	-0.5	10.8	18.3	5.4	8.1	13.0	Korea, Republic of
13.1	15.7	14.3	10.7	12.8	7.1	3.5	-12.4	-17.2	0.2	Malaysia
2.8	-1.9	10.4	4.7	-1.1	-0.6	1.7	-6.6	9.3	-0.9	Papua New Guinea
8.5	4.7	3.1	-1.5	2.4	0.3	-17.3	-13.9	5.5	15.6	Philippines
13.1	8.9	10.9	12.1	8.2	3.1	3.4	-6.3	6.2	8.8	Singapore
..	..	..	..	..	..	..	..	..	..	Solomon Islands
4.8	9.1	8.8	3.4	-11.2	24.9	4.4	-10.4	1.6	35.4	Thailand
..	..	..	..	..	..	..	..	..	..	Vanuatu

Table 19. Growth of merchandise imports (cont'd)

Average annual growth (percent)	1967	1968	1969	1970	1971	1972	1973	1974	1975	1976	1977
LATIN AMERICA AND CARIBBEAN	*2.3*	*9.5*	*4.4*	*9.6*	*4.7*	*2.9*	*4.2*	*4.7*	*1.9*	*-0.1*	*3.3*
Argentina	-6.3	1.1	25.9	-6.0	8.2	-17.1	7.1	10.4	2.5	-24.8	24.4
Bahamas	..	..	..	..	..	..	..	..	..	..	..
Barbados	-6.1	15.5	2.1	5.2	-2.7	2.3	-5.2	-13.9	3.4	3.3	5.2
Belize	..	..	..	..	..	..	..	..	..	..	..
Bolivia	6.5	4.3	1.9	-7.9	0.2	-4.6	2.8	23.0	50.0	4.3	-8.6
Brazil	8.4	25.3	-3.0	20.9	18.4	14.6	15.2	13.3	-4.8	0.3	-9.3
Chile	4.9	3.6	8.5	1.0	6.9	1.1	-5.6	-13.3	-15.3	-10.8	42.9
Colombia	-24.0	26.3	2.8	16.0	4.9	-17.4	-1.4	20.0	-10.1	12.9	11.9
Costa Rica	1.8	12.8	8.7	21.0	3.5	1.0	2.1	4.6	-7.9	11.7	21.8
Dominican Republic	5.0	9.1	15.0	12.5	11.3	4.7	0.4	-3.7	11.6	-15.6	15.1
Ecuador	66.5	-9.0	5.1	6.6	20.4	-18.6	-15.4	56.4	7.3	-4.8	12.3
El Salvador	2.3	0.0	-12.1	-11.7	21.5	-1.3	11.9	4.2	-1.6	17.8	23.1
Grenada	..	..	..	..	..	..	..	..	..	..	..
Guatemala	10.2	-5.2	-5.5	8.4	13.0	9.3	1.7	9.1	-1.4	11.8	16.3
Guyana	9.9	-9.5	2.2	8.0	-11.1	-2.1	6.4	-17.5	27.7	2.3	-17.9
Haiti	1.8	0.6	27.1	6.5	3.9	11.5	2.5	6.0	8.8	26.2	8.9
Honduras	9.0	22.9	-2.6	14.1	-15.1	-6.8	6.2	-9.8	0.2	6.0	18.7
Jamaica	6.5	7.5	8.3	13.2	5.1	0.0	-13.7	-14.2	16.8	-18.8	-9.0
Mexico	10.4	11.2	4.8	10.6	-5.4	12.9	20.6	-6.7	1.7	-8.3	-15.9
Nicaragua	13.9	-5.5	-5.4	9.6	-0.5	-6.3	17.0	14.0	-11.3	0.1	32.3
Panama	2.6	10.0	10.6	7.8	-9.7	-2.8	-4.9	-14.6	8.3	-12.1	-7.6
Paraguay	..	..	..	..	..	..	..	..	..	..	..
Peru	-0.3	-19.5	-10.9	-6.4	22.2	4.9	-0.6	6.8	41.2	-23.0	-15.5
St. Vincent and the Grenadines	..	..	..	..	..	..	..	..	..	..	..
Trinidad and Tobago	-9.6	6.2	13.3	12.7	-8.5	-0.1	-22.0	-16.0	-27.4	31.2	-19.9
Uruguay	-7.2	-3.8	-1.4	26.4	-19.4	-18.5	-1.0	4.8	9.4	12.3	-2.9
Venezuela	5.7	18.0	-2.1	8.0	1.4	6.3	-6.2	19.1	33.0	26.5	32.1
MIDDLE EAST AND NORTH AFRICA	*-15.0*	*2.0*	*9.3*	*9.2*	*-6.8*	*1.1*	*-3.9*	*31.6*	*37.2*	*3.1*	*23.0*
Algeria	4.3	29.6	18.4	22.5	-2.7	3.7	18.0	35.6	36.1	-10.5	24.2
Egypt, Arab Republic of	-25.2	-12.2	-2.3	22.6	-4.7	-14.2	-35.4	79.5	72.3	0.2	17.4
Jordan	-17.3	5.3	14.3	-6.1	7.6	5.1	-8.2	6.4	50.0	39.2	25.2
Morocco	9.9	14.4	-2.2	15.4	-3.8	1.8	4.6	14.6	26.8	5.0	14.3
Oman	..	..	..	..	..	..	..	..	..	..	..
Syrian Arab Republic	-27.4	26.6	12.3	-15.9	2.2	3.6	-10.8	42.0	29.6	18.5	29.3
Tunisia	-7.1	-19.9	17.0	21.9	0.1	37.8	0.6	20.8	20.2	8.3	10.6
Yemen Arab Republic	10.8	7.1	9.5	-18.4	5.5	95.7	19.9	4.2	46.9	42.3	139.7
Yemen, PDR	-23.3	-18.6	21.2	-11.1	-38.0	-13.8	-17.4	-9.3	-19.8	27.8	16.2
EUROPE AND MEDITERRANEAN	..	..	..	..	..	..	..	..	..	..	..
Cyprus	9.0	7.2	10.8	9.1	0.4	6.7	4.1	-35.1	-27.4	37.2	32.5
Greece	0.1	14.3	11.4	15.9	-3.7	10.9	22.5	-22.4	18.3	6.7	0.6
Hungary	..	..	..	..	..	..	..	..	..	..	..
Israel	9.5	28.8	14.6	19.0	3.5	-6.4	45.6	-13.5	6.8	-7.4	-6.6
Malta	5.2	10.2	9.4	4.7	-7.7	7.4	1.4	5.2	-1.3	8.9	11.5
Poland	..	..	..	..	..	..	..	..	..	..	..
Portugal	9.3	11.3	7.3	22.5	-2.4	4.0	1.6	3.4	-17.3	8.5	6.9
Turkey	-3.3	14.8	-6.6	11.4	24.4	21.6	7.7	8.5	18.1	9.8	6.5
Yugoslavia	8.4	8.5	10.7	29.5	9.2	-12.9	24.3	3.7	-0.7	-7.5	10.7
Developing countries	3.7	8.9	4.7	12.2	2.9	0.4	7.5	5.1	6.3	3.9	7.1
Highly indebted	5.8	10.4	4.0	12.1	6.7	0.0	8.8	8.0	8.8	1.9	7.3
Low-income countries	0.1	3.9	-2.9	7.8	-1.3	-5.0	9.7	-1.5	9.5	-7.4	6.4
Low-income Africa	-0.8	11.5	11.3	2.7	2.2	-3.9	-7.4	-7.8	6.4	-4.3	7.6
China and India	-1.8	-0.8	-16.9	17.5	3.6	0.6	17.1	7.5	8.7	-12.1	6.7
Other low-income Asia	6.5	0.5	-1.3	1.1	-20.6	-25.3	43.5	-12.6	19.5	-0.9	3.4
Middle-income countries	4.5	10.1	6.4	13.1	3.7	1.4	7.1	6.2	5.7	5.8	7.1
High-income oil exporters	13.4	8.0	8.5	-8.8	8.7	26.8	22.5	23.4	23.2	40.7	37.2
Industrialized countries	9.6	12.1	7.8	12.9	1.5	5.5	6.6	-12.5	-2.3	10.0	3.5
Japan	25.1	15.8	12.4	24.8	-4.9	8.2	9.0	-15.5	-2.4	3.9	2.0
United States	4.0	18.9	5.5	6.8	3.3	12.6	7.3	-8.9	-6.1	19.5	11.8
European Community	10.3	10.0	8.8	13.0	3.2	3.7	5.9	-15.5	-0.7	9.2	1.8

NOTE: See glossary for definitions.

1978	1979	1980	1981	1982	1983	1984	1985	1986	1987	
2.9	*4.6*	*11.9*	*4.3*	*-15.8*	*-22.1*	*2.2*	*1.8*	*1.4*	*-0.5*	*LATIN AMERICA AND CARIBBEAN*
-18.6	47.4	33.6	-10.7	-41.4	-13.4	4.1	-16.7	22.7	20.9	Argentina
..	..	..	..	..	..	..	..	..	..	Bahamas
2.6	13.1	6.8	8.6	0.5	15.4	7.5	-6.8	-2.5	-17.6	Barbados
..	..	..	..	..	..	..	..	..	..	Belize
15.2	-5.2	-27.9	37.7	-34.8	3.2	-15.6	12.2	16.5	3.4	Bolivia
4.7	4.4	-2.6	-8.1	-5.7	-16.3	-7.2	-4.1	25.7	-7.8	Brazil
21.5	22.8	6.6	1.1	-31.9	-21.0	20.2	-9.7	2.3	24.0	Chile
25.6	-1.8	25.7	10.4	9.7	-7.4	-7.8	-6.5	-13.9	-1.6	Colombia
2.2	2.3	-5.1	-21.2	-22.7	7.4	11.7	1.6	-0.8	7.8	Costa Rica
-6.2	1.0	9.4	-1.8	-7.9	5.5	-0.4	5.3	0.6	10.6	Dominican Republic
12.6	15.6	1.5	-15.1	-4.8	-13.1	9.9	10.0	-9.2	13.6	Ecuador
-4.7	-15.9	-15.3	4.6	-5.9	-2.3	12.5	-0.6	-13.3	4.4	El Salvador
..	..	..	..	..	..	..	..	..	..	Grenada
6.7	-10.5	-0.7	23.3	-26.9	-15.0	14.5	-7.2	-18.2	35.9	Guatemala
-15.6	-9.5	-4.8	16.3	-32.8	-9.9	-11.6	17.4	-10.7	-1.3	Guyana
-2.9	1.0	17.5	5.6	-11.4	-5.5	6.4	5.9	-11.8	-7.7	Haiti
7.2	0.5	4.9	-8.4	-24.4	23.4	10.9	-0.2	-4.1	-6.1	Honduras
-8.8	-6.4	-6.5	21.6	-0.9	15.0	-24.6	1.9	-10.3	14.1	Jamaica
27.6	33.6	40.0	23.9	-35.6	-46.1	49.7	22.8	-23.2	-4.6	Mexico
-7.0	-54.7	76.2	9.9	-18.3	6.3	8.4	15.1	-14.3	-0.1	Nicaragua
-4.2	3.1	-1.8	4.8	5.1	-6.7	1.8	-1.2	-14.2	-13.6	Panama
..	..	-4.3	12.1	-5.8	-22.6	15.1	-15.2	45.2	36.5	Paraguay
-19.8	3.4	41.0	23.6	2.0	-25.4	-17.1	-7.4	27.0	11.1	Peru
..	..	..	..	..	..	..	..	..	..	St. Vincent and the Grenadines
-1.3	-22.2	21.3	-5.5	25.9	-28.2	-24.0	-17.9	-19.7	-15.2	Trinidad and Tobago
3.6	23.4	15.8	-3.9	-28.7	-26.0	0.2	-6.5	33.3	11.3	Uruguay
-5.7	-21.0	0.8	11.0	2.3	-32.5	-11.6	11.7	3.4	-15.6	Venezuela
3.6	*-3.1*	*5.7*	*9.8*	*1.5*	*3.1*	*2.9*	*-3.5*	*-15.6*	*2.0*	*MIDDLE EAST AND NORTH AFRICA*
8.4	-15.8	13.3	7.5	-2.3	-1.6	0.6	-0.7	-14.1	-4.9	Algeria
21.0	-10.8	3.1	11.0	8.3	18.8	8.6	-6.0	-24.7	16.9	Egypt, Arab Republic of
-2.6	9.6	5.7	29.3	7.3	-3.8	-5.4	1.4	-15.7	-10.0	Jordan
-13.7	4.5	-5.3	3.0	5.9	-14.6	9.4	3.9	6.3	3.6	Morocco
..	..	..	32.2	21.8	-6.1	11.3	14.0	-36.0	..	Oman
-19.0	18.6	0.7	19.0	-15.0	18.2	-8.7	-13.0	-25.4	-4.0	Syrian Arab Republic
6.1	13.2	3.7	6.4	-4.6	-7.3	2.2	-8.5	2.8	-4.1	Tunisia
14.0	-5.8	12.0	-5.8	-9.0	5.5	0.4	-10.8	-35.3	1.9	Yemen Arab Republic
1.1	26.3	17.1	-13.0	21.8	-0.5	6.1	-12.8	24.5	-0.9	Yemen, PDR
..	..	..	..	..	..	..	..	..	..	*EUROPE AND MEDITERRANEAN*
7.4	11.8	3.3	-9.3	15.0	2.3	13.9	-6.6	1.8	7.7	Cyprus
3.7	2.9	-8.5	-17.5	19.0	-1.9	3.2	8.0	13.3	4.9	Greece
..	..	..	..	..	..	..	..	..	..	Hungary
11.7	-0.6	-5.2	4.8	-1.0	1.4	2.3	5.1	1.2	11.1	Israel
-1.1	10.9	11.1	-9.1	-3.8	-5.3	-0.5	7.8	9.3	4.7	Malta
..	..	..	..	..	..	..	..	..	..	Poland
-2.5	3.9	18.8	5.5	2.9	-12.7	-1.3	1.3	29.8	21.8	Portugal
-24.4	-11.5	18.6	12.3	5.5	10.2	18.3	10.6	3.5	7.3	Turkey
0.2	19.3	-8.6	3.6	-5.7	-11.5	0.4	4.1	6.0	3.0	Yugoslavia
6.7	3.7	11.1	7.3	-4.0	-3.4	4.9	2.8	0.1	6.0	Developing countries
2.8	3.1	11.2	5.7	-14.7	-20.0	-2.7	1.4	1.1	-0.9	Highly indebted
15.3	7.7	10.5	0.1	0.7	2.7	13.0	29.0	-5.7	-3.6	Low-income countries
4.4	-6.5	9.8	-8.4	-1.7	-9.8	1.1	2.3	3.4	0.7	Low-income Africa
21.9	16.1	10.9	4.0	1.0	9.3	19.9	42.2	-8.4	-5.3	China and India
19.9	9.3	9.9	1.0	3.8	-0.8	4.0	9.2	-1.6	-0.6	Other low-income Asia
5.5	3.1	11.2	8.5	-4.7	-4.4	3.5	-2.2	1.5	8.3	Middle-income countries
4.5	7.1	10.4	18.1	7.8	-6.4	-9.5	-24.9	-19.0	-3.1	High-income oil exporters
6.0	4.5	-1.0	-4.7	-0.8	1.8	10.4	6.6	11.6	..	Industrialized countries
4.7	4.8	-3.4	-1.4	-0.1	0.2	9.7	-0.5	7.7	..	Japan
10.6	-3.0	-7.6	4.5	-5.6	9.5	26.1	9.6	13.1	..	United States
6.9	7.3	1.2	-10.3	1.9	-0.8	4.4	7.2	12.3	..	European Community

Table 20. Growth of merchandise exports

Average annual growth (percent)	1967	1968	1969	1970	1971	1972	1973	1974	1975	1976	1977
SUB-SAHARAN AFRICA	*-0.1*	*-6.2*	*37.1*	*32.9*	*21.4*	*13.0*	*5.3*	*-18.6*	*-2.2*	*9.2*	*1.0*
Excluding Nigeria	*12.3*	*7.3*	*6.6*	*6.9*	*3.2*	*8.1*	*-4.8*	*-1.6*	*7.2*	*-4.9*	*1.6*
Benin	16.5	45.2	21.3	31.4	26.7	-15.6	-22.0	5.3	42.2	-47.3	7.2
Botswana	-9.9	-21.3	69.6	18.6	26.3	85.9	9.8	19.1	15.7	22.8	2.2
Burkina Faso	16.2	17.7	-4.9	-11.2	-8.6	33.0	-22.6	47.9	30.1	-9.5	14.5
Burundi	52.3	16.5	-28.4	66.2	-15.6	22.9	-4.9	-9.7	9.7	-17.7	7.6
Cameroon	5.4	18.9	14.5	-1.0	2.8	-3.3	9.6	11.0	7.3	-25.1	-4.2
Cape Verde	..	..	..	..	..	..	..	..	..	..	..
Central African Republic	-2.1	16.0	-15.1	3.0	4.5	-13.2	-2.3	-6.5	0.9	-9.2	22.0
Chad	..	..	..	..	..	..	..	..	..	..	..
Comoros	..	..	..	..	..	..	..	..	..	..	..
Congo, People's Republic	5.2	-2.7	-14.3	-22.7	21.0	37.6	135.0	22.2	-12.2	-3.2	-23.4
Cote d'Ivoire	8.3	20.8	3.0	-2.9	5.5	16.7	5.4	6.9	9.0	-5.8	-5.4
Ethiopia	3.2	5.4	12.7	-15.3	12.9	10.6	3.9	12.8	-12.0	-32.3	-20.5
Gabon	61.9	14.0	14.0	-0.6	13.7	13.7	24.3	-8.8	28.3	9.5	9.8
Gambia, The	6.7	-14.2	9.8	-16.9	-8.0	11.2	-35.1	89.4	49.9	-32.8	11.7
Ghana	2.3	-5.5	-13.9	67.6	-4.8	-0.3	-9.7	-16.4	35.0	-27.6	-18.3
Guinea-Bissau	..	..	..	..	..	..	..	..	..	..	..
Kenya	1.6	6.3	11.1	9.4	-4.6	-1.2	-0.9	-20.6	4.2	3.2	9.1
Lesotho	9.8	-1.0	3.2	-6.9	-31.2	70.0	9.4	-5.2	-4.5	23.8	-19.3
Liberia	12.2	6.9	17.4	1.2	11.3	9.6	-6.5	11.7	-8.5	12.1	-7.0
Madagascar	30.4	8.3	-6.5	16.4	-2.6	2.7	-4.4	-8.0	29.8	-33.1	-10.9
Malawi	14.8	-8.1	7.0	2.6	13.7	8.9	0.3	-16.7	2.3	28.6	2.8
Mali	23.8	-34.9	55.4	88.9	-5.2	-2.0	1.6	0.2	-32.7	90.3	48.3
Mauritania	24.6	7.0	7.3	-11.4	17.9	33.4	29.0	-24.1	-15.9	4.8	-12.5
Mauritius	-9.2	2.8	-40.2	3.7	-22.0	7.6	-0.7	-18.9	32.4	38.5	43.9
Mozambique	..	..	..	..	..	..	..	..	..	..	..
Niger	-28.2	8.9	-21.8	21.9	27.7	26.5	3.2	-39.7	60.0	41.4	14.5
Nigeria	-19.9	-36.4	153.3	74.5	39.2	16.6	12.2	-28.2	-9.6	22.2	0.6
Rwanda	28.3	5.7	-9.8	49.6	-1.4	-17.2	94.7	-8.7	10.3	25.6	-17.4
Senegal	0.7	6.4	-11.7	23.3	-20.0	46.5	-27.0	6.7	28.1	10.6	18.9
Seychelles	14.1	16.9	27.4	11.1	-32.6	65.8	-5.1	-22.2	13.3	21.4	17.6
Sierra Leone	-31.0	38.4	14.2	1.5	-0.6	13.3	-20.8	-22.4	11.9	-35.7	-5.3
Somalia	-4.5	3.3	-1.4	-10.9	19.9	26.8	-22.0	35.4	51.5	-22.6	-9.8
Sudan	2.1	7.3	14.6	16.7	-1.1	-2.1	-29.5	-3.0	15.0	4.1	16.4
Swaziland	..	..	..	..	..	..	..	..	..	..	..
Tanzania	54.4	8.0	-11.4	-1.6	0.6	14.5	-25.4	-28.5	-3.3	10.3	-8.1
Togo	1.7	19.1	17.5	-8.3	13.5	-6.0	-23.1	8.1	-29.3	31.9	20.6
Uganda	3.4	0.8	7.7	4.3	3.1	1.0	-11.1	0.8	-11.7	-32.3	13.4
Zaire	12.5	6.4	23.3	4.6	23.8	10.4	-27.5	10.7	-7.1	-17.7	26.3
Zambia	26.6	4.5	18.0	-1.3	-10.2	9.6	-10.0	5.4	-5.6	14.6	-8.0
Zimbabwe	2.8	-5.9	15.6	4.2	9.8	22.7	0.2	6.2	9.7	-8.2	-13.3
SOUTH ASIA	*5.3*	*1.6*	*-0.1*	*-6.7*	*-3.0*	*5.3*	*-5.5*	*-3.0*	*6.4*	*22.3*	*1.5*
Bangladesh	..	..	..	..	..	..	-16.3	-15.4	-24.2	36.9	10.6
India	6.5	14.2	3.9	-14.5	1.1	12.6	0.7	-0.6	4.6	29.8	4.8
Nepal	-3.4	-23.8	25.6	-6.5	-5.8	-1.1	-16.6	-10.9	62.8	4.9	-26.9
Pakistan	11.0	-21.0	-3.3	7.1	-13.9	-6.4	-21.6	-5.3	17.8	9.0	-14.0
Sri Lanka	13.8	7.0	-24.4	16.7	-8.1	-3.6	7.5	-5.4	10.6	-7.7	0.5
EAST ASIA AND PACIFIC	*6.2*	*9.4*	*10.3*	*5.6*	*6.5*	*19.1*	*15.7*	*-8.4*	*5.2*	*17.5*	*11.6*
China	-13.6	-7.3	-4.7	3.2	10.8	10.3	21.4	3.9	15.2	-11.0	2.1
Fiji	28.5	3.8	-9.0	11.2	-13.5	-15.6	-23.3	-32.6	40.0	10.5	55.3
Hong Kong	16.1	6.3	24.1	7.9	-0.9	-0.1	2.5	-6.8	12.0	32.1	12.3
Indonesia	11.0	19.7	19.9	1.5	6.4	58.3	22.5	-16.7	3.9	7.1	12.9
Korea, Republic of	28.2	30.7	45.3	23.5	21.4	49.2	60.4	10.8	10.6	35.5	27.9
Malaysia	5.3	14.4	4.7	4.8	1.0	-9.2	6.1	-5.9	7.5	24.1	-1.9
Papua New Guinea	7.7	-14.3	5.8	4.4	53.6	18.7	21.4	95.1	-10.9	-33.3	7.5
Philippines	-1.5	-3.9	-3.5	23.1	9.0	-13.5	11.6	-4.3	3.6	11.8	17.4
Singapore	12.6	16.2	9.3	-2.1	-0.1	11.8	2.6	-19.5	-1.7	10.6	18.1
Solomon Islands	..	..	..	..	..	..	..	..	..	..	..
Thailand	6.8	-5.5	-0.7	8.0	19.4	21.5	-0.5	2.9	-0.2	40.0	8.4
Vanuatu	..	..	..	..	..	..	..	..	..	..	..

NOTE: See glossary for definitions.

1978	1979	1980	1981	1982	1983	1984	1985	1986	1987	
-9.0	*17.7*	*-2.2*	*-21.4*	*-11.0*	*-5.3*	*10.5*	*12.6*	*0.3*	*-3.0*	*SUB-SAHARAN AFRICA*
-3.9	*15.0*	*5.3*	*-3.1*	*-0.9*	*-4.1*	*7.3*	*10.3*	*3.0*	*0.1*	*Excluding Nigeria*
7.3	1.8	71.5	-18.4	-11.9	-14.1	20.9	13.8	-17.8	7.1	Benin
13.4	59.9	6.6	-16.3	16.6	42.7	7.6	12.8	2.8	7.7	Botswana
4.5	4.8	12.8	7.9	-11.1	-6.3	7.9	-1.6	37.5	18.3	Burkina Faso
4.3	40.2	-31.3	30.4	15.3	-7.8	14.9	16.4	17.1	23.2	Burundi
24.0	44.9	11.9	22.6	-2.5	18.1	12.8	16.5	-0.8	-2.5	Cameroon
..	..	..	..	..	..	..	..	..	..	Cape Verde
14.4	-10.1	12.8	-9.6	-4.7	5.7	-4.1	24.6	-2.8	2.7	Central African Republic
..	..	..	..	..	..	..	..	..	..	Chad
..	..	..	..	..	..	..	..	..	..	Comoros
-7.0	151.4	21.3	5.9	19.0	-1.9	16.6	-5.8	-2.1	1.6	Congo, People's Republic
13.8	-5.2	27.5	-6.8	-6.0	-10.5	20.2	16.7	2.3	-4.5	Cote d'Ivoire
21.9	17.8	11.0	2.8	4.1	1.3	-5.9	-17.1	10.1	21.1	Ethiopia
-22.7	22.3	-22.5	-5.1	6.2	0.1	3.3	3.4	-19.9	-4.1	Gabon
-21.9	51.8	-50.4	-17.2	120.7	0.1	2.8	-0.5	18.2	-24.1	Gambia, The
6.7	-5.7	32.5	-0.7	-2.3	-47.5	1.3	24.6	48.3	6.0	Ghana
..	..	..	..	..	..	..	..	..	..	Guinea-Bissau
-3.7	-8.2	10.0	-11.5	-4.2	-6.8	0.1	6.4	11.1	-7.3	Kenya
100.0	14.3	20.8	-8.6	-26.4	0.6	-16.5	-28.1	1.4	8.7	Lesotho
15.0	-11.8	-3.4	0.2	-10.6	-4.1	6.3	2.0	-6.1	-10.9	Liberia
29.6	-9.7	-0.1	-9.4	0.0	-4.1	4.6	-16.6	-2.5	16.8	Madagascar
-6.5	3.7	12.7	-0.6	-6.3	-6.0	37.0	6.8	-11.9	12.2	Malawi
-17.1	23.5	21.9	-21.2	11.7	4.1	15.5	12.1	15.7	-8.6	Mali
-13.7	0.2	17.3	45.6	-5.7	34.5	-9.8	40.6	42.7	-0.9	Mauritania
5.4	-4.6	-49.8	5.3	96.2	1.0	31.4	35.5	17.2	13.5	Mauritius
..	..	..	..	..	..	..	..	..	..	Mozambique
54.8	48.9	31.5	-23.2	-24.2	-8.2	-21.2	1.0	1.5	4.2	Niger
-12.8	19.9	-7.9	-37.5	-24.7	-7.5	16.4	16.4	-4.1	-8.5	Nigeria
10.2	51.7	-30.4	-1.6	-8.1	19.3	5.9	-11.9	2.1	21.6	Rwanda
-30.9	8.6	-25.0	-1.6	28.3	7.9	3.1	-10.8	29.8	8.7	Senegal
39.5	1.0	-32.9	-23.3	-1.3	35.6	18.3	27.2	19.6	-2.2	Seychelles
45.8	-8.8	2.7	-30.5	-13.9	9.1	8.4	-0.1	-6.1	-6.9	Sierra Leone
16.5	-28.0	30.8	42.4	1.5	-44.7	-40.4	72.8	17.6	1.6	Somalia
-29.6	9.9	-10.9	-6.7	-6.7	21.7	29.3	-12.7	-1.7	-10.3	Sudan
..	..	..	..	..	..	..	..	..	..	Swaziland
-9.5	-7.1	2.6	18.9	-24.8	-8.0	-13.8	-18.8	17.4	8.6	Tanzania
48.7	0.7	28.8	-21.3	5.2	-14.7	-16.8	14.4	16.0	-8.6	Togo
-21.4	14.7	-15.5	-15.3	35.5	10.2	-2.0	3.3	-13.4	-22.0	Uganda
-21.1	71.7	19.0	-4.3	-16.9	-12.6	13.9	1.5	-7.1	3.9	Zaire
-6.5	9.7	-10.2	-9.7	9.6	0.1	0.5	-9.7	-13.1	-1.0	Zambia
-0.7	18.6	4.2	3.9	1.4	-15.8	3.3	11.5	14.0	-8.6	Zimbabwe
1.8	*8.0*	*-5.3*	*-0.1*	*6.6*	*6.6*	*0.4*	*3.8*	*8.9*	*10.2*	*SOUTH ASIA*
20.3	4.1	11.5	-9.4	1.7	6.2	7.3	7.5	2.2	22.0	Bangladesh
-2.4	4.4	-9.7	-5.0	12.1	2.1	1.8	-1.5	7.4	12.1	India
-9.6	-0.4	1.1	29.2	-2.9	-18.9	42.4	21.2	-19.7	0.4	Nepal
13.5	23.1	12.9	6.4	-6.0	28.6	-12.7	15.3	25.3	9.8	Pakistan
19.9	4.0	-21.3	31.4	3.7	-3.3	20.4	14.1	-4.4	7.0	Sri Lanka
13.2	*5.9*	*9.9*	*5.7*	*5.0*	*11.0*	*14.8*	*4.5*	*12.4*	*14.3*	*EAST ASIA AND PACIFIC*
15.7	17.4	22.7	16.3	7.9	7.1	16.9	10.9	13.5	11.6	China
20.1	-6.4	-28.2	10.3	41.8	-13.5	26.4	11.8	16.2	0.2	Fiji
25.4	11.3	19.3	0.6	1.2	15.4	31.4	5.0	0.9	24.3	Hong Kong
6.3	-6.5	-2.8	-4.5	-3.1	16.3	-1.6	-2.7	5.5	7.0	Indonesia
13.8	-0.3	10.0	19.0	7.0	15.0	18.3	7.0	16.5	25.3	Korea, Republic of
6.8	15.8	-0.8	-0.4	13.6	15.6	12.7	4.5	8.8	7.3	Malaysia
32.5	0.1	19.1	-5.2	-7.7	3.1	3.5	13.7	14.2	12.9	Papua New Guinea
1.7	9.2	18.8	3.6	-4.8	-5.2	2.3	0.0	7.7	-1.2	Philippines
12.7	9.5	9.8	2.7	6.6	8.0	11.3	-2.1	9.6	5.5	Singapore
..	..	..	..	..	..	..	..	..	..	Solomon Islands
-1.2	11.1	12.4	17.0	17.6	-10.3	21.4	8.9	12.5	24.2	Thailand
..	..	..	..	..	..	..	..	..	..	Vanuatu

Table 20. Growth of merchandise exports (cont'd)

Average annual growth (percent)	1967	1968	1969	1970	1971	1972	1973	1974	1975	1976	1977
LATIN AMERICA AND CARIBBEAN	*7.2*	*1.0*	*2.0*	*1.9*	*-16.5*	*-1.6*	*6.1*	*-21.6*	*-6.3*	*2.5*	*2.7*
Argentina	-1.8	-0.2	7.5	2.7	-4.6	0.1	16.8	2.2	-22.7	31.5	42.2
Bahamas	..	..	..	..	..	..	..	..	..	..	..
Barbados	79.7	-1.5	-29.8	-8.3	8.3	-18.9	-22.7	-20.6	32.0	-6.5	11.8
Belize	..	..	..	..	..	..	..	..	..	..	..
Bolivia	75.6	3.6	-1.0	-17.9	14.3	15.3	-9.1	3.1	-7.2	16.8	-7.6
Brazil	1.2	11.2	18.1	9.0	6.6	26.3	6.7	8.1	8.2	2.9	-0.9
Chile	31.9	-2.2	-1.2	15.2	-0.7	-13.8	-15.0	71.3	0.4	24.1	2.2
Colombia	0.0	-8.4	18.8	-1.6	-11.4	1.7	-1.7	-13.0	4.2	-20.5	2.1
Costa Rica	6.1	19.2	6.2	12.0	7.4	9.2	-1.7	9.5	6.4	-3.3	11.1
Dominican Republic	12.0	1.0	-10.1	8.0	8.8	8.9	-2.6	-23.7	65.4	-15.3	-6.1
Ecuador	12.7	17.6	-29.8	24.6	16.8	147.1	115.6	-28.0	-12.4	11.3	-8.9
El Salvador	15.9	-1.3	-8.0	-7.3	8.2	14.6	-9.0	9.5	18.2	-11.8	-2.6
Grenada	..	..	..	..	..	..	..	..	..	..	..
Guatemala	-10.9	11.9	10.4	-1.5	1.7	5.4	0.2	10.8	7.3	-8.3	23.1
Guyana	-1.7	-5.7	11.5	14.7	12.5	-9.7	-25.5	2.3	24.6	-13.4	-3.7
Haiti	-3.6	10.6	-0.3	7.7	16.8	6.7	16.0	2.5	12.4	2.4	2.1
Honduras	5.7	22.0	-0.6	-0.3	-1.2	-5.0	1.0	-13.8	9.1	6.5	9.7
Jamaica	5.1	-15.9	10.6	21.7	-2.4	3.5	-14.9	12.1	1.1	-18.0	11.2
Mexico	0.3	-0.3	10.7	-12.8	-1.7	24.1	6.2	-8.3	10.1	3.4	20.1
Nicaragua	7.0	5.1	-1.1	1.5	4.9	25.0	-22.3	19.2	6.8	6.7	2.1
Panama	-4.8	-7.8	19.5	-7.7	-4.0	-16.5	-15.1	-13.9	31.1	-29.9	-1.6
Paraguay	..	..	..	..	..	..	..	..	..	..	..
Peru	15.8	14.8	-15.6	15.2	-4.9	-7.8	-32.9	21.9	19.9	-14.4	17.0
St. Vincent and the Grenadines	..	..	..	..	..	..	..	..	..	..	..
Trinidad and Tobago	1.2	5.7	0.9	1.2	-16.2	-3.7	-8.0	-21.2	-12.5	19.4	-9.8
Uruguay	-6.0	28.0	4.3	13.2	-12.9	-19.5	-1.3	20.4	13.9	28.2	6.8
Venezuela	8.8	-0.6	0.0	1.3	-24.9	-10.3	9.3	-40.9	-18.1	-2.3	-6.6
MIDDLE EAST AND NORTH AFRICA	*19.9*	*3.6*	*15.0*	*11.6*	*-23.8*	*27.6*	*-1.0*	*-26.7*	*4.0*	*6.6*	*5.9*
Algeria	46.7	5.9	7.8	11.4	-30.4	48.0	3.5	-39.3	3.0	9.1	8.3
Egypt, Arab Republic of	-20.2	8.8	50.4	23.0	-8.0	-11.5	-17.8	-18.3	13.2	9.7	10.0
Jordan	9.9	7.5	-6.2	-4.7	-4.5	41.0	2.1	47.1	-10.2	67.7	16.8
Morocco	2.3	4.3	0.8	0.2	-1.1	18.6	12.3	6.2	-19.0	9.1	1.2
Oman	..	..	..	..	..	..	..	..	..	..	..
Syrian Arab Republic	-12.0	41.6	54.4	20.3	-4.2	14.6	-7.5	27.4	35.0	0.1	-7.6
Tunisia	42.3	14.4	14.8	6.1	4.4	30.3	-0.1	15.2	6.5	-10.1	5.6
Yemen Arab Republic	12.5	37.5	-9.6	-31.1	65.0	-32.6	30.8	134.6	-33.0	-42.1	38.4
Yemen, PDR	-13.3	-23.6	20.2	1.6	-45.0	-10.5	-26.6	-34.4	-22.9	-6.4	-8.8
EUROPE AND MEDITERRANEAN	..	..	..	..	..	..	..	..	..	..	..
Cyprus	1.5	7.9	0.3	22.0	12.7	12.6	7.0	-24.1	-9.5	71.8	12.0
Greece	29.1	-4.7	11.8	13.5	3.1	23.9	47.9	0.8	12.5	3.9	-0.4
Hungary	..	..	..	..	..	..	..	..	..	..	..
Israel	10.4	13.1	10.0	-8.5	15.9	23.7	11.8	-14.4	1.1	26.2	23.1
Malta	-14.2	29.0	1.1	-2.3	-4.3	25.0	8.7	-4.2	25.2	30.9	16.5
Poland	..	..	..	..	..	..	..	..	..	..	..
Portugal	8.9	5.0	10.0	12.0	1.2	5.9	7.8	1.7	-22.8	-0.7	9.0
Turkey	-0.1	-7.5	11.0	6.7	11.1	36.9	12.6	-12.2	-8.0	27.6	-13.5
Yugoslavia	4.1	-4.3	8.3	6.9	7.9	13.2	6.4	-1.3	2.7	19.1	-2.4
Developing countries	7.8	2.6	7.5	7.5	-4.6	9.9	7.0	-14.7	0.0	9.9	6.6
Highly indebted	5.2	-1.8	8.0	9.2	-7.3	3.0	9.4	-21.9	-6.9	8.2	2.5
Low-income countries	1.1	0.6	0.8	2.6	2.0	7.3	-1.8	-1.8	9.2	0.2	2.4
Low-income Africa	10.9	5.5	5.3	10.9	0.9	6.9	-13.7	-6.0	5.7	-7.9	3.7
China and India	-6.1	1.8	-0.6	-5.6	6.4	11.3	12.4	2.1	11.2	3.5	3.3
Other low-income Asia	4.0	-13.6	-6.5	7.1	-8.8	-6.1	-15.1	-7.4	10.0	8.0	-5.9
Middle-income countries	8.8	2.8	8.4	8.1	-5.3	10.3	8.1	-16.2	-1.2	11.3	7.1
High-income oil exporters	8.5	21.1	6.1	21.4	9.9	14.0	7.3	-15.7	-11.0	18.3	1.6
Industrialized countries	6.8	14.0	13.9	10.1	7.0	6.3	3.5	10.8	-1.5	9.1	7.1
Japan	4.5	23.6	18.5	13.7	20.0	6.7	4.0	17.6	3.2	20.6	9.9
United States	5.2	6.9	6.2	8.0	-0.8	7.4	13.9	14.8	0.6	2.1	1.4
European Community	6.5	16.1	15.8	10.3	8.0	5.7	2.5	8.6	-1.3	9.4	8.6

NOTE: See glossary for definitions.

1978	1979	1980	1981	1982	1983	1984	1985	1986	1987	
8.1	*7.2*	*5.2*	*5.5*	*-1.3*	*4.6*	*11.2*	*1.7*	*-4.5*	*6.1*	*LATIN AMERICA AND CARIBBEAN*
5.2	0.7	-5.8	15.0	-6.7	-1.3	4.8	14.0	-15.0	-8.9	Argentina
..	..	..	..	..	..	..	..	..	..	Bahamas
30.0	-9.4	-8.5	32.4	32.9	25.8	28.6	-7.6	-22.6	-44.5	Barbados
..	..	..	..	..	..	..	..	..	..	Belize
-9.9	-7.1	7.8	1.6	-2.2	-6.6	-2.9	-7.6	12.4	13.8	Bolivia
10.7	8.9	25.1	17.7	-5.6	13.2	21.9	4.8	-18.3	14.9	Brazil
9.9	14.8	8.5	-6.2	10.4	-1.9	2.9	8.2	9.5	4.2	Chile
44.5	-3.9	18.9	-13.5	-3.9	2.6	9.7	6.5	31.3	25.7	Colombia
19.5	-11.0	6.1	3.4	-10.2	-2.3	14.5	3.7	0.4	13.1	Costa Rica
-7.2	13.0	-17.7	56.9	-16.5	-0.8	17.7	-16.1	-15.3	16.4	Dominican Republic
8.0	10.1	-12.7	-15.9	13.3	6.6	14.2	14.6	3.5	-16.8	Ecuador
1.7	23.2	-4.1	-15.2	-14.8	8.6	-5.9	-2.9	-12.0	3.9	El Salvador
..	..	..	..	..	..	..	..	..	..	Grenada
7.9	-5.4	20.2	-17.5	4.2	3.1	-3.4	3.7	-11.9	11.4	Guatemala
-0.3	-6.8	7.3	-7.5	-16.7	-20.9	14.7	30.1	2.4	3.2	Guyana
14.3	-11.1	33.4	-16.0	11.9	-4.5	13.9	7.1	-27.5	-6.3	Haiti
23.3	0.4	8.6	-7.0	-6.0	-0.1	12.4	10.4	0.9	7.0	Honduras
-8.5	-4.0	3.5	4.4	-18.9	-1.6	2.0	-21.5	-1.1	8.4	Jamaica
53.0	16.4	34.8	23.0	13.1	13.5	9.7	-5.9	-3.3	9.8	Mexico
9.0	-14.3	-38.0	27.9	-15.7	8.7	-10.3	-11.5	-27.0	35.5	Nicaragua
-5.6	-5.9	-2.1	-9.0	-0.3	-5.0	-6.4	35.7	13.5	1.9	Panama
..	..	-1.6	6.3	10.1	-25.0	11.7	9.7	88.3	49.6	Paraguay
19.6	44.7	-2.7	-9.8	16.0	-6.8	6.4	-1.0	-8.6	-5.9	Peru
..	..	..	..	..	..	..	..	..	..	St. Vincent and the Grenadines
-8.0	-9.6	-1.8	-16.7	-9.3	-17.0	-5.4	2.9	4.9	-13.3	Trinidad and Tobago
-0.4	-5.3	24.7	18.2	-6.4	6.5	-11.1	-2.3	23.5	-6.7	Uruguay
-4.5	7.9	-16.6	-5.8	-9.6	-3.5	11.5	-3.5	8.4	-1.7	Venezuela
9.2	*14.2*	*-9.1*	*-1.8*	*3.2*	*5.4*	*5.0*	*5.7*	*5.5*	*1.1*	*MIDDLE EAST AND NORTH AFRICA*
4.4	12.8	-13.5	-7.2	0.9	5.2	4.3	4.8	9.1	1.1	Algeria
27.8	11.5	-4.3	-1.1	9.9	20.7	7.0	2.7	-1.9	2.4	Egypt, Arab Republic of
13.0	13.0	25.0	33.5	8.7	-17.8	30.5	8.4	-18.0	-18.3	Jordan
8.9	9.6	6.6	1.1	-0.2	6.8	4.4	3.8	4.5	3.4	Morocco
..	..	..	12.0	4.2	-2.4	6.6	17.3	8.5	..	Oman
-0.8	18.2	-14.5	-7.9	7.1	1.0	-1.1	0.5	7.9	-1.9	Syrian Arab Republic
21.7	16.6	-9.2	9.4	-9.0	2.8	1.2	1.1	7.3	9.7	Tunisia
-38.3	60.5	53.9	112.3	-9.2	-31.7	21.3	74.3	-61.0	-22.8	Yemen Arab Republic
7.0	82.0	3.7	-29.4	44.2	-7.6	-2.2	9.4	8.4	-16.6	Yemen, PDR
..	..	..	..	..	..	..	..	..	..	*EUROPE AND MEDITERRANEAN*
-0.2	11.3	7.6	6.7	4.2	-8.6	20.5	-17.4	-3.4	8.0	Cyprus
18.6	-6.9	14.9	-12.3	4.1	8.1	16.6	-2.6	20.5	4.9	Greece
..	..	..	..	..	..	..	..	..	..	Hungary
18.4	-2.0	6.5	6.3	-2.5	-0.8	17.1	9.9	11.7	13.8	Israel
10.1	3.4	1.2	2.4	-16.8	6.0	12.1	2.8	4.5	2.7	Malta
..	..	..	..	..	..	..	..	..	..	Poland
3.0	21.4	10.1	0.7	14.7	18.1	12.9	8.2	18.4	4.6	Portugal
19.9	-18.2	19.2	67.4	37.7	0.7	29.7	16.6	-15.3	21.8	Turkey
0.6	1.7	15.5	18.0	-4.4	-7.5	6.9	7.3	-5.9	1.0	Yugoslavia
8.3	8.7	5.7	0.8	1.4	6.7	11.4	4.9	5.3	9.3	Developing countries
3.2	10.4	4.3	-1.8	-4.3	2.4	11.7	3.5	-3.9	3.6	Highly indebted
3.9	12.3	8.8	5.1	4.6	3.5	9.2	7.0	10.8	9.4	Low-income countries
-5.9	12.0	7.2	-5.4	-4.0	-7.6	3.6	0.7	6.7	1.7	Low-income Africa
7.5	12.0	10.3	9.6	9.0	5.7	12.8	7.9	12.1	11.7	China and India
12.4	16.0	3.4	8.4	-1.7	14.3	-1.6	12.4	11.1	7.6	Other low-income Asia
8.8	8.3	5.3	0.3	0.9	7.2	11.7	4.6	4.5	9.3	Middle-income countries
-5.2	8.9	-4.9	-10.7	-20.9	-22.6	-4.9	-14.7	25.9	-11.3	High-income oil exporters
4.3	3.4	6.2	3.3	-1.8	1.8	9.7	4.2	1.5	..	Industrialized countries
0.2	-2.7	18.0	11.4	-3.9	9.9	16.0	5.8	-1.1	..	Japan
10.4	5.8	12.9	-0.6	-7.8	-8.5	7.9	-1.0	-5.4	..	United States
3.8	3.4	3.3	2.9	-0.2	1.8	8.2	5.5	3.0	..	European Community

Country Pages

ALGERIA	1967	1968	1969	1970	1971	1972	1973	1974	1975	1976	1977
CURRENT GNP PER CAPITA (US $)	260	300	330	360	350	460	530	680	880	1,070	1,190
POPULATION (thousands)	12,586	12,953	13,340	13,746	14,169	14,608	15,063	15,533	16,018	16,518	17,033
ORIGIN AND USE OF RESOURCES	*(Billions of current Algerian Dinars)*										
Gross National Product (GNP)	16.18	18.64	20.88	23.92	25.78	31.44	35.62	55.18	60.87	72.77	85.66
Net Factor Income from Abroad	-0.48	-0.44	-0.17	-0.15	0.86	1.03	1.02	-0.38	-0.70	-1.31	-1.58
Gross Domestic Product (GDP)	16.67	19.08	21.04	24.07	24.92	30.41	34.59	55.56	61.57	74.08	87.24
Indirect Taxes, net	..	..	..	..	..	..	..	..	..	..	..
GDP at factor cost	..	..	..	..	..	..	..	..	..	..	..
Agriculture	1.97	2.41	2.22	2.43	2.62	2.83	2.73	3.87	5.82	6.68	6.74
Industry	6.37	7.25	8.12	9.27	8.95	12.61	15.94	29.35	29.06	37.27	44.37
Manufacturing	2.05	2.38	2.78	3.37	3.47	4.09	4.78	4.56	5.06	6.57	7.51
Services, etc.	8.33	9.42	10.70	12.37	13.36	14.98	15.92	22.34	26.69	30.12	36.12
Resource Balance	0.26	-0.28	-0.92	-1.70	-2.34	-1.59	-2.17	1.83	-5.69	-3.03	-9.78
Exports of Goods & NFServices	3.88	4.36	4.95	5.31	4.57	6.16	8.75	21.40	20.71	24.36	26.55
Imports of Goods & NFServices	3.62	4.65	5.87	7.01	6.91	7.76	10.92	19.57	26.40	27.39	36.33
Domestic Absorption	16.40	19.37	21.97	25.77	27.27	32.01	36.76	53.73	67.26	77.11	97.02
Private Consumption, etc.	9.82	11.14	12.04	13.44	14.28	16.81	17.79	25.72	31.45	35.93	44.62
General Gov't Consumption	2.66	2.92	3.17	3.58	4.10	4.76	5.03	5.93	7.97	9.25	11.59
Gross Domestic Investment	3.92	5.31	6.76	8.75	8.89	10.44	13.94	22.08	27.84	31.93	40.81
Fixed Investment	..	..	6.17	8.16	8.34	9.81	12.42	16.96	23.97	31.36	38.43
Memo Items:											
Gross Domestic Saving	4.18	5.03	5.84	7.05	6.54	8.85	11.77	23.91	22.15	28.90	31.03
Gross National Saving	4.53	5.42	6.64	7.86	8.50	10.88	14.13	24.86	22.85	29.19	30.60
	(Billions of 1980 Algerian Dinars)										
Gross National Product	62.05	69.04	75.57	82.95	78.49	98.43	101.52	108.39	113.68	123.37	131.20
GDP at factor cost	..	..	..	..	..	..	..	..	..	..	..
Agriculture	5.72	6.65	6.17	6.66	6.95	6.97	6.42	8.01	10.31	10.09	9.37
Industry	34.90	38.28	41.73	45.68	38.39	54.79	58.05	58.22	59.98	67.69	73.73
Manufacturing	3.99	4.48	5.11	6.05	5.95	6.85	7.86	7.12	7.25	8.98	9.76
Services, etc.	22.61	25.11	28.02	30.95	31.30	34.46	34.95	42.84	44.50	47.49	50.17
Resource Balance	29.43	32.65	33.98	33.73	21.34	33.68	31.32	18.56	12.90	16.74	8.05
Exports of Goods & NFServices	41.66	47.20	51.48	53.41	39.55	54.89	57.96	54.52	54.79	56.72	56.36
Imports of Goods & NFServices	12.23	14.56	17.49	19.69	18.21	21.21	26.64	35.96	41.89	39.98	48.31
Domestic Absorption	33.79	37.39	41.93	49.56	55.30	62.53	68.12	90.51	101.88	108.52	125.23
Private Consumption, etc.	15.13	16.17	17.43	20.22	25.42	28.04	27.59	38.14	42.99	46.30	54.09
General Gov't Consumption	7.26	7.46	7.42	7.86	8.76	9.82	9.78	10.06	12.97	13.99	15.46
Gross Domestic Investment	11.40	13.76	17.09	21.48	21.12	24.67	30.75	42.30	45.92	48.23	55.67
Fixed Investment	..	..	14.51	18.74	18.62	21.71	25.64	30.89	37.57	45.03	49.85
Memo Items:											
Capacity to Import	13.12	13.66	14.74	14.91	12.03	16.85	21.35	39.33	32.87	35.56	35.31
Terms of Trade Adjustment	-28.54	-33.54	-36.73	-38.51	-27.52	-38.04	-36.61	-15.19	-21.92	-21.17	-21.05
Gross Domestic Income	34.69	36.50	39.19	44.78	49.12	58.18	62.82	93.87	92.86	104.10	112.22
Gross National Income	33.51	35.50	38.84	44.45	50.98	60.39	64.91	93.20	91.76	102.20	110.14
DOMESTIC PRICES (DEFLATORS)	*(Index 1980 = 100)*										
Overall (GDP)	26.4	27.2	27.7	28.9	32.5	31.6	34.8	50.9	53.6	59.1	65.5
Domestic Absorption	48.5	51.8	52.4	52.0	49.3	51.2	54.0	59.4	66.0	71.0	77.5
Agriculture	34.4	36.2	35.9	36.5	37.6	40.5	42.5	48.4	56.5	66.2	71.9
Industry	18.2	18.9	19.5	20.3	23.3	23.0	27.5	50.4	48.5	55.1	60.2
Manufacturing	51.5	53.2	54.4	55.7	58.2	59.7	60.8	64.1	69.8	73.2	76.9
MANUFACTURING ACTIVITY											
Employment (1980=100)	..	..	34.9	37.7	40.3	42.6	46.8	51.5	57.9	67.1	71.5
Real Earnings per Empl. (1980=100)	..	..	..	97.9	97.5	100.1	99.4	105.7	102.6	103.9	97.5
Real Output per Empl. (1980=100)	..	..	..	101.2	94.7	98.6	103.7	102.1	95.8	90.4	89.0
Earnings as % of Value Added	..	..	..	44.6	47.4	45.6	44.6	50.0	53.6	52.9	52.7
MONETARY HOLDINGS	*(Billions of current Algerian Dinars)*										
Money Supply, Broadly Defined	7.59	10.30	12.37	13.45	14.59	19.00	21.42	27.12	35.44	45.88	55.25
Money as Means of Payment	7.01	9.31	11.01	11.63	12.95	16.75	18.92	24.25	31.97	41.07	48.55
Currency Ouside Banks	3.23	3.70	4.16	4.74	5.70	7.05	8.82	10.45	12.74	17.24	20.58
Demand Deposits	3.79	5.61	6.85	6.89	7.25	9.70	10.11	13.80	19.23	23.83	27.98
Quasi-Monetary Liabilities	0.58	0.99	1.36	1.83	1.64	2.25	2.50	2.87	3.47	4.81	6.69
GOVERNMENT DEFICIT (-) OR SURPLUS	*(Millions of current Algerian Dinars)*										
	..	..	..	..	..	..	..	..	..	..	..
Current Revenue	..	..	..	..	..	..	..	..	..	..	..
Current Expenditure	..	..	..	..	..	..	..	..	..	..	..
Current Budget Balance	..	..	..	..	..	..	..	..	..	..	..
Capital Receipts	..	..	..	..	..	..	..	..	..	..	..
Capital Payments	..	..	..	..	..	..	..	..	..	..	..

1978	1979	1980	1981	1982	1983	1984	1985	1986	1987 estimate	NOTES	ALGERIA
1,380	1,680	1,960	2,250	2,400	2,430	2,420	2,550	2,590	2,680	..	**CURRENT GNP PER CAPITA (US $)**
17,562	18,106	18,666	19,241	19,833	20,443	21,071	21,718	22,402	23,101	..	**POPULATION (thousands)**
				(Billions of current Algerian Dinars)							**ORIGIN AND USE OF RESOURCES**
102.38	123.90	157.84	185.84	201.50	227.81	253.15	282.35	279.90	305.90	..	Gross National Product (GNP)
-2.45	-4.33	-4.67	-5.63	-6.06	-5.94	-6.70	-6.80	-6.70	-7.40	..	Net Factor Income from Abroad
104.83	128.22	162.51	191.47	207.55	233.75	259.85	289.15	286.60	313.30	..	Gross Domestic Product (GDP)
..	..	..	..	..	..	..	..	..	..	..	Indirect Taxes, net
..	..	..	..	..	..	..	..	..	..	..	GDP at factor cost
8.42	10.78	12.92	16.25	16.11	16.61	19.48	27.07	33.30	38.90	..	Agriculture
50.95	65.22	87.36	100.71	107.71	119.63	130.49	140.50	121.20	132.60	..	Industry
9.84	12.16	13.98	16.42	18.60	22.38	26.56	30.95	34.80	38.20	..	Manufacturing
45.46	52.22	62.22	74.51	83.74	97.51	109.88	121.58	132.10	141.80	..	Services, etc.
-15.26	-2.17	6.46	7.10	3.99	5.06	6.13	9.68	-9.90	-0.70	..	Resource Balance
26.69	39.91	55.80	66.18	64.22	65.34	67.69	69.20	41.00	42.80	..	Exports of Goods & NFServices
41.95	42.08	49.34	59.08	60.23	60.29	61.56	59.53	50.90	43.50	..	Imports of Goods & NFServices
120.10	130.40	156.04	184.37	203.56	228.70	253.72	279.48	296.50	314.00	..	Domestic Absorption
50.88	58.48	70.18	87.18	95.56	106.18	118.30	136.20	157.00	171.80	..	Private Consumption, etc.
14.60	17.48	22.35	26.35	30.66	34.69	40.81	45.83	46.70	51.60	..	General Gov't Consumption
54.62	54.43	63.51	70.84	77.34	87.82	94.61	97.45	92.80	90.60	..	Gross Domestic Investment
50.79	50.37	54.88	63.04	71.49	80.32	87.36	92.65	95.40	93.00	..	Fixed Investment
											Memo Items:
39.36	52.26	69.98	77.94	81.33	92.88	100.74	107.13	82.90	89.90	..	Gross Domestic Saving
38.08	49.14	66.24	73.60	76.48	87.79	94.80	101.99	79.60	84.79	..	Gross National Saving
				(Billions of 1980 Algerian Dinars)							
142.76	155.74	157.84	163.34	169.93	179.83	187.03	197.44	199.24	201.74	..	Gross National Product
..	..	..	..	..	..	..	..	..	..	..	GDP at factor cost
10.45	11.42	12.92	13.02	11.92	11.72	12.82	16.33	17.23	18.33	..	Agriculture
81.79	88.86	87.36	87.66	95.26	100.76	106.06	109.95	111.75	113.45	..	Industry
11.53	13.38	13.98	15.38	16.58	18.08	19.87	21.97	22.27	..	..	Manufacturing
53.36	60.82	62.22	67.72	67.92	72.02	73.22	75.83	75.03	74.83	..	Services, etc.
7.70	16.37	6.46	-2.64	-3.24	-1.04	-0.44	1.56	11.27	19.27	..	Resource Balance
59.79	64.10	55.80	52.40	52.30	53.90	55.60	57.30	57.90	58.30	..	Exports of Goods & NFServices
52.09	47.74	49.34	55.04	55.54	54.94	56.04	55.74	46.64	39.03	..	Imports of Goods & NFServices
137.90	144.74	156.04	171.05	178.36	185.56	192.56	200.56	192.76	187.37	..	Domestic Absorption
53.73	63.67	70.18	82.09	84.40	85.50	91.00	96.91	96.51	97.71	..	Private Consumption, etc.
17.33	18.96	22.35	23.35	24.25	24.75	26.74	29.24	26.44	26.14	..	General Gov't Consumption
66.84	62.11	63.51	65.61	69.71	75.31	74.81	74.41	69.81	63.51	..	Gross Domestic Investment
59.48	54.78	54.88	56.28	61.78	65.78	66.18	67.38	64.58	59.98	..	Fixed Investment
											Memo Items:
33.13	45.27	55.80	61.66	59.22	59.55	61.62	64.80	37.57	38.40	..	Capacity to Import
-26.65	-18.83	0.00	9.26	6.92	5.65	6.02	7.50	-20.34	-19.90	..	Terms of Trade Adjustment
118.95	142.27	162.51	177.67	182.02	190.15	198.12	209.61	183.67	186.71	..	Gross Domestic Income
116.10	136.90	157.84	172.60	176.86	185.48	193.05	204.94	178.90	181.84	..	Gross National Income
				(Index 1980 = 100)							**DOMESTIC PRICES (DEFLATORS)**
72.0	79.6	100.0	113.7	118.5	126.7	135.3	143.1	140.5	151.6	..	Overall (GDP)
87.1	90.1	100.0	107.8	114.1	123.2	131.8	139.3	153.8	167.6	..	Domestic Absorption
80.6	94.4	100.0	124.8	135.1	141.7	151.9	165.8	193.3	212.2	..	Agriculture
62.3	73.4	100.0	114.9	113.1	118.7	123.0	127.8	108.5	116.9	..	Industry
85.3	90.8	100.0	106.8	112.2	123.8	133.7	140.9	156.3	..	..	Manufacturing
											MANUFACTURING ACTIVITY
76.4	89.6	100.0	111.3	123.8	137.8	153.3	170.6			..	Employment (1980=100)
99.5	101.6	100.0	92.1	87.8	88.1	87.7	83.0			..	Real Earnings per Empl. (1980=100)
93.5	102.4	100.0	99.2	95.8	93.9	93.0	90.9			..	Real Output per Empl. (1980=100)
53.4	53.0	53.0	53.0	53.0	53.0	53.0	53.0			..	Earnings as % of Value Added
				(Billions of current Algerian Dinars)							**MONETARY HOLDINGS**
71.99	86.29	103.84	121.24	154.12	185.02	216.63	249.46	257.62	..	D	Money Supply, Broadly Defined
62.13	72.21	84.43	97.92	125.30	152.76	180.43	202.23	204.75	223.87	..	Money as Means of Payment
27.29	35.40	42.34	48.06	49.16	60.02	67.46	76.64	89.36	96.89	..	Currency Ouside Banks
34.84	36.81	42.09	49.87	76.14	92.74	112.97	125.59	115.39	126.98	..	Demand Deposits
9.86	14.08	19.41	23.32	28.82	32.26	36.20	47.23	52.87	..	..	Quasi-Monetary Liabilities
				(Millions of current Algerian Dinars)							**GOVERNMENT DEFICIT (-) OR SURPLUS**
..	..	..	..	..	..	..	..	..	..	..	Current Revenue
..	..	..	..	..	..	..	..	..	..	..	Current Expenditure
..	..	..	..	..	..	..	..	..	..	..	Current Budget Balance
..	..	..	..	..	..	..	..	..	..	..	Capital Receipts
..	..	..	..	..	..	..	..	..	..	..	Capital Payments

ALGERIA	1967	1968	1969	1970	1971	1972	1973	1974	1975	1976	1977
FOREIGN TRADE (CUSTOMS BASIS)				*(Millions of current US dollars)*							
Value of Exports, fob	724	830	934	1,009	857	1,306	1,906	4,260	4,291	4,972	5,809
Nonfuel Primary Products	134	177	230	232	142	159	262	234	233	274	195
Fuels	563	589	632	709	641	1,080	1,582	3,941	3,964	4,654	5,577
Manufactures	26	64	72	68	73	67	62	86	94	44	38
Value of Imports, cif	639	815	1,009	1,257	1,227	1,492	2,259	4,036	5,974	5,307	7,102
Nonfuel Primary Products	214	195	208	233	263	347	473	1,040	1,520	1,109	1,520
Fuels	10	13	16	27	43	34	36	60	102	84	97
Manufactures	415	607	785	998	922	1,112	1,749	2,936	4,353	4,113	5,485
				(Index 1980 = 100)							
Terms of Trade	18.4	20.3	20.2	19.3	23.5	20.6	22.7	63.4	57.0	61.0	61.1
Export Prices, fob	5.3	5.8	6.0	5.8	7.1	7.3	10.3	38.1	37.2	39.5	42.7
Nonfuel Primary Products	42.4	42.6	46.6	48.9	43.6	47.2	60.8	66.3	72.8	71.9	71.0
Fuels	4.3	4.3	4.3	4.3	5.6	6.2	8.9	36.7	35.7	38.4	42.0
Manufactures	33.1	38.7	33.9	36.3	37.9	46.8	50.0	82.3	80.6	74.2	75.8
Import Prices, cif	28.9	28.4	29.7	30.2	30.3	35.6	45.6	60.1	65.3	64.8	69.8
Trade at Constant 1980 Prices				*(Millions of 1980 US dollars)*							
Exports, fob	13,611	14,414	15,535	17,299	12,032	17,811	18,427	11,188	11,528	12,573	13,612
Imports, cif	2,212	2,868	3,396	4,159	4,045	4,196	4,953	6,717	9,145	8,188	10,173
BALANCE OF PAYMENTS				*(Millions of current US dollars)*							
Exports of Goods & Services	..	..	..	1,129	935	1,349	2,135	5,270	4,857	5,537	6,384
Merchandise, fob	..	..	..	1,010	816	1,224	1,950	4,944	4,501	5,221	6,009
Nonfactor Services	..	..	..	98	107	107	162	178	284	237	288
Factor Services	..	..	..	21	11	18	23	148	73	80	87
Imports of Goods & Services	..	..	..	1,487	1,339	1,741	2,917	5,017	6,894	6,827	8,988
Merchandise, fob	..	..	..	1,078	996	1,303	2,141	3,667	5,452	4,693	6,198
Nonfactor Services	..	..	..	245	246	327	517	716	1,015	1,305	1,532
Factor Services	..	..	..	164	97	111	259	633	427	830	1,259
Long-Term Interest	..	..	..	10	16	51	67	218	209	340	405
Current Transfers, net	..	..	..	195	225	224	337	319	355	386	278
Workers' Remittances	..	..	..	211	238	272	332	351	412	433	349
Total to be Financed	..	..	..	-163	-180	-167	-446	572	-1,681	-904	-2,327
Official Capital Grants	..	..	..	37	222	41	-2	-396	20	18	4
Current Account Balance	..	..	..	-125	42	-127	-447	176	-1,661	-886	-2,323
Long-Term Capital, net	..	..	..	96	226	240	1,013	133	1,393	1,758	1,901
Direct Investment	..	..	..	45	-151	17	50	358	85	184	173
Long-Term Loans	..	..	..	274	237	247	1,368	298	1,260	1,578	2,436
Disbursements	..	..	..	308	290	385	1,601	790	1,507	2,011	3,075
Repayments	..	..	..	34	52	138	233	491	247	433	639
Other Long-Term Capital	..	..	..	-260	-82	-66	-404	-128	28	-22	-711
Other Capital, net	..	..	..	-54	-268	-38	17	255	-66	-250	75
Change in Reserves	..	..	..	83	0	-75	-583	-564	334	-622	347
Memo Item:				*(Algerian Dinars per US dollar)*							
Conversion Factor (Annual Avg)	4.940	4.940	4.940	4.940	4.910	4.480	3.960	4.180	3.950	4.160	4.150
EXTERNAL DEBT, ETC.				*(Millions of US dollars, outstanding at end of year)*							
Public/Publicly Guar. Long-Term	..	..	..	937	1,233	1,488	2,932	3,305	4,477	5,934	8,632
Official Creditors	..	..	..	483	579	584	782	1,000	1,079	1,159	1,471
IBRD and IDA	..	..	..	16	13	10	8	6	36	85	138
Private Creditors	..	..	..	454	654	904	2,150	2,304	3,399	4,775	7,161
Private Non-guaranteed Long-Term	..	..	..	..	..	..	..	..	..	..	..
Use of Fund Credit	..	..	..	0	0	0	0	0	0	0	0
Short-Term Debt	..	..	..	..	..	..	..	..	..	..	1,686
Memo Items:				*(Millions of US dollars)*							
Int'l Reserves Excluding Gold	286.7	286.4	204.3	147.9	298.6	285.1	912.2	1,454.3	1,128.3	1,764.9	1,683.6
Gold Holdings (at market price)	156.3	245.7	206.4	204.4	238.7	355.1	614.1	1,020.4	767.4	737.3	907.1
SOCIAL INDICATORS											
Total Fertility Rate	7.5	7.5	7.4	7.4	7.4	7.4	7.3	7.3	7.3	7.2	7.2
Crude Birth Rate	49.8	49.4	49.1	48.7	48.4	48.0	47.4	46.8	46.2	45.6	45.0
Infant Mortality Rate	150.0	146.4	142.8	139.2	135.6	132.0	128.0	124.0	120.0	116.0	112.0
Life Expectancy at Birth	51.4	51.8	52.2	52.7	53.1	53.5	54.0	54.5	55.0	55.5	56.0
Food Production, p.c. ('79-81 = 100)	107.2	121.5	114.8	113.2	115.0	114.7	105.8	108.2	124.2	110.8	93.3
Labor Force, Agriculture (%)	53.1	51.2	49.2	47.3	45.7	44.1	42.4	40.8	39.2	37.6	36.0
Labor Force, Female (%)	5.2	5.5	5.8	6.0	6.3	6.5	6.7	6.9	7.1	7.3	7.5
School Enroll. Ratio, primary	..	..	..	76.0	..	..	..	..	93.0	..	95.0
School Enroll. Ratio, secondary	..	..	..	11.0	..	..	..	..	20.0	23.0	27.0

1978	1979	1980	1981	1982	1983	1984	1985	1986	1987 estimate	NOTES	ALGERIA
				(Millions of current US dollars)							**FOREIGN TRADE (CUSTOMS BASIS)**
6,126	9,863	13,871	14,396	13,144	12,583	12,795	13,034	8,066	9,242	f	Value of Exports, fob
206	184	194	204	148	104	118	105	89	119	..	Nonfuel Primary Products
5,885	9,640	13,628	14,124	12,908	12,365	12,549	12,827	7,870	8,986	..	Fuels
35	40	49	68	89	114	129	103	107	137	..	Manufactures
8,667	8,407	10,525	11,302	10,679	10,332	10,263	9,814	9,234	9,263	f	Value of Imports, cif
1,726	1,878	2,718	2,865	2,658	2,728	2,536	3,050	2,540	2,292	..	Nonfuel Primary Products
130	165	259	230	165	214	211	184	274	347	..	Fuels
6,812	6,364	7,548	8,207	7,856	7,390	7,516	6,579	6,420	6,625	..	Manufactures
				(Index 1980 = 100)							
54.9	68.0	100.0	112.0	104.9	97.1	95.8	96.8	50.1	53.9	..	Terms of Trade
43.1	61.5	100.0	111.9	101.3	92.2	89.8	87.4	49.5	56.2	..	Export Prices, fob
82.4	100.1	100.0	88.6	87.6	85.3	81.7	77.5	86.8	91.4	..	Nonfuel Primary Products
42.3	61.0	100.0	112.5	101.6	92.5	90.2	87.5	49.0	55.5	..	Fuels
74.2	108.1	100.0	88.7	79.8	73.4	72.1	78.1	92.3	101.3	..	Manufactures
78.6	90.5	100.0	99.9	96.6	94.9	93.7	90.3	98.9	104.3	..	Import Prices, cif
				(Millions of 1980 US dollars)							Trade at Constant 1980 Prices
14,209	16,028	13,871	12,866	12,979	13,651	14,241	14,919	16,280	16,457	..	Exports, fob
11,028	9,285	10,525	11,318	11,061	10,887	10,948	10,871	9,342	8,883	..	Imports, cif
				(Millions of current US dollars)							**BALANCE OF PAYMENTS**
6,735	10,109	14,631	15,157	14,473	13,696	13,628	13,846	8,869	10,045	..	Exports of Goods & Services
6,340	9,484	13,653	14,116	13,510	12,742	12,793	13,040	8,066	9,033	..	Merchandise, fob
326	468	606	558	635	767	655	616	631	657	..	Nonfactor Services
69	156	372	482	328	187	179	190	172	355	..	Factor Services
10,588	12,061	14,649	15,376	14,901	13,960	13,701	13,172	11,816	11,219	..	Imports of Goods & Services
7,293	7,805	9,597	10,088	9,889	9,517	9,241	8,820	7,889	7,916	..	Merchandise, fob
1,959	2,391	3,465	3,503	3,388	3,031	2,949	2,803	2,332	2,338	..	Nonfactor Services
1,336	1,865	1,587	1,786	1,624	1,413	1,511	1,549	1,595	964	..	Factor Services
594	1,234	1,394	1,314	1,371	1,222	1,292	1,296	1,433	1,377	..	Long-Term Interest
295	313	243	300	263	178	153	331	724	..	..	Current Transfers, net
393	417	334	424	427	327	291	263	310	434	..	Workers' Remittances
-3,558	-1,639	225	80	-165	-87	80	1,004	-2,224	..	K	Total to be Financed
20	8	24	5	-18	1	-5	11	0	..	K	Official Capital Grants
-3,538	-1,631	249	85	-183	-85	74	1,015	-2,224	-406	..	Current Account Balance
3,667	2,440	921	11	-974	-851	-410	-25	363	-1,565	..	Long-Term Capital, net
135	10	315	-1	-65	-14	-14	-2	11	-20	..	Direct Investment
4,146	2,634	776	115	-695	-185	-95	-85	1,010	653	..	Long-Term Loans
5,040	4,190	3,236	2,640	2,203	3,055	3,180	3,201	4,257	4,196	..	Disbursements
894	1,556	2,460	2,524	2,898	3,240	3,274	3,286	3,248	3,543	..	Repayments
-633	-212	-193	-108	-196	-653	-296	51	-657	-2,198	..	Other Long-Term Capital
-36	-325	171	24	88	516	2	30	664	1,899	..	Other Capital, net
-93	-484	-1,341	-120	1,069	421	333	-1,020	1,198	71	..	Change in Reserves
				(Algerian Dinars per US dollar)							Memo Item:
3.970	3.850	3.840	4.320	4.590	4.790	4.980	5.030	4.700	4.850	..	Conversion Factor (Annual Avg)
				(Millions of US dollars, outstanding at end of year)							**EXTERNAL DEBT, ETC.**
13,416	16,029	16,361	15,307	13,932	12,945	12,106	13,468	16,148	19,240	..	Public/Publicly Guar. Long-Term
1,949	2,479	2,803	2,879	2,894	2,810	2,555	2,792	3,008	3,268	..	Official Creditors
187	227	253	275	280	323	284	480	693	924	..	IBRD and IDA
11,467	13,549	13,558	12,428	11,037	10,135	9,551	10,675	13,140	15,972	..	Private Creditors
..	..	..	..	..	..	..	..	..	..	..	Private Non-guaranteed Long-Term
0	0	0	0	0	0	0	0	0	..	..	Use of Fund Credit
1,985	1,933	2,325	2,307	2,751	1,957	1,759	1,862	3,152	3,641	..	Short-Term Debt
				(Millions of US dollars)							Memo Items:
1,980.5	2,658.8	3,772.6	3,695.4	2,422.0	1,880.4	1,464.2	2,819.0	1,660.2	1,640.5	..	Int'l Reserves Excluding Gold
1,249.1	2,858.5	3,291.2	2,219.2	2,550.8	2,129.8	1,721.2	1,825.6	2,182.3	2,702.6	..	Gold Holdings (at market price)
											SOCIAL INDICATORS
7.1	7.0	6.9	6.8	6.7	6.5	6.4	6.2	6.1	5.9	..	Total Fertility Rate
44.5	44.1	43.6	43.2	42.7	42.0	41.3	40.6	39.9	39.3	..	Crude Birth Rate
107.2	102.4	97.6	92.8	88.0	82.8	77.6	72.4	67.2	62.0	..	Infant Mortality Rate
56.8	57.6	58.4	59.2	60.0	60.4	60.8	61.2	61.6	61.9	..	Life Expectancy at Birth
93.2	95.6	107.4	97.1	90.4	92.9	96.1	103.7	102.0	103.5	..	Food Production, p.c. ('79-81 = 100)
34.3	32.7	31.1	..	..	..	..	..	..	..	..	Labor Force, Agriculture (%)
7.8	8.0	8.2	8.3	8.5	8.6	8.8	8.9	9.0	9.2	..	Labor Force, Female (%)
95.0	95.0	95.0	94.0	94.0	94.0	94.0	94.0	92.0	..	..	School Enroll. Ratio, primary
29.0	31.0	33.0	36.0	..	43.0	47.0	51.0	..	..	..	School Enroll. Ratio, secondary

ANTIGUA AND BARBUDA	1967	1968	1969	1970	1971	1972	1973	1974	1975	1976	1977
CURRENT GNP PER CAPITA (US $)	700	680	690	700	710	800	840	880	940	820	870
POPULATION (thousands)	62	63	65	66	67	68	69	70	71	71	73

ORIGIN AND USE OF RESOURCES *(Millions of current East Caribbean dollar)*

	1967	1968	1969	1970	1971	1972	1973	1974	1975	1976	1977
Gross National Product (GNP)	74.6	81.3	85.5	91.7	95.9	109.3	118.2	131.5	142.1	135.1	162.2
Net Factor Income from Abroad	0.0	0.0	0.0	0.0	0.0	0.0	0.0	0.0	0.0	0.0	-0.3
Gross Domestic Product (GDP)	74.6	81.3	85.5	91.7	95.9	109.3	118.2	131.5	142.1	135.1	162.5
Indirect Taxes, net	..	..	..	..	..	..	..	..	..	..	20.5
GDP at factor cost	..	..	..	..	..	..	..	..	..	..	142.0
Agriculture	..	..	..	..	..	..	..	..	..	..	..
Industry	..	..	..	..	..	..	..	..	..	..	..
Manufacturing	..	..	..	..	..	..	..	..	..	..	..
Services, etc.	..	..	..	..	..	..	..	..	..	..	..
Resource Balance	..	..	..	..	..	..	..	..	..	..	-19.8
Exports of Goods & NFServices	..	..	..	..	..	..	..	..	..	..	90.8
Imports of Goods & NFServices	..	..	..	..	..	..	..	..	..	..	110.6
Domestic Absorption	..	..	..	..	..	..	..	..	..	..	182.3
Private Consumption, etc.	..	..	..	..	..	..	..	..	..	..	90.4
General Gov't Consumption	..	..	..	..	..	..	..	..	..	..	36.8
Gross Domestic Investment	..	..	..	..	..	..	..	..	..	..	55.1
Fixed Investment	..	..	..	..	..	..	..	..	..	..	..
Memo Items:											
Gross Domestic Saving	..	..	..	..	..	..	..	..	..	..	35.3
Gross National Saving	..	..	..	..	..	..	..	..	..	..	38.2

(Millions of 1980 East Caribbean dollar)

	1967	1968	1969	1970	1971	1972	1973	1974	1975	1976	1977
Gross National Product	244.68	238.24	242.75	244.43	238.11	253.71	244.43	228.56	221.60	194.25	209.17
GDP at factor cost	..	..	..	..	..	..	..	..	..	..	183.67
Agriculture	..	..	..	..	..	..	..	..	..	..	..
Industry	..	..	..	..	..	..	..	..	..	..	..
Manufacturing	..	..	..	..	..	..	..	..	..	..	..
Services, etc.	..	..	..	..	..	..	..	..	..	..	..
Resource Balance	..	..	..	..	..	..	..	..	..	..	-45.83
Exports of Goods & NFServices	..	..	..	..	..	..	..	..	..	..	123.49
Imports of Goods & NFServices	..	..	..	..	..	..	..	..	..	..	169.32
Domestic Absorption	..	..	..	..	..	..	..	..	..	..	255.43
Private Consumption, etc.	..	..	..	..	..	..	..	..	..	..	136.86
General Gov't Consumption	..	..	..	..	..	..	..	..	..	..	47.43
Gross Domestic Investment	..	..	..	..	..	..	..	..	..	..	71.13
Fixed Investment	..	..	..	..	..	..	..	..	..	..	..
Memo Items:											
Capacity to Import	..	..	..	..	..	..	..	..	..	..	139.01
Terms of Trade Adjustment	..	..	..	..	..	..	..	..	..	..	15.51
Gross Domestic Income	..	..	..	..	..	..	..	..	..	..	225.12
Gross National Income	..	..	..	..	..	..	..	..	..	..	224.69

DOMESTIC PRICES (DEFLATORS) *(Index 1980 = 100)*

	1967	1968	1969	1970	1971	1972	1973	1974	1975	1976	1977
Overall (GDP)	30.5	34.1	35.2	37.5	40.3	43.1	48.4	57.5	64.1	69.5	77.5
Domestic Absorption	..	..	..	..	..	..	..	..	..	..	71.4
Agriculture	..	..	..	..	..	..	..	..	..	..	..
Industry	..	..	..	..	..	..	..	..	..	..	..
Manufacturing	..	..	..	..	..	..	..	..	..	..	..

MANUFACTURING ACTIVITY

	1967	1968	1969	1970	1971	1972	1973	1974	1975	1976	1977
Employment (1980=100)	..	..	..	..	..	..	..	..	..	..	..
Real Earnings per Empl. (1980=100)	..	..	..	..	..	..	..	..	..	..	..
Real Output per Empl. (1980=100)	..	..	..	..	..	..	..	..	..	..	..
Earnings as % of Value Added	..	..	..	..	..	..	..	..	..	..	..

MONETARY HOLDINGS *(Millions of current East Caribbean dollar)*

	1967	1968	1969	1970	1971	1972	1973	1974	1975	1976	1977
Money Supply, Broadly Defined	..	..	..	..	..	..	..	..	98.0	106.5	102.5
Money as Means of Payment	..	..	..	..	..	..	..	..	19.0	24.3	30.9
Currency Ouside Banks	..	..	..	..	..	..	..	..	10.8	11.9	13.2
Demand Deposits	..	..	..	..	..	..	..	..	8.2	12.4	17.8
Quasi-Monetary Liabilities	..	..	..	..	..	..	..	..	79.1	82.2	71.6

(Millions of current East Caribbean dollar)

	1967	1968	1969	1970	1971	1972	1973	1974	1975	1976	1977
GOVERNMENT DEFICIT (-) OR SURPLUS	..	..	..	..	..	..	..	..	..	..	..
Current Revenue	..	..	..	..	..	..	..	..	..	..	..
Current Expenditure	..	..	..	..	..	..	..	..	..	..	..
Current Budget Balance	..	..	..	..	..	..	..	..	..	..	..
Capital Receipts	..	..	..	..	..	..	..	..	..	..	..
Capital Payments	..	..	..	..	..	..	..	..	..	..	..

1978	1979	1980	1981	1982	1983	1984	1985	1986	1987 estimate	NOTES	ANTIGUA AND BARBUDA
970	1,110	1,310	1,530	1,670	1,700	1,920	2,160	2,380	2,570	..	CURRENT GNP PER CAPITA (US $)
73	74	75	76	77	78	79	80	81	82	..	POPULATION (thousands)
			(Millions of current East Caribbean dollar)								ORIGIN AND USE OF RESOURCES
182.7	229.3	275.1	313.1	344.3	351.0	440.3	479.0	522.9	565.1	..	Gross National Product (GNP)
-2.2	-4.9	-5.7	-5.7	-1.6	-2.9	-3.4	-3.7	-4.3	..	..	Net Factor Income from Abroad
184.9	234.2	280.8	318.8	345.9	353.9	443.7	482.7	527.2	..	..	Gross Domestic Product (GDP)
25.6	32.7	40.6	48.6	54.7	54.3	..	..	..	..	..	Indirect Taxes, net
159.3	201.5	240.2	270.2	291.2	299.6	..	..	..	..	..	GDP at factor cost
..	..	..	..	..	..	..	..	..	..	..	Agriculture
..	..	..	..	..	..	..	..	..	..	..	Industry
..	..	..	..	..	..	..	..	..	..	..	Manufacturing
..	..	..	..	..	..	..	..	..	..	..	Services, etc.
-17.8	-69.1	-95.3	-146.1	-163.9	-53.0	..	..	..	..	..	Resource Balance
113.7	136.9	196.8	229.5	223.8	229.2	..	..	..	..	..	Exports of Goods & NFServices
131.5	206.0	292.1	375.6	387.7	282.2	..	..	..	..	..	Imports of Goods & NFServices
202.7	303.3	376.1	464.9	509.8	406.9	..	..	..	..	..	Domestic Absorption
101.3	167.0	203.2	227.3	255.3	221.0	..	..	..	..	..	Private Consumption, etc.
36.0	47.8	57.1	68.2	78.7	84.1	..	..	..	..	..	General Gov't Consumption
65.4	88.5	115.8	169.4	175.8	101.8	..	..	..	..	..	Gross Domestic Investment
..	..	..	..	..	..	..	..	..	..	..	Fixed Investment
											Memo Items:
47.6	19.4	20.5	23.3	11.9	48.8	..	..	..	..	..	Gross Domestic Saving
55.4	36.6	33.7	38.7	33.8	74.5	..	..	..	..	..	Gross National Saving
			(Millions of 1980 East Caribbean dollar)								
239.24	252.53	275.10	288.43	302.08	303.49	328.24	351.43	368.57	394.35	..	Gross National Product
208.90	222.74	240.20	249.90	256.76	259.47	268.14	..	..	..	..	GDP at factor cost
..	..	..	..	..	..	..	..	..	..	..	Agriculture
..	..	..	..	..	..	..	..	..	..	..	Industry
..	..	..	..	..	..	..	..	..	..	..	Manufacturing
..	..	..	..	..	..	..	..	..	..	..	Services, etc.
-34.01	-88.46	-95.30	-140.11	-165.32	-79.41	..	..	..	..	..	Resource Balance
156.13	161.85	196.80	200.06	183.88	170.28	..	..	..	..	..	Exports of Goods & NFServices
190.14	250.31	292.10	340.17	349.20	249.69	..	..	..	..	..	Imports of Goods & NFServices
276.24	346.69	376.10	433.81	468.83	385.75	..	..	..	..	..	Domestic Absorption
137.29	196.37	203.20	214.83	245.47	224.76	..	..	..	..	..	Private Consumption, etc.
54.26	52.72	57.10	62.77	68.96	72.83	..	..	..	..	..	General Gov't Consumption
84.69	97.60	115.80	156.21	154.40	88.17	..	..	..	..	..	Gross Domestic Investment
..	..	..	..	..	..	..	..	..	..	..	Fixed Investment
											Memo Items:
164.40	166.34	196.80	207.85	201.58	202.80	..	..	..	..	..	Capacity to Import
8.27	4.50	0.00	7.79	17.70	32.52	..	..	..	..	..	Terms of Trade Adjustment
250.50	262.73	280.80	301.49	321.20	338.86	..	..	..	..	..	Gross Domestic Income
247.51	257.03	275.10	296.21	319.77	336.01	..	..	..	..	..	Gross National Income
			(Index 1980 = 100)								DOMESTIC PRICES (DEFLATORS)
76.3	90.7	100.0	108.5	114.0	115.5	134.0	136.1	141.7	..	..	Overall (GDP)
73.4	87.5	100.0	107.2	108.7	105.5	..	..	..	..	..	Domestic Absorption
..	..	..	..	..	..	..	..	..	..	..	Agriculture
..	..	..	..	..	..	..	..	..	..	..	Industry
..	..	..	..	..	..	..	..	..	..	..	Manufacturing
											MANUFACTURING ACTIVITY
..	..	..	..	..	..	..	..	..	..	..	Employment (1980=100)
..	..	..	..	..	..	..	..	..	..	..	Real Earnings per Empl. (1980=100)
..	..	..	..	..	..	..	..	..	..	..	Real Output per Empl. (1980=100)
..	..	..	..	..	..	..	..	..	..	..	Earnings as % of Value Added
			(Millions of current East Caribbean dollar)								MONETARY HOLDINGS
113.2	130.7	145.4	168.5	191.3	236.8	286.1	319.9	378.1	449.4	..	Money Supply, Broadly Defined
29.4	34.8	38.6	43.2	41.7	51.1	58.4	68.5	87.4	110.6	..	Money as Means of Payment
15.1	17.1	15.3	16.3	16.3	17.5	24.0	26.5	32.5	42.2	..	Currency Ouside Banks
14.2	17.7	23.3	26.9	25.4	33.6	34.4	42.0	54.9	68.4	..	Demand Deposits
83.8	95.9	106.8	125.4	149.6	185.7	227.7	251.5	290.8	338.8	..	Quasi-Monetary Liabilities
			(Millions of current East Caribbean dollar)								
..	..	..	..	..	..	..	..	..	..	..	GOVERNMENT DEFICIT (-) OR SURPLUS
..	..	..	..	..	..	..	..	..	..	..	Current Revenue
..	..	..	..	..	..	..	..	..	..	..	Current Expenditure
..	..	..	..	..	..	..	..	..	..	..	Current Budget Balance
..	..	..	..	..	..	..	..	..	..	..	Capital Receipts
..	..	..	..	..	..	..	..	..	..	..	Capital Payments

ANTIGUA AND BARBUDA	1967	1968	1969	1970	1971	1972	1973	1974	1975	1976	1977
FOREIGN TRADE (CUSTOMS BASIS)				*(Millions of current US dollars)*							
Value of Exports, fob	..	..	..	..	..	..	..	..	..	..	..
Nonfuel Primary Products	..	..	..	..	..	..	..	..	..	..	..
Fuels	..	..	..	..	..	..	..	..	..	..	..
Manufactures	..	..	..	..	..	..	..	..	..	..	..
Value of Imports, cif	..	..	..	..	..	..	..	..	..	..	..
Nonfuel Primary Products	..	..	..	..	..	..	..	..	..	..	..
Fuels	..	..	..	..	..	..	..	..	..	..	..
Manufactures	..	..	..	..	..	..	..	..	..	..	..
					(Index 1980 = 100)						
Terms of Trade	..	..	..	..	..	..	..	..	..	..	..
Export Prices, fob	..	..	..	..	..	..	..	..	..	..	..
Nonfuel Primary Products	..	..	..	..	..	..	..	..	..	..	..
Fuels	..	..	..	..	..	..	..	..	..	..	..
Manufactures	..	..	..	..	..	..	..	..	..	..	..
Import Prices, cif	..	..	..	..	..	..	..	..	..	..	..
Trade at Constant 1980 Prices				*(Millions of 1980 US dollars)*							
Exports, fob	..	..	..	..	..	..	..	..	..	..	..
Imports, cif	..	..	..	..	..	..	..	..	..	..	..
BALANCE OF PAYMENTS				*(Millions of current US dollars)*							
Exports of Goods & Services	..	..	..	..	..	..	..	..	..	..	32
Merchandise, fob	..	..	..	..	..	..	..	..	..	..	7
Nonfactor Services	..	..	..	..	..	..	..	..	..	..	25
Factor Services	..	..	..	..	..	..	..	..	..	..	0
Imports of Goods & Services	..	..	..	..	..	..	..	..	..	..	44
Merchandise, fob	..	..	..	..	..	..	..	..	..	..	37
Nonfactor Services	..	..	..	..	..	..	..	..	..	..	6
Factor Services	..	..	..	..	..	..	..	..	..	..	0
Long-Term Interest	..	..	..	..	..	..	..	..	..	..	..
Current Transfers, net	..	..	..	..	..	..	..	..	..	..	1
Workers' Remittances	..	..	..	..	..	..	..	..	..	..	0
Total to be Financed	..	..	..	..	..	..	..	..	..	..	-11
Official Capital Grants	..	..	..	..	..	..	..	..	..	..	2
Current Account Balance	..	..	..	..	..	..	..	..	..	..	-10
Long-Term Capital, net	..	..	..	..	..	..	..	..	..	..	10
Direct Investment	..	..	..	..	..	..	..	..	..	..	2
Long-Term Loans	..	..	..	..	..	..	..	..	..	..	..
Disbursements	..	..	..	..	..	..	..	..	..	..	..
Repayments	..	..	..	..	..	..	..	..	..	..	..
Other Long-Term Capital	..	..	..	..	..	..	..	..	..	..	7
Other Capital, net	..	..	..	..	..	..	..	..	..	..	-5
Change in Reserves	..	..	..	0	0	0	0	0	0	2	4
Memo Item:				*(East Caribbean dollar per US dollar)*							
Conversion Factor (Annual Avg)	1.740	2.000	2.000	2.000	1.970	1.920	1.960	2.050	2.170	2.610	2.700
EXTERNAL DEBT, ETC.				*(Millions of US dollars, outstanding at end of year)*							
Public/Publicly Guar. Long-Term	..	..	..	..	..	..	..	..	..	..	..
Official Creditors	..	..	..	..	..	..	..	..	..	..	..
IBRD and IDA	..	..	..	..	..	..	..	..	..	..	..
Private Creditors	..	..	..	..	..	..	..	..	..	..	..
Private Non-guaranteed Long-Term	..	..	..	..	..	..	..	..	..	..	..
Use of Fund Credit	..	..	..	..	..	..	..	..	..	..	..
Short-Term Debt	..	..	..	..	..	..	..	..	..	..	..
Memo Items:				*(Thousands of US dollars)*							
Int'l Reserves Excluding Gold	..	..	..	..	..	..	..	..	7,316	9,666	5,104
Gold Holdings (at market price)	..	..	..	..	..	..	..	..	..	..	..
SOCIAL INDICATORS											
Total Fertility Rate	..	..	..	..	..	..	..	..	..	..	..
Crude Birth Rate	27.6	27.2	26.7	26.2	24.6	23.1	21.5	20.0	18.4	19.1	19.9
Infant Mortality Rate	..	..	..	..	..	..	..	..	..	..	..
Life Expectancy at Birth	..	..	..	..	..	..	..	..	..	..	..
Food Production, p.c. ('79-81 = 100)	95.3	90.2	98.5	99.2	124.4	101.8	100.7	98.9	98.3	101.6	102.0
Labor Force, Agriculture (%)	..	..	..	..	..	..	..	..	..	..	..
Labor Force, Female (%)	..	..	..	..	..	..	..	..	..	..	..
School Enroll. Ratio, primary	..	..	..	..	..	..	..	..	..	..	..
School Enroll. Ratio, secondary	..	..	..	..	..	..	..	..	..	..	..

1978	1979	1980	1981	1982	1983	1984	1985	1986	1987 estimate	NOTES	ANTIGUA AND BARBUDA
				(Millions of current US dollars)							**FOREIGN TRADE (CUSTOMS BASIS)**
..	..	60	51	49	38	41	45	53	..	..	Value of Exports, fob
..	..	..	..	..	..	..	..	..	..	..	Nonfuel Primary Products
..	..	..	..	..	..	..	..	..	..	..	Fuels
..	..	..	..	..	..	..	..	..	..	..	Manufactures
..	..	192	184	181	209	224	232		..	..	Value of Imports, cif
..	..	..	..	..	..	..	..	..	..	..	Nonfuel Primary Products
..	..	..	..	..	..	..	..	..	..	..	Fuels
..	..	..	..	..	..	..	..	..	..	..	Manufactures
				(Index 1980 = 100)							
..	..	..	..	..	..	..	..	..	..	..	Terms of Trade
..	..	..	..	..	..	..	..	..	..	..	Export Prices, fob
..	..	..	..	..	..	..	..	..	..	..	Nonfuel Primary Products
..	..	..	..	..	..	..	..	..	..	..	Fuels
..	..	..	..	..	..	..	..	..	..	..	Manufactures
..	..	..	..	..	..	..	..	..	..	..	Import Prices, cif
				(Millions of 1980 US dollars)							Trade at Constant 1980 Prices
..	..	..	..	..	..	..	..	..	..	..	Exports, fob
..	..	..	..	..	..	..	..	..	..	..	Imports, cif
				(Millions of current US dollars)							**BALANCE OF PAYMENTS**
42	51	107	107	108	104	132	134	144	..	..	Exports of Goods & Services
13	12	59	51	49	37	35	28	25	..	..	Merchandise, fob
30	38	45	53	52	63	91	103	116	..	..	Nonfactor Services
0	1	3	3	7	5	5	2	3	..	..	Factor Services
50	81	136	150	159	124	152	172	237	..	..	Imports of Goods & Services
42	68	115	125	127	100	119	133	181	..	..	Merchandise, fob
7	10	17	20	22	18	28	34	47	..	..	Nonfactor Services
0	3	4	5	10	6	5	5	9	..	..	Factor Services
..	..	..	..	..	..	..	..	..	..	..	Long-Term Interest
4	8	7	8	9	11	12	14	14	..	..	Current Transfers, net
0	0	0	0	0	0	0	0	0	..	..	Workers' Remittances
-4	-21	-22	-35	-42	-9	-8	-25	-79	..	K	Total to be Financed
2	2	3	3	1	..	2	1	3	..	K	Official Capital Grants
-2	-20	-19	-33	-42	-9	-6	-24	-76	..	..	Current Account Balance
5	19	27	39	31	-1	2	4	78	..	..	Long-Term Capital, net
-7	8	20	22	23	5	9	10	13	..	..	Direct Investment
..	..	..	..	..	..	..	..	..	..	..	Long-Term Loans
..	..	..	..	..	..	..	..	..	..	..	Disbursements
..	..	..	..	..	..	..	..	..	..	..	Repayments
10	9	4	14	7	-6	-9	-7	62	..	..	Other Long-Term Capital
-2	6	-11	-7	11	12	11	21	2	..	..	Other Capital, net
-1	-5	3	0	-1	-1	-7	-1	-4	3	..	Change in Reserves
				(East Caribbean dollar per US dollar)							Memo Item:
2.700	2.700	2.700	2.700	2.700	2.700	2.700	2.700	2.700	2.700	..	Conversion Factor (Annual Avg)
				(Millions of US dollars, outstanding at end of year)							**EXTERNAL DEBT, ETC.**
..	..	..	..	..	..	..	..	..	..	..	Public/Publicly Guar. Long-Term
..	..	..	..	..	..	..	..	..	..	..	Official Creditors
..	..	..	..	..	..	..	..	..	..	..	IBRD and IDA
..	..	..	..	..	..	..	..	..	..	..	Private Creditors
..	..	..	..	..	..	..	..	..	..	..	Private Non-guaranteed Long-Term
..	..	..	..	..	..	..	..	..	..	..	Use of Fund Credit
..	..	..	..	..	..	..	..	..	..	..	Short-Term Debt
				(Thousands of US dollars)							Memo Items:
5,973	11,233	7,817	7,336	8,520	9,930	15,440	16,580	28,260	25,600	..	Int'l Reserves Excluding Gold
..	..	..	..	..	..	..	..	..	..	..	Gold Holdings (at market price)
											SOCIAL INDICATORS
..	..	..	..	0.3	0.6	0.9	1.3	1.6	1.9	..	Total Fertility Rate
18.3	18.8	16.5	16.4	16.2	15.9	15.7	15.4	15.2	15.0	..	Crude Birth Rate
..	..	31.5	29.6	27.5	25.5	23.6	21.9	20.4	18.9	..	Infant Mortality Rate
..	..	..	..	24.2	48.4	72.6	72.7	72.8	73.0	..	Life Expectancy at Birth
116.2	93.8	103.3	102.9	101.4	102.6	95.6	96.8	95.8	95.6	..	Food Production, p.c. ('79-81=100)
..	..	..	..	..	..	..	..	..	..	..	Labor Force, Agriculture (%)
..	..	..	..	..	..	..	..	..	..	..	Labor Force, Female (%)
..	..	..	..	..	..	..	..	..	..	..	School Enroll. Ratio, primary
..	..	..	..	..	..	..	..	..	..	..	School Enroll. Ratio, secondary

ARGENTINA	1967	1968	1969	1970	1971	1972	1973	1974	1975	1976	1977
CURRENT GNP PER CAPITA (US $)	910	900	900	910	1,000	1,040	1,240	1,630	1,810	1,680	1,610
POPULATION (thousands)	22,915	23,247	23,596	23,962	24,349	24,759	25,186	25,620	26,052	26,485	26,920

ORIGIN AND USE OF RESOURCES	*(Billions of current Argentine Australes*										
Gross National Product (GNP)	5.90E-06	6.80E-06	8.00E-06	8.70E-06	1.20E-05	2.00E-05	3.50E-05	4.80E-05	1.40E-04	7.50E-04	2.10E-03
Net Factor Income from Abroad	-1.00E-07	-1.00E-07	-1.00E-07	-1.00E-07	-1.00E-07	-3.00E-07	-4.00E-07	-4.00E-07	-1.30E-06	-9.50E-06	-2.70E-05
Gross Domestic Product (GDP)	6.00E-06	6.90E-06	8.10E-06	8.80E-06	1.20E-05	2.10E-05	3.50E-05	4.90E-05	1.40E-04	7.60E-04	2.10E-03
Indirect Taxes, net	7.00E-07	8.00E-07	9.00E-07	1.00E-06	1.40E-06	2.40E-06	4.00E-06	5.50E-06	1.60E-05	8.70E-05	2.40E-04
GDP at factor cost	5.30E-06	6.10E-06	7.20E-06	7.80E-06	1.10E-05	1.80E-05	3.10E-05	4.30E-05	1.30E-04	6.70E-04	1.90E-03
Agriculture	7.10E-07	8.00E-07	8.90E-07	9.50E-07	1.60E-06	2.70E-06	5.00E-06	5.90E-06	1.10E-05	7.40E-05	2.00E-04
Industry	2.20E-06	2.60E-06	3.00E-06	3.10E-06	4.50E-06	7.40E-06	1.20E-05	1.70E-05	5.60E-05	3.00E-04	7.80E-04
Manufacturing	1.70E-06	1.90E-06	2.20E-06	2.30E-06	3.30E-06	5.60E-06	9.00E-06	1.20E-05	4.10E-05	2.20E-04	5.80E-04
Services, etc.	2.40E-06	2.70E-06	3.30E-06	3.70E-06	5.00E-06	8.20E-06	1.50E-05	2.10E-05	6.00E-05	3.00E-04	8.70E-04
Resource Balance	2.00E-07	1.00E-07	0.00E+00	0.00E+00	-1.00E-07	1.00E-07	9.00E-07	4.00E-07	-3.00E-07	3.30E-05	6.50E-05
Exports of Goods & NFServices	6.00E-07	6.00E-07	7.00E-07	8.00E-07	1.00E-06	2.00E-06	3.60E-06	4.50E-06	1.10E-05	9.40E-05	2.70E-04
Imports of Goods & NFServices	4.00E-07	5.00E-07	7.00E-07	8.00E-07	1.10E-06	1.90E-06	2.70E-06	4.10E-06	1.20E-05	6.10E-05	2.10E-04
Domestic Absorption	5.80E-06	6.80E-06	8.10E-06	8.80E-06	1.20E-05	2.00E-05	3.40E-05	4.80E-05	1.40E-04	7.30E-04	2.00E-03
Private Consumption, etc.	4.10E-06	4.80E-06	5.60E-06	6.00E-06	8.50E-06	1.40E-05	2.40E-05	3.20E-05	8.80E-05	4.50E-04	1.30E-03
General Gov't Consumption	6.00E-07	7.00E-07	8.00E-07	9.00E-07	1.30E-06	2.00E-06	4.20E-06	6.60E-06	1.90E-05	7.40E-05	2.00E-04
Gross Domestic Investment	1.10E-06	1.30E-06	1.70E-06	1.90E-06	2.60E-06	4.30E-06	6.40E-06	9.40E-06	3.70E-05	2.00E-04	5.70E-04
Fixed Investment	1.10E-06	1.30E-06	1.70E-06	1.90E-06	2.60E-06	4.30E-06	6.40E-06	9.40E-06	3.70E-05	2.00E-04	5.70E-04
Memo Items:											
Gross Domestic Saving	1.30E-06	1.40E-06	1.70E-06	1.90E-06	2.70E-06	4.50E-06	7.40E-06	9.90E-06	3.60E-05	2.40E-04	6.30E-04
Gross National Saving	1.20E-06	1.30E-06	1.60E-06	1.80E-06	2.60E-06	4.20E-06	7.00E-06	..	3.50E-05	2.30E-04	6.10E-04

	(Thousands of 1980 Argentine Australes)										
Gross National Product	18,064	18,820	20,573	21,833	22,728	23,103	23,896	25,249	25,120	25,060	26,687
GDP at factor cost	15,927	16,731	18,243	19,434	20,168	20,575	21,292	22,471	22,366	22,305	23,764
Agriculture	1,938	1,856	1,938	1,758	1,786	1,820	2,015	2,070	2,013	2,109	2,160
Industry	6,114	6,751	7,480	7,393	7,871	8,144	8,246	8,737	8,662	8,706	9,435
Manufacturing	4,653	4,915	5,483	5,342	5,672	5,898	6,132	6,492	6,326	6,135	6,614
Services, etc.	7,875	8,125	8,824	10,282	10,511	10,611	11,031	11,663	11,690	11,490	12,169
Resource Balance	186	31	1	242	5	80	251	198	59	657	751
Exports of Goods & NFServices	1,069	994	1,205	1,219	1,098	1,120	1,277	1,280	1,161	1,527	1,944
Imports of Goods & NFServices	883	963	1,203	977	1,093	1,040	1,026	1,082	1,102	870	1,193
Domestic Absorption	17,998	18,910	20,682	21,698	22,766	23,155	23,800	25,179	25,196	24,537	26,085
Private Consumption, etc.	12,422	12,791	13,764	14,693	15,165	15,552	16,251	17,356	17,285	16,154	16,487
General Gov't Consumption	2,381	2,642	2,642	2,637	2,756	2,706	2,881	3,104	3,108	3,248	3,387
Gross Domestic Investment	3,194	3,476	4,275	4,369	4,846	4,897	4,667	4,719	4,803	5,134	6,210
Fixed Investment	3,220	3,634	4,463	4,540	4,907	4,963	4,602	4,788	4,798	5,298	6,405
Memo Items:											
Capacity to Import	1,324	1,155	1,203	977	994	1,095	1,368	1,187	1,073	1,348	1,566
Terms of Trade Adjustment	255	162	-1	-242	-104	-26	91	-92	-88	-178	-378
Gross Domestic Income	18,439	19,102	20,682	21,698	22,667	23,210	24,142	25,285	25,167	25,015	26,458
Gross National Income	18,319	18,982	20,572	21,591	22,623	23,078	23,987	25,156	25,033	24,882	26,309

DOMESTIC PRICES (DEFLATORS)	*(Index 1980 = 100)*										
Overall (GDP)	0.033	0.036	0.039	0.040	0.055	0.089	0.150	0.190	0.570	3.000	7.800
Domestic Absorption	0.032	0.036	0.039	0.041	0.054	0.089	0.140	0.190	0.570	3.000	7.800
Agriculture	0.036	0.043	0.046	0.054	0.089	0.150	0.250	0.280	0.560	3.500	9.400
Industry	0.036	0.038	0.040	0.042	0.057	0.091	0.140	0.190	0.640	3.500	8.300
Manufacturing	0.036	0.038	0.041	0.044	0.059	0.095	0.150	0.190	0.650	3.600	8.700

MANUFACTURING ACTIVITY											
Employment (1980=100)	..	..	..	113.0	116.3	118.9	123.3	130.1	135.0	130.2	122.2
Real Earnings per Empl. (1980=100)	..	..	..		53.2	81.0	141.7	101.8	117.6	82.2	79.7
Real Output per Empl. (1980=100)	..	..	..	79.2	75.3	92.3	82.9	76.9	93.9	69.8	106.6
Earnings as % of Value Added	..	..	..	29.6	30.5	28.2	32.4	35.7	32.0	20.0	19.6

MONETARY HOLDINGS	*(Billions of current Argentine Australes)*										
Money Supply, Broadly Defined	1.00E-06	2.00E-06	2.40E-06	2.80E-06	3.00E-06	5.30E-06	9.70E-06	1.60E-05	3.60E-05	1.60E-04	5.20E-04
Money as Means of Payment	1.00E-06	1.00E-06	1.70E-06	1.80E-06	2.00E-06	3.30E-06	5.70E-06	9.80E-06	2.80E-05	1.00E-04	2.30E-04
Currency Ouside Banks	1.00E-06	1.00E-06	7.00E-07	8.00E-07	1.00E-06	1.30E-06	2.70E-06	4.30E-06	1.20E-05	4.10E-05	1.10E-04
Demand Deposits	0.00E+00	0.00E+00	1.00E-06	1.00E-06	1.00E-06	2.00E-06	3.00E-06	5.50E-06	1.60E-05	6.40E-05	1.20E-04
Quasi-Monetary Liabilities	0.00E+00	1.00E-06	7.00E-07	1.00E-06	1.00E-06	2.00E-06	4.00E-06	6.00E-06	8.00E-06	5.40E-05	2.90E-04

GOVERNMENT DEFICIT (-) OR SURPLUS	*(Millions of current Argentine Australes)*										
	..	..	..	..	..	-0.001	-0.002	-0.003	-0.015	-0.054	-0.058
Current Revenue	..	..	..	0.001	0.002	0.003	0.004	0.008	0.016	0.093	0.300
Current Expenditure	..	..	..	0.001	0.002	0.004	0.005	0.008	0.024	0.097	0.200
Current Budget Balance	..	..	..			-0.001	-0.001	..	-0.008	-0.004	0.100
Capital Receipts	..	..	..					..	..	..	..
Capital Payments	..	..	..				0.001	0.003	0.007	0.003	0.200

1978	1979	1980	1981	1982	1983	1984	1985	1986	1987 estimate	NOTES	ARGENTINA
1,570	1,790	1,960	1,930	1,860	1,950	2,140	2,120	2,360	2,360	A	CURRENT GNP PER CAPITA (US $)
27,357	27,795	28,237	28,683	29,136	29,598	30,073	30,564	31,030	31,436	..	POPULATION (thousands)
				(Billions of current Argentine Australes							ORIGIN AND USE OF RESOURCES
0.0052	0.0140	0.0280	0.0530	0.1300	0.6300	4.89	36.40	70.14	162.15	..	Gross National Product (GNP)
-0.0001	-0.0001	-0.0003	-0.0016	-0.0130	-0.0570	-0.39	-3.19	-4.17	-10.96	..	Net Factor Income from Abroad
0.0052	0.0140	0.0280	0.0550	0.1500	0.6800	5.28	39.59	74.31	173.11	..	Gross Domestic Product (GDP)
0.0006	0.0013	0.0032	0.0062	0.0170	0.0780	0.60	4.52	8.47	19.73	..	Indirect Taxes, net
0.0046	0.0130	0.0250	0.0490	0.1300	0.6000	4.68	35.08	65.84	153.38	..	GDP at factor cost
0.0005	0.0013	0.0022	0.0044	0.0150	0.0770	0.59	4.44	8.33	19.41	..	Agriculture
0.0019	0.0050	0.0093	0.0170	0.0500	0.2600	2.03	15.24	28.61	66.66	..	Industry
0.0013	0.0035	0.0063	0.0120	0.0370	0.1900	1.44	10.80	20.27	47.23	..	Manufacturing
0.0023	0.0067	0.0140	0.0270	0.0660	0.2700	2.05	15.39	28.89	67.31	..	Services, etc.
0.0002	0.0000	-0.0006	-0.0002	0.0048	0.0330	0.22	2.63	1.67	0.31	..	Resource Balance
0.0006	0.0013	0.0019	0.0052	0.0200	0.1000	0.68	6.19	8.30	17.64	..	Exports of Goods & NFServices
0.0004	0.0012	0.0026	0.0054	0.0150	0.0670	0.45	3.56	6.62	17.32	..	Imports of Goods & NFServices
0.0050	0.0140	0.0290	0.0550	0.1400	0.6500	5.06	36.96	72.64	172.80	..	Domestic Absorption
0.0031	0.0093	0.0190	0.0370	0.1000	0.4400	3.88	29.22	57.32	145.87	..	Private Consumption, etc.
0.0006	0.0016	0.0037	0.0075	0.0160	0.0890	0.58	4.38	8.76	9.71	..	General Gov't Consumption
0.0013	0.0032	0.0063	0.0100	0.0230	0.1200	0.60	3.37	6.56	17.22	..	Gross Domestic Investment
0.0013	0.0032	0.0063	0.0100	0.0230	0.7200	0.60	3.37	6.56	17.22	..	Fixed Investment
											Memo Items:
0.0015	0.0033	0.0057	0.0100	0.0280	0.1500	0.82	6.00	8.23	17.53	..	Gross Domestic Saving
0.0014	0.0031	0.0053	0.0084	0.0160	0.0940	0.43	2.81	4.07	6.56	..	Gross National Saving
				(Thousands of 1980 Argentine Australes)							
25,752	27,576	28,004	25,694	24,094	24,721	25,347	24,334	25,925	26,318	..	Gross National Product
22,956	24,615	25,107	23,366	22,064	22,707	23,267	22,225	23,509	23,888	I f	GDP at factor cost
2,220	2,284	2,158	2,200	2,353	2,395	2,473	2,440	2,373	2,416	..	Agriculture
8,707	9,389	9,276	8,087	7,528	7,960	8,034	7,403	8,187	8,319	..	Industry
5,918	6,506	6,275	5,285	5,015	5,524	5,735	5,142	5,804	5,771	..	Manufacturing
12,029	12,943	13,673	13,079	12,184	12,352	12,759	12,382	12,949	13,152	..	Services, etc.
986	300	-616	-263	790	1,017	923	1,403	986	741	..	Resource Balance
2,117	2,049	1,944	2,051	2,123	2,287	2,271	2,557	2,349	2,223	..	Exports of Goods & NFServices
1,131	1,749	2,560	2,314	1,333	1,270	1,348	1,154	1,363	1,482	..	Imports of Goods & NFServices
24,945	27,497	28,953	26,640	24,138	24,637	25,363	23,711	25,567	26,241	..	Domestic Absorption
16,227	18,052	18,930	18,029	16,746	17,560	18,596	17,618	18,839	18,619	..	Private Consumption, etc.
3,496	3,691	3,740	3,798	3,528	3,624	3,699	3,635	3,827	4,254	..	General Gov't Consumption
5,221	5,755	6,283	4,813	3,864	3,453	3,068	2,457	2,901	3,368	..	Gross Domestic Investment
5,585	5,963	6,283	5,176	3,751	3,556	3,234	2,856	3,081	3,534	..	Fixed Investment
											Memo Items:
1,704	1,799	1,944	2,212	1,753	1,902	2,015	2,006	1,707	1,509	..	Capacity to Import
-413	-250	0	161	-370	-386	-256	-551	-642	-714	..	Terms of Trade Adjustment
25,517	27,547	28,337	26,539	24,558	25,268	26,030	24,563	25,911	26,268	..	Gross Domestic Income
25,339	27,326	28,004	25,855	23,724	24,335	25,091	23,783	25,282	25,604	..	Gross National Income
				(Index 1980 = 100)							DOMESTIC PRICES (DEFLATORS)
20.2	51.3	100.0	207.6	592.2	2,661.0	20,090.8	160,000.0	280,000.0	640,000.0	..	Overall (GDP)
20.2	51.7	100.0	206.4	591.8	2,635.8	19,936.4	160,000.0	280,000.0	660,000.0	..	Domestic Absorption
21.2	58.3	100.0	199.0	654.7	3,194.9	23,942.8	180,000.0	350,000.0	800,000.0	..	Agriculture
21.8	52.7	100.0	215.6	660.9	3,302.5	25,310.3	210,000.0	350,000.0	800,000.0	..	Industry
22.7	53.8	100.0	220.4	740.6	3,370.9	25,120.4	210,000.0	350,000.0	820,000.0	..	Manufacturing
											MANUFACTURING ACTIVITY
110.7	108.3	100.0	87.7	82.9	85.6	87.9	84.6	..	..	J	Employment (1980=100)
79.1	89.5	100.0	90.8	80.3	103.4	125.6	103.7	..	..	J	Real Earnings per Empl. (1980=100)
92.4	81.6	100.0	104.2	140.2	104.5	109.8	102.4	..	..	J	Real Output per Empl. (1980=100)
20.8	22.9	26.4	23.2	16.1	20.2	22.8	18.7	..	..	J	Earnings as % of Value Added
				(Billions of current Argentine Australes)							MONETARY HOLDINGS
0.0015	0.0044	0.0082	0.0170	0.0460	0.2300	1.6400	8.4300	17.7600	48.4200	..	Money Supply, Broadly Defined
0.0006	0.0014	0.0027	0.0047	0.0160	0.0750	0.4500	3.0300	5.6000	12.5800	..	Money as Means of Payment
0.0003	0.0008	0.0016	0.0030	0.0087	0.0460	0.3100	2.0200	3.9900	9.2600	..	Currency Ouside Banks
0.0003	0.0006	0.0011	0.0017	0.0075	0.0290	0.1400	1.0100	1.6100	3.3200	..	Demand Deposits
0.0009	0.0030	0.0055	0.0130	0.0300	0.1600	1.1900	5.4000	12.1600	35.8400	..	Quasi-Monetary Liabilities
				(Millions of current Argentine Australes)							
-0.2	-0.4	-1.0	-4.5	-10.6	-87.0	-267.4	-2,923.8	..	..	..	GOVERNMENT DEFICIT (-) OR SURPLUS
0.8	2.3	4.9	9.2	23.2	106.2	782.2	8,284.9	..	..	..	Current Revenue
0.8	2.1	4.7	10.3	26.5	133.6	804.7	8,583.1	..	..	..	Current Expenditure
..	0.2	0.2	-1.1	-3.3	-27.4	-22.5	-298.2	..	..	..	Current Budget Balance
..	0.001	0.004	0.002	0.013	0.057	0.100	1.400	..	..	..	Capital Receipts
0.2	0.6	1.2	3.3	7.3	59.7	245.0	2,627.0	..	..	..	Capital Payments

ARGENTINA	1967	1968	1969	1970	1971	1972	1973	1974	1975	1976	1977
FOREIGN TRADE (CUSTOMS BASIS)				*(Millions of current US dollars)*							
Value of Exports, fob	1,465	1,368	1,612	1,773	1,740	1,941	3,266	3,931	2,961	3,912	5,642
Nonfuel Primary Products	1,338	1,189	1,386	1,519	1,468	1,540	2,527	2,958	2,224	2,919	4,265
Fuels	8	12	6	8	9	6	6	12	15	20	28
Manufactures	119	166	220	247	264	395	733	961	723	973	1,349
Value of Imports, cif	1,096	1,169	1,571	1,689	1,846	1,904	2,235	3,635	3,945	3,029	4,158
Nonfuel Primary Products	249	271	367	374	390	411	552	780	763	511	615
Fuels	94	85	106	80	123	72	170	528	522	536	682
Manufactures	753	814	1,098	1,234	1,332	1,422	1,514	2,327	2,660	1,982	2,861
Terms of Trade	163.1	144.6	148.6	139.3	141.8	*(Index 1980 = 100)* 127.0	166.9	133.4	122.8	120.9	111.0
Export Prices, fob	33.1	31.0	33.9	36.4	37.4	41.7	60.1	70.7	68.9	69.2	70.2
Nonfuel Primary Products	36.2	34.8	37.1	39.2	40.8	45.2	76.2	84.0	73.7	73.9	73.1
Fuels	4.3	4.3	4.3	4.3	5.6	6.2	8.9	36.7	35.7	38.4	42.0
Manufactures	21.6	23.4	25.1	29.8	29.2	34.5	35.7	48.0	58.5	59.1	63.2
Import Prices, cif	20.3	21.4	22.8	26.1	26.4	32.8	36.0	53.0	56.1	57.3	63.2
Trade at Constant 1980 Prices				*(Millions of 1980 US dollars)*							
Exports, fob	4,428	4,417	4,749	4,876	4,651	4,654	5,437	5,557	4,296	5,650	8,037
Imports, cif	5,402	5,461	6,876	6,466	6,995	5,798	6,211	6,855	7,029	5,287	6,577
BALANCE OF PAYMENTS				*(Millions of current US dollars)*							
Exports of Goods & Services	..	..	..	2,148	2,131	2,318	3,747	4,714	3,587	4,691	6,738
Merchandise, fob	..	..	..	1,773	1,740	1,941	3,266	3,930	2,961	3,918	5,651
Nonfactor Services	..	..	..	331	360	359	433	623	537	691	937
Factor Services	..	..	..	44	31	17	48	161	89	82	150
Imports of Goods & Services	..	..	..	2,308	2,518	2,541	3,047	4,596	4,878	4,058	5,643
Merchandise, fob	..	..	..	1,499	1,653	1,685	1,978	3,216	3,510	2,765	3,799
Nonfactor Services	..	..	..	487	512	460	564	808	813	703	913
Factor Services	..	..	..	322	354	395	505	572	555	590	931
Long-Term Interest	..	..	..	338	343	367	461	511	723	470	551
Current Transfers, net	..	..	..	0	0	0	0	..	6	24	32
Workers' Remittances	..	..	..	0	0	0	0	0	0	0	0
Total to be Financed	..	..	..	-160	-387	-223	700	118	-1,286	657	1,127
Official Capital Grants	..	..	..	-3	-3	-4	11	..	-1	-6	-1
Current Account Balance	..	..	..	-163	-390	-227	711	118	-1,287	651	1,126
Long-Term Capital, net	..	..	..	117	335	182	52	-2	-287	799	472
Direct Investment	..	..	..	11	11	10	10	10	..	143	
Long-Term Loans	..	..	..	135	342	265	320	464	-502	1,119	426
Disbursements	..	..	..	907	1,068	1,027	1,234	1,460	920	2,173	1,577
Repayments	..	..	..	772	726	761	915	996	1,422	1,053	1,151
Other Long-Term Capital	..	..	..	-26	-15	-88	-288	-476	217	-314	-96
Other Capital, net	..	..	..	122	-390	-20	79	-41	496	-531	247
Change in Reserves	..	..	..	-76	445	65	-842	-75	1,078	-919	-1,845
Memo Item:				*(Argentine Australes per US dollar)*							
Conversion Factor (Annual Avg)	3.00E-07	4.00E-07	4.00E-07	4.00E-07	5.00E-07	8.00E-07	9.00E-07	9.00E-07	3.70E-06	2.00E-05	5.00E-05
Additional Conversion Factor	3.00E-07	3.00E-07	3.00E-07	4.00E-07	5.00E-07	5.00E-07	5.00E-07	5.00E-07	3.70E-06	1.40E-05	4.08E-05
EXTERNAL DEBT, ETC.				*(Millions of US dollars, outstanding at end of year)*							
Public/Publicly Guar. Long-Term	..	..	..	1,880	2,154	2,378	2,787	3,251	3,124	4,431	5,036
Official Creditors	..	..	..	637	713	743	893	1,087	1,125	1,190	1,268
IBRD and IDA	..	..	..	181	219	256	302	340	341	343	343
Private Creditors	..	..	..	1,243	1,442	1,635	1,893	2,164	1,999	3,241	3,768
Private Non-guaranteed Long-Term	..	..	..	3,291	3,410	3,461	3,432	3,460	3,457	3,298	3,267
Use of Fund Credit	..	..	..	0	0	189	210	78	293	529	419
Short-Term Debt	..	..	..	..	..	..	..	..	..	..	2,724
Memo Items:				*(Millions of US dollars)*							
Int'l Reserves Excluding Gold	643.0	650.7	402.8	532.6	192.5	313.3	1,148.7	1,144.3	287.8	1,444.9	3,153.8
Gold Holdings (at market price)	84.2	130.2	135.5	149.2	111.9	259.1	448.6	745.3	560.4	538.5	690.2
SOCIAL INDICATORS											
Total Fertility Rate	3.0	3.1	3.1	3.1	3.1	3.1	3.2	3.2	3.3	3.3	3.4
Crude Birth Rate	21.1	21.3	24.8	22.9	23.5	22.9	22.7	23.6	24.6	25.5	25.4
Infant Mortality Rate	56.0	54.6	53.2	51.8	50.4	49.0	47.4	45.8	44.2	42.6	41.0
Life Expectancy at Birth	66.0	66.2	66.5	66.8	67.0	67.3	67.6	67.9	68.1	68.4	68.7
Food Production, p.c. ('79-81=100)	99.5	89.3	95.0	94.3	90.3	84.7	87.7	97.6	94.1	100.1	97.5
Labor Force, Agriculture (%)	17.4	16.9	16.5	16.0	15.7	15.4	15.1	14.8	14.5	14.2	13.9
Labor Force, Female (%)	23.7	24.1	24.5	24.9	25.1	25.3	25.5	25.7	26.0	26.1	26.3
School Enroll. Ratio, primary	..	..	..	105.0	..	..	..	..	106.0	109.0	111.0
School Enroll. Ratio, secondary	..	..	..	44.0	..	..	..	..	54.0	57.0	56.0

1978	1979	1980	1981	1982	1983	1984	1985	1986	1987 estimate	NOTES	ARGENTINA
				(Millions of current US dollars)							**FOREIGN TRADE (CUSTOMS BASIS)**
6,394	7,808	8,021	9,143	7,625	7,836	8,107	8,396	6,852	6,360	f	Value of Exports, fob
4,669	5,871	5,883	6,724	5,228	6,219	6,353	5,988	4,895	4,368	..	Nonfuel Primary Products
52	49	278	620	548	334	332	614	153	312	..	Fuels
1,674	1,888	1,861	1,800	1,849	1,283	1,423	1,795	1,804	1,681	..	Manufactures
3,832	6,692	10,539	9,430	5,337	4,504	4,585	3,814	4,724	6,119	f	Value of Imports, cif
566	1,096	1,299	1,039	722	626	694	492	771	1,335	..	Nonfuel Primary Products
477	1,109	1,086	1,022	682	463	478	462	466	706	..	Fuels
2,790	4,488	8,155	7,370	3,933	3,414	3,413	2,860	3,488	4,078	..	Manufactures
				(Index 1980 = 100)							
105.6	108.0	100.0	98.9	91.5	97.8	98.8	89.8	85.4	81.2	..	Terms of Trade
75.6	91.7	100.0	99.1	88.6	92.3	91.1	82.8	79.4	81.0	..	Export Prices, fob
80.5	95.2	100.0	97.0	84.3	93.5	91.7	78.9	72.9	73.1	..	Nonfuel Primary Products
42.3	61.0	100.0	112.5	101.6	92.5	90.2	87.5	44.9	56.9	..	Fuels
66.1	83.0	100.0	103.0	99.0	87.0	89.0	97.0	114.6	125.8	..	Manufactures
71.6	84.8	100.0	100.2	96.8	94.4	92.3	92.2	93.0	99.7	..	Import Prices, cif
				(Millions of 1980 US dollars)							Trade at Constant 1980 Prices
8,458	8,519	8,024	9,228	8,606	8,491	8,895	10,144	8,626	7,856	..	Exports, fob
5,353	7,888	10,539	9,413	5,514	4,774	4,968	4,137	5,077	6,137	..	Imports, cif
				(Millions of current US dollars)							**BALANCE OF PAYMENTS**
7,836	9,916	11,993	12,430	10,048	9,953	10,292	10,495	9,074	8,599	..	Exports of Goods & Services
6,401	7,810	8,021	9,143	7,624	7,836	8,107	8,396	6,852	6,356	..	Merchandise, fob
1,083	1,366	2,744	2,402	1,901	1,677	1,921	1,846	1,865	2,025	..	Nonfactor Services
352	740	1,228	885	523	440	264	253	357	218	..	Factor Services
6,048	10,486	16,783	17,122	12,437	12,429	12,686	11,448	11,935	12,876	..	Imports of Goods & Services
3,488	6,028	10,541	9,430	5,337	4,564	4,585	3,814	4,724	5,800	..	Merchandise, fob
1,473	2,745	3,483	3,107	1,858	2,017	2,125	2,077	2,438	2,373	..	Nonfactor Services
1,087	1,713	2,759	4,585	5,242	5,848	5,976	5,557	4,773	4,703	..	Factor Services
754	957	1,337	2,044	2,435	2,417	3,277	4,389	3,707	3,775	..	Long-Term Interest
48	35	23	-22	34	16	2	0	2	-8	..	Current Transfers, net
0	0	0	0	0	0	0	0	0	0	..	Workers' Remittances
1,836	-535	-4,767	-4,714	-2,355	-2,460	-2,392	-953	-2,859	-4,285	K	Total to be Financed
20	22	0	0	0	0	0	0	0	..	K	Official Capital Grants
1,856	-513	-4,767	-4,714	-2,355	-2,460	-2,392	-953	-2,859	-4,285	..	Current Account Balance
1,410	3,016	4,188	9,484	7,658	2,793	-479	3,705	6,337	2,244	..	Long-Term Capital, net
227	147	788	927	257	183	269	919	574	-19	..	Direct Investment
1,278	4,566	2,855	6,396	5,759	1,469	8	2,772	560	2,421	..	Long-Term Loans
3,316	5,865	4,708	8,346	7,054	2,833	820	3,790	2,602	3,116	..	Disbursements
2,038	1,299	1,853	1,950	1,294	1,364	812	1,018	2,043	695	..	Repayments
-115	-1,719	545	2,161	1,642	1,141	-756	14	5,203	-158	..	Other Long-Term Capital
-1,131	1,657	-2,217	-8,228	-6,058	-61	3,078	-1,888	-4,186	-332	..	Other Capital, net
-2,135	-4,160	2,796	3,458	755	-272	-207	-864	708	2,373	..	Change in Reserves
				(Argentine Australes per US dollar)							Memo Item:
1.20E-04	2.80E-04	5.00E-04	9.50E-04	2.60E-03	1.10E-02	0.068	0.600	0.940	2.140	..	Conversion Factor (Annual Avg)
7.96E-05	1.32E-04	1.84E-04	4.40E-04	2.59E-03	1.05E-02	0.068	0.602	0.943	2.144	..	Additional Conversion Factor
				(Millions of US dollars, outstanding at end of year)							**EXTERNAL DEBT, ETC.**
6,746	8,600	10,181	10,570	15,886	25,445	26,700	35,706	38,774	47,451	..	Public/Publicly Guar. Long-Term
1,653	1,871	1,903	1,913	1,959	2,908	2,618	4,168	4,979	7,124	..	Official Creditors
352	367	404	477	504	533	503	700	1,140	2,147	..	IBRD and IDA
5,093	6,729	8,278	8,657	13,927	22,537	24,082	31,538	33,795	40,327	..	Private Creditors
3,102	5,439	6,593	12,166	11,227	10,393	10,340	4,575	4,559	2,858	..	Private Non-guaranteed Long-Term
0	0	0	0	0	1,173	1,098	2,312	2,741	3,853	..	Use of Fund Credit
3,428	6,911	10,383	12,921	16,521	8,913	10,718	6,730	3,641	2,651	..	Short-Term Debt
				(Millions of US dollars)							Memo Items:
4,966.5	9,388.0	6,719.5	3,268.4	2,506.4	1,172.4	1,242.6	3,273.0	2,718.0	1,617.0	..	Int'l Reserves Excluding Gold
966.8	2,238.5	2,577.3	1,737.9	1,997.6	1,667.9	1,347.9	1,429.6	1,709.4	2,117.0	..	Gold Holdings (at market price)
											SOCIAL INDICATORS
3.4	3.4	3.4	3.4	3.3	3.3	3.2	3.2	3.2	3.1	..	Total Fertility Rate
25.2	24.2	24.3	24.5	24.1	23.6	23.2	22.8	22.5	22.1	..	Crude Birth Rate
40.0	39.0	38.0	37.0	35.2	33.4	31.7	30.1	28.6	27.2	..	Infant Mortality Rate
68.9	69.1	69.3	69.5	69.7	69.9	70.1	70.3	70.5	70.7	..	Life Expectancy at Birth
101.4	103.8	95.6	100.7	104.1	99.6	101.1	97.6	98.1	97.2	..	Food Production, p.c. ('79-81 = 100)
13.6	13.3	13.0	..	..	..	..	..	..	..	..	Labor Force, Agriculture (%)
26.5	26.7	26.9	27.0	27.1	27.2	27.4	27.5	27.6	27.7	..	Labor Force, Female (%)
..	112.0	..	106.0	108.0	107.0	107.0	108.0	..	..	..	School Enroll. Ratio, primary
56.0	56.0	..	57.0	..	60.0	65.0	70.0	..	..	..	School Enroll. Ratio, secondary

AUSTRALIA	1967	1968	1969	1970	1971	1972	1973	1974	1975	1976	1977
CURRENT GNP PER CAPITA (US $)	2,250	2,530	2,730	2,960	3,190	3,480	4,150	5,130	6,410	7,460	7,650
POPULATION (thousands)	11,799	12,009	12,263	12,507	12,937	13,177	13,380	13,723	13,893	14,033	14,192
ORIGIN AND USE OF RESOURCES					*(Billions of current Australian Dollars)*						
Gross National Product (GNP)	23.79	26.82	29.63	32.78	36.79	41.08	49.05	58.43	69.46	81.72	89.96
Net Factor Income from Abroad	-0.34	-0.36	-0.48	-0.50	-0.53	-0.58	-0.44	-0.54	-0.88	-1.10	-1.26
Gross Domestic Product (GDP)	24.12	27.19	30.11	33.29	37.33	41.65	49.49	58.98	70.34	82.82	91.22
Indirect Taxes, net	2.31	2.55	2.81	3.03	3.32	3.87	4.54	5.69	6.95	8.68	9.33
GDP at factor cost	21.81	24.64	27.30	30.25	34.00	37.79	44.95	53.28	63.40	74.14	81.89
Agriculture	1.82	2.31	2.14	1.94	2.23	2.99	4.05	3.55	3.68	4.23	4.09
Industry	9.34	10.40	11.69	12.98	14.36	15.49	18.09	21.56	25.57	29.73	32.61
Manufacturing	6.08	6.70	7.34	8.09	8.77	9.32	10.92	12.58	14.54	16.81	18.25
Services, etc.	12.97	14.47	16.29	18.36	20.74	23.17	27.35	33.87	41.09	48.87	54.52
Resource Balance	-0.49	-0.31	0.08	0.04	0.03	1.18	1.30	-1.10	0.53	0.16	-0.88
Exports of Goods & NFServices	3.47	3.79	4.63	4.94	5.32	6.22	7.53	8.74	10.50	12.36	13.86
Imports of Goods & NFServices	3.96	4.09	4.55	4.90	5.29	5.04	6.23	9.83	9.96	12.19	14.74
Domestic Absorption	24.61	27.49	30.03	33.24	37.30	40.47	48.19	60.07	69.81	82.66	92.10
Private Consumption, etc.	16.56	18.33	20.19	19.58	21.98	24.66	28.87	35.06	41.57	47.74	53.62
General Gov't Consumption	3.46	3.62	4.00	4.56	5.26	5.88	7.10	9.15	11.97	14.39	16.16
Gross Domestic Investment	4.59	5.54	5.84	9.10	10.07	9.92	12.21	15.87	16.27	20.53	22.32
Fixed Investment	6.56	7.38	8.05	8.96	10.17	10.56	12.08	13.80	16.59	19.85	21.74
Memo Items:											
Gross Domestic Saving	4.10	5.23	5.92	9.14	10.09	11.11	13.51	14.77	16.80	20.70	21.44
Gross National Saving	3.83	4.96	5.51	8.71	9.61	10.59	13.07	14.21	15.86	19.54	20.16
					(Billions of 1980 Australian Dollars)						
Gross National Product	76.11	83.31	87.70	92.88	98.49	102.33	108.87	110.34	112.88	116.89	117.76
GDP at factor cost	68.97	75.58	79.85	85.59	91.02	94.09	99.45	100.29	102.89	105.88	107.19
Agriculture	..	..	6.08	6.01	6.69	5.86	6.28	6.91	7.46	7.75	7.59
Industry	..	..	33.57	35.81	37.44	39.01	42.09	41.20	41.08	43.31	43.17
Manufacturing	..	..	19.91	20.30	20.81	21.68	23.52	22.44	22.17	22.94	22.81
Services, etc.	..	..	49.57	52.51	55.79	58.90	61.50	63.29	65.80	67.43	68.66
Resource Balance	-3.79	-3.60	-2.88	-1.39	0.05	2.53	-0.58	-5.60	-0.69	-1.46	-1.23
Exports of Goods & NFServices	11.03	11.75	13.68	14.93	16.36	17.36	17.31	16.52	17.95	19.43	19.62
Imports of Goods & NFServices	14.82	15.35	16.56	16.32	16.31	14.83	17.89	22.11	18.63	20.89	20.85
Domestic Absorption	81.04	88.11	92.10	95.72	99.87	101.24	110.45	116.99	115.03	119.94	120.64
Private Consumption, etc.	46.44	49.08	52.02	53.85	56.57	60.02	64.50	67.74	68.69	69.24	70.31
General Gov't Consumption	13.29	13.41	13.94	14.47	14.99	15.53	16.69	17.77	19.29	20.45	20.94
Gross Domestic Investment	21.30	25.62	26.14	27.41	28.30	25.68	29.26	31.49	27.05	30.25	29.39
Fixed Investment	21.30	23.03	24.26	25.48	27.16	26.71	28.06	27.31	27.40	28.73	28.42
Memo Items:											
Capacity to Import	12.99	14.20	16.85	16.47	16.39	18.31	21.63	19.65	19.63	21.17	19.61
Terms of Trade Adjustment	1.96	2.45	3.17	1.54	0.03	0.95	4.32	3.13	1.69	1.74	-0.02
Gross Domestic Income	79.20	86.97	92.39	95.87	99.95	104.73	114.20	114.52	116.03	120.22	119.40
Gross National Income	78.07	85.76	90.87	94.42	98.52	103.29	113.19	113.47	114.57	118.63	117.74
DOMESTIC PRICES (DEFLATORS)					*(Index 1980 = 100)*						
Overall (GDP)	31.2	32.2	33.7	35.3	37.4	40.1	45.0	52.9	61.5	69.9	76.4
Domestic Absorption	30.4	31.2	32.6	34.7	37.4	40.0	43.6	51.3	60.7	68.9	76.3
Agriculture	..	..	35.2	32.4	33.4	51.1	64.4	51.3	49.3	54.5	53.9
Industry	..	..	34.8	36.3	38.4	39.7	43.0	52.3	62.2	68.6	75.5
Manufacturing	..	..	36.8	39.8	42.2	43.0	46.4	56.1	65.6	73.3	80.0
MANUFACTURING ACTIVITY											
Employment (1980=100)	103.3	105.4	108.9	112.1	112.5	112.4	111.8	115.4	108.0	104.0	101.8
Real Earnings per Empl. (1980=100)	67.9	69.9	75.0	77.7	79.9	83.8	85.1	88.4	97.6	99.0	100.3
Real Output per Empl. (1980=100)	..	..	..	71.5	70.3	73.7	74.2	70.8	73.9	77.9	84.4
Earnings as % of Value Added	50.4	50.3	52.3	52.5	53.9	54.1	54.3	54.5	56.0	55.9	54.8
MONETARY HOLDINGS					*(Billions of current Australian Dollars)*						
Money Supply, Broadly Defined	12.74	13.63	14.93	15.70	17.07	20.42	24.72	26.97	32.55	36.55	38.71
Money as Means of Payment	4.49	4.75	5.21	5.45	5.75	6.90	8.05	7.99	9.81	10.68	11.39
Currency Ouside Banks	0.99	1.07	1.19	1.33	1.48	1.66	1.96	2.36	2.76	3.13	3.55
Demand Deposits	3.50	3.68	4.01	4.12	4.27	5.23	6.09	5.64	7.05	7.55	7.84
Quasi-Monetary Liabilities	8.24	8.88	9.73	10.25	11.32	13.53	16.67	18.97	22.75	25.87	27.32
GOVERNMENT DEFICIT (-) OR SURPLUS					*(Millions of current Australian Dollars)*						
	..	..	..	-260	-78	130	-409	135	-2,505	-3,646	-2,748
Current Revenue	..	..	..	7,308	8,219	9,087	9,691	12,270	15,693	18,780	21,958
Current Expenditure	..	..	..	5,844	6,597	7,374	8,431	9,992	13,651	18,014	21,019
Current Budget Balance	..	..	..	1,464	1,622	1,713	1,260	2,278	2,042	766	939
Capital Receipts	..	..	..	15	22	44	49	29	20	34	33
Capital Payments	..	..	..	1,739	1,722	1,627	1,718	2,172	4,567	4,446	3,720

1978	1979	1980	1981	1982	1983	1984	1985	1986	1987 estimate	NOTES	AUSTRALIA
8,100	8,780	9,650	10,910	11,550	11,450	11,890	11,630	11,190	11,150	..	CURRENT GNP PER CAPITA (US $)
14,358	14,514	14,692	14,927	15,178	15,369	15,544	15,758	15,974	16,180	..	POPULATION (thousands)
			(Billions of current Australian Dollars)								ORIGIN AND USE OF RESOURCES
99.57	112.53	128.17	144.41	159.96	172.81	195.60	217.58	237.06	266.49	..	Gross National Product (GNP)
-1.55	-1.97	-2.20	-2.99	-3.69	-4.57	-6.27	-7.78	-9.34	-10.50	..	Net Factor Income from Abroad
101.11	114.50	130.37	147.41	163.65	177.38	201.87	225.35	246.41	276.99	..	Gross Domestic Product (GDP)
10.21	11.95	14.24	15.82	18.26	20.46	23.93	28.22	29.38	34.19	..	Indirect Taxes, net
90.90	102.55	116.13	131.59	145.39	156.92	177.94	197.13	217.03	242.79	B	GDP at factor cost
6.62	7.34	7.16	7.42	5.99	8.97	9.23	9.21	10.17	..	..	Agriculture
35.22	40.53	46.96	51.95	57.23	61.02	69.53	76.89	80.02	..	..	Industry
19.36	21.87	25.05	28.07	29.51	31.61	35.49	38.71	41.88	..	..	Manufacturing
59.28	66.64	76.25	88.04	100.43	107.39	123.11	139.25	156.21	..	..	Services, etc.
-1.83	0.38	-0.98	-4.33	-5.40	-2.64	-4.48	-6.02	-7.04	-3.91	..	Resource Balance
14.57	19.46	22.20	22.12	24.34	25.75	30.49	37.21	39.32	45.00	..	Exports of Goods & NFServices
16.40	19.08	23.18	26.44	29.74	28.39	34.97	43.23	46.36	48.91	..	Imports of Goods & NFServices
102.94	114.12	131.34	151.74	169.05	180.02	206.35	231.37	253.45	280.89	..	Domestic Absorption
59.98	66.09	75.64	85.92	96.67	106.58	119.66	131.79	147.90	165.13	..	Private Consumption, etc.
18.18	19.85	23.10	26.58	30.26	33.98	38.20	42.65	46.71	50.82	..	General Gov't Consumption
24.79	28.18	32.61	39.24	42.12	39.46	48.49	56.93	58.84	64.94	..	Gross Domestic Investment
24.18	27.18	31.59	37.62	41.42	40.45	46.37	54.01	57.21	62.12	..	Fixed Investment
											Memo Items:
22.96	28.56	31.63	34.91	36.72	36.82	44.01	50.91	51.80	61.04	..	Gross Domestic Saving
21.43	26.68	29.68	32.18	33.33	32.71	38.25	44.12	43.68	52.13	..	Gross National Saving
			(Billions of 1980 Australian Dollars)								
121.65	125.55	128.17	131.62	131.62	131.38	139.41	146.66	148.83	155.44	..	Gross National Product
111.14	114.44	116.13	119.92	119.69	119.32	126.82	133.07	136.64	141.97	B	GDP at factor cost
8.81	8.00	7.16	8.30	6.79	8.81	9.13	9.01	..	..	..	Agriculture
43.90	45.71	46.96	48.54	46.65	45.69	49.34	53.35	..	..	..	Industry
23.37	24.79	25.05	25.89	24.16	23.43	24.98	26.44	..	..	..	Manufacturing
70.82	74.04	76.25	77.50	81.20	80.35	85.42	89.48	..	..	..	Services, etc.
-1.41	0.64	-0.98	-3.99	-3.81	-1.83	-3.22	-2.02	0.42	2.64	..	Resource Balance
20.14	22.59	22.20	21.33	22.80	22.09	25.61	28.16	29.42	32.08	..	Exports of Goods & NFServices
21.55	21.95	23.18	25.32	26.62	23.91	28.83	30.18	29.00	29.44	..	Imports of Goods & NFServices
124.94	127.11	131.34	138.33	138.45	136.67	147.10	153.85	154.08	158.65	..	Domestic Absorption
72.52	72.83	75.64	78.62	80.34	81.27	85.81	88.52	91.28	94.46	..	Private Consumption, etc.
22.04	22.26	23.10	23.87	23.81	25.09	26.31	27.87	28.41	29.19	..	General Gov't Consumption
30.39	32.02	32.61	35.85	34.30	30.30	34.98	37.46	34.39	35.00	..	Gross Domestic Investment
29.32	30.21	31.59	34.33	34.04	30.85	33.08	36.04	34.75	35.05	..	Fixed Investment
											Memo Items:
19.15	22.39	22.20	21.17	21.78	21.69	25.14	25.98	24.59	27.09	..	Capacity to Import
-0.99	-0.20	0.00	-0.15	-1.02	-0.40	-0.47	-2.18	-4.83	-4.99	..	Terms of Trade Adjustment
122.54	127.55	130.37	134.19	133.62	134.44	143.41	149.65	149.68	156.30	..	Gross Domestic Income
120.66	125.35	128.17	131.46	130.60	130.98	138.94	144.48	144.00	..	..	Gross National Income
			(Index 1980 = 100)								DOMESTIC PRICES (DEFLATORS)
81.9	89.6	100.0	109.7	121.5	131.5	140.3	148.4	159.5	171.7	..	Overall (GDP)
82.4	89.8	100.0	109.7	122.1	131.7	140.3	150.4	164.5	177.1	..	Domestic Absorption
75.1	91.7	100.0	89.3	88.3	101.8	101.1	102.3	..	..	..	Agriculture
80.2	88.7	100.0	107.0	122.7	133.5	140.9	144.1	..	..	..	Industry
82.9	88.3	100.0	108.4	122.1	134.9	142.1	146.4	..	..	..	Manufacturing
											MANUFACTURING ACTIVITY
99.3	99.3	100.0	100.2	100.1	91.5	87.8	88.9	..	..	..	Employment (1980=100)
100.8	99.4	100.0	101.6	104.3	106.1	106.7	106.1	..	..	..	Real Earnings per Empl. (1980=100)
89.1	94.3	100.0	102.7	101.8	101.4	108.6	111.3	..	..	..	Real Output per Empl. (1980=100)
54.9	54.0	52.1	50.6	53.8	55.9	51.1	48.1	..	..	..	Earnings as % of Value Added
			(Billions of current Australian Dollars)								MONETARY HOLDINGS
50.25	56.76	65.55	72.46	80.47	91.44	102.17	120.47	131.99	..	D	Money Supply, Broadly Defined
12.71	14.66	17.22	18.06	18.03	20.80	22.49	23.30	25.95	31.22	..	Money as Means of Payment
3.96	4.37	4.98	5.53	6.02	6.88	7.86	8.63	9.54	10.84	..	Currency Ouside Banks
8.75	10.29	12.25	12.53	12.01	13.91	14.64	14.67	16.41	20.38	..	Demand Deposits
37.54	42.10	48.33	54.39	62.44	70.64	79.68	97.17	106.04	..	..	Quasi-Monetary Liabilities
			(Millions of current Australian Dollars)								GOVERNMENT DEFICIT (-) OR SURPLUS
-3,325	-3,400	-2,049	-1,042	-563	-4,526	-8,003	-6,777	-5,832	-3,312	C	
24,146	26,291	30,551	35,957	41,684	45,520	49,832	58,585	65,751	73,749	..	Current Revenue
24,114	26,743	29,735	34,042	38,832	45,578	52,948	60,100	66,228	71,953	..	Current Expenditure
32	-452	816	1,915	2,852	-58	-3,116	-1,515	-477	1,796	..	Current Budget Balance
103	59	62	285	41	64	95	138	233	343	..	Capital Receipts
3,460	3,007	2,927	3,242	3,456	4,532	4,982	5,400	5,588	5,451	..	Capital Payments

AUSTRALIA	1967	1968	1969	1970	1971	1972	1973	1974	1975	1976	1977
FOREIGN TRADE (CUSTOMS BASIS)					*(Millions of current US dollars)*						
Value of Exports, fob	3,368	3,401	4,045	4,621	5,073	6,306	9,596	11,000	11,877	13,126	13,306
Nonfuel Primary Products	2,694	2,583	3,018	3,393	3,651	4,567	6,764	7,513	8,088	8,630	8,561
Fuels	123	154	191	251	317	389	589	954	1,384	1,648	1,783
Manufactures	550	665	836	978	1,105	1,350	2,243	2,534	2,405	2,847	2,963
Value of Imports, cif	3,924	4,366	4,538	5,056	5,228	5,028	7,393	11,982	10,697	12,232	13,511
Nonfuel Primary Products	516	570	569	614	611	625	977	1,452	1,065	1,286	1,525
Fuels	307	312	335	276	249	236	303	1,008	1,045	1,170	1,364
Manufactures	3,102	3,484	3,633	4,167	4,368	4,167	6,113	9,522	8,588	9,776	10,623
Terms of Trade	141.1	130.9	125.2	120.6	105.7	119.7	143.6	131.3	108.8	112.7	102.8
					(Index 1980 = 100)						
Export Prices, fob	29.3	27.9	27.4	29.9	30.2	37.6	56.8	71.0	63.0	67.2	67.6
Nonfuel Primary Products	34.3	32.2	34.5	36.2	36.8	44.3	70.4	77.8	71.1	71.8	67.8
Fuels	13.8	13.4	13.1	16.3	19.2	21.6	37.8	66.9	78.7	74.8	78.7
Manufactures	21.7	23.6	18.8	22.3	21.7	29.3	39.4	54.6	44.9	58.0	68.1
Import Prices, cif	20.8	21.3	21.9	24.8	28.6	31.4	39.6	54.1	57.9	59.6	65.7
Trade at Constant 1980 Prices					*(Millions of 1980 US dollars)*						
Exports, fob	11,497	12,187	14,782	15,436	16,799	16,765	16,880	15,501	18,853	19,526	19,691
Imports, cif	18,907	20,479	20,762	20,372	18,306	15,998	18,669	22,165	18,481	20,510	20,560
BALANCE OF PAYMENTS					*(Millions of current US dollars)*						
Exports of Goods & Services	..	..	..	5,705	6,266	7,751	11,322	13,535	14,461	15,579	15,845
Merchandise, fob	..	..	..	4,623	5,063	6,276	9,266	10,767	11,688	12,981	13,193
Nonfactor Services	..	..	..	902	967	1,107	1,399	2,001	2,175	2,087	2,125
Factor Services	..	..	..	180	236	369	657	767	598	511	528
Imports of Goods & Services	..	..	..	6,467	7,073	7,205	10,754	16,093	15,133	17,330	18,733
Merchandise, fob	..	..	..	4,108	4,475	4,294	6,450	10,635	9,487	10,946	12,176
Nonfactor Services	..	..	..	1,357	1,516	1,697	2,446	3,493	3,502	3,922	4,108
Factor Services	..	..	..	1,001	1,081	1,214	1,858	1,966	2,144	2,463	2,449
Long-Term Interest	..	..	..	..	..	..	..	..	..	..	..
Current Transfers, net	..	..	..	80	56	66	-11	-32	-84	-72	-21
Workers' Remittances	..	..	..	0	0	0	0	0	0	0	0
Total to be Financed	..	..	..	-682	-751	612	557	-2,590	-756	-1,823	-2,908
Official Capital Grants	..	..	..	-95	-82	-150	-140	-248	-332	-186	-232
Current Account Balance	..	..	..	-777	-833	462	418	-2,838	-1,088	-2,009	-3,140
Long-Term Capital, net	..	..	..	950	1,800	1,571	-1,111	617	461	1,319	1,954
Direct Investment	..	..	..	778	1,046	926	-117	1,087	289	776	874
Long-Term Loans	..	..	..	..	..	..	..	..	..	..	..
Disbursements	..	..	..	..	..	..	..	..	..	..	..
Repayments	..	..	..	..	..	..	..	..	..	..	..
Other Long-Term Capital	..	..	..	267	836	795	-854	-222	505	728	1,312
Other Capital, net	..	..	..	172	402	684	375	588	-247	306	32
Change in Reserves	..	..	..	-345	-1,369	-2,717	318	1,633	874	384	1,154
Memo Item:					*(Australian Dollars per US dollar)*						
Conversion Factor (Annual Avg)	0.890	0.890	0.890	0.890	0.890	0.860	0.780	0.680	0.730	0.790	0.870
EXTERNAL DEBT, ETC.					*(Millions US dollars, outstanding at end of year)*						
Public/Publicly Guar. Long-Term	..	..	..	..	..	..	..	..	..	..	..
Official Creditors	..	..	..	..	..	..	..	..	..	..	..
IBRD and IDA	..	..	..	..	..	..	..	..	..	..	..
Private Creditors	..	..	..	..	..	..	..	..	..	..	..
Private Non-guaranteed Long-Term	..	..	..	..	..	..	..	..	..	..	..
Use of Fund Credit	..	..	..	..	..	..	..	..	..	..	..
Short-Term Debt	..	..	..	..	..	..	..	..	..	..	..
Memo Items:					*(Millions of US dollars)*						
Int'l Reserves Excluding Gold	1,133.4	1,185.5	998.4	1,453.5	3,033.9	5,859.5	5,386.4	3,953.5	2,954.3	2,870.0	2,057.5
Gold Holdings (at market price)	232.3	307.7	264.5	255.2	324.1	480.3	827.4	1,376.0	1,035.0	992.3	1,261.7
SOCIAL INDICATORS											
Total Fertility Rate	2.8	2.9	2.9	2.9	3.0	2.7	2.5	2.4	2.2	2.1	2.0
Crude Birth Rate	19.4	20.1	20.4	20.6	21.4	20.1	18.5	17.8	16.7	16.2	15.9
Infant Mortality Rate	18.3	17.8	17.9	17.9	17.3	16.7	16.5	16.1	14.2	13.8	12.4
Life Expectancy at Birth	70.9	71.0	71.2	71.4	71.5	71.7	72.1	72.4	72.8	73.1	73.5
Food Production, p.c. ('79-81 = 100)	79.0	96.9	86.8	86.5	90.3	86.5	99.2	92.0	98.5	102.2	97.2
Labor Force, Agriculture (%)	9.0	8.7	8.4	8.1	8.0	7.8	7.7	7.5	7.4	7.3	7.2
Labor Force, Female (%)	29.8	30.3	30.8	31.2	31.9	32.5	33.2	33.8	34.4	35.0	35.7
School Enroll. Ratio, primary	..	..	..	115.0	..	..	..	..	107.0	108.0	109.0
School Enroll. Ratio, secondary	..	..	..	82.0	..	..	..	..	87.0	88.0	88.0

1978	1979	1980	1981	1982	1983	1984	1985	1986	1987 estimate	NOTES	AUSTRALIA
											FOREIGN TRADE (CUSTOMS BASIS)
				(Millions of current US dollars)							
14,522	18,888	21,985	22,160	21,415	19,841	22,720	22,292	21,838	25,283	..	Value of Exports, fob
8,310	11,393	13,346	12,913	12,835	10,782	12,709	11,876	12,244	13,920	..	Nonfuel Primary Products
2,013	2,269	2,419	3,376	3,845	4,454	5,392	5,869	4,810	5,064	..	Fuels
4,198	5,226	6,220	5,871	4,736	4,605	4,618	4,548	4,784	6,300	..	Manufactures
15,567	18,190	22,399	26,215	26,667	21,458	25,919	25,889	26,104	29,318	..	Value of Imports, cif
1,630	1,878	2,282	2,303	2,144	2,014	2,389	2,250	2,253	2,676	..	Nonfuel Primary Products
1,415	1,937	3,098	3,563	3,849	2,324	2,336	1,749	1,207	1,427	..	Fuels
12,522	14,375	17,018	20,348	20,674	17,120	21,194	21,889	22,644	25,214	..	Manufactures
				(Index 1980 = 100)							
95.3	98.7	100.0	95.8	90.6	91.5	93.5	88.6	79.8	71.9	..	Terms of Trade
70.4	85.7	100.0	93.3	84.3	81.8	81.8	76.4	73.6	72.8	..	Export Prices, fob
75.6	95.6	100.0	91.8	82.1	82.1	80.1	76.0	74.7	79.6	..	Nonfuel Primary Products
83.5	88.2	100.0	98.4	91.3	84.3	83.3	83.3	72.5	61.7	..	Fuels
61.9	72.7	100.0	94.0	85.0	79.0	85.0	70.0	72.0	69.8	..	Manufactures
73.9	86.8	100.0	97.4	93.1	89.5	87.5	86.3	92.2	101.2	..	Import Prices, cif
				(Millions of 1980 US dollars)							Trade at Constant 1980 Prices
20,632	22,041	21,985	23,749	25,410	24,246	27,779	29,177	29,675	34,726	..	Exports, fob
21,075	20,945	22,399	26,916	28,659	23,983	29,620	30,006	28,306	28,969	..	Imports, cif
				(Millions of current US dollars)							**BALANCE OF PAYMENTS**
17,336	22,472	26,330	26,318	25,901	24,644	28,515	27,466	27,839	33,514	..	Exports of Goods & Services
14,099	18,571	21,564	21,225	20,787	19,517	22,778	22,280	22,144	26,275	..	Merchandise, fob
2,616	3,140	3,737	4,068	4,012	3,763	3,997	3,673	4,139	5,085	..	Nonfactor Services
621	760	1,029	1,025	1,101	1,364	1,739	1,513	1,556	2,153	..	Factor Services
21,616	24,885	30,514	34,816	34,295	30,697	37,217	36,664	38,297	43,243	..	Imports of Goods & Services
14,002	16,057	20,192	23,580	23,378	19,482	23,680	23,510	24,235	26,795	..	Merchandise, fob
4,721	5,223	6,201	6,815	6,749	6,090	6,990	6,647	6,671	7,436	..	Nonfactor Services
2,894	3,606	4,120	4,420	4,168	5,125	6,546	6,506	7,391	9,011	..	Factor Services
..	..	..	..	..	..	..	..	..	..	..	Long-Term Interest
24	104	275	298	301	415	451	683	813	1,118	..	Current Transfers, net
0	0	0	0	0	0	0	0	0	0	..	Workers' Remittances
-4,256	-2,310	-3,909	-8,200	-8,093	-5,638	-8,251	-8,515	-9,645	-8,611	K	Total to be Financed
-295	-354	-336	-347	-389	-335	-332	-157	-148	-77	K	Official Capital Grants
-4,551	-2,664	-4,245	-8,547	-8,482	-5,973	-8,584	-8,672	-9,793	-8,688	..	Current Account Balance
2,999	2,125	3,823	6,048	11,399	7,039	5,989	7,260	9,877	6,827	..	Long-Term Capital, net
1,437	1,110	1,366	1,532	1,397	2,298	-1,127	327	605	57	..	Direct Investment
..	..	..	..	..	..	..	..	..	..	..	Long-Term Loans
..	..	..	..	..	..	..	..	..	..	..	Disbursements
..	..	..	..	..	..	..	..	..	..	..	Repayments
1,857	1,369	2,794	4,864	10,391	5,076	7,448	7,090	9,420	6,848	..	Other Long-Term Capital
1,465	-159	1,022	2,353	1,983	1,965	1,288	-869	621	2,231	..	Other Capital, net
87	698	-601	146	-4,900	-3,031	1,307	2,281	-705	-371	..	Change in Reserves
				(Australian Dollars per US dollar)							Memo Item:
0.890	0.880	0.900	0.860	0.910	1.070	1.100	1.300	1.430	1.510	..	Conversion Factor (Annual Avg)
				(Millions US dollars, outstanding at end of year)							**EXTERNAL DEBT, ETC.**
..	..	..	..	..	..	..	..	..	..	..	Public/Publicly Guar. Long-Term
..	..	..	..	..	..	..	..	..	..	..	Official Creditors
..	..	..	..	..	..	..	..	..	..	..	IBRD and IDA
..	..	..	..	..	..	..	..	..	..	..	Private Creditors
..	..	..	..	..	..	..	..	..	..	..	Private Non-guaranteed Long-Term
..	..	..	..	..	..	..	..	..	..	..	Use of Fund Credit
..	..	..	..	..	..	..	..	..	..	..	Short-Term Debt
				(Millions of US dollars)							Memo Items:
2,061.7	1,424.1	1,690.4	1,671.1	6,370.8	8,962.3	7,441.1	5,767.6	7,246.4	8,743.9	..	Int'l Reserves Excluding Gold
1,760.8	4,061.7	4,676.5	3,153.4	3,624.1	3,025.7	2,445.1	2,593.4	3,100.2	3,839.9	..	Gold Holdings (at market price)
											SOCIAL INDICATORS
2.0	1.9	1.9	2.0	2.0	2.0	2.0	2.0	2.0	1.9	..	Total Fertility Rate
15.6	15.3	15.3	15.7	15.7	15.7	15.7	15.7	15.6	15.6	..	Crude Birth Rate
12.1	11.3	10.7	9.9	10.0	10.1	10.1	10.2	10.3	10.4	..	Infant Mortality Rate
73.9	74.4	74.8	75.2	75.7	75.7	75.7	77.5	76.9	76.3	..	Life Expectancy at Birth
117.0	109.8	90.9	99.3	84.9	111.0	104.4	99.4	98.5	93.0	..	Food Production, p.c. ('79-81=100)
7.1	7.0	6.9	..	..	..	..	..	..	..	..	Labor Force, Agriculture (%)
36.3	36.9	37.5	37.6	37.6	37.7	37.7	37.8	37.9	37.9	..	Labor Force, Female (%)
111.0	111.0	109.0	..	..	108.0	107.0	106.0	106.0	..	..	School Enroll. Ratio, primary
87.0	86.0	84.0	..	..	90.0	94.0	95.0	96.0	..	..	School Enroll. Ratio, secondary

AUSTRIA	1967	1968	1969	1970	1971	1972	1973	1974	1975	1976	1977
CURRENT GNP PER CAPITA (US $)	1,490	1,610	1,780	1,960	2,180	2,520	3,120	3,960	4,760	5,380	6,120
POPULATION (thousands)	7,338	7,362	7,384	7,426	7,456	7,495	7,525	7,533	7,537	7,540	7,544
ORIGIN AND USE OF RESOURCES				*(Billions of current Austrian Schillings)*							
Gross National Product (GNP)	284.3	305.0	333.3	373.9	417.7	476.8	540.1	615.6	652.5	719.7	789.3
Net Factor Income from Abroad	-1.3	-1.9	-1.7	-2.0	-1.9	-2.7	-3.4	-3.0	-3.6	-5.0	-6.9
Gross Domestic Product (GDP)	285.6	306.8	335.0	375.9	419.6	479.5	543.5	618.6	656.1	724.7	796.2
Indirect Taxes, net	39.6	44.9	48.7	55.2	62.4	74.6	88.0	94.0	92.6	98.7	112.6
GDP at factor cost	246.0	262.0	286.3	320.7	357.2	405.0	455.4	524.6	563.6	626.0	683.6
Agriculture	22.7	21.6	23.3	25.8	25.1	28.2	31.4	33.1	33.1	36.4	36.1
Industry	130.5	138.6	150.8	170.5	190.8	219.2	231.4	265.1	268.4	294.0	318.9
Manufacturing	93.7	100.3	111.5	126.7	140.9	158.3	166.7	190.6	187.7	208.7	223.0
Services, etc.	132.4	146.6	160.9	179.6	203.7	232.1	280.7	320.3	354.7	394.3	441.3
Resource Balance	-2.3	-0.8	4.0	3.6	3.0	2.7	2.1	-2.1	4.2	-11.0	-21.4
Exports of Goods & NFServices	71.5	78.8	95.1	116.8	128.7	146.4	166.0	204.3	208.9	236.3	257.0
Imports of Goods & NFServices	73.8	79.6	91.1	113.1	125.6	143.7	163.9	206.3	204.8	247.3	278.4
Domestic Absorption	287.9	307.6	331.0	372.2	416.6	476.9	541.3	620.7	651.9	735.8	817.6
Private Consumption, etc.	154.3	166.6	181.3	205.1	229.7	259.6	291.9	330.8	368.1	410.2	456.6
General Gov't Consumption	41.6	45.2	50.5	55.2	62.0	70.1	81.9	97.4	113.1	127.8	138.7
Gross Domestic Investment	92.0	95.8	99.2	111.9	124.9	147.2	167.6	192.4	170.8	197.8	222.2
Fixed Investment	73.4	76.1	81.0	93.8	113.6	136.6	146.8	170.0	171.6	188.7	209.0
Memo Items:											
Gross Domestic Saving	89.7	95.0	103.2	115.5	127.9	149.9	169.7	190.3	175.0	186.8	200.8
Gross National Saving	89.7	94.2	102.6	113.8	126.2	146.5	164.7	184.4	168.9	179.9	191.9
				(Billions of 1980 Austrian Schillings)							
Gross National Product	580.4	605.1	644.0	689.5	725.3	769.5	806.5	839.6	836.0	873.0	909.5
GDP at factor cost	502.8	520.0	552.9	589.5	618.4	651.1	676.3	713.9	720.4	758.0	786.3
Agriculture	34.9	34.7	35.1	36.4	33.6	33.9	36.0	37.5	39.3	40.7	39.2
Industry	223.9	236.4	256.4	275.4	293.6	314.8	332.3	344.5	329.8	345.3	362.3
Manufacturing	149.9	159.7	176.9	190.9	202.0	217.4	228.6	236.5	222.4	236.4	249.2
Services, etc.	324.0	337.8	355.7	381.4	401.4	425.2	443.4	461.7	471.6	493.2	516.0
Resource Balance	-14.4	-13.9	-3.3	-4.6	-4.6	-9.2	-19.3	-11.7	-5.3	-22.3	-33.8
Exports of Goods & NFServices	127.9	138.7	163.2	190.1	202.3	222.8	234.9	260.1	253.8	281.9	294.8
Imports of Goods & NFServices	142.4	152.7	166.4	194.6	206.9	232.0	254.2	271.7	259.1	304.2	328.6
Domestic Absorption	597.3	622.8	650.4	697.8	733.3	783.1	831.0	855.4	846.0	901.5	951.3
Private Consumption, etc.	327.3	342.0	356.4	375.6	401.0	425.5	448.4	462.3	477.5	499.6	528.2
General Gov't Consumption	113.6	117.1	119.8	123.7	127.8	133.0	137.0	144.8	150.6	157.0	162.7
Gross Domestic Investment	156.4	163.7	174.3	198.5	204.5	224.6	245.6	248.3	217.8	244.9	260.5
Fixed Investment	148.7	153.0	160.5	176.3	200.6	224.9	225.7	234.6	223.0	231.5	243.5
Memo Items:											
Capacity to Import	138.0	151.1	173.7	200.9	211.9	236.3	257.5	269.0	264.4	290.6	303.3
Terms of Trade Adjustment	10.0	12.4	10.5	10.8	9.6	13.5	22.6	8.9	10.6	8.7	8.6
Gross Domestic Income	592.9	621.3	657.7	704.1	738.3	787.4	834.3	852.7	851.2	887.9	926.0
Gross National Income	590.5	617.5	654.6	700.3	734.9	783.0	829.1	848.6	846.5	881.7	918.0
DOMESTIC PRICES (DEFLATORS)				*(Index 1980 = 100)*							
Overall (GDP)	49.0	50.4	51.8	54.2	57.6	62.0	67.0	73.3	78.0	82.4	86.8
Domestic Absorption	48.2	49.4	50.9	53.3	56.8	60.9	65.1	72.6	77.1	81.6	85.9
Agriculture	65.0	62.1	66.5	70.8	74.7	83.3	87.1	88.2	84.2	89.6	92.0
Industry	58.3	58.6	58.8	61.9	65.0	69.6	69.6	77.0	81.4	85.1	88.0
Manufacturing	62.5	62.8	63.1	66.4	69.7	72.8	72.9	80.6	84.4	88.3	89.5
MANUFACTURING ACTIVITY											
Employment (1980=100)	85.1	83.8	86.6	93.9	98.4	100.7	102.7	101.7	96.4	100.4	100.5
Real Earnings per Empl. (1980=100)	63.8	66.5	68.8	71.8	74.8	78.4	82.7	87.1	90.7	89.9	93.1
Real Output per Empl. (1980=100)	51.1	55.0	59.3	63.6	64.9	68.1	72.2	80.7	81.6	83.5	87.8
Earnings as % of Value Added	48.2	47.4	47.1	47.1	47.7	48.5	52.0	51.5	56.2	55.9	58.4
MONETARY HOLDINGS				*(Billions of current Austrian Schillings)*							
Money Supply, Broadly Defined	154.7	169.9	190.6	215.1	247.1	285.9	323.2	363.6	430.5	503.2	548.9
Money as Means of Payment	52.9	56.2	60.8	64.8	73.7	89.2	95.9	101.4	115.9	125.5	127.3
Currency Ouside Banks	30.3	31.2	32.7	34.0	37.3	42.9	46.1	48.7	52.3	55.2	58.7
Demand Deposits	22.6	25.0	28.1	30.8	36.4	46.4	49.8	52.7	63.6	70.4	68.6
Quasi-Monetary Liabilities	101.8	113.7	129.8	150.4	173.3	196.6	227.3	262.3	314.6	377.6	421.6
GOVERNMENT DEFICIT (-) OR SURPLUS	..	..	..	*(Billions of current Austrian Schillings)*							
	..	..	..	-1.80	0.24	-0.85	-8.96	-9.76	-26.34	-34.06	-30.21
Current Revenue	..	..	..	110.06	124.63	142.93	163.80	189.51	205.49	226.70	254.14
Current Expenditure	..	..	..	99.94	111.72	125.64	144.73	171.27	203.15	230.63	255.66
Current Budget Balance	..	..	..	10.12	12.91	17.29	19.07	18.24	2.34	-3.93	-1.52
Capital Receipts	..	..	..	0.27	0.14	0.17	0.17	0.21	0.27	0.37	0.42
Capital Payments	..	..	..	12.19	12.81	18.31	28.20	28.21	28.95	30.50	29.11

1978	1979	1980	1981	1982	1983	1984	1985	1986	1987 estimate	NOTES	AUSTRIA
6,890	8,460	9,990	10,150	9,800	9,320	9,100	9,100	9,970	12,010	..	**CURRENT GNP PER CAPITA (US $)**
7,547	7,551	7,554	7,558	7,571	7,549	7,552	7,555	7,565	7,557	..	POPULATION (thousands)
				(Billions of current Austrian Schillings)							**ORIGIN AND USE OF RESOURCES**
834.0	910.7	986.4	1,047.2	1,125.1	1,192.4	1,270.2	1,347.0	1,420.8	1,474.9	..	Gross National Product (GNP)
-8.4	-7.8	-8.4	-8.8	-8.4	-8.8	-8.6	-7.1	-11.8	-12.6	..	Net Factor Income from Abroad
842.3	918.5	994.7	1,056.0	1,133.5	1,201.2	1,278.7	1,354.1	1,432.6	1,487.5	..	Gross Domestic Product (GDP)
112.7	124.3	132.8	142.3	150.7	161.7	180.2	186.7	190.3	197.3	..	Indirect Taxes, net
729.6	794.3	861.9	913.7	982.8	1,039.5	1,098.6	1,167.3	1,242.3	1,290.2	B	GDP at factor cost
39.3	40.3	44.3	43.4	43.7	44.1	48.7	44.8	47.0	..	..	Agriculture
334.8	368.2	393.4	410.4	436.6	454.7	472.4	502.6	534.2	..	..	Industry
234.7	260.2	276.7	288.9	308.1	322.8	341.3	367.7	388.7	..	..	Manufacturing
468.2	510.1	557.0	602.1	653.2	702.4	757.7	806.7	851.4	..	..	Services, etc.
0.2	-3.8	-19.3	-13.9	18.9	15.7	2.0	2.3	12.5	12.2	..	Resource Balance
280.7	327.6	366.3	404.6	431.3	449.7	497.7	549.0	527.0	526.3	..	Exports of Goods & NFServices
280.5	331.5	385.5	418.5	412.4	434.0	495.7	546.7	514.5	514.1	..	Imports of Goods & NFServices
842.1	922.4	1,014.0	1,069.9	1,114.7	1,185.5	1,276.7	1,351.7	1,420.0	1,475.3	..	Domestic Absorption
468.8	511.6	552.6	596.3	640.4	694.8	733.3	774.9	802.0	830.1	..	Private Consumption, etc.
154.1	166.0	178.7	195.2	214.3	226.9	237.8	255.0	272.5	283.4	..	General Gov't Consumption
219.2	244.8	282.7	278.4	260.0	263.8	305.6	321.8	345.5	361.9	..	Gross Domestic Investment
211.1	228.0	251.6	265.5	259.3	264.7	284.8	308.5	326.0	336.7	..	Fixed Investment
											Memo Items:
219.4	241.0	263.4	264.5	278.9	279.5	307.6	324.2	358.1	374.1	..	Gross Domestic Saving
211.0	233.9	254.3	254.8	269.8	270.0	298.6	315.9	346.3	362.3	..	Gross National Saving
				(Billions of 1980 Austrian Schillings)							
912.9	957.3	986.4	985.1	996.8	1,018.6	1,033.1	1,063.8	1,078.2	1,090.6	..	Gross National Product
797.1	833.3	861.9	861.3	871.4	888.1	893.6	922.0	939.7	951.0	B	GDP at factor cost
41.7	42.5	44.3	42.2	48.5	46.5	47.9	45.5	46.3	..	..	Agriculture
363.1	383.4	393.4	387.8	384.6	388.5	394.3	409.9	417.9	..	..	Industry
251.4	268.6	276.7	274.5	276.1	280.7	287.8	301.6	307.0	..	..	Manufacturing
517.4	539.7	557.0	563.2	571.1	591.2	597.9	614.0	623.1	..	..	Services, etc.
-11.9	-16.2	-19.3	0.4	20.4	12.4	0.5	0.2	-16.3	-27.8	..	Resource Balance
312.4	346.2	366.3	380.1	387.4	400.4	426.8	456.1	445.6	449.5	..	Exports of Goods & NFServices
324.3	362.4	385.5	379.6	367.0	387.9	426.3	455.9	461.9	477.3	..	Imports of Goods & NFServices
934.1	981.9	1,014.0	992.8	983.9	1,013.7	1,039.6	1,069.1	1,103.7	1,129.7	..	Domestic Absorption
520.1	544.4	552.6	555.2	562.7	592.8	591.4	604.8	613.3	629.4	..	Private Consumption, etc.
168.9	174.4	178.7	182.2	186.5	190.7	191.8	196.2	200.5	202.9	..	General Gov't Consumption
245.2	263.1	282.7	255.4	234.7	230.3	256.3	268.2	289.9	297.3	..	Gross Domestic Investment
234.4	242.8	251.6	246.2	228.7	226.2	231.6	243.6	253.0	257.6	..	Fixed Investment
											Memo Items:
324.5	358.2	366.3	367.0	383.8	402.0	428.1	457.8	473.2	488.6	..	Capacity to Import
12.1	12.0	0.0	-13.1	-3.6	1.6	1.2	1.7	27.6	39.1	..	Terms of Trade Adjustment
934.3	977.7	994.7	980.2	1,000.7	1,027.8	1,041.3	1,071.1	1,114.9	1,141.0	..	Gross Domestic Income
925.0	969.3	986.4	972.0	993.2	1,020.2	1,034.4	1,065.5	1,105.8	..	..	Gross National Income
				(Index 1980 = 100)							**DOMESTIC PRICES (DEFLATORS)**
91.3	95.1	100.0	106.3	112.9	117.1	122.9	126.6	131.8	135.0	..	Overall (GDP)
90.2	93.9	100.0	107.8	113.3	116.9	122.8	126.4	128.7	130.6	..	Domestic Absorption
94.3	94.6	100.0	102.9	90.2	95.0	101.6	98.4	101.4	..	..	Agriculture
92.2	96.0	100.0	105.8	113.5	117.0	119.8	122.6	127.8	..	..	Industry
93.3	96.9	100.0	105.3	111.6	115.0	118.6	121.9	126.6	..	..	Manufacturing
											MANUFACTURING ACTIVITY
99.4	100.1	100.0	97.2	93.3	93.5	93.0	93.0	..	..	..	Employment (1980=100)
97.2	98.6	100.0	101.8	104.2	104.2	103.2	110.9	..	..	..	Real Earnings per Empl. (1980=100)
89.0	93.3	100.0	105.2	107.9	110.4	114.7	119.8	..	..	..	Real Output per Empl. (1980=100)
58.4	57.3	58.3	58.2	57.7	55.8	54.6	56.1	..	..	..	Earnings as % of Value Added
				(Billions of current Austrian Schillings)							**MONETARY HOLDINGS**
627.4	678.3	765.0	843.6	936.1	984.7	1,047.3	1,110.2	1,205.9	1,296.5	..	Money Supply, Broadly Defined
137.9	125.4	145.1	141.6	153.3	170.4	176.4	181.9	193.5	213.5	..	Money as Means of Payment
63.2	67.0	71.6	73.2	75.7	84.1	83.9	84.5	87.9	93.0	..	Currency Ouside Banks
74.7	58.4	73.4	68.5	77.5	86.3	92.5	97.4	105.7	120.5	..	Demand Deposits
489.5	552.8	620.0	702.0	782.8	814.3	870.9	928.3	1,012.4	1,083.0	..	Quasi-Monetary Liabilities
				(Billions of current Austrian Schillings)							
-34.92	-35.31	-33.57	-32.52	-54.53	-71.84	-58.22	-63.33	-84.22	-78.76	E	**GOVERNMENT DEFICIT (-) OR SURPLUS**
290.01	317.76	346.98	382.46	397.57	417.64	456.97	476.12	497.45	516.05	..	Current Revenue
291.79	314.92	338.12	370.13	405.34	433.57	461.64	484.60	520.81	549.69	..	Current Expenditure
-1.78	2.84	8.86	12.33	-7.77	-15.93	-4.67	-8.48	-23.36	-33.64	..	Current Budget Balance
0.40	0.40	0.43	0.47	0.51	0.53	0.46	0.58	0.84	0.80	..	Capital Receipts
33.54	38.55	42.86	45.32	47.27	56.44	54.01	55.43	61.70	45.92	..	Capital Payments

113

AUSTRIA	1967	1968	1969	1970	1971	1972	1973	1974	1975	1976	1977
FOREIGN TRADE (CUSTOMS BASIS)					*(Millions of current US dollars)*						
Value of Exports, fob	1,809	1,989	2,412	2,857	3,169	3,883	5,285	7,161	7,518	8,507	9,808
Nonfuel Primary Products	352	366	442	493	519	619	868	1,068	943	1,232	1,378
Fuels	57	59	60	72	64	77	113	144	155	159	189
Manufactures	1,400	1,564	1,911	2,291	2,586	3,187	4,303	5,950	6,420	7,116	8,241
Value of Imports, cif	2,309	2,496	2,825	3,549	4,189	5,216	7,121	9,023	9,392	11,523	14,248
Nonfuel Primary Products	547	572	642	777	819	950	1,392	1,743	1,622	1,898	2,339
Fuels	161	187	204	294	340	370	533	1,096	1,187	1,409	1,465
Manufactures	1,601	1,737	1,980	2,479	3,029	3,896	5,196	6,184	6,584	8,216	10,444
Terms of Trade	132.9	134.0	126.9	135.9	126.6	123.2	127.0	104.4	112.8	103.1	100.5
					(Index 1980 = 100)						
Export Prices, fob	29.0	28.5	28.4	29.9	31.7	35.4	46.1	56.4	63.8	61.1	66.1
Nonfuel Primary Products	26.4	27.2	29.6	30.3	29.7	34.6	47.9	48.6	50.5	52.1	49.8
Fuels	8.5	7.3	5.4	5.9	7.0	8.5	24.5	36.6	33.6	38.5	36.8
Manufactures	33.1	32.4	32.4	34.2	35.2	38.5	46.8	58.8	68.0	63.9	71.4
Import Prices, cif	21.8	21.3	22.4	22.0	25.0	28.7	36.3	54.1	56.5	59.2	65.8
Trade at Constant 1980 Prices					*(Millions of 1980 US dollars)*						
Exports, fob	6,240	6,985	8,507	9,566	10,011	10,971	11,465	12,692	11,785	13,933	14,842
Imports, cif	10,585	11,744	12,641	16,150	16,752	18,159	19,622	16,690	16,612	19,464	21,658
BALANCE OF PAYMENTS					*(Millions of current US dollars)*						
Exports of Goods & Services	..	..	..	4,331	5,036	6,173	8,514	11,273	12,456	13,754	16,264
Merchandise, fob	..	..	..	2,836	3,142	3,842	5,218	7,505	7,553	8,472	9,721
Nonfactor Services	..	..	..	1,358	1,723	2,155	3,031	3,224	4,292	4,649	5,794
Factor Services	..	..	..	136	170	177	265	544	611	634	749
Imports of Goods & Services	..	..	..	4,415	5,130	6,296	8,688	11,308	12,540	14,760	18,317
Merchandise, fob	..	..	..	3,516	4,073	4,969	6,787	8,813	9,505	11,053	13,541
Nonfactor Services	..	..	..	687	810	1,033	1,461	1,788	2,216	2,791	3,611
Factor Services	..	..	..	212	246	294	439	707	819	916	1,165
Long-Term Interest	..	..	..	..	..	..	..	..	..	..	..
Current Transfers, net	..	..	..	11	9	-30	-80	-158	-143	-103	-123
Workers' Remittances	..	..	..	13	19	27	40	66	74	90	103
Total to be Financed	..	..	..	-73	-85	-152	-254	-192	-227	-1,108	-2,176
Official Capital Grants	..	..	..	-2	-4	-5	-5	-10	-3	-11	-17
Current Account Balance	..	..	..	-75	-89	-157	-259	-202	-229	-1,119	-2,193
Long-Term Capital, net	..	..	..	12	-22	73	-227	393	1,051	-84	584
Direct Investment	..	..	..	104	107	107	121	157	53	35	10
Long-Term Loans	..	..	..	..	..	..	..	..	..	..	..
Disbursements	..	..	..	..	..	..	..	..	..	..	..
Repayments	..	..	..	..	..	..	..	..	..	..	..
Other Long-Term Capital	..	..	..	-90	-125	-29	-344	246	1,001	-108	591
Other Capital, net	..	..	..	244	484	450	316	197	350	1,087	1,265
Change in Reserves	..	..	..	-181	-373	-366	171	-388	-1,172	116	344
Memo Item:					*(Austrian Schillings per US dollar)*						
Conversion Factor (Annual Avg)	26.000	26.000	26.000	26.000	24.960	23.120	19.580	18.690	17.420	17.940	16.530
EXTERNAL DEBT, ETC.					*(Millions US dollars, outstanding at end of year)*						
Public/Publicly Guar. Long-Term	..	..	..	..	..	..	..	..	..	..	..
Official Creditors	..	..	..	..	..	..	..	..	..	..	..
IBRD and IDA	..	..	..	..	..	..	..	..	..	..	..
Private Creditors	..	..	..	..	..	..	..	..	..	..	..
Private Non-guaranteed Long-Term	..	..	..	..	..	..	..	..	..	..	..
Use of Fund Credit	..	..	..	..	..	..	..	..	..	..	..
Short-Term Debt	..	..	..	..	..	..	..	..	..	..	..
Memo Items:					*(Millions of US dollars)*						
Int'l Reserves Excluding Gold	783	796	822	1,044	1,547	1,927	1,992	2,535	3,583	3,561	3,351
Gold Holdings (at market price)	705	855	719	762	909	1,353	2,344	3,894	2,928	2,814	3,463
SOCIAL INDICATORS											
Total Fertility Rate	2.6	2.6	2.5	2.3	2.2	2.1	2.0	1.9	1.8	1.7	1.6
Crude Birth Rate	17.4	17.1	16.4	15.1	14.6	13.9	13.0	12.9	12.5	11.6	11.4
Infant Mortality Rate	26.4	25.5	25.4	25.9	26.1	25.2	23.8	23.5	20.5	18.2	16.9
Life Expectancy at Birth	69.9	70.0	70.1	70.3	70.4	70.6	70.8	71.1	71.4	71.7	72.0
Food Production, p.c. ('79-81 = 100)	88.1	90.5	91.2	87.2	88.0	84.2	89.3	93.8	94.8	94.5	93.6
Labor Force, Agriculture (%)	17.5	16.6	15.7	14.8	14.2	13.6	13.1	12.5	11.9	11.3	10.7
Labor Force, Female (%)	39.3	39.1	38.9	38.7	38.9	39.1	39.3	39.5	39.6	39.8	39.9
School Enroll. Ratio, primary	..	..	..	104.0	..	..	..	..	102.0	102.0	100.0
School Enroll. Ratio, secondary	..	..	..	72.0	..	..	..	..	74.0	75.0	72.0

1978	1979	1980	1981	1982	1983	1984	1985	1986	1987 estimate	NOTES	AUSTRIA
					(Millions of current US dollars)						**FOREIGN TRADE (CUSTOMS BASIS)**
12,174	15,478	17,478	15,840	15,690	15,423	15,712	17,102	22,517	27,163	..	Value of Exports, fob
1,745	2,327	2,723	2,310	2,116	2,137	2,175	2,128	2,621	3,100	..	Nonfuel Primary Products
191	223	274	276	241	217	234	346	273	479	..	Fuels
10,238	12,928	14,480	13,255	13,333	13,070	13,304	14,628	19,622	23,583	..	Manufactures
15,975	20,231	24,415	21,013	19,514	19,322	19,573	20,803	26,793	32,638	..	Value of Imports, cif
2,567	3,266	3,781	3,229	3,002	2,866	3,072	3,139	3,796	4,340	..	Nonfuel Primary Products
1,715	2,499	3,781	3,920	3,150	2,667	2,955	3,090	2,320	2,358	..	Fuels
11,694	14,465	16,852	13,863	13,362	13,788	13,546	14,574	20,678	25,940	..	Manufactures
					(Index 1980 = 100)						
108.2	108.7	100.0	89.3	93.1	90.8	87.3	90.2	103.0	108.2	..	Terms of Trade
78.5	94.0	100.0	87.2	86.0	81.4	76.6	77.5	95.3	110.4	..	Export Prices, fob
62.1	92.0	100.0	86.4	79.0	83.6	79.4	73.7	75.9	83.6	..	Nonfuel Primary Products
44.7	93.6	100.0	104.0	97.0	85.4	83.6	81.1	76.2	65.3	..	Fuels
83.7	94.6	100.0	87.0	87.0	81.0	76.0	78.0	99.0	117.0	..	Manufactures
72.6	86.5	100.0	97.7	92.3	89.6	87.6	85.9	92.5	102.0	..	Import Prices, cif
					(Millions of 1980 US dollars)						Trade at Constant 1980 Prices
15,502	16,463	17,478	18,172	18,252	18,944	20,525	22,068	23,631	24,599	..	Exports, fob
22,017	23,395	24,415	21,517	21,136	21,559	22,333	24,221	28,959	31,987	..	Imports, cif
					(Millions of current US dollars)						**BALANCE OF PAYMENTS**
20,570	25,987	31,289	28,764	28,637	27,837	27,926	30,043	38,343	45,464	..	Exports of Goods & Services
12,163	15,446	17,023	15,609	15,371	15,160	15,465	16,634	21,482	26,614	..	Merchandise, fob
7,406	8,958	11,728	9,951	10,083	10,046	9,549	10,183	13,043	14,341	..	Nonfactor Services
1,000	1,583	2,538	3,203	3,183	2,631	2,913	3,226	3,819	4,508	..	Factor Services
21,252	27,160	32,931	30,139	27,928	27,524	28,138	30,190	38,186	45,684	..	Imports of Goods & Services
15,443	19,716	23,633	20,660	18,826	18,815	19,491	21,019	26,079	31,137	..	Merchandise, fob
4,232	5,274	6,116	5,727	5,425	5,590	5,307	5,606	7,496	9,042	..	Nonfactor Services
1,577	2,170	3,182	3,752	3,677	3,118	3,340	3,565	4,611	5,505	..	Factor Services
..	..	..	..	..	..	..	..	..	..	..	Long-Term Interest
2	53	-58	-56	-37	-38	-22	-55	2	65	..	Current Transfers, net
135	177	215	198	201	189	176	183	267	364	..	Workers' Remittances
-681	-1,120	-1,700	-1,431	672	275	-234	-203	159	-155	K	Total to be Financed
-11	-23	-31	-30	-36	-44	-39	-39	-45	-71	K	Official Capital Grants
-692	-1,144	-1,731	-1,460	636	231	-272	-241	115	-226	..	Current Account Balance
1,394	-548	514	808	-608	-1,384	-366	-338	-1,758	636	..	Long-Term Capital, net
53	103	138	123	125	106	93	191	-41	134	..	Direct Investment
..	..	..	..	..	..	..	..	..	..	..	Long-Term Loans
..	..	..	..	..	..	..	..	..	..	..	Disbursements
..	..	..	..	..	..	..	..	..	..	..	Repayments
1,352	-628	407	715	-698	-1,446	-421	-491	-1,673	573	..	Other Long-Term Capital
648	670	2,557	1,031	180	662	708	668	2,372	-8	..	Other Capital, net
-1,350	1,022	-1,341	-379	-208	491	-70	-88	-728	-402	..	Change in Reserves
					(Austrian Schillings per US dollar)						Memo Item:
14.520	13.370	12.940	15.930	17.060	17.960	20.010	20.690	15.270	12.640		Conversion Factor (Annual Avg)
					(Millions US dollars, outstanding at end of year)						**EXTERNAL DEBT, ETC.**
..	..	..	..	..	..	..	..	..	..	..	Public/Publicly Guar. Long-Term
..	..	..	..	..	..	..	..	..	..	..	Official Creditors
..	..	..	..	..	..	..	..	..	..	..	IBRD and IDA
..	..	..	..	..	..	..	..	..	..	..	Private Creditors
..	..	..	..	..	..	..	..	..	..	..	Private Non-guaranteed Long-Term
..	..	..	..	..	..	..	..	..	..	..	Use of Fund Credit
..	..	..	..	..	..	..	..	..	..	..	Short-Term Debt
					(Millions of US dollars)						Memo Items:
5,047	4,075	5,281	5,285	5,300	4,515	4,244	4,767	6,162	7,532	..	Int'l Reserves Excluding Gold
4,758	10,808	12,445	8,392	9,649	8,060	6,516	6,912	8,265	10,237	..	Gold Holdings (at market price)
											SOCIAL INDICATORS
1.6	1.7	1.7	1.6	1.6	1.6	1.6	1.5	1.5	1.5	..	Total Fertility Rate
11.4	11.5	12.1	12.4	12.2	12.1	11.9	11.7	11.6	11.4	..	Crude Birth Rate
14.9	14.8	13.9	12.6	13.2	13.9	14.6	15.4	16.2	17.0	..	Infant Mortality Rate
72.1	72.3	72.5	72.7	72.8	73.0	73.2	73.5	73.7	74.1	..	Life Expectancy at Birth
97.0	97.9	103.5	98.6	112.1	106.5	111.1	108.7	108.0	109.1	..	Food Production, p.c. ('79-81=100)
10.2	9.6	9.0	..	..	..	..	..	..	..	..	Labor Force, Agriculture (%)
40.1	40.2	40.4	40.3	40.3	40.3	40.2	40.2	40.2	40.2	..	Labor Force, Female (%)
100.0	99.0	98.0	..	98.0	98.0	97.0	99.0	100.0	..	..	School Enroll. Ratio, primary
72.0	72.0	74.0	..	74.0	74.0	76.0	79.0	79.0	..	..	School Enroll. Ratio, secondary

BAHAMAS	1967	1968	1969	1970	1971	1972	1973	1974	1975	1976	1977
CURRENT GNP PER CAPITA (US $)	2,240	2,480	2,780	2,700	2,900	2,900	3,260	2,950	2,730	3,000	3,330
POPULATION (thousands)	151	157	164	171	174	178	181	185	189	193	197

ORIGIN AND USE OF RESOURCES *(Millions of current Bahamian Dollars)*

	1967	1968	1969	1970	1971	1972	1973	1974	1975	1976	1977
Gross National Product (GNP)	350.7	399.8	474.7	475.3	505.2	520.6	591.1	557.2	525.3	565.7	628.2
Net Factor Income from Abroad	-47.3	-54.0	-64.0	-64.2	-68.2	-70.3	-79.8	-75.2	-70.9	-76.4	-84.8
Gross Domestic Product (GDP)	398.0	453.8	538.7	539.5	573.4	590.9	670.9	632.4	596.2	642.1	713.0
Indirect Taxes, net	..	..	..	..	..	..	..	..	..	..	..
GDP at factor cost											
Agriculture	..	..	..	..	..	..	..	..	..	..	..
Industry	..	..	..	..	..	..	..	..	..	..	..
Manufacturing	..	..	..	..	..	..	..	..	..	..	..
Services, etc.	..	..	..	..	..	..	..	..	..	..	..
Resource Balance	..	..	..	..	..	..	..	..	..	..	117.7
Exports of Goods & NFServices	..	..	..	..	..	..	..	..	..	..	589.3
Imports of Goods & NFServices	..	..	..	..	..	..	..	..	..	..	471.6
Domestic Absorption	..	..	..	..	..	..	..	..	..	..	595.3
Private Consumption, etc.	..	..	..	..	..	..	..	..	..	..	396.8
General Gov't Consumption	..	..	..	..	..	..	..	..	..	..	109.3
Gross Domestic Investment	..	..	..	..	..	..	..	..	..	..	89.2
Fixed Investment	..	..	..	..	..	..	..	..	..	..	81.3
Memo Items:											
Gross Domestic Saving	..	..	..	..	..	..	..	..	..	..	206.9
Gross National Saving	..	..	..	..	..	..	..	..	..	..	103.5

(Millions of 1980 Bahamian Dollars)

	1967	1968	1969	1970	1971	1972	1973	1974	1975	1976	1977
Gross National Product	782.4	848.4	924.7	872.4	886.6	855.6	920.4	766.3	652.7	686.5	749.4
GDP at factor cost											
Agriculture	..	..	..	..	..	..	..	..	..	..	..
Industry	..	..	..	..	..	..	..	..	..	..	..
Manufacturing	..	..	..	..	..	..	..	..	..	..	..
Services, etc.	..	..	..	..	..	..	..	..	..	..	..
Resource Balance	..	..	..	..	..	..	..	..	..	..	92.0
Exports of Goods & NFServices	..	..	..	..	..	..	..	..	..	..	746.6
Imports of Goods & NFServices	..	..	..	..	..	..	..	..	..	..	654.6
Domestic Absorption	..	..	..	..	..	..	..	..	..	..	778.6
Private Consumption, etc.	..	..	..	..	..	..	..	..	..	..	520.0
General Gov't Consumption	..	..	..	..	..	..	..	..	..	..	141.8
Gross Domestic Investment	..	..	..	..	..	..	..	..	..	..	116.7
Fixed Investment	..	..	..	..	..	..	..	..	..	..	106.5
Memo Items:											
Capacity to Import	..	..	..	..	..	..	..	..	..	..	817.9
Terms of Trade Adjustment	..	..	..	..	..	..	..	..	..	..	71.4
Gross Domestic Income	..	..	..	..	..	..	..	..	..	..	941.9
Gross National Income	..	..	..	..	..	..	..	..	..	..	820.8

DOMESTIC PRICES (DEFLATORS) *(Index 1980 = 100)*

	1967	1968	1969	1970	1971	1972	1973	1974	1975	1976	1977
Overall (GDP)	43.8	46.0	50.2	53.2	55.7	59.5	62.8	71.1	78.6	80.5	81.9
Domestic Absorption	..	..	..	..	..	..	..	..	..	..	76.5
Agriculture	..	..	..	..	..	..	..	..	..	..	..
Industry	..	..	..	..	..	..	..	..	..	..	..
Manufacturing	..	..	..	..	..	..	..	..	..	..	..

MANUFACTURING ACTIVITY

	1967	1968	1969	1970	1971	1972	1973	1974	1975	1976	1977
Employment (1980=100)	..	..	..	..	..	..	..	..	..	..	77.9
Real Earnings per Empl. (1980=100)	..	..	..	..	..	..	..	..	..	..	45.2
Real Output per Empl. (1980=100)	..	..	..	..	..	..	..	..	..	..	..
Earnings as % of Value Added	..	..	..	..	..	..	..	..	..	..	39.8

MONETARY HOLDINGS *(Millions of current Bahamian Dollars)*

	1967	1968	1969	1970	1971	1972	1973	1974	1975	1976	1977
Money Supply, Broadly Defined	..	..	200.8	183.6	189.2	204.5	252.9	254.2	276.9	356.8	338.3
Money as Means of Payment	..	..	88.0	80.5	68.8	84.9	80.4	77.8	73.7	77.5	89.5
Currency Ouside Banks	..	..	17.4	17.8	15.3	21.3	20.2	19.3	20.1	20.5	23.3
Demand Deposits	..	..	70.6	62.7	53.5	63.6	60.2	58.5	53.6	57.0	66.2
Quasi-Monetary Liabilities	..	..	112.8	103.1	120.4	119.6	172.5	176.4	203.2	279.3	248.8

(Millions of current Bahamian Dollars)

	1967	1968	1969	1970	1971	1972	1973	1974	1975	1976	1977
GOVERNMENT DEFICIT (-) OR SURPLUS	..	..	..	..	..	..	1.9	-40.1	-0.4	-6.9	-9.8
Current Revenue	..	..	..	..	..	..	107.6	118.0	137.2	152.1	156.0
Current Expenditure	..	..	..	..	..	..	97.3	103.0	113.4	127.7	140.7
Current Budget Balance	..	..	..	..	..	..	10.3	15.0	23.8	24.4	15.3
Capital Receipts	..	..	..	..	..	..	..	..	..	..	..
Capital Payments	..	..	..	..	..	..	8.4	55.1	24.2	31.3	25.1

1978	1979	1980	1981	1982	1983	1984	1985	1986	1987 estimate	NOTES	BAHAMAS
3,680	5,070	5,790	5,660	6,620	7,170	8,300	9,000	9,550	10,320	..	**CURRENT GNP PER CAPITA (US $)**
201	206	210	214	218	222	226	231	236	241	..	**POPULATION (thousands)**
				(Millions of current Bahamian Dollars)							**ORIGIN AND USE OF RESOURCES**
702.7	1,016.4	1,221.7	1,287.7	1,463.4	1,609.5	1,892.6	2,158.1	2,309.2	2,551.9	..	Gross National Product (GNP)
-129.7	-123.4	-113.6	-138.8	-114.9	-123.3	-148.5	-162.6	-163.3	-162.1	..	Net Factor Income from Abroad
832.4	1,139.8	1,335.3	1,426.5	1,578.3	1,732.8	2,041.1	2,320.7	2,472.5	2,714.0	..	Gross Domestic Product (GDP)
..	125.9	146.6	152.6	157.0	179.0	254.0	303.1	322.8	335.4	..	Indirect Taxes, net
..	1,013.9	1,188.7	1,273.9	1,421.3	1,553.8	1,787.1	2,017.6	2,149.7	2,378.6	..	GDP at factor cost
..	..	..	..	..	..	..	..	..	..	..	Agriculture
..	..	..	..	..	..	..	..	..	..	..	Industry
..	..	..	..	..	..	..	..	..	..	..	Manufacturing
..	..	..	..	..	..	..	..	..	..	..	Services, etc.
126.1	98.8	100.2	71.5	98.4	148.2	107.8	128.1	146.2	149.8	..	Resource Balance
688.1	785.0	939.8	961.2	1,038.3	1,153.5	1,297.9	1,503.6	1,577.9	1,664.3	..	Exports of Goods & NFServices
562.0	686.2	839.6	889.7	939.9	1,005.3	1,190.1	1,375.5	1,431.7	1,514.5	..	Imports of Goods & NFServices
706.3	1,041.0	1,235.1	1,355.0	1,479.9	1,584.5	1,933.3	2,192.6	2,326.3	2,564.2	..	Domestic Absorption
483.5	751.0	823.9	888.9	932.2	994.3	1,335.6	1,458.6	1,570.7	1,727.7	..	Private Consumption, etc.
128.4	143.8	167.0	197.3	208.9	238.0	264.0	289.1	300.1	326.4	..	General Gov't Consumption
94.4	146.2	244.2	268.8	338.8	352.2	333.7	444.9	455.5	510.1	..	Gross Domestic Investment
85.9	135.3	212.9	238.9	298.7	310.7	326.5	407.8	425.2	508.1	..	Fixed Investment
											Memo Items:
220.5	245.0	344.4	340.3	437.2	500.5	441.5	573.0	601.7	659.9	..	Gross Domestic Saving
68.6	105.8	211.1	187.7	304.7	366.9	278.3	395.9	424.4	..	..	Gross National Saving
				(Millions of 1980 Bahamian Dollars)							
820.1	1,118.4	1,221.7	1,073.9	1,177.4	1,212.3	1,373.3	1,433.9	1,488.3	1,546.2	..	Gross National Product
..	1,112.8	1,188.7	1,074.9	1,159.5	1,193.8	1,332.6	1,380.9	1,402.6	1,451.0	..	GDP at factor cost
..	..	..	..	..	..	..	..	..	..	..	Agriculture
..	..	..	..	..	..	..	..	..	..	..	Industry
..	..	..	..	..	..	..	..	..	..	..	Manufacturing
..	..	..	..	..	..	..	..	..	..	..	Services, etc.
131.0	93.6	100.2	-11.5	23.8	31.5	-21.6	-97.6	-102.0	-108.2	..	Resource Balance
823.2	853.6	939.8	860.6	934.2	1,015.6	1,145.0	1,212.9	1,244.6	1,228.3	..	Exports of Goods & NFServices
692.2	760.1	839.6	872.1	910.4	984.1	1,166.6	1,310.5	1,346.6	1,336.5	..	Imports of Goods & NFServices
863.0	1,160.2	1,235.1	1,223.6	1,269.6	1,308.5	1,551.5	1,701.2	1,734.4	1,789.4	..	Domestic Absorption
592.5	838.2	823.9	799.5	793.6	813.9	1,069.2	1,131.1	1,170.1	1,194.3	..	Private Consumption, etc.
156.9	161.2	167.0	177.5	177.3	194.0	207.1	217.2	216.2	220.6	..	General Gov't Consumption
113.6	160.8	244.2	246.6	298.8	300.6	275.2	352.8	348.1	374.5	..	Gross Domestic Investment
103.4	148.9	212.9	219.2	263.6	265.1	269.3	323.5	325.1	373.3	..	Fixed Investment
											Memo Items:
847.5	869.5	939.8	942.2	1,005.7	1,129.2	1,272.3	1,432.6	1,484.1	1,468.7	..	Capacity to Import
24.3	15.9	0.0	81.5	71.5	113.6	127.2	219.7	239.5	240.4	..	Terms of Trade Adjustment
1,018.3	1,269.6	1,335.3	1,293.6	1,364.9	1,453.5	1,657.2	1,823.2	1,871.9	1,921.6	..	Gross Domestic Income
844.4	1,134.3	1,221.7	1,155.5	1,248.9	1,325.9	1,500.6	1,653.6	1,727.9	1,786.6	..	Gross National Income
				(Index 1980 = 100)							**DOMESTIC PRICES (DEFLATORS)**
83.7	90.9	100.0	117.7	122.0	129.3	133.4	144.7	151.5	161.4	..	Overall (GDP)
81.8	89.7	100.0	110.7	116.6	121.1	124.6	128.9	134.1	143.3	..	Domestic Absorption
..	..	..	..	..	..	..	..	..	..	..	Agriculture
..	..	..	..	..	..	..	..	..	..	..	Industry
..	..	..	..	..	..	..	..	..	..	..	Manufacturing
											MANUFACTURING ACTIVITY
101.7	96.8	100.0	..	..	..	..	..	..	..	..	Employment (1980=100)
104.1	100.8	100.0	..	..	..	..	..	..	..	..	Real Earnings per Empl. (1980=100)
..	..	..	..	..	..	..	..	..	..	..	Real Output per Empl. (1980=100)
93.8	45.3	42.6	..	..	..	..	..	..	..	..	Earnings as % of Value Added
				(Millions of current Bahamian Dollars)							**MONETARY HOLDINGS**
388.2	456.9	520.3	583.4	684.7	792.6	857.9	937.9	1,065.5	1,241.2	D	Money Supply, Broadly Defined
107.1	138.3	136.3	143.5	159.7	176.9	187.7	208.1	249.2	277.7	..	Money as Means of Payment
26.5	29.8	33.4	37.4	41.1	45.5	51.0	57.8	65.0	74.8	..	Currency Ouside Banks
80.6	108.5	102.9	106.1	118.6	131.4	136.7	150.3	184.2	202.9	..	Demand Deposits
281.1	318.6	384.0	439.9	525.0	615.7	670.2	729.8	816.3	963.5	..	Quasi-Monetary Liabilities
				(Millions of current Bahamian Dollars)							
-13.6	6.7	18.9	-33.9	-45.1	-52.2	14.4	4.5	-17.4	..	..	**GOVERNMENT DEFICIT (-) OR SURPLUS**
185.2	226.5	270.4	314.2	311.3	337.9	376.7	430.6	451.2	..	..	Current Revenue
149.8	180.4	228.1	261.4	289.0	318.1	343.8	385.0	436.0	..	..	Current Expenditure
35.4	46.1	42.3	52.8	22.3	19.8	32.9	45.6	15.2	..	..	Current Budget Balance
..	..	..	..	..	..	..	..	..	..	..	Capital Receipts
49.0	39.4	23.4	86.7	67.4	72.0	18.5	41.1	32.6	..	..	Capital Payments

117

BAHAMAS	1967	1968	1969	1970	1971	1972	1973	1974	1975	1976	1977
FOREIGN TRADE (CUSTOMS BASIS)					*(Millions of current US dollars)*						
Value of Exports, fob	31.7	50.8	53.2	87.5	266.5	343.4	529.8	1,795.4	2,216.1	2,616.0	2,408.7
Nonfuel Primary Products	..	..	..	..	..	..	..	..	..	..	..
Fuels	..	..	..	..	..	..	..	..	..	..	..
Manufactures	..	..	..	..	..	..	..	..	..	..	..
Value of Imports, cif	162.2	176.4	296.3	331.9	511.3	484.9	764.2	2,111.5	2,482.1	2,892.6	2,871.3
Nonfuel Primary Products	..	..	..	..	..	..	..	..	..	..	..
Fuels	..	..	..	..	..	..	..	..	..	..	..
Manufactures	..	..	..	..	..	..	..	..	..	..	..
					(Index 1980 = 100)						
Terms of Trade	..	..	..	..	..	..	..	..	..	..	..
Export Prices, fob	..	..	..	..	..	..	..	..	..	..	..
Nonfuel Primary Products	..	..	..	..	..	..	.. ..	..	..	..	..
Fuels	..	..	..	..	..	..	..	..	..	..	..
Manufactures	..	..	..	..	..	..	..	..	..	..	..
Import Prices, cif	..	..	..	..	..	..	..	..	..	..	..
Trade at Constant 1980 Prices					*(Millions of 1980 US dollars)*						
Exports, fob	..	..	..	..	..	..	..	..	..	..	..
Imports, cif	..	..	..	..	..	..	..	..	..	..	..
BALANCE OF PAYMENTS					*(Millions of current US dollars)*						
Exports of Goods & Services	..	..	..	..	..	..	593.7	696.0	711.9	693.6	712.2
Merchandise, fob	..	..	..	..	..	..	95.0	122.8	116.2	149.4	135.8
Nonfactor Services	..	..	..	..	..	..	374.8	444.0	470.1	539.8	572.4
Factor Services	..	..	..	..	..	..	123.9	129.2	125.5	4.5	4.0
Imports of Goods & Services	..	..	..	..	..	..	721.5	800.8	658.3	597.7	632.7
Merchandise, fob	..	..	..	..	..	..	388.8	411.1	339.0	389.5	390.1
Nonfactor Services	..	..	..	..	..	..	120.6	140.2	129.9	127.0	152.7
Factor Services	..	..	..	..	..	..	212.1	249.5	189.4	81.2	89.9
Long-Term Interest	..	..	..	2.7	2.4	2.2	2.7	3.8	4.8	5.5	4.6
Current Transfers, net	..	..	..	..	..	..	-20.3	-20.1	-14.9	-16.5	-18.6
Workers' Remittances	..	..	..	..	..	..	0.0	0.0	0.0	0.0	0.0
Total to be Financed	..	..	..	..	..	..	-148.1	-125.0	38.6	79.4	60.9
Official Capital Grants	..	..	..	..	..	..	3.0	2.4	4.7	4.9	5.4
Current Account Balance	..	..	..	..	..	..	-145.1	-122.5	43.3	84.3	66.3
Long-Term Capital, net	..	..	..	..	..	..	93.6	115.5	42.3	14.5	46.1
Direct Investment	..	..	..	..	..	..	79.4	111.1	48.6	14.6	31.4
Long-Term Loans	..	..	..	-4.6	-4.6	1.5	18.6	18.9	16.0	-7.2	0.5
Disbursements	..	..	..	..	..	6.2	23.3	23.6	23.3	0.6	30.8
Repayments	..	..	..	4.6	4.6	4.7	4.7	4.7	7.3	7.8	30.3
Other Long-Term Capital	..	..	..	..	..	..	-7.4	-17.0	-27.0	2.2	8.7
Other Capital, net	..	..	..	..	..	..	58.7	14.2	-84.0	-104.7	-93.3
Change in Reserves	..	..	..	4.4	-7.8	-7.6	-7.2	-7.1	-1.6	5.9	-19.1
Memo Item:					*(Bahamian Dollars per US dollar)*						
Conversion Factor (Annual Avg)	1.020	1.020	1.020	1.000	1.000	1.000	1.000	1.000	1.000	1.000	1.000
EXTERNAL DEBT, ETC.					*(Millions of US dollars, outstanding at end of year)*						
Public/Publicly Guar. Long-Term	..	..	..	44.20	39.70	41.20	59.80	78.70	94.60	87.30	87.70
Official Creditors	..	..	..	27.90	24.60	21.30	18.00	16.40	36.00	33.90	13.50
IBRD and IDA	..	..	..	0.00	0.00	0.00	0.00	0.00	0.00	0.00	0.40
Private Creditors	..	..	..	16.40	15.00	19.90	41.80	62.30	58.60	53.40	74.20
Private Non-guaranteed Long-Term	..	..	..	..	..	..	..	..	..	..	..
Use of Fund Credit	..	..	..	0.00	0.00	0.00	0.00	0.00	0.00	0.00	0.00
Short-Term Debt	..	..	..	..	..	..	..	..	..	..	..
Memo Items:					*(Millions of US dollars)*						
Int'l Reserves Excluding Gold	..	44.4	26.1	21.7	29.5	37.1	43.2	49.8	53.3	47.4	67.1
Gold Holdings (at market price)	..	..	..	..	..	..	..	..	..	..	1.4
SOCIAL INDICATORS											
Total Fertility Rate	..	..	..	3.4	3.3	3.1	2.9	2.7	2.5	2.7	2.8
Crude Birth Rate	26.7	26.2	25.8	25.3	24.9	24.5	24.0	23.6	21.3	27.4	24.7
Infant Mortality Rate	40.1	38.4	36.7	35.0	32.6	30.3	27.9	25.5	34.2	24.7	27.7
Life Expectancy at Birth	64.9	65.3	65.6	65.7	66.2	66.7	66.9	67.1	67.4	67.6	67.8
Food Production, p.c. ('79-81 = 100)	75.2	76.0	108.6	130.9	76.9	87.0	71.8	81.3	84.6	95.7	91.4
Labor Force, Agriculture (%)	..	..	..	..	..	..	..	..	..	..	..
Labor Force, Female (%)	..	..	..	..	..	..	..	..	..	..	..
School Enroll. Ratio, primary	..	..	..	..	..	..	..	..	..	..	..
School Enroll. Ratio, secondary	..	..	..	..	..	..	..	..	..	..	..

1978	1979	1980	1981	1982	1983	1984	1985	1986	1987 estimate	NOTES	BAHAMAS
				(Millions of current US dollars)							**FOREIGN TRADE (CUSTOMS BASIS)**
2,118.1	3,783.5	4,906.1	6,178.2	4,520.0	3,938.0	3,911.2	3,945.2	3,974.2	..	..	Value of Exports, fob
..	..	..	..	..	..	..	..	..	..	..	Nonfuel Primary Products
..	..	..	..	..	..	..	..	..	..	..	Fuels
..	..	..	..	..	..	..	..	..	..	..	Manufactures
2,482.2	3,985.0	3,479.2	4,124.0	3,615.4	3,342.0	3,523.6	3,265.2	2,615.1	..	..	Value of Imports, cif
..	..	..	..	..	..	..	..	..	..	..	Nonfuel Primary Products
..	..	..	..	..	..	..	..	..	..	..	Fuels
..	..	..	..	..	..	..	..	..	..	..	Manufactures
				(Index 1980 = 100)							
..	..	..	..	..	..	..	..	..	..	..	Terms of Trade
..	..	..	..	..	..	..	..	..	..	..	Export Prices, fob
..	..	..	..	..	..	..	..	..	..	..	Nonfuel Primary Products
..	..	..	..	..	..	..	..	..	..	..	Fuels
..	..	..	..	..	..	..	..	..	..	..	Manufactures
..	..	..	..	..	..	..	..	..	..	..	Import Prices, cif
				(Millions of 1980 US dollars)							Trade at Constant 1980 Prices
..	..	..	..	..	..	..	..	..	..	..	Exports, fob
..	..	..	..	..	..	..	..	..	..	..	Imports, cif
				(Millions of current US dollars)							**BALANCE OF PAYMENTS**
836.1	952.4	1,159.8	1,157.0	1,156.6	1,262.0	1,326.1	1,542.0	1,614.5	..	..	Exports of Goods & Services
145.8	170.5	200.4	176.9	212.6	225.1	261.9	296.9	293.1	..	..	Merchandise, fob
683.5	771.1	946.5	963.1	920.1	1,021.6	1,045.4	1,224.7	1,302.0	..	..	Nonfactor Services
6.9	10.9	13.0	17.1	23.8	15.3	18.8	20.4	19.4	..	..	Factor Services
778.1	932.2	1,175.0	1,242.2	1,227.4	1,306.2	1,376.8	1,580.7	1,633.2	..	..	Imports of Goods & Services
466.4	597.0	801.3	799.1	756.0	823.6	868.6	996.7	1,022.4	..	..	Merchandise, fob
175.1	187.8	223.0	267.4	312.6	325.5	320.8	374.9	411.3	..	..	Nonfactor Services
136.6	147.4	150.7	175.7	158.8	157.1	187.4	209.1	199.5	..	..	Factor Services
4.9	6.8	7.6	9.0	25.2	27.0	25.9	20.1	21.8	13.6	..	Long-Term Interest
-22.2	-15.8	-19.7	-13.8	-17.6	-10.3	-14.7	-14.5	-14.0	..	..	Current Transfers, net
0.0	0.0	0.0	0.0	0.0	0.0	0.0	0.0	0.0	..	..	Workers' Remittances
35.8	4.4	-34.8	-99.0	-88.4	-54.5	-65.3	-53.3	-32.6	..	K	Total to be Financed
7.5	11.8	17.9	11.1	21.4	17.4	15.4	21.4	20.7	..	K	Official Capital Grants
43.3	16.3	-17.0	-88.0	-67.0	-37.1	-49.9	-31.9	-11.9	..	..	Current Account Balance
-17.4	20.8	27.3	159.9	96.3	36.6	9.1	-13.3	21.0	..	..	Long-Term Capital, net
-0.7	9.7	3.8	34.4	3.0	-6.0	-4.7	-29.4	-12.7	..	..	Direct Investment
12.1	-4.4	-6.5	71.4	66.4	6.2	-27.5	-24.6	12.2	-31.9	..	Long-Term Loans
23.5	15.6	11.2	102.3	84.5	23.4	1.7	6.2	33.3	4.2	..	Disbursements
11.4	20.0	17.7	30.9	18.1	17.2	29.2	30.8	21.1	36.1	..	Repayments
-36.3	3.7	12.1	43.0	5.5	18.9	25.9	19.3	0.9	..	..	Other Long-Term Capital
-34.8	-20.9	-2.1	-66.6	-21.8	11.1	80.0	64.5	39.7	..	..	Other Capital, net
9.0	-16.1	-8.2	-5.3	-7.4	-10.6	-39.2	-19.3	-48.8	63.6	..	Change in Reserves
				(Bahamian Dollars per US dollar)							**Memo Item:**
1.000	1.000	1.000	1.000	1.000	1.000	1.000	1.000	1.000	1.000	..	Conversion Factor (Annual Avg)
				(Millions of US dollars, outstanding at end of year)							**EXTERNAL DEBT, ETC.**
99.60	95.20	89.70	158.70	230.30	236.70	206.70	186.50	202.10	174.70	..	Public/Publicly Guar. Long-Term
11.10	8.50	8.90	13.50	22.40	23.90	21.80	23.00	26.80	32.30	..	Official Creditors
0.60	0.90	2.80	3.90	7.80	9.20	7.60	11.00	14.30	20.60	..	IBRD and IDA
88.50	86.70	80.70	145.20	207.90	212.80	184.80	163.50	175.40	142.40	..	Private Creditors
..	..	..	..	..	..	..	..	..	..	..	Private Non-guaranteed Long-Term
0.00	0.00	0.00	0.00	0.00	0.00	0.00	0.00	0.00	..	..	Use of Fund Credit
81.00	83.00	61.00	68.00	73.00	27.00	20.00	29.00	50.00	58.00	..	Short-Term Debt
				(Millions of US dollars)							**Memo Items:**
58.1	77.5	92.3	100.2	113.5	122.0	161.1	182.5	231.5	170.1	..	Int'l Reserves Excluding Gold
2.9	8.8	..	..	..	..	..	..	..	..	..	Gold Holdings (at market price)
											SOCIAL INDICATORS
3.0	3.1	3.3	3.3	3.2	3.1	3.1	3.0	3.0	2.9	..	Total Fertility Rate
21.6	23.7	24.2	25.0	25.0	25.0	25.0	25.0	25.0	25.1	..	Crude Birth Rate
32.5	25.9	30.0	22.3	23.3	24.4	25.5	26.7	27.9	29.2	..	Infant Mortality Rate
68.0	68.2	68.4	68.6	68.8	69.1	69.3	69.5	69.7	69.9	..	Life Expectancy at Birth
94.2	94.5	98.5	107.0	103.6	103.6	103.5	102.5	103.9	104.5	..	Food Production, p.c. ('79-81 = 100)
..	..	..	..	..	..	..	..	..	..	..	Labor Force, Agriculture (%)
..	..	..	..	..	..	..	..	..	..	..	Labor Force, Female (%)
..	..	..	..	..	..	..	..	..	..	..	School Enroll. Ratio, primary
..	..	..	..	..	..	..	..	..	..	..	School Enroll. Ratio, secondary

BAHRAIN	1967	1968	1969	1970	1971	1972	1973	1974	1975	1976	1977
CURRENT GNP PER CAPITA (US $)	..	..	..	..	..	..	..	..	..	..	300
POPULATION (thousands)	202	208	214	220	230	239	250	261	272	285	300

ORIGIN AND USE OF RESOURCES *(Millions of current Bahrain Dinars)*

	1967	1968	1969	1970	1971	1972	1973	1974	1975	1976	1977
Gross National Product (GNP)	..	..	..	..	..	..	..	..	..	..	..
Net Factor Income from Abroad	..	..	..	..	..	..	..	..	..	..	..
Gross Domestic Product (GDP)	..	..	..	..	..	..	..	..	..	..	777.2
Indirect Taxes, net	..	..	..	..	..	..	..	..	..	..	..
GDP at factor cost	..	..	..	..	..	..	..	..	..	..	..
Agriculture	..	..	..	..	..	..	..	..	..	..	..
Industry	..	..	..	..	..	..	..	..	..	..	..
Manufacturing	..	..	..	..	..	..	..	..	..	..	..
Services, etc.	..	..	..	..	..	..	..	..	..	..	..
Resource Balance	..	..	..	..	..	..	..	..	..	..	..
Exports of Goods & NFServices	..	..	..	..	..	..	..	..	..	..	..
Imports of Goods & NFServices	..	..	..	..	..	..	..	..	..	..	..
Domestic Absorption	..	..	..	..	..	..	..	..	..	..	..
Private Consumption, etc.	..	..	..	..	..	..	..	..	..	..	..
General Gov't Consumption	..	..	..	..	..	..	..	..	..	..	..
Gross Domestic Investment	..	..	..	..	..	..	..	..	..	..	..
Fixed Investment	..	..	..	..	..	..	..	..	..	..	..
Memo Items:											
Gross Domestic Saving	..	..	..	..	..	..	..	..	..	..	351.9
Gross National Saving	..	..	..	..	..	..	..	..	..	..	..

(Millions of 1980 Bahrain Dinars)

	1967	1968	1969	1970	1971	1972	1973	1974	1975	1976	1977
Gross National Product	..	..	..	..	..	..	..	..	..	..	..
GDP at factor cost	..	..	..	..	..	..	..	..	..	..	..
Agriculture	..	..	..	..	..	..	..	..	..	..	..
Industry	..	..	..	..	..	..	..	..	..	..	..
Manufacturing	..	..	..	..	..	..	..	..	..	..	..
Services, etc.	..	..	..	..	..	..	..	..	..	..	..
Resource Balance	..	..	..	..	..	..	..	..	..	..	..
Exports of Goods & NFServices	..	..	..	..	..	..	..	..	..	..	..
Imports of Goods & NFServices	..	..	..	..	..	..	..	..	..	..	..
Domestic Absorption	..	..	..	..	..	..	..	..	..	..	..
Private Consumption, etc.	..	..	..	..	..	..	..	..	..	..	..
General Gov't Consumption	..	..	..	..	..	..	..	..	..	..	..
Gross Domestic Investment	..	..	..	..	..	..	..	..	..	..	..
Fixed Investment	..	..	..	..	..	..	..	..	..	..	..
Memo Items:											
Capacity to Import	..	..	..	..	..	..	..	..	..	..	..
Terms of Trade Adjustment	..	..	..	..	..	..	..	..	..	..	..
Gross Domestic Income	..	..	..	..	..	..	..	..	..	..	..
Gross National Income	..	..	..	..	..	..	..	..	..	..	..

DOMESTIC PRICES (DEFLATORS) *(Index 1980 = 100)*

	1967	1968	1969	1970	1971	1972	1973	1974	1975	1976	1977
Overall (GDP)	..	..	..	..	..	..	..	..	..	..	..
Domestic Absorption	..	..	..	..	..	..	..	..	..	..	..
Agriculture	..	..	..	..	..	..	..	..	..	..	..
Industry	..	..	..	..	..	..	..	..	..	..	..
Manufacturing	..	..	..	..	..	..	..	..	..	..	..

MANUFACTURING ACTIVITY

	1967	1968	1969	1970	1971	1972	1973	1974	1975	1976	1977
Employment (1980=100)	..	..	..	..	..	..	..	..	..	..	..
Real Earnings per Empl. (1980=100)	..	..	..	..	..	..	..	..	..	..	..
Real Output per Empl. (1980=100)	..	..	..	..	..	..	..	..	..	..	..
Earnings as % of Value Added	..	..	..	..	..	..	..	..	..	..	..

MONETARY HOLDINGS *(Millions of current Bahrain Dinars)*

	1967	1968	1969	1970	1971	1972	1973	1974	1975	1976	1977
Money Supply, Broadly Defined	37.1	43.9	48.4	56.4	73.1	87.9	99.6	145.8	184.2	303.8	355.5
Money as Means of Payment	28.4	34.8	34.4	38.0	48.6	55.2	59.4	61.3	77.8	127.9	152.5
Currency Ouside Banks	12.7	15.4	17.2	18.9	21.2	23.8	14.9	16.9	24.0	34.2	43.8
Demand Deposits	15.7	19.4	17.2	19.1	27.4	31.4	44.5	44.4	53.8	93.7	108.7
Quasi-Monetary Liabilities	8.7	9.1	14.0	18.5	24.6	32.7	40.3	84.5	106.5	175.9	203.0

(Millions of current Bahrain Dinars)

	1967	1968	1969	1970	1971	1972	1973	1974	1975	1976	1977
GOVERNMENT DEFICIT (-) OR SURPLUS	..	..	..	..	..	..	..	40	6	-36	-13
Current Revenue	..	..	..	..	..	..	..	118	129	187	255
Current Expenditure	..	..	..	..	..	..	..	38	62	78	100
Current Budget Balance	..	..	..	..	..	..	..	79	67	108	155
Capital Receipts	..	..	..	..	..	..	..	..	..	..	..
Capital Payments	..	..	..	..	..	..	..	40	62	145	168

1978	1979	1980	1981	1982	1983	1984	1985	1986	1987 estimate	NOTES	BAHRAIN
..	..	..	..	..	..	..	..	..	..	..	**CURRENT GNP PER CAPITA** (US $)
315	331	347	360	373	387	402	417	431	445	..	**POPULATION** (thousands)
				(Millions of current Bahrain Dinars)							**ORIGIN AND USE OF RESOURCES**
..	..	..	..	..	..	..	..	..	..	..	Gross National Product (GNP)
..	..	..	..	..	..	..	..	..	..	..	Net Factor Income from Abroad
917.8	1,029.2	1,344.1	1,541.5	1,686.3	1,711.0	1,734.0	1,602.9	1,383.0	1,314.3	..	Gross Domestic Product (GDP)
..	..	42.7	53.8	60.4	66.2	51.2	41.2	12.6	..	..	Indirect Taxes, net
..	..	1,301.4	1,487.7	1,625.9	1,644.8	1,682.8	1,561.7	1,370.4	..	B	GDP at factor cost
..	..	14.9	18.2	18.9	19.2	19.4	19.9	20.6	19.3	..	Agriculture
..	..	725.5	776.3	723.6	704.4	719.8	690.7	522.0	465.2	..	Industry
..	..	198.4	222.3	173.8	167.4	177.4	140.7	170.8	146.6	..	Manufacturing
..	..	603.7	747.0	943.8	987.4	994.8	892.3	840.4	829.8	..	Services, etc.
..	..	126.4	208.2	243.7	97.9	118.0	215.8	221.0	..	..	Resource Balance
..	..	1,378.7	1,694.8	1,567.1	1,324.3	1,426.3	1,383.9	1,126.3	..	..	Exports of Goods & NFServices
..	..	1,252.3	1,486.6	1,323.4	1,226.4	1,308.3	1,168.1	905.3	..	..	Imports of Goods & NFServices
..	..	1,217.7	1,333.3	1,442.6	1,613.0	1,616.4	1,387.1	1,162.0	..	..	Domestic Absorption
..	..	470.3	539.3	584.1	627.9	673.8	646.8	568.4	..	..	Private Consumption, etc.
..	..	153.0	189.5	242.2	259.2	275.2	287.4	281.5	..	..	General Gov't Consumption
..	..	594.4	604.5	616.3	725.9	667.4	452.9	312.1	..	..	Gross Domestic Investment
..	..	323.2	511.4	496.1	677.2	649.2	504.7	482.0	..	..	Fixed Investment
											Memo Items:
420.7	503.0	720.8	812.7	860.0	823.9	785.0	668.7	533.1	..	..	Gross Domestic Saving
..	..	..	..	..	..	..	..	..	..	..	Gross National Saving
				(Millions of 1980 Bahrain Dinars)							
..	..	..	..	..	..	..	..	..	..	..	Gross National Product
..	..	1,301.4	1,298.9	1,410.4	1,408.7	1,493.6	1,460.1	1,522.3	..	B H	GDP at factor cost
..	..	14.9	18.3	19.6	18.6	17.9	17.5	17.6	17.0	..	Agriculture
..	..	725.5	677.3	625.7	652.4	732.7	714.7	724.6	691.1	..	Industry
..	..	198.4	263.1	230.4	212.9	228.2	209.7	306.9	305.0	..	Manufacturing
..	..	603.7	655.0	823.1	801.1	792.0	768.8	792.6	790.5	..	Services, etc.
..	..	126.4	167.4	271.2	85.7	241.5	334.7	406.0	..	..	Resource Balance
..	..	1,378.7	1,583.4	1,541.7	1,364.2	1,588.8	1,575.9	1,978.9	..	..	Exports of Goods & NFServices
..	..	1,252.3	1,416.0	1,270.5	1,278.5	1,347.3	1,241.2	1,572.9	..	..	Imports of Goods & NFServices
..	..	1,217.7	1,244.6	1,304.7	1,451.6	1,408.4	1,303.9	1,279.2	..	..	Domestic Absorption
..	..	470.3	577.7	658.8	722.4	747.2	746.3	680.3	..	..	Private Consumption, etc.
..	..	153.0	168.5	200.3	200.0	197.4	205.8	202.0	..	..	General Gov't Consumption
..	..	594.4	498.3	445.6	529.2	463.8	351.8	396.9	..	..	Gross Domestic Investment
..	..	323.2	455.2	439.2	598.9	596.2	449.5	449.6	..	..	Fixed Investment
											Memo Items:
..	..	1,378.7	1,614.4	1,504.4	1,380.6	1,468.8	1,470.6	1,956.9	..	..	Capacity to Import
..	..	0.0	30.9	-37.2	16.4	-120.0	-105.4	-22.1	..	..	Terms of Trade Adjustment
..	..	1,344.1	1,381.5	1,431.2	1,488.5	1,422.6	1,395.7	1,512.8	..	..	Gross Domestic Income
..	..	..	..	..	..	..	..	..	..	..	Gross National Income
				(Index 1980 = 100)							**DOMESTIC PRICES (DEFLATORS)**
73.3	76.3	100.0	114.1	114.8	116.2	112.4	106.8	90.1	87.7	..	Overall (GDP)
..	..	100.0	107.1	110.6	111.1	114.8	106.4	90.8	..	..	Domestic Absorption
..	..	100.0	99.2	96.6	103.2	108.4	113.4	116.8	113.4	..	Agriculture
..	..	100.0	114.6	115.6	108.0	98.2	96.6	72.0	67.3	..	Industry
..	..	100.0	84.5	75.4	78.6	77.7	67.1	55.7	48.1	..	Manufacturing
											MANUFACTURING ACTIVITY
..	..	..	..	..	..	..	..	..	..	..	Employment (1980=100)
..	..	..	..	..	..	..	..	..	..	B	Real Earnings per Empl. (1980=100)
..	..	..	..	..	..	..	..	..	..	..	Real Output per Empl. (1980=100)
..	..	..	..	..	..	..	..	..	..	..	Earnings as % of Value Added
				(Millions of current Bahrain Dinars)							**MONETARY HOLDINGS**
402.4	411.9	524.1	730.5	780.2	843.1	827.8	903.3	885.2	968.5	..	Money Supply, Broadly Defined
171.3	186.1	192.2	248.8	266.9	250.0	239.3	243.1	235.7	246.7	..	Money as Means of Payment
44.1	49.9	58.3	63.3	70.5	73.5	78.2	79.0	80.0	84.1	..	Currency Ouside Banks
127.2	136.2	133.9	185.4	196.5	176.5	161.1	164.2	155.7	162.6	..	Demand Deposits
231.0	225.8	331.8	481.8	513.3	593.1	588.5	660.2	649.5	721.8	..	Quasi-Monetary Liabilities
				(Millions of current Bahrain Dinars)							
-31	22	68	111	32	-96	-23	14	-52	..	..	**GOVERNMENT DEFICIT (-) OR SURPLUS**
274	304	446	537	554	491	523	541	460	..	..	Current Revenue
137	157	192	231	299	313	329	344	355	..	..	Current Expenditure
137	148	254	306	256	177	194	197	105	..	..	Current Budget Balance
..	..	..	..	..	..	..	..	..	..	..	Capital Receipts
168	125	186	195	223	274	217	184	157	..	..	Capital Payments

BAHRAIN	1967	1968	1969	1970	1971	1972	1973	1974	1975	1976	1977
FOREIGN TRADE (CUSTOMS BASIS)					*(Millions of current US dollars)*						
Value of Exports, fob	211	211	226	236	318	323	407	1,163	1,147	1,634	1,859
Nonfuel Primary Products	8	18	22	13	12	16	23	105	101	144	139
Fuels	191	176	179	184	259	253	328	982	935	1,290	1,462
Manufactures	13	18	26	39	47	54	56	76	112	200	259
Value of Imports, cif	206	193	208	247	302	378	533	1,199	1,158	1,670	2,031
Nonfuel Primary Products	23	24	23	38	42	51	65	85	86	154	163
Fuels	96	87	92	80	84	152	218	764	589	708	925
Manufactures	87	82	93	130	176	175	249	350	484	807	943
					(Index 1980 = 100)						
Terms of Trade	58.5	61.2	62.8	50.1	46.0	60.6	61.7	92.4	83.8	83.7	85.4
Export Prices, fob	4.6	5.0	5.2	5.3	6.6	7.6	10.5	39.2	38.0	41.5	45.6
Nonfuel Primary Products	38.1	36.3	38.6	40.8	39.3	37.7	54.2	58.9	43.4	52.5	59.6
Fuels	4.3	4.3	4.3	4.3	5.6	6.2	8.9	36.7	35.7	38.4	42.0
Manufactures	30.1	31.7	32.4	37.5	36.5	38.5	49.5	67.2	66.1	66.7	71.7
Import Prices, cif	7.9	8.2	8.3	10.6	14.4	12.6	17.1	42.4	45.4	49.6	53.4
Trade at Constant 1980 Prices					*(Millions of 1980 US dollars)*						
Exports, fob	4,542	4,224	4,334	4,449	4,810	4,242	3,861	2,966	3,017	3,936	4,077
Imports, cif	2,594	2,363	2,506	2,336	2,103	3,006	3,118	2,826	2,550	3,365	3,801
BALANCE OF PAYMENTS					*(Millions of current US dollars)*						
Exports of Goods & Services	..	..	..	246	300	392	541	1,436	1,323	1,696	2,074
Merchandise, fob	..	..	..	218	265	348	479	1,272	1,203	1,518	1,849
Nonfactor Services	..	..	..	15	19	24	34	89	75	111	149
Factor Services	..	..	..	13	16	20	28	75	46	67	76
Imports of Goods & Services	..	..	..	291	356	425	607	1,324	1,299	1,805	2,196
Merchandise, fob	..	..	..	247	303	361	516	1,126	1,090	1,510	1,837
Nonfactor Services	..	..	..	40	48	58	83	180	179	243	296
Factor Services	..	..	..	4	5	6	8	18	30	51	63
Long-Term Interest	..	..	..	..	..	..	..	..	..	..	..
Current Transfers, net	..	..	..	-44	-54	-63	-92	-198	-228	-253	-301
Workers' Remittances	..	..	..	..	..	..	..	..	0	0	0
Total to be Financed	..	..	..	-89	-110	-96`	-158	-86	-203	-362	-423
Official Capital Grants	..	..	..	0	0	0	0	0	..	1	99
Current Account Balance	..	..	..	-89	-110	-96	-158	-86	-203	-360	-324
Long-Term Capital, net	..	..	..	73	57	45	-14	56	88	281	501
Direct Investment	..	..	..	..	..	..	..	..	..	..	..
Long-Term Loans	..	..	..	..	..	..	..	..	..	..	..
Disbursements	..	..	..	..	..	..	..	..	..	..	..
Repayments	..	..	..	..	..	..	..	..	..	..	..
Other Long-Term Capital	..	..	..	73	57	45	-14	56	88	280	402
Other Capital, net	..	..	..	19	76	49	152	97	284	226	-109
Change in Reserves	..	..	..	-3	-23	3	20	-67	-169	-147	-68
Memo Item:					*(Bahrain Dinars per US dollar)*						
Conversion Factor (Annual Avg)	0.480	0.480	0.480	0.480	0.470	0.440	0.400	0.390	0.400	0.400	0.400
EXTERNAL DEBT, ETC.					*(Millions of US dollars, outstanding at end of year)*						
Public/Publicly Guar. Long-Term	..	..	..	..	..	..	..	..	..	..	..
Official Creditors	..	..	..	..	..	..	..	..	..	..	..
IBRD and IDA	..	..	..	..	..	..	..	..	..	..	..
Private Creditors	..	..	..	..	..	..	..	..	..	..	..
Private Non-guaranteed Long-Term	..	..	..	..	..	..	..	..	..	..	..
Use of Fund Credit	..	..	..	..	..	..	..	..	..	..	..
Short-Term Debt	..	..	..	..	..	..	..	..	..	..	..
Memo Items:					*(Millions of US dollars)*						
Int'l Reserves Excluding Gold	20.0	41.7	59.6	62.8	86.0	83.5	64.0	131.3	289.5	436.4	503.9
Gold Holdings (at market price)	8.3	9.9	8.3	8.9	10.3	15.4	26.6	44.4	21.0	20.2	24.7
SOCIAL INDICATORS											
Total Fertility Rate	7.0	6.8	6.6	6.4	6.1	5.9	5.8	5.7	5.5	5.4	5.2
Crude Birth Rate	43.4	41.9	40.4	39.0	37.5	36.0	35.7	35.4	35.0	34.7	34.4
Infant Mortality Rate	78.0	73.4	68.8	64.2	59.6	55.0	51.6	48.2	44.8	41.4	38.0
Life Expectancy at Birth	60.0	60.7	61.4	62.1	62.8	63.5	64.3	65.1	65.9	66.7	67.5
Food Production, p.c. ('79-81 = 100)	..	..	..	..	..	..	..	..	..	..	..
Labor Force, Agriculture (%)	9.1	8.4	7.8	7.2	6.8	6.3	5.9	5.4	5.0	4.6	4.2
Labor Force, Female (%)	6.0	5.6	5.3	4.9	5.9	6.9	7.8	8.8	9.8	10.0	10.2
School Enroll. Ratio, primary	..	..	..	99.0	..	..	..	..	96.0	..	97.0
School Enroll. Ratio, secondary	..	..	..	51.0	..	..	..	..	52.0	..	50.0

1978	1979	1980	1981	1982	1983	1984	1985	1986	1987 estimate	NOTES	BAHRAIN
											FOREIGN TRADE (CUSTOMS BASIS)
			(Millions of current US dollars)								
1,867	2,471	3,795	4,347	3,791	3,200	3,139	2,820	2,344	2,344	..	Value of Exports, fob
203	148	129	177	245	270	224	148	209	162	..	Nonfuel Primary Products
1,487	2,160	3,553	3,882	2,985	2,586	2,712	2,442	1,970	1,967	..	Fuels
177	163	113	288	561	344	203	230	165	215	..	Manufactures
2,033	2,478	3,479	4,124	3,615	3,342	3,524	3,159	2,427	2,613	..	Value of Imports, cif
194	232	311	254	265	243	252	214	229	249	..	Nonfuel Primary Products
880	1,277	2,029	2,517	1,900	1,832	1,941	1,816	987	1,025	..	Fuels
959	969	1,139	1,353	1,450	1,267	1,331	1,129	1,211	1,339	..	Manufactures
			(Index 1980 = 100)								
79.5	87.2	100.0	102.4	97.5	98.0	97.6	96.2	71.4	74.4	..	Terms of Trade
45.8	63.5	100.0	109.5	96.7	91.7	89.2	86.0	48.5	61.3	..	Export Prices, fob
62.4	88.2	100.0	78.3	62.5	86.4	79.3	64.2	72.9	92.9	..	Nonfuel Primary Products
42.3	61.0	100.0	112.5	101.6	92.5	90.2	87.5	44.9	56.9	..	Fuels
76.1	90.0	100.0	99.1	94.8	90.7	89.3	89.0	103.2	113.0	..	Manufactures
57.6	72.8	100.0	107.0	99.1	93.6	91.4	89.4	68.0	82.5	..	Import Prices, cif
											Trade at Constant 1980 Prices
			(Millions of 1980 US dollars)								
4,074	3,890	3,795	3,968	3,921	3,489	3,518	3,278	4,829	3,822	..	Exports, fob
3,527	3,403	3,479	3,854	3,648	3,572	3,855	3,534	3,568	3,169	..	Imports, cif
			(Millions of current US dollars)								**BALANCE OF PAYMENTS**
2,235	2,774	4,085	4,943	4,635	3,801	4,266	3,843	3,235	..	..	Exports of Goods & Services
1,903	2,447	3,540	4,177	3,693	3,119	3,204	2,897	2,338	..	..	Merchandise, fob
226	214	231	248	452	412	693	634	629	..	..	Nonfactor Services
107	112	314	518	490	269	369	312	268	..	..	Factor Services
2,314	2,866	3,727	4,390	3,923	3,464	3,840	3,303	2,648	..	..	Imports of Goods & Services
1,873	2,093	3,020	3,559	3,036	2,867	3,132	2,797	2,161	..	..	Merchandise, fob
343	370	476	610	715	536	551	446	426	..	..	Nonfactor Services
98	403	231	221	172	61	156	60	61	..	..	Factor Services
..	..	..	..	..	..	..	..	..	..	..	Long-Term Interest
-388	-279	-283	-318	-331	-300	-346	-396	-672	..	..	Current Transfers, net
0	0	0	0	0	0	0	0	0	..	..	Workers' Remittances
-466	-371	75	235	381	36	80	145	-85	..	K	Total to be Financed
91	98	181	194	190	143	125	120	120	..	K	Official Capital Grants
-375	-273	257	430	570	179	205	265	35	..	..	Current Account Balance
237	269	192	192	95	184	276	157	53	..	..	Long-Term Capital, net
23	145	-418	..	29	64	141	105	-46	..	..	Direct Investment
..	..	..	..	..	..	..	..	..	..	..	Long-Term Loans
..	..	..	..	..	..	..	..	..	..	..	Disbursements
..	..	..	..	..	..	..	..	..	..	..	Repayments
123	25	428	-3	-123	-23	11	-68	-22	..	..	Other Long-Term Capital
114	113	-11	-30	-669	-470	-436	-245	-312	..	..	Other Capital, net
24	-109	-437	-591	4	107	-45	-177	224	349	..	Change in Reserves
											Memo Item:
			(Bahrain Dinars per US dollar)								
0.390	0.380	0.380	0.380	0.380	0.380	0.380	0.380	0.380	0.380	..	Conversion Factor (Annual Avg)
											EXTERNAL DEBT, ETC.
			(Millions of US dollars, outstanding at end of year)								
..	..	..	..	..	..	..	..	..	..	..	Public/Publicly Guar. Long-Term
..	..	..	..	..	..	..	..	..	..	..	Official Creditors
..	..	..	..	..	..	..	..	..	..	..	IBRD and IDA
..	..	..	..	..	..	..	..	..	..	..	Private Creditors
..	..	..	..	..	..	..	..	..	..	..	Private Non-guaranteed Long-Term
..	..	..	..	..	..	..	..	..	..	..	Use of Fund Credit
..	..	..	..	..	..	..	..	..	..	..	Short-Term Debt
			(Millions of US dollars)								Memo Items:
493.4	613.9	953.4	1,544.1	1,534.8	1,426.4	1,302.4	1,659.7	1,489.4	1,148.5	..	Int'l Reserves Excluding Gold
33.9	76.8	88.4	59.6	68.5	57.2	46.2	49.1	58.6	72.6	..	Gold Holdings (at market price)
											SOCIAL INDICATORS
5.1	5.0	4.9	4.8	4.6	4.5	4.4	4.3	4.2	4.1	..	Total Fertility Rate
33.7	33.0	32.4	31.7	31.0	30.5	29.9	29.4	28.8	28.3	..	Crude Birth Rate
36.8	35.6	34.4	33.2	32.0	31.6	31.2	30.8	30.5	30.1	..	Infant Mortality Rate
67.8	68.2	68.5	68.9	69.2	69.3	69.4	69.5	69.6	69.7	..	Life Expectancy at Birth
..	..	..	..	..	..	..	..	..	..	..	Food Production, p.c. ('79-81=100)
3.8	3.4	3.0	..	..	..	..	..	..	..	..	Labor Force, Agriculture (%)
10.4	10.6	10.9	10.8	10.7	10.7	10.6	10.6	10.5	10.5	..	Labor Force, Female (%)
99.0	102.0	104.0	104.0	104.0	108.0	110.0	111.0	..	..	..	School Enroll. Ratio, primary
52.0	54.0	64.0	68.0	78.0	77.0	82.0	86.0	..	..	..	School Enroll. Ratio, secondary

BANGLADESH	1967	1968	1969	1970	1971	1972	1973	1974	1975	1976	1977
CURRENT GNP PER CAPITA (US $)	80	90	100	100	100	70	80	90	130	130	110
POPULATION (millions)	63	65	66	68	71	73	75	77	79	81	83
ORIGIN AND USE OF RESOURCES					*(Billions of current Bangladesh Taka)*						
Gross National Product (GNP)	25.66	26.52	30.14	32.01	30.99	27.70	45.23	71.15	125.69	107.33	104.88
Net Factor Income from Abroad	0.06	0.11	0.25	0.27	0.11	-0.04	0.12	0.06	-0.05	-0.13	-0.48
Gross Domestic Product (GDP)	25.60	26.41	29.89	31.73	30.88	27.74	45.11	71.09	125.74	107.46	105.36
Indirect Taxes, net	0.82	0.91	1.00	1.10	1.07	0.96	1.07	2.11	1.79	6.07	5.49
GDP at factor cost	24.77	25.50	28.89	30.63	29.81	26.78	44.04	68.98	123.95	101.39	99.87
Agriculture	14.30	14.14	16.56	17.31	15.76	16.53	26.10	41.50	78.62	57.34	53.67
Industry	2.07	2.45	2.75	2.77	2.37	1.68	4.56	7.28	14.17	13.88	14.70
Manufacturing	1.43	1.62	1.75	1.84	1.69	1.10	2.90	4.27	8.38	8.17	8.66
Services, etc.	9.23	9.82	10.58	11.65	12.75	9.52	14.46	22.30	32.94	36.24	36.99
Resource Balance	-0.84	-0.86	-1.05	-1.33	-1.37	-2.24	0.45	-4.98	-6.57	-13.96	-5.35
Exports of Goods & NFServices	2.38	2.20	2.44	2.64	1.94	1.57	2.71	2.41	3.66	4.98	7.14
Imports of Goods & NFServices	3.22	3.06	3.49	3.97	3.31	3.81	2.26	7.39	10.22	18.94	12.49
Domestic Absorption	26.44	27.27	30.94	33.06	32.25	29.97	44.66	76.06	132.31	121.42	110.71
Private Consumption, etc.	20.83	20.37	23.24	25.22	29.71	28.67	38.81	67.26	120.49	105.98	92.48
General Gov't Consumption	2.87	3.16	3.96	4.24	..	..	1.82	3.40	4.02	4.49	5.59
Gross Domestic Investment	2.74	3.74	3.74	3.60	2.54	1.30	4.03	5.40	7.80	10.95	12.65
Fixed Investment	..	..	..	..	..	..	..	..	..	..	..
Memo Items:											
Gross Domestic Saving	0.89	2.87	2.69	2.27	1.17	-0.93	4.48	0.43	1.23	-3.01	7.30
Gross National Saving	0.95	2.99	2.94	2.59	1.33	-0.91	4.87	0.64	1.49	-2.71	7.75
					(Billions of 1980 Bangladesh Taka)						
Gross National Product	135.91	149.11	151.55	160.04	150.51	128.93	133.79	149.64	154.53	173.26	174.85
GDP at factor cost	131.26	143.46	145.40	153.33	144.88	124.60	130.26	145.07	152.40	163.70	166.47
Agriculture	78.09	86.11	86.95	91.66	87.53	78.16	78.37	86.56	85.69	95.69	92.80
Industry	14.72	17.65	18.92	18.86	14.91	7.88	13.50	15.45	21.94	22.99	24.72
Manufacturing	8.40	9.54	9.76	10.20	8.65	4.74	7.85	9.23	14.87	15.94	16.59
Services, etc.	42.81	44.80	44.53	48.29	47.62	43.02	41.54	47.50	46.96	54.79	58.12
Resource Balance	-14.64	-13.19	-14.49	-16.32	-13.28	-14.61	-1.04	-13.91	-11.67	-15.28	-6.77
Exports of Goods & NFServices	6.80	7.07	7.39	8.17	5.91	4.56	8.47	7.04	8.41	7.89	10.36
Imports of Goods & NFServices	21.43	20.25	21.89	24.48	19.19	19.17	9.51	20.96	20.08	23.17	17.14
Domestic Absorption	150.26	161.75	164.90	175.13	163.34	143.68	134.46	163.42	166.26	188.75	182.41
Private Consumption, etc.	..	..	..	..	..	..	..	..	..	..	..
General Gov't Consumption	..	..	..	..	..	..	..	..	..	..	..
Gross Domestic Investment	19.74	26.63	24.53	22.80	17.31	7.68	15.08	14.31	12.71	13.45	16.05
Fixed Investment	..	..	..	..	..	..	..	..	..	..	..
Memo Items:											
Capacity to Import	15.83	14.54	15.28	16.27	11.26	7.92	11.40	6.84	7.18	6.10	9.80
Terms of Trade Adjustment	9.04	7.47	7.89	8.11	5.34	3.36	2.94	-0.20	-1.23	-1.79	-0.56
Gross Domestic Income	144.66	156.04	158.29	166.92	155.41	132.43	136.35	149.31	153.36	171.67	175.07
Gross National Income	144.95	156.58	159.44	168.15	155.86	132.30	136.73	149.44	153.29	171.47	174.29
DOMESTIC PRICES (DEFLATORS)					*(Index 1980 = 100)*						
Overall (GDP)	18.9	17.8	19.9	20.0	20.6	21.5	33.8	47.5	81.3	61.9	60.0
Domestic Absorption	17.6	16.9	18.8	18.9	19.7	20.9	33.2	46.5	79.6	64.3	60.7
Agriculture	18.3	16.4	19.0	18.9	18.0	21.2	33.3	47.9	91.8	59.9	57.8
Industry	14.1	13.9	14.5	14.7	15.9	21.3	33.7	47.2	64.6	60.4	59.5
Manufacturing	17.0	17.0	17.9	18.1	19.6	23.3	36.9	46.2	56.3	51.3	52.2
MANUFACTURING ACTIVITY											
Employment (1980=100)	55.8	57.0	64.1	50.1	57.2	47.9	44.5	82.2	86.6	89.3	87.3
Real Earnings per Empl. (1980=100)	92.2	98.5	97.0	120.0	120.3	110.7	87.4	89.5	59.6	71.3	73.2
Real Output per Empl. (1980=100)	98.5	103.7	102.0	116.2	89.5	64.6	58.5	58.5	69.8	101.9	122.1
Earnings as % of Value Added	28.0	26.4	24.3	26.4	33.8	35.0	31.4	35.5	32.7	32.4	27.0
MONETARY HOLDINGS					*(Billions of current Bangladesh Taka)*						
Money Supply, Broadly Defined	..	..	..	..	..	..	..	13.99	14.74	17.47	21.61
Money as Means of Payment	..	..	..	..	..	..	..	8.39	8.28	9.22	11.67
Currency Ouside Banks	..	..	..	..	..	..	3.21	4.10	3.62	3.82	4.90
Demand Deposits	..	..	..	..	..	..	..	4.29	4.66	5.40	6.76
Quasi-Monetary Liabilities	..	..	..	..	..	..	..	5.60	6.46	8.25	9.94
					(Millions of current Bangladesh Taka)						
GOVERNMENT DEFICIT (-) OR SURPLUS	..	..	..	..	..	..	-864.0	-330.0	1,444.0	-3,728.0	209.0
Current Revenue	..	..	..	..	..	..	4,022.0	5,970.0	11,863.0	10,763.0	15,912.0
Current Expenditure	..	..	..	..	..	..	..	..	..	..	..
Current Budget Balance	..	..	..	..	..	..	..	..	..	..	..
Capital Receipts	..	..	..	..	..	..	..	..	..	..	..
Capital Payments	..	..	..	..	..	..	..	..	..	..	..

1978	1979	1980	1981	1982	1983	1984	1985	1986	1987 estimate	NOTES	BANGLADESH
100	120	140	160	160	150	140	150	160	160	..	CURRENT GNP PER CAPITA (US $)
85	86	89	91	93	95	98	101	103	106	..	POPULATION (millions)
				(Billions of current Bangladesh Taka)							ORIGIN AND USE OF RESOURCES
145.88	172.48	198.21	232.90	263.14	285.82	348.32	414.56	458.31	535.45	C	Gross National Product (GNP)
-0.48	-0.33	0.22	-0.37	-2.00	-2.60	-1.60	-2.40	-3.70	-3.72	f	Net Factor Income from Abroad
146.37	172.82	197.98	233.26	265.14	288.42	349.92	416.96	462.01	539.17	C	Gross Domestic Product (GDP)
7.17	8.91	10.35	13.46	13.82	15.47	18.85	21.79	25.52	30.08	..	Indirect Taxes, net
139.20	163.91	187.63	219.80	251.32	272.95	331.07	395.17	436.50	509.09	B C	GDP at factor cost
80.09	91.35	99.50	108.95	121.84	135.87	169.33	207.98	216.78	255.06	..	Agriculture
19.87	25.98	29.49	36.70	42.53	44.71	50.98	59.50	66.11	72.60	..	Industry
13.36	15.83	19.56	22.86	25.70	28.07	30.95	34.63	37.33	40.21	f	Manufacturing
46.41	55.49	68.99	87.61	100.77	107.84	129.61	149.48	179.12	211.51	..	Services, etc.
-13.87	-16.87	-25.57	-29.73	-38.82	-38.33	-38.89	-42.52	-43.46	-49.90	..	Resource Balance
7.55	8.99	11.34	13.95	15.09	19.26	21.24	25.71	27.12	29.67	..	Exports of Goods & NFServices
21.42	25.86	36.91	43.69	53.91	57.59	60.13	68.24	70.57	79.57	..	Imports of Goods & NFServices
160.23	189.69	223.56	262.99	303.96	326.75	388.81	459.49	505.47	589.07	..	Domestic Absorption
135.99	159.24	181.08	210.12	248.21	270.77	322.19	377.68	415.03	484.24	..	Private Consumption, etc.
7.20	10.94	12.68	15.64	15.91	16.77	23.87	29.79	35.76	42.91	..	General Gov't Consumption
17.05	19.51	29.80	37.23	39.84	39.21	42.75	52.01	54.68	61.92	..	Gross Domestic Investment
..	..	29.80	37.23	39.84	39.21	42.75	52.01	54.68	61.92	..	Fixed Investment
											Memo Items:
3.18	2.64	4.23	7.50	1.02	0.88	3.87	9.49	11.22	12.02	..	Gross Domestic Saving
4.41	4.48	7.71	13.33	7.51	13.20	17.91	19.49	25.04	30.68	..	Gross National Saving
				(Billions of 1980 Bangladesh Taka)							
186.50	195.15	198.21	211.03	211.56	218.91	228.97	236.89	247.93	258.29	C	Gross National Product
177.97	185.50	187.63	199.18	202.00	208.97	217.62	225.82	236.10	245.40	B C	GDP at factor cost
100.81	99.34	99.50	104.82	105.77	110.66	112.43	113.49	118.07	118.21	..	Agriculture
25.21	30.98	29.49	31.71	32.67	32.87	35.60	37.81	38.67	41.45	..	Industry
16.84	19.16	19.56	20.62	20.95	20.61	21.36	22.06	22.47	23.90	f	Manufacturing
61.10	65.21	68.99	74.82	74.65	77.28	81.97	86.98	93.16	100.23	..	Services, etc.
-17.39	-16.67	-25.57	-22.23	-20.56	-19.30	-22.90	-31.12	-22.97	-26.20	..	Resource Balance
11.07	11.00	11.34	13.03	14.08	14.63	14.70	13.46	15.83	17.50	..	Exports of Goods & NFServices
28.46	27.66	36.91	35.26	34.65	33.93	37.60	44.58	38.80	43.70	..	Imports of Goods & NFServices
204.51	212.20	223.56	233.59	233.66	240.11	252.90	269.39	272.87	286.09	..	Domestic Absorption
..	..	..	..	..	..	..	..	..	..	..	Private Consumption, etc.
..	..	..	..	..	..	..	..	..	..	..	General Gov't Consumption
20.40	21.30	29.80	30.73	40.42	38.07	35.97	37.77	37.56	37.92	..	Gross Domestic Investment
..	..	..	..	..	..	..	..	..	..	..	Fixed Investment
											Memo Items:
10.04	9.61	11.34	11.26	9.70	11.35	13.28	16.80	14.91	16.29	..	Capacity to Import
-1.04	-1.38	0.00	-1.76	-4.39	-3.28	-1.42	3.34	-0.92	-1.21	..	Terms of Trade Adjustment
186.08	194.15	197.98	209.59	208.71	217.53	228.58	241.61	248.98	258.68	..	Gross Domestic Income
185.46	193.77	198.21	209.27	207.17	215.63	227.55	240.23	247.01	257.08	..	Gross National Income
				(Index 1980 = 100)							DOMESTIC PRICES (DEFLATORS)
78.2	88.4	100.0	110.4	124.4	130.6	152.1	175.0	184.9	207.5	..	Overall (GDP)
78.4	89.4	100.0	112.6	130.1	136.1	153.7	170.6	185.2	205.9	..	Domestic Absorption
79.4	92.0	100.0	103.9	115.2	122.8	150.6	183.3	183.6	215.8	..	Agriculture
78.8	83.9	100.0	115.7	130.2	136.0	143.2	157.4	171.0	175.2	..	Industry
79.3	82.6	100.0	110.9	122.7	136.2	144.9	157.0	166.2	168.2	..	Manufacturing
											MANUFACTURING ACTIVITY
95.7	97.3	100.0	106.4	110.7	108.9	113.1	117.4	..	..	J	Employment (1980=100)
79.7	98.9	100.0	98.2	87.1	84.2	85.3	82.9	..	..	J	Real Earnings per Empl. (1980=100)
86.4	97.0	100.0	99.1	100.7	98.1	98.3	97.8	..	..	J	Real Output per Empl. (1980=100)
29.6	34.6	31.3	33.5	32.1	30.7	32.1	32.1	..	..	J	Earnings as % of Value Added
				(Billions of current Bangladesh Taka)							MONETARY HOLDINGS
27.30	33.41	40.38	47.16	52.78	73.90	100.58	114.28	132.79	138.71	..	Money Supply, Broadly Defined
14.64	18.36	20.17	22.72	23.34	31.64	42.27	45.95	50.00	..	..	Money as Means of Payment
6.33	7.11	8.27	9.14	9.74	13.44	17.25	17.67	19.03	22.49	..	Currency Ouside Banks
8.31	11.24	11.90	13.57	13.59	18.19	25.02	28.28	30.97	..	..	Demand Deposits
12.67	15.05	20.21	24.44	29.44	42.26	58.31	68.32	82.79	..	..	Quasi-Monetary Liabilities
				(Millions of current Bangladesh Taka)							
4,274.0	874.0	4,976.0	-7,396.0	3,135.0	9,003.0	2,873.0	-717.0	..	..	C	GOVERNMENT DEFICIT (-) OR SURPLUS
21,672.0	24,098.0	28,068.0	30,078.0	41,232.0	44,257.0	44,355.0	47,851.0	..	..	..	Current Revenue
..	..	..	..	..	..	..	..	..	..	..	Current Expenditure
..	..	..	..	..	..	..	..	..	..	..	Current Budget Balance
..	..	..	..	..	..	..	..	..	..	..	Capital Receipts
..	..	..	..	..	..	..	..	..	..	..	Capital Payments

BANGLADESH	1967	1968	1969	1970	1971	1972	1973	1974	1975	1976	1977
FOREIGN TRADE (CUSTOMS BASIS)					*(Millions of current US dollars)*						
Value of Exports, fob	..	..	..	..	..	321.0	326.0	367.6	264.8	361.7	435.0
Nonfuel Primary Products	..	..	..	..	..	124.1	127.0	151.1	84.3	137.9	177.1
Fuels	..	..	..	..	..	0.2	0.3	0.2	0.2	1.4	6.7
Manufactures	..	..	..	..	..	196.6	198.7	216.4	180.2	222.5	251.3
Value of Imports, cif	..	..	..	..	..	611.0	834.5	1,032.0	1,199.5	1,136.5	1,057.5
Nonfuel Primary Products	..	..	..	..	..	..	384.7	544.3	631.0	469.9	403.8
Fuels	..	..	..	..	..	..	94.0	92.0	97.0	120.0	172.0
Manufactures	..	..	..	..	..	..	355.8	395.7	471.5	546.6	481.7
					(Index 1980 = 100)						
Terms of Trade	..	..	..	..	..	156.1	166.5	112.8	109.4	117.7	126.2
Export Prices, fob	..	..	..	..	..	48.0	58.2	77.6	73.8	73.6	80.0
Nonfuel Primary Products	..	..	..	..	..	80.2	81.8	99.7	98.3	89.2	100.0
Fuels	4.3	4.3	4.3	4.3	5.6	6.2	8.9	36.7	35.7	38.4	42.0
Manufactures	30.1	31.7	32.4	37.5	36.5	38.5	49.5	67.2	66.1	66.7	71.7
Import Prices, cif	..	..	..	..	..	30.7	35.0	68.8	67.4	62.5	63.4
Trade at Constant 1980 Prices					*(Millions of 1980 US dollars)*						
Exports, fob	..	..	..	..	..	668.9	559.8	473.9	359.0	491.6	543.7
Imports, cif	..	..	..	..	..	1,986.0	2,386.7	1,500.6	1,779.5	1,817.6	1,667.9
BALANCE OF PAYMENTS					*(Millions of current US dollars)*						
Exports of Goods & Services	..	..	..	526	344	355	412	418	440	451	489
Merchandise, fob	..	..	..	420	295	260	354	370	344	372	405
Nonfactor Services	..	..	..	61	40	76	43	34	83	63	59
Factor Services	..	..	..	45	9	19	14	13	13	16	25
Imports of Goods & Services	..	..	..	770	665	801	817	984	1,478	1,361	988
Merchandise, fob	..	..	..	660	570	683	780	925	1,403	1,275	875
Nonfactor Services	..	..	..	105	94	113	35	49	56	51	58
Factor Services	..	..	..	5	1	5	2	10	19	35	56
Long-Term Interest	..	..	..	0	0	0	4	9	17	28	30
Current Transfers, net	..	..	..	10	10	10	34	19	35	29	60
Workers' Remittances	..	..	..	0	0	0	0	0	9	16	45
Total to be Financed	..	..	..	-234	-311	-436	-371	-547	-1,003	-882	-439
Official Capital Grants	..	..	..	120	160	210	431	174	382	245	265
Current Account Balance	..	..	..	-114	-151	-226	60	-373	-621	-637	-174
Long-Term Capital, net	..	..	..	280	300	305	527	365	888	750	522
Direct Investment	..	..	..	..	..	..	0	0	0	0	0
Long-Term Loans	..	..	..	..	..	60	309	502	550	257	290
Disbursements	..	..	..	..	..	60	315	518	605	297	332
Repayments	..	..	..	..	..	..	6	16	55	40	42
Other Long-Term Capital	..	..	..	160	140	35	-213	-311	-44	248	-33
Other Capital, net	..	..	..	-166	-149	-145	-683	-84	-222	-246	-217
Change in Reserves	..	..	..	0	0	66	97	93	-45	134	-130
Memo Item:					*(Bangladesh Taka per US dollar)*						
Conversion Factor (Annual Avg)	4.760	4.760	4.760	4.760	4.760	6.030	7.780	7.970	8.880	14.850	15.470
EXTERNAL DEBT, ETC.					*(Millions of US dollars, outstanding at end of year)*						
Public/Publicly Guar. Long-Term	..	..	..	..	..	77.8	416.6	1,085.8	1,604.8	1,877.1	2,181.3
Official Creditors	..	..	..	..	..	75.8	401.1	1,068.9	1,581.0	1,854.2	2,159.8
IBRD and IDA	..	..	..	0.0	0.0	3.0	91.2	258.9	349.7	471.3	559.5
Private Creditors	..	..	..	..	..	2.0	15.5	16.9	23.8	22.9	21.4
Private Non-guaranteed Long-Term	..	..	..	..	..	..	..	..	..	..	..
Use of Fund Credit	..	..	..	0.0	0.0	67.9	75.4	164.6	200.9	275.2	237.5
Short-Term Debt	..	..	..	..	..	..	..	..	..	..	23.0
Memo Items:					*(Millions of US dollars)*						
Int'l Reserves Excluding Gold	..	..	..	..	..	270.5	143.2	138.2	148.3	288.9	232.7
Gold Holdings (at market price)	..	..	..	..	..	..	..	..	..	..	8.8
SOCIAL INDICATORS											
Total Fertility Rate	6.9	6.9	7.0	7.0	7.0	7.0	6.9	6.7	6.6	6.4	6.3
Crude Birth Rate	47.5	47.7	47.9	48.1	48.3	48.5	48.3	48.0	47.8	47.5	47.3
Infant Mortality Rate	140.0	140.0	140.0	140.0	140.0	140.0	149.7	138.8	138.2	137.6	137.0
Life Expectancy at Birth	44.9	44.9	44.9	44.9	44.9	44.9	45.2	45.6	46.0	46.4	46.8
Food Production, p.c. ('79-81 = 100)	123.1	122.5	123.9	114.2	102.7	97.8	104.5	98.0	105.3	97.4	101.7
Labor Force, Agriculture (%)	82.8	82.3	81.9	81.4	80.7	80.1	79.4	78.8	78.1	77.4	76.8
Labor Force, Female (%)	5.1	5.2	5.3	5.4	5.5	5.6	5.6	5.7	5.8	5.9	6.0
School Enroll. Ratio, primary	..	..	..	54.0	..	..	..	..	73.0	79.0	72.0
School Enroll. Ratio, secondary	..	..	..	..	..	..	..	..	26.0	21.0	21.0

1978	1979	1980	1981	1982	1983	1984	1985	1986	1987 estimate	NOTES	BANGLADESH
				(Millions of current US dollars)							**FOREIGN TRADE (CUSTOMS BASIS)**
552.8	661.6	758.9	662.0	671.1	704.9	811.2	934.0	819.0	1,074.0	C	Value of Exports, fob
199.8	223.6	249.2	214.8	239.0	245.3	269.8	291.6	266.2	356.3	..	Nonfuel Primary Products
6.7	4.0	0.0	0.0	15.0	26.2	15.2	23.4	7.2	10.3	..	Fuels
346.4	434.1	509.7	447.2	417.0	433.4	526.2	619.0	545.6	707.4	..	Manufactures
1,338.0	1,710.5	2,253.5	2,649.0	2,581.0	2,246.0	2,353.0	2,647.0	2,364.0	2,620.0	C	Value of Imports, cif
463.6	509.7	642.5	571.0	584.3	637.9	700.7	872.8	519.9	667.1	..	Nonfuel Primary Products
193.9	247.0	416.0	509.0	485.7	370.1	300.0	359.0	355.0	233.7	f	Fuels
680.4	953.8	1,195.0	1,569.0	1,511.0	1,238.0	1,352.3	1,415.2	1,489.1	1,719.2	f	Manufactures
				(Index 1980 = 100)							
120.0	114.1	100.0	94.5	100.2	101.3	109.1	124.1	112.9	102.9	..	Terms of Trade
84.5	97.2	100.0	96.3	96.0	94.9	101.8	109.0	93.5	100.5	..	Export Prices, fob
109.2	116.7	100.0	91.0	97.8	103.8	141.6	116.2	96.1	98.0	..	Nonfuel Primary Products
42.3	61.0	100.0	112.5	101.6	92.5	90.2	87.5	44.9	56.9	..	Fuels
76.1	90.0	100.0	99.1	94.8	90.7	89.3	106.9	93.6	103.0	..	Manufactures
70.5	85.2	100.0	101.9	95.9	93.7	93.4	87.8	82.9	97.7	..	Import Prices, cif
				(Millions of 1980 US dollars)							Trade at Constant 1980 Prices
654.0	680.6	758.9	687.2	699.0	742.4	796.6	856.7	875.9	1,068.7	..	Exports, fob
1,899.1	2,007.1	2,253.7	2,599.1	2,692.6	2,396.7	2,520.2	3,013.9	2,853.1	2,681.8	..	Imports, cif
				(Millions of current US dollars)							**BALANCE OF PAYMENTS**
599	765	976	985	874	916	1,090	1,220	1,079	1,336	C f	Exports of Goods & Services
490	610	722	711	626	686	811	934	819	1,074	..	Merchandise, fob
82	114	163	222	214	213	222	228	224	227	..	Nonfactor Services
27	42	92	52	34	17	57	58	36	35	..	Factor Services
1,490	1,739	2,622	2,792	2,890	2,651	2,664	3,011	2,749	3,033	C f	Imports of Goods & Services
1,349	1,556	2,372	2,533	2,572	2,309	2,353	2,647	2,364	2,620	..	Merchandise, fob
82	119	173	185	187	156	190	217	223	256	..	Nonfactor Services
59	64	77	75	131	186	121	148	162	157	..	Factor Services
37	40	47	56	60	67	74	91	108	132	..	Long-Term Interest
113	143	210	379	424	628	627	477	586	731	f	Current Transfers, net
102	127	197	362	368	576	527	364	497	617	..	Workers' Remittances
-778	-831	-1,436	-1,428	-1,592	-1,107	-947	-1,314	-1,084	-966	K	Total to be Financed
393	553	592	584	697	720	708	701	963	657	K	Official Capital Grants
-385	-278	-844	-844	-895	-387	-239	-614	-122	-309	..	Current Account Balance
849	1,027	1,275	1,137	1,197	1,272	1,178	1,247	1,115	1,338	C f	Long-Term Capital, net
0	0	0	0	0	0	0	0	2	2	..	Direct Investment
443	555	668	499	606	474	437	468	813	733	..	Long-Term Loans
495	598	699	548	669	541	533	600	975	923	..	Disbursements
52	43	31	48	62	67	96	132	162	191	..	Repayments
13	-81	15	54	-106	78	33	79	-663	-54	..	Other Long-Term Capital
-512	-603	-635	-692	-810	-684	-584	-710	-866	-826	C f	Other Capital, net
48	-147	204	399	508	-200	-355	77	-128	-203	..	Change in Reserves
				(Bangladesh Taka per US dollar)							Memo Item:
15.120	15.220	15.480	16.340	20.040	23.760	24.950	26.000	29.890	30.630	..	Conversion Factor (Annual Avg)
				(Millions of US dollars, outstanding at end of year)							**EXTERNAL DEBT, ETC.**
2,668.6	2,793.1	3,549.3	3,911.5	4,434.2	4,865.6	5,123.5	5,978.0	7,272.4	8,851.1	..	Public/Publicly Guar. Long-Term
2,632.3	2,745.5	3,491.4	3,844.7	4,350.2	4,741.3	4,979.5	5,830.5	7,063.2	8,632.1	..	Official Creditors
662.6	825.2	980.9	1,139.7	1,324.9	1,513.7	1,727.0	2,075.7	2,510.8	3,055.7	..	IBRD and IDA
36.2	47.6	58.0	66.9	84.0	124.4	144.0	147.5	209.2	219.0	..	Private Creditors
..	..	..	..	..	..	..	..	..	..	..	Private Non-guaranteed Long-Term
231.6	226.3	269.0	322.2	412.6	441.1	356.1	423.5	461.2	581.4	..	Use of Fund Credit
67.0	110.0	212.0	228.0	209.0	170.0	133.0	135.0	125.0	73.9	..	Short-Term Debt
				(Millions of US dollars)							Memo Items:
315.2	386.3	299.7	138.4	182.6	524.1	389.9	336.5	409.1	843.2	..	Int'l Reserves Excluding Gold
6.0	27.4	31.5	21.3	24.4	21.4	16.5	17.5	20.9	25.9	..	Gold Holdings (at market price)
											SOCIAL INDICATORS
6.2	6.2	6.3	6.1	6.0	5.9	5.8	5.7	5.6	5.5	..	Total Fertility Rate
46.2	45.1	43.3	42.8	41.7	41.4	41.1	39.0	40.5	40.2	..	Crude Birth Rate
135.2	133.4	140.0	129.8	128.0	126.0	123.9	125.0	119.8	117.8	..	Infant Mortality Rate
47.1	47.5	47.9	48.3	48.6	49.0	49.4	49.8	50.1	50.5	..	Life Expectancy at Birth
103.7	99.6	101.7	98.8	99.6	99.1	98.1	98.5	96.5	89.2	..	Food Production, p.c. ('79-81 = 100)
76.1	75.5	74.8	..	..	..	..	..	..	..	..	Labor Force, Agriculture (%)
6.1	6.2	6.3	6.4	6.5	6.6	6.7	6.8	6.9	7.0	..	Labor Force, Female (%)
67.0	65.0	62.0	62.0	61.0	62.0	62.0	60.0	..	..	..	School Enroll. Ratio, primary
18.0	15.0	18.0	15.0	..	18.0	19.0	18.0	..	..	..	School Enroll. Ratio, secondary

BARBADOS	1967	1968	1969	1970	1971	1972	1973	1974	1975	1976	1977
CURRENT GNP PER CAPITA (US $)	590	630	680	750	830	900	1,040	1,220	1,520	1,740	1,960
POPULATION (thousands)	237	238	238	239	240	242	243	245	246	246	247
ORIGIN AND USE OF RESOURCES					*(Millions of current Barbados Dollars)*						
Gross National Product (GNP)	240.8	275.2	308.7	361.6	402.1	444.6	554.6	704.2	809.8	877.8	989.9
Net Factor Income from Abroad	2.5	0.6	-0.1	-1.4	-0.8	-3.8	-3.3	5.1	-2.6	4.4	-3.7
Gross Domestic Product (GDP)	238.3	274.6	308.8	363.0	402.9	448.4	557.9	699.1	812.4	873.4	993.6
Indirect Taxes, net	23.6	27.1	35.5	43.1	47.7	53.3	59.0	58.7	111.8	85.4	103.5
GDP at factor cost	214.7	247.5	273.3	319.9	355.2	395.1	498.9	640.4	700.6	788.0	890.1
Agriculture	41.2	36.2	32.7	35.1	33.4	39.0	51.7	68.5	93.4	76.8	91.8
Industry	41.7	48.9	55.1	62.5	69.7	76.5	92.2	125.8	130.1	153.9	181.2
Manufacturing	..	..	..	25.4	29.6	34.2	43.8	62.6	71.9	84.8	102.6
Services, etc.	131.8	162.4	185.5	222.3	252.1	279.6	355.0	446.1	477.1	557.3	617.1
Resource Balance	-22.8	-32.1	-50.5	-69.4	-51.0	-59.5	-64.7	-49.0	-81.5	-163.8	-146.4
Exports of Goods & NFServices	143.8	173.7	190.2	216.0	246.7	290.8	336.5	435.6	409.3	395.1	503.0
Imports of Goods & NFServices	166.6	205.8	240.7	285.4	297.7	350.3	401.2	484.6	490.8	558.9	649.4
Domestic Absorption	261.1	306.7	359.3	432.4	453.9	507.9	622.6	748.1	893.9	1,037.2	1,140.0
Private Consumption, etc.	177.7	208.2	243.4	289.9	306.7	349.3	417.9	505.2	602.9	649.9	773.8
General Gov't Consumption	34.5	33.4	41.7	47.7	54.2	58.8	80.8	83.4	134.8	151.4	172.1
Gross Domestic Investment	48.9	65.1	74.2	94.8	93.0	99.8	123.9	159.5	156.2	235.9	194.1
Fixed Investment	48.9	65.1	74.2	94.8	93.0	99.8	123.9	159.5	156.2	235.9	194.1
Memo Items:											
Gross Domestic Saving	26.1	33.0	23.7	25.4	42.0	40.3	59.2	110.5	74.7	72.1	47.7
Gross National Saving	34.9	40.4	30.8	31.8	48.8	45.1	67.2	128.7	85.8	98.2	69.6
					(Millions of 1980 Barbados Dollars)						
Gross National Product	1,020.8	1,082.2	1,160.5	1,265.7	1,317.3	1,325.5	1,347.3	1,301.5	1,339.5	1,350.9	1,396.5
GDP at factor cost	910.1	973.3	1,027.4	1,119.9	1,163.5	1,177.9	1,212.1	1,183.4	1,158.7	1,212.7	1,255.7
Agriculture	212.6	176.2	158.3	167.7	151.1	127.1	133.4	126.2	122.0	130.6	133.4
Industry	184.1	200.7	214.7	225.3	227.0	238.4	254.8	242.3	239.6	278.3	264.4
Manufacturing	..	..	..	90.7	98.9	109.3	117.1	111.6	123.5	144.4	141.7
Services, etc.	513.3	596.4	654.4	726.9	785.3	812.4	823.9	814.9	797.2	803.8	857.9
Resource Balance	..	..	..	..	..	..	..	..	..	..	..
Exports of Goods & NFServices	..	..	..	..	..	..	..	..	..	..	..
Imports of Goods & NFServices	..	..	..	..	..	..	..	..	..	..	..
Domestic Absorption	..	..	..	..	..	..	..	..	..	..	..
Private Consumption, etc.	..	..	..	..	..	..	..	..	..	..	..
General Gov't Consumption	..	..	..	..	..	..	..	..	..	..	..
Gross Domestic Investment	..	..	..	..	..	..	..	..	..	..	..
Fixed Investment	..	..	..	..	..	..	..	..	..	..	..
Memo Items:											
Capacity to Import	..	..	..	..	..	..	..	..	..	..	..
Terms of Trade Adjustment	..	..	..	..	..	..	..	..	..	..	..
Gross Domestic Income	..	..	..	..	..	..	..	..	..	..	..
Gross National Income	..	..	..	..	..	..	..	..	..	..	..
DOMESTIC PRICES (DEFLATORS)					*(Index 1980 = 100)*						
Overall (GDP)	23.6	25.4	26.6	28.6	30.5	33.5	41.2	54.1	60.5	65.0	70.9
Domestic Absorption	..	..	..	..	..	..	..	..	..	..	..
Agriculture	19.4	20.6	20.7	20.9	22.1	30.7	38.8	54.3	76.6	58.8	68.8
Industry	22.7	24.4	25.7	27.7	30.7	32.1	36.2	51.9	54.3	55.3	68.5
Manufacturing	..	..	..	28.0	29.9	31.3	37.4	56.1	58.2	58.7	72.4
MANUFACTURING ACTIVITY											
Employment (1980=100)	..	..	..	91.6	104.1	109.1	116.2	103.1	107.2	114.4	129.7
Real Earnings per Empl. (1980=100)	..	..	..	99.6	92.2	99.3	95.0	86.3	75.9	78.9	78.6
Real Output per Empl. (1980=100)	..	..	..	78.6	73.4	78.9	77.5	76.7	77.5	81.4	66.1
Earnings as % of Value Added	..	..	..	56.7	54.9	51.9	51.1	46.9	53.1	56.2	61.7
MONETARY HOLDINGS					*(Millions of current Barbados Dollars)*						
Money Supply, Broadly Defined	124.8	158.8	169.5	193.9	228.6	240.2	253.8	311.2	362.4	413.0	455.5
Money as Means of Payment	42.1	51.9	56.8	55.7	62.5	67.8	74.0	88.3	105.5	113.5	135.2
Currency Ouside Banks	12.7	14.3	15.7	21.2	21.4	23.7	26.8	33.9	40.8	46.7	55.2
Demand Deposits	29.4	37.6	41.1	34.5	41.1	44.1	47.2	54.4	64.7	66.8	80.0
Quasi-Monetary Liabilities	82.7	106.9	112.7	138.2	166.1	172.4	179.7	222.9	257.0	299.5	320.3
					(Millions of current Barbados Dollars)						
GOVERNMENT DEFICIT (-) OR SURPLUS	..	..	..	..	..	-8.0	-33.8	-22.0	-20.0	-58.5	-62.8
Current Revenue	..	..	..	..	..	119.5	138.6	177.5	214.2	230.3	267.0
Current Expenditure	..	..	..	..	..	102.0	139.1	161.0	185.3	227.2	244.6
Current Budget Balance	..	..	..	..	..	17.5	-0.5	16.5	28.9	3.1	22.4
Capital Receipts	..	..	..	..	..	..	..	..	..	..	..
Capital Payments	..	..	..	..	..	25.5	33.3	38.5	48.9	61.6	85.2

1978	1979	1980	1981	1982	1983	1984	1985	1986	1987 estimate	NOTES	BARBADOS
2,250	2,670	3,130	3,520	3,760	4,010	4,410	4,640	5,140	5,350	..	**CURRENT GNP PER CAPITA (US $)**
248	247	249	250	251	252	253	253	254	254	..	POPULATION (thousands)
				(Millions of current Barbados Dollars)							**ORIGIN AND USE OF RESOURCES**
1,115.4	1,345.3	1,683.2	1,904.6	1,984.8	2,082.6	2,280.5	2,395.9	2,617.3	2,712.1	..	Gross National Product (GNP)
3.3	-3.1	4.7	0.0	-5.2	-30.1	-22.3	-25.2	-59.6	-70.8	..	Net Factor Income from Abroad
1,112.1	1,348.4	1,678.5	1,904.6	1,990.0	2,112.7	2,302.8	2,421.1	2,676.9	2,782.9	..	Gross Domestic Product (GDP)
127.6	152.2	194.7	198.4	205.8	213.7	228.2	229.3	348.7	349.0	..	Indirect Taxes, net
984.5	1,196.2	1,483.8	1,706.2	1,784.2	1,899.0	2,074.6	2,191.8	2,328.2	2,433.9	B	GDP at factor cost
91.7	109.8	152.2	128.9	122.0	135.6	139.2	150.1	149.5	..	..	Agriculture
210.1	252.1	331.8	379.7	388.4	440.4	493.2	449.2	463.6		..	Industry
112.4	136.4	169.0	189.7	205.5	238.7	264.1	231.7	229.3		f	Manufacturing
682.7	834.3	999.8	1,197.6	1,273.8	1,323.0	1,442.2	1,592.5	1,715.1		..	Services, etc.
-87.8	-130.0	-100.2	-285.5	-124.7	-112.6	15.9	95.2	5.0	5.4	..	Resource Balance
645.6	867.8	1,129.4	1,072.5	1,215.5	1,353.7	1,573.8	1,558.7	1,483.8	1,251.4	..	Exports of Goods & NFServices
733.4	997.8	1,229.6	1,358.0	1,340.2	1,466.3	1,557.9	1,463.5	1,478.8	1,246.0	..	Imports of Goods & NFServices
1,199.9	1,478.4	1,778.7	2,190.1	2,114.7	2,225.2	2,286.9	2,325.9	2,671.9	2,777.5	..	Domestic Absorption
755.7	950.9	1,062.0	1,343.9	1,335.2	1,457.9	1,525.4	1,497.8	1,779.3	1,841.5	f	Private Consumption, etc.
189.7	210.3	292.4	321.2	329.0	346.1	387.6	456.3	468.8	525.0	..	General Gov't Consumption
254.5	317.2	424.3	525.0	450.5	421.2	373.9	371.8	423.8	411.0	f	Gross Domestic Investment
254.5	317.2	424.3	525.0	450.5	421.2	373.9	371.8	423.8	..	..	Fixed Investment
											Memo Items:
166.7	187.2	324.1	239.5	325.8	308.7	389.8	467.0	428.8	416.4	..	Gross Domestic Saving
199.1	218.1	372.0	287.9	355.0	312.2	400.7	471.3	406.5	321.5	..	Gross National Saving
				(Millions of 1980 Barbados Dollars)							
1,492.3	1,598.9	1,683.2	1,624.6	1,538.8	1,523.8	1,582.3	1,588.3	1,709.6	1,728.2	..	Gross National Product
1,316.9	1,421.8	1,483.8	1,455.4	1,383.6	1,389.8	1,439.8	1,453.2	1,521.3	1,551.5	B	GDP at factor cost
133.4	144.3	152.2	126.0	123.1	127.7	139.9	139.1	145.4	120.9	..	Agriculture
280.0	317.5	331.8	328.3	306.2	316.0	324.1	307.7	326.0	324.1	..	Industry
145.1	165.4	169.0	162.9	154.2	157.8	161.2	145.8	153.3	145.6	f	Manufacturing
903.5	960.0	999.8	1,001.1	954.3	946.1	975.9	1,006.4	1,049.9	1,106.5	..	Services, etc.
..	..		..	..	..	..	..	..	..	..	Resource Balance
..	..		..	..	..	..	..	..	..	..	Exports of Goods & NFServices
..	..		..	..	..	..	..	..	..	..	Imports of Goods & NFServices
..	..		..	..	..	..	..	..	..	..	Domestic Absorption
..	..		..	..	..	..	..	..	..	..	Private Consumption, etc.
..	..		..	..	..	..	..	..	..	..	General Gov't Consumption
..	..		..	..	..	..	..	..	..	..	Gross Domestic Investment
..	..		..	..	..	..	..	..	..	..	Fixed Investment
											Memo Items:
..	..		..	..	..	..	..	..	..	..	Capacity to Import
..	..		..	..	..	..	..	..	..	..	Terms of Trade Adjustment
..	..		..	..	..	..	..	..	..	..	Gross Domestic Income
..	..		..	..	..	..	..	..	..	..	Gross National Income
				(Index 1980 = 100)							**DOMESTIC PRICES (DEFLATORS)**
74.7	84.1	100.0	117.2	129.0	136.6	144.1	150.8	153.0	156.9	..	Overall (GDP)
..	..		..	..	..	..	..	..	..	..	Domestic Absorption
68.7	76.1	100.0	102.3	99.1	106.2	99.5	107.9	102.8	..	..	Agriculture
75.0	79.4	100.0	115.6	126.9	139.4	152.2	146.0	142.2	..	..	Industry
77.5	82.4	100.0	116.4	133.3	151.3	163.9	158.9	149.6	..	..	Manufacturing
											MANUFACTURING ACTIVITY
104.9	115.1	100.0	107.5	140.4	150.0	142.9	..	..	..	G	Employment (1980=100)
87.0	76.9	100.0	95.3	98.4	104.0	115.9	..	..	..	G	Real Earnings per Empl. (1980=100)
80.0	82.0	100.0	91.5	76.8	71.8	77.8	..	..	..	G	Real Output per Empl. (1980=100)
51.6	53.7	63.0	60.2	69.8	64.0	61.3	..	..	..	..	Earnings as % of Value Added
				(Millions of current Barbados Dollars)							**MONETARY HOLDINGS**
512.6	645.9	741.3	856.3	926.5	1,029.9	1,111.0	1,208.3	1,315.5	1,567.6	D	Money Supply, Broadly Defined
160.7	222.2	244.8	249.6	251.5	312.9	305.7	353.5	395.4	466.9	..	Money as Means of Payment
65.9	80.2	101.6	111.2	110.6	114.1	118.1	123.5	137.4	156.6	..	Currency Ouside Banks
94.9	142.1	143.3	138.3	140.9	198.7	187.6	230.0	258.0	310.3	..	Demand Deposits
351.9	423.6	496.5	606.7	675.0	717.0	805.3	854.8	920.2	1,100.7	..	Quasi-Monetary Liabilities
				(Millions of current Barbados Dollars)							**GOVERNMENT DEFICIT (-) OR SURPLUS**
-4.3	-20.9	-60.9	-128.8	-76.1	-25.7	-91.0	-71.4	..	..	C F	
330.0	377.5	476.0	512.2	572.5	636.9	666.1	748.5	..	..	..	Current Revenue
261.0	..	393.1	484.9	513.9	532.3	613.6	674.9	..	..	..	Current Expenditure
69.0	..	82.9	27.3	58.6	104.6	52.5	73.6	..	..	..	Current Budget Balance
..	..		0.7	..	..	..	..	..	..	..	Capital Receipts
73.3	..	143.8	156.8	134.7	130.3	143.5	145.0	..	..	..	Capital Payments

BARBADOS	1967	1968	1969	1970	1971	1972	1973	1974	1975	1976	1977
FOREIGN TRADE (CUSTOMS BASIS)					*(Millions of current US dollars)*						
Value of Exports, fob	41.48	40.11	37.13	39.04	41.88	44.73	53.49	86.05	107.27	86.09	96.18
Nonfuel Primary Products	30.01	26.93	22.58	23.23	21.98	22.74	26.47	40.53	62.11	36.90	38.24
Fuels	6.27	6.54	4.56	4.07	7.26	6.01	4.97	18.02	14.44	11.36	12.62
Manufactures	5.20	6.64	9.99	11.74	12.64	15.98	22.04	27.49	30.72	37.84	45.33
Value of Imports, cif	76.93	84.01	97.28	117.27	121.84	141.66	167.62	203.96	216.39	236.58	271.61
Nonfuel Primary Products	24.04	25.10	27.82	32.65	35.13	42.41	51.84	60.90	59.73	65.93	71.06
Fuels	6.66	8.31	7.40	6.50	7.97	8.29	11.13	32.35	36.54	30.79	36.01
Manufactures	46.23	50.60	62.06	78.12	78.74	90.96	104.65	110.71	120.12	139.86	164.54
					(Index 1980 = 100)						
Terms of Trade	52.0	53.9	62.7	62.7	58.2	67.4	83.6	119.8	110.3	89.4	81.9
Export Prices, fob	10.8	10.6	13.9	15.9	15.8	20.8	32.2	65.2	61.6	52.9	52.8
Nonfuel Primary Products	13.5	13.1	17.5	19.7	21.9	29.5	40.4	96.6	71.2	48.2	43.1
Fuels	4.3	4.3	4.3	4.3	5.6	6.2	8.9	36.7	35.7	38.4	42.0
Manufactures	30.1	31.7	32.4	37.5	36.5	38.5	49.5	67.2	66.1	66.7	71.7
Import Prices, cif	20.7	19.6	22.2	25.4	27.2	30.9	38.5	54.4	55.9	59.1	64.5
Trade at Constant 1980 Prices					*(Millions of 1980 US dollars)*						
Exports, fob	385.98	380.15	266.90	244.87	265.16	214.94	166.24	131.94	174.11	162.86	182.03
Imports, cif	371.84	429.42	438.40	461.14	448.77	458.98	435.21	374.74	387.35	400.28	421.19
BALANCE OF PAYMENTS					*(Millions of current US dollars)*						
Exports of Goods & Services	..	..	..	101.65	110.74	125.23	149.35	189.80	220.78	210.21	260.01
Merchandise, fob	..	..	..	35.75	33.27	37.68	47.93	67.31	94.45	76.11	85.36
Nonfactor Services	..	..	..	59.15	71.65	81.04	92.38	112.46	114.05	123.82	166.73
Factor Services	..	..	..	6.75	5.82	6.51	9.03	10.03	12.28	10.28	7.92
Imports of Goods & Services	..	..	..	148.65	151.25	174.99	208.96	244.64	269.39	287.01	327.34
Merchandise, fob	..	..	..	106.90	112.16	128.04	152.56	185.51	197.02	219.09	250.20
Nonfactor Services	..	..	..	34.65	31.04	37.11	44.25	46.80	56.09	59.19	64.93
Factor Services	..	..	..	7.10	8.05	9.84	12.15	12.32	16.29	8.73	12.21
Long-Term Interest	..	..	..	0.80	0.70	0.70	0.30	2.60	2.40	2.00	1.70
Current Transfers, net	..	..	..	..	..	..	..	..	..	..	..
Workers' Remittances	..	..	..	0.00	0.00	0.00	0.00	0.00	0.00	0.00	0.00
Total to be Financed	..	..	..	..	..	..	..	..	..	..	..
Official Capital Grants	..	..	..	..	..	..	..	..	..	..	..
Current Account Balance	..	..	..	-41.80	-35.14	-43.25	-52.32	-47.83	-41.33	-64.18	-51.38
Long-Term Capital, net	..	..	..	12.40	17.87	21.70	26.34	2.48	21.43	24.69	38.20
Direct Investment	..	..	..	8.45	14.79	17.28	4.80	2.24	22.13	5.94	4.48
Long-Term Loans	..	..	..	..	0.30	-5.60	17.90	1.30	-0.30	4.80	19.10
Disbursements	..	..	..	..	0.30	0.90	21.10	2.50	1.40	6.80	26.10
Repayments	..	..	..	..	..	6.50	3.20	1.20	1.70	2.00	7.00
Other Long-Term Capital	..	..	..	3.95	2.79	10.02	3.64	-1.06	-0.39	13.95	14.62
Other Capital, net	..	..	..	23.55	28.96	24.62	26.94	38.65	28.51	27.46	14.83
Change in Reserves	..	..	..	5.85	-11.70	-3.07	-0.96	6.70	-8.61	12.03	-1.66
Memo Item:					*(Barbados Dollars per US dollar)*						
Conversion Factor (Annual Avg)	1.740	2.000	2.000	2.000	1.970	1.920	1.960	2.050	2.020	2.000	2.010
EXTERNAL DEBT, ETC.					*(Millions of US dollars, outstanding at end of year)*						
Public/Publicly Guar. Long-Term	..	..	..	12.9	14.3	7.8	25.8	27.0	26.6	31.2	49.4
Official Creditors	..	..	..	..	0.3	1.2	2.4	4.8	5.2	10.3	30.0
IBRD and IDA	..	..	..	0.0	0.0	0.0	0.0	0.0	0.0	0.0	0.0
Private Creditors	..	..	..	12.9	14.0	6.6	23.4	22.2	21.4	20.9	19.5
Private Non-guaranteed Long-Term	..	..	..	..	..	..	..	..	..	..	..
Use of Fund Credit	..	..	..	0.0	0.0	0.0	0.0	0.0	0.0	0.0	7.9
Short-Term Debt	..	..	..	..	..	..	..	..	..	..	..
Memo Items:					*(Millions of US dollars)*						
Int'l Reserves Excluding Gold	9.25	15.52	12.47	16.58	18.86	27.99	32.37	39.15	39.58	27.98	37.01
Gold Holdings (at market price)	..	..	..	..	..	..	..	..	..	..	..
SOCIAL INDICATORS											
Total Fertility Rate	3.5	3.3	3.2	3.0	2.9	2.8	2.7	2.6	2.4	2.3	2.2
Crude Birth Rate	21.9	21.7	20.5	20.6	21.6	22.0	20.8	20.3	19.2	18.2	17.6
Infant Mortality Rate	33.0	31.8	30.6	29.4	28.2	27.0	24.4	21.8	30.0	16.6	14.0
Life Expectancy at Birth	67.6	67.9	68.3	68.7	69.0	69.4	69.8	70.1	70.5	70.9	71.2
Food Production, p.c. ('79-81 = 100)	..	..	..	..	102.9	92.3	91.1	88.9	80.3	84.8	93.5
Labor Force, Agriculture (%)	20.6	19.8	19.0	18.2	17.4	16.5	15.7	14.8	14.0	13.2	12.4
Labor Force, Female (%)	40.3	40.2	40.1	40.0	40.7	41.4	42.1	42.8	43.5	44.6	44.2
School Enroll. Ratio, primary	..	..	..	102.0	..	..	..	..	103.0	109.0	116.0
School Enroll. Ratio, secondary	..	..	..	69.0	..	..	..	..	77.0	77.0	81.0

1978	1979	1980	1981	1982	1983	1984	1985	1986	1987 estimate	NOTES	BARBADOS
											FOREIGN TRADE (CUSTOMS BASIS)
				(Millions of current US dollars)							
129.81	151.74	196.00	244.00	263.00	321.00	391.00	352.00	274.70	155.89	f	Value of Exports, fob
38.65	45.29	72.59	45.85	55.35	71.25	78.07	57.97	40.16	14.88	..	Nonfuel Primary Products
25.24	13.52	27.02	70.30	47.33	63.35	60.93	95.03	46.54	63.94	..	Fuels
65.92	92.93	96.39	127.85	160.32	186.40	252.00	199.00	188.00	77.07	..	Manufactures
312.35	420.75	522.00	572.00	551.00	621.00	658.60	601.90	593.00	520.00	f	Value of Imports, cif
82.90	93.94	112.06	115.64	109.29	124.67	107.36	105.22	112.02	88.55	..	Nonfuel Primary Products
36.23	59.17	80.17	91.50	89.23	99.80	90.24	106.68	60.98	78.37	..	Fuels
193.22	267.64	˙329.77	364.86	352.48	396.53	461.00	390.00	420.00	353.08	..	Manufactures
				(Index 1980 = 100)							
75.9	82.3	100.0	93.2	78.8	78.2	75.1	74.5	74.3	71.4	..	Terms of Trade
54.9	70.8	100.0	94.0	76.2	74.0	70.0	68.2	68.8	70.3	..	Export Prices, fob
42.8	51.0	100.0	67.5	42.8	44.5	37.4	31.5	35.3	36.2	..	Nonfuel Primary Products
42.3	61.0	100.0	112.5	101.6	92.5	90.2	87.5	45.0	56.9	..	Fuels
76.1	90.0	100.0	99.1	94.8	90.7	89.3	89.0	103.2	113.0	..	Manufactures
72.3	86.1	100.0	100.9	96.8	94.5	93.3	91.5	92.5	98.5	..	Import Prices, cif
				(Millions of 1980 US dollars)							Trade at Constant 1980 Prices
236.58	214.23	196.00	259.49	344.98	434.09	558.28	515.96	399.49	221.72	..	Exports, fob
432.12	488.72	522.00	566.65	569.39	656.86	705.88	657.67	641.11	528.08	..	Imports, cif
				(Millions of current US dollars)							**BALANCE OF PAYMENTS**
329.54	441.56	572.66	555.41	626.31	698.35	827.28	796.58	768.86	..	..	Exports of Goods & Services
110.97	131.51	180.78	162.73	208.22	272.16	339.73	300.50	243.71	..	..	Merchandise, fob
206.78	294.39	371.35	370.16	396.11	405.01	458.66	466.64	486.08	..	..	Nonfactor Services
11.78	15.66	20.53	22.52	21.98	21.18	28.89	29.44	39.07	..	..	Factor Services
377.47	497.64	619.75	691.99	685.13	761.00	822.80	761.82	785.82	..	..	Imports of Goods & Services
288.02	379.11	480.78	527.12	507.03	571.47	606.23	559.28	521.19	..	..	Merchandise, fob
75.97	95.71	118.33	133.50	144.68	148.66	165.47	150.82	194.32	..	..	Nonfactor Services
13.47	22.82	20.63	31.37	33.41	40.87	51.11	51.72	70.31	..	..	Factor Services
3.40	5.30	5.80	8.80	15.70	17.10	14.90	20.10	31.10	32.30	..	Long-Term Interest
..	..	21.48	24.06	17.10	16.71	16.51	14.67	18.54	..	..	Current Transfers, net
0.00	0.00	0.00	0.00	0.00	0.00	0.00	0.00	0.00	..	..	Workers' Remittances
..	..	-23.36	-109.63	-38.45	-42.16	25.14	53.58	7.94	..	..	Total to be Financed
..	..	-2.30	-9.00	2.80	0.20	-6.00	-13.30	-13.30	..	..	Official Capital Grants
-31.27	-34.21	-25.66	-118.63	-35.65	-41.96	19.14	40.28	-5.36	-82.80	..	Current Account Balance
14.17	-11.78	15.46	90.94	18.10	37.34	-26.45	14.72	-3.17	..	..	Long-Term Capital, net
8.95	4.97	0.94	7.21	4.03	1.49	-1.54	-5.17	4.96	..	..	Direct Investment
15.00	4.10	29.70	79.70	53.20	68.90	26.20	48.20	88.50	-8.20	..	Long-Term Loans
19.40	10.90	38.40	87.10	60.40	80.20	39.20	72.00	114.10	30.20	..	Disbursements
4.40	6.80	8.70	7.40	7.20	11.30	13.00	23.80	25.60	38.40	..	Repayments
-9.78	-20.86	-12.88	13.03	-41.93	-33.25	-45.11	-15.01	-83.33	..	..	Other Long-Term Capital
38.13	50.77	27.83	48.94	19.42	0.19	-6.28	-32.73	28.69	97.44	..	Other Capital, net
-21.03	-4.78	-17.64	-21.24	-1.87	4.44	13.59	-22.27	-20.16	-14.64	..	Change in Reserves
				(Barbados Dollars per US dollar)							Memo Item:
2.010	2.010	2.010	2.010	2.010	2.010	2.010	2.010	2.010	2.010	..	Conversion Factor (Annual Avg)
				(Millions of US dollars, outstanding at end of year)							**EXTERNAL DEBT, ETC.**
64.2	68.7	97.8	175.5	226.0	292.9	305.4	360.8	465.5	500.5	..	Public/Publicly Guar. Long-Term
38.6	47.6	60.6	94.0	120.7	164.5	167.7	184.5	204.3	236.7	..	Official Creditors
0.0	0.0	0.9	4.5	12.6	17.9	19.6	31.4	39.3	47.5	..	IBRD and IDA
25.6	21.1	37.2	81.5	105.4	128.4	137.7	176.3	261.1	263.8	..	Private Creditors
..	..	..	..	..	..	..	..	..	..	..	Private Non-guaranteed Long-Term
8.5	8.6	2.9	0.9	24.5	37.4	42.7	47.9	39.7	22.1	..	Use of Fund Credit
26.0	79.0	65.0	57.0	85.0	249.0	42.0	48.0	107.0	98.0	..	Short-Term Debt
				(Millions of US dollars)							Memo Items:
59.84	66.12	78.92	100.56	121.60	123.28	132.52	139.77	151.71	145.21	..	Int'l Reserves Excluding Gold
..	..	1.65	2.41	2.77	2.31	1.87	1.98	2.37	2.93	..	Gold Holdings (at market price)
											SOCIAL INDICATORS
2.2	2.1	2.1	2.0	1.9	1.9	1.9	1.8	1.8	1.8	..	Total Fertility Rate
16.3	16.7	16.6	18.0	18.0	17.8	16.7	16.7	16.6	16.6	..	Crude Birth Rate
13.9	13.7	13.6	13.5	13.4	13.2	13.1	13.4	13.6	13.9	..	Infant Mortality Rate
71.5	71.8	72.1	72.4	72.6	73.1	73.5	74.0	74.4	74.9	..	Life Expectancy at Birth
90.7	97.5	108.8	93.7	85.2	79.3	83.4	81.0	86.2	72.5	..	Food Production, p.c. ('79-81 = 100)
11.5	10.7	9.9	8.4	8.4	7.8	8.5	7.9	7.9	..	..	Labor Force, Agriculture (%)
43.6	45.3	45.8	44.8	45.6	45.9	45.6	46.6	47.3	47.3	..	Labor Force, Female (%)
122.0	115.0	109.0	107.0	102.0	108.0	110.0	..	..	..	..	School Enroll. Ratio, primary
81.0	85.0	88.0	88.0	89.0	89.0	93.0	..	..	..	..	School Enroll. Ratio, secondary

BELGIUM	1967	1968	1969	1970	1971	1972	1973	1974	1975	1976	1977
CURRENT GNP PER CAPITA (US $)	2,030	2,210	2,450	2,710	2,950	3,390	4,140	5,170	6,050	6,980	7,780
POPULATION (thousands)	9,557	9,590	9,613	9,638	9,673	9,709	9,738	9,768	9,795	9,811	9,822

ORIGIN AND USE OF RESOURCES
(Billions of current Belgian Franc)

	1967	1968	1969	1970	1971	1972	1973	1974	1975	1976	1977
Gross National Product (GNP)	977.3	1,045.9	1,159.7	1,291.8	1,411.9	1,580.7	1,791.6	2,102.7	2,325.8	2,649.9	2,858.6
Net Factor Income from Abroad	7.6	8.4	8.4	10.9	9.5	12.2	9.3	11.8	12.7	17.3	12.1
Gross Domestic Product (GDP)	969.7	1,037.5	1,151.3	1,280.9	1,402.4	1,568.5	1,782.3	2,090.9	2,313.1	2,632.5	2,846.5
Indirect Taxes, net	117.2	119.9	126.6	148.2	159.7	165.1	181.2	215.6	233.6	271.1	294.0
GDP at factor cost	852.5	917.6	1,024.7	1,132.7	1,242.7	1,403.4	1,601.2	1,875.3	2,079.6	2,361.4	2,552.6
Agriculture	42.5	46.7	51.2	46.0	49.3	63.5	69.3	62.2	68.2	78.6	71.4
Industry	396.2	420.6	479.0	541.5	570.2	631.8	719.8	859.4	889.8	1,007.4	1,075.9
Manufacturing	291.3	317.1	368.0	411.3	424.8	475.7	544.2	644.0	634.2	717.1	758.4
Services, etc.	531.0	570.2	621.1	693.4	782.9	873.3	993.3	1,169.3	1,355.1	1,546.6	1,699.2
Resource Balance	1.4	1.4	9.3	29.3	28.6	50.7	35.7	7.2	4.1	4.4	-25.7
Exports of Goods & NFServices	352.1	401.5	482.6	561.8	608.9	683.6	847.0	1,116.1	1,065.6	1,265.4	1,474.0
Imports of Goods & NFServices	350.7	400.1	473.3	532.4	580.4	632.9	811.3	1,108.8	1,061.4	1,261.0	1,499.6
Domestic Absorption	968.3	1,036.1	1,142.0	1,251.6	1,373.8	1,517.8	1,746.6	2,083.6	2,309.0	2,628.2	2,872.2
Private Consumption, etc.	610.7	661.9	717.5	767.9	846.5	948.1	1,083.6	1,256.2	1,421.4	1,611.2	1,767.5
General Gov't Consumption	133.4	144.0	159.8	175.3	202.0	232.4	264.3	314.3	388.4	441.0	489.1
Gross Domestic Investment	224.2	230.2	264.7	308.4	325.4	337.3	398.7	513.1	499.2	575.9	615.7
Fixed Investment	218.2	218.4	240.6	286.5	298.7	324.6	368.9	452.1	501.5	561.5	601.9
Memo Items:											
Gross Domestic Saving	225.6	231.6	274.1	337.7	353.9	388.0	434.4	520.3	503.3	580.3	590.0
Gross National Saving	233.2	240.0	282.5	..	..	..	..	..	..	..	..

(Billions of 1980 Belgian Franc)

	1967	1968	1969	1970	1971	1972	1973	1974	1975	1976	1977
Gross National Product	2,174.3	2,267.4	2,414.7	2,571.2	2,661.7	2,807.8	2,967.4	3,092.2	3,048.3	3,224.6	3,235.7
GDP at factor cost	1,888.6	1,982.2	2,125.5	2,244.0	2,334.4	2,482.7	2,639.0	2,748.4	2,719.2	2,867.5	2,881.7
Agriculture	70.7	74.8	73.3	72.0	75.9	75.4	78.4	81.8	68.1	65.3	68.0
Industry	734.1	758.3	825.7	906.5	942.5	1,004.6	1,082.5	1,126.9	1,080.6	1,157.1	1,175.6
Manufacturing	478.9	513.3	573.2	628.8	654.0	705.5	768.1	796.8	745.2	800.8	806.0
Services, etc.	1,352.2	1,415.7	1,497.7	1,570.2	1,625.0	1,705.4	1,790.5	1,865.7	1,882.6	1,980.8	1,978.2
Resource Balance	-41.3	-43.5	-50.0	-24.5	-14.3	2.6	-60.1	-75.3	-54.4	-51.0	-100.6
Exports of Goods & NFServices	753.8	858.2	983.8	1,082.4	1,151.1	1,265.9	1,447.8	1,545.8	1,407.6	1,571.9	1,771.6
Imports of Goods & NFServices	795.1	901.6	1,033.8	1,106.9	1,165.4	1,263.3	1,507.9	1,621.1	1,462.0	1,622.9	1,872.2
Domestic Absorption	2,198.4	2,292.3	2,446.7	2,573.3	2,657.6	2,782.8	3,011.5	3,149.6	3,085.7	3,254.2	3,322.3
Private Consumption, etc.	1,297.7	1,378.1	1,461.7	1,528.4	1,600.6	1,699.6	1,831.3	1,881.0	1,892.2	1,985.5	2,033.7
General Gov't Consumption	374.6	387.9	411.6	425.1	450.1	476.7	502.2	521.2	545.6	567.5	582.7
Gross Domestic Investment	526.1	526.3	573.3	619.9	607.0	606.5	678.0	747.5	647.9	701.2	706.0
Fixed Investment	520.7	512.9	541.6	589.4	577.8	594.4	636.0	683.0	670.4	695.8	696.6
Memo Items:											
Capacity to Import	798.2	904.8	1,054.1	1,167.9	1,222.8	1,364.4	1,574.3	1,631.7	1,467.7	1,628.5	1,840.2
Terms of Trade Adjustment	44.4	46.6	70.4	85.5	71.7	98.5	126.5	85.9	60.1	56.6	68.6
Gross Domestic Income	2,201.5	2,295.4	2,467.1	2,634.3	2,715.0	2,884.0	3,077.9	3,160.2	3,091.4	3,259.9	3,290.3
Gross National Income	2,218.7	2,314.0	2,485.1	2,656.7	2,733.4	2,906.4	3,093.9	3,178.1	3,108.4	3,281.3	3,304.3

DOMESTIC PRICES (DEFLATORS)
(Index 1980 = 100)

	1967	1968	1969	1970	1971	1972	1973	1974	1975	1976	1977
Overall (GDP)	45.0	46.1	48.0	50.3	53.1	56.3	60.4	68.0	76.3	82.2	88.4
Domestic Absorption	44.0	45.2	46.7	48.6	51.7	54.5	58.0	66.2	74.8	80.8	86.5
Agriculture	60.1	62.5	69.8	63.8	65.0	84.1	88.5	76.1	100.2	120.3	105.0
Industry	54.0	55.5	58.0	59.7	60.5	62.9	66.5	76.3	82.3	87.1	91.5
Manufacturing	60.8	61.8	64.2	65.4	65.0	67.4	70.9	80.8	85.1	89.5	94.1

MANUFACTURING ACTIVITY

	1967	1968	1969	1970	1971	1972	1973	1974	1975	1976	1977
Employment (1980=100)	122.6	121.1	125.5	131.1	131.6	130.0	125.5	126.8	119.0	114.2	109.8
Real Earnings per Empl. (1980=100)	56.4	58.8	61.1	63.1	67.7	72.3	77.9	82.6	85.2	89.6	92.1
Real Output per Empl. (1980=100)	32.5	37.8	44.3	50.6	55.1	57.8	68.9	76.3	71.9	82.2	83.8
Earnings as % of Value Added	51.3	49.3	46.7	46.4	48.4	47.5	46.1	46.9	49.7	49.1	48.8

MONETARY HOLDINGS
(Billions of current Belgian Franc)

	1967	1968	1969	1970	1971	1972	1973	1974	1975	1976	1977
Money Supply, Broadly Defined	597.8	652.7	689.2	741.2	838.6	971.0	1,101.0	1,200.2	1,384.4	1,554.1	1,686.8
Money as Means of Payment	350.5	375.9	353.5	378.1	420.3	484.2	520.3	552.6	639.4	684.1	741.0
Currency Ouside Banks	173.6	178.8	178.1	183.2	196.4	216.8	231.5	248.9	281.2	299.5	327.4
Demand Deposits	176.9	197.1	175.4	194.9	223.9	267.4	288.8	303.7	358.2	384.6	413.6
Quasi-Monetary Liabilities	247.3	276.8	335.7	363.1	418.3	486.8	580.7	647.6	745.0	870.0	945.8

GOVERNMENT DEFICIT (-) OR SURPLUS
(Billions of current Belgian Franc)

	1967	1968	1969	1970	1971	1972	1973	1974	1975	1976	1977
GOVERNMENT DEFICIT (-) OR SURPLUS	..	..	..	-21.2	-39.6	-67.9	-61.8	-46.6	-109.0	-147.5	-168.3
Current Revenue	..	..	..	448.9	495.1	555.7	649.6	780.5	927.3	1,059.6	1,183.0
Current Expenditure	..	..	..	416.7	463.0	545.3	634.5	742.4	939.8	1,091.4	1,225.5
Current Budget Balance	..	..	..	32.2	32.1	10.4	15.1	38.1	-12.5	-31.8	-42.5
Capital Receipts	..	..	..	0.5	0.6	0.3	0.5	1.1	0.6	0.4	0.5
Capital Payments	..	..	..	53.9	72.3	78.6	77.4	85.8	97.1	116.1	126.3

1978	1979	1980	1981	1982	1983	1984	1985	1986	1987 estimate	NOTES	BELGIUM
8,980	10,690	12,440	12,050	10,780	9,370	8,730	8,540	9,490	11,480	..	CURRENT GNP PER CAPITA (US $)
9,830	9,837	9,847	9,852	9,856	9,856	9,853	9,857	9,859	9,860	..	POPULATION (thousands)
				(Billions of current Belgian Franc)							ORIGIN AND USE OF RESOURCES
3,067.5	3,261.0	3,507.4	3,635.4	3,940.2	4,178.6	4,497.6	4,811.7	5,132.6	5,300.5	..	Gross National Product (GNP)
10.1	-3.5	-18.3	-23.1	-38.3	-36.7	-27.8	-31.3	-14.9	-12.0	..	Net Factor Income from Abroad
3,057.4	3,264.5	3,525.7	3,658.5	3,978.5	4,215.3	4,525.4	4,843.0	5,147.5	5,312.5	..	Gross Domestic Product (GDP)
317.1	331.9	361.8	380.4	422.4	450.4	462.4	491.2	505.1	537.1	..	Indirect Taxes, net
2,740.3	2,932.6	3,163.9	3,278.0	3,556.1	3,764.9	4,063.0	4,351.8	4,642.4	4,775.3	B	GDP at factor cost
78.6	75.6	79.4	88.2	97.8	109.7	112.3	112.4	114.2	..	..	Agriculture
1,138.1	1,197.1	1,253.0	1,201.6	1,318.1	1,378.7	1,445.1	1,541.6	1,637.9	..	..	Industry
798.6	841.5	861.5	830.9	933.6	975.3	1,021.9	1,087.5	1,163.9	..	..	Manufacturing
1,840.6	1,991.7	2,193.3	2,368.7	2,562.6	2,726.9	2,967.9	3,189.0	3,395.4	..	..	Services, etc.
-30.1	-69.6	-98.3	-72.5	-56.5	56.4	56.6	91.6	168.9	154.3	..	Resource Balance
1,540.3	1,796.6	2,026.4	2,284.9	2,637.6	2,919.3	3,333.1	3,467.1	3,360.6	3,339.9	..	Exports of Goods & NFServices
1,570.4	1,866.2	2,124.7	2,357.4	2,694.1	2,862.9	3,276.5	3,375.6	3,191.7	3,185.6	..	Imports of Goods & NFServices
3,087.5	3,334.1	3,624.0	3,731.0	4,035.0	4,158.9	4,468.8	4,751.4	4,978.6	5,158.2	..	Domestic Absorption
1,888.2	2,057.8	2,224.5	2,389.9	2,616.1	2,752.1	2,961.9	3,177.0	3,305.1	3,443.9	..	Private Consumption, etc.
544.0	588.1	643.6	702.8	740.6	761.3	799.7	852.5	881.3	872.5	..	General Gov't Consumption
655.3	688.2	755.8	638.3	678.4	645.5	707.1	722.0	792.3	841.8	..	Gross Domestic Investment
644.9	655.5	728.4	657.1	691.0	696.8	750.0	795.9	859.3	916.1	..	Fixed Investment
											Memo Items:
625.2	618.6	657.5	565.8	621.9	701.9	763.8	813.6	961.2	996.1	..	Gross Domestic Saving
..	..	..	..	..	..	..	..	..	..	..	Gross National Saving
				(Billions of 1980 Belgian Franc)							
3,325.2	3,381.8	3,507.4	3,459.8	3,500.6	3,507.7	3,587.7	3,637.2	3,737.4	3,806.4	..	Gross National Product
2,962.3	3,032.9	3,163.9	3,127.7	3,167.6	3,167.6	3,248.5	3,295.3	3,377.8	3,425.8	B	GDP at factor cost
75.9	76.2	79.4	82.7	87.4	85.3	92.9	91.8	95.6	..	..	Agriculture
1,197.3	1,220.0	1,253.0	1,195.0	1,214.5	1,244.6	1,258.6	1,282.8	1,306.4	..	..	Industry
818.9	840.8	861.5	852.3	885.4	921.3	940.7	959.3	977.5	..	..	Manufacturing
2,040.8	2,089.4	2,193.3	2,203.6	2,231.9	2,208.0	2,257.9	2,285.9	2,346.3	..	..	Services, etc.
-109.7	-156.9	-98.3	13.4	35.8	119.5	121.2	127.3	96.7	76.8	..	Resource Balance
1,831.5	1,961.4	2,026.4	2,088.6	2,134.0	2,201.6	2,327.6	2,355.4	2,487.5	2,587.0	..	Exports of Goods & NFServices
1,941.2	2,118.3	2,124.7	2,075.2	2,098.2	2,082.1	2,206.4	2,228.1	2,390.8	2,510.3	..	Imports of Goods & NFServices
3,423.7	3,542.5	3,624.0	3,467.9	3,498.0	3,418.3	3,488.2	3,533.2	3,651.7	3,738.3	..	Domestic Absorption
2,084.4	2,184.3	2,224.5	2,214.2	2,252.6	2,222.2	2,254.7	2,296.5	2,358.0	2,419.2	..	Private Consumption, etc.
617.0	633.6	643.6	651.3	641.2	640.5	642.8	654.9	660.3	652.4	..	General Gov't Consumption
722.3	724.5	755.8	602.4	604.2	555.7	590.7	581.8	633.3	666.6	..	Gross Domestic Investment
714.9	696.6	728.4	610.7	598.8	573.0	585.5	591.7	630.5	663.8	..	Fixed Investment
											Memo Items:
1,903.9	2,039.3	2,026.4	2,011.4	2,054.2	2,123.1	2,244.5	2,288.5	2,517.3	2,631.9	..	Capacity to Import
72.4	77.9	0.0	-77.2	-79.8	-78.5	-83.1	-66.9	29.8	44.8	..	Terms of Trade Adjustment
3,386.4	3,463.5	3,525.7	3,404.1	3,454.0	3,459.4	3,526.4	3,593.6	3,778.2	3,859.9	..	Gross Domestic Income
3,397.6	3,459.8	3,507.4	3,382.6	3,420.8	3,429.2	3,504.7	3,570.4	3,767.2	..	..	Gross National Income
				(Index 1980 = 100)							DOMESTIC PRICES (DEFLATORS)
92.3	96.4	100.0	105.1	112.6	119.2	125.4	132.3	137.3	139.3	..	Overall (GDP)
90.2	94.1	100.0	107.6	115.4	121.7	128.1	134.5	136.3	138.0	..	Domestic Absorption
103.6	99.3	100.0	106.6	111.9	128.7	120.9	122.5	119.4	..	..	Agriculture
95.1	98.1	100.0	100.6	108.5	110.8	114.8	120.2	125.4	..	..	Industry
97.5	100.1	100.0	97.5	105.4	105.9	108.6	113.4	119.1	..	..	Manufacturing
											MANUFACTURING ACTIVITY
105.1	102.3	100.0	95.0	91.6	89.1	88.4	86.8	..	..	..	Employment (1980=100)
94.3	97.8	100.0	100.6	99.7	97.3	96.0	95.3	..	..	..	Real Earnings per Empl. (1980=100)
85.1	91.3	100.0	102.0	109.3	114.1	116.5	119.6	..	..	..	Real Output per Empl. (1980=100)
48.8	49.1	49.9	53.2	49.4	47.7	46.5	46.2	..	..	..	Earnings as % of Value Added
				(Billions of current Belgian Franc)							MONETARY HOLDINGS
1,818.7	1,935.6	1,996.6	2,110.3	2,237.2	2,411.7	2,527.4	2,699.6	2,931.3	3,150.0	D	Money Supply, Broadly Defined
784.4	804.1	806.1	823.5	855.8	929.8	932.5	962.4	1,037.7	1,088.0	..	Money as Means of Payment
349.7	359.0	364.2	370.1	369.5	383.3	381.9	379.9	400.5	410.7	..	Currency Ouside Banks
434.7	445.1	441.9	453.4	486.3	546.5	550.6	582.5	637.2	677.3	..	Demand Deposits
1,034.3	1,131.5	1,190.5	1,286.8	1,381.4	1,481.9	1,594.9	1,737.2	1,893.6	2,062.0	..	Quasi-Monetary Liabilities
				(Billions of current Belgian Franc)							GOVERNMENT DEFICIT (-) OR SURPLUS
-209.2	-247.7	-269.0	-447.0	-432.8	-513.8	-619.7	-558.9	-543.7	..	..	GOVERNMENT DEFICIT (-) OR SURPLUS
1,300.1	1,397.6	1,523.4	1,597.7	1,781.7	1,860.5	2,018.8	2,181.7	2,234.7	..	..	Current Revenue
1,360.1	1,494.4	1,609.8	1,840.6	2,010.6	2,166.8	2,325.4	2,457.1	2,551.6	..	..	Current Expenditure
-60.0	-96.8	-86.4	-242.9	-228.9	-306.3	-306.6	-275.4	-316.9	..	..	Current Budget Balance
0.6	0.5	0.6	0.7	1.9	1.7	0.7	1.1	0.9	..	..	Capital Receipts
149.8	151.4	183.2	204.8	205.8	209.2	313.8	284.6	227.7	..	..	Capital Payments

BELGIUM	1967	1968	1969	1970	1971	1972	1973	1974	1975	1976	1977
FOREIGN TRADE (CUSTOMS BASIS)					*(Millions of current US dollars)*						
Value of Exports, fob	7,032	8,164	10,065	11,609	12,391	16,044	22,393	28,126	28,760	32,783	37,450
Nonfuel Primary Products	1,573	1,816	2,160	2,451	2,330	2,914	4,366	5,286	4,873	5,559	6,447
Fuels	195	255	332	310	332	466	639	938	1,393	1,572	2,012
Manufactures	5,264	6,093	7,572	8,849	9,729	12,663	17,388	21,902	22,495	25,652	28,991
Value of Imports, cif	7,176	8,333	9,989	11,362	12,856	15,589	21,916	29,446	30,191	35,161	40,140
Nonfuel Primary Products	2,471	2,849	3,180	3,577	3,530	4,125	6,086	7,720	7,215	8,514	9,346
Fuels	615	793	893	1,035	1,223	1,567	1,900	4,219	4,323	4,982	5,637
Manufactures	4,090	4,691	5,915	6,751	8,103	9,897	13,931	17,507	18,652	21,666	25,157
					(Index 1980 = 100)						
Terms of Trade	146.9	148.0	127.7	146.0	129.6	139.6	134.9	108.3	107.1	105.0	101.6
Export Prices, fob	30.5	29.6	27.0	31.5	31.2	36.3	48.3	59.9	60.9	61.8	64.8
Nonfuel Primary Products	37.1	38.2	42.2	42.5	39.0	43.1	65.3	76.3	64.5	65.6	64.9
Fuels	8.5	7.3	5.4	5.9	7.0	8.5	24.5	36.6	33.6	38.5	36.8
Manufactures	31.8	31.5	29.1	34.2	33.5	39.7	46.9	58.5	63.3	63.3	68.4
Import Prices, cif	20.7	20.0	21.1	21.6	24.1	26.0	35.8	55.4	56.9	58.8	63.8
Trade at Constant 1980 Prices					*(Millions of 1980 US dollars)*						
Exports, fob	23,079	27,603	37,310	36,880	39,744	44,158	46,396	46,928	47,242	53,088	57,810
Imports, cif	34,589	41,694	47,269	52,702	53,444	59,915	61,274	53,192	53,096	59,773	62,948
BALANCE OF PAYMENTS					*(Millions of current US dollars)*						
Exports of Goods & Services	..	..	..	12,358	13,596	17,173	24,700	33,734	33,811	37,337	46,227
Merchandise, fob	..	..	..	9,062	9,800	12,789	18,392	24,215	23,067	26,066	31,219
Nonfactor Services	..	..	..	2,176	2,501	2,920	3,937	5,266	6,176	6,736	9,619
Factor Services	..	..	..	1,120	1,295	1,463	2,371	4,252	4,568	4,535	5,389
Imports of Goods & Services	..	..	..	11,487	12,791	15,665	23,028	32,523	33,025	36,385	46,164
Merchandise, fob	..	..	..	8,598	9,589	11,998	17,545	24,193	23,905	27,310	34,200
Nonfactor Services	..	..	..	1,857	2,040	2,431	3,366	4,439	5,063	5,209	7,158
Factor Services	..	..	..	1,032	1,161	1,236	2,117	3,891	4,057	3,865	4,805
Long-Term Interest	..	..	..	..	..	..	..	..	..	..	..
Current Transfers, net	..	..	..	33	44	34	62	10	-109	-196	-163
Workers' Remittances	..	..	..	154	186	199	280	294	286	264	322
Total to be Financed	..	..	..	904	849	1,542	1,734	1,221	677	757	-100
Official Capital Grants	..	..	..	-187	-206	-234	-370	-389	-504	-324	-459
Current Account Balance	..	..	..	717	643	1,308	1,363	831	173	433	-558
Long-Term Capital, net	..	..	..	-550	-752	-1,132	-1,100	-601	-772	-177	-758
Direct Investment	..	..	..	140	221	209	451	658	723	520	811
Long-Term Loans	..	..	..	..	..	..	..	..	..	..	..
Disbursements	..	..	..	..	..	..	..	..	..	..	..
Repayments	..	..	..	..	..	..	..	..	..	..	..
Other Long-Term Capital	..	..	..	-503	-767	-1,107	-1,180	-869	-990	-372	-1,110
Other Capital, net	..	..	..	64	489	412	738	-782	516	-1,494	1,070
Change in Reserves	..	..	..	-230	-380	-587	-1,002	551	83	1,238	246
Memo Item:					*(Belgian Franc per US dollar)*						
Conversion Factor (Annual Avg)	50.000	50.000	50.000	50.000	48.870	44.010	38.980	38.950	36.780	38.610	35.840
EXTERNAL DEBT, ETC.					*(Millions of US dollars, outstanding at end of year)*						
Public/Publicly Guar. Long-Term	..	..	..	..	..	..	..	..	..	..	..
Official Creditors	..	..	..	..	..	..	..	..	..	..	..
IBRD and IDA	..	..	..	..	..	..	..	..	..	..	..
Private Creditors	..	..	..	..	..	..	..	..	..	..	..
Private Non-guaranteed Long-Term	..	..	..	..	..	..	..	..	..	..	..
Use of Fund Credit	..	..	..	..	..	..	..	..	..	..	..
Short-Term Debt	..	..	..	..	..	..	..	..	..	..	..
Memo Items:					*(Millions of US dollars)*						
Int'l Reserves Excluding Gold	1,110	663	868	1,377	1,797	2,232	3,319	3,538	4,069	3,491	3,956
Gold Holdings (at market price)	1,488	1,824	1,528	1,570	1,925	2,796	4,734	7,865	5,915	5,683	7,003
SOCIAL INDICATORS											
Total Fertility Rate	2.4	2.3	2.2	2.2	2.2	2.1	1.9	1.8	1.7	1.7	1.7
Crude Birth Rate	15.3	14.8	14.8	14.8	14.6	14.0	13.3	12.7	12.2	12.3	12.4
Infant Mortality Rate	22.9	21.7	21.2	21.1	20.4	18.8	17.7	17.4	16.1	15.3	13.6
Life Expectancy at Birth	71.0	71.0	71.1	71.2	71.2	71.3	71.4	71.8	71.9	72.1	72.3
Food Production, p.c. ('79-81=100)	..	..	..	..	..	..	..	..	..	..	..
Labor Force, Agriculture (%)	5.8	5.4	5.1	4.8	4.6	4.4	4.2	4.0	3.8	3.6	3.4
Labor Force, Female (%)	29.3	29.7	30.1	30.5	30.8	31.2	31.5	31.9	32.2	32.6	32.9
School Enroll. Ratio, primary	..	..	..	103.0	..	..	..	..	102.0	101.0	102.0
School Enroll. Ratio, secondary	..	..	..	81.0	..	..	..	..	84.0	85.0	85.0

1978	1979	1980	1981	1982	1983	1984	1985	1986	1987 estimate	NOTES	BELGIUM
											FOREIGN TRADE (CUSTOMS BASIS)
				(Millions of current US dollars)							
44,793	56,083	63,967	55,228	51,695	51,676	51,416	53,316	68,649	82,951	f	Value of Exports, fob
7,382	9,201	11,217	9,693	8,994	8,707	9,108	9,017	11,013	13,175	..	Nonfuel Primary Products
1,912	3,526	5,310	4,961	4,440	4,292	4,016	3,439	3,295	3,058	..	Fuels
35,498	43,357	47,440	40,574	38,261	38,676	38,293	40,860	54,342	66,718	..	Manufactures
48,268	60,186	71,192	61,417	57,213	53,653	54,746	55,561	68,025	82,598	f	Value of Imports, cif
10,733	13,536	15,764	13,361	12,253	11,828	12,148	11,933	13,667	15,997	..	Nonfuel Primary Products
5,986	8,555	12,361	12,523	11,886	9,450	10,242	9,257	7,233	7,688	..	Fuels
31,549	38,095	43,067	35,533	33,074	32,376	32,357	34,371	47,126	58,913	..	Manufactures
				(Index 1980 = 100)							
105.4	106.3	100.0	89.4	87.8	87.5	84.6	86.7	101.3	84.9	f	Terms of Trade
75.1	91.3	100.0	87.5	80.8	78.2	73.6	73.7	84.9	104.9	f	Export Prices, fob
73.0	94.2	100.0	91.5	82.2	85.4	82.0	74.6	76.3	81.6	..	Nonfuel Primary Products
44.7	93.6	100.0	104.0	97.0	85.4	83.6	81.1	41.3	..	..	Fuels
78.5	90.5	100.0	85.0	79.0	76.0	71.0	73.0	93.0	106.0	..	Manufactures
71.3	85.9	100.0	97.9	92.0	89.4	87.1	85.0	83.8	123.6	f	Import Prices, cif
				(Millions of 1980 US dollars)							Trade at Constant 1980 Prices
59,625	61,436	63,967	63,093	63,950	66,107	69,840	72,302	80,830	79,087	f	Exports, fob
67,733	70,099	71,192	62,732	62,168	60,038	62,885	65,355	81,131	66,842	f	Imports, cif
				(Millions of current US dollars)							**BALANCE OF PAYMENTS**
56,390	73,936	88,942	87,239	82,093	76,668	77,354	80,684	101,509	..	f	Exports of Goods & Services
38,170	49,180	55,173	48,529	46,070	45,641	46,176	47,257	60,283	..	..	Merchandise, fob
10,709	13,061	15,157	14,269	12,973	12,491	11,999	12,399	17,305	..	..	Nonfactor Services
7,511	11,696	18,613	24,441	23,051	18,537	19,180	21,028	23,921	..	..	Factor Services
56,467	76,109	92,653	90,185	83,453	76,047	76,575	79,351	97,043	..	f	Imports of Goods & Services
40,824	53,804	60,323	53,128	49,475	47,449	47,404	47,691	59,176	..	..	Merchandise, fob
8,784	10,957	13,501	12,381	10,525	9,772	9,818	10,372	13,853	..	..	Nonfactor Services
6,859	11,347	18,829	24,676	23,453	18,826	19,354	21,289	24,015	..	..	Factor Services
..	..	..	..	..	..	..	..	..	..	..	Long-Term Interest
-283	-344	-379	-411	-248	-181	-174	-127	-103	..	f	Current Transfers, net
460	471	530	455	388	394	358	384	479	..	..	Workers' Remittances
-360	-2,517	-4,089	-3,357	-1,608	440	604	1,206	4,363	..	K	Total to be Financed
-465	-547	-856	-822	-926	-915	-654	-523	-777	..	K	Official Capital Grants
-825	-3,063	-4,945	-4,178	-2,534	-476	-50	683	3,586	..	..	Current Account Balance
-740	-22	3,132	2,891	558	-3,282	-2,949	-4,945	-7,607	..	f	Long-Term Capital, net
874	-201	1,345	1,268	1,423	941	105	766	-990	..	..	Direct Investment
..	..	..	..	..	..	..	..	..	..	..	Long-Term Loans
..	..	..	..	..	..	..	..	..	..	..	Disbursements
..	..	..	..	..	..	..	..	..	..	..	Repayments
-1,150	726	2,643	2,444	61	-3,308	-2,400	-5,189	-5,840	..	..	Other Long-Term Capital
1,841	-142	1,714	-354	1,199	3,347	3,288	4,641	4,158	..	..	Other Capital, net
-276	3,228	98	1,641	777	411	-290	-378	-138	-3,899	..	Change in Reserves
				(Belgian Franc per US dollar)							Memo Item:
31.490	29.320	29.240	37.130	45.690	51.130	57.780	59.380	44.670	37.330	..	Conversion Factor (Annual Avg)
				(Millions of US dollars, outstanding at end of year)							**EXTERNAL DEBT, ETC.**
..	..	..	..	..	..	..	..	..	..	..	Public/Publicly Guar. Long-Term
..	..	..	..	..	..	..	..	..	..	..	Official Creditors
..	..	..	..	..	..	..	..	..	..	..	IBRD and IDA
..	..	..	..	..	..	..	..	..	..	..	Private Creditors
..	..	..	..	..	..	..	..	..	..	..	Private Non-guaranteed Long-Term
..	..	..	..	..	..	..	..	..	..	..	Use of Fund Credit
..	..	..	..	..	..	..	..	..	..	..	Short-Term Debt
				(Millions of US dollars)							Memo Items:
3,966	5,443	7,823	4,952	3,927	4,714	4,564	4,849	5,538	9,620	..	Int'l Reserves Excluding Gold
9,625	17,516	20,151	13,588	15,618	13,041	10,539	11,178	13,362	16,279	..	Gold Holdings (at market price)
											SOCIAL INDICATORS
1.7	1.7	1.7	1.7	1.6	1.6	1.6	1.5	1.5	1.5	..	Total Fertility Rate
12.5	12.6	12.6	12.6	12.4	12.1	11.9	11.7	11.5	11.3	..	Crude Birth Rate
13.3	11.1	11.0	11.7	11.9	12.1	12.4	12.6	12.9	13.1	..	Infant Mortality Rate
72.5	72.7	72.9	73.1	73.3	73.7	74.1	74.4	74.8	75.2	..	Life Expectancy at Birth
..	..	..	..	..	..	..	..	..	..		Food Production, p.c. ('79-81 = 100)
3.2	3.0	2.8	..	..	..	..	..	..	..	..	Labor Force, Agriculture (%)
33.2	33.6	33.9	33.9	33.9	33.9	33.8	33.8	33.8	33.8	..	Labor Force, Female (%)
101.0	101.0	101.0	..	98.0	97.0	95.0	95.0	96.0	..	..	School Enroll. Ratio, primary
87.0	87.0	88.0	..	89.0	92.0	94.0	96.0	96.0	..	..	School Enroll. Ratio, secondary

135

BELIZE	1967	1968	1969	1970	1971	1972	1973	1974	1975	1976	1977
CURRENT GNP PER CAPITA (US $)	420	430	430	440	470	530	580	670	800	840	870
POPULATION (thousands)	113	115	118	120	122	124	127	129	132	134	137
ORIGIN AND USE OF RESOURCES					*(Millions of current Belize Dollars)*						
Gross National Product (GNP)	68.4	74.2	78.1	88.1	96.0	103.7	124.1	162.0	196.7	202.8	230.2
Net Factor Income from Abroad	-0.3	-1.8	-2.1	-2.4	-2.8	-3.9	-5.9	-17.6	-20.8	-17.1	-9.3
Gross Domestic Product (GDP)	68.7	76.0	80.2	90.5	98.8	107.6	130.0	179.6	217.5	219.9	239.5
Indirect Taxes, net	..	..	..	..	..	..	..	..	..	..	..
GDP at factor cost	..	..	..	..	..	..	..	..	..	..	..
Agriculture	..	..	..	..	..	..	..	..	..	..	..
Industry	..	..	..	..	..	..	..	..	..	..	..
Manufacturing	..	..	..	..	..	..	..	..	..	..	..
Services, etc.	..	..	..	..	..	..	..	..	..	..	..
Resource Balance	..	..	..	..	..	..	..	..	..	..	..
Exports of Goods & NFServices	..	..	..	..	..	..	..	..	..	..	..
Imports of Goods & NFServices	..	..	..	..	..	..	..	..	..	..	..
Domestic Absorption	..	..	..	..	..	..	..	..	..	..	..
Private Consumption, etc.	..	..	..	..	..	..	..	..	..	..	..
General Gov't Consumption	..	..	..	..	..	..	..	..	..	..	..
Gross Domestic Investment	..	..	..	..	..	..	..	..	..	..	..
Fixed Investment	..	..	..	..	..	..	..	..	..	..	..
Memo Items:											
Gross Domestic Saving	..	..	..	..	..	..	..	..	..	..	..
Gross National Saving	..	..	..	..	..	..	..	..	..	..	..
					(Millions of 1980 Belize Dollars)						
Gross National Product	166.34	174.45	183.32	193.42	202.63	210.14	229.56	239.47	244.77	256.79	283.92
GDP at factor cost	..	..	..	195.73	205.39	216.07	226.62	243.07	253.06	245.28	282.11
Agriculture	..	..	..	55.96	59.53	61.58	64.77	70.26	72.69	65.67	87.38
Industry	..	..	..	30.84	32.51	36.50	39.54	45.10	47.93	49.09	54.65
Manufacturing	..	..	..	20.73	23.89	26.35	29.63	34.55	35.26	34.20	41.35
Services, etc.	..	..	..	108.94	113.34	117.99	122.31	127.71	132.44	130.53	140.08
Resource Balance	..	..	..	..	..	..	..	..	..	..	..
Exports of Goods & NFServices	..	..	..	..	..	..	..	..	..	..	..
Imports of Goods & NFServices	..	..	..	..	..	..	..	..	..	..	..
Domestic Absorption	..	..	..	..	..	..	..	..	..	..	..
Private Consumption, etc.	..	..	..	..	..	..	..	..	..	..	..
General Gov't Consumption	..	..	..	..	..	..	..	..	..	..	..
Gross Domestic Investment	..	..	..	..	..	..	..	..	..	..	..
Fixed Investment	..	..	..	..	..	..	..	..	..	..	..
Memo Items:											
Capacity to Import	..	..	..	..	..	..	..	..	..	..	..
Terms of Trade Adjustment	..	..	..	..	..	..	..	..	..	..	..
Gross Domestic Income	..	..	..	..	..	..	..	..	..	..	..
Gross National Income	..	..	..	..	..	..	..	..	..	..	..
DOMESTIC PRICES (DEFLATORS)					*(Index 1980 = 100)*						
Overall (GDP)	39.5	41.6	41.5	44.7	46.6	48.8	53.5	70.1	74.8	73.0	79.6
Domestic Absorption	..	..	..	..	..	..	..	..	..	..	..
Agriculture	..	..	..	..	..	..	..	..	..	..	..
Industry	..	..	..	..	..	..	..	..	..	..	..
Manufacturing	..	..	..	..	..	..	..	..	..	..	..
MANUFACTURING ACTIVITY											
Employment (1980=100)	..	..	..	..	..	..	..	..	..	..	..
Real Earnings per Empl. (1980=100)	..	..	..	..	..	..	..	..	..	..	..
Real Output per Empl. (1980=100)	..	..	..	..	..	..	..	..	..	..	..
Earnings as % of Value Added	..	..	..	..	..	..	..	..	..	..	..
MONETARY HOLDINGS					*(Millions of current Belize Dollars)*						
Money Supply, Broadly Defined	..	..	..	..	..	..	..	..	..	71.6	73.4
Money as Means of Payment	..	..	..	..	..	..	..	..	..	21.3	24.7
Currency Ouside Banks	..	..	..	..	..	..	..	..	..	11.3	12.6
Demand Deposits	..	..	..	..	..	..	..	..	..	10.0	12.1
Quasi-Monetary Liabilities	..	..	..	..	..	..	..	..	..	50.3	48.7
					(Thousands of current Belize Dollars)						
GOVERNMENT DEFICIT (-) OR SURPLUS	..	..	..	..	..	..	..	..	..	..	-12,100
Current Revenue	..	..	..	..	..	..	..	..	..	..	49,900
Current Expenditure	..	..	..	..	..	..	..	..	..	..	38,600
Current Budget Balance	..	..	..	..	..	..	..	..	..	..	11,300
Capital Receipts	..	..	..	..	..	..	..	..	..	..	200
Capital Payments	..	..	..	..	..	..	..	..	..	..	23,600

1978	1979	1980	1981	1982	1983	1984	1985	1986	1987 estimate	NOTES	BELIZE
920	1,090	1,190	1,250	1,170	1,100	1,140	1,150	1,200	1,250	..	**CURRENT GNP PER CAPITA (US $)**
140	143	146	150	154	158	162	166	170	175	..	**POPULATION (thousands)**
				(Millions of current Belize Dollars)							**ORIGIN AND USE OF RESOURCES**
266.0	300.9	342.9	360.2	326.7	335.4	370.2	370.8	410.3	440.0	..	Gross National Product (GNP)
-11.4	-8.8	-11.3	-0.5	-21.5	-19.5	-16.2	-17.0	-5.3	..	..	Net Factor Income from Abroad
277.4	309.7	354.2	360.7	348.2	354.9	386.4	387.8	415.6		..	Gross Domestic Product (GDP)
25.5	32.8	46.1	47.2	45.6	46.5	50.6	50.8	54.4		..	Indirect Taxes, net
251.9	276.9	308.1	313.5	302.6	308.4	335.8	337.0	361.2		..	GDP at factor cost
79.0	85.2	91.6	86.0	67.4	67.1	72.9	70.3	74.7		..	Agriculture
62.1	60.7	68.7	62.7	55.1	55.3	65.3	64.3	75.4		..	Industry
45.1	41.9	46.5	39.3	30.4	37.7	36.8	34.0	39.5		..	Manufacturing
110.8	131.0	147.8	164.8	180.1	186.0	197.6	202.4	211.1		..	Services, etc.
..	..	..	..	..	..	..	..	..	..	..	Resource Balance
..	..	..	..	..	..	..	..	..	..	..	Exports of Goods & NFServices
..	..	..	..	..	..	..	..	..	..	..	Imports of Goods & NFServices
..	..	..	..	..	..	..	..	..	..	..	Domestic Absorption
..	..	..	..	..	..	..	..	..	..	..	Private Consumption, etc.
..	..	..	..	..	..	..	..	..	..	..	General Gov't Consumption
..	..	..	..	..	..	..	..	..	..	..	Gross Domestic Investment
..	..	..	..	..	..	..	..	..	..	..	Fixed Investment
											Memo Items:
..	43.0	40.4	30.4	11.2	10.7	23.9	13.8		..	..	Gross Domestic Saving
..	..	..	..	..	..	39.5	36.0		..	..	Gross National Saving
				(Millions of 1980 Belize Dollars)							
302.14	334.17	342.90	346.81	321.19	305.19	321.56	327.27	341.62	358.74	..	Gross National Product
286.93	296.62	308.10	314.04	320.98	313.75	325.70	331.94	334.30		..	GDP at factor cost
92.24	90.83	91.60	90.45	96.07	90.32	93.13	91.47	87.51		..	Agriculture
59.68	62.30	68.70	67.65	66.81	68.49	68.49	68.18	71.01		..	Industry
43.22	40.99	46.50	44.27	44.16	46.97	43.10	42.63	43.57		..	Manufacturing
135.01	143.48	147.80	155.94	158.10	154.94	164.07	172.30	175.78		..	Services, etc.
..	..	..	..	..	..	..	..	..	..	..	Resource Balance
..	..	..	..	..	..	..	..	..	..	..	Exports of Goods & NFServices
..	..	..	..	..	..	..	..	..	..	..	Imports of Goods & NFServices
..	..	..	..	..	..	..	..	..	..	..	Domestic Absorption
..	..	..	..	..	..	..	..	..	..	..	Private Consumption, etc.
..	..	..	..	..	..	..	..	..	..	..	General Gov't Consumption
..	..	..	..	..	..	..	..	..	..	..	Gross Domestic Investment
..	..	..	..	..	..	..	..	..	..	..	Fixed Investment
											Memo Items:
..	..	..	..	..	..	..	..	..	..	..	Capacity to Import
..	..	..	..	..	..	..	..	..	..	..	Terms of Trade Adjustment
..	..	..	..	..	..	..	..	..	..	..	Gross Domestic Income
..	..	..	..	..	..	..	..	..	..	..	Gross National Income
				(Index 1980 = 100)							**DOMESTIC PRICES (DEFLATORS)**
86.5	90.6	100.0	100.7	98.9	101.1	104.9	103.4	107.0	..	..	Overall (GDP)
..	..	..	..	..	..	..	..	..	..	..	Domestic Absorption
85.6	93.8	100.0	95.1	70.2	74.3	78.3	76.9	85.4	..	..	Agriculture
104.1	97.4	100.0	92.7	82.5	80.7	95.3	94.3	106.2	..	..	Industry
104.3	102.2	100.0	88.8	68.8	80.3	85.4	79.7	90.7	..	..	Manufacturing
											MANUFACTURING ACTIVITY
..	..	..	..	..	..	..	..	..	..	..	Employment (1980=100)
..	..	..	..	..	..	..	..	..	..	..	Real Earnings per Empl. (1980=100)
..	..	..	..	..	..	..	..	..	..	..	Real Output per Empl. (1980=100)
..	..	..	..	..	..	..	..	..	..	..	Earnings as % of Value Added
				(Millions of current Belize Dollars)							**MONETARY HOLDINGS**
92.9	95.9	108.2	117.8	125.1	148.2	154.6	167.7	199.1	241.7	..	Money Supply, Broadly Defined
36.0	36.9	41.7	39.6	39.4	42.4	50.0	58.6	71.0	84.3	..	Money as Means of Payment
16.7	16.7	17.5	19.0	20.6	21.5	22.8	22.6	25.9	29.6	..	Currency Ouside Banks
19.3	20.2	24.1	20.6	18.8	20.9	27.2	36.0	45.1	54.7	..	Demand Deposits
56.9	59.0	66.6	78.3	85.7	105.8	104.6	109.1	128.1	157.4	..	Quasi-Monetary Liabilities
				(Thousands of current Belize Dollars)							
-2,600	-5,000	-5,900	..	..	..	-5,200	-11,000	..	..	C F	**GOVERNMENT DEFICIT (-) OR SURPLUS**
62,000	71,000	85,000	88,800	97,600	95,600	97,900	103,100	..	..	..	Current Revenue
43,700	51,600	65,400	77,900	84,000	84,000	93,600	99,900	..	..	..	Current Expenditure
18,300	19,400	19,600	10,900	13,600	11,600	4,300	3,200	..	..	..	Current Budget Balance
500	700	300	300	300	300	500	200	..	..	..	Capital Receipts
21,400	25,100	25,800	-77,900	-84,000	-84,000	10,000	14,400	..	..	..	Capital Payments

BELIZE	1967	1968	1969	1970	1971	1972	1973	1974	1975	1976	1977
FOREIGN TRADE (CUSTOMS BASIS)					*(Millions of current US dollars)*						
Value of Exports, fob	..	..	..	..	..	..	..	..	..	..	..
Nonfuel Primary Products	..	..	..	..	..	..	..	..	..	..	..
Fuels	..	..	..	..	..	..	..	..	..	..	..
Manufactures	..	..	..	..	..	..	..	..	..	..	..
Value of Imports, cif	..	..	..	..	..	..	..	..	..	..	..
Nonfuel Primary Products	..	..	..	..	..	..	..	..	..	..	..
Fuels	..	..	..	..	..	..	..	..	..	..	..
Manufactures	..	..	..	..	..	..	..	..	..	..	..
					(Index 1980 = 100)						
Terms of Trade	..	..	..	..	..	..	..	..	..	..	..
Export Prices, fob	..	..	..	..	..	..	..	..	..	..	..
Nonfuel Primary Products	..	..	..	..	..	..	..	..	..	..	..
Fuels	..	..	..	..	..	..	..	..	..	..	..
Manufactures	30.0	30.0	32.0	33.0	36.0	40.0	51.0	58.0	64.0	64.0	71.0
Import Prices, cif	..	..	..	..	..	..	..	..	..	..	..
Trade at Constant 1980 Prices					*(Millions of 1980 US dollars)*						
Exports, fob	..	..	..	..	..	..	..	..	..	..	..
Imports, cif	..	..	..	..	..	..	..	..	..	..	..
BALANCE OF PAYMENTS					*(Millions of current US dollars)*						
Exports of Goods & Services	..	..	..	..	..	..	..	..	..	..	..
Merchandise, fob	..	..	..	..	..	..	..	..	..	..	..
Nonfactor Services	..	..	..	..	..	..	..	..	..	..	..
Factor Services	..	..	..	..	..	..	..	..	..	..	..
Imports of Goods & Services	..	..	..	..	..	..	..	..	..	..	..
Merchandise, fob	..	..	..	..	..	..	..	..	..	..	..
Nonfactor Services	..	..	..	..	..	..	..	..	..	..	..
Factor Services	..	..	..	..	..	..	..	..	..	..	..
Long-Term Interest	..	..	..	0	0	0	0	0	0	0	0
Current Transfers, net	..	..	..	..	..	..	..	..	..	..	..
Workers' Remittances	..	..	..	..	..	..	..	..	..	..	..
Total to be Financed	..	..	..	..	..	..	..	..	..	..	..
Official Capital Grants	..	..	..	..	..	..	..	..	..	..	..
Current Account Balance	..	..	..	..	..	..	..	..	..	..	..
Long-Term Capital, net	..	..	..	..	..	..	..	..	..	..	..
Direct Investment	..	..	..	..	..	..	..	..	..	..	..
Long-Term Loans	..	..	..	4	..	..	0	0	0	5	4
Disbursements	..	..	..	4	..	..	..	1	0	5	5
Repayments	..	..	..	..	..	..	0	0	0	0	0
Other Long-Term Capital	..	..	..	..	..	..	..	..	..	..	..
Other Capital, net	..	..	..	..	..	..	..	..	..	..	..
Change in Reserves	..	..	..	0	0	0	0	0	0	0	-3
Memo Item:					*(Belize Dollars per US dollar)*						
Conversion Factor (Annual Avg)	1.450	1.670	1.670	1.670	1.640	1.600	1.630	1.710	1.810	2.230	2.000
EXTERNAL DEBT, ETC.					*(Millions of US dollars, outstanding at end of year)*						
Public/Publicly Guar. Long-Term	..	..	..	4	4	5	5	5	5	9	14
Official Creditors	..	..	..	4	4	5	5	5	5	9	14
IBRD and IDA	..	..	..	0	0	0	0	0	0	0	0
Private Creditors	..	..	..	..	..	..	..	..	..	..	..
Private Non-guaranteed Long-Term	..	..	..	..	..	..	..	..	..	..	..
Use of Fund Credit	..	..	..	0	0	0	0	0	0	0	0
Short-Term Debt	..	..	..	..	..	..	..	..	..	..	7
Memo Items:					*(Thousands of US dollars)*						
Int'l Reserves Excluding Gold	..	..	..	..	..	..	..	..	..	5,491	8,021
Gold Holdings (at market price)	..	..	..	..	..	..	..	..	..	..	..
SOCIAL INDICATORS											
Total Fertility Rate	6.2	5.8	5.6	6.9	6.4	6.3	6.2	6.2	6.1	6.0	5.9
Crude Birth Rate	..	..	..	..	..	38.7	38.7	38.9	39.2	39.4	39.2
Infant Mortality Rate	..	..	..	..	33.7	35.1	36.6	38.1	39.5	40.9	42.4
Life Expectancy at Birth	..	..	..	..	..	..	..	..	..	..	46.0
Food Production, p.c. ('79-81 = 100)	73.0	81.4	73.4	81.9	82.1	83.3	87.6	93.4	85.5	75.6	97.3
Labor Force, Agriculture (%)	..	..	..	..	..	..	..	..	..	..	..
Labor Force, Female (%)	..	..	..	..	..	..	..	..	..	..	..
School Enroll. Ratio, primary	..	..	..	..	..	..	..	..	..	..	..
School Enroll. Ratio, secondary	..	..	..	..	..	..	..	..	..	..	..

1978	1979	1980	1981	1982	1983	1984	1985	1986	1987 estimate	NOTES	BELIZE
				(Millions of current US dollars)							**FOREIGN TRADE (CUSTOMS BASIS)**
..	..	82	75	60	65	73	64	73	..	..	Value of Exports, fob
..	..	..	..	..	..	..	..	..	..	..	Nonfuel Primary Products
..	..	..	..	..	..	..	..	..	..	..	Fuels
..	..	..	..	..	..	..	..	..	..	..	Manufactures
..	..	228	240	234	237	252	240	252	..	..	Value of Imports, cif
..	..	..	..	..	..	..	..	..	..	..	Nonfuel Primary Products
..	..	..	..	..	..	..	..	..	..	..	Fuels
..	..	..	..	..	..	..	..	..	..	..	Manufactures
				(Index 1980 = 100)							
..	..	..	..	..	..	..	..	..	..	..	Terms of Trade
..	..	..	..	..	..	..	..	..	..	..	Export Prices, fob
..	..	..	..	..	..	..	..	..	..	..	Nonfuel Primary Products
..	..	..	..	..	..	..	..	..	..	..	Fuels
83.0	94.0	..	..	..	..	..	..	..	..	..	Manufactures
..	..	..	..	..	..	..	..	..	..	..	Import Prices, cif
				(Millions of 1980 US dollars)							Trade at Constant 1980 Prices
..	..	..	..	..	..	..	..	..	..	..	Exports, fob
..	..	..	..	..	..	..	..	..	..	..	Imports, cif
				(Millions of current US dollars)							**BALANCE OF PAYMENTS**
..	..	..	..	..	..	127	128	131	152	..	Exports of Goods & Services
..	..	..	..	..	..	90	87	75	87	..	Merchandise, fob
..	..	..	..	..	..	35	38	47	57	..	Nonfactor Services
..	..	..	..	..	..	1	3	10	9	..	Factor Services
..	..	..	..	..	..	160	155	160	176	..	Imports of Goods & Services
..	..	..	..	..	..	115	114	117	138	..	Merchandise, fob
..	..	..	..	..	..	30	28	30	24	..	Nonfactor Services
..	..	..	..	..	..	15	13	14	15	..	Factor Services
0	1	1	2	2	2	3	6	5	4	..	Long-Term Interest
..	..	..	..	..	..	16	20	15	18	..	Current Transfers, net
..	..	..	..	..	..	0	0	0	0	..	Workers' Remittances
..	..	..	..	..	..	-17	-8	-14	-7	K	Total to be Financed
..	..	..	..	..	..	9	13	9	12	K	Official Capital Grants
..	..	..	..	..	..	-8	5	-4	6	..	Current Account Balance
..	..	..	..	..	..	9	23	13	15	..	Long-Term Capital, net
..	..	..	..	..	..	-4	4	1	1	..	Direct Investment
9	11	13	11	9	16	6	13	1	5	..	Long-Term Loans
9	11	14	13	12	19	8	22	9	12	..	Disbursements
0	0	1	1	2	3	2	9	8	7	..	Repayments
..	..	..	..	..	..	-3	-6	2	-4	..	Other Long-Term Capital
..	..	..	..	..	..	-6	-26	5	-8	..	Other Capital, net
-6	4	-2	2	1	4	4	-2	-14	-12	..	Change in Reserves
				(Belize Dollars per US dollar)							Memo Item:
2.000	2.000	2.000	2.000	2.000	2.000	2.000	2.000	2.000	2.000	..	Conversion Factor (Annual Avg)
				(Millions of US dollars, outstanding at end of year)							**EXTERNAL DEBT, ETC.**
22	34	47	56	62	76	76	95	99	113	..	Public/Publicly Guar. Long-Term
22	33	40	46	48	58	57	79	87	100	..	Official Creditors
0	0	0	0	0	1	2	5	7	9	..	IBRD and IDA
..	1	7	10	15	18	18	16	12	13	..	Private Creditors
..	..	..	..	..	..	..	..	..	..	..	Private Non-guaranteed Long-Term
0	0	0	0	0	4	5	11	12	11	..	Use of Fund Credit
11	38	16	4	7	21	16	13	11	14	..	Short-Term Debt
				(Thousands of US dollars)							Memo Items:
13,978	10,463	12,676	10,335	9,838	9,306	6,072	14,813	26,897	36,415	..	Int'l Reserves Excluding Gold
..	..	..	..	..	..	..	..	..	..	..	Gold Holdings (at market price)
											SOCIAL INDICATORS
5.8	5.7	5.8	5.4	5.4	5.2	5.1	5.0	4.9	5.0	..	Total Fertility Rate
39.1	38.9	38.8	38.6	38.5	38.3	38.2	38.0	36.1	36.8	..	Crude Birth Rate
43.8	45.3	46.8	48.2	49.6	51.1	52.6	54.0	..	..	..	Infant Mortality Rate
48.1	50.1	52.2	54.3	56.4	58.4	60.5	62.6	64.7	66.7	..	Life Expectancy at Birth
106.7	95.3	103.2	101.5	106.8	103.6	98.2	94.6	90.8	90.3	..	Food Production, p.c. ('79-81 = 100)
..	..	..	..	..	..	..	..	..	..	..	Labor Force, Agriculture (%)
..	..	22.8	..	..	32.5	32.5	32.6	32.5	..	..	Labor Force, Female (%)
..	..	..	..	..	..	..	..	..	..	..	School Enroll. Ratio, primary
..	..	..	..	..	..	..	..	..	..	..	School Enroll. Ratio, secondary

BENIN	1967	1968	1969	1970	1971	1972	1973	1974	1975	1976	1977
CURRENT GNP PER CAPITA (US $)	100	110	110	110	100	110	130	150	170	180	200
POPULATION (thousands)	2,457	2,522	2,589	2,657	2,728	2,800	2,874	2,951	3,029	3,109	3,192
ORIGIN AND USE OF RESOURCES				*(Billions of current CFA Francs)*							
Gross National Product (GNP)	61.4	65.6	69.5	74.6	74.4	83.9	91.8	108.4	118.1	136.2	150.3
Net Factor Income from Abroad	-0.1	-0.4	-0.7	-0.7	-0.9	-0.6	0.0	-0.6	-0.3	0.0	-0.1
Gross Domestic Product (GDP)	61.5	65.9	70.2	75.3	75.3	84.4	91.8	109.0	118.4	136.2	150.4
Indirect Taxes, net	4.9	7.5	7.5	7.5	8.3	8.7	9.2	9.1	10.2	13.3	15.9
GDP at factor cost	56.6	58.4	62.6	67.7	67.0	75.7	82.5	99.9	108.2	122.9	134.5
Agriculture	31.7	31.4	33.1	33.5	31.7	35.4	37.9	42.2	44.3	55.4	58.7
Industry	4.8	5.2	5.9	7.7	7.4	9.1	10.2	14.8	15.0	14.4	15.5
Manufacturing	..	..	..	5.3	5.7	6.3	6.8	9.2	9.6	8.3	9.0
Services, etc.	20.1	21.9	23.6	26.5	27.8	31.2	34.5	43.0	48.9	53.0	60.3
Resource Balance	-8.9	-6.9	-6.7	-5.8	-5.9	-10.5	-8.2	-17.9	-25.6	-28.5	-31.0
Exports of Goods & NFServices	8.4	11.0	14.7	20.3	23.9	21.6	25.6	27.8	30.6	31.5	43.3
Imports of Goods & NFServices	17.3	17.9	21.4	26.2	29.8	32.1	33.8	45.7	56.2	60.0	74.3
Domestic Absorption	70.4	72.8	76.8	81.1	81.2	94.9	100.0	126.8	144.0	164.7	181.4
Private Consumption, etc.	54.5	55.6	58.3	61.4	62.5	70.6	74.1	91.0	102.5	122.6	133.3
General Gov't Consumption	6.9	7.3	7.7	9.0	9.3	10.5	11.2	11.5	13.4	14.3	15.3
Gross Domestic Investment	8.9	10.0	10.9	10.7	9.5	13.8	14.7	24.3	28.0	27.8	32.7
Fixed Investment	..	..	..	..	..	..	..	..	..	..	..
Memo Items:											
Gross Domestic Saving	0.0	3.1	4.2	4.9	3.6	3.3	6.5	6.5	2.4	-0.7	1.7
Gross National Saving	-0.5	2.7	3.4	4.2	3.2	3.6	7.2	7.0	5.4	5.7	11.0
				(Billions of 1980 CFA Francs)							
Gross National Product	172.8	174.5	180.5	184.9	181.4	193.9	202.2	207.1	197.4	200.7	210.0
GDP at factor cost	164.3	163.7	162.4	168.5	164.6	175.6	183.2	193.5	181.9	182.1	189.5
Agriculture	..	..	..	86.6	83.5	86.4	87.9	84.2	79.6	88.6	88.2
Industry	..	..	..	22.1	20.0	23.3	22.8	29.7	25.4	22.8	25.6
Manufacturing	..	..	..	14.6	14.7	15.4	14.5	17.4	15.3	12.7	15.9
Services, etc.	..	..	..	59.8	61.1	65.9	72.5	79.6	76.8	70.7	75.8
Resource Balance	-28.5	-28.8	-24.9	-25.5	-27.8	-31.3	-33.4	-34.7	-39.0	-35.2	-39.5
Exports of Goods & NFServices	18.1	20.3	28.4	32.2	41.2	43.2	40.1	39.8	40.0	38.6	41.3
Imports of Goods & NFServices	46.5	49.1	53.3	57.7	69.0	74.5	73.5	74.5	79.0	73.8	80.8
Domestic Absorption	194.4	204.0	211.1	212.3	211.5	226.6	235.6	242.9	236.8	235.8	249.6
Private Consumption, etc.	165.4	169.2	174.9	175.0	174.5	181.9	189.2	190.2	185.1	185.9	205.0
General Gov't Consumption	21.6	21.9	22.4	24.9	25.0	27.2	28.6	26.8	25.8	23.4	20.3
Gross Domestic Investment	11.7	12.7	13.3	12.5	12.1	17.5	17.7	26.0	26.0	26.5	24.4
Fixed Investment	..	..	..	..	..	..	..	..	..	..	..
Memo Items:											
Capacity to Import	22.6	30.1	36.7	44.9	55.3	50.1	55.6	45.3	43.1	38.7	47.1
Terms of Trade Adjustment	4.6	9.8	8.3	12.7	14.1	7.0	15.5	5.6	3.0	0.1	5.8
Gross Domestic Income	180.2	183.3	195.7	199.5	197.8	202.3	217.7	213.8	200.9	200.8	215.9
Gross National Income	180.1	183.0	194.6	197.6	195.4	200.9	217.7	212.7	200.4	200.8	215.9
DOMESTIC PRICES (DEFLATORS)				*(Index 1980 = 100)*							
Overall (GDP)	35.5	37.7	38.6	40.3	41.0	43.2	45.4	52.3	59.8	67.9	71.6
Domestic Absorption	36.2	35.7	36.4	38.2	38.4	41.9	42.4	52.2	60.8	69.8	72.7
Agriculture	..	..	..	38.6	38.0	41.0	43.1	50.1	55.6	62.6	66.6
Industry	..	..	..	35.0	37.1	39.0	44.5	49.8	59.0	63.4	60.5
Manufacturing	..	..	..	36.2	38.5	40.6	47.2	52.9	62.7	65.1	56.5
MANUFACTURING ACTIVITY											
Employment (1980=100)	..	..	..	..	..	..	..	..	95.2	96.1	100.6
Real Earnings per Empl. (1980=100)	..	..	..	..	..	..	..	..	101.6	99.8	98.2
Real Output per Empl. (1980=100)	..	..	..	..	..	..	..	..	95.6	99.8	123.7
Earnings as % of Value Added	..	..	..	..	..	..	..	19.2	20.9	24.3	23.9
MONETARY HOLDINGS				*(Billions of current CFA Francs)*							
Money Supply, Broadly Defined	6.3	7.3	8.6	10.0	12.3	13.9	14.8	18.5	31.9	30.6	34.9
Money as Means of Payment	6.1	6.9	8.2	9.6	11.5	12.7	12.3	15.7	26.9	25.6	29.3
Currency Ouside Banks	3.2	3.3	4.0	4.5	5.0	5.8	6.0	6.1	8.5	9.4	9.9
Demand Deposits	2.9	3.6	4.2	5.1	6.6	6.9	6.3	9.7	18.4	16.3	19,4
Quasi-Monetary Liabilities	0.3	0.4	0.4	0.4	0.7	1.2	2.5	2.7	5.0	5.0	5.6
GOVERNMENT DEFICIT (-) OR SURPLUS				*(Millions of current CFA Francs)*							
Current Revenue	..	..	..	..	..	..	..	..	..	..	1,434
Current Expenditure	..	..	..	..	..	..	..	..	..	..	41,982
Current Budget Balance	..	..	..	..	..	..	..	..	..	..	22,416
Capital Receipts	..	..	..	..	..	..	..	..	..	..	19,566
Capital Payments	..	..	..	..	..	..	..	..	..	15	13
	..	..	..	..	..	..	..	..	..	..	18,145

1978	1979	1980	1981	1982	1983	1984	1985	1986	1987 estimate	NOTES	BENIN
210	260	320	350	330	280	260	260	270	300	..	**CURRENT GNP PER CAPITA (US $)**
3,276	3,363	3,464	3,573	3,685	3,801	3,920	4,043	4,177	4,315	..	**POPULATION (thousands)**
				(Billions of current CFA Francs)							**ORIGIN AND USE OF RESOURCES**
171.1	206.0	241.7	283.2	338.1	368.4	409.9	457.0	478.8	494.6	f	Gross National Product (GNP)
0.1	0.1	-0.6	-3.2	-2.0	-5.1	-7.3	-4.0	-10.1	-10.4	..	Net Factor Income from Abroad
171.1	205.9	242.3	286.3	340.0	373.4	417.2	461.0	488.8	505.0	f	Gross Domestic Product (GDP)
15.9	18.2	22.3	32.7	41.1	31.6	27.6	27.4	26.8	33.8	..	Indirect Taxes, net
155.2	187.7	220.1	253.6	298.9	341.8	389.6	433.6	462.0	471.2	..	GDP at factor cost
72.5	88.3	105.2	114.6	135.3	160.2	176.7	208.7	226.1	218.2	..	Agriculture
19.4	24.2	26.9	36.6	47.6	51.0	59.6	67.8	61.1	66.9	..	Industry
12.2	12.1	12.8	12.6	20.5	18.5	16.3	19.1	16.7	..	..	Manufacturing
63.3	75.3	88.0	102.5	116.1	130.6	153.3	157.1	174.7	186.1	..	Services, etc.
-33.6	-44.1	-60.2	-97.6	-125.0	-43.0	-57.8	-66.7	-60.2	-48.3	..	Resource Balance
50.3	64.2	68.3	81.2	105.0	92.0	102.6	108.9	69.7	73.8	..	Exports of Goods & NFServices
83.9	108.3	128.5	178.8	230.0	135.0	160.4	175.6	129.9	122.1	..	Imports of Goods & NFServices
204.7	250.0	302.5	383.9	465.0	416.4	475.0	527.7	549.0	553.2	..	Domestic Absorption
153.6	181.8	231.9	296.8	313.3	332.2	401.5	424.8	441.4	434.6	..	Private Consumption, etc.
17.4	20.3	25.7	32.1	35.6	41.0	42.5	40.0	45.7	48.3	..	General Gov't Consumption
33.7	48.0	45.0	55.0	116.1	43.2	30.9	62.8	62.0	70.4	..	Gross Domestic Investment
..	..	..	..	..	..	..	..	..	..	..	Fixed Investment
											Memo Items:
0.1	3.9	-15.2	-42.6	-8.9	0.2	-26.8	-3.9	1.8	22.1	..	Gross Domestic Saving
12.5	19.1	-6.5	-33.4	0.5	4.2	-25.2	1.2	3.9	22.9	..	Gross National Saving
				(Billions of 1980 CFA Francs)							
213.8	227.6	241.7	263.4	280.5	272.8	278.0	299.0	298.4	290.9	..	Gross National Product
194.7	208.5	220.1	236.5	253.2	248.1	254.2	274.1	276.0	265.7	I	GDP at factor cost
97.4	102.1	105.2	97.7	99.5	100.5	102.0	115.8	122.7	112.1	..	Agriculture
24.1	26.9	26.9	32.7	42.8	43.0	47.1	52.4	47.2	48.9	..	Industry
14.6	13.0	12.8	12.1	19.7	17.7	15.6	18.3	16.0	..	..	Manufacturing
73.2	79.5	88.0	106.1	110.9	104.6	105.1	105.9	106.1	104.8	..	Services, etc.
-40.1	-47.3	-60.2	-66.1	-76.1	-16.7	-22.1	-23.5	-21.4	-21.1	..	Resource Balance
45.3	61.6	68.3	74.5	88.4	71.1	72.7	70.8	61.9	63.2	..	Exports of Goods & NFServices
85.4	108.9	128.5	140.6	164.5	87.8	94.8	94.3	83.3	84.3	..	Imports of Goods & NFServices
253.8	274.8	302.5	330.3	358.2	293.2	304.9	325.0	322.9	314.9	..	Domestic Absorption
210.6	220.0	231.9	251.1	273.4	237.9	253.2	268.7	268.8	259.2	..	Private Consumption, etc.
20.2	21.8	25.7	28.0	30.1	29.8	31.5	33.3	31.8	31.8	..	General Gov't Consumption
23.0	32.9	45.0	51.2	54.8	25.5	20.2	23.0	22.3	23.9	..	Gross Domestic Investment
..	..	..	..	..	..	..	..	..	..	..	Fixed Investment
											Memo Items:
51.2	64.6	68.3	63.9	75.1	59.8	60.6	58.5	44.7	51.0	..	Capacity to Import
5.9	3.0	0.0	-10.7	-13.3	-11.3	-12.1	-12.3	-17.2	-12.3	..	Terms of Trade Adjustment
219.6	230.5	242.3	253.6	268.8	265.2	270.8	289.2	284.3	281.6	..	Gross Domestic Income
219.7	230.6	241.7	252.8	267.2	261.5	265.9	286.7	281.2	278.6	..	Gross National Income
				(Index 1980 = 100)							**DOMESTIC PRICES (DEFLATORS)**
80.0	90.5	100.0	108.4	120.5	135.1	147.5	152.9	162.1	171.8	..	Overall (GDP)
80.6	91.0	100.0	116.2	129.8	142.0	155.8	162.4	170.0	175.7	..	Domestic Absorption
74.4	86.5	100.0	117.2	135.9	159.4	173.1	180.2	184.2	194.7	..	Agriculture
80.5	89.9	100.0	112.0	111.1	118.6	126.7	129.4	129.5	136.9	..	Industry
83.7	93.0	100.0	104.3	104.3	104.3	104.3	104.3	104.3	..	..	Manufacturing
											MANUFACTURING ACTIVITY
100.8	101.8	100.0	113.4	115.9	118.5	121.1	123.8	..	..	..	Employment (1980=100)
97.2	105.3	100.0	88.0	..	..	..	..	..	..	..	Real Earnings per Empl. (1980=100)
85.9	97.6	100.0	103.4	111.5	121.5	134.0	134.3	..	..	..	Real Output per Empl. (1980=100)
25.1	25.4	24.8	25.0	25.0	25.0	25.0	25.0	..	..	..	Earnings as % of Value Added
				(Billions of current CFA Francs)							**MONETARY HOLDINGS**
39.0	40.7	61.4	75.4	98.4	97.0	110.6	111.6	109.1	96.8	..	Money Supply, Broadly Defined
31.0	33.7	45.4	60.4	83.4	82.0	89.5	87.1	78.9	61.8	..	Money as Means of Payment
8.0	13.3	17.1	28.1	28.8	22.5	27.0	20.3	26.2	19.6	..	Currency Ouside Banks
23.1	20.5	28.4	32.3	54.6	59.5	62.5	66.8	52.8	42.2	..	Demand Deposits
8.0	7.0	16.0	15.0	15.0	15.0	21.1	24.5	30.2	35.0	..	Quasi-Monetary Liabilities
				(Millions of current CFA Francs)							**GOVERNMENT DEFICIT (-) OR SURPLUS**
5,052	-1,010	..	..	..	..	..	..	..	..	..	Current Revenue
40,695	45,898	..	..	..	..	..	..	..	..	..	Current Expenditure
18,853	26,315	..	..	..	..	..	..	..	..	..	Current Budget Balance
21,842	19,583	..	..	..	..	..	..	..	..	..	Capital Receipts
12	7	..	..	..	..	..	..	..	..	..	Capital Payments
16,802	20,600	..	..	..	..	..	..	..	..	..	Capital Payments

BENIN	1967	1968	1969	1970	1971	1972	1973	1974	1975	1976	1977
FOREIGN TRADE (CUSTOMS BASIS)				*(Millions of current US dollars)*							
Value of Exports, fob	23.0	33.4	42.8	58.0	69.7	67.3	93.0	93.0	116.1	85.9	129.1
Nonfuel Primary Products	20.9	30.3	39.5	51.8	65.4	61.5	85.8	72.0	98.6	74.2	119.4
Fuels	0.0	0.0	0.0	0.1	0.0	0.0	0.0	4.5	0.3	0.3	0.5
Manufactures	2.1	3.1	3.3	6.1	4.3	5.7	7.2	16.5	17.2	11.4	9.2
Value of Imports, cif	43.4	49.4	54.7	63.6	76.3	93.2	111.7	164.3	197.0	218.8	246.2
Nonfuel Primary Products	10.3	11.9	12.2	13.0	16.7	23.7	28.0	32.6	43.2	52.6	67.8
Fuels	1.9	2.3	1.9	2.4	2.8	4.0	6.4	16.1	15.2	12.4	12.3
Manufactures	31.1	35.2	40.6	48.1	56.9	65.5	77.3	115.6	138.6	153.8	166.1
				(Index 1980 = 100)							
Terms of Trade	129.1	130.9	126.0	125.2	108.1	117.1	176.5	116.0	93.8	123.9	152.0
Export Prices, fob	31.7	31.7	33.4	34.5	32.7	37.4	66.3	63.0	55.3	77.6	108.8
Nonfuel Primary Products	32.0	31.9	33.7	34.6	32.5	37.4	68.4	64.9	53.8	80.0	114.1
Fuels	4.3	4.3	4.3	4.3	5.6	6.2	8.9	36.7	35.7	38.4	42.0
Manufactures	30.1	31.7	32.4	37.5	36.5	38.5	49.5	67.2	66.1	66.7	71.7
Import Prices, cif	24.5	24.2	26.5	27.6	30.3	32.0	37.6	54.3	59.0	62.7	71.6
Trade at Constant 1980 Prices				*(Millions of 1980 US dollars)*							
Exports, fob	72.7	105.5	128.0	168.1	213.0	179.8	140.3	147.7	210.0	110.7	118.7
Imports, cif	176.9	204.5	206.1	230.7	252.0	291.5	297.4	302.6	334.2	349.2	343.9
BALANCE OF PAYMENTS				*(Millions of current US dollars)*							
Exports of Goods & Services	..	..	..	72	84	85	115	119	148	120	165
Merchandise, fob	..	..	..	58	70	67	93	93	116	86	129
Nonfactor Services	..	..	..	12	12	15	19	22	28	31	33
Factor Services	..	..	..	2	2	2	3	4	4	3	3
Imports of Goods & Services	..	..	..	95	109	132	155	191	268	273	330
Merchandise, fob	..	..	..	66	78	94	116	148	206	209	256
Nonfactor Services	..	..	..	25	26	33	36	41	58	61	71
Factor Services	..	..	..	4	5	4	3	2	5	3	4
Long-Term Interest	..	..	..	0	1	1	0	1	1	1	1
Current Transfers, net	..	..	..	..	..	..	..	..	..	..	..
Workers' Remittances	..	..	..	2	4	6	7	7	18	28	39
Total to be Financed	..	..	..	..	..	..	..	..	..	..	..
Official Capital Grants	..	..	..	..	..	..	..	..	..	..	..
Current Account Balance	..	..	..	-3	-2	-9	-14	-29	-53	-65	-57
Long-Term Capital, net	..	..	..	28	30	44	34	57	70	79	88
Direct Investment	..	..	..	7	3	5	3	-2	2	2	3
Long-Term Loans	..	..	..	1	5	4	9	12	11	24	30
Disbursements	..	..	..	2	8	7	11	17	17	27	33
Repayments	..	..	..	1	3	2	2	5	5	3	3
Other Long-Term Capital	..	..	..	20	22	35	21	48	57	53	55
Other Capital, net	..	..	..	-18	-23	-33	-18	-27	-36	-9	-31
Change in Reserves	..	..	..	-6	-6	-2	-2	0	20	-5	0
Memo Item:				*(CFA Francs per US dollar)*							
Conversion Factor (Annual Avg)	246.850	246.850	259.710	277.710	277.130	252.210	222.700	240.500	214.320	238.980	245.670
EXTERNAL DEBT, ETC.				*(Millions of US dollars, outstanding at end of year)*							
Public/Publicly Guar. Long-Term	..	..	..	41	48	51	65	79	89	110	143
Official Creditors	..	..	..	29	37	40	52	62	73	94	127
IBRD and IDA	..	..	..	0	3	5	11	15	19	25	32
Private Creditors	..	..	..	11	11	11	13	17	17	16	16
Private Non-guaranteed Long-Term	..	..	..	..	..	..	..	..	..	..	..
Use of Fund Credit	..	..	..	0	0	0	0	0	0	0	0
Short-Term Debt	..	..	..	..	..	..	..	..	..	..	27
Memo Items:				*(Thousands of US dollars)*							
Int'l Reserves Excluding Gold	8,000.0	10,020.0	7,470.0	15,510.0	24,596.0	28,417.0	33,081.0	34,603.0	15,035.0	19,231.0	20,360.0
Gold Holdings (at market price)	..	..	..	..	..	..	..	..	..	..	923.7
SOCIAL INDICATORS											
Total Fertility Rate	6.9	6.9	6.9	6.9	6.9	6.9	6.8	6.7	6.6	6.6	6.5
Crude Birth Rate	49.5	49.6	49.6	49.7	49.7	49.8	49.7	49.6	49.4	49.3	49.2
Infant Mortality Rate	160.0	158.2	156.4	154.6	152.8	151.0	146.8	142.6	138.4	134.2	130.0
Life Expectancy at Birth	43.2	43.4	43.7	44.0	44.2	44.5	44.8	45.0	45.3	45.6	45.8
Food Production, p.c. ('79-81=100)	99.4	102.4	99.9	101.6	100.0	99.4	101.1	95.6	92.8	100.2	97.5
Labor Force, Agriculture (%)	82.2	81.7	81.3	80.9	79.8	78.8	77.7	76.7	75.6	74.5	73.4
Labor Force, Female (%)	47.7	47.8	47.8	47.9	48.0	48.1	48.2	48.3	48.4	48.5	48.6
School Enroll. Ratio, primary	..	..	..	36.0	..	..	..	..	50.0	..	52.0
School Enroll. Ratio, secondary	..	..	..	5.0	..	..	..	..	9.0	..	11.0

1978	1979	1980	1981	1982	1983	1984	1985	1986	1987 estimate	NOTES	BENIN
											FOREIGN TRADE (CUSTOMS BASIS)
				(Millions of current US dollars)							
125.6	132.9	222.5	184.0	143.6	121.9	146.7	152.0	93.9	121.1	f	Value of Exports, fob
114.3	120.8	101.3	71.4	63.4	80.2	64.4	68.6	38.5	52.4	..	Nonfuel Primary Products
0.4	0.4	111.6	105.8	74.1	34.1	68.2	67.1	35.0	44.0	..	Fuels
11.0	11.7	9.6	6.8	6.1	7.6	14.1	16.3	20.4	24.7	..	Manufactures
312.0	320.0	331.0	543.0	464.0	283.0	285.0	312.0	370.0	346.1	f	Value of Imports, cif
82.0	82.9	89.0	144.8	126.0	76.2	29.8	32.6	68.0	45.6	..	Nonfuel Primary Products
16.9	21.8	12.1	23.5	12.1	9.9	187.2	203.8	89.5	117.0	..	Fuels
213.1	215.3	229.9	374.6	325.9	197.0	68.0	75.6	212.6	183.4	..	Manufactures
				(Index 1980 = 100)							
128.1	114.7	100.0	102.5	92.4	94.7	97.3	90.4	67.8	87.5	..	Terms of Trade
98.6	102.5	100.0	101.3	89.7	88.7	88.3	80.4	60.5	72.8	..	Export Prices, fob
102.0 .	104.1	100.0	88.5	78.6	87.1	86.3	73.0	66.8	78.1	..	Nonfuel Primary Products
42.3	61.0	100.0	112.5	101.6	92.5	90.2	87.5	44.9	56.9	..	Fuels
76.1	90.0	100.0	99.1	94.8	90.7	89.3	89.0	103.2	113.0	..	Manufactures
77.0	89.3	100.0	98.8	97.1	93.7	90.7	89.0	89.2	83.2	..	Import Prices, cif
				(Millions of 1980 US dollars)							Trade at Constant 1980 Prices
127.4	129.7	222.5	181.6	160.0	137.4	166.1	189.0	155.3	166.2	..	Exports, fob
405.4	358.3	331.0	549.4	477.9	302.0	314.1	350.6	414.7	416.0	..	Imports, cif
				(Millions of current US dollars)							**BALANCE OF PAYMENTS**
168	182	306	256	215	242	235	243	200	211	f	Exports of Goods & Services
126	133	222	184	144	176	170	177	181	168	..	Merchandise, fob
39	43	84	72	71	66	65	66	18	22	..	Nonfactor Services
4	6	0	0	0	0	0	1	1	21	..	Factor Services
374	389	611	661	706	368	384	400	386	471	f	Imports of Goods & Services
285	289	499	522	576	251	266	276	321	390	..	Merchandise, fob
85	95	109	136	124	104	101	115	50	67	..	Nonfactor Services
4	5	3	3	6	14	17	9	15	14	..	Factor Services
2	3	3	3	6	13	17	14	22	15	..	Long-Term Interest
..	..	44	45	35	24	21	20	35	37	f	Current Transfers, net
56	73	44	45	35	24	21	20	35	..	..	Workers' Remittances
..	..	-261	-360	-457	-103	-128	-137	-151	-223	..	Total to be Financed
..	..	63	82	65	55	55	51	26	15	..	Official Capital Grants
-90	-52	-197	-278	-392	-48	-74	-87	-125	-208	..	Current Account Balance
87	115	129	182	278	215	127	125	90	79	f	Long-Term Capital, net
1	3	4	2	-1	-1	0	..	..	..	..	Direct Investment
41	145	66	101	213	105	17	11	33	49	..	Long-Term Loans
44	148	72	107	222	116	39	38	69	68	..	Disbursements
3	3	6	7	9	11	22	27	36	19	..	Repayments
46	-33	-4	-2	1	56	56	64	31	15	..	Other Long-Term Capital
-3	-67	49	180	-35	-168	-54	-37	34	128	f	Other Capital, net
6	4	19	-84	150	1	1	-1	1	1	..	Change in Reserves
				(CFA Francs per US dollar)							Memo Item:
225.640	212.720	211.300	271.730	328.620	381.070	436.960	449.260	346.300	300.540	..	Conversion Factor (Annual Avg)
				(Millions of US dollars, outstanding at end of year)							**EXTERNAL DEBT, ETC.**
181	306	348	403	583	632	595	677	775	929	..	Public/Publicly Guar. Long-Term
158	174	230	233	261	306	307	359	434	534	..	Official Creditors
34	40	52	62	73	89	100	125	158	196	..	IBRD and IDA
23	132	118	170	322	326	288	318	342	395	..	Private Creditors
..	..	..	..	..	..	..	..	..	..	..	Private Non-guaranteed Long-Term
0	0	0	0	0	0	0	0	0	..	..	Use of Fund Credit
56	92	68	85	79	83	87	142	166	204	..	Short-Term Debt
				(Thousands of US dollars)							Memo Items:
15,515.0	14,241.0	8,145.8	57,642.0	4,943.4	3,674.7	2,452.3	4,102.6	3,899.0	3,595.0	..	Int'l Reserves Excluding Gold
1,898.4	5,683.2	6,543.5	4,412.3	5,071.6	4,234.7	3,422.1	3,629.7	4,339.0	5,373.5	..	Gold Holdings (at market price)
											SOCIAL INDICATORS
6.5	6.5	6.5	6.5	6.5	6.5	6.5	6.5	6.5	6.5	..	Total Fertility Rate
49.2	49.2	49.3	49.3	49.3	49.2	49.2	49.1	49.1	49.0	..	Crude Birth Rate
128.0	126.0	124.0	122.0	120.0	117.2	114.5	111.7	108.9	106.2	..	Infant Mortality Rate
46.2	46.7	47.1	47.6	48.0	48.5	48.9	49.4	49.9	50.4	..	Life Expectancy at Birth
106.3	107.5	98.1	94.4	93.6	94.3	111.3	113.5	115.7	111.8	..	Food Production, p.c. ('79-81 = 100)
72.4	71.3	70.2	..	..	..	..	..	..	..	..	Labor Force, Agriculture (%)
48.7	48.9	49.0	48.8	48.7	48.5	48.4	48.3	48.1	47.9	..	Labor Force, Female (%)
58.0	60.0	64.0	66.0	70.0	67.0	67.0	65.0	..	..	..	School Enroll. Ratio, primary
11.0	12.0	16.0	19.0	22.0	21.0	20.0	..	..	..	..	School Enroll. Ratio, secondary

BHUTAN	1967	1968	1969	1970	1971	1972	1973	1974	1975	1976	1977
CURRENT GNP PER CAPITA (US $)	..	..	..	..	..	..	..	..	..	..	..
POPULATION (thousands)	940	951	963	976	983	996	1,011	1,028	1,047	1,068	1,091
ORIGIN AND USE OF RESOURCES				*(Millions of current Bhutanese Ngultrum)*							
Gross National Product (GNP)	..	..	..	..	..	..	..	..	..	..	..
Net Factor Income from Abroad	..	..	..	..	..	..	..	..	..	..	..
Gross Domestic Product (GDP)	..	..	..	..	..	..	..	..	..	..	..
Indirect Taxes, net	..	..	..	..	..	..	..	..	..	..	..
GDP at factor cost	..	..	..	..	..	..	..	..	..	..	..
Agriculture	..	..	..	..	..	..	..	..	..	..	..
Industry	..	..	..	..	..	..	..	..	..	..	..
Manufacturing	..	..	..	..	..	..	..	..	..	..	..
Services, etc.	..	..	..	..	..	..	..	..	..	..	..
Resource Balance	..	..	..	..	..	..	..	..	..	..	..
Exports of Goods & NFServices	..	..	..	..	..	..	..	..	..	..	..
Imports of Goods & NFServices	..	..	..	..	..	..	..	..	..	..	..
Domestic Absorption	..	..	..	..	..	..	..	..	..	..	..
Private Consumption, etc.	..	..	..	..	..	..	..	..	..	..	..
General Gov't Consumption	..	..	..	..	..	..	..	..	..	..	..
Gross Domestic Investment	..	..	..	..	..	..	..	..	..	..	..
Fixed Investment	..	..	..	..	..	..	..	..	..	..	..
Memo Items:											
Gross Domestic Saving	..	..	..	..	..	..	..	..	..	..	..
Gross National Saving	..	..	..	..	..	..	..	..	..	..	..
				(Millions of 1983 Bhutanese Ngultrum)							
Gross National Product	..	..	..	..	..	..	..	..	..	..	..
GDP at factor cost	..	..	..	..	..	..	..	..	..	..	..
Agriculture	..	..	..	..	..	..	..	..	..	..	..
Industry	..	..	..	..	..	..	..	..	..	..	..
Manufacturing	..	..	..	..	..	..	..	..	..	..	..
Services, etc.	..	..	..	..	..	..	..	..	..	..	..
Resource Balance	..	..	..	..	..	..	..	..	..	..	..
Exports of Goods & NFServices	..	..	..	..	..	..	..	..	..	..	..
Imports of Goods & NFServices	..	..	..	..	..	..	..	..	..	..	..
Domestic Absorption	..	..	..	..	..	..	..	..	..	..	..
Private Consumption, etc.	..	..	..	..	..	..	..	..	..	..	..
General Gov't Consumption	..	..	..	..	..	..	..	..	..	..	..
Gross Domestic Investment	..	..	..	..	..	..	..	..	..	..	..
Fixed Investment	..	..	..	..	..	..	..	..	..	..	..
Memo Items:											
Capacity to Import	..	..	..	..	..	..	..	..	..	..	..
Terms of Trade Adjustment	..	..	..	..	..	..	..	..	..	..	..
Gross Domestic Income	..	..	..	..	..	..	..	..	..	..	..
Gross National Income	..	..	..	..	..	..	..	..	..	..	..
DOMESTIC PRICES (DEFLATORS)				*(Index 1983 = 100)*							
Overall (GDP)	..	..	..	..	..	..	..	..	..	..	..
Domestic Absorption	..	..	..	..	..	..	..	..	..	..	..
Agriculture	..	..	..	..	..	..	..	..	..	..	..
Industry	..	..	..	..	..	..	..	..	..	..	..
Manufacturing	..	..	..	..	..	..	..	..	..	..	..
MANUFACTURING ACTIVITY											
Employment (1980=100)	..	..	..	..	..	..	..	..	..	..	..
Real Earnings per Empl. (1980=100)	..	..	..	..	..	..	..	..	..	..	..
Real Output per Empl. (1980=100)	..	..	..	..	..	..	..	..	..	..	..
Earnings as % of Value Added	..	..	..	..	..	..	..	..	..	..	..
MONETARY HOLDINGS				*(Millions of current Bhutanese Ngultrum)*							
Money Supply, Broadly Defined	..	..	..	..	..	..	..	..	..	..	..
Money as Means of Payment	..	..	..	..	..	..	..	..	..	..	..
Currency Ouside Banks	..	..	..	..	..	..	..	..	..	..	..
Demand Deposits	..	..	..	..	..	..	..	..	..	..	..
Quasi-Monetary Liabilities	..	..	..	..	..	..	..	..	..	..	..
GOVERNMENT DEFICIT (-) OR SURPLUS				*(Millions of current Bhutanese Ngultrum)*							
Current Revenue	..	..	..	..	..	..	..	..	..	..	..
Current Expenditure	..	..	..	..	..	..	..	..	..	..	..
Current Budget Balance	..	..	..	..	..	..	..	..	..	..	..
Capital Receipts	..	..	..	..	..	..	..	..	..	..	..
Capital Payments	..	..	..	..	..	..	..	..	..	..	..

1978	1979	1980	1981	1982	1983	1984	1985	1986	1987 estimate	NOTES	BHUTAN
..	..	..	..	..	110	120	120	130	150	..	**CURRENT GNP PER CAPITA (US $)**
1,115	1,140	1,165	1,188	1,212	1,237	1,262	1,287	1,313	1,345	..	**POPULATION (thousands)**
			(Millions of current Bhutanese Ngultrum)								**ORIGIN AND USE OF RESOURCES**
..	..	..	1,016.2	1,200.9	1,357.9	1,630.7	1,919.2	2,171.6	2,648.1	..	Gross National Product (GNP)
..	..	..	-274.4	-358.4	-446.2	-382.1	-381.1	-506.7	-618.9	..	Net Factor Income from Abroad
..	..	..	1,290.6	1,559.3	1,804.1	2,012.8	2,300.3	2,678.3	3,267.0	..	Gross Domestic Product (GDP)
..	..	..	..	..	..	..	..	..	..	..	Indirect Taxes, net
											GDP at factor cost
..	..	..	624.0	764.3	897.5	1,016.2	1,160.1	1,373.8	..	..	Agriculture
..	..	..	222.0	290.7	338.4	357.5	351.7	441.2	..	..	Industry
..	..	..	48.1	52.2	69.2	79.1	94.9	96.0	..	..	Manufacturing
..	..	..	444.6	504.3	568.2	639.1	788.5	863.3	..	..	Services, etc.
..	..	..	..	..	..	..	..	..	..	..	Resource Balance
..	..	..	..	..	..	..	..	..	..	..	Exports of Goods & NFServices
..	..	..	..	..	..	..	..	..	..	..	Imports of Goods & NFServices
..	..	..	..	..	..	..	..	..	..	..	Domestic Absorption
..	..	..	..	..	..	..	..	..	..	..	Private Consumption, etc.
..	..	..	..	..	..	..	..	..	..	..	General Gov't Consumption
..	..	..	..	..	..	..	..	..	..	..	Gross Domestic Investment
..	..	..	..	..	..	..	..	..	..	..	Fixed Investment
											Memo Items:
..	..	..	..	..	..	..	..	..	..	..	Gross Domestic Saving
..	..	..	..	..	..	..	..	..	..	..	Gross National Saving
			(Millions of 1983 Bhutanese Ngultrum)								
..	..	..	1,209.0	1,310.0	1,357.9	1,499.0	1,594.4	1,691.8	1,928.7	..	Gross National Product
..	..	..	..	..	..	..	..	..	..	..	GDP at factor cost
..	..	..	750.7	836.8	897.5	929.3	991.3	1,084.8	..	..	Agriculture
..	..	..	271.5	321.1	338.4	325.9	292.6	353.9	..	..	Industry
..	..	..	60.5	59.6	69.2	72.5	81.0	76.3	..	..	Manufacturing
..	..	..	513.2	543.0	568.2	595.0	627.1	647.9	..	..	Services, etc.
..	..	..	..	..	..	..	..	..	..	..	Resource Balance
..	..	..	..	..	..	..	..	..	..	..	Exports of Goods & NFServices
..	..	..	..	..	..	..	..	..	..	..	Imports of Goods & NFServices
..	..	..	..	..	..	..	..	..	..	..	Domestic Absorption
..	..	..	..	..	..	..	..	..	..	..	Private Consumption, etc.
..	..	..	..	..	..	..	..	..	..	..	General Gov't Consumption
..	..	..	..	..	..	..	..	..	..	..	Gross Domestic Investment
..	..	..	..	..	..	..	..	..	..	..	Fixed Investment
											Memo Items:
..	..	..	..	..	..	..	..	..	..	..	Capacity to Import
..	..	..	..	..	..	..	..	..	..	..	Terms of Trade Adjustment
..	..	..	..	..	..	..	..	..	..	..	Gross Domestic Income
..	..	..	..	..	..	..	..	..	..	..	Gross National Income
			(Index 1983 = 100)								**DOMESTIC PRICES (DEFLATORS)**
..	..	..	84.1	91.7	100.0	108.8	120.4	128.4	137.3	..	Overall (GDP)
..	..	..	..	..	..	..	..	..	..	..	Domestic Absorption
..	..	..	83.1	91.3	100.0	109.4	117.0	126.6	..	..	Agriculture
..	..	..	81.8	90.5	100.0	109.7	120.2	124.7	..	..	Industry
..	..	..	79.5	87.6	100.0	109.1	117.2	125.8	..	..	Manufacturing
											MANUFACTURING ACTIVITY
..	..	..	..	..	..	..	..	..	..	..	Employment (1980 = 100)
..	..	..	..	..	..	..	..	..	..	..	Real Earnings per Empl. (1980 = 100)
..	..	..	..	..	..	..	..	..	..	..	Real Output per Empl. (1980 = 100)
..	..	..	..	..	..	..	..	..	..	..	Earnings as % of Value Added
			(Millions of current Bhutanese Ngultrum)								**MONETARY HOLDINGS**
..	..	65.3	74.3	85.7	103.8	157.4	202.9	231.6	316.8	..	Money Supply, Broadly Defined
..	..	..	..	..	..	..	..	..	..	..	Money as Means of Payment
..	..	..	..	..	..	..	..	..	..	..	Currency Ouside Banks
..	..	..	..	..	..	..	..	..	..	..	Demand Deposits
..	..	..	..	..	..	..	..	..	..	..	Quasi-Monetary Liabilities
			(Millions of current Bhutanese Ngultrum)								**GOVERNMENT DEFICIT (-) OR SURPLUS**
..	..	..	..	..	..	..	..	..	..	..	Current Revenue
..	..	..	..	..	..	..	..	..	..	..	Current Expenditure
..	..	..	..	..	..	..	..	..	..	..	Current Budget Balance
..	..	..	..	..	..	..	..	..	..	..	Capital Receipts
..	..	..	..	..	..	..	..	..	..	..	Capital Payments

BHUTAN	1967	1968	1969	1970	1971	1972	1973	1974	1975	1976	1977
FOREIGN TRADE (CUSTOMS BASIS)					*(Thousands of current US dollars)*						
Value of Exports, fob	..	..	..	..	..	..	..	..	..	..	..
Nonfuel Primary Products	..	..	..	..	..	..	..	..	..	..	..
Fuels	..	..	..	..	..	..	..	..	..	..	..
Manufactures	..	..	..	..	..	..	..	..	..	..	..
Value of Imports, cif	..	..	..	..	..	..	..	..	..	..	..
Nonfuel Primary Products	..	..	..	..	..	..	..	..	..	..	..
Fuels	..	..	..	..	..	..	..	..	..	..	..
Manufactures	..	..	..	..	..	..	..	..	..	..	..
					(Index 1980 = 100)						
Terms of Trade	..	..	..	..	..	..	..	..	..	..	..
Export Prices, fob	..	..	..	..	..	..	..	..	..	..	..
Nonfuel Primary Products	..	..	..	..	..	..	..	..	..	..	..
Fuels	..	..	..	..	..	..	..	..	..	..	..
Manufactures	..	..	..	..	..	..	..	..	..	..	..
Import Prices, cif	..	..	..	..	..	..	..	..	..	..	..
Trade at Constant 1980 Prices					*(Thousands of 1980 US dollars)*						
Exports, fob	..	..	..	..	..	..	..	..	..	..	..
Imports, cif	..	..	..	..	..	..	..	..	..	..	..
BALANCE OF PAYMENTS					*(Millions of current US dollars)*						
Exports of Goods & Services	..	..	..	..	..	..	..	..	..	..	..
Merchandise, fob	..	..	..	..	..	..	..	..	..	..	..
Nonfactor Services	..	..	..	..	..	..	..	..	..	..	..
Factor Services	..	..	..	..	..	..	..	..	..	..	..
Imports of Goods & Services	..	..	..	..	..	..	..	..	..	..	..
Merchandise, fob	..	..	..	..	..	..	..	..	..	..	..
Nonfactor Services	..	..	..	..	..	..	..	..	..	..	..
Factor Services	..	..	..	..	..	..	..	..	..	..	..
Long-Term Interest	..	..	..	..	..	..	..	..	..	..	..
Current Transfers, net	..	..	..	..	..	..	..	..	..	..	..
Workers' Remittances	..	..	..	..	..	..	..	..	..	..	..
Total to be Financed	..	..	..	..	..	..	..	..	..	..	..
Official Capital Grants	..	..	..	..	..	..	..	..	..	..	..
Current Account Balance	..	..	..	..	..	..	..	..	..	..	..
Long-Term Capital, net	..	..	..	..	..	..	..	..	..	..	..
Direct Investment	..	..	..	..	..	..	..	..	..	..	..
Long-Term Loans	..	..	..	..	..	..	..	..	..	..	..
Disbursements	..	..	..	..	..	..	..	..	..	..	..
Repayments	..	..	..	..	..	..	..	..	..	..	..
Other Long-Term Capital	..	..	..	..	..	..	..	..	..	..	..
Other Capital, net	..	..	..	..	..	..	..	..	..	..	..
Change in Reserves	..	..	..	..	..	..	..	..	..	..	..
Memo Item:					*(Bhutanese Ngultrum per US dollar)*						
Conversion Factor (Annual Avg)	7.560	7.600	7.570	7.560	7.470	7.730	7.860	7.980	8.650	8.940	8.560
EXTERNAL DEBT, ETC.					*(Thousands of US dollars, outstanding at end of year)*						
Public/Publicly Guar. Long-Term	..	..	..	..	..	..	..	..	..	..	..
Official Creditors	..	..	..	..	..	..	..	..	..	..	..
IBRD and IDA	..	..	..	..	..	..	..	..	..	..	..
Private Creditors	..	..	..	..	..	..	..	..	..	..	..
Private Non-guaranteed Long-Term	..	..	..	..	..	..	..	..	..	..	..
Use of Fund Credit	..	..	..	0.0	0.0	0.0	0.0	0.0	0.0	0.0	0.0
Short-Term Debt	..	..	..	..	..	..	..	..	..	..	..
Memo Items:					*(Millions US dollars)*						
Int'l Reserves Excluding Gold	..	..	..	..	..	..	..	..	..	..	..
Gold Holdings (at market price)	..	..	..	..	..	..	..	..	..	..	..
SOCIAL INDICATORS											
Total Fertility Rate	6.0	6.0	6.0	6.0	6.0	6.0	6.0	6.0	6.0	6.0	6.0
Crude Birth Rate	43.0	43.0	43.0	43.0	43.0	43.0	43.0	43.0	43.0	43.0	43.0
Infant Mortality Rate	..	..	..	..	..	..	..	..	..	..	..
Life Expectancy at Birth	33.3	33.5	33.7	33.9	34.1	34.3	35.3	36.3	37.3	38.3	39.3
Food Production, p.c. ('79-81 = 100)	95.4	95.7	96.3	96.7	97.0	97.5	97.6	98.0	98.5	98.7	98.9
Labor Force, Agriculture (%)	94.4	94.3	94.2	94.1	93.9	93.8	93.6	93.5	93.3	93.1	92.9
Labor Force, Female (%)	35.2	35.1	34.9	34.8	34.7	34.6	34.5	34.4	34.3	34.2	34.0
School Enroll. Ratio, primary	..	..	..	6.0	..	..	..	..	..	9.0	10.0
School Enroll. Ratio, secondary	..	..	..	1.0	..	..	..	..	1.0	1.0	..

1978	1979	1980	1981	1982	1983	1984	1985	1986	1987 estimate	NOTES	BHUTAN
				(Thousands of current US dollars)							**FOREIGN TRADE (CUSTOMS BASIS)**
..	3.0	17.0	19.0	17.0	16.0	17.0	22.0	25.0	..	..	Value of Exports, fob
..	..	..	..	..	..	..	..	..	..	..	Nonfuel Primary Products
..	..	..	..	..	..	..	..	..	..	..	Fuels
..	..	..	..	..	..	..	..	..	..	..	Manufactures
..	15.0	57.0	66.0	67.0	71.0	69.0	76.0	88.0	..	..	Value of Imports, cif
..	..	..	..	..	..	..	..	..	..	..	Nonfuel Primary Products
..	..	..	..	..	..	..	..	..	..	..	Fuels
..	..	..	..	..	..	..	..	..	..	..	Manufactures
				(Index 1980 = 100)							Terms of Trade
..	..	..	..	..	..	..	..	..	..	..	Export Prices, fob
..	..	..	..	..	..	..	..	..	..	..	Nonfuel Primary Products
..	• ..	..	..	..	..	..	..	..	..	..	Fuels
..	..	..	..	..	..	..	..	..	..	..	Manufactures
..	..	..	..	..	..	..	..	..	..	..	Import Prices, cif
				(Thousands of 1980 US dollars)							Trade at Constant 1980 Prices
..	..	..	..	..	..	..	..	..	..	..	Exports, fob
..	..	..	..	..	..	..	..	..	..	..	Imports, cif
				(Millions of current US dollars)							**BALANCE OF PAYMENTS**
..	..	..	19.2	16.5	15.6	17.4	22.2	33.4	60.8	..	Exports of Goods & Services
..	..	..	..	..	..	..	..	..	..	..	Merchandise, fob
..	..	..	..	..	..	..	..	..	..	..	Nonfactor Services
..	..	..	..	..	..	..	..	..	..	..	Factor Services
..	..	..	65.6	67.1	70.8	69.4	85.1	90.4	102.8	..	Imports of Goods & Services
..	..	..	..	..	..	..	..	..	..	..	Merchandise, fob
..	..	..	..	..	..	..	..	..	..	..	Nonfactor Services
..	..	..	..	..	..	..	..	..	..	..	Factor Services
..	..	..	..	0.0	0.0	0.0	0.0	0.2	0.5	..	Long-Term Interest
..	..	..	..	..	..	..	..	..	..	..	Current Transfers, net
..	..	..	..	..	..	..	..	..	..	..	Workers' Remittances
..	..	..	-72.1	-78.4	-94.1	-78.2	-93.6	-78.2	-56.2	K	Total to be Financed
..	..	0.0	0.0	0.0	0.0	0.0	0.0	0.0	..	K	Official Capital Grants
..	..	..	-72.1	-78.4	-94.1	-78.2	-93.6	-78.2	-56.2	..	Current Account Balance
..	..	..	..	..	..	..	..	..	..	..	Long-Term Capital, net
..	..	..	..	..	..	..	..	..	..	..	Direct Investment
..	..	..	..	0.8	0.7	1.1	5.5	11.4	16.2	..	Long-Term Loans
..	..	..	..	0.8	0.7	1.1	5.5	11.4	16.2	..	Disbursements
..	..	..	..	..	..	..	..	..	..	..	Repayments
..	..	..	..	..	..	..	..	..	..	..	Other Long-Term Capital
..	..	..	78.9	84.6	102.0	89.4	99.1	92.6	83.6	..	Other Capital, net
..	..	..	-6.8	-6.2	-7.9	-11.2	-5.5	-14.4	-27.4	..	Change in Reserves
				(Bhutanese Ngultrum per US dollar)							Memo Item:
8.210	8.080	7.860	8.660	9.460	10.100	11.360	12.370	12.610	12.960	..	Conversion Factor (Annual Avg)
				(Thousands of US dollars, outstanding at end of year)							**EXTERNAL DEBT, ETC.**
..	..	..	..	1,100.0	1,800.0	2,700.0	8,800.0	21,000.0	40,700.0	..	Public/Publicly Guar. Long-Term
..	..	..	..	1,100.0	1,800.0	2,700.0	8,800.0	21,000.0	40,700.0	..	Official Creditors
..	..	..	..	0.0	0.0	0.0	500.0	2,300.0	8,500.0	..	IBRD and IDA
..	..	..	..	..	..	..	..	..	..	..	Private Creditors
..	..	..	..	..	..	..	..	..	..	..	Private Non-guaranteed Long-Term
0.0	0.0	0.0	0.0	0.0	0.0	0.0	0.0	0.0	..	..	Use of Fund Credit
..	..	..	..	..	..	..	..	..	..	..	Short-Term Debt
				(Millions US dollars)							Memo Items:
..	..	..	..	..	..	..	..	..	..	..	Int'l Reserves Excluding Gold
..	..	..	..	..	..	..	..	..	..	..	Gold Holdings (at market price)
											SOCIAL INDICATORS
6.0	6.1	6.1	6.2	6.2	6.1	5.9	5.8	5.7	5.5	..	Total Fertility Rate
43.0	43.0	42.9	42.9	42.9	42.1	41.3	40.4	39.6	38.8	..	Crude Birth Rate
..	..	..	..	163.0	155.0	147.0	139.0	142.7	146.3	..	Infant Mortality Rate
40.0	40.7	41.5	42.2	42.9	43.4	44.0	44.5	45.0	45.6	..	Life Expectancy at Birth
99.5	99.6	100.3	100.2	99.8	100.9	107.3	106.2	108.2	121.7	..	Food Production, p.c. ('79-81 = 100)
92.8	92.6	92.4	..	..	..	..	..	..	..	..	Labor Force, Agriculture (%)
33.9	33.8	33.6	33.5	33.3	33.2	33.1	32.9	32.8	32.6	..	Labor Force, Female (%)
10.0	11.0	15.0	16.0	22.0	24.0	25.0	25.0	..	..	..	School Enroll. Ratio, primary
1.0	1.0	..	..	3.0	4.0	4.0	4.0	..	..	..	School Enroll. Ratio, secondary

BOLIVIA	1967	1968	1969	1970	1971	1972	1973	1974	1975	1976	1977
CURRENT GNP PER CAPITA (US $)	210	220	230	230	250	270	290	320	360	400	410
POPULATION (thousands)	4,028	4,125	4,225	4,328	4,435	4,545	4,659	4,777	4,899	5,025	5,156

ORIGIN AND USE OF RESOURCES *(Millions of current Bolivian Pesos)*

	1967	1968	1969	1970	1971	1972	1973	1974	1975	1976	1977
Gross National Product (GNP)	9.9E-03	1.1E-02	1.1E-02	1.2E-02	1.3E-02	1.7E-02	2.5E-02	4.2E-02	4.7E-02	5.4E-02	6.2E-02
Net Factor Income from Abroad	-1.7E-04	-2.8E-04	-3.6E-04	-2.8E-04	-1.8E-04	-3.1E-04	-4.9E-04	-8.2E-04	-7.9E-04	-9.0E-04	-1.7E-03
Gross Domestic Product (GDP)	1.0E-02	1.1E-02	1.2E-02	1.2E-02	1.3E-02	1.7E-02	2.6E-02	4.3E-02	4.8E-02	5.5E-02	6.4E-02
Indirect Taxes, net	6.7E-04	7.0E-04	7.5E-04	1.0E-03	1.1E-03	1.4E-03	2.6E-03	5.6E-03	5.8E-03	6.5E-03	7.7E-03
GDP at factor cost	9.4E-03	1.0E-02	1.1E-02	1.1E-02	1.2E-02	1.6E-02	2.3E-02	3.7E-02	4.2E-02	4.9E-02	5.6E-02
Agriculture	2.0E-03	2.2E-03	2.3E-03	2.4E-03	2.7E-03	3.3E-03	5.1E-03	8.9E-03	9.8E-03	1.1E-02	1.2E-02
Industry	3.0E-03	3.2E-03	3.5E-03	3.9E-03	4.0E-03	5.8E-03	9.8E-03	1.7E-02	1.6E-02	1.8E-02	2.1E-02
Manufacturing	1.4E-03	1.4E-03	1.5E-03	1.6E-03	1.8E-03	2.2E-03	3.3E-03	5.2E-03	5.9E-03	7.0E-03	8.2E-03
Services, etc.	5.0E-03	5.5E-03	5.7E-03	5.8E-03	6.6E-03	7.8E-03	1.1E-02	1.7E-02	2.3E-02	2.6E-02	3.0E-02
Resource Balance	-3.0E-04	-3.9E-04	-2.7E-04	5.1E-05	-3.7E-04	-4.6E-04	-1.0E-04	3.7E-03	-2.9E-03	-1.1E-03	-1.7E-03
Exports of Goods & NFServices	2.4E-03	2.4E-03	2.7E-03	3.0E-03	2.8E-03	3.6E-03	7.1E-03	1.5E-02	1.3E-02	1.5E-02	1.7E-02
Imports of Goods & NFServices	2.7E-03	2.8E-03	3.0E-03	2.9E-03	3.2E-03	4.0E-03	7.2E-03	1.1E-02	1.6E-02	1.6E-02	1.9E-02
Domestic Absorption	1.0E-02	1.1E-02	1.2E-02	1.2E-02	1.4E-02	1.7E-02	2.6E-02	3.9E-02	5.1E-02	5.6E-02	6.6E-02
Private Consumption, etc.	7.6E-03	7.6E-03	8.0E-03	8.0E-03	9.1E-03	1.1E-02	1.6E-02	2.6E-02	2.8E-02	3.5E-02	4.1E-02
General Gov't Consumption	8.6E-04	1.0E-03	1.1E-03	1.2E-03	1.3E-03	1.7E-03	2.5E-03	3.9E-03	5.0E-03	5.9E-03	7.6E-03
Gross Domestic Investment	1.9E-03	2.6E-03	2.7E-03	2.9E-03	3.2E-03	5.0E-03	7.4E-03	9.1E-03	1.8E-02	1.6E-02	1.7E-02
Fixed Investment	1.5E-03	2.1E-03	2.1E-03	2.2E-03	2.4E-03	3.6E-03	5.6E-03	7.4E-03	1.3E-02	1.2E-02	1.4E-02
Memo Items:											
Gross Domestic Saving	1.6E-03	2.2E-03	2.5E-03	2.9E-03	2.8E-03	4.5E-03	7.3E-03	1.3E-02	1.5E-02	1.5E-02	1.6E-02
Gross National Saving	1.5E-03	1.9E-03	2.1E-03	2.7E-03	2.7E-03	4.3E-03	7.0E-03	1.2E-02	1.4E-02	1.4E-02	1.4E-02

(Thousands of 1980 Bolivian Pesos)

	1967	1968	1969	1970	1971	1972	1973	1974	1975	1976	1977
Gross National Product	74.27	77.95	79.92	80.03	84.78	89.32	95.21	99.93	107.05	113.52	117.17
GDP at factor cost	..	..	..	..	..	..	..	..	..	..	..
Agriculture	14.01	15.01	14.87	15.41	16.33	17.28	18.08	18.75	20.21	21.22	21.09
Industry	30.11	32.07	33.78	33.01	34.63	37.57	43.56	43.35	43.72	45.49	46.59
Manufacturing	9.77	10.12	10.31	10.83	11.22	12.13	12.74	14.18	15.04	16.29	17.43
Services, etc.	31.29	32.71	33.60	33.42	34.89	35.98	35.26	39.79	44.68	48.52	52.41
Resource Balance	6.86	6.78	7.14	3.25	3.11	3.32	6.14	8.52	2.33	5.85	5.40
Exports of Goods & NFServices	28.50	29.19	29.93	24.28	27.08	29.91	33.27	32.94	31.48	34.87	34.17
Imports of Goods & NFServices	21.65	22.41	22.79	21.03	23.98	26.59	27.12	24.41	29.15	29.02	28.77
Domestic Absorption	68.56	72.99	75.11	78.59	82.74	87.51	90.75	93.36	106.28	109.39	114.68
Private Consumption, etc.	49.74	48.44	50.63	53.88	55.48	55.17	59.80	61.55	64.65	69.63	73.72
General Gov't Consumption	6.30	7.32	7.54	8.24	8.97	9.96	11.13	12.14	13.48	14.29	14.73
Gross Domestic Investment	12.52	17.24	16.93	16.48	18.29	22.39	19.83	19.67	28.16	25.46	26.23
Fixed Investment	9.57	14.11	13.04	12.40	13.65	15.25	14.88	16.21	19.48	20.39	21.55
Memo Items:											
Capacity to Import	19.21	19.29	20.73	21.40	21.17	23.55	26.74	32.40	23.65	27.08	26.18
Terms of Trade Adjustment	-9.29	-9.90	-9.21	-2.88	-5.91	-6.35	-6.52	-0.54	-7.84	-7.79	-7.99
Gross Domestic Income	66.13	69.88	73.04	78.96	79.94	84.48	90.37	101.35	100.78	107.44	112.10
Gross National Income	64.98	68.05	70.71	77.15	78.87	82.96	88.69	99.39	99.22	105.73	109.18

DOMESTIC PRICES (DEFLATORS) *(Index 1980 = 100)*

	1967	1968	1969	1970	1971	1972	1973	1974	1975	1976	1977
Overall (GDP)	13.3	13.6	14.0	14.8	15.5	18.6	26.4	41.7	44.4	48.1	53.3
Domestic Absorption	15.1	15.4	15.7	15.4	16.5	19.9	28.3	41.6	48.2	51.6	57.3
Agriculture	14.5	14.4	15.2	15.6	16.5	19.3	28.2	47.4	48.6	52.0	59.0
Industry	9.9	10.0	10.4	11.8	11.5	15.5	22.5	38.9	35.5	39.9	45.3
Manufacturing	13.9	14.2	14.5	14.9	15.8	17.8	25.6	36.6	39.5	43.2	46.8

MANUFACTURING ACTIVITY

	1967	1968	1969	1970	1971	1972	1973	1974	1975	1976	1977
Employment (1980=100)	..	..	..	87.8	83.8	85.1	95.5	118.1	119.1	117.4	113.5
Real Earnings per Empl. (1980=100)	..	..	..	77.9	85.7	94.3	92.1	76.6	80.1	86.1	93.7
Real Output per Empl. (1980=100)	..	..	..	67.8	74.3	81.7	79.4	73.3	75.4	83.5	95.0
Earnings as % of Value Added	..	..	..	43.5	43.8	42.4	40.7	42.7	42.1	39.4	38.6

MONETARY HOLDINGS *(Millions of current Bolivian Pesos)*

	1967	1968	1969	1970	1971	1972	1973	1974	1975	1976	1977
Money Supply, Broadly Defined	1.3E-03	1.5E-03	1.7E-03	1.9E-03	2.3E-03	2.8E-03	3.8E-03	5.4E-03	6.7E-03	9.9E-03	1.3E-02
Money as Means of Payment	1.2E-03	1.3E-03	1.4E-03	1.5E-03	1.8E-03	2.2E-03	3.0E-03	4.3E-03	4.8E-03	6.5E-03	7.9E-03
Currency Ouside Banks	9.0E-04	9.5E-04	1.0E-03	1.2E-03	1.3E-03	1.6E-03	2.1E-03	2.7E-03	3.1E-03	4.0E-03	4.9E-03
Demand Deposits	2.9E-04	3.4E-04	3.2E-04	3.8E-04	4.8E-04	6.1E-04	9.0E-04	1.5E-03	1.7E-03	2.5E-03	3.0E-03
Quasi-Monetary Liabilities	1.5E-04	2.3E-04	3.1E-04	3.8E-04	4.9E-04	6.3E-04	8.1E-04	1.2E-03	2.0E-03	3.4E-03	5.0E-03

(Thousands of current Bolivian Pesos)

	1967	1968	1969	1970	1971	1972	1973	1974	1975	1976	1977
GOVERNMENT DEFICIT (-) OR SURPLUS	..	..	..	..	..	-0.3	..	0.5	0.1	-0.3	-0.6
Current Revenue	..	..	..	..	..	1.3	2.5	5.1	5.7	6.8	7.6
Current Expenditure	..	..	..	..	..	1.5	2.3	4.1	4.8	6.0	6.8
Current Budget Balance	..	..	..	..	..	-0.2	0.2	1.0	0.9	0.8	0.8
Capital Receipts	..	..	..	..	..	..	..	..	..	..	..
Capital Payments	..	..	..	..	..	0.1	0.2	0.5	0.8	1.1	1.4

1978	1979	1980	1981	1982	1983	1984	1985	1986	1987 estimate	NOTES	BOLIVIA
440	460	490	550	530	490	490	490	510	570	A	CURRENT GNP PER CAPITA (US $)
5,291	5,431	5,576	5,726	5,881	6,041	6,207	6,378	6,562	6,749	..	POPULATION (thousands)
				(Millions of current Bolivian Pesos)							ORIGIN AND USE OF RESOURCES
7.3E-02	8.6E-02	1.2E-01	1.5E-01	3.6E-01	1.4E+00	2.2E+01	2.5E+03	7.4E+03	8.5E+03	..	Gross National Product (GNP)
-2.3E-03	-3.9E-03	-7.2E-03	-9.9E-03	-3.0E-02	-9.8E-02	-1.3E+00	-1.8E+02	-6.0E+02	-6.6E+02	..	Net Factor Income from Abroad
7.5E-02	8.9E-02	1.2E-01	1.6E-01	3.9E-01	1.5E+00	2.3E+01	2.6E+03	8.0E+03	9.2E+03	..	Gross Domestic Product (GDP)
8.5E-03	1.0E-02	1.3E-02	1.0E-02	1.0E-02	1.0E-02	1.7E-01	2.8E+01	1.3E+02	2.0E+02	..	Indirect Taxes, net
6.7E-02	7.9E-02	1.1E-01	1.5E-01	3.8E-01	1.5E+00	2.3E+01	2.6E+03	7.9E+03	9.0E+03	B	GDP at factor cost
1.4E-02	1.6E-02	2.3E-02	2.9E-02	7.7E-02	3.2E-01	5.1E+00	7.3E+02	2.0E+03	2.2E+03	..	Agriculture
2.4E-02	2.8E-02	4.3E-02	5.1E-02	1.3E-01	4.6E-01	6.5E+00	7.4E+02	1.9E+03	2.2E+03	..	Industry
9.4E-03	1.2E-02	1.8E-02	2.4E-02	4.1E-02	2.2E-01	3.6E+00	3.6E+02	1.0E+03	1.2E+03	..	Manufacturing
3.7E-02	4.5E-02	5.8E-02	7.8E-02	1.9E-01	6.9E-01	1.1E+01	1.2E+03	4.2E+03	4.8E+03	..	Services, etc.
-5.6E-03	-2.8E-03	6.3E-03	-9.6E-04	1.7E-02	4.2E-02	6.5E-01	2.4E+01	-4.0E+02	-7.4E+02	..	Resource Balance
1.7E-02	2.2E-02	2.6E-02	2.5E-02	6.1E-02	2.0E-01	2.4E+00	3.7E+02	1.4E+03	1.3E+03	..	Exports of Goods & NFServices
2.3E-02	2.5E-02	2.0E-02	2.6E-02	4.4E-02	1.5E-01	1.8E+00	3.5E+02	1.8E+03	2.1E+03	..	Imports of Goods & NFServices
8.1E-02	9.2E-02	1.2E-01	1.6E-01	3.8E-01	1.4E+00	2.2E+01	2.6E+03	8.4E+03	9.9E+03	..	Domestic Absorption
5.3E-02	6.1E-02	8.2E-02	1.2E-01	2.8E-01	1.1E+00	1.6E+01	2.2E+03	6.9E+03	7.8E+03	..	Private Consumption, etc.
9.5E-03	1.4E-02	1.7E-02	1.9E-02	6.1E-02	2.5E-01	4.7E+00	4.1E+02	9.0E+02	1.3E+03	..	General Gov't Consumption
1.8E-02	1.8E-02	1.7E-02	1.9E-02	3.8E-02	5.1E-02	1.3E+00	-2.1E+01	6.4E+02	8.7E+02	..	Gross Domestic Investment
1.6E-02	1.7E-02	1.8E-02	1.8E-02	4.5E-02	9.8E-02	1.1E+00	1.9E+02	7.7E+02	9.5E+02	..	Fixed Investment
											Memo Items:
1.2E-02	1.5E-02	2.4E-02	1.8E-02	5.5E-02	9.3E-02	2.0E+00	2.6E+00	2.4E+02	1.4E+02	..	Gross Domestic Saving
1.0E-02	1.1E-02	1.7E-02	8.6E-03	2.7E-02	3.9E-03	7.2E-01	-1.7E+02	-3.2E+02	-4.8E+02	..	Gross National Saving
				(Thousands of 1980 Bolivian Pesos)							
120.55	118.88	115.69	116.59	108.59	102.07	102.20	100.45	98.55	101.88	..	Gross National Product
109.71	110.78	109.68	108.94	114.27	109.24	109.27	106.69	102.20	103.72	B	GDP at factor cost
21.53	22.23	22.56	22.35	23.90	19.98	24.55	26.79	25.53	25.48	..	Agriculture
45.48	42.58	42.71	41.66	38.91	37.14	32.70	29.60	27.75	28.65	..	Industry
18.22	17.87	17.97	16.58	14.53	13.86	11.93	10.82	11.04	11.83	..	Manufacturing
57.10	59.28	57.67	60.06	55.85	53.82	53.35	54.05	53.92	55.61	..	Services, etc.
3.23	3.91	6.34	-2.09	8.59	7.75	9.27	4.71	3.63	2.47	..	Resource Balance
33.15	33.63	25.90	24.85	25.39	25.32	24.66	22.25	24.70	23.95	..	Exports of Goods & NFServices
29.92	29.71	19.56	26.93	16.80	17.58	15.39	17.54	21.06	21.48	..	Imports of Goods & NFServices
120.88	120.18	116.60	126.16	110.07	103.19	101.34	105.74	103.57	107.26	..	Domestic Absorption
79.25	79.65	81.98	95.39	80.12	80.10	80.97	86.65	89.41	89.35	..	Private Consumption, etc.
15.16	17.68	17.26	13.63	17.03	15.63	14.60	12.91	10.00	12.90	..	General Gov't Consumption
26.48	22.86	17.36	17.14	12.93	7.46	5.77	6.18	4.16	5.02	..	Gross Domestic Investment
23.61	21.73	17.56	14.21	13.47	7.74	6.61	4.95	5.34	5.69	..	Fixed Investment
											Memo Items:
22.58	26.32	25.90	25.95	23.30	22.37	21.07	18.74	16.34	13.81	..	Capacity to Import
-10.57	-7.31	0.00	1.10	-2.09	-2.95	-3.58	-3.51	-8.36	-10.14	..	Terms of Trade Adjustment
113.55	116.78	122.94	125.18	116.57	107.99	107.02	106.93	98.85	99.59	..	Gross Domestic Income
109.98	111.57	115.69	117.70	106.49	99.12	98.61	96.93	90.19	91.73	..	Gross National Income
				(Index 1980 = 100)							DOMESTIC PRICES (DEFLATORS)
60.4	72.1	100.0	127.0	330.9	1,325.8	20,689	2,400,000	7,500,000	8,400,000	..	Overall (GDP)
66.7	76.8	100.0	125.7	341.3	1,384.4	21,935	2,500,000	8,100,000	9,300,000	..	Domestic Absorption
64.6	73.2	100.0	127.7	320.8	1,586.1	20,856	2,700,000	7,600,000	8,500,000	..	Agriculture
53.6	66.5	100.0	123.2	334.9	1,237.7	19,958	2,500,000	6,700,000	7,600,000	..	Industry
51.4	66.0	100.0	146.8	279.0	1,611.7	30,214	3,300,000	9,200,000	10,000,000	..	Manufacturing
											MANUFACTURING ACTIVITY
108.3	95.6	100.0	104.4	106.5	121.2	108.4	..	..	..	G	Employment (1980=100)
103.1	112.3	100.0	96.0	110.9	98.9	122.2	..	..	..	G	Real Earnings per Empl. (1980=100)
107.2	111.4	100.0	86.8	115.9	65.8	61.4	..	..	..	G	Real Output per Empl. (1980=100)
39.0	37.8	34.6	34.6	34.6	34.6	34.6	..	..	..	..	Earnings as % of Value Added
				(Millions of current Bolivian Pesos)							MONETARY HOLDINGS
1.4E-02	1.7E-02	2.3E-02	2.9E-02	9.7E-02	2.7E-01	4.1E+00	3.0E+02	1.9E+03	2.1E+03	..	Money Supply, Broadly Defined
8.8E-03	1.0E-02	1.5E-02	1.8E-02	5.8E-02	1.8E-01	3.4E+00	2.2E+02	3.8E+02	5.2E+02	..	Money as Means of Payment
5.8E-03	7.2E-03	9.5E-03	1.1E-02	3.9E-02	1.2E-01	2.9E+00	..	..	..	..	Currency Ouside Banks
3.0E-03	3.1E-03	5.2E-03	6.7E-03	1.9E-02	5.3E-02	4.8E-01	..	..	..	..	Demand Deposits
5.7E-03	6.3E-03	8.4E-03	1.2E-02	3.9E-02	8.8E-02	6.8E-01	8.9E+01	1.5E+03	1.6E+03	..	Quasi-Monetary Liabilities
				(Thousands of current Bolivian Pesos)							
-1.5	-3.4	-6.6	-6.9	..	-95.9	-6,100.0	..	..	..	C F	GOVERNMENT DEFICIT (-) OR SURPLUS
8.5	8.4	11.1	14.3	19.2	63.5	800.0	..	..	..	..	Current Revenue
8.4	10.2	16.7	19.7	..	100.0	6,800.0	..	..	..	..	Current Expenditure
0.1	-1.8	-5.6	-5.4	..	-36.5	-6,000.0	..	..	..	..	Current Budget Balance
..	..	..	..	..	..	..	..	..	..	..	Capital Receipts
1.6	1.6	1.0	1.5	..	100.0	100.0	..	..	..	..	Capital Payments

BOLIVIA	1967	1968	1969	1970	1971	1972	1973	1974	1975	1976	1977
FOREIGN TRADE (CUSTOMS BASIS)					*(Millions of current US dollars)*						
Value of Exports, fob	150	152	172	190	181	201	261	557	444	568	632
Nonfuel Primary Products	125	130	151	176	156	164	205	378	299	404	481
Fuels	21	22	20	9	20	34	48	166	129	149	117
Manufactures	5	1	1	6	5	2	7	13	16	14	34
Value of Imports, cif	151	153	165	159	170	179	230	366	574	593	586
Nonfuel Primary Products	35	30	31	35	40	42	55	88	109	98	93
Fuels	2	2	2	2	2	2	2	4	13	11	8
Manufactures	114	121	132	122	128	135	173	274	452	485	485
					(Index 1980 = 100)						
Terms of Trade	49.50	49.90	53.90	69.20	54.20	47.30	53.80	86.10	70.90	78.30	87.20
Export Prices, fob	15.00	14.60	16.80	22.50	18.80	18.10	25.80	53.40	45.90	50.30	60.50
Nonfuel Primary Products	25.20	24.60	27.50	28.60	26.70	29.90	46.40	66.80	52.10	56.90	67.40
Fuels	4.30	4.30	4.30	4.30	5.60	6.20	8.90	36.70	35.70	38.40	42.00
Manufactures	19.40	18.40	20.40	21.80	24.30	25.70	33.00	50.50	48.50	49.00	66.00
Import Prices, cif	30.20	29.30	31.10	32.60	34.70	38.20	47.90	62.00	64.80	64.20	69.40
Trade at Constant 1980 Prices					*(Millions of 1980 US dollars)*						
Exports, fob	1,002	1,038	1,028	844	965	1,112	1,011	1,042	968	1,130	1,044
Imports, cif	499	521	531	489	489	467	480	591	886	924	845
BALANCE OF PAYMENTS					*(Millions of current US dollars)*						
Exports of Goods & Services	..	..	..	207	202	223	294	598	495	636	701
Merchandise, fob	..	..	..	190	182	201	261	557	445	563	634
Nonfactor Services	..	..	..	14	16	21	26	37	41	60	61
Factor Services	..	..	..	3	5	1	7	4	9	13	6
Imports of Goods & Services	..	..	..	207	212	240	297	474	668	707	834
Merchandise, fob	..	..	..	135	144	153	193	324	470	512	579
Nonfactor Services	..	..	..	44	46	64	71	118	148	146	180
Factor Services	..	..	..	28	22	23	32	32	50	49	75
Long-Term Interest	..	..	..	7	8	13	17	19	26	39	61
Current Transfers, net	..	..	..	2	2	5	5	3	3	3	2
Workers' Remittances	..	..	..	0	0	0	0	0	0	0	0
Total to be Financed	..	..	..	2	-7	-12	2	126	-140	-64	-131
Official Capital Grants	..	..	..	2	5	9	11	11	10	11	13
Current Account Balance	..	..	..	4	-2	-4	13	137	-130	-54	-118
Long-Term Capital, net	..	..	..	36	61	101	47	111	169	226	338
Direct Investment	..	..	..	-76	2	-11	5	26	53	-8	-1
Long-Term Loans	..	..	..	39	48	96	8	73	116	248	345
Disbursements	..	..	..	58	67	127	41	127	170	322	451
Repayments	..	..	..	19	19	31	33	54	54	74	106
Other Long-Term Capital	..	..	..	70	6	8	25	1	-11	-24	-19
Other Capital, net	..	..	..	-41	-60	-97	-63	-124	-79	-117	-158
Change in Reserves	..	..	..	1	1	-1	3	-124	40	-56	-62
Memo Item:					*(Bolivian Pesos per US dollar)*						
Conversion Factor (Annual Avg)	1.20E-05	1.20E-05	1.20E-05	1.20E-05	1.20E-05	1.30E-05	2.00E-05	2.60E-05	2.80E-05	2.80E-05	2.90E-05
Additional Conversion Factor	1.19E-05	1.19E-05	1.19E-05	1.19E-05	1.19E-05	1.33E-05	2.00E-05	2.00E-05	2.00E-05	2.00E-05	2.00E-05
EXTERNAL DEBT, ETC.					*(Millions of US dollars, outstanding at end of year)*						
Public/Publicly Guar. Long-Term	..	..	..	480	528	622	639	715	824	1,065	1,428
Official Creditors	..	..	..	272	311	393	416	459	495	589	712
IBRD and IDA	..	..	..	18	37	49	53	62	70	85	109
Private Creditors	..	..	..	208	217	230	223	256	329	476	716
Private Non-guaranteed Long-Term	..	..	..	11	12	14	13	15	16	25	20
Use of Fund Credit	..	..	..	6	7	9	22	18	16	0	0
Short-Term Debt	..	..	..	..	..	..	..	..	..	..	285
Memo Items:					*(Thousands of US dollars)*						
Int'l Reserves Excluding Gold	28,590	28,600	30,300	32,820	39,649	44,257	54,930	176,220	139,530	151,060	211,090
Gold Holdings (at market price)	9,353	13,049	11,767	13,539	16,755	26,383	45,691	76,053	57,434	55,766	99,277
SOCIAL INDICATORS											
Total Fertility Rate	6.6	6.5	6.5	6.5	6.5	6.5	6.5	6.5	6.4	6.4	6.4
Crude Birth Rate	45.6	45.6	45.5	45.5	45.5	45.4	45.3	45.2	45.1	45.0	44.8
Infant Mortality Rate	157.0	155.8	154.6	153.4	152.2	151.0	148.4	145.8	143.2	140.6	138.0
Life Expectancy at Birth	45.1	45.4	45.7	46.1	46.4	46.7	47.1	47.5	47.9	48.3	48.6
Food Production, p.c. ('79-81=100)	88.5	93.3	91.5	93.9	94.5	99.5	101.0	103.0	108.6	109.7	101.2
Labor Force, Agriculture (%)	53.4	52.9	52.5	52.1	51.5	50.9	50.4	49.8	49.2	48.6	48.1
Labor Force, Female (%)	21.2	21.3	21.4	21.4	21.6	21.7	21.8	21.9	22.0	22.1	22.2
School Enroll. Ratio, primary	..	..	..	76.0	..	..	..	..	85.0	89.0	89.0
School Enroll. Ratio, secondary	..	..	..	24.0	..	..	..	..	31.0	30.0	32.0

1978	1979	1980	1981	1982	1983	1984	1985	1986	1987 estimate	NOTES	BOLIVIA
				(Millions of current US dollars)							**FOREIGN TRADE (CUSTOMS BASIS)**
629	760	942	912	828	755	725	623	564	566	f	Value of Exports, fob
511	619	692	573	436	360	329	257	230	222	..	Nonfuel Primary Products
107	106	222	314	372	388	391	362	323	332	..	Fuels
11	35	28	25	20	7	4	4	11	12	..	Manufactures
769	842	665	917	578	589	492	552	716	776	f	Value of Imports, cif
119	131	104	143	90	92	75	80	104	136	..	Nonfuel Primary Products
11	7	7	9	6	6	10	10	10	12	..	Fuels
639	703	554	765	482	491	407	461	602	629	..	Manufactures
				(Index 1980 = 100)							
84.50	95.30	100.00	95.20	91.40	90.40	90.40	84.10	60.80	51.20	..	Terms of Trade
66.80	86.90	100.00	95.30	88.40	86.40	85.40	79.50	64.00	56.40	..	Export Prices, fob
75.70	93.40	100.00	88.40	79.80	80.80	80.40	70.50	51.00	54.50	..	Nonfuel Primary Products
42.30	61.00	100.00	112.50	101.60	92.50	90.20	87.50	77.20	56.90	..	Fuels
81.10	92.70	100.00	85.90	82.00	80.00	78.00	74.00	87.50	96.10	..	Manufactures
79.10	91.20	100.00	100.20	96.70	95.60	94.50	94.50	105.20	110.30	..	Import Prices, cif
				(Millions of 1980 US dollars)							Trade at Constant 1980 Prices
941	874	942	957	936	874	849	784	882	1,003	..	Exports, fob
973	923	665	916	597	616	520	584	681	704	..	Imports, cif
				(Millions of current US dollars)							**BALANCE OF PAYMENTS**
706	873	1,039	1,016	912	893	842	728	681	618	..	Exports of Goods & Services
627	760	942	912	828	755	725	612	547	473	..	Merchandise, fob
76	105	81	87	76	97	88	98	117	129	..	Nonfactor Services
3	8	16	16	8	41	30	18	18	16	..	Factor Services
1,065	1,322	1,105	1,520	1,131	1,137	1,105	1,090	1,165	1,235	..	Imports of Goods & Services
724	738	574	828	496	496	412	463	597	658	..	Merchandise, fob
222	391	250	311	213	236	245	234	241	271	..	Nonfactor Services
119	193	281	381	422	406	448	393	328	306	..	Factor Services
90	129	173	180	191	204	213	156	100	62	..	Long-Term Interest
6	11	13	13	17	40	22	20	19	20	..	Current Transfers, net
0	0	0	1	1	1	1	0	1	1	..	Workers' Remittances
-353	-438	-53	-491	-202	-204	-241	-342	-466	-597	K	Total to be Financed
22	41	47	26	29	66	67	60	82	112	K	Official Capital Grants
-332	-397	-6	-465	-173	-138	-174	-282	-384	-485	..	Current Account Balance
314	290	351	486	55	354	-81	-175	19	130	..	Long-Term Capital, net
12	35	47	76	31	7	7	10	10	22	..	Direct Investment
303	223	339	251	169	48	52	-43	239	134	..	Long-Term Loans
582	384	485	380	291	152	205	133	345	209	..	Disbursements
279	162	145	129	122	105	153	177	106	74	..	Repayments
-22	-9	-82	132	-174	234	-206	-201	-311	-139	..	Other Long-Term Capital
-54	66	-490	-417	-139	286	446	542	589	216	..	Other Capital, net
71	42	146	397	257	-502	-191	-86	-225	139	..	Change in Reserves
				(Bolivian Pesos per US dollar)							**Memo Item:**
3.20E-05	3.30E-05	4.00E-05	4.70E-05	1.20E-04	4.70E-04	7.10E-03	7.60E-01	1.92E+00	2.06E+00	..	Conversion Factor (Annual Avg)
2.00E-05	2.04E-05	2.45E-05	2.45E-05	6.41E-05	2.32E-04	3.14E-03	4.40E-01	1.92E+00	2.05E+00	..	Additional Conversion Factor
				(Millions of US dollars, outstanding at end of year)							**EXTERNAL DEBT, ETC.**
1,718	1,908	2,228	2,765	2,861	3,279	3,386	3,484	4,064	4,599	..	Public/Publicly Guar. Long-Term
867	977	1,147	1,424	1,541	1,959	2,011	2,087	2,758	3,515	..	Official Creditors
134	168	239	281	302	305	269	301	332	411	..	IBRD and IDA
851	931	1,081	1,341	1,320	1,321	1,375	1,398	1,306	1,084	..	Private Creditors
60	95	92	80	129	376	340	555	555	200	..	Private Non-guaranteed Long-Term
20	20	80	71	86	89	64	51	145	141	..	Use of Fund Credit
364	528	300	303	237	311	479	650	765	608	..	Short-Term Debt
				(Thousands of US dollars)							**Memo Items:**
169,770	178,200	106,100	99,840	155,910	160,140	251,640	200,010	163,670	97,314	..	Int'l Reserves Excluding Gold
145,673	349,640	447,260	329,428	406,659	348,680	281,463	292,482	349,640	433,003	..	Gold Holdings (at market price)
											SOCIAL INDICATORS
6.4	6.3	6.3	6.3	6.3	6.2	6.2	6.1	6.1	6.1	..	Total Fertility Rate
44.7	44.5	44.3	44.2	44.0	43.7	43.5	43.2	43.0	42.7	..	Crude Birth Rate
135.2	132.4	129.6	126.8	124.0	122.5	120.9	119.4	117.9	116.3	..	Infant Mortality Rate
49.1	49.5	49.9	50.3	50.7	51.4	52.1	52.7	53.4	54.0	..	Life Expectancy at Birth
99.2	97.3	99.5	103.3	105.1	77.0	93.0	97.5	88.9	96.0	..	Food Production, p.c. ('79-81 = 100)
47.5	47.0	46.4	..	..	..	..	..	..	..	..	Labor Force, Agriculture (%)
22.3	22.4	22.5	22.9	23.2	23.5	23.8	24.2	24.5	24.8	..	Labor Force, Female (%)
..	..	84.0	86.0	85.0	87.0	91.0	..	..	..	..	School Enroll. Ratio, primary
..	..	36.0	34.0	35.0	35.0	37.0	..	..	..	..	School Enroll. Ratio, secondary

BOTSWANA	1967	1968	1969	1970	1971	1972	1973	1974	1975	1976	1977
CURRENT GNP PER CAPITA (US $)	100	100	120	130	150	180	240	340	350	400	420
POPULATION (thousands)	578	593	608	624	648	673	699	727	755	782	811
ORIGIN AND USE OF RESOURCES					*(Millions of current Botswana Pula)*						
Gross National Product (GNP)	39.1	42.5	49.9	58.6	71.9	88.3	129.5	180.3	188.1	248.0	290.7
Net Factor Income from Abroad	-1.2	-1.3	-1.3	-1.3	-6.1	-15.3	-21.1	-8.0	-24.9	-25.9	-24.4
Gross Domestic Product (GDP)	40.3	43.8	51.2	59.9	78.0	103.6	150.6	188.3	213.0	273.9	315.1
Indirect Taxes, net	2.2	2.6	2.1	4.0	5.6	11.2	12.0	14.2	18.1	22.9	29.2
GDP at factor cost	38.1	41.2	49.1	55.9	72.4	92.4	138.6	174.1	194.9	251.0	285.9
Agriculture	16.4	18.3	23.2	19.8	25.7	34.2	49.7	62.4	61.2	65.7	74.4
Industry	6.0	6.6	5.2	16.5	21.5	28.6	41.6	49.5	60.5	84.4	91.8
Manufacturing	3.3	3.6	2.8	3.5	4.6	6.1	8.9	10.1	15.5	20.9	25.3
Services, etc.	17.9	18.9	22.8	23.6	30.8	40.8	59.3	76.4	91.3	123.8	148.9
Resource Balance	-12.8	-17.6	-20.3	-24.3	-25.7	-25.9	-33.0	-45.4	-53.8	-52.3	-53.5
Exports of Goods & NFServices	10.5	10.2	11.7	13.6	20.4	39.8	55.2	76.4	93.8	135.2	155.5
Imports of Goods & NFServices	23.3	27.8	32.0	37.9	46.1	65.7	88.2	121.8	147.6	187.5	209.0
Domestic Absorption	53.1	61.4	71.5	84.2	103.7	129.5	183.6	233.7	266.8	326.2	368.6
Private Consumption, etc.	33.2	37.8	43.0	46.7	52.2	59.1	89.9	104.6	123.6	156.5	194.2
General Gov't Consumption	10.5	11.6	11.3	12.2	14.3	16.0	18.6	31.7	41.5	54.7	75.7
Gross Domestic Investment	9.4	12.0	17.2	25.3	37.2	54.4	75.1	97.4	101.7	115.0	98.7
Fixed Investment	9.0	9.9	9.9	17.3	30.3	54.5	64.3	79.6	57.3	79.1	77.8
Memo Items:											
Gross Domestic Saving	-3.4	-5.6	-3.1	1.0	11.5	28.5	42.1	52.0	47.9	62.7	45.2
Gross National Saving	-4.6	-6.9	-4.4	0.4	8.3	20.9	34.2	72.5	17.5	41.6	26.0
					(Millions of 1980 Botswana Pula)						
Gross National Product	124.4	132.2	153.0	172.2	190.4	229.4	282.5	383.5	344.6	410.6	429.3
GDP at factor cost	121.5	128.7	150.8	164.6	194.0	246.6	310.1	373.3	364.8	420.2	425.6
Agriculture	33.3	36.4	45.5	37.3	44.1	58.4	71.1	98.7	87.7	88.9	92.8
Industry	18.4	20.0	15.0	60.8	69.2	91.5	111.6	131.7	134.3	166.7	160.6
Manufacturing	8.2	8.6	6.6	6.6	7.9	9.5	12.5	15.4	17.4	24.5	31.4	34.6
Services, etc.	76.8	80.2	96.6	78.0	95.5	126.2	153.9	173.0	176.3	202.5	215.1
Resource Balance	-41.0	-93.7	-98.8	-110.5	-92.7	-71.1	-98.8	-96.6	-96.3	-58.5	-42.9
Exports of Goods & NFServices	37.1	37.8	43.3	50.3	75.4	137.5	156.5	188.9	194.2	261.0	264.8
Imports of Goods & NFServices	78.1	131.5	142.1	160.8	168.1	208.6	255.3	285.5	290.5	319.5	307.7
Domestic Absorption	169.5	230.3	255.9	286.6	301.5	347.2	435.4	500.0	494.6	516.6	511.4
Private Consumption, etc.	103.9	146.2	153.6	156.5	135.2	132.2	179.8	202.0	222.3	239.2	261.8
General Gov't Consumption	35.9	38.8	37.4	38.8	41.1	46.0	44.8	83.1	83.6	95.7	115.1
Gross Domestic Investment	29.7	45.3	64.9	91.3	125.2	169.0	210.8	214.9	188.7	181.7	134.5
Fixed Investment	..	..	..	..	..	..	..	182.9	112.5	128.3	108.1
Memo Items:											
Capacity to Import	35.2	48.2	52.0	57.7	74.4	126.4	159.8	179.1	184.6	230.4	228.9
Terms of Trade Adjustment	-1.9	10.4	8.7	7.4	-1.0	-11.1	3.3	-9.8	-9.6	-30.6	-35.9
Gross Domestic Income	126.6	147.1	165.8	183.5	207.8	265.0	339.9	393.6	388.7	427.5	432.7
Gross National Income	122.5	142.6	161.7	179.6	189.4	218.2	285.8	373.7	335.0	379.9	393.4
DOMESTIC PRICES (DEFLATORS)					*(Index 1980 = 100)*						
Overall (GDP)	31.4	32.1	32.6	34.0	37.4	37.5	44.7	46.7	53.5	59.8	67.3
Domestic Absorption	31.3	26.7	27.9	29.4	34.4	37.3	42.2	46.7	53.9	63.1	72.1
Agriculture	49.2	50.3	51.0	53.1	58.3	58.6	69.9	63.2	69.8	73.9	80.2
Industry	32.6	33.0	34.7	27.1	31.1	31.3	37.3	37.6	45.0	50.6	57.2
Manufacturing	40.2	41.9	42.4	44.3	48.4	48.8	57.8	58.0	63.3	66.6	73.1
MANUFACTURING ACTIVITY											
Employment (1980=100)	..	..	..	..	42.2	48.4	52.0	60.2	70.3	78.0	75.7
Real Earnings per Empl. (1980=100)	..	..	..	..	47.5	44.4	47.9	84.7	127.9	54.1	104.3
Real Output per Empl. (1980=100)	..	..	..	..	..	101.7	..	137.3	125.6	119.0	131.3
Earnings as % of Value Added	..	17.5	..	..	..	27.6	..	39.6	51.6	20.6	37.1
MONETARY HOLDINGS					*(Millions of current Botswana Pula)*						
Money Supply, Broadly Defined	..	..	..	..	..	..	..	..	..	86.2	108.6
Money as Means of Payment	..	..	..	..	..	..	..	..	..	43.5	58.2
Currency Ouside Banks	..	..	..	..	..	..	..	..	..	10.4	12.4
Demand Deposits	..	..	..	..	..	..	..	..	..	33.1	45.8
Quasi-Monetary Liabilities	..	..	..	..	..	..	..	..	..	42.7	50.4
GOVERNMENT DEFICIT (-) OR SURPLUS					*(Millions of current Botswana Pula)*						
	..	..	..	..	-11.2	-21.0	-14.0	-5.7	1.2	-20.8	-4.6
Current Revenue	..	..	..	..	19.2	28.8	45.6	65.9	89.3	84.1	115.2
Current Expenditure	..	..	..	..	18.1	20.2	26.0	36.0	46.8	63.5	75.8
Current Budget Balance	..	..	..	..	1.1	8.6	19.6	29.9	42.5	20.6	39.4
Capital Receipts	..	..	..	..	0.0	..	..	..	..	0.8	0.3
Capital Payments	..	..	..	..	12.3	29.6	33.7	35.6	41.2	42.1	44.3

1978	1979	1980	1981	1982	1983	1984	1985	1986	1987 estimate	NOTES	BOTSWANA
490	600	790	950	1,000	1,040	980	870	910	1,050	..	**CURRENT GNP PER CAPITA (US $)**
840	870	902	933	966	999	1,034	1,070	1,107	1,146	..	**POPULATION (thousands)**
				(Millions of current Botswana Pula)							**ORIGIN AND USE OF RESOURCES**
328.4	470.3	655.0	771.1	781.8	996.3	1,162.0	1,431.8	1,915.2	2,436.0	C	Gross National Product (GNP)
-31.9	-45.8	-54.5	-19.1	-11.1	-53.4	-140.1	-228.9	-278.3	-311.4	..	Net Factor Income from Abroad
360.3	516.1	709.5	790.2	792.9	1,049.7	1,302.1	1,660.7	2,193.5	2,747.4	C f	Gross Domestic Product (GDP)
40.9	61.6	102.0	120.6	118.8	131.9	165.4	155.0	..	..	..	Indirect Taxes, net
319.4	454.5	607.5	669.6	674.1	917.8	1,136.7	1,505.7	2,037.4	..	B C f	GDP at factor cost
71.7	81.7	83.3	90.5	87.8	77.2	76.1	82.9	93.1	87.4	..	Agriculture
107.3	193.1	291.3	310.6	270.7	440.4	590.7	814.2	1,182.3	1,559.5	..	Industry
24.4	42.8	29.2	49.3	71.2	78.7	82.0	88.0	127.1	169.2	..	Manufacturing
181.3	241.3	334.9	389.1	434.4	532.1	635.3	763.6	918.1	1,100.5	..	Services, etc.
-98.1	-81.0	-88.4	-167.7	-288.5	-131.5	-7.8	-27.9	343.9	..	..	Resource Balance
161.1	275.9	357.8	398.0	349.9	618.3	772.3	970.3	1,472.0	..	..	Exports of Goods & NFServices
259.2	356.9	446.2	565.7	638.4	749.8	780.1	998.2	1,128.1	..	..	Imports of Goods & NFServices
458.4	597.1	797.9	957.9	1,081.4	1,181.2	1,309.9	1,688.6	1,849.6	..	..	Domestic Absorption
224.6	295.2	366.3	409.8	488.4	571.2	629.3	746.3	817.0	..	..	Private Consumption, etc.
91.3	104.3	140.7	203.2	242.6	301.9	362.8	443.1	526.7	..	..	General Gov't Consumption
142.5	197.6	290.9	344.9	350.4	308.1	317.8	499.2	505.9	..	..	Gross Domestic Investment
110.1	162.9	248.8	306.6	304.6	320.3	337.6	484.0	511.9	..	..	Fixed Investment
											Memo Items:
44.4	116.6	202.5	177.2	61.9	176.6	310.0	471.3	849.8	..	..	Gross Domestic Saving
17.8	72.5	146.9	156.7	50.6	122.8	160.5	236.4	566.6	..	..	Gross National Saving
				(Millions of 1980 Botswana Pula)							
507.3	561.0	655.0	657.8	749.7	888.3	990.5	1,029.1	1,187.7	1,386.5	C	Gross National Product
492.4	548.4	607.5	652.3	638.9	812.2	973.6	1,056.9	..	..	B C	GDP at factor cost
86.8	86.6	83.3	75.0	71.8	60.1	50.9	46.8	54.6	48.8	..	Agriculture
240.3	254.2	291.3	345.0	321.3	477.9	636.1	655.1	782.5	942.7	..	Industry
32.4	43.6	29.2	37.0	45.8	42.4	44.0	35.8	43.7	48.5	..	Manufacturing
227.7	281.0	334.9	350.2	369.2	405.7	444.1	522.1	556.8	606.9	..	Services, etc.
-33.4	-62.0	-88.4	-55.8	-115.4	129.9	291.4	151.8	359.8	..	..	Resource Balance
316.5	369.7	357.8	435.2	388.0	639.5	790.0	766.4	975.4	..	..	Exports of Goods & NFServices
349.9	431.7	446.2	491.0	503.4	509.6	498.6	614.6	615.6	..	..	Imports of Goods & NFServices
588.2	683.8	797.9	826.0	877.7	813.8	839.7	1,072.2	1,034.1	..	..	Domestic Absorption
278.5	339.0	366.3	366.2	397.6	399.2	405.2	481.8	451.1	..	..	Private Consumption, etc.
125.3	119.3	140.7	163.6	194.0	216.4	240.0	280.5	305.1	..	..	General Gov't Consumption
184.4	225.5	290.9	296.2	286.1	198.2	194.5	309.9	277.9	..	..	Gross Domestic Investment
145.0	187.3	248.8	264.7	247.4	211.7	210.1	299.6	282.0	..	..	Fixed Investment
											Memo Items:
217.5	333.7	357.8	345.4	275.9	420.2	493.6	597.4	803.3	..	..	Capacity to Import
-99.0	-36.0	0.0	-89.8	-112.1	-219.3	-296.4	-169.0	-172.1	..	..	Terms of Trade Adjustment
455.8	585.8	709.5	680.4	650.2	724.4	834.7	1,055.0	1,221.8	..	..	Gross Domestic Income
408.2	525.0	655.0	568.0	637.6	669.0	694.1	860.1	1,015.6	..	..	Gross National Income
				(Index 1980 = 100)							**DOMESTIC PRICES (DEFLATORS)**
64.9	83.0	100.0	102.6	104.0	111.2	115.1	135.7	157.4	171.9	..	Overall (GDP)
77.9	87.3	100.0	116.0	123.2	145.1	156.0	157.5	178.9	..	..	Domestic Absorption
82.6	94.3	100.0	120.7	122.3	128.5	149.5	177.1	170.5	179.1	..	Agriculture
44.7	76.0	100.0	90.0	84.3	92.2	92.9	124.3	151.1	165.4	..	Industry
75.3	98.2	100.0	133.2	155.5	185.6	186.4	245.8	290.8	348.9	..	Manufacturing
											MANUFACTURING ACTIVITY
81.1	99.9	100.0	120.4	135.9	177.0	177.0	182.5	..	..	G	Employment (1980=100)
123.5	92.8	100.0	96.0	96.9	80.3	81.2	84.8	..	..	G	Real Earnings per Empl. (1980=100)
110.9	114.0	100.0	89.2	86.5	70.0	68.5	..	..	..	G	Real Output per Empl. (1980=100)
50.4	29.7	51.1	48.1	36.6	38.5	40.3	..	..	..	..	Earnings as % of Value Added
				(Millions of current Botswana Pula)							**MONETARY HOLDINGS**
121.0	199.0	236.9	226.5	245.6	316.3	368.9	557.4	606.2	1,013.4	..	Money Supply, Broadly Defined
60.9	82.1	90.6	114.8	127.4	137.3	150.6	188.2	243.5	312.1	..	Money as Means of Payment
15.8	17.9	24.4	29.6	29.0	30.2	35.2	43.4	58.5	68.6	..	Currency Ouside Banks
45.2	64.2	66.2	85.2	98.4	107.2	115.4	144.9	184.9	243.6	..	Demand Deposits
60.0	116.9	146.3	111.7	118.2	178.9	218.3	369.2	362.7	701.3	..	Quasi-Monetary Liabilities
				(Millions of current Botswana Pula)							
-6.9	21.4	-1.3	-18.3	-20.1	103.2	188.3	413.8	..	..	C	**GOVERNMENT DEFICIT (-) OR SURPLUS**
160.1	242.3	299.9	311.4	385.7	554.5	787.9	1,116.3	..	..	..	Current Revenue
100.0	137.0	178.6	211.9	270.1	316.0	416.3	497.6	..	..	..	Current Expenditure
60.1	105.3	121.3	99.5	115.6	238.5	371.6	618.7	..	..	..	Current Budget Balance
0.7	0.9	1.8	3.7	1.9	1.7	3.2	4.4	..	..	..	Capital Receipts
67.7	84.8	124.4	121.5	137.6	136.9	186.5	209.2	..	..	..	Capital Payments

BOTSWANA	1967	1968	1969	1970	1971	1972	1973	1974	1975	1976	1977
FOREIGN TRADE (CUSTOMS BASIS)				*(Millions of current US dollars)*							
Value of Exports, fob	13.0	10.0	18.0	22.0	28.0	58.0	85.0	121.0	142.0	176.0	180.0
Nonfuel Primary Products	12.4	9.5	17.1	16.5	18.6	31.9	55.3	70.1	92.2	126.2	115.0
Fuels	0.0	0.0	0.0	0.0	0.0	0.1	0.1	0.1	0.2	0.2	0.2
Manufactures	0.6	0.5	0.9	5.5	9.3	26.0	29.7	50.8	49.6	49.6	64.9
Value of Imports, cif	31.0	33.0	43.0	49.0	49.0	109.0	166.0	185.0	218.0	209.0	276.0
Nonfuel Primary Products	8.4	8.9	11.6	13.3	13.3	29.5	45.0	50.1	58.8	56.4	74.3
Fuels	3.0	3.1	4.1	4.7	4.7	10.4	15.8	17.6	21.6	20.7	27.9
Manufactures	19.6	20.9	27.3	31.1	31.1	69.1	105.2	117.3	137.6	131.9	173.8
				(Index 1980 = 100)							
Terms of Trade	177.0	174.4	179.2	179.6	158.5	158.4	163.0	119.8	115.7	112.5	104.0
Export Prices, fob	35.3	34.5	36.6	37.7	38.0	42.4	56.5	67.6	68.5	69.2	69.2
Nonfuei Primary Products	36.0	35.2	37.4	38.4	39.2	46.7	61.7	68.0	70.0	70.2	67.9
Fuels	4.3	4.3	4.3	4.3	5.6	6.2	8.9	36.7	35.7	38.4	42.0
Manufactures	30.1	31.7	32.4	37.5	36.5	38.5	49.5	67.2	66.1	66.7	71.7
Import Prices, cif	19.9	19.8	20.4	21.0	24.0	26.8	34.7	56.4	59.2	61.5	66.5
Trade at Constant 1980 Prices				*(Millions of 1980 US dollars)*							
Exports, fob	36.8	29.0	49.1	58.3	73.6	136.9	150.3	179.0	207.2	254.5	260.2
Imports, cif	155.4	166.8	210.3	233.1	204.1	407.3	478.5	327.8	368.1	340.0	415.1
BALANCE OF PAYMENTS				*(Millions of current US dollars)*							
Exports of Goods & Services	..	..	..	58.0	85.0	102.0	157.0	159.0	219.9	275.9	311.3
Merchandise, fob	..	..	..	26.0	38.0	57.0	92.0	120.0	142.0	169.7	191.6
Nonfactor Services	..	..	..	31.0	46.0	43.0	63.0	36.0	30.8	46.8	51.4
Factor Services	..	..	..	1.0	1.0	2.0	2.0	3.0	47.1	59.3	68.3
Imports of Goods & Services	..	..	..	94.0	130.0	161.0	225.0	179.7	273.6	310.5	370.7
Merchandise, fob	..	..	..	57.0	78.0	98.0	137.0	164.7	181.2	180.2	226.4
Nonfactor Services	..	..	..	34.0	46.0	50.0	70.0	41.0	46.7	78.4	89.9
Factor Services	..	..	..	3.0	6.0	13.0	18.0	15.0	45.7	51.9	54.4
Long-Term Interest	..	..	..	0.4	0.6	1.2	2.2	3.0	3.4	3.2	3.5
Current Transfers, net	..	..	..	1.0	4.0	10.0	19.0	42.0	-7.4	5.5	6.2
Workers' Remittances	..	..	..	..	..	..	..	..	0.0	0.0	0.0
Total to be Financed	..	..	..	-35.0	-41.0	-49.0	-49.0	-21.6	-37.9	-29.1	-53.2
Official Capital Grants	..	..	..	5.0	4.0	10.0	1.0	2.0	28.3	42.7	78.4
Current Account Balance	..	..	..	-30.0	-37.0	-39.0	-48.0	-19.6	-9.6	13.6	25.2
Long-Term Capital, net	..	..	..	20.0	96.0	166.0	143.0	126.0	78.4	62.4	67.9
Direct Investment	..	..	..	6.0	38.0	60.0	53.0	46.0	-38.3	11.2	12.0
Long-Term Loans	..	..	..	5.5	15.4	44.9	41.4	18.2	21.0	24.8	13.4
Disbursements	..	..	..	5.7	15.7	45.3	41.7	18.5	24.7	25.9	15.8
Repayments	..	..	..	0.2	0.3	0.4	0.3	0.3	3.7	1.1	2.4
Other Long-Term Capital	..	..	..	3.5	38.6	51.1	47.6	59.8	67.4	-16.2	-35.8
Other Capital, net	..	..	..	10.5	-59.0	-74.0	-53.0	-106.4	..	-4.1	-75.4
Change in Reserves	..	..	..	-0.5	0.0	-53.0	-42.0	0.0	..	-71.9	-17.7
Memo Item:				*(Botswana Pula per US dollar)*							
Conversion Factor (Annual Avg)	0.710	0.710	0.710	0.710	0.720	0.770	0.690	0.680	0.740	0.870	0.840
EXTERNAL DEBT, ETC.				*(Millions of US dollars, outstanding at end of year)*							
Public/Publicly Guar. Long-Term	..	..	..	17.4	33.2	74.8	114.9	134.1	147.3	165.2	180.9
Official Creditors	..	..	..	14.2	30.3	72.0	111.8	131.1	147.3	163.0	178.7
IBRD and IDA	..	..	..	5.2	6.0	18.9	34.9	42.9	46.4	50.0	53.7
Private Creditors	..	..	..	3.2	2.9	2.8	3.1	2.9	0.1	2.2	2.2
Private Non-guaranteed Long-Term	..	..	..	..	..	..	..	..	..	..	..
Use of Fund Credit	..	..	..	0.0	0.0	0.0	0.0	0.0	0.0	0.0	0.0
Short-Term Debt	..	..	..	..	..	..	..	..	..	..	3.0
Memo Items:				*(Millions of US dollars)*							
Int'l Reserves Excluding Gold	..	..	..	..	..	..	..	..	..	74.9	100.1
Gold Holdings (at market price)	..	..	..	..	..	..	..	..	..	..	..
SOCIAL INDICATORS											
Total Fertility Rate	6.9	6.9	6.9	6.9	6.9	6.9	6.9	6.8	6.8	6.7	6.7
Crude Birth Rate	53.7	53.4	53.1	52.7	52.4	52.1	51.8	51.5	51.2	50.9	50.6
Infant Mortality Rate	110.0	107.0	104.0	101.0	98.0	95.0	92.4	89.8	87.2	84.6	82.0
Life Expectancy at Birth	48.5	48.9	49.3	49.7	50.1	50.5	50.9	51.3	51.7	52.1	52.5
Food Production, p.c. ('79-81 = 100)	167.8	155.9	159.6	156.6	174.3	163.9	145.2	151.4	129.0	144.7	127.2
Labor Force, Agriculture (%)	87.4	86.8	86.1	85.5	84.0	82.5	81.1	79.6	78.1	76.5	75.0
Labor Force, Female (%)	44.3	44.1	43.9	43.7	43.1	42.5	41.9	41.3	40.6	40.1	39.5
School Enroll. Ratio, primary	..	..	..	65.0	..	..	..	..	72.0	82.0	89.0
School Enroll. Ratio, secondary	..	..	..	7.0	..	..	..	..	16.0	18.0	20.0

1978	1979	1980	1981	1982	1983	1984	1985	1986	1987 estimate	NOTES	BOTSWANA
											FOREIGN TRADE (CUSTOMS BASIS)
				(Millions of current US dollars)							
222.0	436.0	503.0	413.0	456.0	637.0	673.6	744.5	857.7	1,010.1	f	Value of Exports, fob
114.9	176.9	150.1	191.9	166.4	240.9	266.3	281.6	297.9	368.4	..	Nonfuel Primary Products
0.3	0.4	0.1	0.4	0.4	0.5	0.5	0.5	0.2	0.3	..	Fuels
106.9	258.6	352.8	220.7	289.2	395.6	406.8	462.4	559.6	641.3	..	Manufactures
353.0	521.0	691.0	799.0	686.0	735.0	706.8	583.1	684.1	783.3	f	Value of Imports, cif
96.6	134.9	171.1	199.4	174.7	185.9	168.6	127.5	155.7	193.9	..	Nonfuel Primary Products
30.4	72.4	90.4	104.8	103.5	106.5	92.2	93.2	41.5	59.7	..	Fuels
226.0	313.6	429.5	494.9	407.8	442.6	446.0	362.5	486.9	529.7	..	Manufactures
				(Index 1980 = 100)							
102.1	107.5	100.0	98.2	97.8	97.7	97.2	98.0	104.7	106.1	..	Terms of Trade
75.2	92.4	100.0	98.2	93.0	91.0	89.5	87.6	98.2	107.4	..	Export Prices, fob
74.6	96.3	100.0	97.1	90.0	91.5	89.7	85.5	90.1	98.8	..	Nonfuel Primary Products
42.3	61.0	100.0	112.5	101.6	92.5	90.2	87.5	44.9	56.9	..	Fuels
76.1	90.0	100.0	99.1	94.8	90.7	89.3	89.0	103.2	113.0	..	Manufactures
73.7	86.0	100.0	100.0	95.0	93.1	92.0	89.4	93.8	101.2	..	Import Prices, cif
											Trade at Constant 1980 Prices
				(Millions of 1980 US dollars)							
295.1	471.8	503.0	420.8	490.4	700.0	753.0	849.6	873.5	940.9	..	Exports, fob
479.1	605.8	691.1	799.0	721.7	789.2	768.1	652.2	729.2	773.9	..	Imports, cif
				(Millions of current US dollars)							**BALANCE OF PAYMENTS**
332.8	575.6	747.5	604.4	645.5	836.0	877.6	889.0	1,066.2	1,866.3	..	Exports of Goods & Services
223.4	442.2	544.5	401.3	460.6	640.3	677.7	727.9	850.2	1,582.2	..	Merchandise, fob
50.5	61.1	100.9	97.0	102.5	113.0	107.2	76.3	99.8	126.9	..	Nonfactor Services
58.9	72.3	102.2	106.0	82.4	82.7	92.7	84.9	116.2	157.2	..	Factor Services
465.4	666.6	953.5	948.6	835.4	989.0	997.6	875.2	1,038.1	1,406.2	..	Imports of Goods & Services
295.0	442.1	602.5	687.1	579.8	615.3	583.4	494.2	607.0	852.3	..	Merchandise, fob
101.8	128.3	177.0	176.5	177.9	210.7	195.7	147.4	169.4	188.8	..	Nonfactor Services
68.6	96.2	174.0	85.0	77.7	163.1	218.5	233.6	261.7	365.1	..	Factor Services
4.6	5.8	6.6	6.7	9.7	12.8	15.1	20.4	27.0	31.9	..	Long-Term Interest
6.4	2.1	-1.4	-1.7	-0.2	-0.4	-7.3	-3.2	-2.6	-1.8	..	Current Transfers, net
0.0	0.0	0.0	0.0	0.0	0.0	0.0	0.0	0.0	0.0	..	Workers' Remittances
-126.2	-89.0	-207.4	-345.9	-190.2	-153.4	-127.4	10.7	25.5	458.3	K	Total to be Financed
65.7	98.9	132.4	128.1	98.7	128.5	110.9	76.9	114.2	138.9	K	Official Capital Grants
-60.5	9.9	-75.0	-217.8	-91.5	-24.9	-16.5	87.6	139.6	597.2	..	Current Account Balance
133.4	198.8	268.0	249.8	204.4	192.5	225.7	190.6	255.9	264.0	..	Long-Term Capital, net
40.8	127.9	109.2	88.3	21.1	22.5	61.9	52.1	90.2	125.1	..	Direct Investment
7.7	12.1	18.4	22.6	52.9	24.7	59.6	42.4	10.7	63.9	..	Long-Term Loans
11.3	15.6	24.4	24.6	56.6	35.7	75.6	68.3	28.8	101.6	..	Disbursements
3.6	3.5	6.0	2.0	3.7	11.0	16.0	25.9	18.1	37.7	..	Repayments
19.2	-40.1	8.0	10.7	31.8	16.8	-6.7	19.2	40.8	-63.9	..	Other Long-Term Capital
-33.6	-92.7	-102.5	-105.8	-58.1	-44.0	-84.9	-23.9	-93.9	-285.6	..	Other Capital, net
-39.4	-116.0	-90.4	73.8	-54.9	-123.6	-124.3	-254.3	-301.6	-575.6	..	Change in Reserves
											Memo Item:
				(Botswana Pula per US dollar)							
0.830	0.810	0.780	0.830	1.020	1.100	1.280	1.890	1.890	1.810	..	Conversion Factor (Annual Avg)
				(Millions of US dollars, outstanding at end of year)							**EXTERNAL DEBT, ETC.**
120.6	134.3	151.6	164.6	210.6	230.4	262.6	341.3	387.4	514.4	..	Public/Publicly Guar. Long-Term
117.7	129.6	146.7	160.2	176.5	199.4	225.9	306.5	347.9	470.9	..	Official Creditors
56.0	61.4	65.6	76.2	87.0	101.4	103.7	137.0	162.8	194.8	..	IBRD and IDA
2.9	4.7	4.9	4.4	34.2	30.9	36.7	34.9	39.5	43.5	..	Private Creditors
..	..	..	..	..	..	..	..	..	..	..	Private Non-guaranteed Long-Term
0.0	0.0	0.0	0.0	0.0	0.0	0.0	0.0	0.0	..	..	Use of Fund Credit
3.0	17.0	4.0	8.0	3.0	4.0	5.0	2.0	3.0	3.2	..	Short-Term Debt
				(Millions of US dollars)							Memo Items:
150.6	267.3	343.7	253.4	293.0	395.7	474.3	783.2	1,197.7	2,057.1	..	Int'l Reserves Excluding Gold
..	..	..	..	..	..	..	..	..	..	..	Gold Holdings (at market price)
											SOCIAL INDICATORS
6.7	6.8	6.8	6.9	6.9	6.8	6.7	6.7	6.6	6.5	..	Total Fertility Rate
49.8	49.1	48.3	47.6	46.8	46.3	45.9	45.4	44.9	44.4	..	Crude Birth Rate
80.8	79.6	78.4	77.2	76.0	74.6	73.2	71.9	70.5	69.1	..	Infant Mortality Rate
53.4	54.3	55.1	56.0	56.9	57.3	57.8	58.3	58.7	59.2	..	Life Expectancy at Birth
101.3	110.7	88.3	101.0	96.9	87.5	79.1	73.9	75.5	76.6	..	Food Production, p.c. ('79-81=100)
73.4	71.9	70.3	..	..	..	..	..	..	..	..	Labor Force, Agriculture (%)
38.9	38.4	37.8	37.5	37.2	36.9	36.6	36.3	36.0	35.8	..	Labor Force, Female (%)
91.0	95.0	91.0	102.0	96.0	99.0	101.0	104.0	..	..	..	School Enroll. Ratio, primary
21.0	21.0	19.0	23.0	21.0	21.0	25.0	29.0	..	..	..	School Enroll. Ratio, secondary

BRAZIL	1967	1968	1969	1970	1971	1972	1973	1974	1975	1976	1977
CURRENT GNP PER CAPITA (US $)	310	370	410	450	500	570	700	880	1,070	1,300	1,470
POPULATION (millions)	89	91	93	96	98	101	103	106	108	111	113

ORIGIN AND USE OF RESOURCES

(Billions of current Brazilian Cruzados)

	1967	1968	1969	1970	1971	1972	1973	1974	1975	1976	1977
Gross National Product (GNP)	7.9E-02	1.1E-01	1.5E-01	1.9E-01	2.6E-01	3.4E-01	4.8E-01	7.1E-01	9.9E-01	2.0E+00	2.0E+00
Net Factor Income from Abroad	-1.8E-03	-2.4E-03	-3.2E-03	-1.8E-03	-2.5E-03	-3.3E-03	-4.5E-03	-6.2E-03	-1.4E-02	-2.5E-02	-4.0E-02
Gross Domestic Product (GDP)	8.1E-02	1.2E-01	1.5E-01	1.9E-01	2.6E-01	3.5E-01	4.9E-01	7.1E-01	1.0E+00	2.0E+00	2.0E+00
Indirect Taxes, net	1.1E-02	1.8E-02	2.5E-02	3.1E-02	3.8E-02	5.1E-02	6.9E-02	9.3E-02	1.2E-01	2.0E-01	3.0E-01
GDP at factor cost	7.0E-02	9.7E-02	1.3E-01	1.6E-01	2.2E-01	3.0E-01	4.2E-01	6.2E-01	8.9E-01	1.0E+00	2.0E+00
Agriculture	1.1E-02	1.4E-02	1.7E-02	2.0E-02	2.9E-02	3.8E-02	5.6E-02	8.0E-02	1.1E-01	1.9E-01	3.2E-01
Industry	2.4E-02	3.5E-02	4.8E-02	6.3E-02	8.5E-02	1.1E-01	1.6E-01	2.5E-01	3.6E-01	5.7E-01	8.5E-01
Manufacturing	1.8E-02	2.7E-02	3.7E-02	4.8E-02	6.5E-02	8.7E-02	1.2E-01	1.9E-01	2.7E-01	4.4E-01	6.4E-01
Services, etc.	3.5E-02	4.8E-02	6.3E-02	8.1E-02	1.1E-01	1.4E-01	2.0E-01	2.9E-01	4.2E-01	6.8E-01	1.0E+00
Resource Balance	0.0E+00	-7.0E-04	1.0E-04	-8.2E-04	-4.5E-03	-5.5E-03	-6.0E-03	-4.2E-02	-4.0E-02	-3.9E-02	-1.7E-02
Exports of Goods & NFServices	4.7E-03	6.9E-03	1.0E-02	1.4E-02	1.7E-02	2.5E-02	4.0E-02	5.7E-02	7.6E-02	1.1E-01	1.8E-01
Imports of Goods & NFServices	4.7E-03	7.6E-03	1.0E-02	1.4E-02	2.1E-02	3.1E-02	4.6E-02	9.9E-02	1.2E-01	1.5E-01	2.0E-01
Domestic Absorption	8.1E-02	1.2E-01	1.5E-01	2.0E-01	2.6E-01	3.5E-01	4.9E-01	7.6E-01	1.0E+00	2.0E+00	3.0E+00
Private Consumption, etc.	5.8E-02	8.1E-02	1.0E-01	1.3E-01	1.8E-01	2.4E-01	3.3E-01	5.0E-01	6.7E-01	1.0E+00	2.0E+00
General Gov't Consumption	9.4E-03	1.3E-02	1.6E-02	2.2E-02	2.9E-02	3.7E-02	5.1E-02	7.0E-02	1.1E-01	1.7E-01	2.3E-01
Gross Domestic Investment	1.4E-02	2.2E-02	3.4E-02	4.0E-02	5.5E-02	7.4E-02	1.1E-01	1.8E-01	2.7E-01	3.8E-01	5.5E-01
Fixed Investment	..	..	..	3.7E-02	5.1E-02	7.0E-02	1.0E-01	1.6E-01	2.4E-01	3.7E-01	5.3E-01
Memo Items:											
Gross Domestic Saving	1.3E-02	2.1E-02	3.4E-02	3.9E-02	5.0E-02	6.8E-02	1.1E-01	1.4E-01	2.3E-01	3.4E-01	5.3E-01
Gross National Saving	1.2E-02	1.9E-02	3.1E-02	3.7E-02	4.8E-02	6.5E-02	1.0E-01	1.3E-01	2.2E-01	3.1E-01	..

(Millions of 1980 Brazilian Cruzados)

	1967	1968	1969	1970	1971	1972	1973	1974	1975	1976	1977
Gross National Product	4,044	4,510	4,947	5,446	6,061	6,791	7,743	8,448	8,838	9,693	10,130
GDP at factor cost	3,706	4,122	4,529	4,959	5,519	6,184	7,048	7,685	8,080	8,873	9,283
Agriculture	706	740	767	775	854	888	888	897	961	985	1,104
Industry	1,393	1,582	1,771	1,928	2,157	2,466	2,889	3,130	3,279	3,664	3,777
Manufacturing	1,094	1,248	1,389	1,518	1,698	1,935	2,256	2,432	2,524	2,830	2,895
Services, etc.	1,607	1,799	1,990	2,255	2,509	2,830	3,272	3,659	3,840	4,224	4,401
Resource Balance	-62	-99	-82	-175	-300	-341	-446	-744	-603	-588	-490
Exports of Goods & NFServices	295	343	410	435	459	570	651	666	744	741	739
Imports of Goods & NFServices	358	442	492	610	759	911	1,097	1,410	1,346	1,329	1,229
Domestic Absorption	4,196	4,704	5,137	5,673	6,419	7,198	8,261	9,265	9,568	10,431	10,787
Private Consumption, etc.	2,975	3,316	3,373	3,871	4,396	4,897	5,487	6,172	6,120	6,891	7,305
General Gov't Consumption	526	556	599	639	696	756	849	850	979	1,054	997
Gross Domestic Investment	696	832	1,165	1,162	1,326	1,545	1,925	2,242	2,469	2,486	2,485
Fixed Investment	..	..	..	1,115	1,286	1,501	1,816	2,056	2,256	2,415	2,387
Memo Items:											
Capacity to Import	358	401	497	576	598	748	955	814	882	992	1,126
Terms of Trade Adjustment	62	58	87	140	139	178	304	147	138	250	387
Gross Domestic Income	4,196	4,664	5,142	5,638	6,258	7,034	8,119	8,669	9,104	10,093	10,683
Gross National Income	4,106	4,568	5,035	5,586	6,200	6,969	8,047	8,595	8,976	9,943	10,517

DOMESTIC PRICES (DEFLATORS)

(Index 1980 = 100)

	1967	1968	1969	1970	1971	1972	1973	1974	1975	1976	1977
Overall (GDP)	2.0	2.5	3.0	3.5	4.3	5.1	6.2	8.4	11.2	16.5	24.2
Domestic Absorption	1.9	2.5	3.0	3.4	4.1	4.9	5.9	8.2	10.9	16.0	23.2
Agriculture	1.5	1.8	2.2	2.6	3.4	4.3	6.3	8.9	11.2	18.9	29.2
Industry	1.7	2.2	2.7	3.2	3.9	4.6	5.6	7.9	10.9	15.6	22.5
Manufacturing	1.7	2.2	2.7	3.2	3.8	4.5	5.5	7.8	10.6	15.4	22.2

MANUFACTURING ACTIVITY

	1967	1968	1969	1970	1971	1972	1973	1974	1975	1976	1977
Employment (1980=100)	41.3	44.1	44.5	46.5	49.4	53.7	69.8	73.9	77.1	82.1	86.7
Real Earnings per Empl. (1980=100)	44.2	46.4	49.9	67.2	75.0	86.1	87.5	93.5	89.2	97.7	99.6
Real Output per Empl. (1980=100)	57.0	59.0	63.7	71.0	72.1	79.4	85.4	94.1	101.5	99.3	97.0
Earnings as % of Value Added	17.3	16.5	16.7	22.3	23.6	25.4	23.3	21.9	18.9	20.1	20.9

MONETARY HOLDINGS

(Billions of current Brazilian Cruzados)

	1967	1968	1969	1970	1971	1972	1973	1974	1975	1976	1977
Money Supply, Broadly Defined	2.1E-02	2.9E-02	3.9E-02	5.0E-02	5.5E-02	8.3E-02	1.3E-01	1.7E-01	2.8E-01	4.2E-01	6.4E-01
Money as Means of Payment	1.5E-02	2.1E-02	2.7E-02	3.5E-02	4.2E-02	5.9E-02	8.7E-02	1.2E-01	1.7E-01	2.3E-01	3.2E-01
Currency Ouside Banks	3.0E-03	4.0E-03	5.0E-03	7.0E-03	9.0E-03	1.2E-02	1.6E-02	2.2E-02	3.1E-02	4.6E-02	6.5E-02
Demand Deposits	1.2E-02	1.7E-02	2.2E-02	2.8E-02	3.3E-02	4.7E-02	7.1E-02	9.5E-02	1.4E-01	1.8E-01	2.5E-01
Quasi-Monetary Liabilities	5.8E-03	8.0E-03	1.2E-02	1.5E-02	1.3E-02	2.4E-02	4.0E-02	5.8E-02	1.1E-01	1.8E-01	3.2E-01

(Billions of current Brazilian Cruzados)

	1967	1968	1969	1970	1971	1972	1973	1974	1975	1976	1977
GOVERNMENT DEFICIT (-) OR SURPLUS	..	..	..	-1.00E-03	-2.00E-03	-1.00E-03	2.00E-03	9.00E-03	-5.00E-03	-3.00E-03	-2.10E-02
Current Revenue	..	..	..	3.40E-02	4.60E-02	6.50E-02	9.30E-02	1.40E-01	2.00E-01	3.20E-01	5.60E-01
Current Expenditure	..	..	..	3.10E-02	3.90E-02	5.10E-02	7.20E-02	1.00E-01	1.60E-01	2.50E-01	4.30E-01
Current Budget Balance	..	..	..	3.00E-03	7.00E-03	1.40E-02	2.10E-02	3.40E-02	3.50E-02	6.90E-02	1.30E-01
Capital Receipts	..	..	..	1.00E-03	2.00E-03	1.00E-03	2.00E-03	1.00E-03	1.00E-03	4.00E-03	4.00E-03
Capital Payments	..	..	..	5.00E-03	1.10E-02	1.60E-02	2.10E-02	2.60E-02	4.10E-02	7.60E-02	1.50E-01

1978	1979	1980	1981	1982	1983	1984	1985	1986	1987 estimate	NOTES	BRAZIL
1,660	1,890	2,080	2,060	2,060	1,820	1,710	1,670	1,830	2,020	..	**CURRENT GNP PER CAPITA** (US $)
116	119	121	124	127	130	133	136	138	141	..	**POPULATION** (millions)

(Billions of current Brazilian Cruzados) — **ORIGIN AND USE OF RESOURCES**

1978	1979	1980	1981	1982	1983	1984	1985	1986	1987	NOTES	
4.0E+00	6.0E+00	1.2E+01	2.4E+01	4.6E+01	1.1E+02	3.7E+02	1.3E+03	3.7E+03	1.2E+04	f	Gross National Product (GNP)
-8.4E-02	-1.6E-01	-4.0E-01	-1.0E+00	-3.0E+00	-7.0E+00	-2.2E+01	-7.4E+01	-1.6E+02	-4.5E+02		Net Factor Income from Abroad
4.0E+00	6.0E+00	1.3E+01	2.5E+01	4.9E+01	1.2E+02	3.9E+02	1.4E+03	3.8E+03	1.3E+04	f	Gross Domestic Product (GDP)
4.2E-01	6.1E-01	1.0E+00	3.0E+00	5.0E+00	1.2E+01	3.4E+01	1.2E+02	4.1E+02	1.1E+03	..	Indirect Taxes, net
3.0E+00	5.0E+00	1.1E+01	2.2E+01	4.4E+01	1.1E+02	3.6E+02	1.3E+03	3.4E+03	1.2E+04	f	GDP at factor cost
3.7E-01	6.0E-01	1.0E+00	2.0E+00	4.0E+00	1.2E+01	4.1E+01	1.4E+02	3.8E+02	..	..	Agriculture
1.0E+00	2.0E+00	5.0E+00	9.0E+00	1.7E+01	4.1E+01	1.4E+02	5.0E+02	1.3E+03	..	..	Industry
9.9E-01	2.0E+00	4.0E+00	7.0E+00	1.3E+01	3.1E+01	1.1E+02	3.8E+02	9.5E+02	..	..	Manufacturing
2.0E+00	3.0E+00	6.0E+00	1.1E+01	2.2E+01	5.5E+01	1.8E+02	6.5E+02	1.7E+03	..	..	Services, etc.
-4.3E-02	-1.2E-01	-2.8E-01	-9.3E-02	-3.4E-01	3.0E+00	2.2E+01	7.1E+01	9.6E+01	3.9E+02		Resource Balance
2.4E-01	4.3E-01	1.0E+00	2.0E+00	4.0E+00	1.3E+01	5.2E+01	1.7E+02	3.3E+02	1.1E+03		Exports of Goods & NFServices
2.9E-01	5.6E-01	1.0E+00	2.0E+00	4.0E+00	1.1E+01	3.1E+01	9.8E+01	2.3E+02	7.1E+02		Imports of Goods & NFServices
4.0E+00	6.0E+00	1.3E+01	2.5E+01	4.9E+01	1.2E+02	3.7E+02	1.3E+03	3.7E+03	1.2E+04		Domestic Absorption
2.0E+00	4.0E+00	9.0E+00	1.7E+01	3.4E+01	8.7E+01	2.8E+02	9.7E+02	2.6E+03	8.3E+03		Private Consumption, etc.
3.5E-01	5.9E-01	1.0E+00	2.0E+00	5.0E+00	1.1E+01	3.2E+01	1.4E+02	3.9E+02	1.6E+03		General Gov't Consumption
8.3E-01	1.0E+00	3.0E+00	5.0E+00	1.0E+01	1.8E+01	6.0E+01	2.4E+02	7.1E+02	2.5E+03		Gross Domestic Investment
8.1E-01	1.0E+00	3.0E+00	5.0E+00	1.0E+01	1.9E+01	6.0E+01	2.4E+02	7.1E+02	2.5E+03		Fixed Investment

Memo Items:

1978	1979	1980	1981	1982	1983	1984	1985	1986	1987	NOTES	
7.9E-01	1.0E+00	3.0E+00	5.0E+00	9.0E+00	2.0E+01	8.2E+01	3.1E+02	8.1E+02	2.9E+03		Gross Domestic Saving
7.1E-01	1.0E+00	2.0E+00	4.0E+00	7.0E+00	1.4E+01	6.1E+01	2.4E+02	6.5E+02	2.5E+03		Gross National Saving

(Millions of 1980 Brazilian Cruzados)

1978	1979	1980	1981	1982	1983	1984	1985	1986	1987	NOTES	
10,543	11,260	12,222	11,725	11,707	11,324	11,982	13,044	14,250	14,822	..	Gross National Product
9,727	10,432	11,412	11,043	11,165	10,857	11,473	12,439	13,440	13,837	..	GDP at factor cost
1,072	1,124	1,232	1,333	1,328	1,323	1,363	1,501	1,380	1,548	..	Agriculture
4,018	4,284	4,678	4,249	4,245	3,975	4,231	4,610	5,157	5,192	..	Industry
3,072	3,282	3,581	3,210	3,195	3,000	3,183	3,448	3,836	..	..	Manufacturing
4,638	5,024	5,502	5,461	5,591	5,558	5,879	6,328	6,902	7,097	..	Services, etc.
-449	-476	-278	134	83	462	800	919	488	597	..	Resource Balance
836	914	1,121	1,360	1,235	1,412	1,724	1,843	1,655	1,540	..	Exports of Goods & NFServices
1,285	1,391	1,400	1,227	1,153	950	924	924	1,167	942	..	Imports of Goods & NFServices
11,241	12,047	12,905	12,097	12,281	11,552	11,895	12,839	14,377	14,711	..	Domestic Absorption
7,606	8,312	8,875	8,482	8,710	8,676	8,872	9,304	10,169	10,934	..	Private Consumption, etc.
1,071	1,164	1,139	1,129	1,284	1,124	1,027	1,305	1,511	1,340	..	General Gov't Consumption
2,563	2,571	2,890	2,486	2,287	1,752	1,996	2,230	2,697	2,436	..	Gross Domestic Investment
2,500	2,597	2,835	2,460	2,320	1,921	1,996	2,230	2,697	2,436	..	Fixed Investment

Memo Items:

1978	1979	1980	1981	1982	1983	1984	1985	1986	1987	NOTES	
1,091	1,080	1,121	1,179	1,060	1,204	1,579	1,595	1,659	1,457	..	Capacity to Import
254	165	0	-181	-175	-207	-145	-248	4	-83	..	Terms of Trade Adjustment
11,047	11,736	12,626	12,050	12,188	11,806	12,550	13,510	14,869	15,225	..	Gross Domestic Income
10,797	11,425	12,222	11,544	11,532	11,117	11,837	12,795	14,254	14,739	..	Gross National Income

(Index 1980 = 100) — **DOMESTIC PRICES (DEFLATORS)**

1978	1979	1980	1981	1982	1983	1984	1985	1986	1987	NOTES	
33.6	52.4	100.0	200.7	394.5	991.4	3076.7	10307.2	25740.2	83541.8	..	Overall (GDP)
32.6	51.3	100.0	203.7	399.9	1006.5	3101.1	10490.3	25944.0	84319.4	..	Domestic Absorption
34.8	53.4	100.0	168.2	285.3	891.9	2976.0	9564.3	27669.7	..	..	Agriculture
32.1	51.7	100.0	205.9	411.2	1021.2	3271.6	10938.4	25339.9	..	..	Industry
32.1	51.5	100.0	209.6	412.4	1044.6	3349.9	10900.5	24701.6	..	..	Manufacturing

MANUFACTURING ACTIVITY

1978	1979	1980	1981	1982	1983	1984	1985	1986	1987	NOTES	
91.9	95.4	100.0	105.3	110.9	116.7	122.9	129.4	..	..	G	Employment (1980=100)
104.2	108.7	100.0	99.7	92.3	84.0	90.8	92.9	..	..	G	Real Earnings per Empl. (1980=100)
94.8	94.3	100.0	84.4	79.8	71.1	71.7	73.7	..	..	G	Real Output per Empl. (1980=100)
21.2	20.7	17.1	19.7	19.7	19.7	19.7	19.7	..	..	..	Earnings as % of Value Added

(Billions of current Brazilian Cruzados) — **MONETARY HOLDINGS**

1978	1979	1980	1981	1982	1983	1984	1985	1986	1987	NOTES	
9.6E-01	2.0E+00	3.0E+00	6.0E+00	1.3E+01	3.6E+01	1.3E+02	4.7E+02	..	..	D	Money Supply, Broadly Defined
4.3E-01	7.6E-01	1.0E+00	2.0E+00	4.0E+00	8.0E+00	2.4E+01	1.1E+02	..	..	..	Money as Means of Payment
9.3E-02	1.7E-01	2.9E-01	5.2E-01	1.0E+00	2.0E+00	6.0E+00	2.4E+01	..	..	..	Currency Ouside Banks
3.4E-01	5.9E-01	1.0E+00	2.0E+00	3.0E+00	6.0E+00	1.8E+01	8.2E+01	..	..	..	Demand Deposits
5.2E-01	9.2E-01	2.0E+00	4.0E+00	9.0E+00	2.8E+01	1.0E+02	3.6E+02	..	..	..	Quasi-Monetary Liabilities

(Billions of current Brazilian Cruzados) — **GOVERNMENT DEFICIT (-) OR SURPLUS**

1978	1979	1980	1981	1982	1983	1984	1985	1986	1987	NOTES	
-6.30E-02	-3.50E-02	-3.00E-01	-6.00E-01	-1.29E+00	-4.84E+00	-1.88E+01	-1.54E+02	..	..	E	GOVERNMENT DEFICIT (-) OR SURPLUS
8.30E-01	1.34E+00	2.74E+00	6.00E+00	1.26E+01	3.00E+01	9.05E+01	3.63E+02	..	..	..	Current Revenue
6.50E-01	1.04E+00	2.34E+00	4.56E+00	9.74E+00	2.37E+01	7.84E+01	3.49E+02	..	..	..	Current Expenditure
1.80E-01	3.00E-01	4.10E-01	1.44E+00	2.85E+00	6.29E+00	1.21E+01	1.38E+01	..	..	..	Current Budget Balance
5.00E-03	2.40E-02	5.40E-02	7.40E-02	1.80E-01	1.30E+00	1.35E+00	4.82E+00	..	..	..	Capital Receipts
2.50E-01	3.60E-01	7.60E-01	2.12E+00	4.32E+00	1.24E+01	3.22E+01	1.73E+02	..	..	..	Capital Payments

BRAZIL	1967	1968	1969	1970	1971	1972	1973	1974	1975	1976	1977
FOREIGN TRADE (CUSTOMS BASIS)				*(Millions of current US dollars)*							
Value of Exports, fob	1,654	1,881	2,311	2,739	2,904	3,991	6,199	7,951	8,669	10,128	12,120
Nonfuel Primary Products	1,480	1,716	2,062	2,335	2,350	3,167	4,779	5,769	6,098	7,377	8,763
Fuels	1	1	3	16	24	40	84	111	201	251	217
Manufactures	173	164	245	388	530	784	1,337	2,071	2,371	2,500	3,141
Value of Imports, cif	1,667	2,129	2,263	2,845	3,696	4,776	6,992	14,163	13,578	13,713	13,254
Nonfuel Primary Products	470	537	512	531	591	690	1,287	2,199	1,607	1,922	1,832
Fuels	260	316	292	351	523	643	1,054	3,372	3,551	4,332	4,502
Manufactures	936	1,276	1,459	1,962	2,581	3,443	4,652	8,591	8,421	7,459	6,920
Terms of Trade				*(Index 1980 = 100)*							
Terms of Trade	189.5	190.2	180.4	188.6	171.0	165.1	189.0	125.5	125.5	141.5	160.4
Export Prices, fob	30.5	31.2	32.4	35.2	35.1	38.2	55.5	65.9	66.4	75.3	91.0
Nonfuel Primary Products	31.1	31.3	32.7	36.7	35.4	38.9	61.1	64.6	64.6	76.4	96.1
Fuels	4.3	4.3	4.3	4.3	5.6	6.2	8.9	36.7	35.7	38.4	42.0
Manufactures	26.4	31.0	32.6	37.2	43.4	46.5	55.8	72.9	77.5	79.8	85.3
Import Prices, cif	16.1	16.4	18.0	18.7	20.5	23.1	29.4	52.5	52.9	53.2	56.7
Trade at Constant 1980 Prices				*(Millions of 1980 US dollars)*							
Exports, fob	5,431	6,037	7,129	7,772	8,284	10,461	11,164	12,073	13,066	13,443	13,320
Imports, cif	10,369	12,989	12,595	15,227	18,033	20,662	23,804	26,978	25,681	25,764	23,359
BALANCE OF PAYMENTS				*(Millions of current US dollars)*							
Exports of Goods & Services	..	..	..	3,117	3,334	4,508	7,037	9,371	9,939	11,283	13,510
Merchandise, fob	..	..	..	2,739	2,891	3,941	6,093	7,814	8,492	9,961	11,923
Nonfactor Services	..	..	..	320	388	421	603	824	926	887	1,080
Factor Services	..	..	..	58	55	145	341	734	521	435	507
Imports of Goods & Services	..	..	..	3,975	4,985	6,204	9,222	16,934	16,950	17,841	18,622
Merchandise, fob	..	..	..	2,507	3,256	4,193	6,154	12,562	12,042	12,347	12,023
Nonfactor Services	..	..	..	788	939	1,151	1,626	2,354	2,281	2,340	2,623
Factor Services	..	..	..	680	790	860	1,442	2,019	2,627	3,154	3,976
Long-Term Interest	..	..	..	224	297	425	747	1,472	2,055	1,719	2,026
Current Transfers, net	..	..	..	-3	-7	-17	11	4	13	7	..
Workers' Remittances	..	..	..	0	0	0	0	0	0	0	0
Total to be Financed	..	..	..	-861	-1,658	-1,713	-2,174	-7,560	-6,998	-6,551	-5,116
Official Capital Grants	..	..	..	24	20	23	17	-2	-10	-3	4
Current Account Balance	..	..	..	-837	-1,638	-1,690	-2,158	-7,562	-7,008	-6,554	-5,112
Long-Term Capital, net	..	..	..	1,241	1,688	3,565	4,116	6,228	4,926	6,105	6,044
Direct Investment	..	..	..	407	536	570	1,341	1,268	1,190	1,372	1,687
Long-Term Loans	..	..	..	1,336	1,430	3,552	2,662	6,290	4,808	5,638	5,827
Disbursements	..	..	..	1,792	1,995	4,311	3,812	8,210	7,050	8,161	9,494
Repayments	..	..	..	456	564	758	1,150	1,920	2,242	2,523	3,668
Other Long-Term Capital	..	..	..	-527	-298	-580	96	-1,327	-1,062	-902	-1,474
Other Capital, net	..	..	..	90	433	556	338	349	1,017	3,121	-411
Change in Reserves	..	..	..	-494	-483	-2,431	-2,296	984	1,065	-2,672	-521
Memo Item:				*(Brazilian Cruzados per US dollar)*							
Conversion Factor (Annual Avg)	0.003	0.003	0.004	0.005	0.005	0.006	0.006	0.007	0.008	0.011	0.014
EXTERNAL DEBT, ETC.				*(Millions of US dollars, outstanding at end of year)*							
Public/Publicly Guar. Long-Term	..	..	..	3,421	4,201	5,927	7,789	11,234	14,144	17,898	22,400
Official Creditors	..	..	..	1,886	2,102	2,386	2,712	3,464	4,067	4,531	5,074
IBRD and IDA	..	..	..	258	348	485	647	871	1,093	1,217	1,413
Private Creditors	..	..	..	1,535	2,099	3,541	5,077	7,770	10,076	13,366	17,326
Private Non-guaranteed Long-Term	..	..	..	1,706	2,427	4,239	5,150	8,182	9,593	11,133	13,006
Use of Fund Credit	..	..	..	0	0	0	0	0	0	0	0
Short-Term Debt	..	..	..	..	..	..	..	..	..	..	5,991
Memo Items:				*(Millions of US dollars)*							
Int'l Reserves Excluding Gold	154.5	212.3	611.3	1,141.7	1,696.2	4,132.8	6,359.9	5,215.8	3,980.4	6,488.0	7,192.0
Gold Holdings (at market price)	45.7	54.1	45.4	48.2	57.7	86.1	149.0	247.5	186.1	178.8	249.9
SOCIAL INDICATORS											
Total Fertility Rate	5.3	5.2	5.1	4.9	4.8	4.7	4.6	4.5	4.4	4.3	4.2
Crude Birth Rate	36.4	35.9	35.3	34.8	34.2	33.6	33.3	33.0	32.7	32.3	32.0
Infant Mortality Rate	100.0	98.2	96.4	94.6	92.8	91.0	88.6	86.2	83.8	81.4	79.0
Life Expectancy at Birth	57.9	58.3	58.7	59.0	59.4	59.8	60.2	60.6	61.0	61.4	61.8
Food Production, p.c. ('79-81 = 100)	85.0	84.6	84.0	87.3	87.7	89.1	87.7	92.4	94.1	97.7	100.3
Labor Force, Agriculture (%)	47.1	46.3	45.6	44.9	43.5	42.1	40.7	39.3	37.9	36.6	35.2
Labor Force, Female (%)	20.5	20.9	21.3	21.7	22.3	22.8	23.3	23.9	24.4	24.9	25.4
School Enroll. Ratio, primary	..	..	..	..	82.0	..	..	..	88.0	..	89.0
School Enroll. Ratio, secondary	..	..	..	26.0	17.0	..	..	..	26.0	30.0	32.0

1978	1979	1980	1981	1982	1983	1984	1985	1986	1987 estimate	NOTES	BRAZIL
											FOREIGN TRADE (CUSTOMS BASIS)
				(Millions of current US dollars)							
12,659	15,244	20,100	23,300	20,200	21,900	27,000	25,600	22,396	26,225	..	Value of Exports, fob
8,128	9,140	11,973	12,658	10,785	11,831	13,729	12,511	11,328	13,321	..	Nonfuel Primary Products
196	228	358	1,178	1,444	1,158	1,824	1,625	906	1,154	..	Fuels
4,335	5,876	7,770	9,465	7,971	8,911	11,447	11,464	10,162	11,750	..	Manufactures
15,016	19,731	24,900	24,100	21,100	16,800	15,200	14,300	15,555	16,581	..	Value of Imports, cif
2,435	3,642	3,978	3,209	2,664	2,047	2,166	2,098	3,505	2,798	..	Nonfuel Primary Products
4,937	7,314	10,732	12,175	11,296	9,393	8,029	6,755	4,160	4,472	..	Fuels
7,645	8,776	10,190	8,716	7,140	5,360	5,005	5,447	7,890	9,311	..	Manufactures
				(Index 1980 = 100)							
139.9	122.9	100.0	93.5	92.5	93.1	96.7	89.1	110.2	97.1	..	Terms of Trade
85.8	94.9	100.0	98.5	90.5	86.6	87.6	79.3	84.9	86.5	..	Export Prices, fob
85.9	97.0	100.0	90.4	86.1	88.2	87.9	77.7	83.1	74.8	..	Nonfuel Primary Products
42.3	61.0	100.0	112.5	101.6	92.5	90.2	87.5	44.9	60.0	..	Fuels
89.9	93.8	100.0	110.0	95.0	84.0	87.0	80.0	94.6	111.0	..	Manufactures
61.4	77.2	100.0	105.3	97.8	93.0	90.7	89.0	77.0	89.0	..	Import Prices, cif
				(Millions of 1980 US dollars)							Trade at Constant 1980 Prices
14,745	16,062	20,101	23,660	22,331	25,280	30,806	32,294	26,394	30,329	..	Exports, fob
24,468	25,551	24,900	22,888	21,579	18,056	16,762	16,070	20,205	18,627	..	Imports, cif
				(Millions of current US dollars)							**BALANCE OF PAYMENTS**
14,489	17,998	23,275	26,923	23,469	24,341	30,205	29,309	25,302	28,673	..	Exports of Goods & Services
12,473	15,244	20,132	23,276	20,173	21,898	27,002	25,634	22,392	26,206	..	Merchandise, fob
1,191	1,464	1,725	2,246	1,794	1,713	1,936	2,079	1,894	1,867	..	Nonfactor Services
825	1,290	1,418	1,401	1,502	730	1,267	1,596	1,016	600	..	Factor Services
21,597	28,493	36,250	38,873	39,773	31,286	30,334	29,737	29,865	30,048	..	Imports of Goods & Services
13,631	17,961	22,955	22,091	19,395	15,429	13,916	13,168	14,044	15,052	..	Merchandise, fob
2,863	3,763	4,833	5,109	5,366	4,105	3,679	3,760	4,002	2,697	..	Nonfactor Services
5,103	6,769	8,462	11,673	15,012	11,752	12,739	12,809	11,819	12,299	..	Factor Services
3,128	4,754	6,324	7,943	9,356	7,664	7,209	7,331	7,499	5,834	..	Long-Term Interest
69	12	127	189	-10	106	161	139	95	100	..	Current Transfers, net
0	5	4	14	6	2	4	2	0	..	..	Workers' Remittances
-7,039	-10,483	-12,848	-11,761	-16,314	-6,839	32	-289	-4,468	-1,275	K	Total to be Financed
3	5	42	10	2	2	10	16	-9	..	K	Official Capital Grants
-7,036	-10,478	-12,806	-11,751	-16,312	-6,837	42	-273	-4,477	-1,275	..	Current Account Balance
10,091	6,465	6,249	11,657	8,013	7,746	8,090	1,121	-188	3,895	..	Long-Term Capital, net
1,882	2,223	1,544	2,313	2,534	1,373	1,556	1,267	331	582	..	Direct Investment
9,418	5,729	4,495	8,686	7,854	4,383	5,121	-469	-212	-2,128	..	Long-Term Loans
14,617	12,241	11,323	16,095	15,380	7,964	9,587	2,466	3,100	1,555	..	Disbursements
5,199	6,512	6,827	7,410	7,526	3,582	4,466	2,935	3,312	3,682	..	Repayments
-1,212	-1,492	168	648	-2,377	1,988	1,403	307	-298	5,441	..	Other Long-Term Capital
1,573	1,226	2,850	647	1,464	-1,494	-3,255	-974	1,037	-974	..	Other Capital, net
-4,628	2,787	3,707	-553	6,835	585	-4,877	126	3,628	-1,646	..	Change in Reserves
				(Brazilian Cruzados per US dollar)							Memo Item:
0.018	0.027	0.053	0.093	0.180	0.580	1.850	6.200	13.650	39.230	..	Conversion Factor (Annual Avg)
				(Millions of US dollars, outstanding at end of year)							**EXTERNAL DEBT, ETC.**
30,394	35,921	40,434	45,260	50,797	59,123	69,936	73,672	84,349	91,653	..	Public/Publicly Guar. Long-Term
5,842	6,219	6,870	7,618	8,508	10,094	12,372	15,432	20,447	25,201	..	Official Creditors
1,602	1,827	2,069	2,319	2,724	3,655	3,997	5,305	7,576	9,411	..	IBRD and IDA
24,553	29,702	33,564	37,642	42,289	49,029	57,564	58,240	63,902	66,452	..	Private Creditors
16,152	15,864	16,605	19,792	23,124	21,512	19,304	17,176	14,641	14,434	..	Private Non-guaranteed Long-Term
0	0	0	0	550	2,644	4,185	4,619	4,501	3,976	..	Use of Fund Credit
7,068	8,634	13,526	15,321	17,451	14,204	11,500	11,017	9,286	13,868	..	Short-Term Debt
				(Millions of US dollars)							Memo Items:
11,826.0	8,966.3	5,769.3	6,603.5	3,927.9	4,355.1	11,508.0	10,605.0	5,803.0	6,299.2	..	Int'l Reserves Excluding Gold
363.6	872.4	1,105.9	876.1	69.4	206.4	452.9	1,013.4	950.7	1,177.8	..	Gold Holdings (at market price)
											SOCIAL INDICATORS
4.1	4.1	4.0	3.9	3.8	3.7	3.7	3.6	3.5	3.4	..	Total Fertility Rate
31.7	31.5	31.2	30.9	30.6	30.1	29.6	29.0	28.5	28.0	..	Crude Birth Rate
77.4	75.8	74.2	72.6	71.0	66.4	61.8	57.2	52.6	47.9	..	Infant Mortality Rate
62.1	62.5	62.8	63.1	63.4	63.8	64.1	64.5	64.9	65.3	..	Life Expectancy at Birth
93.0	95.0	103.1	101.9	107.6	100.7	105.0	111.7	102.5	108.1	..	Food Production, p.c. ('79-81=100)
33.9	32.5	31.2	..	..	..	..	..	..	..	..	Labor Force, Agriculture (%)
25.9	26.4	26.9	27.0	27.0	27.1	27.1	27.2	27.2	27.3	..	Labor Force, Female (%)
92.0	97.0	99.0	96.0	102.0	103.0	103.0	104.0	105.0	..	..	School Enroll. Ratio, primary
32.0	32.0	34.0	37.0	34.0	35.0	35.0	..	..	..	..	School Enroll. Ratio, secondary

BURKINA FASO	1967	1968	1969	1970	1971	1972	1973	1974	1975	1976	1977
CURRENT GNP PER CAPITA (US $)	60	60	60	60	60	60	70	90	100	110	120
POPULATION (thousands)	5,313	5,419	5,526	5,633	5,741	5,848	5,959	6,076	6,202	6,337	6,481
ORIGIN AND USE OF RESOURCES					*(Billions of current CFA Francs)*						
Gross National Product (GNP)	75.6	77.7	84.8	88.0	92.6	101.2	106.2	125.8	137.3	161.1	190.6
Net Factor Income from Abroad	2.3	2.4	2.6	4.3	4.8	4.8	6.8	6.3	4.3	6.9	6.9
Gross Domestic Product (GDP)	73.3	75.3	82.2	83.7	87.8	96.4	99.4	119.5	133.0	154.2	183.7
Indirect Taxes, net	3.4	3.5	3.9	4.3	6.5	6.5	7.7	9.0	10.5	14.3	18.4
GDP at factor cost	69.9	71.8	78.3	79.4	81.3	89.9	91.7	110.5	122.6	140.0	165.2
Agriculture	33.5	32.6	36.0	34.9	36.7	42.0	39.9	50.3	52.5	61.2	71.4
Industry	15.9	16.9	18.2	18.3	20.3	21.3	24.8	26.0	32.9	38.9	37.1
Manufacturing	..	..	..	..	..	..	..	..	..	..	..
Services, etc.	20.5	22.4	24.1	26.1	24.3	26.6	27.0	34.1	37.2	39.9	56.7
Resource Balance	-8.2	-8.2	-11.0	-11.3	-16.7	-18.6	-21.8	-24.7	-40.6	-37.6	-54.8
Exports of Goods & NFServices	6.6	8.5	8.7	8.4	8.8	11.3	11.5	18.2	17.9	23.1	26.7
Imports of Goods & NFServices	14.8	16.7	19.7	19.6	25.5	29.9	33.2	42.8	58.5	60.7	81.5
Domestic Absorption	81.5	83.5	93.1	95.0	104.5	115.0	121.1	144.1	173.7	191.8	238.5
Private Consumption, etc.	67.7	66.7	73.5	72.7	75.3	79.7	78.7	84.8	100.7	115.5	156.1
General Gov't Consumption	5.9	8.4	9.0	10.8	11.6	12.9	14.8	18.2	31.6	28.9	34.0
Gross Domestic Investment	7.9	8.4	10.6	11.6	17.6	22.4	27.6	41.1	41.4	47.5	48.4
Fixed Investment	..	..	..	..	..	..	..	..	..	..	..
Memo Items:											
Gross Domestic Saving	-0.3	0.2	-0.3	0.3	0.9	3.8	5.9	16.4	0.8	9.9	-6.4
Gross National Saving	2.0	2.6	6.1	9.1	11.7	14.3	20.6	29.7	12.1	25.5	10.3
					(Billions of 1980 CFA Francs)						
Gross National Product	178.2	195.7	201.9	205.9	209.4	212.9	217.4	232.3	234.0	256.9	256.3
GDP at factor cost	..	..	..	185.9	188.5	194.3	192.8	209.6	214.7	229.6	228.5
Agriculture	..	..	..	97.3	98.5	98.2	92.7	106.6	105.7	109.8	103.4
Industry	..	..	..	39.5	41.1	41.9	47.4	41.8	50.6	56.1	46.6
Manufacturing	..	..	..	..	..	..	..	..	..	..	..
Services, etc.	..	..	..	49.1	49.0	54.2	52.7	61.1	58.4	63.8	78.5
Resource Balance	-40.9	-44.4	-49.0	-38.7	-50.0	-47.6	-52.8	-53.0	-68.4	-62.0	-83.6
Exports of Goods & NFServices	15.0	19.3	18.4	20.8	21.0	28.3	26.1	28.0	27.5	25.4	26.6
Imports of Goods & NFServices	55.9	63.7	67.4	59.5	71.0	75.9	78.9	81.0	95.9	87.4	110.2
Domestic Absorption	193.9	217.1	225.9	234.1	248.1	250.1	255.9	273.4	295.0	307.9	330.3
Private Consumption, etc.	149.6	172.9	174.0	177.4	178.7	164.8	159.5	168.3	177.1	189.7	221.6
General Gov't Consumption	16.2	23.1	23.1	27.3	28.0	32.5	34.5	36.6	50.1	49.7	46.3
Gross Domestic Investment	22.2	23.4	28.2	29.4	41.4	52.9	61.9	68.5	67.9	68.5	62.5
Fixed Investment	..	..	..	..	..	..	..	..	..	..	..
Memo Items:											
Capacity to Import	24.9	32.6	29.9	25.3	24.6	28.6	27.2	34.4	29.3	33.3	36.1
Terms of Trade Adjustment	9.9	13.3	11.5	4.6	3.6	0.3	1.1	6.3	1.7	7.9	9.5
Gross Domestic Income	175.2	189.1	197.9	199.9	201.7	202.8	204.2	226.7	228.4	253.7	256.2
Gross National Income	182.8	199.0	208.3	210.5	213.0	213.2	218.6	238.6	235.7	264.8	265.8
DOMESTIC PRICES (DEFLATORS)					*(Index 1980 = 100)*						
Overall (GDP)	42.9	40.5	42.9	42.9	44.3	47.6	48.9	54.2	58.7	62.7	74.4
Domestic Absorption	42.0	38.5	41.2	40.6	42.1	46.0	47.3	52.7	58.9	62.3	72.2
Agriculture	..	..	..	35.9	37.3	42.8	43.1	47.2	49.7	55.7	69.0
Industry	..	..	..	46.4	49.3	50.9	52.3	62.2	65.0	69.4	79.7
Manufacturing	..	..	..	..	..	..	..	..	..	..	..
MANUFACTURING ACTIVITY											
Employment (1980=100)	..	..	..	..	..	..	..	91.8	93.1	94.5	95.9
Real Earnings per Empl. (1980=100)	..	..	..	..	..	..	..	55.9	60.4	85.3	81.1
Real Output per Empl. (1980=100)	..	..	..	..	..	..	..	84.0	89.0	93.3	90.4
Earnings as % of Value Added	..	..	..	..	..	..	..	11.1	13.3	15.4	17.5
MONETARY HOLDINGS					*(Billions of current CFA Francs)*						
Money Supply, Broadly Defined	6.65	7.59	8.71	9.36	9.93	10.59	14.49	17.53	24.32	31.55	35.57
Money as Means of Payment	6.59	7.45	8.51	9.14	9.47	9.89	13.50	16.38	22.50	27.55	30.78
Currency Ouside Banks	4.60	4.77	5.54	5.74	5.65	5.77	7.13	8.51	10.65	12.85	14.75
Demand Deposits	1.99	2.68	2.97	3.40	3.82	4.11	6.36	7.86	11.85	14.71	16.03
Quasi-Monetary Liabilities	0.06	0.14	0.20	0.23	0.46	0.71	0.99	1.16	1.82	4.00	4.79
					(Millions of current CFA Francs)						
GOVERNMENT DEFICIT (-) OR SURPLUS	..	..	..	..	..	..	354	2,212	-1,664	-1,272	4,313
Current Revenue	..	..	..	..	..	..	12,850	15,971	16,644	21,728	31,384
Current Expenditure	..	..	..	..	..	..	10,255	11,457	14,487	17,503	22,343
Current Budget Balance	..	..	..	..	..	..	2,595	4,514	2,157	4,225	9,041
Capital Receipts	..	..	..	..	..	..	12	2	18	37	30
Capital Payments	..	..	..	..	..	..	2,253	2,304	3,839	5,534	4,758

1978	1979	1980	1981	1982	1983	1984	1985	1986	1987 estimate	NOTES	BURKINA FASO
140	170	200	210	200	170	150	150	170	200	..	**CURRENT GNP PER CAPITA (US $)**
6,633	6,794	6,962	7,137	7,318	7,504	7,692	7,881	8,101	8,330	..	**POPULATION (thousands)**
				(Billions of current CFA Francs)							**ORIGIN AND USE OF RESOURCES**
233.1	269.1	303.6	361.0	397.1	417.9	428.5	518.9	549.2	579.0	f	Gross National Product (GNP)
13.0	23.3	34.3	42.1	38.7	37.0	38.0	39.0	41.4	39.9	f	Net Factor Income from Abroad
220.1	245.8	269.3	318.9	358.4	380.9	390.5	479.9	507.8	539.1	..	Gross Domestic Product (GDP)
20.0	13.5	16.0	17.7	20.5	18.1	16.2	21.2	29.6	31.4	..	Indirect Taxes, net
200.0	232.3	253.3	301.2	337.9	362.8	374.3	458.7	478.2	507.7	..	GDP at factor cost
90.6	98.7	104.7	129.9	143.2	152.0	164.2	214.0	199.3	203.4	..	Agriculture
44.9	48.2	50.7	56.8	50.2	60.0	56.9	83.4	102.0	96.9	..	Industry
..	..	..	..	39.9	48.1	47.5	50.9	54.9	58.5	..	Manufacturing
64.5	85.4	97.9	114.5	144.5	150.8	153.2	161.3	176.9	207.4	..	Services, etc.
-61.0	-68.7	-79.1	-92.5	-119.8	-118.1	-88.7	-137.2	-98.9	-126.0	..	Resource Balance
28.1	36.2	43.6	53.6	56.2	55.9	88.1	86.1	100.2	120.8	..	Exports of Goods & NFServices
89.1	104.9	122.7	146.1	176.0	174.0	176.8	223.3	199.1	246.8	..	Imports of Goods & NFServices
281.0	314.5	348.4	411.4	478.2	499.0	479.2	617.1	606.7	665.1	..	Domestic Absorption
185.9	211.2	247.7	296.2	316.7	324.6	303.1	439.9	413.8	468.5	..	Private Consumption, etc.
40.8	49.0	45.9	54.5	77.4	84.1	85.2	81.9	96.4	111.9	..	General Gov't Consumption
54.3	54.3	54.8	60.7	84.1	90.3	90.9	95.3	96.5	84.7	..	Gross Domestic Investment
..	48.8	48.4	56.6	77.6	87.7	87.6	70.0	66.2	61.5	..	Fixed Investment
											Memo Items:
-6.7	-14.4	-24.3	-31.8	-35.7	-27.8	2.2	-41.9	-2.4	-41.3	..	Gross Domestic Saving
18.7	26.6	33.6	42.9	32.2	43.1	71.4	39.5	140.8	..	..	Gross National Saving
				(Billions of 1980 CFA Francs)							
273.3	294.7	303.6	318.1	330.2	323.8	328.2	372.4	413.7	427.3	I	Gross National Product
241.4	254.5	253.3	265.5	285.1	285.1	291.7	334.5	374.1	386.2	..	GDP at factor cost
109.2	111.2	104.7	113.2	115.9	115.7	115.6	142.2	161.8	157.7	..	Agriculture
50.1	50.9	50.7	50.0	46.4	42.0	42.0	47.1	51.6	54.9	..	Industry
..	..	..	..	..	..	..	..	..	..	..	Manufacturing
82.1	92.4	97.9	102.3	122.9	127.3	134.0	145.2	160.8	173.6	..	Services, etc.
-72.1	-74.0	-79.1	-82.2	-87.5	-79.8	-60.7	-76.2	-58.7	-56.8	..	Resource Balance
38.4	41.3	43.6	49.4	41.4	32.3	45.8	52.1	63.6	70.4	..	Exports of Goods & NFServices
110.5	115.3	122.7	131.6	129.0	112.2	106.4	128.2	122.3	127.2	..	Imports of Goods & NFServices
330.1	343.3	348.4	363.2	387.7	376.6	362.1	422.3	445.4	456.6	..	Domestic Absorption
218.6	232.5	247.7	258.0	262.1	247.9	230.5	275.4	295.1	303.1	..	Private Consumption, etc.
47.0	51.1	45.9	50.5	59.4	64.8	70.4	74.7	86.2	92.6	..	General Gov't Consumption
64.6	59.8	54.8	54.7	66.2	63.9	61.2	72.2	64.2	60.9	..	Gross Domestic Investment
..	53.7	48.4	50.9	60.5	62.3	59.2	62.3	45.0	44.4	..	Fixed Investment
											Memo Items:
34.9	39.8	43.6	48.3	41.2	36.0	53.0	49.4	61.5	62.3	..	Capacity to Import
-3.5	-1.5	0.0	-1.1	-0.3	3.7	7.3	-2.6	-2.0	-8.2	..	Terms of Trade Adjustment
254.6	267.8	269.3	279.9	299.9	300.5	308.7	343.5	384.7	391.7	..	Gross Domestic Income
269.8	293.2	303.6	317.0	330.0	327.5	335.4	369.8	411.6	419.1	..	Gross National Income
				(Index 1980 = 100)							**DOMESTIC PRICES (DEFLATORS)**
85.3	91.3	100.0	113.5	119.4	128.4	129.5	138.6	131.3	134.8	..	Overall (GDP)
85.1	91.6	100.0	113.3	123.4	132.5	132.3	146.1	136.2	145.7	..	Domestic Absorption
82.9	88.7	100.0	114.8	123.6	131.3	142.0	150.5	123.1	129.0	..	Agriculture
89.6	94.7	100.0	113.7	108.3	142.7	135.4	177.0	197.9	176.4	..	Industry
..	..	..	..	..	..	..	..	..	..	..	Manufacturing
											MANUFACTURING ACTIVITY
97.3	98.9	100.0	101.0	102.7	104.2	105.6	107.0	..	..	..	Employment (1980=100)
96.4	112.5	100.0	99.5	96.9	94.1	104.9	106.9	..	..	..	Real Earnings per Empl. (1980=100)
89.5	94.2	100.0	105.6	91.5	91.4	97.4	105.6	..	..	..	Real Output per Empl. (1980=100)
20.7	22.8	23.0	21.2	19.8	18.3	19.7	19.7	..	..	..	Earnings as % of Value Added
				(Billions of current CFA Francs)							**MONETARY HOLDINGS**
42.52	46.24	53.22	63.72	71.38	80.06	92.77	93.32	114.19	127.51	..	Money Supply, Broadly Defined
34.40	34.76	41.67	48.83	54.36	60.33	66.50	69.54	85.43	91.18	..	Money as Means of Payment
13.50	17.36	19.90	24.83	27.03	31.74	31.24	30.97	43.43	43.71	..	Currency Ouside Banks
20.90	17.41	21.77	24.00	27.34	28.60	35.26	38.57	42.00	47.47	..	Demand Deposits
8.12	11.48	11.55	14.89	17.01	19.72	26.26	23.78	28.76	36.33	..	Quasi-Monetary Liabilities
				(Millions of current CFA Francs)							
1,495	-5,622	879	-4,674	-6,185	562	-3,350	7,705	..	..	F	**GOVERNMENT DEFICIT (-) OR SURPLUS**
32,542	35,550	43,109	48,674	54,527	54,540	63,071	73,409	..	..	..	Current Revenue
27,203	34,661	35,305	41,140	55,354	47,987	60,854	57,903	..	75,874	..	Current Expenditure
5,339	889	7,804	7,534	-827	6,553	2,217	15,506	..	..	..	Current Budget Balance
..	1	30	46	33	..	17	99	..	64	..	Capital Receipts
3,844	6,512	6,955	12,254	5,391	5,991	5,584	7,900	..	11,781	..	Capital Payments

BURKINA FASO	1967	1968	1969	1970	1971	1972	1973	1974	1975	1976	1977
FOREIGN TRADE (CUSTOMS BASIS)				*(Millions of current US dollars)*							
Value of Exports, fob	23.6	28.2	26.5	24.6	24.6	36.0	44.2	66.0	73.5	83.1	94.8
Nonfuel Primary Products	22.6	27.1	25.5	23.4	23.0	32.8	41.6	60.6	68.7	78.1	89.4
Fuels	0.0	0.0	0.1	0.1	0.1	0.1	0.2	0.0	0.1	0.1	0.1
Manufactures	1.0	1.1	0.9	1.1	1.5	3.1	2.3	5.4	4.8	4.9	5.3
Value of Imports, cif	44.9	51.0	62.0	56.5	73.0	94.0	130.0	180.0	216.0	195.0	260.0
Nonfuel Primary Products	16.4	16.2	16.6	15.3	20.7	28.0	39.0	66.3	52.5	43.7	60.9
Fuels	2.9	3.4	4.1	4.6	6.5	8.1	9.2	11.7	19.1	15.3	22.1
Manufactures	25.5	31.4	41.3	36.5	45.9	58.0	81.8	102.0	144.4	136.0	177.1
				(Index 1980 = 100)							
Terms of Trade	146.7	152.7	147.5	161.0	158.0	154.6	178.0	108.4	94.1	118.6	109.7
Export Prices, fob	33.8	34.3	33.9	35.5	38.8	42.7	67.8	68.4	58.5	73.2	72.9
Nonfuel Primary Products	34.1	34.6	34.6	36.5	39.9	44.3	71.8	68.5	58.1	73.7	73.0
Fuels	4.3	4.3	4.3	4.3	5.6	6.2	8.9	36.7	35.7	38.4	42.0
Manufactures	30.1	31.7	32.4	37.5	36.5	38.5	49.5	67.2	66.1	66.7	71.7
Import Prices, cif	23.0	22.5	23.0	22.0	24.6	27.6	38.1	63.1	62.2	61.7	66.5
Trade at Constant 1980 Prices				*(Millions of 1980 US dollars)*							
Exports, fob	69.8	82.2	78.1	69.3	63.4	84.3	65.2	96.5	125.6	113.6	130.1
Imports, cif	194.9	226.9	269.5	256.3	297.1	340.2	341.5	285.2	347.1	316.2	391.3
BALANCE OF PAYMENTS				*(Millions of current US dollars)*							
Exports of Goods & Services	..	..	..	31	33	45	58	84	95	106	119
Merchandise, fob	..	..	..	25	25	36	44	66	73	83	95
Nonfactor Services	..	..	..	4	6	6	8	9	14	17	17
Factor Services	..	..	..	2	3	3	6	9	7	6	6
Imports of Goods & Services	..	..	..	68	89	112	155	211	281	260	342
Merchandise, fob	..	..	..	45	58	74	104	148	188	167	221
Nonfactor Services	..	..	..	21	28	34	45	51	74	77	95
Factor Services	..	..	..	2	3	4	6	11	19	16	26
Long-Term Interest	..	..	..	0	1	1	1	1	2	2	2
Current Transfers, net	..	..	..	16	22	22	36	29	32	37	40
Workers' Remittances	..	..	..	18	21	26	36	34	46	50	60
Total to be Financed	..	..	..	-21	-34	-45	-61	-97	-154	-117	-183
Official Capital Grants	..	..	..	29	35	48	63	92	99	85	100
Current Account Balance	..	..	..	9	1	3	3	-5	-54	-33	-83
Long-Term Capital, net	..	..	..	31	37	43	81	105	118	104	141
Direct Investment	..	..	..	0	1	-1	4	2	0	2	5
Long-Term Loans	..	..	..	0	1	4	9	14	18	24	42
Disbursements	..	..	..	2	3	6	12	17	22	27	45
Repayments	..	..	..	2	2	2	2	3	4	3	4
Other Long-Term Capital	..	..	..	1	0	-9	4	-3	1	-8	-6
Other Capital, net	..	..	..	-27	-36	-44	-72	-83	-70	-70	-76
Change in Reserves	..	..	..	-12	-2	-2	-11	-16	6	-1	18
Memo Item:				*(CFA Francs per US dollar)*							
Conversion Factor (Annual Avg)	246.850	246.850	259.710	277.710	277.130	252.210	222.700	240.500	214.320	238.980	245.670
EXTERNAL DEBT, ETC.				*(Millions of US dollars, outstanding at end of year)*							
Public/Publicly Guar. Long-Term	..	..	..	21	24	20	31	47	63	87	134
Official Creditors	..	..	..	21	22	19	30	46	61	83	130
IBRD and IDA	..	..	..	0	0	2	3	6	11	18	31
Private Creditors	..	..	..	0	1	1	1	1	2	3	4
Private Non-guaranteed Long-Term	..	..	..	..	..	..	..	..	..	..	..
Use of Fund Credit	..	..	..	0	0	0	0	0	0	0	0
Short-Term Debt	..	..	..	..	..	..	..	..	..	..	30
Memo Items:				*(Millions of US dollars)*							
Int'l Reserves Excluding Gold	18.4	23.4	25.8	36.4	43.0	47.5	62.6	83.6	76.5	71.4	56.2
Gold Holdings (at market price)	..	..	..	..	..	..	..	..	..	..	0.9
SOCIAL INDICATORS											
Total Fertility Rate	6.3	6.4	6.4	6.4	6.4	6.4	6.4	6.4	6.5	6.5	6.5
Crude Birth Rate	48.1	48.0	48.0	47.9	47.9	47.8	47.7	47.5	47.4	47.2	47.1
Infant Mortality Rate	183.0	178.8	174.6	170.4	166.2	162.0	161.0	160.0	159.0	158.0	157.0
Life Expectancy at Birth	39.1	39.5	40.0	40.4	40.8	41.2	41.6	42.0	42.4	42.8	43.2
Food Production, p.c. ('79-81 = 100)	116.4	118.6	115.6	116.7	108.7	102.5	90.4	99.8	109.9	95.4	97.5
Labor Force, Agriculture (%)	88.9	88.7	88.5	88.3	88.1	88.0	87.8	87.7	87.5	87.3	87.2
Labor Force, Female (%)	48.5	48.4	48.4	48.3	48.3	48.2	48.2	48.2	48.1	48.1	48.0
School Enroll. Ratio, primary	..	..	..	13.0	..	..	..	..	16.0	15.0	16.0
School Enroll. Ratio, secondary	..	..	..	1.0	..	..	..	..	2.0	2.0	2.0

1978	1979	1980	1981	1982	1983	1984	1985	1986	1987 estimate	NOTES	BURKINA FASO
											FOREIGN TRADE (CUSTOMS BASIS)
				(Millions of current US dollars)							
107.8	132.7	160.6	159.4	126.4	130.0	138.0	112.0	142.5	202.3	f	Value of Exports, fob
99.6	115.5	143.1	136.0	108.9	116.4	118.3	92.6	124.5	198.2	..	Nonfuel Primary Products
0.2	0.8	0.2	0.3	0.0	0.0	0.0	0.0	0.0	..	..	Fuels
8.1	16.4	17.3	23.2	17.5	13.6	19.7	19.4	18.0	4.1	..	Manufactures
296.0	370.0	446.0	422.0	436.0	377.0	319.0	334.0	426.0	513.0	f	Value of Imports, cif
92.2	92.9	103.5	117.1	123.3	108.6	74.5	72.1	104.1	108.2	..	Nonfuel Primary Products
25.2	41.9	58.9	65.6	71.8	64.5	54.5	60.0	9.3	13.5	..	Fuels
178.6	235.2	283.5	239.3	240.9	203.9	190.0	201.9	312.6	391.3	..	Manufactures
				(Index 1980 = 100)							
106.7	108.3	100.0	88.7	83.4	94.5	95.1	80.0	66.2	73.6	..	Terms of Trade
79.3	93.2	100.0	92.0	82.0	90.0	88.6	73.1	67.6	81.1	..	Export Prices, fob
79.7	94.0	100.0	90.8	80.3	90.0	88.5	70.4	64.4	80.7	..	Nonfuel Primary Products
42.3	61.0	100.0	112.5	101.6	92.5	90.2	87.5	44.9	56.9	..	Fuels
76.1	90.0	100.0	99.1	94.8	90.7	89.3	89.0	103.2	113.0	..	Manufactures
74.3	86.0	100.0	103.6	98.3	95.3	93.2	91.3	102.0	110.2	..	Import Prices, cif
				(Millions of 1980 US dollars)							Trade at Constant 1980 Prices
135.9	142.4	160.6	173.4	154.1	144.4	155.8	153.3	210.8	249.4	..	Exports, fob
398.2	430.2	446.0	407.1	443.3	395.8	342.3	365.9	417.5	465.6	..	Imports, cif
				(Millions of current US dollars)							**BALANCE OF PAYMENTS**
136	181	225	209	184	157	175	159	232	..	..	Exports of Goods & Services
108	133	161	159	126	113	141	112	162	..	..	Merchandise, fob
24	42	49	41	48	38	27	33	48	..	..	Nonfactor Services
5	7	16	9	9	6	7	13	22	..	..	Factor Services
406	507	596	552	553	472	409	439	650	..	..	Imports of Goods & Services
255	312	368	348	360	309	270	276	415	..	f	Merchandise, fob
135	178	208	185	174	147	124	142	215	..	f	Nonfactor Services
16	16	20	19	19	16	14	21	20	..	..	Factor Services
3	5	6	6	8	8	6	10	12	14	..	Long-Term Interest
55	83	112	120	89	89	71	94	294	..	..	Current Transfers, net
78	110	150	153	110	113	90	..	150	..	..	Workers' Remittances
-215	-242	-259	-223	-281	-226	-162	-186	-124	..	K	Total to be Financed
156	186	211	181	189	166	159	119	0	..	K	Official Capital Grants
-60	-57	-49	-42	-92	-60	-3	-66	-124	..	..	Current Account Balance
190	255	268	218	245	245	203	165	94	..	f	Long-Term Capital, net
0	1	0	2	2	2	2	..	..	..	..	Direct Investment
40	68	62	78	74	82	43	39	72	95	..	Long-Term Loans
44	72	73	86	81	89	55	56	94	112	..	Disbursements
5	5	11	8	8	8	12	17	22	17	..	Repayments
-5	0	-5	-43	-19	-5	-1	7	22	..	..	Other Long-Term Capital
-155	-178	-209	-162	-153	-148	-163	-67	122	..	f	Other Capital, net
25	-21	-11	-14	0	-36	-37	-32	-92	-87	..	Change in Reserves
				(CFA Francs per US dollar)							Memo Item:
225.640	212.720	211.300	271.730	328.620	381.070	436.960	449.260	346.300	300.540	..	Conversion Factor (Annual Avg)
			(Millions of US dollars, outstanding at end of year)								**EXTERNAL DEBT, ETC.**
185	257	299	313	346	395	407	497	620	794	..	Public/Publicly Guar. Long-Term
181	239	278	294	310	350	367	460	584	756	..	Official Creditors
47	65	77	84	95	113	124	149	186	221	..	IBRD and IDA
4	19	20	19	36	45	41	38	35	38	..	Private Creditors
..	..	..	..	..	..	..	..	..	..	..	Private Non-guaranteed Long-Term
0	0	0	0	0	0	0	0	0	..	..	Use of Fund Credit
78	38	35	35	32	27	26	43	49	67	..	Short-Term Debt
				(Millions of US dollars)							Memo Items:
36.3	61.6	68.2	70.8	61.8	85.0	106.3	139.5	233.5	322.6	..	Int'l Reserves Excluding Gold
1.9	5.7	6.6	4.5	5.1	4.2	3.4	3.6	4.3	5.4	..	Gold Holdings (at market price)
											SOCIAL INDICATORS
6.5	6.5	6.5	6.5	6.5	6.5	6.5	6.5	6.5	6.5	..	Total Fertility Rate
47.1	47.0	47.0	46.9	46.9	47.0	47.1	47.2	47.3	47.4	..	Crude Birth Rate
155.6	154.2	152.8	151.4	150.0	144.0	138.0	132.0	126.0	120.0	..	Infant Mortality Rate
43.6	44.0	44.4	44.8	45.3	45.7	46.1	46.6	47.0	47.4	..	Life Expectancy at Birth
101.2	103.2	94.9	101.9	100.1	99.3	96.1	114.3	124.6	113.9	..	Food Production, p.c. ('79-81 = 100)
87.0	86.9	86.7	..	..	..	..	..	..	..	..	Labor Force, Agriculture (%)
48.0	47.9	47.9	47.7	47.5	47.4	47.2	47.0	46.9	46.7	..	Labor Force, Female (%)
17.0	18.0	21.0	23.0	24.0	26.0	29.0	32.0	..	..	..	School Enroll. Ratio, primary
2.0	3.0	3.0	3.0	3.0	4.0	4.0	5.0	..	..	..	School Enroll. Ratio, secondary

BURUNDI	1967	1968	1969	1970	1971	1972	1973	1974	1975	1976	1977
CURRENT GNP PER CAPITA (US $)	50	50	50	70	80	70	80	90	100	120	140
POPULATION (thousands)	3,217	3,261	3,305	3,350	3,366	3,407	3,576	3,647	3,720	3,794	3,870

ORIGIN AND USE OF RESOURCES

(Billions of current Burundi Francs)

	1967	1968	1969	1970	1971	1972	1973	1974	1975	1976	1977
Gross National Product (GNP)	14.6	14.9	15.6	20.6	21.3	20.6	23.7	26.5	32.5	37.7	48.4
Net Factor Income from Abroad	-1.0	-1.2	-1.1	-0.7	-0.8	-1.0	-0.7	-0.7	-0.7	-1.0	-1.2
Gross Domestic Product (GDP)	15.6	16.0	16.6	21.2	22.1	21.6	24.4	27.2	33.2	38.7	49.6
Indirect Taxes, net	1.0	1.2	1.2	1.6	1.4	1.5	1.7	2.0	2.2	3.2	6.4
GDP at factor cost	14.6	14.9	15.5	19.7	20.7	20.1	22.7	25.2	31.0	35.5	43.2
Agriculture	..	..	..	13.9	14.3	13.1	15.4	16.6	20.3	23.0	27.4
Industry	..	..	..	2.0	2.2	2.3	2.8	3.3	4.2	4.5	6.0
Manufacturing	..	..	..	1.4	1.6	1.8	2.3	2.6	2.8	3.2	3.9
Services, etc.	..	..	..	3.8	4.2	4.6	4.4	5.3	6.5	8.0	9.8
Resource Balance	-0.2	-0.5	-0.4	-0.2	-1.3	-0.7	-0.6	-1.6	-3.6	-1.1	0.2
Exports of Goods & NFServices	1.7	1.7	1.6	2.3	1.9	2.5	2.7	2.7	2.7	5.3	8.7
Imports of Goods & NFServices	1.9	2.1	2.0	2.5	3.1	3.3	3.3	4.2	6.3	6.4	8.5
Domestic Absorption	15.8	16.5	17.1	21.4	23.4	22.3	24.9	28.8	36.7	39.8	49.4
Private Consumption, etc.	13.6	14.0	14.3	18.4	19.4	18.8	20.8	24.2	30.4	32.9	38.4
General Gov't Consumption	1.1	1.3	1.4	2.1	2.3	2.8	2.8	3.4	3.8	3.4	5.5
Gross Domestic Investment	1.1	1.3	1.4	1.0	1.7	0.7	1.3	1.1	2.5	3.5	5.5
Fixed Investment	1.1	1.3	1.4	0.9	1.1	1.1	1.4	1.8	3.1	3.5	5.5
Memo Items:											
Gross Domestic Saving	0.9	0.8	1.0	0.8	0.4	0.0	0.7	-0.5	-1.0	2.4	5.7
Gross National Saving	-0.1	-0.4	-0.1	-0.3	-0.9	-1.8	-0.5	-1.7	-2.0	1.0	5.0

(Billions of 1980 Burundi Francs)

	1967	1968	1969	1970	1971	1972	1973	1974	1975	1976	1977
Gross National Product	50.5	49.4	49.3	63.4	64.9	60.1	65.5	65.1	66.1	70.9	79.1
GDP at factor cost	51.1	50.0	49.5	61.1	63.4	59.2	63.3	62.5	63.4	67.4	71.7
Agriculture	31.8	30.3	30.2	42.2	43.6	39.3	44.0	42.2	44.2	45.8	47.6
Industry	4.7	4.7	4.7	5.2	5.4	5.3	5.4	5.6	5.6	6.2	7.8
Manufacturing	3.6	3.7	3.6	3.8	4.2	4.1	4.2	4.3	4.3	4.6	4.8
Services, etc.	14.6	14.9	14.5	13.6	14.4	14.6	13.9	14.7	13.6	15.4	16.2
Resource Balance	-0.5	-2.1	-0.6	-2.6	-4.8	-2.5	-2.7	-4.0	-5.8	-6.2	-10.4
Exports of Goods & NFServices	7.7	7.5	8.0	7.5	7.3	9.0	8.2	7.6	9.0	7.7	6.6
Imports of Goods & NFServices	8.2	9.5	8.6	10.1	12.1	11.6	10.9	11.6	14.8	13.8	17.0
Domestic Absorption	54.6	55.5	53.4	68.1	72.2	65.7	70.1	71.0	73.3	79.0	91.6
Private Consumption, etc.	45.8	45.8	43.5	56.9	57.9	55.0	57.8	55.9	60.0	65.2	71.0
General Gov't Consumption	4.6	4.9	5.1	7.9	8.4	8.9	8.6	8.7	8.8	6.9	10.6
Gross Domestic Investment	4.1	4.9	4.9	3.3	6.0	1.8	3.8	6.3	4.6	6.8	10.1
Fixed Investment	..	..	..	2.9	3.7	3.5	4.1	4.3	6.5	6.8	10.1
Memo Items:											
Capacity to Import	7.4	7.3	6.9	9.3	7.2	9.0	9.0	7.3	6.4	11.4	17.3
Terms of Trade Adjustment	-0.3	-0.1	-1.1	1.8	-0.1	0.0	0.8	-0.3	-2.5	3.7	10.7
Gross Domestic Income	53.8	53.3	51.7	67.3	67.3	63.1	68.3	66.7	65.0	76.5	91.9
Gross National Income	50.2	49.2	48.2	65.2	64.8	60.1	66.3	64.8	63.5	74.6	89.8

DOMESTIC PRICES (DEFLATORS)

(Index 1980 = 100)

	1967	1968	1969	1970	1971	1972	1973	1974	1975	1976	1977
Overall (GDP)	28.9	30.0	31.5	32.4	32.8	34.2	36.1	40.6	49.1	53.1	61.0
Domestic Absorption	28.9	29.8	31.9	31.5	32.4	34.0	35.5	40.5	50.1	50.4	53.9
Agriculture	..	..	..	32.9	32.9	33.4	35.1	39.3	46.0	50.3	57.5
Industry	..	..	..	38.3	39.7	43.9	52.4	58.8	73.9	72.8	76.9
Manufacturing	..	..	..	37.7	39.1	43.6	54.4	59.7	66.0	69.1	81.1

MANUFACTURING ACTIVITY

	1967	1968	1969	1970	1971	1972	1973	1974	1975	1976	1977
Employment (1980=100)	..	..	55.4	60.0	66.4	64.7	72.3	71.8	85.4	68.9	82.8
Real Earnings per Empl. (1980=100)	..	..	..	..	203.9	199.2	186.3	189.7	171.9	183.2	215.9
Real Output per Empl. (1980=100)	..	..	..	..	117.9	107.1	91.4	97.0	84.4	101.3	108.2
Earnings as % of Value Added	..	..	..	..	27.6	27.5	25.0	25.4	28.1	24.9	25.0

MONETARY HOLDINGS

(Billions of current Burundi Francs)

	1967	1968	1969	1970	1971	1972	1973	1974	1975	1976	1977
Money Supply, Broadly Defined	1.66	1.66	1.73	2.14	2.44	2.44	2.99	3.45	3.35	4.71	6.55
Money as Means of Payment	1.61	1.61	1.65	2.00	2.32	2.31	2.83	3.31	3.21	4.55	6.15
Currency Ouside Banks	1.01	1.04	1.03	1.19	1.33	1.37	1.55	1.87	1.71	2.41	3.22
Demand Deposits	0.60	0.57	0.62	0.81	0.99	0.94	1.28	1.43	1.50	2.14	2.93
Quasi-Monetary Liabilities	0.05	0.05	0.08	0.14	0.12	0.13	0.16	0.14	0.14	0.17	0.40

(Millions of current Burundi Francs)

	1967	1968	1969	1970	1971	1972	1973	1974	1975	1976	1977
GOVERNMENT DEFICIT (-) OR SURPLUS	..	..	..	..	..	..	..	-86	-1,081	-237	-791
Current Revenue	..	..	..	..	..	..	4,719	5,658	5,729	8,435	9,821
Current Expenditure	..	..	..	..	..	..	..	..	..	..	..
Current Budget Balance	..	..	..	..	..	..	..	..	..	..	..
Capital Receipts	..	..	..	..	..	..	1	2	17	2	21
Capital Payments	..	..	..	..	..	..	..	..	..	..	..

1978	1979	1980	1981	1982	1983	1984	1985	1986	1987 estimate	NOTES	BURUNDI
150	170	200	250	240	230	230	240	240	250	..	CURRENT GNP PER CAPITA (US $)
3,948	4,022	4,114	4,224	4,338	4,454	4,573	4,696	4,834	4,978	..	POPULATION (thousands)
				(Billions of current Burundi Francs)							ORIGIN AND USE OF RESOURCES
53.6	69.5	82.2	85.4	88.8	98.3	115.0	134.9	142.8	147.2	..	Gross National Product (GNP)
-1.3	-0.9	-0.6	-1.8	-2.4	-2.5	-3.2	-4.0	-6.0	-8.2	..	Net Factor Income from Abroad
54.9	70.4	82.8	87.2	91.2	100.7	118.2	138.8	148.8	155.4	..	Gross Domestic Product (GDP)
6.1	7.9	6.2	4.5	6.8	6.6	10.8	12.6	13.4	12.8	..	Indirect Taxes, net
48.8	62.5	76.6	82.7	84.4	94.1	107.4	126.2	135.4	142.6	..	GDP at factor cost
29.6	37.9	47.9	50.4	48.1	53.7	64.5	77.7	82.1	84.1	..	Agriculture
7.9	10.4	9.7	11.1	13.0	14.6	14.9	16.4	18.5	20.3	..	Industry
4.5	6.6	5.7	6.4	7.5	8.4	9.3	10.2	11.6	12.4	..	Manufacturing
11.3	14.3	19.0	21.3	23.4	25.8	28.0	32.1	34.8	38.2	..	Services, etc.
-4.7	-7.4	-11.2	-10.0	-14.0	-14.8	-16.2	-12.3	-5.8	-19.0	..	Resource Balance
6.4	10.0	7.3	7.9	9.2	9.0	13.4	14.6	22.0	13.3	..	Exports of Goods & NFServices
11.1	17.4	18.5	17.9	23.2	23.8	29.7	26.9	27.8	32.3	..	Imports of Goods & NFServices
59.6	77.9	94.0	97.2	105.2	115.5	134.4	151.1	154.6	174.4	..	Domestic Absorption
44.3	58.3	71.6	68.4	78.3	79.2	97.8	114.3	117.4	117.4	..	Private Consumption, etc.
7.6	9.1	10.9	14.0	13.7	13.3	15.4	17.6	21.2	26.1	..	General Gov't Consumption
7.7	10.5	11.5	14.8	13.2	23.0	21.2	19.2	16.0	30.9	..	Gross Domestic Investment
7.7	10.5	11.5	11.8	13.8	19.4	20.8	19.9	19.2	28.2	..	Fixed Investment
											Memo Items:
3.0	3.1	0.3	4.9	-0.8	8.2	5.0	6.9	10.2	11.9	..	Gross Domestic Saving
2.3	2.3	-0.1	3.6	-2.5	6.3	2.5	3.9	4.7	4.2	..	Gross National Saving
				(Billions of 1980 Burundi Francs)							
78.5	80.8	82.2	92.1	88.9	89.9	92.0	95.6	99.6	101.9	..	Gross National Product
72.3	73.7	76.6	87.7	83.3	85.4	84.6	89.4	92.4	97.1	..	GDP at factor cost
46.4	47.1	47.9	56.2	50.7	52.7	50.0	54.0	55.7	57.7	f	Agriculture
8.7	8.8	9.7	10.4	10.5	11.0	11.8	12.4	12.8	13.7	f	Industry
5.1	5.3	5.7	6.5	6.3	6.7	7.4	8.2	8.4	9.0	..	Manufacturing
17.2	17.9	19.0	21.1	22.0	21.6	22.8	23.0	23.8	25.7	..	Services, etc.
-8.9	-11.1	-11.2	-5.8	-8.9	-13.0	-12.5	-8.3	-8.6	-5.9	..	Resource Balance
9.3	10.3	7.3	11.6	12.5	10.5	12.9	14.3	13.9	15.6	..	Exports of Goods & NFServices
18.2	21.5	18.5	17.4	21.4	23.6	25.4	22.6	22.5	21.5	..	Imports of Goods & NFServices
89.4	93.0	94.0	99.7	100.1	105.1	107.1	106.8	111.6	111.3	..	Domestic Absorption
65.9	69.9	71.6	71.6	75.8	71.0	78.1	80.0	79.7	..	..	Private Consumption, etc.
11.8	11.1	10.9	12.3	12.7	12.4	12.2	12.5	14.2	..	..	General Gov't Consumption
11.8	12.0	11.5	15.8	11.6	21.7	16.8	14.3	17.7	..	..	Gross Domestic Investment
11.8	12.0	11.5	11.2	12.4	17.4	16.4	14.9	16.7	..	..	Fixed Investment
											Memo Items:
10.5	12.3	7.3	7.7	8.5	8.9	11.5	12.3	17.8	8.9	..	Capacity to Import
1.3	2.0	0.0	-3.9	-4.0	-1.6	-1.4	-2.1	3.9	-6.8	..	Terms of Trade Adjustment
81.7	83.8	82.8	90.0	87.2	90.5	93.2	96.5	106.9	98.6	..	Gross Domestic Income
79.7	82.7	82.2	88.1	84.9	88.2	90.7	93.5	103.5	95.2	..	Gross National Income
				(Index 1980 = 100)							DOMESTIC PRICES (DEFLATORS)
68.2	86.0	100.0	92.8	100.0	109.3	125.0	140.8	144.5	147.5	..	Overall (GDP)
66.7	83.7	100.0	97.5	105.1	109.8	125.5	141.4	138.6	156.7	..	Domestic Absorption
63.9	80.6	100.0	89.6	94.8	101.9	129.1	143.9	147.3	145.8	..	Agriculture
90.2	118.2	100.0	106.0	123.4	132.7	126.4	132.7	144.4	148.4	..	Industry
88.5	123.0	100.0	98.4	118.8	124.8	125.4	123.9	138.8	137.6	..	Manufacturing
											MANUFACTURING ACTIVITY
79.2	80.3	100.0	..	..	138.9	..	..	..	..	..	Employment (1980 = 100)
172.6	112.1	100.0	..	..	132.9	..	..	..	..	..	Real Earnings per Empl. (1980 = 100)
122.1	119.9	100.0	..	..	135.3	..	..	..	..	..	Real Output per Empl. (1980 = 100)
18.8	12.7	13.8	..	..	18.1	..	..	..	..	..	Earnings as % of Value Added
				(Billions of current Burundi Francs)							MONETARY HOLDINGS
8.90	9.48	12.49	15.64	15.30	19.24	19.97	23.91	24.10	24.43	..	Money Supply, Broadly Defined
8.41	9.06	9.64	12.23	10.83	13.65	14.53	18.18	19.66	19.45	..	Money as Means of Payment
4.54	4.88	4.97	7.06	6.42	7.26	7.50	7.25	8.01	8.73	..	Currency Ouside Banks
3.86	4.18	4.67	5.17	4.41	6.38	7.03	10.93	11.65	10.72	..	Demand Deposits
0.50	0.42	2.84	3.41	4.48	5.59	5.44	5.73	4.44	4.98	..	Quasi-Monetary Liabilities
				(Millions of current Burundi Francs)							
-1,610	-3,758	-3,234	-4,933	..	..	..	..	..	..	..	GOVERNMENT DEFICIT (-) OR SURPLUS
10,970	12,471	14,708	16,096	..	..	..	..	..	..	..	Current Revenue
..	..	..	10,060	..	..	..	..	..	..	..	Current Expenditure
..	..	..	6,036	..	..	..	..	..	..	..	Current Budget Balance
5	9	12	22	..	..	..	..	..	..	..	Capital Receipts
..	..	..	10,991	..	..	..	..	..	..	..	Capital Payments

BURUNDI	1967	1968	1969	1970	1971	1972	1973	1974	1975	1976	1977
FOREIGN TRADE (CUSTOMS BASIS)					*(Millions of current US dollars)*						
Value of Exports, fob	14.0	16.3	11.9	24.4	18.5	26.0	30.1	29.6	31.6	53.6	89.0
Nonfuel Primary Products	13.3	15.4	11.4	24.0	17.9	25.4	29.3	28.9	30.6	52.6	87.1
Fuels	0.0	0.0	0.0	0.0	0.1	0.0	0.2	0.2	0.2	0.3	0.5
Manufactures	0.7	1.0	0.5	0.3	0.6	0.6	0.6	0.5	0.9	0.7	1.4
Value of Imports, cif	20.0	23.0	21.7	22.4	29.9	31.3	31.5	40.3	62.7	58.0	90.9
Nonfuel Primary Products	4.8	5.5	4.6	7.1	8.5	9.5	10.0	12.0	14.6	14.6	21.0
Fuels	1.7	1.9	1.6	2.5	2.0	1.9	2.0	3.0	3.7	5.1	6.2
Manufactures	13.5	15.6	15.6	12.7	19.4	19.9	19.6	25.3	44.4	38.3	63.7
					(Index 1980 = 100)						
Terms of Trade	124.8	125.0	120.1	169.3	109.5	111.4	104.8	74.5	68.7	143.0	204.5
Export Prices, fob	25.9	25.9	26.4	32.5	29.3	33.4	40.6	44.3	43.2	88.8	137.0
Nonfuel Primary Products	25.9	25.8	26.3	32.6	29.5	33.4	41.6	44.0	42.8	89.8	140.8
Fuels	4.3	4.3	4.3	4.3	5.6	6.2	8.9	36.7	35.7	38.4	42.0
Manufactures	30.1	31.7	32.4	37.5	36.5	38.5	49.5	67.2	66.1	66.7	71.7
Import Prices, cif	20.7	20.7	21.9	19.2	26.8	30.0	38.8	59.4	62.8	62.1	67.0
Trade at Constant 1980 Prices					*(Millions of 1980 US dollars)*						
Exports, fob	54.1	63.0	45.1	74.9	63.3	77.8	74.0	66.8	73.3	60.3	65.0
Imports, cif	96.4	110.9	99.1	116.5	111.8	104.2	81.2	67.9	99.9	93.4	135.6
BALANCE OF PAYMENTS					*(Millions of current US dollars)*						
Exports of Goods & Services	..	..	..	26.5	22.0	29.7	34.8	34.9	35.9	62.8	98.9
Merchandise, fob	..	..	..	24.5	19.6	26.6	31.1	31.3	32.3	56.9	93.2
Nonfactor Services	..	..	..	1.4	1.7	2.4	2.7	2.4	2.5	4.6	3.1
Factor Services	..	..	..	0.6	0.7	0.7	1.0	1.2	1.1	1.3	2.6
Imports of Goods & Services	..	..	..	29.5	37.4	39.2	42.0	55.4	81.9	76.4	111.6
Merchandise, fob	..	..	..	22.4	29.9	31.3	31.4	43.1	61.7	58.2	76.3
Nonfactor Services	..	..	..	5.8	5.9	5.9	9.5	10.5	18.3	16.6	18.4
Factor Services	..	..	..	1.3	1.6	2.0	1.1	1.8	1.9	1.6	16.9
Long-Term Interest	..	..	..	0.2	0.2	0.3	0.2	0.2	0.4	0.5	0.6
Current Transfers, net	..	..	..	..	..	..	..	..	..	..	..
Workers' Remittances	..	..	..	..	..	..	..	..	..	..	..
Total to be Financed	..	..	..	..	..	..	..	..	..	..	..
Official Capital Grants	..	..	..	..	..	..	..	..	..	..	..
Current Account Balance	..	..	..	2.3	-9.2	-4.2	-0.6	-10.7	-32.6	6.9	16.2
Long-Term Capital, net	..	..	..	17.8	19.3	19.8	24.5	32.2	48.8	41.2	54.1
Direct Investment	..	..	..	0.2	..	..	..	0.2	..	..	..
Long-Term Loans	..	..	..	1.0	0.1	-1.1	0.2	1.5	9.3	2.8	17.9
Disbursements	..	..	..	1.4	0.5	0.5	0.9	2.2	10.9	5.1	20.3
Repayments	..	..	..	0.4	0.4	1.6	0.7	0.7	1.6	2.3	2.4
Other Long-Term Capital	..	..	..	16.6	19.2	20.9	24.3	30.5	39.5	38.4	36.2
Other Capital, net	..	..	..	..	..	..	..	..	-2.9	-32.7	-24.2
Change in Reserves	..	..	..	..	..	..	..	..	-13.3	-15.4	-46.1
Memo Item:					*(Burundi Francs per US dollar)*						
Conversion Factor (Annual Avg)	87.500	87.500	87.500	87.500	87.500	87.500	80.030	78.750	78.750	86.250	90.000
EXTERNAL DEBT, ETC.				*(Millions of US dollars, outstanding at end of year)*							
Public/Publicly Guar. Long-Term	..	..	..	7.30	7.80	6.90	7.80	9.40	18.00	21.00	40.70
Official Creditors	..	..	..	5.80	6.30	5.70	6.70	8.10	11.60	14.70	34.10
IBRD and IDA	..	..	..	3.60	3.80	4.20	4.30	4.50	4.40	5.00	7.10
Private Creditors	..	..	..	1.50	1.60	1.20	1.10	1.30	6.40	6.30	6.60
Private Non-guaranteed Long-Term	..	..	..	..	..	..	..	..	..	..	..
Use of Fund Credit	..	..	..	7.70	5.80	0.00	0.00	0.00	1.40	0.00	0.00
Short-Term Debt	..	..	..	..	..	..	..	..	..	..	5.00
Memo Items:					*(Thousands of US dollars)*						
Int'l Reserves Excluding Gold	5,050.0	2,910.0	7,420.0	15,360.0	17,685.0	18,487.0	21,483.0	14,189.0	30,713.0	49,030.0	94,406.0
Gold Holdings (at market price)	..	40.0	34.0	36.0	42.0	62.0	108.0	179.0	135.0	129.0	1,500.0
SOCIAL INDICATORS											
Total Fertility Rate	6.4	6.4	6.4	6.4	6.4	6.4	6.4	6.4	6.5	6.5	6.5
Crude Birth Rate	46.5	46.4	46.4	46.3	46.3	46.2	46.6	47.0	47.5	47.9	48.3
Infant Mortality Rate	140.0	139.0	138.0	137.0	136.0	135.0	134.0	133.0	132.0	131.0	130.0
Life Expectancy at Birth	44.4	44.5	44.6	44.7	44.8	44.9	45.0	45.2	45.4	45.5	45.7
Food Production, p.c. ('79-81 = 100)	102.6	104.3	101.6	101.4	101.1	94.4	102.3	91.9	102.9	103.1	102.6
Labor Force, Agriculture (%)	93.9	93.7	93.6	93.5	93.4	93.3	93.3	93.2	93.1	93.0	92.9
Labor Force, Female (%)	50.0	49.9	49.8	49.8	49.7	49.6	49.6	49.5	49.5	49.4	49.3
School Enroll. Ratio, primary	..	..	..	30.0	..	..	..	..	22.0	..	23.0
School Enroll. Ratio, secondary	..	..	..	2.0	..	..	..	..	3.0	2.0	2.0

1978	1979	1980	1981	1982	1983	1984	1985	1986	1987 estimate	NOTES	BURUNDI
				(Millions of current US dollars)							**FOREIGN TRADE (CUSTOMS BASIS)**
69.0	104.9	65.3	71.0	88.0	77.1	98.0	111.6	167.0	139.5	..	Value of Exports, fob
67.8	103.5	60.6	66.8	82.9	72.5	93.4	103.2	147.4	118.8	..	Nonfuel Primary Products
0.4	0.6	0.3	0.4	0.5	0.4	0.4	0.4	0.2	0.3	..	Fuels
0.8	0.9	4.3	3.8	4.7	4.2	4.3	8.1	19.4	20.5	..	Manufactures
98.0	152.7	167.2	161.0	214.0	194.0	186.9	188.8	207.0	225.0	..	Value of Imports, cif
19.3	25.1	29.4	28.3	37.3	33.9	31.2	30.6	33.4	36.8	..	Nonfuel Primary Products
7.9	14.2	26.5	22.1	29.4	27.0	23.7	23.2	9.7	11.7	..	Fuels
70.7	113.4	111.4	110.6	147.3	133.1	132.0	135.0	164.0	176.5	..	Manufactures
				(Index 1980 = 100)							
137.4	127.0	100.0	81.8	92.1	89.8	100.9	99.8	117.8	74.9	..	Terms of Trade
101.8	110.4	100.0	83.4	89.7	85.3	94.3	92.3	117.9	79.9	..	Export Prices, fob
103.0	111.0	100.0	82.6	89.4	84.9	94.6	92.6	120.4	76.1	..	Nonfuel Primary Products
42.3	61.0	100.0	112.5	101.6	92.5	90.2	87.5	44.9	56.9	..	Fuels
76.1	90.0	100.0	99.1	94.8	90.7	89.3	89.0	103.2	113.0	..	Manufactures
74.1	86.9	100.0	102.0	97.4	95.0	93.5	92.5	100.1	106.6	..	Import Prices, cif
				(Millions of 1980 US dollars)							Trade at Constant 1980 Prices
67.8	95.0	65.3	85.1	98.1	90.4	103.9	120.9	141.6	174.6	..	Exports, fob
132.3	175.7	167.2	157.9	219.7	204.3	199.9	204.2	206.8	211.0	..	Imports, cif
				(Millions of current US dollars)							**BALANCE OF PAYMENTS**
78.7	114.9	92.0	97.0	108.5	98.6	102.5	122.5	151.0	110.1	f	Exports of Goods & Services
69.5	102.0	66.3	74.9	87.8	80.6	87.8	112.0	126.0	96.0	..	Merchandise, fob
4.4	6.6	14.7	12.8	14.9	16.1	13.4	9.3	23.6	11.6	..	Nonfactor Services
4.8	6.9	11.0	9.3	5.5	1.9	1.5	1.5	2.0	2.1	..	Factor Services
142.2	195.1	224.0	228.4	289.7	284.3	244.4	238.2	268.4	299.5	f	Imports of Goods & Services
96.5	145.0	168.0	161.0	214.0	184.0	184.0	171.0	179.0	190.0	..	Merchandise, fob
26.9	32.8	37.7	37.4	43.9	72.1	31.9	33.3	34.8	43.3	..	Nonfactor Services
19.2	17.1	18.0	29.4	31.7	28.3	28.4	34.2	54.4	66.5	..	Factor Services
1.0	1.4	2.0	1.8	2.4	3.4	8.1	8.6	12.1	15.1	..	Long-Term Interest
..	..	3.1	6.2	7.4	6.0	6.2	7.8	4.6	4.1	..	Current Transfers, net
..	..	..	..	..	..	..	..	..	..	..	Workers' Remittances
..	..	-119.6	-105.8	-166.1	-173.9	-137.5	-108.9	-94.2	-169.2	..	Total to be Financed
..	..	35.2	38.5	40.1	39.9	39.9	39.9	43.3	37.2	..	Official Capital Grants
-31.3	-45.4	-84.4	-67.3	-126.0	-134.0	-97.6	-69.0	-50.9	-132.0	..	Current Account Balance
72.1	96.5	121.3	122.6	141.5	188.6	149.1	133.5	191.7	205.4	f	Long-Term Capital, net
..	..	1.1	0.6	1.5	0.4	0.9	0.5	3.0	1.5	..	Direct Investment
24.3	38.9	40.8	26.1	51.6	102.4	70.4	54.6	84.1	113.0	..	Long-Term Loans
26.2	41.7	45.0	29.7	55.0	106.6	79.6	69.2	103.2	140.3	..	Disbursements
1.9	2.8	4.2	3.6	3.4	4.2	9.2	14.6	19.1	27.3	..	Repayments
47.8	57.6	44.2	57.4	48.3	45.9	37.9	38.5	61.3	53.7	..	Other Long-Term Capital
-50.0	-67.0	-42.8	-86.3	-50.9	-40.1	-53.2	-55.4	-124.1	-82.7	..	Other Capital, net
9.2	15.9	5.9	31.0	35.4	-14.5	1.7	-9.1	-16.7	9.3	..	Change in Reserves
				(Burundi Francs per US dollar)							Memo Item:
90.000	90.000	90.000	90.000	90.000	92.950	119.710	120.690	114.170	123.560	..	Conversion Factor (Annual Avg)
				(Millions of US dollars, outstanding at end of year)							**EXTERNAL DEBT, ETC.**
69.60	110.40	141.30	157.20	200.80	292.10	334.00	415.30	529.40	718.20	..	Public/Publicly Guar. Long-Term
59.10	98.10	133.00	152.60	196.20	275.30	309.40	387.70	502.60	698.60	..	Official Creditors
14.10	25.40	37.00	47.40	65.50	92.30	116.80	136.70	185.30	251.80	..	IBRD and IDA
10.60	12.20	8.30	4.60	4.60	16.90	24.60	27.60	26.90	19.50	..	Private Creditors
..	..	..	..	..	..	..	..	..	..	..	Private Non-guaranteed Long-Term
0.00	0.00	12.10	11.10	10.50	5.00	0.00	0.00	0.00	..	..	Use of Fund Credit
5.00	13.00	12.00	10.00	16.00	10.00	12.00	31.00	23.00	37.00	..	Short-Term Debt
				(Thousands of US dollars)							Memo Items:
81,304.0	89,987.0	94,502.0	61,300.0	29,489.0	26,939.0	19,731.0	29,472.0	69,073.0	60,732.0	..	Int'l Reserves Excluding Gold
2,055.0	6,738.0	10,152.0	6,846.0	7,869.0	6,570.0	5,310.0	5,632.0	6,732.0	8,337.0	..	Gold Holdings (at market price)
											SOCIAL INDICATORS
6.5	6.5	6.5	6.5	6.5	6.5	6.5	6.5	6.5	6.5	..	Total Fertility Rate
47.9	47.6	47.2	46.9	46.5	46.7	46.8	47.0	47.1	47.3	..	Crude Birth Rate
128.8	127.6	126.4	125.2	124.0	121.8	119.6	117.4	115.2	113.1	..	Infant Mortality Rate
45.9	46.1	46.3	46.4	46.6	47.0	47.5	47.9	48.4	48.9	..	Life Expectancy at Birth
105.7	98.6	96.9	104.5	101.4	97.7	92.2	100.0	100.3	98.2	..	Food Production, p.c. ('79-81 = 100)
92.9	92.8	92.7	..	..	..	..	..	..	..	..	Labor Force, Agriculture (%)
49.3	49.2	49.1	49.0	48.8	48.6	48.5	48.3	48.1	47.9	..	Labor Force, Female (%)
23.0	25.0	29.0	33.0	40.0	44.0	49.0	53.0	..	..	..	School Enroll. Ratio, primary
3.0	3.0	3.0	3.0	..	3.0	4.0	4.0	..	..	..	School Enroll. Ratio, secondary

CAMEROON	1967	1968	1969	1970	1971	1972	1973	1974	1975	1976	1977
CURRENT GNP PER CAPITA (US $)	130	150	170	180	180	180	210	270	310	360	410
POPULATION (thousands)	6,088	6,225	6,364	6,506	6,683	6,864	7,051	7,242	7,439	7,676	7,920
ORIGIN AND USE OF RESOURCES					*(Billions of current CFA Francs)*						
Gross National Product (GNP)	215.9	244.4	272.8	304.8	316.5	347.2	390.9	477.2	563.7	639.7	773.9
Net Factor Income from Abroad	-14.4	-14.6	-13.3	-13.4	-24.1	-30.4	-32.9	-42.1	-48.6	-52.8	-60.1
Gross Domestic Product (GDP)	230.3	259.0	286.1	318.2	340.6	377.6	423.8	519.3	612.3	692.5	834.0
Indirect Taxes, net	21.6	24.4	22.6	20.4	18.4	29.6	46.9	70.7	96.5	116.0	80.4
GDP at factor cost	208.7	234.6	263.5	297.8	322.2	348.0	376.9	448.6	515.8	576.5	753.6
Agriculture	71.7	81.6	87.9	99.8	105.6	120.7	130.5	153.4	178.3	191.2	280.6
Industry	46.3	50.5	55.6	59.3	65.5	71.2	78.9	92.8	111.7	131.3	153.4
Manufacturing	22.5	24.0	28.1	32.6	35.2	39.2	43.4	51.0	63.0	73.0	74.7
Services, etc.	112.3	126.9	142.6	159.1	169.5	185.7	214.4	273.1	322.3	370.0	400.0
Resource Balance	-4.5	-6.5	1.5	4.9	-14.1	-21.5	-14.0	13.7	-17.5	-32.7	-22.9
Exports of Goods & NFServices	47.5	57.3	66.9	83.4	78.4	77.0	87.7	132.1	138.9	156.8	209.0
Imports of Goods & NFServices	52.0	63.8	65.4	78.5	92.5	98.5	101.7	118.4	156.4	189.5	231.9
Domestic Absorption	234.8	265.5	284.6	313.3	354.7	399.1	437.8	505.6	629.8	725.2	856.9
Private Consumption, etc.	171.2	196.6	215.2	224.2	256.5	284.2	304.5	359.6	440.8	528.9	593.2
General Gov't Consumption	31.9	34.3	38.8	38.1	41.5	46.0	49.0	57.3	66.7	74.5	81.9
Gross Domestic Investment	31.7	34.6	30.6	51.0	56.7	68.9	84.3	88.7	122.3	121.8	181.8
Fixed Investment	..	..	..	..	..	..	..	..	100.3	118.8	163.5
Memo Items:											
Gross Domestic Saving	27.2	28.1	32.1	55.9	42.6	47.4	70.3	102.4	104.8	89.1	158.9
Gross National Saving	12.8	13.5	18.8	40.5	16.4	14.4	31.7	55.6	51.5	32.3	98.4
					(Billions of 1980 CFA Francs)						
Gross National Product	615.2	656.6	695.3	719.6	716.6	729.4	772.0	849.4	846.1	887.6	962.2
GDP at factor cost	591.1	628.3	670.2	702.1	734.4	734.6	746.8	803.3	777.5	801.5	943.6
Agriculture	239.4	252.0	265.7	278.7	282.3	298.7	310.4	325.4	329.4	333.8	338.2
Industry	103.6	116.1	123.7	122.1	136.8	132.1	137.6	145.1	150.4	162.0	191.6
Manufacturing	46.5	52.9	60.0	61.8	70.2	71.1	71.3	72.6	78.5	83.6	94.8
Services, etc.	309.3	325.6	338.3	349.4	357.2	366.2	391.7	459.4	443.1	467.1	514.4
Resource Balance	-39.3	-61.6	-44.8	-50.9	-102.3	-91.3	-61.8	-37.3	-72.2	-69.9	-126.2
Exports of Goods & NFServices	138.3	144.3	160.2	174.5	171.5	180.0	184.7	210.0	188.4	218.7	192.6
Imports of Goods & NFServices	177.6	205.9	205.0	225.4	273.7	271.3	246.5	247.2	260.5	288.5	318.8
Domestic Absorption	691.6	755.3	772.5	801.1	878.6	888.3	901.5	967.2	995.1	1,032.7	1,170.5
Private Consumption, etc.	543.4	602.9	613.2	621.1	678.4	667.9	664.1	726.1	738.3	773.6	876.2
General Gov't Consumption	70.7	74.3	84.8	81.5	79.6	91.3	99.6	114.6	109.0	122.1	106.1
Gross Domestic Investment	77.5	78.1	74.6	98.6	120.5	129.1	137.8	126.4	147.8	137.1	188.2
Fixed Investment	..	..	..	..	..	..	..	..	119.9	133.4	171.2
Memo Items:											
Capacity to Import	162.3	184.9	209.7	239.5	232.0	212.1	212.5	275.8	231.4	238.7	287.3
Terms of Trade Adjustment	23.9	40.6	49.5	65.0	60.5	32.1	27.8	65.9	43.0	20.1	94.7
Gross Domestic Income	676.2	734.3	777.2	815.2	836.8	829.1	867.6	995.8	965.9	982.9	1,139.0
Gross National Income	639.1	697.2	744.9	784.6	777.2	761.5	799.8	915.2	889.2	907.7	1,056.9
DOMESTIC PRICES (DEFLATORS)					*(Index 1980 = 100)*						
Overall (GDP)	35.3	37.3	39.3	42.4	43.9	47.4	50.5	55.8	66.3	71.9	79.9
Domestic Absorption	34.0	35.2	36.8	39.1	40.4	44.9	48.6	52.3	63.3	70.2	73.2
Agriculture	29.9	32.4	33.1	35.8	37.4	40.4	42.0	47.1	54.1	57.3	83.0
Industry	44.7	43.5	44.9	48.6	47.9	53.9	57.3	64.0	74.3	81.1	80.0
Manufacturing	48.4	45.4	46.8	52.7	50.1	55.1	60.9	70.3	80.2	87.3	78.8
MANUFACTURING ACTIVITY											
Employment (1980=100)	..	..	..	80.5	95.0	82.5	..	83.6	103.7	91.5	95.8
Real Earnings per Empl. (1980=100)	..	..	..	55.6	61.5	65.8	..	69.2	69.2	76.2	73.8
Real Output per Empl. (1980=100)	..	..	..	59.1	69.0	68.1	..	60.8	51.5	60.9	78.6
Earnings as % of Value Added	..	..	..	29.4	31.2	26.2	..	33.5	38.6	38.0	36.5
MONETARY HOLDINGS					*(Billions of current CFA Francs)*						
Money Supply, Broadly Defined	30.0	35.1	40.4	45.5	51.6	57.4	70.3	94.4	105.7	132.5	183.5
Money as Means of Payment	27.1	31.7	35.2	38.4	43.6	47.2	55.4	72.9	76.0	94.6	126.6
Currency Ouside Banks	14.6	17.1	17.9	19.6	20.9	21.9	25.9	31.0	32.2	36.9	48.9
Demand Deposits	12.4	14.6	17.3	18.9	22.7	25.3	29.5	41.9	43.8	57.7	77.7
Quasi-Monetary Liabilities	2.9	3.5	5.3	7.1	8.0	10.2	14.9	21.6	29.6	37.9	56.9
GOVERNMENT DEFICIT (-) OR SURPLUS					*(Billions of current CFA Francs)*						
	..	..	..	..	..	..	..	..	-13.0	-15.9	-3.0
Current Revenue	..	..	..	..	..	..	..	..	91.0	109.6	132.8
Current Expenditure	..	..	..	..	..	..	..	..	73.4	77.7	95.9
Current Budget Balance	..	..	..	..	..	..	..	..	17.6	31.9	36.8
Capital Receipts	..	..	..	..	..	..	..	..	..	..	..
Capital Payments	..	..	..	..	..	..	..	..	30.6	47.7	39.9

1978	1979	1980	1981	1982	1983	1984	1985	1986	1987 estimate	NOTES	CAMEROON
500	600	760	900	900	870	820	840	920	960	..	**CURRENT GNP PER CAPITA (US $)**
8,172	8,433	8,701	8,980	9,269	9,567	9,874	10,191	10,548	10,927	..	**POPULATION (thousands)**
				(Billions of current CFA Francs)							**ORIGIN AND USE OF RESOURCES**
987.5	1,167.1	1,427.3	1,800.4	2,158.1	2,609.5	3,059.6	3,757.9	3,994.7	3,883.3	C	Gross National Product (GNP)
-64.7	-92.0	-141.6	-179.6	-145.3	-175.0	-213.0	-243.0	-216.0	-151.0	..	Net Factor Income from Abroad
1,052.2	1,259.1	1,568.9	1,980.0	2,303.4	2,784.5	3,272.6	4,000.9	4,210.7	4,034.3	C	Gross Domestic Product (GDP)
108.2	83.5	118.2	..	..	..	..	..	..	..		Indirect Taxes, net
944.0	1,175.6	1,450.7	..	..	..	..	..	..	..	B C	GDP at factor cost
329.9	388.1	437.0	527.4	633.9	656.0	758.7	853.7	926.0	958.7	..	Agriculture
167.9	259.9	406.0	617.9	810.4	1,085.9	1,333.4	1,629.8	1,435.0	1,240.8	..	Industry
86.7	101.9	124.1	173.7	247.0	290.9	358.5	422.5	510.7	519.9	..	Manufacturing
554.4	611.1	725.9	834.7	859.1	1,042.6	1,180.5	1,517.4	1,849.7	1,834.8	..	Services, etc.
-47.0	-90.2	-49.1	-45.9	-6.0	121.8	390.3	428.0	11.0	-143.0	..	Resource Balance
248.9	265.4	378.0	516.0	599.0	797.0	1,085.0	1,318.0	964.0	660.0	..	Exports of Goods & NFServices
295.9	355.6	427.1	561.9	605.0	675.2	694.7	890.0	953.0	803.0	..	Imports of Goods & NFServices
1,099.2	1,349.3	1,618.0	2,025.9	2,309.4	2,662.7	2,882.3	3,572.9	4,199.7	4,177.3	..	Domestic Absorption
757.6	960.4	1,185.3	1,365.2	1,571.4	1,734.3	1,894.7	2,446.9	2,927.0	2,996.7	..	Private Consumption, etc.
102.8	116.3	136.8	172.3	199.3	248.3	306.4	345.3	437.9	446.8	..	General Gov't Consumption
238.8	272.6	295.9	488.4	538.7	680.1	681.2	780.7	834.8	733.8	..	Gross Domestic Investment
204.7	251.7	282.4	441.4	507.2	654.5	661.8	764.4	..	..	..	Fixed Investment
											Memo Items:
191.8	182.4	246.8	442.5	532.7	801.9	1,071.5	1,208.7	845.8	590.8	..	Gross Domestic Saving
124.3	83.5	84.0	247.2	363.4	595.6	830.7	950.9	611.8	408.2	..	Gross National Saving
				(Billions of 1980 CFA Francs)							
1,121.7	1,258.5	1,427.3	1,612.0	1,703.6	1,837.4	1,940.2	2,171.7	2,401.3	2,371.5	C	Gross National Product
1,074.5	1,266.8	1,450.7	..	..	..	..	..	..	..	B C	GDP at factor cost
351.0	403.4	437.0	494.5	513.4	474.5	516.4	494.6	534.9	552.8	..	Agriculture
211.7	304.1	406.0	530.6	617.3	745.4	777.3	866.4	884.3	848.3	..	Industry
109.4	113.3	124.1	164.5	221.5	236.2	232.9	237.1	250.6	238.7	..	Manufacturing
635.0	649.3	725.9	746.4	687.7	739.4	780.1	955.0	1,107.0	1,055.9	..	Services, etc.
-92.7	-85.0	-49.1	-27.8	-5.2	104.4	265.9	194.4	97.4	91.2	..	Resource Balance
253.8	300.9	378.0	461.5	440.5	529.3	703.0	799.1	795.9	691.2	..	Exports of Goods & NFServices
346.5	385.9	427.1	489.3	445.6	424.9	437.1	604.7	698.5	600.0	..	Imports of Goods & NFServices
1,290.4	1,441.8	1,618.0	1,799.2	1,823.5	1,855.0	1,807.9	2,121.6	2,428.8	2,365.8	..	Domestic Absorption
889.8	1,045.8	1,185.3	1,223.4	1,233.9	1,221.4	1,187.1	1,457.1	1,690.3	1,707.1	..	Private Consumption, etc.
122.4	129.5	136.8	158.1	163.8	178.6	194.2	199.9	262.7	272.1	..	General Gov't Consumption
278.2	266.4	295.9	417.7	425.8	455.0	426.6	464.6	475.8	386.6	..	Gross Domestic Investment
238.5	246.0	282.4	377.5	400.9	437.9	414.9	451.9	..	..	..	Fixed Investment
											Memo Items:
291.5	288.0	378.0	449.3	441.2	501.5	682.7	895.5	706.6	493.2	..	Capacity to Import
37.7	-12.9	0.0	-12.2	0.7	-27.7	-20.3	96.4	-89.3	-198.0	..	Terms of Trade Adjustment
1,235.4	1,343.9	1,568.9	1,759.3	1,819.1	1,931.6	2,053.5	2,412.4	2,436.9	2,259.0	..	Gross Domestic Income
1,159.4	1,245.6	1,427.3	1,599.8	1,704.4	1,809.7	1,919.9	2,268.1	2,311.9	2,173.4	..	Gross National Income
				(Index 1980 = 100)							**DOMESTIC PRICES (DEFLATORS)**
87.9	92.8	100.0	111.8	126.7	142.1	157.8	172.8	166.7	164.2	..	Overall (GDP)
85.2	93.6	100.0	112.6	126.6	143.5	159.4	168.4	172.9	176.6	..	Domestic Absorption
94.0	96.2	100.0	106.7	123.5	138.3	146.9	172.6	173.1	173.4	..	Agriculture
79.3	85.5	100.0	116.5	131.3	145.7	171.5	188.1	162.3	146.3	..	Industry
79.3	89.9	100.0	105.6	111.5	123.2	153.9	178.2	203.8	217.8	..	Manufacturing
											MANUFACTURING ACTIVITY
100.5	97.9	100.0	102.1	104.3	106.6	108.8	111.2	..	..	..	Employment (1980=100)
75.6	85.8	100.0	123.8	152.1	150.4	162.9	197.0	..	..	..	Real Earnings per Empl. (1980=100)
82.1	80.9	100.0	129.8	171.1	178.6	199.6	215.0	..	..	..	Real Output per Empl. (1980=100)
39.0	34.3	36.6	36.6	36.6	36.6	36.6	36.6	..	..	..	Earnings as % of Value Added
				(Billions of current CFA Francs)							**MONETARY HOLDINGS**
212.3	260.1	315.4	405.6	483.4	612.4	736.2	864.5	831.0	677.7	..	Money Supply, Broadly Defined
146.9	184.3	208.2	258.9	298.5	377.1	410.8	426.7	448.0	387.0	..	Money as Means of Payment
58.5	68.2	78.1	101.8	107.6	127.6	134.4	148.3	167.7	171.1	..	Currency Ouside Banks
88.5	116.0	130.1	157.1	190.9	249.4	276.3	278.4	280.3	215.9	..	Demand Deposits
65.4	75.8	107.2	146.7	184.9	235.3	325.5	437.8	383.1	290.7	..	Quasi-Monetary Liabilities
				(Billions of current CFA Francs)							
4.1	31.7	7.2	-58.5	-55.2	33.9	52.9	30.5	..	..	C	**GOVERNMENT DEFICIT (-) OR SURPLUS**
178.8	222.4	230.7	314.5	389.2	668.1	795.3	885.7	..	..	..	Current Revenue
124.4	133.9	148.7	248.4	256.4	344.6	428.8	466.2	..	..	..	Current Expenditure
54.4	88.5	82.1	66.1	132.8	323.6	366.5	419.5	..	..	..	Current Budget Balance
..	..	..	..	1.2	3.7	0.1	..	..	..	..	Capital Receipts
50.3	56.8	74.8	124.6	189.2	293.4	313.7	389.0	..	..	..	Capital Payments

CAMEROON	1967	1968	1969	1970	1971	1972	1973	1974	1975	1976	1977
FOREIGN TRADE (CUSTOMS BASIS)				*(Millions of current US dollars)*							
Value of Exports, fob	146.3	185.7	232.1	230.9	212.5	225.3	361.6	490.9	469.9	534.8	704.0
Nonfuel Primary Products	142.3	170.4	214.1	211.5	192.3	203.2	328.3	447.2	421.3	483.3	660.9
Fuels	0.0	0.0	0.1	0.0	0.1	0.4	0.6	1.0	1.4	2.0	13.2
Manufactures	4.0	15.3	17.9	19.3	20.0	21.7	32.8	42.6	47.2	49.6	29.9
Value of Imports, cif	178.3	178.3	207.2	242.1	249.5	298.8	334.7	437.3	598.3	594.7	746.9
Nonfuel Primary Products	28.2	29.4	27.0	32.9	37.2	39.7	47.6	67.3	76.4	67.2	102.3
Fuels	10.1	11.0	11.9	13.1	15.3	17.8	21.4	42.4	60.2	53.2	67.2
Manufactures	140.1	137.9	168.3	196.0	197.0	241.3	265.7	327.6	461.7	474.3	577.4
					(Index 1980 = 100)						
Terms of Trade	109.0	120.3	125.1	119.3	98.4	97.0	114.3	93.1	78.6	117.2	147.2
Export Prices, fob	25.5	27.2	29.7	29.9	26.7	29.3	43.0	52.6	46.9	71.3	97.9
Nonfuel Primary Products	25.4	26.9	29.6	29.4	26.1	28.8	42.7	51.5	45.4	72.0	102.3
Fuels	4.3	4.3	4.3	4.3	5.6	6.2	8.9	36.7	35.7	38.4	42.0
Manufactures	30.1	31.7	32.4	37.5	36.5	38.5	49.5	67.2	66.1	66.7	71.7
Import Prices, cif	23.4	22.6	23.8	25.0	27.2	30.2	37.6	56.5	59.7	60.8	66.5
Trade at Constant 1980 Prices				*(Millions of 1980 US dollars)*							
Exports, fob	573.6	681.9	780.5	772.7	794.5	768.1	841.7	934.0	1,002.5	750.6	719.3
Imports, cif	761.6	787.9	871.8	966.9	918.0	988.2	890.5	774.7	1,002.8	978.4	1,123.0
BALANCE OF PAYMENTS				*(Millions of current US dollars)*							
Exports of Goods & Services	..	..	..	279.3	294.3	320.0	521.2	583.8	672.3	721.2	979.0
Merchandise, fob	..	..	..	218.7	235.7	239.3	409.5	493.2	512.0	584.2	809.1
Nonfactor Services	..	..	..	56.7	51.7	75.0	108.1	86.8	147.9	133.7	161.8
Factor Services	..	..	..	3.9	6.9	5.6	3.6	3.7	12.4	3.3	8.1
Imports of Goods & Services	..	..	..	319.4	347.6	417.8	531.5	601.9	848.4	851.2	1,103.6
Merchandise, fob	..	..	..	190.8	223.1	257.6	310.6	389.9	540.3	554.9	719.2
Nonfactor Services	..	..	..	122.4	114.2	146.4	194.1	175.7	249.4	251.9	335.1
Factor Services	..	..	..	6.2	10.3	13.8	26.8	36.3	58.8	44.4	49.3
Long-Term Interest	..	..	..	4.9	6.1	7.1	11.1	13.1	17.2	20.9	31.1
Current Transfers, net	..	..	..	-7.2	-7.4	-10.2	-25.6	-19.4	-22.1	-16.7	-1.5
Workers' Remittances	..	..	..	0.0	0.0	0.0	0.0	0.0	0.0	0.0	0.0
Total to be Financed	..	..	..	-47.3	-60.8	-108.0	-35.9	-37.5	-198.3	-146.7	-126.1
Official Capital Grants	..	..	..	17.5	16.3	16.4	19.2	20.6	45.8	54.7	40.5
Current Account Balance	..	..	..	-29.8	-44.4	-91.6	-16.7	-17.0	-152.5	-92.1	-85.6
Long-Term Capital, net	..	..	..	59.9	53.2	33.4	70.7	52.7	105.1	167.6	151.7
Direct Investment	..	..	..	16.0	1.7	3.3	-0.7	13.5	25.3	8.2	4.3
Long-Term Loans	..	..	..	33.1	26.0	69.8	31.1	41.8	104.6	162.6	320.8
Disbursements	..	..	..	39.5	35.3	81.2	50.0	60.3	132.3	189.9	357.4
Repayments	..	..	..	6.4	9.3	11.4	18.9	18.5	27.7	27.3	36.6
Other Long-Term Capital	..	..	..	-6.7	9.1	-56.0	21.1	-23.2	-70.5	-57.9	-213.9
Other Capital, net	..	..	..	-0.7	-25.4	23.8	-51.3	-17.7	-12.1	-83.0	-68.2
Change in Reserves	..	..	..	-29.4	16.6	34.4	-2.7	-18.0	59.5	7.5	2.1
Memo Item:				*(CFA Francs per US dollar)*							
Conversion Factor (Annual Avg)	246.850	246.850	246.850	274.330	277.710	265.390	240.960	230.050	222.420	225.090	247.780
EXTERNAL DEBT, ETC.				*(Millions of US dollars, outstanding at end of year)*							
Public/Publicly Guar. Long-Term	..	..	..	131.2	162.2	195.4	236.8	274.9	371.6	513.3	861.3
Official Creditors	..	..	..	119.6	148.6	164.1	209.8	242.2	292.5	345.3	537.4
IBRD and IDA	..	..	..	12.3	18.7	32.2	53.2	62.8	99.6	125.3	160.0
Private Creditors	..	..	..	11.6	13.6	31.2	27.1	32.7	79.1	168.0	323.9
Private Non-guaranteed Long-Term	..	..	..	9.2	11.9	20.1	22.2	34.4	33.6	37.9	42.8
Use of Fund Credit	..	..	..	0.0	0.0	0.0	0.0	5.7	14.2	39.4	41.2
Short-Term Debt	..	..	..	..	..	..	..	..	..	..	111.4
Memo Items:				*(Thousands of US dollars)*							
Int'l Reserves Excluding Gold	24,350.0	40,851.0	48,086.0	80,807.0	73,603.0	43,638.0	51,153.0	78,527.0	28,828.0	43,804.0	42,389.0
Gold Holdings (at market price)	..	..	..	..	..	..	..	..	..	..	2,474.0
SOCIAL INDICATORS											
Total Fertility Rate	5.4	5.5	5.6	5.8	5.9	6.0	6.1	6.2	6.3	6.4	6.5
Crude Birth Rate	41.3	42.0	42.6	43.3	43.9	44.6	45.3	45.9	46.6	47.2	47.9
Infant Mortality Rate	136.0	132.6	129.2	125.8	122.4	119.0	117.4	115.8	114.2	112.6	111.0
Life Expectancy at Birth	46.4	47.2	48.1	48.9	49.8	50.6	50.9	51.2	51.5	51.8	52.1
Food Production, p.c. ('79-81 = 100)	96.2	97.3	99.0	103.7	105.8	108.2	109.1	114.4	115.9	109.7	110.7
Labor Force, Agriculture (%)	85.1	84.6	84.0	83.4	82.1	80.7	79.4	78.0	76.7	75.3	74.0
Labor Force, Female (%)	37.8	37.7	37.6	37.5	37.3	37.1	36.9	36.7	36.5	36.3	36.1
School Enroll. Ratio, primary	..	..	..	89.0	..	..	..	..	97.0	99.0	101.0
School Enroll. Ratio, secondary	..	..	..	7.0	..	..	..	..	13.0	15.0	16.0

1978	1979	1980	1981	1982	1983	1984	1985	1986	1987 estimate	NOTES	CAMEROON
				(Millions of current US dollars)							**FOREIGN TRADE (CUSTOMS BASIS)**
805.0	1,170.0	1,447.0	1,764.0	1,616.2	1,830.0	2,080.0	2,322.0	1,790.7	1,770.4	f	Value of Exports, fob
749.2	836.2	893.1	619.2	480.9	561.1	647.0	723.9	677.3	779.4	..	Nonfuel Primary Products
25.3	266.4	503.9	1,058.6	1,051.1	1,183.3	1,342.8	1,489.2	992.5	858.0	..	Fuels
30.6	67.4	50.1	86.2	84.2	85.6	90.2	108.9	120.9	133.0	..	Manufactures
1,008.5	1,270.6	1,538.4	1,421.0	1,243.2	1,249.7	1,213.0	1,163.0	1,704.0	1,921.7	f	Value of Imports, cif
113.2	152.7	157.0	155.8	145.8	142.8	126.4	116.1	33.4	14.1	..	Nonfuel Primary Products
72.4	137.1	179.3	108.7	45.3	65.6	103.8	109.1	46.9	35.4	..	Fuels
822.9	980.9	1,202.1	1,156.5	1,052.1	1,041.4	982.8	937.8	1,623.7	1,872.2	..	Manufactures
				(Index 1980 = 100)							
119.5	104.4	100.0	98.3	95.6	93.8	96.4	92.3	62.2	56.1	..	Terms of Trade
90.3	90.5	100.0	99.4	93.4	89.6	90.3	86.5	67.2	68.2	..	Export Prices, fob
94.6	107.1	100.0	83.0	79.2	83.9	90.6	84.0	90.7	80.3	..	Nonfuel Primary Products
42.3	'61.0	100.0	112.5	101.6	92.5	90.2	87.5	55.2	56.9	..	Fuels
76.1	90.0	100.0	99.1	94.8	90.7	89.3	89.0	103.2	113.0	..	Manufactures
75.5	86.7	100.0	101.2	97.7	95.5	93.6	93.6	108.1	121.5	..	Import Prices, cif
				(Millions of 1980 US dollars)							Trade at Constant 1980 Prices
891.9	1,292.6	1,447.0	1,774.3	1,729.9	2,043.1	2,304.5	2,685.3	2,663.2	2,595.8	..	Exports, fob
1,335.6	1,465.8	1,538.3	1,404.3	1,271.8	1,308.8	1,295.7	1,241.9	1,576.1	1,581.8	..	Imports, cif
				(Millions of current US dollars)							**BALANCE OF PAYMENTS**
1,318.6	1,718.0	1,827.7	2,218.3	2,032.7	2,259.6	2,588.3	2,819.0	2,545.0	2,117.0	C f	Exports of Goods & Services
1,095.9	1,354.1	1,418.1	1,763.1	1,617.1	1,829.6	2,080.2	2,337.0	1,995.0	1,714.0	..	Merchandise, fob
206.5	351.5	388.6	430.5	402.4	417.3	461.3	461.0	496.0	362.0	..	Nonfactor Services
16.3	12.4	21.0	24.7	13.1	12.7	46.9	21.0	54.0	41.0	..	Factor Services
1,503.5	1,843.7	2,226.2	2,676.5	2,368.6	2,311.2	2,380.0	2,458.0	3,074.0	3,124.0	C f	Imports of Goods & Services
949.0	1,270.8	1,452.3	1,611.4	1,285.4	1,166.2	1,194.7	1,088.0	1,475.0	1,739.0	..	Merchandise, fob
464.1	464.7	589.4	776.5	753.6	737.5	718.9	810.0	996.0	913.0	..	Nonfactor Services
90.4	108.3	184.5	288.6	329.6	407.4	466.4	560.0	603.0	472.0	..	Factor Services
48.3	71.4	119.4	137.4	150.7	139.2	166.1	136.0	187.7	176.4	..	Long-Term Interest
-12.2	-32.7	-100.3	-57.9	-72.9	-82.2	-63.7	-33.0	-52.0	-105.0	..	Current Transfers, net
0.0	3.5	11.0	21.1	17.3	25.6	14.2	1.0	2.0	3.0	..	Workers' Remittances
-197.1	-158.4	-498.8	-516.1	-408.8	-133.8	144.6	328.0	-581.0	-1,112.0	K	Total to be Financed
21.8	46.0	103.6	68.5	57.0	78.8	56.5	0.0	0.0	..	K	Official Capital Grants
-175.3	-112.4	-395.1	-447.6	-351.8	-55.0	201.1	328.0	-581.0	-1,112.0	..	Current Account Balance
161.6	339.4	718.8	509.6	319.9	376.5	130.4	26.0	-70.0	156.0	C f	Long-Term Capital, net
33.7	59.9	105.2	30.5	27.6	57.8	92.1	21.0	0.0	31.0	..	Direct Investment
299.4	521.2	509.6	328.2	166.9	339.5	198.9	-175.2	59.6	106.5	..	Long-Term Loans
371.4	613.1	623.5	452.7	362.8	508.9	398.9	295.0	491.7	519.6	..	Disbursements
72.0	91.9	113.9	124.5	195.9	169.4	200.0	470.2	432.1	413.1	..	Repayments
-193.3	-287.7	0.4	82.3	68.4	-99.6	-217.1	180.2	-129.6	18.5	..	Other Long-Term Capital
19.4	-153.0	-178.5	-163.2	-19.4	-114.3	-548.0	-315.4	829.0	437.0	C f	Other Capital, net
-5.7	-74.1	-145.2	101.2	51.3	-207.2	216.5	-38.6	-178.0	519.0	..	Change in Reserves
				(CFA Francs per US dollar)							Memo Item:
238.590	216.650	209.210	235.270	296.680	354.660	409.520	471.130	386.600	318.660	..	Conversion Factor (Annual Avg)
				(Millions of US dollars, outstanding at end of year)							**EXTERNAL DEBT, ETC.**
1,184.9	1,684.7	2,048.6	2,048.8	1,958.7	1,876.4	1,720.7	2,007.1	2,377.7	2,785.3	..	Public/Publicly Guar. Long-Term
753.6	1,047.7	1,219.0	1,332.0	1,369.2	1,381.2	1,332.8	1,584.2	1,909.2	2,229.8	..	Official Creditors
210.0	256.7	298.0	339.9	386.2	433.6	430.2	513.6	632.5	783.7	..	IBRD and IDA
431.3	637.0	829.7	716.8	589.5	495.2	387.9	422.9	468.6	555.6	..	Private Creditors
86.2	160.0	178.0	250.0	361.0	532.8	609.2	381.1	504.9	520.2	..	Private Non-guaranteed Long-Term
42.7	32.7	15.3	4.1	0.6	0.0	0.0	0.0	0.0	..	..	Use of Fund Credit
164.1	238.9	270.5	245.3	396.4	327.9	382.1	528.7	783.6	722.4	..	Short-Term Debt
				(Thousands of US dollars)							Memo Items:
52,284.0	125,700.0	188,860.0	85,187.0	67,227.0	159,090.0	53,853.0	132,470.0	59,016.0	63,760.0	..	Int'l Reserves Excluding Gold
3,390.0	15,206.0	17,508.0	11,806.0	13,570.0	11,331.0	9,249.0	9,810.0	11,727.0	14,523.0	..	Gold Holdings (at market price)
											SOCIAL INDICATORS
6.5	6.5	6.5	6.5	6.5	6.6	6.7	6.8	6.9	7.0	..	Total Fertility Rate
47.6	47.3	46.9	46.6	46.3	46.7	47.0	47.4	47.8	48.1	..	Crude Birth Rate
109.4	107.8	106.2	104.6	103.0	98.6	94.2	89.8	85.4	81.0	..	Infant Mortality Rate
52.4	52.7	53.0	53.3	53.6	54.1	54.6	55.1	55.6	56.2	..	Life Expectancy at Birth
102.1	101.3	98.9	99.8	98.3	97.3	93.6	95.4	94.3	91.6	..	Food Production, p.c. ('79-81 = 100)
72.6	71.3	69.9	..	..	..	..	..	..	..	..	Labor Force, Agriculture (%)
35.9	35.7	35.5	35.2	35.0	34.8	34.6	34.4	34.2	34.0	..	Labor Force, Female (%)
103.0	104.0	104.0	106.0	107.0	106.0	107.0	..	..	..	..	School Enroll. Ratio, primary
17.0	18.0	19.0	19.0	21.0	22.0	23.0	..	..	..	..	School Enroll. Ratio, secondary

CANADA	1967	1968	1969	1970	1971	1972	1973	1974	1975	1976	1977
CURRENT GNP PER CAPITA (US $)	3,030	3,320	3,640	3,880	4,320	4,850	5,590	6,510	7,330	8,390	8,920
POPULATION (thousands)	20,412	20,744	21,028	21,324	21,592	21,822	22,072	22,364	22,697	22,993	23,273
ORIGIN AND USE OF RESOURCES					*(Billions of current Canadian Dollars)*						
Gross National Product (GNP)	67.88	74.17	81.83	87.76	95.78	107.17	125.64	149.87	169.00	194.39	213.31
Net Factor Income from Abroad	-1.24	-1.22	-1.21	-1.35	-1.51	-1.46	-1.73	-2.24	-2.54	-3.54	-4.57
Gross Domestic Product (GDP)	69.12	75.39	83.03	89.12	97.29	108.63	127.37	152.11	171.54	197.92	217.88
Indirect Taxes, net	8.46	9.27	9.94	11.09	12.05	13.63	15.31	17.87	17.09	20.99	23.19
GDP at factor cost	60.66	66.13	73.09	78.02	85.24	95.00	112.06	134.24	154.45	176.93	194.69
Agriculture	3.10	3.32	3.59	3.44	3.59	3.78	5.89	7.03	7.68	7.81	7.75
Industry	23.80	25.53	27.74	28.45	31.14	34.41	41.71	50.14	55.30	62.70	70.44
Manufacturing	15.11	16.34	18.01	17.81	19.49	21.69	25.48	29.88	31.83	35.54	39.21
Services, etc.	33.76	37.27	41.76	46.13	50.50	56.81	64.46	77.08	91.48	106.42	116.50
Resource Balance	0.69	0.97	0.13	2.24	1.62	1.01	1.74	0.48	-2.40	-0.99	-0.09
Exports of Goods & NFServices	14.16	16.16	17.84	20.07	21.18	23.77	29.79	37.82	38.97	44.28	51.17
Imports of Goods & NFServices	13.46	15.19	17.71	17.83	19.55	22.75	28.05	37.35	41.37	45.28	51.26
Domestic Absorption	68.43	74.42	82.90	86.88	95.67	107.62	125.63	151.63	173.94	198.92	217.97
Private Consumption, etc.	40.84	44.46	49.06	51.14	55.49	62.76	71.99	84.83	97.40	111.74	122.56
General Gov't Consumption	11.10	12.70	14.19	16.45	18.23	20.14	22.85	27.48	33.27	38.27	43.41
Gross Domestic Investment	16.48	17.26	19.66	19.28	21.94	24.72	30.79	39.32	43.27	48.90	52.00
Fixed Investment	15.03	15.25	16.78	17.58	19.69	21.57	25.58	30.60	36.13	40.86	44.71
Memo Items:											
Gross Domestic Saving	17.18	18.23	19.79	21.52	23.57	25.73	32.53	39.80	40.88	47.91	51.91
Gross National Saving	16.00	17.10	18.62	20.24	22.23	24.46	31.03	37.98	38.72	44.80	47.69
					(Billions of 1980 Canadian Dollars)						
Gross National Product	170.76	180.09	190.17	194.97	206.14	218.60	235.43	245.58	251.83	266.55	275.07
GDP at factor cost	153.23	161.02	171.21	172.50	182.62	194.44	210.67	221.08	230.39	243.05	250.45
Agriculture	9.85	10.47	11.15	10.14	11.55	10.35	11.38	10.29	11.03	12.02	12.15
Industry	68.41	72.63	75.74	75.47	78.77	84.82	94.77	95.18	89.47	95.76	98.57
Manufacturing	35.89	38.27	40.97	39.23	41.73	45.19	50.06	51.59	48.14	51.67	53.56
Services, etc.	74.97	77.91	84.33	86.89	92.30	99.27	104.52	115.60	129.89	135.28	139.73
Resource Balance	6.97	8.74	7.67	12.49	12.37	10.82	9.99	2.62	0.31	1.50	6.20
Exports of Goods & NFServices	39.16	44.10	47.64	51.77	54.47	58.74	64.94	63.65	59.34	65.63	71.45
Imports of Goods & NFServices	32.19	35.35	39.97	39.29	42.10	47.91	54.94	61.03	59.03	64.13	65.25
Domestic Absorption	167.02	174.42	185.42	185.60	197.12	210.68	228.59	246.47	255.25	269.81	274.92
Private Consumption, etc.	95.86	99.54	106.14	104.69	110.83	120.82	130.17	138.94	144.55	154.23	157.33
General Gov't Consumption	34.11	36.73	37.88	41.42	43.24	44.41	46.99	49.63	52.84	53.87	56.35
Gross Domestic Investment	37.04	38.14	41.40	39.49	43.05	45.45	51.43	57.90	57.87	61.71	61.24
Fixed Investment	35.81	35.99	37.85	37.97	40.95	42.72	46.96	50.08	52.99	55.44	56.59
Memo Items:											
Capacity to Import	33.85	37.62	40.27	44.22	45.60	50.05	58.35	61.81	55.61	62.72	65.13
Terms of Trade Adjustment	-5.31	-6.48	-7.37	-7.56	-8.87	-8.69	-6.59	-1.84	-3.73	-2.91	-6.32
Gross Domestic Income	168.68	176.68	185.71	190.53	200.62	212.81	232.00	247.25	251.84	268.41	274.80
Gross National Income	165.45	173.61	182.79	187.42	197.26	209.91	228.84	243.74	248.10	263.65	268.76
DOMESTIC PRICES (DEFLATORS)					*(Index 1980 = 100)*						
Overall (GDP)	39.7	41.2	43.0	45.0	46.4	49.0	53.4	61.1	67.1	73.0	77.5
Domestic Absorption	41.0	42.7	44.7	46.8	48.5	51.1	55.0	61.5	68.1	73.7	79.3
Agriculture	31.5	31.7	32.2	33.9	31.1	36.6	51.8	68.3	69.6	65.0	63.8
Industry	34.8	35.1	36.6	37.7	39.5	40.6	44.0	52.7	61.8	65.5	71.5
Manufacturing	42.1	42.7	44.0	45.4	46.7	48.0	50.9	57.9	66.1	68.8	73.2
MANUFACTURING ACTIVITY											
Employment (1980=100)	89.1	88.6	90.4	88.4	88.0	90.5	94.5	96.4	94.1	94.1	92.0
Real Earnings per Empl. (1980=100)	76.3	79.1	81.0	84.2	87.6	90.0	90.9	92.7	93.5	98.9	101.4
Real Output per Empl. (1980=100)	61.2	65.5	68.5	68.6	72.7	76.5	82.8	88.6	84.9	90.5	97.6
Earnings as % of Value Added	51.3	50.8	50.6	53.0	52.3	51.6	49.5	46.6	49.5	51.2	50.4
MONETARY HOLDINGS					*(Billions of current Canadian Dollars)*						
Money Supply, Broadly Defined	37.17	39.93	43.23	46.31	49.92	55.83	70.32	83.72	97.62	116.83	135.48
Money as Means of Payment	11.70	15.97	15.30	15.57	17.61	19.77	21.50	21.83	25.97	26.36	29.10
Currency Ouside Banks	2.82	3.05	3.33	3.56	3.99	4.56	5.20	5.86	6.78	7.32	8.08
Demand Deposits	8.88	12.92	11.97	12.01	13.63	15.21	16.30	15.97	19.19	19.04	21.01
Quasi-Monetary Liabilities	25.47	23.95	27.93	30.74	32.30	36.05	48.82	61.90	71.65	90.47	106.38
GOVERNMENT DEFICIT (-) OR SURPLUS	..	..	..	..	*(Billions of current Canadian Dollars)* -1.72	..	..	-1.97	-5.70	-6.30	-9.41
Current Revenue	..	..	..	..	18.14	..	..	31.62	34.89	37.72	38.24
Current Expenditure	..	..	..	..	..	..	..	28.87	34.85	38.41	43.71
Current Budget Balance	..	..	..	..	..	..	..	2.75	0.04	-0.69	-5.47
Capital Receipts	..	..	..	..	..	..	..	..	..	..	..
Capital Payments	..	..	..	..	..	..	..	4.72	5.74	5.61	3.94

1978	1979	1980	1981	1982	1983	1984	1985	1986	1987 estimate	NOTES	CANADA
9,510	10,040	10,680	11,860	12,020	12,770	13,680	14,140	14,240	15,160	..	**CURRENT GNP PER CAPITA (US $)**
23,517	23,747	24,043	24,343	24,632	24,886	25,150	25,379	25,612	25,861	..	**POPULATION (thousands)**
				(Billions of current Canadian Dollars)							**ORIGIN AND USE OF RESOURCES**
235.65	268.94	302.06	344.66	361.77	394.11	432.09	465.11	493.01	536.86	..	Gross National Product (GNP)
-5.95	-7.16	-7.83	-11.34	-12.67	-11.60	-13.51	-14.33	-16.89	-17.00	..	Net Factor Income from Abroad
241.60	276.10	309.89	355.99	374.44	405.72	445.60	479.45	509.90	553.87	..	Gross Domestic Product (GDP)
24.82	26.64	27.27	36.46	38.91	40.14	43.02	47.46	53.92	58.36	..	Indirect Taxes, net
216.78	249.46	282.62	319.54	335.53	365.58	402.58	431.99	455.97	495.51	B	GDP at factor cost
9.29	10.96	11.79	13.33	12.74	12.27	13.35	..	..	..	..	Agriculture
76.52	90.80	102.35	114.42	113.32	125.80	141.35	..	..	..	..	Industry
44.07	51.60	55.48	62.18	58.35	66.30	77.20	..	..	..	..	Manufacturing
130.98	147.70	168.48	191.79	209.47	227.51	247.88	..	..	..	..	Services, etc.
1.12	1.82	5.68	3.88	14.06	13.67	16.14	12.38	6.21	4.93	..	Resource Balance
61.16	75.07	87.60	96.88	96.67	103.46	126.32	135.41	138.33	143.94	..	Exports of Goods & NFServices
60.04	73.25	81.92	93.00	82.61	89.79	110.17	123.03	132.12	139.01	..	Imports of Goods & NFServices
240.48	274.27	304.21	352.12	360.38	392.05	429.46	467.07	503.69	548.93	..	Domestic Absorption
137.34	153.55	172.56	196.02	210.15	229.37	250.28	273.91	296.94	323.44	..	Private Consumption, etc.
47.39	52.29	59.25	68.79	78.65	84.57	89.99	96.37	101.20	107.75	..	General Gov't Consumption
55.75	68.44	72.40	87.31	71.58	78.10	89.19	96.78	105.55	117.74	..	Gross Domestic Investment
49.39	58.15	69.72	86.12	84.09	87.60	92.72	104.00	113.43	129.43	..	Fixed Investment
											Memo Items:
56.87	70.26	78.08	91.18	85.64	91.77	105.33	109.16	111.76	122.68	..	Gross Domestic Saving
51.20	63.54	70.98	80.79	74.09	81.17	92.86	95.94	96.16	108.39	..	Gross National Saving
				(Billions of 1980 Canadian Dollars)							
286.99	297.63	302.06	310.97	300.42	311.18	330.37	344.55	355.00	372.04	..	Gross National Product
264.82	276.57	282.62	288.13	278.65	287.60	307.34	319.46	328.15	340.62	B	GDP at factor cost
12.05	11.24	11.79	12.58	12.94	13.12	13.00	13.38	14.55		..	Agriculture
99.26	104.13	102.35	104.77	96.72	101.11	111.14	116.45	118.36		..	Industry
56.00	58.06	55.48	57.50	50.11	53.34	61.46	64.54	66.08		..	Manufacturing
153.50	161.20	168.48	170.78	169.00	173.37	183.21	189.63	195.24		..	Services, etc.
11.12	7.21	5.68	2.55	14.04	12.97	17.30	16.10	14.28	11.32	..	Resource Balance
81.20	85.28	87.60	91.45	89.46	95.19	113.13	119.91	125.56	132.57	..	Exports of Goods & NFServices
70.08	78.07	81.92	88.91	75.43	82.22	95.84	103.81	111.28	121.25	..	Imports of Goods & NFServices
282.86	298.15	304.21	318.73	296.90	307.81	323.61	339.30	352.96	370.07	..	Domestic Absorption
164.10	168.93	172.56	176.33	171.57	175.91	183.82	193.21	201.06	210.18	..	Private Consumption, etc.
57.29	57.62	59.25	60.75	62.19	63.09	64.01	65.75	66.38	67.98	..	General Gov't Consumption
61.47	71.60	72.40	81.65	63.13	68.81	75.78	80.34	85.53	91.91	..	Gross Domestic Investment
58.35	63.33	69.72	77.94	69.37	68.88	69.98	75.66	79.52	87.29	..	Fixed Investment
											Memo Items:
71.39	80.01	87.60	92.62	88.27	94.74	109.88	114.25	116.51	125.56	..	Capacity to Import
-9.81	-5.27	0.00	1.16	-1.20	-0.45	-3.25	-5.66	-9.05	-7.01	..	Terms of Trade Adjustment
284.17	300.09	309.89	322.43	309.74	320.32	337.66	349.75	358.19	374.37	..	Gross Domestic Income
277.18	292.36	302.06	312.13	299.23	310.73	327.11	338.89	345.95		..	Gross National Income
				(Index 1980 = 100)							**DOMESTIC PRICES (DEFLATORS)**
82.2	90.4	100.0	110.8	120.4	126.5	130.7	134.9	138.8	145.2	..	Overall (GDP)
85.0	92.0	100.0	110.5	121.4	127.4	132.7	137.7	142.7	148.3	..	Domestic Absorption
77.1	97.5	100.0	106.0	98.4	93.5	102.7	..	..	..	..	Agriculture
77.1	87.2	100.0	109.2	117.2	124.4	127.2	..	..	..	..	Industry
78.7	88.9	100.0	108.1	116.5	124.3	125.6	..	..	..	..	Manufacturing
											MANUFACTURING ACTIVITY
96.6	100.1	100.0	100.1	92.0	90.3	92.9	94.8	..	..	..	Employment (1980 = 100)
99.8	100.1	100.0	99.5	99.1	100.5	101.8	116.9	..	..	..	Real Earnings per Empl. (1980 = 100)
101.0	103.1	100.0	104.8	102.0	106.6	117.7	130.4	..	..	..	Real Output per Empl. (1980 = 100)
48.8	46.8	47.4	47.3	51.3	48.5	45.8	48.5	..	..	..	Earnings as % of Value Added
				(Billions of current Canadian Dollars)							**MONETARY HOLDINGS**
159.57	188.43	211.37	253.66	256.36	264.79	282.76	303.94	331.45	362.60	D	Money Supply, Broadly Defined
31.12	31.56	34.75	34.64	38.87	43.71	52.44	69.85	80.22	85.07	..	Money as Means of Payment
8.95	9.45	10.40	10.71	11.62	12.80	13.50	14.61	15.59	16.82	..	Currency Ouside Banks
22.17	22.11	24.35	23.93	27.26	30.92	38.94	55.24	64.62	68.25	..	Demand Deposits
128.45	156.87	176.62	219.02	217.49	221.08	230.32	234.09	251.23	277.53	..	Quasi-Monetary Liabilities
				(Billions of current Canadian Dollars)							
-11.95	-10.56	-10.73	-8.43	-20.81	-25.16	-28.87	-28.68	-20.11	..	C E	**GOVERNMENT DEFICIT (-) OR SURPLUS**
41.64	48.22	57.49	72.28	73.24	77.33	85.30	91.10	99.12	..	..	Current Revenue
49.15	54.62	64.75	74.84	88.02	97.02	108.78	115.06	117.03	..	..	Current Expenditure
-7.52	-6.40	-7.25	-2.56	-14.78	-19.69	-23.48	-23.96	-17.91	..	..	Current Budget Balance
..	..	..	..	..	..	..	..	..	..	..	Capital Receipts
4.44	4.16	3.48	5.87	6.03	5.47	5.39	4.72	2.20	..	..	Capital Payments

CANADA	1967	1968	1969	1970	1971	1972	1973	1974	1975	1976	1977
FOREIGN TRADE (CUSTOMS BASIS)				*(Millions of current US dollars)*							
Value of Exports, fob	10,555	12,556	13,754	16,185	17,675	20,178	25,207	32,783	32,300	38,370	41,293
Nonfuel Primary Products	5,083	5,666	5,551	6,918	7,214	7,981	10,627	13,045	11,904	14,250	15,076
Fuels	558	645	744	978	1,262	1,734	2,466	5,179	5,334	5,300	5,171
Manufactures	4,914	6,245	7,460	8,288	9,199	10,463	12,115	14,560	15,063	18,821	21,046
Value of Imports, cif	10,858	12,296	14,004	14,253	16,541	20,042	24,746	34,414	36,212	40,356	42,257
Nonfuel Primary Products	1,691	1,812	1,990	2,113	2,293	2,697	3,675	4,950	4,827	5,332	5,552
Fuels	706	785	707	800	971	1,149	1,405	3,613	4,362	4,365	4,195
Manufactures	8,462	9,698	11,307	11,340	13,278	16,196	19,666	25,851	27,023	30,659	32,510
				(Index 1980 = 100)							
Terms of Trade	143.4	144.4	125.8	129.9	117.8	112.6	135.4	102.8	112.5	115.8	103.7
Export Prices, fob	31.9	32.2	31.1	32.4	32.5	34.4	51.7	56.0	64.5	69.0	68.1
Nonfuel Primary Products	31.7	32.6	34.7	36.9	34.9	38.3	58.7	65.2	63.0	62.5	58.5
Fuels	9.7	8.3	6.2	6.8	8.0	9.6	27.9	31.1	54.6	64.5	69.9
Manufactures	43.6	45.1	46.1	49.6	51.1	52.6	55.5	66.5	70.4	76.8	76.8
Import Prices, cif	22.3	22.3	24.7	25.0	27.5	30.5	38.2	54.5	57.3	59.6	65.7
Trade at Constant 1980 Prices				*(Millions of 1980 US dollars)*							
Exports, fob	33,080	38,995	44,187	49,885	54,452	58,731	48,741	58,540	50,075	55,579	60,603
Imports, cif	48,787	55,129	56,601	57,080	60,051	65,666	64,787	63,182	63,162	67,677	64,292
BALANCE OF PAYMENTS				*(Billions of current US dollars)*							
Exports of Goods & Services	..	..	..	20.23	21.96	24.82	30.67	39.89	39.79	46.48	49.79
Merchandise, fob	..	..	..	16.82	18.41	21.09	26.34	34.34	33.94	40.03	43.18
Nonfactor Services	..	..	..	2.45	2.60	2.74	3.15	4.13	4.21	4.81	4.91
Factor Services	..	..	..	0.97	0.94	0.99	1.19	1.42	1.64	1.64	1.70
Imports of Goods & Services	..	..	..	19.56	22.07	25.49	30.92	41.94	44.83	51.16	54.20
Merchandise, fob	..	..	..	13.66	15.62	19.01	23.35	32.40	34.29	38.14	40.06
Nonfactor Services	..	..	..	3.35	3.66	3.93	4.56	5.72	6.27	7.64	8.17
Factor Services	..	..	..	2.55	2.79	2.55	3.01	3.82	4.28	5.38	5.97
Long-Term Interest	..	..	..	..	..	..	..	..	..	..	..
Current Transfers, net	..	..	..	0.06	0.16	0.20	0.24	0.43	0.37	0.43	0.33
Workers' Remittances	..	..	..	0.00	0.00	0.00	0.00	0.00	0.00	0.00	0.00
Total to be Financed	..	..	..	0.74	0.05	-0.48	-0.02	-1.63	-4.67	-4.24	-4.08
Official Capital Grants	..	..	..	0.32	0.35	0.37	0.42	0.56	0.58	0.11	-0.03
Current Account Balance	..	..	..	1.06	0.40	-0.10	0.41	-1.07	-4.09	-4.14	-4.11
Long-Term Capital, net	..	..	..	1.10	0.57	1.48	0.85	1.49	4.44	7.84	3.33
Direct Investment	..	..	..	0.57	0.69	0.22	0.06	0.03	-0.19	-1.06	-0.47
Long-Term Loans	..	..	..	..	..	..	..	..	..	..	..
Disbursements	..	..	..	..	..	..	..	..	..	..	..
Repayments	..	..	..	..	..	..	..	..	..	..	..
Other Long-Term Capital	..	..	..	0.22	-0.46	0.88	0.36	0.90	4.05	8.80	3.84
Other Capital, net	..	..	..	-0.72	-0.20	-1.16	-1.72	-0.39	-0.75	-3.18	-0.57
Change in Reserves	..	..	..	-1.44	-0.77	-0.21	0.47	-0.03	0.40	-0.53	1.35
Memo Item:				*(Canadian Dollars per US dollar)*							
Conversion Factor (Annual Avg)	1.080	1.080	1.080	1.050	1.010	0.990	1.000	0.980	1.020	0.990	1.060
EXTERNAL DEBT, ETC.				*(Millions US dollars, outstanding at end of year)*							
Public/Publicly Guar. Long-Term	..	..	..	..	..	..	..	..	..	..	..
Official Creditors	..	..	..	..	..	..	..	..	..	..	..
IBRD and IDA	..	..	..	..	..	..	..	..	..	..	..
Private Creditors	..	..	..	..	..	..	..	..	..	..	..
Private Non-guaranteed Long-Term	..	..	..	..	..	..	..	..	..	..	..
Use of Fund Credit	..	..	..	..	..	..	..	..	..	..	..
Short-Term Debt	..	..	..	..	..	..	..	..	..	..	..
Memo Items:				*(Millions of US dollars)*							
Int'l Reserves Excluding Gold	1,702	2,183	2,234	3,888	4,839	5,216	4,841	4,885	4,385	4,924	3,653
Gold Holdings (at market price)	1,021	1,033	877	844	990	1,425	2,464	4,094	3,079	2,913	3,630
SOCIAL INDICATORS											
Total Fertility Rate	2.5	2.4	2.3	2.3	2.1	2.0	1.9	1.8	1.8	1.8	1.8
Crude Birth Rate	18.2	17.6	17.6	17.4	16.8	15.9	15.6	15.4	15.8	15.6	15.5
Infant Mortality Rate	22.0	20.8	19.3	18.8	17.5	17.1	15.5	15.0	14.2	13.4	12.3
Life Expectancy at Birth	72.0	72.2	72.4	72.6	72.8	73.0	73.2	73.4	73.6	73.8	74.1
Food Production, p.c. ('79-81 = 100)	93.5	99.5	95.9	85.5	97.1	94.4	93.3	85.6	94.2	105.2	102.7
Labor Force, Agriculture (%)	9.3	8.8	8.3	7.8	7.5	7.3	7.0	6.8	6.5	6.3	6.0
Labor Force, Female (%)	30.5	31.2	31.8	32.5	33.3	34.0	34.8	35.6	36.4	37.1	37.8
School Enroll. Ratio, primary	..	..	..	101.0	..	..	..	..	99.0	100.0	100.0
School Enroll. Ratio, secondary	..	..	..	65.0	..	..	..	..	91.0	91.0	89.0

1978	1979	1980	1981	1982	1983	1984	1985	1986	1987 estimate	NOTES	CANADA
				(Millions of current US dollars)							**FOREIGN TRADE (CUSTOMS BASIS)**
44,080	55,117	63,105	68,281	66,977	72,420	84,844	85,737	84,381	92,886	..	Value of Exports, fob
15,342	19,717	23,500	23,258	21,311	22,486	24,281	22,191	22,676	26,851	..	Nonfuel Primary Products
4,816	7,281	9,010	9,451	9,601	10,017	11,117	12,047	8,146	9,382	..	Fuels
23,923	28,119	30,595	35,573	36,065	39,917	49,446	51,499	53,559	56,653	..	Manufactures
46,548	56,953	62,838	70,342	58,360	65,077	77,789	80,640	85,068	92,594	..	Value of Imports, cif
6,044	7,497	9,063	9,069	7,652	8,211	9,078	8,460	9,163	9,751	..	Nonfuel Primary Products
4,209	5,286	7,778	8,620	5,887	4,488	5,016	4,886	4,020	4,531	..	Fuels
36,295	44,169	45,996	52,653	44,821	52,378	63,694	67,294	71,884	78,312	..	Manufactures
				(Index 1980 = 100)							
99.7	99.4	100.0	107.2	111.6	116.5	121.4	121.6	112.3	101.5	..	Terms of Trade
73.5	86.9	100.0	103.6	102.3	103.9	105.5	104.4	98.0	96.4	..	Export Prices, fob
67.2	92.7	100.0	92.5	84.2	86.5	85.0	79.8	77.1	79.4	..	Nonfuel Primary Products
68.3	71.6	100.0	137.2	148.6	142.1	161.1	170.1	114.0	103.2	..	Fuels
79.6	· 88.0	100.0	105.0	107.0	109.0	110.0	109.0	108.0	106.0	..	Manufactures
73.7	87.5	100.0	96.6	91.7	89.3	86.9	85.8	87.2	95.0	..	Import Prices, cif
				(Millions of 1980 US dollars)							Trade at Constant 1980 Prices
59,978	63,404	63,105	65,919	65,478	69,678	80,433	82,140	86,147	96,371	..	Exports, fob
63,133	65,111	62,838	72,791	63,674	72,915	89,518	93,942	97,571	97,497	..	Imports, cif
				(Billions of current US dollars)							**BALANCE OF PAYMENTS**
55.59	66.35	78.02	84.25	82.78	88.54	102.23	104.42	103.82	113.96	..	Exports of Goods & Services
48.13	57.64	67.51	72.55	70.50	75.74	88.67	89.70	88.50	97.77	..	Merchandise, fob
5.39	6.46	7.39	8.35	7.83	8.26	8.65	9.16	10.15	10.93	..	Nonfactor Services
2.06	2.26	3.11	3.34	4.45	4.55	4.92	5.56	5.16	5.25	..	Factor Services
59.80	70.93	79.89	90.45	81.63	86.90	100.75	106.45	112.47	123.50	..	Imports of Goods & Services
43.95	53.47	59.51	66.00	55.45	60.76	72.64	77.09	80.94	89.02	..	Merchandise, fob
8.64	9.09	10.58	11.63	11.45	12.18	12.78	13.31	14.48	16.72	..	Nonfactor Services
7.21	8.36	9.81	12.83	14.72	13.96	15.34	16.05	17.05	17.76	..	Factor Services
..	..	..	..	..	..	..	..	..	..	..	Long-Term Interest
0.24	0.37	0.62	0.79	0.91	0.81	0.80	0.81	0.92	2.04	..	Current Transfers, net
0.00	0.00	0.00	0.00	0.00	0.00	0.00	0.00	0.00	0.00	..	Workers' Remittances
-3.98	-4.21	-1.26	-5.41	2.06	2.45	2.28	-1.22	-7.73	-7.50	K	Total to be Financed
-0.30	0.05	0.28	0.33	0.25	0.05	-0.20	-0.23	0.13	-0.47	K	Official Capital Grants
-4.28	-4.15	-0.98	-5.09	2.31	2.51	2.08	-1.44	-7.60	-7.96	..	Current Account Balance
2.87	2.49	-0.05	1.31	5.96	0.21	2.75	2.09	13.67	9.19	..	Long-Term Capital, net
-1.41	-0.74	-3.17	-8.74	-2.24	-3.78	-0.92	-5.96	-1.82	-0.92	..	Direct Investment
..	..	..	..	..	..	..	..	..	..	..	Long-Term Loans
..	..	..	..	..	..	..	..	..	..	..	Disbursements
..	..	..	..	..	..	..	..	..	..	..	Repayments
4.58	3.18	2.85	9.72	7.95	3.94	3.87	8.28	15.36	10.58	..	Other Long-Term Capital
1.22	0.72	0.37	3.93	-8.85	-2.27	-5.70	-0.72	-5.59	2.12	..	Other Capital, net
0.19	0.94	0.66	-0.15	0.58	-0.45	0.87	0.07	-0.48	-3.34	..	Change in Reserves
				(Canadian Dollars per US dollar)							Memo Item:
1.140	1.170	1.170	1.200	1.230	1.230	1.300	1.370	1.390	1.330	..	Conversion Factor (Annual Avg)
				(Millions US dollars, outstanding at end of year)							**EXTERNAL DEBT, ETC.**
..	..	..	..	..	..	..	..	..	..	..	Public/Publicly Guar. Long-Term
..	..	..	..	..	..	..	..	..	..	..	Official Creditors
..	..	..	..	..	..	..	..	..	..	..	IBRD and IDA
..	..	..	..	..	..	..	..	..	..	..	Private Creditors
..	..	..	..	..	..	..	..	..	..	..	Private Non-guaranteed Long-Term
..	..	..	..	..	..	..	..	..	..	..	Use of Fund Credit
..	..	..	..	..	..	..	..	..	..	..	Short-Term Debt
				(Millions of US dollars)							Memo Items:
3,544	2,856	3,041	3,492	3,000	3,466	2,491	2,503	3,251	7,277	..	Int'l Reserves Excluding Gold
5,001	11,356	12,369	8,134	9,258	7,695	6,208	6,575	7,710	8,965	..	Gold Holdings (at market price)
											SOCIAL INDICATORS
1.8	1.8	1.8	1.7	1.7	1.7	1.7	1.7	1.7	1.7	..	Total Fertility Rate
15.2	15.4	15.4	15.3	15.2	15.0	14.9	14.7	14.6	14.5	..	Crude Birth Rate
11.9	10.9	10.4	9.6	9.8	10.0	10.2	10.4	10.6	10.8	..	Infant Mortality Rate
74.4	74.7	75.0	75.4	75.6	75.8	76.0	76.2	76.4	76.6	..	Life Expectancy at Birth
104.5	93.6	99.1	107.4	113.9	105.5	103.8	106.2	114.3	109.5	..	Food Production, p.c. ('79-81 = 100)
5.8	5.5	5.3	..	..	..	..	..	..	..	..	Labor Force, Agriculture (%)
38.4	39.1	39.8	39.8	39.8	39.8	39.8	39.8	39.8	39.8	..	Labor Force, Female (%)
101.0	101.0	102.0	..	107.0	106.0	105.0	105.0	105.0	..	..	School Enroll. Ratio, primary
88.0	88.0	92.0	..	99.0	102.0	103.0	103.0	103.0	..	..	School Enroll. Ratio, secondary

175

CAPE VERDE	1967	1968	1969	1970	1971	1972	1973	1974	1975	1976	1977
CURRENT GNP PER CAPITA (US $)	..	..	..	..	..	..	..	..	270	320	360
POPULATION (thousands)	247	254	262	269	272	272	271	271	276	281	283

ORIGIN AND USE OF RESOURCES *(Millions of current Cape Verde Escudos)*

	1967	1968	1969	1970	1971	1972	1973	1974	1975	1976	1977
Gross National Product (GNP)	..	..	..	..	..	..	1,678	1,759	1,991	2,524	2,994
Net Factor Income from Abroad	..	..	..	..	..	..	197	314	179	514	814
Gross Domestic Product (GDP)	..	..	..	1,124	1,218	1,271	1,481	1,445	1,812	2,010	2,180
Indirect Taxes, net	..	..	..	90	100	104	110	117	122	120	137
GDP at factor cost	..	..	..	1,034	1,118	1,167	1,371	1,328	1,690	1,890	2,043
Agriculture	..	..	..	..	..	..	236	245	334	445	471
Industry	..	..	..	..	..	..	394	367	406	459	509
Manufacturing	..	..	..	..	..	..	..	..	..	..	..
Services, etc.	..	..	..	..	..	..	851	833	1,073	1,106	1,200
Resource Balance	..	..	..	..	..	..	-696	-597	-639	-1,107	-1,586
Exports of Goods & NFServices	..	..	..	..	..	..	319	279	234	173	77
Imports of Goods & NFServices	..	..	..	..	..	..	1,016	876	873	1,280	1,663
Domestic Absorption	..	..	..	..	..	..	2,177	2,042	2,451	3,117	3,766
Private Consumption, etc.	..	..	..	..	..	..	1,711	1,461	1,755	2,072	2,206
General Gov't Consumption	..	..	..	..	..	..	120	213	305	379	532
Gross Domestic Investment	..	..	..	..	..	..	346	369	391	667	1,028
Fixed Investment	..	..	..	..	..	..	..	..	..	..	..
Memo Items:											
Gross Domestic Saving	..	..	..	..	..	..	-350	-229	-248	-440	-558
Gross National Saving	..	..	..	..	..	..	..	..	..	..	..

(Millions of 1980 Cape Verde Escudos)

	1967	1968	1969	1970	1971	1972	1973	1974	1975	1976	1977
Gross National Product	..	..	..	..	..	..	2,922	2,992	2,756	3,343	3,736
GDP at factor cost	..	..	..	..	..	..	..	..	..	..	..
Agriculture	..	..	..	..	..	..	..	..	..	..	..
Industry	..	..	..	..	..	..	..	..	..	..	..
Manufacturing	..	..	..	..	..	..	..	..	..	..	..
Services, etc.	..	..	..	..	..	..	..	..	..	..	..
Resource Balance	..	..	..	..	..	..	-1,139	-795	-801	-1,535	-2,298
Exports of Goods & NFServices	..	..	..	..	..	..	1,212	848	617	648	185
Imports of Goods & NFServices	..	..	..	..	..	..	2,351	1,643	1,418	2,183	2,482
Domestic Absorption	..	..	..	..	..	..	3,887	3,467	3,557	4,290	5,070
Private Consumption, etc.	..	..	..	..	..	..	2,835	2,194	2,325	2,592	2,707
General Gov't Consumption	..	..	..	..	..	..	224	394	465	520	678
Gross Domestic Investment	..	..	..	..	..	..	828	879	767	1,178	1,685
Fixed Investment	..	..	..	..	..	..	..	..	..	..	..
Memo Items:											
Capacity to Import	..	..	..	..	..	..	739	522	380	295	115
Terms of Trade Adjustment	..	..	..	..	..	..	-473	-326	-237	-353	-70
Gross Domestic Income	..	..	..	..	..	..	2,274	2,346	2,519	2,402	2,703
Gross National Income	..	..	..	..	..	..	2,448	2,667	2,519	2,990	3,666

DOMESTIC PRICES (DEFLATORS) *(Index 1980 = 100)*

	1967	1968	1969	1970	1971	1972	1973	1974	1975	1976	1977
Overall (GDP)	..	..	..	36.0	42.0	45.9	53.9	54.1	65.7	73.0	78.6
Domestic Absorption	..	..	..	..	..	..	56.0	58.9	68.9	72.7	74.3
Agriculture	..	..	..	..	..	..	..	..	..	..	..
Industry	..	..	..	..	..	..	..	..	..	..	..
Manufacturing	..	..	..	..	..	..	..	..	..	..	..

MANUFACTURING ACTIVITY

	1967	1968	1969	1970	1971	1972	1973	1974	1975	1976	1977
Employment (1980=100)	..	..	..	..	..	..	..	..	..	..	..
Real Earnings per Empl. (1980=100)	..	..	..	..	..	..	..	..	..	..	..
Real Output per Empl. (1980=100)	..	..	..	..	..	..	..	..	..	..	..
Earnings as % of Value Added	..	..	..	..	..	..	..	..	..	..	..

MONETARY HOLDINGS *(Millions of current Cape Verde Escudos)*

	1967	1968	1969	1970	1971	1972	1973	1974	1975	1976	1977
Money Supply, Broadly Defined	..	..	..	..	..	..	..	..	..	1,022	1,441
Money as Means of Payment	..	..	..	..	..	..	..	..	..	1,000	1,411
Currency Ouside Banks	..	..	..	..	..	..	..	..	..	465	537
Demand Deposits	..	..	..	..	..	..	..	..	..	536	874
Quasi-Monetary Liabilities	..	..	..	..	..	..	..	..	..	21	30

GOVERNMENT DEFICIT (-) OR SURPLUS *(Millions of current Cape Verde Escudos)*

	1967	1968	1969	1970	1971	1972	1973	1974	1975	1976	1977
Current Revenue	..	..	..	..	..	..	..	..	..	..	..
Current Expenditure	..	..	..	..	..	..	..	..	..	..	..
Current Budget Balance	..	..	..	..	..	..	..	..	..	..	..
Capital Receipts	..	..	..	..	..	..	..	..	..	..	..
Capital Payments	..	..	..	..	..	..	..	..	..	..	..

1978	1979	1980	1981	1982	1983	1984	1985	1986	1987 estimate	NOTES	CAPE VERDE
360	380	500	460	450	430	430	420	460	500	..	CURRENT GNP PER CAPITA (US $)
285	290	296	303	310	315	320	327	335	343	..	POPULATION (thousands)
			(Millions of current Cape Verde Escudos)								ORIGIN AND USE OF RESOURCES
3,474	3,999	5,322	6,080	7,565	9,139	10,948	12,378	13,660	15,188	..	Gross National Product (GNP)
857	898	1,746	1,565	1,302	1,129	1,495	1,787	1,259	1,399	..	Net Factor Income from Abroad
2,617	3,101	3,576	4,515	6,263	8,010	9,453	10,591	12,401	13,789	..	Gross Domestic Product (GDP)
204	228	280	329	399	459	566	..	..	..	..	Indirect Taxes, net
2,413	2,873	3,296	4,186	5,864	7,551	8,887	..	..	..	B	GDP at factor cost
650	901	1,137	1,069	1,347	1,603	1,629	2,055	..	..	..	Agriculture
608	709	643	1,139	1,772	2,458	1,830	2,162	..	..	..	Industry
..	..	..	181	250	360	283	378	..	..	..	Manufacturing
1,358	1,490	1,796	2,307	3,144	3,949	5,994	6,374	..	..	..	Services, etc.
-1,942	-2,408	-2,741	-3,866	-4,557	-5,832	-5,647	-6,702			..	Resource Balance
277	464	771	1,155	1,816	2,540	2,240	2,541			..	Exports of Goods & NFServices
2,220	2,872	3,512	5,021	6,373	8,372	7,887	9,243			..	Imports of Goods & NFServices
4,559	5,509	6,316	8,381	10,820	13,842	15,100	17,293	..	..	..	Domestic Absorption
2,899	3,067	3,949	4,492	5,950	7,732	8,351	9,502	..	..	..	Private Consumption, etc.
558	680	856	1,048	1,284	1,515	2,012	2,331	..	..	..	General Gov't Consumption
1,102	1,763	1,511	2,841	3,586	4,595	4,737	5,460	..	..	..	Gross Domestic Investment
..	..	..	..	..	..	..	..	..	..	..	Fixed Investment
											Memo Items:
-840	-646	-1,229	-1,025	-971	-1,237	-910	-1,242	..	..	..	Gross Domestic Saving
883	1,207	2,112	2,298	2,196	2,380	2,418	2,582	..	..	..	Gross National Saving
			(Millions of 1980 Cape Verde Escudos)								
3,915	4,050	5,322	5,067	5,210	5,201	5,466	5,618	6,201	6,573	..	Gross National Product
..	..	..	..	..	..	..	..	..	..	B	GDP at factor cost
..	..	..	..	..	..	..	..	..	..	..	Agriculture
..	..	..	..	..	..	..	..	..	..	..	Industry
..	..	..	..	..	..	..	..	..	..	..	Manufacturing
..	..	..	..	..	..	..	..	..	..	..	Services, etc.
-2,360	-2,521	-2,741	-2,562	-2,693	-3,789	-3,320	-3,985			..	Resource Balance
581	603	771	1,050	1,892	2,066	1,672	1,650			..	Exports of Goods & NFServices
2,940	3,124	3,512	3,612	4,585	5,855	4,992	5,636			..	Imports of Goods & NFServices
5,428	5,926	6,317	6,324	7,006	8,348	8,039	9,002	..	..	..	Domestic Absorption
3,375	3,411	3,949	3,083	3,652	4,870	4,670	5,466	..	..	..	Private Consumption, etc.
580	580	856	873	885	862	1,005	1,058	..	..	..	General Gov't Consumption
1,473	1,936	1,511	2,368	2,470	2,615	2,364	2,478	..	..	..	Gross Domestic Investment
..	..	..	..	..	..	..	..	..	..	..	Fixed Investment
											Memo Items:
368	505	771	831	1,307	1,776	1,418	1,549	..	..	..	Capacity to Import
-213	-98	0	-219	-586	-289	-254	-101	..	..	..	Terms of Trade Adjustment
2,855	3,307	3,576	3,543	3,728	4,269	4,465	4,915	..	..	..	Gross Domestic Income
3,702	3,952	5,322	4,847	4,624	4,912	5,211	5,517	..	..	..	Gross National Income
			(Index 1980 = 100)								DOMESTIC PRICES (DEFLATORS)
85.3	91.1	100.0	120.0	145.2	175.7	200.3	211.1	234.5	246.0	..	Overall (GDP)
84.0	93.0	100.0	132.5	154.4	165.8	187.8	192.1	..	..	..	Domestic Absorption
..	..	..	..	..	..	..	..	..	..	..	Agriculture
..	..	..	..	..	..	..	..	..	..	..	Industry
..	..	..	..	..	..	..	..	..	..	..	Manufacturing
											MANUFACTURING ACTIVITY
..	..	..	..	..	..	..	..	..	..	..	Employment (1980=100)
..	..	..	..	..	..	..	..	..	..	..	Real Earnings per Empl. (1980=100)
..	..	..	..	..	..	..	..	..	..	..	Real Output per Empl. (1980=100)
..	..	..	..	..	..	..	..	..	..	..	Earnings as % of Value Added
			(Millions of current Cape Verde Escudos)								MONETARY HOLDINGS
1,674	1,966	2,563	3,047	..	..	..	..	..	..	..	Money Supply, Broadly Defined
1,615	1,823	2,302	2,553	3,228	3,956	4,202	4,843	5,621	6,889	..	Money as Means of Payment
638	736	872	1,040	1,282	1,333	1,434	1,628	1,826	1,956	..	Currency Ouside Banks
977	1,086	1,430	1,512	1,946	2,623	2,768	3,215	3,794	4,933	..	Demand Deposits
59	143	261	494	..	..	..	..	..	..	..	Quasi-Monetary Liabilities
			(Millions of current Cape Verde Escudos)								GOVERNMENT DEFICIT (-) OR SURPLUS
..	..	..	..	..	..	..	..	..	..	..	Current Revenue
..	..	..	..	..	..	..	..	..	..	..	Current Expenditure
..	..	..	..	..	..	..	..	..	..	..	Current Budget Balance
..	..	..	..	..	..	..	..	..	..	..	Capital Receipts
..	..	..	..	..	..	..	..	..	..	..	Capital Payments

CAPE VERDE	1967	1968	1969	1970	1971	1972	1973	1974	1975	1976	1977
FOREIGN TRADE (CUSTOMS BASIS)					*(Thousands of current US dollars)*						
Value of Exports, fob	..	..	..	..	..	..	..	..	..	..	..
Nonfuel Primary Products	..	..	..	..	..	..	..	..	..	..	..
Fuels	..	..	..	..	..	..	..	..	..	..	..
Manufactures	..	..	..	..	..	..	..	..	..	..	..
Value of Imports, cif	..	..	..	..	..	..	..	..	..	..	..
Nonfuel Primary Products	..	..	..	..	..	..	..	..	..	..	..
Fuels	..	..	..	..	..	..	..	..	..	..	..
Manufactures	..	..	..	..	..	..	..	..	..	..	..
					(Index 1980 = 100)						
Terms of Trade	..	..	..	..	..	..	..	..	..	..	..
Export Prices, fob	..	..	..	..	..	..	..	..	..	..	..
Nonfuel Primary Products	..	..	..	..	..	..	..	..	..	..	..
Fuels	..	..	..	..	..	..	..	..	..	..	..
Manufactures	..	..	..	..	..	..	..	..	..	..	..
Import Prices, cif	..	..	..	..	..	..	..	..	..	..	..
Trade at Constant 1980 Prices					*(Thousands of 1980 US dollars)*						
Exports, fob	..	..	..	..	..	..	..	..	..	..	..
Imports, cif	..	..	..	..	..	..	..	..	..	..	..
BALANCE OF PAYMENTS					*(Thousands of current US dollars)*						
Exports of Goods & Services	..	..	..	..	..	..	..	..	..	..	..
Merchandise, fob	..	..	..	..	..	..	..	..	..	..	..
Nonfactor Services	..	..	..	..	..	..	..	..	..	..	..
Factor Services	..	..	..	..	..	..	..	..	..	..	..
Imports of Goods & Services	..	..	..	..	..	..	..	..	..	..	..
Merchandise, fob	..	..	..	..	..	..	..	..	..	..	..
Nonfactor Services	..	..	..	..	..	..	..	..	..	..	..
Factor Services	..	..	..	..	..	..	..	..	..	..	..
Long-Term Interest	..	..	..	..	..	..	..	..	0	100	100
Current Transfers, net	..	..	..	..	..	..	..	..	..	..	..
Workers' Remittances	..	..	..	..	..	..	..	..	..	..	..
Total to be Financed	..	..	..	..	..	..	..	..	..	..	..
Official Capital Grants	..	..	..	..	..	..	..	..	..	..	..
Current Account Balance	..	..	..	..	..	..	..	..	..	..	..
Long-Term Capital, net	..	..	..	..	..	..	..	..	..	..	..
Direct Investment	..	..	..	..	..	..	..	..	..	..	..
Long-Term Loans	..	..	..	..	..	..	..	..	500	11,300	2,300
Disbursements	..	..	..	..	..	..	..	..	500	11,300	2,300
Repayments	..	..	..	..	..	..	..	..	..	..	..
Other Long-Term Capital	..	..	..	..	..	..	..	..	..	..	..
Other Capital, net	..	..	..	..	..	..	..	..	..	..	..
Change in Reserves	..	..	..	..	..	..	..	..	..	..	..
Memo Item:					*(Cape Verde Escudos per US dollar)*						
Conversion Factor (Annual Avg)	28.750	28.750	28.750	28.750	28.320	27.010	24.670	25.410	25.550	30.140	34.050
EXTERNAL DEBT, ETC.					*(Millions of US dollars, outstanding at end of year)*						
Public/Publicly Guar. Long-Term	..	..	..	..	..	..	..	..	0.5	11.7	13.7
Official Creditors	..	..	..	..	..	..	..	..	0.5	11.7	13.7
IBRD and IDA	..	..	..	..	..	..	..	..	0.0	0.0	0.0
Private Creditors	..	..	..	..	..	..	..	..	..	..	..
Private Non-guaranteed Long-Term	..	..	..	..	..	..	..	..	..	..	..
Use of Fund Credit	..	..	..	0.0	0.0	0.0	0.0	0.0	0.0	0.0	0.0
Short-Term Debt	..	..	..	..	..	..	..	..	..	..	..
Memo Items:					*(Millions US dollars)*						
Int'l Reserves Excluding Gold	..	..	..	..	..	..	..	..	..	..	..
Gold Holdings (at market price)	..	..	..	..	..	..	..	..	..	..	..
SOCIAL INDICATORS											
Total Fertility Rate	7.3	7.4	7.4	7.5	7.4	7.2	7.1	7.0	6.9	6.7	6.6
Crude Birth Rate	41.0	39.6	39.8	36.1	41.1	38.9	36.7	36.3	36.9	41.4	39.8
Infant Mortality Rate	..	..	..	..	129.0	92.0	112.0	79.0	109.0	100.0	85.0
Life Expectancy at Birth	54.6	55.1	55.6	56.1	56.5	57.0	57.4	57.8	58.3	58.7	59.2
Food Production, p.c. ('79-81=100)	154.8	119.6	116.0	101.8	86.4	83.3	81.1	79.3	82.0	90.9	84.2
Labor Force, Agriculture (%)	65.9	65.3	64.7	64.1	62.9	61.7	60.4	59.2	58.0	56.7	55.5
Labor Force, Female (%)	23.9	23.8	23.7	23.7	24.1	24.5	24.9	25.3	25.8	26.1	26.4
School Enroll. Ratio, primary	..	..	..	66.0	..	..	..	..	145.0	..	..
School Enroll. Ratio, secondary	..	..	..	..	..	..	..	..	8.0	..	..

1978	1979	1980	1981	1982	1983	1984	1985	1986	1987 estimate	NOTES	CAPE VERDE
				(Thousands of current US dollars)							**FOREIGN TRADE (CUSTOMS BASIS)**
..	..	9	6	3	2	3	3	3	..	..	Value of Exports, fob
..	..	..	..	..	..	..	..	..	..	..	Nonfuel Primary Products
..	..	..	..	..	..	..	..	..	..	..	Fuels
..	..	..	..	..	..	..	..	..	..	..	Manufactures
..	..	68	..	..	..	..	..	..	..	..	Value of Imports, cif
..	..	..	..	..	..	..	..	..	..	..	Nonfuel Primary Products
..	..	..	..	..	..	..	..	..	..	..	Fuels
..	..	..	..	..	..	..	..	..	..	..	Manufactures
				(Index 1980 = 100)							
..	..	..	..	..	..	..	..	..	..	..	Terms of Trade
..	..	..	..	..	..	..	..	..	..	..	Export Prices, fob
..	..	..	..	..	..	..	..	..	..	..	Nonfuel Primary Products
..	..	..	..	..	..	..	..	..	..	..	Fuels
..	..	..	..	..	..	..	..	..	..	..	Manufactures
..	..	..	..	..	..	..	..	..	..	..	Import Prices, cif
				(Thousands of 1980 US dollars)							Trade at Constant 1980 Prices
..	..	..	..	..	..	..	..	..	..	..	Exports, fob
..	..	..	..	..	..	..	..	..	..	..	Imports, cif
				(Thousands of current US dollars)							**BALANCE OF PAYMENTS**
..	..	..	..	..	..	..	..	..	..	..	Exports of Goods & Services
..	..	9,110	6,250	3,750	1,900	2,600	4,700	3,990	4,910	..	Merchandise, fob
..	..	..	..	..	..	..	..	..	..	..	Nonfactor Services
..	..	..	..	..	..	..	..	..	..	..	Factor Services
..	..	..	..	..	..	..	..	..	..	..	Imports of Goods & Services
..	..	..	..	96,700	104,500	82,000	81,300	91,600	111,700	..	Merchandise, fob
..	..	..	..	..	..	..	..	..	..	..	Nonfactor Services
..	..	..	..	..	..	..	..	..	..	..	Factor Services
100	100	100	200	900	2,000	3,000	2,600	1,900	3,400	..	Long-Term Interest
24,400	25,510	39,700	36,100	32,000	34,710	21,630	21,900	0	..	..	Current Transfers, net
..	..	..	..	..	..	..	..	..	..	..	Workers' Remittances
-29,660	-37,530	-25,060	-42,730	-49,680	-51,460	-46,200	-51,290	..	..	K	Total to be Financed
0	0	0	0	35,300	38,800	38,700	37,300	43,000	53,900	K	Official Capital Grants
-29,660	-37,530	-25,060	-42,730	-14,380	-12,660	-7,500	-13,990	..	..	..	Current Account Balance
22,540	37,220	31,790	39,820	55,410	53,840	49,000	50,660	..	..	..	Long-Term Capital, net
..	..	..	..	..	..	..	..	..	..	..	Direct Investment
1,400	2,900	2,800	20,400	21,800	16,100	13,900	19,300	8,400	3,900	..	Long-Term Loans
1,500	3,000	2,900	20,500	22,400	17,100	16,100	21,400	10,800	7,400	..	Disbursements
100	100	100	100	600	1,000	2,200	2,100	2,400	3,500	..	Repayments
21,140	34,320	28,990	19,420	-1,690	-1,060	-3,600	-5,940	..	..	..	Other Long-Term Capital
4,660	6,190	-1,000	3,480	-29,520	-32,970	-35,970	-34,560	..	..	..	Other Capital, net
2,460	-5,880	-5,730	-570	-11,510	-8,210	-5,530	-2,110	..	..	..	Change in Reserves
				(Cape Verde Escudos per US dollar)							Memo Item:
35.500	37.430	40.170	48.690	58.290	71.690	84.760	93.000	87.400	87.000	..	Conversion Factor (Annual Avg)
				(Millions of US dollars, outstanding at end of year)							**EXTERNAL DEBT, ETC.**
14.9	17.6	20.2	38.5	57.7	70.6	75.5	96.0	107.9	120.6	..	Public/Publicly Guar. Long-Term
14.9	17.6	20.2	38.5	57.7	70.6	74.1	93.6	104.9	117.7	..	Official Creditors
0.0	0.0	0.0	0.0	0.0	0.2	1.2	3.0	4.6	6.3	..	IBRD and IDA
..	..	..	..	..	..	1.4	2.4	3.0	2.9	..	Private Creditors
..	..	..	..	..	..	..	..	..	..	..	Private Non-guaranteed Long-Term
0.0	0.0	0.0	0.0	0.0	0.0	0.0	0.0	0.0	..	..	Use of Fund Credit
5.0	1.0	..	..	..	..	..	1.0	6.9	10.6	..	Short-Term Debt
				(Millions US dollars)							Memo Items:
..	..	..	..	..	..	..	..	..	..	..	Int'l Reserves Excluding Gold
..	..	..	..	..	..	..	..	..	..	..	Gold Holdings (at market price)
											SOCIAL INDICATORS
6.5	6.3	6.2	6.0	5.8	5.7	5.6	5.4	5.3	5.2	..	Total Fertility Rate
35.4	35.2	38.5	40.5	40.3	40.0	39.8	39.6	39.3	39.1	..	Crude Birth Rate
94.0	60.0	93.8	75.7	70.1	65.0	60.2	55.7	51.6	47.8	..	Infant Mortality Rate
59.7	60.2	60.7	62.0	63.4	63.7	64.1	64.5	64.9	65.3	..	Life Expectancy at Birth
94.5	87.1	121.9	91.0	88.0	77.9	90.6	80.8	104.6	110.9	..	Food Production, p.c. ('79-81=100)
54.2	53.0	51.7	..	..	..	..	..	..	..	..	Labor Force, Agriculture (%)
26.8	27.1	27.5	27.7	28.0	28.3	28.6	28.9	29.0	29.0	..	Labor Force, Female (%)
..	..	131.0	..	106.0	110.0	110.0	108.0	..	..	..	School Enroll. Ratio, primary
..	..	10.0	11.0	..	11.0	11.0	13.0	..	..	..	School Enroll. Ratio, secondary

CENTRAL AFRICAN REPUBLIC	1967	1968	1969	1970	1971	1972	1973	1974	1975	1976	1977
CURRENT GNP PER CAPITA (US $)	90	100	110	100	100	110	120	150	170	200	230
POPULATION (thousands)	1,791	1,820	1,849	1,879	1,909	1,939	1,970	2,002	2,034	2,082	2,131
ORIGIN AND USE OF RESOURCES					*(Billions of current CFA Francs)*						
Gross National Product (GNP)	40.2	44.2	46.8	49.6	52.6	55.5	57.0	71.9	80.5	107.1	123.6
Net Factor Income from Abroad	-0.1	-0.1	-0.1	-0.1	-0.1	-0.1	-0.1	-0.1	-0.1	0.1	0.1
Gross Domestic Product (GDP)	40.3	44.3	46.9	49.7	52.7	55.6	57.1	72.0	80.6	107.0	123.5
Indirect Taxes, net	2.6	3.1	3.3	2.9	3.3	3.9	4.5	4.9	5.1	5.8	7.2
GDP at factor cost	37.7	41.2	43.6	46.8	49.4	51.7	52.6	67.1	75.5	101.2	116.3
Agriculture	14.9	16.1	16.5	16.6	18.4	19.9	20.6	26.5	28.5	40.6	46.8
Industry	9.1	11.3	12.1	12.2	13.3	14.1	14.1	15.1	17.3	20.5	22.9
Manufacturing	2.1	2.3	2.3	3.2	3.3	3.8	4.1	5.2	5.7	6.2	7.6
Services, etc.	13.7	13.8	15.0	18.0	17.7	17.7	17.9	25.5	29.7	40.1	46.6
Resource Balance	-8.3	-5.7	-7.0	-7.6	-7.0	-7.6	-8.1	-6.3	-15.8	-8.5	-12.5
Exports of Goods & NFServices	10.6	13.5	11.2	13.9	13.9	12.1	14.4	19.0	18.4	25.4	32.7
Imports of Goods & NFServices	18.9	19.2	18.2	21.5	20.9	19.7	22.5	25.3	34.2	33.9	45.2
Domestic Absorption	48.6	50.0	53.9	57.3	59.7	63.2	65.2	78.3	96.4	115.5	136.0
Private Consumption, etc.	30.9	32.8	36.1	37.4	38.8	41.5	40.8	54.3	71.3	88.0	106.2
General Gov't Consumption	8.9	9.5	9.8	10.5	10.5	11.0	11.8	13.5	13.9	14.6	15.4
Gross Domestic Investment	8.8	7.7	8.0	9.4	10.4	10.7	12.6	10.5	11.2	12.9	14.4
Fixed Investment	..	..	..	..	..	..	..	..	..	..	14.5
Memo Items:											
Gross Domestic Saving	0.5	2.0	1.0	1.8	3.4	3.1	4.5	4.2	-4.6	4.4	1.9
Gross National Saving	0.4	1.3	0.2	0.6	1.9	1.9	3.2	2.3	-7.3	2.0	-0.5
					(Billions of 1980 CFA Francs)						
Gross National Product	123.8	132.8	143.5	146.6	148.1	149.9	153.2	162.3	162.7	171.0	177.1
GDP at factor cost	120.7	122.5	130.6	135.6	136.5	136.6	137.9	148.2	149.6	155.3	163.5
Agriculture	50.4	52.5	53.1	54.1	55.8	56.1	56.5	58.8	60.2	62.2	63.3
Industry	16.9	19.1	20.2	22.0	21.8	24.8	25.7	28.1	28.0	26.5	29.1
Manufacturing	..	..	..	..	..	..	..	10.2	9.5	9.7	10.6
Services, etc.	53.4	50.9	57.2	59.5	58.9	55.7	55.7	61.3	61.5	66.7	71.1
Resource Balance	-20.4	-26.8	-25.8	-28.5	-26.0	-25.2	-22.5	-2.7	-3.3	-7.7	-21.7
Exports of Goods & NFServices	34.6	27.7	22.2	25.1	25.8	23.1	27.1	38.5	46.6	38.4	36.5
Imports of Goods & NFServices	55.0	54.5	48.0	53.6	51.8	48.3	49.6	41.2	49.9	46.1	58.2
Domestic Absorption	152.1	161.6	169.4	175.2	174.6	175.4	176.0	165.3	166.3	178.5	198.6
Private Consumption, etc.	100.3	101.6	108.3	109.8	109.4	110.8	108.6	117.1	123.6	135.2	156.3
General Gov't Consumption	24.1	29.9	30.3	31.5	29.9	29.3	30.6	25.2	23.7	23.0	21.8
Gross Domestic Investment	27.7	30.1	30.9	34.0	35.3	35.3	36.8	23.0	19.1	20.3	20.5
Fixed Investment	..	..	..	..	..	..	..	..	..	..	20.7
Memo Items:											
Capacity to Import	30.9	38.3	29.6	34.6	34.4	29.7	31.7	31.0	26.8	34.6	42.1
Terms of Trade Adjustment	-3.8	10.6	7.3	9.5	8.6	6.6	4.6	-7.6	-19.8	-3.9	5.6
Gross Domestic Income	128.0	145.4	151.0	156.3	157.2	156.8	158.2	155.0	143.3	166.9	182.5
Gross National Income	120.0	143.4	150.8	156.1	156.7	156.4	157.8	154.7	143.0	167.1	182.7
DOMESTIC PRICES (DEFLATORS)					*(Index 1980 = 100)*						
Overall (GDP)	30.6	32.9	32.7	33.9	35.5	37.0	37.2	44.3	49.4	62.6	69.8
Domestic Absorption	31.9	30.9	31.8	32.7	34.2	36.0	37.0	47.4	58.0	64.7	68.5
Agriculture	29.6	30.7	31.1	30.7	33.0	35.5	36.5	45.1	47.4	65.3	74.0
Industry	53.8	59.2	59.8	55.4	61.1	56.8	54.9	53.7	61.8	77.5	78.6
Manufacturing	..	..	..	..	..	..	..	51.2	60.3	63.7	71.9
MANUFACTURING ACTIVITY											
Employment (1980=100)	..	..	..	..	..	..	229.0	235.2	159.0	143.1	123.5
Real Earnings per Empl. (1980=100)	..	..	..	..	..	..	56.6	61.1	72.2	74.9	93.9
Real Output per Empl. (1980=100)	..	..	..	..	..	..	68.5	77.0	94.1	87.7	127.0
Earnings as % of Value Added	..	..	..	..	..	..	30.1	27.3	31.4	31.6	39.3
MONETARY HOLDINGS					*(Billions of current CFA Francs)*						
Money Supply, Broadly Defined	6.34	7.78	7.56	8.37	8.36	9.77	10.43	13.63	13.68	19.35	20.49
Money as Means of Payment	5.87	7.20	6.70	7.54	7.54	8.92	9.21	12.48	12.27	17.07	18.20
Currency Ouside Banks	3.69	4.22	4.66	4.96	4.90	5.53	5.75	7.38	7.48	8.96	11.16
Demand Deposits	2.18	2.98	2.03	2.58	2.64	3.39	3.46	5.10	4.79	8.12	7.04
Quasi-Monetary Liabilities	0.47	0.58	0.86	0.83	0.82	0.85	1.21	1.15	1.41	2.27	2.29
GOVERNMENT DEFICIT (-) OR SURPLUS	..	..	..	..	*(Millions of current CFA Francs)*	..	..	..	..	..	..
Current Revenue	..	..	..	..	..	..	..	..	..	..	..
Current Expenditure	..	..	..	..	..	..	..	..	..	..	..
Current Budget Balance	..	..	..	..	..	..	..	..	..	..	..
Capital Receipts	..	..	..	..	..	..	..	..	..	..	..
Capital Payments	..	..	..	..	..	..	..	..	..	..	..

1978	1979	1980	1981	1982	1983	1984	1985	1986	1987 estimate	NOTES	CENTRAL AFRICAN REPUBLIC
260	290	320	330	340	290	280	280	290	330	..	CURRENT GNP PER CAPITA (US $)
2,182	2,233	2,286	2,343	2,400	2,460	2,521	2,583	2,654	2,727	..	POPULATION (thousands)
				(Billions of current CFA Francs)							ORIGIN AND USE OF RESOURCES
136.4	151.1	168.9	190.1	244.6	247.5	274.5	311.2	326.6	318.1	..	Gross National Product (GNP)
0.1	0.2	0.5	1.3	-1.3	-3.5	-4.2	-5.0	-4.3	-5.6	..	Net Factor Income from Abroad
136.3	150.9	168.4	188.8	245.9	251.0	278.7	316.2	330.9	323.7	..	Gross Domestic Product (GDP)
7.2	8.1	9.9	11.1	14.1	13.7	18.7	22.4	22.8	18.9	..	Indirect Taxes, net
129.1	142.8	158.5	177.7	231.8	237.3	260.0	293.8	308.1	304.8	..	GDP at factor cost
51.3	55.8	63.4	70.5	96.6	95.8	109.3	117.3	124.5	124.6	..	Agriculture
28.8	32.2	31.8	33.6	30.4	34.8	39.0	40.3	42.8	40.4	..	Industry
10.8	11.8	11.4	13.2	17.9	20.6	21.7	21.5	20.5	22.9	..	Manufacturing
49.0	54.8	63.3	73.6	104.9	106.7	111.7	136.2	140.8	139.8	..	Services, etc.
-15.9	-17.1	-28.0	-20.0	-33.2	-32.8	-35.0	-46.5	-47.7	-53.0	..	Resource Balance
33.7	35.2	44.0	48.5	54.6	60.5	65.4	79.8	64.3	54.5	..	Exports of Goods & NFServices
49.6	52.3	72.0	68.5	87.8	93.3	100.4	126.3	112.0	107.5	..	Imports of Goods & NFServices
152.2	168.0	196.4	208.8	279.1	283.8	313.7	362.7	378.6	376.7	..	Domestic Absorption
117.8	130.1	159.1	164.0	223.6	211.7	239.7	273.4	284.0	288.2	f	Private Consumption, etc.
19.4	23.3	25.5	28.3	39.1	42.4	39.7	43.6	43.8	43.5	..	General Gov't Consumption
15.0	14.6	11.8	16.5	16.4	29.7	34.3	45.7	50.8	45.0	f	Gross Domestic Investment
12.5	16.7	11.7	15.2	15.7	26.3	33.8	49.1	48.3	42.2	..	Fixed Investment
											Memo Items:
-0.9	-2.5	-16.2	-3.5	-16.8	-3.1	-0.7	-0.8	3.1	-8.0	..	Gross Domestic Saving
-3.8	-4.1	-19.2	-6.0	-23.7	-12.3	-9.8	-11.6	-5.7	-19.3	..	Gross National Saving
				(Billions of 1980 CFA Francs)							
181.3	176.8	168.9	166.0	175.9	162.6	177.2	183.2	186.1	186.8	..	Gross National Product
165.4	161.4	158.5	156.4	163.5	152.4	166.0	172.4	175.1	179.8	..	GDP at factor cost
65.3	64.4	63.4	62.3	63.3	60.7	66.4	68.8	70.8	73.4	..	Agriculture
33.3	31.4	31.8	29.3	29.1	31.8	32.7	33.8	34.9	34.1	..	Industry
12.8	11.7	11.4	10.3	10.6	10.7	10.8	11.3	11.4	10.7	..	Manufacturing
66.8	65.6	63.3	64.8	71.1	59.9	67.0	69.7	69.4	72.3	..	Services, etc.
-24.5	-20.3	-28.0	-16.0	-21.7	-19.1	-21.2	-22.1	-29.3	-27.6	..	Resource Balance
35.6	36.6	44.0	45.7	36.5	37.4	37.6	39.1	34.3	32.2	..	Exports of Goods & NFServices
60.2	56.9	72.0	61.7	58.2	56.5	58.9	61.2	63.6	59.8	..	Imports of Goods & NFServices
205.8	196.8	196.4	180.6	198.6	184.3	200.9	208.6	218.4	218.6	..	Domestic Absorption
160.4	152.8	159.1	141.6	156.3	135.4	148.1	153.7	167.5	170.2	f	Private Consumption, etc.
25.5	27.1	25.5	24.7	21.8	18.6	19.4	20.3	20.1	20.7	..	General Gov't Consumption
19.9	16.9	11.8	14.4	20.5	30.3	33.4	34.7	30.9	27.7	f	Gross Domestic Investment
16.7	19.4	11.7	13.0	19.7	27.0	28.4	30.0	30.7	27.5	..	Fixed Investment
											Memo Items:
40.9	38.3	44.0	43.7	36.2	36.7	38.3	38.7	36.5	30.3	..	Capacity to Import
5.2	1.7	0.0	-2.0	-0.3	-0.8	0.7	-0.4	2.2	-1.9	..	Terms of Trade Adjustment
186.6	178.2	168.4	162.6	176.6	164.4	180.4	186.1	191.3	189.1	..	Gross Domestic Income
186.6	178.5	168.9	164.0	175.6	161.9	177.9	182.8	188.3	185.0	..	Gross National Income
				(Index 1980 = 100)							DOMESTIC PRICES (DEFLATORS)
75.2	85.5	100.0	114.7	139.0	152.0	155.1	169.5	175.0	169.5	..	Overall (GDP)
73.9	85.4	100.0	115.6	140.5	154.0	156.2	173.8	173.3	172.3	..	Domestic Absorption
78.6	86.7	100.0	113.1	152.7	157.8	164.6	170.5	175.8	169.7	..	Agriculture
86.4	102.5	100.0	114.8	104.2	109.5	119.3	119.1	122.8	118.5	..	Industry
84.4	101.0	100.0	128.3	169.4	192.4	200.1	191.1	179.9	213.9	..	Manufacturing
											MANUFACTURING ACTIVITY
112.9	106.3	100.0	94.2	101.1	168.7	172.3	176.0	..	..	J	Employment (1980=100)
98.1	102.3	100.0	107.9	99.6	101.1	105.2	103.2	..	..	J	Real Earnings per Empl. (1980=100)
97.7	89.9	100.0	97.6	76.2	69.8	81.4	80.0	..	..	J	Real Output per Empl. (1980=100)
49.8	48.2	46.7	45.4	50.5	55.7	50.5	50.5	..	..	J	Earnings as % of Value Added
				(Billions of current CFA Francs)							MONETARY HOLDINGS
22.51	27.10	36.62	45.33	43.29	47.93	52.11	56.55	58.36	60.02	..	Money Supply, Broadly Defined
20.52	26.13	34.61	42.32	40.56	44.78	48.24	51.54	52.19	53.55	..	Money as Means of Payment
13.36	17.94	24.37	31.96	31.22	34.31	37.24	37.15	40.61	41.46	..	Currency Ouside Banks
7.16	8.19	10.24	10.37	9.34	10.47	11.01	14.39	11.58	12.09	..	Demand Deposits
1.99	0.97	2.01	3.01	2.73	3.15	3.87	5.01	6.17	6.48	..	Quasi-Monetary Liabilities
				(Millions of current CFA Francs)							
..	..	..	-6,671	..	..	..	..	..	..	..	GOVERNMENT DEFICIT (-) OR SURPLUS
..	..	..	35,036	..	..	..	..	..	..	..	Current Revenue
..	..	..	34,977	..	..	..	..	..	..	..	Current Expenditure
..	..	..	59	..	..	..	..	..	..	..	Current Budget Balance
..	..	..	15	..	..	..	..	..	..	..	Capital Receipts
..	..	..	7,179	..	..	..	..	..	..	..	Capital Payments

CENTRAL AFRICAN REPUBLIC	1967	1968	1969	1970	1971	1972	1973	1974	1975	1976	1977
FOREIGN TRADE (CUSTOMS BASIS)					*(Millions of current US dollars)*						
Value of Exports, fob	35.8	44.1	36.9	43.4	43.9	40.5	55.0	57.4	54.0	69.5	104.5
Nonfuel Primary Products	18.7	20.4	19.1	24.1	27.2	27.3	36.6	45.0	41.2	57.2	79.6
Fuels	0.2	0.1	0.1	0.1	0.1	0.0	0.3	0.0	0.0	0.0	0.0
Manufactures	16.9	23.6	17.7	19.2	16.6	13.1	18.2	12.4	12.8	12.3	25.0
Value of Imports, cif	60.4	53.5	52.7	56.6	50.7	56.4	69.0	83.5	109.8	99.6	136.3
Nonfuel Primary Products	9.4	8.8	8.9	10.6	10.3	10.7	12.6	15.0	19.7	19.8	24.3
Fuels	2.9	2.6	4.4	0.8	0.8	1.1	0.7	1.1	0.9	1.4	1.3
Manufactures	48.1	42.1	39.4	45.2	39.5	44.6	55.8	67.4	89.2	78.4	110.7
					(Index 1980 = 100)						
Terms of Trade	109.6	117.6	132.4	100.2	92.2	90.0	98.0	86.4	73.2	102.5	115.1
Export Prices, fob	26.4	28.1	27.7	31.6	30.6	32.5	45.2	50.5	47.1	66.8	82.3
Nonfuel Primary Products	24.8	25.5	25.1	28.6	28.1	30.4	44.6	47.2	43.2	66.8	86.3
Fuels	4.3	4.3	4.3	4.3	5.6	6.2	8.9	36.7	35.7	38.4	42.0
Manufactures	30.1	31.7	32.4	37.5	36.5	38.5	49.5	67.2	66.1	66.7	71.7
Import Prices, cif	24.1	23.9	20.9	31.5	33.2	36.1	46.1	58.5	64.4	65.1	71.5
Trade at Constant 1980 Prices					*(Millions of 1980 US dollars)*						
Exports, fob	135.4	157.0	133.3	137.4	143.5	124.5	121.6	113.7	114.7	104.1	127.0
Imports, cif	250.2	224.0	252.1	179.6	152.8	156.2	149.5	142.8	170.6	152.9	190.8
BALANCE OF PAYMENTS					*(Millions of current US dollars)*						
Exports of Goods & Services	..	..	..	60.8	60.4	58.0	78.1	78.9	81.0	101.6	130.6
Merchandise, fob	..	..	..	43.4	43.9	40.5	55.0	57.4	54.0	69.5	104.5
Nonfactor Services	..	..	..	17.0	16.1	16.8	22.7	20.7	26.0	30.6	23.4
Factor Services	..	..	..	0.4	0.4	0.7	0.5	0.8	1.0	1.5	2.8
Imports of Goods & Services	..	..	..	80.8	78.2	80.5	102.5	125.2	154.1	134.5	177.5
Merchandise, fob	..	..	..	39.6	44.8	46.3	50.7	62.9	82.9	75.9	103.9
Nonfactor Services	..	..	..	39.0	30.5	31.5	48.3	59.1	68.2	56.5	70.2
Factor Services	..	..	..	2.2	2.9	2.7	3.6	3.3	2.9	2.2	3.4
Long-Term Interest	..	..	..	0.7	0.6	0.6	1.2	1.2	2.1	0.9	1.2
Current Transfers, net	..	..	..	-3.9	-5.2	-4.5	-5.3	-7.7	-12.3	-10.4	-10.1
Workers' Remittances	..	..	..	0.0	0.0	0.0	0.0	0.0	0.0	0.0	0.0
Total to be Financed	..	..	..	-23.9	-23.1	-26.9	-29.7	-54.0	-85.4	-43.3	-57.0
Official Capital Grants	..	..	..	12.4	14.1	28.0	26.7	39.5	48.1	48.8	38.1
Current Account Balance	..	..	..	-11.5	-8.9	1.1	-3.0	-14.6	-37.3	5.5	-18.9
Long-Term Capital, net	..	..	..	12.3	21.0	25.0	26.4	57.5	68.1	65.4	47.9
Direct Investment	..	..	..	0.9	0.3	1.2	-0.5	5.7	4.3	3.5	-3.0
Long-Term Loans	..	..	..	-0.6	2.5	14.9	15.8	6.1	9.8	9.3	20.3
Disbursements	..	..	..	1.8	4.1	16.0	19.3	9.6	13.4	11.2	23.4
Repayments	..	..	..	2.4	1.6	1.1	3.5	3.5	3.6	1.9	3.1
Other Long-Term Capital	..	..	..	-0.4	4.0	-19.1	-15.7	6.3	6.0	3.9	-7.6
Other Capital, net	..	..	..	-2.2	-15.1	-26.0	-23.4	-45.4	-31.3	-59.7	-24.2
Change in Reserves	..	..	..	1.4	3.1	-0.1	0.0	2.5	0.5	-11.2	-4.9
Memo Item:					*(CFA Francs per US dollar)*						
Conversion Factor (Annual Avg)	246.850	246.850	259.710	277.710	277.130	252.210	222.700	240.500	214.320	238.980	245.670
EXTERNAL DEBT, ETC.					*(Millions of US dollars, outstanding at end of year)*						
Public/Publicly Guar. Long-Term	..	..	..	24.1	28.2	36.3	55.3	63.9	71.3	78.9	103.3
Official Creditors	..	..	..	17.9	21.7	22.3	26.7	30.1	38.6	47.0	59.8
IBRD and IDA	..	..	..	0.2	2.4	6.1	8.1	9.1	9.7	9.9	10.9
Private Creditors	..	..	..	6.1	6.5	14.0	28.6	33.8	32.7	31.8	43.6
Private Non-guaranteed Long-Term	..	..	..	..	..	..	..	..	..	..	..
Use of Fund Credit	..	..	..	0.0	0.0	0.0	0.0	0.0	5.7	0.0	0.0
Short-Term Debt	..	..	..	..	..	..	..	..	..	..	10.0
Memo Items:					*(Thousands of US dollars)*						
Int'l Reserves Excluding Gold	4,753.0	4,697.0	1,196.0	1,388.0	211.0	1,719.0	1,783.0	1,743.0	3,829.0	18,829.0	25,351.0
Gold Holdings (at market price)	..	..	..	..	..	..	..	..	..	..	924.0
SOCIAL INDICATORS											
Total Fertility Rate	4.7	4.8	4.8	4.9	4.9	5.0	5.1	5.2	5.3	5.4	5.5
Crude Birth Rate	35.6	36.0	36.3	36.7	37.0	37.4	38.2	39.0	39.8	40.6	41.4
Infant Mortality Rate	160.0	157.6	155.2	152.8	150.4	148.0	147.4	146.8	146.2	145.6	145.0
Life Expectancy at Birth	40.9	41.4	41.9	42.4	42.9	43.4	43.8	44.3	44.7	45.1	45.5
Food Production, p.c. ('79-81 = 100)	89.6	90.5	94.1	95.1	97.8	99.6	102.3	102.6	101.8	99.4	101.5
Labor Force, Agriculture (%)	86.1	85.0	84.0	82.9	81.9	80.8	79.8	78.7	77.7	76.6	75.6
Labor Force, Female (%)	49.4	49.3	49.2	49.2	49.0	48.9	48.8	48.7	48.6	48.5	48.4
School Enroll. Ratio, primary	..	..	..	64.0	..	..	..	..	73.0	..	70.0
School Enroll. Ratio, secondary	..	..	..	4.0	..	..	..	..	8.0	9.0	10.0

1978	1979	1980	1981	1982	1983	1984	1985	1986	1987 estimate	NOTES	CENTRAL AFRICAN REPUBLIC
				(Millions of current US dollars)							**FOREIGN TRADE (CUSTOMS BASIS)**
110.3	122.2	147.2	117.7	111.5	114.6	114.7	131.7	145.1	153.8	f	Value of Exports, fob
68.9	67.9	108.7	81.3	76.6	79.4	83.2	93.3	97.7	102.2	..	Nonfuel Primary Products
0.0	0.0	0.0	0.0	0.0	0.0	0.0	0.0	0.0	..	..	Fuels
41.5	54.3	38.5	36.5	35.0	35.3	31.5	38.4	47.4	51.6	..	Manufactures
156.1	158.7	221.0	170.8	165.7	149.7	152.6	170.1	182.0	192.5	f	Value of Imports, cif
29.6	28.1	52.2	37.5	36.2	33.0	31.9	34.7	35.8	32.7	..	Nonfuel Primary Products
2.5	3.9	3.9	3.2	3.2	2.8	3.0	3.4	1.2	1.7	..	Fuels
124.0	126.7	164.9	130.1	126.3	113.8	117.7	132.0	145.0	158.1	..	Manufactures
				(Index 1980 = 100)							
96.0	103.5	100.0	88.1	89.7	89.3	94.6	87.5	85.7	84.3	..	Terms of Trade
75.9	93.6	100.0	88.4	87.9	85.5	89.2	82.2	93.2	96.2	..	Export Prices, fob
75.8	96.7	100.0	84.4	85.1	83.4	89.2	79.7	89.0	89.5	..	Nonfuel Primary Products
42.3	61.0	100.0	112.5	101.6	92.5	90.2	87.5	44.9	56.9	..	Fuels
76.1	90.0	100.0	99.1	94.8	90.7	89.3	89.0	103.2	113.0	..	Manufactures
79.1	90.4	100.0	100.4	98.0	95.7	94.3	93.9	108.7	114.1	..	Import Prices, cif
				(Millions of 1980 US dollars)							Trade at Constant 1980 Prices
145.3	130.6	147.2	133.1	126.8	134.0	128.6	160.1	155.7	159.8	..	Exports, fob
197.4	175.5	221.0	170.1	169.1	156.4	161.8	181.1	167.5	168.7	..	Imports, cif
				(Millions of current US dollars)							**BALANCE OF PAYMENTS**
146.7	159.8	205.4	177.4	170.8	162.1	151.9	180.7	179.9	184.2	f	Exports of Goods & Services
110.3	122.2	147.2	117.7	124.4	123.4	114.4	130.7	129.8	130.4	..	Merchandise, fob
32.7	35.0	53.9	51.7	41.5	35.6	34.8	46.9	47.3	50.9	..	Nonfactor Services
3.7	2.6	4.3	8.0	4.9	3.1	2.7	3.1	2.8	2.9	..	Factor Services
209.5	254.2	329.2	241.3	266.9	252.3	238.1	295.3	338.5	379.1	f	Imports of Goods & Services
119.0	132.9	185.1	144.6	149.7	137.5	140.1	168.0	191.1	204.2	f	Merchandise, fob
87.9	113.3	142.3	92.1	104.9	101.0	85.7	113.1	132.3	153.5	..	Nonfactor Services
2.5	8.0	1.8	4.6	12.3	13.8	12.4	14.2	15.1	21.4	..	Factor Services
1.4	0.4	0.5	1.0	2.4	6.7	5.7	6.7	9.1	9.1	..	Long-Term Interest
-13.3	-8.3	-16.7	-13.9	-17.0	-14.8	-11.2	-12.9	-13.0	-19.0	..	Current Transfers, net
0.0	0.0	0.0	0.0	0.0	0.0	0.0	0.0	0.0	..	..	Workers' Remittances
-76.1	-102.7	-140.5	-77.8	-113.1	-105.1	-97.4	-127.5	-171.6	-213.9	K	Total to be Financed
51.8	86.9	97.5	73.6	71.8	78.3	64.6	72.1	95.3	117.9	K	Official Capital Grants
-24.3	-15.8	-43.1	-4.2	-41.3	-26.8	-32.7	-55.4	-76.3	-96.0	..	Current Account Balance
69.7	124.8	153.2	90.7	74.0	114.1	114.4	132.1	170.6	215.2	f	Long-Term Capital, net
4.8	22.4	5.3	5.8	8.8	4.0	4.9	14.0	14.0	20.0	..	Direct Investment
25.8	12.8	39.6	31.6	16.7	25.2	24.1	40.4	67.9	63.0	..	Long-Term Loans
28.4	13.1	40.7	34.3	19.0	36.4	33.7	47.0	76.9	76.1	..	Disbursements
2.6	0.3	1.1	2.7	2.3	11.2	9.6	6.6	9.0	13.1	..	Repayments
-12.6	2.8	10.8	-20.3	-23.3	6.5	20.8	5.6	-6.6	14.3	..	Other Long-Term Capital
-51.0	-90.6	-99.6	-83.1	-48.3	-89.4	-74.5	-84.4	-86.6	-109.7	..	Other Capital, net
5.5	-18.4	-10.6	-3.4	15.6	2.0	-7.3	7.7	-7.7	-9.5	..	Change in Reserves
				(CFA Francs per US dollar)							Memo Item:
225.640	212.720	211.300	271.730	328.620	381.070	436.960	449.260	346.300	300.540	..	Conversion Factor (Annual Avg)
				(Millions of US dollars, outstanding at end of year)							**EXTERNAL DEBT, ETC.**
110.1	120.3	159.8	189.0	208.0	215.4	224.2	297.0	394.5	520.3	..	Public/Publicly Guar. Long-Term
53.1	65.0	114.4	155.5	175.2	185.3	198.5	271.7	369.6	494.6	..	Official Creditors
11.5	18.6	28.8	33.1	34.9	38.2	49.1	62.7	86.2	125.9	..	IBRD and IDA
57.0	55.3	45.3	33.5	32.9	30.0	25.7	25.3	24.9	25.6	..	Private Creditors
..	..	..	..	..	..	..	..	..	..	..	Private Non-guaranteed Long-Term
13.4	0.0	7.4	22.9	23.7	26.9	23.9	28.4	33.1	36.8	..	Use of Fund Credit
9.0	16.0	23.9	19.0	18.9	14.7	13.3	18.1	28.3	28.3	..	Short-Term Debt
				(Thousands of US dollars)							Memo Items:
24,134.0	44,109.0	54,978.0	69,274.0	46,370.0	46,792.0	52,676.0	49,624.0	65,355.0	96,727.0	..	Int'l Reserves Excluding Gold
1,921.0	5,837.0	6,720.0	4,532.0	5,209.0	4,349.0	3,422.0	3,630.0	4,339.0	5,374.0	..	Gold Holdings (at market price)
											SOCIAL INDICATORS
5.5	5.5	5.5	5.5	5.5	5.6	5.6	5.6	5.7	5.8	..	Total Fertility Rate
41.4	41.4	41.3	41.3	41.3	41.6	41.9	42.2	42.5	42.9	..	Crude Birth Rate
144.4	143.8	143.2	142.6	142.0	135.2	128.3	121.5	114.7	107.8	..	Infant Mortality Rate
45.9	46.4	46.8	47.3	47.7	48.2	48.6	49.1	49.5	50.0	..	Life Expectancy at Birth
101.2	99.5	100.2	100.3	98.9	95.8	91.5	94.7	94.7	93.2	..	Food Production, p.c. ('79-81=100)
74.5	73.5	72.4	..	..	..	..	..	..	..	..	Labor Force, Agriculture (%)
48.3	48.2	48.0	47.8	47.6	47.4	47.2	47.0	46.7	46.5	..	Labor Force, Female (%)
69.0	70.0	72.0	73.0	74.0	77.0	79.0	73.0	..	..	..	School Enroll. Ratio, primary
10.0	11.0	14.0	15.0	16.0	..	16.0	13.0	..	..	..	School Enroll. Ratio, secondary

CHAD	1967	1968	1969	1970	1971	1972	1973	1974	1975	1976	1977
CURRENT GNP PER CAPITA (US $)	90	90	100	100	90	100	100	120	150	150	160
POPULATION (thousands)	3,458	3,521	3,586	3,652	3,725	3,799	3,874	3,951	4,030	4,116	4,203

ORIGIN AND USE OF RESOURCES					*(Billions of current CFA Francs)*						
Gross National Product (GNP)	77.5	78.7	85.8	91.3	97.3	104.9	101.8	110.9	129.7	145.1	160.4
Net Factor Income from Abroad	-0.4	-0.4	-0.5	0.0	0.0	1.0	0.3	0.3	-0.7	-0.6	-1.4
Gross Domestic Product (GDP)	77.9	79.1	86.3	91.3	97.3	103.9	101.5	110.6	130.4	145.7	161.8
Indirect Taxes, net	5.3	4.8	6.6	7.4	7.6	6.2	5.8	7.1	6.7	7.3	1.3
GDP at factor cost	72.6	74.3	79.7	83.9	89.7	97.7	95.7	103.5	123.7	138.4	160.5
Agriculture	36.3	36.5	38.7	39.5	42.2	45.8	43.2	47.3	56.2	60.8	67.0
Industry	13.1	14.5	14.6	15.2	16.2	17.7	19.8	21.9	25.8	26.7	29.2
Manufacturing	12.4	13.7	13.7	14.2	15.2	16.5	18.3	20.1	23.7	24.2	26.3
Services, etc.	28.5	28.1	33.0	36.6	38.8	40.3	38.5	41.4	48.4	58.2	65.5
Resource Balance	-4.2	-4.2	-5.5	-7.5	-7.1	-7.1	-11.2	-13.2	-21.2	-21.3	-29.9
Exports of Goods & NFServices	15.9	16.8	16.2	21.1	22.0	21.5	22.7	28.9	26.7	33.3	35.3
Imports of Goods & NFServices	20.1	21.0	21.7	28.6	29.1	28.6	33.9	42.1	47.9	54.6	65.2
Domestic Absorption	82.1	83.3	91.8	98.8	104.4	111.0	112.7	123.8	151.6	167.0	191.7
Private Consumption, etc.	51.0	51.6	53.7	58.1	60.4	65.0	85.3	70.4	88.6	104.6	119.2
General Gov't Consumption	17.4	18.4	21.8	24.3	26.6	28.8	31.3	30.5	33.0	34.5	38.8
Gross Domestic Investment	13.7	13.4	16.3	16.5	17.3	17.2	-3.9	22.9	30.0	27.9	33.6
Fixed Investment	..	..	..	..	..	..	..	..	..	..	..
Memo Items:											
Gross Domestic Saving	9.5	9.2	10.8	9.0	10.2	10.1	-15.1	9.7	8.8	6.6	3.7
Gross National Saving	9.1	7.8	8.7	7.3	8.3	8.9	-16.8	7.9	6.4	5.0	-1.2

					(Billions of 1980 CFA Francs)						
Gross National Product	175.7	174.9	186.8	191.5	187.2	191.5	174.1	182.7	197.3	203.5	207.1
GDP at factor cost	..	..	..	..	..	..	..	..	..	..	207.3
Agriculture	84.6	83.4	88.5	88.0	85.8	88.7	77.0	79.4	86.3	85.4	86.4
Industry	27.7	28.5	28.6	29.4	29.8	30.3	30.5	32.1	34.5	37.5	37.8
Manufacturing	26.0	26.6	26.6	27.2	27.7	27.7	27.7	28.9	31.1	34.0	34.0
Services, etc.	64.4	64.0	70.9	74.1	71.5	70.4	66.0	70.7	77.7	81.6	84.8
Resource Balance	-37.3	-35.3	-34.1	-41.7	-40.0	-45.0	-59.5	-60.8	-55.1	-50.3	-54.9
Exports of Goods & NFServices	43.8	44.7	41.6	51.8	55.1	47.4	48.8	35.8	40.8	52.9	52.9
Imports of Goods & NFServices	81.2	80.0	75.7	93.5	95.1	92.4	108.3	96.7	95.9	103.2	107.9
Domestic Absorption	214.0	211.2	222.1	233.2	227.2	234.3	232.9	242.9	253.6	254.7	263.9
Private Consumption, etc.	128.0	125.4	122.3	131.4	125.1	130.7	181.8	137.6	142.1	147.3	149.3
General Gov't Consumption	45.1	46.3	53.0	55.9	54.3	59.7	61.8	58.0	55.5	56.8	57.2
Gross Domestic Investment	41.0	39.6	46.8	45.8	47.8	43.9	-10.6	47.3	56.0	50.7	57.4
Fixed Investment	..	..	..	..	..	..	..	..	..	..	..
Memo Items:											
Capacity to Import	64.2	64.0	56.6	69.0	71.9	69.5	72.5	66.4	53.4	63.0	58.4
Terms of Trade Adjustment	20.4	19.3	14.9	17.2	16.8	22.0	23.7	30.5	12.6	10.0	5.5
Gross Domestic Income	197.1	195.2	202.9	208.7	204.0	211.4	197.2	212.6	211.2	214.4	214.5
Gross National Income	196.0	194.2	201.7	208.7	204.0	213.5	197.8	213.2	210.0	213.5	212.5

DOMESTIC PRICES (DEFLATORS)					*(Index 1980 = 100)*						
Overall (GDP)	44.1	45.0	45.9	47.7	52.0	54.8	58.5	60.7	65.7	71.3	77.4
Domestic Absorption	38.4	39.4	41.3	42.4	45.9	47.3	48.4	51.0	59.8	65.6	72.6
Agriculture	42.8	43.7	43.7	44.9	49.2	51.6	56.1	59.6	65.2	71.2	77.5
Industry	47.4	50.9	51.1	51.6	54.3	58.6	65.0	68.2	74.6	71.3	77.4
Manufacturing	47.5	51.3	51.3	52.2	54.8	59.5	66.0	69.6	76.2	71.3	77.4

MANUFACTURING ACTIVITY											
Employment (1980=100)	..	..	..	..	..	..	..	..	..	..	..
Real Earnings per Empl. (1980=100)	..	..	..	..	..	..	..	..	..	..	..
Real Output per Empl. (1980=100)	..	..	..	..	..	..	..	..	..	..	..
Earnings as % of Value Added	..	..	..	..	..	..	..	..	13.5	13.5	13.5

MONETARY HOLDINGS					*(Billions of current CFA Francs)*						
Money Supply, Broadly Defined	6.7	7.4	8.0	9.2	9.8	10.1	10.7	15.1	16.7	20.8	23.7
Money as Means of Payment	6.5	7.3	7.7	8.6	9.3	9.6	9.8	14.1	15.5	19.5	22.5
Currency Ouside Banks	4.2	4.5	4.9	5.3	6.2	6.9	6.5	8.5	10.9	13.0	14.4
Demand Deposits	2.4	2.8	2.8	3.3	3.1	2.7	3.3	5.6	4.6	6.5	8.1
Quasi-Monetary Liabilities	0.2	0.1	0.4	0.6	0.5	0.5	0.9	1.0	1.2	1.3	1.2

					(Millions of current CFA Francs)						
GOVERNMENT DEFICIT (-) OR SURPLUS	..	..	..	..	..	-2,783	-3,910	-2,543	-2,130	-2,848	..
Current Revenue	..	..	..	..	..	12,852	13,450	16,707	18,191	20,292	..
Current Expenditure	..	..	..	..	..	12,554	14,164	15,571	15,371	19,085	..
Current Budget Balance	..	..	..	..	..	298	-714	1,136	2,820	1,207	..
Capital Receipts	..	..	..	..	..	..	..	..	..	..	..
Capital Payments	..	..	..	..	..	3,081	3,196	3,679	4,950	4,055	..

1978	1979	1980	1981	1982	1983	1984	1985	1986	1987 estimate	NOTES	CHAD
170	150	160	160	150	130	110	140	140	150	..	**CURRENT GNP PER CAPITA (US $)**
4,293	4,384	4,477	4,577	4,681	4,789	4,902	5,018	5,146	5,273	..	**POPULATION (thousands)**
				(Billions of current CFA Francs)							**ORIGIN AND USE OF RESOURCES**
175.0	151.1	153.7	167.8	188.0	221.5	221.5	296.9	279.8	288.7	..	Gross National Product (GNP)
-1.9	0.7	0.1	0.1	0.4	-2.5	-2.5	-3.1	-3.2	-4.5	f	Net Factor Income from Abroad
176.9	150.4	153.6	167.7	187.6	224.0	224.0	300.0	283.0	293.2	..	Gross Domestic Product (GDP)
1.5	1.0	0.5	0.7	0.9	1.2	1.5	1.7	1.7	2.9	..	Indirect Taxes, net
175.5	149.4	153.1	167.0	186.7	222.8	222.5	298.3	281.3	290.3	B	GDP at factor cost
74.5	72.4	82.0	78.8	87.9	99.9	91.2	137.9	130.4	125.7	f	Agriculture
30.8	20.9	18.2	26.4	31.0	37.6	51.8	54.3	50.4	53.0	..	Industry
27.4	19.7	17.2	25.3	29.6	35.6	48.4	50.4	45.6	45.4	..	Manufacturing
71.6	57.1	53.5	62.6	68.7	86.5	81.0	107.8	102.2	114.5	..	Services, etc.
-29.0	-30.5	-37.8	-26.8	-27.8	-22.8	-21.1	-18.4	-82.2	-90.1	..	Resource Balance
41.3	43.8	43.1	44.7	45.7	53.7	59.9	65.2	49.3	49.9	..	Exports of Goods & NFServices
70.3	74.3	80.9	71.5	73.5	76.4	81.0	83.7	131.5	140.0	..	Imports of Goods & NFServices
205.9	180.9	191.4	194.6	215.4	246.8	245.1	318.4	..	383.3	..	Domestic Absorption
128.8	103.1	111.5	121.9	146.8	176.5	173.1	239.3	..	305.1	..	Private Consumption, etc.
44.9	47.1	52.1	53.8	51.6	53.8	55.1	60.1	..	24.3	..	General Gov't Consumption
32.2	30.7	27.9	18.9	17.0	16.5	16.9	19.1	..	53.9	..	Gross Domestic Investment
..	..	..	..	..	..	..	..	..	..	..	Fixed Investment
											Memo Items:
3.2	0.2	-9.9	-7.9	-10.8	-6.3	-4.2	0.7	..	-36.2	..	Gross Domestic Saving
-1.2	-1.1	-10.6	-8.0	-10.7	-9.9	-7.4	0.7	..	-44.3	..	Gross National Saving
				(Billions of 1980 CFA Francs)							
205.6	164.2	153.7	155.2	163.8	170.5	160.3	213.3	201.8	202.0	..	Gross National Product
206.3	162.4	153.1	154.5	162.7	171.5	167.0	196.5	203.6	204.7	B	GDP at factor cost
87.6	78.7	82.0	72.9	76.6	80.4	66.2	92.5	94.3	88.0	f	Agriculture
36.2	22.7	18.2	24.4	27.0	30.3	37.5	36.5	35.3	37.1	..	Industry
32.2	21.4	17.2	23.4	25.8	28.6	35.1	33.8	31.9	31.8	..	Manufacturing
84.2	62.0	53.5	57.9	59.9	61.9	58.5	86.7	74.7	80.2	..	Services, etc.
-48.6	-40.0	-37.8	-24.4	-19.2	-20.5	-19.4	-17.5	-77.9	-85.1	..	Resource Balance
50.9	47.8	43.1	38.9	37.8	33.1	31.6	32.8	40.0	40.5	..	Exports of Goods & NFServices
99.5	87.9	80.9	63.3	57.0	53.6	51.0	50.3	118.0	125.6	..	Imports of Goods & NFServices
256.6	203.5	191.4	179.5	182.6	193.0	181.6	233.2	282.2	290.4	..	Domestic Absorption
151.9	114.0	111.5	115.0	128.9	144.3	134.8	184.9	..	221.0	..	Private Consumption, etc.
59.5	53.9	52.1	47.7	40.5	37.3	36.0	36.7	..	19.4	..	General Gov't Consumption
45.2	35.5	27.9	16.8	13.3	11.5	10.8	11.6	..	49.9	..	Gross Domestic Investment
..	..	..	..	..	..	..	..	..	..	..	Fixed Investment
											Memo Items:
58.5	51.8	43.1	39.6	35.4	37.7	37.7	39.2	44.2	44.8	..	Capacity to Import
7.6	4.0	0.0	0.7	-2.4	4.5	6.1	6.4	4.2	4.3	..	Terms of Trade Adjustment
215.6	167.4	153.6	155.8	161.1	177.1	168.3	222.1	208.4	209.5	..	Gross Domestic Income
213.2	168.2	153.7	155.9	161.4	175.0	166.4	219.7	206.0	206.2	..	Gross National Income
				(Index 1980 = 100)							**DOMESTIC PRICES (DEFLATORS)**
85.1	92.0	100.0	108.1	114.8	129.8	138.1	139.1	138.6	142.8	..	Overall (GDP)
80.2	88.9	100.0	108.4	118.0	127.8	135.0	136.6	..	132.0	..	Domestic Absorption
85.1	92.0	100.0	108.1	114.8	124.2	137.9	149.0	138.3	142.9	..	Agriculture
85.1	92.0	100.0	108.1	114.8	124.2	137.9	149.0	142.9	142.9	..	Industry
85.1	92.0	100.0	108.1	114.8	124.2	137.9	149.0	142.9	142.8	..	Manufacturing
											MANUFACTURING ACTIVITY
..	..	..	..	..	..	..	..	..	..	..	Employment (1980=100)
..	..	..	..	..	..	..	..	..	..	..	Real Earnings per Empl. (1980=100)
..	..	..	..	..	..	..	..	..	..	..	Real Output per Empl. (1980=100)
13.5	13.5	13.5	13.5	13.5	13.5	13.5	13.5	..	..	..	Earnings as % of Value Added
				(Billions of current CFA Francs)							**MONETARY HOLDINGS**
29.4	33.2	28.1	33.2	34.7	42.4	67.8	71.9	72.7	75.4	..	Money Supply, Broadly Defined
27.6	31.6	26.5	31.7	33.2	40.8	65.1	68.3	69.1	70.6	..	Money as Means of Payment
16.5	22.5	17.4	22.1	23.6	29.2	44.9	47.4	46.7	46.7	..	Currency Ouside Banks
11.1	9.0	9.0	9.5	9.6	11.6	20.2	21.0	22.5	23.9	..	Demand Deposits
1.9	1.7	1.7	1.5	1.5	1.6	2.6	3.6	3.6	4.8	..	Quasi-Monetary Liabilities
				(Millions of current CFA Francs)							
..	..	..	..	..	..	..	..	-3,757	..	..	**GOVERNMENT DEFICIT (-) OR SURPLUS**
..	..	..	..	..	..	..	..	21,535	..	..	Current Revenue
..	..	..	..	..	..	..	..	..	..	..	Current Expenditure
..	..	..	..	..	..	..	..	..	..	..	Current Budget Balance
..	..	..	..	..	..	..	..	..	..	..	Capital Receipts
..	..	..	..	..	..	..	..	..	..	..	Capital Payments

185

CHAD	1967	1968	1969	1970	1971	1972	1973	1974	1975	1976	1977
FOREIGN TRADE (CUSTOMS BASIS)					*(Millions of current US dollars)*						
Value of Exports, fob	26.9	27.6	31.1	24.7	26.4	34.2	36.0	35.1	40.0	63.0	107.0
Nonfuel Primary Products	..	..	..	..	..	..	..	..	..	..	..
Fuels	0.1	0.0	0.0	0.0	0.0	2.6	3.8	3.4	3.2	5.0	8.5
Manufactures	0.4	0.9	0.7	1.2	1.5	1.9	2.0	1.4	3.3	5.1	8.7
Value of Imports, cif	59.0	54.0	54.0	62.0	62.0	61.4	82.4	87.1	133.0	116.0	189.0
Nonfuel Primary Products	..	..	..	..	..	..	..	..	..	..	..
Fuels	10.1	10.2	9.3	9.9	10.3	8.9	13.5	13.8	20.3	..	..
Manufactures	37.4	34.7	31.4	37.4	37.9	35.8	43.7	50.8	89.5	..	..
					(Index 1980 = 100)						
Terms of Trade	..	..	..	..	..	..	..	..	..	..	..
Export Prices, fob	..	..	..	..	..	..	..	..	..	..	..
Nonfuel Primary Products	..	..	..	..	..	..	..	..	..	..	..
Fuels	..	..	..	..	..	..	..	..	..	..	..
Manufactures	..	..	..	..	..	..	..	..	..	..	..
Import Prices, cif	..	..	..	..	..	..	..	..	..	..	..
Trade at Constant 1980 Prices					*(Millions of 1980 US dollars)*						
Exports, fob	..	..	..	..	..	..	..	..	..	..	..
Imports, cif	..	..	..	..	..	..	..	..	..	..	..
BALANCE OF PAYMENTS					*(Millions of current US dollars)*						
Exports of Goods & Services	..	..	..	69.8	73.2	78.4	93.8	109.4	104.4	128.6	132.4
Merchandise, fob	..	..	..	39.8	42.3	39.9	48.8	70.5	57.9	100.8	106.6
Nonfactor Services	..	..	..	28.8	30.0	37.2	43.3	37.2	44.3	25.9	24.2
Factor Services	..	..	..	1.2	0.9	1.3	1.8	1.8	2.2	2.0	1.6
Imports of Goods & Services	..	..	..	96.5	100.1	107.1	145.8	163.3	230.0	215.2	249.6
Merchandise, fob	..	..	..	52.5	53.1	58.5	73.6	82.7	126.2	115.3	142.2
Nonfactor Services	..	..	..	41.7	44.5	45.8	68.6	77.0	100.9	97.6	103.4
Factor Services	..	..	..	2.3	2.5	2.7	3.7	3.6	2.9	2.3	3.9
Long-Term Interest	..	..	..	0.4	0.9	0.7	0.5	0.7	1.4	0.8	0.6
Current Transfers, net	..	..	..	-6.2	-6.9	-8.6	-9.2	-9.0	-8.1	-4.2	-14.0
Workers' Remittances	..	..	..	0.0	0.0	0.0	0.0	0.0	0.0	0.0	0.0
Total to be Financed	..	..	..	-32.9	-33.8	-37.2	-61.2	-62.9	-133.7	-90.8	-131.2
Official Capital Grants	..	..	..	34.5	37.1	37.4	54.4	58.7	73.3	85.6	102.7
Current Account Balance	..	..	..	1.6	3.3	0.1	-6.8	-4.2	-60.3	-5.2	-28.5
Long-Term Capital, net	..	..	..	38.1	38.0	24.2	58.5	71.9	100.2	122.0	135.7
Direct Investment	..	..	..	0.6	0.3	-0.1	6.1	13.8	20.3	26.8	21.1
Long-Term Loans	..	..	..	3.3	9.0	-1.2	10.6	23.5	18.0	6.5	37.0
Disbursements	..	..	..	5.8	14.8	2.4	13.4	26.1	23.1	10.2	39.8
Repayments	..	..	..	2.5	5.8	3.6	2.8	2.6	5.1	3.7	2.8
Other Long-Term Capital	..	..	..	-0.3	-8.4	-11.8	-12.5	-24.1	-11.4	3.2	-25.1
Other Capital, net	..	..	..	-43.0	-34.3	-26.8	-60.4	-57.3	-52.2	-103.1	-111.2
Change in Reserves	..	..	..	3.3	-7.0	2.5	8.7	-10.4	12.4	-13.7	4.0
Memo Item:					*(CFA Francs per US dollar)*						
Conversion Factor (Annual Avg)	246.850	246.850	259.710	277.710	277.130	252.210	222.700	240.500	214.320	238.980	245.670
EXTERNAL DEBT, ETC.				*(Millions of US dollars, outstanding at end of year)*							
Public/Publicly Guar. Long-Term	..	..	..	32.5	44.1	33.5	45.6	71.0	87.3	89.2	130.0
Official Creditors	..	..	..	24.6	26.0	17.0	29.2	46.2	63.5	66.0	99.2
IBRD and IDA	..	..	..	0.3	1.9	2.6	4.5	6.9	11.4	16.0	23.8
Private Creditors	..	..	..	7.8	18.1	16.5	16.4	24.9	23.8	23.3	30.9
Private Non-guaranteed Long-Term	..	..	..	..	..	..	..	..	..	..	..
Use of Fund Credit	..	..	..	2.5	2.5	2.4	1.2	6.2	3.5	13.4	14.0
Short-Term Debt	..	..	..	..	..	..	..	..	..	..	4.0
Memo Items:					*(Thousands of US dollars)*						
Int'l Reserves Excluding Gold	983.0	-2,246.0	-4,811.0	2,309.0	11,224.0	10,075.0	1,471.0	15,267.0	3,059.0	23,275.0	18,783.0
Gold Holdings (at market price)	..	..	..	..	..	..	..	..	..	..	924.0
SOCIAL INDICATORS											
Total Fertility Rate	6.1	6.0	6.0	6.0	6.0	6.0	6.0	5.9	5.9	5.9	5.9
Crude Birth Rate	45.2	45.1	45.0	44.8	44.7	44.6	44.5	44.4	44.3	44.2	44.1
Infant Mortality Rate	179.0	176.4	173.8	171.2	168.6	166.0	163.6	161.2	158.8	156.4	154.0
Life Expectancy at Birth	37.0	37.4	37.8	38.2	38.6	39.0	39.4	39.8	40.2	40.6	41.0
Food Production, p.c. ('79-81 = 100)	116.0	118.6	117.6	111.2	108.5	96.8	90.4	96.2	98.0	98.3	99.2
Labor Force, Agriculture (%)	91.4	91.0	90.6	90.2	89.5	88.8	88.1	87.4	86.7	86.0	85.3
Labor Force, Female (%)	23.1	23.1	23.1	23.0	23.0	22.9	22.8	22.7	22.7	22.6	22.6
School Enroll. Ratio, primary	..	..	..	35.0	..	..	..	..	35.0	35.0	..
School Enroll. Ratio, secondary	..	..	..	2.0	..	..	..	..	3.0	3.0	..

1978	1979	1980	1981	1982	1983	1984	1985	1986	1987 estimate	NOTES	CHAD
				(Millions of current US dollars)							**FOREIGN TRADE (CUSTOMS BASIS)**
99.0	72.2	72.4	83.4	57.7	73.9	111.0	..	..	..	..	Value of Exports, fob
..	..	..	..	..	..	..	..	..	..	..	Nonfuel Primary Products
7.9	..	..	..	..	..	..	..	..	..	..	Fuels
8.1	5.7	..	..	..	..	..	..	..	..	..	Manufactures
217.0	85.0	74.0	108.0	109.0	157.0	171.0	..	..	..	..	Value of Imports, cif
..	..	..	..	..	..	..	..	..	..	..	Nonfuel Primary Products
..	2.0	..	..	..	..	..	..	..	..	..	Fuels
..	58.7	..	..	..	..	..	..	..	..	..	Manufactures
				(Index 1980 = 100)							
..	..	..	..	..	..	..	..	..	..	..	Terms of Trade
..	..	..	..	..	..	..	..	..	..	..	Export Prices, fob
..	..	..	..	..	..	..	..	..	..	..	Nonfuel Primary Products
..	..	..	..	..	..	..	..	..	..	..	Fuels
..	..	..	..	..	..	..	..	..	..	..	Manufactures
..	..	..	..	..	..	..	..	..	..	..	Import Prices, cif
				(Millions of 1980 US dollars)							Trade at Constant 1980 Prices
..	..	..	..	..	..	..	..	..	..	..	Exports, fob
..	..	..	..	..	..	..	..	..	..	..	Imports, cif
				(Millions of current US dollars)							**BALANCE OF PAYMENTS**
119.5	92.8	71.4	87.5	62.0	106.8	147.9	99.5	146.2	168.4	..	Exports of Goods & Services
99.0	88.3	71.0	83.4	57.7	78.2	109.7	61.9	98.3	111.4	..	Merchandise, fob
20.6	4.6	0.4	4.1	2.3	24.2	36.8	32.6	44.4	54.6	..	Nonfactor Services
0.0	0.0	0.0	0.0	1.9	4.4	1.3	5.0	3.5	2.3	..	Factor Services
271.2	127.1	83.2	107.6	104.7	180.1	224.7	327.5	389.1	480.1	..	Imports of Goods & Services
163.4	64.1	55.3	81.2	81.7	99.2	128.3	166.3	211.5	260.9	..	Merchandise, fob
104.1	54.4	24.2	25.3	22.1	77.6	90.1	153.9	165.3	204.9	..	Nonfactor Services
3.7	8.6	3.7	1.1	1.0	3.3	6.3	7.3	12.3	14.3	..	Factor Services
0.7	0.2	0.1	0.1	0.2	0.2	0.3	1.8	1.7	3.2	..	Long-Term Interest
-11.4	-9.3	-4.1	-0.6	-0.7	-2.8	-1.7	6.8	-5.4	-11.9	..	Current Transfers, net
0.0	0.0	0.0	0.0	0.0	0.0	0.0	..	1.2	1.7	..	Workers' Remittances
-163.0	-43.5	-15.9	-20.7	-43.4	-76.1	-78.5	-221.1	-248.2	-323.6	K	Total to be Financed
116.6	41.9	28.2	44.1	61.9	114.1	87.6	133.9	189.0	240.4	K	Official Capital Grants
-46.4	-1.6	12.3	23.4	18.5	38.0	9.1	-87.3	-59.2	-83.3	..	Current Account Balance
157.9	38.0	24.1	42.2	60.8	96.9	91.9	201.4	238.2	325.0	..	Long-Term Capital, net
33.1	-1.3	-0.4	-0.1	-0.1	-0.1	9.2	53.4	27.7	4.0	..	Direct Investment
45.4	22.6	2.7	9.2	3.6	13.5	4.9	1.9	28.6	47.7	..	Long-Term Loans
47.8	24.4	5.2	9.7	4.6	13.9	9.0	8.8	31.5	51.0	..	Disbursements
2.4	1.8	2.5	0.5	1.0	0.4	4.1	6.9	2.9	3.3	..	Repayments
-37.2	-25.3	-6.4	-11.0	-4.6	-30.5	-9.7	12.2	-7.2	33.0	..	Other Long-Term Capital
-117.5	-38.7	-41.3	-67.1	-76.8	-117.6	-81.9	-136.3	-192.5	-220.2	..	Other Capital, net
6.1	2.3	4.9	1.5	-2.6	-17.3	-19.1	22.1	13.6	-21.6	..	Change in Reserves
				(CFA Francs per US dollar)							Memo Item:
225.640	212.720	211.300	271.730	328.620	381.070	436.960	449.260	346.300	300.540	..	Conversion Factor (Annual Avg)
			(Millions of US dollars, outstanding at end of year)								**EXTERNAL DEBT, ETC.**
186.0	213.5	200.7	182.2	152.7	155.7	149.1	154.6	195.6	269.6	..	Public/Publicly Guar. Long-Term
137.5	154.1	147.0	132.6	106.2	112.8	112.7	125.0	161.6	227.0	..	Official Creditors
30.8	35.8	36.1	36.1	36.1	36.4	39.9	39.9	48.0	61.4	..	IBRD and IDA
48.5	59.4	53.7	49.6	46.6	42.9	36.4	29.6	34.0	42.7	..	Private Creditors
..	..	..	..	..	..	..	..	..	..	..	Private Non-guaranteed Long-Term
11.6	10.1	6.9	8.3	7.8	7.4	4.4	8.7	8.6	9.9	..	Use of Fund Credit
14.0	13.0	10.6	10.7	7.4	5.3	2.7	20.0	31.0	38.3	..	Short-Term Debt
				(Thousands of US dollars)							Memo Items:
11,790.0	11,267.0	5,053.0	7,315.0	12,405.0	27,999.0	44,162.0	33,460.0	15,910.0	52,108.0	..	Int'l Reserves Excluding Gold
1,921.0	5,837.0	6,720.0	4,532.0	5,209.0	4,349.0	3,422.0	3,630.0	4,339.0	5,374.0	..	Gold Holdings (at market price)
											SOCIAL INDICATORS
5.9	5.9	5.9	5.9	5.9	5.9	5.9	5.9	5.7	5.9	..	Total Fertility Rate
44.1	44.1	44.2	44.2	44.2	44.3	44.3	44.4	44.4	44.5	..	Crude Birth Rate
151.8	149.6	147.4	145.2	143.0	144.8	146.5	148.3	139.0	151.8	..	Infant Mortality Rate
41.4	41.8	42.2	42.6	43.0	43.5	44.0	44.5	45.0	45.5	..	Life Expectancy at Birth
103.6	103.5	102.1	94.4	95.6	98.7	87.5	106.3	105.7	100.1	..	Food Production, p.c. ('79-81=100)
84.7	84.0	83.3	..	..	..	..	..	..	..	..	Labor Force, Agriculture (%)
22.5	22.4	22.4	22.3	22.1	22.0	21.9	21.7	21.6	21.5	..	Labor Force, Female (%)
..	..	..	..	..	..	38.0	..	..	..	..	School Enroll. Ratio, primary
..	..	..	..	..	..	6.0	..	..	..	..	School Enroll. Ratio, secondary

CHILE	1967	1968	1969	1970	1971	1972	1973	1974	1975	1976	1977
CURRENT GNP PER CAPITA (US $)	770	790	820	850	1,000	1,090	1,090	1,160	870	890	1,040
POPULATION (thousands)	8,853	9,025	9,197	9,368	9,545	9,722	9,899	10,026	10,196	10,372	10,551
ORIGIN AND USE OF RESOURCES					*(Billions of current Chilean Pesos)*						
Gross National Product (GNP)	0.033	0.046	0.066	0.096	0.130	0.230	1.100	9.100	34.100	124.500	281.900
Net Factor Income from Abroad	-0.001	-0.001	-0.002	-0.002	-0.002	-0.003	-0.012	-0.140	-1.400	-4.200	-5.900
Gross Domestic Product (GDP)	0.034	0.047	0.069	0.098	0.130	0.230	1.100	9.200	35.400	128.700	287.800
Indirect Taxes, net	0.004	0.005	0.007	0.010	0.012	0.020	0.120	1.400	5.000	17.700	41.300
GDP at factor cost	0.031	0.043	0.062	0.088	0.110	0.210	1.000	7.800	30.500	111.000	246.400
Agriculture	0.003	0.004	0.005	0.007	0.010	0.019	0.076	0.520	2.300	10.900	28.300
Industry	0.014	0.019	0.029	0.041	0.049	0.089	0.470	4.500	13.500	51.400	103.900
Manufacturing	0.008	0.012	0.017	0.025	0.031	0.055	0.310	2.700	7.200	29.900	62.600
Services, etc.	0.018	0.024	0.034	0.051	0.068	0.130	0.600	4.200	19.600	66.400	155.600
Resource Balance	0.000	0.000	0.002	0.001	-0.001	-0.008	-0.021	0.062	-0.700	5.600	-5.200
Exports of Goods & NFServices	0.005	0.007	0.012	0.015	0.014	0.024	0.160	1.900	9.000	32.300	59.300
Imports of Goods & NFServices	0.005	0.006	0.010	0.014	0.016	0.032	0.180	1.800	9.700	26.800	64.500
Domestic Absorption	0.034	0.047	0.067	0.098	0.130	0.240	1.200	9.100	36.100	123.100	293.000
Private Consumption, etc.	0.024	0.034	0.049	0.069	0.090	0.180	0.930	5.700	25.900	88.700	209.500
General Gov't Consumption	0.004	0.005	0.008	0.013	0.019	0.038	0.150	1.400	5.600	18.000	41.900
Gross Domestic Investment	0.006	0.008	0.010	0.016	0.018	0.029	0.091	1.900	4.600	16.400	41.500
Fixed Investment	0.005	0.007	0.010	0.015	0.018	0.031	0.150	1.600	6.300	17.100	38.300
Memo Items:											
Gross Domestic Saving	0.006	0.008	0.012	0.017	0.017	0.020	0.069	2.000	3.900	22.000	36.300
Gross National Saving	0.005	0.007	0.010	0.014	0.016	0.018	0.058	1.900	2.600	18.200	32.200
					(Billions of 1980 Chilean Pesos)						
Gross National Product	740.6	768.2	797.6	819.5	904.4	898.9	844.9	847.9	719.0	749.6	833.1
GDP at factor cost	685.1	713.8	741.5	749.9	825.4	827.0	764.4	726.7	642.3	667.4	727.4
Agriculture	65.1	68.4	60.4	62.3	61.6	56.5	50.8	64.4	67.3	66.2	73.2
Industry	317.0	329.9	342.4	348.9	383.5	375.2	348.5	368.2	287.9	299.2	316.8
Manufacturing	190.3	196.5	201.7	205.7	233.7	238.9	220.4	214.8	160.1	169.7	184.1
Services, etc.	381.4	392.4	418.9	427.3	469.4	472.3	454.5	428.1	391.9	408.4	459.7
Resource Balance	-56.5	-67.6	-86.3	-86.1	-100.5	-120.8	-107.7	-76.7	1.7	26.2	-0.3
Exports of Goods & NFServices	85.8	87.5	90.6	92.5	93.3	79.2	81.4	118.8	121.6	151.2	169.2
Imports of Goods & NFServices	142.2	155.1	176.9	178.6	193.8	199.9	189.1	195.5	119.9	125.0	169.5
Domestic Absorption	819.9	858.3	907.9	924.5	1,015.0	1,024.7	961.5	937.4	745.3	747.7	850.0
Private Consumption, etc.	598.0	620.7	653.5	651.2	734.6	783.8	731.7	608.5	532.6	534.9	617.1
General Gov't Consumption	85.7	89.5	95.7	101.4	114.0	120.5	122.6	134.3	120.5	120.4	125.0
Gross Domestic Investment	136.2	148.1	158.7	171.9	166.4	120.3	107.2	194.6	92.2	92.4	107.9
Fixed Investment	131.8	144.2	151.5	161.3	157.6	125.9	118.3	140.9	108.8	92.7	107.0
Memo Items:											
Capacity to Import	154.6	164.8	205.2	186.1	177.8	148.4	166.9	202.2	111.3	151.1	155.8
Terms of Trade Adjustment	68.8	77.3	114.6	93.6	84.5	69.2	85.5	83.4	-10.3	-0.1	-13.3
Gross Domestic Income	832.3	868.0	936.2	932.0	999.0	973.1	939.3	944.1	736.7	773.7	836.4
Gross National Income	809.4	845.5	912.2	913.1	988.9	968.1	930.5	931.3	708.6	749.5	819.7
DOMESTIC PRICES (DEFLATORS)					*(Index 1980 = 100)*						
Overall (GDP)	0.0	0.0	0.0	0.0	0.0	0.0	0.1	1.1	4.7	16.6	33.9
Domestic Absorption	0.0	0.0	0.0	0.0	0.0	0.0	0.1	1.0	4.8	16.5	34.5
Agriculture	0.0	0.0	0.0	0.0	0.0	0.0	0.2	0.8	3.5	16.4	38.6
Industry	0.0	0.0	0.0	0.0	0.0	0.0	0.1	1.2	4.7	17.2	32.8
Manufacturing	0.0	0.0	0.0	0.0	0.0	0.0	0.1	1.3	4.5	17.6	34.0
MANUFACTURING ACTIVITY											
Employment (1980=100)	126.2	116.9	114.8	117.9	119.5	125.5	128.0	122.7	114.2	106.3	109.5
Real Earnings per Empl. (1980=100)	..	..	..	..	..	29.2	47.8	51.8	51.3	60.7	65.2
Real Output per Empl. (1980=100)	49.6	55.4	62.0	60.2	71.5	61.6	49.9	61.3	87.4	74.3	71.7
Earnings as % of Value Added	25.1	21.1	18.3	18.6	22.8	29.6	16.3	12.1	12.3	14.6	17.5
MONETARY HOLDINGS					*(Billions of current Chilean Pesos)*						
Money Supply, Broadly Defined	0.01	0.01	0.01	0.02	0.03	0.08	0.44	2.00	7.50	20.00	45.90
Money as Means of Payment	..	..	..	..	..	0.05	0.22	0.84	3.00	8.80	18.30
Currency Ouside Banks	..	..	..	..	..	0.03	0.10	0.35	1.40	4.50	9.30
Demand Deposits	..	..	..	..	..	0.03	0.13	0.49	1.60	4.30	9.00
Quasi-Monetary Liabilities	..	..	..	..	..	0.03	0.22	1.20	4.50	11.20	27.60
					(Billions of current Chilean Pesos)						
GOVERNMENT DEFICIT (-) OR SURPLUS	..	..	..	..	..	-0.03	-0.08	-0.49	0.05	1.76	-3.19
Current Revenue	..	..	..	..	..	0.07	0.32	2.57	12.34	39.65	89.13
Current Expenditure	..	..	..	..	..	0.08	0.31	2.10	9.74	33.11	81.39
Current Budget Balance	..	..	..	..	..	-0.01	0.01	0.47	2.60	6.54	7.74
Capital Receipts	..	..	..	..	..	..	0.01	0.04	0.12	1.40	2.41
Capital Payments	..	..	..	..	..	0.02	0.10	1.00	2.67	6.18	13.34

1978	1979	1980	1981	1982	1983	1984	1985	1986	1987 estimate	NOTES	CHILE
1,340	1,690	2,110	2,610	2,230	1,930	1,700	1,450	1,310	1,310	..	CURRENT GNP PER CAPITA (US $)
10,733	10,917	11,104	11,294	11,487	11,682	11,878	12,074	12,249	12,423	..	POPULATION (thousands)
				(Billions of current Chilean Pesos)							ORIGIN AND USE OF RESOURCES
476.500	752.700	1,043.400	1,220.300	1,149.000	1,430.700	1,710.600	2,291.200	2,882.000	3,814.400	..	Gross National Product (GNP)
-11.000	-19.500	-31.900	-52.800	-90.100	-127.000	-182.800	-285.400	-364.000	-345.400	..	Net Factor Income from Abroad
487.500	772.200	1,075.300	1,273.100	1,239.100	1,557.700	1,893.400	2,576.600	3,246.100	4,159.800	f	Gross Domestic Product (GDP)
67.400	95.800	129.800	181.500	168.200	..	..	..	..	..	..	Indirect Taxes, net
420.100	676.400	945.500	1,091.700	1,071.000	..	..	..	..	..	B f	GDP at factor cost
37.100	56.100	77.700	80.700	69.400	88.800	..	..	..	..	..	Agriculture
175.500	288.000	401.100	466.400	438.500	604.200	..	..	..	..	..	Industry
109.200	164.000	230.500	284.200	233.500	320.800	..	..	..	..	..	Manufacturing
274.900	428.100	596.400	726.000	731.300	864.700	..	..	..	..	..	Services, etc.
-16.300	-21.900	-44.700	-131.600	-23.500	42.400	-20.100	71.100	124.000	170.600	..	Resource Balance
100.400	179.700	245.400	209.000	239.900	374.500	459.500	749.200	994.200	1,394.300	..	Exports of Goods & NFServices
116.700	201.600	290.100	340.600	263.400	332.100	479.600	678.100	870.200	1,223.700	..	Imports of Goods & NFServices
503.800	794.100	1,120.000	1,404.700	1,262.600	1,515.300	1,913.500	2,505.600	3,122.100	3,989.200	..	Domestic Absorption
346.600	546.300	760.500	948.200	932.700	1,141.900	1,381.700	1,785.200	2,237.300	2,811.000	..	Private Consumption, etc.
70.300	110.400	133.900	167.400	190.100	220.700	273.800	367.100	410.700	475.100	..	General Gov't Consumption
86.800	137.400	225.600	289.000	139.900	152.800	258.000	353.200	474.100	703.100	..	Gross Domestic Investment
71.600	115.000	178.900	236.800	181.500	186.500	233.800	366.400	472.700	666.800	..	Fixed Investment
											Memo Items:
70.500	115.500	180.900	157.400	116.400	195.200	237.900	424.300	598.100	873.700	..	Gross Domestic Saving
61.900	99.300	151.500	106.000	28.300	72.400	59.100	146.400	-741.200	540.600	..	Gross National Saving
				(Billions of 1980 Chilean Pesos)							
900.8	970.5	1,043.4	1,085.3	905.2	899.1	935.3	978.0	1,033.6	1,114.4	..	Gross National Product
793.3	873.5	945.5	973.3	841.3	..	..	..	..	..	B f	GDP at factor cost
70.5	74.8	77.7	80.6	79.7	77.7	83.5	88.2	96.0	99.1	..	Agriculture
339.8	370.8	401.1	426.4	362.8	365.4	392.8	407.0	428.7	449.7	..	Industry
201.2	217.1	230.5	236.4	186.9	192.6	211.5	213.9	231.0	243.7	..	Manufacturing
510.4	551.6	596.4	628.0	531.0	523.9	552.0	558.0	587.7	627.6	..	Services, etc.
-11.1	-29.7	-44.7	-112.2	20.2	54.0	40.1	80.6	88.8	79.6	..	Resource Balance
188.1	214.7	245.4	223.4	234.0	235.4	251.4	268.7	295.1	321.0	..	Exports of Goods & NFServices
199.2	244.3	290.1	335.7	213.7	181.4	211.3	188.1	206.3	241.4	..	Imports of Goods & NFServices
931.8	1,026.9	1,120.0	1,247.3	953.3	913.0	988.2	972.7	1,023.7	1,096.8	..	Domestic Absorption
665.2	709.2	760.5	842.3	729.9	707.1	721.0	714.9	750.4	788.3	..	Private Consumption, etc.
133.5	145.7	133.9	129.7	127.9	127.2	129.1	128.8	126.1	123.4	..	General Gov't Consumption
133.2	172.0	225.6	275.3	95.5	78.7	138.1	129.0	147.2	185.0	..	Gross Domestic Investment
125.6	146.8	178.9	208.9	138.0	117.5	128.0	147.0	157.4	182.8	..	Fixed Investment
											Memo Items:
171.4	217.8	245.4	206.0	194.7	204.5	202.5	207.8	235.7	275.0	..	Capacity to Import
-16.7	3.2	0.0	-17.4	-39.3	-30.9	-48.9	-60.9	-59.4	-46.0	..	Terms of Trade Adjustment
903.9	1,000.4	1,075.3	1,117.6	934.3	936.1	979.4	992.4	1,053.1	1,130.4	..	Gross Domestic Income
884.1	973.7	1,043.4	1,067.9	865.9	868.2	886.4	917.2	974.2	1,068.4	..	Gross National Income
				(Index 1980 = 100)							DOMESTIC PRICES (DEFLATORS)
53.0	77.4	100.0	112.2	127.3	161.1	184.1	244.6	291.8	353.6	..	Overall (GDP)
54.1	77.3	100.0	112.6	132.4	166.0	193.6	257.6	305.0	363.7	..	Domestic Absorption
52.6	75.0	100.0	100.1	87.0	114.3	..	..	..	..	..	Agriculture
51.6	77.7	100.0	109.4	120.8	165.3	..	..	..	..	..	Industry
54.3	75.5	100.0	120.2	124.9	166.5	..	..	..	..	..	Manufacturing
											MANUFACTURING ACTIVITY
107.4	105.7	100.0	93.1	73.5	73.2	81.6	85.7	..	..	..	Employment (1980=100)
77.2	89.8	100.0	120.6	124.4	112.4	105.3	110.6	..	..	..	Real Earnings per Empl. (1980=100)
78.0	93.0	100.0	95.1	112.3	123.1	..	..	..	..	..	Real Output per Empl. (1980=100)
17.9	18.2	18.4	23.0	20.3	17.1	15.4	18.0	..	..	..	Earnings as % of Value Added
				(Billions of current Chilean Pesos)							MONETARY HOLDINGS
87.70	146.90	231.00	311.20	392.40	411.70	557.20	..	..	..	..	Money Supply, Broadly Defined
30.60	50.30	78.90	74.10	81.10	102.70	116.20	..	..	..	..	Money as Means of Payment
16.40	24.90	35.60	44.70	43.00	51.90	64.20	..	..	..	..	Currency Ouside Banks
14.20	25.40	43.20	29.40	38.20	50.80	52.00	..	..	..	..	Demand Deposits
57.10	96.60	152.10	237.10	311.30	309.00	441.00	..	..	..	..	Quasi-Monetary Liabilities
				(Billions of current Chilean Pesos)							
-0.52	37.22	58.21	32.99	-12.19	-40.89	-56.16	-60.73	-31.41	..	..	GOVERNMENT DEFICIT (-) OR SURPLUS
154.21	252.43	358.73	406.35	368.30	429.04	544.11	745.90	914.33	..	..	Current Revenue
135.60	195.09	272.26	341.41	395.00	462.99	563.03	725.81	861.52	..	..	Current Expenditure
18.61	57.34	86.47	64.94	-26.70	-33.95	-18.92	20.09	52.81	..	..	Current Budget Balance
3.19	11.05	8.37	12.96	6.09	3.11	3.30	6.18	6.63	..	..	Capital Receipts
22.32	31.17	36.63	44.91	-8.42	10.05	40.54	87.00	90.85	..	..	Capital Payments

CHILE	1967	1968	1969	1970	1971	1972	1973	1974	1975	1976	1977
FOREIGN TRADE (CUSTOMS BASIS)					*(Millions of current US dollars)*						
Value of Exports, fob	907.7	935.9	1,067.9	1,233.6	961.2	855.4	1,249.4	2,480.6	1,648.7	2,208.5	2,138.4
Nonfuel Primary Products	873.2	906.0	1,026.0	1,178.4	909.8	806.9	1,200.9	2,351.3	1,469.1	1,943.3	1,842.8
Fuels	1.2	0.9	0.7	0.4	0.9	2.8	2.6	20.0	14.3	34.6	58.6
Manufactures	33.3	28.9	41.2	54.8	50.5	45.7	46.0	109.3	165.4	230.5	237.0
Value of Imports, cif	738.0	833.0	891.0	980.0	1,049.0	1,168.0	1,562.0	2,137.0	1,733.0	1,672.0	2,418.0
Nonfuel Primary Products	188.6	206.8	213.2	210.0	271.1	384.3	518.4	847.8	405.7	544.7	430.8
Fuels	62.3	52.1	64.8	60.8	95.5	106.2	114.2	305.3	342.9	196.7	485.2
Manufactures	487.1	574.1	613.0	709.2	682.4	677.5	929.4	983.9	984.4	930.6	1,502.0
					(Index 1980 = 100)						
Terms of Trade	227.3	220.0	257.5	237.1	185.8	174.1	211.1	155.0	107.1	106.9	100.1
Export Prices, fob	46.7	49.2	56.8	57.0	44.7	46.1	79.2	91.9	60.8	65.6	62.2
Nonfuel Primary Products	47.2	49.7	57.2	56.9	44.8	47.4	81.1	93.1	61.5	66.9	63.3
Fuels	4.3	4.3	4.3	4.3	5.6	6.2	8.9	36.7	35.7	38.4	42.0
Manufactures	49.1	51.5	59.1	63.7	48.5	42.7	68.4	90.1	58.5	62.0	60.8
Import Prices, cif	20.5	22.4	22.1	24.0	24.1	26.5	37.5	59.3	56.8	61.4	62.1
Trade at Constant 1980 Prices					*(Millions of 1980 US dollars)*						
Exports, fob	1,944.6	1,902.4	1,880.1	2,165.7	2,149.9	1,854.1	1,576.7	2,700.6	2,711.3	3,364.8	3,439.3
Imports, cif	3,593.8	3,724.8	4,039.8	4,078.6	4,358.9	4,406.1	4,161.4	3,606.0	3,053.1	2,724.2	3,892.3
BALANCE OF PAYMENTS					*(Millions of current US dollars)*						
Exports of Goods & Services	..	..	..	1,274	1,142	985	1,468	2,351	1,842	2,425	2,621
Merchandise, fob	..	..	..	1,113	1,000	851	1,316	2,152	1,590	2,116	2,186
Nonfactor Services	..	..	..	134	130	133	147	176	248	297	417
Factor Services	..	..	..	27	12	1	5	24	4	12	18
Imports of Goods & Services	..	..	..	1,371	1,347	1,464	1,761	2,658	2,344	2,325	3,268
Merchandise, fob	..	..	..	867	927	1,012	1,329	1,901	1,520	1,473	2,151
Nonfactor Services	..	..	..	281	290	303	317	462	529	507	720
Factor Services	..	..	..	223	130	149	114	295	295	345	397
Long-Term Interest	..	..	..	104	109	50	65	114	195	254	265
Current Transfers, net	..	..	..	2	3	5	5	6	4	32	80
Workers' Remittances	..	..	..	0	0	0	0	0	0	0	0
Total to be Financed	..	..	..	-95	-202	-473	-289	-301	-498	132	-567
Official Capital Grants	..	..	..	4	4	2	10	8	8	16	16
Current Account Balance	..	..	..	-91	-198	-471	-279	-292	-490	148	-551
Long-Term Capital, net	..	..	..	144	-110	110	-54	25	180	62	65
Direct Investment	..	..	..	-79	-66	-1	-5	-557	50	-1	16
Long-Term Loans	..	..	..	448	-57	165	186	510	45	14	196
Disbursements	..	..	..	655	223	292	358	761	465	654	995
Repayments	..	..	..	207	280	127	172	251	420	640	800
Other Long-Term Capital	..	..	..	-229	9	-57	-245	64	77	33	-163
Other Capital, net	..	..	..	33	49	225	379	146	85	68	617
Change in Reserves	..	..	..	-86	259	137	-46	121	225	-278	-131
Memo Item:					*(Chilean Pesos per US dollar)*						
Conversion Factor (Annual Avg)	0.005	0.007	0.009	0.012	0.012	0.019	0.110	0.830	4.910	13.050	21.530
EXTERNAL DEBT, ETC.					*(Millions of US dollars, outstanding at end of year)*						
Public/Publicly Guar. Long-Term	..	..	..	2,067	2,181	2,590	2,814	3,792	3,733	3,609	3,675
Official Creditors	..	..	..	1,181	1,204	1,428	1,550	2,009	2,130	2,045	1,950
IBRD and IDA	..	..	..	130	141	145	151	145	151	157	163
Private Creditors	..	..	..	886	977	1,163	1,264	1,783	1,603	1,564	1,725
Private Non-guaranteed Long-Term	..	..	..	501	394	374	365	534	641	772	980
Use of Fund Credit	..	..	..	2	43	86	95	196	387	467	365
Short-Term Debt	..	..	..	..	..	..	..	..	..	..	864
Memo Items:					*(Millions of US dollars)*						
Int'l Reserves Excluding Gold	81.3	162.0	296.0	341.8	170.1	96.8	121.6	41.1	55.9	405.1	426.5
Gold Holdings (at market price)	45.3	55.4	47.7	49.9	58.7	87.9	154.5	268.2	181.9	180.0	225.0
SOCIAL INDICATORS											
Total Fertility Rate	4.4	4.3	4.1	4.0	3.8	3.6	3.5	3.3	3.2	3.0	2.9
Crude Birth Rate	29.3	27.6	26.0	26.8	27.3	26.3	27.1	26.2	24.5	23.2	21.6
Infant Mortality Rate	99.7	91.6	87.4	82.2	73.9	76.5	65.8	65.2	57.5	56.6	50.1
Life Expectancy at Birth	60.6	61.2	61.8	62.4	63.0	63.6	64.3	65.0	65.7	66.4	67.2
Food Production, p.c. ('79-81 = 100)	102.3	106.1	97.1	101.5	99.8	92.4	81.6	92.7	95.8	92.1	99.4
Labor Force, Agriculture (%)	25.2	24.6	23.9	23.2	22.5	21.8	21.1	20.4	19.7	19.1	18.4
Labor Force, Female (%)	22.2	22.2	22.3	22.4	22.8	23.3	23.8	24.3	24.8	25.3	25.8
School Enroll. Ratio, primary	..	..	..	107.0	..	..	..	..	118.0	116.0	117.0
School Enroll. Ratio, secondary	..	..	..	39.0	..	..	..	..	48.0	49.0	50.0

1978	1979	1980	1981	1982	1983	1984	1985	1986	1987 estimate	NOTES	CHILE
											FOREIGN TRADE (CUSTOMS BASIS)
				(Millions of current US dollars)							
2,462.1	3,894.0	4,705.0	3,836.0	3,710.0	3,836.0	3,650.0	3,822.9	4,222.4	5,090.9	..	Value of Exports, fob
2,154.4	3,510.4	4,197.1	3,474.6	3,339.6	3,450.4	3,240.5	3,437.7	3,831.2	4,598.2	..	Nonfuel Primary Products
41.8	54.9	59.8	66.3	61.2	63.0	59.5	61.2	34.2	44.7	..	Fuels
265.9	328.7	448.1	295.1	309.2	322.6	350.0	324.0	357.0	448.0	..	Manufactures
3,276.0	4,777.0	6,185.0	6,331.0	4,117.0	3,186.0	3,770.0	3,300.0	3,471.0	4,621.9	..	Value of Imports, cif
714.6	802.5	1,144.3	962.4	659.6	513.1	616.2	494.1	515.2	736.7	..	Nonfuel Primary Products
550.1	1,014.4	1,140.7	925.0	660.4	509.6	679.7	582.3	308.7	458.4	..	Fuels
2,011.2	2,960.1	3,900.1	4,443.6	2,797.1	2,163.3	2,474.1	2,223.6	2,647.1	3,426.8	..	Manufactures
				(Index 1980 = 100)							
94.0	109.1	100.0	85.8	78.7	84.7	79.5	79.4	77.8	83.9	..	Terms of Trade
65.1	89.8	100.0	86.9	76.1	80.2	74.2	71.8	72.4	83.8	..	Export Prices, fob
66.3	90.8	100.0	86.9	76.3	80.8	74.9	72.2	72.3	84.0	..	Nonfuel Primary Products
42.3	61.0	100.0	112.5	101.6	92.5	90.2	87.5	44.9	56.9	..	Fuels
61.4	'86.0	100.0	83.0	71.0	73.0	66.0	66.0	78.0	85.6	..	Manufactures
69.3	82.3	100.0	101.3	96.8	94.7	93.3	90.4	93.0	99.9	..	Import Prices, cif
				(Millions of 1980 US dollars)							Trade at Constant 1980 Prices
3,779.2	4,337.7	4,705.2	4,414.4	4,874.2	4,781.5	4,920.4	5,324.5	5,830.0	6,074.9	..	Exports, fob
4,727.4	5,803.7	6,184.9	6,250.3	4,255.1	3,363.2	4,041.9	3,649.6	3,732.2	4,627.4	..	Imports, cif
				(Millions of current US dollars)							**BALANCE OF PAYMENTS**
2,984	4,746	6,276	5,614	5,154	4,831	4,814	4,669	5,349	6,488	..	Exports of Goods & Services
2,460	3,835	4,705	3,836	3,706	3,831	3,650	3,804	4,199	5,224	..	Merchandise, fob
481	785	1,263	1,172	936	797	843	664	922	1,082	..	Nonfactor Services
43	126	308	606	512	203	321	201	228	182	..	Factor Services
4,169	6,040	8,360	10,455	7,567	6,045	6,973	6,058	6,570	7,415	..	Imports of Goods & Services
2,886	4,190	5,469	6,513	3,643	2,845	3,357	2,954	3,099	3,994	..	Merchandise, fob
735	1,027	1,554	1,741	1,377	1,204	1,238	967	1,318	1,495	..	Nonfactor Services
548	823	1,337	2,201	2,547	1,996	2,378	2,137	2,153	1,926	..	Factor Services
384	583	918	1,420	1,942	1,381	1,998	1,636	1,407	1,420	..	Long-Term Interest
75	88	64	36	41	54	41	47	40	56	..	Current Transfers, net
0	0	0	0	0	0	0	0	0	0	..	Workers' Remittances
-1,110	-1,206	-2,020	-4,805	-2,372	-1,160	-2,118	-1,342	-1,181	-871	K	Total to be Financed
22	17	49	72	68	43	58	14	44	60	K	Official Capital Grants
-1,088	-1,189	-1,971	-4,733	-2,304	-1,117	-2,060	-1,328	-1,137	-811	..	Current Account Balance
1,532	1,702	2,291	3,651	1,727	40	3,595	914	848	893	..	Long-Term Capital, net
177	233	170	362	384	132	67	62	57	97	..	Direct Investment
1,137	1,466	2,089	3,399	1,428	1,052	1,149	1,026	733	483	..	Long-Term Loans
2,233	2,785	3,551	5,198	2,678	1,962	1,643	1,460	1,182	777	..	Disbursements
1,096	1,318	1,462	1,799	1,250	910	493	434	449	294	..	Repayments
197	-14	-17	-182	-153	-1,187	2,321	-188	14	253	..	Other Long-Term Capital
286	534	925	1,149	-581	536	-1,516	316	62	-24	..	Other Capital, net
-730	-1,047	-1,245	-67	1,158	541	-19	98	227	-58	..	Change in Reserves
				(Chilean Pesos per US dollar)							Memo Item:
31.660	37.250	39.000	39.000	50.910	78.840	98.660	161.080	193.020	219.540	..	Conversion Factor (Annual Avg)
				(Millions of US dollars, outstanding at end of year)							**EXTERNAL DEBT, ETC.**
4,357	4,811	4,705	4,487	5,243	6,765	10,723	12,904	14,604	15,536	..	Public/Publicly Guar. Long-Term
1,836	1,578	1,369	1,268	1,180	1,448	1,608	2,124	3,045	3,957	..	Official Creditors
173	179	184	201	220	232	232	505	979	1,494	..	IBRD and IDA
2,521	3,233	3,336	3,219	4,063	5,317	9,115	10,780	11,558	11,579	..	Private Creditors
1,569	2,736	4,693	8,138	8,726	8,125	6,427	4,731	2,821	2,466	..	Private Non-guaranteed Long-Term
347	179	123	49	6	606	779	1,088	1,331	1,465	..	Use of Fund Credit
1,101	1,635	2,560	2,989	3,338	2,599	1,914	1,668	1,480	1,772	..	Short-Term Debt
				(Millions of US dollars)							Memo Items:
1,090.1	1,938.3	3,123.2	3,213.3	1,815.0	2,036.3	2,302.9	2,449.9	2,351.3	2,503.8	..	Int'l Reserves Excluding Gold
314.1	780.3	1,004.5	676.5	782.2	583.3	471.4	500.0	597.7	740.2	..	Gold Holdings (at market price)
											SOCIAL INDICATORS
2.9	2.9	2.8	2.8	2.8	2.7	2.7	2.6	2.5	2.4	..	Total Fertility Rate
21.4	21.5	22.2	23.4	23.0	22.5	22.1	21.7	21.3	20.9	..	Crude Birth Rate
40.0	37.8	33.0	26.9	26.4	25.9	25.4	24.9	24.4	23.9	..	Infant Mortality Rate
67.9	68.7	69.5	70.2	71.0	71.1	71.2	71.3	71.5	71.6	..	Life Expectancy at Birth
92.1	98.1	97.7	104.2	100.7	94.8	98.2	100.3	104.8	105.8	..	Food Production, p.c. ('79-81 = 100)
17.8	17.1	16.5	..	..	..	..	..	..	..	..	Labor Force, Agriculture (%)
26.3	26.8	27.3	27.4	27.5	27.7	27.8	27.9	28.1	28.2	..	Labor Force, Female (%)
118.0	119.0	117.0	115.0	108.0	108.0	107.0	109.0	..	..	..	School Enroll. Ratio, primary
52.0	55.0	55.0	57.0	58.0	63.0	66.0	69.0	..	..	..	School Enroll. Ratio, secondary

CHINA /1	1967	1968	1969	1970	1971	1972	1973	1974	1975	1976	1977
CURRENT GNP PER CAPITA (US $)	100	90	100	120	120	130	150	160	180	170	180
POPULATION (millions)	755	775	796	818	841	862	882	900	916	931	944
ORIGIN AND USE OF RESOURCES					*(Billions of current Chinese Yuan)*						
Gross National Product (GNP)	173.00	164.60	188.10	224.00	241.50	248.40	269.50	272.90	290.90	282.00	307.20
Net Factor Income from Abroad	0.00	0.00	0.00	0.00	0.00	0.00	0.00	0.00	0.00	0.00	0.00
Gross Domestic Product (GDP)	173.00	164.60	188.10	224.00	241.50	248.40	269.50	272.90	290.90	282.00	307.20
Indirect Taxes, net	..	..	..	..	..	..	..	..	..	..	..
GDP at factor cost					..						..
Agriculture	69.20	70.20	71.00	78.30	81.20	81.40	89.20	92.70	95.30	96.20	94.10
Industry	61.30	54.50	70.50	91.90	103.80	108.70	117.50	117.80	133.30	127.70	143.90
Manufacturing	46.50	39.60	52.80	70.90	80.10	83.50	89.60	87.10	98.40	90.80	104.00
Services, etc.	42.50	39.90	46.60	53.80	56.50	58.30	62.80	62.40	62.40	58.20	69.20
Resource Balance	0.60	0.70	1.40	0.10	1.80	2.00	1.30	-1.60	-0.50	0.60	0.70
Exports of Goods & NFServices	6.40	6.30	6.50	6.20	7.50	9.00	12.60	15.00	15.40	14.60	15.10
Imports of Goods & NFServices	5.80	5.50	5.10	6.10	5.70	6.90	11.30	16.60	16.00	14.00	14.40
Domestic Absorption	172.40	163.90	186.80	224.00	239.80	246.40	268.10	274.60	291.40	281.50	306.50
Private Consumption, etc.	111.40	105.20	117.20	123.90	129.40	137.60	146.90	152.20	156.40	155.40	167.60
General Gov't Consumption	26.50	25.70	31.10	34.70	38.60	39.80	41.10	41.40	44.70	43.90	47.70
Gross Domestic Investment	34.50	33.00	38.50	65.40	71.70	69.00	80.10	81.00	90.40	82.20	91.20
Fixed Investment	..	..	..	..	..	..	..	..	..	..	..
Memo Items:											
Gross Domestic Saving	35.10	33.70	39.80	65.50	73.50	71.00	81.40	79.30	89.90	82.80	91.90
Gross National Saving	35.10	33.70	39.80	..	..	..	..	..	..	..	92.80
					(Billions of 1980 Chinese Yuan)						
Gross National Product	182.92	170.95	203.81	251.44	268.98	276.90	300.09	303.26	328.34	310.61	335.02
GDP at factor cost	..	..	..	..	..	..	..	..	..	..	..
Agriculture	95.45	94.02	94.10	103.61	105.92	104.47	113.58	118.13	120.59	121.75	119.14
Industry	54.36	49.37	66.22	90.32	102.57	108.03	117.36	118.54	134.89	129.48	147.27
Manufacturing	41.20	35.90	49.60	69.63	79.18	82.97	89.55	87.60	99.56	92.06	106.47
Services, etc.	33.11	27.56	43.49	57.51	60.49	64.40	69.15	66.59	72.86	59.38	68.61
Resource Balance	3.98	3.91	4.72	2.19	6.22	6.70	5.78	1.87	5.06	4.15	2.64
Exports of Goods & NFServices	15.14	15.18	15.04	13.76	16.42	19.34	21.71	20.65	24.14	20.15	19.75
Imports of Goods & NFServices	11.16	11.27	10.32	11.57	10.20	12.64	15.93	18.78	19.08	16.00	17.11
Domestic Absorption	178.94	167.04	199.09	249.25	262.76	270.20	294.31	301.39	323.28	306.46	332.38
Private Consumption, etc.	117.55	107.32	127.25	145.15	147.43	155.89	167.09	172.75	181.00	173.33	186.37
General Gov't Consumption	30.82	29.85	35.73	39.86	44.48	45.72	47.19	47.20	50.79	49.74	52.62
Gross Domestic Investment	30.57	29.87	36.11	64.24	70.85	68.59	80.03	81.44	91.49	83.39	93.39
Fixed Investment	..	..	..	..	..	..	..	..	..	..	..
Memo Items:											
Capacity to Import	12.34	12.73	13.09	11.70	13.45	16.30	17.81	16.95	18.45	16.64	17.90
Terms of Trade Adjustment	-2.80	-2.45	-1.95	-2.06	-2.97	-3.04	-3.90	-3.70	-5.69	-3.51	-1.85
Gross Domestic Income	180.12	168.50	201.86	249.38	266.01	273.86	296.19	299.56	322.65	307.10	333.17
Gross National Income	180.12	168.50	201.86	249.38	266.01	273.86	296.19	299.56	322.65	307.10	333.17
DOMESTIC PRICES (DEFLATORS)					*(Index 1980 = 100)*						
Overall (GDP)	94.6	96.3	92.3	89.1	89.8	89.7	89.8	90.0	88.6	90.8	91.7
Domestic Absorption	96.4	98.1	93.8	89.9	91.2	91.2	91.1	91.1	90.2	91.8	92.2
Agriculture	72.5	74.7	75.5	75.6	76.7	77.9	78.5	78.5	79.0	79.0	79.0
Industry	112.8	110.4	106.5	101.8	101.2	100.6	100.1	99.4	98.8	98.6	97.7
Manufacturing	112.8	110.4	106.5	101.8	101.2	100.6	100.1	99.4	98.8	98.6	97.7
MANUFACTURING ACTIVITY											
Employment (1980=100)	..	..	..	..	..	..	..	..	..	..	94.1
Real Earnings per Empl. (1980=100)	..	..	..	..	..	..	..	..	..	..	80.7
Real Output per Empl. (1980=100)	..	..	..	..	..	..	..	..	..	..	81.4
Earnings as % of Value Added	..	..	..	..	..	..	..	..	..	..	
MONETARY HOLDINGS					*(Billions of current Chinese Yuan)*						
Money Supply, Broadly Defined	..	..	..	..	..	..	..	..	..	..	85.8
Money as Means of Payment	..	..	..	..	..	..	..	..	..	..	58.0
Currency Ouside Banks	..	..	..	..	..	..	..	..	..	..	19.5
Demand Deposits	..	..	..	..	..	..	..	..	..	..	38.5
Quasi-Monetary Liabilities	..	..	..	..	..	..	..	..	..	..	27.8
GOVERNMENT DEFICIT (-) OR SURPLUS					*(Millions of current Chinese Yuan)*						
Current Revenue	..	..	..	..	..	..	..	..	..	..	..
Current Expenditure	..	..	..	..	..	..	..	..	..	..	..
Current Budget Balance	..	..	..	..	..	..	..	..	..	..	..
Capital Receipts	..	..	..	..	..	..	..	..	..	..	..
Capital Payments	..	..	..	..	..	..	..	..	..	..	..

1978	1979	1980	1981	1982	1983	1984	1985	1986	1987 estimate	NOTES	CHINA /1
210	250	290	310	310	310	320	320	300	290	..	**CURRENT GNP PER CAPITA (US $)**
956	969	981	994	1,008	1,020	1,030	1,040	1,054	1,069	..	**POPULATION (millions)**
				(Billions of current Chinese Yuan)							**ORIGIN AND USE OF RESOURCES**
350.20	389.10	429.30	457.50	504.70	565.20	679.90	832.20	937.20	1,085.90	..	Gross National Product (GNP)
0.00	-0.10	-0.20	-1.20	0.90	2.50	3.80	1.60	-0.80	-6.10	..	Net Factor Income from Abroad
350.20	389.20	429.50	458.70	503.80	562.70	676.10	830.60	938.00	1,092.00	f	Gross Domestic Product (GDP)
..	..	..	..	..	..	..	..	..	..		Indirect Taxes, net
..	..	..	..	..	..	..	..	..	..	B f	GDP at factor cost
101.90	126.40	137.60	156.40	177.80	197.00	230.50	259.60	287.90	335.40	..	Agriculture
167.60	182.60	206.00	209.00	225.10	250.30	295.10	376.60	433.80	534.90	..	Industry
124.20	136.40	151.40	151.50	163.60	180.50	216.60	278.70	315.80	..	..	Manufacturing
80.70	80.20	85.90	93.30	100.90	115.40	150.50	194.50	216.30	221.80	..	Services, etc.
-2.10	-4.30	-5.50	1.60	9.30	5.90	1.10	-36.90	-26.20	0.90	..	Resource Balance
17.50	23.30	30.50	41.60	45.00	46.80	62.10	82.50	102.10	145.60	..	Exports of Goods & NFServices
19.60	27.60	36.00	40.00	35.80	40.90	60.90	119.40	128.20	144.70	..	Imports of Goods & NFServices
352.30	393.50	435.00	457.10	494.50	556.80	675.00	867.50	964.20	1,091.10	..	Domestic Absorption
178.30	200.20	236.30	258.70	274.10	307.40	360.40	429.90	472.70	536.20	..	Private Consumption, etc.
54.20	66.70	69.90	72.90	73.80	82.90	97.00	115.70	127.30	144.30	..	General Gov't Consumption
119.80	126.60	128.80	125.50	146.70	166.50	217.60	321.90	364.20	410.60	..	Gross Domestic Investment
..	..	103.00	101.10	120.00	136.90	183.30	254.30	302.00	340.40		Fixed Investment
											Memo Items:
117.70	122.30	123.30	127.10	155.90	172.40	218.70	285.00	338.00	411.50	..	Gross Domestic Saving
118.70	123.20	124.10	126.70	157.80	175.70	223.20	287.10	338.10	405.70	..	Gross National Saving
				(Billions of 1980 Chinese Yuan)							
377.00	402.90	429.30	448.80	491.11	542.69	622.12	698.52	751.56	818.32	..	Gross National Product
..	..	..	..	..	..	..	..	..	..	B f	GDP at factor cost
125.60	140.00	137.60	146.40	162.65	175.50	197.09	203.79	210.72	223.15	..	Agriculture
171.20	183.50	206.00	210.70	227.13	252.57	293.74	357.48	398.24	468.73	..	Industry
126.86	137.09	151.41	152.75	165.13	182.11	215.61	264.54	289.92	..	..	Manufacturing
80.20	79.40	85.90	93.50	100.47	112.19	127.77	135.89	143.28	131.07	..	Services, etc.
-3.90	-5.10	-5.50	2.00	4.28	-0.98	-5.97	-25.01	-7.53	8.24	..	Resource Balance
20.40	25.90	30.50	35.60	36.05	39.54	44.27	59.03	65.38	75.39	..	Exports of Goods & NFServices
24.30	31.00	36.00	33.60	31.77	40.52	50.24	84.03	72.91	67.15	..	Imports of Goods & NFServices
380.90	408.00	435.00	448.60	485.98	541.24	624.57	722.17	759.76	814.71	..	Domestic Absorption
199.09	209.12	236.30	250.96	265.32	291.89	318.57	329.95	341.03	346.97	..	Private Consumption, etc.
59.41	71.68	69.90	71.14	71.42	78.58	85.76	88.82	91.80	93.40	..	General Gov't Consumption
122.40	127.20	128.80	126.50	149.24	170.77	220.24	303.39	326.93	374.34	..	Gross Domestic Investment
..	..	..	..	..	..	..	..	..	..	..	Fixed Investment
											Memo Items:
21.70	26.17	30.50	34.94	39.99	46.35	51.16	58.07	58.02	67.56	..	Capacity to Import
1.30	0.27	0.00	-0.66	3.94	6.81	6.89	-0.95	-7.36	-7.83	..	Terms of Trade Adjustment
378.30	403.17	429.50	449.94	494.19	547.07	625.48	696.20	744.87	815.12	..	Gross Domestic Income
378.30	403.17	429.30	448.14	495.05	549.50	629.01	697.56	744.20	810.50	..	Gross National Income
				(Index 1980 = 100)							**DOMESTIC PRICES (DEFLATORS)**
92.9	96.6	100.0	101.8	102.8	104.2	109.3	119.1	124.7	132.7	..	Overall (GDP)
92.5	96.4	100.0	101.9	101.8	102.9	108.1	120.1	126.9	133.9	..	Domestic Absorption
81.1	90.3	100.0	106.8	109.3	112.3	117.0	127.4	136.6	150.3	..	Agriculture
97.9	99.5	100.0	99.2	99.1	99.1	100.5	105.3	108.9	114.1	..	Industry
97.9	99.5	100.0	99.2	99.1	99.1	100.5	105.3	108.9	..	..	Manufacturing
											MANUFACTURING ACTIVITY
94.9	96.5	100.0	104.5	108.9	..	..	..	..	..	..	Employment (1980=100)
86.6	94.1	100.0	97.4	96.8	..	..	..	..	..	..	Real Earnings per Empl. (1980=100)
91.3	96.1	100.0	98.6	102.5	..	..	..	..	..	..	Real Output per Empl. (1980=100)
..	..	..	..	..	..	..	..	..	..	..	Earnings as % of Value Added
				(Billions of current Chinese Yuan)							**MONETARY HOLDINGS**
89.0	132.8	167.1	197.8	226.6	271.3	359.8	487.5	634.9	795.7	..	Money Supply, Broadly Defined
58.0	92.1	114.9	134.5	148.8	174.9	251.5	301.7	385.9	457.4	..	Money as Means of Payment
21.2	26.8	34.6	39.6	43.9	53.0	79.2	98.8	121.8	145.5	..	Currency Ouside Banks
36.8	65.4	80.3	94.9	104.9	121.9	172.3	202.9	264.1	311.9	..	Demand Deposits
30.9	40.6	52.2	63.3	77.7	96.4	108.4	185.8	249.0	338.3	..	Quasi-Monetary Liabilities
				(Millions of current Chinese Yuan)							
..	..	..	..	..	..	..	..	..	..	..	**GOVERNMENT DEFICIT (-) OR SURPLUS**
..	..	..	..	..	..	..	..	..	..	..	Current Revenue
..	..	..	..	..	..	..	..	..	..	..	Current Expenditure
..	..	..	..	..	..	..	..	..	..	..	Current Budget Balance
..	..	..	..	..	..	..	..	..	..	..	Capital Receipts
..	..	..	..	..	..	..	..	..	..	..	Capital Payments

1/ Not including Taiwan, China (see Footnote on next page)

CHINA	1967	1968	1969	1970	1971	1972	1973	1974	1975	1976	1977
FOREIGN TRADE (CUSTOMS BASIS)					*(Millions of current US dollars)*						
Value of Exports, fob	2,140	2,100	2,200	2,260	2,640	3,440	5,820	6,950	7,260	6,860	7,590
Nonfuel Primary Products	1,193	1,176	1,177	1,220	1,380	1,768	2,878	3,181	3,170	2,724	2,843
Fuels	48	22	18	21	26	26	78	617	1,033	826	1,068
Manufactures	899	902	1,004	1,019	1,233	1,647	2,864	3,152	3,056	3,310	3,680
Value of Imports, cif	2,020	1,950	1,830	2,330	2,210	2,860	5,160	7,620	7,486	6,578	7,214
Nonfuel Primary Products	921	901	912	939	814	1,288	2,497	3,331	2,337	1,965	2,931
Fuels	1	2	4	11	6	12	5	48	50	14	59
Manufactures	1,099	1,047	915	1,381	1,390	1,560	2,658	4,241	5,099	4,599	4,224
					(Index 1980 = 100)						
Terms of Trade	88.3	94.5	100.3	97.8	99.2	108.2	105.0	100.9	94.2	96.6	101.5
Export Prices, fob	28.7	30.4	33.4	33.3	35.1	41.5	57.8	66.4	60.2	63.9	69.3
Nonfuel Primary Products	35.6	34.6	36.9	37.1	38.0	47.1	69.9	67.4	62.2	67.5	73.3
Fuels	4.3	4.3	4.3	4.3	5.6	6.2	8.9	36.7	35.7	38.4	42.0
Manufactures	30.1	30.1	33.8	33.8	36.1	39.8	56.4	77.4	75.2	72.9	81.2
Import Prices, cif	32.5	32.2	33.3	34.0	35.4	38.3	55.0	65.8	64.0	66.2	68.3
Trade at Constant 1980 Prices					*(Millions of 1980 US dollars)*						
Exports, fob	7,456	6,912	6,585	6,794	7,524	8,298	10,075	10,466	12,056	10,728	10,955
Imports, cif	6,212	6,063	5,495	6,854	6,246	7,461	9,382	11,574	11,706	9,934	10,570
BALANCE OF PAYMENTS					*(Millions of current US dollars)*						
Exports of Goods & Services 1/	..	..	..	..	..	..	..	..	..	..	8,770
Merchandise, fob	..	..	..	2,309	2,803	3,652	5,677	7,108	7,689	6,943	8,050
Nonfactor Services	..	..	..	128	81	132	103	128	139	440	500
Factor Services	..	..	..	..	..	..	..	..	..	..	220
Imports of Goods & Services	..	..	..	..	..	..	..	..	..	..	8,286
Merchandise, fob	..	..	..	2,280	2,144	2,819	5,031	7,791	7,926	6,660	7,627
Nonfactor Services	..	..	..	238	862	1,099	220	168	171	465	521
Factor Services	..	..	..	..	..	..	..	..	..	..	138
Long-Term Interest	..	..	..	..	..	..	..	..	..	..	..
Current Transfers, net	..	..	..	..	..	..	..	..	..	..	497
Workers' Remittances	..	..	..	..	..	..	..	..	..	..	..
Total to be Financed	..	..	..	-81	-122	-134	529	-723	-269	258	981
Official Capital Grants	..	..	..	..	..	..	..	..	..	..	-70
Current Account Balance /2	..	..	..	-81	-122	-134	529	-723	-269	258	911
Long-Term Capital, net	..	..	..	..	..	..	..	..	..	..	-1,060
Direct Investment	..	..	..	..	..	..	..	..	..	..	..
Long-Term Loans	..	..	..	..	..	..	..	..	..	..	..
Disbursements	..	..	..	..	..	..	..	..	..	..	..
Repayments	..	..	..	..	..	..	..	..	..	..	..
Other Long-Term Capital	..	..	..	..	..	..	..	..	..	..	-990
Other Capital, net	..	..	..	81	122	134	-529	723	269	-258	149
Change in Reserves	..	..	..	0	0	0	0	0	0	0	0
Memo Item:					*(Chinese Yuan per US dollar)*						
Conversion Factor (Annual Avg)	2	2	2	2	2	2	2	2	2	2	2
EXTERNAL DEBT, ETC.					*(Millions of US dollars, outstanding at end of year)*						
Public/Publicly Guar. Long-Term	..	..	..	..	..	..	..	..	..	..	..
Official Creditors	..	..	..	..	..	..	..	..	..	..	..
IBRD and IDA	..	..	..	..	..	..	..	..	..	..	..
Private Creditors	..	..	..	..	..	..	..	..	..	..	..
Private Non-guaranteed Long-Term	..	..	..	..	..	..	..	..	..	..	..
Use of Fund Credit	..	..	..	0	0	0	0	0	0	0	0
Short-Term Debt	..	..	..	..	..	..	..	..	..	..	..
Memo Items:					*(Millions of US dollars)*						
Int'l Reserves Excluding Gold /3	..	..	..	..	..	..	..	..	..	..	2,345
Gold Holdings (at market price)	..	..	..	..	..	..	..	..	..	..	2,111
SOCIAL INDICATORS											
Total Fertility Rate	5.7	6.4	5.8	5.8	5.1	4.9	4.4	3.8	3.4	2.9	2.7
Crude Birth Rate	34.0	35.6	34.1	33.4	30.6	29.8	27.9	24.8	23.0	19.9	18.9
Infant Mortality Rate	81.0	77.0	73.0	69.0	66.0	61.0	56.0	51.0	46.0	41.0	39.0
Life Expectancy at Birth	55.3	56.6	57.9	59.2	60.5	61.8	63.0	64.3	64.8	65.2	65.6
Food Production, p.c. ('79-81 = 100)	86.4	82.3	80.5	86.5	88.5	85.7	91.2	89.6	90.3	89.1	86.9
Labor Force, Agriculture (%)	79.8	79.3	78.8	78.3	77.9	77.5	77.1	76.7	76.3	75.9	75.5
Labor Force, Female (%)	41.3	41.5	41.6	41.7	41.8	42.0	42.1	42.3	42.4	42.6	42.7
School Enroll. Ratio, primary	..	..	..	89.0	..	..	..	..	126.0	..	106.0
School Enroll. Ratio, secondary	..	..	..	24.0	..	..	..	..	47.0	..	83.0
					(Millions of current US dollars)						
1/ *Taiwan, China: Exports of Goods & Services*	..	..	..	1,733	2,384	3,388	5,149	6,496	6,274	8,967	10,869
2/ *Taiwan, China: Current Account Balance*	..	..	..	1	171	513	566	-1,113	-589	292	911
3/ *Taiwan, China: Int'l Reserves Excl. Gold*	335	302	361	540	617	952	1,026	1,092	1,074	1,516	1,345

1978	1979	1980	1981	1982	1983	1984	1985	1986	1987 estimate	NOTES	CHINA
											FOREIGN TRADE (CUSTOMS BASIS)
colspan over *(Millions of current US dollars)*											
9,115	12,525	18,265	21,554	21,820	22,175	25,922	27,326	31,148	39,542	..	Value of Exports, fob
2,999	3,391	5,088	5,127	4,864	4,950	5,749	6,762	7,753	6,700	..	Nonfuel Primary Products
1,409	2,654	4,388	4,677	5,314	4,656	6,027	7,144	3,670	5,220	..	Fuels
4,708	6,480	8,789	11,750	11,642	12,569	14,146	13,420	19,725	27,622	..	Manufactures
10,893	15,675	19,550	19,482	18,546	21,346	27,350	42,526	43,172	43,392	..	Value of Imports, cif
3,673	4,991	6,910	6,863	7,275	5,689	6,123	7,040	4,928	6,018	..	Nonfuel Primary Products
66	21	177	108	180	111	139	172	492	689	..	Fuels
7,155	10,662	12,463	12,512	11,091	15,546	21,088	35,314	37,752	36,685	..	Manufactures

(Index 1980 = 100)

1978	1979	1980	1981	1982	1983	1984	1985	1986	1987	NOTES	CHINA
93.3	93.3	100.0	103.4	104.4	96.5	98.4	94.7	83.5	86.6	..	Terms of Trade
71.9	84.2	100.0	101.5	95.2	90.3	90.3	85.8	86.2	98.1	..	Export Prices, fob
82.0	99.5	100.0	94.4	84.1	86.6	89.5	76.7	81.7	88.5	..	Nonfuel Primary Products
42.3	61.0	100.0	112.5	101.6	92.5	90.2	87.5	44.9	56.9	..	Fuels
82.7	91.0	100.0	100.8	97.7	91.0	90.7	90.3	106.8	117.3	..	Manufactures
77.0	90.2	100.0	98.1	91.2	93.6	91.8	90.6	103.3	113.3	..	Import Prices, cif

| | | | | | | | | | | | Trade at Constant 1980 Prices |

(Millions of 1980 US dollars)

1978	1979	1980	1981	1982	1983	1984	1985	1986	1987	NOTES	CHINA
12,680	14,884	18,265	21,245	22,926	24,561	28,703	31,840	36,125	40,297	..	Exports, fob
14,144	17,385	19,550	19,851	20,337	22,813	29,804	46,943	41,785	38,286	..	Imports, cif

(Millions of current US dollars)

1978	1979	1980	1981	1982	1983	1984	1985	1986	1987	NOTES	CHINA
											BALANCE OF PAYMENTS
10,606	15,351	20,324	25,157	24,878	24,982	28,724	29,641	30,683	40,147	..	Exports of Goods & Services 1/
9,607	13,658	18,492	22,027	21,125	20,707	23,905	25,108	25,756	34,734	..	Merchandise, fob
763	1,325	1,320	2,433	2,661	2,726	2,811	3,055	3,827	4,386	..	Nonfactor Services
236	368	512	697	1,092	1,549	2,008	1,478	1,100	1,027	..	Factor Services
11,912	17,582	24,058	24,380	19,541	21,006	26,657	41,301	38,096	40,071	..	Imports of Goods & Services
10,745	15,619	22,049	21,047	16,876	18,717	23,891	38,231	34,896	36,395	..	Merchandise, fob
923	1,518	1,386	2,459	2,024	1,994	2,378	2,524	2,276	2,486	..	Nonfactor Services
244	445	623	874	641	295	388	546	924	1,191	..	Factor Services
..	..	317	519	543	525	611	588	643	1,069	..	Long-Term Interest
597	656	640	464	530	436	305	171	255	95	..	Current Transfers, net
..	..	0	..	541	446	317	180	208	166	..	Workers' Remittances
-709	-1,575	-3,094	1,241	5,867	4,412	2,372	-11,489	-7,158	171	K	Total to be Financed
-69	-30	-70	108	-44	75	137	72	124	129	K	Official Capital Grants
-778	-1,605	-3,164	1,349	5,823	4,487	2,509	-11,417	-7,034	300	..	Current Account Balance /2
-899	792	271	739	365	1,183	1,745	4,511	7,381	4,868	..	Long-Term Capital, net
..	..	57	265	386	543	1,124	1,031	1,425	1,668	..	Direct Investment
..	..	1,927	961	535	986	1,070	4,006	4,315	3,930	..	Long-Term Loans
..	..	2,540	2,165	1,837	2,375	2,357	5,302	6,125	5,704	..	Disbursements
..	..	613	1,204	1,302	1,389	1,287	1,297	1,810	1,774	..	Repayments
-830	822	-1,643	-595	-512	-421	-586	-598	1,517	-860	..	Other Long-Term Capital
889	1,410	3,137	-378	137	-887	-2,395	2,232	-2,392	-315	..	Other Capital, net
788	-597	-244	-1,710	-6,325	-4,783	-1,859	4,674	2,045	-4,852	..	Change in Reserves

(Chinese Yuan per US dollar)

1978	1979	1980	1981	1982	1983	1984	1985	1986	1987	NOTES	CHINA
											Memo Item:
2	2	2	2	2	2	2	3	3	4	..	Conversion Factor (Annual Avg)

(Millions of US dollars, outstanding at end of year)

1978	1979	1980	1981	1982	1983	1984	1985	1986	1987	NOTES	CHINA
											EXTERNAL DEBT, ETC.
..	..	4,503	5,274	5,562	5,625	6,482	10,303	16,439	23,659	..	Public/Publicly Guar. Long-Term
..	..	446	1,278	1,790	2,346	2,924	4,806	7,057	9,527	..	Official Creditors
..	..	0	0	1	71	254	930	1,739	2,757	..	IBRD and IDA
..	..	4,057	3,996	3,773	3,279	3,558	5,497	9,382	14,132	..	Private Creditors
..	..	..	..	..	..	..	..	..	..	..	Private Non-guaranteed Long-Term
0	0	0	524	496	0	0	0	731	848	..	Use of Fund Credit
..	..	..	..	2,300	3,984	5,600	6,419	4,769	5,720	..	Short-Term Debt

(Millions of US dollars)

1978	1979	1980	1981	1982	1983	1984	1985	1986	1987	NOTES	CHINA
											Memo Items:
1,557	2,154	2,545	5,058	11,349	14,987	17,366	12,728	11,453	16,305	..	Int'l Reserves Excluding Gold /3
2,893	6,554	7,546	5,048	5,803	4,845	3,915	4,153	4,964	6,148	..	Gold Holdings (at market price)
											SOCIAL INDICATORS
2.6	2.3	2.5	3.0	2.3	2.3	2.2	2.3	2.3	2.4	..	Total Fertility Rate
18.3	17.8	18.2	20.9	21.1	18.6	17.5	17.8	20.8	21.1	..	Crude Birth Rate
38.0	43.0	41.0	41.0	40.0	39.0	..	..	..	33.0	..	Infant Mortality Rate
66.1	66.5	66.9	67.4	67.8	68.1	68.5	68.8	69.1	69.5	..	Life Expectancy at Birth
94.9	100.2	98.8	101.0	108.6	115.6	122.4	120.7	125.2	127.2	..	Food Production, p.c. ('79-81 = 100)
75.0	74.6	74.2	..	..	..	..	..	61.0	..	..	Labor Force, Agriculture (%)
42.9	43.0	43.2	43.2	43.2	43.2	43.2	43.2	43.2	43.2	..	Labor Force, Female (%)
124.0	122.0	105.0	118.0	112.0	113.0	118.0	124.0	..	..	..	School Enroll. Ratio, primary
63.0	56.0	47.0	35.0	36.0	35.0	37.0	39.0	..	..	..	School Enroll. Ratio, secondary

(Millions of current US dollars)

1978	1979	1980	1981	1982	1983	1984	1985	1986	1987	NOTES	CHINA
14,426	18,414	22,627	26,080	25,691	28,832	34,735	35,421	46,236	61,403	..	1/ *Taiwan, China: Exports of Goods & Services*
1,639	181	-913	519	2,248	4,412	6,976	9,195	16,217	17,925	..	2/ *Taiwan, China: Current Account Balance*
1,406	1,392	2,205	7,235	8,532	11,859	15,664	22,556	46,310	..	..	3/ *Taiwan, China: Int'l Reserves Excl. Gold*

COLOMBIA	1967	1968	1969	1970	1971	1972	1973	1974	1975	1976	1977
CURRENT GNP PER CAPITA (US $)	310	310	320	340	360	390	430	510	560	620	710
POPULATION (thousands)	19,655	20,216	20,753	21,266	21,750	22,199	22,628	23,058	23,502	23,958	24,425
ORIGIN AND USE OF RESOURCES	*(Billions of current Colombian Pesos)*										
Gross National Product (GNP)	83.1	96.2	110.3	129.4	152.4	185.3	238.0	317.2	396.9	521.3	708.3
Net Factor Income from Abroad	-1.5	-2.0	-2.7	-3.3	-3.5	-4.3	-5.1	-5.2	-8.2	-10.9	-7.7
Gross Domestic Product (GDP)	84.6	98.2	113.0	132.8	155.9	189.6	243.2	322.4	405.1	532.3	716.0
Indirect Taxes, net	5.7	7.2	8.6	9.9	11.3	13.1	16.6	22.3	31.7	49.7	75.0
GDP at factor cost	78.9	91.0	104.3	122.9	144.6	176.5	226.6	300.1	373.4	482.6	641.0
Agriculture	22.8	26.3	29.6	33.5	36.9	46.0	59.0	79.0	97.3	126.1	179.6
Industry	19.1	22.1	25.7	30.6	36.3	45.9	63.2	82.1	97.2	122.4	161.5
Manufacturing	13.5	15.1	17.4	21.3	25.1	32.2	44.4	60.5	73.2	93.4	116.5
Services, etc.	37.0	42.6	49.1	58.8	71.5	84.6	104.4	139.1	178.9	234.2	300.0
Resource Balance	0.4	-1.3	-1.3	-2.5	-6.3	-0.9	2.5	-3.5	7.3	16.8	26.3
Exports of Goods & NFServices	10.0	12.5	14.7	18.6	19.7	26.9	37.1	46.9	64.1	90.7	120.8
Imports of Goods & NFServices	9.5	13.8	15.9	21.0	26.0	27.7	34.6	50.4	56.8	74.0	94.5
Domestic Absorption	84.2	99.5	114.2	135.3	162.2	190.5	240.7	325.9	397.8	515.5	689.8
Private Consumption, etc.	62.7	72.2	83.3	96.1	114.8	138.0	173.3	228.5	292.8	378.3	500.3
General Gov't Consumption	7.1	8.1	9.7	12.3	17.1	18.1	23.0	28.2	36.2	43.7	55.2
Gross Domestic Investment	14.4	19.1	21.3	26.9	30.3	34.4	44.4	69.2	68.8	93.5	134.3
Fixed Investment	13.3	17.0	19.2	23.9	27.3	30.5	38.4	52.8	62.1	84.6	104.0
Memo Items:											
Gross Domestic Saving	14.8	17.9	20.0	24.4	24.0	33.5	46.9	65.7	76.2	110.3	160.5
Gross National Saving	13.3	15.8	17.2	21.0	20.4	29.4	42.0	60.9	68.9	100.7	154.3
	(Billions of 1980 Colombian Pesos)										
Gross National Product	747.1	793.4	843.5	908.8	966.4	1,041.8	1,116.9	1,182.8	1,204.8	1,261.3	1,317.2
GDP at factor cost	698.2	741.2	789.0	842.4	896.3	974.1	1,054.2	1,103.3	1,124.9	1,173.0	1,216.7
Agriculture	177.4	189.6	196.1	200.8	202.3	218.3	223.2	235.4	249.4	256.1	264.2
Industry	200.3	211.2	230.2	246.8	262.6	293.1	331.3	336.9	334.2	345.7	346.2
Manufacturing	126.4	134.2	143.9	155.9	170.7	193.6	217.0	226.9	227.1	234.3	234.3
Services, etc.	320.5	340.4	362.7	394.7	431.4	462.7	499.8	530.9	541.3	571.2	606.3
Resource Balance	38.7	26.7	26.4	4.9	-9.2	18.7	20.8	14.4	56.0	32.8	9.2
Exports of Goods & NFServices	128.2	138.9	145.4	146.9	154.3	174.4	179.5	169.8	194.3	188.1	179.7
Imports of Goods & NFServices	89.4	112.1	119.0	142.0	163.5	155.7	158.7	155.4	138.3	155.3	170.5
Domestic Absorption	719.9	780.3	833.2	916.9	988.0	1,036.4	1,109.6	1,179.1	1,162.4	1,243.5	1,318.9
Private Consumption, etc.	527.9	560.2	605.9	652.8	695.1	753.0	799.3	830.8	855.8	914.2	948.8
General Gov't Consumption	62.9	65.1	70.0	80.1	101.1	96.3	106.1	103.7	106.1	110.1	115.3
Gross Domestic Investment	129.0	155.0	157.4	184.0	191.8	187.1	204.1	244.6	200.5	219.2	254.8
Fixed Investment	117.4	135.0	138.3	160.1	167.9	164.6	178.9	194.4	187.0	204.8	206.2
Memo Items:											
Capacity to Import	93.5	101.9	109.6	125.2	123.9	150.8	170.0	144.5	156.1	190.6	217.8
Terms of Trade Adjustment	-34.7	-37.0	-35.9	-21.7	-30.3	-23.7	-9.5	-25.3	-38.2	2.4	38.2
Gross Domestic Income	723.9	770.0	823.7	900.0	948.4	1,031.5	1,120.8	1,168.3	1,180.2	1,278.7	1,366.2
Gross National Income	712.4	756.5	807.7	887.1	936.0	1,018.1	1,107.4	1,157.5	1,166.6	1,263.7	1,355.3
DOMESTIC PRICES (DEFLATORS)	*(Index 1980 = 100)*										
Overall (GDP)	11.1	12.2	13.1	14.4	15.9	18.0	21.5	27.0	33.2	41.7	53.9
Domestic Absorption	11.7	12.7	13.7	14.8	16.4	18.4	21.7	27.6	34.2	41.5	52.3
Agriculture	12.8	13.9	15.1	16.7	18.2	21.1	26.4	33.5	39.0	49.2	68.0
Industry	9.5	10.5	11.1	12.4	13.8	15.6	19.1	24.4	29.1	35.4	46.7
Manufacturing	10.7	11.3	12.1	13.6	14.7	16.6	20.5	26.7	32.2	39.9	49.7
MANUFACTURING ACTIVITY											
Employment (1980=100)	55.7	57.3	62.8	66.6	68.6	74.4	82.0	86.9	88.6	91.1	94.5
Real Earnings per Empl. (1980=100)	95.2	96.4	93.9	102.9	107.5	105.2	96.1	93.0	91.4	94.1	91.9
Real Output per Empl. (1980=100)	81.0	87.1	84.2	83.9	90.8	91.0	89.1	94.3	91.2	97.3	96.4
Earnings as % of Value Added	25.2	23.4	23.2	24.9	24.0	23.3	23.0	19.9	20.6	19.8	19.6
MONETARY HOLDINGS	*(Billions of current Colombian Pesos)*										
Money Supply, Broadly Defined	17.9	20.7	25.6	28.9	32.0	40.2	53.1	65.6	97.4	135.1	181.8
Money as Means of Payment	13.7	15.9	19.4	22.4	25.1	31.9	41.6	49.1	58.9	79.4	103.5
Currency Ouside Banks	4.7	5.5	6.5	7.8	8.5	10.7	12.4	15.9	20.8	28.8	40.5
Demand Deposits	9.0	10.4	12.9	14.6	16.5	21.1	29.2	33.2	38.1	50.6	63.0
Quasi-Monetary Liabilities	4.2	4.9	6.2	6.5	7.0	8.4	11.4	16.5	38.5	55.8	78.3
	(Billions of current Colombian Pesos)										
GOVERNMENT DEFICIT (-) OR SURPLUS	..	..	..	..	-3	-5	-6	-4	-1	5	5
Current Revenue	..	..	..	..	17	20	25	34	51	64	84
Current Expenditure	..	..	..	..	11	13	15	22	34	38	52
Current Budget Balance	..	..	..	..	6	7	10	12	17	25	32
Capital Receipts	..	..	..	..	0	0	0	0	0	0	0
Capital Payments	..	..	..	..	8	12	16	16	18	20	28

1978	1979	1980	1981	1982	1983	1984	1985	1986	1987 estimate	NOTES	COLOMBIA
870	1,050	1,220	1,370	1,440	1,440	1,430	1,350	1,280	1,230	..	**CURRENT GNP PER CAPITA (US $)**
24,903	25,392	25,892	26,399	26,911	27,422	27,927	28,418	28,961	29,498	..	**POPULATION (thousands)**
				(Billions of current Colombian Pesos)							**ORIGIN AND USE OF RESOURCES**
901.6	1,181.6	1,573.4	1,972.3	2,459.8	2,991.0	3,757.5	4,824.1	6,548.9	8,217.4	..	Gross National Product (GNP)
-7.9	-7.2	-5.7	-10.5	-37.5	-63.1	-99.1	-141.8	-152.5	-562.0	..	Net Factor Income from Abroad
909.5	1,188.8	1,579.1	1,982.8	2,497.3	3,054.1	3,856.6	4,965.9	6,701.4	8,779.4	..	Gross Domestic Product (GDP)
99.0	122.9	158.4	167.4	214.5	253.7	360.7	516.8	797.3	1,030.9	..	Indirect Taxes, net
810.5	1,065.9	1,420.7	1,815.4	2,282.8	2,800.5	3,495.9	4,449.1	5,904.1	7,748.5	..	GDP at factor cost
210.0	255.9	305.3	381.3	468.8	571.4	673.9	846.8	1,164.9	1,503.7	..	Agriculture
197.6	262.1	395.6	506.5	644.2	808.7	1,065.2	1,429.1	1,975.1	2,710.1	..	Industry
138.7	180.4	267.9	327.8	406.7	495.7	647.4	809.1	1,130.1	1,499.5	..	Manufacturing
402.9	547.9	719.8	927.6	1,169.8	1,420.4	1,756.8	2,173.3	2,764.1	3,534.7	..	Services, etc.
25.7	21.1	9.8	-70.7	-106.8	-84.9	-22.3	63.7	451.6	588.7	..	Resource Balance
151.2	180.9	256.1	235.0	272.5	319.4	458.3	685.7	1,259.7	1,674.7	..	Exports of Goods & NFServices
125.5	159.8	246.3	305.7	379.4	404.4	480.7	622.0	808.1	1,086.0	..	Imports of Goods & NFServices
883.8	1,167.8	1,569.3	2,053.5	2,604.1	3,139.1	3,878.9	4,902.2	6,249.8	8,190.8	..	Domestic Absorption
639.7	841.2	1,108.8	1,437.7	1,819.7	2,196.9	2,721.9	3,425.4	4,379.4	5,672.2	..	Private Consumption, etc.
77.8	110.8	159.4	206.9	272.8	334.6	425.6	531.3	667.4	846.4	..	General Gov't Consumption
166.3	215.8	301.1	408.9	511.6	607.6	731.4	945.5	1,203.0	1,672.2	..	Gross Domestic Investment
139.9	183.3	264.9	350.0	436.1	524.8	654.5	870.5	1,183.7	1,647.8	..	Fixed Investment
											Memo Items:
192.0	236.8	310.9	338.2	404.8	522.6	709.1	1,009.2	1,654.7	2,260.9	..	Gross Domestic Saving
185.8	233.8	312.9	340.9	378.0	470.9	639.1	932.2	1,656.2	1,941.7	..	Gross National Saving
				(Billions of 1980 Colombian Pesos)							
1,425.7	1,503.8	1,573.4	1,603.3	1,611.0	1,633.6	1,681.7	1,734.6	1,815.5	1,912.9	..	Gross National Product
1,288.7	1,341.8	1,420.7	1,460.1	1,468.7	1,496.3	1,541.4	1,589.1	1,671.4	1,756.8	..	GDP at factor cost
285.8	299.9	305.3	315.2	309.6	318.1	325.0	330.3	340.9	360.8	..	Agriculture
346.8	350.1	395.6	398.8	395.0	410.8	438.0	469.2	521.2	553.1	..	Industry
236.8	237.7	267.9	264.6	257.8	262.2	276.6	284.4	311.6	330.1	..	Manufacturing
656.1	691.8	719.8	746.1	764.0	767.4	778.3	789.5	809.2	842.9	..	Services, etc.
19.1	36.3	9.8	-32.6	-56.7	-33.3	-0.6	50.4	84.8	111.4	..	Resource Balance
224.9	243.7	256.1	225.8	222.3	220.3	243.0	277.9	318.8	357.0	..	Exports of Goods & NFServices
205.8	207.4	246.3	258.4	278.9	253.6	243.6	227.5	233.9	245.6	..	Imports of Goods & NFServices
1,417.5	1,475.6	1,569.3	1,646.6	1,686.5	1,689.4	1,714.3	1,719.5	1,779.8	1,853.0	..	Domestic Absorption
1,019.2	1,062.6	1,108.8	1,140.3	1,154.4	1,166.1	1,204.5	1,233.5	1,275.1	1,323.8	..	Private Consumption, etc.
125.9	141.5	159.4	165.3	173.0	171.9	179.0	187.1	190.5	198.1	..	General Gov't Consumption
272.4	271.6	301.1	341.0	359.2	351.4	330.7	298.8	314.2	331.1	..	Gross Domestic Investment
225.5	234.1	264.9	281.5	289.8	293.3	296.9	281.4	304.1	320.8	..	Fixed Investment
											Memo Items:
248.0	234.7	256.1	198.6	200.4	200.4	232.3	250.8	364.7	378.8	..	Capacity to Import
23.1	-8.9	0.0	-27.2	-21.9	-19.9	-10.7	-27.1	45.9	21.7	..	Terms of Trade Adjustment
1,459.6	1,502.9	1,579.1	1,586.8	1,608.0	1,636.2	1,703.0	1,742.7	1,910.5	1,986.2	..	Gross Domestic Income
1,448.8	1,494.9	1,573.4	1,576.1	1,589.1	1,613.7	1,671.0	1,707.4	1,861.4	1,934.6	..	Gross National Income
				(Index 1980 = 100)							**DOMESTIC PRICES (DEFLATORS)**
63.3	78.6	100.0	122.8	153.2	184.4	225.0	280.6	359.4	446.9	..	Overall (GDP)
62.3	79.1	100.0	124.7	154.4	185.8	226.3	285.1	351.2	442.0	..	Domestic Absorption
73.5	85.3	100.0	121.0	151.4	179.6	207.3	256.3	341.7	416.8	..	Agriculture
57.0	74.9	100.0	127.0	163.1	196.9	243.2	304.6	378.9	490.0	..	Industry
58.6	75.9	100.0	123.9	157.8	189.0	234.0	282.5	362.7	454.3	..	Manufacturing
											MANUFACTURING ACTIVITY
96.9	100.1	100.0	97.1	94.6	91.5	89.6	88.1	..	..	..	Employment (1980=100)
109.3	105.3	100.0	101.3	103.5	108.5	117.3	122.3	..	..	..	Real Earnings per Empl. (1980=100)
100.1	101.5	100.0	100.9	96.9	102.0	109.2	118.9	..	..	..	Real Output per Empl. (1980=100)
22.4	19.7	18.4	19.0	20.5	20.9	19.9	20.5	..	..	..	Earnings as % of Value Added
				(Billions of current Colombian Pesos)							**MONETARY HOLDINGS**
240.0	301.5	447.2	615.4	773.9	1,010.3	1,243.7	1,495.3	..	..	D	Money Supply, Broadly Defined
132.9	165.9	212.0	256.0	321.0	397.0	492.0	545.0	..	..	..	Money as Means of Payment
53.7	67.3	84.1	101.6	130.3	167.7	211.7	186.7	..	..	..	Currency Ouside Banks
79.2	98.6	127.9	154.4	190.7	229.3	280.3	358.3	..	..	..	Demand Deposits
107.1	135.6	235.2	359.4	452.9	613.3	751.7	950.3	..	..	..	Quasi-Monetary Liabilities
				(Billions of current Colombian Pesos)							
6	-12	-29	-59	-41	-33	..	..	..	..	F	**GOVERNMENT DEFICIT (-) OR SURPLUS**
111	138	189	229	335	407	..	..	..	..	..	Current Revenue
74	124	165	218	287	371	439	..	..	..	..	Current Expenditure
37	14	25	11	48	36	..	..	..	..	..	Current Budget Balance
0	0	0	0	0	0	0	..	..	..	..	Capital Receipts
31	26	54	70	89	79	-439	..	..	..	..	Capital Payments

COLOMBIA	1967	1968	1969	1970	1971	1972	1973	1974	1975	1976	1977
FOREIGN TRADE (CUSTOMS BASIS)				*(Millions of current US dollars)*							
Value of Exports, fob	509.9	558.3	602.9	727.7	689.1	863.4	1,175.5	1,416.7	1,464.9	1,744.6	2,443.0
Nonfuel Primary Products	395.9	453.5	465.6	595.8	527.4	625.5	804.6	901.3	1,053.8	1,289.8	1,881.6
Fuels	74.7	50.8	77.0	73.2	70.2	65.3	61.9	117.2	105.6	71.0	95.8
Manufactures	39.3	54.0	60.3	58.7	91.5	172.6	309.0	398.2	305.5	383.8	465.6
Value of Imports, cif	496.9	643.3	685.3	842.9	929.4	858.9	1,061.5	1,597.2	1,494.8	1,708.1	2,028.3
Nonfuel Primary Products	74.4	99.1	112.8	131.8	166.7	147.0	224.4	367.4	257.8	342.1	398.1
Fuels	7.1	5.2	8.1	8.8	11.0	5.5	4.5	4.4	15.3	42.7	137.7
Manufactures	415.4	539.0	564.3	702.4	751.7	706.5	832.7	1,225.4	1,221.7	1,323.4	1,492.5
					(Index 1980 = 100)						
Terms of Trade	51.3	59.8	52.5	60.7	61.8	68.0	75.2	83.0	79.1	117.0	151.3
Export Prices, fob	14.9	17.8	16.2	19.8	21.2	26.1	36.2	50.1	49.7	74.4	102.1
Nonfuel Primary Products	26.4	26.8	27.3	33.5	30.1	33.7	42.5	47.0	48.0	86.3	126.8
Fuels	4.3	4.3	4.3	4.3	5.6	6.2	8.9	36.7	35.7	38.4	42.0
Manufactures	22.1	20.8	27.5	33.6	37.6	43.0	47.0	67.1	67.1	57.7	68.5
Import Prices, cif	29.0	29.7	30.8	32.6	34.3	38.4	48.1	60.3	62.8	63.6	67.5
Trade at Constant 1980 Prices				*(Millions of 1980 US dollars)*							
Exports, fob	3,428.3	3,141.3	3,732.0	3,673.0	3,253.4	3,307.5	3,250.8	2,828.6	2,947.7	2,344.5	2,392.6
Imports, cif	1,713.9	2,165.2	2,226.5	2,583.8	2,710.0	2,238.7	2,207.1	2,647.6	2,379.2	2,685.8	3,004.6
BALANCE OF PAYMENTS				*(Millions of current US dollars)*							
Exports of Goods & Services	..	..	..	1,019	997	1,228	1,588	1,948	2,180	2,850	3,502
Merchandise, fob	..	..	..	788	754	979	1,263	1,495	1,683	2,202	2,660
Nonfactor Services	..	..	..	189	215	224	279	363	422	559	718
Factor Services	..	..	..	42	28	26	47	90	75	89	124
Imports of Goods & Services	..	..	..	1,348	1,485	1,454	1,677	2,351	2,400	2,738	3,173
Merchandise, fob	..	..	..	802	903	850	983	1,511	1,415	1,654	1,970
Nonfactor Services	..	..	..	324	381	377	431	558	592	639	766
Factor Services	..	..	..	222	201	228	263	282	393	445	437
Long-Term Interest	..	..	..	59	67	81	103	133	136	157	154
Current Transfers, net	..	..	..	-4	0	9	12	18	30	39	40
Workers' Remittances	..	..	..	6	8	9	12	16	18	42	42
Total to be Financed	..	..	..	-333	-488	-217	-77	-384	-190	151	369
Official Capital Grants	..	..	..	40	34	26	23	33	18	12	7
Current Account Balance	..	..	..	-293	-454	-191	-55	-352	-172	163	375
Long-Term Capital, net	..	..	..	267	228	290	308	261	320	91	232
Direct Investment	..	..	..	39	40	17	23	35	33	14	43
Long-Term Loans	..	..	..	118	185	283	243	118	327	149	186
Disbursements	..	..	..	253	357	470	450	401	505	366	412
Repayments	..	..	..	134	172	187	206	283	179	217	225
Other Long-Term Capital	..	..	..	70	-31	-36	19	76	-58	-84	-4
Other Capital, net	..	..	..	43	208	62	-92	-3	-86	360	41
Change in Reserves	..	..	..	-17	18	-161	-161	94	-62	-614	-648
Memo Item:				*(Colombian Pesos per US dollar)*							
Conversion Factor (Annual Avg)	14.510	16.290	17.320	18.440	19.930	21.870	23.640	26.060	30.930	34.690	36.770
EXTERNAL DEBT, ETC.				*(Millions of US dollars, outstanding at end of year)*							
Public/Publicly Guar. Long-Term	..	..	..	1,297	1,436	1,675	1,955	2,144	2,379	2,478	2,700
Official Creditors	..	..	..	1,112	1,221	1,353	1,517	1,638	1,737	1,802	1,883
IBRD and IDA	..	..	..	374	410	474	528	584	656	694	738
Private Creditors	..	..	..	184	215	322	438	506	641	676	816
Private Non-guaranteed Long-Term	..	..	..	283	344	392	365	302	367	404	389
Use of Fund Credit	..	..	..	55	58	0	0	0	0	0	0
Short-Term Debt	..	..	..	..	..	..	..	..	..	..	1,964
Memo Items:				*(Millions of US dollars)*							
Int'l Reserves Excluding Gold	52.0	142.0	195.0	189.1	188.0	309.0	516.2	431.1	474.7	1,100.8	1,747.3
Gold Holdings (at market price)	31.2	37.1	26.2	18.2	17.5	27.8	48.2	80.0	157.9	190.4	285.5
SOCIAL INDICATORS											
Total Fertility Rate	5.9	5.7	5.5	5.3	5.0	4.8	4.5	4.3	4.0	4.0	4.0
Crude Birth Rate	43.0	40.7	38.4	36.9	35.3	33.8	32.1	31.7	31.3	30.8	30.4
Infant Mortality Rate	85.0	83.0	81.0	79.0	77.0	75.0	72.0	69.0	66.0	64.5	63.0
Life Expectancy at Birth	57.9	58.3	58.7	59.1	59.4	59.8	60.2	60.6	61.0	61.4	61.8
Food Production, p.c. ('79-81=100)	81.1	82.2	82.2	83.2	85.8	85.8	83.2	86.6	91.6	97.0	94.7
Labor Force, Agriculture (%)	42.5	41.5	40.4	39.3	38.8	38.3	37.7	37.2	36.7	36.2	35.7
Labor Force, Female (%)	20.7	20.9	21.1	21.3	21.4	21.5	21.6	21.7	21.9	22.0	22.1
School Enroll. Ratio, primary	..	..	..	108.0	..	..	..	..	118.0	..	125.0
School Enroll. Ratio, secondary	..	..	..	25.0	..	..	..	..	39.0	..	43.0

1978	1979	1980	1981	1982	1983	1984	1985	1986	1987 estimate	NOTES	COLOMBIA
				(Millions of current US dollars)							**FOREIGN TRADE (CUSTOMS BASIS)**
3,038.3	3,300.4	3,950.0	3,160.0	3,110.0	3,080.0	3,480.0	3,550.0	5,110.0	5,024.4	..	Value of Exports, fob
2,377.5	2,453.7	3,033.9	2,074.4	2,090.8	2,034.0	2,350.3	2,315.1	3,652.0	2,307.2	..	Nonfuel Primary Products
131.0	131.8	112.4	248.0	268.0	451.0	517.8	577.9	665.0	1,634.2	..	Fuels
529.8	714.9	803.7	837.6	751.2	595.0	611.9	657.0	793.0	1,083.0	..	Manufactures
2,836.3	3,233.2	4,660.0	5,200.0	5,460.0	4,970.0	4,490.0	4,130.0	3,850.0	4,230.0	..	Value of Imports, cif
488.5	561.1	823.2	819.9	873.3	801.5	759.6	685.8	627.0	669.0	..	Nonfuel Primary Products
206.3	325.8	567.3	731.0	662.5	649.0	474.3	484.2	153.0	121.0	..	Fuels
2,141.5	2,346.3	3,269.5	3,649.1	3,924.2	3,519.5	3,256.0	2,960.0	3,070.0	3,440.0	..	Manufactures
				(Index 1980 = 100)							
116.9	113.9	100.0	91.5	97.9	96.1	100.9	98.3	99.5	69.7	..	Terms of Trade
87.9	99.4	100.0	92.5	94.7	91.4	94.1	90.2	98.8	77.3	..	Export Prices, fob
97.6	105.2	100.0	85.2	88.8	85.8	90.6	86.6	110.3	73.1	..	Nonfuel Primary Products
42.3	61.0	100.0	112.5	101.6	92.5	90.2	87.5	53.4	63.4	..	Fuels
74.5	92.6	100.0	110.0	113.0	116.0	116.0	109.0	128.8	141.4	..	Manufactures
75.2	87.2	100.0	101.0	96.7	95.1	93.3	91.8	99.3	110.9	..	Import Prices, cif
				(Millions of 1980 US dollars)							Trade at Constant 1980 Prices
3,457.3	3,320.8	3,949.7	3,417.1	3,283.1	3,370.1	3,696.9	3,936.5	5,170.6	6,499.2	..	Exports, fob
3,773.1	3,706.8	4,660.0	5,146.0	5,643.9	5,224.6	4,814.8	4,501.0	3,876.3	3,815.4	..	Imports, cif
				(Millions of current US dollars)							**BALANCE OF PAYMENTS**
4,100	4,851	5,860	5,014	4,974	4,103	5,328	4,616	6,614	7,104	..	Exports of Goods & Services
3,155	3,441	3,986	3,158	3,114	2,970	4,273	3,650	5,331	5,700	..	Merchandise, fob
771	1,091	1,331	1,131	1,309	814	894	826	1,097	1,182	..	Nonfactor Services
173	319	543	725	551	319	161	140	186	222	..	Factor Services
3,914	4,515	6,231	7,217	8,197	7,270	7,028	6,886	7,097	7,850	..	Imports of Goods & Services
2,552	2,978	4,283	4,730	5,358	4,464	4,027	3,673	3,490	3,874	..	Merchandise, fob
852	941	1,160	1,284	1,335	1,290	1,288	1,420	1,677	1,746	..	Nonfactor Services
510	596	788	1,203	1,505	1,516	1,713	1,793	1,930	2,230	..	Factor Services
195	252	310	461	705	625	620	867	978	1,177	..	Long-Term Interest
44	99	164	241	167	145	289	455	793	1,001	..	Current Transfers, net
44	63	68	99	71	63	71	105	393	616	..	Workers' Remittances
229	435	-207	-1,962	-3,056	-3,022	-1,411	-1,815	310	255	K	Total to be Financed
29	3	1	1	2	19	10	6	-8	..	K	Official Capital Grants
258	438	-206	-1,961	-3,054	-3,003	-1,401	-1,809	302	255	..	Current Account Balance
137	758	816	1,581	1,617	1,547	1,832	2,356	2,459	250	..	Long-Term Capital, net
66	103	51	228	337	514	561	1,016	642	349	..	Direct Investment
51	679	808	1,401	1,255	1,049	1,359	1,335	1,635	-107	..	Long-Term Loans
335	1,130	1,071	1,775	1,653	1,662	2,041	2,085	2,756	1,296	..	Disbursements
285	451	263	374	398	614	682	751	1,121	1,403	..	Repayments
-9	-27	-44	-49	23	-35	-98	-1	190	8	..	Other Long-Term Capital
174	300	308	321	587	-388	-816	-392	-1,491	-90	..	Other Capital, net
-569	-1,496	-919	60	850	1,845	385	-155	-1,270	-415	..	Change in Reserves
				(Colombian Pesos per US dollar)							Memo Item:
39.100	42.550	47.280	54.490	64.090	78.850	100.820	142.310	194.260	242.610	..	Conversion Factor (Annual Avg)
				(Millions of US dollars, outstanding at end of year)							**EXTERNAL DEBT, ETC.**
2,812	3,384	4,088	5,076	5,990	6,875	7,732	9,570	12,185	13,828	..	Public/Publicly Guar. Long-Term
2,011	2,152	2,393	2,629	3,018	3,491	3,791	5,276	6,802	8,263	..	Official Creditors
773	860	1,012	1,185	1,366	1,531	1,597	2,417	3,279	4,127	..	IBRD and IDA
801	1,232	1,696	2,448	2,972	3,383	3,941	4,294	5,383	5,566	..	Private Creditors
363	473	515	866	1,192	1,278	1,437	1,568	1,585	1,524	..	Private Non-guaranteed Long-Term
0	0	0	0	0	0	0	0	0	..	..	Use of Fund Credit
1,927	2,012	2,337	2,774	3,124	3,260	2,868	3,099	1,597	1,654	..	Short-Term Debt
				(Millions of US dollars)							Memo Items:
2,366.4	3,843.8	4,830.0	4,740.0	3,860.0	1,900.0	1,360.0	1,600.0	2,700.0	3,090.0	..	Int'l Reserves Excluding Gold
443.2	1,186.3	1,643.3	1,341.3	1,745.7	1,612.0	422.0	601.7	785.9	330.1	..	Gold Holdings (at market price)
											SOCIAL INDICATORS
3.8	3.7	3.5	3.5	3.5	3.4	3.4	3.3	3.2	3.2	..	Total Fertility Rate
30.2	29.9	29.7	28.8	27.9	27.7	27.5	27.2	27.0	26.8	..	Crude Birth Rate
61.3	59.7	58.0	55.5	53.0	51.6	50.2	48.8	47.4	45.9	..	Infant Mortality Rate
62.2	62.6	63.1	63.5	64.0	64.3	64.7	65.0	65.4	65.7	..	Life Expectancy at Birth
98.2	101.1	98.9	100.0	96.4	93.2	95.8	95.0	96.6	99.9	..	Food Production, p.c. ('79-81 = 100)
35.2	34.7	34.2	..	..	..	..	..	..	..	..	Labor Force, Agriculture (%)
22.2	22.3	22.4	22.4	22.4	22.3	22.3	22.2	22.2	22.1	..	Labor Force, Female (%)
129.0	133.0	128.0	130.0	114.0	121.0	119.0	117.0			..	School Enroll. Ratio, primary
45.0	48.0	44.0	45.0	46.0	46.0	49.0	50.0	..	..	..	School Enroll. Ratio, secondary

COMOROS	1967	1968	1969	1970	1971	1972	1973	1974	1975	1976	1977
CURRENT GNP PER CAPITA (US $)	100	90	90	100	110	120	140	150	170	190	200
POPULATION (thousands)	249	254	260	266	272	278	285	291	298	305	312

ORIGIN AND USE OF RESOURCES *(Millions of current CFA Francs)*

	1967	1968	1969	1970	1971	1972	1973	1974	1975	1976	1977
Gross National Product (GNP)	5,631	5,631	6,123	6,856	8,041	8,539	9,447	14,305	12,797	12,361	13,771
Net Factor Income from Abroad	..	..	..	..	..	..	..	..	..	..	..
Gross Domestic Product (GDP)	..	..	..	..	..	..	..	..	..	..	..
Indirect Taxes, net	..	..	..	..	..	..	..	..	..	..	..
GDP at factor cost	..	..	..	..	..	..	..	..	..	..	..
Agriculture	..	..	..	..	..	..	..	..	..	..	..
Industry	..	..	..	..	..	..	..	..	..	..	..
Manufacturing	..	..	..	..	..	..	..	..	..	..	..
Services, etc.	..	..	..	..	..	..	..	..	..	..	..
Resource Balance	..	..	..	..	..	..	..	..	..	..	..
Exports of Goods & NFServices	..	..	..	..	..	..	..	..	..	..	..
Imports of Goods & NFServices	..	..	..	..	..	..	..	..	..	..	..
Domestic Absorption	..	..	..	..	..	..	..	..	..	..	..
Private Consumption, etc.	..	..	..	..	..	..	..	..	..	..	..
General Gov't Consumption	..	..	..	..	..	..	..	..	..	..	..
Gross Domestic Investment	..	..	..	..	..	..	..	..	..	..	..
Fixed Investment	..	..	..	..	..	..	..	..	..	..	..
Memo Items:											
Gross Domestic Saving	..	..	..	..	..	..	..	..	..	..	..
Gross National Saving	..	..	..	..	..	..	..	..	..	..	..

(Millions of 1985 CFA Francs)

	1967	1968	1969	1970	1971	1972	1973	1974	1975	1976	1977
Gross National Product	..	..	..	..	..	..	..	..	..	..	..
GDP at factor cost	..	..	..	..	..	..	..	..	..	..	..
Agriculture	..	..	..	..	..	..	..	..	..	..	..
Industry	..	..	..	..	..	..	..	..	..	..	..
Manufacturing	..	..	..	..	..	..	..	..	..	..	..
Services, etc.	..	..	..	..	..	..	..	..	..	..	..
Resource Balance	..	..	..	..	..	..	..	..	..	..	..
Exports of Goods & NFServices	..	..	..	..	..	..	..	..	..	..	..
Imports of Goods & NFServices	..	..	..	..	..	..	..	..	..	..	..
Domestic Absorption	..	..	..	..	..	..	..	..	..	..	..
Private Consumption, etc.	..	..	..	..	..	..	..	..	..	..	..
General Gov't Consumption	..	..	..	..	..	..	..	..	..	..	..
Gross Domestic Investment	..	..	..	..	..	..	..	..	..	..	..
Fixed Investment	..	..	..	..	..	..	..	..	..	..	..
Memo Items:											
Capacity to Import	..	..	..	..	..	..	..	..	..	..	..
Terms of Trade Adjustment	..	..	..	..	..	..	..	..	..	..	..
Gross Domestic Income	..	..	..	..	..	..	..	..	..	..	..
Gross National Income	..	..	..	..	..	..	..	..	..	..	..

DOMESTIC PRICES (DEFLATORS) *(Index 1985 = 100)*

	1967	1968	1969	1970	1971	1972	1973	1974	1975	1976	1977
Overall (GDP)	..	..	..	..	..	..	..	..	..	..	..
Domestic Absorption	..	..	..	..	..	..	..	..	..	..	..
Agriculture	..	..	..	..	..	..	..	..	..	..	..
Industry	..	..	..	..	..	..	..	..	..	..	..
Manufacturing	..	..	..	..	..	..	..	..	..	..	..

MANUFACTURING ACTIVITY

	1967	1968	1969	1970	1971	1972	1973	1974	1975	1976	1977
Employment (1985 = 100)	..	..	..	..	..	..	..	..	..	..	..
Real Earnings per Empl. (1985 = 100)	..	..	..	..	..	..	..	..	..	..	..
Real Output per Empl. (1985 = 100)	..	..	..	..	..	..	..	..	..	..	..
Earnings as % of Value Added	..	..	..	..	..	..	..	..	..	..	..

MONETARY HOLDINGS *(Millions of current CFA Francs)*

	1967	1968	1969	1970	1971	1972	1973	1974	1975	1976	1977
Money Supply, Broadly Defined	..	..	..	..	..	..	..	..	..	..	..
Money as Means of Payment	..	..	..	..	..	..	..	..	..	..	..
Currency Ouside Banks	..	..	..	..	..	..	..	..	..	..	..
Demand Deposits	..	..	..	..	..	..	..	..	..	..	..
Quasi-Monetary Liabilities	..	..	..	..	..	..	..	..	..	..	..

(Millions of current CFA Francs)

	1967	1968	1969	1970	1971	1972	1973	1974	1975	1976	1977
GOVERNMENT DEFICIT (-) OR SURPLUS	..	..	..	..	..	..	..	..	..	..	..
Current Revenue	..	..	..	..	..	..	..	..	..	..	..
Current Expenditure	..	..	..	..	..	..	..	..	..	..	..
Current Budget Balance	..	..	..	..	..	..	..	..	..	..	..
Capital Receipts	..	..	..	..	..	..	..	..	..	..	..
Capital Payments	..	..	..	..	..	..	..	..	..	..	..

1978	1979	1980	1981	1982	1983	1984	1985	1986	1987 estimate	NOTES	COMOROS
210	260	340	380	350	330	310	300	320	370	..	**CURRENT GNP PER CAPITA (US $)**
319	326	333	345	357	369	382	395	409	424	..	**POPULATION (thousands)**
				(Millions of current CFA Francs)							**ORIGIN AND USE OF RESOURCES**
16,494	19,656	25,551	30,115	36,150	42,321	46,832	51,087	56,117	59,335	..	Gross National Product (GNP)
..	..	..	..	-1,609	-175	-136	-350	-153	-337	..	Net Factor Income from Abroad
..	..	..	..	37,759	42,496	46,968	51,437	56,270	59,672	..	Gross Domestic Product (GDP)
..	..	..	..	..	..	..	..	..	..	..	Indirect Taxes, net
..	..	..	..	..	..	..	..	..	..	..	GDP at factor cost
..	..	..	..	..	13,724	14,861	16,266	18,577	21,049	21,363	..	Agriculture
..	..	..	..	..	5,236	6,038	7,195	7,237	7,286	8,413	..	Industry
..	..	..	..	..	1,431	1,516	1,716	1,887	2,062	2,148	..	Manufacturing
..	..	..	..	..	18,799	21,597	23,507	25,623	27,935	29,896	..	Services, etc.
..	..	..	..	..	-11,135	-26,587	-19,966	-15,255	-13,772	..	Resource Balance
..	..	..	..	..	8,426	4,236	8,943	9,251	6,431	..	Exports of Goods & NFServices
..	..	..	..	..	19,561	30,823	28,909	24,506	20,203	..	Imports of Goods & NFServices
..	..	..	..	..	53,631	73,555	71,403	71,525	73,444	..	Domestic Absorption
..	..	..	..	..	29,207	38,727	40,841	42,700	50,269	..	Private Consumption, etc.
..	..	..	..	..	12,084	13,305	15,310	15,552	15,819	..	General Gov't Consumption
..	..	..	..	..	12,340	21,523	15,252	13,273	7,356	..	Gross Domestic Investment
..	..	..	..	..	11,366	15,740	13,410	11,773	..	..	Fixed Investment
											Memo Items:
..	..	..	..	..	1,205	-5,064	-4,714	-1,982	-6,416	..	Gross Domestic Saving
..	..	..	..	..	208	-6,502	-5,369	-2,833	-6,489	..	Gross National Saving
				(Millions of 1985 CFA Francs)							
..	..	40,563	44,902	44,606	47,880	49,960	51,087	52,389	53,297	..	Gross National Product
..	..	..	..	..	..	..	..	..	..	..	GDP at factor cost
..	..	..	..	16,572	17,073	17,724	18,577	19,186	20,130	..	Agriculture
..	..	..	..	6,758	7,039	7,531	7,237	7,267	7,431	..	Industry
..	..	..	..	1,662	1,683	1,806	1,887	2,040	2,290	..	Manufacturing
..	..	..	..	23,311	23,971	24,847	25,623	26,083	26,040	..	Services, etc.
..	..	..	..	..	-14,062	-26,976	-19,966	-16,384	-12,721	..	Resource Balance
..	..	..	..	..	7,653	5,355	8,943	9,401	8,501	..	Exports of Goods & NFServices
..	..	..	..	..	21,715	32,331	28,909	25,785	21,222	..	Imports of Goods & NFServices
..	..	..	..	..	62,145	77,078	71,403	68,920	66,322	..	Domestic Absorption
..	..	..	..	..	32,762	41,252	40,841	39,781	43,638	..	Private Consumption, etc.
..	..	..	..	..	13,412	14,064	15,310	14,521	14,957	..	General Gov't Consumption
..	..	..	..	..	15,971	21,762	15,252	14,618	7,727	..	Gross Domestic Investment
..	..	..	..	..	14,877	15,867	13,410	12,853	..	..	Fixed Investment
											Memo Items:
..	..	..	..	..	9,354	4,443	8,943	9,734	6,755	..	Capacity to Import
..	..	..	..	..	1,701	-912	0	333	-1,746	..	Terms of Trade Adjustment
..	..	..	..	..	49,784	49,190	51,437	52,869	51,855	..	Gross Domestic Income
..	..	..	..	..	49,581	49,048	51,087	52,721	51,551	..	Gross National Income
				(Index 1985 = 100)							**DOMESTIC PRICES (DEFLATORS)**
..	..	..	..	81.0	88.4	93.7	100.0	107.1	111.3	..	Overall (GDP)
..	..	..	..	..	86.3	95.4	100.0	103.8	110.7	..	Domestic Absorption
..	..	..	..	82.8	87.0	91.8	100.0	109.7	106.1	..	Agriculture
..	..	..	..	77.5	85.8	95.5	100.0	100.3	113.2	..	Industry
..	..	..	..	86.1	90.1	95.0	100.0	101.1	93.8	..	Manufacturing
											MANUFACTURING ACTIVITY
..	..	..	..	..	..	..	..	..	..	..	Employment (1985=100)
..	..	..	..	..	..	..	..	..	..	..	Real Earnings per Empl. (1985=100)
..	..	..	..	..	..	..	..	..	..	..	Real Output per Empl. (1985=100)
..	..	..	..	..	..	..	..	..	..	..	Earnings as % of Value Added
				(Millions of current CFA Francs)							**MONETARY HOLDINGS**
..	..	..	..	6,370	8,521	7,213	8,192	8,664	..	..	Money Supply, Broadly Defined
..	..	..	..	5,202	7,292	6,254	7,177	7,232	..	..	Money as Means of Payment
..	..	..	..	2,434	3,427	3,142	3,448	3,118	..	..	Currency Ouside Banks
..	..	..	..	2,768	3,865	3,112	3,729	4,114	..	..	Demand Deposits
..	..	..	..	1,168	1,229	959	1,015	1,432	..	..	Quasi-Monetary Liabilities
				(Millions of current CFA Francs)							
..	..	..	..	..	..	..	..	..	..	..	**GOVERNMENT DEFICIT (-) OR SURPLUS**
..	..	..	..	..	..	..	..	..	..	..	Current Revenue
..	..	..	..	..	..	..	..	..	..	..	Current Expenditure
..	..	..	..	..	..	..	..	..	..	..	Current Budget Balance
..	..	..	..	..	..	..	..	..	..	..	Capital Receipts
..	..	..	..	..	..	..	..	..	..	..	Capital Payments

COMOROS	1967	1968	1969	1970	1971	1972	1973	1974	1975	1976	1977
FOREIGN TRADE (CUSTOMS BASIS)					*(Thousands of current US dollars)*						
Value of Exports, fob	..	..	..	..	..	..	..	..	..	..	..
Nonfuel Primary Products	..	..	..	..	..	..	..	..	..	..	..
Fuels	..	..	..	..	..	..	..	..	..	..	..
Manufactures	..	..	..	..	..	..	..	..	..	..	..
Value of Imports, cif	..	..	..	..	..	..	..	..	..	..	..
Nonfuel Primary Products	..	..	..	..	..	..	..	..	..	..	..
Fuels	..	..	..	..	..	..	..	..	..	..	..
Manufactures	..	..	..	..	..	..	..	..	..	..	..
Terms of Trade					*(Index 1980 = 100)*						
Export Prices, fob	..	..	..	..	..	..	..	..	..	..	..
Nonfuel Primary Products	..	..	..	..	..	..	..	..	..	..	..
Fuels	..	..	..	..	..	..	..	..	..	..	..
Manufactures	..	..	..	..	..	..	..	..	..	..	..
Import Prices, cif	..	..	..	..	..	..	..	..	..	..	..
Trade at Constant 1980 Prices					*(Thousands of 1980 US dollars)*						
Exports, fob	..	..	..	..	..	..	..	..	..	..	..
Imports, cif	..	..	..	..	..	..	..	..	..	..	..
BALANCE OF PAYMENTS					*(Thousands of current US dollars)*						
Exports of Goods & Services	..	..	..	..	..	..	..	..	..	..	..
Merchandise, fob	..	..	..	..	..	..	..	..	..	..	..
Nonfactor Services	..	..	..	..	..	..	..	..	..	..	..
Factor Services	..	..	..	..	..	..	..	..	..	..	..
Imports of Goods & Services	..	..	..	..	..	..	..	..	..	..	..
Merchandise, fob	..	..	..	..	..	..	..	..	..	..	..
Nonfactor Services	..	..	..	..	..	..	..	..	..	..	..
Factor Services	..	..	..	..	..	..	..	..	..	..	..
Long-Term Interest	..	..	..	0	0	0	0	0	0	0	0
Current Transfers, net	..	..	..	..	..	..	..	..	..	..	..
Workers' Remittances	..	..	..	..	..	..	..	..	..	..	..
Total to be Financed	..	..	..	..	..	..	..	..	..	..	..
Official Capital Grants	..	..	..	..	..	..	..	..	..	..	..
Current Account Balance	..	..	..	..	..	..	..	..	..	..	..
Long-Term Capital, net	..	..	..	..	..	..	..	..	..	..	..
Direct Investment	..	..	..	..	..	..	..	..	..	..	..
Long-Term Loans	..	..	..	..	..	0	..	2	1	14	4
Disbursements	..	..	..	..	..	0	0	2	1	15	5
Repayments	..	..	..	..	..	0	0	0	0	0	0
Other Long-Term Capital	..	..	..	..	..	..	..	..	..	..	..
Other Capital, net	..	..	..	..	..	..	..	..	..	..	..
Change in Reserves	..	..	..	0	0	0	0	0	0	0	0
Memo Item:					*(CFA Francs per US dollar)*						
Conversion Factor (Annual Avg)	246.850	246.850	259.710	277.710	277.130	252.210	222.700	240.500	214.320	238.980	225.670
EXTERNAL DEBT, ETC.					*(Millions of US dollars, outstanding at end of year)*						
Public/Publicly Guar. Long-Term	..	..	..	1	1	2	2	4	5	19	23
Official Creditors	..	..	..	1	1	2	2	4	5	19	23
IBRD and IDA	..	..	..	0	0	0	0	0	0	0	0
Private Creditors	..	..	..	..	..	..	..	..	..	..	..
Private Non-guaranteed Long-Term	..	..	..	..	..	..	..	..	..	..	..
Use of Fund Credit	..	..	..	0	0	0	0	0	0	0	1
Short-Term Debt	..	..	..	..	..	..	..	..	..	..	1
Memo Items:					*(Thousands of US dollars)*						
Int'l Reserves Excluding Gold	..	..	..	..	..	..	..	..	..	..	..
Gold Holdings (at market price)	..	..	..	..	..	..	..	..	..	..	..
SOCIAL INDICATORS											
Total Fertility Rate	7.0	7.0	7.0	7.0	7.0	7.0	7.0	7.0	7.0	7.0	7.0
Crude Birth Rate	49.3	49.3	49.3	49.3	49.3	49.3	49.3	49.3	49.3	49.3	49.3
Infant Mortality Rate	115.0	113.2	111.4	109.6	107.8	106.0	104.2	102.4	100.6	98.8	97.0
Life Expectancy at Birth	46.9	47.4	47.8	48.3	48.7	49.2	49.6	50.1	50.6	51.0	51.5
Food Production, p.c. ('79-81 = 100)	107.6	107.7	107.4	108.1	108.0	106.1	107.1	106.0	105.1	106.0	100.8
Labor Force, Agriculture (%)	87.4	87.2	86.9	86.7	86.3	85.9	85.6	85.2	84.8	84.4	84.1
Labor Force, Female (%)	42.9	42.9	42.9	42.9	42.8	42.7	42.7	42.6	42.6	42.6	42.6
School Enroll. Ratio, primary	..	..	..	34.0	..	..	46.0	..	..	..	..
School Enroll. Ratio, secondary	..	..	..	3.0	..	..	7.0	..	..	..	..

|------|------|------|------|------|------|------|------|------|------|-------|---------|
						FOREIGN TRADE (CUSTOMS BASIS)					
		(Thousands of current US dollars)									
..	..	11,190	16,420	19,580	19,470	7,050	15,230	18,460	..	..	Value of Exports, fob
..	..	..	..	..	..	..	..	..	..	..	Nonfuel Primary Products
..	..	..	..	..	..	..	..	..	..	..	Fuels
..	..	..	..	..	..	..	..	..	..	..	Manufactures
..	..	54,700	52,100	50,100	49,700	60,200	63,600	70,900	..	..	Value of Imports, cif
..	..	..	..	..	..	..	..	..	..	..	Nonfuel Primary Products
..	..	..	..	..	..	..	..	..	..	..	Fuels
..	..	..	..	..	..	..	..	..	..	..	Manufactures
		(Index 1980 = 100)									
..	..	..	..	..	..	..	..	..	..	..	Terms of Trade
..	..	..	..	..	..	..	..	..	..	..	Export Prices, fob
..	..	..	..	..	..	..	..	..	..	..	Nonfuel Primary Products
..	..	..	..	..	..	..	..	..	..	..	Fuels
..	..	..	..	..	..	..	..	..	..	..	Manufactures
..	..	..	..	..	..	..	..	..	..	..	Import Prices, cif
		(Thousands of 1980 US dollars)									
											Trade at Constant 1980 Prices
..	..	..	..	..	..	..	..	..	..	..	Exports, fob
..	..	..	..	..	..	..	..	..	..	..	Imports, cif
		(Thousands of current US dollars)									**BALANCE OF PAYMENTS**
..	..	14	18	23	23	11	20	29	23	..	Exports of Goods & Services
..	..	11	16	20	19	7	16	20	12	..	Merchandise, fob
..	..	2	1	3	3	3	4	7	10	..	Nonfactor Services
..	..	1	1	1	1	1	1	1	2	..	Factor Services
..	..	34	51	51	52	72	66	74	85	..	Imports of Goods & Services
..	..	22	25	25	29	33	28	28	46	..	Merchandise, fob
..	..	12	25	25	22	38	36	42	36	..	Nonfactor Services
..	..	0	1	1	2	2	2	3	3	..	Factor Services
0	0	0	0	1	1	2	2	1	1	..	Long-Term Interest
..	..	0	0	-2	-2	-3	-1	-2	1	..	Current Transfers, net
..	..	2	1	2	2	2	4	5	7	..	Workers' Remittances
..	..	-20	-32	-30	-31	-64	-46	-47	-61	K	Total to be Financed
..	..	11	24	19	20	31	32	32	39	K	Official Capital Grants
..	..	-9	-8	-11	-11	-33	-14	-16	-23	..	Current Account Balance
..	..	-6	13	4	-1	4	12	11	31	..	Long-Term Capital, net
..	..	..	..	..	..	..	..	..	8	..	Direct Investment
4	10	13	11	16	18	23	22	24	13	..	Long-Term Loans
4	10	13	11	16	18	24	22	25	13	..	Disbursements
0	..	..	0	0	0	1	0	1	0	..	Repayments
..	..	-30	-22	-32	-39	-51	-42	-44	-28	..	Other Long-Term Capital
..	..	11	-1	11	15	23	9	8	0	..	Other Capital, net
0	..	4	-3	-4	-3	6	-6	-4	-9	..	Change in Reserves
		(CFA Francs per US dollar)									**Memo Item:**
225.640	212.720	211.300	271.730	328.610	381.060	436.960	449.260	346.300	300.540	..	Conversion Factor (Annual Avg)
		(Millions of US dollars, outstanding at end of year)									**EXTERNAL DEBT, ETC.**
28	38	43	53	68	84	101	129	159	188	..	Public/Publicly Guar. Long-Term
28	38	43	52	67	83	101	129	159	188	..	Official Creditors
0	3	4	6	8	9	13	20	26	33	..	IBRD and IDA
..	..	0	0	1	1	0	0	0	0	..	Private Creditors
..	..	..	..	..	..	..	..	..	..	..	Private Non-guaranteed Long-Term
1	1	0	0	0	0	0	0	0	..	..	Use of Fund Credit
1	1	1	2	1	2	3	4	8	15	..	Short-Term Debt
		(Thousands of US dollars)									**Memo Items:**
..	..	6,373	8,392	10,831	10,794	3,509	11,748	17,546	..	..	Int'l Reserves Excluding Gold
..	..	..	..	..	221	176	186	223	..	..	Gold Holdings (at market price)
											SOCIAL INDICATORS
7.0	7.0	7.0	7.0	7.0	7.0	7.0	7.0	7.0	7.0	..	Total Fertility Rate
49.3	49.3	49.3	49.3	49.3	49.4	49.4	49.5	49.5	49.6	..	Crude Birth Rate
95.2	93.4	91.6	89.8	88.0	86.7	85.5	84.2	82.9	81.7	..	Infant Mortality Rate
52.0	52.4	52.9	53.3	53.8	54.2	54.7	55.1	55.6	56.0	..	Life Expectancy at Birth
100.4	99.9	105.3	94.7	96.5	99.5	95.3	97.5	94.7	94.1	..	Food Production, p.c. ('79-81=100)
83.7	83.4	83.0	..	..	..	..	..	..	..	..	Labor Force, Agriculture (%)
42.6	42.5	42.5	42.3	42.1	41.8	41.6	41.4	41.2	40.9	..	Labor Force, Female (%)
..	..	91.0	..	89.0	..	..	..	..	..	..	School Enroll. Ratio, primary
..	..	23.0	..	69.0	..	..	..	..	..	..	School Enroll. Ratio, secondary

CONGO, PEOPLE'S REPUBLIC OF THE	1967	1968	1969	1970	1971	1972	1973	1974	1975	1976	1977
CURRENT GNP PER CAPITA (US $)	200	220	240	240	260	290	340	430	520	510	500
POPULATION (thousands)	1,120	1,149	1,178	1,208	1,239	1,270	1,303	1,338	1,376	1,417	1,460
ORIGIN AND USE OF RESOURCES				*(Billions of current CFA Francs)*							
Gross National Product (GNP)	57.2	61.3	67.6	74.2	86.7	101.9	114.0	135.7	157.9	170.6	178.7
Net Factor Income from Abroad	-1.2	-0.9	-1.3	-1.8	-2.0	-1.6	-6.8	-5.2	-6.5	-9.7	-9.3
Gross Domestic Product (GDP)	58.4	62.2	68.9	76.0	88.7	103.5	120.8	140.9	164.4	180.3	188.0
Indirect Taxes, net	9.1	10.0	10.9	9.7	11.8	11.9	12.9	14.5	21.6	21.0	26.7
GDP at factor cost	49.3	52.2	58.0	66.3	76.9	91.6	107.9	126.4	142.8	159.3	161.3
Agriculture	11.6	12.2	12.2	13.6	15.2	17.0	19.0	21.3	23.8	24.3	29.0
Industry	11.6	12.0	13.4	18.2	19.4	22.0	26.7	35.8	53.3	67.8	59.5
Manufacturing	..	..	..	..	..	..	..	..	..	..	..
Services, etc.	35.2	38.0	43.3	44.2	54.1	64.5	75.1	83.8	87.3	88.2	99.5
Resource Balance	-14.8	-16.7	-17.1	-17.5	-17.8	-24.3	-14.2	-8.8	-45.8	-47.2	-37.0
Exports of Goods & NFServices	17.8	18.9	22.1	26.4	28.9	29.3	38.2	75.0	59.0	72.8	85.7
Imports of Goods & NFServices	32.6	35.6	39.2	43.9	46.7	53.6	52.4	83.8	104.8	120.0	122.7
Domestic Absorption	73.2	78.9	86.0	93.5	106.5	127.8	135.0	149.7	210.2	227.5	225.0
Private Consumption, etc.	45.4	49.3	55.4	62.2	67.7	79.7	75.2	75.2	116.3	139.7	138.6
General Gov't Consumption	9.4	11.2	12.2	12.9	15.2	18.0	21.2	25.1	29.6	31.9	36.4
Gross Domestic Investment	18.4	18.4	18.4	18.4	23.6	30.1	38.6	49.4	64.3	55.9	50.0
Fixed Investment	..	..	..	..	..	..	..	47.3	61.5	53.6	48.9
Memo Items:											
Gross Domestic Saving	3.6	1.7	1.3	0.9	5.8	5.8	24.4	40.6	18.5	8.7	13.0
Gross National Saving	2.4	0.8	0.0	-2.1	1.6	1.0	15.9	21.6	5.0	-8.7	-4.5
				(Billions of 1980 CFA Francs)							
Gross National Product	142.5	153.7	164.2	178.9	191.9	208.6	216.8	243.7	268.5	269.5	243.1
GDP at factor cost	121.8	129.5	139.3	158.9	168.9	186.2	204.5	226.3	242.4	251.6	218.9
Agriculture	28.3	29.6	28.6	31.1	32.2	33.5	34.7	36.0	37.3	35.5	36.9
Industry	48.2	48.4	50.2	61.7	62.1	66.1	75.5	93.9	118.3	130.1	103.9
Manufacturing	32.8	33.8	34.2	..	..	..	..	..	..	..	..
Services, etc.	69.2	78.1	88.8	90.7	102.2	112.4	120.3	123.5	124.3	119.3	115.3
Resource Balance	-92.9	-93.2	-90.8	-83.9	-80.5	-77.0	-73.4	-69.7	-79.7	-66.3	-38.4
Exports of Goods & NFServices	54.8	70.6	91.1	118.0	122.2	126.6	130.9	135.5	135.8	139.9	155.1
Imports of Goods & NFServices	147.6	163.9	182.0	201.9	202.7	203.5	204.4	205.2	215.5	206.2	193.5
Domestic Absorption	216.3	223.6	229.9	235.9	252.5	271.0	291.7	315.0	349.5	321.5	277.5
Private Consumption, etc.	96.5	106.7	117.1	122.9	124.4	125.7	126.6	127.0	133.8	145.4	126.1
General Gov't Consumption	30.3	32.7	33.8	38.8	41.1	43.5	45.9	48.7	51.3	50.1	49.9
Gross Domestic Investment	89.5	84.1	79.0	74.2	87.0	101.8	119.2	139.4	164.5	126.1	101.5
Fixed Investment	..	..	..	..	..	..	..	..	..	..	..
Memo Items:											
Capacity to Import	80.6	87.0	102.6	121.4	125.5	111.3	149.0	183.6	121.3	125.1	135.1
Terms of Trade Adjustment	25.8	16.4	11.5	3.4	3.3	-15.3	18.1	48.1	-14.5	-14.8	-20.0
Gross Domestic Income	171.5	172.5	179.1	186.9	199.8	196.8	248.5	301.5	265.3	270.1	236.1
Gross National Income	168.4	170.1	175.7	182.3	195.1	193.3	234.8	291.8	254.1	254.7	223.2
DOMESTIC PRICES (DEFLATORS)				*(Index 1980 = 100)*							
Overall (GDP)	40.1	39.8	41.1	41.4	45.1	48.8	52.4	55.6	58.8	63.3	73.4
Domestic Absorption	33.8	35.3	37.4	39.6	42.2	47.2	46.3	47.5	60.1	70.8	81.1
Agriculture	40.9	41.2	42.6	43.7	47.1	50.8	54.7	59.2	63.9	68.4	78.5
Industry	24.1	24.8	26.7	29.5	31.2	33.3	35.4	38.1	45.1	52.1	57.3
Manufacturing	..	..	..	..	..	..	..	..	..	..	..
MANUFACTURING ACTIVITY											
Employment (1980=100)	..	..	..	..	..	..	..	..	..	..	..
Real Earnings per Empl. (1980=100)	..	..	..	..	..	..	..	..	..	..	..
Real Output per Empl. (1980=100)	..	..	..	..	..	..	..	..	..	..	..
Earnings as % of Value Added	..	33.0	34.3	33.8	35.1	40.6	44.3	49.8	53.7	43.8	..
MONETARY HOLDINGS				*(Billions of current CFA Francs)*							
Money Supply, Broadly Defined	10.1	11.1	11.6	13.4	14.9	15.9	18.5	25.6	29.1	33.9	34.5
Money as Means of Payment	9.5	10.1	10.9	12.5	13.7	14.7	17.1	23.7	26.9	31.3	30.6
Currency Ouside Banks	4.2	4.6	5.1	5.4	6.4	7.8	8.9	11.6	14.0	15.1	15.9
Demand Deposits	5.3	5.5	5.8	7.1	7.4	6.9	8.2	12.1	12.9	16.2	14.8
Quasi-Monetary Liabilities	0.6	1.0	0.8	0.9	1.2	1.3	1.4	1.9	2.2	2.6	3.9
GOVERNMENT DEFICIT (-) OR SURPLUS				*(Millions of current CFA Francs)*							
GOVERNMENT DEFICIT (-) OR SURPLUS	..	..	..	-1,090	-2,400	..	..	..	..	..	..
Current Revenue	..	..	..	16,700	17,610	18,800	19,600	43,770	47,830	46,920	..
Current Expenditure	..	..	..	16,110	17,660	..	..	..	..	..	..
Current Budget Balance	..	..	..	590	-50	..	..	..	..	..	..
Capital Receipts	..	..	..	10	..	..	500	350	..	1,380	..
Capital Payments	..	..	..	1,690	2,350	..	..	..	..	..	..

1978	1979	1980	1981	1982	1983	1984	1985	1986	1987 estimate	NOTES	CONGO, PEOPLE'S REPUBLIC OF THE
540	650	880	1,190	1,300	1,210	1,170	1,070	930	870	..	**CURRENT GNP PER CAPITA (US $)**
1,506	1,554	1,605	1,658	1,713	1,769	1,826	1,884	1,950	2,020	..	**POPULATION (thousands)**
				(Billions of current CFA Francs)							**ORIGIN AND USE OF RESOURCES**
185.6	234.8	334.9	501.2	654.1	732.9	871.0	867.8	562.2	567.2	..	Gross National Product (GNP)
-12.7	-20.2	-25.5	-40.5	-55.9	-66.3	-87.5	-103.2	-78.2	-79.6	..	Net Factor Income from Abroad
198.3	255.0	360.4	541.7	710.0	799.2	958.5	970.9	640.4	646.8	..	Gross Domestic Product (GDP)
21.8	35.4	58.1	90.2	166.2	..	..	..	..	..	..	Indirect Taxes, net
176.5	219.6	302.3	451.5	543.8	..	..	..	..	..	B	GDP at factor cost
31.8	36.5	42.1	42.7	55.8	60.6	66.3	72.3	77.4	78.8	..	Agriculture
58.5	92.8	168.0	275.7	373.2	434.2	542.4	523.4	208.5	211.2	..	Industry
18.6	21.5	27.0	34.3	33.9	44.1	46.5	54.6	61.3	53.0	f	Manufacturing
108.0	125.7	150.3	223.3	281.0	304.4	349.8	375.2	354.5	356.8	..	Services, etc.
-34.4	-3.1	-0.4	-97.9	-94.0	49.4	153.4	113.6	-88.5	-15.5	..	Resource Balance
86.7	120.9	216.3	314.3	392.4	463.1	590.7	551.9	255.2	281.1	..	Exports of Goods & NFServices
121.1	124.0	216.7	412.2	486.4	413.7	437.3	438.3	343.7	296.6	..	Imports of Goods & NFServices
232.7	258.1	360.8	639.6	804.0	749.8	805.1	857.3	728.9	662.3	..	Domestic Absorption
133.0	139.9	168.5	305.9	284.1	318.4	372.1	406.2	380.5	376.8	..	Private Consumption, etc.
45.5	52.0	63.4	72.8	95.8	124.2	141.7	159.7	159.8	132.9	..	General Gov't Consumption
54.2	66.2	128.9	260.9	424.1	307.2	291.3	291.4	188.6	152.6	..	Gross Domestic Investment
46.4	56.4	118.7	239.7	404.7	302.9	277.3	274.3	182.9	148.1	..	Fixed Investment
											Memo Items:
19.8	63.1	128.5	163.0	330.1	356.6	444.7	405.0	100.1	137.1	..	Gross Domestic Saving
4.1	36.2	89.4	113.7	258.7	274.7	337.5	285.0	7.7	43.3	..	Gross National Saving
				(Billions of 1980 CFA Francs)							
253.1	279.2	334.9	416.7	487.7	520.1	543.4	531.5	518.9	497.5	..	Gross National Product
240.3	261.0	302.3	375.7	409.1	..	..	..	..	..	B H	GDP at factor cost
36.8	39.1	42.1	40.9	42.8	44.7	44.0	44.3	46.9	45.3	..	Agriculture
107.6	134.3	168.0	220.2	301.8	349.1	359.3	362.5	370.8	367.9	..	Industry
23.2	23.2	27.0	34.9	37.0	48.7	51.7	56.6	53.2	50.1	f	Manufacturing
126.5	130.1	150.3	189.2	182.7	173.9	199.7	191.9	152.5	133.3	..	Services, etc.
-21.8	24.3	-0.4	-83.8	-95.6	25.1	63.6	41.1	90.8	..	..	Resource Balance
170.1	177.6	216.3	236.3	259.1	303.8	333.6	311.9	301.1	..	..	Exports of Goods & NFServices
191.8	153.4	216.7	320.1	354.7	278.8	270.0	270.8	210.3	..	..	Imports of Goods & NFServices
269.4	174.0	360.8	562.6	614.4	532.9	536.2	542.1	456.2	..	..	Domestic Absorption
136.2	31.6	168.5	263.0	228.2	235.5	253.4	263.1	245.2	..	..	Private Consumption, etc.
52.7	56.0	63.4	66.3	67.0	78.0	87.7	93.7	87.6	..	..	General Gov't Consumption
80.6	86.4	128.9	233.3	319.2	219.4	195.1	185.3	123.5	..	..	Gross Domestic Investment
..	..	..	..	..	..	..	..	..	..	..	Fixed Investment
											Memo Items:
137.3	149.5	216.3	244.1	286.1	312.1	364.7	341.0	156.1	..	..	Capacity to Import
-32.7	-28.1	0.0	7.8	27.0	8.2	31.1	29.1	-145.0	..	..	Terms of Trade Adjustment
238.2	275.4	360.4	457.9	554.3	576.0	634.1	627.8	425.2	..	..	Gross Domestic Income
220.4	251.1	334.9	424.5	514.7	528.3	574.5	560.6	374.0	..	..	Gross National Income
				(Index 1980 = 100)							**DOMESTIC PRICES (DEFLATORS)**
73.2	84.0	100.0	120.3	134.6	140.8	159.0	162.2	112.3	118.4	..	Overall (GDP)
86.4	148.3	100.0	113.7	130.9	140.7	150.1	158.2	159.8	..	..	Domestic Absorption
86.5	93.3	100.0	104.5	130.3	135.4	150.7	163.2	165.1	174.0	..	Agriculture
54.4	69.1	100.0	125.2	123.7	124.4	151.0	144.4	56.2	57.4	..	Industry
80.1	92.5	100.0	98.4	91.5	90.5	90.0	96.4	115.3	105.9	..	Manufacturing
											MANUFACTURING ACTIVITY
..	..	..	..	..	..	..	..	..	..	..	Employment (1980 = 100)
..	..	..	..	..	..	..	..	..	..	B	Real Earnings per Empl. (1980 = 100)
..	..	..	..	..	..	..	..	..	..	..	Real Output per Empl. (1980 = 100)
..	..	..	..	..	..	57.0	..	..	..	..	Earnings as % of Value Added
				(Billions of current CFA Francs)							**MONETARY HOLDINGS**
36.9	44.9	61.4	92.1	116.0	113.8	122.3	147.5	130.5	138.5	..	Money Supply, Broadly Defined
33.0	39.6	54.3	75.1	97.6	90.5	101.3	111.9	97.5	102.3	..	Money as Means of Payment
16.3	19.0	22.9	30.6	43.2	42.9	43.7	48.1	49.4	55.3	..	Currency Ouside Banks
16.7	20.7	31.5	44.5	54.4	47.6	57.7	63.7	48.1	47.1	..	Demand Deposits
3.9	5.3	7.1	17.0	18.4	23.3	20.9	35.6	33.0	36.1	..	Quasi-Monetary Liabilities
				(Millions of current CFA Francs)							
..	..	-18,760	..	..	..	..	..	..	..	E F	**GOVERNMENT DEFICIT (-) OR SURPLUS**
..	..	159,930	..	..	..	..	..	..	..	..	Current Revenue
..	..	..	..	..	..	..	..	..	..	..	Current Expenditure
..	..	..	..	..	..	..	..	..	..	..	Current Budget Balance
..	..	70	..	..	..	..	..	..	..	..	Capital Receipts
..	..	..	..	..	..	..	..	..	..	..	Capital Payments

CONGO, PEOPLE'S REPUBLIC OF THE	1967	1968	1969	1970	1971	1972	1973	1974	1975	1976	1977
FOREIGN TRADE (CUSTOMS BASIS)				*(Millions of current US dollars)*							
Value of Exports, fob	40.4	39.9	34.5	28.2	35.0	46.0	78.5	207.8	178.9	189.5	173.6
Nonfuel Primary Products	19.5	16.1	20.2	18.9	21.0	23.1	23.1	31.6	27.0	29.2	45.8
Fuels	0.7	1.0	0.5	0.3	0.5	2.6	28.2	158.0	131.2	136.9	99.1
Manufactures	20.2	22.8	13.7	9.0	13.5	20.2	27.2	18.2	20.7	23.5	28.8
Value of Imports, cif	82.0	83.5	78.6	57.2	78.9	89.7	125.1	123.3	164.8	167.6	183.1
Nonfuel Primary Products	12.2	13.1	14.0	12.1	13.6	17.4	19.2	22.0	28.7	29.2	40.7
Fuels	4.9	2.5	2.5	1.2	2.0	2.4	3.5	10.6	13.0	15.1	2.7
Manufactures	64.8	67.9	62.1	44.0	63.3	70.0	102.4	90.7	123.1	123.3	139.7
Terms of Trade					*(Index 1980 = 100)*						
	102.8	90.3	88.6	83.3	80.7	70.7	42.3	69.7	63.7	69.3	72.4
Export Prices, fob	23.5	23.8	24.0	25.4	26.1	24.9	18.1	39.2	38.4	42.0	50.2
Nonfuel Primary Products	22.0	22.2	22.6	23.8	23.7	25.6	38.0	43.1	39.8	49.2	65.9
Fuels	4.3	4.3	4.3	4.3	5.6	6.2	8.9	36.7	35.7	38.4	42.0
Manufactures	30.1	31.7	32.4	37.5	36.5	38.5	49.5	67.2	66.1	66.7	71.7
Import Prices, cif	22.8	26.4	27.1	30.5	32.3	35.2	42.8	56.2	60.2	60.6	69.3
Trade at Constant 1980 Prices				*(Millions of 1980 US dollars)*							
Exports, fob	172.2	167.6	143.5	111.0	134.2	184.8	434.2	530.6	466.1	451.2	345.7
Imports, cif	359.1	316.3	290.3	187.6	244.2	254.9	292.6	219.4	273.6	276.4	264.1
BALANCE OF PAYMENTS				*(Millions of current US dollars)*							
Exports of Goods & Services	..	..	..	80.2	105.1	115.2	173.0	304.9	276.9	293.1	348.6
Merchandise, fob	..	..	..	58.0	75.3	78.1	122.2	261.2	230.8	221.6	266.7
Nonfactor Services	..	..	..	22.0	28.2	34.7	47.7	41.3	43.0	68.5	77.2
Factor Services	..	..	..	0.2	1.6	2.4	3.1	2.4	3.2	3.1	4.8
Imports of Goods & Services	..	..	..	129.0	175.5	217.8	255.5	363.8	529.2	529.5	535.2
Merchandise, fob	..	..	..	72.0	98.6	119.2	126.8	193.4	272.0	235.1	234.1
Nonfactor Services	..	..	..	50.0	68.3	89.9	105.3	152.1	224.9	254.1	258.6
Factor Services	..	..	..	7.0	8.6	8.7	23.4	18.3	32.4	40.3	42.5
Long-Term Interest	..	..	..	3.4	3.4	3.3	5.1	6.1	10.0	12.4	13.1
Current Transfers, net	..	..	..	-4.2	-8.1	-12.7	-7.6	-57.2	-32.9	-32.3	-33.5
Workers' Remittances	..	..	..	0.8	0.0	0.0	0.0	0.0	0.0	0.0	0.0
Total to be Financed	..	..	..	-53.0	-78.5	-115.3	-90.1	-116.2	-285.2	-268.7	-220.1
Official Capital Grants	..	..	..	8.0	11.8	21.6	11.8	23.1	36.3	41.3	26.7
Current Account Balance	..	..	..	-45.0	-66.7	-93.7	-78.3	-93.1	-248.9	-227.3	-193.3
Long-Term Capital, net	..	..	..	67.0	69.9	107.3	86.2	95.7	195.6	179.9	162.4
Direct Investment	..	..	..	30.0	48.7	66.3	68.3	46.1	15.4	1.4	2.0
Long-Term Loans	..	..	..	14.6	17.0	30.1	33.8	70.3	121.3	66.4	119.1
Disbursements	..	..	..	20.4	22.6	36.3	43.5	85.8	147.0	79.6	144.3
Repayments	..	..	..	5.8	5.6	6.2	9.7	15.5	25.7	13.2	25.2
Other Long-Term Capital	..	..	..	14.4	-7.7	-10.8	-27.7	-43.7	22.6	70.8	14.6
Other Capital, net	..	..	..	-20.2	-3.5	-15.5	-11.6	16.1	43.0	42.9	22.6
Change in Reserves	..	..	..	-1.8	0.3	2.0	3.7	-18.8	10.3	4.5	8.4
Memo Item:				*(CFA Francs per US dollar)*							
Conversion Factor (Annual Avg)	246.850	246.850	259.710	277.710	277.130	252.210	222.700	240.500	214.320	238.980	245.670
EXTERNAL DEBT, ETC.				*(Millions of US dollars, outstanding at end of year)*							
Public/Publicly Guar. Long-Term	..	..	..	124.3	148.8	176.2	211.3	295.0	400.7	446.0	590.3
Official Creditors	..	..	..	107.4	130.0	143.2	158.2	189.1	262.7	285.3	418.1
IBRD and IDA	..	..	..	29.3	27.8	28.9	30.0	32.6	34.7	43.0	44.2
Private Creditors	..	..	..	16.9	18.8	33.0	53.0	105.9	138.1	160.7	172.2
Private Non-guaranteed Long-Term	..	..	..	..	..	..	..	..	..	..	..
Use of Fund Credit	..	..	..	0.0	0.0	0.0	0.0	0.0	0.0	0.0	12.0
Short-Term Debt	..	..	..	..	..	..	..	..	..	..	129.6
Memo Items:				*(Thousands of US dollars)*							
Int'l Reserves Excluding Gold	2,623.0	7,407.4	5,741.7	8,884.8	10,820.0	10,338.0	7,858.1	24,101.0	13,815.0	12,168.0	13,530.0
Gold Holdings (at market price)	..	..	..	..	..	..	..	..	..	..	923.7
SOCIAL INDICATORS											
Total Fertility Rate	5.8	5.8	5.9	5.9	6.0	6.0	6.0	6.0	6.0	6.0	6.0
Crude Birth Rate	42.3	42.4	42.6	42.7	42.9	43.0	43.0	43.0	43.0	43.0	43.0
Infant Mortality Rate	110.0	106.0	102.0	98.0	94.0	90.0	89.0	88.0	87.0	86.0	85.0
Life Expectancy at Birth	50.2	50.6	50.9	51.3	51.7	52.1	52.5	52.9	53.3	53.7	54.2
Food Production, p.c. ('79-81=100)	111.0	108.0	104.3	103.1	103.7	100.8	100.6	100.1	102.2	101.3	99.1
Labor Force, Agriculture (%)	65.6	65.4	65.2	65.0	64.7	64.5	64.2	64.0	63.7	63.4	63.2
Labor Force, Female (%)	40.2	40.2	40.1	40.1	40.1	40.1	40.1	40.1	40.0	40.0	39.9
School Enroll. Ratio, primary	..	..	..	..	..	..	..	133.0	..	..	..
School Enroll. Ratio, secondary	..	..	..	..	..	..	..	33.0	..	..	..

1978	1979	1980	1981	1982	1983	1984	1985	1986	1987 estimate	NOTES	CONGO, PEOPLE'S REPUBLIC OF THE
											(Millions of current US dollars)
											FOREIGN TRADE (CUSTOMS BASIS)
160.9	523.9	979.6	1,114.8	1,212.5	1,096.6	1,265.0	1,145.0	718.0	883.5	..	Value of Exports, fob
35.6	51.4	60.2	106.8	134.6	112.1	131.2	119.2	127.2	155.2	..	Nonfuel Primary Products
93.5	433.7	855.7	907.0	952.0	878.0	1,016.0	909.1	457.1	591.1	..	Fuels
31.8	38.8	63.7	101.0	125.9	106.5	117.8	116.8	133.8	137.2	..	Manufactures
242.0	266.4	500.0	828.0	807.0	746.0	737.0	752.0	629.0	570.0	..	Value of Imports, cif
58.1	75.5	107.5	193.1	190.0	173.8	160.8	162.4	140.6	109.1	..	Nonfuel Primary Products
19.4	17.2	69.3	95.4	92.0	86.9	83.9	68.4	32.9	42.4	..	Fuels
164.6	173.7	323.3	539.5	525.1	485.3	492.3	521.2	455.5	418.4	..	Manufactures
											(Index 1980 = 100)
67.4	73.5	100.0	106.1	101.3	96.4	97.4	94.5	56.7	64.1	..	Terms of Trade
50.1	64.9	100.0	107.5	98.2	90.5	89.5	86.1	55.1	66.7	..	Export Prices, fob
60.9	96.1	100.0	82.7	81.3	77.8	85.2	74.3	81.3	94.9	..	Nonfuel Primary Products
42.3	61.0	100.0	112.5	101.6	92.5	90.2	87.5	44.9	56.9	..	Fuels
76.1	90.0	100.0	99.1	94.8	90.7	89.3	89.0	103.2	113.0	..	Manufactures
74.3	88.3	100.0	101.3	96.9	93.9	92.0	91.1	97.1	104.1	..	Import Prices, cif
											(Millions of 1980 US dollars)
											Trade at Constant 1980 Prices
321.3	807.8	979.6	1,037.5	1,235.1	1,211.1	1,412.8	1,330.1	1,302.8	1,323.8	..	Exports, fob
325.5	301.9	500.0	817.4	832.6	794.3	801.5	825.3	647.5	547.7	..	Imports, cif
											(Millions of current US dollars)
											BALANCE OF PAYMENTS
380.4	568.7	1,029.3	1,173.4	1,205.8	1,159.4	1,355.2	1,228.8	781.7	1,045.2	..	Exports of Goods & Services
308.2	495.7	910.6	1,072.7	1,108.5	1,066.2	1,268.3	1,145.0	670.8	912.1	..	Merchandise, fob
68.8	69.7	110.6	84.0	77.6	87.8	80.3	74.7	102.9	102.8	..	Nonfactor Services
3.4	3.3	8.0	16.6	19.6	5.4	6.5	9.1	8.0	30.3	f	Factor Services
599.3	679.1	1,194.9	1,645.7	1,519.2	1,556.2	1,142.4	1,393.0	1,399.6	1,295.8	..	Imports of Goods & Services
282.1	363.0	545.2	803.6	663.8	649.5	617.6	630.2	511.0	494.4	..	Merchandise, fob
259.2	219.9	480.1	713.3	697.3	727.4	391.3	525.7	655.0	506.1	..	Nonfactor Services
58.0	96.3	169.6	128.8	158.0	179.3	133.5	237.2	233.6	295.3	f	Factor Services
11.2	37.6	42.0	94.6	103.8	110.3	107.0	132.0	78.6	44.9	..	Long-Term Interest
-13.3	-31.7	-64.4	-32.3	-47.2	-41.1	-45.0	-37.5	-40.9	-47.2	..	Current Transfers, net
0.0	0.0	0.8	0.0	0.0	0.0	0.0	0.0	0.0	..	..	Workers' Remittances
-232.2	-142.1	-230.0	-504.6	-360.7	-437.9	167.9	-201.7	-658.8	-297.8	K	Total to be Financed
51.6	43.0	64.3	44.8	29.9	37.7	42.3	40.4	64.1	53.2	K	Official Capital Grants
-180.6	-99.1	-165.7	-459.8	-330.8	-400.2	210.2	-161.3	-594.7	-244.6	f	Current Account Balance
229.0	120.8	310.8	182.4	482.2	212.7	-13.1	68.3	7.9	235.4	..	Long-Term Capital, net
4.1	16.5	40.0	30.8	35.3	56.1	34.9	12.7	22.4	-39.9	..	Direct Investment
171.4	109.5	528.2	335.2	378.3	242.0	174.4	284.6	313.3	382.4	..	Long-Term Loans
186.8	183.1	577.0	401.0	503.8	442.9	408.5	551.8	550.4	532.0	..	Disbursements
15.4	73.6	48.8	65.8	125.5	200.9	234.1	267.2	237.1	149.6	..	Repayments
2.0	-48.2	-321.7	-228.4	38.6	-123.1	-264.8	-269.4	-391.9	-160.3	..	Other Long-Term Capital
-58.6	6.4	-86.0	340.0	-232.9	155.2	-205.5	91.6	577.3	5.2	..	Other Capital, net
10.2	-28.1	-59.1	-62.6	81.5	32.4	8.4	1.5	9.4	4.0	..	Change in Reserves
											(CFA Francs per US dollar)
											Memo Item:
225.640	212.720	211.300	271.730	328.620	381.070	436.960	449.260	346.300	300.540	..	Conversion Factor (Annual Avg)
											(Millions of US dollars, outstanding at end of year)
											EXTERNAL DEBT, ETC.
815.7	953.9	1,426.6	1,559.5	1,756.3	1,826.5	1,831.1	2,397.5	3,001.5	3,679.2	..	Public/Publicly Guar. Long-Term
527.2	606.8	656.3	720.2	750.9	762.8	758.0	860.5	1,235.5	1,594.9	..	Official Creditors
62.0	62.7	61.4	78.7	91.8	101.7	98.7	124.2	152.2	194.0	..	IBRD and IDA
288.5	347.1	770.3	839.3	1,005.4	1,063.7	1,073.2	1,537.0	1,766.0	2,084.3	..	Private Creditors
..	..	..	..	..	..	..	..	..	..	..	Private Non-guaranteed Long-Term
17.2	18.6	6.1	0.0	0.0	0.0	0.0	0.0	11.6	13.5	..	Use of Fund Credit
124.6	192.6	244.0	115.2	184.6	174.6	201.2	679.9	708.0	943.7	..	Short-Term Debt
											(Thousands of US dollars)
											Memo Items:
9,429.1	42,233.0	85,899.0	123,370.0	37,015.0	7,360.1	4,113.8	3,963.4	6,818.5	3,398.1	..	Int'l Reserves Excluding Gold
1,921.0	5,836.8	6,720.3	4,531.5	5,208.7	4,349.1	3,422.1	3,629.7	4,339.0	5,373.5	..	Gold Holdings (at market price)
											SOCIAL INDICATORS
6.0	6.0	6.0	6.0	6.0	6.1	6.2	6.3	6.4	6.5	..	Total Fertility Rate
43.0	43.0	43.0	43.0	43.0	43.9	44.7	45.6	46.4	47.3	..	Crude Birth Rate
84.2	83.4	82.6	81.8	81.0	79.0	77.0	75.0	72.9	70.9	..	Infant Mortality Rate
54.6	55.0	55.4	55.8	56.2	56.7	57.2	57.6	58.1	58.7	..	Life Expectancy at Birth
97.8	98.7	100.2	101.1	101.2	94.0	92.7	93.4	92.5	91.5	..	Food Production, p.c. ('79-81 = 100)
62.9	62.7	62.4	..	..	..	..	..	..	..	..	Labor Force, Agriculture (%)
39.9	39.9	39.8	39.7	39.6	39.5	39.4	39.3	39.2	39.1	..	Labor Force, Female (%)
..	..	..	..	..	163.0	..	..	..	..	..	School Enroll. Ratio, primary
..	..	..	..	..	87.0	..	..	..	..	..	School Enroll. Ratio, secondary

COSTA RICA	1967	1968	1969	1970	1971	1972	1973	1974	1975	1976	1977
CURRENT GNP PER CAPITA (US $)	430	470	500	560	600	650	750	860	950	1,070	1,320
POPULATION (thousands)	1,590	1,634	1,685	1,727	1,798	1,843	1,873	1,922	1,968	2,009	2,066

ORIGIN AND USE OF RESOURCES
(Billions of current Costa Rican Colones)

	1967	1968	1969	1970	1971	1972	1973	1974	1975	1976	1977
Gross National Product (GNP)	4.5	5.0	5.5	6.4	7.0	8.0	9.9	12.9	16.3	20.1	25.7
Net Factor Income from Abroad	-0.1	-0.1	-0.1	-0.1	-0.1	-0.3	-0.3	-0.3	-0.5	-0.6	-0.6
Gross Domestic Product (GDP)	4.6	5.1	5.7	6.5	7.1	8.2	10.2	13.2	16.8	20.7	26.3
Indirect Taxes, net	0.5	0.5	0.6	0.7	0.8	0.7	1.0	1.5	2.1	2.6	3.4
GDP at factor cost	4.2	4.6	5.1	5.8	6.4	7.5	9.2	11.8	14.7	18.1	22.9
Agriculture	1.1	1.2	1.3	1.5	1.4	1.6	2.0	2.5	3.4	4.2	5.8
Industry	1.1	1.2	1.3	1.6	1.8	2.1	2.6	3.6	4.6	5.7	6.9
Manufacturing	..	..	..	..	..	..	..	..	..	..	..
Services, etc.	2.5	2.7	3.0	3.5	3.9	4.5	5.6	7.1	8.8	10.8	13.7
Resource Balance	-0.3	-0.2	-0.3	-0.4	-0.7	-0.5	-0.6	-2.0	-1.4	-1.2	-1.4
Exports of Goods & NFServices	1.2	1.5	1.5	1.8	1.9	2.5	3.2	4.4	5.1	6.0	8.1
Imports of Goods & NFServices	1.5	1.7	1.8	2.3	2.7	3.1	3.8	6.4	6.5	7.2	9.6
Domestic Absorption	4.9	5.4	6.0	7.0	7.9	8.7	10.8	15.2	18.2	21.9	27.8
Private Consumption, etc.	3.4	3.8	4.1	4.8	5.2	5.8	6.9	9.8	12.0	13.7	17.2
General Gov't Consumption	0.6	0.6	0.7	0.8	1.0	1.2	1.4	1.9	2.6	3.3	4.2
Gross Domestic Investment	0.9	0.9	1.1	1.3	1.7	1.8	2.4	3.5	3.6	4.9	6.4
Fixed Investment	0.8	0.9	1.0	1.3	1.6	1.8	2.3	3.2	3.7	4.9	5.9
Memo Items:											
Gross Domestic Saving	0.6	0.7	0.8	0.9	1.0	1.3	1.8	1.6	2.2	3.7	5.0
Gross National Saving	0.5	0.6	0.8	0.8	0.9	1.1	1.6	1.3	1.8	3.2	4.5

(Millions of 1980 Costa Rican Colones)

	1967	1968	1969	1970	1971	1972	1973	1974	1975	1976	1977
Gross National Product	19,235	20,813	22,005	23,761	25,367	27,038	29,201	30,978	31,333	32,917	35,946
GDP at factor cost	..	..	..	..	..	..	..	..	..	..	..
Agriculture	4,553	4,964	5,480	5,705	5,969	6,292	6,648	6,535	6,733	6,767	6,916
Industry	3,975	4,510	4,773	5,197	5,718	6,431	6,979	7,786	8,062	8,766	9,689
Manufacturing	..	..	..	..	..	..	..	..	..	..	..
Services, etc.	11,097	11,770	12,134	13,165	13,987	15,010	16,245	17,157	17,314	18,321	20,271
Resource Balance	-1,606	-1,180	-1,455	-2,308	-2,382	-1,270	-1,264	-1,563	-965	-2,074	-4,173
Exports of Goods & NFServices	3,783	4,883	5,354	6,178	6,705	7,849	8,421	9,033	8,855	9,335	10,096
Imports of Goods & NFServices	5,388	6,063	6,808	8,486	9,087	9,120	9,685	10,596	9,820	11,408	14,269
Domestic Absorption	21,231	22,424	23,841	26,376	28,055	29,004	31,135	33,041	33,075	35,928	41,049
Private Consumption, etc.	14,389	15,488	16,133	17,760	18,094	19,145	20,025	20,985	21,309	22,105	24,938
General Gov't Consumption	3,354	3,507	3,746	4,106	4,367	4,674	4,957	5,376	5,680	6,123	6,659
Gross Domestic Investment	3,488	3,430	3,962	4,510	5,595	5,185	6,153	6,680	6,085	7,699	9,452
Fixed Investment	3,307	3,356	3,670	4,399	5,119	5,359	5,813	6,379	6,301	7,794	8,762
Memo Items:											
Capacity to Import	4,335	5,232	5,715	6,844	6,579	7,538	8,098	7,350	7,658	9,447	12,122
Terms of Trade Adjustment	553	349	361	666	-126	-312	-324	-1,683	-1,197	112	2,027
Gross Domestic Income	20,178	21,593	22,748	24,734	25,547	27,422	29,548	29,795	30,913	33,966	38,903
Gross National Income	19,787	21,162	22,366	24,428	25,241	26,726	28,877	29,295	30,136	33,029	37,972

DOMESTIC PRICES (DEFLATORS)
(Index 1980 = 100)

	1967	1968	1969	1970	1971	1972	1973	1974	1975	1976	1977
Overall (GDP)	23.6	24.1	25.3	27.1	27.8	29.6	34.0	42.0	52.3	61.1	71.4
Domestic Absorption	23.2	23.9	25.0	26.4	28.1	30.1	34.6	45.9	55.1	61.0	67.7
Agriculture	23.4	23.7	23.8	25.8	24.2	25.5	29.5	38.6	50.8	62.3	83.3
Industry	26.6	26.8	28.2	30.4	31.4	32.3	36.8	45.9	57.1	64.7	71.1
Manufacturing	..	..	..	..	..	..	..	..	..	..	..

MANUFACTURING ACTIVITY

	1967	1968	1969	1970	1971	1972	1973	1974	1975	1976	1977
Employment (1980=100)	..	..	..	..	..	..	..	..	..	..	..
Real Earnings per Empl. (1980=100)	..	..	..	..	..	..	..	..	..	..	..
Real Output per Empl. (1980=100)	..	..	..	..	..	..	..	..	..	..	..
Earnings as % of Value Added	..	23.4	41.9	41.4	41.9	41.8	40.0	38.4	39.5	40.1	38.6

MONETARY HOLDINGS
(Billions of current Costa Rican Colones)

	1967	1968	1969	1970	1971	1972	1973	1974	1975	1976	1977
Money Supply, Broadly Defined	1.08	1.07	1.19	1.28	1.81	2.17	2.64	3.45	4.90	6.59	8.66
Money as Means of Payment	0.83	0.85	0.96	1.01	1.32	1.50	1.87	2.15	2.77	3.41	4.50
Currency Ouside Banks	0.28	0.31	0.35	0.38	0.43	0.52	0.64	0.73	0.85	1.12	1.41
Demand Deposits	0.55	0.54	0.61	0.63	0.88	0.98	1.23	1.41	1.92	2.29	3.09
Quasi-Monetary Liabilities	0.25	0.22	0.24	0.27	0.49	0.66	0.77	1.30	2.13	3.18	4.16

GOVERNMENT DEFICIT (-) OR SURPLUS
(Millions of current Costa Rican Colones)

	1967	1968	1969	1970	1971	1972	1973	1974	1975	1976	1977
	..	..	..	..	..	-356	-380	-89	-371	-704	-818
Current Revenue	..	..	..	..	..	1,260	1,620	2,449	3,027	3,648	4,393
Current Expenditure	..	..	..	..	..	1,262	1,565	1,994	2,748	3,500	4,161
Current Budget Balance	..	..	..	..	..	-2	55	455	279	148	232
Capital Receipts	..	..	..	..	..	..	..	..	..	..	..
Capital Payments	..	..	..	..	..	354	435	543	649	852	1,051

1978	1979	1980	1981	1982	1983	1984	1985	1986	1987 estimate	NOTES	COSTA RICA
1,560	1,790	1,950	1,530	1,140	1,060	1,260	1,400	1,500	1,550	..	CURRENT GNP PER CAPITA (US $)
2,115	2,166	2,218	2,271	2,324	2,379	2,434	2,490	2,557	2,626	..	POPULATION (thousands)
				(Billions of current Costa Rican Colones)							ORIGIN AND USE OF RESOURCES
29.3	33.3	39.4	50.7	81.4	115.6	151.1	181.1	230.3	256.6	..	Gross National Product (GNP)
-0.9	-1.3	-2.0	-6.4	-16.1	-13.7	-11.9	-11.3	-11.9	-13.7	..	Net Factor Income from Abroad
30.2	34.6	41.4	57.1	97.5	129.3	163.0	192.4	242.1	270.3	..	Gross Domestic Product (GDP)
4.0	4.2	4.9	6.8	11.3	18.6	23.9	26.7	33.3	23.1	..	Indirect Taxes, net
26.2	30.4	36.5	50.3	86.2	110.7	139.1	165.7	208.9	247.1	B	GDP at factor cost
6.2	6.4	7.4	13.1	23.9	28.5	34.6	38.4	53.1	49.8	..	Agriculture
7.9	9.2	11.2	15.2	25.1	37.0	48.0	56.0	68.8	77.8	..	Industry
..	..	..	..	..	..	..	..	..	..	f	Manufacturing
16.2	19.0	22.9	28.8	48.5	63.9	80.4	98.1	120.3	142.7	..	Services, etc.
-2.4	-3.6	-4.3	-2.8	2.9	-1.0	0.8	-3.4	3.2	-8.7	..	Resource Balance
8.5	9.3	11.0	24.7	44.0	46.6	57.0	61.9	79.2	91.7	..	Exports of Goods & NFServices
10.9	12.9	15.3	27.5	41.1	47.6	56.2	65.2	76.0	100.4	..	Imports of Goods & NFServices
32.6	38.1	45.7	59.9	94.7	130.3	162.2	195.8	238.9	279.0	..	Domestic Absorption
20.4	23.1	27.1	34.3	56.4	79.5	104.0	127.7	151.1	182.0	..	Private Consumption, etc.
5.1	6.2	7.5	9.0	14.2	19.5	23.9	28.2	34.7	39.6	..	General Gov't Consumption
7.1	8.8	11.0	16.6	24.1	31.3	34.4	39.9	53.1	57.4	..	Gross Domestic Investment
7.0	9.1	9.9	13.7	19.8	23.3	32.7	37.3	45.5	56.8	..	Fixed Investment
											Memo Items:
4.7	5.2	6.7	13.8	26.9	30.3	35.2	36.6	56.3	48.7	..	Gross Domestic Saving
4.0	4.1	4.9	7.9	11.9	17.6	24.7	27.4	46.6	37.3	..	Gross National Saving
				(Millions of 1980 Costa Rican Colones)							
37,925	39,536	39,418	37,842	34,064	35,644	38,792	39,315	41,528	42,753	..	Gross National Product
..	36,067	36,543	35,880	33,730	33,940	37,189	36,815	36,586	37,662	B	GDP at factor cost
7,373	7,409	7,372	7,748	7,384	7,679	8,450	8,212	8,454	8,030	..	Agriculture
10,420	11,038	11,166	10,697	9,254	9,632	10,718	10,904	11,578	12,272	..	Industry
..	..	..	..	..	..	..	..	..	..	f	Manufacturing
21,366	22,651	22,867	22,025	21,010	21,383	22,548	23,012	24,017	25,048	..	Services, etc.
-4,240	-4,326	-4,282	953	2,322	717	730	-1,160	-2,748	-2,934	..	Resource Balance
11,099	11,461	10,963	12,182	11,517	11,366	12,545	11,722	11,113	12,728	..	Exports of Goods & NFServices
15,340	15,787	15,245	11,230	9,196	10,649	11,815	12,882	13,861	15,662	..	Imports of Goods & NFServices
43,399	45,424	45,687	39,517	35,326	37,977	40,986	43,289	46,798	48,284	..	Domestic Absorption
27,083	27,705	27,140	25,544	23,278	24,444	26,783	28,767	30,091	32,225	..	Private Consumption, etc.
6,904	7,436	7,544	7,121	6,940	6,736	6,661	6,513	7,087	7,051	..	General Gov't Consumption
9,412	10,283	11,003	6,852	5,108	6,797	7,541	8,009	9,619	9,008	..	Gross Domestic Investment
9,475	10,923	9,894	7,430	5,364	5,807	7,326	7,720	8,358	9,069	..	Fixed Investment
											Memo Items:
11,997	11,428	10,963	10,086	9,832	10,434	11,982	12,221	14,442	14,307	..	Capacity to Import
898	-34	0	-2,097	-1,685	-933	-564	499	3,329	1,579	..	Terms of Trade Adjustment
40,057	41,065	41,406	38,373	35,962	37,761	41,152	42,628	47,379	46,929	..	Gross Domestic Income
38,823	39,502	39,418	35,746	32,379	34,711	38,228	39,814	44,858	44,332	..	Gross National Income
				(Index 1980 = 100)							DOMESTIC PRICES (DEFLATORS)
77.1	84.2	100.0	141.1	259.0	334.2	390.8	456.8	549.7	596.0	..	Overall (GDP)
75.0	84.0	100.0	151.6	268.0	343.0	395.8	452.2	510.6	577.7	..	Domestic Absorption
83.6	86.4	100.0	169.7	323.5	370.5	409.1	467.9	628.1	619.6	..	Agriculture
75.5	82.9	100.0	142.1	271.2	384.2	447.9	513.1	593.8	634.0	..	Industry
..	..	..	..	..	..	..	..	..	..	..	Manufacturing
											MANUFACTURING ACTIVITY
..	..	..	..	..	..	..	..	..	..	G	Employment (1980 = 100)
..	..	..	..	..	..	..	..	..	..	G	Real Earnings per Empl. (1980 = 100)
..	..	..	..	..	..	..	..	..	..	G	Real Output per Empl. (1980 = 100)
40.0	40.5	42.9	..	..	..	..	..	..	..	..	Earnings as % of Value Added
				(Billions of current Costa Rican Colones)							MONETARY HOLDINGS
11.07	14.87	17.24	32.27	40.99	56.14	65.76	76.00	92.16	107.15	..	Money Supply, Broadly Defined
5.62	6.23	7.27	10.83	18.45	25.62	30.13	32.44	42.49	42.61	..	Money as Means of Payment
1.70	1.95	2.26	3.50	5.44	6.94	8.59	9.94	13.24	14.78	..	Currency Ouside Banks
3.92	4.27	5.02	7.33	13.01	18.68	21.54	22.50	29.25	27.83	..	Demand Deposits
5.44	8.64	9.96	21.44	22.55	30.52	35.63	43.56	49.67	64.54	..	Quasi-Monetary Liabilities
				(Millions of current Costa Rican Colones)							
-1,507	-2,341	-3,062	-1,640	-864	-2,591	-1,195	-2,467	-11,034	..	..	GOVERNMENT DEFICIT (-) OR SURPLUS
5,755	6,316	7,373	10,198	17,024	28,115	36,377	41,011	54,564	..	..	Current Revenue
5,867	7,038	8,806	10,631	17,487	27,005	33,077	36,964	53,690	..	..	Current Expenditure
-112	-722	-1,433	-432	-463	1,110	3,300	4,046	874	..	..	Current Budget Balance
..	1	0	..	2	0	5	..	..	..	..	Capital Receipts
1,394	1,620	1,629	1,207	403	3,700	4,500	6,513	11,907	..	..	Capital Payments

COSTA RICA	1967	1968	1969	1970	1971	1972	1973	1974	1975	1976	1977
FOREIGN TRADE (CUSTOMS BASIS)				*(Millions of current US dollars)*							
Value of Exports, fob	144	171	190	231	225	281	345	440	494	600	840
Nonfuel Primary Products	117	137	154	185	172	221	264	322	364	429	636
Fuels	0	0	0	1	2	1	0	1	0	1	1
Manufactures	27	34	36	46	51	59	80	118	130	171	203
Value of Imports, cif	191	214	245	317	350	373	455	720	694	800	1,059
Nonfuel Primary Products	27	33	32	45	52	49	61	109	89	95	115
Fuels	9	10	11	12	16	20	32	65	74	74	102
Manufactures	155	171	203	260	282	304	363	546	531	631	842
				(Index 1980 = 100)							
Terms of Trade	130.8	131.1	130.0	132.5	112.7	121.9	127.1	98.1	98.9	120.4	139.4
Export Prices, fob	32.2	32.1	33.5	36.5	33.1	37.8	47.1	55.0	58.0	72.9	91.8
Nonfuel Primary Products	32.7	32.6	34.0	37.7	34.0	38.2	46.8	51.6	55.7	75.9	101.0
Fuels	4.3	4.3	4.3	4.3	5.6	6.2	8.9	36.7	35.7	38.4	42.0
Manufactures	30.1	31.7	32.4	37.5	36.5	38.5	49.5	67.2	66.1	66.7	71.7
Import Prices, cif	24.6	24.5	25.8	27.5	29.4	31.0	37.1	56.1	58.7	60.6	65.8
Trade at Constant 1980 Prices				*(Millions of 1980 US dollars)*							
Exports, fob	447	533	566	634	681	743	731	800	851	823	915
Imports, cif	775	874	951	1,150	1,190	1,202	1,228	1,284	1,182	1,321	1,609
BALANCE OF PAYMENTS				*(Millions of current US dollars)*							
Exports of Goods & Services	..	..	..	278	281	344	419	540	601	711	969
Merchandise, fob	..	..	..	231	225	279	345	440	493	592	828
Nonfactor Services	..	..	..	46	56	64	71	95	103	113	131
Factor Services	..	..	..	1	1	1	3	4	4	6	10
Imports of Goods & Services	..	..	..	358	403	451	538	815	828	925	1,211
Merchandise, fob	..	..	..	287	317	337	412	649	627	695	925
Nonfactor Services	..	..	..	54	68	75	82	121	130	149	195
Factor Services	..	..	..	17	18	39	43	46	70	81	90
Long-Term Interest	..	..	..	14	15	18	24	31	37	41	52
Current Transfers, net	..	..	..	..	..	..	..	..	..	..	..
Workers' Remittances	..	..	..	0	0	0	0	0	0	0	0
Total to be Financed	..	..	..	..	..	..	..	..	..	..	..
Official Capital Grants	..	..	..	..	..	..	..	..	..	..	..
Current Account Balance	..	..	..	-74	-114	-100	-112	-266	-218	-202	-226
Long-Term Capital, net	..	..	..	46	61	83	92	137	238	219	299
Direct Investment	..	..	..	26	22	26	38	46	69	61	63
Long-Term Loans	..	..	..	19	35	53	61	84	158	156	248
Disbursements	..	..	..	60	78	103	118	158	260	276	399
Repayments	..	..	..	41	43	50	57	74	102	120	151
Other Long-Term Capital	..	..	..	1	4	4	-7	7	11	2	-11
Other Capital, net	..	..	..	13	60	21	46	79	-18	34	33
Change in Reserves	..	..	..	15	-6	-4	-26	51	-2	-52	-106
Memo Item:				*(Costa Rican Colones per US dollar)*							
Conversion Factor (Annual Avg)	6.630	6.630	6.630	6.630	6.630	6.640	6.650	7.930	8.570	8.570	8.570
EXTERNAL DEBT, ETC.				*(Millions of US dollars, outstanding at end of year)*							
Public/Publicly Guar. Long-Term	..	..	..	134	167	207	249	303	421	542	733
Official Creditors	..	..	..	96	112	138	154	178	239	304	385
IBRD and IDA	..	..	..	41	48	60	69	81	90	109	129
Private Creditors	..	..	..	38	55	69	95	125	182	238	347
Private Non-guaranteed Long-Term	..	..	..	112	116	132	153	182	228	263	326
Use of Fund Credit	..	..	..	0	0	0	0	23	35	38	36
Short-Term Debt	..	..	..	..	..	..	..	..	..	..	223
Memo Items:				*(Millions of US dollars)*							
Int'l Reserves Excluding Gold	16.1	18.5	27.1	14.2	27.2	40.6	48.5	42.1	48.8	95.4	190.5
Gold Holdings (at market price)	2.1	2.5	2.1	2.2	2.6	3.9	6.7	11.1	8.4	8.0	12.1
SOCIAL INDICATORS											
Total Fertility Rate	5.8	5.5	5.2	4.9	4.6	4.3	4.2	4.1	3.9	3.8	3.7
Crude Birth Rate	40.2	36.2	34.4	33.4	31.3	31.2	28.5	29.5	29.5	29.8	31.0
Infant Mortality Rate	60.3	59.7	67.1	61.5	56.5	54.4	44.8	37.5	37.8	33.1	27.9
Life Expectancy at Birth	65.7	66.1	66.6	67.1	67.6	68.1	68.4	68.7	69.0	69.4	69.7
Food Production, p.c. ('79-81=100)	75.5	86.7	90.5	96.7	100.9	102.5	103.8	98.2	107.1	107.5	106.4
Labor Force, Agriculture (%)	45.2	44.3	43.5	42.6	41.4	40.2	39.0	37.8	36.6	35.4	34.3
Labor Force, Female (%)	17.3	17.6	17.8	18.1	18.4	18.7	19.0	19.3	19.6	19.9	20.2
School Enroll. Ratio, primary	..	..	..	110.0	..	..	..	..	107.0	108.0	109.0
School Enroll. Ratio, secondary	..	..	..	28.0	..	..	..	..	42.0	43.0	44.0

1978	1979	1980	1981	1982	1983	1984	1985	1986	1987 estimate	NOTES	COSTA RICA
											FOREIGN TRADE (CUSTOMS BASIS)
				(Millions of current US dollars)							
919	934	1,032	1,011	877	867	967	976	1,125	1,155	..	Value of Exports, fob
655	705	671	676	621	602	664	637	716	681	..	Nonfuel Primary Products
1	1	6	13	8	8	8	9	5	6	..	Fuels
263	228	354	322	248	257	294	331	405	467	..	Manufactures
1,212	1,446	1,597	1,274	945	993	1,091	1,098	1,148	1,377	..	Value of Imports, cif
123	153	198	158	114	121	122	112	104	95	..	Nonfuel Primary Products
118	190	246	205	189	186	172	171	93	129	..	Fuels
971	1,104	1,153	911	642	686	798	815	951	1,153	..	Manufactures
				(Index 1980 = 100)							
114.1	111.8	100.0	93.6	94.2	97.5	96.5	94.8	103.4	84.2	..	Terms of Trade
84.1	96.1	100.0	94.7	91.5	92.6	90.2	87.8	100.8	91.5	..	Export Prices, fob
88.0	98.3	100.0	92.5	90.2	93.5	90.6	87.2	100.3	81.3	..	Nonfuel Primary Products
42.3	61.0	100.0	112.5	101.6	92.5	90.2	87.5	44.9	56.9	..	Fuels
76.1	90.0	100.0	99.1	94.8	90.7	89.3	89.0	103.2	113.0	..	Manufactures
73.7	86.0	100.0	101.2	97.1	95.0	93.5	92.6	97.5	108.6	..	Import Prices, cif
											Trade at Constant 1980 Prices
				(Millions of 1980 US dollars)							
1,093	972	1,032	1,067	958	936	1,072	1,112	1,116	1,263	..	Exports, fob
1,644	1,683	1,597	1,259	973	1,045	1,167	1,186	1,177	1,268	..	Imports, cif
				(Millions of current US dollars)							**BALANCE OF PAYMENTS**
1,025	1,111	1,219	1,199	1,143	1,173	1,314	1,270	1,440	1,503	f	Exports of Goods & Services
864	942	1,001	1,003	869	853	998	939	1,085	1,114	..	Merchandise, fob
144	156	197	173	248	280	279	281	310	342	..	Nonfactor Services
17	13	21	24	27	40	38	50	45	48	..	Factor Services
1,405	1,682	1,897	1,639	1,451	1,528	1,610	1,619	1,676	1,916	f	Imports of Goods & Services
1,049	1,257	1,375	1,091	805	898	997	1,005	1,049	1,375	..	Merchandise, fob
225	262	283	213	238	250	255	274	296	208	..	Nonfactor Services
131	163	239	335	409	380	358	340	331	334	..	Factor Services
85	116	171	135	109	529	243	342	217	139	..	Long-Term Interest
..	..	20	27	30	23	32	43	..	37	..	Current Transfers, net
0	0	0	0	0	0	0	0	0	..	..	Workers' Remittances
..	..	-659	-409	-273	-329	-264	-306	..	..	..	Total to be Financed
..	..	-5	0	6	46	109	176	..	..	..	Official Capital Grants
-363	-558	-664	-409	-267	-283	-155	-130	-84	-225	..	Current Account Balance
353	350	397	218	28	1,221	176	520	59	-408	f	Long-Term Capital, net
47	42	48	66	27	55	52	65	57	65	..	Direct Investment
212	199	370	188	123	293	101	170	-14	10	..	Long-Term Loans
515	465	533	366	244	430	239	311	190	86	..	Disbursements
304	266	163	178	121	137	137	141	203	76	..	Repayments
95	109	-16	-37	-127	828	-86	109	15	-483	..	Other Long-Term Capital
-17	125	229	147	352	-866	-26	-259	105	696	f	Other Capital, net
27	83	38	45	-113	-72	5	-130	-79	-62	..	Change in Reserves
											Memo Item:
				(Costa Rican Colones per US dollar)							
8.570	8.570	8.570	21.760	37.410	41.090	44.530	50.450	55.990	62.780	..	Conversion Factor (Annual Avg)
				(Millions of US dollars, outstanding at end of year)							**EXTERNAL DEBT, ETC.**
947	1,301	1,692	2,192	2,378	3,128	3,160	3,505	3,580	3,629	..	Public/Publicly Guar. Long-Term
501	600	776	894	1,063	1,372	1,404	1,711	1,800	1,998	..	Official Creditors
146	161	183	197	204	210	195	313	417	500	..	IBRD and IDA
446	700	916	1,298	1,315	1,757	1,756	1,794	1,780	1,631	..	Private Creditors
417	398	412	372	381	348	317	302	306	290	..	Private Non-guaranteed Long-Term
32	58	57	103	93	192	156	189	172	132	..	Use of Fund Credit
284	353	575	621	778	495	338	379	471	676	..	Short-Term Debt
				(Millions of US dollars)							Memo Items:
193.9	118.6	145.6	131.4	226.1	311.3	405.0	506.4	523.4	488.9	..	Int'l Reserves Excluding Gold
18.1	44.5	51.3	11.5	23.7	33.4	7.1	19.0	27.1	30.3	..	Gold Holdings (at market price)
											SOCIAL INDICATORS
3.7	3.6	3.6	3.6	3.5	3.5	3.4	3.4	3.3	3.3	..	Total Fertility Rate
29.9	30.2	30.3	30.4	30.0	29.6	29.3	28.9	28.5	28.1	..	Crude Birth Rate
23.8	23.3	20.1	17.9	17.8	17.6	17.5	17.4	17.2	17.1	..	Infant Mortality Rate
70.4	71.0	71.7	72.4	73.0	73.2	73.3	73.4	73.5	73.7	..	Life Expectancy at Birth
104.6	105.1	98.6	96.4	89.0	89.2	91.2	91.7	92.4	91.6	..	Food Production, p.c. ('79-81 = 100)
33.1	32.0	30.8	..	..	..	..	..	..	..	..	Labor Force, Agriculture (%)
20.6	20.9	21.2	21.3	21.4	21.5	21.6	21.7	21.7	21.7	..	Labor Force, Female (%)
107.0	107.0	106.0	106.0	103.0	102.0	101.0	101.0	..	..	..	School Enroll. Ratio, primary
..	48.0	47.0	47.0	46.0	44.0	42.0	41.0	..	..	..	School Enroll. Ratio, secondary

COTE D'IVOIRE	1967	1968	1969	1970	1971	1972	1973	1974	1975	1976	1977
CURRENT GNP PER CAPITA (US $)	210	240	260	270	270	280	320	390	500	580	670
POPULATION (thousands)	4,876	5,078	5,290	5,510	5,739	5,976	6,222	6,481	6,754	7,041	7,341
ORIGIN AND USE OF RESOURCES					*(Billions of current CFA Francs)*						
Gross National Product (GNP)	248.6	298.6	334.6	380.3	407.2	433.8	509.5	679.6	767.2	1,006.9	1,406.9
Net Factor Income from Abroad	-17.8	-18.6	-19.3	-22.0	-29.0	-32.3	-49.6	-59.4	-67.3	-107.1	-132.3
Gross Domestic Product (GDP)	266.4	317.2	353.9	402.3	436.2	466.1	559.1	739.0	834.5	1,114.0	1,539.2
Indirect Taxes, net	46.0	55.1	69.5	83.9	79.1	83.8	110.2	164.5	149.8	299.8	501.5
GDP at factor cost	220.4	262.1	284.4	318.4	357.1	382.3	448.9	574.5	684.7	814.2	1,037.7
Agriculture	92.6	112.8	118.8	128.2	134.7	138.9	173.6	188.2	235.7	272.7	373.5
Industry	50.3	51.7	54.6	74.4	85.9	92.0	99.2	118.9	142.2	182.3	233.3
Manufacturing	32.5	30.9	32.7	41.4	45.1	51.7	56.8	75.6	78.4	104.8	117.6
Services, etc.	77.5	97.6	111.0	115.8	136.5	151.4	176.1	267.4	306.8	359.2	430.9
Resource Balance	19.8	40.3	34.3	26.8	16.5	21.0	14.1	54.1	1.2	61.2	96.9
Exports of Goods & NFServices	96.5	129.1	130.1	143.9	140.5	156.2	199.9	337.7	306.5	465.0	656.1
Imports of Goods & NFServices	76.7	88.8	95.8	117.1	124.0	135.2	185.8	283.6	305.3	403.8	559.2
Domestic Absorption	246.6	276.9	319.6	375.5	419.7	445.1	545.0	684.9	833.3	1,052.8	1,442.3
Private Consumption, etc.	164.7	184.8	207.3	230.4	255.1	275.1	327.3	402.5	504.2	616.4	811.9
General Gov't Consumption	29.2	35.6	43.2	54.6	69.5	72.7	88.1	119.7	141.8	180.3	209.7
Gross Domestic Investment	52.7	56.5	69.1	90.5	95.1	97.3	129.6	162.7	187.3	256.1	420.7
Fixed Investment	51.1	55.0	61.3	83.1	91.5	94.2	121.9	143.6	183.9	247.2	397.6
Memo Items:											
Gross Domestic Saving	72.5	96.8	103.4	117.3	111.6	118.3	143.7	216.8	188.5	317.3	517.6
Gross National Saving	54.7	66.2	71.1	79.9	64.3	63.5	66.7	123.9	81.8	141.0	300.6
					(Billions of 1980 CFA Francs)						
Gross National Product	846.9	994.2	1,074.9	1,155.5	1,278.2	1,364.2	1,396.4	1,516.2	1,680.5	1,718.8	1,691.6
GDP at factor cost	771.4	893.3	938.4	998.3	1,157.1	1,231.3	1,278.5	1,376.2	1,568.6	1,506.1	1,379.4
Agriculture	346.5	403.5	405.0	422.4	454.8	457.3	485.6	434.1	474.9	503.5	481.6
Industry	105.7	104.7	108.2	143.2	173.6	185.7	178.9	175.5	200.9	232.1	255.3
Manufacturing	69.2	62.5	64.5	79.5	90.0	103.4	102.1	113.8	116.2	142.8	133.7
Services, etc.	319.2	385.1	425.2	432.8	528.7	588.3	614.0	766.6	892.8	770.5	642.5
Resource Balance	83.4	152.6	128.9	109.1	129.0	179.3	88.0	137.5	125.9	74.1	-135.0
Exports of Goods & NFServices	349.8	447.4	437.9	471.4	498.9	558.5	546.9	627.6	602.6	673.3	620.0
Imports of Goods & NFServices	266.4	294.8	309.0	362.3	369.8	379.2	458.9	490.2	476.8	599.2	755.0
Domestic Absorption	839.3	917.3	1,022.2	1,129.8	1,255.1	1,295.8	1,458.4	1,504.4	1,683.3	1,839.3	2,045.3
Private Consumption, etc.	621.1	684.0	743.6	791.5	891.1	941.5	1,030.7	1,067.7	1,219.7	1,272.7	1,312.5
General Gov't Consumption	68.9	80.4	94.7	115.6	142.4	141.6	160.5	175.0	194.4	226.6	244.2
Gross Domestic Investment	149.2	152.9	183.9	222.7	221.6	212.7	267.2	261.8	269.2	339.9	488.6
Fixed Investment	144.0	148.3	160.5	201.4	211.0	203.7	248.1	226.9	263.2	325.2	465.2
Memo Items:											
Capacity to Import	335.2	428.7	419.6	445.2	419.0	438.1	493.7	583.7	478.6	690.0	885.8
Terms of Trade Adjustment	-14.6	-18.8	-18.3	-26.2	-79.8	-120.4	-53.2	-44.0	-124.0	16.7	265.9
Gross Domestic Income	908.0	1,051.1	1,132.8	1,212.7	1,304.3	1,354.7	1,493.2	1,598.0	1,685.2	1,930.1	2,176.2
Gross National Income	832.3	975.5	1,056.6	1,129.3	1,198.4	1,243.7	1,343.2	1,472.2	1,556.5	1,735.5	1,957.4
DOMESTIC PRICES (DEFLATORS)					*(Index 1980 = 100)*						
Overall (GDP)	28.9	29.6	30.7	32.5	31.5	31.6	36.2	45.0	46.1	58.2	80.6
Domestic Absorption	29.4	30.2	31.3	33.2	33.4	34.3	37.4	45.5	49.5	57.2	70.5
Agriculture	26.7	28.0	29.3	30.4	29.6	30.4	35.8	43.4	49.6	54.2	77.6
Industry	47.6	49.4	50.4	52.0	49.5	49.5	55.4	67.7	70.8	78.5	91.4
Manufacturing	47.0	49.5	50.7	52.1	50.1	50.0	55.6	66.4	67.5	73.4	88.0
MANUFACTURING ACTIVITY											
Employment (1980=100)	34.8	39.2	42.9	49.7	53.8	56.8	61.0	64.4	73.5	80.8	87.4
Real Earnings per Empl. (1980=100)	91.7	88.9	99.9	98.3	103.8	110.4	106.6	105.6	100.2	99.2	89.9
Real Output per Empl. (1980=100)	53.8	51.1	51.5	51.6	56.9	62.3	63.0	76.1	72.4	77.0	82.2
Earnings as % of Value Added	21.7	22.5	26.1	27.1	26.2	25.6	27.6	25.0	29.8	29.7	26.1
MONETARY HOLDINGS					*(Billions of current CFA Francs)*						
Money Supply, Broadly Defined	58.2	72.5	91.9	106.8	117.6	122.8	147.9	223.2	244.6	349.7	524.5
Money as Means of Payment	48.5	59.1	69.8	83.5	92.1	103.2	117.9	162.8	179.8	260.1	383.1
Currency Ouside Banks	27.6	30.6	34.0	39.8	46.9	51.5	57.0	77.5	89.6	106.7	137.3
Demand Deposits	20.9	28.5	35.8	43.7	45.1	51.7	60.9	85.3	90.2	153.4	245.8
Quasi-Monetary Liabilities	9.7	13.4	22.1	23.2	25.5	19.7	30.0	60.4	64.7	89.6	141.4
GOVERNMENT DEFICIT (-) OR SURPLUS					*(Billions of current CFA Francs)*						
Current Revenue	..	..	..	..	..	..	..	..	..	..	..
Current Expenditure	..	..	..	..	..	..	..	..	..	..	..
Current Budget Balance	..	..	..	..	..	..	..	..	..	..	..
Capital Receipts	..	..	..	..	..	..	..	..	..	..	..
Capital Payments	..	..	..	..	..	..	..	..	..	..	..

1978	1979	1980	1981	1982	1983	1984	1985	1986	1987 estimate	NOTES	COTE D'IVOIRE
820	980	1,170	1,130	960	760	660	630	700	750	..	CURRENT GNP PER CAPITA (US $)
7,655	7,984	8,327	8,684	9,056	9,441	9,840	10,252	10,650	11,069	..	POPULATION (thousands)
				(Billions of current CFA Francs)							ORIGIN AND USE OF RESOURCES
1,610.5	1,742.4	2,102.7	2,150.8	2,319.8	2,388.1	2,661.9	2,832.0	2,998.0	2,836.7	..	Gross National Product (GNP)
-172.3	-202.5	-118.9	-140.7	-166.7	-193.8	-207.4	-305.8	-246.3	-217.8	f	Net Factor Income from Abroad
1,782.8	1,944.9	2,221.6	2,291.5	2,486.5	2,581.9	2,869.3	3,137.8	3,244.3	3,054.5	..	Gross Domestic Product (GDP)
448.6	448.4	429.4	373.5	450.1	502.8	728.4	792.9	696.3	756.8	..	Indirect Taxes, net
1,334.2	1,496.5	1,792.2	1,918.0	2,036.4	2,079.1	2,140.9	2,344.9	2,548.0	2,297.7	..	GDP at factor cost
461.4	513.5	597.9	655.8	650.2	633.8	788.4	833.5	916.1	820.0	..	Agriculture
293.8	331.8	357.4	375.8	462.9	487.8	506.8	618.0	613.9	585.3	..	Industry
137.0	158.1	195.6	201.0	263.5	301.6	334.2	399.2	412.3	..	..	Manufacturing
579.0	651.2	836.9	886.4	923.3	957.5	845.7	893.4	1,018.0	892.4	..	Services, etc.
-16.3	-59.5	-134.6	-162.8	-72.2	-43.2	330.2	414.3	389.2	170.8	f	Resource Balance
651.1	673.0	756.0	806.0	905.7	942.8	1,314.1	1,437.9	1,269.2	1,039.4	..	Exports of Goods & NFServices
667.4	732.5	890.6	968.8	977.9	986.0	983.9	1,023.6	880.0	868.6	..	Imports of Goods & NFServices
1,799.1	2,004.4	2,356.2	2,454.3	2,558.7	2,625.1	2,539.1	2,723.5	2,855.1	2,883.7	..	Domestic Absorption
978.1	1,106.7	1,333.6	1,462.0	1,553.9	1,658.5	1,787.8	1,891.7	2,004.8	1,974.4	..	Private Consumption, etc.
290.4	353.8	395.9	398.0	427.8	435.0	438.0	437.1	491.7	511.5	..	General Gov't Consumption
530.6	543.9	626.7	594.3	577.0	531.6	313.3	394.7	358.6	397.8	..	Gross Domestic Investment
528.8	527.1	581.8	558.4	538.7	469.5	352.6	359.3	386.2	393.7	..	Fixed Investment
											Memo Items:
514.3	484.4	492.1	431.5	504.8	488.4	643.5	809.0	747.8	568.6	..	Gross Domestic Saving
238.6	159.3	222.0	156.4	209.5	172.8	308.9	377.9	353.9	235.6	..	Gross National Saving
				(Billions of 1980 CFA Francs)							
1,901.8	1,955.4	2,102.7	2,176.0	2,199.3	2,157.0	2,064.0	2,202.8	2,333.7	2,282.5	..	Gross National Product
1,708.6	1,794.6	1,792.2	1,904.6	1,936.5	1,936.3	1,799.0	2,085.1	2,161.7	2,077.0	..	GDP at factor cost
527.4	529.5	597.9	610.4	602.0	535.5	550.5	602.4	674.8	668.9	..	Agriculture
304.1	328.4	357.4	393.0	334.2	344.5	354.1	340.6	321.3	303.6	..	Industry
143.3	155.4	195.6	212.2	183.4	204.7	259.7	295.5	290.6	..	..	Manufacturing
877.1	936.7	836.9	901.2	1,000.3	1,056.3	894.4	1,142.1	1,165.6	1,104.5	..	Services, etc.
-210.0	-167.7	-134.6	-37.3	67.2	71.7	215.9	252.3	228.7	198.9	f	Resource Balance
655.6	670.5	756.0	808.5	820.2	754.5	831.8	824.4	796.0	754.4	..	Exports of Goods & NFServices
865.6	838.3	890.6	845.8	752.9	682.8	615.9	572.1	567.3	555.5	..	Imports of Goods & NFServices
2,385.5	2,406.9	2,356.2	2,354.3	2,286.4	2,253.7	2,007.0	2,176.1	2,287.6	2,243.3	..	Domestic Absorption
1,525.3	1,562.8	1,333.6	1,523.6	1,538.6	1,554.6	1,558.0	1,695.2	1,841.6	1,681.1	..	Private Consumption, etc.
291.5	315.2	395.9	337.8	322.2	311.3	263.3	206.5	226.9	322.5	..	General Gov't Consumption
568.7	528.9	626.7	492.9	425.6	387.8	185.8	274.4	219.0	239.8	..	Gross Domestic Investment
566.9	511.9	581.8	463.1	397.4	342.5	209.1	249.8	235.8	237.3	..	Fixed Investment
											Memo Items:
844.5	770.2	756.0	703.7	697.4	652.9	822.6	803.6	818.2	664.7	..	Capacity to Import
188.9	99.6	0.0	-104.8	-122.8	-101.6	-9.2	-20.8	22.2	-89.7	..	Terms of Trade Adjustment
2,364.4	2,338.8	2,221.6	2,212.2	2,230.8	2,223.8	2,213.8	2,407.7	2,538.4	2,352.6	..	Gross Domestic Income
2,090.7	2,055.0	2,102.7	2,071.1	2,076.5	2,055.4	2,054.8	2,182.1	2,355.8	2,192.8	..	Gross National Income
				(Index 1980 = 100)							DOMESTIC PRICES (DEFLATORS)
81.9	86.9	100.0	98.9	105.6	111.0	129.1	129.2	128.9	125.1	..	Overall (GDP)
75.4	83.3	100.0	104.2	111.9	116.5	126.5	125.2	124.8	128.5	..	Domestic Absorption
87.5	97.0	100.0	107.4	108.0	118.3	143.2	138.4	135.8	122.6	..	Agriculture
96.6	101.0	100.0	95.6	138.5	141.6	143.1	181.5	191.1	192.8	..	Industry
95.6	101.7	100.0	94.7	143.7	147.4	128.7	135.1	141.9	..	..	Manufacturing
											MANUFACTURING ACTIVITY
89.8	94.5	100.0	96.5	93.9	80.7	..	..	..	..	..	Employment (1980=100)
97.2	96.7	100.0	101.5	104.6	136.0	..	..	..	..	..	Real Earnings per Empl. (1980=100)
85.3	85.7	100.0	117.4	92.4	..	..	..	..	..	..	Real Output per Empl. (1980=100)
29.3	30.5	30.3	33.0	31.0	..	..	..	..	..	..	Earnings as % of Value Added
				(Billions of current CFA Francs)							MONETARY HOLDINGS
581.6	566.2	581.8	639.6	660.3	691.9	826.1	939.4	963.9	930.0	..	Money Supply, Broadly Defined
415.6	433.8	438.7	464.4	460.3	488.0	574.6	620.2	636.4	598.6	..	Money as Means of Payment
164.5	193.7	210.9	229.8	219.1	232.0	278.7	307.1	317.7	304.7	..	Currency Ouside Banks
251.1	240.1	227.8	234.6	241.2	256.0	295.9	313.1	318.8	293.9	..	Demand Deposits
166.1	132.4	143.1	175.3	200.0	203.9	251.5	319.3	327.5	331.4	..	Quasi-Monetary Liabilities
				(Billions of current CFA Francs)							
..	-168.3	-233.2	..	..	..	-90.0	..	..	..	..	GOVERNMENT DEFICIT (-) OR SURPLUS
..	467.1	517.4	..	..	..	816.1	..	..	..	..	Current Revenue
..	..	409.6	..	..	..	748.5	..	..	..	..	Current Expenditure
..	..	107.8	..	..	..	67.6	..	..	..	..	Current Budget Balance
..	..	0.2	..	..	..	..	..	..	..	..	Capital Receipts
..	..	341.2	..	..	..	157.6	..	..	..	..	Capital Payments

COTE D'IVOIRE	1967	1968	1969	1970	1971	1972	1973	1974	1975	1976	1977
FOREIGN TRADE (CUSTOMS BASIS)				*(Millions of current US dollars)*							
Value of Exports, fob	325.1	422.2	458.1	468.8	455.6	552.9	860.6	1,214.3	1,181.6	1,630.8	2,154.8
Nonfuel Primary Products	303.6	395.6	428.8	437.9	424.7	488.7	769.4	1,069.7	974.3	1,446.1	1,912.3
Fuels	3.7	3.8	4.1	2.9	2.5	11.4	19.6	45.3	66.9	63.0	81.9
Manufactures	17.8	22.8	25.1	28.0	28.4	52.8	71.6	99.3	140.4	121.7	160.5
Value of Imports, cif	262.8	306.6	334.1	387.2	398.1	452.8	713.5	966.7	1,126.5	1,295.4	1,751.4
Nonfuel Primary Products	48.7	57.4	55.4	70.8	74.0	90.5	160.6	189.7	191.2	206.8	286.8
Fuels	14.4	16.7	17.4	18.5	19.1	29.0	33.1	137.6	156.6	166.4	199.9
Manufactures	199.7	232.4	261.3	298.0	305.0	333.3	519.8	639.3	778.7	922.2	1,264.7
Terms of Trade	97.5	105.4	106.5	105.9	87.8	88.9	96.2	93.3	80.2	115.2	145.6
Export Prices, fob	23.2	24.9	26.2	27.7	25.5	26.5	39.1	51.6	46.1	67.5	94.3
Nonfuel Primary Products	24.2	25.8	27.3	28.2	25.5	27.7	42.0	51.4	45.0	69.9	102.5
Fuels	4.3	4.3	4.3	4.3	5.6	6.2	8.9	36.7	35.7	38.4	42.0
Manufactures	30.1	31.7	32.4	37.5	36.5	38.5	49.5	67.2	66.1	66.7	71.7
Import Prices, cif	23.8	23.6	24.6	26.1	29.0	29.8	40.7	55.3	57.5	58.6	64.8
Trade at Constant 1980 Prices				*(Millions of 1980 US dollars)*							
Exports, fob	1,402.6	1,695.0	1,745.1	1,695.2	1,788.3	2,087.3	2,199.7	2,352.5	2,563.3	2,414.2	2,284.8
Imports, cif	1,105.2	1,297.7	1,355.6	1,482.2	1,372.0	1,519.0	1,754.3	1,748.0	1,960.6	2,208.9	2,703.7
BALANCE OF PAYMENTS				*(Millions of current US dollars)*							
Exports of Goods & Services	..	..	..	565.7	576.3	696.9	995.3	1,445.7	1,503.4	1,998.4	2,779.2
Merchandise, fob	..	..	..	497.1	496.0	595.7	861.8	1,253.0	1,238.8	1,735.1	2,412.1
Nonfactor Services	..	..	..	53.5	65.4	84.4	116.9	167.3	225.8	235.2	324.8
Factor Services	..	..	..	15.1	14.9	16.8	16.6	25.4	38.7	28.0	42.3
Imports of Goods & Services	..	..	..	583.6	654.8	765.2	1,145.5	1,414.1	1,740.5	1,988.5	2,656.4
Merchandise, fob	..	..	..	375.1	400.4	460.2	701.1	894.4	1,012.1	1,161.3	1,597.2
Nonfactor Services	..	..	..	145.6	173.0	228.3	324.5	397.6	548.8	640.9	822.7
Factor Services	..	..	..	62.9	81.4	76.7	119.9	122.1	179.6	186.2	236.5
Long-Term Interest	..	..	..	12.1	16.0	19.0	28.9	42.5	59.2	71.3	110.1
Current Transfers, net	..	..	..	-55.5	-66.0	-89.4	-123.1	-139.3	-183.8	-289.6	-344.8
Workers' Remittances	..	..	..	0.0	0.0	0.0	0.0	0.0	0.0	0.0	0.0
Total to be Financed	..	..	..	-73.4	-144.5	-157.6	-273.4	-107.7	-421.0	-279.7	-221.9
Official Capital Grants	..	..	..	35.5	39.0	61.0	54.0	46.7	42.0	30.6	45.2
Current Account Balance	..	..	..	-37.9	-105.5	-96.6	-219.3	-61.0	-379.0	-249.2	-176.7
Long-Term Capital, net	..	..	..	104.9	135.8	86.2	274.0	220.3	312.2	306.8	615.8
Direct Investment	..	..	..	30.7	15.7	18.7	51.0	32.6	69.1	44.8	14.7
Long-Term Loans	..	..	..	51.2	79.8	42.5	205.9	105.2	291.6	263.4	738.5
Disbursements	..	..	..	81.7	111.5	84.5	252.6	187.0	375.0	390.5	919.9
Repayments	..	..	..	30.5	31.7	42.0	46.7	81.8	83.4	127.1	181.4
Other Long-Term Capital	..	..	..	-12.5	1.2	-36.0	-37.0	35.9	-90.5	-32.0	-182.5
Other Capital, net	..	..	..	-31.9	-50.8	-61.3	-60.9	-97.7	-25.6	-27.5	-320.5
Change in Reserves	..	..	..	-35.2	20.5	71.8	6.3	-61.6	92.4	-30.0	-118.7
Memo Item:				*(CFA Francs per US dollar)*							
Conversion Factor (Annual Avg)	246.850	246.850	259.710	277.710	277.130	252.210	222.700	240.500	214.320	238.980	245.670
EXTERNAL DEBT, ETC.				*(Millions of US dollars, outstanding at end of year)*							
Public/Publicly Guar. Long-Term	..	..	..	255.4	350.6	398.0	580.2	690.7	942.8	1,170.9	1,897.0
Official Creditors	..	..	..	143.5	179.3	209.3	254.7	315.1	384.5	435.0	591.9
IBRD and IDA	..	..	..	4.5	10.6	17.7	28.9	49.2	73.8	88.4	127.0
Private Creditors	..	..	..	112.0	171.3	188.7	325.5	375.6	558.3	735.8	1,305.1
Private Non-guaranteed Long-Term	..	..	..	11.0	13.0	12.0	32.0	44.0	65.0	74.0	155.0
Use of Fund Credit	..	..	..	0.0	0.0	0.0	0.0	13.7	13.1	27.1	16.3
Short-Term Debt	..	..	..	..	..	..	..	..	..	..	478.0
Memo Items:				*(Thousands of US dollars)*							
Int'l Reserves Excluding Gold	71,100.0	80,700.0	73,900.0	118,840.0	89,433.0	87,178.0	88,373.0	65,667.0	102,760.0	76,403.0	184,810.0
Gold Holdings (at market price)	..	..	..	..	..	..	..	..	..	..	3,694.9
SOCIAL INDICATORS											
Total Fertility Rate	7.4	7.4	7.4	7.4	7.4	7.4	7.4	7.4	7.4	7.4	7.4
Crude Birth Rate	51.7	51.6	51.5	51.4	51.3	51.2	51.2	51.1	51.1	51.1	51.1
Infant Mortality Rate	143.0	140.2	137.4	134.6	131.8	129.0	126.4	123.8	121.2	118.6	116.0
Life Expectancy at Birth	43.0	43.5	44.0	44.5	45.0	45.5	46.0	46.5	47.0	47.5	48.0
Food Production, p.c. ('79-81 = 100)	84.5	82.7	86.2	83.8	88.9	81.6	83.0	87.8	100.3	94.5	94.5
Labor Force, Agriculture (%)	79.0	78.1	77.3	76.5	75.4	74.3	73.1	72.0	70.9	69.8	68.6
Labor Force, Female (%)	38.8	38.5	38.3	38.1	37.8	37.5	37.2	36.9	36.6	36.3	35.9
School Enroll. Ratio, primary	..	..	..	58.0	..	..	..	..	62.0	64.0	66.0
School Enroll. Ratio, secondary	..	..	..	9.0	..	..	..	..	13.0	13.0	14.0

1978	1979	1980	1981	1982	1983	1984	1985	1986	1987 estimate	NOTES	COTE D'IVOIRE
											FOREIGN TRADE (CUSTOMS BASIS)
				(Millions of current US dollars)							
2,322.9	2,506.8	3,142.0	2,535.2	2,287.9	2,092.0	2,707.0	2,972.1	3,179.2	2,960.9	..	Value of Exports, fob
2,081.6	2,182.4	2,657.7	2,081.8	1,743.3	1,615.0	2,187.5	2,446.1	2,782.7	2,556.3	..	Nonfuel Primary Products
86.0	112.4	188.9	191.2	298.0	239.2	319.5	259.0	107.5	122.1	..	Fuels
155.3	212.0	295.4	262.2	246.7	237.7	200.0	267.0	289.0	282.5	..	Manufactures
2,309.6	2,388.9	3,015.0	2,393.1	2,183.7	1,813.5	1,496.9	1,749.2	1,969.1	2,245.4	..	Value of Imports, cif
358.1	417.4	606.8	545.1	469.1	413.6	271.5	301.9	415.9	505.9	..	Nonfuel Primary Products
221.3	272.7	504.4	527.5	469.1	336.3	253.6	378.6	263.0	326.0	..	Fuels
1,730.2	1,698.8	1,903.8	1,320.5	1,245.6	1,063.6	971.8	1,068.7	1,290.2	1,413.5	..	Manufactures
				(Index 1980 = 100)							
121.2	118.3	100.0	84.5	86.6	91.6	99.7	95.9	103.5	91.5	..	Terms of Trade
89.4	101.7	100.0	86.6	83.1	84.9	91.4	86.0	89.9	87.7	..	Export Prices, fob
95.0	106.8	100.0	83.5	79.3	83.1	91.7	85.5	92.3	87.8	..	Nonfuel Primary Products
42.3	61.0	100.0	112.5	101.6	92.5	90.2	87.5	44.9	56.9	..	Fuels
76.1	90.0	100.0	99.1	94.8	90.7	89.3	89.0	103.2	113.0	..	Manufactures
73.7	86.0	100.0	102.5	96.0	92.7	91.6	89.7	86.9	95.8	..	Import Prices, cif
				(Millions of 1980 US dollars)							Trade at Constant 1980 Prices
2,598.9	2,464.0	3,142.1	2,928.2	2,752.4	2,464.2	2,962.6	3,457.2	3,535.5	3,376.8	..	Exports, fob
3,133.3	2,777.6	3,015.0	2,334.4	2,274.4	1,957.0	1,633.3	1,950.7	2,265.8	2,344.0	..	Imports, cif
				(Millions of current US dollars)							**BALANCE OF PAYMENTS**
3,086.6	3,292.5	3,639.8	2,915.3	2,843.9	2,538.1	3,042.2	3,200.0	3,705.2	3,624.0	f	Exports of Goods & Services
2,615.9	2,722.8	3,012.7	2,435.1	2,347.2	2,066.3	2,624.7	2,761.4	3,162.0	2,960.8	..	Merchandise, fob
410.4	510.5	564.2	434.3	450.7	425.1	370.3	398.9	475.5	638.5	..	Nonfactor Services
60.3	59.2	63.0	46.0	46.0	46.7	47.1	39.6	67.7	24.7	..	Factor Services
3,507.2	4,104.0	4,760.5	3,847.6	3,498.4	3,173.7	2,833.2	2,879.2	3,481.7	3,881.2	f	Imports of Goods & Services
2,042.9	2,233.5	2,613.6	2,067.9	1,789.7	1,635.1	1,487.3	1,410.1	1,620.8	1,840.7	..	Merchandise, fob
1,095.2	1,351.1	1,521.2	1,215.9	1,155.5	983.3	823.9	763.2	1,113.3	1,188.5	..	Nonfactor Services
369.1	519.5	625.7	563.8	553.2	555.3	522.0	705.9	747.6	852.0	..	Factor Services
190.8	258.2	384.5	445.0	555.8	520.1	587.0	665.6	767.0	597.4	..	Long-Term Interest
-458.2	-576.3	-715.6	-494.6	-391.4	-319.6	-291.1	-278.9	-426.2	-383.3	..	Current Transfers, net
0.0	0.0	0.0	0.0	0.0	0.0	0.0	0.0	0.0	0.0	..	Workers' Remittances
-879.0	-1,387.8	-1,836.4	-1,426.8	-1,045.9	-955.2	-82.2	41.9	-202.7	-640.5	K	Total to be Financed
40.3	4.7	9.9	15.5	29.8	26.5	28.1	26.0	68.0	16.7	K	Official Capital Grants
-838.6	-1,383.1	-1,826.4	-1,411.3	-1,016.1	-928.7	-54.0	67.9	-134.8	-623.8	..	Current Account Balance
906.1	719.7	1,080.5	936.8	928.5	560.0	306.0	222.4	333.2	305.0	f	Long-Term Capital, net
83.3	74.7	94.7	32.8	47.5	37.5	3.0	29.2	107.1	0.0	..	Direct Investment
860.4	724.6	1,132.8	745.2	1,485.8	511.0	970.5	712.4	467.4	622.6	..	Long-Term Loans
1,114.2	1,138.7	1,691.1	1,357.5	2,098.6	1,127.6	1,435.0	1,282.1	1,240.7	1,502.1	..	Disbursements
253.8	414.1	558.3	612.3	612.8	616.6	464.5	569.7	773.3	879.5	..	Repayments
-77.9	-84.3	-156.9	143.4	-634.6	-15.0	-695.6	-545.2	-309.3	-334.3	..	Other Long-Term Capital
98.6	334.4	634.1	97.4	-62.2	224.0	-278.2	-263.7	-108.5	-48.5	f	Other Capital, net
-166.1	328.9	111.8	377.1	149.9	144.7	26.2	-26.6	-89.9	367.3	..	Change in Reserves
				(CFA Francs per US dollar)							Memo Item:
225.640	212.720	211.300	271.730	328.620	381.070	436.960	449.260	346.300	300.540	..	Conversion Factor (Annual Avg)
				(Millions of US dollars, outstanding at end of year)							**EXTERNAL DEBT, ETC.**
2,822.7	3,687.5	4,328.4	4,470.0	5,071.0	4,915.1	4,968.8	5,921.1	6,777.6	8,449.7	..	Public/Publicly Guar. Long-Term
811.3	1,074.3	1,209.5	1,153.3	1,443.3	1,642.5	1,954.7	2,592.0	3,255.5	4,729.4	..	Official Creditors
182.6	235.4	313.7	348.1	547.9	726.2	775.9	972.2	1,260.6	1,886.3	..	IBRD and IDA
2,011.3	2,613.3	3,118.9	3,316.7	3,627.7	3,272.6	3,014.1	3,329.1	3,522.1	3,720.3	..	Private Creditors
253.0	190.0	414.0	602.0	1,158.0	1,367.0	1,989.0	2,569.0	2,955.0	3,264.0	..	Private Non-guaranteed Long-Term
0.0	0.0	0.0	371.5	479.4	617.2	591.2	621.7	622.9	576.4	..	Use of Fund Credit
729.0	857.0	1,059.0	1,165.0	1,106.0	833.0	630.0	725.0	787.0	1,265.0	..	Short-Term Debt
				(Thousands of US dollars)							Memo Items:
447,970.0	147,030.0	19,704.0	17,847.0	2,195.5	19,721.0	5,367.0	4,732.4	19,556.0	8,901.9	..	Int'l Reserves Excluding Gold
7,571.0	22,784.0	26,232.8	17,688.8	20,332.0	16,976.8	13,719.4	14,551.5	17,395.0	21,542.4	..	Gold Holdings (at market price)
											SOCIAL INDICATORS
7.4	7.4	7.4	7.4	7.4	7.3	7.2	7.2	7.1	7.0		Total Fertility Rate
51.1	51.0	51.0	51.0	51.0	50.4	49.9	49.4	48.8	48.3		Crude Birth Rate
113.8	111.6	109.4	107.2	105.0	103.1	101.2	99.3	97.4	95.5		Infant Mortality Rate
48.5	49.0	49.5	50.0	50.5	50.9	51.4	51.8	52.3	52.8		Life Expectancy at Birth
97.2	101.2	99.0	99.8	93.1	89.3	101.2	109.0	104.3	102.4		Food Production, p.c. ('79-81 = 100)
67.5	66.3	65.2	..	..	..	..	..	..	..		Labor Force, Agriculture (%)
35.6	35.2	34.9	34.9	34.8	34.8	34.7	34.7	34.6	34.5		Labor Force, Female (%)
70.0	72.0	..	..	82.0	81.0	78.0	..	..	..		School Enroll. Ratio, primary
17.0	20.0	..	..	19.0	19.0	20.0	..	..	..		School Enroll. Ratio, secondary

CYPRUS	1967	1968	1969	1970	1971	1972	1973	1974	1975	1976	1977
CURRENT GNP PER CAPITA (US $)	..	..	..	..	..	..	..	..	..	..	..
POPULATION (thousands)	595	602	608	615	614	613	611	610	609	611	614

ORIGIN AND USE OF RESOURCES	*(Millions of current Cyprus Pounds)*										
Gross National Product (GNP)	..	..	..	..	..	..	..	..	259.0	336.2	431.6
Net Factor Income from Abroad	..	..	..	..	..	..	..	..	2.0	2.3	8.5
Gross Domestic Product (GDP)	..	..	..	..	..	..	..	..	257.0	333.9	423.1
Indirect Taxes, net	..	..	..	..	..	..	..	..	17.1	13.7	25.4
GDP at factor cost	..	..	..	..	..	..	..	..	239.9	320.2	397.7
Agriculture	..	..	..	..	..	..	..	..	40.4	53.1	55.9
Industry	..	..	..	..	..	..	..	..	64.7	96.3	131.1
Manufacturing	..	..	..	..	..	..	..	..	36.8	57.6	75.2
Services, etc.	..	..	..	..	..	..	..	..	134.8	170.8	210.7
Resource Balance	..	..	..	..	..	..	..	..	-54.5	-43.0	-84.4
Exports of Goods & NFServices	..	..	..	..	..	..	..	..	91.2	166.1	202.3
Imports of Goods & NFServices	..	..	..	..	..	..	..	..	145.7	209.1	286.7
Domestic Absorption	..	..	..	..	..	..	..	..	311.5	376.9	507.5
Private Consumption, etc.	..	..	..	..	..	..	..	..	208.7	233.7	300.7
General Gov't Consumption	..	..	..	..	..	..	..	..	44.9	54.5	59.2
Gross Domestic Investment	..	..	..	..	..	..	..	..	57.9	88.7	147.6
Fixed Investment	..	..	..	..	..	..	..	..	50.5	70.3	124.6
Memo Items:											
Gross Domestic Saving	..	..	..	..	..	..	..	..	3.4	45.7	63.2
Gross National Saving	..	..	..	..	..	..	..	..	10.5	53.4	77.6

	(Millions of 1980 Cyprus Pounds)										
Gross National Product	..	..	..	..	..	..	..	..	439.3	527.9	619.2
GDP at factor cost	..	..	..	..	..	..	..	..	407.9	488.9	566.1
Agriculture	..	..	..	..	..	..	..	..	63.4	68.8	68.9
Industry	..	..	..	..	..	..	..	..	123.7	162.1	200.3
Manufacturing	..	..	..	..	..	..	..	..	69.3	90.3	104.3
Services, etc.	..	..	..	..	..	..	..	..	220.8	258.0	296.9
Resource Balance	..	..	..	..	..	..	..	..	-86.7	-90.0	-125.2
Exports of Goods & NFServices	..	..	..	..	..	..	..	..	119.9	198.1	242.3
Imports of Goods & NFServices	..	..	..	..	..	..	..	..	206.6	288.1	367.5
Domestic Absorption	..	..	..	..	..	..	..	..	522.6	614.3	732.2
Private Consumption, etc.	..	..	..	..	..	..	..	..	346.6	388.8	433.0
General Gov't Consumption	..	..	..	..	..	..	..	..	79.0	91.6	92.6
Gross Domestic Investment	..	..	..	..	..	..	..	..	97.0	133.9	206.6
Fixed Investment	..	..	..	..	..	..	..	..	81.3	106.5	175.0
Memo Items:											
Capacity to Import	..	..	..	..	..	..	..	..	129.3	228.9	259.3
Terms of Trade Adjustment	..	..	..	..	..	..	..	..	9.4	30.8	17.0
Gross Domestic Income	..	..	..	..	..	..	..	..	445.3	555.1	624.0
Gross National Income	..	..	..	..	..	..	..	..	448.7	558.7	636.2

DOMESTIC PRICES (DEFLATORS)	*(Index 1980 = 100)*										
Overall (GDP)	..	..	..	..	..	..	..	..	59.0	63.7	69.7
Domestic Absorption	..	..	..	..	..	..	..	..	59.6	61.4	69.3
Agriculture	..	..	..	..	..	..	..	..	63.7	77.2	81.1
Industry	..	..	..	..	..	..	..	..	52.3	59.4	65.5
Manufacturing	..	..	..	..	..	..	..	..	53.1	63.8	72.1

MANUFACTURING ACTIVITY											
Employment (1980=100)	56.4	60.7	63.2	66.5	69.5	76.4	78.1	63.2	53.7	62.8	77.9
Real Earnings per Empl. (1980=100)	..	..	..	..	..	..	..	..	67.6	68.9	75.1
Real Output per Empl. (1980=100)	..	..	..	..	..	..	..	..	91.6	100.0	91.9
Earnings as % of Value Added	36.5	38.7	38.6	37.8	35.7	35.1	37.0	40.6	36.2	30.0	32.6

MONETARY HOLDINGS	*(Millions of current Cyprus Pounds)*										
Money Supply, Broadly Defined	84.2	95.3	109.3	120.4	145.0	175.2	197.0	228.5	229.7	275.1	316.0
Money as Means of Payment	29.1	32.7	38.0	42.6	46.7	57.4	60.9	67.0	62.1	80.0	85.9
Currency Ouside Banks	13.6	15.5	17.2	18.4	21.8	26.3	29.7	35.8	33.7	39.3	43.2
Demand Deposits	15.5	17.2	20.8	24.2	25.0	31.2	31.1	31.3	28.4	40.7	42.8
Quasi-Monetary Liabilities	55.1	62.6	71.3	77.8	98.3	117.7	136.1	161.4	167.6	195.1	230.1

GOVERNMENT DEFICIT (-) OR SURPLUS	*(Millions of current Cyprus Pounds)*										
	..	..	..	1.4	0.4	-4.4	-12.1	-20.9	-20.4	-24.9	-12.7
Current Revenue	..	..	..	41.3	46.9	52.4	60.3	61.1	66.8	78.7	106.7
Current Expenditure	..	..	..	31.9	37.3	45.3	60.9	72.2	76.2	83.9	91.8
Current Budget Balance	..	..	..	9.4	9.6	7.1	-0.6	-11.1	-9.4	-5.2	14.9
Capital Receipts	..	..	..	0.1	0.1	0.0	0.1	0.0	0.1	0.1	0.1
Capital Payments	..	..	..	8.0	9.3	11.6	11.7	9.8	11.1	19.8	27.7

1978	1979	1980	1981	1982	1983	1984	1985	1986	1987 estimate	NOTES	CYPRUS
..	..	..	..	..	..	4,070	4,110	4,450	5,200	..	**CURRENT GNP PER CAPITA (US $)**
618	623	629	637	645	655	657	665	672	679	..	**POPULATION (thousands)**
				(Millions of current Cyprus Pounds)							**ORIGIN AND USE OF RESOURCES**
516.0	638.2	772.3	887.4	1,036.7	1,136.2	1,339.2	1,485.6	1,611.7	1,795.5	..	Gross National Product (GNP)
9.5	8.4	12.0	11.5	12.2	-0.4	2.5	4.9	3.8	3.9	..	Net Factor Income from Abroad
506.5	629.8	760.3	875.9	1,024.5	1,136.6	1,336.7	1,480.7	1,607.9	1,791.6	..	Gross Domestic Product (GDP)
32.5	44.0	51.3	55.9	70.8	83.3	94.8	106.0	103.1	115.0	..	Indirect Taxes, net
474.0	585.8	709.0	820.0	953.7	1,053.3	1,241.9	1,374.7	1,504.8	1,676.6	..	GDP at factor cost
55.3	64.5	72.9	81.1	95.0	89.8	119.6	111.0	117.3	135.2	..	Agriculture
167.8	209.2	255.8	283.9	316.0	341.9	387.9	420.1	438.1	478.6	..	Industry
92.6	110.8	133.4	154.2	174.5	187.9	215.2	231.9	240.5	269.2	..	Manufacturing
250.9	312.1	380.3	455.0	542.7	621.6	734.4	843.6	949.4	1,062.8	..	Services, etc.
-104.3	-120.2	-135.4	-114.5	-136.5	-154.2	-166.4	-149.6	-53.9	4.7	..	Resource Balance
214.4	281.4	344.1	440.2	521.9	573.0	731.0	722.4	721.3	843.5	..	Exports of Goods & NFServices
318.7	401.6	479.5	554.7	658.4	727.2	897.4	872.0	775.2	838.8	..	Imports of Goods & NFServices
610.8	750.0	895.7	990.4	1,161.0	1,290.8	1,503.1	1,630.3	1,661.8	1,786.9	..	Domestic Absorption
356.0	425.6	504.3	566.2	683.8	775.9	864.3	971.1	1,019.7	1,103.4	..	Private Consumption, etc.
66.3	80.5	103.9	128.0	152.1	172.6	189.5	209.5	227.1	252.1	..	General Gov't Consumption
188.5	243.9	287.5	296.2	325.1	342.3	449.3	449.7	415.0	431.4	..	Gross Domestic Investment
170.4	219.5	260.0	275.9	304.7	316.0	412.5	403.0	384.0	405.8	..	Fixed Investment
											Memo Items:
84.2	123.7	152.1	181.7	188.6	188.1	282.9	300.1	361.1	436.1	..	Gross Domestic Saving
102.1	141.1	175.6	204.7	212.6	200.0	298.1	318.6	377.3	452.7	..	Gross National Saving
				(Millions of 1980 Cyprus Pounds)							
665.7	727.5	772.3	793.8	842.6	876.3	956.3	1,003.1	1,043.4	1,118.4	..	Gross National Product
609.3	669.5	709.0	730.6	776.5	817.4	890.1	932.3	970.7	1,040.7	..	GDP at factor cost
66.6	69.0	72.9	72.4	72.7	71.9	78.6	78.4	76.9	81.0	..	Agriculture
221.1	243.4	255.8	249.5	255.6	259.7	272.2	276.0	277.5	295.3	..	Industry
113.3	123.6	133.4	141.9	150.4	153.3	163.1	166.8	168.7	182.5	..	Manufacturing
321.6	357.1	380.3	408.7	448.2	485.8	539.3	577.9	616.3	664.4	..	Services, etc.
-131.0	-142.8	-135.4	-98.9	-130.8	-126.8	-153.3	-128.1	-69.0	-35.7	..	Resource Balance
257.2	310.5	344.1	392.6	423.6	458.2	527.4	521.7	520.9	590.1	..	Exports of Goods & NFServices
388.2	453.3	479.5	491.5	554.4	585.0	680.7	649.8	589.9	625.8	..	Imports of Goods & NFServices
784.4	860.7	895.7	882.4	963.5	1,003.4	1,107.8	1,127.9	1,109.9	1,151.7	..	Domestic Absorption
456.9	489.7	504.3	509.7	578.1	611.9	651.6	684.3	691.7	721.5	..	Private Consumption, etc.
91.7	95.7	103.9	111.6	116.0	123.2	127.9	132.8	139.7	147.0	..	General Gov't Consumption
235.8	275.3	287.5	261.1	269.4	268.3	328.3	310.8	278.5	283.2	..	Gross Domestic Investment
212.4	247.1	260.0	243.2	252.4	246.9	300.1	276.9	257.2	265.6	..	Fixed Investment
											Memo Items:
261.2	317.6	344.1	390.0	439.5	461.0	554.5	538.3	548.9	629.3	..	Capacity to Import
4.0	7.1	0.0	-2.6	15.9	2.8	27.1	16.6	28.0	39.2	..	Terms of Trade Adjustment
657.4	725.0	760.3	780.9	848.6	879.4	981.6	1,016.4	1,068.9	1,155.2	..	Gross Domestic Income
669.7	734.6	772.3	791.2	858.5	879.1	983.4	1,019.7	1,071.4	1,157.6	..	Gross National Income
				(Index 1980 = 100)							**DOMESTIC PRICES (DEFLATORS)**
77.5	87.7	100.0	111.8	123.0	129.7	140.0	148.1	154.5	160.5	..	Overall (GDP)
77.9	87.1	100.0	112.2	120.5	128.6	135.7	144.5	149.7	155.2	..	Domestic Absorption
83.0	93.5	100.0	112.0	130.7	124.9	152.2	141.6	152.5	166.9	..	Agriculture
75.9	85.9	100.0	113.8	123.6	131.7	142.5	152.2	157.9	162.1	..	Industry
81.7	89.6	100.0	108.7	116.0	122.6	131.9	139.0	142.6	147.5	..	Manufacturing
											MANUFACTURING ACTIVITY
88.4	94.8	100.0	104.7	106.3	106.9	110.7	113.8	..	..	J	Employment (1980=100)
81.7	92.3	100.0	107.7	116.9	122.3	124.8	128.2	..	..	J	Real Earnings per Empl. (1980=100)
84.1	90.4	100.0	106.3	107.0	109.0	111.2	108.0	..	..	J	Real Output per Empl. (1980=100)
34.8	38.3	37.6	40.4	41.4	43.0	42.2	46.9	..	..	J	Earnings as % of Value Added
				(Millions of current Cyprus Pounds)							**MONETARY HOLDINGS**
363.3	433.5	501.2	601.4	709.5	791.1	899.4	992.3	1,095.5	1,238.4	..	Money Supply, Broadly Defined
100.8	129.5	153.3	188.3	218.5	248.5	259.4	285.3	282.5	314.0	..	Money as Means of Payment
51.2	64.0	76.0	89.5	101.6	115.9	122.2	127.9	130.7	142.6	..	Currency Ouside Banks
49.6	65.5	77.4	98.8	116.8	132.6	137.1	157.4	151.9	171.5	..	Demand Deposits
262.5	304.0	347.9	413.1	491.0	542.7	640.1	707.0	813.0	924.4	..	Quasi-Monetary Liabilities
				(Millions of current Cyprus Pounds)							
-28.8	-41.6	-52.5	-57.3	-54.2	-78.3	-74.0	-58.0	-63.8	..	..	**GOVERNMENT DEFICIT (-) OR SURPLUS**
113.3	136.6	173.7	203.2	252.1	288.6	344.0	389.9	426.4	..	..	Current Revenue
100.9	129.3	165.1	212.4	249.3	301.2	337.2	362.9	381.1	..	..	Current Expenditure
12.4	7.3	8.6	-9.2	2.8	-12.6	6.8	27.0	45.3	..	..	Current Budget Balance
0.2	0.2	0.1	0.1	0.2	0.8	0.3	0.2	0.7	..	..	Capital Receipts
41.4	49.1	61.2	48.2	57.6	66.6	81.2	85.2	109.7	..	..	Capital Payments

CYPRUS	1967	1968	1969	1970	1971	1972	1973	1974	1975	1976	1977
FOREIGN TRADE (CUSTOMS BASIS)				*(Millions of current US dollars)*							
Value of Exports, fob	74.9	81.8	88.2	108.5	114.8	133.8	173.1	152.2	151.2	256.9	317.5
Nonfuel Primary Products	72.5	78.6	84.5	91.1	95.5	110.2	138.0	114.3	91.5	151.4	175.2
Fuels	0.0	0.0	0.0	0.0	0.0	0.5	0.1	0.6	0.6	9.9	5.9
Manufactures	2.4	3.2	3.7	17.3	19.3	23.1	34.9	37.3	59.1	95.5	136.5
Value of Imports, cif	168.1	169.7	202.4	235.1	259.0	315.4	449.0	406.1	305.5	429.8	619.1
Nonfuel Primary Products	34.2	30.4	36.2	45.2	50.1	63.9	119.9	93.3	85.6	114.1	129.9
Fuels	12.4	14.7	15.6	16.9	20.8	22.0	25.1	52.2	48.8	65.1	86.3
Manufactures	121.5	124.6	150.6	173.0	188.2	229.5	304.0	260.6	171.1	250.5	402.8
				(Index 1980 = 100)							
Terms of Trade	184.7	198.6	198.2	187.7	160.7	145.6	128.8	107.0	113.3	109.2	110.8
Export Prices, fob	39.8	40.3	43.3	43.7	41.0	42.5	51.3	59.4	65.3	64.5	71.2
Nonfuel Primary Products	40.2	40.8	44.0	45.3	42.2	44.5	52.0	57.5	65.1	66.1	72.5
Fuels	4.3	4.3	4.3	4.3	5.6	6.2	8.9	36.7	35.7	38.4	42.0
Manufactures	30.1	31.7	32.4	37.5	36.5	38.5	49.5	67.2	66.1	66.7	71.7
Import Prices, cif	21.6	20.3	21.9	23.3	25.5	29.2	39.9	55.6	57.6	59.1	64.2
Trade at Constant 1980 Prices				*(Millions of 1980 US dollars)*							
Exports, fob	188.1	203.0	203.5	248.2	279.8	315.2	337.2	256.1	231.7	398.2	446.1
Imports, cif	779.7	835.7	925.6	1,010.1	1,014.1	1,081.7	1,126.4	730.8	530.5	727.5	963.7
BALANCE OF PAYMENTS				*(Millions of current US dollars)*							
Exports of Goods & Services	..	..	..	229.2	264.6	329.9	423.2	365.5	301.0	459.4	565.5
Merchandise, fob	..	..	..	102.4	109.0	121.9	163.6	142.4	142.3	250.2	304.3
Nonfactor Services	..	..	..	111.3	140.9	191.2	236.4	192.5	133.3	179.8	214.6
Factor Services	..	..	..	15.5	14.6	16.8	23.2	30.5	25.4	29.5	46.5
Imports of Goods & Services	..	..	..	268.0	298.7	373.8	516.3	498.9	410.5	533.5	728.1
Merchandise, fob	..	..	..	206.3	227.4	281.4	402.1	379.1	301.2	397.8	558.8
Nonfactor Services	..	..	..	47.9	58.0	78.4	92.7	99.7	89.2	111.6	143.6
Factor Services	..	..	..	13.8	13.3	14.0	21.5	20.1	20.0	24.1	25.7
Long-Term Interest	..	..	..	2.4	2.8	2.4	3.0	3.7	4.9	5.3	8.6
Current Transfers, net	..	..	..	..	..	..	..	..	..	..	..
Workers' Remittances	..	..	..	0.0	0.0	0.0	0.0	0.0	0.0	0.0	0.0
Total to be Financed	..	..	..	..	..	..	..	..	..	..	..
Official Capital Grants	..	..	..	..	..	..	..	..	..	..	..
Current Account Balance	..	..	..	-21.7	-15.0	-25.2	-78.2	-72.4	-37.3	-13.4	-89.9
Long-Term Capital, net	..	..	..	28.2	29.9	37.5	50.4	89.4	71.9	110.1	136.0
Direct Investment	..	..	..	20.2	29.6	36.4	37.2	32.4	18.1	32.4	41.4
Long-Term Loans	..	..	..	4.3	-17.3	-0.8	16.6	17.1	4.7	18.0	58.4
Disbursements	..	..	..	7.4	6.8	4.3	21.9	21.9	9.8	27.3	67.7
Repayments	..	..	..	3.1	24.1	5.1	5.3	4.8	5.1	9.3	9.3
Other Long-Term Capital	..	..	..	3.7	17.6	1.9	-3.4	39.9	49.1	59.7	36.2
Other Capital, net	..	..	..	22.2	55.9	19.3	11.1	-63.4	-88.6	-15.5	-22.9
Change in Reserves	..	..	..	-28.7	-70.7	-31.6	16.7	46.4	54.0	-81.2	-23.2
Memo Item:				*(Cyprus Pounds per US dollar)*							
Conversion Factor (Annual Avg)	..	..	..	..	..	..	..	..	0.370	0.410	0.410
EXTERNAL DEBT, ETC.				*(Millions of US dollars, outstanding at end of year)*							
Public/Publicly Guar. Long-Term	..	..	..	56.0	41.2	39.9	57.3	76.2	76.3	94.4	162.0
Official Creditors	..	..	..	25.6	30.5	31.5	40.8	50.2	55.3	80.5	102.7
IBRD and IDA	..	..	..	15.2	19.1	21.5	30.3	37.9	38.9	39.7	40.5
Private Creditors	..	..	..	30.5	10.7	8.4	16.5	26.0	21.1	13.9	59.3
Private Non-guaranteed Long-Term	..	..	..	..	..	..	..	..	..	..	..
Use of Fund Credit	..	..	..	0.0	0.0	0.0	0.0	7.8	9.5	50.0	52.3
Short-Term Debt	..	..	..	..	..	..	..	..	..	..	61.0
Memo Items:				*(Millions of US dollars)*							
Int'l Reserves Excluding Gold	103.5	140.9	161.9	194.0	268.7	303.3	288.7	250.1	197.7	274.5	313.5
Gold Holdings (at market price)	..	17.9	15.1	16.0	18.7	27.8	48.0	79.8	60.0	57.7	72.6
SOCIAL INDICATORS											
Total Fertility Rate	2.8	2.7	2.6	2.4	2.3	2.2	2.2	2.2	2.2	2.2	2.2
Crude Birth Rate	21.0	20.4	19.8	19.2	18.6	18.0	18.1	18.3	18.4	18.6	18.7
Infant Mortality Rate	29.0	29.0	29.0	29.0	29.0	29.0	27.8	26.6	25.4	24.2	23.0
Life Expectancy at Birth	70.3	70.5	70.8	71.0	71.2	71.4	71.9	72.3	72.8	73.2	73.7
Food Production, p.c. ('79-81=100)	104.5	98.2	105.1	95.1	117.7	114.1	80.1	102.9	89.8	97.4	100.6
Labor Force, Agriculture (%)	39.5	39.2	38.8	38.5	37.3	36.0	34.8	33.5	32.3	31.0	29.8
Labor Force, Female (%)	33.4	33.4	33.4	33.5	33.6	33.8	33.9	34.0	34.2	34.3	34.4
School Enroll. Ratio, primary	..	..	..	..	..	..	..	..	..	..	..
School Enroll. Ratio, secondary	..	..	..	..	..	..	..	..	..	..	..

1978	1979	1980	1981	1982	1983	1984	1985	1986	1987 estimate	NOTES	CYPRUS
				(Millions of current US dollars)							**FOREIGN TRADE (CUSTOMS BASIS)**
343.7	456.3	532.8	559.0	554.0	494.2	575.0	476.3	500.3	579.1	..	Value of Exports, fob
172.9	205.6	209.6	215.6	228.5	178.8	217.8	169.1	167.2	226.8	..	Nonfuel Primary Products
8.3	23.1	27.9	31.7	38.5	41.4	36.2	38.6	18.6	25.8	..	Fuels
162.5	227.6	295.4	311.7	286.9	274.0	321.0	268.6	314.5	326.5	..	Manufactures
748.2	999.8	1,195.1	1,101.4	1,206.8	1,207.8	1,350.8	1,233.5	1,263.4	1,463.3	..	Value of Imports, cif
153.4	190.6	214.4	212.5	217.3	242.8	242.3	212.9	186.2	250.7	..	Nonfuel Primary Products
83.1	125.5	222.5	238.7	246.7	227.7	247.9	223.6	140.9	182.5	..	Fuels
511.8	683.7	758.2	650.2	742.8	737.3	860.6	797.0	936.3	1,030.1	..	Manufactures
				(Index 1980 = 100)							
106.8	106.6	100.0	96.8	96.7	96.4	94.8	97.2	105.1	104.8	..	Terms of Trade
77.2	92.1	100.0	98.3	93.6	91.3	88.2	88.4	96.2	103.0	..	Export Prices, fob
81.6	100.5	100.0	95.5	90.9	91.9	86.2	87.7	96.0	99.6	..	Nonfuel Primary Products
42.3	61.0	100.0	112.5	101.6	92.5	90.2	87.5	44.9	56.9	..	Fuels
76.1	90.0	100.0	99.1	94.8	90.7	89.3	89.0	103.2	113.0	..	Manufactures
72.3	86.4	100.0	101.6	96.8	94.7	93.0	90.9	91.5	98.3	..	Import Prices, cif
				(Millions of 1980 US dollars)							Trade at Constant 1980 Prices
445.2	495.4	532.9	568.4	592.0	541.3	652.2	538.8	520.3	562.0	..	Exports, fob
1,035.1	1,157.1	1,195.1	1,084.1	1,247.2	1,275.3	1,453.1	1,356.6	1,381.0	1,488.0	..	Imports, cif
				(Millions of current US dollars)							**BALANCE OF PAYMENTS**
658.9	896.5	1,107.2	1,180.4	1,247.3	1,224.3	1,386.0	1,335.4	1,554.7	1,942.6	..	Exports of Goods & Services
326.8	421.5	489.2	508.6	499.5	438.2	524.7	417.5	450.7	566.5	..	Merchandise, fob
271.3	400.9	520.3	569.7	632.5	683.1	751.3	802.0	978.5	1,232.6	..	Nonfactor Services
60.8	74.2	97.7	102.1	115.3	102.9	110.0	115.8	125.4	143.5	..	Factor Services
888.7	1,183.2	1,422.1	1,392.1	1,475.0	1,484.9	1,635.0	1,533.5	1,616.3	1,895.3	..	Imports of Goods & Services
684.1	906.1	1,079.4	1,044.3	1,090.9	1,094.1	1,229.3	1,122.7	1,139.4	1,327.1	..	Merchandise, fob
169.5	226.5	279.0	273.1	294.5	287.2	300.2	303.0	353.1	420.8	..	Nonfactor Services
35.1	50.5	63.7	74.7	89.7	103.6	105.6	107.8	123.8	147.4	..	Factor Services
13.4	24.8	29.3	41.5	49.5	62.6	66.0	67.3	78.3	93.9	..	Long-Term Interest
..	..	32.6	27.4	24.8	23.4	21.6	22.3	..	..	..	Current Transfers, net
0.0	0.0	0.0	0.0	0.0	0.0	0.0	0.0	0.0	0.0	..	Workers' Remittances
..	..	-282.4	-184.8	-202.8	-237.7	-224.1	-173.8	..	..	..	Total to be Financed
..	..	40.8	35.7	46.6	51.0	18.7	16.6	..	..	..	Official Capital Grants
-162.3	-215.0	-241.6	-149.1	-156.2	-186.7	-205.4	-157.2	-16.1	94.0	..	Current Account Balance
188.6	173.8	257.9	223.1	266.9	188.1	228.8	131.8	184.7	60.5	..	Long-Term Capital, net
57.0	70.5	85.0	78.4	71.6	68.3	52.7	58.2	46.3	52.0	..	Direct Investment
65.7	52.2	107.8	104.7	136.1	62.8	151.2	73.7	131.2	27.3	..	Long-Term Loans
78.1	67.7	138.0	139.4	185.8	130.3	212.0	166.8	244.5	193.6	..	Disbursements
12.4	15.5	30.2	34.7	49.7	67.5	60.8	93.1	113.3	166.3	..	Repayments
65.8	51.1	24.3	4.3	12.6	5.9	6.1	-16.7	7.2	-18.8	..	Other Long-Term Capital
-7.4	29.1	19.4	10.1	29.5	38.8	72.6	0.7	-7.0	-154.5	..	Other Capital, net
-18.9	12.1	-35.7	-84.1	-140.2	-40.2	-96.0	24.7	-161.6	0.0	..	Change in Reserves
				(Cyprus Pounds per US dollar)							**Memo Item:**
0.370	0.350	0.350	0.420	0.480	0.530	0.590	0.610	0.520	0.480	..	Conversion Factor (Annual Avg)
				(Millions of US dollars, outstanding at end of year)							**EXTERNAL DEBT, ETC.**
241.3	299.6	398.0	486.1	609.1	650.7	758.6	936.9	1,192.7	1,419.0	..	Public/Publicly Guar. Long-Term
131.7	149.9	163.0	176.6	219.4	259.7	307.0	459.6	649.8	941.0	..	Official Creditors
43.7	50.1	58.1	61.2	68.1	74.4	72.3	95.3	97.5	121.2	..	IBRD and IDA
109.6	149.6	235.0	309.5	389.7	391.0	451.6	477.3	542.8	478.0	..	Private Creditors
..	..	..	..	..	..	..	..	..	..	..	Private Non-guaranteed Long-Term
42.9	50.0	39.0	25.2	13.6	5.7	3.1	0.0	0.0	..	..	Use of Fund Credit
86.0	113.0	108.0	111.0	107.0	200.0	258.0	409.0	421.0	597.2	..	Short-Term Debt
				(Millions of US dollars)							**Memo Items:**
345.7	353.1	368.3	426.4	523.2	519.1	540.6	595.3	752.8	873.5	..	Int'l Reserves Excluding Gold
100.1	235.0	270.6	182.5	209.7	175.1	141.5	150.1	179.4	222.2	..	Gold Holdings (at market price)
											SOCIAL INDICATORS
2.2	2.2	2.3	2.3	2.4	2.3	2.3	2.3	2.3	2.3	..	Total Fertility Rate
19.0	19.4	19.7	20.1	20.4	20.0	19.6	19.1	18.7	18.3	..	Crude Birth Rate
21.8	20.6	19.4	18.2	17.0	16.5	16.0	15.5	15.0	14.5	..	Infant Mortality Rate
73.8	73.8	73.9	73.9	74.0	74.1	74.2	74.3	74.4	74.5	..	Life Expectancy at Birth
97.1	97.3	104.6	98.1	105.0	88.8	97.9	90.2	84.5	87.5	..	Food Production, p.c. ('79-81 = 100)
28.5	27.3	26.0	..	..	..	..	..	..	..	..	Labor Force, Agriculture (%)
34.6	34.7	34.8	34.9	35.0	35.1	35.2	35.3	35.3	35.4	..	Labor Force, Female (%)
..	..	..	..	..	..	..	..	..	..	..	School Enroll. Ratio, primary
..	..	..	..	..	..	..	..	..	..	..	School Enroll. Ratio, secondary

DENMARK	1967	1968	1969	1970	1971	1972	1973	1974	1975	1976	1977
CURRENT GNP PER CAPITA (US $)	2,440	2,660	2,950	3,130	3,450	3,980	4,820	5,760	6,910	8,040	8,930
POPULATION (thousands)	4,839	4,867	4,891	4,929	4,963	4,992	5,022	5,045	5,060	5,073	5,088
ORIGIN AND USE OF RESOURCES					*(Billions of current Danish Kroner)*						
Gross National Product (GNP)	84.75	94.28	107.18	118.41	130.71	149.98	172.03	192.35	214.59	249.26	276.24
Net Factor Income from Abroad	-0.06	-0.08	-0.14	-0.21	-0.41	-0.75	-0.83	-1.28	-1.66	-1.96	-3.07
Gross Domestic Product (GDP)	84.81	94.36	107.32	118.63	131.12	150.73	172.86	193.63	216.26	251.21	279.31
Indirect Taxes, net	11.36	13.54	15.53	17.29	19.25	21.89	23.76	23.98	27.54	33.09	39.11
GDP at factor cost	73.45	80.82	91.79	101.33	111.87	128.84	149.10	169.65	188.71	218.12	240.20
Agriculture	5.87	6.13	7.10	6.61	7.16	8.56	10.53	11.44	10.93	12.30	14.56
Industry	26.49	28.80	32.27	35.06	37.92	43.34	48.52	53.83	58.99	66.29	71.11
Manufacturing	16.35	18.08	20.27	21.97	23.55	26.65	30.91	34.96	39.05	43.97	47.64
Services, etc.	41.09	45.89	52.42	59.66	66.79	76.94	90.06	104.38	118.80	139.53	154.53
Resource Balance	-1.68	-1.29	-2.31	-3.54	-2.40	0.79	-3.28	-5.81	-2.00	-11.68	-10.20
Exports of Goods & NFServices	23.06	25.96	29.41	33.08	36.19	40.71	49.28	61.37	65.07	72.42	80.48
Imports of Goods & NFServices	24.75	27.25	31.72	36.62	38.59	39.92	52.56	67.18	67.07	84.09	90.68
Domestic Absorption	86.50	95.64	109.63	122.17	133.52	149.94	176.14	199.44	218.25	262.89	289.51
Private Consumption, etc.	60.92	66.59	75.29	68.34	73.55	80.91	94.38	107.18	123.57	146.11	161.64
General Gov't Consumption	15.13	17.56	20.25	23.67	27.87	32.08	36.81	45.25	53.18	60.52	66.77
Gross Domestic Investment	10.44	11.50	14.08	30.15	32.11	36.95	44.95	47.00	41.50	56.26	61.10
Fixed Investment	19.94	21.44	25.59	28.43	31.38	37.08	42.74	45.51	44.42	56.94	60.90
Memo Items:											
Gross Domestic Saving	8.76	10.21	11.77	26.61	29.70	37.74	41.67	41.19	39.50	44.58	50.90
Gross National Saving	8.70	10.13	11.61	26.33	29.25	36.87	40.77	39.74	37.64	42.34	47.77
					(Billions of 1980 Danish Kroner)						
Gross National Product	265.48	276.00	293.27	299.02	306.57	322.10	333.85	330.19	327.61	348.80	353.29
GDP at factor cost	223.68	229.65	242.70	253.35	260.10	273.81	286.87	290.38	286.85	303.96	305.98
Agriculture	16.15	16.47	16.45	13.39	15.16	15.66	14.52	17.49	16.23	14.72	16.74
Industry	71.86	74.69	78.54	81.01	82.81	89.66	90.19	88.05	83.45	88.18	88.76
Manufacturing	41.53	44.75	47.18	48.57	49.50	53.88	56.90	57.17	55.38	58.47	58.75
Services, etc.	135.66	138.49	147.71	158.95	162.13	168.49	182.16	184.84	187.17	201.06	200.48
Resource Balance	-17.18	-15.14	-22.10	-26.98	-21.76	-18.61	-25.41	-17.41	-13.56	-26.93	-22.77
Exports of Goods & NFServices	65.40	71.45	75.85	80.10	84.56	89.28	96.26	99.63	97.85	101.85	106.04
Imports of Goods & NFServices	82.57	86.58	97.96	107.09	106.32	107.89	121.67	117.04	111.41	128.78	128.81
Domestic Absorption	282.79	291.31	315.74	326.57	329.33	342.41	360.97	349.85	343.80	378.54	380.09
Private Consumption, etc.	162.97	166.11	176.56	182.75	181.37	184.37	193.31	187.77	194.68	210.00	212.27
General Gov't Consumption	54.33	56.89	60.76	64.94	68.49	72.43	75.32	77.96	79.53	83.08	85.08
Gross Domestic Investment	65.49	68.31	78.42	78.88	79.47	85.62	92.35	84.11	69.59	85.46	82.74
Fixed Investment	65.33	66.54	74.38	76.01	77.48	84.65	87.59	79.84	69.95	81.93	79.92
Memo Items:											
Capacity to Import	76.96	82.50	90.82	96.73	99.70	110.03	114.08	106.92	108.09	110.90	114.32
Terms of Trade Adjustment	11.56	11.05	14.97	16.63	15.14	20.75	17.83	7.29	10.24	9.05	8.28
Gross Domestic Income	277.18	287.23	308.61	316.22	322.71	344.55	353.39	339.72	340.48	360.66	365.60
Gross National Income	277.04	287.06	308.24	315.65	321.71	342.85	351.68	337.48	337.86	357.85	361.58
DOMESTIC PRICES (DEFLATORS)					*(Index 1980 = 100)*						
Overall (GDP)	31.9	34.2	36.5	39.6	42.6	46.5	51.5	58.2	65.5	71.4	78.2
Domestic Absorption	30.6	32.8	34.7	37.4	40.5	43.8	48.8	57.0	63.5	69.4	76.2
Agriculture	36.3	37.2	43.2	49.4	47.2	54.7	72.5	65.4	67.3	83.6	87.0
Industry	36.9	38.6	41.1	43.3	45.8	48.3	53.8	61.1	70.7	75.2	80.1
Manufacturing	39.4	40.4	43.0	45.2	47.6	49.5	54.3	61.1	70.5	75.2	81.1
MANUFACTURING ACTIVITY											
Employment (1980=100)	99.7	99.4	103.7	110.0	106.5	108.0	111.9	108.3	98.3	99.0	101.2
Real Earnings per Empl. (1980=100)	72.2	72.5	77.3	78.6	84.4	86.7	88.4	94.1	99.6	102.3	100.6
Real Output per Empl. (1980=100)	57.5	60.2	62.8	64.4	66.8	70.2	73.4	82.3	78.3	83.7	88.3
Earnings as % of Value Added	55.9	55.1	54.5	56.4	58.1	56.6	57.0	59.1	59.1	57.4	57.5
MONETARY HOLDINGS					*(Billions of current Danish Kroner)*						
Money Supply, Broadly Defined	41.61	46.92	51.88	54.39	59.21	67.12	76.27	82.70	104.99	117.30	128.19
Money as Means of Payment	21.12	24.06	27.13	27.47	29.61	33.64	37.59	39.36	51.27	54.51	58.85
Currency Ouside Banks	4.72	4.85	5.23	4.87	4.92	5.56	5.99	6.04	7.63	8.44	9.91
Demand Deposits	16.40	19.21	21.90	22.60	24.69	28.08	31.60	33.32	43.64	46.07	48.94
Quasi-Monetary Liabilities	20.49	22.86	24.75	26.92	29.60	33.48	38.68	43.34	53.72	62.79	69.34
GOVERNMENT DEFICIT (-) OR SURPLUS	..	..	..	2.95	*(Billions of current Danish Kroner)* 3.57	4.09	6.14	1.34	-4.35	-0.52	-2.73
Current Revenue	..	..	..	41.83	47.63	53.57	58.64	67.27	71.24	83.17	91.70
Current Expenditure	..	..	..	35.39	40.34	45.67	48.38	61.01	69.79	77.44	88.35
Current Budget Balance	..	..	..	6.44	7.29	7.90	10.26	6.25	1.45	5.73	3.35
Capital Receipts	..	..	..	0.11	0.09	0.19	0.15	0.21	0.16	1.00	0.21
Capital Payments	..	..	..	3.60	3.81	4.00	4.28	5.13	5.96	7.25	6.29

1978	1979	1980	1981	1982	1983	1984	1985	1986	1987 estimate	NOTES	DENMARK
10,010	11,890	13,120	12,850	12,120	11,390	11,160	11,310	12,640	15,000	..	CURRENT GNP PER CAPITA (US $)
5,104	5,117	5,123	5,122	5,118	5,114	5,112	5,114	5,121	5,105	..	POPULATION (thousands)
			(Billions of current Danish Kroner)								ORIGIN AND USE OF RESOURCES
306.78	340.31	364.50	395.11	446.65	494.11	541.41	590.54	638.56	669.30	..	Gross National Product (GNP)
-4.60	-6.58	-9.28	-12.69	-17.82	-18.43	-23.86	-26.17	-27.36	-28.00	..	Net Factor Income from Abroad
311.38	346.89	373.79	407.79	464.47	512.54	565.27	616.71	665.92	697.25	..	Gross Domestic Product (GDP)
46.32	54.58	57.80	62.67	67.05	74.67	83.73	94.54	110.55	112.54	..	Indirect Taxes, net
265.06	292.32	315.99	345.12	397.41	437.87	481.54	522.18	555.38	584.72	B	GDP at factor cost
16.78	16.07	17.82	20.96	26.27	24.94	31.27	30.49	30.42	..	..	Agriculture
77.26	83.49	93.59	95.58	110.50	121.59	132.97	146.37	161.00	..	..	Industry
51.41	56.98	64.31	67.15	76.26	86.04	95.48	103.16	112.36	..	..	Manufacturing
171.02	192.76	204.58	228.57	260.64	291.34	317.30	345.32	363.96	..	..	Services, etc.
-6.76	-9.77	-3.95	3.03	2.06	10.13	6.65	1.17	-0.15	15.93	..	Resource Balance
86.48	101.50	122.26	149.08	168.95	186.25	207.57	225.48	212.50	220.73	..	Exports of Goods & NFServices
93.24	111.27	126.21	146.05	166.89	176.12	200.92	224.31	212.65	204.80	..	Imports of Goods & NFServices
318.13	356.66	377.74	404.76	462.41	502.41	558.62	615.54	666.07	681.32	..	Domestic Absorption
176.99	195.97	208.81	227.77	255.38	280.64	306.82	338.65	360.75	372.64	..	Private Consumption, etc.
76.25	86.83	99.73	113.22	131.10	140.54	146.18	155.59	160.14	177.11	..	General Gov't Consumption
64.90	73.86	69.19	63.77	75.94	81.22	105.63	121.31	145.18	131.57	..	Gross Domestic Investment
66.91	72.73	70.31	63.44	75.05	81.89	97.76	113.48	137.60	134.58	..	Fixed Investment
											Memo Items:
58.14	64.09	65.24	66.80	77.99	91.35	112.28	122.48	145.04	147.51	..	Gross Domestic Saving
53.37	..	55.45	53.25	59.85	72.07	88.35	95.73	116.77	119.13	..	Gross National Saving
			(Billions of 1980 Danish Kroner)								
357.04	368.16	364.50	359.11	367.21	377.31	391.37	405.80	420.41	416.72	..	Gross National Product
306.65	315.08	315.99	314.37	327.29	334.76	348.55	359.10	365.02	363.09	B	GDP at factor cost
17.14	17.24	17.82	19.17	21.10	19.31	23.84	23.55	23.97	..	..	Agriculture
89.20	91.31	93.59	86.99	89.02	93.21	97.70	102.49	110.22	..	..	Industry
58.63	62.00	64.31	61.86	62.87	67.18	70.67	72.76	75.70	..	..	Manufacturing
200.30	206.53	204.58	208.21	217.17	222.24	227.01	233.05	230.83	..	..	Services, etc.
-21.67	-19.10	-3.95	8.20	6.82	11.14	8.92	3.14	-4.90	2.16	..	Resource Balance
107.30	116.26	122.26	132.28	135.59	142.18	147.11	153.18	153.32	160.07	..	Exports of Goods & NFServices
128.96	135.37	126.21	124.09	128.77	131.04	138.18	150.04	158.22	157.91	..	Imports of Goods & NFServices
384.26	394.55	377.74	362.26	374.83	380.12	399.51	420.55	443.53	432.34	..	Domestic Absorption
213.86	216.80	208.81	203.99	206.95	212.29	219.47	231.44	240.12	236.52	..	Private Consumption, etc.
90.35	95.65	99.73	102.36	105.51	105.46	105.03	107.60	108.79	110.42	..	General Gov't Consumption
80.05	82.10	69.19	55.91	62.37	62.36	75.01	81.51	94.61	85.39	..	Gross Domestic Investment
80.81	80.45	70.31	56.80	60.82	61.95	69.93	77.51	91.61	85.39	..	Fixed Investment
											Memo Items:
119.61	123.48	122.26	126.66	130.36	138.58	142.76	150.82	158.11	170.19	..	Capacity to Import
12.32	7.21	0.00	-5.62	-5.23	-3.60	-4.35	-2.36	4.79	10.12	..	Terms of Trade Adjustment
374.91	382.66	373.79	364.84	376.42	387.66	404.08	421.33	443.42	444.62	..	Gross Domestic Income
369.36	375.38	364.50	353.49	361.97	373.72	387.02	403.45	425.20	..	..	Gross National Income
			(Index 1980 = 100)								DOMESTIC PRICES (DEFLATORS)
85.9	92.4	100.0	110.1	121.7	131.0	138.4	145.6	151.8	160.5	..	Overall (GDP)
82.8	90.4	100.0	111.7	123.4	132.2	139.8	146.4	150.2	157.6	..	Domestic Absorption
97.9	93.2	100.0	109.4	124.5	129.2	131.2	129.5	126.9	..	..	Agriculture
86.6	91.4	100.0	109.9	124.1	130.4	136.1	142.8	146.1	..	..	Industry
87.7	91.9	100.0	108.6	121.3	128.1	135.1	141.8	148.4	..	..	Manufacturing
											MANUFACTURING ACTIVITY
100.8	102.0	100.0	95.1	94.8	94.5	99.2	106.2			..	Employment (1980 = 100)
99.6	100.7	100.0	97.8	99.0	99.9	98.3	96.5	..	..	..	Real Earnings per Empl. (1980 = 100)
88.3	92.9	100.0	107.5	106.9	111.6	113.1	108.6	..	..	..	Real Output per Empl. (1980 = 100)
57.6	57.7	56.8	55.1	54.4	52.5	51.7	51.8	..	..	..	Earnings as % of Value Added
			(Billions of current Danish Kroner)								MONETARY HOLDINGS
136.44	150.38	167.99	186.21	206.83	247.57	309.79	366.67	401.19	417.79	..	Money Supply, Broadly Defined
68.30	75.05	83.21	93.03	105.24	114.18	153.84	195.77	214.95	235.40	..	Money as Means of Payment
10.75	11.57	12.36	13.57	14.18	15.42	16.37	17.57	18.82	20.45	..	Currency Ouside Banks
57.55	63.47	70.85	79.46	91.06	98.76	137.47	178.20	196.13	214.95	..	Demand Deposits
68.14	75.33	84.78	93.18	101.59	133.39	155.95	170.90	186.24	182.39	..	Quasi-Monetary Liabilities
			(Billions of current Danish Kroner)								
-1.07	-2.58	-10.00	-24.69	-37.52	-35.15	-22.24	-3.73	..	..	E	GOVERNMENT DEFICIT (-) OR SURPLUS
106.80	122.09	135.23	148.11	166.21	192.20	223.31	251.18	282.05	295.16	..	Current Revenue
101.61	118.61	137.09	162.04	191.50	213.75	233.94	242.58	241.33	257.01	..	Current Expenditure
5.19	3.48	-1.86	-13.93	-25.29	-21.55	-10.63	8.60	40.72	38.15	..	Current Budget Balance
0.62	0.61	3.67	0.24	0.27	0.38	0.38	0.33	0.99	0.90	..	Capital Receipts
6.88	6.67	11.81	10.99	12.50	13.99	11.99	12.66	-241.33	-257.01	..	Capital Payments

DENMARK	1967	1968	1969	1970	1971	1972	1973	1974	1975	1976	1977
FOREIGN TRADE (CUSTOMS BASIS)				*(Millions of current US dollars)*							
Value of Exports, fob	2,474	2,582	2,958	3,285	3,601	4,312	6,116	7,683	8,663	8,980	9,911
Nonfuel Primary Products	1,289	1,261	1,328	1,429	1,556	1,831	2,654	3,085	3,479	3,577	4,014
Fuels	33	39	56	82	81	89	132	288	292	335	317
Manufactures	1,152	1,283	1,574	1,774	1,963	2,392	3,330	4,310	4,892	5,068	5,580
Value of Imports, cif	3,134	3,213	3,800	4,385	4,582	5,027	7,714	9,857	10,326	12,404	13,227
Nonfuel Primary Products	692	664	725	869	849	967	1,486	1,761	1,640	2,133	2,539
Fuels	332	372	384	459	557	549	821	1,889	1,920	2,038	2,227
Manufactures	2,110	2,178	2,691	3,057	3,177	3,511	5,407	6,206	6,766	8,232	8,461
				(Index 1980 = 100)							
Terms of Trade	157.7	162.2	149.7	151.0	151.1	149.3	152.3	107.6	107.3	106.9	103.3
Export Prices, fob	29.4	29.1	29.3	30.0	32.8	37.8	50.5	55.9	58.3	61.8	65.7
Nonfuel Primary Products	30.5	30.7	34.2	36.1	38.7	46.5	60.1	57.5	54.6	60.7	62.9
Fuels	8.5	7.3	5.4	5.9	7.0	8.5	24.5	36.6	33.6	38.5	36.8
Manufactures	30.7	30.7	30.7	31.9	34.0	37.2	46.5	56.1	65.4	66.0	71.9
Import Prices, cif	18.6	17.9	19.6	19.9	21.7	25.3	33.2	51.9	54.3	57.8	63.6
Trade at Constant 1980 Prices				*(Millions of 1980 US dollars)*							
Exports, fob	8,413	8,875	10,086	10,952	10,977	11,421	12,108	13,754	14,856	14,538	15,097
Imports, cif	16,806	17,915	19,394	22,078	21,108	19,880	23,259	18,984	19,005	21,464	20,813
BALANCE OF PAYMENTS				*(Millions of current US dollars)*							
Exports of Goods & Services	..	..	..	4,549	5,152	6,230	8,697	10,896	12,127	12,897	14,482
Merchandise, fob	..	..	..	3,317	3,610	4,363	6,181	7,703	8,653	9,054	10,007
Nonfactor Services	..	..	..	1,148	1,462	1,769	2,348	2,969	3,227	3,588	4,112
Factor Services	..	..	..	84	80	98	167	224	248	255	362
Imports of Goods & Services	..	..	..	5,050	5,524	6,203	9,393	12,032	12,704	15,027	16,590
Merchandise, fob	..	..	..	4,077	4,321	4,793	7,367	9,489	9,970	11,931	12,725
Nonfactor Services	..	..	..	844	1,052	1,191	1,700	2,061	2,161	2,481	2,955
Factor Services	..	..	..	129	151	218	325	482	573	616	910
Long-Term Interest	..	..	..	..	..	..	..	..	..	..	..
Current Transfers, net	..	..	..	-9	-6	-17	-11	-29	-35	-47	-11
Workers' Remittances	..	..	..	0	0	0	0	0	0	0	0
Total to be Financed	..	..	..	-510	-378	10	-707	-1,165	-612	-2,177	-2,120
Official Capital Grants	..	..	..	-34	-45	-73	241	186	109	265	397
Current Account Balance	..	..	..	-544	-423	-63	-466	-979	-503	-1,912	-1,723
Long-Term Capital, net	..	..	..	77	333	193	752	530	243	2,128	2,879
Direct Investment	..	..	..	75	73	16	114	..	188	-254	-85
Long-Term Loans	..	..	..	..	..	..	..	..	..	..	..
Disbursements	..	..	..	..	..	..	..	..	..	..	..
Repayments	..	..	..	..	..	..	..	..	..	..	..
Other Long-Term Capital	..	..	..	36	305	250	397	344	-54	2,117	2,568
Other Capital, net	..	..	..	478	269	-40	157	11	217	-197	-440
Change in Reserves	..	..	..	-11	-178	-91	-443	438	42	-19	-717
Memo Item:				*(Danish Kroner per US dollar)*							
Conversion Factor (Annual Avg)	6.960	7.500	7.500	7.500	7.420	6.950	6.050	6.090	5.750	6.050	6.000
EXTERNAL DEBT, ETC.				*(Millions US dollars, outstanding at end of year)*							
Public/Publicly Guar. Long-Term	..	..	..	..	..	..	..	..	..	..	..
Official Creditors	..	..	..	..	..	..	..	..	..	..	..
IBRD and IDA	..	..	..	..	..	..	..	..	..	..	..
Private Creditors	..	..	..	..	..	..	..	..	..	..	..
Private Non-guaranteed Long-Term	..	..	..	..	..	..	..	..	..	..	..
Use of Fund Credit	..	..	..	..	..	..	..	..	..	..	..
Short-Term Debt	..	..	..	..	..	..	..	..	..	..	..
Memo Items:				*(Millions of US dollars)*							
Int'l Reserves Excluding Gold	427	335	357	419	653	786	1,247	858	803	841	1,589
Gold Holdings (at market price)	109	137	89	69	79	118	204	338	254	244	318
SOCIAL INDICATORS											
Total Fertility Rate	2.4	2.1	2.0	1.9	2.0	2.0	1.9	1.9	1.9	1.7	1.7
Crude Birth Rate	16.8	15.3	14.6	14.4	15.2	15.1	14.3	14.1	14.2	12.9	12.2
Infant Mortality Rate	15.8	16.4	14.8	14.2	13.5	12.2	11.5	10.7	10.4	10.2	8.7
Life Expectancy at Birth	72.9	73.0	73.1	73.2	73.4	73.5	73.6	73.8	73.9	74.1	74.1
Food Production, p.c. ('79-81 = 100)	95.0	99.8	93.9	85.9	91.0	87.4	86.8	99.0	91.3	87.1	97.9
Labor Force, Agriculture (%)	13.1	12.5	11.8	11.2	10.8	10.4	10.0	9.6	9.2	8.8	8.4
Labor Force, Female (%)	34.9	35.3	35.7	36.1	36.8	37.6	38.4	39.2	40.0	40.8	41.6
School Enroll. Ratio, primary	..	..	..	96.0	..	..	..	..	104.0	101.0	99.0
School Enroll. Ratio, secondary	..	..	..	78.0	..	..	..	..	80.0	80.0	83.0

1978	1979	1980	1981	1982	1983	1984	1985	1986	1987 estimate	NOTES	DENMARK
											FOREIGN TRADE (CUSTOMS BASIS)
				(Millions of current US dollars)							
11,676	14,342	16,407	15,697	14,953	15,601	15,486	16,469	20,558	24,697	..	Value of Exports, fob
4,961	5,906	6,589	6,304	6,110	5,899	5,748	5,967	7,581	8,924	..	Nonfuel Primary Products
298	562	566	505	385	780	801	903	643	734	..	Fuels
6,417	7,874	9,252	8,888	8,458	8,922	8,936	9,599	12,334	15,039	..	Manufactures
14,777	18,412	19,315	17,521	16,834	16,179	16,536	17,985	22,726	25,334	..	Value of Imports, cif
2,847	3,464	3,664	3,323	3,017	3,066	3,052	3,036	3,997	4,642	..	Nonfuel Primary Products
2,303	3,603	4,328	4,216	3,790	3,163	2,999	3,097	2,009	1,996	..	Fuels
9,627	11,345	11,323	9,982	10,027	9,950	10,485	11,852	16,719	18,696	..	Manufactures
				(Index 1980 = 100)							
109.7	111.7	100.0	93.6	93.5	94.5	92.3	95.5	105.0	106.4	..	Terms of Trade
77.0	94.0	100.0	92.2	86.6	85.0	80.7	81.4	91.0	102.2	..	Export Prices, fob
77.4	97.8	100.0	93.2	86.8	86.6	81.5	78.9	81.6	85.9	..	Nonfuel Primary Products
44.7	93.6	100.0	104.0	97.0	85.4	83.6	81.1	41.3	44.4	..	Fuels
80.4	92.2	100.0	91.0	86.0	84.0	80.0	83.0	105.0	124.0	..	Manufactures
70.2	84.2	100.0	98.5	92.6	89.9	87.4	85.2	86.7	96.1	..	Import Prices, cif
				(Millions of 1980 US dollars)							Trade at Constant 1980 Prices
15,160	15,254	16,407	17,016	17,274	18,350	19,183	20,237	22,593	24,167	..	Exports, fob
21,056	21,880	19,315	17,783	18,182	17,992	18,912	21,117	26,215	26,374	..	Imports, cif
				(Millions of current US dollars)							**BALANCE OF PAYMENTS**
17,107	..	..	23,174	22,238	22,511	22,451	24,016	29,545	36,155	..	Exports of Goods & Services
11,790	14,571	16,789	16,120	15,675	16,207	16,078	17,061	21,204	25,585	..	Merchandise, fob
4,776	..	..	5,867	5,418	5,313	5,214	5,553	6,348	7,948	..	Nonfactor Services
542	..	..	1,188	1,145	991	1,160	1,402	1,993	2,622	..	Factor Services
19,155	..	..	24,885	24,310	23,500	24,152	26,598	33,526	38,897	..	Imports of Goods & Services
14,141	17,966	18,811	17,046	16,469	15,975	16,283	17,822	22,284	24,762	..	Merchandise, fob
3,588	..	..	4,670	4,542	4,514	4,469	4,911	5,814	7,382	..	Nonfactor Services
1,427	..	..	3,169	3,299	3,011	3,400	3,865	5,428	6,753	..	Factor Services
..	..	..	..	..	..	..	..	..	..	..	Long-Term Interest
-32	..	-89	-122	-40	-93	-6	-54	-112	-55	..	Current Transfers, net
0	..	0	0	0	0	0	0	0	0	..	Workers' Remittances
-2,080	-2,983	-2,578	-1,833	-2,111	-1,082	-1,707	-2,635	-4,093	-2,798	K	Total to be Financed
583	..	112	-33	-144	-95	73	-73	-167	-153	K	Official Capital Grants
-1,497	-2,983	-2,466	-1,865	-2,255	-1,177	-1,634	-2,708	-4,261	-2,951	..	Current Account Balance
3,018	931	2,652	1,319	2,265	2,364	1,973	4,384	3,319	8,187	..	Long-Term Capital, net
56	103	-91	-40	55	-96	-86	..	..	..	..	Direct Investment
..	..	..	..	..	..	..	..	..	..	..	Long-Term Loans
..	..	..	..	..	..	..	..	..	..	..	Disbursements
..	..	..	..	..	..	..	..	..	..	..	Repayments
2,379	827	2,631	1,392	2,354	2,555	1,987	4,457	3,486	8,339	..	Other Long-Term Capital
-25	1,981	-109	-169	-308	186	-707	-150	-1,024	-770	..	Other Capital, net
-1,496	71	-76	715	298	-1,373	368	-1,525	1,966	-4,466	..	Change in Reserves
				(Danish Kroner per US dollar)							Memo Item:
5.510	5.260	5.640	7.120	8.330	9.150	10.360	10.600	8.090	6.840	..	Conversion Factor (Annual Avg)
				(Millions US dollars, outstanding at end of year)							**EXTERNAL DEBT, ETC.**
..	..	..	..	..	..	..	..	..	..	..	Public/Publicly Guar. Long-Term
..	..	..	..	..	..	..	..	..	..	..	Official Creditors
..	..	..	..	..	..	..	..	..	..	..	IBRD and IDA
..	..	..	..	..	..	..	..	..	..	..	Private Creditors
..	..	..	..	..	..	..	..	..	..	..	Private Non-guaranteed Long-Term
..	..	..	..	..	..	..	..	..	..	..	Use of Fund Credit
..	..	..	..	..	..	..	..	..	..	..	Short-Term Debt
				(Millions of US dollars)							Memo Items:
3,129	3,236	3,387	2,548	2,266	3,621	3,009	5,429	4,964	10,066	..	Int'l Reserves Excluding Gold
447	840	960	648	744	621	502	533	637	788	..	Gold Holdings (at market price)
											SOCIAL INDICATORS
1.7	1.6	1.5	1.4	1.4	1.4	1.4	1.4	1.4	1.4	..	Total Fertility Rate
12.2	11.6	11.2	10.4	10.4	10.5	10.5	10.5	10.6	10.6	..	Crude Birth Rate
8.8	8.8	8.4	7.9	8.5	9.1	9.8	10.5	11.3	12.1	..	Infant Mortality Rate
74.1	74.1	74.1	74.2	74.3	74.9	75.1	75.2	75.4	75.6	..	Life Expectancy at Birth
96.9	100.0	99.4	100.7	110.7	104.6	126.3	123.0	119.9	121.1	..	Food Production, p.c. ('79-81=100)
8.1	7.7	7.3	..	..	..	..	..	..	..	..	Labor Force, Agriculture (%)
42.3	43.1	43.9	44.0	44.0	44.1	44.2	44.2	44.3	44.4	..	Labor Force, Female (%)
98.0	98.0	96.0	..	98.0	99.0	98.0	98.0	..	..	..	School Enroll. Ratio, primary
99.0	102.0	105.0	..	105.0	103.0	103.0	105.0	..	..	..	School Enroll. Ratio, secondary

DOMINICA	1967	1968	1969	1970	1971	1972	1973	1974	1975	1976	1977
CURRENT GNP PER CAPITA (US $)	240	240	250	290	320	340	350	350	380	420	470
POPULATION (thousands)	66	67	68	68	69	70	70	71	71	71	72
ORIGIN AND USE OF RESOURCES					*(Millions of current East Caribbean Dollars)*						
Gross National Product (GNP)	27.7	30.3	32.3	40.3	43.0	45.6	49.6	56.9	60.2	74.3	98.2
Net Factor Income from Abroad	0.0	0.0	0.0	0.0	0.0	0.0	0.0	0.0	0.0	0.0	0.0
Gross Domestic Product (GDP)	27.7	30.3	32.3	40.3	43.0	45.6	49.6	56.9	60.2	74.3	98.2
Indirect Taxes, net	..	..	..	..	..	..	..	..	..	..	13.4
GDP at factor cost	..	..	..	..	..	..	..	..	..	..	84.8
Agriculture	..	..	..	..	..	..	..	..	..	..	31.9
Industry	..	..	..	..	..	..	..	..	..	..	12.4
Manufacturing	..	..	..	..	..	..	..	..	..	..	4.2
Services, etc.	..	..	..	..	..	..	..	..	..	..	40.5
Resource Balance	..	..	..	..	..	..	..	..	..	..	-19.4
Exports of Goods & NFServices	..	..	..	..	..	..	..	..	..	..	40.5
Imports of Goods & NFServices	..	..	..	..	..	..	..	..	..	..	59.9
Domestic Absorption	..	..	..	..	..	..	..	..	..	..	117.6
Private Consumption, etc.	..	..	..	..	..	..	..	..	..	..	62.7
General Gov't Consumption	..	..	..	..	..	..	..	..	..	..	33.4
Gross Domestic Investment	..	..	..	..	..	..	..	..	..	..	21.5
Fixed Investment	..	..	..	..	..	..	..	..	..	..	21.5
Memo Items:											
Gross Domestic Saving	..	..	..	..	..	..	..	..	..	..	2.1
Gross National Saving	..	..	..	..	..	..	..	..	..	..	6.7
					(Millions of 1980 East Caribbean Dollars)						
Gross National Product	154.16	157.83	154.33	159.84	180.02	190.87	194.04	185.36	161.67	151.66	157.00
GDP at factor cost	..	..	..	..	..	..	..	..	..	..	136.16
Agriculture	..	..	..	..	..	..	..	..	..	..	59.22
Industry	..	..	..	..	..	..	..	..	..	..	18.21
Manufacturing	..	..	..	..	..	..	..	..	..	..	4.67
Services, etc.	..	..	..	..	..	..	..	..	..	..	58.73
Resource Balance	..	..	..	..	..	..	..	..	..	..	..
Exports of Goods & NFServices	..	..	..	..	..	..	..	..	..	..	..
Imports of Goods & NFServices	..	..	..	..	..	..	..	..	..	..	..
Domestic Absorption	..	..	..	..	..	..	..	..	..	..	..
Private Consumption, etc.	..	..	..	..	..	..	..	..	..	..	..
General Gov't Consumption	..	..	..	..	..	..	..	..	..	..	..
Gross Domestic Investment	..	..	..	..	..	..	..	..	..	..	..
Fixed Investment	..	..	..	..	..	..	..	..	..	..	..
Memo Items:											
Capacity to Import	..	..	..	..	..	..	..	..	..	..	..
Terms of Trade Adjustment	..	..	..	..	..	..	..	..	..	..	..
Gross Domestic Income	..	..	..	..	..	..	..	..	..	..	..
Gross National Income	..	..	..	..	..	..	..	..	..	..	..
DOMESTIC PRICES (DEFLATORS)					*(Index 1980 = 100)*						
Overall (GDP)	18.0	19.2	20.9	25.2	23.9	23.9	25.6	30.7	37.2	49.0	62.5
Domestic Absorption	..	..	..	..	..	..	..	..	..	..	..
Agriculture	..	..	..	..	..	..	..	..	..	..	53.9
Industry	..	..	..	..	..	..	..	..	..	..	68.1
Manufacturing	..	..	..	..	..	..	..	..	..	..	89.9
MANUFACTURING ACTIVITY											
Employment (1980=100)	..	..	..	..	..	..	..	..	..	..	..
Real Earnings per Empl. (1980=100)	..	..	..	..	..	..	..	..	..	..	..
Real Output per Empl. (1980=100)	..	..	..	..	..	..	..	..	..	..	..
Earnings as % of Value Added	..	..	..	..	..	..	..	..	..	..	..
MONETARY HOLDINGS					*(Millions of current East Caribbean Dollars)*						
Money Supply, Broadly Defined	..	..	..	..	..	..	..	..	36.65	41.25	45.15
Money as Means of Payment	..	..	..	..	..	..	..	..	8.54	9.36	11.53
Currency Ouside Banks	..	..	..	..	..	..	..	..	2.77	3.65	4.81
Demand Deposits	..	..	..	..	..	..	..	..	5.77	5.70	6.72
Quasi-Monetary Liabilities	..	..	..	..	..	..	..	..	28.11	31.89	33.63
					(Thousands of current East Caribbean Dollars)						
GOVERNMENT DEFICIT (-) OR SURPLUS	..	..	..	..	..	..	..	..	..	-4,500	200
Current Revenue	..	..	..	..	..	..	..	..	..	21,680	30,970
Current Expenditure	..	..	..	..	..	..	..	..	..	19,500	23,700
Current Budget Balance	..	..	..	..	..	..	..	..	..	2,180	7,270
Capital Receipts	..	..	..	..	..	..	..	..	..	20	30
Capital Payments	..	..	..	..	..	..	..	..	..	6,700	7,100

1978	1979	1980	1981	1982	1983	1984	1985	1986	1987 estimate	NOTES	DOMINICA
580	570	730	910	980	1,000	1,100	1,190	1,320	1,440	..	**CURRENT GNP PER CAPITA (US $)**
72	73	73	74	75	77	77	78	79	80	..	**POPULATION (thousands)**
				(Millions of current East Caribbean Dollars)							**ORIGIN AND USE OF RESOURCES**
121.3	121.7	157.5	177.4	192.6	212.1	234.4	259.4	289.3	316.1	..	Gross National Product (GNP)
-0.5	-0.3	-1.6	-1.1	-3.6	-5.7	-7.8	-2.8	-8.1	-9.0	..	Net Factor Income from Abroad
121.8	122.0	159.1	178.5	196.2	217.8	242.2	262.2	297.4	325.1	..	Gross Domestic Product (GDP)
16.0	16.2	15.4	25.6	31.2	35.3	39.8	43.3	..	..	..	Indirect Taxes, net
105.8	105.8	143.7	152.9	165.0	182.5	202.4	218.9	252.9	276.4	..	GDP at factor cost
41.4	34.8	44.0	48.5	49.8	52.5	56.5	62.6	72.3	..	..	Agriculture
14.9	15.5	30.1	31.0	33.5	33.2	38.4	38.5	42.0	..	..	Industry
6.0	5.0	6.9	10.2	13.4	14.1	13.8	15.5	..	..	..	Manufacturing
49.5	55.5	69.6	73.4	81.7	96.8	107.5	117.8	138.6	..	..	Services, etc.
-27.0	-56.7	-104.8	-77.5	-49.7	-38.9	-59.4	-64.5	..	..	..	Resource Balance
52.1	52.1	42.4	62.9	85.6	96.4	99.6	103.4	..	..	..	Exports of Goods & NFServices
79.1	108.8	147.2	140.4	135.3	135.3	159.0	167.9	..	..	..	Imports of Goods & NFServices
148.8	178.7	263.9	256.0	245.5	256.6	302.1	326.7	..	..	..	Domestic Absorption
86.9	96.6	167.2	186.1	174.7	184.1	227.7	246.5	..	..	..	Private Consumption, etc.
34.4	41.1	43.8	45.9	48.8	50.4	54.1	58.2	..	..	..	General Gov't Consumption
27.5	41.0	52.9	24.0	22.0	22.1	20.3	22.0	..	..	..	Gross Domestic Investment
27.5	31.0	45.9	24.0	22.0	20.8	20.0	22.0	..	..	..	Fixed Investment
											Memo Items:
0.5	-15.7	-51.9	-53.5	-27.3	-16.7	-39.6	-42.5	..	..	..	Gross Domestic Saving
9.7	2.4	-36.5	-38.9	-18.5	-8.9	-30.4	-27.8	..	..	..	Gross National Saving
				(Millions of 1980 East Caribbean Dollars)							
174.71	143.40	157.50	175.20	181.32	182.87	192.56	195.78	209.97	219.65	..	Gross National Product
153.02	124.93	143.70	154.36	158.12	161.29	171.18	172.89	184.89	193.37	..	GDP at factor cost
65.91	44.93	44.00	53.65	54.95	55.51	58.67	57.92	63.12	..	..	Agriculture
21.00	20.41	30.10	29.22	29.37	29.07	34.36	33.33	32.30	..	..	Industry
6.57	5.45	6.90	8.12	9.57	9.68	9.68	10.57	..	..	..	Manufacturing
66.12	59.59	69.60	71.48	73.80	76.70	78.15	81.63	89.46	..	..	Services, etc.
..	..	..	..	..	..	..	..	..	..	..	Resource Balance
..	..	..	..	..	..	..	..	..	..	..	Exports of Goods & NFServices
..	..	..	..	..	..	..	..	..	..	..	Imports of Goods & NFServices
..	..	..	..	..	..	..	..	..	..	..	Domestic Absorption
..	..	..	..	..	..	..	..	..	..	..	Private Consumption, etc.
..	..	..	..	..	..	..	..	..	..	..	General Gov't Consumption
..	..	..	..	..	..	..	..	..	..	..	Gross Domestic Investment
..	..	..	..	..	..	..	..	..	..	..	Fixed Investment
											Memo Items:
..	..	..	..	..	..	..	..	..	..	..	Capacity to Import
..	..	..	..	..	..	..	..	..	..	..	Terms of Trade Adjustment
..	..	..	..	..	..	..	..	..	..	..	Gross Domestic Income
..	..	..	..	..	..	..	..	..	..	..	Gross National Income
				(Index 1980 = 100)							**DOMESTIC PRICES (DEFLATORS)**
69.4	84.9	100.0	101.2	106.8	116.5	121.4	129.1	137.1	143.3	..	Overall (GDP)
..	..	..	..	..	..	..	..	..	..	..	Domestic Absorption
62.8	77.5	100.0	90.4	90.6	94.6	96.3	108.1	114.5	..	..	Agriculture
71.0	75.9	100.0	106.1	114.1	114.2	111.8	115.5	130.0	..	..	Industry
91.4	91.7	100.0	125.6	140.0	145.6	142.5	146.6	..	..	..	Manufacturing
											MANUFACTURING ACTIVITY
..	..	..	..	..	..	..	..	..	..	..	Employment (1980 = 100)
..	..	..	..	..	..	..	..	..	..	..	Real Earnings per Empl. (1980 = 100)
..	..	..	..	..	..	..	..	..	..	..	Real Output per Empl. (1980 = 100)
..	..	..	..	..	..	..	..	..	..	..	Earnings as % of Value Added
				(Millions of current East Caribbean Dollars)							**MONETARY HOLDINGS**
53.94	77.60	79.47	81.29	96.26	105.71	122.59	128.20	147.27	190.85	..	Money Supply, Broadly Defined
17.49	30.09	27.11	26.44	25.41	25.40	32.40	31.22	37.03	58.67	..	Money as Means of Payment
5.15	7.33	7.48	7.73	6.58	6.39	12.22	9.64	6.64	20.77	..	Currency Ouside Banks
12.34	22.75	19.63	18.71	18.83	19.01	20.18	21.58	30.39	37.89	..	Demand Deposits
36.45	47.52	52.36	54.84	70.85	80.31	90.19	96.98	110.24	132.18	..	Quasi-Monetary Liabilities
				(Thousands of current East Caribbean Dollars)							
-2,700	-8,100	..	..	..	..	..	..	..	..	C	**GOVERNMENT DEFICIT (-) OR SURPLUS**
43,280	52,050	..	..	..	..	..	..	..	..	..	Current Revenue
35,200	37,800	..	..	..	..	..	..	..	..	..	Current Expenditure
8,080	14,250	..	..	..	..	..	..	..	..	..	Current Budget Balance
20	50	..	..	..	..	..	..	..	..	..	Capital Receipts
10,800	22,400	..	..	..	..	..	..	..	..	..	Capital Payments

DOMINICA	1967	1968	1969	1970	1971	1972	1973	1974	1975	1976	1977
FOREIGN TRADE (CUSTOMS BASIS)				*(Thousands of current US dollars)*							
Value of Exports, fob	..	..	..	..	..	..	..	..	..	..	..
Nonfuel Primary Products	..	..	..	..	..	..	..	..	..	..	..
Fuels	..	..	..	..	..	..	..	..	..	..	..
Manufactures	..	..	..	..	..	..	..	..	..	..	..
Value of Imports, cif	..	..	..	..	..	..	..	..	..	..	..
Nonfuel Primary Products	..	..	..	..	..	..	..	..	..	..	..
Fuels	..	..	..	..	..	..	..	..	..	..	..
Manufactures	..	..	..	..	..	..	..	..	..	..	..
					(Index 1980 = 100)						
Terms of Trade	..	..	..	..	..	..	..	..	..	..	..
Export Prices, fob	..	..	..	..	..	..	..	..	..	..	..
Nonfuel Primary Products	..	..	..	..	..	..	..	..	..	..	..
Fuels	..	..	..	..	..	..	..	..	..	..	..
Manufactures	..	..	..	..	..	..	..	..	..	..	..
Import Prices, cif	..	..	..	..	..	..	..	..	..	..	..
Trade at Constant 1980 Prices				*(Thousands of 1980 US dollars)*							
Exports, fob	..	..	..	..	..	..	..	..	..	..	..
Imports, cif	..	..	..	..	..	..	..	..	..	..	..
BALANCE OF PAYMENTS				*(Thousands of current US dollars)*							
Exports of Goods & Services	..	..	..	..	..	..	..	..	..	14,002	15,300
Merchandise, fob	..	..	..	..	..	..	..	..	..	11,099	12,000
Nonfactor Services	..	..	..	..	..	..	..	..	..	2,501	3,000
Factor Services	..	..	..	..	..	..	..	..	..	402	300
Imports of Goods & Services	..	..	..	..	..	..	..	..	..	19,315	22,500
Merchandise, fob	..	..	..	..	..	..	..	..	..	17,272	19,907
Nonfactor Services	..	..	..	..	..	..	..	..	..	1,844	2,293
Factor Services	..	..	..	..	..	..	..	..	..	199	300
Long-Term Interest	..	..	..	..	..	..	..	..	..	..	..
Current Transfers, net	..	..	..	..	..	..	..	..	0	1,300	1,700
Workers' Remittances	..	..	..	..	..	..	..	..	..	1,300	3,200
Total to be Financed	..	..	..	..	..	..	..	..	..	-3,997	-5,500
Official Capital Grants	..	..	..	..	..	..	..	..	0	2,899	4,000
Current Account Balance	..	..	..	..	..	..	..	..	..	-1,098	-1,500
Long-Term Capital, net	..	..	..	..	..	..	..	..	..	4,398	6,000
Direct Investment	..	..	..	..	..	..	..	..	..	..	..
Long-Term Loans	..	..	..	..	..	..	..	..	..	..	..
Disbursements	..	..	..	..	..	..	..	..	..	..	..
Repayments	..	..	..	..	..	..	..	..	..	..	..
Other Long-Term Capital	..	..	..	..	..	..	..	..	..	1,499	2,000
Other Capital, net	..	..	..	..	..	..	..	..	..	-2,874	-3,541
Change in Reserves	..	..	..	0	0	0	0	0	0	-426	-959
Memo Item:				*(East Caribbean Dollars per US dollar)*							
Conversion Factor (Annual Avg)	1.760	2.000	2.000	2.000	1.970	1.920	1.960	2.050	2.170	2.610	2.700
EXTERNAL DEBT, ETC.				*(Millions of US dollars, outstanding at end of year)*							
Public/Publicly Guar. Long-Term	..	..	..	..	..	..	..	..	..	..	..
Official Creditors	..	..	..	..	..	..	..	..	..	..	..
IBRD and IDA	..	..	..	..	..	..	..	..	..	..	..
Private Creditors	..	..	..	..	..	..	..	..	..	..	..
Private Non-guaranteed Long-Term	..	..	..	..	..	..	..	..	..	..	..
Use of Fund Credit	..	..	..	..	..	..	..	..	..	..	..
Short-Term Debt	..	..	..	..	..	..	..	..	..	..	..
Memo Items:				*(Thousands of US dollars)*							
Int'l Reserves Excluding Gold	..	..	..	..	..	..	..	..	345	1,173	2,230
Gold Holdings (at market price)	..	..	..	..	..	..	..	..	..	..	..
SOCIAL INDICATORS											
Total Fertility Rate	..	..	..	..	..	..	..	..	..	..	..
Crude Birth Rate	..	..	..	..	25.8	31.4	37.0	32.7	29.2	22.9	24.7	24.0
Infant Mortality Rate	..	..	..	..	..	..	..	38.7	28.1	26.9	24.0	27.0
Life Expectancy at Birth	..	..	..	..	..	..	..	..	..	..	..
Food Production, p.c. ('79-81 = 100)	129.7	135.2	139.0	121.7	117.7	119.0	112.3	121.4	126.8	138.5	141.4
Labor Force, Agriculture (%)	..	..	..	..	..	..	..	..	..	..	..
Labor Force, Female (%)	..	..	..	..	..	..	..	..	..	..	..
School Enroll. Ratio, primary	..	..	..	..	..	..	..	..	..	..	..
School Enroll. Ratio, secondary	..	..	..	..	..	..	..	..	..	..	..

1978	1979	1980	1981	1982	1983	1984	1985	1986	1987	NOTES	
											FOREIGN TRADE (CUSTOMS BASIS)
		(Thousands of current US dollars)									
..	..	9,700	19,100	24,500	27,500	25,600	28,400	33,500	..	..	Value of Exports, fob
..	..	..	..	..	..	..	..	..	..	..	Nonfuel Primary Products
..	..	..	..	..	..	..	..	..	..	..	Fuels
..	..	..	..	..	..	..	..	..	..	..	Manufactures
..	..	38,680	29,510	37,860	25,150	30,670	32,560	34,650	..	..	Value of Imports, cif
..	..	..	..	..	..	..	..	..	..	..	Nonfuel Primary Products
..	..	..	..	..	..	..	..	..	..	..	Fuels
..	..	..	..	..	..	..	..	..	..	..	Manufactures
		(Index 1980 = 100)									
..	..	..	..	..	..	..	..	..	..	..	Terms of Trade
..	..	..	..	..	..	..	..	..	..	..	Export Prices, fob
..	..	..	..	..	..	..	..	..	..	..	Nonfuel Primary Products
..	..	..	..	..	..	..	..	..	..	..	Fuels
..	..	..	..	..	..	..	..	..	..	..	Manufactures
..	..	..	..	..	..	..	..	..	..	..	Import Prices, cif
		(Thousands of 1980 US dollars)									
											Trade at Constant 1980 Prices
..	..	..	..	..	..	..	..	..	..	..	Exports, fob
..	..	..	..	..	..	..	..	..	..	..	Imports, cif
		(Thousands of current US dollars)									
											BALANCE OF PAYMENTS
19,200	19,600	16,300	24,400	32,700	37,300	38,141	38,878	50,152	..	..	Exports of Goods & Services
15,900	9,800	10,100	19,700	25,100	27,800	26,100	28,605	39,294	..	..	Merchandise, fob
3,100	9,500	5,600	3,700	6,600	8,200	11,300	9,903	10,489	..	..	Nonfactor Services
200	300	600	1,000	1,000	1,300	741	370	369	..	..	Factor Services
29,700	40,700	55,000	52,600	50,900	51,404	62,200	65,810	65,423	..	..	Imports of Goods & Services
25,907	35,819	48,359	45,178	43,178	42,815	50,722	52,008	50,500	..	..	Merchandise, fob
3,393	4,481	6,341	6,922	6,922	6,885	9,478	11,802	12,828	..	..	Nonfactor Services
400	400	300	500	800	1,704	2,000	2,000	2,094	..	..	Factor Services
..	..	..	..	..	..	..	..	..	..	..	Long-Term Interest
3,600	6,800	6,300	5,800	4,600	5,000	6,300	6,501	6,782	..	..	Current Transfers, net
5,800	8,700	8,600	8,500	7,500	9,400	10,800	11,102	11,569	..	..	Workers' Remittances
-6,900	-14,300	-32,400	-22,400	-13,600	-9,104	-17,759	-20,433	-8,507	..	K	Total to be Financed
5,600	20,600	18,100	9,598	5,799	4,400	14,601	11,903	4,783	..	K	Official Capital Grants
-1,300	6,300	-14,300	-12,802	-7,801	-4,704	-3,158	-8,530	-3,724	..	..	Current Account Balance
6,725	21,229	19,567	12,276	14,172	10,462	21,760	19,430	16,008	..	..	Long-Term Capital, net
..	..	..	..	200	200	2,300	3,000	6,084	..	..	Direct Investment
..	..	..	..	..	..	..	..	..	..	..	Long-Term Loans
..	..	..	..	..	..	..	..	..	..	..	Disbursements
..	..	..	..	..	..	..	..	..	..	..	Repayments
1,125	629	1,467	2,678	8,173	5,862	4,859	4,527	5,142	..	..	Other Long-Term Capital
-5,329	-21,970	-9,229	-7,989	-7,938	-10,117	-12,839	-11,375	-6,187	..	..	Other Capital, net
-96	-5,559	3,963	8,514	1,568	4,359	-5,763	476	-6,098	-9,201	..	Change in Reserves
		(East Caribbean Dollars per US dollar)									
											Memo Item:
2.700	2.700	2.700	2.700	2.700	2.700	2.700	2.700	2.700	2.700	..	Conversion Factor (Annual Avg)
		(Millions of US dollars, outstanding at end of year)									
											EXTERNAL DEBT, ETC.
..	..	..	..	..	..	..	..	..	..	..	Public/Publicly Guar. Long-Term
..	..	..	..	..	..	..	..	..	..	..	Official Creditors
..	..	..	..	..	..	..	..	..	..	..	IBRD and IDA
..	..	..	..	..	..	..	..	..	..	..	Private Creditors
..	..	..	..	..	..	..	..	..	..	..	Private Non-guaranteed Long-Term
..	..	..	..	..	..	..	..	..	..	..	Use of Fund Credit
..	..	..	..	..	..	..	..	..	..	..	Short-Term Debt
		(Thousands of US dollars)									
											Memo Items:
1,925	9,836	5,078	3,057	4,329	1,479	5,250	3,271	9,593	18,427	..	Int'l Reserves Excluding Gold
..	..	..	..	..	..	..	..	..	..	..	Gold Holdings (at market price)
											SOCIAL INDICATORS
				3.5	3.4	3.4	3.3	3.2	3.1	..	Total Fertility Rate
24.0	21.0	24.7	22.5	21.8	22.9	20.8	20.2	..	21.6	..	Crude Birth Rate
21.9	12.3	12.6	10.2	17.1	13.7	..	..	..	..	..	Infant Mortality Rate
..	..	..	..	74.6	74.7	74.8	74.9	75.0	75.1	..	Life Expectancy at Birth
154.0	107.2	86.3	106.5	115.9	124.5	120.0	121.1	135.5	136.8	..	Food Production, p.c. ('79-81=100)
..	..	..	..	..	..	..	..	..	..	..	Labor Force, Agriculture (%)
..	..	..	..	..	..	..	..	..	..	..	Labor Force, Female (%)
..	..	..	..	..	..	..	..	..	..	..	School Enroll. Ratio, primary
..	..	..	..	..	..	..	..	..	..	..	School Enroll. Ratio, secondary

DOMINICAN REPUBLIC	1967	1968	1969	1970	1971	1972	1973	1974	1975	1976	1977
CURRENT GNP PER CAPITA (US $)	260	260	290	320	330	370	440	510	600	690	770
POPULATION (thousands)	4,052	4,175	4,299	4,424	4,549	4,674	4,799	4,925	5,051	5,178	5,306
ORIGIN AND USE OF RESOURCES					*(Millions of current Dominican Pesos)*						
Gross National Product (GNP)	1,015	1,060	1,207	1,353	1,519	1,799	2,100	2,627	3,229	3,545	4,136
Net Factor Income from Abroad	-20	-19	-23	-26	-29	-47	-77	-90	-113	-124	-123
Gross Domestic Product (GDP)	1,035	1,079	1,231	1,379	1,547	1,845	2,177	2,717	3,342	3,669	4,259
Indirect Taxes, net	120	138	151	160	181	205	229	398	432	386	449
GDP at factor cost	915	941	1,080	1,219	1,367	1,640	1,949	2,319	2,910	3,283	3,810
Agriculture	210	219	262	282	290	320	408	507	605	589	720
Industry	282	278	327	388	450	549	658	833	1,139	1,238	1,353
Manufacturing	199	189	230	275	306	347	399	545	752	815	871
Services, etc.	542	582	641	709	808	976	1,112	1,377	1,599	1,842	2,186
Resource Balance	-72	-88	-94	-139	-151	-68	-98	-263	-84	-247	-282
Exports of Goods & NFServices	187	200	227	256	292	411	513	729	1,009	840	918
Imports of Goods & NFServices	259	288	321	394	444	478	611	992	1,093	1,087	1,200
Domestic Absorption	1,107	1,167	1,324	1,518	1,699	1,913	2,276	2,979	3,425	3,916	4,542
Private Consumption, etc.	810	856	945	1,067	1,237	1,349	1,570	2,013	2,328	2,887	3,358
General Gov't Consumption	136	145	156	167	164	172	188	283	216	147	184
Gross Domestic Investment	161	166	223	284	297	392	518	683	882	882	1,000
Fixed Investment	154	171	192	246	294	427	498	644	803	780	939
Memo Items:											
Gross Domestic Saving	89	78	130	146	146	324	420	421	799	635	717
Gross National Saving	69	68	113	150	134	306	372	364	720	634	730
					(Millions of 1980 Dominican Pesos)						
Gross National Product	2,742.7	2,757.6	3,051.0	3,366.9	3,492.6	3,817.8	4,274.4	4,537.4	4,751.0	5,069.1	5,332.2
GDP at factor cost	2,315.6	2,296.0	2,558.3	2,842.9	3,143.9	3,477.0	3,938.3	3,974.0	4,247.3	4,658.2	4,881.5
Agriculture	600.9	630.2	693.1	745.8	819.0	849.9	923.3	923.6	900.4	966.4	983.5
Industry	606.1	586.5	683.7	808.4	952.7	1,113.9	1,343.3	1,412.4	1,525.7	1,641.0	1,736.8
Manufacturing	398.2	369.8	443.2	527.4	595.6	644.4	730.2	764.9	820.6	876.0	925.4
Services, etc.	1,406.5	1,408.3	1,532.3	1,656.0	1,781.3	1,943.5	2,130.6	2,316.1	2,451.3	2,598.2	2,737.5
Resource Balance	-106.2	-159.7	-192.1	-281.6	-260.1	-84.9	-76.4	-499.2	-520.2	-290.4	-291.2
Exports of Goods & NFServices	498.3	499.7	532.1	592.0	691.9	899.9	1,010.6	960.0	995.3	1,192.6	1,272.3
Imports of Goods & NFServices	604.4	659.4	724.3	873.6	952.0	984.8	1,087.0	1,459.2	1,515.5	1,483.0	1,563.5
Domestic Absorption	2,719.7	2,784.6	3,101.2	3,491.9	3,813.1	3,992.2	4,473.6	5,151.3	5,397.6	5,495.9	5,749.0
Private Consumption, etc.	1,974.1	2,070.0	2,256.2	2,524.1	2,759.3	2,846.9	3,095.1	3,453.1	3,680.8	3,953.1	4,070.3
General Gov't Consumption	347.5	317.8	317.2	330.2	306.6	305.1	312.4	428.5	344.3	258.6	290.9
Gross Domestic Investment	398.0	396.9	527.9	637.5	747.2	840.2	1,066.0	1,269.6	1,372.6	1,284.2	1,387.8
Fixed Investment	368.0	398.8	447.6	564.2	755.8	897.8	1,064.8	1,168.7	1,311.7	1,207.5	1,325.9
Memo Items:											
Capacity to Import	437.0	457.5	512.6	566.8	627.2	845.9	912.8	1,072.9	1,399.7	1,146.1	1,196.1
Terms of Trade Adjustment	-61.3	-42.2	-19.5	-25.2	-64.7	-54.1	-97.8	112.8	404.4	-46.5	-76.1
Gross Domestic Income	2,552.2	2,582.8	2,889.6	3,185.1	3,488.3	3,853.2	4,299.4	4,764.9	5,281.8	5,159.0	5,381.6
Gross National Income	2,681.4	2,715.4	3,031.4	3,341.7	3,427.9	3,763.7	4,176.6	4,650.3	5,155.3	5,022.6	5,256.1
DOMESTIC PRICES (DEFLATORS)					*(Index 1980 = 100)*						
Overall (GDP)	39.6	41.1	42.3	43.0	43.5	47.2	49.5	58.4	68.5	70.5	78.0
Domestic Absorption	40.7	41.9	42.7	43.5	44.6	47.9	50.9	57.8	63.5	71.3	79.0
Agriculture	35.0	34.8	37.8	37.8	35.4	37.7	44.2	54.9	67.2	61.0	73.3
Industry	46.6	47.4	47.8	48.0	47.2	49.3	49.0	59.0	74.6	75.5	77.9
Manufacturing	50.0	51.1	51.8	52.2	51.4	53.9	54.6	71.2	91.6	93.0	94.1
MANUFACTURING ACTIVITY											
Employment (1980 = 100)	71.4	65.9	65.7	75.8	78.1	85.1	93.4	95.5	83.8	76.3	81.1
Real Earnings per Empl. (1980 = 100)	105.8	117.9	133.0	124.9	121.5	120.2	100.6	101.6	125.0	131.0	112.8
Real Output per Empl. (1980 = 100)	50.9	54.9	64.1	63.0	70.5	69.8	85.4	86.2	95.5	116.7	98.3
Earnings as % of Value Added	37.5	40.0	37.5	35.3	33.2	34.4	30.5	26.5	24.2	22.7	22.1
MONETARY HOLDINGS					*(Millions of current Dominican Pesos)*						
Money Supply, Broadly Defined	173	211	243	290	356	442	560	799	947	998	1,158
Money as Means of Payment	120	139	149	172	188	222	260	364	380	390	460
Currency Ouside Banks	59	65	72	81	84	99	116	141	158	172	203
Demand Deposits	61	74	77	90	105	124	144	224	222	218	257
Quasi-Monetary Liabilities	53	72	94	118	168	220	300	435	567	607	698
					(Millions of current Dominican Pesos)						
GOVERNMENT DEFICIT (-) OR SURPLUS	..	..	..	..	..	-3	-21	-42	56	-11	0
Current Revenue	..	..	..	..	..	348	386	499	684	617	671
Current Expenditure	..	..	..	..	..	215	234	276	302	344	390
Current Budget Balance	..	..	..	..	..	133	152	223	382	273	281
Capital Receipts	..	..	..	..	..	13	12	14	10	5	7
Capital Payments	..	..	..	..	..	149	184	279	336	289	288

1978	1979	1980	1981	1982	1983	1984	1985	1986	1987 estimate	NOTES	DOMINICAN REPUBLIC
810	890	1,000	1,120	1,170	1,130	950	760	710	730	A	CURRENT GNP PER CAPITA (US $)
5,435	5,567	5,700	5,836	5,975	6,119	6,267	6,420	6,568	6,716	..	POPULATION (thousands)
				(Millions of current Dominican Pesos)							ORIGIN AND USE OF RESOURCES
4,255	4,900	5,964	6,492	7,208	7,744	9,780	12,902	14,430	17,807	..	Gross National Product (GNP)
-136	-206	-210	-293	-254	-297	-328	-926	-918		..	Net Factor Income from Abroad
4,391	5,106	6,174	6,785	7,462	8,041	10,108	13,828	15,348		..	Gross Domestic Product (GDP)
403	456	507	541	473	574	824	..	..	..	..	Indirect Taxes, net
3,988	4,650	5,667	6,244	6,989	7,467	9,285	..	..	..	B	GDP at factor cost
694	804	1,088	1,098	1,149	1,212	1,537	2,476	2,641	..	..	Agriculture
1,383	1,600	1,876	2,008	2,288	2,439	3,173	4,051	4,551	..	..	Industry
874	929	1,015	1,133	1,455	1,528	1,923	2,173	2,443	..	..	Manufacturing
2,314	2,703	3,210	3,679	4,025	4,391	5,399	7,301	8,156	..	..	Services, etc.
-421	-471	-805	-488	-639	-452	-324	-853	-421	..	..	Resource Balance
828	1,135	1,271	1,513	1,142	1,534	2,829	4,085	4,053	..	..	Exports of Goods & NFServices
1,249	1,606	2,076	2,000	1,781	1,986	3,152	4,938	4,474	..	..	Imports of Goods & NFServices
4,811	5,577	6,979	7,272	8,101	8,493	10,432	14,681	15,769	..	..	Domestic Absorption
3,417	3,775	4,826	4,910	5,747	5,939	7,384	11,070	..	..	..	Private Consumption, etc.
263	407	504	660	708	767	876	1,149	..	..	..	General Gov't Consumption
1,130	1,394	1,649	1,702	1,645	1,787	2,172	2,461	..	..	..	Gross Domestic Investment
1,032	1,335	1,567	1,640	1,541	1,705	2,063	..	..	..	..	Fixed Investment
											Memo Items:
710	924	844	1,214	1,006	1,335	1,849	1,609	..	..	..	Gross Domestic Saving
721	895	833	1,104	942	1,233	1,725	1,436	..	..	..	Gross National Saving
				(Millions of 1980 Dominican Pesos)							
5,469.0	5,674.9	5,963.9	6,140.0	6,312.2	6,580.2	6,484.1	6,190.7	6,213.0	6,685.2	..	Gross National Product
5,073.9	5,307.4	5,666.9	5,908.4	6,118.3	6,342.4	6,153.2	..	..	..	B	GDP at factor cost
1,028.2	1,039.5	1,087.8	1,147.7	1,200.4	1,236.0	1,234.6	1,192.1	1,151.8	..	..	Agriculture
1,695.4	1,829.2	1,876.5	1,931.9	1,863.0	2,010.0	2,039.3	1,945.0	1,996.4	..	..	Industry
924.2	966.8	1,015.4	1,042.8	1,096.6	1,115.4	1,082.2	1,038.2	1,061.9	..	..	Manufacturing
2,862.4	2,958.9	3,209.8	3,340.3	3,467.4	3,583.3	3,424.2	3,480.3	3,524.4	..	..	Services, etc.
-233.2	-208.4	-804.7	-524.7	-457.7	-289.3	-101.4	-131.0	-270.6	..	..	Resource Balance
1,255.3	1,555.4	1,271.3	1,350.2	1,135.6	1,273.2	1,325.7	1,333.1	1,354.1	..	..	Exports of Goods & NFServices
1,488.5	1,763.8	2,076.0	1,875.0	1,593.2	1,562.5	1,427.1	1,464.1	1,624.6	..	..	Imports of Goods & NFServices
5,819.1	6,036.1	6,978.8	6,944.6	6,988.5	7,118.6	6,799.6	6,748.5	6,943.3	..	..	Domestic Absorption
4,057.5	4,085.7	4,826.2	4,854.8	5,020.4	5,083.4	4,832.8	4,696.1	..	..	..	Private Consumption, etc.
336.4	408.8	504.0	621.1	605.2	639.7	495.0	486.5	..	..	..	General Gov't Consumption
1,425.3	1,541.6	1,648.6	1,468.8	1,362.9	1,395.4	1,471.9	1,565.9	..	..	..	Gross Domestic Investment
1,318.8	1,499.3	1,566.8	1,432.4	1,272.0	1,340.6	1,412.6	..	..	..	..	Fixed Investment
											Memo Items:
987.1	1,246.4	1,271.3	1,417.9	1,021.5	1,207.3	1,280.6	1,211.2	1,471.7	..	..	Capacity to Import
-268.2	-309.0	0.0	67.7	-114.1	-65.9	-45.1	-121.8	117.7	..	..	Terms of Trade Adjustment
5,317.7	5,518.7	6,174.1	6,487.6	6,416.7	6,763.3	6,653.1	6,495.6	6,790.4	..	..	Gross Domestic Income
5,200.8	5,365.9	5,963.9	6,207.7	6,198.1	6,514.3	6,439.1	6,068.9	6,330.7	..	..	Gross National Income
				(Index 1980 = 100)							DOMESTIC PRICES (DEFLATORS)
78.6	87.6	100.0	105.7	114.3	117.7	150.9	209.0	230.0	..	..	Overall (GDP)
82.7	92.4	100.0	104.7	115.9	119.3	153.4	217.5	227.1	..	..	Domestic Absorption
67.5	77.3	100.0	95.7	95.7	98.0	124.5	207.7	229.3	..	..	Agriculture
81.6	87.4	100.0	103.9	122.8	121.4	155.6	208.3	228.0	..	..	Industry
94.5	96.1	100.0	108.7	132.7	137.0	177.7	209.3	230.0	..	..	Manufacturing
											MANUFACTURING ACTIVITY
83.2	92.0	100.0	100.1	101.5	102.4	106.8	111.4	..	..	..	Employment (1980=100)
113.5	111.1	100.0	99.8	103.2	100.9	100.8	79.4	..	..	..	Real Earnings per Empl. (1980=100)
98.5	106.3	100.0	103.3	105.2	105.7	98.6	90.7	..	..	..	Real Output per Empl. (1980=100)
23.4	25.8	26.9	25.9	23.0	22.7	23.9	23.9	..	..	..	Earnings as % of Value Added
				(Millions of current Dominican Pesos)							MONETARY HOLDINGS
1,184	1,392	1,494	1,724	1,994	2,212	2,768	3,395	5,107	6,021	D	Money Supply, Broadly Defined
458	598	580	660	731	781	1,160	1,355	1,989	2,609	..	Money as Means of Payment
224	274	275	324	358	415	593	677	937	1,313	..	Currency Ouside Banks
234	325	305	337	374	367	567	678	1,051	1,297	..	Demand Deposits
726	794	915	1,063	1,263	1,430	1,609	2,040	3,119	3,411	..	Quasi-Monetary Liabilities
				(Millions of current Dominican Pesos)							
-70	-315	-173	-181	-247	-216	-114	-262	..	..	..	GOVERNMENT DEFICIT (-) OR SURPLUS
680	732	946	985	819	987	1,263	1,718	..	..	..	Current Revenue
460	631	753	809	824	908	1,061	1,468	..	..	..	Current Expenditure
220	101	194	175	-5	79	202	250	..	..	..	Current Budget Balance
16	8	10	10	8	14	19	11	..	..	..	Capital Receipts
306	424	376	367	249	309	335	523	..	..	..	Capital Payments

DOMINICAN REPUBLIC	1967	1968	1969	1970	1971	1972	1973	1974	1975	1976	1977
FOREIGN TRADE (CUSTOMS BASIS)					*(Millions of current US dollars)*						
Value of Exports, fob	156.2	163.5	183.4	214.0	240.8	353.5	443.2	648.7	893.8	716.4	780.4
Nonfuel Primary Products	150.8	157.5	175.6	203.8	228.7	284.9	341.8	535.0	761.4	567.0	653.3
Fuels	0.1	0.0	0.0	0.0	0.0	0.0	0.0	0.0	0.0	0.5	0.0
Manufactures	5.3	6.1	7.9	10.2	12.0	68.6	101.4	113.8	132.4	149.0	127.1
Value of Imports, cif	197.0	222.0	246.0	304.0	358.0	388.0	489.0	808.0	889.0	878.0	975.0
Nonfuel Primary Products	40.2	55.5	48.9	60.2	65.5	65.2	116.6	181.9	154.7	233.2	205.9
Fuels	14.8	14.8	20.1	21.0	30.9	40.2	58.7	134.0	200.8	14.0	202.7
Manufactures	142.0	151.6	177.0	222.8	261.5	282.6	313.7	492.1	533.6	630.8	566.4
					(Index 1980 = 100)						
Terms of Trade	74.2	74.5	96.4	94.8	92.6	120.6	123.8	138.4	116.9	94.4	113.6
Export Prices, fob	15.9	16.5	20.6	22.2	23.0	31.0	39.9	76.6	63.8	60.3	70.0
Nonfuel Primary Products	15.7	16.2	20.3	21.8	22.6	29.6	37.7	78.9	63.4	58.9	69.6
Fuels	4.3	4.3	4.3	4.3	5.6	6.2	8.9	36.7	35.7	38.4	42.0
Manufactures	30.1	31.7	32.4	37.5	36.5	38.5	49.5	67.2	66.1	66.7	71.7
Import Prices, cif	21.5	22.2	21.4	23.5	24.8	25.7	32.3	55.4	54.6	63.9	61.6
Trade at Constant 1980 Prices					*(Millions of 1980 US dollars)*						
Exports, fob	981.1	991.1	890.8	962.3	1,047.2	1,140.2	1,110.5	847.1	1,401.1	1,187.4	1,115.5
Imports, cif	918.1	1,001.7	1,152.4	1,296.0	1,442.3	1,509.9	1,516.5	1,459.7	1,629.0	1,374.6	1,582.5
BALANCE OF PAYMENTS					*(Millions of current US dollars)*						
Exports of Goods & Services	..	..	..	259	292	413	516	735	1,015	853	940
Merchandise, fob	..	..	..	214	241	348	442	637	894	716	781
Nonfactor Services	..	..	..	43	50	64	72	93	116	128	147
Factor Services	..	..	..	2	2	2	3	5	5	9	12
Imports of Goods & Services	..	..	..	392	439	490	644	1,011	1,127	1,108	1,208
Merchandise, fob	..	..	..	278	310	338	422	673	773	764	849
Nonfactor Services	..	..	..	86	99	104	142	243	236	226	248
Factor Services	..	..	..	27	30	48	80	95	118	118	111
Long-Term Interest	..	..	..	13	14	16	20	24	33	34	48
Current Transfers, net	..	..	..	30	16	29	29	33	34	123	136
Workers' Remittances	..	..	..	25	15	24	24	27	28	112	124
Total to be Financed	..	..	..	-103	-131	-49	-98	-243	-78	-132	-132
Official Capital Grants	..	..	..	1	1	2	2	2	5	3	4
Current Account Balance	..	..	..	-102	-129	-47	-97	-241	-73	-129	-129
Long-Term Capital, net	..	..	..	110	83	84	71	163	164	170	222
Direct Investment	..	..	..	72	65	44	35	54	64	60	72
Long-Term Loans	..	..	..	33	38	44	35	108	95	115	137
Disbursements	..	..	..	60	71	76	77	152	158	189	227
Repayments	..	..	..	27	33	32	42	44	62	73	90
Other Long-Term Capital	..	..	..	5	-21	-5	0	0	0	-8	10
Other Capital, net	..	..	..	-13	62	-32	58	78	-98	-64	-51
Change in Reserves	..	..	..	5	-15	-5	-33	0	7	23	-43
Memo Item:					*(Dominican Pesos per US dollar)*						
Conversion Factor (Annual Avg)	1.000	1.000	1.000	1.000	1.000	1.000	1.000	1.000	1.000	1.000	1.000
EXTERNAL DEBT, ETC.					*(Millions of US dollars, outstanding at end of year)*						
Public/Publicly Guar. Long-Term	..	..	..	212	233	278	313	354	411	506	610
Official Creditors	..	..	..	202	223	236	243	252	286	340	393
IBRD and IDA	..	..	..	11	25	25	26	27	27	31	35
Private Creditors	..	..	..	10	9	42	70	102	126	166	217
Private Non-guaranteed Long-Term	..	..	..	141	159	157	157	224	262	281	303
Use of Fund Credit	..	..	..	7	12	4	0	11	0	25	44
Short-Term Debt	..	..	..	..	..	..	..	..	..	..	164
Memo Items:					*(Millions of US dollars)*						
Int'l Reserves Excluding Gold	29.4	32.6	36.8	29.1	52.8	55.3	84.3	87.1	112.6	123.5	180.1
Gold Holdings (at market price)	3.0	3.6	3.0	3.2	3.8	5.6	9.7	16.0	12.1	11.6	17.2
SOCIAL INDICATORS											
Total Fertility Rate	6.7	6.5	6.3	6.0	5.8	5.6	5.4	5.3	5.1	4.9	4.7
Crude Birth Rate	44.9	43.6	42.4	41.2	40.0	38.8	38.0	37.2	36.4	35.6	34.9
Infant Mortality Rate	105.0	102.8	100.6	98.4	96.2	94.0	92.0	90.0	88.0	86.0	84.0
Life Expectancy at Birth	57.0	57.6	58.2	58.7	59.3	59.9	60.3	60.8	61.2	61.6	62.1
Food Production, p.c. ('79-81 = 100)	94.3	94.2	102.0	106.9	110.0	109.4	107.5	109.1	101.4	105.0	105.1
Labor Force, Agriculture (%)	57.4	56.6	55.7	54.8	53.9	53.0	52.0	51.1	50.2	49.3	48.4
Labor Force, Female (%)	10.6	10.7	10.8	10.9	11.0	11.2	11.3	11.5	11.6	11.8	11.9
School Enroll. Ratio, primary	..	..	..	100.0	..	..	..	..	104.0	..	95.0
School Enroll. Ratio, secondary	..	..	..	21.0	..	..	..	..	36.0	..	..

1978	1979	1980	1981	1982	1983	1984	1985	1986	1987	NOTES	DOMINICAN REPUBLIC
				(Millions of current US dollars)							**FOREIGN TRADE (CUSTOMS BASIS)**
675.5	868.6	961.9	1,188.0	767.7	781.7	868.1	735.2	717.6	711.3	..	Value of Exports, fob
548.2	675.0	795.8	1,002.1	665.2	626.9	668.7	535.2	509.8	557.1	..	Nonfuel Primary Products
0.0	0.0	0.0	0.1	0.0	0.0	0.0	0.0	0.0	0.0	..	Fuels
127.3	193.6	166.1	185.9	102.5	154.8	199.4	200.0	207.8	154.2	..	Manufactures
987.0	1,213.0	1,640.0	1,668.0	1,444.0	1,471.0	1,446.0	1,487.0	1,433.0	1,782.5	..	Value of Imports, cif
202.5	255.9	337.1	340.3	268.5	259.8	285.4	279.6	271.9	315.6	..	Nonfuel Primary Products
222.9	321.5	416.5	548.4	492.9	533.5	427.4	407.3	200.7	261.3	..	Fuels
561.6	635.6	886.5	779.4	682.6	677.7	733.3	800.2	960.4	1,205.6	..	Manufactures
				(Index 1980 = 100)							
98.2	91.8	100.0	76.0	62.6	66.5	63.6	65.7	79.1	59.9	..	Terms of Trade
65.3	74.3	100.0	78.7	60.9	62.5	58.9	59.5	68.6	58.4	..	Export Prices, fob
63.2	70.8	100.0	75.8	57.7	58.0	53.5	52.9	60.3	51.5	..	Nonfuel Primary Products
42.3	61.0	100.0	112.5	101.6	92.5	90.2	87.5	45.0	56.9	..	Fuels
76.1	90.0	100.0	99.1	94.8	90.7	89.3	89.0	103.2	113.0	..	Manufactures
66.5	80.9	100.0	103.5	97.3	93.9	92.6	90.5	86.7	97.4	..	Import Prices, cif
				(Millions of 1980 US dollars)							Trade at Constant 1980 Prices
1,035.1	1,169.2	961.9	1,509.7	1,260.8	1,251.1	1,473.1	1,235.8	1,046.6	1,218.3	..	Exports, fob
1,484.7	1,498.8	1,640.0	1,611.2	1,484.4	1,566.6	1,561.0	1,643.5	1,653.8	1,829.7	..	Imports, cif
				(Millions of current US dollars)							**BALANCE OF PAYMENTS**
849	1,167	1,313	1,524	1,146	1,249	1,375	1,344	1,426	..	..	Exports of Goods & Services
676	869	962	1,188	768	785	868	739	722	..	..	Merchandise, fob
153	266	309	325	374	457	502	584	687	..	..	Nonfactor Services
21	32	42	12	4	7	6	22	17	..	..	Factor Services
1,311	1,704	2,238	2,107	1,794	1,882	1,804	1,808	1,816	..	..	Imports of Goods & Services
862	1,138	1,520	1,452	1,257	1,279	1,257	1,286	1,266	..	..	Merchandise, fob
292	347	399	367	277	299	300	275	283	..	..	Nonfactor Services
157	220	319	288	259	304	247	248	267	..	..	Factor Services
61	78	121	144	126	130	111	139	184	106	..	Long-Term Interest
146	177	200	183	190	195	205	242	242	..	..	Current Transfers, net
132	161	183	183	190	195	205	242	242	..	..	Workers' Remittances
-315	-360	-725	-400	-458	-438	-223	-222	-148	..	K	Total to be Financed
4	29	5	10	15	20	60	114	29	..	K	Official Capital Grants
-312	-331	-720	-389	-443	-418	-163	-108	-119	..	..	Current Account Balance
164	189	354	246	298	578	354	300	299	..	..	Long-Term Capital, net
64	17	93	80	-1	48	69	36	50	..	..	Direct Investment
120	124	347	162	274	61	205	165	84	62	..	Long-Term Loans
223	351	482	299	454	248	293	251	187	144	..	Disbursements
103	227	135	136	180	187	89	86	103	82	..	Repayments
-23	19	-90	-6	11	449	21	-16	136	..	..	Other Long-Term Capital
112	98	321	174	-104	-278	-101	-118	-70	..	..	Other Capital, net
36	45	45	-31	248	118	-89	-75	-110	131	..	Change in Reserves
				(Dominican Pesos per US dollar)							**Memo Item:**
1.000	1.000	1.000	1.000	1.090	1.260	2.060	3.110	2.900	3.840	..	Conversion Factor (Annual Avg)
				(Millions of US dollars, outstanding at end of year)							**EXTERNAL DEBT, ETC.**
736	868	1,220	1,400	1,666	2,192	2,344	2,632	2,758	2,938	..	Public/Publicly Guar. Long-Term
438	495	803	1,012	1,332	1,436	1,582	1,852	1,968	2,120	..	Official Creditors
42	47	83	115	138	161	141	174	201	240	..	IBRD and IDA
298	372	417	389	334	757	762	780	790	818	..	Private Creditors
296	295	254	233	252	181	156	151	146	133	..	Private Non-guaranteed Long-Term
48	124	49	23	71	246	221	297	304	284	..	Use of Fund Credit
255	317	480	638	529	298	371	362	311	341	..	Short-Term Debt
				(Millions of US dollars)							**Memo Items:**
154.1	238.6	201.8	225.2	129.0	171.3	253.5	340.1	376.3	182.2	..	Int'l Reserves Excluding Gold
23.5	57.9	77.2	56.5	41.6	29.4	5.6	5.9	7.0	8.7	..	Gold Holdings (at market price)
											SOCIAL INDICATORS
4.6	4.5	4.4	4.3	4.2	4.1	4.0	3.9	3.8	3.8	..	Total Fertility Rate
34.6	34.4	34.1	33.8	33.6	33.1	32.6	32.2	31.7	31.2	..	Crude Birth Rate
82.2	80.4	78.6	76.8	75.0	68.9	62.8	56.8	50.7	44.6	..	Infant Mortality Rate
62.5	62.9	63.3	63.7	64.1	64.5	64.9	65.3	65.6	66.0	..	Life Expectancy at Birth
106.1	103.2	98.5	98.3	100.6	103.4	105.0	99.3	97.0	100.6	..	Food Production, p.c. ('79-81 = 100)
47.5	46.6	45.7	..	..	..	..	..	..	..	..	Labor Force, Agriculture (%)
12.1	12.2	12.4	12.6	12.9	13.2	13.4	13.7	14.0	14.2	..	Labor Force, Female (%)
102.0	104.0	114.0	116.0	109.0	112.0	121.0	124.0	..	..	..	School Enroll. Ratio, primary
38.0	39.0	43.0	45.0	50.0	50.0	50.0	50.0	..	..	..	School Enroll. Ratio, secondary

ECUADOR	1967	1968	1969	1970	1971	1972	1973	1974	1975	1976	1977
CURRENT GNP PER CAPITA (US $)	240	260	280	290	290	300	360	430	550	680	790
POPULATION (thousands)	5,505	5,683	5,865	6,051	6,241	6,434	6,631	6,831	7,035	7,243	7,455
ORIGIN AND USE OF RESOURCES					*(Billions of current Ecuadoran Sucres)*						
Gross National Product (GNP)	24.8	26.9	29.6	34.3	39.1	45.2	58.7	87.1	105.3	128.9	158.9
Net Factor Income from Abroad	-0.5	-0.5	-0.5	-0.7	-0.9	-1.7	-3.5	-5.7	-2.5	-4.1	-7.5
Gross Domestic Product (GDP)	25.2	27.4	30.1	35.0	40.0	46.9	62.2	92.8	107.7	132.9	166.4
Indirect Taxes, net	2.7	3.0	3.1	3.8	4.6	5.5	7.7	10.4	10.7	10.6	13.4
GDP at factor cost	22.5	24.5	27.0	31.3	35.4	41.4	54.5	82.4	97.0	122.4	152.9
Agriculture	6.6	6.8	7.3	8.4	9.2	10.5	12.2	17.4	19.3	22.6	27.7
Industry	5.7	6.2	7.1	8.6	10.7	13.1	20.1	35.4	36.5	47.0	57.3
Manufacturing	4.6	5.0	5.7	6.4	7.5	8.8	10.8	14.3	17.2	22.9	29.9
Services, etc.	12.9	14.4	15.7	18.0	20.1	23.2	29.9	40.0	51.9	63.3	81.4
Resource Balance	-1.0	-1.5	-1.8	-1.6	-3.8	-1.7	2.0	4.8	-7.0	-1.8	-5.0
Exports of Goods & NFServices	3.9	4.1	3.9	4.9	6.0	8.8	15.5	33.6	28.2	34.2	41.3
Imports of Goods & NFServices	4.9	5.6	5.7	6.5	9.8	10.5	13.5	28.8	35.2	36.0	46.3
Domestic Absorption	26.2	29.0	32.0	36.6	43.8	48.5	60.2	88.0	114.7	134.7	171.4
Private Consumption, etc.	19.8	21.5	23.6	26.4	30.4	34.4	41.7	55.5	70.3	84.5	102.6
General Gov't Consumption	2.1	2.5	3.1	3.9	4.1	4.7	6.4	11.6	15.6	18.6	24.7
Gross Domestic Investment	4.4	4.9	5.3	6.4	9.3	9.4	12.1	20.8	28.8	31.6	44.1
Fixed Investment	3.4	3.9	4.8	5.8	8.7	8.4	10.9	16.9	24.9	29.5	39.3
Memo Items:											
Gross Domestic Saving	3.4	3.4	3.5	4.8	5.5	7.7	14.1	25.6	21.8	29.8	39.1
Gross National Saving	3.0	3.0	3.0	4.3	4.8	6.2	10.8	20.3	19.7	25.9	31.7
					(Billions of 1980 Ecuadoran Sucres)						
Gross National Product	104.0	107.5	110.2	118.3	125.6	144.0	180.5	190.2	208.3	226.7	242.0
GDP at factor cost	94.6	98.2	101.2	107.3	113.4	131.6	167.7	179.8	191.9	215.3	228.6
Agriculture	23.8	24.8	24.6	26.4	27.7	28.8	29.1	31.7	32.4	33.4	34.2
Industry	26.3	25.5	27.1	28.9	32.8	50.3	84.2	82.4	81.7	93.8	95.8
Manufacturing	13.7	13.3	14.1	20.9	21.9	23.9	26.1	28.9	33.3	37.6	42.1
Services, etc.	56.5	60.3	61.7	65.5	68.2	70.4	78.1	88.5	99.1	106.6	118.8
Resource Balance	-3.4	-5.7	-7.7	-7.8	-11.3	11.4	43.4	24.7	10.2	17.7	4.2
Exports of Goods & NFServices	21.0	22.1	18.8	20.0	22.3	43.8	77.6	73.9	67.7	73.4	69.7
Imports of Goods & NFServices	24.3	27.8	26.5	27.8	33.5	32.5	34.2	49.3	57.5	55.7	65.5
Domestic Absorption	109.9	116.3	121.1	128.5	139.9	138.1	148.0	177.9	203.0	216.2	244.5
Private Consumption, etc.	73.4	76.8	82.1	86.4	90.6	92.6	97.8	106.2	118.6	132.0	142.7
General Gov't Consumption	9.8	11.2	12.2	13.7	13.4	14.2	15.5	23.7	28.2	30.8	37.1
Gross Domestic Investment	26.7	28.2	26.9	28.4	36.0	31.3	34.7	48.0	56.3	53.3	64.8
Fixed Investment	23.2	24.0	25.8	26.9	34.1	28.0	31.6	40.0	49.4	50.1	57.8
Memo Items:											
Capacity to Import	19.4	20.2	18.0	21.0	20.6	27.3	39.3	57.4	46.1	52.9	58.5
Terms of Trade Adjustment	-1.5	-1.9	-0.8	1.0	-1.7	-16.6	-38.3	-16.5	-21.6	-20.5	-11.3
Gross Domestic Income	105.1	108.7	112.6	121.7	126.9	132.9	153.1	186.0	191.6	213.4	237.5
Gross National Income	102.5	105.6	109.4	119.3	123.9	127.4	142.3	173.7	186.7	206.2	230.8
DOMESTIC PRICES (DEFLATORS)					*(Index 1980 = 100)*						
Overall (GDP)	23.7	24.8	26.6	29.0	31.1	31.4	32.5	45.8	50.5	56.8	66.9
Domestic Absorption	23.8	24.9	26.4	28.5	31.3	35.2	40.7	49.5	56.5	62.3	70.1
Agriculture	27.7	27.7	29.5	31.8	33.2	36.6	42.1	54.8	59.6	67.8	81.0
Industry	21.8	24.3	26.3	29.9	32.7	26.0	23.9	43.0	44.7	50.1	59.8
Manufacturing	34.0	37.7	40.4	30.5	34.4	36.6	41.4	49.5	51.8	60.9	71.1
MANUFACTURING ACTIVITY											
Employment (1980=100)	34.7	37.6	39.5	42.3	44.4	47.3	52.0	58.9	66.2	71.7	77.4
Real Earnings per Empl. (1980=100)	56.3	53.4	58.5	60.8	63.8	66.5	65.3	62.4	65.0	68.9	66.8
Real Output per Empl. (1980=100)	58.0	54.1	57.4	83.1	82.2	83.4	86.0	92.1	90.9	94.6	100.7
Earnings as % of Value Added	28.1	25.6	27.1	26.7	25.7	28.0	27.4	24.8	29.1	26.3	25.4
MONETARY HOLDINGS					*(Billions of current Ecuadoran Sucres)*						
Money Supply, Broadly Defined	4.5	5.6	6.3	7.7	8.9	11.0	14.4	21.0	23.1	28.8	36.0
Money as Means of Payment	3.4	4.2	4.8	6.0	6.7	8.4	11.3	16.9	18.3	22.8	29.9
Currency Ouside Banks	1.4	1.6	1.7	2.3	2.4	2.9	3.6	4.8	5.4	7.6	9.1
Demand Deposits	2.0	2.6	3.0	3.7	4.3	5.5	7.7	12.1	13.0	15.2	20.7
Quasi-Monetary Liabilities	1.0	1.4	1.5	1.7	2.2	2.6	3.1	4.2	4.7	6.0	6.1
					(Millions of current Ecuadoran Sucres)						
GOVERNMENT DEFICIT (-) OR SURPLUS	..	..	..	..	..	..	124	-3	-666	-2,160	-5,388
Current Revenue	..	..	..	..	..	..	7,973	11,390	12,391	14,653	16,452
Current Expenditure	..	..	..	..	..	..	..	..	..	..	..
Current Budget Balance	..	..	..	..	..	..	..	..	..	..	..
Capital Receipts	..	..	..	..	..	..	..	..	..	..	..
Capital Payments	..	..	..	..	..	..	..	..	..	..	..

1978	1979	1980	1981	1982	1983	1984	1985	1986	1987 estimate	NOTES	ECUADOR
930	1,080	1,260	1,490	1,490	1,310	1,150	1,150	1,130	1,040	A	**CURRENT GNP PER CAPITA (US $)**
7,672	7,895	8,123	8,358	8,600	8,850	9,109	9,378	9,638	9,898	..	**POPULATION (thousands)**
				(Billions of current Ecuadoran Sucres)							**ORIGIN AND USE OF RESOURCES**
185.8	224.0	278.8	330.4	384.8	518.5	740.1	1,025.9	1,244.6	1,656.0	..	Gross National Product (GNP)
-5.5	-9.9	-14.5	-18.3	-30.9	-41.8	-72.6	-84.0	-137.5	-152.4	..	Net Factor Income from Abroad
191.3	234.0	293.3	348.7	415.7	560.3	812.6	1,109.9	1,382.1	1,808.4	..	Gross Domestic Product (GDP)
16.5	19.1	24.4	32.6	35.7	46.7	63.9	136.5	..	..	..	Indirect Taxes, net
174.9	214.9	268.9	316.0	380.0	513.6	748.8	973.4	..	..	B	GDP at factor cost
28.5	31.7	35.6	41.6	50.4	73.0	110.0	148.0	209.0	290.9	..	Agriculture
66.1	90.3	111.7	137.0	167.6	227.5	335.6	451.8	482.8	567.3	..	Industry
36.3	44.9	51.8	60.0	73.9	103.6	168.0	210.3	273.9	338.0	..	Manufacturing
96.8	112.0	146.1	170.0	197.8	259.8	367.0	510.2	690.3	950.1	..	Services, etc.
-10.8	1.3	-0.7	3.5	-9.5	22.4	52.4	65.3	4.9	-118.3	..	Resource Balance
40.8	60.6	73.8	75.9	87.6	133.1	209.9	296.9	316.8	421.2	..	Exports of Goods & NFServices
51.6	59.3	74.5	72.4	97.0	110.6	157.4	231.7	311.9	539.6	..	Imports of Goods & NFServices
202.1	232.7	294.1	345.2	425.2	537.8	760.2	1,044.7	1,377.2	1,926.7	..	Domestic Absorption
121.2	143.3	174.9	214.7	262.2	369.3	520.6	715.7	925.5	1,283.8	..	Private Consumption, etc.
26.5	30.1	42.6	49.7	58.1	70.1	99.6	127.3	168.6	220.0	..	General Gov't Consumption
54.4	59.3	76.6	80.8	104.8	98.4	140.0	201.7	283.2	422.9	..	Gross Domestic Investment
50.1	55.4	69.3	77.6	94.2	93.0	125.2	178.3	254.7	416.1	..	Fixed Investment
											Memo Items:
43.7	60.6	75.9	84.3	95.4	120.9	192.4	267.0	288.1	304.5	..	Gross Domestic Saving
38.4	50.7	61.4	65.9	64.5	79.1	119.8	182.9	150.5	152.1	..	Gross National Saving
				(Billions of 1980 Ecuadoran Sucres)							
258.4	268.6	278.8	289.3	285.8	279.8	285.0	298.3	308.5	287.8	..	Gross National Product
243.2	257.7	269.0	276.8	282.2	277.7	289.9	287.7	..	..	B	GDP at factor cost
32.9	33.8	35.6	38.0	38.8	33.4	36.9	40.3	43.5	48.0	..	Agriculture
105.2	112.3	111.7	119.7	120.0	128.1	131.9	136.9	140.0	111.0	..	Industry
45.6	50.0	51.8	56.3	57.2	56.4	55.4	55.5	55.3	54.7	..	Manufacturing
128.0	134.4	146.1	147.6	149.9	140.8	145.6	150.5	154.3	157.1	..	Services, etc.
4.2	7.9	-0.7	9.7	1.2	20.7	31.5	37.7	47.1	22.9	..	Resource Balance
72.0	75.6	73.8	77.3	73.5	75.2	84.7	94.8	103.8	86.0	..	Exports of Goods & NFServices
67.7	67.7	74.5	67.6	72.3	54.5	53.2	57.1	56.7	63.1	..	Imports of Goods & NFServices
261.8	272.6	294.1	295.6	307.5	281.6	283.0	290.1	290.6	293.2	..	Domestic Absorption
152.2	163.1	174.9	185.7	189.0	189.0	192.7	197.4	197.1	200.0	..	Private Consumption, etc.
37.2	39.0	42.6	43.6	43.8	41.2	39.7	38.0	37.9	37.1	..	General Gov't Consumption
72.5	70.4	76.6	66.4	74.7	51.4	50.6	54.7	55.7	56.1	..	Gross Domestic Investment
65.5	65.3	69.3	64.3	64.8	47.8	45.7	48.8	50.2	53.7	..	Fixed Investment
											Memo Items:
53.6	69.2	73.8	70.9	65.2	65.6	70.9	73.2	57.6	49.3	..	Capacity to Import
-18.4	-6.4	0.0	-6.4	-8.2	-9.7	-13.7	-21.6	-46.2	-36.8	..	Terms of Trade Adjustment
247.7	274.1	293.3	298.9	300.5	292.6	300.7	306.1	291.5	279.3	..	Gross Domestic Income
240.0	262.2	278.8	282.9	277.6	270.1	271.3	276.7	262.3	251.0	..	Gross National Income
				(Index 1980 = 100)							**DOMESTIC PRICES (DEFLATORS)**
71.9	83.4	100.0	114.2	134.7	185.3	258.4	338.6	409.2	572.1	..	Overall (GDP)
77.2	85.4	100.0	116.8	138.3	191.0	268.6	360.2	473.9	657.2	..	Domestic Absorption
86.8	93.7	100.0	109.6	129.9	218.7	297.9	366.8	480.4	606.2	..	Agriculture
62.8	80.4	100.0	114.4	139.6	177.6	254.4	329.9	344.9	511.1	..	Industry
79.8	89.8	100.0	106.4	129.2	183.8	303.6	378.6	495.3	618.3	..	Manufacturing
											MANUFACTURING ACTIVITY
87.8	94.6	100.0	92.2	89.8	84.4	85.2	86.1	..	..	J	Employment (1980=100)
67.1	72.1	100.0	107.2	108.7	93.3	143.0	139.8	..	..	J	Real Earnings per Empl. (1980=100)
101.6	99.0	100.0	109.7	113.0	114.5	123.1	122.3	..	..	J	Real Output per Empl. (1980=100)
21.4	24.8	39.4	43.3	52.8	34.8	43.6	43.6	..	..	J	Earnings as % of Value Added
				(Billions of current Ecuadoran Sucres)							**MONETARY HOLDINGS**
39.7	52.2	66.2	75.7	93.6	118.0	169.3	221.9	275.9	403.4	..	Money Supply, Broadly Defined
32.9	42.0	53.6	61.8	73.1	95.1	129.1	158.1	191.1	265.7	..	Money as Means of Payment
10.3	12.3	15.3	17.4	20.5	25.4	35.3	42.7	54.6	74.8	..	Currency Ouside Banks
22.6	29.6	38.3	44.4	52.6	69.7	93.7	115.4	136.5	191.0	..	Demand Deposits
6.8	10.2	12.6	13.9	20.5	22.9	40.2	63.8	84.8	137.7	..	Quasi-Monetary Liabilities
				(Millions of current Ecuadoran Sucres)							**GOVERNMENT DEFICIT (-) OR SURPLUS**
-2,296	-1,520	-4,119	-16,837	-18,480	-14,047	-6,785	21,971	..	..	..	Current Revenue
19,057	23,078	37,549	39,297	45,996	60,187	99,872	189,472	..	..	..	Current Revenue
..	..	34,872	43,135	49,336	..	..	..	..	..	..	Current Expenditure
..	..	2,677	-3,838	-3,340	..	..	..	..	..	..	Current Budget Balance
..	..	..	..	..	..	..	..	..	..	..	Capital Receipts
..	..	6,796	12,999	15,140	..	..	..	..	..	..	Capital Payments

ECUADOR	1967	1968	1969	1970	1971	1972	1973	1974	1975	1976	1977
FOREIGN TRADE (CUSTOMS BASIS)				*(Millions of current US dollars)*							
Value of Exports, fob	157.9	195.2	152.5	189.9	199.1	326.3	532.1	1,123.5	973.9	1,257.6	1,436.3
Nonfuel Primary Products	153.7	190.8	148.1	185.7	193.0	259.0	236.1	405.3	364.3	492.0	692.9
Fuels	1.1	1.1	0.6	0.9	2.0	59.9	282.8	696.7	588.0	740.5	713.3
Manufactures	3.1	3.3	3.8	3.3	4.0	7.4	13.2	21.5	21.6	25.1	30.0
Value of Imports, cif	213.3	252.8	239.4	272.9	338.5	325.0	396.4	781.3	984.6	951.5	1,165.4
Nonfuel Primary Products	32.1	34.0	30.9	27.2	39.1	44.3	57.0	99.1	115.2	112.0	125.5
Fuels	21.6	10.1	16.5	17.1	28.4	21.4	12.1	137.6	20.5	11.3	10.4
Manufactures	159.6	208.7	192.0	228.7	271.0	259.2	327.3	544.7	848.9	828.2	1,029.5
Terms of Trade	140.5	113.5	140.1	130.9	114.1	64.1	33.6	78.2	65.9	75.3	86.5
Export Prices, fob	27.1	28.5	31.7	31.7	28.4	18.9	14.3	41.8	41.3	48.0	60.1
Nonfuel Primary Products	28.0	29.3	32.3	32.6	29.5	34.3	46.1	53.6	53.8	74.7	106.6
Fuels	4.3	4.3	4.3	4.3	5.6	6.2	8.9	36.7	35.7	38.4	42.0
Manufactures	42.0	37.3	45.3	46.7	48.7	52.7	58.7	62.7	66.7	78.0	76.0
Import Prices, cif	19.3	25.1	22.6	24.2	24.9	29.4	42.4	53.5	62.8	63.7	69.5
Trade at Constant 1980 Prices				*(Millions of 1980 US dollars)*							
Exports, fob	582.4	684.9	480.9	599.3	700.2	1,730.3	3,730.1	2,687.3	2,355.2	2,621.3	2,389.0
Imports, cif	1,105.3	1,006.4	1,057.5	1,127.5	1,357.9	1,104.8	934.4	1,461.6	1,568.5	1,492.8	1,675.9
BALANCE OF PAYMENTS				*(Millions of current US dollars)*							
Exports of Goods & Services	..	..	..	259	266	367	633	1,333	1,127	1,432	1,626
Merchandise, fob	..	..	..	235	238	323	585	1,225	1,013	1,307	1,401
Nonfactor Services	..	..	..	24	27	42	42	83	97	111	202
Factor Services	..	..	..	0	1	2	6	25	17	13	24
Imports of Goods & Services	..	..	..	389	438	460	653	1,327	1,379	1,469	2,006
Merchandise, fob	..	..	..	250	307	284	398	875	1,006	1,048	1,361
Nonfactor Services	..	..	..	110	94	108	112	225	289	264	414
Factor Services	..	..	..	30	37	68	144	226	84	157	232
Long-Term Interest	..	..	..	10	11	14	20	25	32	42	69
Current Transfers, net	..	..	..	8	8	8	8	16	14	8	0
Workers' Remittances	..	..	..	0	0	0	0	0	0	0	0
Total to be Financed	..	..	..	-122	-164	-85	-12	22	-239	-29	-379
Official Capital Grants	..	..	..	9	8	8	19	15	19	23	36
Current Account Balance	..	..	..	-113	-156	-77	7	38	-220	-7	-343
Long-Term Capital, net	..	..	..	120	189	167	96	121	219	180	627
Direct Investment	..	..	..	89	162	81	52	77	95	-20	35
Long-Term Loans	..	..	..	21	20	84	29	86	240	160	633
Disbursements	..	..	..	48	54	121	72	176	304	270	761
Repayments	..	..	..	26	34	37	43	90	64	110	128
Other Long-Term Capital	..	..	..	0	-1	-6	-4	-58	-135	17	-77
Other Capital, net	..	..	..	-1	-44	-21	-10	-49	-63	0	-162
Change in Reserves	..	..	..	-6	11	-69	-92	-110	64	-173	-121
Memo Item:				*(Ecuadoran Sucres per US dollar)*							
Conversion Factor (Annual Avg)	18.000	18.000	18.000	20.920	25.000	25.000	25.000	25.000	25.000	25.000	25.000
EXTERNAL DEBT, ETC.				*(Millions of US dollars, outstanding at end of year)*							
Public/Publicly Guar. Long-Term	..	..	..	193	211	287	314	311	435	591	1,111
Official Creditors	..	..	..	132	146	159	173	194	236	303	364
IBRD and IDA	..	..	..	39	40	46	53	59	71	73	79
Private Creditors	..	..	..	61	65	128	141	117	200	288	747
Private Non-guaranteed Long-Term	..	..	..	49	53	62	68	160	274	279	407
Use of Fund Credit	..	..	..	14	6	9	0	0	0	0	0
Short-Term Debt	..	..	..	..	..	..	..	..	..	..	856
Memo Items:				*(Millions of US dollars)*							
Int'l Reserves Excluding Gold	52.0	31.1	42.9	55.2	37.1	121.1	210.4	318.6	253.4	477.4	623.1
Gold Holdings (at market price)	17.2	31.4	22.2	20.4	23.3	23.1	43.3	72.0	54.1	52.0	66.0
SOCIAL INDICATORS											
Total Fertility Rate	6.7	6.6	6.4	6.3	6.2	6.1	5.9	5.8	5.7	5.5	5.4
Crude Birth Rate	44.5	43.8	43.2	42.5	41.9	41.2	40.6	40.0	39.4	38.8	38.2
Infant Mortality Rate	107.0	104.6	102.2	99.8	97.4	95.0	92.4	89.8	87.2	84.6	82.0
Life Expectancy at Birth	56.8	57.2	57.6	58.1	58.5	58.9	59.4	59.9	60.4	60.9	61.4
Food Production, p.c. ('79-81 = 100)	111.9	113.6	105.6	107.1	107.5	100.5	99.5	107.8	106.1	105.0	104.1
Labor Force, Agriculture (%)	53.1	52.2	51.4	50.6	49.4	48.2	46.9	45.7	44.5	43.3	42.1
Labor Force, Female (%)	16.3	16.3	16.3	16.3	16.5	16.8	17.1	17.4	17.7	18.0	18.3
School Enroll. Ratio, primary	..	..	..	97.0	..	..	..	..	104.0	104.0	106.0
School Enroll. Ratio, secondary	..	..	..	22.0	..	..	..	..	40.0	..	..

1978	1979	1980	1981	1982	1983	1984	1985	1986	1987 estimate	NOTES	ECUADOR
				(Millions of current US dollars)							**FOREIGN TRADE (CUSTOMS BASIS)**
1,557.5	2,104.2	2,480.2	2,167.9	2,290.8	2,348.0	2,621.0	2,904.7	2,171.5	2,021.3	..	Value of Exports, fob
800.3	868.6	841.4	753.4	749.7	562.6	726.1	883.0	1,095.1	1,120.9	..	Nonfuel Primary Products
718.4	1,178.2	1,564.5	1,341.7	1,471.9	1,733.4	1,833.7	1,929.7	981.4	815.8	..	Fuels
38.8	57.4	74.3	72.8	69.2	52.0	61.2	92.0	95.0	84.6	..	Manufactures
1,498.6	1,986.0	2,215.3	1,906.8	1,736.7	1,487.0	1,616.0	1,767.0	1,810.0	2,250.0	..	Value of Imports, cif
164.2	228.9	267.9	178.1	182.0	218.3	216.1	229.0	140.4	178.4	..	Nonfuel Primary Products
13.6	21.7	28.3	250.7	28.1	23.7	26.9	27.0	39.6	61.6	..	Fuels
1,320.8	1,735.4	1,919.1	1,478.1	1,526.6	1,245.0	1,373.0	1,511.0	1,630.0	2,010.0	..	Manufactures
				(Index 1980 = 100)							
76.0	81.4	100.0	102.4	99.9	97.4	96.3	93.7	60.0	61.4	..	Terms of Trade
60.4	74.1	100.0	103.9	96.9	93.2	91.0	88.0	63.6	71.1	..	Export Prices, fob
95.8	103.1	100.0	90.8	87.3	93.3	91.9	86.8	93.8	82.8	..	Nonfuel Primary Products
42.3	61.0	100.0	112.5	101.6	92.5	90.2	87.5	45.0	56.9	..	Fuels
83.3	87.3	100.0	114.0	122.0	120.0	108.0	120.0	142.0	156.0	..	Manufactures
79.4	91.0	100.0	101.4	97.0	95.6	94.5	94.0	106.0	116.0	..	Import Prices, cif
				(Millions of 1980 US dollars)							Trade at Constant 1980 Prices
2,580.2	2,840.4	2,480.2	2,086.5	2,363.2	2,520.2	2,879.3	3,299.4	3,414.7	2,841.1	..	Exports, fob
1,886.7	2,181.9	2,215.3	1,879.8	1,789.5	1,554.8	1,709.4	1,880.5	1,707.9	1,940.4	..	Imports, cif
				(Millions of current US dollars)							**BALANCE OF PAYMENTS**
1,738	2,473	2,953	2,981	2,702	2,688	2,972	3,323	2,617	2,383	..	Exports of Goods & Services
1,529	2,151	2,520	2,527	2,327	2,348	2,622	2,905	2,186	2,021	..	Merchandise, fob
174	260	345	384	349	295	273	389	403	346	..	Nonfactor Services
35	62	88	69	26	45	77	29	28	16	..	Factor Services
2,483	3,133	3,653	4,018	3,948	2,716	3,140	3,254	3,275	3,475	..	Imports of Goods & Services
1,704	2,097	2,242	2,353	2,187	1,421	1,567	1,611	1,631	2,054	..	Merchandise, fob
463	556	683	806	725	401	473	564	606	354	..	Nonfactor Services
316	480	728	859	1,036	894	1,100	1,079	1,038	1,067	..	Factor Services
136	269	365	539	715	455	793	737	654	279	..	Long-Term Interest
12	0	0	0	0	0	0	0	0	..	..	Current Transfers, net
0	0	0	0	0	0	0	0	0	0	..	Workers' Remittances
-732	-660	-700	-1,037	-1,246	-28	-168	69	-658	-1,251	K	Total to be Financed
29	29	30	25	20	24	20	80	45	75	K	Official Capital Grants
-703	-630	-670	-1,012	-1,226	-4	-148	149	-613	-1,176	..	Current Account Balance
811	596	793	1,102	184	1,396	380	554	818	1,171	..	Long-Term Capital, net
49	63	70	60	40	50	50	62	70	75	..	Direct Investment
577	559	754	1,449	132	113	615	297	859	409	..	Long-Term Loans
768	1,411	1,289	2,254	1,309	366	830	595	1,101	652	..	Disbursements
191	852	535	805	1,177	253	215	298	243	243	..	Repayments
156	-56	-61	-432	-8	1,209	-305	115	-156	612	..	Other Long-Term Capital
-117	60	123	-457	596	-1,353	-240	-573	-330	-52	..	Other Capital, net
10	-25	-246	368	446	-39	8	-130	125	57	..	Change in Reserves
				(Ecuadoran Sucres per US dollar)							Memo Item:
25.000	25.000	25.000	25.000	33.400	52.900	79.500	91.500	122.790	170.460	..	Conversion Factor (Annual Avg)
				(Millions of US dollars, outstanding at end of year)							**EXTERNAL DEBT, ETC.**
2,217	2,602	3,300	4,349	4,042	5,495	6,553	7,161	8,193	9,026	..	Public/Publicly Guar. Long-Term
1,116	1,089	1,314	1,861	1,695	1,697	1,853	2,053	2,555	3,160	..	Official Creditors
97	122	146	188	210	237	241	304	501	723	..	IBRD and IDA
1,101	1,513	1,986	2,488	2,347	3,798	4,700	5,108	5,638	5,866	..	Private Creditors
561	774	1,122	1,452	1,629	670	177	97	59	30	..	Private Non-guaranteed Long-Term
0	0	0	0	0	213	238	360	486	490	..	Use of Fund Credit
1,198	1,149	1,575	2,021	2,191	1,166	1,283	980	490	891	..	Short-Term Debt
				(Millions of US dollars)							Memo Items:
635.8	722.0	1,013.0	632.4	304.2	644.5	611.2	718.2	644.2	491.1	..	Int'l Reserves Excluding Gold
92.0	212.1	244.2	164.7	189.3	158.0	127.7	135.5	161.9	200.5	..	Gold Holdings (at market price)
											SOCIAL INDICATORS
5.3	5.2	5.2	5.1	5.0	4.9	4.7	4.6	4.5	4.3	..	Total Fertility Rate
37.9	37.6	37.4	37.1	36.8	36.0	35.2	34.4	33.6	32.8	..	Crude Birth Rate
79.6	77.2	74.8	72.4	70.0	64.9	59.9	54.8	49.7	44.7	..	Infant Mortality Rate
62.0	62.6	63.1	63.7	64.3	64.6	65.0	65.3	65.7	66.0	..	Life Expectancy at Birth
99.3	98.2	101.1	100.7	101.4	84.7	92.0	101.4	105.1	96.2	..	Food Production, p.c. ('79-81=100)
41.0	39.8	38.6	..	..	..	..	..	..	..	..	Labor Force, Agriculture (%)
18.6	19.0	19.3	19.3	19.3	19.3	19.3	19.3	19.3	19.3	..	Labor Force, Female (%)
108.0	112.0	115.0	115.0	116.0	117.0	114.0	..	..	..	..	School Enroll. Ratio, primary
..	49.0	52.0	58.0	55.0	52.0	55.0	..	..	..	..	School Enroll. Ratio, secondary

EGYPT, ARAB REPUBLIC OF	1967	1968	1969	1970	1971	1972	1973	1974	1975	1976	1977
CURRENT GNP PER CAPITA (US $)	180	190	210	230	240	250	270	270	310	350	390
POPULATION (thousands)	30,855	31,587	32,319	33,053	33,788	34,520	35,272	36,076	36,953	37,898	38,907
ORIGIN AND USE OF RESOURCES					*(Millions of current Egyptian Pounds)*						
Gross National Product (GNP)	2,512	2,605	2,809	3,033	3,212	3,363	3,757	4,282	5,107	6,555	8,102
Net Factor Income from Abroad	-11	-10	-24	-26	-29	-27	-49	-57	-111	-172	-242
Gross Domestic Product (GDP)	2,523	2,615	2,834	3,058	3,241	3,390	3,806	4,339	5,218	6,727	8,344
Indirect Taxes, net	323	351	388	432	449	388	342	142	162	562	810
GDP at factor cost	2,200	2,264	2,446	2,627	2,792	3,002	3,464	4,197	5,056	6,165	7,534
Agriculture	628	666	730	773	814	933	1,062	1,280	1,468	1,744	2,038
Industry	587	613	679	740	787	804	852	1,052	1,360	1,615	2,051
Manufacturing	..	..	..	..	..	..	..	746	880	993	1,120
Services, etc.	985	984	1,037	1,114	1,191	1,265	1,550	1,865	2,228	2,806	3,445
Resource Balance	-79	-100	-94	-140	-165	-196	-197	-726	-1,101	-789	-894
Exports of Goods & NFServices	369	345	403	434	447	452	532	890	1,053	1,498	1,876
Imports of Goods & NFServices	448	444	496	573	612	649	729	1,616	2,154	2,287	2,770
Domestic Absorption	2,601	2,714	2,928	3,198	3,406	3,586	4,003	5,065	6,319	7,516	9,238
Private Consumption, etc.	1,719	1,783	1,882	2,016	2,139	2,259	2,429	3,191	3,280	3,936	5,176
General Gov't Consumption	519	602	679	756	839	909	1,074	899	1,298	1,670	1,628
Gross Domestic Investment	364	330	367	427	429	418	500	975	1,741	1,910	2,434
Fixed Investment	325	313	342	353	363	378	462	685	1,282	1,471	1,873
Memo Items:											
Gross Domestic Saving	285	231	274	287	263	222	303	249	640	1,121	1,540
Gross National Saving	279	222	253	276	251	243	303	282	707	1,278	1,673
					(Millions of 1980 Egyptian Pounds)						
Gross National Product	7,309	7,513	7,997	8,460	8,757	8,925	8,944	9,180	9,937	11,419	12,919
GDP at factor cost	6,107	6,228	6,645	6,993	7,268	7,605	8,170	8,618	9,508	10,828	11,923
Agriculture	1,967	2,009	2,086	2,137	2,165	2,339	2,409	2,416	2,565	2,600	2,528
Industry	1,969	2,059	2,297	2,446	2,544	2,575	2,528	2,446	2,840	3,282	4,006
Manufacturing	..	..	..	..	..	..	..	1,266	1,348	1,453	1,551
Services, etc.	2,171	2,159	2,261	2,411	2,559	2,690	3,233	3,756	4,104	4,946	5,390
Resource Balance	-71	-381	-351	-503	-502	-523	-811	-1,937	-2,272	-1,102	-992
Exports of Goods & NFServices	2,296	2,036	2,296	2,523	2,488	2,614	2,482	2,581	3,182	4,059	4,449
Imports of Goods & NFServices	2,367	2,417	2,646	3,026	2,989	3,136	3,294	4,518	5,454	5,161	5,441
Domestic Absorption	7,412	7,923	8,419	9,035	9,338	9,519	9,878	11,247	12,434	12,817	14,285
Private Consumption, etc.	..	..	..	..	..	..	..	7,157	6,558	6,782	8,157
General Gov't Consumption	..	..	..	..	..	..	..	1,985	2,583	2,753	2,427
Gross Domestic Investment	1,166	1,050	1,145	1,203	1,055	1,088	1,429	2,104	3,292	3,282	3,701
Fixed Investment	..	..	..	..	..	..	..	1,479	2,424	2,527	2,848
Memo Items:											
Capacity to Import	1,951	1,875	2,147	2,289	2,182	2,188	2,404	2,488	2,666	3,380	3,685
Terms of Trade Adjustment	-344	-161	-148	-234	-305	-425	-79	-93	-516	-678	-764
Gross Domestic Income	6,997	7,381	7,920	8,298	8,531	8,570	8,988	9,217	9,646	11,037	12,529
Gross National Income	6,964	7,352	7,849	8,226	8,452	8,499	8,865	9,087	9,421	10,741	12,155
DOMESTIC PRICES (DEFLATORS)					*(Index 1980 = 100)*						
Overall (GDP)	34.4	34.7	35.1	35.8	36.7	37.7	42.0	46.6	51.3	57.4	62.8
Domestic Absorption	35.1	34.3	34.8	35.4	36.5	37.7	40.5	45.0	50.8	58.6	64.7
Agriculture	31.9	33.2	35.0	36.2	37.6	39.9	44.1	53.0	57.2	67.1	80.6
Industry	29.8	29.8	29.5	30.3	30.9	31.2	33.7	43.0	47.9	49.2	51.2
Manufacturing	..	..	..	..	..	..	..	58.9	65.3	68.3	72.2
MANUFACTURING ACTIVITY											
Employment (1980=100)	63.5	63.2	67.4	68.6	72.4	74.4	78.8	80.1	84.3	86.9	89.9
Real Earnings per Empl. (1980=100)	64.7	63.6	65.4	64.6	63.5	68.5	72.0	72.1	75.1	77.6	78.8
Real Output per Empl. (1980=100)	79.7	80.9	85.1	81.4	85.9	90.1	89.3	69.3	69.1	72.3	81.4
Earnings as % of Value Added	53.4	51.7	55.5	53.7	50.3	50.5	49.6	42.4	50.9	50.2	46.4
MONETARY HOLDINGS					*(Millions of current Egyptian Pounds)*						
Money Supply, Broadly Defined	916	944	999	1,053	1,084	1,255	1,536	2,000	2,430	3,061	4,103
Money as Means of Payment	707	722	746	783	846	989	1,205	1,503	1,863	2,239	2,943
Currency Ouside Banks	450	460	496	525	559	631	777	948	1,156	1,388	1,749
Demand Deposits	257	261	250	258	288	358	428	555	707	851	1,194
Quasi-Monetary Liabilities	209	223	253	270	238	266	331	498	567	822	1,160
					(Millions of current Egyptian Pounds)						
GOVERNMENT DEFICIT (-) OR SURPLUS	..	..	..	..	..	..	..	..	-938	-1,557	-1,114
Current Revenue	..	..	..	..	..	..	..	..	2,235	2,424	3,301
Current Expenditure	..	..	..	..	..	..	..	..	2,418	2,662	3,209
Current Budget Balance	..	..	..	..	..	..	..	..	-183	-238	92
Capital Receipts	..	..	..	..	..	..	..	..	54	105	147
Capital Payments	..	..	..	..	..	..	..	..	809	1,424	1,353

1978	1979	1980	1981	1982	1983	1984	1985	1986	1987 estimate	NOTES	EGYPT, ARAB REPUBLIC OF
390	410	480	530	580	590	610	640	650	670	A	CURRENT GNP PER CAPITA (US $)
39,979	41,108	42,289	43,510	44,760	46,021	47,276	48,503	49,739	50,954	..	POPULATION (thousands)
				(Millions of current Egyptian Pounds)							ORIGIN AND USE OF RESOURCES
9,358	11,957	15,446	16,088	19,332	22,557	26,585	29,995	34,780	39,725	C	Gross National Product (GNP)
-437	-748	-1,051	-1,232	-1,449	-1,613	-1,919	-3,136	-3,576	-3,962	..	Net Factor Income from Abroad
9,795	12,705	16,497	17,320	20,781	24,170	28,504	33,132	38,356	43,687	C	Gross Domestic Product (GDP)
774	604	757	768	684	929	1,103	1,180	2,125	1,879	..	Indirect Taxes, net
9,021	12,101	15,740	16,552	20,097	23,241	27,401	31,952	36,231	41,808	C	GDP at factor cost
2,286	2,530	2,875	3,326	3,932	4,564	5,494	6,386	7,531	8,844	..	Agriculture
2,583	4,337	5,789	6,245	6,521	6,970	8,024	9,125	9,715	10,441	..	Industry
1,319	1,650	1,928	2,144	2,670	3,068	3,624	4,316	4,805	5,806	..	Manufacturing
4,152	5,234	7,076	6,981	9,644	11,707	13,883	16,441	18,985	22,523	..	Services, etc.
-1,496	-2,364	-2,038	-2,673	-3,096	-2,646	-3,837	-4,018	-3,783	-4,766	..	Resource Balance
2,130	3,777	5,034	5,780	5,618	6,159	6,371	6,598	6,034	6,593	..	Exports of Goods & NFServices
3,626	6,141	7,072	8,453	8,714	8,805	10,208	10,616	9,817	11,359	..	Imports of Goods & NFServices
11,291	15,069	18,535	19,993	23,877	26,816	32,341	37,150	42,139	48,453	..	Domestic Absorption
6,178	8,724	11,411	11,588	13,923	15,712	19,367	22,600	26,706	33,710	..	Private Consumption, etc.
2,012	2,172	2,585	3,294	3,704	4,160	5,140	5,712	6,340	6,328	..	General Gov't Consumption
3,101	4,173	4,539	5,111	6,250	6,944	7,834	8,838	9,093	8,415	..	Gross Domestic Investment
2,685	3,763	4,062	4,702	6,150	7,144	7,634	8,338	8,593	8,865	..	Fixed Investment
											Memo Items:
1,605	1,809	2,501	2,438	3,154	4,298	3,997	4,820	5,310	3,649	..	Gross Domestic Saving
1,878	2,834	3,282	3,248	3,096	4,918	4,848	4,149	3,831	1,693	..	Gross National Saving
				(Millions of 1980 Egyptian Pounds)							
13,491	14,100	15,446	15,918	17,612	18,995	20,160	20,899	21,550	22,308	C	Gross National Product
13,005	14,245	15,740	16,363	18,249	19,629	20,853	22,253	22,845	23,424	C	GDP at factor cost
2,668	2,777	2,875	2,925	3,043	3,131	3,197	3,299	3,369	3,439	..	Agriculture
4,630	5,049	5,789	6,003	6,160	6,585	7,194	7,766	7,886	8,063	..	Industry
1,638	1,766	1,928	2,004	2,188	2,329	2,517	2,680	2,774	2,833	..	Manufacturing
5,707	6,418	7,076	7,436	9,046	9,913	10,462	11,188	11,590	11,922	..	Services, etc.
-1,166	-2,237	-2,038	-2,489	-2,025	-1,615	-2,280	-2,290	-1,286	110	..	Resource Balance
4,461	4,302	5,034	4,960	4,444	4,921	5,222	5,436	5,461	5,808	..	Exports of Goods & NFServices
5,627	6,540	7,072	7,450	6,469	6,536	7,503	7,726	6,747	5,697	..	Imports of Goods & NFServices
15,247	17,196	18,535	19,619	20,893	21,908	23,838	25,295	24,902	24,104	..	Domestic Absorption
8,505	10,167	11,411	11,577	12,303	12,800	13,764	14,693	15,127	15,479	..	Private Consumption, etc.
2,621	2,424	2,585	3,131	3,059	3,368	3,747	3,795	3,822	3,733	..	General Gov't Consumption
4,120	4,604	4,539	4,911	5,531	5,741	6,328	6,807	5,953	4,892	..	Gross Domestic Investment
3,568	4,152	4,062	4,518	5,443	5,914	6,164	6,416	5,592	5,162	..	Fixed Investment
											Memo Items:
3,305	4,022	5,034	5,094	4,170	4,572	4,683	4,802	4,147	3,307	..	Capacity to Import
-1,155	-280	0	134	-274	-349	-540	-634	-1,314	-2,501	..	Terms of Trade Adjustment
12,925	14,678	16,497	17,263	18,594	19,944	21,018	22,371	22,302	21,714	..	Gross Domestic Income
12,336	13,820	15,446	16,052	17,339	18,646	19,620	20,265	20,236	19,807	..	Gross National Income
				(Index 1980 = 100)							DOMESTIC PRICES (DEFLATORS)
69.6	84.9	100.0	101.1	110.1	119.1	132.2	144.0	162.4	180.4	..	Overall (GDP)
74.1	87.6	100.0	101.9	114.3	122.4	135.7	146.9	169.2	201.0	..	Domestic Absorption
85.7	91.1	100.0	113.7	129.2	145.8	171.9	193.6	223.6	257.2	..	Agriculture
55.8	85.9	100.0	104.0	105.9	105.8	111.5	117.5	123.2	129.5	..	Industry
80.5	93.5	100.0	107.0	122.0	131.8	144.0	161.0	173.2	205.0	..	Manufacturing
											MANUFACTURING ACTIVITY
95.0	93.0	100.0	100.9	104.0	107.2	110.5	114.0	..	..	..	Employment (1980=100)
82.7	88.0	100.0	118.8	118.5	121.5	116.8	121.4	..	..	..	Real Earnings per Empl. (1980=100)
81.8	83.8	100.0	102.5	117.2	127.6	128.4	140.5	..	..	..	Real Output per Empl. (1980=100)
52.1	54.0	56.8	61.4	57.4	57.4	57.4	57.4	..	..	..	Earnings as % of Value Added
				(Millions of current Egyptian Pounds)							MONETARY HOLDINGS
5,212	6,844	10,364	13,566	17,792	21,817	25,929	30,676	37,102	44,878	..	Money Supply, Broadly Defined
3,553	4,354	6,775	7,646	9,552	10,933	12,443	14,696	15,973	18,241	..	Money as Means of Payment
2,184	2,657	3,398	4,291	5,503	6,475	7,097	8,284	8,803	9,537	..	Currency Ouside Banks
1,369	1,697	3,377	3,355	4,049	4,458	5,346	6,412	7,170	8,704	..	Demand Deposits
1,659	2,490	3,589	5,920	8,240	10,884	13,486	15,980	21,129	26,637	..	Quasi-Monetary Liabilities
				(Millions of current Egyptian Pounds)							
-1,246	-1,964	..	-1,096	-3,554	-2,364	-3,258	-3,439	-4,708	..	F	GOVERNMENT DEFICIT (-) OR SURPLUS
3,778	4,363	..	7,893	9,116	10,714	11,951	13,245	15,126	..	..	Current Revenue
3,488	4,516	..	6,333	9,347	9,637	12,519	12,891	15,042	..	..	Current Expenditure
290	-153	..	1,560	-231	1,077	-568	354	84	..	..	Current Budget Balance
42	323	..	188	601	363	395	655	743	..	..	Capital Receipts
1,578	2,134	..	2,844	3,924	3,804	3,085	4,448	5,535	..	..	Capital Payments

EGYPT, ARAB REPUBLIC OF	1967	1968	1969	1970	1971	1972	1973	1974	1975	1976	1977
FOREIGN TRADE (CUSTOMS BASIS)				*(Millions of current US dollars)*							
Value of Exports, fob	566	656	824	870	927	887	1,117	1,588	1,596	1,785	2,015
Nonfuel Primary Products	417	476	572	585	634	564	721	988	792	761	866
Fuels	21	19	48	79	73	68	113	201	327	644	720
Manufactures	128	161	204	207	220	256	283	399	478	380	429
Value of Imports, cif	792	666	638	787	920	898	914	2,351	3,934	3,862	4,815
Nonfuel Primary Products	401	286	232	274	370	361	380	1,225	1,703	1,344	1,586
Fuels	56	52	56	74	71	60	23	66	272	221	109
Manufactures	334	328	349	439	479	478	511	1,060	1,958	2,296	3,120
					(Index 1980 = 100)						
Terms of Trade	110.3	122.7	104.5	89.1	84.1	79.8	77.7	94.4	86.3	89.8	86.8
Export Prices, fob	24.4	26.0	21.7	18.6	21.6	23.3	35.7	62.2	55.2	56.3	57.8
Nonfuel Primary Products	32.8	34.0	32.7	32.5	36.0	39.0	63.3	69.7	60.3	72.0	71.8
Fuels	4.3	4.3	4.3	4.3	5.6	6.2	8.9	36.7	35.7	38.4	42.0
Manufactures	23.0	23.7	22.3	20.1	18.0	20.1	39.6	67.6	71.9	87.1	75.5
Import Prices, cif	22.1	21.2	20.8	20.9	25.7	29.2	46.0	65.9	63.9	62.7	66.6
Trade at Constant 1980 Prices				*(Millions of 1980 US dollars)*							
Exports, fob	2,320	2,524	3,797	4,670	4,298	3,805	3,126	2,555	2,891	3,172	3,489
Imports, cif	3,578	3,143	3,069	3,762	3,585	3,077	1,988	3,569	6,151	6,163	7,234
BALANCE OF PAYMENTS				*(Millions of current US dollars)*							
Exports of Goods & Services	..	..	..	962	1,005	1,017	1,304	2,338	2,589	3,391	4,001
Merchandise, fob	..	..	..	817	851	813	1,000	1,818	1,875	2,169	2,346
Nonfactor Services	..	..	..	143	153	203	300	433	628	1,150	1,542
Factor Services	..	..	..	2	1	1	4	87	86	72	113
Imports of Goods & Services	..	..	..	1,447	1,518	1,592	1,986	4,165	5,471	5,596	6,417
Merchandise, fob	..	..	..	1,084	1,131	1,170	1,429	3,618	4,608	4,659	5,110
Nonfactor Services	..	..	..	297	312	352	429	341	533	523	769
Factor Services	..	..	..	66	74	69	128	206	330	414	538
Long-Term Interest	..	..	..	..	..	..	..	..	..	..	380
Current Transfers, net	..	..	..	..	..	..	..	231	456	842	960
Workers' Remittances	..	..	..	29	27	104	117	189	366	755	897
Total to be Financed	..	..	..	..	..	..	..	-1,596	-2,426	-1,363	-1,456
Official Capital Grants	..	..	..	..	..	..	..	1,261	986	705	382
Current Account Balance	..	..	..	-148	-207	-174	20	-335	-1,440	-658	-1,074
Long-Term Capital, net	..	..	..	306	288	594	784	2,588	2,927	2,419	3,263
Direct Investment	..	..	..	..	..	..	..	87	225	444	477
Long-Term Loans	..	..	..	..	..	..	..	..	..	..	2,335
Disbursements	..	..	..	..	..	..	..	..	..	..	3,099
Repayments	..	..	..	..	..	..	..	..	..	..	764
Other Long-Term Capital	..	..	..	306	288	594	784	1,240	1,716	1,270	69
Other Capital, net	..	..	..	-170	-140	-405	-633	-1,171	-927	-548	-455
Change in Reserves	..	..	..	12	58	-14	-171	-1,082	-560	-1,213	-1,734
Memo Item:				*(Egyptian Pounds per US dollar)*							
Conversion Factor (Annual Avg)	0.450	0.430	0.410	0.400	0.390	0.390	0.400	0.480	0.460	0.500	0.570
EXTERNAL DEBT, ETC.				*(Millions of US dollars, outstanding at end of year)*							
Public/Publicly Guar. Long-Term	..	..	..	1,713	2,081	2,239	2,423	2,960	4,983	6,018	8,467
Official Creditors	..	..	..	1,227	1,406	1,377	1,748	2,090	4,048	5,017	7,226
IBRD and IDA	..	..	..	22	17	12	30	36	98	176	255
Private Creditors	..	..	..	487	675	862	675	870	935	1,000	1,242
Private Non-guaranteed Long-Term	..	..	..	..	..	..	..	..	..	..	18
Use of Fund Credit	..	..	..	49	76	27	75	114	80	207	310
Short-Term Debt	..	..	..	..	..	..	..	..	..	..	..
Memo Items:				*(Millions of US dollars)*							
Int'l Reserves Excluding Gold	102.0	75.0	51.0	74.1	57.1	51.5	259.6	251.9	193.9	239.6	430.7
Gold Holdings (at market price)	94.0	112.0	94.1	91.0	106.1	157.8	273.0	453.6	341.1	327.7	401.2
SOCIAL INDICATORS											
Total Fertility Rate	6.6	6.4	6.2	5.9	5.7	5.5	5.5	5.4	5.4	5.3	5.3
Crude Birth Rate	41.8	41.1	40.4	39.8	39.1	38.4	38.9	39.4	39.8	40.3	40.8
Infant Mortality Rate	170.0	166.0	162.0	158.0	154.0	150.0	144.0	138.0	132.0	126.0	120.0
Life Expectancy at Birth	49.7	50.2	50.7	51.1	51.6	52.1	52.8	53.5	54.2	54.9	55.6
Food Production, p.c. ('79-81 = 100)	98.8	108.3	107.8	105.4	108.0	108.1	107.0	106.7	108.0	107.2	101.2
Labor Force, Agriculture (%)	53.8	53.2	52.6	52.0	51.4	50.7	50.1	49.4	48.8	48.2	47.6
Labor Force, Female (%)	7.1	7.1	7.1	7.1	7.3	7.4	7.5	7.7	7.8	8.0	8.1
School Enroll. Ratio, primary	..	..	..	72.0	..	..	..	..	71.0	73.0	73.0
School Enroll. Ratio, secondary	..	..	..	35.0	..	..	..	..	43.0	45.0	47.0

1978	1979	1980	1981	1982	1983	1984	1985	1986	1987 estimate	NOTES	EGYPT, ARAB REPUBLIC OF
				(Millions of current US dollars)							**FOREIGN TRADE (CUSTOMS BASIS)**
2,458	3,320	4,759	5,048	5,033	5,936	6,076	5,193	4,040	4,482	C f	Value of Exports, fob
752	702	786	894	809	795	808	696	592	693	..	Nonfuel Primary Products
1,203	2,245	3,585	3,750	3,734	4,485	4,608	3,858	2,658	2,768	..	Fuels
504	373	388	404	490	656	660	639	790	1,021	..	Manufactures
6,727	7,073	8,047	8,839	9,078	10,766	11,594	10,581	8,453	10,586	C f	Value of Imports, cif
2,224	2,039	2,800	3,566	3,292	3,610	3,730	3,432	2,612	2,877	..	Nonfuel Primary Products
101	164	214	265	371	484	469	303	189	207	..	Fuels
4,401	4,870	5,033	5,008	5,415	6,672	7,395	6,846	5,652	7,502	..	Manufactures
				(Index 1980 = 100)							
71.7	73.7	100.0	108.3	103.7	101.5	98.0	84.0	62.8	63.5	..	Terms of Trade
55.1	66.7	100.0	107.2	97.3	95.1	91.0	75.8	60.1	65.1	..	Export Prices, fob
75.8	86.4	100.0	89.5	76.2	87.2	83.5	66.4	60.4	84.5	..	Nonfuel Primary Products
42.3	61.0	100.0	112.5	101.6	95.0	90.2	74.2	52.0	52.0	..	Fuels
80.6	77.7	100.0	107.9	112.2	107.9	110.5	105.6	124.8	137.0	..	Manufactures
76.9	90.6	100.0	99.0	93.8	93.7	92.9	90.2	95.7	102.5	..	Import Prices, cif
				(Millions of 1980 US dollars)							Trade at Constant 1980 Prices
4,459	4,974	4,759	4,708	5,172	6,241	6,676	6,853	6,725	6,888	..	Exports, fob
8,752	7,807	8,047	8,929	9,673	11,494	12,481	11,734	8,834	10,331	..	Imports, cif
				(Millions of current US dollars)							**BALANCE OF PAYMENTS**
4,243	5,707	7,087	8,207	8,430	8,870	9,561	9,754	8,754	8,061	..	Exports of Goods & Services
2,558	3,987	4,686	5,617	5,779	5,248	5,924	6,075	5,193	4,040	..	Merchandise, fob
1,541	1,414	2,116	2,186	2,183	2,604	2,558	2,636	2,647	3,231	..	Nonfactor Services
144	306	284	404	468	1,018	1,079	1,043	914	790	..	Factor Services
7,419	10,156	11,300	13,481	13,917	14,411	16,556	18,011	17,074	14,684	..	Imports of Goods & Services
5,998	7,817	8,577	10,334	10,380	9,619	11,328	11,593	10,581	8,453	..	Merchandise, fob
614	964	980	1,078	1,285	1,896	2,086	2,091	2,300	2,176	..	Nonfactor Services
807	1,375	1,744	2,069	2,252	2,896	3,142	4,327	4,193	4,055	..	Factor Services
447	333	456	658	618	665	746	744	768	806	..	Long-Term Interest
..	..	..	..	..	..	..	..	..	..	..	Current Transfers, net
1,761	2,445	2,696	2,855	1,935	3,165	3,931	3,496	2,973	2,845	..	Workers' Remittances
..	..	..	..	..	..	..	..	..	..	..	Total to be Financed
..	..	..	..	..	..	..	..	..	..	..	Official Capital Grants
-1,070	-1,843	-1,420	-2,077	-3,106	-1,675	-2,278	-3,564	-4,038	-2,705	..	Current Account Balance
2,515	2,337	2,172	2,335	2,315	2,787	4,562	3,599	3,606	355	..	Long-Term Capital, net
387	1,375	541	836	885	966	1,275	1,289	1,275	869	..	Direct Investment
2,107	1,738	1,631	1,844	1,736	1,630	1,549	1,357	681	608	..	Long-Term Loans
2,917	2,525	2,706	3,205	3,228	3,177	3,052	2,894	1,860	1,536	..	Disbursements
810	787	1,075	1,361	1,492	1,548	1,504	1,536	1,179	928	..	Repayments
21	-776	0	-345	-306	191	1,738	953	1,650	-1,122	..	Other Long-Term Capital
-178	190	-54	-55	808	-223	-2,145	365	832	253	..	Other Capital, net
-1,267	-684	-698	-203	-17	-889	-139	-400	-400	2,097	..	Change in Reserves
				(Egyptian Pounds per US dollar)							Memo Item:
0.660	0.700	0.720	0.740	0.810	0.860	0.930	0.960	1.100	1.210	..	Conversion Factor (Annual Avg)
				(Millions of US dollars, outstanding at end of year)							**EXTERNAL DEBT, ETC.**
10,703	12,464	15,785	18,979	21,214	23,184	25,136	29,041	32,093	34,515	..	Public/Publicly Guar. Long-Term
9,249	10,348	13,405	15,937	17,366	18,733	20,328	23,492	26,267	28,389	..	Official Creditors
356	525	728	932	1,125	1,377	1,464	1,850	2,214	2,594	..	IBRD and IDA
1,454	2,116	2,381	3,042	3,848	4,451	4,808	5,549	5,826	6,125	..	Private Creditors
75	185	265	320	455	600	550	750	947	1,098	..	Private Non-guaranteed Long-Term
386	325	177	99	57	52	48	41	31	182	..	Use of Fund Credit
..	..	3,644	3,174	4,442	4,381	4,779	4,966	4,790	4,469	..	Short-Term Debt
				(Millions of US dollars)							Memo Items:
491.7	529.5	1,046.0	716.2	698.1	771.1	736.2	792.1	829.0	1,378.3	..	Int'l Reserves Excluding Gold
558.9	1,265.7	1,433.7	966.7	1,111.2	927.8	749.8	795.3	950.7	1,177.3	..	Gold Holdings (at market price)
											SOCIAL INDICATORS
5.2	5.1	5.1	5.0	4.9	4.9	4.8	4.7	4.6	4.5	..	Total Fertility Rate
40.1	39.5	38.8	38.2	37.5	36.7	35.9	35.2	34.4	33.6	..	Crude Birth Rate
116.0	112.0	108.0	104.0	100.0	92.7	85.4	78.1	70.8	63.5	..	Infant Mortality Rate
56.4	57.2	58.0	58.8	59.7	60.0	60.4	60.8	61.2	61.6	..	Life Expectancy at Birth
101.7	102.1	99.2	98.7	104.6	104.7	103.5	104.8	106.2	108.3	..	Food Production, p.c. ('79-81 = 100)
46.9	46.3	45.7	..	..	..	..	..	..	..	..	Labor Force, Agriculture (%)
8.3	8.4	8.6	8.8	8.9	9.0	9.2	9.3	9.5	9.6	..	Labor Force, Female (%)
74.0	75.0	83.0	87.0	82.0	84.0	..	85.0	..	..	..	School Enroll. Ratio, primary
49.0	50.0	51.0	53.0	58.0	58.0	..	62.0	..	..	..	School Enroll. Ratio, secondary

EL SALVADOR	1967	1968	1969	1970	1971	1972	1973	1974	1975	1976	1977
CURRENT GNP PER CAPITA (US $)	280	280	290	290	300	310	330	380	430	500	590
POPULATION (thousands)	3,242	3,360	3,477	3,590	3,699	3,800	3,897	3,993	4,089	4,184	4,277
ORIGIN AND USE OF RESOURCES				*(Millions of current Salvadoran Colones)*							
Gross National Product (GNP)	2,194	2,274	2,362	2,549	2,679	2,855	3,294	3,891	4,412	5,689	7,095
Net Factor Income from Abroad	-22	-18	-20	-22	-25	-27	-38	-53	-66	-17	-72
Gross Domestic Product (GDP)	2,216	2,292	2,382	2,571	2,704	2,882	3,332	3,944	4,478	5,706	7,167
Indirect Taxes, net	163	148	162	196	199	223	268	329	353	562	848
GDP at factor cost	2,053	2,144	2,220	2,375	2,505	2,659	3,064	3,615	4,125	5,144	6,319
Agriculture	600	603	607	731	729	728	922	999	1,028	1,614	2,374
Industry	525	543	573	600	644	712	769	915	1,115	1,246	1,488
Manufacturing	422	448	466	485	519	563	611	707	831	933	1,047
Services, etc.	1,090	1,146	1,201	1,240	1,331	1,442	1,640	2,030	2,335	2,845	3,304
Resource Balance	-67	-40	-62	8	-50	32	-94	-332	-228	21	62
Exports of Goods & NFServices	572	584	556	643	669	842	1,003	1,284	1,485	2,178	2,767
Imports of Goods & NFServices	638	624	617	635	719	810	1,096	1,615	1,713	2,157	2,705
Domestic Absorption	2,282	2,332	2,444	2,563	2,754	2,850	3,425	4,275	4,706	5,685	7,105
Private Consumption, etc.	1,761	1,861	1,892	1,957	2,057	2,135	2,467	2,954	3,214	3,880	4,622
General Gov't Consumption	203	216	249	276	275	308	349	429	501	686	805
Gross Domestic Investment	319	255	303	331	422	408	609	892	991	1,120	1,679
Fixed Investment	316	248	274	298	359	474	521	719	1,031	1,145	1,520
Memo Items:											
Gross Domestic Saving	252	216	241	339	372	440	515	561	762	1,140	1,740
Gross National Saving	250	212	246	348	387	435	507	551	760	1,185	1,744
				(Millions of 1980 Salvadoran Colones)							
Gross National Product	5,857.6	6,053.3	6,270.0	6,454.2	6,752.3	7,109.0	7,451.5	7,925.4	8,340.5	8,743.1	9,174.0
GDP at factor cost	5,477.3	5,710.8	5,891.5	6,015.0	6,313.0	6,620.8	6,930.9	7,357.5	7,789.5	7,904.4	8,177.7
Agriculture	1,644.5	1,674.0	1,735.6	1,849.5	1,918.8	1,947.1	1,981.6	2,184.4	2,321.6	2,138.4	2,215.4
Industry	1,162.3	1,187.6	1,214.8	1,259.2	1,357.1	1,461.1	1,527.2	1,616.6	1,785.8	1,889.6	2,081.0
Manufacturing	917.8	958.3	965.4	1,001.5	1,071.4	1,112.5	1,192.3	1,261.7	1,320.7	1,436.3	1,511.5
Services, etc.	3,104.4	3,242.3	3,369.6	3,400.9	3,537.4	3,767.1	4,027.8	4,225.9	4,348.7	4,742.9	4,986.0
Resource Balance	-114.9	292.3	256.4	-33.3	-52.0	125.9	-369.9	-309.0	120.7	-609.8	-1,838.3
Exports of Goods & NFServices	1,906.4	2,110.4	2,024.9	1,854.7	2,016.2	2,363.1	2,272.9	2,424.2	2,717.0	2,460.2	2,171.8
Imports of Goods & NFServices	2,021.3	1,818.1	1,768.5	1,888.0	2,068.2	2,237.2	2,642.8	2,733.3	2,596.3	3,070.0	4,010.1
Domestic Absorption	6,026.1	5,811.7	6,063.7	6,542.8	6,865.2	7,049.4	7,906.5	8,336.0	8,335.3	9,380.7	11,120.7
Private Consumption, etc.	4,598.3	4,546.0	4,621.6	4,973.5	5,161.3	5,288.9	5,772.4	5,920.1	5,999.2	6,746.9	7,766.3
General Gov't Consumption	575.6	601.6	696.9	742.1	698.7	819.8	914.3	878.9	955.9	1,106.2	1,161.5
Gross Domestic Investment	852.2	664.1	745.1	827.2	1,005.3	940.7	1,219.8	1,537.0	1,380.3	1,527.6	2,192.9
Fixed Investment	843.9	642.4	663.1	742.2	841.3	1,105.5	1,005.8	1,150.5	1,487.5	1,533.0	1,941.1
Memo Items:											
Capacity to Import	1,810.0	1,701.9	1,591.2	1,911.8	1,923.8	2,325.1	2,417.0	2,172.2	2,250.6	3,099.6	4,101.7
Terms of Trade Adjustment	-96.3	-408.4	-433.7	57.1	-92.4	-38.1	144.0	-252.0	-466.4	639.4	1,929.9
Gross Domestic Income	5,814.8	5,695.5	5,886.4	6,566.6	6,720.8	7,137.2	7,680.6	7,774.9	7,989.7	9,410.3	11,212.3
Gross National Income	5,761.3	5,644.9	5,836.3	6,511.2	6,659.9	7,070.9	7,595.5	7,673.4	7,874.1	9,382.5	11,103.9
DOMESTIC PRICES (DEFLATORS)				*(Index 1980 = 100)*							
Overall (GDP)	37.5	37.5	37.7	39.5	39.7	40.2	44.2	49.1	53.0	65.1	77.2
Domestic Absorption	37.9	40.1	40.3	39.2	40.1	40.4	43.3	51.3	56.5	60.6	63.9
Agriculture	36.5	36.0	35.0	39.5	38.0	37.4	46.5	45.7	44.3	75.5	107.2
Industry	45.2	45.7	47.2	47.7	47.5	48.7	50.3	56.6	62.4	65.9	71.5
Manufacturing	46.0	46.8	48.3	48.4	48.5	50.6	51.2	56.0	62.9	64.9	69.2
MANUFACTURING ACTIVITY											
Employment (1980=100)	96.2	91.5	104.4	115.3	110.7	107.8	116.6	124.0	130.2	132.3	151.0
Real Earnings per Empl. (1980=100)	67.2	72.1	73.7	72.7	76.0	80.2	79.6	78.0	74.0	80.8	69.0
Real Output per Empl. (1980=100)	67.7	56.0	67.2	70.8	78.0	72.1	71.4	93.0	93.4	95.0	95.7
Earnings as % of Value Added	25.2	27.3	26.8	28.0	25.3	23.3	24.6	23.3	22.6	22.5	19.2
MONETARY HOLDINGS				*(Millions of current Salvadoran Colones)*							
Money Supply, Broadly Defined	489	509	562	595	658	807	958	1,116	1,353	1,770	2,004
Money as Means of Payment	253	265	288	295	315	390	466	557	648	917	988
Currency Ouside Banks	122	116	133	136	145	175	201	241	253	380	432
Demand Deposits	130	148	155	159	170	215	265	316	395	537	556
Quasi-Monetary Liabilities	236	245	274	300	343	418	492	560	705	854	1,015
GOVERNMENT DEFICIT (-) OR SURPLUS				*(Millions of current Salvadoran Colones)*							
	..	..	..	-15	-28	-30	-12	-27	-72	-28	91
Current Revenue	..	..	..	278	295	332	396	484	562	807	1,174
Current Expenditure	..	..	..	230	241	273	310	373	442	544	652
Current Budget Balance	..	..	..	48	54	59	85	111	120	263	522
Capital Receipts	..	..	..	9	..	9	4	0	..	0	0
Capital Payments	..	..	..	72	82	98	101	138	192	292	431

1978	1979	1980	1981	1982	1983	1984	1985	1986	1987 estimate	NOTES	EL SALVADOR
700	760	740	740	730	760	810	830	820	850	A	**CURRENT GNP PER CAPITA (US $)**
4,367	4,452	4,532	4,581	4,630	4,680	4,730	4,781	4,876	4,973	..	**POPULATION (thousands)**
			(Millions of current Salvadoran Colones)								**ORIGIN AND USE OF RESOURCES**
7,562	8,547	8,789	8,427	8,660	9,795	11,251	13,834	19,128	23,130	..	Gross National Product (GNP)
-130	-60	-128	-219	-306	-357	-406	-497	-635	-636	..	Net Factor Income from Abroad
7,692	8,607	8,917	8,647	8,966	10,152	11,657	14,331	19,763	23,766	..	Gross Domestic Product (GDP)
645	855	642	661	..	..	..	..	..	..	..	Indirect Taxes, net
7,047	7,752	8,275	7,986							B	GDP at factor cost
2,049	2,508	2,480	2,106	2,075	2,160	2,320	2,611	3,996	3,282	..	Agriculture
1,664	1,851	1,846	1,847	1,896	2,175	2,492	3,139	4,051	5,323	..	Industry
1,205	1,338	1,339	1,359	1,382	1,572	1,837	2,346	3,059	4,017	..	Manufacturing
3,979	4,248	4,591	4,693	4,995	5,817	6,846	8,581	11,716	15,161	..	Services, etc.
-713	-15	82	-527	-454	-550	-791	-1,084	-630	-1,463	..	Resource Balance
2,328	3,182	3,046	2,383	2,188	2,486	2,536	3,199	5,130	4,552	..	Exports of Goods & NFServices
3,041	3,197	2,964	2,909	2,642	3,036	3,327	4,283	5,760	6,015	..	Imports of Goods & NFServices
8,405	8,622	8,835	9,173	9,420	10,702	12,448	15,415	20,393	25,229	..	Domestic Absorption
5,575	5,933	6,405	6,574	6,821	7,871	9,184	11,640	15,392	19,214	..	Private Consumption, etc.
996	1,133	1,247	1,369	1,414	1,607	1,869	2,220	2,382	2,698	..	General Gov't Consumption
1,834	1,556	1,183	1,231	1,185	1,224	1,394	1,554	2,619	3,317	..	Gross Domestic Investment
1,652	1,512	1,210	1,173	1,131	1,180	1,336	1,723	2,619	3,317	..	Fixed Investment
											Memo Items:
1,122	1,541	1,265	704	731	673	604	471	1,989	1,854	..	Gross Domestic Saving
1,104	1,593	1,181	583	554	560	493	297	2,225	2,333	..	Gross National Saving
			(Millions of 1980 Salvadoran Colones)								
9,730.6	9,673.6	8,789.1	7,992.3	7,486.6	7,530.9	7,710.9	7,853.8	7,886.5	8,091.7	..	Gross National Product
9,068.9	8,773.8	8,274.6	7,562.7	..	..	..	..	..	..	B	GDP at factor cost
2,525.9	2,616.4	2,480.2	2,322.1	2,213.3	2,143.2	2,214.5	2,188.0	2,122.5	2,175.6	..	Agriculture
2,146.6	2,085.8	1,845.7	1,658.9	1,543.0	1,580.5	1,589.6	1,653.9	1,697.1	1,767.5	..	Industry
1,580.0	1,500.7	1,339.4	1,199.6	1,098.8	1,120.7	1,135.4	1,177.9	1,207.1	1,243.0	..	Manufacturing
5,229.4	5,041.4	4,590.7	4,206.9	3,979.7	4,062.0	4,166.6	4,275.4	4,337.5	4,424.2	..	Services, etc.
-1,501.8	-302.8	81.7	-52.5	10.0	216.0	55.3	-52.4	-282.0	-70.2	..	Resource Balance
2,617.3	3,563.6	3,046.1	2,593.3	2,291.1	2,563.9	2,451.1	2,356.6	2,059.8	2,259.5	..	Exports of Goods & NFServices
4,119.1	3,866.4	2,964.4	2,645.9	2,281.1	2,347.9	2,395.8	2,409.0	2,341.8	2,329.7	..	Imports of Goods & NFServices
11,403.7	10,046.4	8,834.9	8,240.4	7,726.0	7,569.6	7,915.4	8,169.7	8,439.1	8,437.5	..	Domestic Absorption
7,858.0	6,985.1	6,404.5	5,811.4	5,410.5	5,335.9	5,591.2	5,807.3	5,825.8	5,629.2	..	Private Consumption, etc.
1,289.0	1,320.9	1,247.4	1,291.4	1,287.6	1,298.8	1,361.4	1,453.5	1,507.9	1,528.8	..	General Gov't Consumption
2,256.6	1,740.3	1,183.0	1,137.6	1,027.9	934.9	962.8	908.8	1,105.5	1,279.5	..	Gross Domestic Investment
1,992.7	1,681.2	1,210.1	1,079.4	977.1	898.3	919.3	1,013.0	1,089.2	1,177.7	..	Fixed Investment
											Memo Items:
3,153.8	3,848.6	3,046.1	2,166.8	1,889.1	1,922.4	1,826.4	1,799.4	2,085.7	1,763.0	..	Capacity to Import
536.5	285.0	0.0	-426.6	-402.0	-641.5	-624.8	-557.2	25.9	-496.4	..	Terms of Trade Adjustment
10,438.4	10,028.6	8,916.6	7,761.3	7,334.0	7,144.1	7,345.9	7,560.1	8,183.0	7,870.9	..	Gross Domestic Income
10,267.1	9,958.6	8,789.1	7,565.7	7,084.6	6,889.3	7,086.2	7,296.6	7,912.3	7,595.2	..	Gross National Income
			(Index 1980 = 100)								**DOMESTIC PRICES (DEFLATORS)**
77.7	88.3	100.0	105.6	115.9	130.4	146.3	176.5	242.3	284.0	..	Overall (GDP)
73.7	85.8	100.0	111.3	121.9	141.4	157.3	188.7	241.7	299.0	..	Domestic Absorption
81.1	95.9	100.0	90.7	93.8	100.8	104.8	119.3	188.3	150.8	..	Agriculture
77.5	88.7	100.0	111.4	122.9	137.6	156.8	189.8	238.7	301.1	..	Industry
76.2	89.1	100.0	113.3	125.8	140.3	161.8	199.1	253.4	323.2	..	Manufacturing
											MANUFACTURING ACTIVITY
148.9	104.4	100.0	93.9	87.7	89.5	..	..	..	..	G J	Employment (1980=100)
83.6	101.3	100.0	94.2	92.9	89.7	..	..	..	..	G J	Real Earnings per Empl. (1980=100)
86.3	101.0	100.0	96.2	89.4	92.1	..	..	..	..	G J	Real Output per Empl. (1980=100)
25.2	29.0	31.4	30.1	32.4	28.2	..	..	..	..	J	Earnings as % of Value Added
			(Millions of current Salvadoran Colones)								**MONETARY HOLDINGS**
2,241	2,446	2,563	2,834	3,316	3,637	4,366	5,556	7,194	7,756	..	Money Supply, Broadly Defined
1,087	1,321	1,429	1,437	1,717	1,657	1,961	2,488	3,047	3,147	..	Money as Means of Payment
500	743	719	703	732	724	836	1,080	1,156	1,298	..	Currency Ouside Banks
586	578	710	734	985	933	1,124	1,408	1,891	1,849	..	Demand Deposits
1,155	1,125	1,135	1,397	1,600	1,979	2,405	3,068	4,147	4,609	..	Quasi-Monetary Liabilities
			(Millions of current Salvadoran Colones)								**GOVERNMENT DEFICIT (-) OR SURPLUS**
-107	-86	-512	-648	-680	-543	-521	-107	..	..	..	Current Revenue
1,010	1,188	1,015	1,085	1,086	1,210	1,698	2,439	..	..	..	Current Revenue
755	834	1,046	1,183	1,294	1,298	1,708	1,834	2,465	..	..	Current Expenditure
255	354	-31	-98	-208	-88	-10	606	..	..	..	Current Budget Balance
..	1	1	0	..	3	4	0	2	..	..	Capital Receipts
362	441	483	550	472	458	515	713	269	..	..	Capital Payments

EL SALVADOR	1967	1968	1969	1970	1971	1972	1973	1974	1975	1976	1977
FOREIGN TRADE (CUSTOMS BASIS)					*(Millions of current US dollars)*						
Value of Exports, fob	207.2	211.7	202.1	236.2	243.2	301.7	358.3	462.5	531.4	743.3	972.8
Nonfuel Primary Products	144.5	140.3	135.1	169.7	168.7	217.1	247.0	318.1	386.5	563.4	770.5
Fuels	4.2	3.8	2.9	1.1	1.1	1.6	1.9	3.9	6.3	6.0	6.6
Manufactures	58.5	67.6	64.2	65.5	73.4	83.0	109.3	140.5	138.5	174.0	195.6
Value of Imports, cif	223.9	213.5	209.3	213.6	247.4	278.1	373.8	563.4	598.0	717.9	942.4
Nonfuel Primary Products	41.9	46.7	41.2	38.9	41.0	42.2	63.1	88.6	94.5	110.1	135.5
Fuels	10.9	12.1	8.8	5.2	13.1	12.8	21.5	52.4	50.3	52.6	92.3
Manufactures	171.0	154.7	159.2	169.4	193.4	223.1	289.2	422.4	453.2	555.1	714.7
					(Index 1980 = 100)						
Terms of Trade	101.3	110.1	102.4	111.7	111.6	106.1	115.3	93.9	84.7	131.8	165.9
Export Prices, fob	24.5	25.3	26.3	33.1	31.5	34.1	44.6	52.5	51.1	81.0	108.8
Nonfuel Primary Products	26.1	26.3	26.8	33.1	30.6	33.8	44.0	48.1	47.5	87.8	127.3
Fuels	4.3	4.3	4.3	4.3	5.6	6.2	8.9	36.7	35.7	38.4	42.0
Manufactures	30.1	31.7	32.4	37.5	36.5	38.5	49.5	67.2	66.1	66.7	71.7
Import Prices, cif	24.2	23.0	25.7	29.7	28.3	32.2	38.7	55.9	60.3	61.5	65.6
Trade at Constant 1980 Prices					*(Millions of 1980 US dollars)*						
Exports, fob	846.7	835.4	768.9	712.8	771.3	883.8	803.9	880.4	1,040.3	917.6	893.9
Imports, cif	927.1	927.3	815.3	720.3	875.3	864.1	967.0	1,007.6	991.5	1,167.7	1,436.9
BALANCE OF PAYMENTS					*(Millions of current US dollars)*						
Exports of Goods & Services	..	..	..	261	271	339	402	519	600	900	1,126
Merchandise, fob	..	..	..	236	244	302	358	465	533	745	974
Nonfactor Services	..	..	..	20	24	34	40	49	60	117	115
Factor Services	..	..	..	5	4	4	4	5	7	38	38
Imports of Goods & Services	..	..	..	266	303	339	460	671	720	905	1,135
Merchandise, fob	..	..	..	195	226	250	340	522	551	681	861
Nonfactor Services	..	..	..	57	63	73	100	109	123	171	209
Factor Services	..	..	..	14	14	16	21	40	47	53	66
Long-Term Interest	..	..	..	9	10	11	15	17	19	26	31
Current Transfers, net	..	..	..	12	16	9	12	17	25	24	30
Workers' Remittances	..	..	..	0	0	0	0	0	0	0	34
Total to be Financed	..	..	..	7	-15	9	-46	-135	-95	19	22
Official Capital Grants	..	..	..	2	1	3	2	1	2	5	9
Current Account Balance	..	..	..	9	-14	12	-44	-134	-93	24	31
Long-Term Capital, net	..	..	..	8	22	65	39	194	141	78	48
Direct Investment	..	..	..	4	7	7	6	20	13	13	19
Long-Term Loans	..	..	..	9	19	47	3	94	52	74	42
Disbursements	..	..	..	31	49	74	47	141	132	135	137
Repayments	..	..	..	22	30	27	43	46	80	61	95
Other Long-Term Capital	..	..	..	-7	-5	9	28	79	73	-14	-22
Other Capital, net	..	..	..	-14	-15	-63	-5	-45	-18	-17	-37
Change in Reserves	..	..	..	-2	7	-15	10	-15	-30	-84	-41
Memo Item:					*(Salvadoran Colones per US dollar)*						
Conversion Factor (Annual Avg)	2.500	2.500	2.500	2.500	2.500	2.500	2.500	2.500	2.500	2.500	2.500
EXTERNAL DEBT, ETC.					*(Millions of US dollars, outstanding at end of year)*						
Public/Publicly Guar. Long-Term	..	..	..	88	93	109	107	176	196	263	266
Official Creditors	..	..	..	70	75	88	95	104	145	189	240
IBRD and IDA	..	..	..	33	32	36	40	45	58	67	75
Private Creditors	..	..	..	17	18	22	12	72	51	74	26
Private Non-guaranteed Long-Term	..	..	..	88	102	133	138	163	196	203	241
Use of Fund Credit	..	..	..	7	11	10	0	22	21	15	0
Short-Term Debt	..	..	..	..	..	..	..	..	..	..	216
Memo Items:					*(Millions of US dollars)*						
Int'l Reserves Excluding Gold	37.0	44.2	46.5	45.4	46.2	63.9	41.3	77.6	107.0	185.4	211.3
Gold Holdings (at market price)	17.9	21.3	17.4	18.5	21.5	31.5	54.5	90.6	68.1	65.5	82.6
SOCIAL INDICATORS											
Total Fertility Rate	6.6	6.5	6.4	6.3	6.2	6.1	6.0	5.9	5.9	5.8	5.7
Crude Birth Rate	45.5	44.9	44.4	43.9	43.3	42.8	42.5	42.2	41.9	41.7	41.4
Infant Mortality Rate	112.0	109.0	106.0	103.0	100.0	97.0	94.0	91.0	88.0	85.0	82.0
Life Expectancy at Birth	55.9	56.5	57.1	57.6	58.2	58.7	58.5	58.2	57.9	57.7	57.4
Food Production, p.c. ('79-81 = 100)	91.9	93.6	86.5	95.6	97.4	87.2	99.7	96.7	105.6	102.1	100.5
Labor Force, Agriculture (%)	57.6	57.1	56.5	56.0	54.7	53.4	52.0	50.7	49.4	48.2	46.9
Labor Force, Female (%)	19.3	19.7	20.0	20.4	20.8	21.3	21.7	22.2	22.6	23.1	23.5
School Enroll. Ratio, primary	..	..	..	85.0	..	..	..	..	75.0	..	76.0
School Enroll. Ratio, secondary	..	..	..	22.0	..	..	..	..	19.0	21.0	23.0

1978	1979	1980	1981	1982	1983	1984	1985	1986	1987 estimate	NOTES	EL SALVADOR
											FOREIGN TRADE (CUSTOMS BASIS)
				(Millions of current US dollars)							
801.0	1,131.0	1,074.0	797.0	699.0	736.0	725.4	679.0	756.6	577.0	..	Value of Exports, fob
578.0	871.0	799.1	594.5	524.2	553.8	544.8	490.7	577.7	394.3	..	Nonfuel Primary Products
5.8	8.7	20.1	21.3	12.4	14.2	13.4	13.9	4.4	5.7	..	Fuels
217.2	251.3	254.7	181.2	162.3	168.0	167.2	174.4	174.4	177.0	..	Manufactures
1,023.9	1,012.0	975.9	1,044.5	944.8	892.0	977.4	961.4	902.4	1,035.0	..	Value of Imports, cif
158.9	173.6	210.2	214.2	200.1	188.0	177.6	167.8	166.6	169.6	..	Nonfuel Primary Products
80.1	95.4	173.2	218.0	232.5	206.7	156.7	146.4	56.3	79.0	..	Fuels
785.0	743.0	592.5	612.3	512.3	497.3	643.1	647.2	679.6	786.4	..	Manufactures
				(Index 1980 = 100)							
117.8	114.9	100.0	85.5	91.5	91.8	98.8	96.2	112.6	75.3	..	Terms of Trade
88.1	101.0	100.0	87.5	90.0	87.3	91.4	88.1	111.6	81.9	..	Export Prices, fob
94.8	105.4	100.0	83.8	88.3	86.1	92.1	87.8	115.7	73.4	..	Nonfuel Primary Products
42.3	61.0	100.0	112.5	101.6	92.5	90.2	87.5	44.9	56.9	..	Fuels
76.1	90.0	100.0	99.1	94.8	90.7	89.3	89.0	103.2	113.0	..	Manufactures
74.8	87.9	100.0	102.3	98.4	95.1	92.6	91.6	99.1	108.9	..	Import Prices, cif
											Trade at Constant 1980 Prices
				(Millions of 1980 US dollars)							
908.9	1,120.0	1,074.1	911.3	776.9	843.5	793.6	770.8	678.0	704.2	..	Exports, fob
1,369.2	1,151.7	975.9	1,020.8	960.7	938.4	1,056.0	1,049.4	910.2	950.7	..	Imports, cif
				(Millions of current US dollars)							**BALANCE OF PAYMENTS**
960	1,356	1,271	970	872	908	954	952	1,043	928	f	Exports of Goods & Services
802	1,132	1,075	798	704	735	726	679	756	573	f	Merchandise, fob
121	134	140	126	119	138	168	227	270	338	f	Nonfactor Services
37	90	55	46	50	35	61	46	17	17	f	Factor Services
1,297	1,386	1,289	1,281	1,196	1,217	1,315	1,324	1,296	1,347	f	Imports of Goods & Services
951	955	897	898	826	831	915	895	935	975	f	Merchandise, fob
257	301	274	263	215	230	239	289	217	228	f	Nonfactor Services
89	130	119	120	155	156	162	139	144	144	f	Factor Services
37	41	36	39	44	68	74	76	74	76	..	Long-Term Interest
45	45	17	39	52	97	118	129	174	223	..	Current Transfers, net
45	49	11	42	41	93	114	126	..	..	..	Workers' Remittances
-292	15	-1	-272	-271	-211	-243	-243	-79	-196	K	Total to be Financed
7	7	32	21	119	174	190	214	223	323	K	Official Capital Grants
-286	21	31	-250	-152	-37	-54	-29	144	127	..	Current Account Balance
182	85	206	207	308	491	273	314	240	379	..	Long-Term Capital, net
23	-10	6	-6	-1	28	12	12	-36	-41	f	Direct Investment
74	27	100	199	231	317	127	19	-11	1	f	Long-Term Loans
135	90	135	236	285	425	257	157	126	120	..	Disbursements
61	63	35	38	55	108	129	138	137	119	..	Repayments
78	61	69	-7	-40	-29	-56	68	63	96	..	Other Long-Term Capital
159	-240	-317	-47	-149	-425	-227	-264	-340	-449	..	Other Capital, net
-55	134	81	91	-7	-29	7	-21	-43	-56	..	Change in Reserves
											Memo Item:
				(Salvadoran Colones per US dollar)							
2.500	2.500	2.500	2.500	2.570	2.730	2.820	3.670	5.000	5.000	..	Conversion Factor (Annual Avg)
											EXTERNAL DEBT, ETC.
			(Millions of US dollars, outstanding at end of year)								
330	408	524	724	972	1,340	1,409	1,478	1,497	1,597	..	Public/Publicly Guar. Long-Term
309	391	512	713	881	1,123	1,216	1,328	1,386	1,517	..	Official Creditors
86	103	114	125	132	133	121	145	165	191	..	IBRD and IDA
21	17	12	11	91	217	193	150	111	81	..	Private Creditors
251	201	161	147	134	122	114	104	83	70	..	Private Non-guaranteed Long-Term
0	0	7	44	107	118	105	89	43	6	..	Use of Fund Credit
329	278	220	212	206	94	102	82	90	89	..	Short-Term Debt
											Memo Items:
				(Millions of US dollars)							
268.1	142.6	77.7	71.9	108.5	160.2	165.8	179.6	169.7	186.1	..	Int'l Reserves Excluding Gold
113.2	260.2	304.0	205.0	235.6	178.8	144.7	153.4	183.4	227.1	..	Gold Holdings (at market price)
											SOCIAL INDICATORS
5.6	5.5	5.4	5.3	5.2	5.1	5.1	5.0	4.9	4.9	..	Total Fertility Rate
40.7	40.0	39.3	38.6	38.0	37.7	37.4	37.0	36.7	36.4	..	Crude Birth Rate
79.6	77.2	74.8	72.4	70.0	68.4	66.9	65.3	63.8	62.2	..	Infant Mortality Rate
57.4	57.3	57.3	57.2	57.2	58.2	59.2	60.2	61.2	62.2	..	Life Expectancy at Birth
110.5	109.4	99.9	90.7	82.0	83.9	91.7	88.6	88.7	88.6	..	Food Production, p.c. ('79-81 = 100)
45.7	44.4	43.2	..	..	..	..	..	..	..	..	Labor Force, Agriculture (%)
24.0	24.4	24.9	24.9	25.0	25.0	25.0	25.1	25.1	25.1	..	Labor Force, Female (%)
78.0	79.0	74.0	61.0	68.0	69.0	70.0	70.0	..	..	..	School Enroll. Ratio, primary
25.0	26.0	23.0	20.0	23.0	23.0	24.0	24.0	..	..	..	School Enroll. Ratio, secondary

ETHIOPIA	1967	1968	1969	1970	1971	1972	1973	1974	1975	1976	1977
CURRENT GNP PER CAPITA (US $)	50	60	60	60	70	70	70	80	90	90	90
POPULATION (thousands)	26,765	27,471	28,194	28,937	29,699	30,481	31,284	32,108	32,954	33,856	34,782

ORIGIN AND USE OF RESOURCES *(Millions of current Ethiopian Birr)*

	1967	1968	1969	1970	1971	1972	1973	1974	1975	1976	1977
Gross National Product (GNP)	3,594	3,820	4,030	4,441	4,691	4,712	4,955	5,513	5,516	5,993	6,851
Net Factor Income from Abroad	-11	-18	-26	-20	-19	-32	-50	-39	-35	-3	-4
Gross Domestic Product (GDP)	3,606	3,837	4,056	4,461	4,710	4,744	5,005	5,551	5,551	5,996	6,855
Indirect Taxes, net	232	236	237	288	301	327	376	410	421	474	690
GDP at factor cost	3,373	3,602	3,819	4,173	4,409	4,417	4,629	5,141	5,130	5,522	6,164
Agriculture	1,903	2,011	2,120	2,327	2,405	2,286	2,331	2,605	2,450	2,768	3,227
Industry	527	548	588	602	672	707	748	796	855	824	892
Manufacturing	273	302	337	372	420	440	464	508	569	586	636
Services, etc.	944	1,043	1,112	1,244	1,333	1,424	1,550	1,739	1,825	1,930	2,046
Resource Balance	-95	-73	-87	-14	-88	-93	101	189	-164	-58	-216
Exports of Goods & NFServices	396	399	400	489	468	487	653	827	683	760	840
Imports of Goods & NFServices	492	473	487	504	556	580	552	638	847	818	1,056
Domestic Absorption	3,701	3,911	4,143	4,475	4,798	4,837	4,905	5,362	5,715	6,054	7,071
Private Consumption, etc.	2,800	2,969	3,188	3,519	3,783	3,726	3,798	4,227	4,405	4,610	5,497
General Gov't Consumption	384	403	411	443	461	508	538	586	730	866	968
Gross Domestic Investment	517	538	544	512	554	603	569	549	580	578	606
Fixed Investment	506	527	533	512	554	603	569	549	580	578	606
Memo Items:											
Gross Domestic Saving	421	465	457	498	467	510	670	739	416	520	390
Gross National Saving	400	441	424	472	443	485	644	738	411	561	420

(Millions of 1980 Ethiopian Birr)

	1967	1968	1969	1970	1971	1972	1973	1974	1975	1976	1977
Gross National Product	5,672.4	5,712.9	5,870.1	6,288.0	6,560.3	6,827.6	6,956.6	7,127.5	7,159.5	7,516.1	7,718.4
GDP at factor cost	5,357.8	5,447.4	5,649.8	5,971.7	6,223.1	6,502.0	6,663.5	6,757.0	6,749.9	6,909.8	6,925.1
Agriculture	3,263.1	3,291.1	3,361.0	3,439.6	3,509.0	3,633.8	3,660.0	3,639.9	3,569.6	3,667.7	3,654.9
Industry	765.1	726.1	778.6	871.7	955.0	997.7	1,027.0	1,019.7	1,005.0	966.5	978.7
Manufacturing	426.5	397.2	442.8	567.5	619.7	645.1	677.2	670.4	662.7	652.3	657.4
Services, etc.	1,329.7	1,430.2	1,510.2	1,660.5	1,759.2	1,870.6	1,976.5	2,097.4	2,175.3	2,275.6	2,291.4
Resource Balance	-169.5	-133.8	-154.7	-190.6	-182.3	-202.9	25.1	158.6	-184.7	-25.1	-225.2
Exports of Goods & NFServices	804.1	782.2	781.1	766.3	823.4	815.8	945.4	1,126.7	1,038.6	1,133.6	1,159.1
Imports of Goods & NFServices	973.6	916.0	935.8	956.9	1,005.7	1,018.7	920.2	968.1	1,223.3	1,158.7	1,384.2
Domestic Absorption	5,908.4	5,944.7	6,164.3	6,585.8	6,843.9	7,202.5	7,198.7	7,157.3	7,513.0	7,553.5	7,959.1
Private Consumption, etc.	4,418.3	4,449.3	4,677.7	5,163.5	5,350.3	5,493.7	5,534.2	5,601.7	5,709.7	5,447.5	5,828.1
General Gov't Consumption	605.8	604.3	603.0	630.3	648.1	746.5	774.4	770.8	968.6	1,092.7	1,096.4
Gross Domestic Investment	884.4	891.0	883.5	792.0	845.5	962.3	890.1	784.7	834.7	1,013.3	1,034.6
Fixed Investment	..	..	..	..	..	..	..	..	834.7	1,013.3	1,034.6
Memo Items:											
Capacity to Import	784.7	774.1	768.3	929.6	847.1	855.3	1,088.1	1,255.6	986.3	1,076.6	1,100.7
Terms of Trade Adjustment	-19.3	-8.1	-12.8	163.2	23.6	39.5	142.7	128.9	-52.4	-57.1	-58.4
Gross Domestic Income	5,719.6	5,802.8	5,996.8	6,558.5	6,685.2	7,039.2	7,366.6	7,444.8	7,276.0	7,471.4	7,675.6
Gross National Income	5,653.1	5,704.8	5,857.3	6,451.2	6,584.0	6,867.1	7,099.3	7,256.5	7,107.1	7,459.0	7,660.0

DOMESTIC PRICES (DEFLATORS) *(Index 1980 = 100)*

	1967	1968	1969	1970	1971	1972	1973	1974	1975	1976	1977
Overall (GDP)	62.8	66.0	67.5	69.7	70.7	67.8	69.3	75.9	75.7	79.6	88.6
Domestic Absorption	62.6	65.8	67.2	67.9	70.1	67.2	68.1	74.9	76.1	80.1	88.8
Agriculture	58.3	61.1	63.1	67.7	68.5	62.9	63.7	71.6	68.6	75.5	88.3
Industry	68.9	75.4	75.5	69.0	70.3	70.8	72.9	78.1	85.1	85.3	91.1
Manufacturing	64.1	76.0	76.1	65.6	67.7	68.2	68.5	75.7	85.9	89.8	96.7

MANUFACTURING ACTIVITY

	1967	1968	1969	1970	1971	1972	1973	1974	1975	1976	1977
Employment (1980=100)	61.1	63.1	64.1	65.1	67.4	70.0	71.9	75.2	78.7	77.8	82.2
Real Earnings per Empl. (1980=100)	134.7	138.1	148.1	139.5	147.2	160.8	156.4	151.7	146.8	128.3	119.8
Real Output per Empl. (1980=100)	43.7	39.0	45.5	61.2	65.2	69.6	74.5	74.7	63.9	69.0	66.0
Earnings as % of Value Added	27.5	27.9	26.8	23.8	24.0	23.5	23.6	23.3	21.7	23.0	25.3

MONETARY HOLDINGS *(Millions of current Ethiopian Birr)*

	1967	1968	1969	1970	1971	1972	1973	1974	1975	1976	1977
Money Supply, Broadly Defined	472	525	605	640	651	756	982	1,135	1,244	1,403	1,652
Money as Means of Payment	372	400	455	453	437	491	619	754	942	953	1,179
Currency Ouside Banks	251	274	319	323	304	340	404	533	689	575	769
Demand Deposits	121	126	136	130	133	151	215	221	253	378	409
Quasi-Monetary Liabilities	100	125	150	187	215	265	363	381	302	450	473

GOVERNMENT DEFICIT (-) OR SURPLUS *(Millions of current Ethiopian Birr)*

	1967	1968	1969	1970	1971	1972	1973	1974	1975	1976	1977
GOVERNMENT DEFICIT (-) OR SURPLUS	..	..	..	..	..	-64	-49	-44	-226	-325	-233
Current Revenue	..	..	..	..	..	586	651	716	805	853	1,094
Current Expenditure	..	..	..	..	..	545	581	647	848	967	1,091
Current Budget Balance	..	..	..	..	..	41	70	69	-43	-114	3
Capital Receipts	..	..	..	..	..	1	2	1	2	2	3
Capital Payments	..	..	..	..	..	106	121	115	185	214	238

1978	1979	1980	1981	1982	1983	1984	1985	1986	1987 estimate	NOTES	ETHIOPIA
100	110	120	120	120	120	110	110	120	130	..	**CURRENT GNP PER CAPITA (US $)**
35,734	36,712	37,717	38,784	39,881	41,009	42,169	42,271	43,498	44,788	..	**POPULATION (thousands)**
				(Millions of current Ethiopian Birr)							**ORIGIN AND USE OF RESOURCES**
7,295	8,012	8,547	8,892	9,151	10,003	9,902	9,931	10,763	11,058	C	Gross National Product (GNP)
0	24	49	-11	-16	-28	-99	50	-60	-77	..	Net Factor Income from Abroad
7,295	7,987	8,499	8,903	9,167	10,031	10,001	9,881	10,823	11,134	C	Gross Domestic Product (GDP)
749	901	874	806	871	948	1,057	979	1,115	1,196	..	Indirect Taxes, net
6,546	7,086	7,625	8,097	8,297	9,083	8,944	8,902	9,708	9,938	C	GDP at factor cost
3,497	3,656	3,872	4,072	4,062	4,389	4,070	3,916	4,354	4,204	..	Agriculture
894	1,046	1,189	1,258	1,297	1,402	1,476	1,486	1,586	1,790	..	Industry
636	758	830	871	902	984	1,009	1,023	1,073	1,167	..	Manufacturing
2,154	2,383	2,563	2,767	2,938	3,292	3,397	3,500	3,767	3,944	..	Services, etc.
-405	-425	-440	-448	-785	-827	-985	-856	-1,097	-1,244	..	Resource Balance
866	943	1,178	1,147	1,059	1,142	1,267	1,136	1,371	1,276	..	Exports of Goods & NFServices
1,271	1,368	1,619	1,595	1,844	1,969	2,251	1,992	2,468	2,520	..	Imports of Goods & NFServices
7,700	8,413	8,939	9,351	9,952	10,858	10,986	10,737	11,919	12,379	..	Domestic Absorption
5,915	6,546	6,792	7,030	7,384	8,006	7,911	7,682	8,661	8,626	f	Private Consumption, etc.
1,240	1,168	1,293	1,400	1,487	1,732	1,841	1,924	2,045	2,142	..	General Gov't Consumption
545	699	854	922	1,082	1,119	1,234	1,131	1,213	1,610	f	Gross Domestic Investment
545	699	854	922	1,082	1,119	1,234	1,131	1,213	1,610	..	Fixed Investment
											Memo Items:
140	274	414	474	297	293	249	275	116	366	..	Gross Domestic Saving
171	346	504	514	374	440	372	625	490	626	..	Gross National Saving
				(Millions of 1980 Ethiopian Birr)							
7,648.3	8,122.4	8,547.4	8,636.5	8,757.6	9,168.4	8,735.2	8,579.9	8,791.9	9,445.7	C	Gross National Product
6,832.9	7,219.2	7,624.7	7,858.7	7,951.2	8,375.0	8,068.2	7,497.0	7,999.3	8,640.6	C	GDP at factor cost
3,607.9	3,694.9	3,871.9	3,966.9	3,916.0	4,100.8	3,695.1	3,094.4	3,380.0	3,758.5	..	Agriculture
947.2	1,085.9	1,189.4	1,230.2	1,268.2	1,334.3	1,404.6	1,406.6	1,458.6	1,556.8	..	Industry
657.8	772.2	830.0	866.9	900.6	951.3	983.0	993.2	1,036.8	1,087.4	..	Manufacturing
2,277.8	2,438.4	2,563.4	2,661.6	2,767.0	2,939.9	2,968.6	2,996.0	3,160.7	3,325.3	..	Services, etc.
-400.5	-376.5	-440.3	-373.9	-812.2	-798.6	-805.0	-1,073.8	-1,426.7	..	..	Resource Balance
1,016.7	995.6	1,178.2	1,166.9	1,114.9	1,219.1	1,415.8	1,129.3	1,054.9	..	..	Exports of Goods & NFServices
1,417.2	1,372.1	1,618.5	1,540.8	1,927.1	2,017.7	2,220.8	2,203.1	2,481.6	..	..	Imports of Goods & NFServices
8,049.5	8,517.9	8,939.2	9,049.0	9,626.4	10,061.9	9,864.5	9,492.2	10,409.1	..	..	Domestic Absorption
6,116.8	6,582.3	6,792.0	6,733.4	7,068.5	7,272.4	6,952.9	6,855.2	7,638.6	..	f	Private Consumption, etc.
1,304.8	1,190.4	1,293.2	1,369.8	1,435.5	1,609.6	1,687.7	1,706.3	1,764.3	..	..	General Gov't Consumption
627.9	745.2	854.0	945.7	1,122.4	1,179.8	1,224.0	930.7	1,006.1	..	f	Gross Domestic Investment
627.9	745.2	854.0	945.7	1,122.4	1,179.8	1,224.0	930.7	1,006.1	..	..	Fixed Investment
											Memo Items:
965.5	945.5	1,178.2	1,108.1	1,106.7	1,170.5	1,249.5	1,256.8	1,378.6	..	..	Capacity to Import
-51.2	-50.1	0.0	-58.8	-8.2	-48.6	-166.3	127.5	323.7	..	..	Terms of Trade Adjustment
7,597.8	8,091.3	8,498.9	8,616.3	8,806.0	9,214.7	8,893.1	8,546.0	9,306.1	..	..	Gross Domestic Income
7,597.1	8,072.2	8,547.4	8,577.7	8,749.4	9,119.8	8,568.9	8,707.4	9,115.6	..	..	Gross National Income
				(Index 1980 = 100)							**DOMESTIC PRICES (DEFLATORS)**
95.4	98.1	100.0	102.6	104.0	108.3	110.4	117.4	120.5	114.8	..	Overall (GDP)
95.7	98.8	100.0	103.3	103.4	107.9	111.4	113.1	114.5	..	..	Domestic Absorption
96.9	99.0	100.0	102.6	103.7	107.0	110.2	126.5	128.8	111.8	..	Agriculture
94.4	96.4	100.0	102.3	102.2	105.1	105.1	105.7	108.7	115.0	..	Industry
96.7	98.2	100.0	100.5	100.1	103.5	102.7	103.0	103.5	107.4	..	Manufacturing
											MANUFACTURING ACTIVITY
86.5	98.2	100.0	103.3	107.0	109.6	117.9	117.8	..	..	G J	Employment (1980=100)
107.6	97.0	100.0	96.9	95.2	101.2	93.7	78.6	..	..	G J	Real Earnings per Empl. (1980=100)
63.3	80.6	100.0	105.0	111.9	110.4	108.9	112.6	..	..	G J	Real Output per Empl. (1980=100)
25.3	20.6	18.0	18.7	19.0	19.8	19.1	19.3	..	..	J	Earnings as % of Value Added
				(Millions of current Ethiopian Birr)							**MONETARY HOLDINGS**
1,861	2,108	2,196	2,438	2,689	3,198	3,449	3,994	4,468	4,754	..	Money Supply, Broadly Defined
1,378	1,572	1,568	1,720	1,892	2,142	2,309	2,702	3,273	3,341	..	Money as Means of Payment
895	1,012	1,029	1,039	1,150	1,251	1,272	1,418	1,640	1,744	..	Currency Ouside Banks
484	559	539	681	742	892	1,037	1,285	1,633	1,597	..	Demand Deposits
483	536	628	718	797	1,056	1,140	1,292	1,195	1,413	..	Quasi-Monetary Liabilities
				(Millions of current Ethiopian Birr)							
-422	-255	-381	-334	..	..	..	..	..	..	C	**GOVERNMENT DEFICIT (-) OR SURPLUS**
1,253	1,601	1,764	1,975	..	..	..	..	..	..	..	Current Revenue
1,406	1,596	1,859	1,970	..	..	..	..	..	..	..	Current Expenditure
-153	6	-94	5	..	..	..	..	..	..	..	Current Budget Balance
3	3	7	8	..	..	..	..	..	..	..	Capital Receipts
272	263	293	346	..	..	..	..	..	..	..	Capital Payments

ETHIOPIA	1967	1968	1969	1970	1971	1972	1973	1974	1975	1976	1977
FOREIGN TRADE (CUSTOMS BASIS)				*(Millions of current US dollars)*							
Value of Exports, fob	101.1	103.3	119.4	122.3	125.6	168.4	236.7	266.7	229.4	278.3	333.6
Nonfuel Primary Products	99.8	100.7	114.8	114.6	119.6	161.0	225.8	251.5	202.3	265.6	324.8
Fuels	0.0	0.6	1.4	1.4	2.0	1.1	1.6	4.7	9.9	7.0	6.2
Manufactures	1.3	2.0	3.2	6.3	4.0	6.3	9.3	10.5	17.1	5.8	2.5
Value of Imports, cif	143.1	173.1	155.3	173.1	187.7	189.4	213.1	272.9	294.0	353.2	391.4
Nonfuel Primary Products	18.8	20.3	18.0	21.4	21.9	20.4	24.8	28.7	28.3	32.1	33.2
Fuels	12.6	10.8	11.3	13.4	17.7	15.7	20.0	38.7	51.3	53.3	62.0
Manufactures	111.7	142.0	126.0	138.2	148.1	153.3	168.3	205.6	214.4	267.9	296.2
					(Index 1980 = 100)						
Terms of Trade	129.5	113.3	118.3	141.8	120.2	126.2	139.8	87.5	82.4	142.2	198.0
Export Prices, fob	26.5	25.7	26.3	31.9	29.0	35.1	47.5	47.5	46.4	83.1	125.3
Nonfuel Primary Products	26.5	26.5	27.9	33.8	31.0	36.7	48.6	47.2	45.9	86.2	130.7
Fuels	4.3	4.3	4.3	4.3	5.6	6.2	8.9	36.7	35.7	38.4	42.0
Manufactures	27.2	26.5	35.8	54.3	36.4	28.5	58.9	63.6	66.2	68.9	84.8
Import Prices, cif	20.5	22.7	22.3	22.5	24.1	27.8	34.0	54.2	56.3	58.5	63.3
Trade at Constant 1980 Prices				*(Millions of 1980 US dollars)*							
Exports, fob	381.6	402.1	453.3	384.0	433.5	479.5	498.1	562.0	494.3	334.9	266.3
Imports, cif	699.6	763.8	697.7	770.3	778.6	680.6	627.1	503.4	521.8	604.3	618.7
BALANCE OF PAYMENTS				*(Millions of current US dollars)*							
Exports of Goods & Services	..	..	..	185	194	247	348	390	346	385	422
Merchandise, fob	..	..	..	122	126	166	239	267	230	260	312
Nonfactor Services	..	..	..	57	63	78	100	103	100	107	93
Factor Services	..	..	..	7	5	4	9	20	16	17	17
Imports of Goods & Services	..	..	..	226	249	257	298	387	443	414	529
Merchandise, fob	..	..	..	144	159	158	179	250	276	278	368
Nonfactor Services	..	..	..	67	73	76	91	106	133	117	142
Factor Services	..	..	..	14	16	23	29	32	33	19	19
Long-Term Interest	..	..	..	6	7	8	9	10	11	10	11
Current Transfers, net	..	..	..	-3	-2	3	11	18	15	21	17
Workers' Remittances	..	..	..	0	0	0	0	0	0	0	0
Total to be Financed	..	..	..	-43	-56	-7	61	22	-81	-8	-90
Official Capital Grants	..	..	..	11	11	15	14	34	30	29	43
Current Account Balance	..	..	..	-32	-45	8	75	55	-52	21	-47
Long-Term Capital, net	..	..	..	27	47	47	68	85	98	89	93
Direct Investment	..	..	..	4	6	10	31	29	19	4	6
Long-Term Loans	..	..	..	13	31	20	24	25	61	60	38
Disbursements	..	..	..	28	44	33	37	36	76	74	52
Repayments	..	..	..	15	13	13	13	11	14	14	14
Other Long-Term Capital	..	..	..	0	-1	3	-2	-3	-12	-5	7
Other Capital, net	..	..	..	2	-8	-24	-37	-50	-53	-67	-119
Change in Reserves	..	..	..	3	6	-31	-106	-90	7	-44	74
Memo Item:				*(Ethiopian Birr per US dollar)*							
Conversion Factor (Annual Avg)	2.500	2.500	2.500	2.500	2.490	2.300	2.100	2.070	2.070	2.070	2.070
EXTERNAL DEBT, ETC.				*(Millions of US dollars, outstanding at end of year)*							
Public/Publicly Guar. Long-Term	..	..	..	169	204	226	256	283	344	402	447
Official Creditors	..	..	..	140	179	207	239	265	325	388	435
IBRD and IDA	..	..	..	70	84	97	113	121	143	178	208
Private Creditors	..	..	..	29	25	19	17	18	19	14	12
Private Non-guaranteed Long-Term	..	..	..	..	..	..	..	..	..	..	..
Use of Fund Credit	..	..	..	0	0	0	0	0	0	0	0
Short-Term Debt	..	..	..	..	..	..	..	..	..	..	53
Memo Items:				*(Millions of US dollars)*							
Int'l Reserves Excluding Gold	60.1	58.3	62.4	63.3	59.3	83.2	166.0	263.6	276.7	294.7	213.3
Gold Holdings (at market price)	4.6	9.7	9.5	8.5	10.3	16.2	28.9	51.3	38.6	37.1	47.2
SOCIAL INDICATORS											
Total Fertility Rate	5.8	5.8	5.8	5.8	5.8	5.8	5.8	5.9	5.9	6.0	6.0
Crude Birth Rate	43.4	43.4	43.3	43.3	43.2	43.2	43.2	43.1	43.1	43.0	43.0
Infant Mortality Rate	162.0	160.6	159.2	157.8	156.4	155.0	153.8	152.6	151.4	150.2	149.0
Life Expectancy at Birth	43.0	43.1	43.2	43.3	43.4	43.5	43.7	43.9	44.1	44.3	44.5
Food Production, p.c. ('79-81=100)	107.3	108.8	108.1	108.9	103.5	102.3	101.1	94.1	96.8	94.5	90.0
Labor Force, Agriculture (%)	85.7	85.5	85.2	85.0	84.5	84.0	83.4	82.9	82.4	81.9	81.4
Labor Force, Female (%)	40.1	40.0	40.0	40.0	39.9	39.8	39.8	39.7	39.7	39.6	39.5
School Enroll. Ratio, primary	..	..	..	16.0	..	..	..	..	24.0	25.0	24.0
School Enroll. Ratio, secondary	..	..	..	4.0	..	..	..	..	6.0	8.0	7.0

1978	1979	1980	1981	1982	1983	1984	1985	1986	1987 estimate	NOTES	ETHIOPIA
				(Millions of current US dollars)							**FOREIGN TRADE (CUSTOMS BASIS)**
309.8	422.2	424.8	378.1	404.4	403.0	416.8	333.1	453.0	402.4	..	Value of Exports, fob
294.6	399.4	392.0	343.0	370.0	368.4	388.9	306.8	439.3	385.4	..	Nonfuel Primary Products
12.1	20.4	31.5	29.9	30.9	30.9	25.0	23.7	10.8	13.2	..	Fuels
3.1	2.4	1.2	5.3	3.5	3.7	2.9	2.6	3.0	3.8	..	Manufactures
505.3	567.4	721.4	737.3	785.0	875.0	928.2	993.4	1,214.0	1,201.7	..	Value of Imports, cif
46.8	60.0	80.3	95.3	103.9	113.4	112.3	111.6	102.7	83.2	..	Nonfuel Primary Products
60.3	110.2	178.5	171.0	193.3	213.1	195.1	227.1	165.1	219.6	..	Fuels
398.1	397.2	462.6	470.9	487.8	548.5	620.8	654.7	946.2	898.9	..	Manufactures
				(Index 1980 = 100)							
131.8	132.2	100.0	84.0	89.6	91.6	102.1	99.1	110.2	84.1	..	Terms of Trade
95.4	110.4	100.0	86.6	89.0	87.6	96.2	92.8	114.5	84.0	..	Export Prices, fob
100.9	115.1	100.0	85.0	88.2	87.3	96.9	93.4	119.3	85.3	..	Nonfuel Primary Products
42.3	61.0	100.0	112.5	101.6	92.5	90.2	87.5	44.9	56.9	..	Fuels
74.8	109.9	100.0	80.8	76.2	74.2	72.9	74.9	88.5	97.2	..	Manufactures
72.4	83.5	100.0	103.1	99.3	95.6	94.3	93.6	104.0	99.9	..	Import Prices, cif
											Trade at Constant 1980 Prices
				(Millions of 1980 US dollars)							
324.7	382.5	424.7	436.5	454.5	460.3	433.2	359.0	395.5	478.8	..	Exports, fob
697.8	679.8	721.4	714.8	790.3	915.2	984.6	1,061.7	1,167.6	1,202.7	..	Imports, cif
				(Millions of current US dollars)							**BALANCE OF PAYMENTS**
431	465	591	565	538	578	627	559	679	633	C	Exports of Goods & Services
324	360	459	411	376	392	449	359	446	384	..	Merchandise, fob
94	96	110	143	136	160	163	190	216	232	..	Nonfactor Services
13	10	21	11	26	26	15	10	17	17	..	Factor Services
627	673	797	788	926	991	1,122	1,005	1,238	1,271	C	Imports of Goods & Services
471	496	552	542	683	717	854	719	901	911	..	Merchandise, fob
143	165	230	229	208	235	234	244	292	306	..	Nonfactor Services
13	13	15	18	35	39	34	43	46	54	..	Factor Services
12	13	17	17	22	24	31	35	45	50	..	Long-Term Interest
15	23	20	25	45	85	107	145	210	163	..	Current Transfers, net
0	0	0	0	0	0	0	0	..	..	..	Workers' Remittances
-181	-185	-186	-199	-343	-328	-388	-302	-349	-476	K	Total to be Financed
53	60	60	60	68	93	162	298	293	212	K	Official Capital Grants
-128	-125	-126	-139	-275	-236	-226	-3	-56	-264	..	Current Account Balance
90	130	210	221	377	296	342	461	571	393	C	Long-Term Capital, net
..	..	..	..	..	..	..	..	..	..	..	Direct Investment
62	105	92	287	80	203	201	291	223	273	..	Long-Term Loans
77	120	110	313	113	246	254	363	335	403	..	Disbursements
15	15	17	25	33	43	54	72	112	130	..	Repayments
-26	-35	58	-126	230	1	-21	-129	55	-92	C	Other Long-Term Capital
6	-55	-129	-175	-63	-132	-159	-410	-349	-150	..	Other Capital, net
32	51	45	92	-39	72	42	-48	-165	20	..	Change in Reserves
				(Ethiopian Birr per US dollar)							**Memo Item:**
2.070	2.070	2.070	2.070	2.070	2.070	2.070	2.070	2.070	2.070	..	Conversion Factor (Annual Avg)
				(Millions of US dollars, outstanding at end of year)							**EXTERNAL DEBT, ETC.**
511	616	701	964	1,039	1,225	1,401	1,753	2,038	2,434	..	Public/Publicly Guar. Long-Term
494	592	652	871	937	1,117	1,192	1,464	1,723	2,028	..	Official Creditors
237	280	304	327	350	386	422	486	540	658	..	IBRD and IDA
17	24	49	93	101	109	210	289	314	406	..	Private Creditors
..	..	..	..	..	..	..	..	..	..	..	Private Non-guaranteed Long-Term
0	47	46	114	132	106	75	50	66	63	..	Use of Fund Credit
52	64	56	58	67	62	67	77	83	94	..	Short-Term Debt
				(Millions of US dollars)							**Memo Items:**
152.9	172.7	80.1	266.7	181.8	125.9	44.3	148.0	250.5	144.0	..	Int'l Reserves Excluding Gold
64.6	146.4	182.2	103.4	95.5	79.7	64.4	68.3	81.7	101.2	..	Gold Holdings (at market price)
											SOCIAL INDICATORS
6.0	6.0	5.9	5.9	5.9	6.0	6.1	6.2	6.3	6.4	..	Total Fertility Rate
43.1	43.2	43.3	43.4	43.5	44.4	45.3	46.1	47.0	47.9	..	Crude Birth Rate
151.0	153.0	155.0	157.0	159.0	151.5	144.1	136.6	129.2	121.7	..	Infant Mortality Rate
44.1	43.8	43.5	43.1	42.8	43.6	44.5	45.3	46.2	47.1	..	Life Expectancy at Birth
97.6	105.6	98.9	95.6	101.9	93.0	82.0	89.5	90.7	86.2	..	Food Production, p.c. ('79-81 = 100)
80.8	80.3	79.8	..	..	..	..	..	..	..	..	Labor Force, Agriculture (%)
39.5	39.4	39.3	39.2	39.0	38.8	38.6	38.4	38.2	38.0	..	Labor Force, Female (%)
29.0	37.0	41.0	45.0	40.0	39.0	36.0	36.0	..	..	..	School Enroll. Ratio, primary
9.0	10.0	11.0	12.0	10.0	11.0	11.0	12.0	..	..	..	School Enroll. Ratio, secondary

FIJI	1967	1968	1969	1970	1971	1972	1973	1974	1975	1976	1977
CURRENT GNP PER CAPITA (US $)	340	350	350	400	440	520	660	850	1,030	1,190	1,290
POPULATION (thousands)	486	497	508	520	531	543	554	565	576	588	599

ORIGIN AND USE OF RESOURCES
(Millions of current Fiji Dollars)

	1967	1968	1969	1970	1971	1972	1973	1974	1975	1976	1977
Gross National Product (GNP)	126.6	139.0	151.4	183.6	201.8	252.1	333.7	448.2	556.4	622.5	651.5
Net Factor Income from Abroad	-4.2	-6.8	-7.9	-8.2	-10.1	-9.2	-4.6	-1.8	-6.0	-1.0	-8.6
Gross Domestic Product (GDP)	130.8	145.8	159.3	191.8	211.9	261.3	338.3	450.0	562.4	623.5	660.1
Indirect Taxes, net	13.5	16.2	18.8	22.9	27.3	30.8	37.7	39.5	47.0	52.9	54.4
GDP at factor cost	117.3	129.6	140.5	168.9	184.6	230.5	300.6	410.5	515.4	570.6	605.7
Agriculture	34.7	38.5	38.8	48.2	45.7	57.6	75.7	105.0	132.0	147.0	141.3
Industry	30.2	33.1	33.7	39.8	39.8	52.4	66.6	91.0	115.0	128.0	125.4
Manufacturing	20.0	21.2	20.1	23.7	21.1	27.0	31.6	48.0	60.0	67.0	69.4
Services, etc.	52.4	58.0	68.0	80.9	99.1	120.5	158.3	214.5	268.4	295.6	339.0
Resource Balance	-3.5	-6.5	-6.9	-6.2	-16.5	-24.3	-45.6	-23.5	-3.6	-29.5	-18.2
Exports of Goods & NFServices	58.3	69.1	78.9	92.8	105.5	119.9	153.2	221.1	241.8	235.1	289.9
Imports of Goods & NFServices	61.8	75.6	85.8	99.0	122.0	144.2	198.8	244.6	245.4	264.6	308.1
Domestic Absorption	134.3	152.3	166.2	198.0	228.4	285.6	383.9	473.5	566.0	653.0	678.3
Private Consumption, etc.	89.7	95.1	104.4	128.7	145.1	185.1	266.3	334.4	382.5	433.7	423.0
General Gov't Consumption	18.5	19.4	22.1	26.8	30.7	37.9	42.4	54.0	67.5	85.3	102.3
Gross Domestic Investment	26.1	37.8	39.7	42.5	52.6	62.6	75.2	85.1	116.0	134.0	153.0
Fixed Investment	24.2	34.1	36.0	34.8	45.9	53.1	65.7	74.2	103.4	119.5	128.9
Memo Items:											
Gross Domestic Saving	22.6	31.3	32.8	36.3	36.1	38.3	29.6	61.6	112.4	104.5	134.8
Gross National Saving	18.4	24.5	24.9	27.1	25.7	29.8	24.3	58.0	104.4	99.1	122.6

(Millions of 1980 Fiji Dollars)

	1967	1968	1969	1970	1971	1972	1973	1974	1975	1976	1977
Gross National Product	469.5	498.8	509.2	578.3	618.2	671.1	766.2	794.1	793.1	815.8	862.6
GDP at factor cost	454.5	485.7	492.8	554.6	590.8	637.4	718.6	737.3	737.9	757.8	803.2
Agriculture	140.9	150.7	145.6	158.1	149.5	148.3	157.7	152.2	153.0	158.9	184.5
Industry	126.4	138.8	125.1	137.2	137.2	147.6	161.6	154.7	147.3	151.9	168.8
Manufacturing	79.5	83.0	66.6	75.9	69.4	72.4	74.8	77.5	77.9	84.6	92.4
Services, etc.	187.2	196.3	222.1	259.3	304.2	341.6	399.3	430.4	437.7	447.1	449.9
Resource Balance	50.3	46.6	32.6	41.5	31.5	23.5	9.5	-13.9	-39.0	-65.6	-47.4
Exports of Goods & NFServices	242.3	274.5	282.4	311.8	346.7	374.5	416.6	395.1	362.1	345.2	403.5
Imports of Goods & NFServices	192.0	227.8	249.8	270.3	315.2	351.1	407.1	409.0	401.1	410.8	450.9
Domestic Absorption	436.5	478.8	506.2	565.8	617.8	674.7	770.0	813.6	842.7	891.3	921.7
Private Consumption, etc.	274.8	269.9	297.9	346.9	374.3	400.8	481.3	541.2	545.3	570.6	591.1
General Gov't Consumption	55.7	56.6	62.1	72.3	78.0	87.8	88.5	92.8	102.1	123.5	130.8
Gross Domestic Investment	106.0	152.2	146.2	146.6	165.6	186.1	200.2	179.5	195.3	197.3	199.9
Fixed Investment	96.8	134.7	128.8	114.6	139.8	153.6	168.7	148.6	164.0	165.9	168.4
Memo Items:											
Capacity to Import	181.1	208.3	229.7	253.4	272.6	291.9	313.7	369.7	395.2	365.0	424.2
Terms of Trade Adjustment	-61.2	-66.2	-52.7	-58.4	-74.1	-82.6	-102.8	-25.4	33.1	19.8	20.7
Gross Domestic Income	425.7	459.2	486.2	548.9	575.2	615.5	676.6	774.3	836.8	845.5	895.1
Gross National Income	408.3	432.6	456.6	519.9	544.1	588.5	663.3	768.7	826.2	835.6	883.4

DOMESTIC PRICES (DEFLATORS)
(Index 1980 = 100)

	1967	1968	1969	1970	1971	1972	1973	1974	1975	1976	1977
Overall (GDP)	26.9	27.8	29.6	31.6	32.6	37.4	43.4	56.3	70.0	75.5	75.5
Domestic Absorption	30.8	31.8	32.8	35.0	37.0	42.3	49.9	58.2	67.2	73.3	73.6
Agriculture	24.6	25.6	26.7	30.5	30.6	38.8	48.0	69.0	86.3	92.5	76.6
Industry	23.9	23.8	26.9	29.0	29.0	35.5	41.2	58.8	78.1	84.3	74.3
Manufacturing	25.2	25.6	30.2	31.2	30.4	37.3	42.2	61.9	77.0	79.2	75.1

MANUFACTURING ACTIVITY

	1967	1968	1969	1970	1971	1972	1973	1974	1975	1976	1977
Employment (1980=100)	..	74.2	..	65.9	66.5	66.4	72.3	78.8	74.0	89.4	88.8
Real Earnings per Empl. (1980=100)	..	60.0	..	74.6	88.9	89.3	82.6	89.2	96.5	105.2	115.9
Real Output per Empl. (1980=100)	..	82.3	..	85.4	89.5	85.2	85.0	84.4	85.3	74.5	98.2
Earnings as % of Value Added	..	35.5	..	43.6	59.4	50.1	42.6	37.1	37.5	48.4	46.0

MONETARY HOLDINGS
(Millions of current Fiji Dollars)

	1967	1968	1969	1970	1971	1972	1973	1974	1975	1976	1977
Money Supply, Broadly Defined	43.0	47.4	56.0	66.1	78.6	96.6	110.7	153.4	194.2	200.4	236.2
Money as Means of Payment	24.4	26.3	32.4	38.0	43.4	53.1	58.5	70.8	85.5	88.9	87.4
Currency Ouside Banks	8.5	8.5	9.6	11.2	13.1	14.9	17.2	21.5	27.3	30.7	34.0
Demand Deposits	15.9	17.8	22.8	26.8	30.3	38.1	41.3	49.3	58.2	58.2	53.4
Quasi-Monetary Liabilities	18.6	21.1	23.6	28.1	35.2	43.5	52.2	82.6	108.7	111.4	148.7

GOVERNMENT DEFICIT (-) OR SURPLUS
(Millions of current Fiji Dollars)

	1967	1968	1969	1970	1971	1972	1973	1974	1975	1976	1977
GOVERNMENT DEFICIT (-) OR SURPLUS	..	..	..	-5.3	-3.2	-4.0	-11.2	-16.0	-7.6	-23.6	-35.2
Current Revenue	..	..	..	43.4	50.2	55.1	68.2	80.0	109.3	124.1	134.1
Current Expenditure	..	..	..	39.1	46.1	44.2	55.6	70.0	87.2	108.5	119.4
Current Budget Balance	..	..	..	4.3	4.1	10.9	12.6	10.0	22.1	15.6	14.7
Capital Receipts	..	..	..	..	..	..	..	..	..	..	..
Capital Payments	..	..	..	9.5	7.3	14.9	23.8	26.1	29.8	39.2	49.9

1978	1979	1980	1981	1982	1983	1984	1985	1986	1987 estimate	NOTES	FIJI
1,340	1,580	1,750	2,000	1,870	1,700	1,770	1,630	1,730	1,570	..	**CURRENT GNP PER CAPITA (US $)**
610	622	634	646	659	672	686	700	715	723	..	**POPULATION (thousands)**
				(Millions of current Fiji Dollars)							**ORIGIN AND USE OF RESOURCES**
697.7	839.7	969.1	1,046.9	1,079.5	1,106.3	1,234.2	1,265.2	1,411.4	1,360.1	..	Gross National Product (GNP)
-4.5	-12.5	-14.6	-9.2	-33.8	-35.9	-40.8	-45.8	-44.6	-54.9	..	Net Factor Income from Abroad
702.2	852.2	983.7	1,056.1	1,113.3	1,142.2	1,275.0	1,311.0	1,456.0	1,415.0	f	Gross Domestic Product (GDP)
59.2	72.8	82.7	102.5	92.8	110.4	123.6	136.2	135.0	128.0	..	Indirect Taxes, net
643.0	779.4	901.0	953.6	1,020.5	1,031.8	1,151.4	1,174.8	1,321.0	1,287.0	f	GDP at factor cost
141.1	168.0	199.5	189.7	206.8	189.9	220.2	215.7	276.1	269.1	..	Agriculture
126.1	167.9	198.3	202.6	217.6	203.4	216.4	225.2	277.2	270.1	..	Industry
70.8	98.5	107.6	100.1	108.9	94.4	112.5	107.7	148.6	144.8	..	Manufacturing
375.8	443.5	503.2	561.3	596.1	638.5	714.8	733.9	767.7	747.8	..	Services, etc.
-31.0	-46.3	-48.6	-152.2	-71.3	-62.0	-14.0	-5.0	32.0	64.7	..	Resource Balance
299.5	385.8	470.0	454.4	481.3	498.1	546.0	584.0	609.0	613.7	..	Exports of Goods & NFServices
330.5	432.1	518.6	606.6	552.6	560.1	560.0	589.0	577.0	549.0	..	Imports of Goods & NFServices
733.2	898.5	1,032.3	1,208.4	1,184.4	1,204.2	1,289.0	1,316.0	1,424.0	1,350.3	..	Domestic Absorption
458.2	497.7	562.4	673.1	696.1	731.1	801.0	839.0	857.0	913.3	..	Private Consumption, etc.
115.1	143.9	156.7	173.1	203.8	231.6	245.0	247.0	260.0	260.0	..	General Gov't Consumption
159.9	256.9	313.2	362.2	284.5	241.5	243.0	230.0	307.0	177.0	..	Gross Domestic Investment
149.8	197.3	249.8	280.5	262.6	239.2	216.0	223.0	264.0	199.0	..	Fixed Investment
											Memo Items:
128.9	210.6	264.6	209.9	213.4	179.5	229.0	225.0	339.0	241.7	..	Gross Domestic Saving
121.0	191.0	248.1	193.2	176.8	141.6	184.0	167.0	287.8	163.3	..	Gross National Saving
				(Millions of 1980 Fiji Dollars)							
884.9	984.9	969.1	1,037.4	955.0	913.5	989.9	947.6	1,025.0	928.5	..	Gross National Product
817.7	916.1	901.0	954.6	944.0	907.1	982.5	935.2	1,020.8	939.7	f	GDP at factor cost
182.1	213.3	199.5	225.4	229.3	187.8	236.2	203.8	242.2	227.4	..	Agriculture
168.7	195.6	198.3	210.8	198.4	183.9	187.8	175.4	204.5	173.9	..	Industry
98.9	116.8	107.6	118.4	115.1	103.5	121.2	106.1	126.0	112.5	..	Manufacturing
466.8	507.2	503.2	518.5	516.3	535.5	558.5	556.0	574.1	538.4	..	Services, etc.
-50.3	-45.0	-48.6	-77.1	-9.8	-0.6	42.9	50.0	77.5	31.7	..	Resource Balance
401.1	480.0	470.0	461.5	470.0	461.5	492.4	491.4	496.9	434.2	..	Exports of Goods & NFServices
451.4	525.0	518.6	538.7	479.8	462.1	449.5	441.5	419.4	402.6	..	Imports of Goods & NFServices
941.0	1,044.5	1,032.3	1,123.0	993.4	942.4	978.3	930.4	978.5	940.5	..	Domestic Absorption
592.5	597.8	562.4	639.1	586.1	552.7	605.1	575.7	596.0	648.9	..	Private Consumption, etc.
132.7	158.9	156.7	151.7	168.6	181.9	186.6	175.6	179.3	175.7	..	General Gov't Consumption
215.9	287.9	313.2	332.2	238.7	207.8	186.6	179.1	203.2	115.9	..	Gross Domestic Investment
181.6	219.6	249.8	249.3	215.7	186.8	165.1	158.1	181.5	130.7	..	Fixed Investment
											Memo Items:
409.1	468.8	470.0	403.5	417.9	411.0	438.3	437.7	442.7	450.0	..	Capacity to Import
8.0	-11.2	0.0	-58.0	-52.1	-50.6	-54.1	-53.7	-54.2	15.8	..	Terms of Trade Adjustment
898.7	988.2	983.7	987.8	931.5	891.2	967.0	926.7	1,001.7	988.0	..	Gross Domestic Income
892.9	973.6	969.1	979.4	903.0	863.0	935.8	893.9	970.8	944.3	..	Gross National Income
				(Index 1980 = 100)							**DOMESTIC PRICES (DEFLATORS)**
78.8	85.3	100.0	101.0	113.2	121.3	124.9	133.7	137.9	145.5	..	Overall (GDP)
77.9	86.0	100.0	107.6	119.2	127.8	131.8	141.4	145.5	143.6	..	Domestic Absorption
77.5	78.7	100.0	84.2	90.2	101.1	93.2	105.8	114.0	118.3	..	Agriculture
74.8	85.8	100.0	96.1	109.7	110.6	115.2	128.4	135.6	155.3	..	Industry
71.6	84.3	100.0	84.6	94.6	91.2	92.8	101.5	118.0	128.7	..	Manufacturing
											MANUFACTURING ACTIVITY
88.9	93.7	100.0	92.4	95.9	100.0	101.7	103.5	..	..	..	Employment (1980=100)
117.8	110.5	100.0	102.2	97.7	87.1	99.2	100.2	..	..	..	Real Earnings per Empl. (1980=100)
105.6	111.8	100.0	131.0	120.7	104.4	133.4	128.7	..	..	..	Real Output per Empl. (1980=100)
54.6	41.4	43.6	48.9	47.0	45.0	47.0	47.0	..	..	..	Earnings as % of Value Added
				(Millions of current Fiji Dollars)							**MONETARY HOLDINGS**
255.4	305.9	342.8	364.3	395.3	443.7	490.0	502.5	586.5	607.9	..	Money Supply, Broadly Defined
103.2	116.6	105.3	125.6	130.5	141.6	142.3	146.4	178.7	173.2	..	Money as Means of Payment
38.8	45.2	44.1	48.7	52.8	58.7	61.0	61.8	63.1	64.9	..	Currency Ouside Banks
64.4	71.4	61.3	76.9	77.7	82.9	81.3	84.6	115.5	108.3	..	Demand Deposits
152.2	189.3	237.5	238.7	264.8	302.1	347.7	356.1	407.8	434.7	..	Quasi-Monetary Liabilities
				(Millions of current Fiji Dollars)							
-30.6	-24.8	-29.5	-45.3	-70.3	-43.3	-38.7	-35.4	..	..	..	**GOVERNMENT DEFICIT (-) OR SURPLUS**
157.0	194.0	224.3	262.8	264.8	293.6	329.2	339.7	..	..	..	Current Revenue
140.0	165.3	193.6	213.4	244.6	281.5	316.7	318.5	..	..	..	Current Expenditure
17.0	28.7	30.7	49.4	20.2	12.1	12.5	21.2	..	..	..	Current Budget Balance
..	..	..	..	..	..	..	..	..	..	..	Capital Receipts
47.6	53.5	60.3	94.7	90.5	55.4	51.2	56.6	..	..	..	Capital Payments

FIJI	1967	1968	1969	1970	1971	1972	1973	1974	1975	1976	1977
FOREIGN TRADE (CUSTOMS BASIS)				*(Millions of current US dollars)*							
Value of Exports, fob	49.1	52.5	57.3	67.7	67.8	73.0	78.4	143.7	159.3	126.8	173.2
Nonfuel Primary Products	39.3	41.0	45.8	52.1	51.9	55.5	56.9	109.2	127.1	91.6	131.7
Fuels	4.3	5.6	5.8	7.5	8.2	8.5	8.6	18.6	19.3	22.4	24.7
Manufactures	5.5	5.9	5.7	8.1	7.7	9.1	13.0	15.8	12.9	12.8	16.8
Value of Imports, cif	69.9	78.7	89.4	103.9	128.1	158.2	203.5	273.1	267.3	262.7	306.2
Nonfuel Primary Products	19.0	19.3	21.6	24.9	30.3	37.5	49.7	64.9	57.9	58.2	72.2
Fuels	7.7	8.4	9.6	11.5	13.4	15.7	18.2	42.9	46.6	42.2	59.0
Manufactures	43.2	51.0	58.2	67.5	84.4	104.9	135.6	165.3	162.9	162.3	175.0
				(Index 1980 = 100)							
Terms of Trade	56.1	57.5	67.3	70.2	68.5	77.6	81.0	139.9	108.7	75.9	62.1
Export Prices, fob	10.4	10.8	12.9	13.7	15.9	20.3	28.4	77.0	61.0	44.0	38.7
Nonfuel Primary Products	11.2	12.1	15.8	17.6	20.0	27.6	37.0	97.3	67.8	43.5	36.0
Fuels	4.3	4.3	4.3	4.3	5.6	6.2	8.9	36.7	35.7	38.4	42.0
Manufactures	30.1	31.7	32.4	37.5	36.5	38.5	49.5	67.2	66.1	66.7	71.7
Import Prices, cif	18.6	18.7	19.2	19.5	23.2	26.1	35.0	55.0	56.2	57.9	62.3
Trade at Constant 1980 Prices				*(Millions of 1980 US dollars)*							
Exports, fob	469.8	487.6	443.7	493.4	427.0	360.6	276.6	186.6	261.1	288.4	447.8
Imports, cif	375.8	420.2	466.3	531.9	552.3	606.1	581.2	496.1	476.1	453.5	491.8
BALANCE OF PAYMENTS				*(Millions of current US dollars)*							
Exports of Goods & Services	..	..	..	109.0	126.6	151.1	198.4	286.2	295.2	264.9	325.5
Merchandise, fob	..	..	..	62.1	63.4	71.0	82.7	142.8	157.6	120.9	164.5
Nonfactor Services	..	..	..	44.4	59.4	74.4	108.6	130.1	126.0	137.7	153.1
Factor Services	..	..	..	2.5	3.8	5.8	7.0	13.4	11.5	6.3	8.0
Imports of Goods & Services	..	..	..	125.9	157.6	191.6	262.4	316.8	320.8	312.0	350.6
Merchandise, fob	..	..	..	92.5	113.3	139.2	194.7	234.5	230.9	226.8	258.5
Nonfactor Services	..	..	..	21.2	28.7	35.6	55.0	66.6	62.4	67.5	75.3
Factor Services	..	..	..	12.2	15.6	16.8	12.8	15.6	27.4	17.7	16.8
Long-Term Interest	..	..	..	..	..	..	..	..	..	..	..
Current Transfers, net	..	..	..	-1.2	-0.4	0.9	-0.8	-2.3	-2.4	-4.9	-4.0
Workers' Remittances	..	..	..	0.0	0.0	0.0	0.0	0.0	0.0	0.0	0.0
Total to be Financed	..	..	..	-18.1	-31.4	-39.6	-64.9	-32.8	-28.1	-52.0	-29.0
Official Capital Grants	..	..	..	3.8	3.8	7.7	6.7	4.0	1.7	1.3	2.9
Current Account Balance	..	..	..	-14.3	-27.6	-31.9	-58.2	-28.9	-26.4	-50.6	-26.1
Long-Term Capital, net	..	..	..	19.3	18.0	35.6	49.8	53.0	47.1	16.7	39.2
Direct Investment	..	..	..	6.4	6.5	8.5	13.2	12.0	10.9	..	..
Long-Term Loans	..	..	..	..	..	..	..	..	..	..	..
Disbursements	..	..	..	..	..	..	..	..	..	..	..
Repayments	..	..	..	..	..	..	..	..	..	..	..
Other Long-Term Capital	..	..	..	9.1	7.6	19.4	29.9	37.0	34.5	15.3	36.3
Other Capital, net	..	..	..	-4.5	18.9	20.5	16.6	9.7	29.1	11.6	3.4
Change in Reserves	..	..	..	-0.5	-9.2	-24.2	-8.2	-33.9	-49.9	22.4	-16.5
Memo Item:				*(Fiji Dollars per US dollar)*							
Conversion Factor (Annual Avg)	0.810	0.870	0.870	0.870	0.860	0.820	0.790	0.800	0.820	0.900	0.920
EXTERNAL DEBT, ETC.				*(Millions of US dollars, outstanding at end of year)*							
Public/Publicly Guar. Long-Term	..	..	..	11.7	14.8	20.7	47.2	54.1	59.3	63.9	86.5
Official Creditors	..	..	..	8.2	10.0	10.6	19.1	29.8	38.7	47.7	61.7
IBRD and IDA	..	..	..	0.0	0.7	0.7	5.7	11.0	14.7	18.6	23.0
Private Creditors	..	..	..	3.5	4.8	10.1	28.1	24.3	20.6	16.1	24.8
Private Non-guaranteed Long-Term	..	..	..	..	..	..	..	..	..	..	..
Use of Fund Credit	..	..	..	0.0	0.0	0.0	0.0	0.4	0.0	0.0	7.9
Short-Term Debt	..	..	..	..	..	..	..	..	..	..	26.0
Memo Items:				*(Millions of US dollars)*							
Int'l Reserves Excluding Gold	18.4	18.5	26.8	27.4	39.6	69.4	74.0	109.2	148.6	116.3	147.1
Gold Holdings (at market price)	..	..	..	..	..	..	..	..	..	..	0.9
SOCIAL INDICATORS											
Total Fertility Rate	4.6	4.5	4.3	4.2	4.1	4.0	4.0	3.9	3.9	3.9	3.9
Crude Birth Rate	32.0	31.6	31.2	30.8	30.4	30.0	29.9	29.8	28.9	28.2	26.9
Infant Mortality Rate	55.0	53.0	51.0	49.0	47.0	45.0	43.4	41.8	40.2	38.6	37.0
Life Expectancy at Birth	61.2	61.2	61.2	61.3	61.3	61.3	61.4	61.4	61.4	61.5	62.3
Food Production, p.c. ('79-81=100)	96.3	114.1	104.9	111.4	99.1	86.9	88.0	79.3	76.1	76.7	84.9
Labor Force, Agriculture (%)	53.8	53.0	52.3	51.5	50.9	50.3	49.8	49.2	48.6	48.1	47.6
Labor Force, Female (%)	10.2	10.8	11.3	11.8	12.4	13.0	13.7	14.3	14.9	15.4	15.9
School Enroll. Ratio, primary	..	..	..	105.0	..	..	..	..	115.0	112.0	110.0
School Enroll. Ratio, secondary	..	..	..	52.0	..	..	..	..	66.0	55.0	58.0

1978	1979	1980	1981	1982	1983	1984	1985	1986	1987 estimate	NOTES	FIJI
											FOREIGN TRADE (CUSTOMS BASIS)
				(Millions of current US dollars)							
215.5	249.2	361.4	297.7	284.0	240.0	256.0	236.6	273.5	314.4	..	Value of Exports, fob
154.3	190.3	272.2	206.4	178.9	161.7	169.8	151.1	203.3	229.4	..	Nonfuel Primary Products
38.8	30.3	57.5	60.2	59.8	49.1	56.6	57.9	34.2	43.1	..	Fuels
22.5	28.6	31.7	31.1	45.3	29.2	29.6	27.6	36.1	41.9	..	Manufactures
366.3	469.6	561.3	631.1	513.7	483.2	449.8	441.5	437.4	435.1	..	Value of Imports, cif
89.6	94.8	98.4	112.0	94.3	96.4	74.4	70.3	68.4	57.0	..	Nonfuel Primary Products
63.9	86.6	129.5	162.0	146.7	112.2	105.2	103.1	56.7	69.2	..	Fuels
212.8	288.3	333.4	357.0	272.7	274.6	270.2	268.2	312.2	308.9	..	Manufactures
				(Index 1980 = 100)							
57.7	59.0	100.0	72.8	51.7	52.6	45.2	37.7	37.9	39.6	..	Terms of Trade
40.1	49.5	100.0	74.7	50.2	49.1	41.4	34.2	34.0	39.0	..	Export Prices, fob
37.0	45.1	100.0	65.8	39.0	40.0	32.5	25.4	29.3	33.1	..	Nonfuel Primary Products
42.3	61.0	100.0	112.5	101.6	92.5	90.2	87.5	45.0	56.9	..	Fuels
76.1	90.0	100.0	99.1	94.8	90.7	89.3	89.0	103.2	113.0	..	Manufactures
69.4	83.8	100.0	102.6	97.1	93.3	91.5	90.8	89.7	98.5	..	Import Prices, cif
											Trade at Constant 1980 Prices
				(Millions of 1980 US dollars)							
537.9	503.7	361.4	398.7	565.5	489.3	618.7	691.9	804.1	805.8	..	Exports, fob
527.8	560.3	561.3	615.4	529.0	518.0	491.4	486.3	487.5	441.9	..	Imports, cif
				(Millions of current US dollars)							**BALANCE OF PAYMENTS**
369.5	474.9	600.1	565.0	545.5	512.2	527.4	522.2	550.6	516.3	..	Exports of Goods & Services
184.5	241.8	343.2	281.5	253.3	217.6	229.3	206.4	242.6	273.9	..	Merchandise, fob
176.2	219.0	233.8	252.7	269.7	273.9	277.2	301.5	288.9	229.1	..	Nonfactor Services
8.9	14.1	23.1	30.9	22.5	20.7	21.0	14.3	19.0	13.3	..	Factor Services
403.3	546.1	657.9	754.7	654.4	600.8	568.3	557.6	557.3	508.3	..	Imports of Goods & Services
298.2	412.1	492.9	545.3	440.9	421.5	390.8	381.5	366.6	286.0	..	Merchandise, fob
90.9	106.6	124.7	167.7	154.7	130.5	128.5	130.0	140.9	164.0	..	Nonfactor Services
14.2	27.4	40.3	41.7	58.7	48.8	49.1	46.1	49.8	58.3	..	Factor Services
..	..	..	..	..	..	..	..	..	..	..	Long-Term Interest
-4.0	-8.5	-2.3	-8.8	-3.0	-1.9	-3.8	-10.5	-5.8	-19.2	..	Current Transfers, net
0.0	0.0	0.0	0.0	0.0	0.0	0.0	0.0	0.0	..	..	Workers' Remittances
-37.8	-79.7	-60.1	-198.5	-111.9	-90.5	-44.7	-46.0	-12.5	-11.2	K	Total to be Financed
2.0	13.2	35.3	25.9	20.6	26.7	18.8	33.1	14.7	10.3	K	Official Capital Grants
-35.8	-66.5	-24.8	-172.6	-91.3	-63.9	-26.0	-12.9	2.2	-0.9	..	Current Account Balance
3.2	68.6	109.0	132.8	96.5	85.5	41.2	41.4	36.1	2.3	..	Long-Term Capital, net
..	10.2	34.3	34.7	36.0	31.9	22.9	33.3	34.0	20.8	..	Direct Investment
..	..	..	..	..	..	..	..	..	..	..	Long-Term Loans
..	..	..	..	..	..	..	..	..	..	..	Disbursements
..	..	..	..	..	..	..	..	..	..	..	Repayments
1.2	45.2	39.5	72.2	40.0	26.9	-0.4	-24.9	-12.7	-28.8	..	Other Long-Term Capital
11.0	3.4	-44.0	18.0	-27.8	-28.6	-6.4	-32.9	-10.5	-34.8	..	Other Capital, net
21.5	-5.4	-40.2	21.8	22.6	7.0	-8.9	4.4	-27.7	33.4	..	Change in Reserves
											Memo Item:
				(Fiji Dollars per US dollar)							
0.850	0.840	0.820	0.850	0.930	1.020	1.080	1.150	1.130	1.220	..	Conversion Factor (Annual Avg)
				(Millions of US dollars, outstanding at end of year)							**EXTERNAL DEBT, ETC.**
84.7	109.9	180.0	237.4	265.5	292.3	279.4	302.2	311.5	334.2	..	Public/Publicly Guar. Long-Term
65.9	82.1	123.7	179.1	209.8	229.6	210.2	220.1	241.4	280.2	..	Official Creditors
25.4	28.6	33.5	42.6	53.5	70.3	58.7	63.4	71.0	80.1	..	IBRD and IDA
18.9	27.7	56.3	58.3	55.7	62.7	69.2	82.1	70.1	54.0	..	Private Creditors
..	..	64.9	95.7	99.6	105.0	98.8	108.0	101.5	103.7	..	Private Non-guaranteed Long-Term
8.5	8.6	0.0	0.0	14.9	14.1	13.2	14.5	7.9	6.7	..	Use of Fund Credit
10.0	23.0	36.0	39.0	22.0	26.0	22.0	19.0	20.0	21.5	..	Short-Term Debt
				(Millions of US dollars)							Memo Items:
134.7	136.5	167.5	135.1	126.9	115.8	117.4	130.8	171.1	132.1	..	Int'l Reserves Excluding Gold
1.9	5.7	6.5	4.4	5.1	4.2	3.4	3.6	4.3	0.5	..	Gold Holdings (at market price)
											SOCIAL INDICATORS
3.9	3.9	3.8	3.8	3.8	3.7	3.6	3.5	3.4	3.3	..	Total Fertility Rate
27.8	28.7	29.6	29.9	30.2	29.3	28.4	27.9	27.4	26.9	..	Crude Birth Rate
35.8	34.6	33.4	32.2	31.0	31.9	32.9	33.8	34.7	35.7	..	Infant Mortality Rate
62.6	63.0	63.3	63.6	64.0	64.8	65.6	66.5	67.3	68.2	..	Life Expectancy at Birth
85.5	105.5	93.4	101.1	103.1	73.4	107.9	86.0	102.4	85.3	..	Food Production, p.c. ('79-81 = 100)
47.2	46.7	46.2	..	..	..	..	..	..	..	..	Labor Force, Agriculture (%)
16.5	17.0	17.5	17.8	18.1	18.4	18.7	19.0	19.3	19.5	..	Labor Force, Female (%)
110.0	109.0	110.0	110.0	..	130.0	129.0	129.0	..	..	..	School Enroll. Ratio, primary
60.0	62.0	74.0	74.0	..	56.0	56.0	54.0	..	..	..	School Enroll. Ratio, secondary

FINLAND	1967	1968	1969	1970	1971	1972	1973	1974	1975	1976	1977
CURRENT GNP PER CAPITA (US $)	1,990	2,010	2,200	2,380	2,550	2,900	3,510	4,360	5,410	6,170	6,690
POPULATION (thousands)	4,606	4,627	4,624	4,606	4,616	4,640	4,666	4,691	4,711	4,726	4,739

ORIGIN AND USE OF RESOURCES *(Billions of current Finnish Markkaa)*

	1967	1968	1969	1970	1971	1972	1973	1974	1975	1976	1977
Gross National Product (GNP)	31.08	35.60	40.62	45.32	49.76	57.98	70.55	89.01	102.77	115.79	127.31
Net Factor Income from Abroad	-0.24	-0.31	-0.37	-0.43	-0.50	-0.65	-0.82	-1.04	-1.53	-1.86	-2.48
Gross Domestic Product (GDP)	31.32	35.91	40.99	45.74	50.26	58.62	71.36	90.05	104.29	117.64	129.79
Indirect Taxes, net	3.41	4.09	4.40	4.74	5.43	6.37	7.72	8.37	8.87	10.22	12.57
GDP at factor cost	27.91	31.82	36.59	41.00	44.83	52.25	63.64	81.69	95.42	107.42	117.22
Agriculture	3.75	4.35	4.83	5.06	5.55	5.78	6.80	8.34	10.15	10.69	11.48
Industry	10.23	11.66	14.09	16.34	17.37	20.92	26.05	34.99	38.74	42.04	45.26
Manufacturing	6.39	7.56	9.42	10.87	11.57	13.83	17.18	23.66	25.20	28.45	30.27
Services, etc.	13.93	15.81	17.67	19.60	21.91	25.56	30.79	38.36	46.52	54.69	60.49
Resource Balance	-0.29	0.57	0.38	-0.58	-0.88	0.21	-0.43	-3.31	-6.12	-2.29	2.26
Exports of Goods & NFServices	6.18	8.15	9.91	11.75	12.24	14.97	18.18	24.78	24.78	29.51	36.98
Imports of Goods & NFServices	6.47	7.58	9.53	12.33	13.13	14.77	18.60	28.10	30.90	31.80	34.72
Domestic Absorption	31.61	35.34	40.61	46.32	51.14	58.42	71.79	93.37	110.41	119.93	127.53
Private Consumption, etc.	18.61	20.51	23.74	25.17	27.57	32.81	39.63	45.88	57.53	68.52	71.71
General Gov't Consumption	4.66	5.48	5.93	6.61	7.62	8.96	10.69	13.69	17.80	21.31	24.00
Gross Domestic Investment	8.34	9.35	10.93	14.54	15.95	16.65	21.46	33.80	35.08	30.10	31.82
Fixed Investment	8.11	8.55	10.07	12.38	13.94	16.27	19.97	25.31	31.19	32.08	34.34
Memo Items:											
Gross Domestic Saving	8.05	9.92	11.31	13.96	15.07	16.85	21.04	30.49	28.96	27.81	34.08
Gross National Saving	7.82	9.62	10.95	13.54	14.60	16.21	20.22	29.46	27.43	25.92	31.55

(Billions of 1980 Finnish Markkaa)

	1967	1968	1969	1970	1971	1972	1973	1974	1975	1976	1977
Gross National Product	112.14	114.61	125.54	134.86	137.57	147.85	157.54	162.37	163.57	163.90	163.53
GDP at factor cost	100.38	102.15	111.54	121.55	123.52	132.87	141.63	148.69	151.48	151.63	150.12
Agriculture	14.20	14.92	15.67	15.40	15.25	14.83	14.66	14.05	13.58	13.82	14.02
Industry	37.49	38.60	43.04	47.23	47.52	52.69	56.45	58.56	57.12	56.91	56.74
Manufacturing	24.47	25.83	29.15	32.34	32.87	36.57	38.90	40.64	38.55	39.43	38.79
Services, etc.	48.69	48.63	52.83	58.92	60.75	65.35	70.52	76.09	80.78	80.90	79.36
Resource Balance	-4.54	-0.72	-2.47	-6.86	-7.06	-3.65	-6.49	-10.18	-16.63	-10.73	-3.21
Exports of Goods & NFServices	26.18	28.80	33.62	36.55	36.09	41.32	44.32	44.04	37.89	42.72	49.44
Imports of Goods & NFServices	30.72	29.51	36.09	43.41	43.15	44.97	50.82	54.22	54.53	53.45	52.65
Domestic Absorption	117.55	116.33	129.18	143.03	146.04	153.20	165.91	174.50	182.71	177.36	170.03
Private Consumption, etc.	65.39	62.93	71.12	75.34	77.39	84.77	91.46	88.78	96.47	100.56	93.90
General Gov't Consumption	18.12	19.19	19.83	20.92	22.12	23.84	25.17	26.30	28.12	29.72	30.97
Gross Domestic Investment	34.03	34.22	38.23	46.78	46.53	44.59	49.28	59.42	58.13	47.08	45.15
Fixed Investment	32.13	30.47	34.33	38.61	40.06	42.68	46.31	47.93	50.78	46.30	44.68
Memo Items:											
Capacity to Import	29.34	31.71	37.53	41.39	40.26	45.59	49.65	47.83	43.73	49.61	56.07
Terms of Trade Adjustment	3.16	2.92	3.91	4.83	4.16	4.28	5.33	3.78	5.84	6.88	6.63
Gross Domestic Income	116.18	118.53	130.62	141.01	143.15	153.83	164.75	168.10	171.92	173.51	173.45
Gross National Income	115.30	117.52	129.45	139.69	141.73	152.13	162.87	166.16	169.41	170.79	170.16

DOMESTIC PRICES (DEFLATORS) *(Index 1980 = 100)*

	1967	1968	1969	1970	1971	1972	1973	1974	1975	1976	1977
Overall (GDP)	27.7	31.1	32.3	33.6	36.2	39.2	44.8	54.8	62.8	70.6	77.8
Domestic Absorption	26.9	30.4	31.4	32.4	35.0	38.1	43.3	53.5	60.4	67.6	75.0
Agriculture	26.4	29.1	30.9	32.9	36.4	38.9	46.4	59.4	74.8	77.4	81.9
Industry	27.3	30.2	32.7	34.6	36.6	39.7	46.1	59.8	67.8	73.9	79.8
Manufacturing	26.1	29.3	32.3	33.6	35.2	37.8	44.2	58.2	65.4	72.2	78.0

MANUFACTURING ACTIVITY

	1967	1968	1969	1970	1971	1972	1973	1974	1975	1976	1977
Employment (1980=100)	73.7	74.0	79.1	87.1	89.5	92.0	95.8	98.7	97.8	97.4	94.7
Real Earnings per Empl. (1980=100)	66.6	67.3	70.7	75.4	78.9	85.0	88.7	92.9	96.8	98.0	93.5
Real Output per Empl. (1980=100)	66.4	69.7	69.8	71.6	71.7	76.4	76.5	79.8	76.8	78.7	80.8
Earnings as % of Value Added	52.4	46.8	45.5	47.1	50.3	50.2	48.9	44.1	49.5	50.1	49.2

MONETARY HOLDINGS *(Billions of current Finnish Markkaa)*

	1967	1968	1969	1970	1971	1972	1973	1974	1975	1976	1977
Money Supply, Broadly Defined	13.30	14.96	16.91	19.48	21.98	25.71	29.67	34.90	42.70	46.68	51.99
Money as Means of Payment	2.18	2.67	3.13	3.45	4.03	4.96	6.11	7.27	9.77	9.60	9.87
Currency Ouside Banks	0.96	1.09	1.20	1.29	1.48	1.55	1.78	2.15	2.51	2.54	2.84
Demand Deposits	1.22	1.58	1.92	2.16	2.55	3.41	4.33	5.12	7.26	7.06	7.03
Quasi-Monetary Liabilities	11.12	12.29	13.78	16.04	17.96	20.75	23.56	27.63	32.93	37.07	42.12

(Billions of current Finnish Markkaa)

	1967	1968	1969	1970	1971	1972	1973	1974	1975	1976	1977
GOVERNMENT DEFICIT (-) OR SURPLUS	..	..	..	..	..	1	2	1	-2	0	-2
Current Revenue	..	..	..	..	..	16	20	24	29	37	40
Current Expenditure	..	..	..	..	..	12	14	18	25	29	33
Current Budget Balance	..	..	..	..	..	4	6	5	4	7	7
Capital Receipts	..	..	..	..	..	0	0	0	0	0	0
Capital Payments	..	..	..	..	..	3	4	5	7	7	9

1978	1979	1980	1981	1982	1983	1984	1985	1986	1987 estimate	NOTES	FINLAND
7,280	8,580	10,110	11,080	11,430	10,930	10,830	11,040	12,180	14,420	..	CURRENT GNP PER CAPITA (US $)
4,753	4,765	4,780	4,800	4,824	4,863	4,882	4,908	4,929	4,947	..	POPULATION (thousands)

(Billions of current Finnish Markkaa)

ORIGIN AND USE OF RESOURCES

1978	1979	1980	1981	1982	1983	1984	1985	1986	1987	NOTES	
140.68	164.22	189.58	214.41	240.84	269.44	302.46	330.11	350.79	379.80	..	Gross National Product (GNP)
-2.69	-2.77	-3.24	-4.41	-5.35	-5.79	-7.10	-6.71	-7.34	-8.00	..	Net Factor Income from Abroad
143.38	166.99	192.82	218.82	246.19	275.23	309.57	336.82	358.13	387.75	..	Gross Domestic Product (GDP)
14.88	16.59	19.49	22.46	25.44	28.10	33.74	37.46	41.21	45.34	..	Indirect Taxes, net
128.50	150.40	173.34	196.35	220.74	247.13	275.83	299.36	316.92	342.40	B	GDP at factor cost
11.88	13.91	16.74	17.41	19.41	21.06	23.30	24.12	23.99	..	..	Agriculture
50.10	59.42	67.97	76.08	83.43	92.81	102.31	107.43	109.67	..	..	Industry
34.55	41.76	48.48	53.09	57.23	63.28	70.65	74.59	75.27	..	..	Manufacturing
66.51	77.07	88.63	102.86	117.90	133.26	150.22	167.81	183.26		..	Services, etc.
5.68	2.52	-1.48	3.07	1.99	1.35	8.08	3.28	4.63	1.05	..	Resource Balance
43.07	52.51	63.50	72.35	75.78	82.71	94.18	98.17	94.69	98.01	..	Exports of Goods & NFServices
37.39	49.99	64.98	69.28	73.80	81.36	86.10	94.89	90.07	96.96	..	Imports of Goods & NFServices
137.69	164.47	194.30	215.75	244.20	273.88	301.49	333.54	353.50	386.70	..	Domestic Absorption
78.01	89.95	102.12	113.69	131.38	148.77	163.48	185.71	200.17	219.75	..	Private Consumption, etc.
26.34	29.87	34.89	40.83	46.66	53.33	59.74	68.22	74.03	81.09	..	General Gov't Consumption
33.34	44.64	57.30	61.22	66.17	71.79	78.27	79.61	79.31	85.85	..	Gross Domestic Investment
34.10	38.11	47.31	53.81	61.32	69.45	73.12	80.05	81.83	89.11	..	Fixed Investment

Memo Items:

1978	1979	1980	1981	1982	1983	1984	1985	1986	1987		
39.02	47.16	55.82	64.29	68.15	73.13	86.34	82.89	83.93	86.90	..	Gross Domestic Saving
36.27	44.33	52.50	59.84	62.68	67.21	79.11	76.14	75.80	78.19	..	Gross National Saving

(Billions of 1980 Finnish Markkaa)

1978	1979	1980	1981	1982	1983	1984	1985	1986	1987		
167.19	179.87	189.58	192.05	198.66	204.74	210.95	218.98	224.10	230.80	..	Gross National Product
152.40	164.40	173.34	176.00	182.13	187.89	192.32	198.57	202.22	208.86	B	GDP at factor cost
13.92	15.66	16.74	15.42	15.58	16.37	16.57	16.45	15.86	..	..	Agriculture
58.44	63.24	67.97	69.60	71.08	73.48	75.52	78.31	78.99	..	..	Industry
40.42	44.79	48.48	50.07	50.77	52.39	54.49	56.66	57.29	..	..	Manufacturing
80.04	85.51	88.63	90.98	95.48	98.04	100.23	103.82	107.36	..	..	Services, etc.
3.16	-1.43	-1.48	4.69	2.44	2.18	5.17	1.55	-0.44	-5.30	..	Resource Balance
53.86	58.57	63.50	66.63	65.91	67.53	71.19	72.06	72.93	74.40	..	Exports of Goods & NFServices
50.69	60.00	64.98	61.95	63.46	65.35	66.02	70.51	73.38	79.69	..	Imports of Goods & NFServices
167.29	184.40	194.30	191.27	200.61	206.95	210.75	221.88	229.29	242.42	..	Domestic Absorption
94.30	100.61	102.12	101.53	107.28	111.84	114.58	123.42	129.31	136.79	..	Private Consumption, etc.
32.25	33.47	34.89	36.39	37.68	39.08	40.17	42.24	43.59	45.19	..	General Gov't Consumption
40.74	50.31	57.30	53.35	55.65	56.03	56.00	56.22	56.38	60.45	..	Gross Domestic Investment
41.58	42.84	47.31	48.33	50.48	52.53	51.42	52.91	52.67	55.36	..	Fixed Investment

Memo Items:

1978	1979	1980	1981	1982	1983	1984	1985	1986	1987		
58.40	63.03	63.50	64.69	65.17	66.44	72.21	72.94	77.15	80.56	..	Capacity to Import
4.54	4.45	0.00	-1.94	-0.74	-1.09	1.02	0.88	4.21	6.16	..	Terms of Trade Adjustment
175.00	187.42	192.82	194.02	202.32	208.03	216.95	224.32	233.05	243.28	..	Gross Domestic Income
171.73	184.32	189.58	190.10	197.92	203.65	211.98	219.86	228.31	..	..	Gross National Income

(Index 1980 = 100)

DOMESTIC PRICES (DEFLATORS)

1978	1979	1980	1981	1982	1983	1984	1985	1986	1987		
84.1	91.3	100.0	111.7	121.2	131.6	143.4	150.7	156.5	163.5	..	Overall (GDP)
82.3	89.2	100.0	112.8	121.7	132.3	143.1	150.3	154.2	159.5	..	Domestic Absorption
85.3	88.8	100.0	112.9	124.6	128.7	140.6	146.7	151.3	..	..	Agriculture
85.7	94.0	100.0	109.3	117.4	126.3	135.5	137.2	138.8	..	..	Industry
85.5	93.3	100.0	106.0	112.7	120.8	129.7	131.7	131.4	..	..	Manufacturing

MANUFACTURING ACTIVITY

1978	1979	1980	1981	1982	1983	1984	1985	1986	1987		
92.0	95.2	100.0	99.8	97.9	95.9	94.7	93.6			..	Employment (1980=100)
95.1	99.0	100.0	102.0	103.3	104.6	107.2	110.3	..	..	..	Real Earnings per Empl. (1980=100)
84.8	92.3	100.0	105.0	105.7	109.1	113.3	119.1	..	..	..	Real Output per Empl. (1980=100)
45.3	43.4	44.1	45.4	45.6	44.4	42.6	43.9	..	..	..	Earnings as % of Value Added

(Billions of current Finnish Markkaa)

MONETARY HOLDINGS

1978	1979	1980	1981	1982	1983	1984	1985	1986	1987		
59.92	70.75	81.41	94.38	107.06	121.34	140.25	165.59	178.73	200.25	..	Money Supply, Broadly Defined
11.50	14.09	14.98	17.19	19.92	21.43	24.95	27.69	27.84	30.34	..	Money as Means of Payment
3.48	3.91	4.30	4.82	5.17	5.63	5.88	6.14	6.36	7.26	..	Currency Ouside Banks
8.01	10.18	10.67	12.36	14.75	15.80	19.07	21.55	21.48	23.08	..	Demand Deposits
48.43	56.66	66.43	77.19	87.14	99.91	115.31	137.90	150.89	169.91	..	Quasi-Monetary Liabilities

(Billions of current Finnish Markkaa)

1978	1979	1980	1981	1982	1983	1984	1985	1986	1987		
-3	-4	-4	-2	-5	-8	-3	-3	0	-4	E	GOVERNMENT DEFICIT (-) OR SURPLUS
41	46	53	63	70	78	89	100	113	120	..	Current Revenue
37	43	48	55	65	77	82	93	101	..	..	Current Expenditure
4	3	5	7	5	2	7	8	12	..	..	Current Budget Balance
0	0	0	0	0	0	0	0	0	..	..	Capital Receipts
7	7	9	9	10	10	10	11	12	..	..	Capital Payments

FINLAND	1967	1968	1969	1970	1971	1972	1973	1974	1975	1976	1977
FOREIGN TRADE (CUSTOMS BASIS)					*(Millions of current US dollars)*						
Value of Exports, fob	1,534	1,636	1,985	2,306	2,356	2,947	3,719	5,522	5,489	6,339	7,668
Nonfuel Primary Products	560	576	661	750	727	814	1,057	1,435	1,230	1,497	1,818
Fuels	6	5	12	19	5	6	10	47	23	100	165
Manufactures	968	1,055	1,312	1,538	1,624	2,127	2,652	4,040	4,236	4,741	5,686
Value of Imports, cif	1,698	1,592	2,023	2,637	2,796	3,198	4,210	6,850	7,600	7,391	7,608
Nonfuel Primary Products	365	335	393	488	443	524	720	1,113	1,196	1,098	1,218
Fuels	191	208	229	302	383	420	527	1,504	1,447	1,586	1,794
Manufactures	1,142	1,050	1,401	1,847	1,970	2,254	2,963	4,233	4,957	4,708	4,596
					(Index 1980 = 100)						
Terms of Trade	133.7	134.7	139.7	136.3	142.7	143.6	128.2	99.8	111.1	106.4	102.8
Export Prices, fob	24.3	22.9	26.0	26.0	29.3	33.3	39.4	51.4	60.2	59.7	62.9
Nonfuel Primary Products	23.6	24.9	27.4	28.0	28.5	36.1	49.5	46.4	50.7	53.8	51.9
Fuels	8.5	7.3	5.4	5.9	7.0	8.5	24.5	36.6	33.6	38.5	36.8
Manufactures	25.1	22.2	26.3	26.3	30.1	32.6	36.5	53.6	64.1	62.8	69.2
Import Prices, cif	18.2	17.0	18.6	19.1	20.5	23.2	30.7	51.5	54.2	56.1	61.2
Trade at Constant 1980 Prices					*(Millions of 1980 US dollars)*						
Exports, fob	6,315	7,135	7,620	8,868	8,036	8,860	9,437	10,744	9,116	10,614	12,187
Imports, cif	9,339	9,357	10,851	13,824	13,610	13,807	13,697	13,299	14,017	13,174	12,428
BALANCE OF PAYMENTS					*(Millions of current US dollars)*						
Exports of Goods & Services	..	..	..	2,773	2,907	3,613	4,723	6,576	6,747	7,683	9,205
Merchandise, fob	..	..	..	2,294	2,351	2,929	3,820	5,487	5,511	6,297	7,603
Nonfactor Services	..	..	..	411	464	571	765	845	1,098	1,247	1,452
Factor Services	..	..	..	68	91	113	137	244	138	139	150
Imports of Goods & Services	..	..	..	3,007	3,245	3,740	5,095	7,768	8,845	8,754	9,263
Merchandise, fob	..	..	..	2,470	2,626	2,985	4,070	6,367	7,219	6,961	7,164
Nonfactor Services	..	..	..	371	421	506	684	920	1,073	1,173	1,336
Factor Services	..	..	..	166	198	250	341	481	553	620	762
Long-Term Interest	..	..	..	..	..	..	..	..	..	..	..
Current Transfers, net	..	..	..	2	6	1	1	5	-2	-8	-12
Workers' Remittances	..	..	..	0	0	0	0	0	0	0	0
Total to be Financed	..	..	..	-232	-332	-126	-372	-1,187	-2,100	-1,079	-69
Official Capital Grants	..	..	..	-7	-8	9	-15	-23	-33	-37	-38
Current Account Balance	..	..	..	-239	-340	-117	-387	-1,210	-2,133	-1,116	-107
Long-Term Capital, net	..	..	..	65	356	342	55	213	1,271	934	409
Direct Investment	..	..	..	-41	-35	-24	-2	6	42	27	-24
Long-Term Loans	..	..	..	..	..	..	..	..	..	..	..
Disbursements	..	..	..	..	..	..	..	..	..	..	..
Repayments	..	..	..	..	..	..	..	..	..	..	..
Other Long-Term Capital	..	..	..	113	399	357	73	230	1,261	944	470
Other Capital, net	..	..	..	268	177	-204	146	971	638	74	-258
Change in Reserves	..	..	..	-94	-193	-21	186	27	225	108	-43
Memo Item:					*(Finnish Markkaa per US dollar)*						
Conversion Factor (Annual Avg)	3.450	4.200	4.200	4.200	4.180	4.150	3.820	3.770	3.680	3.860	4.030
EXTERNAL DEBT, ETC.				*(Millions US dollars, outstanding at end of year)*							
Public/Publicly Guar. Long-Term	..	..	..	..	..	..	..	..	..	..	..
Official Creditors	..	..	..	..	..	..	..	..	..	..	..
IBRD and IDA	..	..	..	..	..	..	..	..	..	..	..
Private Creditors	..	..	..	..	..	..	..	..	..	..	..
Private Non-guaranteed Long-Term	..	..	..	..	..	..	..	..	..	..	..
Use of Fund Credit	..	..	..	..	..	..	..	..	..	..	..
Short-Term Debt	..	..	..	..	..	..	..	..	..	..	..
Memo Items:					*(Millions of US dollars)*						
Int'l Reserves Excluding Gold	139.0	293.0	281.9	424.6	623.1	667.7	574.3	595.9	433.2	462.1	531.1
Gold Holdings (at market price)	45.2	54.0	45.3	30.8	60.9	90.6	92.4	153.6	115.5	111.0	149.2
SOCIAL INDICATORS											
Total Fertility Rate	2.2	2.1	1.9	1.8	1.7	1.6	1.5	1.6	1.7	1.7	1.7
Crude Birth Rate	16.8	15.9	14.6	14.0	13.2	12.7	12.2	13.3	13.9	14.1	13.8
Infant Mortality Rate	..	..	..	..	..	..	..	11.0	9.5	9.8	9.0
Life Expectancy at Birth	69.6	69.8	70.0	70.3	70.5	70.7	71.0	71.3	71.6	71.9	72.2
Food Production, p.c. ('79-81 = 100)	87.5	86.9	94.8	95.4	101.6	100.0	93.1	97.0	100.9	110.0	98.0
Labor Force, Agriculture (%)	21.9	21.2	20.4	19.6	18.8	18.1	17.3	16.6	15.8	15.0	14.3
Labor Force, Female (%)	42.8	43.1	43.4	43.7	43.9	44.2	44.5	44.7	45.0	45.3	45.6
School Enroll. Ratio, primary	..	..	..	82.0	..	..	..	..	102.0	101.0	100.0
School Enroll. Ratio, secondary	..	..	..	102.0	..	..	..	..	89.0	91.0	93.0

1978	1979	1980	1981	1982	1983	1984	1985	1986	1987 estimate	NOTES	FINLAND
											FOREIGN TRADE (CUSTOMS BASIS)
				(Millions of current US dollars)							
8,572	11,175	14,140	14,007	13,127	12,510	13,498	13,609	16,325	20,039	..	Value of Exports, fob
1,936	2,838	3,658	3,363	2,529	2,530	2,804	2,530	2,742	3,500	..	Nonfuel Primary Products
223	303	618	591	532	646	743	580	395	444	..	Fuels
6,413	8,034	9,864	10,052	10,066	9,334	9,950	10,499	13,188	16,095	..	Manufactures
7,865	11,390	15,632	14,190	13,380	12,846	12,435	13,226	15,324	19,860	..	Value of Imports, cif
1,276	1,735	2,272	1,925	1,946	1,725	1,762	1,761	2,044	2,539	..	Nonfuel Primary Products
1,746	3,018	4,546	4,357	3,654	3,454	3,101	3,218	2,345	2,672	..	Fuels
4,842	6,637	8,814	7,909	7,781	7,667	7,572	8,247	10,936	14,649	..	Manufactures
				(Index 1980 = 100)							
103.5	109.1	100.0	95.0	98.0	94.7	95.5	95.8	106.7	108.6	..	Terms of Trade
68.6	88.2	100.0	94.4	91.1	84.9	83.7	82.2	94.6	108.1	..	Export Prices, fob
63.5	97.4	100.0	88.6	83.1	84.5	86.2	79.5	80.0	85.2	..	Nonfuel Primary Products
44.7	93.6	100.0	104.0	97.0	85.4	83.6	81.1	60.8	65.4	..	Fuels
71.8	85.3	100.0	96.0	93.0	85.0	83.0	83.0	100.0	117.0	..	Manufactures
66.3	80.8	100.0	99.3	92.9	89.7	87.6	85.9	88.6	99.5	..	Import Prices, cif
											Trade at Constant 1980 Prices
				(Millions of 1980 US dollars)							
12,493	12,674	14,140	14,834	14,415	14,731	16,129	16,547	17,265	18,545	..	Exports, fob
11,865	14,098	15,632	14,284	14,405	14,327	14,192	15,405	17,287	19,953	..	Imports, cif
				(Millions of current US dollars)							
											BALANCE OF PAYMENTS
10,562	13,711	17,335	17,276	16,279	15,357	16,331	16,745	19,633	23,715	..	Exports of Goods & Services
8,482	11,095	14,074	13,651	12,828	12,174	13,088	13,327	15,930	19,007	..	Merchandise, fob
1,854	2,244	2,728	2,917	2,719	2,528	2,434	2,415	2,732	3,448	..	Nonfactor Services
226	371	533	708	731	654	809	1,003	971	1,259	..	Factor Services
9,846	13,765	18,618	17,567	16,929	16,151	16,170	17,312	19,997	25,187	..	Imports of Goods & Services
7,395	10,723	14,724	13,299	12,636	12,019	11,600	12,428	14,312	17,680	..	Merchandise, fob
1,572	1,960	2,492	2,539	2,449	2,438	2,583	2,801	3,220	4,430	..	Nonfactor Services
880	1,081	1,402	1,729	1,844	1,694	1,988	2,083	2,466	3,077	..	Factor Services
..	..	..	..	..	..	..	..	..	..	..	Long-Term Interest
-15	-16	-20	-10	-27	-23	-22	-6	-157	-162	..	Current Transfers, net
0	0	0	0	0	0	0	0	0	0	..	Workers' Remittances
701	-70	-1,302	-301	-678	-817	138	-574	-521	-1,633	K	Total to be Financed
-37	-94	-102	-106	-108	-120	-153	-170	-234	-304	K	Official Capital Grants
664	-164	-1,404	-407	-786	-937	-14	-744	-755	-1,938	..	Current Account Balance
858	108	-59	443	134	222	216	629	333	20	..	Long-Term Capital, net
-28	-98	-109	-128	-246	-251	-370	-284	-429	-809	..	Direct Investment
..	..	..	..	..	..	..	..	..	..	..	Long-Term Loans
..	..	..	..	..	..	..	..	..	..	..	Disbursements
..	..	..	..	..	..	..	..	..	..	..	Repayments
923	300	152	677	487	593	739	1,083	996	1,134	..	Other Long-Term Capital
-812	447	1,743	-260	851	488	1,615	701	-1,858	5,940	..	Other Capital, net
-710	-391	-280	224	-200	228	-1,817	-587	2,280	-4,022	..	Change in Reserves
				(Finnish Markkaa per US dollar)							Memo Item:
4.120	3.900	3.730	4.320	4.820	5.570	6.010	6.200	5.070	4.400	..	Conversion Factor (Annual Avg)
				(Millions US dollars, outstanding at end of year)							**EXTERNAL DEBT, ETC.**
..	..	..	..	..	..	..	..	..	..	..	Public/Publicly Guar. Long-Term
..	..	..	..	..	..	..	..	..	..	..	Official Creditors
..	..	..	..	..	..	..	..	..	..	..	IBRD and IDA
..	..	..	..	..	..	..	..	..	..	..	Private Creditors
..	..	..	..	..	..	..	..	..	..	..	Private Non-guaranteed Long-Term
..	..	..	..	..	..	..	..	..	..	..	Use of Fund Credit
..	..	..	..	..	..	..	..	..	..	..	Short-Term Debt
				(Millions of US dollars)							Memo Items:
1,222.9	1,540.0	1,870.2	1,483.7	1,517.5	1,237.7	2,754.3	3,749.9	1,787.1	6,417.5	..	Int'l Reserves Excluding Gold
213.6	504.9	581.3	504.5	580.2	484.5	391.6	625.2	747.4	946.4	..	Gold Holdings (at market price)
											SOCIAL INDICATORS
1.7	1.6	1.7	1.7	1.7	1.7	1.7	1.7	1.7	1.6	..	Total Fertility Rate
13.4	13.3	13.1	13.2	13.1	13.0	12.9	12.8	12.7	12.6	..	Crude Birth Rate
7.5	7.6	7.6	6.5	7.2	8.0	8.9	9.9	11.0	12.2	..	Infant Mortality Rate
72.5	72.8	73.1	73.4	73.6	74.1	74.5	74.9	75.3	75.7	..	Life Expectancy at Birth
99.6	104.0	103.5	92.5	105.7	114.8	111.6	110.1	107.3	97.5	..	Food Production, p.c. ('79-81=100)
13.5	12.8	12.0	..	..	..	..	..	..	..	..	Labor Force, Agriculture (%)
45.8	46.1	46.4	46.5	46.5	46.5	46.6	46.6	46.7	46.8	..	Labor Force, Female (%)
98.0	97.0	96.0	..	100.0	102.0	103.0	104.0	104.0	..	..	School Enroll. Ratio, primary
96.0	97.0	98.0	..	98.0	101.0	101.0	102.0	..	..	..	School Enroll. Ratio, secondary

FRANCE	1967	1968	1969	1970	1971	1972	1973	1974	1975	1976	1977
CURRENT GNP PER CAPITA (US $)	2,360	2,560	2,820	3,000	3,250	3,540	4,240	5,100	6,010	6,700	7,550
POPULATION (thousands)	49,548	49,915	50,318	50,772	51,251	51,701	52,118	52,460	52,705	52,891	53,077
ORIGIN AND USE OF RESOURCES					*(Billions of current French Francs)*						
Gross National Product (GNP)	578.4	628.3	716.0	799.7	887.1	990.1	1,132.2	1,307.1	1,470.3	1,703.5	1,921.7
Net Factor Income from Abroad	3.1	2.9	2.9	3.3	2.9	2.1	2.3	4.1	2.4	2.9	3.9
Gross Domestic Product (GDP)	575.4	625.4	713.0	796.4	884.2	987.9	1,129.8	1,303.0	1,467.9	1,700.6	1,917.8
Indirect Taxes, net	81.8	83.3	98.7	105.3	115.5	130.7	146.2	160.5	176.4	211.5	223.7
GDP at factor cost	493.5	542.1	614.3	691.0	768.6	857.2	983.7	1,142.5	1,291.5	1,489.0	1,694.1
Agriculture	44.7	45.3	47.2	52.0	54.7	65.2	77.0	75.0	75.4	82.9	91.2
Industry	216.5	231.5	260.8	293.5	324.2	362.2	409.8	466.1	528.5	604.5	679.8
Manufacturing	154.4	165.3	190.6	215.8	239.0	265.7	302.9	342.3	381.3	440.7	496.4
Services, etc.	314.1	348.5	405.1	450.8	505.2	560.6	643.0	761.8	864.0	1,013.1	1,146.8
Resource Balance	1.8	0.0	-3.3	4.5	9.8	10.6	9.5	-13.2	17.3	-12.5	2.2
Exports of Goods & NFServices	74.9	81.6	99.0	123.8	145.2	165.3	198.5	269.5	279.6	332.8	393.0
Imports of Goods & NFServices	73.1	81.6	102.3	119.3	135.4	154.7	188.9	282.7	262.3	345.3	390.8
Domestic Absorption	573.5	625.3	716.4	791.8	874.4	977.4	1,120.3	1,316.2	1,450.5	1,713.1	1,915.6
Private Consumption, etc.	329.3	358.9	407.5	446.0	508.2	567.3	640.7	745.2	857.4	987.8	1,112.6
General Gov't Consumption	88.5	100.3	112.9	126.9	134.6	149.8	171.1	204.2	247.9	292.9	335.3
Gross Domestic Investment	155.8	166.2	196.0	218.9	231.6	260.3	308.5	366.8	345.2	432.4	467.7
Fixed Investment	142.1	151.3	173.3	193.1	216.9	244.9	286.1	334.6	351.9	404.8	436.2
Memo Items:											
Gross Domestic Saving	157.6	166.2	192.7	223.5	241.3	270.9	318.0	353.6	362.6	419.9	469.9
Gross National Saving	160.7	166.4	191.9	223.1	239.5	268.4	315.4	352.2	358.4	416.1	466.9
					(Billions of 1980 French Francs)						
Gross National Product	1,675.2	1,745.0	1,865.7	1,972.7	2,133.2	2,225.1	2,346.0	2,421.5	2,411.0	2,513.5	2,595.2
GDP at factor cost	1,422.0	1,498.4	1,593.6	1,698.6	1,837.5	1,914.0	2,023.5	2,111.5	2,111.6	2,191.0	2,283.7
Agriculture	103.5	106.1	101.4	105.5	107.4	107.9	113.9	113.6	105.4	101.2	101.4
Industry	577.2	600.9	657.5	708.7	744.2	788.1	825.2	852.8	844.2	882.5	907.0
Manufacturing	372.7	394.5	441.7	481.3	512.2	544.4	582.4	601.1	588.7	630.3	653.8
Services, etc.	984.9	1,029.5	1,099.0	1,150.2	1,274.4	1,324.2	1,401.8	1,447.3	1,457.4	1,525.4	1,581.4
Resource Balance	-28.5	-38.8	-54.2	-34.0	-57.3	-68.9	-91.8	-64.8	-22.3	-67.4	-32.5
Exports of Goods & NFServices	189.2	207.1	239.6	278.3	335.3	375.7	416.1	452.9	445.3	481.7	517.1
Imports of Goods & NFServices	217.8	245.9	293.9	312.3	392.6	444.6	507.9	517.7	467.5	549.0	549.6
Domestic Absorption	1,694.1	1,775.3	1,912.1	1,998.4	2,183.2	2,289.1	2,432.8	2,478.5	2,429.3	2,576.5	2,622.4
Private Consumption, etc.	947.6	985.6	1,053.1	1,102.2	1,227.0	1,287.6	1,356.2	1,372.8	1,412.1	1,481.0	1,521.6
General Gov't Consumption	325.3	343.5	357.7	372.6	387.1	400.8	414.4	419.3	437.3	455.4	466.3
Gross Domestic Investment	421.2	446.2	501.4	523.6	569.1	600.7	662.2	686.4	579.9	640.1	634.5
Fixed Investment	404.6	427.0	466.1	487.6	540.9	573.5	622.1	630.0	589.4	608.6	597.6
Memo Items:											
Capacity to Import	223.3	245.9	284.3	324.2	420.9	475.0	533.5	493.5	498.4	529.1	552.8
Terms of Trade Adjustment	34.0	38.8	44.7	45.9	85.5	99.3	117.4	40.6	53.2	47.5	35.6
Gross Domestic Income	1,699.6	1,775.4	1,902.6	2,010.3	2,211.5	2,319.5	2,458.4	2,454.4	2,460.2	2,556.6	2,625.5
Gross National Income	1,709.2	1,783.8	1,910.3	2,018.6	2,218.7	2,324.4	2,463.4	2,462.2	2,464.1	2,561.0	2,630.8
DOMESTIC PRICES (DEFLATORS)					*(Index 1980 = 100)*						
Overall (GDP)	34.5	36.0	38.4	40.5	41.6	44.5	48.3	54.0	61.0	67.8	74.0
Domestic Absorption	33.9	35.2	37.5	39.6	40.1	42.7	46.1	53.1	59.7	66.5	73.0
Agriculture	43.2	42.7	46.5	49.3	51.0	60.4	67.7	66.0	71.5	82.0	89.9
Industry	37.5	38.5	39.7	41.4	43.6	46.0	49.7	54.7	62.6	68.5	74.9
Manufacturing	41.4	41.9	43.2	44.8	46.7	48.8	52.0	56.9	64.8	69.9	75.9
MANUFACTURING ACTIVITY											
Employment (1980=100)	97.9	96.2	98.6	101.1	102.7	104.3	106.8	108.2	105.3	104.4	104.0
Real Earnings per Empl. (1980=100)	..	..	..	..	..	..	..	..	..	..	..
Real Output per Empl. (1980=100)	46.9	52.4	57.1	64.1	66.9	69.8	74.1	83.8	78.0	85.2	87.4
Earnings as % of Value Added	..	..	..	..	..	..	..	..	..	..	..
MONETARY HOLDINGS					*(Billions of current French Francs)*						
Money Supply, Broadly Defined	351.3	393.3	427.4	489.7	577.6	681.4	783.9	916.3	1,075.2	1,228.5	1,523.5
Money as Means of Payment	198.3	214.2	209.7	232.5	259.9	299.0	327.9	377.7	425.2	457.0	587.8
Currency Ouside Banks	70.8	72.7	72.6	75.9	78.0	84.0	89.5	97.8	106.6	116.1	121.5
Demand Deposits	127.5	141.4	137.0	156.6	181.9	215.0	238.4	279.9	318.6	340.9	466.3
Quasi-Monetary Liabilities	153.0	179.1	217.7	257.2	317.7	382.4	456.0	538.6	650.0	771.5	935.7
GOVERNMENT DEFICIT (-) OR SURPLUS					*(Billions of current French Francs)*						
GOVERNMENT DEFICIT (-) OR SURPLUS	..	..	..	..	..	6.8	4.7	5.8	-37.8	-17.1	-22.3
Current Revenue	..	..	..	..	..	333.8	372.9	457.5	516.4	627.8	699.0
Current Expenditure	..	..	..	..	..	299.0	335.3	414.0	502.7	578.0	669.2
Current Budget Balance	..	..	..	..	..	34.8	37.6	43.5	13.7	49.8	29.8
Capital Receipts	..	..	..	..	..	0.8	2.6	2.2	1.6	3.8	2.9
Capital Payments	..	..	..	..	..	30.1	30.7	39.3	52.8	64.4	50.6

1978	1979	1980	1981	1982	1983	1984	1985	1986	1987 estimate	NOTES	FRANCE
8,450	10,030	11,900	12,480	11,960	10,650	9,940	9,750	10,730	12,790	..	CURRENT GNP PER CAPITA (US $)
53,277	53,480	53,714	53,966	54,219	54,652	54,947	55,170	55,389	55,609	..	POPULATION (thousands)
				(Billions of current French Francs)							ORIGIN AND USE OF RESOURCES
2,187.3	2,489.7	2,820.9	3,175.1	3,628.0	3,994.8	4,338.4	4,669.2	4,995.8	5,230.2	..	Gross National Product (GNP)
4.8	8.6	12.6	10.3	2.0	-11.7	-23.5	-25.7	-17.2	-19.3	..	Net Factor Income from Abroad
2,182.6	2,481.1	2,808.3	3,164.8	3,626.0	4,006.5	4,361.9	4,695.0	5,013.0	5,249.5	..	Gross Domestic Product (GDP)
260.2	309.4	356.2	386.8	451.6	499.5	546.6	590.7	616.8	663.7	..	Indirect Taxes, net
1,922.4	2,171.7	2,452.1	2,778.0	3,174.5	3,507.0	3,815.3	4,104.2	4,396.3	4,585.9	B	GDP at factor cost
104.5	118.5	119.0	130.9	166.2	169.6	175.2	180.5	186.9	..	..	Agriculture
766.5	859.8	947.3	1,029.8	1,148.4	1,256.6	1,338.7	1,444.1	1,529.4	..	..	Industry
564.3	635.6	679.5	733.5	821.5	901.0	956.3	1,039.2	1,112.0	..	..	Manufacturing
1,311.6	1,502.9	1,742.0	2,004.0	2,311.4	2,580.3	2,848.0	3,070.3	3,296.8	..	..	Services, etc.
28.7	14.4	-34.4	-30.6	-68.8	-6.8	28.5	30.9	52.0	10.1	..	Resource Balance
445.3	526.7	604.4	714.4	790.4	900.7	1,053.2	1,124.2	1,080.6	1,102.4	..	Exports of Goods & NFServices
416.6	512.3	638.8	745.0	859.2	907.5	1,024.8	1,093.2	1,028.6	1,092.2	..	Imports of Goods & NFServices
2,153.9	2,466.7	2,842.7	3,195.4	3,694.8	4,013.3	4,333.4	4,664.0	4,961.0	5,239.4	..	Domestic Absorption
1,258.7	1,434.4	1,645.1	1,897.5	2,191.8	2,422.9	2,640.5	2,855.5	3,018.9	3,183.4	..	Private Consumption, etc.
390.1	443.9	517.5	604.3	711.8	793.5	866.4	924.2	987.0	1,017.8	..	General Gov't Consumption
505.1	588.5	680.1	693.6	791.3	796.8	826.5	884.3	955.1	1,038.2	..	Gross Domestic Investment
487.5	554.4	645.8	712.8	785.5	828.6	868.1	939.4	979.7	1,032.2	..	Fixed Investment
											Memo Items:
533.8	602.9	645.7	663.0	722.5	790.0	855.0	915.2	1,007.1	1,048.3	..	Gross Domestic Saving
530.4	601.7	647.9	660.9	711.6	765.1	822.7	877.8	979.8	1,018.7	..	Gross National Saving
				(Billions of 1980 French Francs)							
2,682.6	2,773.2	2,820.9	2,850.5	2,915.2	2,925.4	2,956.6	3,005.2	3,076.0	3,134.5	..	Gross National Product
2,351.1	2,412.5	2,452.1	2,497.5	2,554.4	2,570.5	2,602.9	2,642.3	2,699.9	2,739.8	B	GDP at factor cost
112.1	121.4	119.0	118.2	138.0	131.2	137.3	137.4	137.3	..	..	Agriculture
925.1	943.3	947.3	947.3	950.6	956.1	944.6	944.2	944.7	..	..	Industry
667.9	684.0	679.5	674.9	680.8	683.5	671.0	664.6	659.3	..	..	Manufacturing
1,639.4	1,698.7	1,742.0	1,775.8	1,825.0	1,846.6	1,890.5	1,940.0	2,004.9	..	..	Services, etc.
-18.4	-34.8	-34.4	1.1	-25.6	14.2	42.2	25.7	-25.7	-63.2	..	Resource Balance
547.7	588.5	604.4	626.6	616.1	638.8	683.5	694.8	693.1	701.7	..	Exports of Goods & NFServices
566.0	623.3	638.8	625.5	641.7	624.5	641.3	669.0	718.8	764.9	..	Imports of Goods & NFServices
2,695.0	2,798.2	2,842.7	2,840.2	2,939.3	2,919.6	2,930.3	2,996.0	3,112.5	3,209.5	..	Domestic Absorption
1,577.9	1,625.6	1,645.1	1,678.9	1,737.2	1,753.3	1,772.2	1,813.5	1,874.4	1,917.7	..	Private Consumption, etc.
490.1	504.9	517.5	533.5	553.4	564.8	571.2	584.1	605.5	618.2	..	General Gov't Consumption
627.0	667.7	680.1	627.7	648.7	601.6	586.8	598.4	632.7	673.6	..	Gross Domestic Investment
610.1	629.3	645.8	633.6	624.9	602.6	586.9	603.3	614.6	632.5	..	Fixed Investment
											Memo Items:
605.0	640.8	604.4	599.8	590.3	619.9	659.1	688.0	755.1	772.0	..	Capacity to Import
57.3	52.3	0.0	-26.8	-25.8	-18.9	-24.4	-6.8	62.0	70.3	..	Terms of Trade Adjustment
2,733.9	2,815.7	2,808.3	2,814.5	2,887.9	2,915.0	2,948.1	3,014.9	3,148.8	3,216.6	..	Gross Domestic Income
2,739.9	2,825.4	2,820.9	2,823.6	2,889.5	2,906.5	2,932.2	2,998.4	3,138.0	..	..	Gross National Income
				(Index 1980 = 100)							DOMESTIC PRICES (DEFLATORS)
81.5	89.8	100.0	111.4	124.4	136.6	146.7	155.4	162.4	166.8	..	Overall (GDP)
79.9	88.2	100.0	112.5	125.7	137.5	147.9	155.7	159.4	163.2	..	Domestic Absorption
93.2	97.6	100.0	110.8	120.4	129.2	127.6	131.4	136.1	..	..	Agriculture
82.9	91.1	100.0	108.7	120.8	131.4	141.7	152.9	161.9	..	..	Industry
84.5	92.9	100.0	108.7	120.7	131.8	142.5	156.4	168.7	..	..	Manufacturing
											MANUFACTURING ACTIVITY
102.2	100.5	100.0	96.7	95.3	93.1	90.4	87.7	..	..	..	Employment (1980=100)
..	..	..	..	..	..	..	..	..	..	..	Real Earnings per Empl. (1980=100)
89.4	95.6	100.0	104.6	103.9	105.4	110.1	106.3	..	..	..	Real Output per Empl. (1980=100)
..	..	..	..	..	..	..	..	..	..	..	Earnings as % of Value Added
				(Billions of current French Francs)							MONETARY HOLDINGS
1,714.4	1,948.3	2,124.9	2,351.0	2,616.7	2,897.8	3,148.7	3,338.3	3,573.7	3,992.0	D	Money Supply, Broadly Defined
644.1	710.6	750.9	857.8	942.4	1,050.3	1,149.0	1,186.2	1,258.5	1,317.9	..	Money as Means of Payment
131.9	139.1	143.7	160.8	177.1	190.6	198.8	209.0	216.2	226.5	..	Currency Ouside Banks
512.2	571.5	607.1	697.0	765.3	859.7	950.2	977.2	1,042.2	1,091.4	..	Demand Deposits
1,070.3	1,237.7	1,374.0	1,493.2	1,674.3	1,847.5	1,999.7	2,152.1	2,315.2	2,674.1	..	Quasi-Monetary Liabilities
				(Billions of current French Francs)							
-29.6	-37.0	-1.0	-85.6	-111.4	-142.4	-116.2	-125.3	-159.1	-40.4	..	GOVERNMENT DEFICIT (-) OR SURPLUS
793.2	942.1	1,112.1	1,271.2	1,486.0	1,655.4	1,832.3	1,981.5	2,109.6	2,246.4	..	Current Revenue
784.2	905.1	1,044.3	1,257.2	1,495.9	1,691.3	1,859.5	2,015.4	2,144.8	2,256.4	..	Current Expenditure
9.0	37.0	67.8	14.0	-9.9	-35.9	-27.2	-33.9	-35.2	-10.0	..	Current Budget Balance
0.9	1.0	1.4	2.1	3.5	6.5	2.3	1.7	1.9	1.9	..	Capital Receipts
51.2	64.9	68.6	87.5	140.7	104.2	108.9	108.8	127.4	70.9	..	Capital Payments

FRANCE	1967	1968	1969	1970	1971	1972	1973	1974	1975	1976	1977
FOREIGN TRADE (CUSTOMS BASIS)				*(Billions of current US dollars)*							
Value of Exports, fob	11.38	12.67	14.88	17.74	20.42	25.84	35.66	45.14	51.60	55.46	63.36
Nonfuel Primary Products	2.78	3.24	3.79	4.11	4.83	6.38	9.20	11.09	10.82	11.65	12.89
Fuels	0.35	0.33	0.35	0.38	0.45	0.58	0.76	1.21	1.41	1.61	1.89
Manufactures	8.25	9.11	10.74	13.25	15.13	18.88	25.70	32.84	39.38	42.20	48.58
Value of Imports, cif	12.38	13.93	17.22	18.92	21.14	26.72	37.05	52.17	53.61	64.02	70.28
Nonfuel Primary Products	4.12	4.31	5.26	5.60	5.60	6.88	9.97	12.16	12.17	13.95	16.47
Fuels	1.80	1.90	1.96	2.29	2.93	3.53	4.58	11.94	12.26	14.37	15.09
Manufactures	6.46	7.71	9.99	11.04	12.60	16.30	22.50	28.07	29.17	35.70	38.71
				(Index 1980 = 100)							
Terms of Trade	176.9	172.0	153.6	161.1	154.4	153.2	153.8	109.0	119.7	111.8	104.9
Export Prices, fob	29.0	29.0	29.0	30.3	31.8	35.7	48.7	56.2	63.1	62.2	64.4
Nonfuel Primary Products	33.7	33.1	35.7	36.5	36.5	41.5	61.9	73.6	69.2	67.6	63.3
Fuels	8.5	7.3	5.4	5.9	7.0	8.5	24.5	36.6	33.6	38.5	36.8
Manufactures	30.9	31.2	31.6	32.5	34.2	37.7	46.5	52.7	64.1	62.8	67.3
Import Prices, cif	16.4	16.9	18.9	18.8	20.6	23.3	31.7	51.6	52.7	55.6	61.4
Trade at Constant 1980 Prices				*(Billions of 1980 US dollars)*							
Exports, fob	39.24	43.67	51.30	58.57	64.22	72.32	73.25	80.30	81.76	89.17	98.34
Imports, cif	75.54	82.53	91.22	100.62	102.61	114.50	117.06	101.20	101.66	115.09	114.47
BALANCE OF PAYMENTS				*(Billions of current US dollars)*							
Exports of Goods & Services	..	..	..	25.29	28.91	35.78	49.90	63.51	73.25	78.69	90.47
Merchandise, fob	..	..	..	17.95	20.42	25.89	34.82	44.00	49.86	53.92	61.16
Nonfactor Services	..	..	..	5.61	6.71	7.81	11.32	14.18	18.06	19.21	22.69
Factor Services	..	..	..	1.73	1.78	2.08	3.76	5.34	5.33	5.56	6.63
Imports of Goods & Services	..	..	..	24.61	27.48	34.26	46.75	65.09	67.92	79.61	88.08
Merchandise, fob	..	..	..	17.69	19.56	24.85	34.41	48.79	48.76	58.91	64.46
Nonfactor Services	..	..	..	5.38	6.24	7.23	8.99	11.47	13.65	14.99	17.36
Factor Services	..	..	..	1.54	1.69	2.18	3.34	4.82	5.51	5.71	6.26
Long-Term Interest	..	..	..	..	..	..	..	..	..	..	..
Current Transfers, net	..	..	..	-0.66	-0.85	-0.90	-1.10	-1.16	-1.52	-1.41	-1.41
Workers' Remittances	..	..	..	0.13	0.15	0.18	0.20	0.21	0.24	0.24	0.27
Total to be Financed	..	..	..	0.02	0.58	0.61	2.05	-2.73	3.81	-2.33	0.99
Official Capital Grants	..	..	..	-0.22	-0.41	-0.74	-0.64	-1.12	-1.10	-1.04	-1.42
Current Account Balance	..	..	..	-0.20	0.16	-0.12	1.41	-3.86	2.71	-3.37	-0.43
Long-Term Capital, net	..	..	..	-0.16	-0.41	-1.39	-3.09	-1.38	-2.62	-3.23	-0.53
Direct Investment	..	..	..	0.25	0.13	0.09	0.19	1.34	0.23	-0.66	0.89
Long-Term Loans	..	..	..	..	..	..	..	..	..	..	..
Disbursements	..	..	..	..	..	..	..	..	..	..	..
Repayments	..	..	..	..	..	..	..	..	..	..	..
Other Long-Term Capital	..	..	..	-0.18	-0.12	-0.74	-2.65	-1.60	-1.74	-1.53	-0.01
Other Capital, net	..	..	..	2.17	3.53	3.12	-0.20	4.89	3.39	3.59	1.64
Change in Reserves	..	..	..	-1.81	-3.28	-1.61	1.89	0.35	-3.48	3.02	-0.68
Memo Item:				*(French Francs per US dollar)*							
Conversion Factor (Annual Avg)	4.940	4.940	5.190	5.550	5.540	5.040	4.450	4.810	4.290	4.780	4.910
EXTERNAL DEBT, ETC.				*(Millions US dollars, outstanding at end of year)*							
Public/Publicly Guar. Long-Term	..	..	..	..	..	..	..	..	..	..	..
Official Creditors	..	..	..	..	..	..	..	..	..	..	..
IBRD and IDA	..	..	..	..	..	..	..	..	..	..	..
Private Creditors	..	..	..	..	..	..	..	..	..	..	..
Private Non-guaranteed Long-Term	..	..	..	..	..	..	..	..	..	..	..
Use of Fund Credit	..	..	..	..	..	..	..	..	..	..	..
Short-Term Debt	..	..	..	..	..	..	..	..	..	..	..
Memo Items:				*(Millions of US dollars)*							
Int'l Reserves Excluding Gold	1,760	324	286	1,428	4,428	6,189	4,268	4,526	8,457	5,620	5,872
Gold Holdings (at market price)	5,264	4,641	3,567	3,771	4,392	6,535	11,327	18,823	14,155	13,612	16,770
SOCIAL INDICATORS											
Total Fertility Rate	2.6	2.6	2.5	2.5	2.5	2.4	2.3	2.1	1.9	1.8	1.9
Crude Birth Rate	17.0	16.7	16.8	16.8	17.2	17.0	16.4	15.3	14.1	13.6	14.0
Infant Mortality Rate	20.7	20.4	19.6	18.2	17.2	16.0	15.4	14.6	13.6	12.5	11.5
Life Expectancy at Birth	71.6	71.7	71.9	72.0	72.2	72.3	72.6	72.9	73.2	73.5	73.8
Food Production, p.c. ('79-81=100)	87.9	91.2	84.9	90.0	90.7	87.6	94.2	96.9	90.7	91.5	87.0
Labor Force, Agriculture (%)	16.2	15.3	14.5	13.6	13.1	12.6	12.1	11.6	11.1	10.6	10.1
Labor Force, Female (%)	35.3	35.6	35.9	36.2	36.5	36.8	37.1	37.4	37.7	38.0	38.4
School Enroll. Ratio, primary	..	..	..	117.0	..	..	..	..	109.0	109.0	111.0
School Enroll. Ratio, secondary	..	..	..	74.0	..	..	..	..	82.0	83.0	83.0

1978	1979	1980	1981	1982	1983	1984	1985	1986	1987 estimate	NOTES	FRANCE
				(Billions of current US dollars)							**FOREIGN TRADE (CUSTOMS BASIS)**
76.49	97.96	110.87	101.25	92.36	91.14	93.16	97.46	119.07	143.08	..	Value of Exports, fob
16.21	20.25	24.73	22.85	20.08	20.49	20.89	21.46	25.39	30.21	..	Nonfuel Primary Products
2.05	3.49	4.49	4.72	3.66	3.46	3.40	3.75	3.19	3.17	..	Fuels
58.24	74.22	81.65	73.68	68.62	67.19	68.87	72.24	90.49	109.70	..	Manufactures
81.86	106.71	134.33	120.28	115.45	105.27	103.61	107.59	127.85	157.52	..	Value of Imports, cif
18.69	22.52	25.61	21.68	20.28	19.53	19.56	19.57	23.65	27.72	..	Nonfuel Primary Products
15.94	22.95	35.73	34.67	30.83	25.75	24.91	23.95	16.14	16.93	..	Fuels
47.24	61.24	73.00	63.93	64.35	59.99	59.15	64.08	88.06	112.87	..	Manufactures
				(Index 1980 = 100)							
110.0	108.7	100.0	89.6	89.4	89.8	90.2	93.8	97.9	98.9	..	Terms of Trade
74.5	90.5	100.0	88.7	83.1	80.5	78.7	80.0	89.7	99.9	..	Export Prices, fob
73.4	92.8	100.0	91.6	81.1	84.8	80.3	76.6	77.5	78.1	..	Nonfuel Primary Products
44.7	93.6	100.0	104.0	97.0	85.4	83.6	81.1	66.5	65.2	..	Fuels
77.6	90.4	100.0	87.0	83.0	79.0	78.0	81.0	95.0	110.0	..	Manufactures
67.8	83.2	100.0	99.0	92.9	89.6	87.3	85.3	91.6	100.9	..	Import Prices, cif
				(Billions of 1980 US dollars)							Trade at Constant 1980 Prices
102.61	108.29	110.87	114.17	111.21	113.27	118.37	121.83	132.80	143.28	..	Exports, fob
120.81	128.19	134.33	121.46	124.32	117.51	118.75	126.13	139.57	156.07	..	Imports, cif
				(Billions of current US dollars)							**BALANCE OF PAYMENTS**
114.08	144.84	171.87	168.81	153.68	145.26	147.70	153.90	186.62	219.27	..	Exports of Goods & Services
74.50	94.27	107.56	100.76	91.41	89.69	92.19	95.66	117.65	139.39	..	Merchandise, fob
29.97	35.80	42.88	41.32	36.47	34.79	34.65	35.94	44.11	51.46	..	Nonfactor Services
9.61	14.77	21.42	26.73	25.79	20.78	20.86	22.30	24.86	28.41	..	Factor Services
103.76	135.65	171.89	169.35	161.14	146.50	145.62	151.45	179.40	218.58	..	Imports of Goods & Services
74.40	97.47	120.97	110.72	107.20	98.36	96.79	101.00	120.06	148.68	..	Merchandise, fob
20.67	25.26	31.65	33.06	27.79	25.51	25.13	25.39	32.25	39.16	..	Nonfactor Services
8.69	12.92	19.27	25.57	26.14	22.63	23.70	25.05	27.08	30.75	..	Factor Services
..	..	..	..	..	..	..	..	..	..	..	Long-Term Interest
-1.81	-2.30	-2.45	-2.27	-1.96	-1.74	-1.01	-1.31	-1.46	-1.72	..	Current Transfers, net
0.34	0.40	0.45	0.35	0.32	0.34	0.34	0.23	0.32	0.40	..	Workers' Remittances
8.51	6.89	-2.47	-2.81	-9.42	-2.98	1.06	1.14	5.77	-1.03	K	Total to be Financed
-1.47	-1.73	-1.72	-1.97	-2.66	-2.08	-1.88	-1.33	-2.85	-3.06	K	Official Capital Grants
7.04	5.16	-4.19	-4.78	-12.08	-5.06	-0.82	-0.18	2.92	-4.09	..	Current Account Balance
-4.91	-6.93	-10.18	-10.99	-1.77	7.25	2.50	1.80	-9.89	-1.03	..	Long-Term Capital, net
0.56	0.61	0.19	-2.13	-1.25	0.03	0.27	0.36	-2.12	-4.00	..	Direct Investment
..	..	..	..	..	..	..	..	..	..	..	Long-Term Loans
..	..	..	..	..	..	..	..	..	..	..	Disbursements
..	..	..	..	..	..	..	..	..	..	..	Repayments
-4.00	-5.81	-8.66	-6.89	2.14	9.30	4.11	2.77	-4.93	6.03	..	Other Long-Term Capital
0.87	3.52	20.44	11.07	10.28	1.97	1.12	0.75	8.36	-3.02	..	Other Capital, net
-3.00	-1.76	-6.07	4.70	3.56	-4.17	-2.80	-2.37	-1.39	8.14	..	Change in Reserves
				(French Francs per US dollar)							Memo Item:
4.510	4.250	4.230	5.430	6.570	7.620	8.740	8.990	6.930	6.010		Conversion Factor (Annual Avg)
				(Millions US dollars, outstanding at end of year)							**EXTERNAL DEBT, ETC.**
..	..	..	..	..	..	..	..	..	..	..	Public/Publicly Guar. Long-Term
..	..	..	..	..	..	..	..	..	..	..	Official Creditors
..	..	..	..	..	..	..	..	..	..	..	IBRD and IDA
..	..	..	..	..	..	..	..	..	..	..	Private Creditors
..	..	..	..	..	..	..	..	..	..	..	Private Non-guaranteed Long-Term
..	..	..	..	..	..	..	..	..	..	..	Use of Fund Credit
..	..	..	..	..	..	..	..	..	..	..	Short-Term Debt
				(Millions of US dollars)							Memo Items:
9,278	17,579	27,340	22,262	16,531	19,851	20,940	26,589	31,454	33,050	..	Int'l Reserves Excluding Gold
23,050	41,942	48,252	32,537	37,399	31,227	25,235	26,766	31,996	39,625	..	Gold Holdings (at market price)
											SOCIAL INDICATORS
1.8	1.9	2.0	2.0	1.9	1.9	1.9	1.9	1.9	1.8	..	Total Fertility Rate
13.8	14.1	14.9	14.9	14.7	14.6	14.4	14.2	14.1	13.9	..	Crude Birth Rate
10.6	10.1	10.0	9.6	9.4	9.3	9.1	9.0	8.8	8.7	..	Infant Mortality Rate
74.2	74.6	74.9	75.3	75.7	76.0	76.3	76.6	76.9	77.1	..	Life Expectancy at Birth
92.4	100.6	100.8	98.7	103.8	99.6	108.1	107.1	105.0	106.5	..	Food Production, p.c. ('79-81 = 100)
9.6	9.1	8.6	..	..	..	..	..	..	..	..	Labor Force, Agriculture (%)
38.7	39.0	39.3	39.4	39.4	39.5	39.6	39.6	39.7	39.7	..	Labor Force, Female (%)
112.0	112.0	111.0	..	109.0	108.0	107.0	110.0	112.0	..	..	School Enroll. Ratio, primary
83.0	84.0	85.0	..	88.0	90.0	91.0	93.0	95.0	..	..	School Enroll. Ratio, secondary

GABON	1967	1968	1969	1970	1971	1972	1973	1974	1975	1976	1977
CURRENT GNP PER CAPITA (US $)	480	520	610	670	690	780	890	1,710	2,660	4,170	3,750
POPULATION (thousands)	499	500	502	504	528	553	580	608	637	668	699
ORIGIN AND USE OF RESOURCES					*(Billions of current CFA Francs)*						
Gross National Product (GNP)	60.6	66.0	76.8	87.7	98.3	114.2	136.8	344.1	432.2	665.0	637.5
Net Factor Income from Abroad	-6.2	-6.9	-5.9	-7.4	-10.3	-11.5	-24.3	-27.6	-30.2	-54.1	-52.7
Gross Domestic Product (GDP)	66.8	72.9	82.7	95.1	108.6	125.7	161.1	371.7	462.4	719.1	690.2
Indirect Taxes, net	9.0	9.6	11.0	13.5	15.1	24.0	35.4	48.0	57.3	58.6	138.3
GDP at factor cost	57.8	63.3	71.7	81.6	93.5	101.7	125.7	323.7	405.1	660.5	551.9
Agriculture	16.1	13.4	15.4	16.6	19.9	16.3	18.7	30.5	29.6	36.0	37.7
Industry	25.3	33.2	37.7	42.6	48.9	64.4	80.0	232.6	283.5	440.6	397.0
Manufacturing	..	..	..	..	..	..	..	..	..	..	..
Services, etc.	25.4	26.3	29.6	35.9	39.8	45.0	62.4	108.6	149.3	242.5	255.5
Resource Balance	0.0	-0.6	0.5	3.3	5.1	0.2	-8.1	50.7	2.7	40.1	48.0
Exports of Goods & NFServices	32.1	35.3	45.2	51.9	60.9	68.7	..	..	..	..	..
Imports of Goods & NFServices	32.1	35.9	44.7	48.6	55.8	68.5	..	..	..	..	..
Domestic Absorption	66.8	73.5	82.2	91.8	103.5	125.5	169.2	321.0	459.7	679.0	642.2
Private Consumption, etc.	35.1	38.5	42.1	45.9	44.5	62.1	69.1	83.2	94.9	139.7	162.1
General Gov't Consumption	9.1	10.2	11.6	13.0	19.5	20.0	25.1	34.4	56.5	76.0	96.3
Gross Domestic Investment	22.6	24.8	28.5	32.9	39.5	43.4	75.0	203.4	308.3	463.3	383.8
Fixed Investment	..	..	..	32.9	48.0	47.4	60.4	156.3	258.1	435.5	338.1
Memo Items:											
Gross Domestic Saving	22.6	24.2	29.0	36.2	44.6	43.6	66.9	254.1	311.0	503.4	431.8
Gross National Saving	16.4	15.3	21.5	26.8	31.8	29.0	38.3	220.5	272.0	437.2	362.9
					(Billions of 1980 CFA Francs)						
Gross National Product	326.8	334.2	370.6	399.9	432.8	483.1	498.2	757.6	911.6	1,223.2	1,068.3
GDP at factor cost	..	..	..	..	..	..	..	..	..	..	..
Agriculture	..	..	..	..	..	..	..	..	..	..	..
Industry	..	..	..	..	..	..	..	..	..	..	..
Manufacturing	..	..	..	..	..	..	..	..	..	..	..
Services, etc.	..	..	..	..	..	..	..	..	..	..	..
Resource Balance	..	..	..	..	..	..	..	..	..	..	..
Exports of Goods & NFServices	..	..	..	..	..	..	..	..	..	..	..
Imports of Goods & NFServices	..	..	..	..	..	..	..	..	..	..	..
Domestic Absorption	..	..	..	..	..	..	..	..	..	..	..
Private Consumption, etc.	..	..	..	..	..	..	..	..	..	..	..
General Gov't Consumption	37.1	39.2	42.1	44.8	49.2	65.4	55.3	66.3	102.1	114.6	128.3
Gross Domestic Investment	69.1	71.6	78.4	87.3	99.9	119.4	137.6	386.6	478.2	684.2	534.4
Fixed Investment	..	..	..	..	..	..	..	274.1	420.4	601.8	400.8
Memo Items:											
Capacity to Import	..	..	..	..	..	..	..	..	..	..	..
Terms of Trade Adjustment	..	..	..	..	..	..	..	..	..	..	..
Gross Domestic Income	..	..	..	..	..	..	..	..	..	..	..
Gross National Income	..	..	..	..	..	..	..	..	..	..	..
DOMESTIC PRICES (DEFLATORS)					*(Index 1980 = 100)*						
Overall (GDP)	18.5	19.7	20.7	21.9	22.7	23.6	27.5	45.4	47.4	54.4	59.7
Domestic Absorption	..	..	..	..	..	..	..	..	..	..	..
Agriculture	..	..	..	..	..	..	..	..	..	..	..
Industry	..	..	..	..	..	..	..	..	..	..	..
Manufacturing	..	..	..	..	..	..	..	..	..	..	..
MANUFACTURING ACTIVITY											
Employment (1980 = 100)	..	..	..	..	..	..	..	..	..	..	..
Real Earnings per Empl. (1980 = 100)	..	..	..	..	..	..	..	..	..	..	..
Real Output per Empl. (1980 = 100)	..	..	..	..	..	..	..	..	..	..	..
Earnings as % of Value Added	..	..	..	..	..	40.1	39.8	47.8	50.3	44.4	42.2
MONETARY HOLDINGS					*(Billions of current CFA Francs)*						
Money Supply, Broadly Defined	9.8	10.6	12.4	13.5	15.3	19.7	26.1	44.0	71.0	132.5	129.6
Money as Means of Payment	8.9	9.6	11.4	12.5	14.5	18.4	22.8	38.1	59.0	104.1	95.8
Currency Ouside Banks	3.6	3.7	4.3	5.0	5.8	7.0	9.1	15.3	21.7	32.7	30.5
Demand Deposits	5.3	5.9	7.2	7.6	8.7	11.4	13.7	22.8	37.3	71.3	65.3
Quasi-Monetary Liabilities	0.9	1.0	1.0	1.0	0.8	1.3	3.3	5.9	12.0	28.4	33.8
					(Billions of current CFA Francs)						
GOVERNMENT DEFICIT (-) OR SURPLUS	..	..	..	..	..	..	-17.6	-15.5	-37.0	-158.1	..
Current Revenue	..	..	..	..	..	..	38.8	89.2	167.3	188.4	..
Current Expenditure	..	..	..	..	..	..	..	..	..	..	..
Current Budget Balance	..	..	..	..	..	..	..	..	..	..	..
Capital Receipts	..	..	..	..	..	..	0.1	0.0	0.1	0.6	..
Capital Payments	..	..	..	..	..	..	..	..	..	..	..

1978	1979	1980	1981	1982	1983	1984	1985	1986	1987 estimate	NOTES	GABON
2,770	3,110	3,830	4,260	4,410	3,950	3,440	3,330	3,190	2,760	..	CURRENT GNP PER CAPITA (US $)
731	764	797	831	868	907	949	997	1,021	1,047	..	POPULATION (thousands)
				(Billions of current CFA Francs)							ORIGIN AND USE OF RESOURCES
471.2	567.1	802.8	960.4	1,098.8	1,236.4	1,305.0	1,388.0	1,041.0	919.0	..	Gross National Product (GNP)
-68.0	-77.5	-101.7	-89.2	-102.9	-107.1	-231.0	-258.0	-132.0	-133.0	..	Net Factor Income from Abroad
539.2	644.6	904.5	1,049.6	1,201.7	1,343.5	1,536.0	1,646.0	1,173.0	1,052.0	..	Gross Domestic Product (GDP)
104.6	113.8	134.3	110.8	126.6	106.8	87.0	105.0	98.0	..	..	Indirect Taxes, net
434.6	530.8	770.2	938.8	1,075.1	1,236.7	1,449.0	1,541.0	1,075.0	..	B	GDP at factor cost
35.1	41.0	65.4	67.2	75.6	80.7	97.0	102.0	109.0	114.0	..	Agriculture
321.6	410.3	560.0	642.1	790.9	769.0	935.0	911.0	485.0	430.0	..	Industry
..	..	..	..	..	..	..	..	..	..	..	Manufacturing
182.5	193.3	279.1	340.3	335.2	493.8	504.0	633.0	579.0	508.0	..	Services, etc.
80.0	135.9	291.6	216.5	223.5	277.9	306.0	225.0	-183.0	29.0	..	Resource Balance
..	..	..	681.5	765.9	838.7	931.0	898.0	400.0	432.0	..	Exports of Goods & NFServices
..	..	..	465.0	542.4	560.8	625.0	673.0	583.0	403.0	..	Imports of Goods & NFServices
459.2	508.7	612.9	833.1	978.2	1,065.6	1,230.0	1,421.0	1,356.0	1,023.0	..	Domestic Absorption
196.6	218.6	236.7	302.1	391.9	375.5	431.0	501.0	550.0	448.0	..	Private Consumption, etc.
73.8	77.9	119.7	149.5	170.0	238.1	284.0	306.0	268.0	244.0	..	General Gov't Consumption
188.8	212.2	256.5	381.5	416.3	452.0	515.0	614.0	538.0	331.0	..	Gross Domestic Investment
197.2	203.1	241.2	325.1	332.3	453.0	478.0	597.0	560.0	331.0	..	Fixed Investment
											Memo Items:
268.8	348.1	548.1	598.0	639.8	729.9	821.0	839.0	355.0	360.0	..	Gross Domestic Saving
183.0	237.9	413.1	483.5	509.5	586.6	549.6	531.9	164.6	183.0	..	Gross National Saving
				(Billions of 1980 CFA Francs)							
767.5	776.4	802.8	772.3	796.4	808.1	791.7	843.9	876.1	761.7	..	Gross National Product
..	..	..	..	..	..	..	..	..	..	..	GDP at factor cost
..	..	..	..	..	..	..	..	..	..	..	Agriculture
..	..	..	..	..	..	..	..	..	..	..	Industry
..	..	..	..	..	..	..	..	..	..	..	Manufacturing
..	..	..	..	..	..	..	..	..	..	..	Services, etc.
..	..	..	..	..	..	..	..	..	..	..	Resource Balance
..	..	..	..	..	..	..	..	..	..	..	Exports of Goods & NFServices
..	..	..	..	..	..	..	..	..	..	..	Imports of Goods & NFServices
..	..	..	..	..	..	..	..	..	..	..	Domestic Absorption
..	..	..	..	..	..	..	..	..	..	..	Private Consumption, etc.
92.4	90.8	119.7	131.1	148.1	152.1	173.2	187.6	179.5	146.5	..	General Gov't Consumption
203.3	251.1	256.5	364.6	332.6	311.0	332.7	371.6	315.9	187.3	..	Gross Domestic Investment
225.1	224.3	241.2	327.0	293.9	312.0	291.6	338.7	..	..	..	Fixed Investment
											Memo Items:
..	..	..	..	..	..	..	..	..	..	..	Capacity to Import
..	..	..	..	..	..	..	..	..	..	..	Terms of Trade Adjustment
..	..	..	..	..	..	..	..	..	..	..	Gross Domestic Income
..	..	..	..	..	..	..	..	..	..	..	Gross National Income
				(Index 1980 = 100)							DOMESTIC PRICES (DEFLATORS)
61.4	73.0	100.0	120.9	134.7	149.3	160.3	161.6	122.0	124.4	..	Overall (GDP)
..	..	..	..	..	..	..	..	..	..	..	Domestic Absorption
..	..	..	..	..	..	..	..	..	..	..	Agriculture
..	..	..	..	..	..	..	..	..	..	..	Industry
..	..	..	..	..	..	..	..	..	..	..	Manufacturing
											MANUFACTURING ACTIVITY
..	..	..	..	..	..	..	..	..	..	..	Employment (1980=100)
..	..	..	..	..	..	..	..	..	..	B	Real Earnings per Empl. (1980=100)
..	..	..	..	..	..	..	..	..	..	..	Real Output per Empl. (1980=100)
44.3	..	..	..	..	..	..	..	..	..	..	Earnings as % of Value Added
				(Billions of current CFA Francs)							MONETARY HOLDINGS
112.8	122.4	152.1	176.2	199.6	235.3	272.8	306.0	274.9	238.7	..	Money Supply, Broadly Defined
89.8	85.4	94.0	114.5	125.6	142.6	167.8	176.6	152.4	133.3	..	Money as Means of Payment
30.5	29.9	34.9	36.7	44.0	48.4	52.6	55.8	47.4	49.5	..	Currency Ouside Banks
59.4	55.6	59.0	77.8	81.6	94.2	115.1	120.8	105.0	83.8	..	Demand Deposits
22.9	36.9	58.1	61.7	74.1	92.7	105.0	129.4	122.5	105.5	..	Quasi-Monetary Liabilities
				(Billions of current CFA Francs)							
..	-3.5	55.1	8.2	35.1	-16.3	3.1	1.1	..	..	F	GOVERNMENT DEFICIT (-) OR SURPLUS
..	275.0	353.5	422.8	488.1	509.3	606.7	648.4	..	..	..	Current Revenue
..	..	..	..	..	253.1	290.2	298.5	..	..	..	Current Expenditure
..	..	..	..	..	256.2	316.5	349.9	..	..	..	Current Budget Balance
..	..	0.3	0.4	0.3	0.1	0.1	..	..	..	..	Capital Receipts
..	..	..	..	..	272.6	313.5	348.8	..	..	..	Capital Payments

GABON	1967	1968	1969	1970	1971	1972	1973	1974	1975	1976	1977
FOREIGN TRADE (CUSTOMS BASIS)					*(Millions of current US dollars)*						
Value of Exports, fob	125.0	128.6	142.0	140.4	178.4	227.6	330.3	767.9	942.6	1,135.0	1,343.0
Nonfuel Primary Products	79.6	76.3	82.8	77.9	87.2	140.1	169.0	179.9	183.2	226.2	249.0
Fuels	36.1	43.4	50.2	51.6	78.2	68.9	141.9	563.3	749.3	895.8	1,064.9
Manufactures	9.3	8.9	9.1	10.9	13.0	18.6	19.4	24.8	10.2	13.0	29.1
Value of Imports, cif	67.2	64.3	77.9	79.8	90.9	133.3	189.7	331.9	469.2	503.2	716.0
Nonfuel Primary Products	11.4	11.8	12.2	12.3	13.9	20.7	27.8	47.9	66.6	66.5	101.1
Fuels	4.1	1.1	1.2	1.1	1.1	1.6	2.2	4.0	6.0	3.9	7.8
Manufactures	51.7	51.4	64.6	66.5	75.9	111.0	159.8	280.1	396.6	432.8	607.1
					(Index 1980 = 100)						
Terms of Trade	44.7	32.5	29.7	27.6	28.7	35.5	34.5	69.9	61.6	65.6	64.0
Export Prices, fob	10.2	9.2	8.9	8.9	9.9	13.4	15.7	40.0	38.2	42.0	45.3
Nonfuel Primary Products	24.0	23.0	22.0	23.4	25.1	26.4	37.7	52.9	52.1	66.4	65.3
Fuels	4.3	4.3	4.3	4.3	5.6	6.2	8.9	36.7	35.7	38.4	42.0
Manufactures	19.6	18.4	22.3	25.7	25.7	30.7	34.1	52.0	55.9	55.3	64.2
Import Prices, cif	22.8	28.3	30.0	32.0	34.5	37.8	45.4	57.2	62.0	64.0	70.8
Trade at Constant 1980 Prices					*(Millions of 1980 US dollars)*						
Exports, fob	1,227.6	1,399.3	1,595.1	1,585.8	1,802.5	1,695.8	2,108.5	1,921.9	2,466.4	2,699.5	2,963.9
Imports, cif	294.8	227.6	259.8	249.2	263.3	353.0	417.8	580.6	756.3	785.7	1,011.6
BALANCE OF PAYMENTS					*(Millions of current US dollars)*						
Exports of Goods & Services	..	..	..	207.9	251.6	310.9	442.6	1,005.4	1,247.3	1,357.7	1,487.0
Merchandise, fob	..	..	..	173.7	225.1	268.0	396.5	956.9	1,149.2	1,217.3	1,300.4
Nonfactor Services	..	..	..	31.4	24.9	40.0	42.3	39.3	85.8	127.0	172.7
Factor Services	..	..	..	2.8	1.7	3.0	3.8	9.1	12.3	13.4	13.9
Imports of Goods & Services	..	..	..	215.8	233.5	329.3	488.9	784.7	1,175.1	1,299.4	1,349.8
Merchandise, fob	..	..	..	99.3	106.9	155.6	231.0	349.7	516.9	527.3	589.2
Nonfactor Services	..	..	..	97.2	101.1	142.0	196.7	299.7	492.7	580.1	595.3
Factor Services	..	..	..	19.3	25.5	31.7	61.2	135.3	165.5	192.0	165.2
Long-Term Interest	..	..	..	2.6	3.4	4.2	10.9	10.7	15.0	25.3	31.0
Current Transfers, net	..	..	..	-7.2	-9.0	-12.2	-19.4	-25.0	-41.2	-50.6	-65.8
Workers' Remittances	..	..	..	0.0	0.0	0.0	0.0	0.0	0.0	0.0	0.0
Total to be Financed	..	..	..	-15.1	9.1	-30.5	-65.7	195.7	31.1	7.7	71.3
Official Capital Grants	..	..	..	12.5	8.8	26.8	22.7	23.5	42.4	44.6	27.6
Current Account Balance	..	..	..	-2.6	18.0	-3.7	-43.0	219.1	73.5	52.3	98.9
Long-Term Capital, net	..	..	..	18.8	26.9	27.7	64.0	140.0	318.8	242.2	10.0
Direct Investment	..	..	..	-0.7	15.7	17.4	15.9	76.4	159.9	1.0	14.6
Long-Term Loans	..	..	..	16.6	32.0	79.8	138.2	114.6	318.2	333.7	131.8
Disbursements	..	..	..	25.9	47.0	98.0	189.9	144.6	367.6	396.4	228.3
Repayments	..	..	..	9.3	15.0	18.2	51.7	30.0	49.4	62.7	96.5
Other Long-Term Capital	..	..	..	-9.6	-29.7	-96.3	-112.7	-74.4	-201.6	-137.1	-163.9
Other Capital, net	..	..	..	-11.3	-37.8	-27.9	1.8	-308.4	-346.3	-311.7	-220.1
Change in Reserves	..	..	..	-4.9	-7.0	3.9	-22.8	-50.8	-46.0	17.2	111.2
Memo Item:					*(CFA Francs per US dollar)*						
Conversion Factor (Annual Avg)	246.850	246.850	259.710	277.710	277.130	252.210	222.700	240.500	214.320	238.980	245.670
EXTERNAL DEBT, ETC.					*(Millions of US dollars, outstanding at end of year)*						
Public/Publicly Guar. Long-Term	..	..	..	90.8	127.5	200.6	346.2	471.5	771.6	1,079.5	1,255.5
Official Creditors	..	..	..	67.0	74.6	77.3	91.1	102.0	172.7	200.0	232.4
IBRD and IDA	..	..	..	29.3	27.6	24.1	19.9	17.3	19.3	21.0	21.3
Private Creditors	..	..	..	23.9	52.9	123.3	255.1	369.5	598.9	879.5	1,023.1
Private Non-guaranteed Long-Term	..	..	..	..	..	..	..	..	..	..	..
Use of Fund Credit	..	..	..	0.0	0.0	0.0	0.0	0.0	0.0	0.0	0.0
Short-Term Debt	..	..	..	..	..	..	..	..	..	..	221.0
Memo Items:					*(Thousands of US dollars)*						
Int'l Reserves Excluding Gold	9,433.0	5,467.0	8,286.0	14,740.0	25,387.0	23,227.0	47,858.0	103,290.0	146,070.0	116,160.0	9,905.0
Gold Holdings (at market price)	..	..	..	..	..	..	..	..	..	..	1,056.0
SOCIAL INDICATORS											
Total Fertility Rate	4.2	4.2	4.2	4.2	4.2	4.3	4.3	4.3	4.3	4.4	4.4
Crude Birth Rate	30.9	30.9	30.9	30.9	30.9	30.9	30.9	30.9	30.9	30.9	30.9
Infant Mortality Rate	147.0	144.0	141.0	138.0	135.0	132.0	130.0	128.0	126.0	124.0	122.0
Life Expectancy at Birth	43.0	43.4	43.8	44.2	44.6	45.0	45.4	45.8	46.3	46.7	47.1
Food Production, p.c. ('79-81 = 100)	108.3	109.2	107.6	107.1	106.0	103.1	101.6	100.0	95.4	95.4	92.1
Labor Force, Agriculture (%)	81.3	80.8	80.2	79.6	79.2	78.8	78.4	78.0	77.6	77.2	76.7
Labor Force, Female (%)	40.3	40.2	40.1	40.0	40.0	39.9	39.8	39.8	39.7	39.6	39.6
School Enroll. Ratio, primary	..	..	..	85.0	..	..	..	..	102.0	..	..
School Enroll. Ratio, secondary	..	..	..	8.0	..	..	..	..	17.0	..	..

1978	1979	1980	1981	1982	1983	1984	1985	1986	1987 estimate	NOTES	GABON
				(Millions of current US dollars)							**FOREIGN TRADE (CUSTOMS BASIS)**
1,107.2	1,848.4	2,172.8	2,200.7	2,160.6	1,975.3	2,011.4	1,974.4	1,114.0	1,284.6	f	Value of Exports, fob
310.3	389.3	340.0	417.8	412.2	404.1	469.6	485.9	454.4	525.4	..	Nonfuel Primary Products
782.3	1,415.8	1,806.8	1,712.4	1,697.8	1,485.4	1,416.6	1,346.0	519.7	623.8	..	Fuels
14.6	43.4	26.0	70.5	50.7	85.7	125.2	142.5	140.0	135.5	..	Manufactures
682.0	673.0	686.0	842.7	798.5	852.2	723.6	861.4	924.0	662.7	f	Value of Imports, cif
145.7	140.5	141.4	148.4	146.5	167.4	127.1	146.2	223.1	141.8	..	Nonfuel Primary Products
8.4	10.2	9.4	22.5	8.2	15.1	15.6	16.2	6.6	8.9	..	Fuels
527.9	522.3	535.2	671.8	643.8	669.7	581.0	699.0	694.3	512.0	..	Manufactures
				(Index 1980 = 100)							
60.5	71.8	100.0	107.1	101.6	95.3	95.4	90.5	55.9	64.1	..	Terms of Trade
48.3	66.0	100.0	106.8	98.6	90.1	88.9	84.4	59.4	71.4	..	Export Prices, fob
73.8	91.7	100.0	92.5	90.2	86.7	91.2	81.4	82.9	95.2	..	Nonfuel Primary Products
42.3	61.0	100.0	112.5	101.6	92.5	90.2	87.5	44.9	56.9	..	Fuels
69.8	77.7	100.0	81.0	80.4	72.1	70.8	69.5	82.2	90.3	..	Manufactures
79.9	91.8	100.0	99.6	97.1	94.6	93.2	93.3	106.3	111.4	..	Import Prices, cif
				(Millions of 1980 US dollars)							Trade at Constant 1980 Prices
2,290.9	2,801.9	2,172.8	2,061.5	2,190.2	2,191.7	2,262.9	2,339.4	1,874.7	1,798.1	..	Exports, fob
853.5	732.7	686.0	845.8	822.7	901.1	776.7	923.5	869.5	594.7	..	Imports, cif
				(Millions of current US dollars)							**BALANCE OF PAYMENTS**
1,482.8	2,031.2	2,433.9	2,533.0	2,350.7	2,218.2	2,203.4	2,119.4	1,212.6	..	..	Exports of Goods & Services
1,308.6	1,815.0	2,084.4	2,200.2	2,160.4	2,000.1	2,017.8	1,951.7	1,071.3	1,286.4	..	Merchandise, fob
169.8	208.8	324.6	307.6	170.3	200.8	137.8	138.5	123.2	..	..	Nonfactor Services
4.4	7.3	24.9	25.2	20.0	17.3	47.7	29.2	18.1	..	..	Factor Services
1,366.7	1,665.6	1,926.0	2,064.6	1,982.1	2,056.8	2,023.5	2,186.2	2,120.9	..	..	Imports of Goods & Services
557.9	554.8	686.1	841.2	722.6	725.5	733.2	854.9	976.5	731.8	..	Merchandise, fob
577.7	741.9	789.4	868.2	926.1	1,031.5	1,018.5	1,058.5	880.3	..	..	Nonfactor Services
231.1	368.8	450.5	355.2	333.3	299.8	271.8	272.8	264.1	..	..	Factor Services
71.5	101.3	122.8	90.3	96.5	52.3	58.4	52.9	52.8	57.2	..	Long-Term Interest
-79.1	-153.7	-157.5	-93.1	-83.5	-95.1	-92.4	-109.3	-168.6	-146.4	..	Current Transfers, net
0.0	0.1	..	0.2	0.1	0.3	0.1	0.1	0.2	0.0	..	Workers' Remittances
37.1	211.9	350.4	375.3	285.0	66.4	87.4	-176.0	-1,076.9	-231.4	K	Total to be Financed
36.7	35.8	33.4	28.0	24.3	5.6	25.3	13.5	22.2	21.6	K	Official Capital Grants
73.9	247.6	383.9	403.3	309.4	72.0	112.7	-162.5	-1,054.7	-209.7	..	Current Account Balance
136.7	67.3	-2.4	-67.9	117.3	77.4	197.9	226.1	590.7	559.5	..	Long-Term Capital, net
56.5	48.3	23.5	47.5	127.0	106.1	4.8	11.1	103.4	121.4	..	Direct Investment
-55.8	195.7	-115.4	-167.7	-102.7	-73.5	49.8	80.1	141.0	252.0	..	Long-Term Loans
146.1	427.4	170.3	59.4	88.1	85.4	236.9	252.2	256.4	265.3	..	Disbursements
201.9	231.7	285.7	227.1	190.8	158.9	187.1	172.1	115.4	13.3	..	Repayments
99.2	-212.5	56.0	24.4	68.7	39.2	118.0	121.4	324.2	164.5	..	Other Long-Term Capital
-202.4	-331.3	-285.4	-217.6	-280.2	-238.9	-271.4	-112.5	332.6	-549.5	..	Other Capital, net
-8.2	16.3	-96.0	-117.8	-146.5	89.4	-39.3	48.9	131.4	199.7	..	Change in Reserves
				(CFA Francs per US dollar)							Memo Item:
225.640	212.720	211.300	271.730	328.620	381.070	436.960	449.260	346.300	300.540	..	Conversion Factor (Annual Avg)
				(Millions of US dollars, outstanding at end of year)							**EXTERNAL DEBT, ETC.**
1,288.0	1,509.7	1,307.9	997.5	825.3	673.5	658.4	871.3	1,135.2	1,604.8	..	Public/Publicly Guar. Long-Term
288.2	380.1	360.8	327.0	296.6	259.5	239.9	296.8	395.5	639.2	..	Official Creditors
21.3	20.3	18.6	17.0	15.2	14.6	10.2	11.1	12.3	13.8	..	IBRD and IDA
999.8	1,129.6	947.0	670.6	528.7	414.0	418.6	574.5	739.8	965.7	..	Private Creditors
..	..	..	..	..	..	..	..	..	..	..	Private Non-guaranteed Long-Term
9.9	19.8	14.5	13.2	10.3	1.9	0.0	0.0	33.5	60.3	..	Use of Fund Credit
214.0	241.0	228.0	147.0	173.0	233.0	251.0	262.0	439.7	405.5	..	Short-Term Debt
				(Thousands of US dollars)							Memo Items:
22,574.0	20,135.0	107,500.0	198,850.0	311,890.0	186,900.0	199,450.0	192,550.0	126,350.0	11,999.0	..	Int'l Reserves Excluding Gold
2,147.0	6,451.0	7,428.0	5,009.0	5,757.0	4,807.0	3,885.0	4,120.0	4,925.0	6,100.0	..	Gold Holdings (at market price)
											SOCIAL INDICATORS
4.4	4.4	4.5	4.5	4.5	4.7	4.9	5.1	5.3	5.5	..	Total Fertility Rate
31.5	32.1	32.6	33.2	33.8	35.4	37.0	38.6	40.2	41.7	..	Crude Birth Rate
120.0	118.0	116.0	114.0	112.0	109.1	106.2	103.3	100.5	97.6	..	Infant Mortality Rate
47.6	48.0	48.4	48.9	49.3	49.9	50.5	51.1	51.7	52.3	..	Life Expectancy at Birth
96.8	99.2	102.3	98.5	100.5	101.9	100.2	97.0	97.3	97.0	..	Food Production, p.c. ('79-81=100)
76.3	75.8	75.4	..	..	..	..	..	..	..	..	Labor Force, Agriculture (%)
39.5	39.4	39.4	39.2	39.0	38.8	38.6	38.4	38.2	38.0	..	Labor Force, Female (%)
..	..	115.0	117.0	119.0	123.0	..	..	..	..	..	School Enroll. Ratio, primary
..	..	21.0	21.0	23.0	25.0	..	..	..	..	..	School Enroll. Ratio, secondary

GAMBIA, THE	1967	1968	1969	1970	1971	1972	1973	1974	1975	1976	1977
CURRENT GNP PER CAPITA (US $)	110	100	110	100	120	130	140	200	210	230	250
POPULATION (thousands)	426	438	451	463	476	489	502	517	533	551	570

ORIGIN AND USE OF RESOURCES

(Millions of current Gambian Dalasis)

	1967	1968	1969	1970	1971	1972	1973	1974	1975	1976	1977
Gross National Product (GNP)	81.1	82.2	96.7	111.8	108.5	117.7	128.8	196.0	220.5	274.0	347.3
Net Factor Income from Abroad	-4.3	-3.9	-5.4	-4.0	-4.7	-5.9	-3.6	0.6	-0.6	-4.3	-7.7
Gross Domestic Product (GDP)	85.4	86.1	102.1	115.8	113.2	123.6	132.4	195.4	221.1	278.3	355.0
Indirect Taxes, net	8.7	7.9	10.4	14.4	10.6	11.7	12.7	14.7	17.6	33.1	46.5
GDP at factor cost	76.7	78.2	91.7	101.4	102.6	111.9	119.7	180.7	203.5	245.2	308.5
Agriculture	27.6	27.7	32.9	33.0	36.5	39.9	42.6	63.0	70.5	92.5	103.8
Industry	7.1	7.2	8.5	9.4	9.4	10.2	10.9	16.3	20.8	27.2	33.3
Manufacturing	2.5	2.6	3.0	3.4	3.4	3.6	3.9	5.8	6.6	11.8	14.3
Services, etc.	42.0	43.3	50.3	59.0	56.7	61.8	66.2	101.4	112.2	125.5	171.4
Resource Balance	-1.5	-10.3	-1.9	5.6	-7.8	0.2	-8.8	8.9	21.9	-34.7	-21.4
Exports of Goods & NFServices	46.3	40.4	55.7	48.4	45.3	55.8	57.8	103.7	131.7	127.5	166.6
Imports of Goods & NFServices	47.8	50.7	57.6	42.8	53.1	55.6	66.6	94.8	109.8	162.2	188.0
Domestic Absorption	86.9	96.4	104.0	110.2	121.0	123.4	141.2	186.5	199.2	313.0	376.4
Private Consumption, etc.	66.7	74.3	81.4	91.0	98.0	96.5	111.2	151.4	140.0	239.3	269.0
General Gov't Consumption	11.9	13.1	13.6	13.9	16.1	17.5	19.6	26.9	33.5	43.5	57.5
Gross Domestic Investment	8.3	9.0	9.0	5.3	6.9	9.4	10.4	8.2	25.7	30.2	49.9
Fixed Investment	..	..	..	..	..	..	..	..	..	..	..
Memo Items:											
Gross Domestic Saving	6.8	-1.3	7.1	10.9	-0.9	9.6	1.6	17.1	47.6	-4.5	28.5
Gross National Saving	2.5	-5.2	1.7	7.5	-4.9	4.8	-1.3	18.0	46.6	-6.0	23.0

(Millions of 1980 Gambian Dalasis)

	1967	1968	1969	1970	1971	1972	1973	1974	1975	1976	1977
Gross National Product	233.61	238.80	261.96	240.12	265.44	282.76	303.23	375.04	364.64	395.90	405.11
GDP at factor cost	230.81	234.01	259.34	233.61	260.47	279.39	292.18	350.13	328.13	346.69	351.62
Agriculture	79.30	80.51	89.31	80.40	89.64	96.02	100.53	120.44	112.96	114.61	114.17
Industry	42.13	42.69	47.35	42.51	47.54	51.08	53.32	63.95	59.66	60.40	62.08
Manufacturing	-66.82	-68.27	-75.54	-68.27	-75.54	-81.35	-84.98	-101.68	-93.69	-106.77	-103.86
Services, etc.	109.37	110.80	122.67	110.70	123.29	132.29	138.33	165.75	155.51	171.68	175.36
Resource Balance	-17.02	-30.15	-12.25	-6.17	-49.25	-28.31	-27.90	10.09	28.66	-33.46	-55.35
Exports of Goods & NFServices	167.31	156.93	176.35	135.07	114.43	145.70	147.90	166.34	202.12	212.13	203.46
Imports of Goods & NFServices	184.33	187.09	188.60	141.24	163.68	174.01	175.80	156.25	173.46	245.60	258.81
Domestic Absorption	263.26	279.86	289.05	255.48	327.31	326.52	339.84	363.73	336.96	435.37	469.89
Private Consumption, etc.	214.53	227.63	239.10	213.24	278.04	269.53	281.03	310.06	251.39	347.96	361.93
General Gov't Consumption	29.75	32.34	31.75	31.90	36.05	39.32	42.58	36.13	39.24	39.24	42.66
Gross Domestic Investment	18.98	19.89	18.19	10.34	13.22	17.67	16.23	17.54	46.33	48.16	65.31
Fixed Investment	..	..	..	..	..	..	..	..	..	..	..
Memo Items:											
Capacity to Import	178.55	149.08	182.38	159.73	139.64	174.64	152.57	170.92	208.06	193.05	229.35
Terms of Trade Adjustment	11.24	-7.85	6.03	24.65	25.21	28.94	4.67	4.58	5.94	-19.08	25.89
Gross Domestic Income	257.47	241.86	282.83	273.96	303.27	327.14	316.61	378.40	371.56	382.83	440.43
Gross National Income	244.85	230.95	267.99	264.77	290.64	311.70	307.91	379.62	370.58	376.82	430.99

DOMESTIC PRICES (DEFLATORS)

(Index 1980 = 100)

	1967	1968	1969	1970	1971	1972	1973	1974	1975	1976	1977
Overall (GDP)	34.7	34.5	36.9	46.4	40.7	41.4	42.4	52.3	60.5	69.2	85.6
Domestic Absorption	33.0	34.4	36.0	43.1	37.0	37.8	41.5	51.3	59.1	71.9	80.1
Agriculture	34.8	34.4	36.8	41.0	40.7	41.6	42.4	52.3	62.4	80.7	90.9
Industry	16.9	16.9	17.9	22.1	19.8	20.0	20.4	25.5	34.9	45.0	53.6
Manufacturing	-3.7	-3.8	-4.0	-5.0	-4.5	-4.4	-4.6	-5.7	-7.0	-11.1	-13.8

MANUFACTURING ACTIVITY

	1967	1968	1969	1970	1971	1972	1973	1974	1975	1976	1977
Employment (1980=100)	..	..	..	..	..	..	..	..	164.0	177.7	189.1
Real Earnings per Empl. (1980=100)	..	..	..	..	..	..	..	..	45.1	52.3	54.0
Real Output per Empl. (1980=100)	..	..	..	..	..	..	..	..	-622.4	-408.7	-480.7
Earnings as % of Value Added	..	..	..	..	..	..	..	..	43.2	43.7	40.8

MONETARY HOLDINGS

(Millions of current Gambian Dalasis)

	1967	1968	1969	1970	1971	1972	1973	1974	1975	1976	1977
Money Supply, Broadly Defined	14.8	16.0	16.8	19.4	19.4	27.1	38.4	42.2	47.9	69.9	61.5
Money as Means of Payment	12.7	13.7	14.1	16.5	16.4	22.8	31.8	33.0	37.3	49.4	39.8
Currency Ouside Banks	8.9	9.4	10.2	12.1	12.7	14.9	25.2	24.1	27.2	32.1	19.8
Demand Deposits	3.8	4.3	3.9	4.3	3.6	7.8	6.6	9.0	10.2	17.3	20.0
Quasi-Monetary Liabilities	2.1	2.4	2.7	2.9	3.1	4.3	6.6	9.2	10.6	20.5	21.7

GOVERNMENT DEFICIT (-) OR SURPLUS

(Millions of current Gambian Dalasis)

	1967	1968	1969	1970	1971	1972	1973	1974	1975	1976	1977
	..	..	..	..	..	..	0	-3	-7	-7	-27
Current Revenue	..	..	..	..	..	19	23	27	35	47	65
Current Expenditure	..	..	..	..	..	..	18	21	31	40	54
Current Budget Balance	..	..	..	..	..	..	5	6	4	8	10
Capital Receipts	..	..	..	..	..	0	0	0	0	0	0
Capital Payments	..	..	..	..	..	..	5	9	11	14	37

1978	1979	1980	1981	1982	1983	1984	1985	1986	1987 estimate	NOTES	GAMBIA, THE
260	340	380	340	320	290	250	230	210	220	..	CURRENT GNP PER CAPITA (US $)
590	612	634	657	680	704	726	748	773	797	..	POPULATION (thousands)
				(Millions of current Gambian Dalasis)							ORIGIN AND USE OF RESOURCES
355.3	412.9	421.2	406.5	439.9	505.3	644.2	822.9	1,032.4	1,221.1	C	Gross National Product (GNP)
-6.5	-4.9	-3.8	-5.1	-11.5	-22.3	-28.5	-36.4	-45.7	-53.9	..	Net Factor Income from Abroad
361.8	417.8	425.0	411.6	451.4	527.6	672.7	859.3	1,078.1	1,275.0	C	Gross Domestic Product (GDP)
46.9	56.5	65.5	44.3	45.6	66.9	82.8	92.9	100.0	111.7	..	Indirect Taxes, net
314.9	361.3	359.5	367.3	405.8	460.7	589.9	766.4	978.1	1,163.3	C	GDP at factor cost
96.5	126.1	112.3	111.9	154.8	195.9	192.0	263.7	345.5	404.6	..	Agriculture
43.0	52.4	48.1	59.9	48.8	45.7	57.5	76.0	102.5	124.6	..	Industry
13.6	20.1	13.8	26.7	22.7	23.3	35.8	47.1	61.3	73.4	..	Manufacturing
175.4	182.8	199.1	195.5	202.2	219.1	340.4	426.7	530.1	634.1	..	Services, etc.
-156.9	-54.7	-91.4	-132.9	-89.3	-64.1	-162.3	-73.4	-189.0	-160.4	..	Resource Balance
138.6	207.1	241.2	190.6	213.7	259.8	327.2	393.8	767.3	799.8	..	Exports of Goods & NFServices
295.5	261.8	332.6	323.5	303.0	323.9	489.5	467.2	956.3	960.2	..	Imports of Goods & NFServices
518.7	472.5	516.4	544.5	540.7	591.7	834.6	932.6	1,267.2	1,435.3	..	Domestic Absorption
330.2	293.5	311.7	342.4	329.1	371.6	582.9	652.0	766.2	940.7	..	Private Consumption, etc.
79.6	68.4	91.1	94.7	106.5	104.9	122.1	141.1	245.4	252.5	..	General Gov't Consumption
108.9	110.6	113.6	107.4	105.1	115.2	129.6	139.5	255.6	242.1	..	Gross Domestic Investment
..	104.6	103.6	87.5	113.2	109.0	129.1	136.1	219.5	241.3	..	Fixed Investment
											Memo Items:
-48.0	55.9	22.2	-25.5	15.8	51.1	-32.3	66.2	66.5	81.8	..	Gross Domestic Saving
-54.3	49.3	24.6	-24.0	10.7	35.8	-44.7	50.4	71.4	88.7	..	Gross National Saving
				(Millions of 1980 Gambian Dalasis)							
392.68	412.52	421.20	392.54	423.52	472.14	481.08	481.70	507.61	536.59	..	Gross National Product
346.88	355.01	359.50	354.95	394.95	444.89	443.10	438.58	477.91	502.15	C I	GDP at factor cost
98.44	130.34	112.30	99.65	123.08	149.37	140.35	146.51	163.12	168.50	..	Agriculture
80.54	43.07	48.10	71.03	76.25	72.90	71.22	71.59	82.03	88.00	..	Industry
-106.77	-13.80	13.80	-130.01	-176.49	-186.66	-172.86	-172.86	-191.02	-201.92	..	Manufacturing
167.89	181.60	199.10	184.26	195.62	222.63	231.53	220.48	232.76	245.65	..	Services, etc.
-210.37	-62.02	-91.40	-136.39	-67.93	-38.87	..	..	..	..	..	Resource Balance
157.05	206.15	241.20	175.01	197.36	218.97	..	..	..	..	..	Exports of Goods & NFServices
367.43	268.17	332.60	311.40	265.28	257.85	..	..	..	..	..	Imports of Goods & NFServices
610.16	479.69	516.40	533.83	501.99	531.00	..	..	..	..	..	Domestic Absorption
421.87	307.63	311.70	329.60	313.10	340.59	..	..	..	..	..	Private Consumption, etc.
55.19	70.11	91.10	93.25	97.41	91.99	..	..	..	..	..	General Gov't Consumption
133.10	101.95	113.60	110.98	91.48	98.42	..	..	..	..	..	Gross Domestic Investment
..	..	..	..	..	..	..	..	..	..	..	Fixed Investment
											Memo Items:
172.34	212.14	241.20	183.47	187.10	206.82	..	..	..	..	..	Capacity to Import
15.28	5.99	0.00	8.46	-10.26	-12.15	..	..	..	..	..	Terms of Trade Adjustment
415.07	423.66	425.00	405.90	423.80	479.97	..	..	..	..	..	Gross Domestic Income
407.96	418.51	421.20	401.00	413.26	459.99	..	..	..	..	..	Gross National Income
				(Index 1980 = 100)							DOMESTIC PRICES (DEFLATORS)
90.5	100.0	100.0	103.6	104.0	107.2	134.2	171.2	203.8	228.0	..	Overall (GDP)
85.0	98.5	100.0	102.0	107.7	111.4	..	..	..	..	..	Domestic Absorption
98.0	96.7	100.0	112.3	125.8	131.2	136.8	180.0	211.8	240.1	..	Agriculture
53.4	121.7	100.0	84.3	64.0	62.7	80.7	106.2	125.0	141.6	..	Industry
-12.7	-145.7	100.0	-20.5	-12.9	-12.5	-20.7	-27.2	-32.1	-36.4	..	Manufacturing
											MANUFACTURING ACTIVITY
136.4	106.4	100.0	101.5	138.6	127.7	117.6	108.4		..	J	Employment (1980 = 100)
79.1	95.2	100.0	102.9	78.8	87.0	108.5	137.0		..	J	Real Earnings per Empl. (1980 = 100)
-550.6	-64.6	100.0	-708.1	-969.6	-1070.3	-730.9	-846.9		..	J	Real Output per Empl. (1980 = 100)
60.8	40.5	25.3	25.8	31.7	27.6	27.6	27.6		..	J	Earnings as % of Value Added
				(Millions of current Gambian Dalasis)							MONETARY HOLDINGS
89.7	82.2	90.7	109.7	127.0	161.0	169.8	257.0	275.6	343.5	..	Money Supply, Broadly Defined
57.8	57.8	61.3	77.0	87.2	100.3	99.6	162.1	166.5	197.8	..	Money as Means of Payment
34.5	36.5	36.8	42.6	55.9	57.2	58.4	85.7	91.2	95.0	..	Currency Ouside Banks
23.3	21.3	24.5	34.4	31.3	43.1	41.2	76.4	75.3	102.8	..	Demand Deposits
31.9	24.4	29.4	32.7	39.8	60.6	70.2	94.9	109.2	145.7	..	Quasi-Monetary Liabilities
				(Millions of current Gambian Dalasis)							
-36	-38	-18	-50	-34	-39	..	..	..	..	C	GOVERNMENT DEFICIT (-) OR SURPLUS
103	89	109	97	135	109	..	..	..	..	..	Current Revenue
62	63	68	85	120		..	..	..	..	..	Current Expenditure
41	26	41	12	15		..	..	..	..	..	Current Budget Balance
0	0	0	0	0		..	..	..	..	..	Capital Receipts
77	65	59	63	49		..	..	..	..	..	Capital Payments

GAMBIA, THE	1967	1968	1969	1970	1971	1972	1973	1974	1975	1976	1977
FOREIGN TRADE (CUSTOMS BASIS)					*(Millions of current US dollars)*						
Value of Exports, fob	17.5	14.4	17.2	15.6	14.7	18.5	20.7	40.2	49.0	36.0	47.6
Nonfuel Primary Products	17.5	14.4	17.2	15.4	14.4	18.1	19.8	39.2	48.1	34.6	45.2
Fuels	0.0	0.0	0.0	0.0	0.0	0.0	0.2	0.4	0.3	0.3	0.3
Manufactures	0.0	0.0	0.0	0.2	0.3	0.4	0.7	0.7	0.6	1.1	2.0
Value of Imports, cif	19.1	20.8	22.4	16.9	21.1	22.8	31.7	37.6	48.7	65.1	73.1
Nonfuel Primary Products	4.8	5.3	5.9	5.8	6.2	5.9	8.0	12.1	13.3	19.6	21.2
Fuels	0.6	0.7	0.7	0.6	0.7	0.8	1.5	2.0	4.4	4.1	4.9
Manufactures	13.7	14.7	15.8	10.5	14.3	16.2	22.2	23.5	31.1	41.4	47.0
Terms of Trade	134.3	132.8	136.8	149.8	137.6	145.0	208.7	141.0	112.6	123.8	132.3
					(Index 1980 = 100)						
Export Prices, fob	36.6	35.3	38.4	42.0	42.9	48.6	83.6	85.6	69.7	76.1	90.2
Nonfuel Primary Products	36.6	35.2	38.6	42.0	43.5	49.6	94.0	87.1	70.2	77.1	91.9
Fuels	4.3	4.3	4.3	4.3	5.6	6.2	8.9	36.7	35.7	38.4	42.0
Manufactures	30.1	31.7	32.4	37.5	36.5	38.5	49.5	67.2	66.1	66.7	71.7
Import Prices, cif	27.3	26.5	28.1	28.0	31.2	33.5	40.1	60.7	61.9	61.5	68.1
Trade at Constant 1980 Prices					*(Millions of 1980 US dollars)*						
Exports, fob	47.6	40.9	44.9	37.3	34.3	38.1	24.8	46.9	70.3	47.2	52.8
Imports, cif	70.1	78.2	79.8	60.3	67.8	68.2	79.1	61.8	78.8	105.9	107.3
BALANCE OF PAYMENTS					*(Millions of current US dollars)*						
Exports of Goods & Services	..	..	..	20	21	27	29	54	69	56	68
Merchandise, fob	..	..	..	18	18	22	21	44	57	44	53
Nonfactor Services	..	..	..	2	2	4	7	9	9	10	12
Factor Services	..	..	..	1	1	1	2	1	3	3	2
Imports of Goods & Services	..	..	..	20	24	29	35	55	62	76	80
Merchandise, fob	..	..	..	16	19	22	27	41	47	60	63
Nonfactor Services	..	..	..	2	3	4	6	11	12	15	17
Factor Services	..	..	..	2	2	3	3	3	2	1	0
Long-Term Interest	..	..	..	0	0	0	0	0	0	0	0
Current Transfers, net	..	..	..	0	0	1	0	0	0	1	1
Workers' Remittances	..	..	..	0	0	0	0	0	0	0	0
Total to be Financed	..	..	..	0	-3	-1	-6	-1	7	-18	-12
Official Capital Grants	..	..	..	0	2	1	1	7	5	1	4
Current Account Balance	..	..	..	0	-1	0	-4	6	11	-17	-8
Long-Term Capital, net	..	..	..	2	3	5	6	11	6	4	8
Direct Investment	..	..	..	0	2	2	1	1	0	1	..
Long-Term Loans	..	..	..	1	0	2	2	3	2	3	10
Disbursements	..	..	..	1	0	3	2	4	3	4	10
Repayments	..	..	..	0	0	0	0	0	0	0	1
Other Long-Term Capital	..	..	..	0	0	0	2	0	-1	-2	-5
Other Capital, net	..	..	..	-1	-1	-3	8	-12	-4	6	1
Change in Reserves	..	..	..	-2	-2	-2	-10	-5	-14	7	-1
Memo Item:					*(Gambian Dalasis per US dollar)*						
Conversion Factor (Annual Avg)	1.840	2.080	2.080	2.080	1.970	1.920	1.680	1.700	2.010	2.340	2.190
EXTERNAL DEBT, ETC.					*(Millions of US dollars, outstanding at end of year)*						
Public/Publicly Guar. Long-Term	..	..	..	5	6	7	9	12	13	15	27
Official Creditors	..	..	..	5	6	7	9	12	13	14	26
IBRD and IDA	..	..	..	0	0	1	2	4	5	5	7
Private Creditors	..	..	..	..	..	..	..	..	..	1	1
Private Non-guaranteed Long-Term	..	..	..	..	..	..	..	..	..	..	..
Use of Fund Credit	..	..	..	0	0	0	0	0	0	0	6
Short-Term Debt	..	..	..	..	..	..	..	..	..	..	7
Memo Items:					*(Thousands of US dollars)*						
Int'l Reserves Excluding Gold	8,370.0	6,550.0	5,990.0	8,110.0	10,939.0	11,385.0	16,241.0	28,046.0	28,552.0	20,635.0	24,395.0
Gold Holdings (at market price)	..	..	..	..	..	..	..	..	..	..	..
SOCIAL INDICATORS											
Total Fertility Rate	6.5	6.5	6.5	6.5	6.5	6.5	6.5	6.5	6.5	6.5	6.5
Crude Birth Rate	49.8	49.6	49.5	49.4	49.3	49.2	49.1	49.1	49.0	48.9	48.9
Infant Mortality Rate	193.0	190.4	187.8	185.2	182.6	180.0	177.4	174.8	172.2	169.6	167.0
Life Expectancy at Birth	34.9	35.3	35.7	36.1	36.5	36.9	37.3	37.7	38.1	38.5	38.9
Food Production, p.c. ('79-81 = 100)	145.0	158.1	147.1	149.7	154.8	135.4	150.4	150.8	153.6	137.0	104.7
Labor Force, Agriculture (%)	87.2	87.0	86.8	86.6	86.3	86.1	85.8	85.6	85.3	85.0	84.8
Labor Force, Female (%)	42.6	42.6	42.7	42.7	42.7	42.6	42.6	42.6	42.6	42.5	42.4
School Enroll. Ratio, primary	..	..	..	24.0	..	..	..	..	33.0	31.0	33.0
School Enroll. Ratio, secondary	..	..	..	7.0	..	..	..	..	10.0	11.0	11.0

1978	1979	1980	1981	1982	1983	1984	1985	1986	1987 estimate	NOTES	GAMBIA, THE
											FOREIGN TRADE (CUSTOMS BASIS)
				(Millions of current US dollars)							
39.0	58.0	31.0	27.0	44.0	48.0	48.9	43.3	48.0	35.3	..	Value of Exports, fob
37.3	55.5	29.6	25.8	42.1	45.9	46.9	41.1	44.4	32.3	.,	Nonfuel Primary Products
0.3	0.4	0.2	0.2	0.3	0.3	0.3	0.3	0.3	0.4	..	Fuels
1.5	2.2	1.2	1.0	1.7	1.8	1.7	1.9	3.3	2.6	..	Manufactures
100.0	141.0	131.8	145.0	150.6	117.6	100.2	93.3	105.0	113.3	..	Value of Imports, cif
29.1	41.1	18.4	57.0	78.2	35.3	55.3	50.7	58.8	64.6	..	Nonfuel Primary Products
6.8	9.5	8.9	9.8	10.2	8.0	10.9	9.5	3.6	4.3	..	Fuels
64.1	90.4	104.5	78.2	62.2	74.4	33.9	33.2	42.7	44.4	..	Manufactures
				(Index 1980 = 100)							
124.8	106.5	100.0	104.1	86.1	91.5	92.3	97.0	86.6	82.7	..	Terms of Trade
94.7	92.7	100.0	105.1	77.6	84.6	83.8	74.6	70.0	67.8	..	Export Prices, fob
96.4	93.2	100.0	105.3	77.0	84.4	83.6	74.0	68.6	65.8	..	Nonfuel Primary Products
42.3	61.0	100.0	112.5	101.6	92.5	90.2	87.5	44.9	56.9	..	Fuels
76.1	90.0	100.0	99.1	94.8	90.7	89.3	89.0	103.2	113.0	..	Manufactures
75.9	87.1	100.0	101.0	90.2	92.5	90.8	76.9	80.8	81.9	..	Import Prices, cif
											Trade at Constant 1980 Prices
				(Millions of 1980 US dollars)							
41.2	62.5	31.0	25.7	56.7	56.7	58.3	58.0	68.6	52.1	..	Exports, fob
131.8	162.0	131.8	143.6	166.9	127.2	110.4	121.3	130.0	138.3	..	Imports, cif
				(Millions of current US dollars)							**BALANCE OF PAYMENTS**
58	79	66	61	81	81	79	88	104	115	..	Exports of Goods & Services
42	54	48	42	57	54	61	65	70	79	..	Merchandise, fob
14	23	18	19	24	27	18	23	34	36	..	Nonfactor Services
2	2	0	0	1	0	0	0	0	..	..	Factor Services
117	138	182	163	138	135	77	80	99	112	..	Imports of Goods & Services
85	96	138	121	92	89	73	81	100	108	..	Merchandise, fob
27	38	42	39	35	30	24	23	29	31	..	Nonfactor Services
5	4	2	3	11	17	-20	-24	-30	-26	..	Factor Services
0	0	0	2	2	2	1	1	5	5	..	Long-Term Interest
0	-1	4	3	3	3	5	5	7	9	..	Current Transfers, net
0	0	0	0	0	0	0	0	0	..	..	Workers' Remittances
-59	-60	-112	-98	-54	-52	-34	-35	-48	-40	K	Total to be Financed
13	26	38	51	32	19	28	34	47	47	K	Official Capital Grants
-46	-33	-74	-47	-22	-33	-6	-1	-1	6	..	Current Account Balance
26	45	46	65	49	14	40	54	52	49	..	Long-Term Capital, net
2	12	..	2	..	..	1	2	6	3	..	Direct Investment
18	25	53	40	21	9	10	13	33	27	..	Long-Term Loans
18	26	54	40	29	14	14	14	39	37	..	Disbursements
0	0	0	1	8	5	4	1	6	10	..	Repayments
-7	-18	-45	-28	-4	-14	2	5	-34	-28	..	Other Long-Term Capital
7	-27	21	-9	-52	10	-33	-43	-46	-47	..	Other Capital, net
13	16	6	-9	25	9	-1	-10	-5	-9	..	Change in Reserves
											Memo Item:
				(Gambian Dalasis per US dollar)							
2.000	1.790	1.750	2.180	2.480	2.940	4.110	4.990	7.310	6.750	..	Conversion Factor (Annual Avg)
											EXTERNAL DEBT, ETC.
			(Millions of US dollars, outstanding at end of year)								
28	54	106	140	155	159	158	182	221	273	..	Public/Publicly Guar. Long-Term
28	51	82	110	120	123	125	144	202	256	..	Official Creditors
8	11	16	20	24	28	30	35	51	69	..	IBRD and IDA
1	3	24	30	35	35	33	38	19	17	..	Private Creditors
..	..	..	..	..	..	..	..	..	..	..	Private Non-guaranteed Long-Term
14	9	8	17	33	28	27	27	21	23	..	Use of Fund Credit
10	20	23	19	20	25	45	36	29	24	..	Short-Term Debt
				(Thousands of US dollars)							Memo Items:
26,072.0	1,929.0	5,670.0	3,948.0	8,387.0	2,918.0	2,259.0	1,734.0	13,563.0	..	..	Int'l Reserves Excluding Gold
..	..	..	..	..	..	..	..	..	..	..	Gold Holdings (at market price)
											SOCIAL INDICATORS
6.5	6.5	6.5	6.5	6.5	6.5	6.5	6.5	6.5	6.5	..	Total Fertility Rate
48.8	48.8	48.8	48.8	48.8	48.8	48.8	48.7	48.7	48.7	..	Crude Birth Rate
164.4	161.8	159.2	156.6	154.0	155.4	156.8	158.3	159.7	161.1	..	Infant Mortality Rate
39.3	39.8	40.2	40.6	41.0	41.5	41.9	42.4	42.9	43.4	..	Life Expectancy at Birth
130.6	90.5	91.7	117.9	138.1	94.7	102.1	122.9	135.7	118.6	..	Food Production, p.c. ('79-81 = 100)
84.5	84.3	84.0	..	..	..	..	..	..	..	..	Labor Force, Agriculture (%)
42.4	42.3	42.2	42.0	41.9	41.7	41.5	41.4	41.2	40.9	..	Labor Force, Female (%)
37.0	42.0	52.0	58.0	62.0	68.0	73.0	75.0	..	..	..	School Enroll. Ratio, primary
11.0	12.0	13.0	15.0	17.0	19.0	21.0	20.0	..	..	..	School Enroll. Ratio, secondary

GERMANY, FEDERAL REPUBLIC OF	1967	1968	1969	1970	1971	1972	1973	1974	1975	1976	1977
CURRENT GNP PER CAPITA (US $)	2,100	2,290	2,550	2,860	3,240	3,840	4,750	5,710	6,670	7,480	8,250
POPULATION (thousands)	59,286	59,500	60,067	60,651	61,302	61,672	61,976	62,054	61,829	61,531	61,401

ORIGIN AND USE OF RESOURCES

(Billions of current Deutsche Mark)

	1967	1968	1969	1970	1971	1972	1973	1974	1975	1976	1977
Gross National Product (GNP)	493.7	533.7	597.8	675.7	751.8	825.1	918.9	985.7	1,029.4	1,126.2	1,199.3
Net Factor Income from Abroad	-0.6	0.4	0.9	0.4	1.2	1.4	1.6	1.0	2.5	4.5	1.4
Gross Domestic Product (GDP)	494.3	533.3	596.9	675.3	750.6	823.7	917.3	984.7	1,026.9	1,121.7	1,197.9
Indirect Taxes, net	62.4	62.0	76.7	77.3	86.1	94.4	102.4	106.2	109.9	119.8	127.9
GDP at factor cost	431.9	471.3	520.3	598.0	664.5	729.3	814.8	878.5	917.0	1,001.9	1,070.0
Agriculture	20.4	22.2	23.5	21.8	22.8	24.8	26.6	25.9	28.5	31.1	31.8
Industry	252.2	258.0	290.4	333.7	361.3	387.8	430.9	456.0	454.9	500.6	529.9
Manufacturing	193.6	200.7	229.6	259.4	278.0	296.6	333.3	355.7	354.1	389.7	414.5
Services, etc.	221.7	253.1	283.1	319.8	366.5	411.2	459.8	502.8	543.5	590.0	636.2
Resource Balance	18.7	20.0	17.2	13.9	15.1	17.9	28.6	44.2	30.3	28.9	28.8
Exports of Goods & NFServices	108.0	121.7	138.4	152.8	169.9	185.2	217.0	279.1	271.4	308.9	325.7
Imports of Goods & NFServices	89.3	101.7	121.2	139.0	154.8	167.3	188.4	234.9	241.1	280.0	296.8
Domestic Absorption	475.6	513.2	579.8	661.4	735.4	805.9	888.6	940.5	996.6	1,092.8	1,169.1
Private Consumption, etc.	324.6	349.0	389.2	368.3	407.8	450.7	493.9	533.0	583.2	627.9	681.2
General Gov't Consumption	80.1	82.8	93.1	106.5	126.8	141.1	163.3	190.2	210.0	221.8	234.9
Gross Domestic Investment	70.9	81.4	97.5	186.7	200.8	214.0	231.4	217.3	203.5	243.1	252.9
Fixed Investment	113.0	118.1	137.4	170.2	193.7	209.4	222.7	218.0	218.7	235.3	252.9
Memo Items:											
Gross Domestic Saving	89.6	101.5	114.7	200.5	215.9	231.9	260.0	261.5	233.7	272.0	281.7
Gross National Saving	86.0	98.9	111.3	195.1	209.8	225.5	252.9	253.0	227.0	266.8	273.9

(Billions of 1980 Deutsche Mark)

	1967	1968	1969	1970	1971	1972	1973	1974	1975	1976	1977
Gross National Product	948.5	1,004.0	1,079.8	1,134.2	1,168.0	1,217.2	1,274.2	1,276.5	1,258.1	1,328.3	1,363.5
GDP at factor cost	817.0	872.9	921.9	1,001.0	1,028.4	1,071.8	1,126.2	1,135.6	1,118.5	1,179.8	1,214.7
Agriculture	26.9	28.5	27.8	28.6	28.1	27.3	29.4	30.8	29.2	28.7	30.4
Industry	419.6	452.2	496.0	522.7	531.5	551.3	580.9	571.9	543.1	582.2	593.8
Manufacturing	302.1	333.8	373.4	392.8	396.7	409.6	435.6	432.5	412.4	444.0	453.2
Services, etc.	503.8	522.4	554.2	582.1	606.4	636.5	661.6	672.5	682.7	712.0	737.7
Resource Balance	26.3	28.7	17.7	-1.0	-9.8	-8.1	8.1	38.9	17.5	17.3	16.7
Exports of Goods & NFServices	190.7	215.1	234.9	248.5	264.6	281.9	310.5	348.0	324.6	356.7	368.4
Imports of Goods & NFServices	164.4	186.4	217.2	249.5	274.4	290.0	302.4	309.1	307.1	339.4	351.7
Domestic Absorption	924.0	974.5	1,060.2	1,134.5	1,175.8	1,223.2	1,263.8	1,236.2	1,237.5	1,305.6	1,345.2
Private Consumption, etc.	496.6	520.0	562.3	606.2	636.4	664.6	684.8	690.2	710.8	735.1	770.2
General Gov't Consumption	193.4	194.3	202.9	211.8	222.7	232.1	243.5	253.3	262.7	266.6	270.4
Gross Domestic Investment	233.9	260.1	295.0	316.5	316.7	326.5	335.5	292.7	264.0	303.9	304.6
Fixed Investment	234.6	243.0	266.9	292.0	309.5	317.8	316.8	286.4	271.3	281.1	291.0
Memo Items:											
Capacity to Import	198.8	223.2	247.9	274.4	301.2	321.0	348.4	367.2	345.7	374.4	385.9
Terms of Trade Adjustment	8.1	8.1	13.0	25.9	36.6	39.1	37.9	19.2	21.1	17.7	17.5
Gross Domestic Income	958.4	1,011.2	1,091.0	1,159.4	1,202.6	1,254.2	1,309.7	1,294.4	1,276.1	1,340.7	1,379.4
Gross National Income	956.6	1,012.1	1,092.9	1,160.1	1,204.6	1,256.3	1,312.1	1,295.7	1,279.2	1,346.0	1,381.0

DOMESTIC PRICES (DEFLATORS)

(Index 1980 = 100)

	1967	1968	1969	1970	1971	1972	1973	1974	1975	1976	1977
Overall (GDP)	52.0	53.2	55.4	59.6	64.4	67.8	72.1	77.2	81.8	84.8	88.0
Domestic Absorption	51.5	52.7	54.7	58.3	62.5	65.9	70.3	76.1	80.5	83.7	86.9
Agriculture	75.8	77.7	84.6	76.0	80.9	90.5	90.4	84.3	97.6	108.2	104.6
Industry	60.1	57.0	58.5	63.8	68.0	70.3	74.2	79.7	83.8	86.0	89.2
Manufacturing	64.1	60.1	61.5	66.0	70.1	72.4	76.5	82.2	85.9	87.8	91.5

MANUFACTURING ACTIVITY

	1967	1968	1969	1970	1971	1972	1973	1974	1975	1976	1977
Employment (1980=100)	101.9	103.3	109.2	113.5	112.6	110.3	111.0	108.1	100.8	98.4	99.0
Real Earnings per Empl. (1980=100)	55.9	59.3	64.1	71.5	75.3	78.0	81.9	86.0	87.1	90.9	94.1
Real Output per Empl. (1980=100)	49.2	55.2	59.3	59.7	60.4	63.3	66.6	71.5	71.9	80.5	88.8
Earnings as % of Value Added	40.6	42.8	43.6	46.3	47.4	47.7	47.9	49.1	49.4	47.7	48.5

MONETARY HOLDINGS

(Billions of current Deutsche Mark)

	1967	1968	1969	1970	1971	1972	1973	1974	1975	1976	1977
Money Supply, Broadly Defined	284.3	307.8	340.5	373.0	423.4	483.4	529.5	569.1	634.0	683.5	752.2
Money as Means of Payment	81.6	90.6	95.4	103.7	116.9	133.4	135.7	150.2	171.7	177.3	198.6
Currency Ouside Banks	31.5	32.6	34.7	36.9	40.3	45.7	47.4	51.5	56.5	60.6	67.5
Demand Deposits	50.1	58.0	60.7	66.8	76.6	87.7	88.2	98.7	115.2	116.7	131.1
Quasi-Monetary Liabilities	202.7	217.2	245.1	269.4	306.6	350.0	393.8	418.8	462.3	506.2	553.6

GOVERNMENT DEFICIT (-) OR SURPLUS

(Billions of current Deutsche Mark)

	1967	1968	1969	1970	1971	1972	1973	1974	1975	1976	1977
GOVERNMENT DEFICIT (-) OR SURPLUS	..	..	..	6.94	6.37	5.83	12.43	-6.43	-37.16	-31.21	-25.57
Current Revenue	..	..	..	166.10	186.69	208.22	242.26	260.36	272.99	302.23	329.27
Current Expenditure	..	..	..	143.54	161.62	184.78	208.38	241.54	283.61	303.65	325.23
Current Budget Balance	..	..	..	22.56	25.07	23.44	33.88	18.82	-10.62	-1.42	4.04
Capital Receipts	..	..	..	0.12	0.10	0.12	0.13	0.12	0.10	0.32	0.10
Capital Payments	..	..	..	15.74	18.80	17.73	21.58	25.37	26.64	30.11	29.71

1978	1979	1980	1981	1982	1983	1984	1985	1986	1987 estimate	NOTES	GERMANY, FEDERAL REPUBLIC OF
9,510	11,560	13,340	13,290	12,270	11,450	11,230	10,990	12,040	14,480	..	**CURRENT GNP PER CAPITA (US $)**
61,327	61,359	61,566	61,682	61,638	61,421	61,181	61,035	60,909	60,824	..	**POPULATION (thousands)**
				(Billions of current Deutsche Mark)							**ORIGIN AND USE OF RESOURCES**
1,291.6	1,396.5	1,485.3	1,545.1	1,597.0	1,680.4	1,770.0	1,844.3	1,945.2	2,020.1	..	Gross National Product (GNP)
6.3	4.3	6.3	4.2	-0.8	5.6	14.1	13.8	14.0	11.0	..	Net Factor Income from Abroad
1,285.3	1,392.2	1,479.0	1,540.9	1,597.8	1,674.8	1,755.9	1,830.5	1,931.2	2,009.1	..	Gross Domestic Product (GDP)
137.9	152.0	162.9	169.2	172.5	182.7	190.0	192.5	195.0	201.9	..	Indirect Taxes, net
1,147.4	1,240.2	1,316.2	1,371.7	1,425.4	1,492.1	1,565.9	1,638.0	1,736.2	1,807.2	B	GDP at factor cost
32.2	31.0	30.4	31.7	36.3	32.2	34.7	31.9	34.0	..	..	Agriculture
561.6	607.7	632.3	641.6	656.9	685.1	710.6	745.2	738.2	..	..	Industry
437.9	470.4	482.9	489.7	502.8	524.9	547.3	583.6	640.2	..	..	Manufacturing
691.5	753.5	816.4	867.7	904.5	957.6	1,010.7	1,053.4	1,159.1	..	..	Services, etc.
37.4	11.2	-3.1	15.5	37.6	38.0	55.0	80.2	114.6	110.9	..	Resource Balance
344.7	377.3	422.2	482.3	517.7	524.8	590.7	647.3	638.2	638.1	..	Exports of Goods & NFServices
307.3	366.1	425.3	466.7	480.1	486.8	535.7	567.2	523.6	527.3	..	Imports of Goods & NFServices
1,247.9	1,381.0	1,482.1	1,525.4	1,560.2	1,636.8	1,701.0	1,750.3	1,816.6	1,898.2	..	Domestic Absorption
723.2	780.9	834.6	884.0	918.3	958.8	990.1	1,024.0	1,054.2	1,101.7	..	Private Consumption, etc.
252.9	273.4	297.8	318.2	326.2	336.2	350.2	365.6	382.6	397.2	..	General Gov't Consumption
271.9	326.7	349.7	323.2	315.7	341.9	360.7	360.7	379.8	399.3	..	Gross Domestic Investment
274.0	307.3	335.8	337.6	332.9	353.4	364.7	372.0	385.3	396.9	..	Fixed Investment
											Memo Items:
309.2	337.9	346.6	338.7	353.3	379.9	415.7	440.9	494.4	510.2	..	Gross Domestic Saving
305.9	332.4	342.2	331.9	341.7	374.4	418.4	444.4	498.4	511.9	..	Gross National Saving
				(Billions of 1980 Deutsche Mark)							
1,408.0	1,463.7	1,485.3	1,485.4	1,471.1	1,499.0	1,548.2	1,578.2	1,614.8	1,643.2	..	Gross National Product
1,247.6	1,297.5	1,316.2	1,321.2	1,314.9	1,332.5	1,371.9	1,403.6	1,438.2	1,465.5	B	GDP at factor cost
31.4	30.0	30.4	30.6	36.0	32.8	35.7	33.7	36.2	..	..	Agriculture
604.1	632.6	632.3	619.2	602.0	609.5	621.6	634.2	599.0	..	..	Industry
461.4	482.8	482.9	475.5	464.2	469.4	482.2	498.6	509.6	..	..	Manufacturing
765.6	796.6	816.4	831.7	833.9	851.7	878.8	898.5	967.8	..	..	Services, etc.
12.7	-9.0	-3.1	36.6	51.8	46.8	66.4	84.7	68.2	48.9	..	Resource Balance
383.8	401.0	422.2	456.7	471.5	469.0	511.0	545.8	545.9	550.1	..	Exports of Goods & NFServices
371.1	410.0	425.3	420.1	419.7	422.2	444.6	461.1	477.7	501.2	..	Imports of Goods & NFServices
1,388.3	1,468.1	1,482.1	1,444.8	1,420.0	1,447.3	1,469.7	1,481.8	1,534.8	1,585.1	..	Domestic Absorption
794.5	825.9	834.6	832.6	826.1	834.7	840.3	852.4	881.6	915.7	..	Private Consumption, etc.
280.7	290.2	297.8	303.2	300.7	301.4	308.8	315.3	323.1	328.2	..	General Gov't Consumption
313.1	352.0	349.7	309.1	293.3	311.2	320.6	314.1	330.1	341.2	..	Gross Domestic Investment
304.8	326.6	335.8	319.7	302.9	312.5	314.9	315.0	325.4	331.3	..	Fixed Investment
											Memo Items:
416.2	422.6	422.2	434.1	452.5	455.2	490.2	526.3	582.3	606.6	..	Capacity to Import
32.4	21.6	0.0	-22.6	-19.0	-13.8	-20.8	-19.5	36.4	56.5	..	Terms of Trade Adjustment
1,433.5	1,480.7	1,479.0	1,458.8	1,452.9	1,480.3	1,515.3	1,547.0	1,639.4	1,690.5	..	Gross Domestic Income
1,440.4	1,485.3	1,485.3	1,462.8	1,452.1	1,485.2	1,527.4	1,558.7	1,651.2	1,699.7	..	Gross National Income
				(Index 1980 = 100)							**DOMESTIC PRICES (DEFLATORS)**
91.7	95.4	100.0	104.0	108.6	112.1	114.3	116.9	120.5	123.0	..	Overall (GDP)
89.9	94.1	100.0	105.6	109.9	113.1	115.7	118.1	118.4	119.8	..	Domestic Absorption
102.8	103.5	100.0	103.7	100.9	98.1	97.1	94.6	93.7	..	..	Agriculture
93.0	96.1	100.0	103.6	109.1	112.4	114.3	117.5	123.2	..	..	Industry
94.9	97.4	100.0	103.0	108.3	111.8	113.5	117.0	125.6	..	..	Manufacturing
											MANUFACTURING ACTIVITY
98.4	98.8	100.0	97.6	94.1	91.0	90.2	91.5	..	..	G	Employment (1980=100)
96.9	99.0	100.0	99.3	98.5	99.3	100.5	102.3	..	..	G	Real Earnings per Empl. (1980=100)
89.5	96.0	100.0	103.9	104.7	108.0	114.4	117.0	..	..	G	Real Output per Empl. (1980=100)
48.2	47.9	50.3	51.1	50.1	48.3	47.6	46.3	..	..	..	Earnings as % of Value Added
				(Billions of current Deutsche Mark)							**MONETARY HOLDINGS**
827.4	873.0	914.4	950.0	1,012.4	1,067.8	1,120.8	1,199.7	1,268.0	1,333.1	D	Money Supply, Broadly Defined
227.5	234.1	243.4	239.6	256.7	278.2	294.7	314.5	340.2	365.7	..	Money as Means of Payment
76.2	79.9	84.0	84.2	88.6	96.4	99.8	103.9	112.1	124.1	..	Currency Ouside Banks
151.3	154.3	159.4	155.4	168.1	181.8	195.0	210.6	228.1	241.6	..	Demand Deposits
599.9	638.9	671.1	710.4	755.7	789.6	826.0	885.2	927.8	967.4	..	Quasi-Monetary Liabilities
				(Billions of current Deutsche Mark)							
-26.49	-27.63	-26.91	-36.31	-32.02	-32.95	-32.31	-20.00	-16.44	-21.23	E	**GOVERNMENT DEFICIT (-) OR SURPLUS**
352.40	377.62	425.69	451.53	480.43	493.75	517.40	554.19	574.76	591.73	..	Current Revenue
348.84	371.51	415.25	450.59	475.39	489.50	518.86	533.64	553.19	578.44	..	Current Expenditure
3.56	6.11	10.44	0.94	5.04	4.25	-1.46	20.55	21.57	13.29	..	Current Budget Balance
0.12	0.12	0.14	0.17	0.30	0.24	0.21	0.24	0.25	0.18	..	Capital Receipts
30.17	33.86	37.49	36.97	36.20	37.59	38.27	38.35	35.34	34.85	..	Capital Payments

GERMANY, FEDERAL REPUBLIC OF	1967	1968	1969	1970	1971	1972	1973	1974	1975	1976	1977
FOREIGN TRADE (CUSTOMS BASIS)				*(Billions of current US dollars)*							
Value of Exports, fob	21.74	24.84	29.05	34.19	39.04	46.21	67.44	89.17	90.02	102.03	117.93
Nonfuel Primary Products	1.79	2.04	2.34	2.73	3.00	3.62	5.96	8.20	7.57	8.70	10.45
Fuels	0.73	0.83	0.81	1.01	1.16	1.21	1.72	3.08	2.81	2.93	3.12
Manufactures	19.22	21.97	25.90	30.45	34.88	41.38	59.76	77.88	79.64	90.40	104.36
Value of Imports, cif	17.35	20.15	24.93	29.81	34.34	39.76	54.50	68.98	74.21	87.78	100.70
Nonfuel Primary Products	7.33	8.16	9.71	10.84	11.10	12.71	17.72	20.52	20.38	23.64	26.89
Fuels	1.74	2.06	2.21	2.63	3.50	3.67	6.21	13.31	13.10	15.83	17.23
Manufactures	8.28	9.93	13.01	16.34	19.74	23.38	30.57	35.15	40.73	48.31	56.58
Terms of Trade	141.4	137.5	127.9	133.9	134.5	131.7	145.2	106.2	113.0	111.1	108.1
Export Prices, fob	27.6	26.6	27.2	29.2	31.5	35.4	47.9	56.4	62.0	63.8	68.3
Nonfuel Primary Products	34.1	34.2	37.2	37.9	36.4	41.1	58.8	66.9	61.1	63.5	65.5
Fuels	8.5	7.3	5.4	5.9	7.0	8.5	24.5	36.6	33.6	38.5	36.8
Manufactures	29.6	28.9	30.2	32.9	35.2	38.5	48.3	56.5	64.1	65.4	70.5
Import Prices, cif	19.5	19.4	21.2	21.8	23.4	26.8	33.0	53.1	54.8	57.4	63.2
Trade at Constant 1980 Prices				*(Billions of 1980 US dollars)*							
Exports, fob	78.83	93.34	107.00	117.04	124.06	130.67	140.83	158.21	145.28	159.85	172.73
Imports, cif	89.00	104.06	117.45	136.62	146.77	148.13	165.28	129.96	135.34	152.86	159.41
BALANCE OF PAYMENTS				*(Billions of current US dollars)*							
Exports of Goods & Services	..	..	..	42.97	48.00	57.00	80.86	106.78	110.40	123.47	141.10
Merchandise, fob	..	..	..	34.32	37.29	44.79	64.93	86.99	87.63	98.48	113.22
Nonfactor Services	..	..	..	6.21	7.69	8.67	11.12	13.88	16.53	18.02	20.39
Factor Services	..	..	..	2.45	3.02	3.54	4.81	5.91	6.24	6.97	7.49
Imports of Goods & Services	..	..	..	39.48	44.08	52.04	70.66	90.33	99.04	112.55	129.53
Merchandise, fob	..	..	..	28.62	30.64	36.39	49.29	65.14	70.77	82.48	93.81
Nonfactor Services	..	..	..	7.90	9.65	11.28	15.64	18.06	21.47	22.88	26.39
Factor Services	..	..	..	2.96	3.80	4.36	5.73	7.12	6.80	7.19	9.33
Long-Term Interest	..	..	..	..	..	..	..	..	..	..	..
Current Transfers, net	..	..	..	-1.59	-2.10	-2.44	-3.26	-3.66	-3.77	-3.85	-3.93
Workers' Remittances	..	..	..	0.00	0.00	0.00	0.00	0.00	0.00	0.00	0.00
Total to be Financed	..	..	..	1.90	1.82	2.52	6.93	12.80	7.60	7.06	7.63
Official Capital Grants	..	..	..	-1.05	-0.88	-1.36	-1.92	-2.23	-3.25	-3.37	-3.70
Current Account Balance	..	..	..	0.85	0.93	1.16	5.02	10.57	4.35	3.69	3.93
Long-Term Capital, net	..	..	..	-0.80	0.03	-0.34	0.22	-4.49	-10.42	-3.99	-9.10
Direct Investment	..	..	..	-0.30	0.11	0.38	0.38	0.21	-1.33	-1.10	-1.26
Long-Term Loans	..	..	..	..	..	..	..	..	..	..	..
Disbursements	..	..	..	..	..	..	..	..	..	..	..
Repayments	..	..	..	..	..	..	..	..	..	..	..
Other Long-Term Capital	..	..	..	0.55	0.81	0.65	1.76	-2.48	-5.84	0.48	-4.14
Other Capital, net	..	..	..	6.00	3.87	3.79	4.21	-6.77	4.97	3.86	8.04
Change in Reserves	..	..	..	-6.05	-4.83	-4.62	-9.44	0.69	1.10	-3.56	-2.86
Memo Item:				*(Deutsche Mark per US dollar)*							
Conversion Factor (Annual Avg)	4.000	4.000	3.940	3.660	3.490	3.190	2.670	2.590	2.460	2.520	2.320
EXTERNAL DEBT, ETC.				*(Millions US dollars, outstanding at end of year)*							
Public/Publicly Guar. Long-Term	..	..	..	..	..	..	..	..	..	..	..
Official Creditors	..	..	..	..	..	..	..	..	..	..	..
IBRD and IDA	..	..	..	..	..	..	..	..	..	..	..
Private Creditors	..	..	..	..	..	..	..	..	..	..	..
Private Non-guaranteed Long-Term	..	..	..	..	..	..	..	..	..	..	..
Use of Fund Credit	..	..	..	..	..	..	..	..	..	..	..
Short-Term Debt	..	..	..	..	..	..	..	..	..	..	..
Memo Items:				*(Millions of US dollars)*							
Int'l Reserves Excluding Gold	3,925	5,409	3,050	9,630	14,231	19,326	28,206	27,359	26,216	30,019	34,708
Gold Holdings (at market price)	4,252	5,434	4,103	4,249	5,082	7,617	13,202	21,934	16,495	15,848	19,514
SOCIAL INDICATORS											
Total Fertility Rate	2.5	2.4	2.2	2.0	1.9	1.7	1.5	1.5	1.5	1.5	1.4
Crude Birth Rate	17.0	16.1	14.8	13.4	12.7	11.3	10.3	10.1	9.7	9.8	9.5
Infant Mortality Rate	22.8	22.6	23.2	23.4	23.1	22.4	22.7	21.1	19.7	17.4	15.4
Life Expectancy at Birth	70.3	70.4	70.4	70.5	70.5	70.6	70.9	71.2	71.6	71.9	72.3
Food Production, p.c. ('79-81 = 100)	88.5	93.7	90.7	92.3	94.0	89.8	91.5	94.9	92.2	90.7	95.6
Labor Force, Agriculture (%)	9.5	8.8	8.2	7.5	7.3	7.1	7.0	6.8	6.6	6.4	6.3
Labor Force, Female (%)	36.0	36.1	36.1	36.2	36.4	36.6	36.8	37.0	37.2	37.4	37.6
School Enroll. Ratio, primary	..	..	..	..							
School Enroll. Ratio, secondary	..	..	..	36.0	..	..	..	..	42.0	..	..

1978	1979	1980	1981	1982	1983	1984	1985	1986	1987 estimate	NOTES	GERMANY, FEDERAL REPUBLIC OF
				(Billions of current US dollars)							**FOREIGN TRADE (CUSTOMS BASIS)**
142.09	171.44	191.64	175.28	175.46	168.75	171.01	183.33	242.40	293.79	..	Value of Exports, fob
12.33	15.91	19.05	17.27	16.21	16.12	16.71	16.97	21.05	24.68	..	Nonfuel Primary Products
4.51	5.69	7.14	6.97	6.47	5.63	5.55	5.06	3.88	3.89	..	Fuels
125.25	149.84	165.45	151.04	152.77	147.00	148.75	161.30	217.47	265.22	..	Manufactures
120.67	157.68	185.92	162.69	154.05	152.01	152.87	157.60	189.65	227.33	..	Value of Imports, cif
30.02	36.54	41.79	34.59	32.73	32.23	32.82	32.97	39.73	45.16	..	Nonfuel Primary Products
19.50	30.87	41.90	39.80	36.43	32.40	31.17	31.35	21.88	21.96	..	Fuels
71.14	90.27	102.22	88.30	84.90	87.37	88.88	93.28	128.04	160.21	..	Manufactures
				(Index 1980 = 100)							
115.7	111.6	100.0	88.7	92.3	91.9	84.6	88.2	115.6	119.6	..	Terms of Trade
80.5	94.0	100.0	87.0	85.1	82.3	73.9	75.2	96.7	112.3	..	Export Prices, fob
72.8	93.8	100.0	90.0	81.7	84.1	79.2	75.5	79.4	83.9	..	Nonfuel Primary Products
44.7	93.6	100.0	104.0	97.0	85.4	83.6	81.1	58.4	51.7	..	Fuels
84.0	94.2	100.0	86.0	85.0	82.0	73.0	75.0	100.0	118.0	..	Manufactures
69.5	84.3	100.0	98.1	92.2	89.5	87.4	85.2	83.7	93.8	..	Import Prices, cif
											Trade at Constant 1980 Prices
				(Billions of 1980 US dollars)							
176.55	182.34	191.64	201.53	206.24	205.03	231.52	243.80	250.63	261.69	..	Exports, fob
173.53	187.11	185.92	165.85	167.08	169.76	175.00	184.87	226.70	242.26	..	Imports, cif
				(Billions of current US dollars)							**BALANCE OF PAYMENTS**
172.45	206.97	233.94	215.62	215.62	207.92	209.52	223.12	296.82	359.08	..	Exports of Goods & Services
135.77	163.89	183.21	166.82	165.77	159.89	161.35	173.15	230.11	278.19	..	Merchandise, fob
25.96	29.43	34.93	33.33	33.82	31.79	31.03	32.32	41.63	47.99	..	Nonfactor Services
10.72	13.65	15.80	15.47	16.03	16.25	17.14	17.64	25.09	32.90	..	Factor Services
154.35	201.27	234.94	207.96	199.87	192.68	189.41	196.49	244.97	298.32	..	Imports of Goods & Services
111.76	147.37	174.46	150.60	140.99	138.49	139.11	144.86	174.81	208.32	..	Merchandise, fob
32.47	39.62	44.33	40.58	40.17	38.02	35.13	35.72	47.04	57.73	..	Nonfactor Services
10.12	14.28	16.15	16.78	18.71	16.18	15.17	15.91	23.12	32.27	..	Factor Services
..	..	..	..	..	..	..	..	..	..	..	Long-Term Interest
-4.80	-5.32	-5.85	-4.86	-4.46	-4.33	-3.99	-3.49	-4.62	-5.17	..	Current Transfers, net
0.00	0.00	0.00	0.00	0.00	0.00	0.00	0.00	0.00	0.00	..	Workers' Remittances
13.30	0.38	-6.85	2.80	11.29	10.91	16.12	23.13	47.23	55.60	K	Total to be Financed
-4.27	-5.96	-7.12	-6.14	-6.23	-5.55	-6.41	-6.38	-7.80	-10.64	K	Official Capital Grants
9.03	-5.57	-13.96	-3.34	5.06	5.37	9.71	16.76	39.43	44.96	..	Current Account Balance
-5.67	0.66	-4.07	-2.60	-12.11	-8.67	-13.30	-11.10	7.18	-23.55	..	Long-Term Capital, net
-2.00	-2.79	-3.61	-3.55	-1.66	-1.60	-3.79	-4.34	-8.58	-7.06	..	Direct Investment
..	..	..	..	..	..	..	..	..	..	..	Long-Term Loans
..	..	..	..	..	..	..	..	..	..	..	Disbursements
..	..	..	..	..	..	..	..	..	..	..	Repayments
0.60	9.41	6.65	7.09	-4.22	-1.53	-3.09	-0.38	23.56	-5.84	..	Other Long-Term Capital
6.38	1.81	2.30	7.51	9.97	2.06	2.45	-4.78	-45.01	-1.18	..	Other Capital, net
-9.74	3.11	15.73	-1.57	-2.91	1.25	1.15	-0.88	-1.59	-20.23	..	Change in Reserves
				(Deutsche Mark per US dollar)							Memo Item:
2.010	1.830	1.820	2.260	2.430	2.550	2.850	2.940	2.170	1.800	..	Conversion Factor (Annual Avg)
			(Millions US dollars, outstanding at end of year)								**EXTERNAL DEBT, ETC.**
..	..	..	..	..	..	..	..	..	..	..	Public/Publicly Guar. Long-Term
..	..	..	..	..	..	..	..	..	..	..	Official Creditors
..	..	..	..	..	..	..	..	..	..	..	IBRD and IDA
..	..	..	..	..	..	..	..	..	..	..	Private Creditors
..	..	..	..	..	..	..	..	..	..	..	Private Non-guaranteed Long-Term
..	..	..	..	..	..	..	..	..	..	..	Use of Fund Credit
..	..	..	..	..	..	..	..	..	..	..	Short-Term Debt
				(Millions of US dollars)							Memo Items:
48,474	52,549	48,592	43,719	44,762	42,674	40,141	44,380	51,734	78,756	..	Int'l Reserves Excluding Gold
26,813	48,769	56,110	37,835	43,489	36,312	29,345	31,125	37,207	46,078	..	Gold Holdings (at market price)
											SOCIAL INDICATORS
1.4	1.4	1.4	1.4	1.4	1.4	1.4	1.3	1.3	1.3	..	Total Fertility Rate
9.4	9.5	10.1	10.1	10.1	10.0	10.0	10.0	9.9	9.9	..	Crude Birth Rate
14.7	13.6	12.7	11.6	12.1	12.5	13.0	13.6	14.1	14.6	..	Infant Mortality Rate
72.6	72.9	73.2	73.2	73.3	74.5	74.6	74.7	74.8	74.9	..	Life Expectancy at Birth
99.6	98.8	101.0	100.3	109.2	106.3	114.4	109.0	115.6	110.4	..	Food Production, p.c. ('79-81 = 100)
6.1	6.0	5.8	..	..	..	..	..	..	..	..	Labor Force, Agriculture (%)
37.8	38.0	38.1	38.0	38.0	37.9	37.8	37.7	37.6	37.5	..	Labor Force, Female (%)
..	..	99.0	..	100.0	99.0	98.0	97.0	..	..	..	School Enroll. Ratio, primary
..	47.0	72.0	..	74.0	74.0	73.0	72.0	..	..	..	School Enroll. Ratio, secondary

GHANA	1967	1968	1969	1970	1971	1972	1973	1974	1975	1976	1977
CURRENT GNP PER CAPITA (US $)	240	220	230	250	270	260	260	290	280	280	300
POPULATION (thousands)	8,135	8,291	8,451	8,614	8,832	9,088	9,362	9,617	9,835	10,027	10,204

ORIGIN AND USE OF RESOURCES
(Millions of current Ghanaian Cedis)

	1967	1968	1969	1970	1971	1972	1973	1974	1975	1976	1977
Gross National Product (GNP)	1,472	1,648	1,952	2,214	2,454	2,779	3,475	4,629	5,241	6,478	11,123
Net Factor Income from Abroad	-32	-52	-49	-45	-47	-37	-26	-31	-42	-48	-40
Gross Domestic Product (GDP)	1,504	1,700	2,001	2,259	2,500	2,815	3,501	4,660	5,283	6,526	11,163
Indirect Taxes, net	161	194	242	309	301	303	281	418	541	672	867
GDP at factor cost	1,343	1,506	1,758	1,950	2,199	2,512	3,220	4,242	4,742	5,854	10,296
Agriculture	605	710	919	1,051	1,104	1,313	1,715	2,383	2,518	3,300	6,274
Industry	308	346	383	412	457	499	651	845	1,110	1,254	1,768
Manufacturing	179	214	248	258	275	306	409	502	737	857	1,204
Services, etc.	592	644	699	796	940	1,003	1,135	1,432	1,655	1,972	3,121
Resource Balance	-38	-8	-12	-31	-112	154	177	-162	49	-22	-118
Exports of Goods & NFServices	262	345	395	482	394	583	751	854	1,023	1,025	1,172
Imports of Goods & NFServices	300	353	407	513	506	428	574	1,016	973	1,047	1,289
Domestic Absorption	1,542	1,708	2,013	2,290	2,612	2,661	3,324	4,822	5,234	6,548	11,281
Private Consumption, etc.	1,162	1,234	1,492	1,680	1,935	2,106	2,626	3,645	3,873	5,170	8,637
General Gov't Consumption	225	285	285	290	324	355	382	569	688	799	1,409
Gross Domestic Investment	155	189	236	320	353	200	316	608	672	580	1,235
Fixed Investment	132	160	200	271	311	244	268	555	614	642	1,049
Memo Items:											
Gross Domestic Saving	117	181	224	289	241	354	493	446	721	558	1,117
Gross National Saving	72	113	161	233	186	314	459	410	708	505	1,070

(Millions of 1980 Ghanaian Cedis)

	1967	1968	1969	1970	1971	1972	1973	1974	1975	1976	1977
Gross National Product	34,828	34,709	36,827	40,594	42,827	42,143	43,033	46,205	40,011	38,627	39,391
GDP at factor cost	31,366	31,113	32,672	35,286	37,909	37,796	39,664	42,151	35,965	34,690	36,322
Agriculture	19,380	19,137	20,457	22,760	23,841	24,756	24,262	26,400	21,138	20,789	19,664
Industry	4,745	5,010	5,507	5,870	5,951	5,351	6,397	6,361	6,306	6,137	6,386
Manufacturing	2,772	3,249	3,781	3,949	3,706	3,368	4,162	3,925	4,288	4,094	4,228
Services, etc.	11,040	11,034	11,277	12,318	13,363	12,262	12,515	13,579	12,709	11,823	13,406
Resource Balance	-320	808	65	1,298	1,864	4,278	2,846	435	1,128	1,632	-117
Exports of Goods & NFServices	5,906	6,003	5,683	6,802	6,366	7,389	7,354	5,773	5,664	5,946	4,465
Imports of Goods & NFServices	6,226	5,195	5,619	5,504	4,502	3,110	4,508	5,338	4,536	4,314	4,582
Domestic Absorption	35,485	34,374	37,176	39,650	41,290	38,090	40,328	45,905	39,023	37,118	39,573
Private Consumption, etc.	30,498	28,961	31,527	32,580	34,335	32,661	34,954	38,693	32,363	31,078	31,357
General Gov't Consumption	2,672	2,928	2,788	3,348	3,056	3,374	2,698	3,331	3,354	3,432	4,417
Gross Domestic Investment	2,315	2,485	2,860	3,722	3,900	2,054	2,675	3,880	3,307	2,608	3,799
Fixed Investment	2,025	2,173	2,501	3,255	3,585	2,363	2,364	3,661	3,106	2,918	3,496
Memo Items:											
Capacity to Import	5,440	5,078	5,450	5,173	3,507	4,231	5,898	4,486	4,764	4,223	4,163
Terms of Trade Adjustment	-466	-925	-233	-1,630	-2,859	-3,157	-1,456	-1,286	-900	-1,723	-302
Gross Domestic Income	34,699	34,256	37,007	39,318	40,295	39,211	41,718	45,053	39,252	37,027	39,154
Gross National Income	34,362	33,784	36,594	38,964	39,968	38,985	41,577	44,919	39,111	36,904	39,089

DOMESTIC PRICES (DEFLATORS)
(Index 1980 = 100)

	1967	1968	1969	1970	1971	1972	1973	1974	1975	1976	1977
Overall (GDP)	4.3	4.8	5.4	5.5	5.8	6.6	8.1	10.1	13.2	16.8	28.3
Domestic Absorption	4.3	5.0	5.4	5.8	6.3	7.0	8.2	10.5	13.4	17.6	28.5
Agriculture	3.1	3.7	4.5	4.6	4.6	5.3	7.1	9.0	11.9	15.9	31.9
Industry	6.5	6.9	7.0	7.0	7.7	9.3	10.2	13.3	17.6	20.4	27.7
Manufacturing	6.5	6.6	6.6	6.5	7.4	9.1	9.8	12.8	17.2	20.9	28.5

MANUFACTURING ACTIVITY

	1967	1968	1969	1970	1971	1972	1973	1974	1975	1976	1977
Employment (1980=100)	50.1	58.7	63.4	71.9	73.4	75.7	81.9	91.6	96.0	110.9	111.6
Real Earnings per Empl. (1980=100)	338.1	323.1	342.8	397.9	394.3	410.4	382.8	363.1	342.5	246.5	161.3
Real Output per Empl. (1980=100)	109.1	119.0	133.3	193.2	185.0	171.6	189.7	183.9	158.7	150.4	146.6
Earnings as % of Value Added	22.5	23.9	24.1	23.0	22.0	22.1	19.5	21.2	20.5	17.1	15.8

MONETARY HOLDINGS
(Millions of current Ghanaian Cedis)

	1967	1968	1969	1970	1971	1972	1973	1974	1975	1976	1977
Money Supply, Broadly Defined	0.32	0.35	0.39	0.43	0.47	0.67	0.79	1.00	1.39	1.90	3.04
Money as Means of Payment	0.24	0.26	0.29	0.31	0.32	0.46	0.56	0.70	1.01	1.43	2.39
Currency Ouside Banks	0.12	0.13	0.15	0.15	0.16	0.24	0.25	0.34	0.49	0.71	1.16
Demand Deposits	0.12	0.13	0.14	0.16	0.16	0.22	0.32	0.36	0.52	0.72	1.24
Quasi-Monetary Liabilities	0.08	0.09	0.10	0.12	0.15	0.21	0.23	0.31	0.38	0.47	0.65

GOVERNMENT DEFICIT (-) OR SURPLUS
(Millions of current Ghanaian Cedis)

	1967	1968	1969	1970	1971	1972	1973	1974	1975	1976	1977
GOVERNMENT DEFICIT (-) OR SURPLUS	..	..	..	..	..	-161	-187	-196	-401	-736	-1,057
Current Revenue	..	..	..	..	..	422	392	584	810	870	1,171
Current Expenditure	..	..	..	..	..	440	453	624	910	1,102	1,361
Current Budget Balance	..	..	..	..	..	-18	-61	-40	-100	-232	-190
Capital Receipts	..	..	..	..	..	..	..	..	..	..	..
Capital Payments	..	..	..	..	..	143	126	156	301	504	867

1978	1979	1980	1981	1982	1983	1984	1985	1986	1987 estimate	NOTES	GHANA
350	370	400	400	370	340	360	370	390	390	A	**CURRENT GNP PER CAPITA (US $)**
10,374	10,548	10,740	11,116	11,505	11,907	12,324	12,737	13,163	13,599	..	**POPULATION (thousands)**
				(Millions of current Ghanaian Cedis)							**ORIGIN AND USE OF RESOURCES**
20,938	28,123	42,670	72,404	86,226	182,398	266,918	337,280	498,796	725,541	..	Gross National Product (GNP)
-48	-99	-182	-222	-225	-1,640	-3,643	-5,768	-12,576	-20,458	..	Net Factor Income from Abroad
20,986	28,222	42,852	72,626	86,451	184,038	270,561	343,048	511,372	745,999	..	Gross Domestic Product (GDP)
1,398	1,524	1,524	2,867	2,937	..	..	..	..	..	..	Indirect Taxes, net
19,588	26,698	41,328	69,759	83,514	..	..	..	..	..	B	GDP at factor cost
12,742	16,924	24,821	38,553	49,572	109,927	133,232	154,003	244,317	377,481	..	Agriculture
2,524	3,466	5,086	6,654	5,401	12,199	28,631	57,209	87,723	118,947	..	Industry
1,813	2,447	3,346	4,338	3,117	7,101	17,306	39,562	57,021	73,710	..	Manufacturing
5,721	7,832	12,945	27,419	31,478	61,912	108,698	131,836	179,332	249,571	..	Services, etc.
-279	20	-295	-412	308	-5,790	-7,300	-10,100	-19,400	-48,300	..	Resource Balance
1,754	3,170	3,628	3,454	2,886	11,238	21,800	36,500	84,700	146,600	..	Exports of Goods & NFServices
2,033	3,150	3,923	3,866	2,578	17,028	29,100	46,600	104,100	194,900	..	Imports of Goods & NFServices
21,265	28,202	43,147	73,038	86,143	189,828	277,861	353,148	530,772	794,299	..	Domestic Absorption
17,766	23,454	35,953	63,333	77,620	167,147	239,613	288,748	435,672	650,199	..	Private Consumption, etc.
2,371	2,903	4,784	6,384	5,603	15,780	19,641	31,400	46,100	64,100	..	General Gov't Consumption
1,128	1,845	2,410	3,321	2,920	6,901	18,607	33,000	49,000	80,000	..	Gross Domestic Investment
1,062	1,899	2,613	3,430	3,053	6,922	18,542	..	..	..	..	Fixed Investment
											Memo Items:
849	1,865	2,115	2,909	3,228	1,111	11,307	22,900	29,600	31,700	..	Gross Domestic Saving
792	1,759	1,924	2,675	3,000	-535	8,410	18,894	23,461	40,877	..	Gross National Saving
				(Millions of 1980 Ghanaian Cedis)							
43,261	42,536	42,670	41,452	38,761	37,063	40,215	41,955	43,994	45,910	..	Gross National Product
40,421	40,310	41,328	39,961	37,588	..	..	..	..	..	B I	GDP at factor cost
23,410	24,292	24,821	24,183	22,865	21,270	23,335	23,486	24,259	24,267	..	Agriculture
5,908	5,069	5,086	4,273	3,546	3,125	3,404	4,006	4,313	4,802	..	Industry
4,080	3,393	3,346	2,700	2,147	1,909	2,153	2,677	2,974	3,270	..	Manufacturing
14,017	13,252	12,945	13,137	12,484	12,768	13,659	14,719	15,763	17,199	..	Services, etc.
25	611	-295	-80	1,719	193	293	-22	803	151	..	Resource Balance
4,271	4,199	3,628	3,307	3,811	2,068	2,264	2,410	3,412	4,415	..	Exports of Goods & NFServices
4,246	3,588	3,923	3,387	2,092	1,875	1,971	2,432	2,609	4,263	..	Imports of Goods & NFServices
43,310	42,002	43,147	41,674	37,177	36,971	40,105	42,233	43,532	46,116	..	Domestic Absorption
35,571	35,094	35,953	33,941	30,564	30,461	33,973	35,755	37,055	37,919	..	Private Consumption, etc.
5,193	4,268	4,784	5,534	4,920	4,823	4,199	4,068	4,141	5,359	..	General Gov't Consumption
2,545	2,641	2,410	2,199	1,692	1,687	1,933	2,410	2,336	2,838	..	Gross Domestic Investment
2,965	2,770	2,613	2,309	1,783	1,742	1,985	2,476	2,395	2,907	..	Fixed Investment
											Memo Items:
3,664	3,610	3,628	3,026	2,342	1,237	1,476	1,905	2,123	3,207	..	Capacity to Import
-607	-588	0	-281	-1,469	-830	-787	-505	-1,289	-1,208	..	Terms of Trade Adjustment
42,728	42,025	42,852	41,313	37,427	36,334	39,611	41,706	43,045	45,060	..	Gross Domestic Income
42,653	41,948	42,670	41,171	37,292	36,233	39,427	41,450	42,705	44,702	..	Gross National Income
				(Index 1980 = 100)							**DOMESTIC PRICES (DEFLATORS)**
48.4	66.2	100.0	174.6	222.3	495.2	669.7	812.7	1153.4	1612.4	..	Overall (GDP)
49.1	67.1	100.0	175.3	231.7	513.4	692.8	836.2	1219.3	1722.4	..	Domestic Absorption
54.4	69.7	100.0	159.4	216.8	516.8	570.9	655.7	1007.1	1555.5	..	Agriculture
42.7	68.4	100.0	155.7	152.3	390.3	841.1	1428.0	2033.8	2477.1	..	Industry
44.4	72.1	100.0	160.7	145.2	372.0	803.8	1478.0	1917.6	2253.9	..	Manufacturing
											MANUFACTURING ACTIVITY
107.0	99.5	100.0	96.6	84.1	73.6	..	..	..	..	G	Employment (1980 = 100)
133.8	106.2	100.0	64.4	68.2	48.4	..	..	..	..	G	Real Earnings per Empl. (1980 = 100)
113.8	105.6	100.0	77.2	97.2	75.9	..	..	..	..	G	Real Output per Empl. (1980 = 100)
21.8	18.3	19.3	20.5	22.1	17.6	..	..	..	..	..	Earnings as % of Value Added
				(Millions of current Ghanaian Cedis)							**MONETARY HOLDINGS**
5.13	5.94	7.95	12.03	14.84	20.80	31.96	46.72	69.11	105.97	..	Money Supply, Broadly Defined
4.13	4.68	6.09	9.41	11.20	16.72	26.85	38.31	55.16	84.17	..	Money as Means of Payment
2.12	2.46	3.52	6.05	6.96	10.39	17.63	22.56	32.35	48.98	..	Currency Ouside Banks
2.00	2.22	2.56	3.36	4.25	6.33	9.22	15.75	22.81	35.19	..	Demand Deposits
1.01	1.26	1.86	2.62	3.63	4.09	5.11	8.41	13.96	21.80	..	Quasi-Monetary Liabilities
				(Millions of current Ghanaian Cedis)							
-1,897	-1,800	-1,808	-4,707	-4,848	-4,933	-4,843	-7,579	299	..	F	**GOVERNMENT DEFICIT (-) OR SURPLUS**
1,393	2,600	2,951	3,279	4,856	10,242	22,642	40,311	73,626	..	..	Current Revenue
2,541	3,500	4,179	6,330	8,603	13,401	23,326	38,461	60,834	..	..	Current Expenditure
-1,148	-900	-1,228	-3,051	-3,747	-3,159	-684	1,850	12,792	..	..	Current Budget Balance
..	..	..	..	..	..	..	..	..	..	..	Capital Receipts
749	900	580	1,656	1,101	1,774	4,159	9,429	12,493	..	..	Capital Payments

GHANA	1967	1968	1969	1970	1971	1972	1973	1974	1975	1976	1977
FOREIGN TRADE (CUSTOMS BASIS)				*(Millions of current US dollars)*							
Value of Exports, fob	277.9	306.7	301.5	432.9	341.4	393.1	565.1	646.6	737.0	765.2	951.1
Nonfuel Primary Products	267.5	299.2	290.9	422.8	326.4	373.6	519.9	614.9	696.7	722.3	906.1
Fuels	1.9	1.7	2.0	0.7	2.7	3.1	3.7	12.4	20.4	16.4	15.3
Manufactures	8.5	5.9	8.6	9.3	12.3	16.4	41.6	19.4	19.9	26.5	29.6
Value of Imports, cif	307.2	307.7	347.3	410.0	433.6	292.2	447.6	817.7	787.9	862.0	1,143.5
Nonfuel Primary Products	67.0	67.9	70.8	100.3	89.8	73.3	132.3	179.0	139.1	162.2	160.0
Fuels	18.4	21.1	22.4	23.9	26.5	33.7	40.1	135.7	131.2	128.7	179.5
Manufactures	221.8	218.7	254.1	285.8	317.3	185.2	275.3	503.1	517.6	571.2	803.9
				(Index 1980 = 100)							
Terms of Trade	100.5	122.8	133.7	106.8	79.6	102.3	113.4	99.7	81.6	113.5	159.3
Export Prices, fob	23.0	26.9	30.7	26.3	21.8	25.1	40.0	54.8	46.3	66.4	101.0
Nonfuel Primary Products	23.6	27.6	32.1	26.4	22.0	25.6	41.7	55.7	46.5	68.1	105.4
Fuels	4.3	4.3	4.3	4.3	5.6	6.2	8.9	36.7	35.7	38.4	42.0
Manufactures	30.6	31.1	31.6	33.2	33.2	31.1	33.7	45.6	51.8	53.4	64.8
Import Prices, cif	22.9	21.9	23.0	24.6	27.4	24.6	35.3	54.9	56.7	58.5	63.4
Trade at Constant 1980 Prices				*(Millions of 1980 US dollars)*							
Exports, fob	1,207.3	1,141.2	982.2	1,646.5	1,568.3	1,564.0	1,412.2	1,180.5	1,593.1	1,152.7	941.8
Imports, cif	1,340.8	1,405.9	1,512.7	1,665.6	1,585.2	1,188.8	1,268.6	1,489.1	1,390.6	1,474.1	1,803.8
BALANCE OF PAYMENTS				*(Millions of current US dollars)*							
Exports of Goods & Services	..	..	..	477	389	443	654	747	895	894	1,020
Merchandise, fob	..	..	..	427	335	384	585	679	801	779	890
Nonfactor Services	..	..	..	45	52	55	62	63	90	112	128
Factor Services	..	..	..	4	3	4	7	4	4	3	3
Imports of Goods & Services	..	..	..	543	535	348	540	942	922	995	1,159
Merchandise, fob	..	..	..	375	368	223	372	708	651	690	860
Nonfactor Services	..	..	..	128	128	100	145	207	231	260	261
Factor Services	..	..	..	40	39	25	23	27	41	44	37
Long-Term Interest	..	..	..	12	15	7	12	13	19	19	16
Current Transfers, net	..	..	..	..	..	..	..	..	..	..	..
Workers' Remittances	..	..	..	0	0	0	0	0	0	0	0
Total to be Financed	..	..	..	..	..	..	..	..	..	..	..
Official Capital Grants	..	..	..	..	..	..	..	..	..	..	..
Current Account Balance	..	..	..	-68	-146	108	127	-172	18	-74	-80
Long-Term Capital, net	..	..	..	110	74	57	50	36	113	18	154
Direct Investment	..	..	..	68	31	12	14	11	71	-18	19
Long-Term Loans	..	..	..	..	..	..	..	..	..	..	..
Disbursements	..	..	..	..	..	..	..	..	..	..	..
Repayments	..	..	..	14	13	7	11	14	35	34	20
Other Long-Term Capital	..	..	..	43	43	46	36	26	42	36	134
Other Capital, net	..	..	..	-51	78	-100	-102	75	-133	-3	36
Change in Reserves	..	..	..	8	-6	-65	-75	60	3	59	-109
Memo Item:				*(Ghanaian Cedis per US dollar)*							
Conversion Factor (Annual Avg)	0.870	1.020	1.020	1.020	1.030	1.320	1.420	1.610	1.880	2.360	3.500
Additional Conversion	0.867	1.020	1.020	1.020	1.035	1.333	1.165	1.150	1.150	1.150	1.150
EXTERNAL DEBT, ETC.				*(Millions of US dollars, outstanding at end of year)*							
Public/Publicly Guar. Long-Term	..	..	..	488	526	556	702	702	668	659	781
Official Creditors	..	..	..	264	301	341	460	569	562	586	663
IBRD and IDA	..	..	..	53	58	66	70	75	83	95	122
Private Creditors	..	..	..	224	225	215	242	133	106	73	118
Private Non-guaranteed Long-Term	..	..	..	10	10	9	8	6	10	10	10
Use of Fund Credit	..	..	..	46	20	2	0	2	45	60	63
Short-Term Debt	..	..	..	..	..	..	..	..	..	..	222
Memo Items:				*(Millions of US dollars)*							
Int'l Reserves Excluding Gold	77.1	90.3	66.2	36.6	36.1	93.7	176.1	71.5	124.7	91.7	148.6
Gold Holdings (at market price)	5.6	6.7	5.6	6.0	7.0	10.4	18.0	29.8	22.4	21.6	33.0
SOCIAL INDICATORS											
Total Fertility Rate	6.8	6.7	6.7	6.7	6.7	6.6	6.6	6.6	6.6	6.5	6.5
Crude Birth Rate	46.6	46.4	46.3	46.1	45.9	45.8	45.6	45.5	45.4	45.2	45.1
Infant Mortality Rate	114.5	113.0	111.5	110.0	108.5	107.0	106.0	105.0	104.0	103.0	102.0
Life Expectancy at Birth	48.4	48.7	49.0	49.4	49.7	50.0	50.2	50.4	50.6	50.8	51.0
Food Production, p.c. ('79-81 = 100)	131.9	128.6	132.6	142.1	139.8	132.2	135.8	149.0	133.3	116.6	105.3
Labor Force, Agriculture (%)	60.0	59.4	58.9	58.4	58.1	57.9	57.6	57.4	57.1	56.8	56.6
Labor Force, Female (%)	42.7	42.5	42.4	42.2	42.2	42.1	42.0	41.9	41.9	41.8	41.7
School Enroll. Ratio, primary	..	..	..	64.0	..	..	..	..	71.0	71.0	71.0
School Enroll. Ratio, secondary	..	..	..	14.0	..	..	..	..	37.0	36.0	36.0

1978	1979	1980	1981	1982	1983	1984	1985	1986	1987 estimate	NOTES	GHANA
				(Millions of current US dollars)							**FOREIGN TRADE (CUSTOMS BASIS)**
992.4	1,041.0	1,257.0	1,063.0	873.0	536.0	575.0	663.0	914.0	1,056.0	f	Value of Exports, fob
943.8	989.0	1,212.0	931.0	789.0	506.0	529.0	572.0	868.0	987.0	..	Nonfuel Primary Products
23.0	32.0	20.0	99.0	69.0	12.0	38.0	81.0	24.0	41.0	..	Fuels
25.6	20.0	25.0	33.0	15.0	18.0	8.0	10.0	22.0	28.0	..	Manufactures
1,002.6	852.7	1,128.6	1,106.0	705.0	600.0	640.0	731.0	783.0	836.0	..	Value of Imports, cif
155.7	109.5	145.9	123.0	93.0	78.0	75.0	78.0	72.2	82.2	..	Nonfuel Primary Products
138.6	175.9	300.9	350.0	242.0	161.0	161.0	199.0	122.9	143.8	..	Fuels
708.2	567.3	681.8	633.0	370.0	361.0	404.0	454.0	588.0	610.0	..	Manufactures
				(Index 1980 = 100)							
138.8	133.1	100.0	82.3	73.2	89.3	96.2	90.7	87.0	84.6	..	Terms of Trade
98.7	109.8	100.0	85.2	71.6	83.7	88.7	82.0	76.3	83.1	..	Export Prices, fob
103.1	113.5	100.0	82.5	69.4	83.4	88.6	81.3	77.3	84.2	..	Nonfuel Primary Products
42.3	61.0	100.0	112.5	101.6	92.5	90.2	87.5	44.9	56.9	..	Fuels
71.5	81.9	100.0	102.1	96.9	89.6	88.1	82.4	97.4	106.9	..	Manufactures
71.1	82.5	100.0	103.5	97.8	93.8	92.1	90.5	87.7	98.2	..	Import Prices, cif
				(Millions of 1980 US dollars)							Trade at Constant 1980 Prices
1,005.3	948.4	1,257.0	1,248.2	1,219.5	640.1	648.6	808.2	1,198.7	1,270.9	..	Exports, fob
1,409.2	1,034.1	1,128.6	1,068.2	720.9	639.5	694.6	807.8	893.3	851.4	..	Imports, cif
				(Millions of current US dollars)							**BALANCE OF PAYMENTS**
997	1,165	1,213	832	714	478	612	676	819		..	Exports of Goods & Services
893	1,066	1,104	711	607	439	566	632	773	..	..	Merchandise, fob
103	97	107	119	104	38	44	38	40	..	..	Nonfactor Services
1	2	3	2	3	0	2	6	5	4	..	Factor Services
1,102	1,121	1,264	1,336	905	724	813	952	1,056	..	..	Imports of Goods & Services
780	803	908	954	589	500	533	669	713	..	..	Merchandise, fob
293	258	266	292	225	134	163	168	228	..	..	Nonfactor Services
29	60	89	90	92	90	117	116	116	..	..	Factor Services
24	30	30	26	29	45	30	31	43	58	..	Long-Term Interest
..	..	..	..	..	-2	21	33	..	202	..	Current Transfers, net
0	1	1	1	1	0	5	0	1	..	..	Workers' Remittances
..	..	..	..	..	-246	-180	-234	..	-397	..	Total to be Financed
..	..	..	..	..	72	141	100	..	122	..	Official Capital Grants
-46	122	29	-421	-109	-174	-39	-134	-43	-275	..	Current Account Balance
157	201	176	181	212	108	344	147	258	..	..	Long-Term Capital, net
10	-3	16	16	16	2	2	6	4	5	..	Direct Investment
..	..	..	..	..	..	..	..	269	240	..	Long-Term Loans
..	..	..	..	..	..	..	..	366	365	..	Disbursements
41	41	71	34	40	67	72	74	97	125	..	Repayments
147	204	160	165	196	34	201	42	-15	..	..	Other Long-Term Capital
-26	-354	-301	204	-104	-190	-373	47	-245	-20	..	Other Capital, net
-86	31	96	36	1	256	67	-60	30	295	..	Change in Reserves
				(Ghanaian Cedis per US dollar)							**Memo Item:**
5.730	7.020	9.640	17.200	21.420	45.360	61.320	76.160	89.290	147.000	..	Conversion Factor (Annual Avg)
1.764	2.750	2.750	2.750	2.750	8.830	35.986	54.365	89.204	153.730	..	Additional Conversion
			(Millions of US dollars, outstanding at end of year)								**EXTERNAL DEBT, ETC.**
856	990	1,128	1,136	1,153	1,209	1,162	1,296	1,690	2,207	..	Public/Publicly Guar. Long-Term
714	839	991	1,023	1,042	1,068	1,036	1,136	1,454	1,935	..	Official Creditors
157	187	213	239	256	269	289	377	580	851	..	IBRD and IDA
142	151	137	113	111	142	127	160	237	272	..	Private Creditors
10	9	10	10	16	21	32	40	38	30	..	Private Non-guaranteed Long-Term
61	74	43	28	21	280	468	656	748	779	..	Use of Fund Credit
359	202	131	286	206	90	238	183	180	108	..	Short-Term Debt
				(Millions of US dollars)							**Memo Items:**
277.2	289.1	180.4	145.6	138.9	144.8	301.6	478.5	513.0	195.1	..	Int'l Reserves Excluding Gold
49.5	112.1	149.1	122.8	175.5	146.5	135.7	73.6	111.0	136.5	..	Gold Holdings (at market price)
											SOCIAL INDICATORS
6.5	6.5	6.5	6.5	6.5	6.4	6.4	6.3	6.3	6.3	..	Total Fertility Rate
45.1	45.1	45.2	45.2	45.2	45.2	45.2	45.2	45.3	45.3	..	Crude Birth Rate
101.2	100.4	99.6	98.8	97.0	95.3	93.6	91.9	90.3	88.6	..	Infant Mortality Rate
51.2	51.4	51.6	51.8	52.0	52.5	52.9	53.4	53.9	54.4	..	Life Expectancy at Birth
101.0	104.1	100.6	95.3	91.0	80.9	114.7	103.9	107.5	105.5	..	Food Production, p.c. ('79-81 = 100)
56.3	56.1	55.8	..	..	..	..	..	..	..	..	Labor Force, Agriculture (%)
41.6	41.5	41.4	41.2	41.1	40.9	40.8	40.6	40.4	40.2	..	Labor Force, Female (%)
71.0	69.0	73.0	77.0	77.0	78.0	67.0	66.0	..	..	..	School Enroll. Ratio, primary
36.0	36.0	37.0	37.0	38.0	38.0	36.0	39.0	..	..	..	School Enroll. Ratio, secondary

GREECE	1967	1968	1969	1970	1971	1972	1973	1974	1975	1976	1977
CURRENT GNP PER CAPITA (US $)	840	930	1,060	1,170	1,300	1,460	1,720	1,940	2,370	2,660	2,860
POPULATION (thousands)	8,716	8,741	8,773	8,793	8,831	8,889	8,929	8,962	9,047	9,167	9,309
ORIGIN AND USE OF RESOURCES					*(Billions of current Greek Drachmas)*						
Gross National Product (GNP)	220.4	239.6	271.5	304.4	338.2	387.3	497.2	582.1	691.4	849.9	994.0
Net Factor Income from Abroad	4.3	5.0	5.0	5.5	7.9	9.6	13.1	17.9	19.2	24.9	30.3
Gross Domestic Product (GDP)	216.1	234.5	266.5	298.9	330.3	377.7	484.2	564.2	672.2	824.9	963.7
Indirect Taxes, net	27.7	31.9	37.5	40.9	42.9	47.7	55.9	56.9	79.0	96.2	119.1
GDP at factor cost	188.4	202.6	229.0	258.0	287.4	330.0	428.2	507.3	593.2	728.7	844.6
Agriculture	42.5	39.6	43.1	47.1	52.3	61.5	87.3	100.4	111.0	136.2	141.5
Industry	50.6	59.3	70.2	81.0	91.7	106.7	142.0	155.2	178.9	222.8	263.8
Manufacturing	31.2	34.7	40.8	49.3	55.6	61.9	86.2	102.6	118.1	146.5	165.3
Services, etc.	95.3	103.7	115.7	130.0	143.4	161.8	198.9	251.8	303.3	369.8	439.3
Resource Balance	-16.0	-20.6	-23.9	-25.0	-26.8	-31.4	-53.2	-53.9	-67.2	-67.9	-81.0
Exports of Goods & NFServices	23.0	22.5	25.9	30.0	34.1	44.3	68.9	90.8	113.3	145.1	162.3
Imports of Goods & NFServices	39.0	43.1	49.8	55.0	60.9	75.7	122.1	144.7	180.6	213.1	243.3
Domestic Absorption	232.1	255.1	290.4	323.9	357.1	409.1	537.3	618.1	739.4	892.9	1,044.7
Private Consumption, etc.	155.7	170.7	187.4	202.2	223.5	251.5	308.7	374.7	456.0	551.8	636.1
General Gov't Consumption	28.1	30.2	33.9	37.7	41.4	45.9	55.4	78.1	102.0	124.3	153.8
Gross Domestic Investment	48.2	54.2	69.0	84.0	92.2	111.7	173.2	165.4	181.3	216.7	254.7
Fixed Investment	43.9	54.4	65.6	70.7	83.3	104.8	135.7	125.5	140.0	175.0	221.4
Memo Items:											
Gross Domestic Saving	32.3	33.6	45.1	59.0	65.4	80.3	120.0	111.4	114.1	148.8	173.8
Gross National Saving	43.6	45.8	58.4	74.8	87.4	107.0	154.8	148.6	156.8	202.9	238.0
					(Billions of 1980 Greek Drachmas)						
Gross National Product	889.0	943.3	1,032.6	1,114.4	1,200.4	1,306.0	1,402.6	1,353.8	1,426.8	1,516.6	1,569.5
GDP at factor cost	766.3	803.7	877.5	950.8	1,024.4	1,116.4	1,204.9	1,186.4	1,247.4	1,318.8	1,352.2
Agriculture	197.8	180.7	192.3	210.1	217.2	230.1	228.6	239.6	253.2	249.8	231.4
Industry	199.0	229.1	262.3	283.5	317.9	356.9	400.4	356.0	376.6	411.7	431.4
Manufacturing	111.1	124.0	142.1	164.1	181.9	196.2	230.7	224.1	236.4	260.0	263.7
Services, etc.	369.5	393.9	422.8	457.3	489.3	529.4	575.9	590.8	617.6	657.3	689.5
Resource Balance	-92.4	-112.7	-131.0	-132.4	-137.4	-148.5	-211.2	-143.1	-143.2	-128.7	-155.4
Exports of Goods & NFServices	95.0	94.1	107.8	121.1	135.5	166.5	205.4	205.6	227.5	264.7	269.4
Imports of Goods & NFServices	187.4	206.8	238.8	253.5	272.9	315.0	416.5	348.7	370.6	393.4	424.8
Domestic Absorption	959.1	1,030.3	1,137.8	1,220.0	1,301.1	1,414.2	1,566.1	1,454.0	1,532.7	1,602.2	1,673.4
Private Consumption, etc.	618.6	665.8	704.8	745.4	798.8	863.8	900.4	879.4	938.7	988.8	1,040.1
General Gov't Consumption	132.7	134.5	144.8	153.3	160.9	170.0	181.6	203.5	227.8	239.5	255.1
Gross Domestic Investment	207.8	230.0	288.3	321.3	341.4	380.4	484.1	371.1	366.2	373.9	378.1
Fixed Investment	222.1	269.5	319.7	315.3	359.5	414.9	446.7	332.4	333.2	355.9	383.5
Memo Items:											
Capacity to Import	110.7	107.9	124.2	138.2	152.8	184.3	235.1	218.7	232.7	268.0	283.4
Terms of Trade Adjustment	15.7	13.8	16.4	17.1	17.4	17.8	29.7	13.1	5.2	3.3	14.0
Gross Domestic Income	882.3	931.4	1,023.2	1,104.7	1,181.0	1,283.5	1,384.6	1,324.0	1,394.7	1,476.8	1,532.0
Gross National Income	904.7	957.2	1,049.0	1,131.6	1,217.7	1,323.8	1,432.3	1,366.9	1,432.0	1,519.8	1,583.5
DOMESTIC PRICES (DEFLATORS)					*(Index 1980 = 100)*						
Overall (GDP)	24.9	25.6	26.5	27.5	28.4	29.8	35.7	43.0	48.4	56.0	63.5
Domestic Absorption	24.2	24.8	25.5	26.6	27.4	28.9	34.3	42.5	48.2	55.7	62.4
Agriculture	21.5	21.9	22.4	22.4	24.1	26.7	38.2	41.9	43.8	54.5	61.2
Industry	25.4	25.9	26.7	28.6	28.8	29.9	35.5	43.6	47.5	54.1	61.1
Manufacturing	28.1	28.0	28.7	30.0	30.6	31.6	37.3	45.8	50.0	56.4	62.7
MANUFACTURING ACTIVITY											
Employment (1980=100)	68.0	65.7	66.9	72.1	75.3	77.8	82.4	87.1	89.9	95.9	97.5
Real Earnings per Empl. (1980=100)	53.4	58.1	61.7	64.3	67.2	71.4	73.6	72.0	78.4	86.3	92.7
Real Output per Empl. (1980=100)	45.0	51.7	55.5	57.0	58.9	63.2	68.8	72.9	77.3	78.2	80.9
Earnings as % of Value Added	36.7	34.1	32.4	31.5	31.7	32.9	29.8	31.6	35.9	37.7	39.0
MONETARY HOLDINGS					*(Billions of current Greek Drachmas)*						
Money Supply, Broadly Defined	88.6	101.9	117.4	139.0	170.5	212.1	250.0	301.4	375.4	472.2	586.2
Money as Means of Payment	44.0	46.5	48.9	54.6	63.6	75.9	93.7	112.2	130.6	159.6	186.6
Currency Ouside Banks	33.7	33.3	35.7	39.1	43.3	50.8	65.3	80.6	92.2	112.3	133.4
Demand Deposits	10.4	13.2	13.2	15.5	20.4	25.1	28.4	31.6	38.4	47.3	53.2
Quasi-Monetary Liabilities	44.6	55.5	68.5	84.4	106.8	136.2	156.3	189.2	244.8	312.6	399.6
GOVERNMENT DEFICIT (-) OR SURPLUS					*(Billions of current Greek Drachmas)*						
	..	..	..	..	..	-6.5	-13.0	-23.4	-26.1	-32.4	-47.4
Current Revenue	..	..	..	..	..	98.5	123.0	143.7	179.5	239.5	283.9
Current Expenditure	..	..	..	..	..	77.7	100.4	132.4	163.0	215.9	256.5
Current Budget Balance	..	..	..	..	..	20.8	22.6	11.3	16.6	23.7	27.4
Capital Receipts	..	..	..	..	..	1.6	0.5	1.2	0.5	0.6	0.5
Capital Payments	..	..	..	..	..	28.9	36.1	36.0	43.2	56.7	75.4

1978	1979	1980	1981	1982	1983	1984	1985	1986	1987 estimate	NOTES	GREECE
3,270	3,870	4,370	4,470	4,330	3,970	3,810	3,620	3,670	4,020	..	**CURRENT GNP PER CAPITA (US $)**
9,430	9,548	9,643	9,729	9,792	9,840	9,896	9,919	9,966	10,002	..	**POPULATION (thousands)**
				(Billions of current Greek Drachmas)							**ORIGIN AND USE OF RESOURCES**
1,193.8	1,472.2	1,767.6	2,109.1	2,632.4	3,108.2	3,806.6	4,582.6	5,475.4	6,326.2	..	Gross National Product (GNP)
32.4	43.5	56.6	59.0	57.8	30.4	1.9	-33.8	-67.8	-63.3	..	Net Factor Income from Abroad
1,161.4	1,428.8	1,710.9	2,050.1	2,574.7	3,077.8	3,804.7	4,616.4	5,543.2	6,389.5	..	Gross Domestic Product (GDP)
144.7	183.6	187.2	190.1	264.0	345.9	443.1	481.0	648.4	850.4	..	Indirect Taxes, net
1,016.7	1,245.2	1,523.7	1,860.0	2,310.7	2,731.9	3,361.6	4,135.4	4,894.8	5,539.1	..	GDP at factor cost
177.1	198.2	270.1	329.3	424.4	462.8	591.4	713.8	791.9	875.0	..	Agriculture
312.9	401.3	474.3	570.4	672.7	811.5	981.9	1,209.3	1,450.5	1,580.4	..	Industry
191.3	238.5	297.0	361.3	422.6	503.1	614.8	752.5	907.4	971.0	..	Manufacturing
526.7	645.7	779.4	960.2	1,213.5	1,457.7	1,788.4	2,212.3	2,652.4	3,083.7	..	Services, etc.
-81.7	-111.3	-91.2	-133.7	-265.3	-315.9	-314.5	-535.9	-470.2	-568.1	..	Resource Balance
204.4	249.6	357.7	422.4	473.0	609.5	824.6	977.6	1,233.1	1,361.6	..	Exports of Goods & NFServices
286.1	360.8	448.9	556.1	738.3	925.4	1,139.1	1,513.5	1,703.3	1,929.8	f	Imports of Goods & NFServices
1,243.1	1,540.0	1,802.2	2,183.8	2,839.9	3,393.7	4,119.2	5,152.3	6,013.4	6,957.6	f	Domestic Absorption
736.6	875.8	1,093.9	1,294.6	1,825.4	2,141.1	2,601.5	3,228.0	3,876.0	4,616.4	..	Private Consumption, etc.
185.2	233.5	280.1	368.6	471.2	579.4	742.8	942.0	1,075.0	1,246.5	..	General Gov't Consumption
321.4	430.7	428.2	520.6	543.3	673.2	774.8	982.2	1,062.3	1,094.7	f	Gross Domestic Investment
278.0	369.2	413.7	456.3	513.5	624.0	702.9	880.4	1,024.4	1,113.5	f	Fixed Investment
											Memo Items:
239.7	319.4	337.0	386.9	278.0	357.3	460.4	446.3	592.1	526.6	..	Gross Domestic Saving
308.1	406.0	439.9	505.5	405.2	469.7	565.6	522.6	660.8	648.9	..	Gross National Saving
				(Billions of 1980 Greek Drachmas)							
1,669.5	1,727.8	1,767.6	1,762.3	1,759.3	1,739.0	1,771.4	1,806.9	1,812.8	1,808.3	..	Gross National Product
1,440.8	1,487.5	1,523.7	1,525.5	1,535.2	1,536.1	1,583.3	1,636.0	1,656.5	1,647.4	I	GDP at factor cost
255.4	239.3	270.1	265.7	272.0	247.8	265.1	270.0	271.8	259.6	..	Agriculture
458.5	486.0	474.3	467.1	456.4	457.2	462.9	479.9	481.2	475.2	..	Industry
281.0	296.5	297.0	296.1	289.7	284.7	288.1	298.3	298.0	291.5	..	Manufacturing
726.9	762.1	779.4	792.7	806.8	831.1	855.2	886.0	903.5	912.6	..	Services, etc.
-141.6	-153.3	-91.2	-128.8	-185.7	-193.5	-137.5	-200.5	-165.8	-166.4	..	Resource Balance
313.6	334.4	357.7	336.4	312.2	337.2	394.3	399.5	455.9	467.7	..	Exports of Goods & NFServices
455.2	487.8	448.9	465.2	497.9	530.7	531.8	600.0	621.7	634.1	f	Imports of Goods & NFServices
1,762.9	1,829.0	1,802.2	1,839.9	1,904.8	1,914.2	1,908.0	2,025.7	2,012.8	2,004.0	f	Domestic Absorption
1,090.3	1,100.8	1,093.9	1,173.4	1,255.1	1,258.8	1,252.6	1,339.8	1,359.9	1,392.7	..	Private Consumption, etc.
264.0	279.5	280.1	299.1	306.0	314.4	324.0	334.3	334.3	338.6	..	General Gov't Consumption
408.6	448.6	428.2	367.3	343.7	340.9	331.4	351.6	318.7	272.8	f	Gross Domestic Investment
406.5	442.3	413.7	382.6	375.3	370.4	349.4	367.5	346.5	335.2	f	Fixed Investment
											Memo Items:
325.2	337.3	357.7	353.4	319.0	349.5	385.0	387.6	450.1	447.4	..	Capacity to Import
11.6	2.9	0.0	16.9	6.8	12.3	-9.3	-11.9	-5.8	-20.3	..	Terms of Trade Adjustment
1,632.9	1,678.5	1,710.9	1,728.0	1,725.9	1,733.0	1,761.2	1,813.3	1,841.2	1,817.3	..	Gross Domestic Income
1,681.1	1,730.7	1,767.6	1,779.2	1,766.1	1,751.3	1,762.1	1,794.9	1,807.0	1,788.0	..	Gross National Income
				(Index 1980 = 100)							**DOMESTIC PRICES (DEFLATORS)**
71.6	85.3	100.0	119.8	149.8	178.9	214.9	252.9	300.1	347.7	..	Overall (GDP)
70.5	84.2	100.0	118.7	149.1	177.3	215.9	254.3	298.7	347.2	..	Domestic Absorption
69.3	82.8	100.0	123.9	156.0	186.7	223.1	264.3	291.4	337.1	..	Agriculture
68.2	82.6	100.0	122.1	147.4	177.5	212.1	252.0	301.4	332.6	..	Industry
68.1	80.4	100.0	122.0	145.9	176.7	213.4	252.3	304.5	333.1	..	Manufacturing
											MANUFACTURING ACTIVITY
98.3	99.2	100.0	100.5	102.1	103.7	105.4	107.0	..	..	..	Employment (1980=100)
95.5	100.7	100.0	98.9	93.2	90.7	92.2	93.1	..	..	..	Real Earnings per Empl. (1980=100)
89.3	97.9	100.0	103.5	96.2	93.0	92.7	94.6	..	..	..	Real Output per Empl. (1980=100)
39.0	39.0	38.7	39.5	39.1	39.1	39.1	39.1	..	..	..	Earnings as % of Value Added
				(Billions of current Greek Drachmas)							**MONETARY HOLDINGS**
797.3	947.4	1,159.8	1,536.7	1,973.3	2,374.3	3,035.6	3,826.8	4,557.9	..	D	Money Supply, Broadly Defined
228.2	265.4	308.8	377.4	459.1	525.6	608.5	719.4	875.6	1,000.0	..	Money as Means of Payment
161.8	184.7	211.8	263.7	305.2	348.2	408.2	513.5	550.1	641.4	..	Currency Ouside Banks
66.4	80.8	96.9	113.7	153.9	177.4	200.4	205.9	325.6	358.6	..	Demand Deposits
569.1	682.0	851.0	1,159.3	1,514.2	1,848.7	2,427.1	3,107.4	3,682.3	..	..	Quasi-Monetary Liabilities
				(Billions of current Greek Drachmas)							**GOVERNMENT DEFICIT (-) OR SURPLUS**
-49.8	-55.3	-85.5	-223.7	-477.5	-160.9	-534.0	-661.3	..	..	..	**GOVERNMENT DEFICIT (-) OR SURPLUS**
348.9	427.3	524.7	600.8	871.6	1,192.7	1,368.0	1,671.3	..	..	..	Current Revenue
321.7	407.1	532.2	728.4	1,206.4	1,162.2	1,661.0	2,033.7	..	..	..	Current Expenditure
27.2	20.2	-7.5	-127.6	-334.8	30.6	-293.0	-362.4	..	..	..	Current Budget Balance
0.9	1.4	1.6	2.3	3.6	0.2	4.2	3.6	..	..	..	Capital Receipts
77.9	76.9	79.6	98.4	146.3	191.7	245.2	302.5	..	..	..	Capital Payments

GREECE	1967	1968	1969	1970	1971	1972	1973	1974	1975	1976	1977
FOREIGN TRADE (CUSTOMS BASIS)				*(Millions of current US dollars)*							
Value of Exports, fob	495.0	468.0	554.0	643.0	662.0	871.0	1,454.0	2,030.0	2,278.0	2,558.0	2,757.0
Nonfuel Primary Products	421.0	372.0	388.0	414.0	454.0	549.0	713.0	942.0	1,024.0	1,156.0	1,251.0
Fuels	5.0	6.0	6.0	6.0	6.0	11.0	203.0	183.0	252.0	150.0	133.0
Manufactures	70.0	90.0	160.0	222.0	202.0	311.0	537.0	905.0	1,003.0	1,253.0	1,373.0
Value of Imports, cif	1,186.0	1,393.0	1,594.0	1,958.0	2,098.0	2,346.0	3,473.0	4,385.0	5,321.0	6,051.0	6,852.0
Nonfuel Primary Products	310.0	323.0	369.0	399.0	448.0	486.0	813.0	981.0	991.0	982.0	1,134.0
Fuels	94.0	100.0	119.0	135.0	153.0	231.0	427.0	974.0	1,178.0	1,234.0	1,041.0
Manufactures	782.0	970.0	1,106.0	1,425.0	1,496.0	1,629.0	2,233.0	2,429.0	3,152.0	3,835.0	4,678.0
Terms of Trade	156.5	151.0	155.7	150.1	*(Index 1980 = 100)* 135.0	142.1	132.7	113.0	109.9	111.4	107.1
Export Prices, fob	32.8	32.5	34.4	35.2	35.2	37.3	42.1	58.4	58.2	62.9	68.0
Nonfuel Primary Products	35.0	35.8	37.8	38.0	36.4	40.0	51.4	61.1	63.4	64.9	69.4
Fuels	4.3	4.3	4.3	4.3	5.6	6.2	22.0	36.7	35.7	38.4	42.0
Manufactures	35.2	34.6	35.2	37.7	38.4	39.6	47.2	62.9	62.9	66.0	71.1
Import Prices, cif	20.9	21.5	22.1	23.4	26.1	26.3	31.8	51.7	53.0	56.5	63.6
Trade at Constant 1980 Prices				*(Millions of 1980 US dollars)*							
Exports, fob	1,512.0	1,440.0	1,610.0	1,827.0	1,883.0	2,333.0	3,451.0	3,478.0	3,913.0	4,067.0	4,052.0
Imports, cif	5,668.0	6,477.0	7,215.0	8,361.0	8,052.0	8,927.0	10,934.0	8,490.0	10,047.0	10,715.0	10,782.0
BALANCE OF PAYMENTS				*(Millions of current US dollars)*							
Exports of Goods & Services	..	..	..	1,100	1,314	1,692	2,401	3,254	3,647	4,202	4,833
Merchandise, fob	..	..	..	612	627	835	1,230	1,774	1,960	2,228	2,522
Nonfactor Services	..	..	..	471	668	818	1,105	1,370	1,559	1,838	2,193
Factor Services	..	..	..	17	19	39	66	111	129	136	118
Imports of Goods & Services	..	..	..	1,868	2,128	2,664	4,323	5,041	5,272	5,940	6,833
Merchandise, fob	..	..	..	1,509	1,726	2,161	3,582	4,125	4,320	4,922	5,686
Nonfactor Services	..	..	..	267	307	392	591	690	720	738	859
Factor Services	..	..	..	92	95	112	149	226	232	280	288
Long-Term Interest	..	..	..	63	79	93	135	199	229	239	246
Current Transfers, net	..	..	..	344	469	572	732	642	733	800	922
Workers' Remittances	..	..	..	333	457	560	715	625	716	777	899
Total to be Financed	..	..	..	-424	-345	-401	-1,190	-1,145	-891	-938	-1,078
Official Capital Grants	..	..	..	2	2	1	1	2	15	10	..
Current Account Balance	..	..	..	-422	-343	-400	-1,189	-1,143	-877	-928	-1,078
Long-Term Capital, net	..	..	..	282	299	649	796	761	804	554	862
Direct Investment	..	..	..	50	42	55	62	67	24	305	387
Long-Term Loans	..	..	..	208	175	367	349	589	686	124	346
Disbursements	..	..	..	307	321	542	596	878	1,076	565	870
Repayments	..	..	..	99	146	175	247	289	390	441	523
Other Long-Term Capital	..	..	..	22	79	225	384	102	79	115	129
Other Capital, net	..	..	..	117	228	222	381	225	-111	302	387
Change in Reserves	..	..	..	23	-184	-472	11	156	183	72	-171
Memo Item:				*(Greek Drachmas per US dollar)*							
Conversion Factor (Annual Avg)	30.000	30.000	30.000	30.000	30.000	30.000	29.630	30.000	32.050	36.520	36.840
EXTERNAL DEBT, ETC.				*(Millions of US dollars, outstanding at end of year)*							
Public/Publicly Guar. Long-Term	..	..	..	905	1,031	1,339	1,652	2,132	2,633	2,452	2,669
Official Creditors	..	..	..	307	314	327	410	417	488	576	704
IBRD and IDA	..	..	..	10	24	33	49	61	71	82	91
Private Creditors	..	..	..	598	717	1,011	1,242	1,714	2,145	1,876	1,965
Private Non-guaranteed Long-Term	..	..	..	388	481	537	630	786	906	1,095	1,309
Use of Fund Credit	..	..	..	0	0	0	0	44	222	288	214
Short-Term Debt	..	..	..	..	..	..	..	..	..	..	1,720
Memo Items:				*(Millions of US dollars)*							
Int'l Reserves Excluding Gold	156.0	182.2	187.2	193.6	412.4	898.8	898.9	781.7	963.6	880.6	1,048.3
Gold Holdings (at market price)	130.9	167.3	130.6	124.4	122.4	226.8	393.1	673.3	509.0	492.0	615.3
SOCIAL INDICATORS											
Total Fertility Rate	2.5	2.6	2.5	2.3	2.3	2.3	2.3	2.4	2.4	2.4	2.3
Crude Birth Rate	18.7	18.2	17.4	16.5	16.0	15.9	15.4	16.1	15.7	16.0	15.4
Infant Mortality Rate	34.3	34.4	31.8	29.6	26.9	27.3	24.1	23.9	24.0	22.5	20.4
Life Expectancy at Birth	71.0	71.3	71.6	71.8	72.2	72.7	72.7	72.8	72.9	73.0	73.1
Food Production, p.c. ('79-81 = 100)	78.3	74.2	75.4	83.3	80.8	86.2	86.0	94.5	99.1	94.6	91.5
Labor Force, Agriculture (%)	45.2	44.2	43.2	42.2	41.1	40.0	38.8	37.7	36.6	35.5	34.3
Labor Force, Female (%)	25.3	25.4	25.5	25.7	25.7	25.8	25.8	25.9	25.9	25.9	26.0
School Enroll. Ratio, primary	..	..	..	107.0	..	..	..	..	104.0	104.0	103.0
School Enroll. Ratio, secondary	..	..	..	63.0	..	..	..	..	78.0	80.0	81.0

1978	1979	1980	1981	1982	1983	1984	1985	1986	1987 estimate	NOTES	GREECE
											FOREIGN TRADE (CUSTOMS BASIS)
3,375.0	3,877.0	5,142.0	4,249.0	4,297.0	4,412.0	4,864.0	4,536.0	5,660.0	6,489.0	..	Value of Exports, fob
1,512.0	1,646.0	1,901.0	1,580.0	1,680.0	1,912.0	1,998.0	1,749.0	2,241.0	2,579.0	..	Nonfuel Primary Products
321.0	459.0	800.0	404.0	462.0	305.0	492.0	546.0	371.0	436.0	..	Fuels
1,543.0	1,773.0	2,441.0	2,266.0	2,154.0	2,194.0	2,374.0	2,241.0	3,048.0	3,474.0	..	Manufactures
7,655.0	9,594.0	10,531.0	8,781.0	10,012.0	9,500.0	9,611.0	10,138.0	11,241.0	12,908.0	..	Value of Imports, cif
1,231.0	1,620.0	1,725.0	1,613.0	1,916.0	1,880.0	1,978.0	1,950.0	1,999.0	3,307.0	..	Nonfuel Primary Products
1,429.0	2,036.0	2,464.0	1,933.0	2,871.0	2,606.0	2,587.0	2,993.0	1,583.0	1,786.0	..	Fuels
4,995.0	5,938.0	6,343.0	5,234.0	5,225.0	5,013.0	5,045.0	5,195.0	7,659.0	7,815.0	..	Manufactures
				(Index 1980 = 100)							
102.6	103.9	100.0	93.3	94.5	92.8	89.6	87.8	92.9	92.7	..	Terms of Trade
70.2	86.6	100.0	94.2	91.5	86.9	82.2	78.6	81.4	89.0	..	Export Prices, fob
77.5	98.8	100.0	92.1	84.1	90.9	85.7	82.7	91.4	94.8	..	Nonfuel Primary Products
42.3	61.0	100.0	112.5	101.6	92.5	90.2	87.5	44.9	56.9	..	Fuels
73.6	86.2	100.0	93.0	96.0	83.0	78.0	74.0	83.0	91.3	..	Manufactures
68.4	83.4	100.0	101.0	96.8	93.6	91.7	89.6	87.7	96.0	..	Import Prices, cif
											Trade at Constant 1980 Prices
				(Millions of 1980 US dollars)							
4,806.0	4,476.0	5,142.0	4,511.0	4,697.0	5,077.0	5,920.0	5,768.0	6,950.0	7,292.0	..	Exports, fob
11,183.0	11,508.0	10,531.0	8,693.0	10,347.0	10,148.0	10,476.0	11,319.0	12,823.0	13,451.0	..	Imports, cif
				(Millions of current US dollars)							**BALANCE OF PAYMENTS**
5,905	7,680	8,374	9,170	7,891	7,166	7,333	7,111	7,905	10,141	..	Exports of Goods & Services
2,999	3,932	4,093	4,772	4,141	4,106	4,394	4,293	4,513	5,613	..	Merchandise, fob
2,742	3,495	4,029	4,052	3,492	2,930	2,756	2,664	3,286	4,343	..	Nonfactor Services
164	253	252	346	258	130	183	154	106	185	..	Factor Services
7,842	10,732	11,670	12,828	11,382	10,811	11,097	12,053	11,948	14,475	..	Imports of Goods & Services
6,498	8,948	9,650	10,149	8,910	8,400	8,624	9,346	8,936	11,111	..	Merchandise, fob
1,001	1,340	1,475	1,773	1,612	1,446	1,320	1,417	1,589	1,739	..	Nonfactor Services
343	444	545	906	860	965	1,153	1,290	1,423	1,625	..	Factor Services
315	393	477	867	888	882	863	1,065	1,215	1,260	..	Long-Term Interest
981	1,164	1,087	1,076	1,039	931	917	797	975	1,371	..	Current Transfers, net
951	1,137	1,066	1,057	1,016	912	898	775	942	1,334	..	Workers' Remittances
-956	-1,888	-2,209	-2,582	-2,452	-2,714	-2,847	-4,145	-3,068	-2,963	K	Total to be Financed
1	1	0	161	560	836	715	869	1,392	1,665	K	Official Capital Grants
-955	-1,887	-2,209	-2,421	-1,892	-1,878	-2,132	-3,276	-1,676	-1,298	..	Current Account Balance
1,041	1,332	1,994	1,750	1,799	2,946	2,488	3,635	3,543	3,051	..	Long-Term Capital, net
428	613	672	520	436	439	485	447	471	683	..	Direct Investment
379	341	1,526	1,282	1,043	1,686	1,793	2,115	1,417	198	..	Long-Term Loans
928	968	2,203	2,095	1,830	2,456	2,582	3,107	2,714	2,776	..	Disbursements
549	627	677	812	788	770	790	992	1,297	2,579	..	Repayments
233	377	-204	-213	-240	-15	-505	204	263	506	..	Other Long-Term Capital
48	512	78	513	-19	-968	-225	181	-1,365	171	..	Other Capital, net
-134	43	137	158	112	-100	-131	-540	-502	-1,924	..	Change in Reserves
											Memo Item:
				(Greek Drachmas per US dollar)							
36.740	37.040	42.620	55.410	66.800	88.060	112.720	138.120	139.980	135.430	..	Conversion Factor (Annual Avg)
				(Millions of US dollars, outstanding at end of year)							**EXTERNAL DEBT, ETC.**
3,153	3,551	4,674	5,731	6,663	8,104	9,363	12,452	15,201	17,437	..	Public/Publicly Guar. Long-Term
831	968	1,005	1,034	1,328	1,713	1,778	2,399	3,611	4,820	..	Official Creditors
94	97	110	113	122	132	129	163	179	185	..	IBRD and IDA
2,322	2,582	3,669	4,697	5,335	6,391	7,584	10,053	11,590	12,616	..	Private Creditors
1,321	1,321	1,618	1,642	1,636	1,600	1,647	1,657	1,659	1,429	..	Private Non-guaranteed Long-Term
241	194	100	16	0	0	0	0	0	..	..	Use of Fund Credit
1,947	2,245	2,941	3,260	2,859	3,354	3,267	4,530	4,188	4,255	..	Short-Term Debt
				(Millions of US dollars)							**Memo Items:**
1,305.0	1,342.8	1,345.9	1,022.0	861.1	900.5	954.2	868.0	1,518.7	2,681.4	..	Int'l Reserves Excluding Gold
852.0	1,949.7	2,260.7	1,531.6	1,769.1	1,480.2	1,265.9	1,348.2	1,293.1	1,617.9	..	Gold Holdings (at market price)
											SOCIAL INDICATORS
2.3	2.3	2.2	2.1	2.0	2.0	2.0	1.9	1.9	1.8	..	Total Fertility Rate
15.5	15.5	15.4	14.5	14.2	13.9	13.6	13.4	13.1	12.8	..	Crude Birth Rate
19.3	18.7	17.9	16.3	15.2	14.2	13.2	12.3	11.5	10.7	..	Infant Mortality Rate
73.5	73.9	74.3	74.7	75.2	75.4	75.6	75.9	76.1	76.4	..	Life Expectancy at Birth
95.9	91.7	104.2	104.1	106.8	100.2	100.6	108.2	103.7	98.5	..	Food Production, p.c. ('79-81 = 100)
33.2	32.0	30.9	..	..	..	..	..	..	..	..	Labor Force, Agriculture (%)
26.0	26.0	26.1	26.1	26.2	26.3	26.3	26.4	26.4	26.5	..	Labor Force, Female (%)
103.0	100.0	102.0	105.0	106.0	106.0	..	..	..	..	..	School Enroll. Ratio, primary
81.0	81.0	81.0	8.0	84.0	86.0	..	..	..	..	..	School Enroll. Ratio, secondary

GRENADA	1967	1968	1969	1970	1971	1972	1973	1974	1975	1976	1977
CURRENT GNP PER CAPITA (US $)	..	..	..	250	..	..	..	..	..	..	..
POPULATION (thousands)	..	..	..	97	..	..	..	..	..	..	..
ORIGIN AND USE OF RESOURCES				*(Millions of current East Caribbean Dollars)*							
Gross National Product (GNP)	32.8	37.0	44.9	47.7	49.7	51.0	55.5	65.9	79.9	98.3	112.9
Net Factor Income from Abroad	-0.1	-0.1	-0.1	-0.1	-3.5	-8.4	-5.6	-0.8	-0.8	0.9	0.0
Gross Domestic Product (GDP)	32.9	37.1	45.0	47.8	53.2	59.4	61.1	66.7	80.7	97.4	112.9
Indirect Taxes, net	..	..	..	..	..	..	..	..	..	..	..
GDP at factor cost	..	..	..	..	..	..	..	..	..	..	..
Agriculture	..	..	..	..	..	..	..	..	..	..	..
Industry	..	..	..	..	..	..	..	..	..	..	..
Manufacturing	..	..	..	..	..	..	..	..	..	..	..
Services, etc.	..	..	..	..	..	..	..	..	..	..	..
Resource Balance	..	..	..	..	..	..	..	..	..	..	..
Exports of Goods & NFServices	..	..	..	..	..	..	..	..	..	..	..
Imports of Goods & NFServices	..	..	..	..	..	..	..	..	..	..	..
Domestic Absorption	..	..	..	..	..	..	..	..	..	..	..
Private Consumption, etc.	..	..	..	..	..	..	..	..	..	..	..
General Gov't Consumption	..	..	..	..	..	..	..	..	..	..	..
Gross Domestic Investment	..	..	..	..	..	..	..	..	..	..	..
Fixed Investment	..	..	..	..	..	..	..	..	..	..	..
Memo Items:											
Gross Domestic Saving	..	..	..	..	..	..	..	..	..	..	..
Gross National Saving	..	..	..	..	..	..	..	..	..	..	..
				(Millions of 1980 East Caribbean Dollars)							
Gross National Product	141.56	143.17	168.51	169.39	169.50	173.04	157.58	135.61	150.81	167.79	176.07
GDP at factor cost	..	..	..	..	..	..	..	..	..	..	..
Agriculture	..	..	..	..	..	..	..	..	..	..	..
Industry	..	..	..	..	..	..	..	..	..	..	..
Manufacturing	..	..	..	..	..	..	..	..	..	..	..
Services, etc.	..	..	..	..	..	..	..	..	..	..	..
Resource Balance	..	..	..	..	..	..	..	..	..	..	..
Exports of Goods & NFServices	..	..	..	..	..	..	..	..	..	..	..
Imports of Goods & NFServices	..	..	..	..	..	..	..	..	..	..	..
Domestic Absorption	..	..	..	..	..	..	..	..	..	..	..
Private Consumption, etc.	..	..	..	..	..	..	..	..	..	..	..
General Gov't Consumption	..	..	..	..	..	..	..	..	..	..	..
Gross Domestic Investment	..	..	..	..	..	..	..	..	..	..	..
Fixed Investment	..	..	..	..	..	..	..	..	..	..	..
Memo Items:											
Capacity to Import	..	..	..	..	..	..	..	..	..	..	..
Terms of Trade Adjustment	..	..	..	..	..	..	..	..	..	..	..
Gross Domestic Income	..	..	..	..	..	..	..	..	..	..	..
Gross National Income	..	..	..	..	..	..	..	..	..	..	..
DOMESTIC PRICES (DEFLATORS)				*(Index 1980 = 100)*							
Overall (GDP)	23.2	25.8	26.6	28.2	29.3	29.5	35.2	48.6	53.0	58.6	64.1
Domestic Absorption	..	..	..	..	..	..	..	..	..	..	..
Agriculture	..	..	..	..	..	..	..	..	..	..	..
Industry	..	..	..	..	..	..	..	..	..	..	..
Manufacturing	..	..	..	..	..	..	..	..	..	..	..
MANUFACTURING ACTIVITY											
Employment (1980=100)	..	..	..	..	..	..	..	..	..	..	..
Real Earnings per Empl. (1980=100)	..	..	..	..	..	..	..	..	..	..	..
Real Output per Empl. (1980=100)	..	..	..	..	..	..	..	..	..	..	..
Earnings as % of Value Added	..	..	..	..	..	..	..	..	..	..	..
MONETARY HOLDINGS				*(Millions of current East Caribbean Dollars)*							
Money Supply, Broadly Defined	..	..	..	42.7	45.4	53.9	55.5	54.5	66.7	79.3	88.1
Money as Means of Payment	..	..	..	10.8	11.9	14.3	15.6	15.8	19.4	26.7	29.5
Currency Ouside Banks	..	..	..	6.1	6.9	8.3	8.4	9.7	12.7	15.3	18.3
Demand Deposits	..	..	..	4.7	5.0	6.0	7.2	6.1	6.7	11.4	11.2
Quasi-Monetary Liabilities	..	..	..	31.9	33.5	39.6	39.9	38.7	47.3	52.6	58.6
				(Thousands of current East Caribbean Dollars)							
GOVERNMENT DEFICIT (-) OR SURPLUS	..	..	..	..	..	..	..	-5,300	-5,100	-6,400	-1,700
Current Revenue	..	..	..	..	..	..	..	18,620	21,200	29,740	35,100
Current Expenditure	..	..	..	..	..	..	..	19,400	22,800	33,300	31,100
Current Budget Balance	..	..	..	..	..	..	..	-780	-1,600	-3,560	4,000
Capital Receipts	..	..	..	..	..	..	..	80	100	60	100
Capital Payments	..	..	..	..	..	..	..	4,500	3,500	2,900	5,800

1978	1979	1980	1981	1982	1983	1984	1985	1986	1987 estimate	NOTES	GRENADA
..	..	740	870	980	1,030	1,080	1,170	1,240	1,340	..	CURRENT GNP PER CAPITA (US $)
..	..	87	88	91	92	94	96	98	100	..	POPULATION (thousands)
				(Millions of current East Caribbean Dollars)							ORIGIN AND USE OF RESOURCES
144.8	171.1	182.8	209.6	240.3	253.6	272.0	307.0	336.2	363.4	..	Gross National Product (GNP)
0.0	0.9	-5.9	-13.5	0.0	-0.3	-4.1	-4.3	-4.3	..	..	Net Factor Income from Abroad
144.8	170.2	188.7	223.1	240.3	253.9	276.1	311.3	340.5	375.7	..	Gross Domestic Product (GDP)
..	..	..	..	42.8	48.4	51.5	65.0	68.4	..	..	Indirect Taxes, net
..	..	..	..	197.5	205.5	224.6	246.3	272.1	..	..	GDP at factor cost
..	..	..	..	44.4	45.1	48.8	48.2	54.7	..	..	Agriculture
..	..	..	..	35.4	32.7	32.7	39.0	45.0	..	..	Industry
..	..	..	..	10.2	10.5	9.8	12.5	13.3	..	..	Manufacturing
..	..	..	..	117.7	127.7	143.1	159.1	172.4	..	..	Services, etc.
..	..	..	..	..	..	..	..	..	..	..	Resource Balance
..	..	..	..	..	..	..	..	..	..	..	Exports of Goods & NFServices
..	..	..	..	..	..	..	..	..	..	..	Imports of Goods & NFServices
..	..	..	..	..	..	..	..	..	..	..	Domestic Absorption
..	..	..	..	..	..	..	..	..	..	..	Private Consumption, etc.
..	..	..	..	..	..	..	..	..	..	..	General Gov't Consumption
..	..	..	..	..	..	..	..	..	..	..	Gross Domestic Investment
..	..	..	..	..	..	..	..	..	..	..	Fixed Investment
											Memo Items:
..	..	..	..	..	..	..	..	..	..	..	Gross Domestic Saving
..	..	..	..	..	..	..	..	..	..	..	Gross National Saving
				(Millions of 1980 East Caribbean Dollars)							
190.31	195.78	182.80	191.38	204.95	211.86	218.77	234.93	244.27	258.90	..	Gross National Product
..	..	..	..	..	..	..	..	..	..	..	GDP at factor cost
..	..	..	..	..	..	..	..	..	..	..	Agriculture
..	..	..	..	..	..	..	..	..	..	..	Industry
..	..	..	..	..	..	..	..	..	..	..	Manufacturing
..	..	..	..	..	..	..	..	..	..	..	Services, etc.
..	..	..	..	..	..	..	..	..	..	..	Resource Balance
..	..	..	..	..	..	..	..	..	..	..	Exports of Goods & NFServices
..	..	..	..	..	..	..	..	..	..	..	Imports of Goods & NFServices
..	..	..	..	..	..	..	..	..	..	..	Domestic Absorption
..	..	..	..	..	..	..	..	..	..	..	Private Consumption, etc.
..	..	..	..	..	..	..	..	..	..	..	General Gov't Consumption
..	..	..	..	..	..	..	..	..	..	..	Gross Domestic Investment
..	..	..	..	..	..	..	..	..	..	..	Fixed Investment
											Memo Items:
..	..	..	..	..	..	..	..	..	..	..	Capacity to Import
..	..	..	..	..	..	..	..	..	..	..	Terms of Trade Adjustment
..	..	..	..	..	..	..	..	..	..	..	Gross Domestic Income
..	..	..	..	..	..	..	..	..	..	..	Gross National Income
				(Index 1980 = 100)							DOMESTIC PRICES (DEFLATORS)
76.1	87.4	100.0	109.5	117.2	119.7	124.3	130.7	137.6	143.3	..	Overall (GDP)
..	..	..	..	..	..	..	..	..	..	..	Domestic Absorption
..	..	..	..	..	..	..	..	..	..	..	Agriculture
..	..	..	..	..	..	..	..	..	..	..	Industry
..	..	..	..	..	..	..	..	..	..	..	Manufacturing
											MANUFACTURING ACTIVITY
..	..	..	..	..	..	..	..	..	..	..	Employment (1980=100)
..	..	..	..	..	..	..	..	..	..	..	Real Earnings per Empl. (1980=100)
..	..	..	..	..	..	..	..	..	..	..	Real Output per Empl. (1980=100)
..	..	..	..	..	..	..	..	..	..	..	Earnings as % of Value Added
				(Millions of current East Caribbean Dollars)							MONETARY HOLDINGS
104.1	123.4	131.0	141.2	146.6	147.0	148.4	177.2	227.0	254.1	..	Money Supply, Broadly Defined
38.0	45.7	48.5	53.3	59.6	58.1	50.7	54.1	69.3	74.4	..	Money as Means of Payment
23.7	28.5	32.6	37.4	40.0	41.4	20.6	25.1	30.7	33.1	..	Currency Ouside Banks
14.4	17.2	15.9	15.9	19.6	16.7	30.1	29.1	38.7	41.4	..	Demand Deposits
66.1	77.8	82.5	87.8	87.0	88.9	97.7	123.1	157.6	179.6	..	Quasi-Monetary Liabilities
				(Thousands of current East Caribbean Dollars)							
..	..	..	..	..	..	..	..	..	..	..	GOVERNMENT DEFICIT (-) OR SURPLUS
..	..	..	..	..	..	..	..	..	..	..	Current Revenue
..	..	..	..	..	..	..	..	..	..	..	Current Expenditure
..	..	..	..	..	..	..	..	..	..	..	Current Budget Balance
..	..	..	..	..	..	..	..	..	..	..	Capital Receipts
..	..	..	..	..	..	..	..	..	..	..	Capital Payments

GRENADA	1967	1968	1969	1970	1971	1972	1973	1974	1975	1976	1977
FOREIGN TRADE (CUSTOMS BASIS)				*(Thousands of current US dollars)*							
Value of Exports, fob	..	..	..	..	..	..	..	..	..	..	..
Nonfuel Primary Products	..	..	..	..	..	..	..	..	..	..	..
Fuels	..	..	..	..	..	..	..	..	..	..	..
Manufactures	..	..	..	..	..	..	..	..	..	..	..
Value of Imports, cif	..	..	..	..	..	..	..	..	..	..	..
Nonfuel Primary Products	..	..	..	..	..	..	..	..	..	..	..
Fuels	..	..	..	..	..	..	..	..	..	..	..
Manufactures	..	..	..	..	..	..	..	..	..	..	..
					(Index 1980 = 100)						
Terms of Trade	..	..	..	..	..	..	..	..	..	..	..
Export Prices, fob	..	..	..	..	..	..	..	..	..	..	..
Nonfuel Primary Products	..	..	..	..	..	..	..	..	..	..	..
Fuels	..	..	..	..	..	..	..	..	..	..	..
Manufactures	..	..	..	..	..	..	..	..	..	..	..
Import Prices, cif	..	..	..	..	..	..	..	..	..	..	..
Trade at Constant 1980 Prices				*(Thousands of 1980 US dollars)*							
Exports, fob	..	..	..	..	..	..	..	..	..	..	..
Imports, cif	..	..	..	..	..	..	..	..	..	..	..
BALANCE OF PAYMENTS				*(Millions of current US dollars)*							
Exports of Goods & Services	..	..	..	..	..	..	..	..	..	..	29
Merchandise, fob	..	..	..	..	..	..	..	..	..	..	14
Nonfactor Services	..	..	..	..	..	..	..	..	..	..	15
Factor Services	..	..	..	..	..	..	..	..	..	..	0
Imports of Goods & Services	..	..	..	..	..	..	..	..	..	..	33
Merchandise, fob	..	..	..	..	..	..	..	..	..	..	29
Nonfactor Services	..	..	..	..	..	..	..	..	..	..	4
Factor Services	..	..	..	..	..	..	..	..	..	..	0
Long-Term Interest	..	..	..	0	0	1	1	0	0	0	0
Current Transfers, net	..	..	..	..	..	..	..	..	..	..	5
Workers' Remittances	..	..	..	..	..	..	..	..	..	..	0
Total to be Financed	..	..	..	..	..	..	..	..	..	..	1
Official Capital Grants	..	..	..	..	..	..	..	..	..	..	1
Current Account Balance	..	..	..	..	..	..	..	..	..	..	1
Long-Term Capital, net	..	..	..	..	..	..	..	..	..	..	2
Direct Investment	..	..	..	..	..	..	..	..	..	..	0
Long-Term Loans	..	..	..	3	0	0	0	1	1	0	2
Disbursements	..	..	..	3	0	..	0	1	1	1	2
Repayments	..	..	..	..	0	0	0	0	0	1	0
Other Long-Term Capital	..	..	..	..	..	..	..	..	..	..	0
Other Capital, net	..	..	..	..	..	..	..	..	..	..	-4
Change in Reserves	..	..	..	-5	-1	1	0	0	0	-3	0
Memo Item:				*(East Caribbean Dollars per US dollar)*							
Conversion Factor (Annual Avg)	1.760	2.000	2.000	2.000	1.970	1.920	1.960	2.050	2.170	2.610	2.700
EXTERNAL DEBT, ETC.				*(Thousands of US dollars, outstanding at end of year)*							
Public/Publicly Guar. Long-Term	..	..	..	7,600	8,000	7,800	7,800	8,200	8,000	7,400	9,200
Official Creditors	..	..	..	2,500	2,900	2,600	2,700	3,300	3,800	3,800	5,700
IBRD and IDA	..	..	..	0	0	0	0	0	0	0	0
Private Creditors	..	..	..	5,100	5,100	5,200	5,100	4,900	4,200	3,500	3,500
Private Non-guaranteed Long-Term	..	..	..	..	..	..	..	..	..	..	..
Use of Fund Credit	..	..	..	0	0	0	0	0	1,000	0	0
Short-Term Debt	..	..	..	..	..	..	..	..	..	..	1,000
Memo Items:				*(Thousands of US dollars)*							
Int'l Reserves Excluding Gold	..	..	..	5,316	5,821	5,050	5,004	5,363	5,036	8,113	7,650
Gold Holdings (at market price)	..	..	..	..	..	..	..	..	..	..	..
SOCIAL INDICATORS											
Total Fertility Rate	..	..	..	..	..	..	..	..	..	..	..
Crude Birth Rate	..	..	..	..	..	..	..	..	..	..	..
Infant Mortality Rate	..	..	..	..	..	..	..	..	..	..	..
Life Expectancy at Birth	..	..	..	..	..	..	..	..	..	..	..
Food Production, p.c. ('79-81 = 100)	109.4	111.8	114.8	112.8	105.5	98.9	98.0	93.4	93.4	100.8	96.6
Labor Force, Agriculture (%)	..	..	..	..	..	..	..	..	..	..	..
Labor Force, Female (%)	..	..	..	..	..	..	..	..	..	..	..
School Enroll. Ratio, primary	..	..	..	..	..	..	..	..	..	..	..
School Enroll. Ratio, secondary	..	..	..	..	..	..	..	..	..	..	..

1978	1979	1980	1981	1982	1983	1984	1985	1986	1987 estimate	NOTES	GRENADA
				(Thousands of current US dollars)							**FOREIGN TRADE (CUSTOMS BASIS)**
..	..	17,400	19,000	18,600	18,800	18,100	22,100	27,800	..	..	Value of Exports, fob
..	..	..	..	..	..	..	..	..	..	..	Nonfuel Primary Products
..	..	..	..	..	..	..	..	..	..	..	Fuels
..	..	..	..	..	..	..	..	..	..	..	Manufactures
..	..	81,250	65,340	79,820	78,040	89,570	86,060	89,440	..	..	Value of Imports, cif
..	..	..	..	..	..	..	..	..	..	..	Nonfuel Primary Products
..	..	..	..	..	..	..	..	..	..	..	Fuels
..	..	..	..	..	..	..	..	..	..	..	Manufactures
				(Index 1980 = 100)							
..	..	..	..	..	..	..	..	..	..	..	Terms of Trade
..	..	..	..	..	..	..	..	..	..	..	Export Prices, fob
..	..	..	..	..	..	..	..	..	..	..	Nonfuel Primary Products
..	..	..	..	..	..	..	..	..	..	..	Fuels
..	..	..	..	..	..	..	..	..	..	..	Manufactures
..	..	..	..	..	..	..	..	..	..	..	Import Prices, cif
				(Thousands of 1980 US dollars)							Trade at Constant 1980 Prices
..	..	..	..	..	..	..	..	..	..	..	Exports, fob
..	..	..	..	..	..	..	..	..	..	..	Imports, cif
				(Millions of current US dollars)							**BALANCE OF PAYMENTS**
34	43	41	41	40	41	42	53	62	..	..	Exports of Goods & Services
17	21	17	19	19	19	18	22	28	..	..	Merchandise, fob
16	21	23	21	20	20	23	29	32	..	..	Nonfactor Services
0	1	1	1	2	2	1	1	2	..	..	Factor Services
41	56	64	77	84	80	74	91	107	..	..	Imports of Goods & Services
33	43	49	56	59	58	51	62	74	..	..	Merchandise, fob
7	13	12	14	19	16	20	26	30	..	..	Nonfactor Services
0	1	3	6	7	6	3	3	3	..	..	Factor Services
0	0	1	1	1	1	1	1	1	2	..	Long-Term Interest
7	6	12	10	11	11	11	11	11	..	..	Current Transfers, net
0	0	0	0	0	0	0	0	0	..	..	Workers' Remittances
0	-7	-11	-25	-33	-29	-21	-28	-35	..	K	Total to be Financed
1	7	13	13	16	10	24	27	25	..	K	Official Capital Grants
0	0	1	-13	-18	-19	3	0	-10	..	..	Current Account Balance
4	10	15	20	27	27	31	32	35	..	..	Long-Term Capital, net
1	..	..	..	2	3	3	5	7	..	..	Direct Investment
3	2	1	7	12	13	3	1	4	7	..	Long-Term Loans
4	2	2	8	13	16	5	3	7	11	..	Disbursements
1	0	1	1	1	3	2	2	3	4	..	Repayments
-1	1	1	1	-4	1	1	-1	0	..	..	Other Long-Term Capital
-2	-7	-14	-11	-11	-8	-33	-23	-23	..	..	Other Capital, net
-2	-3	-2	4	2	0	-1	-9	-1	-3	..	Change in Reserves
				(East Caribbean Dollars per US dollar)							**Memo Item:**
2.700	2.700	2.700	2.700	2.700	2.700	2.700	2.700	2.700	2.700		Conversion Factor (Annual Avg)
				(Thousands of US dollars, outstanding at end of year)							**EXTERNAL DEBT, ETC.**
12,000	14,400	15,300	21,800	33,100	45,000	43,800	48,200	53,200	66,800	..	Public/Publicly Guar. Long-Term
9,400	11,300	12,800	18,000	26,200	40,600	40,000	44,600	50,100	64,000	..	Official Creditors
0	0	0	0	0	0	0	0	2,100	3,200	..	IBRD and IDA
2,600	3,100	2,600	3,800	6,900	4,300	3,800	3,600	3,100	2,700	..	Private Creditors
..	..	..	..	..	..	..	..	..	..	..	Private Non-guaranteed Long-Term
1,600	1,300	300	6,000	4,800	5,700	4,500	2,700	1,400	1,200	..	Use of Fund Credit
1,000	2,000	1,000	2,000	2,000	2,500	1,800	1,600	2,200	3,700	..	Short-Term Debt
				(Thousands of US dollars)							**Memo Items:**
9,703	12,220	12,909	16,097	13,001	14,142	14,234	20,813	20,565	23,331	..	Int'l Reserves Excluding Gold
..	..	..	..	..	..	..	..	..	..	..	Gold Holdings (at market price)
											SOCIAL INDICATORS
..	..	..	..	3.6	3.6	3.5	3.4	3.4	3.3	..	Total Fertility Rate
..	..	..	..	31.4	31.1	30.9	30.6	30.4	30.1	..	Crude Birth Rate
..	..	..	..	39.0	38.0	37.0	36.0	35.1	34.1	..	Infant Mortality Rate
..	..	..	..	67.4	67.6	67.8	68.1	68.3	68.6	..	Life Expectancy at Birth
99.4	102.5	95.7	101.8	93.0	85.9	92.3	84.9	88.5	88.5	..	Food Production, p.c. ('79-81 = 100)
..	..	..	..	..	..	..	..	..	..	..	Labor Force, Agriculture (%)
..	..	..	..	..	..	..	..	..	..	..	Labor Force, Female (%)
..	..	..	..	..	..	..	..	..	..	..	School Enroll. Ratio, primary
..	..	..	..	..	..	..	..	..	..	..	School Enroll. Ratio, secondary

GUATEMALA	1967	1968	1969	1970	1971	1972	1973	1974	1975	1976	1977
CURRENT GNP PER CAPITA (US $)	300	330	340	360	380	390	430	500	570	660	790
POPULATION (thousands)	4,830	4,966	5,105	5,248	5,395	5,546	5,702	5,862	6,026	6,195	6,369
ORIGIN AND USE OF RESOURCES					*(Millions of current Guatemalan Quetzales)*						
Gross National Product (GNP)	1,426	1,578	1,678	1,862	1,941	2,054	2,521	3,112	3,577	4,292	5,448
Net Factor Income from Abroad	-28	-33	-38	-42	-44	-48	-48	-50	-69	-74	-33
Gross Domestic Product (GDP)	1,453	1,611	1,715	1,904	1,985	2,101	2,569	3,161	3,646	4,365	5,481
Indirect Taxes, net	99	109	121	132	138	144	170	227	252	316	492
GDP at factor cost	1,354	1,501	1,595	1,772	1,847	1,958	2,399	2,935	3,394	4,049	4,988
Agriculture	..	..	..	..	..	..	..	..	..	..	..
Industry	..	..	..	..	..	..	..	..	..	..	..
Manufacturing	..	..	..	..	..	..	..	..	..	..	..
Services, etc.	..	..	..	..	..	..	..	..	..	..	..
Resource Balance	-48	-27	6	15	-28	8	17	-103	-66	-262	-99
Exports of Goods & NFServices	236	269	305	354	343	397	536	708	792	942	1,340
Imports of Goods & NFServices	284	297	299	339	371	389	519	811	858	1,204	1,439
Domestic Absorption	1,501	1,638	1,709	1,889	2,013	2,093	2,552	3,264	3,712	4,628	5,579
Private Consumption, etc.	1,201	1,278	1,379	1,493	1,588	1,682	2,034	2,470	2,875	3,396	4,127
General Gov't Consumption	113	115	135	151	139	157	167	207	250	297	355
Gross Domestic Investment	188	244	196	244	285	255	352	588	587	934	1,098
Fixed Investment	192	221	231	239	264	272	357	468	571	900	1,039
Memo Items:											
Gross Domestic Saving	140	217	202	259	258	263	369	485	521	672	1,000
Gross National Saving	121	193	177	234	240	246	364	492	530	796	1,061
					(Millions of 1980 Guatemalan Quetzales)						
Gross National Product	3,713.0	4,033.3	4,217.7	4,467.7	4,712.0	5,057.4	5,421.6	5,780.2	5,879.0	6,321.6	6,877.0
GDP at factor cost	..	..	..	..	..	..	..	..	..	..	..
Agriculture	..	..	..	..	..	..	..	..	..	..	..
Industry	..	..	..	..	..	..	..	..	..	..	..
Manufacturing	..	..	..	..	..	..	..	..	..	..	..
Services, etc.	..	..	..	..	..	..	..	..	..	..	..
Resource Balance	-440.1	-393.6	-259.4	-376.3	-421.3	-205.0	-230.3	-356.6	-231.2	-610.4	-711.8
Exports of Goods & NFServices	748.5	842.2	950.1	928.9	967.6	1,106.4	1,212.4	1,292.9	1,335.6	1,423.7	1,512.3
Imports of Goods & NFServices	1,188.6	1,235.7	1,209.5	1,305.2	1,388.8	1,311.4	1,442.7	1,649.6	1,566.8	2,034.1	2,224.1
Domestic Absorption	4,215.0	4,499.6	4,560.0	4,922.3	5,221.2	5,356.9	5,731.6	6,208.7	6,197.3	7,017.2	7,619.0
Private Consumption, etc.	3,307.5	3,408.2	3,617.3	3,812.9	4,020.4	4,259.8	4,497.7	4,651.6	4,791.3	5,176.8	5,591.8
General Gov't Consumption	272.0	273.7	317.3	353.6	327.4	362.1	362.3	379.5	418.1	464.5	495.8
Gross Domestic Investment	635.6	817.7	625.4	755.8	873.3	735.1	871.6	1,177.5	988.0	1,375.9	1,531.4
Fixed Investment	640.1	727.2	738.7	727.9	789.7	785.2	874.8	858.5	939.8	1,289.8	1,409.3
Memo Items:											
Capacity to Import	987.4	1,122.8	1,235.4	1,363.4	1,284.0	1,337.6	1,491.0	1,440.2	1,446.3	1,590.8	2,071.5
Terms of Trade Adjustment	238.9	280.6	285.3	434.5	316.5	231.3	278.6	147.2	110.7	167.1	559.2
Gross Domestic Income	4,013.8	4,386.7	4,585.8	4,980.5	5,116.4	5,383.2	5,779.9	5,999.3	6,076.8	6,574.0	7,466.4
Gross National Income	3,951.9	4,314.0	4,503.0	4,902.2	5,028.4	5,288.7	5,700.2	5,927.5	5,989.6	6,488.7	7,436.3
DOMESTIC PRICES (DEFLATORS)					*(Index 1980 = 100)*						
Overall (GDP)	38.5	39.2	39.9	41.9	41.4	40.8	46.7	54.0	61.1	68.1	79.3
Domestic Absorption	35.6	36.4	37.5	38.4	38.6	39.1	44.5	52.6	59.9	65.9	73.2
Agriculture	..	..	..	..	..	..	..	..	..	..	..
Industry	..	..	..	..	..	..	..	..	..	..	..
Manufacturing	..	..	..	..	..	..	..	..	..	..	..
MANUFACTURING ACTIVITY											
Employment (1980 = 100)	..	48.5	..	..	70.1	75.2	71.4	77.3	79.2	84.1	89.3
Real Earnings per Empl. (1980 = 100)	..	127.7	..	..	119.0	122.0	125.4	113.3	113.1	101.3	89.5
Real Output per Empl. (1980 = 100)	..	..	..	..	..	..	..	..	..	..	..
Earnings as % of Value Added	..	28.6	..	..	26.7	29.9	27.0	24.4	24.0	21.0	18.3
MONETARY HOLDINGS					*(Millions of current Guatemalan Quetzales)*						
Money Supply, Broadly Defined	263	277	309	344	383	477	580	668	808	1,052	1,249
Money as Means of Payment	148	151	161	173	179	214	264	305	354	494	594
Currency Ouside Banks	83	84	91	97	99	114	137	158	175	237	284
Demand Deposits	66	67	70	76	80	100	127	147	178	257	310
Quasi-Monetary Liabilities	115	126	148	171	204	263	316	363	455	558	655
					(Millions of current Guatemalan Quetzales)						
GOVERNMENT DEFICIT (-) OR SURPLUS	..	..	..	..	..	-45	-38	-46	-31	-111	-51
Current Revenue	..	..	..	..	..	183	212	280	327	409	589
Current Expenditure	..	..	..	..	..	154	171	213	263	311	393
Current Budget Balance	..	..	..	..	..	30	41	67	65	98	197
Capital Receipts	..	..	..	..	..	..	..	..	..	1	0
Capital Payments	..	..	..	..	..	75	79	112	96	210	248

1978	1979	1980	1981	1982	1983	1984	1985	1986	1987 estimate	NOTES	GUATEMALA
910	1,030	1,120	1,200	1,190	1,170	1,180	1,190	1,050	950	A	CURRENT GNP PER CAPITA (US $)
6,548	6,733	6,923	7,119	7,322	7,532	7,749	7,973	8,195	8,438	..	POPULATION (thousands)
				(Millions of current Guatemalan Quetzales)							ORIGIN AND USE OF RESOURCES
6,044	6,890	7,808	8,505	8,596	8,937	9,263	10,849	15,377	17,170	..	Gross National Product (GNP)
-26	-12	-71	-103	-121	-113	-207	-331	-461	-425	..	Net Factor Income from Abroad
6,071	6,903	7,879	8,607	8,717	9,050	9,470	11,180	15,838	17,595	..	Gross Domestic Product (GDP)
536	543	597	561	539	461	437	586	0	..	..	Indirect Taxes, net
5,535	6,360	7,282	8,047	8,178	8,589	9,033	10,594	15,838	17,595	B	GDP at factor cost
..	..	..	..	..	..	..	..	..	..	..	Agriculture
..	..	..	..	..	..	..	..	..	..	..	Industry
..	..	..	..	..	..	..	..	..	..	..	Manufacturing
..	..	..	..	..	..	..	..	..	..	..	Services, etc.
-351	-311	-215	-560	-340	-141	-208	-179	231	-1,141	..	Resource Balance
1,304	1,474	1,748	1,471	1,289	1,176	1,256	2,068	2,542	2,807	..	Exports of Goods & NFServices
1,655	1,784	1,963	2,032	1,629	1,317	1,464	2,247	2,311	3,948	..	Imports of Goods & NFServices
6,422	7,214	8,094	9,168	9,057	9,191	9,678	11,359	15,607	18,736	..	Domestic Absorption
4,675	5,432	6,217	7,022	7,149	7,501	7,856	9,297	12,846	14,931	..	Private Consumption, etc.
435	488	627	680	675	688	726	777	1,124	1,374	..	General Gov't Consumption
1,312	1,294	1,250	1,466	1,233	1,002	1,096	1,285	1,637	2,431	..	Gross Domestic Investment
1,218	1,286	1,294	1,443	1,309	950	912	1,225	1,593	2,150	..	Fixed Investment
											Memo Items:
961	983	1,035	905	893	861	888	1,106	1,868	1,290	..	Gross Domestic Saving
1,050	1,094	1,073	892	834	778	709	794	1,500	1,122	..	Gross National Saving
				(Millions of 1980 Guatemalan Quetzales)							
7,219.7	7,562.5	7,808.2	7,839.9	7,551.3	7,362.2	7,346.6	7,297.2	7,301.7	7,565.0	..	Gross National Product
..	..	..	..	..	..	..	..	..	..	..	GDP at factor cost
..	..	..	..	..	..	..	..	..	..	..	Agriculture
..	..	..	..	..	..	..	..	..	..	..	Industry
..	..	..	..	..	..	..	..	..	..	..	Manufacturing
..	..	..	..	..	..	..	..	..	..	..	Services, etc.
-810.4	-486.1	-215.3	-386.6	-117.9	28.6	-95.9	106.4	94.7	-285.8	..	Resource Balance
1,510.7	1,662.4	1,748.0	1,496.2	1,369.7	1,220.7	1,181.3	1,218.8	1,047.0	1,111.5	..	Exports of Goods & NFServices
2,321.1	2,148.4	1,963.3	1,882.8	1,487.6	1,192.1	1,277.1	1,112.5	952.3	1,397.3	..	Imports of Goods & NFServices
8,062.7	8,080.2	8,094.0	8,317.3	7,767.6	7,425.9	7,586.8	7,338.9	7,360.7	7,974.6	..	Domestic Absorption
5,846.4	6,054.3	6,217.0	6,221.1	5,955.6	5,810.7	5,895.4	5,858.5	5,843.0	6,154.4	..	Private Consumption, etc.
527.1	560.0	627.0	654.9	646.7	647.3	664.4	650.4	684.2	726.4	..	General Gov't Consumption
1,689.3	1,466.0	1,250.0	1,441.3	1,165.2	967.9	1,027.0	830.1	833.6	1,093.8	..	Gross Domestic Investment
1,513.1	1,435.7	1,294.0	1,394.4	1,241.9	896.7	816.1	764.0	795.3	909.9	..	Fixed Investment
											Memo Items:
1,828.4	1,774.2	1,748.0	1,363.3	1,177.1	1,064.5	1,095.7	1,023.9	1,047.5	993.5	..	Capacity to Import
317.7	111.9	0.0	-132.9	-192.6	-156.2	-85.6	-195.0	0.4	-118.0	..	Terms of Trade Adjustment
7,570.1	7,706.0	7,878.7	7,797.8	7,457.1	7,298.2	7,405.4	7,250.3	7,455.9	7,570.8	..	Gross Domestic Income
7,537.4	7,674.3	7,808.2	7,707.1	7,358.7	7,206.0	7,261.0	7,102.2	7,302.2	7,447.0	..	Gross National Income
				(Index 1980 = 100)							DOMESTIC PRICES (DEFLATORS)
83.7	90.9	100.0	108.5	114.0	121.4	126.4	150.2	212.4	228.8	..	Overall (GDP)
79.6	89.3	100.0	110.2	116.6	123.8	127.6	154.8	212.0	234.9	..	Domestic Absorption
..	..	..	..	..	..	..	..	..	..	..	Agriculture
..	..	..	..	..	..	..	..	..	..	..	Industry
..	..	..	..	..	..	..	..	..	..	..	Manufacturing
											MANUFACTURING ACTIVITY
95.3	96.7	100.0	92.9	92.3	90.0	89.2	87.9	..	..	G	Employment (1980 = 100)
90.7	100.9	100.0	107.2	109.7	106.8	110.2	106.2	..	..	G	Real Earnings per Empl. (1980 = 100)
..	..	..	..	..	..	..	..	..	..	G	Real Output per Empl. (1980 = 100)
18.4	21.5	22.5	21.9	25.9	23.0	24.1	24.3	..	..	..	Earnings as % of Value Added
				(Millions of current Guatemalan Quetzales)							MONETARY HOLDINGS
1,424	1,537	1,692	1,907	2,191	2,155	2,399	3,193	3,875	4,168	..	Money Supply, Broadly Defined
664	735	753	778	787	834	869	1,347	1,608	1,766	..	Money as Means of Payment
325	365	381	405	405	438	461	698	805	931	..	Currency Ouside Banks
339	369	372	373	382	396	408	649	804	834	..	Demand Deposits
760	802	940	1,129	1,404	1,321	1,530	1,846	2,267	2,402	..	Quasi-Monetary Liabilities
				(Millions of current Guatemalan Quetzales)							
-71	-149	-307	-535	-417	-318	..	..	..	..	F	GOVERNMENT DEFICIT (-) OR SURPLUS
661	666	883	896	881	874	..	..	..	..	..	Current Revenue
466	525	729	818	805	888	..	..	..	..	..	Current Expenditure
196	141	154	78	76	-14	..	..	..	..	..	Current Budget Balance
0	0	1	3	1	..	..	..	..	..	..	Capital Receipts
267	290	463	616	494	304	..	..	..	..	..	Capital Payments

GUATEMALA	1967	1968	1969	1970	1971	1972	1973	1974	1975	1976	1977
FOREIGN TRADE (CUSTOMS BASIS)				*(Millions of current US dollars)*							
Value of Exports, fob	197.9	227.5	255.4	290.2	283.1	328.1	436.2	572.1	623.4	760.3	1,160.2
Nonfuel Primary Products	157.2	174.2	191.1	208.9	204.9	241.7	318.8	418.7	468.1	580.8	957.2
Fuels	0.1	0.1	0.1	0.1	0.2	0.1	0.1	0.4	0.4	0.3	0.2
Manufactures	40.6	53.2	64.2	81.2	78.1	86.3	117.2	153.0	154.9	179.2	202.8
Value of Imports, cif	247.1	249.4	250.2	284.3	303.3	327.7	427.4	700.5	732.6	838.4	1,052.5
Nonfuel Primary Products	39.8	35.2	32.7	40.7	41.1	40.0	52.3	86.1	86.8	77.3	89.9
Fuels	9.4	6.0	5.3	6.2	15.3	25.0	31.5	94.0	103.3	106.0	148.4
Manufactures	197.9	208.2	212.2	237.4	246.9	262.7	343.7	520.4	542.5	655.1	814.1
				(Index 1980 = 100)							
Terms of Trade	111.8	107.9	103.3	113.7	115.5	128.5	132.9	104.8	100.4	130.3	149.6
Export Prices, fob	28.7	29.4	29.9	34.5	33.1	36.4	48.3	57.2	58.1	77.3	95.8
Nonfuel Primary Products	28.4	28.9	29.3	33.6	32.1	35.8	48.0	54.3	55.9	81.3	103.2
Fuels	4.3	4.3	4.3	4.3	5.6	6.2	8.9	36.7	35.7	38.4	42.0
Manufactures	30.1	31.7	32.4	37.5	36.5	38.5	49.5	67.2	66.1	66.7	71.7
Import Prices, cif	25.6	27.3	29.0	30.4	28.7	28.4	36.4	54.6	57.9	59.3	64.0
Trade at Constant 1980 Prices				*(Millions of 1980 US dollars)*							
Exports, fob	690.7	772.8	853.4	840.7	854.8	900.5	902.2	999.6	1,072.3	983.4	1,210.9
Imports, cif	964.2	914.2	863.5	936.0	1,057.4	1,155.5	1,175.3	1,282.3	1,264.5	1,413.5	1,643.6
BALANCE OF PAYMENTS				*(Millions of current US dollars)*							
Exports of Goods & Services	..	..	..	354	342	398	542	720	798	1,008	1,372
Merchandise, fob	..	..	..	297	287	336	442	582	641	760	1,162
Nonfactor Services	..	..	..	52	52	58	90	120	142	208	161
Factor Services	..	..	..	4	4	4	10	18	15	39	48
Imports of Goods & Services	..	..	..	379	417	439	576	879	941	1,284	1,499
Merchandise, fob	..	..	..	267	290	295	391	632	672	951	1,086
Nonfactor Services	..	..	..	70	81	95	129	181	188	250	338
Factor Services	..	..	..	42	46	49	56	66	81	83	75
Long-Term Interest	..	..	..	7	9	9	10	11	13	15	20
Current Transfers, net	..	..	..	17	26	31	43	57	78	198	94
Workers' Remittances	..	..	..	0	0	0	0	0	0	0	0
Total to be Financed	..	..	..	-8	-48	-10	9	-102	-65	-79	-34
Official Capital Grants	..	..	..	0	-1	-1	-1	-1	-1	1	2
Current Account Balance	..	..	..	-8	-49	-12	8	-103	-66	-78	-32
Long-Term Capital, net	..	..	..	54	38	24	49	57	146	84	200
Direct Investment	..	..	..	29	29	16	35	47	80	13	97
Long-Term Loans	..	..	..	21	14	6	28	29	50	45	109
Disbursements	..	..	..	43	36	44	47	61	72	74	142
Repayments	..	..	..	22	23	39	19	32	23	29	33
Other Long-Term Capital	..	..	..	3	-3	4	-14	-19	17	26	-9
Other Capital, net	..	..	..	-30	21	26	25	35	24	213	14
Change in Reserves	..	..	..	-15	-10	-39	-81	11	-104	-220	-182
Memo Item:				*(Guatemalan Quetzales per US dollar)*							
Conversion Factor (Annual Avg)	1.000	1.000	1.000	1.000	1.000	1.000	1.000	1.000	1.000	1.000	1.000
EXTERNAL DEBT, ETC.				*(Millions of US dollars, outstanding at end of year)*							
Public/Publicly Guar. Long-Term	..	..	..	106	114	109	116	120	143	163	217
Official Creditors	..	..	..	54	62	72	86	107	129	153	209
IBRD and IDA	..	..	..	14	17	22	27	34	39	42	46
Private Creditors	..	..	..	52	52	37	30	13	15	11	8
Private Non-guaranteed Long-Term	..	..	..	14	21	31	50	75	100	125	180
Use of Fund Credit	..	..	..	0	0	0	0	0	0	0	0
Short-Term Debt	..	..	..	..	..	..	..	..	..	..	255
Memo Items:				*(Millions of US dollars)*							
Int'l Reserves Excluding Gold	45.1	45.4	54.1	60.8	74.7	116.3	191.3	181.3	283.8	491.0	668.9
Gold Holdings (at market price)	20.1	24.1	20.1	18.7	21.6	31.9	55.2	91.7	68.9	66.2	83.6
SOCIAL INDICATORS											
Total Fertility Rate	6.6	6.6	6.5	6.5	6.5	6.5	6.4	6.4	6.4	6.4	6.4
Crude Birth Rate	45.6	45.4	45.2	45.0	44.8	44.6	44.5	44.5	44.4	44.4	44.3
Infant Mortality Rate	108.0	105.4	102.8	100.2	97.6	95.0	92.4	89.8	87.2	84.6	82.0
Life Expectancy at Birth	50.1	50.9	51.7	52.5	53.2	54.0	54.5	55.0	55.4	55.9	56.4
Food Production, p.c. ('79-81 = 100)	85.9	91.4	92.0	93.9	93.6	96.6	99.6	98.2	102.3	103.7	103.7
Labor Force, Agriculture (%)	62.9	62.4	61.8	61.3	60.9	60.4	60.0	59.5	59.1	58.6	58.2
Labor Force, Female (%)	12.9	12.9	13.0	13.1	13.2	13.2	13.3	13.4	13.5	13.5	13.6
School Enroll. Ratio, primary	..	..	..	57.0	..	..	..	..	61.0	61.0	63.0
School Enroll. Ratio, secondary	..	..	..	8.0	..	..	..	..	12.0	13.0	13.0

1978	1979	1980	1981	1982	1983	1984	1985	1986	1987 estimate	NOTES	GUATEMALA
				(Millions of current US dollars)							**FOREIGN TRADE (CUSTOMS BASIS)**
1,111.6	1,160.9	1,486.1	1,114.8	1,120.0	1,159.0	1,127.0	1,060.1	1,103.4	1,084.1	..	Value of Exports, fob
881.8	892.6	1,111.4	768.4	794.0	821.6	777.7	744.7	741.9	689.5	..	Nonfuel Primary Products
0.3	0.4	15.6	21.8	17.7	18.6	19.5	10.4	5.1	6.2	..	Fuels
229.5	268.0	359.1	324.6	308.2	318.8	329.8	305.0	356.3	388.4	..	Manufactures
1,260.7	1,361.8	1,559.1	2,009.3	1,388.0	1,135.0	1,277.4	1,174.9	960.0	1,479.4	..	Value of Imports, cif
119.8	134.9	174.4	178.6	131.2	108.0	124.1	107.3	92.1	138.2	..	Nonfuel Primary Products
160.7	145.9	377.4	760.1	449.6	370.7	319.8	283.7	118.6	176.7	..	Fuels
980.2	1,080.9	1,007.3	1,070.6	807.2	656.3	833.5	783.9	749.3	1,164.5	..	Manufactures
				(Index 1980 = 100)							
118.4	108.3	100.0	87.0	88.7	92.5	94.7	86.7	102.6	79.8	..	Terms of Trade
85.1	93.9	100.0	90.9	87.6	87.9	88.4	80.2	94.8	83.6	..	Export Prices, fob
87.8	95.2	100.0	87.4	84.8	86.7	88.0	77.1	91.9	73.2	..	Nonfuel Primary Products
42.3	61.0	100.0	112.5	101.6	92.5	90.2	87.5	44.9	56.9	..	Fuels
76.1	90.0	100.0	99.1	94.8	90.7	89.3	89.0	103.2	113.0	..	Manufactures
71.9	86.7	100.0	104.5	98.7	95.0	93.4	92.6	92.4	104.8	..	Import Prices, cif
				(Millions of 1980 US dollars)							Trade at Constant 1980 Prices
1,306.5	1,236.2	1,486.0	1,226.5	1,278.6	1,318.7	1,274.4	1,321.1	1,163.7	1,296.9	..	Exports, fob
1,753.9	1,570.5	1,559.1	1,922.6	1,405.8	1,194.8	1,368.3	1,269.4	1,038.8	1,411.9	..	Imports, cif
				(Millions of current US dollars)							**BALANCE OF PAYMENTS**
1,354	1,552	1,833	1,517	1,310	1,204	1,261	1,196	1,202	1,172	..	Exports of Goods & Services
1,098	1,222	1,519	1,282	1,169	1,091	1,131	1,066	1,045	983	..	Merchandise, fob
183	228	211	155	108	80	97	100	122	158	..	Nonfactor Services
73	103	103	80	34	33	33	30	36	31	..	Factor Services
1,734	1,884	2,107	2,196	1,777	1,461	1,672	1,456	1,288	1,830	..	Imports of Goods & Services
1,282	1,402	1,472	1,544	1,286	1,057	1,183	1,078	872	1,362	..	Merchandise, fob
361	382	486	485	342	258	246	180	169	259	..	Nonfactor Services
91	100	149	166	149	146	242	198	247	209	..	Factor Services
33	48	60	64	77	89	95	116	154	153	..	Long-Term Interest
115	123	109	90	62	30	28	19	50	103	..	Current Transfers, net
0	0	0	0	0	0	0	0	0	0	..	Workers' Remittances
-265	-209	-165	-590	-404	-227	-382	-241	-36	-555	K	Total to be Financed
1	3	1	1	1	1	1	1	25	91	K	Official Capital Grants
-265	-205	-164	-588	-404	-226	-382	-241	-11	-464	..	Current Account Balance
265	245	240	402	341	285	196	245	67	199	..	Long-Term Capital, net
127	117	111	128	76	45	38	61	67	152	..	Direct Investment
156	184	93	280	279	225	93	112	44	-25	..	Long-Term Loans
203	249	170	344	337	314	254	264	184	125	..	Disbursements
46	65	78	65	59	89	162	152	140	150	..	Repayments
-20	-60	35	-7	-15	15	65	72	-69	-20	..	Other Long-Term Capital
68	-66	-334	-115	24	-50	202	103	53	196	..	Other Capital, net
-68	26	258	301	38	-9	-17	-107	-109	69	..	Change in Reserves
				(Guatemalan Quetzales per US dollar)							Memo Item:
1.000	1.000	1.000	1.000	1.000	1.000	1.000	1.150	2.190	2.500	..	Conversion Factor (Annual Avg)
				(Millions of US dollars, outstanding at end of year)							**EXTERNAL DEBT, ETC.**
304	427	549	807	1,144	1,386	1,947	2,136	2,264	2,345	..	Public/Publicly Guar. Long-Term
297	421	534	739	997	1,093	1,151	1,347	1,529	1,625	..	Official Creditors
64	108	144	171	186	193	151	225	277	330	..	IBRD and IDA
7	5	15	67	147	293	796	789	735	720	..	Private Creditors
250	312	282	210	168	154	105	106	119	116	..	Private Non-guaranteed Long-Term
0	0	0	111	106	140	150	116	70	59	..	Use of Fund Credit
259	301	335	136	120	119	141	226	306	305	..	Short-Term Debt
				(Millions of US dollars)							Memo Items:
741.5	696.3	444.7	149.7	112.2	210.0	274.4	300.9	362.1	287.8	..	Int'l Reserves Excluding Gold
116.3	267.4	307.9	207.6	238.6	199.3	161.0	170.8	204.2	253.1	..	Gold Holdings (at market price)
											SOCIAL INDICATORS
6.3	6.3	6.2	6.2	6.1	6.1	6.0	5.9	5.8	5.8	..	Total Fertility Rate
44.0	43.7	43.3	43.0	42.7	42.3	41.9	41.5	41.1	40.7	..	Crude Birth Rate
79.6	77.2	74.8	72.4	70.0	69.1	68.2	67.3	66.5	65.6	..	Infant Mortality Rate
56.9	57.4	58.0	58.5	59.0	59.4	59.9	60.3	60.7	61.2	..	Life Expectancy at Birth
101.7	99.1	99.5	101.4	105.4	101.7	99.7	97.3	93.9	90.4	..	Food Production, p.c. ('79-81 = 100)
57.7	57.3	56.8	..	..	..	..	..	..	..	..	Labor Force, Agriculture (%)
13.7	13.8	13.8	14.1	14.3	14.6	14.8	15.1	15.3	15.6	..	Labor Force, Female (%)
65.0	67.0	69.0	72.0	70.0	75.0	76.0	..	..	..	..	School Enroll. Ratio, primary
14.0	16.0	16.0	16.0	17.0	17.0	17.0	..	..	..	..	School Enroll. Ratio, secondary

GUINEA-BISSAU	1967	1968	1969	1970	1971	1972	1973	1974	1975	1976	1977
CURRENT GNP PER CAPITA (US $)	..	..	..	..	..	160	160	170	190	170	160
POPULATION (thousands)	525	526	526	526	545	564	584	605	628	660	693
ORIGIN AND USE OF RESOURCES				*(Millions of current Guinea-Bissau Pesos)*							
Gross National Product (GNP)	..	..	..	2,810	2,749	3,070	3,428	3,904	4,615	4,492	4,368
Net Factor Income from Abroad	..	..	..	0	0	0	0	0	102	-316	-232
Gross Domestic Product (GDP)	..	..	..	2,810	2,749	3,070	3,428	3,904	4,513	4,808	4,600
Indirect Taxes, net	..	..	..	125	128	134	168	178	107	169	159
GDP at factor cost	..	..	..	2,685	2,621	2,936	3,259	3,726	4,406	4,639	4,441
Agriculture	..	..	..	1,333	1,233	1,405	1,470	1,615	2,156	2,312	2,123
Industry	..	..	..	597	607	689	836	995	1,145	747	1,041
Manufacturing	..	..	..	597	..	..	..	..	..	..	..
Services, etc.	..	..	..	880	909	975	1,122	1,293	1,212	1,749	1,436
Resource Balance	..	..	..	-736	-829	-960	-1,188	-1,336	-937	-1,090	-1,072
Exports of Goods & NFServices	..	..	..	112	114	134	208	196	234	254	428
Imports of Goods & NFServices	..	..	..	848	943	1,094	1,396	1,532	1,171	1,343	1,500
Domestic Absorption	..	..	..	3,546	3,578	4,030	4,616	5,240	5,450	5,897	5,672
Private Consumption, etc.	..	..	..	2,155	2,213	2,632	3,061	3,598	3,722	3,918	3,909
General Gov't Consumption	..	..	..	560	595	699	839	985	1,033	1,197	853
Gross Domestic Investment	..	..	..	831	770	698	716	656	695	782	910
Fixed Investment	..	..	..	..	..	..	..	..	..	..	..
Memo Items:											
Gross Domestic Saving	..	..	..	95	-59	-262	-472	-680	-242	-308	-162
Gross National Saving	..	..	..	..	..	..	..	..	..	..	..
				(Millions of 1980 Guinea-Bissau Pesos)							
Gross National Product	..	..	..	5,135.7	4,897.2	5,225.5	5,263.3	5,491.6	6,021.8	5,723.8	5,371.4
GDP at factor cost	..	..	..	4,923.3	4,683.9	5,013.2	5,021.3	5,256.4	5,741.1	5,947.6	5,488.8
Agriculture	..	..	..	3,627.1	3,340.3	3,616.2	3,585.7	3,710.9	3,763.8	4,058.7	3,533.5
Industry	..	..	..	928.4	856.5	923.4	892.7	903.0	1,067.5	662.7	869.8
Manufacturing	..	..	..	..	..	..	..	..	..	..	..
Services, etc.	..	..	..	580.1	700.4	685.8	784.9	877.7	1,041.7	1,431.0	1,272.5
Resource Balance	..	..	..	-3,227.7	-3,272.3	-3,599.3	-3,577.9	-3,367.6	-2,351.7	-2,082.7	-1,995.3
Exports of Goods & NFServices	..	..	..	191.7	184.5	201.2	276.6	244.2	382.9	319.5	390.0
Imports of Goods & NFServices	..	..	..	3,419.4	3,456.8	3,800.4	3,854.5	3,611.8	2,734.6	2,402.2	2,385.3
Domestic Absorption	..	..	..	8,363.3	8,169.5	8,824.7	8,841.2	8,859.2	8,224.7	8,235.1	7,671.1
Private Consumption, etc.	..	..	..	4,682.7	4,841.2	5,629.3	5,907.9	6,156.0	5,640.4	5,348.1	5,114.5
General Gov't Consumption	..	..	..	1,263.8	1,293.3	1,444.4	1,503.4	1,586.8	1,522.0	1,721.2	1,210.0
Gross Domestic Investment	..	..	..	2,416.8	2,034.9	1,751.0	1,429.9	1,116.4	1,062.3	1,165.9	1,346.6
Fixed Investment	..	..	..	..	..	..	..	..	..	..	..
Memo Items:											
Capacity to Import	..	..	..	453.1	418.9	464.6	573.5	461.8	546.5	453.4	680.7
Terms of Trade Adjustment	..	..	..	261.3	234.4	263.4	297.0	217.6	163.5	133.9	290.7
Gross Domestic Income	..	..	..	5,397.0	5,131.6	5,488.9	5,560.2	5,709.3	6,036.6	6,286.3	5,966.6
Gross National Income	..	..	..	5,397.0	5,131.6	5,488.9	5,560.2	5,709.3	6,185.3	5,857.7	5,662.1
DOMESTIC PRICES (DEFLATORS)				*(Index 1980 = 100)*							
Overall (GDP)	..	..	..	54.7	56.1	58.7	65.1	71.1	76.8	78.1	81.0
Domestic Absorption	..	..	..	42.4	43.8	45.7	52.2	59.1	66.3	71.6	73.9
Agriculture	..	..	..	36.8	36.9	38.9	41.0	43.5	57.3	57.0	60.1
Industry	..	..	..	64.3	70.9	74.7	93.7	110.2	107.2	112.6	119.7
Manufacturing	..	..	..	..	..	..	..	..	..	..	..
MANUFACTURING ACTIVITY											
Employment (1980=100)	..	..	..	..	..	..	..	..	..	..	..
Real Earnings per Empl. (1980=100)	..	..	..	..	..	..	..	..	..	..	..
Real Output per Empl. (1980=100)	..	..	..	..	..	..	..	..	..	..	..
Earnings as % of Value Added	..	..	..	..	..	..	..	..	..	..	..
MONETARY HOLDINGS				*(Millions of current Guinea-Bissau Pesos)*							
Money Supply, Broadly Defined	..	..	..	..	..	..	..	..	..	..	..
Money as Means of Payment	..	..	..	..	..	..	..	..	..	..	..
Currency Ouside Banks	..	..	..	..	..	..	..	..	..	..	..
Demand Deposits	..	..	..	..	..	..	..	..	..	..	..
Quasi-Monetary Liabilities	..	..	..	..	..	..	..	..	..	..	..
GOVERNMENT DEFICIT (-) OR SURPLUS				*(Millions of current Guinea-Bissau Pesos)*							
Current Revenue	..	..	..	..	..	..	..	..	..	..	..
Current Expenditure	..	..	..	..	..	..	..	..	..	..	..
Current Budget Balance	..	..	..	..	..	..	..	..	..	..	..
Capital Receipts	..	..	..	..	..	..	..	..	..	..	..
Capital Payments	..	..	..	..	..	..	..	..	..	..	..

1978	1979	1980	1981	1982	1983	1984	1985	1986	1987 estimate	NOTES	GUINEA-BISSAU
170	170	140	170	190	190	180	180	170	160	A	CURRENT GNP PER CAPITA (US $)
728	765	809	824	839	854	869	886	905	924	..	POPULATION (thousands)
				(Millions of current Guinea-Bissau Pesos)							ORIGIN AND USE OF RESOURCES
5,234	5,801	5,223	6,567	7,944	9,432	16,206	24,390	33,113	68,241	..	Gross National Product (GNP)
-299	-84	-33	-71	-117	-170	-494	-848	-1,151	-7,602	..	Net Factor Income from Abroad
5,533	5,885	5,257	6,638	8,061	9,602	16,700	25,238	34,264	75,842	f	Gross Domestic Product (GDP)
202	336	321	499	506	..	..	..	..	..	..	Indirect Taxes, net
5,332	5,549	4,935	6,139	7,555						B f	GDP at factor cost
2,856	3,036	2,328	3,271	3,769	4,075	6,933	10,719	15,317	46,415	..	Agriculture
1,106	1,169	1,033	1,057	1,161	1,100	2,359	3,592	4,086	4,399	..	Industry
..	..	..	..	..	..	..	..	..	..	..	Manufacturing
1,571	1,680	1,895	2,311	3,130	4,427	7,408	10,927	14,861	25,028	..	Services, etc.
-1,523	-1,917	-1,869	-1,628	-2,696	-2,520	-6,398	-10,866	-12,207	-10,305	..	Resource Balance
472	527	439	743	632	577	2,270	2,488	3,373	4,858	..	Exports of Goods & NFServices
1,995	2,444	2,308	2,371	3,328	3,097	8,668	13,354	15,580	15,163	..	Imports of Goods & NFServices
7,056	7,802	7,126	8,266	10,757	12,122	23,098	36,104	46,471	86,148	..	Domestic Absorption
4,686	5,208	4,046	4,666	6,106	7,502	13,965	21,909	27,484	65,443	f	Private Consumption, etc.
1,169	1,309	1,525	1,891	2,373	2,444	4,124	6,116	8,151	6,447	..	General Gov't Consumption
1,200	1,285	1,555	1,709	2,278	2,176	5,009	8,079	10,836	14,258	f	Gross Domestic Investment
..	..	..	..	..	..	..	..	..	..	..	Fixed Investment
											Memo Items:
-322	-632	-314	81	-418	-344	-1,389	-2,787	-1,371	3,953	..	Gross Domestic Saving
..	..	..	..	-1,217	-1,154	-2,474	-4,755	-2,830	-4,770	..	Gross National Saving
				(Millions of 1980 Guinea-Bissau Pesos)							
6,101.8	6,388.8	5,223.3	6,180.2	6,425.5	6,190.5	6,460.6	6,733.5	6,689.8	6,598.0	..	Gross National Product
6,226.2	6,122.3	4,935.1	5,782.7	6,120.7	..	..	..	..	..	B f	GDP at factor cost
4,048.6	3,724.7	2,328.0	2,988.2	3,158.5	3,095.5	3,194.5	3,408.5	3,551.7	3,871.4	..	Agriculture
1,187.9	1,187.5	1,033.2	991.4	1,080.6	1,062.1	1,381.7	1,451.9	1,204.4	1,173.0	..	Industry
..	..	..	..	..	..	..	..	..	..	..	Manufacturing
1,213.0	1,567.2	1,895.4	2,270.2	2,287.9	2,159.6	2,092.6	2,114.2	2,172.3	2,281.0	..	Services, etc.
-2,004.9	-2,056.4	-1,869.1	-1,831.8	-2,697.7	-2,596.4	-2,966.4	-3,161.3	-3,072.3	-2,873.0	..	Resource Balance
354.8	354.8	439.4	314.9	281.9	261.0	356.8	258.3	285.7	317.1	..	Exports of Goods & NFServices
2,359.7	2,411.2	2,308.5	2,146.7	2,979.6	2,857.4	3,323.2	3,419.6	3,358.0	3,190.1	..	Imports of Goods & NFServices
8,454.5	8,535.8	7,125.7	8,081.6	9,224.7	8,913.6	9,635.1	10,135.9	10,000.7	10,198.3	..	Domestic Absorption
5,093.0	4,829.8	4,045.9	5,189.4	5,644.0	5,348.4	6,047.4	6,261.3	6,105.0	7,052.1	f	Private Consumption, etc.
1,578.2	1,498.3	1,525.0	1,593.9	1,917.5	1,952.0	1,885.6	2,000.7	2,022.7	1,385.5	..	General Gov't Consumption
1,783.3	2,207.7	1,554.8	1,298.3	1,663.2	1,613.2	1,702.0	1,874.0	1,873.1	1,760.8	f	Gross Domestic Investment
..	..	..	..	..	..	..	..	..	..	..	Fixed Investment
											Memo Items:
558.9	519.9	439.4	672.7	565.8	532.4	870.3	637.1	726.9	1,022.0	..	Capacity to Import
204.1	165.1	0.0	357.8	284.0	271.4	513.5	378.8	441.2	704.9	..	Terms of Trade Adjustment
6,653.7	6,644.6	5,256.6	6,607.6	6,811.0	6,588.5	7,182.2	7,353.5	7,369.6	8,030.2	..	Gross Domestic Income
6,305.9	6,553.9	5,223.3	6,538.0	6,709.4	6,461.9	6,974.1	7,112.3	7,131.0	7,302.9	..	Gross National Income
				(Index 1980 = 100)							DOMESTIC PRICES (DEFLATORS)
85.8	90.8	100.0	106.2	123.5	152.0	250.4	361.9	494.5	1,035.3	..	Overall (GDP)
83.5	91.4	100.0	102.3	116.6	136.0	239.7	356.2	464.7	844.7	..	Domestic Absorption
70.5	81.5	100.0	109.4	119.3	131.7	217.0	314.5	431.3	1,198.9	..	Agriculture
93.1	98.4	100.0	106.6	107.5	103.6	170.7	247.4	339.2	375.0	..	Industry
..	..	..	..	..	..	..	..	..	..	..	Manufacturing
											MANUFACTURING ACTIVITY
..	..	..	..	..	..	..	..	..	..	..	Employment (1980 = 100)
..	..	..	..	..	..	..	..	..	..	..	Real Earnings per Empl. (1980 = 100)
..	..	..	..	..	..	..	..	..	..	..	Real Output per Empl. (1980 = 100)
..	..	..	..	..	..	..	..	..	..	..	Earnings as % of Value Added
				(Millions of current Guinea-Bissau Pesos)							MONETARY HOLDINGS
..	..	..	..	2,779.0	3,402.0	4,541.0	6,433.0	9,104.0	15,615.0	..	Money Supply, Broadly Defined
..	..	..	..	..	3,257.0	..	..	..	..	..	Money as Means of Payment
..	..	..	..	..	..	..	..	..	..	..	Currency Ouside Banks
..	..	..	..	..	..	..	..	..	..	..	Demand Deposits
..	..	..	..	..	145.0	..	..	..	..	..	Quasi-Monetary Liabilities
				(Millions of current Guinea-Bissau Pesos)							GOVERNMENT DEFICIT (-) OR SURPLUS
..	..	..	..	..	-2,215	-4,615	-7,163	..	..	..	Current Revenue
..	..	..	..	..	2,510	5,141	8,405	..	..	..	Current Revenue
..	..	..	..	..	2,311	3,602	5,581	..	..	..	Current Expenditure
..	..	..	..	..	199	1,539	2,825	..	..	..	Current Budget Balance
..	..	..	..	..	5	7	8	..	..	..	Capital Receipts
..	..	..	..	..	2,419	6,161	9,995	..	..	..	Capital Payments

GUINEA-BISSAU	1967	1968	1969	1970	1971	1972	1973	1974	1975	1976	1977
FOREIGN TRADE (CUSTOMS BASIS)				*(Thousands of current US dollars)*							
Value of Exports, fob	..	..	..	..	..	..	..	..	..	..	..
Nonfuel Primary Products	..	..	..	..	..	..	..	..	..	..	..
Fuels	..	..	..	..	..	..	..	..	..	..	..
Manufactures	..	..	..	..	..	..	..	..	..	..	..
Value of Imports, cif	..	..	..	..	..	..	..	..	..	..	..
Nonfuel Primary Products	..	..	..	..	..	..	..	..	..	..	..
Fuels	..	..	..	..	..	..	..	..	..	..	..
Manufactures	..	..	..	..	..	..	..	..	..	..	..
					(Index 1980 = 100)						
Terms of Trade	..	..	..	..	..	..	..	..	..	..	..
Export Prices, fob	..	..	..	..	..	..	..	..	..	..	..
Nonfuel Primary Products	..	..	..	..	..	..	..	..	..	..	..
Fuels	..	..	..	..	..	..	..	..	..	..	..
Manufactures	..	..	..	..	..	..	..	..	..	..	..
Import Prices, cif	..	..	..	..	..	..	..	..	..	..	..
Trade at Constant 1980 Prices				*(Thousands of 1980 US dollars)*							
Exports, fob	..	..	..	..	..	..	..	..	..	..	..
Imports, cif	..	..	..	..	..	..	..	..	..	..	..
BALANCE OF PAYMENTS				*(Thousands of current US dollars)*							
Exports of Goods & Services	..	..	..	..	..	..	..	..	..	..	..
Merchandise, fob	..	..	..	..	..	..	..	..	..	..	..
Nonfactor Services	..	..	..	..	..	..	..	..	..	..	..
Factor Services	..	..	..	..	..	..	..	..	..	..	..
Imports of Goods & Services	..	..	..	..	..	..	..	..	..	..	..
Merchandise, fob	..	..	..	..	..	..	..	..	..	..	..
Nonfactor Services	..	..	..	..	..	..	..	..	..	..	..
Factor Services	..	..	..	..	..	..	..	..	..	..	..
Long-Term Interest	..	..	..	..	..	..	..	..	0	0	0
Current Transfers, net	..	..	..	..	..	..	..	..	..	..	..
Workers' Remittances	..	..	..	..	..	..	..	..	..	..	..
Total to be Financed	..	..	..	..	..	..	..	..	..	..	..
Official Capital Grants	..	..	..	..	..	..	..	..	..	..	..
Current Account Balance	..	..	..	..	..	..	..	..	..	..	..
Long-Term Capital, net	..	..	..	..	..	..	..	..	..	..	..
Direct Investment	..	..	..	..	..	..	..	..	..	..	..
Long-Term Loans	..	..	..	..	..	..	..	..	7,500	12,700	8,500
Disbursements	..	..	..	..	..	..	..	..	7,500	12,700	8,800
Repayments	..	..	..	..	..	..	..	..	..	..	300
Other Long-Term Capital	..	..	..	..	..	..	..	..	..	..	..
Other Capital, net	..	..	..	..	..	..	..	..	..	..	..
Change in Reserves	..	..	..	0	0	0	0	0	0	0	-929
Memo Item:				*(Guinea-Bissau Pesos per US dollar)*							
Conversion Factor (Annual Avg)	28.750	28.750	28.750	35.690	35.000	35.000	38.350	39.520	41.410	42.780	40.010
Additional Conversion Factor	28.750	28.750	28.750	28.750	28.312	27.053	24.515	25.408	25.543	30.229	33.644
EXTERNAL DEBT, ETC.				*(Millions of US dollars, outstanding at end of year)*							
Public/Publicly Guar. Long-Term	..	..	..	..	..	..	..	..	7	19	26
Official Creditors	..	..	..	..	..	..	..	..	7	17	19
IBRD and IDA	..	..	..	..	..	..	..	..	0	0	0
Private Creditors	..	..	..	..	..	..	..	..	..	2	6
Private Non-guaranteed Long-Term	..	..	..	..	..	..	..	..	..	..	..
Use of Fund Credit	..	..	..	0	0	0	0	0	0	0	0
Short-Term Debt	..	..	..	..	..	..	..	..	..	0	1
Memo Items:				*(Millions US dollars)*							
Int'l Reserves Excluding Gold	..	..	..	..	..	..	..	..	..	..	..
Gold Holdings (at market price)	..	..	..	..	..	..	..	..	..	..	..
SOCIAL INDICATORS											
Total Fertility Rate	5.9	5.9	5.9	5.9	5.9	5.9	5.9	5.9	6.0	6.0	6.0
Crude Birth Rate	46.0	46.0	46.0	46.0	46.0	46.0	46.0	46.0	46.0	46.0	46.0
Infant Mortality Rate	192.0	189.8	187.6	185.4	183.2	181.0	179.0	177.0	175.0	173.0	171.0
Life Expectancy at Birth	35.2	35.3	35.4	35.5	35.5	35.6	35.7	35.8	35.8	35.9	36.0
Food Production, p.c. ('79-81 = 100)	130.1	128.6	119.7	122.2	108.4	109.8	105.4	105.9	124.6	135.6	93.9
Labor Force, Agriculture (%)	85.1	84.7	84.4	84.1	83.9	83.7	83.6	83.4	83.2	83.0	82.8
Labor Force, Female (%)	42.5	42.5	42.5	42.5	42.6	42.6	42.6	42.6	42.6	42.6	42.6
School Enroll. Ratio, primary	..	..	..	39.0	..	..	..	..	64.0	..	..
School Enroll. Ratio, secondary	..	..	..	8.0	..	..	..	..	3.0	..	..

1978	1979	1980	1981	1982	1983	1984	1985	1986	1987 estimate	NOTES	GUINEA-BISSAU
				(Thousands of current US dollars)							**FOREIGN TRADE (CUSTOMS BASIS)**
..	..	11,000	14,000	12,000	9,000	17,400	11,600	..	..	..	Value of Exports, fob
..	..	..	..	..	..	..	..	..	..	..	Nonfuel Primary Products
..	..	..	..	..	..	..	..	..	..	..	Fuels
..	..	..	..	..	..	..	..	..	..	..	Manufactures
..	..	53,200	54,600	52,400	58,100	57,400	63,000	71,600	..	..	Value of Imports, cif
..	..	..	..	..	..	..	..	..	..	..	Nonfuel Primary Products
..	..	..	..	..	..	..	..	..	..	..	Fuels
..	..	..	..	..	..	..	..	..	..	..	Manufactures
				(Index 1980 = 100)							
..	..	..	..	..	..	..	..	..	..	..	Terms of Trade
..	..	..	..	..	..	..	..	..	..	..	Export Prices, fob
..	..	..	..	..	..	..	..	..	..	..	Nonfuel Primary Products
..	..	..	..	..	..	..	..	..	..	..	Fuels
..	..	..	..	..	..	..	..	..	..	..	Manufactures
..	..	..	..	..	..	..	..	..	..	..	Import Prices, cif
				(Thousands of 1980 US dollars)							Trade at Constant 1980 Prices
..	..	..	..	..	..	..	..	..	..	..	Exports, fob
..	..	..	..	..	..	..	..	..	..	..	Imports, cif
				(Thousands of current US dollars)							**BALANCE OF PAYMENTS**
..	..	..	..	17,400	14,900	25,000	13,100	16,500	23,800	..	Exports of Goods & Services
..	..	..	..	11,800	8,600	17,400	11,600	8,700	15,300	..	Merchandise, fob
..	..	..	..	5,600	6,300	7,600	1,500	7,800	8,500	..	Nonfactor Services
..	..	..	..	0	0	0	0	0	..	..	Factor Services
..	..	..	..	82,900	77,300	88,200	83,800	89,800	88,000	..	Imports of Goods & Services
..	..	..	..	61,500	58,400	60,100	59,500	58,500	54,500	..	Merchandise, fob
..	..	..	..	18,200	15,500	21,900	18,900	24,800	27,300	..	Nonfactor Services
..	..	..	..	3,200	3,400	6,200	5,400	6,500	6,200	..	Factor Services
200	1,000	1,000	1,300	1,000	400	500	1,900	2,000	4,100	..	Long-Term Interest
..	..	..	..	-14,000	-10,900	-4,900	-7,000	-1,500	-2,000	..	Current Transfers, net
..	..	..	..	0	0	0	0	0	..	..	Workers' Remittances
..	..	..	..	-79,500	-73,300	-68,100	-77,700	-74,800	-66,200	K	Total to be Financed
..	..	..	..	44,500	43,000	29,300	32,400	43,500	40,400	K	Official Capital Grants
..	..	..	..	-35,000	-30,300	-38,800	-45,300	-31,300	-25,800	..	Current Account Balance
..	..	..	..	68,303	59,660	66,129	65,858	60,700	77,800	..	Long-Term Capital, net
..	..	..	..	..	..	..	..	..	..	..	Direct Investment
20,000	17,400	65,600	16,700	20,700	14,600	27,400	49,100	10,500	47,300	..	Long-Term Loans
20,800	18,800	68,600	19,300	22,300	16,200	30,500	52,500	17,900	52,000	..	Disbursements
800	1,400	3,000	2,600	1,600	1,600	3,100	3,400	7,400	4,700	..	Repayments
..	..	..	..	3,103	2,060	9,429	-15,642	6,700	-9,900	..	Other Long-Term Capital
..	..	..	..	-49,205	-42,407	-33,604	-25,613	-44,700	-38,900	..	Other Capital, net
984	1,864	13	3,114	15,903	13,048	6,275	5,055	15,300	-13,100	..	Change in Reserves
				(Guinea-Bissau Pesos per US dollar)							Memo Item:
45.110	49.650	49.860	42.900	48.700	58.700	120.600	160.000	205.000	561.000	..	Conversion Factor (Annual Avg)
35.039	34.057	33.811	37.339	39.866	42.099	105.290	159.620	203.950	559.330	..	Additional Conversion Factor
				(Millions of US dollars, outstanding at end of year)							**EXTERNAL DEBT, ETC.**
46	63	125	130	143	153	190	253	294	391	..	Public/Publicly Guar. Long-Term
35	43	89	100	111	124	135	175	211	309	..	Official Creditors
0	1	5	10	13	23	26	44	59	87	..	IBRD and IDA
11	21	36	30	32	29	55	78	83	82	..	Private Creditors
..	..	..	..	..	..	..	..	..	..	..	Private Non-guaranteed Long-Term
0	1	1	3	3	2	4	3	2	2	..	Use of Fund Credit
6	4	5	5	15	37	50	41	22	31	..	Short-Term Debt
				(Millions US dollars)							Memo Items:
..	..	..	..	..	..	..	..	..	..	..	Int'l Reserves Excluding Gold
..	..	..	..	..	..	..	..	..	..	..	Gold Holdings (at market price)
											SOCIAL INDICATORS
6.0	6.0	6.0	6.0	6.0	6.0	6.0	6.0	6.0	6.0	..	Total Fertility Rate
46.1	46.2	46.4	46.5	46.6	46.6	46.5	46.5	46.4	46.4	..	Crude Birth Rate
168.6	166.2	163.8	161.4	159.0	163.4	167.7	172.1	176.4	180.8	..	Infant Mortality Rate
36.3	36.7	37.0	37.3	37.7	38.0	38.4	38.7	39.0	39.4	..	Life Expectancy at Birth
100.9	97.8	93.3	108.8	122.2	105.7	120.7	123.4	132.0	132.1	..	Food Production, p.c. ('79-81 = 100)
82.7	82.5	82.3	..	..	..	..	..	..	..	..	Labor Force, Agriculture (%)
42.6	42.7	42.7	42.5	42.3	42.1	42.0	41.8	41.6	41.4	..	Labor Force, Female (%)
106.0	98.0	67.0	66.0	63.0	62.0	60.0	..	..	..	..	School Enroll. Ratio, primary
..	..	6.0	8.0	10.0	11.0	11.0	..	..	..	..	School Enroll. Ratio, secondary

GUYANA	1967	1968	1969	1970	1971	1972	1973	1974	1975	1976	1977
CURRENT GNP PER CAPITA (US $)	..	..	..	380	420	400	410	500	640	660	620
POPULATION (thousands)	..	..	..	670	683	696	710	724	730	735	742
ORIGIN AND USE OF RESOURCES				*(Millions of current Guyana Dollars)*							
Gross National Product (GNP)	393.3	417.5	457.6	493.1	528.1	577.1	613.1	905.4	1,154.5	1,075.3	1,057.5
Net Factor Income from Abroad	-32.0	-42.0	-41.0	-42.5	-36.0	-22.2	-31.7	-49.3	-33.0	-60.9	-67.1
Gross Domestic Product (GDP)	425.3	459.5	498.6	535.6	564.1	599.3	644.8	954.7	1,187.5	1,136.2	1,124.6
Indirect Taxes, net	50.4	54.3	60.7	65.6	65.7	68.6	68.4	84.9	90.0	97.8	105.3
GDP at factor cost	374.9	405.2	437.9	470.0	498.4	530.7	576.4	869.8	1,097.5	1,038.4	1,019.3
Agriculture	80.1	79.8	89.4	90.1	101.7	104.2	106.3	264.1	341.4	236.0	210.8
Industry	138.3	158.4	173.9	189.3	190.6	196.3	191.8	287.8	376.9	364.9	364.0
Manufacturing	46.5	49.0	52.5	57.0	61.3	63.9	64.3	120.3	161.6	134.9	122.9
Services, etc.	156.5	167.0	174.6	190.6	206.1	230.2	278.3	317.9	379.2	437.5	444.5
Resource Balance	-10.0	14.2	19.0	-2.9	21.0	-7.8	-107.7	18.4	-0.3	-283.2	-175.0
Exports of Goods & NFServices	244.3	267.6	295.4	302.4	329.5	344.4	336.5	652.5	889.2	750.7	710.9
Imports of Goods & NFServices	254.3	253.4	276.4	305.3	308.5	352.2	444.2	634.1	889.5	1,033.9	885.9
Domestic Absorption	435.3	445.3	479.6	538.5	543.1	607.1	752.5	936.3	1,187.8	1,419.0	1,299.7
Private Consumption, etc.	258.9	273.1	296.7	325.7	336.3	371.3	417.3	522.0	562.3	673.7	682.7
General Gov't Consumption	66.6	69.9	79.1	90.9	101.7	116.9	159.7	162.2	232.9	320.1	290.0
Gross Domestic Investment	109.8	102.3	103.8	121.9	105.1	118.9	175.5	252.1	392.6	425.2	327.0
Fixed Investment	105.0	96.0	98.0	112.6	102.8	108.3	154.8	198.1	350.3	381.0	290.0
Memo Items:											
Gross Domestic Saving	99.8	116.5	122.8	119.0	126.1	111.1	67.8	270.5	392.3	142.4	151.9
Gross National Saving	64.5	70.9	79.6	75.5	89.7	89.4	34.3	216.1	349.0	70.3	75.8
				(Millions of 1980 Guyana Dollars)							
Gross National Product	1,016.37	980.68	1,080.19	1,143.04	1,293.00	1,245.28	1,234.00	1,343.25	1,497.79	1,481.87	1,439.60
GDP at factor cost	1,038.26	1,043.60	1,114.29	1,160.58	1,202.22	1,155.04	1,178.92	1,275.57	1,392.68	1,414.54	1,372.73
Agriculture	267.24	263.74	285.90	284.30	301.79	274.09	274.09	317.83	306.17	317.83	307.33
Industry	427.26	436.35	470.73	490.58	505.95	475.20	464.02	503.16	549.28	526.92	508.75
Manufacturing	98.06	94.13	98.80	98.18	115.36	111.68	105.55	132.55	146.05	149.73	150.83
Services, etc.	343.75	343.50	357.66	385.71	394.47	405.74	440.81	454.58	537.23	569.79	556.64
Resource Balance	-82.85	-42.34	-45.15	-84.59	34.95	-39.42	-308.32	-195.10	-482.26	-597.88	-350.37
Exports of Goods & NFServices	1,343.96	1,359.27	1,416.56	1,399.50	1,446.88	1,338.86	1,257.80	1,212.61	1,245.56	1,198.03	1,036.36
Imports of Goods & NFServices	1,426.80	1,401.60	1,461.71	1,484.10	1,411.93	1,378.28	1,566.12	1,407.71	1,727.82	1,795.91	1,386.73
Domestic Absorption	1,264.65	1,237.05	1,325.10	1,416.87	1,338.96	1,364.61	1,645.01	1,637.72	2,038.90	2,176.38	1,886.75
Private Consumption, etc.	568.63	624.29	695.28	694.88	680.06	760.93	833.65	883.40	833.66	957.81	958.85
General Gov't Consumption	150.23	164.93	191.23	240.43	248.17	260.55	297.99	292.42	410.47	536.57	448.69
Gross Domestic Investment	545.78	447.83	438.59	481.55	410.73	343.13	513.37	461.90	794.77	682.00	479.21
Fixed Investment	..	..	..	..	..	..	..	..	..	608.26	422.81
Memo Items:											
Capacity to Import	1,370.70	1,480.15	1,562.19	1,470.00	1,508.05	1,347.76	1,186.40	1,448.56	1,727.24	1,303.98	1,112.80
Terms of Trade Adjustment	26.74	120.88	145.63	70.50	61.16	8.90	-71.40	235.95	481.68	105.95	76.44
Gross Domestic Income	1,208.54	1,315.59	1,425.58	1,402.77	1,435.08	1,334.08	1,265.29	1,678.57	2,038.32	1,684.45	1,612.81
Gross National Income	1,043.11	1,101.55	1,225.82	1,213.54	1,354.16	1,254.18	1,162.60	1,579.19	1,979.47	1,587.82	1,516.03
DOMESTIC PRICES (DEFLATORS)				*(Index 1980 = 100)*							
Overall (GDP)	36.0	38.5	39.0	40.2	41.1	45.2	48.2	66.2	76.3	72.0	73.2
Domestic Absorption	34.4	36.0	36.2	38.0	40.6	44.5	45.7	57.2	58.3	65.2	68.9
Agriculture	30.0	30.3	31.3	31.7	33.7	38.0	38.8	83.1	111.5	74.3	68.6
Industry	32.4	36.3	36.9	38.6	37.7	41.3	41.3	57.2	68.6	69.3	71.5
Manufacturing	47.4	52.1	53.1	58.1	53.1	57.2	60.9	90.8	110.7	90.1	81.5
MANUFACTURING ACTIVITY											
Employment (1980=100)	..	..	..	..	..	..	..	..	..	..	..
Real Earnings per Empl. (1980=100)	..	..	..	..	..	..	..	..	..	..	..
Real Output per Empl. (1980=100)	..	..	..	..	..	..	..	..	..	..	..
Earnings as % of Value Added	..	..	..	..	..	..	..	..	..	..	..
MONETARY HOLDINGS				*(Millions of current Guyana Dollars)*							
Money Supply, Broadly Defined	129.2	145.1	160.6	173.9	202.0	246.3	291.8	339.1	475.8	523.8	646.1
Money as Means of Payment	50.7	56.3	60.6	61.4	69.4	87.5	100.6	133.2	207.4	221.5	284.9
Currency Ouside Banks	30.1	33.8	35.7	37.3	40.7	48.1	55.6	63.6	91.6	105.2	142.8
Demand Deposits	20.5	22.5	24.9	24.1	28.7	39.4	44.9	69.5	115.8	116.3	142.1
Quasi-Monetary Liabilities	78.6	88.8	99.9	112.5	132.5	158.8	191.2	205.9	268.4	302.3	361.2
GOVERNMENT DEFICIT (-) OR SURPLUS				*(Millions of current Guyana Dollars)*							
	..	..	..	-24	-33	-39	-105	-22	-78	-313	-133
Current Revenue	..	..	..	143	140	163	177	323	503	399	381
Current Expenditure	..	..	..	..	..	..	200	259	303	389	391
Current Budget Balance	..	..	..	..	..	..	-22	64	200	11	-11
Capital Receipts	..	..	..	1	1	1	2	..	..	..	..
Capital Payments	..	..	..	..	..	..	85	86	278	324	123

1978	1979	1980	1981	1982	1983	1984	1985	1986	1987 estimate	NOTES	GUYANA
610	650	720	730	620	550	500	480	470	380	..	**CURRENT GNP PER CAPITA (US $)**
747	754	760	766	772	779	785	790	799	807	..	**POPULATION (thousands)**
				(Millions of current Guyana Dollars)							**ORIGIN AND USE OF RESOURCES**
1,203.1	1,249.0	1,403.0	1,433.0	1,284.0	1,273.9	1,348.4	1,542.7	1,686.8	2,413.3	..	Gross National Product (GNP)
-64.5	-77.0	-105.0	-164.0	-162.0	-194.1	-314.6	-407.3	-483.2	-1,091.7	..	Net Factor Income from Abroad
1,267.6	1,326.0	1,508.0	1,597.0	1,446.0	1,468.0	1,663.0	1,950.0	2,170.0	3,505.0	..	Gross Domestic Product (GDP)
132.0	147.0	172.0	247.0	196.0	268.0	259.0	314.0	333.0	441.0	..	Indirect Taxes, net
1,135.6	1,179.0	1,336.0	1,350.0	1,250.0	1,200.0	1,404.0	1,636.0	1,837.0	3,064.0	..	GDP at factor cost
256.6	263.5	312.0	300.0	292.0	291.0	347.0	439.0	490.0	894.0	..	Agriculture
391.0	400.5	478.0	412.0	362.0	273.0	348.0	397.0	501.0	..	f	Industry
137.5	146.0	162.0	201.0	179.0	158.0	183.0	227.0	271.0	440.0	..	Manufacturing
488.0	515.0	546.0	638.0	596.0	636.0	709.0	800.0	846.0	..	f	Services, etc.
-0.2	-124.9	-158.4	-350.2	-219.0	-272.7	-152.9	-330.3	-281.5	-359.0	..	Resource Balance
799.7	793.1	1,041.9	1,031.9	794.4	675.3	938.0	1,047.7	1,080.5	2,803.8	..	Exports of Goods & NFServices
799.9	918.0	1,200.3	1,382.1	1,013.4	948.0	1,090.9	1,378.0	1,362.0	3,162.8	..	Imports of Goods & NFServices
1,267.9	1,451.0	1,666.4	1,947.2	1,665.0	1,740.7	1,815.9	2,280.4	2,451.6	3,864.0	..	Domestic Absorption
711.9	692.0	781.4	1,012.2	923.0	973.7	1,058.9	1,346.4	1,317.6	2,193.0	..	Private Consumption, etc.
296.0	348.0	436.0	464.0	392.0	458.0	301.0	352.0	401.0	552.0	..	General Gov't Consumption
260.0	411.0	449.0	471.0	350.0	309.0	456.0	582.0	733.0	1,119.0	..	Gross Domestic Investment
242.0	325.0	404.0	441.0	350.0	309.0	456.0	466.0	595.0	946.0	..	Fixed Investment
											Memo Items:
259.7	286.0	290.6	120.8	131.0	36.3	303.1	251.6	451.4	760.0	..	Gross Domestic Saving
195.8	209.5	188.1	-31.2	-48.6	-170.6	-7.5	-164.3	-4.0	..	..	Gross National Saving
				(Millions of 1980 Guyana Dollars)							
1,432.35	1,396.59	1,403.00	1,386.58	1,187.76	1,087.92	1,032.63	1,018.58	995.97	906.87	..	Gross National Product
1,335.85	1,314.16	1,336.00	1,332.25	1,193.35	1,080.73	1,104.18	1,115.51	1,120.04	..	I	GDP at factor cost
333.87	310.54	312.00	319.29	314.92	303.25	313.46	313.46	325.12	306.17	..	Agriculture
471.01	462.63	478.00	468.22	378.77	299.10	306.09	318.67	310.28	..	f	Industry
154.64	160.77	162.00	171.82	149.73	125.18	117.82	114.14	114.14	105.55	..	Manufacturing
530.97	540.99	546.00	544.75	499.67	478.38	484.64	483.39	484.64	..	f	Services, etc.
-6.47	-71.43	-158.40	-217.09	-80.51	-100.00	63.21	-40.96	-116.84	-40.08	..	Resource Balance
1,127.62	1,013.04	1,041.90	980.23	751.94	702.08	768.71	793.05	748.00	776.87	..	Exports of Goods & NFServices
1,134.09	1,084.46	1,200.30	1,197.33	832.45	802.08	705.50	834.01	864.85	816.95	..	Imports of Goods & NFServices
1,519.73	1,553.26	1,666.40	1,751.51	1,406.87	1,338.49	1,177.40	1,293.05	1,388.00	1,316.42	..	Domestic Absorption
790.40	701.23	781.40	958.74	858.78	826.87	751.21	846.50	901.30	882.67	..	Private Consumption, etc.
397.18	397.01	436.00	379.84	265.96	269.68	141.57	143.89	151.93	162.92	..	General Gov't Consumption
331.40	455.01	449.00	412.93	282.13	241.95	284.62	302.66	334.77	270.84	..	Gross Domestic Investment
306.75	360.12	404.00	387.38	280.51	240.56	282.99	256.16	283.57	221.17	..	Fixed Investment
											Memo Items:
1,133.80	936.92	1,041.90	893.94	652.55	571.35	606.62	634.10	686.10	724.22	..	Capacity to Import
6.18	-76.12	0.00	-86.29	-99.39	-130.73	-162.09	-158.95	-61.90	-52.65	..	Terms of Trade Adjustment
1,519.45	1,405.71	1,508.00	1,448.12	1,226.98	1,107.77	1,078.52	1,093.14	1,209.25	1,223.71	..	Gross Domestic Income
1,438.54	1,320.47	1,403.00	1,300.29	1,088.37	957.19	870.54	859.63	934.06	854.18	..	Gross National Income
				(Index 1980 = 100)							**DOMESTIC PRICES (DEFLATORS)**
83.8	89.5	100.0	104.1	109.0	118.5	134.0	155.7	170.7	274.6	..	Overall (GDP)
83.4	93.4	100.0	111.2	118.3	130.0	154.2	176.4	176.6	293.5	..	Domestic Absorption
76.9	84.9	100.0	94.0	92.7	96.0	110.7	140.1	150.7	292.0	..	Agriculture
83.0	86.6	100.0	88.0	95.6	91.3	113.7	124.6	161.5	..	..	Industry
88.9	90.8	100.0	117.0	119.6	126.2	155.3	198.9	237.4	416.9	..	Manufacturing
											MANUFACTURING ACTIVITY
..	..	..	..	..	..	..	..	..	..	..	Employment (1980 = 100)
..	..	..	..	..	..	..	..	..	..	..	Real Earnings per Empl. (1980 = 100)
..	..	..	..	..	..	..	..	..	..	..	Real Output per Empl. (1980 = 100)
..	..	..	..	..	..	..	..	..	..	f	Earnings as % of Value Added
				(Millions of current Guyana Dollars)							**MONETARY HOLDINGS**
716.3	770.8	929.4	1,085.5	1,384.2	1,671.7	1,991.0	2,431.0	2,906.5	4,331.8	D	Money Supply, Broadly Defined
300.4	288.2	328.1	350.8	436.0	508.9	618.6	739.6	880.8	1,332.6	..	Money as Means of Payment
156.5	148.3	167.0	186.1	231.0	268.9	335.8	421.6	508.8	726.2	..	Currency Ouside Banks
143.9	139.9	161.1	164.7	205.0	240.0	282.8	318.0	372.0	606.4	..	Demand Deposits
416.0	482.6	601.4	734.7	948.2	1,162.8	1,372.4	1,691.4	2,025.7	2,999.2	..	Quasi-Monetary Liabilities
				(Millions of current Guyana Dollars)							
-129	-232	-440	-453	..	..	..	..	..	..		**GOVERNMENT DEFICIT (-) OR SURPLUS**
404	481	540	666	651	..	..	963	..	..	..	Current Revenue
403	450	..	..	..	..	..	..	..	..	..	Current Expenditure
2	30	..	..	..	..	..	..	..	..	..	Current Budget Balance
0	0	..	..	..	..	..	..	..	..	..	Capital Receipts
131	263	..	..	..	..	..	..	..	..	..	Capital Payments

GUYANA	1967	1968	1969	1970	1971	1972	1973	1974	1975	1976	1977
FOREIGN TRADE (CUSTOMS BASIS)					*(Millions of current US dollars)*						
Value of Exports, fob	109.8	104.1	116.8	133.4	149.2	147.0	137.1	270.2	364.2	268.2	261.3
Nonfuel Primary Products	105.0	100.5	112.5	126.4	140.4	137.9	126.9	258.5	348.8	227.0	211.9
Fuels	0.0	0.0	0.0	0.0	0.0	0.0	0.0	0.1	0.2	0.2	0.1
Manufactures	4.8	3.7	4.3	7.0	8.8	9.1	10.2	11.6	15.2	41.0	49.3
Value of Imports, cif	130.0	110.0	118.0	134.1	135.0	143.3	177.0	254.9	343.9	364.0	315.0
Nonfuel Primary Products	26.4	24.5	23.1	22.6	25.5	24.2	31.7	44.3	50.3	57.3	51.6
Fuels	10.5	10.5	10.7	11.5	11.9	13.5	23.0	46.5	57.3	54.2	63.1
Manufactures	93.1	75.0	84.3	100.0	97.6	105.7	122.3	164.1	236.3	252.5	200.4
					(Index 1980 = 100)						
Terms of Trade	128.8	138.6	132.9	125.8	110.4	111.2	119.8	132.2	135.4	111.2	106.7
Export Prices, fob	27.3	27.4	27.6	27.5	27.3	29.8	37.3	71.9	77.8	66.2	66.9
Nonfuel Primary Products	27.2	27.3	27.5	27.1	26.9	29.4	36.6	72.2	78.5	66.1	65.9
Fuels	4.3	4.3	4.3	4.3	5.6	6.2	8.9	36.7	35.7	38.4	42.0
Manufactures	30.1	31.7	32.4	37.5	36.5	38.5	49.5	67.2	66.1	66.7	71.7
Import Prices, cif	21.2	19.8	20.8	21.9	24.8	26.8	31.2	54.4	57.5	59.5	62.7
Trade at Constant 1980 Prices					*(Millions of 1980 US dollars)*						
Exports, fob	402.3	379.4	423.0	485.1	545.9	492.7	367.2	375.7	468.0	405.3	390.4
Imports, cif	614.0	555.5	567.9	613.1	545.3	534.0	567.9	468.5	598.2	611.8	502.5
BALANCE OF PAYMENTS					*(Millions of current US dollars)*						
Exports of Goods & Services	..	..	..	147	163	164	158	293	372	295	275
Merchandise, fob	..	..	..	129	146	144	136	270	351	280	259
Nonfactor Services	..	..	..	17	16	19	21	22	19	14	16
Factor Services	..	..	..	1	1	1	1	1	1	2	0
Imports of Goods & Services	..	..	..	168	171	179	221	300	390	431	369
Merchandise, fob	..	..	..	120	120	129	159	230	306	331	287
Nonfactor Services	..	..	..	31	31	38	49	50	64	75	61
Factor Services	..	..	..	17	19	12	13	20	20	26	22
Long-Term Interest	..	..	..	3	4	6	8	9	10	20	15
Current Transfers, net	..	..	..	-1	0	0	-1	-2	-4	-4	-4
Workers' Remittances	..	..	..	0	0	0	0	0	0	0	0
Total to be Financed	..	..	..	-22	-8	-15	-65	-10	-23	-141	-97
Official Capital Grants	..	..	..	..	1	-1	0	-1	-2	-2	0
Current Account Balance	..	..	..	-22	-7	-16	-64	-11	-25	-143	-98
Long-Term Capital, net	..	..	..	17	8	8	18	42	88	43	41
Direct Investment	..	..	..	9	-56	3	8	1	1	-26	-2
Long-Term Loans	..	..	..	12	14	9	16	30	88	57	42
Disbursements	..	..	..	14	16	11	21	35	95	70	59
Repayments	..	..	..	2	1	2	5	5	7	13	17
Other Long-Term Capital	..	..	..	-4	49	-2	-6	12	1	14	1
Other Capital, net	..	..	..	2	0	17	20	15	-14	12	45
Change in Reserves	..	..	..	2	-2	-8	26	-46	-50	88	11
Memo Item:					*(Guyana Dollars per US dollar)*						
Conversion Factor (Annual Avg)	1.740	2.000	2.000	2.000	1.980	2.090	2.110	2.230	2.360	2.550	2.550
EXTERNAL DEBT, ETC.					*(Millions of US dollars, outstanding at end of year)*						
Public/Publicly Guar. Long-Term	..	..	..	83	154	158	173	214	296	364	416
Official Creditors	..	..	..	66	83	89	95	123	163	183	208
IBRD and IDA	..	..	..	1	3	5	9	14	20	24	27
Private Creditors	..	..	..	17	71	69	78	91	133	181	208
Private Non-guaranteed Long-Term	..	..	..	..	..	..	..	..	..	..	..
Use of Fund Credit	..	..	..	0	2	0	5	6	0	20	21
Short-Term Debt	..	..	..	..	..	..	..	..	..	..	44
Memo Items:					*(Thousands of US dollars)*						
Int'l Reserves Excluding Gold	18,850	23,550	20,550	20,400	26,156	36,750	13,974	62,572	100,500	27,283	22,976
Gold Holdings (at market price)	..	..	..	..	..	..	..	..	..	..	..
SOCIAL INDICATORS											
Total Fertility Rate	5.3	5.1	5.0	4.8	4.7	4.5	4.4	4.3	4.2	4.0	3.9
Crude Birth Rate	35.4	34.9	34.4	33.9	33.8	35.3	33.7	31.4	32.8	31.5	31.4
Infant Mortality Rate	56.0	56.0	56.0	56.0	56.0	56.0	54.6	53.2	51.8	50.4	49.0
Life Expectancy at Birth	62.4	63.2	63.9	64.7	65.4	66.1	65.9	65.6	65.4	65.1	64.9
Food Production, p.c. ('79-81 = 100)	121.1	112.3	128.4	114.7	125.9	107.3	98.7	115.9	102.0	107.7	93.2
Labor Force, Agriculture (%)	33.8	33.1	32.5	31.9	31.4	30.9	30.3	29.8	29.3	28.8	28.3
Labor Force, Female (%)	20.1	20.3	20.4	20.5	20.9	21.4	21.8	22.3	22.7	23.0	23.4
School Enroll. Ratio, primary	..	..	..	98.0	..	..	..	..	95.0	99.0	98.0
School Enroll. Ratio, secondary	..	..	..	55.0	..	..	..	..	54.0	61.0	61.0

1978	1979	1980	1981	1982	1983	1984	1985	1986	1987 estimate	NOTES	GUYANA
											FOREIGN TRADE (CUSTOMS BASIS)
				(Millions of current US dollars)							
291.0	289.9	389.0	346.0	241.0	189.0	210.0	206.4	228.5	243.1	..	Value of Exports, fob
243.2	272.8	352.8	313.7	219.3	171.8	191.1	185.6	191.6	205.0	..	Nonfuel Primary Products
0.5	0.1	0.2	0.2	0.1	0.1	0.1	0.1	0.1	0.1	..	Fuels
47.3	17.1	36.0	32.1	21.6	17.1	18.9	20.7	36.9	38.1	..	Manufactures
279.0	318.0	365.0	438.0	282.0	246.0	214.0	248.0	217.0	225.6	..	Value of Imports, cif
50.9	61.3	68.6	82.5	53.2	46.4	41.2	44.7	73.6	73.0	..	Nonfuel Primary Products
68.8	69.2	80.9	97.8	62.7	54.7	47.1	52.3	25.2	34.6	..	Fuels
159.4	187.5	215.6	257.8	166.1	144.9	125.8	151.0	118.2	117.9	..	Manufactures
				(Index 1980 = 100)							
113.5	96.4	100.0	93.1	81.2	83.2	81.9	62.7	69.1	67.7	..	Terms of Trade
74.7	79.9	100.0	96.1	80.4	79.7	77.2	58.3	63.1	65.0	..	Export Prices, fob
74.6	79.4	100.0	95.8	79.2	78.7	76.2	56.2	58.7	60.3	..	Nonfuel Primary Products
42.3	61.0	100.0	112.5	101.6	92.5	90.2	87.5	44.9	56.9	..	Fuels
76.1	90.0	100.0	99.1	94.8	90.7	89.3	89.0	103.2	113.0	..	Manufactures
65.8	82.9	100.0	103.2	98.9	95.8	94.3	93.1	91.2	96.0	..	Import Prices, cif
											Trade at Constant 1980 Prices
				(Millions of 1980 US dollars)							
389.4	362.7	389.0	360.0	299.9	237.2	272.0	353.9	362.4	373.9	..	Exports, fob
423.9	383.4	365.0	424.4	285.0	256.8	226.9	266.4	238.0	234.9	..	Imports, cif
				(Millions of current US dollars)							**BALANCE OF PAYMENTS**
314	315	411	372	264	225	246	262			..	Exports of Goods & Services
296	293	389	346	241	193	217	214	217	..	..	Merchandise, fob
18	19	20	23	23	32	29	48	30	..	..	Nonfactor Services
0	4	2	3	0	0	0	0	..	..	..	Factor Services
337	398	538	556	398	382	346	353			..	Imports of Goods & Services
253	289	386	400	254	226	202	209	229	..	..	Merchandise, fob
60	71	107	99	94	98	99	104	59	..	..	Nonfactor Services
23	38	44	58	49	58	45	40	..	..	..	Factor Services
17	25	27	36	24	29	21	13	16	16	..	Long-Term Interest
0	0	1	4	-6	-4	1	-2	7	..	..	Current Transfers, net
0	0	0	0	2	1	0	0	..	..	..	Workers' Remittances
-23	-83	-127	-179	-139	-161	-98	-93	-123	-170	K	Total to be Financed
-7	0	-2	-4	-2	3	4	-3	0	..	K	Official Capital Grants
-30	-83	-129	-184	-141	-157	-95	-97	-123	-170	..	Current Account Balance
26	15	73	125	8	-30	-24	-39	-8	..	..	Long-Term Capital, net
..	1	1	-2	4	5	4	2	..	..	..	Direct Investment
27	32	51	109	51	34	10	32	41	12	..	Long-Term Loans
60	98	93	153	74	57	25	44	57	21	..	Disbursements
32	66	43	44	23	23	15	11	16	9	..	Repayments
5	-18	23	22	-45	-72	-42	-70	-49		..	Other Long-Term Capital
20	11	13	41	133	187	143	140	133	169	..	Other Capital, net
-17	57	43	18	0	0	-24	-5	-3	1	..	Change in Reserves
											Memo Item:
				(Guyana Dollars per US dollar)							
2.550	2.550	2.550	2.810	3.000	3.000	3.830	4.250	4.270	9.760	..	Conversion Factor (Annual Avg)
				(Millions of US dollars, outstanding at end of year)							**EXTERNAL DEBT, ETC.**
450	502	560	640	678	695	679	739	808	874	..	Public/Publicly Guar. Long-Term
248	290	339	447	490	519	508	568	614	677	..	Official Creditors
35	50	54	68	81	88	79	92	112	133	..	IBRD and IDA
203	212	221	193	188	176	172	171	195	197	..	Private Creditors
..	..	..	..	..	..	..	..	..	..	..	Private Non-guaranteed Long-Term
39	53	86	86	86	77	71	79	88	102	..	Use of Fund Credit
73	66	110	109	158	199	227	290	344	309	..	Short-Term Debt
				(Thousands of US dollars)							Memo Items:
58,266	17,528	12,700	6,911	10,557	6,490	5,850	6,470	9,000	8,430	..	Int'l Reserves Excluding Gold
..	..	..	..	..	..	..	..	..	..	..	Gold Holdings (at market price)
											SOCIAL INDICATORS
3.8	3.6	3.5	3.4	3.3	3.2	3.2	3.1	3.1	3.1	..	Total Fertility Rate
30.4	30.7	30.3	30.0	29.8	29.5	29.3	28.6	27.8	27.1	..	Crude Birth Rate
48.6	48.2	47.8	47.4	47.0	46.5	46.0	45.5	45.0	44.4	..	Infant Mortality Rate
65.0	65.2	65.3	65.4	65.5	65.7	65.8	70.0	..	66.1	..	Life Expectancy at Birth
114.6	102.3	96.0	101.7	98.5	87.7	81.1	81.0	80.9	78.2	..	Food Production, p.c. ('79-81 = 100)
27.8	27.3	26.8	..	..	..	..	..	..	..	..	Labor Force, Agriculture (%)
23.7	24.1	24.4	24.5	24.6	24.7	24.8	24.9	..	24.9	..	Labor Force, Female (%)
98.0	102.0	100.0	99.0	95.0	99.0	..	..	..	..	..	School Enroll. Ratio, primary
60.0	59.0	60.0	60.0	57.0	60.0	..	..	..	..	..	School Enroll. Ratio, secondary

HAITI	1967	1968	1969	1970	1971	1972	1973	1974	1975	1976	1977
CURRENT GNP PER CAPITA (US $)	80	80	90	90	100	100	110	130	150	170	180
POPULATION (thousands)	4,228	4,319	4,411	4,504	4,597	4,689	4,781	4,872	4,963	5,054	5,145
ORIGIN AND USE OF RESOURCES					*(Millions of current Haitian Gourdes)*						
Gross National Product (GNP)	1,741	1,763	1,872	1,954	2,207	2,354	2,821	3,460	3,573	4,359	4,878
Net Factor Income from Abroad	-16	-17	-18	-18	-20	-23	-22	-30	-35	-36	-61
Gross Domestic Product (GDP)	1,756	1,780	1,890	1,972	2,227	2,377	2,843	3,489	3,608	4,395	4,939
Indirect Taxes, net	..	..	..	..	..	..	..	..	..	..	..
GDP at factor cost	..	..	..	..	..	..	..	..	..	..	..
Agriculture	..	..	..	..	..	..	..	..	..	..	..
Industry	..	..	..	..	..	..	..	..	..	..	..
Manufacturing	..	..	..	..	..	..	..	..	..	..	..
Services, etc.	..	..	..	..	..	..	..	..	..	..	..
Resource Balance	-63	-37	-56	-77	-76	-118	-133	-250	-297	-391	-440
Exports of Goods & NFServices	201	232	242	272	304	296	350	422	528	736	905
Imports of Goods & NFServices	264	269	298	349	380	413	483	672	825	1,128	1,345
Domestic Absorption	1,819	1,816	1,946	2,049	2,303	2,494	2,976	3,739	3,905	4,786	5,379
Private Consumption, etc.	1,535	1,530	1,605	1,636	1,821	1,944	2,283	2,908	3,048	3,721	4,201
General Gov't Consumption	167	165	175	188	227	219	241	257	326	372	411
Gross Domestic Investment	118	121	166	225	254	331	451	574	532	694	767
Fixed Investment	..	..	..	..	..	..	..	..	..	..	..
Memo Items:											
Gross Domestic Saving	55	85	110	148	178	213	319	325	235	303	327
Gross National Saving	39	68	145	168	196	258	331	363	309	432	422
					(Millions of 1980 Haitian Gourdes)						
Gross National Product	4,541.1	4,617.8	4,747.9	4,724.7	5,128.3	5,192.0	5,266.9	5,602.7	5,490.6	5,963.8	5,968.3
GDP at factor cost	..	..	..	..	..	..	..	..	..	..	..
Agriculture	..	..	..	..	..	..	..	..	..	..	..
Industry	..	..	..	..	..	..	..	..	..	..	..
Manufacturing	..	..	..	..	..	..	..	..	..	..	..
Services, etc.	..	..	..	..	..	..	..	..	..	..	..
Resource Balance	-77.9	-79.1	-83.6	-198.6	-187.1	-228.6	-164.4	-188.9	-237.8	-435.6	-608.8
Exports of Goods & NFServices	324.8	392.7	406.5	437.3	521.2	552.0	647.5	688.9	694.5	781.8	771.6
Imports of Goods & NFServices	402.7	471.8	490.2	635.9	708.3	780.6	811.9	877.8	932.3	1,217.3	1,380.3
Domestic Absorption	4,670.1	4,754.8	4,893.4	4,991.9	5,385.3	5,498.7	5,497.3	5,858.8	5,776.2	6,447.7	6,649.7
Private Consumption, etc.	3,992.8	4,041.5	4,121.7	4,061.0	4,370.5	4,460.3	4,423.3	4,766.5	4,549.4	5,033.0	5,186.5
General Gov't Consumption	474.6	491.9	509.2	537.2	613.2	563.9	521.2	478.6	474.8	495.2	510.4
Gross Domestic Investment	202.8	221.4	262.4	393.7	401.6	474.5	552.7	613.7	751.9	919.5	952.8
Fixed Investment	..	..	..	..	..	..	..	..	..	..	..
Memo Items:											
Capacity to Import	306.6	407.5	398.0	495.6	566.4	558.8	588.9	551.6	596.9	795.2	928.8
Terms of Trade Adjustment	-18.2	14.8	-8.6	58.3	45.2	6.8	-58.5	-137.3	-97.6	13.4	157.2
Gross Domestic Income	4,574.0	4,690.5	4,801.2	4,851.6	5,243.5	5,276.9	5,274.3	5,532.7	5,440.8	6,025.6	6,198.1
Gross National Income	4,522.9	4,632.6	4,739.3	4,783.0	5,173.5	5,198.9	5,208.4	5,465.4	5,392.9	5,977.2	6,125.5
DOMESTIC PRICES (DEFLATORS)					*(Index 1980 = 100)*						
Overall (GDP)	38.2	38.1	39.3	41.1	42.8	45.1	53.3	61.5	65.1	73.1	81.8
Domestic Absorption	39.0	38.2	39.8	41.1	42.8	45.4	54.1	63.8	67.6	74.2	80.9
Agriculture	..	..	..	..	..	..	..	..	..	..	..
Industry	..	..	..	..	..	..	..	..	..	..	..
Manufacturing	..	..	..	..	..	..	..	..	..	..	..
MANUFACTURING ACTIVITY											
Employment (1980=100)	..	53.7	39.5	37.1	39.4	47.2	53.5	59.8	69.7	74.0	85.0
Real Earnings per Empl. (1980=100)	..	117.5	136.2	138.1	144.0	126.7	104.5	93.4	79.0	91.3	91.1
Real Output per Empl. (1980=100)	..	..	..	..	..	..	..	..	..	..	..
Earnings as % of Value Added	..	..	..	..	..	..	..	..	..	..	..
MONETARY HOLDINGS					*(Millions of current Haitian Gourdes)*						
Money Supply, Broadly Defined	179	201	224	249	290	382	487	584	735	1,015	1,217
Money as Means of Payment	142	160	176	191	215	271	333	342	403	550	629
Currency Ouside Banks	88	94	105	115	126	148	173	183	190	243	265
Demand Deposits	54	66	71	76	88	124	160	160	213	306	364
Quasi-Monetary Liabilities	36	41	49	58	75	110	155	241	332	466	588
GOVERNMENT DEFICIT (-) OR SURPLUS					*(Millions of current Haitian Gourdes)*						
Current Revenue	..	..	..	..	..	..	..	..	..	..	..
Current Expenditure	..	..	..	..	..	310	314	422	528	642	758
Current Budget Balance	..	..	..	..	..	..	..	..	..	..	..
Capital Receipts	..	..	..	..	..	..	..	..	..	..	..
Capital Payments	..	..	..	..	..	32	29	28	108	174	229

1978	1979	1980	1981	1982	1983	1984	1985	1986	1987 estimate	NOTES	HAITI
200	220	250	260	270	270	290	310	340	360	..	CURRENT GNP PER CAPITA (US $)
5,236	5,329	5,422	5,517	5,615	5,717	5,822	5,933	6,050	6,164	..	POPULATION (thousands)
				(Millions of current Haitian Gourdes)							ORIGIN AND USE OF RESOURCES
4,928	5,531	7,230	7,323	7,341	8,026	8,983	9,974	11,134	11,153	C	Gross National Product (GNP)
-74	-67	-80	-74	-84	-123	-99	-73	-84	-82	..	Net Factor Income from Abroad
5,001	5,597	7,309	7,397	7,425	8,148	9,082	10,047	11,218	11,235	C	Gross Domestic Product (GDP)
..	..	..	..	..	645	716	912	1,025	947	..	Indirect Taxes, net
..	..	..	..	..	7,503	8,366	9,135	10,193	10,288	B C	GDP at factor cost
..	..	..	..	..	..	..	..	..	..	..	Agriculture
..	..	..	..	..	..	..	..	..	..	..	Industry
..	..	..	..	..	..	..	..	..	..	..	Manufacturing
..	..	..	..	..	..	..	..	..	..	..	Services, etc.
-443	-557	-649	-1,124	-755	-848	-821	-850	-658	-860	..	Resource Balance
1,018	1,094	1,580	1,243	1,465	1,427	1,587	1,518	1,458	1,388	..	Exports of Goods & NFServices
1,461	1,651	2,229	2,367	2,220	2,276	2,408	2,368	2,116	2,248	f	Imports of Goods & NFServices
5,444	6,154	7,958	8,521	8,180	8,996	9,903	10,897	11,876	12,096	..	Domestic Absorption
4,135	4,573	5,984	6,443	6,102	6,794	7,488	8,269	9,484	9,599	..	Private Consumption, etc.
469	532	736	826	848	871	974	1,197	1,168	1,096	..	General Gov't Consumption
840	1,049	1,238	1,252	1,230	1,331	1,441	1,431	1,225	1,400	f	Gross Domestic Investment
..	..	..	..	..	1,331	1,441	1,431	1,225	1,400	f	Fixed Investment
											Memo Items:
397	492	589	128	475	483	620	581	567	540	..	Gross Domestic Saving
470	596	769	370	634	591	743	755	750	746	..	Gross National Saving
				(Millions of 1980 Haitian Gourdes)							
6,239.4	6,719.3	7,229.5	7,038.1	6,788.1	6,814.2	6,862.5	6,901.0	6,943.1	6,980.7	C	Gross National Product
..	..	..	..	..	..	..	..	..	..	..	GDP at factor cost
..	..	..	..	..	..	..	..	..	..	..	Agriculture
..	..	..	..	..	..	..	..	..	..	..	Industry
..	..	..	..	..	..	..	..	..	..	..	Manufacturing
..	..	..	..	..	..	..	..	..	..	..	Services, etc.
-572.9	-444.9	-649.0	-777.9	-376.6	-490.6	-414.3	-436.8	-420.8	-539.7	..	Resource Balance
846.8	948.7	1,580.5	915.6	1,123.4	1,016.8	1,079.6	1,013.4	921.8	856.9	..	Exports of Goods & NFServices
1,419.8	1,393.6	2,229.5	1,693.6	1,500.1	1,507.3	1,493.9	1,450.2	1,342.6	1,396.6	f	Imports of Goods & NFServices
6,903.8	7,238.2	7,958.0	7,885.7	7,241.1	7,407.0	7,351.2	7,387.4	7,415.2	7,571.1	..	Domestic Absorption
5,248.3	5,387.3	5,983.6	5,887.0	5,367.7	5,504.3	5,368.0	5,485.1	5,751.8	5,915.0	..	Private Consumption, etc.
598.3	619.4	736.4	751.1	711.3	677.5	701.3	795.2	716.9	666.5	..	General Gov't Consumption
1,057.2	1,231.5	1,238.0	1,247.7	1,162.0	1,225.1	1,281.9	1,107.2	946.4	989.5	f	Gross Domestic Investment
..	..	..	..	..	..	..	..	..	..	f	Fixed Investment
											Memo Items:
989.3	923.5	1,580.5	889.3	989.7	945.4	984.6	929.6	924.9	862.1	..	Capacity to Import
142.4	-25.3	0.0	-26.3	-133.7	-71.3	-95.0	-83.7	3.2	5.2	..	Terms of Trade Adjustment
6,473.3	6,768.0	7,309.0	7,081.5	6,730.7	6,845.1	6,841.9	6,866.9	6,997.5	7,036.5	..	Gross Domestic Income
6,381.9	6,694.0	7,229.5	7,011.8	6,654.4	6,742.8	6,767.5	6,817.2	6,946.3	6,985.9	..	Gross National Income
				(Index 1980 = 100)							DOMESTIC PRICES (DEFLATORS)
79.0	82.4	100.0	104.1	108.2	117.8	130.9	144.5	160.4	159.8	..	Overall (GDP)
78.9	85.0	100.0	108.1	113.0	121.5	134.7	147.5	160.2	159.8	..	Domestic Absorption
..	..	..	..	..	..	..	..	..	..	..	Agriculture
..	..	..	..	..	..	..	..	..	..	..	Industry
..	..	..	..	..	..	..	..	..	..	..	Manufacturing
											MANUFACTURING ACTIVITY
84.2	93.8	100.0	102.5	111.0	122.3	112.6	115.5			..	Employment (1980=100)
109.9	111.7	100.0	112.6	115.7	107.6	107.3	101.7	..	..	..	Real Earnings per Empl. (1980=100)
..	..	..	..	..	..	..	..	..	..	..	Real Output per Empl. (1980=100)
..	..	..	..	..	..	..	..	..	..	..	Earnings as % of Value Added
				(Millions of current Haitian Gourdes)							MONETARY HOLDINGS
1,432	1,873	1,938	2,197	2,267	2,348	2,669	2,818	0	3,694	..	Money Supply, Broadly Defined
718	1,108	945	1,175	1,164	1,175	1,400	1,477	0	2,098	..	Money as Means of Payment
311	419	418	487	566	600	691	763	829	980	..	Currency Ouside Banks
406	689	527	687	599	575	709	714	-829	1,118	..	Demand Deposits
714	765	993	1,023	1,103	1,173	1,269	1,340	0	1,596	..	Quasi-Monetary Liabilities
				(Millions of current Haitian Gourdes)							
..	-226	-342	..	-236	..	..	..	..	..	C	GOVERNMENT DEFICIT (-) OR SURPLUS
	824	927		1,113	1,346	1,591	1,930	1,597	..	..	Current Revenue
667	878	1,013	1,300	1,205	1,407	..	1,799	1,780	..	..	Current Expenditure
..	-54	-86	..	-92	-61	..	131	-184	..	..	Current Budget Balance
..	..	..	..	..	..	..	..	..	..	..	Capital Receipts
255	171	256	160	144	-1,407	..	-1,799	-1,780	..	..	Capital Payments

HAITI	1967	1968	1969	1970	1971	1972	1973	1974	1975	1976	1977
FOREIGN TRADE (CUSTOMS BASIS)					*(Millions of current US dollars)*						
Value of Exports, fob	32.9	36.9	38.8	48.1	52.5	63.2	95.0	130.2	142.9	176.3	216.3
Nonfuel Primary Products	24.8	27.8	29.3	29.9	35.0	30.0	35.6	44.8	48.9	73.4	92.6
Fuels	0.0	0.1	0.0	0.0	0.0	0.0	0.0	0.0	0.0	0.0	0.0
Manufactures	8.1	9.0	9.5	18.2	17.5	33.2	59.4	85.4	94.0	102.9	123.8
Value of Imports, cif	38.4	37.1	45.3	59.0	66.2	84.8	112.7	165.6	196.3	246.1	286.9
Nonfuel Primary Products	12.7	9.4	12.6	12.1	16.2	18.7	22.4	32.1	47.1	70.6	65.9
Fuels	2.3	2.4	3.7	2.9	3.8	3.9	4.4	12.4	12.9	17.1	23.6
Manufactures	23.5	25.4	29.1	44.0	46.3	62.1	86.0	121.1	136.3	158.4	197.5
Terms of Trade	117.3	123.8	135.9	127.9	110.8	108.9	108.8	104.9	94.0	114.0	128.0
					(Index 1980 = 100)						
Export Prices, fob	27.0	27.4	28.9	33.3	31.1	35.1	45.5	60.8	59.4	71.5	85.9
Nonfuel Primary Products	26.1	26.5	27.9	31.1	29.0	31.9	40.0	51.4	49.6	79.5	117.1
Fuels	4.3	4.3	4.3	4.3	5.6	6.2	8.9	36.7	35.7	38.4	42.0
Manufactures	30.1	31.7	32.4	37.5	36.5	38.5	49.5	67.2	66.1	66.7	71.7
Import Prices, cif	23.0	22.1	21.3	26.0	28.1	32.2	41.8	57.9	63.1	62.7	67.2
Trade at Constant 1980 Prices					*(Millions of 1980 US dollars)*						
Exports, fob	121.8	134.7	134.3	144.6	168.9	180.1	209.0	214.2	240.7	246.5	251.7
Imports, cif	166.8	167.7	213.1	226.9	235.9	263.1	269.7	285.9	311.0	392.3	427.2
BALANCE OF PAYMENTS					*(Millions of current US dollars)*						
Exports of Goods & Services	..	..	..	6	9	9	8	92	106	142	175
Merchandise, fob	..	..	..	39	44	40	51	63	71	100	138
Nonfactor Services	..	..	..	6	9	9	8	29	35	41	36
Factor Services	..	..	..	0	0	0	0	0	0	1	1
Imports of Goods & Services	..	..	..	11	11	10	12	137	169	226	277
Merchandise, fob	..	..	..	48	53	58	68	97	122	164	200
Nonfactor Services	..	..	..	8	6	4	7	34	38	53	63
Factor Services	..	..	..	4	5	5	5	6	9	8	13
Long-Term Interest	..	..	..	0	0	0	1	1	1	2	4
Current Transfers, net	..	..	..	..	..	..	..	..	..	..	..
Workers' Remittances	..	..	..	17	17	22	22	22	59	74	76
Total to be Financed	..	..	..	..	..	..	..	..	..	..	..
Official Capital Grants	..	..	..	..	..	..	..	..	..	..	..
Current Account Balance	..	..	..	2	4	6	-1	-20	-27	-19	-38
Long-Term Capital, net	..	..	..	10	8	19	12	22	36	66	103
Direct Investment	..	..	..	3	3	4	6	6	3	8	8
Long-Term Loans	..	..	..	1	-1	1	-2	5	11	35	50
Disbursements	..	..	..	4	3	5	3	10	17	44	65
Repayments	..	..	..	3	4	4	5	6	6	9	15
Other Long-Term Capital	..	..	..	6	6	14	8	11	22	23	45
Other Capital, net	..	..	..	-16	-8	-20	-10	-5	-2	-32	-63
Change in Reserves	..	..	..	4	-5	-6	0	3	-6	-14	-3
Memo Item:					*(Haitian Gourdes per US dollar)*						
Conversion Factor (Annual Avg)	5.000	5.000	5.000	5.000	5.000	5.000	5.000	5.000	5.000	5.000	5.000
EXTERNAL DEBT, ETC.					*(Millions of US dollars, outstanding at end of year)*						
Public/Publicly Guar. Long-Term	..	..	..	40	39	43	41	46	57	88	139
Official Creditors	..	..	..	29	28	31	30	36	48	76	124
IBRD and IDA	..	..	..	0	0	0	0	0	6	19	37
Private Creditors	..	..	..	11	10	13	11	10	9	12	14
Private Non-guaranteed Long-Term	..	..	..	..	..	..	..	..	..	..	..
Use of Fund Credit	..	..	..	2	1	0	4	8	13	14	10
Short-Term Debt	..	..	..	..	..	..	..	..	..	..	8
Memo Items:					*(Thousands of US dollars)*						
Int'l Reserves Excluding Gold	1,900	2,500	3,500	4,300	10,387	17,882	17,032	19,700	12,430	27,941	33,844
Gold Holdings (at market price)	..	..	..	..	..	..	..	289	217	208	926
SOCIAL INDICATORS											
Total Fertility Rate	6.2	6.1	6.0	5.9	5.8	5.8	5.7	5.6	5.5	5.4	5.4
Crude Birth Rate	42.5	41.9	41.3	40.7	40.1	39.4	38.9	38.4	37.8	37.3	36.8
Infant Mortality Rate	172.0	168.6	165.2	161.8	158.4	155.0	151.8	148.6	145.4	142.2	139.0
Life Expectancy at Birth	46.2	46.7	47.1	47.6	48.0	48.5	48.9	49.4	49.8	50.2	50.7
Food Production, p.c. ('79-81=100)	107.3	105.4	106.8	106.9	107.7	108.6	107.1	107.4	105.4	105.8	98.7
Labor Force, Agriculture (%)	76.0	75.5	74.9	74.4	74.0	73.5	73.1	72.6	72.2	71.8	71.3
Labor Force, Female (%)	47.0	46.8	46.7	46.6	46.3	46.1	45.9	45.6	45.4	45.1	44.8
School Enroll. Ratio, primary	..	..	..	..	53.0	..	..	..	60.0	61.0	60.0
School Enroll. Ratio, secondary	..	..	..	..	6.0	..	..	..	8.0	10.0	11.0

1978	1979	1980	1981	1982	1983	1984	1985	1986	1987 estimate	NOTES	HAITI
				(Millions of current US dollars)							**FOREIGN TRADE (CUSTOMS BASIS)**
240.6	239.7	341.0	273.2	296.2	272.4	312.9	332.6	285.0	260.7	C f	Value of Exports, fob
97.2	72.7	141.3	74.7	74.8	72.9	69.4	67.1	72.7	48.8	..	Nonfuel Primary Products
0.0	0.0	0.0	0.0	0.1	2.7	0.0	0.0	0.0	..	..	Fuels
143.4	167.0	199.7	198.5	221.3	196.8	243.5	265.5	212.3	211.9	..	Manufactures
311.4	367.8	492.1	527.2	448.2	412.0	435.0	449.2	367.2	378.2	C f	Value of Imports, cif
67.4	71.3	74.6	93.9	100.6	123.4	128.1	127.3	114.8	120.9	..	Nonfuel Primary Products
24.5	34.5	61.4	64.7	55.6	61.3	61.4	63.6	50.4	41.6	..	Fuels
219.5	261.9	356.1	368.6	292.0	227.3	245.5	258.3	202.0	215.7	..	Manufactures
				(Index 1980 = 100)							
111.4	106.8	100.0	94.0	95.0	94.0	95.6	97.3	124.1	108.6	..	Terms of Trade
83.7	93.8	100.0	95.4	92.5	89.0	89.8	89.1	105.4	102.9	..	Export Prices, fob
98.1	103.9	100.0	86.8	86.2	84.8	91.6	89.6	112.4	74.1	..	Nonfuel Primary Products
42.3	61.0	100.0	112.5	101.6	92.5	90.2	87.5	44.9	56.9	..	Fuels
76.1	90.0	100.0	99.1	94.8	90.7	89.3	89.0	103.2	113.0	..	Manufactures
75.1	87.8	100.0	101.5	97.4	94.7	94.0	91.6	84.9	94.7	..	Import Prices, cif
				(Millions of 1980 US dollars)							Trade at Constant 1980 Prices
287.6	255.6	341.0	286.3	320.3	305.9	348.5	373.2	270.4	253.4	..	Exports, fob
414.8	418.9	492.1	519.6	460.2	435.1	463.0	490.3	432.5	399.4	..	Imports, cif
				(Millions of current US dollars)							**BALANCE OF PAYMENTS**
213	216	309	245	278	294	323	343	297	314	C	Exports of Goods & Services
150	138	216	151	177	187	215	223	191	198	..	Merchandise, fob
61	75	90	90	98	103	104	114	102	111	..	Nonfactor Services
2	3	3	4	4	5	4	5	5	5	..	Factor Services
327	346	501	551	519	547	571	584	495	530	C	Imports of Goods & Services
207	220	319	376	330	354	360	345	303	308	..	Merchandise, fob
102	110	165	159	171	174	188	214	172	196	..	Nonfactor Services
17	16	17	17	17	19	22	25	20	27	..	Factor Services
5	4	5	6	7	6	6	7	8	9	..	Long-Term Interest
..	..	52	63	49	46	44	50	..	58	..	Current Transfers, net
78	85	106	123	95	89	89	98	108	116	..	Workers' Remittances
..	..	-137	-234	-199	-211	-203	-182	..	-153	..	Total to be Financed
..	..	33	66	69	69	78	87	..	122	..	Official Capital Grants
-45	-55	-104	-168	-130	-142	-125	-95	-45	-31	..	Current Account Balance
90	101	104	164	105	123	138	118	118	163	C	Long-Term Capital, net
10	12	13	8	7	8	4	5	5	5	..	Direct Investment
46	42	40	102	57	34	47	52	32	80	..	Long-Term Loans
60	50	55	117	66	42	58	65	43	94	..	Disbursements
14	9	15	15	9	8	11	14	11	14	..	Repayments
35	47	18	-12	-28	12	9	-26	81	-44	..	Other Long-Term Capital
-33	-43	-35	-6	63	41	10	-39	-90	-156	C	Other Capital, net
-13	-4	35	10	-38	-22	-23	16	17	24	..	Change in Reserves
				(Haitian Gourdes per US dollar)							Memo Item:
5.000	5.000	5.000	5.000	5.000	5.000	5.000	5.000	5.000	5.000	..	Conversion Factor (Annual Avg)
				(Millions of US dollars, outstanding at end of year)							**EXTERNAL DEBT, ETC.**
185	227	267	362	416	448	491	534	585	674	..	Public/Publicly Guar. Long-Term
172	211	243	279	324	363	421	467	516	606	..	Official Creditors
45	53	66	82	96	122	150	174	209	263	..	IBRD and IDA
12	16	24	83	91	85	71	67	69	68	..	Private Creditors
..	..	..	..	..	..	..	..	..	..	..	Private Non-guaranteed Long-Term
10	8	22	37	47	76	84	82	67	52	..	Use of Fund Credit
6	19	14	24	73	46	80	88	46	79	..	Short-Term Debt
				(Thousands of US dollars)							Memo Items:
38,638	55,015	16,226	24,047	4,225	8,968	12,978	6,388	15,879	16,999	..	Int'l Reserves Excluding Gold
2,195	9,118	10,498	7,079	8,136	6,794	5,490	5,823	6,961	8,621	..	Gold Holdings (at market price)
											SOCIAL INDICATORS
5.3	5.2	5.2	5.1	5.1	5.0	4.9	4.9	4.8	4.7	..	Total Fertility Rate
36.5	36.2	35.9	35.6	35.4	35.2	35.0	34.8	34.7	34.5	..	Crude Birth Rate
136.8	134.6	132.4	130.2	128.0	125.1	122.3	119.4	116.5	113.7	..	Infant Mortality Rate
51.1	51.5	51.9	52.3	52.7	53.1	53.5	53.9	54.3	54.7	..	Life Expectancy at Birth
103.0	103.9	99.0	97.2	95.3	96.9	96.8	96.6	95.4	95.8	..	Food Production, p.c. ('79-81 = 100)
70.9	70.4	70.0	..	..	..	..	..	..	..	..	Labor Force, Agriculture (%)
44.5	44.3	44.0	43.8	43.5	43.3	43.1	42.8	42.6	42.4	..	Labor Force, Female (%)
60.0	64.0	69.0	69.0	72.0	76.0	78.0	..	..	..	..	School Enroll. Ratio, primary
12.0	12.0	13.0	13.0	14.0	16.0	18.0	..	..	..	..	School Enroll. Ratio, secondary

HONDURAS	1967	1968	1969	1970	1971	1972	1973	1974	1975	1976	1977
CURRENT GNP PER CAPITA (US $)	240	260	260	270	280	290	310	330	360	400	460
POPULATION (thousands)	2,418	2,484	2,553	2,627	2,706	2,792	2,884	2,981	3,082	3,188	3,299
ORIGIN AND USE OF RESOURCES	*(Millions of current Honduran Lempiras)*										
Gross National Product (GNP)	1,153.8	1,247.4	1,298.8	1,400.8	1,413.0	1,551.0	1,759.0	2,042.0	2,190.0	2,580.0	3,201.0
Net Factor Income from Abroad	-42.4	-46.2	-37.2	-45.2	-49.0	-55.0	-66.0	-27.0	-58.0	-116.0	-138.0
Gross Domestic Product (GDP)	1,196.2	1,293.6	1,336.0	1,446.0	1,462.0	1,606.0	1,825.0	2,069.0	2,248.0	2,696.0	3,339.0
Indirect Taxes, net	101.0	113.2	117.2	139.0	143.0	151.0	170.0	199.0	221.0	286.0	407.0
GDP at factor cost	1,095.2	1,180.4	1,218.8	1,307.0	1,319.0	1,455.0	1,655.0	1,870.0	2,027.0	2,410.0	2,932.0
Agriculture	423.1	448.0	437.2	424.0	413.0	450.0	511.0	550.0	554.0	688.0	898.0
Industry	226.2	253.5	268.1	290.0	299.0	329.0	397.0	474.0	505.0	560.0	679.0
Manufacturing	136.2	151.0	164.7	181.0	196.0	221.0	261.0	294.0	319.0	363.0	443.0
Services, etc.	445.9	478.9	513.5	593.0	607.0	676.0	747.0	846.0	968.0	1,162.0	1,355.0
Resource Balance	-24.2	-15.6	-38.6	-89.6	-12.0	17.0	-16.0	-247.0	-202.0	-120.0	-148.0
Exports of Goods & NFServices	341.2	393.4	373.8	403.6	434.0	470.0	588.0	665.0	690.0	909.0	1,163.0
Imports of Goods & NFServices	365.4	409.0	412.4	493.2	446.0	453.0	604.0	912.0	892.0	1,029.0	1,311.0
Domestic Absorption	1,220.4	1,309.2	1,374.6	1,535.6	1,474.0	1,589.0	1,841.0	2,316.0	2,450.0	2,816.0	3,487.0
Private Consumption, etc.	865.1	941.1	967.3	1,067.6	1,049.0	1,140.0	1,307.0	1,532.0	1,746.0	1,950.0	2,300.0
General Gov't Consumption	120.2	128.8	148.5	166.0	175.0	193.0	186.0	242.0	278.0	348.0	417.0
Gross Domestic Investment	235.1	239.3	258.8	302.0	250.0	256.0	348.0	542.0	426.0	518.0	770.0
Fixed Investment	213.3	226.7	244.5	268.0	253.0	245.0	325.0	433.0	476.0	550.0	711.0
Memo Items:											
Gross Domestic Saving	210.9	223.7	220.2	212.4	238.0	273.0	332.0	295.0	224.0	398.0	622.0
Gross National Saving	168.5	183.5	190.0	173.0	195.2	224.5	272.9	295.1	175.8	288.6	491.7
	(Millions of 1980 Honduran Lempiras)										
Gross National Product	2,536.0	2,702.1	2,754.0	2,845.0	2,944.7	3,122.1	3,361.9	3,454.3	3,496.9	3,778.7	4,156.4
GDP at factor cost	2,422.9	2,573.4	2,596.1	2,661.7	2,771.0	2,949.5	3,172.6	3,147.2	3,221.4	3,527.1	3,826.3
Agriculture	852.5	914.0	878.2	839.5	915.4	933.3	986.9	898.7	838.4	932.2	988.0
Industry	530.2	565.4	569.0	593.0	610.7	656.2	746.0	783.9	787.7	825.6	934.4
Manufacturing	286.5	301.6	310.7	366.5	390.7	431.6	486.5	489.1	508.2	544.0	618.0
Services, etc.	1,040.2	1,094.0	1,148.9	1,229.2	1,244.9	1,360.0	1,439.7	1,464.6	1,595.3	1,769.2	1,903.9
Resource Balance	-87.1	-66.0	-90.6	-151.8	147.0	199.6	98.4	-181.2	-21.3	-147.8	-430.0
Exports of Goods & NFServices	1,025.5	1,154.8	1,085.7	1,181.5	1,337.7	1,340.1	1,476.6	1,330.3	1,381.9	1,396.7	1,393.0
Imports of Goods & NFServices	1,112.6	1,220.8	1,176.3	1,333.3	1,190.7	1,140.5	1,378.2	1,511.5	1,403.2	1,544.5	1,823.0
Domestic Absorption	2,737.1	2,890.3	2,939.9	3,114.7	2,927.8	3,058.6	3,417.7	3,665.7	3,593.8	4,094.4	4,789.1
Private Consumption, etc.	1,838.5	1,987.5	1,999.7	2,092.1	2,017.9	2,137.2	2,385.0	2,353.8	2,447.1	2,815.5	3,218.1
General Gov't Consumption	274.4	279.3	310.0	358.9	370.8	393.3	362.8	417.1	446.3	528.4	560.1
Gross Domestic Investment	624.2	623.5	630.2	663.7	539.0	528.0	669.8	894.7	700.4	750.5	1,010.9
Fixed Investment	576.9	601.4	604.7	592.5	543.6	509.4	630.3	714.6	778.1	796.5	934.5
Memo Items:											
Capacity to Import	1,038.9	1,174.2	1,066.2	1,091.1	1,158.7	1,183.3	1,341.6	1,102.1	1,085.5	1,364.4	1,617.2
Terms of Trade Adjustment	13.4	19.4	-19.5	-90.5	-179.0	-156.8	-134.9	-228.2	-296.5	-32.3	224.2
Gross Domestic Income	2,663.4	2,843.7	2,829.8	2,872.5	2,895.7	3,101.4	3,381.2	3,256.3	3,276.0	3,914.3	4,583.3
Gross National Income	2,549.4	2,721.5	2,734.5	2,754.5	2,765.7	2,965.3	3,226.9	3,226.1	3,200.4	3,746.4	4,380.7
DOMESTIC PRICES (DEFLATORS)	*(Index 1980 = 100)*										
Overall (GDP)	45.1	45.8	46.9	48.8	47.5	49.3	51.9	59.4	62.9	68.3	76.6
Domestic Absorption	44.6	45.3	46.8	49.3	50.3	52.0	53.9	63.2	68.2	68.8	72.8
Agriculture	49.6	49.0	49.8	50.5	45.1	48.2	51.8	61.2	66.1	73.8	90.9
Industry	42.7	44.8	47.1	48.9	49.0	50.1	53.2	60.5	64.1	67.8	72.7
Manufacturing	47.5	50.1	53.0	49.4	50.2	51.2	53.6	60.1	62.8	66.7	71.7
MANUFACTURING ACTIVITY											
Employment (1980 = 100)	..	..	..	..	..	..	..	..	..	..	..
Real Earnings per Empl. (1980 = 100)	..	..	..	..	..	..	..	..	..	..	..
Real Output per Empl. (1980 = 100)	..	..	..	..	..	..	..	..	..	..	..
Earnings as % of Value Added	..	28.6	35.3	..	37.5	36.9	37.3	38.0	37.5	37.6	37.6
MONETARY HOLDINGS	*(Millions of current Honduran Lempiras)*										
Money Supply, Broadly Defined	213.0	218.6	263.2	300.3	333.1	378.5	460.6	474.4	532.0	692.6	839.4
Money as Means of Payment	114.3	127.4	148.1	158.9	169.4	192.9	238.4	242.4	262.7	361.0	411.3
Currency Ouside Banks	55.9	61.4	73.4	76.8	80.2	90.1	112.2	108.8	114.8	173.3	193.2
Demand Deposits	58.4	66.0	74.7	82.1	89.2	102.8	126.2	133.6	147.9	187.7	218.1
Quasi-Monetary Liabilities	98.7	91.2	115.1	141.4	163.7	185.6	222.2	232.0	269.3	331.6	428.1
GOVERNMENT DEFICIT (-) OR SURPLUS	*(Millions of current Honduran Lempiras)*										
	..	..	..	..	..	-45	-8	-18	-63	-64	..
Current Revenue	..	..	..	..	..	206	242	278	308	390	477
Current Expenditure	..	..	..	..	..	186	198	202	236	288	..
Current Budget Balance	..	..	..	..	..	19	44	76	73	102	..
Capital Receipts	..	..	..	..	..	1	0	0	1	..	..
Capital Payments	..	..	..	..	..	65	52	95	138	166	..

1978	1979	1980	1981	1982	1983	1984	1985	1986	1987 estimate	NOTES	HONDURAS
530	600	640	690	690	700	720	740	770	810	..	**CURRENT GNP PER CAPITA (US $)**
3,415	3,536	3,663	3,795	3,933	4,077	4,226	4,382	4,528	4,677	..	**POPULATION (thousands)**
				(Millions of current Honduran Lempiras)							**ORIGIN AND USE OF RESOURCES**
3,627.0	4,184.0	4,781.0	5,247.0	5,359.0	5,731.0	6,106.0	6,597.0	7,143.0	7,599.0	..	Gross National Product (GNP)
-171.0	-241.0	-307.0	-306.0	-403.0	-304.0	-356.0	-380.0	-422.0	-444.0	..	Net Factor Income from Abroad
3,798.0	4,425.0	5,088.0	5,553.0	5,762.0	6,035.0	6,462.0	6,977.0	7,565.0	8,043.0	..	Gross Domestic Product (GDP)
426.0	496.0	539.0	597.0	577.0	616.0	705.0	842.0	935.0	983.0	..	Indirect Taxes, net
3,372.0	3,929.0	4,549.0	4,956.0	5,185.0	5,419.0	5,757.0	6,135.0	6,630.0	7,060.0	..	GDP at factor cost
945.0	1,038.0	1,132.0	1,166.0	1,186.0	1,226.0	1,253.0	1,328.0	1,495.0	1,529.0	..	Agriculture
836.0	978.0	1,143.0	1,180.0	1,335.0	1,392.0	1,502.0	1,540.0	1,580.0	1,664.0	..	Industry
520.0	606.0	687.0	726.0	766.0	832.0	913.0	926.0	963.0	1,030.0	..	Manufacturing
1,591.0	1,913.0	2,274.0	2,610.0	2,664.0	2,801.0	3,002.0	3,267.0	3,555.0	3,867.0	..	Services, etc.
-175.0	-184.0	-370.0	-354.0	-105.0	-223.0	-437.0	-319.0	-105.0	-215.0	..	Resource Balance
1,380.0	1,679.0	1,886.0	1,771.0	1,537.0	1,606.0	1,699.0	1,811.0	2,022.0	1,967.0	..	Exports of Goods & NFServices
1,555.0	1,863.0	2,256.0	2,125.0	1,642.0	1,829.0	2,136.0	2,130.0	2,127.0	2,182.0	..	Imports of Goods & NFServices
3,973.0	4,609.0	5,458.0	5,907.0	5,867.0	6,258.0	6,899.0	7,296.0	7,670.0	8,258.0	..	Domestic Absorption
2,497.0	2,915.0	3,532.0	3,998.0	4,291.0	4,484.0	4,716.0	4,986.0	5,340.0	5,731.0	..	Private Consumption, etc.
442.0	520.0	678.0	758.0	800.0	877.0	952.0	1,046.0	1,144.0	1,289.0	..	General Gov't Consumption
1,034.0	1,174.0	1,248.0	1,151.0	776.0	897.0	1,231.0	1,264.0	1,186.0	1,238.0	..	Gross Domestic Investment
941.0	1,004.0	1,235.0	1,051.0	966.0	1,073.0	1,246.0	1,245.0	1,143.0	1,155.0	..	Fixed Investment
											Memo Items:
859.0	990.0	878.0	797.0	671.0	674.0	794.0	945.0	1,081.0	1,023.0	..	Gross Domestic Saving
696.9	762.8	586.0	508.8	286.0	389.4	458.6	589.8	684.9	579.0	..	Gross National Saving
				(Millions of 1980 Honduran Lempiras)							
4,485.3	4,750.0	4,781.0	4,900.7	4,698.3	4,773.1	4,890.9	5,017.5	5,133.3	5,359.0	..	Gross National Product
4,192.5	4,581.3	4,549.0	4,608.9	4,553.4	4,531.7	4,622.4	4,707.2	4,818.9	5,031.2	..	GDP at factor cost
1,055.0	1,140.9	1,132.0	1,175.5	1,168.8	1,161.0	1,175.5	1,210.1	1,230.2	1,318.4	..	Agriculture
1,057.0	1,225.2	1,143.0	1,045.6	1,098.7	1,120.2	1,187.3	1,172.1	1,172.1	1,174.6	..	Industry
664.0	707.4	687.0	670.4	646.1	680.6	738.1	721.5	740.6	753.4	..	Manufacturing
2,080.5	2,215.2	2,274.0	2,387.8	2,285.8	2,250.5	2,259.6	2,325.0	2,416.5	2,538.1	..	Services, etc.
-356.0	-187.6	-370.0	-52.4	269.3	175.5	-77.6	-35.1	60.3	78.0	..	Resource Balance
1,696.7	1,990.5	1,886.0	1,929.0	1,701.6	1,863.9	1,894.6	1,945.0	2,029.8	2,076.6	..	Exports of Goods & NFServices
2,052.7	2,178.1	2,256.0	1,981.4	1,432.3	1,688.4	1,972.2	1,980.1	1,969.5	1,998.6	..	Imports of Goods & NFServices
5,078.8	5,208.3	5,458.0	5,217.8	4,791.9	4,872.8	5,266.4	5,388.4	5,442.0	5,651.5	..	Domestic Absorption
3,229.6	3,281.1	3,532.0	3,477.6	3,495.9	3,435.7	3,492.4	3,584.6	3,675.6	3,842.0	..	Private Consumption, etc.
585.3	614.4	678.0	693.9	671.4	679.3	704.5	741.6	780.0	832.9	..	General Gov't Consumption
1,263.9	1,312.8	1,248.0	1,046.3	624.6	757.8	1,069.5	1,062.2	986.4	976.6	..	Gross Domestic Investment
1,149.5	1,123.8	1,235.0	954.0	787.9	902.7	1,081.1	1,046.9	952.8	916.2	..	Fixed Investment
											Memo Items:
1,821.7	1,963.0	1,886.0	1,651.3	1,340.7	1,482.5	1,568.7	1,683.6	1,872.3	1,801.7	..	Capacity to Import
125.0	-27.5	0.0	-277.7	-360.9	-381.4	-325.9	-261.5	-157.5	-274.9	..	Terms of Trade Adjustment
4,847.8	4,993.2	5,088.0	4,887.7	4,700.3	4,667.0	4,862.9	5,091.8	5,344.7	5,454.6	..	Gross Domestic Income
4,610.4	4,722.5	4,781.0	4,623.0	4,337.4	4,391.7	4,565.0	4,756.1	4,975.7	5,084.1	..	Gross National Income
				(Index 1980 = 100)							**DOMESTIC PRICES (DEFLATORS)**
80.4	88.1	100.0	107.5	113.8	119.5	124.5	130.3	137.5	140.4	..	Overall (GDP)
78.2	88.5	100.0	113.2	122.4	128.4	131.0	135.4	140.9	146.1	..	Domestic Absorption
89.6	91.0	100.0	99.2	101.5	105.6	106.6	109.7	121.5	116.0	..	Agriculture
79.1	79.8	100.0	112.8	121.5	124.3	126.5	131.4	134.8	141.7	..	Industry
78.3	85.7	100.0	108.3	118.6	122.2	123.7	128.3	130.0	136.7	..	Manufacturing
											MANUFACTURING ACTIVITY
..	..	..	..	..	..	..	..	..	..	G	Employment (1980 = 100)
..	..	..	..	..	..	..	..	..	..	G	Real Earnings per Empl. (1980 = 100)
..	..	..	..	..	..	..	..	..	..	G	Real Output per Empl. (1980 = 100)
37.6	37.6	37.6	37.6	37.6	37.6	37.6	37.6	..	..	..	Earnings as % of Value Added
				(Millions of current Honduran Lempiras)							**MONETARY HOLDINGS**
1,008.7	1,107.2	1,212.4	1,321.0	1,579.2	1,851.7	2,056.4	2,009.5	2,203.0	2,707.7	D	Money Supply, Broadly Defined
480.4	545.6	610.3	637.4	716.9	814.9	846.0	855.6	955.3	1,119.1	..	Money as Means of Payment
215.4	270.1	274.6	302.2	313.6	361.9	383.5	410.2	425.8	491.5	..	Currency Ouside Banks
265.0	275.5	335.7	335.2	403.3	453.0	462.5	445.4	529.5	627.6	..	Demand Deposits
528.3	561.6	602.1	683.6	862.3	1,036.8	1,210.4	1,153.9	1,247.7	1,588.6	..	Quasi-Monetary Liabilities
				(Millions of current Honduran Lempiras)							
										F	**GOVERNMENT DEFICIT (-) OR SURPLUS**
519	632	748	741	..	..	..	..	..	..	..	Current Revenue
..	..	..	..	..	..	..	..	..	..	..	Current Expenditure
..	..	..	..	..	..	..	..	..	..	..	Current Budget Balance
..	..	..	..	..	..	..	..	..	..	..	Capital Receipts
..	..	..	..	..	..	..	..	..	..	..	Capital Payments

HONDURAS	1967	1968	1969	1970	1971	1972	1973	1974	1975	1976	1977
FOREIGN TRADE (CUSTOMS BASIS)				*(Millions of current US dollars)*							
Value of Exports, fob	155.9	179.0	165.9	169.7	182.9	193.1	246.8	253.3	293.3	391.8	510.7
Nonfuel Primary Products	144.4	162.2	147.3	149.8	174.8	183.5	228.6	211.0	249.5	351.9	461.0
Fuels	0.0	2.1	4.8	6.3	3.0	3.4	4.1	14.5	12.3	1.1	0.6
Manufactures	11.5	14.6	13.8	13.7	5.1	6.2	14.1	27.8	31.5	38.8	49.1
Value of Imports, cif	164.8	184.7	184.3	220.7	193.4	192.8	262.2	380.1	404.3	453.1	579.4
Nonfuel Primary Products	22.1	25.6	23.3	28.4	21.3	24.3	30.1	46.2	60.5	56.6	64.9
Fuels	8.3	12.5	12.8	14.7	17.5	19.2	26.1	63.4	68.5	48.2	69.3
Manufactures	134.4	146.6	148.2	177.6	154.6	149.4	206.1	270.5	275.3	348.3	445.2
Terms of Trade	134.8	139.0	126.5	123.8	130.6	135.7	134.0	99.3	99.3	117.8	130.0
				(Index 1980 = 100)							
Export Prices, fob	32.5	30.5	28.5	29.2	31.9	35.4	44.8	53.4	56.6	71.0	84.4
Nonfuel Primary Products	32.7	33.1	34.5	37.7	34.5	38.7	48.1	53.6	57.2	71.7	86.2
Fuels	4.3	4.3	4.3	4.3	5.6	6.2	8.9	36.7	35.7	38.4	42.0
Manufactures	30.1	31.7	32.4	37.5	36.5	38.5	49.5	67.2	66.1	66.7	71.7
Import Prices, cif	24.1	22.0	22.5	23.6	24.4	26.1	33.4	53.7	57.0	60.3	65.0
Trade at Constant 1980 Prices				*(Millions of 1980 US dollars)*							
Exports, fob	480.2	585.9	582.5	580.5	573.7	545.1	550.7	474.7	518.0	551.7	604.9
Imports, cif	684.1	840.7	818.5	934.3	792.8	738.7	784.3	707.4	709.0	751.3	892.0
BALANCE OF PAYMENTS				*(Millions of current US dollars)*							
Exports of Goods & Services	..	..	..	199	218	237	298	337	351	463	594
Merchandise, fob	..	..	..	178	195	212	267	300	310	412	530
Nonfactor Services	..	..	..	18	21	23	27	32	35	42	51
Factor Services	..	..	..	2	2	3	4	5	7	10	13
Imports of Goods & Services	..	..	..	269	247	257	339	475	481	581	737
Merchandise, fob	..	..	..	203	178	177	243	388	372	433	550
Nonfactor Services	..	..	..	41	42	50	59	68	73	82	105
Factor Services	..	..	..	25	27	30	37	19	35	67	82
Long-Term Interest	..	..	..	4	5	6	10	12	16	22	31
Current Transfers, net	..	..	..	3	3	3	4	14	5	3	4
Workers' Remittances	..	..	..	0	0	0	0	0	0	0	0
Total to be Financed	..	..	..	-68	-26	-16	-38	-123	-125	-115	-139
Official Capital Grants	..	..	..	4	4	3	4	19	13	10	10
Current Account Balance	..	..	..	-64	-23	-13	-35	-104	-112	-105	-129
Long-Term Capital, net	..	..	..	44	39	27	36	82	144	114	159
Direct Investment	..	..	..	8	7	3	7	-1	7	5	9
Long-Term Loans	..	..	..	32	29	25	4	68	99	98	140
Disbursements	..	..	..	39	38	38	21	89	128	136	200
Repayments	..	..	..	6	9	13	17	22	29	37	60
Other Long-Term Capital	..	..	..	-1	-1	-4	22	-3	26	0	0
Other Capital, net	..	..	..	8	-17	-2	7	5	22	29	36
Change in Reserves	..	..	..	12	1	-12	-9	17	-54	-38	-66
Memo Item:											
Conversion Factor (Annual Avg)	2.000	2.000	2.000	*(Honduran Lempiras per US dollar)*							
	2.000	2.000	2.000	2.000	2.000	2.000	2.000	2.000	2.000	2.000	2.000
EXTERNAL DEBT, ETC.				*(Millions of US dollars, outstanding at end of year)*							
Public/Publicly Guar. Long-Term	..	..	..	90	102	119	134	171	264	344	458
Official Creditors	..	..	..	86	100	115	129	157	244	301	372
IBRD and IDA	..	..	..	45	50	56	66	72	85	98	120
Private Creditors	..	..	..	4	2	5	5	15	20	43	86
Private Non-guaranteed Long-Term	..	..	..	19	36	45	62	89	96	114	141
Use of Fund Credit	..	..	..	0	0	0	0	21	20	20	5
Short-Term Debt	..	..	..	..	..	..	..	..	..	..	149
Memo Items:				*(Millions of US dollars)*							
Int'l Reserves Excluding Gold	25.0	31.4	30.8	20.1	21.8	35.1	41.7	44.3	97.0	130.8	179.8
Gold Holdings (at market price)	0.1	0.1	0.1	0.1	0.1	0.2	0.4	0.6	0.4	0.4	2.3
SOCIAL INDICATORS											
Total Fertility Rate	7.4	7.4	7.4	7.4	7.4	7.4	7.2	7.1	6.9	6.7	6.6
Crude Birth Rate	50.1	49.8	49.6	49.3	49.0	48.7	47.7	46.7	45.7	44.8	43.8
Infant Mortality Rate	123.0	120.4	117.8	115.2	112.6	110.0	107.0	104.0	101.0	98.0	95.0
Life Expectancy at Birth	50.9	51.5	52.1	52.7	53.3	54.0	54.7	55.4	56.2	56.9	57.7
Food Production, p.c. ('79-81 = 100)	108.0	110.7	110.7	108.5	113.0	114.9	108.2	99.2	88.8	94.4	99.1
Labor Force, Agriculture (%)	66.5	66.0	65.4	64.9	64.5	64.0	63.6	63.1	62.7	62.3	61.8
Labor Force, Female (%)	13.6	13.8	14.0	14.2	14.3	14.5	14.6	14.8	14.9	15.1	15.2
School Enroll. Ratio, primary	..	..	..	87.0	..	..	..	..	88.0	88.0	..
School Enroll. Ratio, secondary	..	..	..	14.0	..	..	..	..	16.0	17.0	19.0

1978	1979	1980	1981	1982	1983	1984	1985	1986	1987 estimate	NOTES	HONDURAS
				(Millions of current US dollars)							**FOREIGN TRADE (CUSTOMS BASIS)**
601.8	720.9	813.4	712.5	655.7	671.8	725.4	764.6	854.3	826.7	..	Value of Exports, fob
541.2	640.0	707.9	626.9	597.5	606.4	646.9	675.8	757.5	721.0	..	Nonfuel Primary Products
0.5	0.3	3.9	2.4	0.6	4.2	4.8	5.4	4.5	5.6	..	Fuels
60.1	80.6	101.7	83.2	57.6	61.2	73.7	83.4	92.3	100.1	..	Manufactures
699.2	825.8	1,008.7	944.9	689.9	823.0	893.4	888.1	875.0	895.0	..	Value of Imports, cif
79.2	84.7	119.6	114.5	79.6	96.4	94.7	90.2	63.9	58.2	..	Nonfuel Primary Products
76.0	106.8	160.7	149.4	150.3	184.1	153.5	134.2	91.1	125.1	..	Fuels
544.0	634.3	728.4	681.0	460.0	542.5	645.2	663.7	720.0	711.8	..	Manufactures
				(Index 1980 = 100)							
110.4	112.1	100.0	92.0	93.3	99.0	97.1	93.2	100.4	83.4	..	Terms of Trade
80.7	96.3	100.0	94.2	92.2	94.6	90.9	86.8	96.1	86.9	..	Export Prices, fob
81.3	97.1	100.0	93.5	92.0	95.0	91.1	86.5	96.0	84.5	..	Nonfuel Primary Products
42.3	61.0	100.0	112.5	101.6	92.5	90.2	87.5	44.9	56.9	..	Fuels
76.1	90.0	100.0	99.1	94.8	90.7	89.3	89.0	103.2	113.0	..	Manufactures
73.1	85.9	100.0	102.3	98.8	95.6	93.6	93.2	95.7	104.2	..	Import Prices, cif
				(Millions of 1980 US dollars)							Trade at Constant 1980 Prices
745.6	748.9	813.4	756.8	711.1	710.3	798.1	881.1	888.7	951.4	..	Exports, fob
956.5	961.4	1,008.7	923.6	698.2	861.3	954.8	953.3	914.4	858.6	..	Imports, cif
				(Millions of current US dollars)							**BALANCE OF PAYMENTS**
707	859	967	903	784	815	863	934	1,029	994	f	Exports of Goods & Services
626	757	850	784	677	699	737	805	899	863	..	Merchandise, fob
61	82	91	100	90	102	110	114	118	121	..	Nonfactor Services
19	21	26	20	17	14	16	15	13	10	..	Factor Services
881	1,072	1,306	1,233	1,042	1,079	1,260	1,268	1,313	1,323	f	Imports of Goods & Services
655	783	954	899	681	756	885	879	900	894	..	Merchandise, fob
122	147	173	162	142	156	181	184	189	197	..	Nonfactor Services
104	141	179	173	219	166	194	205	224	232	..	Factor Services
43	60	83	96	112	97	88	102	114	92	..	Long-Term Interest
5	7	8	9	9	10	10	12	13	..	..	Current Transfers, net
0	0	0	0	0	0	0	0	0	..	..	Workers' Remittances
-170	-206	-331	-321	-249	-254	-386	-322	-271	-330	K	Total to be Financed
13	14	14	19	21	35	70	111	116	146	K	Official Capital Grants
-157	-192	-317	-303	-228	-219	-316	-211	-155	-183	..	Current Account Balance
188	194	280	228	189	183	344	364	214	190	f	Long-Term Capital, net
13	28	6	-4	14	21	21	28	30	36	..	Direct Investment
162	159	271	234	166	174	186	266	97	32	..	Long-Term Loans
236	279	359	332	258	257	269	351	205	199	..	Disbursements
75	119	87	97	92	83	84	85	108	167	..	Repayments
1	-8	-11	-21	-11	-48	68	-41	-29	-24	..	Other Long-Term Capital
-22	19	-41	1	-15	-7	-42	-142	-31	34	f	Other Capital, net
-10	-20	78	74	54	43	15	-11	-28	-41	..	Change in Reserves
				(Honduran Lempiras per US dollar)							Memo Item:
2.000	2.000	2.000	2.000	2.000	2.000	2.000	2.000	2.000	2.000	..	Conversion Factor (Annual Avg)
				(Millions of US dollars, outstanding at end of year)							**EXTERNAL DEBT, ETC.**
595	760	997	1,257	1,430	1,649	1,793	2,163	2,399	2,681	..	Public/Publicly Guar. Long-Term
481	578	718	886	1,046	1,230	1,373	1,694	1,882	2,180	..	Official Creditors
150	179	216	253	297	349	355	450	546	655	..	IBRD and IDA
113	182	279	371	385	419	420	470	517	501	..	Private Creditors
165	158	191	172	159	195	162	141	125	115	..	Private Non-guaranteed Long-Term
0	0	15	38	104	147	136	134	98	68	..	Use of Fund Credit
173	264	272	242	152	142	199	304	364	439	..	Short-Term Debt
				(Millions of US dollars)							Memo Items:
184.4	209.2	149.8	101.0	112.2	113.6	128.2	105.8	111.3	106.0	..	Int'l Reserves Excluding Gold
3.1	7.1	9.4	6.4	7.3	6.1	4.9	5.2	6.3	7.8	..	Gold Holdings (at market price)
											SOCIAL INDICATORS
6.5	6.4	6.3	6.2	6.2	6.0	5.9	5.8	5.7	5.6	..	Total Fertility Rate
43.5	43.2	42.9	42.6	42.3	41.9	41.4	41.0	40.6	40.2	..	Crude Birth Rate
92.4	89.8	87.2	84.6	82.0	76.3	70.5	64.8	59.0	53.3	..	Infant Mortality Rate
58.5	59.4	60.2	61.1	61.9	62.3	62.7	63.2	63.6	64.0	..	Life Expectancy at Birth
103.3	94.7	101.6	103.7	95.6	86.7	83.6	84.9	88.4	89.8	..	Food Production, p.c. ('79-81 = 100)
61.4	60.9	60.5	..	..	..	..	..	..	..	..	Labor Force, Agriculture (%)
15.4	15.5	15.7	16.0	16.3	16.6	16.9	17.3	17.6	17.9	..	Labor Force, Female (%)
89.0	89.0	95.0	94.0	99.0	101.0	102.0	102.0	..	..	..	School Enroll. Ratio, primary
23.0	..	30.0	..	32.0	33.0	33.0	36.0	..	..	..	School Enroll. Ratio, secondary

HONG KONG	1967	1968	1969	1970	1971	1972	1973	1974	1975	1976	1977
CURRENT GNP PER CAPITA (US $)	680	700	800	900	1,020	1,220	1,560	1,890	2,160	2,700	3,200
POPULATION (thousands)	3,758	3,827	3,888	3,942	4,024	4,110	4,201	4,296	4,396	4,443	4,510

ORIGIN AND USE OF RESOURCES *(Billions of current Hong Kong Dollars)*

	1967	1968	1969	1970	1971	1972	1973	1974	1975	1976	1977
Gross National Product (GNP)	..	..	..	..	..	..	..	..	..	..	..
Net Factor Income from Abroad	..	..	..	..	..	..	..	..	..	..	..
Gross Domestic Product (GDP)	14.72	15.66	18.38	21.88	25.18	30.38	39.10	44.58	46.46	59.34	68.91
Indirect Taxes, net	0.77	0.84	0.95	1.06	1.18	1.50	1.69	1.59	1.92	2.42	2.86
GDP at factor cost	13.95	14.81	17.43	20.82	23.99	28.88	37.41	42.99	44.54	56.91	66.04
Agriculture	0.27	0.30	0.36	0.38	0.41	0.44	0.53	0.54	0.53	0.64	0.72
Industry	5.37	5.80	6.55	7.56	8.55	9.98	12.07	12.51	13.39	17.36	20.11
Manufacturing	3.71	4.48	5.25	6.14	6.70	7.73	9.37	9.44	10.34	13.70	15.65
Services, etc.	8.32	8.71	10.52	12.89	15.04	18.46	24.81	29.94	30.62	38.91	45.21
Resource Balance	0.10	0.16	0.86	0.78	-0.06	1.17	0.96	0.32	0.48	4.05	1.60
Exports of Goods & NFServices	11.78	14.07	17.45	20.33	22.42	25.48	33.29	38.52	38.51	53.16	57.11
Imports of Goods & NFServices	11.68	13.91	16.58	19.55	22.48	24.31	32.33	38.21	38.02	49.11	55.51
Domestic Absorption	14.62	15.50	17.52	21.10	25.24	29.21	38.14	44.26	45.98	55.29	67.31
Private Consumption, etc.	10.32	11.46	12.95	14.78	17.04	19.62	26.18	29.74	31.20	35.36	43.33
General Gov't Consumption	1.20	1.30	1.41	1.64	1.74	2.11	2.59	3.20	3.53	4.03	4.72
Gross Domestic Investment	3.09	2.74	3.16	4.68	6.45	7.48	9.38	11.33	11.26	15.91	19.26
Fixed Investment	2.98	2.62	3.02	4.51	6.26	7.25	9.08	10.65	10.53	12.93	17.56
Memo Items:											
Gross Domestic Saving	3.20	2.90	4.02	5.46	6.39	8.65	10.34	11.64	11.74	19.95	20.86
Gross National Saving	3.20	2.90	4.02	..	..	..	..	..	..	..	..

(Billions of 1980 Hong Kong Dollars)

	1967	1968	1969	1970	1971	1972	1973	1974	1975	1976	1977
Gross National Product	..	..	..	..	..	..	..	..	..	..	..
GDP at factor cost	41.86	43.15	48.36	53.07	57.07	63.13	71.60	73.79	73.55	86.15	96.89
Agriculture	..	..	..	..	..	..	..	..	..	..	..
Industry	..	..	..	..	..	..	..	..	..	..	..
Manufacturing	..	..	..	..	..	..	..	..	..	..	..
Services, etc.	..	..	..	..	..	..	..	..	..	..	..
Resource Balance	2.39	2.21	4.43	4.16	0.47	2.57	2.95	4.88	3.13	6.25	3.84
Exports of Goods & NFServices	33.39	38.03	44.35	49.05	51.13	55.59	61.51	58.96	59.24	75.91	78.81
Imports of Goods & NFServices	31.00	35.82	39.92	44.89	50.65	53.02	58.56	54.08	56.11	69.66	74.97
Domestic Absorption	41.86	43.45	46.61	51.69	59.45	63.95	72.01	71.72	73.63	83.64	97.29
Private Consumption, etc.	25.19	27.76	30.11	33.17	37.50	40.11	45.50	44.62	46.00	49.30	58.09
General Gov't Consumption	3.39	3.56	3.88	4.12	4.20	4.54	4.98	5.42	5.76	6.14	6.76
Gross Domestic Investment	13.28	12.13	12.63	14.39	17.75	19.30	21.53	21.69	21.87	28.21	32.44
Fixed Investment	12.89	11.73	12.18	13.90	17.22	18.72	20.87	20.56	20.97	23.89	30.07
Memo Items:											
Capacity to Import	31.27	36.24	41.99	46.68	50.52	55.57	60.30	54.52	56.83	75.40	77.13
Terms of Trade Adjustment	-2.11	-1.79	-2.36	-2.37	-0.61	-0.02	-1.22	-4.44	-2.41	-0.51	-1.68
Gross Domestic Income	42.13	43.87	48.69	53.47	59.31	66.49	73.74	72.17	74.35	89.38	99.45
Gross National Income	42.13	43.87	48.69	53.47	59.31	66.49	73.74	72.17	74.35	89.38	99.45

DOMESTIC PRICES (DEFLATORS) *(Index 1980 = 100)*

	1967	1968	1969	1970	1971	1972	1973	1974	1975	1976	1977
Overall (GDP)	33.3	34.3	36.0	39.2	42.0	45.7	52.2	58.2	60.5	66.0	68.1
Domestic Absorption	34.9	35.7	37.6	40.8	42.5	45.7	53.0	61.7	62.4	66.1	69.2
Agriculture	..	..	..	..	..	..	..	..	..	..	..
Industry	..	..	..	..	..	..	..	..	..	..	..
Manufacturing	..	..	..	..	..	..	..	..	..	..	..

MANUFACTURING ACTIVITY

	1967	1968	1969	1970	1971	1972	1973	1974	1975	1976	1977
Employment (1980=100)	44.7	51.2	56.7	59.4	61.1	62.6	70.6	65.7	74.8	87.8	84.8
Real Earnings per Empl. (1980=100)	..	..	..	..	..	..	74.3	67.4	70.6	77.0	83.7
Real Output per Empl. (1980=100)	..	..	..	..	..	..	..	..	..	..	..
Earnings as % of Value Added	..	..	..	..	..	..	53.2	53.5	52.5	51.3	53.2

MONETARY HOLDINGS *(Billions of current Hong Kong Dollars)*

	1967	1968	1969	1970	1971	1972	1973	1974	1975	1976	1977
Money Supply, Broadly Defined	..	..	..	..	..	..	..	..	..	..	..
Money as Means of Payment	..	..	..	..	..	..	..	..	..	..	..
Currency Ouside Banks	..	..	..	..	..	..	..	..	..	..	..
Demand Deposits	..	..	..	..	..	..	..	..	..	..	..
Quasi-Monetary Liabilities	..	..	..	..	..	..	..	..	..	..	..

GOVERNMENT DEFICIT (-) OR SURPLUS *(Millions of current Hong Kong Dollars)*

	1967	1968	1969	1970	1971	1972	1973	1974	1975	1976	1977
Current Revenue	..	..	..	..	..	..	..	..	..	..	..
Current Expenditure	..	..	..	..	..	..	..	..	..	..	..
Current Budget Balance	..	..	..	..	..	..	..	..	..	..	..
Capital Receipts	..	..	..	..	..	..	..	..	..	..	..
Capital Payments	..	..	..	..	..	..	..	..	..	..	..

1978	1979	1980	1981	1982	1983	1984	1985	1986	1987 estimate	NOTES	HONG KONG
3,750	4,260	5,210	6,110	6,390	6,230	6,450	6,230	7,030	8,260	f	CURRENT GNP PER CAPITA (US $)
4,597	4,979	5,039	5,097	5,156	5,216	5,276	5,337	5,410	5,479	..	POPULATION (thousands)
				(Billions of current Hong Kong Dollars)							ORIGIN AND USE OF RESOURCES
..	..	..	..	..	..	..	..	..	..	..	Gross National Product (GNP)
..	..	..	..	..	..	..	..	..	..	..	Net Factor Income from Abroad
81.16	107.05	137.08	164.97	186.33	207.56	248.73	261.19	299.83	360.24	f	Gross Domestic Product (GDP)
3.52	4.06	5.20	6.12	6.23	8.10	9.89	12.33	14.75	..	..	Indirect Taxes, net
77.64	102.99	131.88	158.86	180.10	199.46	238.83	248.86	285.08	..	f	GDP at factor cost
0.86	0.99	1.11	1.12	1.23	1.24	1.27	1.24	1.33	..	..	Agriculture
24.31	32.33	41.04	50.45	52.84	61.47	73.74	72.16	83.74	..	..	Industry
18.33	24.72	30.55	36.05	36.39	44.14	55.54	53.07	62.25	..	..	Manufacturing
52.47	69.67	89.74	107.28	126.02	136.75	163.82	175.47	200.01	..	..	Services, etc.
-2.90	-3.12	-6.30	-9.24	-6.77	-4.18	10.45	14.84	13.72	19.34	..	Resource Balance
68.64	95.19	120.40	149.32	157.99	197.62	265.29	281.00	330.18	444.98	..	Exports of Goods & NFServices
71.54	98.31	126.71	158.56	164.76	201.81	254.84	266.16	316.47	425.64	..	Imports of Goods & NFServices
84.06	110.16	143.38	174.22	193.10	211.75	238.28	246.36	286.11	340.90	..	Domestic Absorption
54.16	66.89	85.26	102.45	118.93	138.83	158.60	169.86	193.20	224.05	..	Private Consumption, etc.
5.55	6.83	8.83	12.40	14.91	16.72	18.29	19.92	22.97	25.72	..	General Gov't Consumption
24.35	36.44	49.29	59.36	59.26	56.20	61.38	56.58	69.94	91.13	..	Gross Domestic Investment
22.29	33.19	45.55	55.41	57.86	51.87	55.58	55.12	66.77	86.70	..	Fixed Investment
											Memo Items:
21.46	33.33	42.99	50.12	52.49	52.01	71.83	71.42	83.66	110.47	..	Gross Domestic Saving
..	..	..	..	..	..	..	..	..	..	..	Gross National Saving
				(Billions of 1980 Hong Kong Dollars)							
..	..	..	..	..	..	..	..	..	..	..	Gross National Product
105.82	118.81	131.88	143.30	149.27	158.19	173.02	..	..	..	f	GDP at factor cost
..	..	..	..	..	..	..	..	..	..	..	Agriculture
..	..	..	..	..	..	..	..	..	..	..	Industry
..	..	..	..	..	..	..	..	..	..	..	Manufacturing
..	..	..	..	..	..	..	..	..	..	..	Services, etc.
-1.77	-1.97	-6.30	-5.91	-5.47	-1.52	6.28	4.76	8.24	13.06	..	Resource Balance
88.97	104.97	120.40	137.07	134.99	153.66	183.97	193.82	222.98	290.77	..	Exports of Goods & NFServices
90.74	106.94	126.71	142.99	140.46	155.17	177.69	189.06	214.74	277.71	..	Imports of Goods & NFServices
112.49	125.61	143.38	155.90	159.98	166.07	173.87	175.18	191.79	214.14	..	Domestic Absorption
69.21	75.65	85.26	91.72	96.99	104.66	110.62	115.08	126.16	139.85	..	Private Consumption, etc.
7.50	8.20	8.83	10.79	11.39	11.99	12.32	12.55	13.33	13.95	..	General Gov't Consumption
35.78	41.76	49.29	53.39	51.60	49.41	50.93	47.56	52.30	60.34	..	Gross Domestic Investment
33.06	38.04	45.55	49.59	50.41	46.20	46.93	46.54	49.97	57.55	..	Fixed Investment
											Memo Items:
87.06	103.55	120.40	134.65	134.69	151.96	184.97	199.60	224.04	290.33	..	Capacity to Import
-1.91	-1.42	0.00	-2.42	-0.30	-1.70	1.01	5.78	1.06	-0.44	..	Terms of Trade Adjustment
108.82	122.22	137.08	147.56	154.21	162.85	181.16	185.73	201.10	226.76	..	Gross Domestic Income
108.82	122.22	137.08	147.56	154.21	162.85	181.16	185.73	201.10	226.76	..	Gross National Income
				(Index 1980 = 100)							DOMESTIC PRICES (DEFLATORS)
73.3	86.6	100.0	110.0	120.6	126.1	138.1	145.2	149.9	158.6	..	Overall (GDP)
74.7	87.7	100.0	111.7	120.7	127.5	137.0	140.6	149.2	159.2	..	Domestic Absorption
..	..	..	..	..	..	..	..	..	..	..	Agriculture
..	..	..	..	..	..	..	..	..	..	..	Industry
..	..	..	..	..	..	..	..	..	..	..	Manufacturing
											MANUFACTURING ACTIVITY
93.6	98.2	100.0	100.3	93.7	94.9	99.9	94.0	..	..	G	Employment (1980=100)
94.9	101.8	100.0	100.4	99.3	102.7	105.6	119.4	..	..	G	Real Earnings per Empl. (1980=100)
..	..	..	..	..	..	..	..	..	..	G	Real Output per Empl. (1980=100)
55.6	51.2	52.0	49.8	50.1	46.9	57.0	51.3	..	..	..	Earnings as % of Value Added
				(Billions of current Hong Kong Dollars)							MONETARY HOLDINGS
70.15	79.73	110.36	135.74	..	..	..	..	..	..	D	Money Supply, Broadly Defined
20.11	20.85	24.12	25.06	..	..	..	..	..	..	..	Money as Means of Payment
..	..	..	..	..	..	..	..	..	..	..	Currency Ouside Banks
..	..	..	..	..	..	..	..	..	..	..	Demand Deposits
50.04	58.88	86.24	110.68	..	..	..	..	..	..	..	Quasi-Monetary Liabilities
				(Millions of current Hong Kong Dollars)							GOVERNMENT DEFICIT (-) OR SURPLUS
..	..	..	..	..	..	..	..	..	..	..	Current Revenue
..	..	..	..	..	..	..	..	..	..	..	Current Expenditure
..	..	..	..	..	..	..	..	..	..	..	Current Budget Balance
..	..	..	..	..	..	..	..	..	..	..	Capital Receipts
..	..	..	..	..	..	..	..	..	..	..	Capital Payments

HONG KONG	1967	1968	1969	1970	1971	1972	1973	1974	1975	1976	1977
FOREIGN TRADE (CUSTOMS BASIS)				*(Millions of current US dollars)*							
Value of Exports, fob	1,524	1,744	2,178	2,514	2,871	3,477	5,051	5,907	6,019	8,522	9,624
Nonfuel Primary Products	129	136	159	171	180	225	344	424	389	615	656
Fuels	6	6	7	7	8	8	11	20	19	25	35
Manufactures	1,389	1,602	2,012	2,336	2,683	3,244	4,696	5,463	5,611	7,882	8,933
Value of Imports, cif	1,814	2,058	2,457	2,905	3,387	3,895	5,631	6,710	6,757	8,909	10,457
Nonfuel Primary Products	650	697	747	835	964	1,076	1,581	1,926	2,005	2,398	2,629
Fuels	66	71	79	85	109	120	153	419	429	552	643
Manufactures	1,097	1,290	1,631	1,985	2,314	2,700	3,897	4,364	4,323	5,960	7,186
Terms of Trade	109.5	116.9	112.1	113.6	122.3	133.5	149.1	145.2	126.2	129.8	120.1
Export Prices, fob	28.8	31.0	31.2	33.4	38.5	46.6	66.1	82.9	75.4	80.9	81.3
Nonfuel Primary Products	33.3	34.2	36.1	37.1	37.2	42.6	58.8	62.9	61.5	68.4	73.2
Fuels	4.3	4.3	4.3	4.3	5.6	6.2	8.9	36.7	35.7	38.4	42.0
Manufactures	29.2	31.5	31.5	33.8	39.2	47.7	67.7	85.4	76.9	82.3	82.3
Import Prices, cif	26.3	26.5	27.8	29.4	31.5	34.9	44.3	57.1	59.8	62.3	67.7
Trade at Constant 1980 Prices				*(Millions of 1980 US dollars)*							
Exports, fob	5,288	5,622	6,978	7,527	7,463	7,459	7,647	7,126	7,979	10,540	11,832
Imports, cif	6,888	7,753	8,824	9,876	10,763	11,154	12,711	11,757	11,308	14,304	15,436
BALANCE OF PAYMENTS				*(Millions of current US dollars)*							
Exports of Goods & Services	..	..	..	3,474	3,888	4,708	6,783	8,201	8,509	11,651	13,241
Merchandise, fob	..	..	..	2,514	2,872	3,442	5,041	5,969	6,027	8,512	9,624
Nonfactor Services	..	..	..	840	879	1,078	1,413	1,687	1,753	2,377	2,637
Factor Services	..	..	..	120	137	188	329	545	729	762	980
Imports of Goods & Services	..	..	..	3,249	3,785	4,354	6,394	7,870	7,940	10,412	12,396
Merchandise, fob	..	..	..	2,910	3,394	3,866	5,633	6,785	6,774	8,915	10,476
Nonfactor Services	..	..	..	316	366	447	635	807	907	1,145	1,443
Factor Services	..	..	..	23	25	41	126	278	259	352	477
Long-Term Interest	..	..	..	..	..	..	..	..	..	..	..
Current Transfers, net	..	..	..	..	..	..	..	..	..	..	..
Workers' Remittances	..	..	..	..	..	..	..	..	..	..	..
Total to be Financed	..	..	..	225	103	354	389	331	569	1,239	845
Official Capital Grants	..	..	..	..	..	..	..	..	..	..	..
Current Account Balance	..	..	..	225	103	354	389	331	569	1,239	845
Long-Term Capital, net	..	..	..	-1	-1	-1	0	16	4	41	88
Direct Investment	..	..	..	..	..	..	..	..	..	..	..
Long-Term Loans	..	..	..	..	..	..	..	..	..	..	..
Disbursements	..	..	..	..	..	..	..	..	..	..	..
Repayments	..	..	..	..	..	..	..	..	..	..	..
Other Long-Term Capital	..	..	..	-1	-1	-1	0	16	4	41	88
Other Capital, net	..	..	..	..	..	..	..	..	..	..	..
Change in Reserves	..	..	..	..	..	..	..	..	..	..	..
Memo Item:				*(Hong Kong Dollars per US dollar)*							
Conversion Factor (Annual Avg)	5.740	6.060	6.060	6.060	5.980	5.640	5.150	5.030	4.940	4.900	4.660
EXTERNAL DEBT, ETC.				*(Millions of US dollars, outstanding at end of year)*							
Public/Publicly Guar. Long-Term	..	..	..	..	..	..	..	..	..	..	..
Official Creditors	..	..	..	..	..	..	..	..	..	..	..
IBRD and IDA	..	..	..	..	..	..	..	..	..	..	..
Private Creditors	..	..	..	..	..	..	..	..	..	..	..
Private Non-guaranteed Long-Term	..	..	..	..	..	..	..	..	..	..	..
Use of Fund Credit	..	..	..	..	..	..	..	..	..	..	..
Short-Term Debt	..	..	..	..	..	..	..	..	..	..	..
Memo Items:				*(Millions US dollars)*							
Int'l Reserves Excluding Gold	..	..	..	..	..	..	..	..	..	..	..
Gold Holdings (at market price)	..	..	..	..	..	..	..	..	..	..	..
SOCIAL INDICATORS											
Total Fertility Rate	4.0	3.8	3.5	3.3	3.1	2.9	2.8	2.6	2.5	2.4	2.3
Crude Birth Rate	23.5	22.7	21.9	21.1	20.3	19.5	19.4	19.3	18.1	17.6	17.7
Infant Mortality Rate	23.0	21.8	20.6	19.4	18.2	17.0	16.9	16.8	14.9	13.7	13.5
Life Expectancy at Birth	68.4	69.9	69.9	70.0	70.0	70.0	71.0	71.3	71.5	71.8	72.1
Food Production, p.c. ('79-81 = 100)	..	..	..	..	..	..	..	..	..	..	..
Labor Force, Agriculture (%)	5.4	5.1	4.7	4.4	4.2	3.9	3.7	3.4	3.2	3.0	2.8
Labor Force, Female (%)	33.4	33.8	34.3	34.7	34.8	35.0	35.1	35.3	35.4	35.4	35.5
School Enroll. Ratio, primary	..	..	..	117.0	..	..	..	..	119.0	114.0	115.0
School Enroll. Ratio, secondary	..	..	..	36.0	..	..	..	..	49.0	52.0	54.0

1978	1979	1980	1981	1982	1983	1984	1985	1986	1987 estimate	NOTES	HONG KONG
				(Millions of current US dollars)							**FOREIGN TRADE (CUSTOMS BASIS)**
11,499	15,156	19,703	21,737	20,964	21,951	28,317	30,185	35,440	48,475	..	Value of Exports, fob
764	1,013	1,410	1,575	1,587	1,755	2,017	2,281	2,587	3,473	..	Nonfuel Primary Products
42	58	84	96	101	107	131	145	207	222	..	Fuels
10,693	14,085	18,208	20,066	19,277	20,089	26,169	27,758	32,645	44,780	..	Manufactures
13,451	17,137	22,027	24,671	23,461	24,009	28,567	29,580	35,365	48,462	..	Value of Imports, cif
3,015	3,398	4,145	4,303	4,383	4,415	4,731	4,841	5,306	6,757	..	Nonfuel Primary Products
666	979	1,217	1,955	1,884	1,595	1,568	1,384	1,135	1,215	..	Fuels
9,770	12,760	16,665	18,414	17,194	17,999	22,268	23,355	28,925	40,491	..	Manufactures
				(Index 1980 = 100)							
100.9	102.6	100.0	109.5	108.2	99.8	99.6	102.9	108.1	105.8	..	Terms of Trade
77.5	91.8	100.0	109.7	104.5	94.8	93.1	94.5	110.0	121.0	..	Export Prices, fob
78.0	98.3	100.0	90.6	87.4	88.8	86.1	77.7	84.3	90.9	..	Nonfuel Primary Products
42.3	61.0	100.0	112.5	101.6	92.5	90.2	87.5	44.9	56.9	..	Fuels
77.7	91.5	100.0	111.5	106.2	95.4	93.7	96.3	113.8	125.0	..	Manufactures
76.8	89.5	100.0	100.1	96.6	95.0	93.5	91.8	101.8	114.4	..	Import Prices, cif
				(Millions of 1980 US dollars)							Trade at Constant 1980 Prices
14,841	16,512	19,702	19,820	20,066	23,149	30,416	31,926	32,216	40,048	..	Exports, fob
17,520	19,151	22,026	24,639	24,290	25,269	30,551	32,206	34,748	42,358	..	Imports, cif
				(Millions of current US dollars)							**BALANCE OF PAYMENTS**
15,986	20,935	24,190	26,646	26,039	27,199	33,968	36,187	42,116	..	..	Exports of Goods & Services
11,517	15,181	19,731	21,791	20,986	22,119	28,325	30,186	35,449	..	..	Merchandise, fob
3,149	3,850	3,686	4,001	4,184	4,265	4,594	4,800	5,372	..	..	Nonfactor Services
1,320	1,904	773	854	869	815	1,049	1,201	1,295	..	..	Factor Services
16,049	20,736	25,448	28,269	27,126	27,756	32,555	34,318	40,564	..	..	Imports of Goods & Services
13,516	17,262	22,453	24,838	23,685	24,304	28,754	29,861	35,564	..	..	Merchandise, fob
1,769	2,393	2,643	3,027	3,036	3,046	3,354	3,928	4,423	..	..	Nonfactor Services
764	1,081	352	404	405	406	447	529	577	..	..	Factor Services
..	..	..	..	..	..	..	..	..	..	..	Long-Term Interest
..	..	..	..	..	..	..	..	..	..	..	Current Transfers, net
..	..	..	..	..	..	..	..	..	..	..	Workers' Remittances
-63	199	-1,258	-1,623	-1,087	-557	1,413	1,869	1,552	..	K	Total to be Financed
..	..	0	0	0	0	0	0	0	..	K	Official Capital Grants
-63	199	-1,258	-1,623	-1,087	-557	1,413	1,869	1,552	..	..	Current Account Balance
29	103	329	125	208	206	224	270	295	..	..	Long-Term Capital, net
..	..	250	300	200	200	243	257	282	..	..	Direct Investment
..	..	..	..	..	..	..	..	..	..	..	Long-Term Loans
..	..	..	..	..	..	..	..	..	..	..	Disbursements
..	..	..	..	..	..	..	..	..	..	..	Repayments
29	103	79	-175	8	6	-19	13	13	..	..	Other Long-Term Capital
..	..	250	300	200	200	243	257	282	..	..	Other Capital, net
..	..	679	1,198	679	151	-1,880	-2,396	-2,129	..	..	Change in Reserves
				(Hong Kong Dollars per US dollar)							**Memo Item:**
4.680	5.000	4.980	5.590	6.070	7.270	7.820	7.790	7.800	7.800	..	Conversion Factor (Annual Avg)
				(Millions of US dollars, outstanding at end of year)							**EXTERNAL DEBT, ETC.**
..	..	..	..	..	..	..	..	..	..	..	Public/Publicly Guar. Long-Term
..	..	..	..	..	..	..	..	..	..	..	Official Creditors
..	..	..	..	..	..	..	..	..	..	..	IBRD and IDA
..	..	..	..	..	..	..	..	..	..	..	Private Creditors
..	..	..	..	..	..	..	..	..	..	..	Private Non-guaranteed Long-Term
..	..	..	..	..	..	..	..	..	..	..	Use of Fund Credit
..	..	..	..	..	..	..	..	..	..	..	Short-Term Debt
				(Millions US dollars)							**Memo Items:**
..	..	..	..	..	..	..	..	..	..	..	Int'l Reserves Excluding Gold
..	..	..	..	..	..	..	..	..	..	..	Gold Holdings (at market price)
											SOCIAL INDICATORS
2.2	2.1	2.1	2.0	2.0	1.9	1.9	1.9	1.9	1.8	..	Total Fertility Rate
17.6	16.4	16.9	16.8	16.8	16.8	16.7	16.7	16.7	16.7	..	Crude Birth Rate
11.8	12.3	11.2	9.7	9.9	10.2	10.5	10.7	11.0	11.3	..	Infant Mortality Rate
72.8	73.4	74.1	74.8	75.5	75.5	75.5	75.7	75.8	75.9	..	Life Expectancy at Birth
..	..	..	..	..	..	..	..	78.5	90.8	..	Food Production, p.c. ('79-81 = 100)
2.5	2.3	2.1	..	..	..	..	..	..	..	..	Labor Force, Agriculture (%)
35.5	35.5	35.5	35.3	35.1	34.9	34.7	34.5	34.3	34.1	..	Labor Force, Female (%)
113.0	111.0	107.0	106.0	106.0	106.0	105.0	..	..	..	..	School Enroll. Ratio, primary
57.0	60.0	64.0	65.0	67.0	68.0	69.0	..	..	..	..	School Enroll. Ratio, secondary

HUNGARY	1967	1968	1969	1970	1971	1972	1973	1974	1975	1976	1977
CURRENT GNP PER CAPITA (US $)	..	..	..	..	..	..	..	..	..	..	1,310
POPULATION (thousands)	10,224	10,264	10,303	10,337	10,365	10,394	10,436	10,471	10,532	10,589	10,637

ORIGIN AND USE OF RESOURCES — *(Billions of current Hungarian Forint)*

	1967	1968	1969	1970	1971	1972	1973	1974	1975	1976	1977
Gross National Product (GNP)	..	..	..	..	..	..	..	..	475.2	524.4	574.9
Net Factor Income from Abroad	..	..	..	..	..	..	..	..	-7.5	-4.5	-7.1
Gross Domestic Product (GDP)	256.8	281.1	312.4	332.5	360.8	391.0	429.0	448.9	482.7	528.9	582.0
Indirect Taxes, net	..	..	..	..	..	..	..	..	..	..	..
GDP at factor cost	..	..	..	..	..	..	..	..	..	..	..
Agriculture	50.9	56.8	64.7	60.6	69.1	70.7	82.3	84.7	86.4	94.7	105.4
Industry	141.9	127.7	139.4	150.5	163.8	178.6	200.5	224.9	247.8	259.1	285.6
Manufacturing	..	..	..	..	..	..	..	..	..	..	..
Services, etc.	64.0	96.6	108.3	121.5	128.0	141.7	146.2	139.3	148.5	175.1	191.0
Resource Balance	-4.8	-2.2	6.9	-7.7	-22.5	4.8	19.6	-19.9	-36.0	-21.3	-26.3
Exports of Goods & NFServices	..	..	..	100.2	108.4	133.4	163.9	186.8	200.2	205.2	240.6
Imports of Goods & NFServices	..	..	..	107.9	130.9	128.6	144.2	206.7	236.2	226.5	266.9
Domestic Absorption	261.6	283.3	305.5	340.2	383.3	386.2	409.4	468.8	518.7	550.2	608.3
Private Consumption, etc.	..	..	..	194.1	209.6	223.5	241.7	261.4	286.0	307.3	334.0
General Gov't Consumption	..	..	..	34.4	37.8	38.6	40.5	46.9	50.3	53.2	57.8
Gross Domestic Investment	77.6	86.9	95.4	111.7	135.9	124.1	127.2	160.5	182.4	189.7	216.5
Fixed Investment	64.8	73.9	83.1	100.3	113.1	116.8	123.1	139.2	161.0	168.2	197.7
Memo Items:											
Gross Domestic Saving	72.8	84.7	102.3	104.0	113.4	128.9	146.8	140.7	146.4	168.4	190.2
Gross National Saving	..	..	..	..	..	..	..	..	142.2	166.8	184.9

(Billions of 1980 Hungarian Forint)

	1967	1968	1969	1970	1971	1972	1973	1974	1975	1976	1977
Gross National Product	375.4	403.7	423.7	452.8	473.9	503.5	534.3	571.0	604.5	630.2	670.1
GDP at factor cost	..	..	..	..	..	..	..	..	..	..	..
Agriculture	89.0	91.8	103.3	91.6	97.5	100.0	105.5	108.4	110.8	106.5	119.6
Industry	145.4	152.3	158.7	169.8	181.5	193.3	208.6	226.0	241.9	255.9	272.2
Manufacturing	..	..	..	..	..	..	..	..	..	..	..
Services, etc.	151.3	160.8	170.4	191.5	202.2	217.4	231.8	243.7	261.3	273.2	286.6
Resource Balance	..	..	..	-43.2	-66.2	-29.7	-12.0	-40.3	-44.0	-36.1	-32.5
Exports of Goods & NFServices	..	..	..	127.6	136.8	163.0	186.8	193.3	201.7	216.9	246.2
Imports of Goods & NFServices	..	..	..	170.7	203.0	192.7	198.8	233.6	245.7	253.0	278.7
Domestic Absorption	..	..	..	496.1	547.4	540.4	557.9	618.4	658.0	671.7	710.8
Private Consumption, etc.	..	..	..	302.4	318.7	333.4	349.7	369.0	386.9	394.3	413.3
General Gov't Consumption	..	..	..	48.4	51.4	51.7	52.5	57.9	60.0	62.1	64.3
Gross Domestic Investment	117.1	118.7	124.8	145.3	177.3	155.3	155.7	191.5	211.1	215.3	233.2
Fixed Investment	103.6	105.3	113.9	133.2	147.3	145.7	150.4	166.9	186.1	186.1	208.7
Memo Items:											
Capacity to Import	..	..	..	158.6	168.1	199.8	225.9	211.2	208.3	229.2	251.2
Terms of Trade Adjustment	..	..	..	31.0	31.3	36.8	39.0	17.9	6.5	12.3	5.1
Gross Domestic Income	..	..	..	483.9	512.5	547.5	584.9	595.9	620.6	647.9	683.4
Gross National Income	..	..	..	..	500.8	530.0	567.1	605.6	611.0	642.5	675.1

DOMESTIC PRICES (DEFLATORS) — *(Index 1980 = 100)*

	1967	1968	1969	1970	1971	1972	1973	1974	1975	1976	1977
Overall (GDP)	66.6	69.4	72.2	73.4	75.0	76.5	78.6	77.7	78.6	83.2	85.8
Domestic Absorption	..	..	..	68.6	70.0	71.5	73.4	75.8	78.8	81.9	85.6
Agriculture	57.2	61.9	62.6	66.1	70.9	70.7	78.0	78.2	78.0	88.9	88.2
Industry	97.6	83.8	87.8	88.6	90.2	92.4	96.1	99.5	102.4	101.2	104.9
Manufacturing	..	..	..	..	..	..	..	..	..	..	..

MANUFACTURING ACTIVITY

	1967	1968	1969	1970	1971	1972	1973	1974	1975	1976	1977
Employment (1980=100)	100.7	107.7	111.2	112.0	110.1	109.6	111.3	112.6	112.2	104.5	104.3
Real Earnings per Empl. (1980=100)	58.6	60.1	62.0	74.4	76.1	79.2	83.7	88.1	90.3	102.4	104.9
Real Output per Empl. (1980=100)	31.7	38.3	38.1	41.2	44.7	47.7	49.7	53.0	59.1	69.5	72.9
Earnings as % of Value Added	32.8	31.2	28.6	27.9	26.0	25.0	24.9	24.0	23.2	24.6	28.5

MONETARY HOLDINGS — *(Billions of current Hungarian Forint)*

	1967	1968	1969	1970	1971	1972	1973	1974	1975	1976	1977
Money Supply, Broadly Defined	..	..	..	..	134.7	146.0	173.8	194.3	216.9	236.7	262.1
Money as Means of Payment	..	..	..	..	69.2	72.9	89.5	94.3	101.2	106.6	117.0
Currency Ouside Banks	..	..	..	..	25.2	26.2	31.3	34.6	41.3	45.1	50.7
Demand Deposits	..	..	..	..	44.0	46.7	58.2	59.7	59.9	61.6	66.4
Quasi-Monetary Liabilities	..	..	..	..	65.5	73.1	84.3	100.0	115.7	130.1	145.0

GOVERNMENT DEFICIT (-) OR SURPLUS — *(Billions of current Hungarian Forint)*

	1967	1968	1969	1970	1971	1972	1973	1974	1975	1976	1977
Current Revenue	..	..	..	..	..	..	..	..	..	..	..
Current Expenditure	..	..	..	..	..	..	..	..	..	..	..
Current Budget Balance	..	..	..	..	..	..	..	..	..	..	..
Capital Receipts	..	..	..	..	..	..	..	..	..	..	..
Capital Payments	..	..	..	..	..	..	..	..	..	..	..

1978	1979	1980	1981	1982	1983	1984	1985	1986	1987 estimate	NOTES	HUNGARY
1,500	1,680	1,930	2,140	2,270	2,160	2,060	1,940	2,020	2,240	..	CURRENT GNP PER CAPITA (US $)
10,673	10,698	10,711	10,712	10,706	10,690	10,667	10,649	10,628	10,617	..	POPULATION (thousands)
					(Billions of current Hungarian Forint)						ORIGIN AND USE OF RESOURCES
619.7	668.7	708.4	751.5	824.7	877.1	949.3	995.9	1,047.8	1,178.7	..	Gross National Product (GNP)
-10.0	-13.6	-12.6	-28.4	-23.2	-19.3	-29.2	-37.8	-41.0	-45.2	..	Net Factor Income from Abroad
629.7	682.3	721.0	779.9	847.9	896.4	978.5	1,033.7	1,088.8	1,223.9	..	Gross Domestic Product (GDP)
..	..	..	..	..	..	..	..	..	..	..	Indirect Taxes, net
..	..	..	..	..	..	..	..	..	..	B	GDP at factor cost
107.8	108.8	123.5	136.8	148.5	152.9	166.1	166.6	182.6	188.9	..	Agriculture
308.5	332.4	296.8	323.8	350.1	369.3	400.8	425.9	440.0	493.4	..	Industry
..	..	..	..	..	..	..	..	..	..	..	Manufacturing
213.4	241.1	300.7	319.3	349.3	374.2	411.6	441.2	466.2	541.6	..	Services, etc.
-57.7	-22.3	-15.6	-8.2	6.8	17.1	30.9	21.4	-15.4	-5.9	..	Resource Balance
243.9	283.3	281.8	308.2	321.8	360.7	402.0	436.2	431.6	464.4	..	Exports of Goods & NFServices
301.6	305.6	297.4	316.4	315.0	343.6	371.1	414.8	447.0	470.3	..	Imports of Goods & NFServices
687.4	704.6	736.6	788.1	841.1	879.3	947.6	1,012.3	1,104.2	1,229.8	..	Domestic Absorption
361.7	401.2	441.2	477.5	515.1	551.2	600.5	649.3	695.5	775.9	..	Private Consumption, etc.
65.8	71.3	74.1	79.1	84.2	90.9	95.3	104.6	116.0	125.8	..	General Gov't Consumption
259.9	232.1	221.3	231.3	241.8	237.2	251.8	258.4	292.7	328.1	..	Gross Domestic Investment
214.4	220.8	207.7	206.7	213.9	220.1	225.4	232.1	261.2	293.3	..	Fixed Investment
											Memo Items:
202.2	209.8	205.7	223.1	248.6	254.3	282.7	279.8	277.3	322.2	..	Gross Domestic Saving
195.9	197.8	194.6	196.4	227.7	237.4	256.6	245.4	239.9	281.9	..	Gross National Saving
					(Billions of 1980 Hungarian Forint)						
698.3	706.4	708.4	714.9	742.0	752.0	765.2	757.8	768.4	795.3	..	Gross National Product
..	..	..	..	..	..	..	..	..	..	B	GDP at factor cost
120.7	119.1	123.5	127.7	142.6	142.7	149.3	143.2	148.3	148.3	..	Agriculture
287.0	301.8	296.8	309.6	321.4	327.7	331.9	323.7	322.4	334.6	..	Industry
..	..	..	..	..	..	..	..	..	..	..	Manufacturing
301.9	299.9	300.7	304.5	298.9	298.2	307.5	319.7	327.8	342.7	..	Services, etc.
-64.2	-20.5	-15.6	-5.3	16.6	34.7	53.3	48.3	31.5	38.9	..	Resource Balance
251.0	280.2	281.8	296.9	307.4	327.9	349.8	367.9	359.7	377.1	..	Exports of Goods & NFServices
315.2	300.7	297.4	302.2	290.8	293.2	296.5	319.7	328.3	338.2	..	Imports of Goods & NFServices
773.8	741.2	736.6	747.1	746.3	733.9	735.5	738.3	766.9	786.6	..	Domestic Absorption
428.3	438.1	441.2	454.4	466.6	470.0	474.8	481.3	490.7	505.8	..	Private Consumption, etc.
70.6	74.1	74.1	76.2	70.8	70.4	71.8	74.8	78.2	79.1	..	General Gov't Consumption
275.0	229.0	221.3	216.5	208.9	193.5	188.9	182.2	198.0	201.7	..	Gross Domestic Investment
218.8	220.4	207.7	198.8	195.6	188.9	181.9	176.4	187.9	198.3	..	Fixed Investment
											Memo Items:
254.9	278.7	281.8	294.3	297.1	307.8	321.2	336.1	316.9	333.9	..	Capacity to Import
3.9	-1.5	0.0	-2.6	-10.3	-20.1	-28.6	-31.8	-42.8	-43.1	..	Terms of Trade Adjustment
713.5	719.3	721.0	739.3	752.5	748.5	760.2	754.8	755.6	782.4	..	Gross Domestic Income
702.2	705.0	708.4	712.3	731.7	731.9	736.6	726.0	725.6	752.1	..	Gross National Income
					(Index 1980 = 100)						DOMESTIC PRICES (DEFLATORS)
88.7	94.7	100.0	105.1	111.1	116.6	124.1	131.4	136.4	148.3	..	Overall (GDP)
88.8	95.1	100.0	105.5	112.7	119.8	128.8	137.1	144.0	156.3	..	Domestic Absorption
89.3	91.4	100.0	107.1	104.2	107.2	111.2	116.3	123.1	127.4	..	Agriculture
107.5	110.1	100.0	104.6	108.9	112.7	120.8	131.6	136.5	147.5	..	Industry
..	..	..	..	..	..	..	..	..	..	..	Manufacturing
											MANUFACTURING ACTIVITY
104.5	102.7	100.0	97.7	95.3	91.7	91.1	92.3	..	..	..	Employment (1980 = 100)
108.1	104.4	100.0	103.8	103.0	100.8	106.1	108.2	..	..	..	Real Earnings per Empl. (1980 = 100)
76.9	80.7	100.0	105.0	108.3	114.3	115.6	110.5	..	..	..	Real Output per Empl. (1980 = 100)
28.9	27.2	33.7	32.9	32.5	32.1	32.6	33.5	..	..	..	Earnings as % of Value Added
					(Billions of current Hungarian Forint)						MONETARY HOLDINGS
292.2	317.6	352.3	380.2	405.6	421.2	443.0	487.8	554.4	590.5	..	Money Supply, Broadly Defined
127.5	140.2	160.7	177.6	196.8	203.1	212.9	243.0	289.1	306.1	..	Money as Means of Payment
55.5	61.9	71.4	79.8	84.9	94.8	105.4	116.7	130.7	153.7	..	Currency Ouside Banks
72.1	78.4	89.3	97.8	111.9	108.3	107.5	126.3	158.4	152.4	..	Demand Deposits
164.7	177.4	191.6	202.6	208.8	218.1	230.1	244.8	265.3	284.4	..	Quasi-Monetary Liabilities
					(Billions of current Hungarian Forint)						
..	..	..	-22.1	-16.1	-6.4	15.7	-10.1	-34.6	..	..	GOVERNMENT DEFICIT (-) OR SURPLUS
..	..	..	416.8	438.7	484.5	535.2	538.7	619.4	..	..	Current Revenue
..	..	..	380.0	399.0	438.6	463.4	491.7	608.5	..	..	Current Expenditure
..	..	..	36.8	39.7	45.9	71.8	47.0	10.9	..	..	Current Budget Balance
..	..	..	1.3	..	1.0	..	0.3	..	..	..	Capital Receipts
..	..	..	60.2	55.8	53.3	56.1	57.4	45.5	..	..	Capital Payments

309

HUNGARY	1967	1968	1969	1970	1971	1972	1973	1974	1975	1976	1977
FOREIGN TRADE (CUSTOMS BASIS)					*(Millions of current US dollars)*						
Value of Exports, fob	1,268.0	1,333.6	1,552.6	1,726.2	1,847.3	2,402.8	3,353.7	3,942.0	4,519.3	4,926.7	5,834.1
Nonfuel Primary Products	..	..	..	..	..	..	..	..	..	..	..
Fuels	..	..	..	..	..	..	..	..	..	..	..
Manufactures	..	..	..	..	..	..	..	..	..	..	..
Value of Imports, cif	1,330.0	1,350.2	1,443.8	1,876.7	2,247.8	2,355.6	3,018.0	4,453.2	5,400.3	5,533.0	6,531.0
Nonfuel Primary Products	..	..	..	..	..	..	..	..	..	..	..
Fuels	..	..	..	..	..	..	..	..	..	..	..
Manufactures	..	..	..	..	..	..	..	..	..	..	..
					(Index 1980 = 100)						
Terms of Trade	..	..	..	..	..	..	..	..	..	..	..
Export Prices, fob	..	..	..	..	..	..	..	..	..	..	..
Nonfuel Primary Products	..	..	..	..	..	..	..	..	..	..	..
Fuels	..	..	..	..	..	..	..	..	..	..	..
Manufactures	..	..	..	..	..	..	..	..	..	..	..
Import Prices, cif	..	..	..	..	..	..	..	..	..	..	..
Trade at Constant 1980 Prices					*(Millions of 1980 US dollars)*						
Exports, fob	..	..	..	..	..	..	..	..	..	..	..
Imports, cif	..	..	..	..	..	..	..	..	..	..	..
BALANCE OF PAYMENTS					*(Millions of current US dollars)*						
Exports of Goods & Services	..	..	..	2,144	2,108	2,707	3,684	4,628	5,333	5,518	6,473
Merchandise, fob	..	..	..	1,819	1,864	2,421	3,327	4,021	4,764	5,042	5,877
Nonfactor Services	..	..	..	307	229	262	313	538	492	386	504
Factor Services	..	..	..	18	15	24	43	69	77	90	92
Imports of Goods & Services	..	..	..	2,204	2,498	2,672	3,501	5,046	6,059	6,208	7,330
Merchandise, fob	..	..	..	1,921	2,206	2,325	2,957	4,421	5,277	5,528	6,496
Nonfactor Services	..	..	..	204	212	253	415	444	534	482	569
Factor Services	..	..	..	79	80	93	129	181	248	198	264
Long-Term Interest	..	..	..	..	..	..	..	..	..	..	..
Current Transfers, net	..	..	..	36	42	60	73	80	75	69	44
Workers' Remittances	..	..	..	0	0	0	0	0	0	0	0
Total to be Financed	..	..	..	-61	-422	35	191	-419	-733	-694	-864
Official Capital Grants	..	..	..	0	0	0	0	0	0	0	0
Current Account Balance	..	..	..	-61	-422	35	191	-419	-733	-694	-864
Long-Term Capital, net	..	..	..	..	217	137	5	143	413	419	637
Direct Investment	..	..	..	..	..	..	..	..	..	..	-6
Long-Term Loans	..	..	..	..	..	..	..	..	..	..	..
Disbursements	..	..	..	..	..	..	..	..	..	..	..
Repayments	..	..	..	..	..	..	..	..	..	..	..
Other Long-Term Capital	..	..	..	..	217	137	5	143	413	419	643
Other Capital, net	..	..	..	61	204	1	94	495	206	210	294
Change in Reserves	..	..	..	0	0	-172	-290	-220	114	66	-67
Memo Item:					*(Hungarian Forint per US dollar)*						
Conversion Factor (Annual Avg)	..	60.000	60.000	60.000	59.820	55.260	48.970	46.750	43.970	41.570	40.960
EXTERNAL DEBT, ETC.					*(Millions of US dollars, outstanding at end of year)*						
Public/Publicly Guar. Long-Term	..	..	..	..	..	..	..	..	..	..	..
Official Creditors	..	..	..	..	..	..	..	..	..	..	..
IBRD and IDA	..	..	..	..	..	..	..	..	..	..	..
Private Creditors	..	..	..	..	..	..	..	..	..	..	..
Private Non-guaranteed Long-Term	..	..	..	..	..	..	..	..	..	..	..
Use of Fund Credit	..	..	..	0	0	0	0	0	0	0	0
Short-Term Debt	..	..	..	..	..	..	..	..	..	..	..
Memo Items:					*(Millions of US dollars)*						
Int'l Reserves Excluding Gold	..	..	..	..	634	793	1,226	1,524	1,584	1,556	1,656
Gold Holdings (at market price)	..	..	..	..	63	108	204	248	133	178	212
SOCIAL INDICATORS											
Total Fertility Rate	2.0	2.1	2.0	2.0	1.9	1.9	2.0	2.3	2.4	2.3	2.2
Crude Birth Rate	14.6	15.1	15.0	14.7	14.5	14.7	15.0	17.8	18.4	17.5	16.7
Infant Mortality Rate	37.0	35.8	35.7	35.9	35.1	33.2	33.8	34.3	32.8	29.8	26.2
Life Expectancy at Birth	69.6	69.6	69.6	69.6	69.7	69.7	69.7	69.8	69.8	69.8	69.7
Food Production, p.c. ('79-81=100)	70.7	75.0	79.7	71.1	80.4	85.9	87.6	90.4	94.6	87.0	95.8
Labor Force, Agriculture (%)	29.1	27.7	26.4	25.1	24.4	23.7	23.1	22.4	21.7	21.0	20.3
Labor Force, Female (%)	38.6	39.0	39.4	39.8	40.3	40.7	41.1	41.6	42.0	42.4	42.8
School Enroll. Ratio, primary	..	..	..	97.0	..	..	..	..	99.0	99.0	98.0
School Enroll. Ratio, secondary	..	..	..	63.0	..	..	..	..	63.0	..	..

1978	1979	1980	1981	1982	1983	1984	1985	1986	1987 estimate	NOTES	HUNGARY
				(Millions of current US dollars)							**FOREIGN TRADE (CUSTOMS BASIS)**
6,408.2	7,929.9	8,671.5	8,706.8	8,773.0	8,701.7	8,562.8	8,538.1	9,645.4	..	..	Value of Exports, fob
..	..	..	..	..	..	..	..	..	..	..	Nonfuel Primary Products
..	..	..	..	..	..	..	..	..	..	..	Fuels
..	..	..	..	..	..	..	..	..	..	..	Manufactures
7,990.2	8,681.8	9,244.6	9,138.6	8,819.3	8,508.9	8,091.4	8,223.7	9,598.7	..	..	Value of Imports, cif
..	..	..	..	..	..	..	..	..	..	..	Nonfuel Primary Products
..	..	..	..	..	..	..	..	..	..	..	Fuels
..	..	..	..	..	..	..	..	..	..	..	Manufactures
				(Index 1980 = 100)							
..	..	..	..	..	..	..	..	..	..	..	Terms of Trade
..	..	..	..	..	..	..	..	..	..	..	Export Prices, fob
..	..	..	..	..	..	..	..	..	..	..	Nonfuel Primary Products
..	..	..	..	..	..	..	..	..	..	..	Fuels
..	..	..	..	..	..	..	..	..	..	..	Manufactures
..	..	..	..	..	..	..	..	..	..	..	Import Prices, cif
				(Millions of 1980 US dollars)							Trade at Constant 1980 Prices
..	..	..	..	..	..	..	..	..	..	..	Exports, fob
..	..	..	..	..	..	..	..	..	..	..	Imports, cif
				(Millions of current US dollars)							**BALANCE OF PAYMENTS**
7,275	8,678	10,176	10,332	10,478	10,288	10,456	10,403	10,963	12,067	..	Exports of Goods & Services
6,565	7,949	8,877	8,894	9,057	8,881	8,836	8,935	9,140	9,826	..	Merchandise, fob
603	625	806	1,192	1,257	1,262	1,433	1,230	1,559	1,995	..	Nonfactor Services
107	103	493	246	164	146	186	238	265	246	..	Factor Services
8,812	9,662	10,801	11,278	10,965	10,311	10,251	10,528	12,333	12,753	..	Imports of Goods & Services
7,752	8,509	9,020	8,855	8,579	8,453	8,024	8,324	9,668	9,659	..	Merchandise, fob
688	669	900	1,058	1,264	1,046	1,279	1,210	1,544	1,885	..	Nonfactor Services
372	484	881	1,365	1,122	812	948	994	1,121	1,208	..	Factor Services
..	..	606	672	808	795	894	907	936	1,130	..	Long-Term Interest
99	42	48	48	63	56	66	68	77	105	..	Current Transfers, net
0	0	0	0	0	0	0	0	0	0	..	Workers' Remittances
-1,637	-985	-631	-943	-482	-19	199	-116	-1,365	-676	K	Total to be Financed
0	0	0	0	0	0	0	0	0	..	K	Official Capital Grants
-1,637	-985	-631	-943	-482	-19	199	-116	-1,365	-676	..	Current Account Balance
1,143	943	778	990	125	83	871	1,728	682	863	..	Long-Term Capital, net
-4	6	0	2	..	..	..	..	..	..	..	Direct Investment
..	..	741	941	225	154	1,016	2,025	1,028	1,070	..	Long-Term Loans
..	..	1,552	1,880	1,203	1,409	2,858	4,214	3,861	3,168	..	Disbursements
..	..	811	939	978	1,255	1,842	2,189	2,833	2,097	..	Repayments
1,147	937	38	48	-100	-71	-145	-297	-346	-207	..	Other Long-Term Capital
626	-124	356	-491	-655	488	-873	-414	540	-1,077	..	Other Capital, net
-132	166	-503	443	1,012	-552	-197	-1,198	143	890	..	Change in Reserves
				(Hungarian Forint per US dollar)							**Memo Item:**
37.910	35.580	32.530	34.310	36.630	42.670	48.040	50.120	45.830	46.970	..	Conversion Factor (Annual Avg)
				(Millions of US dollars, outstanding at end of year)							**EXTERNAL DEBT, ETC.**
..	..	6,409	6,933	6,739	6,541	7,380	10,160	12,636	15,931	..	Public/Publicly Guar. Long-Term
..	..	542	1,027	866	1,015	1,039	1,395	1,524	1,731	..	Official Creditors
..	..	0	0	0	57	169	372	629	977	..	IBRD and IDA
..	..	5,867	5,906	5,874	5,527	6,341	8,765	11,112	14,200	..	Private Creditors
..	..	..	..	..	..	..	..	..	..	..	Private Non-guaranteed Long-Term
0	0	0	0	237	573	953	971	1,031	808	..	Use of Fund Credit
..	..	3,905	3,092	2,013	2,503	1,763	1,881	2,620	2,217	..	Short-Term Debt
				(Millions of US dollars)							**Memo Items:**
1,887	1,828	2,090	1,652	1,154	1,564	2,109	3,119	3,062	2,272	..	Int'l Reserves Excluding Gold
447	910	1,220	670	295	585	636	761	917	794	..	Gold Holdings (at market price)
											SOCIAL INDICATORS
2.1	2.0	1.9	1.9	1.8	1.8	1.8	1.8	1.8	1.8	..	Total Fertility Rate
15.8	15.0	13.9	13.3	12.5	11.9	11.7	12.2	12.1	11.5	..	Crude Birth Rate
24.4	23.7	23.1	20.6	19.7	19.0	20.2	20.4	19.0	26.7	..	Infant Mortality Rate
69.6	69.6	69.5	69.6	69.9	70.1	70.3	70.5	70.7	70.9	..	Life Expectancy at Birth
96.8	96.1	102.6	101.4	111.8	109.1	115.8	108.3	110.3	110.9	..	Food Production, p.c. ('79-81 = 100)
19.6	18.9	18.2	..	..	..	..	..	..	..	..	Labor Force, Agriculture (%)
43.1	43.5	43.9	44.0	44.1	44.2	44.3	44.4	44.5	44.6	..	Labor Force, Female (%)
97.0	96.0	97.0	99.0	99.0	99.0	99.0	98.0	..	..	..	School Enroll. Ratio, primary
..	70.0	70.0	72.0	73.0	73.0	73.0	72.0	..	..	..	School Enroll. Ratio, secondary

ICELAND	1967	1968	1969	1970	1971	1972	1973	1974	1975	1976	1977
CURRENT GNP PER CAPITA (US $)	2,920	2,630	2,370	2,420	2,820	3,380	4,190	5,520	6,370	7,090	8,230
POPULATION (thousands)	199	201	203	204	206	209	212	215	218	220	222
ORIGIN AND USE OF RESOURCES	*(Billions of current Icelandic Kronur)*										
Gross National Product (GNP)	0.26	0.28	0.34	0.43	0.55	0.69	0.96	1.40	1.98	2.80	4.06
Net Factor Income from Abroad	0.00	0.00	-0.01	0.00	-0.01	-0.01	-0.01	-0.02	-0.05	-0.07	-0.09
Gross Domestic Product (GDP)	0.26	0.28	0.35	0.44	0.56	0.70	0.97	1.42	2.03	2.87	4.14
Indirect Taxes, net	0.04	0.05	0.06	0.08	0.10	0.13	0.19	0.30	0.42	0.60	0.84
GDP at factor cost	0.22	0.23	0.29	0.36	0.46	0.56	0.78	1.12	1.60	2.27	3.30
Agriculture	..	..	..	..	..	..	0.10	0.15	0.19	0.26	0.43
Industry	..	..	..	..	..	..	0.30	0.40	0.58	0.83	1.17
Manufacturing	..	..	..	..	..	..	0.19	0.23	0.34	0.50	0.72
Services, etc.	..	..	..	..	..	..	0.38	0.57	0.84	1.19	1.71
Resource Balance	-0.02	-0.02	0.01	0.01	-0.03	-0.01	-0.01	-0.14	-0.16	0.03	-0.01
Exports of Goods & NFServices	0.08	0.09	0.16	0.21	0.22	0.26	0.37	0.47	0.72	1.05	1.44
Imports of Goods & NFServices	0.10	0.12	0.15	0.20	0.25	0.27	0.38	0.61	0.88	1.02	1.45
Domestic Absorption	0.28	0.30	0.34	0.43	0.59	0.71	0.99	1.55	2.19	2.84	4.15
Private Consumption, etc.	0.16	0.17	0.20	0.25	0.33	0.42	0.57	0.88	1.19	1.61	2.34
General Gov't Consumption	0.04	0.04	0.05	0.07	0.08	0.11	0.15	0.23	0.37	0.51	0.67
Gross Domestic Investment	0.08	0.09	0.09	0.11	0.18	0.18	0.26	0.44	0.63	0.72	1.14
Fixed Investment	0.07	0.08	0.08	0.10	0.16	0.18	0.28	0.44	0.59	0.75	1.08
Memo Items:											
Gross Domestic Saving	0.06	0.07	0.10	0.12	0.14	0.17	0.25	0.31	0.46	0.75	1.13
Gross National Saving	0.06	0.06	0.10	0.11	0.14	0.17	0.24	0.29	0.42	0.68	..
	(Millions of 1980 Icelandic Kronur)										
Gross National Product	7,928	7,440	7,651	8,301	9,372	9,948	10,415	11,005	11,079	11,635	12,739
GDP at factor cost	6,653	6,303	6,560	6,909	7,788	8,123	8,353	8,763	9,012	9,397	10,272
Agriculture	..	..	..	..	..	..	957	984	1,033	979	1,099
Industry	..	..	..	..	..	..	3,254	3,214	3,221	3,574	3,842
Manufacturing	..	..	..	..	..	..	1,962	1,973	1,967	2,175	2,492
Services, etc.	..	..	..	..	..	..	4,142	4,566	4,757	4,844	5,331
Resource Balance	-502	-384	282	84	-748	-428	-850	-1,458	-736	-150	-592
Exports of Goods & NFServices	2,510	2,357	2,692	3,164	3,040	3,365	3,651	3,621	3,716	4,147	4,572
Imports of Goods & NFServices	3,012	2,741	2,411	3,080	3,787	3,794	4,501	5,078	4,452	4,297	5,164
Domestic Absorption	8,507	7,933	7,505	8,312	10,208	10,509	11,402	12,618	12,083	12,096	13,620
Private Consumption, etc.	4,518	4,384	4,242	4,956	5,749	6,197	6,543	7,123	6,557	6,851	7,669
General Gov't Consumption	1,094	1,155	1,192	1,297	1,396	1,599	1,734	1,881	2,057	2,160	2,207
Gross Domestic Investment	2,894	2,393	2,071	2,059	3,062	2,713	3,124	3,614	3,469	3,085	3,744
Fixed Investment	2,506	2,295	1,748	1,882	2,724	2,698	3,268	3,574	3,262	3,173	3,546
Memo Items:											
Capacity to Import	2,398	2,239	2,577	3,263	3,267	3,650	4,338	3,953	3,622	4,427	5,141
Terms of Trade Adjustment	-112	-118	-116	99	227	284	687	332	-94	280	569
Gross Domestic Income	7,893	7,430	7,670	8,494	9,687	10,365	11,239	11,492	11,253	12,226	13,597
Gross National Income	7,816	7,322	7,535	8,400	9,599	10,232	11,102	11,337	10,985	11,915	13,309
DOMESTIC PRICES (DEFLATORS)	*(Index 1980 = 100)*										
Overall (GDP)	3.3	3.7	4.5	5.2	5.9	6.9	9.2	12.7	17.9	24.0	31.8
Domestic Absorption	3.3	3.8	4.5	5.1	5.8	6.7	8.6	12.3	18.1	23.5	30.5
Agriculture	..	..	..	..	..	..	10.6	15.3	18.3	26.3	39.3
Industry	..	..	..	..	..	..	9.4	12.6	18.0	23.2	30.4
Manufacturing	..	..	..	..	..	..	9.4	11.9	17.3	22.8	29.1
MANUFACTURING ACTIVITY											
Employment (1980=100)	70.2	69.0	70.8	77.9	81.5	83.2	83.7	81.3	82.9	85.7	90.0
Real Earnings per Empl. (1980=100)	..	..	..	..	..	..	..	..	..	..	..
Real Output per Empl. (1980=100)	..	..	..	..	..	..	..	..	..	..	..
Earnings as % of Value Added	..	..	..	..	..	..	..	..	..	..	..
MONETARY HOLDINGS	*(Billions of current Icelandic Kronur)*										
Money Supply, Broadly Defined	0.100	0.110	0.140	0.170	0.210	0.250	0.330	0.420	0.540	0.720	1.040
Money as Means of Payment	0.030	0.030	0.040	0.050	0.060	0.070	0.100	0.120	0.170	0.210	0.310
Currency Ouside Banks	0.010	0.010	0.010	0.010	0.020	0.020	0.030	0.030	0.040	0.060	0.090
Demand Deposits	0.020	0.020	0.030	0.030	0.040	0.050	0.070	0.090	0.120	0.150	0.220
Quasi-Monetary Liabilities	0.080	0.080	0.100	0.130	0.150	0.180	0.230	0.300	0.380	0.510	0.730
	(Millions of current Icelandic Kronur)										
GOVERNMENT DEFICIT (-) OR SURPLUS	..	..	..	..	..	-18	-30	-66	-122	-69	-173
Current Revenue	..	..	..	..	..	194	285	415	579	763	1,068
Current Expenditure	..	..	..	..	..	154	212	344	496	595	862
Current Budget Balance	..	..	..	..	..	39	74	71	83	168	205
Capital Receipts	..	..	..	..	..	..	0	2	9	10	4
Capital Payments	..	..	..	..	..	57	104	140	213	247	382

1978	1979	1980	1981	1982	1983	1984	1985	1986	1987	NOTES	ICELAND
									estimate		
9,930	11,820	13,580	14,870	14,610	12,410	11,760	11,740	13,400	16,660	..	CURRENT GNP PER CAPITA (US $)
224	226	228	231	234	237	239	241	243	245	..	POPULATION (thousands)
			(Billions of current Icelandic Kronur)								ORIGIN AND USE OF RESOURCES
6.32	9.39	15.08	23.59	36.79	63.03	83.57	114.33	152.63	200.28	..	Gross National Product (GNP)
-0.16	-0.24	-0.41	-0.81	-1.49	-3.07	-4.55	-5.58	-6.23	-6.20	..	Net Factor Income from Abroad
6.48	9.63	15.49	24.40	38.28	66.09	88.13	119.91	158.86	206.48	..	Gross Domestic Product (GDP)
1.24	1.80	3.03	5.02	7.81	12.28	17.67	22.85	29.64	..	..	Indirect Taxes, net
5.24	7.83	12.46	19.39	30.47	53.81	70.45	97.06	129.22	..	..	GDP at factor cost
0.70	1.02	1.57	2.30	2.93	5.13	7.21	11.43	..	..	..	Agriculture
1.78	2.81	4.67	6.97	11.68	19.16	25.29	32.67	..	..	..	Industry
1.13	1.91	3.13	4.57	7.40	11.56	14.93	19.55	..	..	..	Manufacturing
2.76	4.00	6.22	10.12	15.86	29.52	37.94	52.95	..	..	..	Services, etc.
0.24	0.18	0.10	-0.21	-1.61	1.80	0.43	0.77	6.75	-1.21	..	Resource Balance
2.48	3.81	5.75	8.73	12.72	27.08	34.30	49.82	63.10	73.39	..	Exports of Goods & NFServices
2.24	3.63	5.65	8.94	14.33	25.27	33.87	49.05	56.35	74.59	..	Imports of Goods & NFServices
6.24	9.45	15.40	24.61	39.89	64.29	87.70	119.14	152.11	207.68	..	Domestic Absorption
3.71	5.55	8.84	14.30	23.25	38.53	51.07	70.71	92.51	126.58	..	Private Consumption, etc.
1.05	1.62	2.55	4.10	6.71	12.44	16.08	22.52	29.46	37.29	..	General Gov't Consumption
1.47	2.28	4.01	6.21	9.94	13.32	20.55	25.91	30.14	43.81	..	Gross Domestic Investment
1.53	2.23	3.93	5.96	9.12	14.17	19.90	26.51	31.86	44.22	..	Fixed Investment
											Memo Items:
1.71	2.46	4.10	5.99	8.33	15.12	20.97	26.68	36.89	42.60	..	Gross Domestic Saving
1.55	2.21	3.68	5.17	6.79	12.04	16.49	21.15	30.89	..	..	Gross National Saving
			(Millions of 1980 Icelandic Kronur)								
13,536	14,288	15,082	15,620	15,889	15,111	15,546	16,165	17,306	18,608	..	Gross National Product
11,128	11,894	12,463	12,820	13,156	12,860	13,062	13,685	14,569	..	..	GDP at factor cost
1,258	1,307	1,570	1,541	1,435	1,347	1,557	1,740	..	..	..	Agriculture
3,878	4,040	4,674	4,759	4,795	4,527	4,835	4,802	..	..	..	Industry
2,557	2,730	3,126	3,184	3,118	2,845	3,020	3,056	..	..	..	Manufacturing
5,992	6,547	6,219	6,519	6,927	6,985	6,670	7,144	..	..	..	Services, etc.
-87	111	98	-231	-727	154	-198	-144	251	-1,023	..	Resource Balance
5,265	5,595	5,746	5,825	5,261	5,802	5,976	6,631	7,043	7,325	..	Exports of Goods & NFServices
5,352	5,484	5,648	6,056	5,988	5,648	6,174	6,775	6,792	8,347	..	Imports of Goods & NFServices
13,984	14,553	15,395	16,391	17,263	15,706	16,606	17,111	17,783	20,235	..	Domestic Absorption
8,317	8,560	8,842	9,523	10,059	9,413	9,671	10,155	10,815	12,332	..	Private Consumption, etc.
2,364	2,493	2,546	2,733	2,902	3,039	3,045	3,234	3,444	3,633	..	General Gov't Consumption
3,303	3,500	4,007	4,135	4,302	3,254	3,890	3,722	3,524	4,269	..	Gross Domestic Investment
3,433	3,440	3,927	3,965	3,945	3,461	3,768	3,807	3,725	4,308	..	Fixed Investment
											Memo Items:
5,918	5,756	5,746	5,912	5,314	6,051	6,252	6,881	7,606	8,212	..	Capacity to Import
653	161	0	87	53	249	276	250	563	888	..	Terms of Trade Adjustment
14,550	14,825	15,493	16,247	16,589	16,109	16,684	17,217	18,597	20,100	..	Gross Domestic Income
14,190	14,449	15,082	15,707	15,942	15,360	15,821	16,415	17,869	19,496	..	Gross National Income
			(Index 1980 = 100)								DOMESTIC PRICES (DEFLATORS)
46.6	65.7	100.0	151.0	231.5	416.7	537.1	706.7	880.9	1074.7	..	Overall (GDP)
44.6	64.9	100.0	150.2	231.1	409.3	528.1	696.3	855.4	1026.4	..	Domestic Absorption
55.9	78.3	100.0	149.3	204.1	381.4	463.4	657.2	..	..	..	Agriculture
46.0	69.6	100.0	146.4	243.5	423.8	523.1	680.5	..	..	..	Industry
44.4	69.8	100.0	143.7	237.5	406.3	494.2	639.7	..	..	..	Manufacturing
											MANUFACTURING ACTIVITY
91.6	96.0	100.0	102.0	101.9	103.0	105.2	107.4	..	..	G	Employment (1980 = 100)
..	..	100.0	103.6	107.6	88.6	84.5	..	..	..	G	Real Earnings per Empl. (1980 = 100)
..	..	..	..	..	..	..	..	..	..	G	Real Output per Empl. (1980 = 100)
..	..	52.9	57.3	54.2	54.7	52.8	53.9	..	..	..	Earnings as % of Value Added
			(Billions of current Icelandic Kronur)								MONETARY HOLDINGS
1.540	2.420	4.000	6.870	10.860	19.450	26.030	38.450	52.000	70.380	..	Money Supply, Broadly Defined
0.430	0.630	1.010	1.630	2.080	3.700	7.670	15.200	22.590	34.370	..	Money as Means of Payment
0.120	0.160	0.220	0.410	0.530	0.770	0.970	1.250	1.730	2.240	..	Currency Ouside Banks
0.310	0.470	0.790	1.220	1.550	2.930	6.700	13.950	20.860	32.130	..	Demand Deposits
1.110	1.790	2.990	5.240	8.780	15.750	18.360	23.250	29.410	36.000	..	Quasi-Monetary Liabilities
			(Millions of current Icelandic Kronur)								
-156	-193	-193	-172	-958	-1,998	-1,610	-4,719	..	..	E	GOVERNMENT DEFICIT (-) OR SURPLUS
1,717	2,626	4,089	6,623	10,762	17,492	23,630	31,231	..	..	..	Current Revenue
1,435	2,243	3,388	5,342	9,017	16,505	19,665	28,644	..	..	..	Current Expenditure
282	383	700	1,281	1,745	987	3,965	2,587	..	..	..	Current Budget Balance
3	4	2	1	10	21	4	25	..	..	..	Capital Receipts
441	580	895	1,454	2,713	3,006	5,579	7,331	..	..	..	Capital Payments

ICELAND	1967	1968	1969	1970	1971	1972	1973	1974	1975	1976	1977
FOREIGN TRADE (CUSTOMS BASIS)					*(Millions of current US dollars)*						
Value of Exports, fob	97.0	85.4	107.7	146.9	150.1	191.4	290.7	329.3	307.5	404.0	512.6
Nonfuel Primary Products	93.5	82.8	102.1	141.7	143.2	181.6	277.2	310.4	290.8	379.8	479.7
Fuels	0.0	0.0	0.0	0.0	0.0	0.0	0.0	0.0	0.0	0.0	0.0
Manufactures	3.5	2.6	5.6	5.2	6.9	9.8	13.5	19.0	16.7	24.3	32.9
Value of Imports, cif	163.5	137.6	123.2	157.1	220.7	233.4	327.2	524.5	488.3	470.0	606.9
Nonfuel Primary Products	27.0	24.4	29.1	34.0	39.5	42.6	48.4	76.4	73.6	77.5	88.0
Fuels	13.0	15.3	12.2	14.5	17.9	17.9	25.9	63.6	62.0	57.7	77.9
Manufactures	123.5	97.8	81.9	108.7	163.3	172.9	252.9	384.6	352.7	334.8	441.1
					(Index 1980 = 100)						
Terms of Trade	108.5	137.1	139.7	150.9	134.2	126.0	136.1	103.6	79.7	92.9	103.7
Export Prices, fob	22.4	25.0	27.6	31.6	33.7	35.2	47.6	55.6	45.6	54.9	67.1
Nonfuel Primary Products	22.4	25.1	27.5	31.6	33.8	35.4	48.2	56.2	45.6	55.2	67.7
Fuels	8.5	7.3	5.4	5.9	7.0	8.5	24.5	36.6	33.6	38.5	36.8
Manufactures	21.9	22.8	30.1	33.3	32.4	32.0	37.4	46.6	45.7	51.6	59.8
Import Prices, cif	20.7	18.2	19.8	21.0	25.1	28.0	34.9	53.7	57.2	59.1	64.7
Trade at Constant 1980 Prices					*(Millions of 1980 US dollars)*						
Exports, fob	432.7	341.6	390.4	464.4	445.0	543.2	611.3	592.5	674.4	735.6	763.6
Imports, cif	790.8	754.9	623.8	749.6	878.4	834.2	936.4	977.6	853.5	795.0	937.7
BALANCE OF PAYMENTS					*(Millions of current US dollars)*						
Exports of Goods & Services	..	..	..	234.8	251.7	294.9	414.9	476.0	462.7	577.7	729.5
Merchandise, fob	..	..	..	146.2	150.8	191.7	291.0	329.4	306.9	402.0	512.6
Nonfactor Services	..	..	..	85.1	97.2	99.8	118.9	140.7	153.7	172.5	212.1
Factor Services	..	..	..	3.5	3.7	3.4	5.0	5.9	2.1	3.2	4.8
Imports of Goods & Services	..	..	..	232.3	300.7	320.9	447.3	637.0	605.7	600.8	777.8
Merchandise, fob	..	..	..	143.5	200.6	215.5	326.0	476.7	440.7	427.9	564.7
Nonfactor Services	..	..	..	71.3	85.0	84.0	93.2	125.3	118.5	117.2	149.8
Factor Services	..	..	..	17.5	15.1	21.4	28.0	35.0	46.5	55.7	63.3
Long-Term Interest	..	..	..	..	..	..	..	..	..	..	..
Current Transfers, net	..	..	..	0.0	-0.1	0.8	1.7	1.4	0.5	1.1	..
Workers' Remittances	..	..	..	0.0	0.0	0.0	0.0	0.0	0.0	0.0	0.0
Total to be Financed	..	..	..	2.5	-49.0	-25.3	-30.8	-159.6	-142.5	-22.0	-47.6
Official Capital Grants	..	..	..	-0.4	..	-0.5	14.5	-0.8	-1.2	-0.8	-0.7
Current Account Balance	..	..	..	2.1	-49.0	-25.8	-16.2	-160.4	-143.8	-22.8	-48.3
Long-Term Capital, net	..	..	..	-3.3	53.2	23.6	37.4	108.1	133.9	63.9	111.0
Direct Investment	..	..	..	4.5	19.8	2.8	-1.5	13.2	42.4	4.5	4.2
Long-Term Loans	..	..	..	..	..	..	..	..	..	..	..
Disbursements	..	..	..	..	..	..	..	..	..	..	..
Repayments	..	..	..	..	..	..	..	..	..	..	..
Other Long-Term Capital	..	..	..	-7.4	33.4	21.3	24.4	95.7	92.8	60.2	107.5
Other Capital, net	..	..	..	22.0	10.1	13.7	-5.2	-16.0	-8.8	-37.5	-46.5
Change in Reserves	..	..	..	-20.8	-14.2	-11.4	-16.0	68.3	18.6	-3.6	-16.2
Memo Item:					*(Icelandic Kronur per US dollar)*						
Conversion Factor (Annual Avg)	0.440	0.620	0.880	0.880	0.880	0.880	0.900	1.000	1.540	1.820	1.990
EXTERNAL DEBT, ETC.					*(Millions US dollars, outstanding at end of year)*						
Public/Publicly Guar. Long-Term	..	..	..	..	..	..	..	..	..	..	..
Official Creditors	..	..	..	..	..	..	..	..	..	..	..
IBRD and IDA	..	..	..	..	..	..	..	..	..	..	..
Private Creditors	..	..	..	..	..	..	..	..	..	..	..
Private Non-guaranteed Long-Term	..	..	..	..	..	..	..	..	..	..	..
Use of Fund Credit	..	..	..	..	..	..	..	..	..	..	..
Short-Term Debt	..	..	..	..	..	..	..	..	..	..	..
Memo Items:					*(Millions of US dollars)*						
Int'l Reserves Excluding Gold	34.24	26.74	37.50	52.72	68.85	83.05	98.40	47.25	45.60	79.38	98.34
Gold Holdings (at market price)	1.02	1.22	1.02	1.08	1.27	1.88	3.26	5.60	4.21	4.04	6.43
SOCIAL INDICATORS											
Total Fertility Rate	3.3	3.1	3.0	2.8	2.9	3.1	2.9	2.7	2.6	2.5	2.3
Crude Birth Rate	22.2	21.0	20.8	19.7	20.8	22.3	21.7	19.9	20.1	19.5	18.0
Infant Mortality Rate	12.9	14.0	11.6	13.2	12.9	11.3	9.6	11.7	12.5	7.7	9.5
Life Expectancy at Birth	73.4	73.6	73.8	74.0	74.1	74.3	74.7	75.1	75.5	75.9	76.3
Food Production, p.c. ('79-81 = 100)	101.7	85.8	88.7	86.8	89.5	99.6	95.9	102.5	102.1	96.0	92.5
Labor Force, Agriculture (%)	19.4	18.7	17.9	17.2	16.5	15.8	15.0	14.3	13.6	12.9	12.2
Labor Force, Female (%)	32.1	32.8	33.4	34.1	34.8	35.6	36.4	37.1	37.9	38.7	39.5
School Enroll. Ratio, primary	..	..	..	97.0	..	..	..	..	100.0	99.0	..
School Enroll. Ratio, secondary	..	..	..	80.0	..	..	..	..	80.0	83.0	..

1978	1979	1980	1981	1982	1983	1984	1985	1986	1987 estimate	NOTES	ICELAND
				(Millions of current US dollars)							**FOREIGN TRADE (CUSTOMS BASIS)**
640.9	790.8	931.2	894.6	684.9	748.4	744.2	813.8	1,095.8	1,375.1	..	Value of Exports, fob
608.5	738.9	853.1	825.0	619.1	677.1	651.1	733.0	995.2	1,247.3	..	Nonfuel Primary Products
0.0	0.0	0.0	0.0	0.0	0.0	0.0	0.0	..	..	..	Fuels
32.4	51.9	78.1	69.6	65.8	71.3	93.1	80.8	100.6	127.8	..	Manufactures
673.5	825.0	1,000.1	1,021.0	941.6	815.3	843.6	904.0	1,115.3	1,581.4	..	Value of Imports, cif
97.5	114.4	148.2	144.2	134.2	130.5	128.5	128.0	147.8	179.8	..	Nonfuel Primary Products
79.2	161.8	167.3	168.1	146.3	132.5	132.9	142.1	107.1	117.3	..	Fuels
496.9	548.8	684.6	708.7	661.1	552.3	582.2	633.9	860.4	1,284.3	..	Manufactures
				(Index 1980 = 100)							
100.9	95.0	100.0	98.6	90.0	91.0	84.9	79.2	73.8	74.9	..	Terms of Trade
72.3	78.9	100.0	95.8	82.8	81.3	74.0	68.0	71.4	78.8	..	Export Prices, fob
72.7	79.0	100.0	96.5	83.9	81.9	73.4	67.8	70.5	77.6	..	Nonfuel Primary Products
44.7	93.6	100.0	104.0	97.0	85.4	83.6	81.1	41.3	37.2	..	Fuels
65.3	77.6	100.0	88.0	74.0	76.0	78.0	70.0	81.0	92.0	..	Manufactures
71.7	83.1	100.0	97.1	92.1	89.3	87.1	85.9	96.7	105.2	..	Import Prices, cif
				(Millions of 1980 US dollars)							Trade at Constant 1980 Prices
886.1	1,002.0	931.2	933.7	826.8	921.1	1,006.3	1,196.4	1,535.3	1,745.9	..	Exports, fob
939.5	993.2	1,000.1	1,051.1	1,022.5	912.6	969.0	1,052.5	1,153.4	1,503.8	..	Imports, cif
				(Millions of current US dollars)							**BALANCE OF PAYMENTS**
908.2	1,084.3	1,213.5	1,221.6	1,054.3	1,100.0	1,106.0	1,224.4	1,569.7	..	..	Exports of Goods & Services
640.3	779.6	919.8	896.4	685.5	742.0	743.2	814.0	1,096.8	..	..	Merchandise, fob
262.0	292.8	279.8	302.8	340.6	342.5	347.1	394.3	453.2	..	..	Nonfactor Services
5.9	11.9	13.9	22.4	28.2	15.5	15.7	16.1	19.7	..	..	Factor Services
888.2	1,102.6	1,288.4	1,366.2	1,312.4	1,154.9	1,238.1	1,339.7	1,556.8	..	..	Imports of Goods & Services
618.4	754.1	899.9	925.5	837.6	721.6	756.6	814.3	1,024.0	..	..	Merchandise, fob
192.1	253.5	262.8	282.0	308.5	280.7	310.9	362.9	363.9	..	..	Nonfactor Services
77.7	95.0	125.7	158.7	166.3	152.6	170.6	162.5	168.9	..	..	Factor Services
..	..	..	..	..	..	..	..	..	..	..	Long-Term Interest
-0.3	-1.4	-3.0	-1.6	-3.4	-0.6	2.1	1.3	5.7	..	..	Current Transfers, net
0.0	0.0	0.0	0.0	0.0	0.0	0.0	0.0	0.0	..	..	Workers' Remittances
19.7	-19.7	-77.9	-146.2	-261.5	-55.5	-130.0	-114.0	18.6	..	K	Total to be Financed
-0.8	-1.5	-1.2	-1.7	-1.7	-1.4	-1.3	-1.1	-1.7	..	K	Official Capital Grants
18.9	-21.2	-79.1	-147.9	-263.2	-56.9	-131.3	-115.1	16.9	..	..	Current Account Balance
77.3	93.4	151.8	189.6	212.4	92.1	111.7	154.2	155.0	..	..	Long-Term Capital, net
8.1	3.1	22.3	53.0	35.8	-23.4	13.7	23.6	6.4	..	..	Direct Investment
..	..	..	..	..	..	..	..	..	..	..	Long-Term Loans
..	..	..	..	..	..	..	..	..	..	..	Disbursements
..	..	..	..	..	..	..	..	..	..	..	Repayments
70.0	91.8	130.7	138.3	178.3	116.9	99.3	131.7	150.3	..	..	Other Long-Term Capital
-44.9	-35.4	-38.5	30.2	-44.8	-24.3	4.4	25.1	-60.6	..	..	Other Capital, net
-51.3	-36.8	-34.2	-71.9	95.6	-10.9	15.2	-64.2	-111.3	-14.4	..	Change in Reserves
				(Icelandic Kronur per US dollar)							Memo Item:
2.710	3.530	4.800	7.220	12.350	24.840	31.690	41.510	41.100	38.680	..	Conversion Factor (Annual Avg)
				(Millions US dollars, outstanding at end of year)							**EXTERNAL DEBT, ETC.**
..	..	..	..	..	..	..	..	..	..	..	Public/Publicly Guar. Long-Term
..	..	..	..	..	..	..	..	..	..	..	Official Creditors
..	..	..	..	..	..	..	..	..	..	..	IBRD and IDA
..	..	..	..	..	..	..	..	..	..	..	Private Creditors
..	..	..	..	..	..	..	..	..	..	..	Private Non-guaranteed Long-Term
..	..	..	..	..	..	..	..	..	..	..	Use of Fund Credit
..	..	..	..	..	..	..	..	..	..	..	Short-Term Debt
				(Millions of US dollars)							Memo Items:
135.83	161.94	173.81	229.53	145.17	149.29	127.60	205.54	309.84	311.31	..	Int'l Reserves Excluding Gold
9.94	25.09	28.89	19.48	22.39	18.69	15.11	16.02	19.15	23.72	..	Gold Holdings (at market price)
											SOCIAL INDICATORS
2.4	2.4	2.5	2.4	2.3	2.2	2.1	2.1	2.0	1.9	..	Total Fertility Rate
18.6	19.3	19.9	18.8	18.3	17.7	17.2	16.7	16.2	15.8	..	Crude Birth Rate
11.3	9.5	7.7	7.2	7.5	7.8	8.0	8.3	8.7	9.0	..	Infant Mortality Rate
76.4	76.5	76.6	76.7	76.8	76.8	76.9	76.9	77.0	77.0	..	Life Expectancy at Birth
102.0	97.6	101.9	100.5	96.2	87.0	89.4	91.2	89.8	89.3	..	Food Production, p.c. ('79-81 = 100)
11.6	10.9	10.2	..	..	..	..	..	..	..	..	Labor Force, Agriculture (%)
40.3	41.1	41.9	41.9	42.0	42.1	42.1	42.2	42.3	42.4	..	Labor Force, Female (%)
96.0	96.0	99.0	..	99.0	101.0	100.0	98.0	99.0	..	..	School Enroll. Ratio, primary
83.0	80.0	86.0	..	88.0	90.0	93.0	95.0	..	..	..	School Enroll. Ratio, secondary

INDIA	1967	1968	1969	1970	1971	1972	1973	1974	1975	1976	1977
CURRENT GNP PER CAPITA (US $)	90	90	90	100	100	110	120	140	160	160	160
POPULATION (millions)	511	523	535	548	560	573	586	600	613	628	642
ORIGIN AND USE OF RESOURCES	*(Billions of current Indian Rupees)*										
Gross National Product (GNP)	320.3	330.3	365.8	400.8	431.4	476.6	587.5	695.1	741.8	800.8	897.6
Net Factor Income from Abroad	-2.6	-2.5	-2.7	-1.8	-2.3	-2.6	-2.4	-1.6	-1.6	-1.2	-0.9
Gross Domestic Product (GDP)	322.9	332.8	368.5	402.6	433.7	479.2	589.9	696.7	743.4	802.0	898.5
Indirect Taxes, net	24.3	27.3	30.6	35.3	41.0	46.8	52.2	64.1	79.7	85.3	89.2
GDP at factor cost	298.6	305.5	337.9	367.3	392.7	432.4	537.7	632.6	663.7	716.7	809.3
Agriculture	150.7	147.1	161.3	174.2	180.8	199.4	267.9	290.5	278.5	285.7	327.2
Industry	59.7	64.5	74.4	79.7	87.4	96.1	109.9	138.7	153.9	175.3	197.4
Manufacturing	38.9	41.7	48.6	52.2	57.5	63.8	75.7	98.6	103.8	115.2	128.4
Services, etc.	88.2	93.9	102.2	113.4	124.5	136.9	159.9	203.4	231.3	255.7	284.7
Resource Balance	-6.9	-3.0	-1.2	-3.2	-4.2	-2.1	-3.3	-8.1	-3.0	6.4	4.9
Exports of Goods & NFServices	15.1	16.0	16.3	16.1	17.1	21.5	26.5	36.4	48.9	60.8	66.2
Imports of Goods & NFServices	22.0	19.0	17.5	19.3	21.3	23.6	29.8	44.5	51.9	54.4	61.3
Domestic Absorption	329.8	335.8	369.7	405.8	437.8	481.4	593.2	704.8	746.4	795.6	893.7
Private Consumption, etc.	244.8	249.9	270.7	294.4	309.1	348.6	428.7	498.3	508.7	536.4	620.8
General Gov't Consumption	27.9	30.5	34.2	38.0	44.6	47.5	51.0	61.4	73.5	82.1	86.7
Gross Domestic Investment	57.1	55.4	64.8	73.4	84.1	85.3	113.5	145.1	164.2	177.1	186.2
Fixed Investment	50.8	53.8	59.0	63.0	70.7	80.7	90.3	109.3	132.5	153.0	172.2
Memo Items:											
Gross Domestic Saving	50.2	52.4	63.6	70.2	80.0	83.1	110.2	137.0	161.2	183.5	191.0
Gross National Saving	48.6	51.1	61.7	69.0	78.5	81.2	109.0	137.5	163.7	188.5	199.3
	(Billions of 1980 Indian Rupees)										
Gross National Product	841.6	871.5	927.3	984.3	1,006.7	1,000.0	1,038.6	1,042.6	1,147.2	1,160.2	1,255.4
GDP at factor cost	775.6	796.3	845.6	891.9	905.8	898.2	940.6	949.6	1,039.0	1,048.0	1,137.0
Agriculture	341.4	343.9	364.8	395.9	394.1	370.0	397.3	390.2	439.8	413.2	461.8
Industry	191.4	199.7	216.6	218.7	224.6	232.6	236.4	240.0	253.4	276.6	295.9
Manufacturing	124.9	130.5	144.6	146.5	150.8	156.9	164.5	168.4	172.1	187.2	199.3
Services, etc.	242.8	252.7	264.2	277.3	287.1	295.6	306.9	319.4	345.8	358.2	379.3
Resource Balance	-73.5	-50.5	-44.1	-54.4	-64.5	-61.8	-44.3	-30.8	-11.7	-14.7	-30.4
Exports of Goods & NFServices	41.3	44.8	43.7	42.4	44.8	49.5	49.5	55.0	65.0	80.1	76.9
Imports of Goods & NFServices	114.8	95.3	87.8	96.8	109.3	111.3	93.8	85.8	76.7	94.8	107.3
Domestic Absorption	922.4	928.9	978.5	1,043.3	1,076.5	1,067.2	1,087.0	1,075.7	1,161.3	1,176.6	1,287.1
Private Consumption, etc.	702.1	710.3	734.2	788.7	797.8	799.4	787.1	770.3	835.2	830.9	930.8
General Gov't Consumption	54.3	61.0	70.3	69.2	77.6	78.9	79.1	79.3	88.9	96.6	100.4
Gross Domestic Investment	166.0	157.6	174.0	185.4	201.1	188.9	220.8	226.1	237.2	249.1	255.9
Fixed Investment	147.3	150.8	156.5	157.0	166.5	175.7	175.7	170.7	186.9	211.5	232.4
Memo Items:											
Capacity to Import	78.8	80.3	81.8	80.8	87.7	101.4	83.4	70.2	72.3	106.0	115.9
Terms of Trade Adjustment	37.5	35.5	38.1	38.4	42.9	51.9	33.9	15.2	7.3	25.9	39.0
Gross Domestic Income	886.4	913.9	972.5	1,027.3	1,054.9	1,057.3	1,076.6	1,060.1	1,156.9	1,187.8	1,295.7
Gross National Income	879.1	907.0	965.4	1,022.7	1,049.6	1,051.9	1,072.5	1,057.8	1,154.5	1,186.1	1,294.4
DOMESTIC PRICES (DEFLATORS)	*(Index 1980 = 100)*										
Overall (GDP)	38.0	37.9	39.4	40.7	42.9	47.7	56.6	66.7	64.7	69.0	71.5
Domestic Absorption	35.8	36.2	37.8	38.9	40.7	45.1	54.6	65.5	64.3	67.6	69.4
Agriculture	44.1	42.8	44.2	44.0	45.9	53.9	67.4	74.4	63.3	69.1	70.9
Industry	31.2	32.3	34.3	36.4	38.9	41.3	46.5	57.8	60.7	63.4	66.7
Manufacturing	31.1	32.0	33.6	35.6	38.1	40.7	46.0	58.6	60.3	61.5	64.4
MANUFACTURING ACTIVITY											
Employment (1980=100)	61.7	62.5	64.7	67.9	71.9	72.5	75.0	77.4	81.2	83.6	89.1
Real Earnings per Empl. (1980=100)	79.0	87.7	90.4	93.3	122.3	103.6	90.8	84.8	91.7	96.1	95.4
Real Output per Empl. (1980=100)	81.1	86.7	92.8	95.2	95.4	99.5	95.9	96.1	100.8	108.9	111.3
Earnings as % of Value Added	48.9	50.6	45.7	46.5	60.8	54.2	47.8	43.6	47.2	43.3	49.9
MONETARY HOLDINGS	*(Billions of current Indian Rupees)*										
Money Supply, Broadly Defined	78.9	86.4	97.5	109.0	126.9	145.3	172.6	192.8	218.2	271.5	322.1
Money as Means of Payment	51.0	53.9	60.4	67.6	76.5	86.2	101.0	111.3	122.3	152.8	178.5
Currency Ouside Banks	32.1	33.7	37.6	41.6	45.6	49.1	57.8	61.4	64.4	73.2	84.2
Demand Deposits	18.9	20.2	22.7	26.1	30.8	37.1	43.3	49.9	57.9	79.6	94.3
Quasi-Monetary Liabilities	27.9	32.5	37.2	41.4	50.4	59.2	71.5	81.5	95.8	118.8	143.6
	(Billions of current Indian Rupees)										
GOVERNMENT DEFICIT (-) OR SURPLUS	..	..	..	..	..	..	..	-23.6	-32.0	-36.9	-37.9
Current Revenue	..	..	..	..	..	..	..	75.8	93.6	104.2	115.6
Current Expenditure	..	..	..	..	..	..	..	65.6	80.8	92.6	101.7
Current Budget Balance	..	..	..	..	..	..	..	10.3	12.8	11.6	13.8
Capital Receipts	..	..	..	..	..	..	..	0.5	0.9	1.1	1.2
Capital Payments	..	..	..	..	..	..	..	34.3	45.7	49.6	52.9

1978	1979	1980	1981	1982	1983	1984	1985	1986	1987 estimate	NOTES	INDIA
180	190	230	270	280	280	280	290	290	300	..	**CURRENT GNP PER CAPITA** (US $)
657	672	687	703	718	734	750	765	781	797	..	**POPULATION** (millions)
				(Billions of current Indian Rupees)							**ORIGIN AND USE OF RESOURCES**
977.6	1,078.1	1,363.0	1,597.6	1,773.1	2,067.3	2,285.5	2,607.8	2,912.9	3,212.2	C	Gross National Product (GNP)
0.1	2.7	4.8	3.4	-2.8	-5.4	-10.0	-9.5	-15.0	-14.9	..	Net Factor Income from Abroad
977.5	1,075.4	1,358.1	1,594.2	1,775.9	2,072.7	2,295.4	2,617.3	2,927.9	3,227.3	C	Gross Domestic Product (GDP)
105.3	121.8	135.9	165.4	187.4	208.7	228.1	284.3	322.0	356.4	..	Indirect Taxes, net
872.2	953.6	1,222.3	1,428.8	1,588.5	1,864.1	2,067.3	2,333.0	2,605.8	2,870.8	C	GDP at factor cost
337.0	347.4	464.8	523.7	556.2	678.0	710.9	768.6	824.6	848.3	..	Agriculture
221.0	248.8	316.3	381.0	432.6	502.5	576.2	666.1	760.2	870.5	..	Industry
147.4	169.5	216.4	252.6	280.7	330.5	374.1	435.6	489.9	579.5	..	Manufacturing
314.2	357.4	441.1	524.1	599.7	683.5	780.2	898.4	1,021.0	1,152.0	..	Services, etc.
-12.5	-25.4	-48.9	-52.3	-48.8	-48.0	-54.9	-82.3	-63.8	-63.7	..	Resource Balance
68.8	80.6	88.5	101.7	107.9	126.7	156.4	155.4	182.0	220.4	..	Exports of Goods & NFServices
81.3	106.0	137.4	154.0	156.7	174.7	211.3	237.6	245.8	284.1	..	Imports of Goods & NFServices
989.9	1,100.8	1,407.0	1,646.5	1,824.6	2,120.7	2,350.2	2,699.6	2,991.7	3,290.9	..	Domestic Absorption
663.9	729.1	946.9	1,077.2	1,226.3	1,427.5	1,563.8	1,725.2	1,923.9	2,097.6	..	Private Consumption, etc.
96.2	110.3	130.8	153.5	182.7	211.4	243.5	292.6	349.2	410.8	..	General Gov't Consumption
229.8	261.4	329.3	415.7	415.6	481.8	542.9	681.8	718.6	782.5	..	Gross Domestic Investment
188.8	213.1	262.8	314.5	357.7	398.7	448.5	545.5	633.8	701.4	..	Fixed Investment
											Memo Items:
217.4	236.0	280.4	363.5	366.9	433.8	488.1	599.5	654.7	718.9	..	Gross Domestic Saving
226.9	253.7	307.1	387.5	388.2	454.8	508.1	616.9	667.6	729.7	..	Gross National Saving
				(Billions of 1980 Indian Rupees)							
1,340.3	1,278.9	1,363.0	1,451.2	1,503.6	1,617.4	1,668.0	1,770.4	1,845.2	1,892.6	C	Gross National Product
1,202.2	1,145.3	1,222.3	1,297.8	1,338.3	1,443.9	1,489.6	1,560.8	1,623.3	1,663.8	C I	GDP at factor cost
475.5	414.6	464.8	491.4	483.6	536.0	526.4	527.8	513.9	480.5	..	Agriculture
318.4	311.7	316.3	341.4	357.6	386.9	412.3	444.8	481.5	515.2	..	Industry
220.9	217.0	216.4	233.8	249.1	273.8	292.9	318.7	347.2	376.8	..	Manufacturing
408.3	419.0	441.1	465.0	497.2	520.9	550.9	588.3	627.9	668.1	..	Services, etc.
-62.5	-66.0	-48.9	-53.4	-51.2	-63.2	-52.9	-73.4	-64.0	-60.7	..	Resource Balance
77.4	81.4	88.5	89.9	90.2	96.0	105.4	101.5	114.5	125.9	..	Exports of Goods & NFServices
139.9	147.4	137.4	143.3	141.4	159.2	158.3	174.8	178.5	186.6	..	Imports of Goods & NFServices
1,402.7	1,341.8	1,407.0	1,501.6	1,557.2	1,685.0	1,728.2	1,850.3	1,918.9	1,961.9	..	Domestic Absorption
998.0	933.2	946.9	994.1	1,062.5	1,171.7	1,187.6	1,233.7	1,290.7	1,300.9	..	Private Consumption, etc.
110.4	120.6	130.8	136.6	150.7	157.3	169.8	190.0	210.9	236.2	..	General Gov't Consumption
294.3	288.0	329.3	370.8	343.9	356.0	370.8	426.6	417.3	424.8	..	Gross Domestic Investment
234.7	228.5	262.8	280.8	293.0	292.4	301.9	332.1	362.5	371.5	..	Fixed Investment
											Memo Items:
118.4	112.1	88.5	94.7	97.4	115.5	117.2	114.3	132.2	144.8	..	Capacity to Import
41.0	30.7	0.0	4.8	7.1	19.5	11.8	12.8	17.7	18.9	..	Terms of Trade Adjustment
1,381.2	1,306.5	1,358.1	1,452.9	1,513.1	1,641.2	1,687.1	1,789.8	1,872.5	1,920.1	..	Gross Domestic Income
1,381.3	1,309.6	1,363.0	1,456.0	1,510.7	1,636.9	1,679.8	1,783.3	1,862.9	1,911.5	..	Gross National Income
				(Index 1980 = 100)							**DOMESTIC PRICES (DEFLATORS)**
72.9	84.3	100.0	110.1	117.9	127.8	137.0	147.3	157.9	169.8	..	Overall (GDP)
70.6	82.0	100.0	109.7	117.2	125.9	136.0	145.9	155.9	167.7	..	Domestic Absorption
70.9	83.8	100.0	106.6	115.0	126.5	135.0	145.6	160.5	176.5	..	Agriculture
69.4	79.8	100.0	111.6	121.0	129.9	139.8	149.7	157.9	168.9	..	Industry
66.7	78.1	100.0	108.0	112.7	120.7	127.7	136.7	141.1	153.8	..	Manufacturing
											MANUFACTURING ACTIVITY
91.1	97.0	100.0	99.6	102.3	99.4	101.5	103.7	..	..	G	Employment (1980=100)
100.5	102.5	100.0	98.7	105.0	112.7	116.4	122.0	..	..	G	Real Earnings per Empl. (1980=100)
119.1	113.6	100.0	111.1	121.2	124.6	137.7	145.1	..	..	G	Real Output per Empl. (1980=100)
48.6	49.5	50.7	47.9	48.6	48.7	48.4	48.4	..	..	..	Earnings as % of Value Added
				(Billions of current Indian Rupees)							**MONETARY HOLDINGS**
388.3	455.9	527.7	618.0	721.0	840.4	988.4	1,152.5	1,356.0	1,575.4	D	Money Supply, Broadly Defined
157.6	176.9	204.6	232.5	273.7	308.5	365.6	412.4	478.7	543.2	..	Money as Means of Payment
94.6	108.0	126.3	137.4	157.4	181.3	218.1	239.4	268.0	315.6	..	Currency Ouside Banks
63.1	68.9	78.3	95.0	116.3	127.2	147.5	173.0	210.7	227.6	..	Demand Deposits
230.7	279.0	323.1	385.5	447.2	531.9	622.8	740.1	877.3	1,032.2	..	Quasi-Monetary Liabilities
				(Billions of current Indian Rupees)							
-50.8	-63.0	-88.6	-87.3	-107.3	-133.3	-175.8	-231.5	-234.7	..	C E	**GOVERNMENT DEFICIT** (-) OR SURPLUS
130.9	146.2	163.6	196.3	225.9	254.9	297.0	358.5	391.3	..	..	Current Revenue
116.4	140.0	159.1	181.3	213.3	248.5	301.9	372.5	408.0	..	..	Current Expenditure
14.5	6.3	4.5	15.0	12.7	6.4	-4.9	-14.0	-16.7	..	..	Current Budget Balance
1.4	1.6	1.9	3.4	3.1	3.6	3.9	4.9	4.9	..	..	Capital Receipts
66.7	70.8	95.0	105.7	123.1	143.3	174.8	222.4	222.9	..	..	Capital Payments

INDIA	1967	1968	1969	1970	1971	1972	1973	1974	1975	1976	1977
FOREIGN TRADE (CUSTOMS BASIS)				*(Millions of current US dollars)*							
Value of Exports, fob	1,613	1,754	1,834	2,026	2,043	2,422	2,968	3,906	4,365	5,526	6,355
Nonfuel Primary Products	809	846	812	949	932	1,073	1,353	1,839	2,354	2,486	2,590
Fuels	13	16	14	17	11	16	40	26	39	35	34
Manufactures	791	892	1,008	1,061	1,100	1,333	1,576	2,041	1,972	3,004	3,731
Value of Imports, cif	2,722	2,507	2,118	2,094	2,406	2,230	3,146	5,167	6,385	5,710	6,601
Nonfuel Primary Products	1,291	978	825	838	773	593	1,009	1,475	2,120	1,933	1,998
Fuels	101	108	86	162	245	266	436	1,446	1,419	1,390	1,640
Manufactures	1,330	1,421	1,206	1,095	1,387	1,371	1,701	2,247	2,846	2,387	2,962
				(Index 1980 = 100)							
Terms of Trade	111.3	115.2	107.9	158.9	155.7	157.8	149.8	112.2	114.4	113.5	115.3
Export Prices, fob	29.3	27.9	28.0	36.3	36.1	38.1	46.3	61.3	65.5	63.9	70.1
Nonfuel Primary Products	39.1	38.2	38.3	42.5	41.0	42.8	58.4	68.0	70.1	76.1	90.1
Fuels	4.3	4.3	4.3	4.3	5.6	6.2	8.9	36.7	35.7	38.4	42.0
Manufactures	25.3	24.1	24.7	35.8	34.6	37.0	43.2	56.8	61.7	56.8	61.1
Import Prices, cif	26.3	24.2	26.0	22.8	23.2	24.1	30.9	54.6	57.3	56.3	60.8
Trade at Constant 1980 Prices				*(Millions of 1980 US dollars)*							
Exports, fob	5,508	6,291	6,539	5,589	5,652	6,365	6,410	6,370	6,662	8,647	9,061
Imports, cif	10,338	10,358	8,146	9,175	10,362	9,248	10,179	9,456	11,149	10,145	10,851
BALANCE OF PAYMENTS				*(Millions of current US dollars)*							
Exports of Goods & Services	..	..	..	2,142	2,381	2,826	3,455	4,683	5,787	7,008	8,040
Merchandise, fob	..	..	..	1,870	2,089	2,460	3,017	3,987	4,828	5,742	6,345
Nonfactor Services	..	..	..	272	292	328	384	578	825	1,057	1,384
Factor Services	..	..	..	0	0	38	54	118	134	209	311
Imports of Goods & Services	..	..	..	2,817	3,241	3,378	4,139	5,892	6,322	6,425	7,582
Merchandise, fob	..	..	..	2,294	2,678	2,785	3,503	5,212	5,483	5,389	6,471
Nonfactor Services	..	..	..	279	283	271	323	364	514	691	693
Factor Services	..	..	..	244	280	322	313	316	325	345	418
Long-Term Interest	..	..	..	193	219	241	255	268	264	279	303
Current Transfers, net	..	..	..	83	112	97	159	258	470	692	1,077
Workers' Remittances	..	..	..	65	88	108	175	129	489	677	1,079
Total to be Financed	..	..	..	-592	-748	-455	-525	-951	-65	1,275	1,535
Official Capital Grants	..	..	..	206	400	109	91	118	342	377	432
Current Account Balance	..	..	..	-386	-348	-346	-434	-833	277	1,652	1,967
Long-Term Capital, net	..	..	..	789	1,013	582	654	1,079	1,694	1,416	1,190
Direct Investment	..	..	..	0	0	0	0	0	0	0	0
Long-Term Loans	..	..	..	594	622	494	547	866	1,361	1,048	817
Disbursements	..	..	..	908	959	865	959	1,393	1,901	1,585	1,387
Repayments	..	..	..	314	336	371	412	527	540	537	570
Other Long-Term Capital	..	..	..	-11	-9	-21	16	95	-9	-9	-59
Other Capital, net	..	..	..	-387	-626	-191	-345	-817	-1,397	-1,165	-779
Change in Reserves	..	..	..	-16	-39	-45	125	571	-574	-1,903	-2,378
Memo Item:				*(Indian Rupees per US dollar)*							
Conversion Factor (Annual Avg)	7.500	7.500	7.500	7.500	7.440	7.710	7.790	7.980	8.650	8.940	8.560
EXTERNAL DEBT, ETC.				*(Millions of US dollars, outstanding at end of year)*							
Public/Publicly Guar. Long-Term	..	..	..	7,837	8,792	9,705	10,367	11,351	12,171	13,232	14,647
Official Creditors	..	..	..	7,508	8,450	9,385	10,100	11,068	11,871	12,940	14,258
IBRD and IDA	..	..	..	1,562	1,650	2,086	2,392	2,777	3,245	3,789	4,196
Private Creditors	..	..	..	330	342	321	267	283	300	293	389
Private Non-guaranteed Long-Term	..	..	..	100	100	100	139	267	277	295	295
Use of Fund Credit	..	..	..	0	0	0	75	620	808	471	155
Short-Term Debt	..	..	..	..	..	..	..	..	..	..	432
Memo Items:				*(Millions of US dollars)*							
Int'l Reserves Excluding Gold	419.0	439.0	683.0	763.3	942.4	916.3	848.8	1,027.7	1,089.1	2,791.7	4,872.0
Gold Holdings (at market price)	244.8	291.4	244.8	259.9	303.4	451.3	780.5	1,296.9	975.3	937.1	1,213.4
SOCIAL INDICATORS											
Total Fertility Rate	6.0	5.9	5.9	5.8	5.7	5.6	5.5	5.4	5.3	5.3	5.2
Crude Birth Rate	43.2	42.5	41.8	41.1	40.4	39.7	39.0	38.2	37.5	36.7	36.0
Infant Mortality Rate	145.0	143.0	141.0	139.0	137.0	135.0	133.2	131.4	129.6	127.8	126.0
Life Expectancy at Birth	46.1	46.6	47.0	47.5	48.0	48.4	49.1	49.7	50.4	51.1	51.7
Food Production, p.c. ('79-81 = 100)	91.1	95.3	96.2	99.8	99.2	92.4	97.2	90.9	100.8	98.8	104.0
Labor Force, Agriculture (%)	72.4	72.2	71.9	71.7	71.5	71.3	71.1	70.9	70.7	70.5	70.3
Labor Force, Female (%)	30.3	30.1	29.9	29.7	29.4	29.2	28.9	28.7	28.5	28.2	27.9
School Enroll. Ratio, primary	..	..	..	73.0	..	..	..	..	79.0	79.0	78.0
School Enroll. Ratio, secondary	..	..	..	26.0	..	..	..	..	26.0	26.0	27.0

1978	1979	1980	1981	1982	1983	1984	1985	1986	1987 estimate	NOTES	INDIA
				(Millions of current US dollars)							**FOREIGN TRADE (CUSTOMS BASIS)**
6,650	7,850	8,332	8,698	9,225	9,770	10,192	9,465	10,460	12,548	..	Value of Exports, fob
2,429	2,974	3,125	2,586	3,220	2,909	3,079	2,751	3,248	3,386	..	Nonfuel Primary Products
24	20	32	197	657	914	714	648	397	503	f	Fuels
4,197	4,857	5,175	5,916	5,348	5,947	6,399	6,066	6,815	8,658	f	Manufactures
7,854	10,142	14,090	15,654	14,387	14,782	15,424	17,295	16,801	18,985	..	Value of Imports, cif
1,910	1,813	2,302	2,563	2,465	2,257	2,507	2,443	2,549	3,039	..	Nonfuel Primary Products
1,973	3,546	6,168	6,308	4,735	4,684	4,550	4,078	1,679	2,146	..	Fuels
3,970	4,783	5,621	6,783	7,187	7,841	8,367	10,774	12,574	13,800	f	Manufactures
				(Index 1980 = 100)							
114.7	108.5	100.0	106.3	108.5	115.5	118.5	114.2	118.7	114.3	..	Terms of Trade
75.2	85.0	100.0	109.9	104.0	107.8	110.5	104.2	107.2	114.7	..	Export Prices, fob
84.5	95.9	100.0	92.3	90.3	93.1	98.4	87.8	96.1	90.2	..	Nonfuel Primary Products
42.3	61.0	100.0	112.5	101.6	92.5	90.2	87.5	44.9	56.9	..	Fuels
71.0	79.6	100.0	119.8	114.8	120.2	120.7	116.4	124.0	137.4	..	Manufactures
65.6	78.4	100.0	103.4	95.8	93.3	93.3	91.2	90.3	100.3	..	Import Prices, cif
				(Millions of 1980 US dollars)							Trade at Constant 1980 Prices
8,844	9,231	8,332	7,915	8,870	9,060	9,222	9,086	9,760	10,939	..	Exports, fob
11,977	12,943	14,090	15,141	15,012	15,838	16,539	18,964	18,609	18,921	..	Imports, cif
				(Millions of current US dollars)							**BALANCE OF PAYMENTS**
8,854	10,779	12,294	12,306	11,732	12,740	13,651	13,243	14,735	17,459	C f	Exports of Goods & Services
6,769	7,679	8,332	8,697	8,388	9,089	9,769	9,461	10,460	12,548	..	Merchandise, fob
1,607	2,304	2,879	2,697	2,819	3,202	3,389	3,235	3,775	4,406	..	Nonfactor Services
478	796	1,083	912	525	449	493	547	500	505	..	Factor Services
10,369	13,586	17,876	17,779	17,089	17,921	19,105	20,742	20,896	23,501	C f	Imports of Goods & Services
9,015	11,857	15,892	15,552	14,387	14,782	15,424	17,295	16,706	18,985	..	Merchandise, fob
890	1,262	1,516	1,696	1,884	2,164	2,350	2,124	2,520	2,864	..	Nonfactor Services
464	467	468	531	818	975	1,331	1,323	1,670	1,652	..	Factor Services
379	396	420	443	601	780	851	1,129	1,313	1,517	..	Long-Term Interest
1,150	1,852	2,771	2,318	2,505	2,570	2,526	2,198	2,174	1,974	..	Current Transfers, net
1,149	1,825	2,724	2,325	2,523	2,564	2,509	2,220	2,000	2,000	..	Workers' Remittances
-365	-955	-2,811	-3,155	-2,852	-2,611	-2,928	-5,301	-3,987	-4,068	K	Total to be Financed
449	577	643	497	399	367	453	359	270	318	K	Official Capital Grants
84	-378	-2,168	-2,658	-2,453	-2,244	-2,475	-4,942	-3,717	-3,750	..	Current Account Balance
998	1,219	2,776	2,181	2,667	2,656	4,029	3,409	2,615	4,098	C f	Long-Term Capital, net
0	0	8	10	65	63	62	160	208	253	..	Direct Investment
588	644	2,117	1,657	2,194	2,170	3,355	2,838	1,848	3,511	..	Long-Term Loans
1,259	1,362	2,878	2,401	3,104	3,204	4,386	4,421	4,090	6,191	..	Disbursements
671	717	761	744	910	1,034	1,031	1,583	2,242	2,680	..	Repayments
-39	-2	8	17	9	56	159	52	289	16	..	Other Long-Term Capital
521	-617	-1,265	-2,609	-1,677	-836	-1,358	2,289	1,730	182	C f	Other Capital, net
-1,603	-224	657	3,086	1,463	424	-196	-756	-628	-530	..	Change in Reserves
				(Indian Rupees per US dollar)							**Memo Item:**
8.210	8.080	7.890	8.930	9.630	10.310	11.890	12.240	12.790	13.000	..	Conversion Factor (Annual Avg)
				(Millions of US dollars, outstanding at end of year)							**EXTERNAL DEBT, ETC.**
15,483	15,823	17,662	18,065	19,685	21,306	22,986	27,726	32,119	37,325	..	Public/Publicly Guar. Long-Term
15,088	15,427	16,986	17,257	18,399	19,550	20,029	23,561	27,164	30,763	..	Official Creditors
4,618	5,233	5,969	7,087	8,377	9,599	10,233	12,146	14,004	16,276	..	IBRD and IDA
394	396	675	808	1,287	1,756	2,957	4,165	4,955	6,562	..	Private Creditors
348	335	336	873	1,239	1,767	2,611	3,093	2,598	3,442	..	Private Non-guaranteed Long-Term
0	0	327	964	2,876	4,150	3,932	4,290	4,291	3,653	..	Use of Fund Credit
651	698	926	1,204	1,827	1,573	1,743	1,516	2,303	1,950	..	Short-Term Debt
				(Millions of US dollars)							**Memo Items:**
6,426.4	7,432.5	6,943.9	4,692.9	4,315.2	4,937.3	5,842.3	6,420.4	6,395.7	6,453.6	..	Int'l Reserves Excluding Gold
1,889.8	4,382.9	5,065.9	3,416.0	3,926.4	3,278.5	2,693.6	3,072.8	4,084.5	5,058.4	..	Gold Holdings (at market price)
											SOCIAL INDICATORS
5.1	5.0	5.0	4.9	4.8	4.7	4.6	4.5	4.4	4.3	..	Total Fertility Rate
35.6	35.3	34.9	34.6	34.2	33.7	33.1	32.6	32.1	31.6	..	Crude Birth Rate
119.7	113.3	107.0	100.6	94.3	92.2	90.1	88.0	85.9	83.9	..	Infant Mortality Rate
52.4	53.0	53.7	54.3	55.0	55.4	55.8	56.2	56.7	57.1	..	Life Expectancy at Birth
104.9	97.7	98.2	104.0	100.3	111.9	112.2	111.9	110.9	103.7	..	Food Production, p.c. ('79-81 = 100)
70.1	69.9	69.7	..	..	..	..	..	..	..	..	Labor Force, Agriculture (%)
27.7	27.4	27.2	27.0	26.8	26.6	26.4	26.2	26.0	25.8	..	Labor Force, Female (%)
76.0	81.0	81.0	82.0	85.0	90.0	92.0	..	..	..	..	School Enroll. Ratio, primary
28.0	30.0	31.0	33.0	35.0	34.0	35.0	..	..	..	..	School Enroll. Ratio, secondary

INDONESIA	1967	1968	1969	1970	1971	1972	1973	1974	1975	1976	1977
CURRENT GNP PER CAPITA (US $)	50	60	70	90	90	100	120	150	210	280	330
POPULATION (millions)	109	111	114	116	119	121	124	127	130	133	136
ORIGIN AND USE OF RESOURCES					*(Billions of current Indonesian Rupiahs)*						
Gross National Product (GNP)	902	2,225	2,887	3,540	3,878	4,756	7,002	10,984	13,004	16,122	19,749
Net Factor Income from Abroad	7	12	19	15	3	-60	-124	-315	-338	-200	-336
Gross Domestic Product (GDP)	895	2,213	2,868	3,525	3,875	4,816	7,126	11,300	13,341	16,321	20,084
Indirect Taxes, net	31	94	135	188	229	236	328	447	520	635	846
GDP at factor cost	864	2,119	2,734	3,337	3,646	4,580	6,798	10,853	12,821	15,686	19,238
Agriculture	460	1,075	1,347	1,584	1,655	1,848	2,726	3,517	4,026	4,840	5,940
Industry	114	352	514	659	806	1,215	1,891	3,862	4,466	5,560	6,880
Manufacturing	72	208	292	363	357	521	756	1,036	1,308	1,691	2,115
Services, etc.	321	785	1,007	1,282	1,414	1,753	2,509	3,920	4,849	5,921	7,264
Resource Balance	-63	-82	-139	-69	-59	38	122	943	292	357	969
Exports of Goods & NFServices	80	244	262	458	563	815	1,449	3,255	3,097	3,870	4,824
Imports of Goods & NFServices	143	326	402	528	622	776	1,328	2,312	2,805	3,513	3,855
Domestic Absorption	958	2,295	3,008	3,594	3,934	4,778	7,004	10,357	13,050	15,964	19,116
Private Consumption, etc.	814	1,918	2,428	2,754	2,894	3,328	4,833	7,342	8,687	10,500	12,422
General Gov't Consumption	60	150	191	282	328	398	688	809	1,206	1,530	1,997
Gross Domestic Investment	83	227	389	558	712	1,052	1,483	2,206	3,157	3,934	4,696
Fixed Investment	..	..	..	..	..	..	..	..	..	..	..
Memo Items:											
Gross Domestic Saving	20	144	250	489	653	1,090	1,605	3,149	3,449	4,292	5,665
Gross National Saving	27	157	269	504	656	1,030	1,481	2,833	3,111	4,092	5,330
					(Billions of 1980 Indonesian Rupiahs)						
Gross National Product	16,678	19,047	20,714	22,580	24,032	26,153	29,036	31,029	32,603	35,066	38,053
GDP at factor cost	16,539	18,691	20,240	21,924	23,207	25,689	28,692	31,105	32,687	34,948	37,958
Agriculture	6,301	6,999	7,173	7,547	8,006	8,131	8,889	9,220	9,220	9,656	9,778
Industry	4,609	5,948	7,162	8,293	8,642	10,500	12,753	13,656	13,779	15,560	17,646
Manufacturing	1,254	1,362	1,560	1,707	1,761	2,027	2,336	2,714	3,048	3,343	3,803
Services, etc.	6,226	6,577	6,907	7,325	8,021	8,384	8,437	9,512	11,016	11,148	12,204
Resource Balance	3,146	3,502	3,931	4,694	5,123	6,398	7,131	6,575	5,671	6,653	7,627
Exports of Goods & NFServices	4,523	4,974	5,713	6,690	7,741	9,382	11,131	11,862	11,574	13,544	14,825
Imports of Goods & NFServices	1,376	1,473	1,782	1,996	2,618	2,984	4,000	5,287	5,903	6,892	7,198
Domestic Absorption	13,397	15,392	16,663	17,840	18,944	19,902	22,170	25,193	28,023	29,206	31,547
Private Consumption, etc.	10,701	12,203	12,921	13,087	13,361	13,434	14,413	16,610	17,871	18,412	19,019
General Gov't Consumption	1,110	1,257	1,288	1,501	1,623	1,754	2,239	2,005	2,614	2,805	3,265
Gross Domestic Investment	1,585	1,932	2,453	3,252	3,960	4,714	5,518	6,578	7,537	7,989	9,264
Fixed Investment	..	..	..	..	..	..	..	..	..	..	..
Memo Items:											
Capacity to Import	767	1,101	1,163	1,733	2,369	3,132	4,367	7,442	6,517	7,593	9,007
Terms of Trade Adjustment	-3,756	-3,873	-4,550	-4,957	-5,372	-6,250	-6,764	-4,419	-5,057	-5,952	-5,818
Gross Domestic Income	13,380	15,651	16,692	18,208	19,297	20,765	23,315	27,969	28,957	30,412	33,809
Gross National Income	12,923	15,174	16,164	17,624	18,661	19,903	22,272	26,609	27,546	29,115	32,235
DOMESTIC PRICES (DEFLATORS)					*(Index 1980 = 100)*						
Overall (GDP)	5.2	11.3	13.5	15.2	15.7	17.8	23.7	34.9	39.2	44.9	50.7
Domestic Absorption	7.1	14.9	18.1	20.1	20.8	24.0	31.6	41.1	46.6	54.7	60.6
Agriculture	7.3	15.4	18.8	21.0	20.7	22.7	30.7	38.1	43.7	50.1	60.8
Industry	2.5	5.9	7.2	7.9	9.3	11.6	14.8	28.3	32.4	35.7	39.0
Manufacturing	5.8	15.3	18.7	21.3	20.3	25.7	32.4	38.2	42.9	50.6	55.6
MANUFACTURING ACTIVITY											
Employment (1980=100)	..	..	..	50.2	52.5	63.4	64.3	64.1	71.6	82.6	81.5
Real Earnings per Empl. (1980=100)	..	..	..	54.9	63.4	81.7	77.1	78.5	78.4	77.6	83.9
Real Output per Empl. (1980=100)	..	..	..	41.9	52.3	56.1	60.9	68.1	89.9	72.0	78.8
Earnings as % of Value Added	..	..	..	26.4	23.0	23.5	20.9	23.9	20.5	21.6	20.6
MONETARY HOLDINGS					*(Billions of current Indonesian Rupiahs)*						
Money Supply, Broadly Defined	54	128	233	330	468	696	994	1,454	2,022	2,651	3,133
Money as Means of Payment	52	116	183	250	319	474	671	942	1,274	1,601	2,006
Currency Ouside Banks	34	77	116	155	198	269	375	497	650	779	979
Demand Deposits	17	40	68	96	121	205	296	445	625	822	1,027
Quasi-Monetary Liabilities	2	12	50	80	148	222	323	512	747	1,050	1,127
GOVERNMENT DEFICIT (-) OR SURPLUS					*(Billions of current Indonesian Rupiahs)*						
	..	..	..	..	..	-117	-163	-167	-468	-693	-393
Current Revenue	..	..	..	..	..	644	1,020	1,832	2,300	2,968	3,634
Current Expenditure	..	..	..	..	..	413	695	1,282	1,519	1,809	2,113
Current Budget Balance	..	..	..	..	..	231	325	550	781	1,159	1,521
Capital Receipts	..	..	..	..	..	..	..	..	..	..	..
Capital Payments	..	..	..	..	..	348	488	718	1,249	1,852	1,914

1978	1979	1980	1981	1982	1983	1984	1985	1986	1987 estimate	NOTES	INDONESIA
380	410	480	560	620	580	560	520	500	450	..	**CURRENT GNP PER CAPITA (US $)**
140	143	146	150	153	157	160	163	167	170	..	**POPULATION (millions)**
				(Billions of current Indonesian Rupiahs)							**ORIGIN AND USE OF RESOURCES**
23,540	32,860	46,903	56,496	59,611	70,338	82,887	90,561	91,770	108,480	..	Gross National Product (GNP)
-462	-1,484	-2,011	-1,925	-3,036	-3,360	-4,168	-3,932	-4,053	-6,039	..	Net Factor Income from Abroad
24,002	34,345	48,914	58,421	62,646	73,698	87,055	94,493	95,823	114,519	..	Gross Domestic Product (GDP)
1,029	1,305	1,635	1,752	2,133	2,547	..	..	..	..	..	Indirect Taxes, net
22,973	33,040	47,278	56,669	60,514	71,151	..	..	..	..	B	GDP at factor cost
6,745	9,374	11,725	13,649	15,001	17,696	20,334	22,413	24,696	29,208	..	Agriculture
8,578	12,944	20,405	24,076	23,745	27,301	32,479	34,163	30,080	38,103	..	Industry
2,816	4,003	6,353	7,067	7,482	8,211	11,082	12,676	13,585	15,952	..	Manufacturing
8,680	12,027	16,783	20,696	23,901	28,701	34,241	37,917	41,047	47,208	..	Services, etc.
587	2,402	6,277	2,367	253	-787	4,358	1,834	298	3,163	..	Resource Balance
5,317	10,147	16,162	16,401	15,324	20,448	22,985	21,671	21,161	29,776	..	Exports of Goods & NFServices
4,730	7,746	9,886	14,034	15,071	21,235	18,627	19,837	20,863	26,614	..	Imports of Goods & NFServices
23,416	31,943	42,637	56,054	62,393	74,485	82,697	92,660	95,525	111,356	..	Domestic Absorption
15,126	19,516	25,595	32,293	37,924	44,739	51,399	56,858	60,591	69,439	..	Private Consumption, etc.
2,556	3,277	5,148	6,452	7,229	8,077	9,121	10,893	11,329	11,764	..	General Gov't Consumption
5,734	9,150	11,894	17,309	17,241	21,668	22,177	24,908	23,606	30,154	..	Gross Domestic Investment
..	7,668	10,550	14,135	15,822	18,974	19,625	19,618	20,806	24,616	..	Fixed Investment
											Memo Items:
6,321	11,551	18,171	19,676	17,494	20,881	26,534	26,742	23,903	33,316	..	Gross Domestic Saving
5,858	10,067	16,160	17,751	14,458	17,530	22,421	22,877	19,942	27,462	..	Gross National Saving
				(Billions of 1980 Indonesian Rupiahs)							
40,907	43,184	46,903	50,797	51,301	52,880	56,030	57,436	59,843	61,764	..	Gross National Product
40,859	43,595	47,278	50,969	50,578	52,199	..	..	..	..	B H	GDP at factor cost
10,283	10,967	11,725	12,289	12,420	12,653	13,178	13,735	14,091	14,465	..	Agriculture
18,238	18,946	20,405	21,444	19,764	20,224	21,856	21,982	23,014	23,455	..	Industry
4,443	5,177	6,353	6,853	6,935	7,142	8,498	9,211	9,726	10,484	..	Manufacturing
14,168	15,404	16,783	18,812	20,176	21,190	22,305	22,986	24,029	25,409	..	Services, etc.
6,864	6,307	6,277	18	-1,361	-1,499	1,691	374	856	1,192	..	Resource Balance
14,973	15,316	16,162	13,245	12,052	12,622	12,693	11,676	13,356	14,167	..	Exports of Goods & NFServices
8,109	9,008	9,886	13,227	13,413	14,121	11,002	11,302	12,501	12,975	..	Imports of Goods & NFServices
35,696	39,111	42,637	55,634	56,776	58,530	57,600	60,984	62,816	64,493	..	Domestic Absorption
21,199	23,077	25,595	28,196	29,952	31,775	33,309	34,120	35,255	37,014	..	Private Consumption, etc.
3,840	4,301	5,148	5,654	6,163	6,049	6,255	6,721	6,920	6,909	..	General Gov't Consumption
10,657	11,733	11,894	21,784	20,661	20,706	18,036	20,143	20,641	20,570	..	Gross Domestic Investment
..	8,349	10,550	11,907	12,636	12,794	12,034	11,306	11,688	12,206	..	Fixed Investment
											Memo Items:
9,115	11,801	16,162	15,458	13,639	13,598	13,576	12,347	12,679	14,517	..	Capacity to Import
-5,858	-3,514	0	2,213	1,586	975	882	671	-677	349	..	Terms of Trade Adjustment
36,831	41,803	48,914	54,757	53,947	55,043	58,222	59,373	60,456	63,679	..	Gross Domestic Income
35,049	39,670	46,903	53,010	52,888	53,855	56,913	58,107	59,165	62,113	..	Gross National Income
				(Index 1980 = 100)							**DOMESTIC PRICES (DEFLATORS)**
56.2	75.8	100.0	111.2	119.6	136.3	151.8	161.0	156.7	180.8	..	Overall (GDP)
65.6	81.7	100.0	100.8	109.9	127.3	143.6	151.9	152.1	172.7	..	Domestic Absorption
65.6	85.5	100.0	111.1	120.8	139.9	154.3	163.2	175.3	201.9	..	Agriculture
47.0	68.3	100.0	112.3	120.1	135.0	148.6	155.4	130.7	162.4	..	Industry
63.4	71.3	100.0	103.1	107.9	115.0	130.4	137.6	139.7	152.2	..	Manufacturing
											MANUFACTURING ACTIVITY
84.5	89.0	100.0	104.4	110.1	115.5	119.5	126.8	..	..	J	Employment (1980=100)
91.8	93.6	100.0	108.3	123.2	128.4	131.9	153.1	..	..	J	Real Earnings per Empl. (1980=100)
86.1	96.1	100.0	113.1	117.4	128.7	137.6	156.8	..	..	J	Real Output per Empl. (1980=100)
20.3	19.6	16.4	17.5	19.8	21.0	18.3	20.6	..	..	J	Earnings as % of Value Added
				(Billions of current Indonesian Rupiahs)							**MONETARY HOLDINGS**
3,822	5,159	7,707	9,705	11,074	14,670	17,937	23,177	27,615	33,904	..	Money Supply, Broadly Defined
2,488	3,316	5,011	6,474	7,120	7,576	8,581	10,124	11,631	12,705	..	Money as Means of Payment
1,240	1,546	2,169	2,546	2,934	3,340	3,712	4,460	5,338	5,802	..	Currency Ouside Banks
1,248	1,771	2,842	3,929	4,185	4,236	4,869	5,664	6,293	6,903	..	Demand Deposits
1,334	1,843	2,695	3,231	3,954	7,094	9,356	13,053	15,984	21,199	..	Quasi-Monetary Liabilities
				(Billions of current Indonesian Rupiahs)							
-754	-764	-1,102	-1,172	-1,191	-1,862	1,219	-948	-3,621	..	C	**GOVERNMENT DEFICIT (-) OR SURPLUS**
4,378	7,050	10,406	13,763	12,815	15,511	18,724	20,347	21,324	..	..	Current Revenue
2,570	3,959	5,731	6,882	6,996	8,412	9,429	11,426	13,560	..	..	Current Expenditure
1,808	3,091	4,675	6,881	5,819	7,099	9,295	8,921	7,764	..	..	Current Budget Balance
..	..	..	..	..	..	..	..	..	..	..	Capital Receipts
2,562	3,855	5,777	8,053	7,010	8,961	8,076	9,869	11,385	..	..	Capital Payments

INDONESIA	1967	1968	1969	1970	1971	1972	1973	1974	1975	1976	1977
FOREIGN TRADE (CUSTOMS BASIS)				*(Millions of current US dollars)*							
Value of Exports, fob	665	731	800	1,055	1,199	1,778	3,211	7,426	7,130	8,556	10,853
Nonfuel Primary Products	401	421	405	694	700	827	1,533	2,145	1,704	2,419	3,283
Fuels	240	297	379	346	478	913	1,609	5,211	5,339	6,014	7,379
Manufactures	25	12	16	15	22	37	69	70	88	124	191
Value of Imports, cif	891	923	1,122	1,258	1,379	1,611	2,969	5,231	6,196	7,700	8,495
Nonfuel Primary Products	105	255	214	208	180	223	441	1,037	1,069	1,490	1,805
Fuels	19	8	26	19	36	41	54	251	334	598	1,010
Manufactures	767	660	882	1,031	1,163	1,346	2,474	3,943	4,793	5,612	5,681
					(Index 1980 = 100)						
Terms of Trade	33.0	27.9	27.0	32.4	34.1	28.9	33.1	69.4	61.2	69.7	75.1
Export Prices, fob	9.2	8.5	7.7	10.0	10.7	10.0	14.8	41.1	38.0	42.6	47.8
Nonfuel Primary Products	26.9	26.3	29.1	29.9	27.7	28.3	45.9	57.2	46.7	57.9	68.5
Fuels	4.3	4.3	4.3	4.3	5.6	6.2	8.9	36.7	35.7	38.4	42.0
Manufactures	27.3	24.2	21.2	23.2	27.8	31.3	39.9	58.1	50.5	52.0	61.1
Import Prices, cif	27.9	30.3	28.6	31.0	31.5	34.8	44.8	59.2	62.1	61.1	63.7
Trade at Constant 1980 Prices				*(Millions of 1980 US dollars)*							
Exports, fob	7,209	8,633	10,354	10,513	11,182	17,702	21,690	18,065	18,763	20,092	22,684
Imports, cif	3,189	3,043	3,926	4,060	4,380	4,630	6,632	8,837	9,979	12,599	13,333
BALANCE OF PAYMENTS				*(Millions of current US dollars)*							
Exports of Goods & Services	..	..	..	1,189	1,339	1,837	3,306	7,464	7,025	8,776	10,926
Merchandise, fob	..	..	..	1,173	1,311	1,793	3,215	7,265	6,888	8,615	10,760
Nonfactor Services	..	..	..	16	28	45	52	71	93	111	114
Factor Services	..	..	..	0	0	0	38	127	44	50	51
Imports of Goods & Services	..	..	..	1,565	1,757	2,222	3,836	6,915	8,160	9,700	10,997
Merchandise, fob	..	..	..	1,116	1,230	1,445	2,663	4,634	5,469	6,819	7,473
Nonfactor Services	..	..	..	316	355	430	544	952	1,306	1,669	1,839
Factor Services	..	..	..	133	173	347	629	1,329	1,385	1,212	1,685
Long-Term Interest	..	..	..	46	57	86	118	186	322	489	610
Current Transfers, net	..	..	..	0	0	0	0	0	0	0	0
Workers' Remittances	..	..	..	0	0	0	0	0	0	0	0
Total to be Financed	..	..	..	-376	-418	-385	-530	548	-1,135	-925	-71
Official Capital Grants	..	..	..	66	46	51	55	49	27	15	25
Current Account Balance	..	..	..	-310	-372	-334	-476	598	-1,109	-910	-47
Long-Term Capital, net	..	..	..	356	423	552	576	541	2,270	2,279	1,515
Direct Investment	..	..	..	83	139	207	15	-49	476	344	235
Long-Term Loans	..	..	..	517	438	976	1,288	1,508	2,264	2,153	1,356
Disbursements	..	..	..	636	600	1,190	1,615	2,017	2,945	3,004	2,727
Repayments	..	..	..	120	162	214	327	510	681	851	1,371
Other Long-Term Capital	..	..	..	-310	-200	-683	-783	-967	-496	-234	-100
Other Capital, net	..	..	..	-69	-81	160	240	-451	-2,013	-467	-473
Change in Reserves	..	..	..	23	30	-377	-341	-688	851	-902	-996
Memo Item:				*(Indonesian Rupiahs per US dollar)*							
Conversion Factor (Annual Avg)	153.700	300.100	326.000	365.000	393.400	415.000	415.000	415.000	415.000	415.000	415.000
EXTERNAL DEBT, ETC.				*(Millions of US dollars, outstanding at end of year)*							
Public/Publicly Guar. Long-Term	..	..	..	2,443	3,452	4,204	5,249	6,358	7,994	10,002	11,670
Official Creditors	..	..	..	2,160	3,056	3,453	4,025	4,614	5,004	5,909	7,073
IBRD and IDA	..	..	..	5	32	68	135	211	375	631	869
Private Creditors	..	..	..	283	396	752	1,224	1,744	2,991	4,092	4,597
Private Non-guaranteed Long-Term	..	..	..	461	649	913	1,285	1,844	2,369	2,624	2,842
Use of Fund Credit	..	..	..	139	136	116	23	76	0	0	0
Short-Term Debt	..	..	..	..	..	..	..	..	..	..	1,925
Memo Items:				*(Millions of US dollars)*							
Int'l Reserves Excluding Gold	2.0	82.5	117.7	156.0	185.3	572.2	804.7	1,489.5	584.3	1,496.5	2,508.7
Gold Holdings (at market price)	4.0	4.8	4.0	4.3	2.5	7.9	6.4	10.6	8.0	7.7	27.9
SOCIAL INDICATORS											
Total Fertility Rate	5.6	5.5	5.5	5.5	5.4	5.4	5.3	5.1	5.0	4.9	4.8
Crude Birth Rate	42.6	42.4	42.1	41.9	41.6	41.4	40.4	39.4	38.4	37.4	36.4
Infant Mortality Rate	125.2	123.7	122.1	120.6	119.0	117.4	115.9	114.3	112.8	111.2	109.7
Life Expectancy at Birth	45.1	45.9	46.7	47.4	48.2	49.0	49.5	50.0	50.4	50.9	51.4
Food Production, p.c. ('79-81 = 100)	73.2	80.8	79.7	84.4	83.3	84.9	87.8	89.1	88.5	88.2	88.5
Labor Force, Agriculture (%)	68.8	68.0	67.1	66.3	65.4	64.5	63.6	62.7	61.8	60.9	60.0
Labor Force, Female (%)	29.2	29.6	29.9	30.2	30.4	30.5	30.6	30.7	30.8	30.9	31.0
School Enroll. Ratio, primary	..	..	..	80.0	..	..	..	..	86.0	84.0	90.0
School Enroll. Ratio, secondary	..	..	..	16.0	..	..	..	..	20.0	20.0	21.0

1978	1979	1980	1981	1982	1983	1984	1985	1986	1987 estimate	NOTES	INDONESIA
				(Millions of current US dollars)							**FOREIGN TRADE (CUSTOMS BASIS)**
11,643	15,590	21,909	22,260	19,747	20,961	20,345	18,711	13,567	17,651	..	Value of Exports, fob
3,431	4,936	5,633	3,763	2,690	3,350	3,581	3,153	3,452	4,160	..	Nonfuel Primary Products
7,986	10,166	15,743	17,764	16,189	15,993	14,598	13,097	7,154	8,665	..	Fuels
225	488	533	733	868	1,618	2,166	2,461	2,961	4,826	..	Manufactures
9,493	10,364	14,139	18,527	19,996	19,853	16,853	14,230	13,103	14,453	..	Value of Imports, cif
2,210	2,311	2,637	3,194	2,381	2,453	2,181	1,469	1,064	935	..	Nonfuel Primary Products
831	1,150	2,289	2,459	4,297	5,035	3,289	3,807	1,897	2,271	..	Fuels
6,452	6,903	9,213	12,874	13,318	12,365	11,383	8,954	10,142	11,247	..	Manufactures
				(Index 1980 = 100)							
65.7	81.6	100.0	104.9	100.8	95.2	95.9	93.5	64.8	68.8	..	Terms of Trade
48.3	69.2	100.0	106.4	97.4	88.9	87.7	82.9	57.0	69.3	..	Export Prices, fob
69.8	92.4	100.0	86.6	80.6	84.0	92.1	79.8	82.5	87.8	..	Nonfuel Primary Products
42.3	61.0	100.0	112.5	101.6	92.5	90.2	87.5	44.9	56.9	..	Fuels
74.2	93.4	100.0	94.4	86.9	70.7	69.5	67.3	79.6	87.4	..	Manufactures
73.5	84.7	100.0	101.4	96.7	93.4	91.4	88.6	87.9	100.6	..	Import Prices, cif
											Trade at Constant 1980 Prices
				(Millions of 1980 US dollars)							
24,103	22,537	21,909	20,919	20,264	23,573	23,196	22,570	23,819	25,485	..	Exports, fob
12,907	12,232	14,139	18,270	20,684	21,260	18,434	16,053	14,915	14,363	..	Imports, cif
				(Millions of current US dollars)							**BALANCE OF PAYMENTS**
11,309	15,536	22,208	24,878	21,274	19,866	22,152	20,139	15,972	19,542	..	Exports of Goods & Services
11,019	15,138	21,762	23,348	19,747	18,689	20,754	18,527	14,396	17,991	..	Merchandise, fob
233	315	327	449	504	546	570	844	844	842	..	Nonfactor Services
58	83	120	1,081	1,023	631	828	768	732	709	..	Factor Services
12,745	14,591	19,403	25,694	26,732	26,318	24,175	22,150	20,142	21,752	..	Imports of Goods & Services
8,382	9,240	12,599	16,542	17,854	17,726	15,047	12,705	11,938	14,263	..	Merchandise, fob
2,345	2,885	3,478	4,998	4,862	4,311	4,239	5,135	4,256	2,397	..	Nonfactor Services
2,018	2,465	3,327	4,154	4,016	4,281	4,889	4,310	3,948	5,092	..	Factor Services
720	1,049	1,181	1,431	1,568	1,635	1,908	1,920	2,360	2,748	..	Long-Term Interest
0	0	0	0	0	10	53	61	71	112	..	Current Transfers, net
0	0	0	0	0	10	53	61	71	112	..	Workers' Remittances
-1,436	946	2,805	-816	-5,458	-6,442	-1,970	-1,950	-4,099	-2,098	K	Total to be Financed
14	30	201	250	134	104	114	27	188	261	K	Official Capital Grants
-1,422	975	3,006	-566	-5,324	-6,338	-1,856	-1,923	-3,911	-1,837	..	Current Account Balance
1,600	1,349	2,349	2,401	5,230	5,427	3,095	1,907	3,070	3,134	..	Long-Term Capital, net
279	226	180	133	225	292	222	310	258	425	..	Direct Investment
865	659	1,617	2,057	2,710	3,842	2,592	1,278	1,802	2,457	..	Long-Term Loans
2,975	2,670	3,245	3,845	4,650	5,865	4,884	4,385	4,669	6,191	..	Disbursements
2,110	2,011	1,628	1,788	1,941	2,023	2,293	3,107	2,866	3,734	..	Repayments
442	434	350	-39	2,161	1,189	167	292	822	-9	..	Other Long-Term Capital
-7	-880	-2,926	-2,209	-1,759	1,094	-258	526	-162	-153	..	Other Capital, net
-171	-1,444	-2,428	374	1,853	-183	-981	-510	1,003	-1,144	..	Change in Reserves
											Memo Item:
				(Indonesian Rupiahs per US dollar)							
442.000	623.100	627.000	631.800	661.400	909.300	1,025.900	1,110.600	1,282.600	1,643.800	..	Conversion Factor (Annual Avg)
				(Millions of US dollars, outstanding at end of year)							**EXTERNAL DEBT, ETC.**
13,150	13,278	14,971	15,870	18,513	21,654	22,355	26,863	32,851	41,284	..	Public/Publicly Guar. Long-Term
8,388	8,509	9,507	10,058	11,109	11,997	12,318	15,191	18,729	24,941	..	Official Creditors
1,059	1,264	1,605	1,941	2,442	2,899	3,176	4,434	5,915	8,256	..	IBRD and IDA
4,762	4,769	5,464	5,811	7,403	9,657	10,036	11,672	14,122	16,343	..	Private Creditors
3,040	3,140	3,142	3,579	3,200	3,400	3,800	3,810	3,828	4,105	..	Private Non-guaranteed Long-Term
0	0	0	0	0	445	413	46	51	716	..	Use of Fund Credit
1,786	2,108	2,775	3,274	4,787	4,639	5,384	5,280	6,309	6,476	..	Short-Term Debt
				(Millions of US dollars)							**Memo Items:**
2,626.1	4,061.8	5,391.7	5,014.2	3,144.5	3,718.4	4,773.0	4,974.2	4,051.3	5,592.3	..	Int'l Reserves Excluding Gold
50.6	143.4	1,411.3	1,233.8	1,418.2	1,184.2	957.0	1,015.0	1,213.4	1,502.6	..	Gold Holdings (at market price)
											SOCIAL INDICATORS
4.7	4.6	4.5	4.4	4.3	4.2	4.1	4.0	3.8	3.7	..	Total Fertility Rate
35.6	34.7	33.9	33.0	32.2	31.7	31.1	30.6	30.1	29.5	..	Crude Birth Rate
108.1	106.6	105.0	103.6	102.2	98.5	94.7	91.0	87.3	83.5	..	Infant Mortality Rate
51.8	52.2	52.7	53.1	53.6	54.3	55.1	55.8	56.5	57.3	..	Life Expectancy at Birth
92.3	94.1	101.0	104.9	104.2	108.4	117.6	115.7	118.3	118.2	..	Food Production, p.c. ('79-81 = 100)
59.0	58.1	57.2	..	..	..	..	..	..	..	..	Labor Force, Agriculture (%)
31.1	31.2	31.3	31.3	31.3	31.3	31.3	31.3	31.3	31.2	..	Labor Force, Female (%)
99.0	107.0	107.0	111.0	113.0	116.0	118.0	..	..	..	..	School Enroll. Ratio, primary
22.0	24.0	29.0	31.0	35.0	37.0	39.0	..	..	..	..	School Enroll. Ratio, secondary

IRELAND	1967	1968	1969	1970	1971	1972	1973	1974	1975	1976	1977
CURRENT GNP PER CAPITA (US $)	1,070	1,160	1,230	1,300	1,460	1,720	2,020	2,340	2,610	2,660	3,000
POPULATION (thousands)	2,900	2,913	2,926	2,950	2,978	3,024	3,073	3,124	3,177	3,228	3,272
ORIGIN AND USE OF RESOURCES	*(Millions of current Irish Pounds)*										
Gross National Product (GNP)	1,128	1,277	1,467	1,648	1,882	2,284	2,741	3,041	3,796	4,617	5,595
Net Factor Income from Abroad	25	32	28	28	27	30	12	19	4	-36	-108
Gross Domestic Product (GDP)	1,104	1,245	1,438	1,620	1,856	2,255	2,729	3,021	3,792	4,653	5,703
Indirect Taxes, net	149	169	205	236	272	320	375	376	389	593	532
GDP at factor cost	955	1,076	1,233	1,385	1,584	1,935	2,354	2,645	3,403	4,060	5,171
Agriculture	181	205	213	233	258	352	441	424	589	705	958
Industry	325	370	436	509	582	712	883	1,046	1,214	1,414	1,812
Manufacturing	..	..	..	327	370	455	573	689	773	911	1,198
Services, etc.	449	501	584	643	743	871	1,030	1,175	1,599	1,941	2,401
Resource Balance	-35	-81	-131	-131	-133	-118	-182	-436	-230	-369	-523
Exports of Goods & NFServices	417	483	536	598	670	773	1,028	1,272	1,618	2,153	2,816
Imports of Goods & NFServices	452	564	667	729	803	892	1,210	1,709	1,848	2,521	3,339
Domestic Absorption	1,138	1,326	1,569	1,752	1,988	2,373	2,911	3,458	4,022	5,022	6,226
Private Consumption, etc.	741	845	954	1,060	1,200	1,398	1,685	1,979	2,461	2,936	3,623
General Gov't Consumption	162	183	213	260	309	366	440	549	727	880	1,030
Gross Domestic Investment	235	298	402	431	480	608	786	930	834	1,206	1,573
Fixed Investment	230	270	347	383	452	529	703	749	897	1,165	1,399
Memo Items:											
Gross Domestic Saving	200	217	271	300	347	490	604	494	604	837	1,050
Gross National Saving	247	276	329	359	408	555	656	556	654	850	994
	(Millions of 1980 Irish Pounds)										
Gross National Product	5,189	5,629	5,935	6,127	6,338	6,792	7,066	7,367	7,520	7,558	8,089
GDP at factor cost	4,375	4,709	4,928	5,092	5,264	5,661	5,947	6,386	6,721	6,609	7,457
Agriculture	..	..	..	854	884	1,097	1,096	642	1,040	891	1,035
Industry	..	..	..	..	..	..	..	..	2,274	2,342	2,672
Manufacturing	..	..	..	..	..	..	..	..	..	..	..
Services, etc.	..	..	..	..	..	..	..	..	3,400	3,387	3,746
Resource Balance	-413	-610	-878	-853	-905	-986	-1,374	-1,260	-652	-942	-1,042
Exports of Goods & NFServices	1,929	2,097	2,194	2,291	2,385	2,471	2,741	2,760	2,958	3,198	3,647
Imports of Goods & NFServices	2,341	2,708	3,072	3,144	3,290	3,457	4,115	4,020	3,610	4,140	4,689
Domestic Absorption	5,484	6,093	6,692	6,869	7,147	7,681	8,404	8,580	8,163	8,561	9,293
Private Consumption, etc.	3,506	3,838	4,074	4,186	4,344	4,560	4,904	4,966	4,963	5,023	5,408
General Gov't Consumption	854	897	957	1,029	1,118	1,198	1,280	1,375	1,465	1,505	1,535
Gross Domestic Investment	1,123	1,359	1,661	1,654	1,685	1,923	2,221	2,239	1,735	2,033	2,350
Fixed Investment	1,126	1,274	1,507	1,512	1,638	1,730	2,043	1,877	1,808	1,991	2,085
Memo Items:											
Capacity to Import	2,161	2,318	2,469	2,578	2,747	2,998	3,496	2,994	3,161	3,535	3,955
Terms of Trade Adjustment	232	221	275	286	362	527	755	234	202	336	307
Gross Domestic Income	5,303	5,704	6,090	6,303	6,604	7,223	7,785	7,553	7,713	7,956	8,558
Gross National Income	5,422	5,850	6,211	6,414	6,699	7,318	7,821	7,601	7,722	7,894	8,397
DOMESTIC PRICES (DEFLATORS)	*(Index 1980 = 100)*										
Overall (GDP)	21.8	22.7	24.7	26.9	29.7	33.7	38.8	41.3	50.5	61.1	69.1
Domestic Absorption	20.8	21.8	23.4	25.5	27.8	30.9	34.6	40.3	49.3	58.7	67.0
Agriculture	..	..	..	27.3	29.2	32.1	40.2	66.1	56.6	79.1	92.5
Industry	..	..	..	..	..	..	..	..	53.4	60.4	67.8
Manufacturing	..	..	..	..	..	..	..	..	..	..	..
MANUFACTURING ACTIVITY											
Employment (1980=100)	78.5	81.3	86.2	87.6	87.0	87.5	92.5	93.1	86.2	87.4	90.5
Real Earnings per Empl. (1980=100)	59.0	56.5	63.0	66.8	71.0	75.1	79.6	82.9	88.0	87.5	89.1
Real Output per Empl. (1980=100)	..	..	..	12.3	12.8	14.1	14.2	14.8	20.1	22.7	20.4
Earnings as % of Value Added	47.8	47.0	47.1	48.7	48.5	48.5	46.4	45.3	46.0	42.0	42.4
MONETARY HOLDINGS	*(Millions of current Irish Pounds)*										
Money Supply, Broadly Defined	786	899	990	1,084	1,177	1,352	1,682	2,033	2,456	2,769	3,261
Money as Means of Payment	341	364	389	415	440	518	572	624	748	875	1,072
Currency Ouside Banks	124	131	137	155	173	191	219	244	292	342	390
Demand Deposits	217	233	252	260	267	327	354	380	456	533	682
Quasi-Monetary Liabilities	445	535	601	669	737	834	1,109	1,409	1,708	1,894	2,189
GOVERNMENT DEFICIT (-) OR SURPLUS	*(Millions of current Irish Pounds)*										
	..	..	..	-89.0	-98.0	-125.0	-183.0	-357.0	-472.0	-470.0	-534.0
Current Revenue	..	..	..	506.0	598.0	695.0	842.0	956.0	1,209.0	1,629.0	1,933.0
Current Expenditure	..	..	..	481.0	560.0	663.0	810.0	987.0	1,406.0	1,750.0	2,065.0
Current Budget Balance	..	..	..	25.0	38.0	32.0	32.0	-31.0	-197.0	-121.0	-132.0
Capital Receipts	..	..	..	..	..	..	..	..	..	..	..
Capital Payments	..	..	..	114.0	136.0	157.0	215.0	326.0	275.0	349.0	402.0

1978	1979	1980	1981	1982	1983	1984	1985	1986	1987 estimate	NOTES	IRELAND
3,430	4,120	5,040	5,580	5,510	5,090	4,910	4,730	5,000	6,040	..	CURRENT GNP PER CAPITA (US $)
3,314	3,368	3,401	3,443	3,483	3,508	3,535	3,552	3,582	3,611	..	POPULATION (thousands)
				(Millions of current Irish Pounds)							ORIGIN AND USE OF RESOURCES
6,529	7,634	9,003	10,854	12,454	13,499	14,660	15,324	16,207	17,305	..	Gross National Product (GNP)
-228	-283	-358	-505	-928	-1,184	-1,660	-1,992	-2,032	-2,041	..	Net Factor Income from Abroad
6,757	7,917	9,361	11,359	13,381	14,683	16,320	17,316	18,239		..	Gross Domestic Product (GDP)
482	525	830	1,225	1,567	1,780	1,880	1,801	1,959		..	Indirect Taxes, net
6,275	7,392	8,530	10,134	11,814	12,903	14,439	15,515	16,280		B	GDP at factor cost
1,085	1,055	991	1,143	1,386	1,558	1,787	1,666	1,607		..	Agriculture
2,246	2,706	3,182	3,666	4,350	4,773	5,268	5,823	6,086		..	Industry
1,488	1,767	..	..	..	..	..	..	..		..	Manufacturing
2,944	3,631	4,358	5,325	6,078	6,573	7,385	8,025	8,587		..	Services, etc.
-667	-1,294	-1,261	-1,612	-979	-415	-46	391	629	1,310	..	Resource Balance
3,376	3,938	4,639	5,505	6,434	7,751	9,767	10,741	10,334	11,559	..	Exports of Goods & NFServices
4,042	5,233	5,900	7,117	7,412	8,166	9,814	10,350	9,705	10,249	..	Imports of Goods & NFServices
7,423	9,211	10,622	12,971	14,360	15,098	16,366	16,925	17,610	..	..	Domestic Absorption
4,314	5,194	6,158	7,415	7,746	8,351	8,951	9,421	9,812	10,679	..	Private Consumption, etc.
1,216	1,470	1,860	2,212	2,582	2,815	3,044	3,182	3,459	3,462	..	General Gov't Consumption
1,893	2,547	2,604	3,344	4,032	3,932	4,371	4,323	4,338	4,477	..	Gross Domestic Investment
1,806	2,389	2,718	3,457	3,779	3,723	3,931	3,941	4,054	4,158	..	Fixed Investment
											Memo Items:
1,226	1,253	1,343	1,732	3,053	3,517	4,324	4,714	4,968		..	Gross Domestic Saving
1,055	1,015	1,045	1,288	2,190	2,379	2,634	2,702	2,895		..	Gross National Saving
				(Millions of 1980 Irish Pounds)							
8,533	8,768	9,003	9,239	9,173	8,990	9,059	8,987	8,844	9,300	..	Gross National Product
8,186	8,481	8,530	8,631	8,697	8,580	8,919	9,102	8,886	..	B	GDP at factor cost
1,078	943	991	937	1,043	1,024	1,151	1,146	1,028		..	Agriculture
2,990	3,292	3,182	3,315	3,301	3,251	3,460	3,532	3,510		..	Industry
..	..	..	..	..	..	..	..	..		..	Manufacturing
4,118	4,246	4,358	4,379	4,383	5,098	4,308	4,425	4,349		..	Services, etc.
-1,329	-1,816	-1,261	-1,272	-823	-572	-272	-27	-138	137	..	Resource Balance
4,097	4,362	4,639	4,729	4,991	5,513	6,418	6,850	7,035	8,027	..	Exports of Goods & NFServices
5,426	6,178	5,900	6,001	5,814	6,085	6,690	6,877	7,173	7,890	..	Imports of Goods & NFServices
10,175	10,919	10,622	10,935	10,686	10,376	10,385	10,216	10,154	..	..	Domestic Absorption
5,907	6,153	6,158	6,254	5,791	5,786	5,729	5,760	5,740	..	..	Private Consumption, etc.
1,661	1,737	1,860	1,866	1,927	1,940	1,939	1,933	1,995	1,935	..	General Gov't Consumption
2,607	3,029	2,604	2,815	2,968	2,650	2,716	2,523	2,419	2,414	..	Gross Domestic Investment
2,467	2,824	2,718	2,917	2,820	2,566	2,504	2,394	2,338	2,324	..	Fixed Investment
											Memo Items:
4,531	4,650	4,639	4,642	5,046	5,776	6,658	7,137	7,638	8,898	..	Capacity to Import
434	288	0	-88	55	263	241	287	603	871	..	Terms of Trade Adjustment
9,281	9,392	9,361	9,576	9,918	10,067	10,353	10,476	10,619	..	..	Gross Domestic Income
8,968	9,056	9,003	9,151	9,228	9,253	9,300	9,274	9,447	9,989	..	Gross National Income
				(Index 1980 = 100)							DOMESTIC PRICES (DEFLATORS)
76.4	87.0	100.0	117.5	135.7	149.8	161.4	169.9	182.1	..	..	Overall (GDP)
73.0	84.4	100.0	118.6	134.4	145.5	157.6	165.7	173.4	..	..	Domestic Absorption
100.7	111.9	100.0	121.9	132.9	152.1	155.2	145.4	156.4	..	..	Agriculture
75.1	82.2	100.0	110.6	131.8	146.8	152.2	164.8	173.4	..	..	Industry
..	..	..	..	..	..	..	..	..	..	..	Manufacturing
											MANUFACTURING ACTIVITY
91.7	100.7	100.0	98.6	94.9	89.6	86.9	84.2	..	..	G	Employment (1980=100)
97.0	98.6	100.0	96.2	93.8	96.2	119.8	141.8	..	..	G	Real Earnings per Empl. (1980=100)
24.3	26.1	100.0	100.0	99.6	116.9	137.4	143.3	..	..	G	Real Output per Empl. (1980=100)
41.1	42.3	43.6	41.2	39.6	35.9	38.9	38.9	..	..	..	Earnings as % of Value Added
				(Millions of current Irish Pounds)							MONETARY HOLDINGS
4,142	4,914	5,954	6,832	7,043	7,486	8,257	8,750	8,620	9,559	D	Money Supply, Broadly Defined
1,367	1,479	1,686	1,743	1,838	2,048	2,245	2,288	2,382	2,640	..	Money as Means of Payment
476	582	663	739	811	907	923	970	1,024	1,107	..	Currency Ouside Banks
892	896	1,023	1,004	1,027	1,142	1,322	1,318	1,357	1,532	..	Demand Deposits
2,775	3,435	4,268	5,088	5,205	5,438	6,012	6,461	6,238	6,919	..	Quasi-Monetary Liabilities
				(Millions of current Irish Pounds)							
-768.0	-944.0	-1,224.0	-1,662.0	-1,983.0	-1,835.0	-1,741.0	-2,051.0	-2,115.0	..	..	GOVERNMENT DEFICIT (-) OR SURPLUS
2,256.0	2,728.0	3,550.0	4,427.0	5,535.0	6,433.0	7,099.0	7,555.0	8,005.0	..	..	Current Revenue
2,531.0	3,081.0	3,950.0	5,046.0	6,411.0	7,265.0	7,868.0	8,644.0	9,169.0	..	..	Current Expenditure
-275.0	-353.0	-400.0	-619.0	-876.0	-832.0	-769.0	-1,089.0	-1,164.0	..	..	Current Budget Balance
..	..	..	..	..	..	..	..	..	..	..	Capital Receipts
493.0	591.0	824.0	1,043.0	1,107.0	1,003.0	972.0	962.0	951.0	..	..	Capital Payments

IRELAND	1967	1968	1969	1970	1971	1972	1973	1974	1975	1976	1977
FOREIGN TRADE (CUSTOMS BASIS)				*(Millions of current US dollars)*							
Value of Exports, fob	783	798	891	1,035	1,308	1,611	2,131	2,629	3,179	3,313	4,400
Nonfuel Primary Products	492	475	523	583	688	819	1,079	1,243	1,661	1,546	1,951
Fuels	26	17	22	25	16	16	16	35	42	22	29
Manufactures	265	305	346	427	604	776	1,036	1,352	1,475	1,745	2,420
Value of Imports, cif	1,075	1,175	1,413	1,568	1,835	2,102	2,793	3,813	3,769	4,192	5,381
Nonfuel Primary Products	286	308	323	344	355	426	599	802	726	793	1,023
Fuels	103	98	107	127	166	158	190	531	531	562	678
Manufactures	686	768	983	1,097	1,314	1,517	2,004	2,481	2,512	2,837	3,680
Terms of Trade	147.2	140.9	135.2	154.7	163.5	159.6	145.0	110.1	107.8	108.0	102.4
Export Prices, fob	28.8	29.0	29.7	34.3	39.7	45.4	54.4	59.6	61.0	63.3	65.7
Nonfuel Primary Products	33.6	33.3	36.1	38.1	40.9	48.3	62.6	60.0	58.1	60.4	58.2
Fuels	8.5	7.3	5.4	5.9	7.0	8.5	24.5	36.6	33.6	38.5	36.8
Manufactures	28.4	28.4	30.4	40.5	43.9	46.6	48.6	59.5	67.6	67.6	75.0
					(Index 1980 = 100)						
Import Prices, cif	19.6	20.6	22.0	22.2	24.3	28.4	37.5	54.1	56.5	58.6	64.1
Trade at Constant 1980 Prices				*(Millions of 1980 US dollars)*							
Exports, fob	2,714	2,755	2,996	3,017	3,293	3,550	3,920	4,414	5,214	5,234	6,700
Imports, cif	5,486	5,717	6,427	7,071	7,558	7,391	7,449	7,050	6,667	7,155	8,392
BALANCE OF PAYMENTS				*(Millions of current US dollars)*							
Exports of Goods & Services	..	..	..	1,539	1,754	2,101	2,730	3,330	3,689	4,073	5,678
Merchandise, fob	..	..	..	1,092	1,271	1,580	2,090	2,491	2,774	3,151	4,621
Nonfactor Services	..	..	..	306	331	331	397	508	568	561	715
Factor Services	..	..	..	141	152	190	243	331	347	360	343
Imports of Goods & Services	..	..	..	1,842	2,065	2,376	3,196	4,327	4,162	4,785	6,864
Merchandise, fob	..	..	..	1,524	1,711	1,958	2,617	3,578	3,218	3,728	5,517
Nonfactor Services	..	..	..	220	239	269	350	407	550	572	739
Factor Services	..	..	..	98	115	149	229	342	395	485	608
Long-Term Interest	..	..	..	..	..	..	..	..	..	..	..
Current Transfers, net	..	..	..	75	82	89	97	101	102	88	92
Workers' Remittances	..	..	..	0	0	0	0	0	0	0	0
Total to be Financed	..	..	..	-228	-229	-186	-370	-895	-403	-634	-1,066
Official Capital Grants	..	..	..	30	29	36	116	203	290	228	495
Current Account Balance	..	..	..	-198	-200	-150	-254	-692	-113	-405	-571
Long-Term Capital, net	..	..	..	132	14	68	392	421	677	884	691
Direct Investment	..	..	..	32	25	31	52	51	158	173	136
Long-Term Loans	..	..	..	..	..	..	..	..	..	..	..
Disbursements	..	..	..	..	..	..	..	..	..	..	..
Repayments	..	..	..	..	..	..	..	..	..	..	..
Other Long-Term Capital	..	..	..	70	-40	1	224	167	230	483	59
Other Capital, net	..	..	..	58	394	168	-137	403	-214	-149	284
Change in Reserves	..	..	..	8	-209	-87	-1	-132	-350	-329	-404
Memo Item:				*(Irish Pounds per US dollar)*							
Conversion Factor (Annual Avg)	0.360	0.420	0.420	0.420	0.410	0.400	0.410	0.430	0.450	0.560	0.570
EXTERNAL DEBT, ETC.				*(Millions US dollars, outstanding at end of year)*							
Public/Publicly Guar. Long-Term	..	..	..	..	..	..	..	..	..	..	..
Official Creditors	..	..	..	..	..	..	..	..	..	..	..
IBRD and IDA	..	..	..	..	..	..	..	..	..	..	..
Private Creditors	..	..	..	..	..	..	..	..	..	..	..
Private Non-guaranteed Long-Term	..	..	..	..	..	..	..	..	..	..	..
Use of Fund Credit	..	..	..	..	..	..	..	..	..	..	..
Short-Term Debt	..	..	..	..	..	..	..	..	..	..	..
Memo Items:				*(Millions of US dollars)*							
Int'l Reserves Excluding Gold	414.0	465.7	652.1	680.7	978.0	1,109.4	1,007.2	1,247.2	1,512.6	1,818.5	2,351.1
Gold Holdings (at market price)	25.1	94.6	39.2	17.1	19.9	29.7	48.1	83.4	62.7	60.2	78.0
SOCIAL INDICATORS											
Total Fertility Rate	3.8	3.8	3.8	3.9	4.0	3.9	3.7	3.6	3.4	3.3	3.3
Crude Birth Rate	21.1	21.0	21.5	21.9	22.7	22.7	22.4	22.1	21.1	21.0	21.1
Infant Mortality Rate	24.4	21.0	20.6	19.5	18.0	18.0	18.0	17.8	17.5	15.5	15.5
Life Expectancy at Birth	71.1	71.1	71.1	71.1	71.1	71.3	71.4	71.6	71.7	71.9	72.0
Food Production, p.c. ('79-81 = 100)	86.1	84.5	83.6	81.9	91.4	86.4	81.2	93.9	106.4	92.0	101.6
Labor Force, Agriculture (%)	29.4	28.3	27.3	26.3	25.5	24.7	24.0	23.2	22.4	21.6	20.9
Labor Force, Female (%)	26.0	26.1	26.2	26.2	26.5	26.7	27.0	27.2	27.4	27.6	27.7
School Enroll. Ratio, primary	..	..	..	106.0	..	..	..	..	103.0	106.0	105.0
School Enroll. Ratio, secondary	..	..	..	74.0	..	..	..	..	86.0	91.0	92.0

1978	1979	1980	1981	1982	1983	1984	1985	1986	1987 estimate	NOTES	IRELAND
											FOREIGN TRADE (CUSTOMS BASIS)
				(Millions of current US dollars)							
5,681	7,173	8,473	7,784	8,060	8,609	9,627	10,399	12,604	15,970	..	Value of Exports, fob
2,565	3,059	3,509	2,913	2,782	2,772	2,907	3,016	3,732	4,900	..	Nonfuel Primary Products
23	34	55	51	51	99	116	132	99	115	..	Fuels
3,093	4,080	4,909	4,820	5,227	5,737	6,604	7,251	8,773	10,955	..	Manufactures
7,102	9,850	11,133	10,595	9,696	9,169	9,658	10,049	11,564	13,613	..	Value of Imports, cif
1,218	1,662	1,862	1,846	1,611	1,591	1,634	1,700	2,020	2,272	..	Nonfuel Primary Products
718	1,187	1,647	1,556	1,435	1,237	1,198	1,195	981	1,005	..	Fuels
5,166	7,001	7,624	7,193	6,651	6,342	6,825	7,154	8,563	10,336	..	Manufactures
				(Index 1980 = 100)							
108.0	110.6	100.0	106.4	86.5	89.2	91.3	107.3	110.0	106.7	..	Terms of Trade
78.5	95.4	100.0	103.2	79.2	79.7	79.4	91.0	104.3	110.3	..	Export Prices, fob
75.8	98.4	100.0	94.9	90.8	88.4	82.5	81.1	84.9	89.1	..	Nonfuel Primary Products
44.7	93.6	100.0	104.0	97.0	85.4	83.6	81.1	77.0	77.1	..	Fuels
82.4	93.9	100.0	109.0	74.0	76.0	78.0	96.0	116.0	124.0	..	Manufactures
72.7	86.3	100.0	97.0	91.5	89.3	86.9	84.8	94.8	103.3	..	Import Prices, cif
				(Millions of 1980 US dollars)							
7,234	7,520	8,473	7,540	10,181	10,802	12,129	11,434	12,089	14,484	..	Trade at Constant 1980 Prices Exports, fob
9,771	11,416	11,133	10,922	10,600	10,262	11,112	11,851	12,204	13,174	..	Imports, cif
				(Millions of current US dollars)							**BALANCE OF PAYMENTS**
7,283	9,017	9,788	9,634	9,731	9,988	11,129	12,596	14,943	18,944	..	Exports of Goods & Services
5,946	7,282	7,599	7,641	7,797	8,227	9,299	10,544	12,470	15,916	..	Merchandise, fob
896	1,169	1,381	1,218	1,222	1,176	1,167	1,296	1,598	2,024	..	Nonfactor Services
442	565	808	775	712	584	663	756	875	1,004	..	Factor Services
9,084	12,374	12,954	13,059	12,458	12,013	12,991	14,300	16,901	19,798	..	Imports of Goods & Services
7,075	9,714	9,651	9,875	8,920	8,461	9,082	9,890	11,316	13,245	..	Merchandise, fob
1,024	1,372	1,593	1,443	1,376	1,374	1,373	1,463	1,953	2,514	..	Nonfactor Services
984	1,288	1,710	1,740	2,163	2,178	2,535	2,947	3,632	4,039	..	Factor Services
..	..	..	..	..	..	..	..	..	..	..	Long-Term Interest
109	93	123	97	91	57	-33	-21	-55	-159	..	Current Transfers, net
0	0	0	0	0	0	0	0	0	0	..	Workers' Remittances
-1,678	-3,254	-3,049	-3,327	-2,620	-1,941	-1,884	-1,783	-2,033	-1,087	K	Total to be Financed
777	1,053	1,081	749	750	772	836	1,063	1,344	1,478	K	Official Capital Grants
-901	-2,201	-1,969	-2,578	-1,870	-1,168	-1,049	-720	-690	391	..	Current Account Balance
1,741	1,908	2,406	2,462	2,373	1,786	1,621	1,807	2,438	2,834	..	Long-Term Capital, net
375	337	286	204	242	169	120	..	..	..	..	Direct Investment
..	..	..	..	..	..	..	..	..	..	..	Long-Term Loans
..	..	..	..	..	..	..	..	..	..	..	Disbursements
..	..	..	..	..	..	..	..	..	..	..	Repayments
590	518	1,039	1,509	1,382	845	664	744	1,094	1,356	..	Other Long-Term Capital
-698	-293	275	-14	-382	-430	-614	-1,039	-1,845	-2,340	..	Other Capital, net
-142	586	-712	131	-121	-188	43	-48	96	-885	..	Change in Reserves
				(Irish Pounds per US dollar)							**Memo Item:**
0.520	0.490	0.490	0.620	0.700	0.800	0.920	0.950	0.740	0.670	..	Conversion Factor (Annual Avg)
			(Millions US dollars, outstanding at end of year)								**EXTERNAL DEBT, ETC.**
..	..	..	..	..	..	..	..	..	..	..	Public/Publicly Guar. Long-Term
..	..	..	..	..	..	..	..	..	..	..	Official Creditors
..	..	..	..	..	..	..	..	..	..	..	IBRD and IDA
..	..	..	..	..	..	..	..	..	..	..	Private Creditors
..	..	..	..	..	..	..	..	..	..	..	Private Non-guaranteed Long-Term
..	..	..	..	..	..	..	..	..	..	..	Use of Fund Credit
..	..	..	..	..	..	..	..	..	..	..	Short-Term Debt
				(Millions of US dollars)							**Memo Items:**
2,668.4	2,212.0	2,860.3	2,651.3	2,622.1	2,639.6	2,352.4	2,939.6	3,236.3	4,796.2	..	Int'l Reserves Excluding Gold
101.0	196.2	210.7	142.6	164.3	137.1	110.8	117.5	140.5	173.9	..	Gold Holdings (at market price)
											SOCIAL INDICATORS
3.2	3.2	3.2	3.1	3.0	2.9	2.8	2.7	2.6	2.5	..	Total Fertility Rate
21.2	21.5	21.8	21.0	20.5	19.9	19.4	18.9	18.4	17.9	..	Crude Birth Rate
14.9	12.8	11.1	11.6	12.3	13.1	13.8	14.7	15.6	16.5	..	Infant Mortality Rate
72.2	72.4	72.6	72.8	73.0	73.2	73.3	73.5	73.7	73.8	..	Life Expectancy at Birth
108.4	100.0	109.1	91.0	95.5	96.0	107.3	104.5	97.4	92.5	..	Food Production, p.c. ('79-81 = 100)
20.1	19.4	18.6	..	..	..	..	..	..	..	..	Labor Force, Agriculture (%)
27.8	28.0	28.1	28.3	28.4	28.6	28.8	28.9	29.0	29.1	..	Labor Force, Female (%)
104.0	100.0	100.0	99.0	100.0	100.0	100.0	..	..	..	..	School Enroll. Ratio, primary
92.0	90.0	90.0	92.0	93.0	95.0	96.0	..	..	..	..	School Enroll. Ratio, secondary

ISRAEL	1967	1968	1969	1970	1971	1972	1973	1974	1975	1976	1977
CURRENT GNP PER CAPITA (US $)	1,430	1,630	1,770	1,830	2,100	2,450	2,680	3,310	3,890	4,080	4,080
POPULATION (thousands)	2,745	2,803	2,877	2,974	3,069	3,148	3,278	3,377	3,455	3,533	3,613

ORIGIN AND USE OF RESOURCES

(Millions of current Israel New Sheqalims)

	1967	1968	1969	1970	1971	1972	1973	1974	1975	1976	1977
Gross National Product (GNP)	1.0E+00	1.0E+00	2.0E+00	2.0E+00	2.0E+00	3.0E+00	4.0E+00	6.0E+00	8.0E+00	1.1E+01	1.5E+01
Net Factor Income from Abroad	-2.0E-02	-2.4E-02	-2.8E-02	-3.5E-02	-4.3E-02	-5.7E-02	-1.1E-01	-1.6E-01	-2.8E-01	-3.5E-01	-3.6E-01
Gross Domestic Product (GDP)	1.0E+00	1.0E+00	2.0E+00	2.0E+00	2.0E+00	3.0E+00	4.0E+00	6.0E+00	8.0E+00	1.1E+01	1.6E+01
Indirect Taxes, net	..	..	..	..	..	..	..	..	..	..	..
GDP at factor cost	..	..	..	..	..	..	..	..	..	..	..
Agriculture	8.5E-02	9.0E-02	9.5E-02	1.0E-01	1.3E-01	1.5E-01	1.9E-01	2.9E-01	4.0E-01	5.7E-01	7.8E-01
Industry	3.1E-01	3.8E-01	4.7E-01	5.9E-01	7.5E-01	9.7E-01	1.0E+00	2.0E+00	2.0E+00	3.0E+00	4.0E+00
Manufacturing	..	..	..	..	..	..	..	..	..	..	..
Services, etc.	..	..	..	..	..	..	..	..	..	..	..
Resource Balance	-1.5E-01	-2.0E-01	-2.7E-01	-3.8E-01	-4.0E-01	-3.6E-01	-9.6E-01	-1.0E+00	-2.0E+00	-2.0E+00	-2.0E+00
Exports of Goods & NFServices	2.6E-01	3.9E-01	4.4E-01	4.9E-01	6.8E-01	8.7E-01	1.0E+00	2.0E+00	2.0E+00	3.0E+00	6.0E+00
Imports of Goods & NFServices	4.0E-01	5.9E-01	7.1E-01	8.7E-01	1.0E+00	1.0E+00	2.0E+00	3.0E+00	4.0E+00	5.0E+00	7.0E+00
Domestic Absorption	1.0E+00	2.0E+00	2.0E+00	2.0E+00	3.0E+00	4.0E+00	5.0E+00	7.0E+00	1.1E+01	1.3E+01	1.8E+01
Private Consumption, etc.	8.0E-01	9.1E-01	1.0E+00	1.0E+00	1.0E+00	2.0E+00	2.0E+00	3.0E+00	5.0E+00	6.0E+00	9.0E+00
General Gov't Consumption	3.5E-01	4.1E-01	4.9E-01	6.7E-01	7.9E-01	9.3E-01	2.0E+00	2.0E+00	3.0E+00	4.0E+00	5.0E+00
Gross Domestic Investment	2.0E-01	3.2E-01	4.3E-01	5.4E-01	7.4E-01	9.7E-01	1.0E+00	2.0E+00	2.0E+00	3.0E+00	4.0E+00
Fixed Investment	2.0E-01	2.9E-01	4.0E-01	5.0E-01	6.9E-01	9.2E-01	1.0E+00	2.0E+00	2.0E+00	3.0E+00	3.0E+00
Memo Items:											
Gross Domestic Saving	5.7E-02	1.2E-01	1.6E-01	1.5E-01	3.5E-01	6.1E-01	3.3E-01	5.0E-01	4.1E-01	7.8E-01	2.0E+00
Gross National Saving	1.6E-01	2.0E-01	2.5E-01	2.8E-01	5.1E-01	8.5E-01	6.9E-01	6.8E-01	6.2E-01	1.0E+00	2.0E+00

(Millions of 1980 Israel New Sheqalims)

	1967	1968	1969	1970	1971	1972	1973	1974	1975	1976	1977
Gross National Product	45.72	53.37	60.62	64.92	72.22	82.38	84.11	89.93	92.49	93.67	94.49
GDP at factor cost											
Agriculture	..	..	..	..	..	..	..	..	..	..	..
Industry	..	..	..	..	..	..	..	..	..	..	..
Manufacturing	..	..	..	..	..	..	..	..	..	..	..
Services, etc.	..	..	..	..	..	..	..	..	..	..	..
Resource Balance	-8.89	-10.93	-13.70	-18.30	-17.87	-14.74	-29.21	-26.55	-29.19	-22.95	-17.60
Exports of Goods & NFServices	12.33	16.63	18.01	19.63	24.36	27.52	29.03	30.70	31.25	36.10	40.06
Imports of Goods & NFServices	21.22	27.56	31.71	37.93	42.22	42.26	58.24	57.26	60.44	59.06	57.67
Domestic Absorption	55.98	65.67	75.86	85.01	92.06	99.07	116.37	119.67	125.40	120.42	115.07
Private Consumption, etc.	29.41	32.86	36.21	37.29	39.46	43.29	46.87	50.47	50.60	53.04	55.61
General Gov't Consumption	16.73	18.26	21.29	26.89	27.24	26.72	38.82	39.90	43.96	39.68	34.36
Gross Domestic Investment	9.83	14.56	18.37	20.83	25.37	29.06	30.68	29.29	30.83	27.70	25.09
Fixed Investment	..	..	..	19.02	23.12	26.30	29.11	28.20	28.32	25.28	22.40
Memo Items:											
Capacity to Import	13.52	18.17	19.70	21.26	26.70	29.77	30.86	31.39	31.55	37.47	43.22
Terms of Trade Adjustment	1.19	1.54	1.69	1.63	2.35	2.25	1.83	0.68	0.30	1.37	3.16
Gross Domestic Income	48.28	56.28	63.86	68.34	76.54	86.58	88.98	93.80	96.50	98.84	100.62
Gross National Income	46.91	54.91	62.31	66.55	74.56	84.63	85.94	90.61	92.79	95.03	97.66

DOMESTIC PRICES (DEFLATORS)

(Index 1980 = 100)

	1967	1968	1969	1970	1971	1972	1973	1974	1975	1976	1977
Overall (GDP)	2.6	2.6	2.7	2.9	3.3	3.9	4.8	6.5	8.8	11.4	16.2
Domestic Absorption	2.4	2.5	2.6	2.8	3.1	3.6	4.4	6.1	8.4	10.8	15.3
Agriculture	..	..	..	..	..	..	..	..	..	..	..
Industry	..	..	..	..	..	..	..	..	..	..	..
Manufacturing	..	..	..	..	..	..	..	..	..	..	..

MANUFACTURING ACTIVITY

	1967	1968	1969	1970	1971	1972	1973	1974	1975	1976	1977
Employment (1980=100)	62.3	70.7	75.2	79.0	84.0	88.7	93.9	96.2	94.7	95.3	97.0
Real Earnings per Empl. (1980=100)	33.7	34.8	36.7	38.7	39.2	41.7	39.2	38.7	44.3	49.1	51.6
Real Output per Empl. (1980=100)	..	..	..	..	..	..	..	..	..	..	..
Earnings as % of Value Added	40.9	37.0	36.3	36.2	36.7	36.9	36.5	33.4	35.4	37.5	36.4

MONETARY HOLDINGS

(Millions of current Israel New Sheqalims)

	1967	1968	1969	1970	1971	1972	1973	1974	1975	1976	1977
Money Supply, Broadly Defined	2.2E-01	2.5E-01	3.1E-01	2.8E-01	1.6E-01	2.0E+00	3.0E+00	3.0E+00	4.0E+00	7.0E+00	1.1E+01
Money as Means of Payment	1.0E-01	1.1E-01	1.2E-01	1.3E-01	1.6E-01	2.0E-01	1.0E+00	1.0E+00	1.0E+00	2.0E+00	2.0E+00
Currency Ouside Banks	1.0E-01	1.1E-01	1.2E-01	1.3E-01	1.6E-01	2.0E-01	2.8E-01	3.3E-01	4.2E-01	5.1E-01	6.9E-01
Demand Deposits	0.0E+00	0.0E+00	0.0E+00	0.0E+00	0.0E+00	0.0E+00	1.0E+00	1.0E+00	1.0E+00	1.0E+00	1.0E+00
Quasi-Monetary Liabilities	1.3E-01	1.4E-01	1.9E-01	1.5E-01	0.0E+00	2.0E+00	2.0E+00	2.0E+00	3.0E+00	5.0E+00	9.0E+00

GOVERNMENT DEFICIT (-) OR SURPLUS

(Millions of current Israel New Sheqalims)

	1967	1968	1969	1970	1971	1972	1973	1974	1975	1976	1977
Current Revenue	..	..	..	..	..	-1	-1	-1	-2	-2	-3
Current Expenditure	..	..	..	..	..	1	2	3	4	6	9
Current Budget Balance	..	..	..	..	..	1	2	3	5	7	10
Capital Receipts	..	..	..	..	..	0	-1	-1	-1	-1	-1
Capital Payments	..	..	..	..	..	..	..	..	..	..	..
						0	0	1	1	1	1

1978	1979	1980	1981	1982	1983	1984	1985	1986	1987 estimate	NOTES	ISRAEL
4,250	4,730	5,320	6,090	6,250	6,480	6,390	6,210	6,220	6,810	..	**CURRENT GNP PER CAPITA (US $)**
3,690	3,786	3,878	3,955	4,029	4,100	4,168	4,233	4,304	4,374	..	**POPULATION (thousands)**
				(Millions of current Israel New Sheqalims)							**ORIGIN AND USE OF RESOURCES**
25	47	111	263	598	1,533	7,321	27,478	41,873	53,345	f	Gross National Product (GNP)
-1	-2	-4	-8	-21	-56	-384	-1,268	-1,956	-2,461	..	Net Factor Income from Abroad
26	49	115	271	620	1,589	7,705	28,746	43,830	55,806	f	Gross Domestic Product (GDP)
..	..	..	..	..	..	..	..	..	..	..	Indirect Taxes, net
..	..	..	..	..	..	..	..	..	..	f	GDP at factor cost
1	2	5	12	22	45	242	1,034	..	..	..	Agriculture
8	14	27	61	140	355	1,771	5,946	..	..	..	Industry
..	..	..	..	..	..	..	..	..	..	..	Manufacturing
..	..	..	..	..	..	..	..	..	..	..	Services, etc.
-4	-7	-13	-36	-77	-182	-732	-1,965	-2,645	..	..	Resource Balance
11	19	46	104	211	504	2,849	11,914	16,531	..	..	Exports of Goods & NFServices
15	26	59	141	288	686	3,581	13,879	19,177	..	f	Imports of Goods & NFServices
30	56	128	307	697	1,771	8,436	30,711	46,475	..	..	Domestic Absorption
15	28	60	145	338	891	4,030	15,810	25,399	..	..	Private Consumption, etc.
9	15	44	108	225	532	2,848	10,150	13,423	..	..	General Gov't Consumption
6	12	25	54	135	348	1,558	4,751	7,653	..	..	Gross Domestic Investment
6	12	24	57	131	345	1,480	4,964	6,934	..	..	Fixed Investment
											Memo Items:
2	5	11	18	57	166	826	2,786	5,007	..	..	Gross Domestic Saving
3	6	12	22	60	162	685	2,552	4,757	..	..	Gross National Saving
				(Millions of 1980 Israel New Sheqalims)							
98.79	104.35	110.88	116.22	116.40	119.52	118.54	122.17	125.44	131.25	f	Gross National Product
..	..	..	..	..	..	..	..	..	..	f	GDP at factor cost
..	..	..	..	..	..	..	..	..	..	..	Agriculture
..	..	..	..	..	..	..	..	..	..	..	Industry
..	..	..	..	..	..	..	..	..	..	..	Manufacturing
..	..	..	..	..	..	..	..	..	..	..	Services, etc.
-20.91	-19.12	-13.31	-16.76	-20.07	-22.80	-16.78	-11.33	-14.53	..	..	Resource Balance
42.20	43.68	46.02	47.98	46.66	47.70	54.24	58.16	61.34	..	..	Exports of Goods & NFServices
63.11	62.80	59.33	64.74	66.73	70.49	71.02	69.50	75.87	..	f	Imports of Goods & NFServices
123.07	127.71	128.44	136.79	141.53	147.93	142.52	140.04	145.83	..	..	Domestic Absorption
60.08	64.85	59.79	66.96	71.72	77.43	71.82	71.57	80.90	..	..	Private Consumption, etc.
37.24	34.01	43.85	46.47	43.23	41.18	43.62	45.02	39.88	..	..	General Gov't Consumption
25.75	28.84	24.80	23.36	26.58	29.32	27.08	23.44	25.05	..	..	Gross Domestic Investment
23.68	26.55	23.95	24.84	25.87	29.47	25.69	23.09	21.53	..	..	Fixed Investment
											Memo Items:
45.63	45.70	46.02	47.97	48.82	51.77	56.51	59.66	65.40	..	..	Capacity to Import
3.43	2.02	0.00	-0.01	2.16	4.08	2.26	1.49	4.06	..	..	Terms of Trade Adjustment
105.59	110.60	115.13	120.03	123.63	129.21	128.01	130.20	135.36	..	..	Gross Domestic Income
102.23	106.37	110.88	116.21	118.57	123.60	120.80	123.66	129.50	..	..	Gross National Income
				(Index 1980 = 100)							**DOMESTIC PRICES (DEFLATORS)**
25.3	44.7	100.0	225.7	510.3	1269.8	6127.2	22335.7	33381.5	40645.0	..	Overall (GDP)
24.5	43.6	100.0	224.8	492.6	1197.2	5919.4	21931.1	31870.0	..	..	Domestic Absorption
..	..	..	..	..	..	..	..	..	..	..	Agriculture
..	..	..	..	..	..	..	..	..	..	..	Industry
..	..	..	..	..	..	..	..	..	..	..	Manufacturing
											MANUFACTURING ACTIVITY
100.2	104.7	100.0	102.7	106.7	109.2	109.2	107.6	..	..	J	Employment (1980 = 100)
49.6	88.5	100.0	102.7	109.9	129.8	51.7	..	..	..	J	Real Earnings per Empl. (1980 = 100)
..	..	..	..	..	..	..	..	..	..	J	Real Output per Empl. (1980 = 100)
24.0	51.1	53.5	52.0	67.6	68.0	47.7	45.2	..	..	J	Earnings as % of Value Added
				(Millions of current Israel New Sheqalims)							**MONETARY HOLDINGS**
19	38	93	224	543	1,671	10,190	27,361	33,039	42,037	..	Money Supply, Broadly Defined
3	3	8	13	28	68	305	1,051	2,238	3,346	..	Money as Means of Payment
1	1	3	4	9	25	123	481	974	1,365	..	Currency Ouside Banks
2	2	5	9	20	43	182	570	1,264	1,981	..	Demand Deposits
16	35	85	211	515	1,603	9,885	26,310	30,801	38,691	..	Quasi-Monetary Liabilities
				(Millions of current Israel New Sheqalims)							
-3	-7	-18	-58	-101	-412	-1,440	-951	..	..	C	**GOVERNMENT DEFICIT (-) OR SURPLUS**
14	31	70	158	381	1,108	5,791	20,949	..	..	..	Current Revenue
15	32	78	190	424	1,319	6,587	19,267	..	..	..	Current Expenditure
-1	-2	-8	-32	-42	-211	-796	1,682	..	..	..	Current Budget Balance
..	..	..	..	..	..	..	..	..	..	..	Capital Receipts
2	5	10	26	58	202	644	2,634	..	..	..	Capital Payments

ISRAEL	1967	1968	1969	1970	1971	1972	1973	1974	1975	1976	1977
FOREIGN TRADE (CUSTOMS BASIS)					*(Millions of current US dollars)*						
Value of Exports, fob	555	640	724	776	960	1,149	1,509	1,825	1,941	2,416	3,083
Nonfuel Primary Products	169	183	207	229	263	286	320	394	455	535	630
Fuels	0	0	0	0	0	0	0	0	0	0	0
Manufactures	385	456	516	547	697	863	1,189	1,431	1,486	1,880	2,453
Value of Imports, cif	1,045	1,351	1,646	2,084	2,375	2,447	4,235	5,380	5,975	5,645	5,745
Nonfuel Primary Products	307	330	354	468	489	484	840	1,094	1,327	1,158	1,124
Fuels	64	79	88	102	119	122	297	800	914	931	876
Manufactures	675	942	1,204	1,513	1,767	1,841	3,098	3,486	3,734	3,556	3,745
					(Index 1980 = 100)						
Terms of Trade	132.8	134.7	130.3	143.4	139.0	122.3	120.9	116.3	117.7	113.7	108.2
Export Prices, fob	30.4	30.9	31.8	37.2	39.7	38.5	45.2	63.9	67.2	66.3	68.7
Nonfuel Primary Products	38.0	38.5	40.0	40.0	38.2	41.3	51.1	57.8	63.4	65.8	69.7
Fuels	4.3	4.3	4.3	4.3	5.6	6.2	8.9	36.7	35.7	38.4	42.0
Manufactures	28.1	28.8	29.5	36.3	40.4	37.7	43.8	65.8	68.5	66.4	68.5
Import Prices, cif	22.9	23.0	24.4	26.0	28.6	31.5	37.4	54.9	57.1	58.3	63.5
Trade at Constant 1980 Prices					*(Millions of 1980 US dollars)*						
Exports, fob	1,828	2,068	2,276	2,083	2,415	2,986	3,340	2,857	2,888	3,644	4,486
Imports, cif	4,570	5,885	6,744	8,027	8,307	7,779	11,329	9,795	10,458	9,684	9,043
BALANCE OF PAYMENTS					*(Millions of current US dollars)*						
Exports of Goods & Services	..	..	..	1,402	1,874	2,120	2,697	3,563	3,688	4,397	5,459
Merchandise, fob	..	..	..	808	1,002	1,219	1,563	2,004	2,178	2,669	3,403
Nonfactor Services	..	..	..	498	703	709	821	1,109	1,083	1,299	1,557
Factor Services	..	..	..	96	169	192	313	450	427	429	499
Imports of Goods & Services	..	..	..	2,614	3,096	3,219	5,263	6,756	7,537	7,466	7,818
Merchandise, fob	..	..	..	1,931	2,205	2,287	3,965	4,995	5,520	5,248	5,364
Nonfactor Services	..	..	..	438	593	549	759	967	1,079	1,269	1,462
Factor Services	..	..	..	245	298	383	539	794	938	949	992
Long-Term Interest	..	..	..	34	103	143	224	247	274	258	302
Current Transfers, net	..	..	..	446	550	708	1,139	740	759	776	803
Workers' Remittances	..	..	..	0	0	0	0	0	0	0	0
Total to be Financed	..	..	..	-766	-672	-391	-1,427	-2,453	-3,090	-2,293	-1,556
Official Capital Grants	..	..	..	204	240	341	1,050	990	1,332	1,657	1,269
Current Account Balance	..	..	..	-562	-432	-50	-377	-1,463	-1,758	-636	-287
Long-Term Capital, net	..	..	..	881	866	1,072	2,049	1,591	2,359	2,891	2,274
Direct Investment	..	..	..	40	53	114	149	84	43	41	75
Long-Term Loans	..	..	..	472	805	808	1,028	626	772	1,336	864
Disbursements	..	..	..	533	1,010	1,140	1,420	1,156	1,396	1,795	1,394
Repayments	..	..	..	61	206	332	392	529	624	459	529
Other Long-Term Capital	..	..	..	165	-232	-191	-178	-109	212	-143	65
Other Capital, net	..	..	..	-312	-223	-474	-1,124	-970	-757	-2,071	-1,773
Change in Reserves	..	..	..	-7	-211	-548	-548	842	156	-184	-214
Memo Item:					*(Israel New Sheqalims per US dollar)*						
Conversion Factor (Annual Avg)	3.00E-04	3.50E-04	3.50E-04	3.50E-04	3.70E-04	4.20E-04	4.20E-04	4.50E-04	6.30E-04	7.90E-04	1.00E-03
EXTERNAL DEBT, ETC.					*(Millions of US dollars, outstanding at end of year)*						
Public/Publicly Guar. Long-Term	..	..	..	2,274	2,996	3,585	4,527	5,208	5,915	7,184	8,086
Official Creditors	..	..	..	984	1,466	1,815	2,163	2,588	3,120	4,051	4,747
IBRD and IDA	..	..	..	83	88	94	96	101	112	113	110
Private Creditors	..	..	..	1,290	1,530	1,770	2,364	2,620	2,795	3,132	3,339
Private Non-guaranteed Long-Term	..	..	..	361	491	650	862	946	958	1,115	1,203
Use of Fund Credit	..	..	..	13	35	0	0	80	244	369	386
Short-Term Debt	..	..	..	..	..	..	..	..	..	..	1,599
Memo Items:					*(Millions of US dollars)*						
Int'l Reserves Excluding Gold	668.7	616.8	366.7	405.2	689.8	1,178.9	1,768.3	1,153.4	1,137.1	1,328.4	1,521.6
Gold Holdings (at market price)	46.1	55.2	46.0	46.3	54.1	74.2	123.1	205.0	154.4	148.6	192.0
SOCIAL INDICATORS											
Total Fertility Rate	3.8	3.8	3.8	3.8	3.8	3.8	3.7	3.7	3.6	3.6	3.5
Crude Birth Rate	23.7	24.9	25.6	26.1	28.0	27.2	27.0	27.5	27.6	27.9	26.3
Infant Mortality Rate	25.9	24.8	23.5	25.3	23.0	24.2	22.8	23.4	22.9	20.1	18.1
Life Expectancy at Birth	71.5	71.1	71.0	71.2	71.7	71.1	71.7	71.7	72.0	73.0	73.2
Food Production, p.c. ('79-81 = 100)	99.0	100.2	100.7	98.2	106.5	121.1	108.2	118.6	113.8	114.4	115.3
Labor Force, Agriculture (%)	11.1	10.6	10.2	9.7	9.3	9.0	8.6	8.3	7.9	7.6	7.2
Labor Force, Female (%)	28.7	29.1	29.5	30.0	30.3	30.7	31.1	31.5	31.8	32.2	32.5
School Enroll. Ratio, primary	..	..	..	96.0	..	..	..	..	97.0	96.0	97.0
School Enroll. Ratio, secondary	..	..	..	57.0	..	..	..	..	66.0	66.0	67.0

1978	1979	1980	1981	1982	1983	1984	1985	1986	1987 estimate	NOTES	ISRAEL
											FOREIGN TRADE (CUSTOMS BASIS)
				(Millions of current US dollars)							
3,924	4,553	5,540	5,664	5,280	5,112	5,803	6,256	7,135	8,750	..	Value of Exports, fob
730	898	988	1,074	1,034	989	1,065	1,044	1,083	1,375	..	Nonfuel Primary Products
0	1	1	0	0	1	1	1	1	2	..	Fuels
3,195	3,654	4,551	4,590	4,246	4,122	4,738	5,212	6,052	7,373	..	Manufactures
7,162	8,428	9,512	10,145	9,506	9,471	9,493	9,752	10,366	12,704	..	Value of Imports, cif
1,256	1,522	1,597	1,917	1,548	1,392	1,513	1,412	1,506	1,931	..	Nonfuel Primary Products
947	1,515	2,520	2,634	2,204	1,666	1,660	1,617	736	896	..	Fuels
4,959	5,391	5,396	5,595	5,754	6,413	6,320	6,724	8,124	9,877	..	Manufactures
				(Index 1980 = 100)							
104.2	104.2	100.0	94.5	95.4	94.8	93.8	94.1	91.5	89.4	..	Terms of Trade
73.9	87.5	100.0	96.2	91.9	89.7	87.0	85.4	87.2	93.9	..	Export Prices, fob
76.8	92.7	100.0	96.8	87.6	92.9	87.2	82.3	78.3	93.7	..	Nonfuel Primary Products
42.3	61.0	100.0	112.5	101.6	92.5	90.2	87.5	44.9	56.9	..	Fuels
73.3	86.3	100.0	96.0	93.0	89.0	87.0	86.0	89.0	94.0	..	Manufactures
70.9	84.0	100.0	101.8	96.3	94.7	92.8	90.7	95.3	105.1	..	Import Prices, cif
											Trade at Constant 1980 Prices
				(Millions of 1980 US dollars)							
5,310	5,204	5,540	5,891	5,746	5,697	6,668	7,329	8,184	9,314	..	Exports, fob
10,099	10,038	9,512	9,969	9,867	10,006	10,234	10,752	10,878	12,085	..	Imports, cif
				(Millions of current US dollars)							**BALANCE OF PAYMENTS**
6,619	8,030	9,787	10,403	10,227	10,135	10,493	10,892	11,703	13,778	..	Exports of Goods & Services
4,074	4,759	5,798	5,906	5,560	5,541	6,189	6,602	7,675	9,085	..	Merchandise, fob
1,921	2,276	2,711	2,697	2,561	2,729	2,845	3,037	2,931	3,524	..	Nonfactor Services
624	995	1,278	1,800	2,106	1,865	1,459	1,253	1,097	1,169	..	Factor Services
9,712	11,649	13,579	14,679	14,800	14,958	15,257	14,803	15,606	19,602	..	Imports of Goods & Services
6,693	7,884	8,948	9,411	8,802	8,753	8,779	9,017	9,614	12,891	..	Merchandise, fob
1,780	2,064	2,242	2,538	2,818	3,028	3,040	2,644	2,912	3,457	..	Nonfactor Services
1,239	1,701	2,389	2,730	3,180	3,177	3,438	3,142	3,080	3,254	..	Factor Services
360	584	1,011	1,188	1,630	1,466	1,432	1,790	1,802	1,864	..	Long-Term Interest
825	970	1,021	1,141	989	936	827	877	1,147	1,329	..	Current Transfers, net
0	0	0	0	0	0	0	0	0	0	..	Workers' Remittances
-2,268	-2,649	-2,771	-3,135	-3,584	-3,887	-3,937	-3,034	-2,756	-4,495	K	Total to be Financed
1,400	1,836	1,964	1,786	1,617	1,928	2,538	4,197	4,225	3,496	K	Official Capital Grants
-868	-813	-807	-1,349	-1,967	-1,959	-1,399	1,163	1,469	-999	..	Current Account Balance
2,416	3,086	3,179	2,914	2,740	4,269	3,625	4,117	4,532	3,745	..	Long-Term Capital, net
32	10	-13	3	-122	-32	21	47	39	148	..	Direct Investment
1,376	20	1,594	2,111	1,837	2,137	1,302	-19	367	218	..	Long-Term Loans
1,797	1,181	2,496	3,830	3,272	3,438	2,382	1,212	1,858	1,846	..	Disbursements
421	1,161	902	1,719	1,435	1,301	1,080	1,231	1,491	1,628	..	Repayments
-392	1,220	-366	-986	-592	236	-236	-108	-99	-117	..	Other Long-Term Capital
-603	-2,062	-1,831	-1,014	146	-2,877	-2,732	-4,881	-5,010	-2,098	..	Other Capital, net
-945	-211	-541	-551	-919	567	506	-399	-991	-648	..	Change in Reserves
											Memo Item:
				(Israel New Sheqalims per US dollar)							Conversion Factor (Annual Avg)
1.70E-03	2.50E-03	5.10E-03	1.10E-02	2.40E-02	5.60E-02	2.90E-01	1.18E+00	1.49E+00	1.59E+00	..	
				(Millions of US dollars, outstanding at end of year)							**EXTERNAL DEBT, ETC.**
9,208	10,339	11,500	13,664	14,614	14,985	15,836	15,833	16,445	16,767	..	Public/Publicly Guar. Long-Term
5,699	7,162	7,665	9,124	10,031	10,509	11,158	11,246	11,825	12,215	..	Official Creditors
114	113	111	109	98	87	63	60	60	61	..	IBRD and IDA
3,509	3,176	3,835	4,540	4,584	4,476	4,678	4,587	4,621	4,552	..	Private Creditors
1,641	2,119	2,730	2,794	3,246	4,057	4,353	4,735	5,211	5,729	..	Private Non-guaranteed Long-Term
396	295	200	102	30	0	0	0	0	..	..	Use of Fund Credit
2,555	3,016	3,100	4,181	4,354	4,228	3,581	3,529	3,386	3,837	..	Short-Term Debt
				(Millions of US dollars)							Memo Items:
2,625.1	3,063.5	3,351.4	3,496.7	3,839.3	3,651.2	3,060.3	3,680.2	4,659.6	5,876.1	..	Int'l Reserves Excluding Gold
264.6	630.3	703.9	474.2	495.3	387.2	313.5	332.6	397.5	492.3	..	Gold Holdings (at market price)
											SOCIAL INDICATORS
3.4	3.3	3.2	3.2	3.1	3.0	3.0	2.9	2.9	2.8	..	Total Fertility Rate
25.0	24.7	24.3	23.6	23.2	22.8	22.4	22.0	21.6	21.3	..	Crude Birth Rate
16.3	15.9	15.1	15.5	15.1	14.7	14.3	13.9	13.5	13.2	..	Infant Mortality Rate
73.1	73.0	72.9	73.7	74.5	74.6	74.7	74.8	74.9	75.0	..	Life Expectancy at Birth
111.0	107.2	98.7	94.1	102.4	111.0	103.4	113.6	97.5	99.4	..	Food Production, p.c. ('79-81 = 100)
6.9	6.5	6.2	..	..	..	..	..	..	..	..	Labor Force, Agriculture (%)
32.8	33.1	33.5	33.5	33.5	33.5	33.5	33.5	33.5	33.6	..	Labor Force, Female (%)
96.0	96.0	96.0	96.0	97.0	99.0	99.0	99.0	..	..	..	School Enroll. Ratio, primary
68.0	71.0	73.0	75.0	73.0	74.0	76.0	76.0	..	..	..	School Enroll. Ratio, secondary

ITALY	1967	1968	1969	1970	1971	1972	1973	1974	1975	1976	1977
CURRENT GNP PER CAPITA (US $)	1,650	1,810	1,980	2,160	2,320	2,580	3,060	3,560	3,840	4,160	4,470
POPULATION (thousands)	52,667	52,987	53,317	53,661	54,006	54,400	54,779	55,130	55,441	55,701	55,930

ORIGIN AND USE OF RESOURCES
(Billions of current Italian Lire)

	1967	1968	1969	1970	1971	1972	1973	1974	1975	1976	1977
Gross National Product (GNP)	54,044	58,614	64,756	72,810	79,322	86,917	103,708	127,367	143,772	179,753	218,517
Net Factor Income from Abroad	224	277	353	332	358	330	268	-247	-737	-808	-571
Gross Domestic Product (GDP)	53,820	58,337	64,403	72,478	78,964	86,587	103,440	127,614	144,509	180,561	219,088
Indirect Taxes, net	4,574	4,661	4,940	5,596	5,790	5,543	6,622	8,202	6,887	10,063	12,990
GDP at factor cost	49,246	53,677	59,464	66,882	73,174	81,044	96,819	119,412	137,623	170,498	206,097
Agriculture	4,945	4,692	5,171	5,290	5,473	5,581	7,205	8,362	9,961	11,591	13,843
Industry	20,206	22,007	24,498	28,352	30,168	32,805	39,867	50,780	56,010	72,029	86,597
Manufacturing	13,643	14,881	16,432	19,339	20,599	22,742	28,496	36,739	39,443	52,536	62,426
Services, etc.	28,669	31,638	34,735	38,836	43,322	48,202	56,368	68,472	78,538	96,941	118,649
Resource Balance	26	548	180	-502	-147	-348	-2,942	-6,342	-1,648	-4,172	-1,186
Exports of Goods & NFServices	6,814	7,782	8,932	10,026	11,167	12,735	15,272	22,916	26,380	35,984	45,719
Imports of Goods & NFServices	6,787	7,235	8,752	10,528	11,315	13,083	18,214	29,258	28,028	40,156	46,904
Domestic Absorption	53,794	57,790	64,224	72,980	79,111	86,936	106,382	133,956	146,158	184,734	220,274
Private Consumption, etc.	34,389	36,438	40,070	45,488	49,433	54,372	64,800	80,478	93,227	114,084	137,795
General Gov't Consumption	7,061	7,696	8,343	9,095	11,126	12,662	14,532	17,503	20,288	24,363	30,356
Gross Domestic Investment	12,345	13,655	15,811	18,398	18,552	19,902	27,051	35,975	32,642	46,287	52,122
Fixed Investment	13,948	15,742	18,004	20,581	21,394	22,915	27,982	35,365	35,756	43,504	51,301
Memo Items:											
Gross Domestic Saving	12,371	14,203	15,990	17,896	18,405	19,554	24,109	29,633	30,994	42,114	50,936
Gross National Saving	12,861	14,785	16,661	18,544	19,109	20,249	24,742	29,758	30,644	41,741	51,092

(Billions of 1980 Italian Lire)

	1967	1968	1969	1970	1971	1972	1973	1974	1975	1976	1977
Gross National Product	242,082	258,080	274,033	288,322	293,043	302,203	323,032	334,877	321,663	340,769	347,889
GDP at factor cost	219,841	235,598	250,844	263,963	269,499	281,022	301,062	314,142	307,848	323,262	328,067
Agriculture	19,988	19,400	19,725	19,549	19,649	18,194	19,469	19,832	20,492	19,652	19,553
Industry	91,844	100,512	108,021	114,141	113,597	117,647	127,948	133,827	121,873	133,452	135,294
Manufacturing	58,224	63,927	68,400	74,029	74,394	77,481	85,860	91,327	82,465	92,885	94,880
Services, etc.	129,220	136,915	144,742	153,268	158,426	165,171	174,761	181,862	180,954	189,193	193,954
Resource Balance	-6,434	-4,471	-7,848	-12,906	-11,324	-12,575	-17,044	-13,622	-5,159	-7,196	-2,884
Exports of Goods & NFServices	29,312	33,378	37,302	39,473	42,300	47,166	48,950	53,792	55,772	63,130	67,332
Imports of Goods & NFServices	35,746	37,849	45,150	52,380	53,625	59,741	65,993	67,414	60,931	70,326	70,216
Domestic Absorption	247,486	261,298	280,336	299,864	302,996	313,587	339,221	349,143	328,477	349,494	351,685
Private Consumption, etc.	145,705	152,349	163,851	178,244	183,496	189,757	200,641	205,877	203,233	210,228	215,178
General Gov't Consumption	38,773	40,783	41,924	43,001	45,477	47,881	48,902	50,434	52,111	53,461	54,668
Gross Domestic Investment	63,008	68,167	74,561	78,619	74,024	75,949	89,679	92,831	73,133	85,804	81,839
Fixed Investment	68,854	76,286	82,200	84,697	81,968	82,724	89,116	92,097	80,351	82,239	81,950
Memo Items:											
Capacity to Import	35,885	40,713	46,077	49,882	52,926	58,150	55,334	52,801	57,349	63,019	68,441
Terms of Trade Adjustment	6,573	7,336	8,776	10,408	10,625	10,984	6,385	-992	1,576	-111	1,109
Gross Domestic Income	247,624	264,163	281,263	297,366	302,297	311,997	328,562	334,529	324,895	342,187	349,909
Gross National Income	248,654	265,415	282,809	298,730	303,668	313,187	329,417	333,886	323,239	340,658	348,998

DOMESTIC PRICES (DEFLATORS)
(Index 1980 = 100)

	1967	1968	1969	1970	1971	1972	1973	1974	1975	1976	1977
Overall (GDP)	22.3	22.7	23.6	25.3	27.1	28.8	32.1	38.0	44.7	52.7	62.8
Domestic Absorption	21.7	22.1	22.9	24.3	26.1	27.7	31.4	38.4	44.5	52.9	62.6
Agriculture	24.7	24.2	26.2	27.1	27.9	30.7	37.0	42.2	48.6	59.0	70.8
Industry	22.0	21.9	22.7	24.8	26.6	27.9	31.2	37.9	46.0	54.0	64.0
Manufacturing	23.4	23.3	24.0	26.1	27.7	29.4	33.2	40.2	47.8	56.6	65.8

MANUFACTURING ACTIVITY

	1967	1968	1969	1970	1971	1972	1973	1974	1975	1976	1977
Employment (1980=100)	89.3	91.2	96.1	98.7	104.6	104.3	107.3	107.7	107.5	105.7	104.1
Real Earnings per Empl. (1980=100)	54.1	56.1	58.3	66.9	71.7	73.1	78.4	81.7	84.1	88.1	93.8
Real Output per Empl. (1980=100)	49.0	52.1	54.4	56.8	56.3	58.6	68.8	81.9	72.6	83.3	85.1
Earnings as % of Value Added	39.3	39.1	39.1	41.4	43.6	42.7	39.1	36.9	39.7	36.9	39.4

MONETARY HOLDINGS
(Billions of current Italian Lire)

	1967	1968	1969	1970	1971	1972	1973	1974	1975	1976	1977
Money Supply, Broadly Defined	37,648	42,257	47,122	53,126	62,207	73,448	88,795	103,300	127,920	154,690	190,060
Money as Means of Payment	18,866	21,109	24,471	31,185	37,100	43,506	54,089	59,162	67,120	79,776	96,880
Currency Ouside Banks	5,054	5,260	5,966	6,472	7,138	8,009	9,609	10,768	12,538	14,226	16,076
Demand Deposits	13,812	15,849	18,505	24,713	29,962	35,497	44,480	48,394	54,582	65,550	80,804
Quasi-Monetary Liabilities	18,782	21,148	22,651	21,941	25,107	29,942	34,706	44,138	60,800	74,914	93,180

GOVERNMENT DEFICIT (-) OR SURPLUS
(Billions of current Italian Lire)

	1967	1968	1969	1970	1971	1972	1973	1974	1975	1976	1977
GOVERNMENT DEFICIT (-) OR SURPLUS	..	..	..	..	..	..	-8,421	-12,086	-21,492	-19,805	-24,349
Current Revenue	..	..	..	..	..	..	24,431	31,594	36,722	48,554	61,380
Current Expenditure	..	..	..	..	..	..	26,129	32,728	41,838	52,372	63,195
Current Budget Balance	..	..	..	..	..	..	-1,698	-1,134	-5,116	-3,818	-1,815
Capital Receipts	..	..	..	..	..	..	8	2	1	12	16
Capital Payments	..	..	..	..	..	..	6,731	10,954	16,377	15,999	22,550

1978	1979	1980	1981	1982	1983	1984	1985	1986	1987 estimate	NOTES	ITALY
4,940	6,090	7,490	8,090	7,960	7,560	7,580	7,690	8,550	10,360	..	CURRENT GNP PER CAPITA (US $)
56,127	56,292	56,416	56,503	56,640	56,836	56,983	57,141	57,240	57,317	..	POPULATION (thousands)
				(Billions of current Italian Lire)							ORIGIN AND USE OF RESOURCES
255,980	312,232	391,101	465,726	541,502	627,373	715,939	800,386	886,214	965,530	..	Gross National Product (GNP)
-188	804	669	-2,323	-3,622	-4,196	-4,736	-5,360	-5,002	-4,733		Net Factor Income from Abroad
256,168	311,428	390,432	468,049	545,124	631,569	720,675	805,746	891,216	970,266	..	Gross Domestic Product (GDP)
14,145	15,317	22,454	24,999	29,751	39,925	45,089	50,153	54,337	66,956	..	Indirect Taxes, net
242,023	296,111	367,978	443,050	515,373	591,644	675,586	755,593	836,879	903,310	B	GDP at factor cost
16,216	19,222	22,305	24,812	27,944	33,306	33,929	36,582	38,704	..	..	Agriculture
100,039	121,656	152,506	176,307	200,927	225,679	253,255	277,807	306,394		..	Industry
71,585	87,699	109,501	125,932	141,022	155,184	173,802	191,607	208,830		..	Manufacturing
139,913	170,550	215,621	266,930	316,253	372,584	433,492	491,357	546,117		..	Services, etc.
1,332	-2,200	-16,223	-15,154	-14,381	-5,738	-12,764	-14,811	4,084	-1,719	..	Resource Balance
54,259	68,563	77,338	100,072	114,177	126,860	150,693	169,257	167,408	177,335	..	Exports of Goods & NFServices
52,927	70,763	93,561	115,226	128,559	132,598	163,457	184,067	163,325	179,054	..	Imports of Goods & NFServices
254,836	313,628	406,655	483,203	559,505	637,307	733,439	820,557	887,132	971,986	..	Domestic Absorption
160,239	193,872	243,259	290,737	341,351	395,052	449,906	504,826	553,482	604,984	..	Private Consumption, etc.
36,856	45,826	58,055	75,353	88,808	104,830	119,223	134,792	146,780	165,528	..	General Gov't Consumption
57,742	73,930	105,341	117,113	129,347	137,425	164,310	180,939	186,870	201,474	..	Gross Domestic Investment
57,961	71,309	94,780	111,399	121,152	137,531	160,305	178,357	189,395	208,747	..	Fixed Investment
											Memo Items:
59,073	71,730	89,118	101,959	114,965	131,687	151,546	166,128	190,953	199,755	..	Gross Domestic Saving
59,822	73,711	90,960	101,276	113,330	129,626	149,366	163,298	188,123	196,685	..	Gross National Saving
				(Billions of 1980 Italian Lire)							
357,904	376,710	391,101	392,941	393,219	397,326	409,966	421,751	434,313	447,993	..	Gross National Product
338,194	357,116	367,978	374,091	374,404	374,695	386,875	398,132	409,483	418,206	B	GDP at factor cost
20,240	21,450	22,305	22,433	21,842	23,798	22,855	22,966	23,380	..	..	Agriculture
137,991	145,608	152,506	151,605	148,670	147,278	151,338	153,506	157,926	..	..	Industry
96,571	103,016	109,501	109,224	107,580	106,428	110,499	112,330	116,184	..	..	Manufacturing
199,939	208,675	215,621	220,834	225,315	228,907	238,483	248,105	255,525	..	..	Services, etc.
-1,812	-5,549	-16,223	-6,904	-7,200	-3,861	-7,134	-8,219	-9,797	-17,000	..	Resource Balance
74,124	80,853	77,338	83,116	82,201	84,069	90,452	93,927	97,161	100,644	..	Exports of Goods & NFServices
75,936	86,401	93,561	90,020	89,401	87,930	97,586	102,146	106,958	117,644	..	Imports of Goods & NFServices
359,981	381,281	406,655	401,776	403,027	403,844	419,811	432,797	446,627	467,394	..	Domestic Absorption
221,648	233,323	243,259	247,005	249,644	251,332	257,361	264,963	274,376	286,229	..	Private Consumption, etc.
55,918	56,862	58,055	59,600	61,319	63,092	64,722	67,004	69,102	71,432	..	General Gov't Consumption
82,415	91,096	105,341	95,171	92,064	89,420	97,727	100,830	103,149	109,733	..	Gross Domestic Investment
81,866	86,645	94,780	92,601	87,285	87,155	91,760	94,070	95,365	100,359	..	Fixed Investment
											Memo Items:
77,847	83,715	77,338	78,181	79,400	84,125	89,966	93,927	109,632	116,514	..	Capacity to Import
3,722	2,863	0	-4,935	-2,801	56	-486	0	12,471	15,870	..	Terms of Trade Adjustment
361,892	378,595	390,432	389,937	393,026	400,039	412,190	424,577	449,302	466,264	..	Gross Domestic Income
361,626	379,573	391,101	388,006	390,418	397,381	409,480	421,751	446,784	..	..	Gross National Income
				(Index 1980 = 100)							DOMESTIC PRICES (DEFLATORS)
71.5	82.9	100.0	118.5	137.7	157.9	174.6	189.8	204.0	215.4	..	Overall (GDP)
70.8	82.3	100.0	120.3	138.8	157.8	174.7	189.6	198.6	208.0	..	Domestic Absorption
80.1	89.6	100.0	110.6	127.9	140.0	148.5	159.3	165.5	..	..	Agriculture
72.5	83.6	100.0	116.3	135.1	153.2	167.3	181.0	194.0	..	..	Industry
74.1	85.1	100.0	115.3	131.1	145.8	157.3	170.6	179.7	..	..	Manufacturing
											MANUFACTURING ACTIVITY
101.5	101.6	100.0	95.6	90.1	93.5	93.5	91.2	..	..	..	Employment (1980 = 100)
96.6	99.4	100.0	102.4	101.3	97.9	104.4	100.8	..	..	..	Real Earnings per Empl. (1980 = 100)
86.5	95.1	100.0	106.4	105.8	107.4	117.9	113.6	..	..	..	Real Output per Empl. (1980 = 100)
39.7	37.9	37.5	37.7	38.4	45.2	45.7	43.1	..	..	..	Earnings as % of Value Added
				(Billions of current Italian Lire)							MONETARY HOLDINGS
234,160	277,250	315,960	349,710	415,770	467,860	517,410	571,654	618,619	..	D	Money Supply, Broadly Defined
122,680	151,750	171,320	188,080	219,620	248,720	279,460	308,650	342,540	368,260	..	Money as Means of Payment
18,992	21,626	25,301	29,612	33,061	37,787	42,542	46,329	49,863	53,789	..	Currency Ouside Banks
103,688	130,124	146,019	158,468	186,559	210,933	236,918	262,321	292,677	314,471	..	Demand Deposits
111,480	125,500	144,640	161,630	196,150	219,140	237,950	263,004	276,079	..	..	Quasi-Monetary Liabilities
				(Billions of current Italian Lire)							
-21,151	-27,452	-41,480	-52,017	-54,081	-71,513	-95,325	-108,220	-128,060	-158,850	F	GOVERNMENT DEFICIT (-) OR SURPLUS
76,365	94,591	123,672	145,290	186,110	230,750	253,130	278,500	334,359	356,883	..	Current Revenue
80,829	103,940	145,720	..	..	..	..	..	392,320	438,220	..	Current Expenditure
-4,464	-9,349	-22,048	..	..	..	..	..	-57,961	-81,337	..	Current Budget Balance
7	8	8	..	..	..	..	..	41	27	..	Capital Receipts
16,694	18,110	19,440	..	..	..	..	..	70,140	77,540	..	Capital Payments

ITALY	1967	1968	1969	1970	1971	1972	1973	1974	1975	1976	1977
FOREIGN TRADE (CUSTOMS BASIS)					*(Billions of current US dollars)*						
Value of Exports, fob	8.70	10.18	11.73	13.21	15.11	18.55	22.22	30.25	34.83	36.97	45.06
Nonfuel Primary Products	1.22	1.27	1.42	1.56	1.77	2.18	2.52	3.32	3.79	3.78	4.90
Fuels	0.53	0.61	0.60	0.67	0.82	0.82	1.26	2.36	2.03	2.10	2.53
Manufactures	6.95	8.31	9.71	10.98	12.52	15.55	18.44	24.57	29.01	31.08	37.63
Value of Imports, cif	9.70	10.25	12.45	14.94	15.97	19.28	27.79	40.68	37.93	42.79	46.68
Nonfuel Primary Products	4.32	4.43	5.22	5.93	6.13	7.56	11.16	13.82	12.52	13.81	14.86
Fuels	1.59	1.67	1.82	2.09	2.69	2.90	3.92	10.83	10.26	11.05	12.00
Manufactures	3.79	4.15	5.41	6.91	7.15	8.82	12.71	16.03	15.15	17.93	19.82
Terms of Trade	183.1	172.3	156.5	161.0	158.7	154.0	139.3	103.4	117.9	105.3	108.1
Export Prices, fob	28.3	26.7	26.4	28.3	30.1	34.4	43.2	53.1	60.3	57.6	64.0
Nonfuel Primary Products	35.8	35.8	38.5	38.1	36.3	40.2	54.3	64.1	64.8	64.6	66.6
Fuels	8.5	7.3	5.4	5.9	7.0	8.5	24.5	36.6	33.6	38.5	36.8
Manufactures	33.1	31.6	32.7	35.2	37.3	40.1	44.3	54.1	63.3	58.9	67.1
Import Prices, cif	15.5	15.5	16.8	17.6	19.0	22.4	31.0	51.4	51.1	54.7	59.2
Trade at Constant 1980 Prices					*(Billions of 1980 US dollars)*						
Exports, fob	30.73	38.15	44.51	46.70	50.23	53.87	51.42	56.94	57.80	64.21	70.40
Imports, cif	62.70	66.20	73.92	85.05	84.25	86.22	89.60	79.18	74.21	78.28	78.86
BALANCE OF PAYMENTS					*(Billions of current US dollars)*						
Exports of Goods & Services	..	..	..	18.88	21.38	25.99	31.15	40.88	45.59	47.66	58.38
Merchandise, fob	..	..	..	13.12	14.97	18.44	22.01	30.03	34.56	36.85	44.76
Nonfactor Services	..	..	..	4.13	4.58	5.37	6.32	7.17	8.28	8.44	10.75
Factor Services	..	..	..	1.63	1.83	2.18	2.82	3.69	2.74	2.37	2.88
Imports of Goods & Services	..	..	..	18.28	20.04	24.42	33.91	48.88	46.41	50.80	56.15
Merchandise, fob	..	..	..	13.50	14.85	18.39	25.96	38.54	35.72	41.09	44.90
Nonfactor Services	..	..	..	3.44	3.66	4.14	5.29	6.25	7.00	6.57	7.85
Factor Services	..	..	..	1.35	1.53	1.89	2.66	4.10	3.68	3.14	3.40
Long-Term Interest	..	..	..	..	..	..	..	..	..	..	..
Current Transfers, net	..	..	..	0.51	0.56	0.63	0.63	0.57	0.59	0.52	0.82
Workers' Remittances	..	..	..	0.45	0.52	0.58	0.62	0.54	0.52	0.46	0.71
Total to be Financed	..	..	..	1.10	1.90	2.20	-2.13	-7.42	-0.23	-2.61	3.06
Official Capital Grants	..	..	..	-0.32	-0.31	-0.19	-0.38	-0.61	-0.34	-0.24	-0.61
Current Account Balance	..	..	..	0.77	1.58	2.01	-2.51	-8.03	-0.57	-2.85	2.45
Long-Term Capital, net	..	..	..	0.05	-0.39	-1.29	2.68	1.09	-0.93	-0.62	-0.33
Direct Investment	..	..	..	0.50	0.12	0.41	0.36	0.40	0.29	-0.06	0.59
Long-Term Loans	..	..	..	..	..	..	..	..	..	..	..
Disbursements	..	..	..	..	..	..	..	..	..	..	..
Repayments	..	..	..	..	..	..	..	..	..	..	..
Other Long-Term Capital	..	..	..	-0.12	-0.20	-1.51	2.71	1.30	-0.87	-0.32	-0.30
Other Capital, net	..	..	..	-0.44	0.29	-1.56	-0.58	2.11	-1.03	5.32	3.81
Change in Reserves	..	..	..	-0.39	-1.48	0.85	0.41	4.84	2.53	-1.85	-5.93
Memo Item:					*(Italian Lire per US dollar)*						
Conversion Factor (Annual Avg)	625.000	625.000	625.000	625.000	619.900	583.200	583.000	650.300	652.800	832.300	882.400
EXTERNAL DEBT, ETC.					*(Millions US dollars, outstanding at end of year)*						
Public/Publicly Guar. Long-Term	..	..	..	..	..	..	..	..	..	..	..
Official Creditors	..	..	..	..	..	..	..	..	..	..	..
IBRD and IDA	..	..	..	..	..	..	..	..	..	..	..
Private Creditors	..	..	..	..	..	..	..	..	..	..	..
Private Non-guaranteed Long-Term	..	..	..	..	..	..	..	..	..	..	..
Use of Fund Credit	..	..	..	..	..	..	..	..	..	..	..
Short-Term Debt	..	..	..	..	..	..	..	..	..	..	..
Memo Items:					*(Millions of US dollars)*						
Int'l Reserves Excluding Gold	3,063	2,418	2,089	2,465	3,689	2,955	2,953	3,406	1,306	3,223	8,104
Gold Holdings (at market price)	2,414	3,500	2,973	3,082	3,595	5,346	9,259	15,383	11,568	11,114	13,676
SOCIAL INDICATORS											
Total Fertility Rate	2.5	2.5	2.5	2.4	2.4	2.4	2.3	2.3	2.2	2.1	1.9
Crude Birth Rate	18.0	17.6	17.5	16.8	16.8	16.3	16.0	15.8	14.9	14.0	13.3
Infant Mortality Rate	33.2	32.7	30.8	29.6	28.5	27.0	26.2	22.9	21.1	19.5	18.1
Life Expectancy at Birth	71.0	71.2	71.4	71.7	71.9	72.1	72.3	72.4	72.6	72.7	73.6
Food Production, p.c. ('79-81 = 100)	89.2	86.6	89.9	89.9	90.7	83.4	89.7	92.2	94.2	89.8	92.7
Labor Force, Agriculture (%)	22.4	21.2	20.0	18.8	18.1	17.4	16.8	16.1	15.4	14.7	14.0
Labor Force, Female (%)	27.9	28.2	28.4	28.7	29.0	29.2	29.5	29.8	30.1	30.4	30.7
School Enroll. Ratio, primary	..	..	..	110.0	..	..	..	..	106.0	105.0	103.0
School Enroll. Ratio, secondary	..	..	..	61.0	..	..	..	..	70.0	72.0	73.0

1978	1979	1980	1981	1982	1983	1984	1985	1986	1987 estimate	NOTES	ITALY
											FOREIGN TRADE (CUSTOMS BASIS)
				(Billions of current US dollars)							
56.05	72.24	77.64	75.25	73.44	72.67	73.36	78.94	97.81	116.58	..	Value of Exports, fob
5.26	7.36	7.45	7.75	7.13	6.78	7.07	7.96	9.36	10.45	..	Nonfuel Primary Products
3.29	4.75	4.40	4.72	4.99	3.89	3.32	3.69	2.73	2.84	..	Fuels
47.49	60.13	65.80	62.77	61.31	62.00	62.96	67.29	85.72	103.29	..	Manufactures
55.11	76.16	98.12	89.00	83.83	78.32	81.97	88.59	99.77	122.21	..	Value of Imports, cif
17.32	23.48	26.00	20.66	20.54	19.64	20.72	22.69	25.73	30.70	..	Nonfuel Primary Products
13.35	18.35	27.34	30.74	27.15	24.54	23.18	23.72	17.34	16.66	..	Fuels
24.45	34.33	44.78	37.59	36.14	34.14	38.08	42.18	56.71	74.85	..	Manufactures
				(Index 1980 = 100)							
110.1	105.8	100.0	83.4	93.4	93.0	92.5	94.8	112.9	89.9	..	Terms of Trade
71.4	87.7	100.0	83.2	87.0	83.5	81.1	80.7	97.1	88.0	..	Export Prices, fob
75.9	93.8	100.0	92.3	81.5	87.2	80.7	78.0	83.7	88.4	..	Nonfuel Primary Products
44.7	93.6	100.0	104.0	97.0	85.4	83.6	81.1	49.5	52.9	..	Fuels
74.1	86.7	100.0	81.0	87.0	83.0	81.0	81.0	102.0	89.6	..	Manufactures
64.8	82.9	100.0	99.7	93.2	89.8	87.6	85.1	86.0	97.9	..	Import Prices, cif
				(Billions of 1980 US dollars)							Trade at Constant 1980 Prices
78.54	82.36	77.64	90.44	84.38	87.03	90.47	97.84	100.74	132.48	..	Exports, fob
85.02	91.83	98.12	89.23	89.99	87.19	93.54	104.09	115.98	124.79	..	Imports, cif
				(Billions of current US dollars)							**BALANCE OF PAYMENTS**
72.87	95.10	105.27	101.67	98.87	97.09	99.58	103.03	128.36	155.56	..	Exports of Goods & Services
55.41	71.37	77.05	76.24	72.81	72.08	73.83	75.86	96.35	116.09	..	Merchandise, fob
13.39	17.50	19.89	16.80	17.45	18.00	18.16	19.23	22.98	28.88	..	Nonfactor Services
4.08	6.23	8.33	8.63	8.61	7.01	7.58	7.94	9.03	10.60	..	Factor Services
66.29	90.07	116.71	111.98	106.31	97.14	103.68	107.79	124.08	155.63	..	Imports of Goods & Services
52.51	72.34	93.98	88.17	81.70	74.57	79.65	82.09	92.02	116.01	..	Merchandise, fob
9.52	12.29	15.17	13.14	13.05	12.50	13.29	14.47	17.63	23.03	..	Nonfactor Services
4.26	5.44	7.56	10.67	11.56	10.06	10.74	11.23	14.43	16.59	..	Factor Services
..	..	..	..	..	..	..	..	..	..	..	Long-Term Interest
1.10	1.42	1.37	1.44	1.47	1.41	1.45	1.32	1.46	1.28	..	Current Transfers, net
0.92	1.15	1.24	1.17	1.19	1.14	1.12	1.08	1.20	1.21	..	Workers' Remittances
7.68	6.45	-10.07	-8.87	-5.97	1.36	-2.65	-3.43	5.74	1.21	K	Total to be Financed
-1.49	-0.97	0.09	-0.54	-0.36	0.07	0.17	-0.27	-3.08	-2.27	K	Official Capital Grants
6.19	5.48	-9.98	-9.41	-6.32	1.43	-2.48	-3.70	2.67	-1.06	..	Current Account Balance
-0.46	-1.38	3.37	7.78	4.51	0.75	0.75	1.96	-6.40	2.94	..	Long-Term Capital, net
0.34	-0.18	-0.16	-0.25	-0.39	-0.94	-0.69	-0.83	-2.76	1.74	..	Direct Investment
..	..	..	..	..	..	..	..	..	..	..	Long-Term Loans
..	..	..	..	..	..	..	..	..	..	..	Disbursements
..	..	..	..	..	..	..	..	..	..	..	Repayments
0.69	-0.23	3.44	8.58	5.25	1.62	1.28	3.06	-0.56	3.47	..	Other Long-Term Capital
1.29	-0.72	7.32	0.92	-2.45	3.56	4.59	-5.89	5.95	3.55	..	Other Capital, net
-7.02	-3.37	-0.71	0.72	4.26	-5.74	-2.87	7.63	-2.22	-5.43	..	Change in Reserves
				(Italian Lire per US dollar)							Memo Item:
848.700	830.900	856.400	1,136.800	1,352.500	1,518.900	1,757.000	1,909.400	1,490.800	1,296.100	..	Conversion Factor (Annual Avg)
				(Millions US dollars, outstanding at end of year)							**EXTERNAL DEBT, ETC.**
..	..	..	..	..	..	..	..	..	..	..	Public/Publicly Guar. Long-Term
..	..	..	..	..	..	..	..	..	..	..	Official Creditors
..	..	..	..	..	..	..	..	..	..	..	IBRD and IDA
..	..	..	..	..	..	..	..	..	..	..	Private Creditors
..	..	..	..	..	..	..	..	..	..	..	Private Non-guaranteed Long-Term
..	..	..	..	..	..	..	..	..	..	..	Use of Fund Credit
..	..	..	..	..	..	..	..	..	..	..	Short-Term Debt
				(Millions of US dollars)							Memo Items:
11,109	18,198	23,126	20,134	14,091	20,105	20,795	15,595	19,987	30,214	..	Int'l Reserves Excluding Gold
18,786	34,157	39,302	26,501	30,462	25,435	20,555	21,801	26,062	32,275	..	Gold Holdings (at market price)
											SOCIAL INDICATORS
1.8	1.7	1.7	1.6	1.5	1.5	1.5	1.5	1.5	1.5	..	Total Fertility Rate
12.7	11.9	11.4	11.0	11.0	11.0	11.0	10.9	10.9	10.9	..	Crude Birth Rate
16.8	15.5	14.3	14.1	13.1	12.2	11.4	10.6	9.8	9.2	..	Infant Mortality Rate
74.0	74.3	74.7	75.1	75.5	75.7	76.0	76.3	76.5	76.8	..	Life Expectancy at Birth
92.4	96.2	102.4	101.5	98.8	107.3	98.8	100.1	99.1	103.6	..	Food Production, p.c. ('79-81 = 100)
13.4	12.7	12.0	..	..	..	..	..	..	..	..	Labor Force, Agriculture (%)
30.9	31.2	31.5	31.5	31.6	31.6	31.7	31.7	31.8	31.8	..	Labor Force, Female (%)
102.0	102.0	101.0	..	100.0	99.0	99.0	98.0	97.0	..	..	School Enroll. Ratio, primary
73.0	73.0	72.0	..	73.0	73.0	74.0	75.0	76.0	..	..	School Enroll. Ratio, secondary

JAMAICA	1967	1968	1969	1970	1971	1972	1973	1974	1975	1976	1977
CURRENT GNP PER CAPITA (US $)	580	620	650	720	760	900	950	1,080	1,260	1,330	1,450
POPULATION (thousands)	1,797	1,818	1,842	1,869	1,900	1,936	1,974	2,010	2,013	2,041	2,063
ORIGIN AND USE OF RESOURCES					*(Millions of current Jamaica Dollars)*						
Gross National Product (GNP)	776	854	941	1,120	1,208	1,414	1,692	2,211	2,620	2,627	2,856
Net Factor Income from Abroad	-55	-49	-51	-51	-74	-25	-27	52	20	-69	-98
Gross Domestic Product (GDP)	831	903	993	1,171	1,282	1,439	1,720	2,159	2,600	2,696	2,954
Indirect Taxes, net	88	83	87	98	114	125	146	180	239	237	216
GDP at factor cost	744	820	905	1,073	1,169	1,314	1,574	1,980	2,362	2,460	2,739
Agriculture	78	78	77	78	102	107	120	154	191	214	249
Industry	333	373	419	499	511	533	633	820	956	992	1,092
Manufacturing	137	152	160	184	207	242	287	387	444	491	545
Services, etc.	420	452	497	593	670	799	968	1,185	1,453	1,490	1,613
Resource Balance	-6	-47	-67	-49	-91	-119	-165	-221	-269	-239	-44
Exports of Goods & NFServices	300	333	365	389	434	472	543	770	917	783	928
Imports of Goods & NFServices	306	380	431	438	525	591	707	991	1,186	1,022	972
Domestic Absorption	838	951	1,059	1,220	1,374	1,558	1,885	2,380	2,869	2,935	2,998
Private Consumption, etc.	521	550	608	713	803	968	1,063	1,469	1,722	1,882	2,024
General Gov't Consumption	73	85	103	137	159	197	280	386	477	562	612
Gross Domestic Investment	243	315	349	369	412	393	542	525	670	491	361
Fixed Investment	219	283	315	367	356	367	448	478	610	451	349
Memo Items:											
Gross Domestic Saving	237	268	283	320	320	274	377	304	401	253	318
Gross National Saving	181	233	246	292	268	277	381	387	441	185	234
					(Millions of 1980 Jamaica Dollars)						
Gross National Product	4,007.2	4,291.8	4,459.0	5,024.7	5,099.3	5,812.1	5,844.7	5,871.5	5,738.7	5,154.6	4,967.5
GDP at factor cost	3,869.6	4,133.3	4,307.0	4,847.8	4,974.1	5,416.1	5,436.3	5,271.0	5,180.7	4,834.6	4,775.0
Agriculture	396.3	381.9	363.2	380.1	430.7	438.4	377.3	395.5	402.2	405.0	416.9
Industry	1,994.1	2,142.5	2,130.7	2,440.7	2,508.2	2,655.4	2,644.6	2,611.7	2,506.9	2,216.8	2,083.6
Manufacturing	872.8	895.0	899.1	955.3	977.5	1,091.8	1,098.7	1,061.7	1,085.8	1,038.1	961.0
Services, etc.	1,921.2	2,015.9	2,218.9	2,457.1	2,505.9	2,827.7	2,909.7	2,730.5	2,785.1	2,672.2	2,647.6
Resource Balance	-142.4	-317.6	-477.2	-383.7	-501.8	-524.9	-622.8	-729.8	-1,242.1	-932.3	-561.3
Exports of Goods & NFServices	2,237.6	2,412.5	2,587.7	2,637.2	2,882.5	2,988.7	2,953.3	2,933.9	2,610.1	2,262.0	2,295.9
Imports of Goods & NFServices	2,380.0	2,730.0	3,064.9	3,020.9	3,384.3	3,513.6	3,576.1	3,663.7	3,852.2	3,194.3	2,857.2
Domestic Absorption	4,454.0	4,857.9	5,190.0	5,661.6	5,946.6	6,446.4	6,554.4	6,467.6	6,936.3	6,226.3	5,709.3
Private Consumption, etc.	2,795.5	2,857.8	2,959.3	3,269.5	3,539.7	4,191.0	3,569.5	4,003.1	4,282.5	3,997.2	3,919.3
General Gov't Consumption	339.6	374.3	408.8	546.4	554.7	623.4	789.0	821.2	834.3	948.8	958.3
Gross Domestic Investment	1,318.8	1,625.8	1,822.0	1,845.8	1,852.3	1,632.0	2,195.9	1,643.3	1,819.5	1,280.4	831.7
Fixed Investment	1,257.3	1,557.6	1,624.7	1,799.4	1,593.2	1,489.3	1,753.7	1,465.4	1,625.4	1,160.8	790.9
Memo Items:											
Capacity to Import	2,331.0	2,390.2	2,592.5	2,683.0	2,795.7	2,804.4	2,743.0	2,847.2	2,978.2	2,448.6	2,729.3
Terms of Trade Adjustment	93.4	-22.2	4.8	45.8	-86.7	-184.4	-210.3	-86.7	368.1	186.6	433.4
Gross Domestic Income	4,405.0	4,518.1	4,717.6	5,323.7	5,358.1	5,737.2	5,721.3	5,651.1	6,062.3	5,480.6	5,581.4
Gross National Income	4,100.6	4,269.6	4,463.8	5,070.4	5,012.5	5,627.8	5,634.4	5,784.8	6,106.8	5,341.2	5,400.9
DOMESTIC PRICES (DEFLATORS)					*(Index 1980 = 100)*						
Overall (GDP)	19.3	19.9	21.1	22.2	23.6	24.3	29.0	37.6	45.7	50.9	57.4
Domestic Absorption	18.8	19.6	20.4	21.5	23.1	24.2	28.8	36.8	41.4	47.1	52.5
Agriculture	19.7	20.3	21.1	20.4	23.6	24.3	31.7	38.9	47.6	52.9	59.7
Industry	16.7	17.4	19.7	20.5	20.4	20.1	23.9	31.4	38.1	44.8	52.4
Manufacturing	15.7	17.0	17.8	19.3	21.2	22.2	26.1	36.5	40.9	47.3	56.7
MANUFACTURING ACTIVITY											
Employment (1980=100)	90.4	98.8	100.5	104.3	108.1	110.6	113.9	123.6	120.4	118.9	105.5
Real Earnings per Empl. (1980=100)	102.1	99.2	113.1	111.4	111.9	129.5	132.7	118.4	129.6	135.8	147.4
Real Output per Empl. (1980=100)	..	..	..	..	..	..	..	..	..	..	..
Earnings as % of Value Added	40.9	40.7	44.2	42.6	41.5	44.2	46.5	42.5	46.3	47.7	45.9
MONETARY HOLDINGS					*(Millions of current Jamaica Dollars)*						
Money Supply, Broadly Defined	221	285	342	393	492	581	667	868	962	1,030	1,145
Money as Means of Payment	76	95	111	127	160	173	218	258	322	339	474
Currency Ouside Banks	28	32	37	46	58	72	82	102	127	138	182
Demand Deposits	48	63	74	80	102	101	136	156	195	200	292
Quasi-Monetary Liabilities	145	190	231	266	332	408	448	609	640	691	671
GOVERNMENT DEFICIT (-) OR SURPLUS					*(Millions of current Jamaica Dollars)*						
Current Revenue	..	..	..	..	..	..	..	..	-253	-395	-497
Current Expenditure	..	..	..	..	..	..	..	..	682	710	734
Current Budget Balance	..	..	..	..	..	..	..	..	585	740	814
Capital Receipts	..	..	..	..	..	..	..	..	97	-30	-80
Capital Payments	..	..	..	..	..	..	..	..	0	..	..
	..	..	..	..	..	..	..	..	350	365	416

1978	1979	1980	1981	1982	1983	1984	1985	1986	1987 estimate	NOTES	JAMAICA
1,420	1,280	1,160	1,250	1,310	1,370	1,150	920	870	960	A	CURRENT GNP PER CAPITA (US $)
2,088	2,112	2,133	2,162	2,200	2,241	2,280	2,311	2,336	2,365	..	POPULATION (thousands)
											ORIGIN AND USE OF RESOURCES
				(Millions of current Jamaica Dollars)							
3,586	4,018	4,432	4,999	5,578	6,745	8,453	9,685	11,758	13,869	..	Gross National Product (GNP)
-152	-259	-318	-293	-289	-248	-902	-1,466	-1,552	-1,848	..	Net Factor Income from Abroad
3,737	4,277	4,750	5,292	5,867	6,993	9,355	11,151	13,310	15,717	..	Gross Domestic Product (GDP)
272	358	387	517	643	592	934	1,470	1,940	2,467	..	Indirect Taxes, net
3,465	3,919	4,363	4,774	5,225	6,401	8,421	9,681	11,370	13,251	B	GDP at factor cost
297	308	392	398	396	451	544	672	820	957	..	Agriculture
1,485	1,722	1,804	1,866	1,977	2,449	3,525	4,106	5,350	6,422	..	Industry
638	699	771	865	1,054	1,399	1,732	2,237	3,028	3,506	..	Manufacturing
1,956	2,246	2,554	3,028	3,494	4,093	5,286	6,373	7,140	8,338	..	Services, etc.
50	-68	-99	-547	-679	-844	-624	-1,148	293	45	..	Resource Balance
1,575	2,065	2,426	2,510	2,240	2,621	4,956	6,521	7,294	8,618	..	Exports of Goods & NFServices
1,525	2,133	2,525	3,058	2,919	3,465	5,580	7,669	7,001	8,574	..	Imports of Goods & NFServices
3,687	4,344	4,849	5,839	6,546	7,837	9,979	12,299	13,017	15,673	..	Domestic Absorption
2,375	2,697	3,136	3,667	4,034	4,874	6,275	7,761	8,504	9,820	..	Private Consumption, etc.
750	825	966	1,095	1,288	1,406	1,550	1,752	2,045	2,295	..	General Gov't Consumption
562	823	748	1,077	1,224	1,557	2,155	2,786	2,468	3,558	..	Gross Domestic Investment
499	748	690	954	1,168	1,436	1,981	2,494	2,335	3,422	..	Fixed Investment
											Memo Items:
612	755	648	530	545	713	1,530	1,638	2,761	3,603	..	Gross Domestic Saving
483	620	476	457	496	647	945	1,023	1,820	2,391	..	Gross National Saving
				(Millions of 1980 Jamaica Dollars)							
4,948.7	4,761.5	4,431.9	4,617.1	4,675.6	4,856.0	4,488.5	4,106.7	4,223.0	4,446.8	..	Gross National Product
4,796.4	4,642.5	4,363.2	4,392.7	4,366.4	4,597.2	4,471.3	4,100.5	4,120.9	4,284.9	B	GDP at factor cost
456.9	409.7	392.2	400.9	369.3	396.1	435.9	420.5	411.7	422.0	..	Agriculture
2,059.4	1,993.1	1,803.6	1,824.3	1,787.7	1,840.2	1,773.7	1,655.5	1,729.7	1,848.8	..	Industry
916.1	869.2	771.3	779.3	836.0	852.2	816.5	820.7	850.5	895.0	..	Manufacturing
2,653.8	2,662.1	2,554.5	2,644.3	2,749.8	2,788.1	2,761.4	2,653.3	2,687.9	2,816.4	..	Services, etc.
-326.4	-74.0	-99.1	-158.4	-394.2	-345.1	-424.9	-585.2	-254.7	-378.4	..	Resource Balance
2,413.6	2,462.7	2,425.8	2,521.8	2,209.8	2,418.2	2,845.1	3,163.2	3,375.7	3,782.1	..	Exports of Goods & NFServices
2,740.0	2,536.7	2,524.9	2,680.1	2,604.0	2,763.3	3,270.1	3,748.4	3,630.4	4,160.5	..	Imports of Goods & NFServices
5,496.5	5,138.9	4,849.4	5,027.9	5,301.0	5,369.5	5,395.8	5,314.4	5,084.0	5,465.6	..	Domestic Absorption
3,573.9	3,257.2	3,135.7	3,060.7	3,359.1	3,403.2	3,598.3	3,542.7	3,482.8	3,698.6	..	Private Consumption, etc.
995.6	986.8	966.2	987.3	1,002.6	992.6	926.4	900.9	895.4	910.7	..	General Gov't Consumption
927.1	894.9	747.5	980.0	939.3	973.7	871.1	870.8	705.9	856.4	..	Gross Domestic Investment
804.1	791.8	690.1	840.3	863.9	851.3	743.4	705.7	633.7	786.3	..	Fixed Investment
											Memo Items:
2,829.8	2,456.3	2,425.8	2,200.4	1,998.3	2,090.3	2,904.3	3,187.4	3,782.2	4,182.1	..	Capacity to Import
416.2	-6.4	0.0	-321.4	-211.6	-327.9	59.1	24.2	406.5	400.0	..	Terms of Trade Adjustment
5,586.3	5,058.4	4,750.3	4,548.2	4,695.3	4,696.5	5,030.0	4,753.4	5,235.8	5,487.3	..	Gross Domestic Income
5,364.9	4,755.1	4,431.9	4,295.8	4,464.0	4,528.1	4,547.6	4,130.9	4,629.5	4,846.9	..	Gross National Income
				(Index 1980 = 100)							DOMESTIC PRICES (DEFLATORS)
72.3	84.4	100.0	108.7	119.6	139.2	188.2	235.8	275.6	309.0	..	Overall (GDP)
67.1	84.5	100.0	116.1	123.5	146.0	184.9	231.4	256.0	286.8	..	Domestic Absorption
65.0	75.3	100.0	99.2	107.2	113.7	124.9	159.7	199.2	226.7	..	Agriculture
72.1	86.4	100.0	102.3	110.6	133.1	198.7	248.0	309.3	347.4	..	Industry
69.7	80.4	100.0	111.0	126.1	164.1	212.1	272.6	356.1	391.7	..	Manufacturing
											MANUFACTURING ACTIVITY
102.3	105.8	100.0	..	..	..	..	..	..	..	..	Employment (1980 = 100)
132.8	107.8	100.0	..	..	..	..	..	..	..	..	Real Earnings per Empl. (1980 = 100)
..	..	100.0	..	..	..	..	..	..	..	..	Real Output per Empl. (1980 = 100)
46.3	45.9	46.6	..	..	..	..	..	..	..	..	Earnings as % of Value Added
				(Millions of current Jamaica Dollars)							MONETARY HOLDINGS
1,353	1,538	1,842	2,358	3,015	3,897	4,692	6,009	7,623	8,930	D	Money Supply, Broadly Defined
570	629	716	775	876	1,066	1,319	1,520	2,140	2,252	..	Money as Means of Payment
173	220	260	282	316	375	436	540	729	844	..	Currency Ouside Banks
397	410	457	492	560	691	882	980	1,410	1,407	..	Demand Deposits
783	909	1,126	1,584	2,139	2,831	3,373	4,489	5,483	6,679	..	Quasi-Monetary Liabilities
				(Millions of current Jamaica Dollars)							
-514	-543	-739	-721	..	..	..	..	..	..	C	GOVERNMENT DEFICIT (-) OR SURPLUS
1,130	1,149	1,382	1,672	..	..	..	..	..	..	..	Current Revenue
..	..	..	..	..	..	..	..	..	..	..	Current Expenditure
..	..	..	..	..	..	..	..	..	..	..	Current Budget Balance
..	1	..	..	..	..	..	..	..	..	..	Capital Receipts
..	..	..	..	..	..	..	..	..	..	..	Capital Payments

JAMAICA

	1967	1968	1969	1970	1971	1972	1973	1974	1975	1976	1977
FOREIGN TRADE (CUSTOMS BASIS)					*(Millions of current US dollars)*						
Value of Exports, fob	228.9	215.4	247.3	334.9	330.2	366.1	390.2	730.9	783.9	632.6	778.0
Nonfuel Primary Products	140.1	188.7	218.8	171.3	170.2	180.9	184.7	293.2	340.8	257.5	323.1
Fuels	12.2	6.5	7.1	8.7	9.2	10.4	9.4	10.9	12.1	17.6	17.7
Manufactures	76.5	20.2	21.4	155.0	150.8	174.7	196.1	426.8	431.1	357.5	437.2
Value of Imports, cif	348.9	387.3	436.0	525.4	550.5	620.2	664.4	935.5	1,122.5	911.2	858.8
Nonfuel Primary Products	86.0	86.1	90.2	107.9	115.6	155.6	180.9	262.8	271.4	250.5	219.3
Fuels	28.4	28.0	30.8	34.0	53.2	56.1	72.5	195.5	215.3	207.3	249.6
Manufactures	234.5	273.1	314.9	383.5	381.7	408.6	410.9	477.3	635.8	453.4	389.9
Terms of Trade					*(Index 1980 = 100)*						
Terms of Trade	103.5	112.1	112.0	117.1	118.7	112.8	113.8	115.8	119.6	117.8	125.8
Export Prices, fob	21.7	24.3	25.2	28.0	28.3	30.3	38.0	63.5	67.4	66.3	73.3
Nonfuel Primary Products	27.2	28.1	29.3	29.7	29.0	30.8	35.2	60.3	71.3	69.1	79.0
Fuels	4.3	4.3	4.3	4.3	5.6	6.2	8.9	36.7	35.7	38.4	42.0
Manufactures	30.1	31.7	32.4	37.5	36.5	38.5	49.5	67.2	66.1	66.7	71.7
Import Prices, cif	21.0	21.6	22.5	24.0	23.9	26.9	33.4	54.8	56.3	56.3	58.3
Trade at Constant 1980 Prices					*(Millions of 1980 US dollars)*						
Exports, fob	1,055.1	887.4	981.4	1,194.7	1,165.7	1,207.0	1,026.8	1,151.4	1,163.7	954.0	1,061.0
Imports, cif	1,664.3	1,789.1	1,937.3	2,193.6	2,306.4	2,306.0	1,989.2	1,706.4	1,993.0	1,618.7	1,473.8
BALANCE OF PAYMENTS					*(Millions of current US dollars)*						
Exports of Goods & Services	..	..	..	537	559	630	649	1,056	1,122	943	1,007
Merchandise, fob	..	..	..	341	344	377	392	752	809	656	738
Nonfactor Services	..	..	..	158	175	206	199	238	242	222	212
Factor Services	..	..	..	38	40	47	58	66	71	64	57
Imports of Goods & Services	..	..	..	713	752	856	924	1,173	1,433	1,251	1,069
Merchandise, fob	..	..	..	449	476	529	570	811	970	792	667
Nonfactor Services	..	..	..	141	148	172	184	247	315	312	224
Factor Services	..	..	..	123	129	155	170	115	148	148	178
Long-Term Interest	..	..	..	64	61	63	81	102	110	96	99
Current Transfers, net	..	..	..	27	27	36	35	34	23	2	15
Workers' Remittances	..	..	..	29	36	52	51	49	47	48	41
Total to be Financed	..	..	..	-149	-167	-190	-240	-83	-288	-307	-47
Official Capital Grants	..	..	..	-4	-5	-7	-7	-9	5	4	5
Current Account Balance	..	..	..	-153	-172	-197	-248	-92	-283	-303	-42
Long-Term Capital, net	..	..	..	156	181	118	195	217	233	98	57
Direct Investment	..	..	..	161	174	97	72	23	-2	-1	-10
Long-Term Loans	..	..	..	10	26	35	150	229	193	-218	80
Disbursements	..	..	..	180	200	216	335	427	402	346	261
Repayments	..	..	..	170	174	182	185	198	209	564	181
Other Long-Term Capital	..	..	..	-10	-14	-7	-19	-27	37	313	-18
Other Capital, net	..	..	..	11	23	48	39	-52	-9	-2	-49
Change in Reserves	..	..	..	-15	-32	31	14	-73	59	206	34
Memo Item:					*(Jamaica Dollars per US dollar)*						
Conversion Factor (Annual Avg)	0.720	0.830	0.830	0.830	0.820	0.800	0.910	0.910	0.910	0.910	0.910
EXTERNAL DEBT, ETC.				*(Millions of US dollars, outstanding at end of year)*							
Public/Publicly Guar. Long-Term	..	..	..	160	188	218	354	529	695	923	969
Official Creditors	..	..	..	59	80	86	120	151	189	380	416
IBRD and IDA	..	..	..	30	34	35	37	38	44	59	68
Private Creditors	..	..	..	101	108	132	234	378	506	543	552
Private Non-guaranteed Long-Term	..	..	..	822	825	823	836	890	903	447	500
Use of Fund Credit	..	..	..	0	0	13	16	16	16	80	107
Short-Term Debt	..	..	..	..	..	..	..	..	..	..	103
Memo Items:					*(Millions of US dollars)*						
Int'l Reserves Excluding Gold	85.0	120.2	117.9	139.2	179.1	159.7	127.5	190.4	125.6	32.4	47.8
Gold Holdings (at market price)	..	..	..	..	..	..	..	..	..	..	1.9
SOCIAL INDICATORS											
Total Fertility Rate	5.4	5.4	5.4	5.4	5.4	5.4	5.1	4.4	4.3	4.1	3.9
Crude Birth Rate	37.3	36.3	35.4	34.4	33.5	32.5	31.6	30.8	29.9	29.1	28.2
Infant Mortality Rate	45.0	43.2	41.4	39.6	37.8	36.0	33.8	31.6	29.4	27.2	25.0
Life Expectancy at Birth	66.3	66.6	66.9	67.2	67.5	67.8	68.0	68.3	68.5	68.7	69.0
Food Production, p.c. ('79-81 = 100)	114.7	103.7	97.9	101.6	108.8	109.5	101.2	102.1	97.8	100.1	103.2
Labor Force, Agriculture (%)	35.6	34.8	34.0	33.2	33.0	32.8	32.5	32.3	32.1	31.9	31.8
Labor Force, Female (%)	41.6	41.9	42.2	42.5	42.9	43.3	43.7	44.2	44.6	44.9	45.1
School Enroll. Ratio, primary	..	..	..	119.0	..	..	..	..	97.0	97.0	98.0
School Enroll. Ratio, secondary	..	..	..	46.0	..	..	..	..	58.0	58.0	58.0

1978	1979	1980	1981	1982	1983	1984	1985	1986	1987 estimate	NOTES	JAMAICA
											FOREIGN TRADE (CUSTOMS BASIS)
				(Millions of current US dollars)							
746.9	817.3	964.6	985.3	739.2	713.0	714.0	549.0	596.0	649.0	..	Value of Exports, fob
286.7	354.8	333.9	304.2	269.1	230.7	213.1	180.6	192.7	210.7	..	Nonfuel Primary Products
18.8	33.9	19.2	27.8	26.5	23.3	21.9	16.4	9.3	13.3	..	Fuels
441.4	428.5	611.5	653.4	443.6	459.0	479.0	352.0	394.0	425.0	..	Manufactures
879.9	991.5	1,177.7	1,487.0	1,373.3	1,531.0	1,129.9	1,123.7	964.0	1,207.2	..	Value of Imports, cif
240.6	208.7	270.1	328.1	313.9	347.7	260.1	245.8	207.5	238.4	..	Nonfuel Primary Products
212.5	317.4	445.5	494.4	398.8	470.9	316.0	305.6	137.3	200.9	..	Fuels
426.8	465.4	462.1	664.5	660.6	712.4	553.8	572.3	619.2	767.9	..	Manufactures
				(Index 1980 = 100)							
117.6	111.3	100.0	94.3	93.5	94.5	94.8	95.2	109.2	99.9	..	Terms of Trade
76.9	87.7	100.0	97.9	90.5	88.7	87.1	85.3	93.6	94.0	..	Export Prices, fob
82.8	88.6	100.0	94.3	83.4	84.6	82.3	78.7	82.2	72.5	..	Nonfuel Primary Products
42.3	61.0	100.0	112.5	101.6	92.5	90.2	87.5	44.9	56.9	..	Fuels
76.1	90.0	100.0	99.1	94.8	90.7	89.3	89.0	103.2	113.0	..	Manufactures
65.4	78.7	100.0	103.8	96.7	93.8	91.9	89.6	85.7	94.1	..	Import Prices, cif
				(Millions of 1980 US dollars)							Trade at Constant 1980 Prices
970.8	932.3	964.5	1,006.6	816.8	804.0	819.7	643.7	636.9	690.2	..	Exports, fob
1,344.8	1,259.2	1,177.8	1,431.9	1,419.5	1,632.0	1,230.0	1,253.8	1,124.9	1,283.1	..	Imports, cif
				(Millions of current US dollars)							**BALANCE OF PAYMENTS**
1,170	1,221	1,422	1,500	1,371	1,332	1,335	1,266	1,411	1,641	f	Exports of Goods & Services
831	818	963	974	767	686	702	569	590	708	..	Merchandise, fob
282	347	396	426	477	534	562	603	735	839	..	Nonfactor Services
57	56	63	100	126	113	71	94	87	93	..	Factor Services
1,246	1,440	1,678	1,961	1,925	1,789	1,788	1,789	1,598	1,916	f	Imports of Goods & Services
750	883	1,038	1,297	1,209	1,124	1,037	1,004	837	1,067	..	Merchandise, fob
282	315	348	385	394	369	380	373	366	419	..	Nonfactor Services
213	242	292	279	322	295	371	412	394	430	..	Factor Services
98	104	121	110	140	156	191	213	219	231	..	Long-Term Interest
15	70	82	123	135	95	80	153	112	116	..	Current Transfers, net
27	49	51	63	75	42	26	92	54	51	..	Workers' Remittances
-61	-149	-175	-338	-419	-362	-372	-370	-75	-160	K	Total to be Financed
11	10	9	1	16	7	40	68	37	63	K	Official Capital Grants
-50	-139	-166	-337	-403	-355	-332	-302	-38	-96	..	Current Account Balance
-85	-15	240	118	365	144	430	201	-146	227	f	Long-Term Capital, net
-27	-26	28	-12	-16	-19	12	-9	-5	..	..	Direct Investment
-313	74	249	293	442	287	277	222	-3	95	..	Long-Term Loans
263	236	351	465	557	410	368	422	227	316	..	Disbursements
576	162	102	172	115	123	91	200	230	221	..	Repayments
244	-73	-46	-164	-77	-131	101	-80	-175	68	..	Other Long-Term Capital
78	-5	-11	-47	-12	102	-78	141	300	173	f	Other Capital, net
57	159	-63	266	50	109	-20	-40	-115	-303	..	Change in Reserves
				(Jamaica Dollars per US dollar)							Memo Item:
1.450	1.760	1.780	1.930	1.990	2.150	3.940	5.560	5.480	5.490	..	Conversion Factor (Annual Avg)
			(Millions of US dollars, outstanding at end of year)								**EXTERNAL DEBT, ETC.**
1,078	1,190	1,421	1,713	2,111	2,361	2,522	2,939	3,110	3,511	..	Public/Publicly Guar. Long-Term
528	648	911	1,217	1,603	1,897	2,031	2,391	2,584	2,936	..	Official Creditors
112	127	176	212	322	368	326	468	573	735	..	IBRD and IDA
550	542	510	496	509	464	491	548	526	575	..	Private Creditors
85	60	75	25	50	75	70	66	64	58	..	Private Non-guaranteed Long-Term
181	351	309	470	583	627	629	693	678	679	..	Use of Fund Credit
80	106	98	92	98	250	224	169	147	199	..	Short-Term Debt
				(Millions of US dollars)							Memo Items:
58.8	63.8	105.0	85.2	109.0	63.2	96.9	161.3	98.4	174.3	..	Int'l Reserves Excluding Gold
..	5.8	..	..	..	..	..	..	..	..	..	Gold Holdings (at market price)
											SOCIAL INDICATORS
3.8	3.7	3.6	3.5	3.4	3.4	3.3	3.3	..	2.9	..	Total Fertility Rate
28.2	28.2	27.5	27.5	27.9	27.4	25.2	24.3	22.6	25.6	..	Crude Birth Rate
24.2	23.4	22.6	21.8	21.0	20.3	19.6	18.8	18.1	17.4	..	Infant Mortality Rate
69.7	70.5	71.3	72.1	72.9	73.1	73.2	73.3	73.4	73.6	..	Life Expectancy at Birth
117.7	106.9	98.7	94.4	91.1	97.7	104.4	102.1	102.5	101.9	..	Food Production, p.c. ('79-81=100)
31.6	31.5	31.3	..	..	..	..	..	..	..	..	Labor Force, Agriculture (%)
45.4	45.7	46.0	45.9	45.9	45.9	45.8	45.8	45.8	45.8	..	Labor Force, Female (%)
99.0	99.0	101.0	104.0	106.0	106.0	..	..	..	..	..	School Enroll. Ratio, primary
60.0	60.0	59.0	61.0	60.0	58.0	..	..	..	..	..	School Enroll. Ratio, secondary

JAPAN	1967	1968	1969	1970	1971	1972	1973	1974	1975	1976	1977
CURRENT GNP PER CAPITA (US $)	1,190	1,430	1,680	1,930	2,130	2,540	3,230	3,800	4,490	4,970	5,690
POPULATION (millions)	101	101	103	104	106	107	108	111	112	113	114
ORIGIN AND USE OF RESOURCES					*(Billions of current Japanese Yen)*						
Gross National Product (GNP)	44,525	52,772	62,097	73,188	80,592	92,401	112,520	133,997	148,170	166,417	185,530
Net Factor Income from Abroad	-104	-151	-162	-157	-109	7	21	-247	-157	-156	-92
Gross Domestic Product (GDP)	44,629	52,923	62,259	73,345	80,701	92,394	112,498	134,244	148,327	166,573	185,622
Indirect Taxes, net	2,552	3,892	3,712	4,397	4,808	5,425	6,710	7,131	7,529	8,689	10,421
GDP at factor cost	42,077	49,032	58,548	68,948	75,893	86,969	105,789	127,113	140,798	157,884	175,201
Agriculture	4,045	4,242	4,421	4,488	4,274	5,050	6,675	7,506	8,141	8,870	9,402
Industry	19,944	23,498	28,392	34,230	37,282	42,134	52,186	59,956	62,901	70,747	76,862
Manufacturing	15,126	17,846	22,006	26,402	28,430	31,918	39,569	45,137	44,801	51,101	55,412
Services, etc.	20,640	25,183	29,446	34,627	39,145	45,209	53,638	66,782	77,285	86,956	99,358
Resource Balance	-8	490	886	796	2,084	2,150	47	-1,258	-112	1,162	2,931
Exports of Goods & NFServices	4,502	5,585	6,849	8,277	9,889	10,385	12,141	19,439	20,251	23,839	25,554
Imports of Goods & NFServices	4,511	5,095	5,963	7,482	7,805	8,235	12,094	20,697	20,363	22,676	22,624
Domestic Absorption	44,638	52,434	61,374	72,550	78,617	90,244	112,451	135,502	148,439	165,410	182,691
Private Consumption, etc.	27,929	32,122	37,157	37,984	42,734	49,334	59,666	72,370	84,142	94,970	105,982
General Gov't Consumption	3,410	3,934	4,558	5,455	6,421	7,537	9,336	12,240	14,890	16,417	18,243
Gross Domestic Investment	13,298	16,377	19,658	29,110	29,461	33,374	43,449	50,891	49,407	54,024	58,466
Fixed Investment	13,260	16,304	19,899	24,170	26,440	30,694	39,026	43,506	46,830	51,656	56,891
Memo Items:											
Gross Domestic Saving	13,290	16,867	20,544	29,906	31,545	35,523	43,496	49,633	49,295	55,186	61,397
Gross National Saving	13,179	16,705	20,375	29,738	31,422	35,491	43,491	49,358	49,109	54,989	61,289
					(Billions of 1980 Japanese Yen)						
Gross National Product	110,261	124,021	139,040	152,229	158,888	172,450	186,065	183,426	188,333	197,366	207,897
GDP at factor cost	103,363	114,013	129,968	142,784	148,901	161,500	174,049	173,457	178,537	186,818	195,843
Agriculture	..	..	..	9,082	8,567	9,713	10,336	10,112	10,071	9,601	9,402
Industry	..	..	..	61,947	65,896	71,504	79,123	77,152	76,053	80,997	84,078
Manufacturing	..	..	..	39,554	41,773	45,749	52,014	51,911	48,745	53,645	56,577
Services, etc.	..	..	..	81,550	84,657	91,218	96,568	96,519	102,418	106,960	114,524
Resource Balance	-6,805	-6,543	-6,699	-8,762	-7,763	-9,235	-14,386	-12,202	-8,657	-6,864	-5,141
Exports of Goods & NFServices	8,049	9,910	11,852	13,842	16,106	16,997	18,203	22,421	22,576	25,996	28,777
Imports of Goods & NFServices	14,854	16,453	18,551	22,604	23,870	26,232	32,589	34,623	31,234	32,861	33,918
Domestic Absorption	117,334	130,977	146,161	161,341	166,884	181,671	200,414	195,985	197,199	204,423	213,145
Private Consumption, etc.	68,184	73,686	81,014	86,535	91,433	99,688	108,746	108,940	113,767	117,717	122,619
General Gov't Consumption	12,623	13,347	14,023	14,819	15,591	16,461	17,357	17,892	19,096	19,990	20,865
Gross Domestic Investment	36,527	43,944	51,124	59,988	59,859	65,521	74,311	69,153	64,337	66,716	69,662
Fixed Investment	32,342	38,772	45,962	53,712	56,136	61,759	69,565	62,960	62,192	63,851	66,383
Memo Items:											
Capacity to Import	14,826	18,035	21,307	25,008	30,244	33,080	32,716	32,518	31,062	34,545	38,312
Terms of Trade Adjustment	6,777	8,125	9,455	11,166	14,138	16,083	14,513	10,097	8,486	8,549	9,535
Gross Domestic Income	117,306	132,559	148,917	163,745	173,258	188,518	200,541	193,880	197,028	206,108	217,539
Gross National Income	117,038	132,146	148,495	163,395	173,026	188,532	200,578	193,523	196,819	205,915	217,432
DOMESTIC PRICES (DEFLATORS)					*(Index 1980 = 100)*						
Overall (GDP)	40.4	42.5	44.6	48.1	50.7	53.6	60.5	73.0	78.7	84.3	89.2
Domestic Absorption	38.0	40.0	42.0	45.0	47.1	49.7	56.1	69.1	75.3	80.9	85.7
Agriculture	..	..	..	49.4	49.9	52.0	64.6	74.2	80.8	92.4	100.0
Industry	..	..	..	55.3	56.6	58.9	66.0	77.7	82.7	87.3	91.4
Manufacturing	..	..	..	66.7	68.1	69.8	76.1	87.0	91.9	95.3	97.9
MANUFACTURING ACTIVITY											
Employment (1980=100)	95.8	99.0	103.8	106.5	105.2	107.3	109.1	104.7	103.0	102.1	99.4
Real Earnings per Empl. (1980=100)	54.9	59.9	66.1	71.5	77.6	83.6	90.9	94.5	93.6	94.6	96.4
Real Output per Empl. (1980=100)	..	..	..	44.7	46.5	49.2	57.3	66.0	62.1	69.1	74.3
Earnings as % of Value Added	31.7	32.1	32.0	31.9	34.3	34.8	33.5	35.4	40.4	38.7	38.8
MONETARY HOLDINGS					*(Billions of current Japanese Yen)*						
Money Supply, Broadly Defined	47,084	54,063	64,068	74,890	92,738	115,670	138,980	158,620	185,070	213,450	241,890
Money as Means of Payment	13,369	15,155	18,282	21,360	27,693	34,526	40,311	44,951	49,949	56,179	60,786
Currency Ouside Banks	3,114	3,595	4,319	5,098	5,958	7,706	9,113	10,731	11,579	12,858	14,122
Demand Deposits	10,255	11,560	13,963	16,262	21,735	26,820	31,198	34,220	38,370	43,321	46,664
Quasi-Monetary Liabilities	33,715	38,908	45,786	53,530	65,045	81,144	98,669	113,669	135,121	157,271	181,104
GOVERNMENT DEFICIT (-) OR SURPLUS					*(Billions of current Japanese Yen)*						
	..	..	..	-79	-393	-1,740	-382	-3,357	-7,666	-9,417	-11,916
Current Revenue	..	..	..	7,963	8,741	10,409	14,204	15,993	14,690	16,635	18,560
Current Expenditure	..	..	..	6,343	6,861	8,696	11,075	14,958	17,375	20,539	23,630
Current Budget Balance	..	..	..	1,620	1,880	1,713	3,129	1,035	-2,685	-3,904	-5,070
Capital Receipts	..	..	..	227	276	100	102	24	11	20	33
Capital Payments	..	..	..	1,926	2,549	3,553	3,613	4,416	4,992	5,533	6,879

1978	1979	1980	1981	1982	1983	1984	1985	1986	1987 estimate	NOTES	JAPAN
7,020	8,620	9,870	10,390	10,280	10,320	10,580	11,270	12,840	15,760	..	**CURRENT GNP PER CAPITA (US $)**
115	116	117	118	118	119	120	121	121	122	..	**POPULATION (millions)**
				(Billions of current Japanese Yen)							**ORIGIN AND USE OF RESOURCES**
204,475	221,825	240,098	256,817	269,697	280,568	298,453	317,441	331,345	345,000	..	Gross National Product (GNP)
70	278	-78	-546	69	311	505	1,137	1,229	1,280	..	Net Factor Income from Abroad
204,405	221,547	240,176	257,363	269,628	280,257	297,948	316,304	330,117	343,720	..	Gross Domestic Product (GDP)
11,199	13,258	14,095	15,710	16,505	16,663	19,136	21,250	21,515	22,679	..	Indirect Taxes, net
193,206	208,289	226,081	241,653	253,123	263,594	278,812	295,054	308,601	321,041	B	GDP at factor cost
9,441	9,623	8,847	8,957	9,064	9,264	9,626	9,798	9,481	..	..	Agriculture
85,200	92,071	100,681	107,308	111,214	113,610	122,148	129,250	134,426	..	..	Industry
60,545	64,815	70,232	74,540	78,191	81,416	88,845	94,160	96,652	..	..	Manufacturing
109,764	119,852	130,649	141,098	149,350	157,382	166,173	177,255	186,209	..	..	Services, etc.
3,629	-1,740	-2,211	1,508	2,137	5,345	8,661	11,943	14,538	13,054	..	Resource Balance
24,116	27,897	35,724	41,824	44,490	43,486	49,996	52,090	43,437	43,791	..	Exports of Goods & NFServices
20,487	29,637	37,935	40,316	42,353	38,142	41,335	40,147	28,899	30,737	..	Imports of Goods & NFServices
200,775	223,286	242,387	255,855	267,491	274,912	289,287	304,361	315,579	330,666	..	Domestic Absorption
116,511	128,313	139,528	147,801	157,523	165,298	173,118	181,057	187,647	194,670	..	Private Consumption, etc.
19,753	21,486	23,568	25,585	26,796	27,996	29,449	30,685	32,571	33,101	..	General Gov't Consumption
64,512	73,487	79,292	82,469	83,171	81,618	86,720	92,619	95,360	102,895	..	Gross Domestic Investment
64,129	70,541	75,876	80,768	82,996	83,726	88,992	95,561	100,992	110,995	..	Fixed Investment
											Memo Items:
68,141	71,747	77,081	83,977	85,308	86,963	95,381	104,562	109,898	115,949	..	Gross Domestic Saving
68,156	71,974	76,948	83,385	85,355	87,231	95,855	105,632	111,027	117,087	..	Gross National Saving
				(Billions of 1980 Japanese Yen)							
218,689	230,250	240,098	248,916	256,592	264,907	278,353	292,030	299,070	311,880	..	Gross National Product
206,017	215,693	226,078	234,211	240,833	248,948	260,139	271,547	278,201	289,812	B	GDP at factor cost
9,358	9,493	8,847	8,578	9,015	9,123	9,314	9,259	8,870	..	..	Agriculture
90,237	95,217	100,681	104,434	108,440	112,801	122,087	129,222	131,279	..	..	Industry
60,388	64,451	70,232	73,416	77,653	83,873	93,569	100,116	100,717	..	..	Manufacturing
119,016	125,241	130,649	136,434	139,071	142,690	146,484	152,509	157,794	..	..	Services, etc.
-7,073	-10,065	-2,211	1,411	2,216	6,067	9,602	12,580	8,475	6,374	..	Resource Balance
28,573	30,356	35,724	41,287	42,779	44,556	52,352	55,298	52,397	54,331	..	Exports of Goods & NFServices
35,646	40,421	37,935	39,876	40,563	38,489	42,750	42,718	43,922	47,957	..	Imports of Goods & NFServices
225,683	240,016	242,387	248,035	254,310	258,548	268,283	278,410	289,467	304,327	..	Domestic Absorption
129,048	137,138	139,528	141,741	147,036	151,655	155,567	159,203	164,197	170,395	..	Private Consumption, etc.
21,961	22,920	23,568	24,698	25,178	25,919	26,638	27,097	28,862	28,705	..	General Gov't Consumption
74,674	79,959	79,292	81,597	82,096	80,973	86,079	92,110	96,409	105,227	..	Gross Domestic Investment
72,055	75,850	75,876	78,264	78,894	78,690	82,553	87,350	92,568	102,111	..	Fixed Investment
											Memo Items:
41,960	38,049	35,724	41,367	42,610	43,882	51,707	55,426	66,016	68,325	..	Capacity to Import
13,387	7,693	0	80	-169	-673	-644	128	13,619	13,994	..	Terms of Trade Adjustment
231,998	237,644	240,176	249,526	256,357	263,941	277,240	291,118	311,562	324,694	..	Gross Domestic Income
232,076	237,943	240,098	248,997	256,422	264,233	277,709	292,159	312,690	325,873	..	Gross National Income
				(Index 1980 = 100)							**DOMESTIC PRICES (DEFLATORS)**
93.5	96.3	100.0	103.2	105.1	105.9	107.2	108.7	110.8	110.6	..	Overall (GDP)
89.0	93.0	100.0	103.2	105.2	106.3	107.8	109.3	109.0	108.7	..	Domestic Absorption
100.9	101.4	100.0	104.4	100.5	101.5	103.3	105.8	106.9	..	..	Agriculture
94.4	96.7	100.0	102.8	102.6	100.7	100.1	100.0	102.4	..	..	Industry
100.3	100.6	100.0	101.5	100.7	97.1	95.0	94.1	96.0	..	..	Manufacturing
											MANUFACTURING ACTIVITY
99.2	99.1	100.0	100.9	100.2	101.7	102.8	103.4	..	..	..	Employment (1980=100)
98.6	101.6	100.0	101.5	104.0	105.2	106.8	112.8	..	..	..	Real Earnings per Empl. (1980=100)
75.8	85.4	100.0	101.1	104.9	109.6	119.7	130.3	..	..	..	Real Output per Empl. (1980=100)
37.8	35.7	34.8	36.2	36.2	36.4	35.1	35.9	..	..	..	Earnings as % of Value Added
				(Billions of current Japanese Yen)							**MONETARY HOLDINGS**
275,360	306,280	337,200	376,530	410,690	442,710	476,220	518,000	559,500	611,040	D	Money Supply, Broadly Defined
68,929	71,020	69,572	76,510	80,900	80,802	86,375	88,980	98,214	102,970	..	Money as Means of Payment
16,259	17,052	17,475	18,584	19,776	20,575	22,114	23,407	26,198	28,583	..	Currency Ouside Banks
52,670	53,968	52,097	57,926	61,124	60,227	64,261	65,573	72,016	74,387	..	Demand Deposits
206,431	235,260	267,628	300,020	329,790	361,908	389,845	429,020	461,286	508,070	..	Quasi-Monetary Liabilities
				(Billions of current Japanese Yen)							
-15,236	-16,318	-16,872	-16,826	-17,583	-18,843	-17,291	-15,643	..	..	..	**GOVERNMENT DEFICIT (-) OR SURPLUS**
20,719	24,219	28,043	31,607	32,973	33,829	36,751	40,143	..	..	..	Current Revenue
27,768	31,657	35,591	38,938	41,253	43,550	45,096	47,314	..	..	..	Current Expenditure
-7,049	-7,438	-7,548	-7,331	-8,280	-9,721	-8,345	-7,171	..	..	..	Current Budget Balance
21	51	62	34	65	86	84	158	..	..	..	Capital Receipts
8,208	8,931	9,386	9,529	9,368	9,208	9,030	8,630	..	..	..	Capital Payments

JAPAN	1967	1968	1969	1970	1971	1972	1973	1974	1975	1976	1977
FOREIGN TRADE (CUSTOMS BASIS)					*(Billions of current US dollars)*						
Value of Exports, fob	10.44	12.97	15.99	19.32	24.02	28.59	36.93	55.54	55.75	67.20	80.47
Nonfuel Primary Products	0.71	0.85	1.05	1.25	1.34	1.43	1.87	3.18	2.17	2.49	2.80
Fuels	0.03	0.03	0.05	0.05	0.06	0.07	0.09	0.25	0.22	0.12	0.16
Manufactures	9.70	12.09	14.89	18.02	22.61	27.08	34.97	52.11	53.36	64.60	77.51
Value of Imports, cif	11.66	12.99	15.02	18.88	19.71	23.47	38.31	62.09	57.86	64.50	70.56
Nonfuel Primary Products	6.89	7.39	8.46	10.20	10.03	11.72	19.92	24.51	21.75	24.04	26.17
Fuels	2.24	2.68	3.04	3.91	4.75	5.72	8.34	24.92	25.65	28.29	31.15
Manufactures	2.54	2.92	3.52	4.78	4.93	6.04	10.05	12.67	10.46	12.18	13.24
					(Index 1980 = 100)						
Terms of Trade	235.5	246.3	248.6	262.3	247.4	250.7	208.1	138.4	141.9	132.0	134.0
Export Prices, fob	33.1	33.2	34.6	36.7	38.1	42.5	52.8	67.5	65.7	65.6	71.4
Nonfuel Primary Products	29.6	31.3	34.9	34.4	32.4	35.0	55.5	67.7	51.5	58.4	61.6
Fuels	8.5	7.3	5.4	5.9	7.0	8.5	24.5	36.6	33.6	38.5	36.8
Manufactures	33.7	33.7	35.2	37.5	39.0	43.4	52.8	67.8	66.7	66.0	72.0
Import Prices, cif	14.0	13.5	13.9	14.0	15.4	16.9	25.4	48.8	46.3	49.7	53.3
Trade at Constant 1980 Prices					*(Billions of 1980 US dollars)*						
Exports, fob	31.57	39.02	46.24	52.57	63.09	67.31	69.97	82.24	84.93	102.45	112.63
Imports, cif	83.07	96.21	107.99	134.77	128.09	138.55	151.06	127.27	125.04	129.80	132.37
BALANCE OF PAYMENTS					*(Billions of current US dollars)*						
Exports of Goods & Services	..	..	..	22.96	28.39	34.26	44.67	66.53	68.33	80.52	94.76
Merchandise, fob	..	..	..	18.96	23.55	27.99	36.15	54.42	54.73	65.96	79.16
Nonfactor Services	..	..	..	3.24	3.81	4.59	5.78	8.44	9.83	10.92	11.58
Factor Services	..	..	..	0.76	1.03	1.68	2.74	3.67	3.76	3.64	4.02
Imports of Goods & Services	..	..	..	20.77	22.34	27.21	44.45	70.91	68.56	76.46	83.46
Merchandise, fob	..	..	..	15.01	15.78	19.08	32.49	53.05	49.73	56.16	62.00
Nonfactor Services	..	..	..	4.44	5.07	6.33	9.13	13.12	14.25	15.85	16.71
Factor Services	..	..	..	1.32	1.49	1.80	2.84	4.74	4.58	4.46	4.75
Long-Term Interest	..	..	..	..	..	..	..	..	..	..	..
Current Transfers, net	..	..	..	-0.03	-0.04	-0.13	-0.10	-0.10	-0.10	-0.14	-0.06
Workers' Remittances	..	..	..	0.00	0.00	0.00	0.00	0.00	0.00	0.00	0.00
Total to be Financed	..	..	..	2.16	6.02	6.93	0.12	-4.47	-0.33	3.91	11.24
Official Capital Grants	..	..	..	-0.18	-0.22	-0.33	-0.20	-0.20	-0.27	-0.22	-0.33
Current Account Balance	..	..	..	1.98	5.80	6.60	-0.08	-4.68	-0.59	3.69	10.91
Long-Term Capital, net	..	..	..	-1.64	-1.17	-3.37	-8.61	-3.76	-0.70	-1.66	-6.63
Direct Investment	..	..	..	-0.26	-0.15	-0.56	-1.93	-1.67	-1.53	-1.87	-1.63
Long-Term Loans	..	..	..	..	..	..	..	..	..	..	..
Disbursements	..	..	..	..	..	..	..	..	..	..	..
Repayments	..	..	..	..	..	..	..	..	..	..	..
Other Long-Term Capital	..	..	..	-1.20	-0.80	-2.48	-6.47	-1.89	1.09	0.43	-4.67
Other Capital, net	..	..	..	0.72	5.64	-0.38	2.34	9.69	0.71	1.76	2.21
Change in Reserves	..	..	..	-1.06	-10.26	-2.85	6.35	-1.25	0.59	-3.80	-6.49
Memo Item:					*(Japanese Yen per US dollar)*						
Conversion Factor (Annual Avg)	360.000	360.000	360.000	360.000	349.330	303.170	271.700	299.080	296.800	296.550	268.510
EXTERNAL DEBT, ETC.					*(Millions US dollars, outstanding at end of year)*						
Public/Publicly Guar. Long-Term	..	..	..	..	..	..	..	..	..	..	..
Official Creditors	..	..	..	..	..	..	..	..	..	..	..
IBRD and IDA	..	..	..	..	..	..	..	..	..	..	..
Private Creditors	..	..	..	..	..	..	..	..	..	..	..
Private Non-guaranteed Long-Term	..	..	..	..	..	..	..	..	..	..	..
Use of Fund Credit	..	..	..	..	..	..	..	..	..	..	..
Short-Term Debt	..	..	..	..	..	..	..	..	..	..	..
Memo Items:					*(Millions of US dollars)*						
Int'l Reserves Excluding Gold	1,692	2,550	3,241	4,308	14,622	17,564	11,355	12,614	11,950	15,746	22,341
Gold Holdings (at market price)	341	426	416	569	847	1,369	2,369	3,937	2,961	2,845	3,567
SOCIAL INDICATORS											
Total Fertility Rate	2.0	2.1	2.1	2.1	2.2	2.1	2.1	2.1	1.9	1.9	1.8
Crude Birth Rate	19.4	18.7	18.5	18.7	19.1	19.2	19.4	18.4	17.0	16.2	15.4
Infant Mortality Rate	14.9	15.2	14.2	13.1	12.4	11.7	11.3	10.7	10.0	9.3	8.9
Life Expectancy at Birth	71.5	71.6	71.8	71.9	72.8	73.1	73.4	73.7	74.3	74.7	75.1
Food Production, p.c. ('79-81 = 100)	142.3	143.4	135.4	124.5	106.5	113.6	111.1	113.4	119.6	106.4	115.0
Labor Force, Agriculture (%)	23.7	22.3	21.0	19.6	18.8	17.9	17.1	16.2	15.4	14.6	13.7
Labor Force, Female (%)	39.0	39.0	39.0	39.0	38.9	38.7	38.6	38.4	38.3	38.2	38.1
School Enroll. Ratio, primary	..	..	..	99.0	..	..	..	..	99.0	99.0	99.0
School Enroll. Ratio, secondary	..	..	..	86.0	..	..	..	..	91.0	92.0	93.0

1978	1979	1980	1981	1982	1983	1984	1985	1986	1987 estimate	NOTES	JAPAN
				(Billions of current US dollars)							**FOREIGN TRADE (CUSTOMS BASIS)**
97.50	102.96	129.54	151.91	138.58	146.80	170.04	175.86	209.08	229.05	..	Value of Exports, fob
3.29	3.64	5.01	4.72	3.96	4.32	4.38	4.17	4.60	4.94	..	Nonfuel Primary Products
0.26	0.36	0.50	0.55	0.41	0.43	0.50	0.54	0.59	0.78	..	Fuels
93.95	98.96	124.03	146.64	134.21	142.05	165.15	171.14	203.90	223.34	..	Manufactures
78.73	110.11	139.89	140.83	130.32	125.02	134.26	127.51	119.42	146.05	..	Value of Imports, cif
29.14	40.03	42.90	40.13	37.36	37.23	40.25	37.63	40.38	50.07	..	Nonfuel Primary Products
31.34	45.29	69.99	72.56	65.62	58.92	60.34	55.79	36.90	39.14	..	Fuels
18.26	24.80	27.00	28.14	27.34	28.86	33.67	34.09	42.14	56.84	..	Manufactures
				(Index 1980 = 100)							
152.2	123.7	100.0	103.4	104.0	106.7	108.9	111.6	153.6	153.3	..	Terms of Trade
86.4	93.8	100.0	105.3	98.0	96.3	96.1	94.0	112.7	123.4	..	Export Prices, fob
69.6	88.0	100.0	87.3	73.5	78.2	72.1	66.3	67.4	78.4	..	Nonfuel Primary Products
44.7	93.6	100.0	104.0	97.0	85.4	83.6	81.1	41.3	38.0	..	Fuels
87.3	94.0	100.0	106.0	99.0	97.0	97.0	95.0	115.0	126.0	..	Manufactures
56.8	75.8	100.0	101.8	94.2	90.3	88.2	84.2	73.4	80.5	..	Import Prices, cif
				(Billions of 1980 US dollars)							Trade at Constant 1980 Prices
112.90	109.82	129.54	144.28	141.38	152.47	176.94	187.12	185.54	185.61	..	Exports, fob
138.73	145.26	139.89	138.36	138.27	138.50	152.17	151.42	162.75	181.48	..	Imports, cif
				(Billions of current US dollars)							**BALANCE OF PAYMENTS**
114.30	126.26	158.23	189.30	178.75	183.06	210.44	219.53	259.29	304.36	..	Exports of Goods & Services
95.32	101.12	126.74	149.52	137.66	145.47	168.29	174.02	205.59	224.63	..	Merchandise, fob
13.29	15.69	19.89	23.35	22.00	21.24	22.48	22.47	23.54	28.90	..	Nonfactor Services
5.69	9.45	11.60	16.43	19.09	16.35	19.67	23.04	30.16	50.83	..	Factor Services
97.08	133.88	167.45	182.91	170.52	160.71	173.93	168.71	171.40	212.97	..	Imports of Goods & Services
71.02	99.38	124.61	129.56	119.58	114.01	124.03	118.03	112.77	127.48	..	Merchandise, fob
20.35	26.08	31.03	34.88	32.26	31.88	32.80	32.74	35.45	48.52	..	Nonfactor Services
5.71	8.42	11.81	18.47	18.68	14.82	17.10	17.94	23.18	36.97	..	Factor Services
..	..	..	..	..	..	..	..	..	..	..	Long-Term Interest
-0.26	-0.23	-0.24	-0.21	-0.09	-0.18	-0.13	-0.28	-0.59	-0.98	..	Current Transfers, net
0.00	0.00	0.00	0.00	0.00	0.00	0.00	0.00	0.00	0.00	..	Workers' Remittances
16.96	-7.85	-9.46	6.18	8.14	22.17	36.38	50.54	87.30	90.41	K	Total to be Financed
-0.42	-0.89	-1.29	-1.41	-1.29	-1.37	-1.38	-1.37	-1.47	-2.75	K	Official Capital Grants
16.54	-8.74	-10.75	4.77	6.85	20.80	35.00	49.17	85.83	87.66	..	Current Account Balance
-12.81	-13.51	1.10	-7.86	-17.54	-20.10	-51.39	-64.63	-133.55	-144.20	..	Long-Term Capital, net
-2.36	-2.66	-2.11	-4.71	-4.10	-3.20	-5.97	-5.81	-14.25	-18.33	..	Direct Investment
..	..	..	..	..	..	..	..	..	..	..	Long-Term Loans
..	..	..	..	..	..	..	..	..	..	..	Disbursements
..	..	..	..	..	..	..	..	..	..	..	Repayments
-10.03	-9.96	4.50	-1.74	-12.15	-15.53	-44.04	-57.45	-117.83	-123.12	..	Other Long-Term Capital
6.23	9.11	14.68	6.73	5.99	0.85	18.51	15.60	62.56	94.33	..	Other Capital, net
-9.96	13.14	-5.03	-3.64	4.70	-1.55	-2.12	-0.14	-14.84	-37.79	..	Change in Reserves
				(Japanese Yen per US dollar)							**Memo Item:**
210.440	219.140	226.740	220.540	249.050	237.520	237.520	238.540	168.520	144.640	..	Conversion Factor (Annual Avg)
				(Millions US dollars, outstanding at end of year)							**EXTERNAL DEBT, ETC.**
..	..	..	..	..	..	..	..	..	..	..	Public/Publicly Guar. Long-Term
..	..	..	..	..	..	..	..	..	..	..	Official Creditors
..	..	..	..	..	..	..	..	..	..	..	IBRD and IDA
..	..	..	..	..	..	..	..	..	..	..	Private Creditors
..	..	..	..	..	..	..	..	..	..	..	Private Non-guaranteed Long-Term
..	..	..	..	..	..	..	..	..	..	..	Use of Fund Credit
..	..	..	..	..	..	..	..	..	..	..	Short-Term Debt
				(Millions of US dollars)							**Memo Items:**
32,407	19,522	24,637	28,208	23,334	24,602	26,429	26,719	42,257	80,973	..	Int'l Reserves Excluding Gold
5,417	12,405	14,282	9,631	11,070	9,243	7,469	7,923	9,471	11,729	..	Gold Holdings (at market price)
											SOCIAL INDICATORS
1.8	1.8	1.8	1.8	1.8	1.8	1.8	1.8	1.8	1.8	..	Total Fertility Rate
14.8	14.1	13.5	13.0	12.8	12.6	12.5	12.3	12.1	12.0	..	Crude Birth Rate
8.3	7.8	7.5	7.1	7.1	7.1	7.0	7.0	7.0	7.0	..	Infant Mortality Rate
75.6	75.8	76.0	76.3	76.5	76.8	77.1	77.4	77.8	78.1	..	Life Expectancy at Birth
110.2	107.2	95.8	97.0	97.5	97.2	106.7	106.6	111.5	110.2	..	Food Production, p.c. ('79-81 = 100)
12.9	12.0	11.2	..	..	..	..	..	..	..	..	Labor Force, Agriculture (%)
38.0	37.9	37.7	37.8	37.8	37.8	37.8	37.8	37.8	37.8	..	Labor Force, Female (%)
99.0	101.0	101.0	..	101.0	101.0	101.0	102.0	102.0	..	..	School Enroll. Ratio, primary
93.0	92.0	93.0	..	95.0	94.0	95.0	96.0	96.0	..	..	School Enroll. Ratio, secondary

JORDAN	1967	1968	1969	1970	1971	1972	1973	1974	1975	1976	1977
CURRENT GNP PER CAPITA (US $)	..	..	..	..	..	370	380	410	530	690	780
POPULATION (thousands)	2,098	2,166	2,233	2,299	2,363	2,424	2,484	2,542	2,600	2,658	2,717
ORIGIN AND USE OF RESOURCES				*(Millions of current Jordan Dinars)*							
Gross National Product (GNP)	..	..	..	184.9	193.7	208.5	223.3	254.0	328.0	439.6	520.8
Net Factor Income from Abroad	..	..	..	11.6	9.7	3.4	5.0	6.7	15.9	18.0	6.6
Gross Domestic Product (GDP)	..	..	..	173.3	184.0	205.1	218.3	247.3	312.1	421.6	514.2
Indirect Taxes, net	..	..	..	21.2	23.6	26.9	29.4	4.9	9.0	43.2	74.3
GDP at factor cost	..	..	..	152.1	160.4	178.2	188.9	242.4	303.1	378.4	439.9
Agriculture	..	..	..	15.6	23.9	26.6	17.6	30.3	26.0	37.3	41.7
Industry	..	..	..	24.9	24.0	29.0	39.2	60.3	78.3	98.3	120.4
Manufacturing	..	..	..	11.6	12.0	13.9	17.2	29.7	39.7	50.0	58.2
Services, etc.	..	..	..	111.6	112.5	122.6	132.1	151.8	198.8	242.8	277.8
Resource Balance	..	..	..	-59.2	-71.1	-80.8	-84.0	-115.8	-179.8	-247.1	-315.4
Exports of Goods & NFServices	..	..	..	17.6	17.8	37.0	52.4	80.3	121.3	189.7	228.0
Imports of Goods & NFServices	..	..	..	76.8	88.9	117.8	136.4	196.1	301.1	436.8	543.4
Domestic Absorption	..	..	..	232.5	255.1	285.9	302.3	363.1	491.9	668.7	829.6
Private Consumption, etc.	..	..	..	151.7	159.5	175.3	183.1	199.8	293.0	362.6	470.5
General Gov't Consumption	..	..	..	58.7	60.4	68.3	80.0	97.7	110.1	155.9	156.6
Gross Domestic Investment	..	..	..	22.1	35.2	42.3	39.2	65.6	88.8	150.2	202.5
Fixed Investment	..	..	..	25.2	30.7	36.3	47.2	63.2	87.9	138.0	197.0
Memo Items:											
Gross Domestic Saving	..	..	..	-37.1	-35.9	-38.5	-44.8	-50.2	-91.0	-96.9	-112.9
Gross National Saving	..	..	..	-25.5	-26.2	-25.4	-21.6	-17.1	-20.1	54.5	213.7
				(Millions of 1980 Jordan Dinars)							
Gross National Product	..	..	..	460.3	461.4	492.3	486.3	478.8	536.4	657.9	720.9
GDP at factor cost	..	..	..	344.0	347.7	382.1	371.3	413.1	449.3	530.8	556.8
Agriculture	..	..	..	29.9	44.3	47.1	26.5	34.9	26.5	34.8	34.5
Industry	..	..	..	63.2	57.0	67.9	89.0	102.0	118.0	162.1	168.5
Manufacturing	..	..	..	28.2	27.7	31.3	37.0	51.3	63.3	80.3	82.1
Services, etc.	..	..	..	250.9	246.4	267.1	255.8	276.2	304.8	333.9	353.8
Resource Balance	..	..	..	-158.6	-164.8	-178.8	-163.2	-203.6	-237.6	-340.3	-421.2
Exports of Goods & NFServices	..	..	..	46.2	45.3	84.1	110.0	116.2	150.7	238.4	283.6
Imports of Goods & NFServices	..	..	..	204.8	210.1	262.9	273.2	319.8	388.3	578.7	704.8
Domestic Absorption	..	..	..	656.0	667.9	733.1	705.8	736.1	823.2	1,002.9	1,115.9
Private Consumption, etc.	..	..	..	490.8	480.0	510.6	482.1	449.5	512.0	563.9	638.4
General Gov't Consumption	..	..	..	109.9	109.5	131.4	143.8	161.3	164.9	221.3	210.2
Gross Domestic Investment	..	..	..	55.3	78.4	91.1	79.9	125.3	146.3	217.7	267.3
Fixed Investment	..	..	..	64.8	70.2	80.3	98.9	124.0	148.7	200.9	260.2
Memo Items:											
Capacity to Import	..	..	..	46.9	42.1	82.6	105.0	131.0	156.4	251.3	295.7
Terms of Trade Adjustment	..	..	..	0.7	-3.2	-1.5	-5.0	14.8	5.7	12.9	12.1
Gross Domestic Income	..	..	..	431.6	435.3	482.9	470.7	481.2	517.9	644.6	724.0
Gross National Income	..	..	..	461.0	458.2	490.8	481.3	493.6	542.1	670.8	733.0
DOMESTIC PRICES (DEFLATORS)				*(Index 1980 = 100)*							
Overall (GDP)	..	..	..	40.2	42.0	42.3	45.9	53.0	60.9	66.7	72.2
Domestic Absorption	..	..	..	35.4	38.2	39.0	42.8	49.3	59.8	66.7	74.3
Agriculture	..	..	..	52.2	54.0	56.5	66.4	86.8	98.1	107.2	120.9
Industry	..	..	..	39.4	42.1	42.7	44.0	59.1	66.4	60.6	71.5
Manufacturing	..	..	..	41.1	43.3	44.4	46.5	57.9	62.7	62.3	70.9
MANUFACTURING ACTIVITY											
Employment (1980 = 100)	95.0	57.3	68.3	..	46.0	..	..	62.9	77.0	78.4	85.6
Real Earnings per Empl. (1980 = 100)	..	..	63.2	..	73.0	..	..	69.3	62.0	66.7	64.9
Real Output per Empl. (1980 = 100)	..	..	..	..	61.1	..	..	68.2	71.7	79.0	76.5
Earnings as % of Value Added	27.8	27.0	28.2	37.2	36.9	..	..	25.1	25.7	27.0	27.9
MONETARY HOLDINGS				*(Millions of current Jordan Dinars)*							
Money Supply, Broadly Defined	94.1	108.8	118.8	129.3	135.2	146.6	176.4	219.8	288.4	387.5	473.9
Money as Means of Payment	75.2	88.0	96.2	105.5	108.0	115.0	139.3	172.0	224.7	276.7	329.0
Currency Ouside Banks	51.5	63.6	71.3	82.4	83.0	81.5	97.5	115.5	139.0	161.5	188.3
Demand Deposits	23.7	24.4	24.9	23.0	25.0	33.6	41.8	56.5	85.7	115.2	140.8
Quasi-Monetary Liabilities	18.9	20.9	22.6	23.8	27.2	31.6	37.2	47.8	63.8	110.8	144.9
GOVERNMENT DEFICIT (-) OR SURPLUS				*(Millions of current Jordan Dinars)*							
Current Revenue	..	..	..	..	..	..	..	-19.4	-21.8	-40.6	-92.6
Current Revenue	..	..	..	..	..	..	..	125.6	179.4	169.6	259.4
Current Expenditure	..	..	..	..	..	..	..	101.6	121.9	183.0	190.9
Current Budget Balance	..	..	..	..	..	..	..	24.0	57.5	-13.4	68.5
Capital Receipts	..	..	..	..	..	..	..	0.1	0.0	0.0	0.0
Capital Payments	..	..	..	..	..	..	..	43.4	79.2	27.2	161.1

1978	1979	1980	1981	1982	1983	1984	1985	1986	1987 estimate	NOTES	JORDAN
950	1,090	1,430	1,640	1,720	1,630	1,580	1,530	1,550	1,560	..	**CURRENT GNP PER CAPITA (US $)**
2,780	2,848	2,923	3,008	3,106	3,219	3,351	3,506	3,620	3,752	..	POPULATION (thousands)
				(Millions of current Jordan Dinars)							**ORIGIN AND USE OF RESOURCES**
637.5	758.9	976.3	1,175.9	1,330.6	1,407.7	1,426.1	1,519.1	1,546.3	1,581.0	..	Gross National Product (GNP)
5.3	5.9	-8.0	11.7	9.4	-15.0	-72.3	-86.8	-93.6	-105.3	..	Net Factor Income from Abroad
632.2	753.0	984.3	1,164.2	1,321.2	1,422.7	1,498.4	1,605.9	1,639.9	1,686.3	..	Gross Domestic Product (GDP)
81.0	84.4	91.1	123.1	151.6	180.4	183.4	215.3	238.8	239.0	..	Indirect Taxes, net
551.2	668.6	893.2	1,041.1	1,169.6	1,242.3	1,315.0	1,390.6	1,401.1	1,447.3	..	GDP at factor cost
58.6	43.6	69.4	75.1	81.8	110.0	98.6	118.7	111.1	127.2	..	Agriculture
152.5	202.2	281.7	339.9	377.5	369.6	411.3	402.2	397.1	399.1	..	Industry
71.4	94.1	127.2	165.1	184.9	176.6	200.0	190.3	177.7	186.9	..	Manufacturing
340.1	422.8	542.1	626.1	710.3	762.7	805.1	869.7	892.9	921.0	..	Services, etc.
-340.3	-485.2	-493.1	-763.2	-844.7	-783.8	-738.0	-690.9	-533.5	-501.2	..	Resource Balance
265.5	340.5	469.8	630.7	667.0	637.2	743.2	778.1	630.3	753.6	..	Exports of Goods & NFServices
605.8	825.7	962.9	1,393.9	1,511.7	1,421.0	1,481.2	1,469.0	1,163.8	1,254.8	..	Imports of Goods & NFServices
972.5	1,238.2	1,477.4	1,927.4	2,165.8	2,206.5	2,236.4	2,296.8	2,173.4	2,187.5	..	Domestic Absorption
559.5	722.9	829.5	1,053.6	1,220.4	1,348.1	1,373.9	1,401.7	1,238.5	1,281.7	..	Private Consumption, etc.
190.0	235.3	243.8	285.9	326.1	348.3	376.9	405.2	449.9	459.8	..	General Gov't Consumption
223.0	280.0	404.1	587.9	619.3	510.1	485.6	489.9	485.0	446.0	..	Gross Domestic Investment
229.1	294.5	397.8	564.8	597.3	502.8	485.6	473.1	459.7	446.0	..	Fixed Investment
											Memo Items:
-117.3	-205.2	-89.0	-175.3	-225.3	-273.7	-252.4	-201.0	-48.5	-55.2	..	Gross Domestic Saving
138.9	277.5	514.3	574.7	501.4	369.1	381.6	389.8	469.0	325.2	..	Gross National Saving
				(Millions of 1980 Jordan Dinars)							
828.4	834.0	976.3	1,040.5	1,114.5	1,116.4	1,165.6	1,197.0	1,247.8	1,225.9	..	Gross National Product
671.8	739.1	893.2	946.2	1,030.6	1,052.8	1,138.8	1,157.4	1,199.8	1,181.3	H	GDP at factor cost
56.3	39.0	69.4	62.7	63.8	60.4	66.3	72.9	84.4	85.2	..	Agriculture
192.6	240.6	281.7	340.1	374.0	376.3	419.6	408.1	408.5	396.3	..	Industry
97.1	111.3	127.2	153.3	151.3	151.4	176.8	173.9	165.9	159.3	..	Manufacturing
422.9	459.5	542.1	543.4	592.8	616.1	652.9	676.4	706.9	699.8	..	Services, etc.
-459.6	-620.0	-493.1	-595.2	-690.1	-700.8	-652.4	-628.2	-732.6	-605.8	..	Resource Balance
313.8	404.4	469.8	550.1	531.9	556.0	595.5	642.5	604.3	778.9	..	Exports of Goods & NFServices
773.4	1,024.4	962.9	1,145.3	1,222.0	1,256.8	1,247.9	1,270.7	1,336.9	1,384.7	..	Imports of Goods & NFServices
1,242.1	1,447.0	1,477.4	1,718.6	1,870.8	1,872.3	1,892.2	1,903.2	1,923.8	1,833.3	..	Domestic Absorption
711.2	876.6	829.5	888.4	1,012.2	1,117.6	1,154.0	1,157.5	1,179.0	1,127.5	..	Private Consumption, etc.
249.9	261.4	243.8	265.5	281.9	286.8	298.8	311.8	346.2	355.6	..	General Gov't Consumption
281.0	309.0	404.1	564.7	576.7	467.9	439.4	433.9	398.6	350.2	..	Gross Domestic Investment
289.6	325.0	397.8	542.6	556.2	461.2	439.4	419.0	377.8	350.2	..	Fixed Investment
											Memo Items:
339.0	422.4	469.8	518.2	539.2	563.6	626.1	673.1	724.0	831.6	..	Capacity to Import
25.2	18.0	0.0	-31.9	7.3	7.6	30.6	30.6	119.7	52.7	..	Terms of Trade Adjustment
846.6	845.1	984.3	998.7	1,114.0	1,136.4	1,256.9	1,299.3	1,456.6	1,371.2	..	Gross Domestic Income
853.6	852.0	976.3	1,008.6	1,121.8	1,124.0	1,196.2	1,227.6	1,367.5	1,278.6	..	Gross National Income
				(Index 1980 = 100)							**DOMESTIC PRICES (DEFLATORS)**
77.0	91.0	100.0	113.0	119.4	126.0	122.2	126.6	122.7	127.9	..	Overall (GDP)
78.3	85.6	100.0	112.1	115.8	117.8	118.2	120.7	113.0	119.3	..	Domestic Absorption
104.1	111.8	100.0	119.8	128.2	182.1	148.7	162.8	131.6	149.3	..	Agriculture
79.2	84.0	100.0	99.9	100.9	98.2	98.0	98.6	97.2	100.7	..	Industry
73.5	84.5	100.0	107.7	122.2	116.6	113.1	109.4	107.1	117.3	..	Manufacturing
											MANUFACTURING ACTIVITY
97.4	87.3	100.0	114.4	118.1	128.3	155.3	..	..	..	..	Employment (1980=100)
74.7	95.8	100.0	104.7	110.5	109.1	100.7	..	..	..	..	Real Earnings per Empl. (1980=100)
72.1	100.1	100.0	161.1	163.9	156.7	174.4	..	..	..	..	Real Output per Empl. (1980=100)
32.3	28.2	27.0	27.4	28.6	29.5	29.9	..	..	..	..	Earnings as % of Value Added
				(Millions of current Jordan Dinars)							**MONETARY HOLDINGS**
607.6	767.2	980.3	1,183.6	1,406.9	1,617.9	1,762.2	1,879.1	2,078.8	2,404.2	..	Money Supply, Broadly Defined
370.5	465.6	580.7	701.7	787.5	869.4	878.4	848.2	897.1	979.8	..	Money as Means of Payment
219.5	275.4	351.6	412.3	470.0	516.0	530.5	531.8	583.9	655.8	..	Currency Ouside Banks
151.1	190.2	229.1	289.3	317.5	353.4	347.9	316.4	313.2	324.0	..	Demand Deposits
237.0	301.6	399.6	481.9	619.4	748.5	883.8	1,030.9	1,181.7	1,424.4	..	Quasi-Monetary Liabilities
				(Millions of current Jordan Dinars)							
-160.3	-74.8	-109.9	-115.9	-128.2	-108.5	-139.7	-153.1	..	..	..	**GOVERNMENT DEFICIT (-) OR SURPLUS**
230.8	386.4	414.5	498.6	543.0	573.8	498.4	599.2	..	..	..	Current Revenue
203.6	309.8	305.4	356.1	451.1	458.3	474.3	524.0	..	..	..	Current Expenditure
27.2	76.6	109.1	142.5	91.9	115.5	24.1	75.2	..	..	..	Current Budget Balance
0.1	0.1	2.8	0.4	0.6	..	0.3	0.7	..	..	..	Capital Receipts
187.6	151.6	221.9	258.8	220.7	224.0	164.1	229.0	..	..	..	Capital Payments

JORDAN	1967	1968	1969	1970	1971	1972	1973	1974	1975	1976	1977
FOREIGN TRADE (CUSTOMS BASIS)				*(Millions of current US dollars)*							
Value of Exports, fob	31.7	34.1	33.3	34.1	32.0	47.6	57.5	154.7	153.2	209.1	249.3
Nonfuel Primary Products	26.9	30.6	28.8	22.7	19.9	27.1	33.1	100.3	104.1	134.6	147.4
Fuels	0.0	0.1	0.0	0.0	0.1	0.0	0.5	0.4	0.9	2.7	0.5
Manufactures	4.7	3.5	4.5	11.3	12.1	20.6	23.9	54.0	48.2	71.8	101.4
Value of Imports, cif	154.1	160.9	190.0	184.5	214.6	266.9	327.9	486.6	730.8	1,022.1	1,381.1
Nonfuel Primary Products	54.1	56.0	63.2	64.7	70.4	92.6	112.6	156.2	183.2	297.2	292.8
Fuels	8.5	9.0	10.7	10.5	13.7	12.8	12.6	16.2	77.8	111.9	130.9
Manufactures	91.6	95.9	116.1	109.3	130.5	161.6	202.6	314.2	469.9	613.0	957.4
Terms of Trade	129.9	130.9	132.2	137.2	124.8	111.1	98.3	128.9	142.0	115.0	108.8
Export Prices, fob	30.8	30.8	32.1	34.4	33.9	35.7	42.3	77.3	85.2	69.4	70.8
Nonfuel Primary Products	31.2	31.0	32.4	33.2	33.0	33.8	40.4	84.5	99.8	72.1	70.4
Fuels	4.3	4.3	4.3	4.3	5.6	6.2	8.9	36.7	35.7	38.4	42.0
Manufactures	30.1	31.7	32.4	37.5	36.5	38.5	49.5	67.2	66.1	66.7	71.7
Import Prices, cif	23.7	23.5	24.3	25.1	27.2	32.1	43.0	60.0	60.1	60.3	65.1
Trade at Constant 1980 Prices				*(Millions of 1980 US dollars)*							
Exports, fob	103.0	110.7	103.8	98.9	94.5	133.3	136.0	200.2	179.7	301.4	351.9
Imports, cif	649.9	684.3	782.3	734.6	790.2	830.6	762.4	811.2	1,217.0	1,694.1	2,121.5
BALANCE OF PAYMENTS				*(Millions of current US dollars)*							
Exports of Goods & Services	..	..	..	123.5	113.1	124.9	189.0	279.7	414.6	590.4	815.3
Merchandise, fob	..	..	..	34.1	32.0	47.6	73.7	154.3	152.9	206.9	261.5
Nonfactor Services	..	..	..	55.0	51.2	65.0	96.2	98.1	226.4	344.3	462.1
Factor Services	..	..	..	34.4	30.0	12.3	19.2	27.3	35.3	39.2	91.7
Imports of Goods & Services	..	..	..	253.9	275.2	330.1	417.9	609.1	951.2	1,309.4	1,797.3
Merchandise, fob	..	..	..	163.8	190.4	236.6	291.9	430.3	648.2	907.9	1,443.0
Nonfactor Services	..	..	..	88.3	82.1	90.7	122.0	172.3	293.3	387.4	287.5
Factor Services	..	..	..	1.8	2.7	2.8	4.0	6.5	9.7	14.2	66.8
Long-Term Interest	..	..	..	1.8	2.7	2.8	2.9	3.8	6.8	8.1	14.7
Current Transfers, net	..	..	..	0.0	0.0	27.2	55.4	82.0	172.0	401.8	971.8
Workers' Remittances	..	..	..	0.0	0.0	20.7	44.7	74.8	166.7	410.9	443.7
Total to be Financed	..	..	..	-130.4	-162.1	-178.1	-173.5	-247.4	-364.6	-317.2	-10.2
Official Capital Grants	..	..	..	110.8	100.3	184.5	186.0	250.9	409.3	353.3	0.0
Current Account Balance	..	..	..	-19.6	-61.8	6.4	12.5	3.4	44.7	36.1	-10.2
Long-Term Capital, net	..	..	..	114.0	121.8	201.6	205.1	283.8	546.9	308.4	162.9
Direct Investment	..	..	..	..	..	-1.1	-3.3	3.4	19.4	-10.2	15.0
Long-Term Loans	..	..	..	11.7	21.0	27.7	30.1	51.9	87.6	71.3	190.1
Disbursements	..	..	..	14.4	25.8	34.7	38.6	62.4	102.0	90.7	213.7
Repayments	..	..	..	2.7	4.8	7.0	8.5	10.5	14.4	19.4	23.6
Other Long-Term Capital	..	..	..	-8.5	0.5	-9.5	-7.6	-22.4	30.6	-105.9	-42.2
Other Capital, net	..	..	..	-99.6	-99.6	-189.3	-177.5	-261.2	-416.1	-316.4	53.5
Change in Reserves	..	..	..	5.2	39.6	-18.7	-40.1	-26.0	-175.5	-28.2	-206.2
Memo Item:				*(Jordan Dinars per US dollar)*							
Conversion Factor (Annual Avg)	0.360	0.360	0.360	0.360	0.360	0.360	0.330	0.320	0.320	0.330	0.330
EXTERNAL DEBT, ETC.				*(Millions of US dollars, outstanding at end of year)*							
Public/Publicly Guar. Long-Term	..	..	..	118.5	145.8	170.3	208.5	268.8	340.0	410.2	622.8
Official Creditors	..	..	..	108.3	127.0	153.5	192.1	238.2	291.3	361.5	469.1
IBRD and IDA	..	..	..	9.1	9.5	12.4	18.0	22.7	32.7	39.1	47.7
Private Creditors	..	..	..	10.2	18.7	16.8	16.4	30.6	48.7	48.7	153.7
Private Non-guaranteed Long-Term	..	..	..	..	..	..	..	..	..	..	..
Use of Fund Credit	..	..	..	0.0	4.9	4.9	1.9	0.0	0.0	0.0	0.0
Short-Term Debt	..	..	..	..	..	..	..	..	..	..	112.0
Memo Items:				*(Millions of US dollars)*							
Int'l Reserves Excluding Gold	240.7	254.8	232.9	227.8	222.9	241.0	270.7	312.8	458.9	471.5	643.2
Gold Holdings (at market price)	3.4	35.5	29.8	29.8	34.8	51.7	89.5	148.6	111.8	107.4	133.0
SOCIAL INDICATORS											
Total Fertility Rate	..	..	..	..	..	..	..	..	7.7	..	7.7
Crude Birth Rate	..	..	..	..	..	..	..	..	47.0	..	..
Infant Mortality Rate	102.0	98.0	94.0	90.0	86.0	82.0	78.6	75.2	71.8	68.4	65.0
Life Expectancy at Birth	51.7	52.6	53.6	54.6	55.6	56.6	57.3	58.0	58.7	59.4	60.1
Food Production, p.c. ('79-81 = 100)	..	..	..	76.7	101.1	111.5	80.2	106.7	68.7	80.6	82.2
Labor Force, Agriculture (%)	33.1	31.3	29.6	27.8	26.0	24.3	22.5	20.8	19.0	17.2	15.5
Labor Force, Female (%)	6.0	6.1	6.2	6.3	6.5	6.7	6.8	7.0	7.1	7.3	7.4
School Enroll. Ratio, primary	..	..	..	..	..	..	..	..	..	..	..
School Enroll. Ratio, secondary	..	..	..	..	..	..	..	..	..	..	..

1978	1979	1980	1981	1982	1983	1984	1985	1986	1987 estimate	NOTES	JORDAN
											FOREIGN TRADE (CUSTOMS BASIS)
				(Millions of current US dollars)							
296.7	402.3	561.4	731.8	738.9	579.0	751.9	788.6	732.7	669.4	..	Value of Exports, fob
162.2	244.6	358.9	407.2	371.8	301.2	358.5	357.2	297.8	183.2	..	Nonfuel Primary Products
0.3	0.4	1.7	0.3	0.2	0.4	0.3	0.3	0.1	0.1	..	Fuels
134.2	157.3	200.8	324.3	366.9	277.5	393.1	431.1	434.8	486.2	..	Manufactures
1,498.8	1,948.6	2,394.4	3,146.3	3,217.5	3,016.3	2,784.1	2,732.9	2,432.0	2,339.3	..	Value of Imports, cif
369.9	459.6	503.1	612.6	666.3	641.9	648.4	602.0	631.2	609.1	..	Nonfuel Primary Products
153.0	246.7	408.5	546.8	682.0	584.2	576.1	543.4	140.6	174.1	..	Fuels
975.9	1,242.3	1,482.8	1,986.8	1,869.1	1,790.2	1,559.6	1,587.5	1,660.2	1,556.1	..	Manufactures
				(Index 1980 = 100)							
102.9	104.1	100.0	96.1	93.6	91.6	93.4	93.3	100.2	104.8	..	Terms of Trade
74.6	89.6	100.0	97.6	90.7	86.4	86.0	83.2	94.3	105.5	..	Export Prices, fob
73.6	89.4	100.0	96.5	87.0	82.8	82.7	77.1	83.8	89.7	..	Nonfuel Primary Products
42.3	61.0	100.0	112.5	101.6	92.5	90.2	87.5	44.9	56.9	..	Fuels
76.1	90.0	100.0	99.1	94.8	90.7	89.3	89.0	103.2	113.0	..	Manufactures
72.5	86.0	100.0	101.6	96.9	94.4	92.1	89.2	94.1	100.6	..	Import Prices, cif
				(Millions of 1980 US dollars)							Trade at Constant 1980 Prices
397.6	449.2	561.4	749.6	814.5	669.8	874.1	947.8	776.7	634.7	..	Exports, fob
2,066.5	2,264.8	2,394.4	3,096.0	3,320.9	3,195.1	3,022.3	3,064.5	2,583.2	2,324.9	..	Imports, cif
				(Millions of current US dollars)							**BALANCE OF PAYMENTS**
973.0	1,279.5	1,781.5	2,214.6	2,208.9	2,042.4	2,162.2	2,181.7	2,022.5	2,376.4	..	Exports of Goods & Services
297.6	402.7	575.1	735.0	750.8	580.2	756.8	789.7	731.8	932.0	..	Merchandise, fob
568.6	728.0	997.4	1,172.8	1,142.5	1,175.2	1,178.2	1,186.7	1,069.8	1,292.2	..	Nonfactor Services
106.8	148.8	209.0	306.8	315.6	287.0	227.2	205.3	220.9	152.2	..	Factor Services
2,071.8	2,874.2	3,459.1	4,490.5	4,579.7	4,243.2	4,272.0	4,156.7	3,814.8	4,158.5	..	Imports of Goods & Services
1,502.2	1,959.0	2,397.6	3,169.7	3,239.0	3,035.9	2,783.7	2,724.2	2,423.4	2,694.1	..	Merchandise, fob
480.6	786.4	828.4	1,049.3	1,051.7	879.0	1,072.7	1,006.9	903.1	1,001.6	..	Nonfactor Services
89.0	128.8	233.1	271.5	289.0	328.3	415.6	425.6	488.3	462.8	..	Factor Services
24.4	39.5	57.7	66.6	62.0	99.5	89.0	139.9	151.6	183.4	..	Long-Term Interest
818.2	1,587.7	2,052.1	2,234.5	2,035.0	1,811.5	1,838.8	1,721.2	1,746.6	1,432.6	..	Current Transfers, net
469.6	540.6	714.6	929.4	975.5	999.0	1,113.0	921.1	1,066.4	844.2	..	Workers' Remittances
-280.6	-7.0	374.5	-41.4	-335.9	-389.3	-271.0	-253.8	-45.7	-349.5	K	Total to be Financed
0.0	0.0	0.0	0.0	0.0	0.0	0.0	0.0	0.0	..	K	Official Capital Grants
-280.6	-7.0	374.5	-41.4	-335.9	-389.3	-271.0	-253.8	-45.7	-349.5	..	Current Account Balance
291.2	183.5	103.7	135.0	353.4	426.2	167.6	339.2	148.0	248.9	..	Long-Term Capital, net
50.0	15.7	27.4	67.7	90.6	24.3	75.2	14.1	22.6	32.7	..	Direct Investment
195.1	204.1	223.9	317.2	344.5	399.4	220.5	317.2	314.3	14.8	..	Long-Term Loans
228.6	260.6	299.3	453.6	479.9	508.3	343.7	528.0	551.3	349.1	..	Disbursements
33.5	56.5	75.4	136.4	135.4	108.9	123.2	210.8	237.0	334.3	..	Repayments
46.1	-36.3	-147.6	-249.9	-81.7	2.5	-128.1	7.9	-188.9	201.4	..	Other Long-Term Capital
110.4	35.7	-121.4	-47.1	-194.7	101.7	-76.9	-38.5	-49.9	-40.8	..	Other Capital, net
-121.0	-212.2	-356.8	-46.5	177.2	-138.6	180.3	-46.9	-52.4	141.4	..	Change in Reserves
				(Jordan Dinars per US dollar)							**Memo Item:**
0.310	0.300	0.300	0.330	0.350	0.360	0.380	0.390	0.350	0.340	..	Conversion Factor (Annual Avg)
											EXTERNAL DEBT, ETC.
			(Millions of US dollars, outstanding at end of year)								
847.4	1,065.1	1,275.6	1,534.5	1,841.3	2,171.5	2,302.2	2,769.2	3,237.6	3,517.8	..	Public/Publicly Guar. Long-Term
614.2	780.9	1,000.0	1,241.1	1,495.4	1,576.5	1,571.5	1,829.3	2,052.9	2,379.2	..	Official Creditors
53.4	70.3	101.5	123.3	137.7	166.8	168.5	249.7	362.8	507.6	..	IBRD and IDA
233.2	284.2	275.6	293.4	345.9	595.0	730.6	940.0	1,184.6	1,138.5	..	Private Creditors
..	..	..	..	..	..	..	..	..	..	..	Private Non-guaranteed Long-Term
0.0	0.0	0.0	0.0	0.0	0.0	0.0	63.0	70.2	81.4	..	Use of Fund Credit
205.0	284.0	480.0	523.0	572.0	583.0	860.0	917.0	985.0	965.0	..	Short-Term Debt
				(Millions of US dollars)							**Memo Items:**
885.6	1,166.1	1,142.8	1,086.7	884.1	824.2	515.0	422.8	437.1	424.7	..	Int'l Reserves Excluding Gold
183.3	417.8	601.9	424.1	493.5	415.8	326.8	347.0	415.9	485.1	..	Gold Holdings (at market price)
											SOCIAL INDICATORS
7.5	7.3	7.0	6.8	6.6	6.4	6.3	6.1	6.0	5.8	..	Total Fertility Rate
..	..	..	..	40.3	39.9	39.4	39.0	38.6	38.1	..	Crude Birth Rate
62.8	60.6	58.4	56.2	54.0	54.0	54.0	54.0	50.8	47.5	..	Infant Mortality Rate
60.8	61.4	62.1	62.8	63.5	63.8	64.2	64.6	65.0	65.3	..	Life Expectancy at Birth
96.5	80.6	112.0	107.4	104.8	109.9	111.9	112.8	102.6	108.1	..	Food Production, p.c. ('79-81 = 100)
13.7	12.0	10.2	..	..	..	..	..	..	..	..	Labor Force, Agriculture (%)
7.6	7.7	7.9	8.2	8.4	8.6	8.9	9.1	9.4	9.6	..	Labor Force, Female (%)
98.0	102.0	104.0	103.0	100.0	99.0	..	..	..	..	..	School Enroll. Ratio, primary
67.0	74.0	76.0	77.0	78.0	79.0	..	..	..	..	..	School Enroll. Ratio, secondary

KENYA	1967	1968	1969	1970	1971	1972	1973	1974	1975	1976	1977
CURRENT GNP PER CAPITA (US $)	120	120	130	130	160	180	180	210	230	240	270
POPULATION (thousands)	10,400	10,748	11,114	11,498	11,902	12,327	12,775	13,246	13,741	14,262	14,810
ORIGIN AND USE OF RESOURCES					*(Billions of current Kenya Shillings)*						
Gross National Product (GNP)	8.53	9.21	10.05	11.03	12.32	14.62	16.69	20.37	23.01	27.71	35.62
Net Factor Income from Abroad	-0.27	-0.45	-0.37	-0.42	-0.38	-0.43	-0.88	-0.85	-0.93	-1.36	-1.57
Gross Domestic Product (GDP)	8.80	9.67	10.42	11.45	12.70	15.05	17.57	21.21	23.93	29.07	37.20
Indirect Taxes, net	0.69	0.81	0.89	1.07	1.30	1.28	1.78	2.44	2.79	3.51	4.38
GDP at factor cost	8.12	8.86	9.53	10.38	11.40	13.78	15.79	18.78	21.14	25.56	32.81
Agriculture	2.96	3.06	3.22	3.46	3.58	4.85	5.60	6.64	7.22	9.69	13.77
Industry	1.54	1.71	1.88	2.06	2.32	2.81	3.28	3.89	4.28	4.76	5.90
Manufacturing	0.91	1.00	1.14	1.24	1.43	1.56	1.89	2.39	2.54	2.88	3.60
Services, etc.	3.62	4.09	4.43	4.86	5.50	6.12	6.92	8.25	9.64	11.12	13.15
Resource Balance	-0.08	0.01	0.15	-0.10	-0.83	-0.32	-0.22	-1.53	-1.12	0.20	1.25
Exports of Goods & NFServices	2.51	2.84	3.08	3.42	3.64	4.00	4.81	7.14	7.14	9.43	13.00
Imports of Goods & NFServices	2.59	2.83	2.93	3.51	4.47	4.32	5.04	8.68	8.26	9.23	11.75
Domestic Absorption	8.88	9.66	10.27	11.55	13.53	15.37	17.79	22.75	25.06	28.87	35.95
Private Consumption, etc.	5.86	6.25	6.54	6.89	8.21	9.36	10.37	13.67	16.33	17.91	20.75
General Gov't Consumption	1.24	1.47	1.71	1.86	2.28	2.65	2.89	3.61	4.39	5.08	6.40
Gross Domestic Investment	1.78	1.94	2.02	2.79	3.04	3.36	4.53	5.46	4.34	5.88	8.80
Fixed Investment	1.64	1.79	1.87	2.25	2.88	3.28	3.59	4.06	4.84	5.81	7.80
Memo Items:											
Gross Domestic Saving	1.70	1.95	2.17	2.70	2.21	3.04	4.31	3.93	3.22	6.09	10.05
Gross National Saving	1.34	1.42	1.73	2.20	1.73	2.59	3.37	2.99	2.20	4.56	8.51
					(Millions of 1980 Kenya Shillings)						
Gross National Product	21,857	23,233	25,351	24,104	29,529	34,658	35,779	37,546	38,202	38,634	42,311
GDP at factor cost	20,686	22,227	23,923	22,600	27,356	31,728	33,092	33,451	34,815	35,647	38,786
Agriculture	8,455	8,866	9,650	8,890	10,324	11,503	11,816	11,527	12,376	12,612	13,873
Industry	3,054	3,456	3,715	3,504	4,425	6,485	6,996	7,052	7,084	7,058	7,964
Manufacturing	1,635	1,851	2,041	1,953	2,528	3,359	3,823	4,023	4,048	4,003	4,640
Services, etc.	9,176	9,905	10,559	10,206	12,607	13,740	14,280	14,872	15,354	15,978	16,949
Resource Balance	-3,386	-3,096	-2,344	-4,736	-7,085	-5,196	-3,820	-5,231	-2,045	-1,351	-3,822
Exports of Goods & NFServices	10,782	12,125	13,235	13,617	14,241	12,716	13,815	15,909	14,061	14,328	14,728
Imports of Goods & NFServices	14,168	15,221	15,578	18,353	21,325	17,912	17,635	21,141	16,107	15,680	18,550
Domestic Absorption	26,030	27,530	28,712	29,840	37,711	41,022	41,680	44,468	41,798	41,968	48,250
Private Consumption, etc.	17,596	15,627	16,096	13,973	21,001	24,806	23,053	26,025	26,389	25,045	27,076
General Gov't Consumption	2,729	3,892	4,404	4,554	5,239	5,681	5,886	6,388	7,151	7,681	8,618
Gross Domestic Investment	5,705	8,011	8,212	11,312	11,471	10,535	12,741	12,055	8,257	9,243	12,555
Fixed Investment	..	..	..	..	..	..	..	..	..	..	..
Memo Items:											
Capacity to Import	13,730	15,275	16,377	17,851	17,356	16,578	16,851	17,408	13,919	16,023	20,526
Terms of Trade Adjustment	2,947	3,150	3,142	4,234	3,115	3,862	3,035	1,498	-142	1,694	5,798
Gross Domestic Income	25,591	27,584	29,511	29,338	33,742	39,688	40,895	40,735	39,610	42,311	50,226
Gross National Income	24,804	26,383	28,493	28,338	32,644	38,520	38,814	39,044	38,060	40,329	48,109
DOMESTIC PRICES (DEFLATORS)					*(Index 1980 = 100)*						
Overall (GDP)	38.9	39.6	39.5	45.6	41.5	42.0	46.4	54.1	60.2	71.6	83.7
Domestic Absorption	34.1	35.1	35.8	38.7	35.9	37.5	42.7	51.2	59.9	68.8	74.5
Agriculture	35.0	34.6	33.4	38.9	34.7	42.2	47.4	57.6	58.4	76.8	99.2
Industry	50.4	49.4	50.6	58.7	52.5	43.4	46.8	55.1	60.4	67.4	74.1
Manufacturing	55.4	54.1	55.7	63.7	56.7	46.4	49.5	59.4	62.8	72.0	77.6
MANUFACTURING ACTIVITY											
Employment (1980=100)	35.7	37.6	41.2	48.0	51.8	63.7	71.7	74.8	74.2	84.4	96.0
Real Earnings per Empl. (1980=100)	131.2	135.2	142.2	135.9	129.2	119.0	104.8	117.1	105.7	101.3	89.7
Real Output per Empl. (1980=100)	41.3	41.1	41.8	38.1	45.7	57.9	57.3	70.3	75.5	79.1	94.1
Earnings as % of Value Added	56.6	56.4	54.2	52.9	50.0	48.3	44.8	44.4	43.7	40.7	39.9
MONETARY HOLDINGS					*(Billions of current Kenya Shillings)*						
Money Supply, Broadly Defined	2.38	2.65	3.14	4.00	4.31	4.91	6.12	6.73	8.05	9.82	14.36
Money as Means of Payment	1.41	1.63	1.81	2.24	2.38	2.80	3.45	3.76	4.14	5.12	7.33
Currency Ouside Banks	0.45	0.49	0.57	0.70	0.74	0.89	0.98	1.09	1.23	1.63	2.18
Demand Deposits	0.96	1.14	1.24	1.54	1.64	1.91	2.47	2.67	2.91	3.49	5.15
Quasi-Monetary Liabilities	0.97	1.02	1.33	1.76	1.93	2.10	2.67	2.98	3.91	4.70	7.02
GOVERNMENT DEFICIT (-) OR SURPLUS					*(Millions of current Kenya Shillings)*						
	..	..	..	..	..	-566	-902	-587	-1,151	-1,709	-1,327
Current Revenue	..	..	..	..	..	2,712	2,827	3,696	4,522	5,325	6,381
Current Expenditure	..	..	..	..	..	2,352	2,618	3,080	4,006	4,848	5,531
Current Budget Balance	..	..	..	..	..	360	209	616	516	477	850
Capital Receipts	..	..	..	..	..	..	1	..	..	..	2
Capital Payments	..	..	..	..	..	926	1,112	1,203	1,667	2,186	2,179

1978	1979	1980	1981	1982	1983	1984	1985	1986	1987 estimate	NOTES	KENYA
310	360	410	420	390	340	310	300	310	330	..	**CURRENT GNP PER CAPITA (US $)**
15,387	15,994	16,632	17,303	18,010	18,752	19,533	20,353	21,221	22,097	..	**POPULATION (thousands)**
				(Billions of current Kenya Shillings)							**ORIGIN AND USE OF RESOURCES**
39.17	43.76	50.90	58.60	66.50	74.00	84.70	94.60	112.40	127.30	..	Gross National Product (GNP)
-1.82	-1.74	-1.70	-1.90	-2.80	-2.50	-3.10	-3.70	-4.20	-5.00	..	Net Factor Income from Abroad
40.99	45.50	52.60	60.50	69.30	76.50	87.80	98.30	116.60	132.30	f	Gross Domestic Product (GDP)
5.39	5.94	7.94	8.83	9.32	10.19	11.83	12.46	15.20	18.20	..	Indirect Taxes, net
35.60	39.56	44.66	51.67	59.98	66.31	75.97	85.84	101.40	114.10	f	GDP at factor cost
13.14	13.58	14.47	16.74	20.40	21.90	26.20	28.60	33.60	35.20	..	Agriculture
7.15	8.42	9.90	11.19	12.29	13.69	15.23	17.07	19.03	21.70	..	Industry
4.39	5.00	5.90	6.56	7.45	8.17	9.22	10.37	11.50	13.10	..	Manufacturing
15.31	17.56	20.28	23.74	27.28	30.73	34.54	40.17	48.77	57.20	..	Services, etc.
-4.00	-2.73	-5.99	-5.41	-2.60	-0.80	-1.24	-1.10	0.10	-6.70	..	Resource Balance
11.86	12.00	15.07	15.50	17.60	19.60	23.40	25.50	30.30	28.00	..	Exports of Goods & NFServices
15.86	14.73	21.05	20.91	20.20	20.40	24.64	26.60	30.20	34.70	..	Imports of Goods & NFServices
44.99	48.23	58.59	65.91	71.90	77.30	89.04	99.40	116.50	139.00	..	Domestic Absorption
24.79	29.03	32.09	37.21	43.70	46.30	52.94	62.00	69.00	80.70	..	Private Consumption, etc.
8.00	8.90	10.70	11.50	12.90	14.80	15.50	17.60	21.20	25.70	..	General Gov't Consumption
12.20	10.30	15.80	17.20	15.30	16.20	20.60	19.80	26.30	32.60	..	Gross Domestic Investment
10.28	10.81	12.45	14.51	13.37	14.35	16.10	17.10	23.50	26.20	..	Fixed Investment
											Memo Items:
8.20	7.57	9.81	11.79	12.70	15.40	19.36	18.70	26.40	25.90	..	Gross Domestic Saving
6.51	5.91	8.31	10.78	10.81	13.74	17.13	16.34	23.14	22.08	..	Gross National Saving
				(Millions of 1980 Kenya Shillings)							
45,168	47,963	50,900	53,104	54,028	55,015	55,906	58,178	61,587	64,563	..	Gross National Product
41,179	42,969	44,658	47,331	49,624	50,753	51,157	53,611	56,529	59,240	f	GDP at factor cost
14,397	14,319	14,473	15,330	17,004	17,316	16,705	17,364	18,236	18,945	..	Agriculture
8,807	9,423	9,904	10,334	10,269	10,446	10,669	11,319	11,739	12,366	..	Industry
5,220	5,609	5,903	6,115	6,253	6,534	6,816	7,122	7,535	7,966	..	Manufacturing
17,975	19,227	20,281	21,667	22,350	22,991	23,782	24,928	26,554	27,929	..	Services, etc.
-8,666	-4,852	-5,988	-2,190	950	3,218	1,467	3,460	3,462	1,599	..	Resource Balance
14,975	14,290	15,066	14,435	14,894	14,597	14,879	16,608	18,062	18,088	..	Exports of Goods & NFServices
23,642	19,142	21,054	16,625	13,944	11,379	13,412	13,147	14,600	16,489	..	Imports of Goods & NFServices
56,121	54,786	58,588	56,913	55,179	53,474	56,353	56,778	60,248	65,260	..	Domestic Absorption
30,674	32,581	32,088	31,541	33,505	32,933	34,752	36,397	37,351	39,462	..	Private Consumption, etc.
9,896	10,463	10,700	10,123	9,972	10,307	10,201	9,917	10,571	11,419	..	General Gov't Consumption
15,551	11,742	15,800	15,250	11,703	10,234	11,400	10,465	12,326	14,379	..	Gross Domestic Investment
..	12,145	12,450	13,016	10,305	8,883	9,154	9,209	10,436	11,392	..	Fixed Investment
											Memo Items:
17,682	15,595	15,066	12,321	12,149	10,933	12,737	12,604	14,648	13,305	.. ·	Capacity to Import
2,707	1,305	0	-2,114	-2,745	-3,665	-2,142	-4,004	-3,414	-4,783	..	Terms of Trade Adjustment
50,161	51,238	52,600	52,610	53,384	53,027	55,678	56,235	60,296	62,076	..	Gross Domestic Income
47,875	49,268	50,900	50,990	51,283	51,350	53,764	54,174	58,173	59,780	..	Gross National Income
				(Index 1980 = 100)							**DOMESTIC PRICES (DEFLATORS)**
86.4	91.1	100.0	110.6	123.5	134.9	151.8	163.2	183.0	197.9	..	Overall (GDP)
80.2	88.0	100.0	115.8	130.3	144.6	158.0	175.1	193.4	213.0	..	Domestic Absorption
91.3	94.8	100.0	109.2	120.0	126.5	156.8	164.7	184.3	185.8	..	Agriculture
81.1	89.3	100.0	108.3	119.7	131.0	142.7	150.8	162.1	175.5	..	Industry
84.0	89.1	100.0	107.3	119.1	125.0	135.3	145.6	152.6	164.4	..	Manufacturing
											MANUFACTURING ACTIVITY
90.7	99.9	100.0	108.5	109.3	113.8	118.4	123.2	..	..	G	Employment (1980 = 100)
97.5	94.6	100.0	93.9	88.0	80.2	79.0	75.5	..	..	G	Real Earnings per Empl. (1980 = 100)
84.8	78.0	100.0	89.4	88.4	89.8	90.0	90.4	..	..	G	Real Output per Empl. (1980 = 100)
44.5	43.9	43.1	47.2	47.7	46.0	46.0	46.0	..	..	..	Earnings as % of Value Added
				(Billions of current Kenya Shillings)							**MONETARY HOLDINGS**
16.57	19.25	20.43	23.71	27.58	30.26	35.86	39.49	49.98	55.65	D	Money Supply, Broadly Defined
7.88	9.18	8.43	9.41	10.64	11.47	13.09	12.92	17.52	18.92	..	Money as Means of Payment
2.30	2.67	3.03	3.57	3.72	4.08	4.37	5.04	6.37	7.69	..	Currency Ouside Banks
5.57	6.50	5.40	5.84	6.91	7.39	8.72	7.89	11.15	11.23	..	Demand Deposits
8.69	10.08	12.00	14.30	16.95	18.78	22.77	26.57	32.46	36.74	..	Quasi-Monetary Liabilities
				(Millions of current Kenya Shillings)							
-1,627	-3,015	-2,409	-4,002	-5,463	-3,838	-4,281	-6,245	..	..	C	**GOVERNMENT DEFICIT (-) OR SURPLUS**
9,364	10,100	12,206	13,959	15,270	16,620	17,984	20,089	..	..	..	Current Revenue
7,643	9,400	10,455	12,851	16,400	17,002	19,093	21,524	..	..	..	Current Expenditure
1,721	700	1,751	1,108	-1,130	-382	-1,109	-1,435	..	..	..	Current Budget Balance
1	2	2	15	11	2	1	1	..	..	..	Capital Receipts
3,349	3,717	4,162	5,125	4,344	3,458	3,173	4,811	..	..	..	Capital Payments

KENYA	1967	1968	1969	1970	1971	1972	1973	1974	1975	1976	1977
FOREIGN TRADE (CUSTOMS BASIS)				*(Millions of current US dollars)*							
Value of Exports, fob	240.0	250.0	272.0	305.0	314.0	372.0	516.0	661.0	647.0	824.0	1,195.0
Nonfuel Primary Products	195.7	202.5	216.6	230.4	228.7	284.1	402.7	460.1	427.4	483.8	842.6
Fuels	26.3	26.0	31.9	37.6	42.0	43.6	50.7	116.4	132.8	171.0	204.5
Manufactures	18.1	21.5	23.5	37.0	43.3	44.2	62.6	84.4	86.9	169.2	147.9
Value of Imports, cif	336.0	356.0	360.0	442.0	560.0	554.0	655.0	1,026.0	987.0	923.0	1,285.6
Nonfuel Primary Products	26.6	32.1	31.5	43.1	68.0	72.1	89.3	109.4	87.4	106.3	130.4
Fuels	37.9	41.1	40.1	45.3	51.0	63.8	72.7	236.6	278.1	237.5	286.1
Manufactures	271.6	282.9	288.4	353.5	441.0	418.0	493.0	680.0	621.5	579.2	869.1
				(Index 1980 = 100)							
Terms of Trade	102.3	102.0	95.8	91.7	81.6	97.2	104.0	104.4	97.4	114.4	136.1
Export Prices, fob	18.9	18.5	18.1	18.6	20.1	24.1	33.7	54.3	51.0	63.0	83.7
Nonfuel Primary Products	33.4	31.1	32.1	36.0	34.2	40.0	48.7	60.1	56.6	77.7	116.3
Fuels	4.3	4.3	4.3	4.3	5.6	6.2	8.9	36.7	35.7	38.4	42.0
Manufactures	26.4	23.9	30.7	31.3	30.1	32.5	47.2	63.2	61.3	70.6	68.7
Import Prices, cif	18.5	18.1	18.9	20.3	24.6	24.8	32.4	52.1	52.4	55.1	61.5
Trade at Constant 1980 Prices				*(Millions of 1980 US dollars)*							
Exports, fob	1,271.2	1,351.3	1,500.7	1,641.3	1,565.4	1,546.2	1,532.1	1,216.4	1,267.6	1,307.9	1,426.9
Imports, cif	1,820.1	1,962.3	1,902.1	2,181.2	2,278.4	2,238.1	2,021.9	1,970.9	1,884.3	1,675.8	2,089.1
BALANCE OF PAYMENTS				*(Millions of current US dollars)*							
Exports of Goods & Services	..	..	..	506	529	581	706	971	1,011	1,143	1,594
Merchandise, fob	..	..	..	286	294	337	470	581	633	746	1,131
Nonfactor Services	..	..	..	173	201	207	209	352	322	362	422
Factor Services	..	..	..	47	35	36	27	38	56	35	41
Imports of Goods & Services	..	..	..	580	699	687	862	1,311	1,280	1,282	1,633
Merchandise, fob	..	..	..	372	479	454	545	898	847	810	1,113
Nonfactor Services	..	..	..	138	161	162	188	273	284	295	321
Factor Services	..	..	..	70	59	70	129	139	149	177	199
Long-Term Interest	..	..	..	17	18	20	29	41	47	50	63
Current Transfers, net	..	..	..	-11	-13	-1	-9	-14	-13	-20	3
Workers' Remittances	..	..	..	0	0	0	0	0	0	0	0
Total to be Financed	..	..	..	-86	-183	-108	-165	-353	-281	-158	-36
Official Capital Grants	..	..	..	37	71	40	39	45	63	34	63
Current Account Balance	..	..	..	-49	-112	-68	-126	-308	-218	-124	28
Long-Term Capital, net	..	..	..	87	108	111	168	230	219	256	259
Direct Investment	..	..	..	14	12	..	..	..	16	42	54
Long-Term Loans	..	..	..	47	29	87	186	222	114	212	165
Disbursements	..	..	..	76	64	115	221	280	204	319	389
Repayments	..	..	..	29	35	28	36	57	91	107	224
Other Long-Term Capital	..	..	..	-11	-4	-15	-56	-38	27	-32	-23
Other Capital, net	..	..	..	8	-67	-20	-21	-10	-39	-46	-9
Change in Reserves	..	..	..	-45	71	-24	-22	88	38	-86	-277
Memo Item:				*(Kenya Shillings per US dollar)*							
Conversion Factor (Annual Avg)	7.140	7.140	7.140	7.140	7.140	7.140	7.000	7.140	7.340	8.370	8.280
EXTERNAL DEBT, ETC.				*(Millions of US dollars, outstanding at end of year)*							
Public/Publicly Guar. Long-Term	..	..	..	319	340	386	517	609	636	790	1,114
Official Creditors	..	..	..	234	248	287	388	476	525	649	817
IBRD and IDA	..	..	..	37	47	76	113	137	187	264	314
Private Creditors	..	..	..	85	92	99	129	133	111	141	297
Private Non-guaranteed Long-Term	..	..	..	88	81	107	206	352	392	415	271
Use of Fund Credit	..	..	..	0	0	0	0	39	80	99	58
Short-Term Debt	..	..	..	..	..	..	..	..	..	..	303
Memo Items:				*(Millions of US dollars)*							
Int'l Reserves Excluding Gold	75.8	100.1	169.7	219.8	171.0	202.0	233.0	193.3	173.4	275.5	522.4
Gold Holdings (at market price)	..	..	..	..	..	..	0.0	0.0	0.0	0.0	3.4
SOCIAL INDICATORS											
Total Fertility Rate	8.0	8.0	8.0	8.0	8.0	8.0	8.0	8.0	8.0	8.0	8.0
Crude Birth Rate	52.2	52.3	52.5	52.6	52.8	52.9	53.0	53.2	53.3	53.5	53.6
Infant Mortality Rate	108.0	106.0	104.0	102.0	100.0	98.0	96.0	94.0	92.0	90.0	88.0
Life Expectancy at Birth	48.5	49.0	49.5	50.0	50.5	51.0	51.5	52.0	52.5	53.0	53.5
Food Production, p.c. ('79-81 = 100)	121.1	125.9	127.1	124.7	121.1	121.7	121.0	115.9	117.7	116.9	118.3
Labor Force, Agriculture (%)	85.6	85.3	85.1	84.8	84.4	84.0	83.7	83.3	82.9	82.5	82.1
Labor Force, Female (%)	42.5	42.5	42.4	42.4	42.3	42.3	42.2	42.2	42.1	42.1	42.0
School Enroll. Ratio, primary	..	..	..	58.0	..	..	..	..	95.0	97.0	95.0
School Enroll. Ratio, secondary	..	..	..	9.0	..	..	..	..	13.0	15.0	17.0

1978	1979	1980	1981	1982	1983	1984	1985	1986	1987 estimate	NOTES	KENYA
				(Millions of current US dollars)							**FOREIGN TRADE (CUSTOMS BASIS)**
1,023.0	1,107.0	1,389.0	1,183.0	1,045.0	979.0	1,083.0	987.6	1,216.8	984.6	f	Value of Exports, fob
682.6	709.6	734.1	654.7	587.7	645.0	766.4	708.3	931.1	645.2	..	Nonfuel Primary Products
185.4	223.1	444.6	374.8	263.3	206.0	198.8	155.9	134.8	175.0	..	Fuels
155.0	174.2	210.4	153.5	194.0	128.0	117.8	123.4	150.9	164.4	..	Manufactures
1,705.9	1,655.5	2,589.9	2,081.3	1,603.0	1,390.0	1,529.4	1,455.1	1,649.1	1,865.8	f	Value of Imports, cif
172.1	154.2	265.3	184.2	167.2	178.5	218.0	193.3	210.2	235.2	..	Nonfuel Primary Products
307.5	395.8	877.1	776.8	595.3	532.0	499.1	461.4	300.0	387.2	..	Fuels
1,226.3	1,105.5	1,447.6	1,120.4	840.6	679.5	812.3	800.4	1,138.8	1,243.4	..	Manufactures
				(Index 1980 = 100)							
107.0	107.0	100.0	92.3	90.1	94.1	103.8	91.9	111.6	87.1	f	Terms of Trade
74.4	87.7	100.0	96.2	88.7	89.1	98.5	84.4	93.7	81.7	f	Export Prices, fob
94.5	104.4	100.0	87.7	86.6	91.7	108.1	87.1	101.7	79.0	..	Nonfuel Primary Products
42.3	61.0	100.0	112.5	101.6	92.5	90.2	87.5	67.4	85.3	..	Fuels
72.4	80.4	100.0	102.5	80.4	74.2	69.0	69.0	82.1	90.0	..	Manufactures
69.5	82.0	100.0	104.3	98.4	94.7	94.9	91.8	83.9	93.8	f	Import Prices, cif
				(Millions of 1980 US dollars)							Trade at Constant 1980 Prices
1,374.5	1,262.3	1,389.1	1,229.9	1,178.8	1,098.6	1,099.8	1,169.7	1,299.3	1,204.5	f	Exports, fob
2,453.7	2,019.6	2,590.0	1,996.2	1,629.1	1,467.1	1,612.0	1,584.3	1,964.7	1,988.5	f	Imports, cif
				(Millions of current US dollars)							**BALANCE OF PAYMENTS**
1,544	1,629	2,061	1,799	1,630	1,525	1,663	1,595	1,907	1,740	f	Exports of Goods & Services
956	1,031	1,261	1,081	936	927	1,034	943	1,170	909	..	Merchandise, fob
540	538	746	680	673	570	590	608	699	794	..	Nonfactor Services
47	61	54	37	22	29	39	43	37	37	..	Factor Services
2,295	2,219	3,095	2,575	2,068	1,753	1,966	1,883	2,158	2,451	f	Imports of Goods & Services
1,632	1,594	2,345	1,834	1,468	1,198	1,348	1,273	1,457	1,623	..	Merchandise, fob
428	377	502	503	398	342	377	344	406	485	..	Nonfactor Services
235	248	248	238	201	213	241	266	295	343	..	Factor Services
89	122	173	171	185	167	176	186	206	244	..	Long-Term Interest
16	11	27	99	83	63	60	82	58	72	..	Current Transfers, net
0	0	0	0	0	0	0	0	0	..	..	Workers' Remittances
-735	-579	-1,006	-678	-354	-164	-243	-207	-193	-639	K	Total to be Financed
75	81	120	119	52	119	119	110	149	143	K	Official Capital Grants
-661	-498	-886	-558	-302	-45	-123	-97	-44	-497	..	Current Account Balance
510	588	668	422	104	228	241	59	252	427	f	Long-Term Capital, net
32	78	78	8	3	9	4	13	27	..	..	Direct Investment
499	411	418	218	232	231	229	121	275	196	..	Long-Term Loans
634	571	625	464	490	498	546	466	635	539	..	Disbursements
135	160	207	246	258	267	317	345	360	343	..	Repayments
-96	18	52	77	-184	-132	-111	-185	-198	89	..	Other Long-Term Capital
-49	101	16	-53	41	-83	-58	-77	-118	-22	f	Other Capital, net
200	-191	202	189	158	-99	-60	115	-90	92	..	Change in Reserves
				(Kenya Shillings per US dollar)							Memo Item:
7.730	7.480	7.420	9.050	10.920	13.310	14.410	16.430	16.230	16.450	..	Conversion Factor (Annual Avg)
				(Millions of US dollars, outstanding at end of year)							**EXTERNAL DEBT, ETC.**
1,417	1,860	2,238	2,349	2,453	2,441	2,543	2,923	3,672	4,482	..	Public/Publicly Guar. Long-Term
944	1,164	1,365	1,515	1,730	1,864	2,107	2,527	3,038	3,703	..	Official Creditors
368	423	528	586	742	840	946	1,159	1,382	1,681	..	IBRD and IDA
473	696	872	834	723	577	435	396	634	778	..	Private Creditors
433	438	437	366	385	497	428	511	459	496	..	Private Non-guaranteed Long-Term
68	143	194	204	342	417	380	486	431	381	..	Use of Fund Credit
403	429	638	466	334	402	369	470	372	591	..	Short-Term Debt
				(Millions of US dollars)							Memo Items:
352.6	627.7	491.7	231.1	211.7	376.0	389.8	390.7	413.3	255.8	..	Int'l Reserves Excluding Gold
16.3	40.8	47.0	31.7	36.4	30.4	24.6	26.1	31.2	38.6	..	Gold Holdings (at market price)
											SOCIAL INDICATORS
8.0	8.0	8.0	8.0	8.0	7.9	7.9	7.8	7.7	7.7	..	Total Fertility Rate
53.7	53.7	53.8	53.8	53.9	53.6	53.4	53.1	52.9	52.6	..	Crude Birth Rate
86.4	84.8	83.2	81.6	80.0	80.7	81.5	82.2	83.0	83.7	..	Infant Mortality Rate
53.9	54.4	54.9	55.4	55.9	56.3	56.7	57.1	57.5	55.5	..	Life Expectancy at Birth
112.3	107.4	98.4	94.3	104.8	100.7	72.2	92.7	97.0	88.3	..	Food Production, p.c. ('79-81 = 100)
81.8	81.4	81.0	..	..	..	..	..	..	..	..	Labor Force, Agriculture (%)
42.0	41.9	41.9	41.7	41.5	41.3	41.1	40.9	40.7	40.5	..	Labor Force, Female (%)
91.0	108.0	104.0	101.0	107.0	105.0	102.0	94.0	..	..	..	School Enroll. Ratio, primary
18.0	18.0	18.0	17.0	18.0	20.0	20.0	20.0	..	..	..	School Enroll. Ratio, secondary

KOREA, REPUBLIC OF	1967	1968	1969	1970	1971	1972	1973	1974	1975	1976	1977
CURRENT GNP PER CAPITA (US $)	140	170	220	260	300	210	460	570	580	750	910
POPULATION (thousands)	30,131	30,834	31,544	32,241	32,883	33,505	34,103	34,692	35,281	35,849	36,412
ORIGIN AND USE OF RESOURCES					*(Billions of current Korean Won)*						
Gross National Product (GNP)	1,281	1,653	2,155	2,777	3,407	4,177	5,356	7,564	10,065	13,818	17,729
Net Factor Income from Abroad	22	23	25	17	-3	-3	-16	-67	-159	-177	-257
Gross Domestic Product (GDP)	1,259	1,630	2,130	2,760	3,410	4,194	5,397	7,631	10,224	13,996	17,985
Indirect Taxes, net	100	152	202	255	302	332	409	605	986	1,433	1,867
GDP at factor cost	1,159	1,478	1,929	2,505	3,108	3,862	4,988	7,026	9,238	12,563	16,118
Agriculture	386	467	595	718	906	1,096	1,322	1,857	2,504	3,244	3,958
Industry	331	454	616	805	974	1,228	1,728	2,485	3,431	4,856	6,470
Manufacturing	241	327	433	584	725	939	1,356	1,984	2,676	3,869	4,943
Services, etc.	542	709	920	1,237	1,530	1,870	2,346	3,289	4,289	5,896	7,557
Resource Balance	-135	-208	-254	-269	-354	-194	-155	-833	-873	-176	37
Exports of Goods & NFServices	145	209	288	389	524	836	1,602	2,140	2,855	4,446	5,848
Imports of Goods & NFServices	279	417	542	658	878	1,031	1,757	2,973	3,728	4,621	5,811
Domestic Absorption	1,394	1,837	2,384	3,028	3,763	4,388	5,552	8,464	11,097	14,171	17,948
Private Consumption, etc.	983	1,237	1,542	2,082	2,573	3,092	3,779	5,322	7,208	9,105	11,125
General Gov't Consumption	130	173	221	263	334	425	452	734	1,121	1,521	1,919
Gross Domestic Investment	281	428	621	683	857	871	1,321	2,408	2,767	3,545	4,904
Fixed Investment	275	414	556	687	765	853	1,243	1,940	2,550	3,365	4,840
Memo Items:											
Gross Domestic Saving	146	220	367	414	503	677	1,166	1,576	1,894	3,370	4,941
Gross National Saving	193	272	433	461	537	708	1,186	1,570	1,811	3,286	4,767
					(Billions of 1980 Korean Won)						
Gross National Product	12,648	13,954	15,773	17,013	18,564	19,547	22,278	24,177	25,816	29,286	32,408
GDP at factor cost	11,448	12,481	14,110	15,367	16,890	18,049	20,696	22,376	23,628	26,560	29,412
Agriculture	4,518	4,579	5,060	4,990	5,158	5,262	5,636	6,037	6,289	6,928	7,132
Industry	2,404	3,031	3,754	4,321	4,833	5,273	6,630	7,586	8,497	10,223	11,957
Manufacturing	1,409	1,792	2,179	2,613	3,097	3,517	4,534	5,264	5,896	7,334	8,455
Services, etc.	5,517	6,150	6,771	7,574	8,552	9,038	10,113	10,693	11,362	12,454	13,744
Resource Balance	-1,155	-1,694	-2,027	-1,996	-2,369	-1,550	-1,489	-2,663	-1,865	-1,273	-1,295
Exports of Goods & NFServices	884	1,233	1,678	2,006	2,430	3,305	5,056	5,018	5,972	8,449	10,413
Imports of Goods & NFServices	2,038	2,927	3,705	4,003	4,800	4,855	6,545	7,681	7,837	9,721	11,708
Domestic Absorption	13,594	15,453	17,612	18,882	20,911	21,124	23,869	26,979	28,013	30,877	34,128
Private Consumption, etc.	9,701	10,606	11,332	12,543	14,044	14,405	16,038	17,085	18,000	19,732	20,696
General Gov't Consumption	1,686	1,843	2,018	2,152	2,349	2,485	2,469	2,866	3,195	3,312	3,554
Gross Domestic Investment	2,206	3,004	4,262	4,187	4,518	4,233	5,362	7,028	6,818	7,833	9,878
Fixed Investment	2,179	2,994	3,735	3,772	3,881	3,913	4,854	5,564	5,986	7,139	9,148
Memo Items:											
Capacity to Import	1,055	1,470	1,968	2,368	2,866	3,939	5,967	5,529	6,002	9,352	11,783
Terms of Trade Adjustment	171	237	290	361	436	635	911	512	30	903	1,370
Gross Domestic Income	12,610	13,996	15,874	17,247	18,978	20,208	23,291	24,827	26,178	30,507	34,203
Gross National Income	12,819	14,191	16,062	17,374	19,000	20,181	23,189	24,689	25,846	30,189	33,778
DOMESTIC PRICES (DEFLATORS)					*(Index 1980 = 100)*						
Overall (GDP)	10.1	11.8	13.7	16.3	18.4	21.4	24.1	31.4	39.1	47.3	54.8
Domestic Absorption	10.3	11.9	13.5	16.0	18.0	20.8	23.3	31.4	39.6	45.9	52.6
Agriculture	8.5	10.2	11.8	14.4	17.6	20.8	23.5	30.8	39.8	46.8	55.5
Industry	13.8	15.0	16.4	18.6	20.2	23.3	26.1	32.8	40.4	47.5	54.1
Manufacturing	17.1	18.3	19.9	22.3	23.4	26.7	29.9	37.7	45.4	52.8	58.5
MANUFACTURING ACTIVITY											
Employment (1980=100)	30.6	35.5	39.3	41.1	40.4	47.1	55.9	63.3	69.3	83.9	93.9
Real Earnings per Empl. (1980=100)	32.6	35.8	40.7	43.3	45.3	46.2	55.0	57.1	60.0	66.5	78.0
Real Output per Empl. (1980=100)	29.0	32.6	36.9	40.0	48.7	49.2	60.9	66.9	71.3	72.7	77.5
Earnings as % of Value Added	26.1	25.0	25.0	25.0	23.3	23.6	22.5	23.2	23.6	24.8	26.1
MONETARY HOLDINGS					*(Billions of current Korean Won)*						
Money Supply, Broadly Defined	289	497	782	990	1,224	1,632	2,191	2,682	3,379	4,511	6,256
Money as Means of Payment	123	179	252	308	358	519	730	946	1,182	1,544	2,173
Currency Ouside Banks	58	83	111	134	162	218	311	411	507	677	953
Demand Deposits	65	96	141	174	196	302	419	535	675	867	1,219
Quasi-Monetary Liabilities	166	318	530	683	866	1,112	1,461	1,737	2,197	2,967	4,083
					(Billions of current Korean Won)						
GOVERNMENT DEFICIT (-) OR SURPLUS	..	..	..	..	-10	-161	-27	-164	-202	-192	-316
Current Revenue	..	..	..	..	528	580	679	1,026	1,554	2,315	2,938
Current Expenditure	..	..	..	..	413	537	574	895	1,258	1,790	2,362
Current Budget Balance	..	..	..	..	115	43	105	131	296	525	576
Capital Receipts	..	..	..	..	8	9	20	13	13	14	20
Capital Payments	..	..	..	..	138	209	148	309	508	730	912

1978	1979	1980	1981	1982	1983	1984	1985	1986	1987 estimate	NOTES	KOREA, REPUBLIC OF
1,190	1,510	1,620	1,830	1,880	2,020	2,120	2,160	2,370	2,690	..	CURRENT GNP PER CAPITA (US $)
36,969	37,534	38,124	38,723	39,326	39,929	40,513	41,056	41,467	42,031	..	POPULATION (thousands)
				(Billions of current Korean Won)							ORIGIN AND USE OF RESOURCES
23,937	30,741	36,672	45,126	50,725	58,986	66,408	72,850	83,976	97,532	..	Gross National Product (GNP)
-306	-483	-1,243	-1,897	-2,188	-2,017	-2,458	-2,661	-2,677	-2,258	..	Net Factor Income from Abroad
24,243	31,224	37,915	47,024	52,913	61,003	68,867	75,511	86,653	99,790	f	Gross Domestic Product (GDP)
2,620	3,469	4,539	5,658	6,619	8,136	8,720	9,370	10,844	12,465	..	Indirect Taxes, net
21,622	27,755	33,376	41,366	46,294	52,867	60,147	66,140	75,808	87,325	B f	GDP at factor cost
4,890	5,878	5,525	7,442	7,732	8,293	9,181	10,158	10,637	11,365	..	Agriculture
9,297	12,502	15,670	18,886	21,149	24,764	28,618	30,906	36,558	42,822	..	Industry
6,788	8,949	11,214	13,714	14,996	17,302	20,019	21,285	25,912	30,262	..	Manufacturing
10,056	12,845	16,720	20,695	24,032	27,946	31,067	34,447	39,457	45,603	..	Services, etc.
-726	-2,265	-2,964	-2,521	-1,351	-782	-208	404	4,977	8,474	..	Resource Balance
7,336	8,563	12,765	17,191	18,802	22,246	25,830	27,327	35,323	44,832	..	Exports of Goods & NFServices
8,062	10,829	15,729	19,712	20,154	23,028	26,037	26,923	30,346	36,358	..	Imports of Goods & NFServices
24,969	33,489	40,879	49,544	54,264	61,785	69,074	75,107	81,676	91,316	..	Domestic Absorption
14,844	19,356	24,822	30,482	33,644	37,411	40,788	44,568	47,543	51,559	..	Private Consumption, etc.
2,501	3,059	4,268	5,383	6,110	6,753	7,079	7,893	9,150	10,734	..	General Gov't Consumption
7,624	11,074	11,789	13,679	14,510	17,621	21,207	22,645	24,983	29,023	..	Gross Domestic Investment
7,488	10,191	11,836	12,931	15,486	18,480	20,795	22,436	26,364	30,661	..	Fixed Investment
											Memo Items:
6,898	8,809	8,825	11,158	13,158	16,839	21,000	23,049	29,960	37,497	..	Gross Domestic Saving
6,802	8,519	7,825	9,548	11,297	15,261	18,957	20,871	28,189	36,225	..	Gross National Saving
				(Billions of 1980 Korean Won)							
35,981	38,503	36,672	39,089	41,212	46,109	50,003	52,705	59,188	66,320	..	Gross National Product
32,466	34,737	33,376	35,815	37,637	41,367	45,292	47,893	53,421	59,364	B f	GDP at factor cost
6,427	6,899	5,525	6,760	6,981	7,436	7,453	7,809	8,114	7,767	..	Agriculture
14,560	15,789	15,670	16,555	17,662	20,187	22,890	24,045	27,630	31,892	..	Industry
10,259	11,340	11,214	12,059	12,559	14,096	16,188	16,805	19,737	22,964	..	Manufacturing
15,424	16,412	16,720	17,409	18,393	20,120	21,529	22,820	25,318	28,162	..	Services, etc.
-3,176	-5,031	-2,964	-1,748	-1,125	-539	-610	155	1,795	2,909	..	Resource Balance
11,720	11,587	12,765	14,684	15,638	18,054	19,855	20,279	25,648	31,809	..	Exports of Goods & NFServices
14,896	16,618	15,729	16,432	16,763	18,593	20,465	20,124	23,853	28,900	..	Imports of Goods & NFServices
39,587	44,131	40,879	42,471	44,161	48,283	52,483	54,519	59,267	64,911	..	Domestic Absorption
22,773	25,258	24,822	25,403	27,039	28,727	30,155	31,605	33,820	35,815	..	Private Consumption, etc.
3,993	4,009	4,268	4,532	4,569	4,811	4,835	5,145	5,795	6,482	..	General Gov't Consumption
12,821	14,864	11,789	12,536	12,553	14,745	17,492	17,769	19,652	22,614	..	Gross Domestic Investment
12,175	13,218	11,836	11,359	12,820	15,017	16,618	17,356	20,020	22,741	..	Fixed Investment
											Memo Items:
13,554	13,142	12,765	14,331	15,639	17,962	20,302	20,426	27,764	35,636	..	Capacity to Import
1,834	1,555	0	-354	1	-92	447	147	2,116	3,827	..	Terms of Trade Adjustment
38,245	40,655	37,915	40,370	43,037	47,652	52,319	54,821	63,179	71,647	..	Gross Domestic Income
37,815	40,058	36,672	38,735	41,213	46,017	50,450	52,852	61,304	70,146	..	Gross National Income
				(Index 1980 = 100)							DOMESTIC PRICES (DEFLATORS)
66.6	79.9	100.0	115.5	123.0	127.8	132.8	138.1	141.9	147.1	..	Overall (GDP)
63.1	75.9	100.0	116.7	122.9	128.0	131.6	137.8	137.8	140.7	..	Domestic Absorption
76.1	85.2	100.0	110.1	110.8	111.5	123.2	130.1	131.1	146.3	..	Agriculture
63.8	79.2	100.0	114.1	119.7	122.7	125.0	128.5	132.3	134.3	..	Industry
66.2	78.9	100.0	113.7	119.4	122.7	123.7	126.7	131.3	131.8	..	Manufacturing
											MANUFACTURING ACTIVITY
103.3	103.5	100.0	99.8	102.4	108.1	114.3	119.8	..	..	G	Employment (1980=100)
94.3	104.6	100.0	98.3	102.8	109.0	118.9	119.4	..	..	G	Real Earnings per Empl. (1980=100)
85.3	90.1	100.0	113.4	116.5	125.8	139.0	139.2	..	..	G	Real Output per Empl. (1980=100)
27.1	31.2	29.3	26.8	27.5	26.3	26.3	26.7	..	..	..	Earnings as % of Value Added
				(Billions of current Korean Won)							MONETARY HOLDINGS
8,374	10,484	13,623	17,153	21,680	24,972	27,320	32,640	38,892	49,202	D	Money Supply, Broadly Defined
2,714	3,275	3,807	3,982	5,799	6,783	6,821	7,558	8,809	10,107	..	Money as Means of Payment
1,364	1,604	1,856	2,025	2,574	2,874	3,109	3,286	3,679	4,443	..	Currency Ouside Banks
1,349	1,671	1,951	1,957	3,226	3,909	3,711	4,272	5,130	5,664	..	Demand Deposits
5,660	7,210	9,816	13,171	15,881	18,189	20,499	25,082	30,083	39,095	..	Quasi-Monetary Liabilities
				(Billions of current Korean Won)							
-300	-545	-849	-1,585	-1,656	-663	-841	-943	-86	-1,280	..	GOVERNMENT DEFICIT (-) OR SURPLUS
4,084	5,376	6,738	8,534	9,875	11,418	12,511	13,738	15,721	17,057	..	Current Revenue
3,196	4,205	5,641	6,931	8,297	9,145	10,213	11,523	12,829	14,564	..	Current Expenditure
888	1,171	1,097	1,603	1,578	2,273	2,298	2,215	2,892	2,493	..	Current Budget Balance
24	70	96	71	108	120	93	185	119	266	..	Capital Receipts
1,212	1,786	2,042	3,259	3,342	3,056	3,232	3,343	3,097	4,039	..	Capital Payments

KOREA, REPUBLIC OF	1967	1968	1969	1970	1971	1972	1973	1974	1975	1976	1977
FOREIGN TRADE (CUSTOMS BASIS)					*(Millions of current US dollars)*						
Value of Exports, fob	320	455	623	835	1,068	1,624	3,225	4,460	5,081	7,715	10,016
Nonfuel Primary Products	105	116	143	186	183	246	473	559	830	801	1,391
Fuels	2	2	5	9	11	18	35	108	104	145	117
Manufactures	214	337	475	641	873	1,360	2,717	3,794	4,147	6,770	8,509
Value of Imports, cif	996	1,468	1,823	1,983	2,394	2,522	4,240	6,844	7,271	8,764	10,803
Nonfuel Primary Products	323	464	667	761	907	860	1,569	2,243	2,199	2,394	2,960
Fuels	62	76	111	136	189	219	312	1,054	1,387	1,747	2,179
Manufactures	612	928	1,045	1,086	1,298	1,443	2,358	3,547	3,685	4,623	5,664
					(Index 1980 = 100)						
Terms of Trade	148.6	154.4	146.0	160.8	157.5	149.2	130.0	110.2	114.7	124.6	118.1
Export Prices, fob	33.1	36.0	33.9	36.8	38.7	39.5	48.9	61.0	62.8	70.4	71.5
Nonfuel Primary Products	30.3	32.7	34.8	36.2	36.4	39.2	47.3	58.5	60.6	61.8	67.3
Fuels	4.3	4.3	4.3	4.3	5.6	6.2	8.9	36.7	35.7	38.4	42.0
Manufactures	36.8	39.4	36.1	41.3	42.6	42.6	52.3	62.6	64.5	72.9	72.9
Import Prices, cif	22.3	23.3	23.2	22.9	24.6	26.5	37.6	55.4	54.8	56.5	60.5
Trade at Constant 1980 Prices					*(Millions of 1980 US dollars)*						
Exports, fob	968	1,265	1,838	2,270	2,756	4,113	6,598	7,311	8,088	10,959	14,017
Imports, cif	4,475	6,295	7,859	8,665	9,736	9,528	11,273	12,363	13,279	15,507	17,857
BALANCE OF PAYMENTS					*(Millions of current US dollars)*						
Exports of Goods & Services	..	..	..	1,379	1,616	2,226	4,121	5,353	5,907	9,457	13,073
Merchandise, fob	..	..	..	882	1,132	1,676	3,271	4,516	5,028	7,814	10,046
Nonfactor Services	..	..	..	426	429	507	780	723	796	1,525	2,784
Factor Services	..	..	..	71	54	43	70	115	83	118	243
Imports of Goods & Services	..	..	..	2,180	2,633	2,764	4,618	7,602	7,992	10,113	13,284
Merchandise, fob	..	..	..	1,804	2,177	2,251	3,837	6,455	6,671	8,404	10,523
Nonfactor Services	..	..	..	300	336	354	567	818	871	1,190	2,022
Factor Services	..	..	..	76	119	159	214	329	450	519	739
Long-Term Interest	..	..	..	76	101	160	254	240	331	424	517
Current Transfers, net	..	..	..	95	105	119	155	154	158	193	170
Workers' Remittances	..	..	..	0	0	0	0	0	0	0	0
Total to be Financed	..	..	..	-706	-912	-418	-342	-2,094	-1,927	-463	-41
Official Capital Grants	..	..	..	83	63	50	36	67	67	153	53
Current Account Balance	..	..	..	-623	-849	-368	-306	-2,027	-1,860	-310	12
Long-Term Capital, net	..	..	..	579	660	546	486	805	1,411	1,485	1,453
Direct Investment	..	..	..	66	39	63	93	105	53	75	73
Long-Term Loans	..	..	..	271	536	622	662	1,019	1,351	1,361	1,604
Disbursements	..	..	..	476	774	906	1,090	1,398	1,769	1,935	2,416
Repayments	..	..	..	205	239	284	428	379	418	575	812
Other Long-Term Capital	..	..	..	159	22	-188	-304	-387	-60	-104	-277
Other Capital, net	..	..	..	92	149	-37	170	1,051	815	138	-95
Change in Reserves	..	..	..	-48	40	-141	-350	171	-365	-1,313	-1,370
Memo Item:					*(Korean Won per US dollar)*						
Conversion Factor (Annual Avg)	270.510	276.640	288.230	310.570	348.200	392.900	398.320	400.430	484.000	484.000	484.000
EXTERNAL DEBT, ETC.					*(Millions of US dollars, outstanding at end of year)*						
Public/Publicly Guar. Long-Term	..	..	..	1,816	2,340	2,903	3,553	4,457	5,654	6,943	8,679
Official Creditors	..	..	..	588	892	1,329	1,777	2,162	2,639	3,341	4,086
IBRD and IDA	..	..	..	36	80	143	199	300	496	779	960
Private Creditors	..	..	..	1,228	1,448	1,574	1,776	2,295	3,015	3,601	4,592
Private Non-guaranteed Long-Term	..	..	..	175	230	300	370	500	580	690	868
Use of Fund Credit	..	..	..	0	0	0	0	159	254	373	365
Short-Term Debt	..	..	..	..	..	..	..	..	..	..	4,457
Memo Items:					*(Millions of US dollars)*						
Int'l Reserves Excluding Gold	353	388	550	606	434	523	885	277	781	1,970	2,967
Gold Holdings (at market price)	3	4	3	4	4	7	12	21	16	15	24
SOCIAL INDICATORS											
Total Fertility Rate	4.5	4.4	4.3	4.2	4.2	4.1	3.9	3.7	3.5	3.3	3.1
Crude Birth Rate	31.9	31.3	30.7	30.0	29.4	28.8	27.8	26.8	25.9	24.9	23.9
Infant Mortality Rate	58.0	55.8	53.6	51.4	49.2	47.0	44.6	42.2	39.8	37.4	35.0
Life Expectancy at Birth	57.7	58.4	59.2	59.9	60.7	61.4	62.3	63.1	63.9	64.7	65.5
Food Production, p.c. ('79-81 = 100)	81.4	78.4	90.0	85.2	84.6	83.8	84.4	87.9	96.0	104.2	108.9
Labor Force, Agriculture (%)	52.7	51.5	50.3	49.1	47.8	46.6	45.3	44.1	42.8	41.5	40.2
Labor Force, Female (%)	30.3	30.9	31.5	32.1	32.4	32.6	32.8	33.1	33.3	33.5	33.6
School Enroll. Ratio, primary	..	..	..	103.0	..	..	..	..	107.0	108.0	107.0
School Enroll. Ratio, secondary	..	..	..	42.0	..	..	..	..	56.0	61.0	64.0

1978	1979	1980	1981	1982	1983	1984	1985	1986	1987 estimate	NOTES	KOREA, REPUBLIC OF
				(Millions of current US dollars)							**FOREIGN TRADE (CUSTOMS BASIS)**
12,695	15,052	17,483	21,250	21,850	24,437	29,248	30,283	34,702	47,172	..	Value of Exports, fob
1,435	1,624	1,715	1,832	1,610	1,661	1,737	1,685	2,153	2,873	..	Nonfuel Primary Products
41	29	46	181	285	536	805	929	618	720	..	Fuels
11,219	13,398	15,722	19,237	19,955	22,240	26,707	27,669	31,931	43,579	..	Manufactures
14,966	20,296	22,228	26,028	24,236	26,174	30,609	31,119	31,518	40,934	..	Value of Imports, cif
3,748	5,292	5,978	6,965	5,440	5,820	6,319	5,910	6,586	9,461	..	Nonfuel Primary Products
2,453	3,785	6,648	7,775	7,593	6,958	7,274	7,333	5,024	6,272	..	Fuels
8,765	11,219	9,601	11,288	11,204	13,396	17,015	17,876	19,908	25,201	..	Manufactures
				(Index 1980 = 100)							
117.3	114.0	100.0	100.9	103.6	103.4	105.7	106.1	111.4	105.2	..	Terms of Trade
79.6	94.7	100.0	102.1	98.2	95.5	96.6	93.4	91.9	99.7	..	Export Prices, fob
73.9	89.4	100.0	94.8	89.5	86.6	82.1	70.9	78.6	82.3	..	Nonfuel Primary Products
42.3	61.0	100.0	112.5	101.6	92.5	90.2	87.5	44.9	56.9	..	Fuels
80.6	95.5	100.0	102.8	98.9	96.3	97.9	95.5	94.9	102.4	..	Manufactures
67.9	83.1	100.0	101.2	94.7	92.4	91.3	88.1	82.5	94.8	..	Import Prices, cif
				(Millions of 1980 US dollars)							Trade at Constant 1980 Prices
15,950	15,896	17,483	20,806	22,256	25,593	30,288	32,410	37,762	47,314	..	Exports, fob
22,052	24,428	22,229	25,708	25,580	28,336	33,511	35,336	38,213	43,183	..	Imports, cif
				(Millions of current US dollars)							**BALANCE OF PAYMENTS**
17,161	19,530	22,577	27,269	28,356	30,383	33,652	33,106	41,965	56,255	..	Exports of Goods & Services
12,711	14,705	17,214	20,671	20,879	23,204	26,335	26,442	33,913	46,244	..	Merchandise, fob
4,056	4,390	4,707	5,756	6,673	6,470	6,452	5,593	6,866	8,797	..	Nonfactor Services
394	435	656	842	804	709	865	1,071	1,186	1,214	..	Factor Services
18,717	24,120	28,347	32,416	31,505	32,581	35,565	34,571	38,387	47,619	..	Imports of Goods & Services
14,491	19,100	21,598	24,299	23,473	24,967	27,371	26,461	29,707	38,585	..	Merchandise, fob
3,193	3,502	4,057	4,419	4,152	4,151	4,192	4,094	4,598	5,397	..	Nonfactor Services
1,033	1,518	2,692	3,698	3,880	3,463	4,002	4,016	4,082	3,637	..	Factor Services
793	1,123	1,636	2,031	2,438	2,279	2,509	2,776	2,860	2,375	..	Long-Term Interest
434	399	399	422	447	566	516	555	1,028	1,199	..	Current Transfers, net
0	0	0	0	0	0	0	0	0	0	..	Workers' Remittances
-1,122	-4,191	-5,371	-4,725	-2,702	-1,632	-1,397	-910	4,606	9,835	K	Total to be Financed
37	40	50	79	52	26	25	23	11	19	K	Official Capital Grants
-1,085	-4,151	-5,321	-4,646	-2,650	-1,606	-1,372	-887	4,617	9,854	..	Current Account Balance
2,148	3,111	2,037	3,612	1,849	1,809	2,677	2,252	-2,560	-8,543	..	Long-Term Capital, net
61	16	-7	60	-76	-57	73	200	325	418	..	Direct Investment
2,425	3,404	2,426	3,822	2,423	3,397	3,475	2,675	-1,557	-8,703	..	Long-Term Loans
3,687	5,128	3,980	5,837	4,564	6,061	6,326	7,080	5,636	4,390	..	Disbursements
1,262	1,724	1,554	2,015	2,140	2,665	2,851	4,405	7,192	13,094	..	Repayments
-375	-349	-432	-349	-550	-1,556	-896	-646	-1,339	-277	..	Other Long-Term Capital
-332	1,914	3,595	705	806	-439	-745	-1,173	-1,980	793	..	Other Capital, net
-731	-874	-311	329	-5	236	-560	-192	-77	-2,104	..	Change in Reserves
				(Korean Won per US dollar)							Memo Item:
484.000	484.000	607.430	681.030	731.080	775.750	805.980	870.000	881.450	822.600	..	Conversion Factor (Annual Avg)
				(Millions of US dollars, outstanding at end of year)							**EXTERNAL DEBT, ETC.**
11,317	13,766	15,933	18,361	20,191	22,175	23,832	28,304	29,355	24,541	..	Public/Publicly Guar. Long-Term
4,938	5,559	6,331	7,240	8,162	9,179	9,102	10,142	11,008	11,666	..	Official Creditors
1,258	1,648	1,836	2,104	2,670	3,087	2,980	3,701	4,409	5,094	..	IBRD and IDA
6,378	8,207	9,601	11,121	12,029	12,997	14,730	18,162	18,347	12,875	..	Private Creditors
1,197	1,817	2,303	3,156	3,452	4,775	5,274	6,614	6,569	6,102	..	Private Non-guaranteed Long-Term
275	138	683	1,246	1,259	1,353	1,567	1,508	1,549	525	..	Use of Fund Credit
4,525	7,165	10,561	10,226	12,427	12,115	11,425	10,732	9,256	9,291	..	Short-Term Debt
				(Millions of US dollars)							Memo Items:
2,764	2,959	2,925	2,682	2,807	2,347	2,754	2,869	3,320	3,584	..	Int'l Reserves Excluding Gold
62	151	176	120	139	116	95	102	124	155	..	Gold Holdings (at market price)
											SOCIAL INDICATORS
3.0	2.9	2.8	2.7	2.6	2.5	2.4	2.3	2.2	2.1	..	Total Fertility Rate
23.7	23.6	23.4	23.3	23.1	22.4	21.7	21.1	20.4	19.7	..	Crude Birth Rate
34.0	33.0	32.0	31.0	30.0	30.7	31.3	32.0	32.6	33.3	..	Infant Mortality Rate
66.0	66.4	66.8	67.3	67.7	68.1	68.4	68.8	69.1	69.0	..	Life Expectancy at Birth
114.4	113.9	89.1	97.0	97.9	97.5	102.3	103.2	99.7	96.8	..	Food Production, p.c. ('79-81 = 100)
39.0	37.7	36.4	..	..	..	..	..	..	..	..	Labor Force, Agriculture (%)
33.8	33.9	34.1	34.1	34.1	34.0	34.0	34.0	34.0	33.9	..	Labor Force, Female (%)
109.0	109.0	109.0	107.0	109.0	103.0	99.0	96.0	94.0	..	..	School Enroll. Ratio, primary
68.0	75.0	76.0	84.0	82.0	87.0	91.0	94.0	95.0	..	..	School Enroll. Ratio, secondary

KUWAIT	1967	1968	1969	1970	1971	1972	1973	1974	1975	1976	1977
CURRENT GNP PER CAPITA (US $)	3,600	3,620	3,540	3,340	3,620	3,680	4,180	6,010	9,200	14,600	14,050
POPULATION (thousands)	585	640	695	748	783	829	884	945	1,007	1,074	1,147
ORIGIN AND USE OF RESOURCES					*(Millions of current Kuwaiti Dinars)*						
Gross National Product (GNP)	734.0	793.0	840.0	851.3	1,115.8	1,102.4	1,262.1	3,532.0	3,711.0	4,280.7	4,557.6
Net Factor Income from Abroad	-138.0	-158.0	-149.0	-175.0	-266.0	-361.6	-342.0	-281.0	224.0	441.0	506.0
Gross Domestic Product (GDP)	872.0	951.0	989.0	1,026.3	1,381.8	1,464.0	1,604.1	3,813.0	3,487.0	3,839.7	4,051.6
Indirect Taxes, net	..	..	..	..	..	..	..	..	..	..	..
GDP at factor cost	..	..	..	..	..	..	..	..	..	..	..
Agriculture	5.0	5.0	5.0	2.9	3.2	3.8	4.7	5.9	8.8	10.3	7.7
Industry	555.0	613.0	636.0	689.7	995.3	1,018.3	1,118.7	3,254.1	2,727.8	2,876.5	2,882.3
Manufacturing	34.0	37.0	36.0	42.8	54.2	66.4	80.2	171.2	197.5	234.2	241.3
Services, etc.	312.0	333.0	348.0	333.7	383.3	441.9	480.7	553.0	750.4	952.9	1,161.6
Resource Balance	271.0	339.0	342.0	366.4	658.9	701.8	798.3	2,711.7	1,899.0	1,742.4	1,158.0
Exports of Goods & NFServices	519.0	587.0	628.0	613.9	916.8	1,004.3	1,153.9	3,239.5	2,806.0	2,992.4	2,918.0
Imports of Goods & NFServices	248.0	248.0	286.0	247.5	257.9	302.5	355.6	527.8	907.0	1,250.0	1,760.0
Domestic Absorption	601.0	612.0	647.0	659.9	722.9	762.2	805.8	1,101.3	1,588.0	2,097.3	2,893.6
Private Consumption, etc.	280.0	297.0	306.0	396.2	420.0	427.4	438.6	563.8	758.8	1,029.8	1,362.7
General Gov't Consumption	135.0	144.0	152.0	139.3	173.3	198.6	214.6	279.1	385.5	432.3	586.3
Gross Domestic Investment	186.0	171.0	189.0	124.4	129.6	136.2	152.6	258.4	443.7	635.2	944.6
Fixed Investment	163.0	157.0	170.0	126.5	126.6	127.3	145.8	221.7	417.6	563.1	815.2
Memo Items:											
Gross Domestic Saving	457.0	510.0	531.0	490.8	788.5	838.0	950.9	2,970.1	2,342.7	2,377.6	2,102.6
Gross National Saving	319.0	352.0	382.0	253.3	429.9	384.3	388.5	2,327.0	2,486.7	2,726.6	2,502.6
					(Millions of 1980 Kuwaiti Dinars)						
Gross National Product	5,833.0	6,294.7	6,819.3	7,135.5	7,486.2	7,300.4	7,227.4	7,617.0	7,960.8	9,065.8	9,089.7
GDP at factor cost	..	..	..	..	..	..	..	..	..	..	..
Agriculture	..	..	..	7.7	7.9	6.8	7.4	7.6	9.2	10.0	11.3
Industry	..	..	..	7,811.9	8,400.1	8,709.8	8,191.9	7,257.5	6,384.1	6,844.5	6,593.1
Manufacturing	..	..	..	..	..	..	..	..	..	..	380.4
Services, etc.	..	..	..	717.1	781.4	865.1	902.6	937.9	1,097.5	1,294.6	1,489.7
Resource Balance	5,740.8	6,543.9	6,979.5	7,482.4	7,957.9	8,435.2	7,802.3	6,733.4	5,939.4	6,314.0	4,857.2
Exports of Goods & NFServices	6,035.0	6,839.8	7,308.6	7,776.4	8,247.8	8,766.3	8,181.4	7,205.5	6,660.0	7,249.2	6,784.0
Imports of Goods & NFServices	294.2	295.9	329.0	294.0	289.9	331.1	379.1	472.1	720.6	935.2	1,926.8
Domestic Absorption	..	..	..	1,414.1	1,460.8	1,501.1	1,481.4	1,741.6	2,307.4	2,836.5	3,651.9
Private Consumption, etc.	..	..	..	811.2	808.1	794.3	746.1	847.2	1,126.9	1,367.4	1,672.1
General Gov't Consumption	..	..	..	323.2	374.0	420.8	450.3	503.7	571.2	629.3	759.7
Gross Domestic Investment	341.0	317.7	339.0	279.7	278.7	286.0	285.0	390.6	609.2	839.8	1,220.1
Fixed Investment	..	..	..	287.6	275.1	270.2	276.0	337.1	577.9	750.8	1,067.5
Memo Items:											
Capacity to Import	615.7	700.3	722.5	729.3	1,030.5	1,099.1	1,230.2	2,897.9	2,229.3	2,238.7	3,194.5
Terms of Trade Adjustment	-5,419.3	-6,139.5	-6,586.1	-7,047.1	-7,217.2	-7,667.2	-6,951.2	-4,307.6	-4,430.7	-5,010.5	-3,589.5
Gross Domestic Income	1,195.7	1,805.0	1,743.6	1,489.5	1,972.1	1,914.6	2,150.7	3,895.3	3,060.0	3,138.6	4,504.6
Gross National Income	-143.8	272.6	161.5	88.4	269.0	-366.7	276.2	3,309.3	3,530.0	4,055.4	5,500.2
DOMESTIC PRICES (DEFLATORS)					*(Index 1980 = 100)*						
Overall (GDP)	11.7	12.5	11.9	12.0	15.0	15.3	17.6	46.5	46.6	47.1	50.1
Domestic Absorption	..	..	..	46.7	49.5	50.8	54.4	63.2	68.8	73.9	79.2
Agriculture	..	..	..	37.7	40.5	55.6	63.7	77.8	95.7	102.5	68.0
Industry	..	..	..	8.8	11.8	11.7	13.7	44.8	42.7	42.0	43.7
Manufacturing	..	..	..	..	..	..	..	..	..	..	63.4
MANUFACTURING ACTIVITY											
Employment (1980=100)	26.5	27.4	31.0	32.1	56.6	56.9	57.4	61.4	61.8	79.3	83.4
Real Earnings per Empl. (1980=100)	..	..	..	..	..	60.8	70.0	69.6	83.4	79.8	88.8
Real Output per Empl. (1980=100)	..	..	..	96.3	66.9	68.9	89.0	101.8	85.3	89.0	77.1
Earnings as % of Value Added	10.4	11.2	17.5	12.2	14.5	26.3	27.5	18.1	24.8	19.6	24.6
MONETARY HOLDINGS					*(Millions of current Kuwaiti Dinars)*						
Money Supply, Broadly Defined	321.6	386.4	379.1	362.1	418.8	493.6	536.3	684.6	891.1	1,220.1	1,568.7
Money as Means of Payment	134.4	120.0	107.6	95.2	107.7	142.0	172.4	195.6	290.3	393.7	490.7
Currency Ouside Banks	52.5	47.0	44.3	44.8	50.4	57.1	71.1	81.7	101.7	129.1	150.9
Demand Deposits	81.9	73.0	63.3	50.4	57.3	84.9	101.3	113.9	188.6	264.6	339.8
Quasi-Monetary Liabilities	187.2	266.4	271.5	266.9	311.1	351.6	363.9	489.0	600.8	826.4	1,078.0
GOVERNMENT DEFICIT (-) OR SURPLUS					*(Millions of current Kuwaiti Dinars)*						
GOVERNMENT DEFICIT (-) OR SURPLUS	..	..	..	..	..	192	162	1,588	..	..	1,514
Current Revenue	..	..	..	..	..	608	694	2,724	..	..	2,989
Current Expenditure	..	..	..	..	..	303	368	737	..	..	989
Current Budget Balance	..	..	..	..	..	305	326	1,987	..	..	2,000
Capital Receipts	..	..	..	..	..	1	2	5	..	..	7
Capital Payments	..	..	..	..	..	114	166	404	..	..	493

1978	1979	1980	1981	1982	1983	1984	1985	1986	1987 estimate	NOTES	KUWAIT
15,240	18,350	18,040	20,590	19,320	18,540	17,500	15,010	16,600	14,870	..	CURRENT GNP PER CAPITA (US $)
1,223	1,299	1,372	1,444	1,517	1,588	1,653	1,710	1,775	1,837	..	POPULATION (thousands)
				(Millions of current Kuwaiti Dinars)							ORIGIN AND USE OF RESOURCES
4,981.3	7,712.9	9,051.2	9,122.8	7,921.3	7,585.2	7,850.8	7,305.3	7,178.0	7,177.0	..	Gross National Product (GNP)
717.0	873.0	1,310.1	2,137.0	1,709.0	1,451.0	1,470.0	1,387.0	2,180.0	2,180.0	..	Net Factor Income from Abroad
4,264.3	6,839.9	7,741.1	6,985.8	6,212.3	6,134.2	6,380.8	5,918.3	4,998.0	4,997.0	..	Gross Domestic Product (GDP)
..	..	..	..	..	..	..	..	..	..	..	Indirect Taxes, net
..	..	..	..	..	..	..	..	..	..	B	GDP at factor cost
10.0	11.8	14.0	24.0	28.5	28.4	34.9	39.3	51.5	..	..	Agriculture
2,984.3	5,201.1	5,794.6	4,798.7	3,335.6	3,688.4	3,975.5	3,559.4	2,554.4	..	..	Industry
282.6	566.7	427.3	415.1	307.9	374.9	300.3	377.2	555.5	..	..	Manufacturing
1,270.0	1,627.0	1,932.5	2,163.1	2,848.2	2,417.4	2,370.4	2,319.6	2,392.1	..	..	Services, etc.
1,308.0	3,362.0	3,410.0	2,167.0	133.0	533.0	825.0	534.0	-22.0	..	..	Resource Balance
3,008.0	5,333.0	6,065.0	4,855.0	3,386.0	3,596.0	3,862.0	3,463.0	2,442.0	..	..	Exports of Goods & NFServices
1,700.0	1,971.0	2,655.0	2,688.0	3,253.0	3,063.0	3,037.0	2,929.0	2,464.0	..	..	Imports of Goods & NFServices
2,956.3	3,477.8	4,331.0	4,818.9	6,079.3	5,601.2	5,555.8	5,384.2	5,020.0	..	..	Domestic Absorption
1,477.5	1,780.3	2,388.5	2,663.9	3,317.0	2,795.7	2,878.7	2,668.4	2,584.6	..	..	Private Consumption, etc.
615.9	763.8	864.9	993.2	1,197.2	1,298.7	1,356.6	1,457.2	1,445.4	..	..	General Gov't Consumption
862.9	933.7	1,077.6	1,161.8	1,565.1	1,506.8	1,320.5	1,258.6	990.0	..	..	Gross Domestic Investment
793.9	789.6	972.6	1,072.8	1,436.0	1,525.1	1,306.3	1,295.5	990.0	..	..	Fixed Investment
											Memo Items:
2,170.9	4,295.8	4,487.7	3,328.7	1,698.1	2,039.8	2,145.5	1,792.7	968.0	..	..	Gross Domestic Saving
2,768.9	5,021.8	5,610.8	5,273.7	3,155.1	3,238.8	3,330.6	2,865.7	2,832.3	..	..	Gross National Saving
				(Millions of 1980 Kuwaiti Dinars)							
10,141.3	11,190.6	9,051.2	8,072.0	6,879.7	7,347.8	7,701.4	7,188.5	9,273.8	9,142.0	..	Gross National Product
..	..	..	..	..	..	..	..	..	..	B H	GDP at factor cost
13.6	12.4	14.0	22.8	25.6	29.6	37.3	43.3	58.1	..	..	Agriculture
7,140.5	8,307.9	5,794.6	4,092.0	3,154.3	3,900.8	4,188.3	3,805.9	4,465.0	..	..	Industry
436.3	433.6	427.3	415.4	451.2	448.0	436.9	457.6	462.5	..	..	Manufacturing
1,547.7	1,625.6	1,932.5	2,039.7	2,197.2	1,982.0	2,014.4	1,953.1	1,887.1	..	..	Services, etc.
5,380.3	6,706.3	3,410.0	1,989.0	496.6	1,026.3	1,336.4	1,057.9	2,163.6	..	..	Resource Balance
7,276.8	8,775.7	6,065.0	4,464.3	3,176.4	3,761.8	4,133.1	3,767.3	4,230.8	..	..	Exports of Goods & NFServices
1,896.5	2,069.4	2,655.0	2,475.3	2,679.9	2,735.6	2,796.7	2,709.4	2,067.2	..	..	Imports of Goods & NFServices
3,524.0	3,745.8	4,331.0	4,429.4	5,251.8	4,704.9	4,645.3	4,420.4	3,947.2	..	..	Domestic Absorption
1,693.4	1,904.0	2,388.5	2,484.2	2,867.2	2,309.0	2,349.0	2,147.4	2,059.6	..	..	Private Consumption, etc.
787.1	812.3	864.9	858.3	978.9	1,026.6	1,058.4	1,093.1	1,028.2	..	..	General Gov't Consumption
1,043.6	1,029.5	1,077.6	1,086.9	1,405.7	1,369.2	1,237.9	1,180.0	859.5	..	..	Gross Domestic Investment
962.0	869.4	972.6	1,003.6	1,285.3	1,388.6	1,226.6	1,216.9	860.9	..	..	Fixed Investment
											Memo Items:
3,355.6	5,599.3	6,065.0	4,470.9	2,789.4	3,211.6	3,556.5	3,203.4	2,048.7	..	..	Capacity to Import
-3,921.1	-3,176.4	0.0	6.6	-387.0	-550.2	-576.6	-563.9	-2,182.1	..	..	Terms of Trade Adjustment
4,780.7	6,769.4	7,741.1	6,161.1	4,990.1	5,362.2	5,663.4	5,238.3	4,228.2	..	..	Gross Domestic Income
6,220.2	8,014.2	9,051.2	8,078.6	6,492.7	6,797.6	7,124.7	6,624.6	7,091.8	..	..	Gross National Income
				(Index 1980 = 100)							DOMESTIC PRICES (DEFLATORS)
49.0	68.8	100.0	113.5	115.5	103.8	102.3	102.0	78.0	78.7	..	Overall (GDP)
83.9	92.8	100.0	108.8	115.8	119.1	119.6	121.8	127.2	..	..	Domestic Absorption
73.7	95.2	100.0	105.4	111.1	95.9	93.6	90.8	88.6	..	..	Agriculture
41.8	62.6	100.0	117.3	105.7	94.6	94.9	93.5	57.2	..	..	Industry
64.8	130.7	100.0	99.9	68.2	83.7	68.7	82.4	120.1	..	..	Manufacturing
											MANUFACTURING ACTIVITY
87.3	82.2	100.0	102.7	104.6	105.8	106.2	..	..	..	G J	Employment (1980=100)
92.0	101.7	100.0	102.0	111.3	112.5	115.1	..	..	..	G J	Real Earnings per Empl. (1980=100)
82.4	75.8	100.0	121.5	141.9	131.7	..	..	..	..	G J	Real Output per Empl. (1980=100)
24.7	13.3	21.6	27.6	32.4	37.5	45.9	..	..	..	J	Earnings as % of Value Added
				(Millions of current Kuwaiti Dinars)							MONETARY HOLDINGS
1,950.4	2,262.7	2,857.6	3,866.0	4,182.7	4,367.8	4,475.8	4,435.3	4,546.9	4,762.1	..	Money Supply, Broadly Defined
636.4	669.4	720.8	1,290.2	1,247.6	1,179.6	968.1	943.9	979.3	1,035.7	..	Money as Means of Payment
177.0	215.9	251.3	284.7	342.7	340.6	325.1	327.9	337.1	338.3	..	Currency Ouside Banks
459.4	453.5	469.5	1,005.5	904.9	839.0	643.0	616.0	642.2	697.4	..	Demand Deposits
1,314.0	1,593.3	2,136.8	2,575.8	2,935.1	3,188.2	3,507.7	3,491.4	3,567.6	3,726.4	..	Quasi-Monetary Liabilities
				(Millions of current Kuwaiti Dinars)						C	GOVERNMENT DEFICIT (-) OR SURPLUS
1,134	1,753	4,545	3,025	566	492	780	501	1,690	..		
3,045	3,643	6,916	6,338	4,272	4,145	4,358	3,794	4,744	..	..	Current Revenue
1,045	1,117	1,460	1,696	1,915	2,161	2,042	2,158	2,009	..	..	Current Expenditure
2,000	2,526	5,456	4,642	2,357	1,984	2,316	1,636	2,735	..	..	Current Budget Balance
5	4	7	13	8	10	9	12	7	3	..	Capital Receipts
871	777	918	1,630	1,799	1,502	1,545	1,147	1,052	..	..	Capital Payments

KUWAIT	1967	1968	1969	1970	1971	1972	1973	1974	1975	1976	1977
FOREIGN TRADE (CUSTOMS BASIS)				*(Millions of current US dollars)*							
Value of Exports, fob	1,200.1	1,285.2	1,372.0	1,901.4	2,572.7	3,056.4	3,784.7	10,954.1	9,186.0	9,838.2	9,753.9
Nonfuel Primary Products	20.8	22.2	20.5	22.6	23.2	28.4	45.5	48.1	46.4	71.2	76.4
Fuels	1,141.8	1,216.9	1,285.5	1,787.4	2,433.4	2,859.7	3,495.1	10,361.5	8,407.5	8,831.4	8,618.4
Manufactures	37.4	46.2	66.0	91.5	116.1	168.3	244.1	544.6	732.1	935.6	1,059.1
Value of Imports, cif	593.3	611.3	646.8	625.1	650.5	797.0	1,042.2	1,553.5	2,388.2	3,329.5	4,845.0
Nonfuel Primary Products	115.6	129.8	124.3	135.8	148.8	178.5	229.4	314.1	443.2	539.7	680.4
Fuels	5.0	4.8	6.0	4.5	6.0	7.9	9.7	18.7	14.1	24.9	34.6
Manufactures	472.7	476.8	516.6	484.9	495.7	610.6	803.2	1,220.7	1,930.9	2,764.9	4,130.0
					(Index 1980 = 100)						
Terms of Trade	14.7	14.7	14.2	13.3	16.5	17.1	20.2	65.3	59.4	64.4	65.0
Export Prices, fob	4.4	4.5	4.5	4.5	5.8	6.6	9.4	37.6	37.5	40.8	44.8
Nonfuel Primary Products	29.7	32.2	34.8	37.6	38.5	41.4	55.9	61.8	63.2	62.0	65.0
Fuels	4.3	4.3	4.3	4.3	5.6	6.2	8.9	36.7	35.7	38.4	42.0
Manufactures	26.2	24.6	25.4	29.5	30.3	30.3	30.3	62.3	82.0	94.3	95.9
Import Prices, cif	30.3	30.4	31.6	33.8	35.3	38.5	46.4	57.5	63.2	63.3	68.9
Trade at Constant 1980 Prices				*(Millions of 1980 US dollars)*							
Exports, fob	27,016.4	28,821.9	30,494.5	42,326.9	44,131.4	46,526.0	40,378.5	29,169.5	24,490.7	24,129.8	21,756.6
Imports, cif	1,958.3	2,011.3	2,048.4	1,849.6	1,844.8	2,071.3	2,245.4	2,699.5	3,780.3	5,258.6	7,027.3
BALANCE OF PAYMENTS				*(Millions of current US dollars)*							
Exports of Goods & Services	..	..	..	1,809	2,668	3,153	4,660	12,212	10,289	11,865	12,151
Merchandise, fob	..	..	..	1,693	2,272	2,558	3,826	10,959	8,485	9,621	9,562
Nonfactor Services	..	..	..	26	275	185	275	486	521	612	625
Factor Services	..	..	..	90	121	410	559	767	1,283	1,632	1,965
Imports of Goods & Services	..	..	..	781	841	1,509	2,038	2,514	3,289	4,398	6,341
Merchandise, fob	..	..	..	625	652	797	1,052	1,552	2,400	3,300	4,735
Nonfactor Services	..	..	..	69	73	317	477	621	759	975	1,406
Factor Services	..	..	..	87	116	395	509	341	131	123	199
Long-Term Interest	..	..	..	..	..	..	..	..	..	..	..
Current Transfers, net	..	..	..	-175	-260	-280	-743	-1,235	-276	-315	-370
Workers' Remittances	..	..	..	..	..	..	..	..	0	0	0
Total to be Financed	..	..	..	853	1,567	1,364	1,879	8,463	6,723	7,152	5,440
Official Capital Grants	..	..	..	0	0	0	0	0	-793	-222	-879
Current Account Balance	..	..	..	853	1,567	1,364	1,879	8,463	5,930	6,929	4,561
Long-Term Capital, net	..	..	..	-40	-90	27	24	126	-1,876	-1,136	-1,494
Direct Investment	..	..	..	..	-8	-40	-54	-484	-93	-109	-52
Long-Term Loans	..	..	..	..	..	..	..	..	..	..	..
Disbursements	..	..	..	..	..	..	..	..	..	..	..
Repayments	..	..	..	..	..	..	..	..	..	..	..
Other Long-Term Capital	..	..	..	-40	-82	67	78	610	-990	-804	-562
Other Capital, net	..	..	..	-792	-1,402	-1,316	-1,794	-7,728	-3,740	-5,546	-2,125
Change in Reserves	..	..	..	-21	-75	-75	-109	-861	-315	-247	-943
Memo Item:				*(Kuwaiti Dinars per US dollar)*							
Conversion Factor (Annual Avg)	0.360	0.360	0.360	0.360	0.360	0.330	0.300	0.290	0.290	0.290	0.290
EXTERNAL DEBT, ETC.				*(Millions of US dollars, outstanding at end of year)*							
Public/Publicly Guar. Long-Term	..	..	..	..	..	..	..	..	..	..	..
Official Creditors	..	..	..	..	..	..	..	..	..	..	..
IBRD and IDA	..	..	..	..	..	..	..	..	..	..	..
Private Creditors	..	..	..	..	..	..	..	..	..	..	..
Private Non-guaranteed Long-Term	..	..	..	..	..	..	..	..	..	..	..
Use of Fund Credit	..	..	..	..	..	..	..	..	..	..	..
Short-Term Debt	..	..	..	..	..	..	..	..	..	..	..
Memo Items:				*(Millions of US dollars)*							
Int'l Reserves Excluding Gold	47.5	51.1	95.9	117.1	193.7	269.0	380.8	1,249.2	1,491.5	1,701.8	2,883.1
Gold Holdings (at market price)	137.2	145.4	86.7	92.0	108.2	161.0	319.5	653.5	559.3	751.6	414.2
SOCIAL INDICATORS											
Total Fertility Rate	7.5	7.4	7.3	7.2	7.1	7.0	6.8	6.6	6.4	6.2	5.9
Crude Birth Rate	49.7	48.6	47.6	46.5	45.5	44.4	43.5	42.7	41.8	41.0	40.1
Infant Mortality Rate	41.8	41.2	40.6	40.0	39.8	39.6	39.4	39.2	39.0	39.1	39.1
Life Expectancy at Birth	64.4	65.0	65.5	66.1	66.7	67.3	67.7	68.1	68.6	69.0	69.5
Food Production, p.c. ('79-81 = 100)	..	..	..	..	..	..	..	..	..	..	..
Labor Force, Agriculture (%)	1.7	1.7	1.8	1.8	1.8	1.8	1.8	1.8	1.8	1.8	1.8
Labor Force, Female (%)	6.5	7.1	7.6	8.1	9.0	9.8	10.6	11.4	12.3	12.4	12.5
School Enroll. Ratio, primary	..	..	..	89.0	..	..	..	..	92.0	97.0	100.0
School Enroll. Ratio, secondary	..	..	..	63.0	..	..	..	..	66.0	74.0	74.0

1978	1979	1980	1981	1982	1983	1984	1985	1986	1987 estimate	NOTES	KUWAIT
											FOREIGN TRADE (CUSTOMS BASIS)
					(Millions of current US dollars)						
10,427.5	18,415.9	20,434.6	16,299.9	10,861.3	11,504.8	12,279.0	10,487.0	7,383.0	7,943.0	..	Value of Exports, fob
84.1	134.3	153.6	219.8	196.8	117.9	148.0	121.0	155.1	167.0	..	Nonfuel Primary Products
9,233.6	16,528.7	18,156.5	13,626.7	8,216.2	9,947.9	11,051.0	9,475.0	6,378.9	6,840.1	f	Fuels
1,109.7	1,752.9	2,124.6	2,453.4	2,448.3	1,439.0	1,080.0	891.0	849.0	935.9	..	Manufactures
4,598.0	5,203.8	6,554.2	6,969.1	8,283.4	7,320.2	6,896.0	5,934.0	5,759.0	5,198.0	..	Value of Imports, cif
778.8	947.9	1,129.0	1,162.1	1,403.1	1,238.2	1,292.0	9.2	1,085.1	980.0	..	Nonfuel Primary Products
27.2	36.1	49.9	39.7	51.4	45.7	39.0	30.8	26.9	24.0	..	Fuels
3,792.0	4,219.9	5,375.3	5,767.3	6,828.9	6,036.3	5,565.0	5,894.0	4,647.0	4,194.0	..	Manufactures
					(Index 1980 = 100)						
56.8	69.4	100.0	111.4	105.8	96.3	97.2	92.2	46.5	54.3	..	Terms of Trade
45.0	63.5	100.0	111.0	102.3	91.7	90.4	88.3	48.9	61.9	..	Export Prices, fob
75.0	94.0	100.0	93.6	86.4	88.5	86.8	77.6	79.3	83.2	..	Nonfuel Primary Products
42.3	61.0	100.0	112.5	101.6	92.5	90.2	87.5	44.9	56.9	..	Fuels
91.0	99.2	100.0	105.0	106.0	87.0	94.0	99.0	117.1	152.5	..	Manufactures
79.3	91.5	100.0	99.6	96.7	95.3	93.0	95.8	105.0	114.0	..	Import Prices, cif
											Trade at Constant 1980 Prices
					(Millions of 1980 US dollars)						
23,160.8	29,015.4	20,434.6	14,688.2	10,621.1	12,546.4	13,576.5	11,879.5	15,111.7	12,835.6	..	Exports, fob
5,797.1	5,688.1	6,554.2	6,997.8	8,567.7	7,684.5	7,413.2	6,196.3	5,486.7	4,559.2	..	Imports, cif
											BALANCE OF PAYMENTS
					(Millions of current US dollars)						
13,837	22,872	27,344	25,819	18,450	18,029	18,847	16,792	16,352	15,219	..	Exports of Goods & Services
10,234	18,114	20,633	16,023	10,819	11,473	12,156	10,374	7,213	8,315	..	Merchandise, fob
702	1,183	1,225	1,392	941	868	888	1,137	1,050	1,055	..	Nonfactor Services
2,901	3,575	5,487	8,404	6,690	5,688	5,803	5,281	8,089	5,849	..	Factor Services
6,475	7,552	10,463	10,381	12,056	11,191	11,096	10,405	9,761	9,546	..	Imports of Goods & Services
4,326	4,870	6,756	6,736	7,811	6,889	6,553	5,662	5,303	4,769	..	Merchandise, fob
1,854	2,265	3,067	2,906	3,491	3,620	3,705	4,077	3,858	4,296	..	Nonfactor Services
295	416	640	739	754	683	838	665	601	481	..	Factor Services
..	..	..	..	..	..	..	..	..	..	..	Long-Term Interest
-433	-532	-692	-689	-875	-865	-963	-1,044	-1,081	-1,102	..	Current Transfers, net
0	0	0	0	0	0	0	0	0	0	..	Workers' Remittances
6,930	14,788	16,190	14,750	5,519	5,973	6,789	5,343	5,519	4,572	K	Total to be Financed
-800	-756	-888	-972	-646	-686	-416	-529	-182	-158	K	Official Capital Grants
6,130	14,032	15,302	13,778	4,873	5,287	6,374	4,815	5,337	4,414	..	Current Account Balance
-1,305	-1,625	-925	-890	-889	-1,589	-1,490	-1,240	-2,152	-72	..	Long-Term Capital, net
-95	-188	-407	-151	-108	-240	-95	-70	-247	-93	..	Direct Investment
..	..	..	..	..	..	..	..	..	..	..	Long-Term Loans
..	..	..	..	..	..	..	..	..	..	..	Disbursements
..	..	..	..	..	..	..	..	..	..	..	Repayments
-411	-680	370	233	-136	-662	-980	-642	-1,723	179	..	Other Long-Term Capital
-5,267	-12,042	-13,331	-12,602	-2,006	-4,686	-4,916	-3,029	-3,269	-6,189	..	Other Capital, net
443	-366	-1,045	-286	-1,978	988	32	-545	83	1,847	..	Change in Reserves
											Memo Item:
					(Kuwaiti Dinars per US dollar)						
0.280	0.280	0.270	0.280	0.290	0.290	0.300	0.300	0.290	0.280	..	Conversion Factor (Annual Avg)
											EXTERNAL DEBT, ETC.
				(Millions of US dollars, outstanding at end of year)							
..	..	..	..	..	..	..	..	..	..	..	Public/Publicly Guar. Long-Term
..	..	..	..	..	..	..	..	..	..	..	Official Creditors
..	..	..	..	..	..	..	..	..	..	..	IBRD and IDA
..	..	..	..	..	..	..	..	..	..	..	Private Creditors
..	..	..	..	..	..	..	..	..	..	..	Private Non-guaranteed Long-Term
..	..	..	..	..	..	..	..	..	..	..	Use of Fund Credit
..	..	..	..	..	..	..	..	..	..	..	Short-Term Debt
											Memo Items:
					(Millions of US dollars)						
2,500.4	2,870.1	3,928.5	4,067.5	5,913.2	5,192.1	4,590.2	5,470.7	5,501.1	4,141.6	..	Int'l Reserves Excluding Gold
570.7	1,300.0	1,496.7	1,009.3	1,160.1	968.6	782.8	830.3	992.5	1,229.1	..	Gold Holdings (at market price)
											SOCIAL INDICATORS
5.8	5.7	5.5	5.4	5.2	5.1	5.0	4.9	4.8	4.6	..	Total Fertility Rate
39.1	38.2	37.2	35.5	34.7	34.6	34.4	33.1	31.8	30.5	..	Crude Birth Rate
35.1	31.1	27.6	24.1	22.8	20.9	19.0	18.4	20.2	22.0	..	Infant Mortality Rate
69.9	70.3	70.7	71.2	71.6	71.8	72.1	72.3	72.6	72.1	..	Life Expectancy at Birth
..	..	..	..	..	..	..	..	..	..	..	Food Production, p.c. ('79-81=100)
1.9	1.9	1.9	..	..	..	..	..	..	..	..	Labor Force, Agriculture (%)
12.7	12.8	13.0	13.1	13.2	13.3	13.3	13.4	13.7	13.9	..	Labor Force, Female (%)
98.0	98.0	103.0	101.0	103.0	104.0	103.0	101.0	..	..	..	School Enroll. Ratio, primary
73.0	74.0	78.0	80.0	82.0	82.0	82.0	83.0	..	..	..	School Enroll. Ratio, secondary

LESOTHO	1967	1968	1969	1970	1971	1972	1973	1974	1975	1976	1977
CURRENT GNP PER CAPITA (US $)	80	80	80	100	110	110	150	200	230	260	290
POPULATION (thousands)	1,002	1,022	1,043	1,064	1,089	1,114	1,139	1,166	1,193	1,220	1,252
ORIGIN AND USE OF RESOURCES					*(Millions of current Lesotho Maloti)*						
Gross National Product (GNP)	57.2	58.9	61.6	74.8	76.3	95.8	132.4	158.1	199.7	247.7	308.0
Net Factor Income from Abroad	10.7	11.5	11.9	22.3	25.8	31.5	44.6	60.1	89.1	119.3	139.9
Gross Domestic Product (GDP)	46.5	47.4	49.7	52.5	50.5	64.3	87.8	98.0	110.6	128.4	168.1
Indirect Taxes, net	2.2	2.2	3.7	4.7	4.4	5.2	10.9	14.0	13.2	14.3	23.6
GDP at factor cost	44.3	45.2	46.0	47.8	46.1	59.1	76.9	84.0	97.4	114.1	144.5
Agriculture	15.3	16.0	15.6	16.7	12.0	21.0	32.7	31.7	31.7	31.0	49.6
Industry	4.4	4.0	4.7	4.4	4.9	6.0	8.3	11.3	12.9	21.7	22.4
Manufacturing	1.5	1.7	1.8	2.1	2.4	2.8	3.9	4.8	5.6	6.2	7.2
Services, etc.	24.6	25.2	25.7	26.7	29.2	32.1	35.9	41.0	52.8	61.4	72.5
Resource Balance	-18.9	-18.9	-17.9	-22.9	-29.5	-41.9	-56.6	-82.0	-98.6	-153.4	-167.3
Exports of Goods & NFServices	5.1	5.3	6.4	5.7	6.8	10.1	12.8	14.1	14.6	21.3	18.0
Imports of Goods & NFServices	24.0	24.2	24.3	28.6	36.3	52.0	69.4	96.1	113.2	174.7	185.3
Domestic Absorption	65.4	66.3	67.6	75.4	80.0	106.2	144.4	180.0	209.2	281.8	335.4
Private Consumption, etc.	53.7	54.7	56.0	63.2	63.7	85.6	116.0	146.9	167.2	210.5	264.2
General Gov't Consumption	6.8	6.2	6.1	6.2	8.3	10.8	12.1	15.8	21.3	24.9	29.2
Gross Domestic Investment	4.9	5.4	5.5	6.0	8.0	9.8	16.3	17.3	20.7	46.4	42.0
Fixed Investment	4.4	4.9	4.9	5.1	7.1	8.6	10.7	14.5	20.7	45.5	40.3
Memo Items:											
Gross Domestic Saving	-14.0	-13.5	-12.4	-16.9	-21.5	-32.1	-40.3	-64.7	-77.9	-107.0	-125.3
Gross National Saving	-3.3	-2.0	-0.5	26.8	26.5	27.8	50.8	52.5	12.2	13.5	15.9
					(Millions of 1980 Lesotho Maloti)						
Gross National Product	182.1	196.4	197.9	200.5	220.1	222.1	282.1	324.9	348.1	393.6	456.6
GDP at factor cost	122.0	134.8	134.2	132.6	140.7	139.7	171.3	188.5	165.0	184.2	218.9
Agriculture	..	..	..	76.9	79.2	63.1	73.7	90.3	68.6	64.9	81.5
Industry	..	..	..	7.9	5.5	11.4	17.0	20.5	19.2	26.6	42.6
Manufacturing	..	..	..	2.8	1.7	5.5	8.3	8.5	7.9	9.2	8.7
Services, etc.	..	..	..	47.8	56.0	65.2	80.6	77.7	77.2	92.7	94.8
Resource Balance	-56.6	-56.4	-52.0	-61.1	-75.7	-103.3	-122.2	-145.3	-217.3	-260.7	-271.2
Exports of Goods & NFServices	17.3	17.3	19.1	20.2	24.4	28.7	29.7	29.0	21.9	29.7	21.2
Imports of Goods & NFServices	73.9	73.7	71.1	81.3	100.1	132.0	151.9	174.3	239.2	290.4	292.4
Domestic Absorption	190.2	205.1	204.4	201.9	223.7	251.0	308.9	352.5	396.5	459.7	513.6
Private Consumption, etc.	165.0	172.6	174.3	172.3	185.6	205.1	250.0	291.4	315.9	358.5	380.1
General Gov't Consumption	16.0	14.4	14.0	13.7	18.0	22.6	24.8	28.6	37.5	37.0	38.6
Gross Domestic Investment	16.0	15.3	15.3	15.9	20.1	23.3	34.1	32.5	43.1	64.2	94.9
Fixed Investment	..	..	..	..	..	..	..	..	..	..	..
Memo Items:											
Capacity to Import	15.7	16.1	18.7	16.2	18.8	25.6	28.0	25.6	30.9	35.4	28.4
Terms of Trade Adjustment	-1.6	-1.2	-0.4	-4.0	-5.7	-3.1	-1.7	-3.4	9.0	5.7	7.2
Gross Domestic Income	118.7	134.0	133.9	136.8	142.4	144.6	185.0	203.8	188.2	204.7	249.6
Gross National Income	174.8	191.1	193.1	196.5	214.5	219.0	280.4	321.5	357.1	399.3	463.8
DOMESTIC PRICES (DEFLATORS)					*(Index 1980 = 100)*						
Overall (GDP)	37.1	34.1	35.8	37.3	34.1	43.5	47.0	47.3	61.7	64.5	69.3
Domestic Absorption	34.4	32.3	33.1	37.3	35.8	42.3	46.7	51.1	52.8	61.3	65.3
Agriculture	..	..	..	21.7	15.2	33.3	44.4	35.1	46.2	47.8	60.9
Industry	..	..	..	55.7	89.1	52.6	48.8	55.1	67.2	81.6	52.6
Manufacturing	..	..	..	75.0	141.2	50.9	47.0	56.5	70.9	67.4	82.8
MANUFACTURING ACTIVITY											
Employment (1980=100)	..	..	..	..	..	64.1	96.7	96.8	96.6	106.4	97.0
Real Earnings per Empl. (1980=100)	..	..	..	..	..	..	65.8	67.9	73.1	67.6	72.4
Real Output per Empl. (1980=100)	..	..	..	..	..	35.5	54.6	54.8	60.1	59.2	60.2
Earnings as % of Value Added	..	..	..	..	..	..	48.2	49.7	47.2	48.4	48.4
MONETARY HOLDINGS					*(Millions of current Lesotho Maloti)*						
Money Supply, Broadly Defined	..	..	..	..	..	..	..	..	..	..	..
Money as Means of Payment	..	..	..	..	..	..	..	..	..	..	..
Currency Ouside Banks	..	..	..	..	..	..	..	..	..	..	..
Demand Deposits	..	..	..	..	..	..	..	..	..	..	..
Quasi-Monetary Liabilities	..	..	..	..	..	..	..	..	..	..	..
GOVERNMENT DEFICIT (-) OR SURPLUS					*(Millions of current Lesotho Maloti)*		5	9	-9	-13	-3
Current Revenue	..	..	..	..	..	..	24	32	32	40	70
Current Expenditure	..	..	..	..	..	..	16	17	27	32	43
Current Budget Balance	..	..	..	..	..	..	8	15	6	8	27
Capital Receipts	..	..	..	..	..	..	0	0	0	0	0
Capital Payments	..	..	..	..	..	..	4	6	14	21	30

1978	1979	1980	1981	1982	1983	1984	1985	1986	1987 estimate	NOTES	LESOTHO
330	360	420	480	540	510	490	410	360	370	..	**CURRENT GNP PER CAPITA (US $)**
1,286	1,320	1,355	1,391	1,428	1,466	1,505	1,545	1,586	1,629	..	**POPULATION (thousands)**
				(Millions of current Lesotho Maloti)							**ORIGIN AND USE OF RESOURCES**
385.1	422.5	502.3	583.2	742.1	818.2	957.9	1,084.8	1,230.6	1,300.5	C	Gross National Product (GNP)
153.3	178.2	205.0	254.8	372.9	423.1	487.2	514.3	583.4	551.5	..	Net Factor Income from Abroad
231.8	244.3	297.3	328.4	369.2	395.1	470.7	570.5	647.2	749.0	C f	Gross Domestic Product (GDP)
46.3	28.8	32.9	41.8	69.5	86.7	95.5	113.8	130.5	191.0	..	Indirect Taxes, net
185.5	215.5	264.4	286.6	299.7	308.4	375.2	456.7	516.7	558.0	C f	GDP at factor cost
55.2	65.7	70.0	74.1	63.9	75.3	98.0	111.0	108.6	117.0	..	Agriculture
43.5	54.0	64.8	64.3	82.7	63.9	83.3	102.0	123.6	159.0	..	Industry
10.1	11.9	13.9	16.8	22.2	29.1	40.5	45.7	59.2	84.0	..	Manufacturing
86.8	95.8	129.6	148.2	153.1	169.2	193.9	243.7	284.5	282.0	..	Services, etc.
-169.9	-246.4	-297.6	-370.2	-470.2	-531.4	-632.5	-688.7	-774.5	-738.0	..	Resource Balance
36.4	50.0	61.9	63.4	57.5	53.2	61.8	75.4	93.5	78.0	..	Exports of Goods & NFServices
206.3	296.4	359.5	433.6	527.7	584.6	694.3	764.1	868.0	816.0	..	Imports of Goods & NFServices
401.7	490.7	594.9	698.6	839.4	926.5	1,103.2	1,259.2	1,421.5	1,487.0	..	Domestic Absorption
300.7	353.7	395.3	473.7	599.9	694.7	817.6	907.9	1,010.7	1,180.0	..	Private Consumption, etc.
40.9	52.4	98.9	108.6	99.0	111.2	124.9	168.3	195.7	118.0	..	General Gov't Consumption
60.1	84.6	100.7	116.3	140.5	120.6	160.7	183.0	215.1	189.0	..	Gross Domestic Investment
51.0	73.7	93.5	107.4	143.0	100.2	150.6	185.3	213.1	189.0	..	Fixed Investment
											Memo Items:
-109.8	-161.8	-196.9	-253.9	-329.7	-410.8	-471.8	-505.7	-559.2	-549.0	..	Gross Domestic Saving
44.9	17.9	9.7	2.7	46.0	15.2	20.2	12.6	28.9	157.2	..	Gross National Saving
				(Millions of 1980 Lesotho Maloti)							
496.1	503.0	502.3	518.4	577.3	564.1	588.6	570.8	575.3	622.3	C	Gross National Product
252.4	262.0	264.4	257.9	259.2	247.9	270.8	278.2	297.7	300.5	C f	GDP at factor cost
90.7	93.9	70.0	68.3	59.7	66.5	71.6	74.4	65.1	68.7	..	Agriculture
49.6	58.8	64.8	54.8	68.2	43.6	50.0	54.1	66.1	64.6	..	Industry
8.9	12.1	13.9	15.0	17.3	19.8	24.5	24.2	27.7	32.3	..	Manufacturing
112.1	109.3	129.6	134.8	131.3	137.8	149.2	149.7	166.5	167.2	..	Services, etc.
-261.7	-277.3	-297.6	-342.0	-357.6	-370.7	..	..	..	..	..	Resource Balance
44.9	62.3	61.9	53.1	46.0	38.6	..	..	..	..	..	Exports of Goods & NFServices
306.6	339.6	359.5	395.1	403.6	409.3	..	..	..	..	..	Imports of Goods & NFServices
548.5	572.4	594.9	635.1	653.5	653.7	..	..	..	..	..	Domestic Absorption
414.2	418.8	395.3	448.7	489.3	493.4	..	..	..	..	..	Private Consumption, etc.
50.3	53.7	98.9	86.0	63.0	71.8	..	..	..	..	..	General Gov't Consumption
84.0	99.9	100.7	100.4	101.2	88.5	..	..	..	..	..	Gross Domestic Investment
..	..	..	..	..	..	..	..	..	..	..	Fixed Investment
											Memo Items:
54.1	57.3	61.9	57.8	44.0	37.3	..	..	..	..	..	Capacity to Import
9.2	-5.0	0.0	4.7	-2.0	-1.4	..	..	..	..	..	Terms of Trade Adjustment
296.0	290.1	297.3	297.8	293.9	281.7	..	..	..	..	..	Gross Domestic Income
505.3	498.0	502.3	523.1	575.3	562.8	..	..	..	..	..	Gross National Income
				(Index 1980 = 100)							**DOMESTIC PRICES (DEFLATORS)**
80.8	82.8	100.0	112.0	124.8	139.6	153.7	183.3	196.5	225.1	..	Overall (GDP)
73.2	85.7	100.0	110.0	128.4	141.7	..	..	..	..	..	Domestic Absorption
60.9	70.0	100.0	108.5	107.0	113.2	136.9	149.2	166.8	170.3	..	Agriculture
87.7	91.8	100.0	117.3	121.3	146.6	166.6	188.5	187.0	246.1	..	Industry
113.5	98.3	100.0	112.0	128.3	147.0	165.3	188.8	213.7	260.1	..	Manufacturing
											MANUFACTURING ACTIVITY
98.0	99.0	100.0	101.0	102.1	103.1	104.1	105.2	..	..	..	Employment (1980=100)
94.1	94.5	100.0	105.0	128.0	112.0	142.6	161.2	..	..	..	Real Earnings per Empl. (1980=100)
61.8	83.1	100.0	107.7	125.5	109.6	136.7	150.9	..	..	..	Real Output per Empl. (1980=100)
48.4	48.4	48.4	48.4	48.4	48.4	48.4	48.4	..	..	..	Earnings as % of Value Added
				(Millions of current Lesotho Maloti)							**MONETARY HOLDINGS**
..	..	116.6	144.2	185.4	217.0	250.3	310.7	352.1	387.1	..	Money Supply, Broadly Defined
..	..	48.5	58.6	77.2	84.7	103.3	133.1	154.6	156.8	..	Money as Means of Payment
..	..	6.9	10.6	17.8	23.1	23.7	25.2	32.6	33.2	..	Currency Ouside Banks
..	..	41.5	48.0	59.4	61.5	79.6	107.9	122.1	123.6	..	Demand Deposits
..	..	68.2	85.6	108.2	132.4	147.0	177.6	197.5	230.3	..	Quasi-Monetary Liabilities
				(Millions of current Lesotho Maloti)							**GOVERNMENT DEFICIT (-) OR SURPLUS**
..	..	..	..	..	-20	-13	-28	..	..	C	
..	..	..	..	..	174	239	242	..	..	..	Current Revenue
..	..	..	..	..	142	168	212	..	..	..	Current Expenditure
..	..	..	..	..	32	72	31	..	..	..	Current Budget Balance
..	..	..	..	..	1	0	0	..	..	..	Capital Receipts
..	..	..	..	..	53	85	59	..	..	..	Capital Payments

LESOTHO	1967	1968	1969	1970	1971	1972	1973	1974	1975	1976	1977
FOREIGN TRADE (CUSTOMS BASIS)					*(Millions of current US dollars)*						
Value of Exports, fob	6.0	6.0	7.0	6.0	4.0	8.0	13.0	14.0	13.0	17.0	14.0
Nonfuel Primary Products	5.5	5.8	6.2	5.6	3.6	6.7	10.6	11.8	10.0	13.5	11.2
Fuels	0.0	0.0	0.0	0.0	0.0	0.0	0.0	0.0	0.0	0.0	0.0
Manufactures	0.5	0.2	0.8	0.4	0.4	1.3	2.4	2.3	3.0	3.5	2.9
Value of Imports, cif	33.0	34.0	34.0	32.0	39.0	56.0	87.0	120.0	160.0	207.0	229.0
Nonfuel Primary Products	18.4	8.5	10.2	10.7	10.8	15.5	24.1	33.2	44.3	57.3	63.4
Fuels	0.9	1.6	2.0	2.1	2.4	3.5	5.4	7.4	9.9	12.7	14.1
Manufactures	13.7	23.9	21.8	19.2	25.8	37.1	57.6	79.4	105.9	137.0	151.5
Terms of Trade	112.3	131.3	149.1	137.4	117.1	121.1	139.0	107.7	100.0	100.4	95.0
Export Prices, fob	30.8	31.1	35.2	32.4	31.4	37.0	54.9	62.4	61.0	64.4	66.1
Nonfuel Primary Products	30.9	31.1	35.6	32.1	31.0	36.7	56.3	61.6	59.8	63.8	64.5
Fuels	4.3	4.3	4.3	4.3	5.6	6.2	8.9	36.7	35.7	38.4	42.0
Manufactures	30.1	31.7	32.4	37.5	36.5	38.5	49.5	67.2	66.1	66.7	71.7
Import Prices, cif	27.5	23.7	23.6	23.6	26.8	30.5	39.5	57.9	61.0	64.1	69.6
Trade at Constant 1980 Prices					*(Millions of 1980 US dollars)*						
Exports, fob	19.5	19.3	19.9	18.5	12.7	21.6	23.7	22.4	21.0	26.0	21.0
Imports, cif	120.2	143.4	144.1	135.7	145.3	183.5	220.3	207.1	262.5	323.1	329.0
BALANCE OF PAYMENTS					*(Millions of current US dollars)*						
Exports of Goods & Services	..	..	..	11	16	18	19	26	153	173	200
Merchandise, fob	..	..	..	6	4	8	12	14	14	18	15
Nonfactor Services	..	..	..	2	5	5	6	6	12	12	14
Factor Services	..	..	..	3	7	5	1	6	127	143	171
Imports of Goods & Services	..	..	..	42	52	68	102	143	175	214	241
Merchandise, fob	..	..	..	31	37	53	83	113	151	189	211
Nonfactor Services	..	..	..	10	14	14	17	27	19	21	25
Factor Services	..	..	..	1	1	1	2	3	4	4	5
Long-Term Interest	..	..	..	0	0	0	0	0	0	0	0
Current Transfers, net	..	..	..	30	31	37	67	84	1	1	1
Workers' Remittances	..	..	..	29	30	36	66	85	0	0	0
Total to be Financed	..	..	..	-1	-5	-13	-16	-33	-21	-40	-39
Official Capital Grants	..	..	..	19	21	20	30	32	20	6	30
Current Account Balance	..	..	..	18	16	7	14	-1	-1	-34	-9
Long-Term Capital, net	..	..	..	19	21	20	30	34	24	8	40
Direct Investment	..	..	..	..	..	..	..	..	..	..	..
Long-Term Loans	..	..	..	0	0	0	0	2	4	2	8
Disbursements	..	..	..	0	1	1	..	2	5	3	8
Repayments	..	..	..	0	0	0	0	0	0	0	0
Other Long-Term Capital	..	..	..	0	0	0	0	0	0	0	2
Other Capital, net	..	..	..	-38	-37	-27	-44	-33	-22	26	-31
Change in Reserves	..	..	..	1	0	0	0	0	-1	0	0
Memo Item:					*(Lesotho Maloti per US dollar)*						
Conversion Factor (Annual Avg)	0.710	0.710	0.710	0.710	0.720	0.770	0.690	0.680	0.740	0.870	0.870
EXTERNAL DEBT, ETC.					*(Millions of US dollars, outstanding at end of year)*						
Public/Publicly Guar. Long-Term	..	..	..	8	9	10	8	10	14	16	24
Official Creditors	..	..	..	8	8	9	7	9	13	15	24
IBRD and IDA	..	..	..	4	4	5	5	6	8	10	15
Private Creditors	..	..	..	1	1	1	1	1	1	1	1
Private Non-guaranteed Long-Term	..	..	..	..	..	..	..	..	..	..	..
Use of Fund Credit	..	..	..	0	0	0	0	0	0	0	0
Short-Term Debt	..	..	..	..	..	..	..	..	..	..	..
Memo Items:					*(Thousands of US dollars)*						
Int'l Reserves Excluding Gold	..	..	..	..	..	..	..	..	..	..	..
Gold Holdings (at market price)	..	..	..	..	..	..	..	..	..	..	..
SOCIAL INDICATORS											
Total Fertility Rate	5.8	5.8	5.8	5.8	5.8	5.8	5.8	5.8	5.8	5.8	5.8
Crude Birth Rate	42.4	42.4	42.4	42.4	42.4	42.4	42.4	42.4	42.4	42.4	42.4
Infant Mortality Rate	140.0	138.0	136.0	134.0	132.0	130.0	128.6	127.2	125.8	124.4	123.0
Life Expectancy at Birth	48.6	48.7	48.9	49.0	49.1	49.2	49.6	50.0	50.4	50.8	51.2
Food Production, p.c. ('79-81=100)	122.0	114.9	119.1	113.9	125.1	100.3	106.7	139.5	103.3	97.9	120.5
Labor Force, Agriculture (%)	90.9	90.6	90.2	89.9	89.5	89.1	88.8	88.4	88.0	87.6	87.3
Labor Force, Female (%)	47.6	47.7	47.8	47.9	47.7	47.5	47.3	47.1	46.9	46.7	46.4
School Enroll. Ratio, primary	..	..	..	87.0	..	..	..	..	105.0	105.0	105.0
School Enroll. Ratio, secondary	..	..	..	7.0	..	..	..	..	13.0	14.0	14.0

1978	1979	1980	1981	1982	1983	1984	1985	1986	1987 estimate	NOTES	LESOTHO
											(Millions of current US dollars) **FOREIGN TRADE (CUSTOMS BASIS)**
32.0	45.0	58.0	51.0	35.0	35.0	28.0	20.5	24.3	27.8	f	Value of Exports, fob
24.2	34.5	43.8	38.7	25.9	26.4	20.8	15.4	15.5	17.0	..	Nonfuel Primary Products
0.0	0.0	0.0	0.0	0.0	0.0	0.0	0.0	0.0	..	..	Fuels
7.9	10.5	14.2	12.3	9.2	8.6	7.2	5.1	8.9	10.9	..	Manufactures
273.0	373.0	490.0	564.0	514.0	561.0	504.0	376.5	425.0	483.4	f	Value of Imports, cif
75.6	111.9	152.9	177.3	133.7	228.1	212.0	145.6	4.2	4.6	..	Nonfuel Primary Products
16.8	23.0	30.2	34.7	31.7	34.5	25.0	22.1	0.2	0.3	..	Fuels
180.6	238.2	307.0	352.0	348.7	298.4	267.1	208.8	420.5	478.5	..	Manufactures
											(Index 1980 = 100)
99.8	102.8	100.0	97.0	96.1	96.1	90.4	100.6	91.3	87.6	..	Terms of Trade
76.7	93.3	100.0	96.2	90.9	90.3	86.5	88.1	103.0	108.4	..	Export Prices, fob
76.9	94.5	100.0	95.3	89.5	90.2	85.6	87.8	102.9	105.6	..	Nonfuel Primary Products
42.3	61.0	100.0	112.5	101.6	92.5	90.2	87.5	44.9	56.9	..	Fuels
76.1	90.0	100.0	99.1	94.8	90.7	89.3	89.0	103.2	113.0	..	Manufactures
76.9	90.8	100.0	99.2	94.6	93.9	95.7	87.5	112.8	123.7	..	Import Prices, cif
											(Millions of 1980 US dollars) Trade at Constant 1980 Prices
42.0	48.0	58.0	53.0	38.5	38.8	32.4	23.3	23.6	25.7	..	Exports, fob
355.0	410.8	490.0	568.8	543.6	597.3	526.8	430.1	376.8	390.8	..	Imports, cif
											(Millions of current US dollars) **BALANCE OF PAYMENTS**
233	285	360	381	425	454	396	283	313	335	..	Exports of Goods & Services
33	46	60	51	37	31	28	22	25	30	..	Merchandise, fob
17	19	27	28	26	28	24	18	20	8	..	Nonfactor Services
184	220	273	303	361	395	343	242	267	297	..	Factor Services
279	365	479	513	509	545	489	358	397	427	..	Imports of Goods & Services
243	324	426	454	447	483	433	314	340	372	..	Merchandise, fob
30	35	45	50	47	50	45	35	46	29	..	Nonfactor Services
6	6	7	9	15	12	11	9	10	26	..	Factor Services
0	1	2	2	5	6	4	4	4	5	..	Long-Term Interest
2	2	2	2	3	3	3	2	2	76	..	Current Transfers, net
0	0	0	0	0	0	0	0	0	..	..	Workers' Remittances
-44	-79	-117	-129	-81	-89	-90	-74	-82	-16	K	Total to be Financed
52	79	106	81	43	78	91	74	65	4	K	Official Capital Grants
9	0	-11	-49	-38	-10	2	0	-17	-12	..	Current Account Balance
63	103	141	116	92	112	91	103	81	37	..	Long-Term Capital, net
..	..	4	5	3	5	2	5	2	2	..	Direct Investment
8	19	12	18	43	20	9	25	12	31	..	Long-Term Loans
9	21	15	21	48	34	26	39	22	41	..	Disbursements
1	2	3	3	5	15	17	14	10	9	..	Repayments
2	5	20	13	4	9	-12	0	2	0	..	Other Long-Term Capital
-72	-103	-86	-65	-45	-75	-83	-98	-53	-25	..	Other Capital, net
0	0	-44	-2	-9	-26	-9	-6	-10	..	..	Change in Reserves
											(Lesotho Maloti per US dollar) Memo Item:
0.870	0.840	0.780	0.870	1.080	1.110	1.440	2.190	2.270	2.040	..	Conversion Factor (Annual Avg)
											(Millions of US dollars, outstanding at end of year) **EXTERNAL DEBT, ETC.**
33	52	63	77	118	131	131	164	183	237	..	Public/Publicly Guar. Long-Term
32	44	51	62	100	106	117	156	175	220	..	Official Creditors
18	20	24	31	40	48	53	62	69	82	..	IBRD and IDA
1	8	11	16	18	25	14	8	8	17	..	Private Creditors
..	..	..	..	..	..	..	..	..	..	..	Private Non-guaranteed Long-Term
0	0	0	0	0	0	0	0	0	..	..	Use of Fund Credit
3	..	8	6	3	4	4	4	4	4	..	Short-Term Debt
											(Thousands of US dollars) Memo Items:
..	..	50,269	43,396	47,535	66,677	48,576	43,521	60,260	67,531	..	Int'l Reserves Excluding Gold
..	..	..	..	..	..	..	..	..	..	..	Gold Holdings (at market price)
											SOCIAL INDICATORS
5.8	5.8	5.8	5.8	5.8	5.8	5.8	5.8	5.8	5.8	..	Total Fertility Rate
42.0	41.6	41.3	40.9	40.5	40.6	40.7	40.8	40.9	41.0	..	Crude Birth Rate
120.6	118.2	115.8	113.4	111.0	105.5	100.0	94.4	88.9	83.4	..	Infant Mortality Rate
51.5	51.9	52.2	52.6	53.0	53.5	54.0	54.5	55.0	55.6	..	Life Expectancy at Birth
125.3	106.5	97.9	95.7	81.4	80.6	80.1	87.9	79.3	81.1	..	Food Production, p.c. ('79-81 = 100)
86.9	86.6	86.2	..	..	..	..	..	..	..	..	Labor Force, Agriculture (%)
46.2	46.0	45.8	45.5	45.3	45.1	44.8	44.6	44.4	44.1	..	Labor Force, Female (%)
103.0	104.0	102.0	..	110.0	111.0	112.0	115.0	..	..	..	School Enroll. Ratio, primary
14.0	17.0	17.0	20.0	19.0	21.0	22.0	..	..	..	..	School Enroll. Ratio, secondary

LIBERIA	1967	1968	1969	1970	1971	1972	1973	1974	1975	1976	1977
CURRENT GNP PER CAPITA (US $)	270	280	290	310	320	320	320	370	410	470	500
POPULATION (thousands)	1,256	1,292	1,330	1,369	1,410	1,452	1,497	1,542	1,590	1,639	1,690

ORIGIN AND USE OF RESOURCES *(Millions of current Liberian Dollars)*

	1967	1968	1969	1970	1971	1972	1973	1974	1975	1976	1977
Gross National Product (GNP)	325	346	381	402	424	462	488	614	721	757	866
Net Factor Income from Abroad	-5	-5	-6	-6	-6	-5	-5	-4	-5	-5	-7
Gross Domestic Product (GDP)	330	351	387	408	430	467	493	618	726	762	873
Indirect Taxes, net	24	27	27	30	30	35	39	48	50	63	85
GDP at factor cost	306	324	360	378	400	431	453	570	676	699	788
Agriculture	74	75	89	91	98	97	136	184	180	208	251
Industry	126	131	145	157	162	185	174	222	305	276	257
Manufacturing	9	9	13	15	17	18	23	35	36	45	50
Services, etc.	106	118	126	130	140	149	144	164	191	214	280
Resource Balance	57	98	130	81	78	91	89	75	32	21	-63
Exports of Goods & NFServices	206	233	274	240	252	275	330	407	404	467	459
Imports of Goods & NFServices	148	136	144	159	174	184	241	332	372	446	522
Domestic Absorption	272	254	257	327	352	376	404	543	694	741	936
Private Consumption, etc.	155	155	148	192	213	222	276	356	374	412	530
General Gov't Consumption	41	43	45	45	53	55	56	65	73	89	120
Gross Domestic Investment	76	56	63	91	86	99	72	122	247	240	286
Fixed Investment	..	..	..	..	..	..	..	..	..	..	..
Memo Items:											
Gross Domestic Saving	133	153	194	172	164	189	161	197	279	261	223
Gross National Saving	128	148	188	152	145	172	139	173	252	234	191

(Millions of 1980 Liberian Dollars)

	1967	1968	1969	1970	1971	1972	1973	1974	1975	1976	1977
Gross National Product	748	789	840	899	946	988	963	1,009	986	1,045	1,070
GDP at factor cost	696	733	787	842	884	913	887	932	914	961	971
Agriculture	185	197	213	235	245	251	276	292	297	324	332
Industry	229	226	256	276	285	299	300	321	281	284	264
Manufacturing	30	31	39	47	51	54	62	77	68	83	86
Services, etc.	282	310	318	332	353	362	311	319	336	354	375
Resource Balance	-41	18	28	136	80	131	141	117	-68	-133	-191
Exports of Goods & NFServices	512	542	596	676	652	711	746	721	548	592	528
Imports of Goods & NFServices	553	524	568	540	572	581	605	603	616	726	720
Domestic Absorption	804	787	830	778	881	868	832	899	1,061	1,185	1,270
Private Consumption, etc.	493	527	544	409	519	379	561	580	585	695	780
General Gov't Consumption	122	123	129	150	158	159	133	144	146	155	171
Gross Domestic Investment	189	137	157	218	204	330	137	175	330	335	320
Fixed Investment	..	..	..	..	..	..	..	..	..	..	..
Memo Items:											
Capacity to Import	767	901	1,083	815	827	866	829	740	668	760	633
Terms of Trade Adjustment	255	360	488	140	175	155	82	19	120	167	105
Gross Domestic Income	1,017	1,164	1,345	1,053	1,135	1,154	1,055	1,036	1,114	1,219	1,184
Gross National Income	1,003	1,148	1,327	1,038	1,121	1,143	1,045	1,028	1,106	1,212	1,175

DOMESTIC PRICES (DEFLATORS) *(Index 1980 = 100)*

	1967	1968	1969	1970	1971	1972	1973	1974	1975	1976	1977
Overall (GDP)	43.2	43.6	45.1	44.7	44.8	46.7	50.6	60.8	73.1	72.4	80.9
Domestic Absorption	33.9	32.2	30.9	42.1	40.0	43.3	48.5	60.4	65.4	62.5	73.7
Agriculture	39.7	38.1	41.5	38.9	39.8	38.6	49.1	63.0	60.6	64.4	75.7
Industry	55.2	58.0	56.6	57.0	56.9	61.7	57.8	69.3	108.7	97.4	97.4
Manufacturing	30.1	29.6	33.0	32.7	33.8	33.4	36.5	44.9	53.1	54.5	58.1

MANUFACTURING ACTIVITY

	1967	1968	1969	1970	1971	1972	1973	1974	1975	1976	1977
Employment (1980=100)	..	..	..	..	..	..	..	..	..	..	..
Real Earnings per Empl. (1980=100)	..	..	..	..	..	..	..	..	..	..	..
Real Output per Empl. (1980=100)	..	..	..	..	..	..	..	..	..	..	..
Earnings as % of Value Added	..	..	..	..	..	..	..	..	..	..	..

MONETARY HOLDINGS *(Millions of current Liberian Dollars)*

	1967	1968	1969	1970	1971	1972	1973	1974	1975	1976	1977
Money Supply, Broadly Defined	..	..	..	..	..	..	..	75.0	73.9	108.9	122.3
Money as Means of Payment	..	..	..	..	..	..	..			..	..
Currency Ouside Banks	..	..	..	..	..	..	..	8.5	8.0	8.4	9.2
Demand Deposits	..	..	..	..	..	..	..	..	..	..	..
Quasi-Monetary Liabilities	..	..	..	..	..	..	..	..	..	..	..

(Millions of current Liberian Dollars)

	1967	1968	1969	1970	1971	1972	1973	1974	1975	1976	1977
GOVERNMENT DEFICIT (-) OR SURPLUS	..	..	..	..	..	..	..	6.8	3.4	-16.0	-24.2
Current Revenue	..	..	..	..	..	..	..	116.7	129.2	163.7	181.1
Current Expenditure	..	..	..	..	..	..	..	84.6	88.9	111.6	119.3
Current Budget Balance	..	..	..	..	..	..	..	32.1	40.3	52.1	61.8
Capital Receipts	..	..	..	..	..	..	..	0.2	3.1	0.2	0.3
Capital Payments	..	..	..	..	..	..	..	25.5	40.0	68.3	86.3

1978	1979	1980	1981	1982	1983	1984	1985	1986	1987 estimate	NOTES	LIBERIA
540	590	590	600	550	500	470	480	460	450	..	CURRENT GNP PER CAPITA (US $)
1,744	1,799	1,856	1,915	1,977	2,042	2,108	2,178	2,253	2,327	..	POPULATION (thousands)

(Millions of current Liberian Dollars)

1978	1979	1980	1981	1982	1983	1984	1985	1986	1987	NOTES	
											ORIGIN AND USE OF RESOURCES
933	1,054	1,093	1,074	1,050	970	977	1,031	1,012	1,063	..	Gross National Product (GNP)
-11	-14	-24	-21	-72	-98	-117	-64	-73	-77	..	Net Factor Income from Abroad
944	1,068	1,117	1,095	1,122	1,067	1,094	1,095	1,085	1,139	..	Gross Domestic Product (GDP)
104	114	116	104	110	103	100	98	100	..	..	Indirect Taxes, net
840	953	1,001	991	1,012	964	994	997	985	..	..	GDP at factor cost
288	327	359	313	326	326	352	364	368	..	f	Agriculture
247	286	282	327	320	293	283	276	274	..	f	Industry
54	82	77	57	54	50	51	49	47	..	f	Manufacturing
305	341	360	351	366	345	359	356	343	..	f	Services, etc.
-49	-34	-1	27	-20	-19	45	62	95	..	..	Resource Balance
500	554	614	588	539	487	471	467	464	..	..	Exports of Goods & NFServices
549	587	614	561	559	506	426	405	369	..	..	Imports of Goods & NFServices
992	1,101	1,117	1,068	1,142	1,086	1,049	1,033	990	..	..	Domestic Absorption
573	623	630	678	739	756	750	708	700	..	..	Private Consumption, etc.
139	157	182	211	243	204	187	230	185	..	..	General Gov't Consumption
281	322	305	179	160	126	112	95	105	..	..	Gross Domestic Investment
..	..	..	169	130	126	117	97	95	..	..	Fixed Investment
											Memo Items:
232	288	305	206	140	107	157	157	200	..	..	Gross Domestic Saving
191	242	252	155	23	-30	-3	65	94	..	..	Gross National Saving

(Millions of 1980 Liberian Dollars)

1978	1979	1980	1981	1982	1983	1984	1985	1986	1987	NOTES	
1,121	1,155	1,093	1,082	1,017	969	935	979	954	944	..	Gross National Product
1,004	1,042	1,001	987	963	951	941	938	919	..	..	GDP at factor cost
344	354	359	345	343	354	364	373	374	..	f	Agriculture
270	282	282	252	235	222	205	198	195	..	f	Industry
91	98	77	73	66	63	62	59	56	..	f	Manufacturing
389	406	360	389	386	375	372	367	350	..	f	Services, etc.
-115	-69	-1	70	68	13	38	57	94	..	..	Resource Balance
619	611	614	612	580	507	512	505	502	..	..	Exports of Goods & NFServices
734	680	614	543	511	495	473	447	408	..	..	Imports of Goods & NFServices
1,249	1,239	1,117	1,033	1,013	1,052	1,009	983	930	..	..	Domestic Absorption
772	719	630	571	552	661	671	612	606	..	..	Private Consumption, etc.
179	191	182	277	303	263	230	280	223	..	..	General Gov't Consumption
298	329	305	185	158	128	109	91	100	..	..	Gross Domestic Investment
..	..	..	..	..	..	..	..	..	..	..	Fixed Investment
											Memo Items:
669	641	614	569	493	476	523	516	513	..	..	Capacity to Import
50	30	0	-43	-87	-31	12	11	12	..	..	Terms of Trade Adjustment
1,184	1,199	1,117	1,060	995	1,033	1,059	1,052	1,035	..	..	Gross Domestic Income
1,171	1,184	1,093	1,039	930	938	947	991	965	..	..	Gross National Income

(Index 1980 = 100)

1978	1979	1980	1981	1982	1983	1984	1985	1986	1987	NOTES	
											DOMESTIC PRICES (DEFLATORS)
83.2	91.3	100.0	99.3	103.8	100.3	104.5	105.2	106.1	112.5	..	Overall (GDP)
79.4	88.9	100.0	103.3	112.7	103.3	104.0	105.1	106.5	..	..	Domestic Absorption
83.7	92.4	100.0	90.7	95.2	92.3	96.7	97.6	98.3	..	..	Agriculture
91.4	101.4	100.0	129.6	136.1	132.0	138.2	139.5	140.6	..	..	Industry
59.1	83.9	100.0	76.9	80.8	78.3	81.9	82.7	83.3	..	..	Manufacturing
											MANUFACTURING ACTIVITY
..	..	100.0	100.0	102.6	97.1	140.4	149.6	..	..	G	Employment (1980=100)
..	..	100.0	84.7	107.7	102.1	110.5	106.6	..	..	G	Real Earnings per Empl. (1980=100)
..	..	..	..	..	..	..	..	..	..	G	Real Output per Empl. (1980=100)
..	..	..	..	..	..	..	..	..	..	..	Earnings as % of Value Added

(Millions of current Liberian Dollars)

1978	1979	1980	1981	1982	1983	1984	1985	1986	1987	NOTES	
											MONETARY HOLDINGS
151.0	154.7	114.5	100.9	126.7	142.1	144.4	170.8	202.2	233.5	..	Money Supply, Broadly Defined
..	..	..	..	..	..	..	..	..	..	..	Money as Means of Payment
10.2	11.0	11.4	11.6	15.7	19.8	28.6	46.2	66.1	..	..	Currency Ouside Banks
..	..	..	..	..	..	..	..	..	..	..	Demand Deposits
..	..	..	..	..	..	..	..	..	..	..	Quasi-Monetary Liabilities

(Millions of current Liberian Dollars)

1978	1979	1980	1981	1982	1983	1984	1985	1986	1987	NOTES	
-56.2	-141.2	-88.3	-110.3	-116.6	-102.9	-61.0	-87.1	-90.9	-83.9	C	GOVERNMENT DEFICIT (-) OR SURPLUS
197.5	224.1	225.1	242.4	278.2	257.2	259.5	228.8	205.4	198.4	..	Current Revenue
126.5	152.9	179.7	211.8	295.4	268.7	244.7	244.1	205.3	226.1	..	Current Expenditure
71.0	71.2	45.4	30.6	-17.2	-11.5	14.8	-15.3	0.1	-27.7	..	Current Budget Balance
0.2	0.2	0.2	0.1	0.1	5.5	0.2	0.5	0.3	0.2	..	Capital Receipts
127.4	212.6	133.9	141.0	99.5	96.9	76.0	72.3	91.3	56.4	..	Capital Payments

LIBERIA	1967	1968	1969	1970	1971	1972	1973	1974	1975	1976	1977
FOREIGN TRADE (CUSTOMS BASIS)				*(Millions of current US dollars)*							
Value of Exports, fob	158.7	167.5	195.3	217.2	222.4	244.0	323.8	399.8	393.9	463.0	454.4
Nonfuel Primary Products	153.0	160.7	190.1	207.0	216.0	236.9	317.3	393.4	386.3	450.7	439.7
Fuels	0.0	0.1	0.0	0.0	0.0	0.1	0.0	0.2	0.2	0.5	0.0
Manufactures	5.7	6.7	5.2	10.2	6.4	7.1	6.5	6.2	7.5	11.8	14.7
Value of Imports, cif	124.9	106.9	114.1	149.7	162.2	178.6	193.5	288.4	331.2	399.2	463.5
Nonfuel Primary Products	21.9	23.6	18.9	27.4	31.6	32.6	37.5	48.5	50.5	57.3	77.2
Fuels	8.2	8.4	4.7	9.5	11.8	12.0	14.7	56.4	48.3	59.5	68.9
Manufactures	94.8	74.9	90.6	112.8	118.8	134.0	141.2	183.6	232.4	282.5	317.5
Terms of Trade	177.6	186.0	150.5	180.8	153.7	136.3	154.7	115.5	116.4	121.8	117.5
				(Index 1980 = 100)							
Export Prices, fob	40.3	39.8	39.5	43.4	39.9	39.9	56.7	62.6	67.5	70.7	74.6
Nonfuel Primary Products	40.9	40.4	39.8	43.7	40.0	40.1	56.9	62.6	67.5	70.9	74.7
Fuels	4.3	4.3	4.3	4.3	5.6	6.2	8.9	36.7	35.7	38.4	42.0
Manufactures	30.1	31.7	32.4	37.5	36.5	38.5	49.5	67.2	66.1	66.7	71.7
Import Prices, cif	22.7	21.4	26.2	24.0	26.0	29.3	36.6	54.2	58.0	58.1	63.5
Trade at Constant 1980 Prices				*(Millions of 1980 US dollars)*							
Exports, fob	393.9	421.2	494.5	500.6	557.2	610.8	571.4	638.2	583.7	654.6	608.9
Imports, cif	550.6	499.9	434.9	623.8	624.9	609.3	528.0	532.0	571.1	687.3	729.9
BALANCE OF PAYMENTS				*(Millions of current US dollars)*							
Exports of Goods & Services	..	..	..	..	..	..	..	..	403.7	467.0	459.0
Merchandise, fob	..	..	..	213.7	223.9	243.6	324.0	400.3	394.4	457.0	447.4
Nonfactor Services	..	..	..	3.2	4.1	3.7	5.9	6.9	9.3	10.0	11.6
Factor Services	..	..	..	..	..	..	..	..	0.0	0.0	0.0
Imports of Goods & Services	..	..	..	230.7	235.2	252.4	323.4	440.4	336.1	412.1	480.9
Merchandise, fob	..	..	..	149.7	162.4	174.7	193.5	289.4	290.4	358.6	417.3
Nonfactor Services	..	..	..	74.9	67.3	73.1	125.2	146.7	40.8	48.9	56.9
Factor Services	..	..	..	6.1	5.5	4.6	4.7	4.3	4.9	4.6	6.7
Long-Term Interest	..	..	..	6.0	5.9	4.9	5.3	5.0	5.1	5.9	7.2
Current Transfers, net	..	..	..	-13.3	-13.1	-12.8	-17.3	-19.4	-21.7	-22.4	-25.5
Workers' Remittances	..	..	..	..	..	..	..	..	0.0	0.0	0.0
Total to be Financed	..	..	..	-27.1	-20.3	-17.9	-10.8	-52.6	45.9	32.5	-47.4
Official Capital Grants	..	..	..	10.8	13.3	9.3	19.6	22.2	27.8	32.3	28.8
Current Account Balance	..	..	..	-16.3	-7.0	-8.6	8.8	-30.4	73.7	64.8	-18.6
Long-Term Capital, net	..	..	..	46.3	43.3	38.5	76.8	72.3	116.2	101.9	116.2
Direct Investment	..	..	..	28.1	20.7	18.7	49.0	45.0	80.8	39.1	44.7
Long-Term Loans	..	..	..	-4.0	1.5	-1.4	-3.8	0.4	8.1	27.8	54.3
Disbursements	..	..	..	7.4	12.3	10.5	8.1	14.9	35.4	43.3	73.3
Repayments	..	..	..	11.4	10.8	11.9	11.9	14.5	27.3	15.5	19.0
Other Long-Term Capital	..	..	..	11.4	7.8	11.9	12.0	4.7	-0.5	2.7	-11.6
Other Capital, net	..	..	..	-29.1	-37.5	-29.6	-84.1	-40.3	-189.3	-163.4	-91.8
Change in Reserves	..	..	..	-0.9	1.2	-0.3	-1.6	-1.6	-0.6	-3.3	-5.8
Memo Item:				*(Liberian Dollars per US dollar)*							
Conversion Factor (Annual Avg)	1.000	1.000	1.000	1.000	1.000	1.000	1.000	1.000	1.000	1.000	1.000
EXTERNAL DEBT, ETC.				*(Millions of US dollars, outstanding at end of year)*							
Public/Publicly Guar. Long-Term	..	..	..	157.9	160.4	159.3	159.0	161.8	178.6	208.5	267.0
Official Creditors	..	..	..	124.0	129.6	133.6	137.7	140.6	160.3	183.2	206.8
IBRD and IDA	..	..	..	7.4	9.2	12.5	18.6	23.2	26.9	35.2	44.9
Private Creditors	..	..	..	33.9	30.8	25.7	21.3	21.2	18.3	25.3	60.1
Private Non-guaranteed Long-Term	..	..	..	..	..	..	..	..	..	..	..
Use of Fund Credit	..	..	..	4.3	1.6	0.0	0.1	3.2	0.0	5.2	5.5
Short-Term Debt	..	..	..	..	..	..	..	..	..	..	25.1
Memo Items:				*(Thousands of US dollars)*							
Int'l Reserves Excluding Gold	..	..	..	..	..	..	..	18,719	19,189	17,170	27,345
Gold Holdings (at market price)	..	..	..	..	..	..	..	..	..	..	..
SOCIAL INDICATORS											
Total Fertility Rate	6.4	6.4	6.4	6.5	6.5	6.5	6.5	6.5	6.5	6.5	6.5
Crude Birth Rate	45.8	45.8	45.8	45.7	45.7	45.7	45.6	45.5	45.4	45.3	45.2
Infant Mortality Rate	132.0	129.4	126.8	124.2	121.6	119.0	116.6	114.2	111.8	109.4	107.0
Life Expectancy at Birth	45.0	45.5	46.0	46.5	47.0	47.5	48.0	48.5	49.0	49.5	50.0
Food Production, p.c. ('79-81 = 100)	94.3	96.4	96.6	100.8	102.7	102.3	101.6	106.8	102.9	104.3	105.4
Labor Force, Agriculture (%)	78.2	77.9	77.7	77.5	77.2	76.9	76.5	76.2	75.9	75.6	75.2
Labor Force, Female (%)	31.9	31.9	32.0	32.0	32.0	31.9	31.9	31.9	31.9	31.9	31.9
School Enroll. Ratio, primary	..	..	..	56.0	..	..	..	..	62.0	..	..
School Enroll. Ratio, secondary	..	..	..	10.0	..	..	..	..	17.0	..	..

1978	1979	1980	1981	1982	1983	1984	1985	1986	1987 estimate	NOTES	LIBERIA
											FOREIGN TRADE (CUSTOMS BASIS)
				(Millions of current US dollars)							
496.3	547.2	597.0	529.2	477.4	438.0	452.1	435.6	404.4	385.3	..	Value of Exports, fob
476.9	522.3	569.8	510.3	459.5	416.3	433.0	416.0	398.5	380.0	..	Nonfuel Primary Products
0.1	0.0	7.0	0.4	1.3	1.4	0.0	0.0	0.0	0.0	..	Fuels
19.4	24.9	20.3	18.5	16.6	20.3	19.1	19.7	5.9	5.3	..	Manufactures
481.0	506.5	534.1	477.4	422.0	412.0	368.7	284.4	234.8	208.0	..	Value of Imports, cif
87.5	100.5	109.3	114.2	96.3	94.0	95.1	62.7	55.2	44.5	..	Nonfuel Primary Products
84.5	103.2	152.1	129.3	112.6	111.0	73.8	74.6	40.1	44.2	..	Fuels
309.0	302.8	272.7	233.9	213.1	207.0	199.9	147.1	139.5	119.3	..	Manufactures
				(Index 1980 = 100)							
102.1	107.7	100.0	85.0	93.5	93.1	93.0	90.9	95.6	93.2	..	Terms of Trade
70.9	88.5	100.0	88.5	89.3	85.4	82.9	78.4	77.5	82.8	..	Export Prices, fob
70.7	88.5	100.0	88.1	89.1	85.1	82.7	77.9	77.2	82.5	..	Nonfuel Primary Products
42.3	61.0	100.0	112.5	101.6	92.5	90.2	87.5	44.9	56.9	..	Fuels
76.1	90.0	100.0	99.1	94.8	90.7	89.3	89.0	103.2	113.0	..	Manufactures
69.4	82.2	100.0	104.1	95.5	91.8	89.2	86.2	81.0	88.8	..	Import Prices, cif
											Trade at Constant 1980 Prices
				(Millions of 1980 US dollars)							
700.4	617.9	597.0	598.2	534.6	512.8	545.2	555.9	522.1	465.2	..	Exports, fob
692.7	616.0	534.1	458.8	441.8	449.0	413.5	330.0	289.9	234.2	..	Imports, cif
				(Millions of current US dollars)							**BALANCE OF PAYMENTS**
500.0	553.6	613.5	540.7	511.9	461.1	486.2	468.8	433.2	..	..	Exports of Goods & Services
486.4	536.6	600.4	529.2	477.4	420.8	446.7	430.5	406.8	..	..	Merchandise, fob
13.6	17.0	13.1	11.5	32.7	38.5	36.9	34.6	25.5	..	..	Nonfactor Services
0.0	0.0	0.0	0.0	1.8	1.8	2.6	3.7	0.9	..	..	Factor Services
506.9	540.1	574.7	503.7	556.3	638.9	547.3	475.7	411.9	..	..	Imports of Goods & Services
430.6	457.5	478.0	423.9	390.2	374.8	325.4	263.8	258.2	..	..	Merchandise, fob
65.5	68.9	72.7	58.7	92.0	108.7	92.5	80.2	72.2	..	..	Nonfactor Services
10.8	13.7	24.0	21.1	74.1	155.4	129.4	131.6	81.5	..	..	Factor Services
13.0	22.9	23.0	17.6	14.6	19.7	10.7	10.1	14.9	6.1	..	Long-Term Interest
-29.9	-32.0	-29.0	-30.0	-45.0	-39.0	-43.6	-28.0	-32.6	..	..	Current Transfers, net
0.0	0.0	0.0	0.0	0.0	0.0	0.0	0.0	0.0	..	..	Workers' Remittances
-36.8	-18.5	9.8	7.0	-89.4	-216.8	-104.7	-34.9	-11.3	..	K	Total to be Financed
29.6	34.7	36.2	68.5	92.9	114.2	103.9	91.7	62.2	..	K	Official Capital Grants
-7.2	16.2	46.0	75.4	3.5	-102.6	-0.9	56.8	51.0	..	..	Current Account Balance
94.2	149.0	116.3	112.5	270.6	186.4	173.2	-23.7	-67.7	..	..	Long-Term Capital, net
..	..	..	..	34.8	49.1	36.2	-16.2	-5.7	..	..	Direct Investment
74.9	122.4	69.0	70.6	46.0	67.2	88.9	55.2	28.5	26.7	..	Long-Term Loans
88.2	174.3	84.6	79.8	65.3	77.4	100.1	63.4	43.6	31.6	..	Disbursements
13.3	51.9	15.6	9.2	19.3	10.2	11.2	8.2	15.1	4.9	..	Repayments
-10.3	-8.1	11.1	-26.6	96.9	-44.1	-55.8	-154.3	-152.7	..	..	Other Long-Term Capital
-98.9	-204.0	-203.0	-242.2	-342.4	-120.6	-206.4	-27.8	14.1	..	..	Other Capital, net
11.9	38.7	40.7	54.3	68.4	36.8	34.1	-5.3	2.6	2.2	..	Change in Reserves
											Memo Item:
				(Liberian Dollars per US dollar)							
1.000	1.000	1.000	1.000	1.000	1.000	1.000	1.000	1.000	1.000	..	Conversion Factor (Annual Avg)
				(Millions of US dollars, outstanding at end of year)							**EXTERNAL DEBT, ETC.**
349.4	470.1	573.5	647.9	658.1	737.0	789.9	908.0	1,019.7	1,151.9	..	Public/Publicly Guar. Long-Term
246.5	315.0	412.0	493.9	512.1	565.1	623.6	732.7	837.0	955.0	..	Official Creditors
56.9	71.7	92.2	113.3	126.9	141.6	143.7	176.4	208.7	251.8	..	IBRD and IDA
103.0	155.1	161.5	154.0	146.0	172.0	166.3	175.3	182.7	196.9	..	Private Creditors
..	..	..	..	..	..	..	..	..	..	..	Private Non-guaranteed Long-Term
5.9	45.2	53.0	101.2	164.0	205.6	208.1	225.7	251.3	291.4	..	Use of Fund Credit
89.3	75.3	81.1	80.7	81.1	54.9	65.9	95.6	134.5	174.8	..	Short-Term Debt
											Memo Items:
				(Thousands of US dollars)							Int'l Reserves Excluding Gold
18,021	54,979	5,450	8,339	6,471	20,400	3,479	1,522	2,665	513	..	Int'l Reserves Excluding Gold
..	..	..	..	..	..	..	..	..	..	..	Gold Holdings (at market price)
											SOCIAL INDICATORS
6.5	6.5	6.6	6.6	6.6	6.6	6.6	6.6	6.6	6.6	..	Total Fertility Rate
45.1	45.1	45.1	45.0	45.0	45.1	45.2	45.4	45.5	45.7	..	Crude Birth Rate
104.8	102.6	100.4	98.2	96.0	94.4	92.7	91.1	89.4	87.8	..	Infant Mortality Rate
50.5	51.0	51.5	52.0	52.5	52.9	53.3	53.7	54.1	54.6	..	Life Expectancy at Birth
103.1	101.0	98.1	100.9	97.4	101.6	101.2	97.8	96.7	92.8	..	Food Production, p.c. ('79-81 = 100)
74.9	74.5	74.2	..	..	..	..	..	..	..	..	Labor Force, Agriculture (%)
31.9	31.9	31.9	31.8	31.6	31.5	31.3	31.1	31.0	30.8	..	Labor Force, Female (%)
60.0	71.0	76.0	..	..	..	..	..	..	..	..	School Enroll. Ratio, primary
19.0	23.0	23.0	..	..	..	..	..	..	..	..	School Enroll. Ratio, secondary

LIBYA	1967	1968	1969	1970	1971	1972	1973	1974	1975	1976	1977
CURRENT GNP PER CAPITA (US $)	1,110	1,510	1,840	1,890	1,860	1,990	2,440	3,170	4,560	6,770	6,630
POPULATION (thousands)	1,733	1,799	1,873	1,956	2,048	2,151	2,261	2,376	2,492	2,610	2,731
ORIGIN AND USE OF RESOURCES					*(Millions of current Libyan Dinars)*						
Gross National Product (GNP)	746.7	1,009.0	1,195.2	1,259.6	1,469.2	1,762.4	2,152.5	3,910.3	3,781.5	4,899.8	5,403.1
Net Factor Income from Abroad	-98.6	-184.6	-162.8	-166.8	-157.6	-36.1	-93.7	27.1	1.5	-7.2	-359.9
Gross Domestic Product (GDP)	845.3	1,193.6	1,358.0	1,426.4	1,626.8	1,798.5	2,246.2	3,883.2	3,780.0	4,907.0	5,763.0
Indirect Taxes, net	30.8	39.0	51.4	42.0	40.3	45.5	63.9	91.2	105.7	138.9	150.0
GDP at factor cost	814.5	1,154.6	1,306.6	1,384.4	1,586.5	1,753.0	2,182.3	3,792.0	3,674.3	4,768.1	5,613.0
Agriculture	30.9	33.4	37.4	33.1	33.0	43.6	60.0	64.7	82.9	99.7	90.0
Industry	502.7	781.8	886.4	950.3	1,078.9	1,154.2	1,459.6	2,850.8	2,499.6	3,402.3	4,058.0
Manufacturing	19.7	24.0	26.7	29.1	32.7	46.0	62.8	71.5	86.2	114.8	125.0
Services, etc.	280.9	339.4	382.8	401.0	474.6	555.2	662.7	876.5	1,091.8	1,266.1	1,465.0
Resource Balance	178.0	347.0	369.0	467.0	539.1	445.4	413.8	1,062.0	387.5	1,210.0	1,482.0
Exports of Goods & NFServices	431.0	680.0	788.0	870.0	975.1	997.8	1,240.3	2,489.9	2,053.2	2,881.4	3,431.0
Imports of Goods & NFServices	253.0	333.0	419.0	403.0	436.0	552.4	826.5	1,427.9	1,665.7	1,671.4	1,949.0
Domestic Absorption	667.3	846.6	989.0	959.4	1,087.7	1,353.1	1,832.4	2,821.2	3,392.5	3,697.0	4,281.0
Private Consumption, etc.	300.5	335.1	382.6	463.9	468.6	543.4	702.8	927.0	1,193.5	1,336.6	1,483.0
General Gov't Consumption	131.8	193.1	256.8	259.9	318.4	359.1	465.4	864.8	1,044.3	1,184.6	1,400.0
Gross Domestic Investment	235.0	318.4	349.6	235.6	300.7	450.6	664.2	1,029.4	1,154.7	1,175.8	1,398.0
Fixed Investment	227.0	312.0	342.0	232.0	288.0	437.0	636.0	971.0	1,055.0	1,226.0	1,368.0
Memo Items:											
Gross Domestic Saving	413.0	665.4	718.6	702.6	839.8	896.0	1,078.0	2,091.4	1,542.2	2,385.8	2,880.0
Gross National Saving	276.2	433.3	502.9	485.8	620.4	759.2	902.4	2,014.9	1,466.8	2,302.4	2,266.0
					(Millions of 1980 Libyan Dinars)						
Gross National Product	4,802.8	6,144.6	7,227.2	7,598.3	7,310.8	7,051.8	6,972.1	6,270.7	6,694.8	8,222.0	8,369.5
GDP at factor cost	5,557.1	7,438.7	8,382.8	8,787.6	8,281.7	7,351.1	7,411.7	6,282.3	6,744.1	8,270.7	9,014.1
Agriculture	42.9	64.5	66.3	49.9	58.7	78.4	90.2	92.6	110.5	126.2	106.6
Industry	6,813.9	9,312.2	10,604.3	11,137.0	10,124.1	8,534.7	8,307.7	6,350.0	6,463.3	8,198.3	8,835.4
Manufacturing	42.3	50.5	55.2	58.2	64.1	70.4	93.3	85.4	110.2	141.4	154.7
Services, etc.	334.2	373.8	386.3	411.1	512.7	598.9	702.7	832.6	1,013.0	1,151.2	1,311.9
Resource Balance	5,568.6	8,592.2	10,331.3	11,061.3	8,761.8	6,955.4	6,696.6	2,770.8	2,918.1	4,612.9	4,821.9
Exports of Goods & NFServices	6,167.3	9,338.1	11,217.4	11,873.9	9,602.9	7,946.1	8,041.2	5,108.4	5,293.8	6,909.9	7,392.0
Imports of Goods & NFServices	598.7	745.9	886.1	812.5	841.0	990.7	1,344.6	2,337.6	2,375.7	2,296.9	2,570.1
Domestic Absorption	1,523.8	1,841.2	1,935.4	1,647.5	2,007.1	2,312.2	2,621.7	4,246.3	4,504.6	4,747.5	5,300.8
Private Consumption, etc.	307.6	313.1	297.0	303.2	555.0	613.2	775.3	1,164.1	1,330.1	1,456.8	1,529.5
General Gov't Consumption	276.5	395.6	497.9	508.3	604.7	642.7	804.6	1,620.1	1,554.3	1,741.1	2,031.3
Gross Domestic Investment	939.7	1,132.5	1,140.6	836.0	847.4	1,056.4	1,041.8	1,461.5	1,620.2	1,549.6	1,740.0
Fixed Investment	..	..	..	..	..	..	..	..	..	..	..
Memo Items:											
Capacity to Import	1,020.0	1,523.2	1,666.5	1,754.1	1,881.0	1,789.4	2,017.9	4,076.2	2,928.3	3,959.8	4,524.3
Terms of Trade Adjustment	-5,147.3	-7,814.9	-9,550.9	-10,120.0	-7,721.9	-6,156.6	-6,023.3	-1,032.2	-2,365.4	-2,950.1	-2,867.7
Gross Domestic Income	289.5	-545.9	-1,339.0	-1,515.5	373.0	1,039.4	1,252.4	5,195.1	4,326.8	5,284.0	6,059.3
Gross National Income	-344.5	-1,670.3	-2,323.7	-2,521.5	-411.1	895.1	948.7	5,238.4	4,329.4	5,272.0	5,501.8
DOMESTIC PRICES (DEFLATORS)					*(Index 1980 = 100)*						
Overall (GDP)	15.5	16.4	16.5	16.6	20.1	25.0	30.9	62.4	56.5	59.6	64.6
Domestic Absorption	43.8	46.0	51.1	58.2	54.2	58.5	69.9	66.4	75.3	77.9	80.8
Agriculture	72.0	51.8	56.5	66.4	56.3	55.6	66.5	69.8	75.0	79.0	84.4
Industry	7.4	8.4	8.4	8.5	10.7	13.5	17.6	44.9	38.7	41.5	45.9
Manufacturing	46.6	47.5	48.3	50.0	51.1	65.3	67.3	83.7	78.2	81.2	80.8
MANUFACTURING ACTIVITY											
Employment (1980=100)	39.8	39.8	41.3	43.7	48.4	52.7	53.7	64.1	74.9	77.9	48.3
Real Earnings per Empl. (1980=100)	..	..	..	..	..	..	..	..	..	..	..
Real Output per Empl. (1980=100)	42.2	51.6	48.4	45.4	46.0	39.5	47.8	46.4	54.8	59.1	77.0
Earnings as % of Value Added	35.6	29.8	35.5	37.1	33.9	37.3	36.9	31.1	32.6	30.6	35.3
MONETARY HOLDINGS					*(Millions of current Libyan Dinars)*						
Money Supply, Broadly Defined	153.8	194.6	255.5	320.9	463.0	588.6	810.9	1,334.3	1,360.5	1,694.9	2,128.1
Money as Means of Payment	116.8	150.2	201.8	241.1	364.5	413.0	514.0	753.8	867.6	1,139.4	1,443.8
Currency Ouside Banks	61.0	70.4	102.4	112.3	120.7	147.4	202.6	262.2	346.0	436.0	585.0
Demand Deposits	55.8	79.8	99.4	128.8	243.8	265.6	311.4	491.6	521.6	703.4	858.8
Quasi-Monetary Liabilities	37.0	44.4	53.7	79.8	98.5	175.7	296.8	580.5	492.9	555.5	684.3
GOVERNMENT DEFICIT (-) OR SURPLUS					*(Millions of current Libyan Dinars)*						
Current Revenue	..	..	..	..	..	..	..	..	..	..	..
Current Expenditure	..	..	..	..	..	..	..	..	..	..	..
Current Budget Balance	..	..	..	..	..	..	..	..	..	..	..
Capital Receipts	..	..	..	..	..	..	..	..	..	..	..
Capital Payments	..	..	..	..	..	..	..	..	..	..	..

1978	1979	1980	1981	1982	1983	1984	1985	1986	1987 estimate	NOTES	LIBYA
6,840	8,300	9,550	9,150	9,170	8,550	7,240	6,610	5,550	5,500	..	**CURRENT GNP PER CAPITA (US $)**
2,853	2,977	3,103	3,230	3,360	3,491	3,626	3,764	3,908	4,057	..	**POPULATION (thousands)**
				(Millions of current Libyan Dinars)							**ORIGIN AND USE OF RESOURCES**
5,375.9	7,594.7	10,503.9	9,379.0	8,524.1	8,307.2	7,356.3	7,016.5	6,286.4	6,899.4	..	Gross National Product (GNP)
-312.1	-251.7	-19.3	32.5	-150.7	-223.8	-159.3	..	..	..	..	Net Factor Income from Abroad
5,688.0	7,846.4	10,523.2	9,346.5	8,674.8	8,531.0	7,920.7	7,203.0	6,473.0	7,086.0	..	Gross Domestic Product (GDP)
192.0	243.4	297.9	385.9	315.6	390.5	395.0	..	..	..	..	Indirect Taxes, net
5,496.0	7,603.0	10,225.3	8,960.6	8,359.2	8,140.5	7,525.7	..	..	..	..	GDP at factor cost
122.0	140.4	164.9	202.3	220.3	273.8	270.4	..	..	..	..	Agriculture
3,705.0	5,539.3	7,808.1	6,156.8	5,490.0	5,092.1	4,310.5	..	..	..	..	Industry
149.0	185.8	202.0	229.1	273.5	294.2	359.9	..	..	..	..	Manufacturing
1,669.0	1,923.3	2,252.3	2,601.5	2,648.9	2,774.6	2,944.8	..	..	..	..	Services, etc.
779.0	1,979.0	3,658.0	1,468.0	1,596.0	1,092.0	..	..	..	..	..	Resource Balance
2,978.0	4,801.0	6,964.0	4,995.0	4,573.0	3,979.0	..	..	..	..	..	Exports of Goods & NFServices
2,199.0	2,822.0	3,306.0	3,527.0	2,977.0	2,887.0	..	..	..	..	..	Imports of Goods & NFServices
4,909.0	5,867.4	6,865.2	7,878.5	7,078.8	7,439.0	..	..	..	..	..	Domestic Absorption
1,665.0	1,973.4	2,242.2	2,478.5	1,655.8	2,172.0	..	..	..	..	..	Private Consumption, etc.
1,692.0	1,929.0	2,298.0	2,539.0	3,005.0	3,113.0	..	..	..	..	..	General Gov't Consumption
1,552.0	1,965.0	2,325.0	2,861.0	2,418.0	2,154.0	2,163.0	..	..	..	..	Gross Domestic Investment
1,532.0	1,855.0	2,230.0	2,811.0	2,362.0	2,093.0	..	..	..	..	..	Fixed Investment
											Memo Items:
2,331.0	3,944.0	5,983.0	4,329.0	4,014.0	3,246.0	..	..	..	..	..	Gross Domestic Saving
1,731.2	3,438.0	5,641.3	3,896.5	3,390.5	2,417.0	..	..	..	..	..	Gross National Saving
				(Millions of 1980 Libyan Dinars)							
8,726.1	10,124.8	10,503.9	8,574.5	8,119.1	7,957.5	7,235.4	6,986.6	6,359.7	6,172.2	..	Gross National Product
9,249.0	10,194.3	10,225.3	8,307.2	8,036.7	7,944.3	7,651.0	..	..	..	H	GDP at factor cost
136.0	150.6	164.9	186.0	190.1	222.4	214.1	..	..	..	..	Agriculture
8,765.6	9,453.4	7,808.1	6,051.6	5,764.1	5,496.5	5,160.6	..	..	..	..	Industry
182.8	226.3	202.0	216.1	248.5	262.7	315.7	..	..	..	..	Manufacturing
1,435.3	1,656.2	2,252.3	1,898.3	1,861.9	1,888.8	1,866.8	..	..	..	..	Services, etc.
4,264.4	4,621.8	3,658.0	1,099.8	1,183.7	1,567.4	..	..	..	..	..	Resource Balance
7,028.5	7,624.0	6,964.0	4,491.4	4,426.9	4,426.9	..	..	..	..	..	Exports of Goods & NFServices
2,764.0	3,002.2	3,306.0	3,391.6	3,243.2	2,859.6	..	..	..	..	..	Imports of Goods & NFServices
5,848.9	6,433.1	6,865.2	6,619.4	6,242.6	5,681.8	..	..	..	..	..	Domestic Absorption
1,872.3	2,076.1	2,242.2	2,182.9	2,308.8	2,201.1	..	..	..	..	..	Private Consumption, etc.
2,135.8	2,211.7	2,298.0	2,128.3	2,269.7	2,100.1	..	..	..	..	..	General Gov't Consumption
1,840.9	2,145.4	2,325.0	2,308.2	1,664.1	1,380.7	1,382.1	..	..	..	..	Gross Domestic Investment
..	..	..	..	..	..	..	..	..	..	..	Fixed Investment
											Memo Items:
3,743.2	5,107.6	6,964.0	4,803.2	4,982.0	3,941.2	..	..	..	..	..	Capacity to Import
-3,285.3	-2,516.5	0.0	311.8	555.0	-485.7	..	..	..	..	..	Terms of Trade Adjustment
5,947.5	7,943.9	10,523.2	8,856.6	8,817.8	7,686.2	..	..	..	..	..	Gross Domestic Income
5,440.8	7,608.4	10,503.9	8,886.3	8,674.2	7,471.8	..	..	..	..	..	Gross National Income
				(Index 1980 = 100)							**DOMESTIC PRICES (DEFLATORS)**
61.6	75.0	100.0	109.4	105.0	104.4	100.7	100.4	98.8	111.8	..	Overall (GDP)
83.9	91.2	100.0	119.0	113.4	130.9	..	..	..	..	..	Domestic Absorption
89.7	93.2	100.0	108.8	115.9	123.1	126.3	..	..	..	..	Agriculture
42.3	58.6	100.0	101.7	95.2	92.6	83.5	..	..	..	..	Industry
81.5	82.1	100.0	106.0	110.0	112.0	114.0	..	..	..	..	Manufacturing
											MANUFACTURING ACTIVITY
43.4	76.8	100.0	..	..	..	..	..	..	..	J	Employment (1980 = 100)
..	..	..	..	..	..	..	..	..	..	J	Real Earnings per Empl. (1980 = 100)
91.1	77.9	100.0	..	..	..	..	..	..	..	J	Real Output per Empl. (1980 = 100)
35.7	47.4	40.9	..	..	..	..	..	..	..	J	Earnings as % of Value Added
				(Millions of current Libyan Dinars)							**MONETARY HOLDINGS**
2,373.7	3,199.8	4,104.6	4,646.8	4,305.4	4,126.8	4,175.3	5,053.6	4,811.4	..	..	Money Supply, Broadly Defined
1,687.8	2,247.3	2,898.9	3,512.2	3,232.3	2,884.5	2,711.3	3,492.2	3,041.4	..	..	Money as Means of Payment
868.5	1,053.7	685.7	791.1	889.9	838.2	767.5	985.0	1,023.7	..	..	Currency Ouside Banks
819.3	1,193.6	2,213.2	2,721.1	2,342.4	2,046.3	1,943.8	2,507.2	2,017.7	..	..	Demand Deposits
685.9	952.5	1,205.7	1,134.6	1,073.1	1,242.3	1,464.0	1,561.4	1,770.0	..	..	Quasi-Monetary Liabilities
				(Millions of current Libyan Dinars)							**GOVERNMENT DEFICIT (-) OR SURPLUS**
..	..	..	..	..	..	..	..	..	..	..	Current Revenue
..	..	..	..	..	..	..	..	..	..	..	Current Expenditure
..	..	..	..	..	..	..	..	..	..	..	Current Budget Balance
..	..	..	..	..	..	..	..	..	..	..	Capital Receipts
..	..	..	..	..	..	..	..	..	..	..	Capital Payments

LIBYA	1967	1968	1969	1970	1971	1972	1973	1974	1975	1976	1977
FOREIGN TRADE (CUSTOMS BASIS)				*(Millions of current US dollars)*							
Value of Exports, fob	1,178.1	1,875.6	2,167.0	2,365.6	2,695.0	2,943.0	3,995.3	8,268.1	6,839.6	9,561.5	11,423.1
Nonfuel Primary Products	2.6	2.4	2.6	2.0	1.6	6.8	9.3	2.7	0.6	0.2	0.5
Fuels	1,166.0	1,868.4	2,161.2	2,355.2	2,686.3	2,930.8	3,983.6	8,262.2	6,833.5	9,554.0	11,410.6
Manufactures	9.4	4.9	3.2	8.4	7.1	5.4	2.4	3.3	5.5	7.3	12.0
Value of Imports, cif	476.4	644.6	675.6	674.0	930.0	1,291.0	2,011.0	3,746.0	4,424.0	4,277.0	5,458.0
Nonfuel Primary Products	101.3	107.3	111.6	139.4	177.8	209.3	413.4	609.9	722.7	555.9	824.0
Fuels	16.0	18.9	21.4	17.6	23.3	22.5	35.5	44.8	68.9	86.3	21.0
Manufactures	359.1	518.4	542.6	517.0	728.9	1,059.3	1,562.1	3,091.3	3,632.4	3,634.7	4,613.0
				(Index 1980 = 100)							
Terms of Trade	27.7	27.1	26.3	24.3	27.2	25.6	29.9	67.4	52.0	54.7	56.8
Export Prices, fob	7.3	7.2	7.2	7.2	8.8	9.4	13.4	38.6	32.3	34.3	38.7
Nonfuel Primary Products	32.2	33.2	35.7	39.0	36.1	37.7	45.4	48.7	56.3	58.5	56.7
Fuels	7.2	7.2	7.2	7.2	8.8	9.4	13.4	38.6	32.3	34.3	38.7
Manufactures	31.3	34.3	34.3	41.0	44.8	47.8	53.7	60.4	74.6	89.6	108.2
Import Prices, cif	26.2	26.6	27.5	29.8	32.5	36.9	44.9	57.3	62.2	62.8	68.2
Trade at Constant 1980 Prices				*(Millions of 1980 US dollars)*							
Exports, fob	16,232.9	25,970.7	30,033.6	32,737.0	30,546.5	31,208.4	29,753.3	21,415.6	21,164.7	27,862.7	29,496.7
Imports, cif	1,819.6	2,422.7	2,459.5	2,262.8	2,862.4	3,502.2	4,474.7	6,539.9	7,117.8	6,813.7	7,998.6
BALANCE OF PAYMENTS				*(Millions of current US dollars)*							
Exports of Goods & Services	..	..	..	2,536	2,886	2,695	3,743	8,237	6,793	9,096	10,786
Merchandise, fob	..	..	..	2,397	2,714	2,470	3,528	7,803	6,418	8,748	10,406
Nonfactor Services	..	..	..	42	42	97	101	143	160	167	136
Factor Services	..	..	..	97	129	128	114	291	215	181	244
Imports of Goods & Services	..	..	..	1,638	1,838	2,049	3,249	5,118	5,977	5,851	7,690
Merchandise, fob	..	..	..	674	930	1,291	2,011	3,746	4,424	4,277	4,929
Nonfactor Services	..	..	..	400	337	520	811	1,173	1,343	1,368	1,301
Factor Services	..	..	..	564	572	238	427	200	210	206	1,459
Long-Term Interest	..	..	..	..	..	..	..	..	..	..	..
Current Transfers, net	..	..	..	-140	-174	-306	-273	-350	-260	-257	-858
Workers' Remittances	..	..	..	0	0	0	0	0	0	0	0
Total to be Financed	..	..	..	758	874	340	222	2,768	556	2,988	2,238
Official Capital Grants	..	..	..	-113	-90	-102	-156	-69	-164	-144	-79
Current Account Balance	..	..	..	645	783	238	66	2,700	392	2,844	2,159
Long-Term Capital, net	..	..	..	26	40	-145	-666	-491	-1,688	-1,652	-1,144
Direct Investment	..	..	..	139	140	-4	-148	-241	-616	-521	-512
Long-Term Loans	..	..	..	..	..	..	..	..	..	..	..
Disbursements	..	..	..	..	..	..	..	..	..	..	..
Repayments	..	..	..	..	..	..	..	..	..	..	..
Other Long-Term Capital	..	..	..	0	-10	-39	-362	-182	-908	-987	-554
Other Capital, net	..	..	..	26	50	341	-381	116	-601	-34	659
Change in Reserves	..	..	..	-697	-874	-434	982	-2,325	1,897	-1,157	-1,674
Memo Item:				*(Libyan Dinars per US dollar)*							
Conversion Factor (Annual Avg)	0.360	0.360	0.360	0.360	0.360	0.330	0.300	0.300	0.300	0.300	0.300
EXTERNAL DEBT, ETC.				*(Millions of US dollars, outstanding at end of year)*							
Public/Publicly Guar. Long-Term	..	..	..	..	..	..	..	..	..	..	..
Official Creditors	..	..	..	..	..	..	..	..	..	..	..
IBRD and IDA	..	..	..	..	..	..	..	..	..	..	..
Private Creditors	..	..	..	..	..	..	..	..	..	..	..
Private Non-guaranteed Long-Term	..	..	..	..	..	..	..	..	..	..	..
Use of Fund Credit	..	..	..	..	..	..	..	..	..	..	..
Short-Term Debt	..	..	..	..	..	..	..	..	..	..	..
Memo Items:				*(Millions of US dollars)*							
Int'l Reserves Excluding Gold	317.4	453.5	832.3	1,504.8	2,572.9	2,832.2	2,023.8	3,511.1	2,095.0	3,106.4	4,786.4
Gold Holdings (at market price)	68.1	102.1	85.8	91.1	106.3	158.2	273.6	454.6	341.9	328.5	403.8
SOCIAL INDICATORS											
Total Fertility Rate	7.5	7.5	7.5	7.5	7.6	7.6	7.5	7.5	7.5	7.4	7.4
Crude Birth Rate	49.5	49.4	49.3	49.2	49.1	49.0	48.7	48.3	48.0	47.6	47.3
Infant Mortality Rate	130.0	127.4	124.8	122.2	119.6	117.0	115.0	113.0	111.0	109.0	107.0
Life Expectancy at Birth	50.4	50.9	51.4	51.9	52.4	52.9	53.5	54.1	54.6	55.2	55.8
Food Production, p.c. ('79-81 = 100)	92.4	97.4	99.7	87.5	66.1	105.2	102.7	110.2	122.1	127.0	88.3
Labor Force, Agriculture (%)	36.1	33.7	31.3	28.9	27.8	26.7	25.7	24.6	23.5	22.4	21.4
Labor Force, Female (%)	6.0	6.1	6.3	6.4	6.4	6.5	6.6	6.6	6.7	6.8	6.9
School Enroll. Ratio, primary	..	..	..	..	..	..	..	..	..	..	..
School Enroll. Ratio, secondary	..	..	..	..	..	..	..	..	..	..	..

1978	1979	1980	1981	1982	1983	1984	1985	1986	1987 estimate	NOTES	LIBYA
											FOREIGN TRADE (CUSTOMS BASIS)
9,907.0	16,084.9	21,919.1	15,575.6	13,948.0	10,957.0	11,136.0	10,350.0	5,680.0	5,830.0	..	Value of Exports, fob
0.2	0.1	0.0	0.0	0.0	0.0	0.0	8.4	1.9	..	..	Nonfuel Primary Products
9,866.9	16,006.7	21,909.7	15,513.3	13,948.0	10,957.0	11,136.0	10,341.6	5,678.1	5,830.0	..	Fuels
40.0	78.2	9.4	62.3	0.0	0.0	0.0	0.0	0.0	..	..	Manufactures
6,386.0	8,634.0	10,370.0	14,563.0	10,975.0	8,977.0	8,463.0	5,705.0	4,440.4	5,020.0	..	Value of Imports, cif
881.3	1,012.7	1,467.6	1,687.6	1,701.6	1,519.8	1,429.0	955.0	748.4	846.0	..	Nonfuel Primary Products
35.7	35.0	44.1	83.7	74.9	66.3	63.0	42.0	33.0	37.0	..	Fuels
5,469.0	7,586.3	8,858.3	12,791.8	9,198.5	7,390.9	6,971.0	4,708.0	3,659.0	4,137.0	..	Manufactures
					(Index 1980 = 100)						
48.4	64.3	100.0	111.2	102.2	91.4	90.7	91.5	39.2	46.6	..	Terms of Trade
38.3	58.8	100.0	110.9	98.9	86.9	84.1	84.1	40.7	52.5	..	Export Prices, fob
65.2	96.4	0.0	0.0	0.0	0.0	0.0	80.9	82.7	..	..	Nonfuel Primary Products
38.2	58.7	100.0	111.0	98.9	86.9	84.1	84.1	40.7	52.5	..	Fuels
106.7	94.0	100.0	95.0	83.0	82.0	83.0	76.0	89.9	118.1	..	Manufactures
79.1	91.4	100.0	99.8	96.7	95.0	92.7	91.9	104.0	112.6	..	Import Prices, cif
											Trade at Constant 1980 Prices
				(Millions of 1980 US dollars)							
25,867.2	27,351.8	21,919.1	14,041.5	14,103.1	12,608.7	13,241.4	12,307.2	13,953.4	11,113.2	..	Exports, fob
8,069.1	9,445.1	10,370.0	14,594.7	11,344.7	9,446.9	9,127.6	6,209.6	4,271.3	4,457.4	..	Imports, cif
				(Millions of current US dollars)							**BALANCE OF PAYMENTS**
10,370	16,547	23,366	16,521	14,732	13,185	11,680	10,890	6,157	..	..	Exports of Goods & Services
9,900	15,981	21,919	14,731	13,701	12,348	11,028	10,355	5,667	..	..	Merchandise, fob
153	143	164	163	163	161	170	122	95	..	..	Nonfactor Services
317	424	1,282	1,627	868	676	482	414	396	..	..	Factor Services
8,582	11,874	14,018	18,838	14,617	12,726	11,884	7,999	5,724	..	..	Imports of Goods & Services
5,764	8,647	10,368	14,563	10,976	8,978	8,464	5,706	4,422	..	..	Merchandise, fob
1,448	1,953	2,303	2,757	2,265	2,315	2,202	1,596	985	..	..	Nonfactor Services
1,370	1,273	1,347	1,517	1,376	1,433	1,218	698	317	..	..	Factor Services
..	..	..	..	..	..	..	..	..	..	..	Long-Term Interest
-972	-859	-1,089	-1,570	-1,597	-2,045	-1,240	-767	-446	..	..	Current Transfers, net
0	0	0	0	0	0	0	0	0	..	..	Workers' Remittances
816	3,814	8,259	-3,888	-1,482	-1,586	-1,444	2,125	-13	..	K	Total to be Financed
-78	-43	-46	-76	-78	-58	-81	-41	-41	..	K	Official Capital Grants
738	3,771	8,214	-3,964	-1,560	-1,643	-1,524	2,084	-54	..	..	Current Account Balance
-1,146	-973	-1,418	-1,781	-1,104	-615	-224	-47	-332	..	..	Long-Term Capital, net
-720	-609	-1,136	-769	-411	-327	-46	31	-80	..	..	Direct Investment
..	..	..	..	..	..	..	..	..	..	..	Long-Term Loans
..	..	..	..	..	..	..	..	..	..	..	Disbursements
..	..	..	..	..	..	..	..	..	..	..	Repayments
-348	-321	-236	-935	-615	-231	-97	-37	-211	..	..	Other Long-Term Capital
-111	-578	-390	1,608	663	480	27	591	385	..	..	Other Capital, net
519	-2,220	-6,407	4,136	2,001	1,778	1,721	-2,628	2	200	..	Change in Reserves
				(Libyan Dinars per US dollar)							**Memo Item:**
0.300	0.300	0.300	0.300	0.300	0.300	0.300	0.300	0.310	0.270	..	Conversion Factor (Annual Avg)
											EXTERNAL DEBT, ETC.
			(Millions of US dollars, outstanding at end of year)								
..	..	..	..	..	..	..	..	..	..	..	Public/Publicly Guar. Long-Term
..	..	..	..	..	..	..	..	..	..	..	Official Creditors
..	..	..	..	..	..	..	..	..	..	..	IBRD and IDA
..	..	..	..	..	..	..	..	..	..	..	Private Creditors
..	..	..	..	..	..	..	..	..	..	..	Private Non-guaranteed Long-Term
..	..	..	..	..	..	..	..	..	..	..	Use of Fund Credit
..	..	..	..	..	..	..	..	..	..	..	Short-Term Debt
											Memo Items:
				(Millions of US dollars)							
4,104.8	6,344.4	13,091.0	9,002.9	7,059.5	5,218.7	3,634.2	5,903.9	5,953.2	5,838.0	..	Int'l Reserves Excluding Gold
553.2	1,261.6	1,814.2	1,422.4	1,635.0	1,365.2	1,124.7	1,177.2	1,407.2	1,742.8	..	Gold Holdings (at market price)
											SOCIAL INDICATORS
7.3	7.3	7.3	7.2	7.2	7.1	7.1	7.0	6.9	6.9	..	Total Fertility Rate
47.0	46.6	46.3	45.9	45.6	45.2	44.8	44.3	43.9	43.5	..	Crude Birth Rate
105.0	103.0	101.0	99.0	97.0	89.7	82.4	75.2	67.9	60.6	..	Infant Mortality Rate
56.2	56.7	57.2	57.7	58.2	59.0	59.8	60.7	61.5	62.3	..	Life Expectancy at Birth
78.6	108.2	99.4	92.4	131.4	126.1	117.7	..	112.4	115.4	..	Food Production, p.c. ('79-81 = 100)
20.3	19.3	18.2	..	..	..	..	..	..	..	..	Labor Force, Agriculture (%)
7.0	7.2	7.3	7.4	7.6	7.8	7.9	8.1	8.3	8.5	..	Labor Force, Female (%)
..	..	..	..	..	..	..	127.0	..	..	..	School Enroll. Ratio, primary
..	..	..	..	..	..	..	87.0	..	..	..	School Enroll. Ratio, secondary

LUXEMBOURG	1967	1968	1969	1970	1971	1972	1973	1974	1975	1976	1977
CURRENT GNP PER CAPITA (US $)	2,260	2,440	2,800	3,170	3,450	4,070	4,980	6,420	7,460	8,460	9,270
POPULATION (thousands)	335	337	339	340	345	348	353	357	361	361	362

ORIGIN AND USE OF RESOURCES
(Billions of current Luxembourg Francs)

	1967	1968	1969	1970	1971	1972	1973	1974	1975	1976	1977
Gross National Product (GNP)	37.32	40.88	47.43	56.94	58.22	66.55	80.25	99.23	98.44	116.30	122.38
Net Factor Income from Abroad	0.20	0.27	0.40	1.90	2.17	3.34	3.43	5.59	11.70	16.69	19.82
Gross Domestic Product (GDP)	37.12	40.61	47.02	55.04	56.05	63.21	76.82	93.64	86.74	99.60	102.56
Indirect Taxes, net	2.94	3.17	3.76	4.72	5.36	6.45	7.68	8.27	8.91	9.44	9.59
GDP at factor cost	34.18	37.44	43.26	50.33	50.69	56.76	69.14	85.38	77.84	90.17	92.97
Agriculture	..	..	..	2.12	2.12	2.53	2.95	2.80	2.95	2.81	3.05
Industry	..	..	..	29.44	27.16	30.11	37.92	48.88	36.00	42.34	40.61
Manufacturing	..	..	..	23.79	20.68	22.35	29.39	39.11	25.36	31.56	29.52
Services, etc.	..	..	..	23.49	26.77	30.57	35.95	41.96	47.80	54.44	58.89
Resource Balance	2.94	4.09	6.83	7.05	1.97	3.74	9.54	19.58	3.57	5.54	3.89
Exports of Goods & NFServices	29.13	32.68	39.62	48.87	49.41	52.40	68.60	96.06	80.25	87.97	89.08
Imports of Goods & NFServices	26.19	28.60	32.79	41.82	47.44	48.66	59.06	76.48	76.68	82.43	85.20
Domestic Absorption	34.18	36.52	40.19	47.99	54.08	59.47	67.28	74.06	83.17	94.06	98.67
Private Consumption, etc.	19.89	21.23	22.70	24.86	28.49	31.41	35.60	40.99	48.07	54.82	58.72
General Gov't Consumption	5.99	6.55	6.92	7.72	8.63	9.48	10.50	12.01	13.82	15.75	16.98
Gross Domestic Investment	8.31	8.75	10.58	15.41	16.96	18.58	21.17	21.06	21.28	23.49	22.97
Fixed Investment	9.09	9.17	10.70	13.00	15.62	17.61	21.16	21.91	22.71	24.22	25.09
Memo Items:											
Gross Domestic Saving	11.24	12.84	17.41	22.46	18.93	22.31	30.71	40.64	24.85	29.04	26.86
Gross National Saving	11.44	13.10	17.81	..	..	..	..	..	..	..	..

(Billions of 1980 Luxembourg Francs)

	1967	1968	1969	1970	1971	1972	1973	1974	1975	1976	1977
Gross National Product	87.74	91.54	100.91	105.75	108.76	117.56	126.87	134.80	134.68	142.24	147.33
GDP at factor cost	79.96	83.31	91.16	91.67	93.99	99.35	107.60	113.00	105.54	108.91	111.55
Agriculture	..	..	..	3.80	3.37	3.59	3.73	3.38	3.64	3.15	3.51
Industry	..	..	..	44.29	45.38	49.02	53.64	56.37	47.57	48.22	48.14
Manufacturing	..	..	..	33.04	32.97	35.04	38.79	41.41	33.71	35.14	35.32
Services, etc.	..	..	..	53.61	55.75	58.73	63.55	66.26	66.93	69.57	71.56
Resource Balance	2.71	3.94	6.14	-0.40	-1.49	0.71	4.15	9.53	0.49	0.97	3.13
Exports of Goods & NFServices	61.34	67.93	77.28	84.26	88.70	93.40	106.69	118.01	99.57	100.89	104.68
Imports of Goods & NFServices	58.63	63.99	71.15	84.66	90.19	92.68	102.54	108.49	99.08	99.92	101.55
Domestic Absorption	84.57	86.99	93.87	102.11	105.99	110.62	116.77	116.48	117.65	119.97	120.08
Private Consumption, etc.	45.41	47.27	49.80	52.85	55.79	58.44	61.75	64.52	67.95	69.99	71.47
General Gov't Consumption	14.47	15.28	15.79	16.44	16.92	17.63	18.23	18.92	19.54	20.09	20.68
Gross Domestic Investment	24.68	24.44	28.28	32.82	33.27	34.55	36.78	33.05	30.15	29.89	27.93
Fixed Investment	24.29	23.28	25.72	27.65	30.63	32.73	36.74	34.50	32.12	30.89	30.56
Memo Items:											
Capacity to Import	65.20	73.13	85.97	98.93	93.93	99.80	119.10	136.26	103.69	106.64	106.18
Terms of Trade Adjustment	3.86	5.20	8.69	14.67	5.23	6.40	12.41	18.25	4.12	5.75	1.50
Gross Domestic Income	91.14	96.13	108.69	116.37	109.73	117.74	133.33	144.25	122.26	126.69	124.71
Gross National Income	91.60	96.74	109.60	120.42	113.99	123.96	139.28	153.04	138.80	147.98	148.83

DOMESTIC PRICES (DEFLATORS)
(Index 1980 = 100)

	1967	1968	1969	1970	1971	1972	1973	1974	1975	1976	1977
Overall (GDP)	42.5	44.7	47.0	54.1	53.6	56.8	63.5	74.3	73.4	82.4	83.2
Domestic Absorption	40.4	42.0	42.8	47.0	51.0	53.8	57.6	63.6	70.7	78.4	82.2
Agriculture	..	..	..	55.7	62.9	70.4	79.0	82.7	80.9	89.3	87.0
Industry	..	..	..	66.5	59.8	61.4	70.7	86.7	75.7	87.8	84.4
Manufacturing	..	..	..	72.0	62.7	63.8	75.8	94.5	75.2	89.8	83.6

MANUFACTURING ACTIVITY

	1967	1968	1969	1970	1971	1972	1973	1974	1975	1976	1977
Employment (1980 = 100)	102.3	101.6	104.1	106.4	109.5	111.7	115.9	120.3	117.9	115.0	112.1
Real Earnings per Empl. (1980 = 100)	60.3	64.5	67.0	76.0	77.9	79.8	82.3	92.6	93.4	91.1	92.8
Real Output per Empl. (1980 = 100)	..	..	..	63.3	72.3	72.3	74.2	79.6	87.4	77.9	80.6
Earnings as % of Value Added	57.3	52.6	46.4	42.9	55.2	56.8	48.5	46.2	63.6	66.3	76.9

MONETARY HOLDINGS
(Billions of current Luxembourg Francs)

	1967	1968	1969	1970	1971	1972	1973	1974	1975	1976	1977
Money Supply, Broadly Defined	..	..	..	..	..	..	..	..	..	..	..
Money as Means of Payment	..	..	..	..	..	..	..	..	..	..	..
Currency Ouside Banks	..	..	..	..	..	..	..	..	..	..	..
Demand Deposits	..	..	..	..	..	..	..	..	..	..	..
Quasi-Monetary Liabilities	..	..	..	..	..	..	..	..	..	..	..

(Millions of current Luxembourg Francs)

	1967	1968	1969	1970	1971	1972	1973	1974	1975	1976	1977
GOVERNMENT DEFICIT (-) OR SURPLUS	..	..	..	1,024	1,129	969	2,125	4,067	1,008	314	706
Current Revenue	..	..	..	17,557	19,841	22,617	27,509	34,341	38,557	44,757	50,523
Current Expenditure	..	..	..	14,296	16,083	18,313	21,187	25,321	32,323	38,013	42,887
Current Budget Balance	..	..	..	3,261	3,758	4,304	6,322	9,020	6,234	6,744	7,636
Capital Receipts	..	..	..	12	3	5	5	16	6	29	9
Capital Payments	..	..	..	2,249	2,632	3,340	4,202	4,969	5,232	6,459	6,939

1978	1979	1980	1981	1982	1983	1984	1985	1986	1987 estimate	NOTES	LUXEMBOURG
10,820	12,830	15,120	15,820	15,900	14,760	14,260	14,280	15,790	18,760	..	**CURRENT GNP PER CAPITA (US $)**
362	364	365	366	366	365	366	366	366	366	..	**POPULATION (thousands)**
				(Billions of current Luxembourg Francs)							**ORIGIN AND USE OF RESOURCES**
134.27	146.65	162.91	181.35	220.70	247.70	274.76	294.98	312.24	317.49	..	Gross National Product (GNP)
22.05	24.51	29.98	39.66	61.92	72.96	78.25	84.07	89.69	85.88	..	Net Factor Income from Abroad
112.21	122.15	132.93	141.69	158.79	174.74	196.51	210.91	222.54	231.61	..	Gross Domestic Product (GDP)
11.26	11.57	14.03	13.86	16.23	20.56	23.75	27.01	24.86	..	..	Indirect Taxes, net
100.96	110.57	118.90	127.83	142.56	154.18	172.76	183.90	197.68	..	B	GDP at factor cost
3.48	3.62	3.46	3.89	5.41	4.98	5.29	..	..	..	..	Agriculture
44.82	48.77	51.42	52.78	61.03	66.45	77.34	..	..	..	..	Industry
33.54	36.85	37.81	38.59	46.69	50.50	60.52	..	..	..	..	Manufacturing
63.92	69.76	78.05	85.01	92.34	103.31	113.88	..	..	..	..	Services, etc.
1.24	4.83	-1.54	-4.40	-3.02	0.00	3.76	11.46	18.06	28.79	..	Resource Balance
93.95	111.09	117.31	122.49	141.10	157.37	196.40	223.51	226.41	243.08	..	Exports of Goods & NFServices
92.71	106.26	118.85	126.90	144.12	157.37	192.64	212.05	208.35	214.29	..	Imports of Goods & NFServices
110.98	117.31	134.47	146.10	161.81	174.74	192.75	199.45	204.48	202.82	..	Domestic Absorption
64.06	69.61	78.09	85.88	94.43	101.49	110.21	115.18	123.12	122.07	..	Private Consumption, etc.
18.35	19.73	22.18	24.34	27.12	29.75	32.04	33.37	35.16	34.18	..	General Gov't Consumption
28.56	27.97	34.20	35.88	40.26	43.50	50.50	50.90	46.20	46.57	..	Gross Domestic Investment
27.01	29.57	35.92	36.43	40.17	38.53	39.88	39.48	47.30	46.88	..	Fixed Investment
											Memo Items:
29.80	32.81	32.66	31.48	37.24	43.50	54.26	62.36	64.26	75.36	..	Gross Domestic Saving
..	..	..	..	..	..	..	..	..	..	..	Gross National Saving
				(Billions of 1980 Luxembourg Francs)							
153.28	158.13	162.91	169.34	186.79	195.76	205.46	213.68	219.97	223.16	..	Gross National Product
115.10	118.78	118.90	119.84	121.05	122.66	130.22	133.93	140.47	..	B	GDP at factor cost
3.77	3.71	3.46	3.55	4.34	3.80	4.10	..	..	..	..	Agriculture
49.71	51.18	51.42	49.44	49.46	52.00	56.25	..	..	..	..	Industry
36.79	38.21	37.81	35.92	36.97	38.95	43.29	..	..	..	..	Manufacturing
74.52	76.51	78.05	79.66	80.91	82.93	87.38	..	..	..	..	Services, etc.
0.81	3.44	-1.54	-2.47	-1.41	2.15	5.55	10.10	15.82	13.96	..	Resource Balance
108.36	118.68	117.31	113.11	114.06	119.86	141.40	154.36	160.81	164.02	..	Exports of Goods & NFServices
107.55	115.24	118.85	115.57	115.48	117.70	135.85	144.26	144.99	150.06	..	Imports of Goods & NFServices
127.20	127.96	134.47	135.13	136.12	136.58	142.19	143.21	141.90	146.97	..	Domestic Absorption
73.52	75.97	78.09	79.36	79.51	79.41	81.32	82.73	85.44	88.43	..	Private Consumption, etc.
21.05	21.52	22.18	22.49	22.83	23.26	23.65	23.96	24.40	24.77	..	General Gov't Consumption
32.63	30.47	34.20	33.27	33.78	33.92	37.23	36.53	32.06	33.77	..	Gross Domestic Investment
30.97	32.24	35.92	33.67	33.81	30.12	29.43	28.34	32.82	33.97	..	Fixed Investment
											Memo Items:
108.99	120.49	117.31	111.56	113.06	117.70	138.50	152.05	157.56	170.22	..	Capacity to Import
0.63	1.80	0.00	-1.55	-1.01	-2.15	-2.89	-2.31	-3.25	6.20	..	Terms of Trade Adjustment
128.64	133.20	132.93	131.11	133.70	136.58	144.85	151.01	154.47	167.13	..	Gross Domestic Income
153.91	159.94	162.91	167.80	185.79	193.61	202.57	211.37	216.72	..	..	Gross National Income
				(Index 1980 = 100)							**DOMESTIC PRICES (DEFLATORS)**
87.7	93.0	100.0	106.8	117.9	126.0	133.0	137.6	141.1	143.9	..	Overall (GDP)
87.2	91.7	100.0	108.1	118.9	127.9	135.6	139.3	144.1	138.0	..	Domestic Absorption
92.3	97.5	100.0	109.6	124.8	131.2	128.9	..	..	..	..	Agriculture
90.1	95.3	100.0	106.8	123.4	127.8	137.5	..	..	..	..	Industry
91.2	96.4	100.0	107.4	126.3	129.7	139.8	..	..	..	..	Manufacturing
											MANUFACTURING ACTIVITY
105.5	102.9	100.0	98.9	97.4	94.8	93.1	92.8	..	..	..	Employment (1980 = 100)
96.7	98.3	100.0	97.4	96.1	91.9	111.9	118.2	..	..	..	Real Earnings per Empl. (1980 = 100)
85.3	97.1	100.0	96.8	98.5	104.5	116.6	123.1	..	..	..	Real Output per Empl. (1980 = 100)
67.4	62.7	65.0	65.6	58.5	57.4	60.5	61.0	..	..	..	Earnings as % of Value Added
				(Billions of current Luxembourg Francs)							**MONETARY HOLDINGS**
..	..	..	..	..	..	..	..	..	..	..	Money Supply, Broadly Defined
..	..	..	..	..	..	..	..	..	..	..	Money as Means of Payment
..	..	..	..	..	..	..	..	..	..	..	Currency Ouside Banks
..	..	..	..	..	..	..	..	..	..	..	Demand Deposits
..	..	..	..	..	..	..	..	..	..	..	Quasi-Monetary Liabilities
				(Millions of current Luxembourg Francs)							**GOVERNMENT DEFICIT (-) OR SURPLUS**
3,801	-294	1,673	-3,016	1,152	-3,560	10,253	21,038	..	..	..	Current Revenue
56,137	58,143	66,339	72,133	79,548	90,293	96,889	104,365	..	..	..	Current Revenue
45,439	50,397	56,580	62,783	71,537	76,746	80,805	84,645	..	..	..	Current Expenditure
10,698	7,746	9,759	9,350	8,011	13,547	16,084	19,720	..	..	..	Current Budget Balance
12	107	13	98	51	96	171	65	..	..	..	Capital Receipts
6,909	8,147	8,571	10,005	10,491	16,423	11,413	11,657	..	..	..	Capital Payments

LUXEMBOURG	1967	1968	1969	1970	1971	1972	1973	1974	1975	1976	1977
FOREIGN TRADE (CUSTOMS BASIS)					*(Millions of current US dollars)*						
Value of Exports, fob	..	..	..	..	..	..	..	..	..	..	..
Nonfuel Primary Products	..	..	..	..	..	..	..	..	..	..	..
Fuels	..	..	..	..	..	..	..	..	..	..	..
Manufactures	..	..	..	..	..	..	..	..	..	..	..
Value of Imports, cif	..	..	..	..	..	..	..	..	..	..	..
Nonfuel Primary Products	..	..	..	..	..	..	..	..	..	..	..
Fuels	..	..	..	..	..	..	..	..	..	..	..
Manufactures	..	..	..	..	..	..	..	..	..	..	..
Terms of Trade					*(Index 1980 = 100)*						
Export Prices, fob	..	..	..	..	..	..	..	..	..	..	..
Nonfuel Primary Products	..	..	..	..	..	..	..	..	..	..	..
Fuels	..	..	..	..	..	..	..	..	..	..	..
Manufactures	..	..	..	..	..	..	..	..	..	..	..
Import Prices, cif	..	..	..	..	..	..	..	..	..	..	..
Trade at Constant 1980 Prices					*(Millions of 1980 US dollars)*						
Exports, fob	..	..	..	..	..	..	..	..	..	..	..
Imports, cif	..	..	..	..	..	..	..	..	..	..	..
BALANCE OF PAYMENTS					*(Millions of current US dollars)*						
Exports of Goods & Services	..	..	..	..	..	..	..	..	..	..	..
Merchandise, fob	..	..	..	..	..	..	..	..	..	..	..
Nonfactor Services	..	..	..	..	..	..	..	..	..	..	..
Factor Services	..	..	..	..	..	..	..	..	..	..	..
Imports of Goods & Services	..	..	..	..	..	..	..	..	..	..	..
Merchandise, fob	..	..	..	..	..	..	..	..	..	..	..
Nonfactor Services	..	..	..	..	..	..	..	..	..	..	..
Factor Services	..	..	..	..	..	..	..	..	..	..	..
Long-Term Interest	..	..	..	..	..	..	..	..	..	..	..
Current Transfers, net	..	..	..	..	..	..	..	..	..	..	..
Workers' Remittances	..	..	..	..	..	..	..	..	..	..	..
Total to be Financed	..	..	..	..	..	..	..	..	..	..	..
Official Capital Grants	..	..	..	..	..	..	..	..	..	..	..
Current Account Balance	..	..	..	..	..	..	..	..	..	..	..
Long-Term Capital, net	..	..	..	..	..	..	..	..	..	..	..
Direct Investment	..	..	..	..	..	..	..	..	..	..	..
Long-Term Loans	..	..	..	..	..	..	..	..	..	..	..
Disbursements	..	..	..	..	..	..	..	..	..	..	..
Repayments	..	..	..	..	..	..	..	..	..	..	..
Other Long-Term Capital	..	..	..	..	..	..	..	..	..	..	..
Other Capital, net	..	..	..	..	..	..	..	..	..	..	..
Change in Reserves	..	..	..	..	..	..	..	..	..	..	..
Memo Item:					*(Luxembourg Francs per US dollar)*						
Conversion Factor (Annual Avg)	50.000	50.000	50.000	50.000	48.870	44.010	38.980	38.950	36.780	38.610	35.840
EXTERNAL DEBT, ETC.					*(Millions of US dollars, outstanding at end of year)*						
Public/Publicly Guar. Long-Term	..	..	..	..	..	..	..	..	..	..	..
Official Creditors	..	..	..	..	..	..	..	..	..	..	..
IBRD and IDA	..	..	..	..	..	..	..	..	..	..	..
Private Creditors	..	..	..	..	..	..	..	..	..	..	..
Private Non-guaranteed Long-Term	..	..	..	..	..	..	..	..	..	..	..
Use of Fund Credit	..	..	..	..	..	..	..	..	..	..	..
Short-Term Debt	..	..	..	..	..	..	..	..	..	..	..
Memo Items:					*(Millions of US dollars)*						
Int'l Reserves Excluding Gold	..	..	..	..	..	..	..	..	..	..	..
Gold Holdings (at market price)	12.08	18.42	15.48	16.43	19.30	28.70	49.65	82.49	62.03	59.60	74.37
SOCIAL INDICATORS											
Total Fertility Rate	2.2	2.1	2.0	2.0	1.9	1.7	1.5	1.6	1.5	1.5	1.4
Crude Birth Rate	14.8	14.0	13.3	13.0	12.9	11.7	10.8	11.0	11.0	10.8	11.2
Infant Mortality Rate	20.4	17.0	17.5	24.9	20.5	14.0	15.3	13.5	14.8	17.9	10.6
Life Expectancy at Birth	69.9	70.1	70.2	70.3	70.5	70.6	70.9	71.1	71.4	71.6	71.9
Food Production, p.c. ('79-81 = 100)	..	..	..	..	..	..	..	..	..	..	..
Labor Force, Agriculture (%)	10.2	9.4	8.7	7.9	7.6	7.4	7.1	6.9	6.6	6.4	6.1
Labor Force, Female (%)	27.1	27.1	27.0	26.9	27.4	27.8	28.3	28.7	29.2	29.8	30.4
School Enroll. Ratio, primary	..	..	..	112.0	..	..	..	..	..	98.0	98.0
School Enroll. Ratio, secondary	..	..	..	48.0	..	..	..	..	59.0	62.0	64.0

1978	1979	1980	1981	1982	1983	1984	1985	1986	1987 estimate	NOTES	LUXEMBOURG
					(Millions of current US dollars)						**FOREIGN TRADE (CUSTOMS BASIS)**
..	..	..	..	..	..	..	..	..	..	f	Value of Exports, fob
..	..	..	..	..	..	..	..	..	..	..	Nonfuel Primary Products
..	..	..	..	..	..	..	..	..	..	..	Fuels
..	..	..	..	..	..	..	..	..	..	..	Manufactures
..	..	..	..	..	..	..	..	..	..	f	Value of Imports, cif
..	..	..	..	..	..	..	..	..	..	..	Nonfuel Primary Products
..	..	..	..	..	..	..	..	..	..	..	Fuels
..	..	..	..	..	..	..	..	..	..	..	Manufactures
					(Index 1980 = 100)						
..	..	..	..	..	..	..	..	..	..	f	Terms of Trade
..	..	..	..	..	..	..	..	..	..	f	Export Prices, fob
..	..	..	..	..	..	..	..	..	..	..	Nonfuel Primary Products
..	..	..	..	..	..	..	..	..	..	..	Fuels
..	..	..	..	..	..	..	..	..	..	..	Manufactures
..	..	..	..	..	..	..	..	..	..	f	Import Prices, cif
					(Millions of 1980 US dollars)						Trade at Constant 1980 Prices
..	..	..	..	..	..	..	..	..	..	f	Exports, fob
..	..	..	..	..	..	..	..	..	..	f	Imports, cif
					(Millions of current US dollars)						**BALANCE OF PAYMENTS**
..	..	..	..	..	..	..	..	..	..	f	Exports of Goods & Services
..	..	..	..	..	..	..	..	..	..	..	Merchandise, fob
..	..	..	..	..	..	..	..	..	..	..	Nonfactor Services
..	..	..	..	..	..	..	..	..	..	..	Factor Services
..	..	..	..	..	..	..	..	..	..	f	Imports of Goods & Services
..	..	..	..	..	..	..	..	..	..	..	Merchandise, fob
..	..	..	..	..	..	..	..	..	..	..	Nonfactor Services
..	..	..	..	..	..	..	..	..	..	..	Factor Services
..	..	..	..	..	..	..	..	..	..	..	Long-Term Interest
..	..	..	..	..	..	..	..	..	..	f	Current Transfers, net
..	..	..	..	..	..	..	..	..	..	..	Workers' Remittances
..	..	..	..	..	..	..	..	..	..	K	Total to be Financed
..	..	..	..	..	..	..	..	..	..	K	Official Capital Grants
..	..	..	..	..	..	..	..	..	..	..	Current Account Balance
..	..	..	..	..	..	..	..	..	..	f	Long-Term Capital, net
..	..	..	..	..	..	..	..	..	..	..	Direct Investment
..	..	..	..	..	..	..	..	..	..	..	Long-Term Loans
..	..	..	..	..	..	..	..	..	..	..	Disbursements
..	..	..	..	..	..	..	..	..	..	..	Repayments
..	..	..	..	..	..	..	..	..	..	..	Other Long-Term Capital
..	..	..	..	..	..	..	..	..	..	..	Other Capital, net
..	..	..	..	..	..	..	..	..	..	..	Change in Reserves
					(Luxembourg Francs per US dollar)						**Memo Item:**
31.490	29.320	29.240	37.130	45.690	51.130	57.780	59.380	44.670	37.330	..	Conversion Factor (Annual Avg)
					(Millions of US dollars, outstanding at end of year)						**EXTERNAL DEBT, ETC.**
..	..	..	..	..	..	..	..	..	..	..	Public/Publicly Guar. Long-Term
..	..	..	..	..	..	..	..	..	..	..	Official Creditors
..	..	..	..	..	..	..	..	..	..	..	IBRD and IDA
..	..	..	..	..	..	..	..	..	..	..	Private Creditors
..	..	..	..	..	..	..	..	..	..	..	Private Non-guaranteed Long-Term
..	..	..	..	..	..	..	..	..	..	..	Use of Fund Credit
..	..	..	..	..	..	..	..	..	..	..	Short-Term Debt
					(Millions of US dollars)						**Memo Items:**
..	..	..	..	..	..	..	..	..	..	..	Int'l Reserves Excluding Gold
102.86	233.03	268.30	180.91	207.95	173.63	132.29	140.31	167.68	207.65	..	Gold Holdings (at market price)
											SOCIAL INDICATORS
1.5	1.5	1.5	1.5	1.5	1.5	1.5	1.4	1.4	1.4	..	Total Fertility Rate
11.2	11.2	11.4	12.1	11.9	11.6	11.4	11.1	10.9	10.7	..	Crude Birth Rate
10.6	13.0	11.5	13.8	14.3	14.9	15.4	16.0	16.6	17.3	..	Infant Mortality Rate
72.1	72.3	72.5	72.7	72.9	73.1	73.3	73.5	73.7	73.9	..	Life Expectancy at Birth
..	..	..	..	..	..	..	..	..	..	..	Food Production, p.c. ('79-81 = 100)
5.9	5.6	5.4	..	..	..	..	..	..	..	..	Labor Force, Agriculture (%)
31.0	31.6	32.2	32.2	32.2	32.2	32.3	32.3	32.1	32.0	..	Labor Force, Female (%)
98.0	94.0	..	..	..	100.0	..	..	..	..	..	School Enroll. Ratio, primary
65.0	66.0	62.0	..	64.0	68.0	69.0	74.0	..	..	..	School Enroll. Ratio, secondary

MADAGASCAR	1967	1968	1969	1970	1971	1972	1973	1974	1975	1976	1977
CURRENT GNP PER CAPITA (US $)	120	130	130	130	140	140	160	190	230	230	240
POPULATION (thousands)	6,307	6,452	6,600	6,752	6,909	7,071	7,240	7,417	7,604	7,801	8,010
ORIGIN AND USE OF RESOURCES				*(Billions of current Malagasy Francs)*							
Gross National Product (GNP)	183.1	199.1	215.3	239.1	257.7	265.0	291.1	367.7	389.3	415.6	460.7
Net Factor Income from Abroad	-7.6	-6.6	-6.3	-10.3	-10.8	-8.1	-6.5	-5.1	-5.9	-5.5	-7.4
Gross Domestic Product (GDP)	190.7	205.7	221.6	249.4	268.5	273.1	297.6	372.8	395.2	421.1	468.1
Indirect Taxes, net	21.6	25.3	28.9	32.5	38.2	36.6	39.9	50.0	53.0	56.9	61.7
GDP at factor cost	169.1	180.4	192.7	216.9	230.3	236.5	257.7	322.8	342.2	364.2	406.4
Agriculture	59.0	60.9	64.2	73.8	78.2	82.4	96.2	154.0	162.4	169.2	185.7
Industry	31.5	34.5	38.2	46.0	49.5	52.3	60.2	66.7	70.6	76.2	89.9
Manufacturing	21.1	23.3	25.8	32.9	..	..	..	..	..	..	..
Services, etc.	100.2	110.3	119.2	129.6	140.8	138.4	141.2	152.1	162.2	175.7	192.5
Resource Balance	-5.7	-8.5	-9.0	-2.8	-16.4	-9.1	-11.8	-14.6	-16.9	-5.8	-12.4
Exports of Goods & NFServices	39.7	43.4	47.7	56.8	46.1	46.3	50.4	68.8	74.5	75.9	91.5
Imports of Goods & NFServices	45.4	51.9	56.7	59.6	62.5	55.4	62.2	83.4	91.4	81.7	103.9
Domestic Absorption	196.4	214.2	230.6	252.2	284.9	282.2	309.4	387.4	412.1	426.9	480.5
Private Consumption, etc.	127.3	135.1	144.7	167.2	181.4	191.7	215.4	277.7	301.1	306.4	348.2
General Gov't Consumption	41.3	45.7	48.3	46.1	56.1	52.6	51.4	58.9	60.4	66.6	72.8
Gross Domestic Investment	27.8	33.4	37.6	38.9	47.4	37.9	42.6	50.8	50.6	53.9	59.5
Fixed Investment	..	..	..	36.4	42.7	36.0	40.1	44.0	48.4	52.1	59.5
Memo Items:											
Gross Domestic Saving	22.1	24.9	28.6	36.1	31.0	28.8	30.8	36.2	33.7	48.1	47.1
Gross National Saving	14.5	18.3	16.6	18.9	13.5	13.0	19.3	29.0	22.5	38.3	35.1
Gross National Product				*(Billions of 1980 Malagasy Francs)*							
	504.30	533.12	574.41	598.04	618.15	620.26	612.07	637.23	642.84	620.70	635.32
GDP at factor cost	..	..	..	..	..	..	..	..	..	..	..
Agriculture	..	..	..	234.78	229.69	234.47	237.01	258.64	257.37	235.10	242.74
Industry	..	..	..	103.96	110.06	111.19	108.93	116.16	116.16	105.54	108.71
Manufacturing	..	..	..	..	..	..	..	..	..	..	..
Services, etc.	..	..	..	288.77	308.16	296.35	281.86	270.72	278.74	287.21	291.89
Resource Balance	-80.62	-102.67	-109.20	-107.45	-131.12	-103.27	-103.01	-75.83	-70.50	-46.14	-53.92
Exports of Goods & NFServices	103.11	114.51	118.63	111.56	89.37	88.19	92.12	97.62	114.70	87.99	92.71
Imports of Goods & NFServices	183.74	217.18	227.83	219.02	220.49	191.45	195.13	173.45	185.21	134.13	146.62
Domestic Absorption	632.17	644.55	705.87	734.97	779.03	745.27	730.82	721.36	722.78	673.99	697.25
Private Consumption, etc.	471.06	467.49	497.29	515.45	520.40	530.78	518.89	505.60	510.52	477.52	506.45
General Gov't Consumption	80.84	88.35	89.93	91.12	107.13	97.64	90.13	86.37	88.15	92.50	93.09
Gross Domestic Investment	106.29	130.05	139.62	128.40	151.51	116.85	121.80	129.39	124.11	103.98	97.70
Fixed Investment	..	..	..	..	..	..	..	..	..	..	..
Memo Items:											
Capacity to Import	160.67	181.61	191.67	208.73	162.63	160.01	158.11	143.08	150.96	124.61	129.12
Terms of Trade Adjustment	57.55	67.10	73.04	97.17	73.26	71.82	66.00	45.47	36.26	36.62	36.42
Gross Domestic Income	614.56	644.93	695.07	724.68	721.18	713.82	693.80	691.00	688.54	664.47	679.75
Gross National Income	582.05	611.65	664.47	695.21	691.42	692.08	678.06	682.70	679.10	657.32	671.74
DOMESTIC PRICES (DEFLATORS)				*(Index 1980 = 100)*							
Overall (GDP)	35.5	36.3	36.6	39.7	41.4	42.5	47.4	57.8	60.6	67.1	72.8
Domestic Absorption	31.1	33.2	32.7	34.3	36.6	37.9	42.3	53.7	57.0	63.3	68.9
Agriculture	..	..	..	31.4	34.0	35.1	40.6	59.5	63.1	72.0	76.5
Industry	..	..	..	44.2	45.0	47.0	55.3	57.4	60.8	72.2	82.7
Manufacturing	..	..	..	..	..	..	..	..	..	..	..
MANUFACTURING ACTIVITY											
Employment (1980=100)	70.0	73.3	81.8	88.6	87.9	91.2	102.9	114.4	104.7	106.6	107.6
Real Earnings per Empl. (1980=100)	91.3	93.8	98.4	101.6	108.5	110.8	112.5	102.9	98.5	96.2	96.3
Real Output per Empl. (1980=100)	..	..	..	91.2	97.8	95.0	82.6	87.3	98.8	86.0	82.4
Earnings as % of Value Added	34.0	32.4	37.3	35.6	38.5	42.1	41.9	42.5	40.7	40.3	37.5
MONETARY HOLDINGS				*(Billions of current Malagasy Francs)*							
Money Supply, Broadly Defined	42.1	45.3	49.5	57.1	60.6	67.5	70.2	85.0	79.2	100.2	108.8
Money as Means of Payment	36.5	39.4	41.5	46.2	47.0	53.3	57.3	67.9	69.4	79.7	100.0
Currency Ouside Banks	18.1	18.1	19.0	22.5	22.2	25.4	27.0	31.9	34.0	35.5	42.0
Demand Deposits	18.4	21.3	22.6	23.7	24.7	28.0	30.3	36.1	35.4	44.2	57.9
Quasi-Monetary Liabilities	5.6	5.9	8.0	10.9	13.7	14.2	12.9	17.1	9.8	20.5	8.8
GOVERNMENT DEFICIT (-) OR SURPLUS				*(Millions of current Malagasy Francs)*							
	..	..	..	..	..	-6,729	-7,309	-9,035	..	..	..
Current Revenue	..	..	..	..	..	49,837	54,234	58,246	..	..	..
Current Expenditure	..	..	..	..	..	41,432	46,951	..	..	..	..
Current Budget Balance	..	..	..	..	..	8,405	7,283	..	..	..	..
Capital Receipts	..	..	..	..	..	..	..	..	..	..	..
Capital Payments	..	..	..	..	..	15,134	14,592	..	..	..	..

1978	1979	1980	1981	1982	1983	1984	1985	1986	1987 estimate	NOTES	MADAGASCAR
250	300	350	330	320	300	270	240	230	210	..	CURRENT GNP PER CAPITA (US $)
8,231	8,465	8,714	8,978	9,260	9,558	9,875	10,212	10,551	10,894	..	POPULATION (thousands)
				(Billions of current Malagasy Francs)							**ORIGIN AND USE OF RESOURCES**
484.6	592.5	680.1	764.1	961.5	1,172.5	1,292.3	1,471.0	1,686.8	2,040.0	..	Gross National Product (GNP)
-2.0	-2.6	-9.7	-24.9	-34.6	-48.4	-76.8	-82.4	-120.1	-170.0	..	Net Factor Income from Abroad
486.6	595.1	689.8	789.0	996.1	1,220.9	1,369.1	1,553.4	1,806.9	2,210.0	..	Gross Domestic Product (GDP)
70.1	83.9	..	..	..	..	..	..	..	..	..	Indirect Taxes, net
416.5	511.2	..	..	..	..	..	..	..	..	B	GDP at factor cost
187.6	212.5	249.1	313.6	409.7	525.0	580.6	653.2	776.0	940.0	..	Agriculture
93.6	114.5	124.3	125.3	149.9	185.0	213.9	254.1	284.4	350.0	..	Industry
..	..	..	..	..	..	..	..	..	..	..	Manufacturing
205.4	268.1	316.4	350.1	436.5	510.9	574.6	646.1	746.5	920.0	..	Services, etc.
-30.4	-97.1	-117.7	-91.4	-95.0	-83.6	-59.7	-82.6	-89.0	-145.0	..	Resource Balance
99.2	103.1	109.4	107.9	132.9	153.4	214.2	225.0	274.0	436.0	..	Exports of Goods & NFServices
129.6	200.2	227.1	199.3	227.9	237.0	273.9	307.6	363.0	581.0	..	Imports of Goods & NFServices
517.0	692.2	807.5	880.4	1,091.1	1,304.5	1,428.8	1,636.0	1,895.9	2,355.0	..	Domestic Absorption
362.0	438.5	527.3	608.8	808.6	978.5	1,058.1	1,208.6	1,409.8	1,743.1	..	Private Consumption, etc.
81.6	103.0	117.8	129.1	149.5	165.3	185.0	209.7	237.5	306.9	..	General Gov't Consumption
73.4	150.7	162.4	142.5	133.0	160.7	185.7	217.7	248.6	305.0	..	Gross Domestic Investment
71.2	144.7	157.6	148.3	126.0	..	..	..	..	..	..	Fixed Investment
											Memo Items:
43.0	53.6	44.7	51.1	38.0	77.1	126.0	135.1	159.6	160.0	..	Gross Domestic Saving
37.6	47.4	30.5	22.9	-1.1	24.4	50.4	77.1	71.9	47.8	..	Gross National Saving
				(Billions of 1980 Malagasy Francs)							
620.90	680.27	680.10	606.29	596.75	599.72	601.79	616.31	613.72	620.14	..	Gross National Product
..	..	..	..	..	..	..	..	..	..	B	GDP at factor cost
226.83	243.06	249.10	237.97	247.51	253.87	261.83	266.92	274.87	280.91	..	Agriculture
113.23	127.69	124.30	96.28	82.49	83.62	87.91	91.76	89.04	97.41	..	Industry
..	..	..	..	..	..	..	..	..	..	..	Manufacturing
283.42	312.39	316.40	292.34	289.22	288.55	288.10	293.67	295.23	295.43	..	Services, etc.
-63.57	-118.14	-117.70	-67.00	-67.01	-55.30	-43.68	-40.14	-36.54	-36.76	..	Resource Balance
102.53	111.17	109.40	80.72	74.83	65.60	69.14	72.08	74.44	60.81	..	Exports of Goods & NFServices
166.10	229.30	227.10	147.73	141.85	120.90	112.82	112.23	110.98	97.56	..	Imports of Goods & NFServices
687.05	801.27	807.50	693.58	686.23	681.34	681.52	692.49	695.68	710.51	..	Domestic Absorption
490.26	518.16	527.30	462.76	479.43	475.14	476.09	483.08	486.47	498.89	..	Private Consumption, etc.
98.43	112.46	117.80	117.60	114.04	114.44	109.04	111.14	111.61	110.84	..	General Gov't Consumption
98.36	170.65	162.40	113.22	92.75	91.76	96.38	98.27	97.60	100.77	..	Gross Domestic Investment
..	..	..	..	..	..	..	..	..	..	..	Fixed Investment
											Memo Items:
127.14	118.09	109.40	79.98	82.72	78.25	88.23	82.09	83.77	73.22	..	Capacity to Import
24.61	6.92	0.00	-0.75	7.89	12.65	19.09	10.01	9.33	12.41	..	Terms of Trade Adjustment
648.09	690.06	689.80	625.83	627.10	638.69	656.93	662.35	668.47	686.16	..	Gross Domestic Income
645.52	687.19	680.10	605.54	604.64	612.37	620.88	626.32	623.05	632.54	..	Gross National Income
				(Index 1980 = 100)							**DOMESTIC PRICES (DEFLATORS)**
78.0	87.1	100.0	125.9	160.9	195.0	214.6	238.1	274.1	328.0	..	Overall (GDP)
75.2	86.4	100.0	126.9	159.0	191.5	209.6	236.2	272.5	331.5	..	Domestic Absorption
82.7	87.4	100.0	131.8	165.5	206.8	221.8	244.7	282.3	334.6	..	Agriculture
82.7	89.7	100.0	130.1	181.7	221.2	243.3	276.9	319.4	359.3	..	Industry
..	..	..	..	..	..	..	..	..	..	..	Manufacturing
											MANUFACTURING ACTIVITY
113.0	110.5	100.0	110.7	106.9	118.1	121.9	..	..	..	J	Employment (1980=100)
103.5	99.5	100.0	91.8	76.4	60.1	62.0	..	..	..	J	Real Earnings per Empl. (1980=100)
89.8	95.0	100.0	71.0	54.6	50.0	56.5	..	..	..	J	Real Output per Empl. (1980=100)
38.5	36.7	38.2	49.0	48.5	40.2	35.5	..	..	..	J	Earnings as % of Value Added
				(Billions of current Malagasy Francs)							**MONETARY HOLDINGS**
144.8	169.0	212.2	267.0	299.8	293.2	363.3	419.3	510.9	625.4	D	Money Supply, Broadly Defined
112.8	124.3	151.3	193.8	208.0	192.7	239.9	238.6	289.6	371.8	..	Money as Means of Payment
48.2	53.5	70.2	83.1	90.4	75.8	89.9	96.2	113.2	140.3	..	Currency Ouside Banks
64.6	70.7	81.1	110.7	117.6	116.8	150.0	142.4	176.3	231.5	..	Demand Deposits
32.0	44.8	60.9	73.1	91.8	100.6	123.4	180.7	221.4	253.6	..	Quasi-Monetary Liabilities
				(Millions of current Malagasy Francs)							
..	..	..	..	..					..	..	**GOVERNMENT DEFICIT (-) OR SURPLUS**
129,940	168,361	133,025	142,722	145,898	..	..	..	..	..	..	Current Revenue
..	..	..	..	..	..	..	..	..	..	..	Current Expenditure
..	..	..	..	..	..	..	..	..	..	..	Current Budget Balance
..	..	..	..	..	..	..	..	..	..	..	Capital Receipts
..	..	..	..	..	..	..	..	..	..	..	Capital Payments

MADAGASCAR	1967	1968	1969	1970	1971	1972	1973	1974	1975	1976	1977
FOREIGN TRADE (CUSTOMS BASIS)					*(Millions of current US dollars)*						
Value of Exports, fob	104.2	115.9	113.0	144.8	146.8	163.8	202.7	244.2	301.4	282.6	347.3
Nonfuel Primary Products	92.5	104.9	98.4	128.7	128.7	148.4	174.9	198.1	263.2	237.0	310.3
Fuels	4.0	4.0	4.0	5.6	5.3	6.5	10.3	23.6	25.9	23.5	13.2
Manufactures	7.7	7.0	10.5	10.5	12.8	8.9	17.4	22.6	12.3	22.1	23.7
Value of Imports, cif	145.3	170.0	182.8	170.4	213.2	202.2	202.9	281.0	366.9	289.9	353.0
Nonfuel Primary Products	19.9	23.4	29.0	25.7	37.3	34.1	43.9	69.9	63.5	48.6	61.3
Fuels	8.5	10.6	12.1	12.6	14.0	17.4	19.4	50.7	73.8	57.7	53.5
Manufactures	116.9	136.1	141.7	132.2	161.9	150.7	139.7	160.4	229.6	183.7	238.2
					(Index 1980 = 100)						
Terms of Trade	104.0	109.7	113.0	125.3	111.4	119.4	120.4	95.5	93.0	128.5	158.6
Export Prices, fob	24.3	24.9	26.0	28.6	29.8	32.4	41.9	54.9	52.2	73.2	100.9
Nonfuel Primary Products	29.8	30.1	32.0	37.2	35.5	39.3	52.7	57.1	54.1	81.2	111.0
Fuels	4.3	4.3	4.3	4.3	5.6	6.2	8.9	36.7	35.7	38.4	42.0
Manufactures	30.1	31.7	32.4	37.5	36.5	38.5	49.5	67.2	66.1	66.7	71.7
Import Prices, cif	23.3	22.7	23.0	22.9	26.8	27.1	34.8	57.5	56.1	56.9	63.6
Trade at Constant 1980 Prices					*(Millions of 1980 US dollars)*						
Exports, fob	429.2	464.6	434.3	505.6	492.6	505.9	483.6	444.9	577.6	386.2	344.2
Imports, cif	622.8	747.9	794.3	745.2	796.9	745.5	583.0	489.0	653.5	509.2	554.7
BALANCE OF PAYMENTS					*(Millions of current US dollars)*						
Exports of Goods & Services	..	..	..	196	206	228	243	288	387	334	388
Merchandise, fob	..	..	..	145	147	166	200	240	320	289	351
Nonfactor Services	..	..	..	49	55	58	37	40	62	41	35
Factor Services	..	..	..	2	3	4	6	7	5	4	2
Imports of Goods & Services	..	..	..	213	261	249	287	371	502	414	456
Merchandise, fob	..	..	..	142	178	168	178	238	332	262	312
Nonfactor Services	..	..	..	64	71	72	95	107	148	132	135
Factor Services	..	..	..	7	12	9	14	25	22	21	9
Long-Term Interest	..	..	..	2	2	2	3	4	4	5	5
Current Transfers, net	..	..	..	-25	-24	-30	-23	-9	-25	-18	-19
Workers' Remittances	..	..	..	0	0	0	0	0	0	0	0
Total to be Financed	..	..	..	-42	-79	-51	-67	-92	-140	-98	-87
Official Capital Grants	..	..	..	52	52	86	56	52	84	70	71
Current Account Balance	..	..	..	10	-27	35	-11	-40	-56	-28	-16
Long-Term Capital, net	..	..	..	67	59	77	83	77	113	84	94
Direct Investment	..	..	..	10	-2	12	11	14	5	1	-3
Long-Term Loans	..	..	..	5	9	11	21	15	32	18	35
Disbursements	..	..	..	11	14	16	30	22	39	26	44
Repayments	..	..	..	5	6	6	9	7	8	8	8
Other Long-Term Capital	..	..	..	0	1	-31	-4	-5	-8	-5	-10
Other Capital, net	..	..	..	-62	-31	-109	-61	-79	-83	-55	-94
Change in Reserves	..	..	..	-15	-1	-3	-12	42	26	-1	16
Memo Item:					*(Malagasy Francs per US dollar)*						
Conversion Factor (Annual Avg)	246.900	246.900	259.700	277.700	277.100	252.200	222.700	240.500	214.300	239.000	245.700
EXTERNAL DEBT, ETC.					*(Millions of US dollars, outstanding at end of year)*						
Public/Publicly Guar. Long-Term	..	..	..	89	103	90	118	139	167	183	225
Official Creditors	..	..	..	85	97	78	106	128	158	175	194
IBRD and IDA	..	..	..	12	17	23	39	51	78	94	102
Private Creditors	..	..	..	5	6	12	12	11	9	8	30
Private Non-guaranteed Long-Term	..	..	..	..	..	..	..	..	..	..	..
Use of Fund Credit	..	..	..	0	0	0	0	10	17	22	25
Short-Term Debt	..	..	..	..	..	..	..	..	..	..	46
Memo Items:					*(Millions of US dollars)*						
Int'l Reserves Excluding Gold	42.9	30.6	19.4	37.1	46.4	52.2	67.9	49.4	35.6	42.2	68.9
Gold Holdings (at market price)	..	..	..	..	..	..	..	..	..	..	..
SOCIAL INDICATORS											
Total Fertility Rate	6.6	6.6	6.6	6.6	6.6	6.6	6.6	6.5	6.5	6.5	6.5
Crude Birth Rate	47.0	46.8	46.6	46.4	46.2	46.0	45.9	45.8	45.8	45.7	45.6
Infant Mortality Rate	195.0	191.0	187.0	183.0	179.0	175.0	170.8	166.6	162.4	158.2	154.0
Life Expectancy at Birth	43.8	44.3	44.9	45.4	45.9	46.5	47.1	47.7	48.3	48.9	49.5
Food Production, p.c. ('79-81=100)	107.2	107.5	111.1	112.8	110.5	109.4	106.9	117.4	112.2	111.8	104.1
Labor Force, Agriculture (%)	84.5	84.3	84.0	83.7	83.4	83.1	82.9	82.6	82.3	82.0	81.7
Labor Force, Female (%)	41.6	41.6	41.6	41.6	41.6	41.5	41.5	41.5	41.5	41.4	41.4
School Enroll. Ratio, primary	..	..	..	90.0	..	..	..	..	95.0	..	..
School Enroll. Ratio, secondary	..	..	..	12.0	..	..	..	..	12.0	..	..

1978	1979	1980	1981	1982	1983	1984	1985	1986	1987 estimate	NOTES	MADAGASCAR
											FOREIGN TRADE (CUSTOMS BASIS)
				(Millions of current US dollars)							
399.3	407.5	402.2	324.4	329.5	310.0	340.0	287.0	320.0	310.0	..	Value of Exports, fob
362.7	358.3	354.7	274.2	280.1	266.0	300.0	246.0	285.0	259.0	..	Nonfuel Primary Products
9.6	19.1	23.1	25.5	25.4	21.0	8.0	11.0	7.0	17.0	..	Fuels
27.0	30.0	24.4	24.6	24.0	23.0	32.0	30.0	28.0	34.0	..	Manufactures
460.1	698.4	676.5	473.0	439.0	411.0	412.0	465.0	374.0	386.0	..	Value of Imports, cif
90.3	130.7	83.8	82.8	84.0	95.0	63.4	71.0	66.4	48.9	..	Nonfuel Primary Products
63.9	112.1	99.0	50.6	106.4	79.0	114.6	92.0	83.6	77.6	..	Fuels
305.9	455.6	493.7	339.7	248.7	237.0	234.0	302.0	224.0	259.4	..	Manufactures
				(Index 1980 = 100)							
124.8	119.5	100.0	87.4	94.2	95.5	100.4	103.5	138.5	95.8	..	Terms of Trade
89.5	101.2	100.0	89.0	90.4	88.7	93.0	94.2	107.7	89.3	..	Export Prices, fob
93.5	106.1	100.0	86.5	89.2	88.3	93.5	95.2	112.0	90.1	..	Nonfuel Primary Products
42.3	61.0	100.0	112.5	101.6	92.5	90.2	87.5	44.9	56.9	..	Fuels
76.1	90.0	100.0	99.1	94.8	90.7	89.3	89.0	103.2	113.0	..	Manufactures
71.7	84.7	100.0	101.8	96.0	92.9	92.7	90.9	77.8	93.2	..	Import Prices, cif
				(Millions of 1980 US dollars)							Trade at Constant 1980 Prices
446.1	402.6	402.2	364.6	364.5	349.4	365.4	304.8	297.2	347.3	..	Exports, fob
641.6	824.4	676.5	464.7	457.3	442.2	444.5	511.4	481.0	414.0	..	Imports, cif
				(Millions of current US dollars)							**BALANCE OF PAYMENTS**
444	490	518	398	382	361	393	336	410	417	..	Exports of Goods & Services
405	414	437	332	327	313	335	270	329	311	..	Merchandise, fob
35	74	79	62	49	45	54	61	76	97	..	Nonfactor Services
4	2	3	4	6	3	5	5	5	9	..	Factor Services
585	959	1,133	878	756	654	660	622	697	712	..	Imports of Goods & Services
404	662	764	546	464	374	353	326	331	315	..	Merchandise, fob
168	265	320	239	192	171	156	167	206	229	..	Nonfactor Services
13	31	48	93	100	109	151	129	160	168	..	Factor Services
8	16	27	32	45	28	32	53	60	83	..	Long-Term Interest
-15	-17	-21	-12	-13	-10	2	37	48	54	..	Current Transfers, net
0	0	0	4	4	4	8	..	..	..	..	Workers' Remittances
-156	-485	-635	-492	-387	-303	-265	-249	-239	-241	K	Total to be Financed
76	59	67	82	88	77	82	59	104	106	K	Official Capital Grants
-80	-427	-568	-410	-299	-226	-183	-190	-135	-135	..	Current Account Balance
106	284	442	363	209	258	253	204	322	317	..	Long-Term Capital, net
-4	-7	..	..	..	..	..	0	0	0	..	Direct Investment
57	314	361	251	213	188	101	118	148	165	..	Long-Term Loans
68	331	392	285	246	207	113	167	199	229	..	Disbursements
11	17	31	34	33	19	13	49	52	64	..	Repayments
-23	-82	14	30	-91	-7	70	28	71	46	..	Other Long-Term Capital
-23	12	209	15	44	-23	-39	-30	-122	-166	..	Other Capital, net
-3	131	-83	32	46	-9	-31	16	-66	-16	..	Change in Reserves
				(Malagasy Francs per US dollar)							Memo Item:
225.600	212.700	211.300	271.700	349.700	430.500	576.600	662.500	676.300	1,069.200	..	Conversion Factor (Annual Avg)
				(Millions of US dollars, outstanding at end of year)							**EXTERNAL DEBT, ETC.**
297	611	956	1,433	1,679	1,787	1,876	2,182	2,673	3,114	..	Public/Publicly Guar. Long-Term
242	380	610	941	1,232	1,375	1,515	1,849	2,340	2,839	..	Official Creditors
111	126	152	185	218	253	276	343	454	586	..	IBRD and IDA
55	231	346	492	447	412	360	333	333	275	..	Private Creditors
..	..	..	..	..	..	..	..	..	..	..	Private Non-guaranteed Long-Term
24	19	55	87	140	139	148	162	184	144	..	Use of Fund Credit
45	163	244	90	101	182	150	120	155	119	..	Short-Term Debt
				(Millions of US dollars)							Memo Items:
59.2	5.0	9.1	26.5	20.0	29.2	58.9	48.4	114.5	185.2	..	Int'l Reserves Excluding Gold
..	..	..	..	..	..	..	..	..	..	..	Gold Holdings (at market price)
											SOCIAL INDICATORS
6.5	6.5	6.5	6.5	6.5	6.4	6.4	6.4	6.4	6.4	..	Total Fertility Rate
45.6	45.7	45.7	45.8	45.8	45.8	45.9	45.9	45.9	45.9	..	Crude Birth Rate
151.2	148.4	145.6	142.8	140.0	132.5	125.0	117.5	109.9	102.4	..	Infant Mortality Rate
49.9	50.3	50.7	51.1	51.5	51.9	52.3	52.7	53.1	53.5	..	Life Expectancy at Birth
100.2	96.6	102.5	100.9	98.5	100.9	99.6	98.8	95.9	97.1	..	Food Production, p.c. ('79-81 = 100)
81.5	81.2	80.9	..	..	..	..	..	..	..	..	Labor Force, Agriculture (%)
41.4	41.3	41.3	41.1	40.9	40.7	40.5	40.4	40.1	39.9	..	Labor Force, Female (%)
100.0	..	..	..	136.0	128.0	121.0	..	..	..	..	School Enroll. Ratio, primary
..	..	..	..	..	37.0	36.0	..	..	..	..	School Enroll. Ratio, secondary

MALAWI	1967	1968	1969	1970	1971	1972	1973	1974	1975	1976	1977
CURRENT GNP PER CAPITA (US $)	70	60	60	60	80	80	90	110	120	120	130
POPULATION (thousands)	4,172	4,280	4,395	4,518	4,649	4,789	4,936	5,089	5,244	5,403	5,567

ORIGIN AND USE OF RESOURCES
(Millions of current Malawi Kwacha)

	1967	1968	1969	1970	1971	1972	1973	1974	1975	1976	1977
Gross National Product (GNP)	187.6	197.3	215.7	236.1	300.3	321.8	364.8	473.9	537.7	594.0	705.0
Net Factor Income from Abroad	-7.8	-7.0	-5.8	-6.0	-3.3	-3.7	0.8	12.4	8.0	-18.0	-23.0
Gross Domestic Product (GDP)	195.4	204.3	221.5	242.1	303.6	325.5	364.0	461.5	529.7	612.0	728.0
Indirect Taxes, net	10.2	10.4	12.8	16.5	21.4	22.2	23.2	28.2	35.0	33.7	44.1
GDP at factor cost	185.2	193.9	208.7	225.6	282.2	303.3	340.8	433.3	494.7	578.3	683.9
Agriculture	86.0	86.6	92.4	99.2	125.1	138.4	141.7	178.4	193.7	233.1	298.0
Industry	28.8	32.1	34.8	39.4	44.6	52.9	57.3	73.9	91.3	103.1	113.5
Manufacturing	..	..	..	..	..	..	..	..	..	..	..
Services, etc.	70.4	75.2	81.5	87.0	112.5	112.0	141.8	181.0	209.7	242.1	272.4
Resource Balance	-21.2	-31.3	-35.7	-36.1	-36.5	-47.7	-36.2	-50.5	-88.8	-51.6	-33.7
Exports of Goods & NFServices	47.0	48.7	52.0	58.7	71.2	75.8	100.6	129.3	154.3	186.3	218.4
Imports of Goods & NFServices	68.2	80.0	87.7	94.8	107.7	123.5	136.8	179.8	243.1	237.9	252.1
Domestic Absorption	216.6	235.6	257.2	278.2	340.1	373.2	400.2	512.0	618.5	663.6	761.7
Private Consumption, etc.	156.9	167.2	179.9	176.3	236.9	248.6	270.0	320.2	365.2	416.6	483.4
General Gov't Consumption	33.3	35.5	36.6	39.6	45.0	45.2	48.7	65.7	74.7	86.3	98.6
Gross Domestic Investment	26.4	32.9	40.7	62.3	58.2	79.4	81.5	126.1	178.6	160.7	179.7
Fixed Investment	..	..	..	..	..	..	74.3	87.3	131.8	135.3	161.6
Memo Items:											
Gross Domestic Saving	5.2	1.6	5.0	26.2	21.7	31.7	45.3	75.6	89.8	109.1	146.0
Gross National Saving	-2.4	-5.2	-1.7	18.5	18.7	28.4	47.8	89.3	100.0	91.6	122.5

(Millions of 1980 Malawi Kwacha)

	1967	1968	1969	1970	1971	1972	1973	1974	1975	1976	1977
Gross National Product	502.0	495.5	530.5	533.6	633.4	669.9	701.0	775.4	810.7	805.5	839.4
GDP at factor cost	496.1	487.0	512.1	508.4	590.7	626.1	647.8	697.8	736.0	781.7	812.6
Agriculture	217.3	203.5	212.0	209.2	244.5	267.9	251.1	256.3	258.6	289.3	321.7
Industry	78.8	82.6	87.6	91.0	95.7	112.6	111.2	119.8	141.0	136.4	143.4
Manufacturing	..	..	..	..	..	..	..	..	..	..	..
Services, etc.	200.0	200.9	212.5	208.2	250.5	245.6	285.5	321.7	336.4	356.0	347.5
Resource Balance	-92.6	-144.1	-168.0	-195.4	-198.5	-210.9	-173.1	-169.2	-206.1	-134.2	-140.8
Exports of Goods & NFServices	128.4	135.8	135.6	134.6	147.6	164.2	187.7	183.5	186.7	201.6	179.4
Imports of Goods & NFServices	220.9	279.9	303.6	330.0	346.1	375.1	360.8	352.7	392.8	335.8	320.1
Domestic Absorption	619.4	661.4	716.5	747.1	841.1	890.6	872.3	920.0	1,002.7	968.8	1,014.6
Private Consumption, etc.	439.8	452.0	480.0	434.5	553.6	548.9	558.2	559.4	594.8	617.7	661.2
General Gov't Consumption	92.9	92.8	93.6	92.9	97.9	97.4	96.0	109.6	115.1	120.9	121.8
Gross Domestic Investment	86.7	116.5	142.9	219.7	189.6	244.3	218.1	251.0	292.8	230.2	231.5
Fixed Investment	..	..	..	..	..	..	..	..	..	..	..
Memo Items:											
Capacity to Import	152.3	170.4	180.0	204.3	228.8	230.2	265.3	253.6	249.3	262.9	277.3
Terms of Trade Adjustment	23.9	34.6	44.4	69.7	81.2	66.0	77.7	70.2	62.6	61.3	98.0
Gross Domestic Income	550.7	551.9	592.9	621.5	723.8	745.8	776.8	820.9	859.3	896.0	971.8
Gross National Income	525.9	530.1	574.9	603.3	714.7	735.9	778.7	845.6	873.3	866.8	937.4

DOMESTIC PRICES (DEFLATORS)
(Index 1980 = 100)

	1967	1968	1969	1970	1971	1972	1973	1974	1975	1976	1977
Overall (GDP)	37.1	39.5	40.4	43.9	47.2	47.9	52.1	61.5	66.5	73.3	83.3
Domestic Absorption	35.0	35.6	35.9	37.2	40.4	41.9	45.9	55.7	61.7	68.5	75.1
Agriculture	39.6	42.6	43.6	47.4	51.2	51.7	56.4	69.6	74.9	80.6	92.6
Industry	36.5	38.8	39.7	43.3	46.6	47.0	51.5	61.7	64.7	75.6	79.2
Manufacturing	..	..	..	..	..	..	..	..	..	..	..

MANUFACTURING ACTIVITY

	1967	1968	1969	1970	1971	1972	1973	1974	1975	1976	1977
Employment (1980=100)	37.9	42.6	45.6	47.4	52.5	61.5	64.2	66.0	71.0	69.0	78.8
Real Earnings per Empl. (1980=100)	109.5	102.4	104.7	113.8	104.2	100.8	99.6	102.7	97.3	..	..
Real Output per Empl. (1980=100)	92.0	103.4	104.9	120.7	123.3	120.6	125.9	121.1	138.3	..	..
Earnings as % of Value Added	48.2	35.9	36.6	36.2	34.6	36.9	40.0	39.6	36.9	..	..

MONETARY HOLDINGS
(Millions of current Malawi Kwacha)

	1967	1968	1969	1970	1971	1972	1973	1974	1975	1976	1977
Money Supply, Broadly Defined	38.7	42.4	48.6	56.0	66.9	74.3	98.8	133.4	139.1	137.4	178.2
Money as Means of Payment	24.5	25.1	28.5	32.7	38.8	40.6	55.1	73.4	73.6	72.8	100.1
Currency Ouside Banks	11.8	11.6	12.5	13.3	14.8	17.3	21.3	28.3	27.8	23.1	24.6
Demand Deposits	12.7	13.5	15.9	19.4	24.0	23.3	33.8	45.1	45.8	49.7	75.5
Quasi-Monetary Liabilities	14.2	17.3	20.1	23.3	28.1	33.7	43.7	60.0	65.5	64.6	78.2

GOVERNMENT DEFICIT (-) OR SURPLUS
(Millions of current Malawi Kwacha)

	1967	1968	1969	1970	1971	1972	1973	1974	1975	1976	1977
GOVERNMENT DEFICIT (-) OR SURPLUS	..	..	..	..	-24.9	-20.0	-20.6	-29.5	-48.5	-37.6	-45.2
Current Revenue	..	..	..	..	51.0	55.0	62.2	72.6	93.4	97.4	125.2
Current Expenditure	..	..	..	..	52.0	56.8	57.9	67.4	79.0	84.5	97.7
Current Budget Balance	..	..	..	..	-1.0	-1.8	4.3	5.2	14.4	12.9	27.5
Capital Receipts	..	..	..	..	0.2	0.3	0.3	0.3	0.3	0.3	0.3
Capital Payments	..	..	..	..	24.1	18.5	25.2	35.0	63.3	50.8	73.0

1978	1979	1980	1981	1982	1983	1984	1985	1986	1987 estimate	NOTES	MALAWI
160	170	180	180	190	190	180	170	160	170	..	CURRENT GNP PER CAPITA (US $)
5,736	5,910	6,091	6,280	6,477	6,685	6,906	7,141	7,380	7,629	..	POPULATION (thousands)
				(Millions of current Malawi Kwacha)							ORIGIN AND USE OF RESOURCES
796.7	797.1	933.6	1,029.5	1,168.4	1,359.4	1,628.9	1,852.2	2,117.8	2,594.4	..	Gross National Product (GNP)
-4.0	-38.6	-81.3	-74.3	-75.6	-75.5	-78.8	-90.9	-112.9	-125.7	..	Net Factor Income from Abroad
800.7	835.7	1,014.9	1,103.8	1,244.0	1,434.9	1,707.7	1,943.1	2,230.7	2,720.1	..	Gross Domestic Product (GDP)
58.2	77.9	103.4	107.8	116.1	139.6	178.1	216.0	216.4	268.3	..	Indirect Taxes, net
742.5	757.8	911.5	996.0	1,127.9	1,295.3	1,529.6	1,727.1	2,014.3	2,451.8	..	GDP at factor cost
294.9	299.6	318.8	345.2	406.6	469.6	581.3	639.0	745.3	907.2	..	Agriculture
143.5	142.4	162.0	186.1	203.4	235.5	275.3	328.1	402.8	441.3	..	Industry
..	..	..	..	..	..	..	..	..	..	..	Manufacturing
304.1	315.8	430.7	464.7	517.9	590.2	673.0	760.0	866.2	1,103.3	..	Services, etc.
-143.5	-170.7	-140.4	-64.2	-79.1	-108.9	33.1	-93.2	-47.1	-66.7	..	Resource Balance
185.7	209.7	249.7	284.4	280.2	298.2	484.4	475.0	504.7	658.2	..	Exports of Goods & NFServices
329.2	380.4	390.1	348.6	359.3	407.1	451.3	568.2	551.8	724.9	..	Imports of Goods & NFServices
944.2	1,006.4	1,155.7	1,168.0	1,323.1	1,543.6	1,674.5	2,036.3	2,277.9	2,786.7	..	Domestic Absorption
502.2	580.8	714.7	773.5	838.2	980.1	1,186.6	1,330.6	1,595.9	1,916.0	..	Private Consumption, etc.
134.2	164.2	192.3	199.2	218.3	235.9	268.0	344.0	433.8	488.6	..	General Gov't Consumption
307.8	261.4	248.7	195.3	266.6	327.6	219.9	361.7	248.2	382.1	..	Gross Domestic Investment
247.1	231.9	223.1	167.8	181.7	197.3	222.7	259.5	242.9	332.8	..	Fixed Investment
											Memo Items:
164.3	90.7	107.9	131.1	187.5	218.9	253.1	268.5	201.0	315.5	..	Gross Domestic Saving
166.7	57.9	37.4	67.5	124.4	153.1	196.5	195.5	114.7	268.0	..	Gross National Saving
				(Millions of 1980 Malawi Kwacha)							
956.6	970.3	933.6	883.5	913.9	960.1	1,018.7	1,011.8	1,034.9	1,242.5	..	Gross National Product
881.8	911.9	911.5	864.1	886.9	920.4	959.8	1,001.7	1,029.0	1,037.6	..	GDP at factor cost
331.2	341.5	318.8	292.7	311.8	310.2	344.2	345.9	349.8	356.3	..	Agriculture
160.9	161.9	162.0	159.5	159.8	164.7	164.0	175.8	188.1	174.3	..	Industry
..	..	..	..	..	..	..	..	..	..	..	Manufacturing
389.8	408.5	430.7	411.9	415.3	445.6	451.6	480.1	491.1	507.0	..	Services, etc.
-246.5	-154.7	-140.4	-97.8	-104.4	-103.4	-21.8	-87.8	-17.6	-20.6	..	Resource Balance
169.4	187.7	249.7	205.0	184.6	190.7	253.1	239.4	228.9	242.0	..	Exports of Goods & NFServices
415.9	342.3	390.1	302.8	288.9	294.1	274.9	327.2	246.5	262.6	..	Imports of Goods & NFServices
1,208.9	1,170.7	1,155.3	1,059.5	1,087.7	1,126.3	1,099.3	1,215.0	1,154.8	1,162.2	..	Domestic Absorption
648.8	698.4	714.3	697.1	682.1	698.5	757.5	795.8	808.9	804.9	..	Private Consumption, etc.
165.4	186.0	192.3	186.5	189.7	189.5	204.5	226.3	248.3	243.8	..	General Gov't Consumption
394.6	286.3	248.7	175.9	215.9	238.3	137.3	192.9	97.7	113.5	..	Gross Domestic Investment
316.8	..	223.1	151.2	147.2	143.6	139.1	138.5	95.7	106.0	..	Fixed Investment
											Memo Items:
234.6	188.7	249.7	247.0	225.3	215.4	295.1	273.5	225.4	238.5	..	Capacity to Import
65.2	1.1	0.0	42.0	40.7	24.7	42.0	34.1	-3.5	-3.6	..	Terms of Trade Adjustment
1,027.6	1,017.0	1,014.9	1,003.8	1,024.1	1,047.6	1,119.4	1,161.4	1,133.8	1,138.0	..	Gross Domestic Income
1,021.8	971.3	933.6	925.5	954.6	984.9	1,060.6	1,046.0	1,031.4	1,238.9	..	Gross National Income
				(Index 1980 = 100)							DOMESTIC PRICES (DEFLATORS)
83.2	82.3	100.0	114.8	126.5	140.3	158.5	172.4	196.2	238.3	..	Overall (GDP)
78.1	86.0	100.0	110.2	121.6	137.1	152.3	167.6	197.3	239.8	..	Domestic Absorption
89.1	87.7	100.0	117.9	130.4	151.4	168.9	184.8	213.1	254.6	..	Agriculture
89.2	88.0	100.0	116.7	127.3	143.0	167.8	186.6	214.1	253.1	..	Industry
..	..	..	..	..	..	..	..	..	..	..	Manufacturing
											MANUFACTURING ACTIVITY
82.7	91.1	100.0	96.0	93.9	91.9	..	..	..	..	G	Employment (1980=100)
..	111.9	100.0	110.0	109.1	105.0	..	..	..	..	G	Real Earnings per Empl. (1980=100)
..	106.3	100.0	110.6	102.5	92.4	..	..	..	..	G	Real Output per Empl. (1980=100)
..	43.3	33.1	34.0	41.7	37.6	..	..	..	..	..	Earnings as % of Value Added
				(Millions of current Malawi Kwacha)							MONETARY HOLDINGS
190.2	192.9	219.0	273.2	317.4	348.0	453.6	407.7	580.7	791.2	D	Money Supply, Broadly Defined
93.8	90.6	97.2	114.8	130.8	127.8	154.3	166.9	220.8	298.0	..	Money as Means of Payment
29.8	32.3	35.3	39.4	49.5	50.0	56.8	66.0	79.3	107.6	..	Currency Ouside Banks
64.0	58.3	61.9	75.4	81.3	77.7	97.4	100.9	141.5	190.4	..	Demand Deposits
96.4	102.3	121.8	158.5	186.6	220.3	299.4	240.8	359.9	493.2	..	Quasi-Monetary Liabilities
				(Millions of current Malawi Kwacha)							
-74.3	-75.5	-160.3	-137.7	-95.0	-101.8	-88.3	-162.6	..	..	C	GOVERNMENT DEFICIT (-) OR SURPLUS
169.0	211.2	235.5	257.2	271.6	310.7	380.4	460.9	..	..	..	Current Revenue
139.1	148.9	180.9	263.9	256.2	293.4	320.2	422.5	..	..	..	Current Expenditure
29.9	62.3	54.6	-6.7	15.4	17.3	60.2	38.4	..	..	..	Current Budget Balance
0.3	0.4	0.6	0.4	0.1	0.2	0.4	0.4	..	..	..	Capital Receipts
104.6	138.2	215.5	131.4	110.6	119.3	148.9	201.4	..	..	..	Capital Payments

MALAWI	1967	1968	1969	1970	1971	1972	1973	1974	1975	1976	1977
FOREIGN TRADE (CUSTOMS BASIS)				*(Millions of current US dollars)*							
Value of Exports, fob	56.5	48.1	52.8	59.2	72.1	80.6	99.3	120.7	139.1	165.6	200.0
Nonfuel Primary Products	51.6	43.3	47.4	50.8	65.0	72.3	89.7	108.0	121.8	153.8	185.5
Fuels	1.3	1.7	1.6	1.6	1.5	2.2	2.2	1.0	0.1	0.1	0.1
Manufactures	3.5	3.1	3.8	6.8	5.6	6.1	7.4	11.8	17.3	11.6	14.4
Value of Imports, cif	70.3	69.7	73.7	85.6	109.0	130.0	142.2	187.6	250.5	205.6	232.4
Nonfuel Primary Products	13.5	11.1	11.1	17.4	15.9	18.6	24.4	28.4	29.6	26.1	24.4
Fuels	3.7	4.1	4.3	4.7	10.2	11.6	12.7	19.5	24.9	27.8	30.2
Manufactures	53.1	54.5	58.4	63.5	83.0	99.8	105.0	139.7	196.0	151.7	177.8
Terms of Trade	127.8	123.3	122.4	125.6	140.0	128.7	124.2	112.5	118.1	109.8	118.4
				(Index 1980 = 100)							
Export Prices, fob	30.6	28.3	29.0	31.7	34.0	34.9	42.9	62.6	70.5	65.2	76.7
Nonfuel Primary Products	36.3	35.8	35.7	38.6	38.4	40.2	46.8	62.5	71.3	65.2	77.1
Fuels	4.3	4.3	4.3	4.3	5.6	6.2	8.9	36.7	35.7	38.4	42.0
Manufactures	30.1	31.7	32.4	37.5	36.5	38.5	49.5	67.2	66.1	66.7	71.7
Import Prices, cif	23.9	22.9	23.7	25.3	24.3	27.1	34.5	55.7	59.7	59.4	64.8
Trade at Constant 1980 Prices				*(Millions of 1980 US dollars)*							
Exports, fob	184.8	169.8	181.7	186.4	212.1	230.8	231.5	192.9	197.3	253.8	260.9
Imports, cif	294.1	303.8	310.5	339.1	449.0	479.1	411.6	337.1	419.6	346.0	358.8
BALANCE OF PAYMENTS				*(Millions of current US dollars)*							
Exports of Goods & Services	..	..	..	82.3	100.8	111.8	145.7	185.1	212.6	196.2	213.2
Merchandise, fob	..	..	..	58.9	71.7	78.6	97.9	119.3	138.7	165.3	199.8
Nonfactor Services	..	..	..	11.5	13.9	16.0	22.3	28.3	35.3	23.4	13.4
Factor Services	..	..	..	11.9	15.2	17.3	25.5	37.5	38.6	7.4	0.0
Imports of Goods & Services	..	..	..	126.2	142.0	169.6	184.9	228.9	301.6	266.0	300.5
Merchandise, fob	..	..	..	82.8	93.1	111.5	123.4	166.1	225.0	182.7	183.0
Nonfactor Services	..	..	..	31.0	36.5	42.6	45.4	46.3	59.3	66.2	117.5
Factor Services	..	..	..	12.4	12.4	15.5	16.1	16.5	17.4	17.2	0.0
Long-Term Interest	..	..	..	3.5	3.8	4.2	6.0	7.7	8.5	8.8	7.2
Current Transfers, net	..	..	..	..	..	..	..	..	..	..	..
Workers' Remittances	..	..	..	0.0	0.0	0.0	0.0	0.0	0.0	0.0	0.0
Total to be Financed	..	..	..	..	..	..	..	..	..	..	..
Official Capital Grants	..	..	..	..	..	..	..	..	..	..	..
Current Account Balance	..	..	..	-34.8	-33.1	-49.1	-27.8	-35.7	-79.7	-42.7	-61.8
Long-Term Capital, net	..	..	..	46.7	39.8	46.3	57.9	76.0	64.0	73.3	85.6
Direct Investment	..	..	..	8.6	9.6	10.1	7.8	22.7	8.6	9.7	5.5
Long-Term Loans	..	..	..	36.7	24.8	30.3	27.8	25.4	52.0	48.8	85.3
Disbursements	..	..	..	39.6	28.4	34.7	32.7	32.1	60.6	57.8	99.8
Repayments	..	..	..	2.9	3.6	4.4	4.9	6.7	8.6	9.0	14.5
Other Long-Term Capital	..	..	..	1.4	5.4	5.9	22.4	27.9	3.4	14.8	-5.2
Other Capital, net	..	..	..	-5.6	-8.1	5.3	3.8	-25.7	-2.7	-66.7	27.6
Change in Reserves	..	..	..	-6.3	1.4	-2.5	-33.9	-14.6	18.4	36.1	-51.4
Memo Item:				*(Malawi Kwacha per US dollar)*							
Conversion Factor (Annual Avg)	0.720	0.830	0.830	0.830	0.830	0.800	0.820	0.840	0.860	0.910	0.900
EXTERNAL DEBT, ETC.				*(Millions of US dollars, outstanding at end of year)*							
Public/Publicly Guar. Long-Term	..	..	..	122.4	140.8	164.5	201.8	229.6	257.1	294.8	368.8
Official Creditors	..	..	..	102.8	121.8	143.2	176.5	203.1	224.9	244.0	284.1
IBRD and IDA	..	..	..	16.9	24.2	34.2	42.8	50.2	61.7	73.7	86.9
Private Creditors	..	..	..	19.6	19.0	21.3	25.3	26.5	32.2	50.7	84.7
Private Non-guaranteed Long-Term	..	..	..	..	..	..	..	..	..	..	..
Use of Fund Credit	..	..	..	0.0	0.0	0.0	0.0	0.0	2.8	8.6	15.5
Short-Term Debt	..	..	..	..	..	..	..	..	..	..	69.0
Memo Items:				*(Thousands of US dollars)*							
Int'l Reserves Excluding Gold	22,500.0	22,510.0	21,010.0	29,200.0	31,920.0	36,237.0	66,643.0	81,791.0	61,458.0	26,218.0	87,491.0
Gold Holdings (at market price)	..	..	..	..	..	..	..	..	..	..	1,059.0
SOCIAL INDICATORS											
Total Fertility Rate	7.8	7.8	7.8	7.8	7.8	7.8	7.7	7.7	7.7	7.6	7.6
Crude Birth Rate	56.0	56.0	56.0	56.0	56.0	56.0	55.9	55.7	55.6	55.4	55.3
Infant Mortality Rate	197.0	195.8	194.6	193.4	192.2	191.0	188.2	185.4	182.6	179.8	177.0
Life Expectancy at Birth	39.5	39.8	40.1	40.4	40.7	41.0	41.3	41.7	42.0	42.3	42.7
Food Production, p.c. ('79-81 = 100)	102.4	93.6	95.1	84.3	103.6	103.6	106.5	102.5	96.9	100.9	105.0
Labor Force, Agriculture (%)	91.4	91.1	90.8	90.5	89.8	89.1	88.4	87.7	87.0	86.3	85.5
Labor Force, Female (%)	45.4	45.3	45.3	45.3	45.2	45.0	44.9	44.8	44.7	44.5	44.4
School Enroll. Ratio, primary	..	..	..	..	..	..	..	..	..	56.0	55.0
School Enroll. Ratio, secondary	..	..	..	..	..	..	..	..	..	4.0	4.0

1978	1979	1980	1981	1982	1983	1984	1985	1986	1987 estimate	NOTES	MALAWI
				(Millions of current US dollars)							**FOREIGN TRADE (CUSTOMS BASIS)**
185.3	222.8	285.2	273.9	246.0	229.0	314.4	250.1	242.7	264.3	..	Value of Exports, fob
171.1	208.1	253.3	242.5	219.2	203.7	279.1	216.2	203.4	221.9	..	Nonfuel Primary Products
0.1	0.2	0.6	0.1	0.2	0.2	0.4	0.4	0.2	0.3	..	Fuels
14.1	14.6	31.3	31.2	26.6	25.1	35.0	33.5	39.1	42.2	..	Manufactures
338.2	397.6	440.2	350.1	311.0	311.0	269.9	286.7	260.4	280.9	..	Value of Imports, cif
23.5	33.6	43.7	46.1	37.5	37.5	28.1	28.4	24.4	24.4	..	Nonfuel Primary Products
40.3	58.0	67.5	59.3	50.9	51.0	41.9	39.8	18.9	25.3	..	Fuels
274.5	306.1	329.1	244.8	222.6	222.6	199.9	218.5	217.1	231.3	..	Manufactures
				(Index 1980 = 100)							
104.9	103.6	100.0	93.9	93.2	95.1	97.4	72.8	75.5	67.0	..	Terms of Trade
75.9	88.1	100.0	96.6	92.6	91.7	91.9	68.4	75.4	73.2	..	Export Prices, fob
76.0	88.0	100.0	96.3	92.4	91.9	92.2	66.0	71.7	68.6	..	Nonfuel Primary Products
42.3	61.0	100.0	112.5	101.6	92.5	90.2	87.5	44.9	56.9	..	Fuels
76.1	90.0	100.0	99.1	94.8	90.7	89.3	89.0	103.2	113.0	..	Manufactures
72.4	85.0	100.0	102.9	99.4	96.4	94.4	94.0	99.9	109.2	..	Import Prices, cif
				(Millions of 1980 US dollars)							Trade at Constant 1980 Prices
244.0	253.0	285.2	283.5	265.6	249.7	342.1	365.5	321.9	361.3	..	Exports, fob
467.4	467.7	440.2	340.2	312.9	322.5	286.1	305.1	260.7	257.3	..	Imports, cif
				(Millions of current US dollars)							**BALANCE OF PAYMENTS**
211.2	258.6	314.7	317.6	274.3	283.5	346.5	281.9	273.7	304.0	..	Exports of Goods & Services
184.5	222.4	280.8	272.5	239.7	246.2	315.7	250.0	248.1	275.8	..	Merchandise, fob
25.6	34.8	31.8	43.7	33.1	36.0	27.1	26.4	22.3	25.4	..	Nonfactor Services
1.1	1.4	2.2	1.3	1.5	1.3	3.7	5.5	3.3	2.8	..	Factor Services
432.7	577.9	642.7	528.8	445.0	466.7	378.7	389.1	363.1	392.7	..	Imports of Goods & Services
263.9	317.6	338.9	299.7	272.0	241.9	162.0	172.0	154.1	176.2	..	Merchandise, fob
119.3	133.2	152.5	97.6	80.1	123.1	157.2	158.7	146.6	155.9	..	Nonfactor Services
49.4	127.2	151.3	131.5	92.9	101.7	59.5	58.4	62.4	60.6	..	Factor Services
16.6	24.3	34.6	49.4	32.7	31.5	30.9	30.2	36.2	26.2	..	Long-Term Interest
..	..	13.3	12.0	11.8	8.3	..	..	..	..	..	Current Transfers, net
0.0	0.0	0.0	0.0	0.0	0.0	0.0	0.0	0.0	..	..	Workers' Remittances
..	..	-314.5	-198.1	-157.9	-174.0	..	..	..	..	..	Total to be Financed
..	..	50.1	46.3	36.3	29.5	..	..	..	..	..	Official Capital Grants
-175.1	-265.7	-264.4	-151.8	-121.6	-144.5	7.9	-72.3	-45.9	-23.7	..	Current Account Balance
140.8	151.3	236.1	98.8	46.7	126.2	61.2	42.5	70.4	96.5	..	Long-Term Capital, net
9.1	-1.2	9.5	1.1	..	2.6	..	0.5	..	..	..	Direct Investment
115.6	103.2	125.1	83.7	43.2	46.9	70.1	16.9	44.7	87.2	..	Long-Term Loans
135.5	122.9	158.4	122.9	73.3	74.6	111.4	68.9	119.6	131.8	..	Disbursements
19.9	19.7	33.3	39.2	30.1	27.7	41.3	52.0	74.9	44.6	..	Repayments
16.0	49.3	51.4	-32.3	-32.8	47.3	-8.9	25.1	25.7	9.3	..	Other Long-Term Capital
20.7	78.0	-0.2	5.6	57.3	-12.4	-2.6	-9.9	-48.9	-27.7	..	Other Capital, net
13.6	36.3	28.5	47.5	17.6	30.6	-66.5	39.7	24.4	-45.1	..	Change in Reserves
				(Malawi Kwacha per US dollar)							Memo Item:
0.840	0.820	0.810	0.900	1.060	1.170	1.410	1.720	1.860	2.210	..	Conversion Factor (Annual Avg)
				(Millions of US dollars, outstanding at end of year)							**EXTERNAL DEBT, ETC.**
503.9	510.2	643.6	674.8	703.4	712.6	730.0	807.7	927.9	1,154.8	..	Public/Publicly Guar. Long-Term
365.8	338.4	450.6	486.8	541.5	576.0	603.4	711.3	864.5	1,106.0	..	Official Creditors
107.4	134.1	156.0	202.9	249.0	277.2	335.8	391.3	506.2	613.6	..	IBRD and IDA
138.1	171.8	193.0	188.0	161.8	136.6	126.6	96.4	63.4	48.8	..	Private Creditors
..	..	..	..	..	..	..	..	..	..	..	Private Non-guaranteed Long-Term
11.9	40.2	61.1	87.6	81.2	102.1	112.6	133.9	124.0	110.3	..	Use of Fund Credit
75.0	115.0	116.1	50.0	85.8	73.0	42.0	80.6	80.1	98.2	..	Short-Term Debt
				(Thousands of US dollars)							Memo Items:
74,796.0	69,508.0	68,388.0	49,055.0	22,661.0	15,401.0	56,600.0	44,949.0	24,604.0	51,831.0	..	Int'l Reserves Excluding Gold
2,176.0	6,573.0	7,567.0	5,103.0	5,865.0	4,897.0	3,958.0	4,198.0	5,018.0	6,214.0	..	Gold Holdings (at market price)
											SOCIAL INDICATORS
7.6	7.6	7.6	7.6	7.6	7.6	7.6	7.6	7.6	7.6	..	Total Fertility Rate
55.0	54.7	54.4	54.1	53.8	53.7	53.7	53.6	53.5	53.4	..	Crude Birth Rate
174.2	171.4	168.6	165.8	163.0	163.2	163.3	163.5	163.7	163.9	..	Infant Mortality Rate
43.0	43.4	43.7	44.1	44.4	44.7	45.1	45.4	45.7	45.8	..	Life Expectancy at Birth
105.5	101.8	98.5	99.7	100.2	94.4	92.4	89.6	88.2	84.4	..	Food Production, p.c. ('79-81 = 100)
84.8	84.0	83.3	..	..	..	..	..	..	..	..	Labor Force, Agriculture (%)
44.2	44.0	43.9	43.6	43.4	43.1	42.9	42.6	42.4	42.1	..	Labor Force, Female (%)
55.0	59.0	61.0	64.0	62.0	60.0	62.0	..	..	..	..	School Enroll. Ratio, primary
4.0	4.0	4.0	4.0	4.0	4.0	4.0	..	..	..	..	School Enroll. Ratio, secondary

MALAYSIA	1967	1968	1969	1970	1971	1972	1973	1974	1975	1976	1977
CURRENT GNP PER CAPITA (US $)	350	370	380	390	410	450	550	700	820	920	1,000
POPULATION (thousands)	10,053	10,318	10,588	10,863	11,144	11,434	11,729	12,021	12,307	12,591	12,875
ORIGIN AND USE OF RESOURCES					*(Millions of current Malaysian Ringgit)*						
Gross National Product (GNP)	10,475	10,926	11,919	12,517	13,186	14,495	18,916	22,892	22,625	28,261	32,529
Net Factor Income from Abroad	-94	-98	-258	-339	-345	-358	-640	-984	-700	-1,074	-1,249
Gross Domestic Product (GDP)	10,569	11,024	12,177	12,856	13,531	14,853	19,556	23,876	23,325	29,335	33,778
Indirect Taxes, net	1,128	1,229	1,258	1,257	1,867	2,115	2,851	3,730	3,343	4,390	5,447
GDP at factor cost	9,441	9,795	10,919	11,599	11,664	12,738	16,705	20,146	19,982	24,945	28,331
Agriculture	2,935	3,040	3,537	3,667	3,504	3,824	5,155	7,061	6,527	7,857	8,682
Industry	2,590	2,575	2,965	3,246	3,571	4,077	5,296	7,418	7,303	9,477	11,236
Manufacturing	1,072	1,135	1,383	1,531	1,720	1,962	2,941	4,024	3,931	5,203	6,212
Services, etc.	5,044	5,409	5,675	5,943	6,456	6,952	9,105	9,397	9,495	12,001	13,860
Resource Balance	249	367	1,149	539	197	-180	1,071	78	123	2,964	2,461
Exports of Goods & NFServices	4,006	4,414	5,303	5,404	5,250	5,129	7,779	11,060	10,187	14,576	16,240
Imports of Goods & NFServices	3,757	4,047	4,154	4,865	5,053	5,309	6,708	10,982	10,064	11,612	13,779
Domestic Absorption	10,320	10,657	11,028	12,317	13,334	15,033	18,485	23,798	23,202	26,371	31,317
Private Consumption, etc.	6,530	6,738	7,162	7,417	8,016	8,696	10,401	12,746	13,172	14,919	16,941
General Gov't Consumption	1,644	1,683	1,762	2,018	2,284	2,882	3,088	3,701	4,130	4,527	5,671
Gross Domestic Investment	2,146	2,236	2,104	2,882	3,034	3,455	4,996	7,351	5,900	6,925	8,705
Fixed Investment	1,830	1,876	1,890	2,430	2,989	3,553	4,669	6,416	6,199	6,868	8,261
Memo Items:											
Gross Domestic Saving	2,395	2,603	3,253	3,421	3,231	3,275	6,067	7,429	6,023	9,889	11,166
Gross National Saving	2,117	2,327	2,787	2,883	2,696	2,743	5,241	6,320	5,208	8,694	9,804
					(Millions of 1980 Malaysian Ringgit)						
Gross National Product	20,522	21,598	23,333	24,152	25,631	28,130	31,067	33,417	34,183	37,897	40,770
GDP at factor cost	19,602	20,777	22,326	23,623	23,971	26,074	29,017	31,030	31,737	35,209	37,473
Agriculture	..	..	..	7,381	7,488	8,059	9,008	9,630	9,338	10,481	10,729
Industry	..	..	..	8,669	9,549	10,345	11,410	11,939	11,979	14,095	15,191
Manufacturing	..	..	..	3,659	4,119	4,538	5,560	6,137	6,318	7,487	8,280
Services, etc.	..	..	..	8,934	9,421	10,531	11,895	13,398	13,933	14,806	16,528
Resource Balance	2,282	3,768	4,321	3,387	3,639	4,270	4,611	2,734	5,062	7,022	5,516
Exports of Goods & NFServices	11,280	13,068	13,771	14,453	14,678	14,975	17,100	19,824	19,230	22,491	23,427
Imports of Goods & NFServices	8,997	9,301	9,450	11,066	11,039	10,705	12,489	17,089	14,168	15,470	17,911
Domestic Absorption	20,379	20,781	20,984	21,598	22,819	24,665	27,702	32,233	30,188	32,360	36,931
Private Consumption, etc.	11,736	11,799	11,791	12,549	13,420	14,177	15,327	16,645	16,545	17,306	19,256
General Gov't Consumption	2,899	2,906	2,983	3,377	3,660	4,278	4,474	5,141	5,490	5,956	6,542
Gross Domestic Investment	4,239	4,421	4,157	5,672	5,739	6,209	7,901	10,447	8,154	9,099	11,133
Fixed Investment	3,627	3,714	3,742	4,794	5,696	6,539	7,618	9,283	8,593	9,023	10,340
Memo Items:											
Capacity to Import	9,594	10,144	12,064	12,292	11,470	10,342	14,483	17,211	14,341	19,418	21,111
Terms of Trade Adjustment	-1,686	-2,924	-1,707	-2,161	-3,209	-4,633	-2,617	-2,613	-4,889	-3,073	-2,317
Gross Domestic Income	19,962	19,967	20,412	22,824	23,249	24,302	29,696	32,354	30,362	36,309	40,131
Gross National Income	19,240	19,501	19,928	21,991	22,423	23,497	28,451	30,804	29,295	34,825	38,453
DOMESTIC PRICES (DEFLATORS)					*(Index 1980 = 100)*						
Overall (GDP)	49.8	50.0	51.1	51.5	51.1	51.3	60.5	68.3	66.2	74.5	79.6
Domestic Absorption	50.6	51.3	52.6	57.0	58.4	60.9	66.7	73.8	76.9	81.5	84.8
Agriculture	..	..	..	49.7	46.8	47.5	57.2	73.3	69.9	75.0	80.9
Industry	..	..	..	37.4	37.4	39.4	46.4	62.1	61.0	67.2	74.0
Manufacturing	..	..	..	41.8	41.8	43.2	52.9	65.6	62.2	69.5	75.0
MANUFACTURING ACTIVITY											
Employment (1980=100)	..	29.2	..	34.2	37.7	44.1	57.9	58.9	61.4	69.4	73.0
Real Earnings per Empl. (1980=100)	..	82.5	..	83.7	83.6	80.1	73.4	76.5	78.3	83.7	88.6
Real Output per Empl. (1980=100)	..	..	..	96.2	92.3	91.9	84.1	87.9	94.2	94.6	95.7
Earnings as % of Value Added	..	31.3	..	28.5	30.3	29.3	25.9	26.9	27.4	27.9	27.1
MONETARY HOLDINGS					*(Millions of current Malaysian Ringgit)*						
Money Supply, Broadly Defined	3,041	3,499	4,180	4,654	5,263	6,564	8,605	13,027	15,199	19,222	22,204
Money as Means of Payment	1,529	1,717	1,913	2,071	2,172	2,715	3,735	4,055	4,349	5,257	6,127
Currency Ouside Banks	772	806	930	1,000	1,061	1,269	1,718	2,030	2,239	2,628	3,112
Demand Deposits	757	911	983	1,071	1,112	1,446	2,017	2,026	2,110	2,629	3,015
Quasi-Monetary Liabilities	1,512	1,782	2,267	2,582	3,091	3,849	4,870	8,972	10,850	13,965	16,077
					(Millions of current Malaysian Ringgit)						
GOVERNMENT DEFICIT (-) OR SURPLUS	..	..	..	..	..	-1,359	-1,094	-1,257	-1,827	-1,976	-2,407
Current Revenue	..	..	..	..	..	2,950	3,378	4,748	5,076	6,125	7,766
Current Expenditure	..	..	..	..	..	3,234	3,399	4,446	4,944	5,984	7,252
Current Budget Balance	..	..	..	..	..	-284	-21	302	132	141	514
Capital Receipts	..	..	..	..	..	..	..	11	13	13	28
Capital Payments	..	..	..	..	..	1,075	1,073	1,570	1,972	2,130	2,949

1978	1979	1980	1981	1982	1983	1984	1985	1986	1987 estimate	NOTES	MALAYSIA
1,140	1,400	1,680	1,890	1,910	1,900	2,040	1,970	1,830	1,810	..	**CURRENT GNP PER CAPITA (US $)**
13,163	13,457	13,763	14,085	14,430	14,805	15,217	15,676	16,110	16,560	..	POPULATION (thousands)
				(Millions of current Malaysian Ringgit)							**ORIGIN AND USE OF RESOURCES**
36,186	44,354	51,390	55,602	59,690	65,154	74,182	71,838	65,851	73,478	..	Gross National Product (GNP)
-1,700	-2,070	-1,918	-2,011	-2,889	-4,411	-5,368	-5,709	-5,293	-5,212	..	Net Factor Income from Abroad
37,886	46,424	53,308	57,613	62,579	69,565	79,550	77,547	71,144	78,690	..	Gross Domestic Product (GDP)
6,099	7,670	9,139	8,834	8,759	10,310	..	..	..	..	..	Indirect Taxes, net
31,787	38,754	44,169	48,779	53,820	59,255	..	..	..	..	B	GDP at factor cost
9,513	10,988	11,680	11,962	12,807	13,555	..	..	..	..	..	Agriculture
13,203	16,736	20,164	21,409	21,851	24,835	..	..	..	..	..	Industry
7,189	8,992	11,002	11,542	11,419	12,935	..	..	..	..	..	Manufacturing
15,170	18,700	21,464	24,242	27,921	31,175	..	..	..	..	..	Services, etc.
2,108	4,120	1,334	-3,563	-5,454	-3,698	1,518	4,028	4,574	10,671	..	Resource Balance
18,585	26,004	30,676	30,154	31,846	36,298	43,171	42,537	40,722	50,517	..	Exports of Goods & NFServices
16,477	21,884	29,342	33,717	37,300	39,996	41,653	38,509	36,148	39,846	..	Imports of Goods & NFServices
35,778	42,304	51,974	61,176	68,033	73,263	78,032	73,519	66,570	68,019	..	Domestic Absorption
19,584	22,406	26,946	30,594	33,226	35,998	39,594	40,360	36,574	37,362	..	Private Consumption, etc.
6,090	6,475	8,811	10,425	11,469	12,156	11,741	11,844	12,127	12,310	..	General Gov't Consumption
10,104	13,423	16,217	20,157	23,338	25,109	26,697	21,315	17,869	18,347	..	Gross Domestic Investment
9,381	12,250	16,597	20,759	22,745	24,534	25,391	23,124	18,865	17,783	f	Fixed Investment
											Memo Items:
12,212	17,543	17,551	16,594	17,884	21,411	28,215	25,343	22,443	29,018	..	Gross Domestic Saving
10,355	15,394	15,539	14,457	14,871	16,919	22,700	19,521	17,101	23,744	..	Gross National Saving
				(Millions of 1980 Malaysian Ringgit)							
43,232	47,231	51,390	55,208	57,732	60,011	64,093	63,099	64,501	68,556	..	Gross National Product
38,414	41,633	44,169	48,190	51,583	53,780	57,612	57,193	60,122	63,979	B	GDP at factor cost
10,905	11,532	11,680	12,247	13,040	12,956	13,324	13,657	14,202	15,273	..	Agriculture
16,707	18,899	20,164	20,879	22,276	24,594	27,411	26,490	27,745	29,635	..	Industry
9,048	10,073	11,002	11,522	12,167	13,125	14,739	14,175	15,242	17,185	..	Manufacturing
17,708	19,182	21,464	23,834	25,039	26,665	28,547	28,367	27,483	28,324	..	Services, etc.
4,988	5,385	1,334	-561	-1,575	-594	2,120	6,386	14,991	18,494	..	Resource Balance
25,205	29,733	30,676	30,421	33,669	37,823	43,036	43,229	50,839	56,562	..	Exports of Goods & NFServices
20,217	24,348	29,342	30,982	35,244	38,417	40,916	36,843	35,847	38,068	..	Imports of Goods & NFServices
40,332	44,227	51,974	57,521	61,929	64,809	67,161	62,128	54,438	54,739	..	Domestic Absorption
21,370	23,537	26,946	28,530	29,333	30,056	31,684	31,662	27,294	27,075	..	Private Consumption, etc.
6,924	7,043	8,811	9,987	10,860	11,357	10,801	10,706	10,842	11,001	..	General Gov't Consumption
12,038	13,647	16,217	19,005	21,737	23,396	24,677	19,760	16,303	16,663	..	Gross Domestic Investment
11,176	13,308	16,597	19,598	21,167	22,866	23,543	21,311	17,395	16,235	f	Fixed Investment
											Memo Items:
22,803	28,932	30,676	27,708	30,090	34,865	42,407	40,696	40,383	48,263	..	Capacity to Import
-2,402	-801	0	-2,713	-3,579	-2,958	-629	-2,533	-10,455	-8,299	..	Terms of Trade Adjustment
42,918	48,811	53,308	54,247	56,776	61,257	68,653	65,982	58,974	64,933	..	Gross Domestic Income
40,831	46,429	51,390	52,495	54,153	57,053	63,463	60,567	54,045	60,258	..	Gross National Income
				(Index 1980 = 100)							**DOMESTIC PRICES (DEFLATORS)**
83.6	93.6	100.0	101.1	103.7	108.3	114.8	113.2	102.5	107.5	..	Overall (GDP)
88.7	95.7	100.0	106.4	109.9	113.0	116.2	118.3	122.3	124.3	..	Domestic Absorption
87.2	95.3	100.0	97.7	98.2	104.6	..	..	..	..	..	Agriculture
79.0	88.6	100.0	102.5	98.1	101.0	..	..	..	..	..	Industry
79.5	89.3	100.0	100.2	93.8	98.6	..	..	..	..	..	Manufacturing
											MANUFACTURING ACTIVITY
80.7	89.5	100.0	111.8	103.2	98.1	99.3	102.4	..	..	..	Employment (1980=100)
88.3	95.3	100.0	102.1	111.2	119.1	125.0	152.8	..	..	..	Real Earnings per Empl. (1980=100)
96.5	103.0	100.0	108.6	122.9	136.2	..	..	..	..	..	Real Output per Empl. (1980=100)
26.2	25.6	27.6	29.8	32.5	30.0	28.5	30.4	..	..	..	Earnings as % of Value Added
				(Millions of current Malaysian Ringgit)							**MONETARY HOLDINGS**
26,558	33,075	41,296	49,359	60,904	66,644	77,070	85,026	96,344	..	D	Money Supply, Broadly Defined
7,243	8,487	9,757	11,015	12,477	13,432	13,357	14,132	14,523	16,375	..	Money as Means of Payment
3,578	4,094	4,758	5,100	5,727	6,025	5,974	6,773	7,146	7,965	..	Currency Ouside Banks
3,664	4,393	4,999	5,915	6,750	7,407	7,383	7,359	7,377	8,410	..	Demand Deposits
19,315	24,588	31,539	38,344	48,427	53,212	63,713	70,894	81,821	..	..	Quasi-Monetary Liabilities
				(Millions of current Malaysian Ringgit)							
-2,356	-1,401	-3,185	-8,572	..	-6,770	-5,139	..	..	..	..	**GOVERNMENT DEFICIT (-) OR SURPLUS**
8,797	10,436	14,016	15,781	16,595	18,542	20,532	20,775	19,273	..	..	Current Revenue
7,610	8,332	10,217	13,686	15,921	16,124	17,506	..	..	..	..	Current Expenditure
1,187	2,104	3,799	2,095	674	2,418	3,026	..	..	..	..	Current Budget Balance
11	30	37	61	33	39	38	105	76	80	..	Capital Receipts
3,554	3,535	7,021	10,728	10,796	9,227	8,203	..	..	..	..	Capital Payments

MALAYSIA	1967	1968	1969	1970	1971	1972	1973	1974	1975	1976	1977
FOREIGN TRADE (CUSTOMS BASIS)					*(Millions of current US dollars)*						
Value of Exports, fob	1,216	1,347	1,650	1,687	1,639	1,722	3,040	4,234	3,847	5,295	6,079
Nonfuel Primary Products	1,038	1,136	1,426	1,438	1,326	1,412	2,511	3,301	2,739	3,706	4,275
Fuels	100	118	115	124	169	118	164	359	419	765	868
Manufactures	78	92	108	125	143	191	364	574	689	824	937
Value of Imports, cif	1,106	1,184	1,162	1,390	1,485	1,674	2,517	4,297	3,851	4,177	5,004
Nonfuel Primary Products	352	395	379	410	396	469	700	1,051	982	969	1,178
Fuels	148	168	157	168	194	136	167	436	463	564	638
Manufactures	605	620	625	812	895	1,069	1,650	2,810	2,406	2,644	3,188
					(Index 1980 = 100)						
Terms of Trade	101.4	101.7	113.8	102.9	87.9	77.2	93.6	92.5	75.2	83.0	89.4
Export Prices, fob	17.4	16.8	19.7	19.2	18.4	21.3	35.5	52.5	44.4	49.2	57.6
Nonfuel Primary Products	24.1	23.7	27.5	26.7	25.4	26.1	43.4	55.8	46.3	52.3	60.7
Fuels	4.3	4.3	4.3	4.3	5.6	6.2	8.9	36.7	35.7	38.4	42.0
Manufactures	21.8	20.5	21.8	24.5	23.1	25.3	39.7	49.3	43.7	49.3	65.1
Import Prices, cif	17.1	16.5	17.3	18.6	21.0	27.6	37.9	56.8	59.1	59.3	64.5
Trade at Constant 1980 Prices					*(Millions of 1980 US dollars)*						
Exports, fob	7,007	8,015	8,395	8,799	8,891	8,071	8,562	8,059	8,664	10,755	10,546
Imports, cif	6,466	7,163	6,728	7,465	7,078	6,062	6,633	7,566	6,519	7,041	7,757
BALANCE OF PAYMENTS					*(Millions of current US dollars)*						
Exports of Goods & Services	..	..	..	1,830	1,793	1,876	3,270	4,733	4,392	5,866	6,841
Merchandise, fob	..	..	..	1,640	1,600	1,680	2,972	4,164	3,784	5,245	6,035
Nonfactor Services	..	..	..	113	106	123	193	408	452	446	549
Factor Services	..	..	..	77	87	74	105	162	156	175	257
Imports of Goods & Services	..	..	..	1,763	1,857	2,068	3,103	5,233	4,855	5,246	6,373
Merchandise, fob	..	..	..	1,291	1,375	1,550	2,320	3,939	3,527	3,780	4,516
Nonfactor Services	..	..	..	279	276	309	409	718	872	864	1,086
Factor Services	..	..	..	193	206	208	374	576	456	602	771
Long-Term Interest	..	..	..	25	27	37	54	71	85	156	179
Current Transfers, net	..	..	..	-65	-62	-62	-76	-52	-48	-48	-46
Workers' Remittances	..	..	..	0	0	0	0	0	0	0	0
Total to be Financed	..	..	..	2	-125	-254	91	-552	-511	572	422
Official Capital Grants	..	..	..	6	17	7	14	9	15	8	14
Current Account Balance	..	..	..	8	-108	-248	105	-543	-496	580	436
Long-Term Capital, net	..	..	..	102	236	303	232	676	737	575	669
Direct Investment	..	..	..	94	100	114	172	571	350	381	406
Long-Term Loans	..	..	..	1	159	221	58	221	756	428	476
Disbursements	..	..	..	58	194	269	146	334	893	676	927
Repayments	..	..	..	57	35	48	89	113	137	248	451
Other Long-Term Capital	..	..	..	1	-41	-39	-11	-125	-384	-243	-227
Other Capital, net	..	..	..	-96	-64	-28	-113	65	-176	-357	-797
Change in Reserves	..	..	..	-14	-64	-27	-224	-198	-65	-798	-308
Memo Item:					*(Malaysian Ringgit per US dollar)*						
Conversion Factor (Annual Avg)	3.060	3.060	3.060	3.060	3.050	2.820	2.440	2.410	2.400	2.540	2.460
EXTERNAL DEBT, ETC.					*(Millions of US dollars, outstanding at end of year)*						
Public/Publicly Guar. Long-Term	..	..	..	390	534	693	729	873	1,342	1,607	2,007
Official Creditors	..	..	..	265	303	357	395	502	619	733	959
IBRD and IDA	..	..	..	141	152	174	188	226	271	305	335
Private Creditors	..	..	..	125	231	336	333	371	723	874	1,048
Private Non-guaranteed Long-Term	..	..	..	50	80	130	164	247	502	661	823
Use of Fund Credit	..	..	..	0	0	0	0	0	0	0	104
Short-Term Debt	..	..	..	..	..	..	..	..	..	..	..
Memo Items:					*(Millions of US dollars)*						
Int'l Reserves Excluding Gold	425.7	450.0	556.6	616.3	754.7	907.1	1,275.2	1,547.1	1,456.1	2,404.1	2,783.9
Gold Holdings (at market price)	31.3	79.2	63.4	51.2	72.4	107.7	186.3	309.6	232.8	223.7	287.0
SOCIAL INDICATORS											
Total Fertility Rate	5.9	5.8	5.6	5.5	5.3	5.1	5.0	4.8	4.6	4.4	4.2
Crude Birth Rate	38.5	37.7	37.0	36.2	35.5	34.7	33.8	33.0	32.1	31.3	30.4
Infant Mortality Rate	50.0	48.4	46.8	45.2	43.6	42.0	40.4	38.8	37.2	35.6	34.0
Life Expectancy at Birth	59.4	60.1	60.8	61.6	62.3	63.0	63.5	63.9	64.4	64.8	65.3
Food Production, p.c. ('79-81 = 100)	63.1	65.7	68.4	72.2	76.5	79.1	81.7	86.1	88.4	91.4	90.4
Labor Force, Agriculture (%)	56.7	55.7	54.8	53.8	52.6	51.4	50.1	48.9	47.7	46.5	45.3
Labor Force, Female (%)	30.2	30.5	30.9	31.2	31.6	31.9	32.3	32.7	33.0	33.3	33.7
School Enroll. Ratio, primary	..	..	..	87.0	..	..	..	..	91.0	..	91.0
School Enroll. Ratio, secondary	..	..	..	34.0	..	..	..	..	42.0	47.0	48.0

1978	1979	1980	1981	1982	1983	1984	1985	1986	1987 estimate	NOTES	MALAYSIA
											FOREIGN TRADE (CUSTOMS BASIS)
				(Millions of current US dollars)							
7,387	11,075	12,939	11,734	12,027	14,100	16,484	15,632	13,830	17,864	..	Value of Exports, fob
4,939	7,088	7,277	6,251	5,798	6,534	7,146	6,432	5,518	6,801	..	Nonfuel Primary Products
1,037	2,020	3,199	3,124	3,448	4,048	4,933	4,930	3,160	3,995	..	Fuels
1,410	1,966	2,464	2,359	2,781	3,518	4,404	4,271	5,152	7,069	..	Manufactures
6,364	8,708	11,602	13,052	14,042	14,669	14,849	12,746	11,447	12,506	..	Value of Imports, cif
1,471	1,773	2,055	2,367	2,328	2,383	2,364	1,853	1,579	1,744	..	Nonfuel Primary Products
683	1,050	1,758	2,243	2,123	2,023	1,511	1,488	608	782	..	Fuels
4,210	5,885	7,789	8,441	9,591	10,263	10,974	9,405	9,260	9,980	..	Manufactures
				(Index 1980 = 100)							
90.4	98.9	100.0	89.6	84.7	88.0	93.3	86.5	64.8	71.6	..	Terms of Trade
65.6	84.9	100.0	91.0	82.1	83.3	86.4	78.4	63.8	76.8	..	Export Prices, fob
70.3	93.3	100.0	85.0	74.4	80.9	90.5	73.7	61.6	74.5	..	Nonfuel Primary Products
42.3	61.0	100.0	112.5	101.6	92.5	90.2	87.5	44.9	56.9	..	Fuels
79.0	92.1	100.0	85.6	80.3	78.6	77.2	76.6	90.5	99.4	..	Manufactures
72.5	85.8	100.0	101.6	97.0	94.6	92.6	90.7	98.4	107.3	..	Import Prices, cif
				(Millions of 1980 US dollars)							Trade at Constant 1980 Prices
11,266	13,045	12,940	12,892	14,645	16,926	19,075	19,933	21,684	23,259	..	Exports, fob
8,773	10,149	11,602	12,842	14,481	15,502	16,039	14,052	11,631	11,654	..	Imports, cif
				(Millions of current US dollars)							**BALANCE OF PAYMENTS**
8,386	12,418	14,836	13,879	14,299	16,218	19,067	17,779	16,303	20,795	..	Exports of Goods & Services
7,311	10,994	12,868	11,675	11,966	13,683	16,407	15,135	13,666	17,668	..	Merchandise, fob
697	870	1,229	1,412	1,683	1,972	2,046	2,052	2,102	2,446	..	Nonfactor Services
378	555	739	792	650	562	614	591	535	681	..	Factor Services
8,233	11,482	15,100	16,331	17,867	19,705	20,700	18,467	16,306	18,600	..	Imports of Goods & Services
5,718	7,838	10,462	11,780	12,719	13,251	13,426	11,557	10,273	11,843	..	Merchandise, fob
1,402	2,162	3,026	2,920	3,351	4,079	4,418	4,049	3,818	4,017	..	Nonfactor Services
1,114	1,482	1,612	1,630	1,797	2,375	2,856	2,860	2,214	2,740	..	Factor Services
190	344	338	515	724	940	1,316	1,492	1,398	1,461	..	Long-Term Interest
-68	-36	-43	-55	-53	-35	-63	-46	-19	-25	..	Current Transfers, net
0	0	0	0	0	0	0	0	0	0	..	Workers' Remittances
85	901	-307	-2,507	-3,622	-3,523	-1,696	-733	-21	2,170	K	Total to be Financed
23	28	23	21	21	26	24	40	56	167	K	Official Capital Grants
108	929	-285	-2,486	-3,601	-3,497	-1,671	-694	35	2,336	..	Current Account Balance
712	954	1,043	2,595	3,626	3,992	3,168	1,622	1,332	-275	..	Long-Term Capital, net
500	573	934	1,265	1,397	1,261	797	695	553	575	..	Direct Investment
402	723	1,111	2,332	3,916	3,177	2,386	739	408	-738	..	Long-Term Loans
1,266	1,198	1,456	2,766	4,510	3,890	3,508	4,389	2,226	1,959	..	Disbursements
864	475	345	434	594	713	1,122	3,650	1,818	2,697	..	Repayments
-212	-371	-1,025	-1,023	-1,708	-471	-40	148	315	-280	..	Other Long-Term Capital
-541	-1,081	-290	-561	-288	-511	-1,010	223	88	-943	..	Other Capital, net
-279	-802	-468	452	262	15	-486	-1,151	-1,455	-1,119	..	Change in Reserves
				(Malaysian Ringgit per US dollar)							**Memo Item:**
2.320	2.190	2.180	2.300	2.340	2.320	2.340	2.480	2.580	2.520	..	Conversion Factor (Annual Avg)
				(Millions of US dollars, outstanding at end of year)							**EXTERNAL DEBT, ETC.**
2,542	3,026	3,950	5,686	8,144	11,835	13,215	15,190	17,108	19,065	..	Public/Publicly Guar. Long-Term
1,123	1,252	1,444	1,663	1,778	2,068	2,685	3,321	3,860	4,559	..	Official Creditors
383	448	504	577	660	721	635	776	908	1,116	..	IBRD and IDA
1,419	1,774	2,506	4,022	6,366	9,767	10,530	11,869	13,248	14,505	..	Private Creditors
813	998	1,248	1,637	3,198	2,732	2,775	2,960	2,891	2,610	..	Private Non-guaranteed Long-Term
0	0	0	221	274	330	258	118	0	..	..	Use of Fund Credit
..	..	..	..	..	..	..	..	..	..	..	Short-Term Debt
				(Millions of US dollars)							**Memo Items:**
3,243.3	3,915.0	4,387.4	4,098.0	3,767.8	3,783.7	3,723.3	4,911.8	6,027.4	7,435.3	..	Int'l Reserves Excluding Gold
427.1	1,090.6	1,367.6	926.2	1,064.6	888.9	718.3	765.2	914.7	1,137.6	..	Gold Holdings (at market price)
											SOCIAL INDICATORS
4.1	4.1	4.0	4.0	3.9	3.8	3.7	3.6	3.5	3.4	..	Total Fertility Rate
30.5	30.6	30.7	30.8	30.9	30.4	29.9	29.4	28.9	28.4	..	Crude Birth Rate
33.2	32.4	31.6	30.8	30.0	31.1	32.3	33.4	34.5	35.6	..	Infant Mortality Rate
65.8	66.3	66.9	67.4	68.0	68.3	68.6	68.9	69.2	68.2	..	Life Expectancy at Birth
87.1	96.5	101.0	102.6	108.6	100.5	113.4	124.7	125.8	127.9	..	Food Production, p.c. ('79-81 = 100)
44.0	42.8	41.6	..	..	..	..	..	..	..	..	Labor Force, Agriculture (%)
34.0	34.3	34.6	34.7	34.7	34.8	34.8	34.9	34.9	35.0	..	Labor Force, Female (%)
92.0	92.0	95.0	97.0	94.0	96.0	97.0	99.0	..	..	..	School Enroll. Ratio, primary
49.0	51.0	49.0	51.0	49.0	51.0	53.0	53.0	..	..	..	School Enroll. Ratio, secondary

MALI	1967	1968	1969	1970	1971	1972	1973	1974	1975	1976	1977
CURRENT GNP PER CAPITA (US $)	..	..	60	70	70	80	80	90	120	140	170
POPULATION (thousands)	5,004	5,113	5,224	5,335	5,446	5,557	5,668	5,784	5,905	6,032	6,163

ORIGIN AND USE OF RESOURCES *(Billions of current CFA Francs)*

	1967	1968	1969	1970	1971	1972	1973	1974	1975	1976	1977
Gross National Product (GNP)	63.8	79.9	82.7	92.4	102.3	115.6	118.1	121.3	165.8	210.1	243.5
Net Factor Income from Abroad	-0.1	-0.4	-0.6	-1.4	-0.3	-0.1	-0.4	-1.0	-2.1	-1.6	0.2
Gross Domestic Product (GDP)	63.9	80.3	83.3	93.8	102.6	115.7	118.5	122.3	167.9	211.7	243.3
Indirect Taxes, net	3.5	4.6	5.7	6.9	7.9	5.2	5.9	6.0	6.5	10.1	11.5
GDP at factor cost	60.4	75.7	77.6	86.9	94.7	110.5	112.6	116.3	161.4	201.6	231.8
Agriculture	41.8	49.3	50.2	57.5	62.1	68.4	64.8	59.7	102.4	123.7	142.3
Industry	6.0	8.8	8.9	10.2	11.8	13.9	14.7	16.1	18.0	22.1	25.4
Manufacturing	3.5	5.6	5.4	6.9	7.2	9.5	9.6	10.0	11.4	12.1	13.4
Services, etc.	16.1	22.2	24.3	26.1	28.7	33.4	38.9	46.5	47.5	65.9	75.5
Resource Balance	-8.4	-6.7	-6.9	-5.6	-7.4	-11.7	-15.9	-36.5	-35.0	-22.5	-17.6
Exports of Goods & NFServices	7.7	8.7	9.9	12.5	14.6	12.3	12.8	16.5	17.2	24.7	32.9
Imports of Goods & NFServices	16.2	15.4	16.8	18.1	22.0	24.0	28.7	53.0	52.1	47.2	50.5
Domestic Absorption	72.4	86.9	90.2	99.4	110.1	127.4	134.4	158.8	202.9	234.2	260.9
Private Consumption, etc.	54.0	64.2	66.2	75.0	82.3	93.0	103.2	127.0	158.1	187.7	198.2
General Gov't Consumption	6.6	8.2	8.7	9.3	10.4	13.5	10.0	14.3	17.2	16.5	22.4
Gross Domestic Investment	11.8	14.5	15.3	15.2	17.4	21.0	21.2	17.5	27.6	30.0	40.2
Fixed Investment	9.6	12.0	12.6	13.1	15.3	19.1	19.8	15.6	24.0	27.6	37.2
Memo Items:											
Gross Domestic Saving	3.4	7.9	8.4	9.6	10.0	9.2	5.4	-19.0	-7.4	7.5	22.6
Gross National Saving	4.2	7.1	7.8	8.3	11.1	10.7	6.4	-19.0	-6.9	8.3	27.8

(Billions of 1980 CFA Francs)

	1967	1968	1969	1970	1971	1972	1973	1974	1975	1976	1977
Gross National Product	211.6	217.9	215.4	227.7	235.1	247.6	238.8	232.0	262.9	300.2	323.3
GDP at factor cost	..	..	..	..	..	..	..	..	..	..	..
Agriculture	139.1	142.4	134.9	145.5	148.8	151.1	134.5	125.9	156.8	179.4	193.7
Industry	29.1	29.4	30.2	33.2	26.1	29.6	30.3	30.3	33.0	34.9	39.0
Manufacturing	..	..	..	..	..	..	..	..	..	..	..
Services, etc.	43.8	47.2	52.0	52.7	60.9	67.0	74.8	77.7	76.5	88.3	90.3
Resource Balance	-37.8	-36.0	-37.4	-27.4	-24.0	-34.3	-34.6	-55.2	-50.5	-37.7	-27.7
Exports of Goods & NFServices	22.4	20.9	23.7	23.5	31.9	28.7	26.2	28.7	29.2	33.8	41.8
Imports of Goods & NFServices	60.2	56.8	61.1	50.9	56.0	63.0	60.8	83.9	79.7	71.5	69.5
Domestic Absorption	249.7	255.0	254.4	258.7	259.9	282.1	274.2	289.0	316.8	340.2	350.7
Private Consumption, etc.	183.4	180.2	177.1	185.8	182.0	205.7	200.6	226.5	242.5	270.8	263.2
General Gov't Consumption	25.3	24.8	26.1	27.9	30.8	28.8	28.3	27.1	27.2	23.5	29.7
Gross Domestic Investment	41.1	50.0	51.2	45.0	47.1	47.7	45.3	35.5	47.1	45.9	57.8
Fixed Investment	..	..	..	..	..	..	..	..	..	..	..
Memo Items:											
Capacity to Import	28.8	32.2	36.0	35.1	37.1	32.2	27.2	26.1	26.2	37.4	45.3
Terms of Trade Adjustment	6.4	11.4	12.4	11.6	5.2	3.5	1.0	-2.6	-2.9	3.7	3.5
Gross Domestic Income	218.4	230.4	229.4	242.9	241.0	251.3	240.6	231.2	263.3	306.2	326.5
Gross National Income	218.0	229.2	227.7	239.2	240.3	251.1	239.8	229.3	260.0	303.8	326.8

DOMESTIC PRICES (DEFLATORS) *(Index 1980 = 100)*

	1967	1968	1969	1970	1971	1972	1973	1974	1975	1976	1977
Overall (GDP)	30.2	36.6	38.4	40.5	43.5	46.7	49.5	52.3	63.1	70.0	75.3
Domestic Absorption	29.0	34.1	35.5	38.4	42.4	45.2	49.0	55.0	64.1	68.8	74.4
Agriculture	30.1	34.6	37.2	39.5	41.7	45.3	48.2	47.5	65.3	69.0	73.5
Industry	20.5	30.0	29.4	30.6	45.4	46.8	48.6	53.1	54.7	63.3	65.2
Manufacturing	..	..	..	..	..	..	..	..	..	..	..

MANUFACTURING ACTIVITY

	1967	1968	1969	1970	1971	1972	1973	1974	1975	1976	1977
Employment (1980=100)	..	..	18.1	36.3	39.2	55.1	66.4	56.4	68.4	75.2	83.1
Real Earnings per Empl. (1980=100)	..	..	269.1	250.2	209.3	184.0	160.1	160.7	151.5	151.4	123.5
Real Output per Empl. (1980=100)	..	..	207.0	139.1	97.8	81.5	78.3	90.1	120.0	117.7	139.6
Earnings as % of Value Added	..	..	38.9	45.6	47.8	46.1	49.0	44.2	31.9	30.8	26.1

MONETARY HOLDINGS *(Billions of current CFA Francs)*

	1967	1968	1969	1970	1971	1972	1973	1974	1975	1976	1977
Money Supply, Broadly Defined	12.3	11.8	12.2	13.6	14.7	16.4	17.9	27.1	32.5	36.3	41.8
Money as Means of Payment	12.3	11.3	11.1	13.0	14.0	15.6	17.5	27.0	32.0	35.6	40.7
Currency Ouside Banks	7.4	6.9	7.5	8.9	9.7	10.5	11.1	15.1	18.7	22.5	27.4
Demand Deposits	4.9	4.4	3.6	4.1	4.3	5.1	6.4	11.9	13.3	13.1	13.3
Quasi-Monetary Liabilities	0.1	0.5	1.1	0.7	0.7	0.8	0.3	0.1	0.6	0.6	1.2

GOVERNMENT DEFICIT (-) OR SURPLUS *(Millions of current CFA Francs)*

	1967	1968	1969	1970	1971	1972	1973	1974	1975	1976	1977
GOVERNMENT DEFICIT (-) OR SURPLUS	..	..	..	..	..	..	..	..	..	23	244
Current Revenue	..	..	..	..	..	..	..	..	24,152	27,083	34,700
Current Expenditure	..	..	..	..	..	..	..	..	..	25,252	29,249
Current Budget Balance	..	..	..	..	..	..	..	..	..	1,831	5,450
Capital Receipts	..	..	..	..	..	..	..	..	9	16	36
Capital Payments	..	..	..	..	..	..	..	..	..	1,824	5,242

1978	1979	1980	1981	1982	1983	1984	1985	1986	1987 estimate	NOTES	MALI
170	210	240	250	230	180	160	150	180	210	..	**CURRENT GNP PER CAPITA (US $)**
6,300	6,443	6,590	6,742	6,899	7,060	7,223	7,389	7,576	7,768	..	**POPULATION (thousands)**
				(Billions of current CFA Francs)							**ORIGIN AND USE OF RESOURCES**
262.7	322.4	346.5	370.6	403.9	410.0	459.0	466.6	534.3	580.4	..	Gross National Product (GNP)
2.5	2.3	2.3	-0.2	0.3	-1.3	-4.5	-8.4	-8.6	-8.4	..	Net Factor Income from Abroad
260.2	320.1	344.2	370.8	403.6	411.3	463.5	475.0	542.9	588.8	..	Gross Domestic Product (GDP)
13.6	16.9	14.5	18.4	22.9	23.5	26.2	20.8	23.3	20.4	..	Indirect Taxes, net
246.6	303.2	329.7	352.4	380.7	387.8	437.4	454.2	519.6	568.4	B	GDP at factor cost
143.9	186.7	200.9	207.0	232.0	219.9	233.4	225.1	285.7	315.9	..	Agriculture
27.1	30.0	32.1	38.3	41.7	48.4	63.0	70.2	67.5	68.4	..	Industry
13.3	13.8	14.8	17.9	19.1	22.0	33.9	40.3	34.8	..	..	Manufacturing
89.3	103.4	111.2	125.5	129.9	143.0	167.1	179.7	189.7	204.5	..	Services, etc.
-50.4	-59.0	-64.8	-65.9	-69.1	-79.8	-86.4	-160.6	-131.7	-99.2	..	Resource Balance
32.0	40.3	55.6	54.5	61.8	78.6	101.7	98.9	89.0	97.7	..	Exports of Goods & NFServices
82.4	99.3	120.4	120.4	130.9	158.4	188.1	259.5	220.7	196.9	..	Imports of Goods & NFServices
310.6	379.1	409.0	436.7	472.7	491.1	549.9	635.6	674.6	688.0	..	Domestic Absorption
240.1	294.8	314.8	335.3	360.0	371.4	409.7	486.5	496.0	529.3	..	Private Consumption, etc.
26.7	31.8	35.8	36.6	39.6	44.9	50.1	56.2	58.1	61.6	..	General Gov't Consumption
43.9	52.5	58.4	64.8	73.1	74.8	90.1	92.9	120.5	97.1	..	Gross Domestic Investment
39.9	48.5	58.4	64.8	73.1	74.8	90.1	104.4	106.3	89.1	..	Fixed Investment
											Memo Items:
-6.5	-6.6	-6.4	-1.1	4.0	-5.0	3.7	-67.7	-11.2	-2.1	..	Gross Domestic Saving
3.3	3.1	4.4	7.5	12.5	2.6	8.3	-55.0	-3.9	3.0	..	Gross National Saving
				(Billions of 1980 CFA Francs)							
317.4	351.2	346.5	359.9	384.4	365.7	369.7	365.7	430.9	447.8	..	Gross National Product
..	..	..	..	..	..	..	..	..	..	B	GDP at factor cost
179.0	207.6	200.9	197.5	212.8	186.4	171.4	161.2	212.9	225.8	..	Agriculture
38.4	32.4	32.1	41.4	43.4	46.1	55.0	58.2	64.2	62.0	..	Industry
..	..	..	..	..	..	..	..	..	..	..	Manufacturing
96.9	108.6	111.2	121.3	127.9	134.5	147.0	152.8	160.6	166.4	..	Services, etc.
-62.9	-65.3	-64.8	-63.6	-61.2	-71.3	-78.9	-125.2	-100.4	-78.1	..	Resource Balance
40.0	44.3	55.6	52.6	53.0	58.6	62.1	64.1	67.4	79.3	..	Exports of Goods & NFServices
102.9	109.7	120.4	116.2	114.2	130.0	141.0	189.3	167.8	157.3	..	Imports of Goods & NFServices
377.2	414.0	409.0	423.7	445.3	438.2	452.2	497.4	538.1	532.3	..	Domestic Absorption
288.7	318.6	314.8	327.6	343.5	335.7	337.7	381.9	404.7	417.8	..	Private Consumption, etc.
32.2	34.6	35.8	32.9	34.1	38.3	41.7	43.0	42.7	44.0	..	General Gov't Consumption
56.3	60.8	58.4	63.1	67.8	64.2	72.8	72.5	90.8	70.6	..	Gross Domestic Investment
..	..	..	..	..	..	..	..	..	..	..	Fixed Investment
											Memo Items:
39.9	44.5	55.6	52.6	53.9	64.5	76.2	72.2	67.7	78.1	..	Capacity to Import
0.0	0.2	0.0	0.0	0.9	5.9	14.2	8.0	0.3	-1.2	..	Terms of Trade Adjustment
314.3	348.8	344.2	360.1	385.1	372.8	387.5	380.3	438.0	453.1	..	Gross Domestic Income
317.3	351.3	346.5	359.9	385.3	371.6	383.8	373.8	431.2	446.7	..	Gross National Income
				(Index 1980 = 100)							**DOMESTIC PRICES (DEFLATORS)**
82.8	91.8	100.0	103.0	105.1	112.1	124.2	127.6	124.0	129.6	..	Overall (GDP)
82.3	91.6	100.0	103.1	106.1	112.1	121.6	127.8	125.4	129.2	..	Domestic Absorption
80.4	89.9	100.0	104.8	109.0	118.0	136.2	139.6	134.2	139.9	..	Agriculture
70.5	92.6	100.0	92.5	96.0	105.0	114.6	120.6	105.1	110.3	..	Industry
..	..	..	..	..	..	..	..	..	..	..	Manufacturing
											MANUFACTURING ACTIVITY
89.3	100.6	100.0	98.4	..	..	..	..	..	..	..	Employment (1980=100)
108.4	101.7	100.0	97.8	..	..	..	..	..	..	..	Real Earnings per Empl. (1980=100)
125.8	95.1	100.0	126.5	..	..	..	..	..	..	..	Real Output per Empl. (1980=100)
29.5	29.8	27.6	29.5	..	..	..	..	..	..	..	Earnings as % of Value Added
				(Billions of current CFA Francs)							**MONETARY HOLDINGS**
51.5	60.1	62.8	64.7	72.3	87.3	117.4	127.6	135.3	129.9	..	Money Supply, Broadly Defined
48.5	56.8	59.5	60.2	68.3	81.2	107.1	113.8	117.5	109.3	..	Money as Means of Payment
30.7	38.2	40.6	40.9	45.0	50.2	50.5	59.8	66.9	60.8	..	Currency Ouside Banks
17.8	18.6	18.9	19.3	23.3	31.1	56.6	54.0	50.6	48.5	..	Demand Deposits
3.0	3.3	3.3	4.5	4.0	6.0	10.3	13.8	17.8	20.6	..	Quasi-Monetary Liabilities
				(Millions of current CFA Francs)							
-4,028	-9,060	-15,972	-14,644	-31,549	-34,542	-35,000	-46,700	..	..	F	**GOVERNMENT DEFICIT (-) OR SURPLUS**
35,107	36,107	61,425	72,043	83,247	95,098	107,900	119,300	..	..	..	Current Revenue
33,087	39,595	42,305	46,009	49,707	52,809	59,300	76,100	..	..	..	Current Expenditure
2,020	-3,489	19,120	26,034	33,541	42,289	48,600	43,200	..	..	..	Current Budget Balance
14	39	20	28	12	19	..	..	..	..	..	Capital Receipts
6,062	5,609	35,110	40,706	65,101	76,850	83,600	89,920	..	..	..	Capital Payments

MALI	1967	1968	1969	1970	1971	1972	1973	1974	1975	1976	1977
FOREIGN TRADE (CUSTOMS BASIS)				*(Millions of current US dollars)*							
Value of Exports, fob	16.5	10.7	17.3	35.5	36.0	36.7	58.0	64.2	36.5	84.5	124.6
Nonfuel Primary Products	16.3	10.6	16.8	31.7	32.6	33.2	51.9	56.9	32.3	83.3	123.1
Fuels	0.0	0.0	0.2	0.0	0.1	0.1	0.1	0.1	0.0	0.1	0.2
Manufactures	0.2	0.2	0.4	3.7	3.4	3.5	6.0	7.2	4.3	1.2	1.3
Value of Imports, cif	51.7	34.3	38.9	44.8	59.3	73.7	126.3	179.4	190.1	150.2	158.7
Nonfuel Primary Products	11.9	10.2	8.6	15.5	21.2	19.9	65.0	104.1	57.9	32.0	23.6
Fuels	3.9	2.7	3.4	4.0	5.7	7.9	10.9	15.8	20.0	21.5	28.4
Manufactures	35.9	21.5	26.9	25.3	32.5	45.9	50.3	59.6	112.1	96.7	106.7
Terms of Trade	157.1	158.7	170.7	180.2	*(Index 1980 = 100)* 172.2	168.8	182.2	111.9	100.6	128.6	119.6
Export Prices, fob	34.4	34.4	35.7	38.7	41.5	43.3	67.1	74.1	62.7	76.3	75.8
Nonfuel Primary Products	34.9	34.9	38.9	39.1	42.5	44.1	70.6	75.2	62.3	76.5	76.0
Fuels	4.3	4.3	4.3	4.3	5.6	6.2	8.9	36.7	35.7	38.4	42.0
Manufactures	30.1	31.7	32.4	37.5	36.5	38.5	49.5	67.2	66.1	66.7	71.7
Import Prices, cif	21.9	21.7	20.9	21.5	24.1	25.6	36.8	66.2	62.4	59.3	63.4
Trade at Constant 1980 Prices				*(Millions of 1980 US dollars)*							
Exports, fob	47.9	31.2	48.5	91.6	86.9	85.1	86.5	86.6	58.2	110.8	164.3
Imports, cif	236.0	158.4	186.1	208.6	246.1	288.4	343.0	271.0	304.8	253.1	250.4
BALANCE OF PAYMENTS				*(Millions of current US dollars)*							
Exports of Goods & Services	..	..	..	50.0	55.6	67.9	77.4	81.7	94.8	112.2	148.5
Merchandise, fob	..	..	..	32.8	39.5	45.1	58.5	64.1	71.9	94.4	124.6
Nonfactor Services	..	..	..	17.2	16.1	22.8	18.8	17.6	22.9	17.9	23.9
Factor Services	..	..	..	0.0	0.0	0.0	0.0	0.0	0.0	0.0	0.0
Imports of Goods & Services	..	..	..	72.0	88.8	112.2	166.4	227.8	260.4	206.6	226.4
Merchandise, fob	..	..	..	37.5	49.2	63.4	106.3	129.2	136.2	111.3	111.1
Nonfactor Services	..	..	..	29.2	33.9	43.4	51.6	90.2	102.5	78.7	95.5
Factor Services	..	..	..	5.3	5.7	5.3	8.5	8.4	21.7	16.6	19.8
Long-Term Interest	..	..	..	0.4	0.5	0.5	0.6	0.8	0.8	0.8	2.3
Current Transfers, net	..	..	..	..	..	..	..	..	..	..	..
Workers' Remittances	..	..	..	5.7	6.7	8.1	9.9	14.2	23.3	17.7	26.5
Total to be Financed	..	..	..	..	..	..	..	..	..	..	..
Official Capital Grants	..	..	..	..	..	..	..	..	..	..	..
Current Account Balance	..	..	..	-1.9	-9.0	-11.3	-28.4	-36.1	-56.6	-42.2	5.4
Long-Term Capital, net	..	..	..	23.2	22.1	46.7	64.0	118.5	114.5	67.5	99.6
Direct Investment	..	..	..	..	-0.6	3.4	0.6	..	2.0	-5.0	-5.0
Long-Term Loans	..	..	..	22.6	5.7	17.5	15.8	30.9	28.4	29.8	54.0
Disbursements	..	..	..	22.9	6.1	17.9	20.2	32.7	30.8	33.0	58.7
Repayments	..	..	..	0.3	0.4	0.4	4.4	1.8	2.4	3.2	4.7
Other Long-Term Capital	..	..	..	0.6	17.0	25.8	47.6	87.6	84.1	42.7	50.6
Other Capital, net	..	..	..	-22.8	-13.5	-34.3	-33.9	-78.6	-66.0	-23.1	-103.6
Change in Reserves	..	..	..	1.5	0.4	-1.1	-1.7	-3.7	8.1	-2.1	-1.4
Memo Item:				*(CFA Francs per US dollar)*							
Conversion Factor (Annual Avg)	246.850	246.850	259.710	277.710	277.030	252.210	222.700	240.500	214.320	238.980	245.670
EXTERNAL DEBT, ETC.				*(Millions of US dollars, outstanding at end of year)*							
Public/Publicly Guar. Long-Term	..	..	..	237.6	260.6	256.9	290.3	330.1	338.1	353.4	428.8
Official Creditors	..	..	..	231.5	253.8	248.1	275.7	313.4	320.5	336.5	409.3
IBRD and IDA	..	..	..	6.0	9.5	15.4	21.7	28.2	41.0	54.4	67.1
Private Creditors	..	..	..	6.1	6.8	8.8	14.6	16.7	17.6	16.9	19.5
Private Non-guaranteed Long-Term	..	..	..	..	..	..	..	..	..	..	..
Use of Fund Credit	..	..	..	8.6	8.1	9.0	8.8	11.5	12.4	12.7	10.8
Short-Term Debt	..	..	..	..	..	..	..	..	..	..	20.4
Memo Items:				*(Thousands of US dollars)*							
Int'l Reserves Excluding Gold	1,000.0	600.0	1,100.0	900.0	2,068.0	3,749.0	4,226.0	6,110.0	4,196.0	6,895.0	5,380.0
Gold Holdings (at market price)	..	..	..	..	..	..	..	..	..	..	1,569.0
SOCIAL INDICATORS											
Total Fertility Rate	6.5	6.5	6.5	6.5	6.5	6.5	6.5	6.5	7.7	6.5	6.5
Crude Birth Rate	50.0	50.0	50.0	50.0	50.0	50.0	49.9	49.8	49.7	49.6	49.5
Infant Mortality Rate	206.0	205.4	204.8	204.2	203.6	203.0	200.6	198.2	195.8	193.4	191.0
Life Expectancy at Birth	39.1	39.5	40.0	40.4	40.8	41.3	41.7	42.2	42.6	43.1	43.5
Food Production, p.c. ('79-81 = 100)	106.6	101.4	108.7	104.3	100.5	85.6	77.3	87.7	96.6	95.8	93.9
Labor Force, Agriculture (%)	89.9	89.7	89.4	89.2	88.8	88.4	88.1	87.7	87.3	86.9	86.6
Labor Force, Female (%)	17.2	17.3	17.3	17.3	17.3	17.4	17.4	17.4	17.5	17.5	17.5
School Enroll. Ratio, primary	..	..	..	22.0	..	..	..	..	24.0	26.0	27.0
School Enroll. Ratio, secondary	..	..	..	5.0	..	..	..	..	7.0	8.0	9.0

1978	1979	1980	1981	1982	1983	1984	1985	1986	1987 estimate	NOTES	MALI
											FOREIGN TRADE (CUSTOMS BASIS)
				(Millions of current US dollars)							
112.0	148.0	205.0	154.0	146.0	167.0	192.0	180.5	191.7	216.5	f	Value of Exports, fob
108.2	114.5	170.7	128.4	120.9	138.5	158.5	142.1	134.2	154.4	..	Nonfuel Primary Products
0.4	0.4	0.6	0.4	0.4	0.5	0.6	0.6	0.5	0.6	..	Fuels
3.5	33.1	33.8	25.1	24.7	28.0	33.0	37.8	57.0	61.5	..	Manufactures
285.7	360.6	438.6	364.6	332.0	353.0	368.0	432.0	438.0	447.1	f	Value of Imports, cif
71.0	61.2	80.4	68.2	61.3	65.3	71.0	69.7	62.2	59.2	..	Nonfuel Primary Products
53.3	59.3	74.7	62.2	56.4	60.1	64.8	72.3	51.6	69.5	..	Fuels
161.4	240.1	283.5	234.1	214.3	227.6	232.2	290.1	324.2	318.4	..	Manufactures
				(Index 1980 = 100)							
119.6	105.3	100.0	92.8	82.6	92.9	92.6	82.0	74.7	85.8	..	Terms of Trade
82.3	88.0	100.0	95.4	80.9	88.9	88.5	74.2	68.1	84.2	..	Export Prices, fob
82.7	87.6	100.0	94.6	78.5	88.5	88.3	71.0	59.6	76.6	..	Nonfuel Primary Products
42.3	61.0	100.0	112.5	101.6	92.5	90.2	87.5	44.9	56.9	..	Fuels
76.1	90.0	100.0	99.1	94.8	90.7	89.3	89.0	103.2	113.0	..	Manufactures
68.8	83.6	100.0	102.8	98.0	95.7	95.5	90.5	91.1	98.1	..	Import Prices, cif
											Trade at Constant 1980 Prices
				(Millions of 1980 US dollars)							
136.2	168.2	205.0	161.5	180.4	187.9	217.1	243.3	281.5	257.2	..	Exports, fob
415.6	431.4	438.6	354.7	338.7	368.9	385.2	477.6	480.7	455.8	..	Imports, cif
											BALANCE OF PAYMENTS
				(Millions of current US dollars)							
131.4	191.3	262.7	200.4	190.2	208.8	232.8	220.0	257.0	..	f	Exports of Goods & Services
94.2	145.7	204.9	154.2	145.8	166.8	192.0	176.0	206.0	260.0	..	Merchandise, fob
37.2	45.6	57.7	46.2	44.4	42.0	40.7	44.0	51.4	65.0	..	Nonfactor Services
0.0	0.0	0.0	0.0	0.0	0.0	0.0	0.0	0.0	..	..	Factor Services
363.4	445.9	537.0	470.5	419.5	435.3	458.2	599.8	662.3	..	f	Imports of Goods & Services
199.4	270.3	308.4	269.0	232.7	240.9	257.8	328.0	347.0	346.0	..	Merchandise, fob
150.7	163.8	211.8	168.9	162.7	166.9	172.7	249.0	290.0	310.0	..	Nonfactor Services
13.3	11.8	16.9	32.6	24.2	27.5	27.7	21.8	25.3	..	..	Factor Services
3.0	3.1	3.4	3.5	5.7	6.8	8.2	12.6	12.7	13.3	..	Long-Term Interest
..	..	..	..	25.0	23.4	20.8	..	..	..	..	Current Transfers, net
44.3	54.1	59.4	47.7	39.4	36.5	32.5	35.2	44.9	..	..	Workers' Remittances
..	..	..	..	-208.3	-211.6	-247.0	..	..	..	..	Total to be Financed
..	..	..	..	93.1	108.9	132.5	..	..	..	..	Official Capital Grants
-98.8	-113.3	-124.3	-140.0	-115.2	-102.7	-114.5	-133.0	-173.0	-111.0	..	Current Account Balance
154.0	197.2	227.4	193.4	164.5	186.3	185.6	294.9	312.0	310.0	f	Long-Term Capital, net
-9.8	-3.1	2.4	3.7	1.5	3.2	4.1	4.5	4.3	..	..	Direct Investment
60.7	73.4	96.7	105.9	112.7	149.0	110.9	81.3	145.3	98.5	..	Long-Term Loans
64.8	78.4	102.8	112.0	116.1	156.7	122.6	106.9	167.9	117.3	..	Disbursements
4.1	5.0	6.1	6.1	3.4	7.7	11.7	25.6	22.6	18.8	..	Repayments
103.1	126.9	128.4	83.8	-42.9	-74.8	-62.0	209.2	162.4	211.5	..	Other Long-Term Capital
-59.3	-77.9	-111.1	-60.2	-75.2	-96.1	-70.9	-211.3	-190.1	-206.3	f	Other Capital, net
4.2	-6.0	8.0	6.8	25.9	12.5	-0.3	49.4	51.1	7.3	..	Change in Reserves
											Memo Item:
				(CFA Francs per US dollar)							
225.660	212.720	211.280	271.730	328.610	381.060	436.960	449.260	346.300	300.540	..	Conversion Factor (Annual Avg)
											EXTERNAL DEBT, ETC.
				(Millions of US dollars, outstanding at end of year)							
521.2	523.8	684.7	751.1	823.0	942.8	1,153.5	1,334.2	1,574.1	1,847.3	..	Public/Publicly Guar. Long-Term
491.8	492.3	649.2	721.4	793.7	896.0	1,107.3	1,284.5	1,518.5	1,786.2	..	Official Creditors
81.6	102.0	120.8	139.8	153.8	171.9	190.9	224.4	276.8	339.7	..	IBRD and IDA
29.4	31.5	35.5	29.6	29.3	46.8	46.2	49.7	55.7	61.2	..	Private Creditors
..	..	..	..	..	..	..	..	..	..	..	Private Non-guaranteed Long-Term
10.0	7.9	11.4	7.8	34.2	46.3	63.7	80.8	84.9	75.1	..	Use of Fund Credit
18.8	24.8	24.1	72.6	19.3	21.4	70.3	68.1	82.6	94.0	..	Short-Term Debt
											Memo Items:
				(Thousands of US dollars)							
8,161.0	5,986.0	14,536.0	17,386.0	16,727.0	16,234.0	26,641.0	22,517.0	23,311.0	15,849.0	..	Int'l Reserves Excluding Gold
3,214.0	9,639.0	11,098.0	7,483.0	8,602.0	7,182.0	5,804.0	6,156.0	7,359.0	9,114.0	..	Gold Holdings (at market price)
											SOCIAL INDICATORS
6.5	6.5	6.5	6.5	6.5	6.5	6.5	6.5	6.5	6.5	..	Total Fertility Rate
49.2	48.9	48.6	48.3	48.0	47.9	47.9	47.8	47.7	47.7	..	Crude Birth Rate
188.6	186.2	183.8	181.4	179.0	167.4	155.8	144.3	132.7	121.1	..	Infant Mortality Rate
43.8	44.1	44.4	44.7	45.0	45.4	45.9	46.3	46.8	47.2	..	Life Expectancy at Birth
99.9	98.4	97.1	104.5	110.2	111.6	99.0	100.5	103.2	99.2	..	Food Production, p.c. ('79-81 = 100)
86.2	85.9	85.5	..	..	..	..	..	..	..	..	Labor Force, Agriculture (%)
17.4	17.4	17.4	17.3	17.2	17.1	16.9	16.8	16.7	16.6	..	Labor Force, Female (%)
27.0	27.0	25.0	24.0	26.0	26.0	25.0	25.0	24.0	..	..	School Enroll. Ratio, primary
9.0	..	8.4	8.0	7.7	7.2	7.2	6.3	6.1	..	..	School Enroll. Ratio, secondary

MALTA	1967	1968	1969	1970	1971	1972	1973	1974	1975	1976	1977
CURRENT GNP PER CAPITA (US $)	610	650	680	760	820	910	1,030	1,220	1,540	1,730	1,930
POPULATION (thousands)	319	319	323	326	325	319	322	324	328	329	332

ORIGIN AND USE OF RESOURCES *(Millions of current Maltese Liri)*

	1967	1968	1969	1970	1971	1972	1973	1974	1975	1976	1977
Gross National Product (GNP)	68.5	77.1	88.5	102.1	105.3	110.5	123.3	144.1	184.1	221.8	258.9
Net Factor Income from Abroad	4.8	5.8	6.5	7.3	7.5	8.3	7.5	12.5	18.3	18.1	19.1
Gross Domestic Product (GDP)	63.7	71.3	82.0	94.8	97.8	102.2	115.8	131.6	165.8	203.7	239.8
Indirect Taxes, net	8.0	9.2	11.2	12.6	12.6	12.6	14.9	13.0	12.9	14.3	19.9
GDP at factor cost	55.7	62.1	70.8	82.2	85.2	89.6	100.9	118.6	152.9	189.4	219.9
Agriculture	3.9	4.5	5.2	5.8	6.1	6.7	7.3	8.3	9.2	11.4	12.9
Industry	17.0	19.2	23.9	28.4	26.6	30.3	34.8	44.4	61.8	78.5	91.4
Manufacturing	11.0	12.6	15.9	17.9	17.2	22.0	26.7	33.7	46.7	61.5	72.6
Services, etc.	34.8	38.4	41.7	48.0	52.5	52.6	58.8	65.9	81.9	99.5	115.6
Resource Balance	-12.0	-17.1	-22.4	-28.4	-25.2	-23.5	-22.7	-38.0	-22.4	-23.9	-34.6
Exports of Goods & NFServices	33.4	40.8	46.6	47.1	50.0	53.5	75.3	110.4	137.3	172.7	207.4
Imports of Goods & NFServices	45.4	57.9	69.0	75.5	75.2	77.0	98.0	148.4	159.7	196.6	242.0
Domestic Absorption	75.7	88.4	104.4	123.2	123.0	125.7	138.5	169.6	188.2	227.6	274.4
Private Consumption, etc.	45.9	53.2	63.9	73.8	75.1	80.4	90.1	107.0	118.7	135.7	172.4
General Gov't Consumption	11.2	12.2	13.7	18.4	19.5	19.8	22.8	26.9	30.5	35.9	39.7
Gross Domestic Investment	18.6	23.0	26.8	31.0	28.4	25.5	25.6	35.7	39.0	56.0	62.3
Fixed Investment	15.6	19.9	25.5	27.8	25.5	22.5	22.3	31.2	37.5	54.1	60.0
Memo Items:											
Gross Domestic Saving	6.6	5.9	4.4	2.6	3.2	2.0	2.9	-2.3	16.6	32.1	27.7
Gross National Saving	15.8	18.4	21.4	18.9	19.0	16.8	17.8	18.0	44.1	63.2	63.2

(Millions of 1980 Maltese Liri)

	1967	1968	1969	1970	1971	1972	1973	1974	1975	1976	1977
Gross National Product	119.9	132.6	139.2	157.5	162.4	171.2	185.2	207.2	251.2	288.0	320.4
GDP at factor cost	..	..	..	..	..	..	..	..	..	..	..
Agriculture	..	..	..	..	..	..	..	..	..	..	..
Industry	..	..	..	..	..	..	..	..	..	..	..
Manufacturing	..	..	..	..	..	..	..	..	..	..	..
Services, etc.	..	..	..	..	..	..	..	..	..	..	..
Resource Balance	-78.2	-97.3	-125.1	-131.5	-117.9	-104.2	-83.0	-84.7	-56.1	-57.1	-60.5
Exports of Goods & NFServices	86.0	102.9	107.2	106.4	111.6	108.3	139.9	163.6	196.1	227.7	260.4
Imports of Goods & NFServices	164.2	200.1	232.2	237.9	229.5	212.4	222.9	248.4	252.2	284.8	320.9
Domestic Absorption	188.2	218.4	254.0	276.7	266.6	261.6	255.8	274.7	283.2	322.8	358.7
Private Consumption, etc.	109.7	127.9	155.4	168.7	167.1	167.8	168.2	179.8	182.7	201.2	231.1
General Gov't Consumption	26.2	26.6	27.5	36.2	36.2	36.5	37.2	38.3	43.5	50.5	51.2
Gross Domestic Investment	52.3	63.9	71.0	71.8	63.4	57.2	50.5	56.6	57.0	71.0	76.4
Fixed Investment	52.7	67.0	84.2	76.5	67.0	59.1	49.2	53.4	60.4	75.2	80.3
Memo Items:											
Capacity to Import	120.8	141.0	156.8	148.4	152.6	147.6	171.3	184.8	216.9	250.1	275.0
Terms of Trade Adjustment	34.8	38.1	49.7	42.0	41.0	39.3	31.4	21.1	20.7	22.4	14.6
Gross Domestic Income	144.8	159.3	178.6	187.2	189.7	196.7	204.2	211.1	247.8	288.2	312.8
Gross National Income	154.7	170.7	188.9	199.5	203.3	210.5	216.6	228.3	271.9	310.4	335.0

DOMESTIC PRICES (DEFLATORS) *(Index 1980 = 100)*

	1967	1968	1969	1970	1971	1972	1973	1974	1975	1976	1977
Overall (GDP)	57.9	58.8	63.6	65.3	65.7	64.9	67.0	69.3	73.0	76.6	80.4
Domestic Absorption	40.2	40.5	41.1	44.5	46.1	48.1	54.1	61.7	66.5	70.5	76.5
Agriculture	..	..	..	..	..	..	..	..	..	..	..
Industry	..	..	..	..	..	..	..	..	..	..	..
Manufacturing	..	..	..	..	..	..	..	..	..	..	..

MANUFACTURING ACTIVITY

	1967	1968	1969	1970	1971	1972	1973	1974	1975	1976	1977
Employment (1980=100)	46.0	53.3	60.5	62.2	66.5	71.7	79.0	78.7	78.1	85.0	92.8
Real Earnings per Empl. (1980=100)	33.1	34.7	37.8	42.9	46.9	48.8	49.0	55.0	69.1	77.6	78.2
Real Output per Empl. (1980=100)	..	..	..	..	..	..	..	..	..	..	..
Earnings as % of Value Added	42.8	41.7	42.1	47.3	50.7	50.6	48.0	49.6	49.8	44.7	44.4

MONETARY HOLDINGS *(Millions of current Maltese Liri)*

	1967	1968	1969	1970	1971	1972	1973	1974	1975	1976	1977
Money Supply, Broadly Defined	95.0	107.6	123.9	136.3	157.9	174.8	186.6	205.2	246.8	305.4	336.0
Money as Means of Payment	43.7	50.1	56.6	57.7	67.9	80.1	87.2	99.0	123.8	148.3	165.6
Currency Ouside Banks	33.1	36.2	40.3	45.9	55.7	62.3	72.7	79.6	98.9	119.6	137.8
Demand Deposits	10.6	13.9	16.3	11.8	12.2	17.8	14.5	19.5	24.9	28.7	27.8
Quasi-Monetary Liabilities	51.3	57.6	67.3	78.6	90.0	94.7	99.5	106.2	123.1	157.1	170.3

GOVERNMENT DEFICIT (-) OR SURPLUS *(Millions of current Maltese Liri)*

	1967	1968	1969	1970	1971	1972	1973	1974	1975	1976	1977
Current Revenue	..	..	..	..	..	-4.8	3.1	4.8	-6.2	4.5	3.5
Current Expenditure	..	..	..	..	..	40.8	57.6	69.5	91.3	102.5	104.9
Current Budget Balance	..	..	..	..	..	37.9	40.9	56.4	63.9	70.6	81.5
Capital Receipts	..	..	..	..	..	2.9	16.7	13.1	27.4	31.9	23.4
Capital Payments	..	..	..	..	..	0.1	0.0	4.8	0.9	2.6	0.7
	..	..	..	..	..	7.7	13.6	13.0	34.5	30.0	20.6

1978	1979	1980	1981	1982	1983	1984	1985	1986	1987 estimate	NOTES	MALTA
2,170	2,600	3,150	3,590	3,800	3,480	3,360	3,250	3,440	4,020	..	CURRENT GNP PER CAPITA (US $)
340	347	364	364	360	360	359	360	360	360	..	POPULATION (thousands)
				(Millions of current Maltese Liri)							ORIGIN AND USE OF RESOURCES
294.6	338.7	422.5	477.7	513.8	495.7	506.5	514.8	539.9	579.5	..	Gross National Product (GNP)
16.9	12.9	30.5	41.2	52.0	38.1	45.4	38.8	28.0	33.0	..	Net Factor Income from Abroad
277.7	325.8	392.0	436.5	461.8	457.6	461.1	476.0	511.9	546.5	..	Gross Domestic Product (GDP)
26.4	32.1	43.4	46.0	44.1	40.5	39.7	45.4	50.1	53.7	..	Indirect Taxes, net
251.3	293.7	348.6	390.5	417.7	417.1	421.4	430.6	461.8	492.8	..	GDP at factor cost
11.4	11.5	13.3	15.0	16.8	18.7	19.3	19.4	20.4	21.4	..	Agriculture
106.4	125.2	146.7	162.1	170.4	164.6	169.4	170.0	188.3	200.0	..	Industry
84.4	100.0	115.4	121.4	125.0	120.0	124.7	126.9	134.7	136.2	..	Manufacturing
133.5	157.0	188.6	213.4	230.5	233.8	232.7	241.2	253.1	271.4	..	Services, etc.
-19.9	-16.9	-21.4	-36.5	-74.8	-68.5	-70.0	-75.3	-56.0	-59.1	..	Resource Balance
229.6	290.8	356.6	355.9	319.8	307.6	323.5	345.2	365.7	420.7	..	Exports of Goods & NFServices
249.5	307.7	378.0	392.4	394.6	376.1	393.5	420.5	421.7	479.8	..	Imports of Goods & NFServices
297.6	342.7	413.4	473.0	536.6	526.1	531.1	551.3	567.9	605.6	..	Domestic Absorption
186.4	206.0	253.5	279.4	305.8	306.7	317.5	333.2	347.9	365.7	..	Private Consumption, etc.
46.1	53.7	63.4	75.4	85.2	82.3	80.3	84.3	89.5	98.2	..	General Gov't Consumption
65.1	83.0	96.5	118.2	145.6	137.1	133.3	133.8	130.5	141.7	..	Gross Domestic Investment
60.3	78.2	87.1	105.6	120.1	131.6	126.5	125.9	122.3	146.9	..	Fixed Investment
											Memo Items:
45.2	66.1	75.1	81.7	70.8	68.6	63.3	58.5	74.5	82.6	..	Gross Domestic Saving
73.2	91.5	117.1	135.5	138.1	120.9	123.1	114.2	121.3	137.4	..	Gross National Saving
				(Millions of 1980 Maltese Liri)							
351.3	383.1	422.5	440.6	455.3	443.2	451.9	457.8	467.5	488.5	I	Gross National Product
..	..	..	..	..	..	..	..	..	..	..	GDP at factor cost
..	..	..	..	..	..	..	..	..	..	..	Agriculture
..	..	..	..	..	..	..	..	..	..	..	Industry
..	..	..	..	..	..	..	..	..	..	..	Manufacturing
..	..	..	..	..	..	..	..	..	..	..	Services, etc.
-35.8	-19.7	-21.4	-33.0	-69.8	-50.0	-51.8	-61.6	-44.9	-51.5	..	Resource Balance
272.7	318.5	356.6	315.9	272.3	267.3	278.0	298.7	315.6	352.7	..	Exports of Goods & NFServices
308.4	338.2	378.0	348.9	342.1	317.3	329.8	360.3	360.5	404.2	..	Imports of Goods & NFServices
367.2	385.9	413.4	438.0	484.0	461.7	467.3	487.9	488.3	514.8	..	Domestic Absorption
231.8	236.8	253.5	253.6	256.7	258.1	271.4	290.5	296.5	317.4	..	Private Consumption, etc.
56.4	62.3	63.4	68.1	72.2	71.4	69.6	73.5	76.8	83.9	..	General Gov't Consumption
78.9	86.8	96.5	116.2	155.1	132.2	126.3	123.9	115.1	113.5	..	Gross Domestic Investment
77.6	86.4	87.1	102.1	117.1	135.6	126.1	121.1	110.5	134.7	..	Fixed Investment
											Memo Items:
283.8	319.6	356.6	316.4	277.2	259.5	271.1	295.8	312.6	354.4	..	Capacity to Import
11.2	1.1	0.0	0.5	4.9	-7.8	-6.9	-2.9	-2.9	1.7	..	Terms of Trade Adjustment
342.6	367.3	392.0	405.5	419.2	403.9	408.7	423.4	440.4	465.0	..	Gross Domestic Income
362.5	384.2	422.5	441.1	460.3	435.4	445.0	454.9	464.6	490.2	..	Gross National Income
				(Index 1980 = 100)							DOMESTIC PRICES (DEFLATORS)
83.8	89.0	100.0	107.8	111.5	111.1	111.0	111.7	115.5	118.0	..	Overall (GDP)
81.1	88.8	100.0	108.0	110.9	113.9	113.6	113.0	116.3	117.6	..	Domestic Absorption
..	..	..	..	..	..	..	..	..	..	..	Agriculture
..	..	..	..	..	..	..	..	..	..	..	Industry
..	..	..	..	..	..	..	..	..	..	..	Manufacturing
											MANUFACTURING ACTIVITY
98.3	103.0	100.0	96.8	86.9	82.9	82.9	82.4	..	..	..	Employment (1980=100)
90.3	94.2	100.0	102.0	113.1	114.4	120.8	124.0	..	..	..	Real Earnings per Empl. (1980=100)
..	..	..	..	..	..	..	..	..	..	..	Real Output per Empl. (1980=100)
50.3	49.0	50.4	53.7	54.0	52.7	53.5	53.5	..	..	..	Earnings as % of Value Added
				(Millions of current Maltese Liri)							MONETARY HOLDINGS
381.9	429.7	463.9	513.8	544.9	587.5	629.5	651.1	685.4	748.1	..	Money Supply, Broadly Defined
194.6	231.6	258.8	293.7	306.3	324.6	325.9	317.6	318.7	351.1	..	Money as Means of Payment
155.0	176.3	206.1	239.2	259.7	279.6	283.7	273.3	273.8	300.2	..	Currency Ouside Banks
39.6	55.4	52.7	54.6	46.6	45.0	42.2	44.2	44.9	50.8	..	Demand Deposits
187.3	198.1	205.1	220.0	238.7	262.9	303.6	333.6	366.7	397.0	..	Quasi-Monetary Liabilities
				(Millions of current Maltese Liri)							
11.8	..	4.3	5.5	-7.6	7.3	2.1	-19.2	..	..	F	GOVERNMENT DEFICIT (-) OR SURPLUS
125.4	..	142.3	179.1	192.5	197.5	193.7	197.0	..	..	..	Current Revenue
87.6	..	106.6	134.7	153.0	149.3	151.0	152.4	..	..	..	Current Expenditure
37.8	..	35.7	44.4	39.5	48.2	42.7	44.6	..	..	..	Current Budget Balance
0.1	..	1.1	4.0	0.9	2.2	0.1	0.1	..	..	..	Capital Receipts
26.2	..	32.5	42.9	48.0	43.0	40.7	63.9	..	..	..	Capital Payments

MALTA	1967	1968	1969	1970	1971	1972	1973	1974	1975	1976	1977
FOREIGN TRADE (CUSTOMS BASIS)				*(Millions of current US dollars)*							
Value of Exports, fob	27.1	34.0	38.3	38.6	45.2	67.2	98.1	134.0	166.7	228.4	288.6
Nonfuel Primary Products	6.5	10.5	11.2	8.1	7.3	7.5	14.1	15.5	18.1	23.2	33.2
Fuels	2.6	3.8	3.4	3.7	3.1	2.9	3.7	14.7	19.0	16.3	16.5
Manufactures	17.9	19.6	23.7	26.8	34.7	56.8	80.4	103.7	129.7	188.9	239.0
Value of Imports, cif	111.3	122.9	147.2	160.5	156.1	174.3	239.6	357.3	374.9	420.9	512.7
Nonfuel Primary Products	42.0	42.5	48.4	48.6	52.3	55.6	81.0	109.9	111.5	114.1	132.9
Fuels	6.7	7.2	7.1	7.8	9.6	13.0	14.4	40.4	35.7	37.0	41.4
Manufactures	62.5	73.2	91.7	104.1	94.3	105.6	144.1	207.0	227.7	269.7	338.3
Terms of Trade	84.6	82.0	83.5	82.7	96.0	110.1	109.1	109.7	102.5	104.1	103.4
Export Prices, fob	19.6	19.0	21.2	21.9	26.8	31.9	42.8	61.0	60.6	63.5	68.9
Nonfuel Primary Products	38.7	40.0	41.3	43.3	42.2	45.5	56.2	61.5	70.3	67.7	71.4
Fuels	4.3	4.3	4.3	4.3	5.6	6.2	8.9	36.7	35.7	38.4	42.0
Manufactures	30.1	31.7	32.4	37.5	36.5	38.5	49.5	67.2	66.1	66.7	71.7
Import Prices, cif	23.1	23.2	25.4	26.4	27.9	28.9	39.2	55.6	59.1	61.0	66.6
Trade at Constant 1980 Prices				*(Millions of 1980 US dollars)*							
Exports, fob	138.4	178.5	180.5	176.3	168.8	211.0	229.3	219.6	275.0	359.8	419.1
Imports, cif	480.9	529.8	579.5	607.0	560.5	602.0	610.7	642.4	633.8	690.1	769.6
BALANCE OF PAYMENTS				*(Millions of current US dollars)*							
Exports of Goods & Services	..	..	..	138.9	162.1	204.2	282.5	369.9	451.9	495.1	586.1
Merchandise, fob	..	..	..	34.0	47.0	74.4	109.3	154.9	179.8	237.6	300.7
Nonfactor Services	..	..	..	79.0	86.5	97.5	138.3	167.1	208.4	195.9	215.2
Factor Services	..	..	..	25.9	28.6	32.2	35.0	48.0	63.7	61.6	70.1
Imports of Goods & Services	..	..	..	189.8	194.3	211.2	277.9	401.8	427.0	478.4	591.3
Merchandise, fob	..	..	..	143.9	142.4	155.9	212.9	313.6	332.4	376.6	456.8
Nonfactor Services	..	..	..	37.5	41.9	44.6	51.5	72.9	79.5	85.4	113.4
Factor Services	..	..	..	8.4	10.1	10.7	13.6	15.3	15.1	16.5	21.1
Long-Term Interest	..	..	..	1.7	2.3	0.4	0.5	1.3	1.3	1.0	1.0
Current Transfers, net	..	..	..	21.6	20.4	17.0	20.1	20.2	23.9	30.6	38.8
Workers' Remittances	..	..	..	16.8	13.5	15.2	11.9	10.6	13.5	21.6	24.6
Total to be Financed	..	..	..	-29.3	-11.8	10.0	24.7	-11.7	48.9	47.3	33.6
Official Capital Grants	..	..	..	23.7	17.7	15.2	11.9	24.4	17.2	15.3	12.8
Current Account Balance	..	..	..	-5.6	5.9	25.2	36.6	12.7	66.0	62.6	46.4
Long-Term Capital, net	..	..	..	25.8	-32.0	8.1	-29.0	45.9	48.6	58.6	14.7
Direct Investment	..	..	..	11.6	11.6	4.5	5.2	10.6	15.9	14.1	18.5
Long-Term Loans	..	..	..	-0.7	-8.0	3.7	1.8	3.7	6.9	17.2	1.5
Disbursements	..	..	..	..	7.1	4.5	2.6	5.0	8.3	18.2	2.6
Repayments	..	..	..	0.7	15.1	0.8	0.8	1.3	1.4	1.0	1.1
Other Long-Term Capital	..	..	..	-8.8	-53.3	-15.2	-47.9	7.2	8.7	12.0	-18.1
Other Capital, net	..	..	..	-3.7	43.4	38.8	31.7	-16.6	-1.8	-26.9	-1.0
Change in Reserves	..	..	..	-16.5	-17.2	-72.1	-39.3	-42.0	-112.8	-94.3	-60.1
Memo Item:				*(Maltese Liri per US dollar)*							
Conversion Factor (Annual Avg)	0.360	0.420	0.420	0.420	0.410	0.380	0.370	0.390	0.380	0.430	0.420
EXTERNAL DEBT, ETC.				*(Millions of US dollars, outstanding at end of year)*							
Public/Publicly Guar. Long-Term	..	..	..	25.0	18.1	21.1	23.0	26.9	31.7	48.0	52.8
Official Creditors	..	..	..	25.0	18.1	21.1	20.6	24.6	29.6	46.3	51.2
IBRD and IDA	..	..	..	4.6	4.2	3.9	3.5	3.1	2.6	2.2	1.8
Private Creditors	..	..	..	..	..	..	2.5	2.3	2.2	1.7	1.5
Private Non-guaranteed Long-Term	..	..	..	..	..	..	..	..	..	..	..
Use of Fund Credit	..	..	..	0.0	0.0	0.0	0.0	0.0	0.0	0.0	0.0
Short-Term Debt	..	..	..	..	..	..	..	..	..	..	21.0
Memo Items:				*(Millions of US dollars)*							
Int'l Reserves Excluding Gold	89.5	138.7	127.0	148.2	184.8	261.7	310.4	386.7	485.8	607.8	719.5
Gold Holdings (at market price)	..	14.6	12.3	10.4	15.4	22.9	39.6	65.8	49.5	47.6	59.4
SOCIAL INDICATORS											
Total Fertility Rate	2.2	2.1	2.0	2.0	2.1	2.0	1.6	1.6	2.3	2.2	2.2
Crude Birth Rate	16.7	16.1	15.8	16.3	17.1	16.9	17.5	17.6	18.3	18.0	17.9
Infant Mortality Rate	27.3	27.6	24.3	27.9	23.9	16.7	23.1	19.9	17.5	15.0	13.5
Life Expectancy at Birth	69.4	69.6	69.9	70.1	70.4	70.6	70.6	70.6	70.6	70.6	70.9
Food Production, p.c. ('79-81 = 100)	74.9	85.3	88.6	91.0	86.8	94.0	89.1	84.2	80.0	95.9	105.3
Labor Force, Agriculture (%)	7.8	7.5	7.2	6.9	6.7	6.5	6.4	6.2	6.0	5.8	5.7
Labor Force, Female (%)	21.0	21.1	21.2	21.3	21.4	21.4	21.5	21.6	21.7	21.7	21.7
School Enroll. Ratio, primary	..	..	..	106.0	..	..	..	..	94.0	108.0	110.0
School Enroll. Ratio, secondary	..	..	..	50.0	..	..	..	..	75.0	72.0	71.0

1978	1979	1980	1981	1982	1983	1984	1985	1986	1987 estimate	NOTES	MALTA
											FOREIGN TRADE (CUSTOMS BASIS)
				(Millions of current US dollars)							
341.9	424.4	482.9	447.6	410.8	362.7	394.5	399.5	496.8	557.0	..	Value of Exports, fob
31.2	38.2	32.0	30.1	25.2	25.3	26.7	23.9	25.0	23.8	..	Nonfuel Primary Products
13.1	17.5	23.9	26.0	23.4	15.8	14.1	12.3	6.0	7.6	..	Fuels
297.6	368.7	427.0	391.4	362.1	321.6	353.7	363.3	465.8	525.6	..	Manufactures
569.1	753.3	938.0	855.0	789.0	733.0	717.0	759.0	887.4	1,001.7	..	Value of Imports, cif
146.1	181.4	224.3	189.5	173.4	153.9	145.4	145.1	165.8	188.6	..	Nonfuel Primary Products
41.8	49.9	96.5	115.0	115.3	89.2	92.6	95.0	48.7	64.0	..	Fuels
381.2	522.0	617.2	550.4	500.2	489.9	479.0	518.9	672.9	749.1	..	Manufactures
				(Index 1980 = 100)							
99.1	99.7	100.0	90.3	103.8	88.1	87.0	87.3	97.1	98.3	..	Terms of Trade
74.1	88.9	100.0	90.5	99.9	83.2	80.7	79.5	94.6	103.3	..	Export Prices, fob
79.0	98.5	100.0	96.0	96.5	94.4	85.8	83.8	94.1	92.1	..	Nonfuel Primary Products
42.3	61.0	100.0	112.5	101.6	92.5	90.2	87.5	44.9	56.9	..	Fuels
76.1	90.0	100.0	89.0	100.0	82.0	80.0	79.0	96.0	105.1	..	Manufactures
74.7	89.2	100.0	100.3	96.2	94.4	92.8	91.1	97.5	105.0	..	Import Prices, cif
				(Millions of 1980 US dollars)							Trade at Constant 1980 Prices
461.6	477.3	482.9	494.3	411.3	436.1	488.9	502.4	525.1	539.2	..	Exports, fob
761.4	844.2	938.0	852.6	820.4	776.7	772.9	833.0	910.5	953.5	..	Imports, cif
											BALANCE OF PAYMENTS
				(Millions of current US dollars)							
705.3	909.1	1,171.9	1,073.4	925.4	821.6	815.6	835.9	1,036.7	1,324.2	..	Exports of Goods & Services
346.4	442.4	509.8	470.1	421.0	388.5	410.8	421.7	520.2	631.6	..	Merchandise, fob
280.0	364.6	508.9	441.7	330.5	300.2	266.1	287.6	370.3	559.5	..	Nonfactor Services
78.9	102.1	153.1	161.6	174.0	133.0	138.6	126.6	146.2	133.1	..	Factor Services
679.1	919.7	1,185.2	1,072.7	983.9	889.6	868.9	913.5	1,108.0	1,399.7	..	Imports of Goods & Services
507.3	674.3	884.7	778.1	709.2	659.5	639.2	672.2	784.4	1,024.6	..	Merchandise, fob
138.0	181.9	240.3	240.8	243.6	204.9	208.4	217.0	277.9	354.0	..	Nonfactor Services
33.7	63.6	60.2	53.8	31.1	25.2	21.3	24.3	45.7	21.1	..	Factor Services
0.9	1.9	1.4	1.1	1.0	1.1	1.3	1.5	2.2	2.2	..	Long-Term Interest
28.8	34.9	33.3	32.6	37.1	32.8	31.2	36.0	48.0	63.1	..	Current Transfers, net
16.3	18.1	22.3	19.4	16.3	17.6	13.9	10.7	10.4	16.8	..	Workers' Remittances
55.0	24.3	20.0	33.4	-21.4	-35.1	-22.1	-41.6	-23.3	-12.4	K	Total to be Financed
24.7	28.5	24.6	57.7	37.6	33.5	33.4	18.8	23.1	23.4	K	Official Capital Grants
79.7	52.7	44.6	91.0	16.3	-1.6	11.3	-22.8	-0.3	11.0	..	Current Account Balance
25.2	27.9	48.3	76.0	45.4	36.8	64.4	3.8	-23.6	..	..	Long-Term Capital, net
21.5	16.2	26.6	39.0	20.9	24.5	26.2	19.0	21.8	..	..	Direct Investment
11.7	3.3	55.5	21.4	2.3	23.4	4.9	-2.3	-11.3	-10.4	..	Long-Term Loans
13.1	5.3	57.5	24.2	4.8	32.1	12.2	7.7	3.0	0.3	..	Disbursements
1.4	2.0	2.0	2.8	2.5	8.7	7.3	10.0	14.3	10.7	..	Repayments
-32.7	-20.0	-58.4	-42.1	-15.4	-44.7	-0.1	-31.6	-57.2	..	..	Other Long-Term Capital
20.4	-25.2	-41.5	-19.2	-9.5	69.1	-43.8	-48.1	19.3	-15.5	..	Other Capital, net
-125.2	-55.4	-51.4	-147.8	-52.1	-104.2	-31.9	67.1	4.6	4.5	..	Change in Reserves
				(Maltese Liri per US dollar)							Memo Item:
0.390	0.360	0.350	0.390	0.410	0.430	0.460	0.470	0.390	0.350	..	Conversion Factor (Annual Avg)
			(Millions of US dollars, outstanding at end of year)								**EXTERNAL DEBT, ETC.**
70.7	76.6	127.7	131.1	121.0	136.4	121.1	123.6	114.0	112.3	..	Public/Publicly Guar. Long-Term
69.3	75.3	126.8	130.5	120.5	136.2	121.0	123.5	114.0	112.3	..	Official Creditors
1.3	0.8	0.3	0.0	0.0	0.0	0.0	0.0	0.0	..	..	IBRD and IDA
1.4	1.3	1.0	0.6	0.4	0.3	0.2	0.1	..	..	..	Private Creditors
..	..	..	..	..	..	..	..	..	..	..	Private Non-guaranteed Long-Term
0.0	0.0	0.0	0.0	0.0	0.0	0.0	0.0	0.0	..	..	Use of Fund Credit
19.0	33.0	22.0	28.0	46.0	42.0	85.0	86.0	127.0	182.3	..	Short-Term Debt
				(Millions of US dollars)							Memo Items:
925.1	1,012.7	990.1	1,073.8	1,083.6	1,112.3	990.2	986.9	1,145.4	1,414.6	..	Int'l Reserves Excluding Gold
81.4	187.4	255.8	181.3	211.1	180.1	143.7	152.4	182.2	225.6	..	Gold Holdings (at market price)
											SOCIAL INDICATORS
2.1	2.0	1.9	1.8	1.7	1.7	1.7	1.7	1.7	1.7	..	Total Fertility Rate
17.4	16.9	15.4	15.0	14.4	15.0	14.8	14.2	13.8	13.4	..	Crude Birth Rate
14.4	15.7	15.1	13.1	13.9	14.1	11.6	13.6	..	14.6	..	Infant Mortality Rate
71.2	71.4	71.7	72.0	72.3	72.8	73.4	73.8	74.2	74.6	..	Life Expectancy at Birth
116.6	93.1	106.6	100.3	113.2	111.8	104.9	108.4	115.2	119.1	..	Food Production, p.c. ('79-81 = 100)
5.5	5.4	5.2	..	..	..	..	..	..	..	..	Labor Force, Agriculture (%)
21.7	21.8	21.8	21.9	22.0	22.1	22.2	22.3	22.5	22.7	..	Labor Force, Female (%)
111.0	110.0	100.0	98.0	95.0	95.0	97.0	..	..	..	..	School Enroll. Ratio, primary
70.0	74.0	66.0	71.0	74.0	76.0	76.0	..	..	..	..	School Enroll. Ratio, secondary

MAURITANIA	1967	1968	1969	1970	1971	1972	1973	1974	1975	1976	1977
CURRENT GNP PER CAPITA (US $)	160	170	160	180	170	190	200	260	300	350	360
POPULATION (thousands)	1,142	1,167	1,193	1,220	1,248	1,276	1,306	1,337	1,369	1,402	1,437
ORIGIN AND USE OF RESOURCES	*(Millions of current Mauritanian Ouguiyas)*										
Gross National Product (GNP)	8,670	9,546	9,428	10,878	11,513	12,646	13,913	17,903	19,419	22,089	22,890
Net Factor Income from Abroad	-771	-856	-942	-750	-1,071	-736	-964	-900	-1,095	-1,521	-1,756
Gross Domestic Product (GDP)	9,441	10,402	10,370	11,628	12,585	13,382	14,877	18,803	20,514	23,610	24,646
Indirect Taxes, net	507	845	1,268	700	752	953	1,400	2,317	2,041	2,461	2,217
GDP at factor cost	8,934	9,557	9,102	10,928	11,833	12,429	13,477	16,486	18,473	21,149	22,429
Agriculture	2,873	3,008	2,532	3,199	3,317	3,463	5,201	5,872	5,467	6,023	6,556
Industry	3,296	3,744	3,777	4,204	4,674	4,420	4,068	5,488	6,296	7,163	6,687
Manufacturing	424	517	552	539	579	630	..	..	..	..	..
Services, etc.	2,764	2,804	2,794	3,524	3,842	4,546	4,208	5,126	6,710	7,963	9,186
Resource Balance	1,008	922	791	892	950	-379	170	-1,941	-4,395	-9,308	-9,942
Exports of Goods & NFServices	3,928	3,953	4,028	4,746	5,062	8,365	7,064	9,207	7,995	9,083	8,191
Imports of Goods & NFServices	2,921	3,031	3,237	3,854	4,112	8,744	6,894	11,148	12,390	18,391	18,133
Domestic Absorption	8,433	9,480	9,579	10,736	11,635	13,761	14,707	20,744	24,909	32,918	34,588
Private Consumption, etc.	4,975	5,264	5,708	6,537	6,929	7,114	10,638	14,457	13,807	15,403	16,088
General Gov't Consumption	1,435	1,357	1,544	1,640	1,832	2,581	2,785	3,264	4,033	7,512	8,870
Gross Domestic Investment	2,023	2,860	2,327	2,559	2,873	4,065	1,284	3,023	7,069	10,003	9,630
Fixed Investment	..	..	..	..	..	..	2,415	2,050	4,141	9,319	8,499
Memo Items:											
Gross Domestic Saving	3,031	3,782	3,119	3,451	3,823	3,686	1,454	1,082	2,674	695	-312
Gross National Saving	2,260	2,758	1,995	2,373	2,480	2,753	-142	-396	599	-2,105	-3,114
	(Millions of 1980 Mauritanian Ouguiyas)										
Gross National Product	22,054	24,023	23,329	26,334	25,716	27,722	25,498	28,530	26,421	28,023	27,571
GDP at factor cost	22,637	23,961	23,357	26,259	26,280	26,943	24,769	26,544	25,576	27,576	27,572
Agriculture	11,098	11,549	9,559	9,186	7,985	11,312	9,236	9,206	7,749	7,951	8,412
Industry	6,152	6,930	7,001	7,673	8,122	7,324	7,094	8,184	7,394	8,264	7,362
Manufacturing	..	..	..	..	..	..	..	..	..	..	..
Services, etc.	5,387	5,481	6,798	9,400	10,173	8,306	8,439	9,154	10,434	11,361	11,798
Resource Balance	893	878	1,174	909	911	-1,091	-442	-7,137	-12,555	-23,145	-19,457
Exports of Goods & NFServices	8,470	8,760	8,965	9,361	9,508	15,725	11,360	12,438	9,080	9,890	9,377
Imports of Goods & NFServices	7,576	7,881	7,791	8,452	8,597	16,815	11,802	19,576	21,635	33,035	28,834
Domestic Absorption	23,302	25,458	24,603	27,373	27,377	30,421	27,770	37,272	40,757	53,602	49,514
Private Consumption, etc.	17,999	19,802	19,151	20,392	20,139	21,815	19,355	22,464	20,897	25,093	22,529
General Gov't Consumption	3,368	2,931	3,376	4,596	4,698	5,010	5,762	6,376	7,208	12,372	12,628
Gross Domestic Investment	1,935	2,725	2,076	2,385	2,541	3,596	2,653	8,432	12,653	16,136	14,357
Fixed Investment	..	..	..	..	..	..	5,582	6,525	7,154	15,508	13,056
Memo Items:											
Capacity to Import	10,190	10,279	9,695	10,407	10,583	16,087	12,093	16,167	13,961	16,316	13,025
Terms of Trade Adjustment	1,720	1,519	730	1,047	1,075	362	733	3,729	4,880	6,425	3,648
Gross Domestic Income	25,915	27,856	26,507	29,329	29,363	29,692	28,061	33,864	33,083	36,882	33,705
Gross National Income	23,774	25,542	24,059	27,381	26,791	28,084	26,232	32,259	31,302	34,448	31,219
DOMESTIC PRICES (DEFLATORS)	*(Index 1980 = 100)*										
Overall (GDP)	39.0	39.5	40.2	41.1	44.5	45.6	54.4	62.4	72.7	77.5	82.0
Domestic Absorption	36.2	37.2	38.9	39.2	42.5	45.2	53.0	55.7	61.1	61.4	69.9
Agriculture	25.9	26.0	26.5	34.8	41.5	30.6	56.3	63.8	70.6	75.7	77.9
Industry	53.6	54.0	54.0	54.8	57.5	60.3	57.3	67.1	85.1	86.7	90.8
Manufacturing	..	..	..	..	..	..	..	..	..	..	..
MANUFACTURING ACTIVITY											
Employment (1980=100)	..	..	..	..	..	..	..	..	..	..	..
Real Earnings per Empl. (1980=100)	..	..	..	..	..	..	..	..	..	..	..
Real Output per Empl. (1980=100)	..	..	..	..	..	..	..	..	..	..	..
Earnings as % of Value Added	..	..	..	..	..	..	..	..	..	..	..
MONETARY HOLDINGS	*(Millions of current Mauritanian Ouguiyas)*										
Money Supply, Broadly Defined	749	879	947	1,252	1,257	1,545	1,881	2,958	3,979	4,813	5,058
Money as Means of Payment	693	799	822	1,110	1,132	1,428	1,493	2,451	2,926	3,683	4,095
Currency Ouside Banks	285	312	358	445	461	618	629	954	1,214	1,464	1,529
Demand Deposits	408	487	464	665	671	810	864	1,497	1,712	2,219	2,566
Quasi-Monetary Liabilities	56	80	124	143	125	117	388	507	1,052	1,131	964
GOVERNMENT DEFICIT (-) OR SURPLUS	*(Millions of current Mauritanian Ouguiyas)*										
	..	..	..	..	..	..	..	..	-532	-2,321	-1,685
Current Revenue	..	..	..	..	..	..	..	..	7,639	12,059	9,078
Current Expenditure	..	..	..	..	..	..	..	..	..	..	..
Current Budget Balance	..	..	..	..	..	..	..	..	..	..	..
Capital Receipts	..	..	..	..	..	..	..	..	3	1	327
Capital Payments	..	..	..	..	..	..	..	..	..	..	..

1978	1979	1980	1981	1982	1983	1984	1985	1986	1987 estimate	NOTES	MAURITANIA
360	390	440	480	450	460	420	410	420	440	..	**CURRENT GNP PER CAPITA** (US $)
1,473	1,510	1,549	1,589	1,631	1,674	1,718	1,764	1,809	1,858	..	**POPULATION** (thousands)
			(Millions of current Mauritanian Ouguiyas)								**ORIGIN AND USE OF RESOURCES**
23,332	27,464	30,841	34,897	36,333	40,904	43,611	51,293	57,976	64,169	..	Gross National Product (GNP)
-1,800	-2,094	-1,714	-1,228	-2,505	-2,308	-2,770	-3,805	-4,723	-5,002	..	Net Factor Income from Abroad
25,132	29,558	32,555	36,125	38,838	43,212	46,381	55,098	62,699	69,171	..	Gross Domestic Product (GDP)
1,915	2,100	1,996	2,875	3,637	4,643	4,897	6,522	6,814	7,250	..	Indirect Taxes, net
23,217	27,458	30,559	33,250	35,201	38,569	41,484	48,576	55,885	61,921	..	GDP at factor cost
7,158	7,811	9,285	10,601	11,657	13,035	11,854	13,994	18,865	22,924	f	Agriculture
6,082	7,916	7,941	8,712	8,228	8,188	10,553	13,126	13,539	13,347	f	Industry
..	..	..	..	..	..	..	..	..	..	f	Manufacturing
9,977	11,731	13,333	13,937	15,316	17,346	19,077	21,456	23,481	25,650	..	Services, etc.
-7,795	-7,727	-9,546	-10,568	-17,121	-12,934	-12,485	-8,553	-4,727	-4,671	..	Resource Balance
8,502	9,945	12,154	16,506	15,776	20,038	21,437	31,527	34,782	34,689	..	Exports of Goods & NFServices
16,297	17,672	21,700	27,074	32,897	32,972	33,922	40,080	39,509	39,360	..	Imports of Goods & NFServices
32,927	37,285	42,101	46,693	55,959	56,146	58,866	63,651	67,426	73,842	..	Domestic Absorption
16,413	20,954	22,073	23,742	30,705	40,481	39,279	42,102	44,492	50,558	..	Private Consumption, etc.
8,731	8,793	8,242	7,828	6,960	7,970	7,965	8,172	8,537	9,137	..	General Gov't Consumption
7,783	7,538	11,786	15,123	18,294	7,695	11,622	13,377	14,397	14,147	..	Gross Domestic Investment
6,673	7,375	10,187	13,507	15,774	13,540	11,022	13,540	13,576	13,975	..	Fixed Investment
											Memo Items:
-12	-189	2,240	4,555	1,173	-5,239	-863	4,824	9,670	9,476	..	Gross Domestic Saving
-2,769	-3,756	-762	2,458	-2,777	-9,026	-4,936	-588	2,865	3,144	..	Gross National Saving
			(Millions of 1980 Mauritanian Ouguiyas)								
27,602	28,999	30,841	32,605	30,840	32,754	31,179	31,767	33,264	34,262	..	Gross National Product
27,902	28,973	30,559	31,215	30,096	31,699	30,448	31,349	33,162	34,001	..	GDP at factor cost
8,177	8,326	9,285	10,113	10,282	11,831	9,649	9,887	10,793	11,103	f	Agriculture
7,677	8,494	7,941	7,937	7,615	7,175	9,587	9,982	10,336	10,235	f	Industry
..	..	..	..	..	..	..	..	..	..	f	Manufacturing
12,047	12,153	13,333	13,165	12,199	12,693	11,212	11,480	12,032	12,662	..	Services, etc.
-12,778	-9,891	-9,546	-11,839	-18,185	-13,245	-12,681	-11,254	-9,669	-9,286	..	Resource Balance
10,133	11,536	12,154	14,959	13,725	17,807	16,637	18,356	19,299	19,572	..	Exports of Goods & NFServices
22,911	21,427	21,700	26,798	31,910	31,052	29,318	29,610	28,967	28,858	..	Imports of Goods & NFServices
42,703	41,197	42,101	45,633	51,285	47,963	45,950	45,513	45,794	46,373	..	Domestic Absorption
20,726	21,534	22,073	24,305	27,081	35,284	31,519	30,567	30,785	31,327	..	Private Consumption, etc.
12,023	11,191	8,242	6,779	5,846	5,464	4,351	4,471	4,784	5,681	..	General Gov't Consumption
9,954	8,471	11,786	14,550	18,358	7,215	10,079	10,475	10,225	9,364	..	Gross Domestic Investment
8,643	8,551	10,187	13,252	16,405	13,157	9,906	10,986	9,993	9,587	..	Fixed Investment
											Memo Items:
11,952	12,058	12,154	16,338	15,303	18,871	18,528	23,292	25,502	25,434	..	Capacity to Import
1,819	522	0	1,379	1,578	1,064	1,891	4,935	6,203	5,862	..	Terms of Trade Adjustment
31,744	31,828	32,555	35,173	34,678	35,782	35,159	39,195	42,328	42,948	..	Gross Domestic Income
29,421	29,521	30,841	33,984	32,418	33,818	33,069	36,703	39,466	40,124	..	Gross National Income
			(Index 1980 = 100)								**DOMESTIC PRICES (DEFLATORS)**
84.0	94.4	100.0	106.9	117.3	124.5	139.4	160.8	173.6	186.5	..	Overall (GDP)
77.1	90.5	100.0	102.3	109.1	117.1	128.1	139.9	147.2	159.2	..	Domestic Absorption
87.5	93.8	100.0	104.8	113.4	110.2	122.8	141.5	174.8	206.5	..	Agriculture
79.2	93.2	100.0	109.8	108.0	114.1	110.1	131.5	131.0	130.4	..	Industry
..	..	..	..	..	..	..	..	..	..	..	Manufacturing
											MANUFACTURING ACTIVITY
..	..	..	..	..	..	..	..	..	..	..	Employment (1980=100)
..	..	..	..	..	..	..	..	..	..	..	Real Earnings per Empl. (1980=100)
..	..	..	..	..	..	..	..	..	..	f	Real Output per Empl. (1980=100)
..	..	..	..	..	..	..	..	..	..	..	Earnings as % of Value Added
			(Millions of current Mauritanian Ouguiyas)								**MONETARY HOLDINGS**
5,160	6,293	7,080	9,429	9,572	10,085	11,322	13,848	14,864	..	..	Money Supply, Broadly Defined
4,134	5,081	5,677	7,653	7,135	8,090	9,641	12,173	11,393	..	..	Money as Means of Payment
1,728	2,311	2,376	2,678	2,950	3,024	3,658	4,700	4,418	..	..	Currency Ouside Banks
2,406	2,770	3,301	4,975	4,185	5,065	5,983	7,473	6,975	..	..	Demand Deposits
1,025	1,212	1,403	1,777	2,437	1,995	1,681	1,675	3,471	..	..	Quasi-Monetary Liabilities
			(Millions of current Mauritanian Ouguiyas)								
-871	-1,464	..	..	..	..	..	..	..	..	F	**GOVERNMENT DEFICIT (-) OR SURPLUS**
8,908	8,984	..	..	..	..	..	..	..	..		Current Revenue
8,648	8,479	..	..	..	..	..	..	..	..		Current Expenditure
260	505	..	..	..	..	..	..	..	..		Current Budget Balance
862	679	..	..	..	..	..	..	..	..		Capital Receipts
1,993	2,648	..	..	..	..	..	..	..	..		Capital Payments

MAURITANIA	1967	1968	1969	1970	1971	1972	1973	1974	1975	1976	1977
FOREIGN TRADE (CUSTOMS BASIS)				*(Millions of current US dollars)*							
Value of Exports, fob	72.0	71.8	77.8	88.9	93.9	119.2	155.3	181.5	174.3	178.0	157.0
Nonfuel Primary Products	70.0	70.1	74.3	88.0	86.3	113.0	126.2	178.0	171.2	171.7	152.2
Fuels	0.3	0.2	0.0	0.1	0.0	0.0	8.3	1.7	1.2	2.2	1.7
Manufactures	1.8	1.5	3.5	0.8	7.6	6.2	20.8	1.8	1.9	4.1	3.0
Value of Imports, cif	36.9	35.3	45.2	55.9	63.0	85.2	135.0	188.0	235.0	304.0	334.0
Nonfuel Primary Products	6.8	9.7	9.0	13.9	17.4	20.2	40.4	52.5	76.1	95.0	104.0
Fuels	1.4	2.6	2.5	4.3	4.1	5.2	6.7	22.6	18.9	25.9	29.4
Manufactures	28.8	23.0	33.7	37.6	41.6	59.8	87.9	112.9	140.0	183.0	200.6
Terms of Trade	182.2	206.5	183.9	253.0	188.9	160.0	122.3	130.8	132.6	132.5	126.2
Export Prices, fob	45.0	41.9	42.3	54.6	48.9	46.5	47.0	72.4	82.6	80.5	81.2
Nonfuel Primary Products	47.2	43.4	43.0	55.3	50.4	47.1	64.8	73.1	83.7	82.0	82.3
Fuels	4.3	4.3	4.3	4.3	5.6	6.2	8.9	36.7	35.7	38.4	42.0
Manufactures	30.1	31.7	32.4	37.5	36.5	38.5	49.5	67.2	66.1	66.7	71.7
Import Prices, cif	24.7	20.3	23.0	21.6	25.9	29.1	38.4	55.3	62.3	60.8	64.3
Trade at Constant 1980 Prices				*(Millions of 1980 US dollars)*							
Exports, fob	160.0	171.2	183.7	162.8	192.0	256.2	330.5	250.7	210.9	221.1	193.4
Imports, cif	149.3	173.9	196.2	259.0	243.4	292.9	351.2	339.7	377.0	500.2	519.2
BALANCE OF PAYMENTS				*(Millions of current US dollars)*							
Exports of Goods & Services	..	..	..	104.8	113.6	129.3	142.0	210.2	190.7	205.7	183.0
Merchandise, fob	..	..	..	97.2	102.8	117.3	131.4	187.0	167.3	181.9	157.2
Nonfactor Services	..	..	..	5.7	9.2	10.5	9.7	16.6	18.4	19.5	21.5
Factor Services	..	..	..	1.9	1.6	1.5	1.0	6.6	5.0	4.4	4.3
Imports of Goods & Services	..	..	..	112.3	119.7	148.4	158.6	219.6	312.3	413.6	410.4
Merchandise, fob	..	..	..	72.0	91.6	115.6	119.7	166.6	208.1	269.8	295.5
Nonfactor Services	..	..	..	27.1	12.7	14.8	27.9	41.0	69.2	80.9	67.2
Factor Services	..	..	..	13.2	15.4	18.0	11.0	12.0	35.1	62.9	47.7
Long-Term Interest	..	..	..	0.4	0.8	1.8	2.0	4.3	5.4	7.8	8.9
Current Transfers, net	..	..	..	..	..	..	..	..	..	..	..
Workers' Remittances	..	..	..	0.7	0.8	0.6	0.6	4.3	0.4	0.4	0.3
Total to be Financed	..	..	..								
Official Capital Grants	..	..	..	..	..	..	..	..	..	..	..
Current Account Balance	..	..	..	-4.9	-1.2	-9.9	14.3	47.3	-62.5	-83.8	-122.2
Long-Term Capital, net	..	..	..	23.2	24.6	22.1	44.8	96.1	91.8	242.9	185.0
Direct Investment	..	..	..	0.8	7.5	9.8	9.8	1.9	-122.7	1.6	4.1
Long-Term Loans	..	..	..	1.3	10.6	-3.5	44.0	48.0	21.7	114.8	55.3
Disbursements	..	..	..	4.5	13.6	5.6	54.9	57.9	55.9	184.5	87.4
Repayments	..	..	..	3.2	3.0	9.1	10.9	9.9	34.2	69.7	32.1
Other Long-Term Capital	..	..	..	21.1	6.5	15.8	-9.0	46.2	192.8	126.6	125.5
Other Capital, net	..	..	..	-20.1	-20.9	-7.8	-52.5	-100.4	-84.6	-146.5	-113.2
Change in Reserves	..	..	..	1.8	-2.6	-4.5	-6.6	-43.0	55.2	-12.6	50.4
Memo Item:											
Conversion Factor (Annual Avg)	49.370	49.370	51.940	55.540	55.430	50.440	44.540	45.330	43.100	45.020	45.590
				(Mauritanian Ouguiyas per US dollar)							
EXTERNAL DEBT, ETC.				*(Millions of US dollars, outstanding at end of year)*							
Public/Publicly Guar. Long-Term	..	..	..	27.3	39.8	56.7	105.8	168.5	188.7	393.6	461.5
Official Creditors	..	..	..	19.4	28.3	50.2	83.9	143.9	156.5	268.9	313.2
IBRD and IDA	..	..	..	4.9	8.5	9.6	11.3	20.8	14.0	18.6	24.3
Private Creditors	..	..	..	7.9	11.6	6.5	21.9	24.6	32.1	124.7	148.3
Private Non-guaranteed Long-Term	..	..	..	..	..	..	..	..	..	..	..
Use of Fund Credit	..	..	..	0.0	0.0	0.0	0.0	2.4	0.0	16.1	22.6
Short-Term Debt	..	..	..	..	..	..	..	..	..	..	41.5
Memo Items:				*(Thousands of US dollars)*							
Int'l Reserves Excluding Gold	8,800.0	7,420.0	3,570.0	3,200.0	7,505.0	13,459.0	42,243.0	103,810.0	47,725.0	81,983.0	50,017.0
Gold Holdings (at market price)	..	..	..	..	..	..	..	..	..	..	924.0
SOCIAL INDICATORS											
Total Fertility Rate	6.5	6.5	6.5	6.5	6.5	6.5	6.5	6.5	6.5	6.5	6.5
Crude Birth Rate	47.3	47.3	47.2	47.1	47.1	47.0	46.9	46.9	46.8	46.8	46.7
Infant Mortality Rate	173.0	170.6	168.2	165.8	163.4	161.0	158.6	156.2	153.8	151.4	149.0
Life Expectancy at Birth	37.9	38.3	38.7	39.1	39.5	40.0	40.4	40.8	41.2	41.6	42.0
Food Production, p.c. ('79-81=100)	141.9	141.0	138.8	133.0	119.6	111.2	90.7	87.9	90.6	95.2	95.9
Labor Force, Agriculture (%)	87.3	86.4	85.6	84.8	83.3	81.7	80.2	78.6	77.1	75.5	74.0
Labor Force, Female (%)	21.6	21.6	21.5	21.5	21.3	21.2	21.0	20.9	20.8	20.6	20.4
School Enroll. Ratio, primary	..	..	..	14.0	..	..	..	..	19.0	23.0	26.0
School Enroll. Ratio, secondary	..	..	..	2.0	..	..	..	..	4.0	4.0	6.0

1978	1979	1980	1981	1982	1983	1984	1985	1986	1987 estimate	NOTES	MAURITANIA
											FOREIGN TRADE (CUSTOMS BASIS)
				(Millions of current US dollars)							
123.0	147.0	196.0	270.0	240.0	315.0	294.0	374.3	422.0	403.0	..	Value of Exports, fob
119.3	142.5	190.1	215.0	235.4	305.7	285.0	363.3	412.0	391.0	..	Nonfuel Primary Products
1.3	1.6	2.1	48.0	0.1	3.3	3.0	3.0	1.0	2.0	..	Fuels
2.4	2.9	3.8	7.0	4.6	6.0	6.0	8.0	9.0	10.0	..	Manufactures
302.0	323.0	363.0	436.0	482.0	427.0	341.0	358.0	345.0	353.0	..	Value of Imports, cif
94.5	101.0	113.5	136.3	150.7	139.9	83.1	85.2	78.9	100.0	..	Nonfuel Primary Products
26.2	28.0	31.5	37.8	41.8	54.2	47.9	42.8	24.1	34.0	..	Fuels
181.3	194.0	218.0	261.9	289.4	233.0	210.0	230.0	242.0	219.0	..	Manufactures
				(Index 1980 = 100)							
102.2	103.5	100.0	95.5	96.1	96.2	99.8	97.7	73.1	72.7	..	Terms of Trade
73.7	88.0	100.0	94.6	89.1	87.0	90.0	81.5	64.4	62.1	..	Export Prices, fob
74.3	88.4	100.0	91.2	89.0	86.9	90.0	81.3	64.0	61.4	..	Nonfuel Primary Products
42.3	61.0	100.0	112.5	101.6	92.5	90.2	87.5	44.9	56.9	..	Fuels
76.1	90.0	100.0	99.1	94.8	90.7	89.3	89.0	103.2	113.0	..	Manufactures
72.1	85.0	100.0	99.0	92.8	90.5	90.2	83.4	88.1	85.4	..	Import Prices, cif
											Trade at Constant 1980 Prices
				(Millions of 1980 US dollars)							
166.8	167.1	196.0	285.4	269.2	362.0	326.6	459.1	655.0	648.8	..	Exports, fob
418.6	380.0	363.0	440.2	519.6	471.8	377.9	429.1	391.4	413.4	..	Imports, cif
				(Millions of current US dollars)							**BALANCE OF PAYMENTS**
155.6	204.0	269.7	338.9	305.8	354.7	330.3	402.6	449.0	473.0	f	Exports of Goods & Services
118.6	147.2	196.3	269.9	240.0	315.4	293.8	371.6	423.0	405.0	..	Merchandise, fob
33.8	45.5	56.5	48.9	47.3	30.3	28.0	27.1	23.7	61.0	..	Nonfactor Services
3.1	11.3	17.0	20.1	18.5	9.1	8.5	4.0	2.3	7.0	..	Factor Services
372.6	411.9	492.9	585.0	648.2	626.8	527.5	631.9	581.3	619.0	f	Imports of Goods & Services
267.1	286.0	321.3	386.2	426.6	378.2	302.1	334.0	303.0	317.0	..	Merchandise, fob
75.3	83.2	127.8	128.1	157.6	177.2	177.8	201.8	209.0	227.0	..	Nonfactor Services
30.2	42.8	43.8	70.7	64.1	71.4	47.6	96.1	69.3	75.0	..	Factor Services
9.6	15.4	12.8	18.3	22.6	22.8	23.4	27.2	31.7	28.0	..	Long-Term Interest
..	..	..	..	..	..	..	..	-28.0	-18.0	..	Current Transfers, net
0.6	2.4	5.6	3.7	2.3	1.2	1.0	0.8	1.9	..	..	Workers' Remittances
..	..	..	..	..	..	..	..	-160.3	-164.0	..	Total to be Financed
..	..	..	..	..	..	..	..	75.0	91.0	..	Official Capital Grants
-78.8	-97.6	-134.1	-147.1	-276.6	-213.0	-110.7	-116.1	-85.3	-73.0	..	Current Account Balance
253.0	259.4	249.5	245.9	260.5	252.5	189.6	232.1	142.0	188.0	f	Long-Term Capital, net
2.9	63.2	27.1	12.4	15.0	1.4	8.5	7.0	3.1	5.0	..	Direct Investment
112.2	41.3	116.8	124.0	196.1	165.7	94.9	44.0	164.3	81.8	..	Long-Term Loans
128.8	91.9	133.9	159.7	211.9	180.0	114.5	94.6	210.6	139.7	..	Disbursements
16.6	50.6	17.1	35.7	15.8	14.3	19.6	50.6	46.3	57.9	..	Repayments
138.0	154.8	105.6	109.4	49.4	85.5	86.2	181.1	-100.4	10.2	..	Other Long-Term Capital
-158.4	-133.2	-153.4	-103.7	-32.3	-71.2	-108.0	-141.1	-56.9	-85.0	f	Other Capital, net
-15.8	-28.6	38.0	5.0	48.4	31.7	29.1	25.1	0.2	-30.0	..	Change in Reserves
											Memo Item:
				(Mauritanian Ouguiyas per US dollar)							
46.160	45.890	45.910	48.300	51.770	54.810	63.800	77.090	74.380	73.880	..	Conversion Factor (Annual Avg)
				(Millions of US dollars, outstanding at end of year)							**EXTERNAL DEBT, ETC.**
596.0	633.8	734.2	829.2	1,000.7	1,140.3	1,177.2	1,388.1	1,668.6	1,868.3	..	Public/Publicly Guar. Long-Term
414.6	492.5	600.3	729.6	895.4	1,011.2	1,049.0	1,251.8	1,521.2	1,741.3	..	Official Creditors
29.9	34.1	37.9	54.5	82.1	110.8	105.5	116.3	147.5	216.7	..	IBRD and IDA
181.5	141.3	133.8	99.6	105.4	129.1	128.2	136.2	147.4	127.0	..	Private Creditors
..	..	..	..	..	..	..	..	..	..	..	Private Non-guaranteed Long-Term
24.2	23.4	33.7	34.7	48.2	41.2	29.7	30.3	36.1	47.2	..	Use of Fund Credit
53.5	51.3	65.0	98.3	93.0	107.4	127.1	108.6	117.5	119.3	..	Short-Term Debt
											Memo Items:
				(Thousands of US dollars)							
79,499.0	113,670.0	139,900.0	161,800.0	139,110.0	105,930.0	77,529.0	59,217.0	48,219.0	71,808.0	..	Int'l Reserves Excluding Gold
1,898.0	4,301.0	6,543.0	4,412.0	5,072.0	4,235.0	3,515.0	3,630.0	4,495.0	5,567.0	..	Gold Holdings (at market price)
											SOCIAL INDICATORS
6.5	6.5	6.5	6.5	6.5	6.5	6.5	6.5	6.5	6.5	..	Total Fertility Rate
46.7	46.6	46.6	46.5	46.5	46.7	47.0	47.2	47.5	47.7	..	Crude Birth Rate
146.6	144.2	141.8	139.4	137.0	133.4	129.9	126.3	122.8	119.2	..	Infant Mortality Rate
42.4	42.8	43.2	43.6	44.0	44.7	45.4	46.1	46.8	47.6	..	Life Expectancy at Birth
98.9	98.3	99.4	102.3	93.4	86.9	84.7	85.2	93.2	90.8	..	Food Production, p.c. ('79-81 = 100)
72.4	70.9	69.3	..	..	..	..	..	..	..	..	Labor Force, Agriculture (%)
20.3	20.1	20.0	20.2	20.4	20.6	20.8	21.1	21.3	21.5	..	Labor Force, Female (%)
32.0	32.0	34.0	..	37.0	..	..	..	..	55.0	..	School Enroll. Ratio, primary
9.0	9.0	10.0	..	12.0	..	..	..	..	16.0	..	School Enroll. Ratio, secondary

MAURITIUS	1967	1968	1969	1970	1971	1972	1973	1974	1975	1976	1977
CURRENT GNP PER CAPITA (US $)	300	280	280	280	300	340	430	590	720	860	930
POPULATION (thousands)	788	803	816	829	840	850	860	870	883	898	914
ORIGIN AND USE OF RESOURCES					*(Millions of current Mauritian Rupees)*						
Gross National Product (GNP)	1,131	1,129	1,216	1,236	1,368	1,681	2,187	3,780	4,021	4,657	5,440
Net Factor Income from Abroad	-5	-3	1	7	7	2	16	10	17	-47	-2
Gross Domestic Product (GDP)	1,136	1,132	1,215	1,229	1,361	1,679	2,171	3,770	4,004	4,704	5,442
Indirect Taxes, net	195	207	222	208	224	247	325	479	546	539	666
GDP at factor cost	941	925	993	1,020	1,137	1,431	1,846	3,291	3,458	4,165	4,776
Agriculture	151	144	167	165	195	263	368	985	770	938	939
Industry	218	202	217	223	262	346	436	712	855	1,041	1,213
Manufacturing	130	124	142	146	168	233	277	505	564	631	699
Services, etc.	572	580	609	632	680	823	1,042	1,594	1,832	2,186	2,624
Resource Balance	-57	-27	26	16	-60	37	-46	222	42	-324	-579
Exports of Goods & NFServices	378	452	475	531	523	759	991	2,124	2,269	2,388	2,656
Imports of Goods & NFServices	435	479	449	515	583	722	1,037	1,902	2,227	2,712	3,235
Domestic Absorption	1,193	1,159	1,188	1,213	1,421	1,642	2,217	3,548	3,962	5,028	6,021
Private Consumption, etc.	843	870	864	925	1,042	1,166	1,441	2,238	2,466	2,956	3,658
General Gov't Consumption	163	154	156	166	182	219	235	360	443	575	733
Gross Domestic Investment	187	135	169	122	197	257	541	950	1,053	1,497	1,630
Fixed Investment	145	141	144	145	184	229	480	750	1,138	1,287	1,510
Memo Items:											
Gross Domestic Saving	130	108	195	138	137	294	495	1,172	1,095	1,173	1,051
Gross National Saving	118	100	198	153	153	316	537	1,212	1,149	1,153	1,076
					(Millions of 1980 Mauritian Rupees)						
Gross National Product	4,857	5,054	4,715	4,963	5,167	5,570	6,296	6,877	7,082	7,845	8,562
GDP at factor cost	4,069	4,252	3,904	4,093	4,285	4,727	5,310	5,970	6,092	7,015	7,526
Agriculture	..	..	..	1,371	1,476	1,662	1,736	1,712	1,205	1,333	1,333
Industry	..	..	..	873	920	1,030	1,197	1,502	1,484	1,790	1,952
Manufacturing	..	..	..	620	641	722	794	1,007	926	1,014	1,071
Services, etc.	..	..	..	1,849	1,888	2,035	2,377	2,757	3,403	3,892	4,241
Resource Balance	-260	467	648	408	28	170	241	-1,037	-1,865	-1,554	-1,551
Exports of Goods & NFServices	2,546	3,097	3,199	3,360	3,051	3,733	4,169	3,587	2,918	3,747	4,251
Imports of Goods & NFServices	2,806	2,630	2,551	2,952	3,023	3,564	3,928	4,624	4,784	5,301	5,801
Domestic Absorption	5,010	5,570	4,474	4,537	5,118	5,394	6,019	7,895	8,915	9,473	10,116
Private Consumption, etc.	3,979	4,083	3,399	3,192	3,310	3,328	3,081	4,384	5,241	5,598	6,216
General Gov't Consumption	613	498	498	544	593	678	642	760	816	1,115	1,172
Gross Domestic Investment	1,295	928	1,175	801	1,215	1,389	2,297	2,751	2,858	2,760	2,728
Fixed Investment	..	..	..	..	..	..	..	..	..	2,330	2,481
Memo Items:											
Capacity to Import	2,438	2,482	2,699	3,044	2,712	3,746	3,754	5,164	4,874	4,668	4,763
Terms of Trade Adjustment	-107	-615	-500	-316	-339	13	-415	1,576	1,956	921	512
Gross Domestic Income	4,884	5,139	4,274	4,629	4,807	5,577	5,845	8,435	9,005	8,840	9,078
Gross National Income	4,894	5,133	4,275	4,647	4,828	5,583	5,881	8,454	9,038	8,766	9,075
DOMESTIC PRICES (DEFLATORS)					*(Index 1980 = 100)*						
Overall (GDP)	23.4	22.4	25.8	24.8	26.4	30.2	34.7	55.0	56.8	59.4	63.5
Domestic Absorption	23.8	20.8	26.6	26.7	27.8	30.4	36.8	44.9	44.4	53.1	59.5
Agriculture	..	..	..	12.1	13.2	15.8	21.2	57.6	63.9	70.4	70.4
Industry	..	..	..	25.5	28.4	33.6	36.4	47.4	57.6	58.2	62.2
Manufacturing	..	..	..	23.5	26.2	32.3	34.9	50.2	60.9	62.2	65.3
MANUFACTURING ACTIVITY											
Employment (1980=100)	..	30.8	31.2	32.0	36.3	40.8	51.4	64.1	70.1	88.4	97.3
Real Earnings per Empl. (1980=100)	..	101.4	103.8	101.0	101.8	107.6	96.6	90.7	105.5	116.3	119.3
Real Output per Empl. (1980=100)	..	..	..	138.7	131.5	120.8	179.2	157.3	116.1	100.9	100.7
Earnings as % of Value Added	..	32.2	32.7	34.3	33.9	32.6	26.3	26.2	34.3	44.7	54.7
MONETARY HOLDINGS					*(Millions of current Mauritian Rupees)*						
Money Supply, Broadly Defined	324.0	341.0	432.0	489.0	566.0	713.0	921.0	1,634.0	2,009.0	2,163.0	2,445.0
Money as Means of Payment	224.0	191.0	218.0	230.0	262.0	376.0	467.0	784.0	993.0	1,099.0	1,219.0
Currency Ouside Banks	93.0	89.0	97.0	105.0	126.0	156.0	201.0	314.0	438.0	588.0	694.0
Demand Deposits	130.0	102.0	120.0	125.0	135.0	220.0	266.0	469.0	555.0	511.0	526.0
Quasi-Monetary Liabilities	101.0	150.0	214.0	259.0	304.0	336.0	455.0	851.0	1,015.0	1,064.0	1,226.0
GOVERNMENT DEFICIT (-) OR SURPLUS					*(Millions of current Mauritian Rupees)*						
	..	..	..	..	..	..	-27	-202	-186	-209	-457
Current Revenue	..	..	..	..	..	..	355	460	722	1,058	1,173
Current Expenditure	..	..	..	..	..	..	286	496	669	912	1,176
Current Budget Balance	..	..	..	..	..	..	69	-36	53	147	-3
Capital Receipts	..	..	..	..	..	..	1	0	1	7	..
Capital Payments	..	..	..	..	..	..	97	166	240	363	454

1978	1979	1980	1981	1982	1983	1984	1985	1986	1987 estimate	NOTES	MAURITIUS
1,030	1,210	1,190	1,240	1,230	1,140	1,080	1,100	1,230	1,490	..	CURRENT GNP PER CAPITA (US $)
931	948	966	981	992	1,002	1,011	1,020	1,029	1,042	..	POPULATION (thousands)
				(Millions of current Mauritian Rupees)							ORIGIN AND USE OF RESOURCES
6,258	7,596	8,634	9,949	11,227	12,278	13,734	15,918	18,861	22,332	..	Gross National Product (GNP)
0	-44	-63	-260	-498	-485	-626	-700	-729	-533	..	Net Factor Income from Abroad
6,258	7,640	8,697	10,209	11,725	12,763	14,360	16,618	19,590	22,865	..	Gross Domestic Product (GDP)
764	1,100	1,308	1,444	1,705	2,150	2,310	2,738	3,140	3,780	..	Indirect Taxes, net
5,494	6,540	7,389	8,765	10,020	10,613	12,050	13,880	16,450	19,085	..	GDP at factor cost
977	1,224	914	1,257	1,530	1,465	1,736	2,123	2,510	2,830	..	Agriculture
1,436	1,697	1,912	2,169	2,462	2,596	3,188	4,056	5,194	6,120	..	Industry
801	972	1,127	1,377	1,560	1,678	2,183	2,864	3,830	4,605	..	Manufacturing
3,081	3,619	4,563	5,339	6,028	6,552	7,126	7,701	8,746	10,135	..	Services, etc.
-772	-898	-892	-1,068	-330	-46	-481	-315	1,312	605	..	Resource Balance
2,705	3,260	4,450	4,566	5,529	5,953	6,989	8,895	11,919	15,745	..	Exports of Goods & NFServices
3,477	4,158	5,342	5,634	5,859	5,999	7,470	9,210	10,607	15,140	..	Imports of Goods & NFServices
7,030	8,538	9,589	11,277	12,055	12,809	14,841	16,933	18,278	22,260	..	Domestic Absorption
4,249	5,144	6,562	7,277	8,301	8,874	9,841	11,118	11,890	13,810	..	Private Consumption, etc.
858	1,009	1,224	1,422	1,624	1,706	1,835	1,915	2,068	2,480	..	General Gov't Consumption
1,923	2,385	1,803	2,578	2,130	2,229	3,165	3,900	4,320	5,970	..	Gross Domestic Investment
1,770	1,965	2,028	2,240	2,100	2,300	2,595	3,100	3,890	5,090	..	Fixed Investment
											Memo Items:
1,151	1,487	911	1,510	1,800	2,183	2,684	3,585	5,632	6,575	..	Gross Domestic Saving
1,183	1,475	923	1,354	1,459	1,888	2,325	3,221	5,307	6,570	..	Gross National Saving
				(Millions of 1980 Mauritian Rupees)							
9,070	9,538	8,634	8,734	9,278	9,383	9,789	10,472	11,533	12,961	..	Gross National Product
7,837	8,124	7,389	7,827	8,256	8,337	8,741	9,322	10,136	10,992	..	GDP at factor cost
1,335	1,389	914	1,114	1,335	1,161	1,170	1,302	1,441	1,408	..	Agriculture
2,086	2,111	1,912	1,991	2,082	2,091	2,291	2,608	3,053	3,428	..	Industry
1,153	1,212	1,127	1,225	1,313	1,327	1,489	1,716	2,063	2,353	..	Manufacturing
4,417	4,624	4,563	4,721	4,839	5,085	5,280	5,412	5,641	6,157	..	Services, etc.
-1,647	-1,507	-892	-648	339	254	50	298	396	-855	..	Resource Balance
4,237	4,376	4,450	4,127	4,602	4,645	4,836	5,414	6,874	8,038	..	Exports of Goods & NFServices
5,883	5,883	5,342	4,775	4,263	4,391	4,786	5,116	6,478	8,894	..	Imports of Goods & NFServices
10,717	11,093	9,589	9,767	9,258	9,426	10,093	10,535	11,511	14,088	..	Domestic Absorption
6,519	6,769	6,562	6,332	6,177	6,342	6,596	6,892	7,091	8,367	..	Private Consumption, etc.
1,232	1,216	1,224	1,228	1,243	1,275	1,322	1,322	1,344	1,439	..	General Gov't Consumption
2,966	3,108	1,803	2,207	1,838	1,810	2,174	2,321	3,076	4,282	..	Gross Domestic Investment
2,664	2,495	2,028	1,883	1,619	1,670	1,773	1,950	2,339	2,918	..	Fixed Investment
											Memo Items:
4,577	4,613	4,450	3,870	4,023	4,357	4,478	4,941	7,279	9,249	..	Capacity to Import
341	237	0	-257	-579	-287	-358	-473	405	1,211	..	Terms of Trade Adjustment
9,411	9,823	8,697	8,861	9,018	9,392	9,784	10,360	12,312	14,443	..	Gross Domestic Income
9,411	9,775	8,634	8,477	8,698	9,096	9,431	10,000	11,938	14,172	..	Gross National Income
				(Index 1980 = 100)							DOMESTIC PRICES (DEFLATORS)
69.0	79.7	100.0	112.0	122.2	131.9	141.6	153.4	164.5	172.8	..	Overall (GDP)
65.6	77.0	100.0	115.5	130.2	135.9	147.0	160.7	158.8	158.0	..	Domestic Absorption
73.2	88.1	100.0	112.8	114.6	126.2	148.4	163.1	174.2	201.0	..	Agriculture
68.9	80.4	100.0	108.9	118.2	124.2	139.2	155.5	170.1	178.6	..	Industry
69.5	80.2	100.0	112.4	118.8	126.5	146.7	166.9	185.7	195.7	..	Manufacturing
											MANUFACTURING ACTIVITY
94.4	97.8	100.0	101.3	104.1	108.7	128.7	134.2	..	..	G J	Employment (1980=100)
127.9	122.7	100.0	99.0	94.6	95.5	91.8	111.8	..	..	G J	Real Earnings per Empl. (1980=100)
108.7	115.4	100.0	108.7	113.2	106.6	90.0	104.2	..	..	G J	Real Output per Empl. (1980=100)
53.9	50.5	52.3	48.9	47.8	49.9	46.6	48.1	..	..	J	Earnings as % of Value Added
				(Millions of current Mauritian Rupees)							MONETARY HOLDINGS
2,968.0	3,218.0	3,939.0	4,093.0	5,048.0	5,556.0	6,342.0	7,569.0	9,354.0	13,515.0	D	Money Supply, Broadly Defined
1,449.0	1,426.0	1,720.0	1,533.0	1,741.0	1,804.0	2,050.0	2,373.0	2,670.0	3,241.0	..	Money as Means of Payment
824.0	725.0	735.0	791.0	875.0	922.0	958.0	1,096.0	1,305.0	1,663.0	..	Currency Ouside Banks
625.0	701.0	985.0	742.0	866.0	881.0	1,092.0	1,277.0	1,365.0	1,577.0	..	Demand Deposits
1,519.0	1,792.0	2,218.0	2,560.0	3,307.0	3,752.0	4,292.0	5,196.0	6,684.0	10,274.0	..	Quasi-Monetary Liabilities
				(Millions of current Mauritian Rupees)							GOVERNMENT DEFICIT (-) OR SURPLUS
-727	-882	-897	-1,293	-1,388	-1,160	-857	-823	-640	-939	C	
1,234	1,418	1,813	2,073	2,289	2,825	3,123	3,562	4,131	4,550	..	Current Revenue
1,404	1,758	1,972	2,471	2,892	3,223	3,394	3,691	3,938	4,440	..	Current Expenditure
-170	-340	-159	-398	-604	-397	-271	-129	193	110	..	Current Budget Balance
..	..	..	..	..	..	..	..	..	..	..	Capital Receipts
558	542	737	895	785	763	586	694	833	1,049	..	Capital Payments

MAURITIUS	1967	1968	1969	1970	1971	1972	1973	1974	1975	1976	1977
FOREIGN TRADE (CUSTOMS BASIS)					*(Millions of current US dollars)*						
Value of Exports, fob	62.7	62.8	57.9	69.0	64.9	107.0	137.8	312.6	298.0	265.0	309.8
Nonfuel Primary Products	62.4	62.3	56.8	67.0	62.0	102.2	125.9	285.1	261.4	212.5	237.4
Fuels	0.0	0.0	0.0	0.0	0.0	0.0	0.0	0.0	0.0	0.0	0.1
Manufactures	0.4	0.5	1.1	2.0	2.9	4.8	11.9	27.4	36.5	52.5	72.4
Value of Imports, cif	77.9	75.7	75.2	75.6	83.1	119.9	169.6	308.5	330.7	357.5	444.9
Nonfuel Primary Products	30.6	29.9	28.2	29.3	30.8	39.0	49.7	111.0	97.4	97.3	121.9
Fuels	5.5	6.6	5.5	5.4	5.3	9.6	11.8	28.4	32.4	31.3	41.7
Manufactures	41.9	39.2	41.5	41.0	46.9	71.3	108.1	169.1	200.9	229.0	281.3
					(Index 1980 = 100)						
Terms of Trade	36.9	39.2	55.6	61.9	64.5	94.2	89.6	157.3	113.5	73.6	55.7
Export Prices, fob	8.4	8.1	12.6	14.4	17.4	26.7	34.6	96.8	69.7	44.7	36.3
Nonfuel Primary Products	8.3	8.1	12.4	14.2	17.0	26.3	33.7	101.0	70.2	41.4	31.6
Fuels	4.3	4.3	4.3	4.3	5.6	6.2	8.9	36.7	35.7	38.4	42.0
Manufactures	30.1	31.7	32.4	37.5	36.5	38.5	49.5	67.2	66.1	66.7	71.7
Import Prices, cif	22.7	20.8	22.6	23.3	27.0	28.3	38.6	61.5	61.4	60.8	65.3
Trade at Constant 1980 Prices					*(Millions of 1980 US dollars)*						
Exports, fob	750.4	771.2	461.1	478.4	372.9	401.2	398.3	323.1	427.7	592.5	852.5
Imports, cif	343.8	364.2	333.3	324.3	307.7	423.6	439.4	501.7	539.0	587.9	681.9
BALANCE OF PAYMENTS					*(Millions of current US dollars)*						
Exports of Goods & Services	..	..	..	98	97	145	186	376	379	340	405
Merchandise, fob	..	..	..	70	66	108	139	315	303	265	307
Nonfactor Services	..	..	..	25	28	34	42	57	66	61	93
Factor Services	..	..	..	3	3	3	5	5	10	15	5
Imports of Goods & Services	..	..	..	94	107	138	193	337	373	406	496
Merchandise, fob	..	..	..	65	74	99	143	268	279	308	368
Nonfactor Services	..	..	..	27	30	36	46	66	87	90	121
Factor Services	..	..	..	2	3	3	3	4	8	9	8
Long-Term Interest	..	..	..	2	2	1	1	2	2	3	3
Current Transfers, net	..	..	..	..	..	..	..	..	..	..	..
Workers' Remittances	..	..	..	0	0	0	0	0	0	0	0
Total to be Financed	..	..	..	..	..	..	..	..	..	..	..
Official Capital Grants	..	..	..	..	..	..	..	..	..	..	..
Current Account Balance	..	..	..	8	-5	16	0	54	18	-36	-79
Long-Term Capital, net	..	..	..	4	3	7	3	15	12	11	22
Direct Investment	..	..	..	2	1	-1	-3	1	2	3	2
Long-Term Loans	..	..	..	1	0	1	7	6	9	7	15
Disbursements	..	..	..	2	3	3	9	8	17	13	23
Repayments	..	..	..	1	3	2	2	2	9	6	8
Other Long-Term Capital	..	..	..	2	2	7	-1	9	2	1	6
Other Capital, net	..	..	..	1	0	-1	-7	-6	22	-50	19
Change in Reserves	..	..	..	-14	2	-21	3	-63	-52	76	38
Memo Item:					*(Mauritian Rupees per US dollar)*						
Conversion Factor (Annual Avg)	4.830	5.560	5.560	5.560	5.480	5.340	5.440	5.700	6.030	6.680	6.610
EXTERNAL DEBT, ETC.					*(Millions of US dollars, outstanding at end of year)*						
Public/Publicly Guar. Long-Term	..	..	..	32	33	33	35	41	46	51	71
Official Creditors	..	..	..	21	25	25	28	34	43	48	71
IBRD and IDA	..	..	..	6	6	7	8	9	16	21	30
Private Creditors	..	..	..	10	9	8	7	7	3	3	..
Private Non-guaranteed Long-Term	..	..	..	..	..	3	8	7	7	14	12
Use of Fund Credit	..	..	..	0	0	0	0	0	0	0	13
Short-Term Debt	..	..	..	..	..	..	..	..	..	..	41
Memo Items:					*(Millions of US dollars)*						
Int'l Reserves Excluding Gold	13.3	16.3	33.1	46.2	51.7	70.1	66.8	131.1	166.0	89.5	66.3
Gold Holdings (at market price)	..	..	..	..	..	..	..	..	..	..	1.6
SOCIAL INDICATORS											
Total Fertility Rate	4.3	4.1	3.9	3.6	3.4	3.3	3.2	3.2	3.1	3.1	3.1
Crude Birth Rate	32.2	31.0	29.8	28.5	27.3	26.1	26.2	26.3	26.4	26.5	26.6
Infant Mortality Rate	65.0	63.8	62.6	61.4	60.2	59.0	57.2	55.4	53.6	51.8	50.0
Life Expectancy at Birth	61.6	61.8	62.1	62.4	62.7	62.9	63.3	63.7	64.1	64.5	64.9
Food Production, p.c. ('79-81 = 100)	124.7	110.0	119.1	103.0	106.8	121.0	119.7	118.0	90.4	123.9	115.3
Labor Force, Agriculture (%)	35.7	35.1	34.6	34.0	33.4	32.8	32.2	31.6	31.0	30.4	29.8
Labor Force, Female (%)	20.0	20.0	20.1	20.2	20.5	20.8	21.1	21.5	21.8	22.0	22.2
School Enroll. Ratio, primary	..	..	..	94.0	..	..	..	..	107.0	..	109.0
School Enroll. Ratio, secondary	..	..	..	30.0	..	..	..	..	39.0	43.0	46.0

1978	1979	1980	1981	1982	1983	1984	1985	1986	1987 estimate	NOTES	MAURITIUS
											FOREIGN TRADE (CUSTOMS BASIS)
					(Millions of current US dollars)						
325.8	376.0	430.6	324.0	367.0	368.0	372.3	434.5	662.4	817.4	..	Value of Exports, fob
238.6	271.3	305.9	202.5	248.5	252.1	235.9	244.8	385.8	484.6	..	Nonfuel Primary Products
0.0	0.0	0.0	0.0	0.0	0.0	0.0	0.0	0.0	0.0	..	Fuels
87.1	104.7	124.7	121.6	118.5	115.9	136.4	189.8	276.6	332.9	..	Manufactures
498.4	566.0	619.3	554.0	464.0	440.0	471.0	528.5	675.6	831.3	..	Value of Imports, cif
150.8	174.4	194.2	181.0	149.5	141.8	135.0	146.5	181.6	201.7	..	Nonfuel Primary Products
45.7	66.0	87.5	100.0	77.2	73.3	73.8	77.5	41.6	58.8	..	Fuels
301.8	325.7	337.7	273.0	237.3	224.9	262.3	304.5	452.5	570.7	..	Manufactures
					(Index 1980 = 100)						
49.0	50.9	100.0	69.5	43.1	44.2	34.8	30.5	37.8	37.5	..	Terms of Trade
36.2	43.9	100.0	71.5	41.3	41.0	31.5	27.2	35.3	38.4	..	Export Prices, fob
30.4	36.6	100.0	61.2	32.5	32.7	23.0	17.7	24.0	26.5	..	Nonfuel Primary Products
42.3	61.0	100.0	112.5	101.6	92.5	90.2	87.5	45.0	56.9	..	Fuels
76.1	90.0	100.0	99.1	94.8	90.7	89.3	89.0	103.2	113.0	..	Manufactures
74.0	86.1	100.0	102.8	95.8	92.7	90.6	89.2	93.4	102.6	..	Import Prices, cif
					(Millions of 1980 US dollars)						Trade at Constant 1980 Prices
898.8	857.5	430.6	453.5	889.7	898.4	1,180.3	1,598.8	1,874.2	2,126.6	..	Exports, fob
673.9	657.4	619.3	539.2	484.4	474.8	520.2	592.8	723.3	810.6	..	Imports, cif
					(Millions of current US dollars)						**BALANCE OF PAYMENTS**
443	515	579	510	511	509	508	578	892	1,229	..	Exports of Goods & Services
320	382	434	330	367	370	375	430	673	901	..	Merchandise, fob
118	130	140	174	141	138	130	146	213	322	..	Nonfactor Services
6	4	5	7	4	3	3	2	6	6	..	Factor Services
576	676	718	680	588	557	590	644	848	1,223	..	Imports of Goods & Services
419	483	516	477	397	386	416	458	616	908	..	Merchandise, fob
144	173	174	150	141	127	127	139	172	267	..	Nonfactor Services
14	20	28	53	50	44	48	47	60	48	..	Factor Services
9	15	22	35	35	28	27	28	29	31	..	Long-Term Interest
..	..	..	..	..	..	..	..	30	41	..	Current Transfers, net
0	0	0	0	0	0	0	0	0	..	..	Workers' Remittances
..	..	..	..	..	..	..	..	74	47	..	Total to be Financed
..	..	..	..	..	..	..	..	20	25	..	Official Capital Grants
-119	-149	-119	-153	-42	-22	-54	-30	94	72	..	Current Account Balance
87	85	79	61	68	-9	53	39	22	92	..	Long-Term Capital, net
4	-1	1	1	2	2	5	8	27	44	..	Direct Investment
79	75	73	52	47	-22	42	10	12	44	..	Long-Term Loans
87	86	91	73	79	38	95	54	52	92	..	Disbursements
8	11	18	22	32	60	53	44	40	48	..	Repayments
3	11	5	9	19	11	6	20	-37	-21	..	Other Long-Term Capital
11	17	58	-23	-54	1	19	8	11	2	..	Other Capital, net
21	47	-18	115	28	30	-19	-17	-127	-166	..	Change in Reserves
					(Mauritian Rupees per US dollar)						Memo Item:
6.160	6.310	7.680	8.940	10.870	11.710	13.800	15.440	13.470	12.880	..	Conversion Factor (Annual Avg)
				(Millions of US dollars, outstanding at end of year)							**EXTERNAL DEBT, ETC.**
156	226	296	329	363	328	341	393	445	545	..	Public/Publicly Guar. Long-Term
110	131	161	204	217	226	234	303	360	455	..	Official Creditors
41	49	55	76	81	82	92	137	168	212	..	IBRD and IDA
46	95	135	126	146	102	107	90	85	90	..	Private Creditors
15	23	24	24	18	14	13	15	22	46	..	Private Non-guaranteed Long-Term
15	52	90	149	166	172	154	159	158	150	..	Use of Fund Credit
61	67	47	39	32	49	39	51	38	34	..	Short-Term Debt
					(Millions of US dollars)						Memo Items:
45.8	29.2	90.7	35.1	38.0	17.9	23.6	30.0	136.0	343.5	..	Int'l Reserves Excluding Gold
7.4	19.3	22.2	15.0	17.2	14.3	11.6	12.3	14.7	18.2	..	Gold Holdings (at market price)
											SOCIAL INDICATORS
2.9	2.8	2.7	2.5	2.4	2.4	2.3	2.3	2.2	2.2	..	Total Fertility Rate
25.8	25.0	24.1	23.3	22.5	21.8	21.1	19.0	18.6	20.6	..	Crude Birth Rate
48.4	46.8	45.2	43.6	42.0	42.1	42.2	42.3	42.4	42.4	..	Infant Mortality Rate
65.1	65.2	65.4	65.5	65.7	65.9	66.1	66.3	66.5	66.7	..	Life Expectancy at Birth
116.0	115.1	86.8	98.1	116.5	96.6	93.5	101.0	104.6	104.1	..	Food Production, p.c. ('79-81 = 100)
29.1	28.5	27.9	..	..	..	..	..	..	..	..	Labor Force, Agriculture (%)
22.4	22.7	22.9	23.3	23.7	24.1	24.5	24.9	25.2	25.6	..	Labor Force, Female (%)
109.0	103.0	108.0	114.0	114.0	112.0	106.0	106.0	..	..	..	School Enroll. Ratio, primary
48.0	50.0	48.0	48.0	49.0	51.0	51.0	51.0	..	..	..	School Enroll. Ratio, secondary

MEXICO	1967	1968	1969	1970	1971	1972	1973	1974	1975	1976	1977
CURRENT GNP PER CAPITA (US $)	530	580	620	710	750	820	940	1,120	1,360	1,500	1,490
POPULATION (thousands)	47,792	49,398	51,065	52,789	54,569	56,408	58,277	60,132	61,945	63,720	65,458
ORIGIN AND USE OF RESOURCES					*(Billions of current Mexican Pesos)*						
Gross National Product (GNP)	319	353	390	461	508	586	715	930	1,137	1,410	1,899
Net Factor Income from Abroad	-10	-11	-12	-18	-21	-24	-31	-37	-42	-60	-73
Gross Domestic Product (GDP)	329	364	403	479	529	610	746	966	1,179	1,471	1,972
Indirect Taxes, net	15	17	19	23	26	30	38	45	67	78	114
GDP at factor cost	314	347	384	456	503	580	708	922	1,112	1,393	1,857
Agriculture	41	42	45	56	61	64	83	107	127	151	201
Industry	94	106	119	141	153	178	216	289	352	438	605
Manufacturing	67	76	85	106	119	135	165	217	258	317	443
Services, etc.	194	216	239	282	314	368	447	570	700	882	1,166
Resource Balance	-7	-8	-7	-12	-10	-10	-14	-28	-39	-32	-20
Exports of Goods & NFServices	23	26	31	31	33	41	52	67	68	104	170
Imports of Goods & NFServices	30	34	38	43	43	50	66	96	106	136	190
Domestic Absorption	336	372	410	491	538	620	760	994	1,218	1,503	1,991
Private Consumption, etc.	247	276	302	358	402	456	549	704	845	1,048	1,372
General Gov't Consumption	22	25	28	31	36	47	61	80	110	146	193
Gross Domestic Investment	67	71	80	102	100	116	149	210	263	308	426
Fixed Investment	63	70	78	95	94	115	143	192	252	309	389
Memo Items:											
Gross Domestic Saving	60	63	73	90	91	106	135	182	224	276	407
Gross National Saving	50	52	61	72	70	82	105	147	183	217	336
					(Billions of 1980 Mexican Pesos)						
Gross National Product	1,806.0	1,949.5	2,075.3	2,344.7	2,440.8	2,647.7	2,866.3	3,034.5	3,203.4	3,325.7	3,446.5
GDP at factor cost	1,830.7	1,974.2	2,097.4	2,248.3	2,341.9	2,538.9	2,745.9	2,925.3	3,056.4	3,200.6	3,286.4
Agriculture	240.6	248.1	250.8	263.1	278.2	280.1	291.5	298.9	305.0	308.0	331.2
Industry	551.9	602.4	656.4	718.8	732.1	805.6	894.5	958.5	1,009.0	1,061.9	1,085.5
Manufacturing	381.4	421.3	456.6	496.2	515.3	565.8	625.1	664.8	698.3	733.4	759.5
Services, etc.	1,123.3	1,220.7	1,294.3	1,381.5	1,450.4	1,585.0	1,705.9	1,809.5	1,927.9	2,009.1	2,071.8
Resource Balance	21.6	19.6	41.2	8.0	25.9	42.4	41.7	-9.2	-36.8	4.9	82.5
Exports of Goods & NFServices	180.1	196.1	227.9	214.7	223.1	259.8	295.3	295.9	269.5	314.3	360.4
Imports of Goods & NFServices	158.5	176.5	186.7	206.7	197.2	217.4	253.6	305.1	306.4	309.4	277.9
Domestic Absorption	1,894.3	2,051.5	2,160.2	2,355.5	2,434.8	2,628.3	2,850.2	3,076.1	3,278.7	3,374.1	3,406.0
Private Consumption, etc.	1,364.8	1,486.7	1,553.7	1,643.1	1,727.2	1,840.0	1,954.4	2,053.9	2,179.3	2,271.8	2,310.0
General Gov't Consumption	156.1	172.3	180.0	193.0	213.5	242.2	266.5	283.3	323.4	343.9	340.0
Gross Domestic Investment	373.4	392.5	426.6	519.4	494.1	546.1	629.3	738.9	776.1	758.3	755.9
Fixed Investment	364.0	398.6	427.8	497.2	488.7	548.5	629.4	679.1	742.0	745.3	695.3
Memo Items:											
Capacity to Import	120.9	132.8	152.0	147.3	153.3	176.1	200.0	215.3	194.8	236.2	248.9
Terms of Trade Adjustment	-59.3	-63.3	-75.9	-67.4	-69.8	-83.7	-95.3	-80.6	-74.7	-78.1	-111.5
Gross Domestic Income	1,856.6	2,007.8	2,125.5	2,296.0	2,390.9	2,587.0	2,796.6	2,986.3	3,167.2	3,300.9	3,377.0
Gross National Income	1,746.7	1,886.2	1,999.3	2,277.3	2,371.0	2,564.0	2,771.0	2,953.9	3,128.7	3,247.6	3,335.0
DOMESTIC PRICES (DEFLATORS)					*(Index 1980 = 100)*						
Overall (GDP)	17.2	17.6	18.3	20.3	21.5	22.8	25.8	31.5	36.4	43.5	56.5
Domestic Absorption	17.7	18.1	19.0	20.9	22.1	23.6	26.7	32.3	37.2	44.5	58.5
Agriculture	17.0	16.9	17.8	21.2	22.0	23.0	28.4	35.9	41.6	48.9	60.6
Industry	17.0	17.5	18.1	19.6	20.9	22.1	24.2	30.2	34.9	41.2	55.7
Manufacturing	17.4	18.0	18.7	21.3	23.0	23.9	26.3	32.6	36.9	43.3	58.3
MANUFACTURING ACTIVITY											
Employment (1980=100)	..	..	..	71.4	73.3	75.8	79.6	82.6	82.8	84.7	84.9
Real Earnings per Empl. (1980=100)	..	..	..	92.6	93.4	95.0	96.4	95.5	98.1	104.9	106.3
Real Output per Empl. (1980=100)	..	..	..	77.1	76.2	80.3	85.2	90.2	93.3	94.9	95.1
Earnings as % of Value Added	..	..	..	44.0	42.7	41.4	40.6	39.2	39.1	40.1	37.6
MONETARY HOLDINGS					*(Billions of current Mexican Pesos)*						
Money Supply, Broadly Defined	87	97	115	135	150	168	209	245	296	530	566
Money as Means of Payment	37	42	49	54	58	68	84	101	122	158	208
Currency Ouside Banks	15	17	19	20	22	27	34	43	53	80	89
Demand Deposits	22	26	30	34	36	41	49	58	70	78	119
Quasi-Monetary Liabilities	50	55	67	81	92	100	126	144	173	372	358
					(Billions of current Mexican Pesos)						
GOVERNMENT DEFICIT (-) OR SURPLUS	..	..	..	..	..	-17	-28	-34	-54	-64	-61
Current Revenue	..	..	..	..	..	58	70	96	134	169	241
Current Expenditure	..	..	..	..	..	47	62	93	124	159	222
Current Budget Balance	..	..	..	..	..	11	7	3	10	10	19
Capital Receipts	..	..	..	..	..	..	..	..	..	..	..
Capital Payments	..	..	..	..	..	28	35	37	64	74	80

1978	1979	1980	1981	1982	1983	1984	1985	1986	1987 estimate	NOTES	MEXICO
1,580	1,820	2,320	3,000	2,820	2,290	2,130	2,200	1,900	1,820	..	**CURRENT GNP PER CAPITA (US $)**
67,157	68,819	70,450	72,057	73,652	75,249	76,866	78,524	80,247	81,950	..	**POPULATION (thousands)**
				(Billions of current Mexican Pesos)							**ORIGIN AND USE OF RESOURCES**
2,402	3,142	4,341	5,905	9,240	16,824	27,853	45,307	74,977	191,880	..	Gross National Product (GNP)
-90	-113	-129	-223	-558	-1,055	-1,618	-2,096	-4,377	-3,734	..	Net Factor Income from Abroad
2,491	3,255	4,470	6,128	9,798	17,879	29,472	47,403	79,353	195,615	..	Gross Domestic Product (GDP)
147	220	371	497	974	1,383	2,435	4,627	6,764	20,363	..	Indirect Taxes, net
2,344	3,036	4,099	5,631	8,824	16,496	27,037	42,775	72,589	175,251	B	GDP at factor cost
247	290	368	503	720	1,392	2,533	4,307	7,466	..	..	Agriculture
763	1,017	1,464	1,956	3,057	6,016	9,860	15,804	26,820	..	..	Industry
553	717	989	1,326	2,033	3,772	6,618	11,069	19,557	..	f	Manufacturing
1,481	1,948	2,638	3,668	6,021	10,471	17,078	27,292	45,067	..	..	Services, etc.
-41	-78	-101	-155	491	1,713	2,307	2,304	3,772	3,929	..	Resource Balance
218	306	479	638	1,502	3,397	5,122	7,294	13,734	14,269	..	Exports of Goods & NFServices
259	383	580	793	1,011	1,684	2,815	4,990	9,962	10,341	..	Imports of Goods & NFServices
2,532	3,333	4,572	6,283	9,306	16,166	27,164	45,098	75,582	191,686	..	Domestic Absorption
1,728	2,205	2,909	3,945	6,036	10,882	18,590	30,349	53,692	142,273	..	Private Consumption, etc.
247	324	449	660	1,026	1,574	2,722	4,374	7,235	19,181	..	General Gov't Consumption
557	803	1,214	1,678	2,244	3,710	5,853	10,375	14,655	30,233	..	Gross Domestic Investment
528	770	1,107	1,617	2,249	3,137	5,287	9,076	15,400		..	Fixed Investment
											Memo Items:
516	726	1,113	1,523	2,736	5,423	8,160	12,680	18,426	34,161	..	Gross Domestic Saving
428	616	987	1,303	2,183	4,383	6,580	10,664	14,208	30,830	..	Gross National Saving
				(Billions of 1980 Mexican Pesos)							
3,730.0	4,055.8	4,341.3	4,684.8	4,555.6	4,327.0	4,510.3	4,708.6	4,477.1	4,709.0	..	Gross National Product
3,552.1	3,846.5	4,099.2	4,467.9	4,351.6	4,270.8	4,399.8	4,439.6	4,322.5	4,294.0	B	GDP at factor cost
351.0	343.7	368.0	390.6	382.9	390.6	401.1	416.2	404.8	406.9	..	Agriculture
1,200.5	1,337.6	1,464.4	1,595.8	1,562.8	1,423.0	1,490.2	1,560.6	1,467.9	1,497.3	..	Industry
833.9	922.6	988.9	1,052.7	1,023.8	943.5	990.9	1,050.2	990.5	1,015.2	f	Manufacturing
2,223.9	2,443.5	2,637.6	2,875.9	2,886.0	2,815.3	2,904.8	2,943.1	2,852.6	2,888.7	..	Services, etc.
63.6	11.2	-101.5	-148.8	226.0	457.6	449.7	374.0	439.7	521.2	..	Resource Balance
402.1	451.0	478.5	533.9	650.3	738.5	780.6	749.4	760.9	853.8	..	Exports of Goods & NFServices
338.6	439.8	580.0	682.7	424.3	280.9	330.9	375.4	321.2	332.6	..	Imports of Goods & NFServices
3,711.9	4,113.6	4,571.6	5,011.0	4,605.7	4,171.3	4,346.4	4,545.9	4,285.6	4,271.7	..	Domestic Absorption
2,491.9	2,708.9	2,908.9	3,123.2	3,045.9	2,882.6	2,976.7	3,073.4	2,988.0	2,946.2	..	Private Consumption, etc.
373.8	409.6	448.7	494.8	504.9	518.6	552.8	557.8	569.6	564.1	..	General Gov't Consumption
846.1	995.1	1,214.0	1,393.0	1,054.9	770.1	816.9	914.7	728.0	761.4	..	Gross Domestic Investment
800.8	963.0	1,106.8	1,286.4	1,070.4	767.8	817.0	883.6	777.2	771.4	..	Fixed Investment
											Memo Items:
285.0	350.7	478.5	549.3	630.6	566.6	602.1	548.8	442.8	459.0	..	Capacity to Import
-117.2	-100.3	0.0	15.4	-19.7	-171.9	-178.5	-200.6	-318.1	-394.8	..	Terms of Trade Adjustment
3,658.2	4,024.5	4,470.1	4,877.6	4,812.0	4,457.0	4,617.5	4,719.3	4,407.2	4,398.1	..	Gross Domestic Income
3,612.8	3,955.6	4,341.3	4,700.2	4,535.9	4,155.1	4,331.8	4,508.0	4,159.0	4,314.2	..	Gross National Income
				(Index 1980 = 100)							**DOMESTIC PRICES (DEFLATORS)**
66.0	78.9	100.0	126.0	202.8	386.2	614.5	963.5	1679.3	4081.3	..	Overall (GDP)
68.2	81.0	100.0	125.4	202.1	387.5	625.0	992.1	1763.6	4487.3	..	Domestic Absorption
70.3	84.3	100.0	128.8	187.9	356.4	631.6	1034.8	1844.3	..	..	Agriculture
63.6	76.0	100.0	122.6	195.6	422.7	661.7	1012.6	1827.1	..	..	Industry
66.3	77.8	100.0	126.0	198.6	399.8	667.9	1054.0	1974.6	..	..	Manufacturing
											MANUFACTURING ACTIVITY
88.2	94.8	100.0	105.2	102.8	93.1	91.4	91.7	..	..	J	Employment (1980=100)
103.4	102.4	100.0	103.0	100.6	74.5	72.6	85.6	..	..	J	Real Earnings per Empl. (1980=100)
99.8	102.1	100.0	100.6	98.7	100.9	107.8	106.6	..	..	J	Real Output per Empl. (1980=100)
35.8	34.7	32.9	34.2	34.0	23.8	21.1	26.0	..	..	J	Earnings as % of Value Added
				(Billions of current Mexican Pesos)							**MONETARY HOLDINGS**
759	1,029	1,426	2,141	3,330	5,443	9,185	13,255	23,612	58,555	D	Money Supply, Broadly Defined
270	361	477	635	1,031	1,447	2,315	3,462	5,790	12,627	..	Money as Means of Payment
115	150	195	283	505	681	1,122	1,738	3,067	7,339	..	Currency Ouside Banks
155	211	282	352	526	766	1,193	1,724	2,723	5,288	..	Demand Deposits
489	668	949	1,506	2,299	3,996	6,870	9,793	17,822	45,928	..	Quasi-Monetary Liabilities
				(Billions of current Mexican Pesos)							**GOVERNMENT DEFICIT (-) OR SURPLUS**
-63	-102	-134	-392	-1,454	-1,363	-2,094	-3,978	..	..	..	
323	439	675	895	1,520	3,222	4,773	7,820	..	..	..	Current Revenue
276	349	507	843	2,191	3,660	5,451	9,669	..	..	..	Current Expenditure
47	90	169	52	-671	-438	-678	-1,849	..	..	..	Current Budget Balance
..	..	0	0	0	0	1	0	..	..	..	Capital Receipts
110	192	302	445	783	925	1,417	2,129	..	..	..	Capital Payments

MEXICO	1967	1968	1969	1970	1971	1972	1973	1974	1975	1976	1977
FOREIGN TRADE (CUSTOMS BASIS)				*(Millions of current US dollars)*							
Value of Exports, fob	1,026	1,110	1,254	1,205	1,320	1,845	2,632	2,957	2,993	3,468	4,284
Nonfuel Primary Products	792	850	892	775	785	1,174	1,503	1,713	1,599	1,902	2,089
Fuels	39	34	40	38	31	22	25	124	463	554	1,014
Manufactures	194	226	321	392	504	649	1,103	1,120	931	1,012	1,182
Value of Imports, cif	1,746	1,962	2,078	2,461	2,407	2,935	4,146	6,057	6,572	6,033	5,589
Nonfuel Primary Products	236	226	259	397	339	421	793	1,529	1,347	887	1,142
Fuels	57	57	75	78	104	143	293	426	361	343	166
Manufactures	1,453	1,679	1,744	1,986	1,964	2,371	3,060	4,102	4,864	4,802	4,281
				(Index 1980 = 100)							
Terms of Trade	97.2	104.4	105.4	108.4	116.9	121.8	139.6	109.3	94.1	105.3	98.3
Export Prices, fob	25.2	27.3	27.9	30.7	34.3	38.5	51.8	63.4	58.3	65.3	67.2
Nonfuel Primary Products	31.4	33.1	34.4	36.4	36.9	40.4	57.2	63.5	61.2	76.0	85.8
Fuels	4.3	4.3	4.3	4.3	5.6	6.2	8.9	36.7	35.7	38.4	42.0
Manufactures	31.1	32.6	33.3	43.9	43.2	42.4	50.8	68.9	75.8	74.2	77.3
Import Prices, cif	25.9	26.2	26.5	28.3	29.3	31.6	37.1	58.1	62.0	62.0	68.3
Trade at Constant 1980 Prices				*(Millions of 1980 US dollars)*							
Exports, fob	4,074	4,061	4,495	3,921	3,855	4,786	5,083	4,662	5,135	5,309	6,378
Imports, cif	6,742	7,494	7,851	8,682	8,213	9,274	11,180	10,433	10,606	9,727	8,179
BALANCE OF PAYMENTS				*(Millions of current US dollars)*							
Exports of Goods & Services	..	..	..	2,935	3,171	3,816	4,840	6,368	6,365	7,204	8,210
Merchandise, fob	..	..	..	1,348	1,409	1,717	2,141	2,999	3,009	3,476	4,604
Nonfactor Services	..	..	..	1,397	1,586	1,892	2,463	3,057	3,063	3,364	3,187
Factor Services	..	..	..	190	177	207	236	311	293	364	419
Imports of Goods & Services	..	..	..	4,058	4,064	4,797	6,329	9,365	10,559	10,771	10,229
Merchandise, fob	..	..	..	2,236	2,158	2,610	3,656	5,791	6,292	5,773	5,620
Nonfactor Services	..	..	..	1,181	1,238	1,406	1,580	1,817	2,193	2,519	2,040
Factor Services	..	..	..	641	668	781	1,093	1,757	2,075	2,479	2,569
Long-Term Interest	..	..	..	283	312	353	516	813	1,104	1,360	1,588
Current Transfers, net	..	..	..	25	26	25	41	57	59	73	88
Workers' Remittances	..	..	..	0	0	0	0	0	0	0	0
Total to be Financed	..	..	..	-1,098	-867	-955	-1,448	-2,940	-4,135	-3,494	-1,931
Official Capital Grants	..	..	..	30	31	39	33	65	81	83	82
Current Account Balance	..	..	..	-1,068	-835	-916	-1,415	-2,876	-4,054	-3,410	-1,849
Long-Term Capital, net	..	..	..	656	777	866	1,853	3,111	4,787	5,079	4,669
Direct Investment	..	..	..	323	307	301	457	678	610	628	556
Long-Term Loans	..	..	..	358	453	602	1,819	2,669	3,746	4,504	4,709
Disbursements	..	..	..	1,375	1,502	1,780	3,173	3,979	5,255	6,489	7,819
Repayments	..	..	..	1,017	1,049	1,178	1,354	1,310	1,509	1,985	3,110
Other Long-Term Capital	..	..	..	-55	-14	-76	-456	-301	351	-136	-678
Other Capital, net	..	..	..	441	188	228	-280	-162	-530	-2,530	-2,196
Change in Reserves	..	..	..	-29	-130	-178	-157	-74	-204	862	-624
Memo Item:				*(Mexican Pesos per US dollar)*							
Conversion Factor (Annual Avg)	12.500	12.500	12.500	12.500	12.500	12.500	12.500	12.500	12.500	15.400	22.600
EXTERNAL DEBT, ETC.				*(Millions of US dollars, outstanding at end of year)*							
Public/Publicly Guar. Long-Term	..	..	..	3,196	3,482	3,907	5,554	8,175	11,414	15,806	20,703
Official Creditors	..	..	..	1,149	1,243	1,388	1,592	1,929	2,293	2,613	2,999
IBRD and IDA	..	..	..	582	660	724	808	973	1,123	1,223	1,374
Private Creditors	..	..	..	2,047	2,239	2,519	3,962	6,245	9,120	13,192	17,704
Private Non-guaranteed Long-Term	..	..	..	2,770	2,934	3,121	3,445	3,771	4,195	4,343	4,524
Use of Fund Credit	..	..	..	0	0	0	0	0	0	371	509
Short-Term Debt	..	..	..	..	..	..	..	..	..	..	5,453
Memo Items:				*(Millions of US dollars)*							
Int'l Reserves Excluding Gold	420	492	493	568	752	976	1,160	1,238	1,384	1,188	1,649
Gold Holdings (at market price)	167	198	170	188	229	321	520	683	513	216	289
SOCIAL INDICATORS											
Total Fertility Rate	6.7	6.6	6.6	6.5	6.4	6.4	6.1	5.8	5.5	5.2	4.9
Crude Birth Rate	44.5	44.1	43.8	43.4	43.0	42.6	41.0	39.3	37.7	36.1	34.4
Infant Mortality Rate	79.0	77.0	75.0	73.0	71.0	69.0	67.2	65.4	63.6	61.8	60.0
Life Expectancy at Birth	60.3	60.8	61.2	61.7	62.2	62.6	63.2	63.7	64.3	64.8	65.4
Food Production, p.c. ('79-81=100)	92.0	90.9	88.0	88.7	89.6	92.5	91.8	90.2	92.4	91.3	96.7
Labor Force, Agriculture (%)	47.4	46.3	45.2	44.1	43.3	42.6	41.8	41.1	40.3	39.5	38.8
Labor Force, Female (%)	17.0	17.2	17.5	17.8	18.7	19.7	20.7	21.6	22.6	23.5	24.4
School Enroll. Ratio, primary	..	..	..	104.0	..	..	..	..	109.0	..	113.0
School Enroll. Ratio, secondary	..	..	..	22.0	..	..	..	..	34.0	37.0	39.0

1978	1979	1980	1981	1982	1983	1984	1985	1986	1987 estimate	NOTES	MEXICO
				(Millions of current US dollars)							**FOREIGN TRADE (CUSTOMS BASIS)**
6,301	8,817	15,308	20,041	21,230	22,312	24,196	22,108	16,237	20,887	f	Value of Exports, fob
2,793	3,063	2,767	3,575	2,516	3,163	3,544	2,828	3,221	3,310	..	Nonfuel Primary Products
1,804	3,859	10,306	13,802	16,910	15,862	16,403	14,640	5,890	7,803	..	Fuels
1,704	1,894	2,234	2,664	1,804	3,287	4,249	4,640	7,126	9,774	..	Manufactures
8,053	12,587	19,517	24,161	15,041	8,023	11,788	13,994	12,032	12,731	f	Value of Imports, cif
1,548	2,204	3,490	4,079	2,128	1,732	2,803	2,893	2,286	2,441	..	Nonfuel Primary Products
263	313	318	486	310	161	397	611	124	120	..	Fuels
6,242	10,069	15,709	19,596	12,603	6,130	8,588	10,489	9,622	10,170	..	Manufactures
				(Index 1980 = 100)							
83.7	86.0	100.0	106.5	103.2	96.5	97.2	97.7	66.2	70.0	..	Terms of Trade
64.6	77.6	100.0	106.5	99.8	92.3	91.3	88.7	67.4	78.9	..	Export Prices, fob
83.3	102.9	100.0	89.5	88.3	89.0	89.2	81.8	93.5	91.1	..	Nonfuel Primary Products
42.3	61.0	100.0	112.5	101.6	92.5	90.2	87.5	40.6	51.3	..	Fuels
79.5	92.4	100.0	104.0	101.0	95.0	98.0	98.0	115.8	128.1	..	Manufactures
77.1	90.3	100.0	99.9	96.6	95.7	93.9	90.8	101.7	112.7	..	Import Prices, cif
				(Millions of 1980 US dollars)							Trade at Constant 1980 Prices
9,760	11,356	15,308	18,824	21,280	24,162	26,494	24,922	24,107	26,461	..	Exports, fob
10,440	13,943	19,517	24,176	15,564	8,386	12,552	15,411	11,833	11,293	..	Imports, cif
				(Millions of current US dollars)							**BALANCE OF PAYMENTS**
11,401	16,003	24,628	30,453	27,674	28,611	32,469	29,744	23,682	29,759	..	Exports of Goods & Services
6,226	9,301	16,066	19,938	21,230	22,312	24,196	21,663	16,031	20,655	..	Merchandise, fob
4,495	5,830	7,390	8,945	4,937	4,867	5,960	5,961	5,860	6,879	..	Nonfactor Services
680	872	1,172	1,570	1,507	1,432	2,313	2,120	1,791	2,225	..	Factor Services
14,755	21,687	33,065	44,641	34,188	23,494	28,640	29,509	25,815	26,542	..	Imports of Goods & Services
7,971	12,131	18,896	24,037	14,435	8,550	11,255	13,212	11,432	12,222	..	Merchandise, fob
3,337	4,573	6,788	9,505	5,432	4,140	4,825	5,156	4,740	4,774	..	Nonfactor Services
3,447	4,983	7,381	11,099	14,321	10,804	12,560	11,141	9,643	9,546	..	Factor Services
2,258	3,364	4,590	6,133	7,784	8,151	10,262	9,393	7,737	7,091	..	Long-Term Interest
104	131	132	114	98	122	230	312	260	292	..	Current Transfers, net
0	0	0	0	0	0	0	0	0	0	..	Workers' Remittances
-3,250	-5,553	-8,305	-14,074	-6,416	5,239	4,059	547	-1,873	3,509	K	Total to be Financed
88	94	143	175	198	180	181	690	204	375	K	Official Capital Grants
-3,163	-5,459	-8,162	-13,899	-6,218	5,419	4,240	1,237	-1,669	3,884	..	Current Account Balance
5,151	5,238	7,897	13,220	14,793	7,474	2,632	280	638	4,303	..	Long-Term Capital, net
829	1,332	2,186	2,537	1,655	461	390	491	1,523	3,248	..	Direct Investment
4,557	4,115	6,839	12,508	7,978	2,359	1,618	-96	899	4,217	..	Long-Term Loans
9,462	12,174	11,599	17,016	12,509	7,196	7,276	5,026	5,461	8,550	..	Disbursements
4,905	8,059	4,760	4,508	4,531	4,837	5,657	5,122	4,562	4,333	..	Repayments
-323	-303	-1,271	-2,000	4,962	4,474	443	-805	-1,988	-3,537	..	Other Long-Term Capital
-1,602	536	1,223	1,755	-13,365	-9,643	-4,722	-4,248	1,557	-1,083	..	Other Capital, net
-386	-315	-958	-1,076	4,790	-3,250	-2,150	2,731	-526	-7,104	..	Change in Reserves
				(Mexican Pesos per US dollar)							**Memo Item:**
22.800	22.800	23.000	24.500	56.400	120.100	167.800	256.900	611.800	1,378.200	..	Conversion Factor (Annual Avg)
			(Millions of US dollars, outstanding at end of year)								**EXTERNAL DEBT, ETC.**
25,533	29,068	33,987	43,114	51,642	66,857	69,812	72,711	75,990	82,771	..	Public/Publicly Guar. Long-Term
3,395	3,639	4,481	5,385	6,959	6,713	6,997	8,836	11,736	15,940	..	Official Creditors
1,481	1,731	2,063	2,417	2,692	2,870	2,852	4,034	5,566	7,346	..	IBRD and IDA
22,138	25,429	29,507	37,729	44,684	60,144	62,814	63,875	64,255	66,831	..	Private Creditors
4,955	5,600	7,300	10,200	8,100	14,800	16,296	15,745	15,103	14,148	..	Private Non-guaranteed Long-Term
299	136	0	0	221	1,260	2,360	2,969	4,060	5,163	..	Use of Fund Credit
4,946	8,024	16,163	24,983	26,147	10,139	6,440	5,450	5,900	5,800	..	Short-Term Debt
				(Millions of US dollars)							**Memo Items:**
1,841	2,072	2,960	4,074	834	3,913	7,272	4,906	5,670	12,464	..	Int'l Reserves Excluding Gold
428	1,016	1,215	897	944	881	747	772	1,004	1,228	..	Gold Holdings (at market price)
											SOCIAL INDICATORS
4.8	4.6	4.5	4.3	4.2	4.1	4.0	3.8	3.7	3.6	..	Total Fertility Rate
33.9	33.3	32.8	32.2	31.6	30.9	30.2	29.5	28.8	28.1	..	Crude Birth Rate
58.6	57.2	55.8	54.4	53.0	49.3	45.7	42.0	38.4	34.7	..	Infant Mortality Rate
65.8	66.2	66.6	67.0	67.4	67.6	67.9	68.1	68.4	68.6	..	Life Expectancy at Birth
102.7	96.2	100.1	103.7	97.2	100.9	99.2	95.6	95.0	100.1	..	Food Production, p.c. ('79-81 = 100)
38.0	37.3	36.5	..	..	..	..	..	..	..	..	Labor Force, Agriculture (%)
25.2	26.1	27.0	27.0	27.0	27.0	27.0	27.0	27.1	27.1	..	Labor Force, Female (%)
118.0	120.0	120.0	119.0	119.0	119.0	116.0	115.0	..	..	..	School Enroll. Ratio, primary
42.0	44.0	47.0	51.0	53.0	55.0	55.0	55.0	..	..	..	School Enroll. Ratio, secondary

MOROCCO	1967	1968	1969	1970	1971	1972	1973	1974	1975	1976	1977
CURRENT GNP PER CAPITA (US $)	220	240	250	260	280	300	340	410	500	570	610
POPULATION (thousands)	14,107	14,506	14,908	15,310	15,711	16,111	16,509	16,907	17,305	17,706	18,111

ORIGIN AND USE OF RESOURCES

(Millions of current Moroccan Dirhams)

	1967	1968	1969	1970	1971	1972	1973	1974	1975	1976	1977
Gross National Product (GNP)	15,234	16,317	18,261	19,793	21,742	23,050	25,313	33,334	36,034	40,451	48,894
Net Factor Income from Abroad	-182	-238	-218	-228	-259	-295	-323	-268	-359	-561	-867
Gross Domestic Product (GDP)	15,416	16,555	18,479	20,021	22,001	23,345	25,636	33,602	36,393	41,012	49,761
Indirect Taxes, net	..	..	..	..	..	..	..	..	..	..	..
GDP at factor cost	..	..	..	..	..	..	..	..	..	..	..
Agriculture	3,351	3,712	3,572	3,992	4,811	4,962	5,339	6,872	6,298	7,860	8,153
Industry	4,399	4,652	5,108	5,399	6,000	6,431	7,170	11,742	12,642	13,336	16,198
Manufacturing	2,477	2,667	3,078	3,245	3,584	3,922	4,447	5,246	6,031	6,764	8,241
Services, etc.	7,666	8,191	9,799	10,630	11,190	11,952	13,127	14,988	17,453	19,816	25,410
Resource Balance	-340	-283	-211	-790	-610	-152	-325	-211	-3,958	-7,969	-10,164
Exports of Goods & NFServices	2,799	3,056	3,376	3,531	3,730	4,336	5,342	9,242	8,184	7,588	8,407
Imports of Goods & NFServices	3,139	3,339	3,587	4,321	4,340	4,488	5,667	9,453	12,142	15,557	18,571
Domestic Absorption	15,756	16,838	18,690	20,811	22,611	23,497	25,961	33,813	40,351	48,981	59,925
Private Consumption, etc.	11,835	12,684	13,859	14,703	15,954	17,039	18,647	22,853	25,265	28,238	32,646
General Gov't Consumption	1,878	2,070	2,208	2,407	2,709	2,891	2,991	4,036	5,921	9,211	10,249
Gross Domestic Investment	2,043	2,084	2,623	3,701	3,948	3,567	4,323	6,924	9,165	11,532	17,030
Fixed Investment	2,104	2,193	2,427	2,988	3,269	3,177	3,471	4,932	9,037	12,176	15,901
Memo Items:											
Gross Domestic Saving	1,703	1,801	2,412	2,911	3,338	3,415	3,998	6,713	5,207	3,563	6,866
Gross National Saving	1,521	1,553	2,321	2,865	3,454	3,614	4,541	7,754	6,801	5,208	8,456

(Millions of 1980 Moroccan Dirhams)

	1967	1968	1969	1970	1971	1972	1973	1974	1975	1976	1977
Gross National Product	34,379	38,136	40,901	42,786	45,396	46,389	47,577	49,778	52,591	58,553	62,003
GDP at factor cost	..	..	..	..	..	..	..	..	..	..	..
Agriculture	8,477	10,986	10,517	10,956	12,051	11,784	10,685	11,898	10,158	11,348	10,329
Industry	11,217	11,652	12,499	12,895	13,893	14,762	15,655	16,362	18,173	19,950	21,763
Manufacturing	5,733	6,014	6,697	6,973	7,395	7,802	8,459	8,551	9,108	9,745	10,715
Services, etc.	15,109	16,056	18,380	19,431	19,999	20,438	21,850	21,923	24,801	28,096	31,009
Resource Balance	-2,451	-1,717	-1,473	-2,629	-2,779	-607	-1,015	-3,264	-8,852	-12,608	-14,885
Exports of Goods & NFServices	7,788	9,100	9,869	10,123	9,787	11,576	13,044	12,003	10,120	10,810	11,927
Imports of Goods & NFServices	10,239	10,818	11,343	12,753	12,566	12,184	14,059	15,267	18,973	23,419	26,812
Domestic Absorption	37,254	40,411	42,870	45,911	48,723	47,591	49,206	53,448	61,984	72,002	77,986
Private Consumption, etc.	27,780	29,993	30,794	31,971	33,516	34,276	35,471	37,391	37,000	40,441	43,011
General Gov't Consumption	3,874	4,254	4,446	4,830	5,116	5,146	5,247	5,899	8,188	11,845	12,189
Gross Domestic Investment	5,600	6,164	7,630	9,111	10,091	8,168	8,488	10,158	16,796	19,716	22,785
Fixed Investment	5,787	6,472	7,072	8,188	8,989	7,969	7,707	9,417	16,559	19,625	22,367
Memo Items:											
Capacity to Import	9,130	9,901	10,675	10,421	10,800	11,771	13,253	14,926	12,788	11,423	12,137
Terms of Trade Adjustment	1,342	801	806	298	1,013	195	209	2,923	2,668	612	210
Gross Domestic Income	36,145	39,494	42,203	43,580	46,957	47,178	48,400	53,107	55,800	60,006	63,311
Gross National Income	35,721	38,936	41,707	43,084	46,409	46,584	47,786	52,700	55,259	59,166	62,214

DOMESTIC PRICES (DEFLATORS)

(Index 1980 = 100)

	1967	1968	1969	1970	1971	1972	1973	1974	1975	1976	1977
Overall (GDP)	44.3	42.8	44.6	46.3	47.9	49.7	53.2	67.0	68.5	69.1	78.9
Domestic Absorption	42.3	41.7	43.6	45.3	46.4	49.4	52.8	63.3	65.1	68.0	76.8
Agriculture	39.5	33.8	34.0	36.4	39.9	42.1	50.0	57.8	62.0	69.3	78.9
Industry	39.2	39.9	40.9	41.9	43.2	43.6	45.8	71.8	69.6	66.8	74.4
Manufacturing	43.2	44.3	46.0	46.5	48.5	50.3	52.6	61.4	66.2	69.4	76.9

MANUFACTURING ACTIVITY

	1967	1968	1969	1970	1971	1972	1973	1974	1975	1976	1977
Employment (1980=100)	60.0	..	62.2	..	..	..	..	..	..	81.3	88.0
Real Earnings per Empl. (1980=100)	93.7	..	87.2	..	..	..	..	..	..	93.7	97.5
Real Output per Empl. (1980=100)	65.0	..	79.2	..	..	..	..	..	..	90.6	91.3
Earnings as % of Value Added	60.8	..	45.7	..	..	..	..	..	..	44.0	52.0

MONETARY HOLDINGS

(Millions of current Moroccan Dirhams)

	1967	1968	1969	1970	1971	1972	1973	1974	1975	1976	1977
Money Supply, Broadly Defined	4.9	5.5	6.1	6.6	7.4	8.8	10.2	13.1	15.7	18.5	22.2
Money as Means of Payment	4.1	4.6	5.1	5.5	6.2	7.3	8.6	10.9	12.8	15.1	18.0
Currency Ouside Banks	1.6	1.9	2.1	2.3	2.5	2.9	3.4	4.1	4.7	5.7	6.7
Demand Deposits	2.5	2.7	2.9	3.3	3.7	4.4	5.2	6.8	8.2	9.4	11.4
Quasi-Monetary Liabilities	0.8	0.9	1.0	1.1	1.2	1.4	1.6	2.2	2.9	3.3	4.2

GOVERNMENT DEFICIT (-) OR SURPLUS

(Millions of current Moroccan Dirhams)

	1967	1968	1969	1970	1971	1972	1973	1974	1975	1976	1977
GOVERNMENT DEFICIT (-) OR SURPLUS	..	..	..	-619	-658	-897	-526	-1,348	-3,341	-7,217	-7,647
Current Revenue	..	..	..	3,857	3,884	4,330	5,151	8,449	9,441	9,424	12,219
Current Expenditure	..	..	..	3,326	3,472	4,035	4,373	7,582	8,025	8,782	10,080
Current Budget Balance	..	..	..	531	412	295	778	867	1,416	642	2,139
Capital Receipts	..	..	..	5	6	5	13	19	88	177	114
Capital Payments	..	..	..	1,155	1,076	1,197	1,317	2,234	4,845	8,036	9,900

1978	1979	1980	1981	1982	1983	1984	1985	1986	1987 estimate	NOTES	MOROCCO
650	760	880	840	820	710	620	560	580	620	..	CURRENT GNP PER CAPITA (US $)
18,523	18,946	19,382	19,837	20,314	20,820	21,360	21,941	22,466	22,968	..	POPULATION (thousands)
				(Millions of current Moroccan Dirhams)							ORIGIN AND USE OF RESOURCES
53,850	60,336	67,808	72,986	86,094	90,150	99,664	111,556	127,956	133,592	..	Gross National Product (GNP)
-1,304	-1,707	-2,353	-3,751	-3,994	-4,485	-5,176	-7,754	-6,378	-6,408	..	Net Factor Income from Abroad
55,154	62,043	70,161	76,737	90,088	94,635	104,840	119,310	134,334	140,000	..	Gross Domestic Product (GDP)
..	..	..	..	..	..	..	..	..	..	..	Indirect Taxes, net
..	..	..	..	..	..	..	..	..	..	B	GDP at factor cost
10,435	11,116	12,711	11,422	16,256	16,130	17,547	21,996	28,589	26,000	..	Agriculture
17,287	20,268	22,646	25,762	28,457	30,093	33,358	38,324	40,374	43,430	..	Industry
9,367	10,436	12,010	13,416	14,570	15,954	17,360	20,217	23,502	25,860	..	Manufacturing
27,432	30,659	34,804	39,553	45,375	48,412	53,935	58,990	65,371	70,570	..	Services, etc.
-7,555	-7,974	-7,772	-11,525	-13,429	-8,741	-11,911	-11,637	-9,797	-6,922	..	Resource Balance
9,029	10,554	12,883	15,941	17,882	21,290	26,765	32,048	32,920	35,396	..	Exports of Goods & NFServices
16,584	18,528	20,655	27,466	31,311	30,031	38,676	43,685	42,717	42,318	..	Imports of Goods & NFServices
62,709	70,017	77,933	88,262	103,517	103,376	116,751	130,947	144,131	146,922	..	Domestic Absorption
37,212	41,591	47,782	54,306	63,307	64,910	74,899	82,782	93,546	95,722	..	Private Consumption, etc.
11,469	13,231	14,283	16,769	19,239	18,707	19,032	20,789	23,325	24,500	..	General Gov't Consumption
14,028	15,195	15,868	17,187	20,971	19,759	22,820	27,376	27,260	26,700	..	Gross Domestic Investment
13,732	14,875	14,811	16,825	21,090	20,544	22,460	24,964	26,264	28,150	..	Fixed Investment
											Memo Items:
6,473	7,221	8,096	5,662	7,542	11,018	10,909	15,739	17,463	19,778	..	Gross Domestic Saving
8,094	8,989	9,696	7,019	8,606	12,847	13,198	17,701	23,741	28,071	..	Gross National Saving
				(Millions of 1980 Moroccan Dirhams)							
63,406	66,025	67,808	65,037	70,037	71,125	72,267	74,405	80,510	80,735	..	Gross National Product
..	..	..	..	..	..	..	..	..	..	B	GDP at factor cost
12,184	11,975	12,711	9,799	11,751	11,318	11,286	12,649	15,555	13,538	..	Agriculture
21,557	23,219	22,646	22,622	23,130	23,791	23,527	24,043	23,948	24,692	..	Industry
11,292	11,470	12,010	12,010	12,010	12,478	12,267	12,480	12,911	13,429	..	Manufacturing
31,197	32,716	34,804	35,899	38,325	39,542	41,144	42,825	45,079	46,497	..	Services, etc.
-9,835	-10,461	-7,772	-8,141	-8,158	-4,599	-5,021	-4,400	-5,542	-3,799	..	Resource Balance
12,316	12,430	12,883	12,912	13,541	14,643	15,055	15,698	16,011	17,353	..	Exports of Goods & NFServices
22,151	22,891	20,655	21,053	21,698	19,242	20,076	20,098	21,553	21,151	..	Imports of Goods & NFServices
74,773	78,371	77,933	76,461	81,364	79,250	80,978	83,917	90,124	88,526	..	Domestic Absorption
44,838	47,142	47,782	46,803	49,457	49,788	51,297	52,110	57,531	55,565	..	Private Consumption, etc.
13,043	14,110	14,283	15,588	16,894	16,427	16,437	17,512	19,335	20,045	..	General Gov't Consumption
16,892	17,119	15,868	14,070	15,013	13,035	13,244	14,294	13,259	12,915	..	Gross Domestic Investment
16,472	16,699	14,811	13,809	15,123	13,584	13,074	13,165	12,736	13,576	..	Fixed Investment
											Memo Items:
12,060	13,039	12,883	12,219	12,392	13,641	13,893	14,745	16,610	17,692	..	Capacity to Import
-256	609	0	-693	-1,149	-1,002	-1,162	-954	599	339	..	Terms of Trade Adjustment
64,682	68,519	70,161	67,627	72,058	73,650	74,795	78,563	85,181	85,066	..	Gross Domestic Income
63,150	66,634	67,808	64,344	68,888	70,123	71,105	73,452	81,109	81,074	..	Gross National Income
				(Index 1980 = 100)							DOMESTIC PRICES (DEFLATORS)
84.9	91.4	100.0	112.3	123.1	126.8	138.0	150.0	158.8	165.2	..	Overall (GDP)
83.9	89.3	100.0	115.4	127.2	130.4	144.2	156.0	159.9	166.0	..	Domestic Absorption
85.6	92.8	100.0	116.6	138.3	142.5	155.5	173.9	183.8	192.0	..	Agriculture
80.2	87.3	100.0	113.9	123.0	126.5	141.8	159.4	168.6	175.9	..	Industry
83.0	91.0	100.0	111.7	121.3	127.9	141.5	162.0	182.0	192.6	..	Manufacturing
											MANUFACTURING ACTIVITY
82.9	91.0	100.0	104.6	109.4	114.4	119.6	125.1	..	..	..	Employment (1980=100)
102.1	101.5	100.0	95.4	89.6	88.4	81.8	84.6	..	..	..	Real Earnings per Empl. (1980=100)
97.8	99.0	100.0	93.3	89.2	88.6	83.3	81.0	..	..	..	Real Output per Empl. (1980=100)
51.1	50.8	50.5	50.8	50.8	50.8	50.8	50.8	..	..	..	Earnings as % of Value Added
				(Millions of current Moroccan Dirhams)							MONETARY HOLDINGS
26.3	30.1	33.6	38.9	43.4	51.3	56.8	66.8	77.4	84.9	D	Money Supply, Broadly Defined
20.8	23.4	25.3	29.0	30.1	34.2	36.8	42.8	50.0	54.5	..	Money as Means of Payment
7.7	9.0	9.8	11.1	12.0	13.6	14.8	16.2	18.7	20.0	..	Currency Ouside Banks
13.1	14.3	15.5	17.9	18.0	20.6	22.0	26.6	31.3	34.5	..	Demand Deposits
5.6	6.7	8.3	9.8	13.3	17.1	20.0	23.9	27.4	30.4	..	Quasi-Monetary Liabilities
				(Millions of current Moroccan Dirhams)							
-5,773	-6,039	-7,184	-10,557	-10,630	-7,680	-6,762	-9,424	..	..	..	GOVERNMENT DEFICIT (-) OR SURPLUS
13,183	15,562	17,235	20,066	24,080	24,220	26,383	29,896	..	..	..	Current Revenue
11,403	13,383	16,877	20,755	22,902	24,171	26,156	31,748	..	..	..	Current Expenditure
1,780	2,179	358	-689	1,178	49	227	-1,852	..	..	..	Current Budget Balance
163	241	267	352	308	296	301	317	..	..	..	Capital Receipts
7,716	8,459	7,809	10,220	12,116	8,025	7,290	7,889	..	..	..	Capital Payments

MOROCCO	1967	1968	1969	1970	1971	1972	1973	1974	1975	1976	1977
FOREIGN TRADE (CUSTOMS BASIS)				*(Millions of current US dollars)*							
Value of Exports, fob	424.1	450.1	485.1	487.9	499.1	633.6	876.6	1,706.4	1,543.0	1,262.1	1,299.9
Nonfuel Primary Products	390.1	408.3	437.3	438.7	423.6	541.7	741.7	1,501.5	1,336.1	1,041.8	1,007.1
Fuels	2.1	3.2	1.6	1.9	1.8	1.3	4.7	11.9	14.2	17.7	20.3
Manufactures	31.9	38.6	46.2	47.3	73.7	90.6	130.2	193.0	192.8	202.6	272.4
Value of Imports, cif	517.0	550.4	560.3	684.3	697.0	764.4	1,098.2	1,900.8	2,547.3	2,592.9	3,168.5
Nonfuel Primary Products	207.1	202.0	166.0	211.1	241.3	248.4	416.0	711.1	914.9	710.5	778.7
Fuels	25.1	32.8	32.6	37.4	46.6	54.7	71.2	258.7	276.4	295.2	370.2
Manufactures	284.9	315.6	361.7	435.8	409.2	461.3	611.0	931.0	1,356.0	1,587.2	2,019.5
				(Index 1980 = 100)							
Terms of Trade	121.3	132.7	136.4	129.4	126.3	125.5	112.6	136.7	144.5	111.8	106.4
Export Prices, fob	29.5	30.1	32.2	32.3	33.4	35.7	44.0	80.7	90.1	67.6	68.8
Nonfuel Primary Products	30.4	31.3	32.3	32.5	32.7	34.8	43.6	83.0	94.5	67.8	67.0
Fuels	4.3	4.3	4.3	4.3	5.6	6.2	8.9	36.7	35.7	38.4	42.0
Manufactures	30.6	32.8	38.8	40.3	44.0	46.3	54.5	70.9	74.6	70.9	80.6
Import Prices, cif	24.3	22.6	23.6	24.9	26.4	28.4	39.1	59.0	62.4	60.5	64.7
Trade at Constant 1980 Prices				*(Millions of 1980 US dollars)*							
Exports, fob	1,436.1	1,497.4	1,508.8	1,512.3	1,495.7	1,774.1	1,992.0	2,114.6	1,712.2	1,867.7	1,889.7
Imports, cif	2,124.4	2,430.1	2,377.3	2,744.3	2,639.1	2,686.9	2,810.0	3,219.0	4,082.9	4,288.1	4,900.2
BALANCE OF PAYMENTS				*(Millions of current US dollars)*							
Exports of Goods & Services	..	..	..	705	745	946	1,302	2,131	2,026	1,720	1,871
Merchandise, fob	..	..	..	487	500	642	913	1,704	1,530	1,247	1,284
Nonfactor Services	..	..	..	203	232	291	374	399	468	449	555
Factor Services	..	..	..	15	14	13	14	28	29	24	32
Imports of Goods & Services	..	..	..	902	914	1,036	1,450	2,235	3,070	3,622	4,286
Merchandise, fob	..	..	..	624	637	709	1,037	1,692	2,266	2,308	2,821
Nonfactor Services	..	..	..	218	212	250	320	453	688	1,164	1,241
Factor Services	..	..	..	60	65	77	93	89	117	151	224
Long-Term Interest	..	..	..	25	29	36	44	46	57	77	171
Current Transfers, net	..	..	..	36	74	108	211	300	482	499	546
Workers' Remittances	..	..	..	63	95	140	250	360	533	547	589
Total to be Financed	..	..	..	-161	-94	17	63	196	-562	-1,403	-1,870
Official Capital Grants	..	..	..	37	35	30	33	30	34	6	15
Current Account Balance	..	..	..	-124	-59	48	97	226	-528	-1,397	-1,855
Long-Term Capital, net	..	..	..	151	144	75	35	71	415	1,295	1,765
Direct Investment	..	..	..	20	23	13	-1	-20	0	38	57
Long-Term Loans	..	..	..	136	123	59	30	163	515	661	1,839
Disbursements	..	..	..	175	184	132	122	265	620	776	1,984
Repayments	..	..	..	40	62	72	92	103	106	115	145
Other Long-Term Capital	..	..	..	-42	-37	-28	-28	-101	-134	590	-145
Other Capital, net	..	..	..	-11	-46	-67	-129	-175	85	85	88
Change in Reserves	..	..	..	-16	-40	-56	-2	-122	28	18	2
Memo Item:				*(Moroccan Dirhams per US dollar)*							
Conversion Factor (Annual Avg)	5.060	5.060	5.060	5.060	5.050	4.600	4.110	4.370	4.050	4.420	4.500
EXTERNAL DEBT, ETC.				*(Millions of US dollars, outstanding at end of year)*							
Public/Publicly Guar. Long-Term	..	..	..	711	873	920	1,008	1,206	1,672	2,261	4,100
Official Creditors	..	..	..	582	696	743	817	911	1,118	1,250	1,983
IBRD and IDA	..	..	..	59	79	100	129	182	275	324	378
Private Creditors	..	..	..	129	177	177	190	296	553	1,011	2,117
Private Non-guaranteed Long-Term	..	..	..	15	15	20	30	40	55	75	150
Use of Fund Credit	..	..	..	28	0	0	0	0	0	167	175
Short-Term Debt	..	..	..	..	..	..	..	..	..	..	565
Memo Items:				*(Millions of US dollars)*							
Int'l Reserves Excluding Gold	54.6	64.0	93.0	119.3	150.7	213.7	240.6	391.0	352.4	467.1	505.3
Gold Holdings (at market price)	21.1	25.1	21.1	22.4	26.2	38.9	67.4	113.4	85.3	81.9	104.3
SOCIAL INDICATORS											
Total Fertility Rate	7.1	7.1	7.0	7.0	6.9	6.9	6.7	6.5	6.3	6.1	5.9
Crude Birth Rate	48.2	47.7	47.2	46.6	46.1	45.6	44.4	43.1	41.9	40.6	39.4
Infant Mortality Rate	138.0	134.8	131.6	128.4	125.2	122.0	119.6	117.2	114.8	112.4	110.0
Life Expectancy at Birth	50.4	50.9	51.4	51.9	52.4	52.9	53.5	54.1	54.6	55.2	55.8
Food Production, p.c. ('79-81 = 100)	106.1	139.2	109.9	112.2	121.3	116.8	101.5	114.2	96.2	107.8	90.4
Labor Force, Agriculture (%)	59.8	59.1	58.3	57.6	56.4	55.2	53.9	52.7	51.5	50.3	49.1
Labor Force, Female (%)	12.4	12.9	13.4	13.9	14.4	15.0	15.5	16.0	16.5	16.9	17.2
School Enroll. Ratio, primary	..	..	..	52.0	..	..	..	..	62.0	66.0	69.0
School Enroll. Ratio, secondary	..	..	..	13.0	..	..	..	..	16.0	17.0	19.0

1978	1979	1980	1981	1982	1983	1984	1985	1986	1987 estimate	NOTES	MOROCCO
											FOREIGN TRADE (CUSTOMS BASIS)
					(Millions of current US dollars)						
1,511.3	1,958.6	2,403.4	2,320.3	2,058.6	2,062.0	2,171.9	2,165.1	2,427.5	2,807.0	..	Value of Exports, fob
1,145.1	1,427.3	1,721.5	1,560.7	1,264.3	1,181.8	1,200.7	1,204.8	1,308.7	1,351.0	..	Nonfuel Primary Products
21.2	70.9	116.5	105.0	87.6	81.8	85.9	84.2	61.9	84.0	..	Fuels
345.0	460.5	565.5	654.6	706.7	798.4	885.3	876.1	1,056.9	1,372.0	..	Manufactures
2,950.3	3,673.5	4,182.4	4,352.6	4,315.3	3,596.6	3,906.7	3,849.6	3,790.2	4,229.0	..	Value of Imports, cif
811.6	1,027.1	1,248.7	1,411.0	1,127.1	1,023.5	1,219.3	1,153.4	1,131.2	1,109.1	..	Nonfuel Primary Products
428.2	704.1	986.9	1,187.8	1,173.2	988.3	1,021.2	1,075.9	595.2	910.8	..	Fuels
1,710.6	1,942.3	1,946.8	1,753.8	2,015.0	1,584.8	1,666.2	1,620.3	2,063.8	2,209.1	..	Manufactures
					(Index 1980 = 100)						
105.2	104.5	100.0	94.5	89.8	86.2	87.7	88.7	102.8	106.7	..	Terms of Trade
73.5	86.8	100.0	95.5	84.9	79.6	80.4	77.2	82.8	92.6	..	Export Prices, fob
70.6	87.6	100.0	99.2	87.5	81.5	80.8	76.7	81.7	92.3	..	Nonfuel Primary Products
42.3	61.0	100.0	112.5	101.6	92.5	90.2	87.5	44.9	56.9	..	Fuels
89.6	90.3	100.0	85.8	79.1	76.0	79.0	77.0	88.6	96.6	..	Manufactures
69.8	83.1	100.0	101.1	94.6	92.3	91.7	87.0	80.6	86.8	..	Import Prices, cif
					(Millions of 1980 US dollars)						Trade at Constant 1980 Prices
2,057.5	2,255.5	2,403.4	2,428.9	2,423.8	2,589.5	2,702.4	2,805.8	2,931.5	3,031.6	..	Exports, fob
4,227.2	4,418.9	4,182.5	4,306.5	4,560.3	3,895.5	4,262.4	4,426.5	4,704.3	4,873.3	..	Imports, cif
					(Millions of current US dollars)						**BALANCE OF PAYMENTS**
2,164	2,711	3,270	3,084	2,944	2,931	3,016	3,161	3,566	4,250	..	Exports of Goods & Services
1,488	1,937	2,415	2,283	2,043	2,058	2,161	2,145	2,404	2,782	..	Merchandise, fob
650	735	817	762	876	861	837	993	1,145	1,453	..	Nonfactor Services
27	40	39	39	26	12	18	22	17	16	..	Factor Services
4,262	5,167	5,807	6,014	5,824	4,809	4,943	5,126	5,326	5,845	..	Imports of Goods & Services
2,629	3,245	3,770	3,840	3,815	3,301	3,569	3,514	3,468	3,850	..	Merchandise, fob
1,293	1,444	1,401	1,411	1,320	865	769	820	1,141	1,212	..	Nonfactor Services
340	478	637	764	689	643	605	793	718	782	..	Factor Services
293	449	619	672	615	562	531	492	752	624	..	Long-Term Interest
702	891	1,004	988	840	888	847	966	1,390	1,759	..	Current Transfers, net
762	948	1,054	1,014	849	916	872	967	1,395	1,587	..	Workers' Remittances
-1,395	-1,564	-1,533	-1,943	-2,040	-990	-1,080	-1,000	-370	164	K	Total to be Financed
58	43	114	104	165	101	94	111	160	..	K	Official Capital Grants
-1,338	-1,521	-1,419	-1,839	-1,875	-889	-987	-889	-210	164	..	Current Account Balance
1,499	1,466	1,511	1,396	1,463	866	925	732	576	604	..	Long-Term Capital, net
48	39	89	59	80	46	47	20	1	57	..	Direct Investment
1,038	1,036	1,205	1,217	1,420	401	966	675	712	656	..	Long-Term Loans
1,369	1,480	1,795	1,863	2,245	1,123	1,222	1,209	1,437	1,342	..	Disbursements
331	444	590	646	825	722	256	535	725	686	..	Repayments
355	349	103	17	-201	319	-182	-73	-297	-109	..	Other Long-Term Capital
-190	-20	-341	28	-6	-206	-40	69	-47	-503	..	Other Capital, net
29	74	249	416	418	229	102	88	-318	-265	..	Change in Reserves
					(Moroccan Dirhams per US dollar)						Memo Item:
4.170	3.900	3.940	5.170	6.020	7.110	8.810	10.060	9.100	8.360	..	Conversion Factor (Annual Avg)
					(Millions of US dollars, outstanding at end of year)						**EXTERNAL DEBT, ETC.**
5,348	7,461	8,436	9,185	10,251	10,911	11,736	13,815	16,436	18,468	..	Public/Publicly Guar. Long-Term
2,509	3,913	4,557	5,460	6,122	7,087	7,881	9,702	11,607	13,552	..	Official Creditors
430	542	578	641	738	863	910	1,331	1,901	2,600	..	IBRD and IDA
2,839	3,548	3,879	3,725	4,129	3,824	3,855	4,114	4,829	4,916	..	Private Creditors
125	100	150	210	250	225	169	200	321	372	..	Private Non-guaranteed Long-Term
260	233	317	450	869	920	991	1,190	1,026	1,071	..	Use of Fund Credit
549	735	775	786	1,031	1,131	1,073	1,066	1,063	795	..	Short-Term Debt
					(Millions of US dollars)						Memo Items:
618.4	557.0	398.6	229.7	217.6	106.7	48.7	115.1	211.4	411.1	..	Int'l Reserves Excluding Gold
153.7	360.5	415.0	279.8	321.7	268.6	217.0	230.2	275.2	340.8	..	Gold Holdings (at market price)
											SOCIAL INDICATORS
5.7	5.6	5.4	5.3	5.1	5.0	4.8	4.6	4.5	4.3	..	Total Fertility Rate
38.8	38.2	37.6	37.0	36.4	35.6	34.8	34.0	33.2	32.4	..	Crude Birth Rate
107.4	104.8	102.2	99.6	97.0	96.1	95.2	94.3	93.3	92.4	..	Infant Mortality Rate
56.2	56.7	57.2	57.7	58.2	58.6	59.0	59.5	59.9	60.4	..	Life Expectancy at Birth
108.9	107.9	106.4	85.7	110.1	98.7	95.9	104.8	125.1	98.0	..	Food Production, p.c. ('79-81 = 100)
48.0	46.8	45.6	..	..	..	..	..	..	..	..	Labor Force, Agriculture (%)
17.6	18.0	18.4	18.6	18.9	19.2	19.4	19.7	19.9	20.1	..	Labor Force, Female (%)
72.0	75.0	77.0	79.0	85.0	86.0	80.0	81.0	..	..	..	School Enroll. Ratio, primary
20.0	22.0	24.0	26.0	28.0	29.0	31.0	31.0	..	..	..	School Enroll. Ratio, secondary

MOZAMBIQUE	1967	1968	1969	1970	1971	1972	1973	1974	1975	1976	1977
CURRENT GNP PER CAPITA (US $)	..	..	..	..	..	..	..	..	..	..	..
POPULATION (thousands)	8,786	8,983	9,184	9,390	9,621	9,859	10,102	10,351	10,606	10,890	11,181

ORIGIN AND USE OF RESOURCES *(Billions of current Mozambique Meticais)*

	1967	1968	1969	1970	1971	1972	1973	1974	1975	1976	1977
Gross National Product (GNP)	..	..	..	..	..	..	..	..	..	..	..
Net Factor Income from Abroad	..	..	..	..	..	..	..	..	..	..	..
Gross Domestic Product (GDP)	..	..	..	..	..	..	..	..	..	..	..
Indirect Taxes, net	..	..	..	..	..	..	..	..	..	..	..
GDP at factor cost	..	..	..	..	..	..	..	..	..	..	..
Agriculture	..	..	..	..	..	..	..	..	..	..	..
Industry	..	..	..	..	..	..	..	..	..	..	..
Manufacturing	..	..	..	..	..	..	..	..	..	..	..
Services, etc.	..	..	..	..	..	..	..	..	..	..	..
Resource Balance	..	..	..	..	..	..	..	..	..	..	..
Exports of Goods & NFServices	..	..	..	..	..	..	..	..	..	..	..
Imports of Goods & NFServices	..	..	..	..	..	..	..	..	..	..	..
Domestic Absorption	..	..	..	..	..	..	..	..	..	..	..
Private Consumption, etc.	..	..	..	..	..	..	..	..	..	..	..
General Gov't Consumption	..	..	..	..	..	..	..	..	..	..	..
Gross Domestic Investment	..	..	..	..	..	..	..	..	..	..	..
Fixed Investment	..	..	..	..	..	..	..	..	..	..	..
Memo Items:											
Gross Domestic Saving	..	..	..	..	..	..	..	..	..	..	..
Gross National Saving	..	..	..	..	..	..	..	..	..	..	..

(Millions of 1980 Mozambique Meticais)

	1967	1968	1969	1970	1971	1972	1973	1974	1975	1976	1977
Gross National Product	..	..	..	..	..	..	..	..	..	..	..
GDP at factor cost	..	..	..	..	..	..	..	..	..	..	..
Agriculture	..	..	..	..	..	..	..	..	..	..	..
Industry	..	..	..	..	..	..	..	..	..	..	..
Manufacturing	..	..	..	..	..	..	..	..	..	..	..
Services, etc.	..	..	..	..	..	..	..	..	..	..	..
Resource Balance	..	..	..	..	..	..	..	..	..	..	..
Exports of Goods & NFServices	..	..	..	..	..	..	..	..	..	..	..
Imports of Goods & NFServices	..	..	..	..	..	..	..	..	..	..	..
Domestic Absorption	..	..	..	..	..	..	..	..	..	..	..
Private Consumption, etc.	..	..	..	..	..	..	..	..	..	..	..
General Gov't Consumption	..	..	..	..	..	..	..	..	..	..	..
Gross Domestic Investment	..	..	..	..	..	..	..	..	..	..	..
Fixed Investment	..	..	..	..	..	..	..	..	..	..	..
Memo Items:											
Capacity to Import	..	..	..	..	..	..	..	..	..	..	..
Terms of Trade Adjustment	..	..	..	..	..	..	..	..	..	..	..
Gross Domestic Income	..	..	..	..	..	..	..	..	..	..	..
Gross National Income	..	..	..	..	..	..	..	..	..	..	..

DOMESTIC PRICES (DEFLATORS) *(Index 1980 = 100)*

	1967	1968	1969	1970	1971	1972	1973	1974	1975	1976	1977
Overall (GDP)	..	..	..	..	..	..	..	..	..	..	..
Domestic Absorption	..	..	..	..	..	..	..	..	..	..	..
Agriculture	..	..	..	..	..	..	..	..	..	..	..
Industry	..	..	..	..	..	..	..	..	..	..	..
Manufacturing	..	..	..	..	..	..	..	..	..	..	..

MANUFACTURING ACTIVITY

	1967	1968	1969	1970	1971	1972	1973	1974	1975	1976	1977
Employment (1980 = 100)	..	..	..	..	..	..	..	..	..	..	..
Real Earnings per Empl. (1980 = 100)	..	..	..	..	..	..	..	..	..	..	..
Real Output per Empl. (1980 = 100)	..	..	..	..	..	..	..	..	..	..	..
Earnings as % of Value Added	34.9	32.1	30.4	28.8	40.9	41.2	41.4	..	..	..	..

MONETARY HOLDINGS *(Billions of current Mozambique Meticais)*

	1967	1968	1969	1970	1971	1972	1973	1974	1975	1976	1977
Money Supply, Broadly Defined	..	..	..	..	..	..	..	..	..	..	..
Money as Means of Payment	..	..	..	..	..	..	..	..	..	..	..
Currency Ouside Banks	..	..	..	..	..	..	..	..	..	..	..
Demand Deposits	..	..	..	..	..	..	..	..	..	..	..
Quasi-Monetary Liabilities	..	..	..	..	..	..	..	..	..	..	..

GOVERNMENT DEFICIT (-) OR SURPLUS *(Millions of current Mozambique Meticais)*

	1967	1968	1969	1970	1971	1972	1973	1974	1975	1976	1977
Current Revenue	..	..	..	..	..	..	..	..	..	..	..
Current Expenditure	..	..	..	..	..	..	..	..	..	..	..
Current Budget Balance	..	..	..	..	..	..	..	..	..	..	..
Capital Receipts	..	..	..	..	..	..	..	..	..	..	..
Capital Payments	..	..	..	..	..	..	..	..	..	..	..

1978	1979	1980	1981	1982	1983	1984	1985	1986	1987 estimate	NOTES	MOZAMBIQUE
..	..	..	..	190	160	170	180	220	170	..	**CURRENT GNP PER CAPITA (US $)**
11,478	11,788	12,103	12,423	12,752	13,089	13,436	13,791	14,186	14,591	..	**POPULATION (thousands)**

(Billions of current Mozambique Meticais) — **ORIGIN AND USE OF RESOURCES**

1978	1979	1980	1981	1982	1983	1984	1985	1986	1987	NOTES	
..	..	77.18	79.19	89.24	87.06	104.58	141.45	160.09	396.20	..	Gross National Product (GNP)
..	..	-1.02	-2.31	-3.16	-4.34	-4.52	-6.15	-7.11	-58.20	..	Net Factor Income from Abroad
..	..	78.20	81.50	92.40	91.40	109.10	147.60	167.20	454.40	..	Gross Domestic Product (GDP)
..	..	9.81	9.93	10.50	11.20	11.40	8.73	10.85	28.20	..	Indirect Taxes, net
..	..	68.39	71.57	81.90	80.20	97.70	138.87	156.35	426.20	..	GDP at factor cost
..	..	36.58	37.04	40.97	38.79	41.13	49.18	..	213.60	..	Agriculture
..	..	9.90	10.87	11.61	11.31	12.50	15.96	..	52.50	..	Industry
..	..	7.83	8.60	8.82	8.08	..	..	..	..	..	Manufacturing
..	..	21.91	23.67	29.32	30.10	44.07	73.73	..	160.10	..	Services, etc.
..	..	-14.41	-16.39	-20.95	-18.97	-18.11	-14.62	-19.78	-145.10	..	Resource Balance
..	..	12.92	13.96	12.70	8.93	6.65	6.17	5.92	50.40	..	Exports of Goods & NFServices
..	..	27.33	30.35	33.65	27.90	24.76	20.79	25.70	195.50	..	Imports of Goods & NFServices
..	..	92.61	97.89	113.35	110.37	127.21	162.22	186.98	599.50	..	Domestic Absorption
..	..	64.03	64.93	76.57	79.95	93.29	127.91	143.79	407.10	..	Private Consumption, etc.
..	..	13.81	16.64	18.93	21.27	22.49	24.16	26.91	90.60	..	General Gov't Consumption
..	..	14.77	16.32	17.85	9.15	11.43	10.15	16.28	101.80	..	Gross Domestic Investment
..	..	13.00	15.20	17.00	13.80	12.10	10.70	16.00	97.40	..	Fixed Investment

Memo Items:

1978	1979	1980	1981	1982	1983	1984	1985	1986	1987	NOTES	
..	..	0.36	-0.07	-3.10	-9.82	-6.68	-4.47	-3.50	-43.30	..	Gross Domestic Saving
..	..	0.25	-1.14	-4.77	-11.94	-9.89	-9.94	-9.53	-92.06	..	Gross National Saving

(Millions of 1980 Mozambique Meticais)

1978	1979	1980	1981	1982	1983	1984	1985	1986	1987	NOTES	
..	..	77,183	76,152	73,156	62,978	64,391	58,445	59,045	66,924	..	Gross National Product
..	..	68,390	68,817	67,131	58,116	60,309	57,622	58,123	59,953	..	GDP at factor cost
..	..	36,580	35,612	33,581	28,108	25,389	20,407	..	..	..	Agriculture
..	..	9,900	10,448	9,519	8,199	7,716	6,624	..	..	..	Industry
..	..	7,830	8,272	7,227	5,856	..	..	..	..	..	Manufacturing
..	..	21,910	22,758	24,031	21,809	27,204	30,592	..	..	..	Services, etc.
..	..	-14,410	-15,500	-16,908	-16,436	-15,742	-12,596	-19,649	..	..	Resource Balance
..	..	12,920	12,066	13,298	7,910	4,978	4,802	5,134	..	..	Exports of Goods & NFServices
..	..	27,330	27,566	30,207	24,346	20,720	17,398	24,783	..	..	Imports of Goods & NFServices
..	..	92,610	93,866	92,646	82,668	83,088	73,841	81,805	..	..	Domestic Absorption
..	..	64,030	62,342	60,118	52,779	61,024	61,833	69,664	..	..	Private Consumption, etc.
..	..	13,810	16,314	14,976	13,723	11,301	8,763	7,940	..	..	General Gov't Consumption
..	..	14,770	15,210	17,552	16,166	10,763	3,245	4,200	..	..	Gross Domestic Investment
..	..	13,000	13,523	15,483	14,227	9,498	3,202	3,432	..	..	Fixed Investment

Memo Items:

1978	1979	1980	1981	1982	1983	1984	1985	1986	1987	NOTES	
..	..	12,920	12,679	11,400	7,792	5,565	5,163	5,709	..	..	Capacity to Import
..	..	0	614	-1,898	-117	587	362	574	..	..	Terms of Trade Adjustment
..	..	78,200	78,979	73,840	66,115	67,933	61,606	62,730	..	..	Gross Domestic Income
..	..	77,183	76,766	71,258	62,861	64,978	58,807	59,620	..	..	Gross National Income

(Index 1980 = 100) — **DOMESTIC PRICES (DEFLATORS)**

1978	1979	1980	1981	1982	1983	1984	1985	1986	1987	NOTES	
..	..	100.0	104.0	122.0	138.0	162.0	241.0	269.0	645.0	..	Overall (GDP)
..	..	100.0	104.3	122.3	133.5	153.1	219.7	228.6	..	..	Domestic Absorption
..	..	100.0	104.0	122.0	138.0	162.0	241.0	..	..	..	Agriculture
..	..	100.0	104.0	122.0	138.0	162.0	241.0	..	..	..	Industry
..	..	100.0	104.0	122.0	138.0	..	..	..	..	..	Manufacturing

MANUFACTURING ACTIVITY

1978	1979	1980	1981	1982	1983	1984	1985	1986	1987	NOTES	
..	..	..	..	..	..	..	..	..	..	..	Employment (1980=100)
..	..	..	..	..	..	..	..	..	..	..	Real Earnings per Empl. (1980=100)
..	..	..	..	..	..	..	..	..	..	..	Real Output per Empl. (1980=100)
..	..	..	..	..	..	..	..	..	..	..	Earnings as % of Value Added

(Billions of current Mozambique Meticais) — **MONETARY HOLDINGS**

1978	1979	1980	1981	1982	1983	1984	1985	1986	1987	NOTES	
..	..	..	..	..	..	..	..	..	..	..	Money Supply, Broadly Defined
..	..	..	..	..	..	..	..	..	..	..	Money as Means of Payment
..	..	..	..	..	..	..	..	..	..	..	Currency Ouside Banks
..	..	..	..	..	..	..	..	..	..	..	Demand Deposits
..	..	..	..	..	..	..	..	..	..	..	Quasi-Monetary Liabilities

(Millions of current Mozambique Meticais) — **GOVERNMENT DEFICIT (-) OR SURPLUS**

1978	1979	1980	1981	1982	1983	1984	1985	1986	1987	NOTES	
..	..	..	..	..	..	..	..	..	..	..	Current Revenue
..	..	..	..	..	..	..	..	..	..	..	Current Expenditure
..	..	..	..	..	..	..	..	..	..	..	Current Budget Balance
..	..	..	..	..	..	..	..	..	..	..	Capital Receipts
..	..	..	..	..	..	..	..	..	..	..	Capital Payments

MOZAMBIQUE	1967	1968	1969	1970	1971	1972	1973	1974	1975	1976	1977
FOREIGN TRADE (CUSTOMS BASIS)					*(Millions of current US dollars)*						
Value of Exports, fob	122	155	142	157	160	175	227	296	202	150	149
Nonfuel Primary Products	..	..	..	..	..	..	..	..	..	..	..
Fuels	13	13	12	13	12	10	12	17	22	..	6
Manufactures	3	9	7	15	11	8	12	14	7	..	8
Value of Imports, cif	199	234	260	323	335	327	465	464	417	300	328
Nonfuel Primary Products	..	..	..	..	..	..	..	..	..	..	..
Fuels	17	21	22	27	31	27	31	44	54	39	12
Manufactures	137	167	192	245	241	220	350	302	214	154	225
					(Index 1980 = 100)						
Terms of Trade	..	..	..	..	..	..	..	..	..	..	..
Export Prices, fob	..	..	..	..	..	..	..	..	..	..	..
Nonfuel Primary Products	..	..	..	..	..	..	..	..	..	..	..
Fuels	..	..	..	..	..	..	..	..	..	..	..
Manufactures	..	..	..	..	..	..	..	..	..	..	..
Import Prices, cif	..	..	..	..	..	..	..	..	..	..	..
Trade at Constant 1980 Prices					*(Millions of 1980 US dollars)*						
Exports, fob	..	..	..	..	..	..	..	..	..	..	..
Imports, cif	..	..	..	..	..	..	..	..	..	..	..
BALANCE OF PAYMENTS					*(Millions of current US dollars)*						
Exports of Goods & Services	..	..	..	..	..	..	..	..	..	..	..
Merchandise, fob	..	..	..	..	..	..	..	..	..	..	..
Nonfactor Services	..	..	..	..	..	..	..	..	..	..	..
Factor Services	..	..	..	..	..	..	..	..	..	..	..
Imports of Goods & Services	..	..	..	..	..	..	..	..	..	..	..
Merchandise, fob	..	..	..	..	..	..	..	..	..	..	..
Nonfactor Services	..	..	..	..	..	..	..	..	..	..	..
Factor Services	..	..	..	..	..	..	..	..	..	..	..
Long-Term Interest	..	..	..	..	..	..	..	..	..	..	..
Current Transfers, net	..	..	..	..	..	..	..	..	..	..	..
Workers' Remittances	..	..	..	..	..	..	..	..	..	..	..
Total to be Financed	..	..	..	..	..	..	..	..	..	..	..
Official Capital Grants	..	..	..	..	..	..	..	..	..	..	..
Current Account Balance	..	..	..	..	..	..	..	..	..	..	..
Long-Term Capital, net	..	..	..	..	..	..	..	..	..	..	..
Direct Investment	..	..	..	..	..	..	..	..	..	..	..
Long-Term Loans	..	..	..	..	..	..	..	..	..	..	..
Disbursements	..	..	..	..	..	..	..	..	..	..	..
Repayments	..	..	..	..	..	..	..	..	..	..	..
Other Long-Term Capital	..	..	..	..	..	..	..	..	..	..	..
Other Capital, net	..	..	..	..	..	..	..	..	..	..	..
Change in Reserves	..	..	..	0	0	0	0	0	0	0	0
Memo Item:					*(Mozambique Meticais per US dollar)*						
Conversion Factor (Annual Avg)	28.750	28.750	28.750	28.750	28.320	27.010	24.670	25.410	25.550	30.220	32.930
EXTERNAL DEBT, ETC.					*(Millions of US dollars, outstanding at end of year)*						
Public/Publicly Guar. Long-Term	..	..	..	..	..	..	..	..	..	..	..
Official Creditors	..	..	..	..	..	..	..	..	..	..	..
IBRD and IDA	..	..	..	..	..	..	..	..	..	..	..
Private Creditors	..	..	..	..	..	..	..	..	..	..	..
Private Non-guaranteed Long-Term	..	..	..	..	..	..	..	..	..	..	..
Use of Fund Credit	..	..	..	..	..	..	..	..	..	..	..
Short-Term Debt	..	..	..	..	..	..	..	..	..	..	..
Memo Items:					*(Millions of US dollars)*						
Int'l Reserves Excluding Gold	..	..	..	..	..	..	..	..	..	..	..
Gold Holdings (at market price)	..	..	..	..	..	..	..	..	..	..	..
SOCIAL INDICATORS											
Total Fertility Rate	6.7	6.7	6.7	6.6	6.6	6.6	6.6	6.6	6.5	6.5	6.5
Crude Birth Rate	48.2	48.0	47.8	47.6	47.4	47.3	47.3	47.3	47.2	47.2	47.1
Infant Mortality Rate	163.0	160.0	157.0	154.0	151.0	149.2	147.3	145.5	143.7	141.8	140.0
Life Expectancy at Birth	38.7	39.3	39.9	40.5	41.1	41.5	41.9	42.3	42.7	43.1	43.5
Food Production, p.c. ('79-81 = 100)	133.6	140.9	144.2	143.1	141.4	139.6	139.5	133.4	117.8	115.2	110.0
Labor Force, Agriculture (%)	87.0	86.8	86.6	86.4	86.2	86.0	85.9	85.7	85.5	85.3	85.1
Labor Force, Female (%)	50.2	50.2	50.2	50.1	50.1	50.1	50.1	50.1	50.0	50.0	50.0
School Enroll. Ratio, primary	..	..	..	47.0	..	..	..	..	..	87.0	100.0
School Enroll. Ratio, secondary	..	..	..	5.0	..	..	..	..	..	3.0	4.0

1978	1979	1980	1981	1982	1983	1984	1985	1986	1987 estimate	NOTES	MOZAMBIQUE
				(Millions of current US dollars)							**FOREIGN TRADE (CUSTOMS BASIS)**
163	254	281	359	306	240	..	..	..	..	..	Value of Exports, fob
..	..	..	..	..	..	..	..	..	..	..	Nonfuel Primary Products
..	8	..	..	..	..	..	..	..	..	..	Fuels
..	116	..	..	..	..	..	..	..	..	..	Manufactures
521	571	800	801	836	636	540	424	543	..	..	Value of Imports, cif
..	..	..	..	..	..	..	..	..	..	..	Nonfuel Primary Products
19	37	..	..	..	..	..	..	..	..	..	Fuels
358	371	..	..	..	..	..	..	..	..	..	Manufactures
				(Index 1980 = 100)							
..	..	..	..	..	..	..	..	..	..	..	Terms of Trade
..	..	..	..	..	..	..	..	..	..	..	Export Prices, fob
..	..	..	..	..	..	..	..	..	..	..	Nonfuel Primary Products
..	..	..	..	..	..	..	..	..	..	..	Fuels
..	..	..	..	..	..	..	..	..	..	..	Manufactures
..	..	..	..	..	..	..	..	..	..	..	Import Prices, cif
				(Millions of 1980 US dollars)							Trade at Constant 1980 Prices
..	..	..	..	..	..	..	..	..	..	..	Exports, fob
..	..	..	..	..	..	..	..	..	..	..	Imports, cif
				(Millions of current US dollars)							**BALANCE OF PAYMENTS**
..	..	399	395	337	222	157	143	148	176	..	Exports of Goods & Services
..	..	281	281	229	132	96	77	79	97	..	Merchandise, fob
..	..	118	114	108	91	61	66	69	79	..	Nonfactor Services
..	..	0	0	0	0	0	0	0	..	..	Factor Services
..	..	850	895	953	782	664	599	798	885	..	Imports of Goods & Services
..	..	720	721	752	573	486	381	493	575	..	Merchandise, fob
..	..	124	138	140	121	98	100	100	107	..	Nonfactor Services
..	..	6	36	60	88	81	117	205	203	..	Factor Services
..	..	..	..	..	..	..	..	..	..	..	Long-Term Interest
..	..	28	35	40	55	31	16	27	33	..	Current Transfers, net
..	..	53	65	64	75	57	41	50	58	..	Workers' Remittances
..	..	-423	-464	-575	-505	-476	-440	-623	-676	..	Total to be Financed
..	..	56	57	79	90	168	139	213	304	..	Official Capital Grants
..	..	-367	-407	-496	-415	-308	-301	-410	-372	..	Current Account Balance
..	..	420	466	474	132	308	282	161	221	..	Long-Term Capital, net
..	..	0	0	0	0	0	0	0	..	..	Direct Investment
..	..	..	..	..	..	..	..	..	..	..	Long-Term Loans
..	..	..	..	..	..	..	..	..	..	..	Disbursements
..	..	..	..	..	..	..	..	..	..	..	Repayments
..	..	364	409	395	43	140	143	-52	-83	..	Other Long-Term Capital
..	..	-85	-126	-119	267	23	-1	243	198	..	Other Capital, net
0	0	32	67	141	15	-23	21	6	-47	..	Change in Reserves
				(Mozambique Meticais per US dollar)							Memo Item:
33.000	32.560	32.400	35.350	37.700	40.180	42.440	43.180	40.000	286.000	..	Conversion Factor (Annual Avg)
				(Millions of US dollars, outstanding at end of year)							**EXTERNAL DEBT, ETC.**
..	..	..	..	..	..	..	..	..	..	..	Public/Publicly Guar. Long-Term
..	..	..	..	..	..	..	..	..	..	..	Official Creditors
..	..	..	..	..	..	..	..	..	..	..	IBRD and IDA
..	..	..	..	..	..	..	..	..	..	..	Private Creditors
..	..	..	..	..	..	..	..	..	..	..	Private Non-guaranteed Long-Term
..	..	..	..	..	..	..	..	..	..	..	Use of Fund Credit
..	..	..	..	..	..	..	..	..	..	..	Short-Term Debt
				(Millions of US dollars)							Memo Items:
..	..	..	..	..	..	..	..	..	..	..	Int'l Reserves Excluding Gold
..	..	..	..	..	..	..	..	..	..	..	Gold Holdings (at market price)
											SOCIAL INDICATORS
6.5	6.4	6.4	6.3	6.3	6.3	6.3	6.3	6.3	6.3	..	Total Fertility Rate
46.7	46.3	45.9	45.5	45.1	45.1	45.2	45.2	45.3	45.3	..	Crude Birth Rate
137.6	135.2	132.8	130.4	128.0	126.0	124.0	122.0	120.0	116.7	..	Infant Mortality Rate
43.9	44.3	44.7	45.0	45.4	45.9	46.5	47.0	47.6	48.1	..	Life Expectancy at Birth
104.3	101.5	100.0	98.5	95.6	88.6	85.9	85.0	84.9	83.2	..	Food Production, p.c. ('79-81=100)
84.9	84.7	84.5	..	..	..	..	..	..	..	..	Labor Force, Agriculture (%)
49.9	49.9	49.8	49.6	49.4	49.1	48.9	48.7	48.4	48.2	..	Labor Force, Female (%)
102.0	104.0	75.0	71.0	92.0	83.0	..	84.0	..	..	..	School Enroll. Ratio, primary
5.0	6.0	5.0	5.0	5.0	6.0	..	7.0	..	..	..	School Enroll. Ratio, secondary

NEPAL	1967	1968	1969	1970	1971	1972	1973	1974	1975	1976	1977
CURRENT GNP PER CAPITA (US $)	80	80	80	80	80	80	80	100	110	120	120
POPULATION (thousands)	10,702	10,900	11,115	11,350	11,608	11,890	12,194	12,512	12,841	13,181	13,532
ORIGIN AND USE OF RESOURCES					*(Millions of current Nepalese Rupees)*						
Gross National Product (GNP)	6,415	7,173	7,985	8,768	8,938	10,369	9,969	12,808	16,838	17,671	17,599
Net Factor Income from Abroad	0	0	0	0	0	0	0	0	267	277	319
Gross Domestic Product (GDP)	6,415	7,173	7,985	8,768	8,938	10,369	9,969	12,808	16,571	17,394	17,280
Indirect Taxes, net	91	108	161	55	163	239	204	383	635	805	1,025
GDP at factor cost	6,324	7,065	7,824	8,713	8,775	10,130	9,765	12,425	15,936	16,589	16,255
Agriculture	4,249	4,834	5,304	5,863	5,974	7,035	6,513	8,763	11,435	11,495	10,389
Industry	630	723	827	1,005	817	943	956	1,115	1,303	1,469	1,821
Manufacturing	219	256	307	323	336	409	399	526	664	690	736
Services, etc.	1,445	1,508	1,693	1,845	1,984	2,152	2,296	2,547	3,198	3,625	4,045
Resource Balance	-5	-47	-256	-298	-333	-234	-375	-653	-740	-592	-437
Exports of Goods & NFServices	441	491	560	430	483	587	659	698	1,475	1,874	2,037
Imports of Goods & NFServices	446	538	816	728	816	821	1,034	1,351	2,215	2,466	2,474
Domestic Absorption	6,420	7,220	8,241	9,066	9,271	10,603	10,344	13,461	17,311	17,986	17,717
Private Consumption, etc.	6,100	6,812	7,854	8,543	8,578	9,843	9,429	12,339	13,652	14,060	13,688
General Gov't Consumption	..	..	..	..	..	..	..	..	1,257	1,294	1,260
Gross Domestic Investment	320	408	387	523	693	760	915	1,122	2,402	2,632	2,769
Fixed Investment	..	..	..	..	..	..	..	..	2,223	2,443	2,581
Memo Items:											
Gross Domestic Saving	315	361	131	225	360	526	540	469	1,662	2,040	2,332
Gross National Saving	315	361	131	316	461	617	602	596	2,109	2,645	3,019
Gross National Product	..	..	..	..	..	..	..	..	..	..	..
GDP at factor cost	..	..	..	..	..	..	..	..	..	..	..
Agriculture	12,365	12,431	12,804	13,247	13,341	13,439	13,325	14,074	14,311	14,390	13,788
Industry	..	..	..	..	..	..	..	..	..	..	..
Manufacturing	..	..	..	..	..	..	..	..	..	..	..
Services, etc.	..	..	..	..	..	..	..	..	..	..	..
Resource Balance	..	..	..	..	..	..	..	..	..	..	..
Exports of Goods & NFServices	..	..	..	..	..	..	..	..	..	..	..
Imports of Goods & NFServices	..	..	..	..	..	..	..	..	..	..	..
Domestic Absorption	..	..	..	..	..	..	..	..	..	..	..
Private Consumption, etc.	..	..	..	..	..	..	..	..	..	..	..
General Gov't Consumption	..	..	..	..	..	..	..	..	..	..	..
Gross Domestic Investment	..	..	..	..	..	..	..	..	..	..	..
Fixed Investment	..	..	..	..	..	..	..	..	..	..	..
Memo Items:											
Capacity to Import	..	..	..	..	..	..	..	..	..	..	..
Terms of Trade Adjustment	..	..	..	..	..	..	..	..	..	..	..
Gross Domestic Income	..	..	..	..	..	..	..	..	..	..	..
Gross National Income	..	..	..	..	..	..	..	..	..	..	..
DOMESTIC PRICES (DEFLATORS)					*(Index 1980 = 100)*						
Overall (GDP)	..	..	..	..	..	..	..	..	..	..	..
Domestic Absorption	..	..	..	..	..	..	..	..	..	..	..
Agriculture	34.4	38.9	41.4	44.3	44.8	52.3	48.9	62.3	79.9	79.9	75.3
Industry	..	..	..	..	..	..	..	..	..	..	..
Manufacturing	..	..	..	..	..	..	..	..	..	..	..
MANUFACTURING ACTIVITY											
Employment (1980=100)	..	..	..	..	..	..	..	..	..	..	..
Real Earnings per Empl. (1980=100)	..	..	..	..	..	..	..	..	..	..	..
Real Output per Empl. (1980=100)	..	..	..	..	..	..	..	..	..	..	..
Earnings as % of Value Added	..	..	..	..	..	..	..	..	..	..	24.8
MONETARY HOLDINGS					*(Millions of current Nepalese Rupees)*						
Money Supply, Broadly Defined	602	718	922	931	1,108	1,293	1,663	1,948	2,180	2,811	3,401
Money as Means of Payment	532	596	739	699	784	842	1,090	1,290	1,333	1,636	1,932
Currency Ouside Banks	350	420	487	525	549	596	747	882	882	996	1,212
Demand Deposits	182	176	252	174	235	245	343	408	451	640	720
Quasi-Monetary Liabilities	70	122	183	232	324	451	573	658	847	1,175	1,468
GOVERNMENT DEFICIT (-) OR SURPLUS					*(Millions of current Nepalese Rupees)*						
	..	..	..	..	..	-126	-223	-248	-236	-422	-576
Current Revenue	..	..	..	..	..	783	783	975	1,277	1,447	1,683
Current Expenditure	..	..	..	..	..	..	..	..	..	..	..
Current Budget Balance	..	..	..	..	..	..	..	..	..	..	..
Capital Receipts	..	..	..	..	..	..	..	..	..	..	..
Capital Payments	..	..	..	..	..	..	..	..	..	..	..

1978	1979	1980	1981	1982	1983	1984	1985	1986	1987 estimate	NOTES	NEPAL
120	130	140	160	160	160	170	170	170	160	..	**CURRENT GNP PER CAPITA (US $)**
13,892	14,262	14,640	15,025	15,417	15,814	16,217	16,625	17,038	17,444	..	**POPULATION (thousands)**
				(Millions of current Nepalese Rupees)							**ORIGIN AND USE OF RESOURCES**
20,023	22,605	23,845	27,894	31,603	34,458	40,015	44,920	52,468	59,673	C	Gross National Product (GNP)
291	390	494	587	615	697	625	661	709	1,169	..	Net Factor Income from Abroad
19,732	22,215	23,351	27,307	30,988	33,761	39,390	44,259	51,759	58,504	C f	Gross Domestic Product (GDP)
1,306	1,436	1,465	1,841	1,951	2,117	2,386	2,861	3,300	3,472	..	Indirect Taxes, net
18,426	20,779	21,886	25,466	29,037	31,644	37,004	41,398	48,459	55,032	C f	GDP at factor cost
11,616	13,365	13,520	15,510	17,715	19,082	22,570	23,927	27,713	30,375	..	Agriculture
2,199	2,489	2,608	3,148	3,733	4,049	4,661	5,759	6,839	..	..	Industry
794	848	936	1,049	1,243	1,460	1,816	1,840	2,185	..	..	Manufacturing
4,611	4,925	5,758	6,808	7,589	8,513	9,773	11,712	13,907	..	..	Services, etc.
-967	-929	-1,679	-1,834	-2,236	-3,741	-3,465	-3,945	-4,712	-5,585	..	Resource Balance
2,086	2,618	2,695	3,523	3,592	3,455	4,196	5,372	6,506	7,563	..	Exports of Goods & NFServices
3,053	3,547	4,374	5,357	5,828	7,196	7,661	9,317	11,218	13,148	..	Imports of Goods & NFServices
20,699	23,144	25,030	29,141	33,224	37,502	42,855	48,204	56,471	64,089	..	Domestic Absorption
15,621	17,741	19,195	22,411	25,488	27,287	31,809	34,778	40,415	45,600	..	Private Consumption, etc.
1,471	1,889	1,565	1,922	2,638	3,416	3,644	4,371	5,197	6,225	..	General Gov't Consumption
3,607	3,514	4,270	4,808	5,098	6,799	7,402	9,055	10,859	12,264	..	Gross Domestic Investment
3,294	3,263	3,681	4,299	5,249	6,747	6,958	8,257	9,691	11,191	..	Fixed Investment
											Memo Items:
2,640	2,585	2,591	2,974	2,862	3,058	3,937	5,110	6,147	6,679	..	Gross Domestic Saving
3,231	3,333	3,511	4,118	4,002	4,329	5,250	6,530	7,722	9,224	..	Gross National Saving
				..							
..	..	..	..	..	..	..	..	..	..	C	Gross National Product
..	..	..	..	..	..	..	..	..	..	C f	GDP at factor cost
13,783	14,200	13,520	14,938	15,623	15,454	16,954	17,332	18,080	18,261	..	Agriculture
..	..	..	..	..	..	..	..	..	..	..	Industry
..	..	..	..	..	..	..	..	..	..	..	Manufacturing
..	..	..	..	..	..	..	..	..	..	..	Services, etc.
..	..	..	..	..	..	..	..	..	..	..	Resource Balance
..	..	..	..	..	..	..	..	..	..	..	Exports of Goods & NFServices
..	..	..	..	..	..	..	..	..	..	..	Imports of Goods & NFServices
..	..	..	..	..	..	..	..	..	..	..	Domestic Absorption
..	..	..	..	..	..	..	..	..	..	..	Private Consumption, etc.
..	..	..	..	..	..	..	..	..	..	..	General Gov't Consumption
..	..	..	..	..	..	..	..	..	..	..	Gross Domestic Investment
..	..	..	..	..	..	..	..	..	..	..	Fixed Investment
											Memo Items:
..	..	..	..	..	..	..	..	..	..	..	Capacity to Import
..	..	..	..	..	..	..	..	..	..	..	Terms of Trade Adjustment
..	..	..	..	..	..	..	..	..	..	..	Gross Domestic Income
..	..	..	..	..	..	..	..	..	..	..	Gross National Income
				(Index 1980 = 100)							**DOMESTIC PRICES (DEFLATORS)**
..	..	..	..	..	..	..	..	..	..	..	Overall (GDP)
..	..	..	..	..	..	..	..	..	..	..	Domestic Absorption
84.3	94.1	100.0	103.8	113.4	123.5	133.1	138.1	153.3	166.3	..	Agriculture
..	..	..	..	..	..	..	..	..	..	..	Industry
..	..	..	..	..	..	..	..	..	..	..	Manufacturing
											MANUFACTURING ACTIVITY
..	..	..	..	..	..	..	..	..	..	..	Employment (1980=100)
..	..	..	..	..	..	..	..	..	..	..	Real Earnings per Empl. (1980=100)
..	..	..	..	..	..	..	..	..	..	..	Real Output per Empl. (1980=100)
..	..	..	..	..	..	..	..	..	..	..	Earnings as % of Value Added
				(Millions of current Nepalese Rupees)							**MONETARY HOLDINGS**
4,070	4,712	5,526	6,580	7,987	9,594	10,841	13,014	15,543	19,024	D	Money Supply, Broadly Defined
2,200	2,534	2,864	3,205	3,705	4,366	4,942	5,616	6,951	8,682	..	Money as Means of Payment
1,379	1,627	1,814	2,147	2,408	2,783	3,302	3,797	4,787	5,826	..	Currency Ouside Banks
821	907	1,050	1,058	1,297	1,583	1,641	1,819	2,164	2,855	..	Demand Deposits
1,870	2,178	2,661	3,375	4,283	5,228	5,899	7,398	8,592	10,342	..	Quasi-Monetary Liabilities
				(Millions of current Nepalese Rupees)							
-582	-588	-705	-728	-1,591	-2,954	-2,985	-3,380	..	..	C F	**GOVERNMENT DEFICIT (-) OR SURPLUS**
1,989	2,323	2,620	3,233	3,626	3,867	4,186	4,759	..	..	..	Current Revenue
..	..	..	..	..	..	..	..	..	..	..	Current Expenditure
..	..	..	..	..	..	..	..	..	..	..	Current Budget Balance
..	..	..	..	..	..	..	14	..	..	..	Capital Receipts
..	..	..	..	..	..	..	..	..	..	..	Capital Payments

NEPAL	1967	1968	1969	1970	1971	1972	1973	1974	1975	1976	1977
FOREIGN TRADE (CUSTOMS BASIS)				*(Millions of current US dollars)*							
Value of Exports, fob	55.0	40.0	53.0	48.0	48.0	58.0	63.0	66.0	100.0	98.0	81.0
Nonfuel Primary Products	43.9	26.1	41.8	41.0	42.4	49.7	53.9	46.5	90.0	85.6	67.3
Fuels	0.0	0.1	0.0	0.0	0.0	0.0	0.0	0.0	0.0	0.1	0.0
Manufactures	11.1	13.8	11.2	7.0	5.6	8.3	9.1	19.5	10.0	12.3	13.7
Value of Imports, cif	58.0	50.0	74.0	85.0	70.0	88.0	96.0	110.0	171.0	214.0	244.0
Nonfuel Primary Products	27.5	19.5	28.8	29.0	23.0	27.0	26.0	27.0	55.0	76.5	75.9
Fuels	2.2	2.2	3.0	6.5	6.5	9.4	11.2	23.2	38.6	55.0	62.3
Manufactures	28.4	28.4	42.2	50.0	40.0	52.0	59.0	60.0	77.4	82.4	105.8
Terms of Trade	127.2	126.6	128.5	145.6	*(Index 1980 = 100)* 153.7	183.9	185.9	124.2	118.3	109.5	111.9
Export Prices, fob	35.4	33.8	35.6	34.5	36.6	44.7	58.3	68.5	63.8	59.6	67.4
Nonfuel Primary Products	37.0	35.8	36.6	34.0	36.6	46.0	60.2	69.1	63.5	58.7	66.6
Fuels	4.3	4.3	4.3	4.3	5.6	6.2	8.9	36.7	35.7	38.4	42.0
Manufactures	30.1	31.7	32.4	37.5	36.5	38.5	49.5	67.2	66.1	66.7	71.7
Import Prices, cif	27.8	26.7	27.7	23.7	23.8	24.3	31.3	55.2	53.9	54.4	60.2
Trade at Constant 1980 Prices				*(Millions of 1980 US dollars)*							
Exports, fob	155.5	118.5	148.8	139.2	131.1	129.6	108.1	96.3	156.8	164.5	120.2
Imports, cif	208.6	187.5	267.0	362.0	281.0	347.0	289.0	195.0	317.2	393.4	405.1
BALANCE OF PAYMENTS				*(Millions of current US dollars)*							
Exports of Goods & Services	..	..	..	71.0	81.0	97.0	106.0	114.0	166.0	163.0	170.0
Merchandise, fob	..	..	..	42.0	48.0	58.0	63.0	66.0	100.0	101.0	95.0
Nonfactor Services	..	..	..	25.0	29.0	35.0	38.0	40.0	60.0	55.0	68.0
Factor Services	..	..	..	4.0	4.0	4.0	5.0	8.0	6.0	7.0	7.0
Imports of Goods & Services	..	..	..	96.0	109.0	109.0	132.0	171.0	220.0	207.0	199.0
Merchandise, fob	..	..	..	75.0	85.0	85.0	103.0	134.0	171.0	168.0	164.0
Nonfactor Services	..	..	..	21.0	24.0	24.0	29.0	37.0	48.0	37.0	34.0
Factor Services	..	..	..	0.0	0.0	0.0	0.0	0.0	1.0	2.0	1.0
Long-Term Interest	..	..	..	0.0	0.0	0.0	0.0	0.0	0.0	1.0	1.0
Current Transfers, net	..	..	..	9.0	10.0	9.0	6.0	12.0	17.0	27.0	29.0
Workers' Remittances	..	..	..	..	..	..	..	..	..	0.0	0.0
Total to be Financed	..	..	..	-16.0	-18.0	-3.0	-20.0	-45.0	-37.0	-17.0	0.0
Official Capital Grants	..	..	..	24.0	27.0	24.0	17.0	32.0	36.0	22.0	20.0
Current Account Balance	..	..	..	8.0	9.0	21.0	-3.0	-13.0	-1.0	5.0	21.0
Long-Term Capital, net	..	..	..	22.0	32.0	28.0	25.0	40.0	44.0	34.0	37.0
Direct Investment	..	..	..	..	..	..	..	..	..	0.0	0.0
Long-Term Loans	..	..	..	-2.0	5.0	4.0	8.0	7.0	8.0	11.0	29.0
Disbursements	..	..	..	1.0	5.0	4.0	8.0	8.0	9.0	12.0	31.0
Repayments	..	..	..	2.0	0.0	0.0	0.0	1.0	1.0	1.0	2.0
Other Long-Term Capital	..	..	..	0.0	0.0	0.0	0.0	1.0	0.0	1.0	-11.0
Other Capital, net	..	..	..	-17.0	-35.0	-48.0	-14.0	-25.0	-65.0	-9.0	-35.0
Change in Reserves	..	..	..	-13.0	-6.0	-1.0	-8.0	-2.0	21.0	-30.0	-22.0
Memo Item:				*(Nepalese Rupees per US dollar)*							
Conversion Factor (Annual Avg)	7.620	9.290	10.130	10.130	10.130	10.130	10.280	10.560	10.560	11.970	12.500
EXTERNAL DEBT, ETC.				*(Millions of US dollars, outstanding at end of year)*							
Public/Publicly Guar. Long-Term	..	..	..	2.8	8.3	11.6	19.5	27.1	33.7	44.3	72.1
Official Creditors	..	..	..	2.0	7.4	10.8	18.7	26.3	33.0	43.7	72.1
IBRD and IDA	..	..	..	0.1	0.3	0.4	2.0	3.2	5.5	9.3	21.5
Private Creditors	..	..	..	0.9	0.9	0.8	0.8	0.8	0.7	0.6	..
Private Non-guaranteed Long-Term	..	..	..	..	..	..	..	..	..	..	..
Use of Fund Credit	..	..	..	0.0	0.0	0.0	0.0	0.0	0.0	5.2	5.4
Short-Term Debt	..	..	..	..	..	..	..	..	..	..	23.0
Memo Items:				*(Millions of US dollars)*							
Int'l Reserves Excluding Gold	43.1	54.2	75.5	88.7	96.1	98.4	117.5	121.3	95.7	127.5	139.5
Gold Holdings (at market price)	3.3	11.1	8.2	5.5	6.2	8.8	14.6	24.3	18.3	17.5	21.9
SOCIAL INDICATORS											
Total Fertility Rate	6.2	6.2	6.3	6.4	6.5	6.5	6.5	6.5	6.5	6.5	6.5
Crude Birth Rate	45.5	45.8	46.1	46.5	46.8	47.1	46.6	46.1	45.6	45.1	44.6
Infant Mortality Rate	164.0	161.8	159.6	157.4	155.2	153.0	151.8	150.6	149.4	148.2	147.0
Life Expectancy at Birth	40.6	40.9	41.3	41.6	41.9	42.3	42.6	42.9	43.2	43.5	43.9
Food Production, p.c. ('79-81 = 100)	110.0	109.9	111.5	113.0	108.8	104.7	110.8	111.1	110.4	108.5	102.1
Labor Force, Agriculture (%)	93.8	93.8	93.7	93.6	93.5	93.5	93.4	93.4	93.3	93.2	93.2
Labor Force, Female (%)	35.2	35.2	35.1	35.1	35.0	35.0	35.0	34.9	34.9	34.9	34.9
School Enroll. Ratio, primary	..	..	..	26.0	..	..	..	..	51.0	59.0	69.0
School Enroll. Ratio, secondary	..	..	..	10.0	..	..	..	..	13.0	12.0	14.0

1978	1979	1980	1981	1982	1983	1984	1985	1986	1987 estimate	NOTES	NEPAL
				(Millions of current US dollars)							**FOREIGN TRADE (CUSTOMS BASIS)**
91.0	108.0	109.3	134.0	115.0	94.0	128.2	159.7	141.8	150.8	C f	Value of Exports, fob
76.3	84.3	76.0	100.3	92.2	53.4	51.4	57.9	43.7	38.9	..	Nonfuel Primary Products
0.0	0.0	0.0	0.0	0.1	4.9	7.3	8.0	2.6	3.2	..	Fuels
14.7	23.7	33.3	33.8	22.8	35.8	69.6	93.9	95.5	108.7	..	Manufactures
339.0	379.0	342.0	369.0	395.0	446.6	416.0	453.1	459.3	569.2	C f	Value of Imports, cif
85.7	72.6	57.0	62.8	67.8	84.6	71.6	60.7	53.5	75.1	..	Nonfuel Primary Products
88.5	125.7	152.6	161.0	169.0	49.1	52.8	54.2	36.1	46.7	..	Fuels
164.9	180.7	132.4	145.2	158.2	312.9	291.7	338.2	369.8	447.4	..	Manufactures
				(Index 1980 = 100)							
129.0	125.1	100.0	92.6	88.9	88.8	84.9	91.1	98.7	92.6	..	Terms of Trade
83.8	99.9	100.0	94.9	83.9	84.6	81.0	83.3	92.1	97.6	..	Export Prices, fob
85.4	103.0	100.0	93.6	81.6	80.4	71.1	75.1	78.6	73.7	..	Nonfuel Primary Products
42.3	61.0	100.0	112.5	101.6	92.5	90.2	87.5	44.9	56.9	..	Fuels
76.1	90.0	100.0	99.1	94.8	90.7	89.3	89.0	103.2	113.0	..	Manufactures
65.0	79.8	100.0	102.5	94.4	95.3	95.4	91.4	93.4	105.4	..	Import Prices, cif
				(Millions of 1980 US dollars)							Trade at Constant 1980 Prices
108.6	108.2	109.3	141.2	137.0	111.1	158.2	191.7	153.9	154.6	..	Exports, fob
521.9	474.7	342.0	360.1	418.3	468.5	436.0	495.5	492.0	540.1	..	Imports, cif
				(Millions of current US dollars)							**BALANCE OF PAYMENTS**
177.0	229.0	241.0	307.0	293.0	264.0	279.0	307.0	332.0	352.0	C f	Exports of Goods & Services
86.0	109.0	97.0	134.0	116.0	82.0	111.0	154.0	156.0	139.0	..	Merchandise, fob
83.0	110.0	127.0	159.0	162.0	167.0	162.0	147.0	173.0	209.0	..	Nonfactor Services
8.0	10.0	16.0	13.0	15.0	15.0	6.0	5.0	3.0	5.0	..	Factor Services
248.0	298.0	368.0	450.0	453.0	522.0	503.0	528.0	571.0	610.0	C f	Imports of Goods & Services
204.0	243.0	297.0	370.0	383.0	457.0	426.0	436.0	474.0	507.0	..	Merchandise, fob
43.0	53.0	67.0	76.0	68.0	62.0	74.0	87.0	85.0	89.0	..	Nonfactor Services
1.0	2.0	3.0	3.0	3.0	3.0	4.0	5.0	12.0	15.0	..	Factor Services
1.0	2.0	2.0	3.0	3.0	4.0	5.0	8.0	13.0	14.0	..	Long-Term Interest
24.0	30.0	36.0	46.0	41.0	42.0	45.0	43.0	45.0	64.0	..	Current Transfers, net
0.0	0.0	0.0	0.0	0.0	0.0	0.0	0.0	0.0	..	..	Workers' Remittances
-47.0	-39.0	-92.0	-96.0	-120.0	-217.0	-179.0	-179.0	-194.0	-194.0	K	Total to be Financed
23.0	43.0	64.0	72.0	89.0	95.0	90.0	75.0	69.0	61.0	K	Official Capital Grants
-24.0	3.0	-28.0	-25.0	-30.0	-122.0	-89.0	-104.0	-125.0	-133.0	..	Current Account Balance
47.0	78.0	112.0	125.0	149.0	162.0	169.0	146.0	160.0	171.0	C f	Long-Term Capital, net
0.0	0.0	0.0	0.0	0.0	0.0	0.0	0.0	0.0	..	..	Direct Investment
37.0	37.0	53.0	64.0	67.0	69.0	88.0	97.0	124.0	133.0	..	Long-Term Loans
39.0	38.0	55.0	66.0	69.0	73.0	94.0	105.0	142.0	152.0	..	Disbursements
2.0	2.0	2.0	2.0	3.0	4.0	6.0	8.0	18.0	20.0	..	Repayments
-13.0	-1.0	-5.0	-11.0	-7.0	-2.0	-10.0	-25.0	-33.0	-23.0	..	Other Long-Term Capital
-20.0	-33.0	-81.0	-84.0	-80.0	-94.0	-99.0	-90.0	-37.0	-29.0	C f	Other Capital, net
-4.0	-49.0	-2.0	-16.0	-39.0	54.0	19.0	48.0	1.0	-9.0	..	Change in Reserves
				(Nepalese Rupees per US dollar)							Memo Item:
12.360	12.000	12.000	12.000	12.940	13.800	15.260	17.780	19.330	21.530	..	Conversion Factor (Annual Avg)
				(Millions of US dollars, outstanding at end of year)							**EXTERNAL DEBT, ETC.**
87.0	123.1	173.3	234.5	298.4	363.7	445.0	561.4	711.1	902.0	..	Public/Publicly Guar. Long-Term
87.0	123.1	173.3	234.5	297.8	361.7	431.4	538.4	676.4	867.4	..	Official Creditors
32.5	51.1	76.1	108.9	142.2	172.1	200.0	236.0	295.4	391.9	..	IBRD and IDA
..	..	..	..	0.6	2.1	13.6	23.1	34.7	34.6	..	Private Creditors
..	..	..	..	..	..	..	..	..	..	..	Private Non-guaranteed Long-Term
12.4	11.2	24.2	21.9	15.5	9.6	3.9	11.3	15.1	26.5	..	Use of Fund Credit
16.0	12.0	7.0	22.0	39.0	80.0	24.0	23.0	20.5	18.7	..	Short-Term Debt
				(Millions of US dollars)							Memo Items:
145.1	159.2	182.8	201.9	199.3	133.4	82.0	56.0	86.8	178.2	..	Int'l Reserves Excluding Gold
33.0	76.1	89.2	60.1	69.1	57.7	46.7	49.5	59.1	73.2	..	Gold Holdings (at market price)
											SOCIAL INDICATORS
6.5	6.4	6.4	6.3	6.3	6.2	6.1	6.0	5.9	5.9	..	Total Fertility Rate
44.3	43.9	43.6	43.2	42.9	42.5	42.1	41.6	41.2	40.8	..	Crude Birth Rate
145.4	143.8	142.2	140.6	139.0	137.9	136.7	135.6	134.4	133.3	..	Infant Mortality Rate
44.3	44.7	45.1	45.5	45.9	46.2	46.6	47.0	47.4	47.8	..	Life Expectancy at Birth
103.2	95.8	101.5	102.7	94.4	107.9	105.2	103.5	97.2	97.3	..	Food Production, p.c. ('79-81 = 100)
93.1	93.1	93.0	..	..	..	..	..	..	..	..	Labor Force, Agriculture (%)
34.9	34.9	34.9	34.8	34.7	34.6	34.5	34.3	34.2	34.0	..	Labor Force, Female (%)
77.0	88.0	83.0	65.0	67.0	73.0	77.0	79.0	..	..	..	School Enroll. Ratio, primary
16.0	19.0	21.0	18.0	21.0	22.0	23.0	25.0	..	..	..	School Enroll. Ratio, secondary

NETHERLANDS, THE	1967	1968	1969	1970	1971	1972	1973	1974	1975	1976	1977
CURRENT GNP PER CAPITA (US $)	1,850	2,060	2,300	2,530	2,840	3,210	4,010	5,210	6,070	6,840	7,670
POPULATION (thousands)	12,598	12,730	12,878	13,039	13,195	13,329	13,439	13,545	13,666	13,774	13,856
ORIGIN AND USE OF RESOURCES					*(Billions of current Netherlands Guilders)*						
Gross National Product (GNP)	85.78	94.77	107.44	120.25	136.61	155.03	177.35	201.45	219.95	252.04	275.42
Net Factor Income from Abroad	0.84	0.59	0.79	0.56	0.39	0.70	1.30	1.67	-0.02	0.12	0.49
Gross Domestic Product (GDP)	84.93	94.18	106.66	119.69	136.22	154.33	176.05	199.78	219.97	251.92	274.93
Indirect Taxes, net	8.05	9.43	9.68	11.55	13.79	15.77	17.02	17.96	20.21	22.77	26.73
GDP at factor cost	76.89	84.75	96.98	108.13	122.43	138.56	159.03	181.82	199.76	229.15	248.20
Agriculture	..	..	6.55	6.61	6.98	8.02	9.10	8.29	9.75	11.29	11.42
Industry	..	..	40.72	44.65	50.43	56.96	64.37	73.09	76.96	87.97	92.62
Manufacturing	..	..	28.74	30.93	34.25	38.48	43.64	49.99	48.24	54.37	55.06
Services, etc.	..	..	59.39	68.42	78.81	89.36	102.59	118.39	133.25	152.66	170.89
Resource Balance	-0.77	-0.02	-0.22	-2.08	-0.31	4.28	5.74	5.38	7.48	8.48	3.39
Exports of Goods & NFServices	34.68	38.90	45.70	54.06	61.92	69.40	83.44	107.75	109.76	128.42	130.72
Imports of Goods & NFServices	35.46	38.93	45.92	56.14	62.23	65.12	77.70	102.37	102.28	119.94	127.33
Domestic Absorption	85.71	94.20	106.88	121.76	136.53	150.05	170.31	194.40	212.49	243.44	271.54
Private Consumption, etc.	58.80	64.53	73.63	69.60	77.58	87.33	99.73	113.30	128.98	148.05	164.32
General Gov't Consumption	13.11	14.20	16.20	18.63	21.59	24.37	27.42	32.45	38.25	43.47	47.85
Gross Domestic Investment	13.80	15.46	17.05	33.53	37.36	38.35	43.16	48.64	45.26	51.92	59.37
Fixed Investment	22.78	25.75	26.67	31.52	36.00	38.21	42.85	45.94	48.28	51.23	60.08
Memo Items:											
Gross Domestic Saving	13.03	15.44	16.83	31.45	37.05	42.63	48.90	54.02	52.74	60.40	62.75
Gross National Saving	13.87	16.03	17.62	31.77	37.18	42.97	49.68	55.02	51.99	59.65	62.32
					(Billions of 1980 Netherlands Guilders)						
Gross National Product	207.63	219.91	234.38	246.80	259.45	264.61	278.72	299.93	296.38	310.70	319.13
GDP at factor cost	182.15	192.04	207.29	220.53	231.15	234.31	247.82	270.28	268.66	281.91	286.97
Agriculture	..	..	7.33	7.70	8.07	8.14	8.92	9.81	9.68	9.75	10.17
Industry	..	..	..	..	..	..	..	..	..	..	..
Manufacturing	..	..	..	..	..	..	..	..	..	..	..
Services, etc.	..	..	..	..	..	..	..	..	..	..	..
Resource Balance	-11.09	-12.71	-13.84	-18.56	-13.88	-12.81	-10.75	3.76	4.63	4.31	-2.29
Exports of Goods & NFServices	73.53	82.92	95.32	106.62	118.40	131.44	146.39	150.28	145.57	159.93	157.11
Imports of Goods & NFServices	84.62	95.63	109.15	125.18	132.27	144.24	157.14	146.51	140.93	155.61	159.40
Domestic Absorption	216.25	231.03	246.19	264.14	272.55	276.13	287.28	293.65	291.77	306.24	320.84
Private Consumption, etc.	115.19	122.72	134.94	145.43	150.30	155.43	162.80	168.85	174.33	183.50	191.92
General Gov't Consumption	41.67	42.58	44.50	47.18	48.57	49.57	49.37	50.40	52.31	54.33	56.07
Gross Domestic Investment	59.39	65.73	66.75	71.54	73.67	71.13	75.11	74.40	65.13	68.41	72.85
Fixed Investment	58.50	65.03	63.57	68.36	71.85	70.36	72.27	69.50	66.32	64.52	71.02
Memo Items:											
Capacity to Import	82.77	95.58	108.63	120.54	131.62	153.72	168.74	154.21	151.24	166.61	163.64
Terms of Trade Adjustment	9.25	12.65	13.31	13.92	13.22	22.29	22.35	3.94	5.67	6.69	6.53
Gross Domestic Income	214.40	230.97	245.67	259.51	271.89	285.61	298.88	301.35	302.08	317.24	325.08
Gross National Income	216.88	232.57	247.69	260.72	272.67	286.90	301.07	303.87	302.05	317.39	325.66
DOMESTIC PRICES (DEFLATORS)					*(Index 1980 = 100)*						
Overall (GDP)	41.4	43.1	45.9	48.7	52.7	58.6	63.7	67.2	74.2	81.1	86.3
Domestic Absorption	39.6	40.8	43.4	46.1	50.1	54.3	59.3	66.2	72.8	79.5	84.6
Agriculture	..	..	89.4	85.8	86.4	98.5	102.0	84.5	100.7	115.7	112.3
Industry	..	..	..	..	..	..	..	..	..	..	..
Manufacturing	..	..	..	..	..	..	..	..	..	..	..
MANUFACTURING ACTIVITY											
Employment (1980=100)	126.6	126.4	121.1	121.1	119.2	115.5	113.6	112.7	108.8	104.2	104.8
Real Earnings per Empl. (1980=100)	63.1	65.2	71.4	78.3	81.9	85.1	90.5	94.7	96.1	98.5	96.0
Real Output per Empl. (1980=100)	..	..	62.1	68.9	72.2	74.3	82.6	93.7	88.4	96.4	91.6
Earnings as % of Value Added	52.5	51.8	50.3	52.2	52.8	51.6	51.2	50.7	57.1	53.5	55.7
MONETARY HOLDINGS					*(Billions of current Netherlands Guilders)*						
Money Supply, Broadly Defined	49.95	56.70	63.20	70.57	80.12	109.68	124.88	141.20	159.51	183.33	207.80
Money as Means of Payment	19.29	21.49	23.23	25.95	29.85	35.12	35.14	39.43	47.20	51.05	57.77
Currency Ouside Banks	8.72	8.84	9.43	9.95	10.49	11.41	11.92	12.85	14.49	15.89	17.41
Demand Deposits	10.57	12.64	13.79	16.00	19.36	23.71	23.22	26.58	32.70	35.16	40.35
Quasi-Monetary Liabilities	30.67	35.21	39.98	44.61	50.27	74.55	89.74	101.77	112.31	132.28	150.03
GOVERNMENT DEFICIT (-) OR SURPLUS					*(Billions of current Netherlands Guilders)*						
GOVERNMENT DEFICIT (-) OR SURPLUS	..	..	..	..	..	..	-0.04	-0.04	-6.35	-6.12	-8.36
Current Revenue	..	..	..	..	..	..	77.22	89.71	103.72	119.49	132.96
Current Expenditure	..	..	..	..	..	..	66.75	80.26	97.30	112.61	125.91
Current Budget Balance	..	..	..	..	..	..	10.47	9.45	6.42	6.88	7.05
Capital Receipts	..	..	..	..	..	..	0.05	0.21	0.21	0.23	0.22
Capital Payments	..	..	..	..	..	..	10.56	9.74	12.39	12.40	12.37

1978	1979	1980	1981	1982	1983	1984	1985	1986	1987 estimate	NOTES	NETHERLANDS, THE
8,780	10,400	12,020	11,810	10,940	10,080	9,680	9,360	10,030	11,890	..	**CURRENT GNP PER CAPITA (US $)**
13,942	14,038	14,150	14,247	14,313	14,362	14,420	14,491	14,563	14,616	..	**POPULATION (thousands)**
				(Billions of current Netherlands Guilders)							**ORIGIN AND USE OF RESOURCES**
296.33	315.23	336.12	351.89	368.49	381.58	399.75	418.02	428.75	434.34	..	Gross National Product (GNP)
-0.68	-0.73	-0.62	-0.96	-0.38	0.56	-0.50	1.43	-0.38	-0.01	..	Net Factor Income from Abroad
297.01	315.96	336.74	352.85	368.87	381.02	400.25	416.59	429.13	434.35	..	Gross Domestic Product (GDP)
29.30	29.80	32.02	32.90	32.70	33.92	35.38	36.93	40.19	40.36	..	Indirect Taxes, net
267.71	286.16	304.72	319.95	336.17	347.10	364.87	379.66	388.94	393.99	B	GDP at factor cost
11.77	11.33	11.68	14.56	15.93	16.12	17.26	17.17	18.57	..	..	Agriculture
97.70	102.78	110.45	116.00	121.25	124.62	133.38	139.00	139.75	..	..	Industry
57.95	60.08	60.37	59.78	64.67	67.46	72.57	74.41	84.99	..	..	Manufacturing
187.53	201.85	214.62	222.29	231.69	240.28	249.61	260.33	270.81	..	..	Services, etc.
0.13	-1.69	-1.68	12.44	15.75	14.71	20.66	20.38	17.30	11.39	..	Resource Balance
133.35	155.06	176.87	204.63	212.55	219.84	248.49	266.11	232.05	225.92	..	Exports of Goods & NFServices
133.22	156.75	178.56	192.19	196.80	205.13	227.83	245.73	214.75	214.53	..	Imports of Goods & NFServices
296.88	317.65	338.43	340.41	353.11	366.31	379.59	396.21	411.83	422.96	..	Domestic Absorption
179.21	192.41	205.77	213.35	221.73	229.86	236.74	247.40	255.05	263.51	..	Private Consumption, etc.
52.61	57.17	60.26	62.75	65.12	66.58	66.39	67.55	68.24	69.43	..	General Gov't Consumption
65.06	68.07	72.39	64.31	66.26	69.87	76.47	81.26	88.54	90.02	..	Gross Domestic Investment
64.77	66.96	70.84	66.79	67.18	70.15	75.29	80.80	86.59	89.41	..	Fixed Investment
											Memo Items:
65.19	66.38	70.71	76.75	82.02	84.58	97.12	101.64	105.84	101.41	..	Gross Domestic Saving
63.38	64.47	68.47	73.98	79.75	83.53	95.12	101.58	103.62	99.29	..	Gross National Saving
				(Billions of 1980 Netherlands Guilders)							
325.57	332.18	336.12	333.43	329.38	335.01	344.32	353.82	360.72	370.13	..	Gross National Product
293.43	301.14	304.72	303.14	300.09	304.43	313.95	321.02	326.68	335.59	B	GDP at factor cost
10.92	11.34	11.68	13.29	14.27	14.75	15.65	15.32	16.89	..	..	Agriculture
..	..	..	..	..	..	..	..	..	..	..	Industry
..	..	..	..	..	..	..	..	..	..	..	Manufacturing
..	..	..	..	..	..	..	..	..	..	..	Services, etc.
-7.10	-6.18	-1.68	11.53	9.75	9.64	14.24	13.82	8.95	8.49	..	Resource Balance
162.22	174.23	176.87	179.65	179.67	185.99	199.59	210.37	211.73	222.80	..	Exports of Goods & NFServices
169.32	180.41	178.56	168.12	169.92	176.36	185.34	196.55	202.78	214.31	..	Imports of Goods & NFServices
333.43	339.14	338.43	322.81	319.98	324.88	330.51	338.78	352.10	361.61	..	Domestic Absorption
200.18	205.91	205.77	200.49	198.15	199.83	201.76	205.46	211.84	219.16	..	Private Consumption, etc.
58.26	59.90	60.26	61.51	61.90	62.62	62.29	63.12	64.30	64.81	..	General Gov't Consumption
74.99	73.32	72.39	60.81	59.93	62.43	66.46	70.20	75.95	77.65	..	Gross Domestic Investment
72.78	71.54	70.84	63.37	60.91	62.25	65.64	69.18	74.08	76.48	..	Fixed Investment
											Memo Items:
169.48	178.47	176.87	179.00	183.53	189.01	202.15	212.85	219.11	225.69	..	Capacity to Import
7.26	4.24	0.00	-0.65	3.85	3.01	2.56	2.48	7.39	2.89	..	Terms of Trade Adjustment
333.59	337.20	336.74	333.69	333.58	337.53	347.31	355.08	368.43	372.99	..	Gross Domestic Income
332.83	336.42	336.12	332.78	333.24	338.03	346.88	356.30	368.10	..	..	Gross National Income
				(Index 1980 = 100)							**DOMESTIC PRICES (DEFLATORS)**
91.0	94.9	100.0	105.5	111.9	113.9	116.1	118.2	118.9	117.4	..	Overall (GDP)
89.0	93.7	100.0	105.5	110.4	112.8	114.9	117.0	117.0	117.0	..	Domestic Absorption
107.8	99.9	100.0	109.5	111.6	109.3	110.3	112.1	109.9	..	..	Agriculture
..	..	..	..	..	..	..	..	..	..	..	Industry
..	..	..	..	..	..	..	..	..	..	..	Manufacturing
											MANUFACTURING ACTIVITY
102.1	101.2	100.0	96.7	92.6	88.5	86.7	88.0	..	..	..	Employment (1980 = 100)
99.3	101.0	100.0	98.7	100.0	100.7	111.4	113.9	..	..	..	Real Earnings per Empl. (1980 = 100)
90.8	98.7	100.0	101.6	99.3	107.4	114.6	115.7	..	..	..	Real Output per Empl. (1980 = 100)
56.8	57.7	57.9	58.8	56.8	54.0	56.5	56.5	..	..	..	Earnings as % of Value Added
				(Billions of current Netherlands Guilders)							**MONETARY HOLDINGS**
233.78	258.59	273.69	293.69	312.03	327.30	346.85	366.97	386.50	..	D	Money Supply, Broadly Defined
60.19	61.87	65.58	64.03	72.30	79.66	85.00	90.77	97.21	103.71	..	Money as Means of Payment
18.71	19.98	22.00	22.34	23.39	26.35	27.80	28.60	29.77	32.85	..	Currency Ouside Banks
41.47	41.90	43.59	41.70	48.91	53.31	57.20	62.17	67.44	70.86	..	Demand Deposits
173.59	196.71	208.11	229.66	239.73	247.64	261.85	276.20	289.29	..	..	Quasi-Monetary Liabilities
				(Billions of current Netherlands Guilders)							**GOVERNMENT DEFICIT (-) OR SURPLUS**
-9.21	-14.44	-15.57	-23.06	-28.02	-29.12	-29.22	-22.77	-7.22	-13.77	..	GOVERNMENT DEFICIT (-) OR SURPLUS
143.58	154.52	169.24	179.33	189.75	200.83	206.71	217.48	223.87	225.17	..	Current Revenue
139.89	152.45	164.76	178.65	195.95	206.46	212.41	216.60	222.39	228.85	..	Current Expenditure
3.69	2.07	4.48	0.68	-6.20	-5.63	-5.70	0.88	1.48	-3.68	..	Current Budget Balance
0.15	0.13	0.17	0.19	0.19	0.09	0.18	0.65	0.60	0.29	..	Capital Receipts
12.46	14.96	19.76	23.10	22.35	21.50	23.17	20.56	8.98	11.40	..	Capital Payments

NETHERLANDS, THE	1967	1968	1969	1970	1971	1972	1973	1974	1975	1976	1977
FOREIGN TRADE (CUSTOMS BASIS)					*(Millions of current US dollars)*						
Value of Exports, fob	7,288	8,342	9,965	11,766	13,934	17,351	24,012	32,734	34,957	40,147	43,605
Nonfuel Primary Products	2,502	2,859	3,310	3,842	4,221	5,234	7,385	9,084	10,145	11,298	12,550
Fuels	573	668	827	1,256	1,737	2,077	3,127	5,234	5,960	7,112	7,932
Manufactures	4,213	4,814	5,829	6,668	7,975	10,040	13,501	18,416	18,852	21,737	23,123
Value of Imports, cif	8,337	9,293	10,994	13,393	14,901	17,226	23,753	32,508	34,394	39,452	45,500
Nonfuel Primary Products	2,275	2,515	2,961	3,354	3,484	4,114	6,036	7,825	8,031	9,141	10,679
Fuels	867	943	1,072	1,458	1,937	2,267	3,151	5,890	6,145	7,682	8,380
Manufactures	5,195	5,834	6,960	8,581	9,481	10,845	14,566	18,792	20,219	22,630	26,440
					(Index 1980 = 100)						
Terms of Trade	141.0	133.1	117.7	116.4	111.8	120.1	143.7	103.2	99.6	99.7	94.9
Export Prices, fob	26.8	25.5	23.6	22.9	23.7	28.0	44.4	55.3	55.6	56.9	59.7
Nonfuel Primary Products	33.5	33.9	37.1	38.0	38.3	43.4	59.7	64.3	62.9	65.6	70.1
Fuels	8.5	7.3	5.4	5.9	7.0	8.5	24.5	36.6	33.6	38.5	36.8
Manufactures	32.5	31.8	32.5	33.5	34.9	39.6	46.6	59.7	65.4	62.7	69.3
Import Prices, cif	19.0	19.1	20.0	19.7	21.2	23.4	30.9	53.6	55.8	57.1	62.9
Trade at Constant 1980 Prices					*(Millions of 1980 US dollars)*						
Exports, fob	27,239	32,770	42,218	51,364	58,757	61,888	54,115	59,153	62,920	70,538	72,992
Imports, cif	43,936	48,573	54,840	68,039	70,273	73,764	76,914	60,597	61,659	69,127	72,303
BALANCE OF PAYMENTS					*(Billions of current US dollars)*						
Exports of Goods & Services	..	..	..	15.26	17.85	21.79	30.58	43.07	45.25	50.56	55.64
Merchandise, fob	..	..	..	10.88	12.54	15.58	21.92	30.51	32.01	36.36	39.90
Nonfactor Services	..	..	..	3.07	3.77	4.53	6.06	7.94	9.18	9.63	10.65
Factor Services	..	..	..	1.31	1.54	1.69	2.60	4.61	4.07	4.58	5.10
Imports of Goods & Services	..	..	..	15.70	17.89	20.30	28.10	39.71	42.22	46.72	53.73
Merchandise, fob	..	..	..	11.78	13.15	15.14	20.93	29.92	31.11	35.02	40.14
Nonfactor Services	..	..	..	2.76	3.32	3.69	5.04	6.55	7.28	7.96	9.27
Factor Services	..	..	..	1.16	1.42	1.47	2.13	3.24	3.83	3.74	4.32
Long-Term Interest	..	..	..	..	..	..	..	..	..	..	..
Current Transfers, net	..	..	..	-0.07	-0.07	-0.11	-0.19	-0.25	-0.29	-0.33	-0.38
Workers' Remittances	..	..	..	0.00	0.00	0.00	0.00	0.00	0.00	0.00	0.00
Total to be Financed	..	..	..	-0.51	-0.12	1.37	2.29	3.10	2.75	3.51	1.53
Official Capital Grants	..	..	..	0.03	0.01	-0.02	0.12	-0.06	-0.38	-0.07	-0.31
Current Account Balance	..	..	..	-0.48	-0.11	1.35	2.41	3.04	2.37	3.44	1.22
Long-Term Capital, net	..	..	..	0.60	0.53	-0.96	-1.69	-2.21	-2.20	-4.16	-2.17
Direct Investment	..	..	..	-0.01	0.09	-0.13	-0.06	-1.58	-1.05	-1.49	-2.38
Long-Term Loans	..	..	..	..	..	..	..	..	..	..	..
Disbursements	..	..	..	..	..	..	..	..	..	..	..
Repayments	..	..	..	..	..	..	..	..	..	..	..
Other Long-Term Capital	..	..	..	0.59	0.42	-0.81	-1.76	-0.57	-0.78	-2.61	0.52
Other Capital, net	..	..	..	0.50	-0.26	0.48	0.24	0.15	0.14	1.04	1.24
Change in Reserves	..	..	..	-0.62	-0.16	-0.87	-0.96	-0.98	-0.31	-0.33	-0.30
Memo Item:					*(Netherlands Guilders per US dollar)*						
Conversion Factor (Annual Avg)	3.620	3.620	3.620	3.620	3.500	3.210	2.800	2.690	2.530	2.640	2.550
EXTERNAL DEBT, ETC.					*(Millions US dollars, outstanding at end of year)*						
Public/Publicly Guar. Long-Term	..	..	..	..	..	..	..	..	..	..	..
Official Creditors	..	..	..	..	..	..	..	..	..	..	..
IBRD and IDA	..	..	..	..	..	..	..	..	..	..	..
Private Creditors	..	..	..	..	..	..	..	..	..	..	..
Private Non-guaranteed Long-Term	..	..	..	..	..	..	..	..	..	..	..
Use of Fund Credit	..	..	..	..	..	..	..	..	..	..	..
Short-Term Debt	..	..	..	..	..	..	..	..	..	..	..
Memo Items:					*(Millions of US dollars)*						
Int'l Reserves Excluding Gold	908	766	809	1,454	1,724	2,726	4,253	4,630	4,884	5,178	5,742
Gold Holdings (at market price)	1,722	2,032	1,731	1,908	2,379	3,516	6,098	10,132	7,620	7,321	9,011
SOCIAL INDICATORS											
Total Fertility Rate	2.8	2.7	2.7	2.6	2.4	2.2	1.9	1.8	1.7	1.6	1.6
Crude Birth Rate	18.9	18.6	19.2	18.3	17.2	16.1	14.5	13.7	13.0	12.9	12.5
Infant Mortality Rate	13.4	13.6	13.2	12.7	12.1	11.7	11.5	11.3	10.6	10.7	9.5
Life Expectancy at Birth	73.6	73.6	73.6	73.5	73.8	74.0	74.1	74.3	74.4	74.8	75.3
Food Production, p.c. ('79-81 = 100)	78.3	79.8	77.6	84.7	86.2	80.2	84.3	90.9	92.9	89.2	90.2
Labor Force, Agriculture (%)	7.9	7.6	7.2	6.8	6.7	6.6	6.4	6.3	6.2	6.1	5.9
Labor Force, Female (%)	25.4	25.7	26.0	26.4	26.8	27.3	27.7	28.1	28.6	29.1	29.6
School Enroll. Ratio, primary	..	..	..	102.0	..	..	..	..	98.0	100.0	100.0
School Enroll. Ratio, secondary	..	..	..	75.0	..	..	..	..	88.0	91.0	92.0

1978	1979	1980	1981	1982	1983	1984	1985	1986	1987 estimate	NOTES	NETHERLANDS, THE
				(Millions of current US dollars)							**FOREIGN TRADE (CUSTOMS BASIS)**
50,149	63,667	73,871	68,758	66,404	65,676	65,874	68,282	80,555	92,882	..	Value of Exports, fob
14,693	17,852	19,930	18,789	18,001	17,819	17,720	17,638	22,145	26,560	..	Nonfuel Primary Products
8,023	11,966	16,114	16,231	15,669	15,212	14,878	15,495	12,214	10,089	..	Fuels
27,434	33,849	37,827	33,738	32,734	32,645	33,275	35,149	46,197	56,232	..	Manufactures
53,041	67,281	76,889	66,109	62,583	61,585	62,136	65,212	75,580	91,316	..	Value of Imports, cif
11,942	14,649	16,602	14,092	13,381	13,163	13,271	13,341	15,440	18,460	..	Nonfuel Primary Products
8,253	13,358	18,252	17,258	15,986	15,129	14,470	14,181	8,738	9,936	..	Fuels
32,846	39,274	42,036	34,759	33,216	33,294	34,395	37,690	51,402	62,921	..	Manufactures
				(Index 1980 = 100)							
101.4	111.6	100.0	92.6	96.5	92.0	90.7	91.3	91.2	93.1	..	Terms of Trade
71.2	93.7	100.0	91.9	89.7	83.1	79.9	78.1	84.9	94.2	..	Export Prices, fob
79.5	98.9	100.0	92.0	85.3	87.2	85.1	80.0	85.6	88.3	..	Nonfuel Primary Products
44.7	93.6	100.0	104.0	97.0	85.4	83.6	81.1	60.0	56.4	..	Fuels
81.0	91.5	100.0	87.0	89.0	80.0	76.0	76.0	95.0	111.0	..	Manufactures
70.2	83.9	100.0	99.2	92.9	90.3	88.1	85.6	93.1	101.2	..	Import Prices, cif
				(Millions of 1980 US dollars)							Trade at Constant 1980 Prices
70,480	67,979	73,871	74,817	74,036	79,057	82,417	87,402	94,867	98,636	..	Exports, fob
75,603	80,147	76,889	66,640	67,364	68,198	70,537	76,182	81,176	90,241	..	Imports, cif
				(Billions of current US dollars)							**BALANCE OF PAYMENTS**
63.94	83.60	97.74	92.15	88.87	84.83	86.54	87.08	102.61	123.35	..	Exports of Goods & Services
45.44	58.60	67.49	63.14	60.55	59.27	60.05	62.44	73.56	86.71	..	Merchandise, fob
12.66	14.40	17.08	15.93	15.62	14.03	14.38	15.05	17.44	21.96	..	Nonfactor Services
5.84	10.60	13.17	13.09	12.70	11.53	12.11	9.59	11.60	14.68	..	Factor Services
64.14	82.58	97.54	86.98	82.99	78.91	78.99	82.06	96.43	117.88	..	Imports of Goods & Services
46.91	60.02	68.90	59.27	55.94	55.05	54.41	56.98	66.36	81.49	..	Merchandise, fob
11.46	14.06	16.92	14.94	14.62	13.40	13.71	14.64	18.15	21.83	..	Nonfactor Services
5.77	8.51	11.73	12.78	12.42	10.47	10.87	10.44	11.92	14.56	..	Factor Services
..	..	..	..	..	..	..	..	..	..	..	Long-Term Interest
-0.52	-0.59	-0.81	-0.72	-0.71	-0.57	-0.47	-0.45	-0.75	-1.04	..	Current Transfers, net
0.00	0.00	0.00	0.00	0.00	0.00	0.00	0.00	0.00	0.00	..	Workers' Remittances
-0.72	0.43	-0.62	4.44	5.18	5.35	7.08	4.57	5.43	4.43	K	Total to be Financed
-0.50	-0.20	-0.41	-0.79	-0.75	-0.42	-0.53	-0.52	-0.98	-1.05	K	Official Capital Grants
-1.22	0.23	-1.02	3.66	4.43	4.93	6.55	4.06	4.45	3.37	..	Current Account Balance
-3.14	-5.22	-2.66	-3.80	-4.76	-3.79	-6.52	-3.06	-9.59	-2.80	..	Long-Term Capital, net
-2.32	-4.43	-3.72	-2.98	-2.20	-2.36	-3.40	-1.80	-2.02	-5.50	..	Direct Investment
..	..	..	..	..	..	..	..	..	..	..	Long-Term Loans
..	..	..	..	..	..	..	..	..	..	..	Disbursements
..	..	..	..	..	..	..	..	..	..	..	Repayments
-0.32	-0.58	1.47	-0.03	-1.82	-1.00	-2.59	-0.74	-6.59	3.76	..	Other Long-Term Capital
3.44	4.40	4.86	-0.69	2.12	-1.36	0.01	-0.25	4.73	2.29	..	Other Capital, net
0.92	0.59	-1.18	0.83	-1.78	0.22	-0.04	-0.76	0.41	-2.87	..	Change in Reserves
				(Netherlands Guilders per US dollar)							Memo Item:
2.160	2.010	1.990	2.500	2.670	2.850	3.210	3.320	2.450	2.030	..	Conversion Factor (Annual Avg)
				(Millions US dollars, outstanding at end of year)							**EXTERNAL DEBT, ETC.**
..	..	..	..	..	..	..	..	..	..	..	Public/Publicly Guar. Long-Term
..	..	..	..	..	..	..	..	..	..	..	Official Creditors
..	..	..	..	..	..	..	..	..	..	..	IBRD and IDA
..	..	..	..	..	..	..	..	..	..	..	Private Creditors
..	..	..	..	..	..	..	..	..	..	..	Private Non-guaranteed Long-Term
..	..	..	..	..	..	..	..	..	..	..	Use of Fund Credit
..	..	..	..	..	..	..	..	..	..	..	Short-Term Debt
				(Millions of US dollars)							Memo Items:
5,088	7,591	11,645	9,339	10,133	10,171	9,237	10,782	11,191	16,003	..	Int'l Reserves Excluding Gold
12,380	22,514	25,904	17,467	20,077	16,764	13,547	14,369	17,177	21,272	..	Gold Holdings (at market price)
											SOCIAL INDICATORS
1.6	1.6	1.6	1.6	1.6	1.5	1.5	1.5	1.5	1.5	..	Total Fertility Rate
12.6	12.5	12.8	12.5	12.5	12.4	12.4	12.4	12.4	12.3	..	Crude Birth Rate
9.6	8.7	8.6	8.3	8.5	8.8	9.0	9.3	9.5	9.8	..	Infant Mortality Rate
75.4	75.6	75.7	75.9	75.9	76.4	76.5	76.6	76.7	76.8	..	Life Expectancy at Birth
95.2	96.7	96.3	107.0	108.0	105.6	107.9	105.6	112.4	110.8	..	Food Production, p.c. ('79-81 = 100)
5.8	5.6	5.5	..	..	..	..	..	..	..	..	Labor Force, Agriculture (%)
30.0	30.5	31.0	31.0	31.0	31.0	31.0	31.0	31.0	31.0	..	Labor Force, Female (%)
101.0	101.0	100.0	99.0	97.0	95.0	95.0	114.0	..	..	..	School Enroll. Ratio, primary
92.0	93.0	92.0	94.0	97.0	101.0	102.0	104.0	..	..	..	School Enroll. Ratio, secondary

NEW ZEALAND	1967	1968	1969	1970	1971	1972	1973	1974	1975	1976	1977
CURRENT GNP PER CAPITA (US $)	2,100	2,150	2,220	2,230	2,380	2,690	3,280	4,100	4,520	4,850	4,780
POPULATION (thousands)	2,724	2,748	2,804	2,811	2,854	2,902	2,956	3,024	3,083	3,111	3,120
ORIGIN AND USE OF RESOURCES				*(Millions of current New Zealand Dollars)*							
Gross National Product (GNP)	4,135	4,390	4,877	5,453	6,447	7,370	8,693	9,610	10,310	12,546	14,438
Net Factor Income from Abroad	-46	-32	-41	-41	-46	-54	-37	-82	-165	-265	-336
Gross Domestic Product (GDP)	4,181	4,422	4,918	5,494	6,493	7,424	8,730	9,692	10,475	12,811	14,774
Indirect Taxes, net	387	421	456	518	564	639	710	698	731	1,084	1,215
GDP at factor cost	3,794	4,001	4,462	4,976	5,929	6,785	8,020	8,994	9,744	11,727	13,559
Agriculture	..	..	..	..	776	1,030	1,094	849	1,104	1,468	1,485
Industry	..	..	..	..	2,138	2,385	2,723	3,306	3,524	4,331	4,715
Manufacturing	..	..	..	..	1,537	1,692	1,911	2,341	2,466	3,083	3,257
Services, etc.	..	..	..	..	3,579	4,009	4,914	5,536	5,846	7,012	8,574
Resource Balance	48	140	200	-76	6	62	229	-958	-848	-431	-388
Exports of Goods & NFServices	868	1,073	1,273	1,306	1,497	1,691	2,278	2,226	2,386	3,533	4,080
Imports of Goods & NFServices	820	933	1,072	1,382	1,491	1,630	2,048	3,185	3,235	3,964	4,468
Domestic Absorption	4,133	4,282	4,717	5,570	6,487	7,362	8,501	10,650	11,323	13,243	15,162
Private Consumption, etc.	2,687	2,853	3,033	3,565	3,940	4,411	5,190	6,230	6,938	7,975	8,831
General Gov't Consumption	541	583	628	752	880	985	1,090	1,257	1,573	1,869	2,235
Gross Domestic Investment	905	846	1,056	1,253	1,666	1,966	2,221	3,163	2,812	3,399	4,096
Fixed Investment	984	941	1,116	1,354	1,621	1,985	2,355	2,822	3,119	3,514	3,685
Memo Items:											
Gross Domestic Saving	953	985	1,257	1,177	1,673	2,027	2,451	2,205	1,963	2,967	3,709
Gross National Saving	898	946	1,206	1,132	1,641	2,019	2,470	..	1,847	2,735	3,422
				(Millions of 1980 New Zealand Dollars)							
Gross National Product	16,347	16,538	17,884	18,167	18,933	19,505	20,951	22,665	22,036	22,679	22,056
GDP at factor cost	14,956	15,038	16,338	16,544	17,365	17,845	19,130	21,112	20,840	21,212	20,708
Agriculture	..	..	..	..	..	..	..	..	..	..	2,052
Industry	..	..	..	..	..	..	..	..	..	..	7,190
Manufacturing	..	..	..	..	..	..	..	..	..	..	4,871
Services, etc.	..	..	..	..	..	..	..	..	..	..	13,330
Resource Balance	-402	104	214	-654	-530	-1,141	-1,777	-3,518	-1,429	-510	-612
Exports of Goods & NFServices	3,729	4,230	4,788	4,823	5,062	4,803	5,233	5,116	5,304	6,090	6,146
Imports of Goods & NFServices	4,130	4,125	4,574	5,477	5,592	5,945	7,010	8,634	6,734	6,600	6,758
Domestic Absorption	16,938	16,557	17,824	18,961	19,603	20,799	22,828	26,390	23,814	23,664	23,184
Private Consumption, etc.	10,005	10,307	10,785	11,575	11,512	12,054	13,041	13,802	13,734	13,561	13,098
General Gov't Consumption	2,388	2,443	2,459	2,561	2,660	2,789	2,935	3,135	3,314	3,339	3,418
Gross Domestic Investment	4,546	3,807	4,581	4,825	5,431	5,956	6,851	9,454	6,765	6,763	6,668
Fixed Investment	4,326	3,686	4,239	4,612	4,896	5,621	6,343	7,035	6,571	6,277	5,636
Memo Items:											
Capacity to Import	4,372	4,744	5,429	5,175	5,616	6,169	7,794	6,036	4,967	5,882	6,172
Terms of Trade Adjustment	643	514	641	353	554	1,366	2,561	920	-337	-208	26
Gross Domestic Income	17,180	17,176	18,679	18,659	19,627	21,023	23,612	23,792	22,047	22,945	22,597
Gross National Income	16,990	17,052	18,525	18,520	19,487	20,870	23,512	23,585	21,699	22,471	22,082
DOMESTIC PRICES (DEFLATORS)				*(Index 1980 = 100)*							
Overall (GDP)	25.3	26.5	27.3	30.0	34.0	37.8	41.5	42.4	46.8	55.3	65.5
Domestic Absorption	24.4	25.9	26.5	29.4	33.1	35.4	37.2	40.4	47.5	56.0	65.4
Agriculture	..	..	..	..	..	..	..	..	..	..	72.4
Industry	..	..	..	..	..	..	..	..	..	..	65.6
Manufacturing	..	..	..	..	..	..	..	..	..	..	66.9
MANUFACTURING ACTIVITY											
Employment (1980=100)	72.3	73.4	77.8	81.3	82.4	83.6	87.0	91.9	94.3	96.9	99.5
Real Earnings per Empl. (1980=100)	74.4	74.8	77.7	84.9	87.3	90.5	95.6	104.2	100.5	95.5	95.6
Real Output per Empl. (1980=100)	..	..	..	..	..	..	..	..	..	..	99.0
Earnings as % of Value Added	58.5	58.1	58.9	61.8	63.2	61.6	63.4	65.1	66.5	62.7	66.6
MONETARY HOLDINGS				*(Millions of current New Zealand Dollars)*							
Money Supply, Broadly Defined	2,445	2,579	2,819	3,066	3,351	4,060	5,011	5,236	5,833	6,792	7,807
Money as Means of Payment	781	780	801	861	949	1,219	1,545	1,601	1,749	1,909	1,945
Currency Ouside Banks	153	157	168	195	212	242	290	336	352	418	460
Demand Deposits	628	623	633	665	737	978	1,255	1,265	1,397	1,491	1,486
Quasi-Monetary Liabilities	1,664	1,799	2,018	2,205	2,402	2,841	3,466	3,635	4,083	4,883	5,862
GOVERNMENT DEFICIT (-) OR SURPLUS				*(Millions of current New Zealand Dollars)*							
	..	..	..	-108	-93	-298	-225	-416	-1,194	-614	-789
Current Revenue	..	..	..	1,614	1,901	2,141	2,626	3,121	3,496	4,277	5,150
Current Expenditure	..	..	..	1,380	1,610	1,978	2,326	2,712	3,474	3,870	4,777
Current Budget Balance	..	..	..	234	291	163	300	409	22	407	373
Capital Receipts	..	..	..	6	7	5	6	3	7	6	3
Capital Payments	..	..	..	348	391	466	531	828	1,223	1,027	1,165

1978	1979	1980	1981	1982	1983	1984	1985	1986	1987 estimate	NOTES	NEW ZEALAND
4,770	5,550	6,610	7,940	8,050	8,090	8,070	7,550	7,560	7,850	..	CURRENT GNP PER CAPITA (US $)
3,121	3,109	3,113	3,125	3,183	3,203	3,233	3,254	3,277	3,298	..	POPULATION (thousands)

(Millions of current New Zealand Dollars) — **ORIGIN AND USE OF RESOURCES**

1978	1979	1980	1981	1982	1983	1984	1985	1986	1987	NOTES	
15,613	18,525	21,584	26,159	29,466	32,523	36,389	41,286	50,449	57,804	..	Gross National Product (GNP)
-409	-460	-511	-615	-860	-1,298	-1,904	-2,044	-2,238	-2,500	..	Net Factor Income from Abroad
16,022	18,985	22,095	26,774	30,326	33,821	38,293	43,330	52,687	60,304	..	Gross Domestic Product (GDP)
1,321	1,674	2,032	2,391	2,736	3,254	4,026	4,581	6,144	..	..	Indirect Taxes, net
14,701	17,311	20,063	24,383	27,590	30,567	34,268	38,749	46,543	..	B	GDP at factor cost
1,527	2,308	2,420	2,561	2,494	2,905	3,677	4,022	..	..	..	Agriculture
5,044	5,991	6,855	8,749	10,004	10,918	12,648	13,994	..	..	..	Industry
3,583	4,282	4,884	6,270	7,042	7,734	9,077	9,714	..	..	..	Manufacturing
9,451	10,685	12,820	15,465	17,827	19,998	21,969	25,314	..	..	..	Services, etc.
50	-97	-426	-606	-1,344	-553	-2,138	-840	-734	-1	..	Resource Balance
4,434	5,651	6,780	7,996	8,903	9,885	11,734	14,007	13,953	15,570	..	Exports of Goods & NFServices
4,383	5,747	7,206	8,601	10,247	10,438	13,872	14,847	14,688	15,571	..	Imports of Goods & NFServices
15,972	19,081	22,522	27,380	31,669	34,374	40,431	44,170	53,421	60,305	..	Domestic Absorption
9,867	11,306	13,189	15,463	17,888	19,309	21,590	25,024	29,153	33,849	..	Private Consumption, etc.
2,633	3,040	3,593	4,272	4,840	5,414	5,815	6,511	8,050	8,998	..	General Gov't Consumption
3,473	4,736	5,740	7,645	8,942	9,652	13,027	12,635	16,219	17,457	..	Gross Domestic Investment
3,851	4,097	4,881	6,550	8,313	9,095	10,459	12,029	13,195	15,223	..	Fixed Investment

Memo Items:

3,523	4,639	5,314	7,040	7,598	9,099	10,889	11,795	15,484	17,457	..	Gross Domestic Saving
3,139	4,247	4,914	6,500	6,951	8,049	9,331	10,043	13,585	15,351	..	Gross National Saving

(Millions of 1980 New Zealand Dollars)

20,952	21,186	21,584	22,440	22,145	22,758	24,107	24,852	25,292	24,685	..	Gross National Product
19,718	19,746	20,064	20,912	20,722	21,393	22,707	23,345	23,341	..	B	GDP at factor cost
1,887	2,163	2,420	2,414	2,499	2,349	2,437	2,999	..	..	..	Agriculture
6,793	6,975	6,855	7,429	7,562	7,652	8,716	8,740	..	..	..	Industry
4,701	4,980	4,884	5,311	5,314	5,343	6,167	5,897	..	..	..	Manufacturing
12,825	12,593	12,820	13,127	12,735	13,662	14,210	14,328	..	..	..	Services, etc.
-196	-879	-426	-537	-1,079	-263	-1,062	-34	-155	-459	..	Resource Balance
6,199	6,585	6,780	7,003	6,979	7,311	7,797	8,522	8,489	8,922	..	Exports of Goods & NFServices
6,396	7,463	7,206	7,540	8,058	7,574	8,860	8,556	8,645	9,381	..	Imports of Goods & NFServices
21,701	22,610	22,522	23,506	23,876	23,926	26,425	26,101	26,566	26,930	..	Domestic Absorption
13,242	13,225	13,189	13,340	13,207	13,355	13,885	14,051	14,481	14,485	..	Private Consumption, etc.
3,573	3,609	3,593	3,670	3,657	3,767	3,808	3,846	4,002	4,021	..	General Gov't Consumption
4,885	5,776	5,740	6,497	7,012	6,804	8,733	8,204	8,083	8,424	..	Gross Domestic Investment
5,227	4,864	4,881	5,627	6,282	6,328	6,849	7,106	6,559	6,803	..	Fixed Investment

Memo Items:

6,469	7,338	6,780	7,009	7,001	7,173	7,494	8,072	8,212	9,381	..	Capacity to Import
270	753	0	6	22	-139	-303	-450	-277	459	..	Terms of Trade Adjustment
21,774	22,485	22,095	22,975	22,819	23,525	25,060	25,617	26,133	26,929	..	Gross Domestic Income
21,221	21,939	21,584	22,446	22,167	22,619	23,804	24,403	25,015	..	..	Gross National Income

(Index 1980 = 100) — **DOMESTIC PRICES (DEFLATORS)**

74.5	87.4	100.0	116.6	133.0	142.9	151.0	166.2	199.5	227.8	..	Overall (GDP)
73.6	84.4	100.0	116.5	132.6	143.7	153.0	169.2	201.1	223.9	..	Domestic Absorption
80.9	106.7	100.0	106.1	99.8	123.7	150.9	134.1	..	..	..	Agriculture
74.2	85.9	100.0	117.8	132.3	142.7	145.1	160.1	..	..	..	Industry
76.2	86.0	100.0	118.0	132.5	144.7	147.2	164.7	..	..	..	Manufacturing

MANUFACTURING ACTIVITY

102.2	101.1	100.0	99.0	100.3	101.6	101.6	103.8	..	..	G	Employment (1980=100)
90.2	100.0	100.0	101.1	98.4	93.6	92.1	87.5	..	..	G	Real Earnings per Empl. (1980=100)
91.3	99.8	100.0	105.5	104.2	106.4	116.0	107.3	..	..	G	Real Output per Empl. (1980=100)
65.8	65.4	66.1	64.5	66.5	61.8	55.3	56.5	..	..	..	Earnings as % of Value Added

(Millions of current New Zealand Dollars) — **MONETARY HOLDINGS**

9,506	11,108	12,400	14,456	16,254	18,147	21,478	26,601	31,019	..	..	Money Supply, Broadly Defined
2,379	2,458	2,535	2,926	3,030	3,426	3,761	4,104	4,668	6,667	..	Money as Means of Payment
536	590	577	683	714	739	867	940	1,007	1,059	..	Currency Ouside Banks
1,842	1,868	1,958	2,243	2,316	2,687	2,894	3,164	3,661	5,608	..	Demand Deposits
7,128	8,650	9,865	11,530	13,224	14,721	17,717	22,497	26,351	..	..	Quasi-Monetary Liabilities

(Millions of current New Zealand Dollars)

-1,502	-1,127	-1,541	-2,111	-2,389	-3,209	-3,234	-2,086	-1,985	375	C	GOVERNMENT DEFICIT (-) OR SURPLUS
5,648	6,823	7,871	9,747	11,204	11,711	13,686	16,918	20,683	25,837	..	Current Revenue
5,914	6,858	8,246	10,432	12,171	13,313	15,239	17,298	21,551	25,911	..	Current Expenditure
-266	-35	-375	-685	-967	-1,602	-1,553	-380	-868	-74	..	Current Budget Balance
3	4	6	6	3	13	19	30	161	130	..	Capital Receipts
1,239	1,096	1,172	1,432	1,425	1,620	1,700	1,736	1,278	-319	..	Capital Payments

NEW ZEALAND	1967	1968	1969	1970	1971	1972	1973	1974	1975	1976	1977
FOREIGN TRADE (CUSTOMS BASIS)					*(Millions of current US dollars)*						
Value of Exports, fob	968.7	989.7	1,182.0	1,226.9	1,361.6	1,831.0	2,607.9	2,437.3	2,170.7	2,820.0	3,203.9
Nonfuel Primary Products	898.9	900.4	1,072.7	1,066.8	1,184.6	1,586.4	2,304.7	2,075.8	1,773.2	2,230.4	2,531.1
Fuels	5.9	6.0	7.2	9.1	13.6	18.2	16.5	23.6	37.0	50.8	53.9
Manufactures	63.9	83.4	102.1	151.0	163.5	226.4	286.7	337.9	360.6	538.8	618.9
Value of Imports, cif	955.1	895.9	1,002.5	1,237.6	1,370.8	1,524.3	2,186.4	3,665.7	3,182.7	3,295.1	3,280.2
Nonfuel Primary Products	160.0	150.9	170.6	210.3	206.1	219.4	331.6	505.3	427.9	414.2	426.0
Fuels	74.4	82.1	81.3	83.2	102.0	112.7	157.4	429.6	455.2	481.7	503.0
Manufactures	720.8	662.9	750.6	944.1	1,062.7	1,192.2	1,697.4	2,730.8	2,299.6	2,399.2	2,351.2
Terms of Trade	156.0	153.7	151.0	140.7	136.6	158.8 *(Index 1980 = 100)*	173.2	106.4	96.2	105.6	100.2
Export Prices, fob	32.7	30.3	32.2	33.0	34.9	44.8	62.3	58.5	55.2	61.5	63.6
Nonfuel Primary Products	32.9	30.6	33.0	33.7	35.8	46.8	64.3	58.0	54.6	62.4	63.1
Fuels	8.5	7.3	5.4	5.9	7.0	8.5	24.5	36.6	33.6	38.5	36.8
Manufactures	40.3	34.6	37.1	38.4	40.9	46.5	53.5	64.2	62.9	61.6	71.1
Import Prices, cif	20.9	19.7	21.4	23.4	25.6	28.2	35.9	54.9	57.3	58.2	63.5
Trade at Constant 1980 Prices					*(Millions of 1980 US dollars)*						
Exports, fob	2,966.1	3,270.1	3,666.4	3,721.6	3,901.3	4,088.6	4,188.6	4,168.4	3,935.9	4,586.3	5,033.9
Imports, cif	4,561.8	4,548.5	4,694.1	5,280.5	5,364.1	5,405.2	6,081.9	6,672.0	5,551.7	5,661.3	5,164.8
BALANCE OF PAYMENTS					*(Millions of current US dollars)*						
Exports of Goods & Services	..	..	..	1,471	1,782	2,365	3,166	2,986	3,142	3,738	4,149
Merchandise, fob	..	..	..	1,235	1,479	1,961	2,544	2,228	2,345	2,952	3,239
Nonfactor Services	..	..	..	188	251	329	498	639	698	685	790
Factor Services	..	..	..	48	52	75	124	119	100	102	120
Imports of Goods & Services	..	..	..	1,688	1,810	2,237	3,336	4,789	4,313	4,526	4,834
Merchandise, fob	..	..	..	1,167	1,213	1,461	2,177	3,518	2,963	3,069	3,128
Nonfactor Services	..	..	..	389	460	549	849	1,038	965	944	1,149
Factor Services	..	..	..	132	137	226	310	233	385	513	557
Long-Term Interest	..	..	..	..	..	..	..	..	..	..	..
Current Transfers, net	..	..	..	-5	16	54	76	..	58	32	48
Workers' Remittances	..	..	..	40	56	66	80	83	91	89	112
Total to be Financed	..	..	..	-222	-12	182	-94	-1,755	-1,113	-755	-639
Official Capital Grants	..	..	..	-10	-8	-18	-29	-48	-55	-47	-54
Current Account Balance	..	..	..	-232	-20	164	-123	-1,803	-1,168	-802	-692
Long-Term Capital, net	..	..	..	210	194	139	178	1,063	953	569	650
Direct Investment	..	..	..	137	98	124	191	226	112	238	123
Long-Term Loans	..	..	..	..	..	..	..	..	..	..	..
Disbursements	..	..	..	..	..	..	..	..	..	..	..
Repayments	..	..	..	..	..	..	..	..	..	..	..
Other Long-Term Capital	..	..	..	83	103	34	15	885	896	379	581
Other Capital, net	..	..	..	40	14	187	-25	252	59	48	246
Change in Reserves	..	..	..	-18	-187	-490	-30	488	156	185	-204
Memo Item:					*(New Zealand Dollars per US dollar)*						
Conversion Factor (Annual Avg)	0.720	0.780	0.890	0.890	0.890	0.870	0.820	0.720	0.730	0.890	1.030
EXTERNAL DEBT, ETC.					*(Millions US dollars, outstanding at end of year)*						
Public/Publicly Guar. Long-Term	..	..	..	..	..	..	..	..	..	..	..
Official Creditors	..	..	..	..	..	..	..	..	..	..	..
IBRD and IDA	..	..	..	..	..	..	..	..	..	..	..
Private Creditors	..	..	..	..	..	..	..	..	..	..	..
Private Non-guaranteed Long-Term	..	..	..	..	..	..	..	..	..	..	..
Use of Fund Credit	..	..	..	..	..	..	..	..	..	..	..
Short-Term Debt	..	..	..	..	..	..	..	..	..	..	..
Memo Items:					*(Millions of US dollars)*						
Int'l Reserves Excluding Gold	218.0	161.0	209.0	256.9	492.0	832.2	1,045.1	638.6	426.9	490.8	442.5
Gold Holdings (at market price)	0.6	0.7	0.8	0.8	1.0	1.5	2.5	4.2	3.2	3.0	7.3
SOCIAL INDICATORS											
Total Fertility Rate	3.4	3.3	3.3	3.2	3.2	2.9	2.7	2.6	2.3	2.3	2.2
Crude Birth Rate	22.5	22.7	22.3	22.1	22.6	21.8	20.5	19.6	18.3	17.7	17.3
Infant Mortality Rate	18.0	18.7	16.9	16.7	16.5	15.6	16.2	15.5	15.9	14.0	14.2
Life Expectancy at Birth	71.3	71.4	71.4	71.5	71.6	71.7	71.8	72.0	72.1	72.3	72.4
Food Production, p.c. ('79-81 = 100)	86.5	92.6	97.2	93.8	93.9	96.8	96.4	87.0	93.2	104.6	99.2
Labor Force, Agriculture (%)	12.7	12.4	12.2	11.9	11.8	11.7	11.7	11.6	11.5	11.4	11.4
Labor Force, Female (%)	28.3	28.6	29.0	29.4	29.8	30.3	30.7	31.1	31.6	32.0	32.5
School Enroll. Ratio, primary	..	..	..	110.0	..	..	..	..	106.0	108.0	108.0
School Enroll. Ratio, secondary	..	..	..	77.0	..	..	..	..	81.0	82.0	81.0

1978	1979	1980	1981	1982	1983	1984	1985	1986	1987 estimate	NOTES	NEW ZEALAND
											FOREIGN TRADE (CUSTOMS BASIS)
				(Millions of current US dollars)							
3,985.6	4,668.7	5,453.7	5,567.2	5,479.3	5,269.9	5,508.1	5,731.1	5,937.0	7,179.4	..	Value of Exports, fob
3,074.0	3,624.5	4,151.5	4,176.1	4,142.4	4,102.7	4,006.4	4,148.0	4,293.8	5,335.6	..	Nonfuel Primary Products
61.8	96.3	128.2	155.7	14.6	14.3	22.5	95.3	48.3	64.4	..	Fuels
849.8	947.9	1,174.0	1,235.4	1,322.3	1,152.9	1,479.2	1,487.8	1,594.9	1,779.4	..	Manufactures
3,660.0	4,561.7	5,514.7	5,731.5	5,900.5	5,326.9	6,180.5	5,981.7	6,131.4	7,254.5	..	Value of Imports, cif
497.6	561.3	644.4	620.1	697.9	604.8	699.7	654.1	612.3	762.1	..	Nonfuel Primary Products
513.7	727.0	1,238.6	1,112.9	978.6	964.4	829.7	759.8	530.0	484.0	..	Fuels
2,648.8	3,273.3	3,631.7	3,998.5	4,224.0	3,757.7	4,651.1	4,567.8	4,989.1	6,008.4	..	Manufactures
				(Index 1980 = 100)							
108.4	115.1	100.0	97.3	98.9	100.4	98.8	96.5	97.1	98.0	..	Terms of Trade
76.3	96.8	100.0	95.3	90.8	89.2	85.7	82.3	86.3	94.2	..	Export Prices, fob
75.9	98.7	100.0	93.5	88.5	87.3	82.8	79.6	83.6	90.7	..	Nonfuel Primary Products
44.7	93.6	100.0	104.0	97.0	85.4	83.6	81.1	41.3	34.6	..	Fuels
82.4	90.6	100.0	101.0	99.0	97.0	95.0	91.0	98.0	115.0	..	Manufactures
70.3	84.0	100.0	98.0	91.8	88.9	86.7	85.2	88.9	96.2	..	Import Prices, cif
											Trade at Constant 1980 Prices
				(Millions of 1980 US dollars)							
5,225.9	4,825.3	5,453.7	5,841.3	6,033.9	5,906.0	6,424.2	6,963.3	6,879.8	7,617.6	..	Exports, fob
5,203.3	5,428.5	5,514.7	5,851.2	6,429.1	5,994.6	7,125.3	7,016.9	6,899.5	7,542.3	..	Imports, cif
				(Millions of current US dollars)							**BALANCE OF PAYMENTS**
4,962	6,017	6,505	7,065	6,785	6,920	7,158	7,335	7,789	9,783	..	Exports of Goods & Services
3,998	4,875	5,395	5,599	5,301	5,283	5,490	5,615	5,820	7,279	..	Merchandise, fob
848	1,012	951	1,248	1,261	1,396	1,463	1,493	1,696	2,109	..	Nonfactor Services
116	130	159	218	222	241	205	227	273	395	..	Factor Services
5,426	6,862	7,487	8,229	8,500	7,975	9,029	8,666	9,205	11,320	..	Imports of Goods & Services
3,398	4,590	5,091	5,361	5,522	4,942	5,745	5,524	5,607	6,663	..	Merchandise, fob
1,332	1,638	1,700	2,006	2,016	1,909	2,014	1,837	2,160	2,685	..	Nonfactor Services
696	633	696	862	962	1,124	1,270	1,304	1,439	1,971	..	Factor Services
..	..	..	..	..	..	..	..	..	..	..	Long-Term Interest
26	70	108	65	160	166	196	144	177	233	..	Current Transfers, net
121	156	201	197	215	230	286	318	345	433	..	Workers' Remittances
-439	-778	-874	-1,099	-1,556	-889	-1,675	-1,186	-1,239	-1,304	K	Total to be Financed
-46	-47	-50	-43	-61	-87	-83	-64	-60	-63	K	Official Capital Grants
-486	-824	-924	-1,141	-1,617	-976	-1,758	-1,251	-1,299	-1,368	..	Current Account Balance
527	695	633	1,846	2,193	740	1,282	212	1,984	-604	..	Long-Term Capital, net
217	270	71	173	-82	47	102	93	101	104	..	Direct Investment
..	..	..	..	..	..	..	..	..	..	..	Long-Term Loans
..	..	..	..	..	..	..	..	..	..	..	Disbursements
..	..	..	..	..	..	..	..	..	..	..	Repayments
357	472	612	1,716	2,336	779	1,263	183	1,943	-645	..	Other Long-Term Capital
-241	87	218	-261	-539	448	1,504	651	1,359	1,583	..	Other Capital, net
200	43	72	-444	-37	-212	-1,029	388	-2,044	389	..	Change in Reserves
											Memo Item:
				(New Zealand Dollars per US dollar)							
1.010	0.960	1.000	1.040	1.200	1.370	1.520	1.930	1.960	1.890	..	Conversion Factor (Annual Avg)
											EXTERNAL DEBT, ETC.
				(Millions US dollars, outstanding at end of year)							Public/Publicly Guar. Long-Term
..	..	..	..	..	..	..	..	..	..	..	Official Creditors
..	..	..	..	..	..	..	..	..	..	..	IBRD and IDA
..	..	..	..	..	..	..	..	..	..	..	Private Creditors
..	..	..	..	..	..	..	..	..	..	..	Private Non-guaranteed Long-Term
..	..	..	..	..	..	..	..	..	..	..	Use of Fund Credit
..	..	..	..	..	..	..	..	..	..	..	Short-Term Debt
				(Millions of US dollars)							Memo Items:
451.0	450.7	352.1	673.9	635.9	777.6	1,786.7	1,595.7	3,771.2	3,259.6	..	Int'l Reserves Excluding Gold
14.8	23.2	13.2	8.9	10.2	8.5	6.9	7.3	8.7	10.8	..	Gold Holdings (at market price)
											SOCIAL INDICATORS
2.1	2.2	2.1	2.1	2.0	2.0	2.0	2.0	1.9	1.9	..	Total Fertility Rate
16.3	16.8	16.2	16.2	16.2	16.1	16.1	16.1	16.1	16.0	..	Crude Birth Rate
13.7	12.4	12.8	11.6	12.2	12.9	13.6	14.3	15.1	15.9	..	Infant Mortality Rate
72.8	73.1	73.4	73.8	74.1	74.3	74.4	74.3	74.2	74.1	..	Life Expectancy at Birth
99.1	97.8	99.3	102.9	103.6	104.5	102.5	111.2	106.8	111.0	..	Food Production, p.c. ('79-81=100)
11.3	11.3	11.2	..	..	..	..	..	..	..	..	Labor Force, Agriculture (%)
33.0	33.5	34.0	34.1	34.2	34.3	34.4	34.5	34.6	34.6	..	Labor Force, Female (%)
108.0	107.0	109.0	..	107.0	106.0	106.0	106.0	105.0	..	..	School Enroll. Ratio, primary
80.0	81.0	82.0	..	82.0	85.0	85.0	85.0	84.0	..	..	School Enroll. Ratio, secondary

NICARAGUA	1967	1968	1969	1970	1971	1972	1973	1974	1975	1976	1977
CURRENT GNP PER CAPITA (US $)	350	350	370	380	390	390	430	550	650	710	750
POPULATION (thousands)	1,865	1,926	1,989	2,054	2,122	2,193	2,266	2,339	2,410	2,481	2,552

ORIGIN AND USE OF RESOURCES *(Billions of current Nicaraguan Cordobas)*

	1967	1968	1969	1970	1971	1972	1973	1974	1975	1976	1977
Gross National Product (GNP)	4.5	4.7	5.1	5.3	5.6	5.9	7.3	10.2	10.7	12.4	14.2
Net Factor Income from Abroad	-0.1	-0.2	-0.2	-0.2	-0.2	-0.2	-0.4	-0.5	-0.4	-0.5	-0.5
Gross Domestic Product (GDP)	4.6	4.9	5.2	5.4	5.8	6.2	7.7	10.6	11.1	12.9	14.8
Indirect Taxes, net	0.4	0.4	0.4	0.5	0.5	0.5	0.8	1.1	1.0	1.2	1.5
GDP at factor cost	4.2	4.5	4.8	5.0	5.3	5.6	6.9	9.6	10.1	11.7	13.3
Agriculture	1.1	1.1	1.3	1.4	1.4	1.5	1.9	2.6	2.5	2.9	3.6
Industry	1.1	1.2	1.3	1.4	1.5	1.7	2.1	2.9	3.3	3.7	4.2
Manufacturing	0.8	0.9	1.0	1.1	1.2	1.3	1.6	2.2	2.5	2.7	3.1
Services, etc.	2.4	2.6	2.7	2.7	2.8	3.0	3.7	5.1	5.4	6.4	7.0
Resource Balance	-0.4	-0.2	-0.1	-0.1	-0.1	0.4	-0.7	-1.4	-1.0	0.2	-0.8
Exports of Goods & NFServices	1.3	1.4	1.3	1.5	1.5	2.2	2.2	3.1	3.1	4.3	5.0
Imports of Goods & NFServices	1.6	1.5	1.5	1.6	1.7	1.8	2.9	4.5	4.1	4.1	5.9
Domestic Absorption	5.0	5.0	5.4	5.6	5.9	5.8	8.4	12.1	12.1	12.8	15.6
Private Consumption, etc.	3.6	3.7	3.9	4.0	4.3	4.4	5.9	7.9	8.7	9.3	10.2
General Gov't Consumption	0.4	0.4	0.5	0.5	0.6	0.6	0.6	0.8	1.0	1.2	1.4
Gross Domestic Investment	1.0	0.9	1.0	1.0	1.0	0.8	1.8	3.4	2.4	2.2	4.0
Fixed Investment	0.9	0.8	0.9	0.9	0.9	0.9	1.5	2.5	2.5	2.6	3.6
Memo Items:											
Gross Domestic Saving	0.6	0.7	0.9	0.9	0.9	1.1	1.1	1.9	1.4	2.4	3.2
Gross National Saving	0.5	0.6	0.7	0.7	0.7	0.9	1.0	1.5	1.0	1.9	2.7

(Millions of 1980 Nicaraguan Cordobas)

	1967	1968	1969	1970	1971	1972	1973	1974	1975	1976	1977
Gross National Product	18,022	18,042	19,211	19,382	20,287	20,834	21,437	24,587	25,584	26,633	26,823
GDP at factor cost	17,006	17,344	18,409	18,399	19,257	19,903	20,595	23,189	23,899	25,092	25,110
Agriculture	4,037	3,879	4,391	4,018	4,340	4,381	4,638	5,043	5,372	5,463	5,673
Industry	4,829	4,922	5,182	5,472	5,694	6,068	6,316	7,463	7,540	8,014	9,011
Manufacturing	3,332	3,493	3,753	4,102	4,308	4,535	4,720	5,319	5,489	5,737	6,329
Services, etc.	9,702	10,030	10,404	10,609	11,056	11,312	11,915	13,269	13,429	14,193	13,238
Resource Balance	-1,746	-1,380	-765	-763	-492	1,170	-2,736	-3,123	39	215	-3,508
Exports of Goods & NFServices	5,581	5,959	6,059	6,182	6,412	8,419	7,547	9,102	9,102	9,447	9,003
Imports of Goods & NFServices	7,326	7,339	6,825	6,945	6,903	7,249	10,283	12,225	9,063	9,232	12,511
Domestic Absorption	20,314	20,212	20,743	20,862	21,582	20,591	25,604	28,898	26,302	27,454	31,429
Private Consumption, etc.	15,278	15,761	15,733	15,665	16,218	15,959	19,130	20,753	19,650	20,352	21,845
General Gov't Consumption	1,522	1,465	1,555	1,682	1,731	1,900	1,767	1,966	2,333	2,588	2,576
Gross Domestic Investment	3,513	2,986	3,455	3,515	3,633	2,731	4,708	6,178	4,320	4,514	7,008
Fixed Investment	..	..	..	..	..	..	..	..	..	..	..
Memo Items:											
Capacity to Import	5,719	6,605	6,243	6,359	6,361	8,654	7,674	8,342	6,880	9,567	10,727
Terms of Trade Adjustment	138	646	184	178	-50	234	126	-760	-2,222	119	1,724
Gross Domestic Income	18,706	19,478	20,161	20,277	21,040	21,995	22,995	25,015	24,119	27,789	29,645
Gross National Income	18,160	18,688	19,395	19,559	20,237	21,069	21,564	23,827	23,362	26,753	28,548

DOMESTIC PRICES (DEFLATORS) *(Index 1980 = 100)*

	1967	1968	1969	1970	1971	1972	1973	1974	1975	1976	1977
Overall (GDP)	24.8	25.9	26.2	27.0	27.4	28.4	33.5	41.3	42.3	46.7	52.9
Domestic Absorption	24.4	24.8	25.8	26.7	27.4	28.3	32.8	41.8	46.1	46.6	49.6
Agriculture	27.9	29.1	29.2	33.7	33.0	35.0	40.8	51.0	46.4	53.4	63.3
Industry	22.8	23.7	24.7	25.6	26.6	27.5	32.5	39.4	43.5	45.6	46.7
Manufacturing	24.2	25.1	25.7	27.1	28.0	28.9	34.2	40.6	44.8	46.9	48.7

MANUFACTURING ACTIVITY

	1967	1968	1969	1970	1971	1972	1973	1974	1975	1976	1977
Employment (1980=100)	43.9	50.0	51.9	52.1	53.8	68.1	65.4	75.2	78.2	82.9	89.8
Real Earnings per Empl. (1980=100)	..	..	..	..	..	128.9	110.3	110.6	112.8	118.6	112.2
Real Output per Empl. (1980=100)	198.8	186.9	193.3	205.8	211.6	182.6	191.3	183.9	186.0	183.3	180.9
Earnings as % of Value Added	15.4	16.7	16.8	15.7	15.3	18.4	15.8	15.7	15.5	16.2	16.7

MONETARY HOLDINGS *(Millions of current Nicaraguan Cordobas)*

	1967	1968	1969	1970	1971	1972	1973	1974	1975	1976	1977
Money Supply, Broadly Defined	0.8	0.7	0.7	0.8	1.0	1.2	1.7	2.0	2.0	2.7	2.9
Money as Means of Payment	0.5	0.5	0.5	0.6	0.6	0.8	1.1	1.3	1.3	1.6	1.7
Currency Ouside Banks	0.2	0.2	0.2	0.3	0.3	0.3	0.4	0.5	0.5	0.6	0.7
Demand Deposits	0.3	0.3	0.3	0.3	0.4	0.4	0.7	0.9	0.8	1.0	1.0
Quasi-Monetary Liabilities	0.2	0.2	0.2	0.3	0.3	0.5	0.6	0.7	0.8	1.1	1.2

(Millions of current Nicaraguan Cordobas)

	1967	1968	1969	1970	1971	1972	1973	1974	1975	1976	1977
GOVERNMENT DEFICIT (-) OR SURPLUS	..	..	..	-53	-140	-230	-130	-620	-620	-490	-1,000
Current Revenue	..	..	..	650	720	740	1,040	1,460	1,480	1,730	2,050
Current Expenditure	..	..	..	540	570	620	670	1,090	1,240	1,360	1,770
Current Budget Balance	..	..	..	110	150	130	370	370	240	370	280
Capital Receipts	..	..	..	18	16	15	20	23	50	38	12
Capital Payments	..	..	..	180	300	380	520	1,010	910	900	1,290

1978	1979	1980	1981	1982	1983	1984	1985	1986	1987 estimate	NOTES	NICARAGUA
730	580	690	760	800	820	790	770	730	830	A	CURRENT GNP PER CAPITA (US $)
2,624	2,698	2,775	2,857	2,946	3,044	3,153	3,276	3,388	3,501	..	POPULATION (thousands)

(Billions of current Nicaraguan Cordobas) — ORIGIN AND USE OF RESOURCES

1978	1979	1980	1981	1982	1983	1984	1985	1986	1987	NOTES	
13.4	13.8	20.5	23.1	26.6	31.1	41.0	104.5	393.2	2,327.2	..	Gross National Product (GNP)
-0.7	-0.7	-1.3	-1.4	-1.8	-1.8	-4.0	-11.0	-42.6	-192.1	..	Net Factor Income from Abroad
14.1	14.5	21.9	24.5	28.3	32.9	45.0	115.4	435.7	2,519.3	..	Gross Domestic Product (GDP)
1.3	..	..	..	..	..	..	..	..	..	..	Indirect Taxes, net
12.7	..	..	..	..	..	..	..	..	..	B	GDP at factor cost
3.7	4.1	4.9	5.0	6.1	7.6	11.2	27.3	90.6	..	..	Agriculture
4.1	4.2	6.8	8.1	9.1	10.1	14.3	40.4	146.2	..	..	Industry
3.2	3.6	5.5	6.5	7.4	8.0	11.4	31.9	120.7	..	..	Manufacturing
6.3	6.3	10.1	11.5	13.2	15.2	19.5	47.7	199.0	..	..	Services, etc.
0.5	2.0	-4.0	-5.0	-4.3	-5.2	-7.8	-28.1	-88.9	..	..	Resource Balance
5.2	6.1	5.0	5.5	5.0	5.6	6.9	14.0	44.6	..	..	Exports of Goods & NFServices
4.7	4.1	9.0	10.5	9.2	10.8	14.7	42.1	133.5	..	..	Imports of Goods & NFServices
13.6	12.5	25.9	29.5	32.6	38.1	52.9	143.5	524.6	..	..	Domestic Absorption
10.0	10.7	18.4	18.2	20.1	20.6	26.9	65.0	271.3	..	..	Private Consumption, etc.
1.8	2.6	4.1	5.4	7.2	10.1	15.9	51.8	173.8	..	..	General Gov't Consumption
1.9	-0.8	3.4	5.8	5.3	7.4	10.0	26.7	79.6	..	..	Gross Domestic Investment
2.1	..	..	..	..	..	..	..	..	..	..	Fixed Investment

Memo Items:

1978	1979	1980	1981	1982	1983	1984	1985	1986	1987	NOTES	
2.3	1.2	-0.6	0.8	1.1	2.2	2.2	-1.4	-9.3	..	..	Gross Domestic Saving
1.6	0.5	-1.9	-0.5	-0.6	0.4	-1.8	-11.9	-51.3	..	..	Gross National Saving

(Millions of 1980 Nicaraguan Cordobas)

1978	1979	1980	1981	1982	1983	1984	1985	1986	1987	NOTES	
24,586	18,532	20,544	20,720	20,487	21,602	20,402	19,430	17,637	20,089	..	Gross National Product
23,524	..	..	..	..	..	..	..	..	..	B	GDP at factor cost
6,167	5,922	4,947	5,286	5,436	5,749	5,442	5,181	4,901	..	..	Agriculture
8,271	5,635	6,805	6,723	6,521	6,898	6,963	6,760	6,901	..	..	Industry
6,484	4,650	5,492	5,493	5,496	5,806	5,829	5,555	5,644	..	..	Manufacturing
11,540	8,346	10,141	9,906	9,780	10,092	9,978	9,529	9,534	..	..	Services, etc.
-71	3,528	-3,960	-2,872	-1,297	-1,171	-2,319	-3,562	-3,380	..	..	Resource Balance
9,674	8,484	5,039	5,907	5,235	5,793	4,649	3,901	3,195	..	..	Exports of Goods & NFServices
9,745	4,956	8,999	8,779	6,532	6,964	6,968	7,463	6,575	..	..	Imports of Goods & NFServices
26,049	16,375	25,853	24,787	23,034	23,910	24,702	25,032	24,716	..	..	Domestic Absorption
20,026	14,597	18,381	15,024	13,328	12,690	12,056	10,792	10,733	..	..	Private Consumption, etc.
3,147	3,045	4,107	4,738	5,307	6,437	7,821	9,361	9,164	..	..	General Gov't Consumption
2,876	-1,267	3,365	5,024	4,399	4,783	4,825	4,878	4,819	..	..	Gross Domestic Investment
..	..	..	..	..	..	..	..	..	..	..	Fixed Investment

Memo Items:

1978	1979	1980	1981	1982	1983	1984	1985	1986	1987	NOTES	
10,731	7,404	5,039	4,615	3,521	3,627	3,258	2,477	2,196	..	..	Capacity to Import
1,056	-1,080	0	-1,292	-1,714	-2,166	-1,391	-1,424	-999	..	..	Terms of Trade Adjustment
27,035	18,823	21,893	20,623	20,023	20,573	20,992	20,046	20,337	..	..	Gross Domestic Income
25,642	17,452	20,544	19,427	18,773	19,437	19,011	18,006	16,638	..	..	Gross National Income

(Index 1980 = 100) — DOMESTIC PRICES (DEFLATORS)

1978	1979	1980	1981	1982	1983	1984	1985	1986	1987	NOTES	
54.2	72.9	100.0	111.7	130.4	144.8	201.2	537.5	2042.3	11608.8	..	Overall (GDP)
52.2	76.3	100.0	118.8	141.6	159.4	214.0	573.3	2122.6	..	..	Domestic Absorption
60.1	69.4	100.0	93.7	112.1	132.4	205.8	526.8	1848.9	..	..	Agriculture
49.2	73.7	100.0	119.8	139.4	146.5	205.8	597.6	2117.9	..	..	Industry
49.9	76.8	100.0	118.8	135.3	137.3	196.0	574.3	2139.2	..	..	Manufacturing

MANUFACTURING ACTIVITY

1978	1979	1980	1981	1982	1983	1984	1985	1986	1987	NOTES	
86.1	90.4	100.0	117.1	120.4	115.1	129.8	138.9	..	..	J	Employment (1980 = 100)
112.4	101.3	100.0	98.0	85.3	76.1	70.8	63.1	..	..	J	Real Earnings per Empl. (1980 = 100)
193.6	126.5	100.0	95.5	106.5	121.6	107.2	104.2	..	..	J	Real Output per Empl. (1980 = 100)
15.9	21.8	29.1	30.6	22.9	21.8	19.7	21.5	..	..	J	Earnings as % of Value Added

(Millions of current Nicaraguan Cordobas) — MONETARY HOLDINGS

1978	1979	1980	1981	1982	1983	1984	1985	1986	1987	NOTES	
2.7	3.4	5.7	7.9	9.9	15.0	..	..	..	..	..	Money Supply, Broadly Defined
1.6	2.7	4.1	5.2	6.5	11.0	..	..	..	..	..	Money as Means of Payment
0.9	1.6	2.0	2.4	3.1	5.5	..	..	..	..	..	Currency Ouside Banks
0.7	1.1	2.1	2.8	3.5	5.5	..	..	..	..	..	Demand Deposits
1.1	0.8	1.6	2.7	3.3	4.3	..	..	..	..	..	Quasi-Monetary Liabilities

(Millions of current Nicaraguan Cordobas)

1978	1979	1980	1981	1982	1983	1984	1985	1986	1987	NOTES	
-840	-940	-1,420	-2,620	-5,740	-10,300	-10,610	-25,320	-64,180	..	..	GOVERNMENT DEFICIT (-) OR SURPLUS
1,880	2,140	4,970	6,930	8,160	11,890	18,080	43,220	163,090	..	..	Current Revenue
2,220	2,800	5,150	7,550	10,840	13,930	21,560	60,790	209,450	..	..	Current Expenditure
-350	-660	-180	-620	-2,670	-2,040	-3,480	-17,570	-46,360	..	..	Current Budget Balance
18	0	1	89	94	78	85	210	770	..	..	Capital Receipts
510	280	1,240	2,090	3,160	8,340	7,210	7,960	18,590	..	..	Capital Payments

NICARAGUA	1967	1968	1969	1970	1971	1972	1973	1974	1975	1976	1977
FOREIGN TRADE (CUSTOMS BASIS)				*(Millions of current US dollars)*							
Value of Exports, fob	146.5	157.3	154.6	174.8	183.4	246.3	274.7	377.0	371.5	538.5	633.0
Nonfuel Primary Products	134.5	142.6	135.8	146.4	155.0	210.3	229.6	304.9	306.9	449.4	526.7
Fuels	0.0	0.0	0.0	0.2	0.4	1.8	1.4	1.6	1.4	2.1	1.2
Manufactures	12.0	14.6	18.8	28.1	28.0	34.2	43.7	70.5	63.2	87.0	105.1
Value of Imports, cif	202.4	183.3	175.6	197.9	209.6	218.0	326.4	559.0	516.9	532.1	761.9
Nonfuel Primary Products	22.7	23.6	20.8	25.1	26.2	29.7	47.4	58.1	52.3	57.8	72.8
Fuels	9.6	10.0	10.5	12.1	15.7	16.0	23.7	60.6	73.8	69.4	105.1
Manufactures	170.1	149.7	144.3	160.8	167.8	172.4	255.3	440.4	390.7	405.0	584.1
					(Index 1980 = 100)						
Terms of Trade	125.2	133.4	131.0	141.9	133.4	129.1	144.7	110.9	98.2	129.7	138.0
Export Prices, fob	30.6	31.2	31.0	34.6	34.6	37.1	53.3	61.3	56.6	76.9	88.5
Nonfuel Primary Products	30.6	31.2	30.9	34.4	34.7	38.5	55.8	60.3	55.1	79.6	93.1
Fuels	4.3	4.3	4.3	4.3	5.6	6.2	8.9	36.7	35.7	38.4	42.0
Manufactures	30.1	31.7	32.4	37.5	36.5	38.5	49.5	67.2	66.1	66.7	71.7
Import Prices, cif	24.4	23.4	23.7	24.4	25.9	28.8	36.8	55.3	57.6	59.3	64.2
Trade at Constant 1980 Prices				*(Millions of 1980 US dollars)*							
Exports, fob	479.3	503.8	498.3	505.7	530.4	663.1	515.5	614.7	656.7	700.6	715.0
Imports, cif	828.9	783.2	741.2	812.2	808.3	757.8	886.5	1,010.7	896.8	898.0	1,187.6
BALANCE OF PAYMENTS				*(Millions of current US dollars)*							
Exports of Goods & Services	..	..	..	218	224	291	323	448	458	630	733
Merchandise, fob	..	..	..	179	187	249	278	381	375	542	636
Nonfactor Services	..	..	..	35	33	37	37	56	72	74	83
Factor Services	..	..	..	5	4	5	7	11	11	14	14
Imports of Goods & Services	..	..	..	263	274	297	446	720	660	678	926
Merchandise, fob	..	..	..	179	190	206	328	542	482	485	704
Nonfactor Services	..	..	..	50	48	50	63	105	106	107	137
Factor Services	..	..	..	35	36	42	56	74	71	86	85
Long-Term Interest	..	..	..	7	8	11	18	29	36	41	52
Current Transfers, net	..	..	..	3	3	3	28	3	4	1	1
Workers' Remittances	..	..	..	0	0	0	0	0	0	0	0
Total to be Financed	..	..	..	-43	-47	17	-96	-269	-198	-48	-192
Official Capital Grants	..	..	..	4	2	5	30	12	13	8	11
Current Account Balance	..	..	..	-40	-45	22	-66	-257	-185	-39	-182
Long-Term Capital, net	..	..	..	50	54	55	142	186	191	45	222
Direct Investment	..	..	..	15	13	10	13	14	11	13	10
Long-Term Loans	..	..	..	28	33	46	98	128	141	57	193
Disbursements	..	..	..	44	55	66	148	147	160	91	240
Repayments	..	..	..	16	23	20	50	20	19	34	47
Other Long-Term Capital	..	..	..	3	6	-6	1	32	26	-34	8
Other Capital, net	..	..	..	-1	-4	-53	-11	38	31	-9	-36
Change in Reserves	..	..	..	-9	-5	-23	-65	34	-37	4	-3
Memo Item:				*(Nicaraguan Cordobas per US dollar)*							
Conversion Factor (Annual Avg)	7.000	7.000	7.000	7.000	7.000	7.000	7.000	7.010	7.030	7.030	7.030
EXTERNAL DEBT, ETC.				*(Millions of US dollars, outstanding at end of year)*							
Public/Publicly Guar. Long-Term	..	..	..	147	179	225	324	453	593	652	845
Official Creditors	..	..	..	102	122	132	156	203	247	310	429
IBRD and IDA	..	..	..	30	34	36	39	48	66	83	98
Private Creditors	..	..	..	45	57	94	168	251	346	342	416
Private Non-guaranteed Long-Term	..	..	..	..	..	..	..	..	..	..	..
Use of Fund Credit	..	..	..	8	12	9	15	12	18	10	2
Short-Term Debt	..	..	..	..	..	..	..	..	..	..	427
Memo Items:				*(Millions of US dollars)*							
Int'l Reserves Excluding Gold	31.3	47.7	43.7	48.6	58.1	80.1	116.3	104.5	121.6	146.1	148.3
Gold Holdings (at market price)	0.8	1.0	0.5	0.6	0.8	0.6	1.7	3.5	2.5	2.3	4.3
SOCIAL INDICATORS											
Total Fertility Rate	7.1	7.0	6.9	6.9	6.8	6.7	6.6	6.6	6.5	6.4	6.3
Crude Birth Rate	48.4	48.1	47.8	47.4	47.1	46.8	46.6	46.3	46.1	45.8	45.6
Infant Mortality Rate	115.0	112.0	109.0	106.0	103.0	100.0	98.6	97.2	95.8	94.4	93.0
Life Expectancy at Birth	51.6	52.2	52.9	53.5	54.1	54.7	55.0	55.3	55.7	56.0	56.3
Food Production, p.c. ('79-81 = 100)	103.1	109.0	109.5	113.8	112.0	107.6	102.6	104.3	113.0	115.0	115.3
Labor Force, Agriculture (%)	54.6	53.6	52.5	51.5	51.0	50.5	50.1	49.6	49.1	48.6	48.1
Labor Force, Female (%)	19.2	19.4	19.5	19.7	19.9	20.1	20.2	20.4	20.6	20.8	21.0
School Enroll. Ratio, primary	..	..	..	80.0	..	..	..	..	82.0	88.0	87.0
School Enroll. Ratio, secondary	..	..	..	18.0	..	..	..	..	24.0	28.0	30.0

1978	1979	1980	1981	1982	1983	1984	1985	1986	1987 estimate	NOTES	NICARAGUA
											FOREIGN TRADE (CUSTOMS BASIS)
				(Millions of current US dollars)							
646.0	633.2	413.8	475.9	390.7	434.7	386.7	301.5	247.2	299.9	..	Value of Exports, fob
535.9	556.0	346.5	419.4	355.5	392.3	340.0	261.8	219.5	265.9	..	Nonfuel Primary Products
1.1	5.4	10.3	9.3	5.0	6.3	7.5	6.1	3.1	3.9	..	Fuels
109.0	71.9	57.1	47.2	30.3	36.1	39.2	33.6	24.5	30.2	..	Manufactures
780.0	414.9	881.9	994.2	774.9	799.0	848.4	964.3	856.8	922.6	..	Value of Imports, cif
93.7	64.3	148.9	184.6	103.1	117.6	141.3	155.3	135.5	152.3	..	Nonfuel Primary Products
118.4	88.3	175.8	199.1	179.6	177.6	164.8	164.1	71.5	99.9	..	Fuels
567.8	262.3	557.2	610.5	492.2	503.9	542.4	644.8	649.8	670.4	..	Manufactures
				(Index 1980 = 100)							
117.4	114.4	100.0	87.7	89.5	94.5	95.6	85.2	92.3	76.7	..	Terms of Trade
82.9	94.8	100.0	89.9	87.6	89.7	88.9	78.3	87.9	78.7	..	Export Prices, fob
84.6	96.0	100.0	88.6	86.9	89.5	88.8	76.9	87.6	76.5	..	Nonfuel Primary Products
42.3	61.0	100.0	112.5	101.6	92.5	90.2	87.5	44.9	56.9	..	Fuels
76.1	90.0	100.0	99.1	94.8	90.7	89.3	89.0	103.2	113.0	..	Manufactures
70.6	82.9	100.0	102.6	97.9	94.9	93.0	91.8	95.2	102.6	..	Import Prices, cif
				(Millions of 1980 US dollars)							Trade at Constant 1980 Prices
779.3	667.9	413.8	529.2	446.0	484.8	435.0	385.1	281.3	381.1	..	Exports, fob
1,104.4	500.5	881.9	969.4	791.7	841.8	912.4	1,050.1	900.4	899.2	..	Imports, cif
				(Millions of current US dollars)							**BALANCE OF PAYMENTS**
732	683	514	582	456	478	435	353	295	..	f	Exports of Goods & Services
646	616	450	508	406	429	386	302	247	..	..	Merchandise, fob
74	56	44	45	41	42	45	50	47	..	..	Nonfactor Services
12	11	19	28	9	7	5	2	1	..	..	Factor Services
774	595	1,049	1,244	1,021	1,117	1,164	1,196	1,102	..	f	Imports of Goods & Services
553	389	803	922	724	778	800	800	727	..	..	Merchandise, fob
112	122	103	109	103	123	122	138	130	..	..	Nonfactor Services
109	84	144	212	195	216	243	258	246	..	..	Factor Services
51	37	38	91	109	37	34	28	21	12	..	Long-Term Interest
0	1	2	13	8	4	2	16	9	..	..	Current Transfers, net
0	0	0	0	0	0	0	2	3	..	..	Workers' Remittances
-34	90	-534	-649	-557	-636	-727	-827	-799	..	K	Total to be Financed
9	90	122	57	44	76	88	68	106	..	K	Official Capital Grants
-25	180	-411	-592	-514	-560	-639	-759	-693	..	..	Current Account Balance
144	213	680	733	517	736	569	858	711	..	f	Long-Term Capital, net
7	3	..	..	..	8	2	..	..	..	..	Direct Investment
87	97	222	295	240	272	349	582	592	473	..	Long-Term Loans
135	112	266	365	294	317	379	601	603	495	..	Disbursements
47	16	44	70	54	45	30	19	11	22	..	Repayments
41	23	336	381	233	381	131	208	13	..	..	Other Long-Term Capital
-193	-396	-470	-174	-44	-34	375	-87	-138	..	f	Other Capital, net
74	3	202	33	41	-143	-304	-11	121	0	..	Change in Reserves
				(Nicaraguan Cordobas per US dollar)							**Memo Item:**
7.030	9.260	10.050	10.050	11.700	13.200	15.000	40.000	159.000	786.500	..	Conversion Factor (Annual Avg)
				(Millions of US dollars, outstanding at end of year)							**EXTERNAL DEBT, ETC.**
942	1,085	1,662	2,077	2,488	3,383	4,088	4,484	5,228	6,150	..	Public/Publicly Guar. Long-Term
515	661	898	1,244	1,495	2,332	2,787	3,126	3,900	4,816	..	Official Creditors
103	108	135	171	185	203	193	222	260	308	..	IBRD and IDA
426	424	764	833	992	1,051	1,301	1,358	1,327	1,333	..	Private Creditors
..	..	..	..	..	..	..	..	..	..	..	Private Non-guaranteed Long-Term
3	57	49	25	19	14	9	0	0	..	..	Use of Fund Credit
485	345	460	471	824	781	1,016	1,207	977	1,141	..	Short-Term Debt
				(Millions of US dollars)							**Memo Items:**
50.8	146.6	64.5	111.4	171.2	174.7	..	..	..	..	..	Int'l Reserves Excluding Gold
6.1	9.2	10.6	7.2	8.2	45.8	..	..	..	..	..	Gold Holdings (at market price)
											SOCIAL INDICATORS
6.2	6.2	6.1	6.0	5.9	5.9	5.8	5.7	5.6	5.5	..	Total Fertility Rate
45.3	45.0	44.8	44.5	44.2	43.7	43.2	42.7	42.2	41.7	..	Crude Birth Rate
89.6	86.2	82.8	79.4	76.0	73.6	71.3	68.9	66.5	64.2	..	Infant Mortality Rate
57.0	57.7	58.4	59.1	59.8	60.1	60.5	60.8	61.1	61.5	..	Life Expectancy at Birth
122.6	126.8	84.7	88.5	87.3	84.7	78.6	77.2	72.5	72.5	..	Food Production, p.c. ('79-81 = 100)
47.6	47.1	46.6	..	..	..	..	..	..	..	..	Labor Force, Agriculture (%)
21.2	21.4	21.6	21.9	22.3	22.7	23.1	23.4	23.8	24.1	..	Labor Force, Female (%)
84.0	90.0	99.0	104.0	100.0	103.0	99.0	101.0	..	..	..	School Enroll. Ratio, primary
28.0	30.0	43.0	41.0	39.0	44.0	43.0	39.0	..	..	..	School Enroll. Ratio, secondary

NIGER	1967	1968	1969	1970	1971	1972	1973	1974	1975	1976	1977
CURRENT GNP PER CAPITA (US $)	180	170	160	160	170	160	150	200	230	220	240
POPULATION (thousands)	3,895	3,977	4,061	4,146	4,263	4,384	4,508	4,635	4,767	4,901	5,040

ORIGIN AND USE OF RESOURCES *(Billions of current CFA Francs)*

	1967	1968	1969	1970	1971	1972	1973	1974	1975	1976	1977
Gross National Product (GNP)	162.2	157.0	160.6	177.1	188.3	184.2	206.7	240.6	214.6	247.9	310.1
Net Factor Income from Abroad	-1.5	-1.7	-2.1	-2.5	-2.7	-3.0	-4.2	-6.4	-10.2	-6.5	-7.2
Gross Domestic Product (GDP)	163.7	158.7	162.7	179.6	191.0	187.2	210.9	247.0	224.8	254.4	317.3
Indirect Taxes, net	4.6	4.5	5.5	5.5	6.0	6.5	6.9	9.1	9.6	11.9	19.0
GDP at factor cost	159.1	154.2	157.2	174.1	185.0	180.7	204.0	237.9	215.2	242.5	298.3
Agriculture	115.8	109.8	108.1	116.5	125.0	116.9	127.1	157.6	113.1	119.6	164.3
Industry	7.3	7.7	8.5	12.5	14.1	15.2	19.8	19.1	24.8	32.8	44.4
Manufacturing	4.7	4.9	4.8	8.3	9.2	9.9	13.3	12.1	12.5	14.9	16.3
Services, etc.	40.7	41.2	46.1	50.6	51.8	55.1	64.0	70.3	86.9	102.0	108.6
Resource Balance	-9.7	-10.1	-11.6	-13.2	-6.2	-11.8	-17.3	-41.3	-26.4	-30.7	-30.7
Exports of Goods & NFServices	13.8	12.7	15.2	19.4	23.6	26.6	32.7	30.3	43.3	56.6	62.3
Imports of Goods & NFServices	23.5	22.7	26.8	32.5	29.8	38.4	50.1	71.6	69.7	87.3	93.1
Domestic Absorption	173.5	168.8	174.3	192.8	197.2	199.0	228.3	288.3	251.2	285.1	348.0
Private Consumption, etc.	146.1	139.9	149.9	159.5	166.6	160.8	178.6	230.9	194.7	215.9	240.2
General Gov't Consumption	13.1	13.5	14.2	15.6	15.7	17.4	18.2	20.2	24.3	33.0	45.4
Gross Domestic Investment	14.2	15.4	10.1	17.7	14.9	20.7	31.5	37.2	32.2	36.1	62.5
Fixed Investment	..	..	..	..	..	..	..	..	..	..	..
Memo Items:											
Gross Domestic Saving	4.5	5.4	-1.5	4.5	8.7	9.0	14.2	-4.1	5.8	5.4	31.7
Gross National Saving	3.0	3.7	-5.4	1.0	4.4	3.8	6.4	-14.5	-8.5	-5.8	19.1

(Billions of 1980 CFA Francs)

	1967	1968	1969	1970	1971	1972	1973	1974	1975	1976	1977
Gross National Product	450.4	450.1	431.7	443.1	469.7	441.9	359.3	388.1	370.5	380.5	411.1
GDP at factor cost	442.0	442.2	422.8	435.7	461.6	433.7	354.7	384.2	370.9	371.7	395.2
Agriculture	324.3	317.1	292.8	309.8	325.8	294.7	198.1	240.3	192.8	175.8	203.0
Industry	23.7	26.0	47.7	40.1	47.3	47.7	55.5	48.3	56.9	65.1	72.0
Manufacturing	..	..	..	..	..	..	..	..	..	..	..
Services, etc.	106.8	112.2	97.0	99.5	103.3	106.8	113.1	110.3	138.1	149.5	145.7
Resource Balance	-46.0	-48.1	-40.8	-27.5	-5.8	-25.9	-31.6	-105.7	-18.2	-25.6	-28.0
Exports of Goods & NFServices	74.3	69.1	81.6	85.3	92.1	97.8	112.0	60.7	118.8	129.8	120.3
Imports of Goods & NFServices	120.3	117.2	122.5	112.8	97.9	123.8	143.5	166.4	137.0	155.4	148.4
Domestic Absorption	459.4	466.5	418.3	450.9	440.8	452.8	430.4	489.3	423.9	435.8	469.9
Private Consumption, etc.	318.6	316.5	307.4	293.5	311.0	293.6	222.8	284.9	242.4	246.2	256.5
General Gov't Consumption	43.4	46.4	45.9	49.3	47.6	48.0	45.1	48.5	53.5	58.6	66.2
Gross Domestic Investment	97.5	103.6	65.0	108.2	82.3	111.2	162.5	155.9	128.0	131.0	147.3
Fixed Investment	..	..	..	..	..	..	..	..	..	..	..
Memo Items:											
Capacity to Import	70.5	65.3	69.6	67.1	77.6	85.9	93.8	70.5	85.0	100.8	99.4
Terms of Trade Adjustment	-3.8	-3.7	-12.1	-18.2	-14.5	-12.0	-18.1	9.8	-33.7	-29.0	-21.0
Gross Domestic Income	451.0	451.4	425.3	431.2	461.9	437.1	348.6	408.7	354.0	361.4	399.7
Gross National Income	446.7	446.3	419.6	425.0	455.2	429.9	341.1	397.8	336.8	351.5	390.1

DOMESTIC PRICES (DEFLATORS) *(Index 1980 = 100)*

	1967	1968	1969	1970	1971	1972	1973	1974	1975	1976	1977
Overall (GDP)	36.0	34.9	37.2	40.0	40.1	41.7	57.5	61.9	58.0	65.2	75.4
Domestic Absorption	37.8	36.2	41.7	42.8	44.7	43.9	53.0	58.9	59.3	65.4	74.1
Agriculture	35.7	34.6	36.9	37.6	38.4	39.7	64.2	65.6	58.7	68.1	81.0
Industry	30.7	29.7	17.9	31.1	29.9	31.8	35.6	39.5	43.5	50.3	61.7
Manufacturing	..	..	..	..	..	..	..	..	..	..	..

MANUFACTURING ACTIVITY

	1967	1968	1969	1970	1971	1972	1973	1974	1975	1976	1977
Employment (1980=100)	..	..	..	..	..	..	..	..	..	..	76.6
Real Earnings per Empl. (1980=100)	..	..	..	..	..	..	..	..	..	..	78.9
Real Output per Empl. (1980=100)	..	..	..	..	..	..	..	..	..	..	104.5
Earnings as % of Value Added	..	..	..	..	..	..	..	..	..	..	..

MONETARY HOLDINGS *(Billions of current CFA Francs)*

	1967	1968	1969	1970	1971	1972	1973	1974	1975	1976	1977
Money Supply, Broadly Defined	7.9	8.0	9.2	9.6	12.1	12.9	15.5	20.3	22.3	29.2	37.6
Money as Means of Payment	7.4	7.2	8.5	8.8	10.6	11.4	13.6	17.6	20.1	24.8	32.3
Currency Ouside Banks	4.3	4.3	4.8	4.9	6.0	6.2	6.7	9.4	9.4	13.4	14.9
Demand Deposits	3.0	3.0	3.8	4.0	4.6	5.2	7.0	8.3	10.7	11.4	17.5
Quasi-Monetary Liabilities	0.5	0.7	0.7	0.8	1.6	1.5	1.9	2.7	2.2	4.5	5.3

GOVERNMENT DEFICIT (-) OR SURPLUS *(Millions of current CFA Francs)*

	1967	1968	1969	1970	1971	1972	1973	1974	1975	1976	1977
GOVERNMENT DEFICIT (-) OR SURPLUS	..	..	..	..	..	..	..	..	..	-5,468	-5,261
Current Revenue	..	..	..	..	..	..	..	..	..	30,485	37,644
Current Expenditure	..	..	..	..	..	..	..	..	..	24,820	27,113
Current Budget Balance	..	..	..	..	..	..	..	..	..	5,665	10,531
Capital Receipts	..	..	..	..	..	..	..	..	..	4	11
Capital Payments	..	..	..	..	..	..	..	..	..	11,137	15,803

1978	1979	1980	1981	1982	1983	1984	1985	1986	1987 estimate	NOTES	NIGER
300	360	430	440	380	320	250	240	250	260	..	**CURRENT GNP PER CAPITA (US $)**
5,193	5,352	5,515	5,680	5,850	6,025	6,205	6,391	6,592	6,798	..	**POPULATION (thousands)**
				(Billions of current CFA Francs)							**ORIGIN AND USE OF RESOURCES**
389.3	432.0	528.0	582.1	618.5	649.2	621.1	629.4	622.6	626.5	..	Gross National Product (GNP)
-11.1	-16.7	-8.2	-7.8	-26.9	-28.1	-17.2	-17.7	-20.8	-23.4	..	Net Factor Income from Abroad
400.4	448.7	536.2	589.9	645.4	677.3	638.3	647.1	643.4	649.9	..	Gross Domestic Product (GDP)
23.9	26.0	27.0	27.8	31.3	25.8	26.2	25.6	24.7	21.0	..	Indirect Taxes, net
376.5	422.7	509.2	562.1	614.1	651.5	612.1	621.5	618.7	628.9	B	GDP at factor cost
210.9	218.4	228.2	269.3	293.7	311.1	228.3	237.8	237.9	219.1	..	Agriculture
59.9	84.1	121.6	113.0	114.4	114.5	137.9	135.5	141.8	157.2	..	Industry
18.5	19.0	19.8	23.2	25.6	20.5	44.8	46.2	49.3	56.7	..	Manufacturing
129.6	146.2	186.4	207.6	237.3	251.7	272.1	273.8	263.7	273.6	..	Services, etc.
-21.2	-37.4	-75.5	-68.4	-67.2	-36.0	-30.3	-67.1	-28.9	-30.6	..	Resource Balance
90.2	128.1	130.8	146.7	136.6	152.5	132.1	112.6	126.6	120.8	..	Exports of Goods & NFServices
111.4	165.4	206.3	215.1	203.8	188.5	162.4	179.7	155.5	151.4	..	Imports of Goods & NFServices
421.6	486.1	611.7	658.3	712.6	713.3	668.6	714.2	672.3	680.5	..	Domestic Absorption
283.1	319.0	360.0	474.1	519.8	549.5	583.1	561.7	533.2	544.3	..	Private Consumption, etc.
46.5	51.2	55.2	64.5	75.2	77.2	65.2	70.1	74.0	76.1	..	General Gov't Consumption
92.1	115.9	196.5	119.7	117.6	86.6	20.3	82.4	65.1	60.1	..	Gross Domestic Investment
..	..	152.0	137.6	87.1	99.0	62.9	64.1	63.1	63.1		Fixed Investment
											Memo Items:
70.9	78.5	121.0	51.3	50.4	50.6	-10.0	15.3	36.2	29.5	..	Gross Domestic Saving
51.7	51.2	100.8	29.4	6.1	3.7	-46.1	-16.9	13.9	-6.7	..	Gross National Saving
				(Billions of 1980 CFA Francs)							
463.9	491.4	528.0	535.2	513.1	503.4	435.0	460.5	490.6	466.5	..	Gross National Product
448.7	481.8	509.2	517.0	509.9	506.2	431.1	457.0	487.4	..	B H	GDP at factor cost
230.6	217.0	228.2	226.3	220.5	245.6	221.0	253.7	273.5	..	..	Agriculture
86.1	117.0	121.6	126.9	122.3	109.2	102.5	99.8	100.8	..	..	Industry
..	..	..	..	..	..	..	..	..	..	..	Manufacturing
160.6	177.4	186.4	189.5	193.2	171.4	125.6	121.9	132.1	..	..	Services, etc.
-12.8	-71.1	-75.5	-83.4	-117.2	-102.8	-80.6	-116.0	-79.6	-80.4	..	Resource Balance
160.6	123.0	130.8	128.2	99.1	105.2	95.3	85.8	81.9	81.5	..	Exports of Goods & NFServices
173.4	194.0	206.3	211.6	216.3	208.0	176.0	201.8	161.5	162.0	..	Imports of Goods & NFServices
530.0	607.0	611.7	626.2	619.8	597.0	418.2	530.0	504.8	495.8	..	Domestic Absorption
282.7	326.0	360.0	364.4	387.8	370.9	346.3	346.6	366.4	339.8	..	Private Consumption, etc.
61.9	60.7	55.2	58.2	59.8	63.8	60.7	59.8	57.5	64.8	..	General Gov't Consumption
185.3	220.4	196.5	203.6	172.2	162.3	11.2	123.5	81.0	91.2	..	Gross Domestic Investment
..	..	..	..	..	..	..	..	..	..	..	Fixed Investment
											Memo Items:
140.4	150.2	130.8	144.3	145.0	168.3	143.1	126.5	131.5	129.2	..	Capacity to Import
-20.3	27.2	0.0	16.1	45.9	63.1	47.8	40.7	49.6	47.7	..	Terms of Trade Adjustment
457.1	538.7	536.2	558.7	581.8	589.3	496.9	516.1	556.1	529.4	..	Gross Domestic Income
443.7	518.7	528.0	551.3	558.9	566.4	482.8	501.2	540.2	514.2	..	Gross National Income
				(Index 1980 = 100)							**DOMESTIC PRICES (DEFLATORS)**
83.9	87.7	100.0	108.7	120.4	128.7	142.1	136.1	127.0	134.9	..	Overall (GDP)
79.5	80.1	100.0	105.1	115.0	119.5	159.9	134.8	133.2	137.3	..	Domestic Absorption
91.5	100.6	100.0	119.0	133.2	126.7	103.3	93.7	87.0	..	..	Agriculture
69.6	71.9	100.0	89.1	93.6	104.8	134.6	135.7	140.7	..	..	Industry
..	..	..	..	..	..	..	..	..	..	..	Manufacturing
											MANUFACTURING ACTIVITY
88.9	97.3	100.0	..	125.3	..	..	..	..	..	G	Employment (1980=100)
88.1	96.1	100.0	..	66.6	..	..	..	..	..	G	Real Earnings per Empl. (1980=100)
118.0	133.8	100.0	..	103.8	..	..	..	..	..	G	Real Output per Empl. (1980=100)
43.1	41.2	43.8	..	40.5	..	..	..	..	..	..	Earnings as % of Value Added
				(Billions of current CFA Francs)							**MONETARY HOLDINGS**
54.2	64.5	77.9	94.1	83.0	82.7	101.0	108.1	121.1	114.4	..	Money Supply, Broadly Defined
46.4	57.3	64.6	74.8	70.9	66.6	78.4	80.6	83.0	73.2	..	Money as Means of Payment
19.7	27.3	31.1	34.8	35.3	31.5	30.7	33.4	40.5	35.6	..	Currency Ouside Banks
26.7	30.0	33.5	39.9	35.7	35.0	47.7	47.3	42.5	37.6	..	Demand Deposits
7.8	7.2	13.3	19.3	12.1	16.1	22.6	27.5	38.1	41.3	..	Quasi-Monetary Liabilities
				(Millions of current CFA Francs)							
-12,622	-12,105	-25,241	..	..	..	..	..	..	..	C	**GOVERNMENT DEFICIT (-) OR SURPLUS**
50,382	62,553	77,430	..	..	..	..	..	..	..	..	Current Revenue
34,937	39,163	50,581	..	..	..	..	..	..	..	..	Current Expenditure
15,445	23,390	26,849	..	..	..	..	..	..	..	..	Current Budget Balance
7	6	6	..	..	..	..	..	..	..	..	Capital Receipts
28,074	35,501	52,096	..	..	..	..	..	..	..	..	Capital Payments

NIGER	1967	1968	1969	1970	1971	1972	1973	1974	1975	1976	1977
FOREIGN TRADE (CUSTOMS BASIS)					*(Millions of current US dollars)*						
Value of Exports, fob	25.5	28.9	24.2	31.6	38.4	54.3	62.8	52.6	91.2	133.9	160.1
Nonfuel Primary Products	24.8	27.9	22.7	30.4	36.7	50.4	57.8	46.2	83.2	130.5	155.6
Fuels	0.0	0.0	0.1	0.0	0.0	0.0	0.0	0.0	0.0	0.1	0.1
Manufactures	0.7	1.0	1.4	1.2	1.6	3.9	5.0	6.3	8.0	3.3	4.5
Value of Imports, cif	46.0	41.5	48.7	58.4	53.0	65.7	86.8	96.4	98.9	127.1	196.3
Nonfuel Primary Products	9.6	7.4	8.3	10.7	8.3	13.0	20.0	26.0	28.0	15.7	32.4
Fuels	2.7	2.3	2.1	2.3	4.5	5.8	8.3	13.1	12.4	14.7	14.4
Manufactures	33.7	31.7	38.4	45.3	40.2	46.9	58.6	57.3	58.5	96.7	149.5
Terms of Trade	154.2	159.0	154.6	157.0	163.6	169.5	149.5	124.6	128.4	134.4	124.8
Export Prices, fob	36.1	37.4	40.1	43.0	40.9	45.8	51.3	71.2	77.2	80.2	83.7
Nonfuel Primary Products	36.7	37.9	41.5	43.8	41.2	46.5	51.5	71.8	78.5	80.6	84.2
Fuels	4.3	4.3	4.3	4.3	5.6	6.2	8.9	36.7	35.7	38.4	42.0
Manufactures	30.1	31.7	32.4	37.5	36.5	38.5	49.5	67.2	66.1	66.7	71.7
Import Prices, cif	23.4	23.5	26.0	27.4	25.0	27.0	34.3	57.1	60.2	59.6	67.1
Trade at Constant 1980 Prices					*(Millions of 1980 US dollars)*						
Exports, fob	70.8	77.1	60.3	73.5	93.9	118.7	122.5	73.8	118.1	167.0	191.2
Imports, cif	196.6	176.2	187.6	213.0	212.1	243.4	253.1	168.6	164.5	213.1	292.6
BALANCE OF PAYMENTS					*(Millions of current US dollars)*						
Exports of Goods & Services	..	..	..	60.2	73.5	91.3	126.8	107.6	172.1	203.1	231.7
Merchandise, fob	..	..	..	46.7	58.4	71.0	99.5	81.5	138.5	171.6	196.6
Nonfactor Services	..	..	..	12.3	13.4	18.5	23.3	17.8	23.9	23.4	26.0
Factor Services	..	..	..	1.2	1.7	1.9	4.1	8.2	9.7	8.1	9.0
Imports of Goods & Services	..	..	..	88.8	86.0	115.5	170.6	246.8	274.6	315.4	380.1
Merchandise, fob	..	..	..	55.1	53.1	67.9	112.1	145.0	147.9	198.4	241.4
Nonfactor Services	..	..	..	31.0	29.9	44.7	54.8	87.6	107.6	105.2	129.2
Factor Services	..	..	..	2.7	3.0	2.9	3.7	14.3	19.1	11.8	9.5
Long-Term Interest	..	..	..	..	..	..	..	..	..	2.4	3.1
Current Transfers, net	..	..	..	-3.8	-5.8	-8.8	-16.0	-16.7	-19.0	-19.8	-22.2
Workers' Remittances	..	..	..	0.0	0.6	1.9	1.8	2.1	3.0	3.4	3.3
Total to be Financed	..	..	..	-32.4	-13.6	-33.0	-59.7	-133.4	-91.2	-132.2	-170.3
Official Capital Grants	..	..	..	32.3	30.1	45.1	84.3	119.9	82.7	110.7	74.7
Current Account Balance	..	..	..	-0.1	16.5	12.1	24.6	-13.6	-8.4	-21.5	-95.6
Long-Term Capital, net	..	..	..	53.1	37.2	52.0	94.4	143.1	130.8	163.1	165.0
Direct Investment	..	..	..	0.4	-5.7	0.8	0.8	6.7	16.4	8.6	9.2
Long-Term Loans	..	..	..	..	..	..	..	..	..	20.3	22.9
Disbursements	..	..	..	..	..	..	..	..	..	27.0	29.0
Repayments	..	..	..	..	..	..	..	..	..	6.7	6.1
Other Long-Term Capital	..	..	..	20.4	12.8	6.1	9.3	16.5	31.7	23.6	58.3
Other Capital, net	..	..	..	-42.7	-42.8	-57.8	-114.0	-135.8	-116.6	-103.9	-56.0
Change in Reserves	..	..	..	-10.3	-10.8	-6.3	-5.0	6.3	-5.7	-37.8	-13.4
Memo Item:					*(CFA Francs per US dollar)*						
Conversion Factor (Annual Avg)	246.850	246.850	259.710	277.710	277.130	252.210	222.700	240.500	214.320	238.980	245.670
EXTERNAL DEBT, ETC.					*(Millions of US dollars, outstanding at end of year)*						
Public/Publicly Guar. Long-Term	..	..	..	31.7	41.6	53.9	64.1	91.8	111.6	129.8	118.2
Official Creditors	..	..	..	31.2	41.3	50.2	60.6	80.3	101.0	122.7	113.5
IBRD and IDA	..	..	..	4.2	6.0	9.6	14.0	15.9	18.2	21.8	28.0
Private Creditors	..	..	..	0.5	0.2	3.7	3.5	11.5	10.6	7.1	4.8
Private Non-guaranteed Long-Term	..	..	..	..	..	..	..	..	..	..	..
Use of Fund Credit	..	..	..	0.0	0.0	0.0	0.0	0.0	0.0	0.0	0.0
Short-Term Debt	..	..	..	..	..	..	..	..	..	..	52.0
Memo Items:					*(Millions of US dollars)*						
Int'l Reserves Excluding Gold	1.1	2.7	6.7	18.7	33.6	41.4	50.8	45.5	50.3	82.5	101.1
Gold Holdings (at market price)	..	..	..	..	..	..	..	..	..	..	0.9
SOCIAL INDICATORS											
Total Fertility Rate	6.8	6.8	6.8	6.9	6.9	6.9	6.9	6.9	7.0	7.0	7.0
Crude Birth Rate	49.4	49.6	49.8	50.0	50.2	50.4	50.5	50.6	50.7	50.8	50.9
Infant Mortality Rate	176.0	174.0	172.0	170.0	168.0	166.0	164.2	162.4	160.6	158.8	157.0
Life Expectancy at Birth	37.5	37.8	38.1	38.4	38.7	39.0	39.3	39.6	39.9	40.2	40.5
Food Production, p.c. ('79-81 = 100)	113.6	97.0	108.4	95.0	95.9	90.2	62.4	75.8	69.6	86.2	95.2
Labor Force, Agriculture (%)	94.8	94.7	94.5	94.3	94.0	93.7	93.3	93.0	92.7	92.4	92.1
Labor Force, Female (%)	49.0	49.0	49.0	49.1	49.0	48.9	48.8	48.7	48.6	48.4	48.3
School Enroll. Ratio, primary	..	..	..	14.0	..	..	..	..	20.0	21.0	22.0
School Enroll. Ratio, secondary	..	..	..	1.0	..	..	..	..	2.0	3.0	3.0

1978	1979	1980	1981	1982	1983	1984	1985	1986	1987 estimate	NOTES	NIGER
											FOREIGN TRADE (CUSTOMS BASIS)
				(Millions of current US dollars)							
282.9	448.0	579.7	454.8	333.0	310.7	227.6	222.7	227.5	214.0	f	Value of Exports, fob
277.2	442.6	560.8	440.8	322.7	301.0	74.0	72.0	218.4	206.1	..	Nonfuel Primary Products
0.1	0.1	6.5	3.9	2.9	2.8	151.9	148.8	2.3	2.3	..	Fuels
5.6	5.3	12.4	10.1	7.4	6.9	1.7	1.9	6.9	5.7	..	Manufactures
305.9	461.7	607.7	509.7	442.0	361.3	331.0	388.0	466.0	500.1	f	Value of Imports, cif
61.8	83.9	108.4	138.4	107.8	87.8	71.5	77.7	135.0	148.9	..	Nonfuel Primary Products
43.4	79.2	158.3	75.7	76.5	63.9	50.0	51.0	20.9	31.0	..	Fuels
200.7	298.6	341.0	295.6	257.7	209.6	209.5	259.3	310.1	320.2	..	Manufactures
				(Index 1980 = 100)							
134.7	122.0	100.0	100.0	103.5	107.1	100.0	100.4	96.9	83.1	..	Terms of Trade
95.5	101.6	100.0	102.1	98.7	100.3	93.3	90.4	91.0	82.1	..	Export Prices, fob
96.1	101.8	100.0	102.1	98.7	100.7	100.5	96.9	91.6	81.9	..	Nonfuel Primary Products
42.3	61.0	100.0	112.5	101.6	92.5	90.2	87.5	44.9	56.9	..	Fuels
76.1	90.0	100.0	99.1	94.8	90.7	89.3	89.0	103.2	113.0	..	Manufactures
70.9	83.3	100.0	102.1	95.3	93.7	93.3	90.0	93.9	98.8	..	Import Prices, cif
				(Millions of 1980 US dollars)							Trade at Constant 1980 Prices
296.1	440.8	579.7	445.5	337.5	309.7	244.1	246.4	250.1	260.6	..	Exports, fob
431.3	554.1	607.7	499.2	463.7	385.7	354.9	431.0	496.2	506.1	..	Imports, cif
				(Millions of current US dollars)							**BALANCE OF PAYMENTS**
328.5	533.1	643.7	542.7	440.7	387.5	348.1	250.7	330.0	320.7	..	Exports of Goods & Services
287.7	485.0	576.1	484.6	381.3	335.2	303.3	250.7	330.0	278.8	..	Merchandise, fob
31.5	35.7	40.6	39.8	41.8	40.3	31.4	0.0	0.0	35.1	..	Nonfactor Services
9.3	12.4	27.0	18.2	17.6	11.9	13.4	0.0	0.0	6.9	..	Factor Services
613.1	839.2	1,016.3	873.5	817.6	538.0	462.0	437.9	479.5	479.4	..	Imports of Goods & Services
410.6	527.0	677.4	591.8	515.3	331.6	269.9	309.5	370.6	302.5	..	Merchandise, fob
163.3	242.4	278.9	194.1	213.0	153.7	124.2	73.9	36.9	89.3	..	Nonfactor Services
39.1	69.8	60.0	87.6	89.3	52.7	67.9	54.5	72.0	87.6	..	Factor Services
24.7	42.1	64.9	66.4	71.7	56.9	44.7	41.9	50.0	73.4	..	Long-Term Interest
-35.6	-49.8	-56.7	-51.9	-53.1	-49.3	-43.3	-32.3	-4.3	-42.7	..	Current Transfers, net
4.7	4.6	5.9	4.6	3.4	3.0	4.5	0.0	0.0	..	..	Workers' Remittances
-320.2	-278.8	-429.3	-342.1	-382.8	-197.7	-144.6	-219.5	-153.8	-201.4	K	Total to be Financed
119.3	141.2	153.6	160.6	150.0	135.8	145.8	162.3	148.0	134.4	K	Official Capital Grants
-201.0	-137.6	-275.7	-181.5	-232.8	-61.8	1.2	-57.2	-5.8	-66.9	..	Current Account Balance
282.5	330.5	386.7	385.0	187.4	181.4	176.8	220.6	213.4	187.1	..	Long-Term Capital, net
35.9	36.1	43.9	-6.7	24.9	0.2	-1.1	..	..	..	..	Direct Investment
133.5	191.5	232.6	324.2	23.5	74.6	30.1	42.5	130.8	128.6	..	Long-Term Loans
152.8	224.2	290.1	407.9	180.5	162.1	94.0	106.5	205.3	205.7	..	Disbursements
19.3	32.7	57.5	83.7	157.0	87.5	63.9	64.0	74.5	77.1	..	Repayments
-6.2	-38.3	-43.4	-93.1	-11.1	-29.3	2.0	15.8	-65.4	-75.9	..	Other Long-Term Capital
-67.7	-196.6	-105.2	-200.7	-17.9	-122.5	-146.7	-158.3	-201.2	-66.6	..	Other Capital, net
-13.8	3.6	-5.9	-2.8	63.3	2.9	-31.3	-5.1	-6.5	-53.6	..	Change in Reserves
				(CFA Francs per US dollar)							Memo Item:
225.640	212.720	211.300	271.730	328.620	381.070	436.960	449.260	346.300	300.540	..	Conversion Factor (Annual Avg)
				(Millions of US dollars, outstanding at end of year)							**EXTERNAL DEBT, ETC.**
197.2	262.8	399.0	605.0	604.0	634.3	679.6	837.8	1,013.0	1,258.5	..	Public/Publicly Guar. Long-Term
171.5	199.7	271.3	378.1	390.1	456.5	525.9	681.0	810.2	1,027.8	..	Official Creditors
36.5	48.1	66.1	78.6	91.6	105.2	121.2	146.7	190.9	272.8	..	IBRD and IDA
25.7	63.1	127.7	226.9	213.9	177.7	153.7	156.8	202.8	230.7	..	Private Creditors
158.9	259.7	304.7	303.4	233.8	181.6	162.0	198.5	224.3	254.3	..	Private Non-guaranteed Long-Term
0.0	0.0	0.0	0.0	0.0	32.3	44.3	66.7	88.1	90.8	..	Use of Fund Credit
250.0	106.0	159.3	113.8	123.6	67.2	62.7	98.7	122.0	75.3	..	Short-Term Debt
				(Millions of US dollars)							Memo Items:
128.4	131.7	125.9	105.3	29.6	53.2	88.7	136.4	189.2	248.5	..	Int'l Reserves Excluding Gold
1.9	5.7	6.5	4.4	5.1	4.2	3.4	3.6	4.3	5.4	..	Gold Holdings (at market price)
											SOCIAL INDICATORS
7.0	7.0	7.0	7.0	7.0	7.0	7.0	7.0	7.0	7.0	..	Total Fertility Rate
51.1	51.2	51.4	51.5	51.7	51.6	51.4	51.3	51.1	51.0	..	Crude Birth Rate
154.8	152.6	150.4	148.2	146.0	143.8	141.5	139.3	137.0	134.8	..	Infant Mortality Rate
40.9	41.3	41.7	42.1	42.5	42.9	43.3	43.7	44.1	44.6	..	Life Expectancy at Birth
98.2	100.3	102.4	97.3	95.4	92.1	73.6	90.7	90.6	80.0	..	Food Production, p.c. ('79-81 = 100)
91.7	91.4	91.1	..	..	..	..	..	..	..	..	Labor Force, Agriculture (%)
48.2	48.1	48.0	47.9	47.8	47.6	47.5	47.4	47.3	47.1	..	Labor Force, Female (%)
23.0	..	..	27.0	..	27.0	26.0	28.0	29.0	..	..	School Enroll. Ratio, primary
4.0	..	5.0	6.0	..	6.0	..	6.0	..	..	..	School Enroll. Ratio, secondary

NIGERIA	1967	1968	1969	1970	1971	1972	1973	1974	1975	1976	1977
CURRENT GNP PER CAPITA (US $)	90	90	110	150	190	230	260	360	430	590	700
POPULATION (thousands)	61,453	62,991	64,566	66,182	67,837	69,534	71,274	73,057	74,884	76,758	78,556

ORIGIN AND USE OF RESOURCES					*(Millions of current Nigerian Naira)*						
Gross National Product (GNP)	3,951	3,902	4,981	9,407	10,593	11,237	12,611	21,030	24,806	31,016	36,295
Net Factor Income from Abroad	-64	-113	-152	-332	-751	-854	-866	-680	-529	-659	-868
Gross Domestic Product (GDP)	4,015	4,015	5,132	9,739	11,344	12,091	13,477	21,710	25,335	31,675	37,163
Indirect Taxes, net	281	301	429	680	321	309	296	308	356	477	523
GDP at factor cost	3,734	3,714	4,703	9,059	11,023	11,782	13,181	21,402	24,979	31,198	36,641
Agriculture	2,020	1,872	2,264	3,629	4,273	4,356	4,459	6,518	7,607	8,697	10,378
Industry	468	426	776	1,319	2,017	2,475	3,470	7,868	7,448	10,505	12,017
Manufacturing	241	246	349	388	433	532	615	819	1,449	1,813	1,926
Services, etc.	1,245	1,416	1,664	4,112	4,733	4,951	5,253	7,016	9,924	11,996	14,246
Resource Balance	-106	-100	-27	-22	49	265	635	3,436	237	-49	-271
Exports of Goods & NFServices	515	462	676	943	1,407	1,515	2,441	6,179	5,268	6,524	8,431
Imports of Goods & NFServices	621	561	702	965	1,358	1,250	1,806	2,743	5,031	6,573	8,702
Domestic Absorption	4,121	4,115	5,159	9,761	11,295	11,826	12,842	18,274	25,099	31,723	37,434
Private Consumption, etc.	3,422	3,425	4,189	7,926	8,788	8,773	9,251	13,794	17,349	20,561	23,686
General Gov't Consumption	215	252	420	570	657	837	976	1,312	2,236	2,585	3,826
Gross Domestic Investment	484	438	550	1,265	1,850	2,216	2,615	3,168	5,514	8,577	9,922
Fixed Investment	..	..	..	..	..	..	2,506	2,956	5,020	8,108	9,421
Memo Items:											
Gross Domestic Saving	378	338	524	1,243	1,899	2,481	3,250	6,604	5,751	8,528	9,651
Gross National Saving	304	231	376	925	1,130	1,597	2,348	5,866	5,153	7,768	8,668

					(Millions of 1980 Nigerian Naira)						
Gross National Product	21,727	21,312	26,418	33,202	37,165	38,753	41,104	46,731	45,664	50,028	52,995
GDP at factor cost	21,423	20,976	25,938	33,114	37,895	39,707	42,123	47,303	45,732	50,046	53,135
Agriculture	10,463	10,310	11,860	13,938	14,664	13,596	14,809	16,344	14,647	14,418	15,402
Industry	4,485	3,593	6,444	9,950	13,221	15,774	15,582	18,264	15,770	19,481	20,452
Manufacturing	768	810	1,065	1,363	1,320	1,636	1,821	1,761	2,176	2,684	2,852
Services, etc.	6,475	7,073	7,635	9,225	10,011	10,337	11,732	12,694	15,315	16,147	17,281
Resource Balance	3,125	2,822	4,449	5,451	5,409	6,500	7,164	6,699	3,140	2,407	3,360
Exports of Goods & NFServices	4,616	4,209	6,135	7,663	8,334	9,116	10,367	10,686	9,184	9,163	13,608
Imports of Goods & NFServices	1,491	1,387	1,686	2,212	2,925	2,617	3,203	3,987	6,044	6,756	10,248
Domestic Absorption	18,865	18,963	22,552	28,585	33,625	34,259	35,919	41,280	43,257	48,412	50,546
Private Consumption, etc.	15,505	15,719	17,819	21,873	25,263	25,495	25,384	30,472	27,703	29,092	27,629
General Gov't Consumption	1,059	1,226	2,412	2,345	2,649	2,795	3,418	3,509	4,579	4,631	6,489
Gross Domestic Investment	2,302	2,018	2,321	4,368	5,712	5,969	7,117	7,300	10,974	14,688	16,429
Fixed Investment	..	..	..	..	..	..	6,766	6,918	10,370	13,941	15,602
Memo Items:											
Capacity to Import	1,237	1,141	1,622	2,161	3,031	3,171	4,330	8,982	6,328	6,706	9,929
Terms of Trade Adjustment	-3,379	-3,068	-4,513	-5,502	-5,303	-5,945	-6,037	-1,704	-2,856	-2,457	-3,679
Gross Domestic Income	18,611	18,716	22,488	28,535	33,731	34,814	37,046	46,275	43,541	48,362	50,227
Gross National Income	18,347	18,244	21,905	27,701	31,862	32,808	35,067	45,027	42,808	47,572	49,316

DOMESTIC PRICES (DEFLATORS)					*(Index 1980 = 100)*						
Overall (GDP)	18.3	18.4	19.0	28.6	29.1	29.7	31.3	45.2	54.6	62.3	68.9
Domestic Absorption	21.8	21.7	22.9	34.1	33.6	34.5	35.8	44.3	58.0	65.5	74.1
Agriculture	19.3	18.2	19.1	26.0	29.1	32.0	30.1	39.9	51.9	60.3	67.4
Industry	10.4	11.9	12.0	13.3	15.3	15.7	22.3	43.1	47.2	53.9	58.8
Manufacturing	31.3	30.3	32.8	28.4	32.8	32.6	33.8	46.5	66.6	67.6	67.5

MANUFACTURING ACTIVITY											
Employment (1980=100)	16.2	18.7	23.9	29.8	33.5	38.6	38.7	43.4	55.9	63.3	75.1
Real Earnings per Empl. (1980=100)	104.2	107.8	96.9	89.6	88.1	89.3	94.3	91.3	89.2	91.2	80.1
Real Output per Empl. (1980=100)	78.1	82.6	84.7	104.6	91.0	87.9	98.8	79.2	73.0	92.2	90.1
Earnings as % of Value Added	21.7	21.5	18.1	17.6	20.1	21.6	21.8	20.7	21.8	24.5	23.6

MONETARY HOLDINGS					*(Millions of current Nigerian Naira)*						
Money Supply, Broadly Defined	454	522	663	979	1,042	1,204	1,508	2,730	4,178	5,843	7,813
Money as Means of Payment	323	338	447	643	670	747	926	1,757	2,605	3,864	5,558
Currency Ouside Banks	207	183	253	342	355	385	436	570	1,031	1,351	1,941
Demand Deposits	115	155	195	300	316	362	490	1,187	1,575	2,513	3,617
Quasi-Monetary Liabilities	131	184	215	337	372	457	582	973	1,572	1,979	2,255

GOVERNMENT DEFICIT (-) OR SURPLUS					*(Millions of current Nigerian Naira)*						
	..	..	..	..	..	-82.7	188.9	1,247.6	-1,435.8	-1,870.1	-2,134.2
Current Revenue	..	..	..	..	..	1,053.7	1,830.1	4,308.1	4,740.4	5,449.0	6,659.2
Current Expenditure	..	..	..	..	..	716.0	885.2	..	2,659.5	2,646.4	4,062.8
Current Budget Balance	..	..	..	..	..	337.7	944.9	..	2,080.9	2,802.6	2,596.4
Capital Receipts	..	..	..	..	..	..	..	..	..	..	..
Capital Payments	..	..	..	..	..	420.4	756.0	..	3,516.7	4,672.7	4,730.6

1978	1979	1980	1981	1982	1983	1984	1985	1986	1987 estimate	NOTES	NIGERIA
720	850	1,020	1,090	1,110	980	900	950	700	370	..	**CURRENT GNP PER CAPITA** (US $)
80,563	82,603	84,732	87,529	90,418	93,402	96,485	99,669	103,147	106,736	..	**POPULATION** (thousands)
				(Millions of current Nigerian Naira)							**ORIGIN AND USE OF RESOURCES**
39,067	46,608	54,428	57,421	61,770	63,783	69,671	79,038	80,862	94,042	C	Gross National Product (GNP)
-836	-1,025	-2,063	-1,402	-953	-1,247	-1,636	-1,609	-921	-5,716	..	Net Factor Income from Abroad
39,903	47,633	56,491	58,823	62,723	65,029	71,306	80,647	81,783	99,758	C	Gross Domestic Product (GDP)
691	755	1,731	2,221	2,240	1,736	1,356	1,871	2,043	1,824	..	Indirect Taxes, net
39,212	46,878	54,760	56,602	60,483	63,293	69,950	78,776	79,740	97,934	C	GDP at factor cost
11,442	12,821	14,318	14,448	17,010	20,070	24,999	27,624	28,964	29,633	..	Agriculture
13,637	18,446	22,906	22,030	22,178	19,096	19,454	23,303	21,045	41,699	..	Industry
2,945	4,725	5,038	6,110	7,450	6,323	6,656	8,427	9,116	7,866	..	Manufacturing
14,133	15,611	17,536	20,124	21,295	24,127	25,498	27,848	29,732	26,603	..	Services, etc.
-1,451	2,452	5,114	-2,043	-3,827	-2,235	1,204	2,626	-1,488	4,108	..	Resource Balance
7,114	10,614	14,760	11,362	8,528	7,768	9,418	11,474	9,298	30,946	..	Exports of Goods & NFServices
8,565	8,162	9,645	13,405	12,356	10,003	8,215	8,848	10,785	26,838	..	Imports of Goods & NFServices
41,354	45,181	51,376	60,866	66,550	67,264	70,102	78,021	83,271	95,650	..	Domestic Absorption
26,675	30,717	34,759	42,387	49,230	51,310	56,776	64,162	66,246	68,719	f	Private Consumption, etc.
4,793	4,884	5,051	5,708	6,684	7,941	8,792	7,474	8,876	11,194	..	General Gov't Consumption
9,886	9,580	11,566	12,771	10,636	8,013	4,534	6,386	8,149	15,736	f	Gross Domestic Investment
9,386	9,080	10,976	12,559	10,462	7,876	4,462	5,740	8,017	16,861	..	Fixed Investment
											Memo Items:
8,435	12,032	16,680	10,728	6,809	5,778	5,738	9,011	6,661	19,844	..	Gross Domestic Saving
7,439	10,796	14,393	9,052	5,593	4,260	3,868	7,181	5,515	13,012	..	Gross National Saving
				(Millions of 1980 Nigerian Naira)							
50,170	53,829	54,428	52,398	52,610	49,182	45,312	48,979	50,595	48,077	C	Gross National Product
50,064	53,751	54,760	51,217	51,217	48,482	45,255	48,491	49,494	47,513	C	GDP at factor cost
14,071	13,644	14,318	11,961	12,328	12,339	11,517	13,162	14,376	13,269	..	Agriculture
19,694	23,417	22,906	20,584	20,143	17,160	16,086	17,636	17,148	16,343	..	Industry
3,244	4,766	5,038	5,801	6,424	4,880	4,350	5,029	4,799	5,044	..	Manufacturing
16,299	16,689	17,536	18,672	18,746	18,983	17,652	17,693	17,969	17,900	..	Services, etc.
331	8,390	5,114	-4,933	-4,533	-2,639	-282	929	2,459	2,618	..	Resource Balance
10,932	16,302	14,760	6,578	5,277	4,950	5,719	6,402	6,235	5,450	..	Exports of Goods & NFServices
10,601	7,912	9,645	11,510	9,810	7,588	6,001	5,473	3,776	2,832	..	Imports of Goods & NFServices
50,632	46,231	51,376	58,184	57,679	52,491	46,441	48,755	48,365	45,887	..	Domestic Absorption
31,145	30,037	34,759	39,253	40,808	38,489	37,379	39,629	38,961	35,685	f	Private Consumption, etc.
7,408	6,137	5,051	6,394	6,591	6,356	5,041	4,062	4,577	5,238	..	General Gov't Consumption
12,079	10,057	11,566	12,537	10,280	7,646	4,021	5,065	4,827	4,964	f	Gross Domestic Investment
11,455	9,530	10,976	11,773	9,655	7,176	3,778	4,347	4,534	5,079	..	Fixed Investment
											Memo Items:
8,805	10,290	14,760	9,756	6,771	5,893	6,880	7,097	3,255	3,265	..	Capacity to Import
-2,127	-6,012	0	3,179	1,494	943	1,162	695	-2,980	-2,185	..	Terms of Trade Adjustment
48,836	48,608	56,491	56,430	54,640	50,795	47,320	50,379	47,844	46,320	..	Gross Domestic Income
48,043	47,817	54,428	55,577	54,104	50,125	46,473	49,674	47,615	45,893	..	Gross National Income
				(Index 1980 = 100)							**DOMESTIC PRICES (DEFLATORS)**
78.3	87.2	100.0	110.5	118.0	130.4	154.5	162.3	160.9	205.7	..	Overall (GDP)
81.7	97.7	100.0	104.6	115.4	128.1	151.0	160.0	172.2	208.4	..	Domestic Absorption
81.3	94.0	100.0	120.8	138.0	162.7	217.1	209.9	201.5	223.3	..	Agriculture
69.2	78.8	100.0	107.0	110.1	111.3	120.9	132.1	122.7	255.1	..	Industry
90.8	99.1	100.0	105.3	116.0	129.6	153.0	167.6	190.0	156.0	..	Manufacturing
											MANUFACTURING ACTIVITY
70.7	84.1	100.0	106.8	79.8	75.3	..	..	..	..	..	Employment (1980 = 100)
81.6	89.7	100.0	106.8	103.0	85.8	..	..	..	..	..	Real Earnings per Empl. (1980 = 100)
80.8	86.4	100.0	118.6	104.2	111.7	..	..	..	..	..	Real Output per Empl. (1980 = 100)
22.3	22.1	21.9	30.3	27.2	20.5	..	..	..	..	..	Earnings as % of Value Added
				(Millions of current Nigerian Naira)							**MONETARY HOLDINGS**
7,521	9,849	14,390	15,239	16,694	19,034	21,243	23,153	23,605	28,895	..	Money Supply, Broadly Defined
5,101	6,147	9,227	9,745	10,049	11,283	12,204	13,227	12,663	14,906	..	Money as Means of Payment
2,157	2,351	3,186	3,862	4,223	4,843	4,884	4,910	5,178	6,299	..	Currency Ouside Banks
2,943	3,796	6,041	5,883	5,826	6,440	7,320	8,317	7,485	8,607	..	Demand Deposits
2,420	3,702	5,163	5,494	6,645	7,751	9,039	9,926	10,942	13,989	..	Quasi-Monetary Liabilities
				(Millions of current Nigerian Naira)							
-639.1	..	..	..	..	..	..	..	..	..	C	**GOVERNMENT DEFICIT (-) OR SURPLUS**
5,933.5	..	..	..	..	..	..	..	..	..	..	Current Revenue
2,452.3	..	..	..	..	..	..	..	..	..	..	Current Expenditure
3,481.2	..	..	..	..	..	..	..	..	..	..	Current Budget Balance
..	..	..	..	..	..	..	..	..	..	..	Capital Receipts
4,120.3	..	..	..	..	..	..	..	..	..	..	Capital Payments

437

NIGERIA	1967	1968	1969	1970	1971	1972	1973	1974	1975	1976	1977
FOREIGN TRADE (CUSTOMS BASIS)						*(Millions of current US dollars)*					
Value of Exports, fob	666.8	592.4	905.0	1,227.9	1,793.2	2,161.1	3,448.0	9,195.0	7,992.0	10,568.7	11,848.4
Nonfuel Primary Products	443.1	460.1	482.3	496.4	433.9	339.8	505.8	606.7	496.3	541.2	749.6
Fuels	205.7	107.9	386.6	714.0	1,335.4	1,793.6	2,889.9	8,547.4	7,452.0	9,959.3	11,019.0
Manufactures	18.1	24.4	36.1	17.5	23.9	27.7	52.3	40.9	43.7	68.1	79.7
Value of Imports, cif	625.9	540.9	696.3	1,059.0	1,510.5	1,504.9	1,861.7	2,780.6	6,041.2	8,194.6	11,020.2
Nonfuel Primary Products	86.0	62.0	84.1	118.5	173.0	195.1	255.1	389.7	747.6	1,021.8	1,625.0
Fuels	24.6	40.8	43.8	33.6	12.6	14.9	20.6	88.8	162.8	279.3	199.4
Manufactures	515.3	438.1	568.4	906.9	1,325.0	1,294.9	1,585.9	2,302.1	5,130.9	6,893.6	9,195.8
						(Index 1980 = 100)					
Terms of Trade	39.8	66.6	36.8	23.4	19.8	18.9	22.3	65.3	57.7	62.7	64.0
Export Prices, fob	10.1	14.1	8.5	6.6	6.9	7.2	10.2	37.9	36.4	39.4	43.9
Nonfuel Primary Products	26.0	29.2	34.0	28.6	24.4	27.3	49.3	66.7	49.6	74.7	123.6
Fuels	4.3	4.3	4.3	4.3	5.6	6.2	8.9	36.7	35.7	38.4	42.0
Manufactures	26.1	26.7	26.7	32.4	30.1	33.5	48.9	61.9	56.8	58.5	73.3
Import Prices, cif	25.4	21.2	23.1	28.3	35.0	38.0	45.7	58.1	63.2	62.8	68.7
Trade at Constant 1980 Prices						*(Millions of 1980 US dollars)*					
Exports, fob	6,602.5	4,196.9	10,630.0	18,552.5	25,832.1	30,118.1	33,788.0	24,252.7	21,927.2	26,804.3	26,969.6
Imports, cif	2,468.8	2,550.9	3,008.5	3,736.4	4,313.8	3,956.5	4,069.8	4,787.0	9,561.5	13,040.8	16,041.7
BALANCE OF PAYMENTS						*(Millions of current US dollars)*					
Exports of Goods & Services	..	..	..	1,341.0	2,010.0	2,315.8	3,763.6	10,048.1	9,130.4	10,924.1	13,277.2
Merchandise, fob	..	..	..	1,248.0	1,888.6	2,184.4	3,607.4	9,698.1	8,329.1	10,121.7	12,364.3
Nonfactor Services	..	..	..	83.0	108.3	117.3	131.1	194.8	299.9	359.1	562.6
Factor Services	..	..	..	10.0	13.0	14.1	25.0	155.1	501.4	443.3	350.3
Imports of Goods & Services	..	..	..	1,772.0	2,418.2	2,636.1	3,718.3	5,052.3	8,961.6	11,125.0	14,102.9
Merchandise, fob	..	..	..	939.0	1,393.1	1,365.8	1,714.3	2,479.8	5,484.3	7,477.8	9,665.4
Nonfactor Services	..	..	..	369.0	454.4	575.4	1,020.5	1,852.1	2,695.4	2,915.2	3,552.9
Factor Services	..	..	..	464.0	570.7	694.9	983.5	720.4	781.9	732.0	884.6
Long-Term Interest	..	..	..	27.8	30.1	31.5	38.2	41.7	45.4	44.3	52.1
Current Transfers, net	..	..	..	19.0	-25.1	-46.7	-54.8	-91.4	-111.7	-161.6	-178.6
Workers' Remittances	..	..	..	0.0	0.0	0.0	0.0	0.0	0.0	0.0	0.0
Total to be Financed	..	..	..	-412.0	-433.3	-367.0	-9.5	4,904.4	57.1	-362.5	-1,004.4
Official Capital Grants	..	..	..	44.0	27.1	25.0	1.2	-7.2	-14.6	5.8	-4.7
Current Account Balance	..	..	..	-368.0	-406.2	-342.0	-8.3	4,897.1	42.5	-356.7	-1,009.1
Long-Term Capital, net	..	..	..	249.0	337.0	391.9	307.6	162.4	194.3	-21.9	414.0
Direct Investment	..	..	..	205.0	285.8	305.1	373.1	257.4	417.7	339.4	438.4
Long-Term Loans	..	..	..	13.3	63.2	86.9	-99.0	-41.1	-91.3	-229.8	55.2
Disbursements	..	..	..	81.1	127.6	150.7	91.5	109.2	133.1	126.2	141.3
Repayments	..	..	..	67.8	64.4	63.8	190.5	150.3	224.4	356.0	86.1
Other Long-Term Capital	..	..	..	-13.3	-39.1	-25.0	32.2	-46.7	-117.5	-137.3	-75.0
Other Capital, net	..	..	..	192.0	229.7	-112.3	-97.4	-161.5	-48.1	1.4	-231.9
Change in Reserves	..	..	..	-73.0	-160.5	62.4	-201.9	-4,898.0	-188.6	377.3	827.0
Memo Item:						*(Nigerian Naira per US dollar)*					
Conversion Factor (Annual Avg)	0.710	0.710	0.710	0.710	0.710	0.660	0.660	0.630	0.620	0.630	0.640
EXTERNAL DEBT, ETC.					*(Millions of US dollars, outstanding at end of year)*						
Public/Publicly Guar. Long-Term	..	..	..	452	543	647	1,117	1,181	1,052	803	855
Official Creditors	..	..	..	358	449	544	587	657	704	749	807
IBRD and IDA	..	..	..	182	216	282	300	332	362	402	448
Private Creditors	..	..	..	94	95	103	530	524	349	55	48
Private Non-guaranteed Long-Term	..	..	..	115	108	85	88	93	91	103	130
Use of Fund Credit	..	..	..	0	0	0	0	0	0	0	0
Short-Term Debt	..	..	..	..	..	..	..	..	..	..	2,161
Memo Items:						*(Millions of US dollars)*					
Int'l Reserves Excluding Gold	92.3	97.3	112.4	202.2	408.3	355.5	558.8	5,602.5	5,585.6	5,179.8	4,232.2
Gold Holdings (at market price)	20.1	23.9	20.1	21.3	23.7	35.2	64.1	106.5	80.1	76.9	103.8
SOCIAL INDICATORS											
Total Fertility Rate	6.9	6.9	6.9	6.9	6.9	6.9	6.9	6.9	6.9	6.9	6.9
Crude Birth Rate	51.1	51.0	50.9	50.8	50.7	50.6	50.5	50.4	50.2	50.1	50.0
Infant Mortality Rate	172.0	167.2	162.4	157.6	152.8	148.0	143.2	138.4	133.6	128.8	124.0
Life Expectancy at Birth	42.5	42.9	43.3	43.7	44.1	44.5	44.9	45.3	45.7	46.1	46.5
Food Production, p.c. ('79-81 = 100)	120.3	122.5	129.5	127.1	119.2	105.4	109.7	119.1	106.2	101.4	98.8
Labor Force, Agriculture (%)	71.7	71.4	71.2	71.0	70.7	70.4	70.2	69.9	69.6	69.3	69.0
Labor Force, Female (%)	37.3	37.2	37.2	37.1	37.1	37.1	37.0	37.0	36.9	36.9	36.8
School Enroll. Ratio, primary	..	..	..	37.0	..	..	..	..	51.0	71.0	82.0
School Enroll. Ratio, secondary	..	..	..	4.0	..	..	..	..	8.0	9.0	11.0

1978	1979	1980	1981	1982	1983	1984	1985	1986	1987 est.	NOTES	NIGERIA
				(Millions of current US dollars)							**FOREIGN TRADE (CUSTOMS BASIS)**
10,560.0	17,712.6	25,968.0	17,846.0	12,154.0	10,370.0	11,891.0	12,566.0	6,769.0	7,475.0	f	Value of Exports, fob
822.2	925.1	941.8	991.0	520.6	652.0	621.0	277.2	299.1	579.6	..	Nonfuel Primary Products
9,575.9	16,650.0	24,896.2	16,762.8	11,491.7	9,579.8	11,195.3	12,202.8	6,370.9	6,775.0	..	Fuels
161.9	137.4	130.0	92.1	141.7	138.2	74.6	86.0	99.0	120.4	..	Manufactures
12,762.8	10,274.3	16,642.0	20,877.0	16,060.0	12,254.2	9,364.0	8,890.0	7,465.6	6,133.9	f	Value of Imports, cif
2,079.7	2,038.8	3,243.4	3,873.3	846.5	2,293.6	1,736.4	1,284.5	934.0	680.9	..	Nonfuel Primary Products
246.3	235.4	357.4	453.6	408.7	267.0	236.9	401.8	146.4	189.6	..	Fuels
10,436.8	8,000.1	13,041.2	16,550.1	14,804.8	9,693.6	7,390.7	7,203.7	6,385.2	5,263.4	..	Manufactures
				(Index 1980 = 100)							
57.3	70.7	100.0	108.7	100.9	96.6	97.2	89.8	44.4	49.1	..	Terms of Trade
44.9	62.8	100.0	110.0	99.5	91.7	90.3	82.0	46.1	55.6	..	Export Prices, fob
119.0	121.4	100.0	82.3	69.6	84.3	95.4	83.2	72.1	74.0	..	Nonfuel Primary Products
42.3	61.0	100.0	112.5	101.6	92.5	90.2	82.0	44.9	54.0	..	Fuels
84.1	100.0	100.0	78.4	87.5	81.3	79.8	82.0	96.9	106.4	..	Manufactures
78.4	88.9	100.0	101.1	98.6	95.0	92.9	91.4	103.8	113.3	..	Import Prices, cif
				(Millions of 1980 US dollars)							Trade at Constant 1980 Prices
23,521.6	28,203.8	25,967.9	16,227.1	12,216.5	11,304.1	13,161.8	15,319.5	14,690.2	13,443.0	..	Exports, fob
16,284.6	11,559.4	16,642.3	20,641.5	16,293.5	12,905.0	10,075.2	9,729.6	7,190.8	5,412.4	..	Imports, cif
				(Millions of current US dollars)							**BALANCE OF PAYMENTS**
11,534.5	18,100.4	27,754.4	19,674.8	12,880.4	10,863.5	12,381.1	12,998.7	6,930.7	7,782.0	f	Exports of Goods & Services
10,367.1	16,804.1	25,934.3	18,047.4	12,146.8	10,353.2	11,889.2	12,603.1	6,559.8	7,475.0	..	Merchandise, fob
875.6	1,027.8	1,128.2	904.9	499.1	396.3	426.5	316.2	308.6	262.0	..	Nonfactor Services
291.8	268.4	691.9	722.5	234.5	114.1	65.4	79.4	62.2	45.0	..	Factor Services
15,026.1	16,041.7	22,045.6	25,418.8	19,733.7	14,814.4	11,924.8	11,506.9	6,426.9	7,884.0	f	Imports of Goods & Services
11,540.4	11,879.7	14,754.0	19,006.6	14,869.2	11,438.3	8,879.6	8,346.6	4,039.1	5,541.0	..	Merchandise, fob
2,961.7	3,441.6	5,238.6	4,950.4	3,421.7	2,375.6	1,802.0	1,606.2	1,363.4	898.0	..	Nonfactor Services
524.0	720.5	2,053.1	1,461.7	1,442.7	1,000.5	1,243.3	1,554.1	1,024.4	1,445.0	..	Factor Services
66.2	257.6	531.2	671.0	872.2	1,012.1	1,244.7	1,298.0	326.5	568.5	..	Long-Term Interest
-252.5	-349.9	-410.0	-443.5	-389.5	-375.0	-306.5	-248.0	-128.4	-278.0	..	Current Transfers, net
0.0	0.0	0.0	0.0	0.0	0.0	0.0	0.0	0.0	..	..	Workers' Remittances
-3,744.0	1,708.8	5,298.8	-6,187.5	-7,242.9	-4,325.9	149.8	1,243.9	375.4	-380.0	K	Total to be Financed
-18.9	-39.1	-168.2	-117.6	-39.7	-19.7	-26.1	-2.1	-5.5	..	K	Official Capital Grants
-3,762.9	1,669.7	5,130.6	-6,305.1	-7,282.6	-4,345.6	123.6	1,241.9	369.9	-380.0	..	Current Account Balance
1,601.8	1,198.5	-250.4	1,226.9	1,487.0	1,839.8	-77.5	-372.7	-1,086.1	-186.0	f	Long-Term Capital, net
213.3	304.9	-739.8	546.1	429.8	353.5	188.8	341.4	195.1	386.0	..	Direct Investment
1,626.7	1,284.2	1,482.4	2,454.5	2,942.6	1,897.1	-155.0	-1,215.6	310.6	731.9	..	Long-Term Loans
1,709.3	1,419.2	1,723.2	3,090.2	3,769.2	3,057.8	2,061.4	1,650.2	1,303.2	1,071.3	..	Disbursements
82.6	135.0	240.8	635.7	826.6	1,160.7	2,216.4	2,865.8	992.6	339.4	..	Repayments
-219.3	-351.5	-824.8	-1,656.1	-1,845.7	-391.1	-85.2	503.5	-1,586.4	-1,303.9	..	Other Long-Term Capital
41.9	330.5	-536.4	349.4	3,696.5	2,067.8	435.1	-252.5	122.1	-1.0	f	Other Capital, net
2,119.2	-3,198.6	-4,343.9	4,728.8	2,099.1	437.9	-481.2	-616.6	594.1	567.0	..	Change in Reserves
				(Nigerian Naira per US dollar)							**Memo Item:**
0.640	0.600	0.550	0.620	0.670	0.720	0.770	0.890	1.750	4.020		Conversion Factor (Annual Avg)
			(Millions of US dollars, outstanding at end of year)								**EXTERNAL DEBT, ETC.**
2,308	3,243	4,238	6,268	9,076	12,237	11,524	13,133	20,574	25,707	..	Public/Publicly Guar. Long-Term
870	898	953	1,056	1,257	1,901	2,012	2,359	6,809	11,616	..	Official Creditors
485	516	554	599	711	860	936	1,391	2,170	2,972	..	IBRD and IDA
1,439	2,344	3,285	5,212	7,819	10,335	9,511	10,774	13,765	14,091	..	Private Creditors
336	709	1,097	1,347	1,313	1,300	1,400	1,416	400	350	..	Private Non-guaranteed Long-Term
0	0	0	0	0	0	0	0	0	..	..	Use of Fund Credit
2,446	2,283	3,553	4,424	2,519	5,049	5,740	4,973	3,496	2,657	..	Short-Term Debt
				(Millions of US dollars)							**Memo Items:**
1,886.7	5,547.9	10,235.0	3,895.4	1,612.5	989.9	1,462.3	1,667.2	1,081.4	1,165.3	..	Int'l Reserves Excluding Gold
142.2	351.7	405.0	273.1	313.9	262.1	211.8	224.6	268.5	332.6	..	Gold Holdings (at market price)
											SOCIAL INDICATORS
6.9	6.9	6.9	6.9	6.9	6.9	6.9	6.9	6.9	6.9	..	Total Fertility Rate
49.9	49.8	49.8	49.7	49.6	49.6	49.6	49.5	49.5	49.5	..	Crude Birth Rate
122.0	120.0	118.0	116.0	114.0	111.8	109.5	107.3	105.0	102.8	..	Infant Mortality Rate
46.9	47.3	47.7	48.1	48.5	49.0	49.5	50.1	50.6	51.1	..	Life Expectancy at Birth
94.4	96.1	103.9	100.0	102.0	96.9	98.8	104.7	106.0	103.1	..	Food Production, p.c. ('79-81 = 100)
68.7	68.4	68.1	..	..	..	..	..	..	..	..	Labor Force, Agriculture (%)
36.8	36.7	36.7	36.5	36.3	36.1	35.9	35.7	35.6	35.4	..	Labor Force, Female (%)
92.0	98.0	98.0	..	97.0	92.0	..	..	..	..	..	School Enroll. Ratio, primary
12.0	16.0	..	..	28.0	29.0	..	..	..	..	..	School Enroll. Ratio, secondary

NORWAY	1967	1968	1969	1970	1971	1972	1973	1974	1975	1976	1977
CURRENT GNP PER CAPITA (US $)	2,170	2,320	2,530	2,750	3,090	3,560	4,240	5,290	6,610	7,650	8,550
POPULATION (thousands)	3,785	3,819	3,851	3,877	3,903	3,933	3,961	3,985	4,007	4,026	4,043

ORIGIN AND USE OF RESOURCES					*(Billions of current Norwegian Kroner)*						
Gross National Product (GNP)	59.01	63.07	68.85	79.26	88.36	97.38	110.67	127.88	146.79	167.66	186.80
Net Factor Income from Abroad	-0.69	-0.68	-0.56	-0.62	-0.75	-1.03	-1.19	-1.85	-1.92	-3.05	-4.73
Gross Domestic Product (GDP)	59.70	63.75	69.42	79.88	89.11	98.40	111.85	129.73	148.70	170.71	191.53
Indirect Taxes, net	6.33	6.39	7.19	10.45	11.88	12.96	14.36	15.23	17.20	19.39	22.22
GDP at factor cost	53.37	57.36	62.22	69.43	77.22	85.45	97.50	114.50	131.50	151.32	169.32
Agriculture	3.90	4.06	3.96	4.46	5.02	4.93	5.52	6.46	7.12	8.32	9.47
Industry	19.70	20.87	23.64	25.61	28.13	31.50	35.44	42.72	51.39	57.77	62.74
Manufacturing	12.37	13.36	15.08	17.26	18.81	21.12	24.05	28.38	32.27	34.41	35.96
Services, etc.	36.10	38.82	41.82	49.81	55.96	61.98	70.90	80.55	90.20	104.62	119.33
Resource Balance	-1.09	1.30	1.53	-1.00	-2.89	0.74	-0.53	-3.78	-9.93	-16.27	-20.54
Exports of Goods & NFServices	25.04	27.52	29.40	33.43	35.82	40.04	48.72	60.04	62.21	70.13	76.24
Imports of Goods & NFServices	26.13	26.21	27.86	34.44	38.71	39.30	49.25	63.82	72.14	86.40	96.78
Domestic Absorption	60.79	62.45	67.88	80.88	92.00	97.66	112.38	133.51	158.63	186.98	212.08
Private Consumption, etc.	32.95	35.21	39.24	43.09	47.89	52.50	58.39	66.18	77.52	89.49	103.93
General Gov't Consumption	9.61	10.56	11.67	13.53	15.98	17.86	20.39	23.76	28.70	34.09	38.63
Gross Domestic Investment	18.22	16.68	16.97	24.26	28.13	27.30	33.60	43.57	52.41	63.41	69.52
Fixed Investment	17.98	17.39	17.13	21.52	27.19	28.03	34.33	39.94	50.15	60.49	68.74
Memo Items:											
Gross Domestic Saving	17.14	17.98	18.50	23.26	25.24	28.04	33.07	39.79	42.48	47.14	48.98
Gross National Saving	16.61	17.53	18.15	22.86	..	27.27	32.07	38.06	40.63	44.10	44.15

					(Billions of 1980 Norwegian Kroner)						
Gross National Product	162.69	166.49	174.47	177.99	186.01	195.27	203.23	213.02	222.25	236.30	243.18
GDP at factor cost	147.21	151.19	157.69	155.58	162.15	171.50	178.86	190.70	199.50	214.11	221.53
Agriculture	11.84	11.83	10.62	9.37	9.88	9.94	9.81	10.76	10.39	10.72	10.80
Industry	54.68	55.91	58.91	58.80	61.42	65.78	68.82	72.64	79.00	85.15	86.11
Manufacturing	34.86	35.71	38.31	39.65	41.06	43.03	45.47	47.74	46.70	46.80	45.83
Services, etc.	98.08	100.57	106.36	111.24	116.32	121.60	126.80	132.70	135.72	144.56	152.13
Resource Balance	-1.43	2.34	5.02	-4.99	-9.50	2.72	-2.52	-6.64	-10.97	-13.41	-13.64
Exports of Goods & NFServices	69.97	75.33	79.34	79.42	80.32	91.64	99.23	99.90	102.99	114.59	118.75
Imports of Goods & NFServices	71.40	72.99	74.32	84.41	89.82	88.91	101.75	106.55	113.96	128.00	132.39
Domestic Absorption	166.02	165.97	170.87	184.39	197.12	194.59	207.95	222.74	236.07	253.84	262.69
Private Consumption, etc.	84.15	86.88	94.00	94.85	99.20	102.13	105.09	109.14	114.73	121.71	130.14
General Gov't Consumption	27.48	28.64	30.01	31.90	33.82	35.35	37.28	38.78	41.27	44.32	46.50
Gross Domestic Investment	54.39	50.45	46.85	57.65	64.10	57.11	65.57	74.82	80.07	87.82	86.05
Fixed Investment	48.15	46.68	42.68	49.03	58.22	55.83	63.45	66.69	74.63	82.19	85.18
Memo Items:											
Capacity to Import	68.43	76.62	78.40	81.95	83.11	90.60	100.67	100.24	98.27	103.89	104.29
Terms of Trade Adjustment	-1.54	1.29	-0.94	2.53	2.80	-1.04	1.44	0.33	-4.72	-10.69	-14.46
Gross Domestic Income	163.05	169.60	174.95	181.94	190.41	196.28	206.86	216.43	220.39	229.74	234.59
Gross National Income	161.15	167.78	173.53	180.52	188.81	194.23	204.67	213.35	217.54	225.60	228.72

DOMESTIC PRICES (DEFLATORS)					*(Index 1980 = 100)*						
Overall (GDP)	36.3	37.9	39.5	44.5	47.5	49.9	54.4	60.0	66.1	71.0	76.9
Domestic Absorption	36.6	37.6	39.7	43.9	46.7	50.2	54.0	59.9	67.2	73.7	80.7
Agriculture	32.9	34.3	37.2	47.6	50.8	49.6	56.3	60.1	68.5	77.6	87.6
Industry	36.0	37.3	40.1	43.6	45.8	47.9	51.5	58.8	65.0	67.8	72.9
Manufacturing	35.5	37.4	39.4	43.5	45.8	49.1	52.9	59.4	69.1	73.5	78.5

MANUFACTURING ACTIVITY											
Employment (1980=100)	96.2	96.0	96.6	99.4	100.5	100.7	100.7	103.3	102.8	102.7	103.2
Real Earnings per Empl. (1980=100)	75.5	77.3	80.8	80.1	83.6	85.0	87.3	91.5	98.0	101.6	102.2
Real Output per Empl. (1980=100)	71.2	70.6	74.9	74.7	76.6	75.5	80.8	89.3	87.9	90.6	93.0
Earnings as % of Value Added	53.4	51.2	47.2	50.4	52.2	56.8	55.2	53.4	56.4	58.8	60.6

MONETARY HOLDINGS					*(Billions of current Norwegian Kroner)*						
Money Supply, Broadly Defined	32.78	36.76	40.59	46.70	52.93	59.57	67.61	75.09	86.59	93.71	109.91
Money as Means of Payment	12.26	14.12	15.27	17.20	19.21	22.39	25.81	28.87	33.65	32.42	37.00
Currency Ouside Banks	5.83	6.14	6.52	7.28	7.99	8.75	9.44	10.77	12.39	14.26	16.08
Demand Deposits	6.43	7.98	8.75	9.92	11.22	13.64	16.37	18.10	21.26	18.16	20.92
Quasi-Monetary Liabilities	20.52	22.64	25.32	29.50	33.72	37.18	41.80	46.22	52.94	61.29	72.91

GOVERNMENT DEFICIT (-) OR SURPLUS					*(Billions of current Norwegian Kroner)*						
	..	..	..	..	..	-1.44	-1.03	-1.76	-4.72	-10.01	-13.13
Current Revenue	..	..	..	..	..	37.32	42.91	49.02	55.73	66.12	73.86
Current Expenditure	..	..	..	..	..	32.40	36.66	42.51	49.84	61.28	70.19
Current Budget Balance	..	..	..	..	..	4.93	6.24	6.51	5.90	4.84	3.66
Capital Receipts	..	..	..	..	..	0.00	0.00	0.00	0.00	..	..
Capital Payments	..	..	..	..	..	6.37	7.27	8.27	10.61	14.85	16.80

1978	1979	1980	1981	1982	1983	1984	1985	1986	1987 estimate	NOTES	NORWAY
9,510	10,990	12,890	14,130	14,390	14,160	14,200	14,490	15,440	17,250	..	CURRENT GNP PER CAPITA (US $)
4,059	4,073	4,091	4,100	4,115	4,133	4,140	4,153	4,169	4,175	..	POPULATION (thousands)

(Billions of current Norwegian Kroner)

ORIGIN AND USE OF RESOURCES

1978	1979	1980	1981	1982	1983	1984	1985	1986	1987	NOTES	
205.90	229.34	275.53	317.09	349.57	389.19	439.37	492.34	506.75	550.78	..	Gross National Product (GNP)
-7.18	-9.33	-9.52	-10.58	-12.70	-13.00	-13.14	-9.48	-9.34	-8.96	..	Net Factor Income from Abroad
213.08	238.67	285.05	327.67	362.27	402.20	452.51	501.82	516.09	559.74	..	Gross Domestic Product (GDP)
21.50	24.36	29.06	33.90	38.09	45.29	52.49	65.06	71.40	76.92	..	Indirect Taxes, net
191.58	214.31	255.98	293.77	324.19	356.90	400.02	436.76	444.69	482.82	B	GDP at factor cost
9.81	10.14	10.97	12.96	13.44	13.13	15.04	15.36	16.49	..	..	Agriculture
73.06	88.35	112.90	129.59	142.59	163.11	190.23	207.49	180.54	..	..	Industry
37.53	43.82	45.63	48.57	51.38	56.72	64.52	70.24	79.11	..	..	Manufacturing
130.20	140.18	161.17	185.13	206.24	225.95	247.24	278.97	319.06	..	..	Services, etc.
-1.90	6.24	17.42	25.90	20.39	31.83	41.23	41.08	-17.35	-13.53	..	Resource Balance
87.26	105.36	134.79	156.33	164.97	183.92	214.12	235.34	194.37	199.42	..	Exports of Goods & NFServices
89.17	99.12	117.37	130.43	144.58	152.09	172.89	194.26	211.72	212.95	..	Imports of Goods & NFServices
214.98	232.43	267.62	301.77	341.88	370.36	411.29	460.74	533.44	573.27	..	Domestic Absorption
110.78	119.97	135.24	155.08	175.22	193.08	210.79	246.19	279.76	295.97	..	Private Consumption, etc.
43.54	46.58	53.48	62.62	70.41	78.21	84.10	92.80	102.09	114.63	..	General Gov't Consumption
60.66	65.87	78.90	84.08	96.25	99.07	116.40	121.74	151.59	162.67	..	Gross Domestic Investment
66.10	65.48	70.80	93.34	92.20	104.48	121.42	104.22	140.54	145.69	..	Fixed Investment

Memo Items:

58.75	72.11	96.33	109.98	116.64	130.91	157.63	162.82	134.24	149.14	..	Gross Domestic Saving
51.41	62.52	86.54	99.24	103.57	117.51	144.19	152.78	123.94	138.84	..	Gross National Saving

(Billions of 1980 Norwegian Kroner)

252.13	263.30	275.53	278.08	278.27	292.10	309.77	330.16	345.30	351.80	..	Gross National Product
235.74	246.79	255.98	257.21	257.81	267.81	281.49	293.42	306.68	313.31	B I	GDP at factor cost
10.44	10.75	10.97	11.96	12.45	12.32	13.36	12.69	12.40	..	..	Agriculture
98.03	105.68	112.90	110.88	111.63	119.87	130.54	136.10	139.34	..	..	Industry
44.70	46.14	45.63	44.69	44.79	44.16	47.01	49.04	49.78	..	..	Manufacturing
151.87	157.11	161.17	164.71	164.43	169.69	175.32	187.65	199.37	..	..	Services, etc.
14.24	18.36	17.42	17.54	12.98	23.34	23.66	31.86	19.21	32.06	..	Resource Balance
128.71	132.03	134.79	136.65	136.45	146.79	158.84	175.76	176.86	184.82	..	Exports of Goods & NFServices
114.47	113.67	117.37	119.11	123.47	123.45	135.18	143.90	157.65	152.76	..	Imports of Goods & NFServices
246.11	255.18	267.62	270.01	275.52	278.54	295.55	304.57	331.89	324.83	..	Domestic Absorption
128.03	132.14	135.24	136.78	139.20	141.30	145.14	160.17	169.86	166.13	..	Private Consumption, etc.
48.99	50.73	53.48	56.76	58.98	61.73	63.24	65.36	67.40	69.22	..	General Gov't Consumption
69.10	72.31	78.90	76.46	77.34	75.51	87.17	79.04	94.64	89.49	..	Gross Domestic Investment
75.63	71.87	70.80	83.48	74.30	78.62	87.22	68.88	87.45	82.59	..	Fixed Investment

Memo Items:

112.02	120.82	134.79	142.77	140.88	149.29	167.41	174.33	144.73	143.05	..	Capacity to Import
-16.69	-11.20	0.00	6.11	4.43	2.50	8.57	-1.43	-32.13	-41.76	..	Terms of Trade Adjustment
243.66	262.34	285.05	293.66	292.93	304.38	327.78	335.00	318.97	315.13	..	Gross Domestic Income
235.44	252.10	275.53	284.20	282.69	294.60	318.34	328.73	313.16	..	..	Gross National Income

(Index 1980 = 100)

DOMESTIC PRICES (DEFLATORS)

81.8	87.3	100.0	114.0	125.6	133.2	141.8	149.2	147.0	156.8	..	Overall (GDP)
87.4	91.1	100.0	111.8	124.1	133.0	139.2	151.3	160.7	176.5	..	Domestic Absorption
94.0	94.3	100.0	108.3	108.0	106.6	112.6	121.1	133.0	..	..	Agriculture
74.5	83.6	100.0	116.9	127.7	136.1	145.7	152.5	129.6	..	..	Industry
83.9	95.0	100.0	108.7	114.7	128.5	137.2	143.2	158.9	..	..	Manufacturing

MANUFACTURING ACTIVITY

101.8	100.5	100.0	98.2	95.4	88.7	87.7	88.2	..	..	..	Employment (1980 = 100)
102.1	100.9	100.0	97.9	95.8	98.1	101.1	104.9	..	..	..	Real Earnings per Empl. (1980 = 100)
89.2	91.9	100.0	103.6	105.3	104.4	109.0	117.8	..	..	..	Real Output per Empl. (1980 = 100)
60.9	54.3	57.6	59.4	60.4	57.8	55.2	56.9	..	..	..	Earnings as % of Value Added

(Billions of current Norwegian Kroner)

MONETARY HOLDINGS

123.04	139.67	154.73	175.26	194.60	215.63	256.99	295.47	301.88	359.74	..	Money Supply, Broadly Defined
40.19	43.25	45.56	52.38	58.81	65.92	82.00	98.65	101.76	152.63	..	Money as Means of Payment
17.06	17.73	18.82	20.16	20.93	21.75	22.78	25.05	26.58	28.16	..	Currency Ouside Banks
23.13	25.52	26.74	32.22	37.88	44.17	59.22	73.60	75.18	124.47	..	Demand Deposits
82.85	96.42	109.17	122.88	135.79	149.71	174.99	196.82	200.12	207.11	..	Quasi-Monetary Liabilities

(Billions of current Norwegian Kroner)

-14.47	-14.99	-5.40	6.82	3.89	9.44	9.02	19.04	..	..	..	GOVERNMENT DEFICIT (-) OR SURPLUS
82.69	92.88	117.80	141.24	154.68	177.40	194.94	224.09	246.15	..	..	Current Revenue
81.36	90.22	104.40	119.72	135.15	150.00	163.32	178.85	199.91	..	..	Current Expenditure
1.33	2.67	13.40	21.52	19.53	27.40	31.62	45.24	46.24	..	..	Current Budget Balance
..	..	0.05	0.04	0.10	0.15	0.25	0.39	0.38	..	..	Capital Receipts
15.80	17.66	18.85	14.74	15.74	18.11	22.85	26.59	-199.91	..	..	Capital Payments

NORWAY	1967	1968	1969	1970	1971	1972	1973	1974	1975	1976	1977
FOREIGN TRADE (CUSTOMS BASIS)				*(Millions of current US dollars)*							
Value of Exports, fob	1,736	1,937	2,203	2,457	2,563	3,279	4,680	6,274	7,207	7,917	8,716
Nonfuel Primary Products	764	827	941	1,041	1,061	1,230	1,709	2,061	1,827	2,231	2,424
Fuels	31	41	37	54	58	105	139	401	936	1,501	1,638
Manufactures	942	1,068	1,225	1,361	1,444	1,944	2,831	3,812	4,444	4,185	4,654
Value of Imports, cif	2,746	2,704	2,943	3,702	4,083	4,369	6,219	8,414	9,705	11,105	12,870
Nonfuel Primary Products	544	587	613	791	780	827	1,109	1,550	1,528	1,583	1,647
Fuels	197	213	217	285	308	326	479	1,034	955	1,241	1,424
Manufactures	2,005	1,905	2,113	2,627	2,996	3,216	4,631	5,830	7,222	8,281	9,800
				(Index 1980 = 100)							
Terms of Trade	118.3	121.9	122.0	126.0	114.0	107.4	112.2	100.4	97.4	94.3	94.0
Export Prices, fob	25.3	25.3	26.9	28.4	29.2	30.4	40.0	54.4	56.7	56.2	61.4
Nonfuel Primary Products	26.5	27.9	31.8	33.9	32.8	36.1	51.6	57.3	50.1	58.0	64.8
Fuels	4.3	4.3	4.3	4.3	5.6	6.2	8.9	36.7	35.7	38.4	42.0
Manufactures	28.8	28.8	28.0	31.5	32.2	34.2	41.6	55.7	69.0	66.2	71.0
Import Prices, cif	21.4	20.8	22.0	22.5	25.6	28.3	35.6	54.2	58.2	59.6	65.3
Trade at Constant 1980 Prices				*(Millions of 1980 US dollars)*							
Exports, fob	6,868	7,647	8,200	8,666	8,768	10,779	11,699	11,527	12,709	14,087	14,194
Imports, cif	12,848	13,022	13,368	16,447	15,921	15,432	17,448	15,521	16,665	18,641	19,698
BALANCE OF PAYMENTS				*(Millions of current US dollars)*							
Exports of Goods & Services	..	..	..	4,790	5,203	6,177	8,549	11,156	12,143	13,115	14,557
Merchandise, fob	..	..	..	2,480	2,587	3,307	4,722	6,329	7,287	8,047	9,146
Nonfactor Services	..	..	..	2,166	2,480	2,743	3,624	4,504	4,513	4,745	5,060
Factor Services	..	..	..	144	136	127	203	322	343	323	351
Imports of Goods & Services	..	..	..	5,021	5,727	6,216	8,866	12,172	14,439	16,665	19,299
Merchandise, fob	..	..	..	3,633	4,044	4,322	6,250	8,664	10,132	11,608	13,202
Nonfactor Services	..	..	..	1,138	1,364	1,508	2,030	2,540	3,597	4,174	4,857
Factor Services	..	..	..	250	319	385	585	968	710	882	1,239
Long-Term Interest	..	..	..	..	..	..	..	..	..	..	..
Current Transfers, net	..	..	..	31	..	39	31	20	12	1	-18
Workers' Remittances	..	..	..	0	0	0	0	0	0	0	4
Total to be Financed	..	..	..	-200	-482	0	-286	-996	-2,284	-3,548	-4,760
Official Capital Grants	..	..	..	-42	-41	-59	-79	-121	-171	-199	-274
Current Account Balance	..	..	..	-242	-524	-59	-365	-1,117	-2,455	-3,747	-5,035
Long-Term Capital, net	..	..	..	96	313	227	771	870	2,408	2,297	3,906
Direct Investment	..	..	..	32	62	121	159	198	44	180	640
Long-Term Loans	..	..	..	..	..	..	..	..	..	..	..
Disbursements	..	..	..	..	..	..	..	..	..	..	..
Repayments	..	..	..	..	..	..	..	..	..	..	..
Other Long-Term Capital	..	..	..	106	292	165	691	793	2,535	2,315	3,540
Other Capital, net	..	..	..	217	515	-29	-163	568	388	1,422	1,007
Change in Reserves	..	..	..	-71	-304	-139	-244	-321	-340	28	121
Memo Item:				*(Norwegian Kroner per US dollar)*							
Conversion Factor (Annual Avg)	7.140	7.140	7.140	7.140	7.040	6.590	5.770	5.540	5.230	5.460	5.320
EXTERNAL DEBT, ETC.				*(Millions US dollars, outstanding at end of year)*							
Public/Publicly Guar. Long-Term	..	..	..	..	..	..	..	..	..	..	..
Official Creditors	..	..	..	..	..	..	..	..	..	..	..
IBRD and IDA	..	..	..	..	..	..	..	..	..	..	..
Private Creditors	..	..	..	..	..	..	..	..	..	..	..
Private Non-guaranteed Long-Term	..	..	..	..	..	..	..	..	..	..	..
Use of Fund Credit	..	..	..	..	..	..	..	..	..	..	..
Short-Term Debt	..	..	..	..	..	..	..	..	..	..	..
Memo Items:				*(Millions of US dollars)*							
Int'l Reserves Excluding Gold	659	679	685	788	1,118	1,288	1,533	1,887	2,196	2,189	2,196
Gold Holdings (at market price)	18	28	25	25	41	63	110	183	137	132	178
SOCIAL INDICATORS											
Total Fertility Rate	2.8	2.7	2.7	2.5	2.5	2.4	2.2	2.1	2.0	1.9	1.8
Crude Birth Rate	17.6	17.6	17.6	16.6	16.8	16.3	15.5	14.9	14.0	13.2	12.5
Infant Mortality Rate	14.8	13.7	13.8	12.7	12.8	11.8	11.9	10.4	11.0	10.4	9.1
Life Expectancy at Birth	73.8	74.0	74.1	74.2	74.3	74.4	74.6	74.8	75.0	75.1	75.3
Food Production, p.c. ('79-81 = 100)	89.5	93.5	86.6	85.1	87.1	86.0	85.9	97.5	88.9	87.5	94.8
Labor Force, Agriculture (%)	14.0	13.3	12.5	11.8	11.4	11.1	10.7	10.4	10.0	9.7	9.3
Labor Force, Female (%)	27.5	28.1	28.7	29.3	30.4	31.6	32.7	33.9	35.0	36.0	37.0
School Enroll. Ratio, primary	..	..	..	89.0	..	..	..	..	101.0	101.0	100.0
School Enroll. Ratio, secondary	..	..	..	83.0	..	..	..	..	88.0	89.0	90.0

1978	1979	1980	1981	1982	1983	1984	1985	1986	1987 estimate	NOTES	NORWAY
					(Millions of current US dollars)						**FOREIGN TRADE (CUSTOMS BASIS)**
10,027	13,466	18,481	17,968	17,583	17,972	18,914	18,662	18,230	21,449	..	Value of Exports, fob
2,623	3,175	3,615	3,401	2,877	3,180	3,364	3,032	3,590	4,585	..	Nonfuel Primary Products
2,058	4,863	8,935	9,034	9,135	9,481	10,322	10,012	7,815	8,692	..	Fuels
5,346	5,428	5,931	5,533	5,571	5,311	5,228	5,618	6,825	8,171	..	Manufactures
11,435	13,732	16,952	15,638	15,471	13,494	13,885	14,519	20,298	22,578	..	Value of Imports, cif
1,667	2,094	2,610	2,255	1,930	1,769	1,918	1,886	2,472	2,721	..	Nonfuel Primary Products
1,353	2,071	2,957	2,253	2,030	1,402	1,421	1,265	1,192	1,193	..	Fuels
8,416	9,568	11,384	11,130	11,511	10,323	10,546	11,368	16,634	18,664	..	Manufactures
					(Index 1980 = 100)						
89.5	88.4	100.0	103.6	102.6	97.9	99.0	96.9	69.3	68.1	..	Terms of Trade
64.2	75.1	100.0	100.5	94.2	87.2	86.0	83.0	67.3	73.3	..	Export Prices, fob
68.3	86.8	100.0	88.7	77.3	83.1	78.6	70.8	72.9	83.0	..	Nonfuel Primary Products
42.3	61.0	100.0	112.5	109.0	96.0	93.0	91.0	54.0	54.7	..	Fuels
77.2	86.2	100.0	92.0	85.0	77.0	79.0	78.0	89.0	104.0	..	Manufactures
71.7	84.9	100.0	97.0	91.9	89.1	86.8	85.6	97.2	107.6	..	Import Prices, cif
					(Millions of 1980 US dollars)						Trade at Constant 1980 Prices
15,628	17,930	18,481	17,881	18,659	20,600	21,995	22,488	27,068	29,272	..	Exports, fob
15,955	16,170	16,952	16,120	16,843	15,149	15,993	16,961	20,885	20,982	..	Imports, cif
					(Millions of current US dollars)						**BALANCE OF PAYMENTS**
17,010	21,386	28,263	28,693	27,209	26,553	27,829	29,545	28,924	32,702	..	Exports of Goods & Services
11,032	13,739	18,661	18,484	17,653	18,064	19,112	20,000	18,125	21,142	..	Merchandise, fob
5,511	6,987	8,524	8,686	7,896	7,065	7,068	7,420	8,068	8,441	..	Nonfactor Services
468	659	1,078	1,523	1,660	1,425	1,649	2,126	2,731	3,120	..	Factor Services
18,756	21,992	26,665	26,038	25,999	23,972	24,398	25,899	32,561	35,841	..	Imports of Goods & Services
11,550	13,594	16,759	15,456	15,279	13,700	13,950	15,262	20,203	21,986	..	Merchandise, fob
5,369	5,897	6,901	7,208	7,064	7,065	7,177	7,407	8,382	9,406	..	Nonfactor Services
1,837	2,502	3,005	3,374	3,655	3,207	3,271	3,230	3,975	4,449	..	Factor Services
..	..	..	..	..	..	..	..	..	..	..	Long-Term Interest
-31	-53	-55	-29	-57	-55	-36	-65	-130	-198	..	Current Transfers, net
6	6	13	9	11	9	10	11	11	11	..	Workers' Remittances
-1,777	-660	1,543	2,627	1,154	2,526	3,396	3,580	-3,767	-3,337	K	Total to be Financed
-336	-383	-442	-456	-503	-525	-471	-494	-661	-774	K	Official Capital Grants
-2,113	-1,043	1,101	2,170	650	2,001	2,925	3,086	-4,428	-4,111	..	Current Account Balance
2,874	1,882	-1,372	-1,354	-185	-2,103	-1,128	-1,452	2,178	731	..	Long-Term Capital, net
421	356	-194	503	126	-23	-750	-1,048	-377	-846	..	Direct Investment
..	..	..	..	..	..	..	..	..	..	..	Long-Term Loans
..	..	..	..	..	..	..	..	..	..	..	Disbursements
..	..	..	..	..	..	..	..	..	..	..	Repayments
2,788	1,908	-736	-1,401	193	-1,555	93	91	3,216	2,351	..	Other Long-Term Capital
-124	446	2,153	-386	243	3	1,268	1,818	1,005	6,407	..	Other Capital, net
-637	-1,285	-1,882	-431	-708	98	-3,065	-3,452	1,245	-3,027	..	Change in Reserves
					(Norwegian Kroner per US dollar)						Memo Item:
5.240	5.060	4.940	5.740	6.450	7.300	8.160	8.600	7.390	6.740	..	Conversion Factor (Annual Avg)
					(Millions US dollars, outstanding at end of year)						**EXTERNAL DEBT, ETC.**
..	..	..	..	..	..	..	..	..	..	..	Public/Publicly Guar. Long-Term
..	..	..	..	..	..	..	..	..	..	..	Official Creditors
..	..	..	..	..	..	..	..	..	..	..	IBRD and IDA
..	..	..	..	..	..	..	..	..	..	..	Private Creditors
..	..	..	..	..	..	..	..	..	..	..	Private Non-guaranteed Long-Term
..	..	..	..	..	..	..	..	..	..	..	Use of Fund Credit
..	..	..	..	..	..	..	..	..	..	..	Short-Term Debt
					(Millions of US dollars)						Memo Items:
2,861	4,215	6,048	6,253	6,874	6,629	9,365	13,917	12,525	14,277	..	Int'l Reserves Excluding Gold
256	606	698	471	541	452	365	387	463	573	..	Gold Holdings (at market price)
											SOCIAL INDICATORS
1.8	1.8	1.7	1.7	1.7	1.7	1.7	1.7	1.7	1.7	..	Total Fertility Rate
12.7	12.6	12.4	12.3	12.3	12.3	12.3	12.3	12.3	12.3	..	Crude Birth Rate
8.6	8.7	8.0	7.5	7.7	7.9	8.1	8.3	8.6	8.8	..	Infant Mortality Rate
75.6	75.9	76.2	76.5	76.8	76.9	76.9	77.0	77.0	77.0	..	Life Expectancy at Birth
100.2	96.9	100.2	102.9	107.7	102.2	109.6	105.6	109.3	109.9	..	Food Production, p.c. ('79-81=100)
9.0	8.6	8.3	..	..	..	..	..	..	..	..	Labor Force, Agriculture (%)
38.1	39.1	40.1	40.2	40.3	40.3	40.4	40.5	40.6	40.8	..	Labor Force, Female (%)
100.0	100.0	100.0	..	98.0	98.0	97.0	97.0	98.0	..	..	School Enroll. Ratio, primary
92.0	94.0	94.0	..	97.0	96.0	97.0	97.0	..	..	..	School Enroll. Ratio, secondary

OMAN	1967	1968	1969	1970	1971	1972	1973	1974	1975	1976	1977
CURRENT GNP PER CAPITA (US $)	170	250	310	360	370	390	390	550	1,280	2,850	2,870
POPULATION (thousands)	603	619	636	654	672	691	711	736	766	801	841
ORIGIN AND USE OF RESOURCES					*(Millions of current Omani Rials)*						
Gross National Product (GNP)	35.2	59.5	76.7	88.6	105.1	111.8	106.7	416.5	589.3	736.4	816.6
Net Factor Income from Abroad	-3.6	-19.2	-23.3	-18.2	-20.0	-29.0	-62.7	-152.0	-134.9	-147.9	-130.2
Gross Domestic Product (GDP)	38.8	78.7	100.0	106.8	125.1	140.8	169.4	568.5	724.2	884.3	946.8
Indirect Taxes, net	0.8	1.0	1.2	1.1	1.1	1.6	1.7	2.3	0.5	4.5	4.6
GDP at factor cost	38.0	77.7	98.8	105.7	124.0	139.2	167.7	566.2	723.7	879.8	942.2
Agriculture	14.3	15.1	16.1	16.6	16.8	17.0	16.7	18.4	20.2	18.3	24.1
Industry	20.4	58.1	77.2	82.5	94.8	100.0	120.0	469.1	561.5	616.7	626.7
Manufacturing	0.1	0.1	0.1	0.2	0.2	0.3	0.6	2.0	2.1	4.3	6.7
Services, etc.	4.1	5.5	6.7	7.7	13.5	23.8	32.7	81.0	142.5	249.3	296.0
Resource Balance	4.4	43.6	62.7	57.7	42.1	22.0	21.4	147.6	122.1	144.7	135.0
Exports of Goods & NFServices	13.1	55.2	74.6	78.7	82.3	83.6	114.9	419.1	489.2	551.2	552.0
Imports of Goods & NFServices	8.7	11.6	11.9	21.0	40.2	61.6	93.5	271.5	367.1	406.5	417.0
Domestic Absorption	34.4	35.1	37.3	49.1	83.0	118.8	148.0	420.9	602.1	739.6	811.8
Private Consumption, etc.	20.6	21.4	22.9	20.7	21.4	35.1	40.6	49.6	115.1	181.6	253.4
General Gov't Consumption	1.7	3.8	5.1	13.7	26.0	41.7	63.0	197.2	229.0	240.9	268.5
Gross Domestic Investment	12.1	9.9	9.3	14.7	35.6	42.0	44.4	174.1	258.0	317.1	289.9
Fixed Investment	12.1	9.9	9.3	14.7	35.6	42.0	44.4	174.1	258.0	321.2	310.8
Memo Items:											
Gross Domestic Saving	16.5	53.5	72.0	72.4	77.7	64.0	65.8	321.7	380.1	461.8	424.9
Gross National Saving	12.9	34.3	48.7	..	..	..	..	131.4	173.3	238.0	218.1
					(Millions of 1980 Omani Rials)						
Gross National Product	317.6	478.9	726.0	925.3	1,141.2	1,165.4	1,202.0	1,044.0	1,025.6	1,474.9	1,820.7
GDP at factor cost											
Agriculture	..	..	..	..	..	..	..	..	..	..	..
Industry	..	..	..	..	..	..	..	..	..	..	..
Manufacturing	..	..	..	..	..	..	..	..	..	..	..
Services, etc.	..	..	..	..	..	..	..	..	..	..	..
Resource Balance	..	..	..	..	..	..	..	..	..	..	..
Exports of Goods & NFServices	..	..	..	..	..	..	..	..	..	..	..
Imports of Goods & NFServices	..	..	..	..	..	..	..	..	..	..	..
Domestic Absorption	..	..	..	..	..	..	..	..	..	..	..
Private Consumption, etc.	..	..	..	..	..	..	..	..	..	..	..
General Gov't Consumption	..	..	..	..	..	..	..	..	..	..	..
Gross Domestic Investment	..	..	..	..	..	..	..	..	..	..	..
Fixed Investment	..	..	..	..	..	..	..	..	..	..	..
Memo Items:											
Capacity to Import	..	..	..	..	..	..	..	..	..	..	..
Terms of Trade Adjustment	..	..	..	..	..	..	..	..	..	..	..
Gross Domestic Income	..	..	..	..	..	..	..	..	..	..	..
Gross National Income	..	..	..	..	..	..	..	..	..	..	..
DOMESTIC PRICES (DEFLATORS)					*(Index 1980 = 100)*						
Overall (GDP)	13.7	16.7	11.7	9.9	10.2	11.4	12.5	48.8	55.7	54.7	48.6
Domestic Absorption	..	..	..	..	..	..	..	..	..	..	..
Agriculture	..	..	..	..	..	..	..	..	..	..	..
Industry	..	..	..	..	..	..	..	..	..	..	..
Manufacturing	..	..	..	..	..	..	..	..	..	..	..
MANUFACTURING ACTIVITY											
Employment (1980=100)	..	..	..	..	..	..	..	..	..	..	..
Real Earnings per Empl. (1980=100)	..	..	..	..	..	..	..	..	..	..	..
Real Output per Empl. (1980=100)	..	..	..	..	..	..	..	..	..	..	..
Earnings as % of Value Added	..	..	..	..	..	..	..	..	..	..	..
MONETARY HOLDINGS					*(Millions of current Omani Rials)*						
Money Supply, Broadly Defined	..	..	..	..	..	44.8	46.6	85.2	118.0	164.6	206.6
Money as Means of Payment	..	..	..	..	..	19.0	24.9	48.4	71.6	102.1	111.2
Currency Ouside Banks	..	..	..	..	..	12.3	15.2	28.9	38.5	47.8	55.1
Demand Deposits	..	..	..	..	..	6.7	9.8	19.5	33.1	54.4	56.2
Quasi-Monetary Liabilities	..	..	..	..	..	25.8	21.7	36.8	46.3	62.5	95.3
GOVERNMENT DEFICIT (-) OR SURPLUS					*(Millions of current Omani Rials)*						
	..	..	..	..	..	-17.1	-23.5	-56.4	-37.7	-84.7	58.1
Current Revenue	..	..	..	..	..	53.0	68.5	311.5	429.5	473.5	575.0
Current Expenditure	..	..	..	..	..	..	62.2	186.5	309.9	369.9	367.7
Current Budget Balance	..	..	..	..	..	..	6.3	125.0	119.6	103.6	207.3
Capital Receipts	..	..	..	..	..	..	..	..	0.8	1.8	1.1
Capital Payments	..	..	..	..	..	..	29.8	181.4	158.1	190.1	150.3

1978	1979	1980	1981	1982	1983	1984	1985	1986	1987 estimate	NOTES	OMAN
2,800	3,150	3,140	6,120	7,560	7,090	7,230	7,510	6,510	5,830	..	**CURRENT GNP PER CAPITA (US $)**
885	933	984	1,037	1,091	1,144	1,195	1,242	1,294	1,345	..	**POPULATION (thousands)**
				(Millions of current Omani Rials)							**ORIGIN AND USE OF RESOURCES**
835.9	1,152.2	1,843.6	2,269.4	2,382.7	2,429.9	2,696.7	3,056.9	2,459.7	2,804.5	..	Gross National Product (GNP)
-111.0	-137.3	-222.6	-237.5	-226.7	-310.0	-350.2	-399.0	-338.0	-328.1	..	Net Factor Income from Abroad
946.9	1,289.5	2,066.2	2,506.9	2,609.4	2,739.9	3,046.9	3,455.9	2,797.7	3,132.6	..	Gross Domestic Product (GDP)
4.6	8.1	12.0	..	..	..	..	..	..	..	..	Indirect Taxes, net
942.3	1,281.4	2,054.2							..	B	GDP at factor cost
30.7	40.3	52.6	62.1	66.1	80.5	89.0	95.8	89.3	..	..	Agriculture
584.8	828.1	1,433.1	1,683.1	1,651.7	1,663.1	1,800.5	2,049.0	1,206.5	..	..	Industry
8.5	9.6	13.4	24.0	35.9	67.1	92.3	115.9	178.3	..	..	Manufacturing
331.4	421.1	580.5	761.7	891.6	996.3	1,157.4	1,311.1	1,501.9	..	..	Services, etc.
90.8	262.4	516.6	666.9	401.7	427.0	387.0	437.0	-396.0	..	..	Resource Balance
552.0	787.4	1,294.6	1,621.9	1,527.7	1,470.0	1,532.0	1,722.0	1,060.2	..	..	Exports of Goods & NFServices
461.2	525.0	778.0	955.0	1,126.0	1,043.0	1,145.0	1,285.0	1,456.2	..	..	Imports of Goods & NFServices
855.7	1,027.1	1,549.6	1,840.0	2,207.7	2,312.9	2,659.7	3,018.9	3,193.7	..	..	Domestic Absorption
309.9	337.0	572.3	647.7	832.5	859.5	938.5	1,171.7	1,568.9	..	..	Private Consumption, etc.
272.3	354.7	516.7	616.7	681.2	727.8	808.0	894.1	853.4	..	..	General Gov't Consumption
273.5	335.4	460.6	575.6	694.0	725.6	913.2	953.1	771.4	..	..	Gross Domestic Investment
281.0	318.0	464.3	569.9	709.2					..	..	Fixed Investment
											Memo Items:
364.7	597.8	977.2	1,242.5	1,095.7	1,152.6	1,300.4	1,390.1	375.4	..	..	Gross Domestic Saving
180.4	374.6	629.4	846.5	677.0	602.6	667.2	678.0	-288.9	..	..	Gross National Saving
				(Millions of 1980 Omani Rials)							
1,903.1	1,918.4	1,843.6	2,299.1	2,552.1	3,002.8	3,480.1	4,007.2	4,283.0	4,381.6	..	Gross National Product
..	..	..	..	..	..	..	..	..	..	..	GDP at factor cost
33.0	43.6	52.6	53.4	58.1	68.9	75.6	89.2	79.1	..	..	Agriculture
1,498.0	1,436.2	1,433.1	1,660.5	1,805.2	2,185.2	2,506.7	2,938.0	3,258.0	..	..	Industry
9.1	11.3	13.4	22.1	33.4	51.2	72.4	89.4	81.8	..	..	Manufacturing
438.8	511.2	580.5	698.3	804.9	897.3	1,076.1	1,181.4	1,089.4	..	..	Services, etc.
687.7	675.6	516.6	508.9	341.7	523.3	489.0	678.6	947.4	..	..	Resource Balance
1,319.7	1,291.7	1,294.6	1,452.9	1,504.3	1,606.1	1,701.7	1,972.6	2,180.6	..	..	Exports of Goods & NFServices
632.0	616.2	778.0	944.0	1,162.5	1,082.8	1,212.7	1,294.0	1,233.3	..	..	Imports of Goods & NFServices
1,282.1	1,315.4	1,549.6	1,903.2	2,326.4	2,628.1	3,169.4	3,529.9	3,479.1	..	..	Domestic Absorption
..	..	..	..	..	..	..	..	..	..	..	Private Consumption, etc.
..	..	..	..	..	..	..	..	..	..	..	General Gov't Consumption
370.5	442.4	460.6	577.2	781.4	852.4	1,283.3	1,378.1	1,056.9	..	..	Gross Domestic Investment
..	..	..	..	..	..	..	..	..	..	..	Fixed Investment
											Memo Items:
756.5	924.1	1,294.6	1,603.2	1,577.3	1,526.1	1,622.6	1,734.0	897.9	..	..	Capacity to Import
-563.2	-367.6	0.0	150.3	73.0	-80.0	-79.2	-238.6	-1,282.7	..	..	Terms of Trade Adjustment
1,406.5	1,623.4	2,066.2	2,562.4	2,741.2	3,071.4	3,579.3	3,970.0	3,143.7	..	..	Gross Domestic Income
1,339.9	1,550.8	1,843.6	2,449.4	2,625.1	2,922.7	3,401.0	3,768.6	3,000.3	..	..	Gross National Income
				(Index 1980 = 100)							**DOMESTIC PRICES (DEFLATORS)**
48.1	64.8	100.0	103.9	97.8	86.9	83.3	82.1	63.2	69.2	..	Overall (GDP)
66.7	78.1	100.0	96.7	94.9	88.0	83.9	85.5	91.8	..	..	Domestic Absorption
93.2	92.5	100.0	116.4	113.8	116.8	117.8	107.4	112.9	..	..	Agriculture
39.0	57.7	100.0	101.4	91.5	76.1	71.8	69.7	37.0	..	..	Industry
93.3	85.3	100.0	108.7	107.3	130.9	127.6	129.6	218.0	..	..	Manufacturing
											MANUFACTURING ACTIVITY
..	..	..	..	..	..	..	..	..	..	..	Employment (1980=100)
..	..	..	..	..	..	..	..	..	..	..	Real Earnings per Empl. (1980=100)
..	..	..	..	..	..	..	..	..	..	..	Real Output per Empl. (1980=100)
61.4	61.4	61.4	61.4	61.4	61.4	61.4	61.4	..	..	..	Earnings as % of Value Added
				(Millions of current Omani Rials)							**MONETARY HOLDINGS**
230.6	246.2	325.0	450.7	562.6	672.3	774.0	894.4	829.0	882.0	..	Money Supply, Broadly Defined
114.3	123.1	154.7	212.7	240.5	266.8	283.9	321.5	303.9	324.5	..	Money as Means of Payment
64.4	74.3	94.8	116.2	129.8	140.4	150.0	178.5	168.8	180.4	..	Currency Ouside Banks
49.9	48.9	59.9	96.5	110.7	126.4	134.0	143.0	135.1	144.2	..	Demand Deposits
116.3	123.1	170.3	238.0	322.1	405.5	490.1	572.9	525.0	557.5	..	Quasi-Monetary Liabilities
				(Millions of current Omani Rials)							
-75.3	85.0	9.1	48.5	-222.8	-242.3	-346.8	-364.2	-700.2	..	..	**GOVERNMENT DEFICIT (-) OR SURPLUS**
446.3	650.7	825.2	1,123.4	999.0	1,122.2	1,212.1	1,393.2	888.2	..	..	Current Revenue
410.8	420.5	626.3	787.0	891.2	1,017.4	1,127.1	1,293.4	1,218.1	..	..	Current Expenditure
35.5	230.2	198.9	336.4	107.8	104.8	85.0	99.8	-329.9	..	..	Current Budget Balance
1.3	1.8	3.9	2.5	2.2	2.2	2.3	4.4	6.8	..	..	Capital Receipts
112.1	147.0	193.7	290.4	332.8	349.3	434.1	468.4	377.1	..	..	Capital Payments

OMAN	1967	1968	1969	1970	1971	1972	1973	1974	1975	1976	1977
FOREIGN TRADE (CUSTOMS BASIS)					*(Millions of current US dollars)*						
Value of Exports, fob	35.6	144.0	194.4	206.3	221.1	241.5	344.0	1,211.3	1,436.5	1,566.0	1,573.0
Nonfuel Primary Products	1.6	1.4	1.6	0.9	1.0	1.0	1.7	1.2	3.1	0.0	0.0
Fuels	33.7	142.6	192.7	205.3	220.1	240.5	342.2	1,210.1	1,433.3	1,562.0	1,568.6
Manufactures	0.3	0.1	0.1	0.1	0.0	0.0	0.0	0.0	0.0	4.0	4.3
Value of Imports, cif	9.9	9.6	13.4	18.2	33.1	48.8	116.8	392.5	670.5	667.3	874.5
Nonfuel Primary Products	2.5	1.8	2.5	4.3	12.8	16.1	32.7	57.9	105.0	110.4	152.1
Fuels	1.0	0.9	0.5	0.8	2.4	2.5	5.1	14.8	31.4	48.2	62.3
Manufactures	6.4	6.9	10.4	13.1	17.9	30.2	79.0	319.8	534.2	508.7	660.2
					(Index 1980 = 100)						
Terms of Trade	..	..	..	..	..	..	..	..	..	..	..
Export Prices, fob	..	..	..	..	..	..	..	..	..	..	..
Nonfuel Primary Products	..	..	..	..	..	..	..	..	..	..	..
Fuels	..	..	..	..	..	..	..	..	..	..	..
Manufactures	..	..	..	..	..	..	..	..	..	..	..
Import Prices, cif	..	..	..	..	..	..	..	..	..	..	..
Trade at Constant 1980 Prices					*(Millions of 1980 US dollars)*						
Exports, fob	..	..	..	..	..	..	..	..	..	..	..
Imports, cif	..	..	..	..	..	..	..	..	..	..	..
BALANCE OF PAYMENTS					*(Millions of current US dollars)*						
Exports of Goods & Services	..	..	..	..	..	..	..	1,231	1,433	1,601	1,620
Merchandise, fob	..	..	..	..	..	..	..	1,212	1,416	1,596	1,620
Nonfactor Services	..	..	..	..	..	..	..	0	0	0	0
Factor Services	..	..	..	..	..	..	..	19	17	6	0
Imports of Goods & Services	..	..	..	..	..	..	..	965	1,375	1,397	1,389
Merchandise, fob	..	..	..	..	..	..	..	552	908	1,000	1,044
Nonfactor Services	..	..	..	..	..	..	..	104	155	177	186
Factor Services	..	..	..	..	..	..	..	310	312	220	160
Long-Term Interest	..	..	..	..	..	..	0	4	7	10	25
Current Transfers, net	..	..	..	..	..	..	..	-111	-208	-220	-222
Workers' Remittances	..	..	..	..	..	..	..	0	0	0	0
Total to be Financed	..	..	..	..	..	..	..	155	-150	-16	9
Official Capital Grants	..	..	..	..	..	..	..	24	207	52	268
Current Account Balance	..	..	..	..	..	..	..	179	57	37	277
Long-Term Capital, net	..	..	..	..	..	..	..	41	453	299	401
Direct Investment	..	..	..	..	..	..	..	-61	106	81	48
Long-Term Loans	..	..	..	..	..	..	44	124	121	76	99
Disbursements	..	..	..	..	..	..	44	130	142	106	161
Repayments	..	..	..	..	..	..	..	6	21	30	62
Other Long-Term Capital	..	..	..	..	..	..	..	-47	18	89	-14
Other Capital, net	..	..	..	..	..	..	..	-169	-444	-278	-606
Change in Reserves	..	..	..	0	-14	-14	-5	-51	-66	-57	-72
Memo Item:					*(Omani Rials per US dollar)*						
Conversion Factor (Annual Avg)	0.360	0.420	0.420	0.420	0.420	0.380	0.350	0.350	0.350	0.350	0.350
EXTERNAL DEBT, ETC.					*(Millions of US dollars, outstanding at end of year)*						
Public/Publicly Guar. Long-Term	..	..	..	..	..	..	42	176	287	349	473
Official Creditors	..	..	..	..	..	..	..	70	84	114	207
IBRD and IDA	..	..	..	..	..	..	0	0	0	1	2
Private Creditors	..	..	..	..	..	..	42	106	204	235	266
Private Non-guaranteed Long-Term	..	..	..	..	..	..	..	..	..	..	..
Use of Fund Credit	..	..	..	0	0	0	0	0	0	0	0
Short-Term Debt	..	..	..	..	..	..	..	..	..	..	181
Memo Items:					*(Millions of US dollars)*						
Int'l Reserves Excluding Gold	..	..	..	10.3	24.4	36.4	47.1	92.9	161.3	219.6	289.6
Gold Holdings (at market price)	..	..	..	2.4	0.6	0.9	1.6	5.6	4.2	6.2	16.7
SOCIAL INDICATORS											
Total Fertility Rate	7.2	7.2	7.2	7.2	7.2	7.2	7.2	7.2	7.2	7.2	7.2
Crude Birth Rate	50.0	49.9	49.8	49.8	49.7	49.6	49.5	49.3	49.2	49.0	48.9
Infant Mortality Rate	186.0	180.8	175.6	170.4	165.2	160.0	155.0	150.0	145.0	140.0	135.0
Life Expectancy at Birth	43.8	44.3	44.8	45.4	45.9	46.4	47.0	47.6	48.2	48.8	49.4
Food Production, p.c. ('79-81 = 100)	..	..	..	..	..	..	..	..	..	..	..
Labor Force, Agriculture (%)	59.9	58.9	57.8	56.7	56.0	55.3	54.5	53.8	53.1	52.4	51.8
Labor Force, Female (%)	5.8	5.9	6.0	6.1	6.4	6.6	6.8	7.0	7.2	7.2	7.2
School Enroll. Ratio, primary	..	..	..	3.0	..	..	..	..	44.0	49.0	53.0
School Enroll. Ratio, secondary	..	..	..	..	..	..	..	..	1.0	2.0	5.0

1978	1979	1980	1981	1982	1983	1984	1985	1986	1987 estimate	NOTES	OMAN
				(Millions of current US dollars)							**FOREIGN TRADE (CUSTOMS BASIS)**
1,598.0	2,279.8	3,747.7	4,695.7	4,421.0	4,248.0	4,413.0	4,970.0	2,889.0	..	..	Value of Exports, fob
9.2	44.1	32.0	46.5	34.3	335.1	346.0	336.2	294.0	..	..	Nonfuel Primary Products
1,503.0	2,159.3	3,603.9	4,420.0	4,099.0	3,899.8	4,056.0	4,624.7	2,570.0	..	f	Fuels
85.8	76.4	111.8	229.3	287.7	13.0	11.0	9.0	25.0	..	..	Manufactures
947.4	1,246.4	1,732.0	2,288.2	2,682.5	2,492.3	2,746.0	3,152.6	2,375.0	..	..	Value of Imports, cif
179.6	231.0	294.3	345.9	381.4	395.0	452.0	513.0	426.0	..	..	Nonfuel Primary Products
79.9	84.1	187.1	299.4	279.5	40.3	41.0	42.0	34.0	..	..	Fuels
687.9	931.4	1,250.6	1,642.9	2,021.6	2,057.0	2,253.0	2,597.6	1,915.0	..	..	Manufactures
				(Index 1980 = 100)							
..	..	..	..	..	..	..	..	..	..	..	Terms of Trade
..	..	..	..	..	..	..	..	..	..	..	Export Prices, fob
..	..	..	..	..	..	..	..	..	..	..	Nonfuel Primary Products
..	..	..	..	..	..	..	..	..	..	..	Fuels
..	..	..	..	..	..	..	..	..	..	..	Manufactures
..	..	..	..	..	..	..	..	..	..	..	Import Prices, cif
				(Millions of 1980 US dollars)							Trade at Constant 1980 Prices
..	..	3,715.7	4,161.6	4,336.4	4,232.2	4,511.0	5,293.1	5,741.7	..	..	Exports, fob
..	..	1,437.7	1,900.9	2,314.9	2,173.0	2,419.6	2,758.0	1,765.0	..	..	Imports, cif
				(Millions of current US dollars)							**BALANCE OF PAYMENTS**
1,611	2,287	3,852	4,881	4,753	4,577	4,780	5,348	3,485	..	..	Exports of Goods & Services
1,598	2,280	3,748	4,696	4,423	4,256	4,421	4,972	2,880	..	..	Merchandise, fob
8	8	9	9	12	15	15	15	13	..	..	Nonfactor Services
5	0	96	177	319	307	345	362	593	..	..	Factor Services
1,468	1,669	2,650	3,207	3,737	3,525	3,854	4,332	3,603	..	..	Imports of Goods & Services
1,157	1,285	1,780	2,171	2,583	2,361	2,640	3,028	2,275	..	..	Merchandise, fob
193	235	518	620	701	688	673	690	699	..	..	Nonfactor Services
118	149	352	415	454	476	541	614	629	..	..	Factor Services
20	23	40	35	34	53	85	113	169	177	..	Long-Term Interest
-212	-249	..	..	..	..	..	..	-849	..	..	Current Transfers, net
29	33	35	41	43	43	43	43	39	..	..	Workers' Remittances
-70	370	..	..	..	..	..	..	-966	..	..	Total to be Financed
20	179	..	..	..	..	..	..	0	..	..	Official Capital Grants
-50	549	942	1,360	504	506	319	84	-966	..	..	Current Account Balance
160	182	141	338	295	670	781	282	684	..	..	Long-Term Capital, net
86	118	98	64	180	153	156	159	138	..	..	Direct Investment
-55	31	-59	115	203	428	254	541	538	-94	..	Long-Term Loans
66	216	98	199	287	511	383	688	763	342	..	Disbursements
121	186	157	84	84	83	129	146	225	436	..	Repayments
109	-145	102	160	-87	89	370	-419	7	..	..	Other Long-Term Capital
-146	-567	-293	-462	-53	-824	-781	-242	-346	..	..	Other Capital, net
35	-164	-791	-1,236	-746	-352	-318	-124	628	-426	..	Change in Reserves
				(Omani Rials per US dollar)							Memo Item:
0.350	0.350	0.350	0.350	0.350	0.350	0.350	0.350	0.380	0.380	..	Conversion Factor (Annual Avg)
			(Millions of US dollars, outstanding at end of year)								**EXTERNAL DEBT, ETC.**
449	496	440	537	724	1,132	1,340	1,929	2,483	2,474	..	Public/Publicly Guar. Long-Term
222	356	349	374	361	345	351	406	423	396	..	Official Creditors
6	9	14	17	15	17	32	43	50	66	..	IBRD and IDA
228	141	92	163	363	787	988	1,524	2,060	2,078	..	Private Creditors
..	..	..	..	..	..	..	..	..	..	..	Private Non-guaranteed Long-Term
0	0	0	0	0	0	0	0	0	..	..	Use of Fund Credit
182	146	163	217	234	357	293	422	496	405	..	Short-Term Debt
				(Millions of US dollars)							Memo Items:
254.1	415.6	581.4	744.3	872.4	762.6	900.2	1,090.2	968.0	1,402.2	..	Int'l Reserves Excluding Gold
41.9	95.8	123.1	108.8	127.3	109.9	89.0	94.4	112.9	139.8	..	Gold Holdings (at market price)
											SOCIAL INDICATORS
7.2	7.1	7.1	7.1	7.1	7.0	7.0	6.9	6.9	6.9	..	Total Fertility Rate
48.7	48.4	48.2	47.9	47.7	47.0	46.2	45.5	44.7	44.0	..	Crude Birth Rate
131.4	127.8	124.2	120.6	117.0	116.3	115.6	114.9	114.2	113.6	..	Infant Mortality Rate
50.0	50.5	51.1	51.7	52.3	52.8	53.3	53.8	54.3	54.7	..	Life Expectancy at Birth
..	..	..	..	..	..	..	..	..	..	..	Food Production, p.c. ('79-81=100)
51.1	50.5	49.8	..	..	..	..	..	..	..	..	Labor Force, Agriculture (%)
7.2	7.2	7.1	7.2	7.3	7.3	7.4	7.5	7.7	7.8	..	Labor Force, Female (%)
57.0	..	60.0	66.0	69.0	75.0	83.0	89.0	..	..	..	School Enroll. Ratio, primary
8.0	..	13.0	17.0	22.0	25.0	29.0	32.0	..	..	..	School Enroll. Ratio, secondary

PAKISTAN	1967	1968	1969	1970	1971	1972	1973	1974	1975	1976	1977
CURRENT GNP PER CAPITA (US $)	130	140	150	170	170	160	130	130	140	170	190
POPULATION (thousands)	55,581	57,161	58,827	60,607	62,662	64,710	66,773	68,873	71,033	73,238	75,473
ORIGIN AND USE OF RESOURCES					*(Billions of current Pakistan Rupees)*						
Gross National Product (GNP)	35.09	38.34	40.93	47.75	50.40	54.16	65.99	86.12	111.32	130.63	148.08
Net Factor Income from Abroad	-0.17	-0.19	-0.18	0.00	-0.08	0.10	-0.88	-0.73	-0.94	-1.42	-1.67
Gross Domestic Product (GDP)	35.26	38.53	41.11	47.75	50.49	54.06	66.87	86.85	112.27	132.05	149.75
Indirect Taxes, net	3.05	3.37	3.56	4.40	4.79	4.89	6.08	6.41	7.63	10.63	13.77
GDP at factor cost	32.21	35.15	37.55	43.34	45.70	49.17	60.79	80.44	104.64	121.42	135.98
Agriculture	12.46	13.99	14.04	15.96	16.24	17.93	21.91	28.08	33.53	38.34	43.97
Industry	6.34	6.92	8.21	9.68	10.57	10.63	13.33	17.64	24.53	29.47	31.10
Manufacturing	4.70	5.20	6.01	6.96	7.57	7.77	9.69	12.75	17.48	20.05	20.39
Services, etc.	13.41	14.24	15.30	17.71	18.89	20.61	25.55	34.71	46.58	53.61	60.92
Resource Balance	-2.27	-2.24	-1.11	-3.26	-3.51	-2.66	-1.89	-6.05	-12.71	-12.38	-14.58
Exports of Goods & NFServices	2.90	3.14	3.44	3.71	4.00	4.00	8.93	11.91	12.21	14.17	13.90
Imports of Goods & NFServices	5.17	5.38	4.56	6.97	7.51	6.67	10.83	17.96	24.91	26.54	28.48
Domestic Absorption	37.52	40.76	42.22	51.01	53.99	56.72	68.77	92.90	124.97	144.43	164.33
Private Consumption, etc.	27.71	29.88	32.06	38.62	40.83	42.58	52.39	72.75	94.80	107.32	118.76
General Gov't Consumption	3.93	3.94	4.10	4.85	5.27	6.48	7.72	8.54	11.95	14.34	16.71
Gross Domestic Investment	5.89	6.94	6.07	7.54	7.89	7.66	8.65	11.61	18.22	22.77	28.86
Fixed Investment	5.75	5.74	5.67	6.83	7.05	6.81	7.65	10.61	16.22	22.77	27.86
Memo Items:											
Gross Domestic Saving	3.62	4.70	4.95	4.28	4.39	5.00	6.76	5.57	5.51	10.39	14.27
Gross National Saving	3.65	4.89	5.28	4.67	4.61	5.85	7.39	6.21	6.64	12.22	18.69
					(Billions of 1980 Pakistan Rupees)						
Gross National Product	117.04	125.51	132.48	148.35	148.83	150.48	158.58	164.86	171.75	180.35	187.43
GDP at factor cost	105.32	112.64	119.99	132.44	132.81	134.31	143.79	152.79	160.19	165.62	172.39
Agriculture	38.61	43.14	45.09	49.39	47.88	49.54	50.36	52.47	51.35	53.65	55.01
Industry	21.75	22.87	25.64	29.23	30.19	28.74	32.00	34.95	35.74	37.49	40.11
Manufacturing	15.18	16.14	17.53	19.62	20.12	19.41	21.48	23.08	23.21	23.57	25.37
Services, etc.	44.97	46.63	49.26	53.82	54.74	56.04	61.43	65.37	73.09	74.48	77.27
Resource Balance	-23.89	-25.44	-16.24	-30.97	-25.14	-13.46	-8.41	-16.98	-12.63	-12.04	-19.90
Exports of Goods & NFServices	18.53	19.77	21.56	24.98	25.20	20.87	22.07	18.26	20.37	23.25	19.16
Imports of Goods & NFServices	42.41	45.21	37.81	55.95	50.35	34.33	30.47	35.24	33.00	35.29	39.06
Domestic Absorption	141.54	151.62	149.36	179.31	174.23	163.65	169.25	183.32	185.84	194.39	209.48
Private Consumption, etc.	100.03	106.56	109.80	132.99	127.45	115.81	119.18	137.68	135.01	141.40	151.96
General Gov't Consumption	13.23	13.09	13.26	15.35	15.75	18.50	19.67	16.71	18.60	20.16	20.74
Gross Domestic Investment	28.29	31.98	26.29	30.98	31.03	29.34	30.40	28.93	32.24	32.83	36.78
Fixed Investment	28.30	26.97	25.06	28.48	28.15	26.58	27.61	26.92	28.76	33.31	35.68
Memo Items:											
Capacity to Import	23.81	26.38	28.56	29.78	26.83	20.60	25.15	23.37	16.17	18.84	19.06
Terms of Trade Adjustment	5.29	6.61	7.00	4.80	1.63	-0.27	3.08	5.11	-4.20	-4.41	-0.10
Gross Domestic Income	122.94	132.79	140.11	153.14	150.71	149.93	163.92	171.45	169.01	177.93	189.48
Gross National Income	122.33	132.11	139.47	153.15	150.46	150.21	161.66	169.97	167.55	175.94	187.33
DOMESTIC PRICES (DEFLATORS)					*(Index 1980 = 100)*						
Overall (GDP)	30.0	30.5	30.9	32.2	33.9	36.0	41.6	52.2	64.8	72.4	79.0
Domestic Absorption	26.5	26.9	28.3	28.4	31.0	34.7	40.6	50.7	67.2	74.3	78.4
Agriculture	32.3	32.4	31.1	32.3	33.9	36.2	43.5	53.5	65.3	71.5	79.9
Industry	29.1	30.2	32.0	33.1	35.0	37.0	41.7	50.5	68.6	78.6	77.5
Manufacturing	30.9	32.2	34.3	35.5	37.6	40.1	45.1	55.3	75.3	85.1	80.4
MANUFACTURING ACTIVITY											
Employment (1980=100)	87.3	86.9	95.4	92.6	94.6	95.5	97.4	98.5	99.6	100.8	101.1
Real Earnings per Empl. (1980=100)	55.6	54.6	63.0	65.2	69.8	78.3	89.4	80.5	74.3	78.0	82.9
Real Output per Empl. (1980=100)	45.4	46.2	48.3	50.5	53.1	52.9	58.3	56.1	48.5	50.6	68.4
Earnings as % of Value Added	21.9	19.8	21.1	20.8	21.6	22.0	23.0	24.2	25.5	26.9	23.4
MONETARY HOLDINGS					*(Billions of current Pakistan Rupees)*						
Money Supply, Broadly Defined	15.08	16.96	18.58	20.75	23.61	27.73	31.56	31.18	37.80	49.97	58.94
Money as Means of Payment	10.28	11.04	12.62	14.02	16.49	19.94	22.19	22.52	25.62	34.04	39.97
Currency Ouside Banks	5.93	6.51	7.10	8.06	8.16	9.35	10.99	11.43	11.88	13.85	17.35
Demand Deposits	4.35	4.54	5.52	5.95	8.33	10.59	11.20	11.09	13.74	20.19	22.62
Quasi-Monetary Liabilities	4.80	5.92	5.96	6.74	7.12	7.79	9.37	8.67	12.18	15.92	18.97
					(Millions of current Pakistan Rupees)						
GOVERNMENT DEFICIT (-) OR SURPLUS	..	..	..	..	..	..	-4,554	-5,145	-11,466	-12,239	-12,580
Current Revenue	..	..	..	..	..	..	8,742	12,360	14,637	18,787	21,531
Current Expenditure	..	..	..	..	..	..	8,706	12,328	15,889	17,633	19,310
Current Budget Balance	..	..	..	..	..	..	36	32	-1,252	1,154	2,221
Capital Receipts	..	..	..	..	..	..	..	..	..	..	..
Capital Payments	..	..	..	..	..	..	4,590	5,177	10,214	13,393	14,801

1978	1979	1980	1981	1982	1983	1984	1985	1986	1987 estimate	NOTES	PAKISTAN
220	250	290	330	350	350	350	340	350	350	..	**CURRENT GNP PER CAPITA (US $)**
77,760	80,122	82,581	85,122	87,730	90,427	93,237	96,180	99,215	102,474	..	**POPULATION (thousands)**
				(Billions of current Pakistan Rupees)							**ORIGIN AND USE OF RESOURCES**
174.62	192.81	231.76	275.37	318.41	356.86	412.23	470.33	536.66	596.16	C	Gross National Product (GNP)
-1.80	-2.30	-2.77	-2.59	-3.43	-5.31	-5.97	-7.65	-10.47	-11.99	..	Net Factor Income from Abroad
176.42	195.11	234.53	277.96	321.84	362.16	418.20	477.98	547.13	608.14	C	Gross Domestic Product (GDP)
16.49	17.07	23.93	30.36	32.01	35.98	45.45	47.09	57.75	62.26	..	Indirect Taxes, net
159.93	178.04	210.60	247.60	289.83	326.19	372.75	430.89	489.38	545.89	C	GDP at factor cost
50.57	54.15	62.16	71.70	83.43	90.71	92.16	108.87	120.31	128.16	..	Agriculture
36.46	42.01	52.49	61.50	70.68	81.24	100.50	117.39	133.13	152.63	..	Industry
24.02	27.48	33.55	40.97	48.42	55.20	67.48	75.03	81.83	94.07	..	Manufacturing
72.90	81.88	95.95	114.40	135.73	154.23	180.09	204.63	235.94	265.10	..	Services, etc.
-16.33	-23.54	-27.26	-29.88	-38.25	-40.58	-48.56	-57.91	-51.50	-39.31	..	Resource Balance
16.31	20.86	29.26	34.15	31.30	44.47	47.60	49.89	65.24	81.23	..	Exports of Goods & NFServices
32.64	44.40	56.52	64.02	69.56	85.05	96.16	107.81	116.74	120.53	..	Imports of Goods & NFServices
192.74	218.65	261.79	307.84	360.09	402.74	466.76	535.90	598.63	647.45	..	Domestic Absorption
142.12	163.54	194.91	231.37	268.72	296.80	344.28	397.42	442.61	467.85	..	Private Consumption, etc.
19.12	20.24	23.54	29.00	34.34	42.50	51.55	58.08	66.04	76.24	..	General Gov't Consumption
31.50	34.88	43.34	47.47	57.03	63.44	70.93	80.40	89.98	103.36	..	Gross Domestic Investment
30.51	33.13	41.34	42.97	49.17	56.74	63.44	71.80	80.98	93.86	..	Fixed Investment
											Memo Items:
15.18	11.33	16.08	17.59	18.78	22.87	22.37	22.48	38.48	64.05	..	Gross Domestic Saving
25.55	23.36	31.84	36.30	41.46	57.40	56.12	56.08	74.04	96.17	..	Gross National Saving
				(Billions of 1980 Pakistan Rupees)							
202.76	209.86	231.76	250.84	265.57	281.64	297.55	320.01	341.40	365.28	C	Gross National Product
185.74	196.14	210.60	224.04	240.26	255.65	266.73	290.76	311.20	329.00	C	GDP at factor cost
56.56	58.31	62.16	64.44	66.74	69.28	65.09	73.06	77.53	79.10	..	Agriculture
43.92	47.29	52.49	57.55	63.75	68.86	75.99	81.77	89.43	96.38	..	Industry
28.03	30.36	33.55	37.21	42.41	45.51	49.22	53.34	57.32	61.64	..	Manufacturing
85.25	90.55	95.95	102.05	109.76	117.51	125.64	135.93	144.24	153.53	..	Services, etc.
-20.12	-29.86	-27.26	-12.99	-14.36	-10.69	-16.46	-19.45	-9.99	-3.48	..	Resource Balance
21.60	24.48	29.26	33.08	30.82	40.57	38.82	38.88	53.01	59.34	..	Exports of Goods & NFServices
41.72	54.34	56.52	46.07	45.18	51.26	55.27	58.33	63.01	62.82	..	Imports of Goods & NFServices
225.00	242.27	261.79	266.14	282.70	296.38	318.16	344.49	357.88	375.85	..	Domestic Absorption
164.49	179.94	194.91	198.78	205.98	213.21	230.60	248.80	256.62	264.24	..	Private Consumption, etc.
22.31	22.24	23.54	25.60	27.88	31.04	33.23	36.58	38.59	41.84	..	General Gov't Consumption
38.21	40.10	43.34	41.76	48.84	52.13	54.33	59.12	62.67	69.76	..	Gross Domestic Investment
37.23	38.17	41.34	37.14	41.67	46.09	48.18	52.41	55.93	62.67	..	Fixed Investment
											Memo Items:
20.86	25.53	29.26	24.57	20.33	26.81	27.36	26.99	35.21	42.33	..	Capacity to Import
-0.75	1.05	0.00	-8.51	-10.49	-13.77	-11.46	-11.89	-17.80	-17.01	..	Terms of Trade Adjustment
204.13	213.46	234.53	244.64	257.85	271.92	290.25	313.16	330.08	355.36	..	Gross Domestic Income
202.01	210.91	231.76	242.33	255.09	267.87	286.10	308.12	323.59	348.27	..	Gross National Income
				(Index 1980 = 100)							**DOMESTIC PRICES (DEFLATORS)**
86.1	91.9	100.0	109.8	119.9	126.8	138.6	147.1	157.3	163.3	..	Overall (GDP)
85.7	90.2	100.0	115.7	127.4	135.9	146.7	155.6	167.3	172.3	..	Domestic Absorption
89.4	92.9	100.0	111.3	125.0	130.9	141.6	149.0	155.2	162.0	..	Agriculture
83.0	88.8	100.0	106.8	110.9	118.0	132.2	143.6	148.9	158.4	..	Industry
85.7	90.5	100.0	110.1	114.2	121.3	137.1	140.7	142.7	152.6	..	Manufacturing
											MANUFACTURING ACTIVITY
101.7	99.9	100.0	99.9	104.9	105.3	105.6	105.9	..	..	..	Employment (1980 = 100)
89.4	96.2	100.0	99.8	103.9	115.5	130.0	134.3	..	..	..	Real Earnings per Empl. (1980 = 100)
78.4	86.2	100.0	108.4	122.6	129.3	139.4	151.0	..	..	..	Real Output per Empl. (1980 = 100)
22.7	23.4	20.9	19.6	19.5	20.0	20.0	20.0	..	..	..	Earnings as % of Value Added
				(Billions of current Pakistan Rupees)							**MONETARY HOLDINGS**
70.63	84.12	97.32	108.54	132.23	159.88	167.31	191.99	222.84	259.40	..	Money Supply, Broadly Defined
47.19	56.83	66.67	72.29	87.34	100.57	105.78	123.06	145.25	173.02	..	Money as Means of Payment
21.04	26.45	32.48	34.49	41.15	46.43	52.00	58.68	71.58	81.77	..	Currency Ouside Banks
26.15	30.38	34.19	37.80	46.19	54.14	53.78	64.38	73.67	91.26	..	Demand Deposits
23.43	27.29	30.65	36.25	44.89	59.31	61.53	68.93	77.59	86.38	..	Quasi-Monetary Liabilities
				(Millions of current Pakistan Rupees)							
-13,247	-17,997	-13,344	-16,138	-15,351	-24,784	-25,928	-33,783	-50,121	-49,808	C E	**GOVERNMENT DEFICIT (-) OR SURPLUS**
26,253	30,350	39,928	47,957	52,930	59,939	72,999	79,068	94,812	106,703	..	Current Revenue
23,886	29,864	33,926	45,438	45,966	59,254	72,467	82,508	..	..	..	Current Expenditure
2,367	486	6,002	2,519	6,964	685	532	-3,440	..	..	..	Current Budget Balance
..	..	..	..	..	..	..	..	..	..	..	Capital Receipts
15,614	18,483	19,346	18,657	22,315	25,469	26,460	30,343	..	..	..	Capital Payments

PAKISTAN	1967	1968	1969	1970	1971	1972	1973	1974	1975	1976	1977
FOREIGN TRADE (CUSTOMS BASIS)				*(Millions of current US dollars)*							
Value of Exports, fob	645.1	720.2	681.5	723.3	666.0	698.1	947.1	1,102.4	1,048.8	1,166.8	1,174.4
Nonfuel Primary Products	341.7	346.9	289.8	289.0	288.0	294.8	341.7	485.8	453.1	461.1	422.4
Fuels	3.9	9.1	8.7	8.9	7.7	11.0	7.2	24.9	11.1	24.2	49.4
Manufactures	299.5	364.1	383.0	425.4	370.4	392.3	598.2	591.6	584.6	681.5	702.5
Value of Imports, cif	1,101.1	995.9	1,010.7	1,170.9	925.9	681.9	965.9	1,729.4	2,153.1	2,182.9	2,454.6
Nonfuel Primary Products	305.8	243.5	138.5	321.1	233.2	254.2	332.8	535.7	647.8	589.3	603.0
Fuels	62.8	65.5	63.4	76.0	80.3	53.1	78.2	238.8	385.5	396.9	387.6
Manufactures	732.5	686.9	808.8	773.8	612.4	374.5	554.9	955.0	1,119.7	1,196.6	1,464.1
				(Index 1980 = 100)							
Terms of Trade	87.4	132.8	124.2	117.2	121.8	120.3	159.9	123.6	100.4	103.2	108.2
Export Prices, fob	21.0	29.7	29.1	28.8	30.8	34.5	59.6	73.3	59.2	60.4	70.7
Nonfuel Primary Products	33.4	34.0	31.8	31.6	34.5	37.9	70.7	92.2	67.5	65.6	66.9
Fuels	4.3	4.3	4.3	4.3	5.6	6.2	8.9	36.7	35.7	38.4	42.0
Manufactures	15.3	30.6	31.1	30.6	31.1	36.6	58.5	65.0	54.6	58.5	77.0
Import Prices, cif	24.0	22.4	23.4	24.6	25.3	28.7	37.3	59.3	58.9	58.5	65.3
Trade at Constant 1980 Prices				*(Millions of 1980 US dollars)*							
Exports, fob	3,069.7	2,425.2	2,345.6	2,511.6	2,162.8	2,024.6	1,588.0	1,504.4	1,772.2	1,931.0	1,660.7
Imports, cif	4,580.0	4,453.2	4,319.4	4,763.4	3,661.0	2,378.7	2,589.9	2,917.6	3,654.6	3,728.3	3,756.8
BALANCE OF PAYMENTS				*(Millions of current US dollars)*							
Exports of Goods & Services	..	..	..	805.0	770.3	731.3	923.0	1,236.5	1,267.9	1,459.8	1,436.7
Merchandise, fob	..	..	..	672.0	661.0	615.1	769.0	1,018.2	979.1	1,161.4	1,131.6
Nonfactor Services	..	..	..	107.0	96.3	106.4	136.6	182.2	252.0	267.9	271.6
Factor Services	..	..	..	26.0	13.0	9.8	17.4	36.2	36.8	30.5	33.5
Imports of Goods & Services	..	..	..	1,591.0	1,432.3	1,157.4	1,191.5	1,932.6	2,671.3	2,759.8	3,078.2
Merchandise, fob	..	..	..	1,210.0	1,081.2	859.9	890.5	1,488.7	2,110.5	2,138.7	2,416.9
Nonfactor Services	..	..	..	261.0	274.8	197.6	200.3	334.2	428.8	446.9	458.9
Factor Services	..	..	..	120.0	76.2	99.9	100.6	109.8	132.1	174.2	202.4
Long-Term Interest	..	..	..	77.4	59.3	75.5	86.4	84.4	102.2	130.5	143.6
Current Transfers, net	..	..	..	81.0	65.2	129.2	143.7	139.0	209.6	328.6	614.5
Workers' Remittances	..	..	..	86.0	70.2	130.3	145.8	150.8	220.1	328.4	577.1
Total to be Financed	..	..	..	-705.0	-596.8	-296.8	-124.8	-557.1	-1,193.8	-971.4	-1,027.0
Official Capital Grants	..	..	..	38.0	114.3	44.5	42.9	64.9	125.1	143.9	124.5
Current Account Balance	..	..	..	-667.0	-482.4	-252.3	-81.9	-492.2	-1,068.7	-827.5	-902.5
Long-Term Capital, net	..	..	..	467.0	512.5	245.4	283.4	366.5	504.4	623.0	582.2
Direct Investment	..	..	..	23.0	1.0	17.4	6.3	1.2	11.2	22.3	2.3
Long-Term Loans	..	..	..	379.5	546.7	188.5	241.2	710.2	711.5	747.2	601.8
Disbursements	..	..	..	491.9	639.4	294.9	350.5	829.6	866.8	906.8	797.7
Repayments	..	..	..	112.4	92.7	106.4	109.3	119.4	155.3	159.6	195.9
Other Long-Term Capital	..	..	..	26.5	-149.5	-5.0	-7.0	-409.8	-343.4	-290.4	-146.4
Other Capital, net	..	..	..	82.7	-81.5	-4.1	-47.6	36.4	439.8	212.4	91.3
Change in Reserves	..	..	..	117.3	51.4	11.1	-153.9	89.3	124.5	-7.9	229.0
Memo Item:				*(Pakistan Rupees per US dollar)*							
Conversion Factor (Annual Avg)	4.760	4.760	4.760	4.760	4.760	5.810	10.570	9.900	9.900	9.900	9.900
EXTERNAL DEBT, ETC.				*(Millions of US dollars, outstanding at end of year)*							
Public/Publicly Guar. Long-Term	..	..	..	3,064	3,607	3,797	4,234	4,635	5,095	6,001	6,801
Official Creditors	..	..	..	2,784	3,258	3,466	3,958	4,364	4,828	5,734	6,461
IBRD and IDA	..	..	..	610	659	693	739	789	767	869	952
Private Creditors	..	..	..	280	349	332	276	271	268	266	341
Private Non-guaranteed Long-Term	..	..	..	5	7	10	16	19	19	18	21
Use of Fund Credit	..	..	..	45	46	120	157	291	438	511	532
Short-Term Debt	..	..	..	..	..	..	..	..	..	..	180
Memo Items:				*(Millions of US dollars)*							
Int'l Reserves Excluding Gold	111.0	193.0	278.0	136.2	129.4	220.7	412.0	392.3	340.3	466.2	448.9
Gold Holdings (at market price)	54.5	65.4	54.9	58.3	69.1	103.1	178.3	296.2	222.7	218.0	266.8
SOCIAL INDICATORS											
Total Fertility Rate	7.2	7.2	7.1	7.1	7.0	7.0	7.0	7.0	7.0	7.0	7.0
Crude Birth Rate	47.8	47.3	46.8	46.4	45.9	45.4	45.1	44.8	44.6	44.3	44.0
Infant Mortality Rate	145.0	144.0	143.0	142.0	141.0	140.0	138.0	136.0	134.0	132.0	130.0
Life Expectancy at Birth	45.5	45.7	45.9	46.1	46.3	46.5	46.8	47.1	47.4	47.7	48.0
Food Production, p.c. ('79-81 = 100)	87.5	93.4	93.0	93.9	91.6	91.9	93.4	94.4	94.4	95.8	97.1
Labor Force, Agriculture (%)	59.4	59.3	59.1	58.9	58.5	58.1	57.6	57.2	56.8	56.4	52.2
Labor Force, Female (%)	8.8	8.9	9.0	9.1	9.2	9.4	9.5	9.6	9.8	9.9	10.0
School Enroll. Ratio, primary	..	..	..	40.0	..	..	..	..	46.0	51.0	50.0
School Enroll. Ratio, secondary	..	..	..	13.0	..	..	..	..	15.0	16.0	..

1978	1979	1980	1981	1982	1983	1984	1985	1986	1987 estimate	NOTES	PAKISTAN
				(Millions of current US dollars)							**FOREIGN TRADE (CUSTOMS BASIS)**
1,475.1	2,056.0	2,618.4	2,779.3	2,376.5	3,074.8	2,592.0	2,738.7	3,384.0	4,172.0	..	Value of Exports, fob
567.3	758.2	1,148.7	1,120.6	819.5	1,053.5	774.4	969.0	1,074.9	1,349.8	..	Nonfuel Primary Products
45.4	138.0	184.4	179.4	139.8	57.0	25.3	38.6	23.8	20.7	..	Fuels
862.4	1,159.8	1,285.3	1,479.3	1,417.2	1,964.3	1,792.3	1,731.1	2,285.3	2,801.3	..	Manufactures
3,160.5	4,061.3	5,350.5	5,412.7	5,232.8	5,341.0	5,873.1	5,890.4	5,376.8	5,822.0	..	Value of Imports, cif
840.6	1,056.3	1,020.0	1,205.3	1,051.5	1,094.0	1,471.8	1,463.4	1,324.2	1,361.0	..	Nonfuel Primary Products
597.3	681.9	1,442.1	1,506.2	1,616.8	1,514.1	1,454.6	1,433.2	764.9	1,104.0	..	Fuels
1,722.6	2,323.1	2,888.4	2,701.2	2,564.5	2,733.0	2,946.7	2,993.8	3,287.7	3,357.0	..	Manufactures
				(Index 1980 = 100)							
112.2	103.3	100.0	98.0	94.1	97.0	91.8	88.3	95.1	99.2	..	Terms of Trade
78.3	88.6	100.0	99.8	90.8	91.4	88.3	80.9	79.8	89.6	..	Export Prices, fob
80.6	82.3	100.0	100.7	76.6	78.6	71.5	61.0	56.4	72.1	..	Nonfuel Primary Products
42.3	61.0	100.0	112.5	101.6	92.5	90.2	87.5	44.9	56.9	..	Fuels
80.3	98.9	100.0	97.8	100.5	100.0	98.2	98.7	100.0	101.9	..	Manufactures
69.7	85.8	100.0	101.8	96.5	94.1	96.1	91.6	83.9	90.3	..	Import Prices, cif
				(Millions of 1980 US dollars)							Trade at Constant 1980 Prices
1,885.1	2,320.0	2,618.4	2,785.1	2,617.2	3,365.9	2,937.0	3,386.3	4,243.2	4,657.5	..	Exports, fob
4,532.4	4,733.9	5,350.6	5,317.0	5,420.7	5,673.4	6,109.0	6,428.0	6,409.5	6,449.3	..	Imports, cif
				(Millions of current US dollars)							**BALANCE OF PAYMENTS**
1,682.6	2,151.4	3,007.8	3,556.5	3,185.9	3,541.1	3,643.3	3,403.6	3,968.4	4,529.0	C f	Exports of Goods & Services
1,283.7	1,643.8	2,341.8	2,810.9	2,319.7	2,635.5	2,668.8	2,460.3	2,992.3	3,498.0	..	Merchandise, fob
365.2	461.5	612.8	650.4	735.5	784.5	779.2	787.4	862.9	935.0	..	Nonfactor Services
33.7	46.1	53.3	95.2	130.7	121.2	195.3	155.9	113.2	96.0	..	Factor Services
3,505.6	4,762.8	6,044.4	6,823.3	7,140.5	7,136.6	7,692.3	7,774.4	8,107.2	7,805.0	C f	Imports of Goods & Services
2,750.5	3,815.3	4,856.4	5,563.5	5,774.8	5,620.8	5,993.9	6,015.6	6,096.4	5,792.0	f	Merchandise, fob
539.9	669.4	854.9	902.9	914.3	974.9	1,059.7	1,097.3	1,248.9	1,222.0	..	Nonfactor Services
215.2	278.0	333.1	356.9	451.3	540.9	638.7	661.5	761.9	791.0	..	Factor Services
183.7	214.6	249.2	201.0	254.3	317.3	317.5	312.0	354.2	385.6	..	Long-Term Interest
1,228.7	1,446.7	1,871.6	2,151.4	2,474.9	3,136.9	2,947.4	2,721.2	2,853.3	2,557.0	..	Current Transfers, net
1,155.2	1,397.3	1,747.3	2,097.9	2,226.9	2,888.2	2,738.6	2,457.1	2,631.7	2,172.1	..	Workers' Remittances
-594.4	-1,164.6	-1,164.9	-1,115.4	-1,479.7	-458.7	-1,101.6	-1,649.6	-1,285.5	-719.0	K	Total to be Financed
107.7	341.4	291.8	362.4	369.6	271.1	404.9	381.2	497.1	383.0	K	Official Capital Grants
-486.7	-823.2	-873.1	-753.0	-1,110.1	-187.6	-696.7	-1,268.4	-788.4	-336.0	..	Current Account Balance
752.3	709.4	921.4	862.2	890.6	882.1	798.4	790.9	1,285.7	746.0	C f	Long-Term Capital, net
31.3	33.2	71.6	67.5	122.5	29.3	44.1	78.0	158.7	62.0	..	Direct Investment
560.4	576.8	874.7	409.9	1,068.3	231.1	572.0	243.5	391.7	174.2	..	Long-Term Loans
776.4	884.0	1,227.8	767.8	1,401.9	1,027.6	1,202.5	1,013.7	1,135.8	981.9	..	Disbursements
216.0	307.2	353.1	357.9	333.6	796.5	630.5	770.2	744.1	807.7	..	Repayments
52.9	-242.0	-316.7	22.4	-669.8	350.6	-222.6	88.2	238.2	126.8	..	Other Long-Term Capital
71.8	-146.4	-102.6	-158.7	-303.7	6.8	-219.5	-564.0	-411.2	-146.0	C f	Other Capital, net
-337.4	260.2	54.4	49.5	523.2	-701.3	117.8	1,041.5	-86.1	-264.0	..	Change in Reserves
				(Pakistan Rupees per US dollar)							**Memo Item:**
9.900	9.900	9.900	9.900	10.550	12.700	13.480	15.160	16.130	17.250	..	Conversion Factor (Annual Avg)
				(Millions of US dollars, outstanding at end of year)							**EXTERNAL DEBT, ETC.**
7,583	8,001	8,785	8,835	9,708	9,748	9,979	10,773	11,886	13,150	..	Public/Publicly Guar. Long-Term
7,226	7,614	8,245	8,354	8,809	8,949	9,030	9,986	11,089	12,494	..	Official Creditors
1,011	1,090	1,151	1,219	1,391	1,495	1,593	1,825	2,165	2,687	..	IBRD and IDA
356	387	539	482	900	799	949	787	797	655	..	Private Creditors
21	16	18	25	39	34	26	26	30	56	..	Private Non-guaranteed Long-Term
515	444	382	757	1,165	1,379	1,241	1,266	1,036	804	..	Use of Fund Credit
210	441	737	903	720	797	932	1,310	1,863	2,280	..	Short-Term Debt
				(Millions of US dollars)							**Memo Items:**
407.7	213.0	495.8	721.5	968.5	1,972.5	1,035.5	807.5	709.1	501.9	..	Int'l Reserves Excluding Gold
388.3	931.0	1,072.0	733.8	844.3	710.3	575.0	622.0	755.8	939.0	..	Gold Holdings (at market price)
											SOCIAL INDICATORS
7.0	7.0	7.0	7.0	7.0	6.9	6.9	6.8	6.8	6.7	..	Total Fertility Rate
43.9	43.8	43.7	43.6	43.5	44.2	44.9	45.6	46.4	47.1	..	Crude Birth Rate
128.0	126.0	124.0	122.0	120.0	118.0	116.0	114.0	112.0	110.0	..	Infant Mortality Rate
48.4	48.8	49.2	49.6	50.0	50.4	50.8	51.2	51.5	51.9	..	Life Expectancy at Birth
96.3	98.6	99.0	102.4	102.6	103.0	102.5	102.2	106.2	105.5	..	Food Production, p.c. ('79-81 = 100)
53.9	50.8	50.8	50.8	50.8	50.7	50.7	48.7	52.0	..	..	Labor Force, Agriculture (%)
10.2	10.3	10.4	10.6	10.8	11.0	11.2	11.4	11.7	11.9	..	Labor Force, Female (%)
54.0	42.0	43.0	44.0	42.0	45.0	47.0	..	53.0	..	..	School Enroll. Ratio, primary
15.0	14.0	14.0	14.0	15.0	16.0	17.0	..	24.0	..	..	School Enroll. Ratio, secondary

PANAMA	1967	1968	1969	1970	1971	1972	1973	1974	1975	1976	1977
CURRENT GNP PER CAPITA (US $)	560	590	640	680	740	770	840	920	1,030	1,080	1,120
POPULATION (thousands)	1,405	1,447	1,488	1,531	1,574	1,618	1,662	1,705	1,748	1,790	1,832
ORIGIN AND USE OF RESOURCES					*(Millions of current Panamanian Balboas)*						
Gross National Product (GNP)	778.0	836.1	920.1	994.8	1,120.6	1,231.2	1,404.3	1,599.1	1,820.0	1,902.7	2,008.8
Net Factor Income from Abroad	-22.7	-25.3	-25.3	-26.4	-31.3	-33.7	-42.3	-55.0	-20.8	-53.6	-61.0
Gross Domestic Product (GDP)	800.7	861.4	945.4	1,021.2	1,151.9	1,264.9	1,446.6	1,654.1	1,840.8	1,956.3	2,069.8
Indirect Taxes, net	59.4	61.7	71.4	78.8	90.6	103.1	115.4	136.9	139.6	146.0	181.9
GDP at factor cost	741.3	799.7	874.0	942.4	1,061.3	1,161.8	1,331.2	1,517.2	1,701.2	1,810.3	1,887.9
Agriculture	134.7	146.0	149.2	149.1	164.2	170.9	184.9	184.5	205.6	231.1	263.5
Industry	158.1	174.5	187.7	219.2	247.5	259.8	318.3	364.0	432.1	431.3	424.1
Manufacturing	94.6	105.0	114.8	127.3	135.9	141.0	161.0	202.0	236.0	217.3	234.0
Services, etc.	507.9	540.9	608.5	652.9	740.2	834.2	943.4	1,105.6	1,203.1	1,293.9	1,382.2
Resource Balance	-6.5	8.0	-10.4	-34.2	-43.4	-62.4	-56.8	-155.3	-134.0	-137.4	-75.8
Exports of Goods & NFServices	301.7	330.0	362.9	388.2	426.4	460.7	528.1	761.8	865.4	837.8	921.1
Imports of Goods & NFServices	308.2	322.0	373.3	422.4	469.8	523.1	584.9	917.1	999.4	975.2	996.9
Domestic Absorption	807.2	853.4	955.8	1,055.4	1,195.3	1,327.3	1,503.4	1,809.4	1,974.8	2,093.7	2,145.6
Private Consumption, etc.	536.2	551.8	614.9	618.8	665.2	698.3	767.6	954.0	1,054.1	1,088.8	1,242.6
General Gov't Consumption	102.6	109.7	118.2	152.3	180.2	226.5	250.1	299.3	353.3	386.1	412.1
Gross Domestic Investment	168.4	191.9	222.7	284.3	349.9	402.5	485.7	556.1	567.4	618.8	490.9
Fixed Investment	152.7	173.5	200.7	261.9	306.3	372.2	434.8	465.0	535.5	608.6	445.9
Memo Items:											
Gross Domestic Saving	161.9	199.9	212.3	250.1	306.5	340.1	428.9	400.8	433.4	481.4	415.1
Gross National Saving	139.2	163.0	175.6	212.8	263.0	292.6	363.9	320.2	387.0	400.1	324.2
					(Millions of 1980 Panamanian Balboas)						
Gross National Product	1,612.9	1,719.3	1,871.3	2,011.8	2,205.7	2,314.0	2,439.0	2,501.3	2,590.2	2,597.1	2,622.9
GDP at factor cost	..	..	..	..	..	..	..	..	..	..	..
Agriculture	248.6	262.8	277.1	275.0	296.4	286.6	289.8	272.8	292.5	308.2	323.7
Industry	343.8	374.8	407.0	443.8	498.9	518.1	576.5	559.2	576.5	597.5	551.6
Manufacturing	194.5	213.3	233.2	248.9	264.7	276.8	294.2	298.1	287.4	294.6	298.1
Services, etc.	1,073.3	1,144.6	1,249.7	1,351.8	1,474.8	1,572.3	1,639.2	1,738.4	1,744.2	1,749.3	1,808.6
Resource Balance	-156.1	-132.8	-212.9	-270.7	-350.2	-424.8	-369.7	-458.4	-376.5	-330.5	-182.6
Exports of Goods & NFServices	660.7	710.5	759.1	795.7	829.6	840.0	879.8	941.3	957.3	933.3	1,032.5
Imports of Goods & NFServices	816.7	843.2	972.0	1,066.4	1,179.8	1,264.8	1,249.4	1,399.7	1,333.8	1,263.8	1,215.1
Domestic Absorption	1,821.8	1,915.0	2,146.7	2,341.3	2,620.3	2,801.9	2,875.1	3,028.8	2,989.7	2,985.6	2,866.4
Private Consumption, etc.	1,083.2	1,094.6	1,256.3	1,295.8	1,370.2	1,361.9	1,389.3	1,578.0	1,519.4	1,508.0	1,641.1
General Gov't Consumption	292.5	315.2	312.1	363.9	413.6	482.4	491.7	507.0	559.8	575.8	598.3
Gross Domestic Investment	446.2	505.2	578.3	681.7	836.6	957.6	994.1	943.7	910.4	901.8	627.0
Fixed Investment	399.4	450.9	513.7	621.2	713.9	884.9	886.6	793.4	856.2	873.3	580.1
Memo Items:											
Capacity to Import	799.5	864.2	944.9	980.1	1,070.8	1,114.0	1,128.1	1,162.6	1,154.9	1,085.8	1,122.7
Terms of Trade Adjustment	138.8	153.7	185.9	184.3	241.2	273.9	248.3	221.4	197.7	152.5	90.2
Gross Domestic Income	1,804.6	1,935.9	2,119.6	2,255.0	2,511.3	2,651.0	2,753.7	2,791.8	2,810.8	2,807.6	2,774.0
Gross National Income	1,751.8	1,873.0	2,057.2	2,196.1	2,446.9	2,587.9	2,687.3	2,722.6	2,787.9	2,749.6	2,713.1
DOMESTIC PRICES (DEFLATORS)					*(Index 1980 = 100)*						
Overall (GDP)	48.1	48.3	48.9	49.3	50.7	53.2	57.7	64.4	70.4	73.7	77.1
Domestic Absorption	44.3	44.6	44.5	45.1	45.6	47.4	52.3	59.7	66.1	70.1	74.9
Agriculture	54.2	55.5	53.9	54.2	55.4	59.6	63.8	67.6	70.3	75.0	81.4
Industry	46.0	46.6	46.1	49.4	49.6	50.1	55.2	65.1	75.0	72.2	76.9
Manufacturing	48.6	49.2	49.2	51.2	51.3	50.9	54.7	67.8	82.1	73.8	78.5
MANUFACTURING ACTIVITY											
Employment (1980=100)	56.3	59.1	63.0	69.9	80.8	82.9	87.7	85.9	85.3	84.2	86.2
Real Earnings per Empl. (1980=100)	104.8	106.4	109.6	109.6	107.3	109.0	108.7	105.5	108.3	110.4	119.4
Real Output per Empl. (1980=100)	64.9	66.1	68.0	67.3	71.8	77.7	81.6	100.0	93.7	103.0	99.4
Earnings as % of Value Added	34.4	33.8	36.0	32.0	32.8	32.8	31.3	29.1	28.1	29.5	32.9
MONETARY HOLDINGS					*(Millions of current Panamanian Balboas)*						
Money Supply, Broadly Defined	..	..	..	..	..	..	..	..	635	682	794
Money as Means of Payment	70	81	85	101	105	154	161	196	173	190	213
Currency Ouside Banks	..	..	..	..	..	..	..	..	..	..	..
Demand Deposits	..	..	..	..	..	..	..	..	..	..	..
Quasi-Monetary Liabilities	..	..	..	..	..	..	..	..	462	492	581
GOVERNMENT DEFICIT (-) OR SURPLUS					*(Millions of current Panamanian Balboas)*						
	..	..	..	..	..	..	-91	-119	-150	-203	-119
Current Revenue	..	..	..	..	..	..	306	392	452	440	546
Current Expenditure	..	..	..	..	..	..	304	382	431	479	533
Current Budget Balance	..	..	..	..	..	..	2	9	21	-39	14
Capital Receipts	..	..	..	..	..	..	..	..	..	..	..
Capital Payments	..	..	..	..	..	..	93	129	170	164	133

1978	1979	1980	1981	1982	1983	1984	1985	1986	1987 estimate	NOTES	PANAMA
1,290	1,420	1,680	1,880	2,000	1,980	2,010	2,100	2,180	2,240	..	CURRENT GNP PER CAPITA (US $)
1,873	1,915	1,956	1,998	2,044	2,089	2,134	2,180	2,227	2,272	..	POPULATION (thousands)
				(Millions of current Panamanian Balboas)							ORIGIN AND USE OF RESOURCES
2,403.7	2,721.6	3,331.5	3,658.7	3,962.6	4,124.9	4,288.5	4,555.8	4,815.0	5,128.3	..	Gross National Product (GNP)
-48.8	-78.6	-227.3	-219.3	-316.3	-248.8	-277.0	-345.3	-306.2	-363.0	..	Net Factor Income from Abroad
2,452.5	2,800.2	3,558.8	3,878.0	4,278.9	4,373.7	4,565.5	4,901.1	5,121.2	5,491.3	..	Gross Domestic Product (GDP)
216.0	251.7	268.2	281.6	307.9	331.4	353.7	384.7	427.6	..	..	Indirect Taxes, net
2,236.5	2,548.5	3,290.6	3,596.4	3,971.0	4,042.3	4,211.8	4,516.4	4,693.6	..	B	GDP at factor cost
288.5	304.2	320.4	359.3	371.2	408.4	415.9	450.6	478.6	..	..	Agriculture
511.9	587.1	735.0	821.9	934.3	833.9	836.9	867.6	912.3	..	f	Industry
252.6	293.3	356.0	375.6	394.0	401.0	411.0	420.0	422.1	..	..	Manufacturing
1,652.1	1,908.9	2,503.4	2,696.8	2,973.4	3,131.4	3,312.7	3,582.9	3,730.3	..	..	Services, etc.
-113.8	-246.5	-118.1	-209.5	-179.8	18.1	-74.8	24.3	172.7	..	..	Resource Balance
986.4	1,124.8	1,567.1	1,632.0	1,689.6	1,709.5	1,622.1	1,735.2	1,738.6	..	..	Exports of Goods & NFServices
1,100.2	1,371.3	1,685.2	1,841.5	1,869.4	1,691.4	1,696.9	1,710.9	1,565.9	..	..	Imports of Goods & NFServices
2,566.3	3,046.7	3,676.9	4,087.5	4,458.7	4,355.6	4,640.3	4,876.6	4,948.5	..	..	Domestic Absorption
1,431.7	1,693.8	2,009.5	2,107.4	2,311.5	2,480.0	2,878.0	3,080.0	2,933.8	..	..	Private Consumption, etc.
482.9	567.2	680.5	812.9	962.6	941.5	1,001.3	1,043.6	1,123.7	..	..	General Gov't Consumption
651.7	785.7	986.9	1,167.2	1,184.6	934.1	761.0	753.0	891.0	..	..	Gross Domestic Investment
606.3	661.2	866.4	1,079.6	1,185.4	917.8	779.9	773.1	895.1	..	..	Fixed Investment
											Memo Items:
537.9	539.2	868.8	957.7	1,004.8	952.2	686.2	777.5	1,063.7	..	..	Gross Domestic Saving
455.6	421.3	589.3	690.8	633.5	643.3	377.6	401.4	725.3	..	..	Gross National Saving
				(Millions of 1980 Panamanian Balboas)							
2,905.9	3,024.0	3,331.5	3,497.8	3,652.3	3,639.3	3,668.0	3,800.6	3,959.6	4,016.0	..	Gross National Product
..	..	3,558.8	3,706.6	3,912.4	3,927.9	3,911.4	4,096.8	4,219.3	..	B	GDP at factor cost
348.8	333.9	320.4	347.0	341.6	352.3	358.2	376.1	367.8	397.9	f	Agriculture
620.2	666.6	735.0	737.8	805.7	711.5	669.2	686.0	717.2	732.4	f	Industry
302.8	336.3	356.0	344.3	351.7	345.4	343.7	350.5	359.7	375.9	..	Manufacturing
1,978.1	2,083.1	2,503.4	2,621.8	2,765.1	2,864.1	2,884.0	3,034.7	3,134.3	3,207.2	f	Services, etc.
-291.9	-344.6	-118.1	-162.8	-64.6	96.4	-126.0	-40.8	58.4	..	..	Resource Balance
1,104.5	1,089.7	1,567.1	1,518.3	1,640.1	1,627.4	1,523.6	1,633.3	1,666.1	..	..	Exports of Goods & NFServices
1,396.4	1,434.3	1,685.2	1,681.2	1,704.6	1,530.9	1,649.6	1,674.1	1,607.7	1,755.6	..	Imports of Goods & NFServices
3,239.1	3,428.1	3,676.9	3,869.5	3,977.0	3,831.5	4,037.3	4,137.6	4,160.8	..	..	Domestic Absorption
1,831.1	1,940.9	2,009.5	1,958.2	2,071.2	2,192.6	2,477.6	2,515.1	2,370.3	..	..	Private Consumption, etc.
624.8	643.0	680.5	800.2	873.3	821.5	832.5	845.8	916.8	..	..	General Gov't Consumption
783.1	844.2	986.9	1,111.1	1,032.5	817.4	727.2	776.6	873.7	912.1	..	Gross Domestic Investment
719.1	711.3	866.4	1,010.6	1,020.1	789.8	735.7	786.0	863.1	..	..	Fixed Investment
											Memo Items:
1,251.9	1,176.4	1,567.1	1,489.9	1,540.7	1,547.3	1,576.9	1,697.9	1,785.0	..	..	Capacity to Import
147.5	86.7	0.0	-28.4	-99.4	-80.0	53.2	64.6	118.9	..	..	Terms of Trade Adjustment
3,094.6	3,170.3	3,558.8	3,678.2	3,813.0	3,847.9	3,964.6	4,161.3	4,338.2	..	..	Gross Domestic Income
3,053.4	3,110.7	3,331.5	3,469.4	3,552.9	3,559.2	3,721.2	3,865.1	4,078.5	..	..	Gross National Income
				(Index 1980 = 100)							DOMESTIC PRICES (DEFLATORS)
83.2	90.8	100.0	104.6	109.4	111.3	116.7	119.6	121.4	126.6	..	Overall (GDP)
79.2	88.9	100.0	105.6	112.1	113.7	114.9	117.9	118.9	..	..	Domestic Absorption
82.7	91.1	100.0	103.6	108.7	115.9	116.1	119.8	130.1	..	..	Agriculture
82.5	88.1	100.0	111.4	116.0	117.2	125.1	126.5	127.2	..	..	Industry
83.4	87.2	100.0	109.1	112.0	116.1	119.6	119.8	117.3	..	..	Manufacturing
											MANUFACTURING ACTIVITY
90.0	94.9	100.0	108.6	111.9	115.4	112.6	117.5	..	..	..	Employment (1980=100)
116.1	114.5	100.0	108.7	111.4	116.7	..	..	..	..	..	Real Earnings per Empl. (1980=100)
90.2	104.7	100.0	97.4	94.9	91.6	92.0	92.1	..	..	..	Real Output per Empl. (1980=100)
32.8	30.6	25.1	28.8	30.3	31.9	..	..	..	..	f	Earnings as % of Value Added
				(Millions of current Panamanian Balboas)							MONETARY HOLDINGS
947	1,166	1,458	1,724	1,949	1,971	2,109	2,195	..	..	D	Money Supply, Broadly Defined
246	301	335	360	379	373	381	410	..	..	..	Money as Means of Payment
..	..	..	..	..	..	..	..	..	..	..	Currency Ouside Banks
..	..	..	..	..	..	..	..	..	..	..	Demand Deposits
701	865	1,123	1,364	1,570	1,598	1,728	1,785	..	..	..	Quasi-Monetary Liabilities
				(Millions of current Panamanian Balboas)							
-160	-372	-198	-335	-483	-273	-340	-155	..	..	..	GOVERNMENT DEFICIT (-) OR SURPLUS
598	697	966	1,030	1,204	1,324	1,369	1,393	..	..	..	Current Revenue
615	796	952	1,070	1,401	1,342	1,458	1,468	..	..	..	Current Expenditure
-17	-99	14	-40	-197	-18	-89	-75	..	..	..	Current Budget Balance
..	..	..	..	..	..	5	5	..	..	..	Capital Receipts
142	273	212	295	285	255	256	85	..	..	..	Capital Payments

PANAMA	1967	1968	1969	1970	1971	1972	1973	1974	1975	1976	1977
FOREIGN TRADE (CUSTOMS BASIS)					*(Millions of current US dollars)*						
Value of Exports, fob	85.3	93.8	108.8	109.5	116.5	122.6	137.8	210.5	286.4	236.7	249.5
Nonfuel Primary Products	61.1	73.3	82.3	84.1	88.5	97.7	107.1	110.8	144.7	148.6	158.3
Fuels	23.2	18.8	24.1	21.5	25.2	21.6	24.5	86.3	128.3	66.3	68.3
Manufactures	0.9	1.6	2.5	3.9	2.8	3.3	6.3	13.4	13.5	21.8	22.9
Value of Imports, cif	250.5	266.5	293.6	357.0	395.8	440.5	502.2	822.4	892.1	848.3	858.9
Nonfuel Primary Products	29.2	30.6	32.9	41.7	54.9	53.0	64.8	95.9	86.5	87.9	95.0
Fuels	50.6	56.7	64.8	66.4	70.9	73.3	102.0	293.4	360.5	282.1	283.4
Manufactures	170.8	179.2	196.0	249.0	270.0	314.1	335.4	433.2	445.1	478.4	480.5
					(Index 1980 = 100)						
Terms of Trade	82.1	101.4	98.7	95.5	86.2	94.9	104.6	96.8	100.3	109.2	106.8
Export Prices, fob	11.3	13.5	13.1	14.3	15.9	20.0	26.4	46.9	48.7	57.3	61.4
Nonfuel Primary Products	29.8	29.7	32.0	33.6	32.2	37.8	46.1	57.2	69.1	71.7	74.9
Fuels	4.3	4.3	4.3	4.3	5.6	6.2	8.9	36.7	35.7	38.4	42.0
Manufactures	30.1	31.7	32.4	37.5	36.5	38.5	49.5	67.2	66.1	66.7	71.7
Import Prices, cif	13.8	13.3	13.3	15.0	18.4	21.1	25.3	48.4	48.5	52.5	57.5
Trade at Constant 1980 Prices					*(Millions of 1980 US dollars)*						
Exports, fob	753.2	694.2	829.7	765.8	734.9	613.9	521.4	448.8	588.6	412.8	406.2
Imports, cif	1,816.4	1,998.8	2,210.6	2,383.8	2,152.4	2,091.3	1,988.0	1,697.8	1,839.1	1,616.0	1,493.1
BALANCE OF PAYMENTS					*(Millions of current US dollars)*						
Exports of Goods & Services	..	..	..	395	446	496	604	1,060	1,223	1,231	1,388
Merchandise, fob	..	..	..	130	138	146	162	251	331	269	289
Nonfactor Services	..	..	..	250	279	304	352	494	512	545	608
Factor Services	..	..	..	14	30	46	91	315	379	417	491
Imports of Goods & Services	..	..	..	463	524	599	711	1,301	1,392	1,403	1,541
Merchandise, fob	..	..	..	331	364	409	458	761	823	783	790
Nonfactor Services	..	..	..	89	101	110	121	150	164	166	201
Factor Services	..	..	..	43	60	80	131	390	404	454	550
Long-Term Interest	..	..	..	7	12	17	24	44	41	56	73
Current Transfers, net	..	..	..	-11	-12	-14	-23	-26	-26	-28	-30
Workers' Remittances	..	..	..	0	0	0	0	0	0	0	0
Total to be Financed	..	..	..	-79	-90	-117	-130	-246	-189	-200	-183
Official Capital Grants	..	..	..	15	17	19	19	22	21	24	28
Current Account Balance	..	..	..	-64	-73	-99	-111	-224	-169	-176	-155
Long-Term Capital, net	..	..	..	135	103	147	166	135	205	746	-86
Direct Investment	..	..	..	33	22	13	36	35	8	-11	11
Long-Term Loans	..	..	..	44	51	101	100	104	214	322	239
Disbursements	..	..	..	67	79	136	163	195	245	367	328
Repayments	..	..	..	24	28	35	63	91	31	45	88
Other Long-Term Capital	..	..	..	43	14	14	12	-25	-37	411	-363
Other Capital, net	..	..	..	-71	-31	-30	-58	77	-59	-553	234
Change in Reserves	..	..	..	1	1	-18	3	12	22	-17	7
Memo Item:					*(Panamanian Balboas per US dollar)*						
Conversion Factor (Annual Avg)	1.000	1.000	1.000	1.000	1.000	1.000	1.000	1.000	1.000	1.000	1.000
EXTERNAL DEBT, ETC.					*(Millions of US dollars, outstanding at end of year)*						
Public/Publicly Guar. Long-Term	..	..	..	194	245	345	458	565	771	1,093	1,333
Official Creditors	..	..	..	93	113	133	160	198	269	338	390
IBRD and IDA	..	..	..	7	8	19	26	41	65	88	99
Private Creditors	..	..	..	101	131	212	297	367	502	755	943
Private Non-guaranteed Long-Term	..	..	..	..	..	..	..	..	..	..	..
Use of Fund Credit	..	..	..	0	0	0	0	9	21	50	51
Short-Term Debt	..	..	..	..	..	..	..	..	..	..	300
Memo Items:					*(Thousands of US dollars)*						
Int'l Reserves Excluding Gold	6,730	11,100	14,100	15,750	21,022	43,128	41,715	39,316	34,445	78,942	70,866
Gold Holdings (at market price)	..	..	..	..	..	..	..	..	..	..	..
SOCIAL INDICATORS											
Total Fertility Rate	5.6	5.5	5.3	5.2	5.1	4.9	4.8	4.6	4.4	4.2	4.1
Crude Birth Rate	39.3	38.6	37.9	37.1	36.4	35.7	34.8	33.8	32.9	31.9	31.0
Infant Mortality Rate	52.0	50.2	48.4	46.6	44.8	43.0	40.8	38.6	36.4	34.2	32.0
Life Expectancy at Birth	64.3	64.7	65.1	65.5	65.9	66.4	66.9	67.5	68.1	68.6	69.2
Food Production, p.c. ('79-81 = 100)	92.7	102.0	105.0	91.6	97.7	93.7	95.9	97.1	98.7	95.4	100.7
Labor Force, Agriculture (%)	44.4	43.4	42.5	41.6	40.6	39.6	38.6	37.6	36.6	35.6	34.6
Labor Force, Female (%)	24.0	24.4	24.8	25.2	25.3	25.4	25.5	25.6	25.7	25.8	25.9
School Enroll. Ratio, primary	..	..	..	99.0	..	..	..	..	114.0	121.0	120.0
School Enroll. Ratio, secondary	..	..	..	38.0	..	..	..	..	55.0	60.0	63.0

1978	1979	1980	1981	1982	1983	1984	1985	1986	1987 estimate	NOTES	PANAMA
											FOREIGN TRADE (CUSTOMS BASIS)
				(Millions of current US dollars)							
244.3	291.6	353.4	319.4	310.2	302.6	257.6	335.3	349.6	356.7	..	Value of Exports, fob
163.0	188.8	240.0	229.6	201.2	236.8	18.8	23.6	267.1	262.5	..	Nonfuel Primary Products
60.1	72.4	81.8	58.4	70.3	36.5	236.2	307.8	36.5	45.6	..	Fuels
21.1	30.4	31.5	31.4	38.7	29.3	2.6	3.9	46.0	48.6	..	Manufactures
942.4	1,183.8	1,447.4	1,561.9	1,567.8	1,411.4	1,411.8	1,383.3	1,229.2	1,248.0	..	Value of Imports, cif
100.7	134.5	166.1	162.9	164.2	167.1	165.6	185.7	134.0	52.9	..	Nonfuel Primary Products
230.1	337.6	441.7	439.0	417.5	392.3	367.2	292.9	117.8	90.4	..	Fuels
611.5	711.7	839.6	960.0	986.1	852.0	879.0	904.7	977.5	1,104.7	..	Manufactures
				(Index 1980 = 100)							
96.7	100.6	100.0	96.5	98.4	104.6	96.9	93.7	83.0	70.8	..	Terms of Trade
63.7	80.8	100.0	99.3	96.7	99.3	90.3	86.6	79.6	79.7	..	Export Prices, fob
76.4	90.6	100.0	96.5	95.5	101.7	92.8	76.0	85.2	81.0	..	Nonfuel Primary Products
42.3	61.0	100.0	112.5	101.6	92.5	90.2	87.5	44.9	56.9	..	Fuels
76.1	90.0	100.0	99.1	94.8	90.7	89.3	89.0	103.2	113.0	..	Manufactures
65.9	80.3	100.0	102.9	98.3	94.9	93.3	92.5	95.8	112.7	..	Import Prices, cif
				(Millions of 1980 US dollars)							Trade at Constant 1980 Prices
383.4	360.9	353.4	321.6	320.8	304.7	285.1	387.0	439.3	447.4	..	Exports, fob
1,430.7	1,474.4	1,447.5	1,517.1	1,594.4	1,487.2	1,513.5	1,495.3	1,282.6	1,107.7	..	Imports, cif
				(Millions of current US dollars)							**BALANCE OF PAYMENTS**
1,744	2,583	7,737	9,921	9,407	7,279	6,538	6,317	6,159	5,910	..	Exports of Goods & Services
304	356	2,267	2,540	2,411	1,676	1,686	1,983	2,396	2,508	..	Merchandise, fob
634	782	1,106	1,149	1,169	1,276	1,259	1,326	1,268	1,229	..	Nonfactor Services
806	1,445	4,363	6,232	5,827	4,327	3,593	3,008	2,496	2,173	..	Factor Services
1,949	2,907	8,062	9,897	9,504	6,907	6,432	6,124	5,803	5,648	..	Imports of Goods & Services
862	1,086	2,995	3,315	3,045	2,321	2,509	2,730	2,980	3,035	..	Merchandise, fob
233	280	640	703	659	375	394	422	423	415	..	Nonfactor Services
854	1,541	4,427	5,878	5,800	4,211	3,529	2,972	2,401	2,198	..	Factor Services
122	198	252	281	338	292	304	303	322	226	..	Long-Term Interest
-34	-39	-52	-48	-55	-60	-32	-31	-32	-33	..	Current Transfers, net
0	0	0	0	0	0	0	0	0	0	..	Workers' Remittances
-239	-363	-378	-23	-152	311	74	162	324	229	K	Total to be Financed
31	52	67	79	101	104	144	140	130	112	K	Official Capital Grants
-208	-311	-311	56	-51	416	218	301	454	342	..	Current Account Balance
483	366	-654	295	578	316	249	-148	-3	..	..	Long-Term Capital, net
-3	50	-47	6	3	72	10	59	-72	..	..	Direct Investment
550	219	189	165	493	231	164	46	64	-19	..	Long-Term Loans
992	409	404	379	774	419	397	154	213	139	..	Disbursements
443	190	215	213	281	189	233	107	148	158	..	Repayments
-96	45	-863	45	-19	-90	-68	-393	-125	..	..	Other Long-Term Capital
-189	-82	976	-424	-542	-743	-551	-269	-390	-376	..	Other Capital, net
-86	27	-11	73	16	11	84	115	-61	35	..	Change in Reserves
				(Panamanian Balboas per US dollar)							Memo Item:
1.000	1.000	1.000	1.000	1.000	1.000	1.000	1.000	1.000	1.000	..	Conversion Factor (Annual Avg)
				(Millions of US dollars, outstanding at end of year)							**EXTERNAL DEBT, ETC.**
1,875	2,072	2,271	2,430	2,917	3,145	3,185	3,340	3,572	3,722	..	Public/Publicly Guar. Long-Term
478	529	605	705	854	1,020	1,048	1,181	1,321	1,454	..	Official Creditors
107	118	133	163	191	253	249	310	438	523	..	IBRD and IDA
1,397	1,542	1,666	1,725	2,064	2,126	2,137	2,159	2,251	2,268	..	Private Creditors
..	..	..	..	..	..	..	..	..	..	..	Private Non-guaranteed Long-Term
53	41	23	94	84	193	271	311	353	346	..	Use of Fund Credit
385	491	680	843	922	1,050	912	1,123	1,010	1,256	..	Short-Term Debt
				(Thousands of US dollars)							Memo Items:
150,420	118,720	117,380	119,950	100,970	206,700	215,640	97,963	170,170	77,814	..	Int'l Reserves Excluding Gold
..	..	..	..	..	..	..	..	..	..	..	Gold Holdings (at market price)
											SOCIAL INDICATORS
3.9	3.8	3.7	3.6	3.5	3.4	3.3	3.3	3.2	3.1	..	Total Fertility Rate
30.4	29.8	29.2	28.6	28.0	27.6	27.3	26.9	26.5	26.2	..	Crude Birth Rate
30.8	29.6	28.4	27.2	26.0	25.3	24.6	23.9	23.2	22.5	..	Infant Mortality Rate
69.6	69.9	70.3	70.6	71.0	71.2	71.4	71.5	71.7	71.9	..	Life Expectancy at Birth
101.5	100.5	98.0	101.5	97.7	101.5	98.3	100.4	94.9	91.2	..	Food Production, p.c. ('79-81 = 100)
33.7	32.7	31.7	..	..	..	..	..	..	..	..	Labor Force, Agriculture (%)
26.0	26.1	26.2	26.3	26.4	26.5	26.6	26.7	26.8	26.9	..	Labor Force, Female (%)
118.0	116.0	106.0	111.0	104.0	104.0	105.0	105.0	..	..	..	School Enroll. Ratio, primary
66.0	66.0	61.0	65.0	59.0	59.0	59.0	59.0	..	..	..	School Enroll. Ratio, secondary

PAPUA NEW GUINEA	1967	1968	1969	1970	1971	1972	1973	1974	1975	1976	1977
CURRENT GNP PER CAPITA (US $)	190	200	230	260	280	310	370	460	530	530	550
POPULATION (thousands)	2,256	2,311	2,367	2,422	2,477	2,531	2,585	2,640	2,698	2,758	2,819
ORIGIN AND USE OF RESOURCES	*(Millions of current Papua New Guinea Kina)*										
Gross National Product (GNP)	392.7	429.9	485.9	558.8	602.6	680.8	856.9	947.2	981.0	1,158.2	1,271.8
Net Factor Income from Abroad	-1.7	-3.3	-6.3	-17.6	-31.0	-36.3	-57.8	-75.1	-55.3	-40.4	-26.7
Gross Domestic Product (GDP)	394.4	433.2	492.2	576.4	633.6	717.1	914.7	1,022.3	1,036.3	1,198.6	1,298.5
Indirect Taxes, net	15.9	17.8	21.8	28.5	33.4	35.5	39.3	48.0	57.0	67.6	83.2
GDP at factor cost	378.5	415.4	470.4	547.9	600.2	681.6	875.4	974.3	979.3	1,131.0	1,215.3
Agriculture	165.5	185.8	205.8	214.3	218.5	232.2	258.0	286.2	307.8	380.0	428.3
Industry	78.3	80.3	90.9	127.4	155.0	202.7	330.3	356.4	295.2	301.0	338.7
Manufacturing	..	..	..	31.5	36.7	40.7	51.8	68.8	83.9	101.0	99.9
Services, etc.	150.6	167.1	195.5	234.7	260.1	282.2	326.4	379.7	433.3	517.6	531.5
Resource Balance	-87.8	-91.9	-132.8	-204.4	-233.6	-122.2	92.0	72.0	-64.1	11.9	-16.8
Exports of Goods & NFServices	68.5	81.7	93.1	106.5	135.4	223.8	410.5	479.3	414.3	500.7	584.0
Imports of Goods & NFServices	156.3	173.6	225.9	310.9	369.0	346.0	318.5	407.3	478.4	488.8	600.8
Domestic Absorption	482.2	525.1	625.0	780.8	867.2	839.3	822.7	950.3	1,100.4	1,186.7	1,315.3
Private Consumption, etc.	258.1	285.7	324.8	367.6	401.2	422.1	432.0	473.1	545.0	600.5	697.1
General Gov't Consumption	137.6	145.3	157.2	173.4	195.1	217.6	249.1	299.1	347.7	362.7	337.7
Gross Domestic Investment	86.5	94.1	143.0	239.8	270.9	199.6	141.6	178.1	207.7	223.5	280.5
Fixed Investment	81.1	87.3	136.9	229.3	258.9	185.8	130.2	158.5	175.0	200.6	239.4
Memo Items:											
Gross Domestic Saving	-1.3	2.2	10.2	35.4	37.3	77.4	233.6	250.1	143.6	235.4	263.7
Gross National Saving	-3.0	-1.1	3.9	26.7	8.9	45.3	172.3	165.9	83.0	148.6	192.4
	(Millions of 1980 Papua New Guinea Kina)										
Gross National Product	987.3	1,028.4	1,069.4	1,151.1	1,250.3	1,302.5	1,372.7	1,429.5	1,450.8	1,485.8	1,462.8
GDP at factor cost	956.0	992.6	1,035.1	1,116.9	1,231.4	1,304.5	1,382.7	1,483.1	1,515.4	1,489.4	1,436.8
Agriculture	392.9	401.5	418.6	427.3	435.8	461.5	478.6	504.2	524.8	530.7	536.2
Industry	..	..	..	..	..	..	..	..	..	..	..
Manufacturing	..	..	..	..	..	..	..	..	..	..	..
Services, etc.	..	..	..	..	..	..	..	..	..	..	..
Resource Balance	-262.7	-281.1	-390.2	-557.6	-560.5	-273.3	21.2	31.1	-9.9	-26.0	-78.4
Exports of Goods & NFServices	167.2	190.4	212.6	239.8	301.3	465.9	651.4	740.1	714.3	652.0	678.0
Imports of Goods & NFServices	429.9	471.6	602.7	797.4	861.8	739.2	630.2	709.1	724.2	678.0	756.4
Domestic Absorption	1,296.8	1,357.9	1,424.5	1,629.9	1,934.7	2,023.6	1,811.7	1,610.6	1,643.0	1,670.2	1,630.4
Private Consumption, etc.	614.0	638.8	682.2	734.8	771.6	795.6	822.9	825.7	831.7	853.9	884.6
General Gov't Consumption	428.7	436.8	452.0	453.1	464.1	471.3	458.3	466.2	474.7	438.6	411.0
Gross Domestic Investment	284.0	299.2	434.2	695.7	748.6	507.8	321.5	339.5	336.3	303.9	327.8
Fixed Investment	260.6	270.1	402.8	645.5	692.0	465.8	297.6	306.3	285.9	260.1	278.0
Memo Items:											
Capacity to Import	188.4	221.9	248.4	273.1	316.2	478.1	812.2	834.4	627.2	694.5	735.2
Terms of Trade Adjustment	21.2	31.5	35.8	33.4	14.9	12.2	160.8	94.3	-87.1	42.5	57.2
Gross Domestic Income	965.9	1,009.8	1,062.7	1,152.2	1,271.2	1,332.3	1,404.0	1,633.1	1,609.9	1,428.0	1,497.3
Gross National Income	952.3	996.6	1,045.0	1,126.6	1,218.6	1,251.3	1,315.5	1,505.9	1,465.1	1,331.9	1,429.0
DOMESTIC PRICES (DEFLATORS)	*(Index 1980 = 100)*										
Overall (GDP)	39.4	41.6	45.3	49.0	48.6	51.7	62.5	65.5	64.8	75.6	84.7
Domestic Absorption	37.2	38.7	43.9	47.9	44.8	41.5	45.4	59.0	67.0	71.0	80.7
Agriculture	42.1	46.3	49.2	50.2	50.1	50.3	53.9	56.8	58.7	71.6	79.9
Industry	..	..	..	..	..	..	..	..	..	..	..
Manufacturing	..	..	..	..	..	..	..	..	..	..	..
MANUFACTURING ACTIVITY											
Employment (1980=100)	53.4	55.3	65.5	69.1	74.9	80.5	80.2	85.2	82.4	88.9	90.7
Real Earnings per Empl. (1980=100)	60.4	68.7	68.3	75.0	77.9	83.4	80.6	72.5	81.0	87.6	96.1
Real Output per Empl. (1980=100)	..	..	..	..	..	..	..	..	..	..	..
Earnings as % of Value Added	42.9	43.6	41.9	41.8	40.6	42.8	39.5	39.2	40.4	40.7	37.4
MONETARY HOLDINGS	*(Millions of current Papua New Guinea Kina)*										
Money Supply, Broadly Defined	..	..	..	..	..	..	195.6	320.0	245.5	277.8	428.5
Money as Means of Payment	..	..	..	..	..	..	99.4	140.4	150.8	135.0	169.4
Currency Ouside Banks	..	..	..	..	..	..	57.4	76.6	85.2	47.5	56.7
Demand Deposits	..	..	..	..	..	..	42.0	63.8	65.6	87.5	112.7
Quasi-Monetary Liabilities	..	..	..	..	..	..	96.2	179.6	94.7	142.8	259.1
	(Millions of current Papua New Guinea Kina)										
GOVERNMENT DEFICIT (-) OR SURPLUS	..	..	..	..	..	..	..	..	-57.0	-26.8	-19.4
Current Revenue	..	..	..	..	..	..	..	..	330.1	353.5	389.0
Current Expenditure	..	..	..	..	..	..	..	..	251.7	315.4	353.0
Current Budget Balance	..	..	..	..	..	..	..	..	78.4	38.1	36.0
Capital Receipts	..	..	..	..	..	..	..	..	0.2	0.1	0.0
Capital Payments	..	..	..	..	..	..	..	..	135.5	65.0	55.4

1978	1979	1980	1981	1982	1983	1984	1985	1986	1987 estimate	NOTES	PAPUA NEW GUINEA
660	720	780	810	800	760	710	710	710	740	..	**CURRENT GNP PER CAPITA (US $)**
2,882	2,946	3,011	3,076	3,142	3,206	3,269	3,329	3,411	3,494	..	**POPULATION (thousands)**
				(Millions of current Papua New Guinea Kina)							**ORIGIN AND USE OF RESOURCES**
1,361.5	1,594.6	1,649.4	1,572.7	1,661.1	1,874.3	2,000.4	2,158.5	2,323.0	2,595.0	..	Gross National Product (GNP)
-19.4	-38.0	-58.7	-108.5	-88.0	-99.4	-133.6	-125.5	-129.0	-152.0	..	Net Factor Income from Abroad
1,380.9	1,632.6	1,708.1	1,681.2	1,749.1	1,973.7	2,134.0	2,284.0	2,452.0	2,747.0	f	Gross Domestic Product (GDP)
82.9	95.3	110.0	114.6	134.4	152.0	178.4	189.7	212.0	240.9	..	Indirect Taxes, net
1,298.0	1,537.3	1,598.1	1,566.6	1,614.7	1,821.7	1,955.6	2,094.3	2,240.0	2,506.1	B f	GDP at factor cost
497.5	552.8	565.7	561.4	567.2	662.1	767.0	777.9	833.7	..	..	Agriculture
351.0	474.8	458.5	390.7	406.4	488.8	437.3	563.4	638.4	..	..	Industry
134.0	152.2	162.3	166.6	164.6	178.8	210.9	203.2	221.3	..	..	Manufacturing
532.4	605.0	683.9	729.1	775.5	822.8	929.7	942.7	979.9	..	..	Services, etc.
-48.1	-1.8	-173.2	-344.7	-413.9	-363.5	-315.0	-239.0	-188.0	-149.0	..	Resource Balance
579.1	742.5	737.6	642.9	644.3	766.1	893.0	1,004.0	1,099.0	1,215.0	..	Exports of Goods & NFServices
627.2	744.3	910.8	987.6	1,058.2	1,129.6	1,208.0	1,243.0	1,287.0	1,364.0	..	Imports of Goods & NFServices
1,429.0	1,634.4	1,881.4	2,025.9	2,163.1	2,337.1	2,449.0	2,523.0	2,640.0	2,896.0	..	Domestic Absorption
763.3	879.7	1,039.5	1,113.4	1,132.5	1,239.6	1,339.0	1,485.0	1,563.0	1,691.0	..	Private Consumption, etc.
353.3	371.4	411.2	454.4	468.1	471.3	505.0	537.0	552.0	599.0	..	General Gov't Consumption
312.4	383.3	430.7	458.1	562.5	626.2	605.0	501.0	525.0	606.0	..	Gross Domestic Investment
268.5	326.2	394.2	450.7	576.7	632.4	534.0	467.0	557.0	572.0	..	Fixed Investment
											Memo Items:
264.3	381.5	257.4	113.4	148.5	262.8	290.0	262.0	337.0	457.0	..	Gross Domestic Saving
182.8	275.6	127.9	-80.2	-30.2	82.1	71.4	47.5	131.8	210.5	..	Gross National Saving
				(Millions of 1980 Papua New Guinea Kina)							
1,497.7	1,639.4	1,649.4	1,602.8	1,631.4	1,686.7	1,641.7	1,759.3	1,827.4	1,876.1	..	Gross National Product
1,437.2	1,566.6	1,598.1	1,585.7	1,539.8	1,646.7	..	..	..	..	B f	GDP at factor cost
548.5	552.2	565.7	610.4	601.9	619.9	601.9	618.9	645.5	703.9	..	Agriculture
..	..	458.5	406.3	409.6	444.4	445.5	525.9	569.3	604.1	f	Industry
..	..	162.3	165.3	154.0	158.3	177.1	164.3	170.2	171.2	..	Manufacturing
..	..	683.9	685.1	695.4	702.7	745.9	741.6	758.3	775.8	f	Services, etc.
-30.1	-91.5	-173.2	-134.9	-136.8	-117.6	-56.3	59.7	145.2	162.6	..	Resource Balance
742.0	740.0	737.6	777.4	764.8	778.1	825.9	914.1	986.7	1,025.4	..	Exports of Goods & NFServices
772.1	831.5	910.8	912.3	901.6	895.7	882.2	854.5	841.5	862.8	..	Imports of Goods & NFServices
1,697.5	1,786.3	1,881.3	1,836.7	1,843.6	1,884.6	1,849.5	1,826.6	1,827.9	1,921.3	..	Domestic Absorption
956.5	1,030.0	1,039.4	1,028.7	1,008.6	1,036.1	1,046.2	1,116.4	1,113.8	1,161.0	..	Private Consumption, etc.
418.0	425.6	411.2	395.4	367.7	360.3	364.6	373.3	366.3	384.6	..	General Gov't Consumption
336.8	413.1	430.7	412.7	467.4	488.2	438.7	336.9	347.7	375.7	..	Gross Domestic Investment
302.5	349.5	394.2	404.1	476.6	490.6	388.2	316.5	357.7	355.9	..	Fixed Investment
											Memo Items:
712.8	829.5	737.6	593.9	548.9	607.5	652.1	690.2	718.6	768.5	..	Capacity to Import
-29.1	89.5	0.0	-183.5	-215.9	-170.6	-173.7	-224.0	-268.1	-256.9	..	Terms of Trade Adjustment
1,523.3	1,568.6	1,708.1	1,518.3	1,491.0	1,596.4	1,619.4	1,662.3	1,705.0	1,827.0	..	Gross Domestic Income
1,475.6	1,531.0	1,649.4	1,419.3	1,415.6	1,516.0	1,468.0	1,535.3	1,559.3	1,619.2	..	Gross National Income
				(Index 1980 = 100)							**DOMESTIC PRICES (DEFLATORS)**
89.4	97.3	100.0	98.8	102.5	111.7	119.0	121.1	124.3	131.8	..	Overall (GDP)
84.2	91.5	100.0	110.3	117.3	124.0	132.4	138.1	144.4	150.7	..	Domestic Absorption
90.7	100.1	100.0	92.0	94.2	106.8	127.4	125.7	129.2	..	..	Agriculture
..	..	100.0	96.1	99.2	110.0	98.2	107.1	112.1	..	..	Industry
..	..	100.0	100.8	106.6	112.9	119.1	123.7	130.0	..	..	Manufacturing
											MANUFACTURING ACTIVITY
80.6	91.8	100.0	100.0	109.3	109.6	114.6	117.0	114.6	118.5	J	Employment (1980=100)
95.6	95.4	100.0	86.9	87.0	88.1	88.9	95.9	..	..	J	Real Earnings per Empl. (1980=100)
..	..	..	..	..	..	..	..	..	..	J	Real Output per Empl. (1980=100)
32.8	37.2	37.4	33.6	37.8	37.4	36.3	36.3	..	..	J	Earnings as % of Value Added
				(Millions of current Papua New Guinea Kina)							**MONETARY HOLDINGS**
449.0	571.7	553.1	554.7	567.8	650.0	748.0	820.1	929.0	943.3	..	Money Supply, Broadly Defined
179.0	195.4	201.6	194.4	188.6	206.0	249.0	244.2	256.7	281.3	..	Money as Means of Payment
61.5	67.9	70.5	73.4	72.4	79.9	88.9	94.3	95.7	106.4	..	Currency Ouside Banks
117.4	127.5	131.1	121.0	116.2	126.0	160.1	149.9	160.9	175.0	..	Demand Deposits
270.0	376.3	351.6	360.4	379.2	444.1	499.0	576.0	672.3	662.0	..	Quasi-Monetary Liabilities
				(Millions of current Papua New Guinea Kina)							**GOVERNMENT DEFICIT (-) OR SURPLUS**
-21.5	-62.8	-33.0	-107.7	-97.2	-94.6	-21.1	-56.1	..	..	F	GOVERNMENT DEFICIT (-) OR SURPLUS
431.4	458.8	566.8	555.9	555.7	616.9	723.6	703.8	..	..	..	Current Revenue
386.7	450.3	500.6	570.1	572.5	610.2	649.2	696.3	..	..	..	Current Expenditure
44.7	8.5	66.2	-14.2	-16.8	6.7	74.4	7.5	..	..	..	Current Budget Balance
0.0	0.0	0.1	0.3	0.8	0.3	4.7	0.8	..	..	..	Capital Receipts
66.1	71.3	99.4	93.8	81.2	101.6	100.3	64.4	..	..	..	Capital Payments

PAPUA NEW GUINEA	1967	1968	1969	1970	1971	1972	1973	1974	1975	1976	1977
FOREIGN TRADE (CUSTOMS BASIS)				*(Millions of current US dollars)*							
Value of Exports, fob	58.6	65.0	71.9	79.1	113.4	148.5	294.2	710.3	579.5	455.7	613.2
Nonfuel Primary Products	55.7	61.3	68.0	75.3	81.3	104.7	252.1	656.7	540.9	409.6	596.4
Fuels	0.2	0.3	0.2	0.0	0.0	0.0	0.1	0.1	0.1	0.1	0.0
Manufactures	2.6	3.3	3.7	3.8	32.1	43.8	42.1	53.5	38.5	46.0	16.9
Value of Imports, cif	178.3	178.4	220.1	300.3	355.9	336.4	355.9	502.7	555.7	502.3	642.1
Nonfuel Primary Products	38.0	42.8	54.7	61.0	69.8	70.4	84.6	134.8	123.8	117.2	139.8
Fuels	41.0	19.0	11.4	12.6	15.0	18.8	20.8	61.6	66.8	77.1	132.4
Manufactures	98.9	116.0	154.0	226.7	271.1	247.3	250.5	306.3	365.1	308.0	370.0
					(Index 1980 = 100)						
Terms of Trade	235.9	119.5	127.4	123.7	104.4	111.3	144.0	122.4	108.7	130.0	154.2
Export Prices, fob	23.4	30.3	31.6	33.3	31.1	34.3	56.0	69.3	63.5	74.9	93.7
Nonfuel Primary Products	23.5	31.3	32.2	33.2	29.4	32.9	57.4	69.5	63.3	75.9	94.5
Fuels	4.3	4.3	4.3	4.3	5.6	6.2	8.9	36.7	35.7	38.4	42.0
Manufactures	30.1	31.7	32.4	37.5	36.5	38.5	49.5	67.2	66.1	66.7	71.7
Import Prices, cif	9.9	25.3	24.8	27.0	29.8	30.9	38.9	56.6	58.4	57.6	60.8
Trade at Constant 1980 Prices					*(Millions of 1980 US dollars)*						
Exports, fob	250.8	214.9	227.3	237.3	364.4	432.7	525.1	1,024.3	912.5	608.7	654.3
Imports, cif	1,383.0	947.0	887.1	1,113.7	1,194.4	1,090.2	914.6	887.5	950.9	872.2	1,056.3
BALANCE OF PAYMENTS					*(Millions of current US dollars)*						
Exports of Goods & Services	..	..	..	123.0	159.0	305.0	613.0	718.0	560.0	640.1	751.2
Merchandise, fob	..	..	..	104.0	121.0	222.0	514.0	654.0	456.0	551.4	682.9
Nonfactor Services	..	..	..	15.0	33.0	75.0	81.0	42.0	87.0	68.9	38.0
Factor Services	..	..	..	4.0	5.0	8.0	18.0	22.0	17.0	19.8	30.2
Imports of Goods & Services	..	..	..	372.0	459.0	464.0	577.0	706.0	691.0	685.3	824.2
Merchandise, fob	..	..	..	300.0	356.0	299.0	290.0	428.0	504.0	434.3	559.5
Nonfactor Services	..	..	..	48.0	63.0	114.0	163.0	159.0	122.0	175.5	183.3
Factor Services	..	..	..	24.0	40.0	51.0	124.0	119.0	65.0	75.5	81.4
Long-Term Interest	..	..	..	9.7	17.4	25.8	32.8	35.0	32.0	32.1	33.5
Current Transfers, net	..	..	..	10.0	3.0	5.0	-5.0	-13.0	-7.0	-58.5	-56.3
Workers' Remittances	..	..	..	..	..	..	..	..	..	0.0	0.0
Total to be Financed	..	..	..	-239.0	-297.0	-154.0	31.0	-1.0	-138.0	-103.7	-129.3
Official Capital Grants	..	..	..	150.0	175.0	160.0	184.0	222.0	116.0	145.0	228.4
Current Account Balance	..	..	..	-89.0	-122.0	6.0	215.0	221.0	-22.0	41.3	99.1
Long-Term Capital, net	..	..	..	248.0	392.0	292.0	240.0	280.0	213.0	177.6	293.3
Direct Investment	..	..	..	..	..	19.0	89.0	52.0	34.0	19.0	18.0
Long-Term Loans	..	..	..	133.6	194.1	80.2	10.1	-8.6	9.6	-0.6	38.3
Disbursements	..	..	..	154.0	237.3	159.8	101.5	73.3	79.2	66.9	108.3
Repayments	..	..	..	20.4	43.2	79.6	91.4	81.9	69.6	67.5	70.0
Other Long-Term Capital	..	..	..	-35.6	22.9	32.8	-43.1	14.6	53.4	14.3	8.6
Other Capital, net	..	..	..	-85.0	-198.0	-301.4	-369.8	-401.5	-209.0	-176.1	-254.4
Change in Reserves	..	..	..	-74.0	-72.0	3.4	-85.2	-99.5	18.0	-42.8	-138.0
Memo Item:					*(Papua New Guinea Kina per US dollar)*						
Conversion Factor (Annual Avg)	0.890	0.890	0.890	0.890	0.880	0.830	0.700	0.700	0.760	0.790	0.790
EXTERNAL DEBT, ETC.				*(Millions of US dollars, outstanding at end of year)*							
Public/Publicly Guar. Long-Term	..	..	..	36.2	89.4	130.4	219.6	244.3	275.0	286.0	341.5
Official Creditors	..	..	..	4.0	25.1	51.4	74.2	84.7	99.0	102.8	121.4
IBRD and IDA	..	..	..	1.8	7.8	19.7	30.5	46.4	64.4	71.3	81.6
Private Creditors	..	..	..	32.2	64.2	79.1	145.3	159.6	176.1	183.2	220.1
Private Non-guaranteed Long-Term	..	..	..	172.6	317.8	364.6	306.0	261.0	231.0	206.0	205.3
Use of Fund Credit	..	..	..	0.0	0.0	0.0	0.0	0.0	0.0	34.7	36.3
Short-Term Debt	..	..	..	..	..	..	..	..	..	..	99.0
Memo Items:					*(Millions of US dollars)*						
Int'l Reserves Excluding Gold	..	..	..	..	..	..	31.3	32.5	179.7	257.2	426.6
Gold Holdings (at market price)	..	..	..	..	..	..	..	..	..	..	4.7
SOCIAL INDICATORS											
Total Fertility Rate	6.2	6.1	6.1	6.0	6.0	5.9	6.0	6.1	6.1	6.2	6.3
Crude Birth Rate	42.4	42.0	41.7	41.3	41.0	40.6	41.0	41.4	41.7	42.1	42.5
Infant Mortality Rate	130.0	125.0	120.0	115.0	110.0	105.0	101.0	97.0	93.0	89.0	85.0
Life Expectancy at Birth	45.2	45.7	46.2	46.7	47.2	47.7	48.2	48.7	49.2	49.7	50.3
Food Production, p.c. ('79-81 = 100)	100.5	101.2	102.0	101.4	102.2	102.7	100.6	103.4	104.9	103.2	101.5
Labor Force, Agriculture (%)	85.8	85.3	84.8	84.3	83.5	82.7	81.9	81.1	80.3	79.5	78.7
Labor Force, Female (%)	41.8	41.7	41.7	41.6	41.6	41.5	41.4	41.3	41.2	41.2	41.1
School Enroll. Ratio, primary	..	..	..	52.0	..	..	..	..	56.0	..	57.0
School Enroll. Ratio, secondary	..	..	..	8.0	..	..	..	..	12.0	12.0	11.0

1978	1979	1980	1981	1982	1983	1984	1985	1986	1987 estimate	NOTES	PAPUA NEW GUINEA
				(Millions of current US dollars)							**FOREIGN TRADE (CUSTOMS BASIS)**
756.0	877.3	1,033.0	863.6	791.1	820.0	892.0	911.9	1,033.2	1,172.2	..	Value of Exports, fob
736.7	857.3	972.9	782.9	718.7	747.3	817.4	856.3	971.7	1,099.9	..	Nonfuel Primary Products
0.1	0.1	0.1	0.1	0.5	0.5	0.6	0.6	0.5	0.8	..	Fuels
19.2	19.9	59.9	80.6	71.9	72.2	74.0	55.0	61.0	71.5	..	Manufactures
770.3	902.7	1,176.1	1,261.2	1,169.7	1,120.0	1,109.7	1,008.0	1,130.0	1,222.0	f	Value of Imports, cif
180.2	211.3	253.9	250.5	233.2	224.5	224.8	193.2	238.7	261.2	..	Nonfuel Primary Products
108.5	128.7	222.4	293.0	248.9	242.0	213.0	180.0	98.0	120.9	..	Fuels
481.6	562.7	699.8	717.7	687.7	653.5	671.9	634.8	793.3	840.0	f	Manufactures
				(Index 1980 = 100)							
122.9	119.4	100.0	86.0	91.1	95.1	102.5	94.8	91.7	84.4	..	Terms of Trade
87.2	101.2	100.0	88.2	87.5	88.0	92.4	83.1	82.5	82.9	..	Export Prices, fob
87.6	101.5	100.0	87.2	86.9	87.7	92.7	82.8	81.5	81.5	..	Nonfuel Primary Products
42.3	61.0	100.0	112.5	101.6	92.5	90.2	87.5	44.9	56.9	..	Fuels
76.1	90.0	100.0	99.1	94.8	90.7	89.3	89.0	103.2	113.0	..	Manufactures
71.0	84.7	100.0	102.5	96.0	92.5	90.2	87.7	90.0	98.2	..	Import Prices, cif
				(Millions of 1980 US dollars)							Trade at Constant 1980 Prices
866.7	867.1	1,033.0	979.4	903.7	932.0	964.8	1,097.2	1,252.6	1,414.8	..	Exports, fob
1,085.4	1,065.3	1,176.1	1,230.8	1,217.9	1,210.3	1,230.5	1,149.3	1,256.1	1,244.4	..	Imports, cif
				(Millions of current US dollars)							**BALANCE OF PAYMENTS**
785.8	974.0	1,133.4	972.3	927.8	947.7	1,006.8	1,028.9	1,194.1	1,350.9	..	Exports of Goods & Services
711.8	882.8	1,030.4	839.6	772.8	813.5	891.9	909.3	1,029.4	1,195.5	..	Merchandise, fob
35.7	41.8	43.3	85.3	95.3	75.7	64.3	64.2	72.4	97.2	..	Nonfactor Services
38.4	49.4	59.7	47.4	59.7	58.4	50.6	55.5	92.3	58.2	..	Factor Services
1,007.0	1,178.9	1,561.9	1,651.4	1,546.9	1,488.9	1,518.2	1,323.9	1,428.8	1,777.0	..	Imports of Goods & Services
687.4	782.8	1,020.7	1,098.3	1,018.0	975.0	962.6	875.4	925.3	1,207.5	..	Merchandise, fob
216.5	237.8	301.6	343.3	347.9	333.5	356.1	288.8	301.5	377.2	..	Nonfactor Services
103.1	158.3	239.5	209.9	181.0	180.4	199.5	159.7	202.0	192.3	..	Factor Services
35.9	37.2	51.8	73.3	103.3	121.0	149.6	131.6	134.5	157.1	..	Long-Term Interest
-87.6	-95.4	-105.7	-126.5	-123.0	-97.5	-95.1	-89.0	-78.4	-104.2	..	Current Transfers, net
0.0	0.0	0.0	0.0	0.0	0.0	0.0	0.0	0.0	0.0	..	Workers' Remittances
-308.8	-300.3	-534.1	-805.6	-742.1	-638.7	-606.6	-384.0	-313.1	-530.2	K	Total to be Financed
254.1	251.9	266.6	280.7	260.6	256.9	261.7	218.9	212.4	204.5	K	Official Capital Grants
-54.7	-48.4	-267.5	-525.0	-481.5	-381.9	-344.8	-165.1	-100.7	-325.8	..	Current Account Balance
220.5	296.0	382.7	678.8	742.8	708.6	497.8	338.4	333.2	323.9	..	Long-Term Capital, net
33.9	40.9	59.8	85.7	84.4	138.2	113.6	83.0	82.4	70.8	..	Direct Investment
-52.4	16.6	78.4	344.4	465.2	373.7	168.4	156.0	154.2	96.2	..	Long-Term Loans
30.7	66.9	150.0	424.6	578.1	534.5	394.2	394.9	408.2	444.1	..	Disbursements
83.1	50.3	71.6	80.2	112.9	160.8	225.8	238.9	254.0	347.9	..	Repayments
-15.1	-13.4	-22.1	-31.9	-67.4	-60.2	-46.0	-119.5	-115.8	-47.5	..	Other Long-Term Capital
-158.8	-135.5	-184.2	-204.9	-297.2	-229.5	-105.9	-176.3	-233.7	-4.7	..	Other Capital, net
-7.1	-112.1	68.9	51.0	35.9	-97.2	-47.1	3.0	1.2	6.5	..	Change in Reserves
				(Papua New Guinea Kina per US dollar)							Memo Item:
0.710	0.710	0.670	0.670	0.740	0.830	0.890	1.000	0.970	0.910	..	Conversion Factor (Annual Avg)
				(Millions of US dollars, outstanding at end of year)							**EXTERNAL DEBT, ETC.**
378.8	404.2	510.3	639.3	767.6	956.3	1,010.8	1,098.4	1,238.9	1,471.0	..	Public/Publicly Guar. Long-Term
146.0	159.2	213.1	240.0	275.3	362.1	412.8	500.2	611.0	818.6	..	Official Creditors
87.1	97.3	109.6	118.2	126.6	148.9	159.5	183.6	211.1	264.3	..	IBRD and IDA
232.8	245.0	297.2	399.4	492.2	594.2	598.0	598.2	627.9	652.5	..	Private Creditors
132.2	163.5	138.6	337.9	653.3	820.1	890.0	1,020.4	1,095.3	1,135.0	..	Private Non-guaranteed Long-Term
25.9	16.2	6.1	52.4	49.6	47.1	16.4	11.1	0.0	..	..	Use of Fund Credit
53.0	28.0	64.0	162.0	182.0	67.0	145.0	146.0	62.0	105.0	..	Short-Term Debt
				(Millions of US dollars)							Memo Items:
404.7	503.6	423.4	396.2	452.9	440.1	435.2	442.6	425.5	436.8	..	Int'l Reserves Excluding Gold
10.0	27.3	34.3	24.7	28.5	23.9	19.4	20.6	24.6	30.5	..	Gold Holdings (at market price)
											SOCIAL INDICATORS
6.1	6.0	5.9	5.8	5.6	5.5	5.4	5.3	5.2	5.1	..	Total Fertility Rate
41.8	41.0	40.3	39.5	38.8	38.2	37.6	36.9	36.3	35.7	..	Crude Birth Rate
82.8	80.6	78.4	76.2	74.0	75.7	77.3	79.0	91.0	102.9	..	Infant Mortality Rate
50.6	50.9	51.3	51.6	51.9	52.0	52.1	52.4	52.9	53.4	..	Life Expectancy at Birth
101.1	99.1	99.4	101.6	101.2	98.4	99.3	99.5	98.0	97.3	..	Food Production, p.c. ('79-81=100)
77.9	77.1	76.3	..	..	..	..	..	..	..	..	Labor Force, Agriculture (%)
41.0	41.0	40.9	40.7	40.5	40.2	40.0	39.8	39.6	39.4	..	Labor Force, Female (%)
58.0	60.0	62.0	60.0	60.0	63.5	..	64.4	..	..	..	School Enroll. Ratio, primary
11.0	11.0	11.0	11.0	11.0	13.6	14.0	14.5	..	..	..	School Enroll. Ratio, secondary

PARAGUAY	1967	1968	1969	1970	1971	1972	1973	1974	1975	1976	1977
CURRENT GNP PER CAPITA (US $)	230	240	240	260	270	300	350	440	550	620	720
POPULATION (thousands)	2,172	2,233	2,293	2,354	2,415	2,477	2,541	2,610	2,685	2,767	2,854

ORIGIN AND USE OF RESOURCES *(Billions of current Paraguayan Guaranies)*

	1967	1968	1969	1970	1971	1972	1973	1974	1975	1976	1977
Gross National Product (GNP)	61.1	64.2	68.5	73.7	82.5	94.9	123.3	166.0	188.9	212.1	261.5
Net Factor Income from Abroad	-1.0	-1.1	-1.5	-1.2	-1.3	-2.0	-2.1	-2.0	-1.5	-2.0	-2.1
Gross Domestic Product (GDP)	62.1	65.2	70.1	74.9	83.7	96.9	125.4	168.0	190.4	214.1	263.6
Indirect Taxes, net	4.5	5.1	5.3	5.5	5.9	6.0	6.4	6.8	7.9	8.9	10.3
GDP at factor cost	57.6	60.2	64.8	69.4	77.9	90.9	119.0	161.2	182.6	205.1	253.3
Agriculture	20.4	21.2	22.8	24.0	27.8	33.4	47.3	59.3	70.3	74.0	89.9
Industry	12.6	12.8	14.0	15.5	17.4	19.8	25.6	38.1	40.0	47.5	60.8
Manufacturing	10.1	10.6	11.4	12.5	13.7	15.7	20.0	30.3	29.8	34.2	45.0
Services, etc.	29.1	31.2	33.3	35.4	38.5	43.7	52.6	70.7	80.1	92.6	112.9
Resource Balance	-2.1	-2.9	-3.2	-0.9	-2.2	-0.1	0.9	-3.8	-10.4	-7.2	-7.9
Exports of Goods & NFServices	7.6	8.1	9.5	11.2	11.2	13.3	18.8	26.1	25.2	31.4	51.3
Imports of Goods & NFServices	9.7	10.9	12.8	12.1	13.3	13.4	17.9	29.9	35.6	38.6	59.2
Domestic Absorption	64.2	68.1	73.3	75.8	85.9	97.0	124.6	171.8	200.8	221.3	271.5
Private Consumption, etc.	49.1	52.3	55.8	58.0	66.6	74.6	92.5	127.3	143.0	155.2	190.1
General Gov't Consumption	4.8	5.4	6.3	6.7	7.1	7.8	8.2	9.2	12.0	13.4	16.4
Gross Domestic Investment	10.3	10.3	11.2	11.0	12.2	14.6	23.9	35.3	45.9	52.7	65.1
Fixed Investment	10.1	10.0	10.8	10.9	11.8	13.3	20.4	30.9	39.5	48.7	62.9
Memo Items:											
Gross Domestic Saving	8.1	7.5	8.0	10.1	10.0	14.5	24.7	31.5	35.5	45.5	57.2
Gross National Saving	7.5	6.8	6.8	9.2	9.3	12.9	22.9	29.7	34.1	43.6	55.2

(Billions of 1980 Paraguayan Guaranies)

	1967	1968	1969	1970	1971	1972	1973	1974	1975	1976	1977
Gross National Product	219.6	226.0	233.7	246.6	259.7	274.5	294.9	321.2	343.4	366.0	414.9
GDP at factor cost	207.1	212.1	221.1	232.6	245.3	262.9	285.0	312.1	332.0	354.3	393.1
Agriculture	84.3	85.2	87.0	92.4	93.7	97.3	103.5	113.6	122.9	127.5	135.9
Industry	42.5	43.3	46.0	49.9	53.8	57.6	63.3	68.4	70.1	76.4	91.0
Manufacturing	35.3	36.7	38.6	41.5	43.9	46.9	50.9	54.6	53.6	56.6	66.4
Services, etc.	96.2	101.2	105.8	108.3	116.1	125.2	133.3	143.1	153.1	165.8	182.0
Resource Balance	8.1	3.2	2.9	10.9	5.2	10.1	3.3	-2.2	4.6	-10.0	-23.2
Exports of Goods & NFServices	37.7	37.1	42.8	48.9	47.5	52.7	52.3	59.5	59.0	55.1	68.6
Imports of Goods & NFServices	29.6	33.9	39.9	38.0	42.3	42.6	49.0	61.6	54.5	65.1	91.8
Domestic Absorption	214.9	226.5	235.9	239.8	258.4	270.0	296.8	327.3	341.5	379.6	432.1
Private Consumption, etc.	167.5	178.9	183.5	186.0	201.2	208.1	222.1	248.2	254.0	276.1	315.4
General Gov't Consumption	16.7	18.6	21.2	22.8	22.9	22.9	20.7	18.8	22.9	24.5	28.0
Gross Domestic Investment	30.7	29.0	31.3	31.0	34.3	39.0	54.0	60.3	64.7	79.0	88.7
Fixed Investment	..	..	..	..	..	36.3	48.7	54.7	56.5	71.6	85.1
Memo Items:											
Capacity to Import	23.2	25.0	29.8	35.1	35.5	42.4	51.4	53.8	38.5	52.9	79.6
Terms of Trade Adjustment	-14.6	-12.1	-13.1	-13.7	-12.0	-10.3	-0.9	-5.7	-20.5	-2.2	10.9
Gross Domestic Income	208.4	217.5	225.7	236.9	251.6	269.8	299.2	319.4	325.6	367.4	419.9
Gross National Income	205.1	213.9	220.7	232.9	247.7	264.2	294.0	315.5	322.9	363.8	425.9

DOMESTIC PRICES (DEFLATORS) *(Index 1980 = 100)*

	1967	1968	1969	1970	1971	1972	1973	1974	1975	1976	1977
Overall (GDP)	27.8	28.4	29.4	29.9	31.8	34.6	41.8	51.7	55.0	57.9	64.5
Domestic Absorption	29.9	30.1	31.1	31.6	33.2	35.9	42.0	52.5	58.8	58.3	62.8
Agriculture	24.2	24.9	26.2	26.0	29.7	34.3	45.7	52.2	57.2	58.0	66.2
Industry	29.6	29.7	30.4	31.0	32.3	34.3	40.4	55.7	57.1	62.2	66.8
Manufacturing	28.6	28.8	29.5	30.1	31.3	33.4	39.4	55.5	55.5	60.5	67.7

MANUFACTURING ACTIVITY

	1967	1968	1969	1970	1971	1972	1973	1974	1975	1976	1977
Employment (1980=100)	..	..	..	..	..	..	..	..	..	..	..
Real Earnings per Empl. (1980=100)	..	..	..	..	..	..	..	..	..	..	..
Real Output per Empl. (1980=100)	..	..	..	..	..	..	..	..	..	..	..
Earnings as % of Value Added	..	..	..	..	..	..	..	..	..	..	..

MONETARY HOLDINGS *(Billions of current Paraguayan Guaranies)*

	1967	1968	1969	1970	1971	1972	1973	1974	1975	1976	1977
Money Supply, Broadly Defined	10.3	10.1	11.9	13.5	15.2	18.8	24.3	29.4	37.1	45.7	60.2
Money as Means of Payment	6.7	5.8	6.6	7.3	7.8	9.4	12.5	15.1	17.8	21.6	28.6
Currency Ouside Banks	3.0	3.3	3.5	4.0	4.4	5.1	6.5	7.6	8.9	10.3	13.3
Demand Deposits	3.6	2.5	3.1	3.3	3.4	4.3	6.0	7.6	8.9	11.3	15.2
Quasi-Monetary Liabilities	3.6	4.3	5.3	6.2	7.4	9.4	11.8	14.3	19.3	24.2	31.6

GOVERNMENT DEFICIT (-) OR SURPLUS *(Billions of current Paraguayan Guaranies)*

	1967	1968	1969	1970	1971	1972	1973	1974	1975	1976	1977
	..	..	..	..	..	-2	0	2	-1	-2	2
Current Revenue	..	..	..	..	..	11	13	18	21	24	31
Current Expenditure	..	..	..	..	..	10	11	14	17	19	22
Current Budget Balance	..	..	..	..	..	1	2	5	4	5	10
Capital Receipts	..	..	..	..	..	0	0	0	0	0	0
Capital Payments	..	..	..	..	..	2	2	3	5	7	8

1978	1979	1980	1981	1982	1983	1984	1985	1986	1987 estimate	NOTES	PARAGUAY
830	1,030	1,350	1,680	1,730	1,660	1,500	1,190	1,000	1,000	A	CURRENT GNP PER CAPITA (US $)
2,948	3,047	3,150	3,257	3,367	3,478	3,588	3,696	3,808	3,922	..	POPULATION (thousands)
			(Billions of current Paraguayan Guaranies)								ORIGIN AND USE OF RESOURCES
319.4	428.3	582.2	737.2	762.0	835.8	1,086.9	1,392.8	1,846.8	2,498.3	..	Gross National Product (GNP)
-3.2	-2.2	5.3	8.8	8.3	5.0	-2.4	-12.9	-9.3	-12.6	..	Net Factor Income from Abroad
322.5	430.5	576.9	728.4	753.7	830.8	1,089.3	1,405.7	1,856.0	2,510.9	..	Gross Domestic Product (GDP)
16.3	27.6	34.5	41.9	48.2	45.1	50.7	66.2	89.5	121.7	..	Indirect Taxes, net
306.2	402.9	542.4	686.6	705.5	785.7	1,038.6	1,339.5	1,766.6	2,389.2	B	GDP at factor cost
103.4	135.2	165.1	196.8	190.6	211.6	307.1	403.3	498.9	681.9	..	Agriculture
76.7	102.1	158.3	203.2	208.4	226.1	288.2	357.4	481.5	641.5	..	Industry
54.4	69.6	92.3	118.5	121.0	134.3	172.0	226.1	296.0	404.1	..	Manufacturing
142.4	193.2	253.5	328.5	354.6	393.1	494.0	645.1	875.6	1,187.5	..	Services, etc.
-11.9	-23.7	-77.3	-84.8	-73.3	-40.7	-86.0	-86.2	-215.5	-177.4	..	Resource Balance
59.4	69.1	88.3	83.9	91.1	77.0	146.8	317.1	381.6	541.1	..	Exports of Goods & NFServices
71.3	92.8	165.6	168.7	164.4	117.6	232.8	403.3	597.1	718.6	..	Imports of Goods & NFServices
334.5	454.2	654.2	813.3	827.1	871.5	1,175.4	1,491.9	2,071.5	2,688.4	..	Domestic Absorption
225.2	306.5	436.9	534.7	564.2	621.7	836.0	1,079.9	1,503.1	1,899.8	..	Private Consumption, etc.
21.5	24.7	34.7	48.6	52.3	58.0	69.3	90.2	121.8	162.8	..	General Gov't Consumption
87.7	123.0	182.6	229.9	210.6	191.7	270.0	321.8	446.6	625.8	..	Gross Domestic Investment
81.3	116.1	174.1	219.8	198.6	181.0	255.7	303.3	419.5	591.4	..	Fixed Investment
											Memo Items:
75.8	99.3	105.3	145.1	137.2	151.1	184.0	235.6	231.1	448.4	..	Gross Domestic Saving
72.7	97.5	110.6	154.2	145.7	156.2	182.1	223.3	222.1	436.0	..	Gross National Saving
			(Billions of 1980 Paraguayan Guaranies)								
453.0	506.8	582.2	634.8	624.4	598.8	614.3	629.1	631.7	656.7	..	Gross National Product
430.9	471.8	542.4	590.6	578.1	562.9	587.2	605.2	604.3	628.1	B	GDP at factor cost
142.7	152.2	165.1	181.7	182.5	178.0	188.6	197.2	185.2	198.2	..	Agriculture
105.3	121.1	158.3	169.5	161.8	150.7	155.8	156.1	159.0	161.7	..	Industry
74.2	81.5	92.3	96.3	92.8	88.8	92.8	97.5	96.1	99.5	..	Manufacturing
205.7	230.6	253.5	275.4	273.2	266.5	271.4	281.7	290.6	300.2	..	Services, etc.
-21.8	33.5	-77.3	-73.4	-53.8	-34.0	-46.1	-31.4	-50.8	-43.0	..	Resource Balance
82.4	146.9	88.3	90.6	100.8	72.0	79.7	94.8	114.4	139.3	..	Exports of Goods & NFServices
104.2	113.3	165.6	164.0	154.6	106.0	125.8	126.2	165.2	182.3	..	Imports of Goods & NFServices
475.5	470.4	654.2	700.1	671.4	629.2	661.8	666.4	685.6	703.0	..	Domestic Absorption
329.4	308.2	436.9	449.1	456.6	447.4	484.6	485.1	495.5	504.4	..	Private Consumption, etc.
32.7	30.4	34.7	41.5	42.4	43.8	41.2	42.8	43.8	45.8	..	General Gov't Consumption
113.4	131.8	182.6	209.5	172.4	138.0	136.1	138.5	146.4	152.9	..	Gross Domestic Investment
103.4	123.7	174.1	201.0	162.8	130.5	127.6	129.1	135.9	141.2	..	Fixed Investment
											Memo Items:
86.8	84.4	88.3	81.6	85.6	69.4	79.3	99.2	105.6	137.3	..	Capacity to Import
4.4	-62.4	0.0	-9.0	-15.2	-2.7	-0.4	4.4	-8.8	-2.0	..	Terms of Trade Adjustment
458.1	441.5	576.9	617.6	602.4	592.5	615.4	639.4	626.0	658.0	..	Gross Domestic Income
457.3	444.3	582.2	625.7	609.2	596.2	614.0	633.5	622.9	654.7	..	Gross National Income
			(Index 1980 = 100)								DOMESTIC PRICES (DEFLATORS)
71.1	85.4	100.0	116.2	122.1	139.6	176.9	221.4	292.4	380.4	..	Overall (GDP)
70.3	96.5	100.0	116.2	123.2	138.5	177.6	223.9	302.1	382.4	..	Domestic Absorption
72.5	88.8	100.0	108.3	104.5	118.9	162.9	204.5	269.4	344.1	..	Agriculture
72.8	84.3	100.0	119.9	128.8	150.1	185.0	229.0	302.9	396.9	..	Industry
73.4	85.4	100.0	123.0	130.4	151.1	185.3	232.0	308.0	406.2	..	Manufacturing
											MANUFACTURING ACTIVITY
..	..	..	..	..	..	..	..	..	..	..	Employment (1980=100)
..	..	..	..	..	..	..	..	..	..	..	Real Earnings per Empl. (1980=100)
..	..	..	..	..	..	..	..	..	..	..	Real Output per Empl. (1980=100)
..	..	..	..	..	..	..	..	..	..	..	Earnings as % of Value Added
			(Billions of current Paraguayan Guaranies)								MONETARY HOLDINGS
78.5	97.5	131.3	156.8	165.7	193.4	225.9	272.9	347.6	469.3	..	Money Supply, Broadly Defined
39.8	49.5	62.4	62.4	60.2	75.6	97.8	125.2	158.7	243.7	..	Money as Means of Payment
18.7	24.3	31.2	31.1	33.2	38.5	48.6	62.6	84.5	119.6	..	Currency Ouside Banks
21.1	25.2	31.2	31.3	27.0	37.1	49.2	62.6	74.2	124.1	..	Demand Deposits
38.7	48.0	68.9	94.4	105.5	117.8	128.1	147.7	189.0	225.7	..	Quasi-Monetary Liabilities
			(Billions of current Paraguayan Guaranies)								GOVERNMENT DEFICIT (-) OR SURPLUS
3	4	2	-11	3	-8	-18	..	28	..	..	Current Revenue
41	51	62	73	86	83	102	136	178	..	..	Current Expenditure
26	32	43	57	72	73	90	..	132	..	..	Current Budget Balance
14	19	19	15	13	11	12	..	46	..	..	Capital Receipts
0	0	0	0	0	1	0	0	1	..	..	Capital Payments
11	14	17	26	11	19	31	..	20	..	..	Capital Payments

PARAGUAY	1967	1968	1969	1970	1971	1972	1973	1974	1975	1976	1977
FOREIGN TRADE (CUSTOMS BASIS)					*(Millions of current US dollars)*						
Value of Exports, fob	..	..	..	..	..	..	..	..	..	..	..
Nonfuel Primary Products	..	..	..	..	..	..	..	..	..	..	..
Fuels	..	..	..	..	..	..	..	..	..	..	..
Manufactures	..	..	..	..	..	..	..	..	..	..	..
Value of Imports, cif	..	..	..	..	..	..	..	..	..	..	..
Nonfuel Primary Products	..	..	..	..	..	..	..	..	..	..	..
Fuels	..	..	..	..	..	..	..	..	..	..	..
Manufactures	..	..	..	..	..	..	..	..	..	..	..
Terms of Trade					*(Index 1980 = 100)*						
Export Prices, fob	..	..	..	..	..	..	..	..	..	..	..
Nonfuel Primary Products	..	..	..	..	..	..	..	..	..	..	..
Fuels	..	..	..	..	..	..	..	..	..	..	..
Manufactures	..	..	..	..	..	..	..	..	..	..	..
Import Prices, cif	..	..	..	..	..	..	..	..	..	..	..
Trade at Constant 1980 Prices					*(Millions of 1980 US dollars)*						
Exports, fob	..	..	..	..	..	..	..	..	..	..	..
Imports, cif	..	..	..	..	..	..	..	..	..	..	..
BALANCE OF PAYMENTS					*(Millions of current US dollars)*						
Exports of Goods & Services	..	..	..	90	90	106	151	213	234	258	420
Merchandise, fob	..	..	..	65	67	86	128	173	188	202	327
Nonfactor Services	..	..	..	24	23	20	22	34	34	40	63
Factor Services	..	..	..	0	1	1	2	6	12	16	30
Imports of Goods & Services	..	..	..	111	120	118	171	267	320	331	480
Merchandise, fob	..	..	..	77	83	79	127	198	227	236	360
Nonfactor Services	..	..	..	22	24	27	32	45	55	58	72
Factor Services	..	..	..	13	14	12	12	23	38	36	47
Long-Term Interest	..	..	..	..	..	..	..	7	9	10	13
Current Transfers, net	..	..	..	2	4	3	2	2	1	1	1
Workers' Remittances	..	..	..	0	0	0	0	0	1	0	0
Total to be Financed	..	..	..	-19	-26	-9	-19	-55	-85	-72	-59
Official Capital Grants	..	..	..	3	3	4	4	2	13	3	1
Current Account Balance	..	..	..	-16	-23	-5	-16	-53	-72	-69	-59
Long-Term Capital, net	..	..	..	22	28	24	33	55	99	121	86
Direct Investment	..	..	..	4	7	3	9	21	24	-3	22
Long-Term Loans	..	..	..	..	..	..	..	31	44	54	113
Disbursements	..	..	..	..	..	..	..	47	65	76	143
Repayments	..	..	..	..	..	..	..	16	22	22	30
Other Long-Term Capital	..	..	..	15	18	18	20	1	18	66	-49
Other Capital, net	..	..	..	0	-5	-12	6	28	3	-12	83
Change in Reserves	..	..	..	-6	-1	-7	-24	-30	-30	-41	-110
Memo Item:					*(Paraguayan Guaranies per US dollar)*						
Conversion Factor (Annual Avg)	126.000	126.000	126.000	126.000	126.000	126.000	126.000	126.000	126.000	126.000	126.000
EXTERNAL DEBT, ETC.					*(Millions of US dollars, outstanding at end of year)*						
Public/Publicly Guar. Long-Term	..	..	..	112	123	132	146	164	189	237	336
Official Creditors	..	..	..	82	91	101	113	127	139	178	268
IBRD and IDA	..	..	..	25	28	33	38	41	44	50	62
Private Creditors	..	..	..	30	32	31	33	38	51	59	67
Private Non-guaranteed Long-Term	..	..	..	..	..	..	..	23	39	44	62
Use of Fund Credit	..	..	..	0	0	0	0	0	0	0	0
Short-Term Debt	..	..	..	..	..	..	..	..	..	..	54
Memo Items:					*(Millions of US dollars)*						
Int'l Reserves Excluding Gold	12.2	12.1	10.3	17.5	21.0	31.4	57.0	87.1	115.0	157.5	267.8
Gold Holdings (at market price)	0.1	0.1	0.1	0.1	0.1	0.2	0.3	0.4	0.3	0.3	1.0
SOCIAL INDICATORS											
Total Fertility Rate	6.4	6.3	6.1	6.0	5.8	5.7	5.5	5.4	5.3	5.2	5.1
Crude Birth Rate	39.5	38.9	38.3	37.7	37.2	36.6	36.3	36.1	35.9	35.6	35.4
Infant Mortality Rate	67.0	64.2	61.4	58.6	55.8	53.0	52.2	51.4	50.6	49.8	49.0
Life Expectancy at Birth	65.0	65.1	65.2	65.3	65.5	65.6	65.7	65.8	65.8	65.9	66.0
Food Production, p.c. ('79-81 = 100)	93.8	91.5	90.1	95.7	95.8	92.1	90.4	92.9	89.0	91.8	96.5
Labor Force, Agriculture (%)	53.8	53.4	53.0	52.6	52.2	51.8	51.4	51.0	50.6	50.2	49.8
Labor Force, Female (%)	21.3	21.3	21.3	21.3	21.2	21.2	21.2	21.1	21.1	21.0	21.0
School Enroll. Ratio, primary	..	..	..	109.0	..	..	..	..	102.0	103.0	102.0
School Enroll. Ratio, secondary	..	..	..	17.0	..	..	..	..	20.0	23.0	23.0

1978	1979	1980	1981	1982	1983	1984	1985	1986	1987 estimate	NOTES	PARAGUAY
				(Millions of current US dollars)							**FOREIGN TRADE (CUSTOMS BASIS)**
..	..	310	296	330	258	335	304	233	353	f	Value of Exports, fob
..	..	274	263	299	240	315	287	211	..	..	Nonfuel Primary Products
..	..	0	0	0	0	0	0	0	..	..	Fuels
..	..	37	33	31	18	19	17	21	..	..	Manufactures
..	..	615	600	672	546	586	502	578	616	f	Value of Imports, cif
..	..	78	84	76	59	49	56	64	..	..	Nonfuel Primary Products
..	..	170	127	185	149	172	142	124	..	..	Fuels
..	..	367	389	411	339	364	304	391	..	..	Manufactures
				(Index 1980 = 100)							Terms of Trade
..	..	..	..	..	..	..	..	..	..	..	Terms of Trade
..	..	..	..	..	..	..	..	..	..	..	Export Prices, fob
..	..	..	..	..	..	..	..	..	..	..	Nonfuel Primary Products
..	..	..	..	..	..	..	..	..	..	..	Fuels
..	..	..	..	..	..	..	..	..	..	..	Manufactures
..	..	..	..	..	..	..	..	..	..	..	Import Prices, cif
				(Millions of 1980 US dollars)							Trade at Constant 1980 Prices
..	..	..	..	..	..	..	..	..	..	..	Exports, fob
..	..	..	..	..	..	..	..	..	..	..	Imports, cif
				(Millions of current US dollars)							**BALANCE OF PAYMENTS**
489	594	781	772	792	590	652	650	800	1,044	..	Exports of Goods & Services
356	385	583	561	492	443	478	451	563	793	..	Merchandise, fob
78	129	118	105	178	84	104	118	174	191	..	Nonfactor Services
55	80	80	106	123	63	71	80	62	60	..	Factor Services
608	807	1,399	1,429	1,300	880	1,047	1,021	1,316	1,467	..	Imports of Goods & Services
432	577	1,054	1,070	848	623	761	730	898	935	..	Merchandise, fob
96	154	260	269	361	183	195	185	266	372	..	Nonfactor Services
80	76	85	90	91	74	92	106	152	160	..	Factor Services
18	30	45	36	45	48	61	80	91	96	..	Long-Term Interest
1	3	0	3	2	1	3	2	1	1	..	Current Transfers, net
0	1	2	2	1	0	0	0	0	..	..	Workers' Remittances
-118	-210	-618	-654	-506	-288	-392	-370	-516	-422	K	Total to be Financed
5	4	5	3	4	5	7	6	10	11	K	Official Capital Grants
-113	-206	-613	-651	-503	-283	-386	-364	-505	-411	..	Current Account Balance
172	140	565	606	627	557	403	324	272	237	..	Long-Term Capital, net
20	50	30	26	34	6	1	12	31	9	..	Direct Investment
126	133	125	106	244	229	163	162	104	84	..	Long-Term Loans
158	187	206	201	304	283	240	244	236	214	..	Disbursements
32	54	81	95	61	54	77	82	132	131	..	Repayments
21	-47	405	470	346	317	232	144	128	134	..	Other Long-Term Capital
118	234	199	77	-254	-307	-114	-24	154	212	..	Other Capital, net
-177	-167	-151	-32	130	34	96	64	79	-38	..	Change in Reserves
				(Paraguayan Guaranies per US dollar)							**Memo Item:**
126.000	126.000	126.000	126.000	136.000	146.000	243.700	440.700	517.400	550.000	..	Conversion Factor (Annual Avg)
				(Millions of US dollars, outstanding at end of year)							**EXTERNAL DEBT, ETC.**
445	524	633	709	940	1,145	1,253	1,537	1,832	2,218	..	Public/Publicly Guar. Long-Term
338	369	410	459	559	708	790	1,045	1,263	1,350	..	Official Creditors
74	93	124	157	195	216	225	294	360	416	..	IBRD and IDA
107	155	224	251	381	437	463	492	569	868	..	Private Creditors
86	138	151	133	132	130	110	104	86	28	..	Private Non-guaranteed Long-Term
0	0	0	0	0	0	0	0	0	..	..	Use of Fund Credit
84	145	174	308	226	133	98	151	122	201	..	Short-Term Debt
				(Millions of US dollars)							**Memo Items:**
448.7	609.1	761.9	805.7	739.0	680.2	666.3	533.6	446.7	497.0	..	Int'l Reserves Excluding Gold
2.4	18.0	20.9	14.1	16.2	13.5	10.9	11.6	13.8	17.1	..	Gold Holdings (at market price)
											SOCIAL INDICATORS
5.0	5.0	4.9	4.9	4.8	4.8	4.7	4.7	4.6	4.6	..	Total Fertility Rate
35.4	35.5	35.6	35.7	35.8	35.6	35.4	35.2	35.1	34.9	..	Crude Birth Rate
48.2	47.4	46.6	45.8	45.0	44.2	43.4	42.6	41.9	41.1	..	Infant Mortality Rate
66.1	66.2	66.3	66.4	66.4	66.5	66.6	66.7	66.8	66.9	..	Life Expectancy at Birth
93.2	99.3	98.3	102.4	102.0	102.4	107.6	110.7	100.1	108.8	..	Food Production, p.c. ('79-81 = 100)
49.4	49.0	48.6	..	..	..	..	..	..	..	..	Labor Force, Agriculture (%)
20.9	20.8	20.8	20.8	20.8	20.8	20.8	20.8	20.8	20.7	..	Labor Force, Female (%)
102.0	102.0	..	..	103.0	102.0	101.0	101.0	..	..	..	School Enroll. Ratio, primary
24.0	26.0	..	..	30.0	30.0	31.0	31.0	..	..	..	School Enroll. Ratio, secondary

PERU	1967	1968	1969	1970	1971	1972	1973	1974	1975	1976	1977
CURRENT GNP PER CAPITA (US $)	490	480	490	520	580	620	700	850	1,000	1,040	970
POPULATION (thousands)	12,129	12,474	12,828	13,193	13,568	13,952	14,347	14,750	15,161	15,579	16,004

ORIGIN AND USE OF RESOURCES *(Billions of current Peruvian Intis)*

	1967	1968	1969	1970	1971	1972	1973	1974	1975	1976	1977
Gross National Product (GNP)	1.8E-01	2.1E-01	2.4E-01	2.8E-01	3.1E-01	3.4E-01	4.1E-01	5.2E-01	6.6E-01	8.6E-01	1.1E+00
Net Factor Income from Abroad	0.0E+00	-1.0E-02	-1.0E-02	0.0E+00	0.0E+00	0.0E+00	-1.0E-02	-1.0E-02	-1.0E-02	-2.0E-02	-1.1E-01
Gross Domestic Product (GDP)	1.8E-01	2.2E-01	2.4E-01	2.8E-01	3.1E-01	3.5E-01	4.1E-01	5.2E-01	6.7E-01	8.8E-01	1.2E+00
Indirect Taxes, net	1.0E-02	2.0E-02	2.0E-02	2.0E-02	2.0E-02	3.0E-02	3.0E-02	3.0E-02	5.0E-02	5.0E-02	-9.9E-01
GDP at factor cost	1.7E-01	2.0E-01	2.2E-01	2.6E-01	2.9E-01	3.2E-01	3.8E-01	4.9E-01	6.2E-01	8.2E-01	2.2E+00
Agriculture	3.0E-02	4.0E-02	4.0E-02	5.0E-02	6.0E-02	6.0E-02	7.0E-02	8.0E-02	1.1E-01	1.3E-01	1.9E-01
Industry	6.0E-02	7.0E-02	8.0E-02	9.0E-02	1.0E-01	1.1E-01	1.4E-01	1.8E-01	2.1E-01	3.0E-01	4.1E-01
Manufacturing	3.0E-02	4.0E-02	4.0E-02	6.0E-02	6.0E-02	7.0E-02	9.0E-02	1.1E-01	1.3E-01	1.9E-01	2.5E-01
Services, etc.	9.0E-02	1.1E-01	1.2E-01	1.4E-01	1.6E-01	1.8E-01	2.1E-01	2.6E-01	3.5E-01	4.4E-01	5.9E-01
Resource Balance	-1.0E-02	0.0E+00	0.0E+00	1.0E-02	0.0E+00	0.0E+00	-1.0E-02	-4.0E-02	-7.0E-02	-7.0E-02	-8.0E-02
Exports of Goods & NFServices	3.0E-02	4.0E-02	4.0E-02	5.0E-02	4.0E-02	5.0E-02	6.0E-02	8.0E-02	7.0E-02	1.1E-01	1.9E-01
Imports of Goods & NFServices	4.0E-02	4.0E-02	4.0E-02	4.0E-02	5.0E-02	5.0E-02	7.0E-02	1.1E-01	1.5E-01	1.8E-01	2.7E-01
Domestic Absorption	1.9E-01	2.2E-01	2.4E-01	2.7E-01	3.2E-01	3.5E-01	4.2E-01	5.6E-01	7.4E-01	9.5E-01	1.3E+00
Private Consumption, etc.	1.1E-01	1.4E-01	1.6E-01	2.0E-01	2.2E-01	2.5E-01	2.9E-01	3.7E-01	4.9E-01	6.5E-01	8.8E-01
General Gov't Consumption	2.0E-02	2.0E-02	2.0E-02	3.0E-02	4.0E-02	5.0E-02	5.0E-02	6.0E-02	8.0E-02	1.1E-01	1.7E-01
Gross Domestic Investment	6.0E-02	6.0E-02	6.0E-02	4.0E-02	6.0E-02	6.0E-02	8.0E-02	1.3E-01	1.6E-01	1.9E-01	2.3E-01
Fixed Investment	4.0E-02	4.0E-02	5.0E-02	4.0E-02	5.0E-02	5.0E-02	8.0E-02	1.2E-01	1.5E-01	1.8E-01	2.5E-01
Memo Items:											
Gross Domestic Saving	6.0E-02	5.0E-02	6.0E-02	5.0E-02	5.0E-02	5.0E-02	8.0E-02	1.0E-01	9.0E-02	1.2E-01	1.5E-01
Gross National Saving	5.0E-02	5.0E-02	6.0E-02	5.0E-02	5.0E-02	5.0E-02	7.0E-02	9.0E-02	8.0E-02	1.0E-01	4.0E-02

(Millions of 1980 Peruvian Intis)

	1967	1968	1969	1970	1971	1972	1973	1974	1975	1976	1977
Gross National Product	3,643.7	3,649.8	3,776.3	4,051.4	4,241.3	4,364.6	4,586.1	5,027.2	5,196.9	5,243.7	5,232.0
GDP at factor cost	..	..	..	3,851.8	4,018.2	4,135.6	4,185.3	4,666.1	4,945.9	5,058.6	5,187.8
Agriculture	578.2	561.8	588.2	648.4	636.4	598.0	592.6	621.8	618.4	631.7	627.4
Industry	1,437.1	1,445.9	1,466.5	1,595.9	1,652.5	1,723.0	1,834.4	2,011.4	2,029.1	2,102.8	2,128.7
Manufacturing	779.4	794.2	803.2	872.3	922.0	941.1	1,002.3	1,086.8	1,120.7	1,161.3	1,141.7
Services, etc.	1,721.3	1,742.9	1,840.3	1,873.6	2,003.9	2,096.9	2,229.4	2,454.7	2,618.1	2,632.1	2,631.6
Resource Balance	-164.1	-16.4	-61.1	94.9	25.7	117.2	-190.9	-430.0	-371.5	-177.6	-58.6
Exports of Goods & NFServices	986.1	1,017.0	966.8	986.6	957.7	1,046.7	852.9	897.3	917.9	949.4	1,072.4
Imports of Goods & NFServices	1,150.2	1,033.3	1,028.0	891.7	932.0	929.5	1,043.9	1,327.3	1,289.4	1,127.0	1,131.0
Domestic Absorption	3,900.7	3,766.9	3,956.1	4,023.0	4,267.2	4,300.6	4,847.3	5,518.0	5,637.0	5,544.3	5,446.2
Private Consumption, etc.	2,044.6	2,355.9	2,498.7	2,917.0	3,018.9	3,120.7	3,235.2	3,450.5	3,594.9	3,663.3	3,666.7
General Gov't Consumption	304.3	323.6	334.3	404.2	432.0	460.6	488.3	517.6	575.0	602.6	689.5
Gross Domestic Investment	1,551.7	1,087.4	1,123.0	701.8	816.3	719.3	1,123.8	1,549.8	1,467.1	1,278.3	1,090.0
Fixed Investment	921.8	784.6	809.5	693.0	782.0	811.6	1,109.9	1,405.5	1,456.1	1,253.2	1,149.6
Memo Items:											
Capacity to Import	880.0	993.2	1,069.0	996.4	873.5	888.2	906.4	876.9	637.6	671.3	791.3
Terms of Trade Adjustment	-106.1	-23.7	102.1	9.8	-84.2	-158.5	53.5	-20.4	-280.3	-278.1	-281.1
Gross Domestic Income	3,630.5	3,726.8	3,997.1	4,127.7	4,208.7	4,259.4	4,709.8	5,067.6	4,985.3	5,088.6	5,106.6
Gross National Income	3,537.7	3,626.0	3,878.5	4,061.2	4,157.1	4,206.1	4,639.6	5,006.9	4,916.6	4,965.6	4,950.9

DOMESTIC PRICES (DEFLATORS) *(Index 1980 = 100)*

	1967	1968	1969	1970	1971	1972	1973	1974	1975	1976	1977
Overall (GDP)	4.9	5.8	6.2	6.8	7.3	7.8	8.9	10.3	12.6	16.3	22.1
Domestic Absorption	4.9	5.8	6.1	6.8	7.4	8.1	8.7	10.2	13.1	17.1	23.4
Agriculture	5.9	7.1	7.4	8.1	8.7	9.7	11.5	13.4	17.6	21.2	29.8
Industry	3.9	4.6	5.3	5.5	6.0	6.5	7.6	9.0	10.4	14.2	19.4
Manufacturing	4.0	4.9	5.4	6.3	7.0	7.5	8.5	10.4	11.9	16.3	21.8

MANUFACTURING ACTIVITY

	1967	1968	1969	1970	1971	1972	1973	1974	1975	1976	1977
Employment (1980=100)	64.8	72.0	68.5	71.5	56.8	77.1	77.5	..	..	..	..
Real Earnings per Empl. (1980=100)	82.8	82.5	97.6	..	..	119.6	117.8	..	..	..	..
Real Output per Empl. (1980=100)	62.8	67.6	71.8	83.2	106.2	83.3	88.3	..	..	..	..
Earnings as % of Value Added	22.4	20.3	22.6	..	..	20.5	18.4	16.5	16.5	16.5	16.5

MONETARY HOLDINGS *(Billions of current Peruvian Intis)*

	1967	1968	1969	1970	1971	1972	1973	1974	1975	1976	1977
Money Supply, Broadly Defined	3.0E-02	4.0E-02	4.0E-02	5.0E-02	6.0E-02	7.0E-02	9.0E-02	1.2E-01	1.4E-01	1.8E-01	2.2E-01
Money as Means of Payment	2.0E-02	2.0E-02	3.0E-02	4.0E-02	5.0E-02	6.0E-02	7.0E-02	1.0E-01	1.2E-01	1.5E-01	1.8E-01
Currency Ouside Banks	1.0E-02	1.0E-02	1.0E-02	2.0E-02	2.0E-02	2.0E-02	3.0E-02	3.0E-02	4.0E-02	5.0E-02	6.0E-02
Demand Deposits	1.0E-02	1.0E-02	1.0E-02	2.0E-02	3.0E-02	4.0E-02	5.0E-02	7.0E-02	8.0E-02	1.0E-01	1.2E-01
Quasi-Monetary Liabilities	1.0E-02	1.0E-02	1.0E-02	1.0E-02	2.0E-02	2.0E-02	2.0E-02	2.0E-02	2.0E-02	3.0E-02	4.0E-02

GOVERNMENT DEFICIT (-) OR SURPLUS *(Millions of current Peruvian Intis)*

	1967	1968	1969	1970	1971	1972	1973	1974	1975	1976	1977
GOVERNMENT DEFICIT (-) OR SURPLUS	..	..	..	-1	-6	-3	-12	-10	-18	-32	-34
Current Revenue	..	..	..	39	39	52	52	67	89	114	163
Current Expenditure	..	..	..	36	35	40	51	57	81	111	146
Current Budget Balance	..	..	..	3	4	12	1	10	7	3	17
Capital Receipts	..	..	..	..	..	..	..	0	..	..	..
Capital Payments	..	..	..	4	10	15	14	20	26	35	51

1978	1979	1980	1981	1982	1983	1984	1985	1986	1987 estimate	NOTES	PERU
840	850	990	1,240	1,380	1,140	1,110	1,010	1,150	1,430	..	**CURRENT GNP PER CAPITA (US $)**
16,432	16,863	17,295	17,725	18,152	18,574	18,984	19,383	19,831	20,287	..	**POPULATION (thousands)**
				(Billions of current Peruvian Intis)							**ORIGIN AND USE OF RESOURCES**
1.81	3.28	5.69	10.24	17.27	30.79	68.87	189.62	370.40	750.54	..	Gross National Product (GNP)
-0.09	-0.21	-0.28	-0.41	-0.68	-1.77	-3.98	-10.23	-10.62	-9.62	..	Net Factor Income from Abroad
1.90	3.49	5.97	10.66	17.95	32.56	72.84	199.84	381.02	760.17	..	Gross Domestic Product (GDP)
0.17	0.36	0.57	1.03	1.71	2.75	6.20	19.84	28.67	41.71	..	Indirect Taxes, net
1.74	3.13	5.40	9.63	16.24	29.82	66.65	180.01	352.35	718.46	B	GDP at factor cost
0.25	0.41	0.61	1.10	1.71	3.43	7.98	18.63	45.84	80.36	..	Agriculture
0.73	1.49	2.50	4.06	6.97	12.10	28.13	85.95	137.55	253.55	f	Industry
0.43	0.82	1.21	1.96	3.32	6.07	14.86	50.01	94.10	171.14	..	Manufacturing
0.92	1.59	2.85	5.50	9.27	17.04	36.73	95.26	197.64	426.25	f	Services, etc.
0.01	0.30	0.00	-0.64	-1.02	-0.34	2.69	9.31	-8.03	-12.81	..	Resource Balance
0.39	0.97	1.33	1.74	2.81	5.88	13.49	39.61	43.71	71.89	..	Exports of Goods & NFServices
0.38	0.67	1.33	2.38	3.83	6.22	10.79	30.30	51.74	84.70	..	Imports of Goods & NFServices
1.89	3.19	5.97	11.30	18.97	32.91	70.15	190.54	389.06	772.98	..	Domestic Absorption
1.31	2.13	3.66	6.74	11.07	21.10	45.53	123.41	254.59	506.01	..	Private Consumption, etc.
0.21	0.30	0.67	1.22	2.26	4.08	8.10	22.46	42.34	80.14	..	General Gov't Consumption
0.37	0.76	1.64	3.35	5.64	7.73	16.52	44.67	92.13	186.82	..	Gross Domestic Investment
0.38	0.72	1.40	2.77	4.98	7.56	16.74	44.25	82.65	154.67	..	Fixed Investment
											Memo Items:
0.38	1.06	1.64	2.70	4.62	7.39	19.21	53.98	84.09	174.01	..	Gross Domestic Saving
0.29	0.85	1.37	2.29	3.94	5.62	15.23	43.75	73.47	164.39	..	Gross National Saving
				(Millions of 1980 Peruvian Intis)							
5,150.0	5,364.1	5,690.4	5,993.4	6,014.2	5,185.4	5,429.6	5,576.8	6,256.3	6,792.4	..	Gross National Product
4,930.0	5,056.1	5,395.6	5,682.5	5,687.8	5,026.7	5,248.2	5,285.7	5,962.7	6,496.7	B f	GDP at factor cost
625.1	652.1	610.3	665.4	685.2	611.4	683.7	709.2	747.8	780.7	..	Agriculture
2,194.4	2,374.1	2,504.6	2,531.6	2,539.2	2,148.3	2,244.9	2,306.9	2,545.6	2,757.4	..	Industry
1,095.6	1,141.0	1,206.4	1,214.5	1,202.4	999.3	1,054.2	1,106.0	1,282.2	1,431.4	..	Manufacturing
2,579.7	2,683.7	2,853.6	3,038.8	3,026.0	2,723.9	2,814.4	2,861.3	3,142.2	3,341.3	..	Services, etc.
355.8	445.9	2.8	-247.9	-202.8	122.0	435.6	573.1	186.2	178.5	..	Resource Balance
1,211.1	1,466.3	1,331.7	1,292.6	1,371.7	1,229.9	1,342.0	1,401.0	1,228.1	1,260.0	..	Exports of Goods & NFServices
855.3	1,020.3	1,329.0	1,540.5	1,574.5	1,107.9	906.4	828.0	1,041.9	1,081.5	..	Imports of Goods & NFServices
5,043.4	5,264.0	5,965.6	6,483.7	6,453.1	5,361.6	5,307.4	5,304.3	6,249.5	6,700.9	..	Domestic Absorption
3,417.2	3,514.6	3,658.5	3,850.6	3,874.8	3,543.4	3,633.0	3,713.3	4,188.7	4,459.3	..	Private Consumption, etc.
602.1	545.0	666.7	656.2	743.4	678.2	647.2	670.1	698.7	709.8	..	General Gov't Consumption
1,024.1	1,204.4	1,640.4	1,976.9	1,834.9	1,139.9	1,027.2	920.9	1,362.1	1,531.9	..	Gross Domestic Investment
1,047.4	1,150.0	1,401.7	1,627.6	1,594.7	1,131.8	1,061.9	949.8	1,179.3	1,369.2	..	Fixed Investment
											Memo Items:
884.1	1,482.8	1,331.7	1,124.3	1,153.7	1,046.6	1,132.4	1,082.2	880.2	917.9	..	Capacity to Import
-327.0	16.5	0.0	-168.3	-218.0	-183.3	-209.6	-318.8	-348.0	-342.1	..	Terms of Trade Adjustment
5,072.2	5,726.4	5,968.4	6,067.5	6,032.3	5,300.3	5,533.4	5,558.6	6,087.7	6,537.3	..	Gross Domestic Income
4,823.0	5,380.6	5,690.4	5,825.1	5,796.2	5,002.1	5,220.0	5,258.0	5,908.4	6,450.3	..	Gross National Income
				(Index 1980 = 100)							**DOMESTIC PRICES (DEFLATORS)**
35.2	61.1	100.0	170.9	287.2	593.9	1268.4	3400.2	5920.5	11049.8	..	Overall (GDP)
37.4	60.6	100.0	174.3	294.0	613.8	1321.8	3592.1	6225.4	11535.4	..	Domestic Absorption
39.9	62.5	100.0	165.1	249.5	560.4	1167.5	2627.3	6129.4	10292.7	..	Agriculture
33.2	63.0	100.0	160.4	274.5	563.1	1253.0	3725.9	5403.2	9195.5	..	Industry
39.5	71.8	100.0	161.0	275.8	607.0	1409.4	4521.6	7339.0	11956.5	..	Manufacturing
											MANUFACTURING ACTIVITY
..	96.9	100.0	87.5	106.9	110.0	113.3	116.7	..	..	G	Employment (1980=100)
..	95.9	100.0	109.3	103.3	85.6	87.0	104.0	..	..	G	Real Earnings per Empl. (1980=100)
..	85.5	100.0	90.6	75.5	68.6	66.2	79.4	..	..	G	Real Output per Empl. (1980=100)
16.5	14.8	15.5	21.1	21.1	19.2	19.2	19.2	..	..	..	Earnings as % of Value Added
				(Billions of current Peruvian Intis)							**MONETARY HOLDINGS**
0.36	0.69	1.26	2.12	3.60	7.32	17.01	43.77	66.83	..	..	Money Supply, Broadly Defined
0.27	0.45	0.71	1.04	1.41	2.76	5.97	23.00	42.71	..	..	Money as Means of Payment
0.09	0.16	0.28	0.44	0.63	1.14	2.54	8.28	16.65	..	..	Currency Ouside Banks
0.17	0.29	0.44	0.60	0.78	1.62	3.43	14.72	26.07	..	..	Demand Deposits
0.09	0.23	0.55	1.08	2.19	4.56	11.04	20.76	24.12	..	..	Quasi-Monetary Liabilities
				(Millions of current Peruvian Intis)							
-6	57	-38	-321	..	..	..	..	..	..	..	**GOVERNMENT DEFICIT (-) OR SURPLUS**
283	544	1,008	1,510	..	..	8,666	28,094	44,408	..	..	Current Revenue
222	375	789	1,383	2,013	4,050	7,993	22,576	39,601	..	..	Current Expenditure
62	168	219	127	..	..	673	5,518	4,807	..	..	Current Budget Balance
..	0	0	0	..	..	92	272	..	..	..	Capital Receipts
67	112	258	448	639	-4,050	-7,993	-22,576	-39,601	..	..	Capital Payments

PERU	1967	1968	1969	1970	1971	1972	1973	1974	1975	1976	1977
FOREIGN TRADE (CUSTOMS BASIS)				*(Millions of current US dollars)*							
Value of Exports, fob	755.2	866.1	865.5	1,044.4	892.9	944.4	1,049.5	1,517.4	1,290.9	1,359.5	1,725.6
Nonfuel Primary Products	741.6	847.8	846.6	1,021.0	875.3	919.9	1,006.4	1,446.0	1,210.4	1,245.4	1,553.8
Fuels	8.5	11.1	6.2	7.5	5.6	8.0	12.9	18.2	43.6	53.3	53.5
Manufactures	5.1	7.2	12.7	15.9	12.0	16.5	30.3	53.2	36.9	60.8	118.3
Value of Imports, cif	819.8	629.7	600.0	621.7	749.6	796.3	1,024.2	1,595.3	2,379.6	1,798.0	1,598.3
Nonfuel Primary Products	174.9	169.4	158.5	158.0	177.7	187.3	187.1	311.0	474.0	307.1	294.5
Fuels	26.7	25.4	21.0	14.3	30.3	49.1	61.7	204.7	288.1	299.4	319.3
Manufactures	618.3	434.9	420.5	449.5	541.6	559.9	775.4	1,079.6	1,617.5	1,191.5	984.6
Terms of Trade	119.2	124.8	138.2	130.8	119.1	135.0	172.8	140.5	94.4	118.4	122.1
Export Prices, fob	31.4	31.3	37.1	38.8	34.9	40.1	66.3	78.7	55.8	68.7	74.5
Nonfuel Primary Products	33.7	34.0	39.1	41.0	36.0	41.9	71.9	79.0	56.6	70.7	76.5
Fuels	4.3	4.3	4.3	4.3	5.6	6.2	8.9	36.7	35.7	38.4	42.0
Manufactures	53.6	59.3	57.9	66.4	52.1	52.1	80.7	107.9	71.4	76.4	75.7
Import Prices, cif	26.3	25.1	26.8	29.7	29.3	29.7	38.4	56.0	59.2	58.0	61.1
Trade at Constant 1980 Prices				*(Millions of 1980 US dollars)*							
Exports, fob	2,408.9	2,765.4	2,334.3	2,689.6	2,557.7	2,358.1	1,582.2	1,928.3	2,312.4	1,979.1	2,314.8
Imports, cif	3,117.1	2,509.6	2,236.0	2,093.5	2,558.2	2,684.6	2,667.9	2,848.4	4,022.6	3,097.9	2,617.0
BALANCE OF PAYMENTS				*(Millions of current US dollars)*							
Exports of Goods & Services	..	..	..	1,239	1,086	1,166	1,369	1,880	1,723	1,756	2,144
Merchandise, fob	..	..	..	1,034	890	945	1,112	1,506	1,291	1,361	1,726
Nonfactor Services	..	..	..	190	178	209	231	336	398	385	406
Factor Services	..	..	..	15	19	13	25	39	34	10	13
Imports of Goods & Services	..	..	..	1,119	1,159	1,237	1,673	2,653	3,313	3,008	3,123
Merchandise, fob	..	..	..	699	730	812	1,097	1,909	2,389	2,100	2,164
Nonfactor Services	..	..	..	272	285	291	387	534	648	527	523
Factor Services	..	..	..	148	144	134	188	211	276	381	436
Long-Term Interest	..	..	..	162	163	163	229	266	324	336	411
Current Transfers, net	..	..	..	26	4	7	4	22	17	4	3
Workers' Remittances	..	..	..	0	0	0	0	0	0	0	0
Total to be Financed	..	..	..	146	-69	-64	-300	-752	-1,574	-1,248	-976
Official Capital Grants	..	..	..	56	35	33	38	27	33	55	54
Current Account Balance	..	..	..	202	-34	-32	-262	-725	-1,541	-1,193	-922
Long-Term Capital, net	..	..	..	39	44	139	446	747	1,326	862	1,008
Direct Investment	..	..	..	-70	-58	24	70	58	316	170	54
Long-Term Loans	..	..	..	54	71	96	350	1,119	1,233	1,015	598
Disbursements	..	..	..	387	424	466	888	1,669	1,754	1,538	1,312
Repayments	..	..	..	333	354	370	538	549	521	523	715
Other Long-Term Capital	..	..	..	-1	-3	-13	-13	-457	-256	-378	303
Other Capital, net	..	..	..	5	-61	-112	-97	383	-294	-28	-41
Change in Reserves	..	..	..	-246	51	4	-87	-405	509	358	-46
Memo Item:				*(Peruvian Intis per US dollar)*							
Conversion Factor (Annual Avg)	0.030	0.040	0.040	0.040	0.040	0.040	0.040	0.040	0.040	0.060	0.080
EXTERNAL DEBT, ETC.				*(Millions of US dollars, outstanding at end of year)*							
Public/Publicly Guar. Long-Term	..	..	..	856	901	1,053	1,441	2,221	3,021	3,666	4,711
Official Creditors	..	..	..	373	430	481	559	774	1,074	1,288	1,974
IBRD and IDA	..	..	..	125	129	131	132	132	138	148	171
Private Creditors	..	..	..	483	471	572	882	1,447	1,947	2,378	2,737
Private Non-guaranteed Long-Term	..	..	..	1,799	1,840	1,787	1,771	2,128	2,056	2,436	2,134
Use of Fund Credit	..	..	..	10	4	33	16	0	0	220	243
Short-Term Debt	..	..	..	..	..	..	..	..	..	..	2,100
Memo Items:				*(Millions of US dollars)*							
Int'l Reserves Excluding Gold	105.6	91.4	142.2	296.3	380.9	442.5	526.1	925.2	425.5	289.4	356.8
Gold Holdings (at market price)	20.3	23.7	25.0	42.4	49.4	70.7	112.6	187.1	140.7	135.2	165.4
SOCIAL INDICATORS											
Total Fertility Rate	6.6	6.4	6.2	6.0	5.9	5.7	5.6	5.4	5.3	5.1	5.0
Crude Birth Rate	43.6	42.7	41.8	40.8	39.9	39.0	38.4	37.8	37.2	36.6	36.0
Infant Mortality Rate	126.0	122.8	119.6	116.4	113.2	110.0	109.0	108.0	107.0	106.0	105.0
Life Expectancy at Birth	51.5	52.3	53.1	53.9	54.7	55.6	55.8	56.1	56.4	56.7	57.0
Food Production, p.c. ('79-81 = 100)	135.9	119.7	129.0	136.4	134.8	127.4	125.2	125.2	118.9	117.7	115.2
Labor Force, Agriculture (%)	48.7	48.1	47.6	47.1	46.4	45.7	44.9	44.2	43.5	42.8	42.1
Labor Force, Female (%)	20.5	20.4	20.3	20.3	20.7	21.1	21.5	21.9	22.3	22.7	23.1
School Enroll. Ratio, primary	..	..	..	107.0	..	..	..	..	113.0	112.0	112.0
School Enroll. Ratio, secondary	..	..	..	31.0	..	..	..	..	46.0	47.0	50.0

1978	1979	1980	1981	1982	1983	1984	1985	1986	1987	NOTES	
				(Millions of current US dollars)							**FOREIGN TRADE (CUSTOMS BASIS)**
1,940.7	3,490.9	3,898.3	3,249.0	3,293.3	3,015.2	3,147.0	2,966.0	2,530.6	2,604.5	f	Value of Exports, fob
1,567.6	2,342.3	2,455.4	1,986.7	2,083.8	1,979.7	1,854.6	1,688.5	1,670.1	1,514.7	..	Nonfuel Primary Products
177.7	651.8	776.8	690.8	718.9	544.2	650.3	641.7	287.5	617.7	..	Fuels
195.5	496.8	666.1	571.4	490.6	491.3	642.1	635.8	573.0	472.1	..	Manufactures
1,601.0	1,951.0	3,062.0	3,803.0	3,721.0	2,722.0	2,212.0	2,023.3	2,908.8	3,435.1	..	Value of Imports, cif
297.7	374.5	675.3	795.7	604.6	575.7	424.2	367.0	453.1	552.8	..	Nonfuel Primary Products
77.4	47.2	63.0	40.8	49.1	78.9	43.6	36.0	20.4	29.0	..	Fuels
1,226.0	1,529.3	2,323.7	2,966.4	3,067.3	2,067.4	1,744.2	1,620.3	2,435.3	2,853.4	..	Manufactures
				(Index 1980 = 100)							
91.9	97.0	100.0	91.9	83.7	83.9	83.9	80.8	66.7	68.6	..	Terms of Trade
70.1	87.1	100.0	92.4	80.7	79.3	77.8	74.0	69.1	75.6	..	Export Prices, fob
74.8	97.5	100.0	87.8	75.7	76.6	74.8	69.3	70.1	80.5	..	Nonfuel Primary Products
42.3	61.0	100.0	112.5	101.6	92.5	90.2	87.5	44.9	56.9	..	Fuels
77.1	92.9	100.0	89.0	79.0	78.0	76.0	76.0	89.8	98.6	..	Manufactures
76.3	89.8	100.0	100.5	96.4	94.6	92.7	91.6	103.7	110.2	..	Import Prices, cif
				(Millions of 1980 US dollars)							Trade at Constant 1980 Prices
2,769.6	4,006.3	3,898.5	3,518.0	4,080.3	3,801.9	4,045.3	4,005.6	3,661.2	3,445.4	..	Exports, fob
2,099.7	2,171.7	3,062.0	3,784.6	3,858.5	2,878.1	2,385.6	2,208.7	2,805.6	3,116.2	..	Imports, cif
				(Millions of current US dollars)							**BALANCE OF PAYMENTS**
2,416	4,143	4,832	4,223	4,186	3,842	3,974	3,925	3,398	3,588	..	Exports of Goods & Services
1,941	3,491	3,916	3,249	3,293	3,015	3,147	2,978	2,509	2,559	..	Merchandise, fob
459	594	715	770	784	711	670	814	796	964	..	Nonfactor Services
16	58	202	204	109	116	157	133	93	65	..	Factor Services
2,664	3,536	5,080	6,112	5,962	4,933	4,353	3,934	4,549	..	..	Imports of Goods & Services
1,601	1,951	3,090	3,802	3,721	2,722	2,140	1,806	2,525	..	..	Merchandise, fob
470	560	880	1,087	1,098	965	891	984	1,100	..	..	Nonfactor Services
594	1,025	1,111	1,223	1,143	1,246	1,322	1,144	924	798	..	Factor Services
475	575	669	669	696	539	541	462	270	203	..	Long-Term Interest
3	0	0	0	0	0	0	0	0	..	..	Current Transfers, net
0	0	0	0	0	0	0	0	0	..	..	Workers' Remittances
-245	607	-248	-1,889	-1,776	-1,091	-379	-9	-1,151	-1,419	K	Total to be Financed
53	122	147	161	167	219	158	134	96	-495	K	Official Capital Grants
-192	729	-101	-1,728	-1,609	-872	-221	125	-1,055	-1,914	..	Current Account Balance
317	663	424	517	1,303	1,456	39	-485	-1,111	..	..	Long-Term Capital, net
25	71	27	125	48	38	-89	1	22	..	..	Direct Investment
177	441	353	285	1,404	1,346	1,097	182	195	337	..	Long-Term Loans
890	1,170	1,463	1,814	2,560	1,923	1,572	711	548	597	..	Disbursements
713	729	1,110	1,529	1,155	577	475	529	353	261	..	Repayments
62	28	-103	-54	-316	-147	-1,127	-802	-1,424	..	..	Other Long-Term Capital
-94	-227	329	522	223	-618	432	546	1,871	1,079	..	Other Capital, net
-31	-1,165	-652	689	84	34	-250	-186	295	835	..	Change in Reserves
				(Peruvian Intis per US dollar)							**Memo Item:**
0.160	0.220	0.290	0.420	0.700	1.630	3.470	10.980	13.950	16.840	..	Conversion Factor (Annual Avg)
				(Millions of US dollars, outstanding at end of year)							**EXTERNAL DEBT, ETC.**
5,399	5,932	6,167	6,012	6,956	8,328	9,315	10,351	11,278	12,485	..	Public/Publicly Guar. Long-Term
2,448	2,635	3,062	3,036	3,092	3,627	4,033	4,810	5,528	6,403	..	Official Creditors
186	235	359	416	478	527	508	723	951	1,214	..	IBRD and IDA
2,950	3,297	3,105	2,976	3,864	4,701	5,282	5,541	5,750	6,083	..	Private Creditors
1,830	1,186	1,262	1,405	1,664	1,579	1,465	1,342	1,337	1,433	..	Private Non-guaranteed Long-Term
371	528	489	387	649	698	675	702	728	845	..	Use of Fund Credit
2,098	1,574	2,084	2,479	3,016	1,440	1,704	1,795	2,612	3,295	..	Short-Term Debt
				(Millions of US dollars)							**Memo Items:**
389.7	1,520.7	1,979.9	1,199.5	1,349.7	1,365.1	1,630.1	1,827.0	1,430.0	594.6	..	Int'l Reserves Excluding Gold
226.9	592.9	824.1	555.7	638.8	533.3	431.0	639.0	835.4	724.7	..	Gold Holdings (at market price)
											SOCIAL INDICATORS
4.9	4.8	4.7	4.6	4.5	4.4	4.3	4.2	4.1	4.1	..	Total Fertility Rate
35.5	35.1	34.6	34.2	33.7	33.4	33.0	32.7	32.4	32.0	..	Crude Birth Rate
103.8	102.6	101.4	100.2	99.0	92.5	86.1	79.6	73.1	66.7	..	Infant Mortality Rate
57.3	57.6	57.9	58.3	58.6	59.1	59.5	60.0	60.4	60.9	..	Life Expectancy at Birth
109.8	107.2	93.6	99.3	105.4	95.8	103.5	99.8	96.4	96.7	..	Food Production, p.c. ('79-81 = 100)
41.4	40.7	40.0	..	..	..	..	..	..	..	..	Labor Force, Agriculture (%)
23.4	23.8	24.2	24.2	24.2	24.2	24.2	24.2	24.2	24.2	..	Labor Force, Female (%)
113.0	112.0	114.0	115.0	118.0	..	..	122.0	..	..	..	School Enroll. Ratio, primary
55.0	56.0	59.0	61.0	61.0	..	..	65.0	..	..	..	School Enroll. Ratio, secondary

PHILIPPINES	1967	1968	1969	1970	1971	1972	1973	1974	1975	1976	1977
CURRENT GNP PER CAPITA (US $)	210	230	250	230	220	220	250	300	360	420	450
POPULATION (thousands)	34,056	35,223	36,388	37,542	38,629	39,731	40,844	41,967	43,103	44,288	45,495
ORIGIN AND USE OF RESOURCES					*(Billions of current Philippine Pesos)*						
Gross National Product (GNP)	28.73	31.79	35.01	41.75	49.60	55.90	72.20	99.90	114.40	134.20	153.30
Net Factor Income from Abroad	-0.29	-0.34	-0.28	-0.70	-0.52	-0.60	-0.10	0.40	-0.30	-1.10	-0.90
Gross Domestic Product (GDP)	29.02	32.13	35.30	42.45	50.12	56.50	72.30	99.50	114.70	135.30	154.20
Indirect Taxes, net	2.01	2.17	2.30	3.13	3.94	4.33	6.34	10.16	11.16	11.78	13.60
GDP at factor cost	27.02	29.96	33.00	39.32	46.18	52.17	65.96	89.34	103.54	123.52	140.60
Agriculture	7.54	8.98	10.09	11.78	14.78	16.10	21.20	29.60	33.20	37.60	42.00
Industry	8.17	8.88	9.89	12.58	14.76	18.00	24.00	32.90	38.10	46.00	53.10
Manufacturing	6.16	6.72	7.32	9.57	11.42	14.00	18.20	24.30	28.20	32.90	37.40
Services, etc.	13.31	14.26	15.32	18.08	20.58	22.40	27.10	37.00	43.40	51.70	59.10
Resource Balance	-0.54	-1.04	-1.23	-0.14	-0.39	-0.40	2.50	-3.10	-7.80	-8.60	-5.90
Exports of Goods & NFServices	4.90	4.71	4.58	8.09	9.26	9.90	15.90	22.30	21.30	23.20	28.90
Imports of Goods & NFServices	5.44	5.75	5.81	8.24	9.65	10.30	13.40	25.40	29.10	31.80	34.80
Domestic Absorption	29.56	33.17	36.52	42.59	50.51	56.90	69.80	102.60	122.50	143.90	160.10
Private Consumption, etc.	20.95	23.54	26.20	30.08	35.78	40.70	49.00	68.60	77.60	88.70	101.40
General Gov't Consumption	2.56	2.84	3.12	3.51	4.27	5.30	6.20	8.90	11.10	13.20	14.30
Gross Domestic Investment	6.05	6.79	7.21	8.99	10.45	10.90	14.60	25.10	33.80	42.00	44.40
Fixed Investment	5.26	5.52	5.73	6.70	8.15	9.30	11.40	18.40	27.10	33.70	36.40
Memo Items:											
Gross Domestic Saving	5.52	5.75	5.98	8.85	10.07	10.50	17.10	22.00	26.00	33.40	38.50
Gross National Saving	5.23	5.54	5.87	8.33	9.76	10.44	17.64	23.23	26.90	33.40	38.70
					(Billions of 1980 Philippine Pesos)						
Gross National Product	120.77	126.93	134.44	139.98	149.19	157.69	174.22	184.56	195.00	207.56	221.29
GDP at factor cost	117.06	123.60	129.24	134.69	140.53	148.67	159.42	164.68	176.51	192.51	204.28
Agriculture	34.03	36.46	37.58	38.42	40.31	41.98	44.59	45.89	47.72	51.63	54.24
Industry	37.00	38.83	40.91	43.61	47.01	52.17	58.83	62.02	67.53	74.77	80.86
Manufacturing	27.42	29.18	30.34	32.92	35.12	38.98	44.55	46.50	48.17	50.96	54.85
Services, etc.	54.56	57.22	60.46	63.39	65.23	66.88	71.36	75.54	80.31	84.49	88.97
Resource Balance	-6.54	-12.54	-14.31	-8.90	-8.22	-6.50	-3.67	-13.68	-16.21	-10.81	-7.94
Exports of Goods & NFServices	29.12	25.81	24.74	26.48	27.25	29.98	34.22	30.59	30.89	36.64	42.70
Imports of Goods & NFServices	35.66	38.35	39.05	35.38	35.47	36.47	37.89	44.27	47.10	47.45	50.64
Domestic Absorption	132.14	145.05	153.26	154.32	160.77	167.52	178.45	197.13	211.77	221.71	232.01
Private Consumption, etc.	94.11	103.14	108.83	110.45	114.87	120.71	126.45	134.53	137.26	138.40	148.14
General Gov't Consumption	8.40	9.27	10.15	10.80	11.63	13.54	15.07	17.11	18.65	18.90	19.16
Gross Domestic Investment	29.63	32.64	34.28	33.08	34.27	33.27	36.94	45.48	55.86	64.41	64.72
Fixed Investment	25.17	25.94	26.68	23.72	26.03	27.86	28.46	34.15	44.93	51.52	52.72
Memo Items:											
Capacity to Import	32.15	31.41	30.78	34.77	34.04	35.06	44.96	38.86	34.47	34.62	42.05
Terms of Trade Adjustment	3.03	5.60	6.04	8.29	6.79	5.08	10.74	8.28	3.59	-2.02	-0.64
Gross Domestic Income	128.63	138.11	145.00	153.72	159.34	166.11	185.52	191.73	199.15	208.87	223.43
Gross National Income	123.80	132.53	140.49	148.28	155.98	162.77	184.97	192.84	198.59	205.54	220.65
DOMESTIC PRICES (DEFLATORS)					*(Index 1980 = 100)*						
Overall (GDP)	23.1	24.2	25.4	29.2	32.9	35.1	41.4	54.2	58.7	64.2	68.8
Domestic Absorption	22.4	22.9	23.8	27.6	31.4	34.0	39.1	52.0	57.8	64.9	69.0
Agriculture	22.2	24.6	26.8	30.7	36.7	38.3	47.5	64.5	69.6	72.8	77.4
Industry	22.1	22.9	24.2	28.8	31.4	34.5	40.8	53.0	56.4	61.5	65.7
Manufacturing	22.5	23.0	24.1	29.1	32.5	35.9	40.9	52.3	58.5	64.6	68.2
MANUFACTURING ACTIVITY											
Employment (1980=100)	49.2	40.1	40.6	41.0	42.9	44.3	55.0	54.3	53.2	74.0	84.2
Real Earnings per Empl. (1980=100)	108.5	149.4	153.8	151.4	151.1	145.1	126.0	109.8	122.1	119.5	127.6
Real Output per Empl. (1980=100)	..	93.6	93.7	102.0	107.4	104.2	106.5	128.5	124.8	87.0	97.9
Earnings as % of Value Added	..	24.0	23.6	20.6	20.8	21.6	18.6	15.3	14.8	27.6	26.4
MONETARY HOLDINGS					*(Billions of current Philippine Pesos)*						
Money Supply, Broadly Defined	6.69	7.00	8.14	9.00	10.33	11.87	14.02	16.77	19.25	25.02	32.53
Money as Means of Payment	3.19	3.28	4.02	4.31	5.01	6.47	7.27	9.01	10.32	12.07	14.94
Currency Ouside Banks	1.76	1.78	2.12	2.41	2.65	3.43	3.45	4.31	4.75	5.65	6.73
Demand Deposits	1.43	1.50	1.90	1.90	2.36	3.03	3.81	4.70	5.57	6.42	8.21
Quasi-Monetary Liabilities	3.50	3.72	4.13	4.69	5.31	5.40	6.75	7.76	8.94	12.95	17.59
GOVERNMENT DEFICIT (-) OR SURPLUS					*(Millions of current Philippine Pesos)*						
	..	..	..	..	..	-1,101.0	-843.0	445.0	-1,360.0	-2,352.0	-2,807.0
Current Revenue	..	..	..	..	..	6,970.0	9,484.0	12,151.0	16,822.0	18,282.0	19,944.0
Current Expenditure	..	..	..	..	..	6,546.0	8,506.0	8,762.0	14,651.0	15,798.0	17,719.0
Current Budget Balance	..	..	..	..	..	424.0	978.0	3,389.0	2,171.0	2,484.0	2,225.0
Capital Receipts	..	..	..	..	..	2.0	15.0	6.0	16.0	18.0	15.0
Capital Payments	..	..	..	..	..	1,527.0	1,836.0	2,950.0	3,547.0	4,854.0	5,047.0

1978	1979	1980	1981	1982	1983	1984	1985	1986	1987 estimate	NOTES	PHILIPPINES
500	590	680	770	800	740	630	570	560	590	..	CURRENT GNP PER CAPITA (US $)
46,724	47,975	49,253	50,544	51,848	53,162	54,486	55,819	57,129	58,428	..	POPULATION (thousands)
			(Billions of current Philippine Pesos)								ORIGIN AND USE OF RESOURCES
177.00	218.00	264.50	303.60	335.40	378.80	527.40	597.74	612.00	700.46	..	Gross National Product (GNP)
-0.70	0.50	-0.20	-1.70	-5.20	-5.30	-13.10	-14.94	-12.43	-5.01	..	Net Factor Income from Abroad
177.70	217.50	264.70	305.30	340.60	384.10	540.50	612.68	624.43	705.47	..	Gross Domestic Product (GDP)
16.85	22.38	25.76	25.93	28.66	34.72	43.92	49.35	52.31	70.23	..	Indirect Taxes, net
160.85	195.12	238.94	279.37	311.93	349.38	496.58	563.34	572.12	635.24	B	GDP at factor cost
47.40	55.50	61.80	69.40	76.70	84.60	139.50	162.52	155.99	170.77	..	Agriculture
61.30	76.90	96.80	111.60	122.50	138.20	186.20	200.55	202.28	229.68	..	Industry
43.70	52.10	64.60	75.20	83.10	95.20	137.30	150.53	154.72	173.50	..	Manufacturing
69.00	85.10	106.10	124.30	141.40	161.30	214.80	249.61	266.16	305.01	..	Services, etc.
-8.90	-12.10	-15.10	-16.60	-23.10	-26.20	-0.70	18.06	38.92	6.07	..	Resource Balance
32.40	41.50	53.60	57.80	56.20	75.30	117.70	126.57	155.10	163.47	..	Exports of Goods & NFServices
41.30	53.60	68.70	74.40	79.30	101.50	118.40	108.51	116.19	157.40	..	Imports of Goods & NFServices
186.60	229.60	279.80	321.90	363.70	410.30	541.20	594.62	585.51	699.39	..	Domestic Absorption
119.20	143.60	177.40	203.80	238.00	278.20	413.60	465.17	456.44	533.05	..	Private Consumption, etc.
16.10	18.30	21.20	24.80	29.20	29.50	35.60	44.04	48.44	58.47	..	General Gov't Consumption
51.30	67.70	81.20	93.30	96.50	102.60	92.00	85.40	80.63	107.87	..	Gross Domestic Investment
42.30	56.30	68.00	79.30	86.00	95.30	100.10	89.97	80.82	98.55	..	Fixed Investment
											Memo Items:
42.40	55.60	66.10	76.70	73.40	76.40	91.30	103.47	119.54	113.95	..	Gross Domestic Saving
43.15	57.79	68.15	77.57	70.95	73.73	80.17	91.74	111.91	116.22	..	Gross National Saving
			(Billions of 1980 Philippine Pesos)								
234.79	252.51	264.50	272.08	274.52	278.78	255.15	244.64	250.14	268.41	..	Gross National Product
214.02	225.54	238.94	251.52	259.05	259.92	246.91	235.27	237.87	245.06	B I	GDP at factor cost
56.32	58.93	61.80	64.15	66.23	64.93	66.23	68.45	70.69	69.97	..	Agriculture
85.79	92.45	96.80	101.44	103.47	104.34	93.61	84.05	82.25	88.57	..	Industry
58.75	61.82	64.60	66.83	68.22	69.89	64.88	59.98	60.47	64.51	..	Manufacturing
94.35	100.02	106.10	109.27	113.16	116.74	108.08	103.46	106.51	113.57	..	Services, etc.
-12.50	-19.03	-15.10	-12.72	-15.45	-18.75	-1.16	9.04	15.39	-0.39	..	Resource Balance
44.52	47.54	53.60	54.21	53.60	58.45	63.29	58.60	71.35	70.39	..	Exports of Goods & NFServices
57.01	66.58	68.70	66.93	69.05	77.20	64.45	49.56	55.96	70.78	..	Imports of Goods & NFServices
248.96	270.43	279.80	287.58	298.31	304.76	269.08	246.92	244.07	272.50	..	Domestic Absorption
159.38	172.16	177.40	182.58	194.78	206.22	204.54	191.96	192.50	210.89	..	Private Consumption, etc.
19.67	20.43	21.20	21.97	23.24	22.22	21.20	21.00	20.91	22.41	..	General Gov't Consumption
69.91	77.84	81.20	83.03	80.28	76.32	43.35	33.96	30.65	39.20	..	Gross Domestic Investment
56.92	63.81	68.00	70.40	71.00	68.90	46.73	35.43	30.13	34.83	..	Fixed Investment
											Memo Items:
44.73	51.55	53.60	52.00	48.94	57.27	64.07	57.81	74.70	73.51	..	Capacity to Import
0.21	4.00	0.00	-2.21	-4.66	-1.17	0.78	-0.79	3.36	3.12	..	Terms of Trade Adjustment
236.67	255.41	264.70	272.65	278.19	284.83	268.70	255.17	262.81	275.23	..	Gross Domestic Income
235.00	256.52	264.50	269.87	269.86	277.61	255.93	243.85	253.50	271.53	..	Gross National Income
			(Index 1980 = 100)								DOMESTIC PRICES (DEFLATORS)
75.2	86.5	100.0	111.1	120.4	134.3	201.7	239.4	240.7	259.3	..	Overall (GDP)
75.0	84.9	100.0	111.9	121.9	134.6	201.1	240.8	239.9	256.7	..	Domestic Absorption
84.2	94.2	100.0	108.2	115.8	130.3	210.6	237.4	220.7	244.1	..	Agriculture
71.5	83.2	100.0	110.0	118.4	132.5	198.9	238.6	245.9	259.3	..	Industry
74.4	84.3	100.0	112.5	121.8	136.2	211.6	251.0	255.9	268.9	..	Manufacturing
											MANUFACTURING ACTIVITY
92.1	98.1	100.0	102.3	102.1	72.5	65.8	..	..	..	G J	Employment (1980 = 100)
131.2	106.2	100.0	109.6	111.6	135.1	118.7	..	..	..	G J	Real Earnings per Empl. (1980 = 100)
97.5	100.7	100.0	90.2	95.7	123.3	114.2	..	..	..	G J	Real Output per Empl. (1980 = 100)
25.1	18.5	20.6	24.7	23.0	18.8	18.5	20.1	..	..	J	Earnings as % of Value Added
			(Billions of current Philippine Pesos)								MONETARY HOLDINGS
40.34	45.41	55.44	65.64	78.70	95.86	110.03	124.19	136.52	156.17	..	Money Supply, Broadly Defined
16.95	18.84	22.54	23.52	23.50	32.49	33.63	35.78	42.66	52.38	..	Money as Means of Payment
8.14	9.18	10.18	11.63	12.68	19.61	21.80	24.07	29.31	35.45	..	Currency Ouside Banks
8.81	9.66	12.36	11.90	10.82	12.88	11.84	11.71	13.35	16.93	..	Demand Deposits
23.40	26.57	32.89	42.11	55.21	63.37	76.40	88.41	93.86	103.79	..	Quasi-Monetary Liabilities
			(Millions of current Philippine Pesos)								
-2,171.0	-349.0	-3,385.0	-12,154.0	-14,414.0	-7,468.0	-9,957.0	-11,158.0	..	..	..	GOVERNMENT DEFICIT (-) OR SURPLUS
23,882.0	29,315.0	34,371.0	35,733.0	37,990.0	45,603.0	56,822.0	68,958.0	..	..	..	Current Revenue
19,215.0	20,777.0	24,156.0	26,201.0	30,777.0	34,533.0	42,967.0	55,292.0	..	..	..	Current Expenditure
4,667.0	8,538.0	10,215.0	9,532.0	7,213.0	11,070.0	13,855.0	13,666.0	..	..	..	Current Budget Balance
125.0	4.0	2.0	3.0	3.0	3.0	4.0	3.0	..	..	..	Capital Receipts
6,963.0	8,891.0	13,602.0	21,689.0	21,630.0	18,541.0	23,816.0	24,827.0	..	..	..	Capital Payments

PHILIPPINES	1967	1968	1969	1970	1971	1972	1973	1974	1975	1976	1977
FOREIGN TRADE (CUSTOMS BASIS)				*(Millions of current US dollars)*							
Value of Exports, fob	799.5	816.4	823.0	1,059.7	1,116.3	1,028.8	1,797.3	2,650.7	2,218.1	2,508.4	3,079.6
Nonfuel Primary Products	736.8	743.8	744.8	962.0	1,013.1	923.4	1,492.7	2,290.2	1,786.6	1,843.1	2,282.6
Fuels	12.9	15.2	12.6	17.2	24.2	10.3	16.2	17.5	37.5	33.8	18.9
Manufactures	49.7	57.5	65.7	80.5	79.0	95.1	288.3	343.0	394.0	631.4	778.1
Value of Imports, cif	1,183.8	1,279.5	1,256.1	1,210.4	1,318.7	1,387.7	1,789.5	3,467.6	3,776.2	3,953.3	4,269.8
Nonfuel Primary Products	268.4	276.0	257.6	237.4	284.7	320.3	391.6	626.3	588.8	600.9	703.5
Fuels	114.4	129.7	128.1	144.7	172.7	177.2	231.2	696.0	800.1	927.6	1,040.1
Manufactures	801.0	873.9	870.3	828.3	861.3	890.2	1,166.7	2,145.3	2,387.3	2,424.9	2,526.2
Terms of Trade	144.7	157.8	161.5	176.3	151.6	143.4	170.7	154.4	120.4	119.9	116.4
Export Prices, fob	28.5	30.3	31.6	33.1	32.0	34.1	53.3	82.1	66.3	67.1	70.2
Nonfuel Primary Products	31.0	33.6	34.5	36.9	35.5	35.2	55.5	83.8	64.9	63.3	69.3
Fuels	4.3	4.3	4.3	4.3	5.6	6.2	8.9	36.7	35.7	38.4	42.0
Manufactures	37.9	45.2	44.4	41.1	38.7	40.3	58.1	76.6	80.6	85.5	74.2
Import Prices, cif	19.7	19.2	19.6	18.8	21.1	23.7	31.2	53.2	55.1	55.9	60.3
Trade at Constant 1980 Prices				*(Millions of 1980 US dollars)*							
Exports, fob	2,807.7	2,698.6	2,603.9	3,204.2	3,491.7	3,020.6	3,370.2	3,226.8	3,344.3	3,739.6	4,389.0
Imports, cif	6,015.5	6,672.1	6,418.5	6,451.7	6,252.3	5,844.0	5,728.4	6,519.5	6,854.2	7,066.3	7,085.2
BALANCE OF PAYMENTS				*(Millions of current US dollars)*							
Exports of Goods & Services	..	..	..	1,322	1,402	1,480	2,455	3,527	3,170	3,391	4,236
Merchandise, fob	..	..	..	1,064	1,136	1,136	1,872	2,694	2,263	2,519	3,151
Nonfactor Services	..	..	..	248	250	322	517	663	737	745	732
Factor Services	..	..	..	10	16	22	66	171	170	127	353
Imports of Goods & Services	..	..	..	1,489	1,539	1,662	2,211	4,012	4,412	4,761	5,249
Merchandise, fob	..	..	..	1,090	1,186	1,261	1,596	3,144	3,459	3,632	3,915
Nonfactor Services	..	..	..	259	236	253	435	642	657	748	840
Factor Services	..	..	..	140	117	149	180	226	296	380	494
Long-Term Interest	..	..	..	44	72	84	94	110	118	166	223
Current Transfers, net	..	..	..	29	34	80	94	123	165	148	148
Workers' Remittances	..	..	..	0	0	0	0	0	0	0	125
Total to be Financed	..	..	..	-138	-102	-102	337	-361	-1,076	-1,222	-865
Official Capital Grants	..	..	..	90	100	107	136	154	153	120	112
Current Account Balance	..	..	..	-48	-2	5	473	-208	-923	-1,102	-753
Long-Term Capital, net	..	..	..	220	91	223	268	381	670	1,260	992
Direct Investment	..	..	..	-29	-6	-21	54	4	97	126	211
Long-Term Loans	..	..	..	157	26	196	-14	401	532	1,095	1,015
Disbursements	..	..	..	416	283	452	389	730	831	1,438	1,384
Repayments	..	..	..	260	258	256	403	329	299	343	369
Other Long-Term Capital	..	..	..	2	-29	-61	93	-177	-112	-81	-346
Other Capital, net	..	..	..	-97	10	-47	-30	403	242	-212	-266
Change in Reserves	..	..	..	-75	-99	-181	-712	-577	11	54	27
Memo Item:				*(Philippine Pesos per US dollar)*							
Conversion Factor (Annual Avg)	3.900	3.900	3.900	5.900	6.430	6.670	6.760	6.790	7.250	7.440	7.400
EXTERNAL DEBT, ETC.				*(Millions of US dollars, outstanding at end of year)*							
Public/Publicly Guar. Long-Term	..	..	..	625	700	876	895	1,121	1,448	2,209	3,032
Official Creditors	..	..	..	272	362	532	623	740	912	1,138	1,491
IBRD and IDA	..	..	..	119	133	142	150	172	255	343	432
Private Creditors	..	..	..	353	338	344	273	380	537	1,071	1,542
Private Non-guaranteed Long-Term	..	..	..	919	980	983	1,041	1,213	1,411	1,812	2,165
Use of Fund Credit	..	..	..	69	97	103	92	131	193	450	556
Short-Term Debt	..	..	..	..	..	..	..	..	..	..	2,468
Memo Items:				*(Millions of US dollars)*							
Int'l Reserves Excluding Gold	120.0	99.0	76.0	195.0	309.0	479.8	992.8	1,458.9	1,314.5	1,596.8	1,479.4
Gold Holdings (at market price)	60.3	74.2	45.3	59.8	83.5	120.5	118.7	196.9	148.1	142.3	174.2
SOCIAL INDICATORS											
Total Fertility Rate	6.7	6.6	6.5	6.4	6.3	6.1	5.9	5.7	5.5	5.2	5.0
Crude Birth Rate	40.2	39.5	38.9	38.2	37.6	36.9	36.8	36.7	36.6	36.5	36.4
Infant Mortality Rate	70.0	68.8	67.6	66.4	65.2	64.0	62.0	60.0	58.0	56.0	54.0
Life Expectancy at Birth	56.2	56.5	56.8	57.2	57.5	57.9	58.3	58.7	59.1	59.5	59.9
Food Production, p.c. ('79-81 = 100)	81.4	80.4	84.9	83.1	84.0	81.1	89.1	92.6	93.4	98.6	98.1
Labor Force, Agriculture (%)	56.7	56.1	55.4	54.8	54.5	54.2	53.9	53.6	53.3	53.0	52.7
Labor Force, Female (%)	33.5	33.4	33.3	33.2	33.2	33.1	33.0	32.9	32.8	32.9	32.9
School Enroll. Ratio, primary	..	..	..	108.0	..	..	..	..	107.0	103.0	108.0
School Enroll. Ratio, secondary	..	..	..	46.0	..	..	..	..	54.0	60.0	61.0

1978	1979	1980	1981	1982	1983	1984	1985	1986	1987 estimate	NOTES	PHILIPPINES
				(Millions of current US dollars)							**FOREIGN TRADE (CUSTOMS BASIS)**
3,349.1	4,601.2	5,787.8	5,722.1	5,020.6	5,000.6	5,391.0	4,629.0	4,841.8	5,649.0	..	Value of Exports, fob
2,203.4	2,967.3	3,596.6	3,119.5	2,512.4	2,356.5	2,329.9	1,950.6	1,972.3	2,076.7	..	Nonfuel Primary Products
9.7	11.0	50.0	42.2	16.3	110.5	82.4	34.5	61.5	86.2	..	Fuels
1,136.0	1,622.9	2,141.2	2,560.4	2,491.9	2,533.6	2,978.7	2,643.9	2,808.0	3,486.1	..	Manufactures
5,143.3	6,613.0	8,294.6	8,477.7	8,262.3	7,977.4	6,424.3	5,444.6	5,392.5	7,144.0	..	Value of Imports, cif
750.9	907.4	1,077.0	1,027.4	1,171.3	1,012.7	771.2	792.2	865.2	1,087.9	..	Nonfuel Primary Products
1,088.0	1,466.5	2,354.8	2,553.7	2,187.9	2,193.7	1,696.2	1,507.8	918.4	1,189.8	..	Fuels
3,304.5	4,239.0	4,862.8	4,896.6	4,903.1	4,771.0	3,956.9	3,144.6	3,608.9	4,866.3	..	Manufactures
				(Index 1980 = 100)							
112.2	115.0	100.0	92.0	89.1	97.1	105.1	91.7	94.9	97.8	..	Terms of Trade
75.1	94.4	100.0	95.5	87.9	92.4	97.3	83.6	81.2	95.9	..	Export Prices, fob
74.5	97.6	100.0	89.3	77.5	88.3	98.1	69.0	59.7	71.9	..	Nonfuel Primary Products
42.3	61.0	100.0	112.5	101.6	92.5	90.2	87.5	44.9	56.9	..	Fuels
76.6	89.5	100.0	104.0	101.6	96.5	97.0	99.0	111.4	122.3	..	Manufactures
66.9	82.2	100.0	103.7	98.8	95.1	92.6	91.2	85.6	98.1	..	Import Prices, cif
				(Millions of 1980 US dollars)							Trade at Constant 1980 Prices
4,462.4	4,871.7	5,787.8	5,993.5	5,708.7	5,412.3	5,537.9	5,535.8	5,962.2	5,890.6	..	Exports, fob
7,687.1	8,048.9	8,294.8	8,172.0	8,366.4	8,387.4	6,934.0	5,972.2	6,303.2	7,285.2	..	Imports, cif
				(Millions of current US dollars)							**BALANCE OF PAYMENTS**
4,910	6,256	7,998	8,583	8,004	8,132	8,017	7,917	8,633	9,217	..	Exports of Goods & Services
3,426	4,601	5,788	5,722	5,021	5,005	5,391	4,629	4,842	5,720	f	Merchandise, fob
1,012	1,077	1,448	1,791	1,804	1,808	1,642	2,235	2,860	2,343	..	Nonfactor Services
472	578	762	1,070	1,179	1,319	984	1,053	931	1,154	..	Factor Services
6,323	8,106	10,348	11,151	11,690	11,352	9,671	8,314	8,078	10,310	..	Imports of Goods & Services
4,732	6,142	7,727	7,946	7,667	7,487	6,070	5,111	5,044	6,737	f	Merchandise, fob
983	1,163	1,420	1,608	1,800	1,687	1,142	833	824	1,124	..	Nonfactor Services
608	801	1,201	1,597	2,223	2,178	2,459	2,370	2,210	2,449	..	Factor Services
309	490	573	816	915	938	914	931	1,339	1,497	..	Long-Term Interest
197	229	299	325	322	237	118	172	235	357	..	Current Transfers, net
154	191	205	254	239	180	59	111	163	211	..	Workers' Remittances
-1,216	-1,621	-2,051	-2,243	-3,364	-2,983	-1,536	-225	790	-736	K	Total to be Financed
122	126	148	182	164	235	268	207	206	197	K	Official Capital Grants
-1,094	-1,495	-1,903	-2,061	-3,200	-2,748	-1,268	-18	996	-539	..	Current Account Balance
1,066	1,293	1,077	1,689	1,729	1,694	560	3,275	1,504	652	..	Long-Term Capital, net
101	7	-106	172	16	105	9	12	127	186	..	Direct Investment
1,017	1,188	1,371	1,495	1,521	1,803	944	1,045	660	221	..	Long-Term Loans
1,966	2,148	1,905	2,218	2,478	2,634	1,443	1,636	1,605	1,097	..	Disbursements
949	960	534	723	957	831	499	591	945	876	..	Repayments
-174	-28	-336	-160	28	-449	-661	2,011	511	48	..	Other Long-Term Capital
895	579	1,782	40	753	-992	952	-3,405	-1,369	-381	..	Other Capital, net
-867	-377	-956	332	718	2,046	-244	148	-1,131	268	..	Change in Reserves
				(Philippine Pesos per US dollar)							Memo Item:
7.370	7.380	7.510	7.900	8.540	11.110	16.700	18.700	20.400	20.400	..	Conversion Factor (Annual Avg)
				(Millions of US dollars, outstanding at end of year)							**EXTERNAL DEBT, ETC.**
4,243	5,209	6,527	7,610	8,912	10,653	11,399	13,567	20,409	22,321	..	Public/Publicly Guar. Long-Term
1,933	2,330	2,821	3,523	3,939	4,929	5,360	6,914	8,436	10,437	..	Official Creditors
576	763	961	1,372	1,569	2,109	1,936	2,505	3,109	3,846	..	IBRD and IDA
2,309	2,879	3,706	4,088	4,973	5,724	6,039	6,654	11,973	11,884	..	Private Creditors
2,091	2,071	2,454	2,761	3,229	3,125	2,711	2,998	1,894	1,516	..	Private Non-guaranteed Long-Term
621	712	853	958	833	941	756	1,052	1,173	1,194	..	Use of Fund Credit
3,863	5,315	7,556	9,421	11,325	9,404	9,492	8,573	5,378	4,931	..	Short-Term Debt
				(Millions of US dollars)							Memo Items:
1,763.0	2,249.7	2,846.1	2,065.9	887.8	746.9	602.1	614.9	1,728.2	968.3	..	Int'l Reserves Excluding Gold
341.9	871.4	1,131.8	659.5	852.6	110.3	242.3	483.3	883.0	1,343.9	..	Gold Holdings (at market price)
											SOCIAL INDICATORS
4.9	4.9	4.8	4.8	4.7	4.7	4.6	4.6	4.6	4.6	..	Total Fertility Rate
36.0	35.6	35.3	34.9	34.5	34.5	34.5	34.4	34.4	34.4	..	Crude Birth Rate
53.4	52.8	52.2	51.6	51.0	51.5	52.0	52.5	53.0	53.5	..	Infant Mortality Rate
60.2	60.6	61.0	61.4	61.8	62.1	62.4	62.8	63.1	63.9	..	Life Expectancy at Birth
100.6	98.9	100.1	101.0	96.2	94.7	93.0	94.3	95.6	90.3	..	Food Production, p.c. ('79-81 = 100)
52.4	52.1	51.8	..	..	..	..	..	..	..	..	Labor Force, Agriculture (%)
32.9	32.9	32.9	32.7	32.6	32.4	32.3	32.1	32.0	31.8	..	Labor Force, Female (%)
110.0	107.0	114.0	114.0	109.0	109.0	107.0	106.0	..	..	..	School Enroll. Ratio, primary
63.0	64.0	62.0	61.0	66.0	67.0	68.0	65.0	..	..	..	School Enroll. Ratio, secondary

POLAND	1967	1968	1969	1970	1971	1972	1973	1974	1975	1976	1977
CURRENT GNP PER CAPITA (US $)	..	..	..	..	..	..	..	..	..	..	..
POPULATION (thousands)	31,944	32,426	32,555	32,526	32,805	33,068	33,363	33,691	34,022	34,362	34,698

ORIGIN AND USE OF RESOURCES

(Billions of current Polish Zlotys)

	1967	1968	1969	1970	1971	1972	1973	1974	1975	1976	1977
Gross National Product (GNP)	..	..	..	..	..	..	..	..	..	..	..
Net Factor Income from Abroad	..	..	..	..	..	..	..	..	..	..	..
Gross Domestic Product (GDP)	..	..	..	..	..	..	..	..	..	..	..
Indirect Taxes, net	..	..	..	..	..	..	..	..	..	..	..
GDP at factor cost	..	..	..	..	..	..	..	..	..	..	..
Agriculture	..	..	..	..	..	..	..	..	..	..	..
Industry	..	..	..	..	..	..	..	..	..	..	..
Manufacturing	..	..	..	..	..	..	..	..	..	..	..
Services, etc.	..	..	..	..	..	..	..	..	..	..	..
Resource Balance	..	..	..	..	..	..	..	..	..	..	..
Exports of Goods & NFServices	..	..	..	..	..	..	..	..	..	..	..
Imports of Goods & NFServices	..	..	..	..	..	..	..	..	..	..	..
Domestic Absorption	..	..	..	..	..	..	..	..	..	..	..
Private Consumption, etc.	..	..	..	..	..	..	..	..	..	..	..
General Gov't Consumption	..	..	..	..	..	..	..	..	..	..	..
Gross Domestic Investment	..	..	..	..	..	..	..	..	..	..	..
Fixed Investment	..	..	..	..	..	..	..	..	..	..	..
Memo Items:											
Gross Domestic Saving	..	..	..	..	..	..	..	..	..	..	..
Gross National Saving	..	..	..	..	..	..	..	..	..	..	..

(Billions of 1980 Polish Zlotys)

	1967	1968	1969	1970	1971	1972	1973	1974	1975	1976	1977
Gross National Product	..	..	..	..	..	..	..	..	..	..	..
GDP at factor cost	..	..	..	..	..	..	..	..	..	..	..
Agriculture	..	..	..	..	..	..	..	..	..	..	..
Industry	..	..	..	..	..	..	..	..	..	..	..
Manufacturing	..	..	..	..	..	..	..	..	..	..	..
Services, etc.	..	..	..	..	..	..	..	..	..	..	..
Resource Balance	..	..	..	..	..	..	..	..	..	..	..
Exports of Goods & NFServices	..	..	..	..	..	..	..	..	..	..	..
Imports of Goods & NFServices	..	..	..	..	..	..	..	..	..	..	..
Domestic Absorption	..	..	..	..	..	..	..	..	..	..	..
Private Consumption, etc.	..	..	..	..	..	..	..	..	..	..	..
General Gov't Consumption	..	..	..	..	..	..	..	..	..	..	..
Gross Domestic Investment	..	..	..	..	..	..	..	..	..	..	..
Fixed Investment	..	..	..	..	..	..	..	..	..	..	..
Memo Items:											
Capacity to Import	..	..	..	..	..	..	..	..	..	..	..
Terms of Trade Adjustment	..	..	..	..	..	..	..	..	..	..	..
Gross Domestic Income	..	..	..	..	..	..	..	..	..	..	..
Gross National Income	..	..	..	..	..	..	..	..	..	..	..

DOMESTIC PRICES (DEFLATORS)

(Index 1980 = 100)

	1967	1968	1969	1970	1971	1972	1973	1974	1975	1976	1977
Overall (GDP)	..	..	..	..	..	..	..	..	..	..	..
Domestic Absorption	..	..	..	..	..	..	..	..	..	..	..
Agriculture	..	..	..	..	..	..	..	..	..	..	..
Industry	..	..	..	..	..	..	..	..	..	..	..
Manufacturing	..	..	..	..	..	..	..	..	..	..	..

MANUFACTURING ACTIVITY

	1967	1968	1969	1970	1971	1972	1973	1974	1975	1976	1977
Employment (1980=100)	75.6	79.2	82.3	83.8	86.8	90.5	93.5	96.1	97.9	98.4	99.6
Real Earnings per Empl. (1980=100)	..	..	..	..	..	..	..	..	..	..	..
Real Output per Empl. (1980=100)	..	..	..	..	..	..	..	..	..	..	..
Earnings as % of Value Added	..	..	..	24.1	25.1	25.2	24.7	23.3	24.1	25.7	25.0

MONETARY HOLDINGS

(Billions of current Polish Zlotys)

	1967	1968	1969	1970	1971	1972	1973	1974	1975	1976	1977
Money Supply, Broadly Defined	..	..	..	..	..	..	..	..	..	..	..
Money as Means of Payment	..	..	..	..	..	..	..	..	..	..	..
Currency Ouside Banks	..	..	..	..	..	..	..	..	..	..	..
Demand Deposits	..	..	..	..	..	..	..	..	..	..	..
Quasi-Monetary Liabilities	..	..	..	..	..	..	..	..	..	..	..

(Billions of current Polish Zlotys)

	1967	1968	1969	1970	1971	1972	1973	1974	1975	1976	1977
GOVERNMENT DEFICIT (-) OR SURPLUS	..	..	..	..	..	..	..	..	..	..	..
Current Revenue	..	..	..	..	..	..	..	..	..	..	..
Current Expenditure	..	..	..	..	..	..	..	..	..	..	..
Current Budget Balance	..	..	..	..	..	..	..	..	..	..	..
Capital Receipts	..	..	..	..	..	..	..	..	..	..	..
Capital Payments	..	..	..	..	..	..	..	..	..	..	..

1978	1979	1980	1981	1982	1983	1984	1985	1986	1987 estimate	NOTES	POLAND
..	..	..	..	1,570	1,830	2,100	2,120	2,080	1,920	..	**CURRENT GNP PER CAPITA (US $)**
35,010	35,225	35,578	35,902	36,227	36,571	36,914	37,203	37,503	37,786	..	**POPULATION (thousands)**
				(Billions of current Polish Zlotys)							**ORIGIN AND USE OF RESOURCES**
..	..	2,407.0	2,621.0	5,383.0	6,776.0	8,427.0	10,272.0	12,764.0	16,919.0	..	Gross National Product (GNP)
..	..	-104.0	-131.0	-164.0	-148.0	-149.0	-173.0	-189.0	-218.0	..	Net Factor Income from Abroad
..	..	2,511.0	2,753.0	5,546.0	6,924.0	8,576.0	10,445.0	12,953.0	17,137.0	..	Gross Domestic Product (GDP)
..	..	..	..	..	..	..	..	..	..	..	Indirect Taxes, net
..	..	..	..	..	..	..	..	..	..	..	GDP at factor cost
..	..	..	..	..	..	..	..	..	..	..	Agriculture
..	..	..	..	..	..	..	..	..	..	..	Industry
..	..	..	..	..	..	..	..	..	..	..	Manufacturing
..	..	..	..	..	..	..	..	..	..	..	Services, etc.
..	..	-74.0	-58.0	116.0	122.0	169.0	148.0	221.0		..	Resource Balance
..	..	707.0	638.0	1,078.0	1,192.0	1,516.0	1,901.0	2,357.0		..	Exports of Goods & NFServices
..	..	780.0	697.0	962.0	1,070.0	1,346.0	1,753.0	2,136.0		..	Imports of Goods & NFServices
..	..	2,585.0	2,811.0	5,430.0	6,802.0	8,407.0	10,297.0	12,732.0		..	Domestic Absorption
..	..	1,692.0	2,042.0	3,425.0	4,458.0	5,348.0	6,430.0	..		..	Private Consumption, etc.
..	..	231.0	260.0	455.0	610.0	801.0	979.0	..		..	General Gov't Consumption
..	..	662.0	508.0	1,551.0	1,734.0	2,257.0	2,888.0	3,702.0		..	Gross Domestic Investment
..	..	621.0	514.0	1,117.0	1,395.0	1,781.0	2,211.0	2,833.0		..	Fixed Investment
											Memo Items:
..	..	588.0	450.0	1,667.0	1,856.0	2,426.0	3,036.0	3,923.0		..	Gross Domestic Saving
..	..	513.0	352.0	1,530.0	1,743.0	2,330.0	2,976.0	3,901.0		..	Gross National Saving
				(Billions of 1980 Polish Zlotys)							
..	..	2,406.8	2,152.9	2,089.5	2,224.1	2,359.3	2,470.3	2,597.3	2,652.0	..	Gross National Product
..	..	..	..	..	..	..	..	..	..	..	GDP at factor cost
..	..	..	..	..	..	..	..	..	..	..	Agriculture
..	..	..	..	..	..	..	..	..	..	..	Industry
..	..	..	..	..	..	..	..	..	..	..	Manufacturing
..	..	..	..	..	..	..	..	..	..	..	Services, etc.
..	..	-73.5	-48.4	71.5	96.0	121.5	80.3	..	..	..	Resource Balance
..	..	706.8	577.9	613.8	670.1	749.6	751.5	..	..	..	Exports of Goods & NFServices
..	..	780.3	626.3	542.3	574.0	628.1	671.2	..	..	..	Imports of Goods & NFServices
..	..	2,584.7	2,309.2	2,081.7	2,176.7	2,279.6	2,431.7	..	..	..	Domestic Absorption
..	..	1,692.2	1,574.4	1,359.2	1,420.3	1,472.4	1,570.7	..	..	..	Private Consumption, etc.
..	..	230.8	220.3	225.7	232.7	250.2	269.7	..	..	..	General Gov't Consumption
..	..	661.7	514.5	496.9	523.7	557.0	591.3	..	..	..	Gross Domestic Investment
..	..	621.2	502.8	434.1	472.2	518.6	541.2	..	..	..	Fixed Investment
											Memo Items:
..	..	706.8	574.1	607.8	639.6	707.1	727.9	..	..	..	Capacity to Import
..	..	0.0	-3.9	-6.0	-30.5	-42.5	-23.6	..	..	..	Terms of Trade Adjustment
..	..	2,511.2	2,256.9	2,147.1	2,242.3	2,358.6	2,488.3	..	..	..	Gross Domestic Income
..	..	2,406.8	2,149.1	2,083.4	2,193.6	2,316.8	2,446.7	..	..	..	Gross National Income
				(Index 1980 = 100)							**DOMESTIC PRICES (DEFLATORS)**
..	..	100.0	121.8	257.6	304.7	357.2	415.8	491.4	636.2	..	Overall (GDP)
..	..	100.0	121.7	260.9	312.5	368.8	423.5	..	..	..	Domestic Absorption
..	..	..	..	..	..	..	..	..	..	..	Agriculture
..	..	..	..	..	..	..	..	..	..	..	Industry
..	..	..	..	..	..	..	..	..	..	..	Manufacturing
											MANUFACTURING ACTIVITY
99.1	98.3	100.0	99.2	92.7	91.3	90.5	89.7	..	..	G	Employment (1980 = 100)
..	..	..	..	..	..	..	..	..	..	G	Real Earnings per Empl. (1980 = 100)
..	..	..	..	..	..	..	..	..	..	G	Real Output per Empl. (1980 = 100)
24.0	25.2	28.0	39.1	23.5	24.5	24.0	24.2	..	..	..	Earnings as % of Value Added
				(Billions of current Polish Zlotys)							**MONETARY HOLDINGS**
..	1,385	1,544	1,917	2,660	3,045	3,595	4,465	5,607	7,439	..	Money Supply, Broadly Defined
..	825	910	1,116	1,590	1,752	2,019	2,456	2,989	3,723	..	Money as Means of Payment
..	235	293	402	605	718	824	1,014	1,166	1,312	..	Currency Ouside Banks
..	591	618	713	985	1,034	1,195	1,442	1,823	2,411	..	Demand Deposits
..	559	633	802	1,070	1,292	1,575	2,009	2,618	3,717	..	Quasi-Monetary Liabilities
				(Billions of current Polish Zlotys)							
..	..	..	..	..	..	..	..	-40	..	..	**GOVERNMENT DEFICIT (-) OR SURPLUS**
..	..	..	..	..	..	..	..	5,363	..	..	Current Revenue
..	..	..	..	..	..	..	..	5,070	..	..	Current Expenditure
..	..	..	..	..	..	..	..	293	..	..	Current Budget Balance
..	..	..	..	..	..	..	..	..	..	..	Capital Receipts
..	..	..	..	..	..	..	..	333	..	..	Capital Payments

POLAND	1967	1968	1969	1970	1971	1972	1973	1974	1975	1976	1977
FOREIGN TRADE (CUSTOMS BASIS)				*(Millions of current US dollars)*							
Value of Exports, fob	2,527	2,858	3,320	3,548	3,872	4,927	6,432	8,321	10,289	11,024	12,273
Nonfuel Primary Products	..	..	..	..	..	..	..	..	..	..	..
Fuels	..	..	..	..	..	..	..	..	..	..	..
Manufactures	..	..	..	..	..	..	..	..	..	..	..
Value of Imports, cif	2,353	2,539	2,856	3,210	3,593	4,745	6,953	9,327	11,155	12,898	13,420
Nonfuel Primary Products	..	..	..	..	..	..	..	..	..	..	..
Fuels	..	..	..	..	..	..	..	..	..	..	..
Manufactures	..	..	..	..	..	..	..	..	..	..	..
					(Index 1980 = 100)						
Terms of Trade	..	..	..	..	..	..	..	..	..	..	..
Export Prices, fob	..	..	..	..	..	..	..	..	..	..	..
Nonfuel Primary Products	..	..	..	..	..	..	..	..	..	..	..
Fuels	..	..	..	..	..	..	..	..	..	..	..
Manufactures	..	..	..	..	..	..	..	..	..	..	..
Import Prices, cif	..	..	..	..	..	..	..	..	..	..	..
Trade at Constant 1980 Prices					*(Millions of 1980 US dollars)*						
Exports, fob	..	..	..	..	..	..	..	..	..	..	..
Imports, cif	..	..	..	..	..	..	..	..	..	..	..
BALANCE OF PAYMENTS					*(Millions of current US dollars)*						
Exports of Goods & Services	..	..	..	..	..	..	..	..	..	10,723	11,927
Merchandise, fob	..	..	..	..	..	..	..	..	..	9,506	10,506
Nonfactor Services	..	..	..	..	..	..	..	..	..	1,180	1,385
Factor Services	..	..	..	..	..	..	..	..	..	37	36
Imports of Goods & Services	..	..	..	..	..	..	..	..	..	14,095	14,957
Merchandise, fob	..	..	..	..	..	..	..	..	..	12,263	12,724
Nonfactor Services	..	..	..	..	..	..	..	..	..	1,144	1,309
Factor Services	..	..	..	..	..	..	..	..	..	688	924
Long-Term Interest	..	..	..	..	..	..	..	..	..	..	..
Current Transfers, net	..	..	..	..	..	..	..	..	..	399	463
Workers' Remittances	..	..	..	..	..	..	..	..	..	0	0
Total to be Financed	..	..	..	..	..	..	..	..	..	-2,973	-2,567
Official Capital Grants	..	..	..	..	..	..	..	..	..	180	175
Current Account Balance	..	..	..	..	..	..	..	..	..	-2,793	-2,392
Long-Term Capital, net	..	..	..	..	..	..	..	..	..	2,965	2,794
Direct Investment	..	..	..	..	..	..	..	..	..	-6	-12
Long-Term Loans	..	..	..	..	..	..	..	..	..	..	..
Disbursements	..	..	..	..	..	..	..	..	..	..	..
Repayments	..	..	..	..	..	..	..	..	..	..	..
Other Long-Term Capital	..	..	..	..	..	..	..	..	..	2,791	2,631
Other Capital, net	..	..	..	..	..	..	..	..	..	13	-773
Change in Reserves	..	..	..	0	0	0	0	0	0	-185	371
Memo Item:					*(Polish Zlotys per US dollar)*						
Conversion Factor (Annual Avg)	..	..	..	..	..	..	..	..	..	..	..
EXTERNAL DEBT, ETC.				*(Millions of US dollars, outstanding at end of year)*							
Public/Publicly Guar. Long-Term	..	..	..	..	..	..	..	..	..	..	..
Official Creditors	..	..	..	..	..	..	..	..	..	..	..
IBRD and IDA	..	..	..	..	..	..	..	..	..	..	..
Private Creditors	..	..	..	..	..	..	..	..	..	..	..
Private Non-guaranteed Long-Term	..	..	..	..	..	..	..	..	..	..	..
Use of Fund Credit	..	..	..	..	..	..	..	..	..	..	..
Short-Term Debt	..	..	..	..	..	..	..	..	..	..	..
Memo Items:					*(Millions of US dollars)*						
Int'l Reserves Excluding Gold	..	..	..	..	..	..	..	..	..	..	..
Gold Holdings (at market price)	..	..	..	..	..	..	..	..	..	..	..
SOCIAL INDICATORS											
Total Fertility Rate	2.3	2.2	2.2	2.2	2.3	2.2	2.3	2.3	2.3	2.3	2.3
Crude Birth Rate	16.3	16.2	16.3	16.8	17.1	17.4	17.9	18.4	18.9	19.5	19.0
Infant Mortality Rate	38.1	33.4	34.3	33.2	29.5	28.4	25.8	23.5	24.8	23.7	24.6
Life Expectancy at Birth	69.9	70.0	70.1	70.3	70.4	70.5	71.0	71.0	71.0	71.0	71.0
Food Production, p.c. ('79-81 = 100)	98.9	101.6	93.9	98.7	95.3	100.4	105.7	105.9	106.4	106.6	101.1
Labor Force, Agriculture (%)	41.7	40.8	39.8	38.9	37.9	36.8	35.8	34.7	33.7	32.7	31.6
Labor Force, Female (%)	44.7	44.9	45.2	45.4	45.3	45.3	46.8	45.3	45.3	45.3	45.3
School Enroll. Ratio, primary	..	..	..	101.0	..	..	..	..	100.0	100.0	100.0
School Enroll. Ratio, secondary	..	..	..	62.0	..	..	69.0	..	73.0	..	74.0

1978	1979	1980	1981	1982	1983	1984	1985	1986	1987 estimate	NOTES	POLAND
				(Millions of current US dollars)							**FOREIGN TRADE (CUSTOMS BASIS)**
14,114	16,249	16,997	13,249	11,174	11,572	11,750	11,489	12,074	..	..	Value of Exports, fob
..	..	..	..	..	..	..	..	..	..	..	Nonfuel Primary Products
..	..	..	..	..	..	..	..	..	..	..	Fuels
..	..	..	..	..	..	..	..	..	..	..	Manufactures
14,744	16,142	16,690	12,792	10,648	10,927	10,985	11,855	11,535	..	..	Value of Imports, cif
..	..	..	..	..	..	..	..	..	..	..	Nonfuel Primary Products
..	..	..	..	..	..	..	..	..	..	..	Fuels
..	..	..	..	..	..	..	..	..	..	..	Manufactures
				(Index 1980 = 100)							
..	..	..	..	..	..	..	..	..	..	..	Terms of Trade
..	..	..	..	..	..	..	..	..	..	..	Export Prices, fob
..	..	..	..	..	..	..	..	..	..	..	Nonfuel Primary Products
..	..	..	..	..	..	..	..	..	..	..	Fuels
..	..	..	..	..	..	..	..	..	..	..	Manufactures
..	..	..	..	..	..	..	..	..	..	..	Import Prices, cif
				(Millions of 1980 US dollars)							Trade at Constant 1980 Prices
..	..	..	..	..	..	..	..	..	..	..	Exports, fob
..	..	..	..	..	..	..	..	..	..	..	Imports, cif
				(Millions of current US dollars)							**BALANCE OF PAYMENTS**
13,593	15,221	15,794	11,723	11,533	12,362	13,030	12,152	12,918	13,081	..	Exports of Goods & Services
11,967	13,276	14,170	10,464	10,457	11,000	11,800	10,882	11,558	11,602	..	Merchandise, fob
1,586	1,874	1,485	1,088	984	1,178	1,035	1,094	1,172	1,262	..	Nonfactor Services
40	71	139	171	92	184	195	176	188	217	..	Factor Services
17,024	19,261	19,750	16,611	14,383	14,304	14,776	14,096	14,812	15,072	..	Imports of Goods & Services
14,259	15,660	15,806	10,654	10,582	10,962	10,436	10,914	10,803	11,236	..	Merchandise, fob
1,521	1,863	1,448	2,726	746	566	1,507	452	1,158	704	..	Nonfactor Services
1,244	1,738	2,496	3,231	3,055	2,776	2,833	2,730	2,851	3,132	..	Factor Services
..	..	..	..	..	..	..	1,454	999	960	..	Long-Term Interest
673	672	656	655	323	380	467	768	946	1,413	..	Current Transfers, net
0	0	0	0	0	0	0	0	0	..	..	Workers' Remittances
-2,758	-3,368	-3,300	-4,233	-2,628	-1,760	-1,279	-1,176	-948	-578	K	Total to be Financed
213	3	0	0	0	0	0	0	0	..	K	Official Capital Grants
-2,545	-3,365	-3,300	-4,233	-2,628	-1,760	-1,279	-1,176	-948	-578	..	Current Account Balance
2,408	3,624	2,659	898	-5,211	-3,790	-2,651	-2,070	-3,218	-3,125	..	Long-Term Capital, net
9	17	0	0	0	0	0	0	0	..	..	Direct Investment
..	..	..	..	..	..	..	520	645	-469	..	Long-Term Loans
..	..	..	..	..	..	..	958	1,366	493	..	Disbursements
..	..	..	..	..	..	..	439	721	962	..	Repayments
2,186	3,604	2,659	898	-5,211	-3,790	-2,651	-2,590	-3,863	-2,656	..	Other Long-Term Capital
266	16	239	3,370	8,208	5,669	4,270	3,010	3,993	4,500	..	Other Capital, net
-129	-275	402	-35	-369	-119	-340	236	173	-797	..	Change in Reserves
				(Polish Zlotys per US dollar)							Memo Item:
..	..	44.280	51.250	84.820	91.550	113.240	147.140	175.290	265.080	..	Conversion Factor (Annual Avg)
				(Millions of US dollars, outstanding at end of year)							**EXTERNAL DEBT, ETC.**
..	..	..	..	..	..	..	31,700	31,901	35,569	..	Public/Publicly Guar. Long-Term
..	..	..	..	..	..	..	19,948	20,738	22,914	..	Official Creditors
..	..	..	..	..	..	..	0	0	..	..	IBRD and IDA
..	..	..	..	..	..	..	11,752	11,162	12,656	..	Private Creditors
..	..	..	..	..	..	..	..	..	..	..	Private Non-guaranteed Long-Term
0	..	..	..	..	..	..	0	0	..	..	Use of Fund Credit
..	..	..	..	..	..	..	1,400	4,738	6,565	..	Short-Term Debt
				(Millions of US dollars)							Memo Items:
..	565.1	127.6	277.8	646.8	765.2	1,106.0	870.4	697.8	1,494.7	..	Int'l Reserves Excluding Gold
..	657.9	446.8	186.8	215.2	180.1	145.5	154.3	184.5	228.5	..	Gold Holdings (at market price)
											SOCIAL INDICATORS
2.2	2.2	2.3	2.3	2.3	2.4	2.3	2.3	2.3	2.3	..	Total Fertility Rate
19.0	19.5	19.4	18.9	19.3	19.7	18.9	18.2	16.9	16.7	..	Crude Birth Rate
22.4	21.0	21.2	20.6	20.2	19.2	19.2	18.4	17.5	23.6	..	Infant Mortality Rate
71.0	71.0	71.0	71.0	71.1	71.1	71.3	71.5	71.7	71.9	..	Life Expectancy at Birth
108.9	109.6	95.4	95.0	97.7	101.1	103.1	104.6	111.9	108.6	..	Food Production, p.c. ('79-81 = 100)
30.6	29.5	28.5	..	..	..	..	..	..	..	..	Labor Force, Agriculture (%)
45.3	45.3	45.3	45.3	45.4	45.4	45.4	45.5	45.5	45.5	..	Labor Force, Female (%)
100.0	99.0	99.0	100.0	100.0	101.0	101.0	101.0	..	..	..	School Enroll. Ratio, primary
75.0	76.0	77.0	77.0	75.0	75.0	77.0	77.8	..	..	..	School Enroll. Ratio, secondary

PORTUGAL	1967	1968	1969	1970	1971	1972	1973	1974	1975	1976	1977
CURRENT GNP PER CAPITA (US $)	500	570	610	700	790	910	1,150	1,380	1,480	1,650	1,780
POPULATION (thousands)	9,103	9,115	9,097	9,044	8,991	8,971	8,976	9,009	9,426	9,666	9,736

ORIGIN AND USE OF RESOURCES					*(Billions of current Portuguese Escudos)*						
Gross National Product (GNP)	132.3	146.4	160.9	178.8	199.6	232.5	284.6	343.1	376.8	464.3	617.9
Net Factor Income from Abroad	0.7	0.7	1.1	1.0	0.5	0.7	2.4	3.8	-0.4	-4.6	-7.9
Gross Domestic Product (GDP)	131.6	145.7	159.8	177.8	199.1	231.8	282.2	339.3	377.2	468.9	625.8
Indirect Taxes, net	..	..	..	..	..	..	..	..	..	..	14.8
GDP at factor cost	..	..	..	..	..	..	..	..	..	..	611.0
Agriculture	..	..	..	..	..	..	..	..	..	..	74.6
Industry	..	..	..	..	..	..	..	..	..	..	226.3
Manufacturing	..	..	..	..	..	..	..	..	..	..	..
Services, etc.	..	..	..	..	..	..	..	..	..	..	310.1
Resource Balance	-3.0	-6.9	-6.6	-11.6	-14.0	-10.9	-19.8	-51.9	-46.7	-62.9	-94.1
Exports of Goods & NFServices	35.8	36.5	39.1	43.4	49.9	63.1	75.4	91.2	76.9	81.7	115.3
Imports of Goods & NFServices	38.8	43.4	45.7	55.0	63.9	74.0	95.2	143.1	123.6	144.6	209.4
Domestic Absorption	134.6	152.6	166.4	189.4	213.1	242.7	302.0	391.2	423.9	531.8	719.9
Private Consumption, etc.	82.0	99.0	108.2	118.3	133.9	144.8	182.1	246.5	275.8	345.8	450.4
General Gov't Consumption	17.2	19.1	20.7	24.6	26.9	31.0	36.1	47.9	56.5	64.4	87.8
Gross Domestic Investment	35.4	34.5	37.5	46.5	52.3	66.9	83.8	96.8	91.6	121.6	181.7
Fixed Investment	35.0	32.3	36.1	41.3	49.2	62.8	75.6	88.2	97.8	117.5	167.9
Memo Items:											
Gross Domestic Saving	32.4	27.6	30.9	34.9	38.3	56.0	64.0	44.9	44.9	58.7	87.6
Gross National Saving	33.1	28.3	32.0	49.9	57.1	80.5	93.5	76.9	71.8	83.3	123.1

					(Billions of 1980 Portuguese Escudos)						
Gross National Product	621.6	669.9	729.7	746.0	813.2	864.2	933.9	1,044.3	1,059.0	1,000.8	1,060.4
GDP at factor cost	..	..	..	..	..	..	..	..	..	..	1,058.5
Agriculture	..	..	..	..	..	..	..	..	..	..	98.2
Industry	..	..	..	..	..	..	..	..	..	..	411.8
Manufacturing	..	..	..	..	..	..	..	..	..	..	..
Services, etc.	..	..	..	..	..	..	..	..	..	..	548.4
Resource Balance	11.3	-80.2	-98.0	-105.4	-132.7	-130.1	-174.8	-254.0	-161.9	-175.9	-220.8
Exports of Goods & NFServices	261.3	252.4	259.6	255.5	280.6	332.9	346.7	292.4	246.7	246.7	237.2
Imports of Goods & NFServices	250.0	332.7	357.6	360.9	413.4	463.0	521.4	546.3	408.6	422.6	458.0
Domestic Absorption	655.8	677.0	837.6	872.8	950.7	1,036.0	1,104.5	1,260.8	1,359.3	1,212.1	1,298.7
Private Consumption, etc.	347.2	353.4	507.6	527.7	546.1	613.4	630.3	723.1	828.2	776.9	821.4
General Gov't Consumption	67.2	73.0	75.4	80.8	85.9	93.5	100.7	118.1	125.9	134.8	150.6
Gross Domestic Investment	215.2	208.6	219.2	267.9	276.4	315.6	362.9	334.6	241.0	268.8	326.7
Fixed Investment	224.5	203.7	222.0	247.6	271.7	308.5	337.9	314.3	278.9	281.1	314.9
Memo Items:											
Capacity to Import	230.7	279.8	306.0	284.8	322.8	394.8	413.0	348.2	254.2	238.7	252.2
Terms of Trade Adjustment	-30.6	27.3	46.3	29.2	42.2	61.9	66.3	55.8	7.5	-7.9	15.0
Gross Domestic Income	615.3	652.4	775.8	812.1	863.2	932.4	1,024.7	1,136.7	1,137.7	1,038.6	1,092.9
Gross National Income	613.4	651.7	775.4	812.6	862.8	928.9	1,021.6	1,139.1	1,142.8	1,031.1	1,075.4

DOMESTIC PRICES (DEFLATORS)					*(Index 1980 = 100)*						
Overall (GDP)	21.1	21.7	21.9	23.8	24.5	26.7	30.1	32.6	35.8	46.5	58.1
Domestic Absorption	20.5	22.5	19.9	21.7	22.4	23.4	27.3	31.0	31.2	43.9	55.4
Agriculture	..	..	..	..	..	..	..	..	..	..	76.0
Industry	..	..	..	..	..	..	..	..	..	..	54.9
Manufacturing	..	..	..	..	..	..	..	..	..	..	..

MANUFACTURING ACTIVITY											
Employment (1980=100)	60.9	63.5	65.0	65.1	70.5	80.3	86.1	87.1	88.7	91.3	92.7
Real Earnings per Empl. (1980=100)	70.2	71.0	73.0	79.3	80.9	85.0	86.6	102.0	117.2	114.2	105.2
Real Output per Empl. (1980=100)	..	..	..	..	..	..	..	..	..	..	80.1
Earnings as % of Value Added	31.4	31.6	32.4	33.9	39.2	41.3	40.9	46.6	61.0	60.6	53.0

MONETARY HOLDINGS					*(Billions of current Portuguese Escudos)*						
Money Supply, Broadly Defined	108.5	122.8	144.9	166.5	189.0	234.8	301.3	342.3	385.5	470.7	578.6
Money as Means of Payment	77.9	84.5	93.9	100.4	104.9	122.3	165.6	182.5	227.3	256.2	286.0
Currency Ouside Banks	20.3	19.8	26.7	29.7	31.9	36.1	38.3	69.7	109.8	109.2	113.3
Demand Deposits	57.5	64.7	67.2	70.7	73.0	86.2	127.3	112.8	117.4	147.0	172.7
Quasi-Monetary Liabilities	30.7	38.3	51.0	66.1	84.1	112.5	135.7	159.8	158.2	214.5	292.6

GOVERNMENT DEFICIT (-) OR SURPLUS					*(Billions of current Portuguese Escudos)*						
	..	..	..	..	..	..	..	..	-31.7	-54.2	-40.9
Current Revenue	..	..	..	..	..	..	..	..	95.6	127.4	178.2
Current Expenditure	..	..	..	..	..	..	..	..	108.2	147.5	..
Current Budget Balance	..	..	..	..	..	..	..	..	-12.6	-20.1	..
Capital Receipts	..	..	..	..	..	..	..	0.0	0.5	0.0	0.0
Capital Payments	..	..	..	..	..	..	..	..	19.6	34.1	..

1978	1979	1980	1981	1982	1983	1984	1985	1986	1987 estimate	NOTES	PORTUGAL
1,850	2,050	2,350	2,470	2,480	2,230	1,990	1,970	2,250	2,810	..	**CURRENT GNP PER CAPITA (US $)**
9,796	9,857	9,909	9,957	9,997	10,099	10,132	10,157	10,187	10,212	..	**POPULATION (thousands)**
		(Billions of current Portuguese Escudos)									**ORIGIN AND USE OF RESOURCES**
771.6	971.0	1,202.4	1,412.0	1,753.1	2,170.5	2,648.0	3,330.0	4,187.0	4,913.0	..	Gross National Product (GNP)
-15.7	-22.3	-32.6	-60.7	-103.8	-118.5	-176.0	-197.0	-150.0	-123.0	..	Net Factor Income from Abroad
787.3	993.3	1,235.0	1,472.7	1,856.9	2,289.0	2,824.0	3,527.0	4,337.0	5,036.0	..	Gross Domestic Product (GDP)
13.8	12.5	15.1	20.5	26.6	35.8	34.0	46.0	182.0	206.0	f	Indirect Taxes, net
773.5	980.8	1,219.9	1,452.2	1,830.3	2,253.2	2,790.0	3,481.0	4,155.0	4,830.0	f	GDP at factor cost
94.0	115.0	126.0	123.6	159.8	200.0	265.0	331.0	371.0	448.0	..	Agriculture
291.5	376.3	484.1	579.6	734.6	902.0	1,117.0	1,375.0	1,669.0	1,919.0	..	Industry
..	..	..	..	..	..	..	..	..	..	..	Manufacturing
388.0	489.5	609.8	749.0	935.9	1,151.2	1,408.0	1,775.0	2,115.0	2,463.0	..	Services, etc.
-97.7	-107.6	-186.0	-297.3	-374.0	-282.0	-196.0	-123.0	-117.0	-278.0	..	Resource Balance
158.4	268.7	355.1	403.0	488.7	733.0	1,096.0	1,316.0	1,462.0	1,737.0	..	Exports of Goods & NFServices
256.1	376.3	541.1	700.3	862.7	1,015.0	1,292.0	1,439.0	1,579.0	2,015.0	..	Imports of Goods & NFServices
885.0	1,100.9	1,421.0	1,770.0	2,230.9	2,571.0	3,020.0	3,650.0	4,454.0	5,314.0	..	Domestic Absorption
535.3	670.3	823.1	1,020.3	1,277.5	1,570.0	1,969.0	2,380.0	2,846.0	3,410.0	..	Private Consumption, etc.
109.7	137.6	177.3	218.7	266.6	334.0	406.0	501.0	636.0	696.0	..	General Gov't Consumption
240.0	293.0	420.6	531.0	686.8	667.0	645.0	769.0	972.0	1,208.0	..	Gross Domestic Investment
222.3	270.6	364.2	463.0	587.5	678.0	681.0	777.6	958.0	1,074.0	..	Fixed Investment
											Memo Items:
142.3	185.4	234.6	233.7	312.8	385.0	449.0	646.0	855.0	930.0	..	Gross Domestic Saving
198.1	283.1	351.9	351.8	421.0	502.7	587.0	807.1	1,094.3	1,284.7	..	Gross National Saving
		(Billions of 1980 Portuguese Escudos)									
1,087.2	1,147.3	1,202.4	1,196.7	1,212.1	1,219.0	1,189.0	1,231.2	1,296.1	1,380.4	..	Gross National Product
1,100.6	1,170.3	1,219.9	1,236.3	1,274.9	1,276.6	1,256.4	1,297.2	1,309.9	1,375.2	f	GDP at factor cost
104.1	125.6	126.0	109.4	115.7	109.9	113.0	117.2	..	..	..	Agriculture
440.1	462.3	484.1	493.6	508.7	511.0	498.7	515.0	..	..	..	Industry
..	..	..	..	..	..	..	..	..	..	..	Manufacturing
556.4	582.4	609.8	633.3	650.3	655.6	644.8	665.0		..	..	Services, etc.
-199.1	-177.9	-186.0	-216.8	-226.4	-114.1	-39.0	-2.4	-61.1	-128.1	..	Resource Balance
254.3	323.6	355.1	344.4	365.2	426.1	486.6	540.7	576.9	637.4	..	Exports of Goods & NFServices
453.4	501.5	541.1	561.2	591.6	540.2	525.6	543.1	638.0	765.5	..	Imports of Goods & NFServices
1,314.9	1,359.3	1,421.0	1,465.2	1,513.7	1,400.6	1,303.6	1,308.8	1,423.4	1,559.0	..	Domestic Absorption
809.2	813.6	823.1	851.2	869.0	847.3	811.3	814.3	882.3	957.0	..	Private Consumption, etc.
157.1	167.2	177.3	182.1	187.2	192.2	197.2	200.6	203.7	207.7	..	General Gov't Consumption
348.7	378.5	420.6	431.9	457.5	361.1	295.1	293.8	337.3	394.2	..	Gross Domestic Investment
337.0	333.6	364.2	382.8	393.8	364.4	298.7	289.7	317.3	368.1	..	Fixed Investment
											Memo Items:
280.4	358.1	355.1	323.0	335.1	390.1	445.8	496.6	590.7	659.9	..	Capacity to Import
26.1	34.5	0.0	-21.4	-30.0	-35.9	-40.7	-44.0	13.8	22.5	..	Terms of Trade Adjustment
1,142.0	1,215.9	1,235.0	1,226.9	1,257.3	1,250.5	1,223.9	1,262.4	1,376.1	1,453.1	..	Gross Domestic Income
1,113.4	1,181.7	1,202.4	1,175.3	1,182.1	1,183.1	1,148.2	1,187.2	1,309.8	1,402.9	..	Gross National Income
		(Index 1980 = 100)									**DOMESTIC PRICES (DEFLATORS)**
70.6	84.1	100.0	118.0	144.2	177.9	223.3	270.0	318.4	352.1	..	Overall (GDP)
67.3	81.0	100.0	120.8	147.4	183.6	231.7	278.9	312.9	340.9	..	Domestic Absorption
90.3	91.6	100.0	113.0	138.1	181.9	234.6	282.5	..	..	..	Agriculture
66.2	81.4	100.0	117.4	144.4	176.5	224.0	267.0	..	..	..	Industry
..	..	..	..	..	..	..	..	..	..	..	Manufacturing
											MANUFACTURING ACTIVITY
95.1	98.4	100.0	100.5	99.1	97.7	94.2	96.7	..	..	..	Employment (1980=100)
98.2	93.7	100.0	100.7	98.8	93.9	86.9	97.9	..	..	..	Real Earnings per Empl. (1980=100)
83.0	91.0	100.0	103.3	105.8	114.2	117.2	120.3	..	..	..	Real Output per Empl. (1980=100)
49.8	46.1	43.2	45.1	47.0	43.5	38.4	43.0	..	..	..	Earnings as % of Value Added
		(Billions of current Portuguese Escudos)									**MONETARY HOLDINGS**
738.2	1,012.5	1,365.4	1,755.4	2,218.1	2,661.6	3,400.6	4,226.6	4,992.1	5,553.6	..	Money Supply, Broadly Defined
326.5	414.8	506.0	550.5	638.6	687.3	799.4	1,017.6	1,413.4	1,581.8	..	Money as Means of Payment
121.4	142.1	165.2	188.4	219.5	240.0	267.3	319.0	399.3	457.7	..	Currency Ouside Banks
205.1	272.6	340.8	362.1	419.2	447.3	532.1	698.6	1,014.1	1,124.1	..	Demand Deposits
411.7	597.7	859.4	1,204.9	1,579.5	1,974.3	2,601.2	3,209.0	3,578.7	3,971.8	..	Quasi-Monetary Liabilities
		(Billions of current Portuguese Escudos)									
-92.9	-100.5	-121.4	-179.6	-195.0	-219.7	..	..	..	..	..	**GOVERNMENT DEFICIT (-) OR SURPLUS**
218.0	273.1	380.2	482.5	590.9	790.7	..	..	..	..	..	Current Revenue
..	..	..	..	..	..	..	..	..	..	..	Current Expenditure
..	..	..	..	..	..	..	..	..	..	..	Current Budget Balance
..	..	..	..	..	..	..	..	..	..	..	Capital Receipts
..	..	..	..	..	..	..	..	..	..	..	Capital Payments

PORTUGAL	1967	1968	1969	1970	1971	1972	1973	1974	1975	1976	1977
FOREIGN TRADE (CUSTOMS BASIS)					*(Millions of current US dollars)*						
Value of Exports, fob	701	761	853	949	1,053	1,294	1,862	2,302	1,940	1,820	2,013
Nonfuel Primary Products	239	277	291	321	320	390	543	633	530	551	560
Fuels	10	9	10	22	23	22	22	65	40	39	34
Manufactures	452	475	553	606	709	882	1,297	1,604	1,370	1,231	1,420
Value of Imports, cif	1,059	1,178	1,298	1,590	1,824	2,227	3,073	4,641	3,863	4,316	4,964
Nonfuel Primary Products	348	360	382	427	496	649	934	1,434	1,263	1,352	1,544
Fuels	91	99	110	145	156	154	192	584	588	692	737
Manufactures	620	719	806	1,017	1,172	1,425	1,947	2,623	2,012	2,272	2,683
Terms of Trade	129.0	133.4	132.4	131.6	122.6	*(Index 1980 = 100)* 121.1	119.0	99.1	107.4	98.6	93.0
Export Prices, fob	26.5	27.4	27.9	27.7	30.3	35.2	47.0	57.1	62.3	58.9	59.8
Nonfuel Primary Products	32.2	32.7	34.7	35.5	35.5	40.2	53.7	54.0	57.7	57.7	59.6
Fuels	4.3	4.3	4.3	4.3	5.6	6.2	8.9	36.7	35.7	38.4	42.0
Manufactures	27.0	27.6	27.6	30.3	32.9	37.5	48.0	59.9	65.8	60.5	60.5
Import Prices, cif	20.5	20.5	21.0	21.0	24.7	29.1	39.5	57.6	58.0	59.8	64.3
Trade at Constant 1980 Prices					*(Millions of 1980 US dollars)*						
Exports, fob	2,651	2,783	3,062	3,430	3,472	3,675	3,963	4,029	3,111	3,089	3,365
Imports, cif	5,166	5,747	6,168	7,557	7,374	7,665	7,786	8,051	6,655	7,220	7,720
BALANCE OF PAYMENTS					*(Millions of current US dollars)*						
Exports of Goods & Services	..	..	..	1,349	1,817	2,268	3,091	3,657	2,994	2,628	3,466
Merchandise, fob	..	..	..	956	1,057	1,307	1,843	2,289	1,939	1,790	2,529
Nonfactor Services	..	..	..	310	691	840	1,073	1,136	940	788	867
Factor Services	..	..	..	83	69	121	175	232	114	49	70
Imports of Goods & Services	..	..	..	1,995	2,317	2,795	3,847	5,597	4,788	4,874	5,561
Merchandise, fob	..	..	..	1,448	1,695	2,041	2,753	4,284	3,540	3,960	4,527
Nonfactor Services	..	..	..	495	572	674	1,015	1,211	1,108	714	765
Factor Services	..	..	..	52	50	79	80	102	140	200	268
Long-Term Interest	..	..	..	34	37	34	34	34	44	52	94
Current Transfers, net	..	..	..	488	646	881	1,097	1,110	1,067	966	1,134
Workers' Remittances	..	..	..	523	678	908	1,140	1,145	1,100	908	1,173
Total to be Financed	..	..	..	-158	146	354	341	-830	-727	-1,280	-960
Official Capital Grants	..	..	..	..	..	..	..	..	-28	..	..
Current Account Balance	..	..	..	-158	146	354	341	-830	-755	-1,280	-960
Long-Term Capital, net	..	..	..	278	97	-104	-141	255	-123	60	67
Direct Investment	..	..	..	15	49	66	51	69	106	58	55
Long-Term Loans	..	..	..	-46	14	18	153	94	299	356	788
Disbursements	..	..	..	38	102	117	257	204	404	537	1,066
Repayments	..	..	..	84	88	100	104	110	106	181	279
Other Long-Term Capital	..	..	..	309	34	-188	-345	93	-500	-354	-776
Other Capital, net	..	..	..	-88	97	106	128	-16	-157	187	556
Change in Reserves	..	..	..	-32	-340	-356	-328	590	1,035	1,033	337
Memo Item:					*(Portuguese Escudos per US dollar)*						
Conversion Factor (Annual Avg)	28.750	28.750	28.750	28.750	28.310	27.050	24.670	25.410	25.550	30.230	38.280
EXTERNAL DEBT, ETC.					*(Millions of US dollars, outstanding at end of year)*						
Public/Publicly Guar. Long-Term	..	..	..	485	524	539	723	872	1,093	1,408	2,321
Official Creditors	..	..	..	224	220	262	324	323	316	454	796
IBRD and IDA	..	..	..	52	51	48	45	42	39	36	32
Private Creditors	..	..	..	261	305	277	400	549	777	954	1,526
Private Non-guaranteed Long-Term	..	..	..	268	284	309	338	338	384	413	409
Use of Fund Credit	..	..	..	0	0	0	0	0	0	201	298
Short-Term Debt	..	..	..	..	..	..	..	..	..	..	..
Memo Items:					*(Millions of US dollars)*						
Int'l Reserves Excluding Gold	535	507	570	602	945	1,291	1,676	1,161	398	176	366
Gold Holdings (at market price)	703	1,025	880	963	1,148	1,745	3,091	5,192	3,887	3,729	3,977
SOCIAL INDICATORS											
Total Fertility Rate	3.0	2.9	2.8	2.8	2.8	2.7	2.7	2.6	2.5	2.6	2.4
Crude Birth Rate	22.2	21.4	20.9	20.0	20.2	19.5	19.2	18.9	19.1	19.3	18.6
Infant Mortality Rate	59.2	61.1	55.8	55.1	51.9	41.4	44.8	37.9	38.9	33.4	30.3
Life Expectancy at Birth	66.1	66.6	67.0	67.4	67.7	68.0	68.3	68.6	68.9	69.1	69.4
Food Production, p.c. ('79-81 = 100)	..	..	..	140.3	122.7	119.7	125.9	127.6	119.2	108.2	93.2
Labor Force, Agriculture (%)	35.3	34.2	33.0	31.8	31.1	30.3	29.6	28.8	28.1	27.6	27.2
Labor Force, Female (%)	22.7	23.4	24.1	24.8	26.0	27.2	28.5	29.7	30.9	31.9	32.8
School Enroll. Ratio, primary	..	..	..	98.0	..	..	..	..	113.0	114.0	117.0
School Enroll. Ratio, secondary	..	..	..	57.0	..	..	..	..	53.0	54.0	55.0

1978	1979	1980	1981	1982	1983	1984	1985	1986	1987 estimate	NOTES	PORTUGAL
											FOREIGN TRADE (CUSTOMS BASIS)
				(Millions of current US dollars)							
2,410	3,480	4,629	4,180	4,171	4,601	5,208	5,685	7,160	9,167	..	Value of Exports, fob
609	822	1,052	921	870	906	1,049	1,038	1,236	1,623	..	Nonfuel Primary Products
44	129	255	298	163	231	186	236	217	149	..	Fuels
1,757	2,529	3,322	2,961	3,138	3,464	3,973	4,412	5,707	7,395	..	Manufactures
5,229	6,509	9,293	9,946	9,605	8,257	7,975	7,650	9,393	13,438	..	Value of Imports, cif
1,430	1,845	2,247	2,491	2,149	1,918	2,169	1,974	2,129	2,819	..	Nonfuel Primary Products
826	1,267	2,243	2,396	2,546	2,218	2,286	2,026	1,439	1,563	..	Fuels
2,973	3,397	4,802	5,058	4,911	4,120	3,520	3,650	5,825	9,055	..	Manufactures
				(Index 1980 = 100)							
100.1	99.5	100.0	88.4	81.9	77.7	79.7	84.9	95.5	99.5	..	Terms of Trade
69.6	82.8	100.0	89.7	78.0	72.9	73.1	73.7	78.4	96.0	..	Export Prices, fob
74.6	100.7	100.0	86.0	82.1	81.1	79.1	74.0	81.5	85.6	..	Nonfuel Primary Products
42.3	61.0	100.0	112.5	101.6	92.5	90.2	87.5	45.0	56.9	..	Fuels
69.1	79.6	100.0	89.0	76.0	70.0	71.0	73.0	80.0	100.0	..	Manufactures
69.5	83.2	100.0	101.4	95.2	93.7	91.7	86.8	82.1	96.5	..	Import Prices, cif
				(Millions of 1980 US dollars)							Trade at Constant 1980 Prices
3,465	4,205	4,629	4,662	5,349	6,316	7,129	7,715	9,132	9,552	..	Exports, fob
7,525	7,822	9,293	9,805	10,092	8,810	8,695	8,813	11,441	13,931	..	Imports, cif
				(Millions of current US dollars)							**BALANCE OF PAYMENTS**
3,941	5,328	6,845	6,389	5,901	6,949	7,145	7,940	9,976	12,795	..	Exports of Goods & Services
2,742	3,548	4,584	4,059	4,124	5,229	5,208	5,676	7,179	9,160	..	Merchandise, fob
1,111	1,645	2,086	2,123	1,619	1,547	1,736	1,985	2,494	3,231	..	Nonfactor Services
88	135	176	207	158	173	200	279	303	404	..	Factor Services
6,043	7,859	10,914	11,868	11,797	10,064	9,828	9,770	11,650	15,877	..	Imports of Goods & Services
4,785	6,180	8,610	9,105	8,970	7,612	7,228	7,134	8,812	12,535	..	Merchandise, fob
830	1,104	1,492	1,587	1,421	1,204	1,208	1,230	1,528	2,006	..	Nonfactor Services
427	575	811	1,176	1,406	1,247	1,392	1,406	1,311	1,336	..	Factor Services
184	363	530	870	973	905	1,091	1,156	1,197	1,232	..	Long-Term Interest
1,627	2,468	2,994	2,905	2,668	2,132	2,145	2,102	2,602	3,391	..	Current Transfers, net
1,666	2,451	2,924	2,848	2,607	2,119	2,156	2,075	2,529	3,243	..	Workers' Remittances
-475	-63	-1,073	-2,574	-3,227	-982	-539	271	928	309	K	Total to be Financed
..	6	7	0	0	0	37	108	192	332	K	Official Capital Grants
-475	-57	-1,066	-2,574	-3,227	-982	-502	379	1,121	641	..	Current Account Balance
685	679	708	1,197	2,067	1,217	1,188	1,106	-332	129	..	Long-Term Capital, net
60	70	143	156	136	123	186	231	239	306	..	Direct Investment
1,438	1,489	1,419	1,681	1,628	1,698	1,148	789	-559	-862	..	Long-Term Loans
1,842	2,076	2,099	2,574	2,592	2,869	2,775	2,721	1,893	2,883	..	Disbursements
404	587	681	892	964	1,172	1,627	1,932	2,453	3,744	..	Repayments
-814	-886	-862	-640	303	-604	-182	-22	-204	352	..	Other Long-Term Capital
124	-346	715	1,239	1,283	-799	-681	-572	-900	981	..	Other Capital, net
-334	-276	-357	138	-122	564	-5	-913	111	-1,752	..	Change in Reserves
				(Portuguese Escudos per US dollar)							Memo Item:
43.940	48.620	50.060	61.550	79.470	110.780	146.390	170.390	149.590	140.880	..	Conversion Factor (Annual Avg)
				(Millions of US dollars, outstanding at end of year)							**EXTERNAL DEBT, ETC.**
3,980	5,500	6,703	7,792	8,962	10,072	10,825	12,884	13,661	14,922	..	Public/Publicly Guar. Long-Term
1,763	2,142	2,177	1,961	1,935	2,005	1,835	2,129	2,358	2,783	..	Official Creditors
36	63	109	132	185	260	243	372	491	619	..	IBRD and IDA
2,217	3,358	4,525	5,831	7,028	8,067	8,990	10,755	11,303	12,140	..	Private Creditors
420	504	487	368	766	632	524	552	585	630	..	Private Non-guaranteed Long-Term
264	227	119	55	11	371	561	628	700	529	..	Use of Fund Credit
1,583	1,650	2,395	3,335	3,839	3,422	3,015	2,627	1,429	2,164	..	Short-Term Debt
				(Millions of US dollars)							Memo Items:
871	931	795	534	447	385	516	1,395	1,456	3,327	..	Int'l Reserves Excluding Gold
5,002	11,331	13,068	8,800	10,093	7,796	6,257	6,614	7,880	9,713	..	Gold Holdings (at market price)
											SOCIAL INDICATORS
2.2	2.1	2.1	2.0	2.0	1.9	1.9	1.7	1.7	1.7	..	Total Fertility Rate
17.1	16.3	16.2	15.4	15.4	14.4	14.2	12.8	13.0	13.2	..	Crude Birth Rate
29.1	26.0	23.9	21.9	19.8	19.3	18.6	17.8	16.6	19.9	..	Infant Mortality Rate
70.2	71.0	71.8	72.5	73.3	73.2	73.1	73.0	72.9	72.8	..	Life Expectancy at Birth
93.6	114.2	97.9	87.9	104.9	91.9	101.5	103.8	102.1	103.2	..	Food Production, p.c. ('79-81 = 100)
26.7	26.3	25.8	..	..	..	..	..	..	..	..	Labor Force, Agriculture (%)
33.8	34.8	35.7	35.8	35.9	36.1	36.2	36.3	36.4	36.5	..	Labor Force, Female (%)
103.0	..	119.0	122.0	121.0	120.0	114.0	112.0	..	..	..	School Enroll. Ratio, primary
..	48.0	45.0	38.0	43.0	47.0	..	..	..	..	..	School Enroll. Ratio, secondary

RWANDA	1967	1968	1969	1970	1971	1972	1973	1974	1975	1976	1977
CURRENT GNP PER CAPITA (US $)	50	50	60	60	60	60	70	70	90	120	160
POPULATION (thousands)	3,383	3,484	3,588	3,695	3,819	3,947	4,080	4,216	4,358	4,504	4,655

ORIGIN AND USE OF RESOURCES

					(Billions of current Rwanda Francs)						
Gross National Product (GNP)	16.11	17.28	18.97	22.03	22.38	22.78	24.49	28.48	52.76	61.84	71.82
Net Factor Income from Abroad	0.15	0.06	0.10	0.04	0.15	0.09	0.09	-0.20	-0.01	-0.03	0.19
Gross Domestic Product (GDP)	15.96	17.22	18.87	21.99	22.23	22.70	24.40	28.68	52.77	61.87	71.63
Indirect Taxes, net	0.74	0.79	0.76	1.52	1.29	1.39	1.62	2.24	3.25	5.61	6.44
GDP at factor cost	15.22	16.43	18.10	20.47	20.94	21.31	22.78	26.44	49.52	56.26	65.19
Agriculture	10.51	11.24	12.41	13.55	13.60	13.52	14.89	16.96	25.97	30.45	33.28
Industry	1.56	1.64	1.77	1.89	1.93	2.14	2.13	2.64	10.01	11.69	15.19
Manufacturing	0.39	0.75	0.77	0.79	0.83	0.93	1.00	1.25	6.48	7.60	10.20
Services, etc.	3.89	4.34	4.68	6.55	6.70	7.04	7.38	9.08	16.79	19.73	23.16
Resource Balance	-0.78	-1.10	-1.26	-0.82	-1.48	-1.81	-0.45	-2.63	-4.50	-2.98	-2.43
Exports of Goods & NFServices	1.68	1.50	1.45	2.53	2.27	1.88	2.83	3.49	4.84	9.02	10.22
Imports of Goods & NFServices	2.46	2.61	2.71	3.35	3.75	3.69	3.28	6.12	9.34	12.00	12.65
Domestic Absorption	16.74	18.32	20.13	22.81	23.71	24.51	24.85	31.31	57.27	64.85	74.06
Private Consumption, etc.	13.77	15.44	17.23	19.35	19.48	19.92	19.77	24.86	41.24	45.97	50.86
General Gov't Consumption	1.80	1.47	1.65	1.91	2.20	2.42	2.78	3.45	8.78	10.33	12.41
Gross Domestic Investment	1.17	1.41	1.25	1.55	2.03	2.17	2.30	3.00	7.25	8.55	10.79
Fixed Investment	..	1.22	1.23	1.58	2.01	2.16	..	..	6.96	7.99	9.14
Memo Items:											
Gross Domestic Saving	0.38	0.31	-0.01	0.73	0.55	0.36	1.85	0.37	2.75	5.57	8.36
Gross National Saving	0.54	0.36	-0.10	0.46	0.54	0.15	1.84	-0.27	2.47	5.35	8.30
					(Billions of 1980 Rwanda Francs)						
Gross National Product	53.66	57.33	60.96	67.79	71.63	72.84	72.82	75.31	75.12	77.70	81.85
GDP at factor cost	49.39	52.54	56.24	62.77	64.56	66.11	66.05	67.95	67.64	70.74	74.32
Agriculture	..	..	25.16	27.39	28.12	26.68	27.95	32.96	31.07	39.98	40.28
Industry	..	..	..	..	..	..	..	..	..	15.73	17.23
Manufacturing	..	..	8.95	9.20	9.46	9.72	10.11	10.34	10.56	10.86	12.05
Services, etc.	..	..	..	..	..	..	..	..	..	22.04	24.10
Resource Balance	-3.49	-4.73	-5.43	-3.46	-5.90	-6.95	-2.00	-8.43	-7.97	-5.18	-9.21
Exports of Goods & NFServices	6.61	5.88	5.74	9.11	8.18	6.65	9.64	10.19	7.89	13.42	8.61
Imports of Goods & NFServices	10.10	10.61	11.17	12.56	14.08	13.60	11.64	18.62	15.87	18.60	17.83
Domestic Absorption	56.79	58.16	63.28	70.42	72.26	75.62	76.89	74.17	81.40	82.92	90.83
Private Consumption, etc.	40.97	44.50	50.04	56.77	57.54	57.89	58.07	54.25	59.72	53.42	58.56
General Gov't Consumption	7.64	6.17	7.02	7.39	8.52	9.20	10.17	10.83	15.39	16.52	17.33
Gross Domestic Investment	4.69	5.62	5.05	5.70	7.47	7.84	8.00	8.95	12.07	12.99	14.94
Fixed Investment	..	..	..	..	..	..	..	..	..	12.62	13.13
Memo Items:											
Capacity to Import	6.89	6.13	5.97	9.49	8.52	6.93	10.04	10.62	8.22	13.98	14.40
Terms of Trade Adjustment	0.28	0.25	0.24	0.38	0.34	0.28	0.40	0.42	0.33	0.56	5.79
Gross Domestic Income	53.52	57.25	61.21	67.90	72.11	72.94	73.06	75.70	76.33	78.30	87.41
Gross National Income	54.10	57.78	61.39	68.22	72.23	73.40	73.31	75.94	75.78	78.26	87.64

DOMESTIC PRICES (DEFLATORS)

					(Index 1980 = 100)						
Overall (GDP)	30.1	30.3	31.0	32.6	31.1	31.4	33.6	38.2	69.7	79.6	87.8
Domestic Absorption	29.5	31.5	31.8	32.4	32.8	32.4	32.3	42.2	70.4	78.2	81.5
Agriculture	..	..	49.3	49.5	48.4	50.7	53.3	51.5	83.6	76.2	82.6
Industry	..	..	..	..	..	..	..	..	..	74.3	88.1
Manufacturing	..	..	8.6	8.6	8.8	9.6	9.9	12.1	61.3	70.0	84.7

MANUFACTURING ACTIVITY

Employment (1980 = 100)	..	..	..	..	..	..	..	..	..	..	..
Real Earnings per Empl. (1980 = 100)	..	..	..	..	..	..	..	..	..	..	..
Real Output per Empl. (1980 = 100)	..	..	..	..	..	..	..	..	..	..	..
Earnings as % of Value Added	..	..	39.5	22.0	27.9	34.3	43.8	24.6	..	..	6.1

MONETARY HOLDINGS

					(Billions of current Rwanda Francs)						
Money Supply, Broadly Defined	1.72	1.81	2.12	2.57	2.92	2.97	4.03	5.35	6.00	8.05	10.14
Money as Means of Payment	1.48	1.56	1.82	2.16	2.47	2.52	3.57	4.29	4.85	6.52	8.03
Currency Ouside Banks	1.03	0.99	1.09	1.24	1.41	1.45	2.00	2.55	2.72	3.07	3.95
Demand Deposits	0.45	0.58	0.73	0.93	1.06	1.07	1.56	1.73	2.13	3.45	4.09
Quasi-Monetary Liabilities	0.24	0.25	0.29	0.40	0.45	0.45	0.46	1.06	1.15	1.53	2.10

					(Millions of current Rwanda Francs)						
GOVERNMENT DEFICIT (-) OR SURPLUS	..	..	..	..	..	..	-653	-728	-727	-1,225	-1,026
Current Revenue	..	..	..	..	..	..	2,419	3,369	5,010	6,750	8,703
Current Expenditure	..	..	..	..	..	..	2,679	3,345	4,093	4,777	6,320
Current Budget Balance	..	..	..	..	..	..	-260	24	917	1,973	2,383
Capital Receipts	..	..	..	..	..	..	..	..	..	..	..
Capital Payments	..	..	..	..	..	..	393	752	1,644	3,198	3,409

1978	1979	1980	1981	1982	1983	1984	1985	1986	1987 estimate	NOTES	RWANDA
180	210	240	270	260	270	260	280	300	300	..	**CURRENT GNP PER CAPITA (US $)**
4,811	4,982	5,139	5,305	5,477	5,654	5,837	6,026	6,236	6,454	..	**POPULATION (thousands)**
			(Billions of current Rwanda Francs)								**ORIGIN AND USE OF RESOURCES**
80.57	96.20	108.11	123.41	130.94	141.60	158.62	173.13	168.25	166.24	..	Gross National Product (GNP)
-0.48	0.03	0.12	0.77	-0.02	-0.59	-0.49	-0.56	-0.74	-0.97	..	Net Factor Income from Abroad
81.05	96.17	107.99	122.64	130.96	142.19	159.11	173.70	168.99	167.21	..	Gross Domestic Product (GDP)
5.95	8.24	8.26	7.81	9.24	9.46	11.70	13.03	14.74	..	..	Indirect Taxes, net
75.10	87.93	99.73	114.83	121.72	132.73	147.41	160.67	154.26	..	B	GDP at factor cost
34.38	47.16	49.51	49.96	52.10	53.60	64.40	72.70	62.60	62.50	..	Agriculture
17.93	18.02	23.27	25.58	28.90	35.74	38.46	40.31	40.42	37.80	f	Industry
12.64	12.30	16.48	19.30	18.27	22.55	22.78	23.83	27.13	..	..	Manufacturing
28.74	30.99	35.21	47.10	49.96	52.85	56.25	60.68	65.97	66.91	f	Services, etc.
-7.31	-1.64	-12.90	-14.63	-18.25	-15.12	-13.80	-17.70	-16.15	-19.34	..	Resource Balance
11.98	20.24	15.59	12.05	13.17	14.35	17.49	16.32	19.95	13.32	..	Exports of Goods & NFServices
19.29	21.88	28.49	26.68	31.42	29.47	31.29	34.02	36.10	32.66	..	Imports of Goods & NFServices
88.36	97.81	120.89	137.27	149.21	157.31	172.91	191.40	185.14	186.55	..	Domestic Absorption
62.10	73.62	89.97	96.40	109.02	121.37	131.53	141.78	137.97	138.85	..	Private Consumption, etc.
12.80	12.62	13.49	24.56	16.91	16.71	16.24	19.57	20.13	19.80	..	General Gov't Consumption
13.46	11.57	17.43	16.31	23.28	19.23	25.14	30.05	27.04	27.90	..	Gross Domestic Investment
11.32	14.27	13.19	15.98	18.80	20.70	24.70	27.00	26.80	27.90	..	Fixed Investment
											Memo Items:
6.15	9.93	4.53	1.68	5.03	4.11	11.34	12.35	10.89	8.56	..	Gross Domestic Saving
5.78	10.51	4.36	2.17	5.44	4.04	11.03	12.22	10.76	8.09	..	Gross National Saving
			(Billions of 1980 Rwanda Francs)								
88.54	97.99	108.11	118.24	119.44	126.33	121.02	126.67	131.71	127.90	..	Gross National Product
82.54	89.60	99.73	109.92	110.97	118.39	112.44	117.51	120.70	..	B I f	GDP at factor cost
41.39	50.30	49.51	46.42	53.07	55.34	50.44	51.65	51.93	50.16	..	Agriculture
18.22	18.54	23.27	23.88	23.33	26.55	27.50	28.48	30.28	..	f	Industry
12.78	12.66	16.48	17.44	16.75	17.85	16.38	17.48	20.79	21.37	..	Manufacturing
29.49	29.12	35.21	47.20	43.05	44.98	43.48	46.98	50.12	..	f	Services, etc.
-10.42	-7.34	-12.90	-12.27	-18.12	-15.18	-19.58	-21.82	-22.82	-22.28	..	Resource Balance
12.37	16.75	15.59	15.78	16.07	17.15	17.04	17.58	21.58	18.99	..	Exports of Goods & NFServices
22.79	24.09	28.49	28.06	34.19	32.34	36.63	39.41	44.39	41.26	..	Imports of Goods & NFServices
99.51	105.29	120.89	129.77	137.57	142.05	141.00	148.93	155.15	150.96	..	Domestic Absorption
69.05	79.93	89.97	89.97	92.78	102.28	95.58	97.68	104.86	100.65	..	Private Consumption, etc.
15.88	13.52	13.49	23.02	20.63	19.14	17.64	20.95	21.70	20.53	..	General Gov't Consumption
14.58	11.84	17.43	16.79	24.16	20.63	27.78	30.30	28.59	29.79	..	Gross Domestic Investment
12.68	15.74	13.19	17.08	23.25	25.89	32.48	32.36	33.80	34.88	..	Fixed Investment
											Memo Items:
14.15	22.28	15.59	12.67	14.33	15.75	20.47	18.90	24.53	16.83	..	Capacity to Import
1.78	5.53	0.00	-3.11	-1.74	-1.41	3.43	1.32	2.96	-2.16	..	Terms of Trade Adjustment
90.87	103.49	107.99	114.39	117.71	125.46	124.85	128.43	135.29	126.57	..	Gross Domestic Income
90.33	103.52	108.11	115.13	117.70	124.93	124.44	127.99	134.67	125.79	..	Gross National Income
			(Index 1980 = 100)								**DOMESTIC PRICES (DEFLATORS)**
91.0	98.2	100.0	104.4	109.6	112.1	131.0	136.7	127.7	129.9	..	Overall (GDP)
88.8	92.9	100.0	105.8	108.5	110.7	122.6	128.5	119.3	123.6	..	Domestic Absorption
83.1	93.8	100.0	107.6	98.2	96.9	127.7	140.8	120.5	124.6	..	Agriculture
98.4	97.2	100.0	107.1	123.9	134.6	139.9	141.5	133.5	..	..	Industry
98.9	97.1	100.0	110.6	109.1	126.4	139.1	136.3	130.5	..	..	Manufacturing
											MANUFACTURING ACTIVITY
..	..	..	..	..	..	..	..	..	..	J	Employment (1980 = 100)
..	..	..	..	..	..	..	..	..	..	J	Real Earnings per Empl. (1980 = 100)
..	..	..	..	..	..	..	..	..	..	J	Real Output per Empl. (1980 = 100)
6.6	8.7	..	..	..	..	19.0	..	..	..	J	Earnings as % of Value Added
			(Billions of current Rwanda Francs)								**MONETARY HOLDINGS**
11.18	14.09	15.23	15.98	16.17	18.08	19.89	23.43	26.61	29.34	..	Money Supply, Broadly Defined
8.96	11.26	12.03	11.77	11.46	12.32	13.34	14.70	17.33	17.79	..	Money as Means of Payment
4.44	5.24	5.69	6.09	6.26	6.66	7.03	7.16	7.69	8.20	..	Currency Ouside Banks
4.52	6.01	6.34	5.69	5.20	5.65	6.31	7.54	9.65	9.59	..	Demand Deposits
2.22	2.83	3.20	4.21	4.71	5.76	6.55	8.73	9.27	11.55	..	Quasi-Monetary Liabilities
			(Millions of current Rwanda Francs)								
-1,291	-1,618	-1,875	..	..	..	..	..	..	..	..	**GOVERNMENT DEFICIT (-) OR SURPLUS**
9,180	12,478	13,805	..	..	..	..	..	..	..	..	Current Revenue
6,692	8,443	10,059	..	..	..	..	..	..	..	..	Current Expenditure
2,488	4,035	3,746	..	..	..	..	..	..	..	..	Current Budget Balance
..	..	..	..	..	..	..	..	..	..	..	Capital Receipts
3,779	5,653	5,621	..	..	..	..	..	..	..	..	Capital Payments

RWANDA	1967	1968	1969	1970	1971	1972	1973	1974	1975	1976	1977
FOREIGN TRADE (CUSTOMS BASIS)					*(Millions of current US dollars)*						
Value of Exports, fob	14.0	14.7	14.1	24.5	22.3	20.2	51.3	53.6	57.6	114.2	126.5
Nonfuel Primary Products	14.0	14.7	14.1	24.3	21.7	19.1	50.7	51.1	56.4	112.5	124.8
Fuels	0.0	0.0	0.0	0.0	0.0	0.0	0.0	0.0	0.0	0.0	0.0
Manufactures	0.0	0.1	0.1	0.3	0.7	1.1	0.6	2.5	1.2	1.8	1.7
Value of Imports, cif	20.2	22.5	23.6	29.1	33.1	34.6	41.0	72.0	102.0	128.0	125.0
Nonfuel Primary Products	3.6	5.1	6.2	8.2	6.5	9.2	8.8	19.8	26.2	25.3	23.1
Fuels	1.2	1.4	1.5	1.6	2.0	2.7	3.7	11.4	13.9	14.5	11.7
Manufactures	15.4	16.0	15.9	19.3	24.6	22.7	28.6	40.8	61.9	88.2	90.2
Terms of Trade	125.7	126.6	130.0	140.7	119.0	128.7	136.3	96.3	89.6	138.2	167.4
					(Index 1980 = 100)						
Export Prices, fob	28.9	28.7	30.5	35.5	32.8	35.8	46.7	53.4	52.1	82.2	110.2
Nonfuel Primary Products	28.9	28.7	30.5	35.5	32.7	35.7	46.7	52.9	51.8	82.5	111.0
Fuels	4.3	4.3	4.3	4.3	5.6	6.2	8.9	36.7	35.7	38.4	42.0
Manufactures	30.1	31.7	32.4	37.5	36.5	38.5	49.5	67.2	66.1	66.7	71.7
Import Prices, cif	23.0	22.7	23.5	25.2	27.5	27.8	34.3	55.5	58.1	59.5	65.8
Trade at Constant 1980 Prices					*(Millions of 1980 US dollars)*						
Exports, fob	48.5	51.2	46.2	69.1	68.2	56.4	109.8	100.3	110.6	138.9	114.8
Imports, cif	88.0	99.0	100.6	115.4	120.0	124.3	119.6	129.8	175.5	215.1	189.9
BALANCE OF PAYMENTS					*(Millions of current US dollars)*						
Exports of Goods & Services	..	..	..	26	23	22	56	58	64	123	141
Merchandise, fob	..	..	..	24	21	20	51	54	58	114	126
Nonfactor Services	..	..	..	2	2	2	4	5	6	8	11
Factor Services	..	..	..	0	0	0	0	0	0	1	3
Imports of Goods & Services	..	..	..	35	41	47	62	90	128	159	185
Merchandise, fob	..	..	..	21	23	26	33	57	80	104	102
Nonfactor Services	..	..	..	14	17	21	26	31	44	49	75
Factor Services	..	..	..	0	0	0	3	3	3	5	8
Long-Term Interest	..	..	..	0	0	0	0	0	0	0	1
Current Transfers, net	..	..	..	..	..	..	..	..	..	..	..
Workers' Remittances	..	..	..	1	1	1	1	0	0	0	1
Total to be Financed	..	..	..	..	..	..	..	..	..	..	..
Official Capital Grants	..	..	..	..	..	..	..	..	..	..	..
Current Account Balance	..	..	..	7	0	-5	22	1	-9	18	21
Long-Term Capital, net	..	..	..	19	22	25	33	47	75	79	99
Direct Investment	..	..	..	0	2	1	2	2	3	6	5
Long-Term Loans	..	..	..	0	0	1	4	4	12	23	28
Disbursements	..	..	..	0	0	2	5	5	12	24	28
Repayments	..	..	..	0	0	0	0	0	0	1	1
Other Long-Term Capital	..	..	..	19	20	23	27	41	60	50	66
Other Capital, net	..	..	..	-21	-22	-23	-47	-49	-53	-65	-85
Change in Reserves	..	..	..	-4	0	3	-8	1	-13	-33	-34
Memo Item:					*(Rwanda Francs per US dollar)*						
Conversion Factor (Annual Avg)	100.000	100.000	100.000	100.000	99.740	92.110	84.050	92.840	92.840	92.840	92.840
EXTERNAL DEBT, ETC.					*(Millions of US dollars, outstanding at end of year)*						
Public/Publicly Guar. Long-Term	..	..	..	2	2	3	8	12	24	48	79
Official Creditors	..	..	..	2	2	2	5	10	22	46	72
IBRD and IDA	..	..	..	0	0	1	4	8	13	21	31
Private Creditors	..	..	..	0	0	1	3	3	2	2	6
Private Non-guaranteed Long-Term	..	..	..	..	..	..	..	..	..	..	..
Use of Fund Credit	..	..	..	3	0	0	0	2	2	0	0
Short-Term Debt	..	..	..	..	..	..	..	..	..	..	9
Memo Items:					*(Millions of US dollars)*						
Int'l Reserves Excluding Gold	6.7	5.8	3.0	7.7	5.8	6.4	15.2	13.0	25.6	64.3	82.9
Gold Holdings (at market price)	..	..	..	..	..	..	..	..	..	..	0.7
SOCIAL INDICATORS											
Total Fertility Rate	7.7	7.7	7.8	7.8	7.9	7.9	7.9	7.9	8.0	8.0	8.0
Crude Birth Rate	51.5	51.5	51.5	51.6	51.6	51.6	51.6	51.6	51.7	51.7	51.7
Infant Mortality Rate	137.0	136.4	135.8	135.2	134.6	134.0	133.6	133.2	132.8	132.4	132.0
Life Expectancy at Birth	49.6	49.0	48.3	47.7	47.0	46.4	45.9	45.5	45.0	44.6	44.2
Food Production, p.c. ('79-81 = 100)	95.8	88.9	92.5	96.8	95.2	90.6	91.3	86.3	94.3	96.7	97.7
Labor Force, Agriculture (%)	94.0	93.9	93.8	93.7	93.6	93.5	93.4	93.3	93.2	93.1	93.0
Labor Force, Female (%)	50.2	50.2	50.1	50.0	50.0	49.9	49.9	49.8	49.7	49.7	49.6
School Enroll. Ratio, primary	..	..	..	68.0	..	..	..	..	56.0	64.0	66.0
School Enroll. Ratio, secondary	..	..	..	2.0	..	..	..	..	2.0	2.0	2.0

1978	1979	1980	1981	1982	1983	1984	1985	1986	1987 estimate	NOTES	RWANDA
											FOREIGN TRADE (CUSTOMS BASIS)
111.7	203.0	133.6	113.3	108.5	124.1	142.6	126.1	166.8	132.8	..	Value of Exports, fob
110.3	201.7	132.6	112.2	107.3	122.6	141.3	125.0	165.4	131.3	..	Nonfuel Primary Products
0.0	0.0	0.0	0.0	0.0	0.0	0.0	0.0	0.0	..	..	Fuels
1.4	1.3	1.0	1.1	1.2	1.5	1.3	1.1	1.4	1.5	..	Manufactures
202.0	224.0	277.0	279.0	297.0	277.0	282.0	313.0	348.0	463.0	f	Value of Imports, cif
32.3	37.3	46.5	47.4	50.9	47.5	51.1	54.1	67.1	91.6	..	Nonfuel Primary Products
15.2	18.6	22.8	22.7	24.2	29.8	34.0	39.5	40.4	69.7	f	Fuels
154.5	168.1	207.7	208.8	221.9	199.7	196.9	219.4	240.5	301.7	..	Manufactures
				(Index 1980 = 100)							
118.2	120.4	100.0	85.4	92.3	90.9	101.3	102.5	135.6	84.4	..	Terms of Trade
88.2	105.7	100.0	86.2	89.9	86.2	93.5	93.8	121.5	79.6	..	Export Prices, fob
88.4	105.9	100.0	86.1	89.8	86.1	93.5	93.8	121.7	79.3	..	Nonfuel Primary Products
42.3	61.0	100.0	112.5	101.6	92.5	90.2	87.5	44.9	56.9	..	Fuels
76.1	90.0	100.0	99.1	94.8	90.7	89.3	89.0	103.2	113.0	..	Manufactures
74.7	87.8	100.0	100.9	97.3	94.9	92.3	91.5	89.6	94.2	..	Import Prices, cif
				(Millions of 1980 US dollars)							Trade at Constant 1980 Prices
126.6	192.0	133.6	131.4	120.7	144.0	152.5	134.5	137.3	167.0	..	Exports, fob
270.5	255.2	277.0	276.5	305.1	292.0	305.6	341.9	388.4	491.3	..	Imports, cif
				(Millions of current US dollars)							**BALANCE OF PAYMENTS**
130	235	182	176	158	161	183	170	236	176	..	Exports of Goods & Services
112	203	134	113	108	124	143	126	184	121	..	Merchandise, fob
14	24	32	37	33	28	32	35	43	45	..	Nonfactor Services
4	8	17	25	16	9	8	9	9	10	..	Factor Services
271	319	335	346	354	327	325	351	429	433	..	Imports of Goods & Services
145	159	196	207	215	198	198	219	259	265	..	Merchandise, fob
116	148	123	122	121	113	113	114	148	145	..	Nonfactor Services
10	12	15	17	19	16	15	17	22	23	..	Factor Services
1	1	2	2	2	3	3	4	5	7	..	Long-Term Interest
..	..	-3	-3	5	6	2	4	7	..	..	Current Transfers, net
2	3	1	2	1	2	1	1	2	..	..	Workers' Remittances
..	..	-149	-168	-192	-161	-141	-176	-187	..	..	Total to be Financed
..	..	101	101	105	112	99	112	118	..	..	Official Capital Grants
-47	47	-48	-67	-87	-49	-42	-64	-69	-131	..	Current Account Balance
119	155	158	151	147	149	150	185	200	232	..	Long-Term Capital, net
5	13	16	18	21	11	15	15	18	23	..	Direct Investment
25	39	31	24	29	36	44	54	62	78	..	Long-Term Loans
26	40	34	27	31	41	51	65	74	91	..	Disbursements
1	1	3	2	3	5	6	11	13	13	..	Repayments
89	103	10	8	-8	-10	-8	4	2	131	..	Other Long-Term Capital
-71	-132	-81	-84	-97	-119	-101	-121	-103	-119	..	Other Capital, net
-1	-70	-29	0	37	18	-8	0	-28	18	..	Change in Reserves
				(Rwanda Francs per US dollar)							**Memo Item:**
92.840	92.840	92.840	92.840	92.840	94.340	100.170	101.260	87.640	79.670	..	Conversion Factor (Annual Avg)
			(Millions of US dollars, outstanding at end of year)								**EXTERNAL DEBT, ETC.**
110	135	164	177	197	226	254	328	414	544	..	Public/Publicly Guar. Long-Term
98	124	156	171	189	219	243	316	404	533	..	Official Creditors
37	48	58	65	77	95	118	152	195	251	..	IBRD and IDA
12	11	8	6	8	7	12	12	9	11	..	Private Creditors
..	..	..	..	..	..	..	..	..	..	..	Private Non-guaranteed Long-Term
0	0	0	0	0	0	0	0	0	..	..	Use of Fund Credit
17	21	26	20	21	16	37	27	27	39	..	Short-Term Debt
				(Millions of US dollars)							**Memo Items:**
87.6	152.3	186.6	173.1	128.4	110.9	106.9	113.3	162.3	164.2	..	Int'l Reserves Excluding Gold
0.9	..	..	..	..	..	..	..	..	..	..	Gold Holdings (at market price)
											SOCIAL INDICATORS
8.0	8.0	8.0	8.0	8.0	8.0	8.0	8.0	8.0	8.0	..	Total Fertility Rate
51.7	51.7	51.7	51.7	51.7	51.8	51.9	52.1	52.2	52.3	..	Crude Birth Rate
130.4	128.8	127.2	125.6	124.0	122.1	120.2	118.3	116.4	114.5	..	Infant Mortality Rate
44.6	45.0	45.4	45.8	46.3	46.7	47.2	47.6	48.1	48.5	..	Life Expectancy at Birth
96.4	99.8	97.9	102.3	104.5	98.1	89.2	86.4	86.2	85.2	..	Food Production, p.c. ('79-81 = 100)
93.0	92.9	92.8	..	..	..	..	..	..	..	..	Labor Force, Agriculture (%)
49.6	49.5	49.5	49.3	49.1	49.0	48.8	48.6	48.4	48.3	..	Labor Force, Female (%)
71.0	70.0	64.0	65.0	63.0	62.0	62.0	64.0			..	School Enroll. Ratio, primary
2.0	2.0	2.0	2.0	2.0	2.0	2.0	2.0			..	School Enroll. Ratio, secondary

SAUDI ARABIA	1967	1968	1969	1970	1971	1972	1973	1974	1975	1976	1977
CURRENT GNP PER CAPITA (US $)	490	500	530	570	680	830	980	1,710	2,880	5,520	7,200
POPULATION (thousands)	5,122	5,310	5,517	5,745	5,996	6,268	6,567	6,921	7,251	7,636	8,044

ORIGIN AND USE OF RESOURCES *(Billions of current Saudi Arabian Riyals)*

	1967	1968	1969	1970	1971	1972	1973	1974	1975	1976	1977
Gross National Product (GNP)	11.21	11.70	12.77	13.91	19.10	24.02	30.09	82.35	125.40	165.39	207.72
Net Factor Income from Abroad	-1.94	-2.96	-3.20	-3.49	-3.82	-4.23	-10.46	-16.97	-14.20	0.87	2.67
Gross Domestic Product (GDP)	13.14	14.66	15.98	17.40	22.92	28.26	40.55	99.32	139.60	164.53	205.06
Indirect Taxes, net	0.14	0.14	0.18	0.16	0.35	0.36	0.53	0.51	-0.52	-0.90	-0.67
GDP at factor cost	13.00	14.51	15.80	17.24	22.57	27.90	40.02	98.81	140.12	165.43	205.73
Agriculture	0.85	0.88	0.96	0.98	1.02	1.06	1.14	1.24	1.39	1.59	1.87
Industry	8.16	9.27	9.91	11.03	15.89	20.45	30.93	86.62	120.24	134.27	164.26
Manufacturing	1.07	1.25	1.37	1.67	1.96	1.98	2.43	5.08	7.37	8.17	9.28
Services, etc.	4.13	4.50	5.10	5.38	6.01	6.75	8.48	11.46	17.97	28.67	38.93
Resource Balance	4.11	4.20	4.24	5.31	9.98	13.56	21.74	70.39	87.20	77.42	77.62
Exports of Goods & NFServices	7.65	8.59	9.09	10.30	15.19	19.86	30.01	85.68	114.46	120.28	140.32
Imports of Goods & NFServices	3.54	4.39	4.85	4.99	5.20	6.30	8.27	15.29	27.26	42.86	62.70
Domestic Absorption	9.03	10.46	11.74	12.09	12.94	14.70	18.81	28.93	52.40	87.11	127.43
Private Consumption, etc.	4.00	4.58	5.36	5.86	6.41	6.91	7.89	9.83	18.04	23.90	34.37
General Gov't Consumption	2.67	2.75	3.03	3.42	3.80	4.28	5.33	9.86	15.91	28.88	41.03
Gross Domestic Investment	2.36	3.13	3.35	2.81	2.73	3.50	5.58	9.23	18.45	34.32	52.03
Fixed Investment	2.16	2.39	2.63	2.60	2.93	3.40	5.69	8.40	17.70	33.54	51.19
Memo Items:											
Gross Domestic Saving	6.47	7.32	7.59	8.12	12.71	17.06	27.32	79.62	105.65	111.74	129.65
Gross National Saving	3.93	3.77	3.78	3.80	7.96	11.67	15.31	60.81	89.49	109.12	127.00

(Billions of 1980 Saudi Arabian Riyals)

	1967	1968	1969	1970	1971	1972	1973	1974	1975	1976	1977
Gross National Product	88.73	90.83	97.64	108.41	131.69	158.40	169.60	217.83	233.84	279.57	324.54
GDP at factor cost	103.17	113.03	121.21	134.43	155.29	183.01	222.22	258.81	259.98	279.63	321.53
Agriculture	2.51	2.62	2.71	2.79	2.88	2.98	3.09	3.20	3.33	3.46	3.63
Industry	81.15	89.27	95.33	107.20	126.95	152.67	188.39	218.55	215.96	227.98	262.60
Manufacturing	6.52	7.56	8.38	10.00	11.00	11.05	11.83	12.45	12.09	13.08	14.82
Services, etc.	20.68	22.27	24.54	25.71	27.82	29.63	33.53	38.33	39.76	46.71	54.27
Resource Balance	83.35	89.38	94.80	111.39	133.71	163.15	198.52	214.49	189.67	161.23	157.39
Exports of Goods & NFServices	90.80	98.65	104.77	121.05	143.24	174.43	212.56	238.34	227.49	222.64	242.86
Imports of Goods & NFServices	7.45	9.27	9.96	9.67	9.53	11.27	14.04	23.85	37.82	61.41	85.47
Domestic Absorption	20.98	24.78	27.78	24.32	23.95	22.12	26.49	45.58	69.39	116.92	163.12
Private Consumption, etc.	..	..	..	..	..	..	..	..	..	..	..
General Gov't Consumption	..	..	..	..	..	..	..	..	..	..	..
Gross Domestic Investment	5.01	6.61	6.95	5.55	5.06	6.19	9.52	14.59	29.58	51.30	71.26
Fixed Investment	..	..	..	..	..	..	..	..	..	..	..
Memo Items:											
Capacity to Import	16.11	18.12	18.66	19.96	27.81	35.52	50.95	133.63	158.82	172.33	191.27
Terms of Trade Adjustment	-74.69	-80.53	-86.10	-101.10	-115.43	-138.90	-161.61	-104.71	-68.67	-50.31	-51.58
Gross Domestic Income	29.64	33.63	36.48	34.61	42.23	46.37	63.40	155.36	190.38	227.84	268.93
Gross National Income	14.03	10.31	11.54	7.32	16.26	19.50	7.99	113.12	165.17	229.26	272.96

DOMESTIC PRICES (DEFLATORS) *(Index 1980 = 100)*

	1967	1968	1969	1970	1971	1972	1973	1974	1975	1976	1977
Overall (GDP)	12.6	12.8	13.0	12.8	14.5	15.3	18.0	38.2	53.9	59.1	64.0
Domestic Absorption	43.0	42.2	42.3	49.7	54.0	66.5	71.0	63.5	75.5	74.5	78.1
Agriculture	33.8	33.6	35.3	35.3	35.2	35.6	36.9	38.8	41.8	45.8	51.3
Industry	10.1	10.4	10.4	10.3	12.5	13.4	16.4	39.6	55.7	58.9	62.6
Manufacturing	16.4	16.5	16.4	16.7	17.8	18.0	20.5	40.8	60.9	62.5	62.6

MANUFACTURING ACTIVITY

	1967	1968	1969	1970	1971	1972	1973	1974	1975	1976	1977
Employment (1980=100)	..	..	..	..	..	..	..	..	..	..	..
Real Earnings per Empl. (1980=100)	..	..	..	..	..	..	..	..	..	..	..
Real Output per Empl. (1980=100)	..	..	..	..	..	..	..	..	..	..	..
Earnings as % of Value Added	..	..	..	..	..	..	..	..	..	23.8	..

MONETARY HOLDINGS *(Billions of current Saudi Arabian Riyals)*

	1967	1968	1969	1970	1971	1972	1973	1974	1975	1976	1977
Money Supply, Broadly Defined	2.42	2.80	2.97	3.14	3.59	5.04	6.81	9.78	17.78	29.61	45.56
Money as Means of Payment	1.92	2.20	2.32	2.40	2.65	3.78	5.29	7.33	14.18	24.27	38.41
Currency Ouside Banks	1.20	1.45	1.56	1.63	1.67	2.42	3.05	4.14	6.68	10.59	16.25
Demand Deposits	0.72	0.75	0.77	0.77	0.98	1.36	2.23	3.19	7.50	13.68	22.16
Quasi-Monetary Liabilities	0.50	0.60	0.64	0.74	0.94	1.26	1.52	2.44	3.60	5.34	7.14

GOVERNMENT DEFICIT (-) OR SURPLUS *(Millions of current Saudi Arabian Riyals)*

	1967	1968	1969	1970	1971	1972	1973	1974	1975	1976	1977
GOVERNMENT DEFICIT (-) OR SURPLUS	..	..	..	..	..	..	..	..	..	..	..
Current Revenue	..	..	..	..	..	..	..	..	..	..	..
Current Expenditure	..	..	..	..	..	..	..	..	..	..	..
Current Budget Balance	..	..	..	..	..	..	..	..	..	..	..
Capital Receipts	..	..	..	..	..	..	..	..	..	..	..
Capital Payments	..	..	..	..	..	..	..	..	..	..	..

1978	1979	1980	1981	1982	1983	1984	1985	1986	1987 estimate	NOTES	SAUDI ARABIA
7,760	8,680	10,510	14,360	15,540	12,960	10,460	8,630	6,950	..	..	CURRENT GNP PER CAPITA (US $)
8,473	8,918	9,372	9,829	10,280	10,718	11,131	11,508	11,976	12,459	..	POPULATION (thousands)

(Billions of current Saudi Arabian Riyals) — ORIGIN AND USE OF RESOURCES

1978	1979	1980	1981	1982	1983	1984	1985	1986	1987 estimate	NOTES	SAUDI ARABIA
227.18	256.18	392.45	568.05	536.88	417.05	348.78	315.24	275.18	..	C	Gross National Product (GNP)
1.78	6.64	6.64	47.46	12.16	1.82	-23.24	-15.64	-11.51	..	..	Net Factor Income from Abroad
225.40	249.54	385.81	520.59	524.72	415.23	372.02	330.88	286.69	267.64	C	Gross Domestic Product (GDP)
-0.90	-0.93	..	..	..	..	..	..	..	..		Indirect Taxes, net
226.30	250.47	..	..	..	..	..	..	..	..	B C	GDP at factor cost
3.91	4.20	4.65	5.57	6.74	8.73	9.61	11.14	12.59	..		Agriculture
169.32	179.85	301.25	419.19	405.43	272.68	221.80	180.64	142.81	..		Industry
9.97	12.62	19.29	25.75	22.38	23.97	26.94	27.07	26.20	..		Manufacturing
52.17	65.50	79.91	95.83	112.55	133.82	140.61	139.09	131.28	..		Services, etc.
49.26	39.76	126.14	210.97	167.16	38.18	-19.24	-24.76	-26.11	-26.35		Resource Balance
140.76	147.24	258.49	368.43	354.92	219.45	167.17	134.65	102.61	85.18		Exports of Goods & NFServices
91.50	107.48	132.35	157.46	187.76	181.27	186.41	159.41	128.72	111.53		Imports of Goods & NFServices
176.14	209.78	259.67	309.62	357.55	377.06	391.26	355.64	312.80	293.99		Domestic Absorption
54.61	68.61	102.45	114.90	126.51	137.30	157.37	145.99	122.07	118.89		Private Consumption, etc.
47.03	71.90	77.50	81.92	128.53	126.85	121.33	115.65	113.47	102.45		General Gov't Consumption
74.50	69.27	79.72	112.80	102.52	112.90	112.56	94.00	77.26	72.65		Gross Domestic Investment
66.89	76.65	97.07	106.38	122.32	115.45	103.23	91.21	74.93	..		Fixed Investment

Memo Items:

1978	1979	1980	1981	1982	1983	1984	1985	1986	1987 estimate	NOTES	SAUDI ARABIA
123.76	109.03	205.86	323.77	269.68	151.08	93.32	69.24	51.15	46.30		Gross Domestic Saving
115.62	104.38	198.80	353.40	263.57	134.89	51.64	35.05	22.14	..		Gross National Saving

(Billions of 1980 Saudi Arabian Riyals)

1978	1979	1980	1981	1982	1983	1984	1985	1986	1987 estimate	NOTES	SAUDI ARABIA
336.93	361.91	392.45	451.45	421.88	346.22	313.97	291.97	264.70	..	C	Gross National Product
335.65	353.89	..	..	..	..	..	..	..	..	B C	GDP at factor cost
4.20	4.39	4.65	4.92	5.21	5.73	6.48	7.32	8.27	..		Agriculture
268.38	278.15	301.25	319.30	308.65	231.26	218.97	192.36	169.43	..		Industry
16.12	17.72	19.29	20.67	22.12	24.21	27.38	26.90	26.24	..		Manufacturing
61.75	70.05	79.91	89.27	98.21	107.57	112.10	108.98	99.95	..		Services, etc.
134.42	131.48	126.14	113.18	46.37	-45.40	..	..	..	..		Resource Balance
246.60	249.03	258.49	273.62	241.17	156.42	..	..	..	..		Exports of Goods & NFServices
112.18	117.55	132.35	160.44	194.79	201.81	..	..	..	..		Imports of Goods & NFServices
199.91	221.11	259.67	300.30	365.70	389.96	..	..	..	..		Domestic Absorption
..	..	..	..	..	..	..	..	..	..		Private Consumption, etc.
..	..	..	..	..	..	..	..	..	..		General Gov't Consumption
91.35	75.78	79.72	114.95	106.38	139.49	140.79	..	..	..		Gross Domestic Investment
..	..	..	..	..	..	..	..	..	..		Fixed Investment

Memo Items:

1978	1979	1980	1981	1982	1983	1984	1985	1986	1987 estimate	NOTES	SAUDI ARABIA
172.57	161.03	258.49	375.40	368.23	244.32	..	..	..	..		Capacity to Import
-74.03	-88.00	0.00	101.78	127.06	87.90	..	..	..	..		Terms of Trade Adjustment
260.30	264.59	385.81	515.27	539.13	432.47	..	..	..	..		Gross Domestic Income
262.89	273.91	392.45	553.23	548.94	434.12	..	..	..	..		Gross National Income

(Index 1980 = 100) — DOMESTIC PRICES (DEFLATORS)

1978	1979	1980	1981	1982	1983	1984	1985	1986	1987 estimate	NOTES	SAUDI ARABIA
67.4	70.8	100.0	125.9	127.3	120.5	110.2	107.2	103.3	88.8	..	Overall (GDP)
88.1	94.9	100.0	103.1	97.8	96.7	..	..	..	..	..	Domestic Absorption
93.0	95.5	100.0	113.3	129.3	152.2	148.3	152.2	152.2	..	..	Agriculture
63.1	64.7	100.0	131.3	131.4	117.9	101.3	93.9	84.3	..	..	Industry
61.9	71.2	100.0	124.6	101.2	99.0	98.4	100.7	99.9	..	..	Manufacturing

MANUFACTURING ACTIVITY

1978	1979	1980	1981	1982	1983	1984	1985	1986	1987 estimate	NOTES	SAUDI ARABIA
..	..	..	..	..	..	..	..	..	..	..	Employment (1980 = 100)
..	..	..	..	..	..	..	..	..	..	..	Real Earnings per Empl. (1980 = 100)
..	..	..	..	..	..	..	..	..	..	..	Real Output per Empl. (1980 = 100)
..	..	..	..	..	..	..	..	..	..	..	Earnings as % of Value Added

(Billions of current Saudi Arabian Riyals) — MONETARY HOLDINGS

1978	1979	1980	1981	1982	1983	1984	1985	1986	1987 estimate	NOTES	SAUDI ARABIA
58.03	65.98	77.44	102.99	123.79	137.08	145.35	147.01	160.61	..	..	Money Supply, Broadly Defined
49.21	54.70	58.96	72.98	83.78	84.93	82.98	81.83	86.28	91.77	..	Money as Means of Payment
19.18	23.71	25.68	29.49	34.44	35.42	35.11	35.77	38.81	40.07	..	Currency Ouside Banks
30.03	31.00	33.28	43.49	49.34	49.51	47.86	46.06	47.47	51.70	..	Demand Deposits
8.82	11.28	18.48	30.01	40.01	52.15	62.37	65.18	74.33	..	..	Quasi-Monetary Liabilities

(Millions of current Saudi Arabian Riyals) — GOVERNMENT DEFICIT (-) OR SURPLUS

1978	1979	1980	1981	1982	1983	1984	1985	1986	1987 estimate	NOTES	SAUDI ARABIA
..	..	..	..	..	..	..	..	..	..	..	Current Revenue
..	..	..	..	..	..	..	..	..	..	..	Current Expenditure
..	..	..	..	..	..	..	..	..	..	..	Current Budget Balance
..	..	..	..	..	..	..	..	..	..	..	Capital Receipts
..	..	..	..	..	..	..	..	..	..	..	Capital Payments

SAUDI ARABIA	1967	1968	1969	1970	1971	1972	1973	1974	1975	1976	1977
FOREIGN TRADE (CUSTOMS BASIS)				*(Millions of current US dollars)*							
Value of Exports, fob	1,692	2,026	1,992	2,424	3,845	5,491	9,089	35,562	29,669	38,282	43,458
Nonfuel Primary Products	1	4	4	4	4	9	14	49	31	30	27
Fuels	1,691	2,021	1,987	2,417	3,792	5,391	8,993	35,482	29,466	38,164	43,311
Manufactures	1	2	2	2	49	91	82	31	172	89	120
Value of Imports, cif	502	562	734	692	806	1,125	1,977	2,848	4,141	8,409	14,289
Nonfuel Primary Products	168	191	224	241	256	308	480	639	743	1,233	1,830
Fuels	6	6	10	8	10	11	13	29	28	62	81
Manufactures	328	365	500	443	539	806	1,484	2,180	3,370	7,113	12,378
Terms of Trade	14.1	14.1	13.8	12.9	16.2	16.5	18.8	62.6	56.5	60.5	60.9
				(Index 1980 = 100)							
Export Prices, fob	4.3	4.3	4.3	4.3	5.6	6.3	8.9	36.8	35.9	38.4	42.0
Nonfuel Primary Products	34.0	37.1	37.7	40.8	37.3	42.0	50.9	60.4	70.1	65.3	69.0
Fuels	4.3	4.3	4.3	4.3	5.6	6.2	8.9	36.7	35.7	38.4	42.0
Manufactures	35.4	34.0	35.4	41.7	42.4	45.8	50.0	86.1	69.4	74.3	79.9
Import Prices, cif	30.3	30.2	31.1	33.2	34.8	38.3	47.5	58.7	63.5	63.5	69.0
Trade at Constant 1980 Prices				*(Millions of 1980 US dollars)*							
Exports, fob	39,693	47,461	46,651	56,764	68,211	86,749	101,811	96,745	82,737	99,654	103,385
Imports, cif	1,655	1,859	2,365	2,086	2,315	2,942	4,162	4,849	6,524	13,240	20,700
BALANCE OF PAYMENTS				*(Millions of current US dollars)*							
Exports of Goods & Services	..	..	..	2,372	2,926	4,381	8,264	32,735	30,569	40,272	46,447
Merchandise, fob	..	..	..	2,089	2,581	3,911	7,489	30,132	27,294	35,632	40,351
Nonfactor Services	..	..	..	222	279	364	569	1,383	1,436	1,756	2,108
Factor Services	..	..	..	61	67	106	205	1,220	1,839	2,884	3,989
Imports of Goods & Services	..	..	..	2,037	1,600	1,867	4,856	8,177	12,503	21,601	29,049
Merchandise, fob	..	..	..	829	805	1,197	1,853	3,569	6,004	10,385	14,698
Nonfactor Services	..	..	..	313	374	509	955	2,163	4,375	7,935	10,272
Factor Services	..	..	..	895	421	162	2,048	2,445	2,124	3,280	4,079
Long-Term Interest	..	..	..	..	..	..	..	..	..	..	..
Current Transfers, net	..	..	..	-183	-207	-268	-391	-518	-554	-989	-1,506
Workers' Remittances	..	..	..	0	0	0	0	0	0	0	0
Total to be Financed	..	..	..	152	1,119	2,246	3,016	24,039	17,512	17,683	15,892
Official Capital Grants	..	..	..	-81	-147	-157	-496	-1,014	-3,127	-3,323	-3,901
Current Account Balance	..	..	..	71	972	2,089	2,520	23,025	14,385	14,360	11,991
Long-Term Capital, net	..	..	..	12	-274	-130	-1,408	-9,887	-12,264	-14,479	-11,404
Direct Investment	..	..	..	20	-111	34	-626	-3,732	1,865	-397	783
Long-Term Loans	..	..	..	..	..	..	..	..	..	..	..
Disbursements	..	..	..	..	..	..	..	..	..	..	..
Repayments	..	..	..	..	..	..	..	..	..	..	..
Other Long-Term Capital	..	..	..	73	-16	-7	-286	-5,141	-11,002	-10,759	-8,286
Other Capital, net	..	..	..	4	-25	-902	-194	-2,750	6,938	3,846	2,110
Change in Reserves	..	..	..	-87	-673	-1,057	-917	-10,388	-9,059	-3,727	-2,697
Memo Item:				*(Saudi Arabian Riyals per US dollar)*							
Conversion Factor (Annual Avg)	4.500	4.500	4.500	4.500	4.490	4.320	3.980	3.560	3.530	3.530	3.530
EXTERNAL DEBT, ETC.				*(Millions of US dollars, outstanding at end of year)*							
Public/Publicly Guar. Long-Term	..	..	..	..	..	..	..	..	..	..	..
Official Creditors	..	..	..	..	..	..	..	..	..	..	..
IBRD and IDA	..	..	..	..	..	..	..	..	..	..	..
Private Creditors	..	..	..	..	..	..	..	..	..	..	..
Private Non-guaranteed Long-Term	..	..	..	..	..	..	..	..	..	..	..
Use of Fund Credit	..	..	..	..	..	..	..	..	..	..	..
Short-Term Debt	..	..	..	..	..	..	..	..	..	..	..
Memo Items:				*(Millions of US dollars)*							
Int'l Reserves Excluding Gold	692.0	543.0	488.0	543.0	1,327.0	2,383.0	3,747.0	14,153.0	23,193.0	26,900.0	29,903.0
Gold Holdings (at market price)	69.0	142.0	120.0	127.0	135.0	200.0	346.0	575.0	432.0	415.0	508.0
SOCIAL INDICATORS											
Total Fertility Rate	7.3	7.3	7.3	7.3	7.3	7.3	7.3	7.3	7.3	7.3	7.3
Crude Birth Rate	48.1	48.0	47.9	47.8	47.7	47.6	47.3	46.9	46.6	46.2	45.9
Infant Mortality Rate	135.0	130.0	125.0	120.0	115.5	111.0	106.5	102.0	97.5	93.0	88.5
Life Expectancy at Birth	49.9	50.7	51.5	52.3	53.1	53.9	54.7	55.5	56.3	57.1	57.9
Food Production, p.c. ('79-81 = 100)	118.8	102.5	116.9	123.3	158.4	96.7	108.9	145.7	144.1	126.4	130.1
Labor Force, Agriculture (%)	66.4	65.6	64.9	64.2	62.6	61.0	59.5	57.9	56.3	54.7	53.1
Labor Force, Female (%)	4.6	4.8	4.9	5.0	5.2	5.4	5.5	5.7	5.9	6.0	6.0
School Enroll. Ratio, primary	..	..	..	45.0	..	..	..	..	58.0	59.0	58.0
School Enroll. Ratio, secondary	..	..	..	12.0	..	..	..	..	22.0	24.0	26.0

1978	1979	1980	1981	1982	1983	1984	1985	1986	1987 estimate	NOTES	SAUDI ARABIA
											FOREIGN TRADE (CUSTOMS BASIS)
				(Millions of current US dollars)							
40,715	63,419	109,113	119,913	79,125	45,861	37,530	27,481	20,085	23,138	..	Value of Exports, fob
95	107	181	153	136	153	126	89	225	259	..	Nonfuel Primary Products
40,383	62,847	108,226	119,039	78,165	44,928	36,657	26,677	18,042	20,772	..	Fuels
237	465	705	721	824	780	747	715	1,818	2,107	..	Manufactures
20,177	24,384	29,957	35,042	40,473	39,206	33,696	23,697	19,112	20,465	..	Value of Imports, cif
2,639	3,583	4,965	5,770	5,967	5,732	5,955	3,949	3,651	3,969	..	Nonfuel Primary Products
136	156	191	254	192	247	237	157	150	160	..	Fuels
17,403	20,646	24,801	29,018	34,314	33,227	27,504	19,592	15,311	16,336	..	Manufactures
				(Index 1980 = 100)							
53.6	67.2	100.0	112.5	104.9	96.8	96.9	95.4	46.7	54.3	..	Terms of Trade
42.5	61.2	100.0	112.4	101.8	92.7	90.7	88.1	48.1	60.3	..	Export Prices, fob
67.7	89.7	100.0	94.5	86.4	85.3	83.1	74.5	76.1	79.9	..	Nonfuel Primary Products
42.3	61.0	100.0	112.5	101.6	92.5	90.2	87.5	44.9	56.9	..	Fuels
111.8	95.8	100.0	113.0	118.0	109.0	129.0	117.0	138.4	135.1	..	Manufactures
79.3	91.1	100.0	99.9	97.0	95.7	93.6	92.3	103.1	111.0	..	Import Prices, cif
											Trade at Constant 1980 Prices
				(Millions of 1980 US dollars)							
95,821	103,666	109,113	106,650	77,759	49,487	41,389	31,205	41,747	38,390	..	Exports, fob
25,450	26,769	29,957	35,062	41,709	40,951	36,017	25,683	18,539	18,438	..	Imports, cif
				(Millions of current US dollars)							**BALANCE OF PAYMENTS**
43,461	65,817	111,983	127,855	92,828	65,594	55,023	43,465	33,878	..	..	Exports of Goods & Services
36,993	58,099	100,717	111,831	73,876	45,662	37,423	27,397	19,971	..	..	Merchandise, fob
2,168	2,803	3,824	5,067	4,892	4,064	4,234	3,648	2,658	..	..	Nonfactor Services
4,300	4,915	7,443	10,956	14,060	15,868	13,366	12,420	11,248	..	..	Factor Services
38,929	47,783	59,235	75,678	75,507	73,499	64,541	47,949	38,038	..	..	Imports of Goods & Services
20,020	23,530	25,563	29,889	34,444	33,218	28,557	20,367	16,575	..	..	Merchandise, fob
14,396	22,188	26,756	36,189	34,852	37,259	32,856	25,826	20,806	..	..	Nonfactor Services
4,512	2,065	6,917	9,599	6,210	3,022	3,127	1,756	657	..	..	Factor Services
..	..	..	..	..	..	..	..	..	..	..	Long-Term Interest
-2,845	-3,365	-4,094	-5,348	-5,347	-5,236	-5,284	-5,199	-4,791	..	..	Current Transfers, net
0	0	0	0	0	0	0	0	0	..	..	Workers' Remittances
1,688	14,669	48,654	46,829	11,974	-13,141	-14,802	-9,684	-8,952	..	K	Total to be Financed
-3,900	-3,502	-5,901	-5,700	-4,399	-4,000	-3,598	-3,250	-2,951	..	K	Official Capital Grants
-2,212	11,167	42,754	41,129	7,575	-17,142	-18,401	-12,934	-11,903	..	..	Current Account Balance
-6,633	-6,650	-33,282	-33,425	-7,374	-6,585	-12,154	-11,172	-5,429	..	..	Long-Term Capital, net
556	-1,351	-3,192	6,498	11,128	4,944	4,850	492	964	..	..	Direct Investment
..	..	..	..	..	..	..	..	..	..	..	Long-Term Loans
..	..	..	..	..	..	..	..	..	..	..	Disbursements
..	..	..	..	..	..	..	..	..	..	..	Repayments
-3,289	-1,797	-24,189	-34,223	-14,104	-7,529	-13,406	-8,414	-3,442	..	..	Other Long-Term Capital
2,146	-4,283	-5,535	1,864	-2,509	22,218	29,075	23,396	9,713	..	..	Other Capital, net
6,700	-234	-3,937	-9,568	2,308	1,508	1,480	709	7,619	-2,639	..	Change in Reserves
											Memo Item:
				(Saudi Arabian Riyals per US dollar)							
3.490	3.350	3.350	3.330	3.420	3.440	3.490	3.570	3.650	3.740	..	Conversion Factor (Annual Avg)
				(Millions of US dollars, outstanding at end of year)							**EXTERNAL DEBT, ETC.**
..	..	..	..	..	..	..	..	..	..	..	Public/Publicly Guar. Long-Term
..	..	..	..	..	..	..	..	..	..	..	Official Creditors
..	..	..	..	..	..	..	..	..	..	..	IBRD and IDA
..	..	..	..	..	..	..	..	..	..	..	Private Creditors
..	..	..	..	..	..	..	..	..	..	..	Private Non-guaranteed Long-Term
..	..	..	..	..	..	..	..	..	..	..	Use of Fund Credit
..	..	..	..	..	..	..	..	..	..	..	Short-Term Debt
				(Millions of US dollars)							Memo Items:
19,200.0	19,273.0	23,437.0	32,236.0	29,549.0	27,287.0	24,748.0	25,004.0	18,324.0	22,684.0	..	Int'l Reserves Excluding Gold
1,026.0	2,338.0	2,692.0	1,815.0	2,100.0	1,753.0	1,417.0	1,503.0	1,797.0	2,225.0	..	Gold Holdings (at market price)
											SOCIAL INDICATORS
7.3	7.3	7.3	7.3	7.3	7.2	7.2	7.2	7.1	7.1	..	Total Fertility Rate
45.4	44.9	44.3	43.8	43.3	43.1	42.8	42.6	42.4	42.2	..	Crude Birth Rate
84.0	79.5	75.0	72.2	15.3	30.5	45.8	61.0	59.2	57.3	..	Infant Mortality Rate
58.5	59.1	59.7	60.3	60.9	61.3	61.7	62.2	62.6	63.0	..	Life Expectancy at Birth
124.4	117.3	100.2	82.5	113.3	138.1	132.0	190.0	231.3	205.7	..	Food Production, p.c. ('79-81 = 100)
51.6	50.0	48.4	..	..	..	..	..	..	..	..	Labor Force, Agriculture (%)
6.1	6.2	6.3	6.3	6.4	6.5	6.6	6.7	6.8	6.9	..	Labor Force, Female (%)
60.0	62.0	64.0	66.0	65.0	67.0	..	69.0	..	..	..	School Enroll. Ratio, primary
28.0	30.0	30.0	31.0	33.0	35.0	..	42.0	..	..	..	School Enroll. Ratio, secondary

SENEGAL	1967	1968	1969	1970	1971	1972	1973	1974	1975	1976	1977
CURRENT GNP PER CAPITA (US $)	200	210	200	210	200	220	230	270	340	390	380
POPULATION (thousands)	4,117	4,214	4,313	4,415	4,519	4,625	4,734	4,846	4,960	5,099	5,244
ORIGIN AND USE OF RESOURCES	*(Billions of current CFA Francs)*										
Gross National Product (GNP)	201.4	212.8	211.4	234.3	242.0	267.8	271.8	327.3	394.6	453.3	469.9
Net Factor Income from Abroad	-4.1	-4.4	-5.2	-5.8	-5.2	-5.8	-6.5	-11.5	-11.8	-6.0	-13.7
Gross Domestic Product (GDP)	205.5	217.2	216.6	240.1	247.2	273.6	278.3	338.8	406.4	459.3	483.6
Indirect Taxes, net	25.7	25.2	25.5	27.3	30.3	32.0	32.8	37.4	44.8	52.9	59.0
GDP at factor cost	179.8	192.0	191.1	212.8	216.9	241.6	245.5	301.4	361.6	406.4	424.6
Agriculture	49.1	56.3	50.7	57.8	51.8	67.4	62.8	80.7	122.8	138.0	132.5
Industry	39.9	40.1	44.5	48.6	51.2	54.5	54.7	80.2	92.7	106.9	113.1
Manufacturing	32.8	32.6	36.9	39.2	41.6	44.2	44.8	66.8	74.7	85.5	75.6
Services, etc.	116.5	120.8	121.4	133.7	144.2	151.7	160.8	177.9	190.9	214.4	238.0
Resource Balance	-6.5	-15.9	-17.7	-11.0	-18.0	-10.0	-28.1	-22.4	-21.3	-36.9	-30.0
Exports of Goods & NFServices	52.9	46.6	51.5	65.8	64.0	84.2	79.5	143.7	148.6	166.7	213.0
Imports of Goods & NFServices	59.4	62.5	69.2	76.8	82.0	94.2	107.6	166.1	169.9	203.6	243.0
Domestic Absorption	212.0	233.1	234.3	251.1	265.2	283.6	306.4	361.2	427.7	496.2	513.6
Private Consumption, etc.	152.9	173.1	171.6	177.7	186.4	196.3	210.0	236.1	293.6	348.8	351.1
General Gov't Consumption	36.1	35.0	33.6	35.7	38.8	40.8	43.8	50.0	61.8	71.7	77.9
Gross Domestic Investment	23.0	25.0	29.1	37.7	40.0	46.5	52.6	75.1	72.3	75.7	84.6
Fixed Investment	..	..	..	..	..	..	..	..	..	..	..
Memo Items:											
Gross Domestic Saving	16.5	9.1	11.4	26.7	22.0	36.5	24.5	52.7	51.0	38.8	54.6
Gross National Saving	12.4	5.4	1.9	16.2	11.3	25.3	13.3	35.8	37.2	36.3	44.7
	(Billions of 1980 CFA Francs)										
Gross National Product	480.4	511.0	475.1	516.0	518.4	550.0	519.2	534.2	577.6	638.3	611.5
GDP at factor cost	428.1	460.2	428.5	467.5	463.8	495.2	468.1	490.4	527.9	571.2	552.3
Agriculture	117.6	118.1	120.4	127.5	106.4	131.7	110.5	136.1	141.2	162.4	151.3
Industry	88.8	92.2	94.2	103.0	106.1	108.7	104.3	116.4	124.1	134.1	159.3
Manufacturing	68.2	71.6	73.2	77.2	81.3	82.3	80.4	91.7	97.5	104.2	104.4
Services, etc.	283.2	310.5	271.4	297.3	316.1	320.6	316.0	299.1	328.3	349.5	318.8
Resource Balance	-58.7	-41.9	-28.6	-34.8	-43.7	-21.9	-47.2	-42.0	-27.7	-29.2	-54.2
Exports of Goods & NFServices	117.7	124.1	122.1	133.0	118.2	152.1	117.9	126.8	141.7	177.0	242.2
Imports of Goods & NFServices	176.4	165.9	150.7	167.9	161.9	174.0	165.1	168.8	169.5	206.2	296.4
Domestic Absorption	548.2	562.7	514.6	562.6	572.3	583.0	578.0	593.6	621.3	675.2	683.6
Private Consumption, etc.	390.8	401.1	353.7	382.7	388.2	389.0	382.3	382.8	419.6	457.5	448.9
General Gov't Consumption	91.5	89.4	82.4	84.6	86.3	86.3	85.7	91.5	96.3	95.5	117.8
Gross Domestic Investment	65.9	72.2	78.5	95.3	97.8	107.6	110.0	119.3	105.4	122.2	116.9
Fixed Investment	..	..	..	..	..	..	..	..	..	..	92.6
Memo Items:											
Capacity to Import	157.1	123.7	112.2	143.8	126.4	155.6	122.0	146.1	148.2	168.9	259.8
Terms of Trade Adjustment	39.4	-0.3	-10.0	10.8	8.1	3.4	4.1	19.2	6.5	-8.2	17.6
Gross Domestic Income	528.9	520.5	476.1	538.6	536.8	564.5	534.9	570.8	600.1	637.8	647.0
Gross National Income	519.8	510.6	465.1	526.7	526.5	553.4	523.3	553.4	584.1	630.1	629.1
DOMESTIC PRICES (DEFLATORS)	*(Index 1980 = 100)*										
Overall (GDP)	42.0	41.7	44.6	45.5	46.8	48.8	52.4	61.4	68.5	71.1	76.8
Domestic Absorption	38.7	41.4	45.5	44.6	46.3	48.6	53.0	60.9	68.8	73.5	75.1
Agriculture	41.8	47.7	42.1	45.3	48.7	51.2	56.8	59.3	87.0	85.0	87.6
Industry	44.9	43.5	47.2	47.2	48.3	50.1	52.5	68.9	74.7	79.7	71.0
Manufacturing	48.1	45.5	50.4	50.7	51.2	53.7	55.7	72.9	76.6	82.1	72.4
MANUFACTURING ACTIVITY											
Employment (1980 = 100)	..	..	..	..	..	..	..	69.0	73.7	78.0	84.4
Real Earnings per Empl. (1980 = 100)	..	..	..	..	..	..	..	134.3	122.7	147.6	120.2
Real Output per Empl. (1980 = 100)	..	..	..	..	..	..	..	133.2	137.5	157.0	190.0
Earnings as % of Value Added	..	..	..	..	..	..	..	29.7	30.4	31.3	27.8
MONETARY HOLDINGS	*(Billions of current CFA Francs)*										
Money Supply, Broadly Defined	26.0	29.6	29.7	37.3	38.0	42.8	52.4	77.3	86.1	113.6	131.0
Money as Means of Payment	25.3	28.4	28.0	34.5	35.2	39.1	44.2	67.8	75.2	94.9	109.1
Currency Ouside Banks	10.2	11.9	11.6	15.2	15.9	16.5	19.5	29.0	29.5	33.7	39.5
Demand Deposits	15.1	16.4	16.4	19.3	19.3	22.6	24.8	38.8	45.7	61.1	69.7
Quasi-Monetary Liabilities	0.7	1.3	1.8	2.8	2.8	3.7	8.2	9.5	10.9	18.8	21.9
GOVERNMENT DEFICIT (-) OR SURPLUS	*(Billions of current CFA Francs)*										
	..	..	..	1.2	-1.2	..	-7.5	..	-2.2	..	-15.1
Current Revenue	..	..	..	39.0	41.4	..	46.3	..	77.0	..	86.1
Current Expenditure	..	..	..	37.0	39.5	..	45.3	..	66.4	74.6	77.9
Current Budget Balance	..	..	..	2.0	1.9	..	1.0	..	10.6	..	8.3
Capital Receipts	..	..	..	1.0	0.0	..	1.4	..	0.1	..	0.2
Capital Payments	..	..	..	1.8	3.1	..	9.9	..	12.8	16.8	23.5

1978	1979	1980	1981	1982	1983	1984	1985	1986	1987 estimate	NOTES	SENEGAL
370	450	490	470	490	430	370	370	410	510	..	**CURRENT GNP PER CAPITA (US $)**
5,394	5,548	5,706	5,869	6,036	6,208	6,385	6,567	6,770	6,969	..	**POPULATION (thousands)**
				(Billions of current CFA Francs)							**ORIGIN AND USE OF RESOURCES**
477.3	565.1	606.5	640.1	806.3	893.1	954.9	1,084.8	1,215.0	1,349.9	..	Gross National Product (GNP)
-17.4	-16.8	-21.1	-29.7	-38.0	-46.4	-60.6	-67.2	-80.3	-69.1	..	Net Factor Income from Abroad
494.7	581.9	627.6	669.8	844.3	939.5	1,015.5	1,152.0	1,295.3	1,419.0	..	Gross Domestic Product (GDP)
55.8	63.7	..	..	..	..	..	..	..	..	..	Indirect Taxes, net
438.9	518.2									B	GDP at factor cost
104.6	139.6	120.1	121.0	185.7	204.7	174.1	218.8	290.2	307.8	..	Agriculture
120.4	141.1	156.2	171.6	205.0	235.5	280.4	330.1	344.4	379.3	..	Industry
76.3	89.4	96.3	107.8	133.0	137.8	175.4	212.8	216.7	238.7	..	Manufacturing
269.7	301.2	351.3	377.2	453.6	499.3	561.0	603.1	660.7	731.9	..	Services, etc.
-62.9	-59.3	-100.3	-141.1	-126.6	-133.4	-112.9	-144.3	-100.2	-101.5	..	Resource Balance
154.9	185.3	180.1	270.0	293.2	350.2	402.2	360.3	355.8	393.3	..	Exports of Goods & NFServices
217.8	244.6	280.4	411.1	419.8	483.6	515.1	504.6	456.0	494.8	..	Imports of Goods & NFServices
557.6	641.2	727.9	810.9	970.9	1,072.9	1,128.4	1,296.3	1,395.5	1,520.5	..	Domestic Absorption
380.3	446.6	490.3	550.1	671.3	734.2	760.9	925.5	994.9	1,086.4	..	Private Consumption, etc.
91.1	111.0	140.3	150.7	169.2	186.5	206.8	213.0	222.7	246.9	..	General Gov't Consumption
86.2	83.6	97.3	110.1	130.4	152.2	160.7	157.8	177.9	187.2	..	Gross Domestic Investment
..	82.6	100.2	102.4	124.5	148.3	151.8	161.3	181.4	185.8	..	Fixed Investment
											Memo Items:
23.3	24.3	-3.0	-31.0	3.8	18.8	47.8	13.5	77.7	85.7	..	Gross Domestic Saving
7.7	3.7	-28.3	-63.5	-35.7	-28.2	-12.7	-46.1	6.4	40.3	..	Gross National Saving
				(Billions of 1980 CFA Francs)							
571.0	628.8	606.5	596.3	685.1	700.5	663.0	688.7	715.8	760.0	..	Gross National Product
525.1	576.5	..	..	..	..	..	..	..	..	B	GDP at factor cost
115.6	146.9	120.1	113.3	141.4	148.3	120.3	129.9	161.6	163.7	..	Agriculture
144.9	161.0	156.2	165.0	189.7	194.1	190.5	194.4	207.8	218.4	..	Industry
83.3	100.8	96.3	104.2	121.8	121.8	121.2	122.8	129.2	136.0	..	Manufacturing
331.3	339.6	351.3	345.8	386.4	394.5	394.1	407.0	393.9	416.9	..	Services, etc.
-94.3	-71.5	-100.3	-126.9	-103.4	-94.9	-81.3	-88.1	-64.8	-63.6	..	Resource Balance
167.6	189.2	180.1	196.4	205.8	225.9	209.6	183.0	192.2	202.8	..	Exports of Goods & NFServices
262.0	260.7	280.4	323.4	309.2	320.8	291.0	271.0	257.0	266.3	..	Imports of Goods & NFServices
686.1	719.0	727.9	750.9	820.8	831.7	786.3	819.4	828.1	862.5	..	Domestic Absorption
446.7	484.7	490.3	507.8	565.6	564.7	513.3	551.0	573.2	598.6	..	Private Consumption, etc.
128.3	137.5	140.3	143.0	149.5	155.4	164.2	172.5	145.9	154.4	..	General Gov't Consumption
111.2	96.8	97.3	100.1	105.8	111.7	108.9	95.9	109.0	109.6	..	Gross Domestic Investment
81.2	95.6	100.2	92.8	100.8	108.5	102.5	98.0	111.1	108.7	..	Fixed Investment
											Memo Items:
186.3	197.5	180.1	212.4	215.9	232.3	227.2	193.5	200.5	211.7	..	Capacity to Import
18.7	8.3	0.0	15.9	10.1	6.4	17.6	10.6	8.3	8.9	..	Terms of Trade Adjustment
610.5	655.8	627.6	640.0	727.6	743.2	722.6	741.9	771.6	807.9	..	Gross Domestic Income
589.6	637.1	606.5	612.3	695.2	706.8	680.6	699.2	724.2	769.0	..	Gross National Income
				(Index 1980 = 100)							**DOMESTIC PRICES (DEFLATORS)**
83.6	89.9	100.0	107.3	117.7	127.5	144.0	157.5	169.7	177.6	..	Overall (GDP)
81.3	89.2	100.0	108.0	118.3	129.0	143.5	158.2	168.5	176.3	..	Domestic Absorption
90.5	95.0	100.0	106.8	131.4	138.0	144.7	168.5	179.6	188.1	..	Agriculture
83.1	87.6	100.0	104.0	108.0	121.3	147.2	169.8	165.8	173.7	..	Industry
91.6	88.7	100.0	103.5	109.2	113.2	144.7	173.3	167.7	175.5	..	Manufacturing
											MANUFACTURING ACTIVITY
85.0	98.9	100.0	101.1	116.0	111.2	109.6	114.2	..	..	J	Employment (1980=100)
117.3	107.5	100.0	99.7	85.0	105.0	96.9	101.1	..	..	J	Real Earnings per Empl. (1980=100)
105.1	119.5	100.0	105.7	108.0	124.7	96.2	101.9	..	..	J	Real Output per Empl. (1980=100)
45.9	46.6	43.0	49.2	44.6	43.5	43.2	43.8	..	..	J	Earnings as % of Value Added
				(Billions of current CFA Francs)							**MONETARY HOLDINGS**
158.8	161.1	177.7	216.9	262.3	273.0	287.1	300.1	333.7	332.8	..	Money Supply, Broadly Defined
126.5	121.2	137.9	163.2	189.0	189.2	191.7	193.5	227.0	214.4	..	Money as Means of Payment
46.2	42.9	51.4	73.6	84.5	78.3	77.3	86.2	104.3	100.7	..	Currency Ouside Banks
80.3	78.3	86.6	89.6	104.5	110.9	114.3	107.3	122.7	113.7	..	Demand Deposits
32.3	39.9	39.8	53.7	73.3	83.8	95.5	106.6	106.7	118.4	..	Quasi-Monetary Liabilities
				(Billions of current CFA Francs)							
1.5	-4.3	5.4	-22.6	-52.3	-55.5	..	..	..	..	C F	**GOVERNMENT DEFICIT (-) OR SURPLUS**
98.0	107.7	154.0	152.0	180.0	196.2	..	..	..	..	..	Current Revenue
88.5	99.5	142.3	140.9	169.1	181.1	..	..	..	..	..	Current Expenditure
9.5	8.2	11.7	11.0	11.0	15.1	..	..	..	..	..	Current Budget Balance
0.2	0.1	0.0	0.1	0.5	0.1	..	..	..	..	..	Capital Receipts
8.2	12.6	6.3	33.8	63.7	70.7	..	..	..	..	..	Capital Payments

SENEGAL	1967	1968	1969	1970	1971	1972	1973	1974	1975	1976	1977
FOREIGN TRADE (CUSTOMS BASIS)				*(Millions of current US dollars)*							
Value of Exports, fob	137.3	151.4	123.7	160.6	124.9	215.9	194.7	390.7	462.4	485.1	622.8
Nonfuel Primary Products	130.4	136.9	99.7	125.5	89.0	169.5	140.2	315.3	360.8	383.6	499.1
Fuels	0.0	0.0	3.0	4.7	7.3	8.7	11.4	22.1	32.4	37.3	52.0
Manufactures	6.8	14.5	21.0	30.3	28.6	37.6	43.0	53.3	69.2	64.2	71.6
Value of Imports, cif	157.4	181.0	198.7	192.4	217.8	278.5	359.0	497.4	581.4	643.9	762.3
Nonfuel Primary Products	68.2	75.4	76.2	62.9	77.7	88.8	147.7	186.9	169.2	177.6	198.7
Fuels	1.2	6.2	14.8	9.6	13.6	15.8	21.9	64.7	69.2	78.6	95.3
Manufactures	88.0	99.5	107.6	119.9	126.5	173.9	189.5	245.9	343.0	387.7	468.3
Terms of Trade	90.1	111.5	124.2	113.6	105.8	110.2	102.6	127.6	118.4	114.2	113.0
				(Index 1980 = 100)							
Export Prices, fob	28.8	29.8	27.6	29.1	28.3	33.4	41.2	77.6	71.7	68.0	73.5
Nonfuel Primary Products	28.8	29.7	32.0	34.8	38.4	41.4	54.8	86.6	80.3	73.8	80.0
Fuels	4.3	4.3	4.3	4.3	5.6	6.2	8.9	36.7	35.7	38.4	42.0
Manufactures	30.1	31.7	32.4	37.5	36.5	38.5	49.5	67.2	66.1	66.7	71.7
Import Prices, cif	32.0	26.7	22.2	25.6	26.7	30.3	40.2	60.8	60.6	59.6	65.0
Trade at Constant 1980 Prices				*(Millions of 1980 US dollars)*							
Exports, fob	476.8	507.4	448.0	552.3	441.6	646.9	471.9	503.4	644.8	713.4	847.8
Imports, cif	492.3	676.7	893.7	752.1	814.5	920.0	892.9	817.9	960.2	1,081.3	1,172.6
BALANCE OF PAYMENTS				*(Millions of current US dollars)*							
Exports of Goods & Services	..	..	..	242.6	238.6	343.1	364.4	608.7	696.2	703.7	865.0
Merchandise, fob	..	..	..	158.7	135.5	225.3	214.2	416.9	503.1	513.8	667.2
Nonfactor Services	..	..	..	79.1	96.5	109.7	143.8	183.0	185.1	182.2	188.1
Factor Services	..	..	..	4.8	6.6	8.1	6.4	8.8	8.0	7.7	9.7
Imports of Goods & Services	..	..	..	291.7	312.4	396.9	522.8	737.7	865.7	902.8	1,039.2
Merchandise, fob	..	..	..	203.7	221.9	283.9	374.7	552.5	611.5	659.6	772.5
Nonfactor Services	..	..	..	68.3	67.6	83.0	101.5	126.5	170.3	174.5	202.6
Factor Services	..	..	..	19.7	23.0	30.0	46.6	58.7	83.9	68.7	64.1
Long-Term Interest	..	..	..	1.9	4.5	6.2	8.3	14.4	18.9	18.3	21.9
Current Transfers, net	..	..	..	..	..	..	..	..	..	..	..
Workers' Remittances	..	..	..	2.8	2.8	3.9	5.5	7.8	25.4	45.2	42.1
Total to be Financed	..	..	..	..	..	..	..	..	..	..	..
Official Capital Grants	..	..	..	..	..	..	..	..	..	..	..
Current Account Balance	..	..	..	-16.1	-25.5	10.7	-101.4	-65.6	-85.6	-92.4	-66.9
Long-Term Capital, net	..	..	..	76.2	93.9	85.2	132.0	125.9	142.8	174.9	150.4
Direct Investment	..	..	..	4.7	9.4	12.8	4.8	7.0	15.7	35.5	25.1
Long-Term Loans	..	..	..	12.2	24.0	11.6	67.8	55.4	53.6	64.4	68.9
Disbursements	..	..	..	20.0	36.3	22.7	99.7	79.6	76.4	90.3	105.2
Repayments	..	..	..	7.8	12.3	11.1	31.9	24.2	22.8	25.9	36.3
Other Long-Term Capital	..	..	..	59.3	60.5	60.8	59.4	63.5	73.6	75.0	56.4
Other Capital, net	..	..	..	-50.1	-67.0	-90.5	-62.3	-66.1	-61.7	-101.3	-80.1
Change in Reserves	..	..	..	-10.1	-1.5	-5.4	31.7	5.9	4.5	18.9	-3.4
Memo Item:				*(CFA Francs per US dollar)*							
Conversion Factor (Annual Avg)	246.850	246.850	259.710	277.710	277.130	252.210	222.700	240.500	214.320	238.980	245.670
EXTERNAL DEBT, ETC.				*(Millions of US dollars, outstanding at end of year)*							
Public/Publicly Guar. Long-Term	..	..	..	100.1	122.9	137.3	176.4	243.3	292.8	354.1	436.1
Official Creditors	..	..	..	80.4	86.6	94.7	100.7	129.9	162.0	184.3	245.6
IBRD and IDA	..	..	..	10.9	12.8	18.1	23.5	34.8	53.6	69.5	89.3
Private Creditors	..	..	..	19.6	36.3	42.6	75.8	113.3	130.8	169.8	190.5
Private Non-guaranteed Long-Term	..	..	..	30.8	28.4	23.3	17.7	16.7	17.2	16.0	13.6
Use of Fund Credit	..	..	..	0.0	0.0	0.0	0.0	6.0	29.8	35.2	36.8
Short-Term Debt	..	..	..	..	..	..	..	..	..	..	110.0
Memo Items:				*(Thousands of US dollars)*							
Int'l Reserves Excluding Gold	37,200.0	16,000.0	6,200.0	22,050.0	29,323.0	38,493.0	12,062.0	6,320.0	31,126.0	25,220.0	33,665.0
Gold Holdings (at market price)	..	..	..	..	..	..	..	..	..	..	2,408.3
SOCIAL INDICATORS											
Total Fertility Rate	6.4	6.4	6.5	6.5	6.5	6.5	6.5	6.6	6.6	6.6	6.6
Crude Birth Rate	46.7	46.7	46.7	46.6	46.6	46.6	46.5	46.5	46.4	46.4	46.3
Infant Mortality Rate	168.0	166.8	165.6	164.4	163.2	162.0	160.4	158.8	157.2	155.6	154.0
Life Expectancy at Birth	41.7	42.0	42.3	42.6	42.9	43.2	43.5	43.9	44.2	44.6	44.9
Food Production, p.c. ('79-81 = 100)	183.8	142.9	155.3	116.4	151.6	98.2	112.9	145.0	160.4	139.6	88.0
Labor Force, Agriculture (%)	83.1	82.9	82.8	82.7	82.5	82.3	82.0	81.8	81.6	81.4	81.2
Labor Force, Female (%)	41.3	41.3	41.3	41.3	41.3	41.3	41.3	41.3	41.3	41.3	41.3
School Enroll. Ratio, primary	..	..	..	41.0	..	..	..	..	41.0	..	41.0
School Enroll. Ratio, secondary	..	..	..	10.0	..	..	..	..	10.0	..	10.0

1978	1979	1980	1981	1982	1983	1984	1985	1986	1987 estimate	NOTES	SENEGAL
				(Millions of current US dollars)							**FOREIGN TRADE (CUSTOMS BASIS)**
420.8	533.3	476.7	500.1	547.9	581.0	613.0	498.0	611.0	723.0	f	Value of Exports, fob
299.8	426.4	315.0	214.7	330.9	392.0	421.0	354.0	465.0	542.0	..	Nonfuel Primary Products
60.3	66.2	89.5	187.1	132.0	107.0	100.0	51.0	55.0	73.0	..	Fuels
60.8	40.7	72.1	98.3	85.0	82.0	92.0	93.0	91.0	108.0	..	Manufactures
756.1	931.2	1,037.9	1,075.9	991.9	1,025.0	1,035.0	898.0	933.0	1,073.0	f	Value of Imports, cif
203.8	239.0	268.9	307.1	290.8	310.0	323.0	317.0	323.0	371.9	..	Nonfuel Primary Products
106.2	154.4	263.0	327.0	239.1	234.0	270.0	185.0	138.0	170.1	..	Fuels
446.1	537.8	506.0	441.8	462.0	481.0	442.0	396.0	472.0	531.0	..	Manufactures
				(Index 1980 = 100)							Terms of Trade
100.5	100.4	100.0	102.4	96.5	98.2	102.2	100.1	88.3	94.0	..	Terms of Trade
71.9	83.9	100.0	106.7	91.1	89.5	91.5	83.4	78.8	85.9	..	Export Prices, fob
82.5	88.5	100.0	105.6	86.6	88.4	92.4	81.5	75.9	81.0	..	Nonfuel Primary Products
42.3	61.0	100.0	112.5	101.6	92.5	90.2	87.5	74.4	94.2	..	Fuels
76.1	90.0	100.0	99.1	94.8	90.7	89.3	89.0	103.2	113.0	..	Manufactures
71.5	83.5	100.0	104.2	94.4	91.1	89.5	83.3	89.3	91.4	..	Import Prices, cif
				(Millions of 1980 US dollars)							Trade at Constant 1980 Prices
585.6	635.8	476.7	468.9	601.7	649.5	669.8	597.2	775.0	842.1	..	Exports, fob
1,057.2	1,114.7	1,037.9	1,032.6	1,050.8	1,125.5	1,156.1	1,078.2	1,045.4	1,174.3	..	Imports, cif
				(Millions of current US dollars)							**BALANCE OF PAYMENTS**
686.8	832.6	830.5	1,002.2	903.5	972.3	916.1	815.6	1,047.3	1,288.0	f	Exports of Goods & Services
401.7	478.5	421.7	560.6	501.9	605.8	597.8	481.0	611.0	723.0	..	Merchandise, fob
273.4	337.6	385.0	421.2	380.6	341.0	302.6	321.0	416.0	542.0	..	Nonfactor Services
11.7	16.6	23.7	20.4	21.0	25.5	15.7	13.6	17.3	18.0	..	Factor Services
1,042.6	1,196.9	1,337.0	1,630.8	1,346.7	1,416.1	1,327.3	1,280.0	1,689.0	1,968.0	f	Imports of Goods & Services
744.2	813.3	875.1	1,020.3	815.1	917.2	818.9	790.0	1,010.0	1,170.0	..	Merchandise, fob
210.6	287.7	339.6	494.3	414.2	374.6	369.8	333.0	424.0	555.0	..	Nonfactor Services
87.8	95.9	122.3	116.3	117.4	124.2	138.6	160.0	249.0	248.0	..	Factor Services
31.9	44.9	57.0	42.8	34.2	41.3	51.6	42.9	97.7	115.8	..	Long-Term Interest
..	..	..	-10.1	-4.6	-1.6	0.3	..	..	..	..	Current Transfers, net
53.7	54.4	74.8	64.5	62.2	55.1	50.3	32.2	10.0	10.2	..	Workers' Remittances
..	..	..	-638.5	-447.9	-445.5	-411.0	..	..	..	..	Total to be Financed
..	..	..	177.0	182.3	157.7	138.7	..	..	..	..	Official Capital Grants
-235.1	-263.7	-386.1	-461.5	-265.6	-287.8	-272.3	-317.0	-415.0	-316.0	..	Current Account Balance
253.1	230.0	422.1	418.9	370.3	403.8	371.7	379.0	482.0	378.1	f	Long-Term Capital, net
-5.5	5.0	12.9	19.6	10.1	-36.3	27.2	-3.3	-2.0	-50.0	..	Direct Investment
169.7	143.3	191.7	172.8	271.9	302.1	169.3	163.3	258.4	197.0	..	Long-Term Loans
241.6	222.2	318.4	223.8	283.7	320.1	204.0	212.1	371.7	366.5	..	Disbursements
71.9	78.9	126.7	51.0	11.8	18.0	34.7	48.8	113.3	169.5	..	Repayments
88.9	81.7	217.5	49.5	-94.0	-19.7	36.5	219.0	225.6	231.1	..	Other Long-Term Capital
-59.4	20.6	-82.4	-17.3	-242.7	-135.8	-125.6	-75.1	-73.2	-14.9	f	Other Capital, net
41.5	13.1	46.3	59.9	138.0	19.7	26.2	13.1	6.2	-47.2	..	Change in Reserves
				(CFA Francs per US dollar)							**Memo Item:**
225.640	212.720	211.300	271.730	328.620	381.070	436.960	449.260	346.300	300.540	..	Conversion Factor (Annual Avg)
				(Millions of US dollars, outstanding at end of year)							**EXTERNAL DEBT, ETC.**
642.6	823.8	958.0	1,002.8	1,234.6	1,494.7	1,529.1	1,977.0	2,456.4	3,067.6	..	Public/Publicly Guar. Long-Term
353.4	453.6	584.1	714.6	969.4	1,286.5	1,338.0	1,727.1	2,188.1	2,808.9	..	Official Creditors
104.0	128.5	156.4	222.9	244.2	269.0	274.5	321.0	463.7	632.6	..	IBRD and IDA
289.2	370.2	373.9	288.3	265.1	208.3	191.1	249.8	268.3	258.7	..	Private Creditors
10.9	7.4	9.0	12.8	12.8	10.8	9.7	12.6	39.1	41.7	..	Private Non-guaranteed Long-Term
59.0	61.9	97.9	147.8	184.0	196.6	200.8	241.1	246.6	266.7	..	Use of Fund Credit
141.0	203.0	219.0	237.0	202.0	192.0	260.0	211.0	272.0	319.0	..	Short-Term Debt
				(Thousands of US dollars)							**Memo Items:**
18,816.0	19,150.0	8,069.1	8,664.3	11,422.0	12,207.0	3,687.9	5,050.7	9,379.0	9,202.2	..	Int'l Reserves Excluding Gold
4,949.4	14,848.0	17,095.5	11,527.5	13,250.1	11,063.5	8,940.7	9,483.0	11,336.1	14,038.9	..	Gold Holdings (at market price)
											SOCIAL INDICATORS
6.6	6.6	6.6	6.6	6.6	6.6	6.6	6.6	6.5	6.5	..	Total Fertility Rate
46.2	46.2	46.1	46.1	46.0	45.9	45.7	45.6	45.5	45.4	..	Crude Birth Rate
151.6	149.2	146.8	144.4	142.0	137.4	132.9	128.3	123.8	119.2	..	Infant Mortality Rate
45.0	45.1	45.2	45.2	45.3	45.8	46.2	46.7	47.1	47.6	..	Life Expectancy at Birth
137.6	96.0	85.9	118.2	115.2	81.2	93.5	108.2	104.3	101.5	..	Food Production, p.c. ('79-81 = 100)
81.0	80.8	80.6	..	..	..	..	..	..	..	..	Labor Force, Agriculture (%)
41.3	41.3	41.3	41.1	40.9	40.7	40.5	40.3	40.1	39.9	..	Labor Force, Female (%)
42.0	44.0	46.0	48.0	51.0	53.0	55.0	55.0	..	..	..	School Enroll. Ratio, primary
10.0	10.0	11.0	12.0	12.0	12.0	13.0	13.0	..	..	..	School Enroll. Ratio, secondary

SEYCHELLES	1967	1968	1969	1970	1971	1972	1973	1974	1975	1976	1977
CURRENT GNP PER CAPITA (US $)	340	350	340	350	410	470	580	710	800	870	960
POPULATION (thousands)	50	51	52	53	55	56	57	58	59	60	61
ORIGIN AND USE OF RESOURCES					*(Millions of current Seychelles Rupees)*						
Gross National Product (GNP)	80.0	89.0	91.0	102.0	120.0	163.0	200.0	245.0	287.0	349.8	463.4
Net Factor Income from Abroad	-0.3	-0.3	-0.4	-0.4	-0.5	-0.6	-0.8	-1.0	-1.1	-15.8	-29.6
Gross Domestic Product (GDP)	80.3	89.3	91.4	102.4	120.5	163.6	200.8	246.0	288.1	365.6	493.0
Indirect Taxes, net	..	..	..	..	..	..	..	..	..	36.9	56.8
GDP at factor cost	..	..	..	..	..	..	..	..	..	328.7	436.2
Agriculture	..	..	..	..	..	..	..	..	..	35.2	45.2
Industry	..	..	..	..	..	..	..	..	..	59.3	68.2
Manufacturing	..	..	..	..	..	..	..	..	..	20.1	26.9
Services, etc.	..	..	..	..	..	..	..	..	..	271.1	379.6
Resource Balance	..	..	..	..	..	..	..	..	..	-58.5	-19.9
Exports of Goods & NFServices	..	..	..	..	..	..	..	..	..	271.9	386.3
Imports of Goods & NFServices	..	..	..	..	..	..	..	..	..	330.4	406.2
Domestic Absorption	..	..	..	..	..	..	..	..	..	424.1	512.9
Private Consumption, etc.	..	..	..	..	..	..	..	..	..	200.3	198.5
General Gov't Consumption	..	..	..	..	..	..	..	..	..	81.5	117.1
Gross Domestic Investment	..	..	..	..	..	..	..	..	..	142.3	197.3
Fixed Investment	..	..	..	..	..	..	..	..	..	134.5	190.5
Memo Items:											
Gross Domestic Saving	..	..	..	..	..	..	..	..	..	83.8	177.4
Gross National Saving	..	..	..	..	..	..	..	..	..	65.0	143.8
					(Millions of 1980 Seychelles Rupees)						
Gross National Product	417.0	419.1	450.6	450.6	490.7	568.6	604.4	659.1	667.5	688.6	729.9
GDP at factor cost	..	..	..	..	..	..	..	..	..	647.0	687.1
Agriculture	..	..	..	..	..	..	..	..	..	64.6	59.8
Industry	..	..	..	..	..	..	..	..	..	106.7	107.4
Manufacturing	..	..	..	..	..	..	..	..	..	35.3	42.3
Services, etc.	..	..	..	..	..	..	..	..	..	548.5	609.2
Resource Balance	..	..	..	..	..	..	..	..	..	14.3	-6.8
Exports of Goods & NFServices	..	..	..	..	..	..	..	..	..	547.7	584.9
Imports of Goods & NFServices	..	..	..	..	..	..	..	..	..	533.3	591.7
Domestic Absorption	..	..	..	..	..	..	..	..	..	705.4	783.2
Private Consumption, etc.	..	..	..	..	..	..	..	..	..	297.1	302.8
General Gov't Consumption	..	..	..	..	..	..	..	..	..	155.0	186.1
Gross Domestic Investment	..	..	..	..	..	..	..	..	..	253.4	294.3
Fixed Investment	..	..	..	..	..	..	..	..	..	239.4	284.2
Memo Items:											
Capacity to Import	..	..	..	..	..	..	..	..	..	438.9	562.7
Terms of Trade Adjustment	..	..	..	..	..	..	..	..	..	-108.8	-22.2
Gross Domestic Income	..	..	..	..	..	..	..	..	..	611.0	754.2
Gross National Income	..	..	..	..	..	..	..	..	..	579.8	707.7
DOMESTIC PRICES (DEFLATORS)					*(Index 1980 = 100)*						
Overall (GDP)	18.4	20.4	19.4	21.7	23.5	27.5	31.8	35.7	41.3	50.8	63.5
Domestic Absorption	..	..	..	..	..	..	..	..	..	60.1	65.5
Agriculture	..	..	..	..	..	..	..	..	..	54.5	75.6
Industry	..	..	..	..	..	..	..	..	..	55.6	63.5
Manufacturing	..	..	..	..	..	..	..	..	..	57.0	63.6
MANUFACTURING ACTIVITY											
Employment (1980=100)	..	..	..	..	..	..	..	..	65.2	81.4	85.8
Real Earnings per Empl. (1980=100)	..	..	..	..	..	..	..	..	..	..	..
Real Output per Empl. (1980=100)	..	..	..	..	..	..	..	..	..	..	..
Earnings as % of Value Added	..	..	..	..	..	..	..	..	..	44.1	41.2
MONETARY HOLDINGS					*(Millions of current Seychelles Rupees)*						
Money Supply, Broadly Defined	..	..	..	..	45.3	64.8	72.0	76.3	99.3	147.1	176.7
Money as Means of Payment	..	..	..	..	25.4	37.7	40.3	41.9	47.8	67.6	83.8
Currency Ouside Banks	..	..	..	..	12.4	17.0	18.9	20.1	22.7	31.1	40.1
Demand Deposits	..	..	..	..	13.0	20.7	21.4	21.8	25.1	36.5	43.7
Quasi-Monetary Liabilities	..	..	..	..	19.9	27.1	31.7	34.4	51.5	79.5	92.9
GOVERNMENT DEFICIT (-) OR SURPLUS					*(Millions of current Seychelles Rupees)*						
	..	..	..	..	..	6	-11	-1	-1	-3	20
Current Revenue	..	..	..	..	..	80	68	91	97	126	184
Current Expenditure	..	..	..	..	..	37	49	54	65	87	124
Current Budget Balance	..	..	..	..	..	43	19	37	32	40	61
Capital Receipts	..	..	..	..	..	..	..	..	..	..	0
Capital Payments	..	..	..	..	..	38	30	38	33	43	41

1978	1979	1980	1981	1982	1983	1984	1985	1986	1987 estimate	NOTES	SEYCHELLES
1,130	1,620	2,010	2,290	2,360	2,370	2,360	2,560	2,740	3,170	..	CURRENT GNP PER CAPITA (US $)
61	62	63	63	64	64	65	65	66	66	..	POPULATION (thousands)
				(Millions of current Seychelles Rupees)							ORIGIN AND USE OF RESOURCES
556.6	749.9	907.8	957.2	941.8	960.3	1,026.9	1,162.7	1,259.0	1,355.8	..	Gross National Product (GNP)
-38.3	-56.3	-34.1	-14.6	-26.4	-32.6	-41.2	-42.2	-52.4	-78.2	..	Net Factor Income from Abroad
594.9	806.2	941.9	971.8	968.2	992.9	1,068.1	1,204.9	1,311.4	1,434.0	..	Gross Domestic Product (GDP)
79.0	115.5	152.9	167.5	183.1	186.4	190.6	204.6	213.2	..	..	Indirect Taxes, net
515.9	690.7	789.0	804.3	785.1	806.5	877.5	1,000.3	1,098.2	..	..	GDP at factor cost
50.0	58.8	64.4	76.5	62.0	76.9	68.9	69.3	77.3	..	..	Agriculture
82.1	121.7	147.3	166.7	150.2	156.7	176.5	221.2	227.1	..	..	Industry
36.4	48.6	69.5	83.4	82.6	95.6	100.3	116.4	116.3	..	..	Manufacturing
462.8	625.7	730.2	728.6	756.0	759.3	822.7	914.4	1,007.0	..	..	Services, etc.
-7.4	-48.3	-105.3	-151.5	-227.6	-220.3	-144.3	-186.7	-224.9	..	..	Resource Balance
488.4	582.5	640.1	554.1	507.9	415.0	524.9	581.1	498.7	..	..	Exports of Goods & NFServices
495.8	630.8	745.4	705.6	735.5	635.3	669.2	767.8	723.6	..	..	Imports of Goods & NFServices
602.3	854.5	1,047.2	1,123.3	1,195.8	1,213.2	1,212.4	1,391.6	1,536.3	..	..	Domestic Absorption
195.3	377.6	416.3	497.7	544.6	676.8	653.2	700.7	740.2	..	..	Private Consumption, etc.
149.5	211.4	270.0	308.5	338.0	326.0	327.7	417.4	497.6	..	..	General Gov't Consumption
257.5	265.5	360.9	317.1	313.2	210.4	231.5	273.5	298.5	..	..	Gross Domestic Investment
253.4	256.6	344.2	329.6	302.9	219.9	226.9	273.5	292.7	..	..	Fixed Investment
											Memo Items:
250.1	217.2	255.6	165.6	85.6	-9.9	87.2	86.8	73.6	..	..	Gross Domestic Saving
206.7	154.4	212.2	134.3	38.2	-61.3	29.4	38.9	-10.0	..	..	Gross National Saving
				(Millions of 1980 Seychelles Rupees)							
778.0	899.5	907.8	863.1	839.5	852.4	877.0	962.9	980.8	1,005.5	..	Gross National Product
721.2	828.4	789.0	725.0	700.2	716.3	749.7	828.8	855.9	..	..	GDP at factor cost
55.4	64.4	64.4	82.4	64.2	69.5	62.2	59.4	61.6	..	..	Agriculture
117.8	149.5	147.3	134.4	117.1	117.3	129.0	149.1	153.1	..	..	Industry
49.7	54.6	69.5	54.2	56.2	64.1	64.8	68.3	70.2	..	..	Manufacturing
658.2	753.1	730.2	659.6	681.6	694.6	720.9	789.3	806.9	..	..	Services, etc.
-116.7	-62.4	-105.3	-134.4	-265.3	..	..	..	..	..	..	Resource Balance
625.0	708.4	640.1	544.0	510.8	..	..	..	..	..	..	Exports of Goods & NFServices
741.7	770.7	745.4	678.4	776.1	..	..	..	..	..	..	Imports of Goods & NFServices
948.1	1,029.3	1,047.2	1,010.7	1,128.2	..	..	..	..	..	..	Domestic Absorption
377.1	433.6	416.3	409.3	451.3	..	..	..	..	..	..	Private Consumption, etc.
228.9	267.5	270.0	277.2	291.5	..	..	..	..	..	..	General Gov't Consumption
342.0	328.1	360.9	324.2	385.5	..	..	..	..	..	..	Gross Domestic Investment
335.3	318.7	344.2	314.8	373.4	..	..	..	..	..	..	Fixed Investment
											Memo Items:
730.6	711.7	640.1	532.8	535.9	..	..	..	..	..	..	Capacity to Import
105.6	3.3	0.0	-11.3	25.1	..	..	..	..	..	..	Terms of Trade Adjustment
937.0	970.3	941.9	865.0	888.1	..	..	..	..	..	..	Gross Domestic Income
883.6	902.9	907.8	851.8	864.6	..	..	..	..	..	..	Gross National Income
				(Index 1980 = 100)							DOMESTIC PRICES (DEFLATORS)
71.6	83.4	100.0	110.9	112.2	112.7	117.1	120.8	128.4	134.8	..	Overall (GDP)
63.5	83.0	100.0	111.1	106.0	..	..	..	..	..	..	Domestic Absorption
90.2	91.3	100.0	92.9	96.5	110.6	110.8	116.6	125.4	..	..	Agriculture
69.7	81.4	100.0	124.1	128.3	133.6	136.9	148.4	148.4	..	..	Industry
73.3	89.0	100.0	153.8	147.1	149.2	154.9	170.5	165.7	..	..	Manufacturing
											MANUFACTURING ACTIVITY
98.0	100.0	100.0	99.9	96.7	99.3	96.9	96.7	..	..	..	Employment (1980=100)
..	..	..	..	..	..	..	..	..	..	..	Real Earnings per Empl. (1980=100)
..	..	..	..	..	..	..	..	..	..	..	Real Output per Empl. (1980=100)
36.7	39.0	29.0	30.2	31.5	30.3	30.3	30.7	..	..	..	Earnings as % of Value Added
				(Millions of current Seychelles Rupees)							MONETARY HOLDINGS
186.9	233.3	313.8	307.2	283.2	289.5	322.6	363.4	405.3	441.1	..	Money Supply, Broadly Defined
95.0	114.7	158.7	157.9	143.3	131.3	134.7	155.3	155.0	155.2	..	Money as Means of Payment
43.9	52.9	61.7	65.3	62.6	64.3	69.9	75.8	78.1	82.4	..	Currency Ouside Banks
51.1	61.8	97.0	92.6	80.7	67.0	64.8	79.5	76.9	72.8	..	Demand Deposits
91.9	118.6	155.1	149.3	139.9	158.2	187.9	208.1	250.3	285.9	..	Quasi-Monetary Liabilities
				(Millions of current Seychelles Rupees)							
-29	..	..	..	..	..	..	..	..	..	..	GOVERNMENT DEFICIT (-) OR SURPLUS
197	..	..	..	..	..	..	..	..	..	..	Current Revenue
..	..	..	..	..	..	..	..	..	..	..	Current Expenditure
..	..	..	..	..	..	..	..	..	..	..	Current Budget Balance
..	..	..	..	..	..	..	..	..	..	..	Capital Receipts
..	..	..	..	..	..	..	..	..	..	..	Capital Payments

SEYCHELLES	1967	1968	1969	1970	1971	1972	1973	1974	1975	1976	1977
FOREIGN TRADE (CUSTOMS BASIS)				*(Millions of current US dollars)*							
Value of Exports, fob	2.1	3.1	2.5	2.1	1.9	2.6	3.6	7.3	6.4	8.6	11.1
Nonfuel Primary Products	1.8	2.8	2.0	1.5	1.4	1.7	2.4	3.3	2.1	2.7	2.9
Fuels	0.2	0.2	0.4	0.6	0.4	0.8	1.2	4.0	4.2	5.0	7.4
Manufactures	0.1	0.1	0.1	0.1	0.1	0.1	0.0	0.0	0.0	0.9	0.8
Value of Imports, cif	5.1	6.1	7.2	10.1	15.2	21.0	24.8	28.1	31.8	43.5	45.8
Nonfuel Primary Products	2.0	2.5	2.5	3.0	4.6	5.4	7.4	9.7	9.7	11.1	11.7
Fuels	0.6	0.8	0.8	0.7	1.1	1.2	2.3	4.4	6.0	8.4	8.7
Manufactures	2.5	2.8	3.9	6.4	9.5	14.4	15.2	14.0	16.1	23.9	25.4
Terms of Trade					*(Index 1980 = 100)*						
	113.9	155.0	84.1	54.5	62.0	45.3	58.1	92.7	71.1	79.3	78.9
Export Prices, fob	20.5	26.4	16.6	12.8	16.4	14.0	20.4	52.7	40.5	45.4	49.6
Nonfuel Primary Products	37.9	42.8	38.8	42.1	38.4	32.9	62.8	112.2	55.0	59.0	79.4
Fuels	4.3	4.3	4.3	4.3	5.6	6.2	8.9	36.7	35.7	38.4	42.0
Manufactures	30.1	31.7	32.4	37.5	36.5	38.5	49.5	67.2	66.1	66.7	71.7
Import Prices, cif	18.0	17.1	19.7	23.4	26.4	30.9	35.0	56.8	57.0	57.2	62.8
Trade at Constant 1980 Prices					*(Millions of 1980 US dollars)*						
Exports, fob	10.1	11.8	15.1	16.8	11.3	18.8	17.8	13.8	15.7	19.0	22.4
Imports, cif	28.3	35.8	36.5	43.0	57.5	67.9	70.9	49.5	55.8	76.0	72.9
BALANCE OF PAYMENTS					*(Millions of current US dollars)*						
Exports of Goods & Services	..	..	..	12	14	17	21	29	34	38	55
Merchandise, fob	..	..	..	2	2	2	3	7	6	3	5
Nonfactor Services	..	..	..	10	12	15	18	22	27	34	48
Factor Services	..	..	..	0	0	0	0	0	1	1	2
Imports of Goods & Services	..	..	..	12	18	24	30	35	42	47	65
Merchandise, fob	..	..	..	10	15	21	25	28	32	33	39
Nonfactor Services	..	..	..	1	2	2	3	5	7	10	21
Factor Services	..	..	..	1	1	1	2	2	3	4	6
Long-Term Interest	..	..	..	..	..	..	..	..	..	..	0
Current Transfers, net	..	..	..	0	0	0	0	0	0	0	-1
Workers' Remittances	..	..	..	..	..	..	..	..	..	0	0
Total to be Financed	..	..	..	0	-4	-7	-9	-6	-9	-10	-11
Official Capital Grants	..	..	..	0	0	1	1	2	3	6	9
Current Account Balance	..	..	..	0	-4	-7	-8	-4	-6	-4	-2
Long-Term Capital, net	..	..	..	..	..	..	..	8	10	12	15
Direct Investment	..	..	..	..	..	..	..	6	6	4	5
Long-Term Loans	..	..	..	..	..	..	..	..	..	..	1
Disbursements	..	..	..	..	..	..	..	..	..	..	1
Repayments	..	..	..	..	..	..	..	..	..	..	..
Other Long-Term Capital	..	..	..	..	..	..	..	0	0	2	0
Other Capital, net	..	..	..	0	4	7	8	-3	-3	-8	-10
Change in Reserves	..	..	..	0	0	0	0	-1	-1	-1	-2
Memo Item:					*(Seychelles Rupees per US dollar)*						
Conversion Factor (Annual Avg)	4.830	5.560	5.560	5.560	5.480	5.340	5.440	5.700	6.030	7.420	7.640
EXTERNAL DEBT, ETC.				*(Thousands of US dollars, outstanding at end of year)*							
Public/Publicly Guar. Long-Term	..	..	..							..	1,200
Official Creditors	..	..	..					..	..	..	1,200
IBRD and IDA	..	..	..					..	..	..	0
Private Creditors	..	..	..					..	..	..	..
Private Non-guaranteed Long-Term	..	..	..	..	..	..	..	..	..	..	..
Use of Fund Credit	..	..	..	0	0	0	0	0	0	0	0
Short-Term Debt	..	..	..	..	..	..	..	..	..	..	108,000
Memo Items:					*(Thousands of US dollars)*						
Int'l Reserves Excluding Gold	..	..	..	..	..	..	4,270	5,040	6,340	6,490	11,512
Gold Holdings (at market price)	..	..	..	..	..	..	..	..	..	..	..
SOCIAL INDICATORS											
Total Fertility Rate	..	..	..	..	..	..	..	..	..	4.5	4.3
Crude Birth Rate	..	..	..	..	..	..	..	32.1	30.4	27.1	25.8
Infant Mortality Rate	..	..	..	..	..	..	..	..	..	..	..
Life Expectancy at Birth	..	..	..	..	..	..	..	..	66.0	..	..
Food Production, p.c. ('79-81 = 100)	..	..	..	..	..	..	..	..	..	..	..
Labor Force, Agriculture (%)	..	..	..	..	..	..	..	..	..	..	..
Labor Force, Female (%)	..	..	..	..	..	..	..	..	..	..	..
School Enroll. Ratio, primary	..	..	..	..	..	..	..	..	..	..	..
School Enroll. Ratio, secondary	..	..	..	..	..	..	..	..	..	..	..

1978	1979	1980	1981	1982	1983	1984	1985	1986	1987 estimate	NOTES	SEYCHELLES
											FOREIGN TRADE (CUSTOMS BASIS)
				(Millions of current US dollars)							
15.2	21.9	21.2	17.2	15.2	20.3	25.7	28.0	19.0	24.0	..	Value of Exports, fob
3.1	4.8	5.1	3.9	3.2	3.9	8.6	8.8	4.5	5.8	..	Nonfuel Primary Products
11.4	16.1	15.4	12.4	10.9	14.8	16.3	18.2	13.4	16.4	..	Fuels
0.7	1.0	0.7	0.8	1.1	1.7	0.9	1.0	1.1	1.8	..	Manufactures
52.7	89.5	98.8	93.3	97.9	87.8	87.7	98.8	106.0	137.0	..	Value of Imports, cif
13.4	17.9	22.9	21.5	23.3	18.5	43.5	50.3	30.0	22.0	..	Nonfuel Primary Products
9.7	20.6	24.1	20.7	19.7	22.0	22.2	20.6	13.7	21.4	..	Fuels
29.6	51.0	51.9	51.1	54.9	47.4	22.0	27.9	62.3	93.7	..	Manufactures
				(Index 1980 = 100)							
70.4	84.5	100.0	104.0	99.1	99.9	111.1	102.6	56.4	64.7	..	Terms of Trade
48.6	69.5	100.0	105.8	94.8	93.5	100.0	85.7	48.6	62.8	..	Export Prices, fob
90.1	119.5	100.0	90.0	77.0	99.2	128.3	81.7	55.0	74.4	..	Nonfuel Primary Products
42.3	61.0	100.0	112.5	101.6	92.5	90.2	87.5	44.9	56.9	..	Fuels
76.1	90.0	100.0	99.1	94.8	90.7	89.3	89.0	103.2	113.0	..	Manufactures
69.0	82.3	100.0	101.7	95.6	93.6	90.0	83.5	86.1	97.1	..	Import Prices, cif
											Trade at Constant 1980 Prices
				(Millions of 1980 US dollars)							
31.2	31.5	21.2	16.2	16.0	21.7	25.7	32.7	39.1	38.2	..	Exports, fob
76.3	108.8	98.8	91.7	102.4	93.8	97.4	118.3	123.1	141.0	..	Imports, cif
											BALANCE OF PAYMENTS
				(Millions of current US dollars)							
73	95	103	98	85	84	103	119	130	..	..	Exports of Goods & Services
7	6	6	5	4	5	5	5	5	..	..	Merchandise, fob
64	84	91	89	77	77	95	112	124	..	..	Nonfactor Services
2	5	6	4	4	3	2	2	2	..	..	Factor Services
86	115	132	127	132	122	129	151	173	..	..	Imports of Goods & Services
52	72	84	79	83	74	74	84	91	..	..	Merchandise, fob
27	34	39	40	42	40	46	59	71	..	..	Nonfactor Services
7	9	9	7	7	7	8	8	11	..	..	Factor Services
0	0	0	0	1	1	1	2	3	3	..	Long-Term Interest
-1	-1	-1	-3	-3	-3	-2	-1	-5	..	..	Current Transfers, net
0	0	0	0	0	0	0	0	0	..	..	Workers' Remittances
-14	-21	-30	-32	-50	-40	-28	-33	-48	..	K	Total to be Financed
10	9	15	13	10	14	15	14	16	..	K	Official Capital Grants
-4	-12	-16	-19	-40	-26	-13	-19	-31	..	..	Current Account Balance
18	21	32	23	39	31	33	31	48	..	..	Long-Term Capital, net
4	4	6	3	5	6	6	1	8	..	..	Direct Investment
3	8	12	7	12	8	4	7	8	3	..	Long-Term Loans
3	8	12	7	12	9	6	11	14	11	..	Disbursements
0	0	0	0	0	2	2	4	6	7	..	Repayments
2	0	0	1	12	3	8	8	16	..	..	Other Long-Term Capital
-14	-5	-9	-10	1	-7	-21	-12	-18	..	..	Other Capital, net
1	-4	-8	5	0	3	2	0	1	-6	..	Change in Reserves
											Memo Item:
				(Seychelles Rupees per US dollar)							
6.950	6.330	6.390	6.310	6.550	6.770	7.060	7.130	6.180	5.600	..	Conversion Factor (Annual Avg)
											EXTERNAL DEBT, ETC.
				(Thousands of US dollars, outstanding at end of year)							
4,000	12,900	25,100	27,900	36,400	41,700	42,000	55,000	68,000	84,200	..	Public/Publicly Guar. Long-Term
4,000	12,900	25,100	27,900	30,700	36,300	36,000	46,500	57,000	71,500	..	Official Creditors
0	0	0	0	0	0	0	2,100	4,000	4,900	..	IBRD and IDA
..	..	..	..	5,700	5,400	6,100	8,500	11,000	12,700	..	Private Creditors
..	..	..	..	..	..	..	..	..	..	..	Private Non-guaranteed Long-Term
0	0	0	0	0	0	0	0	0	..	..	Use of Fund Credit
381,000	461,000	59,000	9,000	12,000	8,000	19,000	24,000	38,900	35,000	..	Short-Term Debt
											Memo Items:
				(Thousands of US dollars)							
9,260	12,148	18,439	13,800	13,072	9,972	5,400	8,501	7,749	13,713	..	Int'l Reserves Excluding Gold
..	..	..	..	..	..	..	..	..	..	..	Gold Holdings (at market price)
											SOCIAL INDICATORS
4.2	4.0	3.8	3.7	3.6	3.5	3.4	3.4	3.3	3.2	..	Total Fertility Rate
28.8	27.5	28.9	26.4	26.5	26.5	26.5	26.5	26.5	26.5	..	Crude Birth Rate
..	..	..	..	5.0	10.0	15.0	20.0	25.0	30.0	..	Infant Mortality Rate
..	..	..	..	68.7	68.9	69.1	69.4	69.6	69.8	..	Life Expectancy at Birth
..	..	..	..	..	..	..	..	..	..	..	Food Production, p.c. ('79-81=100)
..	..	..	..	..	..	..	..	..	..	..	Labor Force, Agriculture (%)
..	..	..	..	..	..	..	..	..	..	..	Labor Force, Female (%)
..	..	..	..	..	..	..	..	..	..	..	School Enroll. Ratio, primary
..	..	..	..	..	..	..	..	..	..	..	School Enroll. Ratio, secondary

SIERRA LEONE	1967	1968	1969	1970	1971	1972	1973	1974	1975	1976	1977
CURRENT GNP PER CAPITA (US $)	150	140	150	160	160	160	170	190	220	210	220
POPULATION (thousands)	2,510	2,555	2,602	2,651	2,702	2,754	2,809	2,865	2,924	2,985	3,049

ORIGIN AND USE OF RESOURCES

(Millions of current Sierra Leonean Leones)

	1967	1968	1969	1970	1971	1972	1973	1974	1975	1976	1977
Gross National Product (GNP)	264	262	303	347	345	348	388	471	566	604	733
Net Factor Income from Abroad	-5	-8	-8	-7	-3	-8	-6	-7	-7	-10	-11
Gross Domestic Product (GDP)	270	270	311	353	348	356	393	478	573	613	744
Indirect Taxes, net	24	23	33	34	32	36	40	52	52	55	77
GDP at factor cost	246	247	278	319	316	320	353	426	521	559	667
Agriculture	83	86	90	90	95	97	108	130	186	213	264
Industry	67	62	73	97	91	89	97	119	123	114	130
Manufacturing	15	15	17	18	20	21	21	26	30	31	34
Services, etc.	96	99	115	133	131	134	149	177	212	232	274
Resource Balance	-11	-9	-1	-10	-4	-5	6	-11	-58	-49	-56
Exports of Goods & NFServices	68	65	95	100	109	104	116	134	154	143	156
Imports of Goods & NFServices	79	74	96	110	113	108	110	145	212	192	212
Domestic Absorption	280	278	312	363	352	360	388	489	630	662	800
Private Consumption, etc.	227	224	229	263	268	287	300	364	477	529	631
General Gov't Consumption	20	21	37	42	31	32	42	49	63	64	71
Gross Domestic Investment	34	33	46	59	53	42	46	76	90	69	98
Fixed Investment	..	..	..	..	47	43	44	57	76	72	81
Memo Items:											
Gross Domestic Saving	23	24	45	49	49	37	52	64	33	20	42
Gross National Saving	17	17	37	43	47	29	46	58	27	15	36

(Millions of 1980 Sierra Leonean Leones)

	1967	1968	1969	1970	1971	1972	1973	1974	1975	1976	1977
Gross National Product	764	746	839	935	969	943	980	1,016	1,055	1,018	1,032
GDP at factor cost	725	719	788	871	891	879	912	935	968	947	967
Agriculture	262	258	273	274	280	277	282	280	292	308	324
Industry	242	230	252	298	290	277	278	278	288	258	236
Manufacturing	34	33	38	39	38	41	43	47	51	53	51
Services, etc.	222	231	263	299	321	326	352	378	388	380	407
Resource Balance	-105	-66	16	-41	7	56	92	-62	-8	64	44
Exports of Goods & NFServices	324	351	482	486	466	450	452	370	394	388	329
Imports of Goods & NFServices	429	416	465	527	459	394	360	432	402	324	285
Domestic Absorption	891	844	853	1,001	976	915	907	1,098	1,075	971	1,003
Private Consumption, etc.	..	..	..	..	..	..	..	..	..	..	..
General Gov't Consumption	..	..	..	..	..	..	..	..	..	..	..
Gross Domestic Investment	154	149	178	227	157	115	122	168	173	121	121
Fixed Investment	..	..	..	..	138	119	117	127	147	118	104
Memo Items:											
Capacity to Import	371	367	459	479	443	377	379	399	293	241	210
Terms of Trade Adjustment	47	16	-22	-7	-23	-73	-73	29	-101	-147	-119
Gross Domestic Income	833	795	848	953	960	898	926	1,064	966	888	928
Gross National Income	811	762	817	928	947	870	907	1,044	953	871	913

DOMESTIC PRICES (DEFLATORS)

(Index 1980 = 100)

	1967	1968	1969	1970	1971	1972	1973	1974	1975	1976	1977
Overall (GDP)	34.3	34.7	35.7	36.8	35.5	36.7	39.4	46.1	53.6	59.3	71.1
Domestic Absorption	31.5	33.0	36.6	36.3	36.1	39.4	42.7	44.6	58.6	68.3	79.8
Agriculture	31.8	33.4	33.0	32.8	33.8	35.2	38.1	46.4	63.7	69.1	81.4
Industry	27.7	27.0	29.1	32.4	31.4	32.1	34.9	42.9	42.6	44.0	54.9
Manufacturing	44.8	45.0	44.7	47.7	53.1	50.7	49.2	55.7	59.1	59.1	66.2

MANUFACTURING ACTIVITY

	1967	1968	1969	1970	1971	1972	1973	1974	1975	1976	1977
Employment (1980=100)	..	..	..	..	..	..	..	..	..	..	..
Real Earnings per Empl. (1980=100)	..	..	..	..	..	..	..	..	..	..	..
Real Output per Empl. (1980=100)	..	..	..	..	..	..	..	..	..	..	..
Earnings as % of Value Added	..	..	..	..	..	..	..	..	..	..	..

MONETARY HOLDINGS

(Millions of current Sierra Leonean Leones)

	1967	1968	1969	1970	1971	1972	1973	1974	1975	1976	1977
Money Supply, Broadly Defined	32	39	46	43	48	57	71	85	92	112	136
Money as Means of Payment	22	27	32	29	32	39	48	55	60	72	84
Currency Ouside Banks	14	18	20	19	21	25	30	31	37	41	52
Demand Deposits	8	10	12	10	12	14	19	23	24	31	33
Quasi-Monetary Liabilities	9	12	14	14	16	18	23	30	31	40	52

GOVERNMENT DEFICIT (-) OR SURPLUS

(Millions of current Sierra Leonean Leones)

	1967	1968	1969	1970	1971	1972	1973	1974	1975	1976	1977
Current Revenue	..	..	..	..	..	..	..	-21	-60	-48	-51
Current Expenditure	..	..	..	..	..	..	..	92	97	95	119
Current Budget Balance	..	..	..	..	..	..	..	64	82	83	104
Capital Receipts	..	..	..	..	..	..	..	28	15	12	16
Capital Payments	..	..	..	..	..	..	..	..	..	..	..
								48	75	61	67

1978	1979	1980	1981	1982	1983	1984	1985	1986	1987 estimate	NOTES	SIERRA LEONE
230	270	320	370	380	390	370	350	320	300	..	CURRENT GNP PER CAPITA (US $)
3,115	3,184	3,255	3,329	3,406	3,487	3,570	3,657	3,752	3,845	..	POPULATION (thousands)

(Millions of current Sierra Leonean Leones)

ORIGIN AND USE OF RESOURCES

1978	1979	1980	1981	1982	1983	1984	1985	1986	1987	NOTES	
833	988	1,111	1,269	1,562	1,833	2,690	4,229	6,413	29,339	C f	Gross National Product (GNP)
-17	-41	-45	-23	-43	-43	-40	-81	61	196	..	Net Factor Income from Abroad
850	1,029	1,156	1,292	1,604	1,876	2,729	4,310	6,353	29,143	C f	Gross Domestic Product (GDP)
100	97	93	119	101	77	105	139	220	1,008	..	Indirect Taxes, net
750	932	1,063	1,174	1,504	1,799	2,625	4,171	6,133	28,135	C f	GDP at factor cost
282	327	351	379	539	686	1,053	1,868	2,740	12,571	..	Agriculture
142	204	233	231	256	266	358	566	1,157	5,309	..	Industry
40	55	59	54	100	108	135	192	244	1,118	..	Manufacturing
327	402	479	563	710	847	1,214	1,738	2,236	10,256	..	Services, etc.
-42	-84	-177	-216	-163	-206	-49	-39	-94	272	..	Resource Balance
199	233	264	297	253	208	290	463	855	2,675	..	Exports of Goods & NFServices
242	318	442	513	416	413	339	503	948	2,403	..	Imports of Goods & NFServices
892	1,113	1,333	1,508	1,768	2,082	2,778	4,349	6,446	28,871	..	Domestic Absorption
719	880	1,048	1,172	1,416	1,647	2,242	3,616	5,446	24,284	..	Private Consumption, etc.
77	96	97	90	138	167	189	245	298	2,019	..	General Gov't Consumption
96	138	187	247	215	268	347	488	703	2,569	..	Gross Domestic Investment
100	128	172	236	205	235	332	435	593	2,144	..	Fixed Investment

Memo Items:

1978	1979	1980	1981	1982	1983	1984	1985	1986	1987	NOTES	
54	54	10	31	51	62	298	449	609	2,841	..	Gross Domestic Saving
44	19	-26	19	15	25	279	383	682	..	..	Gross National Saving

(Millions of 1980 Sierra Leonean Leones)

1978	1979	1980	1981	1982	1983	1984	1985	1986	1987	NOTES	
1,023	1,083	1,111	1,199	1,209	1,190	1,219	1,182	1,179	1,184	C f	Gross National Product
963	1,021	1,063	1,101	1,149	1,161	1,155	1,155	1,127	1,132	C f	GDP at factor cost
331	362	351	354	365	362	363	363	402	388	..	Agriculture
197	206	233	227	224	196	213	213	182	209	..	Industry
54	56	59	61	76	68	70	70	61	65	..	Manufacturing
435	453	479	520	560	604	579	579	543	536	..	Services, etc.
-56	-85	-177	-228	-105	-136	2	9	30	25	..	Resource Balance
241	202	264	251	217	169	176	178	188	153	..	Exports of Goods & NFServices
297	287	442	479	321	305	175	169	157	129	..	Imports of Goods & NFServices
1,100	1,205	1,333	1,448	1,347	1,359	1,238	1,201	1,138	1,150	..	Domestic Absorption
..	..	..	..	..	..	..	..	..	..	..	Private Consumption, etc.
..	..	..	..	..	..	..	..	..	..	..	General Gov't Consumption
128	146	187	227	172	164	167	148	126	126	..	Gross Domestic Investment
128	135	172	219	166	149	162	130	107	108	..	Fixed Investment

Memo Items:

1978	1979	1980	1981	1982	1983	1984	1985	1986	1987	NOTES	
245	211	264	277	195	153	150	156	142	143	..	Capacity to Import
5	9	0	26	-22	-16	-27	-22	-46	-10	..	Terms of Trade Adjustment
1,048	1,129	1,156	1,246	1,221	1,207	1,213	1,188	1,123	1,164	..	Gross Domestic Income
1,027	1,092	1,111	1,225	1,188	1,175	1,192	1,160	1,133	1,174	..	Gross National Income

(Index 1980 = 100)

DOMESTIC PRICES (DEFLATORS)

1978	1979	1980	1981	1982	1983	1984	1985	1986	1987	NOTES	
81.5	91.9	100.0	105.9	129.2	153.5	220.2	356.3	543.6	2482.2	..	Overall (GDP)
81.1	92.4	100.0	104.2	131.3	153.2	224.5	362.1	566.3	2511.3	..	Domestic Absorption
85.1	90.4	100.0	107.2	147.5	189.7	290.0	514.5	681.0	3238.3	..	Agriculture
72.1	98.7	100.0	102.1	114.2	135.8	167.9	265.3	637.2	2545.7	..	Industry
74.1	97.2	100.0	88.1	131.9	158.7	192.1	273.8	399.9	1714.7	..	Manufacturing

MANUFACTURING ACTIVITY

1978	1979	1980	1981	1982	1983	1984	1985	1986	1987	NOTES	
..	..	..	..	..	..	..	..	..	..	G	Employment (1980 = 100)
..	..	..	..	..	..	..	..	..	..	G	Real Earnings per Empl. (1980 = 100)
..	..	..	..	..	..	..	..	..	..	G	Real Output per Empl. (1980 = 100)
..	..	..	22.3	..	..	..	..	..	..	..	Earnings as % of Value Added

(Millions of current Sierra Leonean Leones)

MONETARY HOLDINGS

1978	1979	1980	1981	1982	1983	1984	1985	1986	1987	NOTES	
179	215	261	268	420	552	709	1,213	2,285	3,747	..	Money Supply, Broadly Defined
107	128	153	152	253	359	486	900	1,852	2,888	..	Money as Means of Payment
63	72	86	86	121	197	260	442	1,006	1,364	..	Currency Ouside Banks
44	55	66	66	132	162	226	458	846	1,524	..	Demand Deposits
72	87	108	116	166	193	223	313	433	859	..	Quasi-Monetary Liabilities

(Millions of current Sierra Leonean Leones)

GOVERNMENT DEFICIT (-) OR SURPLUS

1978	1979	1980	1981	1982	1983	1984	1985	1986	1987	NOTES	
-79	-119	-148	-121	-167	-271	-206	-376	..	..	C	
170	193	197	237	189	171	243	315	..	..	..	Current Revenue
199	200	..	284	270	253	340	409	..	..	..	Current Expenditure
-30	-7	..	-47	-82	-83	-96	-95	..	..	..	Current Budget Balance
..	..	..	..	3	..	23	12	..	..	..	Capital Receipts
50	112	..	73	89	189	132	293	..	..	..	Capital Payments

SIERRA LEONE	1967	1968	1969	1970	1971	1972	1973	1974	1975	1976	1977
FOREIGN TRADE (CUSTOMS BASIS)				*(Millions of current US dollars)*							
Value of Exports, fob	63.3	90.9	104.6	101.5	98.5	117.5	131.4	146.2	145.8	106.6	130.6
Nonfuel Primary Products	21.5	34.1	29.8	36.2	34.9	39.7	46.6	51.7	56.9	32.4	69.1
Fuels	0.2	0.1	0.2	2.6	3.0	3.3	2.5	2.3	9.9	5.9	6.5
Manufactures	41.6	56.7	74.7	62.6	60.6	74.5	82.3	92.2	79.0	68.3	55.1
Value of Imports, cif	90.4	90.7	111.6	116.9	113.1	121.0	157.7	222.4	159.3	166.3	181.0
Nonfuel Primary Products	23.0	20.4	23.5	30.5	27.8	28.0	46.9	60.8	35.7	39.5	43.1
Fuels	6.7	6.9	6.9	5.5	8.4	9.1	9.4	25.6	21.8	23.7	25.1
Manufactures	60.7	63.4	81.3	80.9	76.9	83.9	101.4	136.1	101.8	103.1	112.8
Terms of Trade	148.1	156.4	143.4	123.2	120.1	114.3	118.8	114.8	98.2	112.2	135.0
Export Prices, fob	31.6	32.8	33.0	31.6	30.8	32.5	45.8	65.7	58.5	66.6	86.2
Nonfuel Primary Products	37.6	35.4	36.2	39.1	34.9	34.4	50.6	65.6	55.9	76.4	116.5
Fuels	4.3	4.3	4.3	4.3	5.6	6.2	8.9	36.7	35.7	38.4	42.0
Manufactures	30.1	31.7	32.4	37.5	36.5	38.5	49.5	67.2	66.1	66.7	71.7
Import Prices, cif	21.3	21.0	23.0	25.6	25.7	28.4	38.6	57.3	59.6	59.3	63.8
Trade at Constant 1980 Prices				*(Millions of 1980 US dollars)*							
Exports, fob	200.4	277.4	316.7	321.5	319.5	362.1	286.7	222.4	248.9	160.1	151.5
Imports, cif	424.1	432.9	484.7	456.2	440.8	426.1	408.8	388.2	267.2	280.2	283.5
BALANCE OF PAYMENTS				*(Millions of current US dollars)*							
Exports of Goods & Services	..	..	..	121.6	113.6	131.5	147.9	170.3	151.6	134.3	162.8
Merchandise, fob	..	..	..	100.0	95.8	113.5	129.1	142.8	129.0	113.9	142.7
Nonfactor Services	..	..	..	18.5	16.9	17.2	18.0	23.3	18.8	18.3	18.3
Factor Services	..	..	..	3.1	1.0	0.9	0.8	4.2	3.9	2.2	1.8
Imports of Goods & Services	..	..	..	142.6	139.3	143.9	183.5	253.2	227.7	203.5	232.5
Merchandise, fob	..	..	..	105.3	102.5	107.7	141.0	199.6	167.6	149.8	165.0
Nonfactor Services	..	..	..	30.1	26.7	28.1	33.5	41.3	45.4	41.5	50.7
Factor Services	..	..	..	7.2	10.1	8.0	8.9	12.3	14.7	12.2	16.8
Long-Term Interest	..	..	..	2.5	2.3	2.6	3.7	4.0	4.4	5.2	4.8
Current Transfers, net	..	..	..	1.2	1.6	-0.3	0.0	0.5	1.7	4.3	4.5
Workers' Remittances	..	..	..	0.0	0.0	0.0	0.0	0.0	0.0	0.0	0.0
Total to be Financed	..	..	..	-19.8	-24.1	-12.7	-35.5	-82.4	-74.4	-64.9	-65.2
Official Capital Grants	..	..	..	3.9	4.2	3.5	6.4	21.5	9.0	6.4	15.8
Current Account Balance	..	..	..	-15.9	-19.9	-9.2	-29.1	-60.9	-65.3	-58.5	-49.4
Long-Term Capital, net	..	..	..	3.1	11.3	12.7	22.4	55.7	44.8	30.9	29.6
Direct Investment	..	..	..	8.2	5.2	3.8	4.1	10.5	10.1	8.5	5.1
Long-Term Loans	..	..	..	-3.0	5.4	5.9	12.5	37.6	22.6	10.5	20.9
Disbursements	..	..	..	7.6	11.7	14.0	21.7	48.1	33.8	27.3	33.2
Repayments	..	..	..	10.6	6.3	8.1	9.2	10.5	11.2	16.8	12.3
Other Long-Term Capital	..	..	..	-6.0	-3.5	-0.5	-0.6	-13.9	3.0	5.4	-12.2
Other Capital, net	..	..	..	12.5	6.7	1.4	15.0	2.7	-1.7	8.3	16.6
Change in Reserves	..	..	..	0.3	1.8	-4.9	-8.3	2.5	22.3	19.4	3.2
Memo Item:				*(Sierra Leonean Leones per US dollar)*							
Conversion Factor (Annual Avg)	0.720	0.830	0.830	0.830	0.830	0.820	0.820	0.840	0.880	1.000	1.140
EXTERNAL DEBT, ETC.				*(Millions of US dollars, outstanding at end of year)*							
Public/Publicly Guar. Long-Term	..	..	..	59.4	68.5	72.8	90.3	135.7	149.4	159.8	189.1
Official Creditors	..	..	..	32.4	43.3	46.6	56.6	70.4	78.9	93.4	114.0
IBRD and IDA	..	..	..	6.3	6.8	7.7	12.4	17.9	22.1	23.6	24.9
Private Creditors	..	..	..	27.0	25.1	26.2	33.7	65.3	70.5	66.4	75.1
Private Non-guaranteed Long-Term	..	..	..	..	..	..	..	..	..	..	..
Use of Fund Credit	..	..	..	0.0	0.0	0.0	0.0	62.9	5.8	33.2	43.2
Short-Term Debt	..	..	..	..	..	..	..	..	..	..	20.5
Memo Items:				*(Thousands of US dollars)*							
Int'l Reserves Excluding Gold	15,900	27,500	35,400	39,360	38,407	46,452	51,775	54,624	28,412	25,204	33,401
Gold Holdings (at market price)	..	..	..	..	..	..	..	..	..	..	..
SOCIAL INDICATORS											
Total Fertility Rate	6.4	6.4	6.4	6.5	6.5	6.5	6.5	6.5	6.5	6.5	6.5
Crude Birth Rate	48.5	48.6	48.7	48.7	48.8	48.9	48.8	48.8	48.7	48.7	48.6
Infant Mortality Rate	204.0	201.8	199.6	197.4	195.2	193.0	190.4	187.8	185.2	182.6	180.0
Life Expectancy at Birth	33.4	33.7	34.0	34.3	34.6	34.9	35.3	35.7	36.1	36.5	36.9
Food Production, p.c. ('79-81 = 100)	104.9	101.2	108.0	103.6	103.1	103.1	99.4	102.4	107.4	106.5	105.5
Labor Force, Agriculture (%)	77.2	76.7	76.1	75.5	74.9	74.3	73.8	73.2	72.6	72.0	71.4
Labor Force, Female (%)	35.8	35.7	35.7	35.6	35.5	35.4	35.3	35.2	35.1	35.0	35.0
School Enroll. Ratio, primary	..	..	..	34.0	..	..	..	..	39.0	37.0	37.0
School Enroll. Ratio, secondary	..	..	..	8.0	..	..	..	..	11.0	12.0	12.0

1978	1979	1980	1981	1982	1983	1984	1985	1986	1987 estimate	NOTES	SIERRA LEONE
											FOREIGN TRADE (CUSTOMS BASIS)
178.3	194.0	207.0	136.0	111.0	120.0	133.0	130.3	127.0	119.9	..	Value of Exports, fob
70.1	82.3	80.5	60.5	56.0	63.1	69.3	57.3	50.6	43.9	..	Nonfuel Primary Products
8.9	9.1	10.3	6.2	4.5	5.7	9.0	11.2	4.8	5.6	..	Fuels
99.3	102.7	116.2	69.4	50.5	51.2	54.8	61.7	71.7	70.4	..	Manufactures
278.0	316.0	425.0	311.0	298.0	160.0	157.1	149.4	141.0	132.1	..	Value of Imports, cif
65.8	74.9	100.8	73.7	70.7	37.9	33.4	30.7	29.7	28.2	..	Nonfuel Primary Products
38.7	44.0	59.1	43.3	41.5	22.3	27.4	25.0	10.6	11.7	..	Fuels
173.5	197.1	265.1	194.0	185.9	99.8	96.3	93.7	100.7	92.2	..	Manufactures
				(Index 1980 = 100)							
114.4	113.0	100.0	93.1	92.1	94.1	98.7	99.8	97.8	93.2	..	Terms of Trade
80.7	96.2	100.0	94.5	89.6	88.8	90.8	89.1	92.5	93.7	..	Export Prices, fob
101.0	113.3	100.0	88.4	84.6	87.0	92.1	89.4	88.3	78.7	..	Nonfuel Primary Products
42.3	61.0	100.0	112.5	101.6	92.5	90.2	87.5	44.9	56.9	..	Fuels
76.1	90.0	100.0	99.1	94.8	90.7	89.3	89.0	103.2	113.0	..	Manufactures
70.5	85.1	100.0	101.5	97.3	94.4	92.0	89.2	94.5	100.6	..	Import Prices, cif
				(Millions of 1980 US dollars)							Trade at Constant 1980 Prices
221.0	201.6	207.0	143.9	123.9	135.2	146.5	146.3	137.4	127.9	..	Exports, fob
394.2	371.2	425.0	306.5	306.3	169.5	170.7	167.5	149.2	131.3	..	Imports, cif
				(Millions of current US dollars)							**BALANCE OF PAYMENTS**
211.6	243.2	278.2	212.5	153.3	142.7	174.2	160.3	152.6	..	..	Exports of Goods & Services
185.1	197.1	213.5	151.6	110.3	107.0	132.6	131.9	125.7	..	..	Merchandise, fob
26.2	45.8	64.1	60.4	42.9	35.4	41.3	28.1	26.8	..	..	Nonfactor Services
0.4	0.3	0.6	0.5	0.2	0.3	0.3	0.3	0.1	..	..	Factor Services
343.1	451.2	495.8	387.5	364.0	196.9	230.2	199.8	163.6	..	..	Imports of Goods & Services
253.0	336.3	385.9	282.0	260.3	133.0	149.7	141.2	111.1	..	..	Merchandise, fob
56.2	72.6	87.5	68.0	68.5	42.8	47.8	43.0	34.2	..	..	Nonfactor Services
33.9	42.3	22.5	37.4	35.2	21.2	32.6	15.6	18.3	..	..	Factor Services
7.7	9.6	7.6	10.0	3.0	4.0	5.0	2.6	4.0	0.9	..	Long-Term Interest
6.8	5.3	8.3	9.4	5.2	3.6	7.9	3.4	1.5	..	..	Current Transfers, net
0.0	0.0	0.0	0.0	0.0	0.0	0.0	0.0	0.0	..	..	Workers' Remittances
-124.6	-202.8	-209.4	-165.6	-205.6	-50.6	-48.1	-36.2	-9.5	..	K	Total to be Financed
12.9	23.9	44.6	34.3	45.2	33.2	25.3	16.5	4.6	..	K	Official Capital Grants
-111.7	-178.8	-164.8	-131.2	-160.4	-17.5	-22.8	-19.7	-4.9	..	..	Current Account Balance
65.0	77.3	102.0	141.4	65.9	-19.2	-31.9	-23.8	-294.2	..	..	Long-Term Capital, net
24.3	16.1	-18.7	7.5	4.7	1.7	5.9	-3.8	-6.4	..	..	Direct Investment
40.7	41.4	63.0	35.1	75.8	16.7	6.7	28.7	14.0	-1.8	..	Long-Term Loans
70.6	73.7	95.1	68.5	83.7	22.8	20.6	35.4	24.9	2.2	..	Disbursements
29.9	32.3	32.1	33.4	7.9	6.1	13.9	6.7	10.9	4.0	..	Repayments
-12.8	-4.1	13.1	64.5	-59.8	-70.8	-69.7	-65.2	-306.4	..	..	Other Long-Term Capital
53.6	104.6	48.0	-56.1	86.0	29.8	33.5	47.3	314.3	..	..	Other Capital, net
-6.9	-3.1	14.9	46.0	8.4	6.9	21.2	-3.8	-15.2	0.3	..	Change in Reserves
											Memo Item:
				(Sierra Leonean Leones per US dollar)							
1.090	1.050	1.050	1.090	1.200	1.260	2.510	3.140	5.210	31.250	..	Conversion Factor (Annual Avg)
				(Millions of US dollars, outstanding at end of year)							**EXTERNAL DEBT, ETC.**
242.8	293.0	350.6	355.1	400.6	391.5	326.3	390.2	462.4	512.8	..	Public/Publicly Guar. Long-Term
149.0	173.8	235.5	234.0	287.0	290.2	239.2	292.1	380.8	425.6	..	Official Creditors
35.0	38.7	42.5	43.9	48.7	53.4	55.5	68.1	80.7	91.3	..	IBRD and IDA
93.9	119.2	115.2	121.1	113.6	101.3	87.2	98.1	81.6	87.2	..	Private Creditors
..	..	..	..	..	..	..	..	..	..	..	Private Non-guaranteed Long-Term
40.9	41.6	28.3	55.6	51.2	68.2	74.3	78.4	71.7	83.1	..	Use of Fund Credit
24.0	44.2	53.6	69.6	59.9	71.7	36.6	71.2	60.7	63.2	..	Short-Term Debt
				(Thousands of US dollars)							Memo Items:
34,787	46,700	30,600	15,950	8,430	16,205	7,749	10,822	13,655	38,028	..	Int'l Reserves Excluding Gold
..	..	..	..	..	..	..	..	..	..	..	Gold Holdings (at market price)
											SOCIAL INDICATORS
6.5	6.5	6.5	6.5	6.5	6.5	6.5	6.5	6.5	6.5	..	Total Fertility Rate
48.6	48.5	48.5	48.4	48.4	48.3	48.3	48.3	48.2	48.2	..	Crude Birth Rate
177.4	174.8	172.2	169.6	167.0	167.8	168.5	169.3	170.1	170.9	..	Infant Mortality Rate
37.3	37.7	38.1	38.5	38.9	39.4	39.9	40.3	40.8	41.3	..	Life Expectancy at Birth
108.9	99.3	100.7	100.0	108.6	109.9	97.9	92.7	100.6	99.4	..	Food Production, p.c. ('79-81 = 100)
70.8	70.2	69.6	..	..	..	..	..	..	..	..	Labor Force, Agriculture (%)
34.9	34.8	34.7	34.5	34.3	34.1	33.9	33.7	33.5	33.3	..	Labor Force, Female (%)
38.0	39.0	45.0	..	58.0	..	..	..	..	..	..	School Enroll. Ratio, primary
12.0	12.0	14.0	..	17.0	..	..	..	..	..	..	School Enroll. Ratio, secondary

SINGAPORE	1967	1968	1969	1970	1971	1972	1973	1974	1975	1976	1977
CURRENT GNP PER CAPITA (US $)	660	740	840	950	1,080	1,270	1,580	2,020	2,540	2,760	2,940
POPULATION (thousands)	1,978	2,012	2,043	2,075	2,113	2,152	2,193	2,230	2,263	2,293	2,325
ORIGIN AND USE OF RESOURCES					*(Millions of current Singapore Dollars)*						
Gross National Product (GNP)	3,844	4,402	5,105	5,861	6,831	8,174	10,033	12,260	13,567	14,570	15,852
Net Factor Income from Abroad	95	87	85	56	-10	-21	-224	-351	124	-81	-187
Gross Domestic Product (GDP)	3,748	4,315	5,020	5,805	6,841	8,195	10,257	12,610	13,443	14,651	16,039
Indirect Taxes, net	302	344	410	485	544	632	767	805	866	989	1,122
GDP at factor cost	3,447	3,971	4,609	5,320	6,297	7,563	9,490	11,805	12,577	13,662	14,917
Agriculture	107	121	129	135	159	161	214	230	254	257	283
Industry	943	1,140	1,373	1,726	2,142	2,756	3,399	4,258	4,585	5,119	5,518
Manufacturing	618	754	936	1,162	1,447	1,853	2,429	3,068	3,209	3,589	3,982
Services, etc.	2,699	3,053	3,518	3,944	4,540	5,278	6,644	8,122	8,603	9,275	10,237
Resource Balance	-316	-284	-532	-1,179	-1,484	-1,378	-1,041	-2,044	-1,416	-1,199	-424
Exports of Goods & NFServices	4,293	4,947	5,907	6,570	7,460	8,520	11,956	18,790	..	..	..
Imports of Goods & NFServices	4,609	5,231	6,439	7,749	8,944	9,899	12,997	20,834	..	..	..
Domestic Absorption	4,064	4,599	5,552	6,984	8,325	9,573	11,298	14,654	14,859	15,850	16,463
Private Consumption, etc.	2,850	3,075	3,555	4,047	4,686	5,190	6,135	7,646	8,066	8,327	8,948
General Gov't Consumption	383	449	560	692	861	990	1,118	1,298	1,423	1,541	1,716
Gross Domestic Investment	831	1,075	1,437	2,244	2,778	3,393	4,045	5,710	5,370	5,982	5,799
Fixed Investment	738	997	1,326	1,888	2,507	3,093	3,606	4,813	4,833	5,288	5,458
Memo Items:											
Gross Domestic Saving	515	792	905	1,065	1,294	2,015	3,004	3,666	3,954	4,782	5,375
Gross National Saving	567	830	941	1,057	1,214	1,988	2,746	3,216	3,987	4,588	5,089
					(Millions of 1980 Singapore Dollars)						
Gross National Product	7,517	8,438	9,528	10,719	11,922	13,490	14,705	15,602	16,857	17,789	19,047
GDP at factor cost	6,691	7,640	8,655	9,807	11,030	12,498	13,892	14,847	15,428	16,524	17,820
Agriculture	232	248	258	265	292	314	297	275	282	309	314
Industry	2,211	2,630	3,130	3,790	4,431	5,204	5,696	5,940	6,090	6,785	7,229
Manufacturing	1,462	1,744	2,140	2,600	3,077	3,632	4,219	4,384	4,296	4,800	5,247
Services, etc.	4,796	5,389	5,977	6,560	7,217	8,008	9,052	9,846	10,327	10,797	11,737
Resource Balance	..	..	..	..	..	..	..	..	..	..	..
Exports of Goods & NFServices	..	..	..	..	..	..	..	..	..	..	..
Imports of Goods & NFServices	..	..	..	..	..	..	..	..	..	..	..
Domestic Absorption	..	..	..	..	..	..	..	..	..	..	..
Private Consumption, etc.	..	..	..	..	..	..	..	..	..	..	..
General Gov't Consumption	697	816	1,004	1,221	1,429	1,620	1,709	1,710	1,757	1,845	2,015
Gross Domestic Investment	2,068	2,569	3,222	4,638	5,437	5,937	6,371	7,662	7,063	7,498	7,222
Fixed Investment	1,623	2,153	2,778	3,698	4,605	5,281	5,728	6,414	6,171	6,470	6,591
Memo Items:											
Capacity to Import	..	..	..	..	..	..	..	..	..	..	..
Terms of Trade Adjustment	..	..	..	..	..	..	..	..	..	..	..
Gross Domestic Income	..	..	..	..	..	..	..	..	..	..	..
Gross National Income	..	..	..	..	..	..	..	..	..	..	..
DOMESTIC PRICES (DEFLATORS)					*(Index 1980 = 100)*						
Overall (GDP)	51.8	52.2	53.6	54.7	57.3	60.6	68.2	78.5	80.5	81.9	83.2
Domestic Absorption	..	..	..	..	..	..	..	..	..	..	..
Agriculture	45.9	49.0	49.8	50.7	54.3	51.1	72.2	83.6	90.3	83.0	90.1
Industry	42.7	43.4	43.9	45.5	48.3	53.0	59.7	71.7	75.3	75.4	76.3
Manufacturing	42.3	43.2	43.7	44.7	47.0	51.0	57.6	70.0	74.7	74.8	75.9
MANUFACTURING ACTIVITY											
Employment (1980=100)	20.8	26.6	35.8	43.1	50.4	60.3	70.0	72.1	67.3	72.7	76.6
Real Earnings per Empl. (1980=100)	61.3	59.1	67.2	69.8	74.0	74.9	74.0	72.7	83.4	87.4	89.9
Real Output per Empl. (1980=100)	78.2	75.3	84.3	74.2	69.9	62.2	66.3	85.9	80.0	89.9	94.3
Earnings as % of Value Added	34.6	33.7	36.2	36.3	36.7	35.4	33.8	30.0	34.6	33.1	32.9
MONETARY HOLDINGS					*(Millions of current Singapore Dollars)*						
Money Supply, Broadly Defined	2,380	2,967	3,566	4,125	4,599	5,803	6,797	7,819	9,390	10,944	12,306
Money as Means of Payment	985	1,196	1,417	1,631	1,760	2,385	2,632	2,858	3,472	4,000	4,412
Currency Ouside Banks	423	501	617	727	806	1,005	1,114	1,306	1,638	1,947	2,243
Demand Deposits	562	695	800	904	954	1,380	1,518	1,552	1,834	2,053	2,169
Quasi-Monetary Liabilities	1,395	1,771	2,149	2,494	2,839	3,418	4,165	4,961	5,918	6,944	7,894
GOVERNMENT DEFICIT (-) OR SURPLUS					*(Millions of current Singapore Dollars)*						
GOVERNMENT DEFICIT (-) OR SURPLUS	..	..	..	..	..	108	-12	197	121	31	164
Current Revenue	..	..	..	..	..	1,777	2,128	2,587	3,146	3,342	3,893
Current Expenditure	..	..	..	..	..	1,213	1,335	1,472	2,005	2,288	2,677
Current Budget Balance	..	..	..	..	..	564	793	1,115	1,141	1,054	1,216
Capital Receipts	..	..	..	..	..	43	156	142	245	155	97
Capital Payments	..	..	..	..	..	1,531	2,266	2,434	1,265	1,178	1,149

1978	1979	1980	1981	1982	1983	1984	1985	1986	1987 estimate	NOTES	SINGAPORE
3,310	3,880	4,550	5,450	6,160	6,930	7,730	7,620	7,450	7,940	..	CURRENT GNP PER CAPITA (US $)
2,353	2,384	2,415	2,443	2,472	2,502	2,529	2,558	2,585	2,610	..	POPULATION (thousands)
				(Millions of current Singapore Dollars)							ORIGIN AND USE OF RESOURCES
17,787	20,444	24,189	28,191	31,776	36,561	40,815	40,330	39,551	43,272	..	Gross National Product (GNP)
-43	-79	-902	-1,148	-894	-172	767	1,407	1,396	1,373	..	Net Factor Income from Abroad
17,830	20,523	25,091	29,339	32,670	36,733	40,048	38,924	38,155	41,899	..	Gross Domestic Product (GDP)
1,276	1,547	1,819	2,173	2,480	3,014	3,496	3,304	2,594	2,780	..	Indirect Taxes, net
16,555	18,976	23,272	27,167	30,190	33,719	36,552	35,619	35,561	39,119	B	GDP at factor cost
274	295	322	356	349	331	340	292	245	221	..	Agriculture
6,083	7,403	9,563	11,107	12,029	13,954	15,712	14,259	14,485	15,810	..	Industry
4,576	5,703	7,313	8,362	8,154	8,908	9,863	9,184	10,185	11,972	..	Manufacturing
11,473	12,825	15,206	17,876	20,292	22,448	23,996	24,372	23,426	25,867	..	Services, etc.
-898	-1,445	-2,216	-1,633	-1,441	-664	-1,113	-946	200	182	..	Resource Balance
..	..	..	..	..	..	..	..	..	..	..	Exports of Goods & NFServices
..	..	..	..	..	..	..	..	..	..	..	Imports of Goods & NFServices
18,728	21,968	27,307	30,973	34,111	37,397	41,161	39,869	37,955	41,717	..	Domestic Absorption
9,806	11,035	13,232	14,597	14,881	15,806	17,411	17,770	18,191	20,013	..	Private Consumption, etc.
1,965	2,034	2,447	2,789	3,570	3,995	4,333	5,548	5,198	5,181	..	General Gov't Consumption
6,957	8,900	11,628	13,587	15,659	17,596	19,417	16,551	14,566	16,523	..	Gross Domestic Investment
6,365	7,520	10,203	12,785	15,506	17,464	19,122	16,425	14,379	15,182	..	Fixed Investment
											Memo Items:
6,060	7,455	9,412	11,954	14,218	16,932	18,304	15,606	14,766	16,705	..	Gross Domestic Saving
5,936	7,309	8,288	10,499	12,909	16,323	18,615	16,561	15,789	17,704	..	Gross National Saving
				(Millions of 1980 Singapore Dollars)							
20,885	22,788	24,188	26,403	28,538	31,623	35,103	35,144	35,775	38,755	..	Gross National Product
19,336	21,146	23,272	25,578	27,231	29,375	31,806	31,480	32,532	35,437	B	GDP at factor cost
312	318	322	316	301	308	323	290	258	232	..	Agriculture
7,741	8,690	9,563	10,600	11,117	12,222	13,421	12,177	12,061	13,257	..	Industry
5,842	6,648	7,313	7,993	7,713	7,929	8,524	7,901	8,565	10,023	..	Manufacturing
12,884	13,869	15,206	16,578	17,966	19,251	20,670	21,411	22,156	24,015	..	Services, etc.
..	..	..	..	..	..	..	..	..	..	..	Resource Balance
..	..	..	..	..	..	..	..	..	..	..	Exports of Goods & NFServices
..	..	..	..	..	..	..	..	..	..	..	Imports of Goods & NFServices
..	..	..	..	..	..	..	..	..	..	..	Domestic Absorption
..	..	..	..	..	..	..	..	..	..	..	Private Consumption, etc.
2,247	2,236	2,447	2,576	2,917	3,198	3,365	4,189	4,184	4,165	..	General Gov't Consumption
8,398	9,983	11,628	12,454	14,301	15,920	17,429	15,222	13,791	15,248	..	Gross Domestic Investment
7,519	8,489	10,203	11,747	14,127	15,651	17,127	15,062	13,402	13,891	..	Fixed Investment
											Memo Items:
..	..	..	..	..	..	..	..	..	..	..	Capacity to Import
..	..	..	..	..	..	..	..	..	..	..	Terms of Trade Adjustment
..	..	..	..	..	..	..	..	..	..	..	Gross Domestic Income
..	..	..	..	..	..	..	..	..	..	..	Gross National Income
				(Index 1980 = 100)							DOMESTIC PRICES (DEFLATORS)
85.2	89.7	100.0	106.7	111.2	115.6	116.4	114.9	110.7	111.7	..	Overall (GDP)
..	..	..	..	..	..	..	..	..	..	..	Domestic Absorption
87.8	92.8	100.0	112.6	116.2	107.4	105.0	101.0	94.7	95.5	..	Agriculture
78.6	85.2	100.0	104.8	108.2	114.2	117.1	117.1	120.1	119.3	..	Industry
78.3	85.8	100.0	104.6	105.7	112.3	115.7	116.2	118.9	119.4	..	Manufacturing
											MANUFACTURING ACTIVITY
84.9	94.5	100.0	98.9	96.6	95.6	96.6	89.2	..	..	..	Employment (1980=100)
90.5	94.8	100.0	108.6	119.1	129.9	141.5	152.1	..	..	..	Real Earnings per Empl. (1980=100)
92.1	98.9	100.0	110.9	108.7	105.5	113.7	115.1	..	..	..	Real Output per Empl. (1980=100)
33.4	30.9	29.7	30.3	35.0	36.3	36.3	37.7	..	..	..	Earnings as % of Value Added
				(Millions of current Singapore Dollars)							MONETARY HOLDINGS
14,014	16,837	20,489	25,198	30,908	35,378	38,948	41,407	35,315	52,492	D	Money Supply, Broadly Defined
4,926	5,706	6,135	7,242	8,157	8,607	8,866	8,785	9,822	11,031	..	Money as Means of Payment
2,583	2,941	3,137	3,382	3,996	4,335	4,619	4,739	5,034	5,440	..	Currency Ouside Banks
2,343	2,765	2,998	3,860	4,161	4,272	4,247	4,046	4,788	5,591	..	Demand Deposits
9,088	11,131	14,354	17,956	22,751	26,771	30,082	32,622	25,493	41,461	..	Quasi-Monetary Liabilities
				(Millions of current Singapore Dollars)							GOVERNMENT DEFICIT (-) OR SURPLUS
146	468	538	213	1,098	661	1,643	816	..	..	C	GOVERNMENT DEFICIT (-) OR SURPLUS
4,195	4,887	6,365	7,797	9,054	10,935	11,451	10,768	..	..	..	Current Revenue
2,797	3,191	3,910	5,377	5,452	6,267	7,802	7,316	..	..	..	Current Expenditure
1,398	1,696	2,455	2,420	3,602	4,668	3,649	3,452	..	..	..	Current Budget Balance
52	294	255	899	1,032	782	246	3,996	..	..	..	Capital Receipts
1,304	1,522	2,172	3,106	3,536	4,789	2,252	6,632	..	..	..	Capital Payments

SINGAPORE	1967	1968	1969	1970	1971	1972	1973	1974	1975	1976	1977
FOREIGN TRADE (CUSTOMS BASIS)				*(Millions of current US dollars)*							
Value of Exports, fob	1,140	1,271	1,549	1,554	1,755	2,181	3,610	5,785	5,377	6,586	8,242
Nonfuel Primary Products	518	582	769	720	661	681	1,220	1,537	1,238	1,610	2,135
Fuels	293	347	378	360	451	536	716	1,857	1,809	1,955	2,481
Manufactures	329	342	402	474	642	964	1,673	2,391	2,331	3,020	3,625
Value of Imports, cif	1,439	1,661	2,040	2,461	2,827	3,383	5,070	8,344	8,135	9,070	10,472
Nonfuel Primary Products	529	544	667	696	708	790	1,272	1,593	1,436	1,809	2,194
Fuels	241	286	321	332	405	491	656	2,003	1,998	2,486	2,677
Manufactures	669	831	1,051	1,433	1,714	2,102	3,142	4,748	4,700	4,774	5,601
					(Index 1980 = 100)						
Terms of Trade	77.2	75.3	77.7	72.7	72.4	73.8	84.6	101.2	94.0	101.2	96.7
Export Prices, fob	11.7	11.3	12.6	12.9	14.5	16.2	26.1	51.9	49.1	54.4	57.6
Nonfuel Primary Products	29.3	28.7	34.3	31.8	29.7	30.5	50.9	59.7	50.9	58.6	66.2
Fuels	4.3	4.3	4.3	4.3	5.6	6.2	8.9	36.7	35.7	38.4	42.0
Manufactures	31.1	31.1	33.1	35.1	37.2	36.5	50.0	68.2	67.6	70.9	70.3
Import Prices, cif	15.2	15.0	16.2	17.7	20.1	21.9	30.8	51.3	52.3	53.8	59.6
Trade at Constant 1980 Prices				*(Millions of 1980 US dollars)*							
Exports, fob	9,707	11,280	12,329	12,067	12,061	13,481	13,838	11,137	10,942	12,104	14,299
Imports, cif	9,459	11,096	12,615	13,905	14,080	15,426	16,442	16,256	15,563	16,872	17,572
BALANCE OF PAYMENTS				*(Millions of current US dollars)*							
Exports of Goods & Services	..	..	..	2,003	2,267	3,243	5,143	8,062	8,276	9,387	11,231
Merchandise, fob	..	..	..	1,447	1,665	2,040	3,426	5,547	5,110	6,218	7,745
Nonfactor Services	..	..	..	489	519	1,090	1,550	2,289	2,787	2,800	3,094
Factor Services	..	..	..	67	83	113	167	225	380	370	391
Imports of Goods & Services	..	..	..	2,567	2,979	3,741	5,658	9,044	8,822	9,905	11,482
Merchandise, fob	..	..	..	2,302	2,655	3,127	4,735	7,764	7,511	8,442	9,729
Nonfactor Services	..	..	..	236	294	484	655	900	966	1,054	1,277
Factor Services	..	..	..	29	30	129	268	380	345	410	475
Long-Term Interest	..	..	..	23	23	25	33	44	48	53	67
Current Transfers, net	..	..	..	-21	-23	-2	-14	-41	-38	-46	-41
Workers' Remittances	..	..	..	0	0	0	0	0	0	0	0
Total to be Financed	..	..	..	-585	-735	-500	-529	-1,023	-584	-564	-291
Official Capital Grants	..	..	..	13	11	5	10	2	0	-3	-4
Current Account Balance	..	..	..	-572	-724	-495	-519	-1,021	-584	-567	-295
Long-Term Capital, net	..	..	..	153	170	218	443	355	272	302	401
Direct Investment	..	..	..	93	116	141	327	280	254	186	206
Long-Term Loans	..	..	..	60	40	74	115	140	86	134	290
Disbursements	..	..	..	114	100	145	177	204	167	229	381
Repayments	..	..	..	55	59	72	61	64	81	95	91
Other Long-Term Capital	..	..	..	-13	2	-1	-9	-66	-67	-15	-91
Other Capital, net	..	..	..	603	874	612	489	961	719	564	207
Change in Reserves	..	..	..	-184	-320	-335	-413	-295	-407	-298	-313
Memo Item:				*(Singapore Dollars per US dollar)*							
Conversion Factor (Annual Avg)	3.060	3.060	3.060	3.060	3.050	2.810	2.460	2.440	2.370	2.470	2.440
EXTERNAL DEBT, ETC.				*(Millions of US dollars, outstanding at end of year)*							
Public/Publicly Guar. Long-Term	..	..	..	151.9	209.0	299.1	422.4	513.2	549.9	705.8	998.2
Official Creditors	..	..	..	139.2	181.1	213.3	285.5	337.0	353.5	386.3	408.2
IBRD and IDA	..	..	..	69.0	81.7	91.4	98.3	98.6	102.2	103.5	104.2
Private Creditors	..	..	..	12.8	27.9	85.8	136.9	176.2	196.4	319.4	590.0
Private Non-guaranteed Long-Term	..	..	..	247.7	236.1	228.6	220.0	275.0	300.0	270.0	340.0
Use of Fund Credit	..	..	..	0.0	0.0	0.0	0.0	0.0	0.0	0.0	0.0
Short-Term Debt	..	..	..	..	..	..	..	..	..	..	55.0
Memo Items:				*(Millions of US dollars)*							
Int'l Reserves Excluding Gold	496.0	712.0	827.0	1,012.0	1,452.0	1,748.0	2,286.0	2,812.0	3,007.0	3,364.0	3,858.0
Gold Holdings (at market price)	..	..	..	..	..	..	..	..	..	..	..
SOCIAL INDICATORS											
Total Fertility Rate	3.9	3.6	3.2	3.1	3.0	3.0	2.8	2.4	2.1	2.1	1.8
Crude Birth Rate	26.8	24.5	22.8	23.0	22.7	23.4	22.2	19.6	17.7	18.7	16.5
Infant Mortality Rate	24.6	22.9	20.7	19.7	19.7	19.0	20.1	16.7	13.9	11.6	12.4
Life Expectancy at Birth	66.5	66.9	67.3	67.7	68.1	68.5	68.9	69.3	69.7	70.1	70.5
Food Production, p.c. ('79-81 = 100)	..	..	..	120.8	137.5	126.3	127.0	119.1	119.5	107.4	135.3
Labor Force, Agriculture (%)	4.7	4.3	3.8	3.4	3.2	3.0	2.9	2.7	2.5	2.3	2.1
Labor Force, Female (%)	23.7	24.4	25.1	25.9	26.8	27.8	28.8	29.7	30.7	31.5	32.2
School Enroll. Ratio, primary	..	..	..	105.0	..	..	..	..	110.0	111.0	110.0
School Enroll. Ratio, secondary	..	..	..	46.0	..	..	..	..	52.0	54.0	55.0

1978	1979	1980	1981	1982	1983	1984	1985	1986	1987 estimate	NOTES	SINGAPORE
											FOREIGN TRADE (CUSTOMS BASIS)
			(Millions of current US dollars)								
10,134	14,233	19,376	20,968	20,788	21,833	24,055	22,815	22,428	28,592	..	Value of Exports, fob
2,588	3,452	4,041	3,530	3,229	3,330	4,102	3,343	3,169	3,591	..	Nonfuel Primary Products
2,867	3,410	4,882	5,726	5,724	6,115	6,162	6,156	4,587	4,523	..	Fuels
4,679	7,372	10,452	11,712	11,834	12,388	13,791	13,317	14,672	20,477	..	Manufactures
13,049	17,638	24,003	27,572	28,167	28,158	28,656	26,250	25,461	32,480	..	Value of Imports, cif
2,469	3,258	3,817	3,565	3,451	3,664	4,290	3,582	3,543	4,108	..	Nonfuel Primary Products
3,122	4,449	6,882	9,295	9,469	8,804	7,964	7,741	5,047	5,956	..	Fuels
7,458	9,931	13,303	14,712	15,248	15,691	16,402	14,927	16,871	22,416	..	Manufactures
			(Index 1980 = 100)								
95.8	99.0	100.0	102.8	101.3	101.6	102.3	101.3	99.4	102.5	..	Terms of Trade
62.9	80.7	100.0	105.4	98.0	95.4	94.4	91.5	82.0	99.1	..	Export Prices, fob
75.8	94.4	100.0	85.9	75.6	83.6	87.3	77.0	70.6	74.6	..	Nonfuel Primary Products
42.3	61.0	100.0	112.5	101.6	92.5	90.2	87.5	44.9	56.9	..	Fuels
79.1	87.8	100.0	109.5	104.7	100.7	98.9	98.1	115.9	127.3	..	Manufactures
65.7	81.5	100.0	102.5	96.8	93.8	92.3	90.3	82.5	96.7	..	Import Prices, cif
			(Millions of 1980 US dollars)								Trade at Constant 1980 Prices
16,110	17,640	19,376	19,896	21,207	22,896	25,480	24,947	27,351	28,854	..	Exports, fob
19,869	21,644	24,003	26,904	29,110	30,013	31,042	29,078	30,879	33,592	..	Imports, cif
			(Millions of current US dollars)								**BALANCE OF PAYMENTS**
13,778	18,649	25,239	29,366	30,386	31,029	31,847	29,735	29,627	37,248	..	Exports of Goods & Services
9,587	13,400	18,200	19,662	19,435	20,429	22,662	21,537	21,279	27,277	..	Merchandise, fob
3,663	4,443	6,085	8,612	9,709	9,299	7,581	6,384	6,258	7,554	..	Nonfactor Services
528	806	953	1,092	1,243	1,300	1,603	1,814	2,090	2,417	..	Factor Services
14,192	19,350	26,695	30,683	31,477	31,425	32,009	29,525	28,895	36,509	..	Imports of Goods & Services
12,090	16,450	22,400	25,785	26,196	26,252	26,734	24,366	23,338	29,817	..	Merchandise, fob
1,547	2,050	2,912	3,254	3,613	3,782	4,024	3,976	4,098	4,919	..	Nonfactor Services
554	849	1,382	1,644	1,668	1,390	1,252	1,183	1,459	1,773	..	Factor Services
97	105	143	184	191	202	218	252	290	305	..	Long-Term Interest
-36	-31	-104	-145	-194	-207	-214	-205	-171	-178	..	Current Transfers, net
0	0	0	0	0	0	0	0	0	0	..	Workers' Remittances
-449	-732	-1,560	-1,462	-1,285	-602	-376	5	560	561	K	Total to be Financed
-3	-4	-3	-8	-11	-8	-9	-8	-19	-22	K	Official Capital Grants
-453	-736	-1,563	-1,470	-1,296	-610	-385	-4	541	539	..	Current Account Balance
296	813	1,460	1,674	1,823	776	764	1,010	320	701	..	Long-Term Capital, net
186	669	1,138	1,645	1,298	1,085	1,210	809	477	982	..	Direct Investment
-150	320	331	212	383	226	599	-81	320	191	..	Long-Term Loans
316	483	685	429	644	712	997	712	808	763	..	Disbursements
466	163	354	217	261	486	398	793	488	572	..	Repayments
262	-172	-6	-176	153	-527	-1,035	289	-458	-450	..	Other Long-Term Capital
821	439	765	705	651	894	1,144	331	-323	-145	..	Other Capital, net
-665	-516	-663	-909	-1,177	-1,059	-1,524	-1,337	-538	-1,095	..	Change in Reserves
			(Singapore Dollars per US dollar)								**Memo Item:**
2.270	2.170	2.140	2.110	2.140	2.110	2.130	2.200	2.180	2.110	..	Conversion Factor (Annual Avg)
			(Millions of US dollars, outstanding at end of year)								**EXTERNAL DEBT, ETC.**
1,133.5	1,323.0	1,320.0	1,318.2	1,460.1	1,495.4	1,900.0	1,798.5	2,137.9	2,542.9	..	Public/Publicly Guar. Long-Term
476.2	527.1	561.8	513.1	472.6	425.6	446.0	392.5	338.5	312.8	..	Official Creditors
121.0	118.1	116.7	109.3	99.0	88.6	73.3	75.5	74.2	66.2	..	IBRD and IDA
657.3	795.8	758.2	805.1	987.5	1,069.9	1,454.0	1,406.0	1,799.4	2,230.2	..	Private Creditors
150.0	250.0	550.0	700.0	900.0	1,050.0	1,182.0	1,345.0	1,469.0	1,643.0	..	Private Non-guaranteed Long-Term
0.0	0.0	0.0	0.0	0.0	0.0	0.0	0.0	0.0	..	..	Use of Fund Credit
93.0	157.0	201.0	245.0	269.0	258.0	208.0	262.0	268.0	305.0	..	Short-Term Debt
			(Millions of US dollars)								**Memo Items:**
5,303.0	5,818.0	6,567.0	7,549.0	8,480.0	9,264.0	10,416.0	12,847.0	12,939.0	15,227.0	..	Int'l Reserves Excluding Gold
..	..	..	..	..	..	..	..	..	..	..	Gold Holdings (at market price)
											SOCIAL INDICATORS
1.8	1.8	1.7	1.7	1.7	1.7	1.7	1.7	1.7	1.6	..	Total Fertility Rate
16.8	17.1	17.1	17.0	16.9	16.7	16.6	16.5	16.3	16.2	..	Crude Birth Rate
12.6	13.2	11.7	10.8	11.9	13.2	14.6	16.2	17.9	19.8	..	Infant Mortality Rate
70.9	71.2	71.5	71.8	72.1	72.2	72.4	72.5	72.7	72.8	..	Life Expectancy at Birth
119.7	100.9	92.3	106.8	80.7	82.7	104.8	93.2	93.3	94.8	..	Food Production, p.c. ('79-81 = 100)
2.0	1.8	1.6	..	..	..	..	..	..	..	..	Labor Force, Agriculture (%)
33.0	33.7	34.5	34.3	34.1	33.8	33.6	33.4	33.1	32.9	..	Labor Force, Female (%)
108.0	106.0	108.0	106.0	109.0	113.0	115.0	..	..	..	..	School Enroll. Ratio, primary
57.0	59.0	58.0	62.0	65.0	69.0	71.0	..	..	..	..	School Enroll. Ratio, secondary

SOLOMON ISLANDS	1967	1968	1969	1970	1971	1972	1973	1974	1975	1976	1977
CURRENT GNP PER CAPITA (US $)	..	..	..	..	..	..	..	270	260	300	320
POPULATION (thousands)	151	155	159	163	168	174	180	186	193	200	206

ORIGIN AND USE OF RESOURCES

(Millions of current Solomon Islands Dollars)

	1967	1968	1969	1970	1971	1972	1973	1974	1975	1976	1977
Gross National Product (GNP)	..	..	..	..	..	23.2	26.7	39.9	37.3	44.5	54.8
Net Factor Income from Abroad	0.0	0.0	0.0	..	..	-0.2	-0.3	-0.5	-2.0	-2.0	-3.0
Gross Domestic Product (GDP)	22.6	25.2	25.7	28.6	30.5	23.4	27.0	40.4	39.3	46.5	57.8
Indirect Taxes, net	..	..	..	2.2	2.3	2.1	2.4	4.1	3.4	4.0	5.4
GDP at factor cost	..	..	..	26.4	28.2	21.3	24.6	36.3	35.8	42.5	52.4
Agriculture	..	..	..	..	..	..	..	..	..	..	..
Industry	..	..	..	..	..	..	..	..	..	..	..
Manufacturing	..	..	..	..	..	..	..	..	..	..	..
Services, etc.	..	..	..	..	..	..	..	..	..	..	..
Resource Balance	..	..	..	..	..	..	..	..	..	..	..
Exports of Goods & NFServices	..	..	..	..	..	..	..	..	..	..	..
Imports of Goods & NFServices	..	..	..	..	..	..	..	..	..	..	..
Domestic Absorption	..	..	..	..	..	..	..	..	..	..	..
Private Consumption, etc.	..	..	..	..	..	..	..	..	..	..	..
General Gov't Consumption	..	..	..	..	..	..	..	..	..	..	..
Gross Domestic Investment	..	..	..	..	..	..	..	..	..	..	..
Fixed Investment	..	..	..	..	..	..	..	..	..	..	..
Memo Items:											
Gross Domestic Saving	..	..	..	1.2	1.9	1.6	0.4	2.1	-5.9	1.7	12.5
Gross National Saving	..	..	..	..	..	..	..	..	-6.5	1.1	11.0

(Millions of 1980 Solomon Islands Dollars)

	1967	1968	1969	1970	1971	1972	1973	1974	1975	1976	1977
Gross National Product	59.4	62.2	60.4	..	..	52.9	58.8	74.2	63.0	72.1	81.8
GDP at factor cost	..	..	..	..	..	..	..	..	..	..	..
Agriculture	..	..	..	..	..	..	..	..	..	..	..
Industry	..	..	..	..	..	..	..	..	..	..	..
Manufacturing	..	..	..	..	..	..	..	..	..	..	..
Services, etc.	..	..	..	..	..	..	..	..	..	..	..
Resource Balance	..	..	..	..	..	..	..	..	..	..	..
Exports of Goods & NFServices	..	..	..	..	..	..	..	..	..	..	..
Imports of Goods & NFServices	..	..	..	..	..	..	..	..	..	..	..
Domestic Absorption	..	..	..	..	..	..	..	..	..	..	..
Private Consumption, etc.	..	..	..	..	..	..	..	..	..	..	..
General Gov't Consumption	..	..	..	..	..	..	..	..	..	..	..
Gross Domestic Investment	..	..	..	..	..	..	..	..	..	..	..
Fixed Investment	..	..	..	..	..	..	..	..	..	..	..
Memo Items:											
Capacity to Import	..	..	..	..	..	..	..	..	..	..	..
Terms of Trade Adjustment	..	..	..	..	..	..	..	..	..	..	..
Gross Domestic Income	..	..	..	..	..	..	..	..	..	..	..
Gross National Income	..	..	..	..	..	..	..	..	..	..	..

DOMESTIC PRICES (DEFLATORS)

(Index 1980 = 100)

	1967	1968	1969	1970	1971	1972	1973	1974	1975	1976	1977
Overall (GDP)	38.0	40.5	42.5	45.0	46.9	50.2	51.9	61.5	67.9	70.6	76.7
Domestic Absorption	..	..	..	..	..	..	..	..	..	..	..
Agriculture	..	..	..	..	..	..	..	..	..	..	..
Industry	..	..	..	..	..	..	..	..	..	..	..
Manufacturing	..	..	..	..	..	..	..	..	..	..	..

MANUFACTURING ACTIVITY

	1967	1968	1969	1970	1971	1972	1973	1974	1975	1976	1977
Employment (1980=100)	..	..	..	..	..	..	..	..	..	..	..
Real Earnings per Empl. (1980=100)	..	..	..	..	..	..	..	..	..	..	..
Real Output per Empl. (1980=100)	..	..	..	..	..	..	..	..	..	..	..
Earnings as % of Value Added	..	..	..	..	..	..	..	..	..	..	..

MONETARY HOLDINGS

(Millions of current Solomon Islands Dollars)

	1967	1968	1969	1970	1971	1972	1973	1974	1975	1976	1977
Money Supply, Broadly Defined	..	..	..	..	..	..	..	..	..	..	..
Money as Means of Payment	..	..	..	..	..	..	..	..	..	..	..
Currency Ouside Banks	..	..	..	..	..	..	..	..	..	..	..
Demand Deposits	..	..	..	..	..	..	..	..	..	..	..
Quasi-Monetary Liabilities	..	..	..	..	..	..	..	..	..	..	..

(Thousands of current Solomon Islands Dollars)

	1967	1968	1969	1970	1971	1972	1973	1974	1975	1976	1977
GOVERNMENT DEFICIT (-) OR SURPLUS	..	..	..	..	..	..	..	..	40	30	-600
Current Revenue	..	..	..	..	..	..	..	..	12,970	16,770	19,800
Current Expenditure	..	..	..	..	..	..	..	..	9,500	10,900	14,100
Current Budget Balance	..	..	..	..	..	..	..	..	3,470	5,870	5,700
Capital Receipts	..	..	..	..	..	..	..	..	30	30	100
Capital Payments	..	..	..	..	..	..	..	..	3,500	5,900	6,500

1978	1979	1980	1981	1982	1983	1984	1985	1986	1987 estimate	NOTES	SOLOMON ISLANDS
350	450	440	550	570	570	560	510	530	420	..	**CURRENT GNP PER CAPITA (US $)**
214	221	228	236	245	254	263	273	283	293	..	**POPULATION (thousands)**
				(Millions of current Solomon Islands Dollars)							**ORIGIN AND USE OF RESOURCES**
64.0	86.2	96.0	112.4	124.4	141.4	193.5	193.1	202.2	220.9	..	Gross National Product (GNP)
-3.0	-4.0	-0.4	0.0	-1.7	-1.1	-0.5	-0.8	-4.9	..	..	Net Factor Income from Abroad
67.0	90.2	96.4	112.4	126.1	142.5	194.0	193.9	207.1	232.7	..	Gross Domestic Product (GDP)
6.2	10.5	10.8	14.4	18.0	18.0	27.3	27.9	27.7	..	..	Indirect Taxes, net
60.8	79.7	85.6	97.9	108.1	124.5	166.7	166.0	179.4	..	..	GDP at factor cost
..	..	..	..	..	..	..	..	..	..	..	Agriculture
..	..	..	..	..	..	..	..	..	..	..	Industry
..	..	..	..	..	..	..	..	..	..	..	Manufacturing
..	..	..	..	..	..	..	..	..	..	..	Services, etc.
..	..	-28.6	-36.2	-18.2	-30.7	-13.0	-42.7	-69.4	..	..	Resource Balance
..	..	67.7	65.3	72.0	86.7	133.7	124.2	146.3	..	..	Exports of Goods & NFServices
..	..	96.3	101.5	90.2	117.4	146.7	166.9	215.7	..	..	Imports of Goods & NFServices
..	..	124.9	148.6	144.3	173.1	207.0	236.7	276.6	..	..	Domestic Absorption
..	..	67.6	84.0	77.2	86.6	115.7	126.1	145.1	..	..	Private Consumption, etc.
..	..	22.2	26.5	29.3	34.1	40.8	47.7	56.3	..	..	General Gov't Consumption
..	..	35.1	38.1	37.8	52.4	50.5	62.9	75.2	..	..	Gross Domestic Investment
..	..	..	..	..	..	..	..	..	..	..	Fixed Investment
											Memo Items:
11.3	31.9	6.6	1.9	19.6	21.8	37.5	20.1	5.7	..	..	Gross Domestic Saving
10.1	29.9	6.8	-3.1	13.5	17.7	34.8	17.3	49.0	..	..	Gross National Saving
				(Millions of 1980 Solomon Islands Dollars)							
89.7	112.1	96.0	109.3	111.1	127.3	129.3	130.3	154.2	146.5	..	Gross National Product
..	..	..	..	..	..	..	..	..	..	..	GDP at factor cost
..	..	..	..	..	..	..	..	..	..	..	Agriculture
..	..	..	..	..	..	..	..	..	..	..	Industry
..	..	..	..	..	..	..	..	..	..	..	Manufacturing
..	..	..	..	..	..	..	..	..	..	..	Services, etc.
..	..	-28.6	-20.2	0.6	3.0	-5.7	-10.0	12.9	..	..	Resource Balance
..	..	67.7	77.0	80.2	95.2	101.4	95.0	119.4	..	..	Exports of Goods & NFServices
..	..	96.3	97.2	79.6	92.2	107.1	105.0	106.5	..	..	Imports of Goods & NFServices
..	..	125.0	129.5	111.9	125.1	135.4	140.8	143.8	..	..	Domestic Absorption
..	..	67.7	72.1	58.7	65.6	74.6	74.2	75.2	..	..	Private Consumption, etc.
..	..	22.2	22.8	22.3	24.3	26.2	28.2	29.8	..	..	General Gov't Consumption
..	..	35.1	34.6	30.9	35.2	34.6	38.4	38.8	..	..	Gross Domestic Investment
..	..	..	..	..	..	..	..	..	..	..	Fixed Investment
											Memo Items:
..	..	67.7	62.5	63.5	68.1	97.6	78.1	72.2	..	..	Capacity to Import
..	..	0.0	-14.5	-16.7	-27.1	-3.8	-16.9	-47.2	..	..	Terms of Trade Adjustment
..	..	96.4	94.8	95.8	101.0	125.9	113.9	109.5	..	..	Gross Domestic Income
..	..	96.0	94.8	94.4	100.2	125.5	113.4	107.0	..	..	Gross National Income
				(Index 1980 = 100)							**DOMESTIC PRICES (DEFLATORS)**
81.6	88.0	100.0	102.8	112.1	111.2	149.6	148.2	132.2	156.3	..	Overall (GDP)
..	..	99.9	114.7	129.0	138.4	152.9	168.1	192.4	..	..	Domestic Absorption
..	..	..	..	..	..	..	..	..	..	..	Agriculture
..	..	..	..	..	..	..	..	..	..	..	Industry
..	..	..	..	..	..	..	..	..	..	..	Manufacturing
											MANUFACTURING ACTIVITY
55.6	80.6	100.0	78.7	82.9	83.3	79.1		..	..	..	Employment (1980 = 100)
..	..	..	..	..	..	..	..	..	..	..	Real Earnings per Empl. (1980 = 100)
..	..	..	..	..	..	..	..	..	..	..	Real Output per Empl. (1980 = 100)
..	..	..	..	..	..	..	..	..	..	..	Earnings as % of Value Added
				(Millions of current Solomon Islands Dollars)							**MONETARY HOLDINGS**
27.3	39.6	39.0	32.2	39.8	47.9	64.6	66.4	72.7	97.6	..	Money Supply, Broadly Defined
7.8	11.0	15.1	14.2	15.8	18.4	28.3	28.3	30.4	37.2	..	Money as Means of Payment
3.9	4.8	5.7	6.4	7.2	9.3	12.8	13.9	13.6	16.8	..	Currency Ouside Banks
4.0	6.2	9.5	7.8	8.6	9.1	15.6	14.5	16.8	20.4	..	Demand Deposits
19.5	28.6	23.9	18.0	24.0	29.5	36.3	38.0	42.3	60.4	..	Quasi-Monetary Liabilities
				(Thousands of current Solomon Islands Dollars)							
700	-2,900	-3,600	-9,200	-10,400	-13,400	-6,500	-19,300	-15,100	-34,800	..	**GOVERNMENT DEFICIT (-) OR SURPLUS**
28,030	31,100	36,200	37,100	39,100	40,300	51,900	54,500	82,500	86,700	..	Current Revenue
16,700	20,100	24,300	30,800	34,500	39,200	46,100	59,500	65,900	66,200	..	Current Expenditure
11,330	11,000	11,900	6,300	4,600	1,100	5,800	-5,000	16,600	20,500	..	Current Budget Balance
70	1,200	600	200	200	100	500	700	..	..	..	Capital Receipts
10,700	15,200	16,100	15,700	15,200	14,600	12,800	15,100	31,600	55,300	..	Capital Payments

SOLOMON ISLANDS	1967	1968	1969	1970	1971	1972	1973	1974	1975	1976	1977
FOREIGN TRADE (CUSTOMS BASIS)				*(Thousands of current US dollars)*							
Value of Exports, fob	..	..	..	..	..	..	..	..	..	..	..
Nonfuel Primary Products	..	..	..	..	..	..	..	..	..	..	..
Fuels	..	..	..	..	..	..	..	..	..	..	..
Manufactures	..	..	..	..	..	..	..	..	..	..	..
Value of Imports, cif	11,000	12,000	11,000	13,000	15,000	17,000	18,000	27,000	33,000	30,000	33,000
Nonfuel Primary Products	..	..	..	..	..	..	..	..	..	..	..
Fuels	..	..	..	..	..	..	..	..	..	..	..
Manufactures	..	..	..	..	..	..	..	..	..	..	..
					(Index 1980 = 100)						
Terms of Trade	..	..	..	..	..	..	..	..	..	..	..
Export Prices, fob	..	..	..	..	..	..	..	..	..	..	..
Nonfuel Primary Products	..	..	..	..	..	..	..	..	..	..	..
Fuels	..	..	..	..	..	..	..	..	..	..	..
Manufactures	..	..	..	..	..	..	..	..	..	..	..
Import Prices, cif	..	..	..	..	..	..	..	..	..	..	..
Trade at Constant 1980 Prices					*(Thousands of 1980 US dollars)*						
Exports, fob	..	..	..	..	..	..	..	..	..	..	..
Imports, cif	..	..	..	..	..	..	..	..	..	..	..
BALANCE OF PAYMENTS					*(Millions of current US dollars)*						
Exports of Goods & Services	..	..	..	..	..	..	..	..	18	27	36
Merchandise, fob	..	..	..	..	..	..	..	..	15	24	33
Nonfactor Services	..	..	..	..	..	..	..	..	2	3	3
Factor Services	..	..	..	..	..	..	..	..	0	0	0
Imports of Goods & Services	..	..	..	..	..	..	..	..	40	39	43
Merchandise, fob	..	..	..	..	..	..	..	..	29	26	29
Nonfactor Services	..	..	..	..	..	..	..	..	10	9	11
Factor Services	..	..	..	..	..	..	..	..	2	4	4
Long-Term Interest	..	..	..	..	..	..	..	..	..	..	..
Current Transfers, net	..	..	..	..	..	..	..	..	2	2	2
Workers' Remittances	..	..	..	..	..	..	..	..	0	0	0
Total to be Financed	..	..	..	..	..	..	..	..	-21	-10	-6
Official Capital Grants	..	..	..	..	..	..	..	..	8	11	11
Current Account Balance	..	..	..	..	..	..	..	..	-13	2	6
Long-Term Capital, net	..	..	..	..	..	..	..	..	16	16	19
Direct Investment	..	..	..	..	..	..	..	..	8	5	4
Long-Term Loans	..	..	..	..	..	..	..	..	..	..	..
Disbursements	..	..	..	..	..	..	..	..	..	..	..
Repayments	..	..	..	..	..	..	..	..	..	..	..
Other Long-Term Capital	..	..	..	..	..	..	..	..	0	0	3
Other Capital, net	..	..	..	..	..	..	..	..	-2	-16	-18
Change in Reserves	..	..	..	0	0	0	0	0	-1	-2	-7
Memo Item:					*(Solomon Islands Dollars per US dollar)*						
Conversion Factor (Annual Avg)	0.890	0.890	0.890	0.890	0.880	0.840	0.700	0.700	0.760	0.820	0.900
EXTERNAL DEBT, ETC.				*(Thousands of US dollars, outstanding at end of year)*							
Public/Publicly Guar. Long-Term	..	..	..	..	..	..	..	..	..	..	..
Official Creditors	..	..	..	..	..	..	..	..	..	..	..
IBRD and IDA	..	..	..	..	..	..	..	..	..	..	..
Private Creditors	..	..	..	..	..	..	..	..	..	..	..
Private Non-guaranteed Long-Term	..	..	..	..	..	..	..	..	..	..	..
Use of Fund Credit	..	..	..	0	0	0	0	0	0	0	0
Short-Term Debt	..	..	..	..	..	..	..	..	..	..	..
Memo Items:					*(Thousands of US dollars)*						
Int'l Reserves Excluding Gold	..	..	..	..	..	..	..	..	..	..	2,871
Gold Holdings (at market price)	..	..	..	..	..	..	..	..	..	..	..
SOCIAL INDICATORS											
Total Fertility Rate	..	..	..	6.4	6.5	6.6	6.8	6.9	7.0	7.1	7.3
Crude Birth Rate	..	..	..	..	..	..	..	..	..	..	45.0
Infant Mortality Rate	..	..	..	..	..	..	..	..	..	..	46.0
Life Expectancy at Birth	..	..	..	40.3	41.5	42.7	43.9	45.2	46.4	47.6	48.8
Food Production, p.c. ('79-81 = 100)	82.5	77.0	79.0	81.2	79.7	77.1	76.5	73.9	73.8	81.3	85.9
Labor Force, Agriculture (%)	..	..	..	..	..	..	..	..	..	..	..
Labor Force, Female (%)	..	..	..	..	..	..	..	..	..	..	..
School Enroll. Ratio, primary	..	..	..	..	..	..	..	..	..	..	..
School Enroll. Ratio, secondary	..	..	..	..	..	..	..	..	..	..	..

1978	1979	1980	1981	1982	1983	1984	1985	1986	1987 estimate	NOTES	SOLOMON ISLANDS
				(Thousands of current US dollars)							**FOREIGN TRADE (CUSTOMS BASIS)**
..	..	74,000	66,190	58,280	61,990	93,100	70,100	66,000	72,000	..	Value of Exports, fob
..	..	..	..	..	..	..	..	..	..	..	Nonfuel Primary Products
..	..	..	..	..	..	..	..	..	..	..	Fuels
..	..	..	..	..	..	..	..	..	..	..	Manufactures
42,000	70,000	89,000	79,000	71,000	74,000	79,000	83,000	80,000	..	..	Value of Imports, cif
..	..	..	..	..	..	..	..	..	..	..	Nonfuel Primary Products
..	..	..	..	..	..	..	..	..	..	..	Fuels
..	..	..	..	..	..	..	..	..	..	..	Manufactures
				(Index 1980 = 100)							
..	..	..	..	..	..	..	..	..	..	..	Terms of Trade
..	..	..	..	..	..	..	..	..	..	..	Export Prices, fob
..	..	..	..	..	..	..	..	..	..	..	Nonfuel Primary Products
..	..	..	..	..	..	..	..	..	..	..	Fuels
..	..	..	..	..	..	..	..	..	..	..	Manufactures
..	..	..	..	..	..	..	..	..	..	..	Import Prices, cif
				(Thousands of 1980 US dollars)							Trade at Constant 1980 Prices
..	..	..	..	..	..	..	..	..	..	..	Exports, fob
..	..	..	..	..	..	..	..	..	..	..	Imports, cif
				(Millions of current US dollars)							**BALANCE OF PAYMENTS**
40	74	85	79	77	85	109	88	..	..	..	Exports of Goods & Services
35	68	73	66	58	62	92	70	66	72	..	Merchandise, fob
5	6	12	13	14	19	13	13	..	..	..	Nonfactor Services
0	0	0	0	5	3	5	5	..	..	..	Factor Services
56	84	117	117	94	103	117	117	..	..	..	Imports of Goods & Services
35	58	74	76	59	61	66	69	67	81	..	Merchandise, fob
12	19	35	32	29	34	41	36	..	..	..	Nonfactor Services
8	7	8	9	6	8	10	12	..	..	..	Factor Services
0	0	0	0	0	0	1	2	2	2	..	Long-Term Interest
2	2	1	-6	-5	-3	-2	-1	28	..	..	Current Transfers, net
0	0	0	0	0	0	0	0	..	..	..	Workers' Remittances
-14	-7	-31	-44	-22	-21	-9	-31	-96	-116	K	Total to be Financed
17	18	19	17	11	15	15	12	28	32	K	Official Capital Grants
3	10	-12	-27	-11	-6	5	-19	-68	-84	..	Current Account Balance
23	23	26	22	25	26	23	20	46	49	..	Long-Term Capital, net
5	3	2	0	1	0	2	1	2	2	..	Direct Investment
1	3	4	4	4	7	13	13	10	9	..	Long-Term Loans
1	3	4	4	4	7	13	14	13	15	..	Disbursements
..	..	..	..	..	..	..	1	4	6	..	Repayments
1	0	0	0	8	3	-6	-6	6	5	..	Other Long-Term Capital
-11	-26	-22	-2	8	-5	-9	-16	16	44	..	Other Capital, net
-15	-8	9	7	-22	-15	-19	15	6	-9	..	Change in Reserves
				(Solomon Islands Dollars per US dollar)							Memo Item:
0.870	0.870	0.830	0.870	0.970	1.150	1.270	1.480	1.740	2.000	..	Conversion Factor (Annual Avg)
				(Thousands of US dollars, outstanding at end of year)							**EXTERNAL DEBT, ETC.**
9,500	12,800	17,400	19,500	22,500	28,800	39,300	55,800	67,800	85,000	..	Public/Publicly Guar. Long-Term
9,500	12,800	17,400	19,500	22,500	28,800	34,300	44,800	52,200	67,600	..	Official Creditors
0	0	0	0	100	1,500	3,100	5,300	7,300	10,000	..	IBRD and IDA
						5,000	11,000	15,500	17,400	..	Private Creditors
..	..	..	..	..	..	..	..	..	..	..	Private Non-guaranteed Long-Term
0	0	0	900	2,700	3,500	3,100	3,000	3,500	2,200	..	Use of Fund Credit
2,000	1,000	2,000	3,000	3,000	3,000	4,000	9,000	3,000	1,400	..	Short-Term Debt
				(Thousands of US dollars)							Memo Items:
29,186	36,957	29,605	21,589	37,230	47,330	44,704	35,606	29,572	36,749	..	Int'l Reserves Excluding Gold
..	..	..	..	..	..	..	..	..	..	..	Gold Holdings (at market price)
											SOCIAL INDICATORS
7.4	7.4	7.4	7.4	7.4	7.3	7.3	7.3	7.3	7.3	..	Total Fertility Rate
45.4	45.8	46.2	46.6	47.0	47.0	47.0	47.0	45.9	44.9	..	Crude Birth Rate
45.1	44.3	43.4	42.5	41.6	40.8	39.9	39.0	..	..	..	Infant Mortality Rate
50.0	51.0	52.0	53.0	54.0	55.0	56.0	57.0	58.1	59.1	..	Life Expectancy at Birth
92.8	100.0	95.1	104.9	101.8	99.3	101.3	99.3	85.0	90.6	..	Food Production, p.c. ('79-81 = 100)
..	..	..	..	..	..	..	..	..	..	..	Labor Force, Agriculture (%)
..	..	..	..	..	..	..	..	..	..	..	Labor Force, Female (%)
..	..	..	..	..	..	..	..	..	..	..	School Enroll. Ratio, primary
..	..	..	..	..	..	..	..	..	..	..	School Enroll. Ratio, secondary

SOMALIA	1967	1968	1969	1970	1971	1972	1973	1974	1975	1976	1977
CURRENT GNP PER CAPITA (US $)	90	90	90	90	100	110	120	110	160	180	220
POPULATION (thousands)	3,317	3,404	3,498	3,595	3,696	3,799	3,903	4,009	4,115	4,222	4,329

ORIGIN AND USE OF RESOURCES *(Billions of current Somali Shillings)*

	1967	1968	1969	1970	1971	1972	1973	1974	1975	1976	1977
Gross National Product (GNP)	1.940	2.040	2.140	2.260	2.360	2.910	3.200	2.960	4.480	5.090	6.860
Net Factor Income from Abroad	-0.003	-0.006	-0.006	0.002	-0.003	0.003	0.010	0.013	0.002	0.008	-0.014
Gross Domestic Product (GDP)	1.940	2.050	2.140	2.260	2.360	2.910	3.180	2.940	4.470	5.080	6.880
Indirect Taxes, net	0.190	0.230	0.240	0.250	0.290	0.330	0.380	0.470	0.450	0.480	0.570
GDP at factor cost	1.750	1.820	1.900	2.000	2.070	2.580	2.800	2.470	4.030	4.600	6.310
Agriculture	1.160	1.120	1.150	1.190	1.190	1.590	1.720	1.130	2.180	2.590	3.850
Industry	0.150	0.240	0.280	0.320	0.350	0.380	0.370	0.500	0.500	0.640	0.760
Manufacturing	0.097	0.110	0.150	0.190	0.210	0.210	0.210	0.200	0.230	0.340	0.320
Services, etc.	0.430	0.460	0.470	0.500	0.530	0.610	0.720	0.840	1.340	1.370	1.700
Resource Balance	-0.17	-0.14	-0.16	-0.12	-0.16	-0.14	-0.41	-0.66	-0.65	-0.70	-2.68
Exports of Goods & NFServices	0.26	0.30	0.34	0.26	0.31	0.46	0.44	0.51	0.56	0.51	0.98
Imports of Goods & NFServices	0.43	0.44	0.51	0.38	0.47	0.60	0.86	1.18	1.20	1.21	3.66
Domestic Absorption	2.11	2.18	2.31	2.38	2.52	3.05	3.60	3.61	5.12	5.78	9.56
Private Consumption, etc.	1.69	1.73	1.83	1.88	2.01	2.06	2.52	2.09	3.20	3.40	5.90
General Gov't Consumption	0.18	0.20	0.22	0.23	0.23	0.60	0.62	0.72	0.91	0.86	0.97
Gross Domestic Investment	0.23	0.25	0.26	0.26	0.28	0.40	0.46	0.79	1.01	1.52	2.69
Fixed Investment	..	..	..	0.20	0.22	0.32	0.36	0.69	0.68	0.82	2.34
Memo Items:											
Gross Domestic Saving	0.066	0.110	0.100	0.150	0.130	0.250	0.044	0.130	0.360	0.820	0.013
Gross National Saving	0.069	0.110	0.100	0.150	0.140	0.270	0.071	0.160	0.380	0.830	0.013

(Millions of 1980 Somali Shillings)

	1967	1968	1969	1970	1971	1972	1973	1974	1975	1976	1977
Gross National Product	10,944	11,614	11,842	11,640	11,831	12,971	12,711	10,031	13,604	13,596	17,020
GDP at factor cost	10,332	10,946	11,025	10,841	10,901	11,812	11,463	8,600	12,374	12,434	15,821
Agriculture	..	..	..	7,135	7,015	7,453	7,124	4,606	7,975	8,262	11,177
Industry	..	..	..	1,304	1,371	1,394	1,176	1,209	1,173	1,298	1,404
Manufacturing	..	..	..	720	770	777	647	575	619	803	701
Services, etc.	..	..	..	2,402	2,515	2,965	3,164	2,785	3,225	2,874	3,240
Resource Balance	-1,208	-1,065	-1,178	-857	-1,451	-911	-1,457	-1,982	-2,024	-2,391	-7,799
Exports of Goods & NFServices	1,018	1,134	1,197	901	1,075	1,522	1,168	1,027	921	771	1,450
Imports of Goods & NFServices	2,227	2,199	2,375	1,758	2,526	2,433	2,625	3,009	2,945	3,162	9,249
Domestic Absorption	11,720	12,482	12,566	12,485	13,295	13,876	14,121	11,954	15,621	15,962	24,859
Private Consumption, etc.	8,566	9,032	8,954	8,867	9,707	7,398	7,759	5,568	8,782	8,656	14,730
General Gov't Consumption	1,098	1,184	1,197	1,278	1,261	3,388	3,320	3,268	3,437	2,837	2,912
Gross Domestic Investment	2,286	2,363	2,357	2,341	2,327	3,090	3,042	3,118	3,403	4,469	7,217
Fixed Investment	..	..	..	1,546	1,429	2,177	2,120	2,243	2,301	2,399	6,246
Memo Items:											
Capacity to Import	1,366	1,522	1,606	1,209	1,682	1,863	1,359	1,314	1,366	1,330	2,484
Terms of Trade Adjustment	348	387	409	308	607	341	190	287	445	560	1,034
Gross Domestic Income	11,208	11,840	12,114	11,937	12,451	13,307	12,854	10,259	14,042	14,131	18,094
Gross National Income	11,208	11,859	12,126	11,948	12,438	13,313	12,901	10,318	14,049	14,155	18,054

DOMESTIC PRICES (DEFLATORS) *(Index 1980 = 100)*

	1967	1968	1969	1970	1971	1972	1973	1974	1975	1976	1977
Overall (GDP)	17.70	17.70	18.10	19.40	19.90	22.40	25.20	29.50	32.90	37.40	40.30
Domestic Absorption	18.00	17.50	18.40	19.00	18.90	22.00	25.50	30.20	32.80	36.20	38.40
Agriculture	..	..	..	16.70	17.00	21.30	24.10	24.40	27.30	31.40	34.40
Industry	..	..	..	24.50	25.90	27.10	31.40	41.70	42.90	49.30	54.10
Manufacturing	..	..	..	25.90	27.30	27.50	32.50	34.80	36.60	42.00	46.10

MANUFACTURING ACTIVITY

	1967	1968	1969	1970	1971	1972	1973	1974	1975	1976	1977
Employment (1980 = 100)	24.30	24.30	32.80	45.30	49.60	43.30	53.10	70.80	79.10	82.90	89.60
Real Earnings per Empl. (1980 = 100)	223.00	216.60	189.40	240.90	222.50	197.50	193.80	169.50	157.80	157.20	146.60
Real Output per Empl. (1980 = 100)	..	..	..	184.90	194.90	225.10	162.60	165.70	163.00	172.00	160.40
Earnings as % of Value Added	32.50	26.50	25.40	27.50	25.50	18.30	26.50	37.10	32.10	26.10	25.20

MONETARY HOLDINGS *(Billions of current Somali Shillings)*

	1967	1968	1969	1970	1971	1972	1973	1974	1975	1976	1977
Money Supply, Broadly Defined	0.26	0.30	0.33	0.47	0.39	0.52	0.61	0.78	1.01	1.20	1.54
Money as Means of Payment	0.23	0.25	0.28	0.38	0.33	0.44	0.51	0.63	0.83	0.99	1.33
Currency Ouside Banks	0.12	0.13	0.14	0.15	0.15	0.22	0.25	0.31	0.39	0.41	0.62
Demand Deposits	0.10	0.12	0.14	0.23	0.18	0.22	0.26	0.32	0.44	0.58	0.70
Quasi-Monetary Liabilities	0.04	0.05	0.05	0.09	0.06	0.08	0.10	0.15	0.18	0.21	0.22

(Millions of current Somali Shillings)

	1967	1968	1969	1970	1971	1972	1973	1974	1975	1976	1977
GOVERNMENT DEFICIT (-) OR SURPLUS	..	..	..	..	..	18.50	-33.40	-59.60	-73.10	-120.30	-367.40
Current Revenue	..	..	..	..	..	412.80	438.00	585.10	622.30	700.20	857.00
Current Expenditure	..	..	..	..	..	347.40	412.30	521.80	586.70	658.90	736.00
Current Budget Balance	..	..	..	..	..	65.40	25.70	63.30	35.60	41.30	121.00
Capital Receipts	..	..	..	..	..	..	..	..	4.60	3.80	7.70
Capital Payments	..	..	..	..	..	46.90	59.10	122.90	113.30	165.40	496.10

1978	1979	1980	1981	1982	1983	1984	1985	1986	1987 estimate	NOTES	SOMALIA
260	260	200	220	240	250	250	270	280	290	A	**CURRENT GNP PER CAPITA (US $)**
4,438	4,552	4,674	4,802	4,938	5,080	5,229	5,384	5,547	5,712	..	**POPULATION (thousands)**
				(Billions of current Somali Shillings)							**ORIGIN AND USE OF RESOURCES**
8.160	9.020	17.320	21.940	28.670	34.920	58.970	82.810	111.150	162.020	..	Gross National Product (GNP)
-0.014	0.032	-0.024	-0.130	-0.430	-0.280	-3.360	-4.620	-7.820	-8.720	..	Net Factor Income from Abroad
8.180	8.990	17.340	22.070	29.100	35.200	62.330	87.420	118.960	170.740	..	Gross Domestic Product (GDP)
1.030	1.160	1.030	1.780	2.050	2.950	2.540	4.100	7.200	7.370	..	Indirect Taxes, net
7.140	7.820	16.310	20.290	27.050	32.250	59.790	83.320	111.770	163.370	..	GDP at factor cost
4.500	4.590	11.160	13.860	18.300	21.410	40.570	55.560	70.420	105.950	..	Agriculture
0.570	0.790	1.300	1.560	2.220	2.460	4.420	6.560	10.070	14.170	..	Industry
0.300	0.450	0.770	0.930	1.420	1.570	2.800	4.150	6.240	8.300	..	Manufacturing
2.080	2.440	3.850	4.860	6.530	8.390	14.800	21.210	31.280	43.260	..	Services, etc.
-2.48	-4.61	-8.98	-8.83	-11.50	-16.06	-26.54	-26.95	-35.19	-58.45	..	Resource Balance
1.59	1.61	5.39	5.03	7.10	8.09	8.79	13.30	15.16	17.96	f	Exports of Goods & NFServices
4.06	6.22	14.37	13.86	18.59	24.15	35.34	40.24	50.35	76.41	..	Imports of Goods & NFServices
10.65	13.60	26.33	30.90	40.60	51.26	88.88	114.37	154.15	229.19	..	Domestic Absorption
4.97	8.08	16.26	22.51	28.44	36.89	59.79	80.34	102.78	151.25	..	Private Consumption, etc.
1.54	2.00	2.39	2.57	3.44	4.72	6.89	8.25	12.09	18.59	..	General Gov't Consumption
4.15	3.52	7.67	5.82	8.72	9.65	22.20	25.79	39.28	59.34	..	Gross Domestic Investment
3.65	3.64	7.79	4.84	7.50	11.03	21.30	23.29	38.08	51.28	..	Fixed Investment
											Memo Items:
1.670	-1.090	-1.310	-3.010	-2.780	-6.400	-4.350	-1.160	4.090	0.890	..	Gross Domestic Saving
2.150	-0.830	-0.980	-2.810	-3.060	-6.380	-4.450	-5.010	-1.990	28.100	..	Gross National Saving
				(Millions of 1980 Somali Shillings)							
18,091	17,031	17,319	18,648	18,887	16,591	16,425	18,099	17,226	19,526	..	Gross National Product
16,095	15,111	16,313	17,550	18,087	15,544	17,084	18,799	17,999	20,244	..	GDP at factor cost
11,746	10,437	11,163	12,826	13,050	10,736	12,254	13,799	12,855	14,827	..	Agriculture
983	1,182	1,305	1,214	1,279	1,108	1,149	1,233	1,342	1,382	..	Industry
592	709	774	644	802	651	603	649	718	746	..	Manufacturing
3,366	3,492	3,846	3,509	3,757	3,699	3,680	3,767	3,802	4,035	..	Services, etc.
-6,919	-9,583	-8,984	-10,679	-11,673	-11,781	-11,515	-11,116	-9,295	-11,357	..	Resource Balance
2,122	2,177	5,391	3,427	2,977	2,847	2,383	2,531	1,354	1,198	f	Exports of Goods & NFServices
9,041	11,759	14,375	14,106	14,650	14,628	13,898	13,647	10,649	12,555	..	Imports of Goods & NFServices
25,047	26,544	26,327	29,459	30,914	28,544	29,145	30,554	28,121	32,261	..	Domestic Absorption
11,489	15,383	16,262	21,127	21,384	19,808	18,372	20,296	18,292	21,331	..	Private Consumption, etc.
4,218	4,537	2,394	2,653	2,912	2,889	2,418	2,355	2,564	3,129	..	General Gov't Consumption
9,340	6,624	7,672	5,680	6,619	5,847	8,355	7,903	7,265	7,801	..	Gross Domestic Investment
8,154	6,885	7,791	4,613	5,765	6,607	7,871	6,911	7,008	6,638	..	Fixed Investment
											Memo Items:
3,529	3,048	5,391	5,120	5,590	4,901	3,459	4,508	3,207	2,950	..	Capacity to Import
1,407	871	0	1,693	2,613	2,055	1,075	1,978	1,853	1,752	..	Terms of Trade Adjustment
19,535	17,832	17,344	20,473	21,855	18,818	18,705	21,416	20,679	22,657	..	Gross Domestic Income
19,498	17,902	17,319	20,341	21,500	18,646	17,501	20,077	19,078	21,278	..	Gross National Income
				(Index 1980 = 100)							**DOMESTIC PRICES (DEFLATORS)**
45.10	53.00	100.00	117.50	151.20	210.00	353.60	449.70	631.90	816.70	..	Overall (GDP)
42.50	51.20	100.00	104.90	131.30	179.60	304.90	374.30	548.20	710.40	..	Domestic Absorption
38.30	44.00	100.00	108.10	140.30	199.40	331.00	402.60	547.80	714.60	..	Agriculture
57.60	67.20	100.00	128.70	173.70	221.80	385.00	532.30	750.60	1,025.50	..	Industry
50.80	62.90	100.00	144.60	177.30	241.60	465.10	639.20	869.10	1,112.80	..	Manufacturing
											MANUFACTURING ACTIVITY
90.40	92.20	100.00	108.40	117.60	127.50	138.30	150.00	..	..	J	Employment (1980=100)
159.00	148.40	100.00	94.60	104.40	90.50	70.90	68.60	..	..	J	Real Earnings per Empl. (1980=100)
134.30	172.00	100.00	94.40	105.10	90.90	70.80	68.50	..	..	J	Real Output per Empl. (1980=100)
31.40	32.80	29.80	29.80	29.80	29.80	29.80	29.80	..	..	J	Earnings as % of Value Added
				(Billions of current Somali Shillings)							**MONETARY HOLDINGS**
2.05	2.81	3.38	4.42	5.12	5.50	6.93	12.56	16.83	..	..	Money Supply, Broadly Defined
1.73	2.34	2.78	3.67	4.03	4.31	5.33	9.77	12.14	..	..	Money as Means of Payment
0.88	1.15	1.51	1.89	1.46	1.36	1.90	3.79	5.21	..	..	Currency Ouside Banks
0.84	1.18	1.28	1.78	2.58	2.95	3.43	5.99	6.93	..	..	Demand Deposits
0.32	0.48	0.60	0.75	1.08	1.19	1.60	2.79	4.69	..	..	Quasi-Monetary Liabilities
				(Millions of current Somali Shillings)							
-700.10	..	..	..	..	..	..	..	..	..	..	**GOVERNMENT DEFICIT (-) OR SURPLUS**
1,347.00	..	..	..	..	..	..	..	..	..	..	Current Revenue
1,288.00	..	..	..	..	..	..	..	..	..	..	Current Expenditure
59.00	..	..	..	..	..	..	..	..	..	..	Current Budget Balance
5.50	..	..	..	..	..	..	..	..	..	..	Capital Receipts
764.60	..	..	..	..	..	..	..	..	..	..	Capital Payments

SOMALIA	1967	1968	1969	1970	1971	1972	1973	1974	1975	1976	1977
FOREIGN TRADE (CUSTOMS BASIS)				*(Millions of current US dollars)*							
Value of Exports, fob	28.20	29.50	33.30	31.10	38.10	56.50	57.10	64.00	88.60	81.00	71.30
Nonfuel Primary Products	25.65	28.67	31.67	29.18	36.07	55.51	55.75	63.13	85.75	78.87	68.64
Fuels	0.05	0.14	0.00	0.00	0.00	0.01	0.03	0.00	0.02	0.05	0.18
Manufactures	2.50	0.70	1.63	1.92	2.03	0.98	1.32	0.87	2.83	2.08	2.48
Value of Imports, cif	44.20	46.60	55.50	45.60	55.80	72.20	112.00	153.60	162.30	175.90	205.70
Nonfuel Primary Products	14.89	13.53	18.59	18.19	25.08	21.98	30.24	39.47	49.08	54.15	55.86
Fuels	2.34	1.97	2.88	2.87	2.36	3.39	4.77	10.31	10.00	12.03	8.97
Manufactures	26.97	31.10	34.03	24.53	28.36	46.83	76.99	103.82	103.22	109.71	140.87
				(Index 1980 = 100)							
Terms of Trade	147.60	144.00	168.20	180.30	151.10	163.10	161.80	93.00	79.10	96.30	86.50
Export Prices, fob	35.80	36.20	41.50	43.50	44.50	52.00	67.40	55.70	51.00	60.20	58.80
Nonfuel Primary Products	37.00	37.70	42.10	43.90	45.00	52.40	68.20	55.60	50.60	60.00	58.40
Fuels	4.30	4.30	4.30	4.30	5.60	6.20	8.90	36.70	35.70	38.40	42.00
Manufactures	30.10	31.70	32.40	37.50	36.50	38.50	49.50	67.20	66.10	66.70	71.70
Import Prices, cif	24.20	25.20	24.70	24.10	29.40	31.90	41.60	59.90	64.50	62.50	67.90
Trade at Constant 1980 Prices				*(Millions of 1980 US dollars)*							
Exports, fob	78.82	81.42	80.29	71.57	85.70	108.72	84.74	114.82	173.81	134.61	121.34
Imports, cif	182.41	185.19	225.04	189.19	189.63	226.55	269.02	256.24	251.76	281.59	302.74
BALANCE OF PAYMENTS				*(Millions of current US dollars)*							
Exports of Goods & Services	..	..	..	43.30	49.25	66.77	73.08	84.55	115.22	111.76	104.05
Merchandise, fob	..	..	..	31.10	38.11	56.46	57.10	63.98	88.63	81.05	71.33
Nonfactor Services	..	..	..	11.20	10.23	9.34	13.59	17.92	23.31	27.59	29.71
Factor Services	..	..	..	1.00	0.90	0.98	2.38	2.65	3.28	3.12	3.02
Imports of Goods & Services	..	..	..	61.90	67.70	90.98	140.43	187.97	217.45	221.78	244.89
Merchandise, fob	..	..	..	40.40	49.55	62.75	97.52	133.73	141.08	153.09	179.05
Nonfactor Services	..	..	..	20.80	16.85	27.47	41.96	53.40	73.33	66.85	65.05
Factor Services	..	..	..	0.70	1.30	0.76	0.95	0.84	3.04	1.85	0.79
Long-Term Interest	..	..	..	0.40	0.60	0.80	0.80	1.10	1.30	0.90	1.20
Current Transfers, net	..	..	..	0.70	2.11	1.85	2.74	3.49	1.94	1.15	2.22
Workers' Remittances	..	..	..	0.00	0.00	0.00	0.00	0.00	0.00	0.00	0.00
Total to be Financed	..	..	..	-17.90	-16.35	-22.37	-64.61	-99.94	-100.29	-108.87	-138.62
Official Capital Grants	..	..	..	12.20	17.15	15.20	25.87	48.23	100.29	39.72	105.85
Current Account Balance	..	..	..	-5.70	0.80	-7.17	-38.74	-51.71	0.00	-69.16	-32.77
Long-Term Capital, net	..	..	..	23.20	21.46	32.90	51.98	109.20	153.10	109.33	170.17
Direct Investment	..	..	..	4.50	1.71	4.45	0.60	0.72	6.68	2.19	7.78
Long-Term Loans	..	..	..	3.70	4.70	15.40	19.20	46.50	58.60	54.70	89.10
Disbursements	..	..	..	4.20	5.30	16.60	21.00	48.90	61.20	56.70	91.60
Repayments	..	..	..	0.50	0.60	1.20	1.80	2.40	2.60	2.00	2.50
Other Long-Term Capital	..	..	..	2.80	-2.09	-2.15	6.31	13.75	-12.46	12.72	-32.56
Other Capital, net	..	..	..	-14.00	-19.42	-9.98	-22.34	-38.44	-110.41	-40.78	-83.21
Change in Reserves	..	..	..	-3.50	-2.85	-15.76	9.11	-19.05	-42.70	0.60	-54.19
Memo Item:				*(Somali Shillings per US dollar)*							
Conversion Factor (Annual Avg)	7.140	7.140	7.140	7.140	7.130	6.980	6.280	6.300	6.300	6.300	13.800
Additional Conversion Factor	7.143	7.143	7.143	7.143	7.129	6.980	6.281	6.295	6.295	6.295	6.295
EXTERNAL DEBT, ETC.				*(Millions of US dollars, outstanding at end of year)*							
Public/Publicly Guar. Long-Term	..	..	..	77.1	83.7	100.3	126.8	176.9	230.2	287.1	387.4
Official Creditors	..	..	..	74.9	81.9	96.3	119.6	170.0	224.4	283.0	384.0
IBRD and IDA	..	..	..	7.1	8.6	10.6	14.5	21.1	30.1	36.7	46.3
Private Creditors	..	..	..	2.2	1.8	4.0	7.2	6.9	5.8	4.1	3.4
Private Non-guaranteed Long-Term	..	..	..	..	..	..	..	..	..	..	..
Use of Fund Credit	..	..	..	0.0	0.0	0.0	0.0	0.0	0.0	0.0	0.0
Short-Term Debt	..	..	..	..	..	..	..	..	..	..	12.0
Memo Items:				*(Thousands of US dollars)*							
Int'l Reserves Excluding Gold	10,000.0	6,750.0	13,420.0	21,140.0	26,659.0	31,331.0	34,997.0	42,323.0	68,368.0	84,876.0	119,980.0
Gold Holdings (at market price)	..	..	810.0	112.0	131.0	195.0	337.0	492.0	370.0	355.0	1,777.0
SOCIAL INDICATORS											
Total Fertility Rate	6.7	6.7	6.7	6.7	6.7	6.7	6.7	6.7	6.7	6.7	6.7
Crude Birth Rate	50.0	50.0	50.0	50.0	50.0	50.0	50.0	50.0	50.1	50.1	50.1
Infant Mortality Rate	162.0	160.6	159.2	157.8	156.4	155.0	153.8	152.6	151.4	150.2	149.0
Life Expectancy at Birth	38.9	39.3	39.7	40.1	40.5	40.9	41.4	41.9	42.3	42.8	43.3
Food Production, p.c. ('79-81 = 100)	136.7	137.4	139.6	139.0	132.9	135.0	123.2	112.6	115.8	110.8	108.3
Labor Force, Agriculture (%)	80.4	80.1	79.7	79.4	79.0	78.6	78.3	77.9	77.5	77.1	76.7
Labor Force, Female (%)	41.3	41.3	41.3	41.2	41.2	41.1	41.0	41.0	40.9	40.8	40.8
School Enroll. Ratio, primary	..	..	..	11.0	..	..	..	..	59.0	45.0	40.0
School Enroll. Ratio, secondary	..	..	..	5.0	..	..	..	..	6.0	4.0	4.0

1978	1979	1980	1981	1982	1983	1984	1985	1986	1987 estimate	NOTES	SOMALIA
											FOREIGN TRADE (CUSTOMS BASIS)
				(Millions of current US dollars)							
109.50	106.00	133.30	175.40	170.80	98.40	54.80	90.60	105.00	118.21	..	Value of Exports, fob
107.95	103.72	125.81	174.39	167.42	96.28	53.44	89.23	103.45	116.36	..	Nonfuel Primary Products
0.32	1.24	6.43	0.37	2.46	1.59	1.03	1.02	0.80	1.01	..	Fuels
1.23	1.04	1.06	0.64	0.92	0.53	0.32	0.35	0.75	0.83	..	Manufactures
275.10	394.00	461.30	425.70	541.60	416.10	535.10	380.00	402.00	425.97	..	Value of Imports, cif
68.90	103.46	170.18	113.50	155.26	121.04	149.36	98.84	79.48	81.16	..	Nonfuel Primary Products
18.25	21.58	3.41	9.33	12.08	8.62	23.99	19.81	8.28	10.94	..	Fuels
187.95	268.96	287.71	302.87	374.27	286.43	361.75	261.34	314.24	333.87	..	Manufactures
				(Index 1980 = 100)							
103.60	117.80	100.00	91.60	92.50	98.20	93.90	90.90	80.20	83.90	..	Terms of Trade
77.40	104.00	100.00	92.30	88.60	92.30	86.40	82.50	81.30	90.10	..	Export Prices, fob
77.60	105.10	100.00	92.20	88.40	92.30	86.30	82.40	81.70	90.40	..	Nonfuel Primary Products
42.30	61.00	100.00	112.50	101.60	92.50	90.20	87.50	44.90	56.90	..	Fuels
76.10	90.00	100.00	99.10	94.80	90.70	89.30	89.00	103.20	113.30	..	Manufactures
74.70	88.30	100.00	100.70	95.80	94.00	92.00	90.70	101.30	107.30	..	Import Prices, cif
				(Millions of 1980 US dollars)							Trade at Constant 1980 Prices
141.39	101.89	133.30	190.10	192.75	106.57	63.45	109.82	129.17	131.24	..	Exports, fob
368.07	446.27	461.30	422.59	565.30	442.43	581.87	418.80	396.66	396.82	..	Imports, cif
				(Millions of current US dollars)							**BALANCE OF PAYMENTS**
151.83	153.25	204.50	255.44	256.14	177.20	106.74	127.62	116.15	107.00	..	Exports of Goods & Services
109.47	106.02	133.33	175.36	170.81	98.44	54.76	90.59	88.27	94.00	..	Merchandise, fob
36.46	37.71	66.20	74.25	83.39	77.03	51.00	35.94	25.36	13.00	..	Nonfactor Services
5.91	9.52	4.97	5.83	1.93	1.73	0.97	1.10	2.52	..	..	Factor Services
322.37	452.95	540.56	519.46	610.28	486.08	595.65	448.08	487.45	507.50	..	Imports of Goods & Services
239.43	342.86	401.52	370.53	471.35	362.09	465.67	330.79	382.00	429.40	..	Merchandise, fob
80.56	103.05	133.19	135.03	130.97	117.60	123.47	96.78	101.64	26.10	..	Nonfactor Services
2.38	7.04	5.85	13.90	7.95	6.39	6.52	20.51	3.81	52.00	..	Factor Services
1.30	1.00	1.90	3.40	6.50	9.50	3.10	2.80	3.60	3.90	..	Long-Term Interest
78.05	35.90	57.25	53.57	13.71	19.13	162.93	19.44	24.02	341.60	..	Current Transfers, net
0.00	36.65	57.25	7.15	19.90	21.70	0.00	0.00	0.00	..	..	Workers' Remittances
-92.42	-263.80	-278.81	-210.45	-340.39	-289.77	-326.03	-301.01	-347.28	-58.90	K	Total to be Financed
27.75	58.08	142.62	127.09	162.94	148.17	194.08	204.35	259.85	306.60	K	Official Capital Grants
-64.67	-205.72	-136.19	-83.37	-177.45	-141.59	-131.95	-96.66	-87.43	247.70	..	Current Account Balance
107.52	144.18	233.30	171.74	301.19	216.45	289.97	281.34	303.09	287.80	..	Long-Term Capital, net
0.30	..	..	..	-0.76	-8.23	-14.94	-0.71	-0.12	..	..	Direct Investment
123.10	112.30	120.90	307.90	159.40	139.60	198.20	106.50	98.30	65.80	..	Long-Term Loans
126.60	115.50	128.00	351.10	169.40	150.00	218.00	108.80	105.50	70.80	..	Disbursements
3.50	3.20	7.10	43.20	10.00	10.40	19.80	2.30	7.20	5.00	..	Repayments
-43.64	-26.20	-30.22	-263.25	-20.39	-63.09	-87.36	-28.80	-54.94	-84.60	..	Other Long-Term Capital
-15.32	-45.73	-120.11	-108.46	-87.75	-152.51	-176.04	-177.35	-239.39	-682.90	..	Other Capital, net
-27.52	107.27	23.00	20.09	-36.00	77.65	18.02	-7.33	23.73	147.40	..	Change in Reserves
				(Somali Shillings per US dollar)							Memo Item:
14.480	15.240	26.910	29.000	35.280	46.480	76.470	94.220	129.590	167.740	..	Conversion Factor (Annual Avg)
6.295	6.295	6.295	6.295	10.750	15.788	20.019	39.487	72.000	105.180	..	Additional Conversion Factor
				(Millions of US dollars, outstanding at end of year)							**EXTERNAL DEBT, ETC.**
525.4	598.6	715.2	1,004.7	1,124.9	1,253.9	1,403.9	1,867.8	2,075.2	2,288.4	..	Public/Publicly Guar. Long-Term
522.6	597.5	715.2	875.5	958.7	1,084.6	1,153.5	1,700.8	1,905.5	2,121.3	..	Official Creditors
55.7	62.5	72.2	91.2	107.7	124.1	147.6	186.0	239.2	297.6	..	IBRD and IDA
2.9	1.1	..	129.3	166.2	169.2	250.4	167.0	169.7	167.1	..	Private Creditors
..	..	..	..	..	..	..	..	..	..	..	Private Non-guaranteed Long-Term
0.0	0.2	4.3	34.1	67.9	111.5	101.6	142.2	145.2	153.7	..	Use of Fund Credit
22.0	26.0	50.6	39.8	55.6	59.9	92.1	59.7	87.8	92.3	..	Short-Term Debt
				(Thousands of US dollars)							Memo Items:
26,340.0	43,791.0	14,564.0	30,712.0	6,505.0	9,166.0	1,049.0	2,500.0	12,800.0	..	..	Int'l Reserves Excluding Gold
3,350.0	7,589.0	11,182.0	7,540.0	8,755.0	7,310.0	5,908.0	6,287.0	7,515.0	9,307.0	..	Gold Holdings (at market price)
											SOCIAL INDICATORS
6.7	6.7	6.8	6.8	6.8	6.8	6.8	6.8	6.8	6.8	..	Total Fertility Rate
50.0	49.9	49.8	49.7	49.6	49.5	49.3	49.2	49.0	48.9	..	Crude Birth Rate
147.8	146.6	145.4	144.2	143.0	138.8	134.6	130.4	126.2	122.0	..	Infant Mortality Rate
43.6	43.9	44.2	44.5	44.8	45.2	45.7	46.1	46.6	47.0	..	Life Expectancy at Birth
107.5	99.9	100.7	99.5	100.1	93.1	93.4	98.9	103.1	103.2	..	Food Production, p.c. ('79-81 = 100)
76.3	75.9	75.5	..	..	..	..	..	..	..	..	Labor Force, Agriculture (%)
40.7	40.7	40.6	40.4	40.2	40.0	39.9	39.7	39.5	39.3	..	Labor Force, Female (%)
43.0	41.0	30.0	25.0	25.0	25.0	..	..	..	..	..	School Enroll. Ratio, primary
5.0	6.0	11.0	14.0	18.0	17.0	..	..	..	..	..	School Enroll. Ratio, secondary

SOUTH AFRICA	1967	1968	1969	1970	1971	1972	1973	1974	1975	1976	1977
CURRENT GNP PER CAPITA (US $)	590	640	710	1,150	890	850	960	1,190	1,410	1,420	1,340
POPULATION (thousands)	20,856	21,381	21,915	22,459	23,013	23,577	24,150	24,727	25,306	25,889	26,476
ORIGIN AND USE OF RESOURCES					*(Millions of current South African Rand)*						
Gross National Product (GNP)	9,389	10,095	11,296	12,399	13,733	15,520	19,208	23,569	26,234	29,386	32,683
Net Factor Income from Abroad	-347	-400	-462	-509	-508	-584	-710	-903	-1,220	-1,414	-1,631
Gross Domestic Product (GDP)	9,736	10,495	11,758	12,908	14,241	16,104	19,918	24,472	27,454	30,800	34,314
Indirect Taxes, net	700	852	800	885	990	1,068	1,243	1,389	1,608	1,880	2,331
GDP at factor cost	9,036	9,643	10,958	12,023	13,251	15,036	18,675	23,083	25,846	28,920	31,983
Agriculture	1,052	949	1,013	973	1,168	1,320	1,531	2,214	2,129	2,275	2,532
Industry	3,667	3,918	4,428	4,817	5,087	5,950	7,802	9,735	11,145	12,515	13,693
Manufacturing	2,067	2,197	2,512	2,796	2,983	3,268	4,092	4,896	5,991	6,834	6,963
Services, etc.	4,317	4,776	5,517	6,233	6,996	7,766	9,342	11,134	12,572	14,130	15,758
Resource Balance	117	374	118	-408	-590	447	644	-179	-731	-323	2,004
Exports of Goods & NFServices	2,483	2,667	2,738	2,757	3,019	3,981	5,015	6,656	7,388	8,477	10,474
Imports of Goods & NFServices	2,366	2,293	2,620	3,165	3,609	3,534	4,371	6,835	8,119	8,800	8,470
Domestic Absorption	9,619	10,121	11,640	13,316	14,831	15,657	19,274	24,651	28,185	31,123	32,310
Private Consumption, etc.	5,715	6,369	7,287	8,068	8,599	9,550	11,582	14,452	15,510	17,562	17,891
General Gov't Consumption	1,061	1,174	1,323	1,564	1,880	1,997	2,291	2,886	3,782	4,537	5,155
Gross Domestic Investment	2,843	2,578	3,030	3,684	4,352	4,110	5,401	7,313	8,893	9,024	9,264
Fixed Investment	..	..	..	3,194	3,741	4,322	5,027	6,158	8,110	9,221	9,571
Memo Items:											
Gross Domestic Saving	2,960	2,952	3,148	3,276	3,762	4,557	6,045	7,134	8,162	8,701	11,268
Gross National Saving	2,639	2,600	2,725	2,789	3,270	3,991	5,318	6,257	7,007	7,317	9,601
					(Millions of 1980 South African Rand)						
Gross National Product	36,415	39,353	40,950	43,336	45,920	46,693	47,942	50,841	51,728	52,716	53,025
GDP at factor cost	34,065	38,188	39,719	42,101	44,334	45,225	46,675	49,831	50,927	51,752	51,879
Agriculture	..	..	..	3,027	3,593	3,607	3,084	3,958	3,678	3,628	3,986
Industry	..	..	..	25,077	25,653	25,583	26,395	26,843	27,071	27,440	27,554
Manufacturing	..	..	..	6,285	6,871	7,307	8,075	8,880	9,308	9,532	8,943
Services, etc.	..	..	..	13,997	15,089	16,035	17,196	19,030	20,178	20,684	20,339
Resource Balance	4,949	6,446	5,270	3,791	3,212	6,296	4,153	746	1,369	3,725	10,097
Exports of Goods & NFServices	13,612	15,261	14,986	15,151	15,677	17,014	16,198	15,442	15,571	16,448	22,820
Imports of Goods & NFServices	8,662	8,815	9,716	11,359	12,465	10,719	12,045	14,696	14,202	12,723	12,723
Domestic Absorption	33,306	37,163	37,328	41,298	44,313	42,061	45,563	52,113	52,770	51,458	45,384
Private Consumption, etc.	19,475	20,774	22,470	24,301	25,546	25,020	26,968	31,044	30,307	30,437	26,310
General Gov't Consumption	3,866	4,107	4,301	4,684	5,068	5,040	5,218	5,572	6,388	6,932	6,270
Gross Domestic Investment	10,482	8,828	10,525	12,313	13,698	12,001	13,377	15,497	16,075	14,089	12,804
Fixed Investment	..	..	..	10,784	11,989	12,644	13,121	13,951	15,377	15,061	13,364
Memo Items:											
Capacity to Import	9,090	10,253	10,154	9,895	10,427	12,074	13,820	14,311	12,923	12,256	15,733
Terms of Trade Adjustment	-4,521	-5,008	-4,832	-5,256	-5,250	-4,940	-2,378	-1,131	-2,648	-4,192	-7,087
Gross Domestic Income	32,824	35,882	37,108	39,834	42,275	43,416	47,337	51,728	51,492	50,991	48,395
Gross National Income	31,470	34,440	35,517	38,080	40,670	41,753	45,564	49,710	49,080	48,523	45,939
DOMESTIC PRICES (DEFLATORS)					*(Index 1980 = 100)*						
Overall (GDP)	25.8	25.7	27.6	28.6	30.0	33.3	40.1	46.3	50.7	55.8	61.8
Domestic Absorption	28.9	27.2	31.2	32.2	33.5	37.2	42.3	47.3	53.4	60.5	71.2
Agriculture	..	..	..	32.1	32.5	36.6	49.6	55.9	57.9	62.7	63.5
Industry	..	..	..	19.2	19.8	23.3	29.6	36.3	41.2	45.6	49.7
Manufacturing	..	..	..	44.5	43.4	44.7	50.7	55.1	64.4	71.7	77.9
MANUFACTURING ACTIVITY											
Employment (1980=100)	..	70.4	..	77.3	..	81.3	84.3	87.9	90.1	97.4	95.3
Real Earnings per Empl. (1980=100)	..	72.2	..	74.4	..	79.5	83.7	86.6	88.8	92.9	93.2
Real Output per Empl. (1980=100)	..	..	..	50.2	..	58.6	66.8	72.0	72.4	68.4	67.7
Earnings as % of Value Added	..	47.5	..	45.6	..	46.8	49.0	51.3	54.1	48.7	47.8
MONETARY HOLDINGS					*(Millions of current South African Rand)*						
Money Supply, Broadly Defined	5.6	6.4	7.1	7.8	8.5	10.0	12.2	14.2	17.0	18.5	20.3
Money as Means of Payment	1.7	2.1	2.2	2.3	2.5	2.8	3.4	4.0	4.3	4.4	4.7
Currency Ouside Banks	0.4	0.4	0.5	0.5	0.6	0.6	0.8	0.9	1.0	1.1	1.2
Demand Deposits	1.3	1.6	1.8	1.8	1.9	2.2	2.6	3.1	3.3	3.3	3.5
Quasi-Monetary Liabilities	3.9	4.3	4.9	5.5	6.0	7.2	8.8	10.2	12.7	14.0	15.7
GOVERNMENT DEFICIT (-) OR SURPLUS					*(Millions of current South African Rand)*						
	..	..	..	..	..	-652	-499	-1,155	-1,438	-2,023	-2,067
Current Revenue	..	..	..	..	..	3,295	4,185	5,114	5,988	6,669	7,599
Current Expenditure	..	..	..	..	..	2,841	3,406	4,419	5,346	6,380	7,075
Current Budget Balance	..	..	..	..	..	454	779	695	642	289	524
Capital Receipts	..	..	..	..	..	30	27	30	37	21	27
Capital Payments	..	..	..	..	..	1,136	1,305	1,880	2,117	2,333	2,618

1978	1979	1980	1981	1982	1983	1984	1985	1986	1987 estimate	NOTES	SOUTH AFRICA
1,460	1,720	2,160	2,570	2,560	2,480	2,450	2,080	1,840	1,870	..	CURRENT GNP PER CAPITA (US $)
27,070	27,672	28,283	28,907	29,546	30,204	30,885	31,593	32,436	33,285	..	POPULATION (thousands)
				(Millions of current South African Rand)							ORIGIN AND USE OF RESOURCES
37,438	44,569	59,272	67,667	75,972	85,711	100,926	113,612	131,873	156,967	f	Gross National Product (GNP)
-1,859	-2,129	-2,735	-3,416	-3,704	-4,164	-4,888	-6,529	-7,822	-7,488	..	Net Factor Income from Abroad
39,297	46,698	62,007	71,083	79,676	89,875	105,814	120,141	139,695	164,455	f	Gross Domestic Product (GDP)
2,779	3,477	4,044	5,081	6,683	6,606	8,386	9,987	11,793	13,371	..	Indirect Taxes, net
36,518	43,221	57,963	66,002	72,993	83,269	97,428	110,154	127,902	151,084	f	GDP at factor cost
2,722	2,885	4,035	4,787	4,581	4,096	5,265	5,907	7,277	8,533	..	Agriculture
16,424	20,698	29,339	30,790	32,840	38,165	43,369	50,468	58,435	66,896	..	Industry
7,886	9,579	12,606	15,646	16,939	19,456	22,704	24,311	27,840	35,103	..	Manufacturing
17,372	19,638	24,589	30,425	35,572	41,008	48,794	53,779	62,190	75,655	..	Services, etc.
3,092	4,846	5,260	-1,043	20	3,723	2,330	12,096	14,633	13,209	..	Resource Balance
12,959	16,724	22,219	20,584	21,833	23,163	28,261	40,395	46,499	48,026	..	Exports of Goods & NFServices
9,867	11,878	16,959	21,627	21,813	19,440	25,931	28,299	31,866	34,817	..	Imports of Goods & NFServices
36,205	41,852	56,747	72,126	79,656	86,152	103,484	108,045	125,062	151,246	..	Domestic Absorption
20,593	22,735	29,458	37,715	45,920	48,137	58,509	61,693	71,404	87,509	..	Private Consumption, etc.
5,673	6,543	8,449	10,378	12,990	14,809	18,787	21,695	26,008	31,613	..	General Gov't Consumption
9,939	12,574	18,840	24,033	20,746	23,206	26,188	24,657	27,650	32,124	..	Gross Domestic Investment
10,342	12,251	16,378	20,132	22,692	24,509	26,131	28,619	28,998	32,068	..	Fixed Investment
											Memo Items:
13,031	17,420	24,100	22,990	20,766	26,929	28,518	36,753	42,283	45,333	..	Gross Domestic Saving
11,186	15,335	21,438	19,652	17,188	22,921	23,809	30,409	34,634	..	..	Gross National Saving
				(Millions of 1980 South African Rand)							
54,535	56,375	59,272	62,384	59,780	59,077	62,264	61,675	62,407	64,603	f	Gross National Product
53,334	55,230	57,963	60,586	59,999	58,537	61,486	61,018	61,713	63,174	f	GDP at factor cost
4,120	4,022	4,035	4,264	3,892	3,015	3,333	3,620	4,203	4,312	..	Agriculture
28,321	29,679	29,339	30,569	30,068	28,935	30,027	29,510	29,374	29,815	..	Industry
9,593	10,385	12,606	13,459	13,052	12,185	12,670	12,204	12,360	12,845	..	Manufacturing
20,894	21,530	24,589	25,754	26,039	26,587	28,127	27,888	28,136	29,047	..	Services, etc.
11,040	11,408	5,260	1,740	4,758	6,383	5,345	9,905	9,940	8,807	..	Resource Balance
23,855	24,049	22,219	20,928	20,948	19,805	21,487	23,665	23,337	22,732	..	Exports of Goods & NFServices
12,815	12,641	16,959	19,188	16,189	13,422	16,142	13,760	13,398	13,925	..	Imports of Goods & NFServices
46,031	47,556	56,747	63,189	59,587	56,421	60,662	55,359	56,019	58,802	..	Domestic Absorption
27,421	27,386	29,458	33,743	34,786	31,399	34,562	32,240	33,987	35,555	..	Private Consumption, etc.
6,325	6,523	8,449	8,738	9,221	9,372	10,097	10,186	10,363	10,958	..	General Gov't Consumption
12,285	13,647	18,840	20,708	15,580	15,650	16,003	12,932	11,668	12,289	..	Gross Domestic Investment
12,987	13,475	16,378	17,784	17,273	16,449	16,117	15,083	12,454	12,281	..	Fixed Investment
											Memo Items:
16,831	17,798	22,219	18,263	16,204	15,993	17,593	19,642	19,550	19,208	..	Capacity to Import
-7,024	-6,251	0	-2,665	-4,744	-3,812	-3,894	-4,024	-3,787	-3,524	..	Terms of Trade Adjustment
50,046	52,713	62,007	62,264	59,602	58,991	62,112	61,240	62,171	64,084	..	Gross Domestic Income
47,511	50,124	59,272	59,719	55,036	55,265	58,370	57,651	58,620	61,079	..	Gross National Income
				(Index 1980 = 100)							DOMESTIC PRICES (DEFLATORS)
68.9	79.2	100.0	109.5	123.8	143.1	160.3	184.1	211.8	243.2	..	Overall (GDP)
78.7	88.0	100.0	114.1	133.7	152.7	170.6	195.2	223.3	257.2	..	Domestic Absorption
66.1	71.7	100.0	112.3	117.7	135.8	158.0	163.2	173.1	197.9	..	Agriculture
58.0	69.7	100.0	100.7	109.2	131.9	144.4	171.0	198.9	224.4	..	Industry
82.2	92.2	100.0	116.3	129.8	159.7	179.2	199.2	225.2	273.3	..	Manufacturing
											MANUFACTURING ACTIVITY
95.2	97.3	100.0	104.2	104.5	98.6	99.2	95.4	..	..	G	Employment (1980=100)
95.1	96.4	100.0	103.2	106.1	108.4	109.1	105.6	..	..	G	Real Earnings per Empl. (1980=100)
72.9	75.6	100.0	99.4	98.5	93.3	95.8	95.3	..	..	G	Real Output per Empl. (1980=100)
48.1	50.7	47.7	49.1	52.3	51.5	49.8	49.5	..	..	..	Earnings as % of Value Added
				(Millions of current South African Rand)							MONETARY HOLDINGS
23.6	27.6	33.8	39.7	45.0	51.2	61.3	70.3	76.5	91.4	D	Money Supply, Broadly Defined
5.1	6.2	8.4	11.3	13.1	16.6	23.4	21.3	23.2	32.0	..	Money as Means of Payment
1.3	1.5	1.9	2.3	2.5	2.8	3.2	3.6	4.2	5.0	..	Currency Ouside Banks
3.9	4.7	6.5	9.0	10.6	13.8	20.2	17.8	19.0	27.0	..	Demand Deposits
18.5	21.4	25.4	28.4	31.9	34.6	37.9	49.0	53.3	59.4	..	Quasi-Monetary Liabilities
				(Millions of current South African Rand)							
-2,095	-1,947	-1,436	-2,810	-3,082	-4,771	-5,060	..	..	..	C F	GOVERNMENT DEFICIT (-) OR SURPLUS
8,956	10,764	14,398	16,034	19,199	21,754	26,474	..	..	..	..	Current Revenue
8,019	9,707	11,709	14,644	18,231	21,936	27,203	..	..	..	..	Current Expenditure
937	1,057	2,689	1,390	968	-182	-729	..	..	..	..	Current Budget Balance
45	38	50	31	40	51	70	..	..	..	..	Capital Receipts
3,077	3,042	4,175	4,231	4,090	4,640	4,401	..	..	..	..	Capital Payments

SOUTH AFRICA	1967	1968	1969	1970	1971	1972	1973	1974	1975	1976	1977
FOREIGN TRADE (CUSTOMS BASIS)				*(Millions of current US dollars)*							
Value of Exports, fob	3,018	3,211	3,356	3,355	3,510	4,196	6,114	8,760	8,959	7,975	9,987
Nonfuel Primary Products	1,809	1,854	2,294	1,807	1,881	2,510	3,349	4,500	4,546	4,004	4,392
Fuels	135	166	26	173	189	159	39	81	114	210	463
Manufactures	1,074	1,191	1,036	1,375	1,440	1,528	2,726	4,179	4,299	3,761	5,132
Value of Imports, cif	2,780	2,789	3,189	3,843	4,364	3,948	5,163	7,856	8,293	7,285	6,270
Nonfuel Primary Products	435	361	282	457	452	421	660	1,010	890	835	817
Fuels	163	182	75	191	286	272	10	21	20	29	33
Manufactures	2,182	2,246	2,832	3,195	3,625	3,255	4,493	6,824	7,383	6,421	5,420
				(Index 1980 = 100)							
Terms of Trade	126.6	128.5	128.0	121.7	120.9	125.4	122.0	128.4	114.8	110.4	100.0
Export Prices, fob	29.1	28.4	36.5	31.2	32.2	36.2	57.5	74.2	71.6	70.2	70.0
Nonfuel Primary Products	35.7	34.6	38.0	39.3	38.2	43.6	63.3	71.5	70.1	72.2	72.5
Fuels	4.3	4.3	4.3	4.3	5.6	6.2	8.9	36.7	35.7	38.4	42.0
Manufactures	50.4	58.6	40.6	66.2	55.6	46.6	55.6	78.9	75.2	71.4	72.2
Import Prices, cif	23.0	22.1	28.5	25.6	26.7	28.9	47.1	57.8	62.3	63.6	70.0
Trade at Constant 1980 Prices				*(Millions of 1980 US dollars)*							
Exports, fob	10,364	11,290	9,201	10,751	10,897	11,585	10,637	11,807	12,521	11,357	14,270
Imports, cif	12,086	12,606	11,194	14,985	16,373	13,664	10,963	13,594	13,302	11,449	8,959
BALANCE OF PAYMENTS				*(Millions of current US dollars)*							
Exports of Goods & Services	..	..	..	4,019	4,402	5,379	7,577	10,073	10,343	10,030	12,288
Merchandise, fob	..	..	..	3,206	3,457	4,386	6,187	8,434	8,436	8,299	10,442
Nonfactor Services	..	..	..	457	527	557	718	921	1,112	966	1,121
Factor Services	..	..	..	356	418	435	672	718	795	765	725
Imports of Goods & Services	..	..	..	5,303	5,938	5,562	7,681	11,655	12,974	12,037	11,807
Merchandise, fob	..	..	..	3,615	4,088	3,698	5,130	8,482	9,164	8,559	7,914
Nonfactor Services	..	..	..	796	942	896	1,156	1,547	1,899	1,549	1,776
Factor Services	..	..	..	892	908	968	1,395	1,626	1,911	1,929	2,117
Long-Term Interest	..	..	..	..	..	..	..	..	..	..	..
Current Transfers, net	..	..	..	31	22	23	-25	38	87	34	-41
Workers' Remittances	..	..	..	0	0	0	0	0	0	0	0
Total to be Financed	..	..	..	-1,253	-1,513	-161	-129	-1,543	-2,544	-1,973	440
Official Capital Grants	..	..	..	38	34	39	45	84	102	100	94
Current Account Balance	..	..	..	-1,215	-1,479	-122	-83	-1,459	-2,442	-1,873	534
Long-Term Capital, net	..	..	..	732	777	752	260	1,716	2,260	1,080	276
Direct Investment	..	..	..	318	231	98	-23	578	63	-14	-189
Long-Term Loans	..	..	..	..	..	..	..	..	..	..	..
Disbursements	..	..	..	..	..	..	..	..	..	..	..
Repayments	..	..	..	..	..	..	..	..	..	..	..
Other Long-Term Capital	..	..	..	376	513	616	237	1,054	2,094	994	371
Other Capital, net	..	..	..	60	317	-102	-381	-409	266	113	-1,316
Change in Reserves	..	..	..	423	385	-529	205	152	-84	680	507
Memo Item:				*(South African Rand per US dollar)*							
Conversion Factor (Annual Avg)	0.710	0.710	0.710	0.710	0.720	0.770	0.690	0.680	0.740	0.870	0.870
EXTERNAL DEBT, ETC.				*(Millions of US dollars, outstanding at end of year)*							
Public/Publicly Guar. Long-Term	..	..	..	..	..	..	..	..	..	..	..
Official Creditors	..	..	..	..	..	..	..	..	..	..	..
IBRD and IDA	..	..	..	..	..	..	..	..	..	..	..
Private Creditors	..	..	..	..	..	..	..	..	..	..	..
Private Non-guaranteed Long-Term	..	..	..	..	..	..	..	..	..	..	..
Use of Fund Credit	..	..	..	..	..	..	..	..	..	..	..
Short-Term Debt	..	..	..	..	..	..	..	..	..	..	..
Memo Items:				*(Millions of US dollars)*							
Int'l Reserves Excluding Gold	196	228	283	346	266	609	449	377	489	425	416
Gold Holdings (at market price)	586	1,488	1,121	711	511	1,164	2,132	3,404	2,489	1,707	1,603
SOCIAL INDICATORS											
Total Fertility Rate	5.9	5.8	5.7	5.7	5.6	5.5	5.4	5.3	5.3	5.2	5.1
Crude Birth Rate	38.1	37.7	37.3	36.9	36.5	36.1	35.7	35.4	35.0	34.7	34.3
Infant Mortality Rate	120.0	118.0	116.0	114.0	112.0	110.0	107.0	104.0	101.0	98.0	95.0
Life Expectancy at Birth	51.9	52.3	52.7	53.1	53.5	53.9	54.3	54.7	55.1	55.5	55.9
Food Production, p.c. ('79-81 = 100)	105.0	86.4	90.1	86.9	99.5	104.6	84.4	103.4	96.3	93.1	97.6
Labor Force, Agriculture (%)	32.3	32.5	32.7	32.9	31.3	29.6	28.0	26.3	24.7	23.1	21.4
Labor Force, Female (%)	30.3	31.2	32.1	33.0	33.1	33.2	33.3	33.5	33.6	33.7	33.8
School Enroll. Ratio, primary	..	..	..	99.0	..	105.0	..	..	..	..	..
School Enroll. Ratio, secondary	..	..	..	18.0	..	20.0	..	..	..	..	..

1978	1979	1980	1981	1982	1983	1984	1985	1986	1987 estimate	NOTES	SOUTH AFRICA
				(Millions of current US dollars)							**FOREIGN TRADE** (CUSTOMS BASIS)
12,875	18,397	25,680	20,895	17,727	18,607	17,348	16,523	18,454	20,066	f	Value of Exports, fob
4,908	6,170	4,663	4,244	3,543	3,697	3,012	3,025	2,956	3,648	..	Nonfuel Primary Products
638	1,098	972	1,274	1,151	1,143	1,033	903	560	692	..	Fuels
7,329	11,129	20,045	15,377	13,033	13,767	13,302	12,595	14,937	15,727	..	Manufactures
7,615	8,989	19,246	22,619	18,575	15,704	16,234	11,469	12,989	14,629	f	Value of Imports, cif
799	1,016	1,444	1,670	1,337	1,141	1,260	785	813	914	..	Nonfuel Primary Products
35	52	76	72	83	65	68	58	33	41	..	Fuels
6,781	7,921	17,726	20,877	17,156	14,498	14,906	10,626	12,142	13,674	..	Manufactures
				(Index 1980 = 100)							
87.7	92.8	100.0	94.8	85.8	82.7	82.1	75.0	71.0	71.2	f	Terms of Trade
69.9	84.7	100.0	94.7	83.7	79.1	77.4	70.6	77.8	85.4	f	Export Prices, fob
76.3	93.7	100.0	95.6	85.3	89.2	87.4	78.1	75.6	81.4	..	Nonfuel Primary Products
42.3	61.0	100.0	112.5	101.6	92.5	90.2	87.5	44.9	56.9	..	Fuels
69.9	83.5	100.0	93.2	82.0	75.9	74.6	68.1	80.5	88.4	..	Manufactures
79.7	91.2	100.0	99.8	97.5	95.7	94.2	94.1	109.6	120.0	f	Import Prices, cif
				(Millions of 1980 US dollars)							Trade at Constant 1980 Prices
18,421	21,719	25,680	22,070	21,180	23,520	22,425	23,401	23,710	23,490	f	Exports, fob
9,556	9,853	19,246	22,654	19,044	16,417	17,233	12,189	11,847	12,191	f	Imports, cif
				(Millions of current US dollars)							**BALANCE OF PAYMENTS**
15,269	20,453	29,246	24,120	20,579	21,427	19,975	18,875	20,978	24,257	..	Exports of Goods & Services
13,034	17,679	25,684	20,596	17,344	18,227	16,931	16,264	18,259	21,088	..	Merchandise, fob
1,327	1,609	2,158	2,207	2,058	1,855	1,772	1,465	1,437	1,687	..	Nonfactor Services
908	1,164	1,405	1,317	1,177	1,345	1,272	1,145	1,282	1,482	..	Factor Services
13,819	17,157	25,987	29,024	23,969	21,646	21,643	16,336	17,941	21,442	..	Imports of Goods & Services
9,212	11,578	18,272	20,646	16,647	14,218	14,747	10,431	11,097	13,903	..	Merchandise, fob
2,108	2,524	3,587	3,978	3,484	3,189	3,028	2,363	2,752	3,172	..	Nonfactor Services
2,498	3,056	4,128	4,399	3,838	4,240	3,868	3,543	4,091	4,367	..	Factor Services
..	..	..	..	..	..	..	..	..	..	..	Long-Term Interest
17	52	94	89	117	140	125	84	76	97	..	Current Transfers, net
0	0	0	0	0	0	0	0	0	0	..	Workers' Remittances
1,467	3,348	3,353	-4,814	-3,273	-79	-1,543	2,622	3,114	2,911	K	Total to be Financed
63	63	145	230	134	56	-25	-11	11	115	K	Official Capital Grants
1,530	3,411	3,498	-4,584	-3,139	-23	-1,568	2,612	3,125	3,027	..	Current Account Balance
-162	-1,565	-860	379	2,414	-338	1,705	-695	-1,280	-998	..	Long-Term Capital, net
-386	-496	-765	-575	308	-92	248	-499	-113	28	..	Direct Investment
..	..	..	..	..	..	..	..	..	..	..	Long-Term Loans
..	..	..	..	..	..	..	..	..	..	..	Disbursements
..	..	..	..	..	..	..	..	..	..	..	Repayments
160	-1,132	-239	724	1,972	-302	1,482	-186	-1,178	-1,142	..	Other Long-Term Capital
-1,135	-1,533	-1,868	3,151	-171	1,342	-662	-2,436	-2,020	-593	..	Other Capital, net
-233	-313	-771	1,055	895	-981	525	519	175	-1,436	..	Change in Reserves
				(South African Rand per US dollar)							Memo Item:
0.870	0.840	0.780	0.880	1.080	1.110	1.440	2.190	2.270	2.030	..	Conversion Factor (Annual Avg)
				(Millions of US dollars, outstanding at end of year)							**EXTERNAL DEBT, ETC.**
..	..	..	..	..	..	..	..	..	..	..	Public/Publicly Guar. Long-Term
..	..	..	..	..	..	..	..	..	..	..	Official Creditors
..	..	..	..	..	..	..	..	..	..	..	IBRD and IDA
..	..	..	..	..	..	..	..	..	..	..	Private Creditors
..	..	..	..	..	..	..	..	..	..	..	Private Non-guaranteed Long-Term
..	..	..	..	..	..	..	..	..	..	..	Use of Fund Credit
..	..	..	..	..	..	..	..	..	..	..	Short-Term Debt
				(Millions of US dollars)							Memo Items:
423	434	726	666	485	823	242	315	370	641	..	Int'l Reserves Excluding Gold
2,213	5,135	7,162	3,693	3,459	2,972	2,269	1,583	1,884	2,822	..	Gold Holdings (at market price)
											SOCIAL INDICATORS
5.0	5.0	4.9	4.8	4.8	4.7	4.7	4.6	4.5	4.5	..	Total Fertility Rate
34.1	33.8	33.6	33.3	33.1	33.4	33.6	33.9	34.2	34.4	..	Crude Birth Rate
92.6	90.2	87.8	85.4	83.0	79.2	75.4	71.7	67.9	64.1	..	Infant Mortality Rate
56.3	56.7	57.1	57.5	57.9	58.6	59.3	60.0	60.7	61.5	..	Life Expectancy at Birth
100.4	95.5	96.4	108.0	91.7	73.7	80.4	83.2	84.3	84.4	..	Food Production, p.c. ('79-81 = 100)
19.8	18.1	16.5	..	..	..	..	..	..	..	..	Labor Force, Agriculture (%)
33.9	34.0	34.1	34.3	34.5	34.6	34.8	34.9	35.1	35.2	..	Labor Force, Female (%)
..	..	..	..	..	..	..	..	..	..	..	School Enroll. Ratio, primary
..	..	..	..	..	..	..	..	..	..	..	School Enroll. Ratio, secondary

SPAIN	1967	1968	1969	1970	1971	1972	1973	1974	1975	1976	1977
CURRENT GNP PER CAPITA (US $)	890	970	1,060	1,100	1,210	1,430	1,790	2,280	2,750	3,060	3,340
POPULATION (thousands)	32,735	33,079	33,427	33,779	34,190	34,498	34,810	35,147	35,515	35,937	36,367
ORIGIN AND USE OF RESOURCES					(Billions of current Spanish Pesetas)						
Gross National Product (GNP)	1,843	2,065	2,344	2,607	2,946	3,459	4,179	5,130	6,005	7,207	9,133
Net Factor Income from Abroad	-9	-11	-16	-17	-16	-16	-12	-1	-19	-40	-62
Gross Domestic Product (GDP)	1,852	2,075	2,360	2,624	2,962	3,476	4,190	5,131	6,023	7,248	9,195
Indirect Taxes, net	128	134	166	182	191	229	293	305	321	386	483
GDP at factor cost	1,724	1,941	2,195	2,443	2,771	3,246	3,898	4,826	5,702	6,862	8,712
Agriculture	..	..	..	..	..	..	..	..	..	..	..
Industry	..	..	..	..	..	..	..	..	..	..	..
Manufacturing	..	..	..	..	..	..	..	..	..	..	..
Services, etc.	..	..	..	..	..	..	..	..	..	..	..
Resource Balance	-46	-38	-49	-25	26	8	-32	-248	-230	-322	-189
Exports of Goods & NFServices	187	241	283	349	422	507	612	740	817	998	1,334
Imports of Goods & NFServices	233	279	332	374	396	499	644	987	1,047	1,321	1,523
Domestic Absorption	1,898	2,113	2,409	2,649	2,936	3,468	4,223	5,379	6,253	7,570	9,384
Private Consumption, etc.	1,228	1,362	1,514	1,670	1,884	2,216	2,667	3,294	3,899	4,770	6,059
General Gov't Consumption	167	182	209	239	275	318	384	489	607	788	1,019
Gross Domestic Investment	504	570	687	739	777	934	1,172	1,596	1,747	2,011	2,306
Fixed Investment	467	537	622	689	719	884	1,119	1,422	1,570	1,812	2,216
Memo Items:											
Gross Domestic Saving	457	532	637	715	803	941	1,140	1,348	1,517	1,689	2,117
Gross National Saving	476	553	660	744	841	982	1,211	1,414	1,566	1,726	2,162
					(Billions of 1980 Spanish Pesetas)						
Gross National Product	8,897	9,494	10,326	10,751	11,257	12,168	13,129	13,864	13,898	14,317	14,728
GDP at factor cost	8,223	8,931	9,451	10,062	10,574	11,399	12,224	13,043	13,193	13,625	14,045
Agriculture	..	..	..	..	..	..	..	..	..	..	..
Industry	..	..	..	..	..	..	..	..	..	..	..
Manufacturing	..	..	..	..	..	..	..	..	..	..	..
Services, etc.	..	..	..	..	..	..	..	..	..	..	..
Resource Balance	-357	-307	-358	-274	-112	-291	-445	-636	-624	-766	-407
Exports of Goods & NFServices	759	899	1,038	1,220	1,393	1,580	1,737	1,720	1,712	1,798	2,015
Imports of Goods & NFServices	1,116	1,206	1,396	1,494	1,505	1,870	2,182	2,356	2,336	2,564	2,422
Domestic Absorption	9,296	9,852	10,756	11,096	11,430	12,518	13,612	14,503	14,565	15,163	15,236
Private Consumption, etc.	5,869	6,221	6,700	6,981	7,333	7,941	8,558	8,991	9,152	9,660	9,806
General Gov't Consumption	1,012	1,030	1,074	1,130	1,179	1,240	1,320	1,443	1,517	1,622	1,686
Gross Domestic Investment	2,415	2,601	2,983	2,986	2,918	3,337	3,734	4,069	3,895	3,880	3,744
Fixed Investment	2,330	2,549	2,799	2,882	2,797	3,193	3,609	3,834	3,661	3,631	3,597
Memo Items:											
Capacity to Import	893	1,043	1,190	1,395	1,605	1,899	2,073	1,765	1,823	1,938	2,121
Terms of Trade Adjustment	134	144	151	176	212	319	336	45	111	140	106
Gross Domestic Income	9,073	9,689	10,549	10,998	11,530	12,547	13,503	13,912	14,052	14,537	14,935
Gross National Income	9,031	9,638	10,477	10,926	11,469	12,487	13,465	13,909	14,009	14,456	14,834
DOMESTIC PRICES (DEFLATORS)					(Index 1980 = 100)						
Overall (GDP)	20.7	21.7	22.7	24.2	26.2	28.4	31.8	37.0	43.2	50.3	62.0
Domestic Absorption	20.4	21.4	22.4	23.9	25.7	27.7	31.0	37.1	42.9	49.9	61.6
Agriculture	..	..	..	..	..	..	..	..	..	..	..
Industry	..	..	..	..	..	..	..	..	..	..	..
Manufacturing	..	..	..	..	..	..	..	..	..	..	..
MANUFACTURING ACTIVITY											
Employment (1980=100)	76.4	77.7	79.7	80.8	82.2	84.6	88.0	91.2	92.0	92.7	91.8
Real Earnings per Empl. (1980=100)	54.5	55.9	63.3	67.4	70.1	76.2	81.0	87.9	95.5	104.7	110.6
Real Output per Empl. (1980=100)	..	..	..	..	..	..	..	..	..	..	..
Earnings as % of Value Added	51.5	50.6	50.1	51.7	52.2	51.3	50.9	51.9	59.6	63.3	65.8
MONETARY HOLDINGS					(Billions of current Spanish Pesetas)						
Money Supply, Broadly Defined	1,204	1,430	1,698	1,955	2,424	2,985	3,710	4,419	5,257	6,270	7,441
Money as Means of Payment	545	612	701	743	920	1,142	1,410	1,654	1,963	2,392	2,836
Currency Ouside Banks	199	218	243	263	294	328	387	447	524	614	777
Demand Deposits	347	394	459	480	626	814	1,023	1,208	1,439	1,778	2,058
Quasi-Monetary Liabilities	658	818	997	1,211	1,505	1,843	2,301	2,764	3,294	3,877	4,605
GOVERNMENT DEFICIT (-) OR SURPLUS					(Billions of current Spanish Pesetas)						
	..	..	..	-16.8	-46.4	-18.7	-10.4	-59.4	-107.9	-65.4	-200.1
Current Revenue	..	..	..	479.0	554.9	683.6	833.6	1,014.1	1,259.4	1,497.2	2,080.4
Current Expenditure	..	..	..	406.7	483.5	569.4	708.0	875.0	1,096.6	1,317.6	1,856.9
Current Budget Balance	..	..	..	72.3	71.4	114.2	125.6	139.1	162.8	179.6	223.5
Capital Receipts	..	..	..	0.5	0.6	1.4	1.9	1.8	1.7	1.7	2.4
Capital Payments	..	..	..	89.6	118.4	134.3	137.9	200.3	272.4	246.7	426.0

1978	1979	1980	1981	1982	1983	1984	1985	1986	1987 estimate	NOTES	SPAIN
3,660	4,380	5,300	5,630	5,320	4,670	4,380	4,290	4,840	6,010	..	**CURRENT GNP PER CAPITA (US $)**
36,778	37,108	37,386	37,654	37,935	38,228	38,415	38,523	38,699	38,832	..	POPULATION (thousands)
				(Billions of current Spanish Pesetas)							**ORIGIN AND USE OF RESOURCES**
11,163	13,079	15,079	16,751	19,283	21,876	24,714	27,583	31,680	35,448	..	Gross National Product (GNP)
-87	-78	-130	-238	-238	-285	-358	-397	-298	-110	..	Net Factor Income from Abroad
11,251	13,158	15,209	16,989	19,567	22,235	25,111	27,914	31,979	35,558	..	Gross Domestic Product (GDP)
480	602	694	909	1,033	1,324	1,568	1,909	2,658	3,158	..	Indirect Taxes, net
10,771	12,556	14,515	16,080	18,534	20,910	23,544	26,005	29,321	32,400	B	GDP at factor cost
..	..	1,073	1,038	1,226	1,370	1,643	1,729	1,759	..	..	Agriculture
..	..	5,877	6,517	7,416	8,373	9,305	10,283	..	..	..	Industry
..	..	4,284	4,762	5,326	6,149	6,904	7,622	..	..	..	Manufacturing
..	..	8,259	9,434	10,925	12,492	14,164	15,902	..	..	..	Services, etc.
90	39	-349	-349	-352	-133	581	603	702	85	..	Resource Balance
1,711	1,974	2,410	3,080	3,673	4,727	5,944	6,525	6,509	7,202	..	Exports of Goods & NFServices
1,621	1,936	2,759	3,429	4,025	4,859	5,363	5,922	5,807	7,118	..	Imports of Goods & NFServices
11,161	13,119	15,558	17,338	19,919	22,368	24,530	27,311	31,277	35,473	..	Domestic Absorption
7,700	9,051	10,080	11,455	13,147	14,812	16,369	18,110	20,427	22,623	..	Private Consumption, etc.
1,292	1,575	1,929	2,242	2,620	3,091	3,448	3,909	4,446	5,132	..	General Gov't Consumption
2,169	2,493	3,548	3,641	4,153	4,464	4,713	5,291	6,404	7,718	..	Gross Domestic Investment
2,568	2,861	3,368	3,718	4,244	4,653	4,838	5,429	6,426	7,667	..	Fixed Investment
											Memo Items:
2,259	2,532	3,199	3,292	3,801	4,331	5,294	5,894	7,106	7,803	..	Gross Domestic Saving
2,299	2,575	3,217	3,210	3,689	4,147	5,088	5,798	7,016	7,775	..	Gross National Saving
				(Billions of 1980 Spanish Pesetas)							
14,925	14,932	15,079	14,963	15,136	15,387	15,668	16,091	16,655	17,530	..	Gross National Product
14,390	14,320	14,515	14,375	14,559	14,725	14,940	15,184	15,400	16,085	B	GDP at factor cost
..	..	1,073	972	957	1,017	1,104	1,131	1,006	..	..	Agriculture
..	..	5,877	5,863	5,842	5,916	5,885	6,010	..	..	..	Industry
..	..	4,284	4,279	4,214	4,287	4,304	4,379	..	..	..	Manufacturing
..	..	8,259	8,336	8,557	8,700	8,925	9,140	..	..	..	Services, etc.
-167	-314	-349	-30	-9	286	667	590	181	-293	..	Resource Balance
2,230	2,356	2,410	2,612	2,737	3,014	3,368	3,458	3,492	3,726	..	Exports of Goods & NFServices
2,397	2,670	2,759	2,642	2,745	2,728	2,700	2,868	3,311	4,019	..	Imports of Goods & NFServices
15,211	15,337	15,558	15,201	15,365	15,347	15,247	15,691	16,633	17,979	..	Domestic Absorption
9,898	10,023	10,080	10,020	10,038	10,073	10,034	10,258	10,634	11,191	..	Private Consumption, etc.
1,778	1,852	1,929	1,966	2,061	2,142	2,203	2,306	2,423	2,642	..	General Gov't Consumption
3,535	3,462	3,548	3,215	3,265	3,133	3,010	3,127	3,576	4,146	..	Gross Domestic Investment
3,499	3,346	3,368	3,259	3,275	3,192	3,007	3,120	3,418	3,886	..	Fixed Investment
											Memo Items:
2,531	2,723	2,410	2,373	2,505	2,654	2,993	3,160	3,711	4,067	..	Capacity to Import
300	367	0	-239	-231	-361	-375	-298	219	341	..	Terms of Trade Adjustment
15,344	15,390	15,209	14,932	15,124	15,273	15,540	15,983	17,033	18,027	..	Gross Domestic Income
15,225	15,299	15,079	14,723	14,905	15,027	15,293	15,793	16,874	..	..	Gross National Income
				(Index 1980 = 100)							**DOMESTIC PRICES (DEFLATORS)**
74.8	87.6	100.0	112.0	127.4	142.2	157.8	171.4	190.2	201.1	..	Overall (GDP)
73.4	85.5	100.0	114.1	129.6	145.7	160.9	174.1	188.0	197.3	..	Domestic Absorption
..	..	100.0	106.8	128.1	134.8	148.8	152.9	174.9	..	..	Agriculture
..	..	100.0	111.2	127.0	141.5	158.1	171.1	..	..	..	Industry
..	..	100.0	111.3	126.4	143.4	160.4	174.1	..	..	..	Manufacturing
											MANUFACTURING ACTIVITY
101.9	99.6	100.0	92.9	86.6	84.3	80.5	77.2	..	..	..	Employment (1980=100)
96.1	97.7	100.0	100.7	101.5	100.9	111.3	119.0	..	..	B	Real Earnings per Empl. (1980=100)
..	..	100.0	106.6	110.8	116.8	121.8	129.3	..	..	..	Real Output per Empl. (1980=100)
44.1	44.6	45.0	45.2	43.4	41.1	43.2	43.2	..	..	..	Earnings as % of Value Added
				(Billions of current Spanish Pesetas)							**MONETARY HOLDINGS**
8,953	10,551	12,319	14,269	16,259	16,083	16,866	18,867	22,027	24,050	..	Money Supply, Broadly Defined
3,326	3,609	4,098	4,630	4,850	5,277	5,746	6,589	7,580	8,899	..	Money as Means of Payment
946	1,039	1,184	1,333	1,531	1,688	1,867	2,083	2,406	2,741	..	Currency Ouside Banks
2,379	2,571	2,913	3,297	3,319	3,589	3,879	4,505	5,174	6,158	..	Demand Deposits
5,628	6,942	8,221	9,639	11,409	10,806	11,120	12,278	14,447	15,151	..	Quasi-Monetary Liabilities
				(Billions of current Spanish Pesetas)							**GOVERNMENT DEFICIT (-) OR SURPLUS**
-264.3	-465.4	-635.1	-873.9	-1,093.3	-1,408.3	-2,113.2	-2,332.0	-1,635.0	..	..	Current Revenue
2,658.8	3,183.8	3,674.1	4,134.6	4,996.7	5,867.3	6,611.4	7,743.1	9,475.2	..	..	Current Revenue
2,557.9	3,170.3	3,612.5	4,016.5	5,178.0	6,078.1	7,129.3	8,447.7	9,730.2	..	..	Current Expenditure
100.9	13.5	61.6	118.1	-181.3	-210.8	-517.9	-704.6	-255.0	..	..	Current Budget Balance
1.6	1.3	1.8	1.3	10.4	15.2	118.1	11.5	3.6	..	..	Capital Receipts
366.8	480.2	698.5	993.3	922.4	1,212.7	1,713.4	1,639.3	1,383.8	..	..	Capital Payments

SPAIN	1967	1968	1969	1970	1971	1972	1973	1974	1975	1976	1977
FOREIGN TRADE (CUSTOMS BASIS)				*(Millions of current US dollars)*							
Value of Exports, fob	1,375	1,589	1,900	2,387	2,938	3,803	5,162	7,059	7,675	8,712	10,218
Nonfuel Primary Products	738	711	756	979	1,039	1,195	1,713	2,011	2,042	2,352	2,625
Fuels	80	137	122	131	126	138	242	478	252	325	378
Manufactures	557	741	1,021	1,277	1,773	2,471	3,206	4,570	5,381	6,035	7,214
Value of Imports, cif	3,453	3,502	4,202	4,714	4,936	6,754	9,536	15,291	16,100	17,288	17,648
Nonfuel Primary Products	1,152	1,174	1,440	1,536	1,657	2,181	3,141	4,529	4,664	4,473	5,000
Fuels	427	541	529	627	812	975	1,243	3,891	4,191	5,093	5,050
Manufactures	1,874	1,787	2,232	2,551	2,467	3,598	5,152	6,871	7,246	7,721	7,597
Terms of Trade	166.7	165.4	144.1	153.1	159.7	155.8	146.4	108.2	117.9	114.0	108.0
					(Index 1980 = 100)						
Export Prices, fob	29.5	26.4	26.0	27.6	30.6	35.1	46.2	56.6	62.4	62.0	64.7
Nonfuel Primary Products	34.8	35.8	38.6	39.5	39.1	41.7	53.8	62.8	66.0	64.9	69.7
Fuels	8.5	7.3	5.4	5.9	7.0	8.5	24.5	36.6	33.6	38.5	36.8
Manufactures	35.0	34.4	33.1	32.5	34.4	38.9	45.9	57.3	63.7	63.1	65.6
Import Prices, cif	17.7	15.9	18.0	18.1	19.1	22.5	31.6	52.3	52.9	54.4	59.9
Trade at Constant 1980 Prices				*(Millions of 1980 US dollars)*							
Exports, fob	4,658	6,029	7,312	8,635	9,608	10,844	11,164	12,466	12,308	14,054	15,803
Imports, cif	19,504	21,969	23,295	26,114	25,787	30,002	30,205	29,226	30,451	31,795	29,483
BALANCE OF PAYMENTS				*(Millions of current US dollars)*							
Exports of Goods & Services	..	..	..	4,900	5,950	7,663	10,463	12,862	13,862	14,545	17,194
Merchandise, fob	..	..	..	2,483	2,979	3,920	5,304	7,211	7,798	8,990	10,546
Nonfactor Services	..	..	..	2,367	2,863	3,577	4,817	4,989	5,453	5,114	6,178
Factor Services	..	..	..	50	108	165	342	663	612	441	470
Imports of Goods & Services	..	..	..	5,480	5,861	7,951	11,287	17,237	18,497	19,985	20,746
Merchandise, fob	..	..	..	4,357	4,578	6,236	8,807	14,258	15,188	16,302	16,738
Nonfactor Services	..	..	..	817	941	1,255	1,894	2,244	2,373	2,649	2,767
Factor Services	..	..	..	306	343	459	585	735	936	1,034	1,240
Long-Term Interest	..	..	..	..	..	..	..	..	..	..	..
Current Transfers, net	..	..	..	659	772	878	1,416	1,151	1,162	1,161	1,416
Workers' Remittances	..	..	..	469	550	599	902	859	972	1,000	1,090
Total to be Financed	..	..	..	79	861	591	592	-3,224	-3,473	-4,280	-2,135
Official Capital Grants	..	..	..	..	-5	-10	-7	-8	-19	-16	-9
Current Account Balance	..	..	..	79	856	581	585	-3,233	-3,492	-4,296	-2,144
Long-Term Capital, net	..	..	..	664	494	838	758	1,759	1,786	1,995	3,004
Direct Investment	..	..	..	179	177	231	337	273	512	284	491
Long-Term Loans	..	..	..	..	..	..	..	..	..	..	..
Disbursements	..	..	..	..	..	..	..	..	..	..	..
Repayments	..	..	..	..	..	..	..	..	..	..	..
Other Long-Term Capital	..	..	..	485	323	617	428	1,495	1,293	1,727	2,523
Other Capital, net	..	..	..	76	104	27	-6	761	877	1,211	286
Change in Reserves	..	..	..	-819	-1,454	-1,446	-1,338	712	829	1,090	-1,146
Memo Item:				*(Spanish Pesetas per US dollar)*							
Conversion Factor (Annual Avg)	60.830	70.000	70.000	70.000	69.470	64.270	58.260	57.690	57.410	66.900	75.960
EXTERNAL DEBT, ETC.				*(Millions US dollars, outstanding at end of year)*							
Public/Publicly Guar. Long-Term	..	..	..	..	..	..	..	..	..	..	..
Official Creditors	..	..	..	..	..	..	..	..	..	..	..
IBRD and IDA	..	..	..	..	..	..	..	..	..	..	..
Private Creditors	..	..	..	..	..	..	..	..	..	..	..
Private Non-guaranteed Long-Term	..	..	..	..	..	..	..	..	..	..	..
Use of Fund Credit	..	..	..	..	..	..	..	..	..	..	..
Short-Term Debt	..	..	..	..	..	..	..	..	..	..	..
Memo Items:				*(Millions of US dollars)*							
Int'l Reserves Excluding Gold	315	364	497	1,320	2,727	4,473	6,170	5,874	5,506	4,705	5,977
Gold Holdings (at market price)	789	940	789	532	621	924	1,601	2,661	2,001	1,922	2,381
SOCIAL INDICATORS											
Total Fertility Rate	2.9	2.9	2.9	2.8	2.9	2.8	2.8	2.9	2.8	2.8	2.6
Crude Birth Rate	20.8	20.2	19.9	19.6	19.7	19.5	19.3	19.6	18.8	18.9	18.0
Infant Mortality Rate	34.0	32.4	30.2	28.1	25.7	22.9	21.5	19.9	18.9	17.1	16.0
Life Expectancy at Birth	71.6	71.9	72.1	72.3	72.6	72.8	72.9	73.1	73.2	73.3	73.5
Food Production, p.c. ('79-81 = 100)	74.1	78.1	76.4	80.1	81.2	85.7	91.9	91.5	94.2	94.4	91.4
Labor Force, Agriculture (%)	30.8	29.2	27.6	26.0	25.1	24.2	23.3	22.4	21.5	20.6	19.7
Labor Force, Female (%)	19.1	19.2	19.3	19.4	19.9	20.3	20.8	21.3	21.7	22.1	22.5
School Enroll. Ratio, primary	..	..	..	123.0	..	..	..	..	111.0	110.0	109.0
School Enroll. Ratio, secondary	..	..	..	56.0	..	..	..	..	73.0	76.0	78.0

1978	1979	1980	1981	1982	1983	1984	1985	1986	1987 estimate	NOTES	SPAIN
											FOREIGN TRADE (CUSTOMS BASIS)
			(Millions of current US dollars)								
13,103	18,196	20,827	20,337	20,271	19,711	23,283	24,307	27,250	34,099	..	Value of Exports, fob
3,153	4,502	5,098	5,009	4,329	4,222	5,009	4,932	5,826	7,827	..	Nonfuel Primary Products
330	347	761	1,008	1,418	1,734	2,027	2,148	1,682	2,001	..	Fuels
9,620	13,347	14,967	14,320	14,525	13,755	16,247	17,227	19,742	24,271	..	Manufactures
18,630	25,370	33,901	32,081	31,282	28,926	28,607	30,066	35,406	49,009	..	Value of Imports, cif
5,397	6,941	7,835	6,730	6,542	6,325	6,384	6,354	7,973	9,648	..	Nonfuel Primary Products
5,306	7,674	13,116	13,641	12,474	11,671	10,826	10,865	6,696	7,993	..	Fuels
7,928	10,755	12,950	11,710	12,266	10,930	11,397	12,848	20,738	31,368	..	Manufactures
			(Index 1980 = 100)								
114.9	114.6	100.0	85.1	91.4	86.8	89.3	89.8	104.5	93.6	..	Terms of Trade
72.9	92.0	100.0	86.5	86.4	79.0	79.3	77.0	88.1	88.2	..	Export Prices, fob
79.5	101.6	100.0	91.3	84.6	87.5	85.8	82.7	89.3	97.3	..	Nonfuel Primary Products
44.7	93.6	100.0	104.0	97.0	85.4	83.6	81.1	41.3	44.4	..	Fuels
72.6	89.2	100.0	84.0	86.0	76.0	77.0	75.0	97.0	93.0	..	Manufactures
63.5	80.2	100.0	101.6	94.5	91.0	88.8	85.7	84.3	94.2	..	Import Prices, cif
											Trade at Constant 1980 Prices
17,967	19,783	20,827	23,504	23,470	24,954	29,366	31,582	30,947	38,652	..	Exports, fob
29,346	31,624	33,901	31,568	33,086	31,786	32,212	35,081	42,024	52,000	..	Imports, cif
			(Millions of current US dollars)								
											BALANCE OF PAYMENTS
22,389	29,975	33,870	34,465	34,798	32,718	36,794	38,367	46,004	56,814	..	Exports of Goods & Services
13,452	18,348	20,564	21,026	21,295	19,865	22,698	23,585	26,504	33,374	..	Merchandise, fob
8,157	10,255	11,547	11,370	11,535	11,499	12,573	12,933	17,805	21,569	..	Nonfactor Services
779	1,373	1,758	2,069	1,968	1,355	1,523	1,848	1,695	1,871	..	Factor Services
22,442	30,625	41,099	41,038	40,606	36,579	35,857	36,705	42,994	59,492	..	Imports of Goods & Services
17,539	24,022	32,282	31,095	30,545	27,541	26,974	27,776	32,758	46,291	..	Merchandise, fob
3,042	4,083	5,561	5,585	5,812	5,217	4,942	5,189	6,446	8,399	..	Nonfactor Services
1,861	2,521	3,256	4,359	4,249	3,821	3,941	3,741	3,790	4,802	..	Factor Services
..	..	..	..	..	..	..	..	..	..	..	Long-Term Interest
1,669	1,804	2,057	1,698	1,577	1,211	1,189	1,377	1,489	2,266	..	Current Transfers, net
1,268	1,426	1,647	1,300	1,124	938	843	1,025	1,180	1,313	..	Workers' Remittances
1,616	1,154	-5,173	-4,876	-4,230	-2,650	2,126	3,038	4,500	-412	K	Total to be Financed
-14	-19	-6	-7	4	-46	-97	-273	-398	361	K	Official Capital Grants
1,602	1,135	-5,178	-4,882	-4,226	-2,696	2,029	2,765	4,102	-51	..	Current Account Balance
1,744	3,148	4,136	4,199	1,771	3,096	3,116	-1,769	-2,508	9,226	..	Long-Term Capital, net
1,067	1,266	1,182	1,440	1,279	1,384	1,525	1,698	3,057	3,814	..	Direct Investment
..	..	..	..	..	..	..	..	..	..	..	Long-Term Loans
..	..	..	..	..	..	..	..	..	..	..	Disbursements
..	..	..	..	..	..	..	..	..	..	..	Repayments
692	1,902	2,960	2,765	488	1,758	1,688	-3,194	-5,167	5,051	..	Other Long-Term Capital
412	-833	247	-44	-649	-665	-328	-3,271	750	3,630	..	Other Capital, net
-3,759	-3,450	795	727	3,104	265	-4,817	2,275	-2,344	-12,805	..	Change in Reserves
											Memo Item:
			(Spanish Pesetas per US dollar)								
76.670	67.130	71.700	92.310	109.860	143.430	160.760	170.040	140.050	123.480	..	Conversion Factor (Annual Avg)
			(Millions US dollars, outstanding at end of year)								**EXTERNAL DEBT, ETC.**
..	..	..	..	..	..	..	..	..	..	..	Public/Publicly Guar. Long-Term
..	..	..	..	..	..	..	..	..	..	..	Official Creditors
..	..	..	..	..	..	..	..	..	..	..	IBRD and IDA
..	..	..	..	..	..	..	..	..	..	..	Private Creditors
..	..	..	..	..	..	..	..	..	..	..	Private Non-guaranteed Long-Term
..	..	..	..	..	..	..	..	..	..	..	Use of Fund Credit
..	..	..	..	..	..	..	..	..	..	..	Short-Term Debt
			(Millions of US dollars)								Memo Items:
10,112	13,224	11,863	10,805	7,655	7,402	11,955	11,175	14,755	30,669	..	Int'l Reserves Excluding Gold
3,282	7,478	8,610	5,806	6,673	5,572	4,510	4,792	5,794	5,770	..	Gold Holdings (at market price)
											SOCIAL INDICATORS
2.5	2.3	2.2	2.0	2.0	1.9	1.9	1.8	1.8	1.8	..	Total Fertility Rate
17.3	16.0	15.1	14.1	13.9	13.8	13.6	13.5	13.3	13.2	..	Crude Birth Rate
15.3	14.4	11.1	10.3	10.3	10.2	10.2	10.2	10.1	10.1	..	Infant Mortality Rate
73.9	74.3	74.7	75.0	75.4	75.6	75.9	76.1	76.3	76.5	..	Life Expectancy at Birth
101.5	100.9	106.0	93.0	101.7	92.9	107.8	103.4	101.6	107.0	..	Food Production, p.c. ('79-81 = 100)
18.9	18.0	17.1	..	..	..	..	..	..	..	..	Labor Force, Agriculture (%)
22.9	23.2	23.6	23.7	23.8	23.9	24.0	24.1	24.2	24.2	..	Labor Force, Female (%)
109.0	109.0	109.0	..	..	108.0	106.0	104.0	101.0	..	..	School Enroll. Ratio, primary
83.0	86.0	87.0	..	..	89.0	92.0	98.0	..	..	..	School Enroll. Ratio, secondary

SRI LANKA	1967	1968	1969	1970	1971	1972	1973	1974	1975	1976	1977
CURRENT GNP PER CAPITA (US $)	170	170	170	170	170	160	170	190	220	230	220
POPULATION (thousands)	11,703	11,992	12,252	12,516	12,608	12,861	13,091	13,284	13,496	13,717	13,942

ORIGIN AND USE OF RESOURCES
(Billions of current Sri Lanka Rupees)

	1967	1968	1969	1970	1971	1972	1973	1974	1975	1976	1977
Gross National Product (GNP)	9.57	11.18	12.33	13.44	13.86	15.07	18.22	23.59	26.36	29.92	36.15
Net Factor Income from Abroad	-0.11	-0.12	-0.18	-0.22	-0.19	-0.18	-0.18	-0.18	-0.21	-0.28	-0.25
Gross Domestic Product (GDP)	9.68	11.30	12.51	13.66	14.05	15.25	18.40	23.77	26.58	30.20	36.41
Indirect Taxes, net	0.30	0.11	0.31	0.48	0.38	0.53	0.48	0.47	0.89	2.17	1.72
GDP at factor cost	9.37	11.19	12.21	13.19	13.67	14.72	17.92	23.30	25.69	28.03	34.68
Agriculture	2.78	3.49	3.52	3.73	3.70	3.88	4.89	7.73	7.80	8.13	10.64
Industry	1.85	2.36	2.80	3.14	3.38	3.56	4.55	5.84	6.79	7.59	9.95
Manufacturing	1.36	1.73	2.02	2.20	2.40	2.59	3.12	4.34	5.16	5.62	8.02
Services, etc.	4.75	5.34	5.88	6.32	6.59	7.28	8.48	9.73	11.10	12.30	14.09
Resource Balance	-0.40	-0.49	-1.13	-0.43	-0.28	-0.24	-0.22	-1.78	-1.98	-0.70	1.33
Exports of Goods & NFServices	2.87	3.42	3.34	3.48	3.46	3.40	4.48	6.28	7.31	8.77	12.31
Imports of Goods & NFServices	3.26	3.92	4.46	3.91	3.74	3.64	4.70	8.06	9.29	9.48	10.98
Domestic Absorption	10.07	11.79	13.64	14.09	14.33	15.49	18.63	25.55	28.56	30.91	35.07
Private Consumption, etc.	7.46	8.68	9.91	9.88	10.17	10.95	14.08	19.07	21.94	22.99	26.70
General Gov't Consumption	1.22	1.39	1.45	1.62	1.76	1.90	2.02	2.74	2.48	3.02	3.12
Gross Domestic Investment	1.40	1.72	2.28	2.59	2.40	2.64	2.53	3.74	4.14	4.90	5.26
Fixed Investment	1.37	1.57	2.28	2.36	2.14	2.21	2.49	2.97	3.70	4.59	5.03
Memo Items:											
Gross Domestic Saving	1.00	1.23	1.16	2.16	2.12	2.40	2.30	1.96	2.16	4.19	6.59
Gross National Saving	0.87	1.10	0.97	1.93	1.90	2.19	2.12	1.77	1.97	3.98	6.48

(Millions of 1980 Sri Lanka Rupees)

	1967	1968	1969	1970	1971	1972	1973	1974	1975	1976	1977
Gross National Product	35,960	37,966	40,946	42,451	43,196	42,773	46,118	47,999	51,137	52,823	55,648
GDP at factor cost	35,497	38,482	40,378	42,167	42,209	43,490	45,088	46,435	47,635	49,101	51,100
Agriculture	12,066	12,772	12,933	13,430	13,106	13,513	13,405	14,186	13,844	14,013	15,470
Industry	9,366	10,814	11,921	12,999	13,039	12,893	13,669	13,174	13,412	14,450	13,894
Manufacturing	7,150	7,842	8,571	9,053	9,391	9,564	9,338	8,913	9,325	9,770	9,713
Services, etc.	14,065	14,897	15,523	15,738	16,064	17,084	18,015	19,075	20,379	20,638	21,735
Resource Balance	-1,875	-4,059	-5,738	-3,563	-1,967	-776	-96	-2,066	1,207	-753	-2,164
Exports of Goods & NFServices	18,088	18,101	17,996	18,767	18,169	17,798	18,004	15,602	18,728	19,150	16,608
Imports of Goods & NFServices	19,962	22,160	23,734	22,330	20,136	18,574	18,100	17,668	17,521	19,903	18,771
Domestic Absorption	38,263	42,458	47,323	46,779	45,803	44,099	46,713	50,452	50,342	54,108	58,259
Private Consumption, etc.	29,122	32,700	36,479	34,963	34,076	31,497	36,062	36,517	37,992	38,789	43,799
General Gov't Consumption	5,059	5,062	5,077	5,263	5,675	6,043	5,457	5,806	4,913	5,653	5,133
Gross Domestic Investment	4,082	4,697	5,767	6,553	6,051	6,559	5,193	8,128	7,437	9,666	9,328
Fixed Investment	3,997	4,277	5,749	5,971	5,400	5,485	5,122	6,468	6,645	9,072	8,930
Memo Items:											
Capacity to Import	17,534	19,377	17,743	19,873	18,623	17,351	17,242	13,776	13,778	18,423	21,049
Terms of Trade Adjustment	-554	1,276	-252	1,106	454	-447	-762	-1,826	-4,951	-727	4,441
Gross Domestic Income	35,834	39,674	41,332	44,322	44,289	42,875	45,855	46,560	46,598	52,628	60,536
Gross National Income	35,406	39,242	40,694	43,557	43,650	42,326	45,356	46,173	46,187	52,096	60,090

DOMESTIC PRICES (DEFLATORS)
(Index 1980 = 100)

	1967	1968	1969	1970	1971	1972	1973	1974	1975	1976	1977
Overall (GDP)	26.6	29.4	30.1	31.6	32.1	35.2	39.5	49.1	51.6	56.6	64.9
Domestic Absorption	26.3	27.8	28.8	30.1	31.3	35.1	39.9	50.6	56.7	57.1	60.2
Agriculture	23.0	27.3	27.3	27.8	28.3	28.7	36.5	54.5	56.3	58.0	68.8
Industry	19.7	21.9	23.5	24.1	25.9	27.6	33.3	44.3	50.6	52.6	71.6
Manufacturing	19.1	22.0	23.6	24.3	25.5	27.1	33.4	48.7	55.3	57.5	82.6

MANUFACTURING ACTIVITY

	1967	1968	1969	1970	1971	1972	1973	1974	1975	1976	1977
Employment (1980=100)	41.9	53.7	56.7	60.4	67.1	68.7	70.9	69.0	78.3	85.2	93.8
Real Earnings per Empl. (1980=100)	..	..	..	..	..	..	..	..	..	..	..
Real Output per Empl. (1980=100)	66.7	62.1	65.8	69.9	71.3	76.3	63.4	69.8	..	..	..
Earnings as % of Value Added	..	..	..	..	..	..	..	..	..	..	..

MONETARY HOLDINGS
(Billions of current Sri Lanka Rupees)

	1967	1968	1969	1970	1971	1972	1973	1974	1975	1976	1977
Money Supply, Broadly Defined	3.06	3.31	3.48	3.81	4.27	4.97	5.26	5.95	6.35	8.21	11.20
Money as Means of Payment	1.79	1.90	1.87	1.95	2.13	2.46	2.76	2.92	3.06	4.13	5.33
Currency Ouside Banks	0.98	1.07	1.08	0.94	1.12	1.20	1.44	1.54	1.61	2.08	2.79
Demand Deposits	0.81	0.83	0.78	1.01	1.01	1.26	1.32	1.38	1.45	2.05	2.54
Quasi-Monetary Liabilities	1.27	1.42	1.61	1.86	2.14	2.51	2.51	3.03	3.29	4.07	5.87

GOVERNMENT DEFICIT (-) OR SURPLUS
(Millions of current Sri Lanka Rupees)

	1967	1968	1969	1970	1971	1972	1973	1974	1975	1976	1977
	..	..	..	-873	-1,023	..	-960	-767	-1,704	-2,518	-1,671
Current Revenue	..	..	..	2,750	2,650	..	3,717	4,608	5,070	5,693	6,754
Current Expenditure	..	..	..	2,781	2,909	..	3,593	4,227	4,851	5,639	6,366
Current Budget Balance	..	..	..	-31	-259	..	124	381	219	54	388
Capital Receipts	..	..	..	2	2	..	2	5	3	28	23
Capital Payments	..	..	..	844	766	..	1,086	1,153	1,926	2,600	2,082

1978	1979	1980	1981	1982	1983	1984	1985	1986	1987 estimate	NOTES	SRI LANKA
220	230	260	300	320	330	360	380	390	400	..	**CURRENT GNP PER CAPITA** (US $)
14,184	14,471	14,738	14,962	15,189	15,416	15,606	15,837	16,101	16,362	..	**POPULATION** (thousands)
			(Billions of current Sri Lanka Rupees)								**ORIGIN AND USE OF RESOURCES**
42.43	52.15	66.10	83.14	97.28	118.39	149.29	158.98	175.61	192.39	..	Gross National Product (GNP)
-0.24	-0.24	-0.43	-1.87	-1.96	-3.21	-4.45	-3.40	-3.86	-4.34	..	Net Factor Income from Abroad
42.67	52.39	66.53	85.00	99.24	121.60	153.75	162.38	179.47	196.72	f	Gross Domestic Product (GDP)
2.19	2.60	4.28	5.67	4.56	7.72	13.71	14.05	15.76	18.99	..	Indirect Taxes, net
40.48	49.78	62.25	79.34	94.68	113.88	140.04	148.32	163.71	177.73	f	GDP at factor cost
12.33	13.41	17.15	21.98	24.96	32.18	40.14	41.07	44.35	47.92	..	Agriculture
11.03	14.05	18.45	22.21	24.89	29.99	36.86	38.86	43.55	48.76	..	Industry
8.09	9.48	11.05	12.88	13.60	15.96	20.89	21.85	24.87	28.47	..	Manufacturing
17.12	22.32	26.64	35.15	44.83	51.71	63.05	68.39	75.81	81.05	..	Services, etc.
-2.04	-6.31	-15.02	-13.67	-18.76	-18.37	-9.13	-19.41	-20.84	-20.66	..	Resource Balance
14.83	17.66	21.43	25.89	27.15	32.02	44.29	42.24	42.57	49.56	..	Exports of Goods & NFServices
16.87	23.97	36.46	39.56	45.90	50.38	53.42	61.65	63.41	70.22	..	Imports of Goods & NFServices
44.70	58.70	81.55	98.67	118.00	139.97	162.88	181.78	200.31	217.39	..	Domestic Absorption
32.11	40.37	53.40	68.75	79.23	94.95	111.23	126.50	139.37	151.95	..	Private Consumption, etc.
4.04	4.80	5.68	6.31	8.24	9.89	11.93	16.60	18.48	19.54	..	General Gov't Consumption
8.55	13.53	22.46	23.61	30.53	35.13	39.71	38.68	42.46	45.90	..	Gross Domestic Investment
8.52	13.25	20.84	23.28	30.28	35.34	39.56	38.46	42.33	45.75	..	Fixed Investment
											Memo Items:
6.52	7.22	7.44	9.94	11.77	16.77	30.58	19.27	21.62	25.24	..	Gross Domestic Saving
6.62	7.73	9.26	11.99	15.31	20.01	33.15	23.10	25.96	30.06	..	Gross National Saving
			(Millions of 1980 Sri Lanka Rupees)								
58,735	62,599	66,095	68,688	71,493	74,997	79,660	80,659	82,387	84,865	..	Gross National Product
55,375	58,886	62,246	65,787	69,029	72,380	75,988	79,783	83,213	84,415	I f	GDP at factor cost
16,309	16,633	17,151	18,342	18,824	19,765	19,691	21,390	21,948	20,668	..	Agriculture
15,960	17,584	18,450	18,903	19,404	19,743	21,135	21,879	23,211	24,673	..	Industry
10,471	10,957	11,048	11,621	12,177	12,275	13,779	14,498	15,720	16,785	..	Manufacturing
23,106	24,669	26,645	28,542	30,801	32,871	35,162	36,514	38,055	39,074	..	Services, etc.
-7,521	-9,882	-15,022	-10,333	-14,233	-16,150	-13,586	-15,712	-17,535	-17,368	..	Resource Balance
18,181	20,689	21,434	23,580	24,649	24,186	27,170	30,052	30,753	32,231	..	Exports of Goods & NFServices
25,702	30,571	36,456	33,914	38,882	40,336	40,756	45,764	48,288	49,599	..	Imports of Goods & NFServices
66,622	72,786	81,549	80,589	87,193	93,309	95,280	98,248	102,126	104,325	..	Domestic Absorption
46,698	48,571	53,399	55,541	59,931	65,827	68,460	74,468	77,250	79,314	..	Private Consumption, etc.
6,037	5,867	5,685	5,202	6,346	6,716	7,180	8,303	8,894	8,966	..	General Gov't Consumption
13,887	18,348	22,465	19,846	20,916	20,766	19,641	15,476	15,981	16,045	..	Gross Domestic Investment
13,834	17,967	20,845	19,568	20,746	20,890	19,566	15,386	15,925	..	..	Fixed Investment
											Memo Items:
22,599	22,524	21,434	22,198	22,995	25,633	33,789	31,355	32,418	35,004	..	Capacity to Import
4,418	1,835	0	-1,383	-1,654	1,447	6,619	1,304	1,665	2,773	..	Terms of Trade Adjustment
63,519	64,739	66,527	68,873	71,306	78,606	88,312	83,840	86,256	89,730	..	Gross Domestic Income
63,153	64,434	66,095	67,305	69,839	76,444	86,278	81,963	84,052	87,637	..	Gross National Income
			(Index 1980 = 100)								**DOMESTIC PRICES (DEFLATORS)**
72.2	83.3	100.0	121.0	136.0	157.6	188.2	196.7	212.2	226.2	..	Overall (GDP)
67.1	80.6	100.0	122.4	135.3	150.0	170.9	185.0	196.1	208.4	..	Domestic Absorption
75.6	80.6	100.0	119.8	132.6	162.8	203.8	192.0	202.1	231.9	..	Agriculture
69.1	79.9	100.0	117.5	128.3	151.9	174.4	177.6	187.6	197.6	..	Industry
77.3	86.6	100.0	110.9	111.7	130.0	151.6	150.7	158.2	169.6	..	Manufacturing
											MANUFACTURING ACTIVITY
90.8	102.8	100.0	99.2	207.0	..	..	..	..	..	..	Employment (1980=100)
..	..	100.0	94.8	83.4	..	..	..	..	..	..	Real Earnings per Empl. (1980=100)
..	69.5	100.0	110.9	101.6	..	..	..	..	..	..	Real Output per Empl. (1980=100)
..	..	25.7	25.2	23.9	..	..	..	..	..	..	Earnings as % of Value Added
			(Billions of current Sri Lanka Rupees)								**MONETARY HOLDINGS**
14.62	20.70	26.26	30.98	39.67	48.22	53.75	63.88	67.00	77.33	D	Money Supply, Broadly Defined
5.89	7.64	9.33	9.95	11.67	14.59	16.65	18.66	21.05	24.90	..	Money as Means of Payment
3.02	3.77	4.18	4.82	5.99	7.20	8.56	9.82	11.57	13.50	..	Currency Ouside Banks
2.88	3.87	5.15	5.13	5.68	7.39	8.09	8.85	9.48	11.41	..	Demand Deposits
8.72	13.06	16.92	21.03	28.00	33.63	37.10	45.22	45.95	52.43	..	Quasi-Monetary Liabilities
			(Millions of current Sri Lanka Rupees)								**GOVERNMENT DEFICIT (-) OR SURPLUS**
-5,290	-6,300	-12,157	-10,640	-14,108	-13,102	-10,251	-15,469	-16,562	-16,697	..	Current Revenue
11,902	13,543	16,057	17,451	19,522	26,711	37,064	39,505	38,466	44,790	..	Current Revenue
11,930	12,276	16,456	16,762	17,687	22,893	27,610	33,574	33,731	37,213	..	Current Expenditure
-28	1,267	-399	689	1,835	3,818	9,454	5,931	4,735	7,577	..	Current Budget Balance
4	5	7	14	20	10	13	15	1,434	20	..	Capital Receipts
5,266	7,572	11,765	11,343	15,963	16,930	19,718	21,415	22,731	24,294	..	Capital Payments

SRI LANKA	1967	1968	1969	1970	1971	1972	1973	1974	1975	1976	1977
FOREIGN TRADE (CUSTOMS BASIS)				*(Millions of current US dollars)*							
Value of Exports, fob	345.0	341.9	321.9	337.7	326.6	325.8	405.4	523.4	559.2	566.8	762.9
Nonfuel Primary Products	330.9	328.1	311.3	327.0	316.0	304.9	349.4	451.8	494.2	485.9	661.8
Fuels	9.1	9.3	0.0	4.9	1.8	1.4	9.1	39.3	0.3	2.8	2.9
Manufactures	5.0	4.5	10.6	5.8	8.8	19.5	46.8	32.3	64.8	78.2	98.2
Value of Imports, cif	359.2	365.1	427.3	386.6	328.1	336.4	422.3	688.2	744.6	552.1	701.1
Nonfuel Primary Products	178.7	177.7	175.6	195.1	165.9	176.9	231.9	344.3	400.2	222.9	301.1
Fuels	25.4	33.7	27.2	10.4	5.1	6.4	45.4	137.3	124.3	137.4	169.1
Manufactures	155.1	153.7	224.5	181.2	157.2	153.1	145.0	206.5	220.1	191.9	231.0
Terms of Trade	159.1	163.9	177.9	132.0	122.4	112.5	129.2	100.5	91.8	115.7	152.9
Export Prices, fob	34.7	32.2	40.1	36.0	37.9	39.2	45.4	61.9	59.8	65.7	88.0
Nonfuel Primary Products	43.3	39.5	40.4	40.6	39.3	40.3	50.3	65.5	59.1	65.8	91.5
Fuels	4.3	4.3	4.3	4.3	5.6	6.2	8.9	36.7	35.7	38.4	42.0
Manufactures	30.1	31.7	32.4	37.5	36.5	38.5	49.5	67.2	66.1	66.7	71.7
Import Prices, cif	21.8	19.6	22.5	27.3	31.0	34.9	35.1	61.6	65.1	56.8	57.5
Trade at Constant 1980 Prices				*(Millions of 1980 US dollars)*							
Exports, fob	993.3	1,062.3	803.4	937.5	861.4	830.4	892.9	845.1	935.0	863.0	867.0
Imports, cif	1,644.8	1,859.6	1,897.4	1,416.3	1,059.3	964.3	1,201.7	1,117.2	1,143.3	972.3	1,218.8
BALANCE OF PAYMENTS				*(Millions of current US dollars)*							
Exports of Goods & Services	..	..	..	378.5	378.1	367.6	426.9	575.9	639.1	636.0	867.4
Merchandise, fob	..	..	..	338.8	325.3	316.5	366.3	509.3	558.4	559.6	762.0
Nonfactor Services	..	..	..	38.1	50.7	49.5	57.9	61.1	73.5	72.5	93.6
Factor Services	..	..	..	1.6	2.2	1.6	2.6	5.5	7.1	3.9	11.8
Imports of Goods & Services	..	..	..	449.0	428.9	412.6	465.2	753.8	829.2	707.1	797.2
Merchandise, fob	..	..	..	353.0	336.3	322.9	371.7	628.7	686.2	579.5	655.9
Nonfactor Services	..	..	..	70.6	70.0	68.8	73.6	102.9	117.5	103.4	113.8
Factor Services	..	..	..	25.4	22.6	20.8	19.9	22.1	25.5	24.1	27.5
Long-Term Interest	..	..	..	..	..	..	..	..	..	..	21.3
Current Transfers, net	..	..	..	-0.9	-3.4	-4.3	0.2	-0.2	2.7	6.7	10.7
Workers' Remittances	..	..	..	3.0	3.4	3.9	7.5	8.2	8.6	13.0	18.5
Total to be Financed	..	..	..	-71.4	-54.2	-49.3	-38.0	-178.1	-187.4	-64.4	80.9
Official Capital Grants	..	..	..	12.6	17.8	16.7	12.9	42.2	77.3	58.5	61.1
Current Account Balance	..	..	..	-58.8	-36.4	-32.6	-25.2	-135.9	-110.0	-5.9	142.0
Long-Term Capital, net	..	..	..	42.8	85.7	65.4	64.6	114.0	146.3	130.5	132.5
Direct Investment	..	..	..	-0.3	0.3	0.4	0.5	1.3	0.1	..	-1.2
Long-Term Loans	..	..	..	..	..	..	..	..	..	..	50.7
Disbursements	..	..	..	..	..	..	..	..	..	..	153.4
Repayments	..	..	..	..	..	..	..	..	..	..	102.7
Other Long-Term Capital	..	..	..	30.5	67.6	48.2	51.3	70.5	68.8	72.0	21.9
Other Capital, net	..	..	..	16.9	-52.6	-31.5	-11.2	-9.3	-62.2	-87.7	-122.0
Change in Reserves	..	..	..	-0.9	3.3	-1.3	-28.2	31.2	25.9	-36.9	-152.5
Memo Item:				*(Sri Lanka Rupees per US dollar)*							
Conversion Factor (Annual Avg)	4.860	5.950	6.600	6.850	7.040	7.640	8.430	8.520	9.000	10.880	13.390
EXTERNAL DEBT, ETC.				*(Millions of US dollars, outstanding at end of year)*							
Public/Publicly Guar. Long-Term	..	..	..	317.0	394.0	421.9	484.8	587.8	597.4	691.6	779.0
Official Creditors	..	..	..	253.3	321.2	352.4	401.3	450.0	490.1	576.2	707.8
IBRD and IDA	..	..	..	27.1	33.0	36.7	45.9	58.8	74.8	79.6	89.0
Private Creditors	..	..	..	63.7	72.8	69.5	83.5	137.8	107.3	115.4	71.2
Private Non-guaranteed Long-Term	..	..	..	..	..	..	..	..	..	..	0.4
Use of Fund Credit	..	..	..	78.5	77.7	80.9	89.4	124.5	146.0	156.0	206.4
Short-Term Debt	..	..	..	..	..	..	..	..	..	..	146.0
Memo Items:				*(Millions of US dollars)*							
Int'l Reserves Excluding Gold	55.0	52.0	40.0	42.7	50.3	59.5	86.6	77.6	57.4	92.3	292.6
Gold Holdings (at market price)	..	..	..	..	..	..	..	..	..	..	..
SOCIAL INDICATORS											
Total Fertility Rate	4.6	4.6	4.5	4.4	4.3	4.2	4.1	4.1	4.0	3.9	3.9
Crude Birth Rate	31.6	32.0	30.4	29.4	30.4	30.0	28.0	27.5	27.8	27.8	27.9
Infant Mortality Rate	..	..	..	..	..	..	..	..	..	..	..
Life Expectancy at Birth	64.2	64.0	63.7	63.5	63.2	63.0	63.7	64.4	65.1	65.8	66.5
Food Production, p.c. ('79-81=100)	74.5	73.6	70.9	74.9	71.9	72.5	68.0	74.4	79.6	80.7	77.3
Labor Force, Agriculture (%)	55.7	55.6	55.4	55.3	55.1	54.9	54.7	54.5	54.3	54.1	53.9
Labor Force, Female (%)	25.2	25.1	25.0	25.0	25.1	25.2	25.4	25.5	25.6	25.9	26.1
School Enroll. Ratio, primary	..	..	..	99.0	..	..	..	..	77.0	79.0	82.0
School Enroll. Ratio, secondary	..	..	..	47.0	..	..	..	..	48.0	48.0	42.0

1978	1979	1980	1981	1982	1983	1984	1985	1986	1987 estimate	NOTES	SRI LANKA
											FOREIGN TRADE (CUSTOMS BASIS)
				(Millions of current US dollars)							
845.9	980.5	1,048.7	1,035.5	1,013.8	1,066.2	1,453.7	1,333.3	1,215.0	1,392.8	..	Value of Exports, fob
730.3	760.8	689.8	659.9	605.7	649.7	929.1	738.1	652.1	760.3	..	Nonfuel Primary Products
50.2	95.3	161.3	134.0	131.1	102.1	126.4	133.6	58.2	72.8	..	Fuels
65.4	124.5	197.6	241.7	276.9	314.4	398.2	461.6	504.7	559.7	..	Manufactures
942.1	1,449.1	2,035.4	1,803.8	1,769.9	1,788.5	1,847.4	1,988.3	1,948.3	2,084.8	..	Value of Imports, cif
318.0	393.7	468.6	403.6	275.0	356.7	331.9	339.2	353.7	403.7	..	Nonfuel Primary Products
155.8	254.3	494.4	450.6	554.9	426.8	475.0	521.1	274.3	344.6	..	Fuels
468.3	801.2	1,072.4	949.6	939.9	1,004.9	1,040.5	1,128.1	1,320.3	1,336.5	..	Manufactures
				(Index 1980 = 100)							
120.0	111.1	123.2	89.7	89.9	101.9	118.4	99.4	96.6	95.7	..	Terms of Trade
81.4	90.7	123.2	92.6	87.4	95.0	107.6	86.5	82.5	88.4	..	Export Prices, fob
87.5	96.7	100.0	87.1	80.6	95.5	120.3	84.7	84.0	83.9	..	Nonfuel Primary Products
42.3	61.0	100.0	112.5	101.6	92.5	90.2	87.5	44.9	56.9	..	Fuels
76.1	90.0	..	100.0	99.1	94.8	90.7	89.3	89.0	103.2	..	Manufactures
67.8	81.6	100.0	103.2	97.2	93.2	90.8	87.1	85.4	92.4	..	Import Prices, cif
											Trade at Constant 1980 Prices
				(Millions of 1980 US dollars)							
1,039.8	1,081.2	851.2	1,118.7	1,160.5	1,122.5	1,351.4	1,541.3	1,473.1	1,576.2	..	Exports, fob
1,389.6	1,775.2	2,035.5	1,747.7	1,820.5	1,918.4	2,033.5	2,283.8	2,281.9	2,256.9	..	Imports, cif
				(Millions of current US dollars)							**BALANCE OF PAYMENTS**
969.0	1,172.8	1,340.1	1,376.9	1,348.3	1,404.5	1,797.9	1,644.5	1,581.8	1,759.0	..	Exports of Goods & Services
844.8	980.3	1,061.8	1,064.5	1,013.7	1,062.4	1,463.4	1,316.1	1,204.0	1,413.0	..	Merchandise, fob
104.1	153.0	231.1	279.4	290.7	297.4	276.1	244.9	309.9	298.0	..	Nonfactor Services
20.0	39.5	47.2	33.0	43.9	44.8	58.3	83.5	67.8	48.0	..	Factor Services
1,113.7	1,591.5	2,269.1	2,182.4	2,325.0	2,316.2	2,273.2	2,499.9	2,466.1	2,642.0	..	Imports of Goods & Services
897.3	1,303.9	1,844.9	1,693.5	1,796.6	1,727.2	1,697.6	1,833.0	1,761.8	2,118.2	..	Merchandise, fob
181.2	232.8	351.4	359.3	390.3	407.4	383.6	456.8	498.7	317.8	..	Nonfactor Services
35.2	54.8	72.8	129.6	138.2	181.7	192.0	210.2	205.6	206.0	..	Factor Services
24.8	28.1	33.4	50.0	69.5	92.7	106.8	116.7	121.5	126.0	..	Long-Term Interest
22.0	48.4	136.1	203.3	264.3	274.3	276.4	266.2	292.7	311.0	f	Current Transfers, net
39.0	60.1	151.7	230.0	289.6	294.3	300.9	292.4	324.5	348.0	..	Workers' Remittances
-122.7	-370.2	-792.9	-602.2	-712.5	-637.5	-198.9	-589.2	-591.6	-572.0	K	Total to be Financed
57.4	143.6	137.7	160.6	162.4	170.5	202.7	174.8	174.6	194.0	K	Official Capital Grants
-65.3	-226.6	-655.2	-441.6	-550.1	-467.0	3.8	-414.4	-417.1	-378.0	..	Current Account Balance
226.1	355.7	378.2	526.4	684.2	581.2	571.8	505.0	519.1	379.0	..	Long-Term Capital, net
1.5	46.8	43.0	49.3	63.6	37.8	32.6	24.8	29.5	..	..	Direct Investment
177.9	151.6	253.3	345.3	395.8	297.1	331.9	323.2	317.5	158.9	..	Long-Term Loans
243.4	201.1	304.5	390.9	470.9	373.0	432.8	451.1	488.2	387.4	..	Disbursements
65.5	49.5	51.2	45.6	75.1	75.9	100.9	127.9	170.7	228.5	..	Repayments
-10.7	13.6	-55.8	-28.8	62.4	75.9	4.6	-17.8	-2.4	26.1	..	Other Long-Term Capital
-107.5	-82.4	-4.5	-122.9	-183.4	-119.3	-266.1	-205.1	-212.9	15.6	..	Other Capital, net
-53.3	-46.7	281.5	38.1	49.3	5.0	-309.4	114.6	110.9	-16.6	..	Change in Reserves
											Memo Item:
				(Sri Lanka Rupees per US dollar)							
15.610	15.570	16.530	19.250	20.810	23.530	25.440	27.160	28.020	29.450	..	Conversion Factor (Annual Avg)
											EXTERNAL DEBT, ETC.
			(Millions of US dollars, outstanding at end of year)								
1,019.4	1,106.8	1,348.4	1,622.7	1,970.3	2,214.3	2,431.9	2,911.9	3,495.8	4,108.8	..	Public/Publicly Guar. Long-Term
968.8	1,046.4	1,208.5	1,305.9	1,438.8	1,632.8	1,788.3	2,242.0	2,790.3	3,389.1	..	Official Creditors
99.2	110.8	128.6	154.5	210.7	280.6	361.8	458.3	570.8	707.4	..	IBRD and IDA
50.6	60.4	139.8	316.8	531.6	581.5	643.6	669.9	705.5	719.7	..	Private Creditors
0.4	1.1	3.3	4.0	2.5	40.2	44.3	98.5	96.0	116.8	..	Private Non-guaranteed Long-Term
242.2	317.7	269.3	403.7	376.7	346.6	321.8	321.3	286.3	233.6	..	Use of Fund Credit
110.0	136.0	219.8	203.9	275.9	283.6	194.7	206.3	184.6	273.3	..	Short-Term Debt
											Memo Items:
				(Millions of US dollars)							
397.6	516.9	245.5	327.4	351.5	297.0	510.8	451.2	352.6	279.1	..	Int'l Reserves Excluding Gold
9.5	32.2	37.1	25.0	28.7	24.0	19.4	20.6	24.6	30.5	..	Gold Holdings (at market price)
											SOCIAL INDICATORS
3.8	3.7	3.6	3.5	3.4	3.3	3.2	3.0	2.9	2.8	..	Total Fertility Rate
28.5	28.9	28.4	28.2	26.8	26.2	24.8	24.3	24.2	24.0	..	Crude Birth Rate
37.1	35.5	33.8	32.2	30.5	30.0	29.5	29.0	28.5	28.0	..	Infant Mortality Rate
67.0	67.6	68.1	68.7	69.2	69.4	69.6	69.8	70.0	70.3	..	Life Expectancy at Birth
81.1	98.1	107.4	94.5	88.2	94.8	83.4	86.8	85.8	77.5	..	Food Production, p.c. ('79-81 = 100)
53.8	53.6	53.4	..	..	..	..	..	..	..	..	Labor Force, Agriculture (%)
26.4	26.6	26.9	26.9	26.9	26.9	26.9	26.9	26.9	26.8	..	Labor Force, Female (%)
89.0	94.0	98.0	103.0	105.0	104.0	103.0	103.0	..	..	..	School Enroll. Ratio, primary
52.0	49.0	51.0	51.0	57.0	59.0	61.0	63.0	..	..	..	School Enroll. Ratio, secondary

ST. KITTS AND NEVIS	1967	1968	1969	1970	1971	1972	1973	1974	1975	1976	1977
CURRENT GNP PER CAPITA (US $)	..	..	..	350	420	470	520	630	720	760	710
POPULATION (thousands)	..	..	..	45	45	44	44	44	44	43	44

ORIGIN AND USE OF RESOURCES *(Millions of current East Caribbean Dollars)*

	1967	1968	1969	1970	1971	1972	1973	1974	1975	1976	1977
Gross National Product (GNP)	28.5	28.0	30.5	31.2	37.3	42.4	45.5	62.2	69.2	75.7	79.2
Net Factor Income from Abroad	-1.0	-1.2	-1.2	-1.4	-1.4	-1.7	-1.9	-2.5	-3.2	-3.0	-2.5
Gross Domestic Product (GDP)	29.5	29.2	31.7	32.6	38.7	44.1	47.4	64.7	72.4	78.7	81.7
Indirect Taxes, net	..	..	..	..	..	..	..	..	..	..	11.5
GDP at factor cost	..	..	..	..	..	..	..	..	..	..	70.2
Agriculture	..	..	..	..	..	..	..	..	..	..	13.1
Industry	..	..	..	..	..	..	..	..	..	..	19.9
Manufacturing	..	..	..	..	..	..	..	..	..	..	12.5
Services, etc.	..	..	..	..	..	..	..	..	..	..	48.7
Resource Balance	..	..	..	..	..	..	..	..	..	..	-12.4
Exports of Goods & NFServices	..	..	..	..	..	..	..	..	..	..	51.8
Imports of Goods & NFServices	..	..	..	..	..	..	..	..	..	..	64.2
Domestic Absorption	..	..	..	..	..	..	..	..	..	..	94.3
Private Consumption, etc.	..	..	..	..	..	..	..	..	..	..	51.1
General Gov't Consumption	..	..	..	..	..	..	..	..	..	..	17.8
Gross Domestic Investment	..	..	..	..	..	..	..	..	..	..	25.4
Fixed Investment	..	..	..	..	..	..	..	..	..	..	30.8
Memo Items:											
Gross Domestic Saving	..	..	..	..	..	..	..	..	..	..	12.8
Gross National Saving	..	..	..	..	..	..	..	..	..	..	..

(Millions of 1980 East Caribbean Dollars)

	1967	1968	1969	1970	1971	1972	1973	1974	1975	1976	1977
Gross National Product	85.77	98.91	84.57	89.08	86.69	96.65	100.77	100.63	103.95	105.55	105.81
GDP at factor cost											
Agriculture	..	..	..	..	..	..	..	..	..	..	..
Industry	..	..	..	..	..	..	..	..	..	..	16.13
Manufacturing	..	..	..	..	..	..	..	..	..	..	24.43
Services, etc.	..	..	..	..	..	..	..	..	..	..	14.43
	..	..	..	..	..	..	..	..	..	..	64.00
Resource Balance	..	..	..	..	..	..	..	..	..	..	..
Exports of Goods & NFServices	..	..	..	..	..	..	..	..	..	..	..
Imports of Goods & NFServices	..	..	..	..	..	..	..	..	..	..	..
Domestic Absorption	..	..	..	..	..	..	..	..	..	..	..
Private Consumption, etc.	..	..	..	..	..	..	..	..	..	..	..
General Gov't Consumption	..	..	..	..	..	..	..	..	..	..	..
Gross Domestic Investment	..	..	..	..	..	..	..	..	..	..	..
Fixed Investment	..	..	..	..	..	..	..	..	..	..	..
Memo Items:											
Capacity to Import	..	..	..	..	..	..	..	..	..	..	..
Terms of Trade Adjustment	..	..	..	..	..	..	..	..	..	..	..
Gross Domestic Income	..	..	..	..	..	..	..	..	..	..	..
Gross National Income	..	..	..	..	..	..	..	..	..	..	..

DOMESTIC PRICES (DEFLATORS) *(Index 1980 = 100)*

	1967	1968	1969	1970	1971	1972	1973	1974	1975	1976	1977
Overall (GDP)	34.9	29.9	38.0	37.0	45.2	46.1	47.5	64.9	70.4	75.4	78.1
Domestic Absorption	..	..	..	..	..	..	..	..	..	..	..
Agriculture	..	..	..	..	..	..	..	..	..	..	81.2
Industry	..	..	..	..	..	..	..	..	..	..	81.5
Manufacturing	..	..	..	..	..	..	..	..	..	..	86.6

MANUFACTURING ACTIVITY

	1967	1968	1969	1970	1971	1972	1973	1974	1975	1976	1977
Employment (1980=100)	..	..	..	..	..	..	..	..	..	..	..
Real Earnings per Empl. (1980=100)	..	..	..	..	..	..	..	..	..	..	..
Real Output per Empl. (1980=100)	..	..	..	..	..	..	..	..	..	..	..
Earnings as % of Value Added	..	..	..	..	..	..	..	..	..	..	..

MONETARY HOLDINGS *(Millions of current East Caribbean Dollars)*

	1967	1968	1969	1970	1971	1972	1973	1974	1975	1976	1977
Money Supply, Broadly Defined	..	..	..	..	..	..	..	..	..	..	..
Money as Means of Payment	..	..	..	..	..	..	..	..	..	..	..
Currency Ouside Banks	..	..	..	..	..	..	..	..	..	..	..
Demand Deposits	..	..	..	..	..	..	..	..	..	..	..
Quasi-Monetary Liabilities	..	..	..	..	..	..	..	..	..	..	..

GOVERNMENT DEFICIT (-) OR SURPLUS *(Millions of current East Caribbean Dollars)*

	1967	1968	1969	1970	1971	1972	1973	1974	1975	1976	1977
	..	..	..	..	..	..	..	..	..	..	..
Current Revenue	..	..	..	..	..	..	..	..	..	..	..
Current Expenditure	..	..	..	..	..	..	..	..	..	..	..
Current Budget Balance	..	..	..	..	..	..	..	..	..	..	..
Capital Receipts	..	..	..	..	..	..	..	..	..	..	..
Capital Payments	..	..	..	..	..	..	..	..	..	..	..

1978	1979	1980	1981	1982	1983	1984	1985	1986	1987 estimate	NOTES	ST. KITTS AND NEVIS
770	910	1,090	1,250	1,320	1,380	1,440	1,450	1,590	1,700	..	CURRENT GNP PER CAPITA (US $)
44	44	44	44	45	45	45	46	46	47	..	POPULATION (thousands)
											(Millions of current East Caribbean Dollars)
											ORIGIN AND USE OF RESOURCES
89.9	105.2	130.6	153.4	161.2	163.4	175.9	181.6	207.2	217.3	..	Gross National Product (GNP)
-3.5	-3.5	-0.2	2.4	1.9	2.7	0.3	-3.8	-4.3	..	..	Net Factor Income from Abroad
93.4	108.7	130.8	151.0	159.3	160.7	175.6	185.4	211.5	..	..	Gross Domestic Product (GDP)
16.0	20.9	27.7	30.5	21.5	24.5	26.7	29.8	..	..	..	Indirect Taxes, net
77.4	87.8	103.1	120.5	137.8	136.2	148.9	155.6	..	..	B	GDP at factor cost
12.2	13.7	16.5	13.8	20.3	16.2	17.7	..	..	..	..	Agriculture
19.6	21.9	27.5	30.2	34.1	30.1	34.4	..	..	..	..	Industry
13.3	12.8	15.7	17.9	18.7	17.5	20.6	..	..	..	..	Manufacturing
61.6	73.1	86.8	107.0	104.9	114.4	123.5	..	..	..	..	Services, etc.
-9.5	-24.7	-37.5	-42.6	-53.7	-69.9	-56.9	-56.4	..	..	..	Resource Balance
58.9	64.7	87.2	104.8	89.4	85.6	104.0	113.7	..	..	..	Exports of Goods & NFServices
68.4	89.3	124.7	147.4	143.1	155.5	160.9	170.1	..	..	..	Imports of Goods & NFServices
103.1	133.8	168.3	193.7	213.0	230.6	232.6	241.8	..	..	..	Domestic Absorption
60.0	78.4	98.3	142.9	161.4	177.9	178.9	176.0	..	..	..	Private Consumption, etc.
22.4	24.7	29.3	34.4	34.8	38.3	40.6	42.9	..	..	..	General Gov't Consumption
20.8	30.6	40.7	16.4	16.8	14.4	13.1	22.9	..	..	..	Gross Domestic Investment
25.2	37.2	49.4	46.1	..	..	..	..	..	..	..	Fixed Investment
											Memo Items:
11.1	5.6	3.2	-26.3	-36.9	-55.5	-43.9	-33.5	..	..	..	Gross Domestic Saving
..	..	25.1	6.1	-10.7	-30.7	-17.1	-12.7	..	..	..	Gross National Saving
											(Millions of 1980 East Caribbean Dollars)
113.69	124.49	130.60	135.52	133.89	134.03	139.96	138.91	144.46	150.02	..	Gross National Product
91.81	99.20	103.10	108.67	116.12	114.65	118.23	119.40	..	..	B	GDP at factor cost
16.87	18.22	16.50	17.36	17.36	14.16	15.15	..	..	..	..	Agriculture
23.20	24.92	27.50	26.03	29.46	25.41	27.13	..	..	..	..	Industry
15.12	15.35	15.70	13.97	15.24	13.51	14.78	..	..	..	..	Manufacturing
71.97	79.80	86.80	92.98	87.76	95.45	97.79	..	..	..	..	Services, etc.
..	..	-37.50	-42.37	-54.13	-72.17	-59.98	-58.71	..	..	..	Resource Balance
..	..	87.20	104.28	90.14	88.74	109.73	118.41	..	..	..	Exports of Goods & NFServices
..	..	124.70	146.65	144.27	160.91	169.72	177.12	..	..	..	Imports of Goods & NFServices
..	..	168.30	178.74	188.72	207.19	200.04	201.69	..	..	..	Domestic Absorption
..	..	98.30	131.71	142.87	161.12	154.08	145.47	..	..	..	Private Consumption, etc.
..	..	29.30	31.20	29.74	32.08	33.11	34.13	..	..	..	General Gov't Consumption
..	..	40.70	15.83	16.11	13.99	12.86	22.05	..	..	..	Gross Domestic Investment
..	..	..	..	..	..	..	..	..	..	..	Fixed Investment
											Memo Items:
..	..	87.20	104.27	90.13	88.58	109.70	118.40	..	..	..	Capacity to Import
..	..	0.00	-0.01	-0.01	-0.16	-0.04	-0.02	..	..	..	Terms of Trade Adjustment
..	..	130.80	136.36	134.58	134.86	140.03	142.97	..	..	..	Gross Domestic Income
..	..	130.60	135.51	133.88	133.86	139.93	138.89	..	..	..	Gross National Income
											(Index 1980 = 100)
											DOMESTIC PRICES (DEFLATORS)
83.4	88.4	100.0	110.7	118.4	119.0	125.4	129.7	142.2	..	..	Overall (GDP)
..	..	100.0	108.4	112.9	111.3	116.3	119.9	..	..	..	Domestic Absorption
72.3	75.2	100.0	79.5	116.9	114.4	116.9	..	..	..	..	Agriculture
84.5	87.9	100.0	116.0	115.7	118.4	126.8	..	..	..	..	Industry
87.9	83.4	100.0	128.1	122.7	129.6	139.4	..	..	..	..	Manufacturing
											MANUFACTURING ACTIVITY
..	..	..	..	..	..	..	..	..	..	..	Employment (1980 = 100)
..	..	..	..	..	..	..	..	..	..	..	Real Earnings per Empl. (1980 = 100)
..	..	..	..	..	..	..	..	..	..	..	Real Output per Empl. (1980 = 100)
..	..	..	..	..	..	..	..	..	..	..	Earnings as % of Value Added
											(Millions of current East Caribbean Dollars)
											MONETARY HOLDINGS
..	84.0	95.3	105.1	117.2	134.1	170.9	204.2	232.6	212.2	..	Money Supply, Broadly Defined
..	9.3	8.0	14.7	11.8	11.0	29.3	32.9	49.8	55.3	..	Money as Means of Payment
..	-1.6	-1.3	-1.4	-2.6	-5.0	12.6	9.4	12.2	13.0	..	Currency Ouside Banks
..	10.8	9.3	16.1	14.4	16.0	16.6	23.5	37.6	42.3	..	Demand Deposits
..	74.7	87.3	90.4	105.4	123.1	141.6	171.3	182.8	157.0	..	Quasi-Monetary Liabilities
											(Millions of current East Caribbean Dollars)
											GOVERNMENT DEFICIT (-) OR SURPLUS
..	..	..	..	..	..	..	..	..	..	..	Current Revenue
..	..	..	..	..	..	..	..	..	..	..	Current Expenditure
..	..	..	..	..	..	..	..	..	..	..	Current Budget Balance
..	..	..	..	..	..	..	..	..	..	..	Capital Receipts
..	..	..	..	..	..	..	..	..	..	..	Capital Payments

ST. KITTS AND NEVIS	1967	1968	1969	1970	1971	1972	1973	1974	1975	1976	1977
FOREIGN TRADE (CUSTOMS BASIS)				*(Millions of current US dollars)*							
Value of Exports, fob	..	..	..	..	..	..	..	..	..	..	..
Nonfuel Primary Products	..	..	..	..	..	..	..	..	..	..	..
Fuels	..	..	..	..	..	..	..	..	..	..	..
Manufactures	..	..	..	..	..	..	..	..	..	..	..
Value of Imports, cif	..	..	..	..	..	..	..	..	..	..	..
Nonfuel Primary Products	..	..	..	..	..	..	..	..	..	..	..
Fuels	..	..	..	..	..	..	..	..	..	..	..
Manufactures	..	..	..	..	..	..	..	..	..	..	..
Terms of Trade				*(Index 1980 = 100)*							
Export Prices, fob	..	..	..	..	..	..	..	..	..	..	..
Nonfuel Primary Products	..	..	..	..	..	..	..	..	..	..	..
Fuels	..	..	..	..	..	..	..	..	..	..	..
Manufactures	..	..	..	..	..	..	..	..	..	..	..
Import Prices, cif	..	..	..	..	..	..	..	..	..	..	..
Trade at Constant 1980 Prices				*(Millions of 1980 US dollars)*							
Exports, fob	..	..	..	..	..	..	..	..	..	..	..
Imports, cif	..	..	..	..	..	..	..	..	..	..	..
BALANCE OF PAYMENTS				*(Thousands of current US dollars)*							
Exports of Goods & Services	..	..	..	..	..	..	..	..	..	..	..
Merchandise, fob	..	..	..	..	..	..	..	..	..	..	..
Nonfactor Services	..	..	..	..	..	..	..	..	..	..	..
Factor Services	..	..	..	..	..	..	..	..	..	..	..
Imports of Goods & Services	..	..	..	..	..	..	..	..	..	..	..
Merchandise, fob	..	..	..	..	..	..	..	..	..	..	..
Nonfactor Services	..	..	..	..	..	..	..	..	..	..	..
Factor Services	..	..	..	..	..	..	..	..	..	..	..
Long-Term Interest	..	..	..	..	..	..	..	..	..	..	..
Current Transfers, net	..	..	..	..	..	..	..	..	..	..	..
Workers' Remittances	..	..	..	..	..	..	..	..	..	..	..
Total to be Financed	..	..	..	..	..	..	..	..	..	..	..
Official Capital Grants	..	..	..	..	..	..	..	..	..	..	..
Current Account Balance	..	..	..	..	..	..	..	..	..	..	..
Long-Term Capital, net	..	..	..	..	..	..	..	..	..	..	..
Direct Investment	..	..	..	..	..	..	..	..	..	..	..
Long-Term Loans	..	..	..	..	..	..	..	..	..	..	..
Disbursements	..	..	..	..	..	..	..	..	..	..	..
Repayments	..	..	..	..	..	..	..	..	..	..	..
Other Long-Term Capital	..	..	..	..	..	..	..	..	..	..	..
Other Capital, net	..	..	..	..	..	..	..	..	..	..	..
Change in Reserves	..	..	..	0	0	0	0	0	0	0	0
Memo Item:											
Conversion Factor (Annual Avg)	1.760	2.000	2.000	*(East Caribbean Dollars per US dollar)*							
				2.000	1.970	1.920	1.960	2.050	2.170	2.610	2.700
EXTERNAL DEBT, ETC.				*(Millions of US dollars, outstanding at end of year)*							
Public/Publicly Guar. Long-Term	..	..	..	..	..	..	..	..	..	..	..
Official Creditors	..	..	..	..	..	..	..	..	..	..	..
IBRD and IDA	..	..	..	..	..	..	..	..	..	..	..
Private Creditors	..	..	..	..	..	..	..	..	..	..	..
Private Non-guaranteed Long-Term	..	..	..	..	..	..	..	..	..	..	..
Use of Fund Credit	..	..	..	..	..	..	..	..	..	..	..
Short-Term Debt	..	..	..	..	..	..	..	..	..	..	..
Memo Items:				*(Thousands of US dollars)*							
Int'l Reserves Excluding Gold	..	..	..	..	..	..	..	..	..	..	..
Gold Holdings (at market price)	..	..	..	..	..	..	..	..	..	..	..
SOCIAL INDICATORS											
Total Fertility Rate	..	..	..	..	..	..	..	..	..	..	..
Crude Birth Rate	..	..	..	25.8	26.0	26.2	26.3	26.5	26.7	26.9	27.8
Infant Mortality Rate	..	..	..	..	..	..	..	..	..	..	..
Life Expectancy at Birth	..	..	..	..	..	..	..	..	..	..	..
Food Production, p.c. ('79-81 = 100)	..	..	..	..	..	..	..	..	..	..	..
Labor Force, Agriculture (%)	..	..	..	..	..	..	..	..	..	..	..
Labor Force, Female (%)	..	..	..	..	..	..	..	..	..	..	..
School Enroll. Ratio, primary	..	..	..	..	..	..	..	..	..	..	..
School Enroll. Ratio, secondary	..	..	..	..	..	..	..	..	..	..	..

1978	1979	1980	1981	1982	1983	1984	1985	1986	1987	NOTES	ST. KITTS AND NEVIS
		(Millions of current US dollars)									**FOREIGN TRADE (CUSTOMS BASIS)**
..	..	24	25	21	18	21	19	20	..	..	Value of Exports, fob
..	..	..	..	..	..	..	..	..	..	..	Nonfuel Primary Products
..	..	..	..	..	..	..	..	..	..	..	Fuels
..	..	..	..	..	..	..	..	..	..	..	Manufactures
..	..	116	115	90	100	113	108	104	..	..	Value of Imports, cif
..	..	..	..	..	..	..	..	..	..	..	Nonfuel Primary Products
..	..	..	..	..	..	..	..	..	..	..	Fuels
..	..	..	..	..	..	..	..	..	..	..	Manufactures
		(Index 1980 = 100)									Terms of Trade
..	..	..	..	..	..	..	..	..	..	..	Export Prices, fob
..	..	..	..	..	..	..	..	..	..	..	Nonfuel Primary Products
..	..	..	..	..	..	..	..	..	..	..	Fuels
..	..	..	..	..	..	..	..	..	..	..	Manufactures
..	..	..	..	..	..	..	..	..	..	..	Import Prices, cif
		(Millions of 1980 US dollars)									Trade at Constant 1980 Prices
											Exports, fob
..	..	..	..	..	..	..	..	..	..	..	Imports, cif
..	..	..	..	..	..	..	..	..	..	..	
		(Thousands of current US dollars)									**BALANCE OF PAYMENTS**
..	..	33	37	35	33	40	45	..	..	..	Exports of Goods & Services
..	..	24	25	21	18	21	22	..	..	..	Merchandise, fob
..	..	8	10	13	13	18	23	..	..	..	Nonfactor Services
..	..	1	1	1	2	1	0	..	..	..	Factor Services
..	..	49	53	53	59	58	65	..	..	..	Imports of Goods & Services
..	..	41	43	40	47	45	49	..	..	..	Merchandise, fob
..	..	6	7	13	11	12	14	..	..	..	Nonfactor Services
..	..	2	2	1	1	1	2	..	..	..	Factor Services
..	..	..	..	..	..	..	..	..	..	..	Long-Term Interest
..	..	8	11	9	8	10	9	..	..	..	Current Transfers, net
..	..	0	0	0	0	0	0	..	..	..	Workers' Remittances
..	..	-7	-5	-9	-18	-8	-11	..	..	K	Total to be Financed
..	..	5	2	2	2	2	2	..	..	K	Official Capital Grants
..	..	-2	-4	-7	-16	-7	-9	..	..	..	Current Account Balance
..	..	10	4	5	16	13	11	..	..	..	Long-Term Capital, net
..	..	1	1	2	13	6	8	..	..	..	Direct Investment
..	..	..	..	..	..	..	..	..	..	..	Long-Term Loans
..	..	..	..	..	..	..	..	..	..	..	Disbursements
..	..	..	..	..	..	..	..	..	..	..	Repayments
..	..	4	1	1	1	5	1	..	..	..	Other Long-Term Capital
..	..	-9	0	-2	0	-5	-1	..	..	..	Other Capital, net
0	0	2	0	4	-1	-2	-2	0	0	..	Change in Reserves
		(East Caribbean Dollars per US dollar)									**Memo Item:**
2.700	2.700	2.700	2.700	2.700	2.700	2.700	2.700	2.700	2.700	..	Conversion Factor (Annual Avg)
		(Millions of US dollars, outstanding at end of year)									**EXTERNAL DEBT, ETC.**
											Public/Publicly Guar. Long-Term
..	..	..	..	..	..	..	..	..	..	..	Official Creditors
..	..	..	..	..	..	..	..	..	..	..	IBRD and IDA
..	..	..	..	..	..	..	..	..	..	..	Private Creditors
..	..	..	..	..	..	..	..	..	..	..	Private Non-guaranteed Long-Term
..	..	..	..	..	..	..	..	..	..	..	Use of Fund Credit
..	..	..	..	..	..	..	..	..	..	..	Short-Term Debt
		(Thousands of US dollars)									**Memo Items:**
..	..	..	..	..	..	..	..	10,232	10,574	..	Int'l Reserves Excluding Gold
..	..	..	..	..	..	..	..	..	..	..	Gold Holdings (at market price)
											SOCIAL INDICATORS
..	..	..	..	0.5	1.0	1.5	2.0	2.5	3.0	..	Total Fertility Rate
24.1	27.7	26.8	25.9	29.1	24.1	25.3	23.7	23.0	25.0	..	Crude Birth Rate
..	..	..	..	9.3	18.5	27.8	28.1	28.4	28.6	..	Infant Mortality Rate
..	..	..	..	63.1	63.4	63.8	65.9	68.0	70.2	..	Life Expectancy at Birth
..	..	..	..	..	..	..	..	..	..	..	Food Production, p.c. ('79-81 = 100)
..	..	..	..	..	..	..	..	..	..	..	Labor Force, Agriculture (%)
..	..	..	..	..	..	..	..	..	..	..	Labor Force, Female (%)
..	..	..	..	..	..	..	..	..	..	..	School Enroll. Ratio, primary
..	..	..	..	..	..	..	..	..	..	..	School Enroll. Ratio, secondary

ST. LUCIA	1967	1968	1969	1970	1971	1972	1973	1974	1975	1976	1977
CURRENT GNP PER CAPITA (US $)	250	240	270	310	340	360	390	420	440	520	590
POPULATION (thousands)	97	99	100	101	103	105	107	110	112	114	117
ORIGIN AND USE OF RESOURCES					*(Millions of current East Caribbean Dollars)*						
Gross National Product (GNP)	41.9	45.5	51.2	65.4	68.2	73.0	84.7	94.4	108.4	147.9	177.1
Net Factor Income from Abroad	-4.3	-4.7	-5.3	-6.8	-7.1	-7.6	-8.9	-9.7	-11.2	-6.9	-7.0
Gross Domestic Product (GDP)	46.2	50.2	56.5	72.2	75.3	80.6	93.6	104.1	119.6	154.8	184.1
Indirect Taxes, net	..	..	..	..	..	..	..	..	..	..	27.4
GDP at factor cost	..	..	..	..	..	..	..	..	..	..	156.7
Agriculture	..	..	..	..	..	..	..	..	..	..	24.2
Industry	..	..	..	..	..	..	..	..	..	..	30.2
Manufacturing	..	..	..	..	..	..	..	..	..	..	13.8
Services, etc.	..	..	..	..	..	..	..	..	..	..	102.3
Resource Balance	..	..	..	..	..	..	..	..	..	..	-61.8
Exports of Goods & NFServices	..	..	..	..	..	..	..	..	..	..	109.1
Imports of Goods & NFServices	..	..	..	..	..	..	..	..	..	..	170.9
Domestic Absorption	..	..	..	..	..	..	..	..	..	..	245.9
Private Consumption, etc.	..	..	..	..	..	..	..	..	..	..	118.1
General Gov't Consumption	..	..	..	..	..	..	..	..	..	..	40.1
Gross Domestic Investment	..	..	..	..	..	..	..	..	..	..	87.7
Fixed Investment	..	..	..	..	..	..	..	..	..	..	..
Memo Items:											
Gross Domestic Saving	..	..	..	..	..	..	..	..	..	..	25.9
Gross National Saving	..	..	..	..	..	..	..	..	..	..	39.4
					(Millions of 1980 East Caribbean Dollars)						
Gross National Product	180.71	175.59	171.30	186.33	212.27	224.49	226.47	232.42	235.06	222.51	258.35
GDP at factor cost	..	..	..	..	..	..	..	..	..	..	226.54
Agriculture	..	..	..	..	..	..	..	..	..	..	33.95
Industry	..	..	..	..	..	..	..	..	..	..	46.18
Manufacturing	..	..	..	..	..	..	..	..	..	..	21.99
Services, etc.	..	..	..	..	..	..	..	..	..	..	146.42
Resource Balance	..	..	..	..	..	..	..	..	..	..	-77.75
Exports of Goods & NFServices	..	..	..	..	..	..	..	..	..	..	153.76
Imports of Goods & NFServices	..	..	..	..	..	..	..	..	..	..	231.51
Domestic Absorption	..	..	..	..	..	..	..	..	..	..	343.26
Private Consumption, etc.	..	..	..	..	..	..	..	..	..	..	160.51
General Gov't Consumption	..	..	..	..	..	..	..	..	..	..	56.82
Gross Domestic Investment	..	..	..	..	..	..	..	..	..	..	125.92
Fixed Investment	..	..	..	..	..	..	..	..	..	..	..
Memo Items:											
Capacity to Import	..	..	..	..	..	..	..	..	..	..	147.80
Terms of Trade Adjustment	..	..	..	..	..	..	..	..	..	..	-5.97
Gross Domestic Income	..	..	..	..	..	..	..	..	..	..	259.54
Gross National Income	..	..	..	..	..	..	..	..	..	..	252.39
DOMESTIC PRICES (DEFLATORS)					*(Index 1980 = 100)*						
Overall (GDP)	25.5	28.5	32.9	38.6	35.3	30.2	41.2	44.6	50.7	69.3	69.3
Domestic Absorption	..	..	..	..	..	..	..	..	..	..	71.6
Agriculture	..	..	..	..	..	..	..	..	..	..	71.3
Industry	..	..	..	..	..	..	..	..	..	..	65.4
Manufacturing	..	..	..	..	..	..	..	..	..	..	62.8
MANUFACTURING ACTIVITY											
Employment (1980 = 100)	..	..	..	..	..	..	..	..	..	..	..
Real Earnings per Empl. (1980 = 100)	..	..	..	..	..	..	..	..	..	..	..
Real Output per Empl. (1980 = 100)	..	..	..	..	..	..	..	..	..	..	..
Earnings as % of Value Added	..	..	..	..	..	..	..	..	..	..	..
MONETARY HOLDINGS					*(Millions of current East Caribbean Dollars)*						
Money Supply, Broadly Defined	..	..	..	..	..	..	..	..	82.1	100.5	106.5
Money as Means of Payment	..	..	..	..	..	..	..	..	20.6	29.1	33.0
Currency Ouside Banks	..	..	..	..	..	..	..	..	9.6	12.1	15.0
Demand Deposits	..	..	..	..	..	..	..	..	11.1	17.0	18.0
Quasi-Monetary Liabilities	..	..	..	..	..	..	..	..	61.5	71.5	73.5
GOVERNMENT DEFICIT (-) OR SURPLUS					*(Millions of current East Caribbean Dollars)*						
Current Revenue	..	..	..	..	..	..	..	..	..	..	..
Current Expenditure	..	..	..	..	..	..	..	..	..	..	..
Current Budget Balance	..	..	..	..	..	..	..	..	..	..	..
Capital Receipts	..	..	..	..	..	..	..	..	..	..	..
Capital Payments	..	..	..	..	..	..	..	..	..	..	..

1978	1979	1980	1981	1982	1983	1984	1985	1986	1987 estimate	NOTES	ST. LUCIA
670	770	840	960	1,040	1,080	1,160	1,250	1,330	1,400	..	**CURRENT GNP PER CAPITA (US $)**
119	121	124	126	129	131	134	137	140	143	..	**POPULATION (thousands)**
											ORIGIN AND USE OF RESOURCES
			(Millions of current East Caribbean Dollars)								
212.9	262.2	298.8	340.7	352.1	381.4	415.6	458.0	514.5	559.4	..	Gross National Product (GNP)
-6.8	-5.4	-5.9	-5.7	-9.2	1.4	2.4	-1.7	5.1	20.9	..	Net Factor Income from Abroad
219.7	267.6	304.7	346.4	361.3	380.0	413.2	459.7	509.4	538.5	..	Gross Domestic Product (GDP)
28.7	38.5	40.5	47.0	48.8	51.6	60.2	70.9	..	..	..	Indirect Taxes, net
191.0	229.1	264.2	299.4	312.5	328.4	353.0	388.8			..	GDP at factor cost
33.2	34.8	31.0	28.9	36.1	47.3	46.2	58.3	70.7	64.1	..	Agriculture
38.4	52.2	65.6	74.6	71.7	63.2	70.2	77.3	84.9	92.3	..	Industry
15.2	18.9	24.7	25.3	26.8	30.2	31.1	33.0	34.0	35.0	..	Manufacturing
119.4	142.1	167.6	195.9	204.7	217.9	236.6	253.2	..	..	..	Services, etc.
-93.9	-122.1	-131.0	-166.1	-125.5	-92.6	-99.1	-79.7	-46.1	-103.9	..	Resource Balance
147.2	176.3	207.6	183.6	198.2	202.0	236.0	260.0	366.7	358.0	..	Exports of Goods & NFServices
241.1	298.4	338.6	349.7	323.7	294.6	335.1	339.7	412.8	461.9	..	Imports of Goods & NFServices
313.7	389.7	435.7	512.5	486.8	472.6	512.5	539.4	555.6	642.3	..	Domestic Absorption
123.7	189.1	191.3	243.9	244.1	249.3	271.2	264.3	238.5	282.7	..	Private Consumption, etc.
42.2	44.8	63.2	69.8	87.2	96.1	96.8	113.1	123.7	123.3	..	General Gov't Consumption
147.8	155.8	181.2	198.8	155.5	127.2	144.5	162.0	193.4	236.3	..	Gross Domestic Investment
..	..	..	..	..	..	..	..	..	..	..	Fixed Investment
											Memo Items:
53.8	33.7	50.2	32.7	30.0	34.6	45.2	82.3	147.2	132.5	..	Gross Domestic Saving
67.8	49.4	74.3	67.2	57.3	73.0	84.6	120.6	152.3	..	..	Gross National Saving
			(Millions of 1980 East Caribbean Dollars)								
287.61	304.44	298.80	300.78	310.87	316.28	338.48	360.32	382.36	390.24	..	Gross National Product
256.18	265.78	264.20	267.54	275.16	285.52	299.79	317.11	..	..	..	GDP at factor cost
41.10	39.14	31.00	26.37	35.07	40.26	43.48	48.81	54.85	52.46	..	Agriculture
57.50	61.62	65.60	67.13	63.77	55.05	57.65	61.62	65.91	70.03	..	Industry
25.50	21.35	24.70	25.18	27.09	29.96	29.96	30.76	31.23	31.55	..	Manufacturing
157.58	165.02	167.60	174.04	176.33	190.21	198.66	206.67	..	..	..	Services, etc.
-125.36	-130.35	-131.00	-212.84	-225.41	-161.06	-175.14	..	..	..	..	Resource Balance
183.92	194.07	207.60	186.32	184.63	222.96	250.73	..	..	..	..	Exports of Goods & NFServices
309.28	324.42	338.60	399.16	410.04	384.02	425.87	..	..	..	..	Imports of Goods & NFServices
419.24	440.55	435.70	518.79	542.33	484.57	518.78	..	..	..	..	Domestic Absorption
163.58	203.47	191.30	243.64	276.46	243.03	258.05	..	..	..	..	Private Consumption, etc.
58.38	52.15	63.20	71.99	100.75	105.85	109.11	..	..	..	..	General Gov't Consumption
197.28	184.93	181.20	203.17	165.12	135.68	151.62	..	..	..	..	Gross Domestic Investment
..	..	..	..	..	..	..	..	..	..	..	Fixed Investment
											Memo Items:
188.83	191.67	207.60	209.57	251.06	263.31	299.92	..	..	..	..	Capacity to Import
4.90	-2.40	0.00	23.25	66.44	40.35	49.20	..	..	..	..	Terms of Trade Adjustment
298.79	307.80	304.70	329.20	383.36	363.86	392.84	..	..	..	..	Gross Domestic Income
292.52	302.05	298.80	324.03	377.31	356.63	387.68	..	..	..	..	Gross National Income
			(Index 1980 = 100)								
74.8	86.3	100.0	113.2	114.0	117.5	120.2	125.7	131.2	135.9	..	**DOMESTIC PRICES (DEFLATORS)** Overall (GDP)
74.8	88.5	100.0	98.8	89.8	97.5	98.8	..	..	..	..	Domestic Absorption
80.8	88.9	100.0	109.6	102.9	117.5	106.2	119.4	128.9	122.2	..	Agriculture
66.8	84.7	100.0	111.1	112.4	114.8	121.8	125.4	128.8	131.8	..	Industry
59.6	88.5	100.0	100.5	98.9	100.8	103.8	107.3	108.9	110.9	..	Manufacturing
											MANUFACTURING ACTIVITY
..	..	..	..	..	..	..	..	..	..	..	Employment (1980 = 100)
..	..	..	..	..	..	..	..	..	..	..	Real Earnings per Empl. (1980 = 100)
..	..	..	..	..	..	..	..	..	..	..	Real Output per Empl. (1980 = 100)
..	..	..	..	..	..	..	..	..	..	..	Earnings as % of Value Added
			(Millions of current East Caribbean Dollars)								
127.4	149.6	167.2	196.4	214.6	245.8	274.0	328.6	412.2	482.3	..	**MONETARY HOLDINGS** Money Supply, Broadly Defined
38.4	45.2	52.7	53.5	57.6	58.4	63.1	71.1	97.6	125.6	..	Money as Means of Payment
17.7	22.1	24.6	27.7	28.0	30.3	30.1	33.0	39.1	52.1	..	Currency Ouside Banks
20.7	23.2	28.1	25.8	29.5	28.1	33.0	38.0	58.6	73.5	..	Demand Deposits
89.0	104.4	114.6	142.9	157.1	187.4	211.0	257.6	314.6	356.8	..	Quasi-Monetary Liabilities
			(Millions of current East Caribbean Dollars)								
-2	5	-13	-8	-16	14	..	..	..	..	C	**GOVERNMENT DEFICIT (-) OR SURPLUS**
66	96	101	120	137	154	..	..	..	..		Current Revenue
51	74	82	92	117	115	..	..	..	..		Current Expenditure
15	22	19	28	20	40	..	..	..	..		Current Budget Balance
..	..	..	..	..	..	..	..	..	..		Capital Receipts
17	18	32	36	36	25	..	..	..	..		Capital Payments

ST. LUCIA	1967	1968	1969	1970	1971	1972	1973	1974	1975	1976	1977
FOREIGN TRADE (CUSTOMS BASIS)					*(Millions of current US dollars)*						
Value of Exports, fob	..	..	..	..	..	..	..	..	..	..	..
Nonfuel Primary Products	..	..	..	..	..	..	..	..	..	..	..
Fuels	..	..	..	..	..	..	..	..	..	..	..
Manufactures	..	..	..	..	..	..	..	..	..	..	..
Value of Imports, cif	..	..	..	..	..	..	..	..	..	..	..
Nonfuel Primary Products	..	..	..	..	..	..	..	..	..	..	..
Fuels	..	..	..	..	..	..	..	..	..	..	..
Manufactures	..	..	..	..	..	..	..	..	..	..	..
Terms of Trade					*(Index 1980 = 100)*						
Export Prices, fob	..	..	..	..	..	..	..	..	..	..	..
Nonfuel Primary Products	..	..	..	..	..	..	..	..	..	..	..
Fuels	..	..	..	..	..	..	..	..	..	..	..
Manufactures	..	..	..	..	..	..	..	..	..	..	..
Import Prices, cif	..	..	..	..	..	..	..	..	..	..	..
Trade at Constant 1980 Prices					*(Millions of 1980 US dollars)*						
Exports, fob	..	..	..	..	..	..	..	..	..	..	..
Imports, cif	..	..	..	..	..	..	..	..	..	..	..
BALANCE OF PAYMENTS					*(Millions of current US dollars)*						
Exports of Goods & Services	..	..	..	..	..	..	..	..	..	..	..
Merchandise, fob	..	..	..	..	..	..	..	..	..	32	41
Nonfactor Services	..	..	..	..	..	..	..	..	..	19	23
Factor Services	..	..	..	..	..	..	..	..	..	12	18
	..	..	..	..	..	..	..	..	..	1	1
Imports of Goods & Services	..	..	..	..	..	..	..	..	..	..	..
Merchandise, fob	..	..	..	..	..	..	..	..	..	50	62
Nonfactor Services	..	..	..	..	..	..	..	..	..	44	54
Factor Services	..	..	..	..	..	..	..	..	..	6	8
Long-Term Interest	..	..	..	..	..	..	..	..	..	0	0
Current Transfers, net	..	..	..	..	..	..	..	..	..	..	..
Workers' Remittances	..	..	..	..	..	..	..	..	..	8	8
	..	..	..	..	..	..	..	..	..	0	0
Total to be Financed	..	..	..	..	..	..	..	..	..	..	..
Official Capital Grants	..	..	..	..	..	..	..	..	..	-10	-13
Current Account Balance	..	..	..	..	..	..	..	..	..	5	3
	..	..	..	..	..	..	..	..	..	-5	-11
Long-Term Capital, net	..	..	..	..	..	..	..	..	..	..	..
Direct Investment	..	..	..	..	..	..	..	..	..	10	17
Long-Term Loans	..	..	..	..	..	..	..	..	..	3	13
Disbursements	..	..	..	..	..	..	..	..	..	..	..
Repayments	..	..	..	..	..	..	..	..	..	..	..
Other Long-Term Capital	..	..	..	..	..	..	..	..	..	3	2
Other Capital, net	..	..	..	..	..	..	..	..	..	-2	-6
Change in Reserves	..	..	..	0	0	0	0	0	0	-3	-1
Memo Item:					*(East Caribbean Dollars per US dollar)*						
Conversion Factor (Annual Avg)	1.760	2.000	2.000	2.000	1.970	1.920	1.960	2.050	2.170	2.610	2.700
EXTERNAL DEBT, ETC.					*(Millions of US dollars, outstanding at end of year)*						
Public/Publicly Guar. Long-Term	..	..	..	..	..	..	..	..	..	..	..
Official Creditors	..	..	..	..	..	..	..	..	..	..	..
IBRD and IDA	..	..	..	..	..	..	..	..	..	..	..
Private Creditors	..	..	..	..	..	..	..	..	..	..	..
Private Non-guaranteed Long-Term	..	..	..	..	..	..	..	..	..	..	..
Use of Fund Credit	..	..	..	..	..	..	..	..	..	..	..
Short-Term Debt	..	..	..	..	..	..	..	..	..	..	..
Memo Items:					*(Thousands of US dollars)*						
Int'l Reserves Excluding Gold	..	..	..	..	..	..	..	..	3,241	5,192	5,653
Gold Holdings (at market price)	..	..	..	..	..	..	..	..	..	..	..
SOCIAL INDICATORS											
Total Fertility Rate	6.1	6.0	5.9	5.8	5.7	5.6	5.5	5.4	5.2	5.1	5.0
Crude Birth Rate	41.6	41.4	41.3	41.2	40.1	38.9	37.8	36.6	35.4	34.3	35.4
Infant Mortality Rate	..	..	..	..	..	..	..	..	..	..	..
Life Expectancy at Birth	61.1	61.8	62.4	63.0	63.6	64.3	64.9	65.5	66.1	66.8	67.4
Food Production, p.c. ('79-81=100)	135.9	132.3	129.0	103.8	105.3	107.7	102.3	111.7	95.3	102.0	100.1
Labor Force, Agriculture (%)	..	..	..	..	..	..	..	..	..	..	..
Labor Force, Female (%)	..	..	..	..	..	..	..	..	..	..	..
School Enroll. Ratio, primary	..	..	..	..	..	..	..	..	..	..	..
School Enroll. Ratio, secondary	..	..	..	..	..	..	..	..	..	..	..

1978	1979	1980	1981	1982	1983	1984	1985	1986	1987 estimate	NOTES	ST. LUCIA
											FOREIGN TRADE (CUSTOMS BASIS)
				(Millions of current US dollars)							
..	..	46	42	42	48	48	52	61	..	..	Value of Exports, fob
..	..	..	..	..	..	..	..	..	..	..	Nonfuel Primary Products
..	..	..	..	..	..	..	..	..	..	..	Fuels
..	..	..	..	..	..	..	..	..	..	..	Manufactures
..	..	176	182	185	235	282	301	..	..	..	Value of Imports, cif
..	..	..	..	..	..	..	..	..	..	..	Nonfuel Primary Products
..	..	..	..	..	..	..	..	..	..	..	Fuels
..	..	..	..	..	..	..	..	..	..	..	Manufactures
				(Index 1980 = 100)							
..	..	..	..	..	..	..	..	..	..	..	Terms of Trade
..	..	..	..	..	..	..	..	..	..	..	Export Prices, fob
..	..	..	..	..	..	..	..	..	..	..	Nonfuel Primary Products
..	..	..	..	..	..	..	..	..	..	..	Fuels
..	..	..	..	..	..	..	..	..	..	..	Manufactures
..	..	..	..	..	..	..	..	..	..	..	Import Prices, cif
				(Millions of 1980 US dollars)							Trade at Constant 1980 Prices
..	..	..	..	..	..	..	..	..	..	..	Exports, fob
..	..	..	..	..	..	..	..	..	..	..	Imports, cif
				(Millions of current US dollars)							**BALANCE OF PAYMENTS**
56	67	88	82	86	102	112	122	..	..	..	Exports of Goods & Services
27	32	46	42	42	48	48	52	..	..	..	Merchandise, fob
28	34	41	39	45	54	64	70	..	..	..	Nonfactor Services
1	1	1	2	0	1	0	0	..	..	..	Factor Services
89	106	135	142	136	129	149	155	..	..	..	Imports of Goods & Services
75	92	113	117	107	97	108	114	..	..	..	Merchandise, fob
13	13	22	24	28	30	38	38	..	..	..	Nonfactor Services
1	1	1	1	1	2	3	3	..	..	..	Factor Services
..	..	..	..	..	..	..	..	..	..	..	Long-Term Interest
8	8	11	15	14	14	14	15	..	..	..	Current Transfers, net
0	0	0	0	0	0	0	0	..	..	..	Workers' Remittances
-25	-32	-37	-45	-37	-13	-24	-19	..	..	K	Total to be Financed
2	3	4	5	6	8	10	6	..	..	K	Official Capital Grants
-23	-28	-33	-40	-31	-5	-13	-13	..	..	..	Current Account Balance
25	32	36	47	38	20	23	25	..	..	..	Long-Term Capital, net
21	26	31	38	27	10	12	17	..	..	..	Direct Investment
..	..	..	..	..	..	..	..	..	..	..	Long-Term Loans
..	..	..	..	..	..	..	..	..	..	..	Disbursements
..	..	..	..	..	..	..	..	..	..	..	Repayments
3	3	2	3	6	2	1	2	..	..	..	Other Long-Term Capital
-1	-3	-5	-12	-5	-15	-8	-8	13	..	..	Other Capital, net
-1	-2	2	5	-2	0	-2	-5	-13	-6	..	Change in Reserves
											Memo Item:
				(East Caribbean Dollars per US dollar)							Conversion Factor (Annual Avg)
2.700	2.700	2.700	2.700	2.700	2.700	2.700	2.700	2.700	2.700	..	
				(Millions of US dollars, outstanding at end of year)							**EXTERNAL DEBT, ETC.**
..	..	..	..	..	..	..	..	..	..	..	Public/Publicly Guar. Long-Term
..	..	..	..	..	..	..	..	..	..	..	Official Creditors
..	..	..	..	..	..	..	..	..	..	..	IBRD and IDA
..	..	..	..	..	..	..	..	..	..	..	Private Creditors
..	..	..	..	..	..	..	..	..	..	..	Private Non-guaranteed Long-Term
..	..	..	..	..	..	..	..	..	..	..	Use of Fund Credit
..	..	..	..	..	..	..	..	..	..	..	Short-Term Debt
				(Thousands of US dollars)							**Memo Items:**
6,757	8,124	8,287	7,612	8,211	8,871	12,380	12,650	25,110	30,800	..	Int'l Reserves Excluding Gold
..	..	..	..	..	..	..	..	..	..	..	Gold Holdings (at market price)
											SOCIAL INDICATORS
4.9	4.8	4.4	4.3	4.2	4.1	4.0	4.0	3.9	3.8	..	Total Fertility Rate
33.9	30.8	30.6	30.6	31.4	31.0	31.0	30.8	28.0	30.6	..	Crude Birth Rate
..	..	..	..	..	..	17.6	23.0	..	22.9	..	Infant Mortality Rate
68.0	68.6	69.3	69.9	70.5	71.2	71.4	71.6	71.8	72.1	..	Life Expectancy at Birth
109.3	106.0	95.3	98.7	98.0	102.0	113.8	117.4	141.2	133.6	..	Food Production, p.c. ('79-81=100)
..	..	..	..	..	..	..	..	..	..	..	Labor Force, Agriculture (%)
..	..	..	..	..	..	..	..	..	..	..	Labor Force, Female (%)
..	..	..	..	..	..	..	..	..	..	..	School Enroll. Ratio, primary
..	..	..	..	..	..	..	..	..	..	..	School Enroll. Ratio, secondary

ST. VINCENT AND THE GRENADINES	1967	1968	1969	1970	1971	1972	1973	1974	1975	1976	1977
CURRENT GNP PER CAPITA (US $)	180	190	200	220	230	290	290	300	340	370	380
POPULATION (thousands)	86	86	87	88	90	92	94	96	93	96	98
ORIGIN AND USE OF RESOURCES				*(Millions of current East Caribbean Dollars)*							
Gross National Product (GNP)	27.90	30.70	33.30	36.90	39.60	53.00	59.10	67.60	71.30	83.00	94.30
Net Factor Income from Abroad	0.00	0.00	0.00	0.00	0.00	0.00	0.00	0.00	0.00	0.00	0.00
Gross Domestic Product (GDP)	27.90	30.70	33.30	36.90	39.60	53.00	59.10	67.60	71.30	83.00	94.30
Indirect Taxes, net	..	..	..	..	..	..	..	..	..	..	14.70
GDP at factor cost	..	..	..	..	..	..	..	..	..	..	79.60
Agriculture	..	..	..	..	..	..	..	..	..	0.00	14.70
Industry	..	..	..	..	..	..	..	..	..	..	18.80
Manufacturing	..	..	..	..	..	..	..	..	..	..	5.90
Services, etc.	..	..	..	..	..	..	..	..	..	..	46.10
Resource Balance	..	..	..	..	..	..	..	..	..	..	-41.10
Exports of Goods & NFServices	..	..	..	..	..	..	..	..	..	..	44.80
Imports of Goods & NFServices	..	..	..	..	..	..	..	..	..	..	85.90
Domestic Absorption	..	..	..	..	..	..	..	..	..	..	135.30
Private Consumption, etc.	..	..	..	..	..	..	..	..	..	..	85.70
General Gov't Consumption	..	..	..	..	..	..	..	..	..	..	23.10
Gross Domestic Investment	..	..	..	..	..	..	..	..	..	..	26.50
Fixed Investment	..	..	..	..	..	..	..	..	..	..	23.90
Memo Items:											
Gross Domestic Saving	..	..	..	..	..	..	..	..	..	..	-14.50
Gross National Saving	..	..	..	..	..	..	..	..	..	..	..
				(Millions of 1980 East Caribbean Dollars)							
Gross National Product	118.59	107.29	114.28	117.55	130.18	134.04	168.67	149.94	136.72	126.32	133.30
GDP at factor cost	..	..	..	..	..	..	..	..	..	..	110.50
Agriculture	..	..	..	..	..	..	..	..	..	..	23.28
Industry	..	..	..	..	..	..	..	..	..	..	26.52
Manufacturing	..	..	..	..	..	..	..	..	..	..	7.55
Services, etc.	..	..	..	..	..	..	..	..	..	..	60.69
Resource Balance	..	..	..	..	..	..	..	..	..	..	-65.30
Exports of Goods & NFServices	..	..	..	..	..	..	..	..	..	..	68.15
Imports of Goods & NFServices	..	..	..	..	..	..	..	..	..	..	133.45
Domestic Absorption	..	..	..	..	..	..	..	..	..	..	198.60
Private Consumption, etc.	..	..	..	..	..	..	..	..	..	..	124.65
General Gov't Consumption	..	..	..	..	..	..	..	..	..	..	32.86
Gross Domestic Investment	..	..	..	..	..	..	..	..	..	..	41.09
Fixed Investment	..	..	..	..	..	..	..	..	..	..	37.01
Memo Items:											
Capacity to Import	..	..	..	..	..	..	..	..	..	..	69.60
Terms of Trade Adjustment	..	..	..	..	..	..	..	..	..	..	1.45
Gross Domestic Income	..	..	..	..	..	..	..	..	..	..	134.75
Gross National Income	..	..	..	..	..	..	..	..	..	..	134.75
DOMESTIC PRICES (DEFLATORS)				*(Index 1980 = 100)*							
Overall (GDP)	23.5	28.6	29.1	31.4	30.4	39.5	35.0	45.1	52.2	65.7	70.7
Domestic Absorption	..	..	..	..	..	..	..	..	..	..	68.1
Agriculture	..	..	..	..	..	..	..	..	..	..	63.1
Industry	..	..	..	..	..	..	..	..	..	..	70.9
Manufacturing	..	..	..	..	..	..	..	..	..	..	78.2
MANUFACTURING ACTIVITY											
Employment (1980=100)	..	..	..	..	..	..	..	..	..	..	..
Real Earnings per Empl. (1980=100)	..	..	..	..	..	..	..	..	..	..	..
Real Output per Empl. (1980=100)	..	..	..	..	..	..	..	..	..	..	..
Earnings as % of Value Added	..	..	..	..	..	..	..	..	..	..	..
MONETARY HOLDINGS				*(Millions of current East Caribbean Dollars)*							
Money Supply, Broadly Defined	..	..	..	..	..	..	..	..	47.11	55.79	61.31
Money as Means of Payment	..	..	..	..	..	..	..	..	13.35	16.08	16.61
Currency Ouside Banks	..	..	..	..	..	..	..	..	8.00	9.52	9.50
Demand Deposits	..	..	..	..	..	..	..	..	5.34	6.55	7.10
Quasi-Monetary Liabilities	..	..	..	..	..	..	..	..	33.76	39.72	44.71
GOVERNMENT DEFICIT (-) OR SURPLUS				*(Thousands of current East Caribbean Dollars)*							
Current Revenue	..	..	..	..	..	..	..	..	..	..	..
Current Expenditure	..	..	..	..	..	..	..	..	..	..	..
Current Budget Balance	..	..	..	..	..	..	..	..	..	..	..
Capital Receipts	..	..	..	..	..	..	..	..	..	..	..
Capital Payments	..	..	..	..	..	..	..	..	..	..	..

1978	1979	1980	1981	1982	1983	1984	1985	1986	1987 estimate	NOTES	ST. VINCENT AND THE GRENADINES
430	480	540	640	740	830	900	970	1,030	1,070	..	CURRENT GNP PER CAPITA (US $)
99	101	103	104	106	107	108	110	111	112	..	POPULATION (thousands)
				(Millions of current East Caribbean Dollars)							ORIGIN AND USE OF RESOURCES
119.50	135.50	153.30	188.20	217.70	240.30	262.00	286.40	307.60	321.50	..	Gross National Product (GNP)
0.00	-3.80	-2.40	-4.60	-7.30	-6.20	-8.10	-7.80	-9.40	-9.50	..	Net Factor Income from Abroad
119.50	139.30	155.70	192.80	225.00	246.50	270.10	294.20	317.00	331.00	..	Gross Domestic Product (GDP)
18.00	21.90	24.20	28.20	36.80	42.20	49.40	55.10	60.30	..	..	Indirect Taxes, net
101.50	117.40	131.50	164.60	188.20	204.30	220.70	239.10	256.70	265.70	..	GDP at factor cost
20.70	19.70	19.80	27.50	31.90	35.50	39.30	45.90	49.10	..	..	Agriculture
24.00	30.90	36.40	44.00	48.30	51.90	51.10	53.40	56.30	..	..	Industry
10.30	13.10	14.20	18.40	20.70	21.30	27.70	26.30	23.70	..	..	Manufacturing
56.80	66.80	75.30	93.10	108.00	116.90	130.30	139.80	151.30	..	..	Services, etc.
-25.20	-51.30	-66.70	-52.30	-69.40	-53.20	-36.40	-18.30	-37.80	..	..	Resource Balance
78.80	88.80	104.50	120.20	129.30	159.00	196.30	222.60	225.10	..	..	Exports of Goods & NFServices
104.00	140.10	171.20	172.50	198.70	212.20	232.70	240.90	262.90	..	..	Imports of Goods & NFServices
144.60	190.40	222.40	245.20	294.40	299.70	306.50	312.60	354.80	..	..	Domestic Absorption
80.70	108.00	121.90	136.10	176.90	180.00	172.00	167.30	205.40	..	..	Private Consumption, etc.
29.50	33.60	37.70	45.00	52.80	56.80	59.10	60.10	53.10	..	..	General Gov't Consumption
34.40	48.80	62.80	64.10	64.70	62.90	75.40	85.20	96.30	..	..	Gross Domestic Investment
32.50	44.40	57.30	58.30	58.50	56.20	70.60	76.10	96.30	..	..	Fixed Investment
											Memo Items:
9.30	-2.30	-3.90	11.70	-4.70	9.70	39.00	66.80	58.50	..	..	Gross Domestic Saving
34.41	26.30	26.10	40.85	25.80	47.78	66.00	86.00	76.57	..	..	Gross National Saving
				(Millions of 1980 East Caribbean Dollars)							
147.67	150.78	153.30	163.67	175.51	187.00	196.72	209.93	218.77	226.58	..	Gross National Product
122.53	126.55	131.50	142.29	150.56	158.14	164.85	173.61	178.87	180.40	..	GDP at factor cost
27.24	23.13	19.80	28.04	29.62	31.05	33.11	37.22	37.38	..	..	Agriculture
29.20	34.71	36.40	37.11	38.66	40.21	40.91	41.34	42.89	..	..	Industry
11.39	13.56	14.20	14.46	15.35	15.74	16.37	15.86	15.10	..	..	Manufacturing
66.08	68.72	75.30	77.14	82.28	86.88	90.83	95.05	98.60	..	..	Services, etc.
-42.53	-58.85	-66.70	-55.64	-80.29	-71.01	-62.27	-60.80	-91.32	..	..	Resource Balance
107.69	107.85	104.50	110.28	115.45	138.88	173.71	184.97	173.56	..	..	Exports of Goods & NFServices
150.23	166.69	171.20	165.92	195.75	209.88	235.98	245.77	264.88	..	..	Imports of Goods & NFServices
190.20	211.46	222.40	223.83	263.43	264.36	267.46	278.78	318.28	..	..	Domestic Absorption
104.12	117.33	121.90	123.75	159.07	160.75	152.07	153.25	190.19	..	..	Private Consumption, etc.
37.70	36.14	37.70	40.69	46.52	49.65	51.50	54.20	48.09	..	..	General Gov't Consumption
48.38	57.99	62.80	59.39	57.84	53.96	63.89	71.33	80.01	..	..	Gross Domestic Investment
45.69	52.65	57.30	54.05	52.34	48.32	59.93	63.96	79.91	..	..	Fixed Investment
											Memo Items:
113.83	105.66	104.50	115.61	127.38	157.26	199.07	227.10	226.79	..	..	Capacity to Import
6.13	-2.19	0.00	5.33	11.93	18.39	25.36	42.13	53.24	..	..	Terms of Trade Adjustment
153.80	150.42	155.70	173.53	195.06	211.74	230.55	260.11	280.20	..	..	Gross Domestic Income
153.80	148.59	153.30	169.01	187.43	205.39	222.08	252.06	272.01	..	..	Gross National Income
				(Index 1980 = 100)							DOMESTIC PRICES (DEFLATORS)
80.9	91.3	100.0	114.6	122.9	127.5	131.6	135.0	139.7	142.0	..	Overall (GDP)
76.0	90.0	100.0	109.5	111.8	113.4	114.6	112.1	111.5	..	..	Domestic Absorption
76.0	85.2	100.0	98.1	107.7	114.3	118.7	123.3	131.3	..	..	Agriculture
82.2	89.0	100.0	118.6	124.9	129.1	124.9	129.2	131.3	..	..	Industry
90.5	96.6	100.0	127.3	134.8	135.4	169.2	165.8	157.0	..	..	Manufacturing
											MANUFACTURING ACTIVITY
..	..	..	..	..	..	..	..	..	..	..	Employment (1980=100)
..	..	..	..	..	..	..	..	..	..	..	Real Earnings per Empl. (1980=100)
..	..	..	..	..	..	..	..	..	..	..	Real Output per Empl. (1980=100)
..	..	..	..	..	..	..	..	..	..	..	Earnings as % of Value Added
				(Millions of current East Caribbean Dollars)							MONETARY HOLDINGS
78.20	96.43	101.10	119.98	135.12	152.05	171.37	196.23	227.13	244.03	..	Money Supply, Broadly Defined
22.62	27.72	28.06	33.28	34.88	40.56	47.79	52.55	63.16	53.16	..	Money as Means of Payment
11.07	12.94	12.85	15.36	17.61	20.14	22.53	32.03	41.42	23.27	..	Currency Ouside Banks
11.55	14.78	15.21	17.93	17.26	20.42	25.27	20.52	21.74	29.89	..	Demand Deposits
55.58	68.71	73.04	86.70	100.24	111.49	123.58	143.68	163.97	190.87	..	Quasi-Monetary Liabilities
				(Thousands of current East Caribbean Dollars)							
..	..	-300	1,600	-7,100	-400	-8,300	6,100	4,400	..	C	GOVERNMENT DEFICIT (-) OR SURPLUS
30,900	39,200	47,000	58,800	68,200	79,600	83,300	96,800	105,700	..	..	Current Revenue
24,900	37,700	45,000	51,200	69,000	73,400	84,300	84,600	90,400	..	..	Current Expenditure
6,000	1,500	2,000	7,600	-800	6,200	-1,000	12,200	15,300	..	..	Current Budget Balance
..	..	..	..	..	..	..	100	100	..	..	Capital Receipts
-24,900	-37,700	2,300	6,000	6,300	6,600	7,300	6,200	11,000	..	..	Capital Payments

ST. VINCENT AND THE GRENADINES	1967	1968	1969	1970	1971	1972	1973	1974	1975	1976	1977
FOREIGN TRADE (CUSTOMS BASIS)					*(Millions of current US dollars)*						
Value of Exports, fob	..	..	..	..	..	..	..	..	..	..	..
Nonfuel Primary Products	..	..	..	..	..	..	..	..	..	..	..
Fuels	..	..	..	..	..	..	..	..	..	..	..
Manufactures	..	..	..	..	..	..	..	..	..	..	..
Value of Imports, cif	..	..	..	..	..	..	..	..	..	..	..
Nonfuel Primary Products	..	..	..	..	..	..	..	..	..	..	..
Fuels	..	..	..	..	..	..	..	..	..	..	..
Manufactures	..	..	..	..	..	..	..	..	..	..	..
					(Index 1980 = 100)						
Terms of Trade	..	..	..	..	..	..	..	..	..	..	..
Export Prices, fob	..	..	..	..	..	..	..	..	..	..	..
Nonfuel Primary Products	..	..	..	..	..	..	..	..	..	..	..
Fuels	..	..	..	..	..	..	..	..	..	..	..
Manufactures	..	..	..	..	..	..	..	..	..	..	..
Import Prices, cif	..	..	..	..	..	..	..	..	..	..	..
Trade at Constant 1980 Prices					*(Millions of 1980 US dollars)*						
Exports, fob	..	..	..	..	..	..	..	..	..	..	..
Imports, cif	..	..	..	..	..	..	..	..	..	..	..
BALANCE OF PAYMENTS					*(Millions of current US dollars)*						
Exports of Goods & Services	..	..	..	..	..	..	..	..	..	..	..
Merchandise, fob	..	..	..	..	..	..	..	..	..	..	..
Nonfactor Services	..	..	..	..	..	..	..	..	..	..	..
Factor Services	..	..	..	..	..	..	..	..	..	..	..
Imports of Goods & Services	..	..	..	..	..	..	..	..	..	..	..
Merchandise, fob	..	..	..	..	..	..	..	..	..	..	..
Nonfactor Services	..	..	..	..	..	..	..	..	..	..	..
Factor Services	..	..	..	..	..	..	..	..	..	..	..
Long-Term Interest	..	..	..	0.00	0.10	0.00	0.10	0.10	0.10	0.10	0.10
Current Transfers, net	..	..	..	..	..	..	..	..	..	..	..
Workers' Remittances	..	..	..	..	..	..	..	..	..	..	..
Total to be Financed	..	..	..	..	..	..	..	..	..	..	..
Official Capital Grants	..	..	..	..	..	..	..	..	..	..	..
Current Account Balance	..	..	..	..	..	..	..	..	..	..	..
Long-Term Capital, net	..	..	..	..	..	..	..	..	..	..	..
Direct Investment	..	..	..	..	..	..	..	..	..	..	..
Long-Term Loans	..	..	..	0.20	..	..	2.10	0.40	0.70	0.60	0.90
Disbursements	..	..	..	0.20	..	..	2.10	0.40	0.70	0.60	0.90
Repayments	..	..	..	..	..	..	..	..	..	..	..
Other Long-Term Capital	..	..	..	..	..	..	..	..	..	..	..
Other Capital, net	..	..	..	..	..	..	..	..	..	..	..
Change in Reserves	..	..	..	0.00	0.00	0.00	0.00	0.00	0.00	0.00	-0.05
Memo Item:					*(East Caribbean Dollars per US dollar)*						
Conversion Factor (Annual Avg)	1.760	2.000	2.000	2.000	1.970	1.920	1.960	2.050	2.170	2.610	2.700
EXTERNAL DEBT, ETC.					*(Thousands of US dollars, outstanding at end of year)*						
Public/Publicly Guar. Long-Term	..	..	..	700	700	800	2,800	3,200	3,400	3,700	4,700
Official Creditors	..	..	..	..	..	..	2,000	2,500	2,800	3,100	4,100
IBRD and IDA	..	..	..	0	0	0	0	0	0	0	0
Private Creditors	..	..	..	700	700	800	700	700	600	600	600
Private Non-guaranteed Long-Term	..	..	..	..	..	..	..	..	..	..	..
Use of Fund Credit	..	..	..	0	0	0	0	0	0	0	0
Short-Term Debt	..	..	..	..	..	..	..	..	..	..	..
Memo Items:					*(Thousands of US dollars)*						
Int'l Reserves Excluding Gold	..	..	..	..	..	..	..	..	..	4,658	4,810
Gold Holdings (at market price)	..	..	..	..	..	..	..	..	..	..	..
SOCIAL INDICATORS											
Total Fertility Rate	6.2	6.0	5.9	5.7	5.6	5.4	5.2	5.1	4.9	4.8	4.6
Crude Birth Rate	..	..	..	..	..	40.2	34.4	35.0	34.5	38.4	30.6
Infant Mortality Rate	..	..	..	..	..	..	..	..	..	..	..
Life Expectancy at Birth	61.7	62.1	62.5	62.8	63.2	63.6	64.0	64.3	64.7	65.1	65.5
Food Production, p.c. ('79-81 = 100)	95.1	96.9	97.3	90.0	95.4	91.0	93.9	88.2	85.4	94.1	92.0
Labor Force, Agriculture (%)	..	..	..	..	..	..	..	..	..	..	..
Labor Force, Female (%)	..	..	..	..	..	..	..	..	..	..	..
School Enroll. Ratio, primary	..	..	..	..	..	..	..	..	..	..	..
School Enroll. Ratio, secondary	..	..	..	..	..	..	..	..	..	..	..

1978	1979	1980	1981	1982	1983	1984	1985	1986	1987 estimate	NOTES	ST. VINCENT AND THE GRENADINES
				(Millions of current US dollars)							**FOREIGN TRADE (CUSTOMS BASIS)**
..	..	22.10	29.80	34.30	41.80	54.20	61.50	68.31	..	..	Value of Exports, fob
..	..	..	..	..	..	..	..	..	..	..	Nonfuel Primary Products
..	..	..	..	..	..	..	..	..	..	..	Fuels
..	..	..	..	..	..	..	..	..	..	..	Manufactures
..	..	98.89	93.44	98.02	87.78	87.40	100.70	117.94	..	..	Value of Imports, cif
..	..	..	..	..	..	..	..	..	..	..	Nonfuel Primary Products
..	..	..	..	..	..	..	..	..	..	..	Fuels
..	..	..	..	..	..	..	..	..	..	..	Manufactures
				(Index 1980 = 100)							
..	..	..	..	..	..	..	..	..	..	..	Terms of Trade
..	..	..	..	..	..	..	..	..	..	..	Export Prices, fob
..	..	..	..	..	..	..	..	..	..	..	Nonfuel Primary Products
..	..	..	..	..	..	..	..	..	..	..	Fuels
..	..	..	..	..	..	..	..	..	..	..	Manufactures
..	..	..	..	..	..	..	..	..	..	..	Import Prices, cif
				(Millions of 1980 US dollars)							Trade at Constant 1980 Prices
..	..	..	..	..	..	..	..	..	..	..	Exports, fob
..	..	..	..	..	..	..	..	..	..	..	Imports, cif
				(Millions of current US dollars)							**BALANCE OF PAYMENTS**
29.40	33.40	39.60	50.10	48.09	59.10	72.70	82.61	87.46	..	..	Exports of Goods & Services
18.10	19.10	21.10	29.80	32.19	41.10	53.60	63.21	67.82	..	..	Merchandise, fob
11.10	13.70	17.70	18.90	15.70	17.80	19.10	19.20	19.45	..	..	Nonfactor Services
0.20	0.60	0.80	1.40	0.20	0.20	0.00	0.20	0.20	..	..	Factor Services
38.73	53.10	65.00	68.30	76.50	81.10	89.20	92.31	100.53	..	..	Imports of Goods & Services
32.91	42.09	51.91	52.91	58.59	63.36	68.94	71.29	78.09	..	..	Merchandise, fob
5.61	9.81	11.39	13.29	15.01	15.24	17.26	17.92	18.75	..	..	Nonfactor Services
0.20	1.20	1.70	2.10	2.90	2.50	3.00	3.10	3.69	..	..	Factor Services
0.10	0.20	0.30	0.40	0.60	0.60	1.10	1.20	1.30	1.30	..	Long-Term Interest
..	..	12.00	12.50	14.00	16.40	13.00	10.00	10.17	..	..	Current Transfers, net
0.00	0.00	0.00	0.00	0.00	0.00	0.00	0.00	0.00	..	..	Workers' Remittances
..	..	-13.20	-4.20	-15.31	-6.60	-4.50	-0.40	-4.81	..	..	Total to be Financed
..	..	3.90	3.40	4.50	4.00	3.50	4.10	8.60	..	..	Official Capital Grants
2.70	-3.60	-9.30	-0.80	-10.81	-2.60	-1.00	3.70	3.79	..	..	Current Account Balance
3.30	6.90	9.60	8.10	7.50	7.70	5.60	7.50	15.36	..	..	Long-Term Capital, net
-0.50	0.60	1.10	0.50	1.50	2.10	1.40	1.80	2.99	..	..	Direct Investment
1.00	1.20	2.90	7.40	3.00	3.10	1.20	1.50	5.50	5.90	..	Long-Term Loans
1.00	1.30	3.00	7.70	3.80	3.80	1.90	2.70	6.90	7.50	..	Disbursements
..	0.10	0.10	0.30	0.80	0.70	0.70	1.20	1.40	1.60	..	Repayments
2.80	5.10	1.70	-3.20	-1.50	-1.50	-0.50	0.10	-1.73	..	..	Other Long-Term Capital
-5.85	0.54	-2.18	-7.16	-0.39	-4.78	-0.10	-4.77	-7.28	..	..	Other Capital, net
-0.15	-3.84	1.88	-0.14	3.70	-0.32	-4.50	-6.43	-11.87	5.60	..	Change in Reserves
				(East Caribbean Dollars per US dollar)							Memo Item:
2.700	2.700	2.700	2.700	2.700	2.700	2.700	2.700	2.700	2.700	..	Conversion Factor (Annual Avg)
				(Thousands of US dollars, outstanding at end of year)							**EXTERNAL DEBT, ETC.**
5,700	7,200	10,300	16,800	19,000	21,500	21,400	23,200	28,800	36,000	..	Public/Publicly Guar. Long-Term
5,100	6,500	9,400	15,100	17,900	20,800	20,900	22,700	28,400	35,700	..	Official Creditors
0	0	0	0	0	0	0	1,100	2,100	3,000	..	IBRD and IDA
500	800	900	1,800	1,100	800	600	500	400	300	..	Private Creditors
..	..	..	..	..	..	..	..	..	..	..	Private Non-guaranteed Long-Term
0	0	300	1,800	1,600	1,600	1,000	300	0	..	..	Use of Fund Credit
..	..	..	..	..	1,000	1,000	2,000	3,000	3,000	..	Short-Term Debt
				(Thousands of US dollars)							Memo Items:
5,256	8,902	7,270	9,005	4,783	5,697	12,820	13,800	25,832	20,220	..	Int'l Reserves Excluding Gold
..	..	..	..	..	..	..	..	..	..	..	Gold Holdings (at market price)
											SOCIAL INDICATORS
4.4	4.3	4.1	4.0	4.2	4.1	3.9	3.8	3.6	3.0	..	Total Fertility Rate
31.2	31.9	28.3	29.3	29.9	28.8	26.2	..	..	26.8	..	Crude Birth Rate
..	..	..	..	..	..	26.5	28.2	29.9	31.6	..	Infant Mortality Rate
65.9	66.2	66.6	67.0	67.4	67.8	68.1	68.5	68.9	69.3	..	Life Expectancy at Birth
103.9	101.9	98.2	99.9	102.8	103.6	129.2	149.6	155.7	157.9	..	Food Production, p.c. ('79-81 = 100)
..	..	..	..	..	..	..	..	..	..	..	Labor Force, Agriculture (%)
..	..	..	..	..	..	..	..	..	..	..	Labor Force, Female (%)
..	..	..	..	..	..	..	..	..	..	..	School Enroll. Ratio, primary
..	..	..	..	..	..	..	..	..	..	..	School Enroll. Ratio, secondary

SUDAN	1967	1968	1969	1970	1971	1972	1973	1974	1975	1976	1977
CURRENT GNP PER CAPITA (US $)	110	120	120	130	150	150	150	190	250	320	380
POPULATION (thousands)	13,082	13,474	13,882	14,302	14,732	15,167	15,612	16,072	16,550	17,044	17,553

ORIGIN AND USE OF RESOURCES *(Millions of current Sudanese Pounds)*

	1967	1968	1969	1970	1971	1972	1973	1974	1975	1976	1977
Gross National Product (GNP)	536	583	585	700	757	828	888	1,237	1,496	1,830	2,322
Net Factor Income from Abroad	0	0	0	-1	-4	-4	-9	-9	-15	-18	-18
Gross Domestic Product (GDP)	536	583	585	701	761	832	897	1,246	1,511	1,848	2,340
Indirect Taxes, net	46	56	79	87	96	110	110	131	169	226	256
GDP at factor cost	490	527	506	614	665	722	787	1,116	1,342	1,622	2,084
Agriculture	190	200	180	264	294	324	342	512	580	623	817
Industry	80	85	83	91	92	98	106	153	185	230	271
Manufacturing	40	42	43	49	51	52	55	71	96	109	132
Services, etc.	220	242	242	259	279	300	338	451	577	770	996
Resource Balance	8	-13	-7	11	-12	-13	2	-48	-173	-184	-148
Exports of Goods & NFServices	79	90	86	116	123	126	151	167	183	223	247
Imports of Goods & NFServices	70	104	93	105	135	139	149	215	356	407	395
Domestic Absorption	528	596	592	691	773	845	895	1,295	1,683	2,032	2,488
Private Consumption, etc.	353	409	384	447	523	623	624	885	1,210	1,374	1,810
General Gov't Consumption	105	112	133	148	160	146	166	181	208	230	278
Gross Domestic Investment	69	75	75	96	90	76	105	229	265	428	399
Fixed Investment	..	..	..	..	..	..	..	..	..	362	312
Memo Items:											
Gross Domestic Saving	78	62	68	106	78	63	107	181	93	244	251
Gross National Saving	77	62	68	105	73	59	99	173	78	279	293

(Millions of 1980 Sudanese Pounds)

	1967	1968	1969	1970	1971	1972	1973	1974	1975	1976	1977
Gross National Product	2,511	2,666	2,612	2,735	2,913	2,852	2,610	2,881	3,237	3,839	4,427
GDP at factor cost	2,294	2,409	2,259	2,398	2,561	2,488	2,311	2,598	2,904	3,399	3,972
Agriculture	891	914	805	1,030	1,133	1,116	1,006	1,191	1,255	1,304	1,557
Industry	375	388	371	357	355	337	311	357	400	481	516
Manufacturing	..	..	..	191	195	178	160	164	207	228	252
Services, etc.	1,029	1,108	1,082	1,012	1,072	1,034	995	1,050	1,249	1,613	1,899
Resource Balance	-129	-161	-92	10	-77	-15	59	-78	-324	-271	-311
Exports of Goods & NFServices	418	463	463	489	430	376	441	343	320	461	405
Imports of Goods & NFServices	547	624	555	480	507	391	382	420	644	732	716
Domestic Absorption	2,640	2,827	2,705	2,730	3,007	2,881	2,577	2,979	3,593	4,142	4,770
Private Consumption, etc.	1,763	1,883	1,786	1,764	2,038	2,105	1,778	2,000	2,533	2,725	3,447
General Gov't Consumption	539	585	567	576	608	502	475	421	461	482	530
Gross Domestic Investment	338	359	352	391	360	275	323	558	599	936	794
Fixed Investment	..	..	..	..	..	..	..	..	..	791	621
Memo Items:											
Capacity to Import	612	544	514	528	463	354	387	326	332	401	448
Terms of Trade Adjustment	194	82	51	39	34	-22	-54	-17	12	-60	43
Gross Domestic Income	2,705	2,748	2,663	2,779	2,963	2,845	2,581	2,885	3,281	3,811	4,502
Gross National Income	2,705	2,748	2,663	2,774	2,946	2,830	2,556	2,864	3,249	3,779	4,470

DOMESTIC PRICES (DEFLATORS) *(Index 1980 = 100)*

	1967	1968	1969	1970	1971	1972	1973	1974	1975	1976	1977
Overall (GDP)	21.4	21.9	22.4	25.6	26.0	29.0	34.0	42.9	46.2	47.7	52.5
Domestic Absorption	20.0	21.1	21.9	25.3	25.7	29.3	34.7	43.5	46.8	49.1	52.1
Agriculture	21.4	21.9	22.4	25.6	26.0	29.0	34.0	42.9	46.2	47.7	52.5
Industry	21.4	21.9	22.4	25.6	26.0	29.0	34.0	43.0	46.2	47.7	52.5
Manufacturing	..	..	..	25.6	26.0	29.0	34.0	42.9	46.2	47.7	52.5

MANUFACTURING ACTIVITY

	1967	1968	1969	1970	1971	1972	1973	1974	1975	1976	1977
Employment (1980 = 100)	..	..	..	..	..	..	..	..	..	..	..
Real Earnings per Empl. (1980 = 100)	..	..	..	..	..	..	..	..	..	..	..
Real Output per Empl. (1980 = 100)	..	..	..	..	..	..	..	..	..	..	..
Earnings as % of Value Added	75.0	67.8	44.8	31.2	29.2	29.5	..	..	..	..	..

MONETARY HOLDINGS *(Millions of current Sudanese Pounds)*

	1967	1968	1969	1970	1971	1972	1973	1974	1975	1976	1977
Money Supply, Broadly Defined	93	98	115	130	140	166	205	278	330	412	589
Money as Means of Payment	82	81	101	116	124	144	176	237	281	350	497
Currency Ouside Banks	42	47	59	67	70	75	93	119	129	153	199
Demand Deposits	39	34	42	49	54	69	83	118	153	197	298
Quasi-Monetary Liabilities	12	17	14	14	16	22	30	41	49	62	92

GOVERNMENT DEFICIT (-) OR SURPLUS *(Millions of current Sudanese Pounds)*

	1967	1968	1969	1970	1971	1972	1973	1974	1975	1976	1977
GOVERNMENT DEFICIT (-) OR SURPLUS	..	..	..	..	..	-6	-16	-10	-75	-59	-168
Current Revenue	..	..	..	..	..	156	157	193	258	299	351
Current Expenditure	..	..	..	..	..	139	167	159	241	256	317
Current Budget Balance	..	..	..	..	..	17	-11	34	17	43	33
Capital Receipts	..	..	..	..	..	0	0	0	1	0	0
Capital Payments	..	..	..	..	..	23	6	44	92	102	202

1978	1979	1980	1981	1982	1983	1984	1985	1986	1987 estimate	NOTES	SUDAN
400	430	430	430	400	380	320	280	300	330	A	**CURRENT GNP PER CAPITA (US $)**
18,075	18,609	19,152	19,703	20,260	20,819	21,377	21,931	22,567	23,214	..	**POPULATION (thousands)**
				(Millions of current Sudanese Pounds)							**ORIGIN AND USE OF RESOURCES**
2,865	3,230	3,931	4,844	6,458	8,637	10,234	13,440	18,606	24,440	C	Gross National Product (GNP)
-18	-24	-41	-136	-261	-390	-743	-960	-1,515	-1,576	..	Net Factor Income from Abroad
2,883	3,254	3,972	4,980	6,719	9,027	10,977	14,400	20,121	26,016	C	Gross Domestic Product (GDP)
305	358	342	460	484	886	985	1,014	1,322	1,218	..	Indirect Taxes, net
2,577	2,896	3,630	4,520	6,235	8,141	9,992	13,386	18,799	24,798	C	GDP at factor cost
1,043	1,089	1,232	1,446	2,396	2,878	3,545	4,336	6,844	9,192	..	Agriculture
306	374	492	659	822	1,237	1,461	2,154	2,922	3,821	..	Industry
148	202	249	359	365	523	652	979	1,463	1,938	..	Manufacturing
1,229	1,433	1,905	2,415	3,017	4,026	4,985	6,896	9,033	11,785	..	Services, etc.
-231	-217	-463	-662	-1,198	-1,209	-1,108	-1,193	-1,733	-1,299	..	Resource Balance
242	244	476	471	593	1,010	1,495	1,606	1,913	2,151	..	Exports of Goods & NFServices
473	460	938	1,134	1,792	2,219	2,603	2,799	3,646	3,450	..	Imports of Goods & NFServices
3,114	3,470	4,435	5,642	7,917	10,236	12,085	15,593	21,854	27,315	..	Domestic Absorption
2,369	2,632	3,200	4,134	5,555	7,401	8,636	10,866	16,306	20,546	..	Private Consumption, etc.
331	407	636	793	1,085	1,264	1,638	2,394	2,913	3,855	..	General Gov't Consumption
414	431	598	715	1,277	1,571	1,811	2,333	2,636	2,914	..	Gross Domestic Investment
324	339	458	507	904	1,538	2,045	2,268	2,376	2,654	..	Fixed Investment
											Memo Items:
183	215	136	52	78	362	703	1,140	903	1,615	..	Gross Domestic Saving
248	292	199	80	146	511	474	1,164	263	742	..	Gross National Saving
				(Millions of 1980 Sudanese Pounds)							
4,362	3,903	3,931	3,923	4,185	4,321	4,012	3,507	3,691	3,800	C	Gross National Product
3,925	3,502	3,630	3,680	4,047	4,065	3,949	3,505	3,712	3,829	C	GDP at factor cost
1,587	1,317	1,232	1,177	1,555	1,437	1,401	1,135	1,351	1,419	..	Agriculture
466	452	493	537	533	618	578	564	577	590	..	Industry
225	244	249	292	237	261	258	251	289	299	..	Manufacturing
1,872	1,733	1,905	1,967	1,958	2,010	1,970	1,806	1,784	1,820	..	Services, etc.
-381	-319	-463	-699	-914	-688	-612	-494	-378	-205	..	Resource Balance
349	283	476	407	292	371	533	275	300	270	..	Exports of Goods & NFServices
730	601	938	1,106	1,206	1,059	1,145	769	678	475	..	Imports of Goods & NFServices
4,771	4,253	4,435	4,754	5,275	5,196	4,951	4,264	4,351	4,223	..	Domestic Absorption
3,609	3,217	3,201	3,502	3,706	3,757	3,562	2,989	3,243	3,169	..	Private Consumption, etc.
504	492	636	646	705	631	647	627	575	595	..	General Gov't Consumption
658	544	598	606	865	807	741	648	533	459	..	Gross Domestic Investment
515	428	458	430	612	790	837	630	480	418	..	Fixed Investment
											Memo Items:
374	318	476	460	399	482	658	441	356	296	..	Capacity to Import
25	36	0	53	107	111	125	166	56	27	..	Terms of Trade Adjustment
4,414	3,970	3,972	4,108	4,468	4,618	4,463	3,936	4,029	4,044	..	Gross Domestic Income
4,387	3,939	3,931	3,976	4,293	4,432	4,137	3,672	3,747	3,827	..	Gross National Income
				(Index 1980 = 100)							**DOMESTIC PRICES (DEFLATORS)**
65.7	82.7	100.0	122.8	154.1	200.3	253.0	382.0	506.5	647.6	..	Overall (GDP)
65.3	81.6	100.0	118.7	150.1	197.0	244.1	365.7	502.3	646.9	..	Domestic Absorption
65.7	82.7	100.0	122.8	154.1	200.3	253.0	382.0	506.5	647.6	..	Agriculture
65.7	82.7	100.0	122.8	154.1	200.3	253.0	382.0	506.5	647.7	..	Industry
65.7	82.7	100.0	122.8	154.1	200.3	253.0	390.9	506.5	647.6	..	Manufacturing
											MANUFACTURING ACTIVITY
..	..	..	..	..	..	..	..	..	..	..	Employment (1980=100)
..	..	..	..	..	..	..	..	..	..	..	Real Earnings per Empl. (1980=100)
..	..	..	..	..	..	..	..	..	..	..	Real Output per Empl. (1980=100)
..	..	..	..	..	..	..	..	..	..	..	Earnings as % of Value Added
				(Millions of current Sudanese Pounds)							**MONETARY HOLDINGS**
751	977	1,264	1,795	2,534	3,110	3,719	6,108	7,813	9,152	..	Money Supply, Broadly Defined
634	837	1,097	1,531	2,091	2,336	2,764	4,145	5,849	..	..	Money as Means of Payment
279	380	508	630	820	1,022	1,247	1,930	2,760	3,625	..	Currency Ouside Banks
354	456	589	901	1,271	1,314	1,517	2,214	3,089	..	..	Demand Deposits
117	141	167	264	443	774	955	1,963	1,964	..	..	Quasi-Monetary Liabilities
				(Millions of current Sudanese Pounds)							**GOVERNMENT DEFICIT (-) OR SURPLUS**
-152	-136	-130	..	-326	..	..	..	..	..	C	
403	544	657	..	1,033	..	..	..	..	..	..	Current Revenue
396	531	627	..	895	..	..	..	..	..	..	Current Expenditure
7	13	30	..	137	..	..	..	..	..	..	Current Budget Balance
0	3	9	..	1	..	..	..	..	..	..	Capital Receipts
159	152	169	..	465	..	..	..	..	..	..	Capital Payments

SUDAN	1967	1968	1969	1970	1971	1972	1973	1974	1975	1976	1977
FOREIGN TRADE (CUSTOMS BASIS)				*(Millions of current US dollars)*							
Value of Exports, fob	214.7	233.4	248.1	294.5	331.7	361.2	416.5	440.5	429.4	577.7	661.0
Nonfuel Primary Products	212.4	231.8	245.0	290.5	325.9	354.7	408.0	421.4	407.6	560.8	645.4
Fuels	0.4	0.4	1.7	2.4	3.5	3.7	6.6	17.1	15.4	10.0	10.0
Manufactures	1.8	1.1	1.4	1.5	2.4	2.9	1.9	2.0	6.4	6.9	5.6
Value of Imports, cif	213.5	257.6	265.6	311.2	355.2	353.5	479.5	655.8	957.0	951.8	982.6
Nonfuel Primary Products	56.0	62.8	43.8	75.4	86.0	90.6	128.1	173.5	192.4	140.6	141.1
Fuels	8.6	13.8	25.5	26.1	25.8	25.4	28.7	44.2	35.3	22.1	22.4
Manufactures	148.9	181.1	196.3	209.6	243.3	237.5	322.6	438.1	729.3	789.1	819.1
				(Index 1980 = 100)							
Terms of Trade	134.1	146.6	157.4	148.4	142.3	142.3	174.7	121.9	93.6	123.0	112.1
Export Prices, fob	33.0	33.4	31.0	31.5	35.9	39.9	65.2	71.2	60.3	78.0	76.7
Nonfuel Primary Products	33.4	33.8	32.4	33.3	38.1	42.3	73.0	74.0	61.9	79.6	77.7
Fuels	4.3	4.3	4.3	4.3	5.6	6.2	8.9	36.7	35.7	38.4	42.0
Manufactures	28.2	28.8	26.3	26.9	30.8	37.2	43.6	71.2	64.1	66.0	71.8
Import Prices, cif	24.6	22.8	19.7	21.2	25.2	28.1	37.4	58.4	64.5	63.4	68.4
Trade at Constant 1980 Prices				*(Millions of 1980 US dollars)*							
Exports, fob	651.5	699.3	801.3	935.5	924.9	905.2	638.4	618.9	711.6	740.6	861.8
Imports, cif	868.5	1,131.6	1,350.3	1,467.1	1,409.1	1,260.0	1,283.4	1,123.3	1,484.3	1,501.0	1,436.5
BALANCE OF PAYMENTS				*(Millions of current US dollars)*							
Exports of Goods & Services	..	..	..	325	353	371	484	448	508	640	713
Merchandise, fob	..	..	..	284	309	325	441	384	412	551	595
Nonfactor Services	..	..	..	38	43	46	41	59	89	89	115
Factor Services	..	..	..	3	1	1	1	5	8	0	4
Imports of Goods & Services	..	..	..	367	393	430	460	742	985	1,219	1,189
Merchandise, fob	..	..	..	268	294	317	334	542	743	1,062	986
Nonfactor Services	..	..	..	81	85	90	103	168	179	106	148
Factor Services	..	..	..	18	14	23	23	33	62	51	55
Long-Term Interest	..	..	..	..	..	..	..	..	..	..	..
Current Transfers, net	..	..	..	..	..	..	..	..	..	151	172
Workers' Remittances	..	..	..	0	0	0	0	0	6	151	172
Total to be Financed	..	..	..	..	..	..	..	..	..	-429	-303
Official Capital Grants	..	..	..	..	..	..	..	..	..	0	11
Current Account Balance	..	..	..	-42	-42	-49	28	-271	-428	-429	-292
Long-Term Capital, net	..	..	..	2	4	21	3	272	80	471	316
Direct Investment	..	..	..	..	-1	0	..	..	..	0	0
Long-Term Loans	..	..	..	..	..	..	..	..	..	..	..
Disbursements	..	..	..	..	..	..	..	..	..	..	..
Repayments	..	..	..	..	..	..	..	..	..	..	..
Other Long-Term Capital	..	..	..	2	5	21	3	272	80	471	305
Other Capital, net	..	..	..	25	52	19	-6	-7	225	-138	-22
Change in Reserves	..	..	..	15	-14	9	-25	5	124	95	-2
Memo Item:				*(Sudanese Pounds per US dollar)*							
Conversion Factor (Annual Avg)	0.350	0.350	0.350	0.350	0.350	0.350	0.350	0.350	0.350	0.350	0.350
EXTERNAL DEBT, ETC.				*(Millions of US dollars, outstanding at end of year)*							
Public/Publicly Guar. Long-Term	..	..	..	307	327	361	453	895	1,231	1,620	1,930
Official Creditors	..	..	..	260	285	310	364	598	774	1,073	1,290
IBRD and IDA	..	..	..	104	105	105	104	99	108	138	156
Private Creditors	..	..	..	46	42	50	90	297	457	547	639
Private Non-guaranteed Long-Term	..	..	..	..	..	..	..	..	..	..	..
Use of Fund Credit	..	..	..	31	..	..	..	..	152	..	..
Short-Term Debt	..	..	..	..	..	..	..	..	..	..	464
Memo Items:				*(Thousands of US dollars)*							
Int'l Reserves Excluding Gold	54,700	47,700	36,400	21,740	27,900	35,587	61,331	124,270	36,369	23,600	23,149
Gold Holdings (at market price)	..	..	..	..	..	..	..	..	..	..	..
SOCIAL INDICATORS											
Total Fertility Rate	6.7	6.7	6.7	6.7	6.7	6.7	6.7	6.7	6.7	6.7	6.7
Crude Birth Rate	47.0	47.0	47.0	47.0	47.0	47.0	47.0	47.0	47.1	47.1	47.1
Infant Mortality Rate	156.0	153.8	151.6	149.4	147.2	145.0	142.2	139.4	136.6	133.8	131.0
Life Expectancy at Birth	40.9	41.3	41.6	41.9	42.3	42.6	43.1	43.6	44.1	44.6	45.1
Food Production, p.c. ('79-81 = 100)	102.7	87.1	95.3	96.7	96.0	92.9	94.0	98.0	101.5	97.2	101.4
Labor Force, Agriculture (%)	79.7	78.8	77.9	77.0	76.4	75.8	75.3	74.7	74.1	73.5	72.9
Labor Force, Female (%)	20.6	20.5	20.4	20.3	20.2	20.2	20.1	20.0	20.0	19.9	19.9
School Enroll. Ratio, primary	..	..	..	38.0	..	..	..	..	47.0	48.0	49.0
School Enroll. Ratio, secondary	..	..	..	7.0	..	..	..	..	14.0	14.0	15.0

1978	1979	1980	1981	1982	1983	1984	1985	1986	1987 estimate	NOTES	SUDAN
											FOREIGN TRADE (CUSTOMS BASIS)
				(Millions of current US dollars)							
483.1	583.3	594.0	538.0	432.0	581.0	722.0	595.0	497.0	482.3	f	Value of Exports, fob
474.5	562.0	573.5	507.1	403.9	544.1	679.4	538.3	440.7	422.2	..	Nonfuel Primary Products
6.3	16.7	5.6	21.3	17.0	22.3	28.0	28.8	24.0	25.5	..	Fuels
2.4	4.6	15.0	9.6	11.1	14.6	14.6	27.9	32.3	34.6	..	Manufactures
878.0	915.8	1,499.3	1,518.7	1,754.0	1,534.0	1,370.0	1,114.0	1,055.0	832.3	f	Value of Imports, cif
183.7	184.0	412.3	336.0	460.7	326.5	250.8	229.2	229.2	165.1	..	Nonfuel Primary Products
12.6	14.8	189.4	289.6	339.0	332.8	351.0	285.8	265.5	184.1	..	Fuels
681.7	717.0	897.6	893.1	954.3	874.7	768.2	599.0	560.3	483.1	..	Manufactures
				(Index 1980 = 100)							
102.8	99.1	100.0	95.8	88.3	99.0	96.2	103.1	82.8	92.8	..	Terms of Trade
79.6	87.5	100.0	97.1	83.6	92.4	88.8	83.9	71.3	77.1	..	Export Prices, fob
80.5	88.5	100.0	96.2	82.4	92.2	88.6	82.9	71.4	76.2	..	Nonfuel Primary Products
42.3	61.0	100.0	112.5	101.6	92.5	90.2	87.5	44.9	56.9	..	Fuels
87.8	100.0	100.0	124.4	117.3	101.3	99.5	101.1	119.5	131.2	..	Manufactures
77.4	88.3	100.0	101.3	94.7	93.3	92.3	81.3	86.1	83.1	..	Import Prices, cif
											Trade at Constant 1980 Prices
				(Millions of 1980 US dollars)							
606.9	666.8	594.0	554.0	516.6	628.8	812.9	709.4	697.3	625.2	..	Exports, fob
1,134.0	1,037.7	1,499.4	1,498.5	1,852.8	1,643.8	1,483.7	1,369.7	1,225.7	1,001.6	..	Imports, cif
				(Millions of current US dollars)							**BALANCE OF PAYMENTS**
705	708	825	763	682	831	952	815	722	712	C f	Exports of Goods & Services
551	527	594	538	432	581	722	595	497	482	..	Merchandise, fob
145	173	216	210	235	240	220	205	205	230	..	Nonfactor Services
9	9	15	15	15	10	10	15	20	..	..	Factor Services
1,420	1,399	1,682	2,029	2,322	2,131	2,118	1,887	1,914	1,664	C f	Imports of Goods & Services
1,188	1,116	1,339	1,540	1,754	1,534	1,370	1,114	1,055	832	..	Merchandise, fob
172	206	258	259	260	270	270	280	283	310	..	Nonfactor Services
60	78	85	230	308	327	478	493	576	522	..	Factor Services
..	..	..	..	..	..	..	..	..	..	..	Long-Term Interest
221	240	209	305	350	415	395	430	350	250	..	Current Transfers, net
221	240	209	305	350	415	395	430	115	..	..	Workers' Remittances
-494	-451	-648	-961	-1,290	-885	-771	-668	-842	-645	..	Total to be Financed
23	17	84	122	174	462	309	314	412	223	..	Official Capital Grants
-471	-434	-564	-839	-1,117	-423	-462	-354	-430	-422	..	Current Account Balance
333	622	230	17	165	135	-81	-34	-65	-107	C f	Long-Term Capital, net
0	0	0	0	0	0	0	0	..	..	..	Direct Investment
..	..	..	..	..	..	..	..	..	..	..	Long-Term Loans
..	..	..	..	..	..	..	..	..	..	..	Disbursements
..	..	..	..	..	..	..	..	..	..	..	Repayments
310	605	146	-105	-9	-327	-390	-348	-477	-330	..	Other Long-Term Capital
60	-179	194	614	613	382	536	288	508	309	C f	Other Capital, net
77	-10	140	208	339	-94	7	100	-13	220	..	Change in Reserves
											Memo Item:
				(Sudanese Pounds per US dollar)							
0.350	0.350	0.590	0.630	0.890	1.230	1.590	2.010	2.720	3.020	..	Conversion Factor (Annual Avg)
											EXTERNAL DEBT, ETC.
			(Millions of US dollars, outstanding at end of year)								
2,314	3,180	3,756	4,534	5,165	5,728	6,151	6,519	6,954	7,876	..	Public/Publicly Guar. Long-Term
1,732	2,729	3,213	3,370	4,008	4,867	5,156	5,426	5,683	6,199	..	Official Creditors
181	202	236	303	387	439	512	560	643	737	..	IBRD and IDA
582	451	544	1,164	1,157	861	995	1,094	1,271	1,677	..	Private Creditors
..	..	325	244	244	244	244	244	244	372	..	Private Non-guaranteed Long-Term
..	..	342	482	501	624	598	665	740	859	..	Use of Fund Credit
554	513	585	909	976	848	1,472	1,501	1,630	2,019	..	Short-Term Debt
											Memo Items:
				(Thousands of US dollars)							
28,356	67,400	48,726	16,970	20,500	16,563	17,200	12,200	58,512	11,714	..	Int'l Reserves Excluding Gold
..	..	..	..	..	..	..	..	..	..	..	Gold Holdings (at market price)
											SOCIAL INDICATORS
6.7	6.6	6.6	6.6	6.6	6.6	6.6	6.6	6.6	6.6	..	Total Fertility Rate
46.8	46.4	46.1	45.7	45.4	45.3	45.2	45.1	45.0	44.9	..	Crude Birth Rate
128.4	125.8	123.2	120.6	118.0	116.5	115.0	113.5	112.0	110.5	..	Infant Mortality Rate
45.5	46.0	46.4	46.8	47.2	47.7	48.1	48.5	49.0	49.4	..	Life Expectancy at Birth
101.0	91.5	98.9	109.5	94.5	94.1	84.5	100.8	103.4	95.9	..	Food Production, p.c. ('79-81 = 100)
72.3	71.7	71.1	..	..	..	..	..	..	..	..	Labor Force, Agriculture (%)
19.8	19.7	19.7	19.9	20.1	20.3	20.5	20.8	21.0	21.2	..	Labor Force, Female (%)
50.0	51.0	50.0	50.0	50.0	49.0	49.0	..	..	..	..	School Enroll. Ratio, primary
16.0	16.0	16.0	17.0	18.0	19.0	19.0	..	..	..	..	School Enroll. Ratio, secondary

SURINAME	1967	1968	1969	1970	1971	1972	1973	1974	1975	1976	1977
CURRENT GNP PER CAPITA (US $)	540	560	570	610	640	700	790	990	1,350	1,480	1,770
POPULATION (thousands)	348	356	364	372	378	382	384	381	364	354	362
ORIGIN AND USE OF RESOURCES					*(Millions of current Suriname Guilders)*						
Gross National Product (GNP)	357.7	369.6	387.5	417.2	452.2	491.3	540.5	693.3	909.0	945.0	1,222.0
Net Factor Income from Abroad	-39.2	-71.6	-79.4	-77.1	-89.1	-83.9	-70.1	-43.9	-21.0	-66.0	-61.0
Gross Domestic Product (GDP)	396.9	441.2	466.9	494.3	541.3	575.2	610.6	737.2	930.0	1,011.0	1,283.0
Indirect Taxes, net	4.6	3.4	6.0	8.1	8.2	9.8	0.1	14.9	195.8	158.6	180.6
GDP at factor cost	392.3	437.8	460.9	486.2	533.1	565.4	610.5	722.3	734.2	852.4	1,102.4
Agriculture	34.5	37.9	35.9	35.6	39.2	41.8	42.9	58.1	58.1	71.6	91.1
Industry	199.7	223.0	235.7	229.0	260.8	265.4	300.1	392.1	299.0	346.0	467.6
Manufacturing	..	..	..	..	..	..	..	..	152.3	181.2	202.9
Services, etc.	158.1	176.9	189.3	221.6	233.1	258.2	267.5	272.1	377.1	434.8	543.7
Resource Balance	5.3	27.0	38.0	29.1	47.4	33.7	25.0	9.0	-53.0	-2.0	-88.0
Exports of Goods & NFServices	234.2	254.5	289.5	297.9	344.1	350.5	367.0	552.0	588.0	618.0	707.0
Imports of Goods & NFServices	228.9	227.5	251.5	268.8	296.7	316.8	342.0	543.0	641.0	620.0	795.0
Domestic Absorption	391.6	414.2	428.9	465.2	493.9	541.5	585.6	728.2	983.0	1,013.0	1,371.0
Private Consumption, etc.	..	..	..	..	..	..	..	..	423.0	243.0	610.0
General Gov't Consumption	..	..	..	..	..	..	..	..	193.0	251.0	282.0
Gross Domestic Investment	111.3	103.5	103.5	105.5	110.5	138.3	205.2	322.2	367.0	519.0	479.0
Fixed Investment	..	..	..	..	..	..	..	..	..	..	479.0
Memo Items:											
Gross Domestic Saving	116.6	130.5	141.5	134.6	157.9	172.0	230.2	331.2	314.0	517.0	391.0
Gross National Saving	76.8	59.1	62.1	56.9	69.7	86.4	158.6	284.5	282.4	451.0	333.7
					(Millions of 1980 Suriname Guilders)						
Gross National Product	932.6	1,037.2	1,047.4	1,053.5	1,098.6	1,135.5	1,182.6	1,312.7	1,523.3	1,487.3	1,639.2
GDP at factor cost	..	..	..	..	..	..	1,332.1	1,366.7	1,230.3	1,341.5	1,478.6
Agriculture	..	..	..	..	..	..	92.9	111.0	97.4	100.6	116.6
Industry	..	..	..	..	..	..	767.7	707.7	530.8	590.0	646.9
Manufacturing	..	..	..	..	..	..	..	..	262.1	281.8	277.3
Services, etc.	..	..	..	..	..	..	471.5	548.1	602.0	650.9	715.1
Resource Balance	..	..	..	..	..	..	..	..	..	..	..
Exports of Goods & NFServices	..	..	..	..	..	..	..	..	..	..	..
Imports of Goods & NFServices	..	..	..	..	..	..	..	..	..	..	..
Domestic Absorption	..	..	..	..	..	..	..	..	..	..	..
Private Consumption, etc.	..	..	..	..	..	..	..	..	..	..	..
General Gov't Consumption	..	..	..	..	..	..	..	..	..	..	..
Gross Domestic Investment	..	..	..	..	..	..	..	..	..	..	..
Fixed Investment	..	..	..	..	..	..	..	..	..	..	..
Memo Items:											
Capacity to Import	..	..	..	..	..	..	..	..	..	..	..
Terms of Trade Adjustment	..	..	..	..	..	..	..	..	..	..	..
Gross Domestic Income	..	..	..	..	..	..	..	..	..	..	..
Gross National Income	..	..	..	..	..	..	..	..	..	..	..
DOMESTIC PRICES (DEFLATORS)					*(Index 1980 = 100)*						
Overall (GDP)	..	..	..	..	38.8	39.7	41.8	48.3	59.7	63.5	74.5
Domestic Absorption	..	..	..	..	..	..	..	..	..	..	..
Agriculture	..	..	..	..	..	..	46.2	52.4	59.6	71.2	78.1
Industry	..	..	..	..	..	..	39.1	55.4	56.3	58.6	72.3
Manufacturing	..	..	..	..	..	..	..	..	58.1	64.3	73.2
MANUFACTURING ACTIVITY											
Employment (1980=100)	..	..	..	..	..	..	..	127.4	137.2	132.2	139.3
Real Earnings per Empl. (1980=100)	..	..	..	..	..	..	..	73.2	75.5	78.3	76.6
Real Output per Empl. (1980=100)	..	..	..	..	..	..	..	..	78.5	96.5	78.7
Earnings as % of Value Added	..	..	..	..	..	..	..	..	..	..	..
MONETARY HOLDINGS					*(Millions of current Suriname Guilders)*						
Money Supply, Broadly Defined	98.4	113.0	132.2	147.8	167.6	189.0	230.8	247.1	293.1	399.5	474.2
Money as Means of Payment	65.4	70.0	80.4	86.1	98.3	103.3	133.1	140.0	168.7	198.2	219.6
Currency Ouside Banks	37.9	39.9	45.4	48.6	53.2	58.2	73.5	78.8	88.6	109.6	125.3
Demand Deposits	27.5	30.1	35.0	37.5	45.1	45.0	59.6	61.2	80.1	88.6	94.3
Quasi-Monetary Liabilities	33.0	43.0	51.8	61.7	69.3	85.8	97.7	107.1	124.5	201.2	254.6
GOVERNMENT DEFICIT (-) OR SURPLUS					*(Millions of current Suriname Guilders)*						
	..	..	..	..	-2.00	-1.80	-19.10	-8.80	-2.10	-21.80	..
Current Revenue	..	..	..	..	164.70	195.50	199.70	228.40	318.10	403.00	..
Current Expenditure	..	..	..	..	146.10	152.80	175.40	193.90	232.00	279.70	..
Current Budget Balance	..	..	..	..	18.60	42.70	24.30	34.50	86.10	123.30	..
Capital Receipts	..	..	..	..	0.40	0.40	0.20	..	..	0.10	..
Capital Payments	..	..	..	..	21.00	44.90	43.60	43.40	88.30	145.20	..

1978	1979	1980	1981	1982	1983	1984	1985	1986	1987 estimate	NOTES	SURINAME
2,100	2,210	2,420	2,950	2,850	2,680	2,630	2,540	2,440	2,360	..	**CURRENT GNP PER CAPITA (US $)**
367	365	356	355	364	374	383	393	402	411	..	**POPULATION (thousands)**
				(Millions of current Suriname Guilders)							**ORIGIN AND USE OF RESOURCES**
1,416.0	1,491.0	1,572.0	1,821.0	1,862.0	1,767.0	1,741.0	1,734.0	1,749.0	1,938.0	..	Gross National Product (GNP)
-55.0	-74.0	-30.0	23.0	13.0	-20.0	-4.0	-7.0	-2.0	-7.0	..	Net Factor Income from Abroad
1,471.0	1,565.0	1,602.0	1,798.0	1,849.0	1,787.0	1,745.0	1,741.0	1,751.0	1,945.0	..	Gross Domestic Product (GDP)
234.9	226.3	247.7	269.6	245.1	226.4	214.1	183.0	146.6	130.5	..	Indirect Taxes, net
1,236.1	1,338.7	1,354.3	1,528.4	1,603.9	1,560.6	1,530.9	1,558.0	1,604.4	1,814.5	..	GDP at factor cost
91.4	114.6	124.0	143.0	144.3	126.9	134.5	142.7	166.0	180.1	..	Agriculture
489.6	529.1	521.6	574.9	538.5	483.6	484.6	464.8	444.0	418.8	..	Industry
225.3	239.6	249.1	271.2	225.8	195.3	197.4	206.6	212.9	180.2	..	Manufacturing
655.1	695.0	708.7	810.5	921.1	950.1	911.8	950.5	994.4	1,215.6	..	Services, etc.
-34.0	-5.0	-86.0	-248.0	-282.0	-278.0	-150.0	-90.0	-45.0	-6.0	..	Resource Balance
814.0	917.0	1,094.0	1,044.0	949.0	803.0	763.0	682.0	648.0	667.0	..	Exports of Goods & NFServices
848.0	922.0	1,180.0	1,292.0	1,231.0	1,081.0	913.0	772.0	693.0	673.0	..	Imports of Goods & NFServices
1,505.0	1,570.0	1,688.0	2,046.0	2,131.0	2,065.0	1,895.0	1,831.0	1,796.0	1,951.0	..	Domestic Absorption
693.0	869.0	929.0	1,060.0	1,104.0	1,353.0	1,193.0	1,106.0	880.0	1,063.0	..	Private Consumption, etc.
335.0	343.0	339.0	431.0	520.0	437.0	499.0	588.0	733.0	746.0	..	General Gov't Consumption
477.0	358.0	420.0	555.0	507.0	275.0	203.0	137.0	183.0	142.0	..	Gross Domestic Investment
477.0	383.0	420.0	555.0	507.0	275.0	203.0	137.0	183.0	142.0	..	Fixed Investment
											Memo Items:
443.0	353.0	334.0	307.0	225.0	-3.0	53.0	47.0	138.0	136.0	..	Gross Domestic Saving
395.0	291.5	315.7	336.6	233.3	-37.3	35.8	32.7	132.4	128.1	..	Gross National Saving
				(Millions of 1980 Suriname Guilders)							
1,786.0	1,633.8	1,572.0	1,732.7	1,620.7	1,526.4	1,503.4	1,487.1	1,448.4	1,331.8	..	Gross National Product
1,559.0	1,467.5	1,354.3	1,454.3	1,396.0	1,348.1	1,321.9	1,336.2	1,328.7	1,246.9	..	GDP at factor cost
111.4	123.1	124.0	140.4	136.1	124.2	129.0	130.8	135.8	125.1	..	Agriculture
639.7	603.8	521.6	543.0	462.1	412.4	426.8	432.9	423.7	329.5	..	Industry
295.0	278.4	249.1	256.1	214.0	193.6	187.5	199.6	196.9	150.2	..	Manufacturing
808.0	740.6	708.7	771.0	797.8	811.5	766.2	772.6	769.2	792.3	..	Services, etc.
..	..	..	..	..	..	..	..	..	..	..	Resource Balance
..	..	..	..	..	..	..	..	..	..	..	Exports of Goods & NFServices
..	..	..	..	..	..	..	..	..	..	..	Imports of Goods & NFServices
..	..	..	..	..	..	..	..	..	..	..	Domestic Absorption
..	..	..	..	..	..	..	..	..	..	..	Private Consumption, etc.
..	..	..	..	..	..	..	..	..	..	..	General Gov't Consumption
..	..	..	..	..	..	..	..	..	..	..	Gross Domestic Investment
..	..	..	..	..	..	..	..	..	..	..	Fixed Investment
											Memo Items:
..	..	..	..	..	..	..	..	..	..	..	Capacity to Import
..	..	..	..	..	..	..	..	..	..	..	Terms of Trade Adjustment
..	..	..	..	..	..	..	..	..	..	..	Gross Domestic Income
..	..	..	..	..	..	..	..	..	..	..	Gross National Income
				(Index 1980 = 100)							**DOMESTIC PRICES (DEFLATORS)**
79.3	91.2	100.0	105.1	114.9	115.8	115.8	116.6	120.8	145.5	..	Overall (GDP)
..	..	..	..	..	..	..	..	..	..	..	Domestic Absorption
82.1	93.1	100.0	101.9	106.0	102.2	104.3	109.1	122.2	144.0	..	Agriculture
76.5	87.6	100.0	105.9	116.5	117.3	113.5	107.4	104.8	127.1	..	Industry
76.4	86.1	100.0	105.9	105.5	100.9	105.3	103.5	108.1	120.0	..	Manufacturing
											MANUFACTURING ACTIVITY
123.4	108.6	100.0	101.6	108.9	108.7	105.8	115.8			G	Employment (1980=100)
85.8	94.8	100.0	103.7	103.6	101.6	101.8	..	..	..	G	Real Earnings per Empl. (1980=100)
102.2	116.2	100.0	115.7	131.7	143.3	155.8	150.9	..	..	G	Real Output per Empl. (1980=100)
..	..	..	..	..	..	..	..	..	..	..	Earnings as % of Value Added
				(Millions of current Suriname Guilders)							**MONETARY HOLDINGS**
554.9	618.9	666.4	795.3	882.6	978.2	1,175.4	1,550.7	1,909.6	2,231.5	..	Money Supply, Broadly Defined
246.1	273.4	294.2	358.0	421.2	455.0	577.3	880.2	1,228.8	1,561.9	..	Money as Means of Payment
145.1	156.3	177.8	197.0	268.1	265.0	305.2	405.4	451.3	637.9	..	Currency Ouside Banks
101.0	117.1	116.3	161.0	153.2	190.1	272.1	474.8	777.5	924.0	..	Demand Deposits
308.8	345.5	372.2	437.3	461.4	523.2	598.1	670.5	680.8	669.6	..	Quasi-Monetary Liabilities
				(Millions of current Suriname Guilders)							**GOVERNMENT DEFICIT (-) OR SURPLUS**
..	..	..	..	..	..	-263.20	-349.20	-445.70	..	..	Current Revenue
..	..	..	..	..	..	518.90	491.60	497.61	..	..	Current Expenditure
..	..	..	..	..	..	697.50	754.00	869.00	..	..	Current Budget Balance
..	..	..	..	..	..	-178.60	-262.40	-371.39	..	..	Capital Receipts
..	..	..	..	..	..	0.10	0.20	0.09	..	..	Capital Payments
..	..	..	..	..	..	84.70	87.00	74.40	..	..	

SURINAME	1967	1968	1969	1970	1971	1972	1973	1974	1975	1976	1977
FOREIGN TRADE (CUSTOMS BASIS)				*(Millions of current US dollars)*							
Value of Exports, fob	..	..	..	..	..	..	..	..	..	..	..
Nonfuel Primary Products	..	..	..	..	..	..	..	..	..	..	..
Fuels	..	..	..	..	..	..	..	..	..	..	..
Manufactures	..	..	..	..	..	..	..	..	..	..	..
Value of Imports, cif	..	..	..	..	..	..	..	..	..	..	..
Nonfuel Primary Products	..	..	..	..	..	..	..	..	..	..	..
Fuels	..	..	..	..	..	..	..	..	..	..	..
Manufactures	..	..	..	..	..	..	..	..	..	..	..
					(Index 1980 = 100)						
Terms of Trade	..	..	..	..	..	..	..	..	..	..	..
Export Prices, fob	..	..	..	..	..	..	..	..	..	..	..
Nonfuel Primary Products	..	..	..	..	..	..	..	..	..	..	..
Fuels	..	..	..	..	..	..	..	..	..	..	..
Manufactures	..	..	..	..	..	..	..	..	..	..	..
Import Prices, cif	..	..	..	..	..	..	..	..	..	..	..
Trade at Constant 1980 Prices				*(Millions of 1980 US dollars)*							
Exports, fob	..	..	..	..	..	..	..	..	..	..	..
Imports, cif	..	..	..	..	..	..	..	..	..	..	..
BALANCE OF PAYMENTS				*(Millions of current US dollars)*							
Exports of Goods & Services	..	..	..	160.70	185.15	206.50	212.32	316.29	338.99	370.37	403.20
Merchandise, fob	..	..	..	136.60	157.97	175.78	180.01	269.63	277.19	303.75	345.85
Nonfactor Services	..	..	..	19.70	23.37	25.08	25.99	34.88	48.20	52.99	50.12
Factor Services	..	..	..	4.40	3.81	5.65	6.32	11.79	13.60	13.62	7.23
Imports of Goods & Services	..	..	..	185.60	205.51	223.87	240.69	336.86	380.76	396.35	486.09
Merchandise, fob	..	..	..	104.10	114.04	128.55	144.01	211.30	242.10	259.19	323.96
Nonfactor Services	..	..	..	36.90	41.72	47.99	54.84	92.48	112.92	88.21	115.03
Factor Services	..	..	..	44.60	49.75	47.34	41.84	33.07	25.74	48.95	47.09
Long-Term Interest	..	..	..	..	..	..	..	..	..	..	..
Current Transfers, net	..	..	..	-0.30	0.50	-0.98	-0.83	-1.56	-5.95	0.00	2.08
Workers' Remittances	..	..	..	0.00	0.00	0.00	0.00	0.00	0.00	0.00	..
Total to be Financed	..	..	..	-25.20	-19.86	-18.35	-29.21	-22.13	-47.72	-25.98	-80.75
Official Capital Grants	..	..	..	12.70	12.84	12.81	13.59	22.85	178.84	89.24	77.30
Current Account Balance	..	..	..	-12.50	-7.02	-5.54	-15.62	0.72	131.13	63.27	-3.45
Long-Term Capital, net	..	..	..	19.40	8.43	28.88	34.21	28.14	81.83	34.64	63.82
Direct Investment	..	..	..	-5.00	-6.82	-1.85	14.19	-0.36	..	..	-12.91
Long-Term Loans	..	..	..	..	..	..	..	..	..	..	..
Disbursements	..	..	..	..	..	..	..	..	..	..	..
Repayments	..	..	..	..	..	..	..	..	..	..	..
Other Long-Term Capital	..	..	..	11.70	2.41	17.91	6.44	5.65	-97.01	-54.61	-0.57
Other Capital, net	..	..	..	0.80	-0.10	-14.40	-0.95	-23.67	-166.10	-70.43	-78.18
Change in Reserves	..	..	..	-7.70	-1.30	-8.94	-17.64	-5.20	-46.87	-27.48	17.82
Memo Item:				*(Suriname Guilders per US dollar)*							
Conversion Factor (Annual Avg)	1.890	1.890	1.890	1.890	1.880	1.780	1.780	1.780	1.780	1.780	1.780
EXTERNAL DEBT, ETC.				*(Millions of US dollars, outstanding at end of year)*							
Public/Publicly Guar. Long-Term	..	..	..	..	..	..	..	..	..	..	..
Official Creditors	..	..	..	..	..	..	..	..	..	..	..
IBRD and IDA	..	..	..	..	..	..	..	..	..	..	..
Private Creditors	..	..	..	..	..	..	..	..	..	..	..
Private Non-guaranteed Long-Term	..	..	..	..	..	..	..	..	..	..	..
Use of Fund Credit	..	..	..	..	..	..	..	..	..	..	..
Short-Term Debt	..	..	..	..	..	..	..	..	..	..	..
Memo Items:				*(Thousands of US dollars)*							
Int'l Reserves Excluding Gold	8,363.0	9,101.0	23,761.0	27,828.0	33,094.0	37,591.0	56,436.0	67,547.0	91,406.0	110,220.0	94,025.0
Gold Holdings (at market price)	18,781.0	22,355.0	8,713.0	9,250.0	10,800.0	16,065.0	16,568.0	27,527.0	20,701.0	19,889.0	24,347.0
SOCIAL INDICATORS											
Total Fertility Rate	5.9	5.8	5.7	5.5	5.4	5.3	5.1	5.0	4.9	4.8	4.6
Crude Birth Rate	40.0	38.6	37.2	35.7	34.3	32.9	30.9	31.0	27.5	31.5	30.7
Infant Mortality Rate	..	..	..	..	..	..	..	..	..	..	..
Life Expectancy at Birth	63.5	63.6	63.7	63.8	63.9	64.0	64.1	64.2	64.3	64.4	64.5
Food Production, p.c. ('79-81 = 100)	121.3	116.7	106.7	101.0	90.5	94.2	88.8	87.7	90.2	86.0	74.5
Labor Force, Agriculture (%)	26.3	25.7	25.2	24.7	24.2	23.7	23.3	22.8	22.3	21.8	21.3
Labor Force, Female (%)	24.6	24.8	25.0	25.3	25.4	25.6	25.7	25.9	26.0	26.4	26.8
School Enroll. Ratio, primary	..	..	..	126.0	..	..	..	..	108.0	106.0	103.0
School Enroll. Ratio, secondary	..	..	..	41.0	..	..	..	..	46.0	49.0	50.0

1978	1979	1980	1981	1982	1983	1984	1985	1986	1987 estimate	NOTES	SURINAME
				(Millions of current US dollars)							**FOREIGN TRADE (CUSTOMS BASIS)**
..	..	514	492	450	381	365	314	..	..	..	Value of Exports, fob
..	..	..	..	..	..	..	..	..	..	..	Nonfuel Primary Products
..	..	..	..	..	..	..	..	..	..	..	Fuels
..	..	..	..	..	..	..	..	..	..	..	Manufactures
..	..	665	726	691	606	512	432	..	..	..	Value of Imports, cif
..	..	..	..	..	..	..	..	..	..	..	Nonfuel Primary Products
..	..	..	..	..	..	..	..	..	..	..	Fuels
..	..	..	..	..	..	..	..	..	..	..	Manufactures
				(Index 1980 = 100)							
..	..	..	..	..	..	..	..	..	..	..	Terms of Trade
..	..	..	..	..	..	..	..	..	..	..	Export Prices, fob
..	..	..	..	..	..	..	..	..	..	..	Nonfuel Primary Products
..	..	..	..	..	..	..	..	..	..	..	Fuels
..	..	..	..	..	..	..	..	..	..	..	Manufactures
..	..	..	..	..	..	..	..	..	..	..	Import Prices, cif
				(Millions of 1980 US dollars)							Trade at Constant 1980 Prices
..	..	..	..	..	..	..	..	..	..	..	Exports, fob
..	..	..	..	..	..	..	..	..	..	..	Imports, cif
				(Millions of current US dollars)							**BALANCE OF PAYMENTS**
481.46	527.35	638.72	602.34	544.40	451.60	434.86	382.60	362.89	419.30	..	Exports of Goods & Services
410.93	443.77	514.46	473.78	427.59	367.34	374.57	335.58	335.69	337.61	..	Merchandise, fob
62.18	69.44	98.65	92.27	80.96	67.82	55.35	44.27	25.64	80.51	..	Nonfactor Services
8.35	14.14	25.61	36.29	35.86	16.44	4.95	2.75	1.57	1.17	..	Factor Services
512.87	571.02	703.53	729.58	696.50	606.57	508.55	392.03	384.07	345.41	..	Imports of Goods & Services
343.28	369.60	454.00	507.09	460.48	402.03	391.97	309.33	302.67	271.60	..	Merchandise, fob
124.26	142.41	203.72	197.87	206.82	176.97	113.01	79.50	76.80	67.99	..	Nonfactor Services
45.32	59.02	45.81	24.62	29.21	27.58	3.56	3.20	4.60	5.83	..	Factor Services
..	..	..	..	..	..	..	..	..	..	..	Long-Term Interest
3.93	7.00	6.56	3.69	-2.64	-8.01	-7.38	-4.08	-2.01	-0.51	..	Current Transfers, net
0.00	0.34	0.78	0.22	0.00	0.00	0.00	0.00	0.00	0.00	..	Workers' Remittances
-27.48	-36.68	-58.24	-123.55	-154.75	-162.98	-81.06	-13.50	-23.18	73.37	K	Total to be Financed
55.85	80.73	74.01	95.72	96.72	2.51	2.57	1.25	0.89	2.32	K	Official Capital Grants
28.37	44.06	15.77	-27.82	-58.03	-160.47	-78.49	-12.25	-22.29	75.69	..	Current Account Balance
70.40	64.13	84.40	131.73	108.01	52.61	-8.38	16.07	-12.99	-55.96	..	Long-Term Capital, net
-7.91	-15.65	10.28	34.39	-5.95	45.40	-40.77	13.17	-32.91	-71.64	..	Direct Investment
..	..	..	..	..	..	..	..	..	..	..	Long-Term Loans
..	..	..	..	..	..	..	..	..	..	..	Disbursements
..	..	..	..	..	..	..	..	..	..	..	Repayments
22.46	-0.95	0.11	1.61	17.24	4.70	29.83	1.64	19.03	13.36	..	Other Long-Term Capital
-57.07	-81.50	-74.24	-90.76	-92.78	-1.23	34.31	-12.30	-4.75	-29.10	..	Other Capital, net
-41.70	-26.69	-25.93	-13.14	42.80	109.09	52.56	8.48	40.03	9.36	..	Change in Reserves
				(Suriname Guilders per US dollar)							Memo Item:
1.780	1.780	1.780	1.780	1.780	1.780	1.780	1.780	1.780	1.780	..	Conversion Factor (Annual Avg)
				(Millions of US dollars, outstanding at end of year)							**EXTERNAL DEBT, ETC.**
..	..	..	..	..	..	..	..	..	..	..	Public/Publicly Guar. Long-Term
..	..	..	..	..	..	..	..	..	..	..	Official Creditors
..	..	..	..	..	..	..	..	..	..	..	IBRD and IDA
..	..	..	..	..	..	..	..	..	..	..	Private Creditors
..	..	..	..	..	..	..	..	..	..	..	Private Non-guaranteed Long-Term
..	..	..	..	..	..	..	..	..	..	..	Use of Fund Credit
..	..	..	..	..	..	..	..	..	..	..	Short-Term Debt
				(Thousands of US dollars)							Memo Items:
132,420.0	169,530.0	189,250.0	207,090.0	175,760.0	59,148.0	24,870.0	23,420.0	20,891.0	15,097.0	..	Int'l Reserves Excluding Gold
12,222.0	27,688.0	31,879.0	21,496.0	24,708.0	20,631.0	16,672.0	17,684.0	21,139.0	26,179.0	..	Gold Holdings (at market price)
											SOCIAL INDICATORS
4.5	4.4	4.3	4.2	4.1	4.0	3.9	3.8	3.8	3.7	..	Total Fertility Rate
29.1	29.0	27.6	28.4	29.0	29.5	30.1	30.7	31.3	32.0	..	Crude Birth Rate
..	..	..	..	48.7	47.4	46.0	44.7	43.4	42.1	..	Infant Mortality Rate
64.6	64.7	64.8	64.9	65.0	65.3	65.7	66.0	66.3	66.6	..	Life Expectancy at Birth
76.8	99.0	98.4	102.6	101.0	99.1	110.6	116.6	101.6	101.5	..	Food Production, p.c. ('79-81 = 100)
20.9	20.4	19.9	..	..	..	..	..	..	..	..	Labor Force, Agriculture (%)
27.1	27.5	27.9	28.1	28.4	28.6	28.8	29.1	29.2	29.3	..	Labor Force, Female (%)
120.0	115.0	..	..	130.0	132.0	5.0	..	..	..	..	School Enroll. Ratio, primary
50.0	51.0	..	..	44.0	44.0	51.0	..	..	..	..	School Enroll. Ratio, secondary

SWAZILAND	1967	1968	1969	1970	1971	1972	1973	1974	1975	1976	1977
CURRENT GNP PER CAPITA (US $)	190	200	260	270	320	330	370	460	590	580	590
POPULATION (thousands)	389	399	409	420	431	443	455	468	482	497	512
ORIGIN AND USE OF RESOURCES					*(Millions of current Swaziland Emalangeni)*						
Gross National Product (GNP)	52.9	56.5	74.8	79.4	100.1	108.9	127.7	162.7	211.6	232.4	260.7
Net Factor Income from Abroad	-6.4	-6.8	-9.7	-10.5	-9.0	-11.6	-20.9	-14.7	-1.8	-2.9	-2.8
Gross Domestic Product (GDP)	59.3	63.3	84.5	89.9	109.1	120.5	148.6	177.4	213.4	235.3	263.5
Indirect Taxes, net	10.1	10.2	18.0	17.3	20.9	24.6	32.0	41.0	52.2	49.0	49.3
GDP at factor cost	49.2	53.1	66.5	72.6	88.2	95.9	116.6	136.4	161.2	186.3	214.2
Agriculture	14.7	13.2	21.9	24.1	32.7	31.7	41.6	47.7	53.5	55.7	64.0
Industry	18.1	17.6	19.3	19.2	21.8	26.7	30.4	35.3	41.0	46.5	53.5
Manufacturing	..	..	..	..	..	..	..	..	..	..	..
Services, etc.	16.4	22.3	25.3	29.3	33.7	37.5	44.6	53.4	66.7	84.1	96.7
Resource Balance	-5.7	-0.2	1.3	5.3	12.6	9.5	21.9	40.0	34.3	19.1	-8.6
Exports of Goods & NFServices	39.3	47.0	53.7	59.9	66.2	75.6	108.4	146.4	155.5	179.4	172.9
Imports of Goods & NFServices	45.0	47.2	52.4	54.6	53.6	66.1	86.5	106.4	121.2	160.3	181.5
Domestic Absorption	65.0	63.5	83.2	84.6	96.5	111.0	126.7	137.4	179.1	216.2	272.1
Private Consumption, etc.	40.7	37.8	54.3	48.7	57.0	59.9	64.4	53.5	105.5	104.6	146.8
General Gov't Consumption	11.8	14.2	14.9	18.6	17.8	24.2	27.6	36.1	35.6	42.0	54.2
Gross Domestic Investment	12.5	11.5	14.0	17.3	21.7	26.9	34.7	47.8	38.0	69.6	71.1
Fixed Investment	..	..	..	..	..	..	..	..	..	..	68.1
Memo Items:											
Gross Domestic Saving	6.8	11.3	15.3	22.6	34.3	36.4	56.6	87.8	72.3	88.7	62.5
Gross National Saving	0.4	4.5	5.6	..	..	..	..	73.4	71.1	85.9	59.7
					(Millions of 1980 Swaziland Emalangeni)						
Gross National Product	184.3	192.2	246.9	250.5	289.4	299.6	315.3	350.8	393.1	382.4	388.5
GDP at factor cost	171.4	180.7	219.5	228.8	257.1	267.1	293.0	296.5	299.6	306.9	318.7
Agriculture	37.9	33.2	51.1	56.0	70.5	66.8	73.3	72.9	70.1	68.3	74.8
Industry	78.6	74.6	79.3	75.7	79.3	88.7	94.8	96.5	92.1	93.6	97.1
Manufacturing	..	..	..	..	49.3	54.5	58.3	58.3	57.9	61.6	65.3
Services, etc.	54.9	72.9	89.1	97.1	107.3	111.6	124.9	127.1	137.4	145.0	146.8
Resource Balance	-55.1	-37.2	-35.7	-22.4	10.1	-9.6	21.1	59.4	43.2	2.9	-20.7
Exports of Goods & NFServices	147.9	172.5	191.3	203.9	208.3	209.3	255.2	294.6	271.3	277.3	241.1
Imports of Goods & NFServices	203.0	209.7	227.0	226.3	198.2	218.9	234.1	235.2	228.1	274.4	261.8
Domestic Absorption	261.7	252.5	314.5	306.0	308.0	345.3	352.3	326.3	353.5	384.7	413.5
Private Consumption, etc.	165.4	154.5	206.8	177.9	176.5	183.3	173.2	116.9	197.1	173.5	202.1
General Gov't Consumption	42.3	49.7	50.5	60.4	53.4	69.4	71.5	80.9	68.6	70.6	82.1
Gross Domestic Investment	54.0	48.3	57.2	67.7	78.1	92.6	107.6	128.5	87.8	140.6	129.3
Fixed Investment	..	..	..	..	..	..	..	..	..	..	..
Memo Items:											
Capacity to Import	177.3	208.8	232.6	248.3	244.8	250.4	293.4	323.6	292.7	307.1	249.4
Terms of Trade Adjustment	29.4	36.3	41.3	44.4	36.5	41.1	38.2	29.0	21.4	29.8	8.3
Gross Domestic Income	236.0	251.6	320.1	328.0	354.6	376.8	411.6	414.7	418.1	417.4	401.1
Gross National Income	213.7	228.5	288.2	294.9	325.9	340.7	353.5	379.8	414.5	412.2	396.8
DOMESTIC PRICES (DEFLATORS)					*(Index 1980 = 100)*						
Overall (GDP)	28.7	29.4	30.3	31.7	34.3	35.9	39.8	46.0	53.8	60.7	67.1
Domestic Absorption	24.8	25.1	26.5	27.6	31.3	32.1	36.0	42.1	50.7	56.2	65.8
Agriculture	38.8	39.8	42.9	43.0	46.4	47.5	56.8	65.4	76.3	81.6	85.6
Industry	23.0	23.6	24.3	25.4	27.5	30.1	32.1	36.6	44.5	49.7	55.1
Manufacturing	..	..	..	..	..	..	..	..	..	..	..
MANUFACTURING ACTIVITY											
Employment (1980=100)	41.3	44.7	..	51.6	56.1	55.7	60.4	..	..	72.9	77.4
Real Earnings per Empl. (1980=100)	77.5	78.4	..	66.8	69.7	79.9	94.4	..	..	79.1	102.9
Real Output per Empl. (1980=100)	..	..	..	..	..	..	..	..	..	..	..
Earnings as % of Value Added	45.9	52.5	..	42.5	50.6	38.7	34.3	..	..	18.2	34.4
MONETARY HOLDINGS					*(Millions of current Swaziland Emalangeni)*						
Money Supply, Broadly Defined	..	..	..	..	..	..	..	57.6	81.7	98.6	106.0
Money as Means of Payment	..	..	..	..	..	..	..	16.9	23.4	28.8	34.2
Currency Ouside Banks	..	..	..	..	..	..	..	3.3	5.1	6.5	7.3
Demand Deposits	..	..	..	..	..	..	..	13.6	18.3	22.3	26.8
Quasi-Monetary Liabilities	..	..	..	..	..	..	..	40.7	58.3	69.8	71.8
					(Millions of current Swaziland Emalangeni)						
GOVERNMENT DEFICIT (-) OR SURPLUS	..	..	..	..	-2	-5	-11	2	17	-9	-9
Current Revenue	..	..	..	..	17	21	29	46	70	55	81
Current Expenditure	..	..	..	..	16	18	23	26	30	40	49
Current Budget Balance	..	..	..	..	1	4	6	21	40	15	33
Capital Receipts	..	..	..	..	0	0	0	0	0	..	0
Capital Payments	..	..	..	..	4	8	17	19	23	23	41

1978	1979	1980	1981	1982	1983	1984	1985	1986	1987 estimate	NOTES	SWAZILAND
600	720	820	1,010	1,010	930	880	740	690	700	..	**CURRENT GNP PER CAPITA (US $)**
529	546	564	583	602	623	644	665	689	713	..	**POPULATION (thousands)**
				(Millions of current Swaziland Emalangeni)							**ORIGIN AND USE OF RESOURCES**
280.6	353.9	416.3	506.8	562.1	609.6	718.6	812.7	1,042.0	1,201.5	C	Gross National Product (GNP)
-42.7	-18.9	-5.8	5.9	15.4	30.0	63.5	70.5	86.0	99.2	..	Net Factor Income from Abroad
323.3	372.8	422.1	500.9	546.7	579.6	655.1	742.2	956.0	1,102.3	C	Gross Domestic Product (GDP)
62.4	73.5	59.6	63.1	61.9	62.9	74.5	91.8	119.5	..	..	Indirect Taxes, net
260.9	299.3	362.5	437.8	484.8	516.7	580.6	650.4	836.5	..	C	GDP at factor cost
78.0	89.5	90.2	104.7	102.1	111.2	127.0	160.3	201.7	..	..	Agriculture
65.1	74.7	114.3	132.8	148.7	140.0	156.5	170.1	247.8	..	..	Industry
..	..	79.6	89.1	100.8	89.7	104.1	105.3	166.3	..	..	Manufacturing
117.8	135.1	158.0	200.3	234.0	265.5	297.1	320.0	387.0	..	..	Services, etc.
-70.3	-123.5	-140.2	-148.6	-165.5	-209.4	-240.0	-263.4	-113.9	..	..	Resource Balance
186.4	221.2	325.7	388.0	416.7	400.6	432.8	474.6	710.7	..	..	Exports of Goods & NFServices
256.7	344.7	465.9	536.6	582.2	610.0	672.8	738.0	824.6	..	..	Imports of Goods & NFServices
393.6	496.3	562.3	649.5	712.2	789.0	895.1	1,005.6	1,069.9	..	..	Domestic Absorption
181.5	273.3	286.8	360.2	395.0	450.4	503.6	590.4	595.3	..	..	Private Consumption, etc.
70.2	74.9	103.9	133.8	141.0	135.8	182.1	177.6	240.7	..	..	General Gov't Consumption
141.9	148.1	171.6	155.5	176.2	202.8	209.4	237.6	233.9	..	..	Gross Domestic Investment
144.9	142.1	147.8	140.2	152.8	208.5	208.5	225.4	228.9	..	..	Fixed Investment
											Memo Items:
71.6	24.6	31.4	6.9	10.7	-6.6	-30.6	-25.8	120.0	..	..	Gross Domestic Saving
27.3	4.3	23.7	13.3	26.6	21.3	33.0	50.2	211.6	..	..	Gross National Saving
				(Millions of 1980 Swaziland Emalangeni)							
387.3	421.3	416.3	455.6	466.6	469.5	504.1	516.6	564.3	578.6	C	Gross National Product
353.5	354.2	362.5	393.6	404.1	402.9	415.9	425.7	463.3	..	C	GDP at factor cost
86.9	83.3	90.2	101.6	96.5	95.2	100.2	109.6	120.9	..	..	Agriculture
116.4	114.8	114.3	126.6	128.9	125.4	124.8	122.9	140.7	..	..	Industry
69.8	71.6	79.6	88.4	93.0	94.0	93.5	92.3	109.1	..	..	Manufacturing
150.2	156.1	158.0	165.4	178.7	182.3	190.9	193.2	201.7	..	..	Services, etc.
-83.9	-133.5	-140.2	-133.7	-50.3	-12.1	-20.8	..	..	..	..	Resource Balance
240.4	245.5	325.7	360.9	389.1	461.5	392.0	..	..	..	..	Exports of Goods & NFServices
324.3	379.0	465.9	494.6	439.4	473.6	412.8	..	..	..	..	Imports of Goods & NFServices
528.5	576.8	562.3	584.0	506.0	464.0	490.1	..	..	..	..	Domestic Absorption
209.2	294.3	286.8	331.8	274.0	245.2	266.7	..	..	..	..	Private Consumption, etc.
95.7	93.2	103.9	122.6	104.4	80.5	95.9	..	..	..	..	General Gov't Consumption
223.6	189.3	171.6	129.6	127.6	138.3	127.5	..	..	..	..	Gross Domestic Investment
..	..	..	..	..	..	..	..	..	..	..	Fixed Investment
											Memo Items:
235.5	243.2	325.7	357.6	314.5	311.0	265.6	..	..	..	..	Capacity to Import
-4.9	-2.3	0.0	-3.3	-74.6	-150.5	-126.5	..	..	..	..	Terms of Trade Adjustment
439.7	441.0	422.1	447.0	381.1	301.4	342.9	..	..	..	..	Gross Domestic Income
382.4	419.0	416.3	452.3	392.0	319.0	377.7	..	..	..	..	Gross National Income
				(Index 1980 = 100)							**DOMESTIC PRICES (DEFLATORS)**
72.7	84.1	100.0	111.2	120.0	128.3	139.6	152.8	180.5	203.1	..	Overall (GDP)
74.5	86.0	100.0	111.2	140.8	170.0	182.6	..	..	..	..	Domestic Absorption
89.8	107.4	100.0	103.1	105.8	116.8	126.7	146.3	166.8	..	..	Agriculture
55.9	65.1	100.0	104.9	115.4	111.6	125.4	138.4	176.1	..	..	Industry
..	..	100.0	100.8	108.4	95.4	111.3	114.1	152.4	..	..	Manufacturing
											MANUFACTURING ACTIVITY
76.7	101.6	100.0	109.1	111.2	104.6	..	..	..	..	G	Employment (1980=100)
108.1	89.4	100.0	91.6	97.3	94.8	..	..	..	..	G	Real Earnings per Empl. (1980=100)
..	..	..	..	..	..	..	..	..	..	G	Real Output per Empl. (1980=100)
42.8	36.5	40.6	43.6	47.0	53.8	..	..	..	..	..	Earnings as % of Value Added
				(Millions of current Swaziland Emalangeni)							**MONETARY HOLDINGS**
132.8	135.7	151.5	165.4	163.7	200.9	242.0	297.4	330.5	..	D	Money Supply, Broadly Defined
37.2	40.9	49.9	51.1	58.1	60.8	67.8	76.9	115.1	125.4	..	Money as Means of Payment
8.9	9.6	11.9	14.4	15.0	15.2	16.8	17.5	25.2	27.2	..	Currency Ouside Banks
28.3	31.3	38.0	36.7	43.1	45.6	51.0	59.5	89.9	98.2	..	Demand Deposits
95.6	94.8	101.6	114.3	105.6	140.1	174.2	220.5	215.4	..	..	Quasi-Monetary Liabilities
				(Millions of current Swaziland Emalangeni)							**GOVERNMENT DEFICIT (-) OR SURPLUS**
-40	4	28	-49	-32	-20	-4	-28	-52	-33	C	GOVERNMENT DEFICIT (-) OR SURPLUS
105	134	156	136	184	186	222	242	250	288	..	Current Revenue
55	60	79	112	124	137	155	171	217	238	..	Current Expenditure
50	74	78	24	60	49	67	72	33	50	..	Current Budget Balance
0	0	0	0	0	..	..	0	0	..	..	Capital Receipts
90	70	50	73	77	68	70	100	85	82	..	Capital Payments

SWAZILAND	1967	1968	1969	1970	1971	1972	1973	1974	1975	1976	1977
FOREIGN TRADE (CUSTOMS BASIS)					*(Millions of current US dollars)*						
Value of Exports, fob	..	..	..	..	..	..	..	..	..	..	..
Nonfuel Primary Products	..	..	..	..	..	..	..	..	..	..	..
Fuels	..	..	..	..	..	..	..	..	..	..	..
Manufactures	..	..	..	..	..	..	..	..	..	..	..
Value of Imports, cif	..	..	..	..	..	..	..	..	..	..	..
Nonfuel Primary Products	..	..	..	..	..	..	..	..	..	..	..
Fuels	..	..	..	..	..	..	..	..	..	..	..
Manufactures	..	..	..	..	..	..	..	..	..	..	..
					(Index 1980 = 100)						
Terms of Trade	..	..	..	..	..	..	..	..	..	..	..
Export Prices, fob	..	..	..	..	..	..	..	..	..	..	..
Nonfuel Primary Products	..	..	..	..	..	..	..	..	..	..	..
Fuels	..	..	..	..	..	..	..	..	..	..	..
Manufactures	..	..	..	..	..	..	..	..	..	..	..
Import Prices, cif	..	..	..	..	..	..	..	..	..	..	..
Trade at Constant 1980 Prices					*(Millions of 1980 US dollars)*						
Exports, fob	..	..	..	..	..	..	..	..	..	..	..
Imports, cif	..	..	..	..	..	..	..	..	..	..	..
BALANCE OF PAYMENTS					*(Millions of current US dollars)*						
Exports of Goods & Services	..	..	..	..	..	..	..	206	235	235	234
Merchandise, fob	..	..	..	..	..	..	..	179	197	193	184
Nonfactor Services	..	..	..	..	..	..	..	23	25	26	29
Factor Services	..	..	..	..	..	..	..	4	13	16	21
Imports of Goods & Services	..	..	..	..	..	..	..	180	203	214	236
Merchandise, fob	..	..	..	..	..	..	..	112	140	156	171
Nonfactor Services	..	..	..	..	..	..	..	30	40	43	47
Factor Services	..	..	..	..	..	..	..	38	23	15	18
Long-Term Interest	..	..	..	2	2	1	2	1	1	1	1
Current Transfers, net	..	..	..	..	..	..	..	0	1	0	0
Workers' Remittances	..	..	..	..	..	..	..	0	0	0	0
Total to be Financed	..	..	..	..	..	..	..	27	33	21	-2
Official Capital Grants	..	..	..	..	..	..	..	16	11	-1	13
Current Account Balance	..	..	..	..	..	..	..	43	44	20	11
Long-Term Capital, net	..	..	..	..	..	..	..	23	31	14	41
Direct Investment	..	..	..	..	..	..	..	4	14	7	20
Long-Term Loans	..	..	..	2	-2	-1	3	-1	1	10	9
Disbursements	..	..	..	4	0	6	11	2	4	11	10
Repayments	..	..	..	2	3	7	8	3	3	1	1
Other Long-Term Capital	..	..	..	..	..	..	..	5	5	-2	-1
Other Capital, net	..	..	..	..	..	..	..	-56	-36	-7	-31
Change in Reserves	..	..	..	1	1	0	0	-10	-39	-28	-22
Memo Item:					*(Swaziland Emalangeni per US dollar)*						
Conversion Factor (Annual Avg)	0.710	0.710	0.710	0.710	0.720	0.770	0.690	0.680	0.740	0.870	0.870
EXTERNAL DEBT, ETC.					*(Millions of US dollars, outstanding at end of year)*						
Public/Publicly Guar. Long-Term	..	..	..	37	35	33	37	36	34	41	52
Official Creditors	..	..	..	21	21	22	32	34	33	40	52
IBRD and IDA	..	..	..	9	9	9	9	8	8	10	15
Private Creditors	..	..	..	16	14	11	5	2	1	1	1
Private Non-guaranteed Long-Term	..	..	..	..	..	..	..	..	..	..	..
Use of Fund Credit	..	..	..	0	0	0	0	0	0	0	0
Short-Term Debt	..	..	..	..	..	..	..	..	..	..	9
Memo Items:					*(Millions of US dollars)*						
Int'l Reserves Excluding Gold	..	..	..	..	..	..	..	13.5	45.6	73.4	94.7
Gold Holdings (at market price)	..	..	..	..	..	..	..	..	..	..	..
SOCIAL INDICATORS											
Total Fertility Rate	6.5	6.5	6.5	6.5	6.5	6.5	6.5	6.5	6.5	6.5	6.5
Crude Birth Rate	47.9	47.8	47.8	47.7	47.6	47.5	47.5	47.4	47.4	47.3	47.3
Infant Mortality Rate	147.0	146.4	145.8	145.2	144.6	144.0	143.2	142.4	141.6	140.8	140.0
Life Expectancy at Birth	44.4	45.0	45.6	46.1	46.7	47.3	47.8	48.3	48.9	49.4	49.9
Food Production, p.c. ('79-81 = 100)	82.5	80.1	83.1	93.3	91.8	96.8	89.2	95.1	94.3	94.9	87.9
Labor Force, Agriculture (%)	83.1	82.2	81.4	80.6	80.0	79.3	78.7	78.0	77.4	76.7	76.1
Labor Force, Female (%)	42.6	42.5	42.3	42.2	42.1	41.9	41.8	41.7	41.6	41.5	41.4
School Enroll. Ratio, primary	..	..	..	87.0	..	..	..	..	99.0	99.0	100.0
School Enroll. Ratio, secondary	..	..	..	18.0	..	..	..	..	32.0	..	36.0

1978	1979	1980	1981	1982	1983	1984	1985	1986	1987	NOTES	SWAZILAND
									estimate		
				(Millions of current US dollars)							**FOREIGN TRADE (CUSTOMS BASIS)**
..	..	368	388	325	304	237	176	267	363	..	Value of Exports, fob
..	..	..	..	..	..	..	..	..	..	..	Nonfuel Primary Products
..	..	..	..	..	..	..	..	..	..	..	Fuels
..	..	..	..	..	..	..	..	..	..	..	Manufactures
..	..	538	506	440	464	381	281	352	425	..	Value of Imports, cif
..	..	..	..	..	..	..	..	..	..	..	Nonfuel Primary Products
..	..	..	..	..	..	..	..	..	..	..	Fuels
..	..	..	..	..	..	..	..	..	..	..	Manufactures
				(Index 1980 = 100)							
..	..	..	..	..	..	..	..	..	..	..	Terms of Trade
..	..	..	..	..	..	..	..	..	..	..	Export Prices, fob
..	..	..	..	..	..	..	..	..	..	..	Nonfuel Primary Products
..	..	..	..	..	..	..	..	..	..	..	Fuels
..	..	..	..	..	..	..	..	..	..	..	Manufactures
..	..	..	..	..	..	..	..	..	..	..	Import Prices, cif
				(Millions of 1980 US dollars)							Trade at Constant 1980 Prices
..	..	..	..	..	..	..	..	..	..	..	Exports, fob
..	..	..	..	..	..	..	..	..	..	..	Imports, cif
				(Millions of current US dollars)							**BALANCE OF PAYMENTS**
247	300	451	496	409	410	344	268	364	507	..	Exports of Goods & Services
199	242	368	388	324	304	231	173	265	363	..	Merchandise, fob
23	31	36	40	31	40	43	27	30	46	..	Nonfactor Services
25	27	46	68	54	67	71	67	69	97	..	Factor Services
341	467	659	659	583	578	488	365	421	514	..	Imports of Goods & Services
247	361	537	502	438	464	371	277	304	370	..	Merchandise, fob
65	85	80	107	104	85	85	55	61	71	..	Nonfactor Services
30	21	42	50	42	29	32	33	56	72	..	Factor Services
2	4	7	8	8	7	8	8	9	12	..	Long-Term Interest
-2	-2	-2	1	0	-2	0	3	2	4	..	Current Transfers, net
0	0	0	0	0	0	0	0	0	0	..	Workers' Remittances
-96	-168	-211	-162	-173	-169	-144	-95	-55	-2	K	Total to be Financed
34	53	76	53	52	78	76	50	49	42	K	Official Capital Grants
-62	-114	-135	-109	-121	-91	-68	-45	-6	40	..	Current Account Balance
103	147	114	99	77	123	87	72	79	33	..	Long-Term Capital, net
22	55	17	33	-12	1	-1	11	12	10	..	Direct Investment
47	38	24	13	20	17	8	-3	5	11	..	Long-Term Loans
49	41	29	21	30	29	17	13	21	30	..	Disbursements
2	4	5	8	10	11	9	15	16	19	..	Repayments
0	0	-4	-1	16	26	4	14	14	-30	..	Other Long-Term Capital
-24	-38	58	-30	38	-15	3	-10	-68	-52	..	Other Capital, net
-16	5	-37	40	6	-17	-23	-18	-5	-21	..	Change in Reserves
				(Swaziland Emalangeni per US dollar)							Memo Item:
0.870	0.840	0.780	0.870	1.080	1.110	1.440	2.190	2.270	2.040	..	Conversion Factor (Annual Avg)
				(Millions of US dollars, outstanding at end of year)							**EXTERNAL DEBT, ETC.**
105	148	166	162	170	175	160	185	219	273	..	Public/Publicly Guar. Long-Term
75	118	142	143	159	170	150	175	210	263	..	Official Creditors
21	24	25	34	44	51	45	51	64	76	..	IBRD and IDA
31	30	24	18	11	6	10	11	10	10	..	Private Creditors
..	..	..	..	..	..	..	..	..	..	..	Private Non-guaranteed Long-Term
0	0	0	0	0	11	10	10	7	3	..	Use of Fund Credit
17	8	15	6	14	37	7	15	17	17	..	Short-Term Debt
				(Millions of US dollars)							Memo Items:
113.6	113.7	158.7	96.4	76.1	92.5	80.1	83.4	96.5	127.2	..	Int'l Reserves Excluding Gold
..	..	..	..	..	..	..	..	..	..	..	Gold Holdings (at market price)
											SOCIAL INDICATORS
6.5	6.5	6.5	6.5	6.5	6.5	6.5	6.5	6.5	6.5	..	Total Fertility Rate
47.2	47.2	47.1	47.1	47.0	47.0	47.0	47.0	47.0	47.0	..	Crude Birth Rate
137.8	135.6	133.4	131.2	129.0	120.1	111.1	102.2	93.3	84.3	..	Infant Mortality Rate
50.5	51.1	51.7	52.3	53.0	53.4	53.9	54.4	54.9	55.4	..	Life Expectancy at Birth
98.4	92.5	103.1	104.4	101.5	100.1	100.9	97.0	113.2	100.6	..	Food Production, p.c. ('79-81 = 100)
75.4	74.8	74.1	..	..	..	..	..	..	..	..	Labor Force, Agriculture (%)
41.4	41.3	41.2	41.0	40.7	40.4	40.2	39.9	39.7	39.5	..	Labor Force, Female (%)
101.0	103.0	106.0	110.0	111.0	111.0	111.0	..	..	..	..	School Enroll. Ratio, primary
37.0	38.0	39.0	40.0	42.0	44.0	44.0	..	..	..	..	School Enroll. Ratio, secondary

SWEDEN	1967	1968	1969	1970	1971	1972	1973	1974	1975	1976	1977
CURRENT GNP PER CAPITA (US $)	3,200	3,490	3,810	4,180	4,430	4,910	5,820	6,930	8,270	9,140	9,930
POPULATION (thousands)	7,868	7,914	7,968	8,043	8,098	8,122	8,137	8,161	8,193	8,222	8,252
ORIGIN AND USE OF RESOURCES				*(Billions of current Swedish Kronor)*							
Gross National Product (GNP)	133.5	141.6	153.7	172.2	186.4	204.1	227.5	256.8	301.6	340.7	369.3
Net Factor Income from Abroad	0.0	0.0	-0.1	-0.1	0.2	0.4	0.8	0.7	0.8	0.5	-0.7
Gross Domestic Product (GDP)	133.5	141.6	153.8	172.2	186.2	203.8	226.7	256.1	300.8	340.2	370.0
Indirect Taxes, net	12.6	14.0	15.1	18.9	24.2	25.1	28.5	28.1	32.4	35.9	41.2
GDP at factor cost	120.9	127.6	138.7	153.3	162.0	178.7	198.2	228.1	268.4	304.3	328.8
Agriculture	..	..	..	7.2	8.1	7.8	8.3	11.5	12.9	14.6	14.5
Industry	..	..	..	63.2	65.0	70.8	80.7	94.8	109.4	120.0	121.6
Manufacturing	..	..	..	43.9	44.6	48.2	55.3	68.6	78.4	82.6	82.3
Services, etc.	..	..	..	82.9	89.0	100.0	109.3	121.8	146.1	169.7	192.8
Resource Balance	-0.1	-0.3	-0.6	-1.0	2.2	3.1	6.3	-1.9	-0.5	-5.7	-6.1
Exports of Goods & NFServices	28.1	30.4	35.0	41.5	45.3	49.3	62.2	82.5	84.7	94.1	101.4
Imports of Goods & NFServices	28.2	30.7	35.6	42.5	43.2	46.2	55.9	84.4	85.2	99.8	107.5
Domestic Absorption	133.6	141.9	154.4	173.2	184.1	200.7	220.5	258.0	301.3	345.9	376.1
Private Consumption, etc.	94.1	99.0	107.3	91.7	99.1	108.9	120.0	137.0	156.0	180.4	197.7
General Gov't Consumption	26.1	29.2	32.0	37.2	42.3	46.8	52.1	60.2	72.4	85.5	102.8
Gross Domestic Investment	13.3	13.7	15.1	44.4	42.7	45.0	48.3	60.8	72.9	79.9	75.6
Fixed Investment	32.6	33.3	35.1	38.1	40.6	45.3	50.1	54.2	62.7	71.6	77.9
Memo Items:											
Gross Domestic Saving	13.2	13.4	14.4	43.4	44.8	48.1	54.6	58.9	72.4	74.3	69.5
Gross National Saving	13.2	13.2	14.2	43.1	44.8	48.1	54.9	59.3	72.6	73.8	67.9
				(Billions of 1980 Swedish Kronor)							
Gross National Product	373.52	386.86	406.00	432.50	437.18	447.56	466.09	480.67	492.92	497.45	487.89
GDP at factor cost	329.94	339.91	357.13	382.87	377.48	389.40	403.76	425.88	436.53	442.28	433.15
Agriculture	..	..	..	18.03	18.42	18.20	18.53	18.71	17.12	17.06	15.90
Industry	..	..	..	149.25	145.33	152.09	160.88	157.86	162.29	163.24	156.78
Manufacturing	..	..	..	102.08	98.73	101.98	109.69	111.63	112.84	112.41	105.84
Services, etc.	..	..	..	215.59	213.73	219.11	224.34	249.32	257.12	261.98	260.47
Resource Balance	-14.55	-16.31	-19.69	-23.43	-13.93	-12.34	-5.10	-11.91	-19.83	-27.71	-19.60
Exports of Goods & NFServices	82.77	89.07	99.29	107.87	113.02	119.64	136.03	143.25	129.95	135.56	137.54
Imports of Goods & NFServices	97.32	105.38	118.97	131.30	126.94	131.98	141.13	155.16	149.78	163.27	157.14
Domestic Absorption	387.92	403.27	426.03	456.08	450.66	459.07	469.56	491.22	511.37	524.45	508.42
Private Consumption, etc.	202.45	211.28	222.03	230.89	230.71	238.41	244.22	252.70	259.90	270.75	267.88
General Gov't Consumption	91.87	98.11	103.25	111.59	114.40	117.43	120.78	124.26	129.91	134.46	138.47
Gross Domestic Investment	93.61	93.88	100.75	113.59	105.55	103.24	104.56	114.26	121.56	119.24	102.08
Fixed Investment	92.43	93.03	97.03	100.19	99.61	103.76	106.52	103.29	106.48	108.46	105.27
Memo Items:											
Capacity to Import	96.98	104.35	116.90	128.18	133.29	140.84	156.98	151.67	148.91	153.98	148.19
Terms of Trade Adjustment	14.21	15.28	17.62	20.31	20.28	21.19	20.95	8.43	18.96	18.42	10.66
Gross Domestic Income	387.58	402.23	423.96	452.96	457.01	467.92	485.41	487.74	510.50	515.16	499.48
Gross National Income	387.72	402.14	423.61	452.81	457.46	468.75	487.04	489.09	511.87	515.86	498.54
DOMESTIC PRICES (DEFLATORS)				*(Index 1980 = 100)*							
Overall (GDP)	35.7	36.6	37.8	39.8	42.6	45.6	48.8	53.4	61.2	68.5	75.7
Domestic Absorption	34.4	35.2	36.2	38.0	40.8	43.7	47.0	52.5	58.9	65.9	74.0
Agriculture	..	..	..	40.0	43.7	43.0	44.9	61.3	75.3	85.7	90.9
Industry	..	..	..	42.3	44.7	46.6	50.1	60.0	67.4	73.5	77.6
Manufacturing	..	..	..	43.0	45.1	47.3	50.4	61.4	69.5	73.5	77.7
MANUFACTURING ACTIVITY											
Employment (1980=100)	103.4	101.6	104.0	106.4	104.4	102.8	104.6	107.3	108.5	108.3	104.5
Real Earnings per Empl. (1980=100)	90.0	93.0	96.1	98.5	99.6	101.9	101.5	102.9	108.7	109.7	105.8
Real Output per Empl. (1980=100)	..	..	..	72.2	73.9	76.5	81.7	85.9	80.7	84.2	85.8
Earnings as % of Value Added	54.9	53.2	52.6	52.2	52.6	51.7	47.6	43.8	46.7	47.8	47.9
MONETARY HOLDINGS				*(Billions of current Swedish Kronor)*							
Money Supply, Broadly Defined	56.97	66.57	70.91	73.58	82.49	93.46	106.48	115.42	127.94	135.04	146.84
Money as Means of Payment	22.39	21.99	22.43	25.30	26.45	29.56	33.46	38.08	43.44	45.00	49.56
Currency Ouside Banks	10.00	10.46	10.94	11.40	12.81	14.14	15.38	17.32	20.13	22.16	24.41
Demand Deposits	12.39	11.54	11.49	13.90	13.64	15.42	18.08	20.76	23.32	22.84	25.15
Quasi-Monetary Liabilities	34.57	44.58	48.48	48.28	56.04	63.90	73.02	77.34	84.50	90.04	97.28
GOVERNMENT DEFICIT (-) OR SURPLUS				*(Billions of current Swedish Kronor)*							
	..	..	..	-3.10	-2.44	-2.47	-3.21	-7.98	-7.55	-1.17	-6.06
Current Revenue	..	..	..	51.29	58.64	66.47	70.50	79.15	93.72	122.88	140.77
Current Expenditure	..	..	..	39.97	44.54	50.93	57.15	68.69	82.01	101.21	122.20
Current Budget Balance	..	..	..	11.32	14.10	15.54	13.35	10.46	11.71	21.67	18.57
Capital Receipts	..	..	..	..	..	..	..	..	..	..	..
Capital Payments	..	..	..	14.42	16.54	18.01	16.56	18.44	19.26	22.84	24.63

1978	1979	1980	1981	1982	1983	1984	1985	1986	1987 estimate	NOTES	SWEDEN
10,870	12,530	14,240	14,870	14,030	12,540	11,970	11,860	13,150	15,630	..	CURRENT GNP PER CAPITA (US $)
8,276	8,294	8,310	8,320	8,325	8,331	8,337	8,350	8,354	8,357	..	POPULATION (thousands)
				(Billions of current Swedish Kronor)							ORIGIN AND USE OF RESOURCES
411.1	461.0	521.1	563.6	613.8	688.2	769.3	838.4	917.5	992.7	..	Gross National Product (GNP)
-1.4	-1.3	-4.0	-9.4	-13.9	-17.2	-20.3	-22.5	-16.2	-16.2	..	Net Factor Income from Abroad
412.5	462.3	525.1	573.0	627.7	705.4	789.6	860.9	933.7	1,009.0	..	Gross Domestic Product (GDP)
40.3	42.3	48.7	57.1	60.7	70.8	87.1	99.1	118.3	136.1	..	Indirect Taxes, net
372.1	420.0	476.4	515.9	567.0	634.6	702.5	761.8	815.4	872.8	B	GDP at factor cost
14.7	14.7	16.9	18.7	20.7	22.7	26.0	26.3	28.6	..	..	Agriculture
131.1	148.7	165.7	176.7	190.3	218.8	248.4	270.6	290.0	..	..	Industry
88.0	100.6	111.3	116.2	126.6	148.7	170.7	186.0	202.2	..	..	Manufacturing
226.4	256.6	293.9	320.5	356.0	393.0	428.1	464.9	496.8	..	..	Services, etc.
4.2	-4.7	-10.0	0.3	-3.8	16.3	30.4	19.7	33.5	28.9	..	Resource Balance
116.4	140.5	156.5	172.5	201.3	249.5	284.7	303.6	310.4	335.7	..	Exports of Goods & NFServices
112.2	145.2	166.5	172.3	205.1	233.2	254.3	283.9	276.9	306.8	..	Imports of Goods & NFServices
408.2	467.0	535.1	572.8	631.5	689.1	759.2	841.2	900.2	980.1	..	Domestic Absorption
219.4	242.4	270.1	298.5	334.1	362.8	398.6	438.7	480.5	525.1	..	Private Consumption, etc.
116.3	132.1	153.2	169.8	185.5	203.7	221.6	239.7	258.1	271.9	..	General Gov't Consumption
72.5	92.5	111.9	104.6	111.9	122.6	139.0	162.9	161.6	183.1	..	Gross Domestic Investment
80.4	91.2	106.0	110.6	119.7	133.6	150.1	169.5	176.3	199.6	..	Fixed Investment
											Memo Items:
76.8	87.8	101.9	104.8	108.1	138.9	169.4	182.6	195.1	212.0	..	Gross Domestic Saving
74.5	85.8	96.6	94.2	91.6	119.5	146.4	156.7	175.5	193.3	..	Gross National Saving
				(Billions of 1980 Swedish Kronor)							
495.70	515.06	521.08	514.95	516.23	527.59	547.57	558.98	570.38	586.93	..	Gross National Product
448.24	468.96	476.41	471.69	477.36	487.15	500.40	508.21	505.96	515.08	B	GDP at factor cost
16.25	16.16	16.87	17.10	18.25	19.18	20.09	19.33	19.44	..	..	Agriculture
155.17	163.38	165.65	161.88	161.57	168.95	180.84	188.15	187.61	..	..	Industry
103.41	109.95	111.30	107.63	107.36	113.46	121.46	125.80	125.14	..	..	Manufacturing
276.82	289.41	293.89	292.71	297.54	299.02	299.46	300.73	298.91	..	..	Services, etc.
-0.27	-8.44	-10.00	3.63	3.90	20.96	25.93	16.55	12.74	9.27	..	Resource Balance
148.27	157.36	156.52	158.28	165.26	182.90	195.23	199.46	205.81	215.62	..	Exports of Goods & NFServices
148.54	165.80	166.52	154.65	161.36	161.94	169.30	182.91	193.07	206.35	..	Imports of Goods & NFServices
497.65	524.92	535.10	519.88	523.83	519.57	535.96	557.35	567.83	587.39	..	Domestic Absorption
266.05	272.38	270.05	268.28	271.80	266.84	270.62	278.66	290.74	302.54	..	Private Consumption, etc.
142.90	149.73	153.16	156.59	157.98	159.32	163.10	166.53	168.70	169.81	..	General Gov't Consumption
88.70	102.80	111.89	95.01	94.05	93.42	102.24	112.16	108.39	115.04	..	Gross Domestic Investment
98.06	102.44	105.99	100.32	99.23	100.79	105.92	112.30	111.23	119.58	..	Fixed Investment
											Memo Items:
154.16	160.42	156.52	154.87	158.37	173.25	189.51	195.60	216.43	225.77	..	Capacity to Import
5.89	3.05	0.00	-3.41	-6.89	-9.65	-5.72	-3.86	10.62	10.15	..	Terms of Trade Adjustment
503.27	519.53	525.10	520.10	520.84	530.88	556.16	570.04	591.18	606.81	..	Gross Domestic Income
501.59	518.11	521.08	511.54	509.34	517.94	541.85	555.12	581.00	..	..	Gross National Income
				(Index 1980 = 100)							DOMESTIC PRICES (DEFLATORS)
82.9	89.5	100.0	109.5	118.9	130.5	140.5	150.0	160.8	169.1	..	Overall (GDP)
82.0	89.0	100.0	110.2	120.5	132.6	141.7	150.9	158.5	166.9	..	Domestic Absorption
90.5	90.7	100.0	109.4	113.4	118.5	129.2	135.9	146.8	..	..	Agriculture
84.5	91.0	100.0	109.2	117.8	129.5	137.4	143.8	154.6	..	..	Industry
85.1	91.5	100.0	108.0	117.9	131.1	140.6	147.9	161.6	..	..	Manufacturing
											MANUFACTURING ACTIVITY
100.9	100.7	100.0	96.8	92.5	89.4	89.8	90.0	..	..	..	Employment (1980=100)
104.7	104.5	100.0	97.2	96.7	95.9	97.4	97.6	..	..	..	Real Earnings per Empl. (1980=100)
88.0	96.9	100.0	101.9	108.4	116.3	120.5	124.0	..	..	..	Real Output per Empl. (1980=100)
47.2	43.3	43.7	43.5	40.4	36.8	36.6	36.7	..	..	..	Earnings as % of Value Added
				(Billions of current Swedish Kronor)							MONETARY HOLDINGS
172.48	202.87	223.54	257.50	276.60	401.80	431.17	..	..		..	Money Supply, Broadly Defined
58.06	67.14	79.41	85.73	80.03	86.75	95.68	102.79	112.15	..	..	Money as Means of Payment
27.57	30.94	33.58	36.06	38.06	41.95	44.86	45.86	53.56	..	..	Currency Ouside Banks
30.49	36.20	45.83	49.67	41.97	44.81	50.82	56.93	58.59	..	..	Demand Deposits
114.42	135.74	144.13	171.77	196.57	315.05	335.49	..	..	..	..	Quasi-Monetary Liabilities
				(Billions of current Swedish Kronor)							
-20.45	-33.22	-43.03	-51.52	-52.59	-60.15	-47.98	-46.39	-26.22	18.39	C E	GOVERNMENT DEFICIT (-) OR SURPLUS
154.65	166.57	186.75	213.68	241.36	272.19	309.71	350.36	377.51	439.00	..	Current Revenue
146.92	172.67	199.22	234.78	262.77	308.92	335.04	376.31	395.73	413.08	..	Current Expenditure
7.73	-6.10	-12.47	-21.10	-21.41	-36.73	-25.33	-25.95	-18.22	25.92	..	Current Budget Balance
..	..	..	0.08	0.05	0.19	0.05	1.56	0.29	1.23	..	Capital Receipts
28.18	27.12	30.56	30.50	31.23	23.61	22.70	22.00	8.29	8.76	..	Capital Payments

SWEDEN	1967	1968	1969	1970	1971	1972	1973	1974	1975	1976	1977
FOREIGN TRADE (CUSTOMS BASIS)					*(Millions of current US dollars)*						
Value of Exports, fob	4,526	4,937	5,687	6,781	7,464	8,749	12,171	15,909	17,434	18,440	19,054
Nonfuel Primary Products	1,249	1,337	1,493	1,639	1,726	1,981	2,851	3,760	3,513	3,632	3,405
Fuels	37	65	57	64	69	90	109	213	272	283	350
Manufactures	3,239	3,535	4,137	5,078	5,670	6,678	9,212	11,936	13,649	14,525	15,299
Value of Imports, cif	4,700	5,121	5,898	7,004	7,082	8,062	10,625	15,820	18,067	19,164	20,140
Nonfuel Primary Products	1,032	1,113	1,220	1,459	1,323	1,432	1,876	2,474	2,738	2,889	3,019
Fuels	527	631	620	745	865	842	1,211	2,848	3,110	3,381	3,513
Manufactures	3,142	3,377	4,058	4,800	4,894	5,788	7,539	10,498	12,219	12,893	13,608
Terms of Trade	138.5	144.7	136.2	142.3	135.6	132.7	134.5	94.8	106.2	105.7	99.4
Export Prices, fob	25.4	25.5	26.5	28.4	29.5	33.8	42.9	49.6	58.3	60.5	62.8
Nonfuel Primary Products	25.8	27.8	30.3	30.8	30.2	35.1	49.7	48.7	52.5	53.4	50.5
Fuels	8.5	7.3	5.4	5.9	7.0	8.5	24.5	36.6	33.6	38.5	36.8
Manufactures	25.8	25.8	26.7	29.0	30.5	34.8	41.5	50.2	61.0	63.4	67.7
Import Prices, cif	18.3	17.6	19.4	19.9	21.7	25.5	31.9	52.4	54.9	57.3	63.2
Trade at Constant 1980 Prices					*(Millions of 1980 US dollars)*						
Exports, fob	17,835	19,384	21,486	23,912	25,322	25,891	28,360	32,049	29,911	30,467	30,322
Imports, cif	25,649	29,089	30,348	35,158	32,586	31,664	33,302	30,206	32,932	33,463	31,867
BALANCE OF PAYMENTS					*(Millions of current US dollars)*						
Exports of Goods & Services	..	..	..	8,256	9,166	10,768	14,840	19,144	21,030	22,407	23,342
Merchandise, fob	..	..	..	6,750	7,415	8,697	12,097	15,797	17,242	18,288	18,924
Nonfactor Services	..	..	..	1,338	1,603	1,896	2,445	3,009	3,394	3,566	3,833
Factor Services	..	..	..	168	149	174	298	338	394	553	585
Imports of Goods & Services	..	..	..	8,368	8,636	9,947	13,100	19,277	20,724	23,308	24,654
Merchandise, fob	..	..	..	6,447	6,520	7,479	10,066	15,405	16,173	18,126	18,657
Nonfactor Services	..	..	..	1,732	1,907	2,222	2,702	3,446	4,033	4,380	4,903
Factor Services	..	..	..	189	209	246	332	426	519	802	1,094
Long-Term Interest	..	..	..	..	..	..	..	..	..	..	..
Current Transfers, net	..	..	..	-48	-55	-69	-96	-90	-152	-203	-208
Workers' Remittances	..	..	..	0	0	0	0	0	0	0	0
Total to be Financed	..	..	..	-160	475	752	1,644	-223	154	-1,104	-1,520
Official Capital Grants	..	..	..	-104	-123	-185	-215	-329	-503	-544	-669
Current Account Balance	..	..	..	-265	352	567	1,429	-552	-349	-1,648	-2,189
Long-Term Capital, net	..	..	..	35	-136	-21	-463	95	1,054	-173	2,412
Direct Investment	..	..	..	-104	-93	-197	-221	-349	-355	-592	-655
Long-Term Loans	..	..	..	..	..	..	..	..	..	..	..
Disbursements	..	..	..	..	..	..	..	..	..	..	..
Repayments	..	..	..	..	..	..	..	..	..	..	..
Other Long-Term Capital	..	..	..	244	80	361	-27	773	1,913	963	3,736
Other Capital, net	..	..	..	259	32	-195	-95	-296	699	1,308	879
Change in Reserves	..	..	..	-29	-248	-351	-870	753	-1,404	513	-1,101
Memo Item:					*(Swedish Kronor per US dollar)*						
Conversion Factor (Annual Avg)	5.170	5.170	5.170	5.170	5.120	4.760	4.370	4.440	4.150	4.360	4.480
EXTERNAL DEBT, ETC.					*(Millions US dollars, outstanding at end of year)*						
Public/Publicly Guar. Long-Term	..	..	..	..	..	..	..	..	..	..	..
Official Creditors	..	..	..	..	..	..	..	..	..	..	..
IBRD and IDA	..	..	..	..	..	..	..	..	..	..	..
Private Creditors	..	..	..	..	..	..	..	..	..	..	..
Private Non-guaranteed Long-Term	..	..	..	..	..	..	..	..	..	..	..
Use of Fund Credit	..	..	..	..	..	..	..	..	..	..	..
Short-Term Debt	..	..	..	..	..	..	..	..	..	..	..
Memo Items:					*(Millions of US dollars)*						
Int'l Reserves Excluding Gold	637.5	590.4	470.0	561.2	892.6	1,357.9	2,284.2	1,487.0	2,838.6	2,255.5	3,414.8
Gold Holdings (at market price)	204.3	269.2	227.0	213.6	252.1	375.2	649.9	1,079.8	812.0	780.2	978.0
SOCIAL INDICATORS											
Total Fertility Rate	2.3	2.1	1.9	1.9	2.0	1.9	1.9	1.9	1.8	1.7	1.6
Crude Birth Rate	15.4	14.3	13.5	13.7	14.1	13.8	13.5	13.4	12.6	11.9	11.6
Infant Mortality Rate	12.9	13.1	11.7	11.0	11.1	10.8	9.9	9.5	8.6	8.3	8.0
Life Expectancy at Birth	74.1	74.3	74.4	74.5	74.6	74.7	74.8	74.9	75.0	75.1	75.2
Food Production, p.c. ('79-81 = 100)	87.0	89.5	79.6	85.6	86.9	88.3	85.1	100.8	91.5	96.1	97.3
Labor Force, Agriculture (%)	10.0	9.4	8.9	8.3	8.0	7.7	7.5	7.2	6.9	6.7	6.4
Labor Force, Female (%)	33.9	34.5	35.1	35.7	36.5	37.3	38.2	39.0	39.8	40.6	41.4
School Enroll. Ratio, primary	..	..	..	94.0	..	..	..	..	101.0	100.0	99.0
School Enroll. Ratio, secondary	..	..	..	86.0	..	..	..	..	78.0	78.0	80.0

1978	1979	1980	1981	1982	1983	1984	1985	1986	1987 estimate	NOTES	SWEDEN
				(Millions of current US dollars)							**FOREIGN TRADE (CUSTOMS BASIS)**
21,768	27,538	30,787	28,493	26,740	27,377	29,258	30,403	37,118	44,313	..	Value of Exports, fob
3,749	4,674	5,118	4,537	4,125	4,443	4,808	4,490	4,883	5,726	..	Nonfuel Primary Products
409	826	1,337	1,262	1,388	1,698	1,630	1,456	1,039	1,268	..	Fuels
17,610	22,039	24,332	22,693	21,227	21,236	22,821	24,457	31,196	37,320	..	Manufactures
20,547	28,579	33,426	28,842	27,533	26,090	26,331	28,538	32,493	40,621	..	Value of Imports, cif
3,197	4,080	4,616	3,904	3,558	3,440	3,652	3,703	4,443	5,109	..	Nonfuel Primary Products
3,347	6,268	8,073	7,160	6,735	5,982	5,124	5,394	3,496	3,627	..	Fuels
14,003	18,231	20,737	17,778	17,240	16,668	17,556	19,441	24,554	31,885	..	Manufactures
				(Index 1980 = 100)							
101.4	106.3	100.0	91.5	88.8	85.6	86.6	87.7	94.5	96.4	..	Terms of Trade
70.6	87.8	100.0	90.1	82.6	76.8	75.8	75.4	88.2	100.5	..	Export Prices, fob
62.0	93.7	100.0	87.5	81.8	83.4	82.6	75.9	74.5	79.0	..	Nonfuel Primary Products
44.7	93.6	100.0	104.0	97.0	85.4	83.6	81.1	52.7	55.4	..	Fuels
73.8	86.6	100.0	90.0	82.0	75.0	74.0	75.0	93.0	108.0	..	Manufactures
69.5	82.7	100.0	98.5	93.1	89.7	87.5	86.0	93.4	104.2	..	Import Prices, cif
				(Millions of 1980 US dollars)							Trade at Constant 1980 Prices
30,853	31,348	30,787	31,613	32,362	35,631	38,608	40,323	42,073	44,094	..	Exports, fob
29,544	34,566	33,426	29,273	29,581	29,072	30,076	33,181	34,803	38,971	..	Imports, cif
				(Millions of current US dollars)							**BALANCE OF PAYMENTS**
27,036	34,993	39,394	37,076	35,132	35,634	37,609	38,720	47,128	56,131	..	Exports of Goods & Services
21,565	27,371	30,671	28,349	26,552	27,201	29,114	30,083	36,732	43,915	..	Merchandise, fob
4,572	6,524	7,395	6,937	6,636	6,711	6,689	6,574	7,984	9,075	..	Nonfactor Services
899	1,098	1,328	1,790	1,944	1,723	1,805	2,064	2,411	3,141	..	Factor Services
26,080	36,277	42,496	38,877	37,556	35,633	36,320	39,020	45,054	55,593	..	Imports of Goods & Services
18,986	28,054	32,861	28,205	26,809	25,319	25,708	27,738	31,682	39,456	..	Merchandise, fob
5,458	6,281	6,811	6,672	6,524	6,305	6,313	6,550	8,028	10,044	..	Nonfactor Services
1,636	1,943	2,824	3,999	4,223	4,009	4,298	4,732	5,344	6,093	..	Factor Services
..	..	..	..	..	..	..	..	..	..	..	Long-Term Interest
-208	-163	-300	-228	-422	-297	-327	-387	-482	-379	..	Current Transfers, net
0	0	0	0	0	0	0	0	0	16	..	Workers' Remittances
748	-1,448	-3,402	-2,029	-2,846	-296	962	-687	1,592	160	K	Total to be Financed
-1,005	-957	-996	-845	-713	-640	-616	-679	-856	-1,013	K	Official Capital Grants
-257	-2,405	-4,398	-2,874	-3,559	-936	346	-1,365	736	-853	..	Current Account Balance
-454	452	3,744	1,043	1,944	1,025	-1,131	727	-2,068	59	..	Long-Term Capital, net
-345	-505	-374	-654	-772	-999	-888	-1,015	-2,245	-2,844	..	Direct Investment
..	..	..	..	..	..	..	..	..	..	..	Long-Term Loans
..	..	..	..	..	..	..	..	..	..	..	Disbursements
..	..	..	..	..	..	..	..	..	..	..	Repayments
896	1,914	5,113	2,542	3,429	2,664	372	2,420	1,033	3,916	..	Other Long-Term Capital
1,364	1,277	549	2,034	1,573	598	743	2,189	1,473	1,676	..	Other Capital, net
-654	676	105	-203	42	-687	42	-1,551	-141	-883	..	Change in Reserves
				(Swedish Kronor per US dollar)							Memo Item:
4.520	4.290	4.230	5.060	6.280	7.670	8.270	8.600	7.120	6.340	..	Conversion Factor (Annual Avg)
				(Millions US dollars, outstanding at end of year)							**EXTERNAL DEBT, ETC.**
..	..	..	..	..	..	..	..	..	..	..	Public/Publicly Guar. Long-Term
..	..	..	..	..	..	..	..	..	..	..	Official Creditors
..	..	..	..	..	..	..	..	..	..	..	IBRD and IDA
..	..	..	..	..	..	..	..	..	..	..	Private Creditors
..	..	..	..	..	..	..	..	..	..	..	Private Non-guaranteed Long-Term
..	..	..	..	..	..	..	..	..	..	..	Use of Fund Credit
..	..	..	..	..	..	..	..	..	..	..	Short-Term Debt
				(Millions of US dollars)							Memo Items:
4,123.8	3,513.9	3,418.5	3,601.1	3,512.5	4,033.8	3,845.0	5,793.5	6,550.6	8,174.3	..	Int'l Reserves Excluding Gold
1,355.8	3,107.1	3,577.4	2,412.3	2,772.7	2,315.2	1,870.9	1,984.4	2,372.2	2,937.8	..	Gold Holdings (at market price)
											SOCIAL INDICATORS
1.6	1.7	1.7	1.7	1.7	1.7	1.7	1.7	1.7	1.7	..	Total Fertility Rate
11.2	11.6	11.6	11.3	11.4	11.4	11.5	11.6	11.6	11.7	..	Crude Birth Rate
7.7	7.4	6.9	6.9	7.3	7.8	8.3	8.8	9.4	10.0	..	Infant Mortality Rate
75.7	76.1	76.5	77.0	77.4	77.3	77.2	77.1	77.0	76.9	..	Life Expectancy at Birth
99.4	97.9	99.3	102.8	105.4	104.1	113.9	107.3	104.5	98.5	..	Food Production, p.c. ('79-81 = 100)
6.2	5.9	5.7	..	..	..	..	..	..	..	..	Labor Force, Agriculture (%)
42.2	43.0	43.8	43.9	43.9	44.0	44.1	44.1	44.2	44.3	..	Labor Force, Female (%)
98.0	98.0	97.0	..	98.0	98.0	98.0	98.0	99.0	..	..	School Enroll. Ratio, primary
84.0	85.0	88.0	..	86.0	85.0	83.0	83.0	..	..	..	School Enroll. Ratio, secondary

SWITZERLAND	1967	1968	1969	1970	1971	1972	1973	1974	1975	1976	1977
CURRENT GNP PER CAPITA (US $)	2,710	2,940	3,210	3,480	3,850	4,400	5,480	6,900	7,960	8,930	10,150
POPULATION (thousands)	5,992	6,068	6,136	6,267	6,324	6,385	6,431	6,443	6,405	6,346	6,327
ORIGIN AND USE OF RESOURCES					*(Billions of current Swiss Francs)*						
Gross National Product (GNP)	72.11	77.39	83.96	93.93	106.48	120.53	134.54	146.49	144.63	147.18	151.90
Net Factor Income from Abroad	1.76	2.27	2.57	3.27	3.49	3.83	4.48	5.39	4.47	5.22	6.11
Gross Domestic Product (GDP)	70.35	75.12	81.39	90.66	103.00	116.71	130.06	141.10	140.16	141.96	145.79
Indirect Taxes, net	4.20	4.40	5.04	5.61	6.01	7.02	7.54	7.40	7.44	7.63	8.02
GDP at factor cost	66.15	70.73	76.35	85.05	96.99	109.70	122.51	133.70	132.72	134.34	137.77
Agriculture	..	..	..	..	..	..	..	..	..	..	..
Industry	..	..	..	..	..	..	..	..	..	..	..
Manufacturing	..	..	..	..	..	..	..	..	..	..	..
Services, etc.	..	..	..	..	..	..	..	..	..	..	..
Resource Balance	0.36	1.19	1.02	-1.54	-1.59	-1.04	-1.45	-2.73	4.00	4.69	3.76
Exports of Goods & NFServices	20.49	23.17	26.63	29.71	32.07	35.78	40.23	45.90	44.02	47.71	53.43
Imports of Goods & NFServices	20.13	21.98	25.61	31.25	33.66	36.82	41.68	48.63	40.02	43.02	49.67
Domestic Absorption	69.99	73.93	80.37	92.21	104.59	117.75	131.51	143.83	136.16	137.27	142.03
Private Consumption, etc.	42.15	44.91	48.67	53.32	59.70	67.73	75.95	83.13	86.01	88.93	92.67
General Gov't Consumption	7.33	7.90	8.68	9.64	11.03	12.34	13.73	15.16	16.13	16.75	17.05
Gross Domestic Investment	20.50	21.11	23.02	29.25	33.86	37.68	41.84	45.55	34.01	31.58	32.31
Fixed Investment	18.33	19.19	20.99	24.95	29.66	33.85	37.85	39.39	35.97	32.54	33.49
Memo Items:											
Gross Domestic Saving	20.86	22.30	24.04	27.70	32.27	36.64	40.38	42.81	38.01	36.28	36.07
Gross National Saving	22.62	24.84	25.43	29.70	34.29	38.80	43.13	46.52	41.03	40.40	41.09
					(Billions of 1980 Swiss Francs)						
Gross National Product	132.46	137.94	145.88	155.82	161.89	166.95	172.28	175.33	161.67	160.25	164.91
GDP at factor cost	121.62	126.09	132.74	141.30	147.54	151.87	156.90	160.27	148.36	146.04	149.47
Agriculture	..	..	..	..	..	..	..	..	..	..	..
Industry	..	..	..	..	..	..	..	..	..	..	..
Manufacturing	..	..	..	..	..	..	..	..	..	..	..
Services, etc.	..	..	..	..	..	..	..	..	..	..	..
Resource Balance	0.33	0.89	1.15	-1.43	-2.49	-3.07	-2.64	-1.62	3.02	1.66	2.05
Exports of Goods & NFServices	31.18	34.30	38.88	41.53	43.15	45.90	49.50	50.01	46.73	51.08	56.04
Imports of Goods & NFServices	30.85	33.41	37.73	42.97	45.64	48.96	52.15	51.64	43.71	49.42	53.99
Domestic Absorption	128.94	133.03	140.31	151.91	159.11	164.69	169.20	170.60	153.65	152.81	156.18
Private Consumption, etc.	78.46	81.52	85.96	90.28	94.39	99.28	102.08	101.88	99.14	99.70	102.67
General Gov't Consumption	13.81	14.34	15.04	15.76	16.67	17.16	17.57	17.86	17.98	18.47	18.56
Gross Domestic Investment	36.68	37.17	39.30	45.87	48.04	48.25	49.55	50.86	36.54	34.64	34.95
Fixed Investment	34.32	35.38	37.48	40.83	44.85	47.08	48.45	46.39	40.09	35.88	36.46
Memo Items:											
Capacity to Import	31.41	35.23	39.24	40.85	43.49	47.59	50.33	48.73	48.08	54.81	58.08
Terms of Trade Adjustment	0.23	0.93	0.36	-0.68	0.34	1.69	0.82	-1.28	1.35	3.73	2.04
Gross Domestic Income	129.50	134.85	141.82	149.80	156.95	163.31	167.38	167.70	158.02	158.21	160.27
Gross National Income	132.69	138.87	146.24	155.13	162.22	168.64	173.11	174.05	163.02	163.98	166.95
DOMESTIC PRICES (DEFLATORS)					*(Index 1980 = 100)*						
Overall (GDP)	54.4	56.1	57.5	60.3	65.8	72.2	78.1	83.5	89.5	91.9	92.1
Domestic Absorption	54.3	55.6	57.3	60.7	65.7	71.5	77.7	84.3	88.6	89.8	90.9
Agriculture	..	..	..	..	..	..	..	..	..	..	..
Industry	..	..	..	..	..	..	..	..	..	..	..
Manufacturing	..	..	..	..	..	..	..	..	..	..	..
MANUFACTURING ACTIVITY											
Employment (1980=100)	123.6	123.2	124.4	124.5	123.9	120.5	117.1	116.0	103.1	98.6	98.5
Real Earnings per Empl. (1980=100)	..	..	..	..	..	..	..	..	..	..	..
Real Output per Empl. (1980=100)	..	..	..	..	..	..	..	..	..	..	..
Earnings as % of Value Added	..	..	..	..	..	..	..	..	..	..	..
MONETARY HOLDINGS					*(Billions of current Swiss Francs)*						
Money Supply, Broadly Defined	73.97	82.78	94.67	104.43	115.59	128.11	135.11	119.72	128.72	140.52	149.80
Money as Means of Payment	28.99	32.59	36.39	40.38	47.51	50.09	49.68	48.03	50.11	55.96	56.43
Currency Ouside Banks	12.09	13.03	13.83	14.54	15.60	17.82	19.08	20.33	20.12	20.78	21.48
Demand Deposits	16.90	19.56	22.56	25.84	31.91	32.27	30.60	27.70	29.99	35.18	34.94
Quasi-Monetary Liabilities	44.98	50.19	58.28	64.05	68.08	78.02	85.43	71.69	78.61	84.56	93.37
GOVERNMENT DEFICIT (-) OR SURPLUS					*(Millions of current Swiss Francs)*						
	..	..	..	487	279	1,065	588	964	-647	-954	-752
Current Revenue	..	..	..	13,983	15,529	17,928	21,696	24,816	26,061	29,068	29,633
Current Expenditure	..	..	..	11,265	12,804	13,909	17,605	20,233	23,415	26,528	28,178
Current Budget Balance	..	..	..	2,718	2,725	4,019	4,091	4,583	2,646	2,540	1,455
Capital Receipts	..	..	..	18	5	8	19	8	21	13	48
Capital Payments	..	..	..	2,249	2,451	2,962	3,522	3,627	3,314	3,507	2,255

1978	1979	1980	1981	1982	1983	1984	1985	1986	1987 estimate	NOTES	SWITZERLAND
11,700	14,140	17,280	17,660	17,030	16,440	16,370	16,290	17,670	21,450	..	**CURRENT GNP PER CAPITA (US $)**
6,337	6,357	6,385	6,429	6,393	6,418	6,442	6,472	6,504	6,501	..	**POPULATION (thousands)**
				(Billions of current Swiss Francs)							**ORIGIN AND USE OF RESOURCES**
157.50	165.19	177.35	193.98	205.17	213.95	226.06	241.35	254.49	265.96	..	Gross National Product (GNP)
5.82	6.65	7.01	9.22	9.19	10.09	12.83	13.40	11.57	11.14	..	Net Factor Income from Abroad
151.68	158.55	170.33	184.76	195.98	203.86	213.23	227.95	242.93	254.82	..	Gross Domestic Product (GDP)
8.59	8.89	9.66	10.51	10.72	11.32	11.85	12.63	13.99	..	..	Indirect Taxes, net
143.09	149.65	160.67	174.25	185.26	192.55	201.38	215.32	228.94	..	..	GDP at factor cost
..	..	..	..	..	..	..	..	..	..	..	Agriculture
..	..	..	..	..	..	..	..	..	..	..	Industry
..	..	..	..	..	..	..	..	..	..	..	Manufacturing
..	..	..	..	..	..	..	..	..	..	..	Services, etc.
3.69	-0.81	-6.02	-1.85	0.89	-0.08	-0.73	0.93	2.55	1.22	..	Resource Balance
53.22	56.02	62.57	69.09	69.53	71.76	80.43	88.99	89.05	89.41	..	Exports of Goods & NFServices
49.52	56.83	68.60	70.93	68.65	71.84	81.16	88.06	86.49	88.18	..	Imports of Goods & NFServices
147.98	159.35	176.35	186.60	195.09	203.95	213.96	227.02	240.37	253.59	..	Domestic Absorption
95.33	100.72	108.01	115.64	122.04	127.36	133.62	140.52	144.96	149.30	..	Private Consumption, etc.
17.53	18.27	19.31	21.19	22.60	23.98	24.89	26.55	27.21	27.82	..	General Gov't Consumption
35.13	40.36	49.03	49.78	50.45	52.61	55.45	59.94	68.21	76.48	..	Gross Domestic Investment
35.82	38.82	44.69	48.98	50.37	53.68	57.37	62.61	67.88	72.76	..	Fixed Investment
											Memo Items:
38.82	39.56	43.00	47.93	51.34	52.53	54.72	60.87	70.76	77.70	..	Gross Domestic Saving
43.59	45.11	48.84	55.42	58.66	60.73	65.61	72.21	80.17	86.53	..	Gross National Saving
				(Billions of 1980 Swiss Francs)							
165.18	169.81	177.35	181.43	179.00	180.75	185.85	193.13	196.49	201.81	..	Gross National Product
149.57	153.50	160.67	162.97	161.35	162.20	165.01	171.86	175.58	..	..	GDP at factor cost
..	..	..	..	..	..	..	..	..	..	..	Agriculture
..	..	..	..	..	..	..	..	..	..	..	Industry
..	..	..	..	..	..	..	..	..	..	..	Manufacturing
..	..	..	..	..	..	..	..	..	..	..	Services, etc.
-1.77	-4.44	-6.02	-2.25	-2.43	-4.74	-5.53	-3.60	-9.18	-13.54	..	Resource Balance
58.12	59.55	62.57	65.48	63.53	64.14	68.23	73.92	74.10	74.71	..	Exports of Goods & NFServices
59.89	63.99	68.60	67.73	65.96	68.88	73.76	77.52	83.28	88.25	..	Imports of Goods & NFServices
160.64	167.27	176.35	175.05	173.28	176.75	180.54	185.78	196.21	205.33	..	Domestic Absorption
104.24	105.28	108.01	109.05	108.93	110.24	111.80	113.87	116.57	117.99	..	Private Consumption, etc.
18.93	19.13	19.31	19.80	19.98	20.74	20.99	21.68	22.27	22.67	..	General Gov't Consumption
37.48	42.86	49.03	46.21	44.36	45.76	47.75	50.23	57.37	64.67	..	Gross Domestic Investment
38.68	40.65	44.69	45.76	44.54	46.44	48.38	51.12	55.55	59.28	..	Fixed Investment
											Memo Items:
64.36	63.08	62.57	65.97	66.81	68.80	73.09	78.34	85.74	89.47	..	Capacity to Import
6.23	3.53	0.00	0.49	3.28	4.66	4.87	4.42	11.64	14.76	..	Terms of Trade Adjustment
165.11	166.37	170.33	173.29	174.13	176.67	179.88	186.60	198.67	206.55	..	Gross Domestic Income
171.42	173.34	177.35	181.91	182.27	185.41	190.72	197.55	208.13	..	..	Gross National Income
				(Index 1980 = 100)							**DOMESTIC PRICES (DEFLATORS)**
95.5	97.4	100.0	106.9	114.7	118.5	121.8	125.1	129.9	132.9	..	Overall (GDP)
92.1	95.3	100.0	106.6	112.6	115.4	118.5	122.2	122.5	123.5	..	Domestic Absorption
..	..	..	..	..	..	..	..	..	..	..	Agriculture
..	..	..	..	..	..	..	..	..	..	..	Industry
..	..	..	..	..	..	..	..	..	..	..	Manufacturing
											MANUFACTURING ACTIVITY
98.8	98.0	100.0	100.2	102.6	98.7	94.1	95.7	..	..	G	Employment (1980 = 100)
..	..	..	..	..	..	..	..	..	..	G	Real Earnings per Empl. (1980 = 100)
..	..	..	..	..	..	..	..	..	..	G	Real Output per Empl. (1980 = 100)
..	..	..	..	..	..	..	..	..	..	..	Earnings as % of Value Added
				(Billions of current Swiss Francs)							**MONETARY HOLDINGS**
166.99	182.55	183.43	195.23	229.88	253.31	273.89	285.24	295.10	323.93	..	Money Supply, Broadly Defined
69.69	68.40	68.33	64.79	69.35	75.82	75.91	73.94	75.91	84.02	..	Money as Means of Payment
23.64	24.99	25.44	24.74	24.19	26.31	28.04	27.42	28.57	28.90	..	Currency Ouside Banks
46.05	43.41	42.89	40.05	45.16	49.51	47.86	46.52	47.33	55.12	..	Demand Deposits
97.30	114.15	115.10	130.44	160.53	177.49	197.98	211.30	219.19	239.91	..	Quasi-Monetary Liabilities
				(Millions of current Swiss Francs)							
-132	-1,087	-344	197	-116	-771	-202	..	..	..	E	**GOVERNMENT DEFICIT (-) OR SURPLUS**
31,519	32,111	34,582	36,988	40,116	42,077	45,662	..	..	..	..	Current Revenue
29,004	30,475	32,206	33,229	37,043	39,144	42,182	..	..	..	..	Current Expenditure
2,515	1,636	2,376	3,759	3,073	2,933	3,480	..	..	..	..	Current Budget Balance
7	7	27	17	46	9	10	..	..	..	..	Capital Receipts
2,654	2,730	2,747	3,579	3,235	3,713	3,692	..	..	..	..	Capital Payments

SWITZERLAND	1967	1968	1969	1970	1971	1972	1973	1974	1975	1976	1977
FOREIGN TRADE (CUSTOMS BASIS)				*(Millions of current US dollars)*							
Value of Exports, fob	3,471	3,951	4,610	5,120	5,768	6,877	9,472	11,838	12,952	14,669	17,325
Nonfuel Primary Products	338	401	461	522	575	664	909	1,151	1,044	1,210	1,480
Fuels	5	6	8	9	5	7	15	26	24	19	24
Manufactures	3,128	3,543	4,141	4,590	5,188	6,207	8,549	10,661	11,884	13,440	15,821
Value of Imports, cif	4,099	4,494	5,266	6,471	7,154	8,471	11,615	14,411	13,272	14,763	17,962
Nonfuel Primary Products	1,041	1,044	1,161	1,407	1,441	1,671	2,329	2,885	2,751	2,896	3,323
Fuels	257	299	299	352	462	462	836	1,439	1,368	1,577	1,702
Manufactures	2,801	3,151	3,807	4,713	5,251	6,338	8,450	10,087	9,153	10,290	12,937
Terms of Trade	112.1	118.3	108.0	107.1	109.8	108.0	114.2	91.8	101.4	100.7	97.2
				(Index 1980 = 100)							
Export Prices, fob	25.4	26.3	26.1	27.2	29.6	33.4	41.8	50.4	58.4	60.1	63.8
Nonfuel Primary Products	30.8	30.4	31.9	32.6	33.8	38.4	46.9	56.6	52.8	58.8	63.6
Fuels	8.5	7.3	5.4	5.9	7.0	8.5	24.5	36.6	33.6	38.5	36.8
Manufactures	25.0	26.1	25.8	26.9	29.3	33.1	41.4	49.8	59.2	60.4	63.9
Import Prices, cif	22.7	22.2	24.2	25.4	27.0	31.0	36.6	54.9	57.7	59.7	65.6
Trade at Constant 1980 Prices				*(Millions of 1980 US dollars)*							
Exports, fob	13,674	15,006	17,640	18,828	19,479	20,570	22,641	23,501	22,159	24,400	27,166
Imports, cif	18,094	20,201	21,762	25,495	26,527	27,360	31,709	26,257	23,015	24,718	27,371
BALANCE OF PAYMENTS				*(Millions of current US dollars)*							
Exports of Goods & Services	..	..	..	7,848	8,938	10,792	14,661	17,985	19,417	21,756	25,520
Merchandise, fob	..	..	..	5,260	5,859	7,044	9,626	12,056	13,109	14,906	17,697
Nonfactor Services	..	..	..	1,767	2,084	2,511	3,283	3,600	3,486	3,808	4,292
Factor Services	..	..	..	821	994	1,237	1,751	2,330	2,822	3,042	3,531
Imports of Goods & Services	..	..	..	7,444	8,478	10,067	13,711	17,108	16,435	18,028	21,459
Merchandise, fob	..	..	..	6,331	7,092	8,361	11,404	14,256	13,118	14,596	17,691
Nonfactor Services	..	..	..	781	931	1,105	1,468	1,780	1,927	2,111	2,398
Factor Services	..	..	..	332	455	600	839	1,072	1,391	1,320	1,371
Long-Term Interest	..	..	..	..	..	..	..	..	..	..	..
Current Transfers, net	..	..	..	-290	-356	-436	-546	-565	-560	-440	-454
Workers' Remittances	..	..	..	0	0	0	0	0	0	0	0
Total to be Financed	..	..	..	114	103	289	404	313	2,422	3,289	3,607
Official Capital Grants	..	..	..	-42	-21	-68	-124	-139	-134	-156	-212
Current Account Balance	..	..	..	72	82	220	280	173	2,288	3,133	3,395
Long-Term Capital, net	..	..	..	-989	-2,213	-2,955	-2,880	-1,723	-3,842	-7,273	-5,969
Direct Investment	..	..	..	..	..	..	..	..	..	..	..
Long-Term Loans	..	..	..	..	..	..	..	..	..	..	..
Disbursements	..	..	..	..	..	..	..	..	..	..	..
Repayments	..	..	..	..	..	..	..	..	..	..	..
Other Long-Term Capital	..	..	..	-947	-2,192	-2,887	-2,756	-1,584	-3,709	-7,117	-5,757
Other Capital, net	..	..	..	1,586	3,327	3,321	2,978	1,980	3,075	7,123	4,032
Change in Reserves	..	..	..	-669	-1,196	-586	-378	-431	-1,521	-2,983	-1,457
Memo Item:				*(Swiss Francs per US dollar)*							
Conversion Factor (Annual Avg)	4.370	4.370	4.370	4.370	4.130	3.820	3.170	2.980	2.580	2.500	2.400
EXTERNAL DEBT, ETC.				*(Millions US dollars, outstanding at end of year)*							
Public/Publicly Guar. Long-Term	..	..	..	..	..	..	..	..	..	..	..
Official Creditors	..	..	..	..	..	..	..	..	..	..	..
IBRD and IDA	..	..	..	..	..	..	..	..	..	..	..
Private Creditors	..	..	..	..	..	..	..	..	..	..	..
Private Non-guaranteed Long-Term	..	..	..	..	..	..	..	..	..	..	..
Use of Fund Credit	..	..	..	..	..	..	..	..	..	..	..
Short-Term Debt	..	..	..	..	..	..	..	..	..	..	..
Memo Items:				*(Millions of US dollars)*							
Int'l Reserves Excluding Gold	607	1,669	1,783	2,401	3,808	4,399	5,007	5,446	7,019	9,607	10,289
Gold Holdings (at market price)	3,107	3,141	2,657	2,916	3,626	5,394	9,339	15,517	11,669	11,222	13,737
SOCIAL INDICATORS											
Total Fertility Rate	2.4	2.3	2.1	2.1	2.0	1.9	1.8	1.7	1.6	1.5	1.5
Crude Birth Rate	17.9	17.3	16.7	15.8	15.2	14.3	13.6	13.1	12.2	11.6	11.5
Infant Mortality Rate	17.5	16.1	15.4	15.4	14.4	13.3	13.2	12.4	10.7	10.7	8.7
Life Expectancy at Birth	72.2	72.5	72.8	73.2	73.5	73.8	74.1	74.4	74.7	74.9	75.2
Food Production, p.c. ('79-81 = 100)	84.6	87.9	84.2	84.9	86.5	84.3	87.0	85.9	87.9	95.5	96.0
Labor Force, Agriculture (%)	8.8	8.4	8.1	7.8	7.6	7.5	7.3	7.2	7.0	6.8	6.7
Labor Force, Female (%)	32.8	32.8	32.8	32.7	33.2	33.7	34.2	34.8	35.3	35.6	35.9
School Enroll. Ratio, primary	..	..	..	..	..	..	..	..	..	..	..
School Enroll. Ratio, secondary	..	..	..	..	..	..	..	..	..	..	..

1978	1979	1980	1981	1982	1983	1984	1985	1986	1987 est.	NOTES	
				(Millions of current US dollars)							**FOREIGN TRADE (CUSTOMS BASIS)**
23,532	26,390	29,471	26,717	25,618	25,308	25,724	27,281	37,534	45,357	..	Value of Exports, fob
1,858	2,203	2,788	1,989	1,812	1,877	2,010	1,961	2,473	3,045	..	Nonfuel Primary Products
21	23	36	31	35	73	90	90	64	61	..	Fuels
21,653	24,163	26,647	24,697	23,770	23,358	23,623	25,230	34,997	42,250	..	Manufactures
23,792	29,306	36,148	30,607	28,577	28,934	29,625	30,626	41,188	50,557	..	Value of Imports, cif
3,950	4,488	6,409	4,541	4,144	4,260	4,260	4,307	5,516	6,586	..	Nonfuel Primary Products
1,924	3,440	4,055	3,742	3,331	3,267	3,040	3,044	2,404	2,245	..	Fuels
17,917	21,378	25,685	22,324	21,102	21,407	22,324	23,275	33,268	41,726	..	Manufactures
				(Index 1980 = 100)							
109.9	105.2	100.0	91.0	94.2	93.8	87.7	87.9	108.6	96.6	..	Terms of Trade
81.4	91.1	100.0	87.9	86.4	83.7	75.9	75.0	98.2	101.1	..	Export Prices, fob
68.7	91.5	100.0	86.9	78.9	79.8	74.5	74.9	80.3	82.7	..	Nonfuel Primary Products
44.7	93.6	100.0	104.0	97.0	85.4	83.6	81.1	41.3	41.4	..	Fuels
82.8	91.1	100.0	88.0	87.0	84.0	76.0	75.0	100.0	103.0	..	Manufactures
74.1	86.6	100.0	96.6	91.7	89.2	86.5	85.4	90.4	104.7	..	Import Prices, cif
											Trade at Constant 1980 Prices
				(Millions of 1980 US dollars)							
28,916	28,968	29,471	30,384	29,656	30,245	33,888	36,369	38,232	44,851	..	Exports, fob
32,122	33,844	36,148	31,690	31,168	32,447	34,241	35,879	45,566	48,308	..	Imports, cif
				(Millions of current US dollars)							**BALANCE OF PAYMENTS**
33,640	37,889	41,925	40,020	45,194	52,495	54,842	56,620	74,023	87,329	..	Exports of Goods & Services
23,499	26,756	29,265	27,097	26,196	33,905	35,756	37,063	48,317	55,161	..	Merchandise, fob
5,461	5,996	6,951	6,881	7,038	7,422	7,298	7,929	10,740	13,581	..	Nonfactor Services
4,681	5,137	5,709	6,042	11,959	11,167	11,788	11,629	14,965	18,587	..	Factor Services
28,994	35,552	42,329	37,561	40,246	50,372	49,409	49,774	68,394	79,951	..	Imports of Goods & Services
23,541	29,209	35,155	30,415	28,303	39,299	38,372	38,624	53,148	60,693	..	Merchandise, fob
3,353	4,036	4,647	4,452	4,511	4,710	4,710	4,978	6,691	8,142	..	Nonfactor Services
2,100	2,306	2,527	2,693	7,433	6,363	6,327	6,172	8,555	11,116	..	Factor Services
..	..	..	..	..	..	..	..	..	..	..	Long-Term Interest
-587	-656	-701	-881	-921	-901	-826	-841	-1,201	-1,544	..	Current Transfers, net
0	0	0	0	84	85	69	66	93	112	..	Workers' Remittances
4,059	1,682	-1,106	1,578	4,026	1,221	4,607	6,005	4,427	5,834	K	Total to be Financed
-336	-400	-439	-122	-99	-13	-9	34	98	46	K	Official Capital Grants
3,723	1,282	-1,544	1,456	3,928	1,209	4,597	6,040	4,525	5,879	..	Current Account Balance
-8,814	-13,985	-11,652	-9,744	-13,703	-3,902	-4,545	-7,326	580	-3,437	..	Long-Term Capital, net
					151	-362	-3,306	383	26	..	Direct Investment
..	..	..	..	..	..	..	..	..	..	..	Long-Term Loans
..	..	..	..	..	..	..	..	..	..	..	Disbursements
..	..	..	..	..	..	..	..	..	..	..	Repayments
-8,478	-13,585	-11,213	-9,622	-13,605	-4,039	-4,174	-4,055	99	-3,508	..	Other Long-Term Capital
11,731	8,769	13,533	7,768	13,274	3,077	1,428	2,511	-4,012	19,883	..	Other Capital, net
-6,641	3,934	-337	520	-3,499	-384	-1,480	-1,225	-1,093	-22,325	..	Change in Reserves
				(Swiss Francs per US dollar)							**Memo Item:**
1.790	1.660	1.670	1.960	2.030	2.100	2.350	2.460	1.800	1.490	..	Conversion Factor (Annual Avg)
											EXTERNAL DEBT, ETC.
				(Millions US dollars, outstanding at end of year)							Public/Publicly Guar. Long-Term
..	..	..	..	..	..	..	..	..	..	..	Official Creditors
..	..	..	..	..	..	..	..	..	..	..	IBRD and IDA
..	..	..	..	..	..	..	..	..	..	..	Private Creditors
..	..	..	..	..	..	..	..	..	..	..	Private Non-guaranteed Long-Term
..	..	..	..	..	..	..	..	..	..	..	Use of Fund Credit
..	..	..	..	..	..	..	..	..	..	..	Short-Term Debt
				(Millions of US dollars)							**Memo Items:**
17,763	16,435	15,656	13,979	15,461	15,034	15,296	18,016	21,786	27,476	..	Int'l Reserves Excluding Gold
18,821	42,638	49,092	33,103	38,049	31,770	25,674	27,232	32,553	40,314	..	Gold Holdings (at market price)
											SOCIAL INDICATORS
1.5	1.5	1.6	1.6	1.5	1.5	1.5	1.5	1.5	1.5	..	Total Fertility Rate
11.2	11.3	11.5	11.4	11.4	11.4	11.4	11.4	11.3	11.3	..	Crude Birth Rate
8.6	8.4	9.0	7.5	8.0	8.5	9.1	9.7	10.3	11.0	..	Infant Mortality Rate
75.8	76.3	76.9	77.4	78.0	77.7	77.4	77.2	76.9	76.7	..	Life Expectancy at Birth
94.9	101.5	100.5	98.0	108.3	103.4	106.4	105.8	106.9	105.8	..	Food Production, p.c. ('79-81 = 100)
6.5	6.4	6.2	..	..	..	..	..	..	..	..	Labor Force, Agriculture (%)
36.1	36.4	36.7	36.7	36.7	36.7	36.7	36.7	36.7	36.7	..	Labor Force, Female (%)
..	..	..	..	..	..	..	..	..	..	..	School Enroll. Ratio, primary
..	..	..	..	..	..	..	..	..	..	..	School Enroll. Ratio, secondary

SYRIAN ARAB REPUBLIC	1967	1968	1969	1970	1971	1972	1973	1974	1975	1976	1977
CURRENT GNP PER CAPITA (US $)	270	290	370	350	390	460	460	650	890	1,020	1,010
POPULATION (thousands)	5,670	5,856	6,052	6,258	6,475	6,703	6,942	7,187	7,438	7,695	7,959

ORIGIN AND USE OF RESOURCES

(Millions of current Syrian Pounds)

	1967	1968	1969	1970	1971	1972	1973	1974	1975	1976	1977
Gross National Product (GNP)	5,495	5,944	6,837	6,842	8,038	9,279	9,956	16,036	20,637	24,703	27,077
Net Factor Income from Abroad	-7	-3	-6	-6	-6	-7	11	85	40	-22	64
Gross Domestic Product (GDP)	5,502	5,947	6,843	6,848	8,044	9,286	9,945	15,951	20,597	24,725	27,013
Indirect Taxes, net	..	..	..	..	..	..	..	..	..	..	..
GDP at factor cost	..	..	..	..	..	..	..	..	..	..	..
Agriculture	1,460	1,411	1,619	1,382	1,610	2,321	1,679	3,224	3,706	4,815	5,001
Industry	1,198	1,325	1,634	1,761	2,054	2,162	2,402	4,170	4,975	6,225	6,602
Manufacturing	..	..	..	..	..	..	..	..	..	..	..
Services, etc.	2,844	3,211	3,590	3,705	4,380	4,803	5,864	8,557	11,916	13,685	15,410
Resource Balance	-186	-187	-311	-242	-400	-513	-366	-1,531	-2,587	-3,651	-6,076
Exports of Goods & NFServices	850	1,014	1,149	1,190	1,390	1,674	2,175	3,816	4,409	4,828	4,908
Imports of Goods & NFServices	1,036	1,201	1,460	1,432	1,790	2,187	2,541	5,347	6,996	8,479	10,984
Domestic Absorption	5,688	6,134	7,154	7,090	8,444	9,799	10,311	17,482	23,184	28,376	33,089
Private Consumption, etc.	4,224	4,296	5,017	4,966	5,823	6,294	6,960	10,704	13,685	15,657	18,199
General Gov't Consumption	834	1,065	1,116	1,187	1,427	1,637	2,122	2,815	4,343	4,960	5,293
Gross Domestic Investment	630	773	1,021	937	1,194	1,868	1,229	3,963	5,156	7,759	9,597
Fixed Investment	630	773	1,021	897	1,150	1,601	1,769	3,067	5,156	7,759	9,597
Memo Items:											
Gross Domestic Saving	444	586	710	695	794	1,355	863	2,432	2,569	4,108	3,521
Gross National Saving	437	625	738	715	818	1,498	1,015	2,683	2,803	4,290	3,948

(Millions of 1980 Syrian Pounds)

	1967	1968	1969	1970	1971	1972	1973	1974	1975	1976	1977
Gross National Product	17,491	18,496	22,182	20,871	22,856	26,651	25,265	32,129	39,269	42,659	42,191
GDP at factor cost	16,411	17,550	20,693	19,449	21,359	24,430	23,015	31,633	38,025	41,289	..
Agriculture	5,230	4,687	5,407	4,150	4,429	6,592	3,953	6,728	6,779	8,084	7,022
Industry	2,968	3,277	4,982	5,471	6,614	7,249	7,214	8,745	11,246	12,249	11,469
Manufacturing	..	..	..	..	..	..	..	..	..	..	
Services, etc.	9,310	10,540	11,808	11,264	11,827	12,824	14,075	16,491	21,170	22,362	23,607
Resource Balance	3,968	4,309	4,479	5,086	4,003	5,265	4,153	-1,893	-3,024	-3,372	-5,872
Exports of Goods & NFServices	7,701	8,682	10,102	9,988	9,476	11,158	10,909	10,006	10,968	10,950	10,163
Imports of Goods & NFServices	3,733	4,373	5,623	4,902	5,473	5,893	6,756	11,899	13,992	14,322	16,035
Domestic Absorption	14,890	15,665	19,358	17,565	20,384	23,112	22,990	34,643	42,219	46,067	47,970
Private Consumption, etc.	10,230	9,841	12,518	11,429	13,095	13,970	15,217	20,997	25,249	26,269	26,090
General Gov't Consumption	2,008	2,555	2,684	2,861	3,366	3,825	4,536	5,767	8,334	8,400	8,567
Gross Domestic Investment	2,652	3,269	4,156	3,275	3,923	5,317	3,237	7,879	8,636	11,398	13,313
Fixed Investment	2,652	3,269	4,156	3,191	3,811	4,595	4,340	6,266	8,636	11,398	13,313
Memo Items:											
Capacity to Import	3,063	3,692	4,425	4,074	4,250	4,511	5,783	8,492	8,818	8,155	7,165
Terms of Trade Adjustment	-4,638	-4,990	-5,677	-5,914	-5,226	-6,647	-5,126	-1,514	-2,150	-2,795	-2,998
Gross Domestic Income	12,870	13,514	16,520	14,971	17,644	20,018	20,116	30,450	37,045	39,900	39,100
Gross National Income	12,853	13,506	16,505	14,956	17,630	20,003	20,138	30,615	37,119	39,864	39,193

DOMESTIC PRICES (DEFLATORS)

(Index 1980 = 100)

	1967	1968	1969	1970	1971	1972	1973	1974	1975	1976	1977
Overall (GDP)	31.4	32.1	30.8	32.8	35.2	34.8	39.4	49.9	52.6	57.9	64.2
Domestic Absorption	38.2	39.2	37.0	40.4	41.4	42.4	44.8	50.5	54.9	61.6	69.0
Agriculture	27.9	30.1	29.9	33.3	36.4	35.2	42.5	47.9	54.7	59.6	71.2
Industry	40.4	40.4	32.8	32.2	31.1	29.8	33.3	47.7	44.2	50.8	57.6
Manufacturing	..	..	..	..	..	..	..	..	..	..	..

MANUFACTURING ACTIVITY

	1967	1968	1969	1970	1971	1972	1973	1974	1975	1976	1977
Employment (1980=100)	55.5	56.7	54.0	56.0	64.4	68.0	72.0	75.8	80.5	86.3	90.4
Real Earnings per Empl. (1980=100)	56.0	58.6	73.2	74.7	72.3	76.3	67.3	61.9	57.2	53.7	63.9
Real Output per Empl. (1980=100)	41.8	42.6	66.0	71.7	89.0	101.0	89.8	79.3	87.2	100.3	98.8
Earnings as % of Value Added	30.2	32.2	34.0	33.1	32.3	25.8	25.1	20.8	21.5	18.3	21.4

MONETARY HOLDINGS

(Millions of current Syrian Pounds)

	1967	1968	1969	1970	1971	1972	1973	1974	1975	1976	1977
Money Supply, Broadly Defined	1,703	1,980	2,215	2,522	2,717	3,428	4,114	5,996	7,585	9,387	12,035
Money as Means of Payment	1,618	1,868	2,088	2,341	2,503	3,151	3,797	5,540	6,966	8,561	10,924
Currency Ouside Banks	1,304	1,491	1,577	1,795	1,846	2,245	2,757	3,413	3,944	5,259	6,797
Demand Deposits	314	377	511	546	657	906	1,040	2,127	3,021	3,302	4,127
Quasi-Monetary Liabilities	85	112	127	181	214	277	317	456	619	826	1,111

GOVERNMENT DEFICIT (-) OR SURPLUS

(Millions of current Syrian Pounds)

	1967	1968	1969	1970	1971	1972	1973	1974	1975	1976	1977
Current Revenue	..	..	..	..	..	-327	-574	-740	-992	-2,332	-2,928
Current Expenditure	..	..	..	..	..	2,333	2,759	4,824	8,688	9,713	10,410
Current Budget Balance	..	..	..	..	..	1,735	2,240	3,203	5,160	6,046	6,634
Capital Receipts	..	..	..	..	..	598	519	1,621	3,528	3,667	3,776
Capital Payments	..	..	..	..	..	13	7	11	10	10	10
	..	..	..	..	..	938	1,100	2,372	4,530	6,009	6,714

1978	1979	1980	1981	1982	1983	1984	1985	1986	1987 estimate	NOTES	SYRIAN ARAB REPUBLIC
1,120	1,240	1,460	1,660	1,670	1,640	1,580	1,610	1,630	1,640	A	**CURRENT GNP PER CAPITA (US $)**
8,230	8,510	8,800	9,101	9,415	9,745	10,091	10,458	10,846	11,248	..	**POPULATION (thousands)**
											ORIGIN AND USE OF RESOURCES
				(Millions of current Syrian Pounds)							
32,556	39,158	51,313	65,698	68,102	72,687	74,885	82,737	99,879	125,156	..	Gross National Product (GNP)
167	184	43	-79	-686	-604	-457	-491	-420	-1,169	..	Net Factor Income from Abroad
32,389	38,974	51,270	65,777	68,788	73,291	75,342	83,227	100,300	126,325	..	Gross Domestic Product (GDP)
..	..	..	..	..	..	..	..	..	..		Indirect Taxes, net
..	..	..	..	..	..	..	..	..	..	B	GDP at factor cost
6,850	6,858	10,369	12,759	13,854	15,627	14,805	17,463	23,816	34,369	..	Agriculture
8,134	10,550	11,947	16,789	15,936	16,473	17,041	18,213	22,613	23,871	..	Industry
..	..	..	..	..	..	..	..	..	..	..	Manufacturing
17,405	21,566	28,954	36,229	38,998	41,191	43,496	47,551	53,871	68,085	..	Services, etc.
-5,306	-6,575	-8,823	-11,397	-7,577	-9,835	-8,723	-10,499	-10,024	-10,911	..	Resource Balance
4,808	7,458	9,345	10,290	9,572	9,714	9,360	10,245	10,134	18,691	..	Exports of Goods & NFServices
10,114	14,033	18,168	21,687	17,149	19,549	18,083	20,744	20,158	29,602	..	Imports of Goods & NFServices
37,695	45,549	60,093	77,174	76,365	83,126	84,065	93,726	110,324	137,236	..	Domestic Absorption
22,338	26,931	34,107	48,256	44,992	49,686	49,121	54,157	65,553	90,818	f	Private Consumption, etc.
6,470	8,424	11,870	13,656	15,103	16,154	17,079	19,785	21,440	22,945	..	General Gov't Consumption
8,887	10,194	14,116	15,262	16,270	17,286	17,865	19,784	23,331	23,473	f	Gross Domestic Investment
8,887	10,194	14,116	15,262	16,270	17,286	17,865	19,784	23,331	23,473	..	Fixed Investment
											Memo Items:
3,581	3,619	5,293	3,865	8,693	7,451	9,142	9,285	13,307	12,562	..	Gross Domestic Saving
6,244	7,341	8,372	6,069	9,757	8,657	9,968	9,945	15,960	12,374	..	Gross National Saving
				(Millions of 1980 Syrian Pounds)							
45,628	47,528	51,313	56,071	57,193	58,345	56,345	58,060	57,467	51,886	..	Gross National Product
..	..	..	..	..	..	..	..	..	..	B H	GDP at factor cost
8,673	7,389	10,369	10,765	10,513	10,458	9,563	10,169	10,762	9,365	..	Agriculture
11,816	12,070	11,947	11,858	12,578	13,265	12,150	13,175	14,773	12,060	..	Industry
..	..	..	..	..	..	..	..	..	..	..	Manufacturing
24,916	27,849	28,954	33,520	34,706	35,132	35,000	35,081	32,182	30,949	..	Services, etc.
-4,821	-7,137	-8,823	-13,561	-9,480	-11,329	-10,947	-11,295	-7,890	-4,969	..	Resource Balance
10,739	10,426	9,345	9,758	10,231	10,636	10,080	9,102	10,092	10,330	..	Exports of Goods & NFServices
15,560	17,563	18,168	23,319	19,711	21,965	21,027	20,397	17,982	15,299	..	Imports of Goods & NFServices
50,226	54,445	60,093	69,704	67,277	70,184	67,660	69,720	65,607	57,343	..	Domestic Absorption
29,329	30,934	34,107	42,838	39,227	40,871	36,883	39,038	37,247	34,844	f	Private Consumption, etc.
9,330	11,771	11,870	12,445	13,154	13,856	15,035	13,913	12,509	10,582	..	General Gov't Consumption
11,567	11,740	14,116	14,421	14,896	15,457	15,742	16,769	15,851	11,917	f	Gross Domestic Investment
11,567	11,740	14,116	14,421	14,896	15,457	15,742	16,769	15,851	11,917	..	Fixed Investment
											Memo Items:
7,397	9,334	9,345	11,064	11,002	10,915	10,884	10,074	9,040	9,660	..	Capacity to Import
-3,342	-1,092	0	1,306	771	279	804	972	-1,052	-670	..	Terms of Trade Adjustment
42,063	46,216	51,270	57,449	58,568	59,134	57,517	59,397	56,665	51,704	..	Gross Domestic Income
42,286	46,436	51,313	57,378	57,964	58,624	57,149	59,032	56,415	51,216	..	Gross National Income
											DOMESTIC PRICES (DEFLATORS)
				(Index 1980 = 100)							
71.3	82.4	100.0	117.2	119.0	124.5	132.8	142.5	173.8	241.2	..	Overall (GDP)
75.1	83.7	100.0	110.7	113.5	118.4	124.2	134.4	168.2	239.3	..	Domestic Absorption
79.0	92.8	100.0	118.5	131.8	149.4	154.8	171.7	221.3	367.0	..	Agriculture
68.8	87.4	100.0	141.6	126.7	124.2	140.3	138.2	153.1	197.9	..	Industry
..	..	..	..	..						..	Manufacturing
											MANUFACTURING ACTIVITY
95.7	95.7	100.0	107.8	108.6	109.3	110.5	113.6	..	..	G	Employment (1980=100)
78.0	94.4	100.0	105.9	99.1	101.2	95.3	94.2	..	..	G	Real Earnings per Empl. (1980=100)
99.7	86.3	100.0	102.4	123.7	135.9	129.1	156.6	..	..	G	Real Output per Empl. (1980=100)
27.6	29.8	27.7	28.4	29.8	30.5	31.0	31.5	..	..	..	Earnings as % of Value Added
											MONETARY HOLDINGS
				(Millions of current Syrian Pounds)							
15,293	17,904	24,030	27,841	33,511	42,182	52,774	63,493	71,121	..	..	Money Supply, Broadly Defined
13,866	16,119	21,854	24,832	29,518	36,978	45,607	54,976	61,214	..	..	Money as Means of Payment
8,459	9,903	13,422	14,046	17,348	20,500	25,155	29,562	36,262	..	..	Currency Ouside Banks
5,407	6,216	8,432	10,786	12,170	16,478	20,452	25,414	24,952	..	..	Demand Deposits
1,427	1,785	2,176	3,009	3,993	5,204	7,167	8,517	9,907	..	..	Quasi-Monetary Liabilities
				(Millions of current Syrian Pounds)							
-2,935	303	-4,976	-4,157	..	..	..	..	..	..	..	**GOVERNMENT DEFICIT (-) OR SURPLUS**
10,397	15,572	19,711	21,202	..	..	..	..	..	..	..	Current Revenue
7,333	9,210	15,546	17,174	..	..	..	..	..	..	..	Current Expenditure
3,064	6,362	4,165	4,028	..	..	..	..	..	..	..	Current Budget Balance
14	13	15	1	..	..	..	..	..	..	..	Capital Receipts
6,013	6,072	9,156	8,186	..	..	..	..	..	..	..	Capital Payments

SYRIAN ARAB REPUBLIC	1967	1968	1969	1970	1971	1972	1973	1974	1975	1976	1977
FOREIGN TRADE (CUSTOMS BASIS)					*(Millions of current US dollars)*						
Value of Exports, fob	154.8	176.2	206.8	203.0	194.6	287.3	351.1	783.7	930.0	1,055.1	1,063.0
Nonfuel Primary Products	129.7	142.8	162.4	146.1	124.9	196.6	218.3	288.6	203.7	268.8	319.4
Fuels	0.7	8.2	21.9	34.0	46.7	52.6	76.3	432.5	653.9	682.6	644.0
Manufactures	24.4	25.2	22.5	22.9	23.0	38.1	56.5	62.7	72.4	103.7	99.6
Value of Imports, cif	264.1	310.8	367.9	350.0	438.3	539.4	613.2	1,229.4	1,669.1	1,959.1	2,655.9
Nonfuel Primary Products	70.6	83.0	82.4	123.5	180.8	165.2	205.7	414.0	432.8	416.7	451.2
Fuels	24.8	34.7	38.2	28.2	26.9	24.1	28.0	74.2	108.7	192.1	445.1
Manufactures	168.6	193.1	247.3	198.3	230.6	350.2	379.6	741.2	1,127.6	1,350.2	1,759.6
Terms of Trade	158.3	136.9	98.7	71.2	58.1	63.1	65.3	81.1	68.1	77.9	81.0
Export Prices, fob	31.2	25.1	19.1	15.6	15.6	20.1	26.5	46.5	40.9	46.3	50.5
Nonfuel Primary Products	32.6	33.2	32.5	32.9	36.0	40.4	63.0	69.6	60.6	78.5	74.3
Fuels	4.3	4.3	4.3	4.3	5.6	6.2	8.9	36.7	35.7	38.4	42.0
Manufactures	30.1	31.7	32.4	37.5	36.5	38.5	49.5	67.2	66.1	66.7	71.7
Import Prices, cif	19.7	18.3	19.4	21.9	26.8	31.9	40.6	57.3	60.1	59.5	62.4
Trade at Constant 1980 Prices					*(Millions of 1980 US dollars)*						
Exports, fob	495.5	701.7	1,083.4	1,303.0	1,248.4	1,430.3	1,322.9	1,685.4	2,274.9	2,277.3	2,103.4
Imports, cif	1,338.0	1,693.8	1,901.3	1,599.8	1,634.2	1,693.7	1,510.0	2,144.9	2,779.4	3,293.0	4,256.2
BALANCE OF PAYMENTS					*(Millions of current US dollars)*						
Exports of Goods & Services	..	..	..	326.0	371.1	471.2	631.8	1,142.5	1,314.9	1,379.7	1,453.0
Merchandise, fob	..	..	..	197.0	195.6	298.6	356.4	782.9	930.0	1,065.6	1,069.8
Nonfactor Services	..	..	..	128.0	174.5	171.5	265.8	304.3	335.1	295.6	329.4
Factor Services	..	..	..	1.0	1.0	1.1	9.5	55.3	49.8	18.5	53.8
Imports of Goods & Services	..	..	..	405.0	458.4	526.6	693.8	1,436.0	1,928.1	2,606.9	2,921.3
Merchandise, fob	..	..	..	332.0	381.1	445.1	568.6	1,039.1	1,425.4	2,102.4	2,402.1
Nonfactor Services	..	..	..	70.0	74.2	78.2	119.2	369.2	466.2	479.1	483.8
Factor Services	..	..	..	3.0	3.0	3.3	6.0	27.7	36.4	25.4	35.4
Long-Term Interest	..	..	..	6.2	6.8	6.8	8.4	10.9	17.3	22.5	28.2
Current Transfers, net	..	..	..	7.0	8.0	39.1	37.0	44.5	52.2	53.1	92.5
Workers' Remittances	..	..	..	7.0	8.0	39.1	37.0	44.5	52.2	53.1	92.5
Total to be Financed	..	..	..	..	..	..	..	..	..	..	..
Official Capital Grants	..	..	..	..	..	..	..	..	..	..	..
Current Account Balance	..	..	..	-69.0	-58.2	28.2	338.6	167.2	93.5	-772.4	-232.6
Long-Term Capital, net	..	..	..	13.0	102.3	58.6	388.6	416.1	644.7	474.5	1,371.0
Direct Investment	..	..	..	..	..	..	..	..	..	..	..
Long-Term Loans	..	..	..	29.4	42.9	42.3	56.6	73.6	190.1	319.4	466.5
Disbursements	..	..	..	60.0	75.7	77.6	94.8	135.1	277.2	404.1	545.2
Repayments	..	..	..	30.6	32.8	35.3	38.2	61.5	87.1	84.7	78.7
Other Long-Term Capital	..	..	..	-16.4	59.4	16.3	332.0	342.5	454.6	155.1	904.5
Other Capital, net	..	..	..	38.6	-24.7	-57.8	-399.5	-472.9	-501.6	-55.6	-920.0
Change in Reserves	..	..	..	17.4	-19.5	-29.1	-327.7	-110.4	-236.6	353.5	-218.4
Memo Item:					*(Syrian Pounds per US dollar)*						
Conversion Factor (Annual Avg)	3.490	3.400	3.060	3.180	3.090	3.020	3.040	3.070	3.020	3.240	3.510
EXTERNAL DEBT, ETC.					*(Millions of US dollars, outstanding at end of year)*						
Public/Publicly Guar. Long-Term	..	..	..	232.5	293.4	336.7	418.7	511.9	685.2	1,003.5	1,495.1
Official Creditors	..	..	..	176.3	242.8	275.3	321.5	331.0	468.2	758.0	1,183.0
IBRD and IDA	..	..	..	4.2	6.4	7.1	7.6	8.2	13.0	47.7	102.4
Private Creditors	..	..	..	56.2	50.7	61.4	97.2	180.9	217.0	245.5	312.1
Private Non-guaranteed Long-Term	..	..	..	..	..	..	..	..	..	..	..
Use of Fund Credit	..	..	..	9.5	5.2	24.6	21.6	21.9	0.0	0.0	15.2
Short-Term Debt	..	..	..	..	..	..	..	..	..	..	316.0
Memo Items:					*(Millions of US dollars)*						
Int'l Reserves Excluding Gold	55.0	39.1	31.0	27.0	58.0	105.3	378.7	466.1	701.9	293.4	483.7
Gold Holdings (at market price)	19.1	33.5	28.2	29.9	34.9	51.9	89.8	147.2	110.7	106.3	133.8
SOCIAL INDICATORS											
Total Fertility Rate	7.8	7.8	7.8	7.7	7.7	7.7	7.6	7.6	7.5	7.5	7.4
Crude Birth Rate	47.6	47.4	47.2	47.0	46.8	46.6	46.5	46.4	46.2	46.1	46.0
Infant Mortality Rate	107.0	103.2	99.4	95.6	91.8	88.0	84.4	80.8	77.2	73.6	70.0
Life Expectancy at Birth	54.0	54.6	55.2	55.8	56.4	57.0	57.6	58.2	58.8	59.4	60.1
Food Production, p.c. ('79-81=100)	73.4	65.1	72.2	52.7	55.8	80.6	45.5	78.4	77.7	88.5	79.4
Labor Force, Agriculture (%)	51.4	51.0	50.6	50.2	48.4	46.6	44.7	42.9	41.1	39.3	37.6
Labor Force, Female (%)	10.8	11.1	11.4	11.6	12.0	12.3	12.6	12.9	13.2	13.5	13.8
School Enroll. Ratio, primary	..	..	..	78.0	..	..	..	..	96.0	90.0	98.0
School Enroll. Ratio, secondary	..	..	..	38.0	..	..	..	..	43.0	..	44.0

1978	1979	1980	1981	1982	1983	1984	1985	1986	1987 estimate	NOTES	SYRIAN ARAB REPUBLIC
				(Millions of current US dollars)							**FOREIGN TRADE (CUSTOMS BASIS)**
1,060.3	1,645.1	2,108.0	2,103.0	2,026.0	1,922.8	1,853.5	1,759.1	1,367.7	1,644.0	..	Value of Exports, fob
298.9	333.5	307.2	260.6	313.6	307.8	454.6	325.1	393.7	519.0	..	Nonfuel Primary Products
665.9	1,186.6	1,662.2	1,661.7	1,513.4	1,323.8	1,187.8	1,165.1	570.7	687.5	..	Fuels
95.5	125.0	138.6	180.7	199.0	291.2	211.1	269.0	403.3	437.5	..	Manufactures
2,443.7	3,324.0	4,118.0	5,040.0	4,014.0	4,542.0	4,115.7	3,487.0	2,325.0	2,464.8	f	Value of Imports, cif
571.5	601.8	736.2	924.4	706.5	1,041.8	695.7	546.7	369.4	410.6	..	Nonfuel Primary Products
324.3	822.6	1,070.0	1,744.7	1,511.3	1,370.7	1,206.0	1,041.3	524.0	650.6	..	Fuels
1,547.9	1,899.5	2,311.8	2,370.9	1,796.2	2,129.5	2,214.0	1,899.1	1,431.6	1,403.6	..	Manufactures
				(Index 1980 = 100)							Terms of Trade
71.7	82.1	100.0	105.4	101.1	99.2	97.5	94.6	76.2	84.6	..	Terms of Trade
50.8	66.7	100.0	108.3	97.5	91.6	89.3	84.3	60.7	74.4	..	Export Prices, fob
77.2	87.5	100.0	92.7	82.6	88.7	87.0	71.6	66.5	84.5	..	Nonfuel Primary Products
42.3	61.0	100.0	112.5	101.6	92.5	90.2	87.5	44.9	56.9	..	Fuels
76.1	90.0	100.0	99.1	94.8	90.7	89.3	89.0	103.2	113.0	..	Manufactures
70.9	81.2	100.0	102.8	96.4	92.3	91.6	89.1	79.6	87.9	..	Import Prices, cif
											Trade at Constant 1980 Prices
				(Millions of 1980 US dollars)							
2,087.0	2,465.9	2,108.0	1,940.9	2,078.6	2,099.7	2,076.3	2,087.0	2,252.6	2,209.9	..	Exports, fob
3,449.0	4,091.0	4,118.0	4,901.5	4,165.4	4,922.3	4,494.1	3,911.5	2,919.1	2,802.8	..	Imports, cif
				(Millions of current US dollars)							**BALANCE OF PAYMENTS**
1,410.2	2,117.8	2,567.7	2,821.1	2,580.1	2,708.8	2,566.9	2,440.4	1,901.0	2,207.0	..	Exports of Goods & Services
1,061.1	1,647.7	2,112.1	2,229.8	2,031.8	1,927.9	1,859.4	1,640.2	1,397.0	1,266.0	..	Merchandise, fob
259.1	376.3	364.8	497.6	522.8	765.1	677.5	774.1	485.0	909.0	..	Nonfactor Services
89.9	93.8	90.7	93.8	25.5	15.8	30.1	26.0	19.0	32.0	..	Factor Services
2,824.9	3,687.2	4,610.4	5,496.5	4,656.6	5,262.7	4,946.7	4,747.2	3,148.0	3,822.0	..	Imports of Goods & Services
2,203.5	3,055.1	4,009.6	4,843.3	3,703.2	4,152.1	3,801.3	3,591.6	2,395.0	2,371.0	..	Merchandise, fob
579.4	590.1	521.0	542.4	776.6	963.1	1,018.1	1,030.7	654.0	1,197.0	..	Nonfactor Services
42.0	42.0	79.7	110.8	176.8	147.5	127.4	124.9	99.0	254.0	..	Factor Services
55.0	71.0	76.8	55.2	73.0	72.9	77.1	79.0	86.2	111.8	..	Long-Term Interest
635.9	901.4	773.5	581.7	445.9	461.1	326.9	293.0	783.0	250.0	..	Current Transfers, net
635.9	901.4	773.5	581.7	445.9	461.1	326.9	293.0	..	250.0	..	Workers' Remittances
..	..	..	..	..	..	..	..	..	-1,365.0	..	Total to be Financed
..	..	..	..	..	..	..	..	..	900.0	..	Official Capital Grants
3.6	958.7	250.7	-275.2	-251.2	-815.0	-851.7	-952.3	-464.0	-465.0	..	Current Account Balance
1,141.7	1,701.7	1,495.0	1,866.8	1,371.7	1,586.6	1,526.9	1,240.7	69.0	1,390.0	..	Long-Term Capital, net
..	..	..	..	..	..	..	..	0.0	..	..	Direct Investment
207.3	253.8	126.4	125.7	58.2	89.2	140.3	262.7	260.8	286.9	..	Long-Term Loans
381.7	448.5	347.1	369.3	300.7	320.7	338.9	478.3	470.7	540.3	..	Disbursements
174.4	194.7	220.7	243.6	242.5	231.5	198.6	215.6	209.9	253.4	..	Repayments
934.4	1,447.9	1,368.6	1,741.1	1,313.5	1,497.4	1,386.6	978.0	-191.8	203.1	..	Other Long-Term Capital
-1,280.2	-2,486.8	-2,004.0	-1,573.0	-1,023.4	-902.3	-688.2	-142.5	456.0	-925.2	..	Other Capital, net
135.0	-173.6	258.3	-18.6	-97.0	130.7	13.0	-146.0	-61.0	0.2	..	Change in Reserves
				(Syrian Pounds per US dollar)							Memo Item:
3.490	3.560	3.930	4.640	4.530	4.590	4.700	4.960	5.260	5.270	..	Conversion Factor (Annual Avg)
				(Millions of US dollars, outstanding at end of year)							**EXTERNAL DEBT, ETC.**
1,737.3	1,998.2	2,107.3	2,193.6	2,229.0	2,288.3	2,319.5	2,712.7	3,121.7	3,647.8	..	Public/Publicly Guar. Long-Term
1,414.1	1,679.4	1,768.6	1,856.5	1,903.4	1,906.1	1,912.3	2,185.9	2,519.9	2,952.2	..	Official Creditors
141.8	198.6	257.4	290.7	302.7	334.1	292.6	356.4	437.3	535.8	..	IBRD and IDA
323.2	318.7	338.7	337.1	325.6	382.1	407.2	526.8	601.9	695.5	..	Private Creditors
..	..	..	..	..	..	..	..	..	..	..	Private Non-guaranteed Long-Term
16.3	16.5	0.0	0.0	0.0	0.0	0.0	0.0	0.0	..	..	Use of Fund Credit
269.0	314.0	631.0	799.0	767.0	751.0	622.0	815.0	1,290.0	1,030.0	..	Short-Term Debt
				(Millions of US dollars)							Memo Items:
381.8	580.8	336.5	291.4	198.1	52.2	268.2	83.1	144.3	..	..	Int'l Reserves Excluding Gold
183.3	426.5	491.1	331.1	380.6	317.8	256.8	272.4	325.6	403.3	..	Gold Holdings (at market price)
											SOCIAL INDICATORS
7.4	7.3	7.3	7.2	7.2	7.1	7.0	7.0	6.9	6.8	..	Total Fertility Rate
45.9	45.8	45.7	45.6	45.5	45.5	45.4	45.4	45.3	45.3	..	Crude Birth Rate
67.8	65.6	63.4	61.2	59.0	57.4	55.8	54.2	52.5	50.9	..	Infant Mortality Rate
60.6	61.1	61.6	62.1	62.6	62.9	63.3	63.7	64.1	64.5	..	Life Expectancy at Birth
90.8	84.2	108.0	107.9	108.1	102.0	88.3	91.3	102.8	95.1	..	Food Production, p.c. ('79-81 = 100)
35.8	34.1	32.3	..	..	..	..	..	..	..	..	Labor Force, Agriculture (%)
14.1	14.4	14.6	14.9	15.2	15.4	15.7	16.0	16.2	16.5	..	Labor Force, Female (%)
98.0	99.0	102.0	103.0	103.0	105.0	107.0	108.0	..	..	..	School Enroll. Ratio, primary
46.0	46.0	47.0	49.0	52.0	56.0	59.0	61.0	..	..	..	School Enroll. Ratio, secondary

TANZANIA	1967	1968	1969	1970	1971	1972	1973	1974	1975	1976	1977
CURRENT GNP PER CAPITA (US $)	90	90	100	100	100	110	120	140	160	180	190
POPULATION (thousands)	12,321	12,706	13,104	13,513	13,967	14,437	14,922	15,424	15,942	16,478	17,032
ORIGIN AND USE OF RESOURCES					*(Billions of current Tanzania Shillings)*						
Gross National Product (GNP)	7.25	7.84	8.23	9.15	9.79	11.16	13.11	16.00	18.99	24.32	28.78
Net Factor Income from Abroad	-0.09	-0.03	-0.04	-0.03	-0.02	-0.01	0.01	0.00	-0.03	-0.10	-0.09
Gross Domestic Product (GDP)	7.34	7.87	8.27	9.17	9.81	11.17	13.10	15.99	19.01	24.42	28.87
Indirect Taxes, net	0.61	0.69	0.81	0.96	0.96	1.14	1.61	1.98	2.02	2.77	3.20
GDP at factor cost	6.73	7.18	7.46	8.21	8.86	10.03	11.49	14.01	16.99	21.65	25.67
Agriculture	2.87	2.99	3.08	3.38	3.49	4.02	4.54	5.44	7.01	9.05	11.13
Industry	1.13	1.17	1.27	1.42	1.56	1.80	2.11	2.41	2.76	4.13	4.89
Manufacturing	0.57	0.65	0.74	0.83	0.85	1.14	1.26	1.48	1.77	2.81	3.29
Services, etc.	2.73	3.02	3.10	3.41	3.80	4.21	4.84	6.16	7.22	8.48	9.64
Resource Balance	0.02	-0.20	0.03	-0.19	-0.46	-0.40	-0.80	-2.34	-2.36	-0.54	-0.94
Exports of Goods & NFServices	1.95	1.91	2.04	2.39	2.67	3.00	3.10	3.47	3.77	5.30	5.63
Imports of Goods & NFServices	1.93	2.10	2.02	2.58	3.13	3.40	3.90	5.80	6.14	5.84	6.57
Domestic Absorption	7.32	8.07	8.25	9.36	10.28	11.57	13.90	18.33	21.38	24.96	29.81
Private Consumption, etc.	5.21	5.79	5.97	6.31	6.55	7.73	9.24	12.05	14.10	15.38	17.98
General Gov't Consumption	0.80	0.88	1.00	0.98	1.14	1.40	1.91	2.76	3.28	3.99	4.31
Gross Domestic Investment	1.31	1.40	1.28	2.07	2.59	2.44	2.76	3.52	4.00	5.60	7.52
Fixed Investment	1.23	1.30	1.21	1.88	2.37	2.36	2.60	3.03	3.54	5.16	6.66
Memo Items:											
Gross Domestic Saving	1.33	1.20	1.31	1.88	2.13	2.04	1.96	1.18	1.64	5.05	6.58
Gross National Saving	1.27	1.23	1.33	1.93	2.14	1.93	1.87	1.10	1.70	5.05	6.65
					(Millions of 1980 Tanzania Shillings)						
Gross National Product	26,527	27,980	28,775	30,593	31,508	33,816	35,595	36,348	37,819	39,077	39,234
GDP at factor cost	24,570	25,814	26,256	27,729	28,764	30,732	31,614	32,225	34,184	34,703	34,858
Agriculture	13,857	14,430	14,486	15,030	14,847	16,062	16,216	15,546	16,864	15,979	16,163
Industry	4,175	4,225	4,423	4,627	5,291	5,549	5,684	5,735	5,647	6,235	6,087
Manufacturing	2,310	2,468	2,714	2,891	3,166	3,433	3,585	3,634	3,647	4,292	4,033
Services, etc.	6,537	7,159	7,347	8,072	8,626	9,121	9,715	10,944	11,673	12,489	12,609
Resource Balance	-1,859	-3,236	-2,387	-3,428	-4,364	-3,317	-4,205	-6,278	-4,828	-3,033	-4,574
Exports of Goods & NFServices	6,621	6,526	6,880	7,767	8,453	9,472	8,236	6,068	6,996	7,267	5,698
Imports of Goods & NFServices	8,481	9,763	9,267	11,195	12,817	12,789	12,441	12,346	11,824	10,300	10,271
Domestic Absorption	28,326	31,195	31,137	34,006	35,861	37,127	39,805	42,626	42,639	42,081	43,794
Private Consumption, etc.	..	..	..	..	..	..	..	..	..	26,207	..
General Gov't Consumption	..	..	..	..	..	..	..	..	..	6,465	..
Gross Domestic Investment	5,578	6,023	5,418	8,113	9,690	7,787	8,148	9,182	8,438	9,409	10,850
Fixed Investment	5,057	5,484	4,936	7,146	8,547	7,278	7,442	7,676	7,067	8,443	9,362
Memo Items:											
Capacity to Import	8,573	8,852	9,382	10,383	10,929	11,287	9,886	7,376	7,267	9,341	8,797
Terms of Trade Adjustment	1,952	2,326	2,502	2,616	2,477	1,815	1,650	1,308	271	2,074	3,100
Gross Domestic Income	28,418	30,284	31,252	33,194	33,973	35,625	37,250	37,656	38,081	41,122	42,320
Gross National Income	28,479	30,306	31,276	33,209	33,984	35,631	37,246	37,655	38,090	41,151	42,333
DOMESTIC PRICES (DEFLATORS)					*(Index 1980 = 100)*						
Overall (GDP)	27.7	28.2	28.8	30.0	31.2	33.0	36.8	44.0	50.3	62.5	73.6
Domestic Absorption	25.8	25.9	26.5	27.5	28.7	31.2	34.9	43.0	50.1	59.3	68.1
Agriculture	20.7	20.7	21.3	22.5	23.5	25.0	28.0	35.0	41.6	56.6	68.9
Industry	27.1	27.8	28.8	30.7	29.5	32.5	37.1	42.0	48.8	66.2	80.4
Manufacturing	24.7	26.3	27.3	28.6	26.8	33.3	35.1	40.8	48.6	65.5	81.5
MANUFACTURING ACTIVITY											
Employment (1980=100)	34.4	42.5	42.8	47.8	53.4	61.5	62.7	69.3	..	..	..
Real Earnings per Empl. (1980=100)	162.9	155.8	181.1	177.3	167.1	167.5	180.0	174.9	..	..	..
Real Output per Empl. (1980=100)	149.1	109.5	123.7	122.4	135.9	118.1	135.7	132.2	..	..	..
Earnings as % of Value Added	42.7	44.4	42.9	42.4	41.6	39.4	37.3	42.3	..	..	..
MONETARY HOLDINGS					*(Billions of current Tanzania Shillings)*						
Money Supply, Broadly Defined	1.54	1.81	1.98	2.22	2.62	3.09	3.65	4.46	5.55	6.95	8.35
Money as Means of Payment	1.19	1.30	1.66	1.68	2.06	2.33	2.77	3.46	4.28	5.33	6.38
Currency Ouside Banks	0.51	0.53	0.60	0.82	0.99	1.20	1.20	1.52	1.76	2.07	2.38
Demand Deposits	0.68	0.77	1.05	0.86	1.07	1.13	1.58	1.94	2.53	3.26	4.00
Quasi-Monetary Liabilities	0.35	0.52	0.33	0.54	0.57	0.76	0.88	1.01	1.27	1.61	1.96
GOVERNMENT DEFICIT (-) OR SURPLUS					*(Millions of current Tanzania Shillings)*						
	..	..	..	..	..	-554	-745	-917	-1,891	-1,527	-1,542
Current Revenue	..	..	..	..	..	1,793	2,294	3,177	4,237	4,393	5,906
Current Expenditure	..	..	..	..	..	1,596	2,087	2,869	4,373	4,300	5,222
Current Budget Balance	..	..	..	..	..	197	207	308	-136	93	684
Capital Receipts	..	..	..	..	..	..	..	..	..	..	..
Capital Payments	..	..	..	..	..	751	952	1,225	1,755	1,620	2,226

1978	1979	1980	1981	1982	1983	1984	1985	1986	1987 estimate	NOTES	TANZANIA
220	250	280	300	320	310	290	300	250	180	f	CURRENT GNP PER CAPITA (US $)
17,605	18,172	18,757	19,407	20,080	20,776	21,497	22,242	23,049	23,884	..	POPULATION (thousands)
											ORIGIN AND USE OF RESOURCES
			(Billions of current Tanzania Shillings)								
32.12	36.21	42.01	48.93	58.00	68.14	85.04	111.13	146.19	181.90	f	Gross National Product (GNP)
-0.05	-0.07	-0.11	-0.18	-0.23	-1.86	-2.06	-9.47	-15.01	-43.20	..	Net Factor Income from Abroad
32.17	36.28	42.12	49.10	58.23	70.00	87.10	120.60	161.20	225.10	f	Gross Domestic Product (GDP)
3.60	4.00	4.70	5.20	5.70	7.30	9.00	12.50	18.10	27.00	..	Indirect Taxes, net
28.57	32.28	37.42	43.90	52.53	62.70	78.10	108.10	143.10	198.10	f	GDP at factor cost
12.51	14.73	16.64	20.34	26.45	32.74	41.29	61.23	84.15	120.94	..	Agriculture
5.40	5.66	6.35	6.84	6.91	6.88	8.48	10.05	13.24	15.52	..	Industry
3.86	3.87	4.10	4.50	4.36	4.87	5.93	6.66	7.42	9.04	..	Manufacturing
10.66	11.90	14.43	16.73	19.17	23.08	28.32	36.82	45.70	61.64	..	Services, etc.
-4.87	-4.63	-5.55	-4.17	-5.90	-4.40	-6.70	-12.00	-23.40	-51.60	..	Resource Balance
4.69	5.13	5.54	5.99	4.90	5.40	7.60	6.80	14.10	28.50	..	Exports of Goods & NFServices
9.56	9.76	11.09	10.16	10.80	9.80	14.30	18.80	37.50	80.10	..	Imports of Goods & NFServices
37.04	40.91	47.67	53.27	64.13	74.40	93.80	132.60	184.60	276.70	..	Domestic Absorption
23.36	25.50	32.49	37.03	43.84	55.37	69.42	102.16	145.75	220.90	..	Private Consumption, etc.
5.58	5.96	5.49	6.11	8.05	9.44	10.78	11.44	12.45	17.50	..	General Gov't Consumption
8.09	9.46	9.69	10.13	12.24	9.59	13.60	19.00	26.40	38.30	..	Gross Domestic Investment
7.33	8.59	8.63	8.63	10.82	7.75	12.00	16.90	23.90	34.60	..	Fixed Investment
											Memo Items:
3.22	4.83	4.14	5.96	6.34	5.19	6.90	7.00	3.00	-13.30	..	Gross Domestic Saving
3.35	5.00	4.20	5.97	6.34	3.54	5.79	1.60	-3.84	-41.72	..	Gross National Saving
			(Millions of 1980 Tanzania Shillings)								
39,994	41,079	42,006	41,908	42,152	41,208	42,824	43,928	45,649	47,705	f	Gross National Product
35,494	36,476	37,418	37,260	37,499	36,737	38,001	39,094	40,580	42,184	f	GDP at factor cost
15,894	16,014	16,636	16,800	17,026	17,512	18,215	19,309	20,414	21,313	..	Agriculture
6,022	6,369	6,348	5,862	5,809	4,891	5,200	5,000	5,216	5,434	..	Industry
4,169	4,308	4,097	3,637	3,518	3,211	3,297	3,169	3,040	3,169	..	Manufacturing
13,578	14,092	14,434	14,598	14,664	14,334	14,586	14,786	14,950	15,436	..	Services, etc.
-8,441	-5,805	-5,547	-3,997	-5,193	-3,318	-4,184	-6,248	-7,061	-6,231	..	Resource Balance
5,441	5,547	5,540	6,856	5,251	4,670	4,529	3,917	3,728	4,434	..	Exports of Goods & NFServices
13,883	11,352	11,087	10,852	10,443	7,989	8,713	10,165	10,789	10,666	..	Imports of Goods & NFServices
48,390	46,866	47,665	45,933	47,395	44,647	46,933	50,215	52,689	53,662	..	Domestic Absorption
..	..	32,486	31,634	31,598	32,689	34,329	37,730	41,765	44,221	..	Private Consumption, etc.
..	..	5,494	5,112	5,726	5,502	5,246	4,168	3,502	3,446	..	General Gov't Consumption
10,462	11,474	9,685	9,187	10,072	6,456	7,358	8,317	7,422	5,995	..	Gross Domestic Investment
9,345	10,474	8,630	8,024	8,969	5,319	6,440	7,419	6,587	5,144	..	Fixed Investment
											Memo Items:
6,810	5,969	5,540	6,401	4,738	4,402	4,631	3,6.7	4,057	3,795	..	Capacity to Import
1,369	421	0	-454	-512	-268	102	-240	329	-639	..	Terms of Trade Adjustment
41,318	41,482	42,118	41,482	41,690	41,060	42,850	43,727	45,957	46,791	..	Gross Domestic Income
41,362	41,500	42,006	41,453	41,640	40,939	42,926	43,688	45,978	47,065	..	Gross National Income
											DOMESTIC PRICES (DEFLATORS)
			(Index 1980 = 100)								
80.5	88.4	100.0	117.1	138.0	169.4	203.8	274.3	353.3	474.6	..	Overall (GDP)
76.5	87.3	100.0	116.0	135.3	166.6	199.9	264.1	350.4	515.6	..	Domestic Absorption
78.7	92.0	100.0	121.1	155.3	186.9	226.7	317.1	412.2	567.4	..	Agriculture
89.7	88.8	100.0	116.6	119.0	140.8	163.1	201.0	253.9	285.7	..	Industry
92.6	89.8	100.0	123.7	124.0	151.6	179.9	210.3	244.0	285.4	..	Manufacturing
											MANUFACTURING ACTIVITY
95.4	96.4	100.0	103.3	99.7	102.6	104.9	107.2	..	..	..	Employment (1980=100)
116.7	121.0	100.0	83.0	69.4	61.4	52.7	45.0	..	..	..	Real Earnings per Empl. (1980=100)
77.1	109.5	100.0	81.6	86.1	77.4	78.3	73.6	..	..	..	Real Output per Empl. (1980=100)
33.0	29.7	32.8	34.1	33.1	34.5	34.0	34.0	..	..	..	Earnings as % of Value Added
											MONETARY HOLDINGS
			(Billions of current Tanzania Shillings)								
9.40	13.81	17.52	20.69	24.73	29.13	30.20	39.39	50.31	66.14	..	Money Supply, Broadly Defined
6.83	10.43	13.35	15.40	18.32	20.56	20.61	25.51	35.81	47.13	..	Money as Means of Payment
2.92	4.06	5.25	6.61	7.99	8.19	10.47	12.72	18.31	24.55	..	Currency Ouside Banks
3.91	6.38	8.10	8.78	10.33	12.37	10.13	12.79	17.50	22.58	..	Demand Deposits
2.57	3.37	4.17	5.29	6.41	8.56	9.59	13.88	14.50	19.01	..	Quasi-Monetary Liabilities
											GOVERNMENT DEFICIT (-) OR SURPLUS
			(Millions of current Tanzania Shillings)								
-2,365	-5,376	-3,537	-3,270	..	..	..	..	..	..	C	GOVERNMENT DEFICIT (-) OR SURPLUS
6,790	7,381	8,539	10,311	..	..	..	..	..	..	..	Current Revenue
5,871	8,214	7,542	9,007	..	..	..	..	..	..	..	Current Expenditure
919	-833	997	1,304	..	..	..	..	..	..	..	Current Budget Balance
..	..	..	..	..	..	..	..	..	..	..	Capital Receipts
3,284	4,543	4,534	4,574	..	..	..	..	..	..	..	Capital Payments

TANZANIA	1967	1968	1969	1970	1971	1972	1973	1974	1975	1976	1977
FOREIGN TRADE (CUSTOMS BASIS)					*(Millions of current US dollars)*						
Value of Exports, fob	216.9	221.2	232.7	238.3	250.9	300.1	343.5	387.8	348.6	490.4	559.0
Nonfuel Primary Products	158.4	170.9	184.6	190.3	184.4	231.7	265.6	292.4	283.1	412.0	491.3
Fuels	19.2	23.3	15.2	16.2	21.6	30.8	26.5	40.8	19.6	21.3	17.9
Manufactures	39.2	26.9	33.0	31.8	44.9	37.6	51.4	54.6	45.9	57.1	49.9
Value of Imports, cif	182.1	214.4	198.6	271.5	337.9	363.4	447.4	760.1	718.2	645.3	730.1
Nonfuel Primary Products	23.4	19.7	21.2	24.4	30.7	47.0	47.7	176.1	159.8	98.1	105.9
Fuels	16.4	19.1	20.1	23.2	30.1	34.0	48.3	140.7	77.4	120.0	96.9
Manufactures	142.3	175.6	157.4	223.9	277.0	282.4	351.3	443.3	481.0	427.2	527.3
Terms of Trade	97.5	93.0	112.3	106.5	98.8	95.3	118.8	107.9	92.8	126.3	140.3
					(Index 1980 = 100)						
Export Prices, fob	19.7	18.6	22.1	23.0	24.1	25.2	38.6	61.0	56.8	72.4	89.8
Nonfuel Primary Products	30.6	30.8	31.0	33.3	34.8	38.6	54.6	66.0	57.8	76.8	96.3
Fuels	4.3	4.3	4.3	4.3	5.6	6.2	8.9	36.7	35.7	38.4	42.0
Manufactures	30.1	31.7	32.4	37.5	36.5	38.5	49.5	67.2	66.1	66.7	71.7
Import Prices, cif	20.2	20.0	19.7	21.6	24.4	26.4	32.5	56.6	61.2	57.3	64.0
Trade at Constant 1980 Prices					*(Millions of 1980 US dollars)*						
Exports, fob	1,099.3	1,187.6	1,052.4	1,035.4	1,041.5	1,192.1	889.1	635.5	614.3	677.3	622.4
Imports, cif	900.1	1,070.6	1,009.1	1,256.5	1,385.9	1,376.1	1,375.2	1,343.5	1,173.6	1,125.4	1,140.4
BALANCE OF PAYMENTS					*(Millions of current US dollars)*						
Exports of Goods & Services	..	..	..	321.8	349.6	411.7	455.8	488.4	491.2	633.2	656.4
Merchandise, fob	..	..	..	245.9	262.0	316.2	363.6	399.2	372.9	490.4	538.5
Nonfactor Services	..	..	..	65.2	78.0	85.4	72.1	71.9	109.4	135.4	106.8
Factor Services	..	..	..	10.7	9.6	10.1	20.0	17.3	9.0	7.4	11.1
Imports of Goods & Services	..	..	..	370.2	455.3	473.3	568.2	822.8	823.6	722.2	842.5
Merchandise, fob	..	..	..	283.5	345.3	359.8	437.8	660.4	670.0	555.6	646.7
Nonfactor Services	..	..	..	72.5	97.1	101.6	111.6	145.3	141.4	141.0	163.9
Factor Services	..	..	..	14.2	12.8	11.8	18.8	17.2	12.1	25.7	31.8
Long-Term Interest	..	..	..	7.5	7.7	8.2	10.9	12.0	14.6	16.0	21.6
Current Transfers, net	..	..	..	..	..	..	..	..	..	..	..
Workers' Remittances	..	..	..	0.0	0.0	0.0	0.0	0.0	0.0	0.0	0.0
Total to be Financed	..	..	..	..	..	..	..	..	..	..	..
Official Capital Grants	..	..	..	..	..	..	..	..	..	..	..
Current Account Balance	..	..	..	-35.6	-99.8	-65.7	-107.5	-285.3	-230.0	-33.7	-70.2
Long-Term Capital, net	..	..	..	73.3	140.0	118.8	158.8	185.3	297.2	146.3	222.9
Direct Investment	..	..	..	..	..	..	..	..	..	..	..
Long-Term Loans	..	..	..	45.3	49.0	72.1	88.1	136.9	222.2	354.4	221.2
Disbursements	..	..	..	58.1	63.4	118.8	120.4	164.4	254.2	382.8	257.8
Repayments	..	..	..	12.8	14.4	46.7	32.3	27.5	32.0	28.4	36.6
Other Long-Term Capital	..	..	..	28.0	91.0	46.7	70.7	48.4	75.0	-208.1	1.7
Other Capital, net	..	..	..	-58.3	-53.5	-3.1	-19.3	-37.5	-82.0	-90.2	7.5
Change in Reserves	..	..	..	20.6	13.3	-50.0	-32.0	137.5	14.7	-22.3	-160.1
Memo Item:					*(Tanzania Shillings per US dollar)*						
Conversion Factor (Annual Avg)	7.140	7.140	7.140	7.140	7.140	7.140	7.020	7.140	7.370	8.380	8.290
EXTERNAL DEBT, ETC.					*(Millions of US dollars, outstanding at end of year)*						
Public/Publicly Guar. Long-Term	..	..	..	250.3	306.7	371.8	464.1	630.5	817.2	1,184.5	1,446.0
Official Creditors	..	..	..	152.9	202.7	286.6	395.1	566.8	755.5	1,112.9	1,308.0
IBRD and IDA	..	..	..	38.3	53.6	73.1	90.5	104.2	161.2	199.0	258.4
Private Creditors	..	..	..	97.4	104.0	85.2	69.0	63.7	61.7	71.5	138.0
Private Non-guaranteed Long-Term	..	..	..	15.0	20.0	30.0	45.5	51.1	54.0	60.0	60.0
Use of Fund Credit	..	..	..	0.0	0.0	0.0	0.0	60.4	73.3	109.3	118.1
Short-Term Debt	..	..	..	..	..	..	..	..	..	..	159.0
Memo Items:					*(Thousands of US dollars)*						
Int'l Reserves Excluding Gold	61,710	77,540	80,260	64,970	60,253	119,590	144,620	50,232	65,446	112,270	281,830
Gold Holdings (at market price)	..	..	..	..	..	..	..	..	..	..	..
SOCIAL INDICATORS											
Total Fertility Rate	6.6	6.5	6.5	6.4	6.4	6.3	6.3	6.4	6.4	6.5	6.5
Crude Birth Rate	49.0	49.1	49.1	49.2	49.2	49.3	49.4	49.5	49.5	49.6	49.7
Infant Mortality Rate	135.0	134.0	133.0	132.0	131.0	130.0	129.0	128.0	127.0	126.0	125.0
Life Expectancy at Birth	43.8	44.3	44.7	45.1	45.6	46.0	46.5	46.9	47.4	47.8	48.3
Food Production, p.c. ('79-81 = 100)	89.8	87.9	86.2	91.3	85.9	83.9	84.9	84.4	92.7	100.0	98.4
Labor Force, Agriculture (%)	..	..	..	..	..	..	..	..	..	..	..
Labor Force, Female (%)	50.7	50.7	50.6	50.6	50.5	50.4	50.3	50.3	50.2	50.1	50.0
School Enroll. Ratio, primary	..	..	..	34.0	..	..	..	..	53.0	70.0	99.0
School Enroll. Ratio, secondary	..	..	..	3.0	..	..	..	..	3.0	4.0	4.0

1978	1979	1980	1981	1982	1983	1984	1985	1986	1987 estimate	NOTES	TANZANIA
											FOREIGN TRADE (CUSTOMS BASIS)
				(Millions of current US dollars)							
472.3	511.4	536.6	564.4	413.0	379.0	339.6	254.9	347.6	347.5	f	Value of Exports, fob
402.9	397.5	427.9	492.4	350.2	331.0	284.6	205.2	287.3	282.0	..	Nonfuel Primary Products
10.8	16.9	26.1	1.0	5.9	6.0	6.0	4.7	2.3	2.9	..	Fuels
58.6	97.0	82.6	70.9	56.9	42.0	49.0	45.0	58.0	62.6	..	Manufactures
1,141.3	1,076.7	1,226.6	1,176.3	1,134.0	795.0	847.3	1,028.0	1,047.5	1,164.6	f	Value of Imports, cif
121.5	83.3	197.1	102.4	118.3	86.1	81.7	88.3	62.6	98.2	..	Nonfuel Primary Products
126.3	149.1	257.2	362.2	297.3	209.7	209.9	246.4	121.7	196.4	..	Fuels
893.5	844.3	772.3	711.7	718.4	499.2	555.6	693.3	863.2	870.0	..	Manufactures
				(Index 1980 = 100)							
115.1	114.9	100.0	85.2	88.4	91.1	96.4	90.5	103.0	89.6	f	Terms of Trade
83.8	97.7	100.0	88.5	86.1	85.9	89.3	82.5	95.8	88.2	f	Export Prices, fob
87.4	102.5	100.0	87.1	84.7	85.3	89.3	81.1	95.3	84.5	..	Nonfuel Primary Products
42.3	61.0	100.0	112.5	101.6	92.5	90.2	87.5	44.9	56.9	..	Fuels
76.1	90.0	100.0	99.1	94.8	90.7	89.3	89.0	103.2	113.0	..	Manufactures
72.8	85.1	100.0	103.8	97.5	94.3	92.7	91.2	93.0	98.4	f	Import Prices, cif
											Trade at Constant 1980 Prices
				(Millions of 1980 US dollars)							
563.5	523.2	536.6	637.8	479.5	441.0	380.2	308.9	362.8	394.1	f	Exports, fob
1,566.7	1,265.5	1,226.6	1,132.7	1,163.5	842.8	914.3	1,127.4	1,126.4	1,183.8	f	Imports, cif
											BALANCE OF PAYMENTS
				(Millions of current US dollars)							
625.3	697.2	684.4	758.8	530.3	486.7	495.7	394.5	444.6	448.6	..	Exports of Goods & Services
476.0	545.7	505.3	563.0	413.0	378.6	388.3	285.6	347.6	347.0	..	Merchandise, fob
130.2	140.0	165.1	184.9	114.5	106.8	105.8	106.4	84.8	96.1	..	Nonfactor Services
19.1	11.6	14.0	10.9	2.8	1.3	1.6	2.5	12.2	5.5	..	Factor Services
1,262.6	1,218.5	1,312.2	1,287.6	1,197.6	909.9	957.8	1,098.7	1,173.2	1,284.0	..	Imports of Goods & Services
992.5	960.7	1,096.2	1,045.0	1,001.5	733.0	786.6	899.3	942.8	1,035.0	..	Merchandise, fob
245.3	237.6	180.5	209.6	162.1	147.0	148.7	178.7	204.0	212.0	..	Nonfactor Services
24.8	20.2	35.5	33.0	34.0	29.9	22.5	20.7	26.4	37.0	..	Factor Services
25.0	32.0	42.0	39.3	39.3	34.5	30.8	23.5	30.0	38.2	..	Long-Term Interest
..	..	21.8	22.6	25.4	18.8	62.1	233.4	249.9	230.0	..	Current Transfers, net
0.0	0.0	0.0	0.0	0.0	0.0	0.0	0.0	0.0	..	..	Workers' Remittances
..	..	-606.0	-506.2	-641.9	-404.4	-400.0	-470.8	-478.7	-605.4	..	Total to be Financed
..	..	104.5	201.9	185.4	145.0	145.6	133.4	363.0	477.0	..	Official Capital Grants
-472.0	-345.2	-501.5	-304.3	-456.5	-259.4	-254.4	-337.4	-115.7	-128.4	..	Current Account Balance
294.4	388.4	411.0	404.2	439.7	415.9	350.1	241.0	364.7	411.0	..	Long-Term Capital, net
..	..	..	..	..	..	..	..	..	..	..	Direct Investment
219.6	297.3	370.8	293.5	339.4	388.9	187.6	131.2	149.7	61.9	..	Long-Term Loans
255.9	347.9	426.0	357.8	383.8	431.1	225.6	172.8	198.3	109.8	..	Disbursements
36.3	50.6	55.2	64.3	44.4	42.2	38.0	41.6	48.6	47.9	..	Repayments
74.8	91.1	-64.3	-91.2	-85.1	-118.0	16.9	-23.6	-148.0	-127.9	..	Other Long-Term Capital
6.4	-100.2	-233.3	-307.3	-387.8	-177.9	-236.3	-355.4	-540.2	-602.2	..	Other Capital, net
171.2	57.0	323.8	207.4	404.6	21.4	140.6	451.8	291.2	319.6	..	Change in Reserves
											Memo Item:
				(Tanzania Shillings per US dollar)							
7.710	8.220	8.200	8.280	9.280	11.140	15.290	17.470	32.700	64.260	..	Conversion Factor (Annual Avg)
											EXTERNAL DEBT, ETC.
			(Millions of US dollars, outstanding at end of year)								
1,574.8	1,731.7	2,044.1	2,209.5	2,404.9	2,749.5	2,711.2	3,057.1	3,641.9	4,067.8	..	Public/Publicly Guar. Long-Term
1,366.9	1,432.3	1,588.7	1,755.6	1,896.1	2,247.3	2,216.3	2,480.3	3,240.3	3,638.2	..	Official Creditors
307.8	376.0	439.8	528.9	625.5	698.5	724.1	833.9	961.3	1,125.6	..	IBRD and IDA
207.9	299.4	455.4	453.9	508.7	502.2	494.9	576.8	401.6	429.7	..	Private Creditors
62.4	68.7	93.1	74.1	76.8	73.3	74.3	18.6	11.4	11.2	..	Private Non-guaranteed Long-Term
89.2	117.8	119.2	98.6	81.1	50.7	23.6	21.1	45.3	64.5	..	Use of Fund Credit
253.0	250.0	307.2	307.9	426.7	531.4	663.5	782.7	367.9	191.8	..	Short-Term Debt
											Memo Items:
				(Thousands of US dollars)							
99,860	68,041	20,300	18,828	4,822	19,415	26,869	16,000	61,090	31,799	..	Int'l Reserves Excluding Gold
..	..	..	..	..	..	..	..	..	..	..	Gold Holdings (at market price)
											SOCIAL INDICATORS
6.6	6.7	6.8	6.9	7.0	7.0	7.0	7.0	7.0	7.0	..	Total Fertility Rate
49.8	49.9	50.1	50.2	50.3	50.2	50.1	50.0	49.9	49.8	..	Crude Birth Rate
123.0	121.0	119.0	117.0	115.0	110.7	106.5	102.2	97.9	93.7	..	Infant Mortality Rate
48.8	49.2	49.7	50.1	50.6	51.1	51.6	52.1	52.6	53.2	..	Life Expectancy at Birth
99.3	100.5	100.1	99.5	93.7	93.7	94.7	90.6	90.0	89.8	..	Food Production, p.c. ('79-81 = 100)
..	..	..	..	..	..	..	..	..	..	..	Labor Force, Agriculture (%)
50.0	49.9	49.8	49.7	49.5	49.3	49.1	48.9	48.7	48.5	..	Labor Force, Female (%)
96.0	100.0	93.0	94.0	90.0	87.0	83.0	72.0	..	..	..	School Enroll. Ratio, primary
4.0	4.0	3.0	3.0	3.0	3.0	3.0	3.0	..	..	..	School Enroll. Ratio, secondary

THAILAND	1967	1968	1969	1970	1971	1972	1973	1974	1975	1976	1977
CURRENT GNP PER CAPITA (US $)	170	190	200	210	210	210	250	300	360	410	460
POPULATION (thousands)	33,019	34,109	35,236	36,370	37,400	38,420	39,410	40,400	41,388	42,385	43,400

ORIGIN AND USE OF RESOURCES

(Billions of current Thai Baht)

	1967	1968	1969	1970	1971	1972	1973	1974	1975	1976	1977
Gross National Product (GNP)	117	127	139	148	153	170	221	279	303	346	402
Net Factor Income from Abroad	0	0	0	0	0	-1	-1	0	0	-1	-1
Gross Domestic Product (GDP)	117	127	139	147	153	170	222	279	303	347	404
Indirect Taxes, net	13	15	16	16	17	18	23	33	33	36	45
GDP at factor cost	104	112	124	131	137	152	199	246	271	310	359
Agriculture	34	36	40	38	37	43	62	75	82	93	100
Industry	29	32	35	37	42	46	60	74	78	96	118
Manufacturing	18	19	22	24	27	31	43	54	57	68	81
Services, etc.	54	59	65	72	75	81	101	129	144	158	185
Resource Balance	-2	-5	-5	-7	-4	-2	-3	-7	-14	-9	-22
Exports of Goods & NFServices	21	21	22	22	25	31	41	60	56	70	81
Imports of Goods & NFServices	23	26	27	29	29	33	45	67	70	79	102
Domestic Absorption	120	131	145	154	158	172	225	286	317	355	425
Private Consumption, etc.	81	87	94	100	103	116	144	185	205	234	274
General Gov't Consumption	11	14	15	17	18	19	22	26	31	38	43
Gross Domestic Investment	27	31	36	38	37	37	60	74	81	83	109
Fixed Investment	27	29	33	35	36	39	50	65	69	79	105
Memo Items:											
Gross Domestic Saving	25	27	31	31	33	35	57	68	67	75	87
Gross National Saving	25	27	31	32	33	35	58	72	68	74	86

(Billions of 1980 Thai Baht)

	1967	1968	1969	1970	1971	1972	1973	1974	1975	1976	1977
Gross National Product	270.29	293.60	316.49	346.37	361.59	374.11	411.43	432.39	451.92	492.63	539.53
GDP at factor cost	240.43	259.60	280.82	307.82	323.72	335.98	370.27	380.72	403.61	442.69	481.88
Agriculture	84.89	93.08	100.19	104.16	108.61	106.63	116.78	120.24	125.44	133.11	136.08
Industry	67.43	71.90	78.49	86.21	93.91	102.05	112.61	117.45	122.94	142.96	165.39
Manufacturing	38.17	40.93	46.22	53.59	59.40	67.37	77.91	82.86	87.60	100.94	115.58
Services, etc.	117.57	127.82	137.41	154.84	159.83	167.74	183.97	193.55	203.54	218.11	240.37
Resource Balance	-27.67	-40.96	-48.66	-44.42	-18.38	-17.86	-46.93	-32.87	-34.87	-23.13	-36.85
Exports of Goods & NFServices	55.85	56.11	56.92	63.99	75.71	88.28	84.28	90.85	86.56	107.42	119.42
Imports of Goods & NFServices	83.52	97.07	105.58	108.41	94.09	106.13	131.20	123.72	121.44	130.55	156.27
Domestic Absorption	297.56	333.76	364.76	389.63	380.74	394.28	460.28	464.10	486.79	517.30	578.69
Private Consumption, etc.	203.70	226.91	240.10	255.38	247.79	266.69	296.96	305.61	319.71	341.56	366.57
General Gov't Consumption	23.83	28.73	31.28	34.83	36.45	37.66	40.50	40.30	45.96	54.87	60.54
Gross Domestic Investment	70.03	78.12	93.38	99.43	96.50	89.92	122.82	118.19	121.12	120.87	151.58
Fixed Investment	62.60	67.97	76.24	89.42	89.89	90.60	99.75	99.51	98.81	111.72	138.47
Memo Items:											
Capacity to Import	75.08	79.31	84.79	83.77	79.76	100.60	121.77	111.51	97.04	116.29	122.85
Terms of Trade Adjustment	19.23	23.19	27.87	19.78	4.06	12.32	37.49	20.66	10.48	8.87	3.43
Gross Domestic Income	289.13	316.00	343.97	364.99	366.41	388.75	450.85	451.90	462.40	503.04	545.27
Gross National Income	289.53	316.80	344.37	366.15	365.64	386.43	448.92	453.05	462.40	501.49	542.96

DOMESTIC PRICES (DEFLATORS)

(Index 1980 = 100)

	1967	1968	1969	1970	1971	1972	1973	1974	1975	1976	1977
Overall (GDP)	43.5	43.2	44.1	42.7	42.3	45.2	53.7	64.7	67.1	70.1	74.5
Domestic Absorption	40.2	39.3	39.6	39.5	41.4	43.6	48.9	61.6	65.2	68.6	73.5
Agriculture	40.4	39.0	39.9	36.7	33.8	40.4	52.7	62.7	65.0	69.5	73.5
Industry	43.4	43.8	44.2	43.3	44.2	45.5	52.8	63.3	63.6	66.9	71.6
Manufacturing	46.9	47.1	46.9	43.8	45.3	46.5	54.7	64.6	64.6	67.6	70.4

MANUFACTURING ACTIVITY

	1967	1968	1969	1970	1971	1972	1973	1974	1975	1976	1977
Employment (1980 = 100)	..	..	..	54.0	52.9	54.1	63.2	80.8	79.7	83.4	88.6
Real Earnings per Empl. (1980 = 100)	..	..	..	83.1	96.3	100.9	96.1	85.1	87.8	94.7	97.4
Real Output per Empl. (1980 = 100)	..	..	..	70.4	77.5	83.2	81.6	71.5	76.3	84.3	88.8
Earnings as % of Value Added	..	..	..	24.5	24.6	24.7	24.7	24.8	24.7	24.7	24.6

MONETARY HOLDINGS

(Billions of current Thai Baht)

	1967	1968	1969	1970	1971	1972	1973	1974	1975	1976	1977
Money Supply, Broadly Defined	32.2	36.8	41.1	46.5	54.0	67.0	82.4	99.1	114.7	135.7	162.9
Money as Means of Payment	15.6	17.2	17.9	19.4	21.3	24.8	30.0	32.7	34.7	41.4	45.4
Currency Ouside Banks	9.8	10.7	11.0	11.9	13.1	15.3	18.7	20.5	22.3	25.8	28.7
Demand Deposits	5.8	6.5	6.9	7.4	8.2	9.5	11.2	12.2	12.4	15.5	16.8
Quasi-Monetary Liabilities	16.6	19.6	23.2	27.1	32.7	42.2	52.5	66.4	80.0	94.3	117.5

GOVERNMENT DEFICIT (-) OR SURPLUS

(Billions of current Thai Baht)

	1967	1968	1969	1970	1971	1972	1973	1974	1975	1976	1977
	..	..	..	..	..	-7.1	-7.0	2.5	-6.3	-13.9	-13.1
Current Revenue	..	..	..	..	..	22.1	26.4	37.8	38.8	42.4	52.1
Current Expenditure	..	..	..	..	..	20.2	25.5	27.8	35.0	40.8	47.1
Current Budget Balance	..	..	..	..	..	1.9	0.9	10.0	3.9	1.6	5.0
Capital Receipts	..	..	..	..	..	0.0	..	0.0	0.0	0.5	0.0
Capital Payments	..	..	..	..	..	9.0	7.9	7.5	10.1	16.0	18.1

1978	1979	1980	1981	1982	1983	1984	1985	1986	1987 estimate	NOTES	THAILAND
530	590	670	750	780	810	840	810	800	840	..	**CURRENT GNP PER CAPITA (US $)**
44,500	45,500	46,700	47,659	48,639	49,639	50,660	51,683	52,642	53,535	..	**POPULATION (thousands)**
				(Billions of current Thai Baht)							**ORIGIN AND USE OF RESOURCES**
485	553	653	748	807	903	962	997	1,077	1,199	..	Gross National Product (GNP)
-4	-6	-5	-12	-13	-7	-11	-18	-22	-24	..	Net Factor Income from Abroad
488	559	659	760	820	910	973	1,014	1,100	1,223	f	Gross Domestic Product (GDP)
53	65	76	85	87	105	116	114	128	151	..	Indirect Taxes, net
435	494	582	675	733	806	858	901	971	1,073	B f	GDP at factor cost
120	134	153	163	157	186	175	170	185	195	..	Agriculture
144	170	203	240	259	286	326	345	372	426	..	Industry
98	118	140	170	176	194	218	225	254	295	..	Manufacturing
224	255	303	357	405	439	473	499	543	602	..	Services, etc.
-21	-38	-42	-48	-14	-66	-42	-29	23	1	..	Resource Balance
97	126	160	181	193	185	216	245	290	355	..	Exports of Goods & NFServices
118	164	201	229	207	251	259	274	267	354	..	Imports of Goods & NFServices
509	596	700	808	834	976	1,016	1,043	1,076	1,222	..	Domestic Absorption
317	378	445	511	535	621	643	656	685	776	..	Private Consumption, etc.
55	67	81	97	110	119	130	143	145	154	..	General Gov't Consumption
138	152	174	200	190	236	243	244	247	292	..	Gross Domestic Investment
123	143	166	188	192	219	239	240	235	276	..	Fixed Investment
											Memo Items:
117	115	133	152	175	170	200	215	270	293	..	Gross Domestic Saving
113	109	129	141	164	167	190	199	249	271	..	Gross National Saving
				(Billions of 1980 Thai Baht)							
592.10	618.38	653.10	687.03	714.45	775.81	824.15	845.68	879.74	938.32	..	Gross National Product
534.40	556.66	582.30	621.67	650.29	690.24	735.85	767.68	798.97	846.81	B f	GDP at factor cost
153.15	150.18	152.90	161.06	166.01	173.44	183.08	194.22	196.94	191.99	..	Agriculture
184.09	195.96	203.00	214.66	221.25	239.29	259.74	259.30	276.24	303.73	..	Industry
125.48	136.03	139.90	148.72	152.38	165.30	176.49	175.41	191.34	210.93	..	Manufacturing
261.81	283.04	302.60	324.04	340.69	367.32	392.51	411.44	431.63	470.76	..	Services, etc.
-32.42	-53.25	-41.50	-22.54	20.25	-35.86	-7.81	18.99	46.07	29.80	..	Resource Balance
134.27	148.27	159.70	173.13	193.13	182.27	214.55	237.12	271.69	315.97	..	Exports of Goods & NFServices
166.69	201.53	201.20	195.67	172.88	218.13	222.36	218.13	225.62	286.17	..	Imports of Goods & NFServices
631.46	682.43	700.00	722.29	707.70	815.91	843.14	845.97	858.74	936.68	..	Domestic Absorption
383.41	431.94	444.60	438.41	441.30	503.71	509.28	514.40	537.93	590.30	..	Private Consumption, etc.
68.44	79.17	81.40	92.33	95.57	102.86	110.15	117.85	118.05	124.12	..	General Gov't Consumption
179.61	171.32	174.00	191.55	170.83	209.34	223.71	213.72	202.76	222.25	..	Gross Domestic Investment
153.73	159.83	165.70	173.21	169.92	191.75	212.64	201.37	190.34	208.18	..	Fixed Investment
											Memo Items:
137.52	155.36	159.70	154.91	160.87	160.82	186.08	195.21	245.13	287.06	..	Capacity to Import
3.24	7.09	0.00	-18.22	-32.26	-21.45	-28.48	-41.91	-26.56	-28.91	..	Terms of Trade Adjustment
602.28	636.27	658.50	681.54	695.69	758.60	806.86	823.05	878.25	937.57	..	Gross Domestic Income
595.34	625.47	653.10	668.81	682.19	754.35	795.67	803.77	853.18	909.41	..	Gross National Income
				(Index 1980 = 100)							**DOMESTIC PRICES (DEFLATORS)**
81.5	88.8	100.0	108.6	112.6	116.7	116.5	117.3	121.5	126.6	..	Overall (GDP)
80.6	87.4	100.0	111.9	117.9	119.6	120.5	123.3	125.3	130.5	..	Domestic Absorption
78.1	89.3	100.0	101.2	94.5	107.0	95.7	87.5	93.8	101.6	..	Agriculture
78.4	86.5	100.0	111.9	116.9	119.4	125.4	133.1	134.7	140.2	..	Industry
77.9	86.5	100.0	114.0	115.8	117.5	123.6	128.0	132.5	139.6	..	Manufacturing
											MANUFACTURING ACTIVITY
94.2	98.5	100.0	99.1	85.5	79.4	77.9	76.5	..	..	..	Employment (1980=100)
99.8	99.9	100.0	105.6	120.4	134.5	150.6	159.6	..	..	..	Real Earnings per Empl. (1980=100)
89.8	97.7	100.0	107.1	126.2	145.5	159.0	163.3	..	..	..	Real Output per Empl. (1980=100)
24.5	23.2	23.2	23.4	23.5	23.6	23.5	23.5	..	..	..	Earnings as % of Value Added
				(Billions of current Thai Baht)							**MONETARY HOLDINGS**
192.0	220.3	271.0	311.6	387.3	480.9	583.4	650.3	733.9	868.2	D	Money Supply, Broadly Defined
54.5	63.5	71.4	73.3	78.3	81.8	93.3	90.1	105.9	132.4	..	Money as Means of Payment
33.2	40.8	45.9	47.8	54.0	59.6	63.5	64.0	71.1	86.7	..	Currency Ouside Banks
21.3	22.7	25.6	25.5	24.3	22.2	29.8	26.1	34.8	45.7	..	Demand Deposits
137.5	156.8	199.5	238.3	309.0	399.1	490.0	560.2	628.1	735.8	..	Quasi-Monetary Liabilities
				(Billions of current Thai Baht)							
-17.8	-20.5	-32.3	-25.7	-53.8	-36.6	-34.0	-55.8	..	..	C	**GOVERNMENT DEFICIT (-) OR SURPLUS**
63.2	76.7	96.7	117.3	120.6	145.8	157.0	166.4	..	..	..	Current Revenue
59.5	74.6	95.6	112.3	131.4	144.1	157.2	175.3	..	..	..	Current Expenditure
3.7	2.1	1.1	5.1	-10.8	1.7	-0.2	-8.9	..	..	..	Current Budget Balance
0.0	0.0	0.0	0.0	0.0	0.0	0.0	0.0	..	..	..	Capital Receipts
21.4	22.6	33.4	30.7	43.0	38.4	33.8	46.9	..	..	..	Capital Payments

THAILAND	1967	1968	1969	1970	1971	1972	1973	1974	1975	1976	1977
FOREIGN TRADE (CUSTOMS BASIS)				*(Millions of current US dollars)*							
Value of Exports, fob	680	654	702	710	827	1,067	1,566	2,449	2,195	2,978	3,490
Nonfuel Primary Products	637	596	629	629	706	864	1,202	2,006	1,785	2,399	2,804
Fuels	7	4	4	5	7	13	21	21	13	7	3
Manufactures	37	54	69	76	113	190	343	422	397	572	683
Value of Imports, cif	1,060	1,150	1,286	1,293	1,287	1,484	2,073	3,156	3,279	3,572	4,613
Nonfuel Primary Products	118	129	154	162	188	223	316	421	402	484	664
Fuels	76	96	88	113	134	152	233	625	708	832	1,026
Manufactures	866	925	1,044	1,019	965	1,110	1,524	2,109	2,170	2,256	2,923
Terms of Trade	135.8	145.9	143.6	141.3	127.6	122.1	143.9	135.1	117.0	110.0	109.0
				(Index 1980 = 100)							
Export Prices, fob	29.4	29.9	32.3	30.3	29.5	31.4	46.2	70.3	63.2	61.2	66.2
Nonfuel Primary Products	31.6	31.7	34.7	32.0	31.1	34.1	55.6	75.0	66.8	63.9	66.2
Fuels	4.3	4.3	4.3	4.3	5.6	6.2	8.9	36.7	35.7	38.4	42.0
Manufactures	25.9	24.4	26.4	29.0	28.5	29.0	34.7	56.0	51.8	52.3	66.3
Import Prices, cif	21.6	20.5	22.5	21.4	23.1	25.7	32.1	52.0	54.0	55.7	60.7
Trade at Constant 1980 Prices				*(Millions of 1980 US dollars)*							
Exports, fob	2,315	2,189	2,173	2,347	2,802	3,403	3,388	3,484	3,476	4,867	5,274
Imports, cif	4,899	5,619	5,718	6,041	5,565	5,777	6,456	6,064	6,074	6,417	7,595
BALANCE OF PAYMENTS				*(Millions of current US dollars)*							
Exports of Goods & Services	..	..	..	1,171	1,278	1,589	2,133	3,173	2,971	3,621	4,179
Merchandise, fob	..	..	..	686	802	1,046	1,515	2,405	2,177	2,959	3,456
Nonfactor Services	..	..	..	406	407	485	547	624	603	508	530
Factor Services	..	..	..	79	68	59	70	143	191	154	193
Imports of Goods & Services	..	..	..	1,470	1,496	1,700	2,323	3,501	3,676	4,108	5,313
Merchandise, fob	..	..	..	1,148	1,152	1,325	1,835	2,793	2,850	3,152	4,236
Nonfactor Services	..	..	..	262	277	302	398	565	616	735	828
Factor Services	..	..	..	60	67	74	91	143	211	221	248
Long-Term Interest	..	..	..	33	40	41	55	85	104	107	125
Current Transfers, net	..	..	..	3	7	30	117	215	56	29	22
Workers' Remittances	..	..	..	0	0	0	0	0	0	24	0
Total to be Financed	..	..	..	-296	-212	-80	-74	-113	-650	-459	-1,112
Official Capital Grants	..	..	..	46	37	29	27	26	24	18	18
Current Account Balance	..	..	..	-250	-175	-51	-46	-87	-625	-441	-1,094
Long-Term Capital, net	..	..	..	156	117	186	79	415	214	315	445
Direct Investment	..	..	..	43	39	68	77	189	22	79	106
Long-Term Loans	..	..	..	90	45	107	-4	253	200	248	346
Disbursements	..	..	..	220	203	246	209	422	454	519	663
Repayments	..	..	..	131	157	139	212	169	254	271	317
Other Long-Term Capital	..	..	..	-23	-4	-19	-23	-53	-32	-30	-25
Other Capital, net	..	..	..	12	47	61	184	157	359	207	658
Change in Reserves	..	..	..	82	10	-196	-216	-485	51	-81	-9
Memo Item:				*(Thai Baht per US dollar)*							
Conversion Factor (Annual Avg)	20.800	20.800	20.800	20.800	20.800	20.800	20.620	20.380	20.380	20.400	20.400
EXTERNAL DEBT, ETC.				*(Millions of US dollars, outstanding at end of year)*							
Public/Publicly Guar. Long-Term	..	..	..	324	357	386	441	513	616	823	1,119
Official Creditors	..	..	..	291	320	349	408	468	555	672	896
IBRD and IDA	..	..	..	159	170	191	219	243	274	309	365
Private Creditors	..	..	..	33	36	37	34	45	62	151	224
Private Non-guaranteed Long-Term	..	..	..	401	425	505	461	648	736	785	880
Use of Fund Credit	..	..	..	0	0	0	0	0	0	78	81
Short-Term Debt	..	..	..	..	..	..	..	..	..	..	1,264
Memo Items:				*(Millions of US dollars)*							
Int'l Reserves Excluding Gold	916.8	928.8	892.8	823.5	788.0	963.4	1,206.8	1,758.2	1,678.8	1,797.5	1,812.7
Gold Holdings (at market price)	92.2	109.8	92.2	87.5	102.1	151.9	262.7	436.4	328.2	315.3	395.4
SOCIAL INDICATORS											
Total Fertility Rate	6.3	6.0	5.7	5.4	5.1	4.9	4.7	4.6	4.5	4.4	4.3
Crude Birth Rate	40.0	42.5	39.6	39.4	38.6	37.9	37.1	36.4	35.6	33.9	32.1
Infant Mortality Rate	84.0	80.2	76.4	72.6	68.8	65.0	63.2	61.4	59.6	57.8	56.0
Life Expectancy at Birth	56.7	57.3	57.9	58.4	59.0	59.6	59.9	60.2	60.6	60.9	61.2
Food Production, p.c. ('79-81 = 100)	75.8	78.6	82.1	83.9	84.1	79.6	93.6	90.2	95.4	97.5	94.5
Labor Force, Agriculture (%)	80.9	80.6	80.2	79.8	78.9	78.0	77.1	76.2	75.3	74.4	73.5
Labor Force, Female (%)	47.5	47.5	47.4	47.3	47.3	47.2	47.2	47.1	47.1	47.1	47.1
School Enroll. Ratio, primary	..	..	..	83.0	..	..	..	..	83.0	83.0	83.0
School Enroll. Ratio, secondary	..	..	..	17.0	..	..	..	..	26.0	27.0	28.0

1978	1979	1980	1981	1982	1983	1984	1985	1986	1987 estimate	NOTES	THAILAND
											FOREIGN TRADE (CUSTOMS BASIS)
			(Millions of current US dollars)								
4,085	5,297	6,505	7,035	6,957	6,368	7,413	7,121	8,876	11,665	..	Value of Exports, fob
3,042	3,883	4,578	4,981	4,915	4,284	4,780	4,222	5,116	7,274	..	Nonfuel Primary Products
5	12	41	35	28	27	50	99	64	91	..	Fuels
1,039	1,402	1,886	2,019	2,014	2,058	2,583	2,800	3,696	4,300	..	Manufactures
5,314	7,132	9,450	10,055	8,527	10,279	10,518	9,239	9,124	13,006	..	Value of Imports, cif
684	1,011	1,169	1,218	1,029	1,218	1,337	1,216	1,313	1,582	..	Nonfuel Primary Products
1,130	1,605	2,876	2,994	2,645	2,488	2,471	2,093	1,266	2,679	..	Fuels
3,501	4,515	5,406	5,843	4,854	6,572	6,710	5,929	6,545	7,716	..	Manufactures
			(Index 1980 = 100)								
117.5	111.5	100.0	89.8	79.1	83.6	81.8	73.6	83.9	84.2	..	Terms of Trade
78.4	91.5	100.0	92.4	77.7	79.3	76.0	67.1	74.3	78.6	..	Export Prices, fob
79.0	91.6	100.0	93.3	76.8	80.5	76.3	63.5	68.3	71.9	..	Nonfuel Primary Products
42.3	61.0	100.0	112.5	101.6	92.5	90.2	87.5	44.9	56.9	..	Fuels
77.2	91.7	100.0	90.2	79.8	76.7	75.3	72.6	85.8	94.2	..	Manufactures
66.8	82.1	100.0	102.9	98.3	94.9	93.0	91.2	88.6	93.3	..	Import Prices, cif
			(Millions of 1980 US dollars)								Trade at Constant 1980 Prices
5,208	5,788	6,506	7,609	8,949	8,031	9,751	10,615	11,943	14,837	..	Exports, fob
7,961	8,687	9,450	9,772	8,677	10,835	11,314	10,135	10,294	13,934	..	Imports, cif
			(Millions of current US dollars)								**BALANCE OF PAYMENTS**
5,126	6,661	8,578	9,252	9,402	9,229	10,416	10,229	12,104	14,301	..	Exports of Goods & Services
4,040	5,234	6,452	6,898	6,820	6,308	7,340	7,065	8,777	10,881	..	Merchandise, fob
817	1,034	1,490	1,612	1,717	1,846	1,963	2,042	2,298	3,160	..	Nonfactor Services
269	394	636	742	865	1,074	1,113	1,122	1,029	260	..	Factor Services
6,321	8,804	10,860	11,995	10,606	12,391	12,697	11,948	12,079	15,091	..	Imports of Goods & Services
4,901	6,781	8,351	8,926	7,569	9,179	9,234	8,411	8,387	11,164	..	Merchandise, fob
984	1,329	1,615	1,784	1,622	1,868	1,860	1,770	1,798	2,046	..	Nonfactor Services
436	694	895	1,285	1,415	1,344	1,604	1,767	1,893	1,880	..	Factor Services
196	322	470	706	711	748	846	911	1,030	1,057	..	Long-Term Interest
6	23	75	50	76	151	59	47	63	67	..	Current Transfers, net
0	0	0	0	0	0	0	0	0	..	..	Workers' Remittances
-1,188	-2,120	-2,208	-2,693	-1,129	-3,012	-2,222	-1,673	88	-723	K	Total to be Financed
34	37	142	119	108	125	115	119	161	137	K	Official Capital Grants
-1,155	-2,083	-2,066	-2,574	-1,021	-2,887	-2,107	-1,554	249	-586	..	Current Account Balance
730	1,562	2,261	2,004	1,482	1,589	1,836	1,694	183	319	..	Long-Term Capital, net
51	51	186	291	191	348	403	160	262	270	..	Direct Investment
678	1,386	1,847	1,594	1,326	1,252	1,491	1,497	-167	-2	..	Long-Term Loans
1,284	2,027	2,623	2,217	2,120	2,265	2,883	3,176	1,889	1,889	..	Disbursements
606	640	776	623	794	1,013	1,392	1,679	2,056	1,891	..	Repayments
-32	87	86	0	-143	-137	-172	-82	-73	-86	..	Other Long-Term Capital
451	486	-356	615	-540	1,138	786	-58	252	1,668	..	Other Capital, net
-27	36	161	-45	79	161	-516	-82	-684	-1,401	..	Change in Reserves
			(Thai Baht per US dollar)								Memo Item:
20.340	20.420	20.480	21.820	23.000	23.000	23.640	27.160	26.290	25.600	..	Conversion Factor (Annual Avg)
			(Millions of US dollars, outstanding at end of year)								**EXTERNAL DEBT, ETC.**
1,819	2,802	4,070	5,127	6,138	7,000	7,266	9,937	11,612	14,023	..	Public/Publicly Guar. Long-Term
1,253	1,666	2,333	2,884	3,468	4,280	4,504	5,844	7,163	8,631	..	Official Creditors
459	580	703	987	1,340	1,738	1,724	2,307	2,890	3,524	..	IBRD and IDA
566	1,136	1,737	2,243	2,670	2,720	2,762	4,093	4,449	5,391	..	Private Creditors
939	1,243	1,703	2,099	2,317	2,655	3,372	3,370	3,108	3,108	..	Private Non-guaranteed Long-Term
177	240	182	706	702	907	791	1,020	988	916	..	Use of Fund Credit
2,094	2,340	2,303	2,878	3,041	3,305	3,551	3,200	2,840	2,664	..	Short-Term Debt
			(Millions of US dollars)								Memo Items:
2,008.6	1,842.8	1,560.2	1,731.8	1,537.6	1,607.1	1,920.6	2,190.1	2,804.4	4,007.1	..	Int'l Reserves Excluding Gold
548.3	1,257.0	1,466.1	988.6	1,136.3	948.8	766.7	813.2	972.2	1,198.6	..	Gold Holdings (at market price)
											SOCIAL INDICATORS
4.1	4.0	3.8	3.7	3.5	3.4	3.2	3.1	3.0	2.8	..	Total Fertility Rate
30.4	30.7	30.1	29.1	28.3	27.5	26.7	26.0	25.2	24.5	..	Crude Birth Rate
54.4	52.8	51.2	49.6	49.9	50.2	50.5	50.8	51.1	51.4	..	Infant Mortality Rate
61.5	61.9	62.2	62.5	62.9	63.2	63.5	63.8	64.1	64.4	..	Life Expectancy at Birth
105.7	97.2	98.8	104.1	104.8	109.1	109.4	111.2	106.5	101.9	..	Food Production, p.c. ('79-81 = 100)
72.7	71.8	70.9	..	..	..	..	..	..	..	..	Labor Force, Agriculture (%)
47.1	47.1	47.1	46.8	46.6	46.4	46.1	45.9	45.6	45.4	..	Labor Force, Female (%)
92.0	95.0	99.0	101.0	..	97.0	97.0	97.0	..	..	..	School Enroll. Ratio, primary
28.0	28.0	29.0	29.0	..	30.0	31.0	30.0	..	..	..	School Enroll. Ratio, secondary

TOGO	1967	1968	1969	1970	1971	1972	1973	1974	1975	1976	1977
CURRENT GNP PER CAPITA (US $)	130	130	140	140	140	150	170	210	260	270	310
POPULATION (thousands)	1,774	1,853	1,934	2,020	2,070	2,121	2,173	2,227	2,282	2,338	2,396

ORIGIN AND USE OF RESOURCES *(Billions of current CFA Francs)*

	1967	1968	1969	1970	1971	1972	1973	1974	1975	1976	1977
Gross National Product (GNP)	55.1	57.7	67.5	69.1	77.8	83.0	89.0	134.5	131.7	147.4	189.7
Net Factor Income from Abroad	-1.9	-2.2	-2.1	-1.1	-1.1	-1.6	-1.6	-0.4	-0.6	-0.6	-1.3
Gross Domestic Product (GDP)	57.0	59.9	69.6	70.2	78.9	84.6	90.6	134.9	132.3	148.0	191.0
Indirect Taxes, net	4.3	4.4	5.6	7.6	8.1	9.2	11.9	12.2	13.9	15.8	18.1
GDP at factor cost	52.7	55.5	64.0	62.6	70.8	75.4	78.7	122.7	118.4	132.2	172.9
Agriculture	25.2	26.7	29.9	23.7	25.0	26.0	29.0	33.2	35.2	47.8	67.7
Industry	13.3	13.1	14.8	14.8	17.8	17.2	18.3	44.7	35.1	30.3	37.2
Manufacturing	5.9	6.9	7.6	7.0	7.0	6.7	7.9	8.2	9.3	10.3	12.0
Services, etc.	18.5	20.1	24.9	31.7	36.1	41.4	43.3	57.0	62.0	69.9	86.1
Resource Balance	0.4	1.2	1.2	-0.9	-3.1	-6.3	-6.1	23.0	-23.1	-13.5	-33.7
Exports of Goods & NFServices	13.4	15.6	19.8	21.6	22.9	20.6	18.6	55.2	35.6	43.4	49.1
Imports of Goods & NFServices	13.0	14.4	18.6	22.5	26.0	26.9	24.7	32.2	58.7	56.9	82.8
Domestic Absorption	56.6	58.7	68.4	71.1	82.0	90.9	96.7	111.9	155.4	161.5	224.7
Private Consumption, etc.	45.5	48.0	54.8	53.1	57.1	61.9	65.9	77.1	99.6	102.5	130.5
General Gov't Consumption	3.9	4.1	4.4	7.4	10.0	11.5	11.1	12.7	19.2	22.2	28.6
Gross Domestic Investment	7.2	6.6	9.2	10.6	14.9	17.5	19.7	22.1	36.6	36.8	65.6
Fixed Investment	..	..	..	..	..	..	..	..	..	..	..
Memo Items:											
Gross Domestic Saving	7.6	7.8	10.4	9.7	11.8	11.2	13.6	45.1	13.5	23.3	31.9
Gross National Saving	5.7	5.6	7.7	7.9	9.7	8.3	12.2	44.5	12.6	23.0	30.5

(Billions of 1980 CFA Francs)

	1967	1968	1969	1970	1971	1972	1973	1974	1975	1976	1977
Gross National Product	126.4	132.1	147.6	153.8	154.2	165.6	172.4	183.6	188.0	184.5	198.0
GDP at factor cost	120.8	126.9	139.7	139.1	140.1	150.1	152.2	167.5	168.8	165.5	180.7
Agriculture	47.4	50.9	54.1	50.3	50.0	52.8	54.5	55.3	56.9	54.0	50.7
Industry	27.9	28.0	30.0	28.0	28.8	33.6	37.0	40.1	41.1	43.1	50.3
Manufacturing	..	..	..	..	..	..	..	..	..	15.5	15.8
Services, etc.	55.5	58.2	67.9	77.9	77.6	82.3	83.9	88.9	90.8	88.2	98.6
Resource Balance	2.8	4.8	6.5	1.9	-3.5	-2.8	-1.3	-2.8	-8.5	-30.1	-67.5
Exports of Goods & NFServices	34.3	38.3	45.5	46.2	49.7	48.0	43.0	41.6	46.1	38.2	50.6
Imports of Goods & NFServices	31.5	33.5	39.0	44.3	53.2	50.8	44.3	44.3	54.5	68.3	118.0
Domestic Absorption	128.0	132.3	145.6	154.3	159.9	171.5	176.8	187.0	197.2	215.5	267.1
Private Consumption, etc.	88.2	94.9	101.8	105.0	103.6	109.0	113.5	121.5	104.6	131.9	137.2
General Gov't Consumption	11.5	12.1	12.1	14.0	17.1	19.4	18.1	19.3	24.9	28.2	31.6
Gross Domestic Investment	28.3	25.3	31.8	35.4	39.1	43.1	45.2	46.2	67.7	55.4	98.3
Fixed Investment	..	..	..	..	..	..	..	..	..	..	..
Memo Items:											
Capacity to Import	32.4	36.3	41.6	42.5	46.9	38.9	33.4	76.0	33.1	52.1	70.0
Terms of Trade Adjustment	-1.8	-2.0	-4.0	-3.7	-2.8	-9.1	-9.6	34.4	-13.0	13.9	19.5
Gross Domestic Income	129.0	135.1	148.1	152.5	153.5	159.6	165.8	218.7	175.7	199.3	219.0
Gross National Income	124.6	130.1	143.7	150.1	151.4	156.5	162.8	218.0	175.0	198.4	217.4

DOMESTIC PRICES (DEFLATORS) *(Index 1980 = 100)*

	1967	1968	1969	1970	1971	1972	1973	1974	1975	1976	1977
Overall (GDP)	43.6	43.7	45.8	44.9	50.5	50.2	51.6	73.2	70.1	79.8	95.7
Domestic Absorption	44.2	44.4	47.0	46.1	51.3	53.0	54.7	59.8	78.8	74.9	84.1
Agriculture	53.2	52.5	55.2	47.1	50.0	49.3	53.2	60.1	61.9	88.5	133.4
Industry	47.7	46.7	49.3	52.8	61.9	51.2	49.4	111.4	85.3	70.3	74.0
Manufacturing	..	..	..	..	..	..	..	..	..	66.7	76.2

MANUFACTURING ACTIVITY

	1967	1968	1969	1970	1971	1972	1973	1974	1975	1976	1977
Employment (1980=100)	..	..	..	67.7	73.4	74.7	..	92.8	97.4	116.6	108.8
Real Earnings per Empl. (1980=100)	..	..	..	..	..	..	..	..	..	..	..
Real Output per Empl. (1980=100)	..	..	..	..	..	..	..	44.9	72.0	79.3	86.2
Earnings as % of Value Added	..	..	..	..	..	..	..	45.5	33.2	31.0	40.5

MONETARY HOLDINGS *(Billions of current CFA Francs)*

	1967	1968	1969	1970	1971	1972	1973	1974	1975	1976	1977
Money Supply, Broadly Defined	7.0	8.6	11.5	12.7	14.1	13.9	15.9	30.5	28.3	41.2	48.1
Money as Means of Payment	6.7	7.6	9.7	10.2	11.9	11.8	11.6	25.1	21.6	32.9	36.5
Currency Ouside Banks	3.2	3.8	4.5	4.6	5.6	5.5	6.0	8.4	10.0	14.2	16.0
Demand Deposits	3.5	3.8	5.2	5.6	6.3	6.4	5.5	16.7	11.6	18.8	20.5
Quasi-Monetary Liabilities	0.3	1.0	1.8	2.5	2.2	2.1	4.4	5.4	6.7	8.3	11.6

GOVERNMENT DEFICIT (-) OR SURPLUS *(Millions of current CFA Francs)*

	1967	1968	1969	1970	1971	1972	1973	1974	1975	1976	1977
GOVERNMENT DEFICIT (-) OR SURPLUS	..	..	..	..	..	..	..	..	..	..	-39,222
Current Revenue	..	..	..	..	..	..	..	..	..	..	47,022
Current Expenditure	..	..	..	..	..	..	..	..	..	..	33,753
Current Budget Balance	..	..	..	..	..	..	..	..	..	..	13,269
Capital Receipts	..	..	..	..	..	..	..	..	..	..	..
Capital Payments	..	..	..	..	..	..	..	..	..	..	52,491

1978	1979	1980	1981	1982	1983	1984	1985	1986	1987 estimate	NOTES	TOGO
360	360	430	400	340	270	240	230	250	290	..	CURRENT GNP PER CAPITA (US $)
2,455	2,516	2,578	2,664	2,753	2,845	2,940	3,038	3,144	3,254	..	POPULATION (thousands)
				(Billions of current CFA Francs)							ORIGIN AND USE OF RESOURCES
199.0	188.5	234.5	248.1	253.7	264.8	276.1	298.5	330.2	345.5	..	Gross National Product (GNP)
-1.9	-1.2	-5.6	-13.4	-16.3	-16.5	-16.9	-16.4	-18.6	-24.2	..	Net Factor Income from Abroad
200.9	189.7	240.1	261.5	270.0	281.3	293.0	314.9	348.8	369.7	f	Gross Domestic Product (GDP)
25.2	26.5	29.0	33.4	36.7	34.0	33.8	38.1	44.5	47.2	..	Indirect Taxes, net
175.7	163.2	211.1	228.1	233.3	247.3	259.2	276.8	304.3	322.5	B	GDP at factor cost
54.4	48.1	66.0	73.7	72.8	90.1	93.4	97.3	98.3	106.4	f	Agriculture
46.3	47.9	59.5	58.8	63.1	62.6	64.2	71.7	62.7	64.8	..	Industry
12.5	13.5	16.5	19.0	20.6	20.1	19.9	22.1	..	..	..	Manufacturing
100.2	93.7	114.6	129.0	134.1	128.6	135.4	145.9	187.8	198.5	..	Services, etc.
-52.6	-60.7	-35.8	-40.4	-37.9	-24.1	-16.8	-29.4	-43.2	-43.6	..	Resource Balance
61.5	58.8	76.0	103.4	110.7	104.6	123.1	127.9	119.5	115.6	..	Exports of Goods & NFServices
114.1	119.5	111.8	143.8	148.6	128.7	139.9	157.3	162.7	159.2	..	Imports of Goods & NFServices
253.5	250.4	275.9	301.9	307.9	305.4	309.8	344.3	392.0	413.3	..	Domestic Absorption
131.1	125.5	168.2	184.1	195.2	201.5	213.1	234.2	236.8	272.9	..	Private Consumption, etc.
29.3	32.3	35.9	38.7	41.8	40.6	42.6	47.2	68.9	76.4	..	General Gov't Consumption
93.1	92.6	71.8	79.1	70.9	63.3	54.1	62.9	86.3	64.0	..	Gross Domestic Investment
..	..	..	67.7	63.6	56.3	64.7	75.8	84.2	63.3	..	Fixed Investment
											Memo Items:
40.5	31.9	36.0	38.7	33.0	39.2	37.3	33.5	43.1	20.4	..	Gross Domestic Saving
38.2	30.3	30.5	25.1	16.1	21.7	19.6	17.2	24.6	-3.6	..	Gross National Saving
				(Billions of 1980 CFA Francs)							
218.3	208.2	234.5	220.7	211.5	196.1	200.7	208.7	217.1	220.3	..	Gross National Product
192.8	180.4	211.1	202.3	193.9	186.0	188.0	193.3	198.2	202.1	B I	GDP at factor cost
59.6	56.6	66.0	65.0	59.2	59.8	67.5	68.8	65.6	66.9	f	Agriculture
52.4	51.7	59.5	56.8	57.2	52.5	48.2	51.6	53.2	54.2	..	Industry
13.4	14.6	16.5	18.0	18.8	17.4	14.2	15.2	..	..	..	Manufacturing
108.5	101.4	114.6	110.2	107.8	99.3	97.0	99.7	108.5	110.7	..	Services, etc.
-72.3	-72.9	-35.8	-31.6	-28.5	-15.9	-12.0	-20.7	-20.8	-18.5	..	Resource Balance
75.3	52.0	76.0	96.6	94.8	81.2	82.8	83.6	88.0	81.6	..	Exports of Goods & NFServices
147.6	124.9	111.8	128.3	123.3	97.1	94.8	104.3	108.8	100.0	..	Imports of Goods & NFServices
292.8	282.6	275.9	263.6	252.8	227.4	224.7	240.8	248.2	250.3	..	Domestic Absorption
145.6	137.9	168.2	160.8	160.9	150.5	145.9	154.3	154.3	170.2	..	Private Consumption, etc.
32.2	35.5	35.9	33.8	34.1	30.0	31.4	33.5	38.6	40.7	..	General Gov't Consumption
115.0	109.1	71.8	69.0	57.8	46.8	47.4	52.9	55.2	39.4	..	Gross Domestic Investment
..	..	..	..	..	..	..	..	..	..	..	Fixed Investment
											Memo Items:
79.6	61.5	76.0	92.2	91.9	78.9	83.4	84.8	79.9	72.6	..	Capacity to Import
4.3	9.5	0.0	-4.4	-3.0	-2.3	0.6	1.2	-8.1	-8.9	..	Terms of Trade Adjustment
224.7	219.1	240.1	227.6	221.3	209.2	213.3	221.3	219.3	222.9	..	Gross Domestic Income
222.5	217.7	234.5	216.3	208.5	193.7	201.3	209.9	209.1	211.3	..	Gross National Income
				(Index 1980 = 100)							DOMESTIC PRICES (DEFLATORS)
91.1	90.5	100.0	112.7	120.4	133.0	137.8	143.1	153.4	159.5	..	Overall (GDP)
86.6	88.6	100.0	114.5	121.8	134.3	137.9	143.0	158.0	165.1	..	Domestic Absorption
91.3	85.0	100.0	113.4	122.9	150.8	138.4	141.5	150.0	159.0	..	Agriculture
88.4	92.7	100.0	103.5	110.3	119.3	133.1	138.9	117.8	119.5	..	Industry
93.6	92.8	100.0	105.6	109.8	115.7	140.2	145.3	..	..	..	Manufacturing
											MANUFACTURING ACTIVITY
130.2	97.7	100.0	98.8	104.4	..	..	..	..	..	G J	Employment (1980=100)
..	..	..	..	..	..	..	..	..	..	G J	Real Earnings per Empl. (1980=100)
71.7	101.8	100.0	115.4	120.9	..	..	..	..	..	G J	Real Output per Empl. (1980=100)
36.0	28.4	..	..	..	..	..	..	..	..	J	Earnings as % of Value Added
				(Billions of current CFA Francs)							MONETARY HOLDINGS
64.9	66.5	72.6	100.6	117.2	117.8	136.1	143.2	165.5	163.8	..	Money Supply, Broadly Defined
48.1	52.7	55.3	80.0	90.1	83.1	90.7	82.7	89.7	91.0	..	Money as Means of Payment
20.8	21.5	27.8	50.7	54.3	45.5	37.0	39.2	46.0	48.3	..	Currency Ouside Banks
27.3	31.1	27.6	29.2	35.8	37.6	53.7	43.5	43.7	42.8	..	Demand Deposits
16.9	13.9	17.2	20.6	27.1	34.7	45.4	60.5	75.8	72.8	..	Quasi-Monetary Liabilities
				(Millions of current CFA Francs)							
-58,836	-17,942	-4,689	-14,821	-4,794	-5,607	-7,862	-6,104	-16,644	..	..	GOVERNMENT DEFICIT (-) OR SURPLUS
56,646	70,008	72,896	70,891	81,649	85,921	102,607	116,151	121,094	..	..	Current Revenue
32,500	38,873	..	61,928	61,505	67,943	85,921	89,204	..	..	..	Current Expenditure
24,146	31,135	..	8,963	20,144	17,978	16,686	26,947	..	..	..	Current Budget Balance
..	..	..	..	..	..	..	97	447	..	..	Capital Receipts
82,982	49,077	..	23,784	24,938	23,585	24,645	33,498	..	..	..	Capital Payments

TOGO	1967	1968	1969	1970	1971	1972	1973	1974	1975	1976	1977
FOREIGN TRADE (CUSTOMS BASIS)				*(Millions of current US dollars)*							
Value of Exports, fob	50.1	60.4	72.7	68.0	73.1	69.9	71.6	215.0	141.0	158.9	199.3
Nonfuel Primary Products	47.8	56.9	67.9	64.9	63.0	67.4	67.3	208.8	133.7	150.4	190.5
Fuels	0.0	0.0	0.0	0.0	0.0	0.0	0.0	0.0	0.0	0.0	0.0
Manufactures	2.3	3.5	4.8	3.1	10.1	2.5	4.3	6.2	7.3	8.5	8.8
Value of Imports, cif	51.1	56.1	67.9	75.8	85.3	93.5	94.5	116.3	269.0	218.2	304.5
Nonfuel Primary Products	15.0	15.6	20.7	22.7	24.5	23.9	20.8	21.2	78.0	49.4	66.6
Fuels	2.1	2.2	2.6	2.8	3.9	4.8	5.2	11.5	13.0	12.8	20.4
Manufactures	34.1	38.4	44.6	50.3	57.0	64.8	68.6	83.6	178.0	156.0	217.5
				(Index 1980 = 100)							
Terms of Trade	103.2	104.0	102.6	100.2	90.2	85.6	92.7	177.4	145.4	124.8	117.0
Export Prices, fob	26.0	26.4	27.0	27.5	26.1	26.5	35.3	98.1	91.0	77.7	80.8
Nonfuel Primary Products	25.9	26.1	26.8	27.2	25.0	26.2	34.7	99.5	92.9	78.4	81.3
Fuels	4.3	4.3	4.3	4.3	5.6	6.2	8.9	36.7	35.7	38.4	42.0
Manufactures	30.1	31.7	32.4	37.5	36.5	38.5	49.5	67.2	66.1	66.7	71.7
Import Prices, cif	25.2	25.3	26.3	27.5	28.9	31.0	38.1	55.3	62.6	62.2	69.1
Trade at Constant 1980 Prices				*(Millions of 1980 US dollars)*							
Exports, fob	192.4	229.1	269.3	246.9	280.3	263.6	202.8	219.1	155.0	204.5	246.6
Imports, cif	202.6	221.3	258.0	275.8	295.0	301.7	248.1	210.3	429.9	350.6	440.8
BALANCE OF PAYMENTS				*(Millions of current US dollars)*							
Exports of Goods & Services	..	..	..	77.0	83.6	81.8	83.6	235.9	175.5	189.5	233.9
Merchandise, fob	..	..	..	68.0	73.1	69.9	71.7	215.1	141.0	158.9	199.3
Nonfactor Services	..	..	..	7.1	6.9	8.6	11.9	14.5	28.0	24.2	27.7
Factor Services	..	..	..	1.9	3.5	3.3	0.0	6.4	6.5	6.5	7.0
Imports of Goods & Services	..	..	..	88.2	104.2	120.8	125.2	142.2	292.0	251.0	363.2
Merchandise, fob	..	..	..	67.0	75.2	81.3	83.7	98.0	211.5	180.6	252.7
Nonfactor Services	..	..	..	14.8	20.9	28.6	28.7	38.2	69.1	60.0	90.3
Factor Services	..	..	..	6.4	8.1	11.0	12.8	6.0	11.3	10.4	20.2
Long-Term Interest	..	..	..	0.7	0.7	1.1	1.2	2.0	4.5	3.5	9.7
Current Transfers, net	..	..	..	-2.4	-3.7	-5.0	0.7	-0.9	-1.4	1.1	-0.3
Workers' Remittances	..	..	..	0.0	0.0	0.0	3.5	4.1	4.8	7.1	8.0
Total to be Financed	..	..	..	-13.6	-24.4	-44.1	-40.9	92.8	-117.9	-60.4	-129.6
Official Capital Grants	..	..	..	16.4	20.9	31.2	30.3	38.9	42.5	36.2	42.4
Current Account Balance	..	..	..	2.8	-3.5	-12.9	-10.6	131.8	-75.4	-24.1	-87.3
Long-Term Capital, net	..	..	..	10.3	26.9	31.9	35.8	15.5	63.1	69.8	91.4
Direct Investment	..	..	..	0.4	3.9	0.2	1.8	-39.1	5.2	5.7	11.6
Long-Term Loans	..	..	..	2.9	3.7	3.2	5.1	24.6	38.9	61.5	123.4
Disbursements	..	..	..	4.5	5.4	7.4	10.5	31.1	51.0	81.0	170.4
Repayments	..	..	..	1.6	1.7	4.2	5.4	6.5	12.1	19.5	47.0
Other Long-Term Capital	..	..	..	-9.4	-1.6	-2.7	-1.4	-9.0	-23.5	-33.6	-86.0
Other Capital, net	..	..	..	-6.2	-22.1	-25.4	-26.9	-113.3	23.8	-46.3	-36.5
Change in Reserves	..	..	..	-6.9	-1.3	6.4	1.7	-34.0	-11.6	0.6	32.5
Memo Item:				*(CFA Francs per US dollar)*							
Conversion Factor (Annual Avg)	246.850	246.850	259.710	277.710	277.130	252.210	222.700	240.500	214.320	238.980	245.670
EXTERNAL DEBT, ETC.				*(Millions of US dollars, outstanding at end of year)*							
Public/Publicly Guar. Long-Term	..	..	..	39.8	46.9	47.0	58.4	89.3	119.5	177.1	320.6
Official Creditors	..	..	..	31.9	38.5	37.8	45.1	55.5	58.4	68.5	109.3
IBRD and IDA	..	..	..	1.9	2.5	3.4	3.9	4.1	6.2	10.6	16.3
Private Creditors	..	..	..	7.8	8.4	9.2	13.3	33.8	61.2	108.6	211.2
Private Non-guaranteed Long-Term	..	..	..	..	..	..	..	..	..	..	..
Use of Fund Credit	..	..	..	0.0	0.0	0.0	0.0	0.0	0.0	8.7	9.1
Short-Term Debt	..	..	..	..	..	..	..	..	..	..	39.0
Memo Items:				*(Millions of US dollars)*							
Int'l Reserves Excluding Gold	22.3	25.5	26.3	35.4	40.5	36.5	37.9	54.5	41.2	66.6	46.1
Gold Holdings (at market price)	..	..	..	..	..	..	..	..	..	..	1.1
SOCIAL INDICATORS											
Total Fertility Rate	6.5	6.5	6.5	6.5	6.5	6.5	6.5	6.5	6.5	6.5	6.5
Crude Birth Rate	50.0	50.0	50.0	50.0	50.0	50.0	50.0	50.0	50.0	50.0	50.0
Infant Mortality Rate	141.0	137.0	133.0	129.0	125.0	121.0	119.0	117.0	115.0	113.0	111.0
Life Expectancy at Birth	43.0	43.5	44.0	44.5	45.0	45.5	46.0	46.5	47.0	47.5	48.0
Food Production, p.c. ('79-81 = 100)	115.0	113.0	109.3	113.8	115.2	105.0	104.2	100.5	101.6	95.3	91.3
Labor Force, Agriculture (%)	77.5	77.3	77.0	76.7	76.3	76.0	75.6	75.3	74.9	74.5	74.1
Labor Force, Female (%)	39.1	39.1	39.1	39.0	39.0	38.9	38.9	38.8	38.8	38.7	38.6
School Enroll. Ratio, primary	..	..	..	71.0	..	..	..	..	99.0	99.0	104.0
School Enroll. Ratio, secondary	..	..	..	7.0	..	..	..	..	19.0	22.0	26.0

1978	1979	1980	1981	1982	1983	1984	1985	1986	1987 estimate	NOTES	TOGO
											FOREIGN TRADE (CUSTOMS BASIS)
				(Millions of current US dollars)							
262.0	290.6	475.8	377.7	344.8	273.8	239.5	242.5	275.0	272.0	f	Value of Exports, fob
213.5	245.4	353.6	351.1	312.0	237.3	204.5	209.5	258.0	250.0	..	Nonfuel Primary Products
31.2	24.8	86.9	2.7	3.4	3.4	0.0	0.0	0.0	..	..	Fuels
17.3	20.5	35.4	23.9	29.4	33.1	35.0	33.0	17.0	22.0	f	Manufactures
497.2	558.1	637.8	503.7	486.6	349.7	310.0	288.0	359.0	350.0	f	Value of Imports, cif
74.0	97.1	144.8	157.7	145.6	109.5	97.1	86.1	101.0	90.9	..	Nonfuel Primary Products
64.8	94.9	125.2	36.7	48.0	30.0	29.9	30.9	17.0	21.1	..	Fuels
358.4	366.1	367.8	309.4	293.0	210.2	183.0	171.0	241.0	238.0	f	Manufactures
				(Index 1980 = 100)							
98.7	93.1	100.0	101.7	93.3	88.5	93.7	87.8	78.6	81.1	..	Terms of Trade
71.4	78.7	100.0	100.9	87.6	81.6	85.7	75.9	74.2	80.3	..	Export Prices, fob
79.0	80.2	100.0	100.9	86.8	80.3	85.2	74.2	72.9	78.3	..	Nonfuel Primary Products
42.3	61.0	100.0	112.5	101.6	92.5	90.2	87.5	44.9	56.9	..	Fuels
76.1	90.0	100.0	99.1	94.8	90.7	89.3	89.0	103.2	113.0	..	Manufactures
72.4	84.5	100.0	99.2	93.9	92.1	91.5	86.4	94.4	99.0	..	Import Prices, cif
				(Millions of 1980 US dollars)							Trade at Constant 1980 Prices
366.7	369.3	475.8	374.4	393.7	335.7	279.3	319.5	370.6	338.7	..	Exports, fob
686.9	660.6	637.8	507.9	518.2	379.6	338.7	333.3	380.4	353.6	..	Imports, cif
				(Millions of current US dollars)							**BALANCE OF PAYMENTS**
305.2	346.0	570.4	490.4	449.8	360.2	384.9	346.6	394.5	442.9	f	Exports of Goods & Services
262.0	290.6	475.8	377.7	344.8	273.5	291.0	244.4	274.2	300.5	..	Merchandise, fob
36.0	45.2	73.9	97.1	88.3	70.6	76.8	78.1	101.1	120.1	..	Nonfactor Services
7.3	10.2	20.7	15.6	16.6	16.1	17.2	24.0	19.3	22.3	..	Factor Services
572.8	641.7	751.5	618.2	603.2	468.7	431.8	427.4	576.3	590.3	f	Imports of Goods & Services
410.9	464.3	524.1	413.9	408.2	291.8	263.2	252.0	354.8	346.7	..	Merchandise, fob
134.2	145.4	166.8	146.0	136.0	117.7	112.6	114.9	153.8	169.7	..	Nonfactor Services
27.7	32.0	60.6	58.3	59.0	59.2	56.0	60.6	67.7	73.9	..	Factor Services
11.1	9.6	19.4	20.0	23.4	27.1	36.7	38.5	43.2	28.6	..	Long-Term Interest
-1.9	-2.1	0.5	-0.7	-2.0	-2.6	-1.9	0.2	0.3	0.7	..	Current Transfers, net
9.3	8.0	9.9	7.9	6.1	6.1	6.6	6.0	8.9	10.7	..	Workers' Remittances
-269.5	-297.9	-180.6	-113.6	-155.4	-111.1	-48.8	-80.6	-181.4	-146.7	K	Total to be Financed
52.1	85.1	85.6	69.5	68.6	67.7	75.0	69.9	76.6	73.5	K	Official Capital Grants
-217.4	-212.8	-95.0	-44.2	-86.8	-43.4	26.2	-10.7	-104.8	-73.2	..	Current Account Balance
268.0	270.0	161.3	116.1	140.7	85.8	87.5	84.2	109.7	160.7	f	Long-Term Capital, net
93.2	52.8	42.3	10.1	16.1	1.5	-9.9	..	13.0	12.0	..	Direct Investment
274.5	186.7	78.1	9.5	31.6	51.0	12.2	1.1	-0.6	15.4	..	Long-Term Loans
311.1	215.5	104.9	35.6	51.2	72.8	52.6	53.2	88.7	49.9	..	Disbursements
36.6	28.8	26.8	26.1	19.6	21.8	40.4	52.1	89.3	34.5	..	Repayments
-151.7	-54.7	-44.8	27.1	24.4	-34.4	10.2	13.2	20.7	59.8	..	Other Long-Term Capital
-22.1	-66.8	-68.2	14.0	-10.5	-24.7	-72.3	-46.9	-30.3	-114.6	..	Other Capital, net
-28.6	9.6	1.9	-86.0	-43.4	-17.8	-41.4	-26.5	25.4	27.1	..	Change in Reserves
				(CFA Francs per US dollar)							Memo Item:
225.640	212.720	211.300	271.730	328.620	381.070	436.960	449.260	346.300	300.540	..	Conversion Factor (Annual Avg)
				(Millions of US dollars, outstanding at end of year)							**EXTERNAL DEBT, ETC.**
644.1	920.0	913.5	833.2	805.2	808.9	692.6	798.5	894.6	1,042.1	..	Public/Publicly Guar. Long-Term
237.7	446.5	496.4	537.7	550.1	647.0	588.5	701.9	813.8	959.9	..	Official Creditors
26.6	34.1	46.6	54.0	64.4	103.7	142.2	186.7	244.2	288.4	..	IBRD and IDA
406.4	473.5	417.1	295.5	255.1	161.9	104.1	96.7	80.7	82.2	..	Private Creditors
..	..	..	..	..	..	..	..	..	..	..	Private Non-guaranteed Long-Term
0.0	0.0	13.9	21.1	20.0	39.3	49.3	62.5	80.8	78.4	..	Use of Fund Credit
90.0	98.0	113.2	110.8	127.5	65.3	63.0	74.0	89.2	102.5	..	Short-Term Debt
				(Millions of US dollars)							Memo Items:
70.0	65.5	77.6	151.5	167.7	172.9	203.3	296.6	332.7	354.9	..	Int'l Reserves Excluding Gold
2.2	6.4	7.4	5.0	5.7	4.8	3.9	4.1	4.9	6.1	..	Gold Holdings (at market price)
											SOCIAL INDICATORS
6.5	6.5	6.5	6.5	6.5	6.5	6.5	6.5	6.5	6.5	..	Total Fertility Rate
49.8	49.7	49.5	49.4	49.2	49.1	49.0	48.9	48.8	48.7	..	Crude Birth Rate
109.2	107.4	105.6	103.8	102.0	100.4	98.8	97.3	95.7	94.1	..	Infant Mortality Rate
48.5	49.0	49.5	50.0	50.5	51.0	51.5	52.0	52.5	53.1	..	Life Expectancy at Birth
98.9	102.6	99.4	98.1	94.0	86.3	93.9	89.7	89.2	89.1	..	Food Production, p.c. ('79-81 = 100)
73.8	73.4	73.0	..	..	..	..	..	..	..	..	Labor Force, Agriculture (%)
38.6	38.5	38.5	38.3	38.1	37.9	37.7	37.5	37.2	37.0	..	Labor Force, Female (%)
110.0	114.0	123.0	118.0	111.0	100.0	97.0	95.0	..	..	..	School Enroll. Ratio, primary
29.0	32.0	34.0	32.0	29.0	24.0	21.0	21.0	..	..	..	School Enroll. Ratio, secondary

TONGA	1967	1968	1969	1970	1971	1972	1973	1974	1975	1976	1977
CURRENT GNP PER CAPITA (US $)	..	..	..	..	..	..	..	..	..	..	..
POPULATION (thousands)	79	81	83	86	87	88	89	89	90	90	90
ORIGIN AND USE OF RESOURCES					*(Millions of current Tongan Pa'anga)*						
Gross National Product (GNP)	..	..	..	..	..	..	..	..	..	..	..
Net Factor Income from Abroad	..	..	..	..	..	..	..	..	..	..	..
Gross Domestic Product (GDP)	..	..	..	..	..	..	..	..	24.83	24.58	30.79
Indirect Taxes, net	..	..	..	..	..	..	..	..	..	..	..
GDP at factor cost	..	..	..	..	..	..	..	..	21.01	21.46	26.68
Agriculture	..	..	..	..	..	..	..	..	10.53	9.82	11.71
Industry	..	..	..	..	..	..	..	..	2.18	2.27	3.03
Manufacturing	..	..	..	..	..	..	..	..	1.11	1.30	1.77
Services, etc.	..	..	..	..	..	..	..	..	8.31	9.37	11.94
Resource Balance	..	..	..	..	..	..	..	..	-7.32	-7.30	-7.18
Exports of Goods & NFServices	..	..	..	..	..	..	..	..	10.26	6.62	10.43
Imports of Goods & NFServices	..	..	..	..	..	..	..	..	17.59	13.92	17.60
Domestic Absorption	..	..	..	..	..	..	..	..	32.15	31.88	37.96
Private Consumption, etc.	..	..	..	..	..	..	..	..	22.08	22.33	27.04
General Gov't Consumption	..	..	..	..	..	..	..	..	3.23	4.71	4.71
Gross Domestic Investment	..	..	..	..	..	..	..	..	6.85	4.83	6.22
Fixed Investment	..	..	..	..	..	..	..	..	5.66	4.86	5.90
Memo Items:											
Gross Domestic Saving	..	..	..	..	0.65	..	-0.90	-0.15	-0.47	-2.46	-0.96
Gross National Saving	..	..	..	..	..	..	..	..	..	..	..
					(Millions of 1980 Tongan Pa'anga)						
Gross National Product	..	..	..	..	..	..	..	..	..	..	..
GDP at factor cost	..	..	..	..	..	..	..	..	33.36	34.83	36.35
Agriculture	..	..	..	..	..	..	..	..	15.47	17.02	16.55
Industry	..	..	..	..	..	..	..	..	3.51	3.12	3.78
Manufacturing	..	..	..	..	..	..	..	..	2.15	2.17	2.59
Services, etc.	..	..	..	..	..	..	..	..	14.38	14.69	16.01
Resource Balance	..	..	..	..	..	..	..	..	..	..	..
Exports of Goods & NFServices	..	..	..	..	..	..	..	..	..	..	..
Imports of Goods & NFServices	..	..	..	..	..	..	..	..	..	..	..
Domestic Absorption	..	..	..	..	..	..	..	..	..	..	..
Private Consumption, etc.	..	..	..	..	..	..	..	..	..	..	..
General Gov't Consumption	..	..	..	..	..	..	..	..	..	..	..
Gross Domestic Investment	..	..	..	..	..	..	..	..	..	..	..
Fixed Investment	..	..	..	..	..	..	..	..	..	..	..
Memo Items:											
Capacity to Import	..	..	..	..	..	..	..	..	..	..	..
Terms of Trade Adjustment	..	..	..	..	..	..	..	..	..	..	..
Gross Domestic Income	..	..	..	..	..	..	..	..	..	..	..
Gross National Income	..	..	..	..	..	..	..	..	..	..	..
DOMESTIC PRICES (DEFLATORS)					*(Index 1980 = 100)*						
Overall (GDP)	..	..	..	..	..	..	..	..	..	..	..
Domestic Absorption	..	..	..	..	..	..	..	..	..	..	..
Agriculture	..	..	..	..	..	..	..	..	68.1	57.7	70.7
Industry	..	..	..	..	..	..	..	..	62.1	72.7	80.2
Manufacturing	..	..	..	..	..	..	..	..	51.7	59.9	68.3
MANUFACTURING ACTIVITY											
Employment (1980=100)	..	..	..	..	..	..	..	..	59.0	65.5	76.1
Real Earnings per Empl. (1980=100)	..	..	..	..	..	..	..	..	..	..	..
Real Output per Empl. (1980=100)	..	..	..	..	..	..	..	..	82.4	66.4	87.8
Earnings as % of Value Added	..	..	..	..	..	..	..	..	..	..	..
MONETARY HOLDINGS					*(Millions of current Tongan Pa'anga)*						
Money Supply, Broadly Defined	..	..	..	..	..	..	..	..	..	..	..
Money as Means of Payment	..	..	..	..	..	..	..	..	..	..	..
Currency Ouside Banks	..	..	..	..	..	..	..	..	..	..	..
Demand Deposits	..	..	..	..	..	..	..	..	..	..	..
Quasi-Monetary Liabilities	..	..	..	..	..	..	..	..	..	..	..
					(Millions of current Tongan Pa'anga)						
GOVERNMENT DEFICIT (-) OR SURPLUS	..	..	..	..	..	..	..	..	..	..	..
Current Revenue	..	..	..	..	..	..	..	..	..	..	..
Current Expenditure	..	..	..	..	..	..	..	..	..	..	..
Current Budget Balance	..	..	..	..	..	..	..	..	..	..	..
Capital Receipts	..	..	..	..	..	..	..	..	..	..	..
Capital Payments	..	..	..	..	..	..	..	..	..	..	..

1978	1979	1980	1981	1982	1983	1984	1985	1986	1987 estimate	NOTES	TONGA
..	..	..	..	..	770	780	710	700	720	..	**CURRENT GNP PER CAPITA** (US $)
90	90	90	92	94	95	96	97	98	99	..	**POPULATION** (thousands)
											ORIGIN AND USE OF RESOURCES
			(Millions of current Tongan Pa'anga)								
..	..	..	57.72	63.69	70.17	80.31	82.60	102.09	113.02	..	Gross National Product (GNP)
..	..	..	3.36	3.01	3.23	5.83	2.60	3.21	3.52	..	Net Factor Income from Abroad
36.32	39.96	46.77	54.36	60.68	66.94	74.48	80.00	98.88	109.50	..	Gross Domestic Product (GDP)
..	..	..	..	8.40	9.31	10.10	10.86	..	..	..	Indirect Taxes, net
31.03	34.71	40.58	47.31	52.29	57.63	64.38	69.14	..	..	..	GDP at factor cost
12.97	14.48	15.62	17.83	24.28	25.45	28.75	28.54	..	..	..	Agriculture
4.34	5.25	5.86	7.17	7.69	9.78	11.57	13.15	..	..	..	Industry
2.30	2.69	2.84	4.47	4.67	5.06	5.80	6.57	..	..	..	Manufacturing
13.72	14.98	19.10	22.31	20.32	22.41	24.06	27.44	..	..	..	Services, etc.
-14.19	-17.47	-16.07	-20.69	-24.75	-27.36	-26.78	-41.60	..	..	..	Resource Balance
10.46	10.42	14.06	14.31	16.45	14.31	19.83	20.10	..	..	..	Exports of Goods & NFServices
24.65	27.89	30.13	35.00	41.20	41.66	46.61	61.70	..	..	..	Imports of Goods & NFServices
50.51	57.43	62.85	75.05	85.43	94.30	101.25	121.60	..	..	..	Domestic Absorption
34.45	39.02	42.56	53.58	59.53	63.10	64.23	78.60	..	..	..	Private Consumption, etc.
5.34	5.96	6.86	7.68	10.82	10.30	11.80	14.10	..	..	..	General Gov't Consumption
10.72	12.45	13.43	13.79	15.09	20.89	25.22	28.90	..	..	..	Gross Domestic Investment
8.96	11.24	12.19	12.56	14.21	20.23	24.67	28.40	..	..	..	Fixed Investment
											Memo Items:
-3.48	-5.02	-2.64	-6.90	-9.66	-6.46	-1.56	-12.70	..	..	..	Gross Domestic Saving
..	..	..	9.38	16.27	18.57	24.37	13.10	..	..	..	Gross National Saving
			(Millions of 1980 Tongan Pa'anga)								
..	..	..	..	..	..	..	..	..	..	..	Gross National Product
36.95	38.29	40.58	40.33	45.39	46.01	47.21	50.50	52.46	..	..	GDP at factor cost
15.35	15.24	15.62	15.04	16.50	14.82	15.09	16.47	16.99	..	..	Agriculture
4.74	5.46	5.86	5.67	5.14	6.07	6.65	7.04	6.95	..	..	Industry
2.73	2.80	2.84	4.95	5.18	5.30	5.56	5.90	6.77	..	..	Manufacturing
16.86	17.58	19.10	19.62	23.75	25.12	25.46	26.99	28.52	..	..	Services, etc.
..	..	..	..	..	..	..	..	..	..	..	Resource Balance
..	..	..	..	..	..	..	..	..	..	..	Exports of Goods & NFServices
..	..	..	..	..	..	..	..	..	..	..	Imports of Goods & NFServices
..	..	..	..	..	..	..	..	..	..	..	Domestic Absorption
..	..	..	..	..	..	..	..	..	..	..	Private Consumption, etc.
..	..	..	..	..	..	..	..	..	..	..	General Gov't Consumption
..	..	..	..	..	..	..	..	..	..	..	Gross Domestic Investment
..	..	..	..	..	..	..	..	..	..	..	Fixed Investment
											Memo Items:
..	..	..	..	..	..	..	..	..	..	..	Capacity to Import
..	..	..	..	..	..	..	..	..	..	..	Terms of Trade Adjustment
..	..	..	..	..	..	..	..	..	..	..	Gross Domestic Income
..	..	..	..	..	..	..	..	..	..	..	Gross National Income
			(Index 1980 = 100)								**DOMESTIC PRICES (DEFLATORS)**
..	..	..	..	..	..	..	..	..	..	..	Overall (GDP)
..	..	..	..	..	..	..	..	..	..	..	Domestic Absorption
84.5	95.0	100.0	118.5	147.1	171.8	190.5	173.3	..	..	..	Agriculture
91.6	96.2	100.0	126.4	149.6	161.1	173.9	186.7	..	..	..	Industry
84.2	96.1	100.0	90.3	90.1	95.5	104.4	111.3	..	..	..	Manufacturing
											MANUFACTURING ACTIVITY
86.8	103.3	100.0	122.1	..	..	..	..	..	..	..	Employment (1980 = 100)
..	107.2	100.0	94.2	..	..	..	..	..	..	..	Real Earnings per Empl. (1980 = 100)
73.0	70.9	100.0	91.9	..	..	..	..	..	..	..	Real Output per Empl. (1980 = 100)
..	..	..	33.8	..	..	..	..	..	..	..	Earnings as % of Value Added
			(Millions of current Tongan Pa'anga)								**MONETARY HOLDINGS**
..	..	..	..	..	..	..	..	27.0	36.1	..	Money Supply, Broadly Defined
..	..	..	..	7.4	9.6	10.1	10.9	13.3	15.2	..	Money as Means of Payment
..	..	..	..	2.5	3.2	3.4	3.6	4.5	4.9	..	Currency Ouside Banks
..	..	..	..	4.9	6.5	6.7	7.4	8.8	10.3	..	Demand Deposits
..	..	..	..	..	..	..	..	13.7	20.9	..	Quasi-Monetary Liabilities
			(Millions of current Tongan Pa'anga)								**GOVERNMENT DEFICIT** (-) OR SURPLUS
..	..	..	..	..	..	..	..	..	..	..	Current Revenue
..	..	..	..	..	..	..	..	..	..	..	Current Expenditure
..	..	..	..	..	..	..	..	..	..	..	Current Budget Balance
..	..	..	..	..	..	..	..	..	..	..	Capital Receipts
..	..	..	..	..	..	..	..	..	..	..	Capital Payments

TONGA	1967	1968	1969	1970	1971	1972	1973	1974	1975	1976	1977
FOREIGN TRADE (CUSTOMS BASIS)					*(Thousands of current US dollars)*						
Value of Exports, fob	..	..	..	..	..	..	..	..	..	..	..
Nonfuel Primary Products	..	..	..	..	..	..	..	..	..	..	..
Fuels	..	..	..	..	..	..	..	..	..	..	..
Manufactures	..	..	..	..	..	..	..	..	..	..	..
Value of Imports, cif	..	..	..	..	..	..	..	..	..	..	..
Nonfuel Primary Products	..	..	..	..	..	..	..	..	..	..	..
Fuels	..	..	..	..	..	..	..	..	..	..	..
Manufactures	..	..	..	..	..	..	..	..	..	..	..
					(Index 1980 = 100)						
Terms of Trade	..	..	..	..	..	..	..	..	..	..	..
Export Prices, fob	..	..	..	..	..	..	..	..	..	..	..
Nonfuel Primary Products	..	..	..	..	..	..	..	..	..	..	..
Fuels	..	..	..	..	..	..	..	..	..	..	..
Manufactures	..	..	..	..	..	..	..	..	..	..	..
Import Prices, cif	..	..	..	..	..	..	..	..	..	..	..
Trade at Constant 1980 Prices					*(Thousands of 1980 US dollars)*						
Exports, fob	..	..	..	..	..	..	..	..	..	..	..
Imports, cif	..	..	..	..	..	..	..	..	..	..	..
BALANCE OF PAYMENTS					*(Thousands of current US dollars)*						
Exports of Goods & Services	..	..	..	..	4	5	6	9	15	10	10
Merchandise, fob	..	..	..	..	3	3	4	5	7	4	4
Nonfactor Services	..	..	..	..	1	1	2	3	8	5	5
Factor Services	..	..	..	..	0	0	0	0	0	1	1
Imports of Goods & Services	..	..	..	..	6	8	10	13	22	19	19
Merchandise, fob	..	..	..	..	5	7	8	11	18	16	15
Nonfactor Services	..	..	..	..	1	1	1	2	4	3	4
Factor Services	..	..	..	..	0	0	0	0	0	0	0
Long-Term Interest	..	..	..	..	..	..	..	..	..	..	..
Current Transfers, net	..	..	..	..	2	2	3	5	7	7	7
Workers' Remittances	..	..	..	..	1	0	1	4	7	5	5
Total to be Financed	..	..	..	..	-1	-1	-1	1	0	-2	-2
Official Capital Grants	..	..	..	..	..	..	..	..	..	..	..
Current Account Balance	..	..	..	..	-1	-1	-1	1	0	-2	-2
Long-Term Capital, net	..	..	..	..	1	1	1	0	2	2	1
Direct Investment	..	..	..	..	..	..	..	..	..	..	..
Long-Term Loans	..	..	..	..	..	..	..	..	..	..	..
Disbursements	..	..	..	..	..	..	..	..	..	..	..
Repayments	..	..	..	..	..	..	..	..	..	..	..
Other Long-Term Capital	..	..	..	..	1	1	1	0	2	2	1
Other Capital, net	..	..	..	..	-1	0	0	1	-1	-2	3
Change in Reserves	..	..	..	..	0	0	0	-2	-2	2	-2
Memo Item:					*(Tongan Pa'anga per US dollar)*						
Conversion Factor (Annual Avg)	0.890	0.890	0.890	0.890	0.890	0.850	0.780	0.680	0.730	0.790	0.870
EXTERNAL DEBT, ETC.					*(Millions of US dollars, outstanding at end of year)*						
Public/Publicly Guar. Long-Term	..	..	..	..	..	..	..	..	..	..	..
Official Creditors	..	..	..	..	..	..	..	..	..	..	..
IBRD and IDA	..	..	..	..	..	..	..	..	..	..	..
Private Creditors	..	..	..	..	..	..	..	..	..	..	..
Private Non-guaranteed Long-Term	..	..	..	..	..	..	..	..	..	..	..
Use of Fund Credit	..	..	..	..	..	..	..	..	..	..	..
Short-Term Debt	..	..	..	..	..	..	..	..	..	..	..
Memo Items:					*(Thousands of US dollars)*						
Int'l Reserves Excluding Gold	..	..	..	..	..	..	..	..	..	..	9,001.0
Gold Holdings (at market price)	..	..	..	..	..	..	..	..	..	..	..
SOCIAL INDICATORS											
Total Fertility Rate	6.9	6.8	6.6	6.4	6.3	6.1	5.9	5.8	5.6	5.4	5.3
Crude Birth Rate	44.4	42.8	41.2	39.6	38.0	36.4	34.8	33.2	31.6	30.0	29.8
Infant Mortality Rate	56.4	56.8	57.2	57.6	58.0	58.4	58.8	59.2	59.6	60.0	..
Life Expectancy at Birth	..	..	..	..	..	..	..	..	..	58.0	58.8
Food Production, p.c. ('79-81 = 100)	103.5	96.7	99.3	90.3	92.8	101.1	95.5	101.3	108.3	110.4	104.1
Labor Force, Agriculture (%)	..	..	..	..	..	..	..	..	..	..	..
Labor Force, Female (%)	..	..	..	..	..	..	..	..	..	..	..
School Enroll. Ratio, primary	..	..	..	..	..	..	..	..	..	..	..
School Enroll. Ratio, secondary	..	..	..	..	..	..	..	..	..	..	..

1978	1979	1980	1981	1982	1983	1984	1985	1986	1987 estimate	NOTES	TONGA
											FOREIGN TRADE (CUSTOMS BASIS)
				(Thousands of current US dollars)							
..	..	9,000	10,000	6,000	6,000	6,000	9,000	..	..	..	Value of Exports, fob
..	..	..	..	..	..	..	..	..	..	..	Nonfuel Primary Products
..	..	..	..	..	..	..	..	..	..	..	Fuels
..	..	..	..	..	..	..	..	..	..	..	Manufactures
..	..	34,300	50,200	50,520	50,660	70,300	62,940	76,160	..	..	Value of Imports, cif
..	..	..	..	..	..	..	..	..	..	..	Nonfuel Primary Products
..	..	..	..	..	..	..	..	..	..	..	Fuels
..	..	..	..	..	..	..	..	..	..	..	Manufactures
				(Index 1980 = 100)							
..	..	..	..	..	..	..	..	..	..	..	Terms of Trade
..	..	..	..	..	..	..	..	..	..	..	Export Prices, fob
..	..	..	..	..	..	..	..	..	..	..	Nonfuel Primary Products
..	..	..	..	..	..	..	..	..	..	..	Fuels
..	..	..	..	..	..	..	..	..	..	..	Manufactures
..	..	..	..	..	..	..	..	..	..	..	Import Prices, cif
				(Thousands of 1980 US dollars)							Trade at Constant 1980 Prices
..	..	..	..	..	..	..	..	..	..	..	Exports, fob
..	..	..	..	..	..	..	..	..	..	..	Imports, cif
				(Thousands of current US dollars)							**BALANCE OF PAYMENTS**
15	14	17	21	23	14	21	26	23	29	..	Exports of Goods & Services
8	6	7	8	8	5	7	8	5	6	..	Merchandise, fob
6	6	7	8	11	5	8	15	15	20	..	Nonfactor Services
1	1	3	5	4	3	6	3	3	4	..	Factor Services
24	27	33	43	45	37	39	47	47	51	..	Imports of Goods & Services
19	23	27	37	39	31	32	32	35	34	..	Merchandise, fob
5	4	5	5	5	6	7	15	12	17	..	Nonfactor Services
0	0	1	1	1	0	1	0	0	..	..	Factor Services
..	..	..	..	..	..	..	..	..	..	..	Long-Term Interest
9	11	12	15	25	20	18	16	23	22	..	Current Transfers, net
7	7	7	13	15	16	14	15	21	20	..	Workers' Remittances
-1	-3	-4	-7	3	-3	0	-5	-1	1	K	Total to be Financed
..	0	0	0	0	0	0	3	1	6	K	Official Capital Grants
-1	-3	-4	-7	3	-3	0	-2	0	7	..	Current Account Balance
3	3	2	3	2	1	5	4	0	7	..	Long-Term Capital, net
..	0	0	0	0	0	0	0	0	0	..	Direct Investment
..	..	..	..	..	..	..	..	..	..	..	Long-Term Loans
..	..	..	..	..	..	..	..	..	..	..	Disbursements
..	..	..	..	..	..	..	..	..	..	..	Repayments
3	3	2	3	2	1	5	1	-1	0	..	Other Long-Term Capital
1	2	5	5	-4	2	1	0	-1	-7	..	Other Capital, net
-3	-2	-4	-1	-1	0	-5	-3	1	-7	..	Change in Reserves
				(Tongan Pa'anga per US dollar)							Memo Item:
0.890	0.880	0.900	0.860	0.910	1.070	1.100	1.430	1.500	1.430	..	Conversion Factor (Annual Avg)
				(Millions of US dollars, outstanding at end of year)							**EXTERNAL DEBT, ETC.**
..	..	..	..	..	..	..	..	..	..	..	Public/Publicly Guar. Long-Term
..	..	..	..	..	..	..	..	..	..	..	Official Creditors
..	..	..	..	..	..	..	..	..	..	..	IBRD and IDA
..	..	..	..	..	..	..	..	..	..	..	Private Creditors
..	..	..	..	..	..	..	..	..	..	..	Private Non-guaranteed Long-Term
..	..	..	..	..	..	..	..	..	..	..	Use of Fund Credit
..	..	..	..	..	..	..	..	..	..	..	Short-Term Debt
				(Thousands of US dollars)							Memo Items:
10,116.0	12,560.0	13,753.0	13,981.0	15,561.0	20,953.0	26,017.0	26,697.0	21,540.0	27,760.0	..	Int'l Reserves Excluding Gold
..	..	..	..	..	..	..	..	..	..	..	Gold Holdings (at market price)
											SOCIAL INDICATORS
5.1	4.9	4.7	4.5	4.3	4.2	4.2	4.1	4.1	4.0	..	Total Fertility Rate
29.7	29.5	29.3	29.2	29.0	29.3	29.6	29.8	30.1	30.4	..	Crude Birth Rate
..	..	..	..	..	..	..	..	..	49.9	..	Infant Mortality Rate
59.6	60.4	61.2	62.0	62.9	63.2	63.6	64.0	64.4	64.8	..	Life Expectancy at Birth
96.3	96.5	101.0	102.5	90.9	86.2	86.0	86.1	88.8	89.0	..	Food Production, p.c. ('79-81 = 100)
..	..	..	..	..	..	..	51.0	..	..	..	Labor Force, Agriculture (%)
..	..	..	..	..	..	..	19.4	..	..	..	Labor Force, Female (%)
..	..	..	..	..	..	..	..	..	..	..	School Enroll. Ratio, primary
..	..	..	..	..	..	..	..	..	..	..	School Enroll. Ratio, secondary

TRINIDAD AND TOBAGO	1967	1968	1969	1970	1971	1972	1973	1974	1975	1976	1977
CURRENT GNP PER CAPITA (US $)	780	790	810	830	850	970	1,110	1,280	2,000	2,470	2,760
POPULATION (thousands)	921	932	944	955	966	975	985	996	1,009	1,024	1,040
ORIGIN AND USE OF RESOURCES					*(Millions of current Trinidad & Tobago Dollars)*						
Gross National Product (GNP)	1,205	1,381	1,415	1,513	1,639	1,945	2,384	3,607	5,119	5,826	7,069
Net Factor Income from Abroad	-119	-137	-143	-130	-132	-136	-180	-586	-181	-265	-464
Gross Domestic Product (GDP)	1,324	1,518	1,558	1,644	1,771	2,081	2,564	4,193	5,300	6,090	7,533
Indirect Taxes, net	65	78	83	94	107	127	133	76	90	62	112
GDP at factor cost	1,259	1,440	1,475	1,550	1,664	1,954	2,432	4,117	5,210	6,029	7,421
Agriculture	61	72	76	80	85	107	111	135	174	238	261
Industry	564	679	670	684	740	885	1,199	2,474	3,159	3,587	4,389
Manufacturing	268	357	367	396	410	482	514	686	792	876	1,027
Services, etc.	634	689	729	786	839	962	1,122	1,508	1,877	2,204	2,771
Resource Balance	113	143	66	18	-91	-139	148	1,028	943	956	953
Exports of Goods & NFServices	551	649	677	703	757	824	1,132	2,378	2,808	3,401	3,733
Imports of Goods & NFServices	438	506	611	685	848	963	984	1,350	1,865	2,445	2,779
Domestic Absorption	1,212	1,375	1,492	1,626	1,862	2,220	2,416	3,165	4,357	5,134	6,579
Private Consumption, etc.	818	900	1,038	986	980	1,234	1,385	1,774	2,256	2,896	3,605
General Gov't Consumption	161	175	187	215	280	334	366	475	652	743	967
Gross Domestic Investment	233	301	268	425	602	652	665	915	1,449	1,496	2,008
Fixed Investment	223	292	272	344	581	615	579	651	1,085	1,398	1,735
Memo Items:											
Gross Domestic Saving	346	443	334	443	511	513	814	1,943	2,392	2,452	2,961
Gross National Saving	230	306	196	318	384	383	630	1,342	2,188	2,162	2,459
					(Millions of 1980 Trinidad & Tobago Dollars)						
Gross National Product	7,818	8,285	8,497	8,876	9,025	9,669	9,732	9,362	10,291	11,041	11,604
GDP at factor cost	7,829	8,299	8,492	8,768	8,898	9,440	9,656	10,337	10,420	11,362	12,072
Agriculture	346	381	358	406	390	431	397	390	406	393	396
Industry	4,907	5,266	5,378	5,547	5,718	6,099	6,221	6,631	6,586	7,432	7,605
Manufacturing	910	995	1,076	1,126	1,120	1,177	1,155	1,204	1,048	1,243	1,207
Services, etc.	2,576	2,651	2,756	2,815	2,789	2,909	3,038	3,315	3,428	3,538	4,071
Resource Balance	3,744	4,411	4,553	4,651	3,341	4,266	4,929	4,588	3,336	3,225	2,838
Exports of Goods & NFServices	4,968	5,664	6,004	6,330	5,408	6,277	6,987	6,882	6,199	6,831	6,465
Imports of Goods & NFServices	1,224	1,252	1,451	1,678	2,067	2,011	2,059	2,294	2,863	3,606	3,627
Domestic Absorption	4,500	4,326	4,399	4,627	6,101	5,759	5,230	5,932	7,258	8,248	9,407
Private Consumption, etc.	3,155	2,817	3,035	2,797	3,576	3,323	2,922	3,039	3,682	4,233	5,369
General Gov't Consumption	559	600	639	733	804	920	941	919	1,183	1,276	1,211
Gross Domestic Investment	786	909	724	1,097	1,721	1,516	1,367	1,975	2,393	2,739	2,828
Fixed Investment	..	..	..	..	..	..	..	..	..	..	..
Memo Items:											
Capacity to Import	1,539	1,606	1,609	1,723	1,845	1,722	2,369	4,042	4,310	5,016	4,870
Terms of Trade Adjustment	-3,429	-4,058	-4,395	-4,607	-3,563	-4,555	-4,618	-2,840	-1,890	-1,815	-1,595
Gross Domestic Income	4,815	4,680	4,556	4,671	5,879	5,469	5,540	7,680	8,705	9,658	10,651
Gross National Income	4,389	4,227	4,102	4,269	5,462	5,114	5,114	6,522	8,402	9,226	10,010
DOMESTIC PRICES (DEFLATORS)					*(Index 1980 = 100)*						
Overall (GDP)	16.1	17.4	17.4	17.7	18.8	20.8	25.2	39.9	50.0	53.1	61.5
Domestic Absorption	26.9	31.8	33.9	35.1	30.5	38.6	46.2	53.3	60.0	62.3	69.9
Agriculture	17.6	18.9	21.1	19.8	21.9	24.8	27.9	34.6	42.8	60.6	65.9
Industry	11.5	12.9	12.5	12.3	12.9	14.5	19.3	37.3	48.0	48.3	57.7
Manufacturing	29.4	35.8	34.1	35.2	36.6	40.9	44.5	57.0	75.6	70.5	85.1
MANUFACTURING ACTIVITY											
Employment (1980 = 100)	42.0	45.5	57.4	62.6	75.1	71.6	82.1	76.7	83.1	91.9	95.0
Real Earnings per Empl. (1980 = 100)	81.2	77.4	..	..	..	..	..	99.6	100.2	103.1	108.6
Real Output per Empl. (1980 = 100)	88.7	75.4	..	..	..	..	..	84.3	73.2	93.8	87.8
Earnings as % of Value Added	40.1	40.5	..	..	..	..	..	38.1	36.8	40.0	38.5
MONETARY HOLDINGS					*(Millions of current Trinidad & Tobago Dollars)*						
Money Supply, Broadly Defined	324	370	424	504	620	742	844	1,100	1,564	2,108	2,695
Money as Means of Payment	130	138	140	151	177	211	213	270	392	572	725
Currency Ouside Banks	42	49	51	57	68	81	80	99	138	177	231
Demand Deposits	87	90	89	95	108	130	133	171	254	395	494
Quasi-Monetary Liabilities	194	232	284	353	444	531	631	831	1,171	1,536	1,969
					(Millions of current Trinidad & Tobago Dollars)						
GOVERNMENT DEFICIT (-) OR SURPLUS	..	..	..	..	..	..	..	..	..	499.9	799.7
Current Revenue	..	..	..	..	..	..	..	..	..	2,340.5	3,037.9
Current Expenditure	..	..	..	..	..	..	..	..	..	948.5	1,069.0
Current Budget Balance	..	..	..	..	..	..	..	..	..	1,392.0	1,968.9
Capital Receipts	..	..	..	..	..	..	..	..	..	0.3	6.0
Capital Payments	..	..	..	..	..	..	..	..	..	892.4	1,175.2

1978	1979	1980	1981	1982	1983	1984	1985	1986	1987 estimate	NOTES	TRINIDAD AND TOBAGO
3,350	3,690	4,690	5,980	6,970	6,450	6,060	6,180	5,360	4,220	..	**CURRENT GNP PER CAPITA (US $)**
1,057	1,076	1,095	1,115	1,134	1,153	1,170	1,185	1,199	1,217	..	POPULATION (thousands)
			(Millions of current Trinidad & Tobago Dollars)								**ORIGIN AND USE OF RESOURCES**
8,363	10,435	14,210	15,957	19,445	18,824	18,356	18,086	17,280	14,977	..	Gross National Product (GNP)
-187	-611	-756	-481	269	-295	-729	-818	-732	-956	..	Net Factor Income from Abroad
8,550	11,046	14,966	16,438	19,176	19,119	19,086	18,905	18,012	15,933	..	Gross Domestic Product (GDP)
49	-231	-588	-537	-856	-664	-141	-17	620	..	..	Indirect Taxes, net
8,501	11,276	15,554	16,975	20,032	19,783	19,226	18,922	17,392	15,323	..	GDP at factor cost
302	322	337	386	432	815	867	877	945	640	..	Agriculture
4,686	6,430	9,358	9,799	9,564	8,364	8,158	7,214	6,157	5,989	..	Industry
1,074	1,567	1,338	1,118	1,349	1,448	1,433	1,265	1,427	1,497	..	Manufacturing
3,513	4,524	5,860	6,791	10,035	10,604	10,201	10,830	10,290	8,694	..	Services, etc.
375	634	1,716	1,590	-1,418	-1,918	-325	743	-678	..	..	Resource Balance
3,766	4,979	7,550	7,542	6,703	5,637	5,870	5,885	5,925	..	..	Exports of Goods & NFServices
3,391	4,345	5,834	5,952	8,121	7,555	6,195	5,143	6,604	..	..	Imports of Goods & NFServices
8,175	10,412	13,249	14,848	20,594	21,037	19,411	18,162	18,690	..	..	Domestic Absorption
4,443	5,663	6,865	8,197	12,041	12,748	11,676	10,810	11,231	..	..	Private Consumption, etc.
1,148	1,536	1,804	2,110	3,136	3,320	3,616	3,464	3,459	..	..	General Gov't Consumption
2,584	3,213	4,580	4,541	5,417	4,969	4,119	3,888	4,000	..	..	Gross Domestic Investment
2,323	2,952	4,204	4,342	5,189	4,770	3,954	3,733	3,736	..	..	Fixed Investment
											Memo Items:
2,959	3,847	6,297	6,131	3,999	3,051	3,794	4,631	3,322	..	..	Gross Domestic Saving
2,722	3,171	5,440	5,484	4,072	2,585	2,893	3,677	2,479	..	..	Gross National Saving
			(Millions of 1980 Trinidad & Tobago Dollars)								
13,346	13,196	14,210	15,118	15,018	13,553	11,971	11,297	11,299	10,828	..	Gross National Product
13,508	14,191	15,554	16,052	15,511	14,267	12,547	11,841	11,165	10,831	..	GDP at factor cost
379	366	337	333	343	401	398	418	429	440	..	Agriculture
8,665	8,920	9,358	9,535	8,419	7,567	6,761	5,940	5,688	5,407	..	Industry
1,273	1,306	1,338	1,250	903	825	717	615	647	778	..	Manufacturing
4,464	4,905	5,860	6,185	6,748	6,299	5,388	5,483	5,049	4,984	..	Services, etc.
2,430	1,553	1,716	2,103	-193	9	1,673	2,470	4,383	..	..	Resource Balance
6,509	6,995	7,550	7,433	6,750	6,568	7,255	7,863	6,442	..	..	Exports of Goods & NFServices
4,079	5,443	5,834	5,330	6,942	6,559	5,582	5,393	2,058	..	..	Imports of Goods & NFServices
11,152	12,356	13,250	13,437	15,008	13,752	10,778	9,359	7,209	..	..	Domestic Absorption
6,174	6,916	6,865	7,548	9,029	8,459	6,307	4,862	4,394	..	..	Private Consumption, etc.
1,358	1,704	1,804	2,071	1,955	1,706	1,822	1,722	1,501	..	..	General Gov't Consumption
3,620	3,736	4,580	3,818	4,023	3,587	2,648	2,775	1,313	..	..	Gross Domestic Investment
..	..	..	..	..	..	..	..	..	..	..	Fixed Investment
											Memo Items:
4,530	6,236	7,550	6,754	5,730	4,894	5,289	6,172	1,847	..	..	Capacity to Import
-1,979	-759	0	-679	-1,020	-1,674	-1,966	-1,691	-4,595	..	..	Terms of Trade Adjustment
11,604	13,149	14,966	14,862	13,795	12,087	10,485	10,138	6,997	..	..	Gross Domestic Income
11,367	12,436	14,210	14,439	13,998	11,879	10,005	9,605	6,705	..	..	Gross National Income
			(Index 1980 = 100)								**DOMESTIC PRICES (DEFLATORS)**
62.9	79.4	100.0	105.8	129.4	138.9	153.3	159.8	155.4	140.7	..	Overall (GDP)
73.3	84.3	100.0	110.5	137.2	153.0	180.1	194.1	259.3	..	..	Domestic Absorption
79.7	88.1	100.0	115.9	126.0	202.9	217.9	209.8	220.4	145.2	..	Agriculture
54.1	72.1	100.0	102.8	113.6	110.5	120.7	121.5	108.2	110.8	..	Industry
84.4	120.0	100.0	89.4	149.4	175.4	199.8	205.8	220.6	192.5	..	Manufacturing
											MANUFACTURING ACTIVITY
97.8	102.7	100.0	97.4	71.5	69.7	68.0	66.4	..	..	..	Employment (1980 = 100)
119.9	134.0	100.0	75.1	110.8	136.3	131.6	120.3	..	..	..	Real Earnings per Empl. (1980 = 100)
97.4	95.0	100.0	95.9	94.5	115.5	111.1	105.4	..	..	..	Real Output per Empl. (1980 = 100)
44.4	41.0	41.0	41.0	41.0	41.0	41.0	41.0	..	..	..	Earnings as % of Value Added
			(Millions of current Trinidad & Tobago Dollars)								**MONETARY HOLDINGS**
3,457	4,426	5,159	6,672	9,027	9,990	10,481	10,791	10,622	..	D	Money Supply, Broadly Defined
936	1,147	1,337	1,855	2,547	2,435	2,294	2,260	2,073	2,187	..	Money as Means of Payment
296	412	467	532	726	758	710	685	723	719	..	Currency Ouside Banks
641	735	870	1,323	1,821	1,676	1,584	1,576	1,350	1,468	..	Demand Deposits
2,521	3,279	3,822	4,817	6,480	7,555	8,187	8,530	8,549	..	..	Quasi-Monetary Liabilities
			(Millions of current Trinidad & Tobago Dollars)								
301.6	-61.0	1,105.6	545.8	..	..	..	..	..	..	..	**GOVERNMENT DEFICIT (-) OR SURPLUS**
3,153.0	4,118.7	6,463.8	7,200.6	..	..	..	..	..	..	..	Current Revenue
1,461.4	2,108.8	2,815.3	3,123.1	..	..	..	..	..	..	..	Current Expenditure
1,691.6	2,009.9	3,648.5	4,077.5	..	..	..	..	..	..	..	Current Budget Balance
2.5	0.1	24.0	32.2	..	..	..	..	..	..	..	Capital Receipts
1,392.5	2,071.0	2,566.9	3,563.9	..	..	..	..	..	..	..	Capital Payments

TRINIDAD AND TOBAGO	1967	1968	1969	1970	1971	1972	1973	1974	1975	1976	1977
FOREIGN TRADE (CUSTOMS BASIS)					*(Millions of current US dollars)*						
Value of Exports, fob	441.2	466.2	474.6	481.5	520.5	557.6	695.7	2,037.7	1,772.7	2,219.3	2,179.8
Nonfuel Primary Products	38.3	42.6	44.0	46.5	44.4	53.1	52.2	88.4	117.1	89.0	77.6
Fuels	342.6	362.8	366.2	371.9	402.8	433.6	571.8	1,838.7	1,543.2	2,007.3	1,997.1
Manufactures	60.4	60.9	64.4	63.2	73.2	70.9	71.8	110.6	112.4	123.0	105.1
Value of Imports, cif	417.5	420.1	482.7	543.4	662.8	762.1	792.2	1,846.5	1,488.5	1,976.3	1,808.5
Nonfuel Primary Products	60.6	56.5	67.3	68.3	74.7	88.3	105.8	157.2	165.1	173.8	215.0
Fuels	207.1	223.1	253.6	287.6	330.4	364.1	402.1	1,327.8	752.6	1,131.4	861.0
Manufactures	149.8	140.4	161.8	187.5	257.7	309.6	284.3	361.5	570.8	671.0	732.5
Terms of Trade	69.4	73.2	72.8	73.1	70.7	68.3	69.5	93.1	83.3	86.4	82.3
					(Index 1980 = 100)						
Export Prices, fob	5.3	5.3	5.3	5.3	6.9	7.6	10.3	38.4	38.2	40.1	43.6
Nonfuel Primary Products	25.5	26.1	28.5	29.2	27.8	31.7	44.5	68.7	76.8	73.3	85.6
Fuels	4.3	4.3	4.3	4.3	5.6	6.2	8.9	36.7	35.7	38.4	42.0
Manufactures	30.1	31.7	32.4	37.5	36.5	38.5	49.5	67.2	66.1	66.7	71.7
Import Prices, cif	7.6	7.2	7.3	7.3	9.7	11.2	14.9	41.3	45.8	46.4	53.0
Trade at Constant 1980 Prices					*(Millions of 1980 US dollars)*						
Exports, fob	8,392.6	8,871.4	8,948.8	9,058.5	7,593.0	7,311.8	6,723.5	5,300.5	4,640.3	5,538.4	4,995.6
Imports, cif	5,509.5	5,851.0	6,630.4	7,470.4	6,831.9	6,824.8	5,324.2	4,472.7	3,247.0	4,261.0	3,411.1
BALANCE OF PAYMENTS					*(Millions of current US dollars)*						
Exports of Goods & Services	..	..	..	349.8	389.1	448.1	572.1	1,172.3	1,333.2	1,451.0	1,625.4
Merchandise, fob	..	..	..	225.3	217.2	254.5	333.3	928.1	987.4	1,054.3	1,175.2
Nonfactor Services	..	..	..	119.1	166.4	187.8	233.8	224.8	302.5	336.4	373.2
Factor Services	..	..	..	5.4	5.4	5.8	5.0	19.5	43.3	60.3	77.0
Imports of Goods & Services	..	..	..	456.7	553.9	568.7	598.8	888.6	980.2	1,165.5	1,420.2
Merchandise, fob	..	..	..	276.2	355.6	384.2	371.5	482.3	658.1	766.0	867.8
Nonfactor Services	..	..	..	108.8	129.2	108.7	131.8	169.3	195.6	233.8	285.1
Factor Services	..	..	..	71.7	69.2	75.8	95.5	237.0	126.5	165.7	267.2
Long-Term Interest	..	..	..	6.1	4.6	6.3	8.1	14.4	11.9	8.9	6.8
Current Transfers, net	..	..	..	2.6	3.0	2.9	-2.0	-7.3	-10.9	-10.5	-15.8
Workers' Remittances	..	..	..	2.7	3.2	3.3	2.6	2.5	2.6	1.6	1.4
Total to be Financed	..	..	..	-104.3	-161.9	-117.7	-28.7	276.4	342.1	275.0	189.4
Official Capital Grants	..	..	..	-4.3	-7.0	-6.2	-6.3	-10.1	-10.3	-19.5	-15.3
Current Account Balance	..	..	..	-108.6	-168.9	-123.9	-35.0	266.3	331.8	255.4	174.1
Long-Term Capital, net	..	..	..	76.0	91.3	97.0	52.5	59.3	32.9	-5.4	215.4
Direct Investment	..	..	..	83.2	103.3	86.1	65.6	120.1	93.0	132.2	83.5
Long-Term Loans	..	..	..	-2.5	-1.2	22.2	28.9	4.5	-3.7	-50.8	151.3
Disbursements	..	..	..	7.5	19.4	27.1	37.3	44.7	11.2	17.1	157.6
Repayments	..	..	..	10.0	20.6	4.9	8.4	40.2	14.9	67.9	6.3
Other Long-Term Capital	..	..	..	-0.4	-3.8	-5.2	-35.7	-55.3	-46.1	-67.2	-4.1
Other Capital, net	..	..	..	13.6	90.1	6.6	-24.2	-37.6	93.5	-49.5	54.4
Change in Reserves	..	..	..	19.0	-12.5	20.3	6.8	-288.0	-458.1	-200.5	-443.9
Memo Item:					*(Trinidad & Tobago Dollars per US dollar)*						
Conversion Factor (Annual Avg)	1.740	2.000	2.000	2.000	1.970	1.920	1.960	2.050	2.170	2.440	2.400
EXTERNAL DEBT, ETC.					*(Millions of US dollars, outstanding at end of year)*						
Public/Publicly Guar. Long-Term	..	..	..	100.6	101.5	122.1	150.8	155.4	148.9	95.9	247.3
Official Creditors	..	..	..	44.8	50.3	55.7	60.9	72.2	78.6	81.4	84.4
IBRD and IDA	..	..	..	23.2	28.8	32.0	36.0	40.2	45.0	51.6	56.3
Private Creditors	..	..	..	55.8	51.1	66.4	89.9	83.2	70.3	14.5	162.9
Private Non-guaranteed Long-Term	..	..	..	..	..	..	..	..	..	..	..
Use of Fund Credit	..	..	..	0.0	0.0	0.0	0.0	0.0	0.0	0.0	0.0
Short-Term Debt	..	..	..	..	..	..	..	..	..	..	92.0
Memo Items:					*(Millions of US dollars)*						
Int'l Reserves Excluding Gold	30.3	49.4	44.7	43.0	69.4	58.3	47.0	390.3	751.0	1,013.6	1,481.7
Gold Holdings (at market price)	..	..	..	..	..	..	..	..	..	..	4.5
SOCIAL INDICATORS											
Total Fertility Rate	3.9	3.8	3.7	3.6	3.6	3.5	3.4	3.4	3.4	3.3	3.3
Crude Birth Rate	30.3	29.6	28.8	28.1	27.4	26.6	26.6	26.5	26.4	26.4	26.3
Infant Mortality Rate	41.0	38.8	36.6	34.4	32.2	30.0	29.2	28.4	27.6	26.8	26.0
Life Expectancy at Birth	65.7	65.8	66.0	66.2	66.3	66.5	66.7	66.9	67.1	67.3	67.5
Food Production, p.c. ('79-81 = 100)	139.7	151.2	141.3	152.3	148.9	157.4	147.4	146.7	141.0	144.9	136.1
Labor Force, Agriculture (%)	19.5	19.2	18.9	18.6	17.8	16.9	16.1	15.2	14.4	13.6	12.7
Labor Force, Female (%)	..	..	..	..	..	..	..	..	29.0	..	..
School Enroll. Ratio, primary	..	..	..	106.0	..	..	..	..	99.0	96.0	94.0
School Enroll. Ratio, secondary	..	..	..	42.0	..	..	..	..	48.0	56.0	58.0

1978	1979	1980	1981	1982	1983	1984	1985	1986	1987 est.	NOTES	TRINIDAD AND TOBAGO
				(Millions of current US dollars)							**FOREIGN TRADE (CUSTOMS BASIS)**
2,042.7	2,610.4	4,080.0	3,760.0	3,090.0	2,352.7	2,173.4	2,160.9	1,385.7	1,462.4	..	Value of Exports, fob
66.2	83.3	86.7	77.1	70.8	58.4	57.1	53.2	65.2	76.1	..	Nonfuel Primary Products
1,829.4	2,370.3	3,787.4	3,368.1	2,697.1	1,963.9	1,760.6	1,710.0	980.4	1,041.4	..	Fuels
147.2	156.9	205.9	314.8	322.1	330.4	355.7	397.7	340.1	344.9	..	Manufactures
1,979.9	2,104.6	3,180.0	3,120.0	3,700.0	2,582.0	1,919.1	1,533.0	1,369.8	1,218.7	..	Value of Imports, cif
262.4	318.3	431.9	487.0	557.7	567.9	532.2	422.1	301.9	330.9	..	Nonfuel Primary Products
798.1	601.9	1,198.1	1,141.8	932.2	82.7	14.4	50.9	38.8	52.4	..	Fuels
919.4	1,184.4	1,550.0	1,491.1	2,210.1	1,931.4	1,372.5	1,060.0	1,029.1	835.5	..	Manufactures
				(Index 1980 = 100)							
75.6	78.3	100.0	106.5	102.4	96.7	96.6	96.0	52.7	61.1	..	Terms of Trade
44.4	62.8	100.0	110.6	100.2	91.9	89.8	86.8	53.1	64.6	..	Export Prices, fob
83.6	89.8	100.0	89.6	78.0	82.7	83.4	60.2	65.9	59.8	..	Nonfuel Primary Products
42.3	61.0	100.0	112.5	101.6	92.5	90.2	87.5	44.9	56.9	..	Fuels
76.1	90.0	100.0	99.1	94.8	90.7	89.3	89.0	103.2	113.0	..	Manufactures
58.8	80.3	100.0	103.9	97.8	95.1	93.0	90.5	100.7	105.7	..	Import Prices, cif
				(Millions of 1980 US dollars)							Trade at Constant 1980 Prices
4,597.4	4,154.1	4,080.0	3,398.7	3,084.1	2,558.9	2,419.5	2,488.6	2,609.6	2,262.7	..	Exports, fob
3,368.3	2,621.9	3,180.0	3,003.8	3,781.4	2,714.3	2,063.0	1,694.5	1,360.0	1,152.8	..	Imports, cif
				(Millions of current US dollars)							**BALANCE OF PAYMENTS**
1,689.7	2,227.6	3,371.2	3,488.2	3,146.2	2,575.9	2,581.2	2,594.7	1,734.0	..	..	Exports of Goods & Services
1,221.9	1,648.9	2,541.6	2,612.2	2,228.7	2,026.5	2,110.7	2,111.0	1,359.4	..	..	Merchandise, fob
342.8	418.8	597.3	531.3	560.6	319.1	329.0	287.8	278.1	..	..	Nonfactor Services
125.0	159.9	232.3	344.7	357.0	230.3	141.5	195.8	96.5	60.0	..	Factor Services
1,609.0	2,215.8	2,972.3	3,021.4	3,651.1	3,498.2	3,020.5	2,625.6	2,124.6	..	..	Imports of Goods & Services
1,057.5	1,334.1	1,789.1	1,763.4	2,486.8	2,233.4	1,704.9	1,354.8	1,206.2	..	..	Merchandise, fob
351.9	469.3	639.8	709.1	873.8	886.6	839.2	713.2	604.6	..	..	Nonfactor Services
199.6	412.4	543.4	548.8	290.5	378.3	476.5	557.6	313.8	..	..	Factor Services
21.1	40.5	50.4	83.2	74.8	122.4	75.2	96.0	118.1	121.3	..	Long-Term Interest
-20.8	-27.1	-42.3	-68.9	-81.5	-71.3	-71.5	-55.4	-30.8	..	..	Current Transfers, net
1.4	1.4	1.4	1.4	1.4	1.4	0.3	0.2	0.2	..	..	Workers' Remittances
59.8	-15.4	356.7	398.0	-586.4	-993.5	-510.8	-86.4	-421.4	-184.0	K	Total to be Financed
-17.1	-18.5	-22.0	-23.5	-58.5	-9.3	-11.7	-4.0	-19.1	..	K	Official Capital Grants
42.7	-33.9	334.7	374.5	-644.9	-1,002.9	-522.5	-90.3	-440.5	-184.0	..	Current Account Balance
245.2	413.4	300.8	300.1	477.7	285.0	1.5	28.4	-93.1	..	..	Long-Term Capital, net
128.8	93.8	184.5	258.1	203.5	114.1	109.8	-7.0	-21.8	..	..	Direct Investment
112.5	156.8	187.0	99.4	126.4	129.1	65.9	143.6	156.4	-133.9	..	Long-Term Loans
121.4	165.8	362.8	128.8	177.5	280.6	176.3	298.7	347.5	128.9	..	Disbursements
8.9	9.0	175.8	29.4	51.1	151.5	110.4	155.1	191.1	262.8	..	Repayments
21.0	181.3	-48.7	-33.9	206.3	51.1	-162.5	-104.2	-208.7	..	..	Other Long-Term Capital
49.6	-35.7	12.7	-348.1	-39.7	-154.9	-171.9	-239.4	-188.0	-174.3	..	Other Capital, net
-337.5	-343.8	-648.3	-326.5	206.9	872.8	692.9	301.3	721.6	358.3	..	Change in Reserves
				(Trinidad & Tobago Dollars per US dollar)							**Memo Item:**
2.400	2.400	2.400	2.400	2.400	2.400	2.400	2.450	3.600	3.600		Conversion Factor (Annual Avg)
				(Millions of US dollars, outstanding at end of year)							**EXTERNAL DEBT, ETC.**
369.7	514.7	712.5	794.8	906.8	1,026.1	1,062.8	1,299.1	1,584.5	1,635.3	..	Public/Publicly Guar. Long-Term
85.1	148.4	262.5	344.5	375.4	350.0	365.2	367.1	392.6	383.8	..	Official Creditors
58.8	57.9	56.5	54.1	51.0	45.9	35.8	38.3	40.4	41.4	..	IBRD and IDA
284.5	366.3	450.0	450.3	531.3	676.1	697.6	932.1	1,191.9	1,251.5	..	Private Creditors
..	..	..	..	..	..	..	..	..	..	..	Private Non-guaranteed Long-Term
0.0	0.0	0.0	0.0	0.0	0.0	0.0	0.0	0.0	..	..	Use of Fund Credit
117.0	166.0	116.0	255.0	296.0	412.0	159.0	149.0	273.0	166.0	..	Short-Term Debt
				(Millions of US dollars)							**Memo Items:**
1,804.8	2,140.0	2,780.8	3,347.5	3,080.5	2,104.5	1,356.7	1,128.5	474.1	187.8	..	Int'l Reserves Excluding Gold
9.1	27.6	31.8	21.4	24.6	20.6	16.6	17.6	21.1	26.1	..	Gold Holdings (at market price)
											SOCIAL INDICATORS
3.3	3.3	3.2	3.2	3.2	3.1	3.0	3.0	2.9	2.8	..	Total Fertility Rate
26.2	26.2	26.1	26.1	26.0	26.0	26.0	25.9	27.0	25.9	..	Crude Birth Rate
25.6	25.2	24.8	24.4	24.0	25.2	26.3	27.5	28.7	29.8	..	Infant Mortality Rate
67.7	68.0	68.2	68.4	68.7	68.9	69.1	69.4	69.6	69.8	..	Life Expectancy at Birth
123.5	107.5	99.2	93.2	86.9	101.0	83.1	96.9	95.3	92.5	..	Food Production, p.c. ('79-81 = 100)
11.9	11.0	10.2	..	..	..	..	..	..	..	..	Labor Force, Agriculture (%)
..	..	..	..	33.1	32.2	33.3	33.0	33.6	..	..	Labor Force, Female (%)
91.0	91.0	101.0	105.0	99.0	96.0	96.0	95.0	..	..	..	School Enroll. Ratio, primary
61.0	..	66.0	69.0	73.0	76.0	76.0	76.0	..	..	..	School Enroll. Ratio, secondary

TUNISIA	1967	1968	1969	1970	1971	1972	1973	1974	1975	1976	1977
CURRENT GNP PER CAPITA (US $)	220	240	260	280	320	390	430	560	710	800	840
POPULATION (thousands)	4,823	4,922	5,024	5,127	5,220	5,315	5,412	5,511	5,611	5,755	5,903

ORIGIN AND USE OF RESOURCES
(Millions of current Tunisian Dinars)

	1967	1968	1969	1970	1971	1972	1973	1974	1975	1976	1977
Gross National Product (GNP)	566.4	618.1	673.2	736.2	866.5	1,042.5	1,110.0	1,506.3	1,694.0	1,862.0	2,106.5
Net Factor Income from Abroad	-10.0	-15.8	-16.8	-19.4	-14.7	-24.9	-41.3	-41.5	-47.4	-71.0	-85.4
Gross Domestic Product (GDP)	576.4	633.9	690.0	755.6	881.2	1,067.4	1,151.3	1,547.8	1,741.4	1,933.0	2,191.9
Indirect Taxes, net	75.7	77.4	93.2	102.4	115.0	136.9	156.7	192.6	205.4	253.6	322.5
GDP at factor cost	500.7	556.5	596.8	653.2	766.2	930.5	994.6	1,355.2	1,536.0	1,679.4	1,869.4
Agriculture	88.7	112.9	108.4	128.7	170.8	229.0	227.3	289.9	321.8	343.2	346.7
Industry	123.9	137.5	147.3	155.5	180.7	214.1	245.5	419.4	451.4	486.4	564.4
Manufacturing	44.3	50.0	57.8	63.5	76.3	101.6	107.2	154.2	158.1	202.3	231.3
Services, etc.	288.1	306.1	341.1	369.0	414.7	487.4	521.8	645.9	762.8	849.8	958.3
Resource Balance	-55.2	-18.8	-30.4	-34.0	-17.8	-12.3	-28.5	5.4	-83.6	-153.7	-224.7
Exports of Goods & NFServices	120.9	133.5	149.9	166.2	212.4	270.6	300.0	547.4	540.6	562.2	648.4
Imports of Goods & NFServices	176.1	152.3	180.3	200.2	230.2	282.9	328.5	542.0	624.2	715.9	873.1
Domestic Absorption	631.6	652.7	720.4	789.6	899.0	1,079.7	1,179.8	1,542.4	1,825.0	2,086.7	2,416.6
Private Consumption, etc.	387.3	396.7	457.1	502.4	571.4	681.7	760.4	938.3	1,082.6	1,201.0	1,392.4
General Gov't Consumption	101.1	110.0	113.7	127.4	138.0	155.6	174.4	204.7	254.5	293.1	354.7
Gross Domestic Investment	143.2	146.0	149.6	159.8	189.6	242.4	245.0	399.4	487.9	592.6	669.5
Fixed Investment	134.3	136.9	145.6	155.0	175.2	211.0	236.0	321.0	448.0	563.0	672.0
Memo Items:											
Gross Domestic Saving	88.0	127.2	119.2	125.8	171.8	230.1	216.5	404.8	404.3	438.9	444.8
Gross National Saving	77.5	117.7	110.3	118.5	180.2	230.6	213.4	409.5	409.7	429.0	426.2

(Millions of 1980 Tunisian Dinars)

	1967	1968	1969	1970	1971	1972	1973	1974	1975	1976	1977
Gross National Product	1,380.6	1,502.7	1,589.2	1,694.9	1,892.4	2,212.7	2,171.5	2,365.9	2,534.4	2,712.6	2,799.2
GDP at factor cost	1,268.5	1,406.2	1,465.4	1,566.7	1,741.7	2,012.6	1,982.3	2,130.6	2,348.4	2,441.1	2,475.3
Agriculture	225.9	280.0	269.7	288.5	351.7	435.8	391.5	448.3	469.2	501.7	451.5
Industry	408.3	505.6	511.9	549.5	582.1	656.0	662.9	687.5	787.7	779.8	837.8
Manufacturing	113.8	123.3	143.9	142.8	167.6	214.2	216.2	191.3	240.5	292.8	305.3
Services, etc.	634.3	620.6	683.8	728.7	807.9	920.8	927.9	994.8	1,091.5	1,159.6	1,186.0
Resource Balance	-130.9	-2.6	6.7	-35.2	61.8	77.8	17.6	-14.3	-94.3	-56.3	-147.8
Exports of Goods & NFServices	409.8	457.9	478.0	518.9	666.3	818.9	750.6	864.6	870.7	1,018.7	1,090.6
Imports of Goods & NFServices	540.7	460.5	471.3	554.1	604.5	741.1	733.0	878.9	965.0	1,075.0	1,238.4
Domestic Absorption	1,535.8	1,542.6	1,620.3	1,773.7	1,861.0	2,185.3	2,232.0	2,446.1	2,698.8	2,866.6	3,053.7
Private Consumption, etc.	722.8	725.5	812.4	961.1	1,031.2	1,205.5	1,286.3	1,364.8	1,520.3	1,599.4	1,736.3
General Gov't Consumption	217.8	230.7	231.6	253.8	261.7	283.4	300.7	323.3	373.7	409.6	442.6
Gross Domestic Investment	595.2	586.4	576.3	558.8	568.1	696.4	645.0	758.0	804.8	857.6	874.8
Fixed Investment	395.1	387.9	396.9	399.3	439.6	502.1	506.1	596.2	744.6	851.1	857.9
Memo Items:											
Capacity to Import	371.2	403.7	391.8	460.0	557.8	708.9	669.4	887.7	835.8	844.2	919.7
Terms of Trade Adjustment	-38.6	-54.2	-86.2	-58.9	-108.5	-110.0	-81.2	23.1	-34.9	-174.5	-170.9
Gross Domestic Income	1,366.3	1,485.8	1,540.8	1,679.6	1,814.3	2,153.1	2,168.4	2,454.9	2,569.6	2,635.8	2,735.0
Gross National Income	1,342.0	1,448.5	1,503.0	1,636.0	1,783.9	2,102.7	2,090.3	2,389.0	2,499.5	2,538.1	2,628.3

DOMESTIC PRICES (DEFLATORS)
(Index 1980 = 100)

	1967	1968	1969	1970	1971	1972	1973	1974	1975	1976	1977
Overall (GDP)	41.0	41.2	42.4	43.5	45.8	47.2	51.2	63.6	66.9	68.8	75.4
Domestic Absorption	41.1	42.3	44.5	44.5	48.3	49.4	52.9	63.1	67.6	72.8	79.1
Agriculture	39.3	40.3	40.2	44.6	48.6	52.5	58.1	64.7	68.6	68.4	76.8
Industry	30.3	27.2	28.8	28.3	31.0	32.6	37.0	61.0	57.3	62.4	67.4
Manufacturing	38.9	40.6	40.2	44.5	45.5	47.4	49.6	80.6	65.7	69.1	75.8

MANUFACTURING ACTIVITY

	1967	1968	1969	1970	1971	1972	1973	1974	1975	1976	1977
Employment (1980=100)	39.2	42.2	40.1	43.6	43.0	46.6	52.7	55.9	62.1	67.0	80.5
Real Earnings per Empl. (1980=100)	67.5	67.0	71.3	70.0	72.2	77.1	73.0	82.3	84.9	89.6	99.1
Real Output per Empl. (1980=100)	97.3	94.5	103.3	94.9	101.7	100.8	98.9	74.4	95.2	94.6	90.6
Earnings as % of Value Added	44.5	44.6	46.8	43.7	43.5	43.4	43.8	41.7	48.3	49.9	50.0

MONETARY HOLDINGS
(Millions of current Tunisian Dinars)

	1967	1968	1969	1970	1971	1972	1973	1974	1975	1976	1977
Money Supply, Broadly Defined	192.6	216.3	233.2	254.2	307.6	365.9	444.0	578.0	721.6	826.7	952.2
Money as Means of Payment	151.3	170.4	180.6	191.9	239.8	277.1	320.8	401.4	467.8	506.3	568.0
Currency Ouside Banks	57.5	62.1	64.5	67.3	80.4	94.2	111.5	139.6	163.0	185.0	213.6
Demand Deposits	93.8	108.3	116.1	124.5	159.4	182.9	209.3	261.9	304.8	321.3	354.4
Quasi-Monetary Liabilities	41.3	46.0	52.5	62.3	67.8	88.9	123.2	176.6	253.8	320.4	384.2

GOVERNMENT DEFICIT (-) OR SURPLUS
(Millions of current Tunisian Dinars)

	1967	1968	1969	1970	1971	1972	1973	1974	1975	1976	1977
GOVERNMENT DEFICIT (-) OR SURPLUS	..	..	..	..	..	-9.5	-17.1	-15.4	-25.2	-62.4	-132.2
Current Revenue	..	..	..	..	..	262.5	293.6	404.3	504.5	533.9	630.6
Current Expenditure	..	..	..	..	..	197.2	226.3	276.8	351.4	385.4	463.4
Current Budget Balance	..	..	..	..	..	65.3	67.3	127.5	153.1	148.5	167.2
Capital Receipts	..	..	..	..	..	0.3	0.4	2.2	1.2	8.4	9.2
Capital Payments	..	..	..	..	..	75.1	84.8	145.1	179.5	219.3	308.6

1978	1979	1980	1981	1982	1983	1984	1985	1986	1987 estimate	NOTES	TUNISIA
930	1,080	1,290	1,390	1,320	1,250	1,220	1,190	1,130	1,210	..	**CURRENT GNP PER CAPITA (US $)**
6,054	6,210	6,369	6,517	6,668	6,823	6,981	7,143	7,311	7,481	..	**POPULATION (thousands)**
											ORIGIN AND USE OF RESOURCES
				(Millions of current Tunisian Dinars)							
2,394.1	2,813.7	3,447.1	4,012.7	4,630.5	5,315.9	6,042.5	6,616.5	6,691.2	7,627.9	..	Gross National Product (GNP)
-89.8	-108.3	-93.4	-149.3	-173.9	-181.5	-197.5	-293.5	-334.8	-387.3	..	Net Factor Income from Abroad
2,483.9	2,922.0	3,540.5	4,162.0	4,804.4	5,497.4	6,240.0	6,910.0	7,026.0	8,015.2	..	Gross Domestic Product (GDP)
357.9	415.0	476.4	521.3	577.7	737.4	821.0	865.8	935.0	1,011.9	..	Indirect Taxes, net
2,126.0	2,507.0	3,064.1	3,640.7	4,226.7	4,760.0	5,419.0	6,044.2	6,091.0	7,003.3	..	GDP at factor cost
374.8	395.7	500.3	568.8	632.0	673.0	863.0	1,048.0	933.0	1,246.0	..	Agriculture
655.0	844.3	1,101.5	1,332.5	1,491.8	1,676.4	1,903.5	2,058.2	2,027.2	2,270.3	..	Industry
274.8	338.8	417.3	493.9	534.1	615.6	733.4	818.0	922.0	1,044.1	..	Manufacturing
1,096.2	1,267.0	1,462.3	1,739.4	2,102.9	2,410.6	2,652.5	2,938.0	3,130.8	3,487.0	..	Services, etc.
-239.4	-146.1	-189.9	-352.4	-505.9	-473.6	-728.8	-423.2	-509.8	-96.4	..	Resource Balance
769.0	1,139.0	1,424.6	1,721.9	1,773.3	1,947.8	2,113.7	2,253.1	2,161.2	2,796.1	..	Exports of Goods & NFServices
1,008.4	1,285.1	1,614.5	2,074.3	2,279.2	2,421.4	2,842.5	2,676.3	2,671.0	2,892.5	..	Imports of Goods & NFServices
2,723.3	3,068.1	3,730.4	4,514.4	5,310.3	5,971.0	6,968.8	7,333.2	7,535.8	8,111.6	..	Domestic Absorption
1,553.2	1,763.8	2,178.7	2,553.2	2,997.2	3,434.2	3,943.8	4,356.2	4,670.0	5,114.0	..	Private Consumption, etc.
406.1	444.4	512.2	615.7	793.8	926.8	1,030.0	1,142.0	1,215.9	1,306.4	..	General Gov't Consumption
764.0	859.9	1,039.5	1,345.5	1,519.3	1,610.0	1,995.0	1,835.0	1,649.9	1,691.2	..	Gross Domestic Investment
771.0	892.0	1,002.0	1,290.0	1,630.0	1,735.0	1,920.0	1,850.0	1,685.0	1,620.0	..	Fixed Investment
											Memo Items:
524.6	713.8	849.6	993.1	1,013.4	1,136.4	1,266.2	1,411.8	1,140.1	1,594.8	..	Gross Domestic Saving
521.1	719.4	897.2	1,025.0	1,066.0	1,200.5	1,315.0	1,343.9	1,090.3	1,608.8	..	Gross National Saving
				(Millions of 1980 Tunisian Dinars)							
2,985.5	3,176.4	3,447.1	3,600.8	3,580.6	3,760.8	3,984.5	4,166.1	4,084.4	4,318.3	..	Gross National Product
2,641.3	2,822.9	3,064.1	3,268.6	3,252.6	3,404.8	3,599.8	3,803.3	3,742.3	3,957.8	..	GDP at factor cost
478.2	455.4	500.3	533.0	478.0	490.0	553.0	649.0	570.0	670.0	..	Agriculture
911.5	1,010.3	1,101.5	1,176.9	1,178.4	1,257.4	1,305.3	1,327.7	1,322.4	1,326.4	..	Industry
326.6	362.1	417.3	468.9	482.0	521.8	556.7	584.3	612.6	639.0	..	Manufacturing
1,251.6	1,357.2	1,462.3	1,558.7	1,596.2	1,657.4	1,741.5	1,826.6	1,849.9	1,961.4	..	Services, etc.
-170.6	-121.1	-189.9	-350.6	-469.2	-415.2	-480.0	-184.8	-74.0	209.9	..	Resource Balance
1,172.1	1,423.9	1,424.6	1,474.0	1,372.0	1,384.8	1,422.2	1,469.2	1,546.0	1,762.4	..	Exports of Goods & NFServices
1,342.7	1,545.0	1,614.5	1,824.6	1,841.2	1,800.0	1,902.2	1,654.0	1,620.0	1,552.5	..	Imports of Goods & NFServices
3,263.9	3,417.4	3,730.4	4,086.6	4,186.9	4,306.9	4,594.7	4,532.3	4,351.7	4,314.4	..	Domestic Absorption
1,833.8	1,921.4	2,178.7	2,335.7	2,402.9	2,513.9	2,663.7	2,732.3	2,760.0	2,795.0	..	Private Consumption, etc.
460.6	465.9	512.2	554.8	596.0	631.0	668.0	696.0	690.0	705.0	..	General Gov't Consumption
969.5	1,030.1	1,039.5	1,196.1	1,188.0	1,162.0	1,263.0	1,104.0	901.7	814.4	..	Gross Domestic Investment
900.0	964.1	1,002.0	1,144.4	1,238.0	1,180.0	1,236.0	1,134.0	926.0	802.0	..	Fixed Investment
											Memo Items:
1,023.9	1,369.4	1,424.6	1,514.6	1,432.5	1,447.9	1,414.5	1,392.5	1,310.8	1,500.8	..	Capacity to Import
-148.2	-54.5	0.0	40.6	60.5	63.1	-7.7	-76.7	-235.2	-261.6	..	Terms of Trade Adjustment
2,945.1	3,241.8	3,540.5	3,776.6	3,778.2	3,954.8	4,107.0	4,270.8	4,042.5	4,262.7	..	Gross Domestic Income
2,837.3	3,121.9	3,447.1	3,641.4	3,641.1	3,823.9	3,976.8	4,089.4	3,849.2	4,056.7	..	Gross National Income
											DOMESTIC PRICES (DEFLATORS)
				(Index 1980 = 100)							
80.3	88.6	100.0	111.4	129.2	141.3	151.7	158.9	164.2	177.2	..	Overall (GDP)
83.4	89.8	100.0	110.5	126.8	138.6	151.7	161.8	173.2	188.0	..	Domestic Absorption
78.4	86.9	100.0	106.7	132.2	137.3	156.1	161.5	163.7	186.0	..	Agriculture
71.9	83.6	100.0	113.2	126.6	133.3	145.8	155.0	153.3	171.2	..	Industry
84.1	93.6	100.0	105.3	110.8	118.0	131.7	140.0	150.5	163.4	..	Manufacturing
											MANUFACTURING ACTIVITY
86.7	96.7	100.0	108.6	119.5	131.4	144.6	159.0	..	..	..	Employment (1980 = 100)
101.0	97.4	100.0	102.0	86.7	83.3	83.4	78.2	..	..	..	Real Earnings per Empl. (1980 = 100)
87.6	90.1	100.0	105.1	95.1	93.6	90.8	86.6	..	..	..	Real Output per Empl. (1980 = 100)
49.7	46.5	46.4	47.7	46.9	46.9	46.9	46.9	..	..	..	Earnings as % of Value Added
											MONETARY HOLDINGS
				(Millions of current Tunisian Dinars)							
1,146.4	1,337.0	1,576.7	1,942.8	2,333.3	2,729.1	3,060.0	3,492.4	3,729.9	..	D	Money Supply, Broadly Defined
679.5	777.3	941.6	1,154.8	1,441.4	1,674.2	1,786.4	2,035.7	2,091.9	2,125.7	..	Money as Means of Payment
249.8	265.1	299.6	342.8	440.0	533.3	573.4	632.6	651.0	704.8	..	Currency Ouside Banks
429.8	512.2	642.0	812.0	1,001.4	1,140.9	1,213.0	1,403.1	1,440.9	1,420.9	..	Demand Deposits
466.9	559.7	635.1	788.0	891.9	1,054.9	1,273.6	1,456.7	1,638.0	..	..	Quasi-Monetary Liabilities
											GOVERNMENT DEFICIT (-) OR SURPLUS
				(Millions of current Tunisian Dinars)							
-101.2	-139.6	-98.9	-105.5	-277.4	-458.8	-307.3	..	..	..	..	
786.7	949.9	1,130.5	1,333.3	1,656.8	1,853.4	2,284.3	..	..	..	..	Current Revenue
559.9	728.2	784.1	909.9	1,273.7	1,514.0	1,735.2	..	..	..	..	Current Expenditure
226.8	221.7	346.4	423.4	383.1	339.4	549.1	..	..	..	..	Current Budget Balance
0.8	2.1	0.7	0.8	0.9	1.1	1.1	..	..	..	..	Capital Receipts
328.8	363.4	446.0	529.7	661.4	799.3	857.5	..	..	..	..	Capital Payments

TUNISIA	1967	1968	1969	1970	1971	1972	1973	1974	1975	1976	1977
FOREIGN TRADE (CUSTOMS BASIS)					*(Millions of current US dollars)*						
Value of Exports, fob	149.3	157.8	165.6	182.5	215.8	310.9	385.6	914.2	856.2	788.8	929.1
Nonfuel Primary Products	95.3	94.1	91.2	97.8	120.3	179.6	178.9	381.0	314.9	251.8	229.2
Fuels	22.2	31.4	43.1	49.7	60.0	84.3	122.7	328.3	373.3	334.1	389.1
Manufactures	31.8	32.3	31.4	35.0	35.5	47.0	83.9	205.0	168.0	202.9	310.8
Value of Imports, cif	260.3	217.2	255.2	304.6	341.9	458.5	605.6	1,120.1	1,417.8	1,525.8	1,820.9
Nonfuel Primary Products	94.6	73.0	86.6	111.5	108.0	128.3	170.4	344.0	346.2	315.7	370.6
Fuels	10.6	6.5	9.3	14.7	15.4	34.3	44.6	137.8	147.5	178.1	206.8
Manufactures	155.1	137.8	159.3	178.4	218.5	295.9	390.5	638.3	924.1	1,031.9	1,243.6
Terms of Trade	74.6	66.2	60.3	63.9	64.6	73.3	69.3	93.1	77.8	80.2	82.9
Export Prices, fob	19.0	17.6	16.1	16.7	18.9	20.9	25.9	53.3	46.9	48.0	53.6
Nonfuel Primary Products	31.8	32.1	33.3	34.6	35.2	40.1	61.5	75.3	80.9	72.6	76.2
Fuels	7.2	7.2	7.2	7.2	8.8	9.4	13.4	38.6	32.3	34.3	38.7
Manufactures	17.9	19.0	19.6	28.6	29.8	32.1	29.8	57.1	59.5	63.1	72.6
Import Prices, cif	25.5	26.5	26.6	26.1	29.2	28.4	37.4	57.2	60.3	59.9	64.6
Trade at Constant 1980 Prices					*(Millions of 1980 US dollars)*						
Exports, fob	785.9	899.0	1,031.9	1,095.1	1,143.1	1,490.1	1,489.0	1,715.2	1,827.1	1,642.6	1,734.4
Imports, cif	1,022.8	819.0	958.3	1,167.7	1,169.4	1,611.7	1,620.6	1,957.5	2,353.1	2,549.0	2,818.6
BALANCE OF PAYMENTS					*(Millions of current US dollars)*						
Exports of Goods & Services	..	..	..	332.0	416.2	564.6	736.7	1,256.8	1,325.9	1,336.5	1,527.0
Merchandise, fob	..	..	..	189.0	213.6	312.7	416.1	871.9	798.9	788.9	928.2
Nonfactor Services	..	..	..	136.0	188.6	238.9	299.2	345.2	486.9	522.2	583.2
Factor Services	..	..	..	7.0	14.0	13.0	21.5	39.7	40.1	25.4	15.6
Imports of Goods & Services	..	..	..	443.0	480.4	654.7	928.7	1,340.9	1,669.5	1,896.3	2,249.9
Merchandise, fob	..	..	..	294.0	334.0	452.7	622.3	971.7	1,238.4	1,531.3	1,730.5
Nonfactor Services	..	..	..	105.0	104.3	136.8	187.2	234.5	273.2	138.3	304.7
Factor Services	..	..	..	44.0	42.1	65.1	119.2	134.7	157.8	226.7	214.7
Long-Term Interest	..	..	..	..	..	..	..	..	..	40.0	63.9
Current Transfers, net	..	..	..	23.0	44.1	53.2	90.6	105.8	131.1	142.6	155.2
Workers' Remittances	..	..	..	29.0	44.1	61.9	98.9	117.9	144.5	143.0	168.3
Total to be Financed	..	..	..	-53.0	6.0	-4.3	-60.8	48.1	-170.0	-417.2	-520.4
Official Capital Grants	..	..	..	..	..	..	..	..	..	53.2	..
Current Account Balance	..	..	..	-53.0	6.0	-4.3	-60.8	48.1	-170.0	-364.0	-520.4
Long-Term Capital, net	..	..	..	86.0	111.3	122.7	174.1	162.4	170.0	234.2	803.0
Direct Investment	..	..	..	16.0	23.1	31.5	57.2	25.3	44.9	102.6	92.3
Long-Term Loans	..	..	..	..	..	..	..	..	..	280.1	663.3
Disbursements	..	..	..	..	..	..	..	..	..	343.4	779.1
Repayments	..	..	..	..	..	..	..	..	..	63.3	115.8
Other Long-Term Capital	..	..	..	70.0	88.3	91.2	116.8	137.1	125.1	-201.7	47.4
Other Capital, net	..	..	..	-13.0	-27.0	-48.5	-35.5	-122.8	-17.4	111.1	-319.2
Change in Reserves	..	..	..	-20.0	-90.3	-69.8	-77.8	-87.7	17.4	18.7	36.6
Memo Item:											
Conversion Factor (Annual Avg)	0.520	0.520	0.520	0.520	0.520	0.480	0.420	0.440	0.400	0.430	0.430

(Tunisian Dinars per US dollar)

	1967	1968	1969	1970	1971	1972	1973	1974	1975	1976	1977
EXTERNAL DEBT, ETC.				*(Millions of US dollars, outstanding at end of year)*							
Public/Publicly Guar. Long-Term	..	..	..	541.4	620.5	695.5	811.4	928.2	1,028.6	1,186.5	1,878.3
Official Creditors	..	..	..	365.3	438.8	535.9	653.6	761.1	848.6	976.3	1,325.7
IBRD and IDA	..	..	..	42.5	60.4	81.1	106.8	131.3	165.1	192.0	234.1
Private Creditors	..	..	..	176.1	181.7	159.6	157.8	167.1	180.0	210.2	552.6
Private Non-guaranteed Long-Term	..	..	..	..	..	..	..	..	..	100.0	125.0
Use of Fund Credit	..	..	..	13.3	2.8	0.0	0.0	0.0	0.0	0.0	29.1
Short-Term Debt	..	..	..	..	..	..	..	..	..	..	201.0
Memo Items:					*(Millions of US dollars)*						
Int'l Reserves Excluding Gold	36.0	31.2	32.6	55.2	142.9	217.8	301.8	412.8	379.9	365.8	351.2
Gold Holdings (at market price)	4.1	5.0	4.3	4.7	5.6	8.3	14.4	24.1	18.2	17.4	24.7
SOCIAL INDICATORS											
Total Fertility Rate	6.8	6.7	6.6	6.4	6.3	6.1	6.0	5.9	5.8	5.7	5.6
Crude Birth Rate	41.8	40.9	39.9	39.0	38.0	37.1	36.8	36.5	36.2	35.9	35.6
Infant Mortality Rate	138.0	134.4	130.8	127.2	123.6	120.0	116.4	112.8	109.2	105.6	102.0
Life Expectancy at Birth	52.1	52.7	53.3	53.9	54.5	55.1	55.9	56.7	57.4	58.2	59.0
Food Production, p.c. ('79-81=100)	94.0	88.8	77.3	82.9	100.6	101.1	105.6	107.1	118.7	110.8	104.8
Labor Force, Agriculture (%)	46.3	45.0	43.6	42.2	41.4	40.6	39.7	38.9	38.1	37.5	36.9
Labor Force, Female (%)	9.8	10.4	11.0	11.6	12.5	13.4	14.2	15.1	16.0	17.0	18.0
School Enroll. Ratio, primary	..	..	..	100.0	..	..	..	..	97.0	99.0	100.0
School Enroll. Ratio, secondary	..	..	..	23.0	..	..	..	..	21.0	22.0	23.0

1978	1979	1980	1981	1982	1983	1984	1985	1986	1987 estimate	NOTES	TUNISIA
											FOREIGN TRADE (CUSTOMS BASIS)
			(Millions of current US dollars)								
1,126.1	1,790.7	2,233.7	2,503.7	1,983.5	1,871.5	1,796.3	1,738.0	1,759.0	2,109.1	..	Value of Exports, fob
264.1	316.2	260.6	315.9	249.0	204.3	244.9	228.4	272.0	316.6	..	Nonfuel Primary Products
433.5	870.0	1,172.5	1,352.4	911.2	851.6	795.5	723.0	427.0	543.1	..	Fuels
428.5	604.6	800.7	835.4	823.4	815.6	755.9	786.6	1,060.0	1,249.4	..	Manufactures
2,157.7	2,842.1	3,508.7	3,770.9	3,395.7	3,099.5	3,114.9	2,756.5	2,890.0	2,955.0	..	Value of Imports, cif
410.1	589.0	756.7	834.0	680.9	741.2	770.1	635.1	727.2	873.0	..	Nonfuel Primary Products
237.9	504.6	727.6	773.9	439.3	366.5	347.8	324.1	197.2	231.4	..	Fuels
1,509.7	1,748.6	2,024.4	2,163.0	2,275.5	1,991.8	1,997.0	1,797.3	1,965.6	1,850.6	..	Manufactures
			(Index 1980 = 100)								
73.9	86.7	100.0	101.4	93.6	87.2	84.1	83.2	76.9	78.9	..	Terms of Trade
53.3	72.8	100.0	102.4	89.2	81.8	77.6	74.2	70.0	76.5	..	Export Prices, fob
76.5	94.1	100.0	96.6	89.3	89.4	86.1	81.1	90.0	90.6	..	Nonfuel Primary Products
38.2	58.7	100.0	111.0	98.9	89.0	81.0	76.0	45.0	56.9	..	Fuels
67.9	94.0	100.0	92.9	80.4	74.0	72.0	71.0	84.0	86.0	..	Manufactures
72.2	84.0	100.0	101.0	95.3	93.8	92.3	89.2	91.0	97.0	..	Import Prices, cif
											Trade at Constant 1980 Prices
			(Millions of 1980 US dollars)								
2,111.4	2,460.9	2,233.7	2,444.7	2,224.4	2,287.7	2,316.2	2,340.9	2,512.9	2,756.5	..	Exports, fob
2,989.4	3,384.7	3,508.8	3,733.7	3,563.3	3,304.8	3,376.0	3,089.3	3,176.9	3,047.0	..	Imports, cif
			(Millions of current US dollars)								**BALANCE OF PAYMENTS**
1,869.4	2,853.4	3,609.1	3,585.4	3,106.1	2,958.2	2,819.7	2,747.8	2,750.0	3,394.3	f	Exports of Goods & Services
1,125.4	1,787.7	2,395.1	2,455.2	1,979.7	1,860.0	1,801.1	1,729.2	1,768.0	2,133.4	..	Merchandise, fob
722.2	1,014.3	1,122.5	1,031.8	1,022.3	1,009.4	919.9	970.8	953.9	1,235.4	..	Nonfactor Services
21.8	51.4	91.6	98.4	104.1	88.7	98.7	47.9	28.1	25.5	..	Factor Services
2,660.5	3,479.2	4,308.6	4,601.4	4,257.0	3,923.2	4,012.3	3,606.8	3,813.8	3,977.1	f	Imports of Goods & Services
2,050.7	2,701.1	3,453.3	3,587.9	3,217.9	2,926.6	3,030.3	2,570.4	2,744.0	2,856.7	..	Merchandise, fob
372.2	460.3	533.1	612.8	640.6	640.5	629.0	636.7	620.0	628.3	..	Nonfactor Services
237.6	317.8	322.2	400.7	398.5	356.1	353.0	399.7	449.8	492.1	..	Factor Services
108.0	174.7	227.5	220.6	209.4	204.0	248.6	253.0	309.8	340.0	..	Long-Term Interest
207.4	280.2	348.1	366.9	383.4	361.8	317.1	270.3	358.9	484.3	..	Current Transfers, net
220.6	284.1	303.2	361.1	371.8	359.2	316.5	270.6	361.6	485.5	..	Workers' Remittances
-551.0	-294.9	-309.9	-628.8	-748.5	-577.9	-846.9	-552.2	-664.3	-62.4	..	Total to be Financed
..	..	..	..	..	..	..	..	..	..	..	Official Capital Grants
-551.0	-294.9	-309.9	-628.8	-748.5	-577.9	-846.9	-552.2	-664.3	-62.4	..	Current Account Balance
628.8	566.4	643.5	684.4	763.1	752.9	656.2	530.7	415.0	322.5	..	Long-Term Capital, net
91.1	50.9	236.0	367.2	400.5	223.8	206.0	139.5	155.0	92.2	..	Direct Investment
505.1	464.6	366.2	296.9	343.6	503.8	421.5	354.7	219.5	191.1	..	Long-Term Loans
655.2	666.7	624.9	657.4	672.6	913.4	858.8	826.8	807.1	849.3	..	Disbursements
150.1	202.1	258.7	360.5	329.0	409.6	437.3	472.1	587.6	658.2	..	Repayments
32.6	50.9	41.3	20.3	19.0	25.3	28.7	36.5	40.5	39.2	..	Other Long-Term Capital
22.6	-123.8	-220.1	-141.6	43.5	-205.4	18.0	-160.2	176.6	-78.6	..	Other Capital, net
-100.4	-147.7	-113.5	86.0	-58.1	30.4	172.7	181.7	72.7	-181.5	..	Change in Reserves
											Memo Item:
			(Tunisian Dinars per US dollar)								
0.420	0.410	0.400	0.490	0.590	0.680	0.780	0.830	0.790	0.830	..	Conversion Factor (Annual Avg)
			(Millions of US dollars, outstanding at end of year)								**EXTERNAL DEBT, ETC.**
2,439.2	3,006.7	3,223.6	3,293.9	3,516.0	3,834.5	3,703.8	4,487.0	5,221.6	6,188.8	..	Public/Publicly Guar. Long-Term
1,574.0	1,818.9	1,989.1	2,101.8	2,302.1	2,420.7	2,497.4	3,030.9	3,598.2	4,418.6	..	Official Creditors
256.7	299.2	336.8	387.0	443.4	501.2	464.3	623.0	853.3	1,115.6	..	IBRD and IDA
865.2	1,187.8	1,234.5	1,192.1	1,213.9	1,413.9	1,206.4	1,456.1	1,623.3	1,770.2	..	Private Creditors
190.0	170.0	180.0	161.4	137.1	112.1	221.5	246.0	250.0	225.7	..	Private Non-guaranteed Long-Term
31.3	31.6	0.0	0.0	0.0	0.0	0.0	0.0	183.1	270.6	..	Use of Fund Credit
281.0	190.0	136.3	133.5	136.3	131.4	169.2	181.9	226.7	223.7	..	Short-Term Debt
			(Millions of US dollars)								Memo Items:
443.0	579.3	590.1	536.1	606.5	567.3	406.3	232.7	305.3	525.5	..	Int'l Reserves Excluding Gold
36.2	87.3	110.0	74.2	85.3	71.2	57.5	61.0	73.0	90.4	..	Gold Holdings (at market price)
											SOCIAL INDICATORS
5.5	5.3	5.1	5.0	4.8	4.7	4.6	4.5	4.4	4.3	..	Total Fertility Rate
35.0	34.3	33.7	33.0	32.4	32.2	32.0	31.8	31.6	31.4	..	Crude Birth Rate
98.6	95.2	91.8	88.4	85.0	79.1	73.2	67.3	61.4	55.5	..	Infant Mortality Rate
59.5	59.9	60.4	60.8	61.3	61.7	62.2	62.6	63.0	63.4	..	Life Expectancy at Birth
102.1	94.6	105.9	99.5	87.6	99.5	96.2	120.9	105.6	116.9	..	Food Production, p.c. ('79-81 = 100)
36.2	35.6	35.0	..	..	..	..	..	..	..	..	Labor Force, Agriculture (%)
19.0	20.0	21.0	21.4	21.8	22.2	22.6	23.0	23.3	23.6	..	Labor Force, Female (%)
100.0	102.0	103.0	106.0	111.0	114.0	116.0	118.0	..	..	..	School Enroll. Ratio, primary
24.0	25.0	27.0	30.0	31.0	34.0	36.0	39.0	..	..	..	School Enroll. Ratio, secondary

TURKEY	1967	1968	1969	1970	1971	1972	1973	1974	1975	1976	1977
CURRENT GNP PER CAPITA (US $)	330	370	400	400	400	410	470	630	830	1,000	1,110
POPULATION (thousands)	32,753	33,585	34,441	35,321	36,429	37,411	38,316	39,189	40,078	40,951	41,776
ORIGIN AND USE OF RESOURCES					*(Billions of current Turkish Liras)*						
Gross National Product (GNP)	101	111	124	145	186	230	294	407	517	659	857
Net Factor Income from Abroad	-1	-1	-1	-1	-1	-2	-2	-3	-2	-5	-6
Gross Domestic Product (GDP)	101	112	124	146	187	232	295	410	519	664	863
Indirect Taxes, net	11	11	13	14	19	26	30	40	51	64	67
GDP at factor cost	90	101	112	131	168	207	266	370	468	600	796
Agriculture	30	31	34	39	50	59	73	105	136	177	220
Industry	24	28	32	36	45	53	68	95	118	146	200
Manufacturing	15	18	20	22	29	35	46	66	80	98	130
Services, etc.	36	42	46	57	73	94	124	169	214	276	376
Resource Balance	-1	-2	-2	-3	-6	-7	-6	-26	-42	-46	-70
Exports of Goods & NFServices	6	6	6	9	13	18	26	29	29	41	41
Imports of Goods & NFServices	7	8	8	12	20	24	33	55	71	86	111
Domestic Absorption	102	114	126	149	194	239	302	436	561	710	933
Private Consumption, etc.	72	80	89	101	135	162	209	300	377	458	599
General Gov't Consumption	13	14	16	19	25	28	37	47	64	85	116
Gross Domestic Investment	18	20	22	29	33	48	56	89	120	167	218
Fixed Investment	17	19	22	27	32	47	59	76	108	154	211
Memo Items:											
Gross Domestic Saving	17	18	20	25	27	42	50	62	78	121	147
Gross National Saving	17	19	21	28	33	51	65	80	96	134	161
					(Billions of 1980 Turkish Liras)						
Gross National Product	2,182	2,334	2,462	2,589	2,819	3,001	3,142	3,413	3,723	4,037	4,227
GDP at factor cost	2,027	2,184	2,310	2,434	2,648	2,811	2,936	3,193	3,475	3,785	3,978
Agriculture	629	639	647	662	749	737	662	731	811	872	862
Industry	575	646	704	731	768	814	902	970	1,056	1,159	1,263
Manufacturing	407	468	516	532	577	616	691	738	799	876	939
Services, etc.	823	899	959	1,042	1,130	1,260	1,373	1,492	1,609	1,754	1,854
Resource Balance	-180	-222	-222	-262	-336	-401	-360	-519	-605	-618	-712
Exports of Goods & NFServices	164	161	174	188	202	262	303	251	256	311	256
Imports of Goods & NFServices	344	382	397	450	537	663	663	770	861	929	968
Domestic Absorption	2,376	2,573	2,705	2,869	3,175	3,426	3,525	3,954	4,346	4,687	4,970
Private Consumption, etc.	1,741	1,881	1,983	2,043	2,288	2,532	2,542	2,785	2,959	3,220	3,377
General Gov't Consumption	253	270	287	293	351	314	328	351	393	429	448
Gross Domestic Investment	382	422	434	532	537	580	656	819	994	1,038	1,144
Fixed Investment	380	431	460	528	508	566	666	753	917	1,032	1,103
Memo Items:											
Capacity to Import	295	286	310	326	360	483	531	405	352	438	357
Terms of Trade Adjustment	131	125	135	138	158	221	228	154	95	128	101
Gross Domestic Income	2,327	2,476	2,618	2,744	2,997	3,246	3,393	3,589	3,836	4,196	4,359
Gross National Income	2,313	2,458	2,598	2,726	2,977	3,222	3,370	3,567	3,818	4,165	4,328
DOMESTIC PRICES (DEFLATORS)					*(Index 1980 = 100)*						
Overall (GDP)	4.6	4.8	5.0	5.6	6.6	7.7	9.3	11.9	13.9	16.3	20.3
Domestic Absorption	4.3	4.4	4.7	5.2	6.1	7.0	8.6	11.0	12.9	15.1	18.8
Agriculture	4.8	4.9	5.2	5.9	6.7	8.0	11.1	14.4	16.8	20.3	25.5
Industry	4.2	4.3	4.5	4.9	5.8	6.5	7.6	9.8	11.2	12.6	15.8
Manufacturing	3.7	3.8	3.8	4.2	5.1	5.6	6.7	9.0	10.0	11.2	13.8
MANUFACTURING ACTIVITY											
Employment (1980 = 100)	52.8	56.5	50.8	63.7	68.2	72.9	81.0	83.6	88.9	92.0	94.9
Real Earnings per Empl. (1980 = 100)	69.3	70.4	78.7	81.9	82.1	81.1	83.1	86.2	95.0	99.0	118.7
Real Output per Empl. (1980 = 100)	91.8	94.9	98.0	107.5	108.9	114.5	116.5	119.6	127.6	132.0	139.6
Earnings as % of Value Added	25.4	26.0	27.0	25.9	26.8	27.4	27.1	26.3	32.0	31.6	37.7
MONETARY HOLDINGS					*(Billions of current Turkish Liras)*						
Money Supply, Broadly Defined	27	32	37	45	57	72	92	116	149	183	245
Money as Means of Payment	23	26	30	36	44	54	71	91	119	152	210
Currency Ouside Banks	9	8	9	12	14	16	21	26	33	42	63
Demand Deposits	14	18	21	24	30	38	50	65	86	109	147
Quasi-Monetary Liabilities	4	5	6	9	13	18	21	25	30	32	35
GOVERNMENT DEFICIT (-) OR SURPLUS					*(Billions of current Turkish Liras)*						
GOVERNMENT DEFICIT (-) OR SURPLUS	..	..	..	-3.6	..	-5.1	-5.5	-7.4	-7.0	-13.3	-52.8
Current Revenue	..	..	..	28.7	38.5	47.4	59.3	71.6	108.8	143.7	186.9
Current Expenditure	..	..	..	23.1	31.3	35.6	48.4	57.1	81.6	110.4	164.0
Current Budget Balance	..	..	..	5.6	7.2	11.8	10.9	14.5	27.2	33.3	22.9
Capital Receipts	..	..	..	0.1	..	0.1	0.1	0.2	0.2	..	..
Capital Payments	..	..	..	9.3	13.6	17.0	16.5	22.1	34.4	46.6	75.7

1978	1979	1980	1981	1982	1983	1984	1985	1986	1987 est.	NOTES	TURKEY
1,210	1,370	1,400	1,450	1,300	1,180	1,100	1,080	1,110	1,200	..	**CURRENT GNP PER CAPITA (US $)**
42,602	43,473	44,438	45,465	46,514	47,642	48,892	50,310	51,416	52,530	..	**POPULATION (thousands)**
				(Billions of current Turkish Liras)							**ORIGIN AND USE OF RESOURCES**
1,264	2,126	4,243	6,255	8,378	11,201	17,684	26,853	38,112	56,055	f	Gross National Product (GNP)
-11	-30	-85	-159	-242	-331	-528	-698	-1,176	-1,711	..	Net Factor Income from Abroad
1,275	2,156	4,328	6,414	8,620	11,532	18,212	27,552	39,288	57,766	..	Gross Domestic Product (GDP)
85	141	230	390	540	714	863	2,026	3,660	5,636	..	Indirect Taxes, net
1,190	2,015	4,098	6,024	8,081	10,817	17,349	25,526	35,628	52,130	..	GDP at factor cost
301	466	925	1,325	1,679	2,118	3,397	4,790	6,586	9,094	..	Agriculture
337	584	1,237	1,858	2,549	3,544	5,807	9,012	12,763	18,938	..	Industry
232	417	865	1,310	1,813	2,583	4,206	6,409	8,998	13,552	..	Manufacturing
551	966	1,936	2,841	3,853	5,155	8,145	11,724	16,279	24,098	..	Services, etc.
-47	-70	-339	-341	-283	-500	-763	-882	-1,170	-1,208	..	Resource Balance
67	91	275	664	1,269	1,768	3,500	5,676	6,933	11,865	..	Exports of Goods & NFServices
114	161	615	1,005	1,552	2,268	4,263	6,558	8,103	13,073	..	Imports of Goods & NFServices
1,322	2,226	4,667	6,754	8,903	12,032	18,975	28,434	40,458	58,974	..	Domestic Absorption
910	1,530	3,175	4,645	6,189	8,597	13,733	20,301	27,296	38,963	..	Private Consumption, etc.
173	294	544	700	939	1,175	1,622	2,332	3,444	5,254	..	General Gov't Consumption
239	402	948	1,409	1,775	2,259	3,621	5,802	9,718	14,757	..	Gross Domestic Investment
280	449	864	1,241	1,647	2,130	3,358	5,561	9,246	14,150	..	Fixed Investment
											Memo Items:
192	332	609	1,068	1,492	1,760	2,858	4,919	8,548	13,549	..	Gross Domestic Saving
208	358	687	1,194	1,606	1,778	3,021	5,141	8,521	13,613	..	Gross National Saving
				(Billions of 1980 Turkish Liras)							
4,362	4,297	4,243	4,399	4,598	4,772	5,053	5,317	5,716	6,131	f	Gross National Product
4,152	4,124	4,098	4,249	4,441	4,618	4,900	5,107	5,481	5,841	..	GDP at factor cost
884	909	925	925	984	982	1,017	1,041	1,125	1,150	..	Agriculture
1,341	1,295	1,237	1,306	1,357	1,442	1,564	1,651	1,796	1,957	..	Industry
974	923	865	945	997	1,082	1,193	1,259	1,381	1,516	..	Manufacturing
1,927	1,921	1,936	2,018	2,100	2,194	2,319	2,414	2,559	2,735	..	Services, etc.
-352	-336	-339	-197	-46	-78	-164	-136	-293	-233	..	Resource Balance
292	264	275	510	714	812	973	1,093	1,077	1,393	..	Exports of Goods & NFServices
644	600	615	707	760	890	1,137	1,229	1,370	1,626	..	Imports of Goods & NFServices
4,752	4,696	4,667	4,706	4,773	4,987	5,362	5,587	6,183	6,548	..	Domestic Absorption
3,391	3,325	3,175	3,145	3,255	3,457	3,816	3,880	4,243	4,507	..	Private Consumption, etc.
492	500	544	542	548	563	563	590	665	711	..	General Gov't Consumption
869	871	948	1,019	971	967	983	1,117	1,276	1,330	..	Gross Domestic Investment
995	959	864	881	910	935	937	1,092	1,228	1,290	..	Fixed Investment
											Memo Items:
379	340	275	467	621	694	934	1,064	1,172	1,476	..	Capacity to Import
87	76	0	-43	-93	-118	-39	-29	95	83	..	Terms of Trade Adjustment
4,487	4,436	4,328	4,466	4,635	4,790	5,159	5,422	5,985	6,398	..	Gross Domestic Income
4,449	4,373	4,243	4,356	4,505	4,654	5,013	5,288	5,811	6,214	..	Gross National Income
				(Index 1980 = 100)							**DOMESTIC PRICES (DEFLATORS)**
29.0	49.4	100.0	142.2	182.3	234.9	350.4	505.4	667.0	914.8	..	Overall (GDP)
27.8	47.4	100.0	143.5	186.5	241.3	353.9	508.9	654.3	900.7	..	Domestic Absorption
34.1	51.3	100.0	143.3	170.6	215.7	334.1	460.0	585.3	791.1	..	Agriculture
25.2	45.1	100.0	142.2	187.8	245.8	371.4	545.8	710.6	967.7	..	Industry
23.8	45.1	100.0	138.7	181.8	238.7	352.5	508.9	651.5	894.0	..	Manufacturing
											MANUFACTURING ACTIVITY
101.6	100.4	100.0	101.4	105.3	101.0	104.3	104.8	110.0	116.5	G	Employment (1980=100)
110.6	110.5	100.0	104.3	97.8	96.0	83.9	89.4	..	..	G	Real Earnings per Empl. (1980=100)
112.0	95.6	100.0	119.2	126.0	127.8	130.7	125.1	..	..	G	Real Output per Empl. (1980=100)
37.7	38.3	30.7	25.3	24.0	24.8	23.8	24.2	..	..	..	Earnings as % of Value Added
				(Billions of current Turkish Liras)							**MONETARY HOLDINGS**
335	542	903	1,706	2,563	3,298	5,220	8,199	11,725	16,613	..	Money Supply, Broadly Defined
289	456	720	969	1,343	1,941	2,292	3,256	5,091	8,381	..	Money as Means of Payment
94	144	218	280	412	548	735	1,011	1,415	2,275	..	Currency Ouside Banks
195	313	502	689	931	1,394	1,557	2,245	3,676	6,106	..	Demand Deposits
46	85	183	737	1,220	1,357	2,928	4,942	6,634	8,232	..	Quasi-Monetary Liabilities
				(Billions of current Turkish Liras)							
-54.6	-136.6	-160.8	-117.0	..	-483.4	-1,815.5	-2,049.7	-1,259.2	..	F	**GOVERNMENT DEFICIT (-) OR SURPLUS**
291.4	496.7	945.3	1,443.5	..	2,303.8	2,711.9	4,820.5	7,016.2	..	..	Current Revenue
248.7	435.1	809.3	1,070.7	..	1,958.5	3,504.1	5,614.2	6,782.1	..	..	Current Expenditure
42.7	61.6	136.0	372.8	..	345.3	-792.2	-793.7	234.1	..	..	Current Budget Balance
..	0.3	11.4	2.5	..	5.1	9.2	15.2	36.0	..	..	Capital Receipts
97.3	198.5	308.2	492.3	..	833.8	1,032.5	1,271.2	1,529.3	..	..	Capital Payments

TURKEY	1967	1968	1969	1970	1971	1972	1973	1974	1975	1976	1977
FOREIGN TRADE (CUSTOMS BASIS)				*(Millions of current US dollars)*							
Value of Exports, fob	522	496	537	589	677	885	1,317	1,538	1,401	1,960	1,753
Nonfuel Primary Products	512	478	500	532	591	739	1,047	1,115	1,039	1,478	1,322
Fuels	0	1	4	4	3	22	49	86	36	16	0
Manufactures	10	17	33	53	83	124	222	337	326	466	431
Value of Imports, cif	685	764	747	886	1,088	1,508	2,049	3,720	4,640	4,993	5,694
Nonfuel Primary Products	78	82	86	139	120	144	205	676	639	427	400
Fuels	54	63	61	67	122	156	222	764	810	1,127	1,471
Manufactures	553	618	601	680	846	1,208	1,622	2,280	3,192	3,439	3,823
Terms of Trade	167.7	177.2	164.9	159.3	166.8	139.9	146.6	116.5	109.3	122.3	118.0
Export Prices, fob	35.1	36.1	35.1	36.1	37.4	35.7	47.2	62.8	62.2	68.2	70.5
Nonfuel Primary Products	35.6	36.9	37.6	38.4	39.6	42.7	58.7	64.6	62.9	69.2	68.8
Fuels	4.3	4.3	4.3	4.3	5.6	6.2	8.9	36.7	35.7	38.4	42.0
Manufactures	26.6	33.8	31.2	33.8	31.2	31.8	48.7	68.8	64.9	66.9	76.0
Import Prices, cif	20.9	20.4	21.3	22.7	22.4	25.5	32.2	53.9	56.9	55.7	59.7
Trade at Constant 1980 Prices				*(Millions of 1980 US dollars)*							
Exports, fob	1,487	1,376	1,527	1,630	1,811	2,478	2,790	2,451	2,254	2,875	2,488
Imports, cif	3,269	3,752	3,505	3,906	4,858	5,906	6,363	6,907	8,160	8,959	9,539
BALANCE OF PAYMENTS				*(Millions of current US dollars)*							
Exports of Goods & Services	..	..	..	776	896	1,266	1,859	2,082	2,018	2,541	2,291
Merchandise, fob	..	..	..	588	677	885	1,320	1,532	1,401	1,960	1,753
Nonfactor Services	..	..	..	183	210	367	526	549	616	580	532
Factor Services	..	..	..	5	9	14	13	1	1	1	6
Imports of Goods & Services	..	..	..	1,150	1,421	1,847	2,452	4,133	5,103	5,690	6,536
Merchandise, fob	..	..	..	830	1,030	1,367	1,829	3,589	4,502	4,872	5,506
Nonfactor Services	..	..	..	237	292	349	476	366	436	517	683
Factor Services	..	..	..	83	99	131	148	178	165	301	347
Long-Term Interest	..	..	..	44	55	66	84	101	118	168	186
Current Transfers, net	..	..	..	317	499	777	1,234	1,466	1,398	1,104	1,068
Workers' Remittances	..	..	..	273	471	740	1,184	1,426	1,312	982	982
Total to be Financed	..	..	..	-57	-26	197	640	-585	-1,687	-2,045	-3,177
Official Capital Grants	..	..	..	13	69	15	20	24	39	16	39
Current Account Balance	..	..	..	-44	43	212	660	-561	-1,648	-2,029	-3,138
Long-Term Capital, net	..	..	..	343	286	712	380	206	1,361	2,575	917
Direct Investment	..	..	..	58	45	43	79	64	114	10	27
Long-Term Loans	..	..	..	198	299	202	334	206	153	523	871
Disbursements	..	..	..	330	411	360	463	362	326	724	1,106
Repayments	..	..	..	131	112	158	129	156	173	201	235
Other Long-Term Capital	..	..	..	74	-128	452	-53	-88	1,055	2,026	-20
Other Capital, net	..	..	..	-155	3	-235	-345	-196	-350	-774	1,854
Change in Reserves	..	..	..	-144	-332	-689	-696	551	637	228	367
Memo Item:				*(Turkish Liras per US dollar)*							
Conversion Factor (Annual Avg)	9.000	9.000	9.000	11.500	14.920	14.150	14.150	13.930	14.440	16.050	18.000
EXTERNAL DEBT, ETC.				*(Millions of US dollars, outstanding at end of year)*							
Public/Publicly Guar. Long-Term	..	..	..	1,844	2,224	2,454	2,866	3,136	3,182	3,619	4,438
Official Creditors	..	..	..	1,768	2,095	2,307	2,696	2,925	3,000	3,279	3,636
IBRD and IDA	..	..	..	137	160	191	255	335	431	555	693
Private Creditors	..	..	..	76	129	147	170	211	182	340	803
Private Non-guaranteed Long-Term	..	..	..	42	49	70	115	146	160	248	479
Use of Fund Credit	..	..	..	74	68	0	0	0	243	391	409
Short-Term Debt	..	..	..	..	..	..	..	..	..	..	6,093
Memo Items:				*(Millions of US dollars)*							
Int'l Reserves Excluding Gold	22.0	26.0	128.0	304.1	631.4	1,262.1	1,985.9	1,561.7	943.8	990.5	638.0
Gold Holdings (at market price)	97.5	116.1	117.7	135.6	149.6	231.8	400.8	665.7	500.6	481.0	599.5
SOCIAL INDICATORS											
Total Fertility Rate	5.6	5.5	5.4	5.3	5.2	5.0	4.9	4.7	4.6	4.5	4.3
Crude Birth Rate	39.0	38.1	37.2	36.3	35.4	34.5	34.0	33.5	33.0	32.5	32.0
Infant Mortality Rate	153.0	150.0	147.0	144.0	141.0	138.0	134.4	130.8	127.2	123.6	120.0
Life Expectancy at Birth	54.4	55.0	55.6	56.2	56.8	57.5	58.1	58.7	59.3	59.9	60.5
Food Production, p.c. ('79-81 = 100)	90.3	92.1	91.1	93.7	94.4	93.1	86.2	92.2	98.1	99.8	99.0
Labor Force, Agriculture (%)	73.1	72.3	71.5	70.7	69.5	68.3	67.2	66.0	64.8	63.5	62.2
Labor Force, Female (%)	38.7	38.5	38.2	37.9	37.6	37.4	37.1	36.8	36.5	36.1	35.7
School Enroll. Ratio, primary	..	..	..	110.0	..	..	..	..	108.0	107.0	105.0
School Enroll. Ratio, secondary	..	..	..	27.0	..	..	..	..	29.0	32.0	34.0

1978	1979	1980	1981	1982	1983	1984	1985	1986	1987 estimate	NOTES	TURKEY
											FOREIGN TRADE (CUSTOMS BASIS)
				(Millions of current US dollars)							
2,288	2,261	2,910	4,702	5,890	5,905	7,389	8,255	7,583	10,322	..	Value of Exports, fob
1,784	1,639	2,086	2,847	2,928	2,795	2,878	2,715	2,924	3,202	..	Nonfuel Primary Products
4	2	42	107	345	233	407	373	181	233	..	Fuels
501	620	782	1,748	2,617	2,877	4,104	5,167	4,478	6,887	..	Manufactures
4,479	4,946	7,573	8,864	8,794	9,235	10,757	11,515	11,027	14,008	..	Value of Imports, cif
262	341	641	816	736	828	1,382	1,582	1,354	981	..	Nonfuel Primary Products
1,440	1,760	3,669	3,919	3,850	3,746	3,794	3,674	1,990	2,496	..	Fuels
2,777	2,845	3,262	4,129	4,208	4,661	5,581	6,259	7,683	10,531	..	Manufactures
				(Index 1980 = 100)							
123.5	119.6	100.0	92.6	89.6	93.6	91.7	90.9	106.6	100.6	..	Terms of Trade
76.7	92.6	100.0	96.5	87.8	87.4	84.4	80.9	87.7	98.0	..	Export Prices, fob
77.2	95.9	100.0	95.5	89.0	92.7	88.8	82.7	84.2	93.1	..	Nonfuel Primary Products
42.3	61.0	100.0	112.5	101.6	92.5	90.2	87.5	44.9	56.9	..	Fuels
75.3	85.1	100.0	97.4	85.1	82.5	81.0	79.5	93.9	103.1	..	Manufactures
62.1	77.5	100.0	104.2	98.0	93.4	92.0	89.0	82.3	97.5	..	Import Prices, cif
											Trade at Constant 1980 Prices
				(Millions of 1980 US dollars)							
2,984	2,441	2,910	4,871	6,705	6,753	8,759	10,208	8,643	10,530	..	Exports, fob
7,213	6,384	7,573	8,507	8,971	9,888	11,695	12,941	13,396	14,372	..	Imports, cif
				(Millions of current US dollars)							**BALANCE OF PAYMENTS**
2,820	2,968	3,671	6,019	7,928	7,946	9,755	11,387	10,833	14,433	..	Exports of Goods & Services
2,288	2,261	2,910	4,703	5,890	5,905	7,389	8,255	7,583	10,322	..	Merchandise, fob
466	673	710	1,264	1,918	1,939	2,157	2,618	2,696	3,520	..	Nonfactor Services
66	34	51	52	120	102	209	514	554	591	..	Factor Services
5,185	6,191	9,251	10,510	11,157	11,629	13,276	14,415	14,310	17,838	..	Imports of Goods & Services
4,369	4,815	7,513	8,567	8,518	8,895	10,331	11,230	10,664	13,556	..	Merchandise, fob
310	366	569	465	1,031	1,166	1,296	1,333	1,349	1,695	..	Nonfactor Services
506	1,010	1,169	1,478	1,608	1,568	1,649	1,852	2,297	2,587	..	Factor Services
206	288	501	980	1,164	1,230	1,182	1,319	1,495	1,885	..	Long-Term Interest
1,086	1,799	2,153	2,559	2,189	1,549	1,885	1,762	1,703	2,070	..	Current Transfers, net
983	1,694	2,071	2,490	2,140	1,513	1,807	1,714	1,634	2,021	..	Workers' Remittances
-1,279	-1,424	-3,427	-1,932	-1,040	-2,134	-1,636	-1,266	-1,774	-1,335	K	Total to be Financed
13	11	18	16	105	236	229	236	246	351	K	Official Capital Grants
-1,266	-1,413	-3,409	-1,916	-935	-1,898	-1,407	-1,030	-1,528	-984	..	Current Account Balance
1,450	1,643	2,934	1,309	1,189	585	1,388	311	896	2,005	f	Long-Term Capital, net
34	75	18	95	55	46	113	99	125	110	..	Direct Investment
1,092	4,038	1,844	1,161	908	469	1,346	425	1,740	1,596	..	Long-Term Loans
1,452	4,516	2,436	1,959	2,094	1,628	2,511	2,788	3,733	4,616	..	Disbursements
360	478	592	797	1,185	1,159	1,165	2,363	1,993	3,020	..	Repayments
310	-2,481	1,054	37	121	-166	-300	-449	-1,215	-52	..	Other Long-Term Capital
-205	-357	548	535	-100	1,303	124	621	1,167	-450	f	Other Capital, net
22	127	-72	72	-154	10	-105	98	-535	-571	..	Change in Reserves
											Memo Item:
				(Turkish Liras per US dollar)							
24.280	31.080	76.040	111.220	162.550	225.460	366.680	521.980	674.510	857.170	..	Conversion Factor (Annual Avg)
											EXTERNAL DEBT, ETC.
				(Millions of US dollars, outstanding at end of year)							
6,464	11,030	14,961	15,225	16,064	16,042	16,536	19,533	24,285	30,490	..	Public/Publicly Guar. Long-Term
5,309	6,830	9,523	10,019	10,510	10,327	10,246	12,389	15,040	18,173	..	Official Creditors
836	1,080	1,347	1,734	2,149	2,520	2,539	3,609	4,836	6,459	..	IBRD and IDA
1,154	4,200	5,439	5,206	5,554	5,714	6,290	7,144	9,245	12,317	..	Private Creditors
557	630	535	440	394	399	425	359	503	866	..	Private Non-guaranteed Long-Term
622	633	1,054	1,322	1,455	1,567	1,426	1,326	1,085	770	..	Use of Fund Credit
7,186	3,596	2,490	2,194	1,764	2,281	3,180	4,759	6,911	8,692	..	Short-Term Debt
											Memo Items:
				(Millions of US dollars)							
801.3	658.2	1,077.0	928.2	1,080.1	1,288.1	1,270.7	1,055.9	1,464.6	1,588.8	..	Int'l Reserves Excluding Gold
828.8	1,927.8	2,221.1	1,497.8	1,722.2	1,440.3	1,171.5	1,261.6	1,500.9	1,854.8	..	Gold Holdings (at market price)
											SOCIAL INDICATORS
4.2	4.1	4.1	4.0	3.9	3.9	3.8	3.8	3.7	3.7	..	Total Fertility Rate
31.6	31.3	30.9	30.6	30.2	30.0	29.8	29.6	29.4	29.1	..	Crude Birth Rate
114.4	108.8	103.2	97.6	92.0	83.5	74.9	66.4	57.8	49.3	..	Infant Mortality Rate
61.0	61.5	62.0	62.5	63.0	63.4	63.8	64.2	64.6	64.9	..	Life Expectancy at Birth
100.6	100.1	99.5	100.4	103.0	99.3	97.7	99.6	102.1	101.3	..	Food Production, p.c. ('79-81 = 100)
61.0	59.7	58.4	..	..	..	..	..	..	..	..	Labor Force, Agriculture (%)
35.2	34.8	34.4	34.3	34.2	34.2	34.1	34.0	34.0	33.9	..	Labor Force, Female (%)
105.0	104.0	97.0	100.0	..	111.0	113.0	116.0	..	..	..	School Enroll. Ratio, primary
34.0	37.0	..	36.0	36.0	38.0	82.0	42.0	..	..	..	School Enroll. Ratio, secondary

UGANDA	1967	1968	1969	1970	1971	1972	1973	1974	1975	1976	1977
CURRENT GNP PER CAPITA (US $)	160	160	180	190	210	210	210	220	220	270	280
POPULATION (thousands)	8,773	9,117	9,446	9,759	10,052	10,323	10,580	10,837	11,105	11,383	11,673

ORIGIN AND USE OF RESOURCES					*(Billions of current Old Uganda Shillings)*						
Gross National Product (GNP)	7.0E+00	7.0E+00	8.0E+00	9.0E+00	1.1E+01	1.1E+01	1.3E+01	1.6E+01	2.2E+01	2.6E+01	5.4E+01
Net Factor Income from Abroad	-1.3E-01	-1.2E-01	-1.3E-01	-1.1E-01	-1.6E-01	-1.2E-01	-1.0E-01	-1.1E-01	-1.6E-02	-5.8E-02	-9.0E-02
Gross Domestic Product (GDP)	7.0E+00	7.0E+00	8.0E+00	9.0E+00	1.1E+01	1.1E+01	1.3E+01	1.6E+01	2.3E+01	2.6E+01	5.4E+01
Indirect Taxes, net	6.7E-01	7.8E-01	8.6E-01	9.2E-01	1.0E+00	1.0E+00	1.0E+00	2.0E+00	2.0E+00	2.0E+00	5.0E+00
GDP at factor cost	6.0E+00	7.0E+00	7.0E+00	9.0E+00	1.0E+01	1.0E+01	1.2E+01	1.4E+01	2.1E+01	2.4E+01	4.9E+01
Agriculture	3.0E+00	3.0E+00	4.0E+00	5.0E+00	5.0E+00	6.0E+00	7.0E+00	9.0E+00	1.5E+01	1.8E+01	3.7E+01
Industry	8.0E-01	9.2E-01	1.0E+00	1.0E+00	1.0E+00	1.0E+00	1.0E+00	2.0E+00	2.0E+00	2.0E+00	3.0E+00
Manufacturing	5.2E-01	6.0E-01	6.9E-01	7.8E-01	8.0E-01	7.9E-01	7.9E-01	1.0E+00	1.0E+00	1.0E+00	3.0E+00
Services, etc.	2.0E+00	2.0E+00	3.0E+00	3.0E+00	3.0E+00	3.0E+00	3.0E+00	4.0E+00	4.0E+00	5.0E+00	9.0E+00
Resource Balance	6.3E-02	9.8E-02	9.9E-02	2.9E-01	-4.2E-01	2.8E-01	4.1E-01	-4.8E-02	-4.8E-01	4.4E-01	6.8E-01
Exports of Goods & NFServices	2.0E+00	2.0E+00	2.0E+00	2.0E+00	2.0E+00	2.0E+00	2.0E+00	2.0E+00	2.0E+00	3.0E+00	5.0E+00
Imports of Goods & NFServices	2.0E+00	2.0E+00	2.0E+00	2.0E+00	2.0E+00	2.0E+00	2.0E+00	2.0E+00	2.0E+00	2.0E+00	4.0E+00
Domestic Absorption	7.0E+00	7.0E+00	8.0E+00	9.0E+00	1.1E+01	1.1E+01	1.3E+01	1.6E+01	2.3E+01	2.6E+01	5.4E+01
Private Consumption, etc.	5.0E+00	6.0E+00	6.0E+00	..	..	..	..	..	..	..	..
General Gov't Consumption	7.4E-01	8.0E-01	8.9E-01	..	..	..	..	..	..	..	..
Gross Domestic Investment	9.2E-01	9.7E-01	1.0E+00	1.0E+00	2.0E+00	1.0E+00	1.0E+00	2.0E+00	2.0E+00	2.0E+00	3.0E+00
Fixed Investment	9.2E-01	9.7E-01	1.0E+00	1.0E+00	2.0E+00	1.0E+00	1.0E+00	2.0E+00	2.0E+00	1.0E+00	..
Memo Items:											
Gross Domestic Saving	9.8E-01	1.0E+00	1.0E+00	2.0E+00	1.0E+00	2.0E+00	1.0E+00	2.0E+00	1.0E+00	2.0E+00	4.0E+00
Gross National Saving	..	..	..	..	..	..	..	..	..	..	..

					(Billions of 1980 Old Uganda Shillings)						
Gross National Product	137.60	142.52	159.35	161.20	163.08	165.59	168.18	170.26	162.61	166.16	169.14
GDP at factor cost	126.74	129.50	145.03	147.08	148.31	150.89	153.10	152.63	150.34	151.25	154.17
Agriculture	88.31	89.54	101.69	102.89	100.49	104.25	109.51	108.45	108.10	108.80	111.67
Industry	11.80	12.54	13.75	14.04	14.21	13.81	12.64	12.51	10.94	10.28	9.86
Manufacturing	10.59	11.03	12.23	12.82	12.96	13.00	12.15	11.89	10.26	10.04	9.75
Services, etc.	26.63	27.43	29.60	30.15	33.60	32.83	30.95	31.67	31.30	32.17	32.64
Resource Balance	-1.66	-2.03	-1.87	-1.85	-5.44	0.09	0.80	-0.08	-0.01	0.22	-3.46
Exports of Goods & NFServices	7.60	7.39	7.82	8.00	7.27	7.61	6.93	5.98	5.28	4.52	3.76
Imports of Goods & NFServices	9.27	9.42	9.69	9.85	12.70	7.53	6.12	6.06	5.29	4.30	7.23
Domestic Absorption	142.03	146.93	163.67	164.91	171.12	167.28	168.64	171.52	162.73	166.30	172.90
Private Consumption, etc.	..	..	..	..	..	..	..	..	..	..	..
General Gov't Consumption	..	..	..	..	..	..	..	..	..	..	..
Gross Domestic Investment	19.02	19.27	20.86	22.02	25.91	18.75	13.93	18.44	12.41	9.67	9.40
Fixed Investment	..	..	..	..	..	..	..	..	..	..	..
Memo Items:											
Capacity to Import	9.61	9.97	10.25	11.43	10.51	8.70	7.70	5.93	4.19	5.09	8.48
Terms of Trade Adjustment	2.01	2.59	2.44	3.43	3.24	1.09	0.78	-0.06	-1.09	0.58	4.72
Gross Domestic Income	142.38	147.48	164.24	166.49	168.92	168.45	170.22	171.39	161.63	167.10	174.15
Gross National Income	139.61	145.10	161.79	164.64	166.32	166.68	168.96	170.21	161.53	166.74	173.86

DOMESTIC PRICES (DEFLATORS)					*(Index 1980 = 100)*						
Overall (GDP)	4.9	5.1	5.2	5.8	6.5	6.7	7.6	9.3	13.8	15.9	32.1
Domestic Absorption	4.8	5.0	5.0	5.6	6.5	6.6	7.4	9.4	14.1	15.6	31.1
Agriculture	3.6	3.7	3.7	4.5	5.4	5.6	6.5	8.2	13.9	16.2	32.8
Industry	6.7	7.3	7.8	8.3	8.5	8.3	8.9	12.5	15.5	17.6	35.3
Manufacturing	4.9	5.4	5.6	6.1	6.2	6.1	6.5	9.5	12.8	14.6	29.6

MANUFACTURING ACTIVITY											
Employment (1980=100)	..	..	..	..	..	..	..	..	..	..	..
Real Earnings per Empl. (1980=100)	..	..	..	..	..	..	..	..	..	..	..
Real Output per Empl. (1980=100)	..	..	..	..	..	..	..	..	..	..	..
Earnings as % of Value Added	45.2	43.9	41.6	..	39.4	..	..	..	..	..	..

MONETARY HOLDINGS					*(Millions of current New Uganda Shillings)*						
Money Supply, Broadly Defined	11.0	13.0	14.0	17.0	17.0	21.0	29.0	39.0	47.0	62.0	74.0
Money as Means of Payment	7.6	9.1	9.7	11.0	11.0	15.0	21.0	30.0	32.0	44.0	58.0
Currency Ouside Banks	3.7	4.4	5.2	5.9	6.0	6.2	8.0	11.0	14.0	22.0	29.0
Demand Deposits	3.9	4.6	4.5	5.2	5.3	9.2	13.0	19.0	19.0	22.0	29.0
Quasi-Monetary Liabilities	3.6	4.2	4.4	5.6	5.3	6.0	7.7	8.5	14.0	17.0	16.0

GOVERNMENT DEFICIT (-) OR SURPLUS					*(Millions of current Old Uganda Shillings)*						
	..	..	..	..	..	-900	-860	-1,530	-1,240	-1,320	-1,280
Current Revenue	..	..	..	..	..	1,540	1,240	1,180	2,090	2,460	3,440
Current Expenditure	..	..	..	..	..	..	..	..	2,700	3,140	3,920
Current Budget Balance	..	..	..	..	..	..	..	..	-610	-680	-480
Capital Receipts	..	..	..	..	..	..	..	..	5	..	..
Capital Payments	..	..	..	..	..	..	..	..	630	640	790

UGANDA	1967	1968	1969	1970	1971	1972	1973	1974	1975	1976	1977
FOREIGN TRADE (CUSTOMS BASIS)				*(Millions of current US dollars)*							
Value of Exports, fob	181.5	183.4	195.7	245.5	235.3	260.5	299.9	315.8	263.6	351.7	588.3
Nonfuel Primary Products	179.7	181.7	194.4	242.9	233.2	258.4	298.6	314.8	262.9	347.3	582.7
Fuels	0.2	0.2	0.2	0.3	0.2	0.3	0.2	0.4	0.5	2.9	3.6
Manufactures	1.6	1.4	1.1	2.2	1.9	1.9	1.2	0.7	0.2	1.5	2.0
Value of Imports, cif	159.3	164.6	174.6	172.1	249.6	162.0	162.1	213.3	205.7	172.3	241.2
Nonfuel Primary Products	14.7	13.9	16.8	15.7	24.8	20.6	22.3	34.0	19.0	17.0	13.5
Fuels	2.4	2.1	2.4	2.7	1.5	1.1	1.0	1.5	3.1	51.1	53.2
Manufactures	142.2	148.7	155.4	153.7	223.3	140.3	138.9	177.8	183.6	104.2	174.5
Terms of Trade	98.9	98.8	94.1	109.3	89.5	89.5	96.6	80.1	70.8	163.9	213.3
Export Prices, fob	28.7	28.7	28.5	34.2	31.8	34.9	45.2	47.2	44.7	88.0	129.9
Nonfuel Primary Products	28.9	28.9	28.7	34.5	31.9	35.0	45.3	47.2	44.7	89.1	132.0
Fuels	4.3	4.3	4.3	4.3	5.6	6.2	8.9	36.7	35.7	38.4	42.0
Manufactures	30.1	31.7	32.4	37.5	36.5	38.5	49.5	67.2	66.1	66.7	71.7
Import Prices, cif	29.0	29.1	30.3	31.3	35.6	39.0	46.8	58.9	63.1	53.7	60.9
Trade at Constant 1980 Prices				*(Millions of 1980 US dollars)*							
Exports, fob	633.2	638.2	687.2	717.1	739.1	746.4	663.2	668.7	590.2	399.6	453.0
Imports, cif	549.6	566.4	576.7	549.2	701.7	415.3	346.1	361.9	326.1	320.8	396.2
BALANCE OF PAYMENTS				*(Millions of current US dollars)*							
Exports of Goods & Services	..	..	..	297.1	284.1	291.5	290.4	307.8	251.8	335.7	556.1
Merchandise, fob	..	..	..	261.6	243.9	263.8	275.1	294.1	237.2	323.6	547.8
Nonfactor Services	..	..	..	32.6	37.1	25.3	11.2	9.4	10.6	9.4	4.9
Factor Services	..	..	..	2.9	3.1	2.4	4.1	4.3	4.0	2.8	3.4
Imports of Goods & Services	..	..	..	271.4	365.1	269.3	246.1	330.6	320.5	294.6	485.3
Merchandise, fob	..	..	..	178.3	247.7	171.9	175.5	236.3	228.4	206.8	366.6
Nonfactor Services	..	..	..	75.1	91.8	78.2	52.2	74.0	84.3	78.2	104.5
Factor Services	..	..	..	18.0	25.6	19.2	18.4	20.3	7.9	9.7	14.2
Long-Term Interest	..	..	..	4.7	4.9	5.1	4.8	4.6	3.2	2.3	4.2
Current Transfers, net	..	..	..	-6.9	-7.4	-9.0	-3.9	-5.1	-3.4	-5.7	-4.7
Workers' Remittances	..	..	..	0.0	0.0	0.0	0.0	0.0	0.0	0.0	0.0
Total to be Financed	..	..	..	18.8	-88.4	13.3	40.4	-27.9	-72.1	35.4	66.1
Official Capital Grants	..	..	..	1.5	2.6	3.2	2.6	3.9	16.0	7.7	2.0
Current Account Balance	..	..	..	20.3	-85.8	16.4	43.0	-24.1	-56.1	43.2	68.1
Long-Term Capital, net	..	..	..	8.2	33.6	32.8	-12.2	20.1	24.0	-8.3	-9.9
Direct Investment	..	..	..	4.2	-1.2	-11.9	5.3	1.7	2.1	1.2	0.8
Long-Term Loans	..	..	..	23.3	20.0	28.2	-3.1	31.2	12.3	33.5	29.5
Disbursements	..	..	..	27.3	24.4	34.0	17.1	42.2	22.0	40.7	48.8
Repayments	..	..	..	4.0	4.4	5.8	20.2	11.0	9.7	7.2	19.3
Other Long-Term Capital	..	..	..	-20.8	12.2	13.4	-16.9	-16.7	-6.4	-50.7	-42.2
Other Capital, net	..	..	..	-30.1	7.8	-44.8	-39.5	-14.2	32.0	-32.5	-57.6
Change in Reserves	..	..	..	1.6	44.4	-4.4	8.6	18.2	0.0	-2.4	-0.5
Memo Item:											
Conversion Factor (Annual Avg)	5.000	4.900	4.700	*(Uganda Shillings per US dollar)*							
	5.000	4.900	4.700	4.900	5.100	5.400	5.900	6.600	9.200	9.600	16.900
EXTERNAL DEBT, ETC.				*(Millions of US dollars, outstanding at end of year)*							
Public/Publicly Guar. Long-Term	..	..	..	138.3	149.9	172.3	176.9	211.9	211.6	239.6	278.0
Official Creditors	..	..	..	108.6	113.2	135.2	154.2	178.6	184.9	214.9	220.5
IBRD and IDA	..	..	..	19.0	26.6	34.6	40.2	40.9	43.1	45.2	45.4
Private Creditors	..	..	..	29.7	36.7	37.1	22.7	33.3	26.7	24.7	57.5
Private Non-guaranteed Long-Term	..	..	..	..	..	..	..	..	..	..	..
Use of Fund Credit	..	..	..	0.0	10.8	10.8	12.1	26.3	28.2	45.5	47.6
Short-Term Debt	..	..	..	..	..	..	..	..	..	..	34.0
Memo Items:				*(Thousands of US dollars)*							
Int'l Reserves Excluding Gold	34,930	49,030	52,270	56,650	26,923	35,953	29,113	16,759	31,023	44,534	47,206
Gold Holdings (at market price)	..	..	..	..	..	..	..	..	..	..	..
SOCIAL INDICATORS											
Total Fertility Rate	6.9	6.9	6.9	6.9	6.9	6.9	6.9	6.9	6.9	6.9	6.9
Crude Birth Rate	49.1	49.3	49.6	49.8	50.1	50.3	50.3	50.3	50.3	50.3	50.3
Infant Mortality Rate	118.0	117.6	117.2	116.8	116.4	116.0	115.6	115.2	114.8	114.4	114.0
Life Expectancy at Birth	46.0	46.2	46.4	46.6	46.8	47.0	46.8	46.6	46.4	46.2	46.0
Food Production, p.c. ('79-81 = 100)	110.7	110.3	115.0	122.4	124.9	124.3	113.6	122.1	127.6	125.7	120.2
Labor Force, Agriculture (%)	90.3	89.9	89.6	89.3	89.0	88.6	88.3	87.9	87.6	87.3	86.9
Labor Force, Female (%)	43.3	43.2	43.2	43.2	43.1	43.1	43.1	43.0	43.0	42.9	42.9
School Enroll. Ratio, primary	..	..	..	38.0	..	..	..	..	44.0	47.0	50.0
School Enroll. Ratio, secondary	..	..	..	4.0	..	..	..	..	4.0	4.0	5.0

1978	1979	1980	1981	1982	1983	1984	1985	1986	1987 estimate	NOTES	UGANDA
											FOREIGN TRADE (CUSTOMS BASIS)
350.0	436.0	345.0	243.0	349.0	372.0	399.0	394.0	443.0	221.5	f	Value of Exports, fob
346.7	431.7	342.0	240.9	346.0	368.8	396.2	391.3	441.3	220.1	..	Nonfuel Primary Products
2.2	2.8	1.9	1.4	2.0	2.1	1.9	1.8	0.7	0.6	..	Fuels
1.2	1.5	1.0	0.7	1.1	1.1	0.9	0.9	1.0	0.8	..	Manufactures
255.0	197.0	293.0	345.2	377.4	377.4	344.1	327.4	349.6	417.3	f	Value of Imports, cif
17.6	13.8	20.1	23.7	25.9	26.2	22.8	19.7	27.7	29.4	..	Nonfuel Primary Products
56.6	45.4	66.7	78.7	86.1	85.8	76.9	90.7	25.7	37.2	..	Fuels
180.8	137.8	206.2	242.8	265.4	265.4	244.4	217.0	296.2	350.8	..	Manufactures
				(Index 1980 = 100)							
147.3	130.5	100.0	80.6	88.6	89.1	99.7	96.1	115.7	67.3	f	Terms of Trade
98.4	106.8	100.0	83.1	88.1	85.2	93.3	89.2	115.8	74.2	f	Export Prices, fob
99.3	107.4	100.0	83.0	88.0	85.2	93.3	89.2	116.1	74.2	..	Nonfuel Primary Products
42.3	61.0	100.0	112.5	101.6	92.5	90.2	87.5	44.9	56.9	..	Fuels
76.1	90.0	100.0	99.1	94.8	90.7	89.3	89.0	103.2	113.0	..	Manufactures
66.8	81.8	100.0	103.2	99.5	95.7	93.5	92.8	100.1	110.2	f	Import Prices, cif
											Trade at Constant 1980 Prices
355.9	408.2	345.0	292.3	396.0	436.5	427.8	441.9	382.6	298.5	f	Exports, fob
381.9	240.7	293.0	334.5	379.2	394.5	367.9	352.9	349.2	378.8	f	Imports, cif
				(Millions of current US dollars)							**BALANCE OF PAYMENTS**
337.4	414.2	330.7	274.4	347.1	384.4	450.4	360.5	421.7	360.4	f	Exports of Goods & Services
323.0	397.2	319.4	229.3	347.1	367.7	410.2	342.9	406.3	335.7	..	Merchandise, fob
8.5	14.9	9.9	44.3	0.0	11.5	33.0	17.1	13.2	23.1	..	Nonfactor Services
5.9	2.0	1.4	0.8	0.0	5.2	7.2	0.5	2.2	1.6	..	Factor Services
471.8	387.1	449.8	391.9	524.3	557.0	518.9	534.3	577.4	678.8	f	Imports of Goods & Services
306.3	265.7	317.6	278.3	337.6	429.6	368.2	410.9	476.9	532.7	..	Merchandise, fob
152.5	108.4	123.4	100.4	160.4	83.5	81.7	73.0	58.4	87.3	..	Nonfactor Services
13.0	13.0	8.8	13.2	26.3	43.9	69.0	50.4	42.1	58.8	..	Factor Services
2.9	3.0	2.4	2.8	10.2	16.9	32.4	20.3	14.7	24.4	..	Long-Term Interest
-8.8	-16.7	-2.2	-0.2	0.0	65.8	27.4	42.6	125.9	118.3	..	Current Transfers, net
0.5	0.8	0.0	0.0	0.0	0.0	0.0	0.0	0.0	..	..	Workers' Remittances
-143.3	10.4	-121.3	-117.7	-177.2	-106.8	-41.1	-131.2	-29.8	-200.1	K	Total to be Financed
5.9	29.1	38.1	149.6	107.3	103.5	85.4	21.7	66.6	93.1	K	Official Capital Grants
-137.4	39.5	-83.2	31.9	-69.9	-3.3	44.3	-109.5	36.8	-107.0	..	Current Account Balance
21.7	122.0	8.1	96.9	143.5	71.5	110.2	139.0	82.1	176.8	f	Long-Term Capital, net
1.0	1.6	..	..	..	0.7	0.9	0.4	0.0	0.5	..	Direct Investment
59.7	136.3	114.3	43.7	57.9	33.1	27.6	95.5	77.3	140.9	..	Long-Term Loans
70.4	148.2	134.1	99.2	102.6	93.4	84.9	138.3	107.5	186.9	..	Disbursements
10.7	11.9	19.8	55.5	44.7	60.3	57.3	42.8	30.2	46.0	..	Repayments
-44.9	-45.0	-144.3	-96.4	-21.7	-65.8	-3.7	21.4	-61.8	-57.7	..	Other Long-Term Capital
122.9	-194.2	19.9	-223.5	-131.0	-116.0	-128.1	-4.4	-63.2	-82.0	f	Other Capital, net
-7.2	32.7	55.2	94.7	57.4	47.8	-26.4	-25.1	-55.7	12.2	..	Change in Reserves
				(Uganda Shillings per US dollar)							**Memo Item:**
22.800	40.200	73.500	156.600	214.000	268.800	359.700	672.000	1,400.000	4,284.100	..	Conversion Factor (Annual Avg)
				(Millions of US dollars, outstanding at end of year)							**EXTERNAL DEBT, ETC.**
362.7	504.0	602.9	536.1	597.3	637.3	686.2	828.3	956.6	1,116.0	..	Public/Publicly Guar. Long-Term
259.5	379.6	455.9	413.3	475.2	541.9	611.8	760.3	889.9	1,058.5	..	Official Creditors
46.6	47.0	47.0	61.2	91.8	123.3	212.6	322.3	412.9	584.1	..	IBRD and IDA
103.2	124.3	147.1	122.8	122.1	95.4	74.4	68.0	66.6	57.6	..	Private Creditors
..	..	..	..	..	..	..	..	..	..	..	Private Non-guaranteed Long-Term
38.0	34.5	60.9	186.2	268.5	353.9	314.9	282.4	229.3	228.9	..	Use of Fund Credit
21.0	30.0	65.7	45.3	56.4	27.7	42.4	46.0	74.1	60.0	..	Short-Term Debt
				(Thousands of US dollars)							**Memo Items:**
52,700	22,839	3,000	30,003	78,324	106,500	67,907	27,288	29,206	54,600	..	Int'l Reserves Excluding Gold
..	..	..	..	..	..	..	..	..	..	..	Gold Holdings (at market price)
											SOCIAL INDICATORS
6.9	6.9	6.9	6.9	6.9	6.9	6.9	6.9	6.9	6.9	..	Total Fertility Rate
50.3	50.3	50.3	50.3	50.3	50.3	50.4	50.4	50.5	50.5	..	Crude Birth Rate
113.6	113.2	112.8	112.4	112.0	112.8	113.7	114.5	115.4	116.2	..	Infant Mortality Rate
46.0	46.0	46.0	46.0	46.0	46.4	46.8	47.3	47.7	48.2	..	Life Expectancy at Birth
122.1	95.0	98.6	106.4	110.0	113.4	82.7	124.1	124.7	119.0	..	Food Production, p.c. ('79-81 = 100)
86.6	86.2	85.9	..	..	..	..	..	..	..	..	Labor Force, Agriculture (%)
42.9	42.8	42.8	42.6	42.5	42.3	42.1	41.9	41.8	41.6	..	Labor Force, Female (%)
51.0	50.0	49.0	52.0	58.0	..	..	..	..	..	..	School Enroll. Ratio, primary
5.0	5.0	5.0	5.0	8.0	..	..	..	..	..	..	School Enroll. Ratio, secondary

UNITED ARAB EMIRATES	1967	1968	1969	1970	1971	1972	1973	1974	1975	1976	1977
CURRENT GNP PER CAPITA (US $)	..	..	..	..	..	..	..	..	9,950	15,950	19,240
POPULATION (thousands)	165	184	203	225	314	401	484	566	646	724	799

ORIGIN AND USE OF RESOURCES
(Millions of current U.A.E. Dirhams)

	1967	1968	1969	1970	1971	1972	1973	1974	1975	1976	1977
Gross National Product (GNP)	..	..	..	..	..	..	9,782	29,999	37,943	48,745	61,311
Net Factor Income from Abroad	..	..	..	..	..	..	-1,619	-1,101	-1,371	-2,083	-1,799
Gross Domestic Product (GDP)	..	..	..	..	..	..	11,400	31,100	39,314	50,828	63,110
Indirect Taxes, net	..	..	..	..	..	..	..	..	-321	-679	-785
GDP at factor cost	..	..	..	..	..	..	..	..	39,635	51,507	63,895
Agriculture	..	..	..	..	..	..	155	208	329	432	491
Industry	..	..	..	..	..	..	6,920	24,511	31,348	39,277	46,556
Manufacturing	..	..	..	..	..	..	225	301	369	593	1,853
Services, etc.	..	..	..	..	..	..	..	..	7,958	11,798	16,848
Resource Balance	..	..	..	..	..	..	5,700	21,400	17,779	21,550	19,174
Exports of Goods & NFServices	..	..	..	..	..	..	9,400	29,400	29,183	36,106	41,265
Imports of Goods & NFServices	..	..	..	..	..	..	3,700	8,000	11,404	14,556	22,091
Domestic Absorption	..	..	..	..	..	..	5,700	9,700	21,535	29,278	43,936
Private Consumption, etc.	..	..	..	..	..	..	1,500	2,200	6,215	7,695	11,557
General Gov't Consumption	..	..	..	..	..	..	1,300	2,700	3,261	4,648	7,413
Gross Domestic Investment	..	..	..	..	..	..	2,900	4,800	12,059	16,935	24,966
Fixed Investment	..	..	..	..	..	..	..	..	12,059	16,585	22,686
Memo Items:											
Gross Domestic Saving	..	..	..	..	..	..	8,600	26,200	29,838	38,485	44,140
Gross National Saving	..	..	..	..	..	..	5,175	21,916	23,631	29,721	34,140

(Millions of 1980 U.A.E. Dirhams)

	1967	1968	1969	1970	1971	1972	1973	1974	1975	1976	1977
Gross National Product	..	..	..	..	..	..	..	39,537	50,863	58,168	69,199
GDP at factor cost	..	..	..	..	..	..	..	..	53,054	61,437	71,956
Agriculture	..	..	..	..	..	..	..	..	367	480	563
Industry	..	..	..	..	..	..	..	..	41,467	45,451	51,713
Manufacturing	..	..	..	..	..	..	..	..	472	702	1,923
Services, etc.	..	..	..	..	..	..	..	..	11,220	15,506	19,680
Resource Balance	..	..	..	..	..	..	..	..	22,687	22,061	17,957
Exports of Goods & NFServices	..	..	..	..	..	..	..	..	42,145	47,713	53,212
Imports of Goods & NFServices	..	..	..	..	..	..	..	..	19,458	25,652	35,255
Domestic Absorption	..	..	..	..	..	..	..	..	30,040	38,597	53,279
Private Consumption, etc.	..	..	..	..	..	..	..	..	9,344	12,231	15,621
General Gov't Consumption	..	..	..	..	..	..	..	..	4,667	6,490	8,280
Gross Domestic Investment	..	..	..	..	..	..	..	..	16,029	19,876	29,378
Fixed Investment	..	..	..	..	..	..	..	..	16,029	19,164	26,078
Memo Items:											
Capacity to Import	..	..	..	..	..	..	..	..	49,793	63,630	65,855
Terms of Trade Adjustment	..	..	..	..	..	..	..	..	7,648	15,916	12,643
Gross Domestic Income	..	..	..	..	..	..	..	..	60,375	76,574	83,879
Gross National Income	..	..	..	..	..	..	..	..	58,511	74,084	81,842

DOMESTIC PRICES (DEFLATORS)
(Index 1980 = 100)

	1967	1968	1969	1970	1971	1972	1973	1974	1975	1976	1977
Overall (GDP)	..	..	..	..	..	..	..	67.5	74.6	83.8	88.6
Domestic Absorption	..	..	..	..	..	..	..	..	71.7	75.9	82.5
Agriculture	..	..	..	..	..	..	..	..	89.6	90.0	87.2
Industry	..	..	..	..	..	..	..	..	75.6	86.4	90.0
Manufacturing	..	..	..	..	..	..	..	..	78.2	84.5	96.4

MANUFACTURING ACTIVITY

	1967	1968	1969	1970	1971	1972	1973	1974	1975	1976	1977
Employment (1980=100)	..	..	..	..	..	..	..	..	..	..	..
Real Earnings per Empl. (1980=100)	..	..	..	..	..	..	..	..	..	..	..
Real Output per Empl. (1980=100)	..	..	..	..	..	..	..	..	..	..	..
Earnings as % of Value Added	..	..	..	..	..	..	..	..	..	..	39.5

MONETARY HOLDINGS
(Millions of current U.A.E. Dirhams)

	1967	1968	1969	1970	1971	1972	1973	1974	1975	1976	1977
Money Supply, Broadly Defined	..	..	..	..	..	..	2,257	6,036	8,820	16,754	15,540
Money as Means of Payment	..	..	..	..	..	..	970	1,536	2,603	4,725	5,215
Currency Ouside Banks	..	..	..	..	..	..	265	429	628	1,077	1,392
Demand Deposits	..	..	..	..	..	..	704	1,107	1,975	3,648	3,822
Quasi-Monetary Liabilities	..	..	..	..	..	..	1,287	4,499	6,217	12,029	10,325

GOVERNMENT DEFICIT (-) OR SURPLUS
(Millions of current U.A.E. Dirhams)

	1967	1968	1969	1970	1971	1972	1973	1974	1975	1976	1977
Current Revenue	..	..	..	..	..	37	30	62	590	597	-211
Current Expenditure	..	..	..	..	..	201	420	800	1,773	3,102	5,995
Current Budget Balance	..	..	..	..	..	148	331	569	886	1,456	4,364
Capital Receipts	..	..	..	..	..	53	89	231	887	1,646	1,631
Capital Payments	..	..	..	..	..	..	..	..	..	..	..
	..	..	..	..	..	16	59	169	297	1,049	1,842

1978	1979	1980	1981	1982	1983	1984	1985	1986	1987 estimate	NOTES	UNITED ARAB EMIRATES
18,530	22,630	28,910	30,520	27,700	24,540	23,510	21,100	15,600	15,770	..	**CURRENT GNP PER CAPITA (US $)**
873	945	1,016	1,085	1,153	1,220	1,285	1,350	1,403	1,456	..	POPULATION (thousands)
											ORIGIN AND USE OF RESOURCES
				(Millions of current U.A.E. Dirhams)							
59,382	78,641	110,434	121,548	113,513	102,617	102,017	99,565	78,542	85,483	..	Gross National Product (GNP)
-1,003	-992	601	709	360	371	374	371	444	483	..	Net Factor Income from Abroad
60,385	79,633	109,833	120,839	113,153	102,246	101,643	99,194	78,098	85,000	..	Gross Domestic Product (GDP)
-885	-1,164	-1,637	-3,215	-2,501	-2,878	-2,861	-2,760	-2,564	..	..	Indirect Taxes, net
61,270	80,797	111,470	124,054	115,654	105,124	104,504	101,954	80,662	87,086	..	GDP at factor cost
604	680	827	1,036	1,144	1,198	1,349	1,440	1,540	..	..	Agriculture
43,982	60,960	86,089	90,170	77,735	68,410	68,639	65,296	45,666	..	..	Industry
2,197	2,533	4,191	8,077	9,436	9,584	9,761	9,255	8,405	..	..	Manufacturing
16,684	19,157	24,554	32,848	36,775	35,516	34,516	35,218	33,456	..	..	Services, etc.
16,042	27,146	47,718	42,617	31,144	23,902	27,707	27,604	7,300	11,800	..	Resource Balance
40,121	56,871	85,592	83,662	71,576	60,874	60,008	58,266	39,396	46,800	..	Exports of Goods & NFServices
24,079	29,725	37,874	41,045	40,432	36,972	32,301	30,662	32,096	35,000	..	Imports of Goods & NFServices
44,343	52,487	62,115	78,222	82,009	78,344	73,936	71,590	70,798	73,200	..	Domestic Absorption
12,501	15,245	18,968	24,946	26,846	26,183	26,744	27,173	29,877	31,000	..	Private Consumption, etc.
8,163	9,600	11,992	21,475	22,000	19,958	17,696	19,484	18,314	19,200	..	General Gov't Consumption
23,679	27,642	31,155	31,801	33,163	32,203	29,496	24,933	22,607	23,000	..	Gross Domestic Investment
25,779	28,442	30,155	30,643	32,683	31,668	29,116	24,458	22,097	..	..	Fixed Investment
											Memo Items:
39,721	54,788	78,873	74,418	64,307	56,105	57,203	52,537	29,907	34,800	..	Gross Domestic Saving
30,322	42,696	64,774	62,730	57,967	68,478	68,545	62,908	38,684	..	..	Gross National Saving
				(Millions of 1980 U.A.E. Dirhams)							
68,413	85,759	110,434	113,596	103,950	98,386	102,944	100,469	78,700	80,659	..	Gross National Product
70,506	88,059	111,470	115,688	106,607	100,481	105,406	102,804	80,395	..	..	GDP at factor cost
649	732	827	1,020	1,079	1,236	1,400	1,525	1,617	..	..	Agriculture
50,639	66,141	86,089	84,618	74,140	66,470	70,657	66,296	45,238	..	..	Industry
2,274	2,542	4,191	7,990	9,251	9,116	9,655	9,443	8,695	..	..	Manufacturing
19,218	21,186	24,554	30,050	31,388	32,775	33,349	34,983	33,540	..	..	Services, etc.
16,432	29,116	47,718	39,278	28,309	23,812	27,429	..	..	..	..	Resource Balance
49,544	66,670	85,592	77,835	66,714	58,377	60,189	..	..	..	..	Exports of Goods & NFServices
33,112	37,554	37,874	38,557	38,405	34,565	32,760	..	..	..	..	Imports of Goods & NFServices
53,134	57,763	62,115	73,676	75,326	74,234	75,138	..	..	..	..	Domestic Absorption
16,785	17,550	18,968	23,262	25,669	24,741	27,125	..	..	..	..	Private Consumption, etc.
8,570	10,204	11,992	20,300	20,042	18,976	18,007	..	..	..	..	General Gov't Consumption
27,779	30,009	31,155	30,114	29,615	30,517	30,006	..	..	..	..	Gross Domestic Investment
30,779	31,009	30,155	29,050	29,160	30,012	29,619	..	..	..	..	Fixed Investment
											Memo Items:
55,172	71,850	85,592	78,591	67,988	56,911	60,860	..	..	..	..	Capacity to Import
5,628	5,180	0	756	1,274	-1,466	671	..	..	..	..	Terms of Trade Adjustment
75,194	92,059	109,833	113,710	104,909	96,580	103,238	..	..	..	..	Gross Domestic Income
74,041	90,939	110,434	114,352	105,223	96,920	103,615	..	..	..	..	Gross National Income
				(Index 1980 = 100)							**DOMESTIC PRICES (DEFLATORS)**
86.8	91.7	100.0	107.0	109.2	104.3	99.1	99.2	99.8	106.0	..	Overall (GDP)
83.5	90.9	100.0	106.2	108.9	105.5	98.4	..	..	..	..	Domestic Absorption
93.1	92.9	100.0	101.6	106.0	96.9	96.4	94.4	95.2	..	..	Agriculture
86.9	92.2	100.0	106.6	104.8	102.9	97.1	98.5	100.9	..	..	Industry
96.6	99.6	100.0	101.1	102.0	105.1	101.1	98.0	96.7	..	..	Manufacturing
											MANUFACTURING ACTIVITY
..	..	..	..	..	..	..	..	..	..	G	Employment (1980=100)
..	..	..	..	..	..	..	..	..	..	G	Real Earnings per Empl. (1980=100)
..	..	..	..	..	..	..	..	..	..	G	Real Output per Empl. (1980=100)
43.1	..	..	38.7	..	..	..	..	..	..	..	Earnings as % of Value Added
				(Millions of current U.A.E. Dirhams)							**MONETARY HOLDINGS**
17,576	18,222	23,527	29,094	33,646	36,342	46,870	49,887	52,076	54,940	..	Money Supply, Broadly Defined
5,776	6,269	7,354	8,969	9,739	9,124	8,892	9,505	9,201	10,096	..	Money as Means of Payment
1,704	1,965	2,143	2,771	2,990	2,879	2,929	3,161	3,246	3,511	..	Currency Ouside Banks
4,072	4,303	5,212	6,198	6,749	6,245	5,963	6,344	5,955	6,585	..	Demand Deposits
11,800	11,953	16,172	20,125	23,907	27,218	37,978	40,382	42,875	44,844	..	Quasi-Monetary Liabilities
				(Millions of current U.A.E. Dirhams)							**GOVERNMENT DEFICIT (-) OR SURPLUS**
-503	202	2,302	2,355	..	..	..	..	..	..	..	Current Revenue
6,984	8,862	17,608	22,460	..	..	..	..	..	..	..	Current Expenditure
6,269	7,498	12,311	17,418	17,648	14,737	14,827	..	..	..	..	Current Budget Balance
715	1,364	5,297	5,042	..	..	..	..	..	..	..	Capital Receipts
..	..	..	..	..	..	..	..	..	..	..	Capital Payments
1,218	1,162	2,995	2,687	-17,648	-14,737	-14,827	..	..	..	..	

UNITED ARAB EMIRATES	1967	1968	1969	1970	1971	1972	1973	1974	1975	1976	1977
FOREIGN TRADE (CUSTOMS BASIS)					*(Millions of current US dollars)*						
Value of Exports, fob	275.0	350.0	450.0	552.4	935.3	1,159.7	2,177.7	7,017.0	6,695.7	8,565.1	9,534.0
Nonfuel Primary Products	0.0	4.6	4.7	17.5	29.1	24.8	43.2	45.0	68.5	99.0	12.7
Fuels	266.0	340.0	431.5	532.1	901.6	1,113.7	2,116.4	6,931.7	6,544.1	8,262.1	9,090.7
Manufactures	9.0	5.5	13.9	2.7	4.7	21.2	18.1	40.4	83.1	204.1	430.6
Value of Imports, cif	240.0	280.0	307.0	271.0	311.5	495.9	841.7	1,781.1	2,754.1	3,443.2	5,179.1
Nonfuel Primary Products	43.4	54.4	58.8	50.0	56.4	75.7	125.9	289.9	342.5	465.0	676.3
Fuels	5.9	8.6	9.8	8.9	9.0	25.7	40.4	136.8	198.2	236.6	317.2
Manufactures	190.7	217.0	238.4	212.1	246.1	394.5	675.5	1,354.5	2,213.4	2,741.6	4,185.7
Terms of Trade	15.9	16.4	16.1	15.5	18.2	*(Index 1980 = 100)* 20.6	23.0	65.6	59.8	63.8	64.0
Export Prices, fob	4.4	4.4	4.4	4.4	5.7	6.4	9.1	36.9	36.1	38.9	42.8
Nonfuel Primary Products	0.0	38.0	37.0	41.7	36.4	38.0	52.7	73.0	65.5	63.2	66.0
Fuels	4.3	4.3	4.3	4.3	5.6	6.2	8.9	36.7	35.7	38.4	42.0
Manufactures	30.1	31.7	32.4	37.5	36.5	38.5	49.5	67.2	66.1	66.7	71.7
Import Prices, cif	27.6	26.6	27.5	28.4	31.6	31.2	39.4	56.3	60.4	61.0	66.9
Trade at Constant 1980 Prices					*(Millions of 1980 US dollars)*						
Exports, fob	6,274.4	8,009.6	10,183.8	12,540.8	16,279.1	17,996.7	24,033.0	18,998.7	18,540.5	22,000.8	22,280.0
Imports, cif	868.3	1,051.3	1,117.7	953.4	986.5	1,587.5	2,136.3	3,161.8	4,563.4	5,646.9	7,746.6
BALANCE OF PAYMENTS					*(Millions of current US dollars)*						
Exports of Goods & Services	..	..	..	625	1,013	1,312	2,428	7,525	7,483	9,278	10,658
Merchandise, fob	..	..	..	510	835	1,082	1,807	6,392	6,970	8,684	9,708
Nonfactor Services	..	..	..	81	133	172	545	1,034	401	448	873
Factor Services	..	..	..	34	45	58	76	99	112	146	77
Imports of Goods & Services	..	..	..	531	664	987	1,407	2,398	3,311	4,316	6,123
Merchandise, fob	..	..	..	267	309	482	821	1,705	2,669	3,327	5,048
Nonfactor Services	..	..	..	49	56	88	105	316	184	316	537
Factor Services	..	..	..	215	299	417	481	377	458	673	538
Long-Term Interest	..	..	..	..	..	..	..	..	..	..	..
Current Transfers, net	..	..	..	-19	-88	-93	-452	-804	-1,221	-1,690	-2,101
Workers' Remittances	..	..	..	..	..	..	..	..	..	..	..
Total to be Financed	..	..	..	75	261	232	569	4,323	2,951	3,272	2,434
Official Capital Grants	..	..	..	0	0	0	0	0	0	0	0
Current Account Balance	..	..	..	75	261	232	569	4,323	2,951	3,272	2,434
Long-Term Capital, net	..	..	..	1	6	-33	204	-247	-325	25	-51
Direct Investment	..	..	..	4	9	-3	-7	-308	-212	-489	-299
Long-Term Loans	..	..	..	..	..	..	..	..	..	..	..
Disbursements	..	..	..	..	..	..	..	..	..	..	..
Repayments	..	..	..	..	..	..	..	..	..	..	..
Other Long-Term Capital	..	..	..	-3	-3	-30	211	61	-113	514	248
Other Capital, net	..	..	..	-30	-76	-14	-605	-2,113	-1,163	-1,052	-1,256
Change in Reserves	..	..	..	-46	-191	-185	-168	-1,963	-1,463	-2,245	-1,127
Memo Item:					*(U.A.E. Dirhams per US dollar)*						
Conversion Factor (Annual Avg)	4.760	4.760	4.760	4.760	4.750	4.390	4.000	3.960	3.960	3.950	3.900
EXTERNAL DEBT, ETC.					*(Millions of US dollars, outstanding at end of year)*						
Public/Publicly Guar. Long-Term	..	..	..	..	..	..	..	..	..	..	..
Official Creditors	..	..	..	..	..	..	..	..	..	..	..
IBRD and IDA	..	..	..	..	..	..	..	..	..	..	..
Private Creditors	..	..	..	..	..	..	..	..	..	..	..
Private Non-guaranteed Long-Term	..	..	..	..	..	..	..	..	..	..	..
Use of Fund Credit	..	..	..	..	..	..	..	..	..	..	..
Short-Term Debt	..	..	..	..	..	..	..	..	..	..	..
Memo Items:					*(Millions of US dollars)*						
Int'l Reserves Excluding Gold	..	..	..	..	..	..	91.7	452.9	987.9	1,906.5	800.3
Gold Holdings (at market price)	..	..	..	..	..	..	..	..	..	73.4	93.8
SOCIAL INDICATORS											
Total Fertility Rate	6.8	6.8	6.8	6.8	6.8	6.8	6.8	6.8	6.8	6.8	6.8
Crude Birth Rate	38.6	37.5	36.4	35.2	34.1	33.0	32.5	32.0	31.5	31.0	30.5
Infant Mortality Rate	85.0	79.4	73.8	68.2	62.6	57.0	54.8	52.6	50.4	48.2	46.0
Life Expectancy at Birth	59.0	59.7	60.4	61.1	61.8	62.5	63.1	63.7	64.3	64.9	65.5
Food Production, p.c. ('79-81 = 100)	..	..	..	..	..	..	..	..	..	..	..
Labor Force, Agriculture (%)	18.2	16.7	15.1	13.6	12.7	11.8	10.9	10.0	9.1	8.2	7.3
Labor Force, Female (%)	3.9	4.0	4.1	4.2	4.2	4.2	4.2	4.1	4.1	4.3	4.5
School Enroll. Ratio, primary	..	..	..	93.0	..	..	..	..	101.0	..	108.0
School Enroll. Ratio, secondary	..	..	..	22.0	..	..	..	..	33.0	..	39.0

1978	1979	1980	1981	1982	1983	1984	1985	1986	1987 estimate	NOTES	UNITED ARAB EMIRATES
				(Millions of current US dollars)							**FOREIGN TRADE (CUSTOMS BASIS)**
9,138.2	13,577.0	20,738.0	20,240.0	17,260.0	15,085.0	16,000.0	14,800.0	10,300.0	12,000.0	f	Value of Exports, fob
130.6	210.2	393.8	344.2	266.1	741.9	743.2	719.8	408.2	516.4	..	Nonfuel Primary Products
8,677.5	12,862.0	19,459.0	18,924.6	16,143.8	13,018.7	13,199.8	12,300.2	8,060.8	9,386.1	f	Fuels
330.1	504.8	885.2	971.2	850.1	1,324.4	2,057.0	1,780.0	1,831.0	2,097.5	..	Manufactures
5,389.0	6,966.0	8,746.0	9,646.0	9,440.0	8,356.0	7,043.2	5,649.0	6,422.0	7,226.0	..	Value of Imports, cif
717.2	965.0	1,227.0	1,291.0	1,183.0	1,213.0	957.2	179.4	346.5	388.3	..	Nonfuel Primary Products
202.0	711.0	999.0	1,262.0	709.0	586.0	507.0	94.6	182.5	205.7	..	Fuels
4,469.8	5,290.0	6,520.0	7,093.0	7,548.0	6,557.0	5,579.0	5,375.0	5,893.0	6,632.0	..	Manufactures
				(Index 1980 = 100)							
55.9	71.5	100.0	109.9	103.9	97.0	96.4	91.7	47.9	53.8	..	Terms of Trade
43.3	62.0	100.0	111.5	100.9	92.0	89.9	87.0	50.8	63.0	..	Export Prices, fob
80.6	90.9	100.0	98.1	80.4	85.8	86.7	74.6	73.1	74.3	..	Nonfuel Primary Products
42.3	61.0	100.0	112.5	101.6	92.5	90.2	87.5	44.9	56.9	..	Fuels
76.1	90.0	100.0	99.1	94.8	90.7	89.3	89.0	103.2	113.0	..	Manufactures
77.4	86.8	100.0	101.4	97.1	94.8	93.3	94.8	106.1	117.1	..	Import Prices, cif
											Trade at Constant 1980 Prices
				(Millions of 1980 US dollars)							
21,110.1	21,884.3	20,738.0	18,158.6	17,111.2	16,404.7	17,801.0	17,016.3	20,265.2	19,046.5	..	Exports, fob
6,963.7	8,023.3	8,746.0	9,510.6	9,723.8	8,816.2	7,550.1	5,955.9	6,052.3	6,172.5	..	Imports, cif
											BALANCE OF PAYMENTS
				(Millions of current US dollars)							
10,720	15,620	23,979	24,073	20,293	18,090	18,476	16,941	11,982	..	..	Exports of Goods & Services
9,154	13,595	20,748	20,240	17,261	15,390	16,072	14,800	9,900	..	..	Merchandise, fob
1,205	1,317	2,341	2,887	2,232	1,990	1,774	1,582	1,411	..	..	Nonfactor Services
361	708	890	946	800	710	630	559	671	..	..	Factor Services
6,768	8,439	10,951	11,867	10,994	8,891	7,883	7,073	7,766	..	..	Imports of Goods & Services
5,364	6,952	8,746	9,648	9,440	7,122	6,325	5,700	6,400	..	..	Merchandise, fob
784	753	1,477	1,466	852	1,160	1,030	915	816	..	..	Nonfactor Services
620	734	728	753	702	609	528	458	550	..	..	Factor Services
..	..	..	..	..	..	..	..	..	..	..	Long-Term Interest
-2,169	-2,909	-3,965	-3,377	-1,825	3,269	2,988	2,724	2,270	..	..	Current Transfers, net
..	..	..	..	..	..	..	..	..	..	..	Workers' Remittances
1,783	4,272	9,063	8,829	7,474	12,469	13,580	12,592	6,486	..	K	Total to be Financed
0	0	0	0	0	0	0	0	0	..	K	Official Capital Grants
1,783	4,272	9,063	8,829	7,474	12,469	13,580	12,592	6,486	..	..	Current Account Balance
336	-52	-620	-1,716	-1,008	-950	-177	200	..	..	..	Long-Term Capital, net
-1,860	-3,355	-4,882	-3,351	-2,833	-4,726	..	..	..	..	..	Direct Investment
..	..	..	..	..	..	..	..	..	..	..	Long-Term Loans
..	..	..	..	..	..	..	..	..	..	..	Disbursements
..	..	..	..	..	..	..	..	..	..	..	Repayments
2,196	3,303	4,262	1,635	1,825	3,776	-177	200	..	..	..	Other Long-Term Capital
-2,083	-3,932	-6,690	-4,770	-4,287	-10,044	-6,559	-7,200	0	..	..	Other Capital, net
-36	-288	-1,753	-2,343	-2,179	-1,475	-6,844	-5,592	-6,486	..	..	Change in Reserves
											Memo Item:
				(U.A.E. Dirhams per US dollar)							
3.870	3.820	3.710	3.670	3.670	3.670	3.670	3.670	3.670	3.670	..	Conversion Factor (Annual Avg)
											EXTERNAL DEBT, ETC.
				(Millions of US dollars, outstanding at end of year)							
..	..	..	..	..	..	..	..	..	..	..	Public/Publicly Guar. Long-Term
..	..	..	..	..	..	..	..	..	..	..	Official Creditors
..	..	..	..	..	..	..	..	..	..	..	IBRD and IDA
..	..	..	..	..	..	..	..	..	..	..	Private Creditors
..	..	..	..	..	..	..	..	..	..	..	Private Non-guaranteed Long-Term
..	..	..	..	..	..	..	..	..	..	..	Use of Fund Credit
..	..	..	..	..	..	..	..	..	..	..	Short-Term Debt
											Memo Items:
				(Millions of US dollars)							
811.8	1,432.3	2,014.7	3,202.2	2,215.5	2,072.4	2,286.9	3,204.3	3,369.9	4,725.3	..	Int'l Reserves Excluding Gold
130.3	295.2	339.8	269.3	373.4	311.8	251.9	267.2	319.4	395.6	..	Gold Holdings (at market price)
											SOCIAL INDICATORS
6.6	6.4	6.3	6.1	5.9	5.9	5.8	5.8	5.7	5.7	..	Total Fertility Rate
30.4	30.2	30.1	29.9	29.8	29.2	28.7	28.1	27.6	27.0	..	Crude Birth Rate
44.4	42.8	41.2	39.6	38.0	36.9	35.7	34.6	33.4	32.3	..	Infant Mortality Rate
65.9	66.3	66.7	67.1	67.5	67.9	68.2	68.5	68.8	69.1	..	Life Expectancy at Birth
..	..	..	..	..	..	..	..	..	..	..	Food Production, p.c. ('79-81 = 100)
6.4	5.5	4.6	..	..	..	..	..	..	..	..	Labor Force, Agriculture (%)
4.7	4.9	5.1	5.2	5.3	5.5	5.6	5.7	5.9	6.0	..	Labor Force, Female (%)
108.0	113.0	86.0	90.0	97.0	97.0	97.0	99.0	..	..	..	School Enroll. Ratio, primary
43.0	47.0	44.0	50.0	58.0	58.0	58.0	58.0	..	..	..	School Enroll. Ratio, secondary

UNITED KINGDOM	1967	1968	1969	1970	1971	1972	1973	1974	1975	1976	1977
CURRENT GNP PER CAPITA (US $)	2,090	2,150	2,180	2,250	2,450	2,750	3,240	3,530	3,910	4,230	4,600
POPULATION (thousands)	54,933	55,157	55,372	55,522	55,942	56,120	56,259	56,272	56,257	56,246	56,220

ORIGIN AND USE OF RESOURCES *(Billions of current Pounds Sterling)*

	1967	1968	1969	1970	1971	1972	1973	1974	1975	1976	1977
Gross National Product (GNP)	41.38	44.81	47.93	52.65	58.57	65.45	75.30	84.93	105.95	125.27	145.73
Net Factor Income from Abroad	0.26	0.19	0.27	0.36	0.38	0.39	0.77	0.82	0.13	0.17	-0.05
Gross Domestic Product (GDP)	41.13	44.62	47.65	52.29	58.19	65.06	74.53	84.12	105.82	125.10	145.78
Indirect Taxes, net	5.90	6.65	6.99	7.47	7.71	8.04	8.57	8.32	10.35	12.85	16.62
GDP at factor cost	35.23	37.97	40.67	44.82	50.47	57.03	65.96	75.80	95.47	112.26	129.15
Agriculture	1.08	1.09	1.20	1.25	1.42	1.55	1.96	2.08	2.49	3.26	3.35
Industry	15.53	16.61	17.89	19.92	21.93	24.93	28.85	31.77	39.10	46.15	55.08
Manufacturing	11.31	12.06	13.16	15.02	16.02	18.21	20.97	23.14	28.11	32.59	38.86
Services, etc.	18.62	20.26	21.58	23.66	27.12	30.55	35.14	41.95	53.88	62.84	70.72
Resource Balance	-0.45	-0.38	0.16	0.37	0.69	-0.13	-1.92	-4.41	-2.00	-1.67	0.76
Exports of Goods & NFServices	7.45	9.04	10.10	11.52	12.91	13.64	17.10	23.00	26.99	35.23	43.34
Imports of Goods & NFServices	7.90	9.43	9.94	11.16	12.21	13.77	19.02	27.41	28.99	36.90	42.58
Domestic Absorption	41.58	45.00	47.50	51.92	57.49	65.20	76.44	88.53	107.82	126.78	145.01
Private Consumption, etc.	25.67	27.60	29.31	31.91	35.83	40.43	46.10	53.18	65.47	75.86	86.85
General Gov't Consumption	7.23	7.68	8.02	9.03	10.30	11.74	13.40	16.72	23.12	27.04	29.48
Gross Domestic Investment	8.68	9.72	10.17	10.98	11.37	13.02	16.95	18.63	19.23	23.88	28.69
Fixed Investment	8.50	9.38	9.74	10.74	11.92	12.86	14.96	17.42	21.35	25.08	27.93
Memo Items:											
Gross Domestic Saving	8.23	9.34	10.33	11.35	12.06	12.89	15.03	14.22	17.23	22.21	29.45
Gross National Saving	8.44	9.47	10.56	11.70	12.45	13.23	15.71	14.91	17.22	22.36	29.35

(Billions of 1980 Pounds Sterling)

	1967	1968	1969	1970	1971	1972	1973	1974	1975	1976	1977
Gross National Product	180.09	187.13	189.91	194.56	197.89	204.09	219.47	215.31	211.30	217.50	222.00
GDP at factor cost	153.07	158.53	161.10	165.43	170.30	177.49	192.50	193.25	190.97	195.46	197.47
Agriculture	5.05	5.66	5.42	5.66	6.06	6.47	3.79	3.88	3.53	3.28	3.65
Industry	49.64	52.01	53.71	54.56	50.00	44.26	92.37	91.20	84.85	86.60	86.65
Manufacturing	37.23	39.93	42.85	43.80	40.45	34.70	62.39	62.36	57.36	58.95	59.15
Services, etc.	98.38	100.85	101.96	105.20	114.24	126.76	96.35	98.18	102.59	105.78	107.18
Resource Balance	-3.49	-2.28	-0.10	0.00	0.75	-3.01	-3.39	-0.47	1.85	4.38	7.43
Exports of Goods & NFServices	31.80	35.62	38.81	40.72	43.61	44.00	49.15	52.76	51.31	55.92	59.61
Imports of Goods & NFServices	35.29	37.90	38.91	40.71	42.85	47.01	52.54	53.22	49.47	51.54	52.18
Domestic Absorption	182.49	188.69	188.97	193.22	195.82	205.88	220.63	213.81	209.21	212.84	214.64
Private Consumption, etc.	103.93	106.92	107.93	111.03	114.49	121.50	127.76	125.91	125.21	125.60	124.99
General Gov't Consumption	38.45	38.60	37.88	38.51	39.66	41.33	43.12	43.93	46.38	46.95	46.17
Gross Domestic Investment	40.12	43.18	43.15	43.67	41.68	43.05	49.75	43.98	37.62	40.29	43.47
Fixed Investment	37.03	39.31	39.10	40.07	40.83	40.70	43.35	42.28	41.54	42.22	41.44
Memo Items:											
Capacity to Import	33.28	36.36	39.53	42.05	45.29	46.55	47.25	44.66	46.06	49.20	53.11
Terms of Trade Adjustment	1.48	0.74	0.72	1.34	1.68	2.55	-1.90	-8.09	-5.25	-6.72	-6.50
Gross Domestic Income	180.48	187.15	189.58	194.56	198.26	205.42	215.34	205.25	205.80	210.51	215.57
Gross National Income	181.57	187.86	190.63	195.90	199.57	206.64	217.57	207.22	206.04	210.79	215.50

DOMESTIC PRICES (DEFLATORS) *(Index 1980 = 100)*

	1967	1968	1969	1970	1971	1972	1973	1974	1975	1976	1977
Overall (GDP)	23.0	23.9	25.2	27.1	29.6	32.1	34.3	39.4	50.1	57.6	65.6
Domestic Absorption	22.8	23.8	25.1	26.9	29.4	31.7	34.6	41.4	51.5	59.6	67.6
Agriculture	21.4	19.3	22.2	22.0	23.4	23.9	51.8	53.6	70.4	99.5	92.0
Industry	31.3	31.9	33.3	36.5	43.9	56.3	31.2	34.8	46.1	53.3	63.6
Manufacturing	30.4	30.2	30.7	34.3	39.6	52.5	33.6	37.1	49.0	55.3	65.7

MANUFACTURING ACTIVITY

	1967	1968	1969	1970	1971	1972	1973	1974	1975	1976	1977
Employment (1980=100)	..	119.9	121.6	123.0	119.9	115.2	116.7	119.0	114.4	111.8	111.5
Real Earnings per Empl. (1980=100)	..	76.3	79.5	82.6	84.3	87.4	90.3	92.9	94.2	93.7	89.8
Real Output per Empl. (1980=100)	..	58.9	63.1	61.9	58.7	49.6	91.5	107.0	94.5	104.1	103.0
Earnings as % of Value Added	..	50.8	51.6	52.2	53.1	51.3	48.4	47.2	51.3	48.3	46.7

MONETARY HOLDINGS *(Billions of current Pounds Sterling)*

	1967	1968	1969	1970	1971	1972	1973	1974	1975	1976	1977
Money Supply, Broadly Defined	21.07	22.83	24.02	26.77	30.93	37.71	45.87	51.27	57.90	64.95	74.52
Money as Means of Payment	8.44	8.78	8.81	9.64	11.09	12.66	13.30	14.74	17.48	19.47	23.52
Currency Ouside Banks	2.82	2.86	3.01	3.32	3.59	4.08	4.38	5.08	5.81	6.58	7.56
Demand Deposits	5.63	5.92	5.81	6.31	7.50	8.58	8.93	9.65	11.68	12.89	15.96
Quasi-Monetary Liabilities	12.63	14.04	15.21	17.13	19.84	25.05	32.57	36.53	40.42	45.48	51.00

GOVERNMENT DEFICIT (-) OR SURPLUS *(Billions of current Pounds Sterling)*

	1967	1968	1969	1970	1971	1972	1973	1974	1975	1976	1977
	..	..	..	0.92	-0.38	-1.74	-2.52	-3.83	-7.80	-7.25	-4.93
Current Revenue	..	..	..	19.07	20.15	21.29	23.14	29.42	37.63	44.14	50.74
Current Expenditure	..	..	..	14.74	16.53	19.16	21.67	28.27	38.60	45.72	51.17
Current Budget Balance	..	..	..	4.33	3.63	2.13	1.47	1.16	-0.97	-1.58	-0.43
Capital Receipts	..	..	..	0.02	0.02	0.02	0.03	0.04	0.08	0.05	0.10
Capital Payments	..	..	..	3.42	4.03	3.89	4.02	5.03	6.90	5.72	4.60

1978	1979	1980	1981	1982	1983	1984	1985	1986	1987 estimate	NOTES	UNITED KINGDOM
5,180	6,450	7,920	9,270	9,760	9,230	8,640	8,470	8,990	10,430	..	CURRENT GNP PER CAPITA (US $)
56,210	56,274	56,360	56,348	56,341	56,377	56,488	56,618	56,734	56,851	..	POPULATION (thousands)
				(Billions of current Pounds Sterling)							ORIGIN AND USE OF RESOURCES
168.18	199.15	230.76	254.61	278.17	303.27	324.46	355.89	381.61	415.34	..	Gross National Product (GNP)
0.37	1.72	0.16	0.29	0.27	1.20	1.38	1.43	2.12	1.11	..	Net Factor Income from Abroad
167.82	197.43	230.60	254.32	277.90	302.08	323.09	354.46	379.49	414.23	..	Gross Domestic Product (GDP)
19.14	25.26	30.94	36.38	40.82	43.11	44.98	49.55	56.26	61.93	..	Indirect Taxes, net
148.68	172.17	199.66	217.94	237.08	258.96	278.10	304.91	323.23	352.31	B	GDP at factor cost
3.60	3.92	4.30	4.81	5.48	5.33	6.18	5.59	5.84	..	..	Agriculture
63.68	74.01	85.25	90.66	98.54	106.98	112.95	123.87	122.41	..	..	Industry
44.70	50.21	54.44	54.84	58.90	62.36	66.69	74.35	80.53	..	..	Manufacturing
81.40	94.25	110.11	122.46	133.06	146.66	158.97	175.45	194.98	..	..	Services, etc.
1.91	0.34	5.13	7.23	4.97	2.87	-0.57	3.73	-2.86	-3.90	..	Resource Balance
47.49	55.01	62.80	67.70	72.98	80.53	92.21	102.65	98.42	107.53	..	Exports of Goods & NFServices
45.58	54.67	57.67	60.47	68.01	77.66	92.78	98.92	101.28	111.43	..	Imports of Goods & NFServices
165.91	197.09	225.47	247.09	272.93	299.21	323.66	350.74	382.35	418.13	..	Domestic Absorption
100.12	118.44	137.47	153.56	168.17	183.81	196.89	215.24	236.39	256.97	..	Private Consumption, etc.
33.41	38.89	49.03	55.47	60.42	65.94	69.88	74.08	79.64	84.93	..	General Gov't Consumption
32.37	39.76	38.98	38.06	44.34	49.46	56.89	61.41	66.31	76.23	..	Gross Domestic Investment
31.70	37.15	41.56	41.89	47.47	52.44	59.36	64.49	67.33	72.51	..	Fixed Investment
											Memo Items:
34.28	40.10	44.11	45.29	49.31	52.33	56.32	65.14	63.45	72.33	..	Gross Domestic Saving
34.52	41.62	44.05	45.54	49.55	53.70	57.95	66.65	65.63	73.30	..	Gross National Saving
				(Billions of 1980 Pounds Sterling)							
230.55	238.39	230.76	228.29	231.80	240.15	246.41	255.25	263.97	275.15	..	Gross National Product
204.27	206.48	199.66	195.39	197.54	204.99	211.29	218.64	223.93	233.59	B	GDP at factor cost
3.95	3.88	4.30	4.38	4.74	4.49	5.30	5.08	5.05	..	..	Agriculture
88.43	91.41	85.25	81.04	82.42	85.48	86.75	90.64	91.81	..	..	Industry
59.95	59.80	54.44	50.86	50.85	52.37	54.41	56.18	56.35	..	..	Manufacturing
111.90	111.19	110.11	109.96	110.38	115.03	119.23	122.92	127.07	..	..	Services, etc.
6.53	3.28	5.13	6.36	4.03	1.79	0.19	2.35	0.22	-1.21	..	Resource Balance
60.74	63.16	62.80	62.39	62.92	64.42	68.92	72.96	75.30	79.53	..	Exports of Goods & NFServices
54.20	59.88	57.67	56.04	58.89	62.63	68.73	70.61	75.08	80.75	..	Imports of Goods & NFServices
223.53	233.07	225.47	221.67	227.54	237.40	245.17	251.87	262.30	275.63	..	Domestic Absorption
131.93	137.49	137.47	137.73	138.81	144.46	147.65	153.38	162.57	171.03	..	Private Consumption, etc.
47.24	48.26	49.03	49.16	49.56	50.55	51.00	51.01	51.59	52.20	..	General Gov't Consumption
44.36	47.32	38.98	34.78	39.17	42.40	46.52	47.49	48.14	52.40	..	Gross Domestic Investment
42.73	43.91	41.56	37.57	39.59	41.62	44.97	46.30	46.18	47.80	..	Fixed Investment
											Memo Items:
56.47	60.25	62.80	62.74	63.19	64.94	68.31	73.27	72.96	77.92	..	Capacity to Import
-4.26	-2.91	0.00	0.35	0.27	0.52	-0.61	0.31	-2.34	-1.61	..	Terms of Trade Adjustment
225.80	233.45	230.60	228.37	231.85	239.72	244.75	254.53	260.18	272.80	..	Gross Domestic Income
226.29	235.48	230.76	228.63	232.08	240.67	245.79	255.56	261.63	..	..	Gross National Income
				(Index 1980 = 100)							DOMESTIC PRICES (DEFLATORS)
72.9	83.5	100.0	111.5	120.0	126.3	131.7	139.4	144.6	151.0	..	Overall (GDP)
74.2	84.6	100.0	111.5	119.9	126.0	132.0	139.3	145.8	151.7	..	Domestic Absorption
91.2	100.9	100.0	109.8	115.7	118.8	116.5	110.2	115.8	..	..	Agriculture
72.0	81.0	100.0	111.9	119.5	125.2	130.2	136.7	133.3	..	..	Industry
74.6	84.0	100.0	107.8	115.8	119.1	122.6	132.3	142.9	..	..	Manufacturing
											MANUFACTURING ACTIVITY
108.8	106.0	100.0	89.0	82.7	78.5	78.4	78.3	..	..	..	Employment (1980 = 100)
95.3	98.2	100.0	100.6	102.2	105.8	108.6	116.8	..	..	..	Real Earnings per Empl. (1980 = 100)
100.4	105.2	100.0	104.8	111.4	120.7	127.7	129.6	..	..	..	Real Output per Empl. (1980 = 100)
46.2	45.7	48.8	47.6	46.3	44.0	43.8	44.7	..	..	..	Earnings as % of Value Added
				(Billions of current Pounds Sterling)							MONETARY HOLDINGS
85.13	97.85	115.09	140.63	157.56	178.23	199.26	221.53	252.49	285.80	D	Money Supply, Broadly Defined
27.36	29.86	31.04	36.53	40.66	45.19	52.16	61.60	75.23	92.40	..	Money as Means of Payment
8.73	9.51	10.24	10.77	11.22	11.87	12.15	12.74	13.36	13.99	..	Currency Ouside Banks
18.63	20.35	20.81	25.77	29.44	33.32	40.02	48.87	61.87	78.41	..	Demand Deposits
57.76	67.99	84.05	104.10	116.90	133.04	147.10	159.93	177.26	193.40	..	Quasi-Monetary Liabilities
				(Billions of current Pounds Sterling)							GOVERNMENT DEFICIT (-) OR SURPLUS
-8.81	-11.16	-10.73	-12.05	-9.57	-13.37	-10.33	-11.27	-6.95	..	..	
55.63	65.04	82.68	93.26	108.45	114.53	122.68	134.41	143.23	..	..	Current Revenue
58.99	69.22	85.07	99.29	109.62	117.58	126.27	135.15	141.27	..	..	Current Expenditure
-3.36	-4.19	-2.39	-6.03	-1.17	-3.05	-3.59	-0.74	1.96	..	..	Current Budget Balance
0.15	0.22	0.23	0.42	0.36	0.37	0.36	0.39	0.47	..	..	Capital Receipts
5.60	7.20	8.57	6.44	8.76	10.69	7.10	10.92	9.38	..	..	Capital Payments

UNITED KINGDOM	1967	1968	1969	1970	1971	1972	1973	1974	1975	1976	1977
FOREIGN TRADE (CUSTOMS BASIS)					*(Billions of current US dollars)*						
Value of Exports, fob	14.37	15.35	17.52	19.35	22.35	24.34	30.53	38.66	43.74	46.03	57.48
Nonfuel Primary Products	2.10	2.27	2.49	2.73	2.83	3.32	4.53	5.46	5.60	5.73	6.96
Fuels	0.37	0.40	0.42	0.50	0.57	0.60	0.91	1.80	1.80	2.25	3.63
Manufactures	11.91	12.67	14.61	16.12	18.95	20.43	25.10	31.40	36.34	38.05	46.88
Value of Imports, cif	17.71	18.96	19.96	21.72	23.94	27.86	38.84	54.15	53.19	55.95	63.62
Nonfuel Primary Products	8.69	8.84	9.11	9.66	9.52	10.32	14.02	17.26	16.31	16.60	18.55
Fuels	2.01	2.17	2.18	2.27	3.04	3.11	4.23	10.84	9.53	10.14	9.13
Manufactures	7.02	7.95	8.66	9.80	11.38	14.43	20.59	26.05	27.34	29.21	35.94
Terms of Trade	132.3	126.9	125.5	126.1	129.5	121.8	103.9	85.0	92.2	86.9	88.4
Export Prices, fob	24.3	23.3	24.4	25.5	27.7	30.5	34.7	45.1	50.2	49.2	55.7
Nonfuel Primary Products	36.0	36.6	40.1	40.8	38.8	44.0	61.6	69.4	63.2	66.1	68.0
Fuels	4.3	4.3	4.3	4.3	5.6	6.2	8.9	36.7	35.7	38.4	42.0
Manufactures	26.7	25.3	26.1	28.1	30.1	32.6	35.7	43.0	49.8	48.3	55.7
Import Prices, cif	18.4	18.4	19.4	20.2	21.4	25.1	33.4	53.1	54.5	56.6	63.0
Trade at Constant 1980 Prices					*(Billions of 1980 US dollars)*						
Exports, fob	59.07	65.84	71.91	75.77	80.63	79.78	87.91	85.69	87.10	93.51	103.17
Imports, cif	96.37	103.24	102.81	107.29	111.80	111.17	116.24	102.06	97.63	98.80	100.97
BALANCE OF PAYMENTS					*(Billions of current US dollars)*						
Exports of Goods & Services	..	..	..	31.26	35.21	42.58	53.94	68.33	74.32	78.31	91.05
Merchandise, fob	..	..	..	19.56	22.02	23.57	29.21	38.37	42.79	45.23	55.38
Nonfactor Services	..	..	..	7.77	9.09	10.08	12.19	14.86	16.37	17.26	19.55
Factor Services	..	..	..	3.93	4.10	8.93	12.54	15.10	15.16	15.81	16.12
Imports of Goods & Services	..	..	..	28.91	32.08	41.39	55.34	75.10	76.83	78.47	89.25
Merchandise, fob	..	..	..	19.64	21.55	25.43	35.51	50.89	50.23	52.25	59.33
Nonfactor Services	..	..	..	6.83	7.83	8.60	10.62	12.75	13.50	13.40	14.41
Factor Services	..	..	..	2.44	2.70	7.36	9.21	11.46	13.11	12.82	15.50
Long-Term Interest	..	..	..	..	..	..	..	..	..	..	..
Current Transfers, net	..	..	..	-0.03	0.00	-0.13	-0.24	-0.28	-0.31	-0.03	-0.08
Workers' Remittances	..	..	..	0.00	0.00	0.00	0.00	0.00	0.00	0.00	0.00
Total to be Financed	..	..	..	2.32	3.13	1.06	-1.64	-7.05	-2.82	-0.20	1.72
Official Capital Grants	..	..	..	-0.40	-0.47	-0.55	-0.85	-0.71	-0.75	-1.40	-1.89
Current Account Balance	..	..	..	1.91	2.66	0.50	-2.48	-7.75	-3.57	-1.60	-0.16
Long-Term Capital, net	..	..	..	-1.47	-1.01	-2.82	-2.55	0.12	-0.04	-0.80	3.13
Direct Investment	..	..	..	-0.19	-0.22	-0.81	-2.26	-0.01	0.36	-1.34	0.26
Long-Term Loans	..	..	..	..	..	..	..	..	..	..	..
Disbursements	..	..	..	..	..	..	..	..	..	..	..
Repayments	..	..	..	..	..	..	..	..	..	..	..
Other Long-Term Capital	..	..	..	-0.88	-0.32	-1.46	0.55	0.83	0.35	1.94	4.76
Other Capital, net	..	..	..	2.19	2.40	0.78	5.36	4.77	3.51	1.46	11.17
Change in Reserves	..	..	..	-2.63	-4.04	1.53	-0.34	2.87	0.10	0.93	-14.13
Memo Item:					*(Pounds Sterling per US dollar)*						
Conversion Factor (Annual Avg)	0.360	0.420	0.420	0.420	0.410	0.400	0.410	0.430	0.450	0.560	0.570
EXTERNAL DEBT, ETC.					*(Millions US dollars, outstanding at end of year)*						
Public/Publicly Guar. Long-Term	..	..	..	..	..	..	..	..	..	..	..
Official Creditors	..	..	..	..	..	..	..	..	..	..	..
IBRD and IDA	..	..	..	..	..	..	..	..	..	..	..
Private Creditors	..	..	..	..	..	..	..	..	..	..	..
Private Non-guaranteed Long-Term	..	..	..	..	..	..	..	..	..	..	..
Use of Fund Credit	..	..	..	..	..	..	..	..	..	..	..
Short-Term Debt	..	..	..	..	..	..	..	..	..	..	..
Memo Items:					*(Millions of US dollars)*						
Int'l Reserves Excluding Gold	1,405	949	1,055	1,479	7,989	4,846	5,589	6,038	4,598	3,375	20,112
Gold Holdings (at market price)	1,297	1,764	1,480	1,440	968	1,368	2,358	3,921	2,949	2,833	3,666
SOCIAL INDICATORS											
Total Fertility Rate	2.7	2.6	2.5	2.4	2.4	2.2	2.0	1.9	1.8	1.7	1.7
Crude Birth Rate	17.6	17.2	16.7	16.3	16.2	14.9	13.9	13.2	12.5	12.1	11.8
Infant Mortality Rate	18.8	18.7	18.6	18.5	17.9	17.5	17.2	16.8	16.0	14.5	14.1
Life Expectancy at Birth	71.4	71.5	71.6	71.7	71.8	72.0	72.1	72.3	72.4	72.6	72.8
Food Production, p.c. ('79-81 = 100)	79.1	76.6	76.9	82.4	82.9	84.1	85.7	92.2	87.3	82.1	92.5
Labor Force, Agriculture (%)	3.2	3.0	2.9	2.8	2.8	2.8	2.7	2.7	2.7	2.7	2.7
Labor Force, Female (%)	34.5	34.9	35.2	35.6	35.9	36.2	36.5	36.8	37.1	37.5	37.8
School Enroll. Ratio, primary	..	..	..	104.0	..	..	..	..	105.0	106.0	105.0
School Enroll. Ratio, secondary	..	..	..	73.0	..	..	..	..	83.0	83.0	83.0

1978	1979	1980	1981	1982	1983	1984	1985	1986	1987 estimate	NOTES	UNITED KINGDOM
				(Billions of current US dollars)							**FOREIGN TRADE (CUSTOMS BASIS)**
71.52	90.49	114.38	102.14	96.58	91.77	94.31	101.17	106.63	131.13	..	Value of Exports, fob
9.15	11.46	15.21	12.66	11.54	11.60	11.29	11.10	13.44	15.40	..	Nonfuel Primary Products
4.50	9.15	14.88	19.36	19.59	19.81	20.48	21.69	12.66	14.36	..	Fuels
57.87	69.88	84.29	70.11	65.45	60.35	62.54	68.39	80.53	101.37	..	Manufactures
78.35	102.41	117.90	101.15	99.10	99.44	105.69	109.41	125.61	154.39	..	Value of Imports, cif
20.29	25.54	28.99	23.98	22.17	22.18	21.92	21.41	24.76	29.08	..	Nonfuel Primary Products
9.17	12.17	15.91	14.28	12.88	10.68	13.80	13.68	9.34	9.98	..	Fuels
48.89	64.70	73.00	62.90	64.04	66.59	69.97	74.32	91.50	115.33	..	Manufactures
				(Index 1980 = 100)							Terms of Trade
93.0	92.9	100.0	99.6	98.8	95.5	93.3	95.7	92.5	98.9	..	Terms of Trade
66.3	80.7	100.0	95.9	89.5	84.4	80.6	81.1	83.8	98.0	..	Export Prices, fob
75.2	96.3	100.0	91.2	83.2	85.6	82.1	77.4	78.4	83.2	..	Nonfuel Primary Products
42.3	61.0	100.0	112.5	109.0	96.0	93.0	91.0	54.0	64.2	..	Fuels
68.2	82.1	100.0	93.0	86.0	81.0	77.0	79.0	93.0	109.0	..	Manufactures
71.2	86.8	100.0	96.3	90.6	88.4	86.4	84.7	90.7	99.0	..	Import Prices, cif
				(Billions of 1980 US dollars)							Trade at Constant 1980 Prices
107.93	112.18	114.38	106.48	107.94	108.70	116.99	124.74	127.19	133.87	..	Exports, fob
109.99	118.01	117.90	105.04	109.44	112.44	122.33	129.13	138.56	155.90	..	Imports, cif
				(Billions of current US dollars)							**BALANCE OF PAYMENTS**
112.53	154.31	201.04	211.01	204.68	186.03	191.29	199.91	213.65	255.27	..	Exports of Goods & Services
67.22	86.46	109.64	102.06	96.67	91.99	93.49	100.62	106.20	130.40	..	Merchandise, fob
22.97	29.55	35.31	33.05	29.91	29.01	28.19	30.89	36.62	44.57	..	Nonfactor Services
22.34	38.30	56.09	75.90	78.10	65.02	69.61	68.41	70.83	80.31	..	Factor Services
107.27	150.75	189.10	193.80	193.76	177.73	186.24	191.26	210.32	252.30	..	Imports of Goods & Services
70.21	93.67	106.28	95.07	92.69	93.33	99.35	103.23	118.58	146.18	..	Merchandise, fob
16.48	21.67	26.71	25.46	25.42	23.63	23.24	23.52	28.54	35.51	..	Nonfactor Services
20.59	35.41	56.11	73.27	75.65	60.77	63.64	64.51	63.20	70.61	..	Factor Services
..	..	..	..	..	..	..	..	..	..	..	Long-Term Interest
-0.25	-0.43	-0.50	-0.09	-0.05	0.26	0.34	0.10	0.09	-0.23	..	Current Transfers, net
0.00	0.00	0.00	0.00	0.00	0.00	0.00	0.00	0.00	0.00	..	Workers' Remittances
5.01	3.13	11.44	17.12	10.86	8.57	5.39	8.75	3.41	2.74	K	Total to be Financed
-3.21	-4.26	-4.11	-3.22	-3.12	-2.94	-2.85	-4.26	-3.25	-5.36	K	Official Capital Grants
1.81	-1.14	7.33	13.89	7.75	5.62	2.55	4.49	0.16	-2.62	..	Current Account Balance
-8.67	-9.83	-16.36	-23.51	-19.21	-17.35	-23.80	-24.86	-31.84	-20.05	..	Long-Term Capital, net
-2.95	-6.06	-1.11	-6.22	-1.81	-2.88	-8.24	-5.73	-10.38	-16.34	..	Direct Investment
..	..	..	..	..	..	..	..	..	..	..	Long-Term Loans
..	..	..	..	..	..	..	..	..	..	..	Disbursements
..	..	..	..	..	..	..	..	..	..	..	Repayments
-2.51	0.49	-11.14	-14.07	-14.29	-11.53	-12.71	-14.87	-18.21	1.66	..	Other Long-Term Capital
4.70	-6.77	7.65	9.02	17.20	10.48	10.38	17.36	37.74	17.56	..	Other Capital, net
2.16	17.73	1.38	0.60	-5.74	1.25	10.87	3.02	-6.07	5.11	..	Change in Reserves
				(Pounds Sterling per US dollar)							Memo Item:
0.520	0.470	0.430	0.500	0.570	0.660	0.750	0.780	0.680	0.610	..	Conversion Factor (Annual Avg)
				(Millions US dollars, outstanding at end of year)							**EXTERNAL DEBT, ETC.**
..	..	..	..	..	..	..	..	..	..	..	Public/Publicly Guar. Long-Term
..	..	..	..	..	..	..	..	..	..	..	Official Creditors
..	..	..	..	..	..	..	..	..	..	..	IBRD and IDA
..	..	..	..	..	..	..	..	..	..	..	Private Creditors
..	..	..	..	..	..	..	..	..	..	..	Private Non-guaranteed Long-Term
..	..	..	..	..	..	..	..	..	..	..	Use of Fund Credit
..	..	..	..	..	..	..	..	..	..	..	Short-Term Debt
				(Millions of US dollars)							Memo Items:
16,026	19,744	20,651	15,238	12,397	11,339	9,440	12,859	18,422	41,715	..	Int'l Reserves Excluding Gold
5,159	9,343	11,104	7,565	8,686	7,253	5,867	6,223	7,431	9,203	..	Gold Holdings (at market price)
											SOCIAL INDICATORS
1.8	1.9	1.9	1.8	1.8	1.8	1.8	1.8	1.8	1.8	..	Total Fertility Rate
12.3	13.1	13.5	13.0	13.1	13.1	13.2	13.2	13.3	13.3	..	Crude Birth Rate
13.3	12.9	12.1	11.2	11.6	12.0	12.4	12.9	13.4	13.8	..	Infant Mortality Rate
73.0	73.4	73.8	73.7	73.7	73.9	74.2	74.5	74.8	75.0	..	Life Expectancy at Birth
94.0	96.9	102.4	100.7	102.7	104.8	114.5	108.2	111.1	106.0	..	Food Production, p.c. ('79-81 = 100)
2.6	2.6	2.6	..	..	..	..	..	..	..	..	Labor Force, Agriculture (%)
38.2	38.5	38.9	38.8	38.8	38.8	38.8	38.7	38.7	38.7	..	Labor Force, Female (%)
105.0	104.0	104.0	..	101.0	102.0	103.0	106.0	..	..	..	School Enroll. Ratio, primary
83.0	82.0	84.0	..	86.0	85.0	85.0	85.0	..	..	..	School Enroll. Ratio, secondary

UNITED STATES	1967	1968	1969	1970	1971	1972	1973	1974	1975	1976	1977
CURRENT GNP PER CAPITA (US $)	4,080	4,450	4,780	4,970	5,320	5,780	6,410	6,890	7,400	8,180	9,040
POPULATION (millions)	199	201	203	205	208	210	212	214	216	218	220
ORIGIN AND USE OF RESOURCES					*(Billions of current US Dollars)*						
Gross National Product (GNP)	817.9	894.0	964.7	1,015.5	1,102.7	1,212.8	1,359.3	1,472.8	1,598.4	1,782.8	1,990.5
Net Factor Income from Abroad	6.0	6.8	6.8	7.3	9.3	11.2	16.2	19.5	17.5	21.1	25.4
Gross Domestic Product (GDP)	811.9	887.2	957.9	1,008.2	1,093.4	1,201.6	1,343.1	1,453.3	1,580.9	1,761.7	1,965.1
Indirect Taxes, net	68.2	75.9	80.5	91.1	100.8	107.4	117.3	127.7	137.7	150.6	162.6
GDP at factor cost	743.7	811.3	877.3	917.1	992.6	1,094.2	1,225.8	1,325.6	1,443.2	1,611.0	1,802.5
Agriculture	23.2	23.9	26.7	27.8	30.0	35.0	53.8	52.1	52.9	51.7	54.2
Industry	298.5	326.0	347.6	348.5	369.8	407.5	454.9	486.9	520.8	592.2	670.3
Manufacturing	224.1	244.8	258.6	253.9	267.0	293.9	328.2	340.5	359.5	411.6	468.0
Services, etc.	490.3	537.3	583.6	631.9	693.6	759.1	834.4	914.3	1,007.2	1,117.8	1,240.6
Resource Balance	6.8	4.7	5.1	8.4	6.3	3.1	16.6	16.3	31.0	18.8	1.9
Exports of Goods & NFServices	50.0	55.1	59.8	68.8	72.4	81.4	114.1	151.6	161.3	177.8	191.5
Imports of Goods & NFServices	43.2	50.4	54.7	60.4	66.1	78.3	97.5	135.3	130.3	159.0	189.6
Domestic Absorption	805.1	882.5	952.8	999.8	1,087.1	1,198.5	1,326.5	1,437.0	1,549.9	1,742.8	1,963.2
Private Consumption, etc.	485.4	541.8	599.4	633.4	681.7	746.2	822.0	896.0	995.5	1,107.4	1,231.1
General Gov't Consumption	176.8	191.9	203.5	218.2	232.4	250.0	266.5	299.1	335.0	356.9	387.3
Gross Domestic Investment	142.8	148.8	149.9	148.2	173.1	202.2	238.0	241.9	219.4	278.5	344.8
Fixed Investment	125.2	138.7	151.4	153.8	173.5	201.9	232.7	239.2	232.1	268.3	326.8
Memo Items:											
Gross Domestic Saving	149.6	153.5	155.0	156.6	179.3	205.4	254.6	258.2	250.3	297.3	346.7
Gross National Saving	155.6	160.6	161.7	162.8	187.5	215.4	269.5	276.6	267.0	317.5	371.3
					(Billions of 1980 US Dollars)						
Gross National Product	1,934.9	2,016.4	2,073.2	2,071.1	2,129.9	2,235.9	2,352.1	2,339.5	2,310.1	2,422.9	2,536.1
GDP at factor cost	1,758.8	1,826.7	1,879.6	1,865.1	1,911.3	2,011.3	2,114.1	2,101.7	2,081.5	2,184.6	2,291.9
Agriculture	62.1	60.2	62.2	65.4	66.7	67.3	68.3	67.1	69.7	67.1	66.7
Industry	734.7	763.9	779.1	745.6	745.2	791.5	855.3	823.3	771.1	822.6	869.8
Manufacturing	435.3	457.2	471.6	445.6	450.0	490.1	544.1	519.2	478.9	524.8	563.9
Services, etc.	1,123.5	1,176.3	1,216.5	1,244.9	1,299.5	1,356.1	1,399.7	1,417.6	1,443.4	1,503.9	1,566.6
Resource Balance	-23.7	-41.7	-45.4	-39.4	-48.7	-59.0	-44.7	-15.1	2.8	-26.8	-50.7
Exports of Goods & NFServices	135.8	140.2	146.9	160.8	161.6	176.1	218.5	242.8	234.2	247.5	254.0
Imports of Goods & NFServices	159.5	181.9	192.3	200.2	210.4	235.1	263.1	257.9	231.4	274.3	304.7
Domestic Absorption	1,944.0	2,042.1	2,103.3	2,095.3	2,160.1	2,273.8	2,367.9	2,323.1	2,281.4	2,420.4	2,553.8
Private Consumption, etc.	1,153.4	1,215.3	1,258.1	1,278.7	1,316.5	1,387.0	1,438.3	1,420.6	1,456.2	1,535.6	1,603.3
General Gov't Consumption	461.2	477.1	488.1	489.4	484.2	487.8	483.1	489.9	496.5	496.0	503.5
Gross Domestic Investment	329.3	349.7	357.2	327.1	359.5	399.0	446.5	412.7	328.7	388.8	447.0
Fixed Investment	308.8	328.5	334.5	322.2	344.9	382.9	414.9	386.6	341.8	372.3	424.7
Memo Items:											
Capacity to Import	184.6	198.8	210.3	228.1	230.4	244.4	307.9	289.0	286.4	306.8	307.8
Terms of Trade Adjustment	48.8	58.6	63.4	67.2	68.7	68.4	89.5	46.2	52.2	59.3	53.7
Gross Domestic Income	1,969.1	2,059.1	2,121.2	2,123.1	2,180.1	2,283.2	2,412.7	2,354.2	2,336.4	2,452.9	2,556.8
Gross National Income	1,983.6	2,075.1	2,136.6	2,138.3	2,198.6	2,304.3	2,441.6	2,385.7	2,362.3	2,482.2	2,589.8
DOMESTIC PRICES (DEFLATORS)					*(Index 1980 = 100)*						
Overall (GDP)	42.3	44.4	46.5	49.0	51.8	54.3	57.8	63.0	69.2	73.6	78.5
Domestic Absorption	41.4	43.2	45.3	47.7	50.3	52.7	56.0	61.9	67.9	72.0	76.9
Agriculture	37.3	39.7	43.0	42.6	44.9	52.1	78.8	77.7	75.9	77.0	81.2
Industry	40.6	42.7	44.6	46.7	49.6	51.5	53.2	59.1	67.5	72.0	77.1
Manufacturing	51.5	53.5	54.8	57.0	59.3	60.0	60.3	65.6	75.1	78.4	83.0
MANUFACTURING ACTIVITY											
Employment (1980=100)	95.9	96.9	99.3	94.8	90.4	93.5	97.8	97.0	89.1	91.7	96.0
Real Earnings per Empl. (1980=100)	100.6	102.6	102.1	100.5	102.6	106.9	107.0	103.8	103.6	106.3	108.3
Real Output per Empl. (1980=100)	59.4	61.9	62.7	62.7	67.3	73.1	80.6	87.2	83.6	89.2	92.2
Earnings as % of Value Added	47.1	46.5	46.9	47.3	45.8	45.3	44.0	42.1	43.1	41.7	41.4
MONETARY HOLDINGS					*(Billions of current US Dollars)*						
Money Supply, Broadly Defined	530.3	571.6	592.2	630.6	714.2	806.5	859.3	907.5	1,023.6	1,162.8	1,285.9
Money as Means of Payment	190.3	204.3	211.0	222.3	236.9	258.9	272.4	284.2	298.1	318.1	343.7
Currency Ouside Banks	41.2	43.8	46.6	50.0	53.4	57.8	61.8	68.1	74.3	81.6	89.9
Demand Deposits	149.1	160.6	164.4	172.3	183.4	201.0	210.6	216.1	223.8	236.5	253.8
Quasi-Monetary Liabilities	339.9	367.2	381.3	408.3	477.3	547.6	587.0	623.3	725.5	844.7	942.2
					(Billions of current US Dollars)						
GOVERNMENT DEFICIT (-) OR SURPLUS	..	..	..	..	..	-18.70	-16.23	-4.48	-53.93	-74.86	-52.23
Current Revenue	..	..	..	..	..	213.89	243.94	277.62	291.63	311.06	371.31
Current Expenditure	..	..	..	..	..	221.10	246.45	267.99	317.02	352.12	388.96
Current Budget Balance	..	..	..	..	..	-7.21	-2.51	9.63	-25.39	-41.06	-17.65
Capital Receipts	..	..	..	..	..	0.18	0.44	1.33	1.07	0.25	0.21
Capital Payments	..	..	..	..	..	11.67	14.16	15.44	29.61	34.05	34.79

1978	1979	1980	1981	1982	1983	1984	1985	1986	1987 estimate	NOTES	UNITED STATES
10,100	11,150	12,000	13,270	13,620	14,510	15,910	16,800	17,560	18,560	..	CURRENT GNP PER CAPITA (US $)
223	225	228	230	233	235	237	239	242	243	..	POPULATION (millions)
(Billions of current US Dollars)											ORIGIN AND USE OF RESOURCES
2,249.7	2,508.1	2,732.0	3,052.6	3,166.0	3,405.7	3,772.2	4,014.9	4,240.3	4,526.7	..	Gross National Product (GNP)
30.5	43.8	47.6	52.1	51.2	49.9	47.4	40.7	34.9	29.5	..	Net Factor Income from Abroad
2,219.2	2,464.3	2,684.4	3,000.5	3,114.8	3,355.9	3,724.8	3,974.1	4,205.4	4,497.2	..	Gross Domestic Product (GDP)
174.2	185.9	207.6	244.7	250.1	268.5	304.0	326.4	335.8	348.0	..	Indirect Taxes, net
2,045.0	2,278.5	2,476.8	2,755.8	2,864.7	3,087.4	3,420.8	3,647.7	3,869.5	4,149.2	B	GDP at factor cost
65.1	77.2	70.2	83.5	81.2	66.4	85.4	84.4	87.5	..	..	Agriculture
754.4	831.0	901.0	1,012.2	1,007.2	1,064.6	1,192.6	1,244.3	1,267.4		..	Industry
522.8	566.5	585.5	647.7	639.7	690.0	778.9	809.4	835.8		..	Manufacturing
1,399.7	1,556.2	1,713.2	1,904.9	2,026.4	2,224.8	2,446.9	2,645.5	2,850.5		..	Services, etc.
4.2	18.7	31.9	34.0	26.4	-5.8	-58.9	-77.9	-104.5	-123.3	..	Resource Balance
227.5	291.3	350.9	382.9	362.0	352.7	383.7	370.9	378.4	427.9	..	Exports of Goods & NFServices
223.3	272.6	319.0	348.8	335.6	358.5	442.6	448.8	482.9	551.1	..	Imports of Goods & NFServices
2,214.9	2,445.6	2,652.5	2,966.5	3,088.5	3,361.7	3,783.7	4,052.1	4,309.9	4,620.5	..	Domestic Absorption
1,373.5	1,521.9	1,685.5	1,861.8	1,999.5	2,183.4	2,384.3	2,590.0	2,773.8	2,981.3	..	Private Consumption, etc.
425.2	467.8	530.3	588.1	641.7	675.1	735.9	820.7	871.2	924.7	..	General Gov't Consumption
416.2	455.9	436.6	516.6	447.3	503.2	663.6	641.3	664.9	714.5	..	Gross Domestic Investment
386.2	437.5	445.5	492.5	471.8	527.8	637.2	690.1	707.2	748.6	..	Fixed Investment
											Memo Items:
420.4	474.6	468.5	550.6	473.7	497.4	604.7	563.4	560.3	591.2	..	Gross Domestic Saving
450.1	517.5	515.1	601.7	523.6	546.3	650.7	602.2	593.9	619.5	..	Gross National Saving
(Billions of 1980 US Dollars)											
2,670.2	2,736.3	2,732.0	2,784.7	2,713.8	2,810.7	3,001.3	3,101.9	3,190.2	3,297.5	..	Gross National Product
2,423.2	2,482.7	2,476.8	2,513.4	2,453.9	2,545.2	2,717.7	2,813.9	2,906.2	3,019.2	B	GDP at factor cost
67.6	71.8	70.2	81.0	82.3	66.7	76.0	87.4	94.6	..	..	Agriculture
918.3	932.2	901.0	905.9	858.2	893.2	987.6	1,022.5	1,034.6		..	Industry
598.7	612.6	585.5	595.1	558.9	591.9	663.1	693.3	712.9		..	Manufacturing
1,647.4	1,684.0	1,713.2	1,750.2	1,729.2	1,809.3	1,899.4	1,960.0	2,034.2		..	Services, etc.
-44.2	-17.7	31.9	24.2	4.0	-39.7	-102.7	-121.8	-154.5	-149.1	..	Resource Balance
281.9	321.6	350.9	354.2	326.5	314.0	335.3	331.2	341.3	385.9	..	Exports of Goods & NFServices
326.1	339.3	319.0	330.0	322.6	353.7	438.0	453.0	495.8	535.0	..	Imports of Goods & NFServices
2,677.6	2,705.6	2,652.5	2,712.8	2,665.7	2,808.8	3,065.7	3,191.7	3,317.9	3,424.9	..	Domestic Absorption
1,666.6	1,691.8	1,685.5	1,706.9	1,733.7	1,822.0	1,921.9	2,020.6	2,116.1	2,179.5	..	Private Consumption, etc.
516.3	520.6	530.3	538.2	548.5	554.7	579.3	624.9	650.0	666.8	..	General Gov't Consumption
494.6	493.2	436.6	467.7	383.5	432.2	564.5	546.2	551.8	578.6	..	Gross Domestic Investment
466.1	483.5	445.5	450.2	407.2	440.5	514.4	541.9	542.1	552.6	..	Fixed Investment
											Memo Items:
332.3	362.6	350.9	362.1	347.9	347.9	379.7	374.4	388.5	415.3	..	Capacity to Import
50.4	41.0	0.0	7.9	21.4	33.9	44.4	43.2	47.2	29.4	..	Terms of Trade Adjustment
2,683.7	2,729.0	2,684.4	2,745.0	2,691.0	2,803.1	3,007.4	3,113.1	3,210.6	3,305.2	..	Gross Domestic Income
2,720.6	2,777.3	2,732.0	2,792.6	2,735.2	2,844.7	3,045.7	3,145.0	3,237.3	3,326.9	..	Gross National Income
(Index 1980 = 100)											DOMESTIC PRICES (DEFLATORS)
84.3	91.7	100.0	109.6	116.7	121.2	125.7	129.5	132.9	137.3	..	Overall (GDP)
82.7	90.4	100.0	109.4	115.9	119.7	123.4	127.0	129.9	134.9	..	Domestic Absorption
96.3	107.5	100.0	103.1	98.7	99.6	112.3	96.5	92.5	..	..	Agriculture
82.1	89.1	100.0	111.7	117.4	119.2	120.8	121.7	122.5	..	..	Industry
87.3	92.5	100.0	108.8	114.5	116.6	117.5	116.7	117.2	..	..	Manufacturing
											MANUFACTURING ACTIVITY
99.8	102.4	100.0	98.0	92.3	90.6	92.6	90.7			..	Employment (1980=100)
108.6	104.6	100.0	99.7	99.8	102.3	104.3	105.9			..	Real Earnings per Empl. (1980=100)
94.6	99.0	100.0	103.6	100.3	107.1	114.4	117.3			..	Real Output per Empl. (1980=100)
41.3	39.9	40.9	40.8	41.4	40.1	39.1	39.7			..	Earnings as % of Value Added
(Billions of current US Dollars)											MONETARY HOLDINGS
1,390.9	1,509.9	1,647.3	1,832.4	2,023.5	2,259.7	2,496.9	2,724.9	2,998.4	3,118.0	D	Money Supply, Broadly Defined
372.2	397.0	424.1	451.4	490.9	538.3	570.3	641.0	746.5	765.9	..	Money as Means of Payment
99.1	107.0	118.8	126.2	136.5	150.6	160.8	173.1	186.2	199.4	..	Currency Ouside Banks
273.1	290.0	305.3	325.2	354.4	387.7	409.4	467.9	560.4	566.6	..	Demand Deposits
1,018.7	1,112.9	1,223.2	1,381.0	1,532.6	1,721.4	1,926.6	2,083.9	2,251.9	2,352.1	..	Quasi-Monetary Liabilities
(Billions of current US Dollars)											
-58.94	-35.95	-76.18	-78.74	-125.69	-202.52	-178.26	-212.11	-212.63	-147.54	E	GOVERNMENT DEFICIT (-) OR SURPLUS
416.57	488.62	545.88	639.79	659.61	653.32	718.32	791.45	823.07	909.82	..	Current Revenue
425.93	473.19	558.12	647.21	722.05	805.70	843.77	931.48	985.26	1,009.40	..	Current Expenditure
-9.36	15.43	-12.24	-7.42	-62.44	-152.38	-125.45	-140.03	-162.19	-99.58	..	Current Budget Balance
0.16	0.14	0.20	0.07	0.31	0.12	0.21	0.23	0.09	0.13	..	Capital Receipts
49.74	51.52	64.14	71.39	63.56	50.26	53.02	72.32	50.54	48.10	..	Capital Payments

UNITED STATES	1967	1968	1969	1970	1971	1972	1973	1974	1975	1976	1977
FOREIGN TRADE (CUSTOMS BASIS)					*(Billions of current US dollars)*						
Value of Exports, fob	31.53	34.39	38.01	43.22	44.13	49.78	71.34	98.51	107.59	114.99	120.13
Nonfuel Primary Products	9.00	9.16	9.33	11.29	10.81	12.95	23.36	29.50	29.23	30.74	32.03
Fuels	1.12	1.07	1.20	1.70	1.59	1.67	1.67	3.45	4.48	4.23	4.20
Manufactures	21.41	24.16	27.47	30.24	31.72	35.16	46.31	65.56	73.88	80.02	83.91
Value of Imports, cif	26.82	33.09	36.04	39.95	45.56	55.56	69.48	101.00	96.90	121.79	147.86
Nonfuel Primary Products	9.35	10.75	10.44	11.35	11.51	13.35	16.97	21.09	18.47	22.67	26.30
Fuels	2.25	2.53	2.79	3.07	3.71	4.80	8.17	25.38	26.40	33.93	44.20
Manufactures	15.22	19.81	22.81	25.53	30.34	37.41	44.33	54.53	52.03	65.20	77.36
Terms of Trade	168.5	165.6	166.6	169.0	157.4	152.4	164.5	123.4	132.6	131.7	124.8
					(Index 1980 = 100)						
Export Prices, fob	34.0	34.7	36.1	38.0	39.2	41.1	51.8	62.3	67.6	70.8	72.9
Nonfuel Primary Products	34.6	34.6	36.3	38.1	37.7	40.8	69.3	80.5	72.5	72.7	71.8
Fuels	13.8	13.4	13.1	16.3	19.2	21.6	37.8	66.9	78.7	74.8	78.7
Manufactures	36.6	37.4	39.1	41.1	41.9	43.2	46.5	56.3	65.4	69.9	73.2
Import Prices, cif	20.2	21.0	21.7	22.5	24.9	27.0	31.5	50.5	51.0	53.7	58.4
Trade at Constant 1980 Prices					*(Billions of 1980 US dollars)*						
Exports, fob	92.74	99.12	105.25	113.63	112.70	120.98	137.84	158.14	159.20	162.48	164.68
Imports, cif	132.86	157.92	166.24	177.51	183.14	205.79	220.83	200.11	190.17	226.65	253.00
BALANCE OF PAYMENTS					*(Billions of current US dollars)*						
Exports of Goods & Services	..	..	..	65.65	68.81	77.50	110.19	146.71	155.73	171.63	184.29
Merchandise, fob	..	..	..	42.45	43.31	49.38	71.41	98.32	107.08	114.75	120.80
Nonfactor Services	..	..	..	9.12	10.24	10.57	13.77	16.96	18.98	23.21	26.18
Factor Services	..	..	..	14.08	15.26	17.55	25.01	31.43	29.67	33.67	37.31
Imports of Goods & Services	..	..	..	59.87	66.42	79.22	98.99	137.33	132.79	162.13	193.79
Merchandise, fob	..	..	..	39.86	45.58	55.80	70.50	103.83	98.18	124.22	151.91
Nonfactor Services	..	..	..	14.28	15.17	16.55	18.45	21.03	21.55	24.11	26.70
Factor Services	..	..	..	5.73	5.67	6.87	10.04	12.47	13.06	13.80	15.18
Long-Term Interest	..	..	..	..	..	..	..	..	..	..	..
Current Transfers, net	..	..	..	-1.10	-1.11	-1.11	-1.25	-1.02	-0.91	-0.91	-0.82
Workers' Remittances	..	..	..	0.00	0.00	0.00	0.00	0.00	0.00	0.00	0.00
Total to be Financed	..	..	..	4.68	1.28	-2.83	9.95	8.36	22.03	8.59	-10.32
Official Capital Grants	..	..	..	-2.35	-2.73	-2.95	-2.87	-6.42	-3.97	-4.41	-4.17
Current Account Balance	..	..	..	2.33	-1.45	-5.78	7.08	1.94	18.06	4.18	-14.49
Long-Term Capital, net	..	..	..	-9.04	-11.82	-8.79	-9.82	-13.94	-23.67	-19.54	-16.51
Direct Investment	..	..	..	-6.13	-7.26	-6.80	-8.53	-4.28	-11.65	-7.60	-8.16
Long-Term Loans	..	..	..	..	..	..	..	..	..	..	..
Disbursements	..	..	..	..	..	..	..	..	..	..	..
Repayments	..	..	..	..	..	..	..	..	..	..	..
Other Long-Term Capital	..	..	..	-0.56	-1.83	0.96	1.58	-3.24	-8.05	-7.53	-4.18
Other Capital, net	..	..	..	-3.99	-17.20	3.51	-2.49	3.19	0.96	4.86	-4.04
Change in Reserves	..	..	..	10.70	30.47	11.06	5.23	8.81	4.65	10.50	35.04
Memo Item:					*(US Dollars per US dollar)*						
Conversion Factor (Annual Avg)	1.000	1.000	1.000	1.000	1.000	1.000	1.000	1.000	1.000	1.000	1.000
EXTERNAL DEBT, ETC.					*(Millions US dollars, outstanding at end of year)*						
Public/Publicly Guar. Long-Term	..	..	..	..	..	..	..	..	..	..	..
Official Creditors	..	..	..	..	..	..	..	..	..	..	..
IBRD and IDA	..	..	..	..	..	..	..	..	..	..	..
Private Creditors	..	..	..	..	..	..	..	..	..	..	..
Private Non-guaranteed Long-Term	..	..	..	..	..	..	..	..	..	..	..
Use of Fund Credit	..	..	..	..	..	..	..	..	..	..	..
Short-Term Debt	..	..	..	..	..	..	..	..	..	..	..
Memo Items:					*(Millions of US dollars)*						
Int'l Reserves Excluding Gold	2,765	4,818	5,105	3,415	2,109	2,663	2,726	4,232	4,627	7,149	7,593
Gold Holdings (at market price)	12,134	13,039	11,927	11,822	12,723	17,910	30,978	51,468	38,528	37,013	45,782
SOCIAL INDICATORS											
Total Fertility Rate	2.6	2.5	2.5	2.5	2.3	2.0	1.9	1.8	1.8	1.7	1.8
Crude Birth Rate	17.8	17.6	17.9	18.4	17.2	15.6	14.8	14.8	14.6	14.6	15.1
Infant Mortality Rate	22.4	21.8	20.9	20.0	19.1	18.5	17.7	16.7	16.1	15.2	14.1
Life Expectancy at Birth	70.6	70.0	70.5	70.8	71.2	71.5	71.9	72.2	72.6	72.9	73.3
Food Production, p.c. ('79-81 = 100)	86.9	86.5	85.7	83.3	89.2	87.2	88.3	88.9	95.0	96.4	99.2
Labor Force, Agriculture (%)	5.0	4.7	4.5	4.3	4.2	4.1	4.1	4.0	3.9	3.8	3.7
Labor Force, Female (%)	35.2	35.6	36.1	36.5	37.0	37.6	38.1	38.6	39.1	39.6	40.1
School Enroll. Ratio, primary	..	..	..	..	..	..	..	..	..	..	..
School Enroll. Ratio, secondary	..	..	..	..	..	..	..	..	..	..	..

1978	1979	1980	1981	1982	1983	1984	1985	1986	1987 estimate	NOTES	UNITED STATES
											FOREIGN TRADE (CUSTOMS BASIS)
				(Billions of current US dollars)							
142.54	176.79	216.92	230.51	210.93	199.14	216.01	211.42	211.90	252.57	..	Value of Exports, fob
39.50	49.98	61.45	59.11	50.32	49.41	51.96	42.80	40.85	47.17	..	Nonfuel Primary Products
3.95	5.69	8.13	10.29	12.78	9.70	9.47	10.10	8.20	7.80	..	Fuels
99.08	121.12	147.34	161.11	147.83	140.04	154.58	158.52	162.84	197.60	..	Manufactures
182.20	217.39	250.28	271.21	253.03	267.97	338.19	358.70	381.36	422.41	..	Value of Imports, cif
32.16	37.15	38.76	38.90	33.88	38.09	43.70	42.91	45.89	47.39	..	Nonfuel Primary Products
44.69	63.67	82.20	84.26	67.42	60.00	63.07	55.68	39.77	46.74	..	Fuels
105.35	116.56	129.31	148.06	151.73	169.88	231.42	260.11	295.71	328.27	..	Manufactures
				(Index 1980 = 100)							
120.5	114.8	100.0	103.4	103.9	110.8	112.0	114.5	123.1	116.3	..	Terms of Trade
78.4	91.9	100.0	106.8	106.0	109.4	110.0	108.6	110.5	113.3	..	Export Prices, fob
76.5	93.6	100.0	96.2	84.2	91.7	90.6	79.8	75.3	77.1	..	Nonfuel Primary Products
83.5	88.2	100.0	98.4	91.3	84.3	83.3	83.3	89.4	80.5	..	Fuels
79.1	91.5	100.0	112.0	118.0	120.0	121.0	123.0	127.0	130.0	..	Manufactures
65.0	80.0	100.0	103.3	102.0	98.7	98.2	94.9	89.8	97.4	..	Import Prices, cif
											Trade at Constant 1980 Prices
				(Billions of 1980 US dollars)							
181.85	192.37	216.92	215.73	199.02	182.07	196.45	194.65	191.68	222.86	..	Exports, fob
280.22	271.62	250.28	262.44	247.98	271.55	344.54	378.08	424.75	433.49	..	Imports, cif
											BALANCE OF PAYMENTS
				(Billions of current US dollars)							
220.01	286.79	342.48	376.52	349.53	334.42	360.74	359.47	375.09	424.84	..	Exports of Goods & Services
142.06	184.47	224.27	237.10	211.20	201.81	219.90	215.94	223.98	249.57	..	Merchandise, fob
29.59	31.76	38.37	45.42	49.30	49.77	48.96	48.84	53.41	62.19	..	Nonfactor Services
48.36	70.56	79.84	94.00	89.03	82.84	91.88	94.69	97.70	113.08	..	Factor Services
229.84	281.65	333.06	362.14	349.30	371.22	455.66	460.59	498.61	565.38	..	Imports of Goods & Services
176.00	212.01	249.77	265.07	247.65	268.89	332.41	338.09	368.52	409.85	..	Merchandise, fob
31.01	35.36	39.90	43.57	45.53	48.57	54.25	58.04	61.36	70.05	..	Nonfactor Services
22.83	34.28	43.39	53.50	56.12	53.76	69.00	64.46	68.73	85.48	..	Factor Services
..	..	..	..	..	..	..	..	..	..	..	Long-Term Interest
-0.86	-0.91	-1.03	-0.99	-1.19	-0.99	-1.43	-1.92	-1.38	-1.22	..	Current Transfers, net
0.00	0.00	0.00	0.00	0.00	0.00	0.00	0.00	0.00	0.00	..	Workers' Remittances
-10.69	4.23	8.39	13.39	-0.96	-37.79	-96.35	-103.04	-124.90	-141.76	K	Total to be Financed
-4.71	-5.20	-6.55	-6.52	-7.70	-8.49	-10.74	-13.39	-13.94	-12.19	K	Official Capital Grants
-15.40	-0.97	1.84	6.87	-8.66	-46.28	-107.09	-116.43	-138.84	-153.95	..	Current Account Balance
-14.84	-25.13	-13.86	-7.24	-19.82	-10.15	27.64	42.76	58.71	12.60	..	Long-Term Capital, net
-8.16	-13.35	-2.30	15.57	11.44	11.58	22.57	1.76	6.28	-2.47	..	Direct Investment
											Long-Term Loans
..	..	..	..	..	..	..	..	..	..	..	Disbursements
..	..	..	..	..	..	..	..	..	..	..	Repayments
-1.97	-6.58	-5.01	-16.29	-23.56	-13.24	15.81	54.39	66.37	27.26	..	Other Long-Term Capital
-1.64	39.73	4.13	-0.87	30.51	52.38	80.17	79.47	46.35	84.41	..	Other Capital, net
31.88	-13.63	7.89	1.24	-2.03	4.05	-0.72	-5.80	33.78	56.94	..	Change in Reserves
				(US Dollars per US dollar)							Memo Item:
1.000	1.000	1.000	1.000	1.000	1.000	1.000	1.000	1.000	1.000	..	Conversion Factor (Annual Avg)
											EXTERNAL DEBT, ETC.
				(Millions US dollars, outstanding at end of year)							
..	..	..	..	..	..	..	..	..	..	..	Public/Publicly Guar. Long-Term
..	..	..	..	..	..	..	..	..	..	..	Official Creditors
..	..	..	..	..	..	..	..	..	..	..	IBRD and IDA
..	..	..	..	..	..	..	..	..	..	..	Private Creditors
..	..	..	..	..	..	..	..	..	..	..	Private Non-guaranteed Long-Term
..	..	..	..	..	..	..	..	..	..	..	Use of Fund Credit
..	..	..	..	..	..	..	..	..	..	..	Short-Term Debt
				(Millions of US dollars)							Memo Items:
6,979	7,784	15,596	18,924	22,809	22,627	23,838	32,096	37,452	34,720	..	Int'l Reserves Excluding Gold
62,469	135,475	155,817	104,984	120,635	100,483	81,018	85,887	102,431	127,018	..	Gold Holdings (at market price)
											SOCIAL INDICATORS
1.8	1.8	1.8	1.8	0.3	0.6	0.9	1.3	1.6	1.9	..	Total Fertility Rate
15.0	15.6	15.9	15.7	2.5	5.1	7.6	10.2	12.7	15.2	..	Crude Birth Rate
13.8	13.1	12.6	11.7	2.4	4.7	7.1	9.4	11.8	14.2	..	Infant Mortality Rate
73.4	73.8	73.7	74.0	74.4	74.5	74.6	74.8	75.0	75.2	..	Life Expectancy at Birth
96.8	100.2	95.8	104.1	102.7	87.0	98.1	102.7	96.8	92.2	..	Food Production, p.c. ('79-81 = 100)
3.7	3.6	3.5	..	..	..	..	..	..	..	..	Labor Force, Agriculture (%)
40.6	41.1	41.6	41.6	41.6	41.6	41.5	41.5	41.5	41.5	..	Labor Force, Female (%)
..	..	98.0	..	101.0	102.0	101.0	101.0	102.0	..	..	School Enroll. Ratio, primary
..	..	90.0	..	94.0	95.0	95.0	99.0	100.0	..	..	School Enroll. Ratio, secondary

URUGUAY	1967	1968	1969	1970	1971	1972	1973	1974	1975	1976	1977
CURRENT GNP PER CAPITA (US $)	470	550	650	780	910	910	990	1,140	1,370	1,440	1,450
POPULATION (thousands)	2,751	2,775	2,794	2,808	2,817	2,821	2,821	2,823	2,830	2,841	2,854

ORIGIN AND USE OF RESOURCES *(Billions of current Uruguayan New Pesos)*

	1967	1968	1969	1970	1971	1972	1973	1974	1975	1976	1977
Gross National Product (GNP)	1.7E-01	3.7E-01	5.0E-01	6.0E-01	7.2E-01	1.2E+00	2.5E+00	4.5E+00	8.0E+00	1.2E+01	2.0E+01
Net Factor Income from Abroad	-3.0E-03	-5.7E-03	-5.6E-03	-6.0E-03	-5.0E-03	-2.0E-02	-2.3E-02	-7.0E-02	-1.9E-01	-2.4E-01	-3.2E-01
Gross Domestic Product (GDP)	1.7E-01	3.7E-01	5.1E-01	6.0E-01	7.2E-01	1.2E+00	2.6E+00	4.5E+00	8.2E+00	1.3E+01	2.0E+01
Indirect Taxes, net	1.8E-02	4.2E-02	5.8E-02	8.1E-02	9.3E-02	2.1E-01	3.4E-01	5.1E-01	1.1E+00	1.8E+00	2.8E+00
GDP at factor cost	1.5E-01	3.3E-01	4.5E-01	5.2E-01	6.3E-01	1.0E+00	2.2E+00	4.0E+00	7.1E+00	1.1E+01	1.7E+01
Agriculture	2.0E-02	4.1E-02	5.5E-02	6.7E-02	8.3E-02	1.8E-01	4.3E-01	6.6E-01	8.5E-01	1.2E+00	2.2E+00
Industry	4.6E-02	1.1E-01	1.4E-01	1.5E-01	1.7E-01	2.7E-01	6.1E-01	1.2E+00	2.2E+00	3.4E+00	5.1E+00
Manufacturing	..	..	..	..	..	..	..	..	..	..	..
Services, etc.	8.7E-02	1.8E-01	2.6E-01	3.0E-01	3.7E-01	5.7E-01	1.2E+00	2.2E+00	4.0E+00	6.3E+00	9.8E+00
Resource Balance	2.1E-03	1.0E-02	5.5E-03	-8.0E-03	-9.0E-03	4.0E-03	3.1E-02	-1.2E-01	-3.0E-01	-9.5E-02	-5.8E-01
Exports of Goods & NFServices	2.4E-02	5.6E-02	6.6E-02	7.3E-02	7.1E-02	1.8E-01	3.5E-01	6.4E-01	1.3E+00	2.4E+00	3.8E+00
Imports of Goods & NFServices	2.2E-02	4.5E-02	6.1E-02	8.1E-02	8.0E-02	1.7E-01	3.2E-01	7.6E-01	1.6E+00	2.4E+00	4.4E+00
Domestic Absorption	1.7E-01	3.6E-01	5.0E-01	6.1E-01	7.3E-01	1.2E+00	2.5E+00	4.7E+00	8.5E+00	1.3E+01	2.1E+01
Private Consumption, etc.	1.2E-01	2.8E-01	3.7E-01	4.5E-01	5.2E-01	9.4E-01	1.8E+00	3.5E+00	6.2E+00	9.1E+00	1.5E+01
General Gov't Consumption	2.4E-02	4.9E-02	7.5E-02	9.2E-02	1.2E-01	1.5E-01	3.6E-01	6.8E-01	1.1E+00	1.8E+00	2.5E+00
Gross Domestic Investment	2.3E-02	3.8E-02	5.5E-02	6.9E-02	9.1E-02	1.5E-01	3.2E-01	5.2E-01	1.1E+00	1.9E+00	3.0E+00
Fixed Investment	2.3E-02	3.8E-02	5.6E-02	6.9E-02	8.3E-02	1.2E-01	2.3E-01	4.6E-01	1.1E+00	2.0E+00	3.0E+00
Memo Items:											
Gross Domestic Saving	2.5E-02	4.8E-02	6.1E-02	6.1E-02	8.2E-02	1.5E-01	3.5E-01	4.0E-01	8.1E-01	1.8E+00	2.4E+00
Gross National Saving	2.3E-02	4.3E-02	5.5E-02	5.5E-02	7.7E-02	1.3E-01	3.3E-01	3.3E-01	6.1E-01	1.5E+00	2.1E+00

(Millions of 1980 Uruguayan New Pesos)

	1967	1968	1969	1970	1971	1972	1973	1974	1975	1976	1977
Gross National Product	58,070	59,654	63,372	67,406	67,507	66,401	66,763	68,998	72,763	75,658	76,802
GDP at factor cost	53,413	54,366	57,521	58,871	58,961	58,126	58,281	60,045	63,679	66,237	67,058
Agriculture	6,671	6,564	7,511	8,311	7,789	6,973	7,155	7,389	7,802	7,950	8,205
Industry	15,874	16,595	17,537	17,746	17,645	17,924	17,502	17,930	19,624	20,409	21,429
Manufacturing	..	..	..	..	..	..	..	..	..	..	..
Services, etc.	30,867	31,207	32,473	32,813	33,528	33,230	33,624	34,727	36,253	37,878	37,424
Resource Balance	-1,917	-479	-2,149	-4,516	-5,210	-4,411	-5,450	-3,440	-2,979	-1,304	-1,538
Exports of Goods & NFServices	5,898	6,753	6,744	6,930	6,592	6,474	6,454	7,789	9,275	11,228	12,031
Imports of Goods & NFServices	7,815	7,233	8,894	11,446	11,802	10,885	11,904	11,229	12,255	12,532	13,569
Domestic Absorption	61,008	61,156	66,451	72,844	73,607	71,805	73,037	73,135	76,833	78,102	79,321
Private Consumption, etc.	48,349	48,487	52,481	57,817	58,110	58,663	58,671	57,925	60,246	59,249	59,215
General Gov't Consumption	7,335	7,946	7,957	8,244	8,298	7,102	8,625	9,247	9,020	9,699	9,374
Gross Domestic Investment	5,325	4,723	6,013	6,783	7,199	6,040	5,741	5,963	7,568	9,154	10,731
Fixed Investment	5,237	4,862	6,213	7,029	6,945	5,888	4,983	5,518	7,744	9,939	10,654
Memo Items:											
Capacity to Import	8,554	8,893	9,701	10,316	10,474	11,135	13,046	9,434	10,006	12,045	11,756
Terms of Trade Adjustment	2,656	2,140	2,956	3,385	3,882	4,661	6,592	1,645	730	817	-275
Gross Domestic Income	61,747	62,817	67,258	71,713	72,279	72,055	74,179	71,340	74,585	77,615	77,508
Gross National Income	60,727	61,793	66,328	70,791	71,389	71,062	73,356	70,643	73,493	76,475	76,527

DOMESTIC PRICES (DEFLATORS) *(Index 1980 = 100)*

	1967	1968	1969	1970	1971	1972	1973	1974	1975	1976	1977
Overall (GDP)	0.29	0.62	0.79	0.88	1.10	1.80	3.80	6.50	11.10	16.50	25.60
Domestic Absorption	0.27	0.60	0.75	0.84	0.99	1.70	3.50	6.40	11.00	16.30	25.80
Agriculture	0.29	0.63	0.74	0.81	1.10	2.70	6.00	8.90	10.90	15.20	26.70
Industry	0.29	0.67	0.78	0.83	0.99	1.50	3.50	6.60	11.40	16.50	23.80
Manufacturing	..	..	..	..	..	..	..	..	..	..	..

MANUFACTURING ACTIVITY

	1967	1968	1969	1970	1971	1972	1973	1974	1975	1976	1977
Employment (1980=100)	..	81.2	..	100.7	..	112.4	119.0	121.6	125.9	132.7	138.6
Real Earnings per Empl. (1980=100)	..	93.0	..	..	..	..	..	..	..	96.1	88.3
Real Output per Empl. (1980=100)	..	73.2	..	..	..	90.1	78.4	79.4	82.5	87.0	94.0
Earnings as % of Value Added	..	34.0	..	..	..	..	..	..	..	35.3	31.6

MONETARY HOLDINGS *(Billions of current Uruguayan New Pesos)*

	1967	1968	1969	1970	1971	1972	1973	1974	1975	1976	1977
Money Supply, Broadly Defined	0.05	0.07	0.11	0.13	0.20	0.32	0.54	0.92	1.60	3.20	5.80
Money as Means of Payment	0.03	0.05	0.08	0.09	0.14	0.20	0.36	0.59	0.83	1.40	1.90
Currency Ouside Banks	0.02	0.03	0.05	0.06	0.08	0.12	0.20	0.32	0.47	0.78	1.10
Demand Deposits	0.01	0.02	0.03	0.03	0.05	0.08	0.16	0.27	0.36	0.60	0.82
Quasi-Monetary Liabilities	0.02	0.02	0.03	0.04	0.06	0.12	0.18	0.34	0.78	1.80	3.90

GOVERNMENT DEFICIT (-) OR SURPLUS *(Billions of current Uruguayan New Pesos)*

	1967	1968	1969	1970	1971	1972	1973	1974	1975	1976	1977
	..	..	..	..	..	-0.03	-0.03	-0.17	-0.36	-0.26	-0.26
Current Revenue	..	..	..	..	..	0.28	0.55	0.91	1.51	2.79	4.51
Current Expenditure	..	..	..	..	..	0.28	0.53	1.06	1.81	3.10	3.92
Current Budget Balance	..	..	..	..	..	0.00	0.02	-0.15	-0.30	-0.31	0.59
Capital Receipts	..	..	..	..	..	0.00	0.00	0.01	0.01	0.01	0.03
Capital Payments	..	..	..	..	..	0.03	0.06	0.03	0.07	-0.04	0.88

1978	1979	1980	1981	1982	1983	1984	1985	1986	1987 estimate	NOTES	URUGUAY
1,660	2,120	2,810	3,600	3,420	2,460	1,960	1,730	1,920	2,160	..	**CURRENT GNP PER CAPITA** (US $)
2,871	2,890	2,910	2,930	2,948	2,963	2,967	2,970	2,983	3,003	..	**POPULATION** (thousands)

(Billions of current Uruguayan New Pesos)

1978	1979	1980	1981	1982	1983	1984	1985	1986	1987	NOTES	
											ORIGIN AND USE OF RESOURCES
30.5	57.2	91.3	121.7	126.1	175.1	274.1	492.6	930.1	1,637.0	..	Gross National Product (GNP)
-0.5	-0.5	-0.9	-0.8	-2.6	-9.9	-20.2	-35.6	-41.6	-63.6	..	Net Factor Income from Abroad
30.9	57.6	92.2	122.5	128.7	185.0	294.4	528.2	971.7	1,700.6	f	Gross Domestic Product (GDP)
4.6	7.9	12.7	16.5	16.7	22.9	38.2	70.5	135.8	246.2	..	Indirect Taxes, net
26.3	49.8	79.5	105.9	112.0	162.1	256.1	457.7	835.9	1,454.3	f	GDP at factor cost
2.9	6.0	8.9	10.0	9.9	16.7	30.4	55.7	109.8	191.9	..	Agriculture
8.1	16.7	26.0	31.9	30.8	48.2	80.9	140.8	259.0	465.7	..	Industry
6.4	13.6	20.6	24.2	21.7	38.3	67.3	118.1	217.8	388.3	..	Manufacturing
15.3	27.0	44.7	64.0	71.2	97.3	144.9	261.1	467.1	796.8	..	Services, etc.
-0.8	-2.6	-5.2	-4.8	-4.0	3.1	12.1	21.1	51.5	34.6	..	Resource Balance
5.5	9.4	13.9	18.0	18.1	44.7	72.1	126.7	227.1	350.4	..	Exports of Goods & NFServices
6.3	12.0	19.0	22.8	22.1	41.6	59.9	105.7	175.6	315.8	..	Imports of Goods & NFServices
31.7	60.2	97.4	127.3	132.7	181.9	282.2	507.1	920.2	1,666.0	..	Domestic Absorption
22.9	43.4	69.9	91.1	94.1	137.8	216.3	394.8	720.3	1,290.8	..	Private Consumption, etc.
3.8	6.8	11.5	17.3	20.1	25.7	36.9	69.2	127.8	216.9	..	General Gov't Consumption
5.0	10.0	16.0	18.8	18.6	18.4	29.1	43.1	72.0	158.2	..	Gross Domestic Investment
4.9	9.3	15.4	19.2	19.4	20.3	27.3	39.2	71.0	144.6	..	Fixed Investment
											Memo Items:
4.2	7.4	10.8	14.0	14.5	21.5	41.2	64.2	123.6	192.8	..	Gross Domestic Saving
3.7	7.0	9.9	13.2	11.9	11.6	21.0	28.6	82.0	129.3	..	Gross National Saving

(Millions of 1980 Uruguayan New Pesos)

1978	1979	1980	1981	1982	1983	1984	1985	1986	1987	NOTES	
80,875	86,393	91,291	93,149	83,081	75,988	73,854	74,055	80,853	85,402	..	Gross National Product
70,687	75,088	79,539	80,946	73,181	68,755	67,790	67,897	72,479	76,164	f	GDP at factor cost
7,659	7,625	8,860	9,349	8,662	8,844	8,244	8,618	9,198	9,268	..	Agriculture
23,415	25,243	25,970	25,246	21,848	19,238	19,319	18,316	20,168	22,342	..	Industry
18,782	20,116	20,603	19,664	16,341	15,195	15,620	15,370	17,229	19,130	..	Manufacturing
39,613	42,219	44,709	46,351	42,670	40,673	40,227	40,963	43,113	44,554	..	Services, etc.
-1,979	-4,299	-5,162	-4,498	-3,433	1,943	4,029	4,643	4,327	1,149	..	Resource Balance
12,553	13,377	13,861	14,717	13,169	15,203	15,309	16,122	18,492	16,742	..	Exports of Goods & NFServices
14,532	17,676	19,023	19,215	16,602	13,260	11,280	11,478	14,165	15,593	..	Imports of Goods & NFServices
83,945	91,361	97,366	98,323	88,333	77,845	74,633	74,141	79,730	87,186	..	Domestic Absorption
61,351	64,542	69,890	71,405	64,369	58,597	55,172	55,762	60,953	66,466	..	Private Consumption, etc.
10,338	11,639	11,482	12,342	12,045	11,693	11,761	12,131	12,564	12,938	..	General Gov't Consumption
12,256	15,180	15,994	14,576	11,919	7,555	7,701	6,248	6,213	7,783	..	Gross Domestic Investment
12,187	14,502	15,422	14,958	12,735	8,558	7,419	5,695	6,161	7,337	..	Fixed Investment
											Memo Items:
12,774	13,870	13,861	15,146	13,572	14,248	13,559	13,76	18,322	17,301	..	Capacity to Import
221	492	0	429	403	-955	-1,750	-2,353	-170	559	..	Terms of Trade Adjustment
82,187	87,554	92,204	94,255	85,303	78,833	76,913	76,432	83,887	88,894	..	Gross Domestic Income
81,095	86,885	91,291	93,579	83,484	75,033	72,104	71,703	80,683	85,961	..	Gross National Income

(Index 1980 = 100)

1978	1979	1980	1981	1982	1983	1984	1985	1986	1987	NOTES	
											DOMESTIC PRICES (DEFLATORS)
37.7	66.2	100.0	130.5	151.6	231.9	374.2	670.4	1156.0	1925.1	..	Overall (GDP)
37.8	65.9	100.0	129.5	150.3	233.7	378.2	683.9	1154.2	1910.8	..	Domestic Absorption
38.5	78.9	100.0	106.8	114.8	188.7	368.9	646.6	1193.7	2070.8	..	Agriculture
34.4	66.2	100.0	126.3	141.0	250.4	418.5	768.6	1284.3	2084.2	..	Industry
33.9	67.6	100.0	122.8	132.9	252.3	430.7	768.4	1264.4	2030.0	..	Manufacturing
											MANUFACTURING ACTIVITY
152.1	160.8	100.0	95.7	77.9	73.3	75.1	76.3	..	..	G J	Employment (1980=100)
93.1	93.3	100.0	108.3	119.4	101.5	78.3	96.2	..	..	G J	Real Earnings per Empl. (1980=100)
89.1	84.1	100.0	106.3	117.9	113.9	115.0	106.8	..	..	G J	Real Output per Empl. (1980=100)
37.3	33.9	32.8	32.0	32.5	28.6	21.0	22.3	..	..	J	Earnings as % of Value Added

(Billions of current Uruguayan New Pesos)

1978	1979	1980	1981	1982	1983	1984	1985	1986	1987	NOTES	
											MONETARY HOLDINGS
11.1	20.6	35.6	53.3	72.5	81.9	132.9	259.8	478.4	703.1	..	Money Supply, Broadly Defined
3.6	6.2	9.1	9.8	13.7	14.9	22.2	46.0	84.2	135.2	..	Money as Means of Payment
1.8	3.2	5.1	6.1	7.9	8.4	12.1	23.3	43.1	76.4	..	Currency Ouside Banks
1.8	3.0	4.0	3.7	5.8	6.5	10.0	22.7	41.1	58.8	..	Demand Deposits
7.5	14.4	26.5	43.5	58.8	66.9	110.7	213.8	394.3	568.0	..	Quasi-Monetary Liabilities

(Billions of current Uruguayan New Pesos)

1978	1979	1980	1981	1982	1983	1984	1985	1986	1987	NOTES	
											GOVERNMENT DEFICIT (-) OR SURPLUS
-0.28	0.00	0.03	-1.84	-11.65	-7.27	-15.31	-11.65	-6.23	..	..	Current Revenue
6.90	12.10	20.52	29.07	27.38	40.06	55.38	109.02	219.63	..	..	Current Expenditure
6.31	10.18	18.54	28.22	35.33	43.02	63.52	111.02	212.35	..	..	Current Budget Balance
0.60	1.92	1.98	0.85	-7.95	-2.96	-8.14	-2.00	7.27	..	..	Capital Receipts
0.02	0.09	0.00	0.01	0.07	0.08	0.14	0.30	0.67	..	..	Capital Payments
0.90	2.01	1.95	2.70	3.78	4.39	7.30	9.95	14.17	..	..	

URUGUAY	1967	1968	1969	1970	1971	1972	1973	1974	1975	1976	1977
FOREIGN TRADE (CUSTOMS BASIS)				*(Millions of current US dollars)*							
Value of Exports, fob	158.7	179.2	200.3	232.5	205.5	214.1	321.5	381.0	381.2	536.0	598.5
Nonfuel Primary Products	148.3	141.6	163.2	185.1	164.2	174.4	267.3	293.5	265.2	354.6	364.6
Fuels	0.2	0.2	0.2	0.1	0.0	0.1	0.9	2.2	1.9	0.1	0.0
Manufactures	10.2	37.4	37.0	47.4	41.2	39.6	53.3	85.2	114.2	181.3	234.0
Value of Imports, cif	170.2	158.7	196.9	232.9	222.1	186.8	284.8	461.1	516.5	599.0	668.8
Nonfuel Primary Products	42.8	42.6	52.2	57.1	51.2	48.5	84.6	104.1	98.2	87.4	114.3
Fuels	29.5	29.1	24.1	34.1	32.0	31.6	33.5	149.6	161.2	207.9	171.1
Manufactures	97.9	86.9	120.6	141.7	138.9	106.7	166.7	207.4	257.1	303.7	383.4
Terms of Trade	232.2	211.3	179.9	197.2	169.2	212.2	209.7	133.6	114.6	121.7	110.7
					(Index 1980 = 100)						
Export Prices, fob	34.5	30.5	32.7	33.5	34.0	44.0	67.0	65.9	57.9	63.5	66.4
Nonfuel Primary Products	35.7	32.3	34.7	34.2	34.6	46.1	75.2	72.1	57.8	63.5	67.2
Fuels	4.3	4.3	4.3	4.3	5.6	6.2	8.9	36.7	35.7	38.4	42.0
Manufactures	25.3	25.9	26.5	31.2	31.8	37.1	46.5	51.8	58.8	63.5	65.3
Import Prices, cif	14.9	14.4	18.1	17.0	20.1	20.7	31.9	49.4	50.5	52.2	60.0
Trade at Constant 1980 Prices					*(Millions of 1980 US dollars)*						
Exports, fob	459.3	588.0	613.5	694.4	604.5	486.5	480.1	577.8	658.0	843.7	900.9
Imports, cif	1,143.7	1,100.3	1,085.0	1,371.8	1,105.4	900.6	891.9	934.4	1,022.0	1,147.7	1,114.1
BALANCE OF PAYMENTS					*(Millions of current US dollars)*						
Exports of Goods & Services	..	..	..	291.5	253.5	352.0	415.5	504.5	554.7	703.0	820.1
Merchandise, fob	..	..	..	224.1	196.8	281.6	327.6	381.4	384.9	565.0	611.5
Nonfactor Services	..	..	..	65.9	55.9	68.9	82.1	118.5	166.1	131.2	197.0
Factor Services	..	..	..	1.5	0.8	1.4	5.7	4.7	3.8	6.8	11.6
Imports of Goods & Services	..	..	..	345.9	325.3	304.5	397.2	639.6	751.1	784.4	993.8
Merchandise, fob	..	..	..	203.1	203.0	178.7	248.6	433.6	494.0	536.6	686.7
Nonfactor Services	..	..	..	116.5	99.9	100.9	117.9	158.7	182.1	168.6	227.5
Factor Services	..	..	..	26.3	22.4	25.0	30.8	47.3	74.9	79.2	79.5
Long-Term Interest	..	..	..	17.4	17.6	21.0	23.9	36.1	51.6	60.6	63.0
Current Transfers, net	..	..	..	-0.9	-0.6	-0.2	-0.1	-1.2	-1.5	-1.0	2.1
Workers' Remittances	..	..	..	0.0	0.0	0.0	0.0	0.0	0.0	0.0	0.0
Total to be Financed	..	..	..	-55.3	-72.4	47.2	18.1	-136.3	-197.8	-82.4	-171.6
Official Capital Grants	..	..	..	10.2	8.9	11.5	19.1	18.6	8.3	8.7	4.6
Current Account Balance	..	..	..	-45.1	-63.5	58.7	37.2	-117.6	-189.5	-73.8	-167.1
Long-Term Capital, net	..	..	..	5.7	61.7	30.1	-13.8	81.7	158.9	87.7	105.8
Direct Investment	..	..	..	..	..	..	..	..	..	..	66.0
Long-Term Loans	..	..	..	-1.1	31.9	28.7	6.4	251.9	36.0	81.4	66.1
Disbursements	..	..	..	50.3	79.3	125.5	86.1	383.5	284.6	240.4	264.8
Repayments	..	..	..	51.4	47.4	96.8	79.7	131.6	248.6	159.0	198.7
Other Long-Term Capital	..	..	..	-3.4	20.9	-10.1	-39.3	-188.9	114.7	-2.3	-30.8
Other Capital, net	..	..	..	10.9	-10.0	-62.5	3.6	7.8	-60.3	80.3	228.5
Change in Reserves	..	..	..	28.5	11.8	-26.4	-27.0	28.2	90.9	-94.2	-167.2
Memo Item:					*(Uruguayan New Pesos per US dollar)*						
Conversion Factor (Annual Avg)	0.120	0.240	0.250	0.250	0.250	0.540	0.870	1.200	2.250	3.340	4.680
EXTERNAL DEBT, ETC.				*(Millions of US dollars, outstanding at end of year)*							
Public/Publicly Guar. Long-Term	..	..	..	269	293	327	347	519	618	694	736
Official Creditors	..	..	..	126	139	176	208	272	264	264	248
IBRD and IDA	..	..	..	49	52	55	59	71	69	79	76
Private Creditors	..	..	..	143	154	151	138	247	354	429	489
Private Non-guaranteed Long-Term	..	..	..	29	40	35	31	119	52	57	83
Use of Fund Credit	..	..	..	18	19	40	39	78	117	145	119
Short-Term Debt	..	..	..	..	..	..	..	..	..	..	166
Memo Items:					*(Millions of US dollars)*						
Int'l Reserves Excluding Gold	22.32	33.60	19.21	13.80	20.10	69.03	100.58	80.55	58.52	176.21	322.38
Gold Holdings (at market price)	140.31	159.60	165.83	172.43	184.51	229.49	396.92	659.46	496.34	477.55	589.86
SOCIAL INDICATORS											
Total Fertility Rate	2.8	2.8	2.9	2.9	3.0	3.0	3.0	3.0	3.0	2.9	2.9
Crude Birth Rate	20.5	20.6	20.7	20.9	21.0	21.1	21.0	20.8	20.6	20.4	20.3
Infant Mortality Rate	48.0	47.8	47.6	47.4	47.2	47.0	46.4	45.8	45.2	44.6	44.0
Life Expectancy at Birth	68.6	68.7	68.7	68.8	68.8	68.8	69.0	69.1	69.3	69.5	69.6
Food Production, p.c. ('79-81=100)	76.1	91.8	97.6	104.4	90.3	85.1	90.5	101.0	99.9	112.3	93.4
Labor Force, Agriculture (%)	19.4	19.1	18.9	18.6	18.3	18.0	17.7	17.4	17.1	16.8	16.5
Labor Force, Female (%)	25.6	25.9	26.1	26.3	26.6	27.0	27.3	27.6	27.9	28.3	28.6
School Enroll. Ratio, primary	..	..	..	112.0	..	..	..	..	107.0	105.0	105.0
School Enroll. Ratio, secondary	..	..	..	59.0	..	..	..	..	60.0	63.0	62.0

1978	1979	1980	1981	1982	1983	1984	1985	1986	1987 estimate	NOTES	URUGUAY
											FOREIGN TRADE (CUSTOMS BASIS)
				(Millions of current US dollars)							
681.9	787.2	1,060.0	1,220.0	1,020.0	1,050.0	925.0	855.0	1,090.0	1,190.0	..	Value of Exports, fob
391.9	411.9	655.4	844.6	684.7	747.7	575.7	550.3	700.0	662.9	..	Nonfuel Primary Products
0.0	0.0	0.0	12.3	2.8	0.8	3.0	1.9	5.0	2.1	..	Fuels
289.9	375.3	404.5	363.1	332.4	301.5	346.3	302.8	385.0	525.0	..	Manufactures
715.7	1,173.2	1,650.0	1,630.0	1,110.0	788.0	776.0	708.0	870.0	1,140.0	..	Value of Imports, cif
113.6	226.6	246.5	189.3	112.2	100.0	118.6	99.8	140.2	171.5	..	Nonfuel Primary Products
231.2	282.3	472.3	515.1	434.6	284.2	281.4	239.7	168.8	178.5	..	Fuels
370.9	664.3	931.2	925.5	563.2	403.8	376.0	368.5	561.0	790.0	..	Manufactures
				(Index 1980 = 100)							
122.5	112.4	100.0	94.7	88.6	89.2	90.0	87.2	97.6	97.0	..	Terms of Trade
76.0	92.6	100.0	97.4	87.0	84.0	83.3	78.8	81.4	95.2	..	Export Prices, fob
79.9	94.5	100.0	95.3	82.1	80.3	76.7	71.4	70.5	79.9	..	Nonfuel Primary Products
42.3	61.0	100.0	112.5	101.6	92.5	90.2	87.5	44.9	56.9	..	Fuels
71.2	90.6	100.0	102.0	99.0	95.0	97.0	97.0	114.6	125.8	..	Manufactures
62.0	82.4	100.0	102.8	98.1	94.2	92.6	90.4	83.3	98.1	..	Import Prices, cif
											Trade at Constant 1980 Prices
				(Millions of 1980 US dollars)							
897.7	850.2	1,059.9	1,253.2	1,173.0	1,249.4	1,110.7	1,085.2	1,339.8	1,250.6	..	Exports, fob
1,154.5	1,424.3	1,650.0	1,585.8	1,131.1	836.8	838.2	783.4	1,044.3	1,161.8	..	Imports, cif
				(Millions of current US dollars)							**BALANCE OF PAYMENTS**
931.3	1,248.4	1,593.7	1,846.5	1,684.5	1,473.9	1,376.6	1,345.5	1,592.3	1,656.7	..	Exports of Goods & Services
686.1	788.1	1,058.5	1,229.7	1,256.4	1,156.4	924.6	853.6	1,087.8	1,189.1	..	Merchandise, fob
226.8	406.1	467.5	471.0	280.9	255.0	364.8	415.4	411.8	364.7	..	Nonfactor Services
18.4	54.2	67.7	145.8	147.2	62.5	87.2	76.5	92.7	102.9	..	Factor Services
1,065.4	1,612.6	2,311.5	2,317.6	1,929.5	1,544.7	1,515.7	1,461.1	1,526.4	1,789.1	..	Imports of Goods & Services
709.8	1,166.2	1,668.2	1,592.1	1,038.4	739.7	732.2	675.4	791.0	1,079.9	..	Merchandise, fob
260.4	337.3	475.5	505.9	547.1	454.7	334.7	357.4	364.7	325.2	..	Nonfactor Services
95.2	109.1	167.8	219.6	344.0	350.3	448.8	428.3	370.7	384.0	..	Factor Services
67.0	80.9	121.3	148.0	179.2	225.6	295.2	288.7	252.9	272.8	..	Long-Term Interest
1.4	1.5	2.0	2.9	0.0	0.0	0.0	0.0	0.0	..	..	Current Transfers, net
0.0	0.0	0.0	0.0	0.0	0.0	0.0	0.0	0.0	0.0	..	Workers' Remittances
-132.7	-362.7	-715.8	-468.2	-245.0	-70.8	-139.1	-115.6	65.9	-132.4	K	Total to be Financed
5.7	5.6	6.7	6.8	10.4	11.0	10.0	10.8	25.3	8.0	K	Official Capital Grants
-127.0	-357.1	-709.1	-461.4	-234.6	-59.8	-129.1	-104.8	91.2	-124.4	..	Current Account Balance
157.9	364.4	411.0	352.4	525.5	637.3	33.4	69.8	162.6	78.1	..	Long-Term Capital, net
128.8	215.5	289.5	48.6	-13.7	-5.6	-3.4	-7.9	-4.5	-4.9	..	Direct Investment
34.6	253.2	226.3	353.3	241.6	381.3	39.5	27.0	102.3	209.0	..	Long-Term Loans
417.2	321.5	356.2	456.0	533.2	502.0	189.4	220.0	207.9	361.8	..	Disbursements
382.6	68.3	129.9	102.7	291.6	120.7	149.9	193.0	105.6	152.8	..	Repayments
-11.2	-109.9	-111.5	-56.3	287.2	250.6	-12.7	39.9	39.5	-134.0	..	Other Long-Term Capital
107.4	66.5	416.3	142.9	-706.9	-647.0	11.9	95.7	33.7	130.0	..	Other Capital, net
-138.3	-73.8	-118.2	-33.9	416.0	69.5	83.8	-60.7	-287.5	-83.7	..	Change in Reserves
				(Uruguayan New Pesos per US dollar)							Memo Item:
6.060	7.860	9.100	10.820	13.910	34.540	56.120	101.430	151.990	226.670	..	Conversion Factor (Annual Avg)
				(Millions of US dollars, outstanding at end of year)							**EXTERNAL DEBT, ETC.**
794	932	1,127	1,348	1,700	2,510	2,528	2,695	2,886	3,048	..	Public/Publicly Guar. Long-Term
256	297	333	325	364	375	385	441	497	649	..	Official Creditors
74	74	72	70	85	92	115	152	211	301	..	IBRD and IDA
538	636	794	1,023	1,336	2,136	2,143	2,255	2,389	2,398	..	Private Creditors
68	185	211	326	206	158	129	60	43	144	..	Private Non-guaranteed Long-Term
0	0	0	0	96	237	222	350	395	392	..	Use of Fund Credit
137	206	322	500	645	386	392	814	584	651	..	Short-Term Debt
				(Millions of US dollars)							Memo Items:
352.41	323.13	383.79	429.99	116.29	206.92	134.42	174.23	481.55	530.00	..	Int'l Reserves Excluding Gold
822.64	1,694.72	2,017.27	1,348.32	1,305.82	992.66	807.13	856.41	1,018.29	1,263.02	..	Gold Holdings (at market price)
											SOCIAL INDICATORS
2.9	2.9	2.8	2.8	2.8	2.8	2.7	2.7	2.6	2.6	..	Total Fertility Rate
20.1	20.0	19.8	19.7	19.5	19.4	19.3	19.1	19.0	18.9	..	Crude Birth Rate
41.2	38.4	35.6	32.8	30.0	29.2	28.3	27.5	26.7	25.8	..	Infant Mortality Rate
69.8	69.9	70.1	70.2	70.3	70.5	70.6	70.8	70.9	71.0	..	Life Expectancy at Birth
91.0	90.2	95.5	114.3	110.8	112.9	102.2	102.9	97.3	101.2	..	Food Production, p.c. ('79-81 = 100)
16.3	16.0	15.7	..	..	..	..	..	..	..	..	Labor Force, Agriculture (%)
29.0	29.3	29.6	29.8	29.9	30.1	30.2	30.4	30.5	30.7	..	Labor Force, Female (%)
105.0	105.0	106.0	122.0	113.0	109.0	108.0	110.0	..	..	..	School Enroll. Ratio, primary
58.0	58.0	60.0	64.0	65.0	67.0	70.0	..	..	..	..	School Enroll. Ratio, secondary

VANUATU	1967	1968	1969	1970	1971	1972	1973	1974	1975	1976	1977
CURRENT GNP PER CAPITA (US $)	..	..	..	..	..	..	..	..	..	..	..
POPULATION (thousands)	78	80	82	84	86	90	95	98	102	105	108

ORIGIN AND USE OF RESOURCES

(Millions of current Vanuatu Vatu)

	1967	1968	1969	1970	1971	1972	1973	1974	1975	1976	1977
Gross National Product (GNP)	..	..	..	..	..	..	..	..	..	..	..
Net Factor Income from Abroad	..	..	..	..	..	..	..	..	..	..	..
Gross Domestic Product (GDP)	..	..	..	..	..	..	..	..	..	..	..
Indirect Taxes, net	..	..	..	..	..	..	..	..	..	..	..
GDP at factor cost	..	..	..	..	..	..	..	..	..	..	..
Agriculture	..	..	..	..	..	..	..	..	..	..	..
Industry	..	..	..	..	..	..	..	..	..	..	..
Manufacturing	..	..	..	..	..	..	..	..	..	..	..
Services, etc.	..	..	..	..	..	..	..	..	..	..	..
Resource Balance	..	..	..	..	..	..	..	..	..	..	..
Exports of Goods & NFServices	..	..	..	..	..	..	..	..	..	..	..
Imports of Goods & NFServices	..	..	..	..	..	..	..	..	..	..	..
Domestic Absorption	..	..	..	..	..	..	..	..	..	..	..
Private Consumption, etc.	..	..	..	..	..	..	..	..	..	..	..
General Gov't Consumption	..	..	..	..	..	..	..	..	..	..	..
Gross Domestic Investment	..	..	..	..	..	..	..	..	..	..	..
Fixed Investment	..	..	..	..	..	..	..	..	..	..	..
Memo Items:											
Gross Domestic Saving	..	..	..	..	..	..	..	..	..	..	..
Gross National Saving	..	..	..	..	..	..	..	..	..	..	..

(Millions of 1980 Vanuatu Vatu)

	1967	1968	1969	1970	1971	1972	1973	1974	1975	1976	1977
Gross National Product	..	..	..	..	..	..	..	..	..	..	..
GDP at factor cost	..	..	..	..	..	..	..	..	..	..	..
Agriculture	..	..	..	..	..	..	..	..	..	..	..
Industry	..	..	..	..	..	..	..	..	..	..	..
Manufacturing	..	..	..	..	..	..	..	..	..	..	..
Services, etc.	..	..	..	..	..	..	..	..	..	..	..
Resource Balance	..	..	..	..	..	..	..	..	..	..	..
Exports of Goods & NFServices	..	..	..	..	..	..	..	..	..	..	..
Imports of Goods & NFServices	..	..	..	..	..	..	..	..	..	..	..
Domestic Absorption	..	..	..	..	..	..	..	..	..	..	..
Private Consumption, etc.	..	..	..	..	..	..	..	..	..	..	..
General Gov't Consumption	..	..	..	..	..	..	..	..	..	..	..
Gross Domestic Investment	..	..	..	..	..	..	..	..	..	..	..
Fixed Investment	..	..	..	..	..	..	..	..	..	..	..
Memo Items:											
Capacity to Import	..	..	..	..	..	..	..	..	..	..	..
Terms of Trade Adjustment	..	..	..	..	..	..	..	..	..	..	..
Gross Domestic Income	..	..	..	..	..	..	..	..	..	..	..
Gross National Income	..	..	..	..	..	..	..	..	..	..	..

DOMESTIC PRICES (DEFLATORS)

(Index 1980 = 100)

	1967	1968	1969	1970	1971	1972	1973	1974	1975	1976	1977
Overall (GDP)	..	..	..	..	..	..	..	..	..	..	..
Domestic Absorption	..	..	..	..	..	..	..	..	..	..	..
Agriculture	..	..	..	..	..	..	..	..	..	..	..
Industry	..	..	..	..	..	..	..	..	..	..	..
Manufacturing	..	..	..	..	..	..	..	..	..	..	..

MANUFACTURING ACTIVITY

	1967	1968	1969	1970	1971	1972	1973	1974	1975	1976	1977
Employment (1980 = 100)	..	..	..	..	..	..	..	..	..	..	..
Real Earnings per Empl. (1980 = 100)	..	..	..	..	..	..	..	..	..	..	..
Real Output per Empl. (1980 = 100)	..	..	..	..	..	..	..	..	..	..	..
Earnings as % of Value Added	..	..	..	..	..	..	..	..	..	..	..

MONETARY HOLDINGS

(Millions of current Vanuatu Vatu)

	1967	1968	1969	1970	1971	1972	1973	1974	1975	1976	1977
Money Supply, Broadly Defined	..	..	..	..	..	..	..	..	..	5,392	5,816
Money as Means of Payment	..	..	..	..	..	..	..	..	..	1,629	1,770
Currency Ouside Banks	..	..	..	..	..	..	..	..	..	494	564
Demand Deposits	..	..	..	..	..	..	..	..	..	1,135	1,206
Quasi-Monetary Liabilities	..	..	..	..	..	..	..	..	..	3,763	4,046

GOVERNMENT DEFICIT (-) OR SURPLUS

(Millions of current Vanuatu Vatu)

	1967	1968	1969	1970	1971	1972	1973	1974	1975	1976	1977
Current Revenue	..	..	..	..	..	..	..	..	..	..	..
Current Expenditure	..	..	..	..	..	..	..	..	..	..	..
Current Budget Balance	..	..	..	..	..	..	..	..	..	..	..
Capital Receipts	..	..	..	..	..	..	..	..	..	..	..
Capital Payments	..	..	..	..	..	..	..	..	..	..	..

1978	1979	1980	1981	1982	1983	1984	1985	1986	1987 estimate	NOTES	VANUATU
..	..	..	..	..	..	..	960	..	..	..	**CURRENT GNP PER CAPITA** (US $)
111	114	117	120	122	125	128	131	135	139	..	**POPULATION** (thousands)
				(Millions of current Vanuatu Vatu)							**ORIGIN AND USE OF RESOURCES**
..	6,880	6,546	7,218	7,942	10,115	12,306	12,986	13,767	..	..	Gross National Product (GNP)
..	-1,320	-1,200	-1,454	-1,501	-75	96	176	496	..	..	Net Factor Income from Abroad
..	8,200	7,746	8,673	9,442	10,190	12,210	12,810	13,270	..	..	Gross Domestic Product (GDP)
..	..	..	..	..	1,551	2,156	2,419	2,360	..	..	Indirect Taxes, net
..	..	..	..	..	8,639	10,054	10,391	10,910	..	B	GDP at factor cost
..	1,800	1,464	1,852	1,886	3,470	4,360	4,335	4,491	..	..	Agriculture
..	1,240	1,255	1,343	1,464	671	865	932	965	..	..	Industry
..	320	327	367	425	312	430	491	509	..	..	Manufacturing
..	5,160	5,027	5,478	6,093	6,049	6,985	7,543	7,814	..	..	Services, etc.
..	..	-668	-205	-239	-1,726	-941	-3,155	..	..	..	Resource Balance
..	..	2,568	3,328	4,038	6,287	8,097	6,615	..	..	..	Exports of Goods & NFServices
..	..	3,236	3,533	4,277	8,013	9,038	9,770	..	..	..	Imports of Goods & NFServices
..	..	..	..	..	11,916	13,151	15,965	..	..	..	Domestic Absorption
..	..	..	..	..	5,130	5,480	7,568	..	..	..	Private Consumption, etc.
..	..	2,092	2,043	2,327	3,806	4,183	4,676	..	..	..	General Gov't Consumption
..	..	..	..	..	2,980	3,488	3,721	..	..	..	Gross Domestic Investment
..	..	..	..	..	..	..	..	..	..	..	Fixed Investment
											Memo Items:
..	..	..	..	..	1,254	2,547	566	..	..	..	Gross Domestic Saving
..	..	..	..	..	1,773	3,330	1,453	..	..	..	Gross National Saving
				(Millions of 1980 Vanuatu Vatu)							
..	..	..	..	..	..	..	..	..	..	..	Gross National Product
..	..	..	..	..	..	..	..	..	..	B	GDP at factor cost
..	1,787	1,464	1,789	1,579	2,859	2,950	2,842	2,949	..	..	Agriculture
..	1,345	1,255	1,190	1,499	676	851	866	916	..	..	Industry
..	..	..	..	..	..	..	..	..	..	..	Manufacturing
..	5,586	5,027	5,046	5,875	5,727	6,066	6,588	6,263	..	..	Services, etc.
..	..	..	..	..	..	..	..	..	..	..	Resource Balance
..	..	..	..	..	..	..	..	..	..	..	Exports of Goods & NFServices
..	..	..	..	..	..	..	..	..	..	..	Imports of Goods & NFServices
..	..	..	..	..	..	..	..	..	..	..	Domestic Absorption
..	..	..	..	..	..	..	..	..	..	..	Private Consumption, etc.
..	..	..	..	..	..	..	..	..	..	..	General Gov't Consumption
..	..	..	..	..	..	..	..	..	..	..	Gross Domestic Investment
..	..	..	..	..	..	..	..	..	..	..	Fixed Investment
											Memo Items:
..	..	..	..	..	..	..	..	..	..	..	Capacity to Import
..	..	..	..	..	..	..	..	..	..	..	Terms of Trade Adjustment
..	..	..	..	..	..	..	..	..	..	..	Gross Domestic Income
..	..	..	..	..	..	..	..	..	..	..	Gross National Income
				(Index 1980 = 100)							**DOMESTIC PRICES (DEFLATORS)**
..	94.1	100.0	108.1	105.5	110.0	123.7	124.4	131.0	..	..	Overall (GDP)
..	..	..	..	..	..	..	..	..	..	..	Domestic Absorption
..	100.7	100.0	103.5	119.4	121.4	147.8	152.5	152.3	..	..	Agriculture
..	92.2	100.0	112.9	97.6	99.3	101.6	107.7	105.4	..	..	Industry
..	..	..	..	..	..	..	..	..	..	..	Manufacturing
											MANUFACTURING ACTIVITY
..	..	..	..	..	..	..	..	..	..	..	Employment (1980=100)
..	..	..	..	..	..	..	..	..	..	..	Real Earnings per Empl. (1980=100)
..	..	..	..	..	..	..	..	..	..	..	Real Output per Empl. (1980=100)
..	..	..	..	..	..	..	..	..	..	..	Earnings as % of Value Added
				(Millions of current Vanuatu Vatu)							**MONETARY HOLDINGS**
5,869	6,043	5,261	5,322	7,935	8,626	10,952	12,391	15,638	14,605	..	Money Supply, Broadly Defined
1,888	2,094	1,882	1,448	1,743	2,272	3,026	2,642	2,810	4,219	..	Money as Means of Payment
699	846	720	597	634	748	922	963	906	1,000	..	Currency Ouside Banks
1,189	1,248	1,162	851	1,110	1,525	2,104	1,679	1,904	3,218	..	Demand Deposits
3,981	3,949	3,379	3,874	6,192	6,353	7,926	9,749	12,828	10,386	..	Quasi-Monetary Liabilities
				(Millions of current Vanuatu Vatu)							
..	..	..	..	..	..	..	..	..	..	..	**GOVERNMENT DEFICIT (-) OR SURPLUS**
..	..	..	..	..	..	..	..	..	..	..	Current Revenue
..	..	..	..	..	..	..	..	..	..	..	Current Expenditure
..	..	..	..	..	..	..	..	..	..	..	Current Budget Balance
..	..	..	..	..	..	..	..	..	..	..	Capital Receipts
..	..	..	..	..	..	..	..	..	..	..	Capital Payments

VANUATU	1967	1968	1969	1970	1971	1972	1973	1974	1975	1976	1977
FOREIGN TRADE (CUSTOMS BASIS)				*(Thousands of current US dollars)*							
Value of Exports, fob	..	..	..	..	..	..	..	..	..	..	..
Nonfuel Primary Products	..	..	..	..	..	..	..	..	..	..	..
Fuels	..	..	..	..	..	..	..	..	..	..	..
Manufactures	..	..	..	..	..	..	..	..	..	..	..
Value of Imports, cif	..	..	..	..	..	..	..	..	..	..	..
Nonfuel Primary Products	..	..	..	..	..	..	..	..	..	..	..
Fuels	..	..	..	..	..	..	..	..	..	..	..
Manufactures	..	..	..	..	..	..	..	..	..	..	..
Terms of Trade					*(Index 1980 = 100)*						
Export Prices, fob	..	..	..	..	..	..	..	..	..	..	..
Nonfuel Primary Products	..	..	..	..	..	..	..	..	..	..	..
Fuels	..	..	..	..	..	..	..	..	..	..	..
Manufactures	..	..	..	..	..	..	..	..	..	..	..
Import Prices, cif	..	..	..	..	..	..	..	..	..	..	..
Trade at Constant 1980 Prices				*(Thousands of 1980 US dollars)*							
Exports, fob	..	..	..	..	..	..	..	..	..	..	..
Imports, cif	..	..	..	..	..	..	..	..	..	..	..
BALANCE OF PAYMENTS				*(Millions of current US dollars)*							
Exports of Goods & Services	..	..	..	..	..	..	..	..	..	..	..
Merchandise, fob	..	..	..	..	..	..	..	..	..	..	..
Nonfactor Services	..	..	..	..	..	..	..	..	..	..	..
Factor Services	..	..	..	..	..	..	..	..	..	..	..
Imports of Goods & Services	..	..	..	..	..	..	..	..	..	..	..
Merchandise, fob	..	..	..	..	..	..	..	..	..	..	..
Nonfactor Services	..	..	..	..	..	..	..	..	..	..	..
Factor Services	..	..	..	..	..	..	..	..	..	..	..
Long-Term Interest	..	..	..	..	..	..	..	..	..	..	..
Current Transfers, net	..	..	..	..	..	..	..	..	..	..	..
Workers' Remittances	..	..	..	..	..	..	..	..	..	..	..
Total to be Financed	..	..	..	..	..	..	..	..	..	..	..
Official Capital Grants	..	..	..	..	..	..	..	..	..	..	..
Current Account Balance	..	..	..	..	..	..	..	..	..	..	..
Long-Term Capital, net	..	..	..	..	..	..	..	..	..	..	..
Direct Investment	..	..	..	..	..	..	..	..	..	..	..
Long-Term Loans	..	..	..	..	..	..	..	..	..	..	..
Disbursements	..	..	..	..	..	..	..	..	..	..	..
Repayments	..	..	..	..	..	..	..	..	..	..	..
Other Long-Term Capital	..	..	..	..	..	..	..	..	..	..	..
Other Capital, net	..	..	..	..	..	..	..	..	..	..	..
Change in Reserves	..	..	..	0	0	0	0	0	0	0	0
Memo Item:				*(Vanuatu Vatu per US dollar)*							
Conversion Factor (Annual Avg)	89.760	89.760	94.440	100.990	99.860	81.610	72.040	77.800	69.270	77.240	79.410
EXTERNAL DEBT, ETC.				*(Thousands of US dollars, outstanding at end of year)*							
Public/Publicly Guar. Long-Term	..	..	..	..	..	..	..	..	..	..	..
Official Creditors	..	..	..	..	..	..	..	..	..	..	..
IBRD and IDA	..	..	..	..	..	..	..	..	..	..	..
Private Creditors	..	..	..	..	..	..	..	..	..	..	..
Private Non-guaranteed Long-Term	..	..	..	..	..	..	..	..	..	..	..
Use of Fund Credit	..	..	..	0	0	0	0	0	0	0	0
Short-Term Debt	..	..	..	..	..	..	..	..	..	..	..
Memo Items:				*(Thousands of US dollars)*							
Int'l Reserves Excluding Gold	..	..	..	..	..	..	..	..	..	..	..
Gold Holdings (at market price)	..	..	..	..	..	..	..	..	..	..	..
SOCIAL INDICATORS											
Total Fertility Rate	..	..	..	..	..	..	..	..	..	..	..
Crude Birth Rate	..	..	..	..	..	..	..	..	..	..	..
Infant Mortality Rate	..	..	..	..	..	..	..	..	..	..	..
Life Expectancy at Birth	..	..	..	..	..	..	..	..	..	..	..
Food Production, p.c. ('79-81 = 100)	134.7	116.4	118.6	105.5	107.7	77.6	87.3	114.1	105.8	104.4	98.6
Labor Force, Agriculture (%)	..	..	..	..	..	..	..	..	..	..	..
Labor Force, Female (%)	..	..	..	..	..	..	..	..	..	..	..
School Enroll. Ratio, primary	..	..	..	..	..	..	..	..	..	..	..
School Enroll. Ratio, secondary	..	..	..	..	..	..	..	..	..	..	..

1978	1979	1980	1981	1982	1983	1984	1985	1986	1987 estimate	NOTES	VANUATU
				(Thousands of current US dollars)							**FOREIGN TRADE (CUSTOMS BASIS)**
..	..	16,100	19,300	10,700	18,000	31,600	17,100	15,300	17,400	..	Value of Exports, fob
..	..	..	..	..	..	..	..	..	..	..	Nonfuel Primary Products
..	..	..	..	..	..	..	..	..	..	..	Fuels
..	..	..	..	..	..	..	..	..	..	..	Manufactures
..	..	69,800	60,100	51,200	51,800	55,200	55,500	56,400	68,100	..	Value of Imports, cif
..	..	..	..	..	..	..	..	..	..	..	Nonfuel Primary Products
..	..	..	..	..	..	..	..	..	..	..	Fuels
..	..	..	..	..	..	..	..	..	..	..	Manufactures
				(Index 1980 = 100)							Terms of Trade
..	..	..	..	..	..	..	..		..	..	Export Prices, fob
..	..	..	..	..	..	..	..		..	..	Nonfuel Primary Products
..	..	..	..	..	..	..	..		..	..	Fuels
..	..	..	..	..	..	..	..		..	..	Manufactures
..	..	..	..	..	..	..	..		..	..	Import Prices, cif
				(Thousands of 1980 US dollars)							Trade at Constant 1980 Prices
..	..	..	..	..	..	..	..		..	..	Exports, fob
..	..	..	..	..	..	..	..		..	..	Imports, cif
				(Millions of current US dollars)							**BALANCE OF PAYMENTS**
..	..	..	..	53	63	84	83	80	..	..	Exports of Goods & Services
..	..	..	..	11	18	33	19	9	11	..	Merchandise, fob
..	..	..	..	37	38	42	38	30	..	..	Nonfactor Services
..	..	..	..	6	7	9	27	41	..	..	Factor Services
..	..	..	..	84	86	100	113	111	..	..	Imports of Goods & Services
..	..	..	..	43	46	52	52	47	67	..	Merchandise, fob
..	..	..	..	30	27	33	33	29	..	..	Nonfactor Services
..	..	..	..	11	13	15	27	36	..	..	Factor Services
..	..	0	0	0	0	8	6	7	7	..	Long-Term Interest
..	..	..	..	8	6	7	7	7	9	..	Current Transfers, net
..	..	..	..	8	6	7	7	7	..	..	Workers' Remittances
..	..	..	..	-23	-18	-9	-23	-24	-37	K	Total to be Financed
..	..	..	..	35	26	32	24	21	35	K	Official Capital Grants
..	..	..	..	12	8	23	1	-3	-2	..	Current Account Balance
..	..	..	..	-3	25	91	-21	35	39	..	Long-Term Capital, net
..	..	..	..	7	6	8	5	2	..	..	Direct Investment
..	..	0	0	2	0	59	19	9	13	..	Long-Term Loans
..	..	..	..	2	1	60	29	26	28	..	Disbursements
..	..	0	0	0	0	0	10	17	15	..	Repayments
..	..	..	..	-46	-7	-8	-68	3	-8	..	Other Long-Term Capital
..	..	..	..	-6	-32	-116	19	-27	-19	..	Other Capital, net
0	0	0	0	-3	-1	3	0	-5	-18	..	Change in Reserves
				(Vanuatu Vatu per US dollar)							**Memo Item:**
72.940	68.760	68.290	87.830	96.210	99.370	99.230	106.030	106.080	109.850	..	Conversion Factor (Annual Avg)
				(Thousands of US dollars, outstanding at end of year)							**EXTERNAL DEBT, ETC.**
..	..	4,100	3,000	4,100	3,600	5,200	6,900	9,000	14,600	..	Public/Publicly Guar. Long-Term
..	..	3,900	2,800	3,900	3,100	3,400	5,200	7,500	13,000	..	Official Creditors
..	..	0	0	0	0	0	200	500	1,500	..	IBRD and IDA
..	..	200	200	100	500	1,800	1,700	1,500	1,600	..	Private Creditors
..	..	..	..	2,000	18,000	75,100	93,000	101,000	110,000	..	Private Non-guaranteed Long-Term
0	0	0	0	0	0	0	0	0	..	..	Use of Fund Credit
..	..	..	..	..	7,000	8,000	9,000	10,000	10,000	..	Short-Term Debt
				(Thousands of US dollars)							**Memo Items:**
..	..	..	8,463.0	5,669.0	6,595.0	8,087.0	10,610.0	21,420.0	40,175.0	..	Int'l Reserves Excluding Gold
..	..	..	..	..	..	..	..	..	..	..	Gold Holdings (at market price)
											SOCIAL INDICATORS
..	..	..	..	..	..	..	..	..	..	..	Total Fertility Rate
..	..	..	..	..	..	..	..	..	..	..	Crude Birth Rate
..	..	..	..	..	..	..	..	..	..	..	Infant Mortality Rate
..	..	..	..	..	..	..	..	..	..	..	Life Expectancy at Birth
112.1	111.9	85.5	102.6	85.6	86.0	94.6	82.4	84.2	80.6	..	Food Production, p.c. ('79-81 = 100)
..	..	..	..	..	..	..	..	..	..	..	Labor Force, Agriculture (%)
..	..	..	..	..	..	..	..	..	..	..	Labor Force, Female (%)
..	..	..	..	..	..	..	..	..	..	..	School Enroll. Ratio, primary
..	..	..	..	..	..	..	..	..	..	..	School Enroll. Ratio, secondary

VENEZUELA	1967	1968	1969	1970	1971	1972	1973	1974	1975	1976	1977
CURRENT GNP PER CAPITA (US $)	1,060	1,140	1,160	1,240	1,270	1,340	1,550	1,920	2,380	2,900	3,160
POPULATION (thousands)	9,574	9,899	10,239	10,597	10,973	11,368	11,781	12,211	12,655	13,112	13,581

ORIGIN AND USE OF RESOURCES

(Billions of current Venezuelan Bolivares)

	1967	1968	1969	1970	1971	1972	1973	1974	1975	1976	1977
Gross National Product (GNP)	45.17	49.52	52.08	57.28	63.03	70.24	83.56	125.96	139.54	161.65	188.20
Net Factor Income from Abroad	-2.98	-3.20	-2.77	-2.49	-3.32	-2.11	-2.96	-2.73	0.43	0.89	0.38
Gross Domestic Product (GDP)	48.15	52.72	54.85	59.77	66.35	72.35	86.52	128.69	139.11	160.75	187.82
Indirect Taxes, net	2.00	2.33	2.48	2.84	2.73	2.65	2.72	3.32	4.57	4.70	5.21
GDP at factor cost	46.15	50.39	52.37	56.93	63.62	69.71	83.81	125.36	134.54	156.06	182.61
Agriculture	2.82	3.06	3.52	3.71	3.84	3.90	4.58	5.65	6.97	7.45	9.27
Industry	19.73	21.70	21.10	23.50	26.55	28.52	37.28	69.28	64.73	73.69	83.74
Manufacturing	..	8.23	8.41	9.63	10.88	11.57	13.91	23.35	21.81	26.32	28.69
Services, etc.	25.59	27.96	30.23	32.56	35.96	39.93	44.66	53.75	67.41	79.62	94.81
Resource Balance	4.20	2.69	2.21	2.43	3.65	2.09	7.22	28.26	9.62	1.19	-12.81
Exports of Goods & NFServices	11.74	11.82	11.70	12.50	14.85	14.79	21.74	49.77	40.07	41.54	43.64
Imports of Goods & NFServices	7.53	9.13	9.48	10.07	11.21	12.70	14.53	21.51	30.46	40.35	56.45
Domestic Absorption	43.94	50.04	52.63	57.34	62.70	70.27	79.31	100.42	129.50	159.56	200.63
Private Consumption, etc.	27.23	29.26	30.28	31.02	33.56	37.76	42.89	54.01	67.98	81.63	96.81
General Gov't Consumption	5.20	5.67	6.14	6.63	7.76	8.50	9.59	12.77	15.94	19.78	22.96
Gross Domestic Investment	11.52	15.11	16.21	19.68	21.38	24.01	26.83	33.64	45.57	58.15	80.86
Fixed Investment	9.76	11.85	13.25	13.35	15.41	18.34	21.55	24.29	35.41	49.50	70.00
Memo Items:											
Gross Domestic Saving	15.72	17.80	18.43	22.12	25.03	26.10	34.04	61.91	55.19	59.34	68.05
Gross National Saving	12.42	14.25	15.25	19.24	21.35	23.59	30.63	58.60	55.00	59.51	67.44

(Billions of 1980 Venezuelan Bolivares)

	1967	1968	1969	1970	1971	1972	1973	1974	1975	1976	1977
Gross National Product	183.96	197.74	201.31	218.87	220.13	227.84	242.77	252.38	264.69	285.73	302.73
GDP at factor cost	187.92	201.17	202.38	217.50	222.16	226.06	243.43	250.00	255.30	276.05	293.77
Agriculture	8.84	9.28	10.21	10.74	10.88	10.73	11.30	11.99	12.83	13.37	13.34
Industry	114.70	124.92	120.08	135.98	134.05	131.01	140.57	134.66	125.16	132.01	137.22
Manufacturing	23.18	24.82	25.54	28.84	29.83	31.17	33.46	35.51	37.11	41.58	43.03
Services, etc.	74.13	78.13	83.20	83.70	88.59	94.42	101.08	111.05	126.17	139.90	151.32
Resource Balance	..	..	..	..	..	..	..	95.58	56.72	38.35	11.86
Exports of Goods & NFServices	..	..	..	..	..	..	..	134.78	107.65	106.73	99.99
Imports of Goods & NFServices	..	..	..	..	..	..	..	39.20	50.92	68.39	88.13
Domestic Absorption	..	..	..	..	..	..	..	190.74	221.12	254.46	294.27
Private Consumption, etc.	..	..	..	..	..	..	..	100.97	115.43	128.74	141.54
General Gov't Consumption	..	..	..	..	..	..	..	26.69	27.50	31.68	33.51
Gross Domestic Investment	..	..	..	..	..	..	..	63.08	78.19	94.03	119.22
Fixed Investment	29.76	34.10	37.06	36.47	40.91	47.07	51.45	50.18	63.05	81.00	104.45
Memo Items:											
Capacity to Import	..	..	..	..	..	..	..	90.71	67.00	70.41	68.13
Terms of Trade Adjustment	..	..	..	..	..	..	..	-44.07	-40.65	-36.33	-31.86
Gross Domestic Income	..	..	..	..	..	..	..	212.56	223.33	248.03	270.30
Gross National Income	..	..	..	..	..	..	..	208.31	224.04	249.40	270.87

DOMESTIC PRICES (DEFLATORS)

(Index 1980 = 100)

	1967	1968	1969	1970	1971	1972	1973	1974	1975	1976	1977
Overall (GDP)	24.6	25.0	25.9	26.2	28.6	30.8	34.4	50.1	52.7	56.5	62.2
Domestic Absorption	..	..	..	..	..	..	..	52.6	58.6	62.7	68.2
Agriculture	31.9	33.0	34.5	34.6	35.3	36.4	40.5	47.1	54.3	60.2	69.5
Industry	17.2	17.4	17.6	17.3	19.8	21.8	26.5	51.4	51.7	55.8	61.0
Manufacturing	..	33.2	32.9	33.4	36.5	37.1	41.6	65.8	58.8	63.3	66.7

MANUFACTURING ACTIVITY

	1967	1968	1969	1970	1971	1972	1973	1974	1975	1976	1977
Employment (1980=100)	41.8	43.1	44.3	46.0	47.8	51.0	54.6	62.5	69.1	82.4	87.9
Real Earnings per Empl. (1980=100)	71.0	76.3	78.9	78.8	78.8	79.3	89.2	95.7	100.2	100.1	103.8
Real Output per Empl. (1980=100)	106.1	113.5	121.9	118.3	114.8	108.1	102.6	94.8	105.2	106.3	106.4
Earnings as % of Value Added	30.1	29.8	28.7	30.6	30.0	31.7	35.0	28.7	27.3	25.3	27.1

MONETARY HOLDINGS

(Billions of current Venezuelan Bolivares)

	1967	1968	1969	1970	1971	1972	1973	1974	1975	1976	1977
Money Supply, Broadly Defined	8.8	9.7	11.2	12.6	14.8	18.3	22.8	29.9	44.3	56.0	71.8
Money as Means of Payment	5.2	5.7	6.2	6.7	7.9	9.5	11.3	16.0	23.3	27.1	34.0
Currency Ouside Banks	1.8	1.9	2.0	2.2	2.3	2.6	2.9	3.8	4.7	5.8	7.4
Demand Deposits	3.5	3.8	4.2	4.6	5.5	6.9	8.5	12.2	18.6	21.3	26.6
Quasi-Monetary Liabilities	3.6	4.0	5.0	5.8	6.9	8.8	11.5	13.9	21.0	28.9	37.7

(Millions of current Venezuelan Bolivares)

	1967	1968	1969	1970	1971	1972	1973	1974	1975	1976	1977
GOVERNMENT DEFICIT (-) OR SURPLUS	..	..	..	-697	233	-165	1,159	4,981	1,916	-3,987	-6,742
Current Revenue	..	..	..	10,360	12,487	13,238	17,007	45,071	42,463	39,985	42,963
Current Expenditure	..	..	..	7,660	8,649	9,634	10,493	15,631	19,705	20,394	24,781
Current Budget Balance	..	..	..	2,700	3,838	3,604	6,514	29,440	22,758	19,591	18,182
Capital Receipts	..	..	..	1	2	290	65	2	18	..	30
Capital Payments	..	..	..	3,398	3,606	4,059	5,419	24,461	20,860	23,578	24,954

1978	1979	1980	1981	1982	1983	1984	1985	1986	1987 estimate	NOTES	VENEZUELA
3,390	3,730	4,070	4,730	4,920	4,790	4,140	3,800	3,550	3,230	..	CURRENT GNP PER CAPITA (US $)
14,058	14,540	15,024	15,485	15,940	16,394	16,851	17,317	17,791	18,265	..	POPULATION (thousands)
											ORIGIN AND USE OF RESOURCES
				(Billions of current Venezuelan Bolivares)							
206.36	247.39	299.21	337.40	333.68	340.00	395.55	448.60	479.20	699.90	..	Gross National Product (GNP)
0.16	-0.01	1.41	2.46	-6.57	-9.08	-13.94	-16.00	-14.60	-19.50	..	Net Factor Income from Abroad
206.20	247.40	297.80	334.93	340.25	349.08	409.49	464.60	493.80	719.40	..	Gross Domestic Product (GDP)
4.54	6.65	8.00	8.42	9.78	20.39	23.90	32.10	17.90	47.30	..	Indirect Taxes, net
201.66	240.75	289.81	326.52	330.47	328.69	385.59	432.50	475.90	672.10	B	GDP at factor cost
10.14	11.94	14.44	16.41	17.68	19.54	21.51	26.90	32.50	42.60	..	Agriculture
88.63	114.37	138.12	148.69	141.35	134.96	171.11	183.40	183.50	273.10	..	Industry
30.71	40.20	47.66	49.85	54.13	57.03	78.36	96.30	113.70	156.30	..	Manufacturing
107.43	121.09	145.24	169.83	181.22	194.58	216.87	254.30	277.80	403.70	..	Services, etc.
-23.04	3.26	20.77	16.46	-8.92	29.02	46.32	44.40	0.30	5.70	..	Resource Balance
42.08	64.22	85.72	89.89	75.37	68.08	116.07	118.70	102.20	160.20	..	Exports of Goods & NFServices
65.12	60.96	64.95	73.42	84.29	39.07	69.75	74.30	101.90	154.50	..	Imports of Goods & NFServices
229.24	244.14	277.03	318.47	349.16	320.06	363.17	420.20	493.50	713.70	..	Domestic Absorption
114.64	134.29	163.40	194.04	212.37	236.03	255.35	291.70	338.90	466.80	..	Private Consumption, etc.
24.06	27.76	35.12	42.64	42.59	41.34	43.59	48.60	54.80	72.30	..	General Gov't Consumption
90.55	82.09	78.51	81.79	94.19	42.70	64.23	79.90	99.80	174.60	..	Gross Domestic Investment
83.14	75.87	74.24	80.77	81.21	64.06	57.37	70.50	93.70	135.60	..	Fixed Investment
											Memo Items:
67.51	85.35	99.28	98.25	85.28	71.72	110.55	124.30	100.10	180.30	..	Gross Domestic Saving
66.08	83.67	98.90	99.07	76.07	61.83	95.77	107.53	84.92	158.94	..	Gross National Saving
				(Billions of 1980 Venezuelan Bolivares)							
309.58	311.80	299.21	299.02	285.36	272.94	267.22	270.55	290.52	300.06	..	Gross National Product
302.50	303.38	289.81	289.38	282.31	263.42	260.44	260.78	288.43	288.09	B H	GDP at factor cost
13.77	14.17	14.44	14.17	14.67	14.73	14.85	16.09	17.42	18.13	..	Agriculture
141.98	145.71	138.12	135.34	130.11	123.22	119.89	120.49	129.46	132.73	..	Industry
45.01	46.49	47.66	46.49	48.37	47.55	49.75	52.20	55.85	57.27	..	Manufacturing
153.40	151.91	145.24	147.16	145.59	141.42	141.33	143.04	151.91	156.97	..	Services, etc.
10.17	37.08	20.77	8.52	-11.02	31.42	20.41	27.47	20.40	35.34	..	Resource Balance
95.05	102.75	85.72	80.93	73.51	71.29	82.65	83.51	87.74	94.44	..	Exports of Goods & NFServices
84.88	65.67	64.95	72.41	84.53	39.88	62.23	56.04	67.33	59.10	..	Imports of Goods & NFServices
306.66	290.62	277.03	282.75	287.16	251.39	252.79	251.20	275.02	273.39	..	Domestic Absorption
156.62	163.17	163.40	167.28	166.84	174.46	168.22	163.99	182.55	167.92	..	Private Consumption, etc.
32.34	33.91	35.12	37.06	36.70	35.70	36.02	35.70	36.31	37.95	..	General Gov't Consumption
117.70	93.54	78.51	78.41	83.62	41.24	48.54	51.51	56.15	67.52	..	Gross Domestic Investment
108.61	86.84	74.24	76.47	73.62	54.22	44.00	46.33	53.46	53.08	..	Fixed Investment
											Memo Items:
54.85	69.18	85.72	88.64	75.59	69.49	103.56	89.52	67.53	61.28	..	Capacity to Import
-40.20	-33.57	0.00	7.71	2.08	-1.80	20.91	6.02	-20.21	-33.16	..	Terms of Trade Adjustment
269.12	278.19	297.80	304.55	292.75	277.96	297.50	286.15	279.08	275.21	..	Gross Domestic Income
269.38	278.24	299.21	306.74	287.44	271.14	288.13	276.56	270.31	266.90	..	Gross National Income
											DOMESTIC PRICES (DEFLATORS)
				(Index 1980 = 100)							
66.7	79.4	100.0	112.8	117.1	124.8	148.1	165.8	165.0	233.3	..	Overall (GDP)
74.8	84.0	100.0	112.6	121.6	127.3	143.7	167.3	179.4	261.1	..	Domestic Absorption
73.6	84.3	100.0	115.9	120.5	132.6	144.8	167.2	186.6	234.9	..	Agriculture
62.4	78.5	100.0	109.9	108.6	109.5	142.7	152.2	141.7	205.8	..	Industry
68.2	86.5	100.0	107.2	111.9	120.0	157.5	184.5	203.6	272.9	..	Manufacturing
											MANUFACTURING ACTIVITY
90.8	98.8	100.0	89.9	91.2	89.7	88.4	89.3	..	..	G	Employment (1980=100)
111.8	114.3	100.0	121.4	118.9	119.2	109.4	109.9	..	..	G	Real Earnings per Empl. (1980=100)
110.4	102.1	100.0	113.8	116.6	115.5	110.6	111.9	..	..	G	Real Output per Empl. (1980=100)
30.5	28.5	26.7	32.2	32.2	31.6	25.8	25.5	..	..	..	Earnings as % of Value Added
											MONETARY HOLDINGS
				(Billions of current Venezuelan Bolivares)							
85.0	95.8	120.5	149.1	171.1	208.2	237.3	255.2	291.3	367.4	D	Money Supply, Broadly Defined
39.0	42.5	50.2	55.0	58.0	70.1	86.7	93.7	98.5	133.8	..	Money as Means of Payment
9.0	10.0	12.3	13.5	13.1	14.7	15.1	16.2	18.7	24.8	..	Currency Ouside Banks
30.0	32.5	37.9	41.4	44.9	55.3	71.6	77.6	79.8	109.0	..	Demand Deposits
46.0	53.3	70.3	94.2	113.1	138.2	150.6	161.5	192.8	233.6	..	Quasi-Monetary Liabilities
				(Millions of current Venezuelan Bolivares)							
-6,892	3,967	113	-3,898	-12,670	-5,102	13,744	10,346	..	..	F	GOVERNMENT DEFICIT (-) OR SURPLUS
42,905	51,113	66,805	97,738	84,019	77,108	103,993	112,595	..	..	..	Current Revenue
27,998	33,582	44,453	60,246	59,011	58,936	71,256	78,843	..	..	..	Current Expenditure
14,907	17,531	22,352	37,492	25,008	18,172	32,737	33,752	..	..	..	Current Budget Balance
..	..	1	89	1	86	9	103	..	..	..	Capital Receipts
21,799	13,564	22,240	41,479	37,679	23,360	19,002	23,509	..	..	..	Capital Payments

VENEZUELA	1967	1968	1969	1970	1971	1972	1973	1974	1975	1976	1977
FOREIGN TRADE (CUSTOMS BASIS)					*(Millions of current US dollars)*						
Value of Exports, fob	3,107.9	3,078.1	3,112.5	3,196.6	3,109.9	3,126.0	4,772.9	11,258.4	9,009.6	9,466.0	9,626.6
Nonfuel Primary Products	177.5	164.6	192.8	239.2	218.0	217.6	263.0	414.7	389.0	440.8	352.2
Fuels	2,891.9	2,876.8	2,874.7	2,909.5	2,846.2	2,847.6	4,432.6	10,700.5	8,508.8	8,860.5	9,132.2
Manufactures	38.5	36.7	45.0	48.0	45.6	60.7	77.2	143.1	111.8	164.7	142.2
Value of Imports, cif	1,445.0	1,665.5	1,718.6	1,869.0	2,078.7	2,430.2	2,806.2	4,186.4	6,003.7	7,662.9	10,938.0
Nonfuel Primary Products	261.7	283.4	275.5	295.8	297.2	340.1	514.8	771.3	981.0	1,221.2	1,791.1
Fuels	9.4	14.0	13.2	26.5	18.8	19.0	21.8	21.9	41.6	38.8	73.4
Manufactures	1,173.9	1,368.1	1,429.8	1,546.7	1,762.7	2,071.1	2,269.6	3,393.2	4,981.1	6,402.9	9,073.5
Terms of Trade	14.8	15.1	14.5	14.6	*(Index 1980 = 100)* 17.3	17.6	20.0	63.6	57.6	61.3	61.8
Export Prices, fob	4.5	4.5	4.6	4.6	6.0	6.7	9.4	37.5	36.6	39.4	42.9
Nonfuel Primary Products	36.2	35.5	38.1	40.5	35.4	35.3	49.3	60.6	59.7	66.4	76.1
Fuels	4.3	4.3	4.3	4.3	5.6	6.2	8.9	36.7	35.7	38.4	42.0
Manufactures	36.4	38.6	33.3	46.2	46.2	47.7	47.7	69.7	75.8	60.6	65.2
Import Prices, cif	30.6	29.9	31.5	31.7	34.8	38.3	47.1	59.0	63.6	64.2	69.4
Trade at Constant 1980 Prices					*(Millions of 1980 US dollars)*						
Exports, fob	68,481.7	68,089.7	68,122.7	68,991.7	51,814.5	46,452.0	50,780.9	30,030.3	24,606.6	24,034.0	22,439.8
Imports, cif	4,725.2	5,576.1	5,458.0	5,893.4	5,976.5	6,351.2	5,958.6	7,097.9	9,436.8	11,935.5	15,763.2
BALANCE OF PAYMENTS					*(Millions of current US dollars)*						
Exports of Goods & Services	..	..	..	2,833	3,339	3,418	5,279	11,971	10,092	10,376	10,947
Merchandise, fob	..	..	..	2,602	3,103	3,152	4,721	11,085	8,853	9,253	9,556
Nonfactor Services	..	..	..	177	197	209	329	530	499	430	609
Factor Services	..	..	..	54	39	57	229	356	740	693	782
Imports of Goods & Services	..	..	..	2,845	3,267	3,424	4,291	6,012	7,748	9,890	13,843
Merchandise, fob	..	..	..	1,713	1,896	2,222	2,626	3,876	5,462	7,337	10,194
Nonfactor Services	..	..	..	525	594	665	748	1,143	1,646	2,068	2,955
Factor Services	..	..	..	607	777	537	917	993	640	485	694
Long-Term Interest	..	..	..	53	62	81	125	144	120	139	266
Current Transfers, net	..	..	..	-86	-79	-89	-105	-135	-143	-169	-231
Workers' Remittances	..	..	..	0	0	0	0	0	0	0	0
Total to be Financed	..	..	..	-98	-7	-95	883	5,824	2,201	317	-3,127
Official Capital Grants	..	..	..	-6	-4	-6	-6	-64	-30	-63	-52
Current Account Balance	..	..	..	-104	-11	-101	877	5,760	2,171	254	-3,179
Long-Term Capital, net	..	..	..	85	370	-269	-59	-874	366	1,403	2,057
Direct Investment	..	..	..	-23	211	-376	-84	-430	418	-889	-3
Long-Term Loans	..	..	..	225	322	405	21	-262	-244	886	1,950
Disbursements	..	..	..	292	432	591	285	180	230	1,201	2,600
Repayments	..	..	..	67	110	185	265	442	474	315	650
Other Long-Term Capital	..	..	..	-111	-159	-292	11	-118	222	1,469	162
Other Capital, net	..	..	..	65	59	540	-207	-418	178	686	1,921
Change in Reserves	..	..	..	-46	-418	-170	-611	-4,468	-2,715	-2,343	-799
Memo Item:					*(Venezuelan Bolivares per US dollar)*						
Conversion Factor (Annual Avg)	4.500	4.500	4.500	4.500	4.500	4.400	4.300	4.280	4.280	4.290	4.290
EXTERNAL DEBT, ETC.					*(Millions of US dollars, outstanding at end of year)*						
Public/Publicly Guar. Long-Term	..	..	..	728	1,021	1,419	1,541	1,493	1,261	2,961	4,426
Official Creditors	..	..	..	365	363	409	414	416	553	536	499
IBRD and IDA	..	..	..	217	216	221	222	226	221	213	196
Private Creditors	..	..	..	362	659	1,010	1,127	1,077	709	2,425	3,928
Private Non-guaranteed Long-Term	..	..	..	236	283	293	350	291	233	350	875
Use of Fund Credit	..	..	..	0	0	0	0	0	0	0	0
Short-Term Debt	..	..	..	..	..	..	..	..	..	..	5,426
Memo Items:					*(Millions of US dollars)*						
Int'l Reserves Excluding Gold	470.5	518.7	529.9	636.6	1,096.8	1,306.8	1,940.1	6,034.2	8,402.6	8,123.8	7,735.2
Gold Holdings (at market price)	403.4	482.3	405.2	409.9	487.3	724.9	1,253.8	2,085.1	1,568.0	1,506.5	1,867.2
SOCIAL INDICATORS											
Total Fertility Rate	5.9	5.7	5.5	5.3	5.1	5.0	4.9	4.8	4.7	4.6	4.4
Crude Birth Rate	40.6	39.7	38.8	37.9	37.0	36.1	35.8	35.4	35.1	34.7	34.4
Infant Mortality Rate	60.0	57.8	55.6	53.4	51.2	49.0	47.8	46.6	45.4	44.2	43.0
Life Expectancy at Birth	63.7	64.2	64.7	65.2	65.7	66.2	66.5	66.8	67.1	67.4	67.7
Food Production, p.c. ('79-81 = 100)	108.3	108.7	112.8	111.9	108.5	104.2	103.2	100.3	109.4	99.5	98.9
Labor Force, Agriculture (%)	28.2	27.4	26.7	26.0	25.0	24.0	22.9	21.9	20.9	19.9	18.9
Labor Force, Female (%)	20.0	20.2	20.5	20.7	21.3	21.8	22.3	22.8	23.3	23.8	24.3
School Enroll. Ratio, primary	..	..	..	94.0	..	..	..	..	97.0	103.0	102.0
School Enroll. Ratio, secondary	..	..	..	33.0	..	..	..	..	43.0	38.0	38.0

1978	1979	1980	1981	1982	1983	1984	1985	1986	1987 estimate	NOTES	VENEZUELA
(Millions of current US dollars)											**FOREIGN TRADE (CUSTOMS BASIS)**
9,270.3	14,317.7	19,292.8	20,125.0	16,499.0	14,571.0	15,841.0	14,660.0	9,122.0	10,567.0	f	Value of Exports, fob
369.2	446.9	715.2	813.8	589.7	770.5	741.3	942.4	1,002.0	1,037.1	..	Nonfuel Primary Products
8,757.5	13,633.0	18,248.0	18,917.2	15,659.0	13,439.1	14,496.8	13,105.1	7,345.3	8,656.9	..	Fuels
143.7	237.8	329.6	394.0	250.3	361.5	602.9	612.5	774.7	873.0	..	Manufactures
11,766.1	10,670.0	11,826.6	13,105.9	12,944.0	8,709.7	7,594.4	8,233.8	9,565.2	8,724.8	f	Value of Imports, cif
1,828.9	1,822.4	2,280.2	2,769.0	2,634.5	1,776.8	1,455.8	1,321.7	1,753.1	1,556.4	..	Nonfuel Primary Products
69.2	122.0	192.7	103.9	128.8	88.7	79.2	68.9	34.5	39.8	..	Fuels
9,868.0	8,725.6	9,353.7	10,233.1	10,180.7	6,844.2	6,059.5	6,843.2	7,777.6	7,128.6	..	Manufactures
(Index 1980 = 100)											
54.7	68.1	100.0	110.9	104.1	95.6	94.5	93.4	47.7	52.0	..	Terms of Trade
43.3	61.9	100.0	110.7	100.4	91.9	89.6	85.9	49.3	58.1	..	Export Prices, fob
73.2	91.6	100.0	84.7	78.8	88.0	83.5	74.8	78.9	87.9	..	Nonfuel Primary Products
42.3	61.0	100.0	112.5	101.6	92.5	90.2	87.5	44.9	53.8	..	Fuels
67.4	86.4	100.0	101.0	89.0	81.0	85.0	74.0	87.5	96.1	..	Manufactures
79.2	90.9	100.0	99.9	96.4	96.1	94.9	92.0	103.4	111.7	..	Import Prices, cif
(Millions of 1980 US dollars)											Trade at Constant 1980 Prices
21,420.5	23,119.5	19,292.9	18,171.8	16,435.8	15,856.7	17,676.6	17,057.4	18,496.1	18,179.3	..	Exports, fob
14,858.2	11,737.5	11,826.5	13,122.6	13,422.5	9,059.8	8,006.6	8,946.5	9,252.8	7,811.8	..	Imports, cif
(Millions of current US dollars)											**BALANCE OF PAYMENTS**
10,855	16,305	22,232	24,519	20,122	17,341	19,045	17,784	12,044	12,690	..	Exports of Goods & Services
9,084	14,159	19,051	19,963	16,332	14,571	15,841	14,660	9,122	10,567	..	Merchandise, fob
719	800	917	975	1,225	1,270	1,069	1,203	1,201	712	..	Nonfactor Services
1,052	1,346	2,264	3,581	2,565	1,500	2,135	1,921	1,721	1,411	..	Factor Services
16,183	15,548	17,065	20,110	23,729	12,703	14,328	13,988	13,504	13,665	..	Imports of Goods & Services
11,234	10,004	10,877	12,123	13,584	6,409	7,260	7,530	7,862	8,832	..	Merchandise, fob
3,935	4,195	4,253	4,980	6,050	2,681	2,902	2,400	2,424	2,048	..	Nonfactor Services
1,014	1,349	1,935	3,007	4,095	3,613	4,166	4,058	3,218	2,785	..	Factor Services
498	821	1,471	1,559	2,010	2,138	1,943	1,769	2,238	2,518	..	Long-Term Interest
-371	-388	-418	-383	-615	-187	-119	-102	-72	-128	..	Current Transfers, net
0	0	0	0	0	0	0	0	0	0	..	Workers' Remittances
-5,699	369	4,749	4,026	-4,222	4,451	4,598	3,694	-1,532	-1,103	K	Total to be Financed
-36	-19	-21	-26	-24	-24	-29	-26	34	-22	K	Official Capital Grants
-5,735	350	4,728	4,000	-4,246	4,427	4,569	3,668	-1,498	-1,125	..	Current Account Balance
3,680	1,412	2,039	784	3,027	260	-2,373	-1,643	-2,453	-2,564	..	Long-Term Capital, net
67	88	55	184	253	86	18	68	16	21	..	Direct Investment
2,173	3,612	1,787	1,362	311	856	-784	-593	-1,800	-1,274	..	Long-Term Loans
2,651	5,898	4,757	3,944	2,732	2,508	1,099	402	263	315	..	Disbursements
479	2,286	2,970	2,582	2,421	1,653	1,883	996	2,063	1,589	..	Repayments
1,476	-2,269	218	-736	2,487	-657	-1,578	-1,092	-703	-1,289	..	Other Long-Term Capital
990	2,336	-3,004	-4,805	-6,941	-4,355	-315	-298	66	2,811	..	Other Capital, net
1,065	-4,098	-3,763	21	8,160	-332	-1,881	-1,727	3,885	878	..	Change in Reserves
(Venezuelan Bolivares per US dollar)											Memo Item:
4.290	4.290	4.290	4.290	4.290	4.300	7.020	7.500	8.080	14.500	..	Conversion Factor (Annual Avg)
(Millions of US dollars, outstanding at end of year)											**EXTERNAL DEBT, ETC.**
6,701	9,603	10,774	11,545	12,342	13,962	18,354	17,098	25,200	25,244	..	Public/Publicly Guar. Long-Term
497	459	509	458	412	408	424	670	851	1,128	..	Official Creditors
179	155	133	106	88	69	54	42	22	11	..	IBRD and IDA
6,205	9,144	10,265	11,087	11,930	13,554	17,930	16,427	24,349	24,116	..	Private Creditors
1,867	2,525	3,181	3,600	5,000	8,959	8,703	8,645	7,934	7,504	..	Private Non-guaranteed Long-Term
0	0	0	0	0	0	0	0	0	..	..	Use of Fund Credit
8,000	11,768	15,535	16,975	14,703	14,510	9,400	8,950	1,575	3,770	..	Short-Term Debt
(Millions of US dollars)											Memo Items:
6,034.6	7,320.3	6,603.9	8,164.4	6,578.5	7,642.8	8,901.1	10,251.0	6,437.4	5,962.7	..	Int'l Reserves Excluding Gold
2,574.1	5,867.5	6,755.7	4,555.4	5,236.1	4,372.0	3,533.1	3,747.4	4,479.7	5,547.8	..	Gold Holdings (at market price)
											SOCIAL INDICATORS
4.4	4.3	4.2	4.2	4.1	4.0	4.0	3.9	3.8	3.8	..	Total Fertility Rate
34.1	33.8	33.6	33.3	33.0	32.5	32.1	31.6	31.1	30.6	..	Crude Birth Rate
42.2	41.4	40.6	39.8	39.0	37.0	35.1	33.1	31.2	29.2	..	Infant Mortality Rate
68.0	68.2	68.5	68.7	69.0	69.2	69.4	69.6	69.8	70.1	..	Life Expectancy at Birth
102.8	103.6	99.7	96.8	92.7	98.4	92.7	89.0	98.0	92.9	..	Food Production, p.c. ('79-81 = 100)
18.0	17.0	16.0	..	13.6	13.6	14.4	14.4	..	..	..	Labor Force, Agriculture (%)
24.8	25.3	25.8	26.0	26.6	26.7	26.8	27.1	26.9	27.1	..	Labor Force, Female (%)
103.0	104.0	105.0	105.0	111.0	108.0	109.0	108.0	..	..	..	School Enroll. Ratio, primary
38.0	39.0	39.0	40.0	43.0	43.0	45.0	45.0	..	..	..	School Enroll. Ratio, secondary

WESTERN SAMOA	1967	1968	1969	1970	1971	1972	1973	1974	1975	1976	1977
CURRENT GNP PER CAPITA (US $)	..	..	..	..	..	..	..	..	..	..	..
POPULATION (thousands)	134	137	139	142	144	146	148	150	151	152	153

ORIGIN AND USE OF RESOURCES — *(Millions of current Western Samoa Tala)*

	1967	1968	1969	1970	1971	1972	1973	1974	1975	1976	1977
Gross National Product (GNP)	..	..	..	..	..	..	..	..	..	..	..
Net Factor Income from Abroad	..	..	..	0.0	..	..	..	..	..	..	..
Gross Domestic Product (GDP)	..	..	..	..	..	..	..	..	..	..	..
Indirect Taxes, net	..	..	..	..	..	..	..	..	..	..	..
GDP at factor cost	..	..	..	..	..	..	..	..	..	..	..
Agriculture	..	..	..	..	..	..	..	..	..	..	..
Industry	..	..	..	..	..	..	..	..	..	..	..
Manufacturing	..	..	..	..	..	..	..	..	..	..	..
Services, etc.	..	..	..	..	..	..	..	..	..	..	..
Resource Balance	..	..	..	..	..	..	..	..	..	..	..
Exports of Goods & NFServices	..	..	..	..	..	..	..	..	..	..	..
Imports of Goods & NFServices	..	..	..	..	..	..	..	..	..	..	..
Domestic Absorption	..	..	..	..	..	..	..	..	..	..	..
Private Consumption, etc.	..	..	..	..	..	..	..	..	..	..	..
General Gov't Consumption	..	..	..	..	..	..	..	..	..	..	..
Gross Domestic Investment	..	..	..	..	..	..	..	..	..	..	..
Fixed Investment	..	..	..	..	..	..	..	..	..	..	..
Memo Items:											
Gross Domestic Saving	..	..	..	..	..	..	..	..	..	..	..
Gross National Saving	..	..	..	..	..	..	..	..	..	..	..

(Millions of 1980 Western Samoa Tala)

	1967	1968	1969	1970	1971	1972	1973	1974	1975	1976	1977
Gross National Product	..	..	..	..	..	..	..	..	..	..	..
GDP at factor cost	..	..	..	..	..	..	..	..	..	..	..
Agriculture	..	..	..	..	..	..	..	..	..	..	..
Industry	..	..	..	..	..	..	..	..	..	..	..
Manufacturing	..	..	..	..	..	..	..	..	..	..	..
Services, etc.	..	..	..	..	..	..	..	..	..	..	..
Resource Balance	..	..	..	..	..	..	..	..	..	..	..
Exports of Goods & NFServices	..	..	..	..	..	..	..	..	..	..	..
Imports of Goods & NFServices	..	..	..	..	..	..	..	..	..	..	..
Domestic Absorption	..	..	..	..	..	..	..	..	..	..	..
Private Consumption, etc.	..	..	..	..	..	..	..	..	..	..	..
General Gov't Consumption	..	..	..	..	..	..	..	..	..	..	..
Gross Domestic Investment	..	..	..	..	..	..	..	..	..	..	..
Fixed Investment	..	..	..	..	..	..	..	..	..	..	..
Memo Items:											
Capacity to Import	..	..	..	..	..	..	..	..	..	..	..
Terms of Trade Adjustment	..	..	..	..	..	..	..	..	..	..	..
Gross Domestic Income	..	..	..	..	..	..	..	..	..	..	..
Gross National Income	..	..	..	..	..	..	..	..	..	..	..

DOMESTIC PRICES (DEFLATORS) — *(Index 1980 = 100)*

	1967	1968	1969	1970	1971	1972	1973	1974	1975	1976	1977
Overall (GDP)	..	..	..	..	..	..	..	..	..	..	..
Domestic Absorption	..	..	..	..	..	..	..	..	..	..	..
Agriculture	..	..	..	..	..	..	..	..	..	..	..
Industry	..	..	..	..	..	..	..	..	..	..	..
Manufacturing	..	..	..	..	..	..	..	..	..	..	..

MANUFACTURING ACTIVITY

	1967	1968	1969	1970	1971	1972	1973	1974	1975	1976	1977
Employment (1980=100)	..	..	..	..	..	..	..	..	..	..	..
Real Earnings per Empl. (1980=100)	..	..	..	..	..	..	..	..	..	..	..
Real Output per Empl. (1980=100)	..	..	..	..	..	..	..	..	..	..	..
Earnings as % of Value Added	..	32.8	..	..	..	..	..	..	..	..	..

MONETARY HOLDINGS — *(Millions of current Western Samoa Tala)*

	1967	1968	1969	1970	1971	1972	1973	1974	1975	1976	1977
Money Supply, Broadly Defined	2.45	2.68	2.95	3.24	4.05	4.54	5.58	6.64	7.11	8.52	10.08
Money as Means of Payment	1.09	1.12	0.90	1.28	1.94	2.31	2.76	3.21	3.48	4.04	4.71
Currency Ouside Banks	0.04	0.06	0.07	0.09	0.10	0.12	0.14	0.17	0.18	0.22	0.23
Demand Deposits	1.05	1.06	0.83	1.19	1.83	2.19	2.62	3.04	3.29	3.83	4.48
Quasi-Monetary Liabilities	1.36	1.56	2.05	1.96	2.11	2.22	2.82	3.43	3.64	4.48	5.38

GOVERNMENT DEFICIT (-) OR SURPLUS — *(Thousands of current Western Samoa Tala)*

	1967	1968	1969	1970	1971	1972	1973	1974	1975	1976	1977
Current Revenue	..	..	..	..	..	..	..	..	..	..	..
Current Expenditure	..	..	..	..	..	..	..	..	..	..	..
Current Budget Balance	..	..	..	..	..	..	..	..	..	..	..
Capital Receipts	..	..	..	..	..	..	..	..	..	..	..
Capital Payments	..	..	..	..	..	..	..	..	..	..	..

1978	1979	1980	1981	1982	1983	1984	1985	1986	1987 estimate	NOTES	WESTERN SAMOA
..	..	..	..	..	..	680	620	560	560	..	**CURRENT GNP PER CAPITA (US $)**
154	154	155	156	157	158	160	163	165	166	..	**POPULATION (thousands)**

(Millions of current Western Samoa Tala) — **ORIGIN AND USE OF RESOURCES**

1978	1979	1980	1981	1982	1983	1984	1985	1986	1987	NOTES	
..	..	..	..	135.5	161.3	177.2	186.0	197.4	210.6	..	Gross National Product (GNP)
..	..	..	..	5.1	6.9	-4.0	-5.2	-3.7	-0.6	..	Net Factor Income from Abroad
71.0	90.0	103.0	108.9	130.4	154.4	181.2	191.2	201.1	211.2	f	Gross Domestic Product (GDP)
..	12.0	13.7	13.0	14.0	14.0	..	..	..	..	..	Indirect Taxes, net
..	78.0	89.3	95.9	116.4	140.4	..	..	..	..	f	GDP at factor cost
..	38.0	41.1	44.9	59.5	..	..	..	..	..	..	Agriculture
..	7.3	10.9	11.4	14.3	..	..	..	..	..	..	Industry
..	3.4	4.9	5.2	6.9	..	..	..	..	..	..	Manufacturing
..	32.7	37.3	39.6	42.6	..	..	..	..	..	..	Services, etc.
..	-57.4	-38.7	-59.9	-44.9	-46.0	-57.5	-70.1	-71.8	-93.9	..	Resource Balance
..	22.1	25.8	17.8	21.9	41.6	50.0	59.4	53.3	57.2	..	Exports of Goods & NFServices
..	79.5	64.5	77.7	66.8	87.6	107.5	129.5	125.1	151.1	..	Imports of Goods & NFServices
..	147.4	141.7	168.8	175.3	200.4	238.7	261.3	272.9	305.1	..	Domestic Absorption
..	88.7	89.4	105.9	118.1	136.2	155.1	172.5	184.4	201.6	..	Private Consumption, etc.
..	16.9	18.2	20.6	24.3	21.9	30.1	34.5	37.3	40.9	..	General Gov't Consumption
..	41.8	34.1	42.3	32.9	42.3	53.5	54.3	51.2	62.6	..	Gross Domestic Investment
..	..	..	..	..	..	..	..	..	..	..	Fixed Investment
											Memo Items:
..	-15.6	-4.6	-17.6	-12.0	-3.7	-4.0	-15.8	-20.6	-31.3	..	Gross Domestic Saving
..	..	..	..	15.6	34.5	29.3	32.0	39.0	45.2	..	Gross National Saving

(Millions of 1980 Western Samoa Tala)

1978	1979	1980	1981	1982	1983	1984	1985	1986	1987	NOTES	
..	..	..	..	..	..	..	..	..	..	..	Gross National Product
..	94.6	89.3	84.2	83.5	85.4	..	..	..	..	f	GDP at factor cost
..	44.5	41.1	39.3	41.5	..	..	..	..	..	..	Agriculture
..	9.9	10.9	10.8	11.7	..	..	..	..	..	..	Industry
..	3.7	4.9	5.2	5.4	..	..	..	..	..	..	Manufacturing
..	40.2	37.3	34.1	30.3	..	..	..	..	..	..	Services, etc.
..	-59.4	-38.7	-48.0	-31.5	-30.8	-36.8	-32.4	-28.9	-38.4	..	Resource Balance
..	22.1	25.8	18.8	18.6	20.8	17.5	25.1	27.6	27.7	..	Exports of Goods & NFServices
..	81.5	64.5	66.8	50.1	51.6	54.3	57.5	56.5	66.1	..	Imports of Goods & NFServices
..	169.2	141.7	141.7	124.3	124.0	131.2	132.5	129.5	140.0	..	Domestic Absorption
..	96.1	89.4	88.7	83.9	87.3	88.6	90.6	89.4	94.9	..	Private Consumption, etc.
..	22.8	18.2	17.1	16.6	14.3	17.6	18.5	19.0	20.1	..	General Gov't Consumption
..	50.3	34.1	35.9	23.8	22.4	25.0	23.4	21.1	25.0	..	Gross Domestic Investment
..	..	..	..	..	..	..	..	..	..	..	Fixed Investment
											Memo Items:
..	22.7	25.8	15.3	16.4	24.5	25.3	26.4	24.1	25.0	..	Capacity to Import
..	0.6	0.0	-3.5	-2.2	3.7	7.8	1.3	-3.5	-2.7	..	Terms of Trade Adjustment
..	110.4	103.0	90.2	90.6	96.9	102.2	101.4	97.1	98.9	..	Gross Domestic Income
..	..	..	..	..	..	..	..	..	..	..	Gross National Income

(Index 1980 = 100) — **DOMESTIC PRICES (DEFLATORS)**

1978	1979	1980	1981	1982	1983	1984	1985	1986	1987	NOTES	
..	82.0	100.0	116.2	140.5	165.7	191.9	191.0	199.9	207.9	..	Overall (GDP)
..	87.1	100.0	119.1	141.0	161.6	181.9	197.2	210.7	217.9	..	Domestic Absorption
..	85.4	100.0	114.2	143.4	..	..	..	..	..	..	Agriculture
..	73.7	100.0	105.6	122.2	..	..	..	..	..	..	Industry
..	91.9	100.0	100.0	127.8	..	..	..	..	..	..	Manufacturing

MANUFACTURING ACTIVITY

1978	1979	1980	1981	1982	1983	1984	1985	1986	1987	NOTES	
..	..	..	..	..	..	..	..	..	..	G	Employment (1980 = 100)
..	..	..	..	..	..	..	..	..	..	G	Real Earnings per Empl. (1980 = 100)
..	..	..	..	..	..	..	..	..	..	G	Real Output per Empl. (1980 = 100)
..	..	..	..	..	..	..	..	..	..	..	Earnings as % of Value Added

(Millions of current Western Samoa Tala) — **MONETARY HOLDINGS**

1978	1979	1980	1981	1982	1983	1984	1985	1986	1987	NOTES	
11.10	15.50	20.14	33.29	44.05	39.98	45.06	54.81	64.98	81.07	..	Money Supply, Broadly Defined
5.18	5.64	8.80	13.82	16.15	16.87	19.64	20.83	21.87	27.39	..	Money as Means of Payment
0.25	0.17	3.56	5.28	6.05	6.02	7.08	8.44	9.18	10.53	..	Currency Ouside Banks
4.93	5.47	5.25	8.54	10.10	10.85	12.56	12.39	12.69	16.86	..	Demand Deposits
5.92	9.86	11.34	19.48	27.90	23.10	25.42	33.98	43.11	53.68	..	Quasi-Monetary Liabilities

(Thousands of current Western Samoa Tala) — **GOVERNMENT DEFICIT (-) OR SURPLUS**

1978	1979	1980	1981	1982	1983	1984	1985	1986	1987	NOTES	
..	..	..	..	..	..	1,000	..	..	..	..	Current Revenue
..	..	..	..	..	..	61,500	..	..	..	..	Current Expenditure
..	..	..	..	..	..	36,800	..	..	..	..	Current Budget Balance
..	..	..	..	..	..	24,700	..	..	..	..	Capital Receipts
..	..	..	..	..	..	23,800	..	..	..	..	Capital Payments

WESTERN SAMOA

	1967	1968	1969	1970	1971	1972	1973	1974	1975	1976	1977
FOREIGN TRADE (CUSTOMS BASIS)				*(Thousands of current US dollars)*							
Value of Exports, fob	..	..	..	..	..	..	..	..	..	..	..
Nonfuel Primary Products	..	..	..	..	..	..	..	..	..	..	..
Fuels	..	..	..	..	..	..	..	..	..	..	..
Manufactures	..	..	..	..	..	..	..	..	..	..	..
Value of Imports, cif	..	..	..	..	..	..	..	..	..	..	..
Nonfuel Primary Products	..	..	..	..	..	..	..	..	..	..	..
Fuels	..	..	..	..	..	..	..	..	..	..	..
Manufactures	..	..	..	..	..	..	..	..	..	..	..
Terms of Trade						*(Index 1980 = 100)*					
Export Prices, fob	..	..	..	..	..	..	..	..	..	..	..
Nonfuel Primary Products	..	..	..	..	..	..	..	..	..	..	..
Fuels	..	..	..	..	..	..	..	..	..	..	..
Manufactures	..	..	..	..	..	..	..	..	..	..	..
Import Prices, cif	..	..	..	..	..	..	..	..	..	..	..
Trade at Constant 1980 Prices				*(Thousands of 1980 US dollars)*							
Exports, fob	..	..	..	..	..	..	..	..	..	..	..
Imports, cif	..	..	..	..	..	..	..	..	..	..	..
BALANCE OF PAYMENTS				*(Thousands of current US dollars)*							
Exports of Goods & Services	..	..	..	7,251	9,438	10,610	11,993	20,076	15,183	12,146	17,148
Merchandise, fob	..	..	..	4,808	6,557	5,168	6,744	12,901	7,476	6,998	14,731
Nonfactor Services	..	..	..	2,343	2,780	5,127	5,130	7,055	7,707	5,033	2,417
Factor Services	..	..	..	100	100	315	119	120	0	115	0
Imports of Goods & Services	..	..	..	14,862	15,209	26,255	27,626	31,233	39,829	29,347	42,543
Merchandise, fob	..	..	..	12,179	12,101	20,979	21,353	23,654	33,212	26,809	37,337
Nonfactor Services	..	..	..	2,541	2,980	5,089	5,942	7,109	6,219	2,177	5,089
Factor Services	..	..	..	142	128	187	331	470	398	360	117
Long-Term Interest	..	..	..	..	..	..	..	600	500	500	1,000
Current Transfers, net	..	..	..	..	..	..	..	..	..	..	..
Workers' Remittances	..	..	..	..	..	..	..	..	..	..	..
Total to be Financed	..	..	..	..	..	..	..	..	..	..	..
Official Capital Grants	..	..	..	..	..	..	..	..	..	..	..
Current Account Balance	..	..	..	..	-2,733	-11,469	-7,620	-1,393	-12,386	-10,233	-9,732
Long-Term Capital, net	..	..	..	3,281	3,728	8,551	10,513	9,949	18,600	13,507	20,358
Direct Investment	..	..	..	..	..	..	..	..	..	..	..
Long-Term Loans	..	..	..	..	..	..	..	2,600	4,000	5,700	7,700
Disbursements	..	..	..	..	..	..	..	3,200	4,600	8,500	8,800
Repayments	..	..	..	..	..	..	..	600	600	2,800	1,100
Other Long-Term Capital	..	..	..	3,281	3,728	8,551	10,513	7,349	14,600	7,807	12,658
Other Capital, net	..	..	..	2,398	-277	176	-3,948	-7,649	-6,791	-5,056	-6,958
Change in Reserves	..	..	..	-607	-718	2,743	1,055	-908	577	1,782	-3,668
Memo Item:											
Conversion Factor (Annual Avg)	0.720	0.720	0.720	0.720	0.720	0.680	0.610	0.610	0.630	0.800	0.790
(Western Samoa Tala per US dollar)											
EXTERNAL DEBT, ETC.				*(Thousands of US dollars, outstanding at end of year)*							
Public/Publicly Guar. Long-Term	..	..	..	..	..	..	..	13,600	15,900	20,400	29,300
Official Creditors	..	..	..	..	..	..	..	3,000	7,100	11,800	18,100
IBRD and IDA	..	..	..	..	..	..	..	0	1,400	2,200	3,300
Private Creditors	..	..	..	..	..	..	..	10,600	8,800	8,600	11,100
Private Non-guaranteed Long-Term	..	..	..	..	..	..	..	..	..	..	..
Use of Fund Credit	..	..	..	0	0	0	0	0	1,500	2,200	2,000
Short-Term Debt	..	..	..	..	..	..	..	..	..	..	..
Memo Items:				*(Thousands of US dollars)*							
Int'l Reserves Excluding Gold	2,160	3,060	5,150	5,220	6,411	4,529	5,078	5,958	6,387	5,240	9,126
Gold Holdings (at market price)	..	..	..	..	..	..	..	..	..	..	..
SOCIAL INDICATORS											
Total Fertility Rate	7.3	7.1	6.9	6.7	6.3	5.9	5.9	5.8	5.8	5.7	5.7
Crude Birth Rate	..	..	..	..	..	..	..	..	..	..	..
Infant Mortality Rate	..	..	..	..	..	..	..	..	..	..	..
Life Expectancy at Birth	..	..	..	..	..	..	..	..	..	62.6	62.9
Food Production, p.c. ('79-81=100)	95.8	94.8	97.5	93.3	97.8	90.6	90.9	91.5	93.9	95.6	98.8
Labor Force, Agriculture (%)	..	..	..	..	..	..	..	..	..	..	..
Labor Force, Female (%)	..	..	..	..	..	..	..	..	..	..	..
School Enroll. Ratio, primary	..	..	..	..	..	..	..	..	..	..	..
School Enroll. Ratio, secondary	..	..	..	..	..	..	..	..	..	..	..

1978	1979	1980	1981	1982	1983	1984	1985	1986	1987 estimate	NOTES	WESTERN SAMOA
											FOREIGN TRADE (CUSTOMS BASIS)
			(Thousands of current US dollars)								
..	..	17,570	11,200	13,140	17,360	19,900	29,800	10,500	..	..	Value of Exports, fob
..	..	..	..	..	..	..	..	..	..	..	Nonfuel Primary Products
..	..	..	..	..	..	..	..	..	..	..	Fuels
..	..	..	..	..	..	..	..	..	..	..	Manufactures
..	..	85,000	100,000	91,000	89,000	87,000	86,000	86,000	..	..	Value of Imports, cif
..	..	..	..	..	..	..	..	..	..	..	Nonfuel Primary Products
..	..	..	..	..	..	..	..	..	..	..	Fuels
..	..	..	..	..	..	..	..	..	..	..	Manufactures
			(Index 1980 = 100)								
..	..	..	..	..	..	..	..	..	..	..	Terms of Trade
..	..	..	..	..	..	..	..	..	..	..	Export Prices, fob
..	..	..	..	..	..	..	..	..	..	..	Nonfuel Primary Products
..	..	..	..	..	..	..	..	..	..	..	Fuels
..	..	..	..	..	..	..	..	..	..	..	Manufactures
..	..	..	..	..	..	..	..	..	..	..	Import Prices, cif
			(Thousands of 1980 US dollars)								
..	..	..	..	..	..	..	..	..	..	..	Trade at Constant 1980 Prices
..	..	..	..	..	..	..	..	..	..	..	Exports, fob
..	..	..	..	..	..	..	..	..	..	..	Imports, cif
											BALANCE OF PAYMENTS
			(Thousands of current US dollars)								
13,649	21,763	25,650	18,345	21,770	26,951	27,203	27,123	24,846	30,807	..	Exports of Goods & Services
9,737	18,132	17,220	10,782	13,459	17,694	18,343	16,125	10,478	11,769	..	Merchandise, fob
3,911	3,631	8,430	7,437	8,200	9,121	8,522	10,317	13,114	16,286	..	Nonfactor Services
0	0	0	126	110	136	338	682	1,253	2,752	..	Factor Services
55,912	78,665	74,230	65,790	60,157	59,395	60,033	60,194	58,149	73,802	..	Imports of Goods & Services
47,655	67,154	56,858	51,379	45,315	44,115	45,551	46,607	42,724	55,751	..	Merchandise, fob
6,763	9,889	14,859	11,585	12,606	12,446	12,184	11,087	13,092	15,600	..	Nonfactor Services
1,494	1,622	2,513	2,826	2,236	2,834	2,298	2,499	2,333	2,451	..	Factor Services
1,200	1,600	2,300	1,800	1,100	1,000	1,400	1,700	1,400	1,300	..	Long-Term Interest
..	..	..	..	..	..	..	..	28,302	..	f	Current Transfers, net
..	..	..	..	..	..	..	..	28,544	..	..	Workers' Remittances
..	..	..	..	..	..	..	..	-1,051	..	..	Total to be Financed
								11,600			Official Capital Grants
-18,103	-22,042	-12,912	-13,094	-6,038	4,803	682	2,009	10,549	8,837	..	Current Account Balance
24,159	30,454	23,344	20,261	13,177	20,021	17,478	11,114	13,195	17,999	..	Long-Term Capital, net
..	..	..	..	..	..	..	..	..	..	..	Direct Investment
8,900	10,700	8,600	3,700	5,300	1,900	7,100	-1,300	-2,200	3,200	..	Long-Term Loans
10,500	12,800	10,900	5,400	7,000	4,700	10,300	2,600	2,200	6,400	..	Disbursements
1,600	2,100	2,300	1,700	1,700	2,800	3,200	3,900	4,400	3,200	..	Repayments
15,259	19,754	14,744	16,561	7,877	18,121	10,378	12,414	3,795	14,799	..	Other Long-Term Capital
-12,045	-8,483	-11,757	-8,866	-5,715	-22,070	-14,946	-9,250	-12,887	-14,825	..	Other Capital, net
5,989	70	1,326	1,699	-1,424	-2,753	-3,214	-3,873	-10,857	-12,011	..	Change in Reserves
											Memo Item:
			(Western Samoa Tala per US dollar)								
0.740	0.820	0.920	1.040	1.210	1.540	1.840	2.240	2.230	2.120	..	Conversion Factor (Annual Avg)
											EXTERNAL DEBT, ETC.
			(Thousands of US dollars, outstanding at end of year)								
37,700	48,100	55,600	56,700	60,300	60,300	64,200	64,800	64,900	71,800	..	Public/Publicly Guar. Long-Term
25,200	37,800	47,300	50,300	55,400	57,400	60,100	61,500	62,500	69,500	..	Official Creditors
4,200	4,300	6,300	7,400	8,500	9,000	9,500	10,100	11,000	12,600	..	IBRD and IDA
12,500	10,300	8,300	6,300	4,900	3,000	4,200	3,200	2,400	2,300	..	Private Creditors
..	..	..	..	..	..	..	..	..	..	..	Private Non-guaranteed Long-Term
3,500	4,400	3,300	5,200	4,200	5,900	8,100	9,400	8,900	7,400	..	Use of Fund Credit
..	..	1,000	1,000	4,000	7,000	2,000	1,000	1,000	1,000	..	Short-Term Debt
											Memo Items:
			(Thousands of US dollars)								
4,782	4,820	2,770	3,282	3,481	7,229	10,557	14,021	23,746	37,198	..	Int'l Reserves Excluding Gold
..	..	..	..	..	..	..	..	..	..	..	Gold Holdings (at market price)
											SOCIAL INDICATORS
5.7	5.6	5.6	5.5	5.5	5.4	5.2	5.1	5.0	4.8	..	Total Fertility Rate
..	..	..	..	37.9	37.0	36.0	35.1	34.1	33.2	..	Crude Birth Rate
..	..	..	..	50.8	50.8	50.8	50.8	50.8	50.8	..	Infant Mortality Rate
63.3	63.6	63.9	64.2	64.5	64.5	64.5	64.5	64.5	64.5	..	Life Expectancy at Birth
93.3	99.0	103.7	97.3	100.3	100.4	94.3	98.7	92.8	81.9	..	Food Production, p.c. ('79-81=100)
..	..	..	..	..	..	..	..	..	..	..	Labor Force, Agriculture (%)
..	..	..	..	..	..	..	..	..	..	..	Labor Force, Female (%)
..	..	..	..	..	..	..	..	..	..	..	School Enroll. Ratio, primary
..	..	..	..	..	..	..	..	..	..	..	School Enroll. Ratio, secondary

YEMEN ARAB REPUBLIC	1967	1968	1969	1970	1971	1972	1973	1974	1975	1976	1977
CURRENT GNP PER CAPITA (US $)	..	..	..	..	..	70	90	120	140	180	220
POPULATION (thousands)	4,876	4,995	5,122	5,258	5,404	5,559	5,723	5,896	6,075	6,260	6,451

ORIGIN AND USE OF RESOURCES *(Millions of current Yemen Rials)*

	1967	1968	1969	1970	1971	1972	1973	1974	1975	1976	1977
Gross National Product (GNP)	..	..	..	1,431	1,753	2,157	2,752	3,436	4,293	5,651	7,467
Net Factor Income from Abroad	..	..	..	200	226	346	488	589	548	763	1,031
Gross Domestic Product (GDP)	..	..	..	1,231	1,527	1,811	2,264	2,847	3,745	4,888	6,436
Indirect Taxes, net	..	..	..	..	..	..	..	..	..	..	..
GDP at factor cost	..	..	..	..	..	..	..	..	..	..	..
Agriculture	..	..	..	642	785	852	1,036	1,202	1,645	2,011	2,313
Industry	..	..	..	127	165	215	272	377	398	542	778
Manufacturing	..	..	..	52	67	87	118	175	211	257	301
Services, etc.	..	..	..	462	577	744	956	1,268	1,702	2,335	3,345
Resource Balance	..	..	..	-324	-362	-451	-622	-801	-1,089	-1,652	-3,227
Exports of Goods & NFServices	..	..	..	23	29	42	62	128	162	214	279
Imports of Goods & NFServices	..	..	..	347	391	493	684	929	1,251	1,866	3,506
Domestic Absorption	..	..	..	1,555	1,889	2,262	2,886	3,648	4,834	6,540	9,663
Private Consumption, etc.	..	..	..	1,434	1,477	1,710	2,188	2,707	3,662	4,939	7,567
General Gov't Consumption	..	..	..	116	161	234	317	365	535	701	886
Gross Domestic Investment	..	..	..	5	251	318	381	576	637	900	1,210
Fixed Investment	..	..	..	58	220	271	334	456	511	745	1,315
Memo Items:											
Gross Domestic Saving	..	..	..	-319	-111	-133	-241	-225	-452	-752	-2,017
Gross National Saving	..	..	..	93	336	544	684	984	1,336	3,098	3,257

(Millions of 1980 Yemen Rials)

	1967	1968	1969	1970	1971	1972	1973	1974	1975	1976	1977
Gross National Product	..	..	..	5,193	6,311	6,995	8,123	8,495	9,034	10,133	11,004
GDP at factor cost	..	..	..	4,260	5,200	5,485	6,250	6,575	7,343	8,048	7,957
Agriculture	..	..	..	2,060	2,665	2,670	3,026	2,861	3,364	3,463	3,186
Industry	..	..	..	459	546	606	691	782	751	982	1,205
Manufacturing	..	..	..	195	224	250	296	347	367	405	431
Services, etc.	..	..	..	1,985	2,329	2,652	3,034	3,462	3,824	4,384	5,026
Resource Balance	..	..	..	-1,422	-1,509	-1,823	-2,358	-3,030	-3,107	-4,168	-5,569
Exports of Goods & NFServices	..	..	..	78	93	130	180	366	349	395	414
Imports of Goods & NFServices	..	..	..	1,500	1,602	1,953	2,538	3,396	3,455	4,563	5,984
Domestic Absorption	..	..	..	5,926	7,049	7,750	9,109	10,134	11,045	12,996	14,986
Private Consumption, etc.	..	..	..	4,753	5,294	5,802	7,106	7,831	8,514	10,008	11,722
General Gov't Consumption	..	..	..	747	887	988	1,065	1,077	1,250	1,388	1,396
Gross Domestic Investment	..	..	..	426	867	961	938	1,226	1,281	1,600	1,868
Fixed Investment	..	..	..	450	759	821	834	959	1,054	1,356	1,959
Memo Items:											
Capacity to Import	..	..	..	99	119	166	230	468	447	523	476
Terms of Trade Adjustment	..	..	..	22	26	37	50	102	99	128	62
Gross Domestic Income	..	..	..	4,526	5,566	5,964	6,801	7,206	8,037	8,956	9,479
Gross National Income	..	..	..	5,215	6,337	7,031	8,173	8,596	9,133	10,261	11,066

DOMESTIC PRICES (DEFLATORS) *(Index 1980 = 100)*

	1967	1968	1969	1970	1971	1972	1973	1974	1975	1976	1977
Overall (GDP)	..	..	..	27.3	27.6	30.6	33.5	40.1	47.2	55.4	68.3
Domestic Absorption	..	..	..	26.2	26.8	29.2	31.7	36.0	43.8	50.3	64.5
Agriculture	..	..	..	31.2	29.5	31.9	34.2	42.0	48.9	58.1	72.6
Industry	..	..	..	27.7	30.2	35.5	39.4	48.2	53.0	55.2	64.6
Manufacturing	..	..	..	26.6	29.9	34.7	39.9	50.4	57.6	63.5	69.8

MANUFACTURING ACTIVITY

	1967	1968	1969	1970	1971	1972	1973	1974	1975	1976	1977
Employment (1980=100)	..	..	..	..	72.2	..	109.0	..	..	..	..
Real Earnings per Empl. (1980=100)	..	..	..	..	..	..	..	..	..	..	..
Real Output per Empl. (1980=100)	..	..	..	..	28.5	..	47.5	..	..	..	..
Earnings as % of Value Added	..	..	..	..	..	..	..	..	..	..	..

MONETARY HOLDINGS *(Millions of current Yemen Rials)*

	1967	1968	1969	1970	1971	1972	1973	1974	1975	1976	1977
Money Supply, Broadly Defined	..	..	..	..	..	..	643	861	1,547	3,478	5,263
Money as Means of Payment	..	..	..	..	..	..	541	712	1,290	2,787	4,479
Currency Ouside Banks	..	..	..	..	..	..	466	621	1,080	2,329	3,819
Demand Deposits	..	..	..	..	..	..	75	92	210	458	660
Quasi-Monetary Liabilities	..	..	..	..	..	..	102	149	257	691	784

GOVERNMENT DEFICIT (-) OR SURPLUS *(Millions of current Yemen Rials)*

	1967	1968	1969	1970	1971	1972	1973	1974	1975	1976	1977
	..	..	..	..	..	..	..	-75.6	162.7	120.0	309.8
Current Revenue	..	..	..	..	..	..	..	384.4	794.0	1,007.2	1,710.7
Current Expenditure	..	..	..	..	..	..	..	322.5	453.0	616.5	841.0
Current Budget Balance	..	..	..	..	..	..	..	61.9	341.0	390.7	869.7
Capital Receipts	..	..	..	..	..	..	..	1.7	1.5	5.8	5.7
Capital Payments	..	..	..	..	..	..	..	139.2	179.8	276.5	637.1

1978	1979	1980	1981	1982	1983	1984	1985	1986	1987 estimate	NOTES	YEMEN ARAB REPUBLIC
290	360	420	470	640	670	690	670	630	590	..	**CURRENT GNP PER CAPITA (US $)**
6,645	6,842	7,039	7,234	7,426	7,612	7,789	7,955	8,191	8,430	..	**POPULATION** (thousands)
											ORIGIN AND USE OF RESOURCES
				(Millions of current Yemen Rials)							
9,608	12,047	13,740	15,064	22,296	24,169	27,101	33,716	41,656	47,062	C	Gross National Product (GNP)
1,389	1,866	2,044	1,953	2,364	2,299	2,345	2,747	3,267	3,603	..	Net Factor Income from Abroad
8,219	10,181	11,696	13,111	19,932	21,870	24,756	30,969	38,389	43,459	C	Gross Domestic Product (GDP)
..	..	..	..	..	..	..	..	..	..	..	Indirect Taxes, net
..	..	..	..	..	..	..	..	..	..	B C	GDP at factor cost
2,409	2,049	3,458	3,685	5,035	5,224	6,236	8,033	10,680	12,163	..	Agriculture
1,299	1,735	1,903	2,191	3,241	3,876	4,486	5,504	6,731	7,562	..	Industry
381	504	655	820	1,720	2,258	2,683	3,465	4,620	5,012	..	Manufacturing
4,511	6,397	6,335	7,235	11,656	12,770	14,034	17,432	20,978	23,734	..	Services, etc.
-4,241	-6,300	-7,662	-8,506	-9,273	-8,486	-7,832	-8,480	-8,413	-14,706	..	Resource Balance
242	474	803	1,130	1,155	1,120	1,125	1,164	1,155	1,599	..	Exports of Goods & NFServices
4,483	6,774	8,465	9,636	10,428	9,606	8,957	9,644	9,568	16,305	..	Imports of Goods & NFServices
12,460	16,481	19,358	21,617	29,205	30,356	32,588	39,449	46,802	58,165	..	Domestic Absorption
7,971	10,648	11,742	12,234	18,627	21,081	23,228	29,434	34,916	43,781	..	Private Consumption, etc.
1,279	1,839	2,255	3,258	4,640	5,030	4,840	5,543	6,898	8,584	..	General Gov't Consumption
3,210	3,994	5,361	6,125	5,938	4,245	4,520	4,472	4,988	5,800	..	Gross Domestic Investment
2,803	3,964	5,006	5,906	5,698	4,393	4,610	4,547	4,938	5,766	..	Fixed Investment
											Memo Items:
-1,031	-2,306	-2,301	-2,381	-3,335	-4,241	-3,312	-4,008	-3,425	-8,906	..	Gross Domestic Saving
4,131	3,844	4,000	3,272	3,368	1,895	2,903	1,895	3,585	-394	..	Gross National Saving
				(Millions of 1980 Yemen Rials)							
12,285	13,237	13,740	14,358	18,674	18,853	19,253	20,040	21,794	22,722	C	Gross National Product
8,675	9,581	10,026	11,319	12,466	12,688	13,024	..	..	..	B C I	GDP at factor cost
2,846	3,317	3,458	3,743	3,915	3,472	3,468	3,763	4,191	4,288	..	Agriculture
1,727	2,005	1,903	1,924	2,718	3,087	3,310	3,439	3,714	4,083	..	Industry
486	568	655	781	1,549	1,909	2,050	2,289	2,534	2,737	..	Manufacturing
5,858	5,793	6,335	6,901	10,085	10,509	10,849	11,237	12,268	12,752	..	Services, etc.
-6,246	-7,611	-7,662	-7,286	-7,582	-6,666	-5,220	-4,863	-3,535	-5,102	..	Resource Balance
307	520	803	1,071	1,073	964	830	739	531	618	..	Exports of Goods & NFServices
6,553	8,130	8,465	8,357	8,655	7,630	6,050	5,602	4,066	5,720	..	Imports of Goods & NFServices
16,678	18,726	19,358	19,854	24,300	23,733	22,847	23,302	23,708	26,224	..	Domestic Absorption
10,963	12,271	11,742	12,342	16,594	17,275	15,950	17,358	17,482	19,311	..	Private Consumption, etc.
1,725	2,033	2,255	2,983	3,485	3,561	3,928	3,471	3,928	4,125	..	General Gov't Consumption
3,990	4,421	5,361	4,530	4,221	2,897	2,969	2,473	2,299	2,789	..	Gross Domestic Investment
3,535	4,329	5,006	4,302	3,987	2,955	2,984	2,467	2,247	2,708	..	Fixed Investment
											Memo Items:
354	569	803	980	959	890	760	676	491	561	..	Capacity to Import
47	49	0	-91	-115	-75	-71	-63	-40	-57	..	Terms of Trade Adjustment
10,479	11,164	11,696	12,477	16,604	16,993	17,557	18,376	20,133	21,065	..	Gross Domestic Income
12,332	13,286	13,740	14,266	18,559	18,778	19,182	19,977	21,754	22,665	..	Gross National Income
				(Index 1980 = 100)							**DOMESTIC PRICES (DEFLATORS)**
78.8	91.6	100.0	104.3	119.2	128.1	140.4	168.0	190.3	205.7	..	Overall (GDP)
74.7	88.0	100.0	108.9	120.2	127.9	142.6	169.3	197.4	221.8	..	Domestic Absorption
84.6	61.8	100.0	98.4	128.6	150.5	179.8	213.5	254.8	283.7	..	Agriculture
75.2	86.5	100.0	113.9	119.2	125.6	135.5	160.1	181.2	185.2	..	Industry
78.3	88.7	100.0	105.0	111.0	118.3	130.9	151.4	182.3	183.1	..	Manufacturing
											MANUFACTURING ACTIVITY
..	..	100.0	..	..	..	..	..	..	..	..	Employment (1980 = 100)
..	..	100.0	..	..	..	..	..	..	..	..	Real Earnings per Empl. (1980 = 100)
..	..	100.0	..	..	..	..	..	..	..	..	Real Output per Empl. (1980 = 100)
..	..	25.7	..	..	..	..	..	..	..	..	Earnings as % of Value Added
											MONETARY HOLDINGS
				(Millions of current Yemen Rials)							
6,957	8,298	9,180	10,301	12,915	15,967	20,365	24,430	30,651	33,743	..	Money Supply, Broadly Defined
5,716	7,037	7,568	8,331	10,669	13,148	16,395	18,823	23,683	26,641	..	Money as Means of Payment
4,963	6,299	6,895	7,439	9,336	10,733	13,314	15,633	19,062	20,159	..	Currency Ouside Banks
753	738	674	892	1,333	2,415	3,081	3,190	4,621	6,482	..	Demand Deposits
1,241	1,261	1,611	1,970	2,246	2,819	3,970	5,607	6,968	7,102	..	Quasi-Monetary Liabilities
				(Millions of current Yemen Rials)							
-283.6	-842.4	-2,333.2	-2,939.5	-5,261.7	-4,807.5	-4,689.4	-5,063.1	-3,793.0	..	F	**GOVERNMENT DEFICIT (-) OR SURPLUS**
2,374.6	3,513.2	3,206.0	4,847.6	5,709.9	5,255.7	5,428.8	5,991.3	9,035.7	..	..	Current Revenue
1,250.4	1,847.2	2,530.9	3,325.0	5,180.3	6,200.3	5,840.1	6,522.7	7,030.3	..	..	Current Expenditure
1,124.2	1,666.0	675.1	1,522.6	529.6	-944.6	-411.3	-531.4	2,005.4	..	..	Current Budget Balance
3.0	14.1	7.5	5.9	3.6	3.8	9.7	10.1	12.1	..	..	Capital Receipts
1,219.0	2,525.5	2,467.3	3,679.9	3,904.6	2,944.3	3,079.6	2,758.3	3,392.1	..	..	Capital Payments

YEMEN ARAB REPUBLIC	1967	1968	1969	1970	1971	1972	1973	1974	1975	1976	1977
FOREIGN TRADE (CUSTOMS BASIS)				*(Millions of current US dollars)*							
Value of Exports, fob	3.0	4.0	4.0	2.8	4.3	4.3	7.9	13.3	10.9	7.7	11.0
Nonfuel Primary Products	3.0	4.0	4.0	2.8	4.2	4.2	7.7	7.3	10.1	6.9	9.0
Fuels	0.0	0.0	0.0	0.0	0.0	0.0	0.0	5.2	0.0	0.0	0.0
Manufactures	0.0	0.0	0.0	0.0	0.1	0.1	0.2	0.8	0.8	0.8	2.0
Value of Imports, cif	30.0	32.0	35.6	31.7	37.0	80.0	122.7	189.7	293.4	410.3	1,040.0
Nonfuel Primary Products	14.2	14.9	14.9	20.5	15.8	41.3	60.9	78.8	135.1	170.9	369.3
Fuels	1.8	1.9	2.2	1.7	1.8	3.9	6.4	1.5	14.8	14.2	32.3
Manufactures	14.0	15.3	18.5	9.5	19.4	34.8	55.4	109.5	143.6	225.2	638.4
Terms of Trade	131.9	128.4	141.3	130.0	109.7	146.2	160.5	78.0	90.6	112.0	109.9
(Index 1980 = 100)											
Export Prices, fob	31.8	30.8	34.4	34.6	32.2	47.5	66.7	48.1	58.8	71.4	74.2
Nonfuel Primary Products	31.8	30.8	34.4	34.6	32.2	47.8	67.2	59.2	58.4	72.0	74.7
Fuels	4.3	4.3	4.3	4.3	5.6	6.2	8.9	36.7	35.7	38.4	42.0
Manufactures	30.1	31.7	32.4	37.5	36.5	38.5	49.5	67.2	66.1	66.7	71.7
Import Prices, cif	24.1	24.0	24.3	26.6	29.4	32.5	41.5	61.7	64.9	63.8	67.5
Trade at Constant 1980 Prices				*(Millions of 1980 US dollars)*							
Exports, fob	9.4	13.0	11.8	8.1	13.4	9.0	11.8	27.6	18.5	10.7	14.8
Imports, cif	124.6	133.5	146.2	119.3	125.8	246.1	295.2	307.7	451.9	643.0	1,541.1
BALANCE OF PAYMENTS				*(Millions of current US dollars)*							
Exports of Goods & Services	..	..	..	20.1	22.2	28.3	37.3	48.0	52.0	96.5	285.0
Merchandise, fob	..	..	..	5.8	6.4	6.8	8.8	11.4	14.2	13.6	15.0
Nonfactor Services	..	..	..	7.3	8.0	8.5	11.0	11.5	20.5	51.6	43.0
Factor Services	..	..	..	7.0	7.8	13.0	17.5	25.0	17.2	31.3	227.0
Imports of Goods & Services	..	..	..	110.7	107.3	147.7	223.5	242.1	310.0	579.8	923.0
Merchandise, fob	..	..	..	88.4	85.6	117.4	178.1	193.3	246.5	475.5	753.0
Nonfactor Services	..	..	..	22.1	21.4	29.4	44.6	47.4	62.9	103.9	169.0
Factor Services	..	..	..	0.2	0.3	0.9	0.8	1.4	0.6	0.4	1.0
Long-Term Interest	..	..	..	..	..	..	..	0.7	0.8	1.0	1.6
Current Transfers, net	..	..	..	38.9	40.8	70.5	94.6	135.5	271.5	676.5	930.0
Workers' Remittances	..	..	..	45.0	47.2	81.0	108.5	156.9	309.7	795.2	1,024.0
Total to be Financed	..	..	..	-51.7	-44.3	-48.9	-91.6	-58.6	13.5	193.3	292.0
Official Capital Grants	..	..	..	17.7	5.7	18.8	23.2	53.3	116.9	102.8	0.0
Current Account Balance	..	..	..	-34.0	-38.6	-30.1	-68.4	-5.3	130.4	296.0	292.0
Long-Term Capital, net	..	..	..	44.4	33.9	28.9	40.4	96.6	135.0	150.1	51.0
Direct Investment	..	..	..	..	..	..	..	..	..	..	0.0
Long-Term Loans	..	..	..	..	..	..	..	44.4	26.2	33.4	48.9
Disbursements	..	..	..	..	..	..	..	53.3	30.2	46.8	57.6
Repayments	..	..	..	..	..	..	..	8.9	4.0	13.4	8.7
Other Long-Term Capital	..	..	..	26.7	28.2	10.1	17.2	-1.1	-8.2	14.0	2.1
Other Capital, net	..	..	..	-8.9	4.7	2.3	45.5	-22.1	-112.4	-61.7	164.0
Change in Reserves	..	..	..	-1.5	0.0	-1.1	-17.5	-69.2	-152.9	-384.5	-507.0
Memo Item:				*(Yemen Rials per US dollar)*							
Conversion Factor (Annual Avg)	..	2.790	4.440	5.460	5.420	4.690	4.620	4.570	4.570	4.560	4.560
EXTERNAL DEBT, ETC.				*(Millions of US dollars, outstanding at end of year)*							
Public/Publicly Guar. Long-Term	..	..	..	..	..	..	..	235.0	250.7	276.7	345.4
Official Creditors	..	..	..	..	..	..	..	229.9	246.9	273.7	343.1
IBRD and IDA	..	..	..	..	..	..	..	2.1	9.2	27.0	43.9
Private Creditors	..	..	..	..	..	..	..	5.1	3.9	3.1	2.4
Private Non-guaranteed Long-Term	..	..	..	..	..	..	..	..	..	..	..
Use of Fund Credit	..	..	..	0.0	0.0	0.0	0.0	0.0	0.0	0.0	0.0
Short-Term Debt	..	..	..	..	..	..	..	..	..	..	86.0
Memo Items:				*(Millions of US dollars)*							
Int'l Reserves Excluding Gold	..	..	..	..	..	..	126.9	198.6	337.5	720.0	1,240.2
Gold Holdings (at market price)	..	..	..	..	..	..	..	..	..	..	0.7
SOCIAL INDICATORS											
Total Fertility Rate	6.8	6.8	6.8	6.8	6.8	6.8	6.8	6.8	6.8	6.8	6.8
Crude Birth Rate	48.8	48.8	48.8	48.7	48.7	48.7	48.7	48.7	48.6	48.6	48.6
Infant Mortality Rate	195.0	192.6	190.2	187.8	185.4	183.0	180.6	178.2	175.8	173.4	171.0
Life Expectancy at Birth	38.0	38.2	38.4	38.6	38.8	39.0	39.5	39.9	40.4	40.8	41.3
Food Production, p.c. ('79-81 = 100)	98.7	94.4	85.1	80.7	97.9	91.6	97.7	95.1	108.8	101.4	97.6
Labor Force, Agriculture (%)	78.0	77.4	76.9	76.4	75.6	74.8	74.1	73.3	72.5	71.8	71.0
Labor Force, Female (%)	6.9	7.0	7.2	7.4	7.9	8.4	9.0	9.5	10.0	10.5	10.9
School Enroll. Ratio, primary	..	..	..	12.0	..	..	..	..	29.0	..	..
School Enroll. Ratio, secondary	..	..	..	1.0	..	..	..	..	4.0	..	..

1978	1979	1980	1981	1982	1983	1984	1985	1986	1987 estimate	NOTES	YEMEN ARAB REPUBLIC
											FOREIGN TRADE (CUSTOMS BASIS)
				(Millions of current US dollars)							
7.3	13.5	22.6	47.5	39.4	26.8	32.0	53.0	25.0	19.1	..	Value of Exports, fob
5.4	8.1	11.1	11.8	13.1	8.9	9.6	15.0	0.9	4.2	..	Nonfuel Primary Products
0.0	0.0	0.0	0.0	0.0	0.0	0.0	0.0	0.0	0.0	..	Fuels
1.9	5.3	11.5	35.7	26.3	17.9	22.4	38.0	24.1	14.9	..	Manufactures
1,357.8	1,491.8	1,853.0	1,758.0	1,521.0	1,593.0	1,565.0	1,359.5	981.9	1,072.0	..	Value of Imports, cif
398.8	408.1	537.8	576.1	478.8	501.6	450.5	379.5	275.8	310.6	..	Nonfuel Primary Products
37.2	35.5	133.5	145.6	113.7	121.2	77.1	59.7	7.3	0.0	..	Fuels
921.7	1,048.2	1,181.7	1,036.3	928.5	970.1	1,037.4	920.3	698.9	761.4	..	Manufactures
				(Index 1980 = 100)							
103.5	101.9	100.0	98.3	94.5	94.7	95.3	92.9	100.7	93.2	..	Terms of Trade
80.0	91.9	100.0	99.0	90.5	90.1	88.7	84.2	101.9	101.1	..	Export Prices, fob
81.5	93.3	100.0	98.7	82.9	88.8	87.3	74.2	76.4	73.7	..	Nonfuel Primary Products
42.3	61.0	100.0	112.5	101.6	92.5	90.2	87.5	44.9	56.9	..	Fuels
76.1	90.0	100.0	99.1	94.8	90.7	89.3	89.0	103.2	113.0	..	Manufactures
77.3	90.2	100.0	100.7	95.8	95.1	93.1	90.7	101.2	108.4	..	Import Prices, cif
				(Millions of 1980 US dollars)							Trade at Constant 1980 Prices
9.1	14.7	22.6	48.0	43.5	29.8	36.1	62.9	24.5	18.9	..	Exports, fob
1,756.5	1,654.3	1,853.0	1,746.1	1,588.2	1,675.4	1,681.4	1,499.2	970.3	989.1	..	Imports, cif
				(Millions of current US dollars)							**BALANCE OF PAYMENTS**
395.0	539.0	634.0	666.0	786.0	764.0	666.0	556.0	518.0	586.0	C f	Exports of Goods & Services
6.0	5.0	13.0	10.0	5.0	10.0	9.0	8.0	16.0	69.0	..	Merchandise, fob
79.0	123.0	164.0	210.0	252.0	239.0	203.0	151.0	113.0	108.0	..	Nonfactor Services
310.0	411.0	457.0	446.0	529.0	515.0	454.0	397.0	389.0	409.0	..	Factor Services
1,201.0	1,720.0	2,253.0	2,137.0	2,328.0	2,148.0	1,703.0	1,367.0	1,117.0	1,739.0	C f	Imports of Goods & Services
944.0	1,412.0	1,869.0	1,733.0	1,952.0	1,796.0	1,414.0	1,106.0	868.0	1,371.0	..	Merchandise, fob
251.0	306.0	375.0	386.0	365.0	339.0	273.0	242.0	207.0	312.0	..	Nonfactor Services
6.0	2.0	9.0	18.0	11.0	13.0	16.0	19.0	42.0	56.0	..	Factor Services
3.7	4.5	5.5	9.9	10.5	13.3	16.4	18.2	42.1	44.9	..	Long-Term Interest
827.0	939.0	933.0	811.0	951.0	838.0	723.0	492.0	506.0	546.0	..	Current Transfers, net
1,012.0	925.0	974.0	624.0	749.0	790.0	653.0	451.0	339.0	428.0	..	Workers' Remittances
21.0	-242.0	-686.0	-660.0	-591.0	-546.0	-314.0	-319.0	-93.0	-607.0	K	Total to be Financed
0.0	0.0	0.0	0.0	0.0	0.0	0.0	0.0	0.0	..	K	Official Capital Grants
21.0	-242.0	-686.0	-660.0	-591.0	-546.0	-314.0	-319.0	-93.0	-607.0	..	Current Account Balance
102.0	92.0	478.0	238.0	192.0	282.0	115.0	137.0	96.0	352.0	C f	Long-Term Capital, net
0.0	22.0	34.0	40.0	5.0	72.0	-48.0	18.0	15.0	-10.0	..	Direct Investment
102.4	102.4	408.9	242.8	217.6	294.3	158.0	132.6	153.8	14.7	..	Long-Term Loans
109.5	110.9	424.7	296.5	263.2	324.1	209.2	188.9	213.9	115.1	..	Disbursements
7.1	8.5	15.8	53.7	45.6	29.8	51.2	56.3	60.1	100.4	..	Repayments
-0.4	-32.4	35.1	-44.8	-30.6	-84.3	5.0	-13.6	-72.8	347.3	..	Other Long-Term Capital
94.0	134.0	9.0	121.0	63.9	28.4	74.2	16.5	183.9	202.0	C f	Other Capital, net
-217.0	16.0	199.0	301.0	335.1	235.6	124.9	165.5	-186.9	53.0	..	Change in Reserves
											Memo Item:
				(Yemen Rials per US dollar)							Conversion Factor (Annual Avg)
4.560	4.560	4.560	4.560	4.560	4.580	5.350	7.270	9.420	10.210	..	
			(Millions of US dollars, outstanding at end of year)								**EXTERNAL DEBT, ETC.**
474.0	496.9	903.2	1,121.3	1,315.8	1,577.4	1,694.6	1,868.6	2,051.8	2,154.5	..	Public/Publicly Guar. Long-Term
469.8	487.4	887.9	1,107.7	1,302.6	1,566.2	1,685.0	1,858.4	1,973.7	2,078.3	..	Official Creditors
66.0	79.8	101.9	119.3	137.5	169.5	193.0	229.3	264.1	309.7	..	IBRD and IDA
4.2	9.5	15.3	13.6	13.2	11.1	9.6	10.2	78.1	76.2	..	Private Creditors
..	..	..	..	..	..	..	..	..	..	..	Private Non-guaranteed Long-Term
0.0	0.0	0.0	0.0	0.0	10.2	9.6	10.7	7.5	1.7	..	Use of Fund Credit
92.0	126.0	86.0	86.0	98.0	155.0	259.0	180.0	249.2	232.3	..	Short-Term Debt
				(Millions of US dollars)							Memo Items:
1,459.4	1,427.4	1,282.6	961.6	554.2	366.0	318.5	296.8	431.7	539.5	..	Int'l Reserves Excluding Gold
1.5	3.4	0.6	0.4	0.5	0.4	0.3	0.3	0.4	0.5	..	Gold Holdings (at market price)
											SOCIAL INDICATORS
6.8	6.8	6.8	6.8	6.8	6.8	6.8	6.8	6.8	6.8	..	Total Fertility Rate
48.5	48.5	48.4	48.4	48.3	48.5	48.6	48.8	49.0	49.1	..	Crude Birth Rate
168.6	166.1	163.7	161.2	158.8	157.0	155.1	153.3	151.5	149.7	..	Infant Mortality Rate
41.8	42.3	42.9	43.4	43.9	44.3	44.7	45.2	45.6	46.0	..	Life Expectancy at Birth
95.8	97.0	99.7	103.3	103.0	94.0	101.0	106.3	120.8	118.7	..	Food Production, p.c. ('79-81 = 100)
70.3	69.5	68.8	..	..	..	..	..	..	..	..	Labor Force, Agriculture (%)
11.3	11.7	12.2	12.3	12.5	12.7	12.8	13.0	13.1	13.2	..	Labor Force, Female (%)
27.0	35.0	47.0	51.0	62.0	67.0	..	..	..	..	..	School Enroll. Ratio, primary
6.0	5.0	5.0	..	7.0	10.0	..	..	..	..	..	School Enroll. Ratio, secondary

YEMEN, PEOPLE'S DEMOCRATIC REP.	1967	1968	1969	1970	1971	1972	1973	1974	1975	1976	1977
CURRENT GNP PER CAPITA (US $)	..	..	..	..	..	..	..	..	..	..	330
POPULATION (thousands)	1,409	1,438	1,467	1,497	1,527	1,558	1,590	1,621	1,653	1,686	1,720

ORIGIN AND USE OF RESOURCES *(Millions of current Yemeni Dinars)*

	1967	1968	1969	1970	1971	1972	1973	1974	1975	1976	1977
Gross National Product (GNP)	..	..	..	..	..	..	97.0	114.7	107.1	142.7	186.6
Net Factor Income from Abroad	..	..	..	..	..	..	13.0	16.6	6.3	15.1	26.4
Gross Domestic Product (GDP)	..	..	..	..	..	..	84.0	98.1	100.8	127.6	160.2
Indirect Taxes, net	..	..	..	..	..	..	8.5	10.9	11.6	15.2	20.2
GDP at factor cost	..	..	..	..	..	..	75.5	87.2	89.2	112.4	140.0
Agriculture	..	..	..	..	..	..	16.4	17.2	17.7	24.5	25.4
Industry	..	..	..	..	..	..	13.2	20.5	17.7	24.5	25.4
Manufacturing	..	..	..	..	..	..	13.2	20.5	17.4	21.9	34.0
Services, etc.	..	..	..	..	..	..	45.9	49.5	54.1	66.0	80.6
Resource Balance	..	..	..	..	..	..	..	..	..	..	..
Exports of Goods & NFServices	..	..	..	..	..	..	..	..	..	..	..
Imports of Goods & NFServices	..	..	..	..	..	..	..	..	..	..	..
Domestic Absorption	..	..	..	..	..	..	..	..	..	..	..
Private Consumption, etc.	..	..	..	..	..	..	..	..	..	..	..
General Gov't Consumption	..	..	..	..	..	..	..	..	..	..	..
Gross Domestic Investment	..	..	..	..	..	..	..	..	..	..	..
Fixed Investment	..	..	..	..	..	..	..	..	..	..	..
Memo Items:											
Gross Domestic Saving	..	..	..	..	..	..	..	..	..	..	..
Gross National Saving	..	..	..	..	..	..	..	..	..	..	..

(Millions of 1980 Yemeni Dinars)

	1967	1968	1969	1970	1971	1972	1973	1974	1975	1976	1977
Gross National Product	..	..	..	..	..	..	..	..	146.7	192.4	243.0
GDP at factor cost	..	..	..	..	..	..	..	..	122.2	151.5	182.3
Agriculture	..	..	..	..	..	..	..	..	21.0	28.2	28.5
Industry	..	..	..	..	..	..	..	..	20.6	26.6	38.9
Manufacturing	..	..	..	..	..	..	..	..	..	..	..
Services, etc.	..	..	..	..	..	..	..	..	80.6	96.7	114.9
Resource Balance	..	..	..	..	..	..	..	..	..	..	..
Exports of Goods & NFServices	..	..	..	..	..	..	..	..	..	..	..
Imports of Goods & NFServices	..	..	..	..	..	..	..	..	..	..	..
Domestic Absorption	..	..	..	..	..	..	..	..	..	..	..
Private Consumption, etc.	..	..	..	..	..	..	..	..	..	..	..
General Gov't Consumption	..	..	..	..	..	..	..	..	..	..	..
Gross Domestic Investment	..	..	..	..	..	..	..	..	..	..	..
Fixed Investment	..	..	..	..	..	..	..	..	..	..	..
Memo Items:											
Capacity to Import	..	..	..	..	..	..	..	..	..	..	..
Terms of Trade Adjustment	..	..	..	..	..	..	..	..	..	..	..
Gross Domestic Income	..	..	..	..	..	..	..	..	..	..	..
Gross National Income	..	..	..	..	..	..	..	..	..	..	..

DOMESTIC PRICES (DEFLATORS) *(Index 1980 = 100)*

	1967	1968	1969	1970	1971	1972	1973	1974	1975	1976	1977
Overall (GDP)	..	..	..	..	..	..	..	..	73.0	74.2	76.8
Domestic Absorption	..	..	..	..	..	..	..	..	..	..	..
Agriculture	..	..	..	..	..	..	..	..	84.3	86.9	89.1
Industry	..	..	..	..	..	..	..	..	84.5	82.3	87.4
Manufacturing	..	..	..	..	..	..	..	..	..	..	..

MANUFACTURING ACTIVITY

	1967	1968	1969	1970	1971	1972	1973	1974	1975	1976	1977
Employment (1980=100)	..	..	..	..	..	..	..	..	..	..	..
Real Earnings per Empl. (1980=100)	..	..	..	..	..	..	..	..	..	..	..
Real Output per Empl. (1980=100)	..	..	..	..	..	..	..	..	..	..	..
Earnings as % of Value Added	..	..	..	..	..	..	..	..	..	..	..

MONETARY HOLDINGS *(Millions of current Yemeni Dinars)*

	1967	1968	1969	1970	1971	1972	1973	1974	1975	1976	1977
Money Supply, Broadly Defined	36.1	33.7	33.3	38.2	39.9	43.7	49.5	56.0	71.8	103.2	151.9
Money as Means of Payment	29.8	28.3	26.8	29.0	31.8	34.4	40.1	46.1	55.4	81.5	115.2
Currency Ouside Banks	20.2	20.6	21.3	24.5	25.9	28.8	32.4	35.8	41.3	63.8	88.2
Demand Deposits	9.6	7.7	5.6	4.6	5.8	5.6	7.7	10.3	14.0	17.7	27.0
Quasi-Monetary Liabilities	6.4	5.4	6.5	9.1	8.2	9.3	9.4	9.9	16.5	21.7	36.6

GOVERNMENT DEFICIT (-) OR SURPLUS *(Millions of current Yemeni Dinars)*

	1967	1968	1969	1970	1971	1972	1973	1974	1975	1976	1977
Current Revenue	..	..	..	..	..	..	..	..	..	..	..
Current Expenditure	..	..	..	..	..	..	..	..	..	..	..
Current Budget Balance	..	..	..	..	..	..	..	..	..	..	..
Capital Receipts	..	..	..	..	..	..	..	..	..	..	..
Capital Payments	..	..	..	..	..	..	..	..	..	..	..

1978	1979	1980	1981	1982	1983	1984	1985	1986	1987 estimate	NOTES	UGANDA
280	290	280	220	240	220	220	230	230	260	A	**CURRENT GNP PER CAPITA (US $)**
11,976	12,297	12,637	12,998	13,382	13,789	14,222	14,680	15,160	15,655	..	**POPULATION (thousands)**
											ORIGIN AND USE OF RESOURCES
			(Billions of current Old Uganda Shillings)								
5.9E+01	8.8E+01	1.3E+02	2.1E+02	403	608	984	2,518	6,163	16,115	..	Gross National Product (GNP)
-3.2E-02	-6.9E-02	-6.7E-02	-6.2E-02	-2	-6	-22	-34	-56	-228	..	Net Factor Income from Abroad
5.9E+01	8.8E+01	1.3E+02	2.1E+02	405	614	1,006	2,552	6,218	16,343	f	Gross Domestic Product (GDP)
4.0E+00	3.0E+00	3.0E+00	1.3E+01	39	72	128	230	352	1,074	..	Indirect Taxes, net
5.5E+01	8.5E+01	1.2E+02	2.0E+02	366	542	878	2,322	5,866	15,269	f	GDP at factor cost
4.1E+01	5.6E+01	8.9E+01	1.4E+02	261	401	657	1,815	4,674	11,609	..	Agriculture
3.0E+00	4.0E+00	6.0E+00	1.8E+01	31	35	45	93	225	793	f	Industry
2.0E+00	3.0E+00	5.0E+00	1.7E+01	30	34	42	87	213	760	..	Manufacturing
1.1E+01	2.6E+01	2.9E+01	4.8E+01	75	106	177	414	967	2,867	f	Services, etc.
-9.8E-01	1.6E-01	-2.0E+00	-1.3E+01	-14	-21	-2	-83	-162	-1,183	..	Resource Balance
3.0E+00	3.0E+00	2.0E+00	1.3E+01	34	58	159	242	587	1,583	..	Exports of Goods & NFServices
4.0E+00	3.0E+00	4.0E+00	2.6E+01	48	79	162	325	749	2,765	..	Imports of Goods & NFServices
6.0E+01	8.8E+01	1.3E+02	2.3E+02	419	634	1,008	2,635	6,381	17,526	..	Domestic Absorption
..	..	..	..	..	563	795	2,280	5,374	14,445	..	Private Consumption, etc.
..	..	..	..	..	18	49	136	503	1,117	..	General Gov't Consumption
5.0E+00	6.0E+00	8.0E+00	1.5E+01	29	53	164	219	504	1,964	..	Gross Domestic Investment
..	..	..	..	..	..	..	..	..	..		Fixed Investment
											Memo Items:
4.0E+00	6.0E+00	6.0E+00	2.0E+00	15	33	162	136	342	781	..	Gross Domestic Saving
..	..	..	..	..	..	149	131	463	1,059	..	Gross National Saving
			(Billions of 1980 Old Uganda Shillings)								
158.11	133.52	126.60	138.61	155.80	166.34	153.95	145.96	137.74	142.53	..	Gross National Product
149.44	129.96	124.00	130.25	141.98	148.54	137.42	134.44	130.94	134.85	f	GDP at factor cost
111.12	95.31	89.32	95.39	105.04	110.39	98.64	95.04	90.43	93.02	..	Agriculture
8.13	5.69	5.56	5.64	6.15	6.19	6.55	5.91	5.75	6.57	f	Industry
7.56	5.00	5.30	5.02	5.73	5.89	6.10	5.42	4.77	5.04	..	Manufacturing
30.19	28.96	29.12	29.22	30.79	31.97	32.23	33.48	34.75	35.26	f	Services, etc.
-2.56	-0.53	-1.78	..	..	..	..	..	..	..	..	Resource Balance
2.91	3.17	2.47	..	..	..	..	..	..	..	..	Exports of Goods & NFServices
5.47	3.70	4.24	..	..	..	..	..	..	..	..	Imports of Goods & NFServices
160.76	134.16	128.45	..	..	..	..	..	..	..	..	Domestic Absorption
..	..	..	..	..	..	..	..	..	..	..	Private Consumption, etc.
..	..	..	..	..	..	..	..	..	..	..	General Gov't Consumption
10.27	8.52	7.66	..	..	..	..	..	..	..	..	Gross Domestic Investment
..	..	..	..	..	..	..	..	..	..		Fixed Investment
											Memo Items:
3.96	3.90	2.47	..	..	..	..	..	..	..	..	Capacity to Import
1.04	0.73	0.00	..	..	..	..	..	..	..	..	Terms of Trade Adjustment
159.24	134.36	126.67	..	..	..	..	..	..	..	..	Gross Domestic Income
159.16	134.26	126.60	..	..	..	..	..	..	..	..	Gross National Income
											DOMESTIC PRICES (DEFLATORS)
			(Index 1980 = 100)								
37.4	66.0	100.0	154.7	258.5	365.5	639.0	1725.0	4473.0	11301.8	..	Overall (GDP)
37.4	65.6	100.0	..	..	..	..	..	..	..	..	Domestic Absorption
36.9	58.5	100.0	141.8	248.3	363.2	665.7	1909.9	5168.3	12479.2	..	Agriculture
35.0	62.1	100.0	311.2	499.2	567.2	681.7	1567.1	3909.3	12074.6	..	Industry
32.3	62.8	100.0	342.9	521.4	575.8	692.9	1607.1	4460.2	15078.1	..	Manufacturing
											MANUFACTURING ACTIVITY
..	..	..	..	..	..	..	..	..	..	G	Employment (1980=100)
..	..	..	..	..	..	..	..	..	..	G	Real Earnings per Empl. (1980=100)
..	..	..	..	..	..	..	..	..	..	G	Real Output per Empl. (1980=100)
..	..	..	..	..	..	..	..	..	..	..	Earnings as % of Value Added
											MONETARY HOLDINGS
			(Millions of current New Uganda Shillings)								
93	140	180	350	390	540	1,000	3,000	7,000	..	..	Money Supply, Broadly Defined
70	110	140	280	300	440	990	2,000	6,000	..	..	Money as Means of Payment
35	58	73	110	130	190	490	1,000	4,000	..	..	Currency Ouside Banks
34	48	67	170	170	250	500	1,000	3,000	..	..	Demand Deposits
23	31	45	62	87	110	180	370	1,000	..	..	Quasi-Monetary Liabilities
											GOVERNMENT DEFICIT (-) OR SURPLUS
			(Millions of current Old Uganda Shillings)								
-160	-3,460	-3,910	-10,050	-14,600	-13,380	-22,120	-64,020	-134,490	..	C	
5,790	2,630	4,010	3,090	27,650	53,680	93,760	166,890	322,820	..	..	Current Revenue
4,550	4,850	6,700	10,600	34,620	57,290	101,680	194,650	335,320	..	..	Current Expenditure
1,240	-2,220	-2,690	-7,510	-6,970	-3,620	-7,920	-27,750	-12,500	..	..	Current Budget Balance
..	..	..	..	10	..	..	..	..	..	..	Capital Receipts
1,400	1,230	1,220	2,540	7,640	9,760	14,200	36,270	121,990	..	..	Capital Payments

1978	1979	1980	1981	1982	1983	1984	1985	1986	1987 estimate	NOTES	YEMEN, PEOPLE'S DEMOCRATIC REP.
370	410	420	490	510	500	540	490	420	420	..	**CURRENT GNP PER CAPITA (US $)**
1,757	1,796	1,838	1,885	1,937	1,996	2,062	2,137	2,205	2,276	..	**POPULATION (thousands)**
				(Millions of current Yemeni Dinars)							**ORIGIN AND USE OF RESOURCES**
214.6	255.0	280.1	313.8	345.2	365.9	388.1	362.8	314.2	324.8	..	Gross National Product (GNP)
38.4	48.9	60.8	68.6	40.9	26.9	14.7	-3.8	-11.7	-16.3	..	Net Factor Income from Abroad
176.2	206.1	219.3	245.2	304.3	339.0	373.4	366.6	325.9	341.1	..	Gross Domestic Product (GDP)
25.2	29.9	31.9	42.6	58.1	63.5	67.0	57.0	52.2	50.1	..	Indirect Taxes, net
151.0	176.2	187.4	202.6	246.2	275.5	306.4	309.6	273.7	291.0	..	GDP at factor cost
19.7	23.4	30.0	28.1	30.0	32.2	37.9	41.3	43.4	45.5	..	Agriculture
48.5	47.3	43.9	52.4	67.0	72.5	82.7	80.1	67.0	68.3	..	Industry
..	..	..	..	..	..	..	..	..	..	..	Manufacturing
82.8	105.5	113.5	122.1	149.2	170.8	185.8	188.2	163.3	177.2	..	Services, etc.
..	..	..	..	..	..	..	..	..	..	..	Resource Balance
..	..	..	..	..	..	..	..	..	..	..	Exports of Goods & NFServices
..	..	..	..	..	..	..	..	..	..	..	Imports of Goods & NFServices
..	..	..	..	..	..	..	..	..	..	..	Domestic Absorption
..	..	..	..	..	..	..	..	..	..	..	Private Consumption, etc.
..	..	..	..	..	..	..	..	..	..	..	General Gov't Consumption
..	..	..	..	..	..	..	..	..	..	..	Gross Domestic Investment
..	..	..	..	..	..	..	..	..	..	..	Fixed Investment
											Memo Items:
..	..	..	..	..	..	..	..	..	..	..	Gross Domestic Saving
..	..	..	..	..	..	..	..	..	..	..	Gross National Saving
				(Millions of 1980 Yemeni Dinars)							
276.6	292.7	280.1	299.1	300.5	287.0	300.8	267.5	229.9	234.3	..	Gross National Product
194.6	202.3	187.4	193.1	..	..	..	..	..	..	..	GDP at factor cost
21.9	24.6	30.0	26.6	..	..	..	..	..	..	..	Agriculture
55.1	47.8	43.9	49.4	..	..	..	..	..	..	..	Industry
..	..	..	..	..	..	..	..	..	..	..	Manufacturing
117.6	129.9	113.5	117.1	..	..	..	..	..	..	..	Services, etc.
..	..	..	..	..	..	..	..	..	..	..	Resource Balance
..	..	..	..	..	..	..	..	..	..	..	Exports of Goods & NFServices
..	..	..	..	..	..	..	..	..	..	..	Imports of Goods & NFServices
..	..	..	..	..	..	..	..	..	..	..	Domestic Absorption
..	..	..	..	..	..	..	..	..	..	..	Private Consumption, etc.
..	..	..	..	..	..	..	..	..	..	..	General Gov't Consumption
..	..	..	..	..	..	..	..	..	..	..	Gross Domestic Investment
..	..	..	..	..	..	..	..	..	..	..	Fixed Investment
											Memo Items:
..	..	..	..	..	..	..	..	..	..	..	Capacity to Import
..	..	..	..	..	..	..	..	..	..	..	Terms of Trade Adjustment
..	..	..	..	..	..	..	..	..	..	..	Gross Domestic Income
..	..	..	..	..	..	..	..	..	..	..	Gross National Income
				(Index 1980 = 100)							**DOMESTIC PRICES (DEFLATORS)**
77.6	87.1	100.0	104.9	114.9	127.5	129.0	135.6	136.7	138.6	..	Overall (GDP)
..	..	..	..	..	..	..	..	..	..	..	Domestic Absorption
90.0	95.1	100.0	105.6	..	..	..	..	..	..	..	Agriculture
88.0	99.0	100.0	106.1	..	..	..	..	..	..	..	Industry
..	..	..	..	..	..	..	..	..	..	..	Manufacturing
											MANUFACTURING ACTIVITY
..	..	..	..	..	..	..	..	..	..	G	Employment (1980 = 100)
..	..	..	..	..	..	..	..	..	..	G	Real Earnings per Empl. (1980 = 100)
..	..	..	..	..	..	..	..	..	..	G	Real Output per Empl. (1980 = 100)
..	..	..	..	..	..	..	..	..	..	..	Earnings as % of Value Added
				(Millions of current Yemeni Dinars)							**MONETARY HOLDINGS**
167.9	217.4	286.1	321.3	382.6	447.2	510.4	551.3	575.6	619.6	..	Money Supply, Broadly Defined
143.6	184.9	234.6	258.6	297.0	331.6	368.7	402.9	419.8	452.3	..	Money as Means of Payment
116.6	143.5	171.1	192.2	223.8	244.3	255.9	276.4	287.9	299.7	..	Currency Ouside Banks
27.0	41.4	63.5	66.5	73.1	87.3	112.8	126.5	131.9	152.6	..	Demand Deposits
24.3	32.5	51.5	62.7	85.6	115.6	141.8	148.5	155.9	167.4	..	Quasi-Monetary Liabilities
				(Millions of current Yemeni Dinars)							
..	..	..	..	..	..	..	..	..	..	..	**GOVERNMENT DEFICIT (-) OR SURPLUS**
..	..	..	..	..	..	..	..	..	..	..	Current Revenue
..	..	..	..	..	..	..	..	..	..	..	Current Expenditure
..	..	..	..	..	..	..	..	..	..	..	Current Budget Balance
..	..	..	..	..	..	..	..	..	..	..	Capital Receipts
..	..	..	..	..	..	..	..	..	..	..	Capital Payments

YEMEN, PEOPLE'S DEMOCRATIC REP.	1967	1968	1969	1970	1971	1972	1973	1974	1975	1976	1977
FOREIGN TRADE (CUSTOMS BASIS)				*(Millions of current US dollars)*							
Value of Exports, fob	137.0	110.0	134.0	135.0	96.0	96.0	100.0	228.0	172.0	177.0	181.0
Nonfuel Primary Products	14.7	12.5	17.8	16.8	11.9	12.0	12.5	16.5	17.8	28.9	34.5
Fuels	108.7	82.2	98.7	101.2	72.0	71.9	74.9	201.1	147.4	147.6	145.9
Manufactures	13.6	15.3	17.5	16.9	12.1	12.1	12.6	10.4	6.7	0.4	0.6
Value of Imports, cif	210.0	203.0	218.2	200.0	156.2	149.0	171.0	419.0	323.0	412.0	544.0
Nonfuel Primary Products	38.0	32.1	59.2	48.0	37.2	35.9	41.1	101.1	67.8	74.0	95.1
Fuels	87.5	66.0	85.0	75.5	58.6	56.4	64.5	157.1	149.7	226.7	256.6
Manufactures	84.5	105.0	74.0	76.5	60.5	56.7	65.4	160.8	105.5	111.3	192.2
Terms of Trade	60.3	53.3	61.0	58.6	60.2	60.8	62.1	79.9	81.3	89.6	88.4
(Index 1980 = 100)											
Export Prices, fob	5.2	5.4	5.5	5.5	7.1	7.9	11.2	39.0	38.2	42.0	47.1
Nonfuel Primary Products	32.1	29.3	32.2	37.5	39.5	41.0	66.2	76.9	64.1	80.4	95.5
Fuels	4.3	4.3	4.3	4.3	5.6	6.2	8.9	36.7	35.7	38.4	42.0
Manufactures	30.1	31.7	32.4	37.5	36.5	38.5	49.5	67.2	66.1	66.7	71.7
Import Prices, cif	8.6	10.2	9.1	9.3	11.8	13.0	18.1	48.8	46.9	46.9	53.2
Trade at Constant 1980 Prices				*(Millions of 1980 US dollars)*							
Exports, fob	2,642.2	2,019.7	2,426.7	2,466.0	1,356.0	1,214.2	890.7	584.6	450.5	421.5	384.6
Imports, cif	2,443.5	1,988.1	2,409.7	2,142.0	1,328.5	1,145.6	946.3	858.3	688.1	879.1	1,021.8
BALANCE OF PAYMENTS				*(Millions of current US dollars)*							
Exports of Goods & Services	..	..	..	105	66	71	66	63	55	92	98
Merchandise, fob	..	..	..	51	29	24	26	17	20	44	47
Nonfactor Services	..	..	..	47	33	44	37	39	31	38	41
Factor Services	..	..	..	7	4	3	4	7	5	9	10
Imports of Goods & Services	..	..	..	161	119	127	151	219	208	307	417
Merchandise, fob	..	..	..	114	93	95	114	180	171	257	344
Nonfactor Services	..	..	..	18	17	17	24	31	36	47	70
Factor Services	..	..	..	29	9	15	13	8	2	3	3
Long-Term Interest	..	..	..	0	0	0	0	0	0	0	0
Current Transfers, net	..	..	..	..	..	..	..	..	..	..	..
Workers' Remittances	..	..	..	60	48	31	34	44	62	121	188
Total to be Financed	..	..	..	..	..	..	..	..	..	..	..
Official Capital Grants	..	..	..	..	..	..	..	..	..	..	..
Current Account Balance	..	..	..	-4	-8	-29	-51	-113	-84	-46	-68
Long-Term Capital, net	..	..	..	12	2	11	26	53	41	108	126
Direct Investment	..	..	..	..	-1	..	..	..	..	..	..
Long-Term Loans	..	..	..	1	2	17	28	25	25	64	92
Disbursements	..	..	..	1	2	17	28	25	25	64	92
Repayments	..	..	..	..	..	..	..	..	0	..	..
Other Long-Term Capital	..	..	..	11	1	-7	-1	28	17	44	34
Other Capital, net	..	..	..	-8	3	18	27	35	16	-48	-44
Change in Reserves	..	..	..	0	3	0	-2	24	27	-14	-14
Memo Item:				*(Yemeni Dinars per US dollar)*							
Conversion Factor (Annual Avg)	0.360	0.420	0.420	0.420	0.420	0.380	0.350	0.350	0.350	0.350	0.350
EXTERNAL DEBT, ETC.				*(Millions of US dollars, outstanding at end of year)*							
Public/Publicly Guar. Long-Term	..	..	..	0.7	2.8	19.9	49.0	76.3	96.8	137.8	238.8
Official Creditors	..	..	..	0.7	2.8	19.9	49.0	76.3	96.8	137.8	238.8
IBRD and IDA	..	..	..	0.0	0.2	0.5	1.1	2.0	3.2	10.7	15.0
Private Creditors	..	..	..	..	..	..	..	..	..	..	..
Private Non-guaranteed Long-Term	..	..	..	..	..	..	..	..	..	..	..
Use of Fund Credit	..	..	..	0.0	0.0	0.0	0.0	11.4	27.7	42.7	44.6
Short-Term Debt	..	..	..	..	..	..	..	..	..	..	46.0
Memo Items:				*(Thousands of US dollars)*							
Int'l Reserves Excluding Gold	61,170	58,510	55,400	58,710	63,681	66,533	75,251	66,813	53,958	81,532	99,324
Gold Holdings (at market price)	1,397	1,663	602	639	746	1,110	1,919	3,133	2,356	2,264	4,817
SOCIAL INDICATORS											
Total Fertility Rate	7.0	7.0	7.0	7.0	7.0	7.0	6.9	6.9	6.9	6.8	6.8
Crude Birth Rate	49.0	48.8	48.7	48.5	48.4	48.2	48.0	47.8	47.6	47.4	47.2
Infant Mortality Rate	191.0	188.2	185.4	182.6	179.8	177.0	174.4	171.8	169.2	166.6	164.0
Life Expectancy at Birth	39.2	39.7	40.1	40.6	41.1	41.5	42.0	42.4	42.9	43.4	43.8
Food Production, p.c. ('79-81 = 100)	105.8	101.8	98.0	87.5	104.9	98.9	107.7	109.4	108.9	112.4	111.1
Labor Force, Agriculture (%)	53.0	52.2	51.5	50.8	49.8	48.9	47.9	47.0	46.0	45.0	44.0
Labor Force, Female (%)	8.5	8.6	8.7	8.8	9.1	9.3	9.5	9.8	10.0	10.2	10.3
School Enroll. Ratio, primary	..	..	..	57.0	..	..	..	..	81.0	81.0	72.0
School Enroll. Ratio, secondary	..	..	..	10.0	..	..	..	..	23.0	26.0	28.0

1978	1979	1980	1981	1982	1983	1984	1985	1986	1987 estimate	NOTES	YEMEN, PEOPLE'S DEMOCRATIC REP.
				(Millions of current US dollars)							**FOREIGN TRADE (CUSTOMS BASIS)**
192.0	466.0	777.0	607.0	795.0	674.0	645.0	681.0	393.0	408.9	..	Value of Exports, fob
32.5	25.1	41.2	43.7	34.4	35.7	33.0	33.8	31.0	33.8	..	Nonfuel Primary Products
158.2	439.8	733.9	561.7	759.4	637.0	610.9	645.7	358.8	371.9	..	Fuels
1.3	1.2	1.9	1.6	1.3	1.3	1.1	1.5	3.2	3.1	..	Manufactures
575.0	925.0	1,527.0	1,419.0	1,599.0	1,483.0	1,543.0	1,311.0	1,105.0	1,450.2	..	Value of Imports, cif
103.2	132.9	190.6	193.2	166.3	149.2	139.7	107.1	139.3	257.3	..	Nonfuel Primary Products
279.4	533.8	968.4	867.2	1,122.5	1,056.2	1,110.2	971.9	606.6	528.1	..	Fuels
192.4	258.3	368.0	358.6	310.2	277.6	293.1	232.1	359.1	664.8	..	Manufactures
				(Index 1980 = 100)							
83.8	87.7	100.0	103.6	101.7	100.1	99.9	98.9	77.8	73.3	..	Terms of Trade
46.7	62.2	100.0	110.7	100.5	92.2	90.2	87.1	46.4	57.8	..	Export Prices, fob
90.7	94.8	100.0	92.2	80.7	87.8	91.8	78.8	66.8	67.1	..	Nonfuel Primary Products
42.3	61.0	100.0	112.5	101.6	92.5	90.2	87.5	44.9	56.9	..	Fuels
76.1	90.0	100.0	99.1	94.8	90.7	89.3	89.0	103.2	113.0	..	Manufactures
55.7	70.9	100.0	106.8	98.8	92.1	90.4	88.1	59.6	79.0	..	Import Prices, cif
				(Millions of 1980 US dollars)							Trade at Constant 1980 Prices
411.5	748.9	777.0	548.5	791.0	731.0	714.7	782.2	847.7	706.9	..	Exports, fob
1,032.7	1,303.9	1,527.1	1,328.8	1,618.5	1,610.1	1,707.7	1,488.8	1,854.2	1,836.7	..	Imports, cif
				(Millions of current US dollars)							**BALANCE OF PAYMENTS**
101	118	182	191	195	182	165	167	135	186	..	Exports of Goods & Services
39	39	60	49	38	40	31	43	30	71	..	Merchandise, fob
49	59	87	91	96	96	100	103	89	100	..	Nonfactor Services
12	20	35	51	61	46	34	22	16	15	..	Factor Services
438	481	733	797	892	895	948	848	633	667	..	Imports of Goods & Services
367	387	598	641	691	684	734	624	447	451	..	Merchandise, fob
66	89	130	149	190	198	197	206	171	202	..	Nonfactor Services
4	5	5	7	12	14	17	18	16	14	..	Factor Services
1	2	7	5	9	12	14	15	18	15	..	Long-Term Interest
..	..	347	404	470	486	499	..	..	..	..	Current Transfers, net
262	322	352	409	475	491	506	429	293	305	..	Workers' Remittances
..	..	-207	-223	-226	-226	-283	..	..	..	..	Total to be Financed
..	..	83	144	126	42	30	..	..	..	..	Official Capital Grants
-29	-15	-124	-79	-100	-184	-253	-222	-175	-122	..	Current Account Balance
141	90	158	298	299	295	176	106	159	202	..	Long-Term Capital, net
..	..	..	..	..	..	..	0	0	..	..	Direct Investment
52	138	83	74	206	159	148	296	145	172	..	Long-Term Loans
53	157	104	92	232	215	211	391	216	228	..	Disbursements
1	19	21	18	26	56	63	95	71	56	..	Repayments
89	-48	-8	80	-33	95	-3	-190	14	30	..	Other Long-Term Capital
-33	-46	7	-176	-165	-97	69	22	-45	-133	..	Other Capital, net
-79	-29	-41	-43	-34	-14	8	94	62	54	..	Change in Reserves
				(Yemeni Dinars per US dollar)							**Memo Item:**
0.350	0.350	0.350	0.350	0.350	0.350	0.350	0.350	0.350	0.350	..	Conversion Factor (Annual Avg)
				(Millions of US dollars, outstanding at end of year)							**EXTERNAL DEBT, ETC.**
304.8	447.5	530.0	553.8	748.4	901.6	1,024.5	1,325.7	1,471.6	1,669.3	..	Public/Publicly Guar. Long-Term
304.8	447.5	530.0	553.8	748.4	901.6	1,024.5	1,325.7	1,471.6	1,669.3	..	Official Creditors
23.0	29.3	35.1	44.1	61.8	83.4	101.0	118.8	142.5	163.7	..	IBRD and IDA
..	..	..	..	..	..	..	..	..	..	..	Private Creditors
..	..	..	..	..	..	..	..	..	..	..	Private Non-guaranteed Long-Term
41.4	24.0	12.4	4.4	18.0	16.1	15.1	14.8	7.0	..	..	Use of Fund Credit
40.0	50.0	71.0	42.0	41.0	70.0	70.0	70.0	125.0	55.0	..	Short-Term Debt
				(Thousands of US dollars)							**Memo Items:**
187,910	209,740	233,770	254,540	286,250	281,610	248,720	186,650	138,020	97,105	..	Int'l Reserves Excluding Gold
6,599	21,299	24,523	16,536	19,007	15,870	12,825	13,603	16,261	20,139	..	Gold Holdings (at market price)
											SOCIAL INDICATORS
6.7	6.7	6.7	6.7	6.7	6.6	6.6	6.6	6.6	6.6	..	Total Fertility Rate
47.2	47.1	47.1	47.0	47.0	47.5	48.0	48.5	49.0	49.6	..	Crude Birth Rate
161.1	158.3	155.4	152.6	149.7	145.7	141.7	137.6	133.6	129.6	..	Infant Mortality Rate
44.3	44.7	45.1	45.6	46.0	47.0	47.9	48.8	49.8	50.7	..	Life Expectancy at Birth
98.9	104.0	98.4	97.6	89.9	91.1	89.2	88.3	88.0	84.9	..	Food Production, p.c. ('79-81 = 100)
43.1	42.1	41.1	..	..	..	..	..	..	..	..	Labor Force, Agriculture (%)
10.5	10.6	10.7	10.9	11.0	11.2	11.3	11.5	11.6	11.7	..	Labor Force, Female (%)
74.0	68.0	65.0	64.0	64.0	66.0	..	..	..	..	..	School Enroll. Ratio, primary
28.0	24.0	18.0	18.0	19.0	19.0	..	..	..	..	..	School Enroll. Ratio, secondary

YUGOSLAVIA	1967	1968	1969	1970	1971	1972	1973	1974	1975	1976	1977
CURRENT GNP PER CAPITA (US $)	520	510	600	650	740	780	870	1,150	1,380	1,620	1,970
POPULATION (thousands)	19,840	20,029	20,209	20,371	20,572	20,772	20,956	21,164	21,365	21,573	21,775

ORIGIN AND USE OF RESOURCES

(Billions of current Yugoslav Dinars)

	1967	1968	1969	1970	1971	1972	1973	1974	1975	1976	1977
Gross National Product (GNP)	118	130	149	171	220	268	330	426	537	677	835
Net Factor Income from Abroad	0	1	1	-1	0	0	0	1	0	-1	0
Gross Domestic Product (GDP)	118	129	148	172	220	268	329	425	537	679	836
Indirect Taxes, net	9	11	12	15	19	20	21	46	47	61	79
GDP at factor cost	109	119	136	157	201	248	308	380	490	618	756
Agriculture	25	23	26	28	34	44	55	61	77	93	106
Industry	45	50	53	64	81	116	125	159	220	259	317
Manufacturing	..	..	..	..	..	..	..	..	..	..	..
Services, etc.	39	45	56	65	86	88	129	159	193	265	334
Resource Balance	-2	-2	-3	-9	-14	-7	-12	-42	-42	-18	-52
Exports of Goods & NFServices	22	23	27	32	43	58	72	91	109	130	145
Imports of Goods & NFServices	24	25	30	41	56	65	84	133	151	148	198
Domestic Absorption	119	131	150	181	234	275	341	467	579	696	888
Private Consumption, etc.	64	70	81	95	125	154	195	264	301	361	442
General Gov't Consumption	20	23	26	30	37	45	53	74	98	119	147
Gross Domestic Investment	35	38	43	56	72	75	93	129	180	217	299
Fixed Investment	30	35	41	52	65	75	86	117	163	207	268
Memo Items:											
Gross Domestic Saving	34	36	41	46	58	69	82	87	138	199	247
Gross National Saving	36	39	46	53	70	86	107	116	169	237	298

(Billions of 1980 Yugoslav Dinars)

	1967	1968	1969	1970	1971	1972	1973	1974	1975	1976	1977
Gross National Product	819	874	980	1,015	1,108	1,129	1,160	1,332	1,337	1,407	1,527
GDP at factor cost	755	800	895	933	1,016	1,045	1,084	1,186	1,220	1,282	1,381
Agriculture	145	140	153	144	153	160	174	183	178	190	201
Industry	268	283	311	334	358	432	448	492	528	549	601
Manufacturing	..	..	..	..	..	..	..	..	..	..	..
Services, etc.	342	376	432	455	505	453	462	511	514	543	580
Resource Balance	-20	-21	-22	-68	-86	-41	-58	-110	-102	-42	-103
Exports of Goods & NFServices	168	178	203	213	231	260	282	267	270	298	292
Imports of Goods & NFServices	188	199	225	282	317	301	340	377	372	340	394
Domestic Absorption	838	891	992	1,087	1,197	1,172	1,216	1,440	1,440	1,452	1,630
Private Consumption, etc.	351	387	478	515	584	580	575	747	693	673	745
General Gov't Consumption	168	180	182	187	191	203	207	222	229	242	255
Gross Domestic Investment	319	323	333	385	422	390	434	471	518	537	631
Fixed Investment	243	262	278	314	329	339	348	380	417	451	493
Memo Items:											
Capacity to Import	175	183	206	219	239	271	292	259	268	299	290
Terms of Trade Adjustment	7	5	3	5	8	11	10	-8	-1	2	-1
Gross Domestic Income	825	875	974	1,024	1,119	1,141	1,168	1,322	1,337	1,411	1,526
Gross National Income	826	879	983	1,020	1,116	1,140	1,170	1,324	1,336	1,409	1,525

DOMESTIC PRICES (DEFLATORS)

(Index 1980 = 100)

	1967	1968	1969	1970	1971	1972	1973	1974	1975	1976	1977
Overall (GDP)	14.4	14.8	15.2	16.9	19.8	23.7	28.4	32.0	40.1	48.2	54.7
Domestic Absorption	14.3	14.7	15.1	16.6	19.6	23.4	28.1	32.4	40.2	48.0	54.5
Agriculture	17.1	16.5	17.0	19.2	22.5	27.5	31.6	33.3	43.2	48.8	52.6
Industry	16.6	17.7	17.2	19.3	22.6	26.8	27.8	32.4	41.6	47.3	52.7
Manufacturing	..	..	..	..	..	..	..	..	..	..	..

MANUFACTURING ACTIVITY

	1967	1968	1969	1970	1971	1972	1973	1974	1975	1976	1977
Employment (1980=100)	56.4	56.7	58.5	61.3	65.5	69.3	72.8	74.9	77.9	86.0	89.6
Real Earnings per Empl. (1980=100)	72.8	78.6	84.1	90.5	95.0	94.5	91.9	95.2	92.8	99.6	105.0
Real Output per Empl. (1980=100)	..	..	..	58.5	62.5	63.3	76.5	100.3	96.9	91.3	98.1
Earnings as % of Value Added	36.4	36.8	38.6	38.8	37.8	38.3	36.8	34.7	35.6	38.9	38.9

MONETARY HOLDINGS

(Billions of current Yugoslav Dinars)

	1967	1968	1969	1970	1971	1972	1973	1974	1975	1976	1977
Money Supply, Broadly Defined	56	70	84	104	127	156	207	255	338	467	571
Money as Means of Payment	22	27	30	36	42	59	81	102	135	217	247
Currency Ouside Banks	8	10	12	15	18	24	29	35	42	49	58
Demand Deposits	14	17	18	21	23	35	52	67	93	168	189
Quasi-Monetary Liabilities	34	43	54	68	85	97	125	153	203	251	324

(Billions of current Yugoslav Dinars)

	1967	1968	1969	1970	1971	1972	1973	1974	1975	1976	1977
GOVERNMENT DEFICIT (-) OR SURPLUS	..	..	..	2.3	1.0	-1.0	-2.6	-5.6	-6.4	..	-9.0
Current Revenue	..	..	..	34.6	42.8	55.5	65.6	93.1	117.1	142.2	80.0
Current Expenditure	..	..	..	..	..	..	..	..	..	..	87.4
Current Budget Balance	..	..	..	..	..	..	..	..	..	..	-7.4
Capital Receipts	..	..	..	..	..	..	..	..	..	..	..
Capital Payments	..	..	..	..	..	..	..	..	..	..	1.6

1978	1979	1980	1981	1982	1983	1984	1985	1986	1987 estimate	NOTES	YUGOSLAVIA
2,400	2,870	3,250	3,450	3,230	2,640	2,270	2,060	2,300	2,480	..	**CURRENT GNP PER CAPITA (US $)**
21,968	22,166	22,304	22,471	22,646	22,800	22,963	23,123	23,280	23,413	..	**POPULATION** (thousands)
											ORIGIN AND USE OF RESOURCES
			(Billions of current Yugoslav Dinars)								
1,020	1,295	1,781	2,464	3,173	4,343	6,766	12,481	24,522	44,552	..	Gross National Product (GNP)
-1	-7	-20	-44	-78	-118	-208	-383	-561	-1,524	..	Net Factor Income from Abroad
1,021	1,302	1,801	2,507	3,251	4,460	6,974	12,863	25,083	46,076	..	Gross Domestic Product (GDP)
97	129	152	218	261	399	599	876	1,709	1,889	..	Indirect Taxes, net
925	1,173	1,648	2,289	2,991	4,061	6,375	11,987	23,374	44,187	..	GDP at factor cost
105	140	197	302	414	598	866	1,385	2,728	5,022	..	Agriculture
389	505	723	1,037	1,297	1,734	2,790	5,531	9,837	19,076	..	Industry
..	..	..	..	..	..	..	..	..	..	..	Manufacturing
431	529	728	950	1,280	1,729	2,718	5,070	10,809	20,089	..	Services, etc.
-49	-90	-76	-62	-2	10	56	229	438	507	..	Resource Balance
167	203	336	497	710	1,216	2,073	3,837	5,982	10,920	..	Exports of Goods & NFServices
216	293	413	559	712	1,206	2,018	3,609	5,544	10,413	..	Imports of Goods & NFServices
1,070	1,392	1,877	2,569	3,253	4,450	6,918	12,635	24,645	45,569	..	Domestic Absorption
542	670	878	1,219	1,607	2,221	3,371	5,932	11,615	21,518	..	Private Consumption, etc.
178	223	281	375	489	639	952	1,713	3,399	6,266	..	General Gov't Consumption
351	499	718	975	1,157	1,591	2,595	4,990	9,630	17,785	..	Gross Domestic Investment
357	448	546	685	855	1,029	1,458	2,567	..	..	..	Fixed Investment
											Memo Items:
302	409	642	914	1,155	1,601	2,651	5,218	10,069	18,292	..	Gross Domestic Saving
361	472	729	1,017	1,298	1,822	2,954	5,722	10,995	17,893	..	Gross National Saving
			(Billions of 1980 Yugoslav Dinars)								
1,665	1,746	1,781	1,791	1,790	1,765	1,792	1,809	1,884	1,854	..	Gross National Product
1,509	1,581	1,648	1,664	1,687	1,651	1,687	1,738	1,796	1,829	..	GDP at factor cost
190	197	197	202	217	216	220	206	220	224	..	Agriculture
656	701	723	739	727	715	740	754	784	801	..	Industry
..	..	..	..	..	..	..	..	..	..	..	Manufacturing
663	683	728	722	743	720	727	778	792	804	..	Services, etc.
-99	-139	-77	-37	-7	1	24	40	14	34	..	Resource Balance
306	314	336	321	302	284	304	319	310	304	..	Exports of Goods & NFServices
404	453	413	358	309	283	280	279	296	270	..	Imports of Goods & NFServices
1,766	1,894	1,877	1,860	1,841	1,813	1,823	1,824	1,914	1,883	..	Domestic Absorption
890	867	878	865	868	850	839	848	938	875	..	Private Consumption, etc.
265	278	281	268	269	263	265	269	281	290	..	General Gov't Consumption
611	750	718	727	705	700	719	706	694	718	..	Gross Domestic Investment
545	580	546	492	465	420	380	364	349	362	..	Fixed Investment
											Memo Items:
312	314	336	319	308	286	288	297	319	283	..	Capacity to Import
7	0	0	-2	6	2	-16	-23	10	-21	..	Terms of Trade Adjustment
1,674	1,755	1,801	1,821	1,840	1,815	1,831	1,842	1,937	1,896	..	Gross Domestic Income
1,672	1,746	1,781	1,789	1,795	1,767	1,776	1,787	1,893	1,833	..	Gross National Income
			(Index 1980 = 100)								**DOMESTIC PRICES (DEFLATORS)**
61.3	74.2	100.0	137.6	177.3	246.0	377.7	690.0	1301.4	2403.6	..	Overall (GDP)
60.6	73.5	100.0	138.1	176.7	245.5	379.6	692.7	1287.9	2420.0	..	Domestic Absorption
55.1	71.0	100.0	149.2	190.2	277.5	393.0	673.8	1239.7	2242.3	..	Agriculture
59.2	72.0	100.0	140.3	178.6	242.4	377.3	733.6	1254.5	2380.4	..	Industry
..	..	..	..	..	..	..	..	..	..	..	Manufacturing
											MANUFACTURING ACTIVITY
93.1	97.5	100.0	104.0	107.5	110.2	113.2	117.1	..	..	G	Employment (1980=100)
104.2	103.7	100.0	105.8	99.5	91.5	86.8	93.2	..	..	G	Real Earnings per Empl. (1980=100)
100.1	102.3	100.0	100.4	99.1	104.7	109.4	100.2	..	..	G	Real Output per Empl. (1980=100)
37.1	36.8	34.3	36.3	35.2	32.5	29.6	29.8	..	..	..	Earnings as % of Value Added
			(Billions of current Yugoslav Dinars)								**MONETARY HOLDINGS**
732	895	1,232	1,620	2,151	2,988	4,360	7,019	12,787	29,600	..	Money Supply, Broadly Defined
291	341	452	568	727	874	1,252	1,820	3,830	7,644	..	Money as Means of Payment
75	91	116	149	196	249	327	551	1,161	2,153	..	Currency Ouside Banks
216	250	336	420	531	625	925	1,269	2,669	5,490	..	Demand Deposits
442	554	781	1,052	1,424	2,114	3,108	5,199	8,957	21,956	..	Quasi-Monetary Liabilities
			(Billions of current Yugoslav Dinars)								
-5.3	-3.8	-20.4	-2.2	2.4	2.1	2.3	-7.0	8.2	..	E F	**GOVERNMENT DEFICIT (-) OR SURPLUS**
90.2	117.9	136.7	198.0	238.1	334.6	497.7	831.1	1.623.0	..	..	Current Revenue
93.6	119.3	155.1	199.1	234.2	331.4	493.9	833.6	1,606.8	..	..	Current Expenditure
-3.4	-1.4	-18.5	-1.1	3.8	3.2	3.9	-2.5	16.2	..	..	Current Budget Balance
..	..	..	..	..	..	..	..	..	..	..	Capital Receipts
1.9	2.4	2.0	1.1	1.5	1.2	1.6	4.5	8.0	..	..	Capital Payments

YUGOSLAVIA	1967	1968	1969	1970	1971	1972	1973	1974	1975	1976	1977
FOREIGN TRADE (CUSTOMS BASIS)				*(Millions of current US dollars)*							
Value of Exports, fob	1,252	1,264	1,474	1,679	1,836	2,237	3,020	3,805	4,072	4,896	4,896
Nonfuel Primary Products	521	500	589	656	665	801	1,062	1,226	1,131	1,452	1,337
Fuels	22	12	14	20	20	18	23	41	30	48	144
Manufactures	708	751	871	1,003	1,151	1,419	1,935	2,538	2,911	3,395	3,415
Value of Imports, cif	1,707	1,797	2,134	2,874	3,297	3,233	4,783	7,520	7,699	7,367	8,973
Nonfuel Primary Products	469	415	522	731	856	866	1,292	2,035	1,452	1,554	1,831
Fuels	85	98	103	138	195	176	380	951	945	1,081	1,207
Manufactures	1,154	1,284	1,508	2,005	2,246	2,190	3,111	4,534	5,302	4,732	5,935
Terms of Trade	116.5	126.7	127.2	130.3	125.6	120.1	128.1	107.9	109.0	106.4	99.1
					(Index 1980 = 100)						
Export Prices, fob	28.0	29.5	31.8	33.8	34.3	36.9	46.8	59.8	62.3	62.9	64.5
Nonfuel Primary Products	34.1	35.1	39.0	40.3	37.5	39.8	55.0	62.6	57.5	59.8	60.9
Fuels	4.3	4.3	4.3	4.3	5.6	6.2	8.9	36.7	35.7	38.4	42.0
Manufactures	29.2	29.2	31.2	35.1	35.7	37.7	45.5	59.1	64.9	64.9	67.5
Import Prices, cif	24.0	23.3	25.0	26.0	27.3	30.7	36.6	55.4	57.1	59.1	65.0
Trade at Constant 1980 Prices				*(Millions of 1980 US dollars)*							
Exports, fob	4,478	4,287	4,643	4,965	5,357	6,062	6,448	6,363	6,535	7,784	7,596
Imports, cif	7,117	7,724	8,547	11,070	12,084	10,521	13,077	13,565	13,472	12,465	13,797
BALANCE OF PAYMENTS				*(Millions of current US dollars)*							
Exports of Goods & Services	..	..	..	2,425	2,689	3,343	4,342	5,690	6,170	6,934	7,547
Merchandise, fob	..	..	..	1,679	1,821	2,238	2,853	3,805	4,073	4,897	5,193
Nonfactor Services	..	..	..	728	851	1,088	1,440	1,785	2,034	1,947	2,218
Factor Services	..	..	..	18	17	17	49	100	63	90	135
Imports of Goods & Services	..	..	..	3,345	3,875	3,961	5,365	8,406	8,611	8,940	11,697
Merchandise, fob	..	..	..	2,637	2,993	2,965	4,137	6,922	7,058	6,762	8,981
Nonfactor Services	..	..	..	566	722	820	994	1,185	1,209	1,809	2,306
Factor Services	..	..	..	142	159	176	234	299	344	369	410
Long-Term Interest	..	..	..	104	110	138	209	246	289	301	367
Current Transfers, net	..	..	..	542	780	1,039	1,510	1,755	1,804	2,187	2,805
Workers' Remittances	..	..	..	441	654	889	1,310	1,500	1,575	1,886	2,507
Total to be Financed	..	..	..	-378	-405	421	488	-962	-636	181	-1,345
Official Capital Grants	..	..	..	6	10	10	14	11	11	-1	-1
Current Account Balance	..	..	..	-372	-395	431	502	-951	-625	180	-1,346
Long-Term Capital, net	..	..	..	194	545	464	581	486	962	1,094	1,401
Direct Investment	..	..	..	..	..	..	..	..	..	..	..
Long-Term Loans	..	..	..	270	484	765	831	586	949	1,311	1,696
Disbursements	..	..	..	645	1,019	1,509	1,676	1,557	2,102	2,450	2,923
Repayments	..	..	..	375	536	744	845	971	1,153	1,139	1,227
Other Long-Term Capital	..	..	..	-82	51	-312	-265	-111	1	-216	-294
Other Capital, net	..	..	..	97	-237	-360	-405	171	-645	-251	233
Change in Reserves	..	..	..	81	87	-534	-677	295	309	-1,023	-287
Memo Item:				*(Yugoslav Dinars per US dollar)*							
Conversion Factor (Annual Avg)	12.500	12.500	12.500	12.500	14.880	17.000	16.240	15.910	17.340	18.180	18.290
EXTERNAL DEBT, ETC.				*(Millions of US dollars, outstanding at end of year)*							
Public/Publicly Guar. Long-Term	..	..	..	1,199	1,363	1,613	1,875	2,091	2,325	2,792	3,087
Official Creditors	..	..	..	857	1,006	1,149	1,371	1,640	1,923	2,371	2,692
IBRD and IDA	..	..	..	244	277	316	352	425	559	655	758
Private Creditors	..	..	..	341	357	464	504	451	401	421	396
Private Non-guaranteed Long-Term	..	..	..	854	1,341	1,825	2,407	2,833	3,493	4,380	5,871
Use of Fund Credit	..	..	..	0	73	91	26	225	182	422	313
Short-Term Debt	..	..	..	..	..	..	..	..	..	..	903
Memo Items:				*(Millions of US dollars)*							
Int'l Reserves Excluding Gold	58	82	202	89	157	675	1,276	1,085	811	1,990	2,044
Gold Holdings (at market price)	22	60	51	54	64	96	164	274	205	198	249
SOCIAL INDICATORS											
Total Fertility Rate	2.6	2.5	2.5	2.3	2.4	2.4	2.3	2.3	2.3	2.2	2.2
Crude Birth Rate	19.6	19.1	18.9	17.8	18.3	18.3	18.1	18.0	18.1	18.1	17.6
Infant Mortality Rate	62.1	58.6	57.3	55.5	49.5	44.4	40.0	40.9	39.7	36.7	35.5
Life Expectancy at Birth	66.6	67.0	67.3	67.7	68.0	68.4	68.5	68.5	68.6	68.7	68.7
Food Production, p.c. ('79-81 = 100)	87.0	86.0	93.6	80.9	88.3	85.0	89.8	96.6	95.1	99.0	101.8
Labor Force, Agriculture (%)	54.0	52.6	51.2	49.8	48.0	46.3	44.5	42.8	41.0	39.3	37.5
Labor Force, Female (%)	36.1	36.2	36.3	36.5	36.6	36.7	36.9	37.0	37.1	37.3	37.5
School Enroll. Ratio, primary	..	..	..	106.0	..	..	..	..	103.0	101.0	99.0
School Enroll. Ratio, secondary	..	..	..	63.0	..	..	..	..	76.0	78.0	79.0

1978	1979	1980	1981	1982	1983	1984	1985	1986	1987 estimate	NOTES	YUGOSLAVIA
											FOREIGN TRADE (CUSTOMS BASIS)
					(Millions of current US dollars)						
5,546	6,799	8,977	10,929	10,752	9,913	10,254	10,641	10,297	11,397	..	Value of Exports, fob
1,433	1,754	2,176	2,136	2,161	2,128	2,023	1,923	1,774	2,235	..	Nonfuel Primary Products
146	204	231	219	199	245	357	297	203	220	..	Fuels
3,967	4,841	6,570	8,574	8,392	7,541	7,874	8,421	8,320	8,942	..	Manufactures
9,769	14,037	15,064	15,757	14,100	12,154	11,996	12,163	11,749	13,114	..	Value of Imports, cif
1,788	2,476	2,973	2,902	2,545	2,238	2,210	2,241	2,264	2,441	..	Nonfuel Primary Products
1,401	2,251	3,549	3,786	3,630	3,304	3,515	3,307	2,605	3,505	..	Fuels
6,580	9,310	8,542	9,069	7,925	6,612	6,271	6,615	6,879	7,168	..	Manufactures
					(Index 1980 = 100)						
102.7	102.7	100.0	102.2	110.9	113.4	111.6	110.8	125.1	126.4	..	Terms of Trade
72.6	87.5	100.0	103.2	106.2	105.9	102.5	99.1	102.0	111.7	..	Export Prices, fob
70.6	95.6	100.0	89.4	82.3	86.5	82.3	75.5	77.8	84.7	..	Nonfuel Primary Products
42.3	61.0	100.0	112.5	101.6	92.5	90.2	87.5	44.9	56.9	..	Fuels
75.3	86.4	100.0	107.1	115.0	113.6	110.1	107.3	113.0	124.6	..	Manufactures
70.7	85.1	100.0	101.0	95.8	93.4	91.8	89.5	81.5	88.4	..	Import Prices, cif
					(Millions of 1980 US dollars)						Trade at Constant 1980 Prices
7,643	7,775	8,978	10,590	10,121	9,363	10,006	10,734	10,097	10,202	..	Exports, fob
13,825	16,488	15,064	15,602	14,712	13,014	13,061	13,593	14,408	14,835	..	Imports, cif
					(Millions of current US dollars)						**BALANCE OF PAYMENTS**
8,554	10,238	13,808	15,753	15,363	13,220	13,448	14,081	15,343	15,969	..	Exports of Goods & Services
5,811	6,802	9,077	10,363	10,460	9,913	10,136	10,623	11,054	12,073	..	Merchandise, fob
2,573	3,247	4,531	5,018	4,621	3,128	3,098	3,236	4,105	3,797	..	Nonfactor Services
170	188	199	372	282	179	213	222	185	99	..	Factor Services
13,098	17,619	20,466	20,935	20,252	16,594	16,310	16,528	18,167	16,676	..	Imports of Goods & Services
9,576	12,871	13,967	13,528	12,484	11,144	10,925	11,212	11,754	13,359	..	Merchandise, fob
3,027	3,926	5,204	5,389	5,676	3,741	3,550	3,430	4,484	1,150	..	Nonfactor Services
495	822	1,295	2,018	2,092	1,709	1,835	1,886	1,929	2,167	..	Factor Services
578	817	1,077	1,463	1,606	1,514	2,339	1,534	1,778	1,717	..	Long-Term Interest
3,263	3,718	4,346	4,224	4,414	3,652	3,342	3,282	3,922	1,526	..	Current Transfers, net
2,951	3,397	4,095	3,964	4,187	3,427	3,168	3,106	3,721	..	..	Workers' Remittances
-1,281	-3,663	-2,311	-958	-475	277	480	835	1,099	819	K	Total to be Financed
-3	-3	-2	0	0	-2	-2	-2	-2	..	K	Official Capital Grants
-1,284	-3,665	-2,313	-958	-475	275	478	833	1,097	819	..	Current Account Balance
1,602	1,300	1,952	576	-93	938	-267	78	-1,404	-1,540	..	Long-Term Capital, net
..	..	..	..	..	..	0	0	..	..	..	Direct Investment
1,753	1,656	2,209	1,227	92	861	-120	-79	-886	-838	..	Long-Term Loans
3,059	3,892	4,589	3,087	2,188	2,517	1,447	1,090	654	546	..	Disbursements
1,306	2,236	2,380	1,859	2,096	1,655	1,568	1,169	1,540	1,384	..	Repayments
-149	-354	-255	-652	-185	79	-145	159	-517	-702	..	Other Long-Term Capital
222	1,066	51	-587	-864	-1,490	-110	-752	1,143	462	..	Other Capital, net
-539	1,299	311	970	1,432	277	-102	-159	-836	259	..	Change in Reserves
											Memo Item:
					(Yugoslav Dinars per US dollar)						
18.640	18.970	24.640	34.970	50.280	92.840	152.820	270.160	379.220	737.000	..	Conversion Factor (Annual Avg)
					(Millions of US dollars, outstanding at end of year)						**EXTERNAL DEBT, ETC.**
3,412	3,669	4,580	5,198	5,461	7,234	8,361	9,349	11,981	14,446	..	Public/Publicly Guar. Long-Term
3,023	3,308	3,600	3,932	3,971	4,059	4,056	4,721	5,675	6,841	..	Official Creditors
898	1,143	1,359	1,483	1,684	1,814	1,731	2,268	2,788	3,311	..	IBRD and IDA
389	362	981	1,266	1,490	3,175	4,305	4,629	6,306	7,605	..	Private Creditors
7,707	10,056	11,005	11,704	10,875	10,033	8,187	7,979	5,831	5,045	..	Private Non-guaranteed Long-Term
238	454	760	1,252	1,754	2,068	1,947	2,108	2,069	1,852	..	Use of Fund Credit
1,171	1,789	2,140	2,492	1,810	1,142	1,026	990	1,340	2,175	..	Short-Term Debt
					(Millions of US dollars)						**Memo Items:**
2,388	1,257	1,384	1,597	775	976	1,158	1,095	1,460	698	..	Int'l Reserves Excluding Gold
369	881	1,094	738	849	710	574	610	729	904	..	Gold Holdings (at market price)
											SOCIAL INDICATORS
2.2	2.2	2.1	2.1	2.1	2.1	2.1	2.0	2.0	1.9	..	Total Fertility Rate
17.3	17.0	17.1	16.7	16.6	16.6	16.4	15.9	15.4	14.9	..	Crude Birth Rate
33.8	32.7	31.4	30.6	29.9	31.7	28.9	28.8	27.3	24.8	..	Infant Mortality Rate
68.8	68.8	68.9	68.9	69.0	69.4	69.9	70.4	70.8	71.3	..	Life Expectancy at Birth
95.2	100.4	99.5	100.1	107.9	101.6	104.1	95.8	98.8	96.6	..	Food Production, p.c. ('79-81 = 100)
35.8	34.0	32.3	..	..	..	..	..	..	..	..	Labor Force, Agriculture (%)
37.6	37.8	37.9	38.0	38.1	38.2	38.3	38.4	38.5	38.6	..	Labor Force, Female (%)
99.0	99.0	99.0	100.0	101.0	100.0	98.0	96.0	..	..	..	School Enroll. Ratio, primary
82.0	82.0	83.0	82.0	83.0	82.0	82.0	82.0	..	..	..	School Enroll. Ratio, secondary

ZAIRE	1967	1968	1969	1970	1971	1972	1973	1974	1975	1976	1977
CURRENT GNP PER CAPITA (US $)	170	170	180	180	200	200	240	270	300	300	330
POPULATION (thousands)	18,222	18,608	19,026	19,481	19,976	20,513	21,094	21,723	22,399	23,121	23,885

ORIGIN AND USE OF RESOURCES

(Billions of current Zaires)

	1967	1968	1969	1970	1971	1972	1973	1974	1975	1976	1977
Gross National Product (GNP)	0.87	1.39	1.77	1.72	1.94	2.14	2.73	3.26	3.44	5.26	7.39
Net Factor Income from Abroad	-0.05	-0.04	-0.05	-0.07	-0.07	-0.07	-0.09	-0.18	-0.22	-0.20	-0.16
Gross Domestic Product (GDP)	0.91	1.43	1.82	1.79	2.00	2.21	2.82	3.43	3.66	5.46	7.55
Indirect Taxes, net	..	..	..	0.10	0.11	0.12	0.15	0.16	0.16	0.16	0.17
GDP at factor cost	..	..	..	1.69	1.89	2.08	2.67	3.27	3.51	5.30	7.39
Agriculture	0.21	0.35	0.37	0.29	0.31	0.35	0.44	0.57	0.69	1.35	1.99
Industry	0.22	0.42	0.61	0.63	0.67	0.67	0.97	1.21	1.12	1.41	1.66
Manufacturing	0.14	0.08	0.09	0.14	0.17	0.18	0.21	0.28	0.35	0.44	0.52
Services, etc.	0.49	0.66	0.84	0.87	1.02	1.19	1.40	1.66	1.86	2.70	3.91
Resource Balance	0.08	0.13	0.15	0.15	0.06	0.03	0.13	0.19	-0.07	-0.05	-0.42
Exports of Goods & NFServices	0.20	0.31	0.36	0.41	0.37	0.37	0.56	0.81	0.52	0.92	1.11
Imports of Goods & NFServices	0.12	0.18	0.21	0.27	0.31	0.35	0.43	0.62	0.59	0.97	1.53
Domestic Absorption	0.83	1.30	1.67	1.64	1.94	2.18	2.69	3.25	3.74	5.51	7.97
Private Consumption, etc.	0.65	0.98	1.25	1.14	1.32	1.54	1.94	2.28	2.69	4.23	5.72
General Gov't Consumption	0.09	0.17	0.21	0.26	0.28	0.27	0.31	0.43	0.45	0.56	0.77
Gross Domestic Investment	0.09	0.14	0.21	0.24	0.34	0.38	0.43	0.54	0.60	0.73	1.49
Fixed Investment	0.09	0.15	0.22	0.20	0.30	0.37	0.37	0.55	0.55	0.66	1.37
Memo Items:											
Gross Domestic Saving	0.18	0.28	0.36	0.38	0.40	0.41	0.56	0.73	0.53	0.68	1.06
Gross National Saving	0.13	0.24	0.27	0.22	0.28	0.29	0.41	0.49	0.27	0.46	0.83

(Millions of 1980 Zaires)

	1967	1968	1969	1970	1971	1972	1973	1974	1975	1976	1977
Gross National Product	25,640	24,587	26,926	26,403	28,247	28,394	30,610	30,870	29,157	28,326	28,988
GDP at factor cost	..	..	..	24,791	26,346	26,379	28,878	30,750	29,274	28,408	28,767
Agriculture	..	8,108	8,148	7,051	7,242	7,353	7,633	7,784	7,668	8,077	7,810
Industry	..	5,190	5,903	8,408	8,917	8,980	9,745	10,237	9,879	9,061	9,161
Manufacturing	..	977	1,081	997	1,054	1,104	1,202	1,301	1,185	1,091	1,068
Services, etc.	..	12,014	13,622	12,146	13,104	12,974	14,315	14,664	13,508	12,269	12,659
Resource Balance	414	1,012	490	330	-12	-392	-939	-713	-546	184	-2,010
Exports of Goods & NFServices	3,159	4,550	4,573	5,181	5,774	6,161	6,743	6,221	5,833	5,467	5,451
Imports of Goods & NFServices	2,745	3,537	4,083	4,851	5,786	6,554	7,682	6,934	6,379	5,283	7,461
Domestic Absorption	29,595	26,714	29,776	28,320	29,862	34,098	38,241	36,460	35,304	33,628	31,641
Private Consumption, etc.	26,153	22,549	24,913	22,800	23,919	28,932	32,344	29,931	29,428	28,746	24,564
General Gov't Consumption	2,168	2,835	2,983	3,581	3,350	2,600	3,108	3,517	3,106	2,725	2,939
Gross Domestic Investment	1,274	1,330	1,880	1,939	2,592	2,567	2,790	3,012	2,770	2,157	4,138
Fixed Investment	..	1,363	1,927	1,646	2,257	2,606	2,425	3,168	2,609	1,987	3,821
Memo Items:											
Capacity to Import	4,748	6,110	6,888	7,493	6,899	7,073	9,967	9,014	5,582	5,026	5,407
Terms of Trade Adjustment	1,589	1,560	2,315	2,312	1,125	912	3,224	2,793	-251	-441	-44
Gross Domestic Income	27,468	26,872	29,988	29,917	30,388	30,219	34,916	35,477	30,805	28,966	29,587
Gross National Income	27,549	26,147	29,241	28,715	29,372	29,306	33,834	33,663	28,906	27,885	28,944

DOMESTIC PRICES (DEFLATORS)

(Index 1980 = 100)

	1967	1968	1969	1970	1971	1972	1973	1974	1975	1976	1977
Overall (GDP)	3.6	5.6	6.6	6.5	6.8	7.5	8.9	10.5	11.8	18.6	25.5
Domestic Absorption	2.8	4.9	5.6	5.8	6.5	6.4	7.0	8.9	10.6	16.4	25.2
Agriculture	..	4.3	4.5	4.1	4.3	4.7	5.8	7.3	9.0	16.8	25.4
Industry	..	8.0	10.3	7.5	7.5	7.5	10.0	11.8	11.3	15.5	18.1
Manufacturing	..	8.2	8.7	14.3	16.3	16.5	17.8	21.3	29.8	40.4	48.3

MANUFACTURING ACTIVITY

	1967	1968	1969	1970	1971	1972	1973	1974	1975	1976	1977
Employment (1980=100)	..	..	..	..	..	..	..	..	..	..	..
Real Earnings per Empl. (1980=100)	..	..	..	..	..	..	..	..	..	..	..
Real Output per Empl. (1980=100)	..	..	..	..	..	..	..	..	..	..	..
Earnings as % of Value Added	..	37.3	40.0	..	..	35.9	..	..	..	..	..

MONETARY HOLDINGS

(Billions of current Zaires)

	1967	1968	1969	1970	1971	1972	1973	1974	1975	1976	1977
Money Supply, Broadly Defined	0.11	0.15	0.17	0.21	0.22	0.27	0.38	0.50	0.55	0.76	1.21
Money as Means of Payment	0.10	0.13	0.15	0.19	0.19	0.23	0.29	0.39	0.46	0.66	1.03
Currency Ouside Banks	0.04	0.05	0.06	0.07	0.08	0.10	0.12	0.16	0.21	0.29	0.47
Demand Deposits	0.06	0.08	0.09	0.11	0.10	0.14	0.17	0.23	0.25	0.37	0.56
Quasi-Monetary Liabilities	0.00	0.02	0.02	0.02	0.04	0.04	0.09	0.11	0.09	0.10	0.18

(Millions of current Zaires)

	1967	1968	1969	1970	1971	1972	1973	1974	1975	1976	1977
GOVERNMENT DEFICIT (-) OR SURPLUS	..	..	..	..	-77	-82	-141	-324	-216	-627	-450
Current Revenue	..	..	..	..	329	344	440	588	491	573	830
Current Expenditure	..	..	..	..	313	320	387	518	553	953	1,003
Current Budget Balance	..	..	..	..	16	24	53	70	-62	-380	-173
Capital Receipts	..	..	..	..	..	..	..	..	..	..	..
Capital Payments	..	..	..	..	92	106	194	394	154	247	276

1978	1979	1980	1981	1982	1983	1984	1985	1986	1987 estimate	NOTES	ZAIRE
360	410	430	400	340	290	200	160	150	150	..	CURRENT GNP PER CAPITA (US $)
24,686	25,519	26,377	27,252	28,134	29,013	29,877	30,712	31,672	32,655	..	POPULATION (thousands)
											ORIGIN AND USE OF RESOURCES
				(Billions of current Zaires)							
9.00	18.22	28.04	38.90	50.03	94.71	146.96	197.69	307.52	590.85	..	Gross National Product (GNP)
-0.18	-0.38	-0.74	-0.94	-2.09	-4.64	-20.34	-24.82	-24.36	-58.11	..	Net Factor Income from Abroad
9.18	18.60	28.79	39.84	52.12	99.35	167.30	222.51	331.88	648.96	..	Gross Domestic Product (GDP)
0.18	0.43	1.01	1.53	1.39	1.88	2.00	7.37	11.25	31.70	..	Indirect Taxes, net
9.00	18.17	27.78	38.31	50.73	97.46	165.30	215.13	320.63	617.26	B	GDP at factor cost
2.78	5.77	8.29	12.65	18.72	34.22	56.38	73.24	103.68	208.74	..	Agriculture
1.96	4.28	8.31	9.32	10.39	23.37	63.44	80.79	130.23	214.48	f	Industry
0.56	0.70	0.87	1.09	1.19	1.48	2.31	2.93	5.74	44.68	..	Manufacturing
4.44	8.56	12.18	17.87	23.01	41.76	47.48	68.48	97.96	225.74	..	Services, etc.
0.08	0.25	-0.34	-2.22	-2.97	-2.69	5.90	10.28	3.28	-20.80	..	Resource Balance
1.71	3.96	6.94	7.77	20.01	41.22	73.44	98.69	119.26	214.28	..	Exports of Goods & NFServices
1.63	3.71	7.28	9.99	22.98	43.91	67.55	88.41	115.98	235.08	..	Imports of Goods & NFServices
9.10	18.35	29.12	42.06	55.09	102.04	161.40	212.23	328.60	669.76	..	Domestic Absorption
7.20	14.32	22.29	32.36	39.22	79.03	113.57	149.40	243.18	472.47	..	Private Consumption, etc.
0.89	1.86	2.56	3.73	8.38	12.18	24.58	32.79	48.25	112.93	..	General Gov't Consumption
1.01	2.18	4.28	5.97	7.49	10.83	23.25	30.04	37.17	84.37	..	Gross Domestic Investment
0.95	1.53	3.44	4.67	6.27	9.82	18.99	27.70	34.67	81.36	..	Fixed Investment
											Memo Items:
1.09	2.43	3.94	3.75	4.52	8.14	29.15	40.32	40.45	63.56	..	Gross Domestic Saving
0.91	2.04	2.98	2.80	2.38	3.54	5.51	12.77	12.40	-2.08	..	Gross National Saving
				(Millions of 1980 Zaires)							
27,516	27,577	28,044	28,950	27,772	27,813	26,675	27,627	29,380	29,726	..	Gross National Product
27,162	27,244	27,776	28,581	28,098	28,531	29,227	29,683	29,705	30,492	B H	GDP at factor cost
7,823	8,063	8,290	8,512	8,685	8,858	9,120	9,413	10,009	10,321	..	Agriculture
8,434	7,881	8,312	8,774	8,503	8,904	9,771	9,999	10,319	10,411	f	Industry
933	879	869	875	780	776	805	829	865	908	..	Manufacturing
11,790	12,223	12,184	12,334	11,666	11,336	10,993	11,213	11,124	11,568	..	Services, etc.
1,138	-51	-335	-448	807	1,773	2,229	2,324	3,494	2,628	..	Resource Balance
6,818	5,711	6,942	6,033	6,605	7,280	7,572	7,723	8,912	7,727	..	Exports of Goods & NFServices
5,680	5,762	7,277	6,481	5,797	5,507	5,342	5,399	5,419	5,099	..	Imports of Goods & NFServices
26,909	28,218	29,121	30,067	28,046	27,325	27,655	28,301	27,958	29,673	..	Domestic Absorption
22,543	22,586	22,289	23,190	22,032	22,745	22,432	23,540	21,857	23,693	..	Private Consumption, etc.
2,422	2,347	2,555	2,524	2,094	1,664	1,255	1,255	1,317	1,314	..	General Gov't Consumption
1,944	3,286	4,278	4,353	3,921	2,916	3,968	3,506	4,784	4,666	..	Gross Domestic Investment
1,834	2,301	3,435	3,411	3,280	2,636	3,293	3,252	4,784	..	..	Fixed Investment
											Memo Items:
5,960	6,150	6,942	5,042	5,048	5,170	5,809	6,027	5,572	4,648	..	Capacity to Import
-858	439	0	-991	-1,557	-2,111	-1,763	-1,696	-3,340	-3,079	..	Terms of Trade Adjustment
27,189	28,607	28,786	28,628	27,297	26,987	28,121	28,929	28,111	29,222	..	Gross Domestic Income
26,658	28,016	28,044	27,959	26,216	25,702	24,912	25,931	26,039	26,647	..	Gross National Income
				(Index 1980 = 100)							**DOMESTIC PRICES (DEFLATORS)**
32.7	66.0	100.0	134.5	180.6	341.4	559.8	726.6	1055.2	2009.1	..	Overall (GDP)
33.8	65.0	100.0	139.9	196.4	373.4	583.6	749.9	1175.3	2257.2	..	Domestic Absorption
35.5	71.5	100.0	148.6	215.6	386.3	618.2	778.1	1035.9	2022.4	..	Agriculture
23.3	54.3	100.0	106.3	122.2	262.5	649.3	808.0	1262.1	2060.1	..	Industry
60.3	79.9	100.0	124.1	152.8	190.5	286.4	353.8	663.7	4923.3	..	Manufacturing
											MANUFACTURING ACTIVITY
..	..	..	..	..	..	..	..	..	..	..	Employment (1980 = 100)
..	..	..	..	..	..	..	..	..	..	..	Real Earnings per Empl. (1980 = 100)
..	..	..	..	..	..	..	..	..	..	..	Real Output per Empl. (1980 = 100)
..	..	..	..	..	..	..	..	..	..	f	Earnings as % of Value Added
				(Billions of current Zaires)							**MONETARY HOLDINGS**
1.86	1.96	3.18	4.84	8.22	13.55	18.78	24.75	38.93	103.30	..	Money Supply, Broadly Defined
1.61	1.57	2.71	4.20	7.42	12.61	17.43	22.89	36.16	96.05	..	Money as Means of Payment
0.80	0.41	1.25	2.09	3.28	6.14	8.80	12.29	18.99	63.92	..	Currency Ouside Banks
0.81	1.16	1.46	2.11	4.14	6.47	8.63	10.60	17.17	32.13	..	Demand Deposits
0.25	0.39	0.47	0.64	0.80	0.94	1.35	1.85	2.77	7.25	..	Quasi-Monetary Liabilities
				(Millions of current Zaires)							
-595	-549	-332	-2,158	-3,484	-1,653	..	..	..	..	..	**GOVERNMENT DEFICIT (-) OR SURPLUS**
910	2,456	4,650	5,905	7,183	13,884	33,503	51,728	..	..	..	Current Revenue
1,239	2,517	4,004	5,783	8,242	12,391	..	..	..	..	..	Current Expenditure
-328	-61	646	122	-1,059	1,492	..	..	..	..	..	Current Budget Balance
..	..	..	..	..	..	..	..	..	..	..	Capital Receipts
267	488	977	2,279	2,425	3,145	..	..	..	..	..	Capital Payments

ZAIRE	1967	1968	1969	1970	1971	1972	1973	1974	1975	1976	1977
FOREIGN TRADE (CUSTOMS BASIS)				*(Millions of current US dollars)*							
Value of Exports, fob	434.8	505.0	679.1	735.4	687.0	737.7	1,001.3	1,381.5	864.8	809.3	1,109.9
Nonfuel Primary Products	406.6	478.3	617.7	684.0	626.9	660.2	927.6	1,310.5	793.5	704.1	1,013.7
Fuels	0.7	0.3	3.2	1.3	10.9	22.0	1.8	2.2	5.8	72.3	40.5
Manufactures	27.6	26.5	58.2	50.1	49.2	55.5	72.0	68.8	65.6	32.9	55.8
Value of Imports, cif	256.1	309.6	451.8	533.0	619.0	766.4	781.9	940.0	932.8	840.3	852.3
Nonfuel Primary Products	61.4	71.8	70.8	95.2	109.8	135.0	160.3	220.9	176.0	198.5	181.3
Fuels	19.4	25.2	58.8	37.8	40.9	47.1	44.6	78.9	91.2	104.6	77.2
Manufactures	175.4	212.6	322.2	400.0	468.3	584.4	577.0	640.2	665.6	537.2	593.8
Terms of Trade	188.9	212.2	274.1	215.0	139.6	121.1	176.3	149.6	96.6	109.6	108.9
Export Prices, fob	40.7	44.4	48.4	50.1	37.8	36.8	68.9	85.9	57.8	65.7	71.4
Nonfuel Primary Products	42.3	45.6	53.8	52.5	42.2	43.8	72.0	87.3	57.5	70.9	73.4
Fuels	4.3	4.3	4.3	4.3	5.6	6.2	8.9	36.7	35.7	38.4	42.0
Manufactures	30.1	31.7	32.4	37.5	36.5	38.5	49.5	67.2	66.1	66.7	71.7
Import Prices, cif	21.5	20.9	17.7	23.3	27.1	30.4	39.1	57.4	59.9	60.0	65.6
Trade at Constant 1980 Prices				*(Millions of 1980 US dollars)*							
Exports, fob	1,069.2	1,137.3	1,402.8	1,467.1	1,815.8	2,004.0	1,453.7	1,609.0	1,495.3	1,231.2	1,554.7
Imports, cif	1,189.8	1,479.1	2,558.3	2,285.8	2,284.0	2,522.1	2,001.3	1,637.4	1,558.5	1,401.4	1,300.1
BALANCE OF PAYMENTS				*(Millions of current US dollars)*							
Exports of Goods & Services	..	..	..	840.8	701.4	695.3	1,059.7	1,556.5	1,024.9	1,177.5	1,282.9
Merchandise, fob	..	..	..	799.5	696.9	690.3	1,038.3	1,520.7	863.4	1,024.2	1,056.4
Nonfactor Services	..	..	..	30.0	4.5	5.0	21.3	35.7	157.7	149.4	218.0
Factor Services	..	..	..	11.3	0.0	0.0	0.0	0.0	3.8	3.9	8.5
Imports of Goods & Services	..	..	..	884.0	901.7	1,009.7	1,300.0	1,910.3	1,669.2	2,095.3	2,799.0
Merchandise, fob	..	..	..	583.2	684.2	752.1	977.3	1,439.3	993.5	1,293.5	1,602.4
Nonfactor Services	..	..	..	255.0	185.8	223.8	274.0	390.6	555.1	602.9	824.9
Factor Services	..	..	..	45.8	31.7	33.9	48.8	80.3	120.6	198.9	371.7
Long-Term Interest	..	..	..	8.8	7.8	17.0	33.8	60.6	62.7	51.0	71.4
Current Transfers, net	..	..	..	..	..	..	..	..	..	..	..
Workers' Remittances	..	..	..	2.1	8.2	8.7	9.7	10.5	26.6	15.5	30.2
Total to be Financed	..	..	..	..	..	..	..	..	..	..	..
Official Capital Grants	..	..	..	..	..	..	..	..	..	..	..
Current Account Balance	..	..	..	-63.7	-102.0	-329.0	-244.9	-372.3	-592.7	-832.8	-1,451.2
Long-Term Capital, net	..	..	..	75.9	179.4	294.2	290.8	241.4	355.9	638.2	1,201.4
Direct Investment	..	..	..	42.2	52.6	104.4	75.8	125.8	15.9	79.8	59.2
Long-Term Loans	..	..	..	3.2	171.1	208.8	320.7	406.9	424.2	546.6	515.2
Disbursements	..	..	..	31.6	200.8	252.7	384.2	533.3	513.0	583.5	560.2
Repayments	..	..	..	28.4	29.7	43.9	63.5	126.4	88.8	36.9	45.0
Other Long-Term Capital	..	..	..	30.5	-44.3	-19.0	-105.7	-291.3	-84.2	11.8	627.0
Other Capital, net	..	..	..	-41.0	-137.9	23.9	-2.5	78.7	109.1	32.8	292.8
Change in Reserves	..	..	..	28.8	60.4	10.8	-43.4	52.3	127.7	161.7	-42.9
Memo Item:				*(Zaires per US dollar)*							
Conversion Factor (Annual Avg)	0.330	0.500	0.500	0.500	0.500	0.500	0.500	0.500	0.500	0.790	0.860
EXTERNAL DEBT, ETC.				*(Millions of US dollars, outstanding at end of year)*							
Public/Publicly Guar. Long-Term	..	..	..	311.1	363.6	573.0	903.7	1,342.7	1,718.4	2,311.8	2,899.5
Official Creditors	..	..	..	221.5	111.2	146.9	174.4	310.9	517.9	909.2	1,149.5
IBRD and IDA	..	..	..	5.7	6.2	10.8	17.8	26.8	49.6	87.0	143.4
Private Creditors	..	..	..	89.6	252.4	426.1	729.3	1,031.8	1,200.5	1,402.6	1,750.0
Private Non-guaranteed Long-Term	..	..	..	..	..	..	..	..	..	..	..
Use of Fund Credit	..	..	..	0.0	0.0	30.6	34.1	34.6	85.8	209.9	253.0
Short-Term Debt	..	..	..	..	..	..	..	..	..	..	277.0
Memo Items:				*(Millions of US dollars)*							
Int'l Reserves Excluding Gold	63.5	125.3	144.2	136.0	90.9	123.2	172.8	118.8	47.9	50.3	133.9
Gold Holdings (at market price)	4.1	15.0	55.2	53.3	62.8	94.4	164.3	93.3	36.5	35.0	42.9
SOCIAL INDICATORS											
Total Fertility Rate	6.0	6.0	6.0	6.0	6.1	6.1	6.1	6.1	6.1	6.1	6.1
Crude Birth Rate	47.0	46.9	46.8	46.8	46.7	46.6	46.4	46.3	46.1	46.0	45.8
Infant Mortality Rate	137.0	135.0	133.0	131.0	129.0	127.0	125.0	123.0	121.0	119.0	117.0
Life Expectancy at Birth	44.0	44.4	44.8	45.2	45.6	46.0	46.4	46.8	47.2	47.6	48.0
Food Production, p.c. ('79-81 = 100)	111.1	112.1	111.2	113.0	110.0	109.2	109.8	110.0	109.8	107.8	106.1
Labor Force, Agriculture (%)	80.8	80.3	79.7	79.1	78.4	77.6	76.9	76.1	75.4	74.6	73.8
Labor Force, Female (%)	43.3	42.9	42.6	42.2	41.7	41.3	40.9	40.4	40.0	39.5	39.0
School Enroll. Ratio, primary	..	..	..	88.0	..	..	..	..	88.0	89.0	89.0
School Enroll. Ratio, secondary	..	..	..	9.0	..	..	..	..	16.0	17.0	20.0

1978	1979	1980	1981	1982	1983	1984	1985	1986	1987 est.	NOTES	ZAIRE
				(Millions of current US dollars)							**FOREIGN TRADE (CUSTOMS BASIS)**
899.4	2,004.1	2,506.9	2,029.9	1,530.2	1,387.9	1,558.3	1,526.2	1,521.0	1,593.5	f	Value of Exports, fob
824.3	1,826.2	2,287.5	1,852.5	1,396.2	1,266.5	1,426.4	1,392.6	1,416.5	1,479.4	..	Nonfuel Primary Products
12.6	52.2	61.8	48.5	37.1	33.6	47.2	45.2	16.3	20.0	..	Fuels
62.5	125.8	157.6	128.9	96.9	87.8	84.7	88.4	88.2	94.1	..	Manufactures
796.7	826.4	1,117.1	1,019.1	912.5	841.7	877.4	918.6	1,008.0	1,149.1	f	Value of Imports, cif
195.1	196.4	264.3	242.2	216.6	199.8	197.0	196.1	205.0	209.6	..	Nonfuel Primary Products
60.2	68.9	92.2	83.6	75.1	69.2	79.5	80.0	29.0	36.7	..	Fuels
541.4	561.1	760.6	693.4	620.8	572.7	600.9	642.4	774.1	902.8	..	Manufactures
				(Index 1980 = 100)							
98.6	108.8	100.0	83.5	78.9	83.5	84.4	82.4	80.6	74.0	..	Terms of Trade
73.3	95.1	100.0	84.6	76.8	79.7	78.6	75.8	81.3	82.0	..	Export Prices, fob
73.9	97.0	100.0	83.2	75.3	78.8	77.7	74.8	81.0	81.1	..	Nonfuel Primary Products
42.3	61.0	100.0	112.5	101.6	92.5	90.2	87.5	44.9	56.9	..	Fuels
76.1	90.0	100.0	99.1	94.8	90.7	89.3	89.0	103.2	113.0	..	Manufactures
74.3	87.4	100.0	101.3	97.3	95.5	93.0	91.9	100.9	110.8	..	Import Prices, cif
				(Millions of 1980 US dollars)							Trade at Constant 1980 Prices
1,227.5	2,107.5	2,507.0	2,398.7	1,992.6	1,741.0	1,983.8	2,013.4	1,870.4	1,943.4	..	Exports, fob
1,071.7	945.1	1,117.1	1,005.6	938.1	881.8	942.9	999.1	998.6	1,036.7	..	Imports, cif
				(Millions of current US dollars)							**BALANCE OF PAYMENTS**
1,923.8	1,915.3	2,403.8	1,828.9	1,684.7	1,801.9	2,059.2	2,006.6	2,027.6	1,930.6	f	Exports of Goods & Services
1,834.2	1,834.6	2,268.6	1,677.9	1,600.8	1,685.8	1,917.8	1,853.3	1,839.2	1,730.0	..	Merchandise, fob
64.7	70.5	102.8	94.3	57.4	101.1	114.8	125.9	155.6	173.0	..	Nonfactor Services
24.9	10.1	32.4	56.6	26.5	15.0	26.7	27.4	32.8	30.6	..	Factor Services
1,613.2	1,757.0	2,883.3	2,676.7	2,426.6	2,327.2	2,467.2	2,369.2	2,545.9	2,711.0	f	Imports of Goods & Services
1,024.5	1,106.9	1,518.9	1,420.9	1,297.2	1,213.3	1,175.7	1,187.1	1,280.0	1,340.0	..	Merchandise, fob
389.2	405.4	834.3	727.5	607.2	689.5	693.9	585.9	659.9	751.0	..	Nonfactor Services
199.4	244.7	530.1	528.3	522.2	424.4	597.6	596.1	606.1	621.0	..	Factor Services
92.1	95.6	197.5	125.7	71.2	94.5	201.4	218.6	120.9	119.3	..	Long-Term Interest
..	..	-79.4	-3.5	-8.8	3.2	-91.2	-54.8	-62.0	..	..	Current Transfers, net
0.0	0.0	0.0	0.0	0.0	0.0	0.0	0.0	0.0	..	..	Workers' Remittances
..	..	-332.1	-651.7	-640.6	-400.9	-376.9	-278.3	-456.6	..	..	Total to be Financed
..	..	40.0	48.0	50.0	52.0	52.0	60.0	60.0	..	..	Official Capital Grants
294.3	146.4	-292.1	-603.7	-590.6	-348.9	-324.9	-218.3	-396.6	-705.0	..	Current Account Balance
142.2	91.1	1,871.5	446.3	125.4	941.3	357.6	330.0	509.7	635.0	f	Long-Term Capital, net
114.9	60.1	6.0	6.0	6.0	6.0	6.0	6.0	5.0	10.0	..	Direct Investment
441.1	122.3	324.7	231.7	108.8	109.1	40.4	32.2	95.8	365.3	..	Long-Term Loans
489.5	195.8	492.2	298.9	170.2	195.5	151.3	151.8	255.2	492.6	..	Disbursements
48.4	73.5	167.5	67.2	61.4	86.4	110.9	119.6	159.4	127.3	..	Repayments
-413.9	-91.2	1,500.8	160.6	-39.4	775.2	259.2	231.8	348.9	259.7	..	Other Long-Term Capital
-443.4	-167.4	-1,544.2	-77.4	306.2	-670.4	-187.7	-148.3	-133.2	-25.6	..	Other Capital, net
6.9	-70.1	-35.2	234.8	159.0	78.0	155.0	36.5	20.1	95.6	..	Change in Reserves
				(Zaires per US dollar)							Memo Item:
0.840	1.730	2.800	4.380	5.750	12.890	36.130	49.870	59.630	112.400	..	Conversion Factor (Annual Avg)
				(Millions of US dollars, outstanding at end of year)							**EXTERNAL DEBT, ETC.**
3,615.6	4,222.0	4,294.4	4,238.5	4,103.8	4,391.1	4,242.2	4,855.3	5,827.3	7,334.3	..	Public/Publicly Guar. Long-Term
1,552.3	2,079.2	2,727.0	2,881.1	2,845.7	3,401.5	3,392.4	4,074.6	4,950.7	6,211.4	..	Official Creditors
188.0	209.9	246.4	256.6	287.0	320.2	345.4	417.6	517.5	788.6	..	IBRD and IDA
2,063.3	2,142.8	1,567.4	1,357.4	1,258.1	989.6	849.8	780.6	876.6	1,122.9	..	Private Creditors
..	..	..	..	..	..	..	..	..	..	..	Private Non-guaranteed Long-Term
260.9	250.9	233.1	345.7	422.6	510.1	579.4	721.0	786.4	833.4	..	Use of Fund Credit
336.0	344.0	296.0	332.0	224.0	210.0	244.0	309.0	318.0	462.0	..	Short-Term Debt
				(Millions of US dollars)							Memo Items:
125.8	206.7	204.1	151.6	38.9	101.6	137.4	189.7	268.6	180.8	..	Int'l Reserves Excluding Gold
69.6	129.0	175.7	142.3	187.3	167.9	143.7	145.5	182.6	236.2	..	Gold Holdings (at market price)
											SOCIAL INDICATORS
6.1	6.1	6.1	6.1	6.1	6.1	6.1	6.1	6.1	6.1	..	Total Fertility Rate
45.7	45.5	45.4	45.2	45.1	45.0	44.9	44.9	44.8	44.7	..	Crude Birth Rate
115.0	113.0	111.0	109.0	107.0	104.9	102.7	100.6	98.4	96.3	..	Infant Mortality Rate
48.4	48.8	49.2	49.6	50.0	50.5	51.0	51.5	52.0	52.6	..	Life Expectancy at Birth
100.1	99.2	100.4	100.4	101.7	101.1	100.6	100.2	98.1	97.6	..	Food Production, p.c. ('79-81 = 100)
73.1	72.3	71.5	..	..	..	..	..	..	..	..	Labor Force, Agriculture (%)
38.6	38.1	37.6	37.4	37.2	37.0	36.8	36.6	36.4	36.2	..	Labor Force, Female (%)
89.0	..	..	..	98.0	98.0	..	..	..	..	..	School Enroll. Ratio, primary
23.0	..	..	..	49.0	57.0	..	..	..	..	..	School Enroll. Ratio, secondary

ZAMBIA	1967	1968	1969	1970	1971	1972	1973	1974	1975	1976	1977
CURRENT GNP PER CAPITA (US $)	320	360	400	450	430	420	420	520	550	540	500
POPULATION (thousands)	3,816	3,923	4,034	4,159	4,288	4,421	4,559	4,700	4,846	4,997	5,152

ORIGIN AND USE OF RESOURCES

(Millions of current Zambian Kwacha)

	1967	1968	1969	1970	1971	1972	1973	1974	1975	1976	1977
Gross National Product (GNP)	907	1,072	1,328	1,244	1,145	1,274	1,465	1,752	1,461	1,748	1,898
Net Factor Income from Abroad	-51	-52	-48	-33	-44	-74	-126	-136	-122	-154	-88
Gross Domestic Product (GDP)	958	1,124	1,376	1,278	1,189	1,348	1,591	1,888	1,583	1,902	1,986
Indirect Taxes, net	92	107	132	122	114	129	152	180	151	182	190
GDP at factor cost	866	1,017	1,244	1,156	1,075	1,219	1,439	1,708	1,432	1,720	1,797
Agriculture	123	128	133	136	154	172	180	199	207	276	326
Industry	499	614	859	699	545	636	845	1,026	670	823	748
Manufacturing	86	109	112	129	150	181	195	240	266	319	353
Services, etc.	335	382	384	442	489	540	566	662	706	803	913
Resource Balance	59	74	437	215	-25	21	251	178	-309	96	-51
Exports of Goods & NFServices	475	544	863	685	501	586	780	944	575	832	781
Imports of Goods & NFServices	416	470	426	471	526	565	529	765	884	736	833
Domestic Absorption	899	1,050	939	1,063	1,214	1,327	1,340	1,709	1,892	1,805	2,038
Private Consumption, etc.	417	499	500	504	493	536	530	660	814	853	1,023
General Gov't Consumption	154	170	176	199	280	315	345	358	436	501	525
Gross Domestic Investment	328	382	264	361	441	476	465	692	642	452	490
Fixed Investment	..	..	..	372	393	445	413	502	602	445	483
Memo Items:											
Gross Domestic Saving	387	456	701	576	416	498	716	870	333	548	439
Gross National Saving	327	380	601	437	263	326	497	648	126	308	274

(Millions of 1980 Zambian Kwacha)

	1967	1968	1969	1970	1971	1972	1973	1974	1975	1976	1977
Gross National Product	2,334	2,381	2,394	2,562	2,540	2,711	2,567	2,778	2,776	2,919	2,912
GDP at factor cost	2,288	2,315	2,313	2,331	2,415	2,643	2,621	2,793	2,728	2,892	2,756
Agriculture	350	353	358	373	382	399	394	412	431	462	466
Industry	1,024	1,018	1,083	1,060	1,081	1,221	1,231	1,344	1,304	1,340	1,250
Manufacturing	339	378	382	413	461	520	529	575	542	573	521
Services, etc.	1,157	1,189	1,117	1,238	1,208	1,302	1,272	1,332	1,280	1,396	1,331
Resource Balance	-1,198	-1,285	-817	-959	-1,180	-1,050	-744	-1,004	-887	-40	-65
Exports of Goods & NFServices	1,072	1,150	1,444	1,330	1,245	1,414	1,293	1,371	1,326	1,611	1,566
Imports of Goods & NFServices	2,270	2,436	2,262	2,289	2,425	2,464	2,037	2,376	2,213	1,651	1,631
Domestic Absorption	3,593	3,673	3,137	3,535	3,758	3,892	3,701	4,247	4,267	3,417	3,136
Private Consumption, etc.	1,904	1,907	1,729	1,586	1,492	1,539	1,421	1,608	1,814	1,719	1,568
General Gov't Consumption	448	477	462	553	711	749	761	722	801	814	772
Gross Domestic Investment	1,241	1,290	946	1,395	1,555	1,604	1,519	1,917	1,653	885	796
Fixed Investment	..	..	..	1,412	1,347	1,477	1,373	1,360	1,493	899	781
Memo Items:											
Capacity to Import	2,592	2,820	4,581	3,334	2,310	2,558	3,005	2,929	1,440	1,867	1,530
Terms of Trade Adjustment	1,520	1,669	3,137	2,004	1,065	1,144	1,712	1,558	114	256	-36
Gross Domestic Income	4,050	4,229	5,694	4,675	3,735	4,066	4,609	4,645	3,130	3,454	3,012
Gross National Income	3,853	4,050	5,531	4,566	3,605	3,855	4,279	4,336	2,890	3,175	2,876

DOMESTIC PRICES (DEFLATORS)

(Index 1980 = 100)

	1967	1968	1969	1970	1971	1972	1973	1974	1975	1976	1977
Overall (GDP)	37.8	43.9	53.8	47.8	44.5	46.1	54.9	61.1	52.5	59.5	65.2
Domestic Absorption	25.0	28.6	29.9	30.1	32.3	34.1	36.2	40.3	44.3	52.8	65.0
Agriculture	35.1	36.3	37.2	36.5	40.3	43.2	45.6	48.4	48.0	59.8	69.8
Industry	48.8	60.3	79.3	66.0	50.5	52.1	68.7	76.4	51.4	61.4	59.9
Manufacturing	25.4	28.9	29.4	31.3	32.5	34.9	36.9	41.8	49.0	55.7	67.7

MANUFACTURING ACTIVITY

	1967	1968	1969	1970	1971	1972	1973	1974	1975	1976	1977
Employment (1980=100)	52.1	56.6	61.9	69.1	73.7	75.7	81.2	91.6	94.7	95.7	96.8
Real Earnings per Empl. (1980=100)	117.7	126.4	125.0	128.4	132.3	133.0	138.8	134.7	132.5	121.4	110.4
Real Output per Empl. (1980=100)	101.8	105.1	113.8	109.7	118.3	126.1	125.9	123.8	114.5	113.1	104.3
Earnings as % of Value Added	44.9	38.9	34.2	33.4	34.1	27.7	30.0	29.2	30.1	28.9	27.7

MONETARY HOLDINGS

(Millions of current Zambian Kwacha)

	1967	1968	1969	1970	1971	1972	1973	1974	1975	1976	1977
Money Supply, Broadly Defined	221	276	342	423	389	414	492	527	589	724	805
Money as Means of Payment	122	161	181	186	199	201	259	266	331	400	393
Currency Ouside Banks	35	40	41	43	58	61	69	80	102	121	118
Demand Deposits	87	120	141	143	140	140	190	186	228	279	274
Quasi-Monetary Liabilities	100	115	161	237	190	212	233	261	258	324	412

GOVERNMENT DEFICIT (-) OR SURPLUS

(Millions of current Zambian Kwacha)

	1967	1968	1969	1970	1971	1972	1973	1974	1975	1976	1977
Current Revenue	..	..	..	..	..	-176	-266	64	-341	-270	-261
Current Expenditure	..	..	..	..	..	297	385	649	462	462	532
Current Budget Balance	..	..	..	..	..	330	372	419	547	562	586
Capital Receipts	..	..	..	..	..	-33	13	230	-85	-100	-54
Capital Payments	..	..	..	..	..	1	86	1	..	..	..
						144	366	167	256	170	207

1978	1979	1980	1981	1982	1983	1984	1985	1986	1987 estimate	NOTES	ZAMBIA
500	510	600	720	660	570	470	410	270	250	..	**CURRENT GNP PER CAPITA (US $)**
5,312	5,477	5,647	5,844	6,048	6,259	6,478	6,704	6,946	7,196	..	**POPULATION (thousands)**
											ORIGIN AND USE OF RESOURCES
			(Millions of current Zambian Kwacha)								
2,109	2,432	2,835	3,388	3,376	3,935	4,556	6,454	10,309	17,013	..	Gross National Product (GNP)
-141	-229	-229	-98	-219	-246	-375	-595	-2,645	-1,067	..	Net Factor Income from Abroad
2,251	2,660	3,064	3,485	3,595	4,181	4,931	7,049	12,954	18,080	f	Gross Domestic Product (GDP)
246	237	221	357	344	632	667	740	1,347	2,210	..	Indirect Taxes, net
2,004	2,423	2,843	3,128	3,251	3,549	4,264	6,309	11,606	15,870	B f	GDP at factor cost
363	397	435	554	492	594	717	925	1,749	2,118	..	Agriculture
886	1,117	1,265	1,351	1,336	1,675	1,907	2,966	5,233	6,472	f	Industry
430	487	566	684	740	830	1,011	1,393	3,368	4,149	..	Manufacturing
1,002	1,146	1,363	1,581	1,767	1,913	2,306	3,158	5,972	9,490	f	Services, etc.
-76	239	-123	-436	-316	-48	188	162	-570	835	..	Resource Balance
755	1,208	1,268	998	995	1,281	1,807	2,628	5,440	8,468	..	Exports of Goods & NFServices
831	969	1,391	1,434	1,312	1,329	1,619	2,466	6,010	7,633	..	Imports of Goods & NFServices
2,326	2,422	3,187	3,921	3,912	4,229	4,743	6,887	13,524	17,245	..	Domestic Absorption
1,252	1,413	1,692	2,262	2,313	2,645	2,779	4,147	8,562	9,957	..	Private Consumption, etc.
538	633	782	986	996	1,009	1,240	1,687	3,057	4,591	..	General Gov't Consumption
537	376	713	673	603	575	724	1,053	1,904	2,697	..	Gross Domestic Investment
437	450	558	610	618	615	623	724	1,127	1,929	..	Fixed Investment
											Memo Items:
461	615	590	237	287	527	912	1,215	1,334	3,532	..	Gross Domestic Saving
252	277	217	3	16	219	455	530	-1,311	2,581	..	Gross National Saving
			(Millions of 1980 Zambian Kwacha)								
2,888	2,696	2,835	3,168	3,001	2,934	2,858	2,868	2,552	2,585	..	Gross National Product
2,736	2,705	2,843	2,909	2,860	2,630	2,667	2,800	2,809	2,805	B H	GDP at factor cost
469	444	435	471	416	451	476	493	537	532	..	Agriculture
1,304	1,237	1,265	1,328	1,320	1,280	1,251	1,262	1,241	1,250	f	Industry
547	580	566	635	613	572	575	619	622	625	..	Manufacturing
1,298	1,293	1,363	1,452	1,428	1,369	1,357	1,374	1,361	1,353	f	Services, etc.
115	76	-123	-62	368	386	326	184	212	252	..	Resource Balance
1,465	1,308	1,268	1,106	1,280	1,155	1,076	1,014	933	1,038	..	Exports of Goods & NFServices
1,351	1,233	1,391	1,168	911	769	750	830	721	786	..	Imports of Goods & NFServices
3,053	2,882	3,187	3,337	2,951	2,729	2,811	3,066	3,279	3,119	..	Domestic Absorption
1,585	1,753	1,692	1,803	1,669	1,692	1,707	1,965	2,110	2,102	..	Private Consumption, etc.
696	696	782	899	797	669	712	677	719	719	..	General Gov't Consumption
771	433	713	634	486	368	393	424	450	298	..	Gross Domestic Investment
622	504	558	570	501	401	343	322	293	344	..	Fixed Investment
											Memo Items:
1,228	1,536	1,268	813	692	741	837	885	652	872	..	Capacity to Import
-238	228	0	-294	-588	-414	-239	-129	-281	-166	..	Terms of Trade Adjustment
2,833	3,202	3,064	2,957	2,575	2,685	2,845	2,999	2,859	2,969	..	Gross Domestic Income
2,650	2,924	2,835	2,874	2,413	2,520	2,619	2,739	2,272	2,419	..	Gross National Income
											DOMESTIC PRICES (DEFLATORS)
			(Index 1980 = 100)								
73.3	89.5	100.0	107.2	113.6	134.9	159.9	225.4	412.6	576.8	..	Overall (GDP)
76.2	84.0	100.0	117.5	132.6	155.0	168.7	224.6	412.4	552.9	..	Domestic Absorption
77.4	89.5	100.0	117.6	118.4	131.7	150.7	187.7	325.6	398.3	..	Agriculture
68.0	90.3	100.0	101.7	101.2	130.9	152.5	235.1	421.6	517.8	..	Industry
78.6	83.9	100.0	107.7	120.8	145.1	175.9	224.9	541.7	663.4	..	Manufacturing
											MANUFACTURING ACTIVITY
97.8	98.9	100.0	101.3	102.7	104.1	105.5	106.9	..	..	..	Employment (1980=100)
103.3	102.6	100.0	109.0	103.5	95.6	95.8	94.8	..	..	..	Real Earnings per Empl. (1980=100)
100.9	106.2	100.0	111.8	106.4	98.0	97.2	103.4	..	..	..	Real Output per Empl. (1980=100)
26.6	25.6	24.6	25.6	25.6	25.6	25.6	25.6	..	..	..	Earnings as % of Value Added
											MONETARY HOLDINGS
			(Millions of current Zambian Kwacha)								
751	950	1,049	1,123	1,458	1,618	1,893	2,336	4,515	..	..	Money Supply, Broadly Defined
397	517	519	564	689	795	870	1,232	2,304	3,225	..	Money as Means of Payment
131	126	151	190	210	239	286	343	593	974	..	Currency Ouside Banks
266	390	368	374	480	556	584	889	1,710	2,251	..	Demand Deposits
354	433	530	560	768	823	1,023	1,105	2,211	..	..	Quasi-Monetary Liabilities
											GOVERNMENT DEFICIT (-) OR SURPLUS
			(Millions of current Zambian Kwacha)								
-325	-241	-568	-450	-668	-327	-414	-1,053	..	..	..	Current Revenue
569	620	791	831	869	1,067	1,112	1,562	..	..	..	Current Expenditure
576	720	1,012	1,142	1,222	1,177	1,258	1,850	..	..	..	Current Budget Balance
-7	-100	-221	-311	-353	-110	-146	-288	..	..	..	Capital Receipts
6	..	..	..	..	1	1	15	..	..	..	Capital Payments
323	141	346	139	316	218	269	780	..	..	..	

ZAMBIA	1967	1968	1969	1970	1971	1972	1973	1974	1975	1976	1977
FOREIGN TRADE (CUSTOMS BASIS)				*(Millions of current US dollars)*							
Value of Exports, fob	658.0	757.1	1,056.2	1,001.0	679.3	758.2	1,141.8	1,406.6	811.5	1,040.5	897.4
Nonfuel Primary Products	652.6	756.2	1,055.1	992.7	669.1	747.1	1,131.9	1,389.2	797.4	1,028.9	885.0
Fuels	0.1	0.0	0.0	0.2	0.3	0.5	0.6	4.0	2.6	3.4	2.7
Manufactures	5.4	0.8	1.1	8.1	9.9	10.6	9.3	13.4	11.6	8.1	9.8
Value of Imports, cif	428.9	455.8	435.6	477.0	558.9	563.4	532.0	787.3	928.7	655.1	671.2
Nonfuel Primary Products	43.2	47.3	56.3	61.5	90.3	73.1	57.1	100.3	91.9	65.6	63.9
Fuels	43.8	46.6	50.0	49.3	45.1	37.1	51.1	95.0	126.1	101.4	102.6
Manufactures	341.9	361.9	329.3	366.2	423.5	453.2	423.9	592.1	710.7	488.1	504.7
Terms of Trade	263.7	292.8	352.8	314.0	188.2	164.7	240.4	171.6	100.6	111.3	95.5
				(Index 1980 = 100)							
Export Prices, fob	50.5	55.6	65.7	63.1	47.7	48.5	81.2	94.8	58.0	64.9	60.8
Nonfuel Primary Products	50.8	55.6	65.8	63.6	48.0	48.9	81.9	95.7	58.0	65.0	60.8
Fuels	4.3	4.3	4.3	4.3	5.6	6.2	8.9	36.7	35.7	38.4	42.0
Manufactures	30.1	31.7	32.4	37.5	36.5	38.5	49.5	67.2	66.1	66.7	71.7
Import Prices, cif	19.1	19.0	18.6	20.1	25.3	29.5	33.8	55.3	57.6	58.3	63.6
Trade at Constant 1980 Prices				*(Millions of 1980 US dollars)*							
Exports, fob	1,303.6	1,362.5	1,608.2	1,587.4	1,425.3	1,562.6	1,406.9	1,483.0	1,400.0	1,604.3	1,476.4
Imports, cif	2,240.8	2,402.0	2,340.3	2,375.3	2,207.2	1,912.3	1,575.9	1,424.5	1,611.9	1,124.0	1,054.7
BALANCE OF PAYMENTS				*(Millions of current US dollars)*							
Exports of Goods & Services	..	..	..	999	727	829	1,203	1,479	880	1,122	974
Merchandise, fob	..	..	..	942	671	760	1,130	1,396	803	1,029	897
Nonfactor Services	..	..	..	17	25	50	57	53	66	83	74
Factor Services	..	..	..	40	31	20	16	30	12	10	4
Imports of Goods & Services	..	..	..	745	823	903	947	1,336	1,474	1,136	1,109
Merchandise, fob	..	..	..	487	562	566	539	791	947	668	683
Nonfactor Services	..	..	..	171	170	215	261	380	398	311	288
Factor Services	..	..	..	87	92	123	147	165	129	157	138
Long-Term Interest	..	..	..	..	..	..	..	..	..	..	..
Current Transfers, net	..	..	..								
Workers' Remittances	..	..	..	0	0	0	0	0	0	0	0
Total to be Financed	..	..	..								
Official Capital Grants	..	..	..	..	..	..	..	..	..	..	..
Current Account Balance	..	..	..	108	-248	-208	130	16	-721	-125	-217
Long-Term Capital, net	..	..	..	-138	-326	104	-13	60	375	111	35
Direct Investment	..	..	..	-297	..	29	32	38	38	31	18
Long-Term Loans	..	..	..								
Disbursements	..	..	..	..	..	..	..	..	..	..	..
Repayments	..	..	..	..	..	..	..	..	..	..	..
Other Long-Term Capital	..	..	..	159	-326	75	-45	22	338	80	18
Other Capital, net	..	..	..	167	310	-42	-130	-50	276	-39	144
Change in Reserves	..	..	..	-137	264	146	14	-26	70	53	38
Memo Item:				*(Zambian Kwacha per US dollar)*							
Conversion Factor (Annual Avg)	0.710	0.710	0.710	0.710	0.710	0.710	0.650	0.640	0.640	0.720	0.790
EXTERNAL DEBT, ETC.				*(Millions of US dollars, outstanding at end of year)*							
Public/Publicly Guar. Long-Term	..	..	..	623	633	676	697	802	1,143	1,299	1,403
Official Creditors	..	..	..	119	147	181	276	423	556	692	791
IBRD and IDA	..	..	..	60	64	73	93	137	183	253	292
Private Creditors	..	..	..	503	486	496	422	380	587	607	611
Private Non-guaranteed Long-Term	..	..	..	30	35	40	56	84	120	115	175
Use of Fund Credit	..	..	..	0	21	41	69	93	89	133	139
Short-Term Debt	..	..	..	..	..	..	..	..	..	..	647
Memo Items:				*(Millions of US dollars)*							
Int'l Reserves Excluding Gold	174.6	193.5	362.9	508.0	277.1	158.4	185.5	164.4	142.0	92.7	66.3
Gold Holdings (at market price)	12.8	7.7	6.4	6.8	8.7	13.0	22.5	31.3	23.6	22.6	27.7
SOCIAL INDICATORS											
Total Fertility Rate	6.6	6.7	6.7	6.7	6.7	6.7	6.7	6.7	6.7	6.7	6.8
Crude Birth Rate	48.9	48.9	49.0	49.0	49.1	49.1	49.1	49.2	49.2	49.3	49.3
Infant Mortality Rate	115.0	112.0	109.0	106.0	103.0	100.0	98.8	97.6	96.4	95.2	94.0
Life Expectancy at Birth	45.3	45.7	46.1	46.5	46.9	47.3	47.7	48.1	48.5	48.9	49.3
Food Production, p.c. ('79-81 = 100)	103.0	100.5	99.9	95.4	103.0	115.1	103.2	109.9	126.2	136.7	128.2
Labor Force, Agriculture (%)	77.9	77.4	77.0	76.6	76.3	75.9	75.6	75.2	74.9	74.5	74.2
Labor Force, Female (%)	27.9	27.9	27.8	27.8	27.7	27.7	27.7	27.6	27.6	27.5	27.5
School Enroll. Ratio, primary	..	..	..	90.0	..	..	..	..	97.0	..	95.0
School Enroll. Ratio, secondary	..	..	..	13.0	..	..	..	..	15.0	16.0	16.0

1978	1979	1980	1981	1982	1983	1984	1985	1986	1987 estimate	NOTES	ZAMBIA
											FOREIGN TRADE (CUSTOMS BASIS)
				(Millions of current US dollars)							
869.2	1,375.5	1,359.5	998.0	932.0	997.0	884.0	815.0	689.2	869.0	..	Value of Exports, fob
846.4	1,345.4	1,331.0	975.7	911.1	976.5	862.9	785.6	666.3	842.1	..	Nonfuel Primary Products
14.3	16.7	15.7	12.5	11.6	11.8	10.6	9.4	3.9	3.9	..	Fuels
8.5	13.4	12.8	9.8	9.4	8.7	10.5	20.0	19.0	23.0	..	Manufactures
628.3	750.2	1,111.0	1,062.0	831.0	690.0	729.6	653.6	662.7	745.4	..	Value of Imports, cif
65.8	83.1	120.0	114.9	90.1	74.8	0.0	32.0	51.7	54.4	..	Nonfuel Primary Products
110.9	134.1	195.2	187.3	146.6	121.7	267.6	146.7	91.0	92.0	..	Fuels
451.6	533.1	795.9	759.8	594.3	493.6	462.0	475.0	520.0	599.0	..	Manufactures
				(Index 1980 = 100)							
91.0	108.3	100.0	79.8	70.9	77.9	70.1	71.6	71.3	79.3	..	Terms of Trade
63.0	90.9	100.0	81.3	69.3	74.0	65.3	66.6	64.9	82.7	..	Export Prices, fob
63.4	91.4	100.0	80.9	68.8	73.7	64.9	66.0	64.4	82.2	..	Nonfuel Primary Products
42.3	61.0	100.0	112.5	101.6	92.5	90.2	87.5	44.9	56.9	..	Fuels
76.1	90.0	100.0	99.1	94.8	90.7	89.3	89.0	103.2	113.0	..	Manufactures
69.2	83.9	100.0	101.9	97.7	95.1	93.1	93.1	91.0	104.2	..	Import Prices, cif
											Trade at Constant 1980 Prices
				(Millions of 1980 US dollars)							
1,380.7	1,514.0	1,359.5	1,227.2	1,345.5	1,346.6	1,353.9	1,222.9	1,062.2	1,051.1	..	Exports, fob
908.0	894.0	1,111.0	1,042.0	850.4	725.7	783.6	701.8	728.0	715.1	..	Imports, cif
											BALANCE OF PAYMENTS
				(Millions of current US dollars)							
952	1,535	1,625	1,169	1,079	1,025	973	867	732	954	..	Exports of Goods & Services
831	1,408	1,457	996	942	923	893	797	681	902	..	Merchandise, fob
113	116	152	153	122	99	75	68	50	51	..	Nonfactor Services
8	11	16	20	15	3	5	2	1	1	..	Factor Services
1,187	1,393	1,986	1,826	1,615	1,286	1,090	1,043	1,166	979	..	Imports of Goods & Services
618	756	1,114	1,065	1,004	711	612	571	549	634	..	Merchandise, fob
407	456	651	584	409	334	291	255	260	224	..	Nonfactor Services
162	181	221	177	203	241	187	217	357	121	..	Factor Services
..	..	..	..	..	..	..	..	..	..	..	Long-Term Interest
..	..	-183	-156	..	..	..	-33	..	..	..	Current Transfers, net
0	0	0	0	0	0	0	0	0	1	..	Workers' Remittances
..	..	-570	-761	..	..	..	-275	..	..	..	Total to be Financed
..	..	33	28	..	..	..	71	..	..	..	Official Capital Grants
-298	37	-537	-734	-565	-271	-153	-204	-391	21	..	Current Account Balance
38	264	195	589	225	165	121	258	468	-113	..	Long-Term Capital, net
39	35	62	-38	39	26	17	0	0	..	..	Direct Investment
..	..	..	..	..	..	..	..	..	..	..	Long-Term Loans
..	..	..	..	..	..	..	..	..	..	..	Disbursements
..	..	..	..	..	..	..	..	..	..	..	Repayments
0	229	101	600	186	139	104	187	468	-113	..	Other Long-Term Capital
71	-409	339	-264	403	37	-107	91	-178	131	..	Other Capital, net
189	108	4	409	-63	69	139	-145	101	-39	..	Change in Reserves
											Memo Item:
				(Zambian Kwacha per US dollar)							
0.800	0.790	0.790	0.870	0.930	1.250	1.790	2.710	7.300	8.890	..	Conversion Factor (Annual Avg)
											EXTERNAL DEBT, ETC.
			(Millions of US dollars, outstanding at end of year)								
1,463	1,820	2,187	2,232	2,379	2,580	2,693	3,203	3,782	4,354	..	Public/Publicly Guar. Long-Term
935	1,251	1,540	1,630	1,730	2,028	2,162	2,591	3,200	3,744	..	Official Creditors
319	335	347	350	359	365	323	475	651	816	..	IBRD and IDA
528	569	647	602	650	552	531	613	582	610	..	Private Creditors
150	125	87	50	44	25	23	..	..	..	..	Private Non-guaranteed Long-Term
343	444	393	731	635	666	698	762	825	957	..	Use of Fund Credit
650	652	586	611	648	514	433	676	1,018	1,089	..	Short-Term Debt
											Memo Items:
				(Millions of US dollars)							
51.1	80.0	78.2	56.2	58.2	54.5	54.2	200.1	70.3	108.8	..	Int'l Reserves Excluding Gold
45.4	111.1	127.9	86.3	99.2	82.8	0.6	1.0	1.2	1.9	..	Gold Holdings (at market price)
											SOCIAL INDICATORS
6.8	6.8	6.8	6.8	6.8	6.8	6.8	6.8	6.8	6.8	..	Total Fertility Rate
49.3	49.4	49.2	49.5	49.5	49.4	49.4	49.3	49.2	49.2	..	Crude Birth Rate
92.8	91.6	90.4	89.2	88.0	89.1	90.3	91.4	92.5	93.7	..	Infant Mortality Rate
49.6	49.8	50.1	50.3	50.6	51.1	51.6	52.1	52.6	53.2	..	Life Expectancy at Birth
116.7	100.4	102.8	96.8	91.5	93.2	90.9	96.9	99.8	94.4	..	Food Production, p.c. ('79-81 = 100)
73.8	73.5	73.1	..	..	..	..	..	..	..	..	Labor Force, Agriculture (%)
27.4	27.4	27.3	27.5	27.7	27.8	28.0	28.2	28.3	28.5	..	Labor Force, Female (%)
95.0	95.0	98.0	97.0	98.0	101.0	103.0	..	..	..	..	School Enroll. Ratio, primary
16.0	16.0	17.0	17.0	17.0	18.0	19.0	..	..	..	..	School Enroll. Ratio, secondary

ZIMBABWE	1967	1968	1969	1970	1971	1972	1973	1974	1975	1976	1977
CURRENT GNP PER CAPITA (US $)	250	240	270	280	320	370	420	530	570	560	540
POPULATION (thousands)	4,721	4,906	5,099	5,249	5,403	5,561	5,724	5,892	6,065	6,243	6,426

ORIGIN AND USE OF RESOURCES
(Millions of current Zimbabwe Dollars)

	1967	1968	1969	1970	1971	1972	1973	1974	1975	1976	1977
Gross National Product (GNP)	786	833	984	1,058	1,214	1,384	1,514	1,821	1,952	2,107	2,150
Net Factor Income from Abroad	-14	-14	-17	-21	-30	-35	-39	-40	-46	-59	-48
Gross Domestic Product (GDP)	800	847	1,001	1,079	1,244	1,419	1,553	1,861	1,998	2,166	2,198
Indirect Taxes, net	55	62	68	68	76	83	103	70	96	102	129
GDP at factor cost	745	786	933	1,011	1,168	1,336	1,450	1,791	1,902	2,064	2,069
Agriculture	152	125	170	153	200	234	215	315	323	350	334
Industry	254	271	322	367	415	485	569	681	722	777	749
Manufacturing	142	152	175	209	251	297	343	421	447	480	460
Services, etc.	339	390	441	491	553	617	666	795	857	937	986
Resource Balance	-3	-34	27	11	-22	40	31	-10	-23	84	52
Exports of Goods & NFServices	..	..	..	..	..	..	..	..	590	617	610
Imports of Goods & NFServices	..	..	..	..	..	..	..	..	613	533	558
Domestic Absorption	803	882	974	1,068	1,266	1,379	1,522	1,871	2,021	2,082	2,146
Private Consumption, etc.	548	573	667	722	844	848	947	1,138	1,240	1,375	1,344
General Gov't Consumption	101	110	119	126	143	157	180	221	256	319	382
Gross Domestic Investment	155	199	188	220	279	374	395	512	525	388	420
Fixed Investment	104	148	142	175	221	256	330	421	468	427	379
Memo Items:											
Gross Domestic Saving	152	165	215	231	257	414	426	502	502	472	472
Gross National Saving	138	151	198	209	225	378	384	452	425	384	414

(Millions of 1980 Zimbabwe Dollars)

	1967	1968	1969	1970	1971	1972	1973	1974	1975	1976	1977
Gross National Product	1,779	2,053	2,089	2,144	2,321	2,681	2,767	3,297	3,225	3,222	3,025
GDP at factor cost	1,814	1,965	2,004	2,438	2,662	2,890	2,959	3,138	3,134	3,110	2,886
Agriculture	..	..	386	357	453	512	414	489	460	512	403
Industry	..	..	875	974	1,033	1,141	1,259	1,288	1,278	1,221	1,129
Manufacturing	..	..	459	513	537	637	687	736	729	687	653
Services, etc.	..	..	743	1,107	1,176	1,237	1,286	1,361	1,396	1,377	1,354
Resource Balance	..	..	..	..	..	..	..	..	..	142	84
Exports of Goods & NFServices	..	..	..	..	..	..	..	..	..	1,041	988
Imports of Goods & NFServices	..	..	..	..	..	..	..	..	..	899	904
Domestic Absorption	..	..	..	..	..	..	..	..	..	3,187	3,008
Private Consumption, etc.	..	..	..	..	..	..	..	..	..	2,084	2,005
General Gov't Consumption	176	185	189	226	248	265	290	326	348	393	425
Gross Domestic Investment	597	666	471	568	670	733	801	1,097	986	710	578
Fixed Investment	..	421	412	508	598	654	797	938	899	728	559
Memo Items:											
Capacity to Import	..	..	..	..	..	..	..	..	..	1,041	988
Terms of Trade Adjustment	..	..	..	..	..	..	..	..	..	0	0
Gross Domestic Income	..	..	..	..	..	..	..	..	..	3,329	3,092
Gross National Income	..	..	..	..	..	..	..	..	..	3,222	3,025

DOMESTIC PRICES (DEFLATORS)
(Index 1980 = 100)

	1967	1968	1969	1970	1971	1972	1973	1974	1975	1976	1977
Overall (GDP)	41.1	40.1	46.5	41.2	43.6	45.9	48.8	54.9	60.2	65.1	71.1
Domestic Absorption	..	..	..	..	..	..	..	..	..	65.3	71.3
Agriculture	..	..	44.0	42.9	44.2	45.7	51.9	64.4	70.2	68.4	82.9
Industry	..	..	36.8	37.7	40.2	42.5	45.2	52.9	56.5	63.6	66.3
Manufacturing	..	..	38.1	40.7	46.7	46.6	49.9	57.2	61.3	69.9	70.4

MANUFACTURING ACTIVITY

	1967	1968	1969	1970	1971	1972	1973	1974	1975	1976	1977
Employment (1980=100)	53.1	57.1	62.2	67.3	75.4	81.8	85.8	92.1	94.6	91.2	87.9
Real Earnings per Empl. (1980=100)	75.3	75.9	79.3	82.0	83.1	85.0	90.1	93.4	95.6	95.0	93.3
Real Output per Empl. (1980=100)	..	..	95.9	97.5	89.2	94.5	98.2	98.9	98.8	92.1	96.2
Earnings as % of Value Added	47.2	46.4	45.4	42.9	42.3	42.6	42.0	40.5	42.3	43.9	46.5

MONETARY HOLDINGS
(Millions of current Zimbabwe Dollars)

	1967	1968	1969	1970	1971	1972	1973	1974	1975	1976	1977
Money Supply, Broadly Defined	..	..	..	..	..	..	..	..	..	..	..
Money as Means of Payment	..	..	..	..	..	..	..	..	324	352	375
Currency Ouside Banks	..	..	..	..	..	..	..	..	67	79	84
Demand Deposits	..	..	..	..	..	..	..	..	257	273	291
Quasi-Monetary Liabilities	..	..	..	..	..	..	..	..	..	..	..

GOVERNMENT DEFICIT (-) OR SURPLUS
(Millions of current Zimbabwe Dollars)

	1967	1968	1969	1970	1971	1972	1973	1974	1975	1976	1977
	..	..	..	..	..	..	..	..	..	-118	-95
Current Revenue	..	..	..	..	..	..	..	..	..	492	575
Current Expenditure	..	..	..	..	..	..	..	..	..	480	622
Current Budget Balance	..	..	..	..	..	..	..	..	..	12	-46
Capital Receipts	..	..	..	..	..	..	..	..	..	1	1
Capital Payments	..	..	..	..	..	..	..	..	..	131	50

1978	1979	1980	1981	1982	1983	1984	1985	1986	1987 estimate	NOTES	ZIMBABWE
530	580	710	860	900	840	730	640	590	590	..	**CURRENT GNP PER CAPITA (US $)**
6,615	6,809	7,009	7,268	7,538	7,817	8,106	8,406	8,705	9,001		**POPULATION (thousands)**
											ORIGIN AND USE OF RESOURCES
(Millions of current Zimbabwe Dollars)											
2,317	2,769	3,394	4,318	4,978	5,976	6,228	7,325	8,496	9,391	..	Gross National Product (GNP)
-46	-53	-47	-115	-194	-248	-195	-284	-384	-334	..	Net Factor Income from Abroad
2,363	2,822	3,441	4,433	5,172	6,224	6,423	7,609	8,880	9,725	..	Gross Domestic Product (GDP)
104	172	217	384	540	874	755	803	971	1,024	..	Indirect Taxes, net
2,259	2,650	3,224	4,049	4,632	5,350	5,668	6,806	7,909	8,701		GDP at factor cost
289	321	451	640	669	544	748	1,039	1,080	947	..	Agriculture
801	1,014	1,248	1,484	1,601	2,287	2,142	2,763	3,297	3,773	..	Industry
515	625	802	1,016	1,121	1,441	1,475	2,043	2,405	2,720	..	Manufacturing
1,169	1,315	1,525	1,925	2,362	2,519	2,778	3,004	3,532	3,981	..	Services, etc.
82	-5	-103	-325	-309	-197	35	86	357	337	..	Resource Balance
675	798	1,043	1,117	1,141	1,345	1,708	2,101	2,559	2,664	..	Exports of Goods & NFServices
593	803	1,146	1,442	1,450	1,542	1,673	2,015	2,202	2,327	..	Imports of Goods & NFServices
2,281	2,827	3,544	4,758	5,481	6,421	6,388	7,523	8,523	9,388	..	Domestic Absorption
1,549	1,932	2,219	2,969	3,406	4,274	3,931	4,362	4,812	5,691	..	Private Consumption, etc.
451	537	677	763	973	1,145	1,341	1,511	1,717	1,901	..	General Gov't Consumption
281	358	648	1,026	1,102	1,002	1,116	1,650	1,994	1,795	..	Gross Domestic Investment
341	395	528	830	1,039	1,238	1,185	1,312	1,580	1,623	..	Fixed Investment
											Memo Items:
363	353	545	701	793	805	1,151	1,736	2,351	2,133	..	Gross Domestic Saving
305	262	421	500	502	441	908	1,363	1,924	1,757	..	Gross National Saving
(Millions of 1980 Zimbabwe Dollars)											
2,946	3,031	3,394	3,772	3,852	3,894	3,867	4,018	4,114	4,148	..	Gross National Product
2,864	2,913	3,224	3,537	3,588	3,459	3,540	3,808	3,897	3,908	..	GDP at factor cost
444	444	451	515	478	403	496	614	583	474	..	Agriculture
1,082	1,142	1,248	1,334	1,325	1,293	1,256	1,333	1,382	1,434	..	Industry
629	697	802	881	877	852	809	902	929	949	..	Manufacturing
1,338	1,327	1,525	1,688	1,785	1,763	1,788	1,861	1,932	2,000	..	Services, etc.
125	-6	-103	-327	-311	-170	23	-8	463	516	..	Resource Balance
1,027	855	1,043	1,123	1,152	1,155	1,079	1,138	1,591	1,616	..	Exports of Goods & NFServices
902	861	1,146	1,450	1,463	1,325	1,056	1,146	1,128	1,100	..	Imports of Goods & NFServices
2,879	3,095	3,544	4,200	4,316	4,227	3,965	4,184	3,823	3,766	..	Domestic Absorption
1,922	2,089	2,219	2,459	2,655	2,697	2,346	2,372	1,929	2,028	..	Private Consumption, etc.
599	615	677	791	845	931	1,020	1,058	1,088	1,099	..	General Gov't Consumption
358	391	648	950	816	599	599	754	806	639	..	Gross Domestic Investment
442	443	528	722	788	765	618	569	603	562	..	Fixed Investment
											Memo Items:
1,027	856	1,043	1,123	1,151	1,156	1,078	1,195	1,311	1,260	..	Capacity to Import
0	1	0	0	-1	1	-1	57	-280	-357	..	Terms of Trade Adjustment
3,004	3,090	3,441	3,873	4,004	4,058	3,987	4,233	4,006	3,926	..	Gross Domestic Income
2,946	3,032	3,394	3,772	3,851	3,894	3,866	4,075	3,834	3,792	..	Gross National Income
											DOMESTIC PRICES (DEFLATORS)
(Index 1980 = 100)											
78.7	91.4	100.0	114.5	129.1	153.4	161.1	182.2	207.2	227.1	..	Overall (GDP)
79.2	91.3	100.0	113.3	127.0	151.9	161.1	179.8	222.9	249.3	..	Domestic Absorption
65.1	72.3	100.0	124.3	140.0	135.0	150.8	169.2	185.2	199.8	..	Agriculture
74.0	88.8	100.0	111.2	120.8	176.9	170.5	207.3	238.6	263.1	..	Industry
81.9	89.7	100.0	115.3	127.8	169.1	182.3	226.5	258.9	286.6	..	Manufacturing
											MANUFACTURING ACTIVITY
85.7	91.7	100.0	107.6	109.5	104.9	100.7	103.1	..	..	..	Employment (1980=100)
95.9	89.6	100.0	108.5	113.7	105.9	114.0	141.9	..	..	..	Real Earnings per Empl. (1980=100)
91.0	93.6	100.0	101.2	100.5	98.2	104.0	113.3	..	..	..	Real Output per Empl. (1980=100)
43.6	42.8	41.8	43.9	48.1	39.9	44.0	44.0	..	..	..	Earnings as % of Value Added
											MONETARY HOLDINGS
(Millions of current Zimbabwe Dollars)											
..	1,637	2,122	2,401	2,839	2,855	3,169	3,742	4,137	5,073	D	Money Supply, Broadly Defined
415	463	633	679	827	751	866	1,005	1,103	1,225	..	Money as Means of Payment
95	108	157	199	238	227	259	321	380	389	..	Currency Ouside Banks
320	356	476	480	589	524	607	684	723	835	..	Demand Deposits
..	1,173	1,489	1,722	2,012	2,103	2,303	2,736	3,034	3,848	..	Quasi-Monetary Liabilities
											GOVERNMENT DEFICIT (-) OR SURPLUS
-254	-293	-376	-262	-545	-394	-647	-602	-613	..	..	Current Revenue
573	611	829	1,130	1,531	1,889	2,110	2,277	2,652	..	..	Current Expenditure
751	835	1,137	1,234	1,716	1,906	2,365	2,413	..	..	..	Current Budget Balance
-179	-224	-308	-104	-184	-17	-255	-136	..	..	..	Capital Receipts
0	1	1	1	0	1	1	17	..	..	..	Capital Payments
75	70	69	159	361	377	393	483	..	..	..	

	1967	1968	1969	1970	1971	1972	1973	1974	1975	1976	1977
FOREIGN TRADE (CUSTOMS BASIS)					*(Millions of current US dollars)*						
Value of Exports, fob	272.0	263.0	325.0	370.0	404.0	515.0	693.0	864.0	936.0	974.0	877.0
Nonfuel Primary Products	230.2	223.3	273.2	310.6	331.2	411.1	562.4	684.3	711.0	723.0	659.0
Fuels	1.0	1.0	1.0	0.4	0.4	2.0	2.8	6.0	13.0	10.0	12.0
Manufactures	40.8	38.7	50.8	59.0	72.4	101.9	127.8	173.7	212.0	241.0	206.0
Value of Imports, cif	261.9	289.9	279.3	329.0	396.7	415.6	549.0	753.0	813.5	611.7	617.8
Nonfuel Primary Products	34.8	39.8	40.5	50.4	68.1	66.9	96.0	124.5	112.1	85.5	89.4
Fuels	22.0	24.1	24.5	25.8	26.6	27.1	37.1	99.8	111.9	84.0	77.2
Manufactures	205.1	226.0	214.3	252.8	301.9	321.7	415.9	528.6	589.6	442.3	451.2
Terms of Trade	160.5	165.6	175.0	178.1	149.4	139.3	147.1	116.5	111.0	123.7	116.6
					(Index 1980 = 100)						
Export Prices, fob	33.3	34.2	36.5	39.9	39.7	41.3	55.4	65.0	64.2	72.8	75.6
Nonfuel Primary Products	34.9	35.8	38.5	40.9	40.8	43.2	58.5	65.0	64.6	76.1	78.1
Fuels	4.3	4.3	4.3	4.3	5.6	6.2	8.9	36.7	35.7	38.4	42.0
Manufactures	30.1	31.7	32.4	37.5	36.5	38.5	49.5	67.2	66.1	66.7	71.7
Import Prices, cif	20.7	20.6	20.9	22.4	26.6	29.6	37.7	55.8	57.8	58.8	64.9
Trade at Constant 1980 Prices					*(Millions of 1980 US dollars)*						
Exports, fob	817.9	769.9	889.9	927.0	1,017.4	1,248.2	1,250.7	1,328.3	1,457.6	1,337.6	1,159.6
Imports, cif	1,264.3	1,405.1	1,338.4	1,468.4	1,492.4	1,403.2	1,457.3	1,348.2	1,406.7	1,039.6	952.7
BALANCE OF PAYMENTS					*(Millions of current US dollars)*						
Exports of Goods & Services	..	..	..	417	454	575	763	960	1,042	1,017	1,005
Merchandise, fob	..	..	..	370	404	516	687	867	922	900	901
Nonfactor Services	..	..	..	32	35	45	60	75	80	78	65
Factor Services	..	..	..	15	15	14	16	18	40	39	40
Imports of Goods & Services	..	..	..	428	508	572	774	1,037	1,191	962	1,004
Merchandise, fob	..	..	..	347	406	458	604	816	844	652	671
Nonfactor Services	..	..	..	46	57	65	115	150	244	198	218
Factor Services	..	..	..	36	45	49	55	71	103	112	115
Long-Term Interest	..	..	..	..	..	..	..	..	..	..	..
Current Transfers, net	..	..	..	-2	-3	-2	-6	-18	-54	-47	-15
Workers' Remittances	..	..	..	..	..	..	..	..	13	13	12
Total to be Financed	..	..	..	-13	-57	2	-16	-94	-203	8	-14
Official Capital Grants	..	..	..	-1	-1	-1	-1	-1	0	..	..
Current Account Balance	..	..	..	-14	-57	1	-17	-96	-203	8	-14
Long-Term Capital, net	..	..	..	25	30	-4	51	61	147	24	-30
Direct Investment	..	..	..	..	..	..	..	..	..	..	-4
Long-Term Loans	..	..	..	..	..	..	..	..	..	..	..
Disbursements	..	..	..	..	..	..	..	..	..	..	..
Repayments	..	..	..	..	..	..	..	..	..	..	..
Other Long-Term Capital	..	..	..	26	31	-2	52	63	147	24	-26
Other Capital, net	..	..	..	-19	14	58	30	-20	65	-35	36
Change in Reserves	..	..	..	8	14	-55	-64	54	-9	3	8
Memo Item:											
Conversion Factor (Annual Avg)	0.710	0.710	0.710	0.710	0.710	0.660	0.590	0.580	0.570	0.630	0.630
					(Zimbabwe Dollars per US dollar)						
EXTERNAL DEBT, ETC.					*(Millions of US dollars, outstanding at end of year)*						
Public/Publicly Guar. Long-Term	..	..	..	229	234	214	218	217	187	142	151
Official Creditors	..	..	..	85	82	74	72	62	53	43	41
IBRD and IDA	..	..	..	41	37	33	28	24	19	15	12
Private Creditors	..	..	..	145	151	140	146	155	134	99	110
Private Non-guaranteed Long-Term	..	..	..	..	..	..	..	..	..	..	..
Use of Fund Credit	..	..	..	0	0	0	0	0	0	0	0
Short-Term Debt	..	..	..	..	..	..	..	..	..	..	46
Memo Items:					*(Millions of US dollars)*						
Int'l Reserves Excluding Gold	58.6	3.6	27.9	20.3	6.1	61.3	124.7	70.8	80.0	76.6	72.6
Gold Holdings (at market price)	30.4	33.6	28.2	39.2	35.7	35.8	84.2	93.6	49.2	35.7	25.1
SOCIAL INDICATORS											
Total Fertility Rate	8.0	7.9	7.8	7.7	7.6	7.5	7.4	7.3	7.2	7.1	7.0
Crude Birth Rate	55.0	54.4	53.9	53.3	52.8	52.2	51.7	51.2	50.6	50.1	49.6
Infant Mortality Rate	101.0	99.4	97.8	96.2	94.6	93.0	91.6	90.2	88.8	87.4	86.0
Life Expectancy at Birth	49.0	49.5	50.0	50.5	51.0	51.5	52.0	52.4	52.9	53.3	53.8
Food Production, p.c. ('79-81 = 100)	116.7	90.7	115.3	99.7	129.2	145.4	99.5	130.3	115.6	124.1	118.9
Labor Force, Agriculture (%)	78.4	78.1	77.7	77.3	76.9	76.4	76.0	75.5	75.1	74.6	74.2
Labor Force, Female (%)	37.8	37.8	37.7	37.6	37.5	37.4	37.3	37.3	37.2	37.1	37.0
School Enroll. Ratio, primary	..	..	..	74.0	..	..	..	..	73.0	72.0	70.0
School Enroll. Ratio, secondary	..	..	..	7.0	..	..	..	..	9.0	9.0	9.0

|------|------|------|------|------|------|------|------|------|------|-------|----------|
											FOREIGN TRADE (CUSTOMS BASIS)
				(Millions of current US dollars)							
891.0	1,217.0	1,423.0	1,406.0	1,273.0	1,128.0	1,154.3	1,112.7	1,254.0	1,358.0	..	Value of Exports, fob
660.0	853.0	1,002.0	1,060.0	944.0	774.0	778.9	713.3	789.7	802.3	..	Nonfuel Primary Products
12.0	17.0	17.0	15.0	16.0	17.0	16.6	15.4	5.2	5.7	..	Fuels
219.0	347.0	404.0	331.0	313.0	337.0	358.8	384.0	459.1	550.0	..	Manufactures
592.0	937.0	1,448.4	1,695.7	1,638.9	1,205.1	955.0	1,030.9	1,099.0	1,054.9	f	Value of Imports, cif
85.4	149.4	255.1	324.0	304.0	235.1	152.2	149.8	155.2	137.4	..	Nonfuel Primary Products
66.9	121.4	245.6	350.8	334.7	220.5	176.7	186.2	71.0	87.3	..	Fuels
439.6	666.2	947.7	1,020.9	1,000.2	749.6	626.1	694.8	872.8	830.2	..	Manufactures
				(Index 1980 = 100)							
106.0	104.2	100.0	92.1	87.3	94.6	95.8	84.3	77.9	84.1	..	Terms of Trade
77.4	89.1	100.0	95.1	84.9	89.3	88.4	76.5	75.6	89.5	..	Export Prices, fob
79.0	89.6	100.0	93.7	81.8	88.6	88.0	70.9	65.7	78.7	..	Nonfuel Primary Products
42.3	61.0	100.0	112.5	101.6	92.5	90.2	87.5	44.9	56.9	..	Fuels
76.1	90.0	100.0	99.1	94.8	90.7	89.3	89.0	103.2	113.0	..	Manufactures
73.0	85.5	100.0	103.2	97.2	94.4	92.3	90.7	97.0	106.5	..	Import Prices, cif
											Trade at Constant 1980 Prices
				(Millions of 1980 US dollars)							
1,151.3	1,365.9	1,423.0	1,478.6	1,499.6	1,263.2	1,305.1	1,455.0	1,658.4	1,516.5	..	Exports, fob
810.7	1,095.8	1,448.4	1,643.0	1,686.3	1,276.5	1,034.8	1,136.9	1,132.5	990.5	..	Imports, cif
											BALANCE OF PAYMENTS
				(Millions of current US dollars)							
1,029	1,231	1,719	1,680	1,576	1,370	1,370	1,301	1,531	1,652	f	Exports of Goods & Services
923	1,080	1,446	1,452	1,314	1,155	1,175	1,134	1,325	1,460	..	Merchandise, fob
68	87	167	132	180	144	139	130	170	160	..	Nonfactor Services
38	64	107	96	82	71	56	37	36	32	..	Factor Services
974	1,283	1,900	2,284	2,200	1,755	1,517	1,403	1,556	1,644	..	Imports of Goods & Services
654	875	1,339	1,535	1,471	1,069	989	922	1,013	1,070	f	Merchandise, fob
223	266	382	522	421	417	350	298	311	338	..	Nonfactor Services
97	142	179	227	308	269	178	183	232	234	..	Factor Services
..	..	..	..	..	..	..	..	..	..	..	Long-Term Interest
-17	-56	-121	-124	-128	-115	-39	-55	-26	-25	..	Current Transfers, net
5	11	9	1	2	0	0	0	0	..	..	Workers' Remittances
37	-109	-302	-728	-752	-500	-185	-157	-51	-22	K	Total to be Financed
..	..	58	91	45	41	88	64	59	72	K	Official Capital Grants
37	-109	-244	-638	-707	-459	-97	-93	8	50	..	Current Account Balance
152	147	4	174	415	201	176	85	128	18	f	Long-Term Capital, net
3	0	2	4	-1	-2	-3	3	7	-24	..	Direct Investment
..	..	..	..	..	..	..	..	..	..	..	Long-Term Loans
..	..	..	..	..	..	..	..	..	..	..	Disbursements
..	..	..	..	..	..	..	..	..	..	..	Repayments
149	146	-55	79	370	162	91	18	62	-30	..	Other Long-Term Capital
-114	82	153	456	269	84	-125	11	-80	95	f	Other Capital, net
-75	-120	87	8	22	174	46	-2	-56	-163	..	Change in Reserves
											Memo Item:
				(Zimbabwe Dollars per US dollar)							
0.680	0.680	0.640	0.690	0.760	1.010	1.240	1.610	1.670	1.660	..	Conversion Factor (Annual Avg)
											EXTERNAL DEBT, ETC.
				(Millions of US dollars, outstanding at end of year)							
416	522	695	791	1,181	1,535	1,390	1,557	1,760	2,044	..	Public/Publicly Guar. Long-Term
39	36	101	165	301	400	471	667	879	1,196	..	Official Creditors
9	6	3	56	69	110	151	234	327	431	..	IBRD and IDA
377	485	594	626	879	1,134	919	891	882	849	..	Private Creditors
..	..	..	13	38	86	78	65	56	51	..	Private Non-guaranteed Long-Term
0	0	0	44	41	200	256	264	234	157	..	Use of Fund Credit
28	35	90	398	582	481	344	308	290	260	..	Short-Term Debt
											Memo Items:
				(Millions of US dollars)							
148.0	298.9	213.5	169.5	140.5	75.4	45.4	93.4	106.4	166.2	..	Int'l Reserves Excluding Gold
36.4	133.1	205.7	185.2	180.0	225.1	214.6	252.1	209.9	204.3	..	Gold Holdings (at market price)
											SOCIAL INDICATORS
7.0	6.9	6.8	6.7	6.6	6.5	6.3	6.2	6.0	5.9	..	Total Fertility Rate
49.5	49.4	49.3	49.2	49.1	48.1	47.1	46.2	45.2	44.2	..	Crude Birth Rate
84.8	83.6	82.4	81.2	80.0	78.5	77.0	75.4	73.9	72.4	..	Infant Mortality Rate
54.2	54.6	55.0	55.4	55.8	56.3	56.8	57.3	57.8	58.3	..	Life Expectancy at Birth
115.5	90.0	91.2	118.8	97.8	71.5	73.9	102.4	101.0	70.0	..	Food Production, p.c. ('79-81=100)
73.7	73.3	72.8	..	..	..	..	..	..	..	..	Labor Force, Agriculture (%)
36.9	36.8	36.7	36.5	36.3	36.1	35.9	35.7	35.4	35.2	..	Labor Force, Female (%)
63.0	61.0	88.0	117.0	126.0	130.0	131.0	131.0	129.0	..	..	School Enroll. Ratio, primary
9.0	8.0	8.0	15.0	23.0	30.0	39.0	43.0	47.0	..	..	School Enroll. Ratio, secondary

CODES FOR GENERAL AND COUNTRY NOTES

GENERAL

Code letters in the *Notes* column of the country pages refer to the following
general footnotes:

A   GNP per capita in US$ is calculated using an alternative conversion factor.

B   GDP by industrial origin data for at least some years are in purchaser values
and not at factor cost.

C   Data are for fiscal years, see *Country Notes*.

D   Money supply reflects total liquid liabilities.

E   State and local government accounts are thought to contribute at least 20
percent of central government tax revenue.

F   Break in comparability of data, see *GFSY* .

G   Data are for *persons engaged*, not *employees*.

H   Partial rebasing into 1980 prices would yield a large rescaling deviation;
therefore GDP has been rescaled directly from the original base year to
1980 prices.

I   Partial rebasing into 1980 prices has resulted in a rescaling deviation of more
than 5 percent of private consumption.

J   Break in comparability of data, see *UN Yearbook of Industrial Statistics*.

K   All unrequited official transfers (net) are included here.

COUNTRY

f   Indicates country-specific notes. These are listed in the pages following,
under main topic headings, and in alphabetical order, by country.

COUNTRY NOTES

National Accounts

BAHRAIN

SOURCES:
All Indicators
1980-87 Ministry of Finance and National Economy of Bahrain

BANGLADESH

BREAKS IN COMPARABILITY:
1960-72 Manufacturing includes mining & quarrying

BARBADOS

SOURCES:
All Indicators
1975 Ministry of Finance and Planning
BREAKS IN COMPARABILITY:
1960-69 Private consumption includes change in stocks (current prices)
1960-69 Gross domestic investment excludes change in stocks (current prices)

BENIN

SOURCES:
All Indicators
1960-67 UN Economic Commission for Africa (UNECA), 12/80

BHUTAN

MISCELLANEOUS:
1981-87 Fiscal year, beginning April l.

BOTSWANA

SOURCES:
All Indicators
1974-84 Botswana Statistical Abstract
1980-87 Bank of Botswana Annual Report, 1987.
All components in GDP by expenditure categories
1970-71 UNECA (current prices)
GDP at mp deflator
1965-72 GDP deflator from Statistical Abstracts for 1965,66,68,69, & 72 (constant prices)
Net factor income from abroad
1974 Bank of Botswana annual report l985 (current prices)
BREAKS IN COMPARABILITY:
1960-66 Data are in old SNA and refer to calender years
1968-87 Data are new SNA, for fiscal years ending June 30
1972-87 Data break in the series in 72 & 74
1965-86 Exports exclude nonfactor services.
1965-86 Imports include goods and net nonfactor services.

BRAZIL

SOURCES:
All components in GDP by expenditure categories
All components in GDP by industrial origin
GNP at market prices
1970-87 Fundacao Getulio Vargas, Conjuntura, June 1988

BURKINA FASO

SOURCES:
All Indicators
1950-69 UNECA
1970-87 Institut National de la Statistique & WB estimates
BREAKS IN COMPARABILITY:
1960-87 Net factor income from abroad includes workers' remittances

BURUNDI

BREAKS IN COMPARABILITY:
1965-66 Agriculture includes construction (constant prices)

CAMEROON

SOURCES:
All Indicators
1960-87 Direction de la Statistique and WB estimates
MISCELLANEOUS:
1950-87 Fiscal year, ending June 30

CENTRAL AFRICAN REPUBLIC

SOURCES:
All Indicators
1960-69 UNECA
BREAKS IN COMPARABILITY:
1960-76 Private consumption includes change in stocks

CHAD

SOURCES:
All Indicators
1960-66 UNECA
1976 UNECA
All components in GDP by expenditure categories
1977-85 UNECA
BREAKS IN COMPARABILITY:
1982-85 Net factor income from abroad includes private transfers (current prices)
1960-85 Agriculture excludes forestry and fishing

CHILE

SOURCES:
All Indicators
1960-69 Central Bank of Chile

CHINA

SOURCES:
All Indicators
1960-77 Statistical Yearbook of China.

CONGO, PEOPLE'S REP.

SOURCES:
All Indicators
1960-66 UNECA
BREAKS IN COMPARABILITY:
1973-87 Manufacturing includes other mining (except petroleum)

COSTA RICA

BREAKS IN COMPARABILITY:
1960-87 Manufacturing includes mining & quarrying

COTE D'IVOIRE

BREAKS IN COMPARABILITY:
1960-73 Exports of goods & nonfactor services excludes nonfactor services.
1960-71 Imports of goods & nonfactor services excludes nonfactor services.
1960-80 Net factor income from abroad includes workers' remittances.

CYPRUS

SOURCES:
All components in GDP by expenditure categories
All components in GDP by industrial origin
1976-87 Dept. of Statistics and Research, Ministry of Finance.
MISCELLANEOUS:
1975-87 Data relate to South Cyprus.

ECUADOR

SOURCES:
All Indicators
1965-69 Central Bank of Ecuador

EGYPT, ARAB REPUBLIC OF

MISCELLANEOUS:
1960-80 Data are for calendar year.
1981-87 Fiscal year, ending June 30.

ETHIOPIA

BREAKS IN COMPARABILITY:
1970-87 Private consumption includes change in stocks
1970-87 Gross domestic investment excludes change in stocks
MISCELLANEOUS:
1965-87 Fiscal year, ending July 7

FIJI

SOURCES:
All components in GDP by expenditure categories
1975-76 Bureau of Statistics, Fiji.
BREAKS IN COMPARABILITY:
1960-76 Break in series from 1976-1977.

GABON

SOURCES:
All Indicators
1950-66 UNECA

GAMBIA, THE

MISCELLANEOUS:
1950-87 Fiscal year, ending June 30

GREECE

SOURCES:
All Indicators
1964-79 OECD
1980-87 Ministry of National Economy.
BREAKS IN COMPARABILITY:
1980-87 Change in stocks includes discrepancy in GDP expenditure estimate
MISCELLANEOUS:
1950-87 Data for imports of goods and nonfactor services exclude ships operating overseas.
1950-87 Data for fixed investment exclude ships operating overseas.

GUINEA-BISSAU

SOURCES:
All Indicators
1970-74 UNECA 06/80
1980-81 Ministry of Planning and WB estimates
BREAKS IN COMPARABILITY:
1970-83 Private consumption includes change in stocks and discrepancy in GDP expenditure estimate

GUYANA

BREAKS IN COMPARABILITY:
1960-85 Other services includes gas, electricity, and water

HAITI

BREAKS IN COMPARABILITY:
1966-87 Data are for fiscal year ending September 30.
1970-75 Imports exclude imports & exports related to assembly industry.
1970-75 Gross domestic fixed investment excludes imports related to assembly industry.

HONG KONG

SOURCES:
All Indicators
1960-87 Estimates of GDP Census & Stat.Dept.,1988 (& various issues).
MISCELLANEOUS:
1960-87 GNP per capita refers to GDP per capita.

INDIA

SOURCES:
All Indicators
1980-86 Central Statistical Organization 1980/81-1985/86 & quick estimates.
MISCELLANEOUS:
1950-86 Data for fiscal year beginning April 1.

ISRAEL

SOURCES:
All components in GDP by expenditure categories
1960-86 Central Bureau of Statistics.
All components in GDP by industrial origin
1972-84 UN Monthly Bulletin of Statistics, various issues. (current prices)
MISCELLANEOUS:
1960-86 Data for exports of goods and nonfactor services excludes subsidies on exports
1980-86 Data relate to new SNA, therefore not comparable to earlier years
1960-84 Data for GDP at factor cost relate to net domestic product at factor cost
1960-86 Data for imports of goods and nonfactor services excludes net taxes on imports

JAMAICA

SOURCES:
All Indicators
1960-87 Statistical Institute, National Income and Product Accounts

JORDAN

MISCELLANEOUS:
1970-87 Data (except population) relate to East Bank only

KENYA

SOURCES:
All Indicators
1972-87 Kenya Economic Survey - Various issues (current prices)
All components in GDP by industrial origin
1979-87 Kenya Economic survey - various issues (constant prices)
MISCELLANEOUS:
1950-87 Data series not strictly comparable; breaks in 1964, & 1972.

KOREA, REPUBLIC OF

SOURCES:
All Indicators
1955-69 National Income Account, Bank of Korea.
1970-87 National Accounts, Bank of Korea.
BREAKS IN COMPARABILITY:
1955-87 Break in series from 1969-1970.

LESOTHO

SOURCES:
All Indicators
1966-87 UNNA
1976, Lesotho National Accounts 1973/74,75/84&76/86, and WB estimates (current prices)
All components in GDP by industrial origin
1980-87 Lesotho National accounts, various issues & WB estimates (constant prices)

GDP at market prices
1980-87 Lesotho National accounts (constant prices)
MISCELLANEOUS:
1982-87 Public administration & defense includes other services
1960-81 Fiscal year, beginning April 1

LIBERIA
BREAKS IN COMPARABILITY:
1960-86 Agriculture includes informal construction and Informal
manufacturing
1981-86 Other services includes ownership of dwellings and gas,
electricity, and water

MALTA
SOURCES:
All Indicators
1955-72 UN Yearbook of National Accounts Statistics, Various
issues.
1973-86 National Accounts of Maltese Islands, various issues.

MAURITANIA
SOURCES:
All Indicators
1960-72 UN Office for Development Research and Policy
Analysis
BREAKS IN COMPARABILITY:
1960-87 Agriculture includes informal manufacturing

NEPAL
BREAKS IN COMPARABILITY:
1960-87 Break in series in 1974 & 1975 especially services
(non-agriculture)
MISCELLANEOUS:
1960-88 Data in fiscal year ending July 15.

NIGER
SOURCES:
All Indicators
1960-66 UNECA
1976-77 UNECA

NIGERIA
SOURCES:
All Indicators
1950-69 Federal Office of Statistics
1973-80 Federal Office of Statistics, 10/85, 02/86, 03/87 & World
Bank estimates
Exports of goods & non-factor services
Imports of goods & non-factor services
1950-80 Central Bank of Nigeria, 12/80, 12/84, & 12/85 (current
prices)
BREAKS IN COMPARABILITY:
1950-80 Private consumption includes change in stocks
MISCELLANEOUS:
1950-79 Data are for fiscal year beginning April 1
1967-69 Data exclude the three Easten states
1980-87 Data are for calendar year

PAKISTAN
MISCELLANEOUS:
1960-87 Fiscal year ending June 30.

PANAMA
SOURCES:
All Indicators
1980-87 Directorate of Statistics and Census
MISCELLANEOUS:
1980-84 Data on transport includes Panama Canal Commission.

PAPUA NEW GUINEA
SOURCES:
All Indicators
1961-76 *PNG Statistical Bulletin* (various issues) & *Compendium
of Statistics*, PNG.

MISCELLANEOUS:
1961-76 Fiscal year data adjusted to calendar year.
1960-76 Data linked to 1977 revised series, as result of new base
year.

PERU
SOURCES:
All Indicators
1979-87 National Institute of Statistics

PORTUGAL
SOURCES:
All components in GDP by expenditure categories
1960-76 *OECD National Accounts of OECD Countries*, various
editions. (current prices)
MISCELLANEOUS:
1977-87 Data for GDP at FC relate to GDP at market prices less
import duties
1977-87 Data for net indirect taxes relate to import duties

RWANDA
SOURCES:
All Indicators
1970-81 *UN Yearbook of National Accounts*, and MBS (current
prices)
All components in GDP by industrial origin
1976-79 *UN Yearbook* (constant prices)
GDP at market prices
1976-79 *UN Yearbook* (constant prices)
Net factor income from abroad
1982-85 *UN Yearbook* (current prices)
Net indirect taxes
1970-74 UNECA (current prices)
BREAKS IN COMPARABILITY:
1969-75 Services include construction
1969-75 Services include mining & quarrying and gas, electricity,
and water (constant prices)

SAUDI ARABIA
MISCELLANEOUS:
1963-87 Fiscal years are based on Islamic (Hijri) year.

SENEGAL
SOURCES:
All Indicators
1975 Ministere du Plan et de la Cooperation
1977-80 Direction de la provision, Ministry of Economy and
Finance, 04/86.

SIERRA LEONE
SOURCES:
All Indicators
1964-73 The Central Statistical Office, *National Accounts of
Sierra Leone* , Freetown
MISCELLANEOUS:
1960-87 Fiscal year, ending June 30.
1964-87 Data on public fixed investment include public entities.

SINGAPORE
SOURCES:
All Indicators
1960-87 Department of Statistics, Singapore.

SOLOMON ISLANDS
SOURCES:
All Indicators
1967-69 UN Monthly Bull. of Stat.(4/74). Data linked to 1980
revised series. (current prices)

SOUTH AFRICA
SOURCES:
All Indicators
1965-76 UN Yearbook
1977-87 South African Quarterly Bulletins March'86 & '88,

Dec.'87.
MISCELLANEOUS:
1960-87 Data, except GNP per capita, includes Namibia .

SRI LANKA
SOURCES:
All Indicators
1960-87 Central BK of Ceylon, Annual Report 1987 & earlier issues. (current prices)

SUDAN
SOURCES:
All components in GDP by industrial origin
1965-69 Components from various issues of Sudan National Accounts (current prices)
MISCELLANEOUS:
1960-87 Fiscal year, ending June 30

SWAZILAND
MISCELLANEOUS:
1960-79 Fiscal year, beginning April 1 through 1976, June 30 from 1976 to 1979

SYRIAN ARAB REPUBLIC
SOURCES:
All Indicators
1963-79 Central Bureau of Statistics.
BREAKS IN COMPARABILITY:
1963-69 1979-87 Private consumption includes change in stocks

1963-69, 1979-87 Gross domestic investment excludes change in stocks.

TANZANIA
SOURCES:
All Indicators
1964-75 National Acounts of Tanzania and *Economic Survey* (various issues). (current prices)
1976-78 Bureau of statistics: National Accounts of Tanzania
MISCELLANEOUS:
1950-87 Data are for mainland Tanzania only.

TOGO
SOURCES:
All Indicators
1984-87 Direction de la Statistique General; WB and IMF staff estimates
MISCELLANEOUS:
1960-80 Data for agriculture represent food crops only

TONGA
BREAKS IN COMPARABILITY:
1975-85 Break in series from 1981-1982.

TURKEY
SOURCES:
All Indicators
1950-69 Turkish State Institute of Statistics.
All components in GDP by expenditure categories
1970-71 OECD Economic Surveys of Nov.1978 (Except exports

& imports of goods & nfs.)

UGANDA
SOURCES:
All components in GDP by industrial origin
1978-87 Central Statistical office and mission estimates
BREAKS IN COMPARABILITY:
1963-87 Other services includes water

MISCELLANEOUS:
1950-87 Data are in old Ugandan shillings.
1978-87 Break in the series from 1978 onwards

URUGUAY
SOURCES:
All Indicators
1966-69 Central Bank of Uruguay, Indicators of Economic and Financial Activity

VANUATU
SOURCES:
All Indicators
1983-85 *The National Income Accounts of Vanuatu, 1983-85*, ADB.
MISCELLANEOUS:
1983-86 Break in series in 1982-1983

WESTERN SAMOA
BREAKS IN COMPARABILITY:
1978-85 Break in series for periods between 1978-83 and 1984-85.

YEMEN, ARAB REPUBLIC OF
BREAKS IN COMPARABILITY:
1970-80 Data are for fiscal year ending June 30.

ZAIRE
BREAKS IN COMPARABILITY:
1960-67 Construction excludes informal construction
1950-87 Data not comparable due to breaks in series in 60,67,70, and 1977.

ZAMBIA
BREAKS IN COMPARABILITY:
1965-85 Public administration & defense includes water
1950-77 Data series has breaks in 1964,1970&1977 - not strictly comparable.

ZIMBABWE
SOURCES:
General government consumption
Resource balance
1960-69 *Monthly Digest of Statistics*, Zwe.Nat.Act&BOP 1974,1978 (current prices)
GDP at market prices
Gross domestic investment
1960-69 Zimbabwe National Income & Expenditure Report 1985 (current prices)

Balance of Payments

ARGENTINA

SOURCES:
All Indicators
1980-87 WB

BANGLADESH

SOURCES:
All Indicators
1973-87 WB

BARBADOS

SOURCES:
All Indicators
1987 WB

BENIN

SOURCES:
All Indicators
1980-87 WB
BREAKS IN COMPARABILITY:
1974-79 Errors & ommissions include smuggling

BOLIVIA

MISCELLANEOUS:
1981-87 Large unrecorded transactions not included.

BURKINA FASO

SOURCES:
All Indicators
1985-86 BCEAO, WB, and IMF estimates
BREAKS IN COMPARABILITY:
1985-86 Imports of goods & nonfactor services include official
transfers deemed capital grants
1985-86 Total short-term capital excludes reserves and lcfar
includes errors & ommissions
MISCELLANEOUS:
1985-86 Data on imports linked to foreign grants, and project
loans.
1985-86 Data for nonmonetary capital used for long-term capital
inflow.

BURUNDI

SOURCES:
All Indicators
1970-87 WB

CAMEROON

SOURCES:
All Indicators
1980-87 WB
MISCELLANEOUS:
1980-87 Fiscal year, ending June 30

CENTRAL AFRICAN REPUBLIC

SOURCES:
All Indicators
1985-87 WB

CHINA

SOURCES:
All Indicators
1980-81, 1987 WB

CONGO, PEOPLE'S REP.

SOURCES:
All Indicators
1987 WB

COSTA RICA

SOURCES:
All Indicators
1980-87 WB

COTE D'IVOIRE

SOURCES:
All Indicators
1987 WB

EGYPT, ARAB REPUBLIC OF

SOURCES:
All Indicators
1974-87 WB
MISCELLANEOUS:
1981-87 Fiscal year, ending June 30.

EL SALVADOR

SOURCES:
All Indicators
1981-85 WB

ETHIOPIA

SOURCES:
All Indicators
1980-87 WB
MISCELLANEOUS:
1980-87 Fiscal year, ending July 7

FIJI

SOURCES:
All Indicators
1987 WB

HAITI

MISCELLANEOUS:
1965-87 Fiscal year, ending September 30.

HONDURAS

SOURCES:
All Indicators
1980-87 WB

INDIA

SOURCES:
All Indicators
1970-87 WB

INDONESIA

SOURCES:
All Indicators
1987 WB

JAMAICA

SOURCES:
All Indicators
1980-87 WB

JORDAN

SOURCES:
All Indicators
1977-87 WB

KENYA

SOURCES:
All Indicators
1986-87 WB

KOREA, REPUBLIC OF
SOURCES:
All Indicators
1986-87 WB

LESOTHO
SOURCES:
All Indicators
1987 WB

MALI
SOURCES:
All Indicators
1985-87 WB
BREAKS IN COMPARABILITY:
1985-87 Data on short-term capital include obligation not serviced, pending settlement

MAURITANIA
SOURCES:
All Indicators
1986-87 WB

NICARAGUA
SOURCES:
All Indicators
1981-87 WB

NIGER
SOURCES:
All Indicators
1987 WB

NIGERIA
SOURCES:
All Indicators
1987 WB

PAKISTAN
SOURCES:
All Indicators
1986-87 WB

PANAMA
MISCELLANEOUS:
1980-87 Exports and imports by enterprises in the Colon Free Zone not included.

PHILIPPINES
SOURCES:
All Indicators
1986-87 WB

RWANDA
BREAKS IN COMPARABILITY:
1987 Official transfers include those deemed capital grants (credit)

SENEGAL
SOURCES:
All Indicators
1985-87 WB

SRI LANKA
SOURCES:
All Indicators
1987 WB

ST. LUCIA
SOURCES:
All Indicators
1980-85 WB

ST. VINCENT
SOURCES:
All Indicators
1981-86 WB

SUDAN
SOURCES:
All Indicators
1979-87 WB
MISCELLANEOUS:
1979-87 Fiscal year, ending June 30

THAILAND
SOURCES:
All Indicators
1987 WB

TUNISIA
SOURCES:
All Indicators
1976-87 WB

UGANDA
SOURCES:
All Indicators
1983-87 WB

WESTERN SAMOA
SOURCES:
All Indicators
1986-87 WB
MISCELLANEOUS:
1980-87 Official transfers treated as official capital grants.

YEMEN, ARAB REPUBLIC OF
SOURCES:
All Indicators
1977-87 WB

ZAIRE
SOURCES:
All Indicators
1987 WB

ZIMBABWE
SOURCES:
All Indicators
1985-87 WB

Trade

ALGERIA
SOURCES:
Exports of all commodities
1965-87 IMF

BANGLADESH
SOURCES:
Imports of fuels
1965-87 IMF
Imports of manufactures
1965-87 IMF

BARBADOS
SOURCES:
All Indicators
1987-88 WB
Exports of all commodities
1983-87 IMF
Imports of all commodities
1980-87 IMF
BREAKS IN COMPARABILITY:
1965-82 data include export of value added from improvement &
repair trade; and exclude military goods.

BENIN
SOURCES:
All Indicators
1984-87 WB

BOLIVIA
SOURCES:
Exports of all commodities
1965-87 IMF

BOTSWANA
SOURCES:
All Indicators
1965-87 IMF

BURKINA FASO
SOURCES:
All Indicators
1984-87 WB
Exp. of food
Exports of other agricultural products
Imports of manufactures
1965-87 IMF

CAMEROON
SOURCES:
Exports of fuel
1965-87 WB

CENTRAL AFRICAN REPUBLIC
SOURCES:
Exports of all commodities
1965-67 WB
1968-87 IMF
Imports of all commodities
1965-67 WB
1968-87 IMF

EGYPT, ARAB REPUBLIC OF
SOURCES:
All Indicators
1965-87 WB
1980-87 WB
MISCELLANEOUS:
1980-87 Fiscal years.

EL SALVADOR
SOURCES:
Exports of beverages
1965-87 IMF .

GABON
SOURCES:
Exports of all commodities
1971-87 IMF
Imports of all commodities
1973-87 IMF

GHANA
SOURCES:
Exports of all commodities
Imports of all commodities
1965-87 WB
MISCELLANEOUS:
1965-87 BOP data plus aluminum trade estimated from partners'
data.

HAITI
SOURCES:
All Indicators
1984-87 WB
Exports of manufactures
Imports of manufactures
1965-87 WB
MISCELLANEOUS:
1965-87 Fiscal year, ending 30 September.

INDIA
SOURCES:
Exports of fuel
1965-87 WB
Exports of manufactures
1976-87 WB
Imports of fats & oil
1979-81 WB.
Imports of manufactures
1975-87 WB
Imports of metals and minerals
1975-87 WB

KENYA
SOURCES:
All Indicators
1965-76 IMF

KUWAIT
SOURCES:
Exports of fuel
1965-69 IMF

LESOTHO
SOURCES:
All Indicators
1965-87 IMF

MALI
SOURCES:
Exports of all commodities
1986-87 WB
Imports of all commodities
1985-87 WB
MISCELLANEOUS:
1984 Extrapolated using BOP data trend.

MEXICO
SOURCES:
All Indicators
1980-87 IMF
MISCELLANEOUS:
1980-87 Trade of in-bond industries excluded.

NEPAL
SOURCES:
All export indicators
1965-87 WB
All import indicators
1965-87 WB. Estimates based on partners' data.

NIGER
SOURCES:
Exp. of food
Exports of metals and minerals
Imports of manufactures
1965-87 IMF

NIGERIA
SOURCES:
Imports of all commodities
1980-85 IMF
1986-87 WB

OMAN
SOURCES:
Exports of fuel
1965-87 IMF

PAKISTAN
SOURCES:
1965-87 Trade in military goods; silver bullion is excluded
Prior to 1971 data include trade with Bangladesh

PAPUA NEW GUINEA
SOURCES:
Imports of manufactures (mach + oman)
1965-87 IMF

PARAGUAY
SOURCES:
All indicators
1980-87 Registered Trade only

PERU
SOURCES:
All export indicators
1975-87 IMF

RWANDA
SOURCES:
Imports of food
Imports of fuels
1973-87 IMF

SENEGAL
SOURCES:
Exports of all commodities
1981-87 IMF
Imports of all commodities
1981-82 IMF
1983-87 WB

SEYCHELLES
SOURCES:
All export indicators
1965-87 WB

SOUTH AFRICA
SOURCES:
All Indicators
1965-87 IMF
Exports of metals and minerals
1973-87 IMF
Imports of all commodities
1965-79 IMF
MISCELLANEOUS:
1965-87 Include trade of Namibia,Lesotho, Botswana, Swaziland
with other countries

SUDAN
SOURCES:
All export indicators
1965-87 WB
Imports of all commodities
1965-87 WB

SYRIAN ARAB REPUBLIC
SOURCES:
Imports of all commodities
1985-87 WB

TANZANIA
SOURCES:
All Indicators
1965-76 IMF
Imports of all commodities
1984-85 WB. 1984 number extrapolated from 1985.

TOGO
SOURCES:
Exports of manufactures
Exports of metals and minerals
Imports of food
Imports of manufactures
1965-87 WB

UGANDA
SOURCES:
All Indicators
1965-76 IMF
Exports of all commodities
1965-87 IMF
All import indicators
1977-87 IMF

UNITED ARAB EMIRATES
SOURCES:
Exports of fuel
1980-82 IMF

VENEZUELA
SOURCES:
Exports of all commodities
1983-87 IMF
All import indicators
1965-87 IMF
Imports of all commodities
1985-87 WB

ZAIRE
SOURCES:
Imports of all commodities
1984-85 WB

ZIMBABWE
SOURCES:
Imports of all commodities
1980-87 IMF

COUNTRY LIST AND
BASE YEARS FOR NATIONAL ACCOUNTS

Algeria	1974	Guatemala	1958	Poland	1982
Antigua	1977	Guinea-Bissau	1979	Portugal	1977
Argentina	1970	Guyana	1977	Rwanda	1976
Australia	1979-80	Haiti	1976	St. Kitts and Nevis	1977
Austria	1976	Honduras	1982	St. Lucia	1977
Bahamas, The	1977	Hong Kong	1980	St. Vincent and the	
Bahrain	1977	Hungary	1981	Grenadines	1977
Bangladesh	1973	Iceland	1980	Saudi Arabia	1970
Barbados	1974	India	1970	Senegal	1979
Belgium	1980	Indonesia	1983	Seychelles	1976
Belize	1984	Ireland	1980	Sierra Leone	1973
Benin	1978	Israel	1980	Singapore	1985
Bhutan	1983	Italy	1980	Solomon Islands	1980
Bolivia	1980	Jamaica	1974	Somalia	1977
Botswana	1980	Japan	1980	South Africa	1980
Brazil	1980	Jordan	1980	Spain	1980
Burkina Faso	1979	Kenya	1982	Sri Lanka	1982
Burundi	1970	Korea, Rep.	1980	Sudan	1982
Cameroon	1980	Kuwait	1972	Suriname	1980
Canada	1981	Lesotho	1980	Swaziland	1980
Cape Verde	1980	Liberia	1971	Sweden	1980
Central African Rep.	1977	Libya	1975	Switzerland	1970
Chad	1977	Luxembourg	1980	Syrian Arab Rep.	1980
Chile	1977	Madagascar	1970	Tanzania	1976
China	1980	Malawi	1978	Thailand	1972
Colombia	1975	Malaysia	1978	Togo	1976
Comoros	1985	Mali	1985	Tonga	1982
Congo, People's Rep.	1975	Malta	1973	Trinidad and Tobago	1970
Costa Rica	1966	Mauritania	1982	Tunisia	1980
Côte d'Ivoire	1984	Mauritius	1976	Turkey	1968
Cyprus	1980	Mexico	1970	Uganda	1966
Denmark	1980	Morocco	1969	United Arab Emirates	1980
Dominica	1977	Mozambique	1980	United Kingdom	1980
Dominican Rep.	1970	Nepal	1975	United States	1982
Ecuador	1975	Netherlands	1980	Uruguay	1978
Egypt, Arab Rep.	1981	New Zealand	1976-77	Vanuatu	1983
El Salvador	1962	Nicaragua	1980	Venezuela	1984
Ethiopia	1981	Niger	1972	Western Samoa	1980
Fiji	1977	Nigeria	1984	Yemen Arab Rep.	1981
Finland	1985	Norway	1980	Yemen, PDR	1980
France	1980	Oman	1976	Yugoslavia	1972
Gabon	1979	Pakistan	1981	Zaire	1980
Gambia, The	1977	Panama	1970	Zambia	1977
Germany, Fed. Rep.	1980	Papua New Guinea	1981	Zimbabwe	1980
Ghana	1975	Paraguay	1982		
Greece	1970	Peru	1979		
Grenada	1984	Philippines	1972		

World Bank publications are sold in local currency through some 70 official publications distributors worldwide. For the name of the World Bank publications distributor in your country, contact World Bank Publications, 1818 H Street, N.W., Washington, D.C., USA, Tel. 202-473-2943, or World Bank Publications, 66, avenue d'Iéna, 75116 Paris, France, Tel. (33) 140.69.30.00